Handbook of ESSENTIAL OILS
Science, Technology, and Applications

SECOND EDITION

Handbook of ESSENTIAL OILS
Science, Technology, and Applications

SECOND EDITION

Edited by
K. Hüsnü Can Başer
Gerhard Buchbauer

CRC Press
Taylor & Francis Group
Boca Raton London New York

CRC Press is an imprint of the
Taylor & Francis Group, an **informa** business

Library
Quest University Canada
3200 University Boulevard
Squamish, BC V8B 0N8

CRC Press
Taylor & Francis Group
6000 Broken Sound Parkway NW, Suite 300
Boca Raton, FL 33487-2742

© 2016 by Taylor & Francis Group, LLC
CRC Press is an imprint of Taylor & Francis Group, an Informa business

No claim to original U.S. Government works

Printed on acid-free paper
Version Date: 20150827

International Standard Book Number-13: 978-1-4665-9046-5 (Hardback)

This book contains information obtained from authentic and highly regarded sources. Reasonable efforts have been made to publish reliable data and information, but the author and publisher cannot assume responsibility for the validity of all materials or the consequences of their use. The authors and publishers have attempted to trace the copyright holders of all material reproduced in this publication and apologize to copyright holders if permission to publish in this form has not been obtained. If any copyright material has not been acknowledged please write and let us know so we may rectify in any future reprint.

Except as permitted under U.S. Copyright Law, no part of this book may be reprinted, reproduced, transmitted, or utilized in any form by any electronic, mechanical, or other means, now known or hereafter invented, including photocopying, microfilming, and recording, or in any information storage or retrieval system, without written permission from the publishers.

For permission to photocopy or use material electronically from this work, please access www.copyright.com (http://www.copyright.com/) or contact the Copyright Clearance Center, Inc. (CCC), 222 Rosewood Drive, Danvers, MA 01923, 978-750-8400. CCC is a not-for-profit organization that provides licenses and registration for a variety of users. For organizations that have been granted a photocopy license by the CCC, a separate system of payment has been arranged.

Trademark Notice: Product or corporate names may be trademarks or registered trademarks, and are used only for identification and explanation without intent to infringe.

Visit the Taylor & Francis Web site at
http://www.taylorandfrancis.com

and the CRC Press Web site at
http://www.crcpress.com

Contents

Preface ix
Editors xi
Contributors xiii

Chapter 1 Introduction 1

K. Hüsnü Can Başer and Gerhard Buchbauer

Chapter 2 History and Sources of Essential Oil Research 5

Karl-Heinz Kubeczka

Chapter 3 Sources of Essential Oils 43

Chlodwig Franz and Johannes Novak

Chapter 4 Natural Variability of Essential Oil Components 87

Éva Németh-Zámboriné

Chapter 5 Production of Essential Oils 127

Erich Schmidt

Chapter 6 Chemistry of Essential Oils 165

Charles Sell

Chapter 7 Analysis of Essential Oils 195

Barbara d'Acampora Zellner, Paola Dugo, Giovanni Dugo, and Luigi Mondello

Chapter 8 Safety Evaluation of Essential Oils: Constituent-Based Approach Utilized
for Flavor Ingredients—An Update 229

Sean V. Taylor

Chapter 9 Metabolism of Terpenoids in Animal Models and Humans 253

Walter Jäger and Martina Höferl

Chapter 10 Biological Activities of Essential Oils: An Update 281

Gerhard Buchbauer and Ramona Bohusch

Chapter 11 Antioxidative Properties of Essential Oils and Single Fragrance Compounds 323

Gerhard Buchbauer and Marina Erkic

Chapter 12 Effects of Essential Oils in the Central Nervous System 345

12.1 Central Nervous System Effects of Essential Oils in Humans 345
Eva Heuberger

12.2 Psychopharmacology of Essential Oils 362
Domingos Sávio Nunes, Viviane de Moura Linck,
Adriana Lourenço da Silva, Micheli Figueiró, and Elaine Elisabetsky

Chapter 13 Phytotherapeutic Uses of Essential Oils 381
Robert Harris

Chapter 14 In Vitro Antimicrobial Activities of Essential Oils Monographed in the *European Pharmacopoeia 8th Edition* 433
Alexander Pauli and Heinz Schilcher

Chapter 15 Aromatherapy with Essential Oils 619
Maria Lis-Balchin

Chapter 16 Essential Oils Used in Veterinary Medicine 655
K. Hüsnü Can Başer and Chlodwig Franz

Chapter 17 Use of Essential Oils in Agriculture 669
Gerhard Buchbauer and Susanne Hemetsberger

Chapter 18 Adulteration of Essential Oils 707
Erich Schmidt and Jürgen Wanner

Chapter 19 Biotransformation of Monoterpenoids by Microorganisms, Insects, and Mammals 747
Yoshiaki Noma and Yoshinori Asakawa

Chapter 20 Biotransformation of Sesquiterpenoids, Ionones, Damascones, Adamantanes, and Aromatic Compounds by Green Algae, Fungi, and Mammals 907
Yoshinori Asakawa and Yoshiaki Noma

Chapter 21 Industrial Uses of Essential Oils 1011
W. S. Brud

Chapter 22 Encapsulation and Other Programmed Release Techniques for Essential Oils and Volatile Terpenes 1023
Jan Karlsen

Contents

Chapter 23 Trade of Essential Oils 1033

    Hugo Bovill

Chapter 24 Storage, Labeling, and Transport of Essential Oils.. 1041

    Jens Jankowski, Jens-Achim Protzen, and Klaus-Dieter Protzen

Chapter 25 Aroma-Vital Cuisine: Healthy and Delightful Consumption by the
Use of Essential Oils 1053

    Maria M. Kettenring and Lara-M. Vucemilovic-Geeganage

Chapter 26 Recent EU Legislation on Flavors and Fragrances and Its Impact on
Essential Oils.. 1071

    Jan C.R. Demyttenaere

Index 1089

Preface

The overwhelming success of the first edition of the *Handbook of Essential Oils: Science, Technology, and Applications* prompted us to edit an updated version. All of our contributors of the first edition were animated by us either to cooperate again or to tell us if their articles of the first edition should be kept as they are. Moreover, we asked experts to write contributions on new topics not covered in the first edition. This was the case in two instances: Prof. Eva Németh-Zámboriné's Chapter 4 report on "Natural Variability of Essential Oils Components," which is based on her excellent plenary lecture at the 44th International Symposium on Essential Oils in Budapest in September 2013, and the important Chapter 18, "Adulteration of Essential Oil," by the perfumers Erich Schmidt and Jürgen Wanner. In the first treatise, the author discusses the reasons why variability of components often occurs in essential oils. The influence of molecular genetics in understanding the background of biodiversity in volatiles formation is elaborated with examples. Thinking on adulteration, the first idea is always a more or less criminal act, such as a fraud or a betrayal to the debit of customers; however, also to a certain extent, these natural products can lose their excellent quality by natural processes, such as storage under inferior conditions or improper access to air thus furnishing oxygenated artifacts. Now, these essential oils are adulterated. Other examples of such processes are described very impressively in this new chapter.

Chapters 3 through 6, 13 through 16, and 19 through 21 remained as they are written in the first edition. Now, they are listed as Chapters 3 (formerly 3), 5 through 7 (formerly 4 through 6), 15 (formerly 13) and 16 (formerly 19), 19 (14), 20 (15), and 23 (20). Especially as to the contribution of Prof. Maria Lis-Balchin (formerly Chapter 13, now Chapter 15), we thought that she has already reported all important facts on aromatherapy that was necessary to tell. Also, the interesting Chapter 25 on "Aroma-Vital Cuisine" by Maria M. Kettenring and Lara M. Vucemilovic, a well done example for the application of essential oils, really, has been kept as it is in the first edition, because an update of the recipes and thus creating a collection of old and new recipes for cooks does not meet the sense of this book. The chapter on "Biological Activities of Essential Oils" (formerly 9) has been updated and divided into two contributions, now Chapters 10 and 11. Chapter 24 "Storage, Labeling, and Transport of Essential Oils" and Chapter 26 "Recent EU Legislation on Flavors and Fragrances and Its Impact on Essential Oils" have been fully revised and updated.

We hope that also this second edition with the alluring mixture of former, updated, and even new contributions will satisfy the curiosity and needs of all of our readers who share with us the vivid interest in this fascinating natural topic, the essential oils.

Editors

K. Hüsnü Can Başer was born on July 15, 1949, in Çankırı, Turkey. He graduated from Eskisehir I.T.I.A. School of Pharmacy with diploma number 1 in 1972 and became a research assistant in the pharmacognosy department of the same school. He did his PhD in pharmacognosy between 1974 and 1978 at Chelsea College of the University of London, London, United Kingdom.

Upon returning home, he worked as a lecturer in pharmacognosy at the school he had earlier graduated from and served as director of Eskisehir I.T.I.A. School of Chemical Engineering between 1978 and 1980. He was promoted to associate professorship in pharmacognosy in 1981.

He served as dean of the faculty of pharmacy at Anadolu University (1993–2001), vice dean of the faculty of pharmacy (1982–1993), head of the department of professional pharmaceutical sciences (1982–1993), head of the pharmacognosy section (1982–present), member of the university board and senate (1982–2001; 2007), and director of the Medicinal and Aromatic Plant and Drug Research Centre (TBAM) (1980–2002) in Anadolu University. He retired from Anadolu University in 2011. Since then, he has been working as a visiting professor in King Saud University, Riyadh, Saudi Arabia.

During 1984–1994, he was appointed as the national project coordinator of Phase I and Phase II of the UNDP/UNIDO projects of the government of Turkey titled "Production of Pharmaceutical Materials from Medicinal and Aromatic Plants," through which TBAM had been strengthened.

He was promoted to full professorship in pharmacognosy in 1987. His major areas of research include essential oils, alkaloids, and biological, chemical, pharmacological, technological, and biological activity research into natural products. He is the 1995 recipient of the Distinguished Service Medal of International Federation of Essential Oils and Aroma Trades (IFEAT) based in London, United Kingdom, and the 2005 recipient of "Science Award" (health sciences) of the Scientific and Technological Research Council of Turkey (TUBITAK) among others. He has published more than 700 papers in international refereed journals (more than 540 in SCI journals), 144 papers in Turkish journals, and 141 papers in conference proceedings. He communicated 940 papers in 262 symposia. He is the author of 52 books or book chapters.

Visit http://www.khcbaser.com for more information.

Gerhard Buchbauer was born in 1943 in Vienna, Austria. He studied pharmacy at the University of Vienna, from where he received his master's degree (Mag. pharm.) in May 1966. In September 1966, he assumed the duties of university assistant at the Institute of Pharmaceutical Chemistry and received his doctorate (PhD) in pharmacy and philosophy in October 1971 with a thesis on synthetic fragrance compounds. Further scientific education was practiced postdoc in the team of Professor C.H. Eugster at the Institute of Organic Chemistry, University of Zurich (1977–1978), followed by the habilitation (postdoctoral lecture qualification) in pharmaceutical chemistry with the inaugural dissertation entitled "Synthesis of Analogies of Drugs and Fragrance Compounds with Contributions to Structure-Activity-Relationships" (1979) and appointment as permanent staff of the University of Vienna and head of the first department of the Institute of Pharmaceutical Chemistry.

In November 1991, he was appointed as a full professor of pharmaceutical chemistry, University of Vienna; in 2002, he was elected as head of this institute. He retired in October 2008. He is married since 1973 and had a son in 1974.

Among others, he is still a member of the permanent scientific committee of International Symposium on Essential Oils (ISEO); a member of the scientific committee of Forum Cosmeticum (1990, 1996, 2002, and 2008); a member of numerous editorial boards (e.g., *Journal of Essential Oil Research*, the *International Journal of Essential Oil Therapeutics*, *Scientia Pharmaceutica*); assistant editor of *Flavour and Fragrance Journal*; regional editor of *Eurocosmetics*; a member of many scientific societies (e.g., *Society of Austrian Chemists*, head of its working group "Food Chemistry, Cosmetics, and Tensides" (2000–2004), *Austrian Pharmaceutical Society*, and *Austrian Phytochemical Society*, vice head of the *Austrian Society of Scientific Aromatherapy*, and so on); technical advisor of IFEAT (1992–2008); and organizer of the 27th ISEO (September 2006, in Vienna) together with Professor Dr. Ch. Franz.

Based on the sound interdisciplinary education of pharmacists, it was possible to establish an almost completely neglected area of fragrance and flavor chemistry as a new research discipline within the pharmaceutical sciences. Our research team is the only one that conducts fragrance research in its entirety and covers synthesis, computer-aided fragrance design, analysis, and pharmaceutical/medicinal aspects. Because of our efforts, it is possible to show and to prove that these small molecules possess more properties than merely emitting a good odor. Now, this research team has gained a worldwide scientific reputation documented by more than 400 scientific publications, about 100 invited lectures, and about 200 contributions to symposia, meetings, and congresses, as short lectures and poster presentations.

Contributors

Yoshinori Asakawa
Faculty of Pharmacy
Tokushima Bunri University
Tokushima, Japan

Barbara d'Acampora Zellner
SCIFAR Department
University of Messina
Messina, Italy

and

L'Oreal Brasil
Rio de Janeiro, Brazil

K. Hüsnü Can Başer
Department of Pharmacognosy
Faculty of Pharmacy
Anadolu University
Eskisehir, Turkey

Ramona Bohusch
Department of Pharmaceutical Chemistry
Division of Clinical Pharmacy & Diagnostics
Center of Pharmacy
University of Vienna
Vienna, Austria

Hugo Bovill
Treatt PLC
Bury St. Edmunds
Suffolk, United Kingdom

W. S. Brud
Academy of Cosmetology and Health Care
and
Pollena-Aroma Ltd
Warsaw, Poland

Gerhard Buchbauer
Department of Pharmaceutical Chemistry
Division of Clinical Pharmacy & Diagnostics
Center of Pharmacy
University of Vienna
Vienna, Austria

Jan C.R. Demyttenaere
EFFA (European Flavour Association)
Brussels, Belgium

Giovanni Dugo
SCIFAR Department
University of Messina
Messina, Italy

Paola Dugo
SCIFAR Department
University of Messina
Messina, Italy

and

Rome Biomedical Campus University
Rome, Italy

Elaine Elisabetsky
Laboratório de Etnofamacologia
ICBS
Universidade Federal do Rio Grande do
Sul, Porto Alegre, Brazil

Marina Erkic
Department of Pharmaceutical Chemistry
Division of Clinical Pharmacy & Diagnostics
Center of Pharmacy
University of Vienna
Vienna, Austria

Micheli Figueiró
Laboratório de Etnofamacologia
ICBS
Universidade Federal do Rio Grande do
Sul, Porto Alegre, Brazil

Chlodwig Franz (professor emeritus)
Department of Farm Animals and Veterinary
   Public Health
University of Veterinary Medicine Vienna
Vienna, Austria

Robert Harris
Quintessential Aromatics
West Sussex, United Kingdom

Susanne Hemetsberger
Department of Pharmaceutical Chemistry
Division of Clinical Pharmacy & Diagnostics
Center of Pharmacy
University of Vienna
Vienna, Austria

Eva Heuberger
Division of Clinical Psychology and Psychotherapy
Department of Psychology
Saarland University
Saarbrücken, Germany

Martina Höferl
Department of Pharmaceutical Chemistry
Division of Clinical Pharmacy & Diagnostics
Center of Pharmacy
University of Vienna
Vienna, Austria

Walter Jäger
Department of Pharmaceutical Chemistry
Division of Clinical Pharmacy & Diagnostics
Center of Pharmacy
University of Vienna
Vienna, Austria

Jens Jankowski
Paul Kaders GmbH
Hamburg, Germany

Jan Karlsen
Department of Pharmaceutics
University of Oslo
Oslo, Norway

Maria M. Kettenring
Villaroma
Neu-Isenburg, Germany

Karl-Heinz Kubeczka
Margetshöcheim, Germany

Maria Lis-Balchin (retired)
South Bank University London
London, United Kingdom

Luigi Mondello
SCIFAR Department
University of Messina
Messina, Italy

and

Rome Biomedical Campus University
Rome, Italy

Viviane de Moura Linck
Laboratório de Etnofamacologia
ICBS
Universidade Federal do Rio Grande do Sul, Porto Alegre, Brazil

Éva Németh-Zámboriné
Department of Medicinal and Aromatic Plants
Corvinus University of Budapest
Budapest, Hungary

Yoshiaki Noma (retired)
Faculty of Human Life Sciences
Tokushima Bunri University
Tokushima, Japan

Johannes Novak
Department of Farm Animals and Veterinary Public Health
University of Veterinary Medicine Vienna
Vienna, Austria

Alexander Pauli (deceased)

Jens-Achim Protzen
Paul Kaders GmbH
Hamburg, Germany

Klaus-Dieter Protzen
Paul Kaders GmbH
Hamburg, Germany

Domingos Sávio Nunes
Departamento de Química
Universidade Estadual de Ponta Grossa
Ponta Grossa, Brazil

Heinz Schilcher (deceased)
Immenstadt, Allgau, Germany

Contributors

Erich Schmidt
Consultant, Essential oils
Nördlingen, Germany

Charles Sell (retired)
Aldington, Kent, England

Adriana Lourenço da Silva
Laboratório de Etnofamacologia
ICBS
Universidade Federal do Rio Grande do Sul, Porto Alegre, Brazil

Sean V. Taylor
Flavor & Extract Manufacturers Association
and
The International Organization of the Flavor Industry
Washington, DC

Lara-M. Vucemilovic-Geeganage
Villaroma
Neu-Isenburg, Germany

Jürgen Wanner
Kurt Kitzing GmbH
Wallerstein, Germany

1 Introduction

K. Hüsnü Can Başer and Gerhard Buchbauer

CONTENTS

Introduction to First Edition 1
Introduction to Second Edition 2

INTRODUCTION TO FIRST EDITION

Essential oils (EOs) are very interesting natural plant products. Among other qualities, they possess various biological properties. The term "biological" comprises all activities that these mixtures of volatile compounds (mainly mono- and sesquiterpenoids, benzenoids, phenylpropanoids, etc.) exert on humans, animals, and other plants. This book intends to make the reader acquainted with all aspects of EOs and their constituent aromachemicals ranging from chemistry, pharmacology, biological activity, production, trade uses, and regulatory aspects. After an overview of research and development activities on EOs with a historical perspective (Chapter 2), Chapter 3 "Sources of Essential Oils" gives an expert insight into vast sources of EOs. The chapter also touches upon agronomic aspects of EO-bearing plants. Traditional and modern production techniques of EOs are illustrated and discussed in Chapter 4. It is followed by two important chapters "Chemistry of Essential Oils" (Chapter 5) and "Analysis of Essential Oils" (Chapter 6) illustrating chemical diversity of EOs, and analytical techniques employed for the analyses of these highly complex mixtures of volatiles.

They are followed by chapters on the biological properties of EOs, starting with "The Toxicology and Safety of Essential Oils: A Constituent-Based Approach" (Chapter 7). On account of the complexity of these natural products, the toxicological or biochemical testing of an EO will always be the sum of its constituents, which either act in a synergistic or in an antagonistic way with one another. Therefore, the chemical characterization of an EO is very important for the understanding of its biological properties. The constituents of these natural mixtures, upon being absorbed into the blood stream of humans or animals, get metabolized and eliminated. This metabolic biotransformation leads, mostly in two steps, to products with high water solubility, which enables the organism to get rid of these "xenobiotics" by renal elimination. This mechanism is thoroughly explained in Chapter 8, "Metabolism of Terpenoids in Animal Models and Humans." In Chapter 9, "Biological Activities of Essential Oils," "uncommon" biological activities of EOs, such as anticancer properties, antinociceptive effects, antiviral activities, antiphlogistic properties, penetration enhancement activities, and antioxidative effects, are reviewed. The psychoactive, particularly stimulating and sedative, effects of fragrances as well as behavioral activities, elucidated, for example, by neurophysiological methods, are the topics of Chapter 10 ("Effects of Essential Oils in the Central Nervous System"), Section 10.2. Here, the emphasis is put on the central nervous system and on psychopharmacology whereas Chapter 10, Section 10.1 mainly deals with reactions of the autonomic nervous system upon contact with EOs and/or their main constituents. The phytotherapeutic uses of EOs is another overview about scientific papers in peer-reviewed journals over the last 30 years, so to say the medical use of these natural plant products excluding aromatherapeutical treatments and single case studies (Chapter 11, "Phytotherapeutic Uses of Essential Oils"). Another contribution only deals with antimicrobial activities of those EOs that are monographed in the European Pharmacopoeia. In Chapter 12, "In Vitro Antimicrobial Activities of Essential Oils Monographed in the European Pharmacopoeia 6th Edition," more than 81 tables show the importance of these valuable properties of EOs. Aromatherapy with EOs covers the other side of the "classical" medical uses. "Aromatherapy with Essential Oils" (Chapter 13) is written by Maria Lis-Balchin, a known expert in this field and far from esoteric quackery. It completes the series of contributions dealing with the biological properties of EO regarded from various sides and standpoints.

Chapters 14 and 15 by the world-renowned experts Y. Asakawa and Y. Noma are concise treatises on the biotransformations of EO constituents. Enzymes in microorganisms and tissues metabolize EO constituents in similar ways by adding mainly oxygen function to molecules to render them water soluble, which facilitates their metabolism. This is also seen as a means of detoxification for these organisms. Many interesting and valuable novel chemicals are biosynthesized by this way. These products are also considered natural since the substrates are natural.

Encapsulation is a technique widely utilized in pharmaceutical, chemical, food, and feed industries to render EOs more manageable in formulations. Classical and modern encapsulation techniques are explained in detail in Chapter 17, "Encapsulation and Other Programmed Release Techniques for EOs and Volatile Terpenes."

EOs and aromachemicals are low-volume high-value products used in perfumery, cosmetics, feed, food, beverages, and pharmaceutical industries. Industrial uses of EOs are covered in an informative chapter from a historical perspective.

"Aroma-Vital Cuisine" (Chapter 18) looks at the possibility of utilizing EOs in the kitchen, where the pleasure of eating, the sensuality, and the enjoyment of lunching and dining of mostly processed food are stressed. Here, the holistic point of view rather than too scientific a way of understanding EOs is the topic, simply to show that these volatile natural plant products can add to the feel-good factor for users.

EOs are not only appealing to humans but also to animals. Applications of EOs as feed additives and for treating diseases in pets and farm animals are illustrated in Chapter 19, "Essential Oils Used in Veterinary Medicine," that comprises a rare collection of information on this subject.

The EO industry is highly complex and fragmented and the trade of EOs is rather conservative and highly specialized. EOs are produced and utilized worldwide in both industrialized and developing countries. The trade situation in the world is summarized in "Trade of Essential Oils" (Chapter 20), authored by a world-renowned expert Hugo Bovill.

Storage and transport of EOs are crucial issues since they are highly sensitive to heat, moisture, and oxygen. Therefore, special precautions and strict regulations apply for their handling in storage and transport. "Storage and Transport of Essential Oils" (Chapter 21) will give the reader necessary guidelines to tackle this problem.

Finally, the regulatory affairs of EOs are dealt with in Chapter 22 to give a better insight to those interested in legislative aspects. "Recent EU Legislation on Flavors and Fragrances and Its Impact on Essential Oils" comprises the most up-to-date regulations and legislative procedures applied on EOs in the European Union.

This book is hoped to satisfy the needs of EO producers, traders, and users as well as researchers, academicians, and legislators who will find the most current information given by selected experts under one cover.

INTRODUCTION TO SECOND EDITION

As mentioned in the Preface, the first new chapter (Chapter 4) deals with "Natural Variability of Essential Oil Components." It is based on Prof. Dr. Eva Nemeth-Zamborini's excellent plenary lecture at the 44th International Symposium on Essential Oils in Budapest in September 2013. In this treatise Prof. Nemeth-Zamborini discusses the reasons why variability of components often occurs in EOs. The influence of molecular genetics in understanding the background of biodiversity in formation of volatiles is elaborated with examples. The second new chapter (Chapter 18) by, the perfumers Erich Schmidt and Jürgen Wanner, is devoted to the subject "Adulteration of Essential Oils," which was not covered in the first edition. On the issue of adulteration, the first idea is always more or less a criminal act, such as fraud or betrayal to the debit of the customers; however, to a certain extent, these natural products can lose their excellent quality by natural processes, such as storage under inferior conditions or improper access to air, thus furnishing oxygenated artefacts. Now, these EOs are adulterated. Other examples of such processes are described very impressively in this chapter.

Chapters 2, 5, 12, 15, 19 through 21, 23, and 25 remained as written in the first edition. The former Chapter 9 has been updated and divided into two chapters, now listed as Chapters 10 and 11. The third new chapter (Chapter 17) is "Use of Essential Oils in Agriculture," which covers the impact of these natural volatile products on the performance of the farmer, especially pointing to benefits he earns using one or more examples of "green chemistry." All other chapters from the first edition have been updated, namely Chapters 8 through 12, 13, 22, and especially 24 and 26, which deal with "Storage, Labeling, and Transport of Essential Oils" and the important news on "Recent EU Legislation on Flavors and Fragrances and Its Impact on Essential Oils," respectively.

2 History and Sources of Essential Oil Research

Karl-Heinz Kubeczka

CONTENTS

2.1 Ancient Historical Background 5
2.2 First Systematic Investigations 7
2.3 Research during the Last Half Century 8
    2.3.1 Essential Oil Preparation Techniques 8
        2.3.1.1 Industrial Processes 8
        2.3.1.2 Laboratory-Scale Techniques 9
        2.3.1.3 Microsampling Techniques 10
    2.3.2 Chromatographic Separation Techniques 14
        2.3.2.1 Thin-Layer Chromatography 15
        2.3.2.2 GC 15
        2.3.2.3 Liquid Column Chromatography 21
        2.3.2.4 Supercritical Fluid Chromatography 23
        2.3.2.5 Countercurrent Chromatography 23
    2.3.3 Hyphenated Techniques 24
        2.3.3.1 Gas Chromatography-Mass Spectrometry 24
        2.3.3.2 High-Resolution GC-FTIR Spectroscopy 25
        2.3.3.3 GC-UV Spectroscopy 27
        2.3.3.4 Gas Chromatography-Atomic Emission Spectroscopy 27
        2.3.3.5 Gas Chromatography-Isotope Ratio Mass Spectrometry 27
        2.3.3.6 High-Performance Liquid Chromatography-Gas Chromatography 27
        2.3.3.7 HPLC-MS, HPLC-NMR Spectroscopy 29
        2.3.3.8 Supercritical Fluid Extraction–Gas Chromatography 29
        2.3.3.9 Supercritical Fluid Chromatography-Gas Chromatography 30
        2.3.3.10 Couplings of SFC-MS and SFC-FTIR Spectroscopy 30
    2.3.4 Identification of Multicomponent Samples without Previous Separation 31
        2.3.4.1 UV Spectroscopy 31
        2.3.4.2 IR Spectroscopy 31
        2.3.4.3 Mass Spectrometry 31
        2.3.4.4 ^{13}C-NMR Spectroscopy 32
References 33

2.1 ANCIENT HISTORICAL BACKGROUND

Plants containing essential oils have been used since furthest antiquities as spices and remedies for the treatment of diseases and in religious ceremonies because of their healing properties and their pleasant odors. In spite of the obscured beginning of the use of aromatic plants in prehistoric times to prevent, palliate, or heal sicknesses, pollen analyses of Stone Age settlements indicate the use of aromatic plants that may be dated to 10,000 BC.

One of the most important medical documents of ancient Egypt is the so-called Papyrus Ebers of about 1550 BC, a 20 m long papyrus, which was purchased in 1872 by the German Egyptologist G. Ebers, for whom it is named, containing some 700 formulas and remedies, including aromatic plants and plant products like anise, fennel, coriander, thyme, frankincense, and myrrh. Much later, the ancient Greek physician Hippocrates (460–377 BC), who is referred to as the father of medicine, mentioned in his treatise *Corpus Hippocratium* approximately 200 medicinal plants inclusive of aromatic plants and described their efficacies.

One of the most important herbal books in history is the five-volume book *De Materia Medica*, written by the Greek physician and botanist Pedanius Dioscorides (ca. 40–90), who practiced in ancient Rome. In the course of his numerous travels all over the Roman and Greek world seeking for medicinal plants, he described more than 500 medicinal plants and respective remedies. His treatise, which may be considered a precursor of modern pharmacopoeias, was later translated into a variety of languages. Dioscorides, as well as his contemporary Pliny the Elder (23–79), a Roman natural historian, mention besides other facts turpentine oil and give some limited information on the methods in its preparation.

Many new medicines and ointments were brought from the east during the Crusades from the eleventh to the thirteenth centuries, and many herbals, whose contents included recipes for the use and manufacture of essential oil, were written during the fourteenth to the sixteenth centuries.

Theophrastus von Hohenheim, known under the name Paracelsus (1493–1541), a physician and alchemist of the fifteenth century, defined the role of alchemy by developing medicines and extracts from healing plants. He believed distillation released the most desirable part of the plant, the *Quinta essentia* or *quintessence* by a means of separating the "essential" part from the "nonessential" containing its subtle and essential constituents. The currently used term "essential oil" still refers to the theory of *Quinta essentia* of Paracelsus.

The roots of distillation methods are attributed to Arabian Alchemists centuries with Avicenna (980–1037) describing the process of steam distillation, who is credited with inventing a coiled cooling pipe to prepare essential oils and aromatic waters. The first description of distilling essential oils is generally attributed to the Spanish physician Arnaldus de Villa Nova (1235–1311) in the thirteenth century. However, in 1975, a perfectly preserved terracotta apparatus was found in the Indus Valley, which is dated to about 3000 BC and which is now displayed in a museum in Taxila, Pakistan. It looks like a primitive still and was presumable used to prepare aromatic waters. Further findings indicate that distillation has also been practiced in ancient Turkey, Persia, and India as far back as 3000 BC.

At the beginning of the sixteenth century appeared a comprehensive treatise on distillation by Hieronymus Brunschwig (ca. 1450–1512), a physician of Strasbourg. He described the process of distillation and the different types of stills in his book *Liber de arte Distillandi de compositis* (Strasbourg 1500 and 1507) with numerous block prints. Although obviously endeavoring to cover the entire field of distillation techniques, he mentions in his book only the four essential oils from rosemary, spike lavender, juniper wood, and the turpentine oil. Just before, until the Middle Ages, the art of distillation was used mainly for the preparation of aromatic waters, and the essential oil appearing on the surface of the distilled water was regarded as an undesirable by-product.

In 1551 appeared at Frankfurt on the Main the *Kräuterbuch*, written by Adam Lonicer (1528–1586), which can be regarded as a significant turning point in the understanding of the nature and the importance of essential oils. He stresses that the art of distillation is a quite recent invention and not an ancient invention and has not been used earlier.

In the *Dispensatorium Pharmacopolarum* of Valerius Cordus, published in Nuremberg in 1546, only three essential were listed; however, the second official edition of the *Dispensatorium Valerii Cordi* issued in 1592, 61 distilled oils were listed illustrating the rapid development and acceptance of essential oils. In that time, the so-called Florentine flask has already been used for separating the essential oil from the water phase.

The German J.R. Glauber (1604–1670), who can be regarded as one of the first great industrial chemists, was born in the little town Karlstadt close to Wuerzburg. His improvements in chemistry, for example, the production of sodium sulfate, as a safe laxative brought him the honor of being named Glauber's salt. In addition, he improved numerous different other chemical processes and especially new distillation devices also for the preparation of essential oils from aromatic plants. However, it lasted until the nineteenth century to get any real understanding of the composition of true essential oils.

2.2 FIRST SYSTEMATIC INVESTIGATIONS

The first systematic investigations of constituents from essential oils may be attributed to the French chemist M.J. Dumas (1800–1884) who analyzed some hydrocarbons and oxygen as well as sulfur- and nitrogen-containing constituents. He published his results in 1833. The French researcher M. Berthelot (1859) characterized several natural substances and their rearrangement products by optical rotation. However, the most important investigations have been performed by O. Wallach, an assistant of Kekule. He realized that several terpenes described under different names according to their botanical sources were often, in fact, chemically identical. He, therefore, tried to isolate the individual oil constituents and to study their basic properties. He employed together with his highly qualified coworkers Hesse, Gildemeister, Betram, Walbaum, Wienhaus, and others fractional distillation to separate essential oils and performed reactions with inorganic reagents to characterize the obtained individual fractions. The reagents he used were hydrochloric acid, oxides of nitrogen, bromine, and nitrosyl chloride—which was used for the first time by W.A. Tilden (1875)—by which frequently crystalline products have been obtained.

At that time, hydrocarbons occurring in essential oils with the molecular formula $C_{10}H_{16}$ were known, which had been named by Kekule *terpenes* because of their occurrence in turpentine oil. Constituents with the molecular formulas $C_{10}H_{16}O$ and $C_{10}H_{18}O$ were also known at that time under the generic name camphor and were obviously related to terpenes. The prototype of this group was camphor itself, which was known since antiquity. In 1891, Wallach characterized the terpenes pinene, camphene, limonene, dipentene, phellandrene, terpinolene, fenchene, and sylvestrene, which has later been recognized to be an artifact.

During 1884–1914, Wallach wrote about 180 articles that are summarized in his book *Terpene und Campher* (Wallach, 1914) compiling all the knowledge on terpenes at that time, and already in 1887, he suggested that the terpenes must be constructed from isoprene units. In 1910, he was honored with the Nobel Prize for Chemistry "in recognition of his outstanding research in organic chemistry and especially in the field of alicyclic compounds" (Laylin, 1993).

In addition to Wallach, the German chemist A. von Baeyer, who also had been trained in Kekule's laboratory, was one of the first chemists to become convinced of the achievements of structural chemistry and who developed and applied it to all of his work covering a broad scope of organic chemistry. Since 1893, he devoted considerable work to the structures and properties of cyclic terpenes (von Baeyer, 1901). Besides his contributions to several dyes, the investigations of polyacetylenes, and so on, his contributions to theoretical chemistry including the strain theory of triple bonds and small carbon cycles have to be mentioned. In 1905, he was awarded the Nobel Prize for Chemistry "in recognition of his contributions to the development of Organic Chemistry and Industrial Chemistry, by his work on organic dyes and hydroaromatic compounds" (Laylin, 1993). The frequently occurring acyclic monoterpenes geraniol, linalool, citral, and so on have been investigated by F.W. Semmler and the Russian chemist G. Wagner (1899), who recognized the importance of rearrangements for the elucidation of chemical constitution, especially the carbon-to-carbon migration of alkyl, aryl, or hydride ions, a type of reaction that was later generalized by H. Meerwein (1914) as Wagner–Meerwein rearrangement.

More recent investigations of J. Read, W. Hückel, H. Schmidt, W. Treibs, and V. Prelog were mainly devoted to disentangle the stereochemical structures of menthols, carvomenthols, borneols, fenchols, and pinocampheols, as well as the related ketones (see Gildemeister and Hoffmann, 1956).

A significant improvement in structure elucidation was the application of dehydrogenation of sesqui- and diterpenes with sulfur and later with selenium to give aromatic compounds as a major method, and the application of the isoprene rule to terpene chemistry, which have been very efficiently used by L. Ruzicka (1953) in Zurich, Switzerland. In 1939, he was honored in recognition of his outstanding investigations with the Nobel Prize in chemistry for his work on "polymethylenes and higher terpenes."

The structure of the frequently occurring bicyclic sesquiterpene ß-caryophyllene was for many years a matter of doubt. After numerous investigations, W. Treibs (1952) has been able to isolate the crystalline caryophyllene epoxide from the autoxidation products of clove oil, and F. Šorm et al. (1950) suggested caryophyllene to have a four- and nine-membered ring on bases of infrared (IR) investigations. This suggestion was later confirmed by the English chemist D.H.R. Barton (Barton and Lindsay, 1951), who was awarded the Nobel Prize in Chemistry in 1969.

The application of ultraviolet (UV) spectroscopy in the elucidation of the structure of terpenes and other natural products was extensively used by R.B. Woodward in the early forties of the last century. On the basis of his large collection of empirical data, he developed a series of rules (later called the Woodward rules), which could be applied to finding out the structures of new natural substances by correlations between the position of UV maximum absorption and the substitution pattern of a diene or an α,β-unsaturated ketone (Woodward, 1941). He was awarded the Nobel Prize in Chemistry in 1965. However, it was not until the introduction of chromatographic separation methods and nuclear magnetic resonance (NMR) spectroscopy into organic chemistry that a lot of further structures of terpenes were elucidated. The almost exponential growth in our knowledge in that field and other essential oil constituents is essentially due to the considerable advances in analytical methods in the course of the last half century.

2.3 RESEARCH DURING THE LAST HALF CENTURY

2.3.1 Essential Oil Preparation Techniques

2.3.1.1 Industrial Processes

The vast majority of essential oils are produced from plant material in which they occur by different kinds of distillation or by cold pressing in the case of the peel oils from citrus fruits.

In water or hydrodistillation, the chopped plant material is submerged and in direct contact with boiling water. In steam distillation, the steam is produced in a boiler separate of the still and blown through a pipe into the bottom of the still, where the plant material rests on a perforated tray or in a basket for quick removal after exhaustive extraction. In addition to the aforementioned distillation at atmospheric pressure, high-pressure steam distillation is most often applied in European and American field stills, and the applied increased temperature significantly reduces the time of distillation. The high-pressure steam-type distillation is often applied for peppermint, spearmint, lavandin, and the like. The condensed distillate, consisting of a mixture of water and oil, is usually separated in a so-called Florentine flask, a glass jar, or more recently in a receptacle made of stainless steel with one outlet near the base and another near the top. There, the distillate separates into two layers from which the oil and the water can be separately withdrawn. Generally, the process of steam distillation is the most widely accepted method for the production of essential oils on a large scale.

Expression or cold pressing is a process in which the oil glands within the peels of citrus fruits are mechanically crushed to release their content. There are several different processes used for the isolation of citrus oils; however, there are four major currently used processes. Those are pellatrice and sfumatrice—most often used in Italy—and the Brown peel shaver as well as the FMC extractor, which are used predominantly in North and South America. For more details, see, for example, Lawrence 1995. All these processes lead to products that are not entirely volatile because they may contain coumarins, plant pigments, and so on; however, they are nevertheless acknowledged as essential oils by the International Organization for Standardization, the different pharmacopoeias, and so on.

History and Sources of Essential Oil Research

In contrast, extracts obtained by solvent extraction with different organic solvents, with liquid carbon dioxide or by supercritical fluid extraction (SFE) may not be considered as true essential oils; however, they possess most often aroma profiles that are almost identical to the raw material from which they have been extracted. They are therefore often used in the flavor and fragrance industry and in addition in food industry, if the chosen solvents are acceptable for food and do not leave any harmful residue in food products.

2.3.1.2 Laboratory-Scale Techniques

The following techniques are used mainly for trapping small amounts of volatiles from aromatic plants in research laboratories and partly for determination of the essential oil content in plant material. The most often used device is the circulatory distillation apparatus, basing on the publication of Clevenger in 1928 and which has later found various modifications. One of those modified apparatus described by Cocking and Middleton (1935) has been introduced in the European pharmacopoeia and several other pharmacopoeias. This device consists of a heated round-bottom flask into which the chopped plant material and water are placed and which is connected to a vertical condenser and a graduated tube, for the volumetric determination of the oil. At the bottom of the tube, a three-way valve permits to direct the water back to the flask, since it is a continuous closed-circuit distillation device, and at the end of the distillation process to separate the essential oil from the water phase for further investigations. The length of distillation depends on the plant material to be investigated; however, it is usually fixed to 3–4 h. For the volumetric determination of the essential oil content in plants according to most of the pharmacopoeias, a certain amount of xylene—usually 0.5 mL—has to be placed over the water before running distillation to separate even small droplets of essential oil during distillation from the water. The volume of essential oil can be determined in the graduated tube after subtracting the volume of the applied xylene.

Improved constructions with regard to the cooling system of the aforementioned distillation apparatus have been published by Stahl (1953) and Sprecher (1963) and, in publications of Kaiser and Lang (1951) and Mechler and Kovar (1977), various apparatus used for the determination of essential oils in plant material are discussed and depicted.

A further improvement was the development of a simultaneous distillation–solvent extraction device by Likens and Nickerson in 1964 (see Nickerson and Likens, 1966). The device permits continuous concentration of volatiles during hydrodistillation in one step using a closed-circuit distillation system. The water distillate is continuously extracted with a small amount of an organic- and water-immiscible solvent. Although there are two versions described, one for high-density and one for low-density solvents, the high-density solvent version using dichloromethane is mostly applied in essential oil research. It has found numerous applications, and several modified versions including different microdistillation devices have been described (e.g., Bicchi et al., 1987; Chaintreau, 2001).

A sample preparation technique basing on Soxhlet extraction in a pressurized container using liquid carbon dioxide as extractant has been published by Jennings (1979). This device produces solvent-free extracts especially suitable for high-resolution gas chromatography (GC). As a less time-consuming alternative, the application of microwave-assisted extraction has been proposed by several researchers, for example, by Craveiro et al. (1989), using a round-bottom flask containing the fresh plant material. This flask was placed into a microwave oven and passed by a flow of air. The oven was heated for 5 min and the obtained mixture of water and oil collected in a small and cooled flask. After extraction with dichloromethane, the solution was submitted to GC–mass spectrometry (GC-MS) analysis. The obtained analytical results have been compared with the results obtained by conventional distillation and exhibited no qualitative differences; however, the percentages of the individual components varied significantly. A different approach yielding solvent-free extracts from aromatic herbs by means of microwave heating has been presented by Lucchesi et al. (2004). The potential of the applied technique has been compared with conventional hydrodistillation showing substantially higher amounts of oxygenated compounds at the expense of monoterpene hydrocarbons.

2.3.1.3 Microsampling Techniques

2.3.1.3.1 Microdistillation

Preparation of very small amounts of essential oils may be necessary if only very small amounts of plant material are available and can be fundamental in chemotaxonomic investigations and control analysis but also for medicinal and spice plant breeding. In the past, numerous attempts have been made to minimize conventional distillation devices. As an example, the modified Marcusson device may be quoted (Bicchi et al., 1983) by which 0.2–3 g plant material suspended in 50 mL water can be distilled and collected in 100 µL analytical grade pentane or hexane. The analytical results proved to be identical with those obtained by conventional distillation.

Microversions of the distillation–extraction apparatus, described by Likens and Nickerson, have also been developed as well for high-density (Godefroot et al., 1981) and low-density solvents (Godefroot et al., 1982). The main advantage of these techniques is that no further enrichment by evaporation is required for subsequent gas chromatographic investigation.

A different approach has been presented by Gießelmann and Kubeczka (1993) and Kubeczka and Gießelmann (1995). By means of a new developed micro-hydrodistillation device, the volatile constituents of very small amounts of plant material have been separated. The microscale hydrodistillation of the sample is performed using a 20 mL crimp-cap glass vial with a Teflon®-lined rubber septum containing 10 mL water and 200–250 mg of the material to be investigated. This vial, which is placed in a heating block, is connected with a cooled receiver vial by a 0.32 mm ID fused silica capillary. By temperature-programmed heating of the sample vial, the water and the volatile constituents are vaporized and passed through the capillary into the cooled receiver vial. There, the volatiles as well as water are condensed and the essential oil collected in pentane for further analysis. The received analytical results have been compared to results from identical samples obtained by conventional hydrodistillation showing a good correlation of the qualitative and quantitative composition. Further applications with the commercially available Eppendorf MicroDistiller® have been published in several papers, for example, by Briechle et al. (1997) and Baser et al. (2001).

A simple device for rapid extraction of volatiles from natural plant drugs and the direct transfer of these substances to the starting point of a thin-layer chromatographic plate has been described by Stahl (1969a) and in his subsequent publications. A small amount of the sample (ca. 100 mg) is introduced into a glass cartridge with a conical tip together with 100 mg silica gel, containing 20% of water, and heated rapidly in a heating block for a short time at a preset temperature. The tip of the glass tube projects ca. 1 mm from the furnace and points to the starting point of the thin-layer plate, which is positioned 1 mm in front of the tip. Before introducing the glass tube, it is sealed with a silicone rubber membrane. This simple technique has proven useful for many years in numerous investigations, especially in quality control, identification of plant drugs, and rapid screening of chemical races. In addition to the aforementioned micro-hydrodistillation with the so-called TAS procedure (T, thermomicro and transfer; A, application; S, substance), several further applications, for example, in structure elucidation of isolated natural compounds such as zinc dust distillation, sulfur and selenium dehydrogenation, and catalytic dehydrogenation with palladium, have been described in the microgram range (Stahl, 1976).

2.3.1.3.2 Direct Sampling from Secretory Structures

The investigation of the essential oils by direct sampling from secretory glands is of fundamental importance in studying the true essential oil composition of aromatic plants, since the usual applied techniques such as hydrodistillation and extraction are known to produce in some cases several artifacts. Therefore, only direct sampling from secretory cavities and glandular trichomes and properly performed successive analysis may furnish reliable results. One of the first investigations with a kind of direct sampling has been performed by Hefendehl (1966), who isolated the glandular hairs from the surfaces of *Mentha piperita* and *Mentha aquatica* leaves by means of a thin film of polyvinyl alcohol, which was removed after drying and extracted with diethyl ether. The composition of this product was in good agreement with the essential oils obtained by hydrodistillation. In contrast to these results, Malingré et al. (1969) observed some qualitative differences in the course of their study on *M. aquatica* leaves after isolation of the essential oil from individual glandular hairs by means of a micromanipulator and a stereomicroscope. In the same year, Amelunxen et al. (1969) published results on *M. piperita*, who separately isolated glandular hairs and glandular trichomes with glass capillaries. They found identical qualitative composition of the oil in both types of hairs, but differing concentrations of the individual components. Further studies have been performed by Henderson et al. (1970) on *Pogostemon cablin* leaves and by Fischer et al. (1987) on *Majorana hortensis* leaves. In the latter study, significant differences regarding the oil composition of the hydrodistilled oil and the oil extracted by means of glass capillaries from the trichomes were observed. Their final conclusion was that the analysis of the respective essential oil is mainly an analysis of artifacts, formed during distillation, and the gas chromatographic analysis. Even if the investigations are performed very carefully and the successive GC has been performed by coldon-column injection to avoid thermal stress in the injection port, significant differences of the GC pattern of directly sampled oils versus the microdistilled samples have been observed in several cases (Bicchi et al., 1985).

2.3.1.3.3   HS Techniques

Headspace (HS) analysis has become one of the very frequently used sampling techniques in the investigation of aromatic plants, fragrances, and spices. It is a means of separating the volatiles from a liquid or solid prior to gas chromatographic analysis and is preferably used for samples that cannot be directly injected into a gas chromatograph. The applied techniques are usually classified according to the different sampling principles in static HS analysis and dynamic HS analysis.

2.3.1.3.3.1   Static HS Methods   In static HS analysis, the liquid or solid sample is placed into a vial, which is heated to a predetermined temperature after sealing. After the sample has reached equilibrium with its vapor (in equilibrium, the distribution of the analytes between the two phases depends on their partition coefficients at the preselected temperature, the time, and the pressure), an aliquot of the vapor phase can be withdrawn with a gas-tight syringe and subjected to gas chromatographic analysis. A simple method for the HS investigation of herbs and spices was described by Chialva et al. (1982), using a blender equipped with a special gas-tight valve. After grinding the herb and until thermodynamic equilibrium is reached, the HS sample can be withdrawn through the valve and injected into a gas chromatograph. Eight of the obtained capillary gas chromatograms are depicted in the paper of Chialva and compared with those of the respective essential oils exhibiting significant higher amounts of the more volatile oil constituents. However, one of the major problems with static HS analyses is the need for sample enrichment with regard to trace components. Therefore, a concentration step such as cryogenic trapping, liquid absorption, or adsorption on a suitable solid has to be inserted for volatiles occurring only in small amounts. A versatile and often-used technique in the last decade is solid-phase microextraction (SPME) for sampling volatiles, which will be discussed in more detail in a separate paragraph. Since different other trapping procedures are a fundamental prerequisite for dynamic HS methods, they will be considered in the succeeding text. A comprehensive treatment of the theoretical basis of static HS analysis including numerous applications has been published by Kolb and Ettre (1997, 2006).

2.3.1.3.3.2   Dynamic HS Methods   The sensitivity of HS analysis can be improved considerably by stripping the volatiles from the material to be investigated with a stream of purified air or inert gas and trapping the released compounds. However, care has to be taken if grinded plant material has to be investigated, since disruption of tissues may initiate enzymatic reactions that may lead to formation of volatile artifacts. After stripping the plant material with gas in a closed vessel, the released volatile compounds are passed through a trap to collect and enrich the sample. This must be done because sample injection of fairly large sample volumes results in band broadening causing peak distortion and poor resolution. The following three techniques are advisable for collecting the highly diluted volatile sample according to Schaefer (1981) and Schreier (1984) with numerous references.

Cryogenic trapping can be achieved by passing the gas containing the stripped volatiles through a cooled vessel or a capillary in which the volatile compounds are condensed (Kolb and Liebhardt, 1986). The most convenient way for trapping the volatiles is to utilize part of the capillary column as a cryogenic trap. A simple device for cryofocusing of HS volatiles by using the first part of capillary column as a cryogenic trap has been shown in the aforementioned reference inclusive of a discussion of the theoretical background of cryogenic trapping. A similar on-column cold trapping device, suitable for extended period vapor sampling, has been published by Jennings (1981).

A different approach can be used if large volumes of stripped volatiles have to be trapped using collection in organic liquid phases. In this case, the volatiles distribute between the gas and the liquid, and efficient collection will be achieved, if the distribution factor K is favorable for solving the stripped compounds in the liquid. A serious drawback, however, is the necessity to concentrate the obtained solution prior to GC with the risk to lose highly volatile compounds. This can be overcome if a short-packed GC column is used containing a solid support coated with a suitable liquid. Novak et al. (1965) have used Celite coated with 30% silicone elastomer E-301 and the absorbed compounds were introduced into a gas chromatograph after thermal desorption. Coating with 15% silicone rubber SE 30 has been successfully used by Kubeczka (1967) with a similar device and the application of a wall-coated tubing with methyl silicone oil SF 96 has been described by Teranishi et al. (1972). A different technique has been used by Bergström (1973) and Bergström et al. (1980). They trapped the scent of flowers on Chromosorb® W coated with 10% silicon high-vacuum grease and filled a small portion of the sorbent containing the volatiles into a precolumn, which was placed in the splitless injection port of a gas chromatograph. There, the volatiles were desorbed under heating and flushed onto the GC column. In 1987, Bichi et al. applied up to 50 cm pieces of thick-film fused silica capillaries coated with a 15 μm dimethyl silicone film for trapping the volatiles in the atmosphere surrounding living plants. The plants under investigation were placed in a glass bell into which the trapping capillary was introduced through a rubber septum, while the other end of the capillary has been connected to pocket sampler. In order to trap even volatile monoterpene hydrocarbons, a capillary length of at least 50 cm and sample volume of maximum 100 mL have to be applied to avoid loss of components through breakthrough. The trapped compounds have been subsequently online thermally desorbed, cold trapped, and analyzed. Finally, a type of *enfleurage* especially designed for field experiments has been described by Joulain (1987) to trap the scents of freshly picked flowers. Around 100 g flowers were spread on the grid of a specially designed stainless steel device and passed by a stream of ambient air, supplied by an unheated portable air drier. The stripped volatiles are trapped on a layer of purified fat placed above the grid. After 2 h, the fat was collected and the volatiles recovered in the laboratory by means of vacuum distillation at low temperature.

With a third often applied procedure, the stripped volatiles from the HS of plant material and especially from flowers are passed through a tube filled with a solid adsorbent on which the volatile compounds are adsorbed. Common adsorbents most often used in investigations of plant volatiles are above all charcoal and different types of synthetic porous polymers. Activated charcoal is an adsorbent with a high adsorption capacity, thermal and chemical stability, and which is not deactivated by water, an important feature, if freshly collected plant material has to be investigated. The adsorbed volatiles can easily be recovered by elution with small amounts (10–50 μL) of carbon disulfide avoiding further concentration of the sample prior to GC analysis. The occasionally observed incomplete recovery of sample components after solvent extraction and artifact formation after thermal desorption has been largely solved by application of small amounts of special type of activated charcoal as described by Grob and Zürcher (1976). Numerous applications have been described using this special type of activated charcoal, for example, by Kaiser (1993) in a great number of field experiments on the scent of orchids. In addition to charcoal, the following synthetic porous polymers have been applied to collect volatile compounds from the HS from flowers and different other plant materials according to Schaefer (1981): Tenax® GC, different Porapak® types (e.g., Porapak P, Q, R, and T), and several Chromosorb types belonging to the 100 series. More recent developed adsorbents are the carbonaceous adsorbents such as Ambersorb®, Carboxen®, and Carbopak®, and their adsorbent properties lie between activated charcoal and the porous polymers. Especially the porous polymers have to be washed repeatedly, for example, with diethyl ether, and conditioned before use in a stream of oxygen-free nitrogen at 200°C–280°C, depending on the sort of adsorbent. The trapped components can be recovered either by thermal desorption or by solvent elution, and the recoveries can be different depending on the applied adsorbent (Cole, 1980). Another very important criterion for the selection of a suitable adsorbent for collecting HS samples is the breakthrough volume limiting the amount of gas passing through the trap.

A comprehensive review concerning HS gas chromatographic analysis of medicinal and aromatic plants and flowers with 137 references, covering the period from 1982 to 1988 has been published by Bicchi and Joulain in 1990, thoroughly describing and explaining the different methodological approaches and applications. Among other things, most of the important contributions of the Finnish research group of Hiltunen and coworkers on the HS of medicinal plants and the optimization of the HS parameters have been cited in the mentioned review.

2.3.1.3.4 Solid-Phase Microextraction

SPME is an easy-to-handle sampling technique, initially developed for the determination of volatile organic compounds in environmental samples (Arthur and Pawliszyn, 1990), and has gained, in the last years, acceptance in numerous fields and has been applied to the analysis of a wide range of analytes in various matrices. Sample preparation is based on sorption of analytes from a sample onto a coated fused silica fiber, which is mounted in a modified GC syringe. After introducing the coated fiber into a liquid or gaseous sample, the compounds to be analyzed are enriched according to their distribution coefficients and can be subsequently thermally desorbed from the coating after introducing the fiber into the hot injector of a gas chromatograph. The commercially available SPME device (Supelco Inc.) consists of a 1 cm length fused silica fiber of ca. 100 µm diameter coated on the outer surface with a stationary phase fixed to a stainless steel plunger and a holder that looks like a modified microliter syringe (Supelco, 2007). The fiber can be drawn into the syringe needle to prevent damage. To use the device, the needle is pierced through the septum that seals the sample vial. Then, the plunger is depressed lowering the coated fiber into the liquid sample or the HS above the sample. After sorption of the sample, which takes some minutes, the fiber has to be drawn back into the needle and withdrawn from the sample vial. By the same procedure, the fiber can be introduced into the gas chromatograph injector where the adsorbed substances are thermally desorbed and flushed by the carrier gas into the capillary GC column.

SPME fibers can be coated with polymer liquid (e.g., polydimethylsiloxane [PDMS]) or a mixed solid and liquid coating (e.g., Carboxen®/PDMS). The selectivity and capacity of the fiber coating can be adjusted by changing the phase type or thickness of the coating on the fiber according to the properties of the compounds to be analyzed. Commercially available are coatings of 7, 30, and 100 µm of PDMS, an 85 µm polyacrylate, and several mixed coatings for different polar components. The influence of fiber coatings on the recovery of plant volatiles was thoroughly investigated by Bicchi et al. (2000). Details concerning the theory of SPME, technology, its application, and specific topics have been described by Pawliszyn (1997) and references cited therein. A number of different applications of SPME in the field of essential oil analysis have been presented by Kubeczka (1997a). An overview on publications of the period 2000–2005 with regard to HS-SPME has been recently published by Belliardo et al. (2006) covering the analysis of volatiles from aromatic and medicinal plants, selection of the most effective fibers and sampling conditions, and discussing its advantages and limitations. The most comprehensive collection of references with regard to the different application of SPME can be obtained from Supelco on CD.

2.3.1.3.5 Stir Bar Sorptive Extraction and HS Sorptive Extraction

Despite the indisputable simplicity and rapidity of SPME, its applicability is limited by the small amount of sorbent on the needle (<0.5 μL), and consequently SPME has no real opportunity to realize quantitative extraction. Parameters governing recovery of analytes from a sample are partitioning constants and the phase ratio between the sorbent and liquid or gaseous sample. Therefore, basing on theoretical considerations, a procedure for sorptive enrichment with the sensitivity of packed PDMS beds (Baltussen et al., 1997) has been developed for the extraction of aqueous samples using modified PDMS-coated stir bars (Baltussen et al., 1999).

The stir bars were incorporated into a narrow glass tube coated with a PDMS layer of 1 mm (corresponding to 55 μL for a 10 mm length) applicable to small sample volumes. Such stir bars are commercially available under the name "Twister" (Gerstel, Germany). After certain stirring time, the stir bar has to be removed, introduced into a glass tube, and transferred to thermal desorption instrument. After desorption and cryofocusing within a cooled programmed temperature vaporization (PTV) injector, the volatiles were transferred onto the analytical GC column. Comparison of SPME and the aforementioned stir bar sorptive extraction (SBSE) technique using identical phases for both techniques exhibited striking differences in the recoveries, which has been attributed to ca. 100 times higher phase ratio in SBSE than in SPME. A comprehensive treatment of SBSE, discussion of the principle, the extraction procedure, and numerous applications was recently been published by David and Sandra (2007).

A further approach for sorptive enrichment of volatiles from the HS of aqueous or solid samples has been described by Tienpont et al. (2000), referred to as HS sorptive extraction (HSSE). This technique implies the sorption of volatiles into PDMS that is chemically bound on the surface of a glass rod support. The device consists of a ca. 5 cm length glass rod of 2 mm diameter and at the last centimeter of 1 mm diameter. This last part is covered with PDMS chemically bound to the glass surface. HS bars with 30, 50, and 100 mg PDMS are commercially available from Gerstel GmbH, Mülheim, Germany. After thermal conditioning at 300°C for 2 h, the glass bar was introduced into the HS of a closed 20 mL HS vial containing the sample to be investigated. After sampling for 45 min, the bar was put into a glass tube for thermal desorption, which was performed with a TDS-2 thermodesorption unit (Gerstel). After desorption and cryofocusing within a PTV injector, the volatiles were transferred onto the analytical GC column. As a result, HSSE exceeded largely the sensitivity attainable with SPME. Several examples referring to the application of HSSE in HS analysis of aromatic and medicinal plants inclusive of details of the sampling procedure were described by Bicchi et al. (2000a).

2.3.2 Chromatographic Separation Techniques

In the course of the last half century, a great number of techniques have been developed and applied to the analysis of essential oils. A part of them has been replaced nowadays by either more effective or easier-to-handle techniques, while other methods maintained their significance and have been permanently improved. Before going into detail, the analytical facilities in the sixties of the last century should be considered briefly. The methods available for the analysis of essential oils have been at that time (Table 2.1) thin-layer chromatography (TLC), various types of liquid column chromatography (LC), and already gas–liquid chromatography (GC). In addition, several spectroscopic techniques such as UV and IR spectroscopy, MS, and ^{1}H-NMR spectroscopy have been available. In the following years, several additional techniques were developed and applied to essential oils analysis, including high-performance liquid chromatography (HPLC); different kinds of countercurrent chromatography (CCC); supercritical fluid chromatography (SFC), including multidimensional coupling techniques, C-13 NMR, near IR (NIR), and Raman spectroscopy; and a multitude of so-called hyphenated techniques, which means online couplings of chromatographic separation devices to spectrometers, yielding valuable structural information of the individual separated components that made their identification feasible.

TABLE 2.1
Techniques Applied to the Analysis of Essential Oils

Chromatographic Techniques Including Two- and Multidimensional Techniques	Spectroscopic and Spectrometric Techniques	Hyphenated Techniques
TLC	UV	GC-MS
GC	IR	GC-UV
LC	MS	HPLC-GC
HPLC	^{1}H-NMR	SFE-GC
CCC	^{13}C-NMR	GC-FTIR
SFC	NIR	GC-AES
	Raman	HPLC-MS
		SFC-GC
		GC-FTIR-MS
		GC-IRMS
		HPLC-NMR

2.3.2.1 Thin-Layer Chromatography

TLC was one of the first chromatographic techniques and has been used for many years for the analysis of essential oils. This method provided valuable information compared to simple measurements of chemical and physical values and has therefore been adopted as a standard laboratory method for characterization of essential oils in numerous pharmacopoeias. Fundamentals of TLC have been described by Geiss (1987) and in a comprehensive handbook by Stahl (1969b), in which numerous applications and examples on investigations of secondary plant metabolites inclusive of essential oils are given. More recently, the third edition of the handbook of TLC from Shema and Fried (2003) appeared. Further approaches in TLC have been the development of high-performance TLC (Kaiser, 1976) and the application of forced flow techniques such as overpressured layer chromatography and rotation planar chromatography described by Tyihák et al. (1979) and Nyiredy (2003).

In spite of its indisputable simplicity and rapidity, this technique is now largely obsolete for analyzing such complex mixtures like essential oils, due to its low resolution. However, for the rapid investigation of the essential oil pattern of chemical races or the differentiation of individual plant species, this method can still be successfully applied (Gaedcke and Steinhoff, 2000). In addition, silver nitrate and silver perchlorate impregnated layers have been used for the separation of olefinic compounds, especially sesquiterpene hydrocarbons (Prasad et al., 1947), and more recently for the isolation of individual sesquiterpenes (Saritas, 2000).

2.3.2.2 GC

However, the separation capability of GC exceeded all the other separation techniques, even if only packed columns have been used. The exiting evolution of this technique in the past can be impressively demonstrated with four examples of the gas chromatographic separation of the essential oil from rue (Kubeczka, 1981a), a medicinal and aromatic plant. This oil was separated by S. Bruno in 1961 into eight constituents and represented one of the first gas chromatographic analyses of that essential oil. Only a few years later in 1964, separation of the same oil has been improved using a Perkin Elmer gas chromatograph equipped with a 2 m packed column and a thermal conductivity detector (TCD) operated under isothermal conditions yielding 20 separated constituents. A further improvement of the separation of the rue oil was obtained after the introduction of temperature programming of the column oven, yielding approximately 80 constituents. The last significant improvements were a result of the development of high-resolution capillary columns and the sensitive flame ionization detector (FID) (Bicchi and Sandra, 1987). By means of a 50 m glass capillary with 0.25 mm ID, the rue oil could be separated into approximately 150 constituents, in 1981. However, the problems associated with the fragility of the glass capillaries and their cumbersome installation lessened the acknowledgment of this column types, despite their outstanding quality. This has changed since flexible fused silica capillaries became commercially available, which are nearly unbreakable in normal usage. In addition, by different cross-linking technologies, the problems associated with wall coating, especially with polar phases, have been overcome, so that all important types of stationary phases used in conventional GC have been commercially available. The most often used stationary phases for the analysis of essential oils have been, and are still today, the polar phases Carbowax® 20M (DB-Wax, Supelcowax-10, HP-20M, Innowax, etc.) and 14% cyanopropylphenyl–86% methyl polysiloxane (DB- 1701, SPB-1701, HP-1701, OV-1701, etc.) and the nonpolar phases PDMS (DB-1, SPB-1, HP-1 and HP-1ms, CPSil-5 CB, OV-1, etc.) and 5% phenyl methyl polysiloxane (DB-5, SPB-5, HP-5, CPSil-8 CB, OV-5, SE-54, etc.). Besides different column diameters of 0.53, 0.32, 0.25, 0.10, and 0.05 mm ID, a variety of film thicknesses can be purchased. Increasing column diameter and film thickness of stationary phase increases the sample capacity at the expense of separation efficiency. However, sample capacity has become important, particularly in trace analysis and with some hyphenated techniques such as GC–Fourier transform IR (GC-FTIR), in which a higher sample capacity is necessary when compared to GC-MS. On the other hand, the application of a narrow bore column with 100 µm ID and a film coating of 0.2 µm have been shown to be highly efficient and theoretical plate numbers of approximately 250,000 were received with a 25 m capillary (Lancas et al., 1988). The most common detector in GC is the FID because of its high sensitivity toward organic compounds. The universal applicable TCD is nowadays used only for fixed-gas detection because of its very low sensitivity as compared to FID, and cannot be used in capillary GC. Nitrogen-containing compounds can be selectively detected with the aid of the selective nitrogen–phosphorus detector and chlorinated compounds by the selective and very sensitive electron-capture detector, which is often used in the analysis of pesticides. Oxygen-containing compounds have been selectively detected with special O-FID analyzer even in very complex samples, which was primarily employed to the analysis of oxygenated compounds in gasoline, utilized as fuel-blending agents (Schneider et al., 1982). The oxygen selectivity of the FID is obtained by two online postcolumn reactions: first, a cracking reaction forming carbon monoxide, which is reduced in a second reactor yielding equimolar quantities of methane, which can be sensitively detected by the FID. Since in total each oxygen atom is converted to one molecule methane, the FID response is proportional to the amount of oxygen in the respective molecule. Application of the O-FID to the analysis of essential oils has been presented by Kubeczka (1991). However, conventional GC using fused silica capillaries with different stationary phases, including chiral phases, and the sensitive FID, is up to now the prime technique for the analysis of essential oils.

2.3.2.2.1 Fast and Ultrafast GC
Due to the demand for faster GC separations in routine work in the field of GC of essential oils, the development of fast and ultrafast GC seems worthy to be mentioned. The various approaches for fast GC have been reviewed in 1999 (Cramers et al., 1999). The most effective way to speed up GC separation without losing separation efficiency is to use shorter columns with narrow inner diameter and thinner coatings, higher carrier gas flow rates, and accelerated temperature ramps. In Figure 2.1, the conventional and fast GC separation of lime oil is shown, indicating virtually the same separation efficiency in the fast GC and a reduction in time from approximately 60 to 13 min (Mondello et al., 2000).

An ultrafast GC separation of the essential oil from lime with an outstanding reduction of time was recently achieved (Mondello et al., 2004) using a 5 m capillary with 50 µm ID and a film thickness of 0.05 µm operated with a high carrier gas velocity of 120 cm/min and an accelerated three-stage temperature program. The analysis of the essential oil was obtained in approximately 90 s, which equates to a speed gain of approximately 33 times in comparison with the conventional GC separation. However, such a separation cannot be performed with conventional GC instruments. In addition, the mass spectrometric identification of the separated components could only be achieved by coupling GC to a time-of-flight mass spectrometer. In Table 2.2, the separation

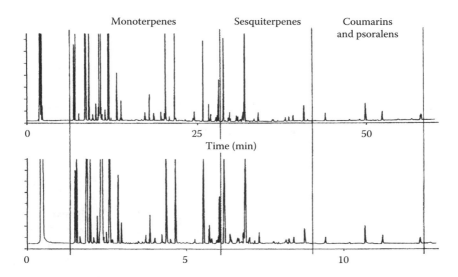

FIGURE 2.1 Comparison of conventional and fast GC separation of lime oil. (From Mondello, L. et al., *LC-GC Eur.*, 13, 495, 2000. With permission.)

TABLE 2.2
Conditions of Conventional, Fast, and Ultrafast GC

	Conventional GC	Fast GC	Ultrafast GC
Column	30 m	10 m	10–15 m
	0.25 mm ID	0.1 mm ID	0.1 mm ID
	0.25 μm film	0.1 μm film	0.1 μm film
Temperature program	50°C–350°C	50°C–350°C	45°C–325°C
	3°C/min	14°C/min	45–200°C/min
Carrier gas	H_2	H_2	H_2
	$u = 36$ cm/s	$u = 57$ cm/s	$u = 120$ cm/s
Sampling frequency	10 Hz	20–50 Hz	50–250 Hz

parameters of conventional, fast, and ultrafast GC separation are given, indicating clearly the relatively low requirements for fast GC, while ultrafast separations can only be realized with modern GC instruments and need a significant higher employment.

2.3.2.2.2 Chiral GC

Besides fast and ultrafast GC separations, one of the most important developments in GC has been the introduction of enantioselective capillary columns in the past with high separation efficiency, so that a great number of chiral substances including many essential oil constituents could be separated and identified. The different approaches of gas chromatographic separation of chiral compounds are briefly summarized in Table 2.3. In the mid-1960s, Gil-Av published results with chiral diamide stationary phases for gas chromatographic separation of chiral compounds, which interacted with the analytes by hydrogen bonding forces (Gil-Av et al., 1965). The ability to separate enantiomers using these phases was therefore limited to substrates with hydrogen bonding donor or acceptor functions.

Diastereomeric association between chiral molecules and chiral transition metal complexes was first described by Schurig (1977). Since hydrogen bonding interaction is not essential for chiral

TABLE 2.3
Different Approaches of Enantioselective GC

1. Chiral diamide stationary phases (Gil-Av et al., 1965)
   Hydrogen bonding interaction
2. Chiral transition metal complexation (Schurig, 1977)
   Complexation gas chromatography
3. Cyclodextrin derivatives (König, 1988 and Schurig, 1988)
   Host–guest interaction, inclusion gas chromatography

recognition in such a system, a number of compounds could be separated, but this method was limited by the nonsufficient thermal stability of the applied metal complexes.

In 1988 König, as well as Schurig, described the use of cyclodextrin derivatives that act enantioselectively by host–guest interaction by partial intrusion of enantiomers into the cyclodextrin cavity. They are cyclic α-(1–4)-bounded glucose oligomers with six-, seven-, or eight-glucose units, which can be prepared by enzymatic degradation of starch with specific cyclodextrin glucosyl transferases from different bacterial strains, yielding α-, β-, and γ-cyclodextrins, and are commercially available. Due to the significant lower reactivity of the 3-hydroxygroups of cyclodextrins, this position can be selectively acylated after alkylation of the two and six positions (Figure 2.2), yielding several nonpolar cyclodextrin derivatives, which are liquid or waxy at room temperature and which proved very useful for gas chromatographic applications.

König and coworkers reported their first results in 1988 with per-O-pentylated and selectively 3-O-acylated-2,6-di-O-pentylated α-, β-, and γ-cyclodextrins, which are highly stable, soluble in nonpolar solvents, and which possess a high enantioselectivity toward many chiral compounds. In the following years, a number of further cyclodextrin derivatives have been synthesized and tested by several groups, allowing the separation of a wide range of chiral compounds, especially due to the improved thermal stability (Table 2.4) (König et al., 1988). With the application of 2,3-pentyl-6-methyl-β- and -γ-cyclodextrin as stationary phases, all monoterpene hydrocarbons commonly occurring in essential oils could be separated (König et al., 1992a). The reason for application of two different columns with complementary properties was that on one column not all enantiomers were satisfactorily resolved. Thus, the simultaneous use of these two columns provided a maximum of information and reliability in peak assignment (König et al., 1992b).

After successful application of enantioselective GC to the analysis of enantiomeric composition of monoterpenoids in many essential oils (e.g., Werkhoff et al., 1993; Bicchi et al., 1995; and references cited therein), the studies have been extended to the sesquiterpene fraction. Standard mixtures of known enantiomeric composition were prepared by isolation of individual enantiomers from numerous essential oils by preparative GC and by preparative enantioselective GC. A gas chromatographic separation of a series of isolated or prepared sesquiterpene hydrocarbon enantiomers, showing the separation of 12 commonly occurring sesquiterpene hydrocarbons on a 2,6-methyl-3-pentyl-β-cyclodextrin capillary column has been presented by König et al. (1995). Further investigations on sesquiterpenes have been published by König et al. (1994). However, due to the complexity

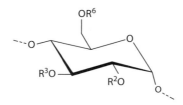

FIGURE 2.2 α-Glucose unit of a cyclodextrin.

TABLE 2.4
Important Cyclodextrin Derivatives

Research Group	Year	Cyclodextrin Derivative
Schurig and Novotny	1988	Per-*O*-methyl-β-CD
König et al.	1988a	Per-*O*-pentyl-(α,β,γ)-CD
König et al.	1988c	3-*O*-acetyl-2,6,-di-*O*-pentyl-(α,β,γ)-CD
König et al.	1989	3-*O*-butyryl-2,6-di-*O*-pentyl-(α,β)-CD
König et al.	1990	6-*O*-methyl-2,3-di-*O*-pentyl-γ-CD
Köng et al.	1990	2,6-Di-*O*-methyl-3-*O*-pentyl-(β,γ)-CD
Dietrich et al.	1992a	2,3-Di-*O*-acetyl-6-*O*-*tert*-butyl-dimethysilyl-β-CD
Dietrich et al.	1992b	2,3-Di-*O*-methyl-6-*O*-*tert*-butyl-dimethylsilyl-(β,γ)-CD
Bicchi et al.	1996	2,3-Di-*O*-ethyl-6-*O*-*tert*-butyl-dimethylsilyl-(β,γ)-CD
Takahisa and Engel	2005a	2,3-Di-*O*-methoxymethyl-6-*O*-*tert*-butyl-dimethylsilyl-β-CD
Takahisa and Engel	2005b	2,3-Di-*O*-methoxymethyl-6-*O*-*tert*-butyl-dimethylsilyl-γ-CD

of the sesquiterpene pattern in many essential oils, it is often impossible to perform directly an enantioselective analysis by coinjection with standard samples on a capillary column with a chiral stationary phase alone. Therefore, in many cases 2D GC had to be performed.

2.3.2.2.3 Two-Dimensional GC

After preseparation of the oil on a nonchiral stationary phase, the peaks of interest have to be transferred to a second capillary column coated with a chiral phase, a technique usually referred to as "heart cutting." In the simplest case, two GC capillaries with different selectivities are serially connected, and the portion of unresolved components from the effluent of the first column is directed into a second column, for example, a capillary with a chiral coating. The basic arrangement used in 2D GC (GC-GC) is shown in Figure 2.3. By means of a valve, the individual fractions of interest eluting from the first column are directed to the second, chiral column, while the rest of the sample may be discarded. With this heart-cutting technique, many separations of chiral oil

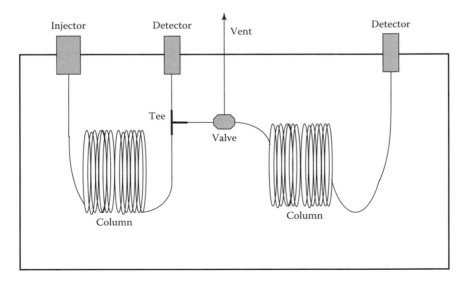

FIGURE 2.3 Basic arrangement used in 2D GC.

constituents have been performed in the past. As an example, the investigation of the chiral sesquiterpene hydrocarbon germacrene D shall be mentioned (Kubeczka, 1996), which was found to be a main constituent of the essential oil from the flowering herb from *Solidago canadensis*. The enantioselective investigation of the germacrene-D fraction from a GC run using a nonchiral DB-Wax capillary transferred to a 2,6-methyl-3-pentyl-β-cyclodextrin capillary exhibited the presence of both enantiomers. This is worthy to be mentioned, since in most of other germacrene D containing higher plants nearly exclusively the (–)-enantiomer can be found.

The previously mentioned 2D GC design, however, in which a valve is used to direct the portion of desired effluent from the first into the second column, has obviously several shortcomings. The sample comes into contact with the metal surface of the valve body, the pressure drop of both connected columns may be significant, and the use of only one-column oven does not permit to adjust the temperature for both columns properly. Therefore, one of the best approaches to overcome these limitations has been realized by a commercially available two-column oven instrument using a Deans-type pressure balancing interface between the two columns called a "live-T connection" (Figure 2.4) providing considerable flexibility (Hener, 1990). By means of that instrument, the enantiomeric composition of several essential oils has been investigated very successfully. As an example, the investigation of the essential oil from *Lavandula angustifolia* shall be mentioned (Kreis and Mosandl, 1992) showing the simultaneous stereoanalysis of a mixture of chiral compounds, which can be found in lavender oils, using the column combination Carbowax 20M as the precolumn and 2,3-di-*O*-acetyl-6-*O*-*tert*-butyldimethylsilyl-β-cyclodextrin as the main column. All the unresolved enantiomeric pairs from the precolumn could be well separated after transferring them to the chiral main column in a single run. As a result, it was found that most of the characteristic and genuine chiral constituents of lavender oil exhibit a high enantiomeric purity.

A different and inexpensive approach for transferring individual GC peaks onto a second column has been presented by Kubeczka (1997), using an SPME device. The highly diluted organic vapor of a fraction eluting from a GC capillary in the carrier gas flow has been absorbed on a coated SPME fiber and introduced onto a second capillary. As could be demonstrated, no modification of the gas chromatograph had to be performed to realize that approach. The eluting fractions were sampled after shutting the valves of the air, of hydrogen and the makeup gas if applied. In

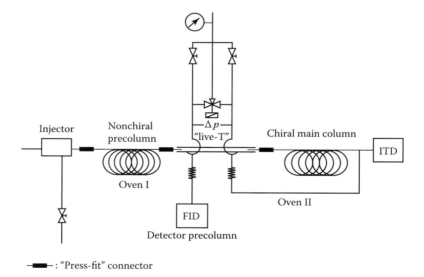

FIGURE 2.4 Scheme of enantioselective multidimensional GC with "live-T" column switching. (From Hener, U., Chirale Aromastoffe—Beiträge zur Struktur, Wirkung und Analytik, Dissertation, Goethe-University of Frankfurt/Main, Frankfurt, Germany, 1990. With permission.)

History and Sources of Essential Oil Research

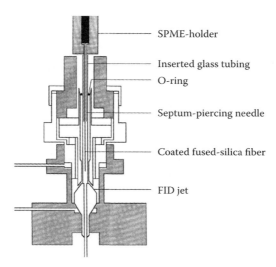

FIGURE 2.5 Cross section of an FID of an HP 5890 gas chromatograph with an inserted SPME fiber. (From Kubeczka, K.-H. *Essential Oil Symposium Proceedings*, 1997b, p. 145. With permission.)

order to minimize the volume of the detector to avoid dilution of the eluting fraction and to direct the gas flow to the fiber surface, a capillary glass tubing of 1.5 mm ID was inserted into the FID and fixed and tightened by an O-ring (Figure 2.5). At the beginning of peak elution, controlled only by time, a 100 μm PDMS fiber was introduced into the mounted glass capillary tubing and withdrawn at the end of peak elution. Afterward, the fiber within the needle was introduced into the injector of a second capillary column with a chiral stationary phase. Two examples concerning the investigation of bergamot oil have been shown. At first, the analysis of an authentic sample of bergamot oil, containing chiral linalool, and the respective chiral acetate is carried out. Both components were cut separately and transferred to an enantioselective cyclodextrin Lipodex® E capillary. The chromatograms clearly have shown that the authentic bergamot oil contains nearly exclusively the (−)-enantiomers of linalool and linalyl acetate, while the respective (+)-enantiomers could only be detected as traces. In contrast to the authentic sample, a commercial sample of bergamot oil, which was analyzed under the same conditions, exhibited the presence of significant amounts of both enantiomers of linalool and linalyl acetate indicating a falsification by admixing the respective racemic alcohol and ester.

2.3.2.2.4 Comprehensive Multidimensional GC

One of the most powerful separation techniques that has been recently applied to the investigation of essential oils is the so-called comprehensive multidimensional GC (GC × GC). This technique is a true multidimensional GC (MDGC) since it combines two directly coupled columns and importantly is able to subject the entire sample to simultaneous two-column separation. Using that technique, the need to select heart cuts, as used in conventional MDGC, is no longer required. Since components now are retained in two different columns, the net capacity is the product of the capacities of the two applied columns increasing considerably the resolution of the total system. Details regarding that technique will be given in Chapter 7.

2.3.2.3 Liquid Column Chromatography

The different types of LC have been mostly used in preparative or semipreparative scale for pre-separation of essential oils or for isolation of individual oil constituents for structure elucidation with spectroscopic methods and were rarely used at that time as an analytical separation tool alone, because GC plays a central role in the study of essential oils.

2.3.2.3.1 Preseparation of Essential Oils

A different approach besides 2D GC, which has often been used in the past to overcome peak overlapping in a single GC run of an essential oil, has been preseparation of the oil with LC. The most common method of fractionation is the separation of hydrocarbons from the oxygenated terpenoids according to Miller and Kirchner (1952), using silica gel as an adsorbent. After elution of the nonpolar components from the column with pentane or hexane, the more polar oxygen-containing constituents are eluted in order of increasing polarity after applying more and more polar eluents.

A very simple and standardized fractionation in terms of speed and simplicity has been published by Kubeczka (1973) using dry-column chromatography. The procedure, which has been proved useful in numerous experiments for prefractionation of an essential oil, allows a preseparation into five fractions of increasing polarity. The preseparation of an essential oil into oxygenated constituents, monoterpene hydrocarbons, and sesquiterpene hydrocarbons, which is—depending on the oil composition—sometimes of higher practical use, can be performed successfully using reversed-phase RP-18 HPLC (Schwanbeck et al., 1982). The HPLC was operated on a semipreparative scale by stepwise elution with methanol–water 82.5:17.5 (solvent A) and pure methanol (solvent B). The elution order of the investigated oil was according to decreasing polarity of the components and within the group of hydrocarbons to increasing molecular weight. Fraction 1 contained all oxygenated mono- and sesquiterpenoids, fraction 2 the monoterpene hydrocarbons, and fraction 3—eluted with pure methanol—the sesquiterpene hydrocarbons. A further alternative to the mentioned separation techniques is flash chromatography, initially developed by Still et al. (1978), which has often been used as a rapid form of preparative LC based on a gas- or air pressure–driven short-column chromatography. This technique, optimized for rapid separation of quantities typically in the range of 0.5–2.0 g, uses dry-packed silica gel in an appropriate column. The separation of the sample generally takes only 5–10 min and can be performed with inexpensive laboratory equipment. However, impurities and active sites on dried silica gel were found to be responsible for isomerization of a number of oil constituents. After deactivation of the dried silica gel by adding 5% water, isomerization processes could be avoided (Scheffer et al., 1976). A different approach using HPLC on silica gel and isocratic elution with a ternary solvent system for the separation of essential oils has been published by Chamblee et al. (1985). In contrast to the aforementioned commonly used offline pretreatment of a sample, the coupling of two or more chromatographic systems in an online mode offers advantages of ease of automation and usually of a shorter analysis time.

2.3.2.3.2 High-Performance Liquid Column Chromatography

The good separations obtained by GC have delayed the application of HPLC to the analysis of essential oils; however, HPLC analysis offers some advantages, if GC analysis of thermolabile compounds is difficult to achieve. Restricting factors for application of HPLC for analyses of terpenoids are the limitations inherent in the commonly available detectors and the relatively small range of k' values of liquid chromatographic systems. Since temperature is an important factor that controls k' values, separation of terpene hydrocarbons was performed at −15°C using a silica gel column and *n*-pentane as a mobile phase. Monitoring has been achieved with UV detection at 220 nm. Under these conditions, mixtures of commonly occurring mono- and sesquiterpene hydrocarbons could be well separated (Schwanbeck et al., 1979; Kubeczka, 1981b). However, the silica gel had to be deactivated by adding 4.8% water prior to separation to avoid irreversible adsorption or alteration of the sample. The investigation of different essential oils by HPLC already has been described in the seventies of the last century (e.g., Komae and Hayashi, 1975; Ross, 1976; Wulf et al., 1978; McKone, 1979; Scott and Kucera, 1979). In the last publication, the authors have used a rather long microbore packed column, which had several hundred thousand theoretical plates. Besides relatively expensive equipment, the HPLC chromatogram of an essential oil, separated on such a column, could only be obtained at the expense of long analysis time. The mentioned separation needed about 20 h and may be only of little value in practical applications.

More recent papers with regard to HPLC separation of essential oils were published, for example, by Debrunner et al. (1995), Bos et al. (1996), and Frérot and Decorzant (2004), and applications using silver ion–impregnated sorbents have been presented by Pettei et al. (1977), Morita et al. (1983), Friedel and Matusch (1987), and van Beek et al. (1994). The literature on the use and theory of silver complexation chromatography has been reviewed by van Beek et al. (1995). HPLC has also been used to separate thermally labile terpenoids at low temperature by Beyer et al. (1986), showing the temperature dependence of the separation efficiency. The investigation of an essential oil fraction from *Cistus ladanifer* using RP-18 reversed-phase HPLC at ambient temperature and an acetonitrile–water gradient was published by Strack et al. (1980). Comparison of the obtained HPLC chromatogram with the respective GC run exhibits a relatively good HPLC separation in the range of sesqui- and diterpenes, while the monoterpenes exhibited, as expected, a significant better resolution by GC. The enantiomeric separation of sesquiterpenes by HPLC with a chiral stationary phase has recently been shown by Nishii et al. (1997), using a Chiralcel® OD column.

2.3.2.4 Supercritical Fluid Chromatography

Supercritical fluids are highly compressed gases above their critical temperature and critical pressure point, representing a hybrid state between a liquid and a gas, which have physical properties intermediate between liquid and gas phases. The diffusion coefficient of a fluid is about two orders of magnitude larger and the viscosity is two orders of magnitude lower than the corresponding properties of a liquid. On the other hand, a supercritical fluid has a significant higher density than a gas. The commonly used carbon dioxide as a mobile phase, however, exhibits a low polarity (comparable to pentane or hexane), limiting the solubility of polar compounds, a problem that has been solved by adding small amounts of polar solvents, for example, methanol or ethanol, to increase mobile-phase polarity, thus permitting separations of more polar compounds (Chester and Innis, 1986). A further strength of SFC lies in the variety of detection systems that can be applied. The intermediate features of SFC between GC and LC can be profitable when used in a variety of detection systems, which can be classified in *LC-* and *GC-like* detectors. In the first case, measurement takes place directly in the supercritical medium or in the liquid phase, whereas GC-like detection proceeds after a decompression stage.

Capillary SFC using carbon dioxide as mobile phase and a FID as detector has been applied to the analysis of several essential oils and seemed to give more reliable quantification than GC, especially for oxygenated compounds. However, the separation efficiency of GC for monoterpene hydrocarbons was, as expected, better than that of SFC. Manninen et al. (1990) published a comparison of a capillary GC versus a chromatogram obtained by capillary SFC from a linalool–methyl chavicol basil oil chemotype exhibiting a fairly good separation by SFC.

2.3.2.5 Countercurrent Chromatography

CCC is according to Conway (1989) a form of liquid–liquid partition chromatography, in which centrifugal or gravitational forces are employed to maintain one liquid phase in a coil or train of chambers stationary, while a stream of a second, immiscible phase is passed through the system in contact with the stationary liquid phase. Retention of the individual components of the sample to be analyzed depends only on their partition coefficients and the volume ratio of the two applied liquid phases. Since there is no porous support, adsorption and catalytic effects encountered with solid supports are avoided.

2.3.2.5.1 Droplet Countercurrent Chromatography

One form of CCC, which has been sporadically applied to separate essential oils into fractions or in the ideal case into individual pure components, is droplet countercurrent chromatography (DCCC). The device, which has been developed by Tanimura et al. (1970), consists of 300–600 glass tubes, which are connected to each other in series with Teflon tubing and filled with a stationary liquid. Separation is achieved by passing droplets of the mobile phase through the columns, thus distributing mixture components at different ratios leading to their separation. With the development of a water-free solvent system, separation of essential oils could be achieved (Becker et al., 1981, 1982). Along with the separation of essential oils, the method allows the concentration of minor components, since relatively large samples can be separated in one analytical run (Kubeczka, 1985).

2.3.2.5.2 Rotation Locular Countercurrent Chromatography

The rotation locular countercurrent chromatography (RLCC) apparatus (Rikakikai Co., Tokyo, Japan) consists of 16 concentrically arranged and serially connected glass tubes. These tubes are divided by Teflon disks with a small hole in the center, thus creating small compartments or locules. After filling the tubes with the stationary liquid, the tubes are inclined to a 30° angle from horizontal. In the ascending mode, the lighter mobile phase is applied to the bottom of the first tube by a constant flow pump, displacing the stationary phase as its volume attains the level of the hole in the disk. The mobile phase passes through this hole and enters into the next compartment, where the process continues until the mobile phase emerges from the uppermost locule. Finally, the two phases fill approximately half of each compartment. The dissolved essential oil subsequently introduced is subjected to a multistage partitioning process that leads to separation of the individual components. Whereas gravity contributes to the phase separation, rotation of the column assembly (60–80 rpm) produces circular stirring of the two liquids to promote partition. If the descending mode is selected for separation, the heavier mobile phase is applied at the top of each column by switching a valve. An overview on applications of RLCC in natural products isolation inclusive of a detailed description of the device and the selection of appropriate solvent systems has been presented by Snyder et al. (1984).

Comparing RLCC to the aforementioned DCCC, one can particularly stress the superior flexibility of RLCC. While DCCC requires under all circumstances a two-phase system able to form droplets in the stationary phase, the choice of solvent systems with RLCC is nearly free. So the limitations of DCCC, when analyzing lipophilic samples, do not apply to RLCC. The separation of a mixture of terpenes has been presented by Kubeczka (1985). A different method, the high-speed centrifugal CCC developed by Ito and coworkers in the mid-1960s (Ito et al., 1966), has been applied to separate a variety of nonvolatile natural compounds; however, separation of volatiles has, strange to say, until now not seriously been evaluated.

2.3.3 Hyphenated Techniques

2.3.3.1 Gas Chromatography-Mass Spectrometry

The advantage of online coupling of a chromatographic device to a spectrometer is that complex mixtures can be analyzed in detail by spectral interpretation of the separated individual components. The coupling of a gas chromatograph with a mass spectrometer is the most often used and a well-established technique for the analysis of essential oils, due to the development of easy-to-handle powerful systems concerning sensitivity, data acquisition and processing, and above all their relatively low cost. The very first application of a GC-MS coupling for the identification of essential oil constituents using a capillary column was already published by Buttery et al. (1963). In those times, mass spectra have been traced on UV recording paper with a five-element galvanometer, and their evaluation was a considerable cumbersome task.

This has changed after the introduction of computerized mass digitizers yielding the mass numbers and the relative mass intensities. The different kinds of GC-MS couplings available at the end of the seventies of the last century have been described in detail by ten Noever de Brauw (1979). In addition, different types of mass spectrometers have been applied in GC-MS investigations such as magnetic sector instruments, quadrupole mass spectrometers, ion-trap analyzers (e.g., ion-trap detector), and time-of-flight mass spectrometers, which are the fastest MS analyzers and therefore used for very fast GC-MS systems (e.g., in comprehensive multidimensional GC-MS). Surprisingly, a time-of-flight mass spectrometer was used in the very first description of a GC-MS investigation of an essential oil mentioned before. From the listed spectrometers, the magnetic sector and quadrupole instruments can also be used for selective ion monitoring, to improve sensitivity for the analysis of target compounds and for discrimination of overlapping GC peaks.

The great majority of today's GC-MS applications utilize 1D capillary GC with quadrupole MS detection and electron ionization. Nevertheless, there are substantial numbers of applications using different types of mass spectrometers and ionization techniques. The proliferation of GC-MS applications is also a result of commercially available easy-to-handle dedicated mass spectral libraries (e.g., NIST/EPA/NIH 2005; WILEY Registry 2006; MassFinder 2007; and diverse printed versions such as Jennings and Shibamoto, 1980; Joulain and König, 1998; Adams, 1989, 1995, 2007 inclusive of retention indices) providing identification of the separated compounds. However, this type of identification has the potential of producing some unreliable results, if no additional information is used, since some compounds, for example, the sesquiterpene hydrocarbons α-cuprenene and β-himachalene, exhibit identical fragmentation pattern and only very small differences of their retention index values. This example demonstrates impressively that even a good library match and the additional use of retention data may lead in some cases to questionable results, and therefore require additional analytical data, for example, from NMR measurements.

2.3.3.1.1 GC-Chemical Ionization-MS and GC-Tandem MS

Although GC-electron impact (EI)-MS is a very useful tool for the analysis of essential oils, this technique can sometimes be not selective enough and requires more sophisticated techniques such as GC-chemical ionization-MS (GC-CI-MS) and GC-tandem MS (GC-MS-MS). The application of CI-MS using different reactant gases is particularly useful, since many terpene alcohols and esters fail to show a molecular ion. The use of OH^- as a reactant ion in negative CI-MS appeared to be an ideal solution to this problem. This technique yielded highly stable quasi-molecular ions M–H, which are often the only ions in the obtained spectra of the aforementioned compounds. As an example, the EI and CI spectra of isobornyl isovalerate—a constituent of valerian oil—shall be quoted (Bos et al., 1982). The respective EI mass spectrum shows only a very small molecular ion at 238. Therefore, the chemical ionization spectra of isobornyl acetate were performed with isobutene as a reactant gas a $[C_{10}H_{17}]^+$ cation and in the negative CI mode with OH^- as a reactant gas two signals with the masses 101, the isovalerate anion, and 237 the quasi-molecular ion $[M-H]^-$. Considering all these obtained data, the correct structure of the oil constituent could be deduced. The application of isobutane and ammonia as reactant gases has been presented by Schultze et al. (1992), who investigated sesquiterpene hydrocarbons by GC-CI-MS. Fundamental aspects of chemical ionization MS have been reviewed by Bruins (1987), discussing the different reactant gases applied in positive and negative ion chemical ionization and their applications in essential oil analysis.

The utilization of GC-MS-MS to the analysis of a complex mixture will be shown in Figure 2.6. In the investigated vetiver oil (Cazaussus et al., 1988), one constituent, the norsesquiterpene ketone khusimone, has been identified by using GC-MS-MS in the collision-activated dissociation mode. The molecular ion at m/z 204 exhibited a lot of daughter ions, but only one of them gave a daughter ion at m/z 108, a fragment rarely occurring in sesquiterpene derivatives so that the presence of khusimone could be undoubtedly identified.

2.3.3.2 High-Resolution GC-FTIR Spectroscopy

A further hyphenated technique, providing valuable analytical information, is the online coupling of a gas chromatograph with a FTIR spectrometer. The capability of IR spectroscopy to provide discrimination between isomers makes the coupling of a gas chromatograph to an FTIR spectrometer suited as a complementary method to GC/MS for the analysis of complex mixtures like essential oils. The GC/FTIR device consists basically of a capillary gas chromatograph and an FTIR spectrometer including a dedicated computer and ancillary equipment. As each GC peak elutes from the GC column, it enters a heated IR measuring cell, the so-called light pipe, usually

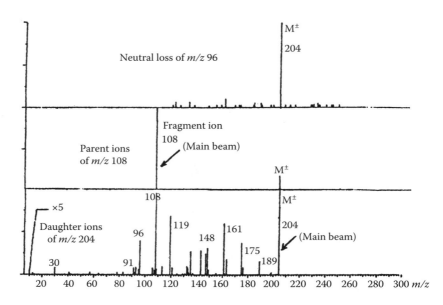

FIGURE 2.6 GC-EIMS-MS of khusimone of vetiver oil. (From Cazaussus, A. et al., *Chromatographia*, 25, 865, 1988. With permission.)

a gold-plated glass tube with IR transparent windows. There, the spectrum is measured as an interferogram from which the familiar absorbance spectrum can be calculated by computerized Fourier transformation. After passing the light pipe, the effluent is directed back into the FID of the gas chromatograph. More detailed information on the experimental setup was given by Herres et al. (1986) and Herres (1987).

In the latter publication, for example, the vapor-phase IR spectra of all the four isomers of pulegol and dihydrocarveol are shown, which have been extracted from a GC/FTIR run. These examples convincingly demonstrate the capability of distinguishing geometrical isomers with the aid of vapor-phase IR spectra, which cannot be achieved by their mass spectra. A broad application of GC-FTIR in the analysis of essential oils, however, is limited by the lack of sufficient vapor-phase spectra of uncommon compounds, which are needed for reference use, since the spectra of isolated molecules in the vapor phase can be significantly different from the corresponding condensed-phase spectra.

A different approach has been published by Reedy et al. in 1985, using a cryogenically freezing of the GC effluent admixed with an inert gas (usually argon) onto a rotating disk maintained at liquid He temperature to form a solid matrix trace. After the separation, reflection absorption spectra can be obtained from the deposited solid trace. A further technique published by Bourne et al. (1990) is the subambient trapping, whereby the GC effluent is cryogenically frozen onto a moving IR transparent window of zinc selenide (ZnSe). An advantage of the latter technique is that the unlike larger libraries of conventional IR spectra can be searched in contrast to the limited number of vapor-phase spectra and those obtained by matrix isolation. A further advantage of both cryogenic techniques is the significant higher sensitivity, which exceeds the detection limits of a light pipe instrument by approximately two orders of magnitude.

Comparing GC/FTIR and GC/MS, advantages and limitations of each technique become visible. The strength of IR lies—as discussed before—in distinguishing isomers, whereas identification of homologues can only be performed successfully by MS. The logical and most sophisticated way to overcome these limitations has been the development of a combined GC/FTIR/MS instrument, whereby simultaneously IR and mass spectra can be obtained.

2.3.3.3 GC-UV Spectroscopy

The instrumental coupling of gas chromatograph with a rapid scanning UV spectrometer has been presented by Kubeczka et al. (1989). In this study, a UV-VIS diode-array spectrometer (Zeiss, Oberkochen, FRG) with an array of 512 diodes was used, which provided continuous monitoring in the range of 200–620 nm. By interfacing the spectrometer via fiber optics to a heated flow cell, which was connected by short heated capillaries to the GC column effluent, interferences of chromatographic resolution could be minimized. With the aid of this device, several terpene hydrocarbons have been investigated. In addition to displaying individual UV spectra, the available software rendered the analyst to define and to display individual window traces, 3D plots, and contour plots, which are valuable tools for discovering and deconvoluting gas chromatographic unresolved peaks.

2.3.3.4 Gas Chromatography-Atomic Emission Spectroscopy

A device for the coupling of capillary GC with atomic emission spectroscopy (GC-AES) has been presented by Wylie and Quimby (1989). By means of this coupling, 23 elements of a compound including all elements of organic substances separated by GC could be selectively detected providing the analyst not only with valuable information on the elemental composition of the individual components of a mixture but also with the percentages of the elemental composition. The device incorporates a microwave-induced helium plasma at the outlet of the column coupled to an optical emission spectrometer. From the 15 most commonly occurring elements in organic compounds, up to 8 could be detected and measured simultaneously, for example, C, O, N, and S, which are of importance with respect to the analysis of essential oils. The examples given in the literature (e.g., Wylie and Quimby, 1989; Bicchi et al., 1992; David and Sandra, 1992; Jirovetz et al., 1992; Schultze, 1993) indicate that the GC-AES coupling can provide the analyst with additional valuable information, which are to some extent complementary to the date obtained by GC-MS and GC-FTIR, making the respective library searches more reliable and more certain.

However, the combined techniques GC-UV and GC-AES have not gained much importance in the field of essential oil research, since UV spectra offer only low information and the coupling of a GC-AES, yielding the exact elemental composition of a component, can to some extent be obtained by precise mass measurement. Nevertheless, the online coupling GC-AES is still today efficiently used in environmental investigations.

2.3.3.5 Gas Chromatography-Isotope Ratio Mass Spectrometry

In addition to enantioselective capillary GC, the online coupling of GC with isotope-ratio MS (GC-IRMS) is an important technique in authentication of food flavors and essential oil constituents. The online combustion of effluents from capillary gas chromatographic separations to determine the isotopic compositions of individual components from complex mixtures was demonstrated by Matthews and Hayes (1978). On the basis of this work, the online interfacing of capillary GC with IRMS was later improved. With the commercially available GC-combustion IRMS device, measurements of the ratios of the stable isotopes $^{13}C/^{12}C$ have been accessible and respective investigations have been reported in several papers (e.g., Bernreuther et al., 1990; Carle et al., 1990; Braunsdorf et al., 1992, 1993; Frank et al., 1995; Mosandl and Juchelka, 1997). A further improvement was the development of the GC-pyrolysis-IRMS (GC-P-IRMS) making measurements of $^{18}O/^{16}O$ ratios and later $^{2}H/^{1}H$ ratios feasible (Juchelka et al., 1998; Ruff et al., 2000; Hör et al., 2001; Mosandl, 2004). Thus, the GC-P-IRMS device (Figure 2.7) appears today as one of the most sophisticated instruments for the appraisal of the genuineness of natural mixtures.

2.3.3.6 High-Performance Liquid Chromatography-Gas Chromatography

The online coupling of an HPLC device to a capillary gas chromatograph offers a number of advantages, above all higher column chromatographic efficiency, simple and rapid method development, simple cleanup of samples from complex matrices, and effective enrichment of the components

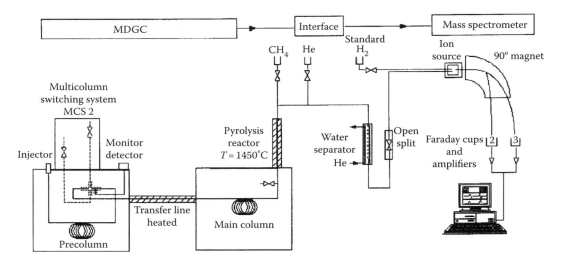

FIGURE 2.7 Scheme of an MDGC-C/P-IRMS device. (From Sewenig, S. et al., *J. Agric. Food Chem.*, 53, 838, 2005. With permission.)

of interest; additionally, the entire analytical procedure can easily be automated, thus increasing accuracy and reproducibility. The commercially available HPLC-GC coupling consists of an HPLC device that is connected with a capillary gas chromatograph via an interface allowing the transfer of HPLC fractions. Two different types of interfaces have been often used. The on-column interface is a modification of the on-column injector for GC; it is particularly suited for the transfer of fairly small fraction containing volatile constituents (Dugo et al., 1994; Mondello et al., 1994a,b, 1995). The second interface uses a sample loop and allows to transfer large sample volumes (up to 1 mL) containing components with limited volatilities. Figure 2.8 gives a schematic view of such an LC-GC

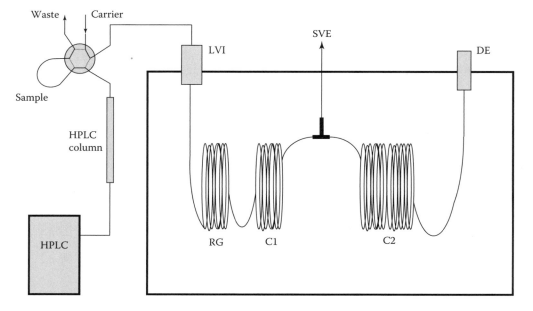

FIGURE 2.8 Basic arrangement of an HPLC-GC device with a sample loop interface. RG, retention gap; C1, retaining column; C2, analytical column; LVI, large volume injector; and SVE, solvent vapor exit.

instrument. In the shown position of the six-port valve, the desired fraction of the HPLC effluent is stored in the sample loop, while the carrier gas is passed through the GC columns. After switching the valve, the content of the sample loop is driven by the carrier gas into the large volume injector and vaporized and enters the precolumns, where the sample components are retained and most of the solvent vapor can be removed through the solvent vapor exit. After closing this valve and increasing the GC-oven temperature, the sample components are volatilized and separated in the main column reaching the detector. The main drawback of this technique, however, may be the loss of highly volatile compounds that are vented together with the solvent. As an example of an HPLC-GC investigation, the preseparation of lemon oil with gradient elution into four fractions is quoted (Munari et al., 1990). The respective gas chromatograms of the individual fractions exhibit good separation into hydrocarbons, esters, carbonyls, and alcohols, facilitating gas chromatographic separation and identification. Due to automation of all analytical steps involved, the manual operations are significantly reduced, and very good reproducibility was obtained. In three excellent review articles, the different kinds of HPLC-GC couplings are discussed in detail, describing their advantages and limitations with numerous references cited therein (Mondello et al., 1996, 1999; Dugo et al., 2003).

2.3.3.7 HPLC-MS, HPLC-NMR Spectroscopy

The online couplings of HPLC with MS and NMR spectroscopy are further important techniques combining high-performance separation with structurally informative spectroscopic techniques, but they are mainly applied to nonvolatile mixtures and shall not be discussed in more detail here, although they are very useful for investigating plant extracts.

Some details concerning the different ionization techniques used in HPLC-MS have been presented among other things by Dugo et al. (2005).

2.3.3.8 Supercritical Fluid Extraction–Gas Chromatography

Although SFE is not a chromatographic technique, separation of mixtures can be obtained during the extraction process by varying the physical properties such as temperature and pressure to obtain fractions of different composition. Detailed reviews on the physical background of SFE and its application to natural products analysis inclusive of numerous applications have been published by Modey et al. (1995) and more recently by Pourmortazavi and Hajimirsadeghi (2007). The different types of couplings (offline and online) have been presented by several authors. Houben et al. (1990) described an online coupling of SFE with capillary GC using a programmed temperature vaporizer as an interface. Similar approaches have been used by Blanch et al. (1994) in their investigations of rosemary leaves and by Ibanez et al. (1997) studying Spanish raspberries. In both the last two papers, an offline procedure was applied. A different device has been used by Hartonen et al. (1992) in a study of the essential oil of *Thymus vulgaris* using a cooled stainless steel capillary for trapping the volatiles connected via a six-port valve to the extraction vessel and the GC column. After sampling of the volatiles within the trap, they have been quickly vaporized and flushed into the GC column by switching the valve. The recoveries of thyme components by SFE-GC were compared with those obtained from hydrodistilled thyme oil by GC exhibiting a good agreement. The SFE-GC analyses of several flavor and fragrance compounds of natural products by transferring the extracted compounds from a small SFE cell directly into a GC capillary has already been presented by Hawthorne et al. (1988). By inserting the extraction cell outlet restrictor (a 20 μm ID capillary) into the GC column through a standard on-column injection port, the volatiles were transferred and focused within the column at 40°C, followed by rapid heating to 70°C (30°C/min) and successive usual temperature programming. The suitability of that approach has been demonstrated with a variety of samples including rosemary, thyme, cinnamon, spruce needles, orange peel, and cedar wood. In a review article from Greibrokk, published in 1995, numerous applications of SFE connected online with GC and other techniques, the different instruments, and interfaces have been discussed, including the main parameters responsible for the quality of the obtained analytical results. In addition, the instrumental setups for SFE-LC and SFE-SFC couplings are given.

2.3.3.9 Supercritical Fluid Chromatography-Gas Chromatography

Online coupling of SFC with GC has sporadically been used for the investigation of volatiles from aromatic herbs and spices. The requirements for instrumentation regarding the pumps, the restrictors, and the detectors are similar to those of SFE-GC. Additional parts of the device are the separation column and the injector, to introduce the sample into the mobile phase and successively into the column. The most common injector type in SFC is the high-pressure valve injector, similar to those used in HPLC. With this valve, the sample is loaded at ambient pressure into a sample loop of defined size and can be swept into the column after switching the valve to the injection position. The separation columns used in SFC may be either packed or open tubular columns with their respective advantages and disadvantages. The latter mentioned open tubular columns for SFC can be compared with the respective GC columns; however, they must have smaller internal diameter. With regard to the detectors used in SFC, the FID is the most common applied detector, presuming that no organic modifiers have been admixed to the mobile phase. In that case, for example, a UV detector with a high-pressure flow cell has to be taken into consideration.

In a paper presented by Yamauchi and Saito (1990), cold-pressed lemon-peel oil has been separated by semipreparative SFC into three fractions (hydrocarbons, aldehydes and alcohols), and esters together with other oil constituents. The obtained fractions were afterward analyzed by capillary GC. SFC has also often been combined with SFE prior to chromatographic separation in plant volatile oil analysis, since in both techniques the same solvents are used, facilitating an online coupling. SFE and online-coupled SFC have been applied to the analysis of turmeric, the rhizomes of *Curcuma longa* L., using modified carbon dioxide as the extractant, yielding fractionation of turmerones curcuminoids in a single run (Sanagi et al., 1993). A multidimensional SFC-GC system was developed by Yarita et al. (1994) to separate online the constituents of citrus essential oils by stepwise pressure programming. The eluting fractions were introduced into a split/splitless injector of a gas chromatograph and analyzed after cryofocusing prior to GC separation. An SFC-GC investigation of cloudberry seed oil extracted with supercritical carbon dioxide was described by Manninen and Kallio (1997), in which SFC was mainly used for the separation of the volatile constituents from the low-boiling compounds, such as triacylglycerols. The volatiles were collected in a trap column and refocused before being separated by GC. Finally, an online technique shall be mentioned by which the compounds eluting from the SFC column can be completely transferred to GC, but also for selective or multistep heart-cutting of various sample peaks as they elute from the SFC column (Levy et al., 2005).

2.3.3.10 Couplings of SFC-MS and SFC-FTIR Spectroscopy

Both coupling techniques such as SFC-MS and SFC-FTIR have nearly exclusively been used for the investigation of low-volatile more polar compounds. Arpino published in 1990 a comprehensive article on the different coupling techniques in SFC-MS, which have been presented up to 1990 including 247 references. A short overview of applications using SFC combined with benchtop mass spectrometers was published by Ramsey and Raynor (1996). However, the only paper concerning the application of SFC-MS in essential oil research was published by Blum et al. (1997). With the aid of a newly developed interface and an injection technique using a retention gap, investigations of thyme extracts have been successfully performed.

The application of SFC-FTIR spectroscopy for the analysis of volatile compounds has also rarely been reported. One publication found in the literature refers to the characterization of varietal differences in essential oil components of hops (Auerbach et al., 2000). In that paper, the IR spectra of the main constituents were taken as films deposited on AgCl disks and compared with spectra obtained after chromatographic separation in a flow cell with IR transparent windows, exhibiting a good correlation.

2.3.4 Identification of Multicomponent Samples without Previous Separation

In addition to chromatographic separation techniques including hyphenated techniques, several spectroscopic techniques have been applied to investigate the composition of essential oils without previous separation.

2.3.4.1 UV Spectroscopy

UV spectroscopy has only little significance for the direct analysis of essential oils due to the inability to provide uniform information on individual oil components. However, for testing the presence of furanocoumarins in various citrus oils, which can cause photodermatosis when applied externally, UV spectroscopy is the method of choice. The presence of those components can be easily determined due to their characteristic UV absorption. In the European pharmacopoeia, for example, quality assessment of lemon oil, which has to be produced by cold pressing, is therefore performed by UV spectroscopy in order to exclude cheaper distilled oils.

2.3.4.2 IR Spectroscopy

Several attempts have also been made to obtain information about the composition of essential oils using IR spectroscopy. One of the first comprehensive investigations of essential oils was published by Bellanato and Hidalgo (1971) in the book *Infrared Analysis of Essential Oils* in which the IR spectra of approximately 200 essential oils and additionally of more than 50 pure reference components have been presented. However, the main disadvantage of this method is the low sensitivity and selectivity of the method in the case of mixtures with a large number of components and, second, the unsolvable problem when attempting to quantitatively measure individual component concentrations.

New approaches to analyze essential oils by vibrational spectroscopy using attenuated reflection (ATR) IR spectroscopy and NIR-FT-Raman spectroscopy have recently been published by Baranska et al. (2005) and numerous papers cited therein. The main components of an essential oil can be identified by both spectroscopic techniques using the spectra of pure oil constituents as references. The spectroscopic analysis is based on characteristic key bands of the individual constituents and made it, for example, possible to discriminate the oil profiles of several eucalyptus species. As can be taken from this paper, valuable information can be obtained as a result of the combined application of ATR-IR and NIR-FT-Raman spectroscopy. Based on reference GC measurements, valuable calibration equations have been developed for numerous essential oil plants and related essential oils in order to quantify the amount of individual oil constituents applying different suitable chemometric algorithms. Main advantages of those techniques are their ability to control the quality of essential oils very fast and easily and, above all, their ability to quantify and analyze the main constituents of essential oils *in situ*, that means in living plant tissues without any isolation process, since both techniques are not destructive.

2.3.4.3 Mass Spectrometry

MS and proton NMR spectroscopy have mainly been used for structure elucidation of isolated compounds. However, there are some reports on mass spectrometric analyses of essential oils. One example has been presented by Grützmacher (1982). The depicted mass spectrum (Figure 2.9) of an essential oil exhibits some characteristic molecular ions of terpenoids with masses at *m/z* 136, 148, 152, and 154. By the application of a double focusing mass spectrometer and special techniques analyzing the decay products of metastable ions, the components anethole, fenchone, borneol, and cineole could be identified, while the assignment of the mass 136 proved to be problematic.

A different approach has been used by Schultze et al. (1986), investigating secondary metabolites in dried plant material by direct mass spectrometric measurement. The small samples (0.1–2 mg, depending on the kind of plant drug) were directly introduced into a mass spectrometer by means

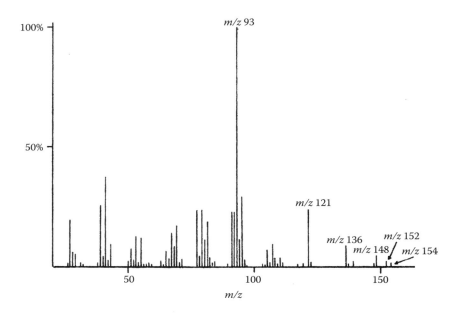

FIGURE 2.9 EI-mass spectrum of an essential oil. (From Grützmacher, H.F., Mixture analysis by new mass spectrometric techniques—A survey, in: Kubeczka, K.H., ed., *Ätherische Öle: Analytik, Physiologie, Zusammensetzung*, Georg Thieme Verlag, Stuttgart, Germany, 1982, pp. 1–24. With permission.)

of a heatable direct probe. By heating the solid sample, stored in a small glass crucible, various substances are released depending on the applied temperature, and subsequently their mass spectra can be taken. With the aid of this technique, numerous medicinal plant drugs have been investigated and their main vaporizable components could be identified.

2.3.4.4 ^{13}C-NMR Spectroscopy

^{13}C-NMR spectroscopy is generally used for the elucidation of molecular structures of isolated chemical species. The application of ^{13}C-NMR spectroscopy to the investigation of complex mixtures is relatively rare. However, the application of ^{13}C-NMR spectroscopy to the analysis of essential oils and similar complex mixtures offers particular advantages, as have been shown in the past (Formaček and Kubeczka, 1979, 1982a; Kubeczka, 2002), to confirm analytical results obtained by GC-MS and for solving certain problems encountered with nonvolatile mixture components or thermally unstable compounds, since analysis is performed at ambient temperature.

The qualitative analysis of an essential oil is based on comparison of the oil spectrum, using broadband decoupling, with spectra of pure oil constituents, which should be recorded under identical conditions regarding solvent, temperature, and so on to ensure that differences in the chemical shifts for individual ^{13}C-NMR lines of the mixture and of the reference substance are negligible. As an example, the identification of the main constituent of celery oil is shown (Figure 2.10). This constituent can be easily identified as limonene by the corresponding reference spectrum. Minor constituents give rise to less intensive signals that can be recognized after a vertical expansion of the spectrum. For recognition of those signals, also a horizontal expansion of the spectrum is advantageous.

The sensitivity of the ^{13}C-NMR technique is limited by diverse factors such as rotational sidebands, ^{13}C–^{13}C couplings, and so on, and at least by the accumulation time. For practical use, the concentration of 0.1% of a component in the entire mixture has to be seen as an interpretable limit. A very pretentious investigation has been presented by Kubeczka (1989). In the investigated essential oil, consisting of more than 80 constituents, approximately 1200 signals were counted after a

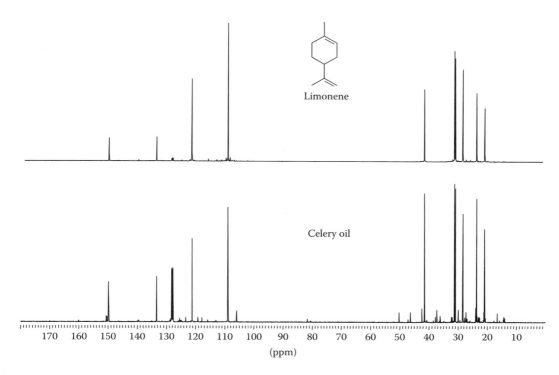

FIGURE 2.10 Identification of limonene in celery oil by ^{13}C-NMR spectroscopy.

horizontal and vertical expansion in the obtained broadband decoupled ^{13}C-NMR spectrum, which reflects impressively the complex composition of that oil. However, the analysis of such a complex mixture is made difficult by the immense density of individual lines, especially in the aliphatic region of the spectrum, making the assignments of lines to individual components ambiguous. Besides, qualitative analysis quantification of the individual sample components is accessible as described by Formaček and Kubeczka (1982b). After elimination of the ^{13}C-NMR signals of non-protonated nuclei and calculation of average signal intensity per carbon atom as a measurement characteristic, it has been possible to obtain satisfactory results as shown by comparison with gas chromatographic analyses.

During the last years, a number of articles have been published by Casanova and coworkers (e.g., Bradesi et al. (1996) and references cited therein). In addition, papers dealing with computer-aided identification of individual components of essential oils after ^{13}C-NMR measurements (e.g., Tomi et al., 1995), and investigations of chiral oil constituents by means of a chiral lanthanide shift reagent by ^{13}C-NMR spectroscopy have been published (Ristorcelli et al., 1997).

REFERENCES

Adams, R. P., 1989. *Identification of Essential Oils by Ion Trap Mass Spectroscopy.* San Diego, CA: Academic Press.

Adams, R. P., 1995. *Identification of Essential Oil Components by Gas Chromatography/Mass Spectroscopy.* Carol Stream, IL: Allured Publishing Corp.

Adams, R. P., 2007. *Identification of Essential Oil Components by Gas Chromatography/Mass Spectrometry*, 4th edn. Carol Stream, IL: Allured Publishing Corp.

Amelunxen, F., T. Wahlig, and H. Arbeiter, 1969. Über den Nachweis des ätherischen Öls in isolierten Drüsenhaaren und Drüsenschuppen von *Mentha piperita* L. *Z. Pflanzenphysiol.*, 61: 68–72.

Arpino, P., 1990. Coupling techniques in LC/MS and SFC/MS. *Fresenius J. Anal. Chem.*, 337: 667–685.

Arthur, C. L. and J. Pawliszyn, 1990. Solid phase microextraction with thermal desorption using fused silica optical fibres. *Anal. Chem.*, 62: 2145–2148.

Auerbach, R. H., D. Kenan, and G. Davidson, 2000. Characterization of varietal differences in essential oil components of hops (*Humulus lupulus*) by SFC-FTIR spectroscopy. *J. AOAC Int.*, 83: 621–626.

Baltussen, E., H. G. Janssen, P. Sandra, and C. A. Cramers, 1997. A novel type of liquid/liquid extraction for the preconcentration of organic micropollutants from aqueous samples: Application to the analysis of PAH's and OCP's in Water. *J. High Resolut. Chromatogr.*, 20: 395–399.

Baltussen, E., P. Sandra, F. David, and C. A. Cramers, 1999. Stir bar sorptive extraction (SBSE), a novel extraction technique for aqueous samples: Theory and principles. *J. Microcol. Sep.*, 11: 737–747.

Baranska, M., H. Schulz, S. Reitzenstein, U. Uhlemann, M. A. Strehle, H. Krüger, R. Quilitzsch, W. Foley, and J. Popp, 2005. Vibrational spectroscopic studies to acquire a quality control method of eucalyptus essential oils. *Biopolymers*, 78: 237–248.

Barton, D. H. R. and A. S. Lindsay, 1951. Sesquiterpenoids. Part I. Evidence for a nine-membered ring in caryophyllene. *J. Chem. Soc.*, 1951: 2988–2991.

Baser, K. H. C., B. Demirci, F. Demirci, N. Kirimer, and I. C. Hedge, 2001. Microdistillation as a useful tool for the analysis of minute amounts of aromatic plant materials. *Chem. Nat. Comp.*, 37: 336–338.

Becker, H., W. C. Hsieh, and C. O. Verelis, 1981. Droplet counter-current chromatography (DCCC). Erste Erfahrungen mit einem wasserfreien Trennsystem. *GIT Fachz. Labor. Suppl. Chromatogr.*, 81: 38–40.

Becker, H., J. Reichling, and W. C. Hsieh, 1982. Water-free solvent system for droplet counter-current chromatography and its suitability for the separation of non-polar substances. *J. Chromatogr.*, 237: 307–310.

Bellanato, J. and A. Hidalgo, 1971. *Infrared Analysis of Essential Oils.* London, U.K.: Heyden & Son Ltd.

Belliardo, F., C. Bicchi, C. Corsero, E. Liberto, P. Rubiolo, and B. Sgorbini, 2006. Headspace-solid-phase microextraction in the analysis of the volatile fraction of aromatic and medicinal plants. *J. Chromatogr. Sci.*, 44: 416–429.

Bergström, G., 1973. Studies on natural odoriferous compounds. *Chem. Scr.*, 4: 135–138.

Bergström, G., M. Appelgren, A. K. Borg-Karlson, I. Groth, S. Strömberg, and St. Strömberg, 1980. Studies on natural odoriferous compounds. *Chem. Scr.*, 16: 173–180.

Bernreuther, A., J. Koziet, P. Brunerie, G. Krammer, N. Christoph, and P. Schreier, 1990. Chirospecific capillary gas chromatography (HRGC) and on-line HRGC-isotope ratio mass spectrometry of γ-decalactone from various sources. *Z. Lebensm. Unters. Forsch.*, 191: 299–301.

Berthelot, M., 1859. Ueber Camphenverbindungen. *Liebigs Ann. Chem.*, 110: 367–368.

Beyer, J., H. Becker, and R. Martin, 1986. Separation of labile terpenoids by low temperature HPLC. *J. Chromatogr.*, 9: 2433–2441.

Bicchi, C., A. D. Amato, F. David, and P. Sandra, 1987. Direct capture of volatiles emitted by living plants. *Flavour Frag. J.*, 2: 49–54.

Bicchi, C., A. D. Amato, C. Frattini, G. M. Nano, E. Cappelletti, and R. Caniato, 1985. Analysis of essential oils by direct sampling from plant secretory structures and capillary gas chromatography. *J. High Resolut. Chromagtogr.*, 8: 431–435.

Bicchi, C., A. D. Amato, V. Manzin, A. Galli, and M. Galli, 1996. Cyclodextrin derivatives in gas chromatographic separation of racemic mixtures of volatile compounds. X. 2,3-di-*O*-ethyl-6-*O*-*tert*-butyl-dimethylsilyl)-β- and -γ-cyclodextrins. *J. Chromatogr. A*, 742: 161–173.

Bicchi, C., A. D. Amato, G. M. Nano, and C. Frattini, 1983. Improved method for the analysis of essential oils by microdistillation followed by capillary gas chromatography. *J. Chromatogr.*, 279: 409–416.

Bicchi, C., C. Cordero, C. Iori, P. Rubiolo, and P. Sandra, 2000a. Headspace sorptive extraction (HSSE) in the headspace analysis of aromatic and medicinal plants. *J. High Resolut. Chromatogr.*, 23: 539–546.

Bicchi, C., C. Cordero, and P. Rubiolo, 2000b. Influence of fibre coating in headspace solid-phase microextraction-gas chromatographic analysis of aromatic and medicinal plants. *J. Chromatogr. A*, 892: 469–485.

Bicchi, C., C. Frattini, G. Pellegrino, P. Rubiolo, V. Raverdino, and G. Tsoupras, 1992. Determination of sulphurated compounds in *Tagetes patula* cv. nana essential oil by gas chromatography with mass spectrometric, Fourier transform infrared and atomic emission spectrometric detection. *J. Chromatogr.*, 609: 305–313.

Bicchi, C. and D. Joulain, 1990. Review: Headspace-gas chromatographic analysis of medicinal and aromatic plants and flowers. *Flavour Frag. J.*, 5: 131–145.

Bicchi, C., V. Manzin, A. D. Amato, and P. Rubiolo, 1995. Cyclodextrin derivatives in GC separation of enantiomers of essential oil, aroma and flavour compounds. *Flavour Frag. J.*, 10: 127–137.

Bicchi, C. and P. Sandra, 1987. Microtechniques in essential oil analysis. In *Capillary Gas Chromatography in Essential Oil Analysis*, P. Sandra and C. Bicchi (eds.), pp. 85–122. Heidelberg, Germany: Alfred Huethig Verlag.

Blanch, G. P., E. Ibanez, M. Herraiz, and G. Reglero, 1994. Use of a programmed temperature vaporizer for off-line SFE/GC analysis in food composition studies. *Anal. Chem.*, 66: 888–892.

Blum, C., K. H. Kubeczka, and K. Becker, 1997. Supercritical fluid chromatography-mass spectrometry of thyme extracts (*Thymus vulgaris* L.). *J. Chromatogr. A*, 773: 377–380.

Bos, R., A. P. Bruins, and H. Hendriks, 1982. Negative ion chemical ionization, a new important tool in the analysis of essential oils. In *Ätherische Öle, Analytik, Physiologie, Zusammensetzung*, K. H. Kubeczka (ed.), pp. 25–32. Stuttgart, Germany: Georg Thieme Verlag.

Bos, R., H. J. Woerdenbag, H. Hendriks, J. H. Zwaving, P.A.G.M. De Smet, G. Tittel, H. V. Wikström, and J. J. C. Scheffer, 1996. Analytical aspects of phytotherapeutic valerian preparations. *Phytochem. Anal.*, 7: 143–151.

Bourne, S., A. M. Haefner, K. L. Norton, and P. R. Griffiths, 1990. Performance characteristics of a real-time direct deposition gas chromatography/Fourier transform infrared system. *Anal. Chem.*, 62: 2448–2452.

Bradesi, P., A. Bighelli, F. Tomi, and J. Casanova, 1996. L'analyse des mélanges complexes par RMN du Carbone-13—Partie I et II. *Cand. J. Appl. Spectrosc.*, 11: 15–24, 41–50.

Braunsdorf, R., U. Hener, and A. Mosandl, 1992. Analytische Differenzierung zwischen natürlich gewachsenen, fermentativ erzeugten und synthetischen (naturidentischen) Aromastoffen II. Mitt.: GC-C-IRMS-Analyse aromarelevanter Aldehyde—Grundlagen und Anwendungsbeispiele. *Z. Lebensm. Unters. Forsch.*, 194: 426–430.

Braunsdorf, R., U. Hener, S. Stein, and A. Mosandl, 1993. Comprehensive cGC-IRMS analysis in the authenticity control of flavours and essential oils. Part I: Lemon oil. *Z. Lebensm. Unters. Forsch.*, 197: 137–141.

Briechle, R., W. Dammertz, R. Guth, and W. Volmer, 1997. Bestimmung ätherischer Öle in Drogen. *GIT Lab. Fachz.*, 41: 749–753.

Bruins, A. P., 1987. Gas chromatography-mass spectrometry of essential oils, Part II: Positive ion and negative ion chemical ionization techniques. In *Capillary Gas Chromatography in Essential Oil Analysis*, P. Sandra and C. Bicchi (eds.), pp. 329–357. Heidelberg, Germany: Dr. A. Huethig Verlag.

Bruno, S., 1961. La chromatografia in phase vapore nell'identificazione di alcuni olii essenziali in materiali biologici. *Farmaco*, 16: 481–186.

Buttery, R. G., W. H. McFadden, R. Teranishi, M. P. Kealy, and T. R. Mon, 1963. Constituents of hop oil. *Nature*, 200: 435–436.

Carle, R., I. Fleischhauer, J. Beyer, and E. Reinhard, 1990. Studies on the origin of (−)-α-bisabolol and chamazulene in chamomile preparations: Part I. Investigations by isotope ratio mass spectrometry (IRMS). *Planta Med.*, 56: 456–460.

Cazaussus, A., R. Pes, N. Sellier, and J. C. Tabet, 1988. GC-MS and GC-MS-MS analysis of a complex essential oil. *Chromatographia*, 25: 865–869.

Chaintreau, A., 2001. Simultaneous distillation–extraction: From birth to maturity—Review. *Flavour Frag. J.*, 16: 136–148.

Chamblee, T. S., B. C. Clark, T. Radford, and G. A. Iacobucci, 1985. General method for the high-performance liquid chromatographic prefractionation of essential oils and flavor mixtures for gas chromatographic-mass spectrometric analysis: Identification of new constituents in cold pressed lime oil. *J. Chromatogr.*, 330: 141–151.

Chester T. L. and D. P. Innis, 1986. Separation of oligo- and polysaccharides by capillary supercritical fluid chromatography, *J. High Resolut. Chromatogr.*, 9: 209–212.

Chialva, F., G. Gabri, P. A. P. Liddle, and F. Ulian, 1982. Qualitative evaluation of aromatic herbs by direct headspace GC analysis. Application of the method and comparison with the traditional analysis of essential oils. *J. High Resolut. Chromatogr.*, 5: 182–188.

Clevenger, J. F., 1928. Apparatus for the determination of volatile oil. *J. Am. Pharm. Assoc.*, 17: 345–349.

Cocking, T. T. and G. Middleton, 1935. Improved method for the estimation of the essential oil content of drugs. *Quart. J. Pharm. Pharmacol.*, 8: 435–442.

Cole, R. A., 1980. The use of porous polymers for the collection of plant volatiles. *J. Sci. Food Agric.*, 31: 1242–1249.

Conway, W. D., 1989. *Countercurrent Chromatography—Apparatus, Theory, and Applications*. New York: VCH Inc.

Cramers, C. A., H. G. Janssen, M. M. van Deursen, and P. A. Leclercq, 1999. High speed gas chromatography: An overview of various concepts. *J. Chromatogr. A*, 856: 315–329.

Craveiro, A. A., F. J. A. Matos, J. Alencar, and M. M. Plumel, 1989. Microwave oven extraction of an essential oil. *Flavour Frag. J.*, 4: 43–44.

David, F. and P. Sandra, 1992. Capillary gas chromatography-spectroscopic techniques in natural product analysis. *Phytochem. Anal.*, 3: 145–152.

David, F. and P. Sandra, 2007. Review: Stir bar sorptive extraction for trace analysis. *J. Chromatogr. A*, 1152: 54–69.

Debrunner, B., M. Neuenschwander, and R. Benneisen, 1995. Sesquiterpenes of *Petasites hybridus* (L.) G.M. et Sch.: Distribution of sesquiterpenes over plant organs. *Pharmaceut. Acta Helv.*, 70: 167–173.

Dietrich, A., B. Maas, V. Karl, P. Kreis, D. Lehmann, B. Weber, and A. Mosandl, 1992a. Stereoisomeric flavour compounds, part LV: Stereodifferentiation of some chiral volatiles on heptakis (2,3-di-*O*-acetyl-6-*O*-tert-butyl-dimethylsilyl)-β-cyclodextrin. *J. High Resolut. Chromatogr.*, 15: 176–179.

Dietrich, A., B. Maas, B. Messer, G. Bruche, V. Karl, A. Kaunzinger, and A. Mosandl, 1992b. Stereoisomeric flavour compounds, part LVIII: The use of heptakis (2,3-di-*O*-methyl-6-*O-tert*-butyl-dimethylsilyl)-β-cyclodextrin as a chiral stationary phase in flavor analysis. *J. High Resolut. Chromatogr.*, 15: 590–593.

Dugo, G., P. Q. Tranchida, A. Cotroneo, P. Dugo, I. Bonaccorsi, P. Marriott, R. Shellie, and L. Mondello, 2005. Advanced and innovative chromatographic techniques for the study of citrus essential oils. *Flavour Frag. J.*, 20: 249–264.

Dugo, G., A. Verzera, A. Cotroneo, I. S. d'Alcontres, L. Mondllo, and K. D. Bartle, 1994. Automated HPLC-HRGC: A powerful method for essential oil analysis. Part II. Determination of the enantiomeric distribution of linalol in sweet orange, bitter orange and mandarin essential oils. *Flavour Frag. J.*, 9: 99–104.

Dugo, P., G. Dugo, and L. Mondello, 2003. On-line coupled LC–GC: Theory and applications. *LC-GC Eur.*, 16(12a): 35–43.

Dumas, M. J., 1833. Ueber die vegetabilischen Substanzen welche sich dem Kampfer nähern, und über einig ätherischen Öle. *Ann. Pharmacie*, 6: 245–258.

Fischer, N., S. Nitz, and F. Drawert, 1987. Original flavour compounds and the essential oil composition of Marjoram (*Majorana hortensis* Moench). *Flavour. Frag. J.*, 2: 55–61.

Formaček, V. and K. H. Kubeczka, 1979. Application of 13C-NMR-spectroscopy in analysis of essential oils. In *Vorkommen und Analytik ätherischer Öle*, K. H. Kubeczka (ed.), pp. 130–138. Stuttgart, Germany: Georg Thieme Verlag.

Formaček, V. and K. H. Kubeczka, 1982a. ^{13}C-NMR analysis of essential oils. In *Aromatic Plants: Basic and Applied Aspects*, N. Margaris, A. Koedam, and D. Vokou (eds.), pp. 177–181. The Hague, the Netherlands: Martinus Nijhoff Publishers.

Formaček, V. and K. H. Kubeczka, 1982b. Quantitative analysis of essential oils by ^{13}C-NMR-spectroscopy. In *Ätherische Öle: Analytik, Physiologie, Zusammensetzung*, K. H. Kubeczka (ed.), pp. 42–53. Stuttgart, Germany: Georg Thieme Verlag.

Frank, C., A. Dietrich, U. Kremer, and A. Mosandl, 1995. GC-IRMS in the authenticy control of the essential oil of *Coriandrum sativum* L. *J. Agric. Food Chem.*, 43: 1634–1637.

Frérot, E. and E. Decorzant 2004. Quantification of total furocoumarins in citrus oils by HPLC couple with UV fluorescence, and mass detection. *J. Agric. Food Chem.*, 52: 6879–6886.

Friedel, H. D. and R. Matusch, 1987. Separation of non-polar sesquiterpene olefins from Tolu balsam by high-performance liquid chromatography: Silver perchlorate impregnation of prepacked preparative silica gel column. *J. Chromatogr.*, 407: 343–348.

Gaedcke, F. and B. Steinhoff, 2000. *Phytopharmaka*. Stuttgart, Germany: Wissenschaftliche Verlagsgesellschaft (Fig. 1.7).

Geiss, F., 1987. *Fundamentals of Thin-Layer Chromatography*. Heidelberg, Germany: Hüthig Verlag.

Gießelmann, G. and K. H. Kubeczka, 1993. A new procedure for the enrichment of headspace constituents versus conventional hydrodistillation. Poster presented at the *24th International Symposium on Essential Oils*, Berlin, Germany.

Gil-Av, E., B. Feibush, and R. Charles-Sigler, 1965. In *Gas Chromatography 1966*, A. B. Littlewood (ed.), 227pp. London, U.K.: Institute of Petroleum.

Gildemeister, E. and F. Hoffmann, 1956. In *Die ätherischen Öle*, W. Treibs (ed.), Vol. 1, p. 14. Berlin, Germany: Akademie-Verlag.

Godefroot, M., P. Sandra, and M. Verzele, 1981. New method for quantitative essential oil analysis. *J. Chromatogr.*, 203: 325–335.

Godefroot, M., M. Stechele, P. Sandra, and M. Verzele, 1982. A new method fort the quantitative analysis of organochlorine pesticides and polychlorinated biphenyls. *J. High Resolut. Chromatogr.*, 5: 75–79.

Greibrokk, T., 1995. Review: Applications of supercritical fluid extraction in multidimensional systems. *J. Chromatogr. A*, 703: 523–536.

Grob, K. and F. Zürcher, 1976. Stripping of trace organic substances from water: Equipment and procedure. *J. Chromatogr.*, 117: 285–294.

Grützmacher, H. F., 1982. Mixture analysis by new mass spectrometric techniques—A survey. In *Ätherische Öle: Analytik, Physiologie, Zusammensetzung*, K. H. Kubeczka (ed.), pp. 1–24. Stuttgart, Germany: Georg Thieme Verlag.

Hartonen, K., M. Jussila, P. Manninen, and M. L. Riekkola, 1992. Volatile oil analysis of *Thymus vulgaris* L. by directly coupled SFE/GC. *J. Microcol. Sep.*, 4: 3–7.

Hawthorne, S. B., M. S. Krieger, and D. J. Miller, 1988. Analysis of flavor and fragrance compounds using supercritical fluid extraction coupled with gas chromatography. *Anal. Chem.*, 60: 472–477.

Hefendehl, F. W., 1966. Isolierung ätherischer Öle aus äußeren Pflanzendrüsen. *Naturw.*, 53: 142.

Henderson, W., J. W. Hart, P. How, and J. Judge, 1970. Chemical and morphological studies on sites of sesquiterpene accumulation in *Pogostemon cablin* (Patchouli). *Phytochemistry*, 9: 1219–1228.

Hener, U., 1990. Chirale Aromastoffe—Beiträge zur Struktur, Wirkung und Analytik. Dissertation, Goethe-University of Frankfurt/Main, Frankfurt, Germany.

Herres, W., 1987. *HRGC-FTIR: Capillary Gas Chromatography-Fourier Transform Infrared Spectroscopy.* Heidelberg, Germany: Alfred Huethig Verlag.

Herres, W., K. H. Kubezka, and W. Schultze, 1986. HRGC-FTIR investigations on volatile terpenes. In *Progress in Essential Oil Research*, E. J. Brunke (ed.), pp. 507–528. Berlin, Germany: W. de Gruyter.

Hör, K., C. Ruff, B. Weckerle, T. König, and P. Schreier, 2001. ^{2}H/^{1}H ratio analysis of flavor compounds by on-line gas chromatography-pyrolysis-isotope ratio mass spectrometry (HRGC-P-IRMS): Citral. *Flavour Frag. J.*, 16: 344–348.

Houben, R. J., H. G. M. Janssen, P. A. Leclercq, J. A. Rijks, and C. A. Cramers, 1990. Supercritical fluid extraction-capillary gas chromatography: On-line coupling with a programmed temperature vaporizer. *J. High Resolut. Chromatogr.*, 13: 669–673.

Ibanez, E., S. Lopez-Sebastian, E. Ramos, J. Tabera, and G. Reglero, 1997. Analysis of highly volatile components of foods by off-line SFE/GC. *J. Agric. Food Chem.*, 45: 3940–3943.

Ito, Y., M. A. Weinstein, I. Aoki, R. Harada, E. Kimura, and K. Nunogaki, 1966. The coil planet centrifuge. *Nature*, 212: 985–987.

Jennings, W., 1981. Recent developments in high resolution gas chromatography. In *Flavour '81*, P. Schreier (ed.), pp. 233–251. Berlin, Germany: Walter de Gruyter & Co.

Jennings, W. and T. Shibamoto, 1980. *Qualitative Analysis of Flavor and Fragrance Volatiles by Glass Capillary Gas Chromatography.* New York: Academic Press.

Jennings, W. G., 1979. Vapor-phase sampling. *J. High Resolut. Chromatogr.*, 2: 221–224.

Jirovetz, L., G. Buchbauer, W. Jäger, A. Woidich, and A. Nikiforov, 1992. Analysis of fragrance compounds in blood samples of mice by gas chromatography, mass spectrometry, GC/FTIR and GC/AES after inhalation of sandalwood oil. *Biomed. Chromatogr.*, 6: 133–134.

Joulain, D., 1987. The composition of the headspace from fragrant flowers: Further results. *Flavour Frag. J.*, 2: 149–155.

Joulain, D. and W. A. König, 1998. *The Atlas of Spectral Data of Sesquiterpene Hydrocarbons.* Hamburg, Germany: E. B. Verlag.

Juchelka, D., T. Beck, U. Hener, F. Dettmar, and A. Mosandl, 1998. Multidimensional gas chromatography coupled on-line with isotope ratio mass spectrometry (MDGC-IRMS): Progress in the analytical authentication of genuine flavor components. *J. High Resolut. Chromatogr.*, 21: 145–151.

Kaiser, H. and W. Lang, 1951. Ueber die Bestimmung des ätherischen Oels in Drogen. *Dtsch. Apoth. Ztg.*, 91: 163–166.

Kaiser, R., 1976. *Einführung in die Hochleistungs-Dünnschicht-Chromatographie.* Bad Dürkheim, Germany: Institut für Chromatographie.

Kaiser, R., 1993. *The Scent of Orchids—Olfactory and Chemical Investigations.* Amsterdam, the Netherlands: Elsevier Science Ltd.

Kolb, B. and L. S. Ettre, 1997. *Static Headspace-Gas Chromatography: Theory and Practice.* New York: Wiley.

Kolb, B. and L. S. Ettre, 2006. *Static Headspace-Gas Chromatography: Theory and Practice*, 2nd edn. New York: Wiley.

Kolb, B. and B. Liebhardt, 1986. Cryofocusing in the combination of gas chromatography with equilibrium headspace sampling. *Chromatographia*, 21: 305–311.

Komae, H. and N. Hayashi, 1975. Separation of essential oils by liquid chromatography. *J. Chromatogr.*, 114: 258–260.
König, W. A., D. Icheln, T. Runge, I. Pforr, and A. Krebs, 1990. Cyclodextrins as chiral stationary phases in capillary gas chromatography. Part VII: Cyclodextrins with an inverse substitution pattern—Synthesis and enantioselectivity. *J. High Resolut. Chromatogr.*, 13: 702–707.
König, W. A., P. Evers, R. Krebber, S. Schulz, C. Fehr, and G. Ohloff, 1989. Determination of the absolute configuration of α-damascenone and α-ionone from black tea by enantioselective capillary gas chromatography. *Tetrahedron*, 45: 7003–7006.
König, W. A., B. Gehrcke, D. Icheln, P. Evers, J. Dönnecke, and W. Wang, 1992a. New, selectively substituted cyclodextrins as stationary phases for the analysis of chiral constituents of essential oils. *J. High Resolut. Chromatogr.*, 15: 367–372.
König, W. A., D. Icheln, T. Runge, P. Evers, B. Gehrcke, and A. Krüger, 1992b. Enantioselective gas chromatography—A new dimension in the analysis of essential oils. In *Proceedings of the 12th International Congress of Flavours, Fragrances and Essential Oils*, H. Woidich and G. Buchbauer (eds.), pp. 177–186. Vienna, Austria: Austrian Association of Flavour and Fragrance Industry.
König, W. A., S. Lutz, P. Mischnick-Lübbecke, B. Brassat, and G. Wenz, 1988a. Cyclodextrins as chiral stationary phases in capillary gas chromatography I. Pentylated α-cyclodextrin. *J. Chromatogr.*, 447: 193–197.
König, W. A., S. Lutz, and G. Wenz, 1988b. Modified cyclodextrins—Novel, highly enantioselective stationary phases for gas chromatography. *Angew. Chem. Int. Ed. Engl.* 27: 979–980.
König, W. A., S. Lutz, G. Wenz, and E. van der Bey, 1988c. Cyclodextrins as chiral stationary phases in capillary gas chromatography II. Heptakis (3-*O*-acetyl-2,6-di-*O*-pentyl)-β-cyclodextrin. *J. High Resolut. Chromatogr. Chromatogr. Commun.*, 11: 506–509.
König, W. A., A. Rieck, C. Fricke, S. Melching, Y. Saritas, and I. H. Hardt, 1995. Enantiomeric composition of sesquiterpenes in essential oils. In *Proceedings of the 13th International Congress of Flavours, Fragrances and Essential Oils*, K. H. C. Baser (ed.), Vol. 2, pp. 169–180. Istanbul, Turkey: AREP Publ.
König, W. A., A. Rieck, I. Hardt, B. Gehrcke, K. H. Kubeczka, and H. Muhle, 1994. Enantiomeric composition of the chiral constituents of essential oils Part 2: Sesquiterpene hydrocarbons. *J. High Resolut. Chromatogr.*, 17: 315–320.
Kreis, P. and A. Mosandl, 1992. Chiral compounds of essential oils XI. Simultaneous stereoanalysis of lavandula oil constituents. *Flavour Frag. J.*, 7: 187–193.
Kubeczka, K. H., 1967. Vorrichtung zur Isolierung, Anreicherung und chemischen Charakterisierung gaschromatographisch getrennter Komponenten im μg-Bereich. *J. Chromatogr.*, 31: 319–325.
Kubeczka, K. H., 1973. Separation of essential oils and similar complex mixtures by means of modifieddry-column chromatography, *Chromatographia*, 6: 106–108.
Kubeczka, K. H., 1981a. Standardization and analysis of essential oils. In *A Perspective of the Perfumes and Flavours Industry in India*, S. Jain (ed.), pp. 105–120. New Delhi, India: Perfumes and Flavours Association of India.
Kubeczka, K. H., 1981b. Application of HPLC for the separation of flavour compounds. In *Flavour 81*, P. Schreier (ed.), pp. 345–359. Berlin, Germany: Walter de Gruyter & Co.
Kubeczka, K. H., 1985. Progress in isolation techniques for essential oil constituents. In *Advances in Medicinal Plant Research*, A. J. Vlietinck and R. A. Dommisse (eds.), pp. 197–224. Stuttgart, Germany: Wissenschaftliche Verlagsgesellschaft mbH.
Kubeczka, K. H., 1989. Studies on complex mixtures: Combined separation techniques versus unprocessed sample analysis. In *Moderne Tecniche in Fitochimica*, C. Bicchi and C. Frattini (eds.), pp. 53–68. Firenze, Tuscany: Società Italiana di Fitochimica.
Kubeczka, K. H., 1991. New methods in essential oil analysis. In *Conferencias Plenarias de la XXIII Reunión Bienal de Quimica*, A. San Feliciano, M. Grande, and J. Casado (eds.), pp. 169–184. Salamanca, Spain: Universidad de Salamanca, Sección local e la R.S.E.Q.
Kubeczka, K. H., 1996. Unpublished results.
Kubeczka, K. H., 1997a. New approaches in essential oil analysis using polymer-coated silica fibers. In *Essential Oils: Basic and Applied Research*, Ch. Franz, A. Máthé, and G. Buchbauer (eds.), pp. 139–146. Carol Stream, IL: Allured Publishing Corp.
Kubeczka, K.-H., 1997b. *Essential Oil Symposium Proceedings*, p. 145.
Kubeczka, K.-H., 2002. *Essential Oils Analysis by Capillary Gas Chromatography and Carbon-13 NMR Spectroscopy*, 2nd completely rev. edn. Baffins Lane, England: Wiley.

Kubeczka, K. H. and G. Gießelmann, 1995. Application of a new micro hydrodistillation device for the investigation of aromatic plant drugs. Poster presented at the *43th Annual Congress on Medicinal Plant Research*, Halle, Germany.

Kubeczka, K. H., W. Schultze, S. Ebel, and M. Weyandt-Spangenberg, 1989. Möglichkeiten und Grenzen der GC-Molekülspektroskopie-Kopplungen. In *Instrumentalized Analytical Chemistry and Computer Technology*, W. Günther and J. P. Matthes (eds.), pp. 131–141. Darmstadt, Germany: GIT Verlag.

Lancas, F., F. David, and P. Sandra, 1988. CGC analysis of the essential oil of citrus fruits on 100 μm i.d. columns. *J. High Resolut. Chromatogr.*, 11: 73–75.

Lawrence, B. M., 1995. The isolation of aromatic materials from natural plant products. In *Manual of the Essential Oil Industry*, K. Tuley De Silva (ed.), pp. 57–154. Vienna, Austria: UNIDO.

Laylin, J. K., 1993. *Nobel Laureates in Chemistry, 1901–1992*. Philadelphia, PA: Chemical Heritage Foundation.

Levy, J. M., J. P. Guzowski, and W. E. Huhak, 2005. On-line multidimensional supercritical fluid chromatography/capillary gas chromatography. *J. High Resolut. Chromatogr.*, 10: 337–341.

Lucchesi, M. E., F. Chemat, and J. Smadja, 2004. Solvent-free microwave extraction of essential oil from aromatic herbs: Comparison with conventional hydro-distillation. *J. Chromatogr. A*, 1043: 323–327.

Malingré, T. M., D. Smith, and S. Batterman, 1969. De Isolering en Gaschromatografische Analyse van de Vluchtige Olie uit Afzonderlijke Klierharen van het Labiatentype. *Pharm. Weekblad*, 104: 429.

Manninen, P. and H. Kallio, 1997. Supercritical fluid chromatography-gas chromatography of volatiles in cloudberry (*Rubus chamaemorus*) oil extracted with supercritical carbon dioxide. *J. Chromatogr. A*, 787: 276–282.

Manninen, P., M. L. Riekkola, Y. Holm, and R. Hiltunen, 1990. SFC in analysis of aromatic plants. *J. High Resolut. Chromatogr.*, 13: 167–169.

MassFinder, 2007. *MassFinder Software*, Version 3.7. Hamburg, Germany: Dr. Hochmuth Scientific Consulting.

Matthews, D. E. and J. M. Hayes, 1978. Isotope-ratio-monitoring gas chromatography-mass spectrometry. *Anal. Chem.*, 50: 1465–1473.

McKone, H. T., 1979. High performance liquid chromatography. *J. Chem. Educ.*, 56: 807–809.

Mechler, E. and K. A. Kovar, 1977. Vergleichende Bestimmungen des ätherischen Öls in Drogen nach dem Europäischen und dem Deutschen Arzneibuch. *Dtsch. Apoth. Ztg.*, 117: 1019–1023.

Meerwein, H., 1914. Über den Reaktionsmechanismus der Umwandlung von Borneol in Camphen. *Liebigs Ann. Chem.*, 405: 129–175.

Miller, J. M. and J. G. Kirchner, 1952. Some improvements in chromatographic techniques for terpenes. *Anal. Chem.*, 24: 1480–1482.

Modey, W. K., D. A. Mulholland, and M. W. Raynor, 1995. Analytical supercritical extraction of natural products. *Phytochem. Anal.*, 7: 1–15.

Mondello, L., K. D. Bartle, G. Dugo, and P. Dugo, 1994a. Automated HPLC-HRGC: A powerful method for essential oil analysis Part III. Aliphatic and terpene aldehydes of orange oil. *J. High Resolut. Chromatogr.*, 17: 312–314.

Mondello, L., K. D. Bartle, P. Dugo, P. Gans, and G. Dugo, 1994b. Automated HPLC-HRGC: A powerful method for essential oils analysis. Part IV. Coupled LC-GC-MS (ITD) for bergamot oil analysis. *J. Microcol. Sep.*, 6: 237–244.

Mondello, L., G. Dugo, and K. D. Bartle, 1996. On-line microbore high performance liquid chromatography-capillary gas chromatography for food and water analyses. A review. *J. Microcol. Sep.*, 8: 275–310.

Mondello, L., P. Dugo, K. D. Bartle, G. Dugo, and A. Cotroneo, 1995. Automated LC-GC: A powerful method for essential oils analysis Part V. Identification of terpene hydrocarbons of bergamot, lemon, mandarin, sweet orange, bitter orange, grapefruit, clementine and Mexican lime oils by coupled HPLC-HRGC-MS (ITD). *Flavour Frag. J.*, 10: 33–42.

Mondello, L., P. Dugo, G. Dugo, A. C. Lewis, and K. D. Bartle, 1999. Review: High-performance liquid chromatography coupled on-line with high resolution gas chromatography, State of the art. *J. Chromatogr. A*, 842: 373–390.

Mondello, L., R. Shellie, A. Casilli, P. Marriott, and G. Dugo, 2004. Ultra-fast essential oil characterization by capillary GC on a 50 μm ID column. *J. Sep. Sci.*, 27: 699–702.

Mondello, L., G. Zappia, G. Errante, P. Dugo, and G. Dugo, 2000. Fast-GC and Fast-GC/MS for the analysis of natural complex matrices. *LC-GC Eur.*, 13: 495–502.

Morita, M., S. Mihashi, H. Itokawa, and S. Hara, 1983. Silver nitrate impregnation of preparative silica gel columns for liquid chromatography. *Anal. Chem.*, 55: 412–414.

Mosandl, A., 2004. Authenticity assessment: A permanent challenge in food flavor and essential oil analysis. *J. Chromatogr. Sci.*, 42: 440–449.

Mosandl, A. and D. Juchelka, 1997. Advances in authenticity assessment of citrus oils. *J. Essent. Oil Res.*, 9: 5–12.

Munari, F., G. Dugo, and A. Cotroneo, 1990. Automated on-line HPLC-HRGC with gradient elution and multiple GC transfer applied to the characterization of citrus essential oils. *J. High Resolut. Chromatogr.*, 13: 56–61.

Nickerson, G. and S. Likens, 1966. Gas chromatographic evidence for the occurrence of hop oil components in beer. *J. Chromatogr.*, 21: 1–5.

Nishii, Y., T. Yoshida, and Y. Tanabe, 1997. Enantiomeric resolution of a germacrene-D derivative by chiral high-performance liquid chromatography. *Biosci. Biotechnol. Biochem.*, 61: 547–548.

NIST/EPA/NIH *Mass Spectral Library 2005*, Version: NIST 05. Gaithersburg, MD: Mass Spectrometry Data Center, National Institute of Standard and Technology.

Novak, J., V. Vašak, and J. Janak, 1965. Chromatographic method for the concentration of trace impurities in the atmosphere and other gases. *Anal. Chem.*, 37: 660–666.

Nyiredy, Sz., 2003. Progress in forced-flow planar chromatography. *J. Chromatogr. A*, 1000: 985–999.

Pawliszyn, J., 1997. *Solid Phase Microextraction Theory and Practice.* New York: Wiley-VCH Inc.

Pettei, M. J., F. G. Pilkiewicz, and K. Nakanishi, 1977. Preparative liquid chromatography applied to difficult separations. *Tetrahedron Lett.*, 24: 2083–2086.

Pourmortazavi, S. M. and S. S. Hajimirsadeghi, 2007. Review: Supercritical fluid extraction in plant essential and volatile oil analysis. *J. Chromatogr. A*, 1163: 2–24.

Prasad, R. S., A. S. Gupta, and S. Dev, 1947. Chromatography of organic compounds III. Improved procedure for the thin-layer chromatography of olefins on silver ion-silica gel layers. *J. Chromatogr.*, 92: 450–453.

Ramsey, E. D. and M. W. Raynor, 1996. Electron ionization and chemical ionization sensitivity studies involving capillary supercritical fluid chromatography combined with benchtop mass spectrometry. *Anal. Commun.*, 33: 95–97.

Reedy, G. T., D. G. Ettinger, J. F. Schneider, and S. Bourne, 1985. High-resolution gas chromatography/matrix isolation infrared spectrometry. *Anal. Chem.*, 57: 1602–1609.

Ristorcelli, D., F. Tomi, and J. Casanova, 1997. Enantiomeric differentiation of oxygenated monoterpenes by carbon-13 NMR in the presence of a chiral lanthanide shift reagent. *J. Magnet. Resonance Anal.*, 1997: 40–46.

Ross, M. S. F., 1976. Analysis of cinnamon oils by high-pressure liquid chromatography. *J. Chromatogr.*, 118: 273–275.

Ruff, C., K. Hör, B. Weckerle, and P. Schreier, 2000. ^{2}H/^{1}H ratio analysis of flavor compounds by on-line gas chromatography pyrolysis isotope ratio mass spectrometry (HRGC-P-IRMS): Benzaldehyde. *J. High Resolut. Chromatogr.*, 23: 357–359.

Ruzicka, L., 1953. The isoprene rule and the biogenesis of terpenic compounds. *Experientia*, 9: 357–396.

Sanagi, M. M., U. K. Ahmad, and R. M. Smith, 1993. Application of supercritical fluid extraction and chromatography to the analysis of turmeric. *J. Chromatogr. Sci.*, 31: 20–25.

Saritas, Y., 2000. Isolierung, Strukturaufklärung und stereochemische Untersuchungen von sesquiterpenoiden Inhaltsstoffen aus ätherischen Ölen von Bryophyta und höheren Pflanzen. PhD dissertation, University of Hamburg, Hamburg, Germany.

Schaefer, J., 1981. Comparison of adsorbents in head space sampling. In *Flavour '81*, P. Schreier (ed.), pp. 301–313. Berlin, Germany: Walter de Gruyter & Co.

Scheffer, J. J. C., A. Koedam, and A. Baerheim Svendsen, 1976. Occurrence and prevention of isomerization of some monoterpene hydrocarbons from essential oils during liquid–solid chromatography on silica gel. *Chromatographia*, 9: 425–432.

Schneider, W., J. C. Frohne, and H. Bruderreck, 1982. Selektive gaschromatographische Messung sauerstoffhaltiger Verbindungen mittels Flammenionisationsdetektor. *J. Chromatogr.*, 245: 71–83.

Schreier, P., 1984. *Chromatographic Studies of Biogenesis of Plant Volatiles.* Heidelberg, Germany: Alfred Hüthig Verlag.

Schultze, W., 1993. Moderne instrumentalanalytische Methoden zur Untersuchung komplexer Gemische. In *Ätherische Öle—Anspruch und Wirklichkeit*, R. Carle (ed.), pp. 135–184. Stuttgart, Germany: Wissenschaftliche Verlagsgesellschaft mbH.

Schultze, W., G. Lange, and G. Heinrich, 1986. Analysis of dried plant material directly introduced into a mass spectrometer. (Part I of investigations on medicinal plants by mass spectrometry). In *Progress in Essential Oil Research*, E. J. Brunke (ed.), pp. 577–596. Berlin, Germany: Walter de Gruyter & Co.

Schultze, W., G. Lange, and G. Schmaus, 1992. Isobutane and ammonia chemical ionization mass spectrometry of sesquiterpene hydrocarbons. *Flavour Frag. J.*, 7: 55–64.

Schurig, V., 1977. Enantiomerentrennung eines chiralen Olefins durch Komplexierungschromatographie an einem optisch aktiven Rhodium(1)-Komplex. *Angew. Chem.*, 89: 113–114.

Schurig, V. and H. P. Nowotny, 1988. Separation of enantiomers on diluted permethylated β-cyclodextrin by high resolution gas chromatography. *J. Chromatogr.*, 441: 155–163.

Schwanbeck, J., V. Koch, and K. H. Kubeczka, 1982. HPLC-separation of essential oils with chemically bonded stationary phases. In *Essential Oils—Analysis, Physiology, Composition*, K. H. Kubeczka (ed.), pp. 70–81. Stuttgart, Germany: Georg Thieme Verlag.

Schwanbeck, J. and K. H. Kubeczka, 1979. Application of HPLC for separation of volatile terpene hydrocarbons. In *Vorkommen und Analytik ätherischer Öle*, K. H. Kubeczka (ed.), pp. 72–76. Stuttgart, Germany: Georg Thieme Verlag.

Scott, R. P. W. and P. Kucera, 1979. Mode of operation and performance characteristics of microbore columns for use in liquid chromatography. *J. Chromatogr.*, 169: 51–72.

Sewenig, S., D. Bullinger, U. Hener, and A. Mosandl, 2005. Comprehensive authentication of (E)-α(β)-ionone from raspberries, using constant flow MDGC-C/P-IRMS and enantio-MDGC-MS. *J. Agric. Food Chem.*, 53: 838–844.

Shema, J. and B. Fried (eds.), 2003. *Handbook of Thin-Layer Chromatography*, 3rd edn. New York: Marcel Dekker.

Snyder, J. K., K. Nakanishi, K. Hostettmann, and M. Hostettmann, 1984. Application of rotation locular countercurrent chromatography in natural products isolation. *J. Liquid Chromatogr.*, 7: 243–256.

Šorm, F., L. Dolejš, and J. Pliva, 1950. *Collect. Czechoslov. Chem. Commun.*, 3: 187.

Sprecher, E., 1963. Rücklaufapparatur zur erschöpfenden Wasserdampfdestillation ätherischen Öls aus voluminösem Destillationsgut. *Dtsch. Apoth. Ztg.*, 103: 213–214.

Stahl, E., 1953. Eine neue Apparatur zur gravimetrischen Erfassung kleinster Mengen ätherischer Öle. *Microchim. Acta*, 40: 367–372.

Stahl, E., 1969a. A thermo micro procedure for rapid extraction and direct application in thin-layer chromatography. *Analyst*, 723–727.

Stahl, E. (ed.), 1969b. *Thin-Layer Chromatography. A Laboratory Handbook*, 2nd edn. Berlin, Germany: Springer.

Stahl, E., 1976. Advances in the field of thermal procedures in direct combination with thin-layer chromatography. *Acc. Chem. Res.*, 9: 75–80.

Still, W. C., M. Kahn, and A. Mitra, 1978. Rapid chromatographic technique for preparative separations with moderate resolution. *J. Org. Chem.*, 43: 2923–2925.

Strack, D., P. Proksch, and P. G. Gülz, 1980. Reversed phase high performance liquid chromatography of essential oils. *Z. Naturforsch.*, 35c: 675–681.

Supelco, 2007. *Solid Phase Microextraction CD*, 6th edn. Bellefonte, PA: Supelco.

Takahisa, E. and K. H. Engel, 2005a. 2,3-Di-O-methoxyethyl-6-O-*tert*-butyl-dimethylsilyl-β-cyclodextrin, a useful stationary phase for gas chromatographic separation of enantiomers. *J. Chromatogr. A*, 1076: 148–154.

Takahisa, E. and K. H. Engel, 2005b. 2,3-Di-O-methoxymethyl-6-O-*tert*-butyl-dimethylsilyl-γ-cyclodextrin: A new class of cyclodextrin derivatives for gas chromatographic separation of enantiomers. *J. Chromatogr. A*, 1063: 181–192.

Tanimura, T., J. J. Pisano, Y. Ito, and R. L. Bowman, 1970. Droplet countercurrent chromatography. *Science*, 169: 54–56.

ten Noever de Brauw, M.C., 1979. Combined gas chromatography-mass spectrometry: A powerful tool in analytical chemistry. *J. Chromatogr.*, 165: 207–233.

Teranishi, R., T. R. Mon, A. B. Robinson, P. Cary, and L. Pauling, 1972. Gas chromatography of volatiles from breath and urine. *Anal. Chem.*, 44: 18–21.

Tienpont, B., F. David, C. Bicchi, and P. Sandra, 2000. High capacity headspace sorptive extraction. *J. Microcol. Sep.*, 12: 577–584.

Tilden, W. A., 1875. On the action of nitrosyl chloride on organic bodies. Part II. On turpentine oil. *J. Chem. Soc.*, 28: 514–518.

Tomi, F., P. Bradesi, A. Bighelli, and J. Casanova, 1995. Computer-aided identification of individual components of essential oils using carbon-13 NMR spectroscopy. *J. Magnet. Resonance Anal.*, 1995: 25–34.

Treibs, W., 1952. Über bi- und polycyclische Azulene. XIII. Das bicyclische Caryophyllen als Azulenbildner. *Liebigs Ann. Chem.*, 576: 125–131.

Tyihák, E., E. Mincsovics, and H. Kalász, 1979. New planar liquid chromatographic technique: Overpressured thin-layer chromatography. *J. Chromatogr.*, 174: 75–81.

van Beek, T. A. and D. Subrtova, 1995. Factors involved in the high pressure liquid chromatographic separation of alkenes by means of argentation chromatography on ion exchangers: Overview of theory and new practical developments. *Phytochem. Anal.*, 6: 1–19.

van Beek, T. A., N. van Dam, A. de Groot, T. A. M. Geelen, and L. H. W. van der Plas, 1994. Determination of the sesquiterpene dialdehyde polygodial by high-pressure liquid chromatography. *Phytochem. Anal.*, 5: 19–23.

von Baeyer, A. and O. Seuffert, 1901. Erschöpfende Bromierung des Menthons. *Ber. Dtsch. Chem. Ges.*, 34: 40–53.

Wagner, G., 1899. *J. Russ. Phys. Chem. Soc.*, 31: 690 (cited in H. Meerwein, 1914. *Liebigs Ann. Chem.*, 405: 129–175).

Wallach, O., 1914. *Terpene und Campher*, 2nd edn., Leipzig, Germany: Veit & Co.

Werkhoff, P., S. Brennecke, W. Bretschneider, M. Güntert, R. Hopp, and H. Surburg, 1993. Chirospecific analysis in essential oil, fragrance and flavor research. *Z. Lebensm. Unters. Forsch.*, 196: 307–328.

WILEY Registry, 2006. *Wiley Registry of Mass Spectral Data*, 8th edn. New York: Wiley.

Woodward, R. B. 1941. Structure and the absorption spectra of α, β-unsaturated ketones. *J. Am. Chem. Soc.*, 63: 1123–1126.

Wulf, L. W., C. W. Nagel, and A. L. Branen, 1978. High-pressure liquid chromatographic separation of the naturally occurring toxicants myristicin, related aromatic ethers and falcarinol. *J. Chromatogr.*, 161: 271–278.

Wylie, P. L. and B. D. Quimby, 1989. Applications of gas chromatography with atomic emission detector. *J. High Resolut. Chromatogr.*, 12: 813–818.

Yamauchi, Y. and M. Saito, 1990. Fractionation of lemon-peel oil by semi-preparative supercritical fluid-chromatography. *J. Chromatogr.*, 505: 237–246.

Yarita, T., A. Nomura, and Y. Horimoto, 1994. Type analysis of citrus essential oils by multidimensional supercritical fluid chromatography/gas chromatography. *Anal. Sci.*, 10: 25–29.

3 Sources of Essential Oils

Chlodwig Franz and Johannes Novak

CONTENTS

3.1 "Essential Oil–Bearing Plants": Attempt of a Definition 43
3.2 Phytochemical Variation 45
    3.2.1 Chemotaxonomy 45
    3.2.2 Inter- and Intraspecific Variation 46
        3.2.2.1 Lamiaceae (Labiatae) and Verbenaceae 46
        3.2.2.2 Asteraceae (Compositae) 52
3.3 Identification of Source Materials 55
3.4 Genetic and Protein Engineering 57
3.5 Resources of Essential Oils: Wild Collection or Cultivation of Plants 59
    3.5.1 Wild Collection and Sustainability 59
    3.5.2 Domestication and Systematic Cultivation 64
    3.5.3 Factors Influencing the Production and Quality of Essential Oil-Bearing Plants 65
        3.5.3.1 Genetic Variation and Plant Breeding 65
        3.5.3.2 Plant Breeding and Intellectual Property Rights 68
        3.5.3.3 Intraindividual Variation between Plant Parts and Depending on the Developmental Stage (*Morpho-* and *Ontogenetic Variation*) 69
        3.5.3.4 Environmental Influences 73
        3.5.3.5 Cultivation Measures, Contaminations, and Harvesting 74
3.6 International Standards for Wild Collection and Cultivation 76
    3.6.1 GA(C)P: Guidelines for Good Agricultural (and Collection) Practice of Medicinal and Aromatic Plants 76
    3.6.2 ISSC-MAP: The International Standard on Sustainable Wild Collection of Medicinal and Aromatic Plants 77
    3.6.3 FairWild 77
3.7 Conclusion 77
References 78

3.1 "ESSENTIAL OIL–BEARING PLANTS": ATTEMPT OF A DEFINITION

Essential oils are complex mixtures of volatile compounds produced by living organisms and isolated by physical means only (pressing and distillation) from a whole plant or plant part of known taxonomic origin. The respective main compounds are mainly derived from three biosynthetic pathways only, the mevalonate pathway leading to sesquiterpenes, the methyl-erythritol pathway leading to mono- and diterpenes, and the shikimic acid pathway *en route* to phenylpropenes. Nevertheless, there are an almost uncountable number of single substances and a tremendous variation in the composition of essential oils. Many of these volatile substances have diverse ecological functions. They can act as internal messengers, as defensive substances against herbivores, or as volatiles not only directing natural enemies to these herbivores but also attracting pollinating insects to their host (Harrewijn et al., 2001).

All plants possess principally the ability to produce volatile compounds, quite often, however, only in traces. "Essential oil plants" in particular are those plant species delivering an essential oil of commercial interest. Two principal circumstances determine a plant to be used as an essential oil plant:

1. A unique blend of volatiles like the flower scents in rose (*Rosa* spp.), jasmine (*Jasminum sambac*), or tuberose (*Polianthes tuberosa*). Such flowers produce and immediately emit the volatiles by the epidermal layers of their petals (Bergougnoux et al., 2007). Therefore, the yield is even in intensive smelling flowers very low, and besides distillation special techniques, as an example, enfleurage has to be applied to recover the volatile fragrance compounds.
2. Secretion and accumulation of volatiles in specialized anatomical structures. These lead to higher concentrations of the essential oil in the plant. Such anatomical storage structures for essential oils can be secretory idioblasts (secretory cells), cavities/ducts, or glandular trichomes (Fahn, 1979, 1988; colorfully documented by Svoboda et al. [2000]).

Secretory idioblasts are individual cells producing an essential oil in large quantities and retaining the oil within the cell like the essential oil idioblasts in the roots of *Vetiveria zizanioides* that occurs within the cortical layer and close to the endodermis (Bertea and Camusso, 2002). Similar structures containing essential oils are also formed in many flowers, for example, *Rosa* sp., *Viola* sp., or *Jasminum* sp.

Cavities or ducts consist of extracellular storage space that originate either by schizogeny (created by the dissolution of the middle lamella between the duct initials and formation of an intercellular space) or by lysogeny (programmed death and dissolution of cells). In both cases, the peripheral cells are becoming epithelial cells highly active in synthesis and secretion of their products into the extracellular cavities (Pickard, 2008). Schizogenic oil ducts are characteristic for the Apiaceae family, for example, *Carum carvi*, *Foeniculum vulgare*, or *Cuminum cyminum*, but also for the Hypericaceae or Pinaceae family. Lysogenic cavities are found in Rutaceae (*Citrus* sp., *Ruta graveolens*), Myrtaceae (e.g., *Syzygium aromaticum*), and others.

Secreting trichomes (glandular trichomes) can be divided into two main categories: peltate and capitate trichomes. Peltate glands consist of a basal epidermal cell, a neck–stalk cell, and a secreting head of 4–16 cells with a large subcuticular space on the apex in which the secretion product is accumulated. The capitate trichomes possess only 1–4 secreting cells with only a small subcuticular space (Werker, 1993; Maleci Bini and Giuliani, 2006). Such structures are typical for Lamiaceae (the mint family), but also for *Pelargonium* sp.

The monoterpene biosynthesis in different species of Lamiaceae, for example, sage (*Salvia officinalis*) and peppermint (*Mentha piperita*), is restricted to a brief period early in leaf development (Croteau et al., 1981; Gershenzon et al., 2000). The monoterpene biosynthesis in peppermint reaches a maximum in 15-day-old leaves; only very low rates were observed in leaves younger than 12 days or older than 20 days. The monoterpene content of the peppermint leaves increased rapidly up to day 21, then leveled off, and kept stable for the remainder of the leaf life (Gershenzon et al., 2000).

The composition of the essential oil often changes between different plant parts. Phytochemical polymorphism is often the case between different plant organs. In *Origanum vulgare* ssp. *hirtum*, a polymorphism within a plant could even be detected on a much lower level, between different oil glands of a leaf (Johnson et al., 2004). This form of polymorphism seems to be not frequently occurring; differences in the composition between oil glands are more often related to the age of the oil glands (Grassi et al., 2004; Johnson et al., 2004; Novak et al., 2006a; Schmiderer et al., 2008).

Such polymorphisms can also be found quite frequently when comparing the essential oil composition of individual plants of a distinct species (intraspecific variation, "chemotypes") and is based on the plants' genetic background.

The differences in the complex composition of two essential oils of one kind may sometimes be difficult to assign to specific chemotypes or to differences arising in the consequence of the reactions of the plants to specific environmental conditions, for example, to different growing locations. In general, the differences due to genetic differences are much bigger than by different environmental conditions. However, many intraspecific polymorphisms are probably not yet detected or have been described only recently even for widely used essential oil crops like sage (Novak et al., 2006b).

3.2 PHYTOCHEMICAL VARIATION

3.2.1 CHEMOTAXONOMY

The ability to accumulate essential oils is not omnipresent in plants but scattered throughout the plant kingdom, in many cases, however, very frequent within—or a typical character of—certain plant families. From the taxonomical and systematic point of view, not the production of essential oils is the distinctive feature since this is a quite heterogeneous group of substances, but either the type of secretory containers (trichomes, oil glands, lysogenic cavities, or schizogenic oil ducts) or the biosynthetically specific group of substances, for example, mono- or sesquiterpenes and phenylpropenes; the more a substance is deduced in the biosynthetic pathway, the more specific it is for certain taxa: monoterpenes are typical for the genus *Mentha*, but menthol is characteristic for *M. piperita* and *Mentha arvensis* ssp. *piperascens* only; sesquiterpenes are common in the *Achillea–millefolium* complex, but only *Achillea roseoalba* (2×) and *Achillea collina* (4×) are able to produce matricine as precursor of (the artifact) chamazulene (Vetter et al., 1997). On the other hand, the phenylpropanoid eugenol, typical for cloves (*S. aromaticum*, Myrtaceae), can also be found in large amounts in distant species, for example, cinnamon (*Cinnamomum zeylanicum*, Lauraceae) or basil (*Ocimum basilicum*, Lamiaceae); as sources for anethole are known not only aniseed (*Pimpinella anisum*) and fennel (*F. vulgare*), which are both Apiaceae, but also star anise (*Illicium verum*, Illiciaceae), *Clausena anisata* (Rutaceae), *Croton zehntneri* (Euphorbiaceae), or *Tagetes lucida* (Asteraceae). Finally, eucalyptol (1,8-cineole)—named after its occurrence in *Eucalyptus* sp. (Myrtaceae)—may also be a main compound of the essential oil of galangal (*Alpinia officinarum*, Zingiberaceae), bay laurel (*Laurus nobilis*, Lauraceae), Japan pepper (*Zanthoxylum piperitum*, Rutaceae), and a number of plants of the mint family, for example, sage (*S. officinalis*, *Salvia fruticosa*, *Salvia lavandulifolia*), rosemary (*Rosmarinus officinalis*), and mints (*Mentha* sp.). Taking the aforementioned facts into consideration, chemotaxonomically relevant are (therefore) common or distinct pathways, typical fingerprints, and either main compounds or very specific even minor or trace substances (e.g., δ-3-carene to separate *Citrus grandis* from other *Citrus* sp. [Gonzalez et al., 2002]).

The plant families comprising species that yield a majority of the most economically important essential oils are not restricted to one specialized taxonomic group but are distributed among all plant classes: gymnosperms, for example, the families Cupressaceae (cedarwood, cedar leaf, juniper oil, etc.) and Pinaceae (pine and fir oils), and angiosperms, and among them within Magnoliopsida, Rosopsida, and Liliopsida. The most important families of dicots are Apiaceae (e.g., fennel, coriander, and other aromatic seed/root oils), Asteraceae or Compositae (chamomile, wormwood, tarragon oil, a.s.o), Geraniaceae (geranium oil), Illiciaceae (star anise oil), Lamiaceae (mint, patchouli, lavender, oregano, and many other herb oils), Lauraceae (litsea, camphor, cinnamon, sassafras oil, etc.), Myristicaceae (nutmeg and mace), Myrtaceae (myrtle, cloves, and allspice), Oleaceae (jasmine oil), Rosaceae (rose oil), and Santalaceae (sandalwood oil). In monocots (Liliopsida), it is substantially restricted to Acoraceae (calamus), Poaceae (vetiver and aromatic grass oils), and Zingiberaceae (e.g., ginger and cardamom).

Apart from the phytochemical group of substances typical for a taxon, the chemical outfit depends, furthermore, on the specific genotype; the stage of plant development, also influenced by environmental factors; and the plant part (see Section 3.2.1). Considering all these influences, chemotaxonomic statements and conclusions have to be based on comparable material, grown and harvested under comparable circumstances.

3.2.2 Inter- and Intraspecific Variation

Knowledge on biochemical systematics and the inheritance of phytochemical characters depends on extensive investigations of taxa (particularly species) and populations on single-plant basis, respectively, and several examples of genera show that the taxa do indeed display different patterns.

3.2.2.1 Lamiaceae (Labiatae) and Verbenaceae

The presumably largest genus among the Lamiaceae is *sage* (*Salvia* L.) consisting of about 900 species widely distributed in the temperate, subtropical, and tropical regions all over the world with major centers of diversity in the Mediterranean, in Central Asia, the Altiplano from Mexico throughout Central and South America, and in southern Africa. Almost 400 species are used in traditional and modern medicine, as aromatic herbs or ornamentals worldwide; among them are *S. officinalis*, *S. fruticosa*, *Salvia sclarea*, *Salvia divinorum*, *Salvia miltiorrhiza*, and *Salvia pomifera*, to name a few. Many applications are based on nonvolatile compounds, for example, diterpenes and polyphenolic acids. Regarding the essential oil, there are a vast number of mono- and sesquiterpenes found in sage but, in contrast to, for example, *Ocimum* sp. and *Perilla* sp. (also Lamiaceae), no phenylpropenes were detected.

To understand species-specific differences within this genus, the Mediterranean *S. officinalis* complex (*S. officinalis*, *S. fruticosa*, and *S. lavandulifolia*) will be confronted with the *Salvia stenophylla* species complex (*S. stenophylla*, *Salvia repens*, and *Salvia runcinata*) indigenous to South Africa: in the *S. officinalis* group, usually α- and β-thujones, 1,8-cineole, camphor, and, in some cases, linalool, β-pinene, limonene, or *cis*-sabinyl acetate are the prevailing substances, whereas in the *S. stenophylla* complex, quite often sesquiterpenes, for example, caryophyllene or α-bisabolol, are main compounds.

Based on taxonomical studies of *Salvia* spp. (Hedge, 1992; Skoula et al., 2000; Reales et al., 2004) and a recent survey concerning the chemotaxonomy of *S. stenophylla* and its allies (Viljoen et al., 2006), Figure 3.1 shows the up-to-now-identified chemotypes within these taxa. Comparing the data of different publications, the picture is, however, not as clear as demonstrated by six *S. officinalis* origins in Figure 3.2 (Chalchat et al., 1998; Asllani, 2000). This might be due to the prevailing chemotype in a population, the variation between single plants, the time of sample collection, and the sample size. This is exemplarily shown by one *S. officinalis* population where the individuals varied in α-thujone, from 9% to 72%; β-thujone, from 2% to 24%; 1,8-cineole, from 4% to 18%; and camphor from 1% to 25%. The variation over 3 years and five harvests of one clone only ranged as follows: α-thujone 35%–72%, β-thujone 1%–7%, 1,8-cineole 8%–15%, and camphor 1%–18% (Bezzi, 1994; Bazina et al., 2002). But also all other (minor) compounds of the essential oil showed respective intraspecific variability (see, e.g., Giannouli and Kintzios, 2000).

S. fruticosa was principally understood to contain 1,8-cineole as main compound but at best traces of thujones, as confirmed by Putievsky et al. (1986) and Kanias et al. (1998). In a comparative study of several origins, Máthé et al. (1996) identified, however, a population with atypically high β-thujone similar to *S. officinalis*. Doubts on if this origin could be true *S. fruticosa* or a spontaneous hybrid of both species were resolved by extensive investigations on the phytochemical and genetic diversities of *S. fruticosa* in Crete (Karousou et al., 1998; Skoula et al., 1999). There, it was shown that all wild populations in western Crete consist of 1,8-cineole chemotypes only, whereas in the eastern part of the island, essential oils with up to 30% thujones, mainly β-thujone, could be observed. In central Crete, finally, mixed populations were found. A cluster analysis based on

Sources of Essential Oils

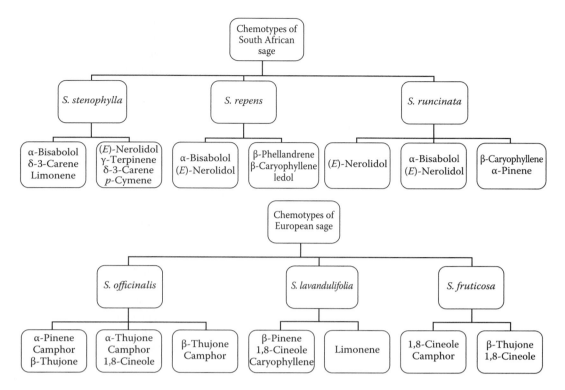

FIGURE 3.1 Chemotypes of some South African and European *Salvia* species.

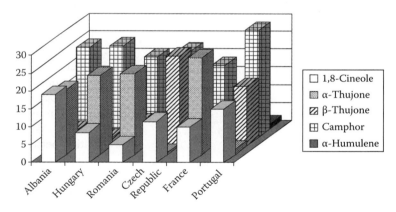

FIGURE 3.2 Composition of the essential oil of six *Salvia officinalis* origins.

random amplification of polymorphic DNA (RAPD) patterns confirmed the genetic differences between the west and east Crete populations of *S. fruticosa* (Skoula et al., 1999).

A rather interesting example of diversity is *oregano*, which counts to the commercially most valued spices worldwide. More than 60 plant species are used under this common name showing similar flavor profiles characterized mainly by cymyl compounds, for example, carvacrol and thymol. With few exemptions, the majority of oregano species belong to the Lamiaceae and Verbenaceae families with the main genera *Origanum* and *Lippia* (Table 3.1). In 1989, almost all of the estimated 15,000 ton/year dried oregano originated from wild collection; today, some 7000 ha of *Origanum onites* are cultivated in Turkey alone (Baser, 2002); *O. onites* and other *Origanum* species are cultivated in Greece, Israel, Italy, Morocco, and other countries.

TABLE 3.1
Species Used Commercially in the World as Oregano

Family/Species	Commercial Name(s) Found in Literature
Labiatae	
Calamintha potosina Schaf.	Oregano de la sierra, oregano, origanum
Coleus amboinicus Lour. (syn. *C. aromaticus* Benth)	Oregano, oregano brujo, oregano de Cartagena, oregano de Espana, oregano Frances
Coleus aromaticus Benth.	Oregano de Espana, oregano, *Origanum*
Hedeoma floribunda Standl.	Oregano, *Origanum*
Hedeoma incona Torr.	Oregano
Hedeoma patens Jones	Oregano, *Origanum*
Hyptis albida HBK.	Oregano, *Origanum*
Hyptis americana (Aubl.) Urb. (*H. gonocephala* Gris.)	Oregano
Hyptis capitata Jacq.	Oregano, *Origanum*
Hyptis pectinata Poit.	Oregano, *Origanum*
Hyptis suaveolens (L.) Poit.	Oregano, oregano cimarron, *Origanum*
Monarda austromontana Epling	Oregano, *Origanum*
Ocimum basilicum L.	Oregano, *Origanum*
Origanum compactum Benth. (syn. *O. glandulosum* Salzm, ex Benth.)	Oregano, *Origanum*
Origanum dictamnus L. (*Majorana dictamnus* L.)	Oregano, *Origanum*
Origanum elongatum (Bonent) Emberger et Maire	Oregano, *Origanum*
Origanum floribundum Munby (*O. cinereum* Noe)	Oregano, *Origanum*
Origanum grosii Pau et Font Quer ex Ietswaart	Oregano, *Origanum*
Origanum majorana L.	Oregano
Origanum microphyllum (Benth) Vogel	Oregano, *Origanum*
Origanum onites L. (syn. *O. smyrneum* L.)	Turkish oregano, oregano, *Origanum*[a]
Origanum scabrum Boiss et Heldr. (syn. *O. pulchrum* Boiss et Heldr.)	Oregano, *Origanum*
Origanum syriacum L. var. *syriacum* (syn. *O. maru* L.)	Oregano, *Origanum*
Origanum vulgare L. ssp. *gracile* (Koch) Ietswaart (syn. *O. gracile* Koch, *O. tyttanthum* Gontscharov)	Oregano, *Origanum*
Origanum vulgare ssp. *hirtum* (Link) Ietswaart (syn. *O. hirtum* Link)	Oregano, *Origanum*
Origanum vulgare ssp. *virens* (Hoffmanns et Link) Ietswaart (syn. *O. virens* Hoffmanns et Link)	Oregano, *Origanum*, oregano verde
Origanum vulgare ssp. *viride* (Boiss.) Hayek (syn. *O. viride*) Halacsy (syn. *O. heracleoticum* L.)	Greek oregano, oregano, *Origanum*[a]
Origanum vulgare L. ssp. *vulgare* (syn. *Thymus origanum* (L.) Kuntze)	Oregano, *Origanum*
Origanum vulgare L.	Oregano, orenga, Oregano de Espana
Poliomintha longiflora Gray	Oregano
Salvia sp.	Oregano
Satureja thymbra L.	Oregano cabruno, oregano, *Origanum*
Thymus capitatus (L.) Hoffmanns et Link (syn. *Coridothymus capitatus* (L.) Rchb.f.)	Spanish oregano, oregano, *Origanum*[a]
Verbenaceae	
Lantana citrosa (Small) Modenke	Oregano xiu, oregano, *Origanum*
Lantana glandulosissima Hayek	Oregano xiu, oregano silvestre, oregano, *Origanum*

(Continued)

TABLE 3.1 (*Continued*)
Species Used Commercially in the World as Oregano

Family/Species	Commercial Name(s) Found in Literature
Lantana hirsuta Mart et Gall.	Oreganillo del monte, oregano, *Origanum*
Lantana involucrata L.	Oregano, *Origanum*
Lantana purpurea (Jacq.) Benth. & Hook. (syn. *Lippia purpurea* Jacq.)	Oregano, *Origanum*
Lantana trifolia L.	Oregano, *Origanum*
Lantana velutina Mart. & Gal.	Oregano xiu, oregano, *Origanum*
Lippia myriocephala Schlecht. & Cham.	Oreganillo
Lippia affinis Schau.	Oregano
Lippia alba (Mill) N.E. Br. (syn. *L. involucrata* L.)	Oregano, *Origanum*
Lippia berlandieri Schau.	Oregano
Lippia cordiostegia Benth.	Oreganillo, oregano montes, oregano, *Origanum*
Lippia formosa T.S. Brandeg.	Oregano, *Origanum*
Lippia geisseana (R.A.Phil.) Soler.	Oregano, *Origanum*
Lippia graveolens HBK	Mexican oregano, oregano cimarron, oregano[a]
Lippia helleri Britton	Oregano del pais, oregano, *Origanum*
Lippia micromera Schau.	Oregano del pais, oregano, *Origanum*
Lippia micromera var. *helleri* (Britton) Moldenke	Oregano
Lippia origanoides HBK	Oregano, origano del pais
Lippia palmeri var. *spicata* Rose	Oregano
Lippia palmeri Wats.	Oregano, *Origanum*
Lippia umbellata Cav.	Oreganillo, oregano montes, oregano, *Origanum*
Lippia velutina Mart. et Galeotti	Oregano, *Origanum*
Rubiaceae	
Borreria sp.	Oreganos, oregano, *Origanum*
Scrophulariaceae	
Limnophila stolonifera (Blanco) Merr.	Oregano, *Origanum*
Apiaceae	
Eryngium foetidum L.	Oregano de Cartagena, oregano, *Origanum*
Asteraceae	
Coleosanthus veronicaefolius HBK	Oregano del cerro, oregano del monte, oregano del campo
Eupatorium macrophyllum L. (syn. *Hebeclinium macrophyllum* DC.)	Oregano, *Origanum*

[a] Oregano species with economic importance according to Lawrence (1984).

In comparison with sage, the genus *Origanum* is much smaller and consists of 43 species and 18 hybrids according to the actual classification (Skoula and Harborne, 2002) with main distribution areas around the Mediterranean. Some subspecies of *O. vulgare* only are also found in the temperate and arid zones of Eurasia up to China. Nevertheless, the genus is characterized by large morphological and phytochemical diversities (Kokkini, 1996; Baser, 2002; Skoula and Harborne, 2002).

The occurrence of several chemotypes is reported, for example, for commercially used *Origanum* species, from Turkey (Baser, 2002). In *O. onites*, two chemotypes are described, a carvacrol type and a linalool type. Additionally, a *mixed type* with both basic types mixed may occur. In Turkey, two chemotypes of *Origanum majorana* are known, one contains *cis*-sabinene hydrate as chemotypical lead compound and is used as marjoram in cooking (*marjoramy*), while the other one contains carvacrol in high amounts and is used to distil *oregano oil* in a commercial scale.

Variability of chemotypes continues also within the *marjoramy O. majorana*. Novak et al. (2002) detected in cultivated marjoram accessions additionally to *cis*-sabinene hydrate the occurrence of polymorphism of *cis*-sabinene hydrate acetate. Since this chemotype did not influence the sensorial impression much, this chemotype was not eliminated in breeding, while an *off-flavor* chemotype would have been certainly eliminated in its cultivation history. In natural populations of *O. majorana* from Cyprus besides the *classical cis*-sabinene hydrate type, a chemotype with α-terpineol as main compound was also detected (Novak et al., 2008). The two extreme *off-flavor* chemotypes in *O. majorana*, carvacrol and α-terpineol chemotypes, are not to be found anywhere in cultivated marjoram, demonstrating one of the advantages of cultivation in delivering homogeneous qualities.

The second *oregano* of commercial value—mainly used in the Americas—is *Mexican oregano* (*Lippia graveolens* HBK., Verbenaceae) endemic to California, Mexico, and throughout Central America (Fischer, 1998). Due to wild harvesting, only a few published data show essential oil contents largely ranging from 0.3% to 3.6%. The total number of up-to-now-identified essential oil compounds comprises almost 70 with the main constituents thymol (3.1%–80.6%), carvacrol (0.5%–71.2%), 1,8-cineole (0.1%–14%), and *p*-cymene (2.7%–28.0%), followed by, for example, myrcene, γ-terpinene, and the sesquiterpene caryophyllene (Lawrence, 1984; Dominguez et al., 1989; Uribe-Hernández et al., 1992; Fischer et al., 1996; Vernin, 2001).

In a comprehensive investigation of wild populations of *L. graveolens* collected from the hilly regions of Guatemala, three different essential oil chemotypes could be identified, a thymol, a carvacrol, and an absolutely irregular type (Fischer et al., 1996). Within the thymol type, contents of up to 85% thymol in the essential oil could be obtained and only traces of carvacrol. The irregular type has shown a very uncommon composition where no compound exceeds 10% of the oil, and also phenylpropenes, for example, eugenol and methyl eugenol, were present (Fischer et al., 1996; Fischer, 1998). In Table 3.2, a comparison of recent data is given including *Lippia alba*, commonly called *oregano* or *oregano del monte*, although carvacrol and thymol are absent from the essential oil of this species. In Guatemala, two different chemotypes were found within *L. alba*: a myrcenone and a citral type (Fischer et al., 2004). Besides it, a linalool, a carvone, a camphor (1,8-cineole), and a limonene–piperitone chemotype have been described (Dellacassa et al., 1990; Pino et al., 1997; Frighetto et al., 1998; Senatore and Rigano, 2001).

Chemical diversity is of special interest if on genus or species level both terpenes and phenylpropenes can be found in the essential oil. Most Lamiaceae preferentially accumulate mono- and sesquiterpenes in their volatile oils, but some genera produce oils also rich in phenylpropenes, among these *Ocimum* sp. and *Perilla* sp.

The genus *Ocimum* comprises over 60 species, of which *Ocimum gratissimum* and *O. basilicum* are of high economic value. Biogenetic studies on the inheritance of *Ocimum* oil constituents were reported by Khosla et al. (1989) and an *O. gratissimum* strain named *clocimum* containing 65% of eugenol in its oil was described by Bradu et al. (1989). A number of different chemotypes of basil (*O. basilicum*) have been identified and classified (Vernin, 1984; Marotti et al., 1996) containing up to 80% linalool, up to 21.5% 1,8-cineole, 0.3%–33.0% eugenol, and also the presumably toxic compounds methyl chavicol (estragole) and methyl eugenol in concentrations close to 50% (Elementi et al., 2006; Macchia et al., 2006).

Perilla frutescens can be classified in several chemotypes as well according to the main monoterpene components perillaldehyde, elsholtzia ketone, or perilla ketones and on the other side phenylpropanoid types containing myristicin, dillapiole, or elemicin (Koezuka et al., 1986). A comprehensive presentation on the chemotypes and the inheritance of the mentioned compounds was given by this author in Hay and Waterman (1993). In the referred last two examples, not only the sensorial but also the toxicological properties of the essential oil compounds are decisive for the (further) commercial use of the respective species' biodiversity.

Although the Labiatae family plays an outstanding role as regards the chemical polymorphism of essential oils, also in other essential oils containing plant families and genera, a comparable phytochemical diversity can be observed.

TABLE 3.2
Main Essential Oil Compounds of *Lippia graveolens* and *L. alba* According to Recent Data

	L. graveolens					*L. alba*			
	Fischer et al. (1996)			Senatore and Rigano (2001)	Vernin et al. (2001)	Fischer (1998)		Senatore and Rigano (2001)	Lorenzo et al. (2001)
	Guatemala			Guatemala	El-Salvador	Guatemala		Guatemala	Uruguay
Compound	Thymol-Type	Carvacrol-Type	Irregular Type			Myrcenone-Type	Cineole-Type		
Myrcene	1.3	1.9	2.7	1.1	t	6.5	1.7	0.2	0.8
p-Cymene	2.7	6.9	2.8	5.5	2.1	t	t	0.7	n.d.
1,8-Cineole	0.1	0.6	5.0	2.1	t	t	**22.8**	**14.2**	1.3
Limonene	0.2	0.3	1.5	0.8	t	1.0	3.2	**43.6**	2.9
Linalool	0.7	1.4	3.8	0.3	t	4.0	2.4	1.2	**55.3**
Myrcenon	n.d.	n.d.	n.d.	n.d.	n.d.	**54.6**	3.2	n.d.	n.d.
Piperitone	n.d.	n.d.	n.d.	**31.6**	n.d.	t	t	**30.6**	n.d.
Thymol	**80.6**	**19.9**	6.8	0.8	7.3	n.d.	n.d.	n.d.	n.d.
Carvacrol	1.3	**45.2**	1.1	4.6	**71.2**	n.d.	n.d.	n.d.	n.d.
β-Caryophyllene	2.8	3.5	8.7	3.0	9.2	2.6	1.2	1.0	9.0
α-Humulene	1.9	2.3	5.7	4.8	5.0	0.7	t	0.6	0.9
Caryophyll.-ox.	0.3	0.8	3.3	n.d.	t	1.8	3.0	1.1	0.6
Z-Dihydrocarvon/Z-Ocimenone	n.d.	n.d.	n.d.	n.d.	n.d.	13.1	0.6	0.1	0.8
E-Dihydrocarvon	n.d.	n.d.	n.d.	n.d.	n.d.	4.9	n.d.	t	1.2

Note: n.d., Not detectable; t, traces. Main compounds in bold.

3.2.2.2 Asteraceae (Compositae)

Only a limited number of genera of the Asteraceae are known as essential oil plants, among them *Tagetes*, *Achillea*, and *Matricaria*. The genus *Tagetes* comprises actually 55 species, all of them endemic to the American continents with the center of biodiversity between 30° northern and 30° southern latitude. One of the species largely used by the indigenous population is *pericon* (*T. lucida* Cav.), widely distributed over the highlands of Mexico and Central America (Stanley and Steyermark, 1976). In contrast to almost all other *Tagetes* species characterized by the content of tagetones, this species contains phenylpropenes and terpenes. A detailed study on its diversity in Guatemala resulted in the identification of several eco- and chemotypes (Table 3.3): anethole, methyl chavicol (estragole), methyl eugenol, and one sesquiterpene type producing higher amounts of nerolidol (Bicchi et al., 1997; Goehler, 2006). The distribution of the three main phenylpropenes in six populations is illustrated in Figure 3.3. In comparison with the plant materials investigated by Ciccio (2004) and Marotti et al. (2004) containing oils with 90%–95% estragole, only the germplasm collection of Guatemaltecan provenances (Goehler, 2006) allows to select individuals with high anethole but low to very low estragole or methyleugenol content—or with interestingly high nerolidol content, as mentioned earlier.

TABLE 3.3
Main Compounds of the Essential Oil of Selected *Tagetes lucida* Types (in% of dm)

Substance	Anethole Type (2)	Estragole Type (8)	Methyleugenol Type (7)	Nerolidol Type (5)	Mixed Type
Linalool	0.26	0.69	1.01	Tr.	3.68
Estragole	11.57	**78.02**	8.68	3.23	**24.28**
Anethole	**73.56**	0.75	0.52	Tr.	**30.17**
Methyleugenol	1.75	5.50	**79.80**	17.76	**17.09**
β-Caryophyllene	0.45	1.66	0.45	2.39	0.88
Germacrene D	2.43	2.89	1.90	Tr.	5.41
Methylisoeugenol	1.42	2.78	2.00	Tr.	3.88
Nerolidol	0.35	0.32	0.31	**40.52**	1.24
Spathulenol	0.10	0.16	0.12	Tr.	0.23
Carophyllene oxide	0.05	0.27	0.45	10.34	0.53

Note: Location of origin in Guatemala: (2) Cabrican/Quetzaltenango, (5) La Fuente/Jalapa, (7) Joyabaj/El Quiche, (8) Sipacapa/S. Marcos, Mixed Type: Taltimiche/San Marcos. Main compounds in bold.

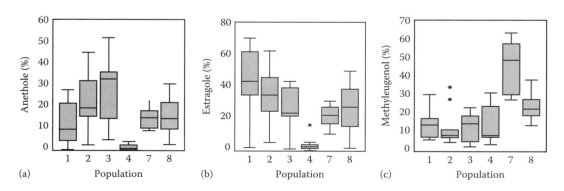

FIGURE 3.3 Variability of (a) anethole, (b) methyl chavicol (estragole), and (c) methyl eugenol in the essential oil of six *Tagetes lucida*—populations from Guatemala. * indicates erratic individuals.

TABLE 3.4
Taxa within the *Achillea-Millefolium*-Group (Yarrow)

Taxon	Ploidy Level	Main Compounds
A. setacea W. et K.	2×	Rupicoline
A. aspleniifolia Vent.	2× (4×)	**7,8-Guajanolide**
		Artabsin-derivatives
		3-Oxa-Guajanolide
A. roseo-alba Ehrend.	2×	Artabsin-derivatives
		3-Oxaguajanolide
		Matricinderivatives
A. collina Becker	4×	Artabsin-derivatives
		3-Oxaguajanolide
		Matricinderivatives
		Matricarinderivatives
A. pratensis Saukel u. Länger	4×	Eudesmanolides
A. distans ssp. Distans W. et K.	6×	Longipinenones
A. distans ssp. styriaca	4×	
A. tanacetifolia (stricta) W. et K.	6×	
A. mill. ssp. sudetica	6×	Guajanolidperoxide
A. mill. ssp. Mill. L.	6×	
A. pannonica Scheele	8× (6×)	Germacrene
		Guajanolidperoxide

Source: Franz, Ch., in *Handbuch des Arznei- u. Gewuerzpflanzenbaus Vol. 5*, Saluplanta, Bernburg, 2013, pp. 453–463.
Note: Substances in bold are proazulenes.

The genus *Achillea* is widely distributed over the northern hemisphere and consists of approximately 120 species, of which the *Achillea millefolium* aggregate (yarrow) represents a polyploid complex of allogamous perennials (Saukel and Länger, 1992; Vetter and Franz, 1996). The different taxa of the recent classification (*minor species* and *subspecies*) are morphologically and chemically to a certain extent distinct and only the diploid taxa *Achillea asplenifolia* and *A. roseoalba* as well as the tetraploids *A. collina* and *Achillea ceretanica* are characterized by proazulens, for example, achillicin, whereas the other taxa, especially 6× and 8×, contain eudesmanolides, longipinenes, germacranolides, and/or guajanolid peroxides (Table 3.4). The intraspecific variation in the proazulene content ranged from traces up to 80%; other essential oil components of the azulenogenic species are, for example, α- and β-pinene, borneol, camphor, sabinene, or caryophyllene (Kastner et al., 1992). The frequency distribution of proazulene individuals among two populations is shown in Figure 3.4.

Crossing experiments resulted in proazulene being a recessive character of di- and tetraploid *Achillea* spp. (Vetter et al., 1997) similar to chamomile (Franz, 1993a,b). Finally, according to Steinlesberger (2002) also a plant-to-plant variation in the enantiomers of, for example, α- and β-pinene as well as sabinene exists in yarrow oils, which makes it even more complicated to use phytochemical characters for taxonomical purposes.

Differences in the essential oil content and composition of chamomile flowers (*Matricaria recutita*) have long been recognized due to the fact that the distilled oil is either dark blue, green, or yellow, depending on the prochamazulene content (matricin as prochamazulene in chamomile is transformed to the blue-colored artifact chamazulene during the distillation process). Recognizing also the great pharmacological potential of the bisabolols, a classification into the chemotypes (–)-α-bisabolol, (–)-α-bisabololoxide A, (–)-α-bisabololoxide B, (–)-α-bisabolonoxide (A), and (pro)

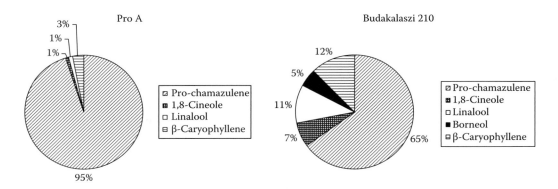

FIGURE 3.4 Frequency distribution of proazulene individuals among two *Achillea* sp. populations.

chamazulene was made by Franz (1982, 1989a). Examining the geographical distribution revealed a regional differentiation, where an α-bisabolol—(pro) chamazulene population was identified on the Iberian peninsula; mixed populations containing chamazulene, bisabolol, and bisabololoxides A/B are most frequent in Central Europe, and prochamazulene—free bisabolonoxide populations are indigenous to southeast Europe and minor Asia. In the meantime, Wogiatzi et al. (1999) have shown for Greece and Taviani et al. (2002) for Italy a higher diversity of chamomile including α-bisabolol types. This classification of populations and chemotypes was extended by analyzing populations at the level of individual plants (Schröder, 1990) resulting in the respective frequency distributions (Figure 3.5).

In addition, the range of essential oil components in the chemotypes of one Central European population is shown in Table 3.5 (Franz, 2000).

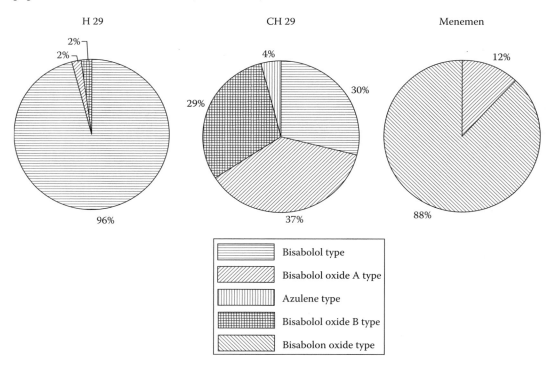

FIGURE 3.5 Frequency distribution of chemotypes in three varieties/populatios of chamomile (*Matricaria recutita* (L.) Rauschert).

TABLE 3.5
Grouping within a European Spontaneous Chamomile, Figures in% of Terpenoids in the Essential Oil of the Flower Heads

	Chamazulen	α-Bisabolol	α-B.-Oxide A	α-B.-Oxide B
α-Bisabolol-type				
Range	2.5–35.2	58.8–92.1	n.d.–1.0	n.d.–3.2
Mean	23.2	**68.8**	n.d.	n.d.
α-Bisabololoxide A-type				
Range	6.6–31.2	0.5–12.3	31.7–66.7	1.9–22.4
Mean	21.3	2.1	**53.9**	11.8
α-Bisabololoxide B-type				
Range	7.6–24.2	0.8–6.5	1.6–4.8	61.6–80.5
Mean	16.8	2.0	2.6	**72.2**
Chamazulene-type				
Range	76.3–79.2	5.8–8.3	n.d.–0.8	n.d.–2.6
Mean	**77.8**	7.1	n.d.	n.d.

Source: Franz, Ch., Biodiversity and random sampling in essential oil plants, Lecture 31st ISEO, Hamburg, Germany, 2000.
Note: Main compounds in bold. n.d., not detected (determined).

Data on inter- and intraspecific variation of essential oils are countless, and recent reviews are known for a number of genera published, for example, in the series "Medicinal and Aromatic Plants—Industrial Profiles" (Harwood Publications, Taylor & Francis, CRC Press, respectively).

The generally observed quantitative and qualitative variations in essential oils draw the attention *i.a.* to appropriate random sampling for getting valid information on the chemical profile of a species or population. As concerns quantitative variations of a certain pattern or substance, Figure 3.6 shows exemplarily the bisabolol content of two chamomile populations depending on the number of individual plants used for sampling. At small numbers, the mean value oscillates strongly, and only after at least 15–20 individuals the range of variation becomes acceptable. Quite different appears the situation at qualitative differences, that is, *either–or variations* within populations or taxa, for example, carvacrol/thymol, α-/β-thujone/1,8-cineole/camphor, or monoterpenes/phenylpropenes. Any random sample may give nonspecific information only on the principal chemical profile of the respective population provided that the sample is representative. This depends on the number of chemotypes, their inheritance, and frequency distribution within the population, and generally speaking, no less than 50 individuals are needed for that purpose, as it can be derived from the comparison of chemotypes in a *Thymus vulgaris* population (Figure 3.7).

The overall high variation in essential oil compositions can be explained by the fact that quite different products might be generated by small changes in the synthase sequences only. On the other hand, different synthases may be able to produce the same substance in systematically distant taxa. The different origin of such substances can be identified by, for example, the $^{12}C/^{13}C$ ratio (Mosandl, 1993). Bazina et al. (2002) stated, "Hence, a simple quantitative analysis of the essential oil composition is not necessarily appropriate for estimating genetic proximity even in closely related taxa."

3.3 IDENTIFICATION OF SOURCE MATERIALS

As illustrated by the previous paragraph, one of the crucial points of using plants as sources for essential oils is their heterogeneity. A first prerequisite for reproducible compositions is therefore an unambiguous botanical identification and characterization of the starting material. The first

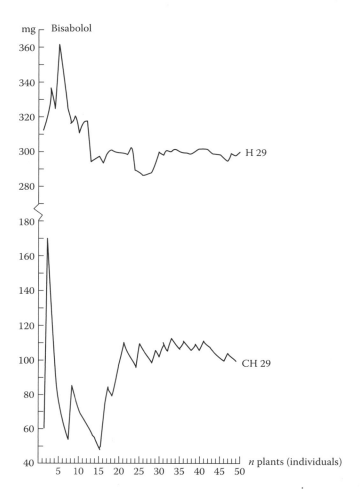

FIGURE 3.6 (−)-α-Bisabolol-content (mg/100 g crude drug) in two chamomile (*Matricaria recutita*) populations: mean value in dependence of the number of individuals used for sampling.

approach is the classical taxonomical identification of plant materials based on macro- and micromorphological features of the plant. The identification is followed by phytochemical analysis that may contribute to species identification as well as to the determination of the quality of the essential oil. This approach is now complemented by DNA-based identification.

DNA is a long polymer of nucleotides, the building units. One of four possible nitrogenous bases is part of each nucleotide, and the sequence of the bases on the polymer strand is characteristic for each living individual. Some regions of the DNA, however, are conserved on the species or family level and can be used to study the relationship of taxa (Taberlet et al., 1991; Wolfe and Liston, 1998). DNA sequences conserved within a taxon but different between taxa can therefore be used to identify a taxon (*DNA barcoding*) (Hebert et al., 2003; Kress et al., 2005). A DNA-barcoding consortium was founded in 2004 with the ambitious goal to build a barcode library for all eukaryotic life in the next 20 years (Ratnasingham and Hebert, 2007). New sequencing technologies (454, Solexa, SOLiD) enable a fast and representative analysis, but will be applied due to their high costs in the moment only in the next phase of DNA barcoding (Frezal and Leblois, 2008). DNA barcoding of animals has already become a routine task. DNA barcoding of plants, however, is still not trivial and a scientific challenge (Pennisi, 2007).

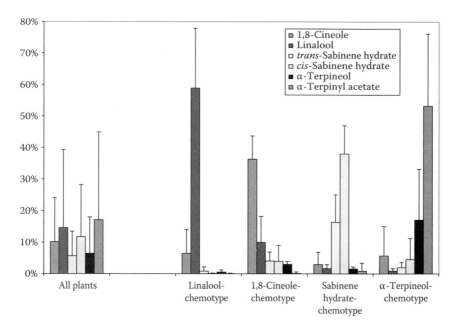

FIGURE 3.7 Mean values of the principal essential compounds of a *Thymus vulgaris* population (left) in comparison to the mean values of the chemotypes within the same population.

Besides sequence information–based approaches, multilocus DNA methods (RAPD, amplified fragment length polymorphism, etc.) are complementing in resolving complicated taxa and can become a barcode for the identification of populations and cultivars (Weising et al., 2005). With multilocus DNA methods, it is furthermore possible to tag a specific feature of a plant of which the genetic basis is still unknown. This approach is called molecular markers (in *sensu strictu*) because they mark the occurrence of a specific trait like a chemotype or flower color. The gene regions visualized, for example, on an agarose gel are not the specific gene responsible for a trait but are located on the genome in the vicinity of this gene and therefore co-occur with the trait and are absent when the trait is absent. An example for such an inexpensive and fast polymerase chain reaction system was developed by Bradbury et al. (2005) to distinguish fragrant from nonfragrant rice cultivars. If markers would be developed for chemotypes in essential oil plants, species identification by DNA and the determination of a chemotype could be performed in one step.

Molecular biological methods to identify species are nowadays routinely used in feed- and foodstuffs to identify microbes, animals, and plants. Especially the discussion about traceability of genetically modified organisms (GMOs) throughout the complete chain ("from the living organism to the supermarket") has sped up research in this area (Auer, 2003; Miraglia et al., 2004). One advantage of molecular biological methods is the possibility to be used in a number of processed materials like fatty oil (Pafundo et al., 2005) or even solvent extracts (Novak et al., 2007). The presence of minor amounts of DNA in an essential oil cannot be excluded *a priori*, although distillation as separation technique would suggest the absence of DNA. However, small plant or DNA fragments could distill over or the essential oil could come in contact with plant material after distillation.

3.4 GENETIC AND PROTEIN ENGINEERING

Genetic engineering is defined as the direct manipulation of the genes of organisms by laboratory techniques, not to be confused with the indirect manipulation of genes in traditional (plant) breeding. *Transgenic or GMOs* are organisms (bacteria, plants, etc.) that have been engineered with single or multiple genes (either from the same species or from a different species), using contemporary molecular biology techniques. These are organisms with improved characteristics, in plants, for example, with resistance or tolerance to biotic or abiotic stresses such as insects, disease, drought, salinity, and temperature. Another important goal in improving agricultural production conditions is to facilitate weed control by transformed plant resistant to broadband herbicides like glufosinate. Peppermint has been successfully transformed with the introduction of the bar gene, which encodes phosphinothricin acetyltransferase, an enzyme inactivating glufosinate ammonium or the ammonium salt of glufosinate, phosphinothricin, making the plant insensitive to the systemic, broad-spectrum herbicide Roundup (*Roundup Ready mint*) (Li et al., 2001).

A first step in genetic engineering is the development and optimization of transformation (gene transfer) protocols for the target species. Such optimized protocols exist for essential oil plants such as lavandin (*Lavandula × intermedia*; Dronne et al., 1999), spike lavender (*Lavandula latifolia*; Nebauer et al., 2000), and peppermint (*M. piperita*; Diemer et al., 1998; Niu et al., 2000).

In spike lavender, an additional copy of the 1-deoxy-D-xylulose-5-phosphate synthase gene, the first enzymatic step in the methylerythritol phosphate (MEP) pathway leading to the precursors of monoterpenes, from *Arabidopsis thaliana*, was introduced and led to an increase of the essential oil of the leaves of up to 360% and of the essential oil of flowers of up to 74% (Munoz-Bertomeu et al., 2006).

In peppermint, many different steps to alter essential oil yield and composition were already targeted (reviewed by Wildung and Croteau 2005; Table 3.6). The overexpression of deoxyxylulose phosphate reductoisomerase (DXR), the second step in the MEP pathway, increased the essential oil yield by approximately 40% tested under field conditions (Mahmoud and Croteau, 2001). The overexpression of geranyl diphosphate synthase leads to a similar increase of the essential oil. Menthofuran, an undesired compound, was downregulated by an antisense method (a method to influence or block the activity of a specific gene). Overexpression of the menthofuran antisense RNA was responsible for an improved oil quality by reducing both menthofuran and pulegone in one transformation step (Mahmoud and Croteau, 2003). The ability to produce a peppermint oil with a new composition was demonstrated by Mahmoud et al. (2004) by upregulating limonene by cosuppression of limonene-3-hydroxylase, the enzyme responsible for the transformation of (−)-limonene to (−)-*trans*-isopiperitenol *en route* to menthol.

Protein engineering is the application of scientific methods (mathematical and laboratory methods) to develop useful or valuable proteins. There are two general strategies for protein engineering, random mutagenesis and rational design. In rational design, detailed knowledge of the structure and function of the protein is necessary to make desired changes by site-directed mutagenesis, a technique already well developed. An impressive example of the rational design of monoterpene synthases was given by Kampranis et al. (2007) who converted a 1,8-cineole synthase from *S. fruticosa* into a synthase producing sabinene, the precursor of α- and β-thujones with a minimum number of substitutions. They went also a step further and converted this monoterpene synthase into a

TABLE 3.6
Essential Oil Composition and Yield of Transgenic Peppermint Transformed with Genes Involved in Monoterpene Biosynthesis

Gene	Method	Limonene	Mentho-Furan	Pulegone	Menthone	Menthol	Oil Yield (lb/acre)
WT	—	1.7	4.3	2.1	20.5	44.5	97.8
DXR	Overexpress	1.6	3.6	1.8	19.6	45.6	137.9
MFS	Antisense	1.7	1.2	0.4	22.7	45.2	109.7
l-3-h	Cosuppress	74.7	0.4	0.1	4.1	3.0	99.6

Source: Wildung, M.R. and R.B. Croteau, *Transgenic Res.*, 14, 365, 2005.
Note: DXR, Deoxyxylulose phosphate reductoisomerase; l-3-h, limonene-3-hydroxylase; MFS, menthofuran synthase; WT, wild type.

sesquiterpene synthase by substituting a single amino acid that enlarged the cavity of the active site enough to accommodate the larger precursor of the sesquiterpenes, farnesyl pyrophosphate.

3.5 RESOURCES OF ESSENTIAL OILS: WILD COLLECTION OR CULTIVATION OF PLANTS

The raw materials for producing essential oil are resourced either from collecting them in nature (*wild collection*) or from cultivating the plants (Table 3.7).

3.5.1 WILD COLLECTION AND SUSTAINABILITY

Since prehistoric times, humans have gathered wild plants for different purposes; among them are aromatic, essential oil–bearing species used as culinary herbs, spices, flavoring agents, and fragrances. With increasing demand of standardized, homogeneous raw material in the industrial societies, more and more wild species have been domesticated and systematically cultivated. Nevertheless, a high number of species are still collected from the wild due to the fact that

- Many plants and plant products are used for the subsistence of the rural population.
- Small quantities of the respective species are requested at the market only which make a systematic cultivation not profitable.
- Some species are difficult to cultivate (slow growth rate and requirement of a special microclimate).
- Market uncertainties or political circumstances do not allow investing in long-term cultivation.
- The market is in favor of *ecologically* or *naturally* labeled wild collected material.

Especially—but not only—in developing countries, parts of the rural population depend economically on gathering high-value plant material. Less than two decades ago, almost all oregano (crude drug as well as essential oil) worldwide came from wild collection (Padulosi, 1996) and even this well-known group of species (*Origanum* sp. and *Lippia* sp.) was counted under "neglected and underutilized crops."

Yarrow (*A. millefolium s.l.*), arnica, and even chamomile originate still partly from wild collection in Central and Eastern Europe, and despite several attempts to cultivate spikenard (*Valeriana celtica*), a tiny European mountain plant with a high content of patchouli alcohol, this species is still wildly gathered in Austria and Italy (Novak et al., 1998, 2000).

To regulate the sustainable use of biodiversity by avoiding overharvesting, genetic erosion, and habitat loss, international organizations such as International Union for Conservation of Nature (IUCN), WWF/TRAFFIC, and World Health Organization (WHO) have launched together the Convention on Biological Diversity (CBD, 2001), the Global Strategy for Plant Conservation (CBD, 2002), and the Guidelines for the Sustainable Use of Biodiversity (CBD, 2004). TRAFFIC is a joint programme of World Wide Fund for Nature (WWF) and the World Conservation Union (IUCN). TRAFFIC also works in close co-operation with the Secretariat of the Convention on International Trade in Endangered Species of Wild Fauna and Flora (CITES). These principles and recommendations address primarily the national and international policy level, but provide also the herbal industry and the collectors with specific guidance on sustainable sourcing practices (Leaman, 2006). A standard for sustainable collection and use of medicinal and aromatic plants (the international standard on sustainable wild collection of medicinal and aromatic plants [ISSC-MAP]) was issued first in 2004, and its principles will be shown at the end of this chapter. This standard certifies wild-crafted plant material insofar as conservation and sustainability are concerned. Phytochemical quality cannot, however, be derived from it, which is the reason for domestication and systematic cultivation of economically important essential oil plants.

TABLE 3.7
Important Essential Oil-Bearing Plants—Common and Botanical Names Including Family, Plant Parts Used, Raw Material Origin, and Trade Quantities of the Essential Oil

Trade Name	Species	Plant Family	Used Plant Part(s)	Wild Collection/ Cultivation	Trade Quantities[a]
Ambrette seed	*Hibiscus abelmoschus* L.	Malvaceae	Seed	Cult	LQ
Amyris	*Amyris balsamifera* L.	Rutaceae	Wood	Wild	LQ
Angelica root	*Angelica archangelica* L.	Apiaceae	Root	Cult	LQ
Anise seed	*Pimpinella anisum* L.	Apiaceae	Fruit	Cult	LQ
Armoise	*Artemisia herba-alba* Asso.	Asteraceae	Herb	Cult/wild	LQ
Asafoetida	*Ferula assa-foetida* L.	Apiaceae	Resin	Wild	LQ
Basil	*Ocimum basilicum* L.	Lamiaceae	Herb	Cult	LQ
Bay	*Pimenta racemosa* Moore	Myrtaceae	Leaf	Cult	LQ
Bergamot	*Citrus aurantium* L. ssp. *bergamia* (Risso et Poit.) Engl.	Rutaceae	Fruit peel	Cult	MQ
Birch tar	*Betula pendula* Roth. (syn. *Betula verrucosa* Erhart. *Betula alba* sensu H.J.Coste. non L.)	Betulaceae	Bark/wood	Wild	LQ
Buchu leaf	*Agathosma betulina* (Bergius) Pillans. *A. crenulata* (L.) Pillans	Rutaceae	Leaf	Wild	LQ
Cade	*Juniperus oxycedrus* L.	Cupressaceae	Wood	Wild	LQ
Cajuput	*Melaleuca leucandendron* L.	Myrtaceae	Leaf	Wild	LQ
Calamus	*Acorus calamus* L.	Araceae	Rhizome	Cult/wild	LQ
Camphor	*Cinnamomum camphora* L. (Sieb.)	Lauraceae	Wood	Cult	LQ
Cananga	*Cananga odorata* Hook. f. et Thoms.	Annonaceae	Flower	Wild	LQ
Caraway	*Carum carvi* L.	Apiaceae	Seed	Cult	LQ
Cardamom	*Elettaria cardamomum* (L.) Maton	Zingiberaceae	Seed	Cult	LQ
Carrot seed	*Daucus carota* L.	Apiaceae	Seed	Cult	LQ
Cascarilla	*Croton eluteria* (L.) W.Wright	Euphorbiaceae	Bark	Wild	LQ
Cedarwood, Chinese	*Cupressus funebris* Endl.	Cupressaceae	Wood	Wild	MQ
Cedarwood, Texas	*Juniperus mexicana* Schiede	Cupressaceae	Wood	Wild	MQ
Cedarwood, Virginia	*Juniperus virginiana* L.	Cupressaceae	Wood	Wild	MQ
Celery seed	*Apium graveolens* L.	Apiaceae	Seed	Cult	LQ
Chamomile	*Matricaria recutita* L.	Asteraceae	Flower	Cult	LQ
Chamomile, Roman	*Anthemis nobilis* L.	Asteraceae	Flower	Cult	LQ
Chenopodium	*Chenopodium ambrosioides* (L.) Gray	Chenopodiaceae	Seed	Cult	LQ
Cinnamon bark, Ceylon	*Cinnamomum zeylanicum* Nees	Lauraceae	Bark	Cult	LQ
Cinnamon bark, Chinese	*Cinnamomum cassia* Blume	Lauraceae	Bark	Cult	LQ
Cinnamon leaf	*Cinnamomum zeylanicum* Nees	Lauraceae	Leaf	Cult	LQ

(Continued)

TABLE 3.7 (*Continued*)
Important Essential Oil-Bearing Plants—Common and Botanical Names Including Family, Plant Parts Used, Raw Material Origin, and Trade Quantities of the Essential Oil

Trade Name	Species	Plant Family	Used Plant Part(s)	Wild Collection/ Cultivation	Trade Quantities[a]
Citronella, Ceylon	*Cymbopogon nardus* (L.) W. Wats.	Poaceae	Leaf	Cult	HQ
Citronella, Java	*Cymbopogon winterianus* Jowitt.	Poaceae	Leaf	Cult	HQ
Clary sage	*Salvia sclarea* L.	Lamiaceae	Flowering herb	Cult	MQ
Clove buds	*Syzygium aromaticum* (L.) Merill et L.M. Perry	Myrtaceae	Leaf/bud	Cult	LQ
Clove leaf	*Syzygium aromaticum* (L.) Merill et L.M. Perry	Myrtaceae	Leaf	Cult	HQ
Coriander	*Coriandrum sativum* L.	Apiaceae	Fruit	Cult	LQ
Cornmint	*Mentha canadensis* L. (syn. *M. arvensis* L. f. *piperascens* Malinv. ex Holmes; *M. arvensis* L. var. *glabrata. M. haplocalyx* Briq.; *M. sachalinensis* [Briq.] Kudo)	Lamiaceae	Leaf	Cult	HQ
Cumin	*Cuminum cyminum* L.	Apiaceae	Fruit	Cult	LQ
Cypress	*Cupressus sempervirens* L.	Cupressaeae	Leaf/twig	Wild	LQ
Davana	*Artemisia pallens* Wall.	Asteraceae	Flowering herb	Cult	LQ
Dill	*Anethum graveolens* L.	Apiaceae	Herb/fruit	Cult	LQ
Dill, India	*Anethum sowa* Roxb.	Apiaceae	Fruit	Cult	LQ
Elemi	*Canarium luzonicum* Miq.	Burseraceae	Resin	Wild	LQ
Eucalyptus	*Eucalyptus globulus* Labill.	Myrtaceae	Leaf	Cult/wild	HQ
Eucalyptus, lemon-scented	*Eucalyptus citriodora* Hook.	Myrtaceae	Leaf	Cult/wild	HQ
Fennel bitter	*Foeniculum vulgare* Mill. ssp. *vulgare* var. *vulgare*	Apiaceae	Fruit	Cult	LQ
Fennel sweet	*Foeniculum vulgare* Mill. ssp. *vulgare* var. *dulce*	Apiaceae	Fruit	Cult	LQ
Fir needle, Canadian	*Abies balsamea* Mill.	Pinaceae	Leaf/twig	Wild	LQ
Fir needle, Siberian	*Abies sibirica* Ledeb.	Pinaceae	Leaf/twig	Wild	LQ
Gaiac	*Guaiacum officinale* L.	Zygophyllaceae	Resin	Wild	LQ
Galbanum	*Ferula galbaniflua* Boiss. *F. rubricaulis* Boiss.	Apiaceae	Resin	Wild	LQ
Garlic	*Allium sativum* L.	Alliaceae	Bulb	Cult	LQ
Geranium	*Pelargonium* spp.	Geraniaceae	Leaf	Cult	MQ
Ginger	*Zingiber officinale* Roscoe	Zingiberaceae	Rhizome	Cult	LQ
Gingergrass	*Cymbopogon martinii* (Roxb.) H. Wats var. *sofia* Burk	Poaceae	Leaf	Cult/wild	
Grapefruit	*Citrus × paradisi* Macfad.	Rutaceae	Fruit peel	Cult	LQ
Guaiacwood	*Bulnesia sarmienti* L.	Zygophyllaceae	Wood	Wild	MQ

(*Continued*)

TABLE 3.7 (*Continued*)
Important Essential Oil-Bearing Plants—Common and Botanical Names Incl. Family, Plant Parts Used, Raw Material Origin, and Trade Quantities of the Essential Oil

Trade Name	Species	Plant Family	Used Plant Part(s)	Wild Collection/ Cultivation	Trade Quantities[a]
Gurjum	*Dipterocarpus* spp.	Dipterocarpaceae	Resin	Wild	LQ
Hop	*Humulus lupulus* L.	Cannabaceae	Flower	Cult	LQ
Hyssop	*Hyssopus officinalis* L.	Lamiaceae	Leaf	Cult	LQ
Juniper berry	*Juniperus communis* L.	Cupressaceae	Fruit	Wild	LQ
Laurel leaf	*Laurus nobilis* L.	Lauraceae	Leaf	Cult/wild	LQ
Lavandin	*Lavandula angustifolia* Mill. × *L. latifolia* Medik.	Lamiaceae	Leaf	Cult	HQ
Lavender	*Lavandula angustifolia* Miller	Lamiaceae	Leaf	Cult	MQ
Lavender, Spike	*Lavandula latifolia* Medik.	Lamiaceae	Flower	Cult	LQ
Lemon	*Citrus limon* (L.) Burman fil.	Rutaceae	Fruit peel	Cult	HQ
Lemongrass, Indian	*Cymbopogon flexuosus* (Nees ex Steud.) H. Wats.	Poaceae	Leaf	Cult	LQ
Lemongrass, West Indian	*Cymbopogon citratus* (DC.) Stapf	Poaceae	Leaf	Cult	LQ
Lime distilled	*Citrus aurantiifolia* (Christm. et Panz.) Swingle	Rutaceae	Fruit	Cult	HQ
Litsea cubeba	*Litsea cubeba* C.H. Persoon	Lauraceae	Fruit/leaf	Cult	MQ
Lovage root	*Levisticum officinale* Koch	Apiaceae	Root	Cult	LQ
Mandarin	*Citrus reticulata* Blanco	Rutaceae	Fruit peel	Cult	MQ
Marjoram	*Origanum majorana* L.	Lamiaceae	Herb	Cult	LQ
Mugwort common	*Artemisia vulgaris* L.	Asteraceae	Herb	Cult/wild	LQ
Mugwort, Roman	*Artemisia pontica* L.	Asteraceae	Herb	Cult/wild	LQ
Myrtle	*Myrtus communis* L.	Myrtaceae	Leaf	Cult/wild	LQ
Neroli	*Citrus aurantium* L. ssp. *aurantium*	Rutaceae	Flower	Cult	LQ
Niaouli	*Melaleuca viridiflora*	Myrtaceae	Leaf	Cult/wild	LQ
Nutmeg	*Myristica fragrans* Houtt.	Myristicaceae	Seed	Cult	LQ
Onion	*Allium cepa* L.	Alliaceae	Bulb	Cult	LQ
Orange	*Citrus sinensis* (L.) Osbeck	Rutaceae	Fruit peel	Cult	HQ
Orange bitter	*Citrus aurantium* L.	Rutaceae	Fruit peel	Cult	LQ
Oregano	*Origanum* spp. *Thymbra spicata* L. *Coridothymus capitatus* Rechb. fil. *Satureja* spp. *Lippia graveolens*	Lamiaceae	Herb	Cult/wild	LQ
Palmarosa	*Cymbopogon martinii* (Roxb.) H. Wats var. *motia* Burk	Poaceae	Leaf	Cult	LQ
Parsley seed	*Petroselinum crispum* (Mill.) Nym. ex A.W. Hill	Apiaceae	Fruit	Cult	LQ
Patchouli	*Pogostemon cablin* (Blanco) Benth.	Lamiaceae	Leaf	Cult	HQ
Pennyroyal	*Mentha pulegium* L.	Lamiaceae	Herb	Cult	LQ
Pepper	*Piper nigrum* L.	Piperaceae	Fruit	Cult	LQ

(*Continued*)

Sources of Essential Oils

TABLE 3.7 (*Continued*)
Important Essential Oil-Bearing Plants—Common and Botanical Names Incl. Family, Plant Parts Used, Raw Material Origin, and Trade Quantities of the Essential Oil

Trade Name	Species	Plant Family	Used Plant Part(s)	Wild Collection/ Cultivation	Trade Quantities[a]
Peppermint	*Mentha* x *piperita* L.	Lamiaceae	Leaf	Cult	HQ
Petitgrain	*Citrus aurantium* L. ssp. *aurantium*	Rutaceae	Leaf	Cult	LQ
Pimento leaf	*Pimenta dioica* (L.) Merr.	Myrtaceae	Fruit	Cult	LQ
Pine needle	*Pinus silvestris* L. *P. nigra* Arnold	Pinaceae	Leaf/twig	Wild	LQ
Pine needle, Dwarf	*Pinus mugo* Turra	Pinaceae	Leaf/twig	Wild	LQ
Pine silvestris	*Pinus silvestris* L.	Pinaceae	Leaf/twig	Wild	LQ
Pine white	*Pinus palustris* Mill.	Pinaceae	Leaf/twig	Wild	LQ
Rose	*Rosa* x *damascena* Miller	Rosaceae	Flower	Cult	LQ
Rosemary	*Rosmarinus officinalis* L.	Lamiaceae	Feaf	Cult/wild	LQ
Rosewood	*Aniba rosaeodora* Ducke	Lauraceae	Wood	Wild	LQ
Rue	*Ruta graveolens* L.	Rutaceae	Herb	Cult	LQ
Sage, Dalmatian	*Salvia officinalis* L.	Lamiaceae	Herb	Cult/wild	LQ
Sage, Spanish	*Salvia lavandulifolia* L.	Lamiaceae	Leaf	Cult	LQ
Sage, three lobed (Greek Turkish)	*Salvia fruticosa* Mill. (syn. *S. triloba* L.)	Lamiaceae	Herb	Cult/wild	LQ
Sandalwood, East Indian	*Santalum album* L.	Santalaceae	Wood	Wild	MQ
Sassafras, Brazilian (Ocotea cymbarum oil)	*Ocotea odorifera* (Vell.) Rohwer (*Ocotea pretiosa* [Nees] Mez.)	Lauraceae	Wood	Wild	HQ
Sassafras, Chinese	*Sassafras albidum* (Nutt.) Nees.	Lauraceae	Root bark	Wild	HQ
Savory	*Satureja hortensis* L. *Satureja montana* L.	Lamiaceae	Leaf	Cult/wild	LQ
Spearmint, Native	*Mentha spicata* L.	Lamiaceae	Leaf	Cult	MQ
Spearmint, Scotch	*Mentha gracilis* Sole	Lamiaceae	Leaf	Cult	HQ
Star anise	*Illicium verum* Hook fil.	Illiciaceae	Fruit	Cult	MQ
Styrax	*Styrax officinalis* L.	Styracaceae	Resin	Wild	LQ
Tansy	*Tanacetum vulgare* L.	Asteraceae	Flowering herb	Cult/wild	LQ
Tarragon	*Artemisia dracunculus* L.	Asteraceae	Herb	Cult	LQ
Tea tree	*Melaleuca* spp.	Myrtaceae	Leaf	Cult	LQ
Thyme	*Thymus vulgaris* L. *T. zygis* Loefl. ex L.	Lamiaceae	Herb	Cult	LQ
Valerian	*Valeriana officinalis* L.	Valerianaceae	Root	Cult	LQ
Vetiver	*Vetiveria zizanoides* (L.) Nash	Poaceae	Root	Cult	MQ
Wintergreen	*Gaultheria procumbens* L.	Ericaceae	Leaf	Wild	LQ
Wormwood	*Artemisia absinthium* L.	Asteraceae	Herb	Cult/wild	LQ
Ylang Ylang	*Cananga odorata* Hook. f. et Thoms.	Annonaceae	Flower	Cult	MQ

[a] HQ, High quantities (>1000 t/a); MQ, medium quantities (100–1000 t/a); LQ, low quantities (<100 t/a).

3.5.2 Domestication and Systematic Cultivation

This offers a number of advantages over wild harvest for the production of essential oils:

- Avoidance of admixtures and adulterations by reliable botanical identification.
- Better control of the harvested volumes.
- Selection of genotypes with desirable traits, especially quality.
- Controlled influence on the history of the plant material and on postharvest handling.

On the other side, it needs arable land and investments in starting material, maintenance, and harvest techniques. On the basis of a number of successful introductions of new crops a scheme and strategy of domestication was developed by this author (Table 3.8).

Recent examples of successful domestication of essential oil-bearing plants are *oregano* (Ceylan et al., 1994; Kitiki, 1997; Putievsky et al., 1997), *Lippia* sp. (Fischer, 1998), *Hyptis suaveolens* (Grassi, 2003), and *T. lucida* (Goehler, 2006). Domesticating a new species starts with studies at the natural habitat. The most important steps are the exact botanical identification and the detailed description of the growing site. National Herbaria are in general helpful in this stage. In the course of collecting seeds and plant material, a first phytochemical screening will be necessary to recognize chemotypes (Fischer et al., 1996; Goehler et al., 1997). The phytosanitary of wild populations should also be observed so as to be informed in advance on specific pests and diseases. The flower heads of wild *Arnica montana*, for instance, are often damaged by the larvae of *Tephritis arnicae* (Fritzsche et al., 2007).

The first phase of domestication results in a germplasm collection. In the next step, the appropriate propagation method has to be developed, which might be derived partly from observations at the natural habitat: while studying wild populations of *T. lucida* in Guatemala we found, besides appropriate seed set, also runners, which could be used for vegetative propagation of selected plants (Goehler et al., 1997). Wherever possible, propagation by seeds and direct sowing is however preferred due to economic reasons.

The appropriate cultivation method depends on the plant type—annual or perennial, herb, vine, or tree—and on the agroecosystem into which the respective species should be introduced.

TABLE 3.8
Domestication Strategy for Plants of the Spontaneous Flora

1. *Studies at the natural habitat*: botany, soil, climate, growing type, natural distribution and propagation, natural enemies, pests and diseases	→ GPS to exactly localize the place
2. *Collection of the wild grown plants and seeds*: establishment of a germplasm collection, ex situ conservation, phytochemical investigation (screening)	
3. *Plant propagation*: vegetatively or by seeds, plantlet cultivation; (biotechnol.: *in vitro* propagation)	→ Biotechnol./*in vitro*
4. *Genetic improvement*: variability, selection, breeding; phytochemical investigation, biotechnology (*in vitro* techniques)	→ Biotechnol./*in vitro*
5. *Cultivation treatments*: growing site, fertilization, crop maintenance, cultivation techniques	
6. *Phytosanitary problems*: pests, diseases	→ Biotechnol./*in vitro*
7. *Duration of the cultivation*: harvest, postharvest handling, phytochemical control of the crop produced	→ Technical processes, solar energy (new techniques)
8. *Economic evaluation and calculation*	→ New techniques

Source: Modified from Franz, Ch., *Plant Res. Dev.*, 37, 101, 1993c; Franz, Ch., Genetic versus climatic factors influencing essential oil formation, *Proceedings of the 12th International Congress of Essential Oils, Fragrances and Flavours*, Vienna, Austria, 1993d, pp. 27–44.

In contrast to large-scale field production of herbal plants in temperate and Mediterranean zones, small-scale sustainable agroforesty and mixed cropping systems adapted to the environment have the preference in tropical regions (Schippmann et al., 2006). Parallel to the cultivation trials dealing with all topics from plant nutrition and maintenance to harvesting and postharvest handling, the evaluation of the genetic resources and the genetic improvement of the plant material must be started to avoid developing of a detailed cultivation scheme with an undesired chemotype.

3.5.3 Factors Influencing the Production and Quality of Essential Oil-Bearing Plants

Since plant material is the product of a predominantly biological process, prerequisite of its productivity is the knowledge on the factors influencing it, of which the most important ones are

1. The already discussed intraspecific chemical polymorphism, derived from it the biosynthesis and inheritance of the chemical features, and as consequence selection and breeding of new cultivars.
2. The intraindividual variation between the plant parts and depending on the developmental stages ("morpho- and ontogenetic variation").
3. The modification due to environmental conditions including infection pressure and immissions.
4. Human influences by cultivation measures, for example, fertilizing, water supply, or pest management.

3.5.3.1 Genetic Variation and Plant Breeding

Phenotypic variation in essential oils was detected very early because of their striking sensorial properties. Due to the high chemical diversity, a continuous selection of the desired chemotypes leads to rather homogenous and reproducible populations, as this is the case with the landraces and common varieties. But Murray and Reitsema (1954) stated already that "a plant breeding program requires a basic knowledge of the inheritance of at least the major essential oil compounds." Such genetic studies have been performed over the last 50 years with a number of species especially of the mint family (e.g., *T. vulgaris*: Vernet, 1976; *Ocimum* sp.: Sobti et al., 1978; Gouyon and Vernet, 1982; *P. frutescens*: Koezuka et al., 1986; *Mentha* sp.: Croteau, 1991), of the Asteraceae/Compositae (*M. recutita*: Horn et al., 1988; Massoud and Franz, 1990), the genus (*Eucalyptus*: Brophy and Southwell, 2002; Doran, 2002), or the *V. zizanioides* (Akhila and Rani, 2002).

The results achieved by inheritance studies have been partly applied in targeted breeding as shown exemplarily in Table 3.9. Apart from the essential oil content and composition there are also other targets to be observed when breeding essential oil plants, as particular morphological characters ensuring high and stable yields of the respective plant part, resistances to pest and diseases as well as abiotic stress, low nutritional requirements to save production costs, appropriate homogeneity, and suitability for technological processes at harvest and postharvest, especially readiness for distillation (Pank, 2007). In general, the following breeding methods are commonly used (Franz, 1999).

3.5.3.1.1 Selection by Exploiting the Natural Variability

Since many essential oil-bearing species are in the transitional phase from wild plants to systematic cultivation, appropriate breeding progress can be achieved by simple selection. Wild collections or accessions of germplasm collections are the basis, and good results were obtained, for example, with *Origanum* sp. (Putievsky et al., 1997) in limited time and at low expenses.

Individual plants showing the desired phenotype will be selected and either generatively or vegetatively propagated (individual selection), or positive or negative mass selection techniques

TABLE 3.9
Some Registered Cultivars of Essential Oil Plant

Species	Cultivar/Variety	Country	Year of Registration	Breeding Method	Specific Characters
Achillea collina	SPAK	CH	1994	Crossing	High in proazulene
Angelica archangelica	VS 2	FR	1996	Recurrent pedigree	Essential oil index of roots: 180
Foeniculum vulgare	Fönicia	HU	1998	Selection	High anethole
Lavandula officinalis	Rapido	FR	1999	Polycross	High essential oil, high linalyl acetate
Levisticum officinale	Amor	PL	2000	Selection	High essential oil
Matricaria recutita	Mabamille	DE	1995	Tetraploid	High α-bisabolol
	Ciclo-1	IT	2000	Line breeding	High chamazulene
	Lutea	SK	1995	Tetraploid	High α-bisabolol
Melissa officinalis	Ildikó	HU	1998	Selection	High essential oil, Citral A + B, linalool
	Landor	CH	1994	Selection	High essential oil
	Lemona	DE	2001	Selection	High essential oil, citral
Mentha piperita	Todd's Mitcham	USA	1972	Mutation	Wilt resistant
	Kubanskaja	RUS	1980s	Crossing and polyploid	High essential oil, high menthol
Mentha spicata	MSH-20	DK	2000	Recurrent pedigree	High menthol, good flavor
Ocimum basilicum	Greco	IT	2000	Synthetic	Flavor
	Perri	ISR	1999	Cross-breeding	Fusarium Resistant
	Cardinal	ISR	2000	Cross-breeding	
Origanum syriacum	Senköy	TR	1992	Selection	5% essential oil, 60% carvacrol
	Carmeli	ISR	1999	Selection	Carvacrol
	Tavor	ISR	1999	Selection	Thymol
Origanum onites		GR	2000	Selfing	Carvacrol
Origanum hirtum		GR	2000	Selfing	Carvacrol
	Vulkan	DE	2002	Crossing	Carvacrol
	Carva	CH	2002	Crossing	Carvacrol
	Darpman	TR	1992	Selection	2.5% essential oil, 55% carvacrol
Origanum majorana (*Majorana hortensis*)	Erfo	DE	1997	Crossing	High essential oil, *cis*-Sabinene-hydrate
	Tetrata	DE	1999	Ployploid	
	G 1	FR	1998	Polycross	
Salvia officinalis	Moran	ISR	1998	Crossing	Herb yield
	Syn 1	IT	2004	Synthetic	α-Thujone
Thymus vulgaris	Varico	CH	1994	Selection	Thymol/carvacrol
	T-16	DK	2000	Recurrent pedigree	Thymol
	Virginia	ISR	2000	Selection	Herb yield

can be applied. Selection is traditionally the most common method of genetic improvement and the majority of varieties and cultivars of essential oil crops have this background. Due to the fact, however, that almost all of the respective plant species are allogamous, a recurrent selection is necessary to maintain the varietal traits, and this has especially to be considered if other varieties or wild populations of the same species are nearby and uncontrolled cross pollination may occur.

The efficacy of selection has been shown by examples of many species, for instance, of the Lamiaceae family, starting from "Mitcham" peppermint and derived varieties (Lawrence, 2007), basil (Elementi et al., 2006), sage (Bezzi, 1994; Bernáth, 2000) to thyme (Rey, 1993). It is a well-known method also in the breeding of caraway (Pank et al., 1996) and fennel (Desmarest, 1992) as well as of tropical and subtropical species such as palmarosa grass (Kulkarni, 1990), tea tree (Taylor, 1996), and eucalyptus (Doran, 2002). At perennial herbs, shrubs, and trees clone breeding, that is, the vegetative propagation of selected high-performance individual plants, is the method of choice, especially in sterile or not type-true hybrids, for example, peppermint (*M. piperita*) or lavandin (*Lavandula* × *hybrida*). But this method is often applied also at sage (Bazina et al., 2002), rosemary (Mulas et al., 2002), lemongrass (Kulkarni and Ramesh, 1992), pepper, cinnamon, and nutmeg (Nair, 1982), and many other species.

3.5.3.1.2 Breeding with Extended Variability (Combination Breeding)

If different desired characters are located in different individuals/genotypes of the same or a closely related crossable species, crossings are made followed by selection of the respective combination products. Artificial crossings are performed by transferring the paternal pollen to the stigma of the female (emasculated) or male sterile maternal flower. In the segregating progenies individuals with the desired combination will be selected and bred to constancy, as exemplarily described for fennel and marjoram by Pank (2002b).

Hybrid breeding—common in large-scale agricultural crops, for example, maize—was introduced into essential oil plants over the last decade only. The advantage of hybrids on the one side is that the F_1 generation exceeds the parent lines in performance due to hybrid vigor and uniformity ("heterosis effect") and on the other side it protects the plant breeder by segregating of the F_2 and following generations in heterogeneous low-value populations. But it needs as precondition separate (inbred) parent lines of which one has to be male sterile and one male fertile with good combining ability.

In addition, a male fertile "maintainer" line is needed to maintain the mother line. Few examples of F_1 hybrid breeding are known especially at Lamiaceae since male sterile individuals are found frequently in these species (Rey, 1994; Langbehn et al., 2002; Novak et al., 2002; Pank et al., 2002a).

Synthetic varieties are based on several (more than two) well-combining parental lines or clones which are grown together in a polycross scheme with open pollination for seed production. The uniformity and performance is not as high as at F_1 hybrids but the method is simpler and cheaper and the seed quality acceptable for crop production until the second or third generation. Synthetic cultivars are known for chamomile (Franz et al., 1985), arnica (Daniel and Bomme, 1991), marjoram (Franz and Novak, 1997), sage (Aiello et al., 2001), or caraway (Pank et al., 2007).

3.5.3.1.3 Breeding with Artificially Generated New Variability

Induced mutations by application of mutagenic chemicals or ionizing radiation open the possibility to find new trait expressions. Although quite often applied, such experiments are confronted with the disadvantages of undirected and incalculable results, and achieving a desired mutation is often like searching for a needle in a haystack. Nevertheless, remarkable achievements are several colchicine-induced polyploid varieties of peppermint (Murray, 1969; Lawrence, 2007), chamomile (Czabajska et al., 1978; Franz et al., 1983; Repčak et al., 1992), and lavender (Slavova et al., 2004).

Further possibilities to obtain mutants are studies of the somaclonal variation of *in vitro* cultures since abiotic stress in cell and tissue cultures induces also mutagenesis. Finally, genetic engineering opens new fields and potentialities to generate new variability and to introduce new traits by gene transfer. Except research on biosynthetic pathways of interesting essential oil compounds genetic engineering, GMO's and transgenic cultivars are until now without practical significance in essential oil crops and also not (yet) accepted by the consumer.

As regards the different traits, besides morphological, technological, and yield characteristics as well as quantity and composition of the essential oil, also stress resistance and resistance to pests and diseases are highly relevant targets in breeding of essential oil plants. Well known in this respect are breeding efforts against mint rust (*Puccinia menthae*) and wilt (*Verticillium dahliae*) resulting in the peppermint varieties "Multimentha," "Prilukskaja," or "Todd's Mitcham" (Murray and Todd, 1972; Lawrence, 2007; Pank, 2007), the development of *Fusarium*-wilt and *Peronospora* resistant cultivars of basil (Dudai, 2006; Minuto et al., 2006), or resistance breeding against *Septoria petroselini* in parsley and related species (Marthe and Scholze, 1996). An overview on this topic is given by Gabler (2002).

3.5.3.2 Plant Breeding and Intellectual Property Rights

Essential oil plants are biological, cultural, and technological resources. They can be found in nature gathered from the wild or developed through domestication and plant breeding. As long as the plant material is wild collected and traditionally used, it is part of the cultural heritage without any individual intellectual property and therefore not possible to protect, for example, by patents. Even finding a new plant or substance is a discovery in the "natural nature" and not an invention since a technical teaching is missing. Intellectual property, however, can be granted to new applications that involve an inventive step. Which consequences can be drawn from these facts for the development of novel essential oil plants and new selections or cultivars?

Selection and genetic improvement of aromatic plants and essential oil crops is not only time consuming but also rather expensive due to the necessity of comprehensive phytochemical and possibly molecular biological investigations. In addition, with few exceptions (e.g., mints, lavender and lavandin, parsley but also *Cymbopogon* sp., black pepper, or cloves) the acreage per species is rather limited in comparison with conventional agricultural and horticultural crops. And finally, there are several "fashion crops" with market uncertainties concerning their longevity or half-life period, respectively. The generally unfavorable cost: benefit ratio to be taken into consideration makes essential oil plant breeding economically risky and there is no incentive for plant breeders unless a sufficiently strong plant intellectual property right (IPR) exists. Questioning "which protection, which property right for which variety?" offers two options (Franz, 2001).

3.5.3.2.1 Plant Variety Protection

By conventional methods bred plant groupings that collectively are distinct from other known varieties and are uniform and stable following repeated reproduction can be protected by way of plant breeder's rights. Basis is the International Convention for the Protection of New Varieties initially issued by UPOV (Union for the Protection of New Varieties of Plants) in 1961 and changed in 1991. A plant breeder's right is a legal title granting its holder the exclusive right to produce reproductive material of his plant variety for commercial purposes and to sell this material within a particular territory for up to 30 years (trees and shrubs) or 25 years (all other plants). A further precondition is the "commercial novelty," that is, it must not have been sold commercially prior to the filing date. Distinctness, uniformity, and stability (DUS) refer to morphological (leaf shape, flower color, etc.) or physiological (winter hardiness, disease resistance, etc.), but not phytochemical characteristics, for example, essential oil content or composition. Such "value for cultivation and use (VCU) characteristics" will not be examined and are therefore not protected by plant breeder's rights (Franz, 2001; Llewelyn, 2002; Van Overwalle, 2006).

3.5.3.2.2 Patent Protection (Plant Patents)

Generally speaking, patentable are inventions (not discoveries!) that are novel, involve an innovative step, and are susceptible to industrial application, including agriculture. Plant varieties or essentially biological processes for the production of plants are explicitly excluded from patenting. But other groupings of plants that fall neither under the term "variety" nor under "natural nature" are possible to be protected by patents. This is especially important for plant groupings with novel phytochemical composition or novel application combined with an inventive step, for example, genetic modification, a technologically new production method or a novel type of isolation (product by process protection).

TABLE 3.10
Advantages and Disadvantages of PVP versus Patent Protection of Specialist Minor Crops (Medicinal and Aromatic Plants)

PVP	Patent
Beginning of protection: registration date	Beginning of protection: application date
Restricted to "varieties"	"Varieties" not patentable, but any other grouping of plants
Requirements: DUS = distinctness, uniformity, stability	Requirements: novelty, inventive step, industrial applicability (= NIA)
Free choice of characters to be used for DUS by PVO (Plant Variety Office)	Repeatability obligatory, product by process option
Phenotypical. Mainly morphological characters (phytochemicals of minor importance)	"Essentially biological process" not patentable
Value for cultivation and use characteristics (VCU) not protected	"Natural nature" not patentable
	Claims (e.g., phytochemical characters) depend on applicant
	Phytochemical characters and use/application (VCU) patentable

Especially for wild plants and essentially allogamous plants not fulfilling DUS for cultivated varieties (cultivars) and plants where the phytochemical characteristics are more important than the morphological ones, plant patents offer an interesting alternative to plant variety protection (PVP) (Table 3.10).

In conformity with the UPOV Convention of 1991 (UPOV, 1991)

- A strong plant IPR is requested.
- Chemical markers (e.g., secondary plant products) must be accepted as protectable characteristics.
- Strong depending rights for essentially derived varieties are needed since it is easy to plagiarize such crops.
- "Double protection" would be very useful (i.e., free decision by the breeder if PVR or patent protection is applied).
- But also researchers exemption and breeders privilege with fair access to genotypes for further development is necessary.

Strong protection does not hinder usage and development; it depends on a fair arrangement only (Le Buanec, 2001).

3.5.3.3 Intraindividual Variation between Plant Parts and Depending on the Developmental Stage (*Morpho-* and *Ontogenetic Variation*)

The formation of essential oils depends on the tissue differentiation (secretory cells and excretion cavities, as discussed in Section 3.1) and on the ontogenetic phase of the respective plant. The knowledge on these facts is necessary to harvest the correct plant parts at the right time.

Regarding the *differences between plant parts*, it is known from cinnamon (*Cinnamomum zeylanicum*) that the root-, stem-, and leaf oils differ significantly (Wijesekera et al., 1974): only the stem bark contains an essential oil with up to 70% cinnamaldehyde, whereas the oil of the root bark consists mainly of camphor and linalool, and the leaves produce oils with eugenol as main compound. In contrast to it, eugenol forms with 70%–90% the main compound in stem, leaf, and bud oils of cloves (*S. aromaticum*) (Lawrence, 1978). This was recently confirmed by Srivastava et al. (2005) for clove oils from India and Madagascar, stating in addition that eugenyl acetate was found in buds up to 8% but in leaves between traces and 1.6% only. The second main substance in leaves as well as buds is β-caryophyllene with up to 20% of the essential oil. In *Aframomum giganteum* (Zingiberaceae), the rhizome essential oil consists of β-caryophyllene, its oxide, and derivatives mainly, whereas in the leaf oil terpentine-4-ol and pinocarvone form the principal components (Agnaniet et al., 2004).

Essential oils of the Rutaceae family, especially citrus oils, are widely used as flavors and fragrances depending on the plant part and species: in lime leaves neral/geranial and nerol/geraniol are prevailing, whereas grapefruit leaf oil consists of sabinene and β-ocimene mainly. The peel of grapefruit contains almost limonene only and some myrcene, but lime peel oil shows a composition of β-pinene, γ-terpinene, and limonene (Gancel et al., 2002). In *Phellodendron* sp., Lis et al. (2004), Lis and Milczarek (2006) found that in flower and fruit oils limonene and myrcene are dominating; in leaf oils, in contrast, α-farnesene, β-elemol, or β-ocimene, are prevailing.

Differences in the essential oil composition between the plant parts of many Umbelliferae (Apiaceae) have exhaustively been studied by the group of Kubeczka, summarized by Kubeczka et al. (1982) and Kubeczka (1997). For instance, the comparison of the essential fruit oil of aniseed (*P. anisum*) with the oils of the herb and the root revealed significant differences (Kubeczka et al., 1986). Contrary to the fruit oil consisting of almost *trans*-anethole only (95%), the essential oil of the herb contains besides anethole, considerable amounts of sesquiterpene hydrocarbons, for example, germacrene D, β-bisabolene, and α-zingiberene. Also pseudoisoeugenyl-2-methylbutyrate and epoxi-pseudoisoeugenyl-2-methylbutyrate together form almost 20% main compounds of the herb oil, but only 8.5% in the root and 1% in the fruit oil. The root essential oil is characterized by a high content of β-bisabolene, geijerene, and pregeijerene and contains only small amounts of *trans*-anethole (3.5%). Recently, Velasco-Neguerela et al. (2002) investigated the essential oil composition in the different plant parts of *Pimpinella cumbrae* from Canary Islands and found in all above-ground parts α-bisabolol as main compound besides of δ-3-carene, limonene, and others, whereas the root oil contains mainly isokessane, geijerene, isogeijerene, dihydroagarofuran, and proazulenes—the latter is also found in *Pimpinella nigra* (Kubeczka et al., 1986). Pseudoisoeugenyl esters, known as chemosystematic characters of the genus *Pimpinella*, have been detected in small concentrations in all organs except leaves.

Finally, Kurowska and Galazka (2006) compared the seed oils of root and leaf parsley cultivars marketed in Poland. Root parsley seeds contained an essential oil with high concentrations of apiole and some lower percentages of myristicin. In leaf parsley seeds, in contrast, the content of myristicin was in general higher than apiole, and a clear differentiation between flat leaved cultivars showing still higher concentrations of apiole and curled cultivars with only traces of apiole could be observed. Allyltetramethoxybenzene as the third marker was found in leaf parsley seeds up to 12.8%, in root parsley seeds, however, in traces only. Much earlier, Franz and Glasl (1976) had published already similar results on parsley seed oils comparing them with the essential oil composition of the other plant parts (Figure 3.8). Leaf oils gave almost the same fingerprint than the seeds with high myristicin in curled leaves, some apiole in flat leaves, and higher apiole concentrations than myristicin in the leaves of root varieties. In all root samples, however, apiole dominated largely over myristicin. It is therefore possible to identify the parsley type by analyzing a small seed sample.

As shown already by Figueiredo et al. (1997), in the major number of essential oil-bearing species the oil composition differs significantly between the plant parts, but there are also plant species—as mentioned before, for example, cloves—which form a rather similar oil composition in each plant organ. Detailed knowledge in this matter is needed to decide, for instance, how exact the separation of plant parts has to be performed before further processing (e.g., distillation) or use.

Another topic to be taken into consideration is the *developmental stage* of the plant and the plant organs, since the formation of essential oils is phase dependent. In most cases, there is a significant increase of the essential oil production throughout the whole vegetative development.

And especially in the generative phase between flower bud formation and full flowering, or until fruit or seed setting, remarkable changes in the oil yield and compositions can be observed.

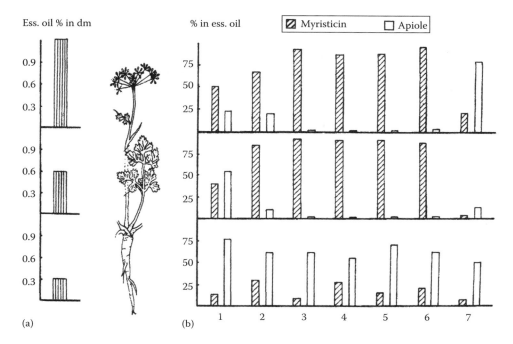

FIGURE 3.8 Differences in the essential oil of fruits, leaves, and roots of parsley cultivars (*Petroselinum crispum* (Mill.) Nyman). (a) Essential oil content, (b) content of myristicin and apiole in the essential oil. 1, 2—flat leaved cv's; 3–7—curled leaves cv's; 7—root parsley).

Obviously, a strong correlation is given between formation of secretory structures (oil glands, ducts, etc.) and essential oil biosynthesis, and different maturation stages, are associated with, for example, higher rates of cyclization or increase of oxygenated compounds (Figueiredo et al., 1997).

Investigations on the ontogenesis of fennel (*F. vulgare* Mill.) revealed that the best time for picking fennel seeds is the phase of full ripeness due to the fact that the anethole content increases from <50% in unripe seeds to over 80% in full maturity (Marotti et al., 1994). In dill weed (*Anethum graveolens* L.) the content on essential oil rises from 0.1% only in young sprouts to more than 1% in herb with milk ripe umbels (Gora et al., 2002). In the herb, oil α-phellandrene prevails until the beginning of flowering with up to 50%, followed by dill ether, *p*-cymene, and limonene. The oil from green as well as ripe umbels contains, on the other hand, mainly (*S*)-carvone and (*R*)-limonene. The flavor of dill oil changes therefore dramatically, which has to be considered when determining the harvest time for distillation.

Among Compositae (Asteraceae) there are not as many results concerning ontogeny due to the fact that in general the flowers or flowering parts of the plants are harvested, for example, chamomile (*M. recutita*), yarrow (*A. millefolium s.l.*), immortelle (*Helichrysum italicum*), or wormwood (*Artemisia* sp.) and therefore the short period between the beginning of flowering and the decay of the flowers is of interest only. In chamomile (*M. recutita*), the flower buds show a relatively high content on essential oil between 0.8% and 1.0%, but the oil yield in this stage is rather low. From the beginning of flowering, the oil content increases until full flowering (all disc florets open) and decreases again with decay of the flower heads. At full bloom there is also the peak of (pro) chamazulene, whereas farnesene and α-bisabolol decrease from the beginning of flowering and the bisabololoxides rise (Franz et al., 1978). This was confirmed by Repčak et al. (1980). The essential oil of *Tagetes minuta* L. at different development stages was investigated by Worku and Bertoldi (1996). Before flower bud formation the oil content was 0.45% only, but it culminated with 1.34% at the immature seed stage. During this period *cis*-ocimene increased from 7.2% to 37.5% and *cis*-ocimenone declined from almost 40%–13.1%. Little variations could be observed at *cis*- and *trans*-tagetone only. Similar results have been reported also by Chalchat et al. (1995).

Also for *Lippia* sp. (Verbenaceae) some results are known concerning development stages (Fischer, 1998; Coronel et al., 2006). The oil content in the aerial parts increases from young buds (<1.0%) to fully blooming (almost 2.0%). But although quantitative variations could be observed for most components of the essential oils, the qualitative composition appeared to be constant throughout the growing season.

A particular situation is given with eucalypts as they develop up to five distinct types of leaves during their lifetime, each corresponding to a certain ontogenetic stage with changing oil concentrations and compositions (Doran, 2002). Usually the oil content increases from young to matured, nonlignified leaves, and is thereafter declining until leaf lignification. Almost the same curve is valid also for the 1,8-cineole concentration in the oil. But comparing the relatively extensive literature on this topic, one may conclude that the concentration at various stages of leaf maturity is determined by a complex pattern of quantitative change in individual or groups of substances, some remaining constant, some increasing, and some decreasing. Tsiri et al. (2003) investigated the volatiles of the leaves of *Eucalyptus camaldulensis* over the course of a year in Greece and found a seasonal variation of the oil concentration with a peak during summer and lowest yields during winter. The constituent with highest concentration was 1,8-cineole (25.3%–44.2%) regardless the time of harvest. The great variation of all oil compounds showed however no clear tendency, neither seasonal nor regarding leaf age or leaf position. Doran (2002) concluded therefore that genotypic differences outweigh any seasonal or environmental effects in eucalypts.

There is an extensive literature on ontogenesis and seasonal variation of Labiatae essential oils. Especially for this plant family, great differences are reported on the essential oil content and composition of young and mature leaves and the flowers may in addition influence the oil quality significantly. Usually, young leaves show higher essential oil contents per area unit compared to old leaves. But the highest oil yield is reached at the flowering period, which is the reason that most of the oils are produced from flowering plants. According to Werker et al. (1993) young basil (*O. basilicum*) leaves contained 0.55% essential oil while the content of mature leaves was only 0.13%. The same is also valid to a smaller extent for *O. sanctum*, where the essential oil decreases from young (0.54%) to senescing leaves (0.38%) (Dey and Choudhuri, 1983). Testing a number of basil cultivars mainly of the linalool chemotype, Macchia et al. (2006) found that only some of the cultivars produce methyl eugenol up to 8% in the vegetative stage. Linalool as main compound is increasing from the vegetative (10%–50%) to the flowering (20%–60%), and postflowering phase (25%–80%), whereas the second important substance eugenol reaches its peak at the beginning of flowering (5%–35%). According to the cultivars, different harvest dates are therefore recommended. In *O. sanctum*, the content of eugenol (60.3%–52.2%) as well as of methyl eugenol (6.6%–2.0%) is decreasing from young to senescent leaves and at the same time β-caryophyllene increases from 20.8% to 30.2% (Dey and Choudhuri, 1983).

As regards oregano (*O. vulgare* ssp. *hirtum*), the early season preponderance of *p*-cymene over carvacrol was reversed as the season progressed and this pattern could also be observed at any time within the plant, from the latest leaves produced (low in cymene) to the earliest (high in cymene) (Johnson et al., 2004; Figure 3.9). Already Kokkini et al. (1996) had shown that oregano contains a higher proportion of *p*-cymene to carvacrol (or thymol) in spring and autumn, whereas carvacrol/thymol prevails in the summer. This is explained by Dudai et al. (1992) as photoperiodic reaction: short days with high *p*-cymene, long days with low *p*-cymene production. But only young plants are capable of making this switch, whereas in older leaves the already produced and stored oil remains almost unchanged (Johnson et al., 2004).

Presumably the most studied essential oil plant is peppermint (*M. piperita* L.). Already in the 1950s Lemli (1955) stated that the proportion of menthol to menthone in peppermint leaves changes in the course of the development toward higher menthol contents. Lawrence (2007) has just recently shown that from immature plants via mature to senescent plants the content of menthol increases

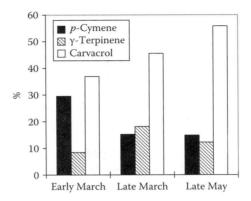

FIGURE 3.9 Average percentages and concentrations of *p*-cymene. γ-Terpinene and carvacrol at the different sampling dates of *Origanum vulgare* ssp. *hirtum*.

(34.8%–39.9%–48.2%) and correspondingly the menthone content decreases dramatically (26.8%–17.4%–4.7%). At the same time, also an increase of menthyl acetate from 8.5% to 23.3% of the oil could be observed. At full flowering, the peppermint herb oil contains only 36.8% menthol but 21.8% menthone, 7.7% menthofuran, and almost 3% pulegone due to the fact that the flower oils are richer in menthone and pulegone and contain a high amount of menthofuran (Hefendehl, 1962). Corresponding differences have been found between young leaves rich in menthone and old leaves with high menthol and menthyl acetate content (Hoeltzel, 1964; Franz, 1972). The developmental stage depends, however, to a large extent from the environmental conditions, especially the day length.

3.5.3.4 Environmental Influences

Essential oil formation in the plants is highly dependent on climatic conditions, especially day length, irradiance, temperature, and water supply. Tropical species follow in their vegetation cycle the dry and rainy season; species of the temperate zones react more on day length, the more distant from the equator their natural distribution area is located.

Peppermint as typical long day plant needs a minimum day length (hours of day light) to switch from the vegetative to the generative phase. This is followed by a change in the essential oil composition from menthone to menthol and menthyl acetate (Hoeltzel, 1964). Franz (1981) tested six peppermint clones at Munich/Germany and at the same time also at Izmir/Turkey. At the development stage "beginning of flowering," all clones contained at the more northern site much more menthol than on the Mediterranean location, which was explained by a maximum day length in Munich of 16 h 45 min, but in Izmir of 14 h 50 min only. Comparable day length reactions have been mentioned already for oregano (Dudai et al., 1992; Kokkini et al., 1996). Also marjoram (*O. majorana* L.) was influenced not only in flower formation by day length, but also in oil composition (Circella et al., 1995). At long day treatment the essential oil contained more *cis*-sabinene hydrate. Terpinene-4-ol prevailed under short day conditions.

Franz et al. (1986) performed ecological experiments with chamomile, growing vegetatively propagated plants at three different sites, in South Finland, Middle Europe, and West Turkey. As regards the oil content, a correlation between flower formation, flowering period, and essential oil synthesis could be observed: the shorter the flowering phase, the less was the time available for oil formation, and thus the lower was the oil content. The composition of the essential oil, on the other hand, showed no qualitative change due to ecological or climatic factors confirming that chemotypes keep their typical pattern. In addition, Massoud and Franz (1990) investigated the genotype–environment interaction of a chamazulene–bisabolol chemotype. The frequency distributions of the essential oil content as well as the content on chamazulene and α-bisabolol have shown that the highest oil- and bisabolol content was reached in Egypt while under German climatic conditions chamazulene was higher. Similar results have been obtained by Letchamo and Marquard (1993). The relatively high heritability coefficients calculated for some essential oil components—informing whether a character is more influenced by genetic or other factors—confirm that the potential to produce a certain chemical pattern is genetically coded, but the gene expression will be induced or repressed by environmental factors also (Franz, 1993b,d).

Other environmental factors, for instance, soil properties, water stress, or temperature, are mainly influencing the productivity of the respective plant species and by this means the oil yield also, but have little effect on the essential oil formation and composition only (Figueiredo et al., 1997; Salamon, 2007).

3.5.3.5 Cultivation Measures, Contaminations, and Harvesting

Essential oil-bearing plants comprise annual, biennial, or perennial herbs, shrubs, and trees, cultivated either in tropical or subtropical areas, in Mediterranean regions, in temperate, or even in arid zones. Surveys in this respect are given, for instance, by Chatterjee (2002) for India, by Carruba et al. (2002) for Mediterranean environments, and by Galambosi and Dragland (2002) for Nordic countries. Nevertheless, some examples should refer to some specific items.

The *cultivation method*—if direct sowing or transplanting—and the timing influence the crop development and by that way also the quality of the product, as mentioned above. Vegetative propagation, necessary for peppermint due to its genetic background as interpecific hybrid, common in *Cymbopogon* sp. and useful to control the ratio between male and female trees in nutmeg (*Myristica fragrans*), results in homogeneous plant populations and fields. A disadvantage could be the easier dispersion of pests and diseases, as known for "yellow rot" of lavandin (*Lavandula × hybrida*) (Fritzsche et al., 2007). Clonal propagation can be performed by leaf or stem cuttings (Goehler et al., 1997; El-Keltawi and Abdel-Rahman, 2006; Nicola et al., 2006) or *in vitro* (e.g., Figueiredo et al., 1997; Mendes and Romano, 1997), the latter method especially for mother plant propagation due to the high costs. *In vitro* essential oil production received increased attention in physiological experiments, but has up to now no practical significance.

As regards *plant nutrition and fertilizing*, a numerous publications have shown its importance for plant growth, development, and biomass yield. The essential oil yield, obviously, depends on the plant biomass; the oil percentage is partly influenced by the plant vigor and metabolic activity. Optimal fertilizing and water supply results in better growth and oil content, for example, in marjoram, oregano, basil, or coriander (Menary, 1994), but also in delay of maturity, which causes quite often "immature" flavors.

Franz (1972) investigated the influence of nitrogen and potassium on the essential oil formation of peppermint. He could show that higher nitrogen supply increased the biomass but retarded the plant development until flowering, whereas higher potassium supply forced the maturity. With increasing nitrogen, a higher oil percentage was observed with lower menthol and higher menthone content; potassium supply resulted in less oil with more menthol and menthyl acetate. Comparable results with *R. officinalis* have been obtained by Martinetti et al. (2006), and Omidbaigi and Arjmandi (2002) have shown for *T. vulgaris* that nitrogen and phosphorus fertilization had significant effect on the herb yield and essential oil content, but did not change the thymol percentage. Also Java citronella (*Cymbopogon winterianus* Jowitt.) responded to nitrogen supply with higher herb and oil yields, but no influence on the geraniol content could be found (Munsi and Mukherjee, 1986).

Extensive pot experiments with chamomile (*M. recutita*) have also shown that high nitrogen and phosphorus nutrition levels resulted in a slightly increased essential oil content of the anthodia, but raising the potassium doses had a respective negative effect (Franz et al., 1983). With nitrogen the flower formation was in delay and lasted longer; with more potassium the flowering phase was reduced, which obviously influenced the period available for essential oil production. This was confirmed by respective ^{14}C-acetate labeling experiments (Franz, 1981).

Almost no effect has been observed on the composition of the essential oil. Also a number of similar pot or field trials came to the same result, as summarized by Salamon (2007).

Salinity and salt stress get an increasing importance in agriculture especially in subtropical and Mediterranean areas. Some essential oil plants, for example, *Artemisia* sp. and *M. recutita* (chamomile) are relatively salt tolerant. Also thyme (*T. vulgaris*) showed a good tolerance to irrigation water salinity up to 2000 ppm, but exceeding concentrations caused severe damages (Massoud et al., 2002). Higher salinity reduced also the oil content, and an increase of *p*-cymene was observed. Recently, Aziz et al. (2008) investigated the influence of salt stress on growth and essential oil in several mint species. In all three mints, salinity reduced the growth severely from 1.5 g/L onward; in peppermint, the menthone content raised and menthol went down to <1.0%, in apple mint, linalool and neryl acetate decreased while myrcene, linalyl acetate, and linalyl propionate increased.

Further problems to be taken into consideration in plant production are *contaminations* with heavy metals, damages caused by pests and diseases, and *residues* of plant protection products. The most important toxic heavy metals Cd, Hg, Pb, and Zn, but also Cu, Ni, and Mn may influence the plant growth severely and by that way also the essential oil, as they may act as cofactors in the plant enzyme system. But as contaminants, they remain in the plant residue after distillation (Zheljazkov and Nielsen, 1996; Zheljazkov et al., 1997). Some plant species, for example, yarrow and chamomile accumulate heavy metals to a greater extent. This is, however, problematic for using the crude drug or for deposition of distillation wastes mainly. The same is valid for the microbial contamination of the plant material. More important in the production of essential oils are pests and diseases that cause damages to the plant material and sometimes alterations in the biosynthesis; but little is known in this respect.

In contrast to organic production, where no use of pesticides is permitted, a small number of insecticides, fungicides, and herbicides are approved for conventional herb production. The number, however, is very restricted (end of 2008 several active substances lost registration at least in Europe), and limits for residues can be found in national law and international regulations, for example, the European Pharmacopoeia. For essential oils, mainly the lipophilic substances are of relevance since they can be enriched over the limits in the oil.

Harvesting and the first steps of *postharvest handling* are the last part of the production chain of starting materials for essential oils. The harvest date is determined by the development stage or maturity of the plant or plant part, Harvesting techniques should keep the quality by avoiding adulterations, admixtures with undesired plant parts, or contaminations, which could cause "off-flavor" in the final product. There are many technical aids at disposal, from simple devices to large-scale harvesters, which will be considered carefully in Chapter 4. From the quality point of view, raising the temperature by fermentation should in general be avoided (except, in vanilla), and during the drying process further contamination with soil, dust, insects, or molds has to be avoided.

Quality and safety of essential oil-bearing plants as raw materials for pharmaceutical products, flavors, and fragrances are of highest priority from the consumer point of view. To meet the respective demands, standards and safety as well as quality assurance measures are needed to ensure that the plants are produced with care, so that negative impacts during wild collection, cultivation, processing, and storage can be limited. To overcome these problems and to guarantee a steady, affordable and sustainable supply of essential oil plants of good quality (Figure 3.10), in recent years guidelines for good agricultural practices (GAP) and standards for Sustainable Wild Collection (ISSC) have been established at the national and international level.

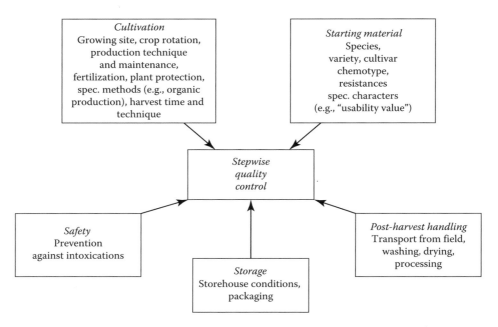

FIGURE 3.10 Main items of "good agricultural practices" (GAP) for medicinal and aromatic plants.

3.6 INTERNATIONAL STANDARDS FOR WILD COLLECTION AND CULTIVATION

3.6.1 GA(C)P: GUIDELINES FOR GOOD AGRICULTURAL (AND COLLECTION) PRACTICE OF MEDICINAL AND AROMATIC PLANTS

First initiatives for the elaboration of such guidelines trace back to a roundtable discussion in Angers, France in 1983, and intensified at an International Symposium in Novi Sad 1988 (Franz, 1989b). A first comprehensive paper was published by Pank et al. (1991) and in 1998 the European Herb Growers Association (EHGA/EUROPAM) released the first version (Máthé and Franz, 1999). The actual version can be downloaded from http://www.europam.net.

In the following it was adopted and slightly modified by the European Agency for the Evaluation of Medicinal Products (EMEA), and finally as Guidelines on good agricultural and collection practices (GACP) by the WHO in 2003.

All these guidelines follow almost the same concept dealing with the following topics:

- Identification and authentication of the plant material, especially botanical identity and deposition of specimens.
- Seeds and other propagation material, respecting the specific standards and certifications.
- Cultivation, including site selection, climate, soil, fertilization, irrigation, crop maintenance, and plant protection with special regard to contaminations and residues.
- Harvest, with specific attention to harvest time and conditions, equipment, damage, contaminations with (toxic) weeds and soil, transport, possible contact with any animals, and cleaning of all equipment and containers.
- Primary processing, that is, washing, drying, distilling; cleanness of the buildings; according to the actual legal situation these processing steps including distillation—if performed by the farmer—is still part of GA(C)P; in all other cases, it is subjected to GMP (good manufacturing practice).

Sources of Essential Oils

- Packaging and labeling, including suitability of the material.
- Storage and transportation, especially storage conditions, protection against pests and animals, fumigation, and transport facilities.
- Equipment: material, design, construction, easy to clean.
- Personnel and facilities, with special regard to education, hygiene, protection against allergens and other toxic compounds, welfare.

In the case of wild collection the standard for sustainable collection should be applied (see Section 3.6.2).

A very important topic is finally the *documentation* of all steps and measurements to be able to trace back the starting material, the exact location of the field, any treatment with agrochemicals, and the special circumstances during the cultivation period. Quality assurance is only possible if the traceability is given and the personnel is educated appropriately. Certification and auditing of the production of essential oil-bearing plants is not yet obligatory, but recommended and often requested by the customer.

3.6.2 ISSC-MAP: THE INTERNATIONAL STANDARD ON SUSTAINABLE WILD COLLECTION OF MEDICINAL AND AROMATIC PLANTS

ISSC-MAP is a joint initiative of the German Bundesamt für Naturschutz (BfN), WWF/TRAFFIC Germany, IUCN Canada, and IUCN Medicinal Plant Specialist Group (MPSG). ISSC-MAP intends to ensure the long-term survival of MAP populations in their habitats by setting principles and criteria for the management of MAP wild collection (Leaman, 2006; Medicinal Plant Specialist Group, 2007). The standard is not intended to address product storage, transport, and processing, or any issues of products, topics covered by the WHO Guidelines on GACP for Medicinal Plants (WHO, 2003). ISSC-MAP includes legal and ethical requirements (legitimacy, customary rights, and transparency), resource assessment, management planning and monitoring, responsible collection, and collection, area practices and responsible business practices. One of the strengths of this standard is that resource management not only includes target MAP resources and their habitats but also social, cultural, and economic issues.

3.6.3 FAIRWILD

The FairWild standard (http://www.fairwild.org) was initiated by the Swiss Import Promotion Organization (SIPPO) and combines principles of FairTrade (Fairtrade Labelling Organizations International, FLO), international labor standards (International Labour Organization, ILO), and sustainability (ISSC-MAP).

3.7 CONCLUSION

This chapter has shown that a number of items concerning the plant raw material have to be taken into consideration when producing essential oils. A quality management has to be established tracing back to the authenticity of the starting material and ensuring that all known influences on the quality are taken into account and documented in an appropriate way. This is necessary to meet the increasing requirements of international standards and regulations. The review also shows that a high number of data and information exist, but sometimes without expected relevance due to the fact that the repeatability of the results is not given by a weak experimental design, an incorrect description of the plant material used, or an inappropriate sampling. On the other side, this opens the chance for many more research work in the field of essential oil-bearing plants.

REFERENCES

Agnaniet, H., C. Menut, and J.M. Bessière, 2004. Aromatic plants of tropical Africa XLIX: Chemical composition of essential oils of the leaf and rhizome of *Aframomum giganteum* K. Schum from Gabun. *Flavour Fragr. J.*, 19: 205–209.

Aiello, N., F. Scartezzini, C. Vender, L. D'Andrea, and A. Albasini, 2001. Caratterisiche morfologiche, produttive e qualitative di una nuova varietà sintetica di salvia confrontata con altre cultivar. *ISAFA Comunicaz. Ric.*, 2001/(1): 5–16.

Akhila, A. and M. Rani, 2002. Chemical constituents and essential oil biogenesis in *Vetiveria zizanioides*. In *Vetiveria—The Genus Vetiveria*, M. Maffei (ed.), pp. 73–109. New York: Taylor & Francis.

Asllani, U., 2000. Chemical composition of albanian sage oil (*Salvia officinalis* L.). *J. Essent. Oil Res.*, 12: 79–84.

Auer, C.A., 2003. Tracking genes from seed to supermarket: Techniques and trends. *Trends Plant Sci.*, 8: 591–597.

Aziz, E.E., H. Al-Amier, and L.E. Craker, 2008. Influence of salt stress on growth and essential oil production in peppermint, pennyroyal and apple mint. *J. Herbs Spices Med. Plants*, 14: 77–87.

Baser, K.H.C., 2002. The Turkish *Origanum* species. In *Oregano: The Genera Origanum and Lippia*, S.E. Kintzios (ed.), pp. 109–126. New York: Taylor & Francis.

Bazina, E., A. Makris, C. Vender, and M. Skoula, 2002. Genetic and chemical relations among selected clones of *Salvia officinalis*. In *Breeding Research on Aromatic and Medicinal Plants*, C.B. Johnson and Ch. Franz (eds.), pp. 269–273. Binghampton, NY: Haworth Press.

Bergougnoux, V., J.C. Caissard, F. Jullien, J.L. Magnard, G. Scalliet, J.M. Cock, P. Hugueney, and S. Baudino, 2007. Both the adaxial and abaxial epidermal layers of the rose petal emit volatile scent compounds. *Planta*, 226: 853–866.

Bernáth, J., 2000. Genetic improvement of cultivated species of the genus Salvia. In *Sage—The Genus Salvia*, S.E. Kintzios (ed.), pp. 109–124. Amsterdam, the Netherlands: Harwood Academic Publishers.

Bernáth, J., 2002. Evaluation of strategies and results concerning genetical improvement of medicinal and aromatic plants. *Acta Hort.*, 576: 116–128.

Bertea, C.M. and W. Camusso, 2002. Anatomy, biochemistry and physiology. In *Vetiveria: Medicinal and Aromatic Plants—Industrial Profiles*, M. Maffei (ed.), Vol. 20, pp. 19–43. London, U.K.: Taylor & Francis.

Bezzi, A., 1994. Selezione clonale e costituzione di varietà di salvia (*Salvia officinalis* L.). *Atti convegno internazionale 'Coltivazione e miglioramento di piante officinali'*, 97–117. Villazzano di Trento, Italy: ISAFA.

Bicchi, C., M. Fresia, P. Rubiolo, D. Monti, Ch. Franz, and I. Goehler, 1997. Constituents of *Tagetes lucida* Cav. ssp. *lucida* essential oil. *Flavour Fragr. J.*, 12: 47–52.

Bradbury, L.M.T., R.J. Henry, Q. Jin, R.F. Reinke, and D.L.E. Waters, 2005. A perfect marker for fragrance genotyping in rice. *Mol. Breed.*, 16: 279–283.

Bradu, B.L., S.N. Sobti, P. Pushpangadan, K.M. Khosla, B.L. Rao, and S.C. Gupta, 1989. Development of superior alternate source of clove oil from 'Clocimum' (*Ocimum gratissimum* Linn.). In *Proceedings of the 11th International Congress of Essential Oils, Fragrances and Flavours*, Vol. 3, pp. 97–103.

Brophy, J.J. and I.A. Southwell, 2002. *Eucalyptus* chemistry. In *Eucalyptus—The Genus Eucalyptus*, J.J.W. Coppen (ed.), pp. 102–160. London, U.K.: Taylor & Francis.

Carruba, A., R. la Torre, and A. Matranga, 2002. Cultivation trials of some aromatic and medicinal plants in a semiarid Mediterranean environment. *Acta Hort.*, 576: 207–214.

CBD, 2001. Convention on biological diversity. In *United Nations Environment Programme, CBD Meeting Nairobi*, Nairobi, Kenya.

CBD, 2002. Global strategy for plant conservation. In *CBD Meeting The Hague*, The Hague, the Netherlands.

CBD, 2004. Sustainable use of biodiversity. In *CBD Meeting Montreal*, Montreal, Quebec, Canada.

Ceylan, A., H. Otan, A.O. Sari, N. Carkaci, E. Bayram, N. Ozay, M. Polat, A. Kitiki, and B. Oguz, 1994. *Origanum onites* L. (Izmir Kekigi) Uzerinde Agroteknik Arastirmalar, Final Report. Izmir, Turkey: AARI.

Chalchat, J., J.C. Gary, and R.P. Muhayimana, 1995. Essential oil of *Tagetes minuta* from Rwanda and France: Chemical composition according to harvesting, location, growing stage and plant part. *J. Essent. Oil Res.*, 7: 375–386.

Chalchat, J., A. Michet, and B. Pasquier, 1998. Study of clones of *Salvia officinalis* L., yields and chemical composition of essential oil. *Flavour Fragr. J.*, 13: 68–70.

Chatterjee, S.K., 2002. Cultivation of medicinal and aromatic plants in India. *Acta Hort.*, 576: 191–202.

Ciccio, J.F., 2004. A source of almost pure methylchavicol: Volatile oil from the aerial parts of *Tagetes lucida* (Asteraceae) cultivated in Costa Rica. *Rev. de Biol. Trop.*, 52: 853–857.

Circella, G., Ch. Franz, J. Novak, and H. Resch, 1995. Influence of day length and leaf insertion on the composition of marjoram essential oil. *Flavour Fragr. J.*, 10: 371–374.

Coronel, A.C., C.M. Cerda-Garcia-Rojas, P. Joseph-Nathan, and C.A.N. Catalán, 2006. Chemical composition, seasonal variation and a new sesquiterpene alcohol from the essential oil of *Lippia integrifolia*. *Flavour Fragr. J.*, 21: 839–847.

Croteau, R., 1991. Metabolism of monoterpenes in mint (*Mentha*) species. *Planta Med.*, 57(Suppl. 1): 10–14.

Croteau, R., M. Felton, F. Karp, and R. Kjonaas, 1981. Relationship of camphor biosynthesis to leaf development in sage (*Salvia officinalis*). *Plant Physiol.*, 67: 820–824.

Czabajska, W., J. Dabrowska, K. Kazmierczak, and E. Ludowicz, 1978. Maintenance breeding of chamomile cultivar 'Zloty Lan'. *Herba Polon.*, 24: 57–64.

Daniel, G. and U. Bomme, 1991. Use of in-vitro culture for arnica (*Arnica montana* L.) breeding. *Landw. Jahrb.*, 68: 249–253.

Dellacassa, E., E. Soler, P. Menéndez, and P. Moyna, 1990. Essentail oils from *Lippia alba* Mill. N.E. Brown and *Aloysia chamaedrifolia* Cham. (Verbenaceae) from Urugay. *Flavour Fragr. J.*, 5: 107–108.

Desmarest, P., 1992. Amelioration du fenoil amier par selection recurrente, clonage et embryogenèse somatique. In *Proceedings of the Second Mediplant Conference*, pp. 19–26. Conthey, Switzerland/CH, P.

Dey, B.B. and M.A. Choudhuri, 1983. Effect of leaf development stage on changes in essential oil of *Ocimum sanctum* L. *Biochem. Physiol. Pflanzen*, 178: 331–335.

Diemer, F., F. Jullien, O. Faure, S. Moja, M. Colson, E. Matthys-Rochon, and J.C. Caissard, 1998. High efficiency transformation of peppermint (*Mentha* x *piperita* L.) with *Agrobacterium tumefaciens*. *Plant Sci.*, 136: 101–108.

Dominguez, X.A., S.H. Sánchez, M. Suárez, X Baldas, J.H., and G. Ma del Rosario, 1989. Chemical constituents of *Lippia graveolens*. *Planta Med.*, 55: 208–209.

Doran, J.C., 2002. Genetic improvement of eucalyptus. In *Eucalyptus—The Genus Eucalyptus, Medicinal and Aromatic Plants—Industrial Profiles*, J.J.W. Coppen (ed.), Vol. 22, pp. 75–101. London, U.K.: Taylor & Francis.

Dronne, S., S. Moja, F. Jullien, F. Berger, and J.C. Caissard, 1999. *Agrobacterium*-mediated transformation of lavandin (*Lavandula* × *intermedia* Emeric ex Loiseleur). *Transgenic Res.*, 8: 335–347.

Dudai, N., 2006. Breeding of high quality basil for the fresh herb market—An overview. In *International Symposium on the Labiatae*, p. 15 San Remo, Italy.

Dudai, N., E. Putievsky, U. Ravid, D. Palevitch, and A.H. Halevy, 1992. Monoterpene content of *Origanum syriacum* L. as affected by environmental conditions and flowering. *Physiol. Plant.*, 84: 453–459.

Elementi, S., R. Nevi, and L.F. D'Antuono, 2006. Biodiversity and selection of 'European' basil (*Ocimum basilicum* L.) types. *Acta Hort.*, 723: 99–104.

El-Keltawi, N.E.M. and S.S.A. Abdel-Rahman, 2006. In vivo propagation of certain sweet basil cultivars. *Acta Hort.*, 723: 297–302.

Fahn, A., 1979. *Secretory Tissues in Plants*. London, U.K.: Academic Press.

Fahn, A., 1988. Secretory tissues in vascular plants. *New Phytologist*, 108: 229–257.

Figueiredo, A.C., J.G. Barroso, L.G. Pedro, and J.J.C. Scheffer, 1997. Physiological aspects of essential oil production. In *Essential Oils: Basic and Applied Research*, Ch. Franz, A. Máthé, and G. Buchbauer (eds.), pp. 95–107. Carol Stream, IL: Allured Publishing.

Fischer, U., 1998. Variabilität Guatemaltekischer Arzneipflanzen der Gattung *Lippia* (Verbenaceae): *Lippia alba*, *L. dulcis*, *L. graveolens*. Dissertation, Veterinärmedizinischen Universität, Wien, Austria.

Fischer, U., Ch. Franz, R. Lopez, and E. Pöll, 1996. Variability of the essential oils of *Lippia graveolens* HBK from Guatemala. In *Essential Oils: Basic and Applied Research*, Ch. Franz, A. Máthé, and A.G. Buchbauer (eds.), pp. 266–269. Carol Stream, IL: Allured Publishing.

Fischer, U., R. Lopez, E. Pöll, S. Vetter, J. Novak, and Ch. Franz, 2004. Two chemotypes within *Lippia alba* populations in Guatemala. *Flavour Fragr. J.*, 19: 333–335.

Franz, C. and H. Glasl, 1976. Comparative investigations of fruit-, leaf- and root-oil of some parsley varieties. *Qual. Plant. Plant Foods Hum. Nutr.*, 25(3/4): 253–262.

Franz, Ch., 1972. Einfluss der Naehrstoffe Stickstoff und Kalium auf die Bildung des aetherischen Oels der Pfefferminze, *Mentha piperita* L. *Planta Med.*, 22: 160–183.

Franz, Ch., 1981. Zur Qualitaet von Arznei- u. Gewuerzpflanzen. Habil.-Schrift. Muenchen, Germany: TUM.

Franz, Ch., 1982. Genetische, ontogenetische und umweltbedingte Variabilität der Bestandteile des ätherischen Öls von Kamille (*Matricaria recutita*(L.) Rauschert). In *Aetherische Oele—Analytik, Physiologie, Zusammensetzung*, K.H. Kubeczka (ed.), pp. 214–224. Stuttgart, Germany: Thieme.

Franz, Ch., 1989a. Biochemical genetics of essential oil compounds. In *Proceedings of the 11th International Congress of Essential Oils, Fragrances and Flavours*, Vol. 3, pp. 17–25. New Delhi, India: Oxford & IBH Publishing.
Franz, Ch., 1989b. Good agricultural practice (GAP) for medicinal and aromatic plant production. *Acta Hort.*, 249: 125–128.
Franz, Ch., 1993a. Probleme bei der Beschaffung pflanzlicher Ausgangsmaterialien. In *Ätherische Öle, Anspruch und Wirklichkeit*, R. Carle (ed.), pp. 33–58. Stuttgart, Germany: Wissenschaftliche Verlagsgesellschaft.
Franz, Ch., 1993b. Genetics. In *Volatile Oil Crops*, R.K.M. Hay and P.G. Waterman (eds.), pp. 63–96. Harlow, U.K.: Longman.
Franz, Ch., 1993c. Domestication of wild growing medicinal plants. *Plant Res. Dev.*, 37: 101–111.
Franz, Ch., 1993d. Genetic versus climatic factors influencing essential oil formation. In *Proceedings of the 12th International Congress of Essential Oils, Fragrances and Flavours*, pp. 27–44. Vienna, Austria.
Franz, Ch., 1999. Gewinnung von biogenen Arzneistoffen und Drogen, In *Biogene Arzneistoffe*, 2nd edn., H. Rimpler (ed.), pp. 1–24. Stuttgart, Germany: Deutscher Apotheker Verlag.
Franz, Ch., 2000. Biodiversity and random sampling in essential oil plants. Lecture 31st ISEO, Hamburg, Germany.
Franz, Ch., 2001. Plant variety rights and specialised plants. In *Proceedings of the PIPWEG 2001, Conference on Plant Intellectual Property within Europe and the Wider Global Community*, pp. 131–137. Sheffield, U.K.: Sheffield Academic Press.
Franz, Ch., 2013. Schafgarbe (*Achillea millefolium* L.). In: *Handbuch des Arznei- u. Gewuerzpflanzenbaus Vol. 5*, pp. 453–463. Saluplanta, Bernburg.
Franz, Ch., K. Hardh, S. Haelvae, E. Mueller H. Pelzmann, and A. Ceylan, 1986. Influence of ecological factors on yield and essential oil of chamomile (*Matricaria recutita* L.). *Acta Hort.*, 188: 157–162.
Franz, Ch., J. Hoelzl, and C. Kirsch, 1983. Influence of nitrogen, phosphorus and potassium fertilization on chamomile (*Chamomilla recutita* (L.) Rauschert). II. Effect on the essential oil. *Gartenbauwiss. Hort. Sci.*, 48: 17–22.
Franz, Ch., J. Hoelzl, and A. Voemel, 1978. Variation in the essential oil of *Matricaria chamomilla* L. depending on plant age and stage of development. *Acta Hort.*, 73: 230–238.
Franz, Ch., C. Kirsch, and O. Isaac, 1983. Process for producing a new tetraploid chamomile variety. German Patent DE3423207.
Franz, Ch., C. Kirsch, and O. Isaac, 1985. Neuere Ergebnisse der Kamillenzüchtung. *Dtsch. Apoth. Ztg.*, 125: 20–23.
Franz, Ch. and Novak, J., 1997. Breeding of *Origanum* sp. In *Proceedings of the IPGRI Workshop*, Padulosi, S. (ed.), pp. 50–57. Oregano.
Frezal, L. and R. Leblois, 2008. Four years of DNA barcoding: Current advances and prospects. *Infect. Genet. Evol.*, 8: 727–736.
Frighetto, N., J.G. de Oliveira, A.C. Siani, and K. Calago das Chagas, 1998. *Lippia alba* Mill (Verbenaceae) as a source of linalool. *J. Essent. Oil Res.*, 10: 578–580.
Fritzsche, R., J. Gabler, H. Kleinhempel, K. Naumann, A. Plescher, G. Proeseler, F. Rabenstein, E. Schliephake, and W. Wradzidlo, 2007. *Handbuch des Arznei- und Gewürzpflanzenbaus: Krankheiten und Schädigungen an Arznei- und Gewürzpflanzen*, Vol. 3. Bernburg, Germany: Saluplanta e.V.
Gabler, J., 2002. Breeding for resistance to biotic and abiotic factors in medicinal and aromatic plants. In *Breeding Research on Aromatic and Medicinal Plants*, C.B. Johnson and Ch. Franz (eds.), pp. 1–12. Binghampton, NY: Haworth Press.
Galambosi, B. and S. Dragland, 2002. Possibilities and limitations for herb production in Nordic countries. *Acta Hort.*, 576: 215–225.
Gancel, A.L., D. Ollé, P. Ollitraut, F. Luro, and J.M. Brillouet, 2002. Leaf and peel volatile compounds of an interspecific citrus somatic hybrid (*Citrus aurantifolia* Swing. × *Citrus paradisi* Macfayden). *Flavour Fragr. J.*, 17: 416–424.
Gershenzon, J., M.E. McConkey, and R.B. Croteau, 2000. Regulation of monoterpene accumulation in leaves of peppermint. *Plant Physiol.*, 122: 205–213.
Giannouli, A.L. and S.E. Kintzios, 2000. Essential oils of *Salvia* spp.: Examples of intraspecific and seasonal variation. In *Sage—The Genus Salvia*, S.E. Kintzios (ed.), pp. 69–80. Amsterdam, the Netherlands: Harwood Academic Publishing.
Goehler, I., 2006. Domestikation von Medizinalpflanzen und Untersuchungen zur Inkulturnahme von *Tagetes lucida* Cav. Dissertation, an der Universität für Bodenkultur Wien, Wein, Austria.

Goehler, I., Ch. Franz, A. Orellana, and C. Rosales, 1997. Propagation of *Tagetes lucida* Cav. Poster WOCMAP II Mendoza, Argentina.
Gonzales de, C.N., A. Quintero, and A. Usubillaga, 2002. Chemotaxonomic value of essential oil compounds in *Citrus* species. *Acta Hort.*, 576: 49–55.
Gora, J., A. Lis, J. Kula, M. Staniszewska, and A. Woloszyn, 2002. Chemical composition variability of essential oils in the ontogenesis of some plants. *Flavour Fragr. J.*, 17: 445–451.
Gouyon, P.H. and P. Vernet, 1982. The consequences of gynodioecy in natural populations of *Thymus vulgaris* L. *Theoret. Appl. Genet.*, 61: 315–320.
Grassi, P., 2003. Botanical and chemical investigations in *Hyptis* spp. (Lamiaceae) in El Salvador. Dissertation, Universität Wien, Wein, Austria.
Grassi, P., J. Novak, H. Steinlesberger, and Ch. Franz, 2004. A direct liquid, non-equilibrium solid-phase micro-extraction application for analysing chemical variation of single peltate trichomes on leaves of *Salvia officinalis*. *Phytochem. Anal.*, 15: 198–203.
Harrewijn, P., van A.M. Oosten, and P.G.M. Piron, 2001. *Natural Terpenoids as Messengers*. Dordrecht, the Netherlands: Kluwer Academic Publishers.
Hay, R.K.M. and P.G. Waterman, 1993. *Volatile Oil Crops*. Burnt Mill, U.K.: Longman Science & Technology Publications.
Hebert, P.D.N., A. Cywinska, S.L. Ball, and J.R. deWaard, 2003. Biological identifications through DNA barcodes. *Proc. R. Soc. Lond. B*, 270: 313–322.
Hedge, I.C., 1992. A global survey of the biography of the Labiatae. In *Advances in Labiatae Science*, R.M. Harley and T. Reynolds (eds.), pp. 7–17. Kew, U.K.: Royal Botanical Gardens.
Hefendehl, F.W., 1962. Zusammensetzung des ätherischen Öls von *Mentha x piperita* im Verlauf der Ontogenese und Versuche zur Beeinflussung der Ölkomposition. *Planta Med.*, 10: 241–266.
Hoeltzel, C., 1964. Über Zusammenhänge zwischen der Biosynthese der ätherischen Öle und dem photoperiodischen Verhalten der Pfefferminze (*Mentha piperita* L.). Dissertation, University of Tübingen, Tübingen, Germany.
Horn, W., Ch. Franz, and I. Wickel, 1988. Zur Genetik der Bisaboloide bei der Kamille. *Plant Breed.*, 101: 307–312.
Johnson, C.B, A. Kazantzis, M. Skoula, U. Miteregger, and J. Novak, 2004. Seasonal, populational and ontogenic variation in the volatile oil content and composition of individuals of *Origanum vulgare* subsp. *hirtum*, assessed by GC headspace analysis and by SPME sampling of individual oil glands. *Phytochem. Anal.*, 15: 286–292.
Kampranis, S.C., D. Ioannidis, A. Purvis, W. Mahrez, E. Ninga, N.A. Katerelos, S. Anssour et al., 2007. Rational conversion of substrate and product specificity in a *Salvia* monoterpene synthase: Structural insights into the evolution of terpene synthase function. *Plant Cell*, 19: 1994–2005.
Kanias, G.D., C. Souleles, A. Loukis, and E. Philotheou-Panou, 1998. Statistical studies of essential oil composition in three cultivated Sage species. *J. Essent. Oil Res.*, 10: 395–403.
Karousou, R., D. Vokou, and Kokkini, 1998. Variation of *Salvia fruticosa* essential oils on the island of Crete (Greece). *Bot. Acta*, 111: 250–254.
Kastner, U., J. Saukel, K. Zitterl-Eglseer, R. Länger, G. Reznicek, J. Jurenitsch, and W. Kubelka, 1992. Ätherisches Öl—ein zusätzliches Merkmal für die Charakterisierung der mitteleuropäischen Taxa der *Achillea-millefolium*-Gruppe. *Sci. Pharm.*, 60: 87–99.
Khosla, M.K., B.L. Bradu, and R.K. Thapa, 1989. Biogenetic studies on the inheritance of different essential oil constituents of *Ocimum* species, their F1 hybrids and synthesized allopolyploids. *Herba Hung.*, 28: 13–19.
Kitiki, A., 1997. Status of cultivation and use of oregano in Turkey. In *Proceedings of the IPGRI Workshop Oregano*, S. Padulosi (ed.), pp. 122–132.
Koezuka, Y., G. Honda, and M. Tabata, 1986. Genetic control of phenylpropanoids in *Perilla frutescens*. *Phytochemistry*, 25: 2085–2087.
Kokkini, S., R. Karousou, A. Dardioti, N. Kirgas, and T. Lanaras, 1996. Autumn essential oils of Greek oregano (*Origanum vulgare* ssp. *hirtum*). *Phytochemistry*, 44: 883–886.
Kress, W.J., K.J. Wurdack, E.A. Zimmer, L.A. Weigt, and D.H. Janzen, 2005. Use of DNA barcodes to identify flowering plants. *PNAS*, 102: 8369–8374.
Kubeczka, K.H., 1997. The essential oil composition of *Pimpinella* species. In *Progress in Essential Oil Research*, K.H.C. Baser and N. Kirimer (eds.), pp. 35–56. Eskisehir, Turkey: ISEO.
Kubeczka, K.H., A. Bartsch, and I. Ullmann, 1982. Neuere Untersuchungen an ätherischen Apiaceen-Ölen. In *Ätherische Öle—Analytik, Physiologie, Zusammensetzung*, K.H. Kubeczka (ed.), pp. 158–187. Stuttgart, Germany: Thieme.

Kubeczka, K.H., I. Bohn, and V. Formacek, 1986. New constituents from the essential oils of *Pimpinella sp.* In *Progress in Essential Oil Research*, E.J. Brunke (ed.), pp. 279–298. Berlin, Germany: W de Gruyter.

Kulkarni, R.N., 1990. Honeycomb and simple mass selection for herb yield and inflorescence-leaf-steam-ratio in palmarose grass. *Euphytica*, 47: 147–151.

Kulkarni, R.N. and S. Ramesh, 1992. Development of lemongrass clones with high oil content through population improvement. *J. Essent. Oil Res.*, 4: 181–186.

Kurowska, A. and I. Galazka, 2006. Essential oil composition of the parsley seed of cultivars marketed in Poland. *Flavour Fragr. J.*, 21: 143–147.

Langbehn, J., F. Pank, J. Novak, and C. Franz, 2002. Influence of Selection and Inbreeding on *Origanum majorana* L. *J. Herbs Spices Med. Plants*, 9: 21–29.

Lawrence, B.M., 1978. *Essential Oils 1976–77*, pp. 84–109. Wheaton, IL: Allured Publishing.

Lawrence, B.M., 1984. The botanical and chemical aspects of Oregano. *Perform. Flavor*, 9(5): 41–51.

Lawrence, B.M., 2007. *Mint: The Genus Mentha*. Boca Raton, FL: CRC Press.

Leaman, D.J., 2006. Sustainable wild collection of medicinal and aromatic plants. In *Medicinal and Aromatic Plants*, R.J. Bogers, L.E. Craker, and D. Lange (eds.), pp. 97–107. Dordrecht, the Netherlands: Springer.

Le Buanec, B., 2001. Development of new plant varieties and protection of intellectual property: An international perspective. In *Proceedings of the PIPWEG Conference on 2001 Angers*, pp. 103–108. Sheffield, U.K.: Sheffield Academic Press.

Lemli, J.A.J.M., 1955. De vluchtige olie van *Mentha piperita* L. gedurende de ontwikkeling van het plant. Dissertation, University of Groningen, Groningen, the Netherlands.

Letchamo, W. and R. Marquard, 1993. The pattern of active substances accumulation in camomile genotypes under different growing conditions and harvesting frequencies. *Acta Hort.*, 331: 357–364.

Li, X., Z. Gong, D. Koiwa, X. Niu, J. Espartero, X. Zhu, P. Veronese et al., 2001. Bar-expressing peppermint (*Mentha* × *piperita* L. var. Black Mitcham) plants are highly resistant to the glufosinate herbicide Liberty. *Mol. Breed.*, 8: 109–118.

Lis, A., E. Boczek, and J. Gora, 2004. Chemical composition of the essential oils from fruits. Leaves and flowers of the Amur cork tree (*Phellodendron amurense* Rupr.). *Flavour Fragr. J.*, 19: 549–553.

Lis, A. and Milczarek, J., 2006. Chemical composition of the essential oils from fruits, leaves and flowers of *Phellodendron sachalinene* (Fr. Schmidt) Sarg. *Flavour Fragr. J.*, 21: 683–686.

Llewelyn, M., 2002. European plant intellectual property. In *Breeding Research on Aromatic and Medicinal Plants*, C.B. Johnson and Ch. Franz (eds.), pp. 389–398. Binghampton, NY: Haworth Press.

Lorenzo, D., D. Paz, P. Davies, R. Vila, S. Canigueral, and E. Dellacassa, 2001. Composition of a new essential oil type of *Lippia alba* (Mill.) N.E. Brown from Uruguay. *Flavour Fragr. J.*, 16: 356–359.

Macchia, M., A. Pagano, L. Ceccarini, S. Benvenuti, P.L. Cioni, and G. Flamini, 2006. Agronomic and phytochimic characteristics in some genotypes of *Ocimum basilicum* L. *Acta Hort.*, 723: 143–149.

Mahmoud, S.S. and R.B. Croteau, 2001. Metabolic engineering of essential oil yield and composition in mint by altering expression of deoxyxylulose phosphate reductoisomerase and menthofuran synthase. *PNAS*, 98: 8915–8920.

Mahmoud, S.S. and R.B. Croteau, 2003. Menthofuran regulates essential oil biosynthesis in peppermint by controlling a downstream monoterpene reductase. *PNAS*, 100: 14481–14486.

Mahmoud, S.S., M. Williams, and R.B. Croteau, 2004. Cosuppression of limonene-3-hydroxylase in peppermint promotes accumulation of limonene in the essential oil. *Phytochemistry*, 65: 547–554.

Maleci Bini, L. and C. Giuliani, 2006. The glandular trichomes of the Labiatae. A review. *Acta Hort.*, 723: 85–90.

Marotti, M., R. Piccaglia, B. Biavati, and I. Marotti, 2004. Characterization and yield evaluation of essential oils from different *Tagetes* species. *J. Essent. Oil Res.*, 16: 440–444.

Marotti, M., R. Piccaglia, and E. Giovanelli, 1994. Effects of variety and ontogenetic stage on the essential oil composition and biological activity of fennel (*Foeniculum vulgare* Mill.). *J. Essent. Oil Res.*, 6: 57–62.

Marotti, M., P. Piccaglia, and E. Giovanelli, 1996. Differences in essential oil composition of basil (*Ocimum basilicum* L.) of Italian cultivars related to morphological characteristics. *J. Agric. Food Chem.*, 44: 3926–3929.

Marthe, F. and P. Scholze, 1996. A screening technique for resistance evaluation to septoria blight (*Septoria petroselini*) in parsley (*Petroselinum crispum*). *Beitr. Züchtungsforsch*, 2: 250–253.

Martinetti, L., E. Quattrini, M. Bononi, and F. Tateo, 2006. Effect of the mineral fertilization and the yield and the oil content of two cultivars of rosemary. *Acta Hort.*, 723: 399–404.

Massoud, H. and C. Franz, 1990. Quantitative genetical aspects of *Chamomilla recutita* (L.) Rauschert. *J. Essent. Oil Res.*, 2: 15–20.

Massoud, H., M. Sharaf El-Din, R. Hassan, and A. Ramadan, 2002. Effect of salinity and some trace elements on growth and leaves essential oil content of thyme (*Thymus vulgaris* L.). *J. Agric. Res. Tanta Univ.*, 28: 856–873.

Máthé, A. and Ch. Franz, 1999. Good agricultural practice and the quality of phytomedicines. *J. Herbs Spices Med. Plants*, 6: 101–113.

Máthé, I., G. Nagy, A. Dobos, V.V. Miklossy, and G. Janicsak, 1996. Comparative studies of the essential oils of some species of Sect. *Salvia*. In *Proceedings of the 27th International Symposium on Essential Oils (ISEO)*, Ch. Franz, A. Máthé, and G. Buchbauer (eds.), pp. 244–247.

Medicinal Plant Specialist Group, 2007. International Standard for Sustainable Wild Collection of Medicinal and Aromatic Plants (ISSC-MAP). Version 1.0. Bundesamt für Naturschutz (BfN), MPSG/SSC/IUCN, WWF Germany, and TRAFFIC, Bonn, Gland, Frankfurt, and Cambridge. *BfN-Skripten* 195.

Menary, R.C., 1994. Factors influencing the yield and composition of essential oils, II: Nutrition, irrigation, plant growth regulators, harvesting and distillation. In *Proceedings of the 4emes Rencontres Internationales*, pp. 116–138. Nyons, France.

Mendes, M.L. and A. Romano, 1997. In vitro cloning of *Thymus mastichina* L. field grown plants. *Acta Hort.*, 502: 303–306.

Minuto, G., A. Minuto, A. Garibaldi, and M.L. Gullino, 2006. Disease control of aromatic crops: Problems and solutions. In *International Symposium on Labiatae*. San Remo, Italy, p. 33.

Miraglia, M., K.G. Berdal, C. Brera, P. Corbisier, A. Holst-Jensen, E.J. Kok, H.J. Marvin et al., 2004. Detection and traceability of genetically modified organisms in the food production chain. *Food Chem. Toxicol.*, 42: 1157–1180.

Mosandl, A., 1993. Neue Methoden zur herkunftsspezifischen Analyse aetherischer Oele. In *Ätherische Öle—Anspruch und Wirklichkeit*, R. Carle (ed.), pp. 103–134. Stuttgart, Germany: Wissenschaftliche Verlagsgesellschaft.

Mulas, M., A.H. Dias Francesconi, B. Perinu, and E. Del Vais, 2002. Selection of Rosemary (*Rosmarinus officinalis* L.) cultivars to optimize biomass yield. In *Breeding Research on Aromatic and Medicinal Plants*, C.B. Johnson and Ch. Franz (eds.), pp. 133–138. Binghampton, NY: Haworth Press.

Munoz-Bertomeu, J., I. Arrillaga, R. Ros, and J. Segura, 2006. Up-regulation of 1-deoxy-D-xylulose-5-phosphate synthase enhances production of essential oils in transgenic spike lavender. *Plant Physiol.*, 142: 890–900.

Munsi, P.S. and Mukherjee, S.K., 1986. Response of Java citronella (*Cymbopogon winterianus* Jowitt.) to harvesting intervals with different nitrogen levels. *Acta Hort.*, 188: 225–229.

Murray, M.J., 1969. *Induced Mutations in Plants*, pp. 345–371. Vienna, Austria: IAEA.

Murray, M.J. and R.H. Reitsema, 1954. The genetic basis of the ketones carvone and menthone in *Mentha crispa* L. *J. Am. Pharm. Assoc. (Sci. Ed.)*, 43: 612–613.

Murray, M.J. and A.W. Todd, 1972. Registration of Todd's Mitcham Peppermint. *Crop Sci.*, 12: 128.

Nair, M.K., 1982. Cultivation of spices. In *Cultivation and Utilization of Aromatic Plants*, C.K. Atal and B.M. Kapur (eds.), pp. 190–214. Jammu-Tawi, India: RRL-CSIR.

Nebauer, S.G., I. Arrillaga, L. del Castillo-Agudo, and J. Segura, 2000. *Agrobacterium tumefaciens*-mediated transformation of the aromatic shrub *Lavandula latifolia*. *Mol. Breed.*, 6: 23–48.

Nicola, S., J. Hoeberechts, and E. Fontana, 2006. Rooting products and cutting timing for peppermint (*Mentha piperita* L.) radication. *Acta Hort.*, 723: 297–302.

Niu, X., X. Li, P. Veronese, R.A. Bressan, S.C. Weller, and P.M. Hasegawa, 2000. Factors affecting *Agrobacterium tumefaciens*-mediated transformation of peppermint. *Plant Cell Rep.*, 19: 304–310.

Novak, J., L. Bahoo, U. Mitteregger, and C. Franz, 2006a. Composition of individual essential oil glands of savory (*Satureja hortensis* L., Lamiaceae) from Syria. *Flavour Fragr. J.*, 21: 731–734.

Novak, J., C. Bitsch, F. Pank, J. Langbehn, and C. Franz, 2002. Distribution of the *cis*-sabinene hydrate acetate chemotype in accessions of marjoram (*Origanum majorana* L.). *Euphytica*, 127: 69–74.

Novak, J., S. Grausgruber-Gröger, and B. Lukas, 2007. DNA-Barcoding of plant extracts. *Food Res. Int.*, 40: 388–392.

Novak, J., B. Lukas, and C. Franz, 2008. The essential oil composition of wild growing sweet marjoram (*Origanum majorana* L., Lamiaceae) from Cyprus—Three chemotypes. *J. Essent. Oil Res.*, 20: 339–341.

Novak, J., M. Marn, and C. Franz, 2006b. An a-pinene chemotype in *Salvia officinalis* L. (Lamiaceae). *J. Essent. Oil Res.*, 18: 239–241.

Novak, J., S. Novak, C. Bitsch, and C. Franz, 2000. Essential oil composition of different populations of *Valeriana celtica* ssp. from Austria and Italy. *Flavour Fragr. J.*, 15: 40–42.

Novak, J., S. Novak, and C. Franz, 1998. Essential oils of rhizomes and rootlets of *Valeriana celtica* L. ssp. *norica* Vierh. from Austria. *J. Essent. Oil Res.*, 10: 637–640.

Omidbaigi, R. and A. Arjmandi, 2002. Effects of NP supply on growth, development, yield and active substances of garden thyme (*Thymus vulgaris* L.). *Acta Hort.*, 576: 263–265.

Padulosi, S. (ed.), 1997. Oregano. Promoting the conservation and use of underutilized and neglected crops. 14. *Proceedings of the IPGRI Internet Workshop on Oregano*, May 8–12, 1996, CIHEAM Valenzano (Bari). IPGRI: Rome.

Pafundo, S., C. Agrimonti, and N. Marmiroli, 2005. Traceability of plant contribution in olive oil by amplified fragment length polymorphisms. *J. Agric. Food Chem.*, 53: 6995–7002.

Pank, F., 2002a. Three approaches to the development of high performance cultivars considering the different biological background of the starting material. *Acta Hort.*, 576: 129–137.

Pank, F. 2002b. Aims and results of current medicinal and aromatic plant breeding projects. *Z. Arznei- u. Gewuerzpfl.* 7(S): 226–236.

Pank, F., 2007. Use of breeding to customise characteristics of medicinal and aromatic plants to postharvest processing requirements. *Stewart Postharvest Rev.*, 4: 1.

Pank, F., E. Herbst, and C. Franz, 1991. Richtlinien für den integrierten Anbau von Arznei- und Gewürzpflanzen. *Drogen Rep.*, 4(S): 45–64.

Pank, F., H. Krüger, and R. Quilitzsch, 1996. Selection of annual caraway (*Carum carvi* L. var. annuum hort.) on essential oil content and carvone in the maturity stage of milky-wax fruits. *Beitr. Züchtungsforsch*, 2: 195–198.

Pank, J., H. Krüger, and R. Quilitzsch, 2007. Results of a polycross-test with annual caraway (*Carum carvi* L. var. annum hort.) *Z. Arznei- u. Gewürzpfl*, 12.

Pennisi, E., 2007. Wanted: A DNA-barcode for plants. *Science*, 318: 190–191.

Pickard, W.F., 2008. Laticifers and secretory ducts: Two other tube systems in plants. *New Phytologist*, 177: 877–888.

Pino, J.A., M. Estarrón, and V. Fuentes, 1997. Essential oil of sage (*Salvia officinalis* L.) grown in Cuba. *J. Essent. Oil Res.*, 9: 221–222.

Putievsky, E., N. Dudai and U. Ravid, 1997. Cultivation, selection and conservation of oregano species in Israel. In: Padulosi, S. (ed.) Oregano. Proc. of the IPGRI Internat. Workshop on Oregano, 8–12 May 1996, CIHEAM Valenzano (Bari). IPGRI: Rome.

Putievsky, E., U. Ravid, and N. Dudai, 1986. The essential oil and yield components from various plant parts of *Salvia fruticosa. J. Nat. Prod.*, 49: 1015–1017.

Ratnasingham, S. and P.D.N. Hebert, 2007. The barcode of life data system (http://www.barcodinglife.org). *Mol. Ecol. Notes*, 7: 355–364.

Reales, A., D. Rivera, J.A. Palazón, and C. Obón, 2004. Numerical taxonomy study of *Salvia* sect. *Salvia* (Labiatae). *Bot. J. Linnean Soc.*, 145: 353–371.

Repčak, M., P. Cernaj, and V. Oravec, 1992. The stability of a high content of a-bisabolol in chamomile. *Acta Hort.*, 306: 324–326.

Repčak, M., J. Halasova, R. Hončariv, and D. Podhradsky, 1980. The content and composition of the essential oil in the course of anthodium development in wild chamomile (*Matricaria chamomilla* L.). *Biol. Plantarum*, 22: 183–191.

Rey, C., 1993. Selection of thyme (*Thymus vulgaris* L.). *Acta Hort.*, 344: 404–407.

Rey, C., 1994. Une variete du thym vulgaire "Varico". *Rev. Suisse Vitic. Arboric. Hortic.*, 26: 249–250.

Salamon, I., 2007. Effect of the internal and external factors on yield and qualitative–quantitative characteristics of chamomile essential oil. *Acta Hort.*, 749: 45–64.

Saukel, J. and R. Länger, 1992. Die *Achillea-millefolium*-Gruppe in Mitteleuropa. *Phyton*, 32: 47–78.

Schippmann, U., D. Leaman, and A.B. Cunningham, 2006. A comparison of cultivation and wild collection of medicinal and aromatic plants under sustainability aspects. In *Medicinal and Aromatic Plants*, R.J. Bogers, L.E. Craker, and D. Lange (eds.), pp. 75–95. Dordrecht, the Netherlands: Springer.

Schmiderer, C., P. Grassi, J. Novak, M. Weber, and C. Franz, 2008. Diversity of essential oil glands of clary sage (*Salvia sclarea* L., Lamiaceae). *Plant Biol.*, 10: 433–440.

Schröder, F.J., 1990. Untersuchungen über die Variabilität des ätherischen Öles in Einzelpflanzen verschiedener Populationen der echten Kamille, *Matricaria chamomilla* L. (syn. *Chamomilla recutita* L.). Dissertation, TU-München-Weihenstephan, Weihenstephan, Germany.

Senatore, F. and D. Rigano, 2001. Essential oil of two *Lippia spp.* (Verbenaceae) growing wild in Guatemala. *Flavour Fragr. J.*, 16: 169–171.

Skoula, M., J.E. Abbes, and C.B. Johnson, 2000. Genetic variation of volatiles and rosmarinic acid in populations of *Salvia fruticosa* Mill. Growing in crete. *Biochem. Syst. Ecol.*, 28: 551–561.
Skoula, M., I. El-Hilalo, and A. Makris, 1999. Evaluation of the genetic diversity of *Salvia fruticosa* Mill. clones using RAPD markers and comparison with the essential oil profiles. *Biochem. Syst. Ecol.*, 27: 559–568.
Skoula, M. and J.B. Harborne, 2002. The taxonomy and chemistry of *Origanum*. In *Oregano-The Genera Origanum and Lippia*, S.E. Kintzios (ed.), pp. 67–108. London, U.K.: Taylor & Francis.
Slavova, Y., F. Zayova, and S. Krastev, 2004. Polyploidization of lavender (*Lavandula vera*) in-vitro. *Bulgarian J. Agric. Sci.*, 10: 329–332.
Sobti, S.N., P. Pushpangadan, R.K. Thapa, S.G. Aggarwal, V.N. Vashist, and C.K. Atal, 1978. Chemical and genetic investigations in essential oils of some *Ocimum* species, their F1 hybrids and synthesized allopolyploids. *Lloydia*, 41: 50–55.
Srivastava, A.K., S.K. Srivastava, and K.V. Syamasundar, 2005. Bud and leaf essential oil composition of *Syzygium aromaticum* from India and Madagascar. *Flavour Fragr. J.*, 20: 51–53.
Stanley, P.C. and J.A. Steyermark, 1976. *Flora of Guatemala: Botany*. Chicago, IL: Field Museum of Natural History.
Steinlesberger, H., 2002. Investigations on progenies of crossing exeriments of Bulgarian and Austrian yarrows (*Achillea millefolium* agg., Compositae) with focus on the enantiomeric ratios of selected Monoterpenes. Dissertation, University of Veterinary Medicine, Wien, Austria.
Svoboda, K.P., T.G. Svoboda, and A.D. Syred, 2000. *Secretory Structures of Aromatic and Medicinal Plants*. Middle Travelly, U.K.: Microscopix Publications.
Taberlet, P., L. Gielly, G. Pautou, and J. Bouvet, 1991. Universal primers for amplification of three non-coding regions of chloroplast DNA. *Plant Mol. Biol.*, 17: 1105–1109.
Taviani, P., D. Rosellini, and F. Veronesi, 2002. Variation for Agronomic and Essential Oil traits among wild populations of *Chamomilla recutita* (L.) Rauschert from Central Italy. In *Breeding Research on Aromatic and Medicinal Plants*, C.B. Johnson and Ch. Franz (eds.), pp. 353–358. Binghampton, NY: Haworth Press.
Taylor, R., 1996. Tea tree—Boosting oil production. *Rural Res.*, 172: 17–18.
Tsiri, D., O. Kretsi, I.B. Chinou, and C.G. Spyropoulos, 2003. Composition of fruit volatiles and annual changes in the volatiles of leaves of *Eucalyptus camaldulensis* Dehn. growing in Greece. *Flavour Fragr. J.*, 18: 244–247.
UPOV, 1991. International Convention for the Protection of New Varieties of Plants, www.upov.int/upovlex/en/conventions/1991/content.html (accessed August 21, 2015).
Uribe-Hernández, C.J., J.B. Hurtado-Ramos, E.R. Olmedo-Arcega, and M.A. Martinez-Sosa, 1992. The essential oil of *Lippia graveolens* HBK from Jalsico, Mexico. *J. Essent. Oil Res.*, 4: 647–649.
Van Overwalle, G., 2006. Intellectual property protection for medicinal and aromatic plants. In *Medicinal and Aromatic Plants*, J. Bogers, L.E. Craker, and D. Lange (eds.), pp. 121–128. Dordrecht, the Netherlands: Springer.
Velasco-Neguerelo, A., J. Pérez-Alonso, P.L. Pérez de Paz, C. García Vallejo, J. Palá-Paúl, and A. Inigo, 2002. Chemical composition of the essential oils from the roots, fruits, leaves and stems of *Pimpinella cumbrae* link growing in the Canary Islands (Spain). *Flavour Fragr. J.*, 17: 468–471.
Vernet, P., 1976. Analyse génétique et écologique de la variabilité de l'essence de *Thymus vulgaris* L. (Labiée). PhD thesis, University of Montpellier, Montpellier, France.
Vernin, G., C. Lageot, E.M. Gaydou, and C. Parkanyi, 2001. Analysis of the essential oil of *Lippia graveolens* HBK from El Salvador. *Flavour Fragr. J.*, 16: 219–226.
Vernin, G., J. Metzger, D. Fraisse, and D. Scharff, 1984. Analysis of basil oils by GC-MS data bank. *Perform. Flavour*, 9: 71–86.
Vetter, S. and C. Franz, 1996. Seed production in selfings of tetraploid *Achillea* species (Asteraceae). *Beitr. Züchtungsforsch*, 2: 124–126.
Vetter, S., C. Franz, S. Glasl, U. Kastner, J. Saukel, and J. Jurenitsch, 1997. Inheritance of sesquiterpene lactone types within the *Achillea millefolium complex* (Compositae). *Plant Breeding*, 116: 79–82.
Viljoen, A.M., A. Gono-Bwalya, G.P.P. Kamatao, K.H.C. Baser, and B. Demirci, 2006. The essential oil composition and chemotaxonomy of *Salvia stenophylla* and its Allies *S. repens* and *S. runcinata*. *J. Essent. Oil Res.*, 18: 37–45.
Weising, K., H. Nybom, K. Wolff, and G. Kahl, 2005. *DNA Fingerpinting in Plants*. Boca Raton, FL: Taylor & Francis.
Werker, E., 1993. Function of essential oil-secreting glandular hairs in aromatic plants of the Lamiaceae—A review. *Flavour Fragr. J.*, 8: 249–255.

Wijesekera R., A.L. Jajewardene, and L.S. Rajapakse, 1974. Composition of the essential oils from leaves, stem bark and root bark of two chemotypes of cinnamom. *J. Sci. Food Agric.*, 25: 1211–1218.

Wildung, M.R. and R.B. Croteau, 2005. Genetic engineering of peppermint for improved essential oil composition and yield. *Transgenic Res.*, 14: 365–372.

WHO., 2003. *Guidelines on Good Agricultural and Collection Practices (GACP) for Medicinal Plants.* Geneva, Switzerland: World Health Organization.

Wogiatzi, E., D. Tassiopoulos, and R. Marquard, 1999. Untersuchungen an Kamillen-Wildsammlungen aus Griechenland. In *Fachtagg. Arznei- u. Gewürzpfl. Gießen*, pp. 186–192. Gießen, Germany: Köhler.

Wolfe, A.D. and A. Liston, 1998. Contributions of PCR-based methods to plant systematics and evolutionary biology. In *Molecular Systematics of Plants II: DNA Sequencing*, D.E. Soltis, P.S. Soltis, and J. Doyle (eds.), pp. 43–86. Dordrecht, the Netherlands: Kluwer Academic Publishers.

Worku, T. and M. Bertoldi, 1996. Essential oils at different development stages of Ethiopian *Tagetes minuta* L. In *Essential Oils: Basic and Applied Research*, Ch. Franz, A. Máthé, and G. Buchbauer (eds.), pp. 339–341. Carol Stream, IL: Allured Publishing.

Zheljazkov, V.D., N. Kovatcheva, S. Stanev, and E. Zheljazkova, 1997. Effect of heavy metal polluted soils on some qualitative and quantitative characters of mint and cornmint. In *Essential Oils: Basic and Applied Research*, Ch. Franz, A. Máthé, and G. Buchbauer (eds.), pp. 128–131. Carol Stream, IL: Allured Publishing.

Zheljazkov, V.D. and N. Nielsen, 1996. Studies on the effect of heavy metals (Cd, Pb, Cu, Mn, Zn and Fe) upon the growth, productivity and quality of lavender (*Lavandula angustifolia* Mill.) production. *J. Essent. Oil Res.*, 8: 259–274.

4 Natural Variability of Essential Oil Components

Éva Németh-Zámboriné

CONTENTS

4.1 Appearance of Variability	87
4.2 Variability at Different Taxonomic Levels	88
4.2.1 Species	88
4.2.2 Populations	95
4.3 Connections of Chemical Diversity with Other Plant Characteristics	99
4.3.1 Propagation and Genetics	99
4.3.2 Morphological Characteristics	101
4.4 Morphogenetic and Ontogenetic Manifestation of the Chemical Variability	102
4.5 Origin of Essential Oil Variability	109
4.6 Chemotaxonomic Considerations	110
4.7 Identification of Natural Variability	115
References	118

4.1 APPEARANCE OF VARIABILITY

It is a long known fact that qualitative and quantitative composition of genuine essential oils is not a standard one. In consequence of this, they possess different quality, value, and price on the market.

As a reflection of this practical experience, in several cases, different qualities are defined for essential oils of the same species. In the standard series of the International Organization for Standardization, lavender oil is published in four different qualities, among which two are lavandin oils. Spearmint oil also has two different standards, and *Eugenia caryophyllus* oils from different organs also represent different qualities. These changing qualities are result of numerous factors: genotype, habitat and environment, and the influence of the special agricultural or technological measures might be all manifested in the products. In the practice, the same species might be utilized for different applications based on the variable composition of its oil like the thyme-odor type, lavender-odor type, and rose-odor type individuals of *Thymus longicaulis* ssp. *longicaulis* (Baser et al., 1993).

In several cases, the real sources of variability are hard to determine. However, for standardization of any product, it is of primary importance that the background of variability and the factors, which influence the composition of the essential oils, are detected, can be managed and controlled.

In the scientific literature, reports on variability of essential oil components are very frequently published. According to a survey on articles in the last volumes (2010–2014) of *Journal of Essential Oil Research* (Taylor & Francis Group), it can be established that approximately one-third of them is evaluating biological variability at generic, specific, or intraspecific taxonomic levels or chemosyndromes due to developmental or morphological differences (Figure 4.1.).

The chemical variability of the essential oils gained from different plant species varies on a large scale. Tétényi (1975) mentioned already 40 years ago that 36 families, 121 genera, and 360 species are polymorphs for essential oil. This number must have increased enormously since that time because of intensive research and highly developed analytical techniques.

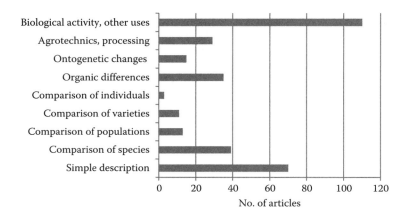

FIGURE 4.1 (See color insert.) Distribution of topics of publications in *Journal of Essential Oil Research* between 2010 and 2014.

The backgrounds of the chemical variability of essential oil composition are usually grouped as abiotic and biotic ones. Abiotic-influencing factors include the effects of the environment (exposure, soil, light intensity and length of illumination, wind, absolute and marginal temperatures, water supply as total and frequency of precipitation) and also those in consequence of human activities (agrotechnical methods, processing, storage). Chapters 3, 5, and 24 deal with these factors in detail.

The biotic/biological factors are the main topic of this chapter. Natural variability should be defined as the phenomenon when diverse quality of essential oils is detectable as result of genetical–biological differences of the source plants. "Natural variability" is, however, a rather complex issue having many aspects as we can see in the succeeding text. In this context, we deal with the essential oil spectrum, quantitative and qualitative composition of the oils, and not discussing other chemical and physical properties.

4.2 VARIABILITY AT DIFFERENT TAXONOMIC LEVELS

4.2.1 SPECIES

Variability in the composition of essential oils was most frequently discussed at the level of plant species and has the highest relevance from practical points of view.

A significant variation in qualitative and quantitative composition might have considerable influence on the recognition and the market value, both of the drug and the essential oil itself. Besides, fluctuations in the composition of the essential oil might have significant effects on the therapeutical efficacy or sensory value of the product. Limonene seems to have a strong influence on the allelopathic property of *Tagetes minuta* (Scrivanti et al., 2003). Similarly, limonene content might contribute to the resistance of *Pinus sylvestris* cultivars against the herbivore *Dioryctria zimmermani* (Sadof and Grant, 1997). In vitro study on the antimicrobial effects of hyssop essential oils proved that higher contents of pinocamphone, isopinocampheol, and linalool increased the efficacy of the oil against several *Fusarium* species (Fraternale et al., 2004). In some phytotherapeutical preparations of chamomile, the antiphlogistic and spasmolytic effect seems to be in closest connection with the content of (−)-α-bisabolol (Schilcher, 2004). On the other hand, adverse effects may be caused by the presence of single compounds like the carcinogen effect of *cis*-isoasarone in the essential oil of calamus (*Acorus calamus*) (Blaschek et al., 1998).

Not each species exhibits a similar size of variability. A huge amount of research data accumulated in the last decades proved that the incidence of diversity is one of the characteristic features of the plant species.

Natural Variability of Essential Oil Components

The well-known caraway (*Carum carvi*) seems to be an essential oil–bearing species of relatively low variability concerning the oil constituents. Caraway is believed to have been cultivated and consumed in Europe longer than any other spice species (Rosengarten, 1969). Nowadays, it is mainly known as a spice and source of essential oil of excellent antimicrobial properties but the spasmolytic and cholagogue effects justify its use in phytotherapy, too. In the oil of caraway, the main components $S(+)$ carvone and $R(+)$ limonene are absolutely the dominating ones, and this is observable in each analysis (Table 4.1). Their ratio in the oil is above 90%, most frequently above 95%. Variability is manifested most frequently only in their proportions compared to each other. Minor constituents have rarely been identified and mentioned. The constituents are almost all monoterpenes with the only sesquiterpene β-caryophyllene and some phenolic compounds. Characteristically, several minor compounds are derivatives of carvone and only rarely detected in other essential oil species.

Biological variability of the oil composition seems to be more pronounced if comparing the two varieties (*Carum carvi* var. *annuum* and *Carum carvi* var. *biennis*) of caraway. In general, biennial types are believed to accumulate higher concentrations of oil and carvone (Table 4.2). Although Bouwmeester and Kuijpers (1993) conclude that the restricted potential of carvone accumulation in annual varieties is the consequence of limited availability of assimilates, the difference is a genetic one without doubt. Based on the majority of available references, carvone content of biennial accessions is regularly higher than one of the annual plants. This fact may explain why biennial caraway is still in cultivation in many countries although production of the annual variety has an economical advantage based on higher yields and more flexible crop rotation.

The Mediterranean species hyssop (*Hyssopus officinalis*) belongs to the Lamiaceae family. It is used for its spicy essential oil in food industry and also as a strong antimicrobial agent.

TABLE 4.1
Variability of Essential Oil Components in Caraway

Compounds (in the Row of their Abundance)	El-Wakeil et al. (1986)	Lawrence (1989)	Puschmann et al. (1992)	Putievsky et al. (1994)	Raal et al. (2012)	Seidler-Łożykowska (2008)
Carvone	80.17	38.8–67.8	47.9–54.5	49.1–62.3	44.3–95.2	55.5–65.1
Limonene	9.8	30.3–48.8	45.0–52.8	33.8–50.2	2.1–50.4	33.0–44.3
Dihydrocarvone	0.7	0.3–0.5	0.1–0.3	0.1–0.7	tr.–0.9	tr.–0.4
Carveol (*cis* + *trans*)	tr.	0.4–1.2	—	0.2–0.6	tr.–0.5	0.1–0.2
β-Myrcene	0.1	0.3–2.4	0.2–0.5	0.2–0.4	0.0–0.4	—
Dihydrocarveol	tr.	0.1–0.2	0.3–1.1	0.1–0.2	—	0.1–0.4
Pinenes (α + β)	0.1	0.1–0.5	—	tr.–0.1	tr.–0.5	0.01
β-Caryophyllene	0.1	1.2–1.7	—	0.1–0.2	tr.–0.3	—
Thujones (α + β)	—	tr.–1.1	—	—	—	—
trans-Anethole	—	0.6–0.8	—	—	—	—
Perillyl-aldehyde	0.2	0.1–0.2	—	—	0.1–0.4	—
Carvyl-acetate	—	—	0.1–0.4	—	—	—
α-Phellandrene	—	0.1–0.3	—	tr.–0.4	—	—
Linalool	—	0.1–0.2	—	tr.–0.1	—	—
γ-Terpinene	0.2	0.1	—	tr.	tr.–0.1	—
p-Cymene	0.1	tr.	—	tr.–0.4	—	—
Sabinene	—	0.1	—	tr.–0.1	—	—
Ocimene	—	tr.–0.1	—	—	—	—
Camphene	—	—	—	tr.	—	0.01
Cumin-aldehyde	0.1	—	—	—	—	—

Note: tr., traces.

TABLE 4.2
Variability of the Main Components Carvone and Limonene in Biennial and Annual Accessions of Caraway

Source of Data	Biennial		Annual	
	Carvone	Limonene	Carvone	Limonene
Argañosa et al. (1998)	54–57	43–45	46–50	49–53
Embong et al. (1977)	39–46	43–49	—	—
Fleischer and Fleischer (1988)	54–68	30–44	—	—
Forwick-Kreutzer et al. (2003)	52–72	—	—	—
Galambosi and Peura (1996)	47–49	39–52	—	—
Pank et al. (2008)	—	—	50–53	45–48
Puschmann et al. (1992)	47–54	—	45–52	—
Putievsky et al. (1994)	53–59	38–44	47–62	3–46
Raal et al. (2012)	44–95[a]	2–50	—	—
Zámboriné-Németh et al. (2005)	51–60	38–44	50–56	43–49

[a] Annual accessions might also be included.

Monoterpenes, which are present as main compounds in the oil of this species (pinocamphone, isopinocamphone), are relatively seldom detected in higher quantities in essential oils of other species. Although as highest number 44 components were detected in hyssop oil (Chalchat et al., 2001), the major ones are relatively uniform and found almost in each examined accession (Table 4.3). Besides the mentioned compounds, the majority of further ones are also monoterpenes, products of related biosynthetic pathways (β-pinene, pinocarvone, myrtenol). In general, it can be observed that besides some mentioned compounds in higher amounts, all the others are present only in minimal concentrations. Thus, the biological variability of the herb oil of hyssop is relatively low as the available data show.

TABLE 4.3
Main Components in the Essential Oil of Hyssop (*Hyssopus officinalis*) According to Different References

Main Compounds	Reference
Pinocamphone, isopinocamphone, β-pinene	Aiello et al. (2001)
Pinocarvone, isopinocamphone, β-pinene	Bernotiené and Butkiené (2010)
Pinocamphone, isopinocamphone	Chalchat et al. (2001)
Isopinocamphone, β-pinene	Danila et al. (2012)
Pinocamphone, isopinocamphone, germacrene D, pinocarvone	Galambosi et al. (1993)
Pinocamphone, isopinocamphone, β-pinene	Fraternale et al. (2004)
Terpineol, bornyl-acetate, linalool	Hodzsimatov and Ramazanova (1974)
Isopinocamphone, β-pinene, pinocamphone	Joulain and Ragaul (1976)
Isopinocamphone, pinocamphone, β-pinene	Koller and Range (1997)
Pinocamphone, β-pinene	Lawrence (1979)
Pinocamphone, isopinocamphone, β-pinene, pinocarvone	Lawrence (1992)
Isopinocamphone, myrtenol, β-pinene, 1,8-cineole, methyl-eugenol, limonene	Piccaglia et al. (1999)
Pinocamphone, camphor, β-pinene	Schulz and Stahl-Biskup (1991)
Isopinocamphone, 1,8-cineole, β-pinene	Tsankova et al. (1993)
1,8-Cineole, β-pinene	Vallejo et al. (1995)

Only samples of the subspecies *aristatus* (Godr.) Briq. collected from three populations of Apennines showed a different character with higher amounts of myrtenol (up to 32%), methyl eugenol (up to 44%), and limonene (up to 15%); however, the characteristic pinane-type compounds have also been found at different quantities (Piccaglia et al., 1999).

Example for a relatively low biological variability in essential oil composition exists also among the species, which provide root drug. Lovage (*Levisticum officinale*) has multiple utilizations: aromatic volatile (essential oil) and nonvolatile (mainly coumarin type) compounds enable the application of each part of the plant as a popular spice. However, the most valuable organ is the root. The main components of the root essential oil are alkylphthalide-type compounds, among which the most abundant ones are usually Z-ligustilide and butylidenephthalide. Besides, references mention 3-butylphthalide, 3-propylidene-dihydrophthalide, termine (validene-dihydrophthalyde), sedanenolide (3-butyl-4,5-dihydrophthalyde), and sedanolide (3-butyl-3α-4,5,6-tetrahydrophthalide) (Venskutonis, 1995; Novák, 2006). Other components of the root oil are monoterpenes, among which the majority is frequently and universally occurring ones like pinenes, camphene, sabinene, and myrcene with the exception of 1-pentil-cyclohexa-1,3-diene, which seems to be characteristic for lovage (Szebeni et al., 1992; Venskutonis, 1995).

Our recent investigations on 10 accessions of lovage originating from different European countries ascertained that the compositional variability is low (Gosztola, 2013). The phthalides are the main components of the distilled oil practically in each accession (Figure 4.2). It is in agreement with the majority of former findings on the root essential oil of lovage (e.g., Perineau et al., 1992; Szebeni et al., 1992). The presence of two isomers E and Z ones make the pictures somewhat more complex; however, their ratios are not significantly different in either of the accessions. In each case, the Z isomer is in multiple concentrations present than the other one.

The seeds of the investigated accessions have been obtained from different countries and regions, but—as in many cases—the real genetic origin is uncertain. Therefore, a common basic source cannot be excluded, either. However, even in this case, it might mean that lovage has a very narrow gene pool and maybe therefore possess a low chemical variability. The connection between the restricted natural distribution and small spectral variance of the oil might support the hypothesis on the development of polychemism as tool in geographical distribution and ecological adaptation (see Section 4.5).

Summarizing of the aforementioned, it seems to be clear that the variation in the oil composition of the aforementioned species is principally a quantitative one. The spectrum seems to be relatively constant; changes are detectable basically in the accumulation proportions of the individual components.

On the other hand, a great number of plant species can be characterized by high intraspecific variability concerning essential oil composition. In these oils, both qualitative and quantitative variations are detectable.

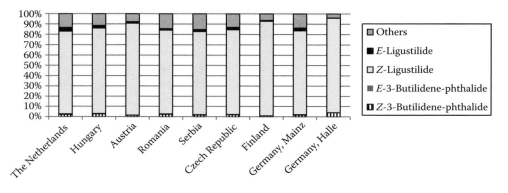

FIGURE 4.2 Proportion of characteristic phthalide components in the root oil of lovage accessions from different European countries. (Németh, E. et al., unpublished data.)

One of the most comprehensively studied genera from this respect is the genus *Achillea*. For the majority of yarrow species, a wide variability in oil composition can be established. Based on comprehensive literature search, in most of the species, 1–3 compounds have been identified as main components (Kindlovits and Németh, 2014). The evaluation is, however, not a simple one because in most cases the different chemical races had been detected separately and mentioned by different authors. Therefore, the comparison of data is always a hard task taking into account the possible role of other influencing factors besides the genetic background.

Chamazulene is till today the most important component of the distilled oil of yarrow. In general, the proazulene accumulation potential of *Achillea collina* (4n) and its relatives, *Achillea asplenifolia* and *Achillea roseoalba* (2n), seems to be widely accepted (Saukel and Länger, 1992; Rauchensteiner et al., 2002; Németh et al., 2007). However, even here, some divergent results can be found in the literature. In plant samples from Yugoslavia, Chalchat et al. (2000) could not identify chamazulene, but 1,8-cineole, chrysanthenone, and camphor are mentioned as main components. Recently, Todorova et al. (2007) presented three chemotypes of *A. collina* (azulene-rich, azulene-poor, sesquiterpene-free ones) based on the analysis of samples from six different populations in Bulgaria.

According to the literature references, the largest intraspecific variability could be devoted to *Achillea millefolium*. For this species, 16 different chemical compounds have already been mentioned in the essential oil as main component (Németh, 2005). Comparing the chamazulene content of the oil, values between 0% and 85% have been detected by different authors (e.g., Hachey et al., 1990; Figueiredo et al., 1992; Michler et al., 1992; Bélanger and Dextraze, 1993; Orth et al., 1999; Orav et al., 2001; Németh et al., 2007).

Taking into consideration only these data, we might assume that *A. millefolium* is an extremely variable species concerning its essential oil spectrum with numerous intraspecific chemical varieties. However, in this case, I would be more cautious because in numerous references, the proper identification of the taxon is not obvious; botanical characterization is missing. The genus *Achillea* is a very complex one with species in a polyploid row, containing intrageneric sections and groups, many spontaneous hybrids, phenocopies, and aneuploid forms. Contradictious results may originate from a false definition of taxa belonging to the *A. millefolium* section only by morphological features or—on the other hand—only by chromosome numbers. Similarly, the lack of representative samples, examining individuals of nonstable spontaneous hybrid or aneuploid character, may lead to invalid information. Detailed morphological and cytological identification of any taxon belonging into the section *A. millefolium* seems to be a prerequisite for reliable chemical characterization, otherwise comparison of the data is not really possible. In the last decade, molecular markers have also been developed for identification of certain taxa (e.g., Guo et al., 2005).

According to the aforementioned, unequivocal definition of the accessions showing diverse essential oil composition in the section *A. millefolium* as chemotypes could be more than questionable, and only references based on comprehensive determination of the investigated plant material can be accepted.

Nevertheless, the high variability concerning the composition of the essential oil of yarrow species is without doubt. Héthelyi et al. (1989) identified at least three chemical varieties according to the spectrum of the main terpenoid components in each of eight examined species. Among others, for *Achillea ochroleuca*, a linalool + borneol + bornyl-acetate + cubebene chemotype, other linalool + borneol + bornyl acetate + elemol + eudesmol chemotype, and an eudesmol chemotype were described. According to Muselli et al. (2009), geographically distinct populations of *A. ligustica* also show chemically distinct characteristics. Corsican samples contain camphor (21%) and santolina alcohol (15%) as main compounds, Sardinian samples have *trans*-sabinyl acetate (18%) and *trans*-sabinol (15%), and those from Sicily can be characterized by high terpinen-4-ol (19%) and carvone (9%) accumulation. Similar results on other species are numerous.

A related plant, wormwood (*Artemisia absinthium*), gained an adverse "reputation" due its thujone content and mutual side effects associated with absinthism. It is widely distributed in Europe and introduced also in other continents. The composition of the essential oil has been studied by several authors and highlighted that large amounts of thujones are representative only for one of the

TABLE 4.4
Chemotypes of *Artemisia absinthium* of Different Origin According to References

Reference	Country	Number of Detected Compounds	Determined Chemotypes (Main Compounds in Area %)
Arino et al. (1999)	Spain	17	Mixed: *cis*-chrysanthenyl acetate (31%–44%) + *cis*-epoxyocimene (34%–42%)
Bagci et al. (2010)	Turkey	31	Chamazulene (29%)
Basta et al. (2007)	Greece	68	Caryophyllene oxide (25%)
Chialva et al. (1983)	Italy	34	*cis*-Epoxyocimene (30%–54%) β-Thujone (41%)
	Romania		β-Thujone (15%)
	France		Sabinyl acetate (32%) Chrysanthenyl acetate (42%)
	Siberia		Sabinyl acetate (85%)
Derwich et al. (2009)	Morocco	—	α-Thujone (40%)
Judzetiene and Budiene (2010)	Lithuania	15	*trans*-Sabinyl acetate (22%–51%), α- and β-Thujones (18%–72%)
Sharopov et al. (2012)	Tajikistan	41	Myrcene (23%) *cis*-Chrysanthenyl acetate (18%)
Simonnet et al. (2012)	Switzerland	6	*cis*-Epoxyocimene (30%–40%)
Pino et al. (1997)	Cuba	40	Bornyl acetate (24%)
Rezaeinodehi and Khangholi (2008)	Iran	28	β-Pinene (24%)

many chemotypes of *A. absinthium*, and other mono- or sometimes sesquiterpenes are more frequently present as major components in the herb oil (Table 4.4). Besides the bicyclic monoterpene *cis*-chrysanthenyl acetate and the irregular monoterpene *cis*-epoxyocimene, mainly terpenoids of thujane skeleton and in some cases those of pinane skeleton are present in highest abundance in the oil. According to the investigation in the last decade, it is obvious that thujone may be not rarely even absent from the oil. Additionally, although wormwood is known among the few proazulene-containing species (Wichtl, 1997), chamazulene is only rarely and in low proportions present in the essential oil of the investigated accessions.

The real source of polychemism in this species seems to be till now unknown. Evaluating the publications, no correlation between main component and geographical distribution of the chemotype could be established. The only conclusion published about this problem was made by Chialva et al. (1983) who mentioned that appearance of different chemotypes might be connected to habitats from different altitudes in the Italian Alps. Unfortunately, detailed study of the data reveals that the plant materials investigated by the authors in this study were not uniform in several respects. Different plant parts, flowers, or leaves have been investigated; some samples have been freshly distilled, others in dried stage; the samples originated, in some cases, from natural habitats and, in other cases, from market; besides, the year of the harvest was also different. Under such conditions, comparison of the results cannot provide reliable conclusions, especially not in chemotaxonomic respect, although the title of the paper is suggesting this.

One of the earliest and most deeply studied plant species from respect of essential oil polymorphism has been tansy (*Tanacetum vulgare*). Formerly—due to the lack of reliable chemical–analytical investigations and systematic evaluation—it has been presented as a characteristic thujone-containing species (Gildemeister and Hofmann, 1968). Although it is true that this is the main component most frequently present in the essential oil, but until today, the number of the detected main compounds in different chemotypes is near to 50. Some of them are summarized

TABLE 4.5
Chemotypes of *Tanacetum vulgare* According to Selected References

Reference	Country	Chemotypes (Main Components)
Collin et al. (1993)	Canada	Camphor-cineole-borneol, β-thujone, chrysanthenone, dihydrocarvone
de Pooter et al. (1989)	Belgium	β-thujone, chrysanthenyl acetate, camphor + thujone
Dragland et al. (2005)	Norway	Thujone, camphor, borneol, bornyl acetate, chrysanthenol, chrysanthenyl acetate, 1,8-cineole, α-terpineol
Forsen and Schantz (1971)	Finland	Chrysanthenyl acetate, isopinocamphone, not identified sesquiterpene
Hendrics et al. (1990)	Nether-land	Artemisia ketone, chrysanthenol + chrysanthenyl acetate, lyratol + lyratyl acetate, β-thujone
Héthelyi et al. (1991)	Hungary	Yomogi alcohol, artemisia alcohol, davanone, lyratol + lyratyl acetate, chrysanthenol, carveol, carvone, dihydrocarvone, terpinene-4-ol, γ-campholenol, myrtenol, β-terpineol, 4-thujene-2-α-yl-acetate, carveyl acetate, β-cubebene, juniper camphor, thymol, β-terpenyl acetate, linalool
Holopainen et al. (1987)	Finland	Sabinene, germacrene D
Mockute and Judzetiene (2004)	Lithuania	1,8-Cineole, artemisia ketone, camphor, α-thujone
Nano et al. (1979)	Italy	Chrysanthenyl acetate
Rohloff et al. (2003)	Norway	β-Thujone, camphor, artemisia ketone, umbellulone, chrysanthenyl acetate, chrysanthenone, chrysanthenol, 1,8-cineole
Sorsa et al. (1968)	Finland	α-Pinene + triciclene, β-pinéne + sabinene, 1,8-cineole, γ-terpinene, artemisia ketone, thujone, camphor, umbellulone, borneol, humulenol
Tétényi (1975)	Hungary	β-Pinene, camphene, chrysanthenyl acetate, 1,8-cineole, γ-terpinene, artemisia ketone, camphor, α- and β-thujone, borneol + bornyl acetate, umbellulon, piperiton

in Table 4.5. The dominant compounds are in most cases monoterpenes, but in some samples also sesquiterpene ones like humulenole, germacrene D, or davanone were detected.

The spectrum of these monoterpenes is very wide. There are representatives of each type of the basic monoterpene skeletons except the carane group. Even if the main component itself is usually not enough for evaluation of the characteristics of the oil, tansy is a good example to illustrate the fact that the main compounds of different chemotypes may not necessarily belong to the same skeleton. It also means that they are not always products of closely related biosynthetic pathways, which might reflect a really heterogeneous genetic structure.

Large intraspecific chemical variability is by no means restricted to Asteraceae species. The genus *Thymus* comprises many species highly polymorphic for essential oil composition. Different chemotypes have been reported in at least 85 cases mainly from the species *Thymus aestivus*, *Thymus herba-barona*, *Thymus hyemalis*, *Thymus mastichina*, *Thymus nitens*, *Thymus vulgaris*, and *Thymus zygis* (Stahl-Biskup and Sáez, 2003). For most of them, 3–6 intraspecific chemotypes have already been described. Different chemotypes are often grouped as ones containing phenolic compounds and chemotypes with nonphenolic ones (Baser et al., 1993).

Common thyme, *Thymus glabrescens* Willd., is a procumbent dwarf shrub, indigenous on sunny hillsides of southeastern and central Europe. Recently, in Hungary, eight populations at different localities have been investigated and new chemotypes identified (Pluhár et al., 2008). Four chemotypes contained thymol as main compound in the oil (15%–34%), but the second and third main compound has been different in each of them. One chemotype contained only monoterpenes as major constituents (*p*-cymene 45.0%, geraniol 13.6%, and linalyl acetate 9.9%) while two other ones only sesquiterpenes (germacrene D 55.4%, β-caryophyllene 14.8%, α-cubebene 50.9%). 1,8-Cineole and thymyl acetate/carvacrol/*p*-cymene chemovarieties were described in Croatia; a terpinyl acetate chemotype was reported in Bosnia; and linalool/thymol/α-terpinyl acetate, geraniol, citronellol, and carvacrol chemovarieties were mentioned in Bulgaria (Pluhár et al., 2008). It can be established that in this species—in contrary to the formerly mentioned ones—the main compounds could be relatively well grouped based on their chemical constitution: acyclic monoterpenes, menthane skeleton group, and sesquiterpene ones. This concludes that intraspecific differences in this species are primarily the results of diversity in biosynthesis at the level of terpene synthases and not in the following transformations.

Within a genus, different species may exhibit different levels of intraspecific chemical variability. The genus *Mentha* is a good example for this. Besides the best known species, *Mentha piperita*, there is only a small variability also in *Mentha pulegium*. While the first one is characterized always by the presence of menthol, the last one contains almost always pulegone as the main compound or one of the main compounds (Lawrence, 2007; Baser et al., 2012). The presence of piperitenone oxide in high percentages has been reported in each of the published studies for the oil composition of *Mentha suaveolens* (Nagell and Hefendehl, 1974; Baser et al., 1999, 2012). Similarly, *Mentha aquatica* seems to be a species of low essential oil variability. According to the available data, menthofuran has been detected in the huge majority of the investigated samples (Guido et al., 1997; Mimica-Ducic et al., 1998; Baser et al., 2012).

On the other hand, numerous species of the genus are really polymorphic concerning their volatile compounds. *Mentha longifolia, Mentha spicata, Mentha arvensis*, and also natural hybrids like *Mentha dumetorum* exhibit a wide spectrum of essential oil compounds, and many chemotypes has been reported (Tétényi, 1970; Lawrence, 2007; Baser et al., 2012).

4.2.2 Populations

During evaluation of the intraspecific essential oil variability of any species, one has to be aware of the fact that in many cases, the investigated plant material is far from a homogenous one. Although representative sampling is a prerequisite for these studies, unfortunately, this is only rarely the fact. It is still quite frequently not taken into account that different populations might reveal significant variability due to the individual differences of single plants. Description of differences among populations without referring to the individual variability within population may lead to significant misinterpretation of data.

This is especially relevant for the wild growing plants because natural populations are often heterogeneous in many respects. A special difficulty is that the size of this diversity is not known either. Therefore, inadequate number of sampled individuals or bulked samples may obscure the real variability that can be demonstrated by several examples.

In natural stands of *Achillea crithmifolia*, considerable variability has been detected, and the level of several essential oil constituents varied on a large scale (Table 4.6). These results of bulked samples, however, are not able to tell us details about the real diversity of the stands. Individual sampling could reveal that three types of plants are present in these populations, based on the ratios of the main components in their essential oil. Chemotype 1 has been characterized by camphor (above 50% in the oil), chemotype 2 showed high concentrations of 1,8-cineole (above 30%), while in the third type of individuals balanced proportions of these components together with borneol had been detected. The abundance of plant individuals belonging to the different chemotypes varied according to habitat (Németh et al., 2000), thus, it seems that the studied natural populations are heterogeneous and the extent of the heterogeneity is also different.

In a similar trial in Bulgaria, analyzing samples from seven habitats, besides camphorous and 1,8-cineole type individuals, an artemisia alcohol chemotype (with 24%–46% artemisia alcohol in the oil) has been described (Konakchiev and Vitkova, 2004). However, in this examination, the populations could not be characterized, and the abundance of the three chemotypes has not been described either, as only a single individual has been sampled from each habitat! Therefore, the

TABLE 4.6
Occurrence Intervals of Main Compounds of *Achillea crithmifolia* Essential Oils in Hungarian Populations

Component	Proportion in the Oil (%)	Component	Proportion in the Oil (%)
α-Pinene	0.0–3.6	Borneol	1.5–24.2
β-Pinene	0.0–5.5	Terpinene-4-ol	0.0–3.5
p-Cymene	0.0–7.0	α-Terpineol	0.0–7.6
1,8-Cineole	0.1–46.0	Ascaridol	0.0–17.6
Linalool	0.0–14.2	Bornyl-acetate	0.0–8.7
Camphor	5.4–77.4	β-Cubebene	0.0–4.7

Source: Németh, É. et al., *J. Essent. Oil Res.*, 12, 53, 2000.

results are only useful to provide data about the existing chemical diversity of the species but not about their distribution in Bulgarian growing areas.

Other data are even more questionable if they could give appropriate information on natural variability of this species. Bulked plant material from a Serbian population "near Niš" was characterized by high (19%) proportions of *trans*-chrysanthenyl acetate (Palić et al., 2003), while another in Greece "from Pelion mountain at the altitude of 700 m" by larger levels of α-terpineol (Tzakou et al., 1993). These data do not tell us anything about the quality of the oil of single individuals where these ratios might be much lower or higher. A single sample from a population might lead to false interpretation if we want to evaluate the practical/pharmaceutical value of these stands because the representativeness is at least questionable.

In the same genus, significant amounts of chamazulene are generally present in the essential oil of *A. collina*. Rarely, we can find, however, any reference about the individual distribution of this compound inside a plant population although collection of bulk samples may again lead to false consequences and is not useful enough for example for strain improvement and breeding. Table 4.7 shows that among 23 Hungarian *A. collina* populations, differences of mean values varied from 33.2% till 67.1%, while the standard deviations show 12-fold differences! A population with 1.8% standard deviation ("Diósd") means in the practice a strongly homogenous stand where the high level of chamazulene manifests itself in almost each individual. On the other hand, a population like "Alsótold" of similar mean value but with a much higher standard deviation can be evaluated as an unstable one, less suiting even for commercial purposes. According to the data, values of chamazulene are more variable than values for the essential oil accumulation level.

A more detailed investigation afterward revealed that the mentioned results could be traced back to individual differences. The plants in the examined wild populations of *A. collina* could be sorted in four groups based on the characteristic spectrum of the essential oil. Individuals, accumulating chamazulene in high proportions as the absolutely main component of the oil, are clearly different from the ones having both β-caryophyllene and chamazulene in higher levels. Individuals of only low levels of chamazulene and having other compounds as major ones form a distinct group, while the plants with essential oil lacking of chamazulene are sorted in the fourth group. The evaluated mean values of the populations most likely reflect the presence and proportion of the individuals belonging to these different chemical types.

As presented in the preceding text, at least 10 chemotypes of wormwood (*A. absinthium*) have been described until recently in the literature. Checking the methods of the published papers, it can, however, be established that in nine of the cited 10 references, the method of sampling has only be described as follows: "aerial parts/leaves/plants were collected…" without providing any information about the number of individuals, replications, or the amount of the sample. Only a single paper

TABLE 4.7
Average Values and Standard Deviations of the Essential Oil Content and Its Chamazulene Level in 23 Spontaneous Hungarian *Achillea collina* Populations

Population (Origin)	Essential Oil Content of Flowers (% d.w.)		Chamazulene Content in Flower Oil (Essential Oil %)	
	Mean	Standard Deviation	Mean	Standard Deviation
Alsótold	0.55	0.49	53.6	25.3
Apc	0.27	0.06	45.7	15.4
Aszód	0.48	0.22	63.1	12.5
Balatonakali	0.36	0.11	67.1	4.2
Balatonudvari	0.29	0.07	61.0	7.2
Bokor	0.35	0.13	64.5	5.8
Csepreg	0.20	0.08	40.0	18.3
Csillebérc	0.33	0.22	60.3	7.2
Diósd	0.42	0.11	61.0	1.8
Jobbágyi	0.33	0.18	33.7	24.6
Kevélynyereg	0.30	0.13	60.7	7.6
Lupasziget	0.31	0.09	52.5	4.8
Makkoshetye	0.18	0.08	40.3	28.1
Mezőnyárád	0.27	0.14	33.2	18.9
Mikóújfalu	0.44	0.07	57.7	8.9
Nagymaros	0.71	0.36	47.8	22.9
Nagymaros	0.53	0.15	64.2	5.0
Oroszlány	0.33	0.11	60.7	2.8
Solymár	0.37	0.08	58.3	5.1
Sopron	0.37	0.29	31.3	26.9
Szigliget	0.35	0.19	47.5	19.2
Tiszavasvári	0.65	0.58	30.5	30.4
Zenta	0.47	0.12	44.7	22.4
Mean	0.39	0.24	51.3	19.3
SD value	0.334	—	25.4	—
P level	0.005	—	0.000	—

Source: Modified from Németh, É. et al., *J. Herbs Spices Med. Plants*, 13, 57, 2007.

mentioned "four different plants," which have been harvested, but in this case, the low number of individual plants is surely not able to represent the population or genotype.

Intrapopulation variability has been studied only in exceptional cases. Llorens-Molina et al. (2012) presented the common occurrence of two well-distinguishable chemotypes (*cis*-beta-epoxyocimene above 70% of essential oil and *cis*-beta-epoxyocimene 60%–70% with *cis*-chrysanthenyl acetate 10%–20%) in a wild habitat in Spain. The two chemically—and presumably also genetically—distinct individuals are distinguishable only by essential oil analysis and do not show any external marker traits. The authors called the attention to the importance of individual monitoring during examination oil composition because of the obvious differences among plants of the same population.

In a plantation at our experimental station originating from commercial seed material, 30 individual plants have been sampled for determination of supposed individual variability. The results

TABLE 4.8
Fruit Characteristics of Selected Individuals in the Stock Plantation of *Foeniculum vulgare* "Foenipharm"

Plant nr.	Essential Oil (% d.w.)	Anethole (% Essential Oil)	Estragole (% Essential Oil)	Fenchone (% Essential Oil)
1	7.91	65.13	2.34	2.14
2	6.10	61.58	2.18	2.38
3	6.34	57.88	2.09	2.07
4	4.73	57.91	2.02	2.51
5	5.59	60.14	2.17	2.41
6	4.94	61.87	2.17	2.26
7	6.33	66.30	2.38	1.87
8	4.49	67.57	2.45	1.76
9	4.91	54.84	1.98	2.53
10	4.90	69.02	2.56	1.79
CV%	18,8	7.5	8.0	12.0

Source: Németh, E. et al., unpublished data.

ascertained the wide diversity of wormwood oil composition not only at the level of populations but also at the level of the individual plants. According to this, β-thujone was the main component in 53% of the plants, but in nine plants, this compound was found only in traces. The second and third most abundant components were β-myrcene and sabinene, respectively, both being main compounds in 13% of the samples. Besides, in 20% of the oils, they were found in approximately equal proportions (Figure 4.3). A single sample contained *trans*-chrysanthenol and another one an unidentified sesquiterpene (retention time [RT]: 44,63; linear retention index [LRI]: 1985; electron ionized mass spectrum [EIMS]: 284 [0,8%], 185 [4%], 159 [5,5%], 145 [33%], 132 [88%], 119 [100%]) as main component (Zámboriné-Németh et al., 2012).

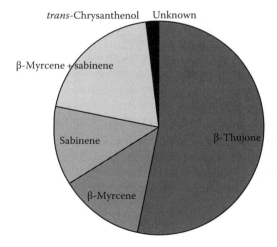

FIGURE 4.3 (See color insert.) Distribution of individuals of different main compounds in their essential oil in a commercial wormwood (*Artemisia absinthium*) population. (From Zámboriné, N.É. et al., Individual variability of wormwood [*Artemisia absinthium* L.] essential oil composition, in: *Program and Book of Abstracts, 43rd ISEO*, Lisbon, Portugal, 2012, p. 105.)

Differences among individuals manifest themselves not only in the main components but also in the total spectrum. The varying ratios of mono- and sesquiterpene compounds to each other demonstrate it very well: individuals with 89% monoterpenes and 11% sesquiterpenes represent one marginal value, while an individual with 20% monoterpenes and 80% sesquiterpenes in the essential oil expressed in gas chromatography (GC) peak area percentages is the contrast.

Sampling of a population of *T. longicaulis* ssp. *longicaulis* in Turkey resulted in distinguishing three different chemotypes: thymol type, geraniol type, and α-terpinyl acetate types. It was shown that individuals belonging to the different chemotypes can be found near to each other even on a 1 m^2 area (Baser et al., 1993).

The aforementioned examples represent quite well that in a chemically diverse species, the populations are heterogeneous too, in consequence of the large individual variability.

4.3 CONNECTIONS OF CHEMICAL DIVERSITY WITH OTHER PLANT CHARACTERISTICS

4.3.1 Propagation and Genetics

The homogeneity or variability of a population often stays in connection with the usual propagation method of the species. Phenotypic manifestation of diverse genetic background and appearance of different chemotypes in a plant stand can be supported by sexual propagation and cross-pollination. To the contrary, vegetative propagation or autogamy enhances uniformity of the population.

Vetter and Franz (1998) proved the large degree of self-incompatibility in five *Achillea* species (*Achillea ceretanica, A. collina, Achillea pratensis, Achillea distans,* and *Achillea monticola*). While the number of seeds in cross-pollinated flowers reached 47–110 pieces, it was solely 0–11 pieces in self-pollinated ones. Our long-term experiences with yarrow ascertain this finding and it is in obvious coincidence with the large intraspecific chemical diversity of these species.

Xenogamy is the preferred way of fertilization in several important medicinal species. As an example, Lamiaceae species are cross-pollinating ones based on the morphological constitution of the flowers and the mechanism of proterandry. Besides xenogamy, also geitonogamy may occur between flowers of the same plant; however, seed-set rates are much lower in this case (Putievsky et al., 1999; Németh and Székely, 2000). In some species of the same genus, both hermaphrodite and male-sterile flowers can be found. In thyme (*T. vulgaris*), it has been described that the latter one occurs primarily in suboptimal environments assuring that outcrossing enhances fitness of the progenies, while the hermaphrodite flower structure enables autogamy. Depending on the type of fertilization, the essential oil pattern varies characteristically (Gouyon et al., 1986).

Species, which are generally propagated by vegetative methods like peppermint and tarragon, do not show any or only a minimum variability among individuals. This fact sometimes is considered as an adverse phenomenon and an obstacle in effective selection and genotype improvement. Therefore, breeders usually try to increase the variability of these plants with specific methods. Mutation breeding proved to be a prosperous tool in producing wilt-resistant strains of peppermint in the United States (Murray et al., 1988). Induction of polyploids by colchicine and the crossing of fertile accessions afterward have been the basics in developing the highly productive variety "Multimentha" in East Germany (Dubiel et al., 1988). Development of new chemical varieties is endeavored today more and more by molecular genetics methods (Croteau et al., 2005; Wagner et al., 2005).

On the other hand, clonal propagation is an optimal way to obtain chemically homogenous populations for commercial production and processing purposes. According to own experiences, seed sowing of tansy results in an enormous segregation of the population, which is not acceptable as raw material for industrial utilization. Therefore, vegetative propagation by young shoots has been developed for the production of selected chemotypes (Zámboriné et al., 1987).

For species, where a large chemical diversity is characteristically present, the uniformity of plant populations seems to be always questionable. It is most frequent that heterogeneity exists also among individuals, and production of drugs of stable quality cannot be carried out from these stands. This phenomenon is an important motivation for introduction of economically important wild species into the agriculture and selection of their stable varieties. Breeding is going on usually parallel with development of technological methods.

Fennel has been cultivated already for many decades, and selected cultivars are registered in numerous countries. The main goals of the breeding have been definitely the increase of essential oil content and stabilization of its composition. During maintenance of our cultivar "Foenipharm," we checked the most important characteristics of individual mother plants. The results show that deviations among the plants are minimal due to the long-term breeding and variety maintenance process (Table 4.8).

Breeding of the polymorph species *A. absinthium* in the Conthey Research Centre (Switzerland) resulted in a uniform variety accumulating *cis*-epoxyocymene as main compound. After screening of more than 800 plants from 24 accessions originating from six countries, the researcher selected and stabilized the desired chemovariety (Simonnet et al., 2012).

Effective breeding necessitates knowledge on the genetic background, but inheritance of volatile compounds is till now only partially detected. The accumulation level of the essential oil is principally a quantitative feature and, thus, target of polygenic inheritance. However, biosynthesis of individual volatile compounds has been explained several times by Mendelian genes and gene interactions.

Classical genetic studies revealed that azulenogenic sesquiterpene lactones are inherited through the recessive allele of a special gene. In tetraploid species, a single homozygote recessive locus assures proazulene accumulation, and in a double homozygote recessive genome, it is manifested at elevated levels (Vetter et al., 1997). Similar mechanism seems to be working in the related chamomile (*Matricaria chamomilla*), and quantitative changes seem to be the result of modifying polygenes (Franz and Wickel, 1985; Wagner et al., 2005).

Multiallelic genetic determination was stipulated for the inheritance of borneol and 1,8-cineole in *Hedeoma drummondii* (Irving and Adams, 1973) or for the inheritance of camphor in *T. vulgare* (Holopainen et al., 1987).

Based on hybridization studies, it has been supposed that at least the majority of the genes regulating biosynthesis of *p*-menthane compounds are universally present in the genus *Mentha*. A single locus may be responsible for the production of anethole and estragole with partial dominance for high-estragole content (Gross et al., 2009). Similarly, existence of chemotypes of different δ-carene levels in scots pine (*P. sylvestris*) are explained by a single gene and inherited in dominant-recessive system (Hiltunen, 1975).

More recently, it has been found that the presence of typical oregano-type "cymyl compounds" (γ-terpinene, *p*-cymene, carvacrol, etc.) in different *Origanum* species is associated with the presence of a specific variant of γ-terpinene synthase gene (Lukas et al., 2010). Six different variants of this gene have been detected, which differ in the presence or absence of specific patterns in intron 3 but not each of them is able to result in the accumulation of the mentioned characteristic compounds of the genus.

According to our recent knowledge, it is obvious that the genetic determination of essential oil compounds should be complex. Besides the direct regulation of the biosynthetic processes, other types of regulation interact with the formation of volatile compounds like intra- and intercellular transportation mechanisms, primarily metabolic processes, or regulation through transcription factors, which are still less known in terpenoid metabolism.

Inherited traits manifest themselves in each plant individual, and this is the background of intraspecific diversity. However, the appearance of variability at population level depends also on the occurrence frequencies of corresponding genes. Complex genetic constitution of a population determines the abundance and diffusion of specific chemotypes in the plant stand and indirectly—the quality of the drug, which can be harvested there.

4.3.2 MORPHOLOGICAL CHARACTERISTICS

Both from theoretical and practical points of view, connection between chemical traits (essential oil composition) and morphological characteristics would be of interest. External features as marker traits for oil composition would be of high importance during cultivation, breeding, or control. However, besides some exemptions, there are no reliable data about this topic.

The leaf form, size of dissections, and color of the leaves show a great variability in *A. crithmifolia*. Our investigations in controlled environment, however, revealed no connection between chemotype and leaf dissection. Plants of light-green leaves accumulated more essential oil, however, showed no correlation with the oil composition (Németh et al., 1999). Similarly, Hofmann (1993) established in several taxa belonging to the section *A. millefolium* that morphological traits may not refer directly to specificities of essential oil composition.

In *A. millefolium*, Gudaityté and Venskutonis (2007) tried to find connection between the color of petals and the azulene accumulation in the flowers; however, it could not be proved. A higher proazulene level was detected to stay in connection with higher number of internode, narrower leaves, and ligulate flowers, but it has not been ascertained by other authors.

The shape of the leaf is also very variable in case of tansy. According to the author's observations (Zámbori-Németh, 1990), some chemotypes can be distinguished from other ones based on this feature. Individuals containing the sesquiterpene davanone have shiny green, oval leaves with dense incisions, while the leaves of the chemotype accumulating thujone as main compound is elongated, leaflets are sparsely incised but lobes are deeper and their color grayish green. However, similar characteristics cannot be generalized as special markers applicable for each chemotype. It is in coincidence with the opinion of Schantz and Forsén (1971) who emphasized that no characteristic connection between essential oil composition and morphology of the examined west European tansy populations could be determined.

In some cases, however, literature references seem to be contradictious in this respect. In a former publication, Hodzsimatov and Ramazanova (1974) declared that the presence of bornyl acetate, terpineol, and linalool in the essential oil of hyssop (*Hyssopus officinalis* L.) is connected to the pink flower color. Chalchat et al. (2001) mentioned that pinocarvone is mostly present in individuals of white petal color, but Galambosi et al. (1993) found the highest pinocarvone proportions in a population of pink flowers. Own (not published) measurements and experiences showed that chemism of white, pink, or blue flowering individuals is independent from flower (petal) color, and there may be larger differences between plants with the same flower color if they originate from different accessions compared to the ones that have different petal color but have the same origin.

The investigations in 48 annual and 18 biennial caraway populations provided data about several significant correlations among oil composition and different morphological and production characteristics like root neck width, number of shoots, number of umbels, and seed biomass. However, no significant correlation could be found between any of these morphological, production features and the carvone content of the oil (Zámboriné, 2005).

In chamomile (*Chamomilla recutita*), Gosztola (2012) carried out a very detailed and comprehensive analysis on Hungarian wild growing populations from different habitats. They studied a wide range of plant characteristics and their connections. As for the oil composition, none of the most important sesquiterpenes showed any significant correlation with morphological features like plant height, diameter of the flowers, and that of the discus (Table 4.9).

By screening of 13 different accessions of fennel (*Foeniculum vulgare* ssp. *capillaceum* var. *vulgare*), it has been established that they represent different chemovarieties of the species (Bernáth et al., 1996). Chemovar 1 represented by a single accession accumulating the largest concentrations of fenchone (above 30% of the oil), chemovar 2 contained three strains characterized as methyl chavicol–rich ones (above 20% of the oil), while the nine accessions belonging to chemovar 3 showed high-anethole contents (above 60% of the oil). Studying the correlations between the main

TABLE 4.9
Correlation Coefficients between Main Essential Oil Components and some Morphological Traits in Chamomile

Chemical Compound	Plant Height	Diameter of Flowers	Discus
β-Farnesene	0.00	0.28	0.38
Bisabolol-oxide B	0.10	0.09	−0.10
α-Bisabolol	−0.34	−0.32	−0.33
Chamazulene	−0.12	−0.04	0.10
Bisabolol-oxide A	0.30	0.26	0.28
cis-Spiroether	0.20	0.13	0.09
trans-Spiroether	−0.02	0.05	0.23

Source: Modified from Gosztola, B., Morphological and chemical diversity of different chamomile (*Matricariarecutita* L.) populations of the Great Hungarian Plain (in Hungarian), PhD dissertation, Corvinus University of Budapest, Budapest, Hungary, 2012.

TABLE 4.10
Correlation Coefficients of Morphological and Chemical Characters of the Accessions Investigated

Morphological Feature	Essential Oil Content (mL/100 g)	Component (in % of the Oil)					
		α-Pinene	β-Pinene	Fenchone	Methyl Chavicol	Anethole	Limonene
Plant height	−0.1002	0.0866	−0.1590	0.2263	0.5927	−0.5805	0.2152
Mass of leaves	0.3531	−0.2713	−0.6458	−0.2800	−0.0768	0.2515	−0.4393
Length of seeds	0.6102	0.2334	0.0219	−0.1940	−0.3321	0.3853	−0.3229
1000 Seed mass	0.4705	0.2186	0.3560	0.1711	−0.4384	0.2624	0.0684

Source: Bernáth, J. et al., *J. Essent. Oil Res.*, 8, 247, 1996.

components of the oil and the morphological features, only loose or medium strength correlations could be determined (Table 4.10).

The connection between the seed size and essential oil content seems to be of the largest practical importance (r = 0.6102). Among the individual components, β-pinene and limonene showed significant negative connection with leaf mass. Similarly, higher plants produced less anethole and more methyl chavicol. Although these results may be interesting, most likely these statistical correlations have hardly any real physiological or genetic background, therefore their universal use as markers is questionable.

4.4 MORPHOGENETIC AND ONTOGENETIC MANIFESTATION OF THE CHEMICAL VARIABILITY

Although until now we discussed about chemical polymorphism of the plants in general, in numerous plant species, there are frequently well-defined deviations between the oil compositions of different plant organs: roots and overground parts, vegetative and generative organs, leaves, and flowers. Rate and pattern of the appearance of divergences are basically characteristics for the species. In some cases, the transition is continuous, and only the proportion of the compounds changes from the basal regions toward the apical parts beyond a relatively standard spectrum. In other species, the spectrum is suddenly changing with the differentiation of new plant organs and generative parts. Besides, there are examples also for uniform composition both of the vegetative and generative organs.

Lovage (*L. officinale*) accumulates in the leaves always α-terpinyl-acetate as main compound (40%–80%) besides a lower level of β-phellandrene (15%–28%). The main components of the roots 3-butylidene-phthalide and Z-ligustilide are only in lower concentrations—not rarely only in traces—present (Novák, 2006). The composition of the fruits is similar to that of the leaves with β-phellandrene as main component up to 60% (Bylaite et al., 1998).

In species, where the root does not provide an official drug, data on essential oil accumulation of the underground parts are obviously much rarer. Existing data, however, show that the composition of the underground parts might be similar or even totally different from that of the shoot system.

In a related Apiaceae species, in fennel, the difference in composition of the roots and that of the overground parts shows the largest deviations, while the difference between the green parts (leaves and shoots) and generative organs (flowers and fruits) is less characteristic (Figure 4.4). In the roots, the absolute main compound is dillapiole; anethole is accumulating in the whole shoot in highest concentration. Composition of the essential oils from stems and leaves of fennel are qualitatively similar to that of the fruits. However, the ratio of anethole varies among organs; the vegetative green parts contain it in higher concentrations than flowers and fruits (Chung and Németh, 1999). At the same time, it can also be established that the majority of the mentioned components are biosynthetically related phenylpropanoids.

Composition of the roots and that of the leaves proved to be surprisingly different both qualitatively and quantitatively in some Asteraceae species.

In the genus *Achillea*, the volatile composition of the root has been studied till now only in a few species. The root oil of *A. distans* contained primarily τ-cadinol, alismol, and α-cadinol (Lazarević et al., 2010), while in *A. millefolium*, epicubenol and the monoterpenic ester neryl isovalerate were detected in highest proportions (Lourenço et al., 1999). The sesquiterpenes (τ-cadinol) and monoterpene esters (neryl isovalerate) were also present in significant amount in *Achillea lingulata* roots (Jovanović et al., 2010). Our recent investigations (Kindlovits et al., 2014) on several *Achillea* accessions showed that n-hexadecanol accumulated in highest concentrations (up to 45%) in the essential

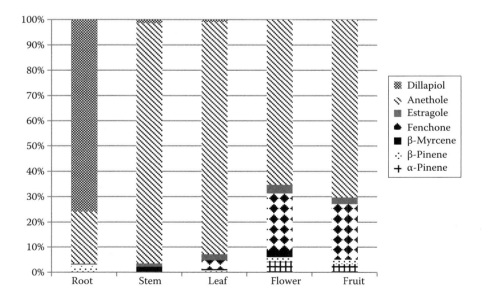

FIGURE 4.4 Composition of the essential oil in different plant organs of the fennel variety "Soroksári." (From Chung, H. and Németh, É., *Int. J. Hortic. Sci.*, 5, 27, 1999.)

oil of the roots of different *A. collina* strains, while fatty acids were predominant in samples of *A. crithmifolia*. It seems that the composition of yarrow root oil is basically dependent on species, but also intraspecific deviations occurred. The composition is basically different from that of the overground parts, and chamazulene is missing.

Intraspecific, moreover, individual differences have been detected in the volatile composition of the root oils also in the related species *A. absinthium*. Major components of the roots and of the overground organs proved to be totally different from each other. Our recent investigations (Németh, E. et al., unpublished data) showed that independent of the major components of the leaves (β-pinene, β-thujone, β-epoxyocimene, β-caryophyllene, etc.), the root oil contained most frequently monoterpenic esters like lavandulyl isovalerate, neryl isovalerate, neryl acetate, and bornyl acetate in highest abundance with hardly any sesquiterpenes, in contrary to yarrow.

The results ascertain that individual variability of essential oil composition might be present both in the overground organs and the underground ones. This shows a rather complex picture because it might mean that based on the essential oil composition of the leaves, a group of individuals is evaluated as belonging to the same chemotype, but concerning the root volatiles, they are frequently not similar to each other. About such data, there are still very few records, which underline the necessity of further investigations.

The presence of significant differences between the composition of the essential oil from the root and that from the shoot is, however, not a universal phenomenon. Schulz and Stahl-Biskup (1991) studied the organic diversity of essential oil spectrum in case of hyssop. They found that pinocamphone can be described as a universal main component in each of the roots, stems, leaves, and flowers at an accumulation proportion of 22%–60%. In case of the roots, the only characteristic difference was the presence of an unidentified—presumably—sesquiterpene type compound in 13.1%–15.6%, while the spectrum of the leaves and flowers proved to be both qualitatively and quantitatively related.

The picture is somewhat similar for peppermint. According to investigations of Murray et al. (1988), components of the essential oil distilled from the stolons are highly comparable with the shoot oil. Major compounds of the stolon oil were menthofuran (46.1%), menthyl acetate (24.5%), and menthol (11.4%), which reflect only quantitative differences compared to the oil distilled from the herb or the leaves. These data refer to a relatively uniform biosynthetic process of these terpenoids in the whole shoot system developing underground or overground.

As a conclusion, we could declare that biological variability is manifested also in the relationship between the volatile compounds of different organs. While some species accumulate qualitatively similar compounds in each organ, some others produce different compounds, however, results of the same biosynthetic route. At the same time, there are also species in which the volatile composition of different plant organs seems to be almost "random," and no closer connection can be established between their biosynthetic origins according to our present knowledge.

The special composition of any plant organ is, however, not a stable phenomenon. Qualitative and quantitative, compositional changes frequently occur during ontogenesis of the plant. These changes are either direct when the same plant organ—leaf, flower—shows an altered character during its development or they may be indirect: variability detected due to morphogenetic changes of the shoot like appearance of buds and fall of the leaves. These changes—onto- and morphogenetic ones—are part of the biological variability, as they are regulated by the metabolomic processes of the plant. Consecutive expression of corresponding genes and changes at translational or enzymatic level might result in different chemosyndromes during the plan life.

In peppermint, it has been detected by in vitro enzyme activity and $^{14}CO_2$ labeling experiments that the background of these ontogenetic changes is a complex process (McConkey et al., 2000). In the first phase ("de novo oil biosynthetic program"), which coincides with leaf expansion and gland filling, the group of enzymes leading from geranyl diphosphate till menthone are extremely active, while about a week later, in the second period ("oil maturation program"), these enzyme activities are strongly diminished and the activity of menthone reductase increases steadily leading to an elevated level of menthol. Early upregulation of menthone reductase in a breeding program may result in transgenic peppermint of elevated menthol accumulation potential.

In Apiaceae species, considerable compositional changes occur during the development of the seeds, which may have influence on the quality of the drug.

In caraway, a timely shift has been proved in the accumulation of the two main components carvone and limonene during seed development, therefore their ratio is primarily depending on the ontogenetic phase. Limonene is formed right after fertilization of the flowers, which is after 5–10 days followed by the accumulation of carvone, the former one being the precursor of carvone. The timely shift is the consequence of changing enzyme activities. Limonene synthase is active at the beginning, while after the mentioned time period, activity of limonene hydroxylase catalyzing the formation of carvone is increasing (Bouwmeester et al., 1998). It has been supposed that this is the rate-limiting step in enhancing carvone accumulation.

A similar phenomenon has been described for the related coriander (*Coriandrum sativum*). In the fruits of this species, linalool is the characteristic main compound, reaching more than 90% of the total essential oil. At the beginning of seed development, besides linalool, the ratio of (*E*)-2-decenal is characteristically high (above 15%–20%), which decreases later, during seed development process (Varga et al., 2012).

In sage (*Salvia officinalis*), it was detected that different types of accumulation structures (peltate glandular trichomes, capitate glandular trichomes, and ambrate resinous droplets) are present in special distribution on the leaves. Each of them has characteristic terpenoid composition (Tirillini et al., 1999). Considerable compositional differences can be found also between older and younger leaves (Grassi et al., 2004). While the younger leaves were rich in β-pinene, bornyl acetate, and sesquiterpenes, the ratio of these compounds was significantly reduced by leaf expansion (Table 4.11). On the other hand, mature leaves contained three times more camphor and camphene than the newly formed ones. In this case, the changes are reflected not only among leaves of different age but also among different segments of the same leaf, too. The relatively immature regions at the basal region of the leaf showed similarity to the composition of young leaves. However, they were characteristic differences observed also between the inside and the marginal regions, therefore it seems to be likely that besides age also other factors may influence the composition of the single oil glands on the leaf surface.

Detailed investigation of Johnson et al. (2004) revealed well-measurable differences in the oil composition of leaves of different age in the related Lamiaceae plant, *Origanum vulgare* ssp. *hirtum*,

TABLE 4.11
Characteristic Main Components (%) of SPME Extracts of Sage Leaves of Different Age

Compound	Young	Old	Intermediary Leaf			
			Base	Margin	Middle	Inside
Camphene	4.5	8.1	4.7	6.5	7.1	5.7
β-Pinene	19.4	4.6	8.1	8.6	9.7	13.4
1,8-Cineole	8.0	15.5	—	—	—	—
α-Thujone	12.1	12.5	16.6	14.6	17.0	21.0
Camphor	9.7	29.1	11.9	20.0	18.2	14.9
Bornyl acetate	2.5	0.0	2.0	0.7	0.4	0.2
α-Humulene	6.0	4.7	7.4	3.5	4.2	3.6
Viridiflorene	4.7	0.5	23.3	14.0	13.6	1.5
Manool	10.4	9.4	6.2	9.2	6.4	5.2

Source: Grassi, P.J. et al., *Phytochem. Anal.*, 15, 198, 2004.

too. Lowest levels of *p*-cymene were always detected in the younger leaves, but to the contrary of sage, the composition of the oil proved to be similar in older and younger parts of the same leaf. It was anticipated that in this case, the background of the differences would be the time of the leaf development and not its relative age. Lengthening days enhance the loss of *p*-cymene compared to carvacrol, but only the young leaves are able to make this switch. In this case, it seems that compositional diversity of leaves of different age may not be considered as a simple biological variability but more a specific environmental response.

The detectable variability of the volatile composition may depend also on the spectrum of individual oil glands even of the same leaf. In oregano, significant differences could be detected in the composition of individual oil glands (Johnson et al., 2004). While the majority of the oil glands yielded carvacrol and virtually no detectable amount of thymol, a small proportion of the glands produced up to 70% thymol. The oil glands of this latter type were randomly distributed on the leaf surface, but the explanation for this interesting phenomenon is still lacking.

The composition of the root oil seems to be varying with the age, too. Although there are hardly any reports on this issue, it seems that this variability may be the result of structural transformations. Stahl-Biskup and Wichtmann (1991) detected a decrease of germacrene B, the characteristic compound (in 51%) of the seedlings parallel with secondary thickening of the roots and formation of secondary oil cavities. The roots of adult plants contained mainly aliphatic aldehydes up to 68% of the oil. In the contrary, the herb oil of seedling and adult plants did not differ significantly from each other.

In many cases, the detectable variability of oil composition is connected to morphological differentiation during shoot development. The emergence of the flowering stem, appearance of flowers, and development of fruits may result in qualitative and quantitative alterations. The shifts are connected not only with aging of the plant but also with changes in its organic structure.

A "classical" example for this phenomenon is the change of the compositional profile of peppermint oil during ontogenesis (Murray et al., 1986). At the beginning of shoot development, the herb contains menthone (above 30%) as main compound, while the ratio of menthol is usually the same or even lower. During shoot growth, the proportion of menthol starts to increase, and at harvest time in a good quality plant material, it reaches more than 40%. By the senescence of the leaves elevated levels of the corresponding ester, menthyl acetate can be measured. Data of Table 4.12 show the mentioned changes during shoot development. On the same day, the older, bottom part (B) of the shoot is more similar to a later phenological phase (more menthol, less menthone) than the upper part (T). Besides, morphogenetic changes are reflected in the elevated values of menthofuran and pulegone on later dates as these are characteristic compounds of the flowering parts of peppermint.

The species-specific behavior in these biogenetic transformations is demonstrated by the fact that the related species *Mentha citrata* (syn. *M. piperita* var. *citrata*) shows an opposite tendency as the compound linalool increases by approximately 30% during flowering, while the corresponding ester, linalyl acetate, is decreasing at the same time (Malizia et al., 1996).

In species possessing different chemical varieties, ontogenetic changes might be characteristic features not only of the species but also of the intraspecific chemotype.

In tansy (*T. vulgare*), Schantz et al. (1966) already described quantitative shift in the essential oil composition from budding till the end of flowering period. Main tendency of the change was an increase of the main components camphor and thujone in these two chemotypes. In own investigation, six different chemotypes were checked in an extended period from early shooting till seed ripening (Németh et al., 1994). In five of the chemotypes, where monoterpene compounds are present as major components of the oil, an increase of the proportion of these components was detected. The increase is slight and continuous in the borneol-, camphor-, and 1,8-cineole-type plants, while it is a sudden and larger one in the thujone- and thujone acetate–type individuals (Figure 4.5). Another dynamics was found in the chemotype accumulating the davanone sesquiterpene lactone in the oil as main component. In these plants, the ratio of davanone is highest right after shooting out in spring and shows a continuous decrease after that.

TABLE 4.12
Main Components (% in the Oil) of the Peppermint Oil from Top (T) and Bottom (B) Parts of the Shoot during the Vegetation Period

Compound	Stem Part	July 21	August 3	August 16	September 7
1,8-Cineole	T	5.3	6.0	5.4	5.4
	B	5.8	4.8	4.5	4.0
Limonene	T	3.2	5.0	3.7	2.8
	B	3.0	2.3	2.7	2.1
Pulegone	T	1.1	3.4	4.6	1.0
	B	0.8	2.0	2.3	1.7
Menthofuran	T	1.4	7.6	9.9	8.2
	B	2.0	2.9	4.6	5.7
Menthone	T	34.1	30.5	28.3	19.2
	B	16.1	17.1	17.1	13.6
Menthol	T	29.9	23.3	27.1	41.5
	B	43.4	39.4	38.8	44.2
Menthyl acetate	T	3.2	3.7	2.7	3.7
	B	8.0	10.3	9.5	9.0

Source: Modified from Murray, M.J.P. et al., Peppermint oil quality differences and the reasons for them, in: Lawrence, B.M., Mookherjee, B.D., and Willis, B.D., eds., *Flavors and Fragrances: A World Perspective*, Elsevier, Amsterdam, the Netherlands, 1988, pp. 189–210.

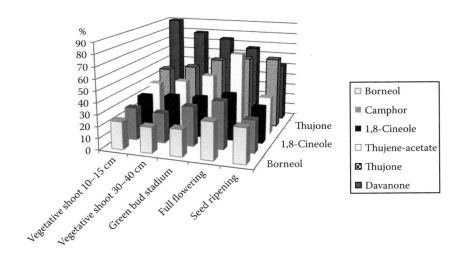

FIGURE 4.5 Proportion of the main components during the vegetation period in different chemotypes of tansy. (From Németh, É. et al., *J. Herbs Spices Med. Plants*, 2, 85, 1994.)

Long ago, Schranz and Hörster (1970) described similar phenomenon in thyme species (*T. vulgaris* and *T. marschallianus*). They established that during the vegetation period, significant changes can be detected in the essential oil composition in both species, which are individually varying. The dynamics of changes of the measured compounds (γ-terpinene, thymol, carvacrol, *p*-cymene, terpinolene, α-pinene, etc.) through plant development was different in each examined individual plant. The large individual variability in the components and in the dynamics of the compositional changes let them conclude that the chemical features of a genus or a species can play only a limited role for systematic characterization.

Less known and investigated are even till today the diurnal changes of volatile components. As a model plant, in *Origanum onites*, the most important compounds carvacrol and thymol fluctuate significantly during the day. Carvacrol has a decrease in early morning and in afternoon, while thymol shows the lowest concentrations at 10 o'clock in the morning and the level increases from early evening till midnight. This dynamics means 1.6–3.0 times differences in the accumulation ratios of the mentioned compounds and do not seem to be influenced severely by the phenological stage from preflowering till postflowering periods (Toncer et al., 2009).

Biological variability of essential oil composition is obviously a complex system manifested in a characteristic row of chemosyndromes in the plant species or in the intraspecific taxon.

The backgrounds of the evaluated changes are until now only partially detected. The spectrum of volatile compounds in the plant depends in a large extent on modifications in varying gene or enzyme activation processes connected to developmental phases.

Changes at transcription level were detected by Grausgruber-Grüger et al. (2012) in garden sage (*S. officinalis*). They investigated the connection between the accumulation of main monoterpenes and transcript levels of their synthases during the vegetation cycle. It has been established that terpene synthase mRNA expression and the level of the respective end products were in significant correlation in case of 1,8-cineole (correlation coefficients $r = 0.51$ and 0.67 for the two investigated cultivars) and camphor ($r = 0.75$ and 0.82), which indicated a transcriptional control of the process. The same correlation, however, could not be proved for α- and β-thujones, which shows the possible role of other regulation levels in accumulation of these compounds.

In several cases, changes of the volatile compounds are presumably connected to differentiation of special anatomical structures like glands or cavities and the enzymatic profile of these organelles. Monoterpenes are formed predominantly in the plastids via the methylerythritol phosphate (MEP) pathway, therefore mesophyll-originated cells rich in plastids may be rich in monoterpenes, while cells of well-developed endoplasmic reticulum may be primary sites of sesquiterpene biosynthesis. Recently, it seems that this statement can only partially be accepted because a metabolic crosstalk between the plastid MEP pathway and cytosolic mevalonic acid (MVA) pathway has been found several times (Dinesh and Nagegowda, 2010). Based on investigations in *Melaleuca alternifolia*, Webb et al. (2011) concluded that both the MVA pathway in the cytosol and the MEP pathway in the plastids may contribute to the sesquiterpene formation, and the intensity of this contribution depends on many factors like the plant species, tissue, and physiological state of the plant. Although data on localization of enzymes and terpenoid biosynthetic processes are continuously accumulating, a general statement about the role of intracellular structures in detected changes of volatile profile of essential oil–bearing plants cannot be established yet.

Looking for the background of biological variations of the essential oil spectrum, synthesis of the volatile compounds is only one side. It cannot be excluded either that the biosynthesized components are specifically translocated and/or further metabolized in the cells. The best documented example for this complex compartmentation system seems to be the formation of different monoterpenic compounds of peppermint oil. Labeling studies revealed that geranyl diphosphate synthase is localized in the leucoplasts, (−)-limonene-6-hydroxylase is associated with the endoplasmic reticulum, (−) *trans*-isopiperitenol dehydrogenase is found in the mitochondria and (+)-pulegone reductase in the cytoplasm. It shows that a well-established subcellular compartmentation and translocation mechanism is needed in fulfilling the total biosynthetic chain. Besides, an active transport is expected in carrying the produced components from the secretory cells into the subcuticular oil storage cavity of the peltate grandular trichomes (Kutchan, 2005). Schratz and Hörster (1970) already many decades ago suggested that compositional changes associated with plant ontogenesis may be present due to the fact that only young oil glands synthesize volatile compounds, and in the fully developed leaves, secondary transformations overdominate. Later, it was proved that monoterpene biosynthesis is going on in peltate glandular trichomes where terpene

4.5 ORIGIN OF ESSENTIAL OIL VARIABILITY

Ontogenetic changes in chemical profile of numerous taxa provide the basis for the "ontogenetic hypothesis" on the genesis of intraspecific chemical taxa. Tétényi (1970) assumed that differentiated intraspecific taxa have emerged by adapting themselves to different life conditions.

According to this, some alterations of ontogenetic metabolistic changes of individual development as acclimatization behavior may have occurred in certain phases characteristic of the given plant species. The special chemical differences of intraspecific taxa seem to have evolved by the stabilization of these chemical features present in consecutive phases of the ontogenesis if the deviation was of durable nature and occurred repeatedly in several progenies. Differences both in intensity and quality of metabolism in single phenological phases might be inherited, becoming a taxonomic characteristics, and thus, establishing the physiological bases of the existence of new chemical taxa.

Individuals of different chemism may show different adaptation capacity to changing environmental circumstances. Thus, plants of the most appropriate compositional profile have a fitness comparing with other ones, therefore can be stabilized and spread through the area (Hegnauer, 1978).

It is without doubt that chemical changes manifested in successive progenies serve frequently as a direct adaptive tool in survival of the population or individual. In other cases, however, adaptation through other—morphological, biological, phenological, propagation—features may also lead to an altered chemical profile as an indirect result.

This general statement for secondary plant compounds may be valid for essential oil compounds as well. Compounds, which have some kind of adaptive value, are presumably distributing to a larger extent inside the population increasing the fitness of the plants. Many examples demonstrate already that in accordance with phylogenesis, appearance of intraspecific chemical variability may be considered as result of adaptive processes (Dudareva and Pichersky, 2008). During natural selection, the presence and accumulation level of volatile compounds have been changed and show now a really colorful spectrum.

In *T. vulgaris,* the broad chemical variability and adaptation potential are manifested during ontogenesis. At the 1–3 months seedling stage, the plants belonging to linalool chemotype accumulate mainly phenolic compounds similarly to the carvacrol and thymol chemotypes and, only in later phenological stages, start to predominate the characteristic linalool in the oil. It was suggested that this behavior of the plant is a chemical defense against herbivores in the young, sensitive age (Linhart and Thompson, 1995).

Seedlings of *Eucalyptus* accumulate essential oil rich in pinenes, while adult plants produce 1,8-cineole (eucalyptol) as main component. Similarly, seedlings of *Cinnamomum camphora* synthesize safrole as major volatile compound of the leaves, but later, different other compounds start to accumulate in abundance corresponding to the different intraspecific chemotype (Tétényi, 1975). Juvenile parts of *Thuja occidentalis* show a multiple difference in the quantitative composition of the volatiles compared with that of the mature shoots. The accumulation rate of sabinene (25.37%) and α-pinene (26.29%) are especially high compared to the leaves of the adult plant (0.76% and 3.54%, respectively) (Gnilka et al., 2010). Interestingly, the adaptive role or other function of these changes is still not adequately declared and is worth of further research.

Several examples demonstrate that adaptation process should be considered here in a wider sense. Chemical properties and production of special volatiles may enhance not only plant survival among adverse ecological conditions and assure a protection against predators but also stimulate competitiveness and distribution.

The important role of the changing volatile profile during ontogenesis in the ecophysiological behavior of the plants has been demonstrated by the example of lavender (Guitton et al., 2010). It was found that in each individual flower, three different group of terpenes appeared sequentially during the flowering period. The authors concluded that the terpenic volatiles in bud stadium and at the end of flowering, when flowers faded and small seeds start to develop, should serve as protective, insect-repellent compounds (e.g., ocimene, limonene, linalool). The ones, however, produced during full flowering are mainly attractive type molecules, presumably enhancing fertilization by engaging pollinators (e.g., linalyl acetate, some sesquiterpenes).

Similarly, Muhlemann et al. (2012) proved that in snapdragon emission of attractive volatiles (primarily myrcene, β-ocimene, linalool, and (E)-nerolidol) follows a diurnal rhythm. Molecular genetics investigations revealed that the synthesis is regulated at the level of transcription. The expression of genes encoding enzymes involved into the formation of these volatiles is increasing after anthesis and reaches its maximum when the flower is ready for pollination. This is specifically detectable in the petals of the flowers, while very few changes were found in sepal expression profiles, which shows a strict organic localization corresponding to the physiological role of the changes. The complex nature of these processes was shown furthermore by the fact that the biosynthesis of secondary compounds and primary metabolic pathways feeding into it are distinctly regulated during development.

Nevertheless, as mentioned earlier, it can be assumed that sometimes an altered chemical profile is not a tool in acclimatization but appearing as consequence of adaptation processes. This fact seems to be an explanation for the existence of variable chemotypes of *T. vulgaris* in different French populations. It has been proposed that the appearance of numerous intraspecific chemotypes and wide range of main components in the essential oil may be the result of natural competition (Gouyon et al., 1986). The species is indigenous thorough the Mediterranean coasts; however, it has a relative weak competition ability, therefore autogamy can assure a survival only in optimal environment. In these habitats, plants are hermaphrodite, and in consequence of the high rate of autogamy, they are of homozygote genetic structure. The constitution of a homozygote genotype could lead to the manifestation of recessively inherited features like the high-thymol accumulation and distribution of this chemotype in the area. However, under less favorable conditions, in order of increased competitiveness, propagation is going on through xenogamy, which results in heterozygotic genetic structure. In such a genotype, the typical oil components are the dominantly inherited ones like geraniol and terpineol, therefore these chemotypes are abundantly found in the population. Based on this hypothesis, the phenotypic manifestation of genetic information, that is, the chemosyndrome is the indirect result of the environmental pressure.

Hybridization represents a further possible way for development of new chemotypes. Interspecific hybridization and polyploidization frequently occur especially in neighboring and overlapping distribution areas of different taxa. Hybrid accessions of elevated fitness—either through a new chemical profile or by other advantageous features—may give rise to the distribution of these genotypes.

One of the most important and widely used medicinal plants, *A. collina*, might have developed by this way. Based on hybridization experiments, it was suggested that this species arose by natural outcrossing and allopolyploidization of *Achillea setacea* and *A. asplenifolia* as their hybrid in neighboring distribution areas. Through this process, new tetraploid species was born with a proazulene producing potential, similarly to the diploid parent *A. asplenifolia*. Hybridization is one way for the formation of new biotypes of better adaptability and chemism (Ehrendorfer, 1963).

Besides the mentioned adaptation processes, interspecific hybridization in overlapping areas with simultaneous blooming and possibilities of crossing was mentioned as another possible evolution mechanism in the development of numerous chemical varieties of *Thymus* species, too (Stahl-Biskup and Sáez, 2003).

4.6 CHEMOTAXONOMIC CONSIDERATIONS

Essential oil variability is target of chemotaxonomy. In connection with the complex regulation of the metabolism of terpenoids and other volatiles, chemotaxonomic aspects should be evaluated already in a more comprehensive way then it has been made for decades.

A big majority of scientific publications on intraspecific chemical variability of essential oil components refer to "chemotypes." Chemotype is in the practice a common term, which can be used in each case when the exact chemical taxon cannot be defined more precisely. Formerly, the term "chemical race" has been a similarly applied category, which, however, has been considered as not unequivocal and a rather indefinitive one (Tétényi, 1975). Therefore, the use of "intraspecific chemical taxon" as a general name and the use of the accepted botanical taxon definitions like *forma, varietas, subspecies*, in case of cultivated species, the term *cultivar* with the prefix "chemo-" as special names for closely defined taxa were suggested. However, this type of definition is rather rare now and thus, from the majority of literature, the range and size of chemical difference is frequently not obvious.

In any case, a special difficulty in evaluation of chemical diversity seems to be that the range of divergence, the altitude of the differences most obviously cannot be universally accepted and applied, although there is a need for some kind of minimum criteria in defining a chemical taxon as a separate one. In case of essential oils and volatile components, this question is maybe even more difficult to answer than in several other secondary compounds, as essential oils consists of a huge number of individual constituents. How many compounds and which of them should be taken into consideration?

Hegnauer (1962) suggested that essential oil constituents that attain at least 1% should be considered for chemotaxonomic evaluation. However, by the quick development of analytical methods, the number of identified components has been increased enormously, thus, 1% is not any more of the same significance as it used to be.

The increase of the number of identified compounds in essential oils can be well demonstrated by comparison of the scientific articles published in the 1990–1991 and 2012–2013 volumes of *Journal of Essential Oil Research*. In the starting volumes, the mean number was 41 component/sample (marginal values: 10–121), while in the last volumes, it reached 65 compound as a mean (marginal values: 22–187), which is a more than 1.5-fold increase during 20 years.

Now, there are no universally accepted criteria for delimitation of a chemical taxon. Evaluation is mostly depending on both the spectrum and relative abundance of the components.

This approach indicates a further question if GC area percentages are appropriate and precise enough for evaluation of the significance of individual components. Area percentage is a relative value, and it is changing by the total number of the identified compounds in the essential oil as a mixture, not directly dependent from the absolute accumulation quantity of the target compound. The majority of literatures, however, publish area percentages, which unfortunately, are less adequate for chemotaxonomic or genetic/biosynthetic conclusions.

The evaluation is sometimes severely aggravated with the—seemingly—obvious question: what is practically an individual component? Even the determination of a chemical compound is to certain extent depending on the goal of the evaluation. For a relatively large group of terpenoid compounds different enantiomers are known in vivo like (±) sabinene, (±) limonene, and (±) pinene, which usually possess also different biological activities. Usually, the plants are producing predominantly only one of the isomers, and in each case, the enantiomeric purity is characteristic for the species. In a recent study, each of the investigated 11 *Thymus* species accumulated mainly (−)-linalool (70%–100%), while the studied 5 *Nepeta* species could be characterized by 100% presence of (+)-linalool in the hydrodistilled oil (Özek et al., 2010). Similarly, (1R)-(+)-camphor overdominated in the oils of both basil species *Ocimum canum* and *Ocimum kilimandscharicum* (Pragadheesh et al., 2013). In some cases, however, racemic appearance could also be found in the genuine oils like in *Salvia microstegia* (52% and 48% of (−) and (+) linalool, respectively) or in *Stachys antalyensis* (60% and 40% of (−) and (+) linalool, respectively) (Özek et al., 2010).

The enantiomeric purity or characteristic ratio of the isomers might be a very valuable marker for authenticity control of essential oils. However, detection of the genuine constitution is often very difficult, because racemization may happen during drying, distillation, and storage and also by nonenzymatic reactions like autoxidation (Kreck et al., 2002). The genetic regulation of producing the corresponding isomer by a species has been less studied. In peppermint, it was proved that in vivo transformation of the not genuine (1S)-pulegone into (S)-isomenthone and (S)-menthofuran was carried out to a high degree (Fuchs et al., 2000), which indicates that the enzymatic processes are less stereoselective, but the end product depends to a large extent on the precursor. Different plant parts may have also specific enantiomeric distribution, which has hardly been investigated yet. (1R)-(+)-α-Pinene is the characteristic isomer in Lithuanian accessions of *Juniperus communis*, but its ratio proved to be 74% ± 13% in leaves and 69% ± 17% in unripe cones compared with (1S)-(−)-α-pinene (Labokas and Ložienė, 2013).

The usage of enantiomeric distribution in chemotaxonomic evaluation is still not widespread. Further research is needed to define its role in systematics in case of different taxonomic units.

Another important aspect in selecting relevant compounds for evaluation is the well-known fact that distilled oils not rarely contain artifacts. The synthesis of these compounds not occurring in the natural oils during water distillation or storage may severely disturb the evaluation both from theoretic and practical points of view. Increased levels of terpinene-4-ol in the distilled oil of marjoram (*Majorana hortensis*) compared to the genuine volatiles (Fischer et al., 1987) or development of spathulenol from bicyclogermacrene during extraction or storage (Toyota et al., 1996) are very good examples for it. The proportions of these compounds are obviously not appropriate ones to reflect the biological variability. Presence of chamazulene in the essential oil may be a sign of the production potential of certain guaianolides; however, it can be used for differentiating genotypes or chemotypes only restrictedly because it can be synthesized from different azulenogenic precursors.

Qualitative and quantitative differences may have of different significance when distinguishing taxonomical units. Denomination of qualitative and quantitative chemotypes, however, raise a further dilemma. The proportion or the absolute quantity of the accumulated volatile compounds is a continuous variable, thus, fixing a borderline between the values would be extremely difficult. A classical approach for this question is illustrated on the example of six Hungarian chemotypes of tansy (*T. vulgare*) in Table 4.13. Among the six chemotypes, the thujone, thujene acetate, and

TABLE 4.13
Proportion of Characteristic Compounds in the Essential Oil of Different Tansy (*Tanacetum vulgare*) Chemotypes

Compound	Thujone	Thujene Acetate	Camphor	Borneol	1,8-Cineole	Davanone
β-Pinene				3–5	7–16	
1,8-Cineole			8–10		23–31	
α-Thujone	55–64					
β-Thujone	26–32					
Thujene acetate		35–45				
Carveyl acetate		22–32				
Artemisia ketone			2–9			
Camphor			38–42	3–4	3–8	
Borneol			2–5	25–30	10–27	
Lyratol			3–5	7–10		
Bornyl acetate				30–43		
Lyratyl acetate				8–11		
Davanone						60–74
Davanol						6–13

Source: Unpublished data from Németh, É. et al.
Note: Only components involved into evaluation, accumulating at least in 5% are indicated.

TABLE 4.14
Chemotypes of *Achillea crithmifolia* in Different European Areas

Area (Country)	Component I	Component II	Component III	Reference
Bulgaria	Camphor 5%–30%	1,8-Cineole 3%–45%	Artemisia alcohol 20%–40%	Konakciev and Vitkova (2004)
Hungary	Camphor 20%–50%	1,8-Cineole 5%–30%	Borneol 5%–10%	Németh et al. (2000)
Serbia	Camphor 30%	1,8-Cineole 30%	Chrysanthenyl acetate 20%	Palić et al. (2003)

davanone types may be accounted as qualitative variants as the main compound is taking the huge majority of the total oil spectrum, and these components are detected in the other types only in traces. The further three chemotypes can be considered as quantitative chemotypes as they have several components common with each other, but definite and stable differences can be detected in the proportions of these compounds.

The example of different chemodemes (chemotype with separate distribution area) of *A. crithmifolia* can only be characterized properly if evaluation is carried out on the basis of three main compounds (Table 4.14). In this case, the complexity of this phenomenon can be clearly seen. Although the first two major compounds camphor and 1,8 cineole are deviating only in quantitative term, the third main compounds seem to be definitive for the characterization of the chemotype. These three compounds are biosynthetically not closely related ones, and it can be supposed that the detectable composition is the result of diverse genetic constitution. These data demonstrate also that chemical variability includes several pathways and products, and changes in any of them may influence also the ratio of the other ones.

We can conclude that biological variability in case of secondary compounds is in the practice a concentrated determination of numerous aspects: appearance of a specific spectrum of chemical compounds in plant populations, in single individuals, in different plant parts, or in different periods of plant life.

For checking and determining the biological variability, the most proper method according to the question should be used. Characteristic requirements, marginal values, limits, or intervals of quantitative features might be different for different reasons. Evaluation of a cultivar checking the drug quality, taxonomic studies, or practical breeding may require different assessments.

Former chemotaxonomic studies often emphasized that not the presence of a compound itself, but the biosynthetic processes, the potential of a plant for the formation of the given compound can be taken into account. According to the present knowledge, this general definition seems to be oversimplified and not fully adequate any more. Today, we have to face two approaches of chemotaxonomic considerations.

A pragmatic evaluation—although often published and mentioned otherwise—has nothing to do with real taxonomic aspects but collects and summarizes the detected chemosyndromes, that is, compositional changes in the essential oil due to internal or external factors.

This approach is quite frequent in the practice if certain quality requirements should be fulfilled. In these cases, a stable composition both qualitatively and quantitatively should be assured, and from this point of view, the proper knowledge on the detectable influencing factors and possibilities for their regulation is of primarily importance.

That means that the range of variability should be evaluated by investigating the row of chemosyndromes, which may appear under different circumstances or due to different treatments. These circumstances or treatments may consist of temperature regimes, illumination or any ecological factor, or even characteristic habitat as a complex background. For this pragmatic approach, it is not necessary that biotic and abiotic factors are strictly distinguished because all of these factors may have an effect on the chemosyndromes. Modifications and control are possible in majority of cases without deeper knowledge about the biosynthetic processes or their genetic regulation. That is why this approach tells nothing about taxonomic relationships of the examined objects, but the results may be directly utilized in cultivation and production of essential oil–bearing species.

However, at a more sophisticated level for a well-established regulation of the desired composition by breeding or evaluating the theoretical connections and taxonomic relationships, we cannot be satisfied by detecting chemosyndromes as result of any detectable factors. This is another approach, when we have to search more detailed for the physiological and genetic background and their less obvious connections.

This is the aspect, which seemed to be simpler some decades ago than today. As cited earlier, Tétényi (1970) declared that determination and separation of an intraspecific chemical taxon should be based on the divergent biosynthetic routes, on the potential of the plant for the synthesis of certain compounds instead on the presence of any compound itself. It is justified by the fact that biosynthesis of a certain compounds might take place by different routes in divergent species therefore simply the presence of any compound may not be a proof for taxonomic relationship.

Really, there are today already several examples for producing the same molecule as active compound by different enzymes and thus likely on the base of different genetic determination. Linalool synthase from *Clarkia breweri* has different constitution than linalool synthase from *Arabidopsis* plants (Dudareva et al., 2006) and only 41% identity to the same enzyme from *M. citrata* (Crowell et al., 2002). The enzyme from *M. citrata* is, however, much more close to other enzymes from the mint (Lamiaceae) family with 62%–72% identity.

The statement that not the compound itself but its biosynthetic route has the taxonomical significance may be still valid. However, we know already that the biosynthetic route consists of complex metabolomic process including the genomic constitution, gene expression, transcriptional and translational regulation, enzyme activities at different biosynthetic levels, interactions with transporters, translocation, and spatial isolation (Dudareva et al., 2006; Dinesh and Nagegowda, 2010). The result of the enzymatic reaction is determined not only by the availability and constitution of the enzyme but also by the availability of precursors and different interactions; therefore, the simple presence or absence of a certain enzyme in general cannot determine the final product.

It has been estimated that about half of all mono- and sesquiterpene synthases act as multiproduct enzymes! Numerous terpene synthases/cyclases are able to produce a wide range of terpenic skeletons. It is usually not possible to predict the product profile of terpene synthases on the basis of their primary structure (Tholl, 2006). Limonene cyclase frequently catalyzes the formation of myrcene and pinenes besides limonene from acyclic precursors (McCaskill and Croteau, 1998). The same synthase converts geranyl-pyrophosphate (GPP) into myrcene and (E)-β-ocimene, while the synthase TPS1 has three acyclic sesquiterpene products (E)-β-farnesene, (3R)-(E)-nerolidol, and (E,E)-farnesol (Dudareva et al., 2006). Taxonomically unrelated species seem to have closely related pathways for the formation of main terpenoid compounds. A 3-hydroxylation of the monoterpene precursor limonene by a P450 enzyme produces *trans*-isopiperitenol, a volatile compound characteristic for mint species (Lupien et al., 1999), while 6-hydroxylation by another P450 enzyme yields *trans*-carveol, which is further oxidized by nonspecific dehydrogenase to carvone, the main volatile compound of caraway, which is not a closely related species taxonomically (Bouwmeester et al., 1998).

According to McConkey et al. (2000), several enzymes of the menthol biosynthetic pathway appear to originate from widely divergent genetic resources in primarily metabolism, which would make their products less significant as chemotaxonomic characteristics.

Although "the biosynthetic route" in the sense of former publications seems to be a quasilinear process resulting in a special compound, the in vivo plant metabolism is a very complex regulatory and operating system nowadays only partially detected. Moreover, it has been supposed that the detectable mechanisms are only a part of the reserves of the plant, and plants have a resource of "hidden" biosynthetic capacities, a practically unlimited potential to produce a large array of different compounds, if they are activated by novel available substances (Lewinsohn and Gijzen, 2009).

Therefore, transgenic operation with single genes would frequently not result in the desired effect concerning the spectrum of volatile compounds. Cloning of a single gene is unlikely to result in a substantial production of the desired volatile compound if the formulation of this compound is the end result of a long metabolic pathway (Dudareva and Pichersky, 2008).

All of these factors contribute to the fact that some volatile components are really characteristic for the species or intraspecific taxon, while others are hardly to join to botanically related taxonomic units. Essential oil components should be individually evaluated if they may serve as chemotaxonomic markers.

A very good example for this is the study of Radulović et al. (2007) about the chemotaxonomic relationships of Balkan *Achillea* species. They evaluated altogether 47 records of 23 yarrow taxa according to their volatile constituents by principal component analysis. If each components detected in the oils at least in 1% were taken into account, except 3 taxa, no clear distinction could be established presumably because of the universal presence of 1,8-cineole, camphor, and borneol in higher amounts. If calculating with classes of compounds like monoterpene and sesquiterpene hydrocarbons, oxygenated monoterpenes and sesquiterpenes, diterpenoids, phenylpropanoids, fatty acid derivatives, and carotenoids, the results were not convincing either. A conclusion could be made that the rate of oxygenation is not an appropriate sign of real taxonomic connections because it may also be influenced by ecological/geographical factors. Finally, choosing the monoterpene structural types (*p*-menthane, bornane, pinane, etc.) as discriminative variables, the best grouping appropriate to accepted taxonomic classification could be set up. It proved to be useful to deal with groups of biosynthetically related compounds instead of individual components.

4.7 IDENTIFICATION OF NATURAL VARIABILITY

Detection of the biological variability is not always an easy job. Unfortunately, in practice, it is rather easy to find irrelevant publications about the chemical variability of essential oil-bearing species as the influencing factors are numerous, and they are usually in connection with each other.

Many articles deal with determination and simple description of the essential oil composition of a given plant material under given conditions (Figure 4.1). This type of articles might be useful to enhance the literature with new information. Mechanical assessment of such articles as a reference about the chemical variability of the given taxon is, however, a bad practice. Comparison of several independent publications about the target species is not an appropriate tool for the evaluation of biological diversity because the genetic material, the habitat, environmental conditions, sampling, processing, and the analytical method itself may strongly influence the measured data, while usually even these circumstances are not adequately provided. Therefore, a sum of publications—even if there is a huge number of them—is not able to reflect the biological variability of a species or any taxon.

Practically, a similar procedure is carried out if analytical results of several samples are compared in the frame of a single publication; however, samples originate from different habitats or other kind of sources. For example, a large pool of samples from different localities of Greece showed that the quantitative composition of essential oil from *M. pulegium* varies greatly (Kokkini et al., 2004). The most variable compound is pulegone, its proportion ranges from traces to 91% of the total oil. Fluctuations in the contents of piperitone (from traces to 97%), menthone (from traces to 53%), isomenthone (from traces to 45%), piperitenone (from traces to 40.0%), and isopiperitenone (from not detected to 23%) has been found, too. The authors emphasize that in localities where the real Mediterranean climate dominates, the total oil content and the amounts of the more reduced products of the *p*-menthane biosynthetic pathway, like menthone and/or isomenthone and their derivatives, were increased. Concerning the market quality of the pennyroyal oil, this conclusion may be enough. However, the changing essential oil profile might be the consequence both of the changing environmental conditions from south to north, might be the manifestation of different genotypes or even both. To conclude about the genetic–biological variability, the plant material of different habitats should be collected and investigated directly on the same plot.

A special care should be taken by choosing and defining a sample. Searching for the biological variability in essential oil composition, only plant material of well-defined origin should be investigated. Rather frequently, we find investigations on commercial samples obtained from any market. It cannot be emphasized enough that these items are not of reproducible quality. Any statement about such samples may provide only general and/or smattering information about the characteristics of the species or taxon.

Twenty commercial seed items were examined, for example, recently by Raal et al. (2012). Although the work includes detailed analytical results on the composition of the samples, a valuable conclusion is impossible because of the inadequate definition of these samples. They originated from pharmacies and other shops of different countries, but it is not enough information about the original genotype. The different age of the samples (2000–2008) also reduces the reliability of the data.

In sampling, the most frequent misunderstanding is caused by the fact that during comparison of populations or bigger taxonomical units, bulk samples are taken, processed, and analyzed. As discussed earlier, individual plant variability is a very frequent and important form of biological variability, which may be hidden in such trials. As mentioned also by Franz and Novak in Chapter 3, representativeness of sample taking in a heterogeneous population is prerequisite for reliable result. This should contain a larger number of individuals (up to 50) even if the analytical methodology would make do with a small quantity of plant material. Random sampling or bulk samples may give bias and does not reflect the characteristic compositional profile of the population. Even if this type of sampling and evaluation may have its role in checking the drug quality of a stand from commercial purpose when harvesting results in bulk material but does not give an answer to the question about biological diversity.

In many cases, a single population is practically a mixture of individuals belonging to different chemotypes and their ratio is most often unknown. As it is reviewed in *A. crithmifolia* in Section 4.2.2, in each habitat, different chemotypes may be found in different proportions (Németh et al., 2000). If this fact is not respected, the published results provide only some kind of "analytical mean" of the different chemotypes instead of characterizing the diversity of the taxon (Tzakou et al., 1993; Chalchat et al., 2000; Konakchiev and Vitkova, 2004).

In summary, relevant data about natural variability should be based on individual plant samples, which have been taken and analyzed in appropriate replications. Figure 4.1 shows the low proportion (approximately 1%) of recent publications dealing with individual differences compared to other topics.

In case of investigating perennial species, the weather conditions of the growth period (vegetation year) and the age of the plantation can hardly be separated. The problem is similar at different harvest times within the same year. Reliable information about the characteristics of the taxon can be obtained only by consequent sampling, keeping other circumstances constant and carrying out the trials for a longer period if necessary.

By taking two samples in October 2007 and in June 2011, Guimarães et al. (2012) wanted to establish the influence of the collection period on the concentrations of the essential oil components in *Mikania glauca*. Even if the sampling would have been representative, there are several problems, which make the evaluation more than questionable. The two random sampling times do not only mean that 4 years passed but also the plants became older, too. Phenological phase of the plants might have been different in October and in June, two different periods of the year. Besides, weather conditions in the growth period or during sampling may also have an effect on the analytical results. The information is useful in showing that quality of the leaves is not stable, but no exact conclusion about the real influencing factors and therefore not about their possible regulation is available from the data.

Unfortunately, in the literature quite frequently, the effect of different factors or treatments is interpreted as direct influencing factor on the essential oil composition; however, the basis of the change is the biological variability of the plants.

Caraway may be a good example for this. Formerly, a negative correlation between wind velocity and loss of essential oil and carvone was anticipated as the consequence of increased volatilization, thus, practically an environmental factor influencing the oil composition. However, Bouwmeester (1998) suggested that the loss of carvone is in connection with seed shedding enhanced by the wind. Shedding results in loss of especially the older/more ripen seeds, while the premature, younger ones remain on the stalks. Due to the delayed activity of limonene synthase, in these premature fruits, carvone usually did not reach its maximum level. In this case, the biological variability connected to the special biological process has been mistaken and evaluated as an environmental effect.

Our investigations in caraway revealed another interesting finding about the effect of row distance on the essential oil components of the seeds (Valkovszky and Németh, 2010). It was established that the wider row distance (48 cm) resulted in 10%–25% loss of carvone in the oil compared to narrower spacing (Figure 4.6). The difference is more pronounced in fertilized plots where optimal nutrient supply is assured for each plant individual. For the farmers, it is worth considering if quality may be improved by this way, but looking for the backgrounds of this phenomenon, the picture seems to be more complex and the studied treatments influence the composition only indirectly. In a wider spacing, the plant develops more umbels, but seeds in higher order umbels ripen several days—sometimes even weeks—later, than older ones. It has been already explained that highest carvone content is detected in the fully ripen seeds. In consequence of large proportion of higher order umbels in plots of wider spacing, most likely a big ratio of seeds still did not reached the phenological stage of maximal carvone content.

Agrotechnical methods in general act indirectly through influencing growth dynamics, developmental characteristics, and organic proportions, which further on determine the accumulation of special compounds.

Propagation method seems to have an effect on essential oil quality in many cases. Zheljazkov et al. (1996) detected differences in the main components of peppermint oil in plantations propagated by different methods. Menthol content of the oil was highest and menthyl acetate level was the lowest in plots established by rhizomes in autumn (Table 4.15). The investigated clones reacted not uniformly, the variety "Zephir," for example, showed the highest pulegone content in plots propagated by rhizomes in summer, while all the other cultivars had the lowest proportions in this treatment. The results, however, should be evaluated as indirect consequence of the propagation method, as it is most unlikely that planting itself has any effect on the biosynthesis of terpenoids in the newly developed shoots several months after transplanting. However, propagation method may have an influence on the phenological phases, appearance of flowers, size, and number of leaves. Besides, accelerated or prolonged development due to the different propagation method and time means different weather conditions for the growing period. All of these factors have been proved to influence the accumulation of volatiles in peppermint, which may reflect itself in the obtained data.

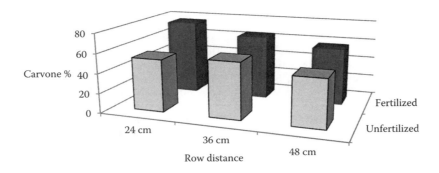

FIGURE 4.6 Carvone content (area %) of caraway seed oil in plots of different row distances. (From Valkovszky, N. and Németh, É., *Acta Aliment. Hung.*, 40, 235, 2010.)

TABLE 4.15
Changes in Essential Oil Composition (%) after Different Propagation Methods of Some Peppermint Cultivars

Variety	Propagation Method	Menthol	Neomenthol	Menthyl Acetate	Pulegone
No. 1	Rooted cuttings	42.3	7.16	10.70	0.53
	Summer rhizomes	44.3	6.41	9.64	0.18
	Autumn rhizomes	47.8	5.84	8.38	0.86
No. 101	Rooted cuttings	53.4	4.24	9.08	0.29
	Summer rhizomes	53.8	4.48	10.40	0.16
	Autumn rhizomes	60.5	4.46	8.63	0.22
Zephir	Rooted cuttings	59.3	1.80	9.81	0.54
	Summer rhizomes	58.9	2.33	9.76	1.06
	Autumn rhizomes	59.1	1.82	8.93	0.61
Mentolna 18	Rooted cuttings	62.4	2.34	7.36	0.17
	Summer rhizomes	62.1	1.95	7.18	0.09
	Autumn rhizomes	63.8	2.10	6.71	0.14

Source: Modified from Zheljazkov, V. et al., *J. Essent. Oil Res.*, 8, 35, 1996.

The results are important in optimalization of agrotechniques under the given circumstances but, obviously, do not represent the primarily background of the detected variability.

Not rarely, essential oil plants are propagated experimentally by in vitro methods. According to Ibrahim et al. (2011), tissue culture provides a fast breeding technology for tarragon (*Artemisia dracunculus*). They report that through callus culture and regeneration estragole, concentrations were effectively reduced in the field grown plants. According to our knowledge, tissue culture from different plant parts as explants is hardly able to change the genotype of the plant, except by spontaneous somatic mutations, which is, however, not reproducible during the technology. On the other hand, volatile concentrations may be influenced by an eventual structural (tissue constitution and essential oil accumulating organelles) change of the plant, which, however, is already an indirect effect and should be evaluated accordingly.

Practically, all of the agrotechnical interventions have effect on the growth and development of the plants. Spacing, irrigation, pruning, and fertilization influence the physiological processes of the plants but not necessarily represent the direct influencing factors of volatile formation.

REFERENCES

Aiello, N., F. Scartezzini, L. D'Andrea, A. Albasini, and P. Rubiolo. 2001. Research on St. John's Wort, Hyssop, Rosemary and Common Golden Rod at Different Planting Density in Trentino. Publications of ISAFA Research Centre, Villanzano di Trento, Nr. 1. pp. 23–31.

Argañosa, G. C., F. W. Sosulski, and A. E. Slinkard. 1998. Seed yields and essential oils of annual and biennial caraway (*Carum carvi* L.) grown in Western Canada. *J. Herbs Spices Med. Plants*, 6: 9–17.

Arino, A., I. Arberas, G. Renobales, S. Arriaga, and J. B. Dominguez. 1999. Seasonal variation in wormwood (*Artemisia absinthium* L.) essential oil composition. *J. Essent. Oil Res.*, 11: 619–622.

Bagci, E., M. Kursat, and S. Civelek. 2010. Essential oil composition of the aerial parts of two *Artemisia* species (*A. vulgaris* and *A. absinthium*) from East Anatolian region. *J. Essent. Oil Bearing Plants*, 13: 66–72.

Baser, K. H. C., M. Kürkçüoğlu, B. Demirci, T. Özek, and G. Tarımcılar. 2012. Essential oils of *Mentha* species from Marmara region of Turkey. *J. Essent. Oil Res.*, 24: 265–272.

Baser, K. H. C., M. Kürkçüoğlu, G. Tarımcılar, and G. Kaynak. 1999. Essential oils of *Mentha* species from Northern Turkey. *J. Essent. Oil Res.*, 11: 579–588.
Baser, K. H. C., T. Özek, N. Kirimer, and G. Tümen. 1993. The occurrence of three chemotypes of *Thymus longicaulis* C. Presl. subsp. *longicaulis* in the same population. *J. Essent. Oil Res.*, 5: 291–295.
Basta, A., O. Tzakou, and M. Couladis. 2007. Chemical composition of *Artemisia absinthium* L. from Greece. *J. Essent. Oil Res.*, 19: 316–318.
Bélanger, A. and Dextraze, L. 1993. Variability of chamazulene within *Achillea millefolium* L. *Acta Hortic.*, 330: 141–145.
Bernáth, J., É. Németh, A. Katta, and É. Héthelyi. 1996. Morphological and chemical evaluation of fennel (*Foeniculum vulgare* Mill.) populations of different origin. *J. Essent. Oil Res.*, 8: 247–253.
Bernotienė, G. and R. Butkienė. 2010. Essential oils of *Hyssopus officinalis* L. cultivated in East Lithuania. *Chemija*, 21: 135–138.
Blaschek, W., R. Hänsel, K. Keller, J. Reichling, H. Rimpler, and G. Schneider. 1998. *Hagers Handbuch der Pharmazeutischen Praxis, Folgeband 2. A-K.* Berlin/Heidelberg, Germany: Springer Verlag.
Bouwmeester, H. J. 1998. Regulation of essential oil formation in caraway. In: *Caraway—The Genus Carum*, É. Németh (ed.), pp. 83–101, Amsterdam, the Netherlands: Harwood Academic Publishers.
Bouwmeester, H. J., J. Gershenzon, M. C. J. M. Koning, and R. Croteau. 1998. Biosynthesis of the monoterpenes limonene and carvone in the fruit of caraway. *Plant Physiol.*, 117: 901–912.
Bouwmeester, H. J. and A. M. Kuijpers. 1993. Relationship between assimilate supply and essential oil accumulation in annual and biennial caraway. *J. Essent. Oil Res.*, 5: 143–152.
Bylaite, E., R. P. Venskutonis, and J. P. Roozen. 1998. Influence of harvesting time of volatile components in different anatomical parts of lovage (*Levisticum officinale* KOCH.). *J. Agric. Food Chem.*, 46: 3735–3740.
Chalchat, J., D. Adamovic, and M. S. Gorunovic. 2001. Composition of oils of three cultivated forms of *Hyssopus officinalis* endemic in Yugoslavia: f. *albus*, f. *cyaneus* and f. *ruber*. *J. Essent. Oil Res.*, 13: 419–421.
Chalchat, J., M. S. Gorunovic, D. Petrovic, and V. V. Zlatkovic. 2000. Aromatic plants of Yugoslavia. II. Chemical composition of oils of *Achillea clavenae*, *A. collina*, *A. lingulata*. *J. Essent. Oil Res.*, 12: 7–10.
Chialva, F., P. A. P. Liddle, and G. Doglia. 1983. Chemotaxonomy of wormwood (*Artemisia absinthium* L.). *Z. Lebensm. Unters. Forsch.*, 176: 363–366.
Chung, H.-G. and Németh, É. 1999. Studies on the essential oil of different fennel (*Foeniculum vulgare* Mill.) populations during ontogeny. *Int. J. Hortic. Sci.*, 5: 27–30.
Collin, G. J., H. Deslauriers, N. Pageau, and M. Gagnon. 1993. Essential oil of tansy (*Tanacetum vulgare* L.) of Canadian origin. *J. Essent. Oil Res.*, 5: 629–638.
Croteau, R. B., R. M. Davis, K. L. Ringer, and M. R. Wildung. 2005. (–)-Menthol biosynthesis and molecular genetics. *Naturwissenschaften*, 92: 562–577.
Crowell, A. L., D. C. Williams, E. M. Davis, M. R. Wildung, and R. Croteau. 2002. Molecular cloning and characterization of a new linalool synthase. *Arch. Biochem. Biophys.*, 405: 112–121.
Danila, D., C. Stefanache, A. Spac, E. Gille, and U. Stanescu. 2012. Studies on the variation of the volatile oil composition in experimental variants of *Hyssopus officinalis*. In: *Program and Book of Abstracts, 43rd ISEO Lisbon*, p. 105, Lisbon, Portugal.
de Pooter, H. L., J. Vermeesch, and N. S. Schamp. 1989. The essential oils of *Tanacetum vulgare* L. and *Tanacetum parthenium* (L.) Schultz-Bip. *J. Essent. Oil Res.*, 1: 9–13.
Derwich, E., Z. Benziane, and A. Boukir. 2009. Chemical compositions and insecticidal activity of essential oils of three plants *Artemisia* sp. (*A. herba-alba*, *A. absinthium* and *A. Pontica* (Morocco). *Electron. J. Environ. Agric. Food Chem.*, 8: 1202–1211.
Dinesh, A. and D. A. Nagegowda. 2010. Plant volatile terpenoid metabolism: Biosynthetic genes, transcriptional regulation and subcellular compartmentation. *FEBS Lett.*, 584: 2965–2973.
Dragland, S., J. Rohloff, R. Mordal, and T. H. Iversen. 2005. Harvest regimen optimization and essential oil production in five tansy (*Tanacetum vulgare* L.) genotypes under Northern climate. *J. Agric. Food Chem.*, 53: 4946–4953.
Dubiel, E., M. Herold, F. Pank, W. Schmidt, and M. Stein. 1988. 30 Jahre "Multimentha" (*Mentha piperita* L.). *Drogenreport*, 1: 31–64.
Dudareva, N., F. Negre, D. A. Nagegowda, and I. Orlova. 2006. Plant volatiles: Recent advances and future perspectives. *Crit. Rev. Plant Sci.*, 25: 417–440.
Dudareva, N. and E. Pichersky. 2008. Metabolic engineering of plant volatiles. *Curr. Opin. Biotechnol.*, 19: 181–189.

Ehrendorfer, F. 1963. Probleme, Methoden und Ergebnisse der experimentellen Systemetik. *Planta Med.*, 11: 224–234.

El-Wakeil, F., M. S. Khairy, R. S. Farga, A. A. Shihata, and A. Z. Badei. 1986. Biochemical studies on the essential oils of some fruits of Umbelliferae family. *Seifen, Öle, Fette, Wachse*, 112: 77–80.

Embong, M. B., D. Hadziyeu, and S. Molnar. 1977. Essential oils from species grown in Alberta, caraway oil (*Carum carvi*). *Can. J. Plant Sci.*, 57: 543–549.

Figueiredo, C., J. M. S. Barroso, M. S. Pais, and J. C. Scheffer. 1992. Composition of the essential oils from leaves and flowers of *Achillea millefolium*. *Flavour Fragr. J.*, 7: 219–222.

Fischer, N., S. Nitz, and F. Drawert. 1987. Original flavour compounds and the essential oil composition of marjoram (*Majorana hortensis*), *Flavour Fragr. J.*, 2: 55–61.

Fleischer, A. and Z. Fleischer. 1988. The essential oil of annual *Carum carvi* L. grown in Israel. In: *Flavours and Fragrances—A World Perspective*, B. M. Lawrence (ed.), pp. 33–40, Amsterdam, the Netherlands: Elsevier Science Publishers B.V.

Forsen, K. and M. Schantz. 1971. Neue Hauptbestandteile im ätherischen Öl des Rainfarns in Finnland. *Arch. Pharm.*, 304: 944–952.

Forwick-Kreutzer, J., B. M. Moseler, R. Wingender, and J. Wunder. 2003. Intraspecific diversity of wild plants and their importance for nature conservation and agriculture. *Mitt. Biol. Bundesanst. Land-Forstwirtsch.*, 393: 210–215.

Franz, Ch. and I. Wickel. 1985. Zur Vererbung der Bestandteile des Kamillenöls—Qualitative Vererbung von Chamazulen und Bisabolol. *Herba Hung.*, 24: 49–59.

Fraternale, D., D. Ricci, F. Epifano, and M. Curini. 2004. Composition and antifungal activity of two essential oils of hyssop. *J. Essent. Oil Res.*, 16: 617–622.

Fuchs, S., T. Beck, and A. Mosandl. 2000. Biogeneseforschung ätherischer Öle mittels SPME-enentio-MDGC/MS. *GIT Labor-Fachzeitschrift*, 8: 358–362.

Galambosi, B. and P. Peura. 1996. Agrobotanical features and oil content of wild and cultivated forms of caraway (*Carum carvi* L.). *J. Essent. Oil Res.*, 8: 389–397.

Galambosi, B., K. P. Svoboda, A. G. Deans, and É. Héthelyi. 1993. Agronomical and phytochemical investigation of *Hyssopus officinalis*. *Agric. Sci. Finland*, 2: 293–302.

Gildemeister, E. and F. Hoffmann. 1968. *Die ätherischen Öle*, Band 1–6. Berlin, Germany: Akademie Verlag.

Gnilka, R., A. Szumny, and Cz. Wawreńczyk. 2010. Efficient method of isolation of pure (−)-α- and (+)-β-thujone from *Thuja occidentalis* essential oil. In: *Program and Book of Abstracts, 41st ISEO Wroclaw*, Poland, Lochynski, S. and Cz. Wawreńczyk (eds.), p. 69.

Gosztola, B. 2012. Morphological and chemical diversity of different chamomile (*Matricaria recutita* L.) populations of the Great Hungarian Plain (in Hungarian), PhD dissertation. Budapest, Hungary: Corvinus University of Budapest.

Gosztola, B., 2013. Results of an in-house experiment of Dr. Beata Gosztola. Department of Medicinal and Aromatic Plants, Corvinus University of Budapest, Hungary.

Gouyon, P. H., Ph. Vernet, J. L. Guillerm, and G. Valdeyron. 1986. Polymorphism and environment: The adaptive value of the oil polymorphism in *Thymus vulgaris* L. *Heredity*, 57: 59–66.

Grassi, P., J. Novak, H. Steinlesberger, and Ch. Franz. 2004. A direct liquid, non-equilibrium solid-phase micro-extraction application for analysing chemical variation of single peltate trichomes on leaves of *Salvia officinalis*. *Phytochem. Anal.*, 15: 198–203.

Grausgruber-Grüger, S., C. Schmiderer, R. Steinborn, and J. Novak. 2012. Seasonal influence on gene expression of monoterpene synthases in *Salvia officinalis* (Lamiaceae). *J. Plant Physiol.*, 169: 353–359.

Gross, M., E. Lewinsohn, Y. Tadmor, E. Bar, N. Dudai, Y. Cohen, and J. Friedman. 2009. The inheritance of volatile phenylpropenes in bitter fennel (*Foeniculum vulgare* Mill. var. vulgare, Apiaceae) chemotypes and their distribution within the plant. *Biochem. Syst. Ecol.*, 37: 308–316.

Gudaityté, O. and P. R. Venskutonis. 2007. Chemotypes of *Achillea millefolium* transferred from 14 different locations in Lithuania to the controlled environment. *Biochem. Syst. Ecol.*, 35: 582–592.

Guido, S., B. Alessandra, F. Guido, and C. Luigi. 1997. Variability of essential oil composition of *Mentha aquatica* collected in two different habitats of North Tuscany, Italy. *J. Essent. Oil. Res.*, 9: 455–457.

Guimarães, L. G. et al. 2012. Chemical analyses of the essential oils from leaves of *Mikania glauca* Mart. ex Baker. *J. Essent. Oil Res.*, 24: 599–604.

Guitton, Y., F. Nicole, S. Moja, T. Bednabdelkader, N. Valot, S. Legrand, F. Julien, and L. Legendre. 2010. Lavender inflorescence—A model to study regulation of terpene synthesis. *Plant Signal. Behav.*, 5–6: 749–775.

Guo, Y. P., J. Saukel, R. Mittermayr, and F. Ehrendorfer. 2005. AFLP analyses demonstrate genetic divergence, hybridization and multiple polyploidization in the evolution of *Achillea* (Asteraceae-Anthemideae). *New Phytol.*, 166: 273–290.

Hachey, J. G. Collin, M. Gagnon, G. Vernin, and D. Fraisse 1990. Extraction and GC/MS analysis of the essential oil of *Achillea millefolium* complex. *J. Essent. Oil Res.*, 2: 317–326.
Hegnauer, R. 1962. *Chemotaxonomie der Pflanzen*, Vol. 1, p. 114. Basel, Switzerland: Birkhäuser Verlag.
Hegnauer, R. 1978. Die systemische Bedeutung der ätherischen Öle. *Dragoco Rep.*, 24: 203–230.
Hendrics, H., D. J. D. van der Elst, F. M. S. van Putten, and R. Bos. 1990. The essential oil of Dutch tansy. *J. Essent. Oil Res.*, 2: 155–162.
Héthelyi, É., B. Dános, and P. Tétényi. 1989. Phytochemical studies on the essential oils of species belonging to the Achillea genus, by GC/MS. *Biomed. Environ. Mass Spectr.*, 18: 629–636.
Héthelyi, É., P. Tétényi, B. Dános, and I. Koczka. 1991. Phytochemical and antimicrobial studies on the essential oils of the *Tanacetum vulgare* clones by GC/MS. *Herba Hung.*, 30: 82–88.
Hiltunen, R. 1975. Variation and inheritance of some monoterpenes in *Pinus sylvestris*. *Planta Med.*, 28: 315–323.
Hodzsimatov, K. H. and N. Ramazanova. 1974. Some biological characteristics of essential oil accumulation and composition in hyssop cultivated in Taskent region (in Russian). *Rastit. Resursi*, 11: 238–242.
Hofmann, L. 1993. Einfluss von Genotyp, Ontogenese und äusseren Faktoren auf pflanzenbauliche Merkmale sowie ätherische Öle und Flavonoide von Klonen der Schafgarbe (*Achillea millefolium* Aggregat), PhD dissertation. München, Germany: Technische Universität.
Holopainen, M., R. Hiltunen, J. Lokki, K. Forsen, and M. Schantz. 1987. Model for the genetic control of thujone, sabinene and umbellulone in tansy (*Tanacetum vulgare*). *Hereditas*, 106: 205–208.
Ibrahim, A. K., A. A. Safwat, S. E. Khattab, and F. M. El Sherif. 2011. Efficient callus induction, plant regeneration and estragole estimation in tarragon (*Artemisia dracunculus* L.). *J. Essent. Oil Res.*, 23: 16–20.
International Organisation for Standardization. 2013. Standards catalogue, ISO/TC54—Essential oils. http://www.iso.org/iso/home/store/catalogue/tc (accessed March 13, 2014).
Irving, S. and R. P. Adams. 1973. Genetic and biosynthetic relationships of monoterpene. In: *Terpenoids, Structure, Biogenesis and Distribution—Recent Advances in Phytochemistry*, Runeckless, E. C. and T. J. Marby (eds.), Vol. 6, pp. 187–214, New York: Academic Press.
Johnson, C. B., A. Kazantzis, M. Skoula, U. Mitteregger, and J. Novak. 2004. Seasonal, populational and otogenetic variation in the volatile oil content and composition of individuals of *Origanum vulgare* subsp. *hirtum* assessed by GC headspace analysis and by SPME sampling of individual oil glands. *Phytochem. Anal.*, 15: 286–292.
Joulain, D. and M. Ragault. 1976. Sur quelques nouvaux constituants de l'huile essentielle d'*Hyssopus officinalis* L. *Rivista Ital.*, 58: 129–131.
Jovanović, O., N. Radulovits, R. Palić, and B. Zlatković. 2010. Root essential oil of *Achillea lingulata* Waldst. & Kit. (Asteraceae). *J. Essent. Oil Res.*, 22: 336–339.
Judzetiene, A. and J. Budiene. 2010. Compositional variation in essential oils of wild *Artemisia absintium* from Lithuania. *J. Essent. Oil Bearing Plants*, 13: 275–285.
Kindlovits, S. and É. Németh. 2012. Sources of variability of yarrow (*Achillea* spp.) essential oil. *Acta Aliment.*, 41: 92–103.
Kindlovits, S., Sz. Sárosi, K. Inotai, and É. Németh. 2014. Volatile constituents in the roots of different yarrow (*Achillea*) accessions. In: *Program and Abstracts, 45th ISEO*, K.H.C. Baser (ed.), p.106, Istanbul, Turkey.
Kokkini, S., E. Hanlidou, R. Karousou, and T. Lanaras. 2004. Clinal variation of *Mentha pulegium* essential oils along the climatic gradient of Greece. *J. Essent. Oil Res.*, 16: 588–593.
Koller, W. D. and P. Range. 1997. Geruchsprägende Inhaltsstoffe von Fenchel und Ysop. *Z. Arzn. Gew. Pfl.*, 2: 73–80.
Konakchiev, A. and A. Vitkova. 2004. Essential oil composition of *Achillea crithmifolia* Waldst. et Kit. *J. Essent. Oil Bearing Plants*, 7: 32–36.
Kreck, M., A. Scharrer, S. Bilke, and A. Mosandl. 2002. Enantioselective analysis of monoterpene compounds in essential oils stir bar sorptive extraction (SBSE)-enantio-MDGC-MS. *Flavour Fragr. J.*, 17: 32–40.
Kutchan, T. M. 2005. A role for intra- and intercellular translocation in natural product biosynthesis. *Curr. Opin. Biotechnol.*, 8: 292–300.
Labokas, J. and K. Ložienė. 2013. Variation of essential oil yield and relative amounts of enantiomers of α-pinene in leaves and unripe cones of *Juniperus communis* L. growing wild in Lithuania. *J. Essent. Oil Res.*, 25: 244–250.
Lawrence, B. M. 1979. *Progress in Essential Oils 1979–80*. Wheaton, IL: Allured Publishing Corporation.
Lawrence, B. M. 1989. Progress in essential oils. *Perfumer Flav.*, 14: 45–48.
Lawrence, B. M. 1992. Progress in essential oils. *Perfumer Flav.*, 17: 54–55.

Lawrence, B. M. 2007. Oil composition of other *Mentha* species. In: *Mint. The Genus Mentha—Medicinal and Aromatic Plants—Industrial Profiles*, B. M. Lawrence (ed.), pp. 217–232, Boca Raton, FL: CRC Press.

Lazarevic J., N. Radulovic, B. Zlatkovic, and R. Palic. 2010. Composition of *Achillea distans* Willd. subsp. *distans* root essential oil. *Nat. Prod. Res.*, 24: 718–731.

Lewinsohn, E. and M. Gijzen. 2009. Phytochemical diversity: The sounds of silent metabolism. *Plant Sci.*, 176: 161–169.

Linhart, Y. B. and J. D. Thompson. 1995. Terpene-based selective herbivory by *Helix aspersa* (Mollusca) on *Thymus vulgaris* (Labiatae). *Oecologia*, 102: 126–132.

Llorens-Molina, J. A., S. Vaces, and H. Boira. 2012. Seasonal variation of essential oil composition in a population of *Artemisia absinthium* L. from Teruel (Spain): Individual sampling vs. individual monitoring. In: *Program and Book of Abstracts, 43rd ISEO*, p. 108, Lisbon, Portugal.

Lourenço, P. M. L., A. C. Figueiredo, J. G. Barroso, L. G. Pedro, M. M. Oliveira, S. G. Deans, and J. J. C. Scheffer. 1999. Essential oil from hairy root cultures and from plant roots of *Achillea millefolium*. *Phytochemistry*, 51: 637–642.

Lukas, B., R. Samuel, and J. Novak. 2010. Oregano or marjoram? The enzyme γ-terpinene synthase affects chemotype formation in the genus *Origanum*. *Isr. J. Plant Sci.*, 58: 211–220.

Lupien, S., F. Karp, M. Wildung, and R. Croteau. 1999. Regiospecific cytochrome P450 limonene hydroxylases from mint (*Mentha*) species: cDNA isolation, characterization, and functional expression of (−)-4S-limonene-3-hydroxylase and (−)-4S-limonene-6-hydroxylase. *Arch. Biochem. Biophys.*, 368: 181–192.

Malizia, R. A., S. Molli, D. A. Cardell, and J. A. Retamar. 1996. Essential oil of *Mentha citrata* grown in Argentina. Variation in the composition and yield at full- and post-flowering. *J. Essent. Oil Res.*, 8: 347–349.

McCaskill, D. and R. Croteau. 1998. Bioengineering terpenoid production: Current problems. *Trends Biotechnol.*, 16: 189–203.

McConkey, M., J. Gershenzon, and R. Croteau, R. 2000. Developmental regulation of monoterpene biosynthesis in the grandular trichomes of peppermint (*Mentha x piperita* L.). *Plant Physiol.*, 122: 215–223.

Michler, B., A. Preitschopf, P. Erhard, and C. Arnold. 1992. *Achillea millefolium*: Zusammenhänge zwischen Standortfaktoren, Ploidiegrad, Vorkommen von Proazulenen und Gehalt an Chamazulen im ätherischen Öl. *PZ-Wissenschaft*, 137: 23–29.

Mimica-Dukic, N, O. Gasic, R. Jancic, and G. Kite. 1998. Essential oil composition of some populations of *Mentha arvensis* L. in Serbia and Montenegro. *J. Essent. Oil Res.*, 10: 502–506.

Mockute D. and Judzetiene, A. 2004. Composition of the essential oils of *Tanacetum vulgare* L. growing wild in Vilnius district (Lithuania). *J. Essent. Oil Res.*, 16: 550–553.

Muhlemann, J. K. et al. 2012. Developmental changes in the metabolic network of snap drag on flowers. *PLoS One*, 7(7): e40381. doi: 10.1371/journal.pone.0040381. Epub July 11, 2012 (accessed December 2, 2013).

Murray, M. J. P. Marble, D. Lincoln, and F. W. Hefendehl. 1986. Peppermint oil quality differences and the reasons for them. In: *Flavors and Fragrances: A World Perspective*, Lawrence, B. M., B. D. Mookherjee, and B. D. Willis (eds.), pp. 189–210, Amsterdam, the Netherlands: Elsevier Publishers.

Murray, M. J. P. et al. 1988. In: *Flavors and Fragrances: A World Perspective*, Lawrence, B. M., B. D. Mookherjee, and B. J Willis (eds.), pp. 189–210, Amsterdam: the Netherlands: Elsevier Publishers.

Muselli A., M. Pau, J. M. Desjobert, M. Foddai, M. Usai, and J. Costa. 2009. Volatile constituents of *Achillea ligustica* all by HS-SPME/GC/GC-MS. Comparison with essential oils obtained by hydrodistillation from Corsica and Sardinia. *Cromatographia*, 69: 575–585.

Nagell, A. and F. W. Hefendehl. 1974. Zusammensetzung des ätherischen Öles von *Mentha rotundifolia*. *Planta Med.*, 26: 4–9.

Nano, G. M., C. Bicchi, C. Frattini, and M. Gallino. 1979. Wild piemontese plants. *Planta Med.*, 35: 270–274.

Németh, É. 2005. Essential oil composition of species in the genus *Achillea*. *J. Essent. Oil Res.*, 17: 501–512.

Németh, É. and G. Székely. 2000. Floral biology of medicinal plants II. Lamiaceae species. *Int. J. Hortic. Sci.*, 6: 137–140.

Németh, É., É. Héthelyi, and J. Bernáth. 1994. Comparison studies on *Tanacetum vulgare* L.chemotypes. *J. Herbs, Spices Med. Plants*, 2: 85–92.

Németh, É., J. Bernáth, and É. Héthelyi. 2000. Chemotypes and their stability in *Achillea crithmifolia* W.et K. populations. *J. Essent. Oil Res.*, 12: 53–58.

Németh, É., J. Bernáth, and Zs. Pluhár. 1999. Variability of selected characters of production biology and essential oil accumulation in populations of *Achillea crithmifolia*. *Plant Breeding*, 118: 263–267.

Németh, É., J. Bernáth, and G. Tarján. 2007. Quantitative and qualitative studies of essential oils of Hungarian *Achillea* populations. *J. Herbs Spices Med. Plants*, 13: 57–69.

Zámbori-Németh, É., 1990. *Biological background of the cultivation of* Mentha piperita, Salvia sclarea and Tanacetum vulgare. PhD Dissertation, Academie of Sciences, Budapest, Hungary, p. 105.

Novák, I. 2006. Possibilities for improving production and quality of lovage (*Levisticum officinale* Koch.) drugs, PhD dissertation. Budapest, Hungary: Corvinus University of Budapest.

Orav, A., T. Kailas, and K. Ivask. 2001. Composition of the essential oil from *Achillea millefolium* from Estonia. *J. Essent. Oil Res.*, 13: 290–294.

Orth, M., F. Czygan, and V. P. Dedkov. 1999. Variation in essential oil composition and chiral monoterpenes of *Achillea millefolium* from Iran. *J. Essent. Oil Res.*, 11: 681–687.

Özek, T., N. Tabanca, F. Demirci, D. E. Wedge, and K. H. C. Baser. 2010. Enantiomeric distribution of some linalool containing essential oils and their biological activities. *Rec. Nat. Prod.*, 4: 180–192.

Palić, R., G. Stojanović, T. Nasković, and N. Ranelović. 2003. Composition and antibacterial activity of *Achillea crithmifolia* and *Achillea nobilis* essential oils. *J. Essent. Oil Res.*, 15: 434–437.

Pank, F., H. Krüger, and R. Quilitzsch. 2008. Ergebnisse zwanzigjähriger rekurrenter Selektion zur Steigerung des Aetherischöl-gehaltes von einjährigem Kümmel (*Carum carvi* L. var. *annuum* hort). *Z. Arzn. Gew. Pfl.*, 13: 24–28.

Perineau, F., L. Ganou, and G. Vilarem. 1992. Studying production of lovage essential oils in a hydrodistillation pilot unit equipped with a cohobation system. *J. Chem. Technol. Biotechnol.*, 53: 165–171.

Piccaglia, R., L. Pace, and E. Tammaro. 1999. Characterisation of essential oils from three Italian ecotypes of hyssop (*Hyssopus officinalis* ssp. *artistatus* Briq.). *J. Essent. Oil Res.*, 11: 693–699.

Pino, J. A., A. Rosado, and V. Fuentes. 1997. Chemical composiition of the essential oil of *Artemisia absinthium* L. from Cuba. *J. Essent. Oil Res.*, 9: 87–89.

Pluhár, Zs., Sz. Sárosi, I. Novák, and G. Kutta. 2008. Essential oil polymorphism of Hungarian common thyme (*Thymus glabrescens* Willd.) populations. *Nat. Prod. Commun.*, 3: 1151–1154.

Pragadheesh, V. S., A. Saroj, A. Yadav, A. Samad, and C. S. Chanotiya. 2013. Compositions, enantiomer characterization and antifungal activity of two *Ocimum* essential oils. *Ind. Crops Prod.*, 50: 333–337.

Puschmann, G., V, Stephani, and D. Fritz. 1992. Investigations on the variability of caraway (*Carum carvi* L.). *Gartenbauwissenschaft*, 57: 275–277.

Putievsky E., A. Paton, E. Lewinsohn, U. Ravid, D. Haimovich, I. Katzir, D. Saadi, and N. Dudai 1999. Crossability and relationship between morphological and chemical varieties of *Ocimum basilicum* L. *J. Herbs Spices Med. Plants*, 6: 11–24.

Putievsky, E., U. Ravid, N. Dudai, and I. Katzir. 1994. A new cultivar of caraway (*Carum carvi* L.) and its essential oil. *J. Herbs Spices Med. Plants*, 2: 85–90.

Raal, A., E. Arak, and A. Orav. 2012. The content and composition of the essential oil found in *Carum carvi* L. commercial fruits obtained from different countries. *J. Essent. Oil Res.*, 24: 53–59.

Radulović, N., B. Zlatković, R. Palić, and G. Stojanović. 2007. Chemotaxonomic significance of the Balkan *Achillea* volatiles. *Nat. Prod. Commun.*, 2: 453–474.

Rauchensteiner F., S. Nejati, I. Werner, S. Glasl, J. Saukel, J. Jurenits, and W. Kubelka. 2002. Determination of taxa of the *Achillea millefolium* group and *Achillea crithmifolia* by morphological and phytochemical methods I. Characterisation of Central European taxa. *Sci. Pharm.*, 70: 199–230.

Rezaeinodehi, A. and S. Khangholi. 2008. Chemical composition of the essential oil of *Artemisia absinthium* growing wild in Iran. *Pak. J. Biol. Sci.*, 11: 946–949.

Rohloff, J., S. Dragland, and R. Mordal. 2003. Chemotypical variation of tansy (*Tanacetum vulgare* L.) from 40 different location in Norway. In: *Program and Book of Abstracts, 34th ISEO*, K.H. Kubeczka (ed.), p. 113, Würzburg, Germany.

Rosengarten, F. 1969. *The Book of spices*. Winnewood, PA: Livingston Publishing Company.

Sadof, C. S. and G. G. Grant. 1997. Monoterpene composition of *Pinus sylvestris* varieties resistant and susceptible to *Dioryctria zimmermani*. *J. Chem. Ecol.*, 23: 1917–1926.

Saukel J. and R. Länger. 1992. Die *Achillea millefolium* Gruppe (Asteraceae) in Mitteleuropa, 1. Problemstellung, Merkmalserhebung und Untersuchungsmaterial. *Phyton-Horn*, 31: 185–207.

Schantz, E. and H. Hörster. 1970. Zusammensetzung des ätherischen Öles von *Thymus vulgaris* and *Thymus marschallianus* in Abhängigkeitt von Blattalter und Jahreszeit. *Planta Med.*, 19: 161–175.

Schantz, M. and K. Forsén. 1971. Begleitstoffe in verschiedenen Chemotypen von *Chrysanthemum vulgare* L. Bernh. I. Reine Thujon- und Campher-Typen. *Farmaseut. Aikakauslehti*, 80: 122–131.

Schantz, M., M. Jarvi, and R. Kaartinen. 1966. Die Veränderungen des ätherischen Öles während der Entwicklung der Blütenkörbchen von *Chrysanthemum vulgare*. *Planta Med.*, 4: 421–435.

Schilcher, H. 2004. *Wirkungsweise und Anwendungsformen der Kamillenblüten*. Berlin, Germany: Berliner Medizinische Verlagsanstalt GmbH.

Schulz, G. and E. Stahl-Biskup. 1991. Essential oils and glycosidic bound volatile from leaves, stems, flowers and roots of *Hyssopus officinalis*. *Flavour Fragr. J.*, 6: 69–73.

Scrivanti, L. R., M. P. Zunino, and J. A. Zygadlo. 2003. *Tagetes minuta* and *Schinus areira* essential oils as allelopathic agents. *Biochem. Syst. Ecol.*, 31: 563–572.

Seidler-Łożykowska, K. 2008. Zmienność morfologiczna genetyczna oraz użytkowa różnych genotypów kminku zwyczajnego (*Carum carvi* L.). *Rozprawy Naukowe*, 390: 47.

Sharopov, S. F., V. Sulaimonova, and W. N. Setzer. 2012. Composition of the essential oil of *Artemisia absinthium* from Tajikistan. *Rec. Nat. Prod.*, 6: 127–134.

Simonnet, X., M. Qennoz, E. Capella, O. Panero, and I. Tonutti. 2012. Agricultural and phytochemical evaluation of *Artemisia absinthium* hybrids. *Acta Hortic.*, 955: 169–172.

Sorsa, M., M. Schantz, J. Lokki, and K. Forsén. 1968. Variability of essential oil components in *Chrysanthemum vulgare* L. in Finland. *Ann. Acad. Sci. Fenn., Series A. IV. Biol.*, 135: 1–12.

Stahl-Biskup, E. and F. Sáez. 2003. *Thyme—The Genus Thymus. Medicinal and Aromatic Plants- Industrial Profiles*. London, U.K.: Taylor and Francis.

Stahl-Biskup, E. and E. M. Wichtmann. 1991. Composition of the essential oils from roots of some *Apiaceae* in relation to the development of their oil duct systems. *Flavour Fragr. J.*, 6: 249–255.

Szebeni, Zs., B. Galambosi, and Y. Holm. 1992. Growth, yield and essential oil content of lovage grown in Finland. *J. Essent. Oil Res.*, 4: 375–380.

Tétényi, P. 1970. *Infraspecific Chemical Taxa of Medicinal Plants*. Budapest, Hungary: Akadémiai Kiadó.

Tétényi, P. 1975. Homology of biosynthetic routes: Base of chemotaxonomy. *Herba Hung.*, 14: 37–42.

Tholl, D. 2006. Terpene synthases and the regulation, diversity and biological roles of terpene metabolism. *Curr. Opin. Biotechnol.*, 9: 297–304.

Tirillini, B., A. Ricci, and R. Pellegrino. 1999. Secretion constituents of leaf glandular trichomes of *Salvia officinalis* L. *J. Essent. Oil Res.*, 11: 565–569.

Todorova M., A. Trendafilova, B. Mikhova, A. Vitkova, and H. Duddeck. 2007. Chemotypes in *Achillea collina* based on sesquiterpene lactone profile. *Phytochemistry*, 68: 1722–1730.

Toncer, O., S. Karaman, S. Kilil, and E. Diraz. 2009. Changes in essential oil composition of Oregano (*Origanum onies* L.) due to diurnal variations at different developmental stages. *Not. Bot. Hort. Agrobot. Cluj.*, 37: 177–181.

Toyota, M., H. Koyama, M. Mizutani, and Y. Asakawa. 1996. (−)-*ent*-spathulenol isolated from liverworts is an artefact. *Phytochemistry*, 41: 1347–1350.

Tsankova, E. T., A. N. Konakchiev, and A. M. Genova. 1993. Chemical composition of the essential oils of two *Hyssopus officinalis* taxa. *J. Essent. Oil Res.*, 5: 609–611.

Tzakou, O., A. Loukis, and N. Argyriadou. 1993. Volatile constituents of *Achillea crithmifolia* flowers from Greece. *J. Essent. Oil Res.*, 5: 345–346.

Valkovszki, N. J. and É. Zámbori-Németh. 2010. Effects of growing conditions on content and composition of the essential oil of annual caraway (*Carum carvi* L. var. *annua*). *Acta Aliment. Hung.*, 40: 235–246.

Vallejo, M. C. G., J. G. Herraiz, M. J. Pérez-Alonso, and A. Velasco-Negueruela. 1995. Volatile oil of *Hyssopus officinalis* L. from Spain. *J. Essent. Oil Res.*, 7: 567–568.

Varga, L., B. Berhardt, B. Gosztola, Sz. Sárosi, and É. Németh-Zámbori. 2012. Essential oil accumulation during ripening process of selected *Apiaceae* species. In: *Proceedings of the Seventh CMAPSEEC Subotica*, Dajic-Stevanovic, Z. and D. Radanovic (eds.), pp. 389–396, Belgrade, Serbia: Inst. Med. Plant Res. Dr. Josif Pančić".

Venskutonis, P. R. 1995. Essential oil composition of some herbs cultivated in Lithuania. In: *Proceedings of the 13th International Congress of Flavours, Fragrances and Essential Oils*, Vol. 2, pp. 108–123, Istanbul.

Vetter, S. and Ch. Franz. 1998. Samenbildung bei Kreuzungen und Selbstungen mit polyploiden *Achillea* Arten. *Z. Arzn. Gew. Pfl.*, 3: 11–14.

Vetter, S., Ch. Franz, S. Glasl, U. Kastner, J. Saukel, and J. Jurenits. 1997. Inheritance of sesquiterpene lactone types within the *Achillea millefolium* complex (Compositae). *Plant Breeding*, 116: 79–82.

Wagner, C., W. Friedt, R. A. Marquard, and F. Ordon. 2005. Molecular analyses on the genetic diversity and inheritance of (−)-α-bisabolol and chamzilene content in tetraploid chamomile (*Chamomilla recutita* (L.) Rausch.). *Plant Sci.*, 169: 917–927.

Webb, H., K. Carsten, L. Rob, J. Hamill, and W. Foley. 2011. The regulation of quantitative variation of foliar terpenes in medicinal tea tree *Melaleuca alternifolia*. *BMC Proc.*, 5 (Suppl. 7): O20.

Wichtl, M. 1997. *Teedrogen und Phytopharmaka*. Stuttgart, Germany: Wissenschaftliche Verlagsgesellschaft.

Yamaura, T., S. Tanaka, M. Tabata. 1992. Localization of the biosynthesis and accumulation of monoterpenoids in glandular trichomes of thyme. *Planta Med.*, 58: 153–158.

Zámboriné, N. É. 2005. Methods of taxonomic investigations on caraway (*Carum carvi* L.), practical importance of single plant characteristics. In: *Widening of the Variety Spectrum in Horticulture*, M. Tóth (ed.), Special issue of *Kertgazdaság*, Corvinus University of Budapest, Hungary, pp. 209–220.

Zámboriné, N. É., D. É. Kertészné, and P. Tétényi. 1987. Propagation of *Tanacetum vulgare* L. *Herba Hung.*, 26: 137–143.

Zámboriné, N. É., J. Bernáth, S. Kindlovits, and Sz. Sárosi. 2012. Individual variability of wormwood (*Artemisia absynthium* L.) essential oil composition. In: *Program and Book of Abstracts, 43rd ISEO*, A.C. Figueiredo, J.G. Baroso, and L.G. Pedro (eds.), p. 105. Centro de Biotechnologia Vegetal, University of Lisbon, Portugal.

Zheljazkov, V., B. Yankov, and V. Topalov. 1996. Comparison of three methods of mint propagation and their effects on the yield of fresh material and essential oil. *J. Essent. Oil Res.*, 8: 35–45.

5 Production of Essential Oils

Erich Schmidt

CONTENTS

5.1 Introduction ... 127
    5.1.1 General Remarks ... 127
    5.1.2 Definition and History ... 129
    5.1.3 Production ... 132
    5.1.4 Climate ... 133
    5.1.5 Soil Quality and Soil Preparation ... 134
    5.1.6 Water Stress and Drought ... 134
    5.1.7 Insect Stress and Microorganisms ... 134
    5.1.8 Location of Oil Cells ... 134
    5.1.9 Types of Biomass Used ... 135
    5.1.10 Timing of the Harvest ... 135
    5.1.11 Agricultural Crop Establishment ... 136
    5.1.12 Propagation from Seed and Clones ... 138
    5.1.13 Commercial Essential Oil Extraction Methods ... 139
    5.1.14 Expression ... 139
    5.1.15 Steam Distillation ... 143
    5.1.16 Concluding Remarks ... 162
Acknowledgments ... 162
References ... 162

5.1 INTRODUCTION

5.1.1 General Remarks

Essential oils have become an integral part of everyday life. They are used in a great variety of ways: as food flavorings, as feed additives, as flavoring agents by the cigarette industry, and in the compounding of cosmetics and perfumes. Furthermore, they are used in air fresheners and deodorizers as well as in all branches of medicine such as in pharmacy, balneology, massage, and homeopathy. A more specialized area will be in the fields of aromatherapy and aromachology. In recent years, the importance of essential oils as biocides and insect repellents has led to a more detailed study of their antimicrobial potential. Essential oils are also good natural sources of substances with commercial potential as starting materials for chemical synthesis.

Essential oils have been known to mankind for hundreds of years, even millenniums. Long before the fragrances themselves were used, the important action of the oils as remedies was recognized. Without the medical care as we enjoy in our time, self-healing was the only option to combat parasites or the suffering of the human body. Later on essential oils were used in the preparation of early cosmetics, powders, and soaps. As the industrial production of synthetic chemicals started and increased during the nineteenth century, the production of essential oils also increased owing to their importance to our way of life.

The quantities of essential oils produced around the world vary widely. The annual output of some essential oils exceeds 35,000 tons, while that of others may reach only a few kilograms. Some production figures, in metric tons, based on the year 2008 are shown in Table 5.1.

TABLE 5.1
Production Figures of Important Essential Oils (2008)

Essential Oil	Production in Metric Tons (2008)	Main Production Countries
Orange oils	51,000	United States, Brazil, Argentina
Cornmint oil	32,000	India, China, Argentina
Lemon oils	9200	Argentina, Italy, Spain
Eucalyptus oils	4000	China, India, Australia, South Africa
Peppermint oil	3300	India, United States, China
Clove leaf oil	1800	Indonesia, Madagascar
Citronella oil	1800	China, Sri Lanka
Spearmint oils	1800	United States, China
Cedarwood oils	1650	United States, China
Litsea cubeba oil	1200	China
Patchouli oil	1200	Indonesia, India
Lavandin oil Grosso	1100	France
Corymbia citriodora	1000	China, Brazil, India, Vietnam

Source: Perfumer & Flavorist, A preliminary report on the world production of some selected essential oils and countries, Vol. 34, January 2009.

Equally wide variations also occur in the monetary value of different essential oils. Prices range from $1.80/kg for orange oil to $120,000.00/kg for orris oil. The total annual value of the world market is of the order of several billions of USD. A large, but variable, labor force is involved in the production of essential oils. While, in some cases, harvesting and oil production will require just a few workers, other cases will require manual harvesting and may require multiple working steps. Essential oil production either from wild-growing or from cultivated plants is possible almost anywhere, excluding the world's coldest, permanently snow-covered regions. It is estimated that the global number of plant species is of the order of 300,000. About 10% of these contain essential oils and could be used as a source for their production. All continents possess their own characteristic flora with many odor-producing species. Occasionally, these plants may be confined to a particular geographical zone such as *Santalum album* to India and Timor in Indonesia, *Pinus mugo* to the European Alps, or *Abies sibirica* to the Commonwealth of Independent States (CIS, former Russia). For many countries, mainly in Africa and Asia, essential oil production is their main source of exports. Essential oil export figures for Indonesia, Sri Lanka, Vietnam, and even India are very high.

Main producer countries are found in every continent. In Europe, the center of production is situated in the countries bordering the Mediterranean Sea: Italy, Spain, Portugal, France, Croatia, Albania, and Greece, as well as middle-eastern Israel, all of which produce essential oils in industrial quantities. Among Central European countries, Bulgaria, Romania, Hungary, and Ukraine should be mentioned. The huge Russian Federation spread over much of Eastern Europe and Northern Asia has not only nearly endless resources of wild-growing plants but also large areas of cultivated land. The Asian continent with its diversity of climates appears to be the most important producer of essential oils. China and India play a major role followed by Indonesia, Sri Lanka, and Vietnam. Many unique and unusual essential oils originate from the huge Australian continent and from neighboring New Zealand and New Caledonia. Major essential oil–producing countries in Africa include Morocco, Tunisia, Egypt, and Algeria with Ivory Coast, South Africa, Ghana, Kenya, Tanzania, Uganda, and Ethiopia playing a minor role. The important spice-producing islands of Madagascar, the Comoros, Mayotte, and Réunion are situated along the eastern coast of the African continent. The American continent is also one of the

Production of Essential Oils

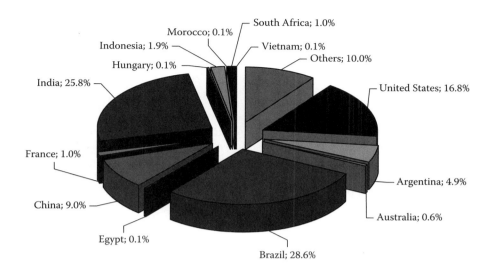

FIGURE 5.1 Production countries and essential oil production worldwide (2008). (Adapted from Perfumer & Flavorist, A preliminary report on the world production of some selected essential oils and countries, Vol. 34, January 2009.)

biggest essential oil producers. The United States, Canada, and Mexico possess a wealth of natural aromatic plant material. In South America, essential oils are produced in Brazil, Argentina, Paraguay, Uruguay, Guatemala, and the Island of Haiti. Apart from the aforementioned major essential oil–producing countries, there are many more, somewhat less important ones, such as Germany, Taiwan, Japan, Jamaica, and the Philippines. Figure 5.1 shows production countries and essential oil production worldwide (2008).

Cultivation of aromatic plants shifted during the last two centuries. From 1850 to 1950, the centers of commercial cultivation of essential oil plants have been the Provence in France, Italy, Spain, and Portugal. With the increase of labor costs, this shifted to the Mediterranean regions of North Africa. As manual harvesting proved too expensive for European conditions, and following improvements in the design of harvesting machinery, only those crops that lend themselves to mechanical harvesting continued to be grown in Europe. In the early 1990s, even North Africa proved too expensive, and the centers of cultivation moved to China and India. At the present time, manual-handling methods are tending to become too costly even in China, and thus India remains as today's center for the cultivation of fragrant plant crops.

5.1.2 Definition and History

Not all odorous extracts of essential oil–bearing plants comply with the International Organization for Standardization (ISO) definition of an "essential oil." An essential oil as defined by the ISO in document ISO 9235.2—aromatic natural raw materials—vocabulary is as follows.

"Product obtained from vegetable raw material—either by distillation with water or steam or—from the epicarp of *Citrus* fruits by a mechanical process, or—by dry distillation" (ISO/DIS 9235.2, 1997, p. 2). Steam distillation can be carried out with or without added water in a still. By contrast, dry distillation of plant material is carried out without the addition of any water or steam to the still (ISO 9235, 1997). Note 2 in Section 3.1.1 of ISO/DIS 9235.2 is of importance. It states that "Essential oils may undergo physical treatments (e.g., re-distillation, aeration) which do not involve significant changes in their composition" (ISO/DIS 9235.2, 1997, p. 2).

An alternative definition of essential oils, established by Professor Dr. Gerhard Buchbauer of the Institute of Pharmaceutical Chemistry, University of Vienna, includes the following suggestion: "Essential oils are more or less volatile substances with more or less odorous impact, produced either by steam distillation or dry distillation or by means of a mechanical treatment from one single species" (Buchbauer et al., 1994). This appears to suggest that mixing several different plant species within the production process is not allowed. As an example, the addition of lavandin plants to lavender plants will yield a natural essential oil but not a natural lavender essential oil. Likewise, wild-growing varieties of *Thymus* will not result in a thyme oil as different chemotypes will totally change the composition of the oil. It follows that blending of different chemotypes of the same botanical species is inadmissible as it will change the chemical composition and properties of the final product. However, in view of the global acceptance of some specific essential oils, there will be exceptions. For example, oil of geranium, ISO/DIS 4730, is obtained from *Pelargonium* × ssp., for example, from hybrids of uncertain parentage rather than from a single botanical species (ISO/DIS 4731, 2005). It is a well established and important article of commerce and may, thus, be considered to be an acceptable exception. In reality, it is impossible to define "one single species" as many essential oils being found on the market come from different plant species. Even in ISO drafts, it is confirmed that various plants are allowed. There are several examples like rosewood oils, distilled from *Aniba rosaeodora* and *Aniba parviflora*, two different plant species. The same happens with the oil of gum turpentine from China, where mainly *Pinus massoniana* will be used, beside other *Pinus* species. Eucalyptus provides another example: oils produced in Portugal have been produced from hybrids such as *Eucalyptus globulus* ssp. *globulus* × *Eucalyptus globulus* ssp. *bicostata* and *Eucalyptus globulus* ssp. *globulus* × *Eucalyptus globulus* ssp. *Eucalyptus globulus* ssp. *pseudoglobulus*. These subspecies were observed from various botanists as separate species. The Chinese eucalyptus oils coming from the Sichuan province are derived from *Cinnamomum longipaniculatum*. Oil of *Melaleuca* (*terpinen-4-ol type*) is produced from *Melaleuca alternifolia* and in smaller amounts also from *Melaleuca linariifolia* and *Melaleuca dissitiflora*. For the future, this definition must be discussed on the level of ISO rules.

Products obtained by other extraction methods, such as solvent extracts, including supercritical carbon dioxide extracts, concretes or pomades, and absolutes as well as resinoids, and oleoresins are *not essential oils* as they do not comply with the earlier mentioned definition. Likewise, products obtained by enzymic treatment of plant material do not meet the requirements of the definition of an essential oil. There exists, though, at least one exception that ought to be mentioned. The well-known "essential oil" of wine yeast, an important flavor and fragrance ingredient, is derived from a microorganism and not from a plant.

In many instances, the commercial terms used to describe perfumery products as essential oils are either wrong or misleading. So-called artificial essential oils, nature-identical essential oils, reconstructed essential oils, and in some cases even essential oils complying with the constants of pharmacopoeias are merely synthetic mixtures of perfumery ingredients and have nothing to do with pure and natural essential oils.

Opinions differ as to the historical origins of essential oil production. According to some, China has been the cradle of hydrodistillation, while others point to the Indus culture (Levey, 1959; Zahn, 1979). On the other hand, some reports also credit the Arabs as being the inventors of distillation. Some literature reports suggest that the earliest practical apparatus for water distillation has been dated from the Indus Culture of some 5000 years ago. However, no written documents have been found to substantiate these claims (Levey, 1955; Zahn, 1979). The earliest documented records of a method and apparatus of what appears to be a kind of distillation procedure were published by Levy from the high culture of Mesopotamia (Levey, 1959). He described a kind of cooking pot from Tepe Gaure in northeastern Mesopotamia, which differed from the design of cooking pots of that period. It was made of brown clay, 53 cm in diameter and 48 cm high. Its special feature was a channel between the raised edges. The total volume of the pot was 37 L and that of the channel was 2.1 L. As the pot was only half-filled when in use, the process appears to represent a true distillation.

FIGURE 5.2 Reconstruction of the distillation plant from Harappa.

While the Arabs appear to be, apart from the existence of the pot discovered in Mesopotamia, the inventors of hydrodistillation, we ought to go back 3000 years BC.

The archaeological museum of Texila in Pakistan has on exhibit a kind of distillation apparatus made of burnt clay. At first sight, it really has the appearance of a typical distillation apparatus, but it is more likely that at that time, it was used for the purification of water (Rovesti, 1977). Apart from that, the assembly resembles an eighteenth-century distillation plant (Figure 5.2). It was again Levy who demonstrated the importance of the distillation culture. Fire was known to be of greatest importance. Initial heating, the intensity of the heat, and its maintenance at a constant level right down to the cooling process were known to be important parameters. The creative ability to produce natural odors points to the fact that the art of distillation was a serious science in ancient Mesopotamia. While the art of distillation had been undergoing improvements right up to the eighth century, it was never mentioned in connection with essential oils, merely with its usefulness for alchemical or medicinal purposes ("Liber servitorius" of Albukasis). In brief, concentration and purification of alcohol appeared to be its main reason for being in existence, its "raison d'être" (Koll and Kowalczyk, 1957).

The Mesopotamian art of distillation had been revived in ancient Egypt as well as being expanded by the expression of citrus oils. The ancient Egyptians improved these processes largely because of their uses in embalming. They also extracted, in addition to myrrh and storax, the exudates of certain East African coastal species of *Boswellia*, none of which are of course essential oils. The thirteenth century Arabian writer Ad-Dimaschki also provided a description of the distillation process, adding descriptions of the production of distilled rose water as well as of the earliest improved cooling systems. It should be understood that the products of these practices were not essential oils in the present accepted sense but merely fragrant distilled water extracts exhibiting the odor of the plant used.

The next important step in the transfer of the practice of distillation to the Occident, from ancient Egypt to the northern hemisphere, was triggered by the crusades of the Middle Ages from the twelfth century onward. Hieronymus Brunschwyk listed in his treatise *The True Art to Distil* about 25 essential oils produced at that time. Once again, one should treat the expression "essential oils" with caution; it would be more accurate to refer to them as "fragrant alcohols" or "aromatic waters." Improvements in the design of equipment led to an enrichment in the diversity of essential oils derived from starting materials such as cinnamon, sandalwood, and also sage and rosemary (Gildemeister and Hoffmann, 1931).

The first evidence capable to discriminate between volatile oils and odorous fatty oils was provided in the sixteenth century. The availability of printed books facilitated "scientists" seeking guidance on the distillation of essential oils. While knowledge of the science of essential oils did not increase during the seventeenth century, the eighteenth century brought about only small progress in the design of equipment and in refinements of the techniques used. The beginning of the nineteenth century brought about progresses in chemistry, including wet analysis, and restarted again, chiefly in France, in an increased development of hydrodistillation methods. Notwithstanding the "industrial" production of lavender already in progress since the mid-eighteenth century, the real breakthrough occurred at the beginning of the following century. While until then the distillation plant was walled in, now the first moveable apparatus appeared. The "Alambique Vial Gattefossé" was easy to transport and placed near the fields. It resulted in improved product quality and reduced the length of transport. These stills were fired with wood or dried plant material. The first swiveling still pots had also been developed that facilitated the emptying of the still residues. These early stills had a capacity of about 50–100 kg of plant material. Later on, their capacity increased to 1000–1200 kg. At the same time, cooling methods were also improved. These improvements spread all over the northern hemisphere to Bulgaria, Turkey, Italy, Spain, Portugal, and even to northern Africa. The final chapter in the history of distillation of plant material came about with the invention of the "alembic à bain-marie," technically speaking a double-walled distillation plant. Steam was not only passed through the biomass, but was also used to heat the wall of the still. This new method improved the speed of the distillation as well as the quality of the top notes of the essential oils thus produced.

The history of the expression of essential oils from the epicarp of citrus fruits is not nearly as interesting as that of hydrodistillation. This can be attributed to the fact that these expressed fragrance concentrates were more readily available in antiquity as expression could be effected by implements made of wood or stone. The chief requirement for this method was manpower, and that was available in unlimited amount. The growth of the industry led to the invention of new mechanical machinery, followed by automation and reduction of manpower. But this topic will be dealt with later on.

5.1.3 PRODUCTION

Before dealing with the basic principles of essential oil production, it is important to be aware of the fact that the essential oil we have in our bottles or drums is not necessarily identical with what is present in the plant. It is wishful thinking, apart for some rare exceptions, to consider an "essential oil" to be the "soul" of the plant and thus an exact replica of what is present in the plant. Only expressed oils that have not come into contact with the fruit juice and that have been protected from aerial oxidation may meet the conditions of a true plant essential oil. The chemical composition of distilled essential oils is not the same as that of the contents of the oil cells present in the plant or with the odor of the plants growing in their natural environment. Headspace technology, a unique method allowing the capture of the volatile constituents of oil cells and thus providing additional information about the plant, has made it possible to detect the volatile components of the plant's "aura." One of the best examples is rose oil. A nonprofessional individual examining pure and natural rose oil on a plotter, even in dilution, will not recognize its plant source. The alteration caused by hydrodistillation is remarkable as plant material in contact with steam undergoes many chemical changes. Hot steam contains more energy than, for example, the surface of the still. Human skin that has come into contact with hot steam suffers tremendous injuries, while short contact with a metal surface at 100°C results merely in a short burning sensation. Hot steam will decompose many aldehydes, and esters may be formed from acids generated during the vaporization of certain essential oil components. Some water-soluble molecules may be lost by solution in the still water, thus altering the fragrance profile of the oil.

Why do so many plants produce essential oils? Certainly neither to regale our nose with pleasant fragrances of rose or lavender nor to heighten the taste (as taste is mostly related to odor) of ginger,

Production of Essential Oils

basil, pepper, thyme, or oregano in our food! Nor to cure diseases of the human body or influence human behavior! Most essential oils contain compounds possessing antimicrobial properties, active against viruses, bacteria, and fungi. Often, different parts of the same plant, such as leaves, roots, and flowers, may contain volatile oils of different chemical composition. Even the height of a plant may play a role. For example, the volatile oil obtained from the gum of the trunk of *Pinus pinaster* at a height of 2 m will contain mainly pinenes and significant car-3-ene, while oil obtained from the gum collected at a height of 4 m will contain very little or no car-3-ene. The reason for this may be protection from deer that browse the bark during the winter months. Some essential oils may act not only as insect repellents but even prevent their reproduction. In many cases, it has been shown that plants attract insects that in turn assist in pollinating the plant. It has also been shown that some plants communicate through the agency of their essential oils. Sometimes, essential oils are considered to be simply metabolic waste products! This may be so in the case of eucalypts as the oil cells present in the mature leaves of *Eucalyptus* species are completely isolated and embedded deeply within the leaf structure. In some cases, essential oils act as germination inhibitors thus reducing competition by other plants (Porter, 2001).

Essential oil yields vary widely and are difficult to predict. The highest oil yields are usually associated with balsams and similar resinous plant exudations, such as gurjun, copaiba, elemi, and Peru balsam, where they can reach 30%–70%. Clove buds and nutmeg can yield between 15% and 17% of essential oil, while other examples worthy of mention are cardamom (about 8%), patchouli (3.5%) and fennel, star anise, caraway seed, and cumin seed (1%–9%). Much lower oil yields are obtained with juniper berries, where 75 kg of berries are required to produce 1 kg of oil, sage (about 0.15%), and other leaf oils such as geranium (also about 0.15%). Rose petals in 700 kg will yield 1 kg of oil, and 1000 kg of bitter orange flowers is required for the production of also just 1 kg of oil. The yields of expressed fruit peel oils, such as bergamot, orange, and lemon, vary from 0.2% to about 0.5%.

A number of important agronomic factors have to be considered before embarking on the production of essential oils, such as climate, soil type, influence of drought and water stress, and stresses caused by insects and microorganisms, propagation (seed or clones), and cultivation practices. Other important factors include precise knowledge on which part of the biomass is to be used, location of the oil cells within the plant, timing of harvest, method of harvesting, storage, and preparation of the biomass prior to essential oil extraction.

5.1.4 CLIMATE

The most important variables include temperature, number of hours of sunshine, and frequency and magnitude of precipitations. Temperature has a profound effect on the yield and quality of the essential oils, as the following example of lavender will show. The last years in the Provence, too cold at the beginning of growth, were followed by very hot weather and a lack of water. As a result, yields decreased by one-third. The relationship between temperature and humidity is an additional important parameter. Humidity coupled with elevated temperatures produces conditions favorable to the proliferation of insect parasites and, most importantly, microorganisms. This sometimes causes plants to increase the production of essential oil for their own protection. Letchamo have studied the relationship between temperature and concentration of daylight on the yield of essential oil and found that the quality of the oil was not influenced (Letchamo et al., 1994). Herbs and spices usually require greater amounts of sunlight. The duration of sunshine in the main areas of herb and spice cultivation, such as the regions bordering the Mediterranean Sea, usually exceeds 8 h/day. In India, Indonesia, and many parts of China, this is well in excess of this figure, and two or even three crops per year can be achieved. Protection against cooling and heavy winds may be required. Windbreaks provided by rows of trees or bushes and even stone walls are particularly common in southern Europe. In China, the *Litsea cubeba* tree is used for the same purpose. In colder countries, the winter snow cover will protect perennials from frost damage. Short periods of frost with temperatures below −10°C will not be too detrimental to plant survival. However, long exposure to heavy frost at very low subzero temperatures will result in permanent damage to the plant ensuing from a lack of water supply.

5.1.5 Soil Quality and Soil Preparation

Every friend of a good wine is aware of the influence of the soil on the grapes and finally on the quality of the wine. The same applies to essential oil–bearing plants. Some crops, such as lavender, thyme, oregano, and clary sage require meager but lime-rich soils. The Jura Chalk of the Haute Provence is destined to produce a good growth of lavender and is the very reason for the good quality and interesting top note of its oils compared with lavender oils of Bulgarian origin growing on different soil types (Meunier, 1985). Soil pH affects significantly oil yield and oil quality. Figueiredo et al. found that the pH value "strongly influences the solubility of certain elements in the soil. Iron, zinc, copper and manganese are less soluble in alkaline than in acidic soils because they precipitate as hydroxides at high pH values" (Figueiredo et al., 2005). It is essential that farmers determine the limits of the elemental profile of the soil. Furthermore, the spacing of plantings should ensure adequate supply with essential trace elements and nutrients. Selection of the optimum site coupled with a suitable climate plays an important role as they will provide a guarantee for optimum crop and essential oil quality.

5.1.6 Water Stress and Drought

It is well known to every gardener that lack of water, as well as too much water, can influence the growth of plants and even kill them. The tolerance of the biomass to soil moisture should be determined in order to identify the most appropriate site for the growing of the desired plant. Since fungal growth is caused by excess water, most plants require well-drained soils to prevent their roots from rotting and the plant from being damaged, thus adversely affecting essential oil production. Lack of water, for example, dryness, exerts a similar deleterious influence. Flowers are smaller than normal and yields drop. Extreme drought can kill the whole plant as its foliage dries closing down its entire metabolism.

5.1.7 Insect Stress and Microorganisms

Plants are living organisms capable of interacting with neighbor plants and warning them of any incipient danger from insect attack. These warning signals are the result of rapid changes occurring in their essential oil composition, which are then transferred to their neighbors who in turn transmit this information on to their neighbors forcing them to change their oil composition as well. In this way, the insect will come into contact with a chemically modified plant material, which may not suit its feeding habits thus obliging it to leave and look elsewhere. Microorganisms can also significantly change the essential oil composition as shown in the case of elderflower fragrance. Headspace gas chromatography coupled with mass spectroscopy (GC/MS) has shown that linalool, the main constituent of elderflowers, was transformed by a fungus present in the leaves, into linalool oxide. The larvae of Cécidomye (*Thomasissiana lavandula*) damage the lavender plant with a concomitant reduction of oil quality. Mycoplasmose and the fungus *Armillaria mellea* can affect the whole plantation and totally spoil the quality of the oil.

5.1.8 Location of Oil Cells

As already mentioned, the cells containing essential oils can be situated in various parts of the plant. Two different types of essential oil cells are known, superficial cells, for example, glandular hairs located on the surface of the plant, common in many herbs such as oregano, mint, and lavender, and cells embedded in plant tissue, occurring as isolated cells containing the secretions (as in citrus fruit and eucalyptus leaves) or as layers of cells surrounding intercellular space (canals or secretory cavities), for example, resin canals of pine. Professor Dr. Johannes Novak (Institute of Applied Botany, Veterinary University, Vienna) has shown impressive pictures and pointed out that the chemical composition of essential oils contained in neighboring cells (oil glands) could be variable but that the typical composition of a particular essential oil was largely due to the averaging of the enormous number of individual cells present in the plant (Novak, 2005). It has been noted in a publication entitled *Physiological Aspects of Essential Oil Production* that individual oil glands do not always secrete the same type of compound and that the process of secretion can be different (Kamatou et al., 2006). Different approaches to distillation are dictated by the location of the oil glands. Preparation of the biomass to be distilled, temperature, and steam pressure affect the quality of the oil produced.

5.1.9 Types of Biomass Used

Essential oils can occur in many different parts of the plant. They can be present in flowers (rose, lavender, magnolia, bitter orange, and blue chamomile) and leaves (cinnamon, patchouli, petitgrain, clove, perilla, and laurel); sometimes the whole aerial part of the plant is distilled (*Melissa officinalis*, basil, thyme, rosemary, marjoram, verbena, and peppermint). The so-called fruit oils are often extracted from seed, which forms part of the fruit, such as caraway, coriander, cardamom, pepper, dill, and pimento. Citrus oils are extracted from the epicarp of species of *Citrus*, such as lemon, lime, bergamot, grapefruit, bitter orange as well as sweet orange, mandarin, clementine, and tangerine. Fruit or perhaps more correctly berry oils are obtained from juniper and *Schinus* species. The well-known bark oils are obtained from birch, cascarilla, cassia, cinnamon, and massoia. Oil of mace is obtained from the aril, a fleshy cover of the seed of nutmeg (*Myristica fragrans*). Flower buds are used for the production of clove oil. Wood and bark exudations yield an important group of essential oils such as galbanum, incense, myrrh, mastix, and storax, to name but a few. The needles of conifers (leaves) are a source of an important group of essential oils derived from species of *Abies*, *Pinus*, and so on. Wood oils are derived mostly from species of *Santalum* (sandalwood), cedar, amyris, cade, rosewood, agarwood, and guaiac. Finally, roots and rhizomes are the source of oils of orris, valerian, calamus, and angelica.

What happens when the plant is cut? Does it immediately start to die as happens in animals and humans? The water content of a plant ranges from 50% to over 80%. The cutting of a plant interrupts its supply of water and minerals. Its life-sustaining processes slow down and finally stop altogether. The production of enzymes stops, and autooxidative processes start, including an increase in bacterial activity leading to rotting and molding. Color and organoleptic properties, such as fragrance, will also change usually to their detriment. As a consequence of this, unless controlled drying or preparation is acceptable options, treatment of the biomass has to be prompt.

5.1.10 Timing of the Harvest

The timing of the harvest of the herbal crop is one of the most important factors affecting the quality of the essential oil. It is a well-documented fact that the chemical composition changes throughout the life of the plant. Occasionally, it can be a matter of days during which the quality of the essential oil reaches its optimum. Knowledge of the precise time of the onset of flowering often has a great influence on the composition of the oil. The chemical changes occurring during the entire life cycle of Vietnamese *Artemisia vulgaris* have shown that 1,8-cineole and β-pinene contents before flowering were below 10% and 1.2%, respectively, whereas at the end of flowering, they reached values above 24% and 10.4% (Nguyen et al., 2004). These are very large variations indeed occurring during the plant's short life span. In the case of the lavender life cycle, the ester value of the oil is the quality-determining factor. It varies within a wide range and influences the value of the oil.

As a rule of thumb, it is held that its maximum value is reached at a time when about two-thirds of the lavender flowers have opened and, thus, that harvesting should commence. In the past, growers knew exactly when to harvest the biomass. These days, the use of a combination of microdistillation and GC techniques enables rapid testing of the quality of the oil and thus the determination of the optimum time for harvesting to start. Oil yields may in some cases be influenced by the time of harvesting. One of the best examples is rose oil. The petals should be collected in the morning between 6 a.m. and 9 a.m. With rising day temperatures, the oil yield will diminish. In the case of oil glands embedded within the leaf structure, such as in the case of eucalypts and pines, oil yield and oil quality are largely unaffected by the time of harvesting.

5.1.11 Agricultural Crop Establishment

The first step is, in most cases, selection of plant seed that suits best the requirements of the product looked for. Preparation of seedbeds, growing from seed, growing and transplanting of seedlings, and so on should follow well-established agricultural practices. The spacing of rows has to be considered (Kassahun et al., 2011; based on example of peppermint leaves). For example, dill prefers wider row spacing than anise, coriander, or caraway (Novak, 2005). The time required before a crop can be obtained depends on the species used and can be very variable. Citronella and lemongrass may take 7–9 months from the time of planting before the first crop can be harvested, while lavender and lavandin require up to 3 years. The most economical way to extract an essential oil is to transport the harvested biomass directly to the distillery. For some plants, this is the only practical option. *M. officinalis* ("lemon balm") is very prone to drying out and thus to loss of oil yield. Some harvested plant material may require special treatment of the biomass before oil extraction, for example, grinding or chipping, breaking or cutting up into smaller fragments, and sometimes just drying. In some cases, fermentation of the biomass should precede oil extraction. Water contained within the plant material can be named as chemically, physicochemically, and mechanically bound water (Grishin et al., 2003). According to these authors, only the mechanically bound water, which is located on the surface and the capillaries of plants, can be reduced. Drying can be achieved simply by spreading the biomass on the ground where wind movement affects the drying process. Drying can also be carried out by the use of appropriate drying equipment. Drying, too, can affect the quality of the essential oil. Until the middle of the 1980s, cut lavender and lavandin have been dried in the field (Figure 5.3), a process requiring about 3 days. The resulting oils exhibited the typical fine, floral odor; however, oil yields were inferior to yields obtained with fresh material. Compared with the present-day procedure with container harvesting and immediate processing (the so-called vert-broyé), this quality of the oil is greener and harsher and requires some time to harmonize. However, yields are better, and one step in the production process has been eliminated. Clary sage is a good example demonstrating the difference between oils distilled from fresh plant material on the one hand and dried plant material on the other. The chemical differences are clearly shown in Table 5.2. Apart from herbal biomass, fruits and seed may also have to be dried before distillation. These include pepper, coriander, cloves, and pimento berries, as well as certain roots such as vetiver, calamus, lovage, and orris. Clary sage is harvested at the beginning of summer but distilled only at the end of the harvesting season.

Seeds and fruits of the families Apiaceae, Piperaceae, and Myristicaceae usually require grinding up prior to steam distillation. In many cases, the seed has to be dried before comminution takes place. Celery, coriander, dill, ambrette, fennel, and anise belong to the Apiaceae. All varieties of pepper belong to the Piperaceae while nutmeg belongs to the Myristicaceae. The finer the material is ground, the better will be the oil yield and, owing to shorter distillation times, also the quality of the oil. In order to reduce losses of volatiles by evaporation during the comminution of the seed or fruit, the grinding can also be carried out under water, preferably in a closed apparatus. Heartwood samples, such as those of *S. album*, *Santalum spicatum*, and *Santalum austrocaledonicum*, have to be reduced to a very fine powder prior to steam distillation in order to achieve complete recovery

FIGURE 5.3 Lavender drying on the field.

TABLE 5.2
Differences in the Composition of the Essential Oil of Clary Sage Manufactured Fresh and Dried

Component	"Vert Broyee" (%)	Traditional (%)
Myrcene	0.9–1.0	0.9–1.1
Limonene	0.2–0.4	0.3–0.5
Ocimene *cis*	0.3–0.5	0.4–0.6
Ocimene *trans*	0.5–0.7	0.8–1.0
Copaene alpha	0.5–0.7	1.4–1.6
Linalool	13.0–24.0	6.5–13.5
Linalyl acetate	56.0–70.5	62.0–78.0
Caryophyllene beta	1.5–1.8	2.5–3.0
Terpineol alpha	1.0–5.0	Max. 2.1
Neryl acetate	0.6–0.8	0.7–1.0
Germacrene D	1.1–7.5	1.5–12
Geranyl acetate	1.4–1.7	2.2–2.5
Geraniol	1.4–1.7	1.2–1.5
Sclareol	0.4–1.8	0.6–2.8
Minor changes		
Middle changes		
Big changes		

of the essential oil. In some cases, coarse chipping of the wood is adequate for efficient essential oil extraction. This includes cedarwood, amyris, rosewood, birch, guaiac, linaloe, cade, and cabreuva.

Plant material containing small branches as well as foliage, which includes pine needles, has to be coarsely chopped up prior to steam distillation. Examples of such material are juniper branches, *M. alternifolia*, *Corymbia citriodora*, *P. mugo*, *P. pinaster*, *Pinus sylvestris*, *Pinus nigra*, *Abies alba*, *A. sibirica*, and *Abies grandis*, as well as mint and peppermint. Present-day mechanized harvesting methods automatically affect the chopping up of the biomass. This also reduces the volume of the biomass, thus increasing the quantity of material that can be packed into the still and making the process more economical.

It appears that the time when the seed is sown influences both oil yield and essential oil composition, for example, whether it is sown in spring or autumn. Important factors affecting production of plant material are application of fertilizers, herbicides, and pesticides and the availability and kind of pollination agents. In many essential oil–producing countries, no artificial nitrogen, potassium, or phosphorus fertilizers are used. Instead, both in Europe and overseas, the biomass left over after steam distillation is spread in the fields as an organic fertilizer. Court et al. reported, from the field tests conducted with peppermint, that an increase in fertilizer affects plant oil yield. However, higher doses did not result in further increases in oil yield or in changes in the oil composition (Court et al., 1993). Herbicides and pesticides do not appear to influence either oil yield or oil composition. The accumulation of pesticide residues in essential oils has a negative influence on their quality and on their uses. The yield and quality of essential oils are also influenced by the timing and type of pollinating agent. If the flower is ready for pollination, the intensity of its fragrance and the amount of volatiles present are at their maximum. If on the other hand the weather is too cold at the time of flowering, pollination will be adversely affected, and transformation to fruit is unlikely to take place. Such an occurrence has a very significant effect on the plant's metabolism and finally on its essential oil. Grapevine cultivators use the following trick to attract pollinators to their vines. A rose flower placed at the end of each grapevine row attracts pollinators who then also pollinate the unattractive flowers of the grapevine.

5.1.12 Propagation from Seed and Clones

Plants can be grown from seed or propagated asexually by cloning. Lavender plants raised from seed are kept for 1 year in pots before transplanting into the field. It then takes another 3 years before the plantation yields enough flowers for commercial harvesting and steam distillation. Plants of any species raised from seed will exhibit wide genetic variations among the progeny, as exist between the members of any species propagated by sexual means, for example, humans. In the case of lavender (*Lavandula angustifolia*), the composition of the essential oil from individual plants varies from plant to plant or, more precisely, from one genotype to the other. Improvement of the crop by selective breeding of those genotypes that yield the most desirable oil is a very slow process requiring years to accomplish. Charles Denny, who initiated the Tasmanian lavender industry in 1921, selected within 11 years 487 genotypes from a source of 2500 genotypes of *L. angustifolia* for closer examination, narrowing them down to just 13 strains exhibiting large yields of superior oil. Finally, four of these genotypes were grown on a large scale and mixed together in what is called "comunelles." The quality of the oils produced was fairly constant from year to year, both in their physicochemical properties and in their olfactory characteristics (Denny, 1995, private information to the author).

Cloning is the preferred method for the replication of plants having particular, usually commercially desirable, characteristics. Clones are obtained from buds or cuttings of the same individual, and the essential oils, for example, obtained from them are the same, or very similar to those of the parent. Cloning procedures are well established but may vary in their detail among different species. One important advantage of clones is that commercial harvesting may be possible after a shorter time as compared with plantations grown from seed. One risk does exist though. If the mother plant is diseased, all clones will also be affected, and the plantation would have to be destroyed.

No field of agriculture requires such a detailed and comprehensive knowledge of botany and soil science as well as of breeding and propagation methods, harvesting methods, and so on as that of the cultivation of essential oil–bearing plants. The importance of this is evident from the very large amount of scientific research carried out in this field by universities as well as by industry.

5.1.13 Commercial Essential Oil Extraction Methods

There are three methods in use. Expression is probably the oldest of these and is used almost exclusively for the production of *Citrus* oils. The second method, hydrodistillation or steam distillation, is the most commonly used one of the three methods, while dry distillation is used only rarely in some very special cases.

5.1.14 Expression

Cold expression, for example, expression at ambient temperature without the involvement of extraneous heat, was practiced long before humans discovered the process of distillation, probably because the necessary tools for it were readily available. Stones or wooden tools were well suited to breaking the oil cells and freeing their fragrant contents. This method was used almost exclusively for the production of *Citrus* peel oils. *Citrus* and the allied genus *Fortunella* belong to the large family Rutaceae. *Citrus* fruits used for the production of the oils are shown in Table 5.3. *Citrus* fruit cultivation is widely spread all over the world with a suitable climate. Oils with the largest production include orange, lemon, grapefruit, and mandarin. Taking world lemon production as an example, the most important lemon-growing areas in Europe are situated in Italy and Spain with Cyprus and Greece being of much lesser importance. Nearly 90%

TABLE 5.3
Important Essential Oil Production from Plants of the Rutaceae Family

Botanical Term	Expressed	Distilled	Used Plant Parts
Citrus aurantifolia (Christm.) Swingle	Lime oil	Lime oil distilled	Pericarp, fruit juice, or crushed fruits
Citrus aurantium L., syn. *Citrus amara* Link, syn. *Citrus bigaradia* Loisel, syn. *Citrus vulgaris* Risso	Bitter orange oil	Neroli oil, bitter orange petitgrain oil	Flower, pericarp, leaf, and twigs with sometimes little green fruits
Citrus bergamia (Risso et Poit.), *Citrus aurantium* L. ssp. *bergamia* (Wight et Arnott) Engler	Bergamot oil	Bergamot petitgrain oil	Pericarp, leaf, and twigs with sometimes little green fruits
Citrus hystrix DC., syn. *Citrus torosa* Blanco	Kaffir lime oil, combava	Kaffir leaves oil	Pericarp, leaves
Citrus latifolia Tanaka	Lime oil Persian type		Pericarp
Citrus limon (L.) Burm. *f.*	Lemon oil	Lemon petitgrain oil	Flower, pericarp, leaf, and twigs with sometimes little green fruits
Citrus reticulate Blanco syn. *Citrus nobilis* Andrews	Mandarin oil	Mandarin petitgrain oil	Flower, pericarp, leaf, and twigs with sometimes little green fruits
Citrus sinensis (L.) Osbeck, *Citrus djalonis* A. Chevalier	Sweet orange oil		Pericarp
Citrus × paradisi Macfad.	Grapefruit oil		Pericarp

of all lemon fruit produced originates from Sicily where the exceptionally favorable climate enables an almost around-the-year production. There is a winter crop from September to April, a spring crop from February to May, and a summer crop from May to September. The Spanish harvest calendar is very similar. Other production areas in the northern hemisphere are in the United States, particularly in Florida, Arizona, and California, and in Mexico. In the southern hemisphere, large-scale lemon producers are Argentina, Uruguay, and Brazil. Lemon production is also being developed in South Africa, Ivory Coast, and Australia. China promises to become a huge producer of lemon in the future.

The reason for extracting citrus oils from fruit peel using mechanical methods is the relative thermal instability of the aldehydes contained in them. Fatty, for example, aliphatic, aldehydes such as heptanal, octanal, nonanal, decanal, and dodecanal are readily oxidized by atmospheric oxygen, which gives rise to the formation of malodorous carboxylic acids. Likewise, terpenoid aldehydes such as neral, geranial, citronellal, and perillaldehyde as well as the α- and β-sinensals are sensitive to oxidation. Hydrodistillation of citrus fruit yields poor quality oils owing to chemical reactions that can be attributed to heat and acid-initiated degradation of some of the unstable fruit volatiles. Furthermore, some of the terpenic hydrocarbons and esters contained in the peel oils are also sensitive to heat and oxygen. One exception to this does exist. Lime oil of commerce can be either cold pressed or steam distilled. The chemical composition of these two types of oil as well as their odors differs significantly from each other. The expressed citrus peel is normally treated with hot steam in order to recover any essential oil still left over in it. The products of this process, consisting mainly of limonene, are used in the solvent industry. The remaining peel and fruit flesh pulp are used as cattle feed.

The oil cells of citrus fruit are situated just under the surface in the epicarp, also called flavedo, in the colored area of the fruit. Figure 5.4 is a cross section of the different parts of the fruit also showing the juice cells present in the fruit. An essential oil is also present in the juice cells. However, the amount of oil present in the juice cells is very much smaller than the amount present in the flavedo; also their composition differs from each other.

Until the beginning of the twentieth century, industrial production of cold-pressed citrus oils was carried out manually. One has to visualize huge halls with hundreds of workers, men and women, seated on small chairs handling the fruit. First of all, the fruit had to be washed and cut into two halves. The pulp was then removed from the fruit using a sharp-edged spoon, called the "rastrello," and after, the peel was soaked in warm water. The fruit peel was now manually turned inside out so that the epicarp was on the inside and squeezed by hand to break the oil glands, and the oil soaked

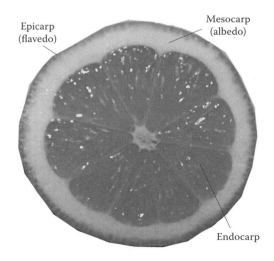

FIGURE 5.4 Parts of a citrus fruit.

up with a sponge. The peel was now turned inside out once again and wiped with the sponge and the sponge squeezed into a terracotta bowl, the "concolina." After decantation, the oil was collected in metal containers. This was an extremely laborious process characterized by substantial oil losses. A later improvement of the fruit peel expression process was the "scodella" method. The apparatus was a metallic hemisphere lined inside with small spikes, with a tube attached at its center. The fruit placed inside the hemisphere was rotated while being squeezed against the spikes thus breaking the oil cells. The oil emulsion, containing some of the wax coating the fruit, flowed into the central tube was collected, and the oil was subsequently separated by centrifugation.

Neither of these methods, even when used simultaneously, was able to satisfy the increased demand for fruit peel oils at the start of the industrial era. The quantity of fruit processed could be increased, but the extraction methods were time wasting and the oil yields too low. With the advent of the twentieth century, the first industrial machinery was developed. Today the only systems of significance in use for the industrial production of peel oils can be classified into four categories: "sfumatrici" machines and "speciale sfumatrici," "pellatrici" machines, "food machinery corporation (FMC) whole fruit process," and "brown oil extractors (BOEs)" (Arnodou, 1991).

It is important to be aware of the fact that the individual oil glands within the epicarp are not connected to neighboring glands. The cell walls of these oil glands are very tough, and it is believed that the oil they contain is either a metabolic waste product or a substance protecting the plant from being browsed by animals.

The machines used in the "sfumatrici" methods consist in principle of two parts, a fixed part and a moveable part. The fruit is cut into two and the flesh is removed. In order to extract the oil, the citrus peel is gently squeezed, by moving it around between the two parts of the device and rinsing off the squeezed-out oil with a jet of water. The oil readily separates from the liquid on standing and is collected by decantation. Since the epicarp may contain organic acids (mainly citric acid followed by malic and oxalic acid), it is occasionally soaked in lime solution in order to neutralize the acids present. Greater concentrations of acid could alter the quality of the oil. Degradation of aldehydes is also an important consideration. In the "special sfumatrici" method, the peel is soaked in the lime solution for 24 h before pressing. By means of a metallic chain drawn by horizontal rollers with ribbed forms, the technical process is finished. The oils obtained by these methods may have to be "wintered," for example, refrigerated in order to freeze out the peel waxes that are then filtered off.

In the "pellatrici" method, the peel oil is removed during the first step and the fruit juice in the second step (Figure 5.5). In the first step, the fruit is fed through a slowly turning Archimedean screw-type valve. The screw is covered with numerous spikes that will bruise the oil cells in the

FIGURE 5.5 "Pellatrici" method. The spiked Archimedes screw with lemons, washed with water.

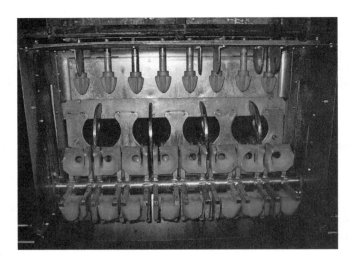

FIGURE 5.6 "Brown" process. A battery of eight juice squeezers waiting for fruits.

epicarp and initiate the flow of oil. The oil is, once again, removed by means of a jet of water. The fruit is finally carried to a fast-rotating, spiked, roller carpet where the remaining oil cells, located deeper within the epicarp, are bruised and their oil content recovered, thus resulting in maximum oil yield. The process involves centrifugation, filtration, and "wintering" as previously mentioned.

The "brown process" (Reeve, 2005) is used mainly in the United States and in South America, but less in Europe. The BOE (Figure 5.6) is somewhat similar to the machinery of the "pellatrici" method. A device at the front end controls the quantity of fruit entering the machine. The machine itself consists of numerous pairs of spiked rollers turning in the same direction, as well as moving horizontally, thus reaching all oil cells. The spiked rollers as well as the fruit are submerged in water for easy transport. Any residual water and oil adhering to the fruit are removed by a special system of rollers and added to the oil emulsion generated on the first set of rollers. Any solid particles are then removed by passing it through a fine sieve. The emulsion is then centrifuged and the aqueous phase recycled. The BOE is manufactured in V4A steel to avoid contact with iron.

The most frequently used type of extractor is the FMC in-line. It is assumed that in the United States, more than 50% of extractors are of the FMC type (Figure 5.7). Other large producer countries,

FIGURE 5.7 Food machinery corporation extractor.

Production of Essential Oils

such as Brazil and Argentina, use exclusively FMC extractors. The reason for this is the design of the machinery, as fruit juice and oil are produced in one step without the two coming into contact with each other. The process requires prior grading of the fruit as the cups used in this process are designed for different sizes of fruit. An optimum fruit size is important as bigger fruit would be over squeezed and some essential oil carried over into the juice making it bitter. On the other hand, if the fruit were too small, the yield of juice would be reduced. Different frame sizes allow treating 3, 5, or 8 fruits at the same time. This technique was revolutionary in its concept and works as follows: the fruit is carried to, and placed into, a fixed cup. Another cup, bearing a mirror image relationship to the fixed cup, is positioned exactly above it. Both cups are built of intermeshing jaws. The moveable cup is lowered toward the fixed cup thus enclosing the fruit. At the same time, a circular knife cuts a hole into the bottom of the fruit. When pressure is applied to the fruit, the expressed juice will exit through the cut hole on to a mesh screen and be transported to the juice manifold, while at the same time, the oil is squeezed out of the surface of the peel. As before, the oil is collected using a jet of water. The oil–water emulsion is then separated by centrifugation.

An examination of the developments in the design of citrus fruit processing machinery shows quite clearly that the quality of the juice was more important than the quality of the oil, the only exception being oil of bergamot. Nevertheless, oil quality improved during the last decades and complies with the requirements of ISO standards. The expressed pulp of the more valuable kind fruit is very often treated with high-pressure steam to recover additional amounts of colorless oils of variable composition. The kinds of fruit treated in this manner are bergamot, lemon, and mandarin.

5.1.15 STEAM DISTILLATION

Steam or water distillation is unquestionably the most frequently used method for the extraction of essential oil from plants. The already mentioned history of steam distillation and the long-standing interest of mankind in extracting the fragrant and useful volatile constituents of plants testify to this. Distillation plants of varying design abound all over the world. While in some developing countries traditional and sometimes rather primitive methods are still being used (Figure 5.8), the essential oils

FIGURE 5.8 Bush distillation device, opened.

produced are often of high quality. Industrialized countries employ technologically more evolved and complex equipment, computer aided with in-process analysis of the final product. Both of these very different ways of commercial essential oil production provide excellent quality oils. One depends on skill and experience, the other on superior technology and expensive equipment. It should be borne in mind that advice by an expert on distillation is a prerequisite for the production of superior quality oils. The term "distillation" is derived from the Latin "distillare," which means "trickling down." In its simplest form, distillation is defined as "evaporation and subsequent condensation of a liquid." All liquids evaporate to a greater or lesser degree, even at room temperature. This is due to thermally induced molecular movements within the liquid resulting in some of the molecules being ejected into the airspace above them (diffusing into the air). As the temperature is increased, these movements increase as well, resulting in more molecules being ejected, for example, in increased evaporation. The definition of an essential oil, ISO 9235, item 3.1.1 is "… product obtained from vegetable raw material—either by distillation with water or steam" and in item 3.1.2 "… obtained with or without added water in the still" (ISO/DIS 9235.2, 1997, p. 2). This means that even "cooking" in the presence of water represents a method suitable for the production of essential oils. The release of the essential oil present in the oil glands (cells) of a plant is due to the bursting of the oil cell walls caused by the increased pressure of the heat-induced expansion of the oil cell contents. The steam flow acts as the carrier of the essential oil molecules. The basic principle of either water or steam distillation is a limit value of a liquid–liquid–vapor system. The theory of hydrodistillation is the following. Two nonmiscible liquids (in our case water and essential oil) A and B form two separate phases. The total vapor pressure of that system is equal to the sum of the partial vapor pressures of the two pure liquids:

$$\rho = \rho_A + \rho_B \quad (\rho \text{ is the total vapor pressure of the system})$$

With complete nonmiscibility of both liquids, r is independent of the composition of the liquid phase. The boiling temperature of the mixture (T_M) lies below the boiling temperatures (T_A and T_B) of liquids A and B. The proportionality between the quantity of each component and the pressure in the vapor phase is given in the formula

$$\frac{N_{oil}}{N_{water}} = \frac{P_{oil}}{P_{water}}$$

where
N_{oil} is the number of moles of the oil in the vapor phase
N_{water} is the number of moles of water in the vapor phase

It is nearly impossible to calculate the proportions as an essential oil is a multicomponent mixture of variable composition.

The simplest method of essential oil extraction is by means of hydrodistillation, for example, by immersion of the biomass in boiling water. The plant material soaks up water during the boiling process, and the oil contained in the oil cells diffuses through the cell walls by means of osmosis. Once the oil has diffused out of the oil cells, it is vaporized and carried away by the stream of steam. The volatility of the oil constituents is not influenced by the rate of vaporization but does depend on the degree of their solubility in water. As a result, the more water-soluble essential components will distil over before the more volatile but less water-soluble ones. The usefulness of hydrodiffusion can be demonstrated by reference to rose oil. It is well known that occasionally some of the essential oil constituents are not present as such in the plant but are artifacts of the extraction process. They can be products of either enzymic splitting or chemical degradation, occurring during the steam distillation, of high-molecular-weight and thus nonvolatile compounds present in the plants. These compounds are often glycosides. The main constituents of rose oil, citronellol, geraniol, and nerol are products of a fermentation that takes place during the water-distillation process.

Hydrolysis of esters to alcohols and acids can occur during steam distillation. This can have serious implications in the case of ester-rich oils, and special precautions have to be taken to prevent or at least to limit the extent of ester degradation. The most important examples of this are lavender or lavandin oils rich in linalyl acetate and cardamom oil rich in α-terpinyl acetate. Chamazulene, a blue bicyclic sesquiterpene, present in the steam-distilled oil of German chamomile, *Chamomilla recutita* (L.) Rauschert, flower heads, is an artifact resulting from matricin by a complex series of chemical reactions: dehydrogenation, dehydration, and ester hydrolysis. As chamazulene is not a particularly stable compound, the deep-blue color of the oil can change to green and even yellow on aging.

The design of a water/steam distillation plant at its simplest, sometimes called "false-bottom apparatus," is as follows: a still pot (a mild steel drum or similar vessel) is fitted with a perforated metal plate or grate, fixed above the intended level of the water, and a lid with a gooseneck outlet. The lid has to be equipped with a gasket or a water seal to prevent steam leaks. The steam outlet is attached to a condenser, for example, a serpentine placed in a drum containing cold water. An oil collector (Florentine flask) placed at the bottom end of the serpentine separates the oil from the distilled water (Figure 5.9). The whole assembly is fixed on a brick fireplace. A separate water inlet is often provided to compensate for water used up during the process. The biomass is placed inside the still pot above the perforated metal plate, and sufficient biomass should be used to completely fill the still pot. The fuel used is firewood. This kind of distillation plant was extensively used at the end of the nineteenth century, mainly for field distillations. A disadvantage of this system was that in some cases excessive heat imparted a burnt smell to the oils. Furthermore, when the water level in the still dropped too much, the plant material could get scorched. Till today, there is a necessity to clean the distillation vessel after two cycles with water to avoid burning notes in the essential oil. In any case, the quality of oils obtained in this type of apparatus was very variable and varied with each distillation. A huge improvement to this process was the introduction of steam generated externally. The early steam generators were very large and unwieldy, and the distillation plant could no longer be transported in the field. The biomass had now to be transported to the distillation plant,

FIGURE 5.9 Old distillation apparatus modernized by electric heating.

unlike with the original type of distillation plant. Originally, the generator was fuelled with dry, extracted biomass. Today, gas or fuel oil is used. The delivery of steam can be carried out in various ways. Most commonly, the steam is led directly into the still through its bottom. Overheating is thus avoided and the biomass is heated rapidly. It also allows regulation of steam quantity and pressure and reduces distillation time and improves oil quality. In another method, the steam is injected in a spiraling motion. This method is more effective as the steam comes into contact with a greater surface of the biomass. The velocity of steam throughput and the duration of the distillation depend on the nature of the biomass. It can vary from 100 kg/h in the case of seed and fruits to 400 kg/h for clary sage. The duration of the distillation can vary from about 20 min for lavender flowers (Denny, 1995, personal communication) to 700 min for dried angelica root. The values quoted are for a 4 m^3 still pot (Omidbaigi, 2005). Specialists on distillation found a formula that distillation can be stopped when the ratio of oil to water coming from condenser will achieve 1:40. In all cases of hydrodistillation, the distillation water is recovered and reused for steam generation. In a cohobation, the aqueous phase of the distillate is continuously reintroduced into the still pot. In this method, any essential oil constituents emulsified or dissolved in the water are captured, thus increasing total oil yield. There is one important exception: in the case of rose oil, the distillation water is collected and redistilled *separately* in a second step. The "floral water" contains increased amounts of β-phenylethyl alcohol, up to 15%, whereas its maximum permissible content in rose oil is 3%. The reason for this is its significant solubility in water, ca. 2%.

The distillation of rose oil is an art in itself as not only quality but also quantity plays an important role. It takes two distillation cycles to produce between 200 and 280 g of rose oil. Jean-François Arnodou describes its manufacture as follows (Arnodou, 1991): the still pot is loaded with 400 kg of rose petals and 1600 L of water. The contents are heated until they boil and steam distilled. Approximately, the quantity of flowers used is then distilled. That action will last about 2–3 h. Specially designed condensers are required in order to obtain a good quality. The condensing system comprises a tubular condenser followed by a second cooler to allow the oil to separate. The oil is collected in Florentine-type oil separators. About 300 L of the oil-saturated still waters are then redistilled in a separate still in order to recover most of the oil contained in them. Both oils are mixed together and constitute the rose oil of commerce. BIOLANDES described in 1991 the whole process, which uses a microprocessor to manage parameters such as pressure and temperature, regulated by servo-controlled pneumatic valves.

A modern distillation plant consists of the biomass container (still pot), a cooling system (condenser), an oil separator, and a high-capacity steam generator. The kettle (still pot) looks like a cylindrical vertical storage tank with steam pipes located at the bottom of the still. Perforated sieve-like plates are often used to separate the plant charge and prevent compaction, thus allowing the steam unimpeded access to the biomass. The outlet for the oil-laden steam is usually incorporated into the design of the usually hemispherical, hinged still pot lid. The steam is then passed through the cooling system, either a plate heat exchanger or a surface heat exchanger, such as a cold-water condenser. The usually liquid condensate is separated into essential oil and distillation water in an appropriate oil separator such as a Florentine flask. The distillation water may, in some cases, be redistilled, and the remaining essential oil is recovered, dried, and stored. Figure 5.10 shows a cross section of such a still.

The following illustrations show different parts of an essential oil production plant. Figure 5.11 shows a battery of four production units in the factory. Each still has a capacity of 3000–5000 L. Owing to their large size, the upper half of the stills is on the level as shown, while the lower half is situated on the lower level. Figure 5.12 shows open stills and displays the steam/oil vapor outlets on the underside of the lids leading to the cooling units. On the right side of the illustration, one can see the perforated plate used to prevent clumping of the biomass. Several such perforated plates, up to 12, depending on the type of biomass, are used to prevent clumping. Spacers on the central upright control the optimum distance between these plates for improved steam penetration. Figure 5.13 shows the unloading of the still. Unloading is much faster than the loading process

Production of Essential Oils

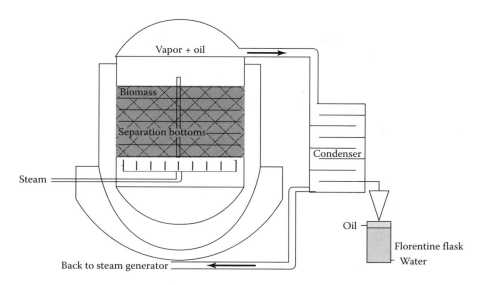

FIGURE 5.10 Cross section of a hydrodistillation plant.

FIGURE 5.11 Battery of four distillation units.

where the biomass is compacted either manually or by means of tractor wheel (Figure 5.14). This type of loading is called "open mouth" loading. Figure 5.15 shows the cooling unit. The cold water enters the tank equipped with a coil condenser. The cooling water is recycled so that no water is wasted. The two main types of industrially used condensers are the following. The earliest was the coil condenser that consisted of a coiled tube fixed in an open vessel of cold water with cold water entering the tank from the bottom and leaving at the top. The oil-rich steam is passed through the coils of the condenser from the top end. The second type of condenser is the pipe bundle condenser where the steam is passed through several vertical tubes immersed in a cold-water tank. The tubes have on the inside walls horizontal protuberances that slow down the rate of the steam flow and thus result in more effective cooling. Figure 5.16 shows the inside of a Florentine flask where the oil is separated from the water. Most essential oils are lighter than water and thus float on top of the water. Some essential oils have a specific gravity >1, for example, they are heavier than water thus

FIGURE 5.12 Open kettle with opening for vapor and oil.

FIGURE 5.13 Unloading a kettle.

FIGURE 5.14 Loading a kettle and pressing by concreted tractor wheel.

FIGURE 5.15 Cooling unit.

collecting at the bottom of the collection vessel. A modified design of the Florentine flask for such oils is shown in Figure 5.17. Figure 5.18 shows oil in the presence of turbid distillation water. The liquid phase is contaminated with biomass matter, and the oil has to be filtered. The capacity of the still pot depends on the biomass. Weights vary from 150 to 650 kg/m^3. Wilted and dried plants are much lighter than seeds and fruits or dried roots that can be very heavy.

A very special case is the production of the essential oils of ylang-ylang from the fresh flowers of *Cananga odorata* (Lam.) Hook. f. et Thomson forma *genuina*. The hydro-distillation process is started and after a certain time the obtained oil is saved. With ongoing distillation, this procedure is repeated three times to achieve at least four separate fractions. The chemical composition of the first

FIGURE 5.16 Inner part of a Florentine flask.

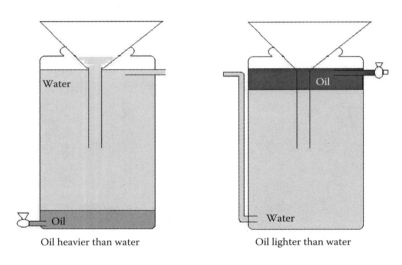

FIGURE 5.17 Two varieties of Florentine flasks.

fraction is characterized by a high concentration of *p*-cresol methylether, methyl benzoate, benzyl acetate, linalool, and *E*-cinnamyl acetate. The second fraction contains less of those volatiles but an increased amount of geraniol, geranyl acetate, and β-caryophyllene. The third fraction contains higher boiling substances such as germacrene-D, (E,E)-α-farnesene, (E,E)-farnesol, benzyl benzoate, (E,E)-farnesyl acetate, and benzyl salicylate. Of course, smaller quantities of the lower boiling components are also present. This kind of fractionation has been practiced for a long time. At the same time, the whole oil, obtained by a single distillation is available as "ylang-ylang complete." This serves as an example of the importance the duration of the distillation can have on the quality of the oil.

Raw materials occurring in the form of hard grains have to be comminuted, for example, ground up before water distillation. This is carried out in the presence of water, such as in a wet-grinding turbine, and the water is used later during the distillation. The stills themselves are equipped with blade stirrers ensuring thorough mixing and particularly dislodging oil particles or biomass articles

FIGURE 5.18 Oil and muddy water in the Florentine flask.

sticking to the walls of the still, the consequence of which can be burning and burnt notes. Dry grinding is likely to result in a significant loss of volatiles. Pepper, coriander, cardamom, celery seed, and angelica seed as well as roots, cumin, caraway, and many other seeds and fruits are treated in this manner. The process used in all these cases is called "turbo distillation." The ratio oil/condensate is very low when this method is used, and it is for that reason that turbo distillation uses a fractionating column to enrich the volatiles. This also assists in preventing small particles of biomass passing into the condenser and contaminating the oil. As in many other distillation and rectification units, cold traps are installed to capture any very volatile oil constituents that may be present. This water-distillation procedure is also used for gums such as myrrh, olibanum, opopanax, and benzoin.

Orris roots are also extracted by water distillation. However, in this case, the distillation has to be carried out under conditions of slightly elevated pressure. This is achieved by means of a reflux column filled with Raschig rings. This is important as the desired constituents, the irones, exhibit very high boiling points. It is noteworthy that in this case, there is no cooling of the vapors, as not only the irones but also the long-chain hydrocarbons will immediately be transported to the top of the column. Figure 5.19 shows a Florentine flask with the condensed oil/water emerging at a temperature of nearly 98°C. Orris oil or orris butter (note that the term orris "concrete" is incorrect, as the process is not a solvent extraction) is one of the few essential oils that are, at least partly, solid at room temperature. Depending on its *trans*-anethole content, rectified star anise oil is another example of this nature.

A relatively new technique that saves time in loading and unloading of the biomass is the "on-site" or "container" distillation. The technique is very simple as the container that is used to pick up the biomass and transport it to the distillery serves itself as the still pot. The first plant crops treated in this way were peppermint and mint, clary sage, lavandin grosso, *L. angustifolia*, *Eucalyptus polybractea*, and tea tree. In its simplest form, the mobile still assembly is composed of the following components: a tractor is coupled to an agricultural harvester that cuts the plant material and delivers it directly into the still pot (or vat) via a chute. The still pot (vat) is permanently fixed onto a trailer that is coupled to the harvester. Once the still pot is completely filled, it is towed by the tractor into the factory where it is uncoupled and attached to the steam supply and condenser and distillation commences. Presupposition for a proper working of the container as vat is a perfect insulation. Every loss of steam and heat will guide to worse quality and diminished quantity. Lids will have to be placed properly and fixed by clamps. The tractor and harvester are attached to an empty still

FIGURE 5.19 Orris distillation, Florentine flask at nearly 98°C.

pot and the process is repeated. The design, shape, and size of the still pot as well as the type of agricultural harvester depend on the type of plant crop, the size of the plantation, the terrain, and so on. The extracted biomass can be used as mulch or, after drying, as fuel for the steam boiler. The unloading is automated using metal chains running over the tubes with steam valves. This method requires less manpower and thus reduces labor costs. Loading and unloading costs are minimal. It lends itself best to fresh biomass, lavender and lavandin, mallee eucalyptus, tea tree, and so on. It may not be as useful for the harvesting and distillation of mint and peppermint as these crops have to be wilted before oil extraction. Figures 5.20 through 5.22 show the harvesting of mallee

FIGURE 5.20 Harvesting blue mallee with distillation container.

Production of Essential Oils

FIGURE 5.21 A battery of containers to be distilled, one opened to show the biomass.

FIGURE 5.22 Distillation plant with container technique.

eucalyptus, containers in processing, and the whole site of container distillation. Figure 5.23 gives a view into the interior of a container.

Another interesting distillation method has been developed by the LBP Freising, Bavaria, Germany. The plant consists of two tubes, each 2 m long and 25 cm in diameter, open at the top. The tubes are attached vertically to a central axis that can be rotated. One tube is connected, hydraulically or mechanically, to the steam generator and on top to a condenser. During the distillation of the contents of the tube, which may take 25–40 min depending on the biomass, the other tube can be loaded. When the distillation of the first tube has been completed, the tubes are rotated around the axis and distillation of the second tube commenced. The first distilled tube can now be emptied and reloaded. The only disadvantage of this type of apparatus is the small size. Only 8.5–21 kg of biomass can be treated. This system has been developed for farmers intending to produce small quantities of essential oil. The apparatus is transportable on a truck and will work satisfactorily provided a supply of power is available.

Most commercially utilized essential oil distillation methods, except the mobile still on-site methods, suffer from high labor costs. Apart from harvesting the biomass, three to four laborers

FIGURE 5.23 View inside the distillation container, showing the steam tubes and the metallic grid.

will be required to load and unload the distillation pots, regulating steam pressure and temperature, and so on. The loading and distribution of the biomass in the distillation vessel may not be homogeneous. This will adversely affect the steam flow through the biomass by channeling, for example, the steam passing through less compacted areas and thus not reaching other more compacted areas. This will result in lower oil yields and perhaps even alter the composition of the oil. In times of high energy costs, the need for consequent recovery has to be considered. Given the demand for greater quantities of essential oils, the question is how to achieve it and at the same time improve the quality of the oils. For this, several considerations have to be taken into account. The first is how to process large quantities of biomass in a given time. Manpower has to be decreased as it still is the most important factor affecting costs. The biomass as a whole has to be treated uniformly to ensure higher oil yields and more constant and thus better oil composition. How can energy costs and water requirements be reduced in an ecologically acceptable way? The answer to this was the development of continuous distillation during the last years of the twentieth century. Until then, all distillation processes were discontinuous. Stills had to be loaded, the distillation stopped, and stills unloaded. The idea was to develop a process where the steam production was continuous with permanent unchanged parameters. This was achieved by the introduction of an endless screw that fed the plant material slowly into the still pot from the top and removed the exhausted plant material from the bottom at the same speed. The plant material moves against the flow of dry steam entering the still from the bottom. In this fashion, all of the biomass comes into contact with the steam ensuring optimum essential oil extraction.

The earliest of these methods is known as the "Padova system." It consists of a still pot 6 m high and about 1.6 m in diameter (Arnodou, 1991). Its total volume is about 8 m^3. The feeding of the still with the plant biomass as well as its subsequent removal is a continuous process. The plant material is delivered via a feed hopper situated at the top of the still. Before entering the still, it is compressed and cut by a rotating knife to ensure a more uniform size. Finally, a horizontal moving cone regulates the quantity of biomass entering the still. The biomass that enters the still moves in the opposite direction to that of the steam. The steam saturated with essential oil vapors is then passed into the cooling system. The exhausted plant material is simultaneously removed by means of an

Archimedes screw. This type of plant was originally designed for the distillation of wine residues. A different system is provided by the "chimie fine aroma process continuous distillation" method. Once again, the plant material is delivered via a hopper to several interconnected tubes. These tubes are slightly inclined and connected to each other. The biomass is carried slowly through the tubes, by means of a worm screw, in a downward direction. Steam is injected at the end of the last tube and is directed upward in the opposite direction to that of the movement of the plant material. The essential oil–laden steam is deflected near the point of entrance of the biomass, into the condenser. The exhausted plant material is unloaded by another worm screw located near the point of the steam entrance to the system.

Texarome, a big producer of cedarwood oil and related products holds a patent on another continuous distillation system. In contrast to other systems, the biomass is conveyed pneumatically within the system. It is a novel system spiked with new technology of that time. Texan cedarwood oil is produced from the whole tree, branches, roots, and stumps. Cedarwood used in Virginia uses exclusively branches, stumps, sawdust, and other waste for oil production; wood is used mainly for furniture making. The wood is passed through a chipper and then through a hammer mill. The dust is collected by means of a cyclone. Any coarse dust is reground to the desired size. The dust is now carried via a plug feeder to the first contactor where superheated steam in reverse flow exhausts it in a first step and following that in a similar second step at the next contactor. The steam and oil vapors are carried into a condenser. The liquid distillate is then separated in Florentine flasks. This process does all transport entirely by pneumatic means. The recycling of cooling water and the use of the dried plant matter as a fuel contribute to environmental requirements (Arnodou, 1991). In the 1990s, the BIOLANDES company designed its own system of continuous distillation. The reason for this was BIOLANDES' engagement in the forests of South West France. The most important area of pine trees (*P. pinaster* Sol.) supplying the paper industry exists between Bordeaux and Biarritz. Twigs and needles have been burnt or left to rot to assist with reforestation with new trees. These needles contain a fine essential oil very similar to that of the dwarf pine oil (*P. mugo* Turra.). Compared to other needle oils, dwarf pine oil is very expensive and greatly appreciated. The oil was produced by a discontinuous distillation but as demand rose, new and improved methods were required. First of all, the collection of the branches had to be improved. A tractor equipped with a crude grinder and a ground wood storage box follows the wood and branch cutters and transports it to the nearby distillation unit where the biomass is exhausted via a continuous distillation process. In contrast to the earlier described methods, the BIOLANDES continuous distillation process operates somewhat differently. The plant material is carried by mechanical means from the storage to the fine cutter and via an Archimedes screw to the top of the distillation pot. The plant material is now compressed by another vertical screw and transported into a chamber that is then hermetically closed on its back but opening at the front. Biomass is falling down allowing the countercurrent passage of hot steam through it. The steam is supplied through numerous nozzles. Endless screws at the bottom of the still continuously dispose of the exhausted biomass. Oil-laden steam is channeled from the top of the still into condenser and then the oil separator.

It is well known that clary sage yields an essential oil on hydrodistillation. However, a very important component of this oil, sclareol, is usually recovered in only very small quantity when this method is used; the reason for this being its very high boiling point. Sclareol can be recovered in very high yield and quality by extraction with volatile solvents. Consequently, BIOLANDES has incorporated an extraction step in its system (Figure 5.24). Any waste biomass, whether of extracted or nonextracted material, is used for energy production or, mostly, for composting. The energy recovery management distinguishes this system from all other earlier described processes. In all of the latter, large amounts of cold water are required to condense the essential oil–laden steam. This results in significant wastage of water as well as in latent energy losses. The BIOLANDES system recovers this latent heat. Hot water from the condenser is carried into an aerodynamic radiator. Air used as the transfer gas takes up the energy of the hot water, cooling it down, so that it can be recycled to the condenser. The hot air is then used to dry about one-third of the biomass waste that

FIGURE 5.24 Scheme of the BIOLANDES continuous production unit. A, biomass; B, distillation vat; C, condenser; D, Florentine flask; E, extraction unit; F, solvent recovery; and G, exhausted biomass.

is used as an energy source for steam and even electricity production. In other words, this system is energetically self-sufficient. Furthermore, since it is fully automated, it results in constant quality products. A unit comprising two stills of 7.5 m^3 capacity can treat per hour 3 ton of pine needles, 1.5 ton of juniper branches, and 0.25 ton of cistus branches (Arnodou, 1991). The advantages are once again short processing time of large amounts of biomass, reduced labor costs, and near-complete energy sufficiency. All operations are automated and water consumption is reduced to a minimum. The system can also operate under slight pressure thus improving the recovery of higher boiling oil constituents.

The following is a controversial method for essential oil extraction by comparison with classical hydrodistillation methods. In this method, the steam enters the distillation chamber from the top passes through the biomass in the still pot (e.g., the distillation chamber) and percolates into the condenser located below it. Separation of the oil from the aqueous phase occurs in a battery of Florentine flasks. It is claimed that this method is very gentle and thus suitable for the treatment of sensitive plants. The biomass is held in the still chamber (e.g., still pot) on a grid that allows easy disposal of the spent plant matter at the completion of the distillation. The whole apparatus is relatively small, distillation times are reduced, and there is less chance of the oil being overheated. It appears that this method is fairly costly and thus likely to be used only for very-high-priced biomass.

Recently, microwave-assisted hydrodistillation methods have been developed, so far mainly in the laboratory or only for small-scale projects. Glass vessels filled with biomass, mainly herbs and fruits or seeds, are heated by microwave power. By controlling the temperature at the center of the vessel, dry heat conditions are established at about 100°C. As the plant material contains enough water, the volatiles are evaporated together with the steam solely generated by the microwave heat and can be collected in a suitably designed condenser/cooling system. In this case, changes in the composition of the oil will be less pronounced than in oil obtained by conventional hydrodistillation. This method has attracted interest owing to the mild heat to which the plant matter is exposed. Kosar reported improvements in the quality of microwave-extracted fennel oil due to increases in the yields of its oxygenated components (Kosar et al., 2007).

Very different products can result from the dry distillation of plant matter. ISO Standard 9235 specifies in Section 3.1.4 that products of dry distillation, for example, "… obtained by distillation without added water or steam" are in fact essential oils (ISO/DIS 9235.2, 1997, p. 2). Dry distillation involves heating in the absence of aerial oxygen, normally in a closed vessel, preventing combustion. The plant material is thus decomposed to new chemical substances. Birch tar from the wood exudate of *Betula pendula* Roth. and cade oil from the wood of *Juniperus oxycedrus* L. are manufactured in this way. Both oils contain phenols, some of which are recognized carcinogens. For this reason, the production of these two oils is no longer of any commercial importance, though very highly rectified and almost phenol-free cade oils do exist.

Some essential oils require rectification. This involves redistillation of the crude oil in order to remove certain undesirable impurities, such as very small amounts of constituents of very low volatility, carried over during the steam or water distillation (such as high-molecular-weight phenols, leaf wax components) as well as small amounts of very volatile compounds exhibiting an undesirable odor, and thus affecting the top note of the oils, such as sulfur compounds (dimethyl sulfide present in crude peppermint oil), isovaleric aldehyde (present in *E. globulus* oil), and certain nitrogenous compounds (low-boiling amines, sulfides, mercaptans, and polysulfides). In some cases, rectification can also be used to enrich the essential oil in a particular component such as 1,8-cineole in low-grade eucalyptus oil. Rectification is usually carried out by redistillation under vacuum to avoid overheating and thus partial decomposition of the oil's constituents. It can also be carried out by steaming. Commonly rectified oils include eucalyptus, clove, mint, turpentine, peppermint, and patchouli. In the case of patchouli and clove oils, rectification improves their, often unacceptably, dark color.

Fractionation of essential oils on a commercial scale is carried out in order to isolate fractions containing a particular compound in very major proportions and occasionally even individual essential oil constituents in a pure state. In order to achieve the required separation, fractionations are conducted under reduced pressure (e.g., under vacuum) to prevent thermal decomposition of the oil constituents, using efficient fractionating columns. A number of different types of fractionating columns are known, but the one most commonly used in laboratory stills or small commercial stills is a glass or stainless steel column filled with Raschig rings. Raschig rings are short, narrow diameter, rings made of glass or any other chemically inert material. Examples of compounds produced on a commercial scale are citral (a mixture of geranial and neral) from *L. cubeba*, 1,8-cineole from eucalyptus oil (mainly *E. polybractea* and other cineole-rich species) as well as from *Cinnamomum camphora* oil, eugenol from clove leaf oil, α-pinene from turpentine, citronellal from citronella oil, linalool from Ho-oil, geraniol from palmarosa oil, and so on. A small-scale high-vacuum plant used for citral production is shown in Figure 5.25. The reflux ratio, for example, the amount of distillate separated and the amount of distillate returned to the still, determines the equilibrium conditions of the vapors near the top of the fractionation column, which are essential for good separation of the oil constituents.

Apart from employing fractional distillation, with or without the application of a vacuum, some essential oil constituents are also obtained on a commercial scale by freezing out from the essential oil, followed by centrifugation at below freezing point of the desired product. Examples are menthol from *Mentha* species (this is usually further purified by recrystallization with a suitable solvent); *trans*-anethole from anise oil, star anise oil, and particularly fennel seed oil; and 1,8-cineole from cineole-rich eucalyptus oils.

Most essential oils are complex mixtures of terpenic and sesquiterpenic hydrocarbons and their oxygenated terpenoid and sesquiterpenoid derivatives (alcohols, aldehydes, ketones, esters, and occasionally carboxylic acids), as well as aromatic (benzenoid) compounds such as phenols, phenolic ethers, and aromatic esters. So-called "terpeneless" and "sesquiterpeneless" essential oils are commonly used in the flavor industry. Many terpenes are bitter in taste, and many, particularly the terpenic hydrocarbons, are poorly soluble or even completely insoluble in water–ethanol mixtures. Since the hydrocarbons rarely contribute anything of importance to their flavoring properties, their removal is a commercial necessity. They are removed by the so-called washing process, a method used mostly for the treatment of citrus oils. This process takes advantage of the different polarities of individual essential oil constituents. The essential oil is added to a carefully selected solvent (usually a water–ethanol solution) and the mixture partitioned by prolonged stirring. This removes some of the more polar oil constituents into the water–ethanol phase (e.g., the solvent phase). Since

FIGURE 5.25 High-vacuum rectification plant in small scale. The distillation assembly is composed of a distillation vessel (1) of glass, placed in an electric heating collar (2). The vessel is surmounted by a jacketed fractionation column (3), packed with glass spirals or Raschig rings, of such a height as to achieve maximum efficiency (e.g., have the maximum number of theoretical plates capable of being achieved for this type of apparatus). The reflux ratio is automatically regulated by a device (4), which also includes the head condenser (5) and a glass tube that leads the product to another condenser (6), from there to both receivers (7). The vacuum pump unit is placed on the right (8).

a single partitioning step is not sufficient to effect complete separation, the whole process has to be repeated several times. The water–ethanol fractions are combined and the solvent removed. The residue contains now very much reduced amounts of hydrocarbons but has been greatly enriched in the desired polar oxygenated flavor constituents, aldehydes such as octanal, nonanal, decanal, hexenal, geranial, and neral; alcohols such as nerol, geraniol, and terpinen-4-ol; oxides such as 1,8-cineole and 1,4-cineole; and esters and sometimes carboxylic acids. Apart from water–ethanol mixtures, hexane or light petroleum fractions (sometimes called "petroleum ether") have sometimes also been added as they will enhance the separation process. However, these are highly flammable liquids, and care has to be taken in their use.

"Folded" or "concentrated" oils are citrus oils from which some of the undesirable components (usually limonene) have been removed by high-vacuum distillation. In order to avoid thermal degradation of the oil, temperatures have to be kept as low as possible. Occasionally, a solvent is used as a "towing" agent to keep the temperature low.

Another, more complex, method for the concentration of citrus oils is a chromatographic separation using packed columns. This method allows a complete elimination of the unwanted hydrocarbons. This method, invented by Erich Ziegler, uses columns packed with either silica or aluminum oxide. The oil is introduced onto the column and the hydrocarbons eluted by means of a suitable nonpolar solvent of very low boiling point. The desirable polar citrus oil components are then washed out using a polar solvent (Ziegler, 1982).

Yet another valuable flavor product of citrus fruits is the "essence oil." The favored method for the transport of citrus juice is in the form of a frozen juice concentrate. The fruit juice is partly dehydrated by distilling off under vacuum the greater part of the water and frozen. Distilling off the water results in significant losses of the desirable volatiles responsible for the aroma of the fruit. These volatiles are captured in several cold traps and constitute the "aqueous essence" or "essence oil" that has the typical fruity and fresh fragrance, but slightly less aldehydic than that of the oil. This oil is used to enhance the flavor of the reconstituted juice obtained by thawing and dilution with water of the frozen concentrate.

Producing essential oils today is, from a marketing point of view, a complex matter. As in the field of other finished products, the requirements of the buyer or producer of the consumer product must be fulfilled. The evaluation of commercial aspects of essential oil production is not an easy task and requires careful consideration. There is no sense in producing oils in oversupply. Areas of short supply, depending on climatic or political circumstances should be identified and acted upon. As in other industries, global trends are an important tool and should continually be monitored. For example, which are the essential oils that cannot be replaced by synthetic substitutes such as patchouli oil or blue chamomile oils? A solution to this problem can lie in the breeding of suitable plants. For example, a producer of a new kind "pastis," the traditional aperitif of France, wants to introduce a new flavor with a rosy note in the fennel component of the flavor. This will require the study and identification of oil constituents with "rosy" notes and help biologists to create new botanical varieties by genetic crossing, for example, by genetic manipulation, of suitable target plant species. Any new lines will first be tested in the laboratory and then in field trials. Test distillations will be carried out and the chemical composition of the oils determined. Agronomists and farmers will be involved in all agricultural aspects of the projects: soil research and harvesting techniques. Variability of all physicochemical aspects of the new strains will be evaluated. At this point, the new types of essential oils will be presented to the client. If the client is satisfied with the quality of the oils, the first larger plantations shall be established, and consumer market research will be initiated. If everything has gone to plan, that is, all technical problems have been successfully resolved and the finished product has met with the approval of the consumers, large-scale production can begin. This example describes the current way of satisfying customer demands.

Global demand for essential oils is on the increase. This also generates some serious problems for which immediate solutions may not easily be found. The first problem is the higher demand for certain essential oils by some of the world's very major producers of cosmetics. They sometimes contract oil quantities that can be of the order of 70% of world production. This will not only raise the price but also restrict consumer access to certain products. From this arises another problem. Our market is to some extent a market of copycats. How can one formulate the fragrance of a competitor's product without having access to the particular essential oil used by him or her, particularly as this oil may have other functions than just being a fragrance, such as certain physiological effects on both the body as well as the mind? Lavender oil from *L. angustifolia* is a calming agent as well as possessing anti-inflammatory activity. No similar or equivalent natural essential oil capable of replacing it is known. Another problem affecting the large global players is ensuring the continuing availability of raw material of the required quality needed to satisfy market demand. This is clearly an almost impossible demand as nobody can assure that climatic conditions required for optimum growth of a particular essential oil crop will remain unchanged. Another problem may be the farmer himself or herself. Sometimes, it may be financially more worthwhile for the farmer to cultivate other than essential oil plant crops. All these factors may have some detrimental effects on the availability of essential oils. Man's responsibility for the continued health of the environment may also be one of the reasons for the disappearance of an essential oil from the market. Sandalwood (species of *Santalum*, but mainly Indian *S. album*) requires in some cases up to 100 years to regenerate to a point where they are large enough to be harvested. This and their uses in religious ceremonies have resulted in significant shortages of Indian oil. Owing to the large monetary value of Indian sandalwood oil, indiscriminate cutting of the wood has just about entirely eliminated it from native forests in Timor (Indonesia). Sandalwood oils of other origins are available, *S. spicatum* from western

Australia and *S. austrocaledonicum* from New Caledonia and Vanuatu. However, their wood oils differ somewhat in odor as well as in chemical composition from genuine Indian oils.

Some essential oils are disappearing from the market owing to the hazardous components they contain and are, therefore, banned from most applications in cosmetics and detergents. These oil components, all of which are labeled as being carcinogenic, include safrole, asarone, methyl eugenol, and elemicin. Plant diseases are another reason for essential oil shortages as they, too, can be affected by a multitude of diseases, some cancerous, which can completely destroy the total crop. For example, French lavender is known to suffer from a condition whereby a particular protein causes a decrease in the growth of the lavender plants. This process could only be slowed down by cultivation at higher altitudes. In the middle of the twentieth century, lavender has been cultivated in the Rhône valley at an altitude of 120 m. Today, lavender is growing only at altitudes around 800 m. The growing shoots of lavender plants are attacked by various pests, in particular the larvae of Cécidomye (*T. lavandula*) that, if unchecked, will defoliate the plants and kill them. Some microorganisms such as *Mycoplasma* and a fungus *A. mellea* can cause serious damage to plantations. At the present time, the use of herbicides and pesticides is an unavoidable necessity. Wild-growing plants are equally prone to attack by insect pests and plant diseases.

The progression from wild-growing plants to essential oil production is an environmental problem. In some developing countries, damage to the natural balance can be traced back to overexploitation of wild-growing plants. Some of these plants are protected worldwide, and their collection, processing, and illegal trading are punishable by law. In some Asian countries, such as in Vietnam, collection from the wild is state controlled and limited to quantities of biomass accruing from natural regeneration.

The state of technical development of the production in the developing countries is very variable and depends largely on the geographical zone they are located in. Areas of particular relevance are Asia, Africa, South America, and Eastern Europe. As a rule, the poorer the country, the more traditional and less technologically sophisticated equipment is used. Generally, standards of the distillation apparatus are those of the 1980s. At that time, the distillation equipment was provided and installed by foreign aid programs with European and American know-how. Most of these units are still in existence and, owing to repairs and improvements by local people, in good working order. Occasionally, primitive equipment has been locally developed, particularly when the state did not provide any financial assistance. Initially, all mastery and expertise of distillation techniques came from Europe, mainly from France. Later on, that knowledge was acquired and transferred to their countries by local people who had studied in Europe. They are no longer dependent on foreign know-how and able to produce oils of constant quality. Conventional hydrodistillation is still the main essential oil extraction method used, one exception being hydrodiffusion often used in Central America, mainly Guatemala and El Salvador, and Brazil in South America. The construction of the equipment is carried out in the country itself and makes the producer independent from higher-priced imports. Steam is generated by oil-burning generators only in the vicinity of cities. In country areas, wood or dried spent biomass is used. As in all other essential oil–producing countries, the distillation plants are close to the cultivation areas. Wild-growing plants are collected, provided the infrastructure exists for their transport to the distillation plant. For certain specific products, permanent fixed distillation plants are used. A forward leap in the technology will be only possible if sufficient investment funds became available in the future. Essential oil quantities produced in those countries are not small, and important specialities such as citral-rich ginger oil from Ecuador play a role on the world market. It should be a compulsory requirement that developing countries treat their wild-growing plant resources with the utmost care. Harvesting has to be controlled to avoid their disappearance from the natural environment and quantities taken adjusted to the ability of the environment to spontaneously regenerate. On the other hand, cultivation will have to be handled with equal care. The avoidance of monoculture will prevent leaching the soil of its nutrients and guard the environment from possible insect propagation. Balanced agricultural practices will lead to a healthy environment and superior quality plants for the production of essential oils.

The following are some pertinent remarks on the now-prevailing views of "green culture" and "organically" grown plants for essential oil production. It is unjustified to suggest that such products are of better quality or greater activity. Comparisons of chemical analyses of "bio-oils," for example, oils from "organically" grown plants, and commercially produced oils show absolutely no differences, qualitative or quantitative, between them. While the concept of pesticide- and fertilizer-free agriculture is desirable and should be supported, the huge worldwide consumption of essential oils could never be satisfied by bio-oils.

Finally, some remarks as to the concept of honesty are attached to the production of natural essential oils. During the last 30 years or so, adulteration of essential oils could be found every day. During the early days, cheap fatty oils (e.g., peanut oil) were used to cut essential oils. Such adulterations were easily revealed by means of placing a drop of the oil on filter paper and allowing it to evaporate (Karg, 1981). While an unadulterated essential oil will evaporate completely or at worst leave only a trace of nonvolatile residue, a greasy patch indicates the presence of a fatty adulterant. As synthetic components of essential oils became available around the turn of the twentieth century, some lavender and lavandin oils have been adulterated by the addition of synthetic linalool and linalyl acetate to the stills before commencing the distillation of the plant material. With the advent of improved analytical methods, such as GC and GC/MS, techniques of adulterating essential oil were also refined. Lavender oil can again serve as an example. Oils distilled from mixtures of lavender and lavandin flowers mimicked the properties of genuine good-quality lavender oils. However, with the introduction of chiral GC techniques, such adulterations were easily identified and the genuineness of the oils guaranteed. This also allowed the verification of the enantiomeric distribution of monoterpenes, monoterpenoid alcohols, and esters present in essential oils. Nuclear magnetic resonance is probably one of the best, but also one of the most expensive, methods available for the authentication of naturalness and will be cost-effective only with large-batch quantities or in the case of very expensive oils. In the future, 2D GC (GC/GC) will provide the next step for the control of naturalness of essential oils.

Another important aspect is the correct botanical source of the essential oil. This can perhaps best be discussed with reference to eucalyptus oil of the 1,8-cineole type. Originally, before commercial eucalyptus oil production commenced in Australia, eucalyptus oil was distilled mainly from *E. globulus* Labill. trees introduced into Europe (mainly Portugal and Spain [ISO Standard 770]). It should be noted that this species exists in several subspecies: *E. globulus* ssp. *bicostata* (Maiden, Blakely, & J. Simm.) Kirkpatr., *E. globulus* Labill. ssp. *globulus.*, *E. globulus* ssp. *pseudoglobulus* (Naudin ex Maiden) Kirkpatr., and *E. globulus* ssp. *maidenii* (F. Muell.) Kirkpatr. It has been shown that the European oils were in fact mixed oils of some of these subspecies and of their hybrids (report by H.H.G. McKern of ISO/TC 54 meeting held in Portugal in 1966). The European Pharmacopoeia Monograph 0390 defines eucalyptus oil as the oil obtained from *E. globulus* Labill., *Eucalyptus fruticetorum* F. von Mueller Syn. *E. polybractea* R.T. Baker (this is the correct botanical name), *Eucalyptus smithii* R.T. Baker, and other species of *Eucalyptus* rich in 1,8-cineole. The Council of Europe's book *Plants in Cosmetics*, Vol. 1, page 127, confuses the matter even further. It entitles the monograph as *E. globulus* Labill. et al. species, for example, and includes any number of unnamed *Eucalyptus* species. The *Pharmacopoeia of the Peoples Republic of China* (English Version, Vol. 1) 1997 goes even further defining eucalyptus oil as the oil obtained from *E. globulus* Labill. and *C. camphora* as well as from other plants of those two families. ISO Standard 3065—Oil of Australian eucalyptus—80%–85% cineole content, simply mentions that the oil is distilled from the appropriate species. The foregoing passage simply shows that Eucalyptus oil does not necessarily have to be distilled from a single species of *Eucalyptus*, for example, *E. globulus*, although suggesting that it is admissible to include 1,8-cineole-rich *Cinnamomum* oils is incorrect and unrealistic. This kind of problem is not unusual or unique. For example, the so-called English lavender oil, considered by many to derive from *L. angustifolia*, is really, in the majority of cases, the hybrid lavandin (Denny, 1995, personal communication).

Another pertinent point is how much twig and leaf material can be used in juniper berry oil. In Indonesia, it is common practice to space individual layers of patchouli leaves in the distillation vessel with twigs of the gurjun tree. Gurjun balsam present in the twigs contains an essential oil that contaminates the patchouli oil. Can this be considered to constitute an adulteration or simply a tool required for the production of the oil?

5.1.16 CONCLUDING REMARKS

As mentioned at the beginning, essential oils do have a future. In spite of regulatory limitations, dangerous substance regulations, and dermatological concerns, and problems with pricing the world production of essentials oil will increase. Essential oils are used in a very large variety of fields. They are an integral constituent of fragrances used in perfumes and cosmetics of all kinds, skin softeners to shower gels and body lotions, and even to "aromatherapy horse care massage oils." They are widely used in the ever-expanding areas of aromatherapy or, better, aromachology. Very large quantities of natural essential oils are used by the food and flavor industries for the flavoring of small goods, fast foods, ice creams, beverages, both alcoholic and nonalcoholic soft drinks, and so on. Their medicinal properties have been known for many years and even centuries. Some possess antibacterial or antifungal activity, while others may assist with the digestion of food. However, as they are multicomponent mixtures of somewhat variable composition, the medicinal use of whole oils has contracted somewhat, the reason being that single essential oil constituents were easier to test for effectiveness and eventual side effects. Despite all that, the use of essential oils is still "number one" on the natural healing scene. With rising health care and medicine costs, self-medication is on the increase and with it a corresponding increase in the consumption of essential oils. Parallel to this, the increase in various esoteric movements is giving rise to further demands for pure natural essential oils.

In the field of agriculture, attempts are being made at the identification of ecologically more friendly natural biocides, including essential oils, to replace synthetic pesticides and herbicides. Essential oils are also used to improve the appetite of farm animals, leading to more rapid increases in body weight as well as to improved digestion.

Finally, some very cheap essential oils or oil components such as limonene, 1,8-cineole, and the pinenes are useful as industrial solvents, while phellandrene-rich eucalyptus oil fractions are marketed as industrial perfumes for detergents and the like.

In conclusion, a "golden future" can be predicted for that useful natural product: the "essential oil"!

ACKNOWLEDGMENTS

The author thanks first of all Dr. Erich Lassak for his tremendous support, for so many detailed information, and also for some pictures; Klaus Dürbeck for some information about production of essential oils in development countries; Dr. Tilmann Miritz, Miritz Citrus Ingredients, for Figures 5.4 through 5.6; Bernhard Mirwald for Figure 5.9; and Tim Denny from Bridestowe Estate, Lilydale, Tasmania, for Figures 5.20 through 5.23.

REFERENCES

Agronomic Characters, Leaf and Essential Oil Yield of Peppermint (Mentha piperita *L.*) *as Influenced by Harvesting Age and Row Spacing, Medicinal and Aromatic Plant Science and Biotechnology.* Global Science Books, p. 1.
Arnodou, J.F., 1991. The taste of nature; industrial methods of natural products extraction. Presented at a Conference organized by the Royal Society of Chemistry in Canterbury, Canterbury, U.K., July 16–19, 1991.

Buchbauer, G., Jäger, W., Jirovetz, L., Nasel, B., Nasel, C., Ilmberger, J., and Diertrich, H., 1994. *25th International Symposium on Essential Oils*. Aromatherapy Research: Studies on the Biological Effects of Fragrance Compounds and Essential Oils upon Inhalation, Grasse, France.

Court, W.A., R.C. Roy, R. Pocs, A.F. More, and P.H. White, 1993. Effects of harvest date on the yield and quality of the essential oil of peppermint. *Can. J. Plant. Sci.*, 73: 815–824.

Denny, T., 1995. Bridestowe estates. Tasmania, Private information to the author.

Dey, D., July 2007. Alberta government, *Agriculture and Food*. Available at http://www1.agric.gov.ab.ca/$department/deptdocs.nsf/all/agdex122.

Figueiredo, A.C., J.G. Barroso, L.G. Pedro, and J.J.C. Scheffer, 2005. Physiological aspects of essential oil production. *Plant Sci.*, 169(6): 1112–1117.

Gildemeister, E. and F. Hoffmann, 1931. *Die Ätherische Öle*. Miltitz, Germany: Verlag Schimmel & Co.

Grishin, A.M., Golovanov, A.N., and Rusakov, S.V., 2003. Evaporation of free water and water bound with forest combustibles under isothermal conditions. *J. Eng. Phys. Thermophys.*, 76(5): 1.

ISO/DIS 4731, 2005. *Oil of Geranium*. Geneva, Switzerland: International Standard Organisation.

ISO/DIS 9235.2, 1997. *Aromatic Natural Raw Materials—Vocabulary*. Geneva, Switzerland: International Standard Organisation.

Kamatou, G.P.P., R.L. van Zyl, S.F. van Vuuren, A.M. Viljoen, A.C. Figueiredo, J.G. Barroso, L.G. Pedro, and P.M. Tilney, 2006. Chemical composition, leaf trichome types and biological activities of the essential oils of four related salvia species indigenous to Southern Africa. *J. Essent. Oil Res.*, 18(Special edition): 72–79.

Karg, J.E., 1981. Das Geschäft mit ätherischen Ölen. *SÖFW, Seifen-Öle-Fette-Wachse*, 107(5/1981): 121–124.

Koll, N. and W. Kowalczyk, 1957. *Fachkunde der Parfümerie und Kosmetik*. Leipzig, Germany: Fachbuchverlag.

Kosar, M., T. Özek, M. Kürkcüoglu, and K.H.C. Baser, 2007. Comparison of microwave-assisted hydrodistillation and hydrodistillation methods for the fruit essential oils of *Foeniculum vulgare*. *J. Essent. Oil Res.*, 19: 426–429.

Letchamo, W., R. Marquard, J. Hölzl, and A. Gosselin, 1994. The selection of Thymus vulgaris cultivars to grow in Canada. *Angewandte Botanik*, 68: 83–88.

Levey, M., 1955. Evidences of ancient distillation, sublimation and extraction in Mesopotamia. *Centaurus*, 4(1): 23–33.

Levey, M., 1959. *Chemistry and Chemical Technology in Ancient Mesopotamia*. Amsterdam, Netherlands: Elsevier.

Meunier, C., 1985. *Lavandes & Lavandins*. Aix-en-Provence, France: ÉDISUD.

Nguyen, T.P.T., T.T. Nguyen, M.H. Tran, H.T. Tran, A. Muselli, A. Bighelli, V. Castola, and J. Casanova, 2004. Artemisia vulgaris L. from Vietnam, chemical variability and composition of the oil along the vegetative life of the plant. *J. Essent. Oil Res.*, 16: 358–361.

Novak, J., 2005. Lecture held on the *35th International Symposium on Essential Oils*, Giardini Naxos, Sicily, Italy.

Omidbaigi, R., 2005. Processing of essential oil plants. In: *Processing, Analysis and Application of Essential Oils*. Har Krishan Bhalla & Sons, Dehradun, India.

Perfumer & Flavorist, 2009. A preliminary report on the world production of some selected essential oils and countries, Vol. 34, January. Carol Stream, IL.

Porter, N., 2001. *Crop and Food Research*. Crop & Foodwatch Research, Christchurch, No. 39, October.

Reeve, D., 2005. A cultivated zest. *Perf. Flav.*, 30(3): 32–35.

Rovesti, P., 1977. Die Destillation ist 5000 Jahre alt. *Dragoco Rep.*, 3: 49–62.

Yanive, Z. and D. Palevitch, 1982. Effect of drought on the secondary metabolites of medicinal and aromatic plants. In: *Cultivation and Utilization of Medicinal Plants*, C.V. Atal and B.M. Kapur (eds.). CSIR, Jammu Tawi, India.

Zahn, J., 1979. *Nichts neues mehr seit Babylon*. Hamburg, Germany: Hoffmann und Campe.

Ziegler, E., 1982. *Die natürlichen und künstlichen Aromen*, pp. 187–188. Heidelberg, Germany: Alfred Hüthig Verlag.

6 Chemistry of Essential Oils

Charles Sell

CONTENTS

6.1 Introduction 165
6.2 Basic Biosynthetic Pathways 165
6.3 Polyketides and Lipids 167
6.4 Shikimic Acid Derivatives... 170
6.5 Terpenoids.. 173
    6.5.1 Hemiterpenoids... 175
    6.5.2 Monoterpenoids 175
    6.5.3 Sesquiterpenoids 179
6.6 Synthesis of Essential Oil Components.. 185
References.. 193

6.1 INTRODUCTION

The term "essential oil" is a contraction of the original "quintessential oil." This stems from the Aristotelian idea that matter is composed of four elements: fire, air, earth, and water. The fifth element, or quintessence, was then considered to be spirit or life force. Distillation and evaporation were thought to be processes of removing the spirit from the plant, and this is also reflected in our language since the term "spirits" is used to describe distilled alcoholic beverages such as brandy, whiskey, and eau de vie. The last of these again shows reference to the concept of removing the life force from the plant. Nowadays, of course, we know that, far from being spirit, essential oils are physical in nature and composed of complex mixtures of chemicals. One thing that we do see from the ancient concepts is that the chemical components of essential oils must be volatile since they are removed by distillation. In order to have boiling points low enough to enable distillation, and atmospheric pressure steam distillation in particular, the essential oil components need to have molecular weights below 300 Da (molecular mass relative to hydrogen = 1) and are usually fairly hydrophobic. Within these constraints, nature has provided an amazingly rich and diverse range of chemicals (Hay and Waterman, 1993; Lawrence, 1985) but there are patterns of molecular structure that give clues to how the molecules were constructed. These synthetic pathways have now been confirmed by experiment and will serve to provide a structure for the contents of this chapter.

6.2 BASIC BIOSYNTHETIC PATHWAYS

The chemicals produced by nature can be classified into two main groups. The primary metabolites are those that are universal across the plant and animal family and constitute the basic building blocks of life. The four subgroups of primary metabolites are proteins, carbohydrates, nucleic acids, and lipids. These families of chemicals contribute little to essential oils although some essential oil components are degradation products of one of these groups, lipids being the most significant. The secondary metabolites are those that occur in some species and not others, and they are usually classified into terpenoids, shikimates, polyketides, and alkaloids. The most important as far as essential oils are concerned are the terpenoids and the shikimates are the second. There are a number of polyketides of importance in essential oils but very few alkaloids. Terpenoids, shikimates, and polyketides will therefore be the main focus of this chapter.

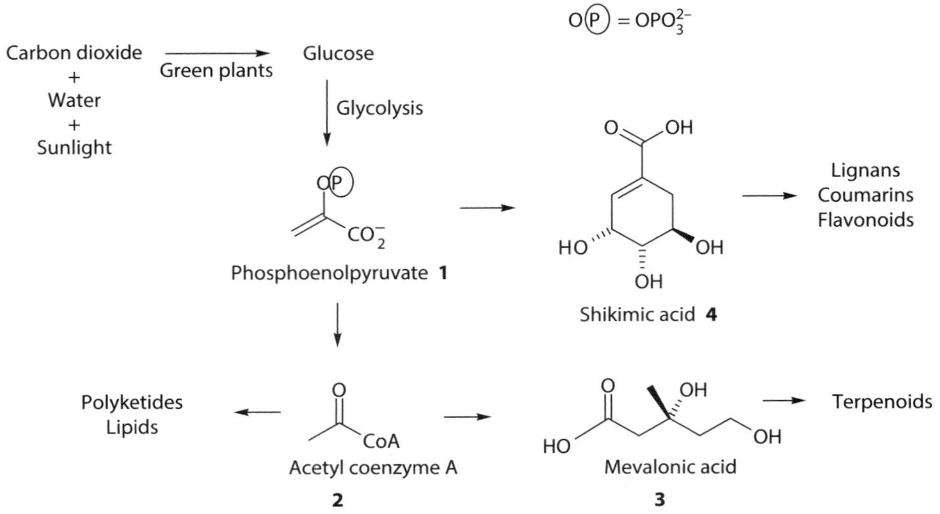

FIGURE 6.1 General pattern of biosynthesis of secondary metabolites.

The general scheme of biosynthetic reactions (Bu'Lock, 1965; Mann et al., 1994) is shown in Figure 6.1. Through photosynthesis, green plants convert carbon dioxide and water into glucose. Cleavage of glucose produces phosphoenolpyruvate (**1**), which is a key building block for the shikimate family of natural products. Decarboxylation of phosphoenolpyruvate gives the two-carbon unit of acetate and this is esterified with coenzyme-A to give acetyl CoA (**2**). Self-condensation of this species leads to the polyketides and lipids. Acetyl CoA is also a starting point for synthesis of mevalonic acid (**3**), which is the key starting material for the terpenoids. In all of these reactions and indeed all the natural chemistry described in this chapter, nature uses the same reactions that chemists do (Sell, 2003). However, nature's reactions tend to be faster and more selective because of the catalysts it uses. These catalysts are called enzymes, and they are globular proteins in which an active site holds the reacting species together. This molecular organization in the active site lowers the activation energy of the reaction and directs its stereochemical course (Lehninger, 1993; Matthews and van Holde, 1990).

Many enzymes need cofactors as reagents or energy providers. Coenzyme-A has already been mentioned earlier. It is a thiol and is used to form thioesters with carboxylic acids. This has two effects on the acid in question. First, the thiolate anion is a better leaving group than alkoxide and so the carbonyl carbon of the thioester is reactive toward nucleophiles. Second, the thioester group increases the acidity of the protons adjacent to the carbonyl group and therefore promotes the formation of the corresponding carbanions. In biosynthesis, a key role of adenosine triphosphate (ATP) is to make phosphate esters of alcohols (phosphorylation). One of the phosphate groups of ATP is added to the alcohol to give the corresponding phosphate ester and adenosine diphosphate. Another group of cofactors of importance to biosynthesis includes pairs such as NADP/NADPH, TPN/TPNH, and DPN/DPNH. These cofactors contain an *N*-alkylated pyridine ring. In each pair, one form comprises an *N*-alkylated pyridinium salt and the other the corresponding *N*-alkyl-1,4-dihydropyridine. The two forms in each pair are interconverted by gain or loss of a hydride anion and therefore constitute redox reagents. In all of the cofactors mentioned here, the reactive part of the molecule is only a small part of the whole. However, the bulk of the molecule has an important role in molecular recognition. The cofactor docks into the active site of the enzyme through recognition, and this holds the cofactor in the optimum spatial configuration relative to the substrate.

Chemistry of Essential Oils

6.3 POLYKETIDES AND LIPIDS

The simplest biosynthetic pathway to appreciate is that of the polyketides and lipids (Bu'Lock, 1965; Mann et al., 1994). The key reaction sequence is shown in Figure 6.2. Acetyl CoA (**2**) is carboxylated to give malonyl CoA (**5**) and the anion of this attacks the CoA ester of a fatty acid. Obviously, the fatty acid could be acetic acid, making this a second molecule of acetyl CoA. After decarboxylation, the product is a b-ketoester with a backbone that is two carbon atoms longer than the first fatty acid. Since this is the route by which fatty acids are produced, it explains why fatty acids are mostly even numbered. If the process is repeated with this new acid as the feedstock, it can be seen that various poly-oxoacids can be built up, each of which will have a carbonyl group on every alternate carbon atom, hence the name polyketides. Alternatively, the ketone function can be reduced to the corresponding alcohol, and then eliminated, and the double bond hydrogenated. This sequence of reactions gives a higher homologue of the starting fatty acid, containing two more carbon atoms in the chain. Long chain fatty acids, whether saturated or unsaturated, are the basis of the lipids.

There are three main paths by which components of essential oils and other natural extracts are formed in this family of metabolites: condensation reactions of polyketides, degradation of lipids, and cyclization of arachidonic acid.

Figure 6.3 shows how condensation of polyketides can lead to phenolic rings. Intramolecular aldol condensation of the tri-keto-octanoic acid and subsequent enolization leads to orsellinic acid (**6**). Polyketide phenols can be distinguished from the phenolic systems of the shikimates by the fact that the former usually retain evidence of oxygenation on alternate carbon atoms, either as acids, ketones, phenols, or as one end of a double bond. The most important natural products containing polyketide phenols are the extracts of oakmoss and tree moss (*Evernia prunastri*). The most significant in odor terms is methyl 3-methylorsellinate (**7**) and ethyl everninate (**8**), which is usually also present in reasonable quantity. Atranol (**9**) and chloratranol (**10**) are minor components but they are skin sensitizers and so limit the usefulness of oakmoss and tree moss extracts, unless they are removed from them. Dimeric esters of orsellinic and everninic acids and analogues also exist in mosses. They are known as depsides and hydrolysis yields the monomers, thus increasing the odor of the sample. However, some depsides, such as atranorin (**11**), are allergens and thus contribute to safety issues with the extracts.

The major metabolic route for fatty acids involves b-oxidation and cleavage giving acetate and a fatty acid with two carbon atoms less than the starting acid, that is, the reverse of the biosynthesis reaction. However, other oxidation routes also exist and these give rise to new metabolites that were

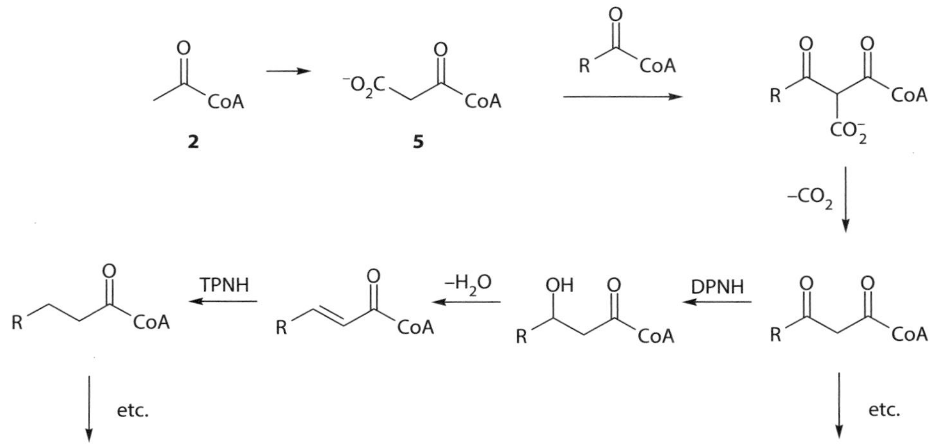

FIGURE 6.2 Polyketide and lipid biosynthesis.

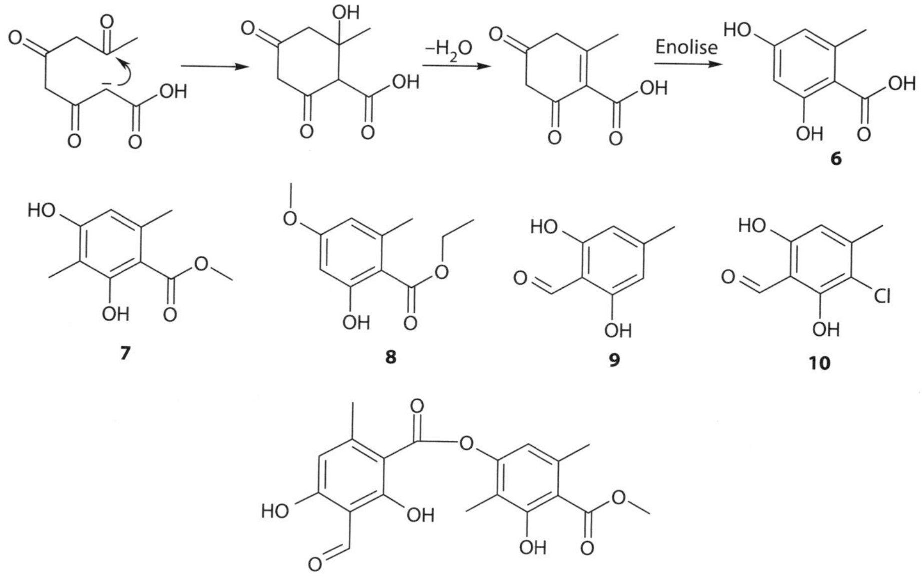

FIGURE 6.3 Polyketide biosynthesis and oakmoss components.

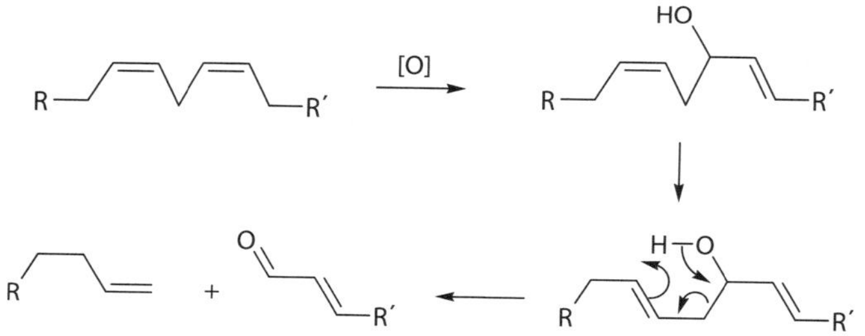

FIGURE 6.4 Fragmentation of polyunsaturated fats to give aldehydes.

not on the biosynthetic pathway. For example, Figure 6.4 shows how allylic oxidation of a dienoic acid and subsequent cleavage can lead to the formation of an aldehyde.

Allylic oxidation followed by lactonization rather than cleavage can, obviously, lead to lactones. Reduction of the acid function to the corresponding alcohols or aldehydes is also possible as are hydrogenation and elimination reactions. Thus, a wide variety of aliphatic entities are made available. Some examples are shown in Figure 6.5 to illustrate the diversity that exists. The hydrocarbon (E,Z)-1,3,5-undecatriene (**12**) is an important contributor to the odor of galbanum. Simple aliphatic alcohols and ethers are found, the occurrence of 1-octanol (**13**) in olibanum and methyl hexyl ether (**14**) in lavender being examples. Aldehydes are often found as significant odor components of oils, for example, decanal (**15**) in orange oil and (E)-4-decenal (**16**) in caraway and cardamom. The ketone 2-nonanone (**17**) that occurs in rue and hexyl propionate (**18**), a component of lavender, is just one of a plethora of esters that are found. The isomeric lactones g-decalactone (**19**) and d-decalactone (**20**) are found in osmanthus (Essential Oils Database, 2006). Acetylenes also occur as essential oil components, often as polyacetylenes such as methyl deca-2-en-4,6,8-triynoate (**21**), which is a component of Artemisia vulgaris.

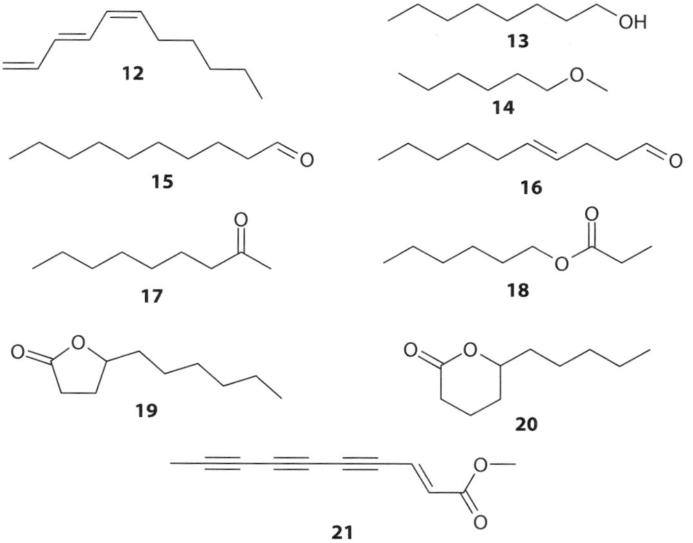

FIGURE 6.5 Some lipid-derived components of essential oils.

Arachidonic acid (**22**) is a polyunsaturated fatty acid that plays a special role as a synthetic intermediate in plants and animals (Mann et al., 1994). As shown in Figure 6.6, allylic oxidation at the 11th carbon of the chain leads to the hydroperoxide (**23**). Further oxidation (at the 15th carbon) with two concomitant cyclization reactions gives the cyclic peroxide (**24**). This is a key intermediate for the biosynthesis of prostaglandins such as 6-ketoprostaglandin F_{1a} (**25**) and

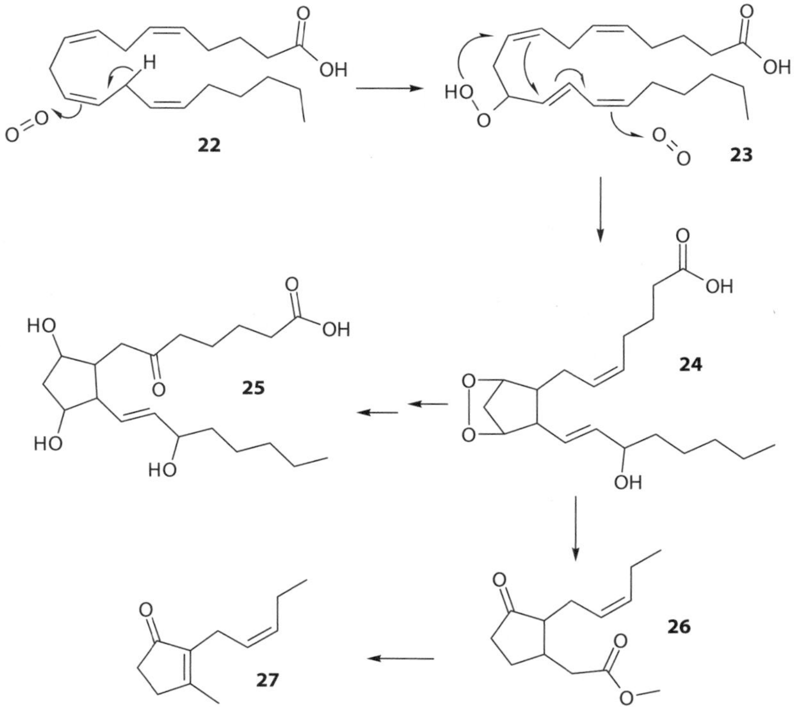

FIGURE 6.6 Biosynthesis of prostaglandins and jasmines.

also for methyl jasmonate (**26**). The latter is the methyl ester of jasmonic acid, a plant hormone, and is a significant odor component of jasmine, as is jasmone (**27**), a product of degradation of jasmonic acid.

6.4 SHIKIMIC ACID DERIVATIVES

Shikimic acid (**4**) is a key synthetic intermediate for plants since it is the key precursor for both the flavonoids and lignin (Bu'Lock, 1965; Mann et al., 1994). The flavonoids are important to plants as antioxidants, colors, protective agents against ultraviolet light, and the like, and lignin is a key component of the structural materials of plants, especially woody tissues. Shikimic acid is synthesized from phosphoenolpyruvate (**1**) and erythrose 4-phosphate (**28**), as shown in Figure 6.7, and thus its biosynthesis starts from the carbohydrate pathway. Its derivatives can usually be recognized by the characteristic shikimate pattern of a six-membered ring with either a one- or three-carbon substituent on position one and oxygenation in the third, and/or fourth, and/or fifth positions. However, the oxygen atoms of the final products are not those of the starting shikimate since these are lost initially and then replaced.

Figure 6.8 shows some of the biosynthetic intermediates stemming from shikimic acid (**4**) and which are of importance in terms of generating materials volatile enough to be essential oil components. Elimination of one of the ring alcohols and reaction with phosphoenolpyruvate (**1**) gives chorismic acid (**29**) that can undergo an oxy-Cope reaction to give prephenic acid (**30**). Decarboxylation and elimination of the ring alcohol now gives the phenylpropionic acid skeleton. Amination and reduction of the ketone function gives the essential amino acid phenylalanine (**31**), whereas reduction and elimination leads to cinnamic acid (**32**). Ring hydroxylation of the latter gives the isomeric *o*- and *p*-coumaric acids, (**33**) and (**34**), respectively. Further hydroxylation gives caffeic acid (**35**) and methylation of this gives ferulic acid (**36**). Oxidation of the methyl ether of the latter and subsequent cyclization gives methylenecaffeic acid (**37**). In shikimate biosynthesis, it is often possible to arrive at a given product by different sequences of the same reactions, and the exact route used will depend on the genetic makeup of the plant.

Aromatization of shikimic acid, without addition of the three additional carbon atoms from phosphoenolpyruvate, gives benzoic acid derivatives. Benzoic acid itself occurs in some oils and its

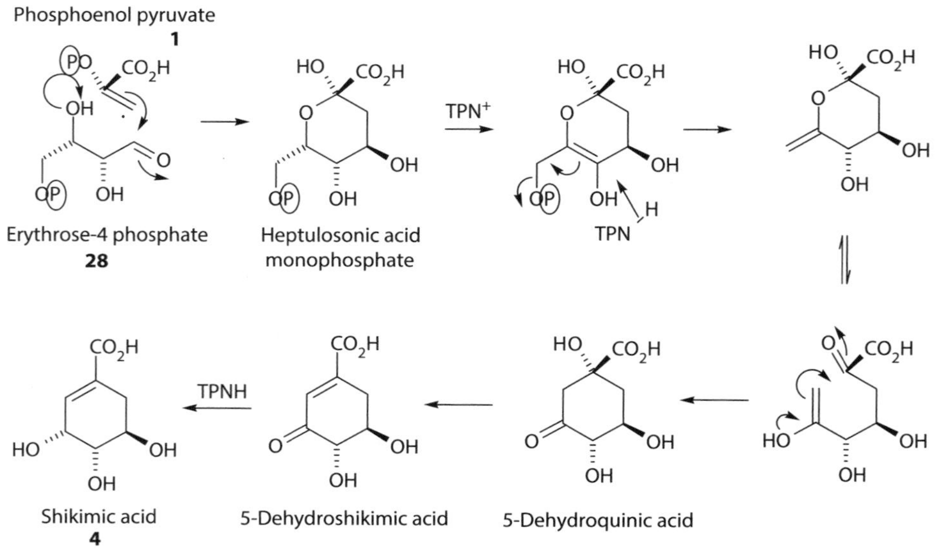

FIGURE 6.7 Biosynthesis of shikimic acid.

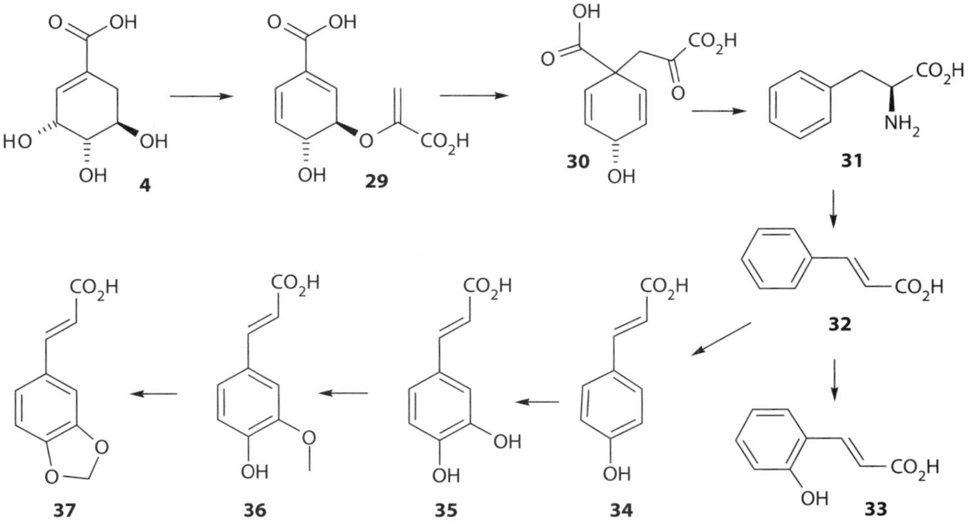

FIGURE 6.8 Key intermediates for shikimic acid.

esters are widespread. For example, methyl benzoate is found in tuberose, ylang ylang, and various lilies. Even more common are benzyl alcohol, benzaldehyde, and their derivatives (Arctander, 1960; Essential Oils Database, 2006; Gildemeister and Hoffmann, 1956; Günther, 1948). Benzyl alcohol occurs in muguet, jasmine, and narcissus, for example, and its acetate is the major component of jasmine oils. The richest sources of benzaldehyde are almond and apricot kernels, but it is also found in a wide range of flowers, including lilac, and other oils such as cassia and cinnamon. Hydroxylation or amination of benzoic acid leads to further series of natural products and some of the most significant, in terms of odors of essential oils, are shown in Figure 6.9. *o*-Hydroxybenzoic acid is known as salicylic acid (**38**) and both it and its esters are widely distributed in nature. For instance, methyl salicylate (**39**) is the major component (about 90% of the volatiles) of wintergreen and makes a significant contribution to the scents of tuberose and ylang ylang although only present at about 10%

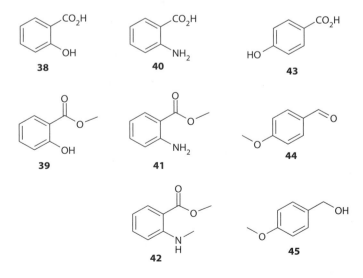

FIGURE 6.9 Hydroxy- and aminobenzoic acid derivatives.

in the former and less than 1% in the latter. o-Aminobenzoic acid is known as anthranilic acid (**40**). Its methyl ester (**41**) has a very powerful odor and is found in such oils as genet, bitter orange flower, tuberose, and jasmine. Dimethyl anthranilate (**42**), in which both the nitrogen and acid functions have been methylated, occurs at low levels in citrus oils. p-Hydroxybenzoic acid has been found in vanilla and orris but much more common is the methyl ester of the corresponding aldehyde, commonly known as anisaldehyde (**44**). As the name suggests, the latter one is an important component of anise and it is also found in oils such as lilac and the smoke of agar wood. The corresponding alcohol, anisyl alcohol (**45**), and its esters are also widespread components of essential oils.

Indole (**46**) and 2-phenylethanol (**47**) are both shikimate derivatives. Indole is particularly associated with jasmine. It usually occurs in jasmine absolute at a level of about 3%–5% and makes a very significant odor contribution to it. However, it does occur in many other essential oils as well. 2-Phenylethanol occurs widely in plants and is especially important for rose where it usually accounts for one-third to three-quarters of the oil. The structures of both are shown in Figure 6.10.

Figure 6.10 also shows some of the commonest cinnamic acid–derived essential oil components. Cinnamic acid (**32**) itself has been found in, for example, cassia and styrax, but its esters, particularly the methyl ester, are more frequently encountered. The corresponding aldehyde, cinnamaldehyde (**48**), is a key component of cinnamon and cassia and also occurs in some other oils. Cinnamyl alcohol (**49**) and its esters are more widely distributed, occurring in narcissus, lilac, and a variety of other oils. Lactonization of o-coumaric acid (**33**) gives coumarin (**50**). This is found in new mown hay to which it gives the characteristic odor. It is also important in the odor profile of lavender and related species and occurs in a number of other oils. Bergapten (**51**) is a more highly oxygenated and substituted coumarin. The commonest source is bergamot oil, but it also occurs in other sources, such as lime and parsley. It is phototoxic and consequently constitutes a safety issue for oils containing it.

Oxygenation in the p-position of cinnamic acid followed by methylation of the phenol and reduction of the acid to alcohol with subsequent elimination of the alcohol gives estragole (also known as methyl chavicol (**52**) and anethole (**53**)). Estragole is found in a variety of oils, mostly herb oils such as basil, tarragon, chervil, fennel, clary sage, anise, and rosemary. Anethole occurs in both the (E)- and (Z)-forms, the more thermodynamically stable (E)-isomer (shown in Figure 6.10) is

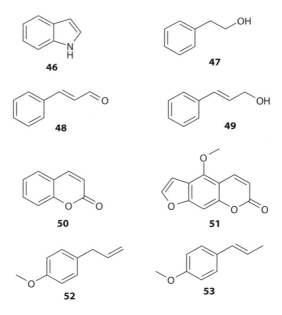

FIGURE 6.10 Some shikimate essential oil components.

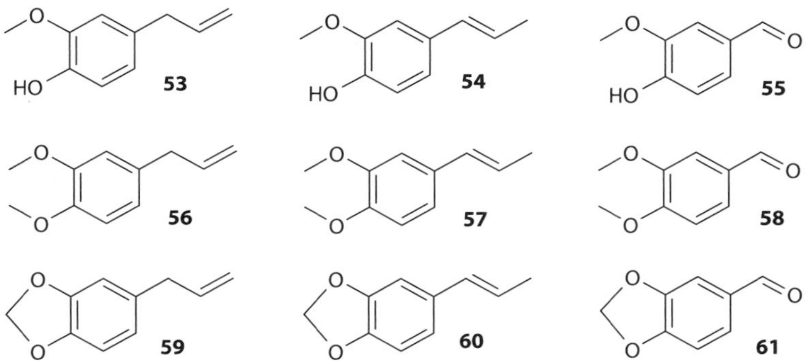

Chemistry of Essential Oils

FIGURE 6.11 Ferulic acid derivatives.

the commoner, the (Z)-isomer is the more toxic of the two. Anethole is found in spices and herbs such as anise, fennel, lemon balm, coriander, and basil and also in flower oils such as ylang ylang and lavender.

Reduction of the side chain of ferulic acid (**36**) leads to an important family of essential oil components, shown in Figure 6.11. The key material is eugenol (**53**), which is widespread in its occurrence. It is found in spices such as clove, cinnamon, and allspice, herbs such as bay and basil, and in flower oils including rose, jasmine, and carnation. Isoeugenol (**54**) is found in basil, cassia, clove, nutmeg, and ylang ylang. Oxidative cleavage of the side chain of shikimates to give benzaldehyde derivatives is common and often significant, as it is in this case, where the product is vanillin (**55**). Vanillin is the key odor component of vanilla and is therefore of considerable commercial importance. It also occurs in other sources such as jasmine, cabreuva, and the smoke of agar wood. The methyl ether of eugenol, methyl eugenol (**56**), is very widespread in nature, which, since it is the subject of some toxicological safety issues, creates difficulties for the essential oils business. The oils of some *Melaleuca* species contain up to 98% methyl eugenol, and it is found in a wide range of species including pimento, bay, tarragon, basil, and rose. The isomer, methyl isoeugenol (**57**), occurs as both (*E*)- and (*Z*)-isomers, the former being slightly commoner. Typical sources include calamus, citronella, and some narcissus species. Oxidative cleavage of the side chain in this set of substances produces veratraldehyde (**58**), a relatively rare natural product. Formation of the methylenedioxy ring, via methylenecaffeic acid (**37**), gives safrole (**59**), the major component of sassafras oil. The toxicity of safrole has led to a ban on the use of sassafras oil by the perfumery industry. Isosafrole (**60**) is found relatively infrequently in nature. The corresponding benzaldehyde derivative, heliotropin (**61**), also known as piperonal, is the major component of heliotrope.

6.5 TERPENOIDS

The terpenoids are, by far, the most important group of natural products as far as essential oils are concerned. Some authors, particularly in older literature, refer to them as terpenes, but this term is nowadays restricted to the monoterpenoid hydrocarbons. They are defined as substances composed of isoprene (2-methylbutadiene) units. Isoprene (**62**) is not often found in essential oils and is not actually an intermediate in biosynthesis, but the 2-methylbutane skeleton is easily discernable in terpenoids. Figure 6.12 shows the structures of some terpenoids. In the case of geraniol (**63**), one end of one isoprene unit is joined to the end of another making a linear structure (2,6-dimethyloctane). In guaiol (**64**), there are three isoprene units joined together to make a molecule with two rings. It is easy to envisage how the three units were first joined together into a chain and then formation of bonds from one point in the chain to another produced the two rings. Similarly, two isoprene units were used to form the bicyclic structure of a-pinene (**65**).

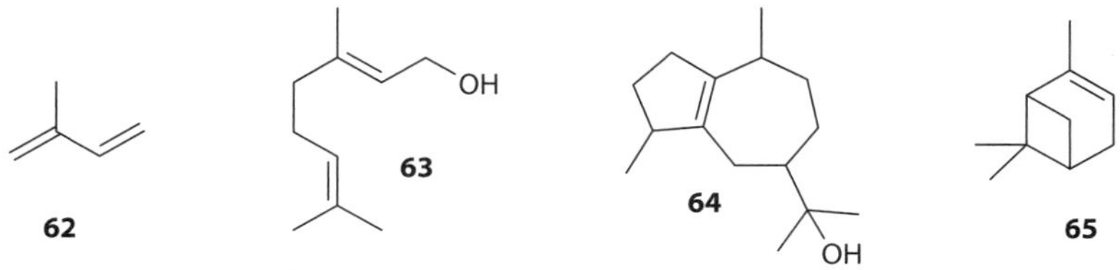

FIGURE 6.12 Isoprene units in some common terpenoids.

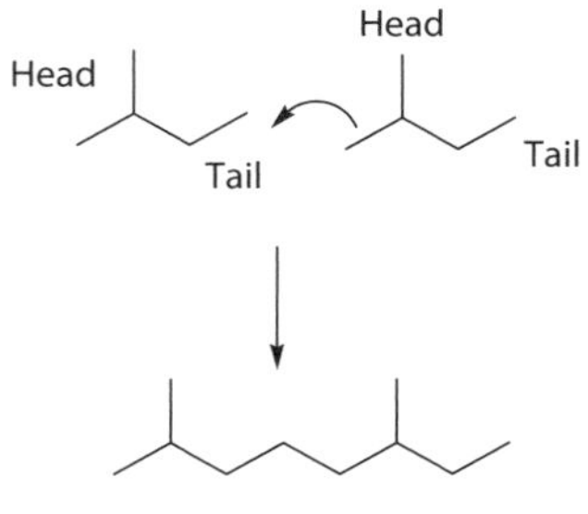

FIGURE 6.13 Head-to-tail coupling of two isoprene units.

The direction of coupling of isoprene units is almost always in one direction, the so-called head-to-tail coupling. This is shown in Figure 6.13. The branched end of the chain is referred to as the head of the molecule and the other as the tail.

This pattern of coupling is explained by the biosynthesis of terpenoids (Bu'Lock, 1965; Croteau, 1987; Mann et al., 1994). The key intermediate is mevalonic acid (**3**), which is made from three molecules of acetyl CoA (**2**). Phosphorylation of mevalonic acid followed by elimination of the tertiary alcohol and concomitant decarboxylation of the adjacent acid group gives isopentenyl pyrophosphate (**66**). This can be isomerized to give prenyl pyrophosphate (**67**). Coupling of these two 5-carbon units gives a 10-carbon unit, geranyl pyrophosphate (**68**), as shown in Figure 6.14, and further additions of isopentenyl pyrophosphate (**66**) lead to 15-, 20-, 25-, and so on carbon units.

It is clear from the mechanism shown in Figure 6.14 that terpenoid structures will always contain a multiple of five carbon atoms when they are first formed. The first terpenoids to be studied contained 10 carbon atoms per molecule and were called monoterpenoids. This nomenclature has remained and so those with 5 carbon atoms are known as hemiterpenoids; those with 15, sesquiterpenoids; those with 20, diterpenoids; and so on. In general, only the hemiterpenoids, monoterpenoids, and

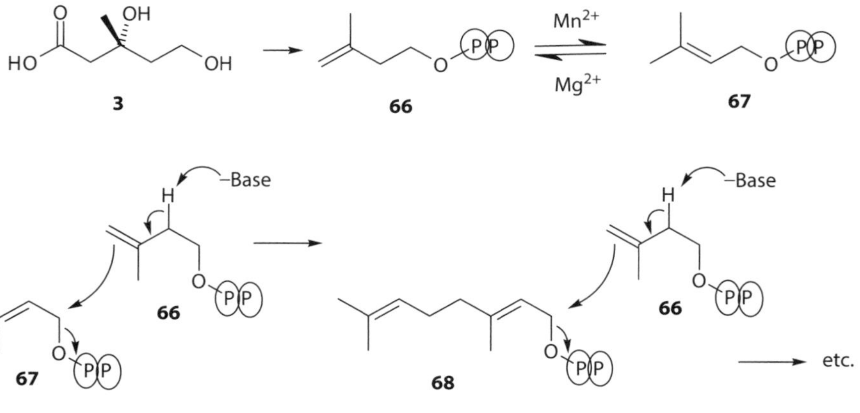

FIGURE 6.14 Coupling of C5 units in terpenoid biosynthesis.

sesquiterpenoids are sufficiently volatile to be components of essential oils. Degradation products of higher terpenoids do occur in essential oils, so they will be included in this chapter.

6.5.1 HEMITERPENOIDS

Many alcohols, aldehydes, and esters, with a 2-methylbutane skeleton, occur as minor components in essential oils. Not surprisingly, in view of the biosynthesis, the commonest oxidation pattern is that of prenol, that is, 3-methylbut-2-ene-1-ol. For example, the acetate of this alcohol occurs in ylang ylang and a number of other oils. However, oxidation has been observed at all positions. Esters such as prenyl acetate give fruity top notes to oils containing them, and the corresponding thioesters contribute to the characteristic odor of galbanum.

6.5.2 MONOTERPENOIDS

Geranyl pyrophosphate (**68**) is the precursor for the monoterpenoids. Heterolysis of its carbon–oxygen bond gives the geranyl carbocation (**69**). In natural systems, this and other carbocations discussed in this chapter do not exist as free ions but rather as incipient carbocations held in enzyme active sites and essentially prompted into cation reactions by the approach of a suitable reagent. For the sake of simplicity, they will be referred to here as carbocations. The reactions are described in chemical terms but all are under enzymic control, and the enzymes present in any given plant will determine the terpenoids it will produce. Thus, essential oil composition can give information about the genetic makeup of the plant. A selection of some of the key biosynthetic routes to monoterpenoids (Devon and Scott, 1972) is shown in Figure 6.15.

Reaction of the geranyl carbocation with water gives geraniol (**63**) that can subsequently be oxidized to citral (**71**). Loss of a proton from (**69**) gives myrcene (**70**) and this can be isomerized to other acyclic hydrocarbons. An intramolecular electrophilic addition reaction of (**69**) gives the monocyclic carbocation (**72**) that can eliminate a proton to give limonene (**73**) or add water to give a-terpineol (**74**). A second intramolecular addition gives the pinyl carbocation (**75**) that can lose a proton to give either a-pinene (**65**) or b-pinene (**76**). The pinyl carbocation (**75**) is also reachable directly from the menthyl carbocation (**72**). Carene (**77**), another bicyclic material, can be produced through similar reactions. Wagner–Meerwein rearrangement of the pinyl carbocation (**75**) gives the bornyl carbocation (**78**). Addition of water to this gives borneol (**79**) and this can be oxidized to camphor (**80**). An alternative Wagner–Meerwein rearrangement of (**75**) gives the fenchyl skeleton (**81**) from which fenchone (**82**) is derived.

Some of the more commonly encountered monoterpenoid hydrocarbons (Arctander, 1960; Essential Oils Database, 2006; Gildemeister and Hoffmann, 1956; Günther, 1948; Sell, 2007) are shown in Figure 6.16. Many of these can be formed by dehydration of alcohols, and so their presence in essential oils could be as artifacts arising from the extraction process. Similarly, p-cymene (**83**) is one of the most stable materials of this class and can be formed from many of the others by appropriate cyclization and/or isomerization and/or oxidation reactions and so its presence in any essential oil could be as an artifact.

Myrcene (**70**) is very widespread in nature. Some sources, such as hops, contain high levels and it is found in most of the common herbs and spices. All isomers of a-ocimene (**84**), b-ocimene (**85**), and allo-ocimene (**86**) are found in essential oils, the isomers of b-ocimene (**85**) being the most frequently encountered. Limonene (**73**) is present in many essential oils, but the major occurrence is in the citrus oils that contain levels up to 90%. These oils contain the dextrorotatory (*R*)-enantiomer, and its antipode is much less common. Both a-phellandrene (**87**) and b-phellandrene (**88**) occur widely in essential oils. For example, (−)-a-phellandrene is found in Eucalyptus dives and (*S*)-(−)-b-phellandrene in the lodgepole pine, Pinus contorta. p-Cymene (**83**) has been identified in many essential oils and plant extracts and thyme and oregano oils are particularly rich in it. a-Pinene (**65**), b-pinene (**76**), and 3-carene (**77**) are all major constituents of turpentine from a wide range of pines, spruces, and firs. The pinenes are often found in other oils, 3-carene less so. Like the pinenes, camphene (**89**) is widespread in nature.

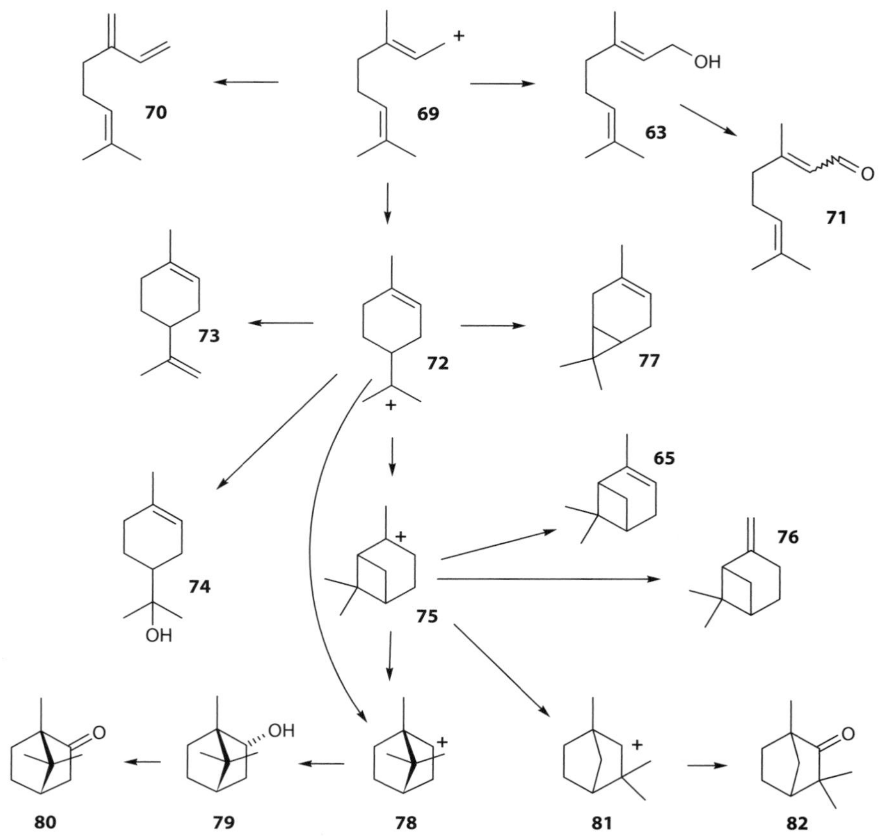

FIGURE 6.15 Formation of monoterpenoid skeletons.

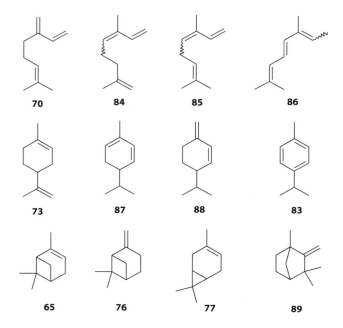

FIGURE 6.16 Some of the more common terpenoid hydrocarbons.

Simple hydrolysis of geranyl pyrophosphate gives geraniol, (E)-3,7-dimethylocta-2,6-dienol (**63**). This is often accompanied in nature by its geometric isomer, nerol (**90**). Synthetic material is usually a mixture of the two isomers and when interconversion is possible, the equilibrium mixture comprises about 60% geraniol (**63**) and 40% nerol (**90**). The name geraniol is often used to describe a mixture of geraniol and nerol. When specifying the geometry of these alcohols, it is better to use the modern (E)/(Z) nomenclature as the terms *cis* and *trans* are somewhat ambiguous in this case and earlier literature is not consistent in their use. Both isomers occur in a wide range of essential oils, geraniol (**63**) being particularly widespread. The oil of Monarda fistulosa contains over 90% geraniol (**63**) and the level in palmarosa is over 80%. Geranium contains about 50% and citronella and lemongrass each contain about 30%. The richest natural sources of nerol include rose, palmarosa, citronella, and davana, although its level in these is usually only in the 10%–15% range. Citronella and related species are used commercially as sources of geraniol, but the price is much higher than that of synthetic material. Citronellol (**91**) is a dihydrogeraniol and occurs widely in nature in both enantiomeric forms. Rose, geranium, and citronella are the oils with the highest levels of citronellol. Geraniol, nerol, and citronellol, together with 2-phenylethanol, are known as the rose alcohols because of their occurrence in rose oils and also because they are the key materials responsible for the rose odor character. Esters (the acetates in particular) of all these alcohols are also commonly encountered in essential oils (Figure 6.17).

Allylic hydrolysis of geranyl pyrophosphate produces linalool (**92**). Like geraniol, linalool occurs widely in nature. The richest source is ho leaf, the oil of which can contain well over 90% linalool. Other rich sources include linaloe, rosewood, coriander, freesia, and honeysuckle. Its acetate is also frequently encountered and is a significant contributor to the odors of lavender and citrus leaf oils.

Figure 6.18 shows a selection of cyclic monoterpenoid alcohols. a-Terpineol (**74**) is found in many essential oils as is its acetate. The isomeric terpinen-4-ol (**93**) is an important component of Ti tree

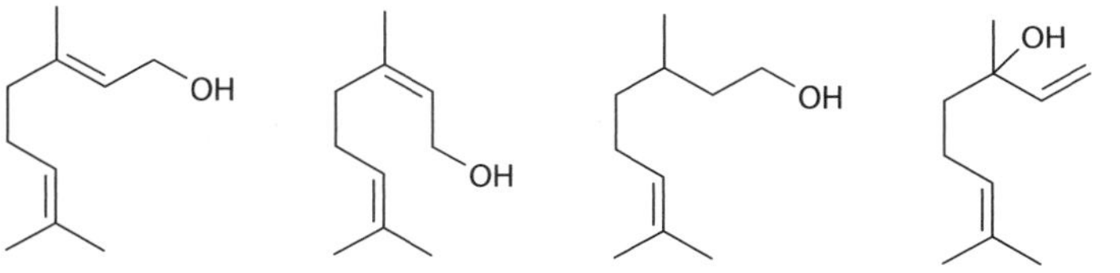

FIGURE 6.17 Key acyclic monoterpenoid alcohols.

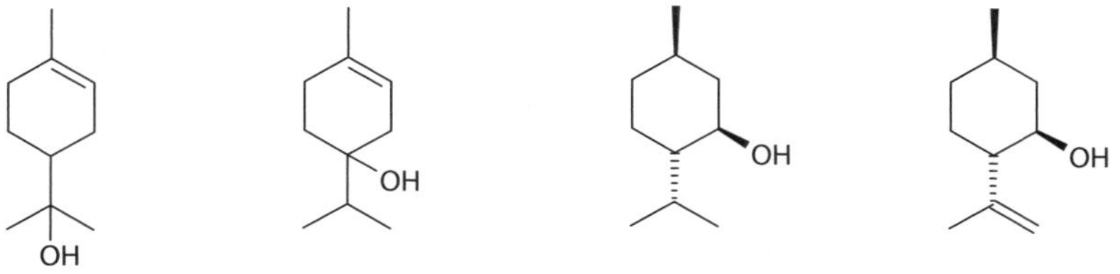

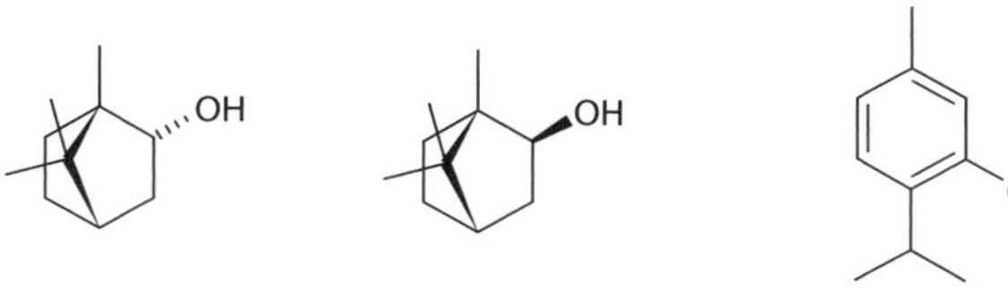

FIGURE 6.18 Some cyclic monoterpenoid alcohol.

oil, but its acetate, surprisingly, is more widely occurring, being found in herbs such as marjoram and rosemary. l-Menthol (**94**) is found in various mints and is responsible for the cooling effect of oils containing it. There are eight stereoisomers of the menthol structure, l-menthol is the commonest in nature and also has the strongest cooling effect. The cooling effect makes menthol and mint oils valuable commodities, the two most important sources being corn mint (*Mentha arvensis*) and peppermint (*Mentha piperita*). Isopulegol (**95**) occurs in some species including *Eucalyptus citriodora* and citronella. Borneol (*endo*-1,7,7-trimethylbicyclo[2.2.1]heptan-2-ol) (**79**) and esters thereof, particularly the acetate, occur in many essential oils. Isoborneol (*exo*-1,7,7-trimethylbicyclo[2.2.1]heptan-2-ol) (**96**) is less common; however, isoborneol and its esters are found in quite a number of oils. Thymol (**97**), being a phenol, possesses antimicrobial properties, and oils, such as thyme and basil, which find appropriate use in herbal remedies. It is also found in various *Ocimum* and *Monarda* species.

Three monoterpenoid ethers are shown in Figure 6.19. 1,8-Cineole (**98**), more commonly referred to simply as cineole, comprises up to 95% of the oil of *Eucalyptus globulus* and about 40%–50% of cajeput oil. It also can be found in an extensive range of other oils and often as a major component. It has antibacterial and decongestant properties and consequently, eucalyptus oil is used in various paramedical applications. Menthofuran (**99**) occurs in mint oils and contributes to the odor of peppermint. It is also found in several other oils. Rose oxide is found predominantly in rose and geranium oils. There are four isomers, the commonest being the levorotatory enantiomer of *cis*-rose oxide (**100**). This is also the isomer with the lowest odor threshold of the four.

The two most significant monoterpene aldehydes are citral (**71**) and its dihydro analogue citronellal (**103**), both of which are shown in Figure 6.20. The word citral is used to describe a mixture of the two geometric isomers geranial (**101**) and neral (**102**) without specifying their relative proportions. Citral occurs widely in nature, both isomers usually being present, the ratio between them usually being in the 40:60 to 60:40 range. Lemongrass contains 70%–90% citral and the fruit of *Litsea cubeba* contains about 60%–75%. Citral also occurs in *Eucalyptus staigeriana*, lemon balm, ginger, basil, rose, and citrus species. It is responsible for the characteristic smell of lemons although lemon oil usually contains only a few percent of it. Citronellal (**103**) also occurs widely in essential oils. *E. citriodora* contains up to 85% citronellal, and significant amounts are also found in some chemotypes of *L. cubeba*, citronella Swangi leaf oil, and *Backhousia citriodora*. Campholenic aldehyde (**104**) occurs in a limited range of species such as olibanum, styrax, and some eucalypts. Material produced from a-pinene (**65**) is important as an intermediate for synthesis.

FIGURE 6.19 Some monoterpenoid ethers.

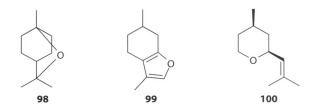

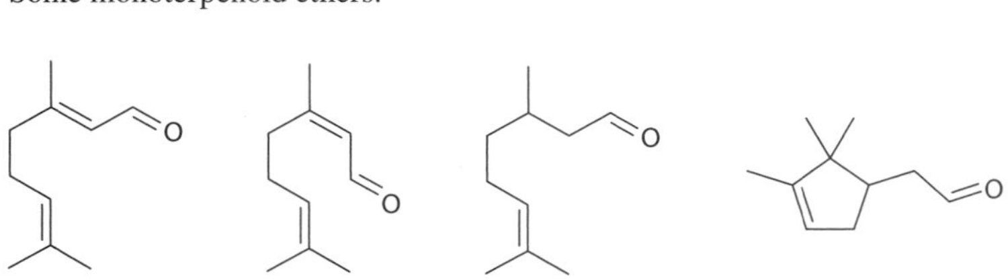

FIGURE 6.20 Some monoterpenoid aldehydes.

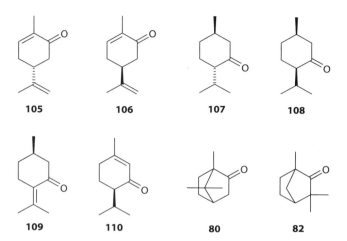

FIGURE 6.21 Some monoterpenoid ketones.

Figure 6.21 shows some of the commoner monoterpenoid ketones found in essential oils. Both enantiomers of carvone are found in nature, the (R)-(−)- (usually referred to as L-carvone) (**105**) being the commoner. This enantiomer provides the characteristic odor of spearmint (*Mentha cardiaca*, *Mentha gracilis*, *Mentha spicata*, and *Mentha viridis*), the oil of which usually contains 55%–75% of L-carvone. The (S)-(+)-enantiomer (**106**) is found in caraway at levels of 30%–65% and in dill at 50%–75%. Menthone is fairly common in essential oils particularly in the mints, pennyroyal, and sages, but lower levels are also found in oils such as rose and geranium. The L-isomer is commoner than the *d*-isomer. Isomenthone is the *cis*-isomer and the two interconvert readily by epimerization. The equilibrium mixture comprises about 70% menthone and 30% isomenthone. The direction of rotation of plane-polarized light reverses on epimerization, and therefore *l*-menthone (**107**) gives D-isomenthone (**108**). (+)-Pulegone (**109**) accounts for about 75% of the oil of pennyroyal and is also found in a variety of other oils. (−)-Piperitone (**110**) also occurs in a variety of oils, the richest source being *E. dives*. Both pulegone and piperitone have strong minty odors. Camphor (**80**) occurs in many essential oils and in both enantiomeric forms. The richest source is the oil of camphor wood, but it is also an important contributor to the odor of lavender and of herbs such as sage and rosemary. Fenchone (**82**) occurs widely, for example, in cedar leaf and lavender. Its laevorotatory enantiomer is an important contributor to the odor of fennel.

6.5.3 Sesquiterpenoids

By definition, sesquiterpenoids contain 15 carbon atoms. This results in their having lower volatilities and hence higher boiling points than monoterpenoids. Therefore, fewer of them (in percentage terms) contribute to the odor of essential oils, but those that do often have low-odor thresholds and contribute significantly as endnotes. They are also important as fixatives for more volatile components.

Just as geraniol (**63**) is the precursor for all the monoterpenoids, farnesol (**111**) is the precursor for all the sesquiterpenoids. Its pyrophosphate is synthesized in nature by the addition of isopentenyl pyrophosphate (**66**) to geranyl pyrophosphate (**68**) as shown in Figure 6.14, and hydrolysis of that gives farnesol. Incipient heterolysis of the carbon–oxygen bond of the phosphate gives the nascent farnesyl carbocation (**112**), and this leads to the other sesquiterpenoids, just as the geranyl carbocation does to monoterpenoids. Starting from farnesyl pyrophosphate, the variety of possible cyclic structures is much greater than that from geranyl pyrophosphate because there are now three double bonds in the molecule. Similarly, there is also a greater scope for further structural variation resulting from rearrangements, oxidations, degradation, and so on (Devon and Scott, 1972).

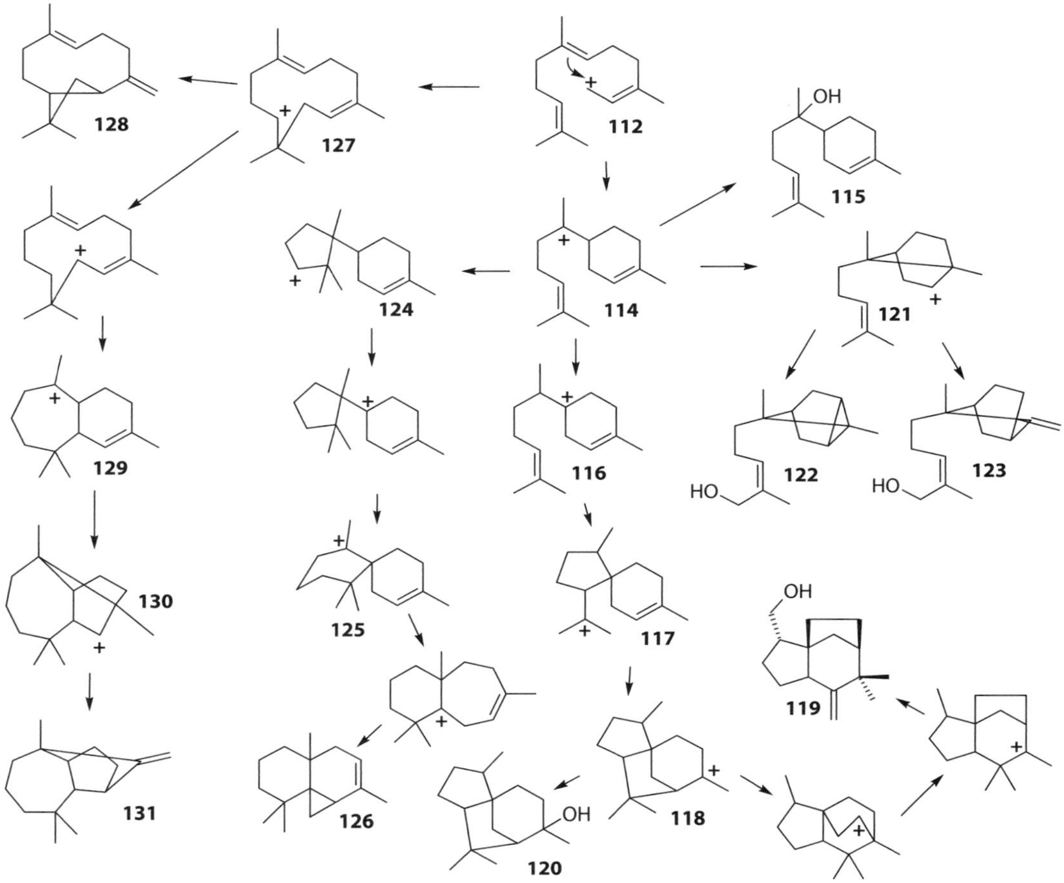

FIGURE 6.22 Some biosynthetic pathways from (Z,E)-farnesol.

The geometry of the double bond in position 2 of farnesol is important in terms of determining the pathway used for subsequent cyclization reactions, and so these are best discussed in two blocks.

Figure 6.22 shows a tiny fraction of the biosynthetic pathways derived from (Z,E)-farnesyl pyrophosphate. Direct hydrolysis leads to acyclic sesquiterpenoids such as farnesol (**111**) and nerolidol (**113**). However, capture of the carbocation (**112**) by the double bond at position 6 gives a cyclic structure that of the bisabolane skeleton (**114**), and quenching of this with water gives bisabolol (**115**). A hydrogen shift in (**114**) leads to the isomeric carbocation (**116**) that still retains the bisabolane skeleton. Further cyclizations and rearrangements take the molecule through various skeletons, including those of the acorane (**117**) and cedrane (**118**) families, to the khusane family, illustrated by khusimol (**119**) in Figure 6.22. Obviously, a wide variety of materials can be generated along this route, an example being cedrol (**120**) formed by reaction of cation (**118**) with water. The bisabolyl carbocation (**114**) can also cyclize to the other double bonds in the molecule leading to, *inter alia*, the campherenane skeleton (**121**) and hence a-santalol (**122**) and b-santalol (**123**), or, via the cuparane (**124**) and chamigrane (**125**) skeletons, to compounds such as thujopsene (**126**). The carbocation function in (**112**) can also add to the double bond at the far end of the chain to give the *cis*-humulane skeleton (**127**). This species can cyclize back to the double bond at carbon 2 before losing a proton, thus giving caryophyllene (**128**). Another alternative is for a series of hydrogen shifts, cyclizations, and rearrangements to lead it through the himachalane (**129**) and longibornane (**130**) skeletons to longifolene (**131**).

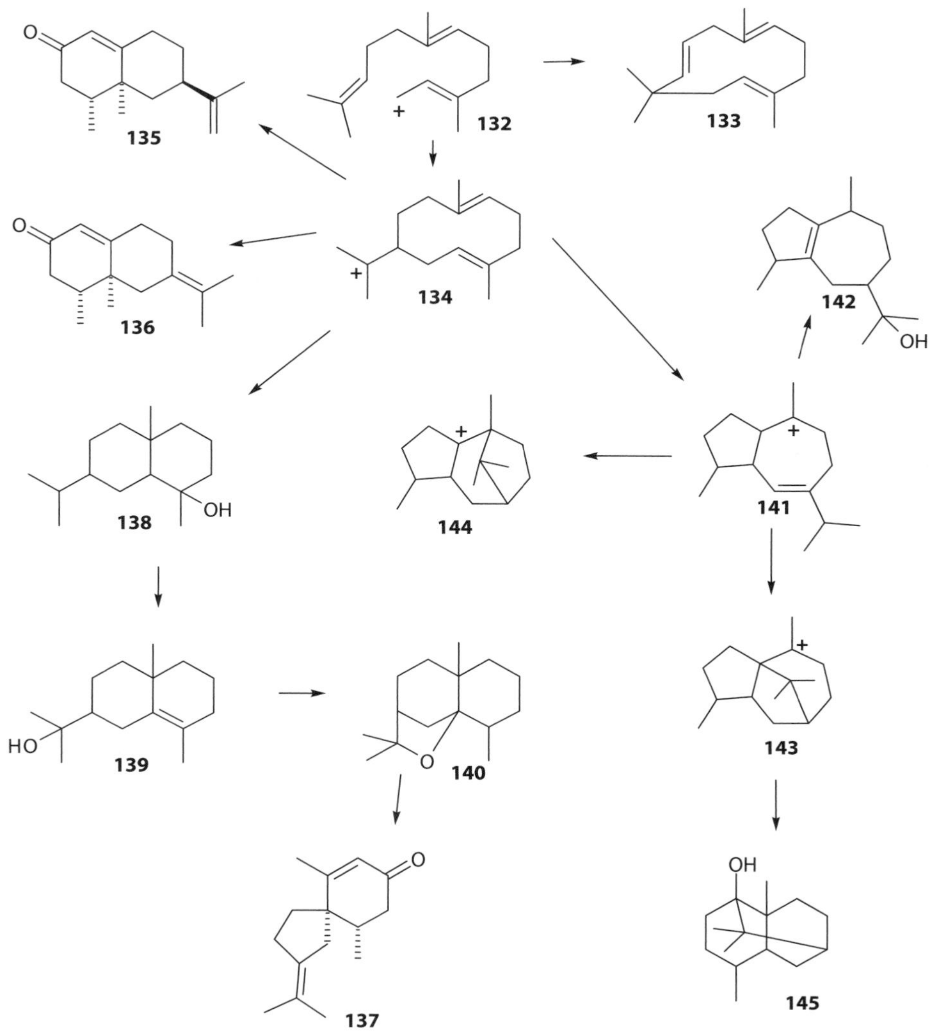

FIGURE 6.23 Some biosynthetic pathways from (*E,E*)-farnesol.

Figure 6.23 shows a few of the many possibilities for biosynthesis of sesquiterpenoids from (*E,E*)-farnesyl pyrophosphate. Cyclization of the cation (**132**) to C-11, followed by loss of a proton gives all *trans*- or a-humulene (**133**), whereas cyclization to the other end of the same double bond gives a carbocation (**134**) with the germacrane skeleton. This is an intermediate in the biosynthesis of odorous sesquiterpenes such as nootkatone (**135**) and a-vetivone (**137**). b-Vetivone (**137**) is synthesized through a route that also produces various alcohols, for example, (**138**) and (**139**), and an ether (**140**) that has the eudesmane skeleton. Rearrangement of the germacrane carbocation (**134**) leads to a carbocation (**141**) with the guaiane skeleton, and this is an intermediate in the synthesis of guaiol (**142**). Carbocation (**141**) is also an intermediate in the biosynthesis of the a-patchoulane (**143**) and b-patchoulane (**144**) skeletons and of patchouli alcohol (**145**).

All four isomers of farnesol (**111**) are found in nature and all have odors in the muguet and linden direction. The commonest is the (*E,E*)-isomer that occurs in, among others, cabreuva and ambrette seed, while the (*Z,E*)-isomer has been found in jasmine and ylang ylang, the (*E,Z*)-isomer in cabreuva, rose, and neroli, and the (*Z,Z*)-isomer in rose. Nerolidol (**113**) is the allylic isomer of farnesol and exists in four isomeric forms: two enantiomers each of two geometric isomers.

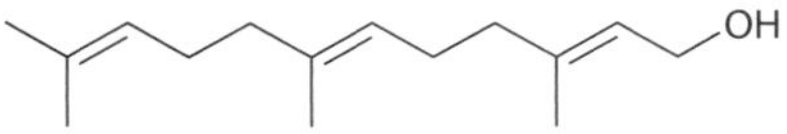

FIGURE 6.24 Some sesquiterpenoid alcohols.

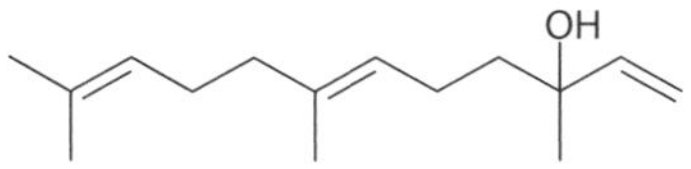

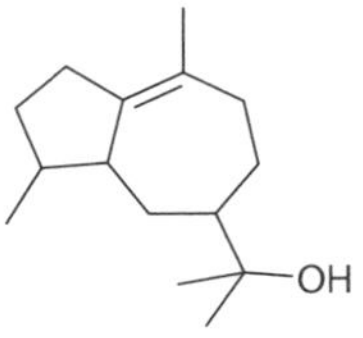

The (*E*)-isomer has been found in cabreuva, niaouli, and neroli oils among others and the (*Z*)-isomer in neroli, jasmine, ho leaf, and so on. Figure 6.24 shows the structures of farnesol and nerolidol with all of the double bonds in the *trans*-configuration.

a-Bisabolol (**115**) is the simplest of the cyclic sesquiterpenoid alcohols. If farnesol is the sesquiterpenoid equivalent of geraniol and nerolidol of linalool, then a-bisabolol is the equivalent of a-terpineol. It has two chiral centers and therefore exists in four stereoisomeric forms, all of which occur in nature. The richest natural source is *Myoporum crassifolium* Forst., a shrub from New Caledonia, but a-bisabolol can be found in many other species including chamomile, lavender, and rosemary. It has a faint floral odor and anti-inflammatory properties and is responsible, at least in part, for the related medicinal properties of chamomile oil.

The santalols (**122**) and (**123**) have more complex structures and are the principal components of sandalwood oil. Cedrol (**120**) is another complex alcohol, but it is more widely occurring in nature than the santalols. It is found in a wide range of species, the most significant being trees of the *Juniperus, Cupressus*, and *Thuja* families. Cedrene (**146**) occurs alongside cedrol in cedarwood oils. Cedrol is dehydrated to cedrene in the presence of acid, and so the latter can be an artifact of the former and the ratio of the two will often depend on the method of isolation. Thujopsene (**126**) also occurs in cedarwood oils, usually at a similar level to that of cedrol/cedrene, and it is found in various other oils also. Caryophyllene (**128**) and a-humulene (the all *trans*-isomer) (**133**) are widespread in nature, cloves being the best-known source of the former and hops of the latter. The ring systems of these two materials are very strained making them quite reactive chemically, and caryophyllene, extracted from clove oil as a by-product of eugenol production, is used as the starting material in the synthesis of several fragrance ingredients. Longifolene (**131**) also possesses a strained ring system. It is a component of Indian turpentine and is therefore readily available as a feedstock for fragrance ingredient manufacture.

Guaiacwood oil is the richest source of guaiol (**142**) and the isomeric bulnesol (**147**), but both are found in other oils, particularly guaiol that occurs in a wide variety of plants. Dehydration and dehydrogenation of these give guaiazulene (**148**), which is used as an anti-inflammatory agent. Guaiazulene is also accessible from a-gurjunene (**149**), the major component of gurjun balsam. Guaiazulene is blue in color as is the related olefin chamazulene (**150**). The latter occurs in a variety of oils, but it is particularly important in chamomile to which it imparts the distinctive blue tint (Figure 6.25).

Vetiver and patchouli are two oils of great importance in perfumery (Williams, 1996, 2004). Both contain complex mixtures of sesquiterpenoids, mostly with complex polycyclic structures (Sell, 2003). The major components of vetiver oil are a-vetivone (**136**), b-vetivone (**137**), and khusimol (**119**), but the most important components as far as odor is concerned are minor constituents such as khusimone (**151**), zizanal (**152**), and methyl zizanoate (**153**). Nootkatone (**154**) is an isomer of a-vetivone and is an important odor component of grapefruit. Patchouli alcohol (**145**) is the

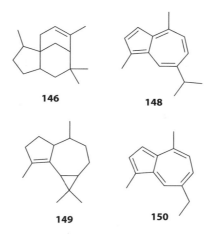

FIGURE 6.25 Some sesquiterpenoid hydrocarbons.

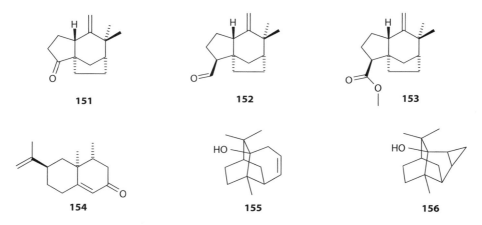

FIGURE 6.26 Components of vetiver, patchouli, and grapefruit.

major constituent of patchouli oil but, as is the case also with vetiver, minor components are more important for the odor profile. These include norpatchoulenol (**155**) and nortetrapatchoulol (**156**) (Figure 6.26).

The molecules of chamazulene (**150**), khusimone (**151**), norpatchoulenol (**155**), and nortetrapatchoulol (**156**) each contain only 14 carbon atoms in place of the normal 15 of sesquiterpenoids. They are all degradation products of sesquiterpenoids. Degradation, either by enzymic action or from environmental chemical processes, can be an important factor in generating essential oil components. Carotenoids are a family of tetraterpenoids characterized by having a tail-to-tail fusion between two diterpenoid fragments. In the case of b-carotene (**157**), both ends of the chain have been cyclized to form cyclohexane rings. Degradation of the central part of the chain leads to a number of fragments that are found in essential oils and the two major families of such are the ionones and damascones. Both have the same carbon skeleton, but in the ionones (Essential Oils Database, 2006; Sell, 2003), the site of oxygenation is three carbon atoms away from the ring, and in damascones oxygenation is found at the chain carbon next to the ring (Figure 6.27).

The ionones occur naturally in a wide variety of flowers, fruits, and leaves, and are materials of major importance in perfumery (Arctander, 1960; Essential Oils Database, 2006; Gildemeister and Hoffmann, 1956; Günther, 1948; Sell, 2007). About 57% of the volatile components of violet flowers

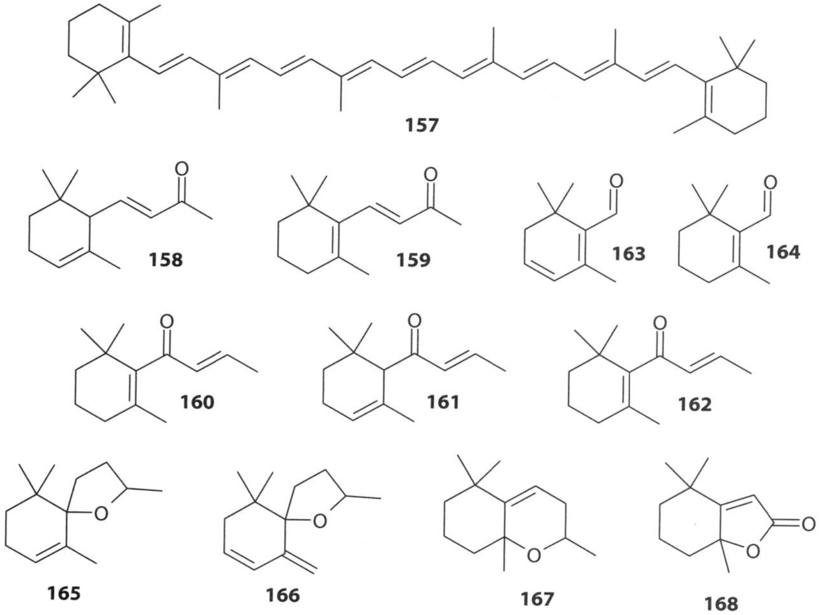

FIGURE 6.27 Carotenoid degradation products.

are a- (**158**) and b-ionones (**159**), and both isomers occur widely in nature. The damascones are also found in a wide range of plants. They usually occur at a very low level, but their very intense odors mean that they still make a significant contribution to the odors of oils containing them. The first to be isolated and characterized was b-damascenone (**160**), which was found at a level of 0.05% in the oil of the Damask rose. Both b-damascenone (**160**) and the a- (**161**) and b-isomers (**162**) have since been found in many different essential oils and extracts. In the cases of safranal (**163**) and cyclocitral (**165**), the side chain is degraded even further leaving only one of its carbon atoms attached to the cyclohexane ring. About 70% of the volatile component of saffron is safranal and it makes a significant contribution to its odor. Other volatile carotenoid degradation products that occur in essential oils and contribute to their odors include the theaspiranes (**165**), vitispiranes (**166**), edulans (**167**), and dihydroactinidiolide (**168**).

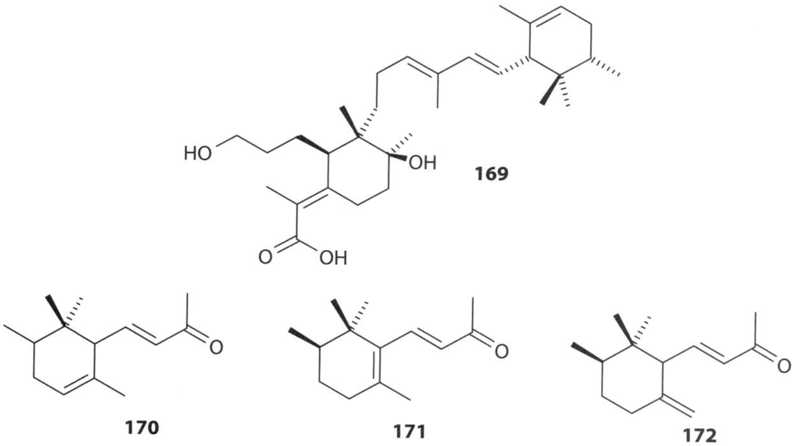

FIGURE 6.28 Iripallidal and the irones.

Chemistry of Essential Oils

The similarity in structure between the ionones and the irones might lead to the belief that the latter are also carotenoid derived. However, this is not the case as the irones are formed by degradation of the triterpenoid iripallidal (**169**), which occurs in the rhizomes of the iris. The three isomers, a- (**170**), b- (**171**), and g-irone (**172**), are all found in iris and the first two in a limited number of other species (Figure 6.28).

6.6 SYNTHESIS OF ESSENTIAL OIL COMPONENTS

It would be impossible, in a volume of this size, to review all of the reported syntheses of essential oil components and so the following discussion will concentrate on some of the more commercially important synthetic routes to selected key substances. In the vast majority of cases, there is a balance between routes using plant extracts as feedstocks and those using petrochemicals. For some materials, plant-derived and petrochemical-derived equivalents might exist in economic competition, while for others, one source is more competitive. The balance will vary over time and the market will respond accordingly. Sustainability of production routes is a complex issue and easy assumptions might be totally incorrect. Production and extraction of plant-derived feedstocks often requires considerable expenditure of energy in fertilizer production, harvesting, and processing, and so it is quite possible that production of a material derived from a plant source would use more mineral oil than the equivalent derived from petrochemical feedstocks.

Figure 6.29 shows some of the plant-derived feedstocks used in the synthesis of lipids and polyketides (Sell, 2006). Rapeseed oil provides erucic acid (**173**) that can be ozonolyzed to give brassylic acid (**174**) and heptanal (**175**), both useful building blocks. The latter can also be obtained,

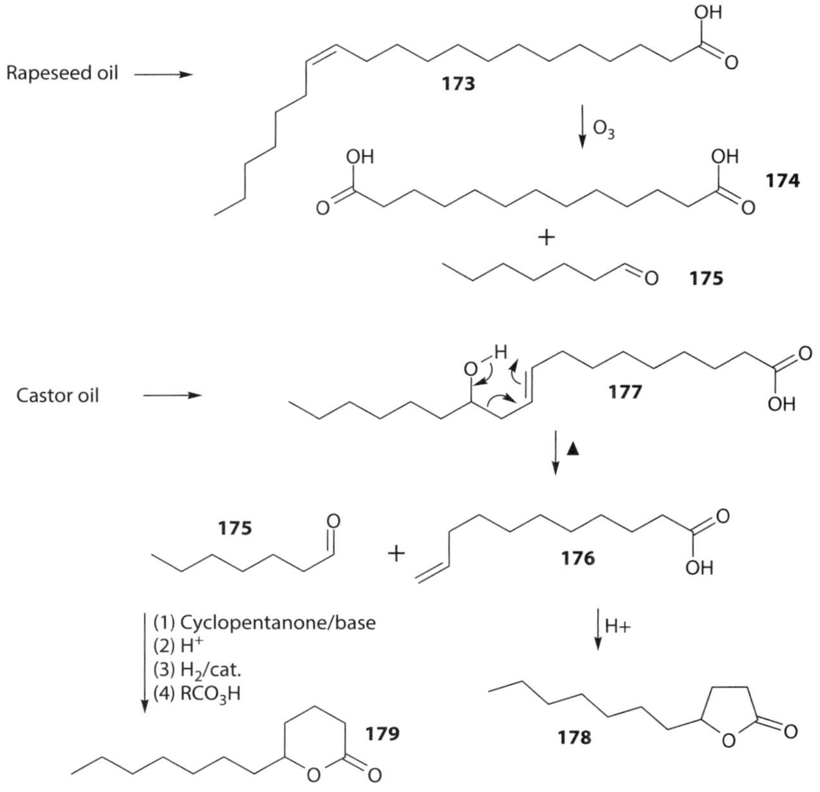

FIGURE 6.29 Some natural feedstocks for synthesis of lipids and polyketides.

together with undecylenic acid (**176**), by pyrolysis of ricinoleic acid (**177**) that is available from castor oil. Treatment of undecylenic acid (**176**) with acid leads to movement of the double bond along the chain and eventual cyclization to give g-undecalactone (**178**), which has been found in narcissus oils. Aldol condensation of heptanal (**175**) with cyclopentanone, followed by Baeyer–Villiger oxidation, gives d-dodecalactone (**179**), identified in the headspace of tuberose. Such aldol reactions, followed by appropriate further conversions, are important in the commercial production of analogues of methyl jasmonate (**26**) and jasmone (**27**).

Ethylene provides a good example of a petrochemical feedstock for the synthesis of lipids and polyketides. It can be oligomerized to provide a variety of alkenes into which functionalization can be introduced by hydration, oxidation, hydroformylation, and so on. Of course, telomerization can be used to provide functionalized materials directly.

Eugenol (**53**) (e.g., clove oil) and safrole (**59**) (e.g., sassafras) are good examples of plant-derived feedstocks that are used in the synthesis of other shikimates. Methylation of eugenol produces methyl eugenol (**56**) and this can be isomerized using acid or metal catalysts to give methyl isoeugenol (**57**). Similarly, isomerization of eugenol gives isoeugenol (**54**), and oxidative cleavage of this, for example, by ozonolysis, gives vanillin (**55**). This last sequence of reactions, when applied to safrole, gives isosafrole (**60**) and heliotropin (**61**). All of these conversions are shown in Figure 6.30.

Production of shikimates from petrochemicals for commercial use mostly involves straightforward chemistry (Arctander, 1969; Bauer and Panten, 2006; Däniker, 1987; Sell, 2006). Nowadays, the major starting materials are benzene (**180**) and toluene (**181**), which are both available in bulk from petroleum fractions. Alkylation of benzene with propylene gives cumene (**182**), the hydroperoxide of which fragments to give phenol (**183**) and acetone. Phenol itself is an important molecular building block and further oxidation gives catechol (**184**). Syntheses using these last two materials will be discussed in the succeeding text. Alkylation of benzene with ethylene gives ethylbenzene, which is converted to styrene (**185**) via autoxidation, reduction, and elimination in a process known as styrene monomer/propylene oxide (SMPO) process. The epoxide (**186**) of styrene serves as an intermediate for 2-phenylethanol (**47**) and phenylacetaldehyde (**187**), both of which occur widely in essential oils. 2-Phenylethanol is also available directly from benzene by Lewis acid–catalyzed addition of ethylene oxide and as a by-product of the SMPO process. Currently, the volume available from the SMPO process provides most of the requirement. All of these processes are illustrated in Figure 6.31.

Phenol (**183**) and related materials, such as guaiacol (**188**), were once isolated from coal tar, but the bulk of their supply is currently produced from benzene via cumene as shown in Figure 6.31.

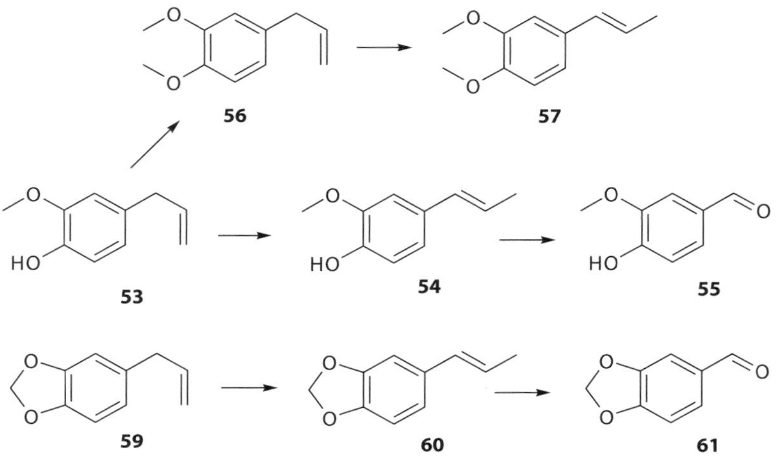

FIGURE 6.30 Shikimates from eugenol and safrole.

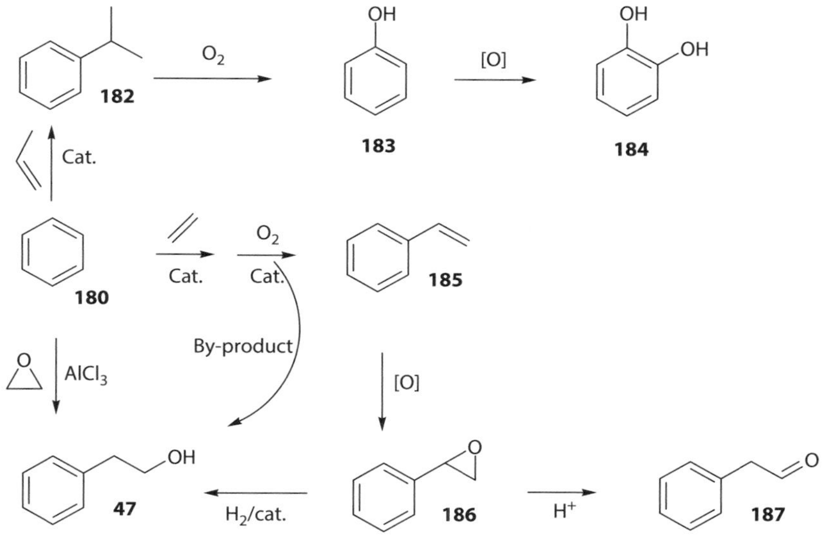

FIGURE 6.31 Benzene as a feedstock for shikimates.

The use of these intermediates to produce shikimates is shown in Figure 6.32. In principle, anethole (**53**) and estragole (methyl chavicol) (**52**) are available from phenol, but in practice, the demand is met by extraction from turpentine. Carboxylation of phenol gives salicylic acid (**38**) and hence serves as a source for the various salicylate esters. Formylation of phenol by formaldehyde, in the presence of a suitable catalyst, has now replaced the Reimer–Tiemann reaction as a route to hydroxybenzaldehydes. The initial products are saligenin (**189**) and *p*-hydroxybenzyl alcohol (**190**), which can be oxidized to salicylaldehyde (**191**) and *p*-hydroxybenzaldehyde (**192**), respectively. Condensation of salicylaldehyde with acetic acid/acetic anhydride gives coumarin (**50**) and *O*-alkylation of *p*-hydroxybenzaldehyde gives anisaldehyde (**44**). As mentioned earlier, oxidation of phenol provides a route to catechol (**184**) and guaiacol (**188**). The latter is a precursor for vanillin, and catechol also provides a route to heliotropin (**61**) via methylenedioxybenzene (**193**).

Oxidation of toluene (**181**) with air or oxygen in the presence of a catalyst gives benzyl alcohol (**194**), benzaldehyde (**195**), or benzoic acid (**196**) depending on the chemistry employed. The demand for benzoic acid far exceeds that for the other two oxidation products and so such processes are usually designed to produce mostly benzoic acid with benzaldehyde as a minor product. For the fragrance industry, benzoic acid is the precursor for the various benzoates of interest, while benzaldehyde, through aldol-type chemistry, serves as the key intermediate for cinnamate esters (such as methyl cinnamate (**197**)) and cinnamaldehyde (**48**). Reduction of the latter gives cinnamyl alcohol (**49**) and hence, through esterification, provides routes to all of the cinnamyl esters. Chlorination of toluene under radical conditions gives benzyl chloride (**198**). Hydrolysis of the chloride gives benzyl alcohol (**194**), which can, in principle, be esterified to give the various benzyl esters (**199**) of interest. However, these are more easily accessible directly from the chloride by reaction with the sodium salt of the corresponding carboxylic acid. All of these conversions are shown in Figure 6.33.

Methyl anthranilate (**41**) is synthesized from either naphthalene (**200**) or *o*-xylene (**201**) as shown in Figure 6.34. Oxidation of either starting material produces phthalic acid (**202**). Conversion of this diacid to its imide, followed by the Hoffmann reaction, gives anthranilic acid, and the methyl ester can then be obtained by reaction with methanol.

In volume terms, the terpenoids represent the largest group of natural and nature identical fragrance ingredients (Däniker, 1987; Sell, 2007). The key materials are the rose alcohols [geraniol (**63**)/nerol (**90**), linalool (**23**), and citronellol (**91**)], citronellal (**103**), and citral (**71**). Interconversion of

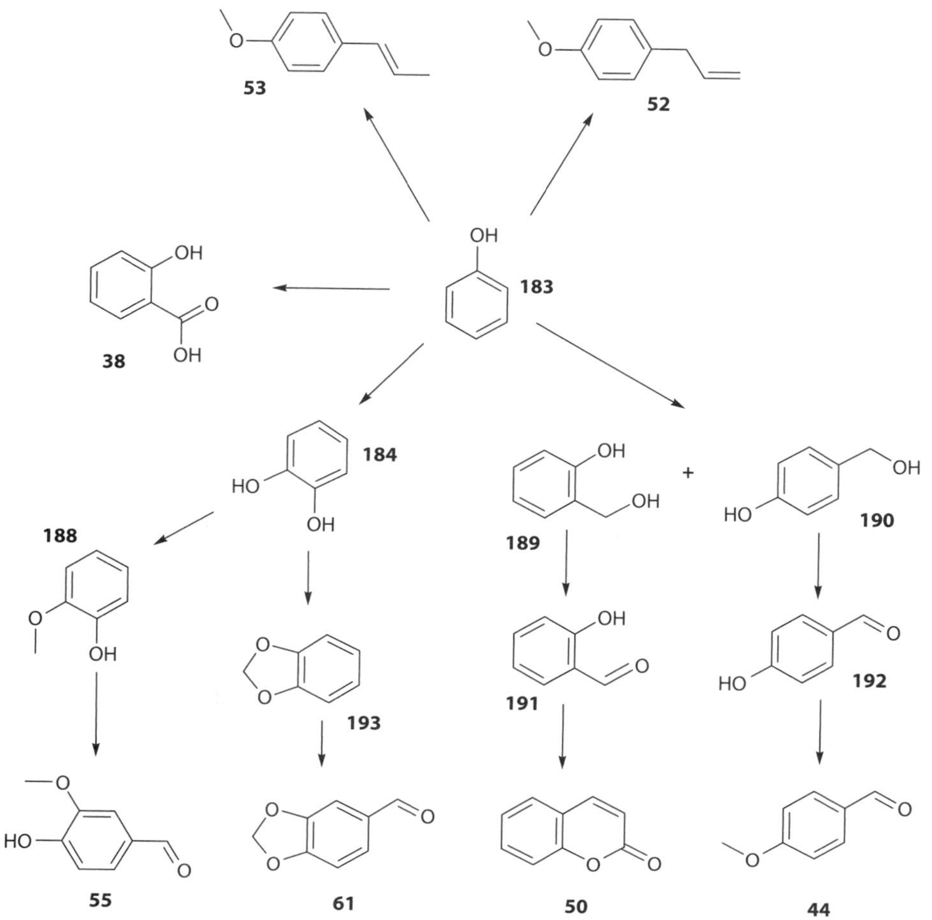

FIGURE 6.32 Synthesis of shikimates from phenol.

these key intermediates is readily achieved by standard functional group manipulation. Materials in this family serve as starting points for the synthesis of a wide range of perfumery materials including esters of the rose alcohols. The ionones are prepared from citral by aldol condensation followed by cyclization of the intermediate y-ionones.

The sources of the aforementioned key substances fall into three main categories: natural extracts, turpentine, and petrochemicals. The balance depends on economics and also on the product in question. For example, while about 10% of geraniol is sourced from natural extracts, it is only about 1% in the case of linalool. Natural grades of geraniol are obtained from the oils of citronella, geranium, and palmarosa (including the variants jamrosa and dhanrosa). Citronella is also used as a source of citronellal. Ho, rosewood, and linaloe were used as sources of linalool, but conservation and economic factors have reduced these sources of supply very considerably. Similarly, citral was once extracted from *L. cubeba* but overharvesting has resulted in loss of that source.

Various other natural extracts are used as feedstocks for the production of terpenoids as shown in Figure 6.35. Two of the most significant ones are clary sage and the citrus oils (obtained as by-products of the fruit juice industry). After distillation of the oil from clary sage, sclareol (**203**) is extracted from the residue, and this serves as a starting material for naphthofuran (**204**), known under trade names such as Ambrofix, Ambrox, and Ambroxan. The conversion is shown in Figure 6.35. Initially, sclareol is oxidized to sclareolide (**205**). This was once effected using oxidants such as

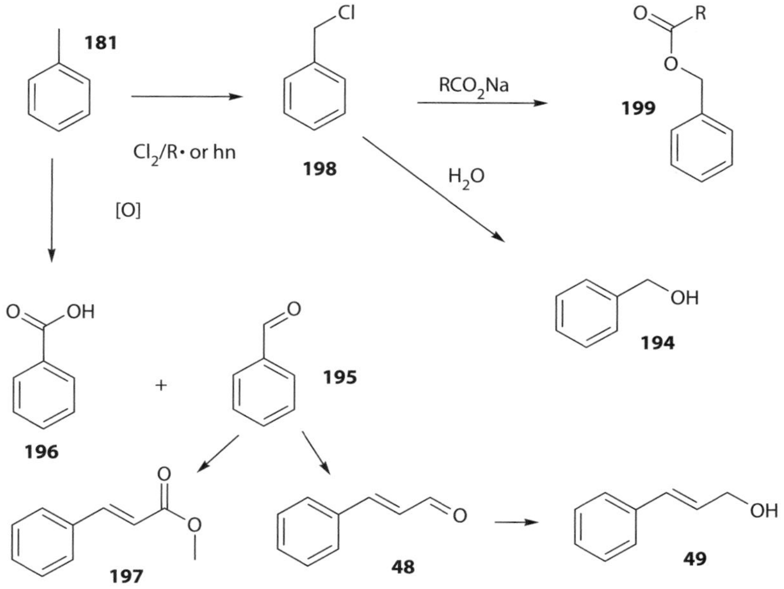

Chemistry of Essential Oils

FIGURE 6.33 Shikimates from toluene.

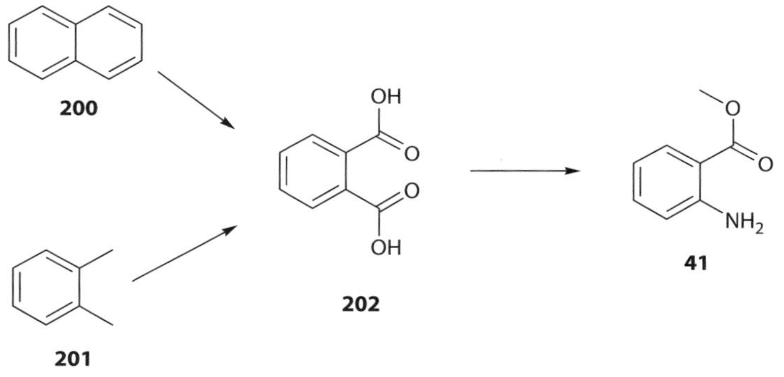

FIGURE 6.34 Synthesis of methyl anthranilate.

permanganate and dichromate, but nowadays, the largest commercial process uses a biotechnological oxidation. Sclareolide is then reduced using lithium aluminum hydride, borane, or similar reagents and the resulting diol is cyclized to the naphthofuran. D-Limonene (**73**) and valencene (**206**) are both extracted from citrus oils. Reaction of D-limonene with nitrosyl chloride gives an adduct that is rearranged to the oxime of L-carvone, and subsequent hydrolysis produces the free ketone (**105**). Selective oxidation of valencene gives nootkatone (**135**).

Turpentine is obtained by tapping of pine trees and this product is known as gum turpentine. However, a much larger commercial source is the so-called crude sulfate turpentine, which is obtained as a by-product of the Kraft paper process. The major components of turpentine are the two pinenes with a-pinene (**65**) predominating. Turpentine also serves as a source of *p*-cymene (**83**) and, as mentioned earlier, the shikimate anethole (**53**) (Zinkel and Russell, 1989).

Figure 6.36 shows some of the major products manufactured from a-pinene (**65**) (Sell, 2003, 2007). Acid-catalyzed hydration of a-pinene gives a-terpineol (**74**), which is the highest tonnage material of all those described here. Acid-catalyzed rearrangement of a-pinene gives camphene (**89**)

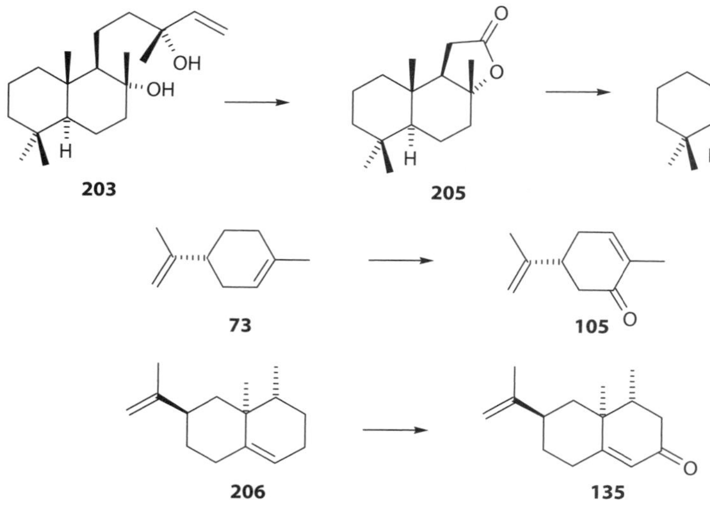

FIGURE 6.35 Partial synthesis of terpenoids from natural extracts.

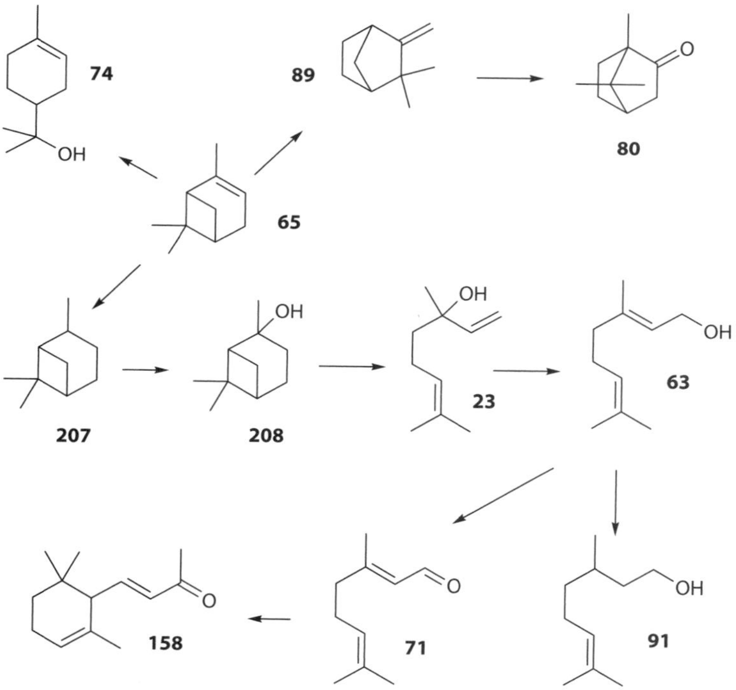

FIGURE 6.36 Products from a-pinene.

and this, in turn, serves as a starting material for production of camphor (**80**). Hydrogenation of a-pinene gives pinane (**207**), which is oxidized to pinanol (**208**) using air as the oxidant. Pyrolysis of pinanol produces linalool (**23**) and this can be rearranged to geraniol (**63**). Hydrogenation of geraniol gives citronellol (**91**), whereas oxidation leads to citral (**71**). The major use of citral is not as a material in its own right, but as a starting material for production of ionones, such as a-ionone (**158**) and vitamins A, E, and K.

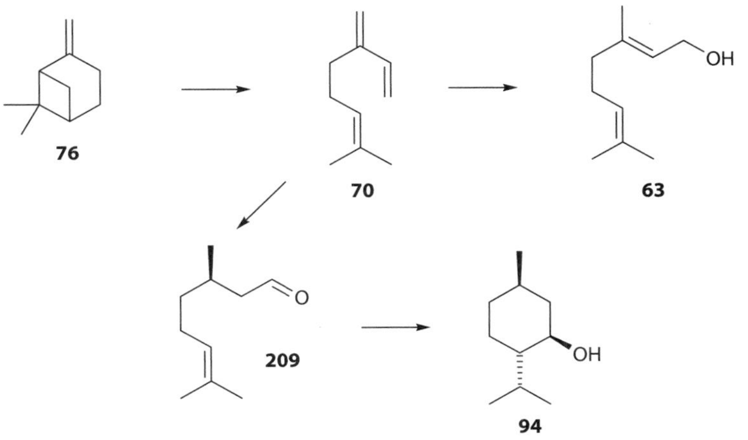

Chemistry of Essential Oils

FIGURE 6.37 Products from b-pinene.

Some of the major products manufactured from b-pinene (**76**) are shown in Figure 6.37. Pyrolysis of b-pinene gives myrcene (**70**) and this can be *hydrated* (not in one step but in a multistage process) to give geraniol (**63**). The downstream products from geraniol are then the same as those described in the preceding paragraph and shown in Figure 6.36. Myrcene is also a starting point for D-citronellol (**209**), which is one of the major feedstocks for the production of L-menthol (**94**) as will be described in the succeeding text.

Currently, there are two major routes to terpenoids that use petrochemical starting materials (Sell, 2003, 2007). The first to be developed is an improved version of a synthetic scheme demonstrated by Arens and van Dorp in 1948. The basic concept is to use two molecules of acetylene (**210**) and two of acetone (**211**) to build the structure of citral (**71**). The route, as it is currently practiced, is shown in Figure 6.38. Addition of acetylene (**210**) to acetone (**211**) in the presence of base gives methylbutynol (**212**), which is hydrogenated, under Lindlar conditions, to methylbutenol (**213**). The second equivalent of acetone is introduced as the methyl ether of its enol form, that is, methoxypropene (**214**). This adds to methylbutenol, and the resultant adduct undergoes a Claisen

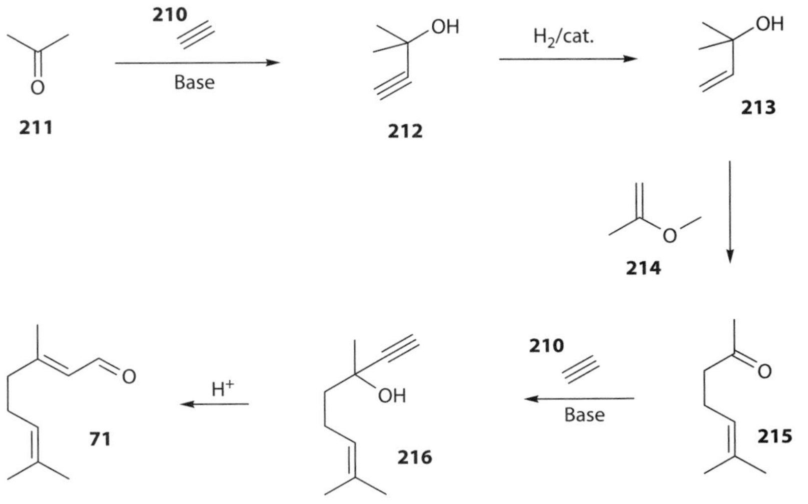

FIGURE 6.38 Citral from acetylene and acetone.

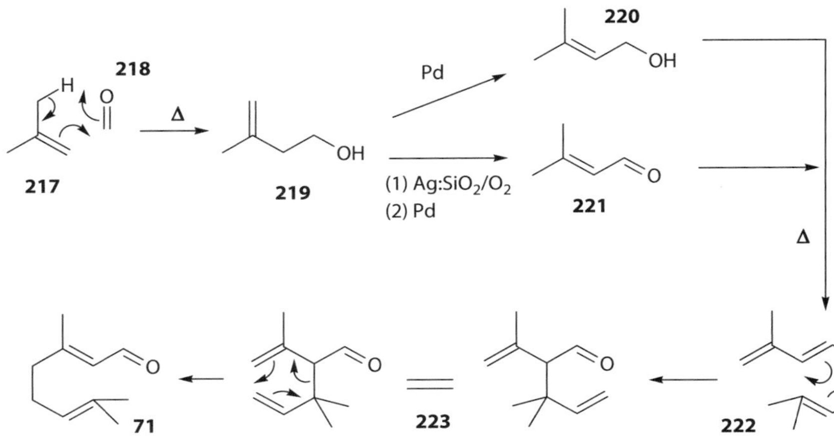

FIGURE 6.39 Citral from isobutylene and acetone.

rearrangement to give methylheptenone (**215**). Base catalyzed addition of the second acetylene to this gives dehydrolinalool (**216**), which can be rearranged under acidic conditions to give citral (**71**). Hydrogenation of dehydrolinalool under Lindlar conditions gives linalool (**23**) and thus opens up all the routes to other terpenoids as described earlier and illustrated in Figure 6.36.

The other major route to citral is shown in Figure 6.39. This starts from isobutene (**217**) and formaldehyde (**218**). The ene reaction between these produces isoprenol (**219**). Isomerization of isoprenol over a palladium catalyst gives prenol (**220**) and aerial oxidation over a silver catalyst gives prenal (senecioaldehyde) (**221**). When heated together, these two add together to form the enol ether (**222**), which then undergoes a Claisen rearrangement to give the aldehyde (**223**). This latter molecule is perfectly set up (after rotation around the central bond) for a Cope rearrangement to give citral (**71**). Development chemists have always striven to produce economic processes with the highest overall yield possible thus minimizing the volume of waste and hence environmental impact. This synthesis is a very good example of the fruits of such work. The reaction scheme uses no reagents, other than oxygen, employs efficient catalysts, and produces only one by-product, water, which is environmentally benign.

The synthesis of L-menthol (**94**) provides an interesting example of different routes operating in economic balance. The three production routes in current use are shown in Figure 6.40. The oldest

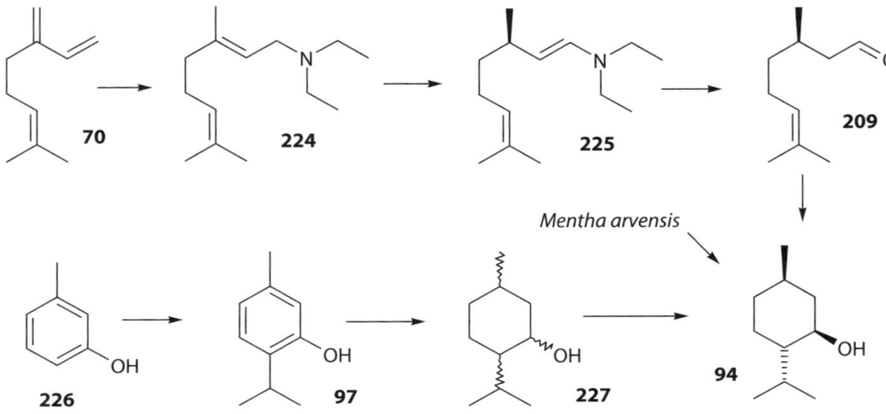

FIGURE 6.40 Competing routes to L-menthol.

and simplest route is extraction from plants of the *Mentha* genus and *M. arvensis* (corn mint) in particular. This is achieved by freezing the oil to force the L-menthol to crystallize out. Diethylamine can be added to myrcene (**70**) in the presence of base and rearrangement of the resultant allyl amine (**224**) using the optically active catalyst ruthenium (*S*)-BINAP perchlorate gives the homochiral enamine (**225**). This can then be hydrolyzed to *d*-citronellol (**209**). The chiral center in this molecule ensures that, on acid-catalyzed cyclization, the two new stereocenters formed possess the correct stereochemistry for conversion, by hydrogenation, to give L-menthol as the final product. Starting from the petrochemically sourced *m*-cresol (**226**), propenylation gives thymol (**97**), which can be hydrogenated to give a mixture of all eight stereoisomers of menthol (**227**). Fractional distillation of this mixture gives racemic menthol. Resolution was originally carried out by fractional crystallization, but recent advances include methods for the enzymic resolution of the racemate to give L-menthol.

Estimation of the long-term sustainability of each of these routes is complex and the final outcome is far from certain. In terms of renewability of feedstocks, *m*-cresol might appear to be at a disadvantage against mint or turpentine. However, as the world's population increases, use of agricultural land will come under pressure for food production, hence increasing pressure on mint cultivation and turpentine, hence, myrcene is a by-product of paper manufacture and is therefore vulnerable to trends in paper recycling and "the paperless office." In terms of energy consumption, and hence current dependence on petrochemicals, the picture is also not as clear as might be imagined. Harvesting and processing of mint requires energy and, if the crop is grown in the same field over time, fertilizer is required and this is produced by the very energy-intensive Haber process. The energy required to turn trees in a forest into pulp at a sawmill is also significant and so turpentine supply will also be affected by energy prices. No doubt, the skills of process chemists will be of increasing importance as we strive to make the best use of natural resources and minimize energy consumption (Baser and Demirci, 2007).

REFERENCES

Arctander, S., 1960. *Perfume and Flavour Materials of Natural Origin*. Elizabeth, NJ: Steffen Arctander. (Currently available from Allured Publishing Corporation.)
Arctander, S., 1969. *Perfume and Flavor Chemicals (Aroma Chemicals)*. Montclair, NJ: Steffen Arctander. (Currently available from Allured Publishing Corporation.)
Baser, K.H.C. and F. Demirci, 2007. Chemistry of essential oils. In *Flavours and Fragrances: Chemistry, Bioprocessing and Sustainability*, R.G. Berger (ed.), pp. 43–86, Berlin, Germany: Springer.
Bauer, K. and J. Panten, 2006. *Common Fragrance and Flavor Materials: Preparation, Properties and Uses*. New York: Wiley-VCH.
Bu'Lock, J.D., 1965. *The Biosynthesis of Natural Products*. New York: McGraw-Hill.
Croteau, R., 1987. Biosynthesis and catabolism of monoterpenoids. *Chem. Rev.*, 87: 929.
Däniker, H.U., 1987. *Flavors and Fragrances (Worldwide)*. Stamford, CA: SRI International.
Devon, T.K. and A.I. Scott, 1972. *Handbook of Naturally Occurring Compounds*, Vol. 2, *The Terpenes*. New York: Academic Press.
Essential Oils Database, 2006. Boelens Aromachemical Information Systems, Leffingwell & Associates, Canton, GA. http://www.leffingwell.com/baciseso.htm (accessed August 20, 2015).
Gildemeister, E. and Fr. Hoffmann, 1956. *Die Ätherischen Öle*. Berlin, Germany: Akademie-Verlag.
Günther, E., 1948. *The Essential Oils*. New York: D van Nostrand.
Hay, R.K.M. and P.G. Waterman (eds.), 1993. *Volatile Oil Crops: Their Biology, Biochemistry and Production*. London, U.K.: Longman.
Lawrence, B.M., 1985. A review of the world production of essential oils. *Perfumer Flavorist*, 10(5): 1.
Lehninger, A.L., 1993. *Principles of Biochemistry*. New York: Worth.
Mann, J., R.S. Davidson, J.B. Hobbs, D.V. Banthorpe, and J.B. Harbourne, 1994. *Natural Products: Their Chemistry and Biological Significance*. London, U.K.: Longman.
Matthews, C.K. and K.E. van Holde. 1990. *Biochemistry*. Redwood City, CA: Benjamin/Cummings.
Sell, C.S., 2003. *A Fragrant Introduction to Terpenoid Chemistry*. Cambridge, U.K.: Royal Society of Chemistry.

Sell, C.S. (ed.), 2006. *The Chemistry of Fragrances from Perfumer to Consumer*, 2nd ed. Cambridge, U.K.: Royal Society of Chemistry.
Sell, C.S., 2007. *Terpenoids*. In *Kirk-Othmer Encyclopedia of Chemical Technology*, 5th ed. New York: Wiley.
Williams, D.G., 1996. *The Chemistry of Essential Oils*. Weymouth, U.K.: Micelle Press.
Williams, D.G., 2004. *Perfumes of Yesterday*. Weymouth, U.K.: Micelle Press.
Zinkel, D.F. and J. Russell (eds.), 1989. *Naval Stores*. New York: Pulp Chemicals Association, Inc.

7 Analysis of Essential Oils

*Barbara d'Acampora Zellner, Paola Dugo,
Giovanni Dugo, and Luigi Mondello*

CONTENTS

7.1 Introduction 195
7.2 Classical Analytical Techniques 196
7.3 Modern Analytical Techniques 200
    7.3.1 Use of GC and Linear Retention Indices in Essential Oil Analysis 201
    7.3.2 GC–MS 202
    7.3.3 Fast GC for Essential Oil Analysis 203
    7.3.4 GC–Olfactometry for the Assessment of Odor-Active Components of Essential Oils 206
    7.3.5 Gas Chromatographic Enantiomer Characterization of Essential Oils 208
    7.3.6 LC and LC Hyphenated to MS in the Analysis of Essential Oils 209
    7.3.7 Multidimensional Gas Chromatographic Techniques 212
    7.3.8 Multidimensional Liquid Chromatographic Techniques 218
    7.3.9 Online Coupled LC–GC 221
7.4 General Considerations on Essential Oil Analysis 221
References 222

7.1 INTRODUCTION

The production of essential oils was industrialized in the first half of the nineteenth century, due to an increased demand for these matrices as perfume and flavor ingredients [1]. As a consequence, the need to perform their systematic investigation also became unprecedented. It is interesting to point out that in the second edition of Parry's monograph, published in 1908, about 90 essential oils were listed, and very little was known about their composition [2]. Further important contributions to the essential oil research field were made by Semmler [3], Gildemeister and Hoffman [4], Finnemore [5], and Guenther [6]. Obviously, it is unfeasible to cite all the researchers involved in the progress of essential oil analysis.

As widely acknowledged, the composition of essential oils is mainly represented by mono- and sesquiterpene hydrocarbons and their oxygenated (hydroxyl and carbonyl) derivatives, along with aliphatic aldehydes, alcohols, and esters. Terpenes can be considered as the most structurally varied class of plant natural products, derived from the repetitive fusion of branched five-carbon units (isoprene units) [7]. In this respect, analytical methods applied in the characterization of essential oils have to account for a great number of molecular species. Moreover, it is also of great importance to highlight that an essential oil chemical profile is closely related to the extraction procedure employed, and, hence, the choice of an appropriate extraction method becomes crucial. On the basis of the properties of the plant material, the following extraction techniques can be applied: steam distillation (SD), possibly followed by rectification and fractionation, solvent extraction (SE), fractionation of solvent extracts, maceration, expression (cold pressing of citrus peels), *enfleurage*, supercritical fluid extraction, pressurized-fluid extraction, simultaneous distillation–extraction, Soxhlet extraction, microwave-assisted hydrodistillation, dynamic and static headspace (HS) techniques, solvent-assisted flavor evaporation, solid-phase microextraction (SPME), and direct thermal desorption, among others.

Apart from the great interest in performing systematic studies on essential oils, there is also the necessity to trace adulterations, mainly in economically important essential oils. As can be observed with almost all commercially available products, market changes occur rapidly, affecting individual plants or industrial processes. In general, market competition, along with the limited interest of consumers with regard to essential oil quality, may induce producers to adulterate their commodities by the addition of products of lower value. Different types of adulterations can be encountered: (1) the simple addition of natural and/or synthetic compounds, with the aim of generating an oil characterized by specific quality values, such as density, optical rotation, residue percentage, and ester value; or (2) refined sophistications in the reconstitution and counterfeiting of commercially valuable oils. In the latter case, natural and/or synthetic compounds are added to enhance the market value of an oil, attempting to maintain the qualitative, or even quantitative, composition of natural essential oils and making adulteration detection a troublesome task. Consequently, the exploitation of modern analytical methodologies, such as gas chromatography (GC) and related hyphenated techniques, is practically unavoidable.

As a consequence of diffused illegal practice in the production of essential oils, there has been an enhanced request for legal standards of commercial purity, while essential oils were included as herbal drugs in pharmacopoeias [8–12] and also in a compendium denominated as *Martindale: the complete drug reference* (formerly named as *Martindale: The Extra Pharmacopoeia*) [13]. In view of the need for standardized methodologies, these pharmacopoeias commonly include the descriptions of several tests, processes, and apparatus. In addition, various international standard regulations have been introduced in which the characteristics of specific essential oils are described, and the botanical source and physicochemical requirements are reported. Such standardized information was created to facilitate the assessment of quality; for example, ISO 3761 (1997) specifies that for Brazilian rosewood essential oil (*Aniba rosaeodora* Ducke), an alcohol content in the 84%–93% range, determined as linalool, is required [14]. Moreover, guidelines for the analysis of essential oils are also available, for example, for the measurement of the refractive index (ISO 280, 1198) and optical rotation (ISO 592, 1998), as also for GC analysis using capillary columns (ISO chromatography [ISO 8432, 1987]) [15]. The French Standards Association (Association Française de Normalisation) also develops norms and standard methods dedicated to the essential oil research field, with the aim of assessing quality in relation to specific physical, organoleptic, chemical, and chromatographic characteristics [16].

The present contribution provides an overview on the classical and modern analytical techniques commonly applied to characterize essential oils. Modern techniques will be focused on chromatographic analyses, including theoretical aspects and applications.

7.2 CLASSICAL ANALYTICAL TECHNIQUES

The thorough study of essential oils is based on the relationship between their physical and chemical properties and is completed by the assessment of organoleptic qualities. The earliest analytical methods applied in the investigation of an essential oil were commonly focused on quality aspects, concerning mainly two properties: identity and purity [17].

The following techniques are commonly applied to assess an essential oil physical properties [6,17]: specific gravity (SG), which is the most frequently reported physicochemical property and is a special case of relative density, $[\rho]^{T(°C)}$, defined as the ratio of the densities of a given oil and of water when both are at identical temperatures. The attained value is characteristic for each essential oil and commonly ranges between 0.696 and 1.118 at 15°C [4]. In cases in which the determinations were made at different temperatures, conversion factors can be used to normalize data.

The measurement of optical rotation, $[\alpha]_D^{20}$, either dextrorotatory or levorotatory, is also widely recognized. Optical activity is determined by using a polarimeter, with the angle of rotation depending on a series of parameters, such as oil nature, the length of the column through which the light passes, the applied wavelength, and the temperature. The degree and direction of rotation are of great importance for purity assessments, since they are related to the structures and the concentration of chiral molecules in the sample. Each optically active substance has its own specific rotation, as defined in Biot's law:

$$[\alpha]_\lambda^T = \frac{[\alpha]_\lambda^T}{c \cdot l},$$

where
   α is the optical rotation at a temperature T expressed in °C
   l is the optical path length in dm
   λ is the wavelength
   c is the concentration in g/100 mL

It is worthy of note that a standard 100 mm tube is commonly used; in cases in which darker or lighter colored oils are analyzed, longer or shorter tubes are used, respectively, and the rotation should be extrapolated for a 100 mm long tube. Moreover, prior to the measurement, the essential oil should be dried out with anhydrous sodium sulfate and filtered.

The determination of the refractive index, $[\eta]_D^{20}$, also represents a characteristic physical constant of an oil, usually ranging from 1.450 to 1.590. This index is represented by the ratio of the sine of the angle of incidence (i) to the sine of the angle of refraction (e) of a beam of light passing from a less dense to a denser medium, such as from air to the essential oil:

$$\frac{\sin i}{\sin e} = \frac{N}{n},$$

where N and n are, respectively, the indices of the more and the less dense medium. The Abbé-type refractometer, equipped with a monochromatic sodium light source, is recommended for routine essential oil analysis; the instrument is calibrated through the analysis of distilled water at 20°C, producing a refractive index of 1.3330. In cases in which the measurement is performed at a temperature above or below 20°C, a correction factor per degree must be added or subtracted, respectively [18].

A further procedure that can be applied for the purity assessment of essential oils is based on water solubility. The test, which reveals the presence of polar substances, such as alcohols, glycols and their esters, and glycerin acetates, is carried out as follows: the oil is added to a saturated solution of sodium chloride, which after homogenization is divided into two phases; the volume of the oil, which is the organic phase, should remain unaltered; and volume reduction indicates the presence of water-soluble substances. On the other hand, the solubility, or immiscibility, of an essential oil in ethanol reveals much on its quality. Considering that essential oils are slightly soluble in water and are miscible with ethanol, it is simple to determine the number of volumes of water-diluted ethanol required for the complete solubility of one volume of oil; the analysis is carried out at 20°C, if the oil is liquid at this temperature. It must be emphasized that oils rich in oxygenated compounds are more readily soluble in dilute ethanol than those richer in hydrocarbons. Moreover, aged or improperly stored oils frequently present decreased solubility [6].

The investigation on the solubility of essential oils in other media is also widely accepted, such as the evaluation of the presence of water by means of a simple procedure: the addition of a volume of essential oil to an equal volume of carbon disulfide or chloroform; in case the oil is rich in oxygenated constituents, it may contain dissolved water, generating turbidity. A further solubility test, in which the oil is dissolved in an aqueous solution of potassium hydroxide, is applied to oils containing molecules with phenolic groups; finally, the incomplete dissolution of oils rich in aldehydes in a dilute bisulfite solution may denote the presence of impurities.

The estimation of melting and congealing points, as well as the boiling range of essential oils, is also of great importance for identity and purity assessments. Melting point evaluations are a valuable modality to control essential oil purity, since a large number of molecules generally comprised in essential oils melt within a range of 0.5°C or, in the case of decomposition, over a narrow temperature range. On the other hand, the determination of the congealing point is usually applied in cases where the essential oil consists mainly of one molecule, such as the oil of cloves that contains about 90% of eugenol. In the latter case, such a test enables the evaluation of the percentage amount of the abundant compound. At congealing point, crystallization occurs accompanied by heat liberation, leading to a rapid increase in temperature that is then stabilized at the so-called congealing point. A further purity evaluation method is represented by the boiling range determination, through which the percentage of oil that distils below a certain temperature or within a temperature range is investigated.

An additional test usually performed in essential oil analysis is the evaporation residue, in which the percentage of the oil that is not released at 100°C is determined. In the specific case of citrus oils, this test enables purity assessment, since a lower amount of residue in an expressed oil may indicate an addition of distilled volatile components to the oil; an increased residue amount reveals the possible presence of terpenes with higher molecular weights, through the addition of single compounds (or other essential oils), or of heavier oils, such as rosin oil and cheaper citrus oils, or by directly using the citrus oil residue. An example consists of the addition of lime oil to sophisticate lemon oils. In oxidized or polymerized oils, the presence of less volatile compounds is common; in this case, a simple test may be carried out by applying a drop of oil on a piece of filter paper; if a transparent spot persists for a period of over 24 h, the oil is most probably degraded. Furthermore, the residue can be subjected to acid and saponification number analyses; for instance, the addition of rosin oil would increase the acid number since this oil, different from other volatile oils, is characterized by the presence of complex acids. By definition, the acid number is the number of milligrams of potassium hydroxide required to neutralize the free acids contained in 1 g of an oil. This number is preserved in cases in which the essential oil has been carefully dried and stored in dark and airtight recipients. As commonly observed, the acid number increases along the aging process of an oil; oxidation of aldehydes and hydrolysis of esters trigger the increase of the acid number.

Classical methodologies have been also widely applied to assess essential oil chemical properties [6,17], such as the determination of the presence of halogenated hydrocarbons and of heavy metals. The former investigation is exploited to reveal the presence of halogenated compounds, commonly added to the oils for adulteration purposes. Several tests have been developed for halogen detection, with the Beilstein method [19] the one most reported. In practice, a copper wire is cleaned and heated in a Bunsen burner flame to form a coating of copper(II) oxide. It is then dipped in the sample to be tested and once again heated in the flame. A positive test is indicated by a green flame caused by the formation of a copper halide. Attention is to be paid to positive or inconclusive results, since they may be induced by trace amounts of organic acids, nitrogen-containing compounds [6], or salts [20]. An alternative to the Beilstein method is the sodium fusion test, in which the oil is first mineralized, and in the case halogenated hydrocarbons are present, a residue of sodium halide is formed, which is soluble in nitric acid, and precipitates as the respective silver halide by the addition of a small amount of silver nitrate solution [17]. With regard to the detection of heavy metals, several tests are described to investigate and ensure the absence especially of copper and lead. One method is based on the extraction of the essential oil with a diluted hydrochloric acid solution, followed by the formation of an aqueous phase to which a buffered thioacetamide solution is added. The latter reagent leads to the formation of sulfite ions that are used in the detection of heavy metals.

The determination of esters derived from phthalic acid is also of great interest for the toxicity evaluation of an essential oil. Considering that esters commonly contained in essential oils are derived from monobasic acids, at first, saponification is carried out through the addition of an ethanolic potassium hydroxide solution. The formed potassium phthalate, which is not soluble in ethanol, generates a crystalline precipitate [17].

The use of qualitative information alone is not sufficient to correctly characterize an essential oil, and quantitative data are of extreme importance. Classical methods are generally focused on chemical groups and the assessment of quantitative information through titration is widely applied, for example, for the acidimetric determination of saponified terpene esters. Saponification can be performed with heat, and in this case, readily saponified esters are to be investigated, in the cold, and afterward, the alkali excess is titrated with aqueous hydrochloric acid; thereafter, the ester number can be calculated. A further test is the determination of terpene alcohols by acetylating with acetic anhydride; part of the acetic anhydride is consumed in the reaction and can be quantified through titration of acetic acid with sodium hydroxide. The percentage of alcohol can then be calculated. The latter method is applied when the alcoholic constituents of an essential oil are not well known; in case these are established, the oil is saponified, and the ester number of the acetylated oil is calculated and used to estimate the free alcohol content.

Other chemical classes worthy of mention are aldehydes and ketones that may be investigated through different tests. The bisulfite method is recommended for essential oils rich in aldehydic compounds, such as lemon grass, bitter almond, and cassia, while the neutral sulfite test is more suitable for ketone-rich oils, such as spearmint, caraway, and dill oils. For essential oils presenting small amounts of aldehydes and ketones, the hydroxylamine method, or its modification, and the Stillman–Reed method are the most indicated ones [20]. In the latter case, the aldehyde and ketone contents are determined through the addition of a neutralized hydroxylamine hydrochloride solution, and subsequent titration with standardized acid (the Stillman–Reed method) [21]; in the former analytical procedure, the aldehyde and ketone content is established through the addition of a hydroxylamine hydrochloride solution, followed by neutralization with the reaction products, that is, alkali of the hydrochloric acid. These methods may be applied in the determination of citral in citrus oils and carvone in caraway oil. With regard to the determination of phenols, such as eugenol in clove oil or thymol and carvacrol in thyme oil, the test is commonly made through the addition of potassium hydroxide solutions, forming water-soluble salts. It has to be pointed out that besides phenols, other constituents are soluble in alkali solutions and in water [6,20].

Essential oils are also often analyzed by means of chromatographic methods. In general, the principle of chromatography is based on the distribution of the constituents to be separated between two immiscible phases; one of these is a stationary bed (a stationary phase) with a large surface area, while the other is a mobile phase that percolates through the stationary bed in a definite direction [22]. Planar chromatography may be referred to as a classical method for essential oil analysis, being well represented by thin-layer chromatography (TLC) and paper chromatography (PC). In both techniques, the stationary phase is distributed as a thin layer on a flat support, in PC being self-supporting, while in TLC coated on a glass, plastic, or metal surface; the mobile phase is allowed to ascend through the layer by capillary forces. TLC is a fast and inexpensive method for identifying substances and testing the purity of compounds, being widely used as a preliminary technique providing valuable information for subsequent analyses [23]. Separations in TLC involve the distribution of one or a mixture of substances between a stationary phase and a mobile phase. The stationary phase is a thin layer of adsorbent (usually silica gel or alumina) coated on a plate. The mobile phase is a developing solvent that travels up the stationary phase, carrying the samples with it. Components of the samples will separate on the stationary phase according to their stationary phase–mobile phase affinities [24]. In practice, a small quantity of the sample is applied near one edge of the plate and its position is marked with a pencil. The plate is then positioned in a developing chamber with one end immersed in the developing solvent, the mobile phase, avoiding the direct contact of the sample with the solvent. When the mobile phase reaches about two-thirds of the plate length, the plate is removed and dried, the solvent front is traced, and the separated components are located. In some cases, the spots are directly visible, but in others, they must be visualized by using methods applicable to almost all organic samples, such as the use of a solution of iodine or sulfuric acid, both of which react with organic compounds yielding dark products. The use of an ultraviolet (UV) lamp is also advisable, especially if a substance that aids in the visualization of compounds is incorporated into the plate, as is the case of many commercially available TLC plates. Data interpretation is made through the calculation of the ratio of fronts (R_f) value for each spot, which is defined as

$$R_f = \frac{Z_S}{Z_{St}},$$

where
Z_S is the distance from the starting point to the center of a specific spot
Z_{St} is the distance from the starting point to the solvent front [24,25]

A concise review on TLC has been made by Sherma [26].

The R_f value is characteristic for any given compound on the same stationary phase using the identical mobile phase. Hence, known R_f values can be compared to those of unknown substances to aid in their identification [24]. On the other hand, separations in PC involve the same principles as those in TLC, differing in the use of a high-quality filter paper as the stationary phase instead of a thin adsorbent layer, by the increased time requirements and poorer resolution. It is worthy to highlight that TLC has largely replaced PC in contemporary laboratory practice [22].

As is well known, essential oils can be characterized by their organoleptic properties, an assessment that involves human subjects as measuring tools. These procedures present an immediate problem, linked to the innate variability between individuals, not only as a result of their previous experiences and expectations, but also to their sensitivity [27]. In this respect, individuals are selected and screened for specific anosmia, as proposed by Friedrich et al. [28]. In case no insensitivities are found, the panelists are introduced to two sensorial properties, quality and intensity. Odor quality is described according to the odor families, while intensity is measured through the rating of a sensation based on an intensity interval scale. The assessment of an essential oil odor can be performed through its addition to filter paper strips and subsequent evaluation by the panelists. Considering that each volatile compound is characterized by a different volatility, the evaluation of the paper strip in different periods of time enables the classification of the odors in top, middle, and bottom notes [29]. In addition, the olfactive assessment during the determination of the evaporation residue is also of significance, since by notes low-boiling adulterants or contaminants may be detected as the oil vaporizes, and the odor of the final hot residue can reveal the addition of high-boiling compounds. Olfactive analyses are also valuable after the determination of phenols in essential oils, by studying the nonphenolic portion [6].

It is noteworthy that the use of the earliest analytical techniques for the systematic study of essential oils, such as SG, relative density, optical activity, and refractive index or melting, congealing, and boiling points determinations, is generally applied for the assessment of pure compounds and may be extended to evaluate essential oils composed of a major compound. Classical methods cannot be used as stand-alone methods and need to be combined with modern analytical techniques, especially GC, for the assessment of essential oil genuineness.

7.3 MODERN ANALYTICAL TECHNIQUES

Most of the methods applied in the analysis of essential oils rely on chromatographic procedures, which enable component separation and identification. However, additional confirmatory evidence is required for reliable identification, avoiding equivocated characterizations.

In the early stages of research in the essential oil field, attention was devoted to the development of methods in order to acquire deeper knowledge on the profiles of volatiles; however, this analytical task was made troublesome due to the complexity of these real-world samples. Over the last decades, the aforementioned research area has benefited from the improvements in instrumental

Analysis of Essential Oils

analytical chemistry, especially in the chromatographic area, and, nowadays, the number of known constituents has drastically increased.

The primary objective in any chromatographic separation is always the complete resolution of the compounds of interest, in the minimum time. To achieve this task, the most suitable analytical column (dimension and stationary phase type) has to be used, and adequate chromatographic parameters must be applied to limit peak enlargement phenomena. A good knowledge of chromatographic theory is, indeed, of great support for the method optimization process, as well as for the development of innovative techniques.

In gas chromatographic analysis, the compounds to be analyzed are vaporized and eluted by the mobile gas phase, the carrier gas, through the column. The analytes are separated on the basis of their relative vapor pressures and affinities for the stationary bed. On the other hand, in liquid chromatographic analysis, the compounds are eluted by a liquid mobile phase consisting of a solvent or a mixture of solvents, the composition of which may vary during the analysis (gradient elution), and are separated according to their affinities for the stationary bed. In general, the volatile fraction of an essential oil is analyzed by GC, while the nonvolatile by liquid chromatography (LC).

At the outlet of the chromatography column, the analytes emerge separated in time. The analytes are then detected, and a signal is recorded generating a chromatogram, which is a signal versus time graphic, and ideally with peaks presenting a Gaussian distribution–curve shape. The peak area and height are a function of the amount of solute present, and its width is a function of band spreading in the column [30], while retention time can be related to the solute's identity. Hence, the information contained in the chromatogram can be used for qualitative and quantitative analysis.

7.3.1 Use of GC and Linear Retention Indices in Essential Oil Analysis

The analysis of essential oils by means of GC began in the 1950s, when Professor Liberti [31] started analyzing citrus essential oils only a few years after James and Martin first described gas–liquid chromatography, commonly referred to as GC [32], a milestone in the evolution of instrumental chromatographic methods.

After its introduction, GC developed at a phenomenal rate, growing from a simple research novelty to a highly sophisticated instrument. Moreover, the current-day requirements for high resolution and trace analysis are satisfied by modern column technology. In particular, inert, thermostable, and efficient open-tubular columns are available, along with associated selective detectors and injection methods, which allow on-column injection of liquid and thermally labile samples. The development of robust fused-silica columns, characterized by superior performances to that of glass columns, brings open-tubular GC columns within the scope of almost every analytical laboratory.

At present, essential oil GC analyses are more frequently performed on capillary columns, which, after their introduction, rapidly replaced packed GC columns. In general, packed columns support larger sample size ranges, from 10 to 20 mL, and thus the dynamic range of the analysis can be enhanced. Trace-level components can be easily separated and quantified without preliminary fractionation or concentration. On the other hand, the use of packed columns leads to lower resolution due to the higher-pressure drop per unit length. Packed columns need to be operated at higher column flow rates, since their low permeability requires high pressures to significantly improve resolution [33]. It is worthy of note that since the introduction of fused-silica capillary columns considerable progress has been made in column technology, a great number of papers regarding GC applications on essential oils have been published.

The choice of the capillary column in an essential oil GC analysis is of great importance for the overall characterization of the matrix; the stationary phase chemical nature and film thickness, as well as the column length and internal diameter, are to be considered. In general, essential oil GC analyses are carried out on 25–50 m columns, with 0.20–0.32 mm internal diameters and 0.25 μm stationary phase film thickness. It must be noted that the degree of separation of two components on two distinct stationary phases can be drastically different. As is well known, nonpolar columns produce boiling-point separations, while on polar stationary phases, compounds are resolved according to their polarity. Considering that essential oil components, such as terpenes and their oxygenated derivatives, frequently present similar boiling points, these elute in a narrow retention time range on a nonpolar column. In order to overcome this limit, the analytical method can be modified by applying a slower oven temperature rate to widen the elution range of the oil or by using a polar stationary phase, as oxygenated compounds are more retained than hydrocarbons. However, choosing different stationary phases may provide little improvement as resolution can be improved for a series of compounds, but new coelutions can also be generated.

Considering gas chromatographic analyses using flame ionization detector (FID), thermal conductivity detector (TCD), or other detectors that do not provide structural information of the analyzed molecules and retention data, more precise retention indices are used as the primary criterion for peak assignment. The retention index system was based on the fact that each analyte is referenced in terms of its position between the two n-paraffins that bracket its retention time. Furthermore, the index calculation is based on a linear interpolation of the carbon chain length of these bracketing paraffins. The most thoroughly studied, diffused, and accepted retention index calculation methods are based on the logarithmic-based equation developed by Kováts [34], for isothermal conditions, and on the equation propounded by van den Dool and Kratz [35], which does not use the logarithmic form and is used in the case of temperature-programmed conditions. Values calculated using the latter approach are commonly denominated in literature as retention index (I), linear retention index (LRI), or programmed-temperature retention index (PTRI or I^T), while the ones derived from the former equation are usually referred to as Kováts index.

In general, retention index systems are based on the incremental structure–retention relationship, that any regular increase in a series of chemical structures should provide a regular increase in the corresponding retention times. This means that the retention index concept is not restricted to the use of n-alkanes as standards. In practice, any homologous series presenting a linear relationship between the adjusted retention time, being logarithmic based or not, and the carbon number can be used.

In the characterization of volatiles, the most commonly applied reference series is n-alkanes. However, the latter commonly present fluctuant behavior on polar stationary phases. In consideration of the fact that retention index values are correlated to retention mechanisms, alternative standard series of intermediate polarity has been introduced, such as 2-alkanones, alkyl ethers, alkyl halides, alkyl acetates, and alkanoic acid methyl esters [22]. Shibamoto [36] suggested the use of polar compound series, such as ethyl esters, as an alternative. The most feasible choice, when analyzing volatiles, is to apply reference series as n-alkanes, fatty acid ethyl esters (FAEEs), or fatty acid methyl esters (FAMEs), employed according to the stationary phase to be used.

Additionally, it is highly advisable to use two analytical columns coated with stationary phases of distinct polarities to obtain two retention index values and enhance confidence in assignments [37–39]. Identifications made on a single column can only be accepted if used in combination with spectroscopic detection systems. When n-alkanes are used, it is accepted that the reproducibility of retention indices between different laboratories are comprised within an acceptable range of ±5 units for methyl silicone stationary phases and ±10 units for polyethylene glycol phases. A further aspect of great importance, which is frequently overseen, is the analytical reproducibility of retention indexes. Moreover, it is worthwhile to highlight that in practice it was found that the use of an initial isothermal hold in the GC oven temperature program does not provide additional resolution [40].

7.3.2 GC–MS

Mass spectrometry (MS) can be defined as the study of systems through the formation of gaseous ions, with or without fragmentation, which are then characterized by their mass-to-charge ratios (m/z) and relative abundances [41]. The analyte may be ionized thermally, by an electric field or by impacting energetic electrons, ions, or photons.

During the past decade, there has been a tremendous growth in popularity of mass spectrometers as a tool for both, routine analytical experiments and fundamental research. This is due to a number of features including relatively low cost, simplicity of design and extremely fast data acquisition rates. Although the sample is destroyed by the mass spectrometer, the technique is very sensitive, and only low amounts of material are used in the analysis.

In addition, the potential of combined GC–mass spectrometry (GC–MS) for determining volatile compounds, contained in very complex flavor and fragrance samples, is well known. The subsequent introduction of powerful data acquisition and processing systems, including automated library search techniques, ensured that the information content of the large quantities of data generated by GC–MS instruments was fully exploited. The most frequent and simple identification method in GC–MS consists of the comparison of the acquired unknown mass spectra with those contained in a reference MS library.

A mass spectrometer produces an enormous amount of data, especially in combination with chromatographic sample inlets [42]. Over the years, many approaches for analysis of GC–MS data have been proposed using various algorithms, many of which are quite sophisticated, in efforts to detect, identify, and quantify all of the chromatographic peaks. Library search algorithms are commonly provided with mass spectrometer data systems with the purpose to assist in the identification of unknown compounds [43].

However, as is well known, compounds such as isomers, when analyzed by means of GC–MS, can be incorrectly identified, a drawback that is often observed in essential oil analysis. As is widely acknowledged, the composition of essential oils is mainly represented by terpenes, which generate very similar mass spectra; hence, a favorable match factor is not sufficient for identification, and peak assignment becomes a difficult, if not impracticable, task (Figure 7.1). In order to increase the reliability of the analytical results and to address the qualitative determination of compositions of complex samples by GC–MS, retention indices can be an effective tool. The use of retention indices in conjunction with the structural information provided by GC–MS is widely accepted and routinely used to confirm the identity of compounds. Besides, retention indices when incorporated to MS libraries can be applied as a filter, thus shortening the search routine for matching results and enhancing the credibility of MS identification [44].

According to Joulain and König [45], provided data contained in mass spectral libraries have been recorded using authentic samples, it can be observed that the mass spectrum of a given sesquiterpene is usually sufficient to ensure its identification when associated with its retention index obtained on methyl silicone stationary phases. Indeed, for the aforecited class of compounds, there would be no need to use a polyethylene glycol phase, which could even lead to misinterpretations caused by possible changes in the retention behavior of sesquiterpene hydrocarbons as a result of column aging or deterioration. Moreover, according to the authors, attention should be paid to the retention index and the mass spectrum registration of each individual sesquiterpene, since many compounds with rather similar mass spectra elute in a narrow range; more than 160 compounds can elute within 100 retention index units on a methyl silicone–based column, for example, 1400–1500.

7.3.3 Fast GC for Essential Oil Analysis

Nowadays in daily routine work, apart from increased analytical sensitivity, demands are also made on the efficiency in terms of speed of the laboratory equipment. Regarding the rapidity of analysis, two aspects need to be considered: (1) the costs in terms of time required, for example, as is the case in quality control analysis, and (2) the efficiency of the utilized analytical equipment.

When compared to conventional GC, the primary objective of fast GC is to maintain sufficient resolving power in a shorter time, by using adequate columns and instrumentation in combination with optimized run conditions to provide 3–10 times faster analysis times [46–48]. The technique can be accomplished by manipulating a number of analysis parameters, such as column length, column internal diameter [ID], stationary phase, film thickness, carrier gas, linear velocity,

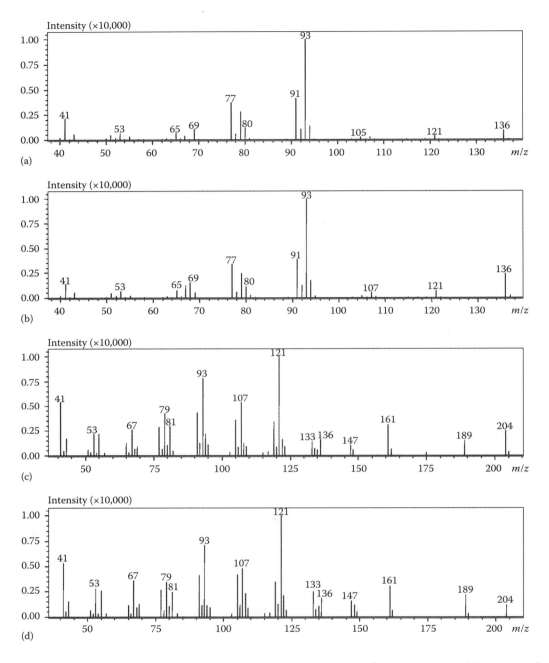

FIGURE 7.1 Representation of the similarity between mass spectra of monoterpenes, sabinene (a) and β-phellandrene (b), and sesquiterpenes, bicyclogermacrene (c) and germacrene B (d).

oven temperature, and ramp rate. Fast GC is typically performed using short, 0.10 or 0.18 mm ID capillary columns with hydrogen carrier gas and rapid oven temperature ramp rates. In general, capillary gas chromatographic analysis may be divided into three groups, based solely on column internal diameter types: (1) conventional GC when 0.25 mm ID columns are applied, (2) fast GC using 0.10–0.18 mm ID columns, and (3) ultrafast GC for columns with an ID of 0.05 mm or less. In addition, GC analyses times between 3 and 12 min can be defined as "fast," between 1 and 3 min as "very fast," and below 1 min as "ultrafast." Fast GC requires instrumentation provided with high split ratio injection systems because of low sample column capacities, increased inlet pressures, rapid oven heating rates, and fast electronics for detection and data collection [49].

The application of two methods, conventional (30 m × 0.25 mm ID, 0.25 μm d_f column) and fast (10 m × 0.10 mm ID, 0.10 μm d_f column), on five different citrus essential oils (bergamot, mandarin, lemon, bitter oranges, and sweet oranges) has been reported [49]. The fast method allowed the separation of almost the same compounds as the conventional analysis, while quantitative data showed good reproducibility. The effectiveness of the fast GC method, through the use of narrow-bore columns, was demonstrated. An ultrafast GC lime essential oil analysis was also performed on a 5 m × 50 μm capillary column with 0.05 μm stationary phase film thickness [50]. The total analysis time of this volatile essential oil was less than 90 s; a chromatogram is presented in Figure 7.2.

Another technique, ultrafast module–GC (UFM–GC) with direct resistively heated narrow-bore columns, has been applied to the routine analysis of four essential oils of differing complexities: chamomile, peppermint, rosemary, and sage [51]. All essential oils were analyzed by conventional GC with columns of different lengths: 5 and 25 m, with a 0.25 mm ID, and by fast GC and UFM–GC with narrow-bore columns (5 m × 0.1 mm ID). Column performances were evaluated and compared through the Grob test, separation numbers, and peak capacities. UFM–GC was successful in the qualitative and quantitative analyses of essential oils of different compositions with analysis times between 40 s and 2 min versus 20–60 min required by conventional GC. UFM–GC allows to drastically reduce the analysis time, although the very high column heating rates may lead to changes in selectivity compared to conventional GC, which are more marked than those of classical fast GC. In a further work, the same researchers [52] stated that in UFM–GC experiments the appropriate flow choice can compensate, in part, the loss of separation capability due to the heating rate increase.

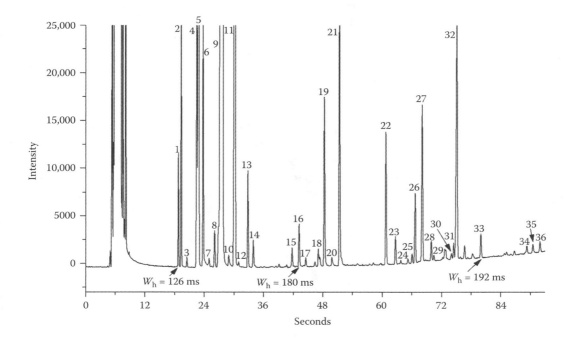

FIGURE 7.2 Fast gas chromatography analysis of a lime essential oil on a 5 m × 5 mm (0.05 μm film thickness) capillary column, applying fast temperature programming. The peak widths of three components are marked to provide an illustration of the high efficiency of the column, even under extreme operating conditions (for peak identification, see Ref. [50]). (From Mondello, L. et al., *J. Sep. Sci.*, 27, 699, 2004. With permission.)

Besides the numerous fast GC applications on citrus essential oils, other oils have also been subjected to analysis, such as rose oil by means of ultrafast GC [53] and very fast GC [54], both using narrow-bore columns. Rosemary and chamomile oils have been investigated by means of fast GC on two short conventional columns of distinct polarity (5 m × 0.25 mm ID) [55]. The latter oil has also been analyzed through fast HS–SPME–GC on a narrow-bore column [56]. Fast and very fast GC analyses on narrow-bore columns have also been carried out on patchouli and peppermint oils [57].

7.3.4 GC–Olfactometry for the Assessment of Odor-Active Components of Essential Oils

The discriminatory capacity of the mammalian olfactory system is such that thousands of volatile chemicals are perceived as having distinct odors. It is accepted that the sensation of odor is triggered by highly complex mixtures of volatile molecules, mostly hydrophobic, and usually occurring in trace-level concentrations (ppm or ppb). These volatiles interact with odorant receptors of the olfactive epithelium located in the nasal cavity. Once the receptor is activated, a cascade of events is triggered to transform the chemical–structural information contained in the odorous stimulus into a membrane potential [58,59], which is projected to the olfactory bulb and then transported to higher regions of the brain [60] where the translation occurs.

It is known that only a small portion of the large number of volatiles occurring in a fragrant matrix contributes to its overall perceived odor [61,62]. Further, these molecules do not contribute equally to the overall flavor profile of a sample; hence, a large GC peak area, generated by a chemical detector does not necessarily correspond to high odor intensities, due to differences in intensity/concentration relationships.

The description of a gas chromatograph modified for the sniffing of its effluent to determine volatile odor activity was first published by Fuller et al. [63]. In general, GC–olfactometry (GC–O) is carried out on a standard GC that has been equipped with a sniffing port, also denominated olfactometry port or transfer line, in substitution of, or in addition to, the conventional detector. When a FID or a mass spectrometer is also used, the analytical column effluent is split and transferred to the conventional detector and to the human nose. GC–O was a breakthrough in analytical aroma research, enabling the differentiation of a multitude of volatiles, previously separated by GC, in odor active and nonodor active, related to their existing concentrations in the matrix under investigation. Moreover, it is a unique analytical technique that associates the resolution power of capillary GC with the selectivity and sensitivity of the human nose.

GC–O systems are often used in addition to either a FID or a mass spectrometer. With regard to detectors, splitting column flow between the olfactory port and a mass spectral detector provides simultaneous identification of odor-active compounds. Another variation is to use an in-line, nondestructive detector such as a TCD [64] or a photoionization detector [65]. Especially when working with GC–O systems equipped with detectors that do not provide structural information, retention indexes are commonly associated to odor description supporting peak assignment.

Over the last decades, GC–O has been extensively used in essential oil analysis in combination with sophisticated olfactometric methods; the latter were developed to collect and process GC–O data and, hence, to estimate the sensory contribution of a single odor-active compound. The odor-active compounds of essential oils extracted from citrus fruits (*Citrus* sp.), such as orange, lime, and lemon, were among the first character impact compounds identified by flavor chemists [66].

GC–O methods are commonly classified in four categories: dilution, time intensity, detection frequency, and posterior intensity methods. Dilution analysis, the most applied method, is based on successive dilutions of an aroma extract until no odor is perceived by the panelists. This procedure, usually performed by a reduced number of assessors, is mainly represented by combined hedonic aroma response method (CHARM) [67], developed by Acree et al., and aroma extraction dilution analysis (AEDA), first presented by Ullrich and Grosch [68]. The former method has been applied to the investigation of two sweet orange oils from different varieties, one Florida Valencia and the other Brazilian Pera [69]. The intensities and qualities of their odor-active components were assessed. CHARM results indicated for both the oils that the most odor-active compounds are associated with the polar fraction compounds: straight chain aldehydes (C_8–C_{14}), β-sinensal, and linalool presented the major CHARM responses. On the other hand, AEDA has been used to investigate the odor-active compounds responsible for the characteristic odors of yuzu oil (*Citrus junos* Sieb. ex Tanaka) [70] and daidai (*Citrus aurantium* L. var. *cyathifera* Y. Tanaka) [71] cold-pressed essential oils.

Time-intensity methods, such as OSME (Greek word for odor), are based on the immediate recording of the intensity as a function of time by moving the cursor of a variable resistor [72]. An interesting application of the time-intensity approach was demonstrated for cold-pressed grapefruit oil [73], in which 38 odor-active compounds were detected and, among these, 22 were considered as aroma-impact compounds. A comparison between the grapefruit oil gas chromatogram and the corresponding time-intensity aromatogram for that sample is shown in Figure 7.3.

A further approach, the detection frequency method [74,75], uses the number of evaluators detecting an odor-active compound in the GC effluent as a measure of its intensity. This GC–O method is performed with a panel composed of numerous and untrained evaluators; 8–10 assessors are a good agreement between low variation of the results and analysis time. It must be added that the results attained are not based on real intensities and are limited by the scale of measurement. An application of the detection frequency method was reported for the evaluation of leaf- and wood-derived essential oils of Brazilian rosewood (*A. roseaodora* Ducke) essential oils by means of enantioselective–GC–O (Es–GC–O) analyses [76].

Another GC–O technique, the posterior intensity method [77], proposes the measurement of a compound odor intensity and its posterior scoring on a previously determined scale. This posterior registration of the perceived intensity may cause a considerable variance between assessors.

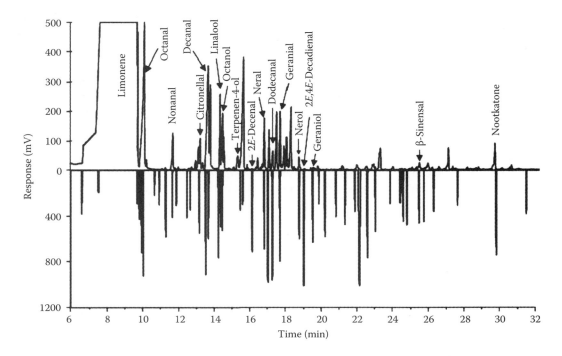

FIGURE 7.3 Gas chromatography–flame ionization detector chromatogram with some components identified by means of mass spectrometry (top) and a time-intensity aromatogram of grapefruit oil (bottom). The separation was performed on a polyethylene glycol column (30 m × 0.32 mm ID, 0.25 µm film thickness). (From Lin, J. and Rouseff, R.L., *Flavour Fragr. J.*, 16, 457, 2001. With permission.)

The attained results may generally be well correlated with detection frequency method results and, to a lesser extent, with dilution methods. In the aforementioned research performed on the essential oils of Brazilian rosewood, this method was also used to give complementary information on the intensity of the linalool enantiomers [76].

Other GC–O applications are also reported in literature using the so-called peak-to-odor impression correlation, the method in which the olfactive quality of an odor-active compound perceived by a panelist is described. The odor-active compounds of the essential oils of black pepper (*Piper nigrum*) and Ashanti pepper (*Piper guineense*) were assessed applying the aforecited correlation method [78]. The odor profile of the essential oils of leaves and flowers of *Hyptis pectinata* (L.) Poit. was also investigated by using the peak-to-odor impression correlation [79].

The choice of the GC–O method is of extreme importance for the correct characterization of a matrix, since the application of different methods to an identical real sample can distinctly select and rank the odor-active compounds according to their odor potency and/or intensity. Commonly, detection frequency and posterior intensity methods result in similar odor intensity/concentration relationships, while dilution analysis investigate and attribute odor potencies.

7.3.5 Gas Chromatographic Enantiomer Characterization of Essential Oils

Capillary GC is currently the method of choice for enantiomer analysis of essential oils, and Es–GC has become an essential tool for stereochemical analysis mainly after the introduction of cyclodextrin (CD) derivatives as chiral stationary phases (CSPs) in 1983 by Sybilska and Koscielski, at the University of Warsaw, for packed columns [80], and is applied to capillary columns in the same decade [81,82]. Moreover, Nowotny et al. first proposed diluting CD derivatives in moderately polar polysiloxane (OV-1701) phases to provide them with good chromatographic properties and a wider range of operative temperatures [83].

The advantage on the application of Es–GC lies mainly in its high separation efficiency and sensitivity, simple detection, unusually high precision and reproducibility, and also the need for a small amount of sample. Moreover, its main use is related with the characterization of the enantiomeric composition and the determination of the enantiomeric excess (ee) and/or ratio (ER) of chiral research chemicals, intermediates, metabolites, flavors and fragrances, drugs, pesticides, fungicides, herbicides, pheromones, and so on. Information on ee or ER is of great importance to characterize natural flavor and fragrance materials, such as essential oils, since the obtained values are useful tools, or even "fingerprints," for the determination of their quality, applied extraction technique, geographic origin, biogenesis, and also authenticity [84].

A great number of essential oils have already been investigated by means of Es–GC using distinct CSPs; unfortunately, a universal chiral selector with widespread potential for enantiomer separation is not available, and thus effective optical separation of all chiral compounds present in a matrix may be unachievable on a single chiral column. In 1997, Bicchi et al. [85] reported the use of columns that addressed particular chiral separations, noting that certain CSPs preferentially resolved certain enantiomers. Thus, a 2,3-di-*O*-ethyl-6-*O*-tert-butyldimethylsilyl-β-CD on polymethylphenylsiloxane (PS086) phase allowed the characterization of lavender and citrus oils containing linalyl oxides, linalool, linalyl acetate, borneol, bornyl acetate, α-terpineol, and *cis*- and *trans*-nerolidol. On the other hand, peppermint oil was better analyzed by using a 2,3-di-*O*-methyl-6-*O*-*tert*-butyldimethylsilyl-β-CD on PS086 phase, and especially for α- and β-pinene, limonene, menthone, isomenthone, menthol, isomenthol, pulegone, and methyl acetate. König [86] performed an exhaustive investigation of the stereochemical correlations of terpenoids, concluding that when using a heptakis (6-*O*-methyl-2,3-di-*O*-penthyl)-β-CD and octakis (6-*O*-methyl-2,3-di-*O*-penthyl)-γ-CD in polysiloxane, the presence of both enantiomers of a single compound is common for monoterpenes, less common for sesquiterpenes, and never observed for diterpenes.

Substantial improvements in chiral separations have been extensively published in the field of chromatography. At present, over 100 stationary phases with immobilized chiral selectors are available [22], presenting increased stability and extended lifetime. It can be affirmed that enantioselective chromatography has now reached a high degree of sophistication. To better characterize an essential oil, it is advisable to perform Es–GC analysis on at least two, or better three, columns coated with different CD derivatives. This procedure enables the separation of more than 85% of the racemates that commonly occur in these matrices [87], while the reversal of enantiomer elution order can take place in several cases. The analyst must be aware of some practical aspects prior to an Es–GC analysis: as is well accepted, variations in linear velocity can affect the separation of enantiomeric pairs; resolution (R_S) can be improved by optimizing the gas linear velocity, a factor of high importance in cases of difficult enantiomer separation. Satisfactory resolution requires $R_S \geq 1$, and baseline resolution is obtained when $R_S \geq 1.5$ [88]. Resolution can be further improved by applying slow temperature ramp rates (1°C–2°C/min is frequently suggested). Moreover, according to the CSP used, the initial GC oven temperature can affect peak width; initial temperatures of 35°C–40°C are recommended for the most column types. Furthermore, attention should be devoted to the column sample capacity, which varies with different compounds, overloading results in broad tailing peaks and reduced enantiomeric resolution. The troublesome separation and identification of enantiomers due to the fact that each chiral molecule splits into two chromatographic signals, for each existing stereochemical center, are also worthy of note. As a consequence, the increase in complexity of certain regions of the chromatogram may lead to imprecise ee and/or ER values. In terms of retention time repeatability, and also reproducibility, it can be affirmed that good results are being achieved with commercially available chiral columns.

The retention index calculation of optically active compounds can be considered as a troublesome issue due to complex inclusion complexation retention mechanisms on CD stationary phases; if a homologous series, such as the *n*-alkanes, is used, the hydrocarbons randomly occupy positions in the chiral cavities. As a consequence, *n*-alkanes can be considered as unsuitable for retention index determinations. Nevertheless, other reference series can be employed on CD stationary phases, such as linear chain FAMEs and FAEEs. However, retention indices are seldom reported for optically active compounds, and publications refer to retention times rather than indices.

The innovations in Es–GC analysis have concerned not only the development and applications of distinct CSPs but also the development of distinct enantioselective analytical techniques, such as Es–GC–MS, Es–GC–O, enantioselective multidimensional GC (Es–MDGC), Es–MDGC–MS, Es–GC hyphenated to isotopic ratio mass spectrometry (Es–GC–IRMS), and Es–MDGC–IRMS.

It is obvious that an enantioselective separation in combination with MS detection presents the additional advantage of qualitative information. Notwithstanding, a difficulty often encountered is that related to peak assignment, due to the similar fragmentation pattern of isomers. The reliability of Es–GC–MS results can be increased by using an effective tool, retention indices. It can be assumed that in the enantioselective recognition of optically active isomers in essential oils, mass spectra can be exploited to locate the two enantiomers in the chromatogram, and the LRI, when possible, enables their identification [89]. In addition, the well-known property of odor activity recognized for several isomers can be assessed by means of Es–GC–MS–O and can represent an outstanding tool for precise enantiomer characterization (Figure 7.4).

As demonstrated by Mosandl and his group [90], Es–GC–O is a valid tool for the simultaneous stereodifferentiation and olfactive evaluation of the volatile optically active components present in essential oils. It is worthwhile to point out that the preponderance of one of the enantiomers, defined by the ee, results in a characteristic aroma [91] and is of great importance for the olfactive characterization of the sample.

7.3.6 LC and LC Hyphenated to MS in the Analysis of Essential Oils

Some natural complex matrices do not need sample preparation prior to GC analysis, for example, essential oils. The latter generally contain only volatile components, since their preparation is performed by SD. Citrus oils, extracted by cold-pressing machines, are an exception, containing

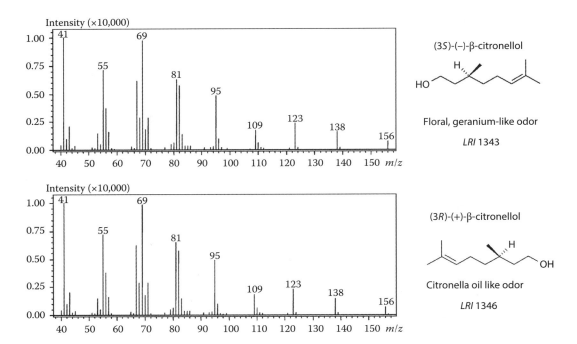

FIGURE 7.4 Representation of the mass spectra similarity of β-citronellol enantiomers.

more than 200 volatile and nonvolatile components. The volatile fraction represents 90%–99% of the entire oil and is represented by mono- and sesquiterpene hydrocarbons and their oxygenated derivatives, along with aliphatic aldehydes, alcohols, and esters; the nonvolatile fraction, constituting 1%–10% of the oil, is represented mainly by hydrocarbons, fatty acids, sterols, carotenoids, waxes, and oxygen heterocyclic compounds (coumarins, psoralens, and polymethoxylated flavones [PMFs]) [92].

In some specific cases, the information attained by means of GC is not sufficient to characterize a citrus essential oil, and the analysis of the nonvolatile fraction can be required. Oxygen heterocyclic compounds, which are a distinct class of flavonoids, can have an important role in the identification of a cold-pressed oil and in the control of both quality and authenticity [92–95]. The analysis of these compounds is usually performed by means of LC, also referred to as high-performance LC (HPLC), in normal (NP-HPLC) or reversed-phase (RP-HPLC) applications. The former method, commonly used when the analytes of interest are slightly polar, separates analytes based on polarity by using a polar stationary phase and a nonpolar mobile phase. The degree of adsorption on the polar stationary phase increases on the basis of analyte polarity, and the extension of this interaction has a great influence on the elution time. In general, the interaction strength is related to the nature of the analyte functional groups and to steric factors. On the other hand, RP-HPLC is based on the use of a nonpolar stationary phase and an aqueous, moderately polar mobile phase. Retention times are therefore shorter for polar molecules, which elute more readily. Moreover, retention times are increased by the addition of a polar solvent to the mobile phase and decreased by the addition of a more hydrophobic solvent.

The online coupling of two columns in the NP-HPLC analysis of bitter orange essential oils with UV detection has been reported: a μ-Porasil (30 cm × 3.9 mm ID, with 10 μm particle size, Waters Corporation, Milford, CT) and a Zorbax silica (25 cm × 4.6 mm ID, with 7 μm particle size, Phenomenex, Bologna, Italy). A large number of cold-pressed Italian and Spanish, commercial, and laboratory-made oils and also mixtures of bitter orange with sweet orange, lemon, lime, and

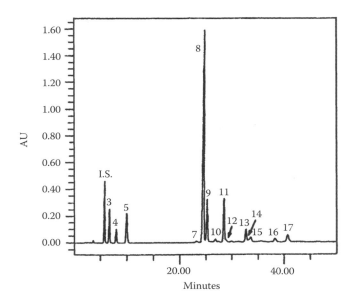

FIGURE 7.5 High-performance liquid chromatography of Italian genuine bitter orange oil. For peak identification, refer to the text (I.S.—Internal Standard). (From Dugo, P. et al., *J. Agric. Food Chem.*, 44, 544, 1996. With permission.)

grapefruit oils were analyzed [93]. A total of four coumarins (osthol [1], meranzin [5], isomeranzin [6], and meranzin hydrate [14]), three psoralens (bergapten [2], epoxybergamottin [3], and epoxybergamottin hydrate [13]), and four PMFs (tangeretin [8], heptamethoxyflavone [9], nobiletin [10], and tetra-*O*-methylscutellarein [11]) were identified. In addition, further three unidentified coumarins (peaks 4, 7, and 12) were detected. The bracketed numbers refer to those in Figure 7.5. In general, Italian essential oils exhibited a higher content of oxygen heterocyclic compounds than the Spanish oils. The use of NP- and RP-HPLC with microbore columns and UV detection has also been reported for lemon and bergamot [96] and bitter orange and grapefruit [97] essential oils. Orange and mandarin essential oils have also been analyzed by NP- and RP-HPLC, but with UV and spectrofluorimetric detection in series [98]. For the identification of chromatographic peaks of all the aforementioned oils, a preparative HPLC was used; the purified fractions were then analyzed by GC–MS and HPLC–MS.

The oxygen heterocyclic compounds present in the nonvolatile residue of citrus essential oils have also been extensively investigated by means of HPLC–atmospheric pressure ionization–mass spectrometry (HPLC–API–MS) [99]. The mass spectra obtained at different voltages of the "sample cone" have been used to build a library. Citrus essential oils have been analyzed with this system, using an optimized NP-HPLC method, and the mass spectra were compared with those of the laboratory-constructed library. This approach allowed the rapid identification and characterization of oxygen heterocyclic compounds of citrus oils, the detection of some minor components for the first time in some oils, and also the detection of authenticity and possible adulteration.

Apart from citrus oils, other essential oils have also been analyzed by means of LC, such as the blackcurrant bud essential oil [100]. The latter was fractionated into hydrocarbons and oxygenated compounds, and the two fractions were submitted for RP-HPLC analysis. Volatile carbonyls consist of some of the most important compounds for the blackcurrant flavor and, hence, were analyzed in detail. The carbonyls were converted into 2,4-dinitrophenylhydrazones and the mixture of 2,4-dinitrophenylhydrazones was separated into derivatives of keto acids and monocarbonyl and dicarbonyl compounds. Each fraction was submitted to chromatographic investigation.

7.3.7 MULTIDIMENSIONAL GAS CHROMATOGRAPHIC TECHNIQUES

In spite of the considerable instrumental advances made, the detection of all the constituents of an essential oil is an extremely difficult task. For example, gas chromatograms relative to complex mixtures are characterized frequently by several overlapping compounds: well-known examples are octanal and α-phellandrene, as well as limonene and 1,8-cineol, on 5% diphenyl–95% dimethylpolysiloxane stationary phases, while insufficient resolution is observed between citronellol and nerol or geraniol and linalyl acetate. On the other hand, the overlapping of monoterpene alcohols and esters with sesquiterpene hydrocarbons is frequently reported on polyethylene glycol stationary phases [101]. Hence, the direct identification of a component in such mixtures can be a cumbersome challenge. The use of MDGC can be of great help in complex sample analysis. In MDGC, key fractions of a sample are selected from the first column and reinjected onto a second one, where ideally, they should be fully resolved. The instrumentation usually involves the use of a switching valve arrangement and two chromatographic columns of differing polarities, but generally of identical dimensions. Furthermore, when heart-cut operations are not carried out, the primary column elutes normally in the first dimension (^{1}D) GC system, while heart-cut fractions are chromatographically resolved on the secondary column [102]. The capillaries employed can be operated in either a single or two distinct GC ovens, with both GC systems commonly equipped with detectors.

The use of MDGC has also been described in a wide range of Es–GC applications [103], involving the use of chiral selectors as the stationary phase for the determination of ee and/or ER as well as for adulteration assessments.

A fully automated, multidimensional, double-oven GC–GC system has been presented by Mondello et al. The system is based on the use of mechanical valves that allow the automatic multiple transfers of different fractions from the precolumn to the analytical one, and the analysis of all transferred fractions during the same gas chromatographic run. A system of pneumatic valves emitted pressure variations in order to maintain constant retention times in the precolumn, even after numerous transfers. In addition, when the system was not applied in the multidimensional configuration, the two gas chromatographs could be operated independently. The system has been used for the determination of the enantiomeric distribution of monoterpene hydrocarbons and monoterpene alcohols in the essential oils of lime [104], mandarin [105], and lemon [106]. In Figure 7.6, the analysis of the latter oil is outlined to illustrate the technique: the chromatogram of the lemon oil obtained on an SE-52 column with the system in standby position is shown in Figure 7.6a, while the one attained with the system in the cut position is illustrated in Figure 7.6b. Figure 7.6c shows the chiral separation of the fractions transferred to the main analytical column. Well-resolved peaks of components present in large amounts, and also of the minor compounds, were attained through the partial transfer of the major concentration components.

MDGC is a useful approach for the fractionation of compounds of particular interest in a specific sample; one of its major application areas is chiral analysis, using a conventional column as ^{1}D and a CSP capillary in the second dimension (^{2}D). Es–MDGC analysis has been used for the direct enantioselective evaluation of limonene in *Rutaceae* and *Gramineae* essential oils [107]. It is noteworthy that (4R)-(+)-limonene of high ee values were found in *Rutaceae* oils, such as bergamot, orange, mandarin, lemon, or lime oils, while the (4S)-(−)-isomer was present in higher amounts in the *Gramineae* oils, such as citronella or lemongrass. The ratios of α-pinene and β-pinene enantiomers were also taken into consideration.

The use of Es–MDGC using a primary polyethylene glycol stationary phase, and a secondary heptakis (2,3-di-*O*-acetyl-6-*O*-*tert*-butyldimethylsylil)-β-cyclodextrin, has been applied to rose oils. The technique proved to be highly efficient for the assessment of origin and quality control of economically important rose oil, using (3S)-(−)-citronellol and (2S,4R)-(−)-*cis*- and *trans*-rose oxides as markers [108]. Buchu leaf oil has also been assessed through Es–MDGC, and (1S)-menthone, isomenthone, (1S)-pulegone, (1S)-thiols, and (1S)-thiolacetates as enantiopure sulfur compounds [109]. A further application was reported on mint and peppermint essential oils, which contain

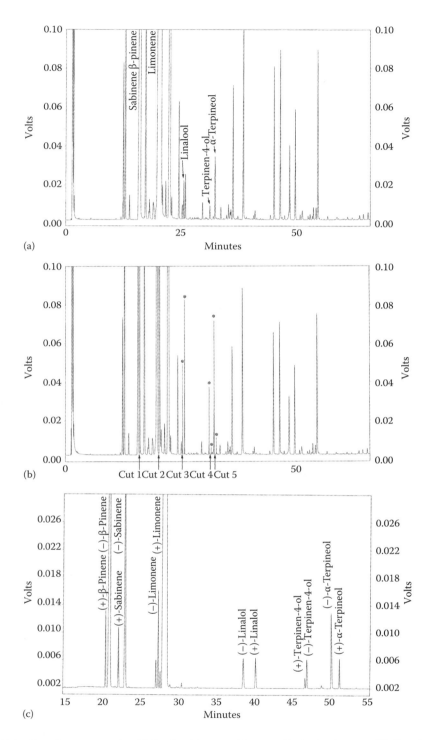

FIGURE 7.6 Gas chromatography (GC) of cold-pressed lemon oil obtained on an SE-52 column (a), GC chromatogram of cold-pressed lemon oil obtained on an SE-52 column with five heart cuts (b), and GC–GC chiral chromatogram of the transferred components (c). (From Mondello, L. et al., *J. High Res. Chromatogr.*, 22, 350, 1999. With permission.)

(1*R*)-configured monoterpenoids [110]. Es–MDGC equipped with a 5% diphenyl–95% dimethylpolysiloxane and a 2,3-di-*O*-methyl-6-*O*-*tert*-butyldimethylsilyl-β-CD, as the ^{1}D and ^{2}D, respectively, enabled the simultaneous stereodifferentiation of menthone, neomenthol, isomenthone, menthol, neoisomenthol, and menthylacetate.

The technique has also been successfully applied to the authenticity assessment of various commercially available rosemary oils [111]. The ER of α-pinene, camphene, β-pinene, limonene, borneol, terpinen-4-ol, α-terpineol, linalool, and camphor were measured; moreover, (1*S*)-(−)-borneol of high enantiomeric purity (higher than 90%) has been defined as a reliable indicator of genuine rosemary oils. A recently created high-performance MDGC system has been recently used in this specific field of essential oil research [111]. A conventional method and a fast MDGC method were developed and applied to the enantioselective analysis of rosemary oil. Prior to "heart cutting," a "standby" analysis was carried out in order to define the retention time cutting windows of the chiral components to be reanalyzed in the ^{2}D; retention time windows of eight peaks were defined. The nine peaks (camphene, β-pinene, sabinene, limonene, camphor, isoborneol, borneol, terpinen-4-ol, and α-terpineol) were then cut and transferred. It must be added that, in some cases, only a portion of the entire peak, that is, limonene, was diverted onto the secondary column. The fast MDGC method was applied on a twin set of 0.1 mm ID microbore columns with the objective of reproducing the conventional analytical result in a much shorter time (Figure 7.7). The overall run time requested for the conventional analysis was 43 min, while it was 8.7 min for the fast MDGC application, with a speed gain of nearly a factor of 5.

MDGC heart-cutting methods, using valve and valveless flow switching interfaces, extend the separation power of capillary GC, although the 2D analysis can only be restricted to a few regions of the chromatogram. In order to attain a complete 2D characterization of a sample, a comprehensive approach, such as comprehensive 2D GC (GC × GC), has to be used.

GC × GC produces an orthogonal two-column separation, with the complete sample transfer achieved by means of a modulator; the latter entraps, refocuses, and releases fractions of the GC effluent from the ^{1}D, onto the ^{2}D column, in a continuous mode; this method enables an accurate screening of complex matrices, offering very high resolution and enhanced detection sensitivity [112,113]. The two columns must possess different separation mechanism, commonly a low polarity or nonpolar column is used in the ^{1}D, and a polar column is used as the fast ^{2}D column. Moreover, a 2D separation can be defined as comprehensive if other two conditions are respected [114,115]: equal percentages (either 100% or less) of all sample components pass through both columns and eventually reach the detector and the resolution obtained in the ^{1}D is essentially maintained.

One of the first applications of GC × GC to essential oils was performed by Dimandja et al. [116], who investigated the separation of peppermint (*Mentha piperita*) and spearmint (*Mentha spicata*) essential oil components. The latter oil is mainly characterized by four major components, that is, carvone, menthol, limonene, and menthone, while the former is mainly represented by menthol, menthone, menthylacetate, and eucalyptol. Both essential oils were considered as being of moderate complexity, containing <100 sample components. The GC × GC system used in this research was composed of a GC equipped with a thermal modulation unit (Zoex Corporation, Lincoln, Nebraska), illustrated in Figure 7.8. The thermal modulation cycle is a three-step process, involving sample accumulation, focusing, and acceleration stages. The column set used was composed of a nonpolar column (1 m × 100 µm ID, 3.5 µm d_f) in the ^{1}D and one of intermediate polarity (2 m × 100 µm ID, 0.5 µm d_f) in the ^{2}D, connected by a press fit. A two- to threefold increase in separation power was obtained for the GC × GC analyses of both mint oils; peppermint essential oil was found to contain 89 identifiable peaks by GC × GC compared to 30 peaks in GC–MS, while in the spearmint oil, 68 peaks were detected by GC × GC and 28 by means of GC–MS. The simple alignment of the ^{2}D retention times of the investigated oils revealed that both have 52 components in common, as opposed to 18 matches by monodimensional GC.

GC × GC analyses of essential oils of high complexity have also been carried out, such as of vetiver oil (*Vetiver zizanioides*) [117]. The work was performed on a gas chromatographic system

Analysis of Essential Oils

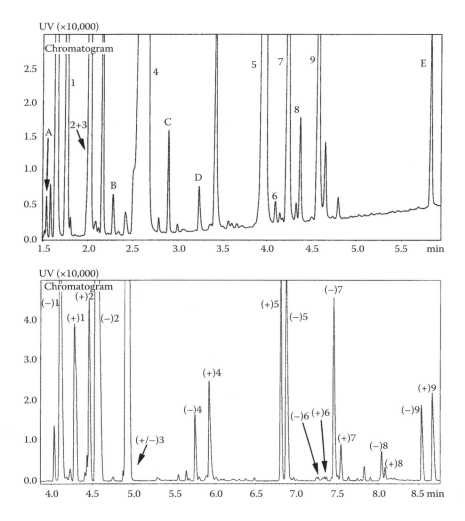

FIGURE 7.7 A 4.5 min chromatogram expansion relative to the rosemary analysis using fast multidimensional gas chromatography (MDGC) with the transfer system in standby position (top); tricyclene (peak A), α-phellandrene (peak B), unknown (peak C), α-terpinolene (peak D), and bornyl acetate (peak E). A 5 min ²D chromatogram expansion relative to the rosemary oil application using fast MDGC (bottom); the peak numbers refer to camphene (1), β-pinene (2), sabinene (3), limonene (4), camphor (5), isoborneol (6), borneol (7), terpinen-4-ol (8), and α-terpineol (9). (From Mondello, L. et al., *J. Chromatogr. A*, 1105, 11, 2006. With permission.)

(6890GC, Agilent Technologies, Santa Clara, CA) retrofitted to a longitudinally modulated cryogenic system (LMCS, Chromatography Concepts, Doncaster, Victoria, Australia). In Figure 7.9, a schematic diagram of the cryotrap is presented; the columns are anchored with retaining nuts to the support frame so that the trap can move up and down along the column, its movement is controlled by a stepper motor. Liquid cryogen is supplied to the inlet of the trap, passing through a narrow restrictor and expanding to cool the body of the trap. A secondary, small flow of nitrogen passing through the center of the body prevents the buildup of ice that would otherwise freeze the column to the trap [118]. Analysis was performed on a 5% phenyl–95% dimethylpolysiloxane ¹D column (25 m × 0.25 mm ID, 0.25 μm d_f) connected to a 50% phenyl–50% dimethylpolysiloxane ²D column (0.8 m × 0.1 mm ID, 0.1 μm d_f), applying a modulation frequency of 4 s cycle. About 200 compounds could be detected by means of GC × GC analysis. The authors reported that

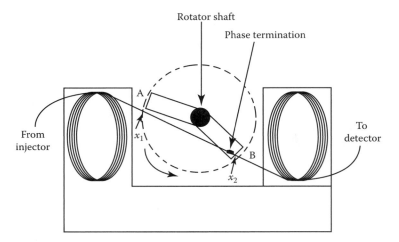

FIGURE 7.8 Top-view scheme of the thermal modulator. The arrows marked as x_1 and x_2 indicate the length of the modulator tube. The rotating heater is in position a priori to each modulation cycle. Its counterclockwise movement (from position A to B) over the modulator tube produces the thermal modulation. (From Dimandja, J.M.D. et al., *J. High Resolut. Chromatogr.*, 23, 208, 2000. With permission.)

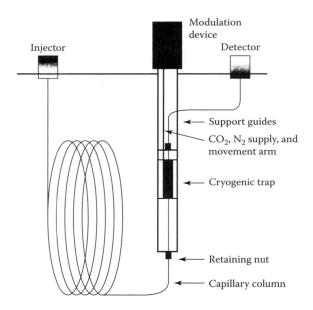

FIGURE 7.9 Scheme of the longitudinally modulated cryogenic system located inside a gas chromatography oven. (From Marriott, P. et al., *Flavour Fragr. J.*, 15, 225, 2000. With permission.)

the GC–MS analysis of vetiver oil, with peak deconvolution, would still not have been sufficient for the identification of coeluting substances.

French lavender (*Lavandula angustifolia*) and tea tree (*Melaleuca alternifolia*) essential oils were also submitted to GC × GC analyses using a nonpolar (5% phenyl–95% dimethylpolysiloxane)–polar (polyethylene glycol) column set [119]. The work, developed using an LMCS, enabled the determination of elution patterns within the ²D space useful for the correlation of component retention behavior with their chemical and structural properties. The GC × GC approach provides higher sensitivity,

Analysis of Essential Oils

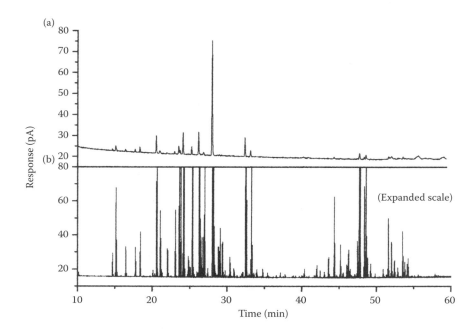

FIGURE 7.10 Comparison of (a) monodimensional gas chromatography (GC) and (b) pulsed GC × GC result for tea tree oil; both chromatograms are shown at identical sensitivity. (From Shellie, R. et al., *J. High Resolut. Chromatogr.*, 23, 554, 2000. With permission.)

greater peak resolution, and capacity, as well as an essential oil fingerprint pattern. The enhanced peak capacity and sensitivity of GC × GC in the tea tree oil application can be observed in the conventional GC and untransformed GC × GC chromatograms illustrated in Figure 7.10.

Lavender essential oil has been further investigated by means of GC × GC retrofitted with an LMCS and hyphenated to time-of-flight mass spectrometry (GC × GC–TOFMS), thus generating a 3D analytical approach [120]. The authors outlined that the vacuum effect on the ^{2}D column in GC × GC–TOFMS may generate differing retention times with respect to an equivalent analysis performed on a GC × GC system at ambient pressure conditions. Lavender essential oil was further investigated through the comparison of GC–MS and GC × GC analyses [121], as illustrated in Figure 7.11. At least 203 components were counted in the ^{2}D separation space, which was characterized by a well-defined monoterpene hydrocarbon region, and a sesquiterpene hydrocarbon area. The oxygenated derivatives of both these groups generally elute closely after the main group in the ^{1}D, but due to their wide range of component polarities, these are found to spread throughout a broader region of the ^{2}D plane. According to the authors, by using GC × GC, the detection of subtle differences in the analyses of lavender essential oils from different cultivars should be simplified, since it could be based on a ^{2}D pictorial representation of the volatile components.

Further relevant essential oil investigations have been performed: the early works using FID, while the more recent ones using, preferably, a TOFMS as detector. Among the essential oils previously studied by means of GC × GC are peppermint [122] and Australian sandalwood [123], with the latter also analyzed through GC × GC–TOFMS in the same work. Essential oils derived from *Thymbra spicata* [124], *Pistacia vera* [125], hop [126], *Teucrium chamaedrys* [127], *Rosa damascena* [128], coriander [129], and *Artemisia annua* [130], as well as tobacco [131], have also been subjected to GC × GC–TOFMS analyses. The references cited herein represent only a fraction of the studies performed by means of GC × GC on essential oils.

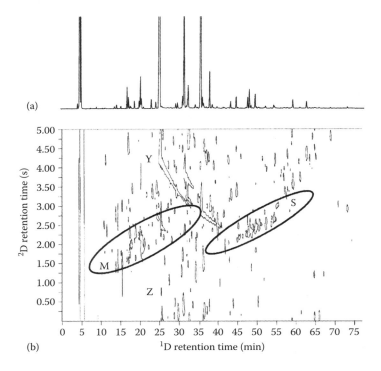

FIGURE 7.11 Reconstructed gas chromatographic trace for a lavender essential oil (a) and the 2D separation space for the 2D gas chromatography analysis of the same sample (b). The minor component Z overlaps completely from the major component Y in the ¹D. M, monoterpene hydrocarbons; S, sesquiterpene hydrocarbons. (From Shellie, R. et al., *J. Chromatogr. A*, 970, 225, 2002. With permission.)

7.3.8 MULTIDIMENSIONAL LIQUID CHROMATOGRAPHIC TECHNIQUES

HPLC has acquired a role of great importance in food analysis, as demonstrated by the wide variety of applications reported. Single–LC column chromatographic processes have been widely applied for sample profile elucidation, providing satisfactory degrees of resolving power; however, whenever highly complex samples require analysis, a monodimensional HPLC system can prove to be inadequate. Moreover, peak overlapping may occur even in the case of relatively simple samples, containing components with similar properties.

The basic principles of MDGC are also valid for multidimensional LC (MDLC). The most common use of MDLC separation is the pretreatment of a complex matrix in an off-line mode. The off-line approach is very easy, but presents several disadvantages: it is time-consuming, operationally intensive, and difficult to automate and to reproduce. Moreover, sample contamination or formation of artifacts can occur. On the other hand, on-line MDLC, though requiring specific interfaces, offers the advantages of ease of automation and greater reproducibility in a shorter analysis time. In the online heart-cutting system, the two columns are connected by means of an interface, usually a switching valve, which allows the transfer of fractions of the first column effluent onto the second column.

In contrast to comprehensive GC (GC × GC), the number of comprehensive LC (LC × LC) applications reported in literature is much less. It can be affirmed that LC × LC presents a greater flexibility when compared to GC × GC since the mobile phase composition can be adjusted in order to obtain enhanced resolution [132]. Comprehensive HPLC systems, developed, and applied to the analysis of food matrixes, have employed the combination of either NP × RP or RP × RP separation modes. However, it is worthy of note that the two separation mechanisms exploited should be as

Analysis of Essential Oils

orthogonal as possible, so that no or little correlation exists between the retention of compounds in both dimensions.

A typical comprehensive 2D HPLC separation is attained through the connection of two columns by means of an interface (usually a high-pressure switching valve), which entraps specific quantities of ^{1}D eluate and directs it onto a secondary column. This means that the first column effluent is divided into "cuts" that are transferred continuously to the ^{2}D by the interface. The type of interface depends on the methods used, although multiport valve arrangements have been the most frequently employed.

Various comprehensive HPLC systems have been developed and proven to be effective for the separation of complex sample components, and in the resolution of a number of practical problems. In fact, the very different selectivities of the various LC modes enable the analysis of complex mixtures with minimal sample preparation. However, comprehensive HPLC techniques are complicated by the operational aspects of switching effectively from one operation step to another, by data acquisition and interpretation issues. Therefore, careful method optimization and several related practical aspects should be considered.

In the most common approach, a microbore LC column in the ^{1}D and a conventional column in the ^{2}D are used. In this case, an 8-, 10-, or 12-port valve equipped with two sample loops (or trapping columns) is used as an interface. A further approach foresees the use of a conventional LC column in the ^{1}D and two conventional columns in the ^{2}D. One or two valves that allow transfers from the first column to two parallel secondary columns (without the use of storage loops) are used as interface.

One of the best examples of the application of comprehensive NPLC × RPLC in essential oil analysis is represented by the analysis of oxygen heterocyclic components in cold-pressed lemon oil, by using a normal phase with a microbore silica column in the ^{1}D and a monolithic C18 column in the ^{2}D with a 10-port switching valve as interface [133]. In Figure 7.12, an NPLC × RPLC separation of the oxygen heterocyclic fraction of a lemon oil sample is presented. Oxygen heterocyclic components (coumarins, psoralens, and PMFs) represent the main part of the nonvolatile fraction of cold-pressed citrus oils. Their structures and substituents have an important role in the characterization of these oils. Positive peak identification of these compounds was obtained both by the relative

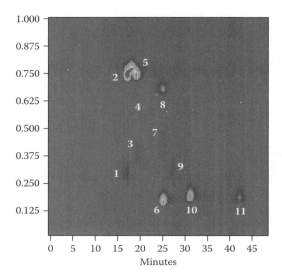

FIGURE 7.12 Comprehensive normal-phase (adsorption) liquid chromatography (LC) × reversed-phase LC separation of the oxygen heterocyclic fraction of a lemon oil sample (for peak identification, see Ref. [133]). (From Dugo, P. et al., *Anal. Chem.*, 73, 2525, 2004. With permission.)

location of the peaks in the 2D plane, which varied in relation to their chemical structure, and by characteristic UV spectra. In a later experiment, a similar setup was used for a citrus oil extract composed of lemon and orange oil [134]. The main difference with respect to the earlier published work [133] was the employment of a bonded-phase (diol) column in the ^{1}D. Under optimized LC conditions, the high degree of orthogonality between the NP and RP systems tested resulted in increased ^{2}D peak capacity.

A novel approach for the analysis of carotenoids, pigments mainly distributed in plant-derived foods, especially in orange and mandarin essential oils, has been recently developed by Dugo et al. [135,136]. In terms of structures, food carotenoids are polyene hydrocarbons, characterized by a C_{40} skeleton that derives from eight isoprene units. They present an extended conjugated double-bond system that is responsible for the yellow, orange, or red colors in plants and are notable for their wide distribution, structural diversity, and various functions. Carotenoids are usually classified in two main groups: hydrocarbon carotenoids, known as carotenes (e.g., β-carotene and lycopene), and oxygenated carotenoids, known as xanthophylls (e.g., β-cryptoxanthin and lutein). The elucidation of carotenoid patterns is particularly challenging, because of the complex composition of carotenoids in natural matrices, their great structural diversity, and their extreme instability. An innovative comprehensive dual-gradient elution HPLC system was employed using an NPLC × RPLC setup, composed of silica and C18 columns in the ^{1}D and ^{2}D, respectively. Free carotenoids in orange essential oil and juice (after saponification) were identified by combining the 2D retention data with UV–visible spectra [136] obtained by using a photodiode array detector (DAD) (Figure 7.13). A recent study of the carotenoid fraction of a saponified mandarin oil has been performed by means of comprehensive LC, in which a ^{1}D microbore silica column was applied for the determination of free carotenoids, and a cyanopropyl column for the separation of esters; a monolithic column was used in the ^{2}D [135]. Detection was performed by connecting a DAD system in parallel with an MS detection system operated in the atmospheric pressure chemical ionization positive-ion mode. Thus, the identification of

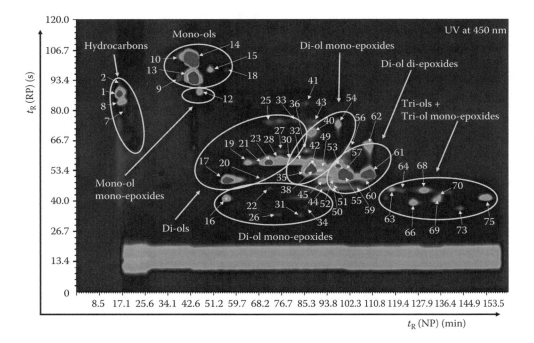

FIGURE 7.13 Contour plot of the comprehensive high-performance liquid chromatography analyses of carotenoids present in sweet orange essential oil with peaks and compound classes indicated (for peak identification, see Ref. [136]). (From Dugo, P. et al., *Anal. Chem.*, 78, 7743, 2006. With permission.)

free carotenoid and carotenoid esters was carried out by combining the information provided by the DAD and MS systems, and the peak positions in the 2D chromatograms.

7.3.9 ONLINE COUPLED LC–GC

The analysis of very complex mixtures is often troublesome due to the variety of chemical classes to which the sample components belong to and to their wide range of concentrations. As such, several compounds cannot be resolved by monodimensional GC. In this respect, less complex and more homogeneous mixtures can be attained by the fractionation of the matrix by means of LC prior to GC separation. The MDLC–GC approach combines the selectivity of the LC separation with the high efficiency and sensitivity of GC separation, enabling the separation of compounds with similar physicochemical properties in samples characterized by a great number of chemical classes.

For the highly volatile components, commonly present in essential oils, the most adequate transfer technique is partially concurrent eluent evaporation [137]. In the latter technique, proposed by Grob, a retention gap is installed, followed by a few meters of precolumn and the analytical capillary GC column, both with identical stationary phase, for the separation of the LC fractionated components. A vapor exit is placed between the precolumn and the analytical column, allowing partial evaporation of the solvent. Hence, column and detector overloading are avoided. This transfer technique can be applied to the analysis of GC components with a boiling point of at least 50°C higher than the solvent.

The composition of citrus essential oils has been greatly exploited by means of LC–GC, and the development of new methods for the study of single classes of components has been well reported. The aldehyde composition in sweet orange oil has been investigated [138], as also industrial citrus oil mono- and sesquiterpene hydrocarbons [139], and the enantiomeric distribution of monoterpene alcohols in lemon, mandarin, sweet orange, and bitter orange oils [140,141].

The hyphenation of LC–GC systems to mass spectrometric detectors has also been reported for the analyses of neroli [142], bitter and sweet oranges, lemon, and petitgrain mandarin oils [143]. It has to be highlighted that the preliminary LC separation, which reduces mutual component interference, greatly simplifies MS identification.

7.4 GENERAL CONSIDERATIONS ON ESSENTIAL OIL ANALYSIS

As evidenced by the numerous techniques described in the present contribution, chromatography, especially GC, has evolved into the dominant method for essential oil analysis. This is to be expected because the complexity of the samples must be unraveled by some type of separation, before the sample constituents can be measured and characterized; in this respect, GC provides the greatest resolving power for most of these volatile mixtures.

In the past, a vast number of investigations have been carried out on essential oils, and many of these natural ingredients have been investigated following the introduction of GC–MS, which marked a real turning point in the study of volatile molecules. Es–GC also represented a landmark in the detection of adulterations, and in the cases where the latter technique could fail, GC correlated to isotope ratio mass spectrometry (GC–IRMS) by means of a combustion interface has proved to be a valuable method to evaluate the genuineness of natural product components. In addition, the introduction of GC–O was a breakthrough in analytical aroma research, enabling the differentiation of a multitude of odor-active and non-odor-active volatiles, according to their existing concentrations in a matrix. The investigation of the nonvolatile fraction of essential oils, by means of LC and its related hyphenated techniques, contributed greatly toward the progress of the knowledge on essential oils. Many extraction techniques have also been developed, boosting the attained results. Moreover, the continuous demand for new synthetic compounds reproducing the sensations elicited by natural flavors triggered analytical investigations toward the attainment of information on scarcely known properties of well-known matrices.

REFERENCES

1. Rowe, D.J., 2005. Introduction. In *Chemistry and Technology of Flavors and Fragrances*, D.J. Rowe (ed.). Oxford, U.K.: The Blackwell Publishing (Chapter 1).
2. Parry, E.J., 1908. *The Chemistry of Essential Oils and Artificial Perfumes*, 2nd edn. London, U.K.: Scott, Greenwood & Son.
3. Semmler, F.W., 1906–1907. *Die Ätherischen Öle*, Vols. I–IV, 3rd edn. Leipzig, Germany: Verlag von Veit.
4. Gildemeister, E. and F.R. Hoffman, 1950. *Die Ätherischen Öle*, Vol. 1, 3rd edn. Leipzig, Germany: Verlag Von Schimmel & Co.
5. Finnemore, H., 1926. *The Essential Oils*, 1st edn. London, U.K.: Ernest Benn.
6. Guenther, E., 1972. *The Essential Oils—Vol. 1: History—Origin in Plants Production—Analysis*, reprint of 1st edn. (1948). Malabar, FL: Krieger Publishing Company.
7. Croteau, R., T.M. Kutchan, and N.G. Lewis, 2000. Natural products (secondary metabolites). In *Biochemistry & Molecular Biology of Plants*, 1st edn., B. Buchanan, W. Gruissen, and R. Jones (eds.). Hoboken, NJ: ASPB and Wiley (Chapter 24).
8. World Health Organisation, 2007. *The International Pharmacopoeia (IntPh)*, 4th edn. Geneva, Switzerland: World Health Organisation Press.
9. Ministry of Health, Labour and Welfare, 2007. *Japanese Pharmacopoeia JP XV, 2007* 15th edn. Tokyo, Japan: Yakuji Nippo, Ltd.
10. The Stationery Office, 2007. *British Pharmacopoeia BP 2008*. Norwich, U.K.: The Stationery Office (TSO).
11. United States Pharmacopoeia Convention, 2007. *The United States Pharmacopoeia USP/NF 2008*. Rockville, MD: United States Pharmacopoeia Convention Inc.
12. European Directorate for the Quality of Medicines & Healthcare (Edqm), 2007. *European Pharmacopoeia*, 6th edn. Strasburg, Germany: European Directorate for the Quality of Medicines & Healthcare (Edqm).
13. Sweetman, S. (ed.), 2007. *Martindale: The Complete Drug Reference*, 35th edn. London, U.K.: Pharmaceutical Press.
14. International Organization for Standardization, 1997. *ISO 3761-1997*. Geneva, Switzerland: International Organization for Standardization, http://www.iso.org/iso/iso_catalogue/catalogue_tc/catalogue_tc_browse. htm?commid=48956 (accessed December 15, 2007).
15. International Organization for Standardization, 2007. *TC 57 Essential Oils*. Geneva, Switzerland: International Organization for Standardization, http://www.iso.org/iso/iso_catalogue/catalogue_tc/catalogue_tc_browse. htm?commid=48956 (accessed December 15, 2007).
16. AFNOR—Association Française de Normalisation, 2004. NF ISO 3515 Huiles essentielles de lavande—(Lavandula angustifolia Mill.). Saint-Denis, http://www.afnor.org (accessed January 11, 2008).
17. Simões, C.M.O. and V. Spitzer, 1999. Óleos voláteis. In *Farmacognosia: da planta ao medicamento*, C.M.O. Simões et al. (eds.). Florianópolis, Brazil: Editora da UFSC and Editora da Universidade/UFRGS (Chapter 18).
18. Bosart, L.W., 1937. Change in refractive index with temperature variation. *Perfumery Essential Oil Record*, 28: 95.
19. Beilstein, F., 1872. Ueber den Nachweis von Chlor, Brom und Jod in organischen Substanzen. *Ber. Dtscg. Chem. Ges.*, 5: 620.
20. Panda, H., 2003. *Essential Oils Handbook*, 1st edn. New Delhi, India: National Institute of Industrial Research (Chapter 110).
21. Stillman, R.C. and R.M. Reed, 1932. Hydroxylamine method for the determination of aldehydes and ketones in essential oils. *Perfumery Essential Oil Record*, 23: 278.
22. Poole, C.F., 2003. *The Essence of Chromatography*, 1st edn. Amsterdam, the Netherlands: Elsevier.
23. Falkenberg, M.B., R.I. Santos, and C.M.O. Simões, 1999. Introdução à Análise Fitoquímica. In *Farmacognosia: da planta ao medicamento*, C.M.O. Simões et al. (eds.). Florianópolis, Brazil: Editora da UFSC and Editora da Universidade/UFRGS (Chapter 10).
24. Wagner, H., S. Bladt, and V. Rickl, 2003. *Plant Drug Analysis: A Thin Layer Chromatography Atlas*, 2nd edn., p. 1. Heidelberg, Germany: Springer.
25. Hahn-Deinstrop, E., 2000. *Applied Thin Layer Chromatography: Best Practice and Avoidance of Mistakes*, 2nd edn. Weinheim, Germany: Wiley-VCH (Chapter 1).
26. Sherma, J., 2000. Thin-layer chromatography in food and agricultural analysis. *J. Chromatogr. A*, 880: 129.
27. Richardson, A., 1999. Measurement of fragrance perception. In *The Chemistry of Fragrances*, D.H. Pybus and C.S. Sell (eds.). Cambridge, U.K.: Royal Society of Chemistry (Chapter 8).

28. Friedrich, J.E., T.E. Acree, and E.H. Lavin, 2001. Selecting standards for gas chromatography-olfactometry. In *Gas Chromatography-Olfactometry: The State of the Art*, J.V. Leland et al. (eds.). Washington, DC: American Chemical Society, pp. 148–155.
29. Curtis, T. and D.G. Williams, 2001. *Introduction to Perfumery*, 2nd edn. New York: Micelle Press.
30. Ettre, L.S. and J.V. Hinshaw, 1993. *Basic Relationships of Gas Chromatography*, 1st edn. Cleveland, OH: Advanstar Data, pp. 15–17.
31. Liberti, A. and G. Conti, 1956. Possibilità di applicazione della cromatografia in fase gassosa allo studio della essenza. In *Proc. 1° Convegno Internazionale di Studi e Ricerche sulle Essenze*, Reggio Calabria, Italy.
32. James, A.T. and A.J.P. Martin, 1952. Gas–liquid partition chromatography: The separation and microestimation of volatile fatty acids from formic acid to dodecanoic acid. *Biochem. J.*, 50: 679.
33. Scott, R.P.W., 2001. *Gas Chromatography*, Chrom Ed. Series, http://www.chromatography-online.org/ (accessed December 15, 2007).
34. Kováts, E., 1958. Gas-chromatographische Charakterisierung organischer Verbindungen. Teil 1: Retentionsindices aliphatischer Halogenide, Alkohole, Aldehyde und Ketone. *Helv. Chim. Acta*, 51: 1915.
35. van den Dool, H. and P.D. Kratz, 1963. A generalization of the retention index system including linear temperature programmed gas–liquid chromatography. *J. Chromatogr.*, 11: 463.
36. Shibamoto, T., 1987. Retention indices in essential oil analysis. In *Capillary Gas Chromatography in Essential Oil Analysis*, 1st edn., S. Sandra and C. Bicchi (eds.). Heidelberg, Germany: Alfred Huethig Verlag, Chapter 8, p. 258.
37. International Organization of the Flavor Industry (I.O.F.I.), 1991. The identification of individual components in flavourings and flavoured food. *Z. Lebensm. Unters. Forsch.*, 192: 530.
38. Davies, N.W., 1990. Gas chromatographic retention indices of monoterpenes and sesquiterpenes on methyl silicone and Carbowax 20 M phases. *J. Chromatogr.*, 503: 1.
39. Royal Society of Chemistry, 1981. Analytical Methods Committee, Application of gas–liquid chromatography to the analysis of essential oils, Part VIII. Fingerprinting of essential oils by temperature-programmed gas–liquid chromatography using methyl silicone stationary phases. *Analyst*, 106: 448.
40. Royal Society of Chemistry, 1980. Analytical Methods Committee, Application of gas–liquid chromatography to the analysis of essential oils, Part VIII. Fingerprinting of essential oils by temperature-programmed gas–liquid chromatography using a Carbowax 20 M stationary phase. *Analyst*, 105: 262.
41. Todd, J.F.J., 1995. Recommendations for nomenclature and symbolism for mass spectroscopy. *Int. J. Mass Spectrom. Ion Process*, 142: 209.
42. Vekey, K., 2001. Mass spectrometry and mass-selective detection in gas chromatography. *J. Chromatogr. A*, 921: 227.
43. McLafferty, F.W., D.A. Stauffer, S.Y. Loh, and C. Wesdemiotis, 1999. Unknown identification using reference mass spectra. Quality evaluation of databases. *J. Am. Soc. Mass Spectrom.*, 10: 1229.
44. Costa, R., M.R. De Fina, M.R. Valentino, P. Dugo, and L. Mondello, 2007. Reliable identification of terpenoids and related compounds by using linear retention indices interactively with mass spectrometry search. *Nat. Prod. Commun.*, 2: 413.
45. Joulain, D. and W.A. König, 1998. *The Atlas of Spectral Data of Sesquiterpene Hydrocarbons*, 1st edn., pp. 1–6. Hamburg, Germany: E.B.-Verlag.
46. Korytar, P., H.G. Janssen, E. Matisova, and U.A.T. Brinkman, 2002. Practical fast gas chromatography: Methods, instrumentation and applications. *TrAC*, 21: 558.
47. Cramers, C.A., H.G. Janssen, M.M. van Deursen, and P.A. Leclercq, 1999. High-speed gas chromatography: An overview of various concepts. *J. Chromatogr. A*, 856: 315.
48. Cramers, C.A. and P.A. Leclercq, 1999. Strategies for speed optimisation in gas chromatography: An overview. *J. Chromatogr. A*, 842: 3.
49. Mondello, L., A. Casilli, P.Q. Tranchida, L. Cicero, P. Dugo, and G. Dugo, 2003. Comparison of fast and conventional GC analysis for citrus essential oils. *J. Agric. Food. Chem.*, 51: 5602.
50. Mondello, L., R. Shellie, A. Casilli, P.Q. Tranchida, P. Marriott, and G. Dugo, 2004. Ultra-fast essential oil characterization by capillary GC on a 50 μm ID column. *J. Sep. Sci.*, 27: 699–702.
51. Bicchi, C., C. Brunelli, C. Cordero, P. Rubiolo, M. Galli, and A. Sironi, 2004. Direct resistively heated column gas chromatography (ultrafast module-GC) for high-speed analysis of essential oils of differing complexities. *J. Chromatogr. A*, 1024: 195.

52. Bicchi, C., C. Brunelli, C. Cordero, P. Rubiolo, M. Galli, and A. Sironi, 2005. High-speed gas chromatography with direct resistively-heated column (ultra fast module-GC)-separation measure (S) and other chromatographic parameters under different analysis conditions for samples of different complexities and volatilities. *J. Chromatogr. A*, 1071: 3.
53. Mondello, L., A. Casilli, P.Q. Tranchida, L. Cicero, P. Dugo, A. Cotroneo, and G. Dugo, 2004. Determinazione della Composizione e Individuazione delle Adulterazioni degli Olii Essenziali mediante Ultrafast-GC. In *Qualità e sicurezza degli Alimenti*, pp. 113–116. Milan, Italy: Morgan Edizioni Scientifiche.
54. Tranchida, P.Q., A. Casilli, G. Dugo, L. Mondello, and P. Dugo, 2005. Fast gas chromatographic analysis with a 0.05 mm ID micro-bore capillary column. *GIT Lab. J.*, 9: 22.
55. Bicchi, C., C. Brunelli, M. Galli, and A. Sironi, 2001. Conventional inner diameter short capillary columns: An approach to speeding up gas chromatographic analysis of medium complexity samples. *J. Chromatogr. A*, 931: 129.
56. Rubiolo, P., F. Belliardo, C. Cordero, E. Liberto, B. Sgorbini, and C. Bicchi, 2006. Headspace-solid-phase microextraction fast GC in combination with principal component analysis as a tool to classify different chemotypes of chamomile flower-heads (*Matricaria recutita* L.). *Phytochem. Anal.*, 17: 217.
57. Proot, M. and P. Sandra, 1986. Resolution of triglycerides in capillary SFC as a function of column temperature. *J. High Res. Chromatogr. Chromatogr. Commun.*, 9: 618.
58. Firestein, S., 1992. Electrical signals in olfactory transduction. *Curr. Opin. Neurobiol.*, 2: 444.
59. Firestein, S., 2001. How the olfactory system makes sense of scents. *Nature*, 413: 211.
60. Malnic, B., J. Hirono, T. Sato, and L.B. Buck, 1999. Combinatorial receptor codes for odors. *Cell*, 96: 713.
61. Grosch, W., 1994. Determination of potent odourants in foods by aroma extract dilution analysis (AEDA) and calculation of odour activity values (OAVs). *Flavour Fragr. J.*, 9: 147.
62. van Ruth, S.M., 2001. Methods for gas chromatography-olfactometry: A review. *Biomol. Eng.*, 17: 121.
63. Fuller, G.H., R. Seltenkamp, and G.A. Tisserand, 1964. The gas chromatograph with human sensor: Perfumer model. *Ann. N. Y. Acad. Sci.*, 116: 711.
64. Nishimura, O., 1995. Identification of the characteristic odorants in fresh rhizomes of ginger (*Zingiber officinale* Roscoe) using aroma extract dilution analysis and modified multidimensional gas chromatography–mass spectroscopy. *J. Agric. Food Chem.*, 43: 2941.
65. Wright, D.W., 1997. Application of multidimensional gas chromatography techniques to aroma analysis. In *Techniques for Analyzing Food Aroma* (*Food Science and Technology*), 1st edn., R. Marsili (ed.). New York: Marcel Dekker, pp. 113–141.
66. McGorrin, R.J., 2002. Character impact compounds: Flavors and off-flavors in foods. In *Flavor, Fragrance, and Odor Analysis*, 1st edn., R. Marsili (ed.). New York: Marcel Dekker, pp. 375–413.
67. Acree, T.E., J. Barnard, and D. Cunningham, 1984. A procedure for the sensory analysis of gas chromatographic effluents. *Food Chem.*, 14: 273.
68. Ullrich, F. and W. Grosch, 1987. Identification of the most intense volatile flavour compounds formed during autoxidation of linoleic acid. *Z. Lebensm. Unters. Forsch.*, 184: 277.
69. Gaffney, B.M., M. Haverkotte, B. Jacobs, and L. Costa, 1996. Charm Analysis of two *Citrus sinensis* peel oil volatiles. *Perf. Flav.*, 21: 1.
70. Song, H.S., M. Sawamura, T. Ito, K. Kawashimo, and H. Ukeda, 2000. Quantitative determination and characteristic flavour of *Citrus junos* (yuzu) peel oil. *Flavour Fragr. J.*, 15: 245.
71. Song, H.S., M. Sawamura, T. Ito, A. Ido, and H. Ukeda, 2000. Quantitative determination and characteristic flavour of daidai (*Citrus aurantium* L. var. *cyathifera* Y. Tanaka) peel oil. *Flavour Fragr. J.*, 15: 323.
72. McDaniel, M.R., R. Miranda-Lopez, B.T. Watson, N.J. Michaels, and L.M. Libbey, 1990. Pinot noir aroma: A sensory/gas chromatographic approach. In *Flavors and Off-Flavors*, Developments in Food Science, Vol. 24, G. Charalambous (ed.), pp. 23–26. Amsterdam, the Netherlands: Elsevier Science Publishers.
73. Lin, J. and R.L. Rouseff, 2001. Characterization of aroma-impact compounds in cold-pressed grapefruit oil using time-intensity GC-olfactometry and GC-MS. *Flavour Fragr. J.*, 16: 457–463.
74. Linssen, J.P.H., J.L.G.M. Janssens, J.P. Roozen, and M.A. Posthumus, 1993. Combined gas chromatography and sniffing port analysis of volatile compounds of mineral water packed in polyethylene laminated packages. *Food Chem.*, 46: 367.
75. Pollien, P., A. Ott, F. Montigon, M. Baumgartner, R. Muñoz-Box, and A. Chaintreau, 1997. Hyphenated headspace-gas chromatography-sniffing technique: Screening of impact odorants and quantitative aromagram comparisons. *J. Agric. Food Chem.*, 45: 2630.

76. d'Acampora Zellner, B., M. Lo Presti, L.E.S. Barata, P. Dugo, G. Dugo, and L. Mondello, 2006. Evaluation of leaf-derived extracts as an environmentally sustainable source of essential oils by using gas chromatography–mass spectrometry and enantioselective gas chromatography-olfactometry. *Anal. Chem.*, 78: 883.
77. Casimir, D.J. and F.B. Whitfield, 1978. Flavour impact values, a new concept for assigning numerical values for the potency of individual flavour components and their contribution to the overall flavour profile. *Ber. Int. Fruchtsaftunion.*, 15: 325.
78. Jirovetz, L., G. Buchbauer, M.B. Ngassoum, and M. Geissler, 2002. Aroma compound analysis of *Piper nigrum* and *Piper guineense* essential oils from Cameroon using solid-phase microextraction-gas chromatography, solid-phase microextraction-gas chromatography–mass spectrometry and olfactometry. *J. Chromatogr. A*, 976: 265.
79. Jirovetz, L. and M.B. Ngassoum, 1999. Olfactory evaluation and CG/MS analysis of the essential oil of leaves and flowers of *Hyptis pectinata* (L.) Poit. From Cameroon. *SoFW J.*, 125: 35.
80. Sybilska, D. and T. Koscielski, 1983. β-Cyclodextrin as a selective agent for the separation of *o*-, *m*- and *p*-xylene and ethylbenzene mixtures in gas–liquid chromatography. *J. Chromatogr.*, 261: 357.
81. Schurig, V. and H.P. Nowotny, 1988. Separation of enantiomers on diluted permethylated β-cyclodextrin by high-resolution gas chromatography. *J. Chromatogr.*, 441: 155.
82. König, W.A., 1991. *Gas Chromatographic Enantiomer Separation with Modified Cyclodextrins*, 1st edn. Heidelberg, Germany: Hüthig.
83. Nowotny, H.P., D. Schmalzing, D. Wistuba, and V. Schurig, 1989. Extending the scope of enantiomer separation on diluted methylated β-cyclodextrin derivatives by high-resolution gas chromatography. *J. High Res. Chromatogr.*, 12: 383.
84. Dugo, G., G. Lamonica, A. Cotroneo, I. Stagno D'Alcontres, A. Verzera, M.G. Donato, P. Dugo, and G. Licandro, 1992. High resolution gas chromatography for detection of adulterations of citrus cold-pressed essential oils. *Perf. Flav.*, 17: 57–74.
85. Bicchi, C., A. D'Amato, V. Manzin, and P. Rubiolo, 1997. Cyclodextrin derivatives in GC separation of racemic mixtures of volatiles. Part XI. Some applications of cyclodextrin derivatives in GC enantioseparations of essential oil components. *Flavour Fragr. J.*, 12: 55.
86. König, W.A., 1998. Enantioselective capillary gas chromatography in the investigation of stereochemical correlations of terpenoids. *Chirality*, 10: 499.
87. Bicchi, C., A. D'Amato, and P. Rubiolo, 1999. Cyclodextrin derivatives as chiral selectors for direct gas chromatographic separation of enantiomers in the essential oil, aroma and flavour fields. *J. Chromatogr. A*, 843: 99.
88. Lee, M.L., F.J. Yang, and K.D. Bartle, 1984. *Open Tubular Column Gas Chromatography*, 1st edn. New York: Wiley.
89. Rubiolo, P., E. Liberto, C. Cagliero, B. Sgorbini, C. Bicchi, B. d'Acampora Zellner, and L. Mondello, 2007. Linear retention indices in enantioselective GC-mass spectrometry (Es-GC-MS) as a tool to identify enantiomers in flavour and fragrance fields. In *Proceedings of the 38th International Symposium on Essential Oils*, Graz, Austria.
90. Lehmann, D., A. Dietrich, U. Hener, and A. Mosandl, 1995. Stereoisomeric flavour compounds LXX, 1-p-menthene-8-thiol: Separation and sensory evaluation of the enantiomers by enantioselective gas chromatography/olfactometry. *Phytochem. Anal.*, 6: 255.
91. Boelens, M.H. and H. Boelens, 1993. Sensory properties of optical isomers. *Perf. Flav.*, 18: 2.
92. Di Giacomo, A. and B. Mincione, 1994. *Gli Olii Essenziali Agrumari in Italia*. Reggio Calabria, Italy: Baruffa editore.
93. Dugo, P., L. Mondello, E. Cogliro, A. Verzera, and G. Dugo, 1996. On the genuineness of citrus essential oils. 51. Oxygen heterocyclic compounds of bitter orange oil (*Citrus aurantium* L.). *J. Agric. Food Chem.*, 44: 544–549.
94. McHale, D. and J.B. Sheridan, 1989. The oxygen heterocyclic compounds of citrus peel oils. *J. Essent. Oil Res.*, 1: 139.
95. McHale, D. and J.B. Sheridan, 1988. Detection of adulteration of cold-pressed lemon oil. *Flavour Fragr. J.*, 3: 127.
96. Benincasa, M., F. Buiarelli, G.P. Cartoni, and F. Coccioli, 1990. Analysis of lemon and bergamot essential oils by HPLC with microbore columns. *Chromatographia*, 30: 271.
97. Buiarelli, F., G.P. Cartoni, F. Coccioli, and T. Leone, 1996. Analysis of bitter essential oils from orange and grapefruit by high-performance liquid chromatography with microbore columns. *J. Chromatogr. A*, 730: 9.
98. Buiarelli, F., G.P. Cartoni, F. Coccioli, and E. Ravazzi, 1991. Analysis of orange and mandarin essential oils by HPLC. *Chromatographia*, 31: 489.

99. Dugo, P., L. Mondello, E. Sebastiani, R. Ottanà, G. Errante, and G. Dogo, 1999. Identification of minor oxygen heterocyclic compounds of citrus essential oils by liquid chromatography-atmospheric pressure chemical ionisation mass spectrometry. *J. Liq. Chromatogr. Rel. Technol.*, 22: 2991.
100. Píry, J. and A. Príbela, 1994. Application of high-performance liquid chromatography to the analysis of the complex volatile mixture of blackcurrant buds (*Ribes nigrum* L.). *J. Chromatogr. A*, 665: 104.
101. Dugo, G., K.D. Bartle, I. Bonaccorsi, M. Catalfamo, A. Cotroneo, P. Dugo, G. Lamonica et al., 1999. Advanced analytical techniques for the analysis of citrus essential oils. Part 1. Volatile fraction: HRGC/MS analysis. *Essenze Derivati Agrumari*, 69: 79.
102. Deans, D.R., 1968. A new technique for heart cutting in gas chromatography. *Chromatographia*, 1: 18.
103. Mosandl, A., 1995. Enantioselective capillary gas chromatography and stable isotope ratio mass spectrometry in the authenticity control of flavours and essential oils. *Food Rev. Int.*, 11: 597.
104. Mondello, L., M. Catalfamo, P. Dugo, and G. Dugo, 1998. Multidimensional capillary GC-GC for the analysis of real complex samples. Part II. Enantiomeric distribution of monoterpene hydrocarbons and monoterpene alcohols of cold-pressed and distilled lime oils. *J. Microcol. Sep.*, 10: 203.
105. Mondello, L., M. Catalfamo, A.R. Proteggente, I. Bonaccorsi, and G. Dugo, 1998. Multidimensional capillary GC-GC for the analysis of real complex samples. 3. Enantiomeric distribution of monoterpene hydrocarbons and monoterpene alcohols of mandarin oils. *J. Agric. Food Chem.*, 46: 54.
106. Mondello, L., M. Catalfamo, G. Dugo, G. Dugo, and H. McNair, 1999. Multidimensional capillary GC-GC for the analysis of real complex samples. Part IV. Enantiomeric distribution of monoterpene hydrocarbons and monoterpene alcohols of lemon oils. *J. High Res. Chromatogr.*, 22: 350–356.
107. Hener, U., P. Kreis, and A. Mosandl, 1990. Enantiomeric distribution of α-pinene, β-pinene and limonene in essential oils and extracts. Part 1. Rutaceae and Gramineae. *Flavour Fragr. J.*, 5: 193.
108. Kreis, P. and A. Mosandl, 1992. Chiral compounds of essential oils. Part XII. Authenticity control of rose oils, using enantioselective multidimensional gas chromatography. *Flavour Fragr. J.*, 7: 199.
109. Köpke, T., A. Dietrich, and A. Mosandl, 1994. Chiral compounds of essential oils XIV: Simultaneous stereoanalysis of buchu leaf oil compounds. *Phytochem. Anal.*, 5: 61.
110. Faber, B., A. Dietrich, and A. Mosandl, 1994. Chiral compounds of essential oils XV: Stereodifferentiation of characteristic compounds of *Mentha* species by multidimensional gas chromatography. *J. Chromatogr.*, 666: 161.
111. Mondello, L., A. Casilli, P.Q. Tranchida, M. Furukawa, K. Komori, K. Miseri, P. Dugo, and G. Dugo, 2006. Fast enantiomeric analysis of a complex essential oil with an innovative multidimensional gas chromatographic system. *J. Chromatogr. A*, 1105: 11–16.
112. Liu, Z. and J.B. Phillips, 1991. Comprehensive 2-dimensional gas chromatography using an on-column thermal modulator interface. *J. Chromatogr. Sci.*, 1067: 227.
113. Phillips, J.B. and J. Beens, 1999. Comprehensive two-dimensional gas chromatography: A hyphenated method with strong coupling between the two dimensions. *J. Chromatogr. A*, 856: 331.
114. Adahchour, M., J. Beens, R.J.J. Vreuls, and U.A.Th. Brinkman, 2006. Recent developments in comprehensive two-dimensional gas chromatography (GC × GC) IV. Further applications, conclusions and perspectives. *TrAC*, 25: 821.
115. Schoenmakers, P., P. Marriott, and J. Beens, 2003. Nomenclature and conventions in comprehensive multidimensional chromatography. *LC–GC Eur.*, 16: 335–339.
116. Dimandja, J.M.D., S.B. Stanfill, J. Grainger, and D.G. Patterson, 2000. Application of comprehensive two-dimensional gas chromatography (GC × GC) to the qualitative analysis of essential oils. *J. High Res. Chromatogr.*, 23: 208–214.
117. Marriott, P., R. Shellie, J. Fergeus, R. Ong, and P. Morrison, 2000. High resolution essential oil analysis by using comprehensive gas chromatographic methodology. *Flavour Fragr. J.*, 15: 225–239.
118. Marriott, P. and R. Kinghorn, 1999. Cryogenic solute manipulation in gas chromatography—The longitudinal modulation approach. *TrAC*, 18: 114.
119. Shellie, R., P. Marriott, and P. Morrison, 2000. Characterization and comparison of tea tree and lavender oils by using comprehensive gas chromatography. *J. High Resolut. Chromatogr.*, 23: 554–560.
120. Shellie, R., P. Marriott, and P. Morrison, 2001. Concepts and preliminary observations on the triple-dimensional analysis of complex volatile samples by using GC × GC-TOFMS. *Anal. Chem.*, 73: 1336.
121. Shellie, R., L. Mondello, P. Marriott, and G. Dugo, 2002. Characterisation of lavender essential oils by using gas chromatography–mass spectrometry with correlation of linear retention indices and comparison with comprehensive two-dimensional gas chromatography. *J. Chromatogr. A*, 970: 225–234.

122. Cordero, C., P. Rubiolo, B. Sgorbini, M. Galli, and C. Bicchi, 2006. Comprehensive two-dimensional gas chromatography in the analysis of volatile samples of natural origin: A multidisciplinary approach to evaluate the influence of second dimension column coated with mixed stationary phases on system orthogonality. *J. Chromatogr. A*, 1132: 268.
123. Shellie, R., P. Marriott, and P. Morrison, 2004. Comprehensive two-dimensional gas chromatography with flame ionization and time-of-flight mass spectrometry detection: Qualitative and quantitative analysis of West Australian sandalwood oil. *J. Chromatogr. Sci.*, 42: 417.
124. Özel, M.Z., F. Gogus, and A.C. Lewis, 2003. Subcritical water extraction of essential oils from *Thymbra spicata*. *Food Chem.*, 82: 381.
125. Özel, M.Z., F. Gogus, J.F. Hamilton, and A.C. Lewis, 2004. The essential oil of *Pistacia vera* L. at various temperatures of direct thermal desorption using comprehensive gas chromatography coupled with time-of-flight mass spectrometry. *Chromatographia*, 60: 79.
126. Roberts, M.T., J.P. Dufour, and A.C. Lewis, 2004. Application of comprehensive multidimensional gas chromatography combined with time-of-flight mass spectrometry (GC × GC-TOFMS) for high resolution analysis of hop essential oil. *J. Sep. Sci.*, 27: 473.
127. Özel, M.Z., F. Göğüs, and A.C. Lewis, 2006. Determination of *Teucrium chamaedrys* volatiles by using direct thermal desorption–comprehensive two-dimensional gas chromatography–time-of-flight mass spectrometry. *J. Chromatogr. A*, 1114: 164.
128. Özel, M.Z., F. Gogus, and A.C. Lewis, 2006. Comparison of direct thermal desorption with water distillation and superheated water extraction for the analysis of volatile components of *Rosa damascena* Mill. Using GC × GC-TOF/MS. *Anal. Chim. Acta*, 566: 172.
129. Eyres, G., P.J. Marriott, and J.P. Dufour, 2007. The combination of gas chromatography–olfactometry and multidimensional gas chromatography for the characterisation of essential oils. *J. Chromatogr. A*, 150: 70.
130. Ma, C., H. Wang, X. Lu, H. Li, B. Liu, and G. Xu, 2007. Analysis of *Artemisia annua* L. volatile oil by comprehensive two-dimensional gas chromatography time-of-flight mass spectrometry. *J. Chromatogr. A*, 1150: 50.
131. Zhu, S., X. Lu, Y. Qiu, T. Pang, H. Kong, C. Wu, and G. Xu, 2007. Determination of retention indices in constant inlet pressure mode and conversion among different column temperature conditions in comprehensive two-dimensional gas chromatography. *J. Chromatogr. A*, 1150: 28.
132. Dugo, P., M.D. Fernandez, A. Cotroneo, and G. Dugo, 2006. Optimization of a comprehensive two-dimensional normal-phase and reversed phase-liquid chromatography system. *J. Chromatogr. Sci.*, 44: 1.
133. Dugo, P., O. Favoino, R. Luppino, G. Dugo, and Mondello, L., 2004. Comprehensive two-dimensional normal-phase (adsorption)-reversed-phase liquid chromatography. *Anal. Chem.*, 73: 2525–2530.
134. François, I.D., A. Villiers, and P. Sandra, 2006. Considerations on the possibilities and limitations of comprehensive normal phase-reversed phase liquid chromatography (NPLC × RPLC). *J. Sep. Sci.*, 29: 492.
135. Dugo, P., M. Herrero, T. Kumm, D. Giuffrida, G. Dugo, and L. Mondello, 2008. Comprehensive normal-phase × reversed-phase liquid chromatography coupled to photodiode array and mass spectrometry detectors for the analysis of free carotenoids and carotenoid esters from mandarin. *J. Chromatogr. A*, 1189: 196–206.
136. Dugo, P., V. Škeříková, T. Kumm, A. Trozzi, P. Jandera, and L. Mondello, 2006. Elucidation of carotenoid patterns in citrus products by means of comprehensive normal-phase × reversed-phase liquid chromatography. *Anal. Chem.*, 78: 7743–7750.
137. Grob, K., 1987. On-line coupled HPLC-HRGC. In *Proceedings of the Eighth International Symposium on Capillary Chromatography*, Riva del Garda, Italy.
138. Mondello, L., K.D. Bartle, G. Dugo, and P. Dugo, 1994. Automated HPLC-HRGC: A powerful method for essential oils analysis. Part III. Aliphatic and terpene aldehydes of orange oil. *J. High Resolut. Chromatogr.*, 17: 312.
139. Mondello, L., P. Dugo, K.D. Bartle, G. Dugo, and A. Cotroneo, 1995. Automated HPLC-HRGC: A powerful method for essential oils analysis. Part V. Identification of terpene hydrocarbons of bergamot, lemon, mandarin, sweet orange, bitter orange, grapefruit, clementine and Mexican lime oils by coupled HPLC-HRGC-MS(ITD). *Flavour Fragr. J.*, 10: 33.
140. Dugo, G., A. Verzera, A. Trozzi, A. Cotroneo, L. Mondello, and K.D. Bartle, 1994. Automated HPLC-HRGC: A powerful method for essential oils analysis. Part I. Investigation on enantiomeric distribution of monoterpene alcohols of lemon and mandarin essential oils. *Essenz. Deriv. Agrum.*, 64: 35.

141. Dugo, G., A. Verzera, A. Cotroneo, I. Stagno d'Alcontres, L. Mondello, and K.D. Bartle, 1994. Automated HPLC-HRGC: A powerful method for essential oil analysis. Part II. Determination of the enantiomeric distribution of linalool in sweet orange, bitter orange and mandarin essential oils. *Flavour Fragr. J.*, 9: 99.
142. Mondello, L., P. Dugo, K.D. Bartle, B. Frere, and G. Dugo, 1994. On-line high performance liquid chromatography coupled with high resolution gas chromatography and mass spectrometry (HPLC-HRGC-MS) for the analysis of complex mixtures containing highly volatile compounds. *Chromatographia*, 39: 529.
143. Mondello, L., P. Dugo, G. Dugo, and K.D. Bartle, 1996. On-line HPLC-HRGC-MS for the analysis of natural complex mixtures. *J. Chromatogr. Sci.*, 34: 174.

8 Safety Evaluation of Essential Oils
Constituent-Based Approach Utilized for Flavor Ingredients—An Update

Sean V. Taylor

CONTENTS

8.1 Introduction 229
8.2 Constituent-Based Evaluation of Essential Oils 231
8.3 Scope of Essential Oils: Used as Flavor Ingredients 231
    8.3.1 Plant Sources 231
    8.3.2 Processing of Essential Oils for Flavor Functions 231
    8.3.3 Chemical Composition and Congeneric Groups 232
    8.3.4 Chemical Assay Requirements and Chemical Description of Essential Oil 234
        8.3.4.1 Intake of the Essential Oil 234
        8.3.4.2 Analytical Limits on Constituent Identification 236
        8.3.4.3 Intake of Congeneric Groups 236
8.4 Safety Considerations for Essential Oils, Constituents, and Congeneric Groups 237
    8.4.1 Essential Oils 237
        8.4.1.1 Safety of Essential Oils: Relationship to Food 237
    8.4.2 Safety of Constituents and Congeneric Groups in Essential Oils 238
8.5 Guide and Example for the Safety Evaluation of Essential Oils 240
    8.5.1 Introduction 240
    8.5.2 Elements of the Guide for the Safety Evaluation of the Essential Oil 240
        8.5.2.1 Introduction 240
        8.5.2.2 Prioritization of Essential Oil According to Presence in Food 240
        8.5.2.3 Organization of Chemical Data: Congeneric Groups and Classes of Toxicity 241
    8.5.3 Summary 248
References 248

8.1 INTRODUCTION

Based on their action on the human senses, plant-derived essential oils have functioned as sources of food, preservatives, medicines, symbolic articles in religious and social ceremonies, and remedies to modify behavior. In many cases, essential oils and extracts gained widespread acceptance as multifunctional agents due to their strong stimulation of the human gustatory (taste) and olfactory (smell) senses. Cinnamon oil exhibits a pleasing warm spicy aftertaste, characteristic spicy aroma, and preservative properties that made it attractive as a food flavoring and fragrance. Four millennia ago, cinnamon oil was the principal ingredient of a holy ointment mentioned in Exodus 32:22–26. Because of its perceived preservative properties, cinnamon and cinnamon oil were sought by Egyptians for embalming. According to Discorides (Discorides, 50 AD), cinnamon was a breath freshener, would aid in digestion, would counteract the bites of venomous beasts, reduced inflammation of the intestines and the kidneys, and acted as a diuretic. Applied to the face, it was purported to remove undesirable spots. It is not surprising, then, that in 1000 BC, cinnamon was more expensive than gold.

It is not unexpected that cultures throughout history ascribed essential oils and extracts with healing and curative powers, and their strong gustatory and olfactory impacts continue to spur desirable emotions in humans, resulting in their often considerable economic value and cultural importance. However, this cultural importance and widespread demand and (when available) use often occurred with only a limited understanding or acknowledgment of the toxic effects associated with high doses of these plant products. The *natural* origin of these products and their long history of use by humans have, in part, mitigated concerns as to whether these products are efficacious or whether they are safe under conditions of intended use (Arctander, 1969). The adverse effects resulting from the human use of pennyroyal oil as an abortifacient or wild germander as a weight control agent are reminders that no substance is inherently safe independent of considerations of dose. In the absence of information concerning efficacy and safety, recommendations for the quantity and quality of natural products, including essential oils, to be consumed as a medicine remain ambiguous. However, when the intended use is, for example, as a flavor that is subject to governmental regulation, effective and safe levels of use are defined by fundamental biological limits and careful risk assessment.

Flavors derived from essential oils (heretofore known as flavors) are complex mixtures that act directly on the gustatory and olfactory receptors in the mouth and nose leading to taste and aroma responses, respectively. Saturation of these receptors by the individual chemicals within the flavors occurs at very low levels in animals. Hence, with few exceptions, the effects of flavors are self-limiting. The evolution of the human diet is tightly tied to the function of these receptors. Taste and aroma not only determine what we eat but often allow us to evaluate the quality of food and, in some cases, identify unwanted contaminants. The principle of self-limitation taken together with the long history of use of essential oils as flavors in food creates initial conditions upon which has been concluded that these complex mixtures are safe under intended conditions of use. In the United States, the conclusion by the U.S. Food and Drug Administration (21 CFR Sec. 182.10, 182.20, 482.40, and 182.50) that certain oils are "generally recognized as safe" (GRAS) for their intended use was based, in large part, on these two considerations. In Europe and Asia, the presumption of "safe under conditions of use" has been bestowed on essential oils based on similar considerations.

For other intended uses such as dietary supplements or direct food additives, a traditional toxicology approach has been used to demonstrate the safety of essential oils. This relies on performing toxicity tests on laboratory animals, assessing intake and intended use, and determining adequate margins of safety between estimated daily intake by humans and toxic levels resulting from animal studies. Given the constantly changing marketplace and the consumer demand for new and interesting products, however, many new intended uses for these complex mixtures are regularly created, and the exact composition of the essential oil or extract may slightly vary based on processing and desired characteristics. The resources necessary to test all complex mixtures for each intended use are simply not economical. Ultimately, for essential oils that are complex mixtures of chemicals being sold into a competitive marketplace, an approach where each mixture is tested is effective only when specifications for the composition and purity are clearly defined and adequate quality controls are in place for the continued commercial use of the oil. In the absence of such specifications, the results of toxicity testing apply specifically and only to the complex mixture tested. Recent safety evaluation approaches (Schilter et al., 2003) suggest that a multifaceted decision tree approach can be applied to prioritize natural products and the extent of data required to demonstrate safety under conditions of use. The latter approach offers many advantages, both economic as well as scientific, over more traditional approaches. Nevertheless, various levels of resource-intensive toxicity testing of an essential oil are required in this approach.

8.2 CONSTITUENT-BASED EVALUATION OF ESSENTIAL OILS

The chemical constitution of a natural product is fundamental to understanding the product's intended use as well as factors that would affect its safety. Recent advances in analytical methodology have made intensive investigation of the chemical composition of a natural product economically feasible and even routine. High-throughput instrumentation necessary to perform extensive qualitative and quantitative analysis of complex chemical mixtures and to evaluate the variation in the composition of the mixture is now a reality. In fact, analytical tools needed to chemically characterize these complex mixtures are becoming more cost effective, while the cost of traditional toxicology is becoming more cost intensive. Based on the wealth of existing chemical and biological data on the constituents of essential oils and similar data on essential oils themselves, it is possible to validate a constituent-based safety evaluation of an essential oil.

As noted earlier, it is scientifically valid to evaluate the safety of a natural mixture based on its chemical composition. Fundamentally, it is the interaction between one or more molecules in the natural product and macromolecules (proteins, enzymes, etc.) that yield the biological response, regardless of whether it is a desired functional effect such as a pleasing taste or a potential toxic effect such as liver necrosis. Many of the advertised beneficial properties of ephedra are based on the presence of the central nervous system stimulant ephedrine. So too, the gustatory and olfactory properties of coriander oil are, in part, based on the binding of linalool, benzyl benzoate, and other molecules to the appropriate receptors. It is these molecular interactions of chemical constituents that ultimately determine conditions of use.

8.3 SCOPE OF ESSENTIAL OILS: USED AS FLAVOR INGREDIENTS

8.3.1 PLANT SOURCES

Essential oils, as products of distillation, are mixtures of mainly low-molecular-weight chemical substances. Sources of essential oils include components (e.g., pulp, bark, peel, leaf, berry, blossom) of fruits, vegetables, spices, and other plants. Essential oils are prepared from food and nonfood sources. Many of the approximately 100 essential oils used as flavoring ingredients in food are derived directly from food (i.e., lemon oil, basil oil, and cardamom oil); far fewer are extracts from plants that are not normally consumed as food (e.g., cedar leaf oil or balsam fir oil).

Whereas an essential oil is typically obtained by steam distillation of the plant or plant part, an oleoresin is produced by extraction of the same with an appropriate organic solvent. The same volatile constituents of the plant isolated in the essential oil are primarily responsible for aroma and taste of the plant as well as the subsequent extract or oleoresin. Hence, borneol, bornyl acetate, camphor, and other volatile constituents in rosemary oil can provide a flavor intensity as potent as the mass of dried rosemary used to produce the oil. A few exceptions include cayenne pepper, black pepper, ginger, paprika, and sesame seeds, which contain key nonvolatile flavor constituents (e.g., gingerol and zingerone in ginger). These nonvolatile constituents are often higher molecular weight, hydrophilic substances that would be lost during distillation in the preparation of an essential oil, but they remain present in the oleoresin or extract. For economic reasons, crude essential oils are often produced via distillation at the source of the plant raw material and subsequently further processed at modern flavor facilities.

8.3.2 PROCESSING OF ESSENTIAL OILS FOR FLAVOR FUNCTIONS

Because essential oils are a product of nature, environmental and genetic factors will impact the chemical composition of the plant. Factors such as species and subspecies, geographical location, harvest time, plant part used, and method of isolation all affect the chemical composition of the crude material separated from the plant. Variability in the composition of the crude essential oil as isolated from nature has been the subject of much research and development since plant yields of essential oils are major economic factors in crop production.

However, the crude essential oil that arrives at the flavor processing plant is not normally used as such. The crude oil is often subjected to a number of processes that are intended to increase purity and to produce a product with the intended flavor characteristics. Some essential oils may be distilled and cooled to remove natural waxes and improve clarity, while others are distilled more than once (i.e., rectified) to remove undesirable fractions or to increase the relative content of certain chemical constituents. Some oils are dry or vacuum distilled. Normally, at some point during processing, the essential oil is evaluated for its technical function as a flavor. This evaluation typically involves analysis (normally by GLC or liquid chromatography) of the composition of the essential oil for chemical constituents that are markers for the desired technical flavor effect. For an essential oil such as cardamom oil, levels of target constituents such as terpinyl acetate, 1,8-cineole, and limonene are markers for technical viability as a flavoring substance. Based on this initial assessment, the crude essential oil may be blended with other sources of the same oil or chemical constituents isolated from the oil to reach target ranges for key constituent markers that reflect flavor function. The mixture may then be further rectified by distillation. Each step of the process is driven by flavor function. Therefore, the chemical composition of product to be marketed may be significantly different from that of the crude oil. Also, the chemical composition of the processed essential oil is more consistent than that of the crude batches of oil isolated from various plant harvests. The range of concentrations for individual constituents and for groups of structurally related constituents in an essential oil are dictated, in large part, by the requirement that target levels of critical flavor imparting constituents—essentially, principal flavor components—must be maintained.

8.3.3 CHEMICAL COMPOSITION AND CONGENERIC GROUPS

In addition to the key chemical constituents that are the principal flavor components within an essential oil or extract and that allow the natural complex mixture to achieve the technical flavor effect, an essential oil found on the market will normally contain many other chemical constituents, some having little or no flavor function. However, the chemical constituents of essential oils are not infinite in structural variation. Because they are derived from higher plants, these constituents are formed via one of four or five major biosynthetic pathways: lipoxygenase oxidation of lipids, shikimic acid, isoprenoid (terpenoid), and photosynthetic pathways. In ripening vegetables, lipoxygenases oxidize polyunsaturated fatty acids, eventually yielding low-molecular-weight aldehydes (2-hexenal), alcohols (2,6-nonadienol), and esters, many exhibiting flavoring properties. Plant amino acids phenylalanine and tyrosine are formed via the shikimic acid pathway and can subsequently be deaminated, oxidized, and reduced to yield important aromatic substances such as cinnamaldehyde and eugenol. The vast majority of constituents detected in commercially viable essential oils are terpenes (e.g., hydrocarbons [limonene], alcohols [menthol], aldehydes [citral], ketones [carvone], acids, and esters [geranyl acetate]) that are formed via the isoprene pathway (Roe and Field, 1965). Since all of these pathways operate in plants, albeit to different extents depending upon the species, season, and growth environment, many of the same chemical constituents are present in a wide variety of essential oils.

A consequence of having a limited number of plant biosynthetic pathways is that structural variation of chemical constituents in an essential oil is limited. Essential oils typically contain 5–10 distinct chemical classes or congeneric groups. Some congeneric groups, such as aliphatic terpene hydrocarbons, contain upward of 100 chemically identified constituents. In some essential oils, a single constituent (e.g., citral in lemongrass oil) or congeneric group of constituents (e.g., hydroxyallylbenzene derivatives, eugenol and eugenyl acetate, in clove bud oil) comprises the majority of the mass of the essential oil. In others, no single congeneric group predominates. For instance, although eight congeneric groups comprise >98% of the composition of oil of *Mentha piperita* (peppermint oil), greater than 95% of the oil is accounted for by three chemical groups: (1) terpene aliphatic and aromatic hydrocarbons; (2) terpene alicyclic secondary alcohols, ketones, and related esters; and (3) terpene 2-isopropylidene–substituted cyclohexanone derivatives and related substances.

The formation and members of a congeneric group are chosen based on a combination of structural features and known biochemical fate. Substances with a common carbon skeletal structure and functional groups that participate in common pathways of metabolism are assigned to the same congeneric group. For instance, menthyl acetate hydrolyzes prior to absorption to yield menthol, which is absorbed and is interconvertible with menthone in fluid compartments (e.g., the blood). Menthol is either conjugated with glucuronic acid and excreted in the urine or undergoes further hydroxylation mainly at C8 to yield a diol that is also excreted, either free or conjugated. Despite the fact that menthyl acetate is an ester, menthol is an alcohol, menthone a ketone, and 3,8-menthanediol a diol, they are structurally and metabolically related (Figure 8.1). Therefore, all are members of the same congeneric group.

In the case of *M. piperita*, the three principal congeneric groups listed earlier have different metabolic options and possess different organ-specific toxic potential. The congeneric group of terpene aliphatic and aromatic hydrocarbons is represented mainly by limonene and myrcene. The second and most predominant congeneric group is the alicyclic secondary alcohols, ketones, and related esters that include d-menthol, menthone, isomenthone, and menthyl acetate. Although the third congeneric group contains alicyclic ketones similar in structure to menthone, it is metabolically quite different in that it contains an exocyclic isopropylidene substituent that undergoes hydroxylation principally at the C9 position, followed by ring closure and dehydration to yield a heteroaromatic furan ring of increased toxic potential. In the absence of a C4–C8 double bond, neither menthone nor isomenthone can participate in this intoxication pathway. Hence, they are assigned to a different congeneric group.

The presence of a limited number of congeneric groups in an essential oil is critical to the organization of constituents and subsequent safety evaluation of the oil itself. Members of each congeneric group exhibit common structural features and participate in common pathways of pharmacokinetics and metabolism and exhibit similar toxicologic potential. Recent guidance on chemical grouping has been published (OECD, 2014). If the mass of the essential oil (>95%) can be adequately characterized chemically and constituents assigned to well-defined congeneric groups, the safety evaluation of the essential oil can be reduced to (1) a safety evaluation of each of the congeneric groups comprising the essential oil and (2) a *sum of the parts* evaluation of the all congeneric groups

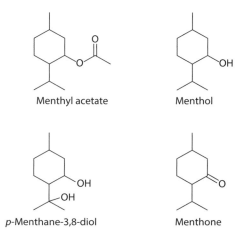

FIGURE 8.1 Congeneric groups are formed by members sharing common structural and metabolic features, such as the group of 2-isopropylidene-substituted cyclohexanone derivatives and related substances.

to account for any chemical or biological interactions between congeneric groups in the essential oil under conditions of intended use. Validation of such an approach lies in the stepwise comparison of the dose and toxic effects for each key congeneric group with similar equivalent doses and toxic effects exhibited by the entire essential oil. Using such an approach, the scientifically independent but industry-sponsored Expert Panel of the U.S. Flavor and Extract Manufacturers Association (FEMA) has conducted safety evaluations on a number of natural flavoring complexes, many of which are essential oils and extracts, under the auspices of the GRAS concept (Smith et al., 2003, 2004). Without question, other approaches have been developed (Meek et al., 2011).

Potential interactions between congeneric groups can, to some extent, be analyzed by an in-depth comparison of the biochemical and toxicologic properties of different congeneric groups in the essential oil. For some representative essential oils that have been the subject of toxicology studies, a comparison of data for the congeneric groups in the essential oil with data on the essential oil itself (congeneric groups together) is a basis for analyzing for the presence or absence of interactions. Therefore, the impact of interaction between congeneric groups is minimal if the levels of and endpoints for toxicity of congeneric groups (e.g., tertiary terpene alcohols) are similar to those of the essential oil (e.g., coriander oil).

Since composition plays such a critical role in the evaluation, analytical identification requirements are also critical to the evaluation. Complete chemical characterization of the essential oil may be difficult or economically unfeasible based on the small volume of essential oil used as a flavor ingredient. In these few cases, mainly for low-volume essential oils, the unknown fraction may be appreciable and a large number of chemical constituents will not be identified. However, if the intake of the essential oil is low or significantly less than its intake from consumption of food (e.g., thyme) from which the essential oil is derived (e.g., thyme oil), there should be no significant concern for safety under conditions of intended use. For those cases in which chemical characterization of the essential oil is limited but the volume of intake is more significant, it may be necessary to perform additional analytical work to decrease the number of unidentified constituents or, in other cases, to perform selected toxicity studies on the essential oil itself. A principal goal of the safety evaluation of essential oils is that no congeneric groups that have significant human intakes should go unevaluated.

8.3.4 Chemical Assay Requirements and Chemical Description of Essential Oil

The safety evaluation of an essential oil initially involves specifying the biological origin, physical and chemical properties, and any other relevant identifying characteristics. An essential oil produced under good manufacturing practices should be of an appropriate purity (quality), and chemical characterization should be complete enough to guarantee a sufficient basis for a thorough safety evaluation of the essential oil under conditions of intended use. Because the evaluation is based primarily on the actual chemical composition of the essential oil, full specifications used in a safety evaluation will necessarily include not only information on the origin of the essential oil (commercial botanical sources, geographical sources, plant parts used, degree of maturity, and methods of isolation) and physical properties (specific gravity, refractive index, optical rotation, solubility, etc.) but also chemical assays for a range of essential oils currently in commerce.

8.3.4.1 Intake of the Essential Oil

Based on current analytical methodology, it is possible to identify literally hundreds of constituents in an essential oil and quantify the constituents to part per million levels. But is this necessary or desirable? From a practical point of view, the level of analysis for constituents should be directly related to the level of exposure to the essential oil. The requirements to identify and quantify constituents for use of 2,000,000 kg of peppermint oil annually should be far greater than that for use of 2000 kg of coriander oil or 50 kg of myrrh oil annually. Also, there is a level at which exposure to each constituent is so low that there is no significant risk associated with intake of that substance.

A conservative no significant risk level of 1.5 µg/day (0.0015 mg/day or 0.000025 mg/kg/day) has been adopted by regulatory authorities as a level at which the human cancer risk is below one in one million—this is commonly referred to as the "threshold of regulation" (FDA, 2005). Therefore, if consumption of an essential oil results in an intake of a constituent that is less than 1.5 µg/day, there should be no requirement to identify and quantify that constituent.

Determining the level to which the constituents of an essential oil should be identified depends upon estimates of intake of food or flavor additives. These estimates are traditionally calculated using a *volume-based* or a *menu–census* approach. A volume-based approach assumes that the total annual volume of use of a substance reported by an industry is distributed over a portion of the population consuming that substance. A menu–census approach is based on the concentration of the substance (essential oil) added to each flavor, the amount of flavor added to each food category, the portion of food consumed daily, and the total of all exposures across all food types. Although the latter is quite accurate for food additives consumed at higher levels in a wide variety of food such as food emulsifiers, the former method provides an efficient and conservative approximation of intake, if a fraction of the total population is assumed to consume all of the substance.

For the World Health Organization (WHO) and the U.S. FDA, intake is calculated using a method known as the *per capita* intake (PCI × 10) method, or alternatively known as the maximized survey-derived intake (MSDI) (Rulis et al., 1984; Woods and Doull, 1991). The MSDI method assumes that only 10% of the population consumes the total annual reported volume of use of a flavor ingredient. This approximation provides a practical and cost-effective approach to the estimation of intake for flavoring substances. The annual volumes of flavoring agents are relatively easy to obtain by industry-wide surveys, which can be performed on a regular basis to account for changes in food trends and flavor consumption. A recent poundage survey of U.S. flavor producers was collected in 2005 and published by FEMA in 2007 (Adams et al., 2007). Similar surveys were conducted in recent years in Europe (EFFA, 2005) and Japan (JFFMA, 2002).

Calculation of intake using the MSDI method has been shown to result in conservative estimates of intake and thus is appropriate for safety evaluation. Over the last three decades, two comprehensive studies of flavor intake have been undertaken. One involved a detailed dietary analysis (DDA) of a panel of 12,000 consumers who recorded all foods that they consumed over a 14-day period, and the flavoring ingredients in each food were estimated by experience flavorists to estimate intake of each flavoring substance (Hall, 1976; Hall and Ford, 1999). The other study utilized a robust full stochastic model (FSM) to estimate intake of flavoring ingredients by typical consumers in the United Kingdom (Lambe et al., 2002). The results of the data-intensive DDA method and the model-based FSM support the use of PCI data as a conservative estimate of intake.

With regard to essential oils, the MSDI method provides overestimates of intake for oils that are widely distributed in food. The large annual volume of use reported for essential oils such as orange oil, lemon oil, and peppermint oil indicate widespread use in a large variety of foods resulting in consumption of these oils by significantly more than 10% of the population. Citrus flavor is pervasive in a multitude of foods and beverages. Therefore, for selected high-volume essential oils, a simple *PCI* rather than a *per capita* × 10 intake may be more appropriate. However, the intake of the congeneric groups and the group of unidentified constituents for these high-volume oils is still estimated by the MSDI method.

An alternative approach to MSDI now employed in the evaluation of flavoring substances by the WHO/UN Food and Agricultural Organization Joint Expert Committee on Food Additives (JECFA) relies on a modification of the traditional menu–census approach. The JECFA single portion exposure technique (SPET) identifies the highest intake from a single food category and assigns it as the overall intake for a flavoring substance. This is calculated from the concentration of the flavoring substance within a food category multiplied by the daily portion size for that food category. At JECFA, the highest intake from either MSDI or SPET is now used to assign an intake to a flavoring substance. JECFA has concluded that both methods, MSDI and SPET, can provide complementary and useful information regarding the uses of flavoring substances. To date, SPET has not been used for the safety evaluation of natural complex mixtures at JECFA, as the flavoring agent evaluations at JECFA have focused thus far on chemically defined substances.

8.3.4.2 Analytical Limits on Constituent Identification

As described earlier, the analytical requirements for detection and identification of the constituents of an essential oil are set by the intake of the oil and by the conservative assumption that constituents with intakes less than 1.5 µg/day will not need to be identified. For instance, if the annual volume of use of coriander oil in the United States is 10,000 kg, then the estimated daily PCI of the oil is

$$\frac{10,000\,\text{kg/year} \times 10^9\,\mu\text{g/kg}}{365\,\text{days/year} \times 31,000,000\,\text{persons}} = 883\,\mu\text{g coriander oil/person/day}$$

Based on the intake of coriander oil (883 µg/day), any constituent present at greater than 0.17% would need to be chemically characterized and quantified:

$$\frac{1.5\,\mu\text{g/day}}{978\,\mu\text{g/day}} \times 100 = 0.17\%$$

For the vast majority of essential oils, meeting these characterization requirements does not require exotic analytical techniques and the identification of the constituents is of a routine nature. However, what would the requirements be for very high-volume essential oils, such as orange oil, cold pressed (567,000 kg), or peppermint oil (1,229,000 kg)? In these cases, a practical limit must be applied and can be justified based on the concept that the intake of these oils is widespread and far exceeds the 10% assumption of MSDI. Based on current analytical capabilities, 0.10% or 0.05% could be used as a reasonable limit of detection, with the lower level used for an essential oil that is known or suspected to contain constituents of higher toxic potential (e.g., methyl eugenol in basil).

8.3.4.3 Intake of Congeneric Groups

Once the analytical limits for identification of constituents have been met, it is key to evaluate the intake of each congeneric group from consumption of the essential oil. A range of concentration of each congeneric group is determined from multiple analyses of different lots of the essential oil used in flavorings. The intake of each congeneric group is determined from mean concentrations (%) of constituents recorded for each congeneric group. For instance, for peppermint oil the alicyclic secondary alcohol/ketone/related ester group may contain (–)-menthol, (–)-menthone, (–)-menthyl acetate, and isomenthone in mean concentrations of 43.0%, 20.3%, 4.4%, and 0.40%, respectively, with the result that that congeneric group accounts for 68.1% of the oil. It should be emphasized that although members in a congeneric group may vary among the different lots of oil, the variation in concentration of congeneric groups in the oil is relatively small.

Routinely, the daily *PCI* of the essential oil is derived from the annual volumes reported in industry surveys (NAS, 1965, 1970, 1975, 1982, 1987; Lucas et al., 1999; JFFMA, 2002; EFFA, 2005). If a conservative estimate of intake of the essential oil is made using a volume-based approach such that a defined group of constituents are set for each essential oil, target constituents can be monitored in an ongoing quality control program, and the composition of the essential oil can become one of the key specifications linking the product that is distributed in the marketplace to the chemically based safety evaluation.

Limited specifications for the chemical composition of some essential oils to be used as food flavorings are currently listed in the Food Chemicals Codex (FCC, 2013). For instance, the chemical assay for cinnamon oil is given as "not less than 80%, by volume, as total aldehydes." Any specification developed related to this safety evaluation procedure should be consistent with already published specifications including FCC and ISO standards. However, based on chemical analyses for the commercially available oil, the chemical specification or assay can and should be expanded to

1. Specify the mean of concentrations for congeneric groups with confidence limits that constitute a sufficient number of commercial lots constituting the vast majority of the oil.
2. Identify key constituents of intake >1.5 µg/day in these groups that can be used to efficiently monitor the quality of the oil placed into commerce over time.
3. Provide information on trace constituents that may be of a safety concern.

For example, given its most recent reported annual volume (1060 kg) (Harman et al., 2013), it is anticipated that a chemical specification for lemongrass oil would include (1) greater than 98.7% of the composition chemically identified; (2) not more than 92% aliphatic terpene primary alcohols, aldehydes, acids, and related esters, typically measured as citral; and (3) not more than 15% aliphatic terpene hydrocarbons, typically measured as myrcene. The principal goal of a chemical specification is to provide sufficient chemical characterization to ensure safety of the essential oil from use as a flavoring. From an industry standpoint, the specification should be sufficiently descriptive as to allow timely quality control monitoring for constituents that are responsible for the technical flavor function. These constituents should also be representative of the major congeneric group or groups in the essential oil. Also, monitored constituents should include those that may be of a safety concern at sufficiently high levels of intake of the essential oil (e.g., pulegone). The scope of a specification should be sufficient to ensure safety in use but not impose an unnecessary burden on industry to perform ongoing analyses for constituents unrelated to the safety or flavor of the essential oil.

8.4 SAFETY CONSIDERATIONS FOR ESSENTIAL OILS, CONSTITUENTS, AND CONGENERIC GROUPS

8.4.1 Essential Oils

8.4.1.1 Safety of Essential Oils: Relationship to Food

The close relationship of natural flavor complexes to food itself has made it difficult to evaluate the safety and regulate the use of essential oils. In the United States, the Federal Food Drug and Cosmetic Act recognizes that a different, lower standard of safety must apply to naturally occurring substances in food than applies to the same ingredient intentionally added to food. For a substance occurring naturally in food, the act applies a realistic standard that the substance must "... not ordinarily render it [the food] injurious to health" (21 CFR 172.30). For added substances, a much higher standard applies. The food is considered to be adulterated if the added substance "... may render it [the food] injurious to health" (21 CFR 172.20). Essential oils used as flavoring substances occupy an intermediate position in that they are composed of naturally occurring substances, many of which are intentionally added to food as individual chemical substances. Because they are considered neither a direct food additive nor a food itself, no current standard can be easily applied to the safety evaluation of essential oils.

The evaluation of the safety of essential oils that have a documented history of use in foods starts with the presumption that they are safe based on their long history of use over a wide range of human exposures without known adverse effects. With a high degree of confidence, one may presume that essential oils derived from food are likely to be safe. Annual surveys of the use of flavoring substances in the United States (NAS, 1965, 1970, 1975, 1981, 1987; Lucas et al., 1999; Harman et al., 2013; 21 CFR 172.510) in part, document the history of use of many essential oils. Conversely, confidence in the presumption of safety decreases for natural complexes that exhibit a significant change in the pattern of use or when novel natural complexes with unique flavor properties enter the food supply. Recent consumer trends that have changed the typical consumer diet have also changed the exposure levels to essential oils in a variety of ways. As one example, changes in the use of cinnamon oil in low-fat cinnamon pastries would alter intake for a specialized population of eaters. Also, increased international trade has coupled with a reduction in cultural cuisine barriers, leading to the introduction of novel plants and plant extracts from previously remote geographical locations. Osmanthus absolute (FEMA No. 3750) and Jambu oleoresin (FEMA No. 3783) are examples of natural complexes recently used as flavoring substances that are derived from plants not indigenous to the United States and not commonly consumed as part of a Western diet. Furthermore, the consumption of some essential oils may not occur solely from intake as flavoring substances; rather, they may be regularly consumed as dietary supplements with advertised functional benefits. These impacts have brought renewed interest in the safety evaluation of essential oils. Although the safety evaluation of essential oil must still rely heavily on knowledge of the history of use, a flexible science-based approach would allow for rigorous safety evaluation of different uses for the same essential oil.

8.4.2 Safety of Constituents and Congeneric Groups in Essential Oils

It is well established that when consumed in high quantities, some plants do indeed exhibit toxicity. Historically, humans have used plants as poisons (e.g., hemlock), and many of the intended medicinal uses of plants (pennyroyal oil as an abortifacient) have produced undesirable toxic side effects. High levels of exposure to selected constituents in the plant or essential oil (i.e., pulegone in pennyroyal oil) have been associated with the observed toxicity. However, with regard to flavor use, experience through long-term use and the predominant self-limiting impact of flavorings on our senses have restricted the amount of a plant or plant part that we use in or on food.

Extensive scientific data on the most commonly occurring major constituents in essential oils have not revealed any results that would give rise to safety concerns at low levels of exposure. Chronic studies have been performed on more 30 major chemical constituents (menthol, carvone, limonene, citral, cinnamaldehyde, benzaldehyde, benzyl acetate, 2-ethyl-1-hexanol, methyl anthranilate, geranyl acetate, furfural, eugenol, isoeugenol, etc.) found in many essential oils. The majority of these studies were hazard determinations that were sponsored by the National Toxicology Program, and they were normally performed at dose levels many orders of magnitude greater than the daily intakes of these constituents from consumption of the essential oil. Even at these high intake levels, the majority of the constituents show no carcinogenic potential (Smith, 2005). In addition to dose/exposure, for some flavor ingredients, the carcinogenic potential that was assessed in the study is related to several additional factors including the mode of administration, species and sex of the animal model, and target organ specificity. In the vast majority of studies, the carcinogenic effect occurs through a nongenotoxic mechanism in which tumors form secondary to preexisting high-dose, chronic organ toxicity, typically to the liver or kidneys. Selected subgroups of structurally related substances (e.g., aldehydes, terpene hydrocarbons) are associated with a single-target organ and tumor type in a specific species and sex of rodent (i.e., male rat kidney tumors secondary to alpha-2u-globulin neoplasms with limonene in male rats) or using a single mode of administration (i.e., forestomach tumors that arise due to high doses of benzaldehyde and hexadienal given by gavage).

Given their long history of use, it appears unlikely that there are essential oils consumed by humans that contain constituents not yet studied that are weak nongenotoxic carcinogens at chronic high-dose levels. Even if there are such cases, because of the relatively low intake (Lucas et al., 1999) as constituents of essential oils, these yet-to-be-discovered constituents would be many orders of magnitude less potent than similar levels of aflatoxins (found in peanut butter), the polycyclic heterocyclic amines (found in cooked foods), or the polynuclear aromatic hydrocarbons (also found in cooked foods). There is nothing to suggest that the major biosynthetic pathways available to higher plants are capable of producing substances such that low levels of exposure to the substance would result in a high level of toxicity or carcinogenicity.

The toxic and carcinogenic potentials exhibited by constituent chemicals in essential oils can largely be equated with the toxic potential of the congeneric group to which that chemical belongs. A comparison of the oral toxicity data (JECFA, 2004) for limonene, myrcene, pinene, and other members of the congeneric group of terpene hydrocarbons shows similar low levels of toxicity with the same high-dose target organ endpoint (kidney) in animal studies, of which the relevance to humans is unlikely. Likewise, dietary toxicity and carcinogenicity data (JECFA, 2001) for cinnamyl alcohol, cinnamaldehyde, cinnamyl acetate, and other members of the congeneric group of 3-phenyl-1-propanol derivatives show similar toxic and carcinogenic endpoints. The safety data for the congeneric chemical groups that are found in vast majority of essential oils have been reviewed (Adams et al., 1996, 1997, 1998, 2004; JECFA 1997, 1998, 1999, 2000a,b, 2001, 2003, 2004; Newberne et al., 1999; Smith et al., 2002a,b). Available data for different representative members in each of these congeneric groups support the conclusion that the toxic and carcinogenic potential of individual constituents adequately represent similar potentials for the corresponding congeneric group.

The second key factor in the determination of safety is the level of intake of the congeneric group from consumption of the essential oil. Intake of the congeneric group will, in turn, depend upon the variability of the chemical composition of the essential oil in the marketplace and on the conditions of use. As discussed earlier, chemical analysis of the different batches of oil obtained from the same and different manufacturers will produce a range of concentrations for individual constituents in each congeneric group of the essential oil. The mean concentration values (%) for constituents are then summed for all members of the congeneric group. The total % determined for the congeneric group is multiplied by the estimated daily intake (PCI × 10) of the essential oil to provide a conservative estimate of exposure to each congeneric group from consumption of the essential oil.

In some essential oils, the intake of one constituent, and therefore, one congeneric group, may account for essentially all of the oil (e.g., linalool in coriander oil, citral in lemongrass oil, benzaldehyde in bitter almond oil). In other oils, exposure to a variety of congeneric groups over a broad concentration range may occur. As noted earlier, cardamom oil is an example of such an essential oil. Ultimately, it is the relative intake and the toxic potential of each congeneric group that is the basis of the congeneric group-based safety evaluation. The combination of relative intake and toxic potential will prioritize congeneric groups for the safety evaluation. Hypothetically, a congeneric group of increased toxic potential that accounts for only 5% of the essential oil may be prioritized higher than a congeneric group of lower toxic potential accounting for 95%.

The following guide and examples therein are intended to more fully illustrate the principles described earlier that are involved in the safety evaluation of essential oils. Fermentation products, process flavors, substances derived from fungi, microorganisms, or animals, and direct food additives are explicitly excluded. The guide is designed primarily for application to essential oils and extracts for use as flavoring substances. The guide is a tool to organize and prioritize the chemical constituents and congeneric groups in an essential oil in such a way as to allow a detailed analysis of their chemical and biological properties. This analysis as well as consideration of other relevant scientific data provides the basis for a safety evaluation of the essential oil under conditions of intended use. Validation of the approach is provided, in large part, by a detailed comparison of the doses and toxic effects exhibited by constituents of the congeneric group with the equivalent doses and effects provided by the essential oil. This methodology, with some variations based on expert judgment as appropriate, has been in use for more than 10 years through the FEMA GRAS program, in safety evaluations conducted by the FEMA Expert Panel (Smith et al., 2004).

8.5 GUIDE AND EXAMPLE FOR THE SAFETY EVALUATION OF ESSENTIAL OILS

8.5.1 INTRODUCTION

The guide is a procedure involving a comprehensive evaluation of the chemical and biological properties of the constituents and congeneric groups of an essential oil. Constituents in, for instance, an essential oil that are of known structure are organized into congeneric groups that exhibit similar metabolic and toxicologic properties. The congeneric groups are further classified according to levels (Structural Classes I, II, and III) of toxicologic concern using a decision tree approach (Cramer et al., 1978; Munro et al., 1996b). Based on intake data for the essential oil and constituent concentrations, the congeneric groups are prioritized according to intake and toxicity potential. The procedure ultimately focuses on those congeneric groups that, due to their structural features and intake, may pose some significant risk from the consumption of the essential oil. Key elements used to evaluate congeneric groups include exposure, structural analogy, metabolism, and toxicology, which includes toxicity, carcinogenicity, and genotoxic potential (Oser and Hall, 1977; Oser and Ford, 1991; Woods and Doull, 1991; Adams et al., 1996, 1997, 1998, 2004; Newberne et al., 1999; Smith et al., 2002a,b, 2004). Throughout the analysis of these data, it is essential that professional judgment and expertise be applied to complete the safety evaluation of the essential oil. As an example of how a typical evaluation process for an essential oil is carried out according to this guide, the safety evaluation for flavor use of corn mint oil (*Mentha arvensis*) is outlined in the following text.

8.5.2 ELEMENTS OF THE GUIDE FOR THE SAFETY EVALUATION OF THE ESSENTIAL OIL

8.5.2.1 Introduction

In Step 1 of the guide, the evaluation procedure estimates intake based on industry survey data for each essential oil. It then organizes the chemically identified constituents that have an intake >1.5 µg/day into congeneric groups that participate in common pathways of metabolism and exhibit a similar toxic potential. In Steps 2 and 3, each identified chemical constituent is broadly classified according to toxic potential (Cramer et al., 1978) and then assigned to a congeneric group of structurally related substances that exhibit similar pathways of metabolism and toxicologic potential.

Before the formal evaluation begins, it is necessary to specify the data (e.g., botanical, physical, chemical) required to completely describe the product being evaluated. In order to effectively evaluate an essential oil, attempted complete analyses must be available for the product intended for the marketplace from a number of flavor manufacturers. Additional quality control data are useful, as they demonstrate consistency in the chemical composition of the product being marketed. A Technical Information Paper drafted for the particular essential oil under consideration organizes and prioritizes these data for efficient sequential evaluation of the essential oil.

In Steps 8 and 9, the safety of the essential oil is evaluated in the context of all congeneric groups and any other related data (e.g., data on the essential oil itself or for an essential oil of similar composition). The procedure organizes the extensive database of information on the essential oil constituents in order to efficiently evaluate the safety of the essential oil under conditions of use. It is important to stress, however, that the guide is not intended to be nor in practice operates as a rigid checklist. Each essential oil that undergoes evaluation is different, and different data will be available for each. The overriding objective of the guide and subsequent evaluation is to ensure that no significant portion of the essential oil should go unevaluated.

8.5.2.2 Prioritization of Essential Oil According to Presence in Food

In Step 1, essential oils are prioritized according to their presence or absence as components of commonly consumed foods (Step 1). This question evaluates the relative intake of the essential oil as an intentionally added flavoring substance versus its intake as a component part of food. Many essential oils are isolated from plants that are commonly consumed as a food. Little or no safety

Safety Evaluation of Essential Oils

concerns should exist for the intentional addition of the essential oil to the diet, if intake of the oil from consumption of traditional foods (garlic) substantially exceeds intake as an intentionally added flavoring substance (garlic oil). In many ways, the first step applies the concept of "long history of safe use" to essential oils. That is, if exposure to the essential oil occurs predominantly from consumption of a normal diet, a conclusion of safety is straightforward. Step 1 of the guide clearly places essential oils that are consumed as part of a traditional diet on a lower level of concern than those oils derived from plants that are either not part of the traditional diet or whose intake is not predominantly from the diet. The first step also mitigates the need to perform comprehensive chemical analysis for essential oils in those cases where intake is low and occurs predominantly from consumption of food. An estimate of the intake of the essential oil is based on the most recent poundage available from flavor industry surveys and the assumption that the essential oil is consumed by only 10% of the population for an oil having a survey volume <50,000 kg/year and 100% of the population for an oil having a survey volume >50,000 kg/year. In addition, the detection limit for constituents is determined based on the daily PCI of the essential oil.

8.5.2.2.1 Corn Mint Oil

To illustrate the type of data considered in Step 1, consider corn mint oil. Corn mint oil is produced by the steam distillation of the flowering herb of *M. arvensis*. The crude oil contains upward of 70% (–)-menthol, some of which is isolated by crystallization at low temperature. The resulting dementholized oil is corn mint oil. Although produced mainly in Brazil during the 1970s and 1980s, corn mint oil is now produced predominantly in China and India. Corn mint has a more stringent taste compared to that of peppermint oil, *M. piperita*, but can be efficiently produced and is used as a more cost-effective substitute. Corn mint oil isolated from various crops undergoes subsequent *clean up*, further distillation, and blending to produce the finished commercial oil. Although there may be significant variability in the concentrations of individual constituents in different samples of crude essential oil, there is far less variability in the concentration of constituents and congeneric groups in the finished commercial oil. The volume of corn mint oil reported in the most recent U.S. poundage survey is 446,000 kg/year (Harman et al., 2013), which is approximately 25% of the potential market of peppermint oil. Because corn mint oil is a high-volume essential oil, it is highly likely that the entire population consumes the annual reported volume, and therefore, the daily PCI is calculated based on 100% of the population (310,000,000). This results in a daily PCI of approximately 3.9 mg/person/day (0.066 mg/kg bw/day) of corn mint oil:

$$\frac{446,000 \text{ kg/year} \times 10^9 \text{ µg/kg}}{365 \text{ days/year} \times 310 \times 10^6 \text{ persons}} = 3942 \text{ µg/person/day}$$

Based on the intake of corn mint oil (3942 µg/day), any constituent present at greater than 0.038% would need to be chemically characterized and quantified:

$$\frac{1.5 \text{ µg/day}}{3942 \text{ µg/day}} \times 100 = 0.038\%$$

8.5.2.3 Organization of Chemical Data: Congeneric Groups and Classes of Toxicity

In Step 2, constituents are assigned to one of three structural classes (I, II, or III) based on toxic potential (Cramer et al., 1978). Class I substances contain structural features that suggest a low order of oral toxicity. Class II substances are clearly less innocuous than Class I substances but do not contain structural features that provide a positive indication of toxicity. Class III substances contain structural features (e.g., an epoxide functional group, unsubstituted heteroaromatic derivatives) that permit no strong presumption of safety and in some cases may even suggest significant toxicity. For instance, the simple aliphatic hydrocarbon, limonene, is assigned to structural Class I, while elemicin, which is an allyl-substituted benzene derivative with a reactive benzylic/allylic position, is assigned to Class III. Likewise, chemically unidentified constituents of the essential oil are automatically placed in Structural Class III, since no presumption of safety can be made.

The toxic potential of each of the three structural classes has been quantified (Munro et al., 1996a). An extensive toxicity database has been compiled for substances in each structural class. The database covers a wide range of chemical structures, including food additives, naturally occurring substances, pesticides, drugs, antioxidants, industrial chemicals, flavors, and fragrances. Conservative no observable effect levels (fifth percentile NOELs) have been determined for each class. These fifth percentile NOELs for each structural class are converted to human exposure threshold levels by applying a 100-fold safety factor and correcting for mean bodyweight (60/100). The human exposure threshold levels are referred to as thresholds of toxicological concern (TTC). With regard to flavoring substances, the TTCs are even more conservative, given that the vast majority of NOELs for flavoring substances are above the 90th percentile. These conservative TTCs have since been adopted by the WHO and Commission of the European Communities for use in the evaluation of chemically identified flavoring agents by JECFA and the European Food Safety Authority (EFSA) (JECFA, 1997; EC, 1999).

Step 3 is a key step in the guide. It organizes the chemical constituents into congeneric groups that exhibit common chemical and biological properties. Based on the well-recognized biochemical pathways operating in plants, essentially all of the volatile constituents found in essential oils, extracts, and oleoresins belong to well-recognized congeneric groups. Recent reports (Maarse et al., 1992, 1994, 2000; Njissen et al., 2003) of the identification of new naturally occurring constituents indicate that newly identified substances fall into existing congeneric groups. The Expert Panel, JECFA, and the EC have acknowledged that individual chemical substances can be evaluated in the context of their respective congeneric group (JECFA, 1997; EC, 1999; Smith et al., 2005a,b). The congeneric group approach provides the basis for understanding the relationship between the biochemical fate of members of a chemical group and their toxicologic potential. Within this framework, the objective is to continuously build a more complete understanding of the absorption, distribution, metabolism, and excretion of members of the congeneric group and their potential to cause systemic toxicity. Within the guidelines, the structural class of each congeneric group is assigned based on the highest structural class of any member of the group. Therefore, if an essential oil contained a group of furanone derivatives that were variously assigned to Structural Classes II and III, in the evaluation of the oil, the congeneric group would, in a conservative manner, be assigned to Class III.

The types and numbers of congeneric groups in a safety evaluation program are, by no means, static. As new scientific data and information become available, some congeneric groups are combined while others are subdivided. This has been the case for the group of alicyclic secondary alcohols and ketones that were the subject of a comprehensive scientific literature review in 1975 (FEMA, 1975). Over the last two decades, experimental data have become available indicating that a few members of this group exhibit biochemical fate and toxicologic potential inconsistent with that for other members of the same group. These inconsistencies, almost without exception, arise at high-dose levels that are irrelevant to the safety evaluation of low levels of exposure to flavor use of the substance. However, given the importance of the congeneric group approach in the safety assessment program, it is critical to resolve these inconsistencies. Additional metabolic and toxicologic studies may be required to distinguish the factors that determine these differences. Often the effect of dose and a unique structural feature results in utilization of a metabolic activation pathway not utilized by other members of a congeneric group. Currently, evaluating bodies including JECFA, EFSA, and the FEMA Expert Panel have classified flavoring substances into the same congeneric groups for the purpose of safety evaluation.

In Steps 5, 6, and 7, each congeneric group in the essential oil is evaluated for safety in use. In Step 5, an evaluation of the metabolism and disposition is performed to determine, under current conditions of intake, whether the group of congeneric constituents is metabolized by well-established detoxication pathways to yield innocuous products. That is, such pathways exist for the congeneric group of

TABLE 8.1
Structural Class Definitions and their Human Intake Thresholds

Class	Description	Fifth Percentile NOEL (mg/kg/day)	Human Exposure Threshold (TTC)[a] (µg/day)
I	Structure and related data suggest a low order of toxicity. If combined with low human exposure, they should enjoy an extremely low priority for investigation. The criteria for adequate evidence of safety would also be minimal. Greater exposures would require proportionately higher priority for more exhaustive study.	3.0	1800
II	Intermediate substances. They are less clearly innocuous than those of Class I, but do not offer the basis either of the positive indication of toxicity or of the lack of knowledge characteristic of those in Class III.	0.91	540
III	Permit no strong initial presumptions of safety or that may even suggest significant toxicity. They thus deserve the highest priority for investigation. Particularly when per capita intake is high of a significant subsection of the population that has a high intake, the implied hazard would then require the most extensive evidence for safety in use.	0.15	90

[a] The human exposure threshold was calculated by multiplying the fifth percentile NOEL by 60 (assuming an individual weighs 60 kg) and dividing by a safety factor of 100.

constituents in an essential oil, and safety concerns will arise only if intake of the congeneric group is sufficient to saturate these pathways potentially leading to toxicity. If a significant intoxication pathway exists (e.g., pulegone), this should be reflected in a higher decision tree class and lower TTC threshold. At Step 6 of the procedure, the intake of the congeneric group relative to the respective TTC for one of the three structural classes (1800 µg/day for Class I; 540 µg/day for Class II; 90 µg/day for Class III; see Table 8.1) is evaluated. If the intake of the congeneric group is less than the threshold for the respective structural class, the intake of the congeneric group presents no significant safety concerns. The group passes the first phase of the evaluation and is then referred to Step 8, the step in which the safety of the congeneric group is evaluated in the context of all congeneric groups in the essential oil.

If, at Step 5, no sufficient metabolic data exist to establish safe excretion of the product or if activation pathways have been identified for a particular congeneric group, then the group moves to Step 7, and toxicity data are required to establish safe use under current conditions of intake. There are examples where low levels of xenobiotic substances can be metabolized to reactive substances. In the event that reactive metabolites are formed at low levels of intake of naturally occurring substances, a detailed analysis of dose-dependent toxicity data must be performed. Also, if the intake of the congeneric group is greater than the human exposure threshold (suggesting metabolic saturation may occur), then toxicity data are also required. If, at Step 7, a database of relevant toxicological data for a representative member or members of the congeneric group indicates that a sufficient margin of safety exists for the intake of the congeneric group, the members of that congeneric group are concluded to be safe under conditions of use of the essential oil. The congeneric group then moves to Step 8.

In the event that insufficient data are available to evaluate a congeneric group at Step 7, or the currently available data result in margins of safety that are not sufficient, the essential oil cannot be further evaluated by this guide and must be set aside for further considerations.

Use of the guide requires scientific judgment at each step of the sequence. For instance, if a congeneric group that accounted for 20% of a high-volume essential oil was previously evaluated and found to be safe under intended conditions of use, the same congeneric group found at less than 2% of a low-volume essential oil does not need to be further evaluated.

Step 8 considers additivity or synergistic interactions between individual substances and between the different congeneric groups in the essential oil. As for all other toxicological concerns, the level of exposure to congeneric groups is relevant to whether additive or synergistic effects present a significant health hazard. The vast majority of essential oils are used in food in extremely low concentrations, which therefore results in very low intake levels of the different congeneric groups within that oil. Moreover, major representative constituents of each congeneric group have been tested individually and pose no toxicological threat even at dose levels that are orders of magnitude greater than normal levels of intake of essential oils from use in traditional foods. Based on the results of toxicity studies both on major constituents of different congeneric groups in the essential oil and on the essential oil itself, it can be concluded that the toxic potential of these major constituents is representative of that of the oil itself, indicating the likely absence of additivity and synergistic interaction. In general, the margin of safety is so wide and the possibility of additivity or synergistic interaction so remote that combined exposure to the different congeneric groups and the unknowns are considered of no health concern, even if expert judgment cannot fully rule out additivity or synergism. However, case-by-case considerations are appropriate. Where possible combined effects might be considered to have toxicological relevance, additional data may be needed for an adequate safety evaluation of the essential oil.

Additivity of toxicologic effect or synergistic interaction is a conservative default assumption that may be applied whenever the available metabolic data do not clearly suggest otherwise. The extensive database of metabolic information on congeneric groups (JECFA, 1997, 1998, 1999, 2000a,b, 2001, 2003, 2004) that are found in essential oils suggests that the potential for additive effects and synergistic interactions among congeneric groups in essential oils is extremely low. Although additivity of effect is the approach recommended by NAS/NRC committees (NRC, 1994, 1988) and regulatory agencies (EPA, 1988), the Presidential Commission of Risk Assessment and Risk Management recommended (Presidential Commission, 1996) that "For risk assessments involving multiple chemical exposures at low concentrations, without information on mechanisms, risks should be added. If the chemicals act through separate mechanisms, their attendant risks should not be added but should be considered separately." Thus, the risks of chemicals that act through different mechanisms, that act on different target systems, or that are toxicologically dissimilar in some other way should be considered to be independent of each other. The congeneric groups in essential oils are therefore considered separately.

Further, the majority of individual constituents that comprise essential oils are themselves used as flavoring substances that pose no toxicological threat at doses that are magnitudes greater than their level of intake from the essential oil. Rulis (1987) reported that "The overwhelming majority of additives present a high likelihood of having safety assurance margins in excess of 10^5." He points out that this is particularly true for additives used in the United States at less than 100,000 lb/year. Because more than 90% of all flavoring ingredients are used at less than 10,000 lb/year (Hall and Oser, 1968), this alone implies intakes commonly many orders of magnitude below the no-effect level. Nonadditivity thus can often be assumed. As is customary in the evaluation of any substance, high-end data for exposure (consumption) are used, and multiple other conservatisms are employed to guard against underestimation of possible risk. All of these apply to complex mixtures as well as to individual substances.

8.5.2.3.1 Corn Mint Oil Congeneric Groups
In corn mint oil, the principal congeneric group is composed of terpene alicyclic secondary alcohols, ketones, and related esters, as represented by the presence of (−)-menthol, (−)-menthone, (+)-isomenthone, (−)-menthyl acetate, and other related substances. Samples of triple-distilled commercial

corn mint oil may contain up to 95% of this congeneric group. The biochemical and biological fate of this group of substances has been previously reviewed (Adams et al., 1996; JECFA, 1999). Key data on metabolism, toxicity, and carcinogenicity are cited in the following text (Table 8.2) in order to complete the evaluation. Although constituents in this group are effectively detoxicated via conjugation of the corresponding alcohol or ω-oxidation followed by conjugation and excretion (Williams, 1940; Madyastha and Srivatsan, 1988; Yamaguchi et al., 1994), the intake of the congeneric group (3745 µg/person/day or 3.75 mg/person/day, see Table 8.2) is higher than the exposure threshold of (540 µg/person/day or 0.540 mg/person/day for Structural Class II. Therefore, toxicity data are required for this congeneric group. In both short- and long-term studies (NCI, 1978; Madsen et al., 1986), menthol, menthone, and other members of the group exhibit NOAELs at least 1000 times the daily PCI (*eaters only*) (3.75 mg/person/day or 0.062 mg/kg bw/day) of this congeneric group resulting from intake of the essential oil. For members of this group, numerous in vitro and in vivo genotoxicity assays are consistently negative (Florin et al., 1980; Heck et al., 1989; Sasaki et al., 1989; Muller, 1993; Zamith et al., 1993; Rivedal et al., 2000, NTP Draft, 2003a). Therefore, the intake of this congeneric group from consumption of *M. arvensis* is not a safety concern.

Although it is a constituent of corn mint oil and is also a terpene alicyclic ketone structurally related to the aforementioned congeneric group, pulegone exhibits a unique structure (i.e., 2-isopropylidenecyclohexanone) that participates in a well-recognized intoxication pathway (see Figure 8.2) (McClanahan et al., 1989; Thomassen et al., 1992; Adams et al., 1996; Chen et al., 2001) that leads to the formation of menthofuran. This metabolite subsequently oxidizes and ring opens to yield a highly reactive 2-ene-1,4-dicarbonyl intermediate that reacts readily with proteins resulting in hepatotoxicity at intake levels at least two orders of magnitude less than no observable effect levels for structurally related alicyclic ketones and secondary alcohols (menthone, carvone, and menthol). Therefore, pulegone and its metabolite (menthofuran), which account for <2% of commercial corn mint oil, are considered separately in the guide. In this case, the daily PCI of 79 µg/person/day (1.3 µg/kg bw/day) does not exceed the 90 µg/day threshold for Class III. However, a 90-day study on pulegone (NTP, 2002) showed a NOAEL (9.375 mg/kg bw/day) that is approximately 7200 times the intake of pulegone and its metabolites as constituents of corn mint oil. Also, in a 28-day study with peppermint oil (*M. piperita*) containing approximately 4% pulegone and menthofuran, a NOAEL of 200 mg/kg bw/day for male rats and a NOAEL of 400 mg/kg bw/day for female rats were established, which corresponds to a NOAEL of 8 mg/kg bw/day for pulegone and menthofuran (Serota, 1990). In a 90-day study with a mixture of *M. piperita* and *M. arvensis* oils (Splindler and Madsen, 1992; Smith et al., 1996), a NOAEL of 100 mg/kg bw/day was established, which corresponds to a NOAEL of 4 mg/kg bw/day for pulegone and menthofuran.

The only other congeneric group that accounts for >2% of the composition of corn mint oil is a congeneric group of terpene hydrocarbons ((+) and (−)-pinene, (+) limonene, etc.). Although these may contribute up to 8% of the oil, upon multiple redistillations during processing, the hydrocarbon content can be significantly reduced (<3%) in the finished commercial oil. Using the 8% figure to determine a conservative estimate of intake, the intake of terpene hydrocarbons is 315 µg/person/day (5.26 µg/kg bw/day). This group is predominantly metabolized by cytochrome P450-catalyzed hydroxylation, conjugation, and excretion (Ishida et al., 1981; Madyastha and Srivatsan, 1987; Crowell et al., 1994; Poon et al., 1996; Vigushin et al., 1998; Miyazawa et al., 2002). The daily PCI of 315 µg/person/day is less than the exposure threshold (1800 µg/person/day) for Structural Class I. Although no additional data would be required to complete the evaluation of this group, NOAELs (300 mg/kg bw/day) from long-term studies (NTP, 1990) on principal members of this group are orders of magnitude greater than the daily *PCI* (*eaters only*) of terpene hydrocarbons (0.088 mg/kg bw/day). Therefore, all congeneric groups in corn mint oil are considered safe for use when consumed in corn mint oil.

Finally, the essential oil itself is evaluated in the context of the combined intake of all congeneric groups and any other related data in Step 8. Interestingly, members of the terpene alicyclic secondary alcohols, ketones, and related esters, multiple members of the monoterpene

TABLE 8.2
Safety Evaluation of Corn Mint Oil, *Mentha arvensis*[a]

Congeneric Group	Step 2. Decision Tree Class (TTC, µg/Person/Day)	Step 3. High% from Multiple Commercial Samples	Step 4. Intake, µg/Person/Day	Step 5. Metabolism Pathways	Step 6. Intake of Congeneric Group or Total of Unidentified Constituents Group <TTC for Class?	Step 7. Relevant Toxicity Data if Intake of Group > TTC
Secondary alicyclic saturated and unsaturated alcohol/ketone/ketal/ester (e.g., menthol, menthone, isomenthone, menthyl acetate)	II (540)	95	3745	1. Glucuronic acid conjugation of the alcohol followed by excretion in the urine. 2. ω-Oxidation of the side-chain substituents to yield various polyols and hydroxyacids and excreted as glucuronic acid conjugates.	No, 3745 > 540 µg/person/day	NOEL of 600,000 µg/kg bw/day for menthol (103-week dietary study in mice) (NCI, 1979) NOEL of 400,000 µg/kg bw/day for menthone (28-day gavage study in rats) (Madsen et al., 1986)
Aliphatic terpene hydrocarbon (e.g., limonene, pinene)	I (1800)	8	315	1. ω-Oxidation to yield polar hydroxy and carboxy metabolites excreted as glucuronic acid conjugates.	Yes, 315 <1800 µg/person/day	Not required
2-Isopropylidene cyclohexanone and metabolites (e.g., pulegone)	III (90)	2	79	1. Reduction to yield menthone or isomenthone, followed by hydroxylation of ring or side-chain positions and then conjugation with glucuronic acid. 2. Conjugation with glutathione in a Michael-type addition leading to mercapturic acid conjugates that are excreted or further hydroxylated and excreted. 3. Hydroxylation catalyzed by cytochrome P-450 to yield a series of ring- and side-chain-hydroxylated pulegone metabolites, one of which is a reactive 2-ene-1,4-dicarbonyl derivative. This intermediate is known to form protein adducts leading to enhanced toxicity. (Austin et al., 1988)	Yes, 79 < 90 µg/person/day	Not required. But NOAEL of 9375 µg/kg bw/day for pulegone (90 day gavage study in rats) (NTP, 2002)

[a] Based on daily per capita intake of 3942 µg/person/day for corn mint oil.

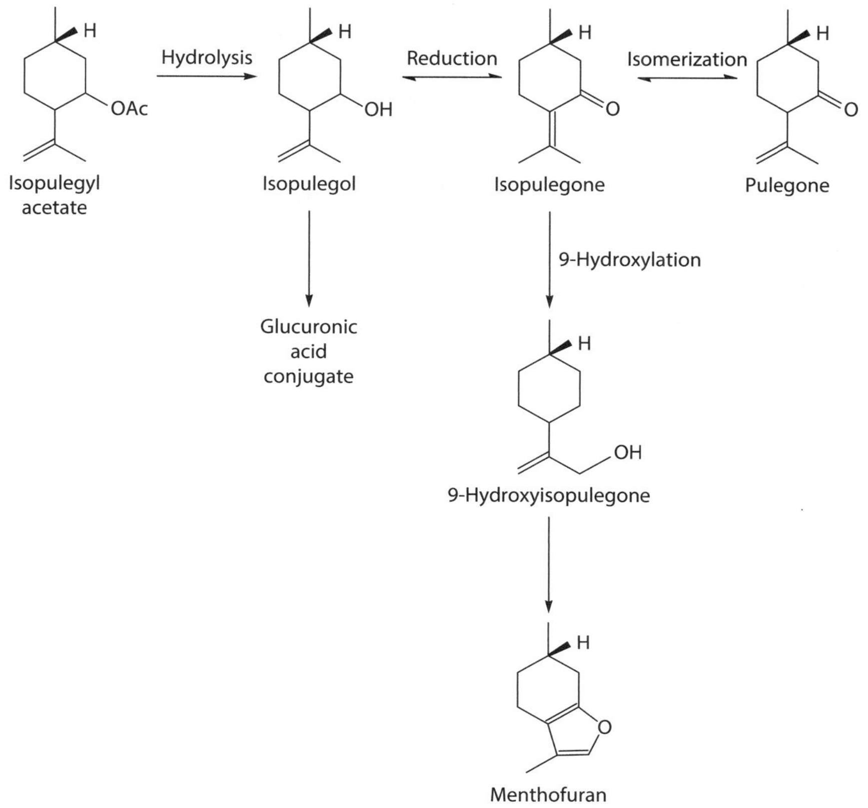

Safety Evaluation of Essential Oils

FIGURE 8.2 Metabolism of isopulegone, pulegone, and isopulegyl acetate.

hydrocarbons, and peppermint oil itself show a common nephrotoxic effect recognized as alpha-2u-globulin nephropathy. The microscopic evidence of histopathology of the kidneys for male rats in the mint oil study is consistent with the presence of alpha-2u-globulin nephropathy. In addition, a standard immunoassay for detecting the presence of alpha-2u-globulin was performed on kidney sections from male and female rats in the mint oil study (Serota, 1990). Results of the assay confirmed the presence of alpha-2u-globulin nephropathy in male rats (Swenberg and Schoonhoven, 2002). This effect is found only in males rats and is not relevant to the human health assessment of corn mint oil. Other toxic interactions between congeneric groups are expected to be minimal given that the NOELs for the congeneric groups and those for finished mint oils are on the same order of magnitude.

Based on the aforementioned assessment and the application of the scientific judgment, corn mint oil is concluded to be *GRAS* under conditions of intended use as a flavoring substance. Given the criteria used in the evaluation, recommended specifications should include the following chemical assay:

1. Less than 95% alicyclic secondary alcohols, ketones, and related esters, typically measured as (−)-menthol
2. Less than 2% 2-isopropylidenecyclohexanones and their metabolites, measured as (−)-pulegone
3. Less than 10% monoterpene hydrocarbons, typically measured as limonene

8.5.3 SUMMARY

The safety evaluation of an essential oil is performed in the context of all available data for congeneric groups of identified constituents and the group of unidentified constituents, data on the essential oil or a related essential oil, and any potential interactions that may occur in the essential oil when consumed as a flavoring substance.

The guide provides a chemically based approach to the safety evaluation of an essential oil. The approach depends on a thorough quantitative analysis of the chemical constituents in the essential oil intended for commerce. The chemical constituents are then assigned to well-defined congeneric groups that are established based on extensive biochemical and toxicologic information, and this is evaluated in the context of intake of the congeneric group resulting from consumption of the essential oil. The intake of unidentified constituents considers the consumption of the essential oil as a food, a highly conservative toxicologic threshold, and toxicity data on the essential oil or an essential oil of similar chemical composition. The flexibility of the guide is reflected in the fact that high intake of major congeneric groups of low toxicologic concern will be evaluated along with low intake of minor congeneric groups of significant toxicological concern (i.e., higher structural class). The guide also provides a comprehensive evaluation of all congeneric groups and constituents that account for the majority of the composition of the essential oil. The overall objective of the guide is to organize and prioritize the chemical constituents of an essential oil in order that no reasonably possible significant risk associated with the intake of essential oil goes unevaluated.

REFERENCES

Adams, T.B., Cohen, S., Doull, J., Feron, V.J., Goodman, J.I., Marnett, L.J., Munro, I.C. et al. (2004) The FEMA GRAS assessment of cinnamyl derivatives used as flavor ingredients. *Food Chem. Toxicol.* 42: 157–185.

Adams, T.B., Doull, J., Goodman, J.I., Munro, I.C., Newberne, P.M., Portoghese, P.S., Smith, R.L. et al. (1997) The FEMA GRAS assessment of furfural used as a flavor ingredient. *Food Chem. Toxicol.* 35: 739–751.

Adams, T.B., Greer, D.B., Doull, J., Munro, I.C., Newberne, P.M., Portoghese, P.S., Smith, R.L. et al. (1998) The FEMA GRAS assessment of lactones used as flavor ingredients. *Food Chem. Toxicol.* 36: 249–278.

Adams, T.B., Hallagan, J.B., Putman, J.M., Gierke, T.L., Doull, J., Munro, I.C., Newberne, P.M. et al. (1996) The FEMA GRAS assessment of alicyclic substances used as flavor ingredients. *Food Chem. Toxicol.* 34: 763–828.

Adams, T.B., McGowen, M.M., Williams, M.C., Cohen, S.M., Feron, V.J., Goodman, J.J., Marnett, L.J. et al. (2007) The FEMA GRAS assessment of aromatic substituted secondary alcohols, ketones and related esters used as flavor ingredients. *Food Chem. Toxicol.* 45: 171–201.

Arctander, S. (1969) *Perfume and Flavor Chemicals,* Vol. 1. Rutgers University, Montclair, NJ (1981).

Austin, C.A., Shephard, E.A., Pike, S.F., Rabin, B.R., and Phillips, I.R. (1988) The effect of terpenoid compounds on cytochrome P-450 levels in rat liver. *Biochem. Pharmacol.* 37(11): 2223–2229.

Chen, L., Lebetkin, E.H., and Burka, L.T. (2001) Metabolism of (R)-(+)-pulegone in F344 rats. *Drug Metabol. Dispos.* 29(12): 1567–1577.

Cramer, G., Ford, R., and Hall, R. (1978) Estimation of toxic hazard—A decision tree approach. *Food Cosmet. Toxicol.* 16: 255–276.

Crowell, P., Elson, C.E., Bailey, H., Elegbede, A., Haag, J., and Gould, M. (1994) Human metabolism of the experimental cancer therapeutic agent D-limonene. *Cancer Chemother. Pharmacol.* 35: 31–37.

Dioscorides (50 AD) *Inquiry into Plants and Growth of Plants—Theophrastus.* De Materia Medica.

EPA (U.S. Environmental Protection Agency). (1988) Technical Support Document on Risk Assessment of Chemical Mixtures. EPA-600/8-90/064. U.S. Environmental Protection Agency, Office of Research and Development, Washington, DC.

European Communities (EC). (1999) Commission of European Communities Regulation No. 2232/96.

European Flavour and Fragrance Association (EFFA). (2005) European inquiry on volume use. Private communication to the Flavor and Extract Manufacturers Association (FEMA), Washington, DC.

Flavor and Extract Manufacturers Association (FEMA). (1975) Scientific literature review of alicyclic substances used as flavor ingredients. U.S. National Technical Information Services, PB86-1558351/LL, FEMA, Wahington, DC.

Florin, I., Rutberg, L., Curvall, M., and Enzell, C.R. (1980) Screening of tobacco smoke constituents for mutagenicity using the Ames test. *Toxicology* 18: 219–232.

Food Chemical Codex (FCC). (2013) *Food Chemicals Codex* (9th edn.). United States Pharmacopeia (USP), Rockville, MD.

Food and Drug Administration. (2005) Threshold of regulation for substances used in food-contact articles. 21 CFR 170.39.

Hall, R.L. (1976) Estimating the distribution of daily intakes of certain GRAS substances. Committee on GRAS list survey—Phase III. National Academy of Sciences/National Research Council, Washington, DC.

Hall, R.L. and Ford, R.A. (1999) Comparison of two methods to assess the intake of flavoring substances. *Food Addit. Contam.* 16: 481–495.

Hall, R.L. and Oser, B.L. (1968) Recent progress in the consideration of flavoring substances under the Food Additives Amendment. *Food Technol.* 19(2): 151.

Harman, C.L., Lipman, M.D., and Hallagan, J.B. (2013) Flavor and Extract Manufacturers Association of the United States 2010 Poundage and Technical Effects Survey. Flavor and Extract Manufacturers Association, Washington, DC.

Heck, J.D., Vollmuth, T.A., Cifone, M.A., Jagannath, D.R., Myhr, B., and Curren, R.D. (1989) An evaluation of food flavoring ingredients in a genetic toxicity screening battery. *Toxicologist* 9(1): 257.

Ishida, T., Asakawa, Y., Takemoto, T., and Aratani, T. (1981) Terpenoids Biotransformation in Mammals III: Biotransformation of alpha-pinene, beta-pinene, 3-carene, carane, myrcene, and p-cymene in rabbits. *J. Pharm. Sci.* 70: 406–415.

Japanese Flavor and Fragrance Manufacturers Association (JFFMA). (2002) Japanese inquiry on volume use. Private communication to the Flavor and Extract Manufacturers Association (FEMA), Washington, DC.

JECFA. (1997) Evaluation of certain food additives and contaminants. Forty-sixth Report of the Joint FAO/WHO Expert Committee on Food Additives. World Health Organization, WHO Technical Report Series 868.

JECFA. (1998) Evaluation of certain food additives and contaminants. Forty-seventh Report of the Joint FAO/WHO Expert Committee on Food Additives. WHO Technical Report Series 876. World Health Organization, Geneva, Switzerland.

JECFA. (1999) Procedure for the Safety Evaluation of Flavouring Agents. Evaluation of certain food additives and contaminants. Forty-ninth Report of the Joint FAO/WHO Expert Committee on Food Additives. World Health Organization, WHO Technical Report Series 884.

JECFA. (2000a) Evaluation of certain food additives and contaminants. Fifty-first report of the Joint FAO/WHO Expert Committee on Food Additives. WHO Technical Report Series No. 891. World Health Organization, Geneva, Switzerland.

JECFA. (2000b) Evaluation of certain food additives and contaminants. Fifty-third report of the Joint FAO/WHO Expert Committee on Food Additives. WHO Technical Report Series No. 896. World Health Organization, Geneva, Switzerland.

JECFA. (2001) Evaluation of certain food additives and contaminants. Fifty-fifth report of the Joint FAO/WHO Expert Committee on Food Additives. WHO Technical Report Series No. 901. World Health Organization, Geneva, Switzerland.

JECFA. (2003) Evaluation of certain food additives and contaminants. Fifty-ninth report of the Joint FAO/WHO Expert Committee on Food Additives. WHO Technical Report Series No. 913. World Health Organization, Geneva, Switzerland.

JECFA. (2004) Evaluation of certain food additives and contaminants. Sixty-first report of the Joint FAO/WHO Expert Committee on Food Additives. World Health Organization, Geneva, Switzerland.

Lambe, J., Cadby, P., and Gibney, M. (2002) Comparison of stochastic modelling of the intakes of intentionally added flavouring substances with theoretical added maximum daily intakes (TAMDI) and maximized survey-derived daily intakes (MSDI). *Food Addit. Contam.* 19(1): 2–14.

Lucas, C.D., Putnam, J.M., and Hallagan, J.B. (1999) Flavor and Extract Manufacturers Association (FEMA) of the United States. 1995 Poundage and Technical Effects Update Survey. Washington, DC.

Maarse, H., Visscher, C.A., Willemsens, L.C., and Boelens, M.H. (1992, 1994, 2000) *Volatile Components in Food-Qualitative and Quantitative Data*. Centraal Instituut Voor Voedingsonderzioek TNO, Zeist, the Netherlands.

Madsen, C., Wurtzen, G., and Carstensen, J. (1986) Short-term toxicity in rats dosed with menthone. *Toxicol. Lett.* 32: 147–152.

Madyastha, K.M. and Srivatsan, V. (1987) Metabolism of beta-myrcene in vivo and in vitro: Its effects on rat-liver microsomal enzymes. *Xenobiotica* 17(5): 539–549.

Madyastha, K.M. and Srivatsan, V. (1988) Studies on the metabolism of l-menthol in rats. *Drug Metab. Dispos.* 16: 765.

McClanahan, R.H., Thomassen, D., Slattery, J.T., and Nelson, S.D. (1989) Metabolic activation of (R)-(+)-pulegone to a reactive enonal that covalently binds to mouse liver proteins. *Chem. Res. Toxicol.* 2: 349–355.

Meek, M.E., Boobis, A.R., Crofton, K.M., Heinemeyer, G., Van Raaij, M., and Vickers, C. (2011) Risk assessment of combined exposure to multiple chemicals: A WHO/IPCS framework. *Reg. Toxicol. Pharmacol.* 60: S1–S14.

Miyazawa, M., Shindo, M., and Shimada, T. (2002) Sex differences in the metabolism of (+)- and (−)-limonene enantiomers to carveol and perillyl alcohol derivatives by cytochrome P450 enzymes in rat liver microsomes. *Chem. Res. Toxicol.* 15(1): 15–20.

Muller, W. (1993) Evaluation of mutagenicity testing with *Salmonella typhimurium* TA102 in three different laboratories. *Environ. Health Perspect. Suppl.* 101: 33–36.

Munro, I., Ford, R., Kennepohl, E., and Sprenger, J. (1996a) Correlation of structural class with no-observed-effect-levels: A proposal for establishing a threshold of concern. *Food Chem. Toxicol.* 34: 829–867.

Munro, I.C., Ford, R.A., Kennepohl, E., and Sprenger, J.G. (1996b) Thresholds of toxicological concern based on structure-activity relationships. *Drug Metab. Rev.* 28(1/2): 209–217.

National Academy of Sciences (NAS). (1965, 1970, 1975, 1981, 1982, 1987) *Evaluating the Safety of Food Chemicals*. National Academy of Sciences, Washington, DC.

National Cancer Institute (NCI). (1979) Bioassay of dl-menthol for possible carcinogenicity. National Technical Report Series No. 98. U.S. Department of Health, Education and Welfare, Bethesda, MD.

National Research Council (NRC). (1988) *Complex Mixtures: Methods for In Vivo Toxicity Testing*. National Academy Press, Washington, DC.

National Research Council (NRC). (1994) *Science and Judgment in Risk Assessment*. National Academy Press, Washington, DC.

National Toxicology Program (NTP). (1990) Carcinogenicity and toxicology studies of d-limonene in F344/N rats and B6C3F1 mice. NTP-TR-347. U.S. Department of Health and Human Services. NIH Publication No. 90-2802. National Toxicology Program, Research Triangle Park, NC.

National Toxicology Program (NTP). (2002) Toxicity studies of pulegone in B6C3F1 mice and rats (Gavage studies). Battelle Research Laboratories, Study No. G004164-X. Unpublished Report. National Toxicology Program, Research Triangle Park, NC.

National Toxicology Program (NTP). (2003a) Draft report on the Initial study results from a 90-day toxicity study on beta-myrcene in mice and rats. Study number C99023 and A06528. National Toxicology Program, Research Triangle Park, NC.

Newberne, P., Smith, R.L., Doull, J., Goodman, J.I., Munro, I.C., Portoghese, P.S., Wagner, B.M. et al. (1999) The FEMA GRAS assessment of *trans*-anethole used as a flavoring substance. *Food Chem. Toxicol.* 37: 789–811.

Nijssen, B., van Ingen-Visscher, K., and Donders, J. (2003) Volatile Compounds in Food 8.1. Centraal Instituut Voor Voedingsonderzioek TNO, Zeist, the Netherlands. http://www.voeding.tno.nl/vcf/VcfNavigate.cfm.

Organization for Economic Cooperation and Development (OECD). (2014) *Guidance on Grouping of Chemicals*, 2nd edn. OECD Environment, Health and Safety Publications, Series on Testing and Assessment, No. 194. OECD, Paris, France.

Oser, B. and Ford, R. (1991) FEMA Expert Panel: 30 years of safety evaluation for the flavor industry. *Food Technol.* 45(11): 84–97.

Oser, B. and Hall, R. (1977) Criteria employed by the Expert Panel of FEMA for the GRAS evaluation of flavoring substances. *Food Cosmet. Toxicol.* 15: 457–466.

Poon, G., Vigushin, D., Griggs, L.J., Rowlands, M.G., Coombes, R.C., and Jarman, M. (1996) Identification and characterization of limonene metabolites in patients with advanced cancer by liquid chromatography/mass spectrometry. *Drug Metab. Dispos.* 24: 565–571.

Presidential Commission on Risk Management and Risk Assessment. (1996) Risk assessment and risk management in regulatory decision making. Final Report, Vols. 1 and 2. Presidential Commission on Risk Management, Washington, DC.

Rivedal, E., Mikalsen, S.O., and Sanner, T. (2000) Morphological transformation and effect on gap junction intercellular communication in Syrian Hamster Embryo Cells Screening Tests for Carcinogens Devoid of Mutagenic Activity. *Toxicol. In Vitro* 14(2): 185–192.

Roe, F. and Field, W. (1965) Chronic toxicity of essential oils and certain other products of natural origin. *Food Cosmet. Toxicol.* 3: 311–324.

Rulis, A.M. (1987) *De Minimis* and the threshold of regulation. In: Felix, C.W. (Ed.), *Food Protection Technology*. Lewis Publishers Inc., Chelsea, MI, pp. 29–37.

Rulis, A.M., Hattan, D.G., and Morgenroth, V.H. (1984) FDA's priority-based assessment of food additives. I. Preliminary results. *Reg. Toxicol. Pharmacol.* 26: 44–51.

Sasaki, Y.F., Imanishi, H., Ohta, T., and Shirasu, Y. (1989) Modifying effects of components of plant essence on the induction of sister-chromatid exchanges in cultured Chinese hamster ovary cells. *Mutat. Res.* 226: 103–110.

Schilter, B., Andersson, C., Anton, R., Constable, A., Kleiner, J., O'Brien, J., Renwick, A.G., Korver, O., Smit, F., and Walker, R. (2003) Guidance for the safety assessment of botanicals and botanical preparations for use in food and food supplements. *Food Chem. Toxicol.* 41: 1625–1649.

Serota, D. (1990) 28-Day toxicity study in rats. Hazelton Laboratories America, HLA Study No. 642-477. Private Communication to FEMA. Unpublished Report.

Smith, R.L., Adams, T.B., Doull, J., Feron, V.J., Goodman, J.I., Marnett, L.J., Portoghese, P.S. et al. (2002a) Safety assessment of allylalkoxybenzene derivatives used as flavoring substances—Methyleugenol and estragole. *Food Chem. Toxicol.* 40: 851–870.

Smith, R.L., Cohen, S.M., Doull, J., Feron, V.J., Goodman, J.I., Marnett, L.J., Munro, I.C. et al. (2005a) Criteria for the safety evaluation of flavoring substances the expert panel of the flavor and extract manufacturers association. *Food Chem. Toxicol.* 43: 1141–1177.

Smith, R.L., Cohen, S., Doull, J., Feron, V.J., Goodman, J.I., Marnett, L.J., Portoghese, P.S., Waddell, W.J., Wagner, B.M., and Adams, T.B. (2003) Recent progress in the consideration of flavor ingredients under the Food Additives Amendment. 21 GRAS Substances. *Food Technol.* 57: 46.

Smith, R.L., Cohen, S., Doull, J., Feron, V.J., Goodman, J.I., Marnett, L.J., Portoghese, P.S., Waddell, W.J., Wagner, B.M., and Adams, T.B. (2004) Safety evaluation of natural flavour complexes. *Toxicol. Lett.* 149: 197–207.

Smith, R.L., Cohen, S.M., Doull, J., Feron, V.J., Goodman, J.I., Marnett, L.J., Portoghese, P.S. et al. (2005b) A procedure for the safety evaluation of natural flavor complexes used as ingredients in food: Essential oils. *Food Chem. Toxicol.* 43: 345–363.

Smith, R.L., Doull, J., Feron, V.J., Goodman, J.I., Marnett, L.J., Munro, I.C., Newberne, P.M. et al. (2002b) The FEMA GRAS assessment of pyrazine derivatives used as flavor ingredients. *Food Chem. Toxicol.* 40: 429–451.

Smith, R.L., Newberne, P., Adams, T.B., Ford, R.A., Hallagan, J.B., and the FEMA Expert Panel. (1996) GRAS flavoring substances 17. *Food Technol.* 50(10): 72–78, 80–81.

Splindler, P. and Madsen, C. (1992) Subchronic toxicity study of peppermint oil in rats. *Toxicol. Lett.* 62: 215–220.

Swenberg, J. and Schoonhoven, R. (2002) Private communication to FEMA.

Thomassen, D., Knebel, N., Slattery, J.T., McClanahan, R.H., and Nelson, S.D. (1992) Reactive intermediates in the oxidation of menthofuran by cytochrome P-450. *Chem. Res. Toxicol.* 5: 123–130.

Vigushin, D., Poon, G.K., Boddy, A., English, J., Halbert, G.W., Pagonis, C., Jarman, M., and Coombes, R.C. (1998) Phase I and pharmacokinetic study of D-limonene in patients with advanced cancer. *Cancer Chemother. Pharmacol.* 42(2): 111–117.

Williams, R.T. (1940) Studies in detoxication. 7. The biological reduction of l-Menthone to d-neomenthol and of d-isomenthone to d-isomenthol in the rabbit. The conjugation of d-neomenthol with glucuronic acid. *Biochem. J.* 34: 690–697.

Woods, L. and Doull, J. (1991) GRAS evaluation of flavoring substances by the Expert Panel of FEMA. *Regul. Toxicol. Pharmacol.* 14(1): 48–58.

Yamaguchi, T., Caldwell, J., and Farmer, P.B. (1994) Metabolic fate of [^{3}H]-l-menthol in the rat. *Drug Metab. Dispos.* 22: 616–624.

Zamith, H.P., Vidal, M.N.P., Speit, G., and Paumgartten, F.J.R. (1993) Absence of genotoxic activity of beta-myrcene in the in vivo cytogenetic bone marrow assay. *Braz. J. Med. Biol. Res.* 26: 93–98.

9 Metabolism of Terpenoids in Animal Models and Humans

Walter Jäger and Martina Höferl

CONTENTS

- 9.1 Introduction 254
- 9.2 Metabolism of Monoterpenes 254
  - 9.2.1 Borneol 254
  - 9.2.2 Camphene 255
  - 9.2.3 Camphor 255
  - 9.2.4 3-Carene 256
  - 9.2.5 Carvacrol 257
  - 9.2.6 Carvone 257
  - 9.2.7 1,4-Cineole 259
  - 9.2.8 1,8-Cineole 259
  - 9.2.9 Citral 260
  - 9.2.10 Citronellal 260
  - 9.2.11 *p*-Cymene 261
  - 9.2.12 Fenchone 262
  - 9.2.13 Geraniol 262
  - 9.2.14 Limonene 262
  - 9.2.15 Linalool 262
  - 9.2.16 Linalyl Acetate 266
  - 9.2.17 Menthofuran 266
  - 9.2.18 Menthol 267
  - 9.2.19 Menthone 267
  - 9.2.20 Myrcene 267
  - 9.2.21 α- and β-Pinene 268
  - 9.2.22 Pulegone 269
  - 9.2.23 α-Terpineol 270
  - 9.2.24 Terpinen-4-ol 272
  - 9.2.25 α- and β-Thujone 272
  - 9.2.26 Thymol 273
- 9.3 Metabolism of Sesquiterpenes 274
  - 9.3.1 Caryophyllene 274
  - 9.3.2 Farnesol 274
  - 9.3.3 Longifolene 275
  - 9.3.4 Patchoulol 275
- References 276

9.1 INTRODUCTION

Terpenoids are main constituents of plant-derived essential oils. Because of their pleasant odor, they are widely used in the food, fragrance, and pharmaceutical industry. Furthermore, in traditional medicine, terpenoids are also well known for their anti-inflammatory, antibacterial, antifungal, antitumor, and sedative activities. Although large amounts are used in the industry, the knowledge about their biotransformation in humans is still scarce. Yet, metabolism of terpenoids can lead to the formation of new biotransformation products with unique structures and often different flavor and biological activities compared to the parent compounds. All terpenoids easily enter the human body by oral absorption, penetration through the skin, or inhalation very often leading to measurable blood concentrations. A number of different enzymes, however, readily metabolize these compounds to more water-soluble molecules. Although nearly every tissue has the ability to metabolize drugs, the liver is the most important organ of drug biotransformation. In general, metabolic biotransformation occurs at two major categories called Phase I and Phase II reactions (Spatzenegger and Jäger, 1995). Phase I concerns mostly cytochrome P450 (CYP)-mediated oxidation as well as reduction and hydrolysis. Phase II is a further step where a Phase I product is completely transformed to high water solubility. This is done by attaching already highly water-soluble endogenous entities such as sugars (glucuronic acids) or salts (sulfates) to the Phase I intermediate and forming a Phase II final product. It is not always necessary for a compound to undergo both Phases I and II; indeed for many terpenoids, one or the other is enough to eliminate these volatile plant constituents. In the following concise review, special emphasis will be put on metabolism of selected mono- and sesquiterpenoids not only in animal and in vitro models but also in humans.

9.2 METABOLISM OF MONOTERPENES

9.2.1 BORNEOL

Borneol is a component in many essential oils, for example, oils of Pinaceae, *Salvia officinalis*, *Rosmarinus officinalis*, and *Artemisia* species (Bornscheuer et al., 2014). *Cinnamomum camphora* chemotype borneol and *Blumea balsamifera*, which are rich in (+)- and (−)-borneol, respectively, are used as sources for preparation of *bing pian*, a drug of traditional Chinese medicine (Zhao et al., 2012). Moreover, borneol is used giving soaps, perfumes, and other products a scent of spruce needles. In vitro studies with rat liver microsomes could provide evidence for four metabolites (Figure 9.1) (Zhang et al., 2008). The main metabolite, camphor, could also be detected in rat plasma (Sun et al., 2014).

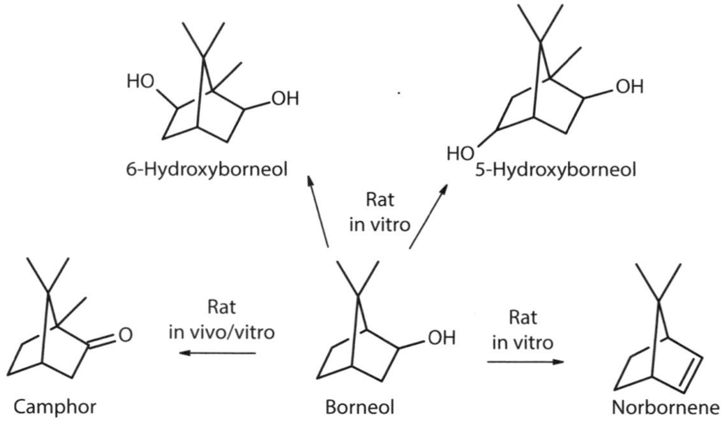

FIGURE 9.1 Proposed metabolism of borneol in rat liver microsomes. (Adapted from Zhang, R. et al., *J. Chromatogr. Sci.*, 46, 419, 2008.)

9.2.2 Camphene

Camphene is found in higher concentrations in the essential oils of common coniferous trees, for example, *Abies alba* or *Tetraclinis articulata*, in the rhizome of *Zingiber officinale* as well as in *S. officinalis* and *R. officinalis* (Bornscheuer et al., 2014). There is only one publication on various biotransformation products in the urine of rabbits after its oral administration. As shown in Figure 9.2, camphene is metabolized into two diastereomeric glycols (camphene-2,10-glycols). Their formation obviously involves two isomeric epoxide intermediates, which are hydrated by epoxide hydrolase. The monohydroxylated camphene and tricyclene derivatives were apparently formed through the nonclassical cation intermediate (Ishida et al., 1979). So far, there are no studies available about the biotransformation of camphene in humans.

9.2.3 Camphor

(+)-Camphor is extracted from the wood of *C. camphora*, a tree endemic to Southeast Asia. Furthermore, it is also one of the major constituents of the essential oils of *S. officinalis* and *R. officinalis*. Camphor is commercially used as a moth repellent and antiseptic in cosmetics (aftershaves, face tonics, mouthwash, etc.) and pharmaceutically in ointments for treatment of rheumatic pains and coughs (Bornscheuer et al., 2014; O'Neil, 2006). In dogs, rabbits, and rats, camphor is extensively metabolized whereat the major hydroxylation products were 5-*endo*- and 5-*exo*-hydroxycamphor. A small amount was also identified as 3-*endo*-hydroxycamphor (Figure 9.3). Both 3- and 5-bornane groups can be further reduced to 2,3- and 2,5-bornanedione. Minor biotransformation steps also involve the reduction of camphor to borneol and isoborneol. Interestingly, all hydroxylated camphor metabolites are further conjugated in a Phase II reaction with glucuronic acid. Camphor is extensively metabolized by human liver microsomes to 5-*exo*-hydroxycamphor (Gyoubu and Miyazawa, 2007; Leibman and Ortiz, 1973). In an in vitro experiment using *Salmonella typhimurium* expressing

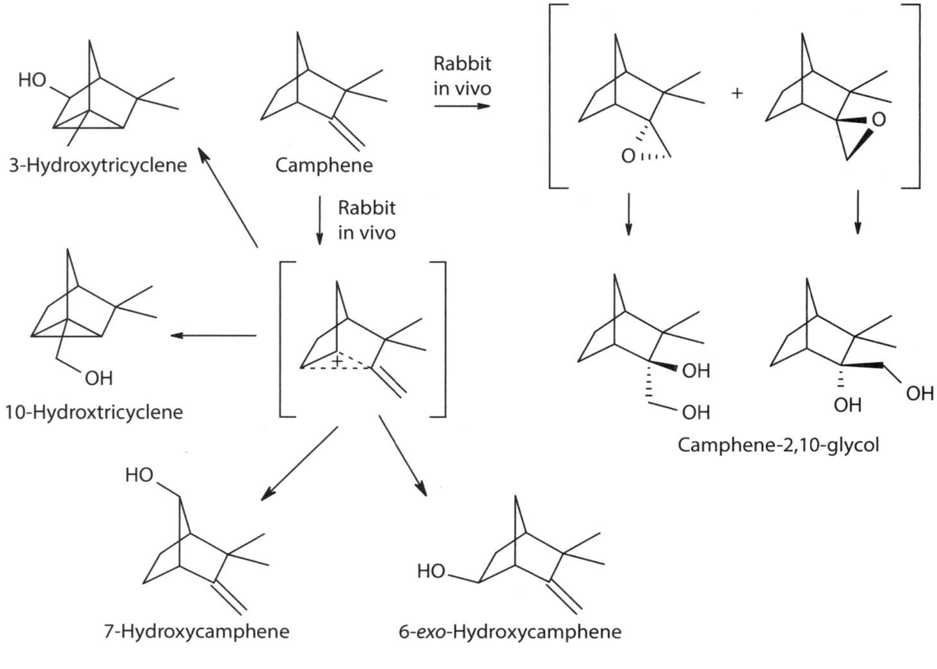

FIGURE 9.2 Urinary excretion of camphene metabolites in rabbits. (Adapted from Ishida, T. et al., *J. Pharm. Sci.*, 68, 928, 1979.)

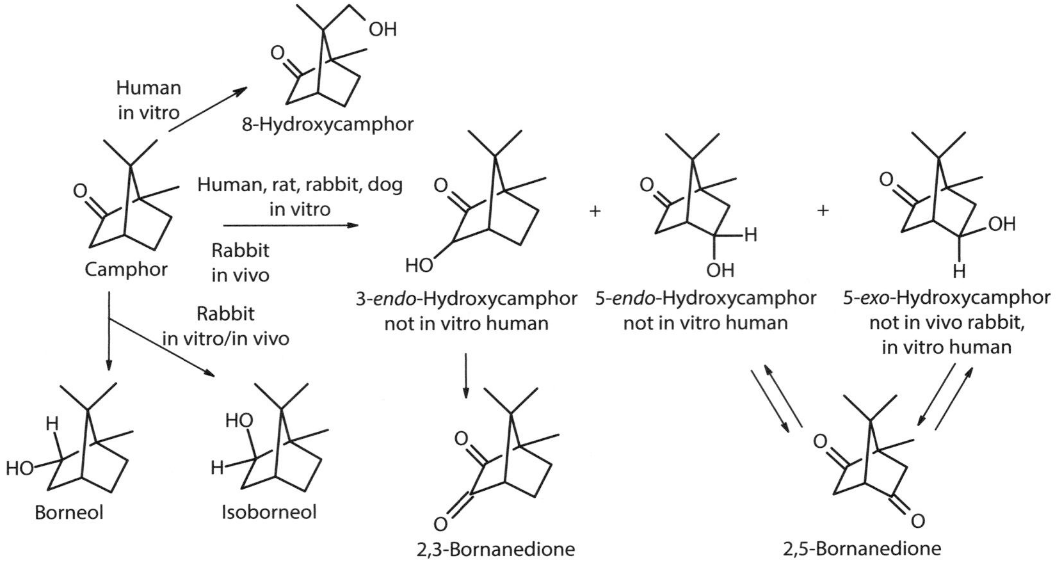

FIGURE 9.3 Metabolisms of camphor in dogs, rabbits, and rats. (Adapted from Leibman, K.C. and Ortiz, E., *Drug Metab. Dispos.*, 1, 543, 1973.)

human CYP2A6 and NADPH-P450 reductase, 5-*exo*-hydroxycamphor was found as a metabolite of camphor together with 8-hydroxycamphor (Nakahashi et al., 2013).

9.2.4 3-Carene

3-Carene is found as in various Pinaceae essential oils: (+)-3-carene is a major compound of *Pinus palustris* essential and (−)-3-carene of *Pinus sylvestris*. It is used as raw material in perfumery (Bornscheuer et al., 2014). In rabbits, 3-carene is metabolized into 3-carene-9-ol and further oxidized into 3-carene-9-carboxylic acid and 3-carene-9,10-dicarboxylic acid (Ishida et al., 1981). In vitro experiments with human liver microsomes revealed 3-carene-10-ol and 3-carene-epoxide as metabolites (Figure 9.4). Hydroxylation was catalyzed by CYP2B6, CYP2C19, and CYP2D6, whereas epoxidation could be attributed to CYP1A2 (Duisken et al., 2005). 3-Carene-10-ol could be detected as metabolite in human urine (Schmidt et al., 2013).

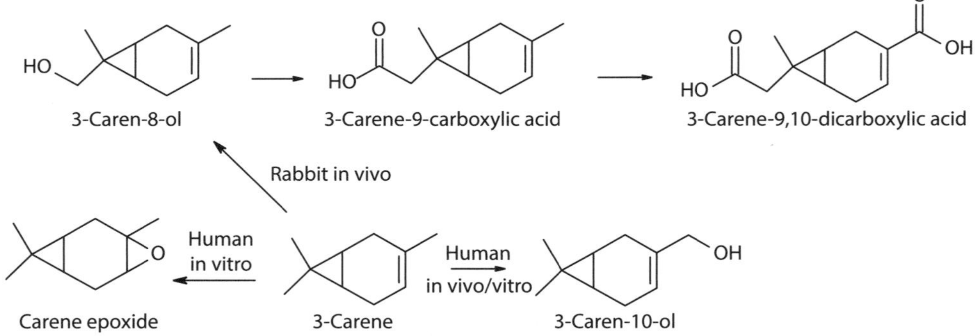

FIGURE 9.4 Proposed metabolism of 3-carene in rabbits and human liver microsome. (Adapted from Ishida, T. et al., *J. Pharm. Sci.*, 70, 406, 1981; Duisken, M. et al., *Curr. Drug Metab.*, 6, 593, 2005.)

9.2.5 Carvacrol

Carvacrol is used as disinfectant and found in high concentrations in the essential oils of, for example, *Thymus vulgaris* chemotype carvacrol, *Origanum vulgare*, *Majorana hortensis*, or *Satureja hortensis*. In rat, only small amounts of unchanged carvacrol were excreted after 24 h after oral application. As β-glucuronidase and sulfatase were used for sample preparation before gas chromatography (GC) analysis, carvacrol might also be excreted as its glucuronide and sulfate, respectively. Both of the aliphatic groups present undergo extensive metabolism, whereas aromatic hydroxylation to 2-hydroxycarvacrol is only a minor important pathway for carvacrol. Further oxidation of 7-hydroxycarvacrol results in isopropylsalicylic acid (Austgulen et al., 1987) (Figure 9.5). Carvacrol is metabolized by recombinant human CYP1A2, CYP2A6, and CYP2B6 (Dong et al., 2012a). An in vitro study with human microsomes demonstrated that recombinant UGT1A9 was mainly responsible for glucuronidation in liver and rUGT1A7 in intestinal microsomes, forming monoglucuronated metabolites (Dong et al., 2012b).

9.2.6 Carvone

The (R)-(−)- and (S)-(+)-enantiomers of monoterpene ketone carvone are found in various plants. While (S)-(+)-carvone is the main constituent of the essential oil *Carum carvi* and *Anethum graveolens*, (R)-(−)-carvone is found in the oil of *Mentha spicata* var. *crispa* (Bornscheuer et al., 2014; O'Neil, 2006). Because of minty odor and taste, large amounts of (R)-(−)-carvone are frequently added to toothpastes, mouth washes, and chewing gums. (S)-(+)-carvone possesses the typical caraway aroma and is mainly used as a flavor compound in food industry. Due to its spasmolytic effect, (S)-(+)-carvone is also used as stomachic and carminative (Jäger et al., 2001). After separate topical

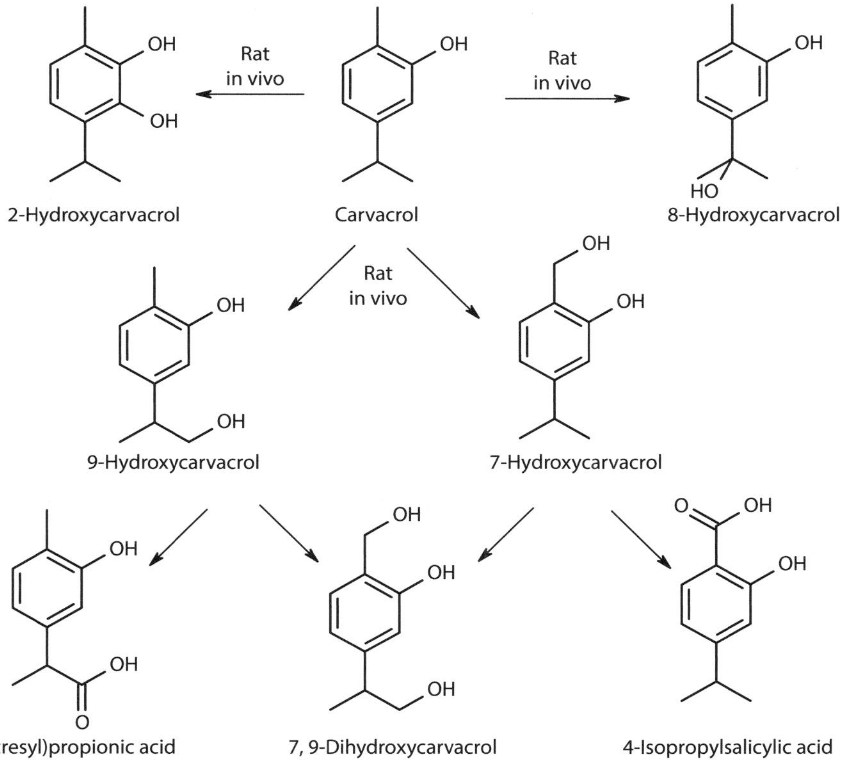

FIGURE 9.5 Metabolism and urinary excretion of carvacrol in rats. (Adapted from Austgulen, L.T. et al., *Pharmacol. Toxicol.*, 61, 98, 1987.)

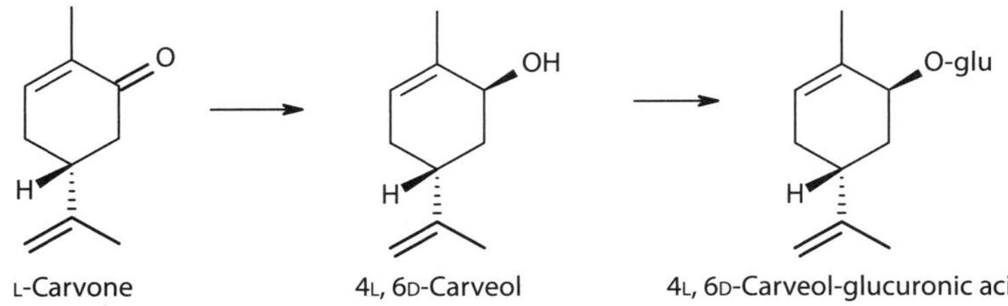

FIGURE 9.6 Metabolic pathway of (R)-(−)-carvone in healthy subjects. (Adapted from Jäger, W. et al., *J. Pharm. Pharmacol.*, 52, 191, 2000.)

applications of (R)-(−)- and (S)-(+)-carvone, both enantiomers are rapidly absorbed resulting in significantly higher maximal plasma concentrations (C_{max}) and areas under the blood concentrations time curves (AUC) for (S)-(+)-compared to (R)-(−)-carvone. As demonstrated in Figure 9.6, analysis of control and ß-glucuronidase pretreated urine samples only revealed stereoselective metabolism of (R)-(−)-carvone but not of (S)-(+)-carvone to (4R,6S)-(−)-carveol and (4R,6S)-(−)-carveol glucuronide indicating that stereoselectivity in Phases I and II metabolism has significant effects on (R)-(−)- and (S)-(+)-carvone pharmacokinetics (Jäger et al., 2000) (Figure 9.6). Contrary to the study of Jäger et al. (2000), Engel could not demonstrate any differences in the formation of metabolites after peroral application of (R)-(−)- and (S)-(+)-carvone to human volunteers, which may be due to the separation of biotransformation products on a nonchiral GC column. As shown in Figure 9.7, besides carveol, several metabolites could be identified in the urine samples (Engel, 2001).

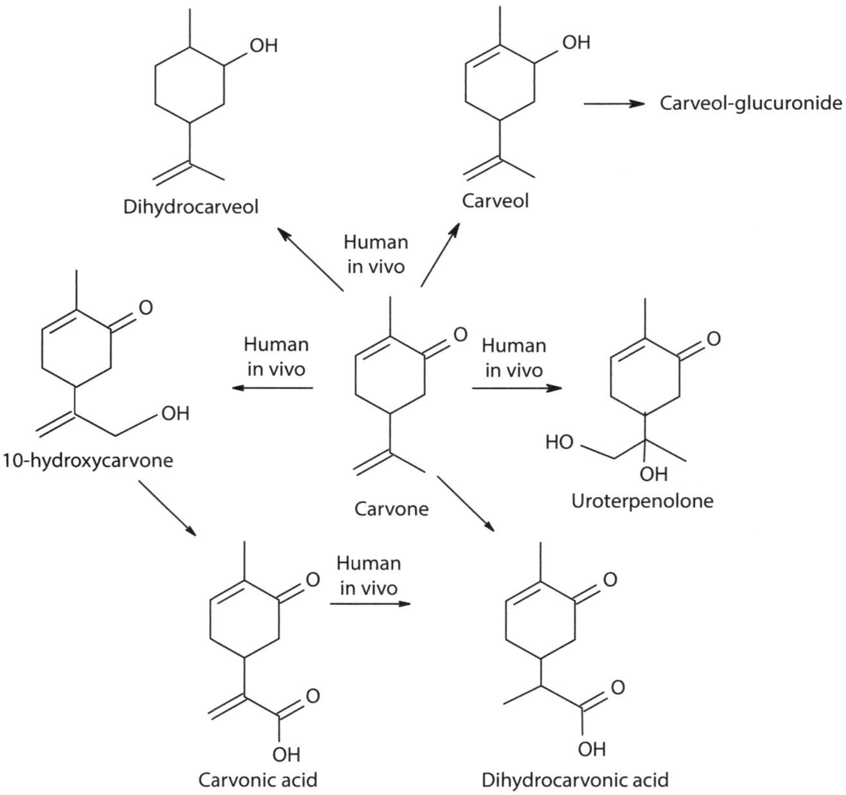

FIGURE 9.7 Proposed metabolic pathway of (R)-(−)- and (S)-(+)-carvone in healthy volunteers. (Adapted from Engel, W., *J. Agric. Food Chem.*, 49, 4069, 2001.)

9.2.7 1,4-Cineole

1,4-Cineole is a flavor constituent of *Citrus aurantifolia* and *Piper cubeba* (Bornscheuer et al., 2014). In vitro and in vivo animal studies demonstrated extensive biotransformation of this monoterpene strongly suggesting biotransformation in the human body too. After oral application to rabbits, four neutral and one acidic metabolite could be isolated from urine (Asakawa et al., 1988). Using rat and human liver microsomes, however, only 2-hydroxylation could be observed indicating species-related differences in 1,4-cineole metabolism (Miyazawa et al., 2001a) (Figure 9.8).

9.2.8 1,8-Cineole

1,8-Cineole is widely distributed in plants and found in high concentrations in the essential oils of *Eucalyptus globulus* and *Laurus nobilis*. It is extensively used in cosmetics, for ointments against cough, muscular pain, and rheumatism (Bornscheuer et al., 2014; O'Neil, 2006). Using rat liver microsomes, 1,8-cineole is predominantly converted to 3-hydroxy-1,8-cineole, followed by 2- and then 9-hydroxycineole (Miyazawa et al., 2001b). In rabbit and koala urine, various 7- and 9-oxylated metabolites were found (Boyle et al., 2001; Shipley et al., 2012). As seen in Figure 9.9, in human liver microsomes, however, only the 2- and 3-hydroxylated products catalyzed by the isoenzyme CYP3A4 were seen (Miyazawa et al., 2001b). Both metabolites could also be identified in the urine of three human volunteers after oral administration of a cold medication containing 1,8-cineole (Miyazawa et al., 2001b) and can therefore be used as urinary markers for the intake of 1,8-cineole in humans. Horst and Rychlik could identify two other hydroxylated metabolites in human urine, 7- and 9-hydroxy-1,8-cineole (Horst and Rychlik, 2010).

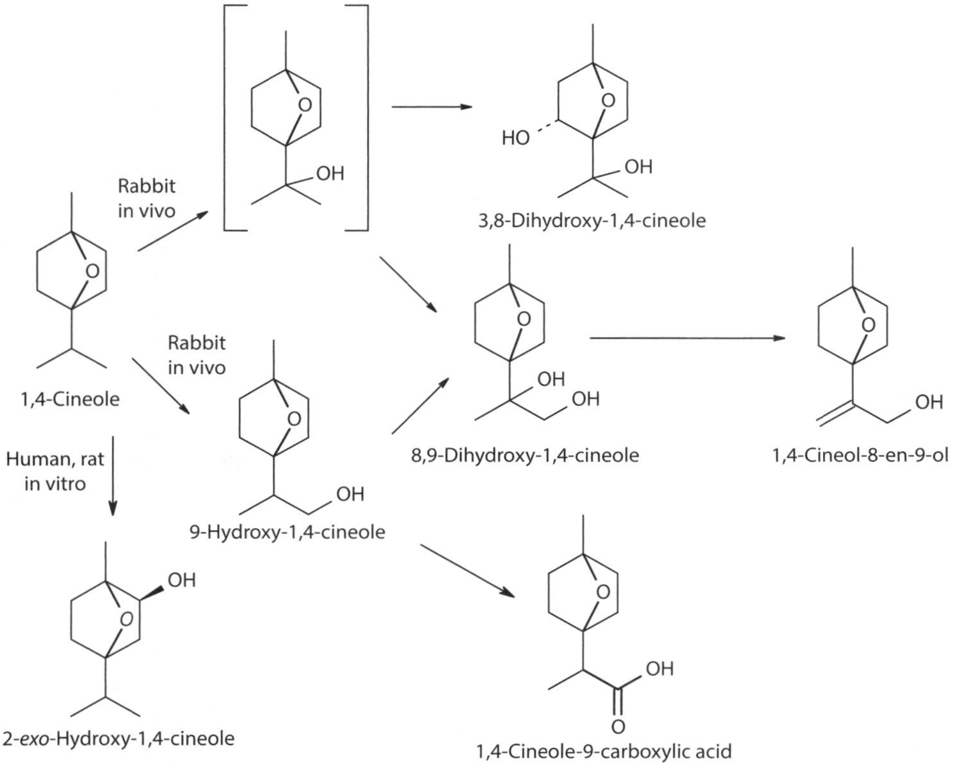

FIGURE 9.8 Proposed metabolism of 1,4-cineole in rabbits and in rat and human liver microsomes. (Adapted from Asakawa, Y. et al., *Xenobiotica*, 18, 1129, 1988; Miyazawa, M. et al., *Xenobiotica*, 31, 713, 2001a.)

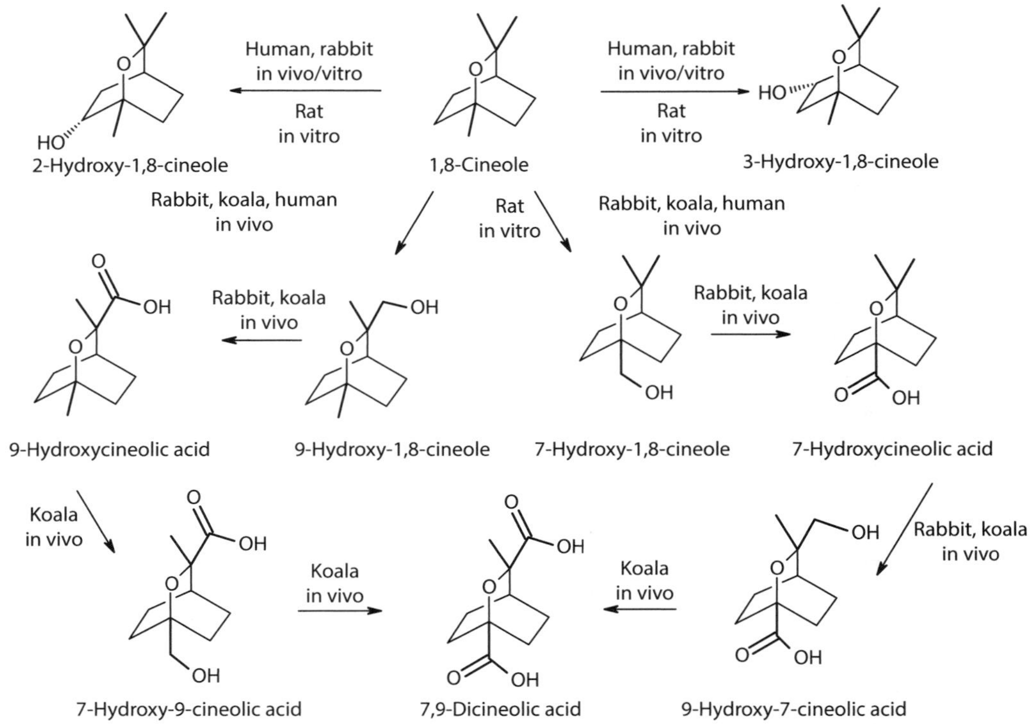

FIGURE 9.9 Proposed metabolism of 1,8-cineole in vitro (rat and human liver microsomes) and in vivo (rabbits, koalas, and humans). (Adapted from Miyazawa, M. et al., *Drug Metab. Dispos.*, 29, 200, 2001b; Boyle, R. et al., *Comp. Biochem. Physiol. C*, 129, 385, 2001; Shipley, L.A. et al., *J. Chem. Ecol.*, 38, 1178, 2012.)

9.2.9 CITRAL

Both natural and synthetic citrals are an isomeric mixture of geranial and neral, in which geranial is usually the predominant isomer. Major amounts are found in the essential oils of *Cymbopogon* sp., *Backhousia citriodora*, *Litsea cubeba*, *Verbena officinalis*, or *Melissa officinalis*. Moreover, it is found in many citrus oils. Because of its intense lemon aroma, citral has been extensively used for flavoring food, cosmetics, and detergents (Bornscheuer et al., 2014; O'Neil, 2006). Studies in rats have shown that citral is rapidly metabolized to several acids and a biliary glucuronide and excreted, with urine as the major route of elimination of citral, followed by expired air, and feces. As demonstrated in Figure 9.10, seven urinary metabolites were isolated and identified (Diliberto et al., 1990). Based on the rat study mentioned earlier, extensive biotransformation of citral in human subjects is highly suggested.

9.2.10 CITRONELLAL

Citronellal is a monocyclic monoterpene aldehyde with high concentrations found in the essential oils of *Corymbia citriodora*, *M. officinalis*, and various *Cymbopogon* species. It is used for perfuming soaps and other products (Bornscheuer et al., 2014; O'Neil, 2006). Only one study described biotransformation of citronellal in rabbits. Ishida et al. could isolate three neutral metabolites of (+)-citronellal in the urine of rabbits (Figure 9.11). An additional acidic metabolite was formed as the result of regioselective oxidation of the aldehyde and dimethyl allyl groups (Ishida et al., 1989). Based on animal data, metabolism of citronellal is also expected in humans.

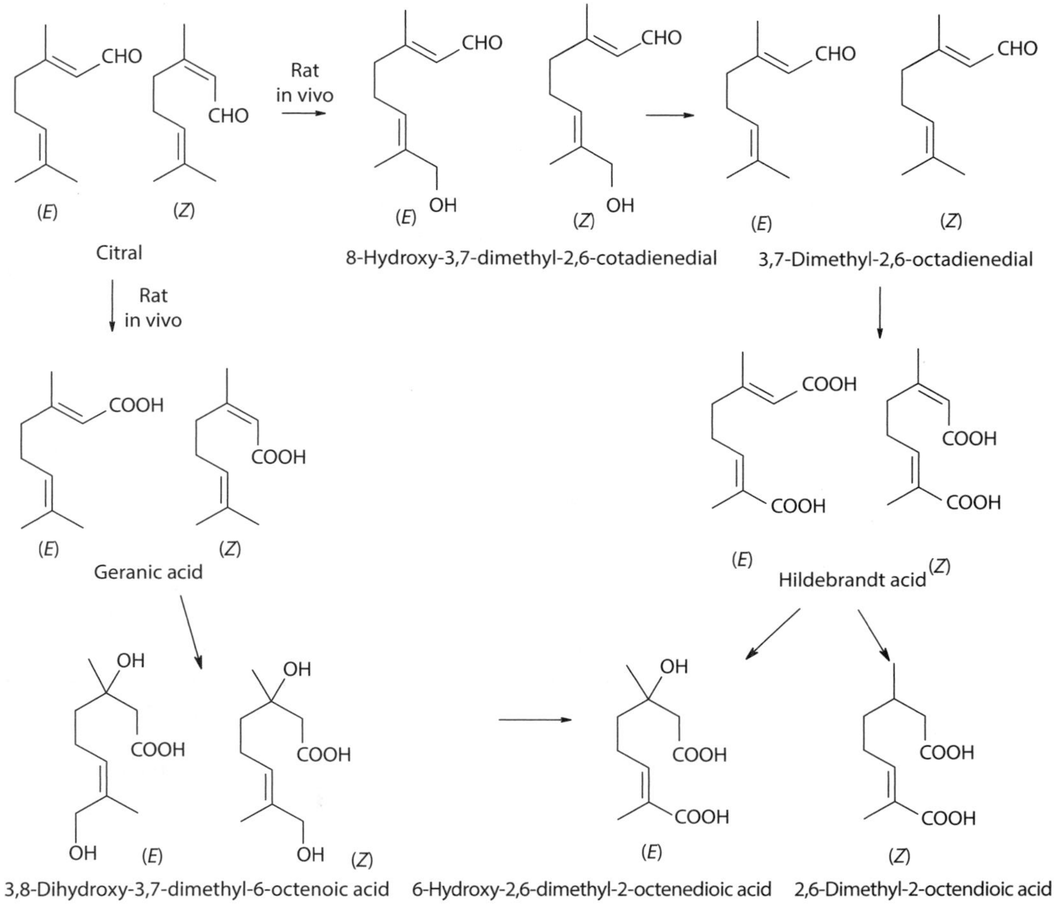

FIGURE 9.10 Proposed metabolism of citral in rats. (Adapted from Diliberto, J.J. et al., *Drug Metab. Dispos.*, 18, 886, 1990.)

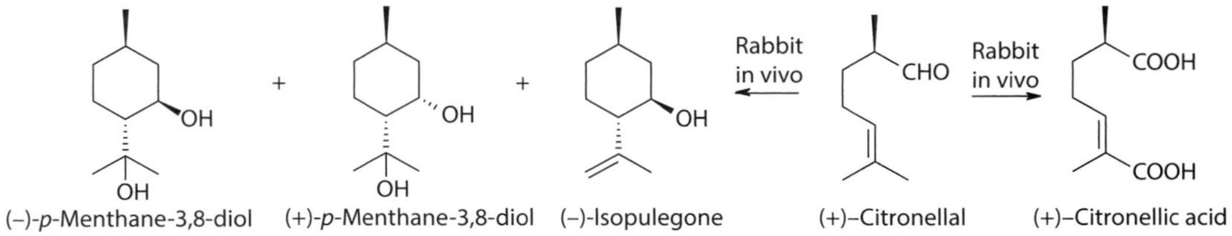

FIGURE 9.11 Proposed metabolism of citronellal in rabbits. (Adapted from Ishida, T. et al., *Xenobiotica*, 19, 843, 1989.)

9.2.11 *p*-Cymene

p-Cymene is found in many essential oils, for example, in *C. carvi* and *T. vulgaris*. It is used as a fragrance compound in perfumery (Bornscheuer et al., 2014). The main *p*-cymene metabolite found in vitro using liver microsomes of brushtail possum, koala, and rat was cuminyl alcohol, accompanied by its oxidation product, cumic acid (Pass et al., 2002). Various hydroxylated and carboxylated *p*-cymene metabolites were found in rabbit (Matsumoto et al., 1992). In vitro assays with human recombinant CYP resulted in several metabolites, thymol, cuminyl alcohol, and cuminaldehyde

(Figure 9.12). In human blood and urine samples, only thymol and its conjugates, thymol glucuronide and sulfate, could be identified (Meesters et al., 2009).

9.2.12 FENCHONE

(+)-Fenchone is found in notable concentrations in the essential oil of *Foeniculum vulgare*, whereas (−)-fenchone is a component of *Thuja occidentalis* essential oil. (+)-Fenchone is used as a food flavor and as carminative (Bornscheuer et al., 2014). A study (Miyazawa and Gyoubu, 2006) investigated the biotransformation of (+)-fenchone in human liver microsomes demonstrating the formation of 6-*exo*-hydroxyfenchone, 6-*endo*-hydroxyfenchone, and 10-hydroxyfenchone (Figure 9.13). Metabolism of (−)-fenchone resulted in similar hydroxylated products (Miyazawa and Gyoubu, 2007). There currently are no data about metabolism of this compound in humans. However, in *S. typhimurium* expressing human CYP2D6 and NADPH-P450 reductase, (+)-fenchone was metabolized into 6-*exo*- and 6-*endo*-hydroxyfenchone (Nakahashi et al., 2013).

9.2.13 GERANIOL

Geraniol is a major component in the essential oils of *Geranium graveolens*, *Cymbopogon martinii* and other *Cymbopogon* species. It has a rose-like odor and is commonly used in perfumes, cosmetics, and as a flavor (Bornscheuer et al., 2014; O'Neil, 2006). Several metabolites could be identified in rat urine after oral administration (Chadha and Madyastha, 1984). Geraniol can be either metabolized to 8-hydroxygeraniol and via 8-carboxygeraniol to Hildebrandt acid or directly oxidized to geranic acid and 3-hydroxycitronellic acid (Figure 9.14). The observed selective oxidation of the C-8 in geraniol also occurs in higher plants as the first step in the biosynthesis of indole alkaloids. When incubated with human CYPs found in skin, geraniol was metabolized not only into the aldehydes neral and geranial but also into epoxides like 2,3-epoxygeraniol, 6,7-epoxygeraniol, and 6,7-epoxygeranial (Hagvall et al., 2008).

9.2.14 LIMONENE

Limonene is one of the most common terpenes found in aromatic plants. The (+)-isomeric form is more abundantly present in plants than the racemic mixture and the (−)-isomeric form. (+)-Limonene has an orange odor and is a major constituent of citrus peel oils such as *Citrus aurantium* sp. *aurantium* and lemon *Citrus limon*, whereas (+)-limonene is found in the essential oil of *Abies procera* and dipentene in turpentine oil. (+)-Limonene is extensively used as fragrances in perfumery and household products (Bornscheuer et al., 2014; O'Neil, 2006). Several research groups have successfully described the biotransformation of (+)-limonene in vitro (rat and human liver microsomes) and in vivo (rat, mice, guinea pigs, dogs, rabbits, human volunteers, and patients). As shown in Figure 9.15, (+)-limonene is extensively biotransformed to several metabolites, whereas in humans, the main biotransformation products are perillyl alcohol; perillic acid; *p*-mentha-1,8-diene-carboxylic acid; *cis*- and *trans*-dihydroperillic acid; carveol; limonene-1,2-diol; limonene-10-ol; uroterpenol; several glucuronides of perillic acid; and limonen-10-ol (Crowell et al., 1992; Miyazawa et al., 2002; Shimada et al., 2002).

9.2.15 LINALOOL

(−)-Linalool is the major compound of the essential oils of *Aniba rosaeodora*, *C. camphora* leaves, *Bursera delpechiana*, and *Lavandula angustifolia*. It has a fresh, light floral odor and is used in large quantities in perfumery as well as soap and detergent products. (+)-Linalool is found in

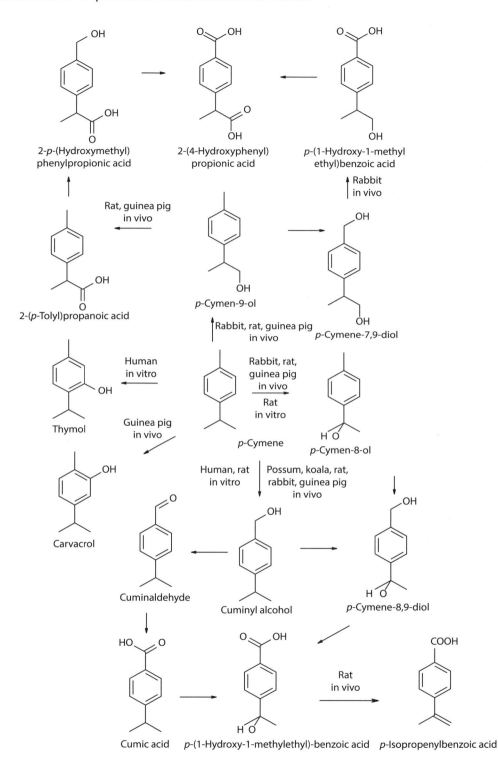

FIGURE 9.12 Proposed metabolism of *p*-cymene in brushtail possum, koala, rabbit, rat, guinea pig, and human volunteers. (Adapted from Matsumoto, T. et al., *Chem. Pharm. Bull.*, 40, 1721, 1992; Meesters, R.J.W. et al., *Xenobiotica*, 39, 663, 2009; Pass, G.J. et al., *Xenobiotica*, 32, 383, 2002.)

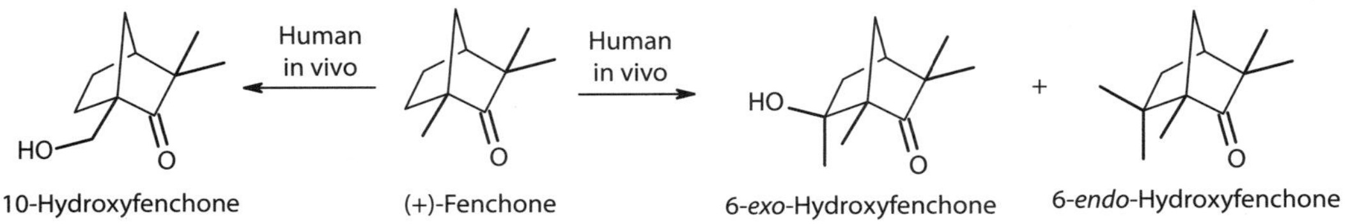

FIGURE 9.13 Proposed metabolism of fenchone in human liver microsomes. (Adapted from Miyazawa, M. and Gyoubu, K., *Biol. Pharm. Bull.*, 29, 2354, 2006; Nakahashi, H. et al., *J. Oleo Sci.*, 62, 293, 2013.)

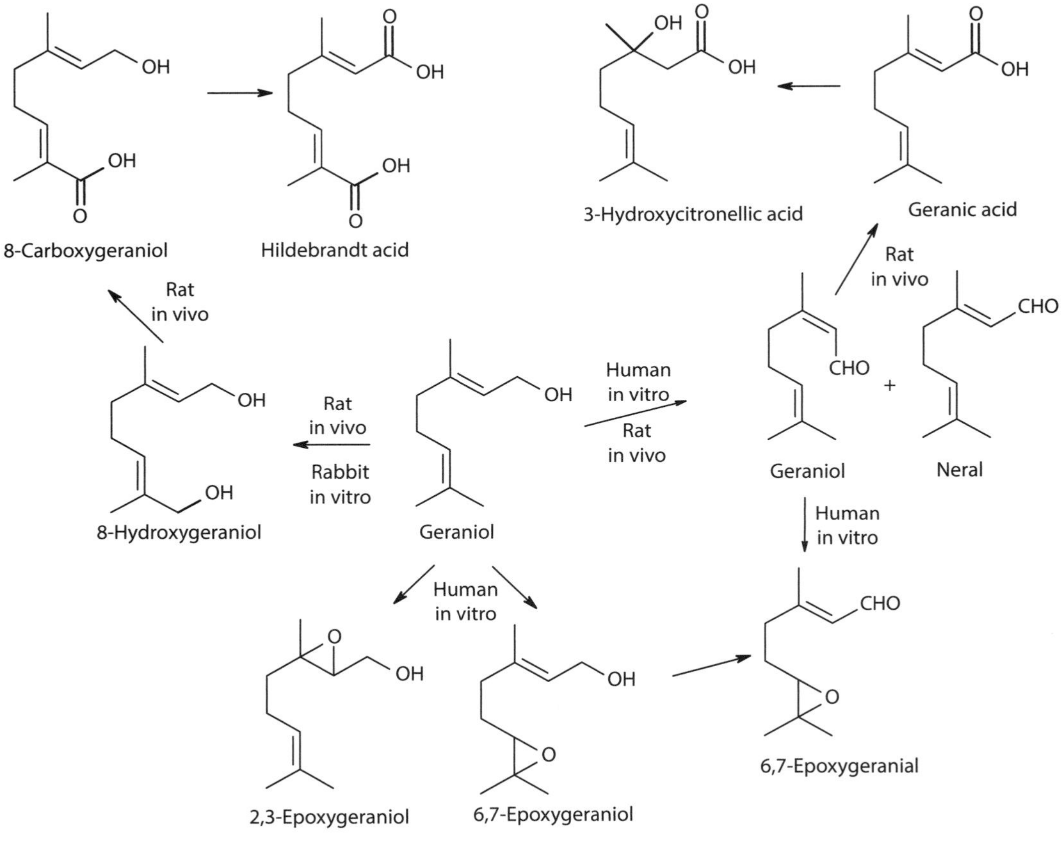

FIGURE 9.14 Proposed metabolism of geraniol in rats. (Adapted from Chadha, A. and Madyastha, M.K., *Xenobiotica*, 14, 365, 1984; Hagvall, L. et al., *Toxicol. Appl. Pharmacol.*, 233, 308, 2008.)

Coriandrum sativum and neroli absolute (*Citrus aurantium* ssp. *aurantium*) (Bornscheuer et al., 2014; O'Neil, 2006). In rat, linalool is metabolized by CYP isoenzymes to dihydro- and tetrahydrolinalool and to 8-hydroxylinalool, which is further oxidized to 8-carboxylinalool (Figure 9.16). CYP-derived metabolites are then converted to glucuronide conjugates (Chadha and Madyastha, 1984). Meesters et al. additionally demonstrated that 6,7-epoxidation by recombinant human CYP2D6 resulted in cyclic ethers, pyranoid- and furanoid-linalool oxide. The proposed intermediate product, 6,7-epoxylinalool, may cause allergic reactions. The enzymatic oxidation of linalool to 8-hydroxylinalool was catalyzed by CYP2C19 and CYP2D6 (Meesters et al., 2007).

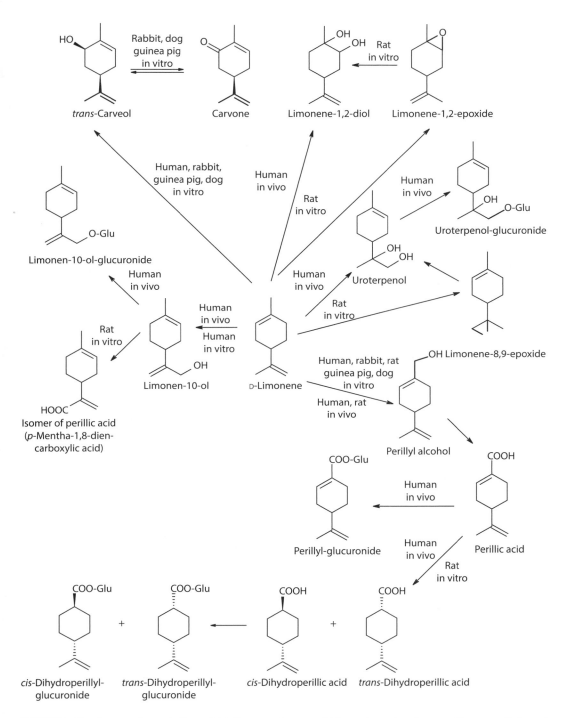

FIGURE 9.15 Proposed metabolism of (+)-limonene in rats, rabbits, guinea pigs, dogs, and humans. (Adapted from Crowell, P.L. et al., *Cancer Chemother. Pharmacol.*, 31, 205, 1992; Miyazawa, M. et al., *Drug Metab. Dispos.*, 30, 602, 2002; Shimada, T. et al., *Drug Metab. Pharmacokin.*, 17, 507, 2002.)

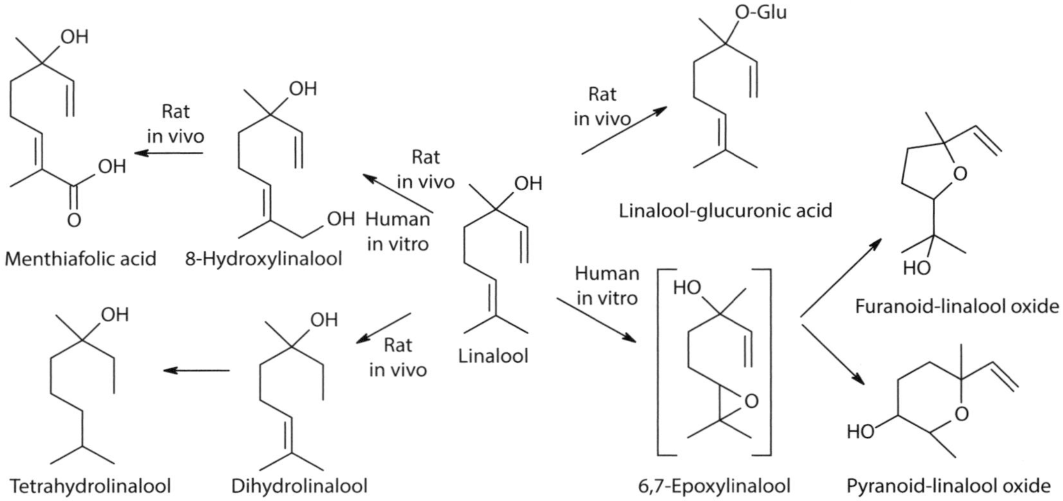

FIGURE 9.16 Proposed metabolism of linalool in rats. (Adapted from Chadha, A. and Madyastha, M.K., *Xenobiotica*, 14, 365, 1984; Meesters, R.J. et al., *Xenobiotica*, 37, 604, 2007.)

9.2.16 LINALYL ACETATE

Due to its pleasant odor, (−)-linalyl acetate is used as ingredient in perfumes and cosmetic products. It is a major compound in the essential oils of petitgrain (*Citrus aurantium* spp. *aurantium*), *Citrus bergamia* and *L. angustifolia* (Bornscheuer et al., 2014; O'Neil, 2006). As an ester, linalyl acetate is hydrolyzed in vivo by carboxylesterases or esterases to linalool (Figure 9.17), which is then further metabolized to numerous oxidized products (see metabolism of linalool) (Bickers et al., 2003).

9.2.17 MENTHOFURAN

(+)-Menthofuran is a constituent of peppermint oils (2%–10%) (Bornscheuer et al., 2014). It could be demonstrated that menthofuran contributes to the nephrotoxicity of pulegone (Khojasteh et al., 2010). Treatment of rat liver slices with toxic concentrations of menthofuran produced several monohydroxylated metabolites, mintlactone and 7-hydroxymintlactone. Glutathione conjugates of 7-hydroxymintlactone and menthofuran could also be identified (Figure 9.18). The metabolites could also be detected in vivo in rat urine.

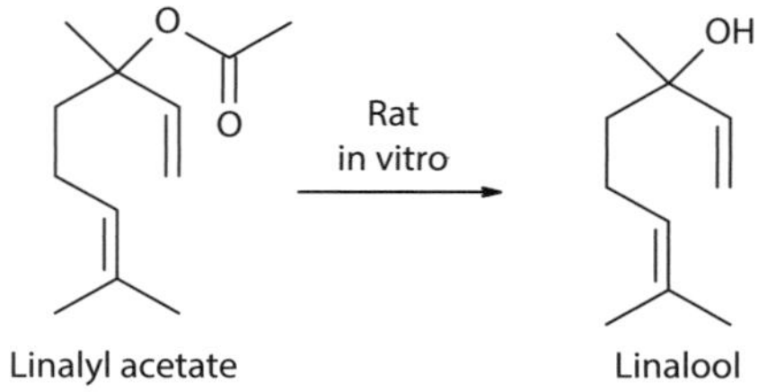

FIGURE 9.17 Proposed metabolism of linalyl acetate in rats. (Adapted from Bickers, D. et al., *Food Chem. Toxicol.*, 41, 919, 2003.)

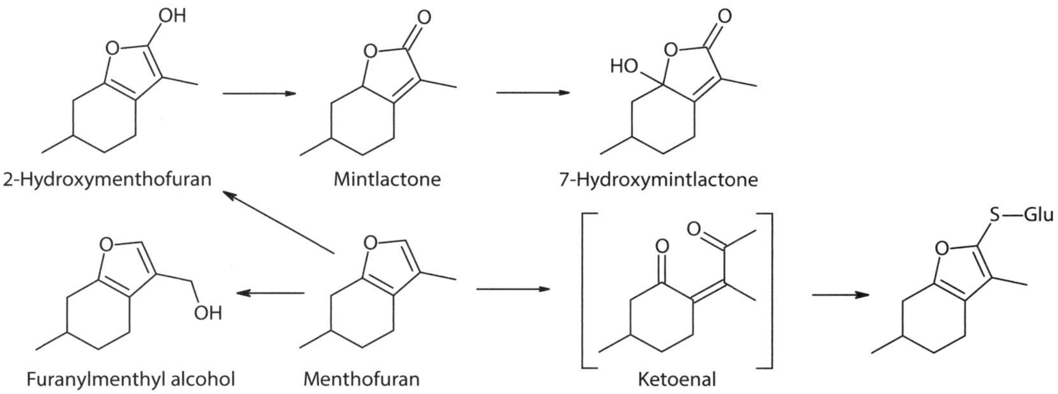

FIGURE 9.18 Proposed metabolism menthofuran in rats. (Adapted from Khojasteh, S.C. et al., *Chem. Res. Toxicol.*, 23, 1824, 2010.)

9.2.18 MENTHOL

(−)-Menthol is a major component of the essential oils of *Mentha piperita* and *Mentha arvensis* var. *piperascens*; besides, it is also found in other mint oils. It has a pleasant typical minty odor and taste and is widely used to flavoring liqueurs and confectionary, in perfumery, tobacco industry, toiletries, oral hygiene products, lotions, and hair tonics. Due to its locally anesthetic, antipruritic, stomachic, and carminative properties, it is used in many indication, for example, topically against coughs, rhinitis, mild burns, insect bites and muscle aches, and internally against dyspetic symptoms (Bornscheuer et al., 2014; O'Neil, 2006). After oral administration to human volunteers, menthol is rapidly metabolized and only menthol glucuronide could be measured in plasma or urine. Interestingly, unconjugated menthol was only detected after a transdermal application. In rats, however, hydroxylation at C-7 and at C-8 and C-9 of the isopropyl moiety form a series a mono- and dihydroxymenthols and carboxylic acids, some of which are excreted in part as glucuronic acid conjugates. Additional metabolites are mono- and or dihydroxylated menthol derivatives (Figure 9.19). Similar to humans, the main metabolite in rats was again menthol glucuronide (Gelal et al., 1999; Madyastha and Srivatsan, 1988a; Spichiger et al., 2006). In vitro experiments using human liver microsomes revealed that the metabolite *p*-menthane-3,8-diol is mainly produced by CYP2A6 (Miyazawa et al., 2011).

9.2.19 MENTHONE

(−)-Menthone is the main compound of *M. arvensis* essential oil and a minor compound of pennyroyal and peppermint oils and used in perfumery (Bornscheuer et al., 2014). In human liver microsomes, (−)-menthone is metabolized into (+)-neomenthol and 7-hydroxymenthone (Figure 9.20) (Miyazawa and Nakanishi, 2006).

9.2.20 MYRCENE

Myrcene is the major constituent of the essential oils of *Humulus lupulus* and *Levisticum officinale* (Bornscheuer et al., 2014). After oral application, several metabolites in the urine of rabbits were identified whereby the formation of the two glycols may be due to the hydration of the corresponding epoxides formed as intermediates (Ishida et al., 1981). The formation of uroterpenol may proceed via limonene, derived from myrcene in the acidic conditions of rabbit stomachs (Figure 9.21).

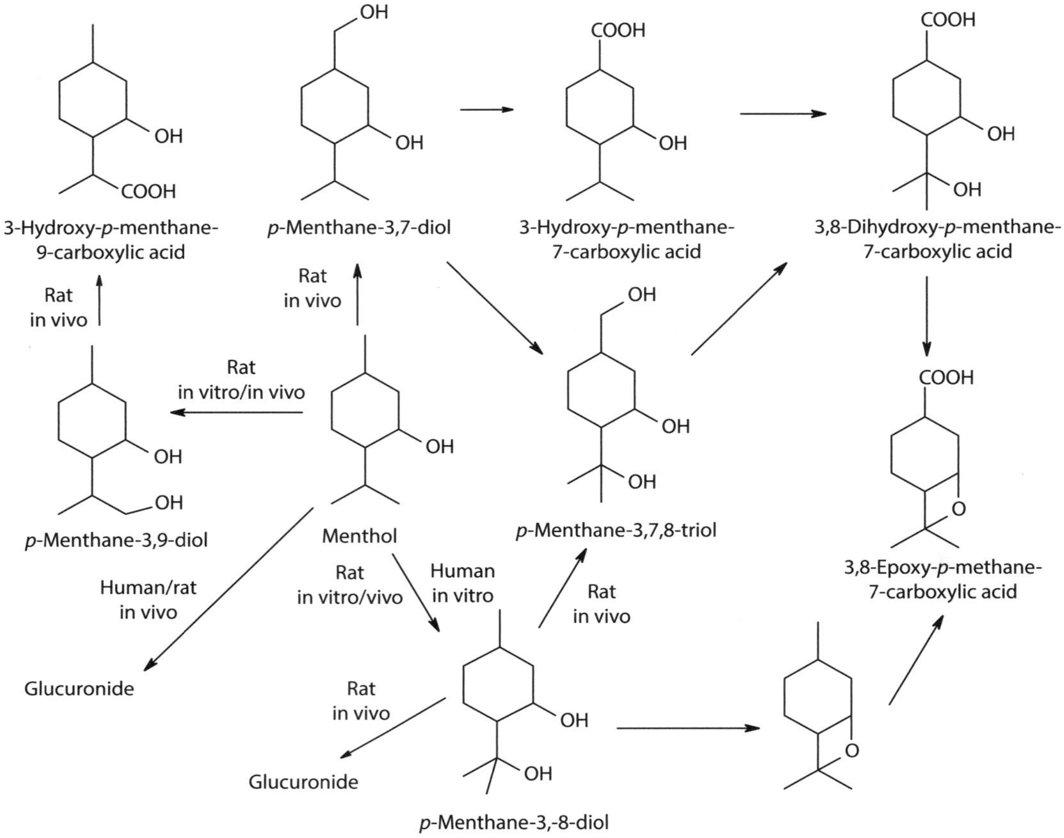

FIGURE 9.19 Proposed metabolism of menthol in rats and humans. (Adapted from Madyastha, K.M. and Srivatsan, V., *Drug Metab. Dispos. Biol. Fate Chem.*, 16, 765, 1988a; Gelal, A. et al., *Clin. Pharmacol. Ther.*, 66, 128, 1999; Spichiger, M. et al., *J. Chromatogr. B*, 799, 111, 2004.)

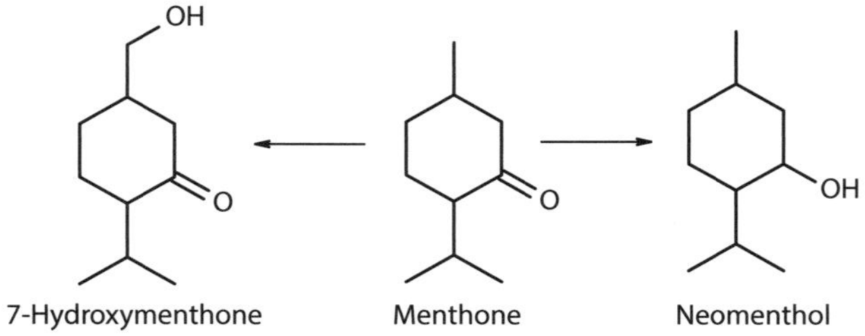

FIGURE 9.20 Proposed metabolism of menthone in rats liver microsomes. (Adapted from Miyazawa, M. and Nakanishi, K., *Biosci. Biotechnol. Biochem.*, 70, 1259, 2006.)

9.2.21 α- AND β-PINENE

The constitutional isomers α- and β-pinene are both major constituents of pine resins and found in many essential oils, for example, *Pinus* species, *Piper nigrum* and *Juniperus communis*. Interestingly, α-pinene is more common in European pines, whereas β-pinene is more common in North America (Bornscheuer et al., 2014; O'Neil, 2006). As shown in Figure 9.22, metabolism of α- and β-pinene in humans leads to the formation of *trans*- and *cis*-verbenol and myrtenol,

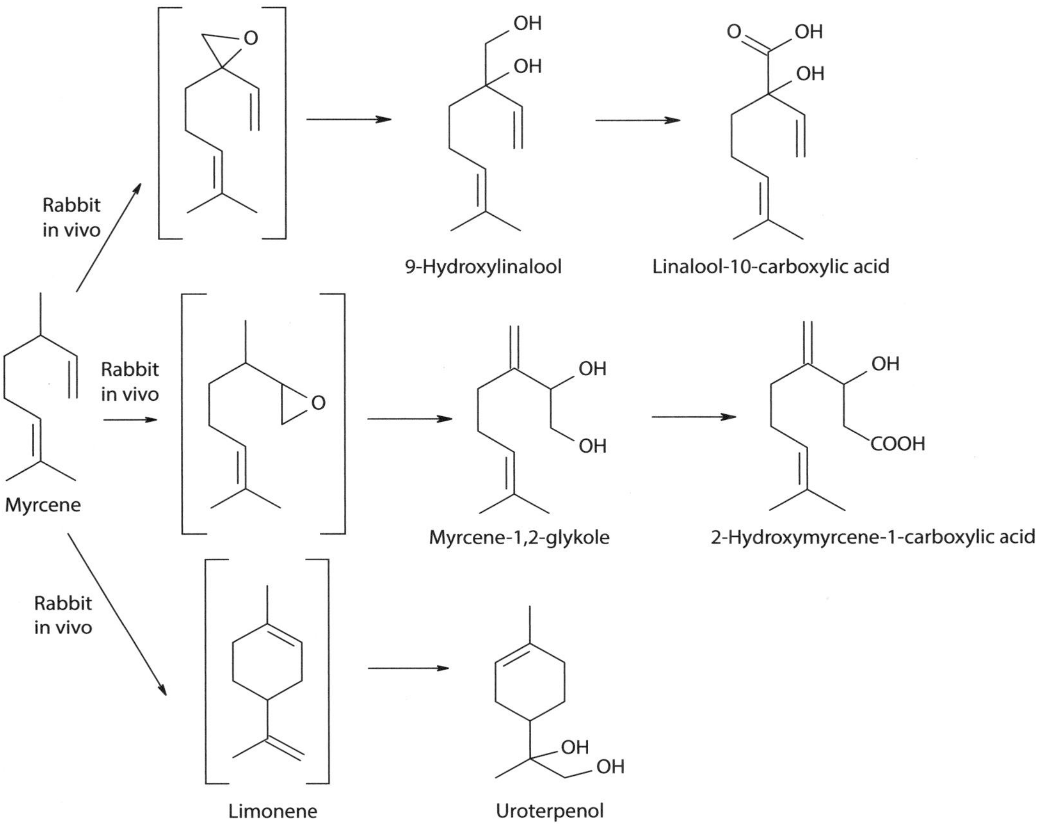

FIGURE 9.21 Proposed metabolism of myrcene in rabbits. (Adapted from Ishida, T. et al., *J. Pharm. Sci.*, 70, 406, 1981.)

respectively (Schmidt et al., 2013). Analysis of the human urine after occupational exposure of sawing fumes also suggest that *cis*- and *trans*-verbenol are being further hydroxylated to diols. The main urinary metabolite of α-pinene in rabbits is *trans*-verbenol; the minor biotransformation products are myrtenol and myrtenic acid. The main urinary metabolite of β-pinene in rabbits, however, is *cis*-verbenol indicating stereoselective hydroxylation (Eriksson and Levin, 1996; Ishida et al., 1981).

9.2.22 Pulegone

(*R*)-(+)-Pulegone is present in essential oils of Lamiaceae. *Hedeoma pulegioides* and *Mentha pulegium*, both commonly called pennyroyal, contain essential oils, which are chiefly, pulegone. Pennyroyal herb had been used for inducing menstruation and abortion. In higher doses, however, it may have resulted in central nervous system toxicity, gastritis, hepatic and renal failure, pulmonary toxicity, and death. Commercially available pennyroyal oils were found to have a pulegone content > 80% and to be both hepatotoxic and pneumotoxic in mice (Engel, 2003). Though pulegone has been used for flavoring food and oral hygiene products, the pulegone content in foods and beverages is restricted by EU law (Bornscheuer et al., 2014; O'Neil, 2006). At nontoxic concentrations, pulegone is oxidized selectively at the 10-position. Alternatively, it may be reduced to menthone, which has been detected in trace levels in urine samples. It might be possible that

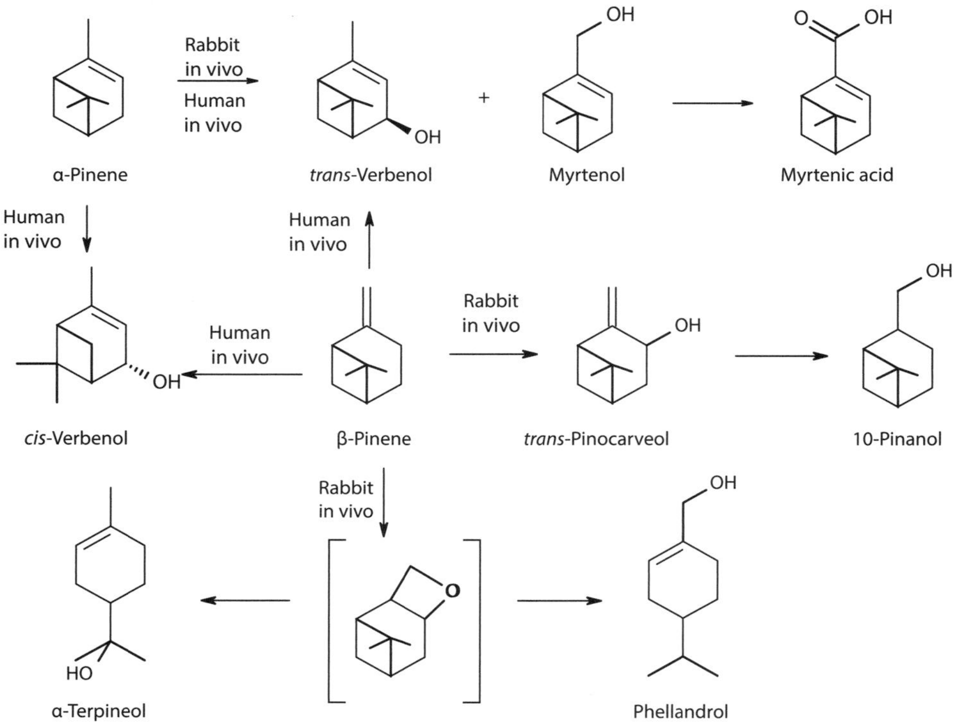

FIGURE 9.22 Proposed metabolism of α- and β-pinene in rabbits and humans. (Adapted from Ishida, T. et al., *J. Pharm. Sci.*, 70, 406, 1981; Eriksson, K. and Levin, J.O., *J. Chromatogr. B*, 677, 85, 1996.)

pulegone is also reduced at the carbonyl group first. Consequently, pulegol is either reduced very efficiently to menthol or rearranged to 3-*p*-menthen-8-ol (Engel, 2003) (Figure 9.23). In rats, three major pathways were identified: (a) hydroxylation followed by glucuronidation, (b) reduction to menthone and hydroxylation, and (c) conjugation with glutathione and further metabolism (Chen et al., 2001).

9.2.23 α-Terpineol

α-Terpineol, a monocyclic monoterpene alcohol, is found in the essential oil sources such as neroli, petitgrain, *Melaleuca alternifolia*, and *O. vulgare* (Bornscheuer et al., 2014). Based on its pleasant lilac-like odor, α-terpineol is widely used in the manufacture of perfumes, cosmetics, soaps, and antiseptic agents. After oral administration to rats (600 mg/kg body weight), α-terpineol is metabolized to *p*-menthane-1,2,8-triol probably formed from the epoxide intermediate. Notably, allylic methyl oxidation and the reduction of the 1,2-double bound are the major routes for the biotransformation of α-terpineol in rat (Figure 9.24). Although allylic oxidation of C-1 methyl seems to be the major pathway, the alcohol *p*-menth-1-ene-7,8-diol could not be isolated from the urine samples. Probably, this compound is accumulated and is readily further oxidized to oleuropeic acid (Madyastha and Srivatsan, 1988b).

Metabolism of Terpenoids in Animal Models and Humans

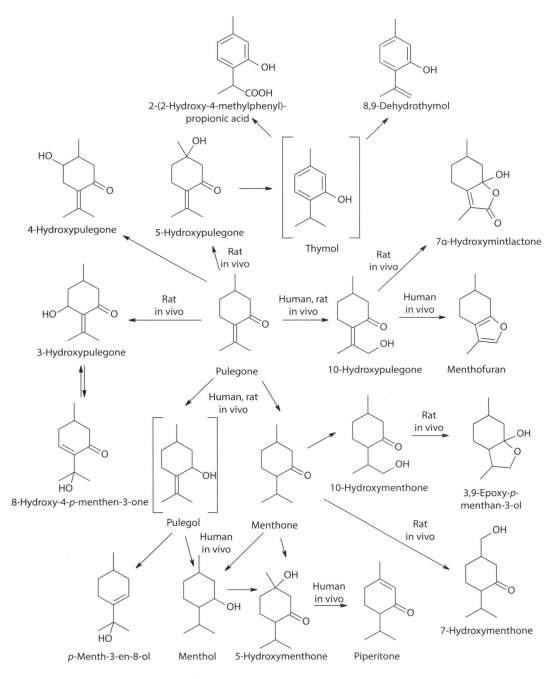

FIGURE 9.23 Proposed metabolism of pulegone in humans and rats. (Adapted from Engel, W., *J. Agric. Food Chem.*, 51, 6589, 2003; Chen, L.-J. et al., *Drug Metab. Dispos.*, 29, 1567, 2001.)

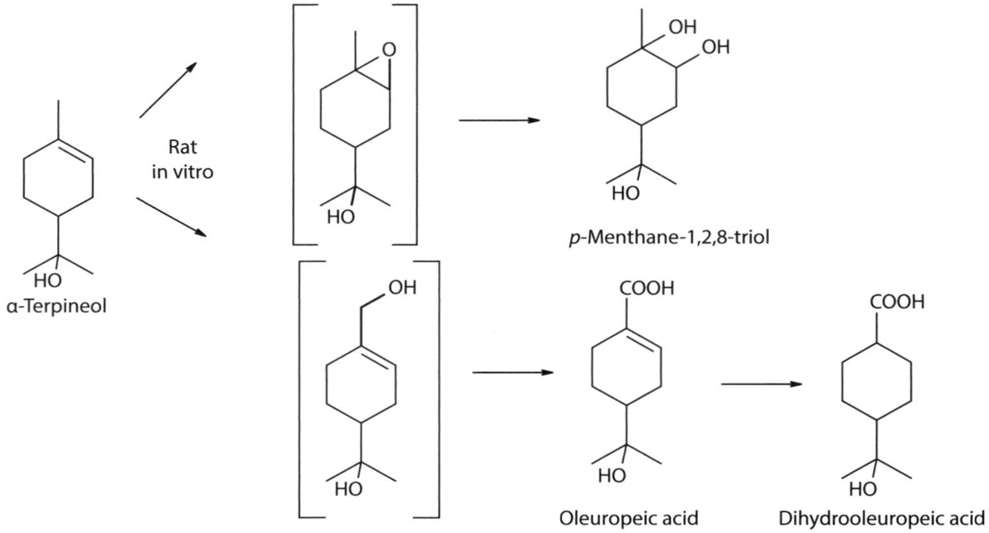

FIGURE 9.24 Proposed metabolism of α-terpineol in rats. (Adapted from Madyastha, K.M. and Srivatsan, V., *Environ. Contam. Toxicol.*, 41, 17, 1988b.)

9.2.24 TERPINEN-4-OL

(−)-Terpinen-4-ol is the main compound of *M. alternifolia* essential oil, which is widely used in cosmetic and pharmaceutical products due to its antiseptic properties (Bornscheuer et al., 2014). Both (+)- and (−)-terpinen-4-ol are epoxidized to 1,2.epoxy-p-menthan-4-ol by human liver microsomes, whereas the formation of an 8-hydroxylated metabolite could only be shown for the (+)-isomer (Haigou and Miyazawa, 2012; Miyazawa and Haigou, 2011). (Figure 9.25).

9.2.25 α- AND β-THUJONE

α- and β-thujone are bicyclic monoterpenes that differ in the stereochemistry of the C-4 methyl group. The isomer ratio depends on the plant source, with high content of α-thujone in *T. occidentalis* and β-thujone in the essential oils of *Tanacetum vulgare*, *S. officinalis*, and *Artemisia absinthium*. Due to their neurotoxicity, maximum permissible amounts for thujones in foods and beverages were been defined in EU (Bornscheuer et al., 2014; O'Neil, 2006). α-Thujone is best known as the active ingredient of the alcoholic beverage absinthe, which was a very popular European drink in the 1800s. In vivo rabbit, rat, and mouse models conformed to the in vitro data using rat, mice, and human liver microsomes where α- and β-thujone are extensively metabolized to six hydroxythujones and three dehydrothujones (Figure 9.26). In Phase II reactions, several metabolites are

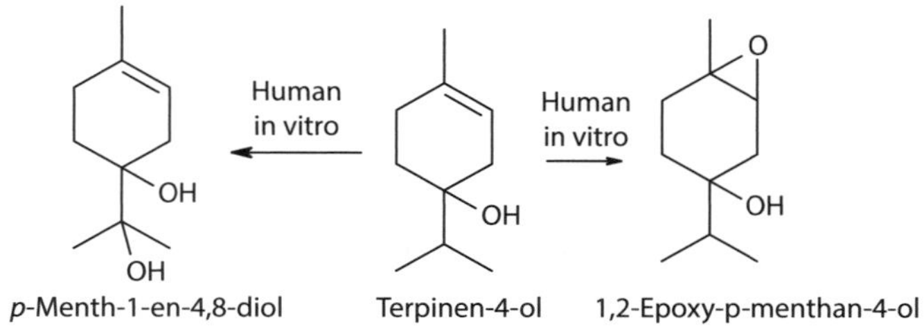

FIGURE 9.25 Proposed metabolism of (−)-terpinen-4-ol in human liver microsomes (Adapted from Miyazawa M. and Haigou R., *Xenobiotica*, 41, 1056, 2011; Haigou R., and Miyazawa, M., *J. Oleo Sci.*, 61, 35, 2012.)

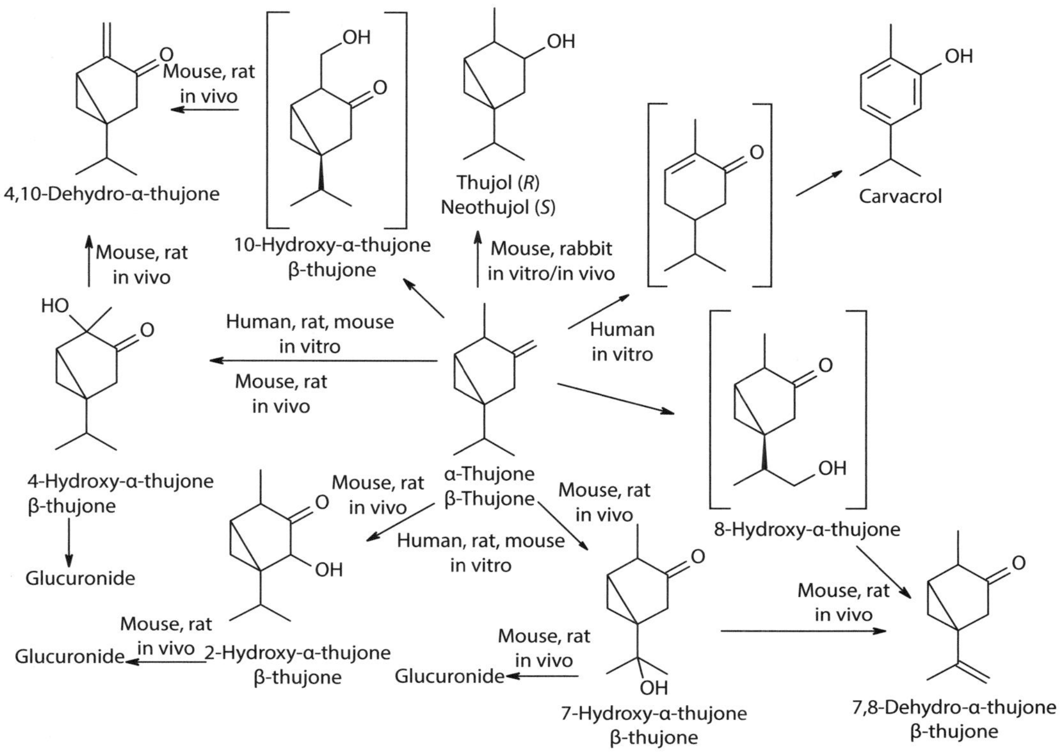

FIGURE 9.26 Proposed in vitro and in vivo metabolism of α- and β-thujone in rats. (Adapted from Ishida, T. et al., *Xenobiotica*, 19, 843, 1989; Höld, K.M. et al., *Environ. Chem. Toxicol.*, 97, 3826, 2000; Höld, K.M. et al., *Chem. Res. Toxicol.*, 14, 589, 2001.)

further conjugated with glucuronic acid (Höld et al., 2000, 2001; Ishida et al., 1989). Recent studies have shown that the main human metabolites 4- and 7-hydroxy-α-thujone are formed by CYP2A, CYP2B6, CYP2D6, and CYP3A4 (Abass et al., 2011). CYP3A4 seems to be involved in the formation of most of the thujone metabolites in human models (Jiang and Ortiz de Montellano, 2009). Based on the in vitro and in vivo data, biotransformation should also be pronounced in humans after the intake of α- and β-thujone.

9.2.26 Thymol

Thymol is the major constituent of the essential oils of *T. vulgaris* and *O. vulgare*. It is medically used in preparations against bronchitis and oral infections. Due to its antiseptic properties, it is also an ingredient of mouthwashes and toothpastes (Bornscheuer et al., 2014; O'Neil, 2006). Although after oral application to rats, large quantities were excreted unchanged or as their glucuronide and sulfate conjugates; extensive oxidation of the methyl and isopropyl groups also occurred (Austgulen et al., 1987) resulting in the formation of derivatives of benzyl alcohol and 2-phenylpropanol and their corresponding carboxylic acids. Ring hydroxylation was only a minor reaction in rats (Figure 9.27), whereas in humans several ring-hydroxylated products were found after oral application. All metabolites could be detected in human urine samples that were treated with glucuronidase prior to analysis (Thalhamer et al., 2011). CYP1A2, CYP2A6, and CYP2D6 were shown to be the main enzymes responsible for metabolizing thymol in human liver microsomes (Dong et al., 2012a). Thymol is excreted as glucuronide and sulfate into human urine, which can be also assumed for its metabolites (Kohlert et al., 2002).

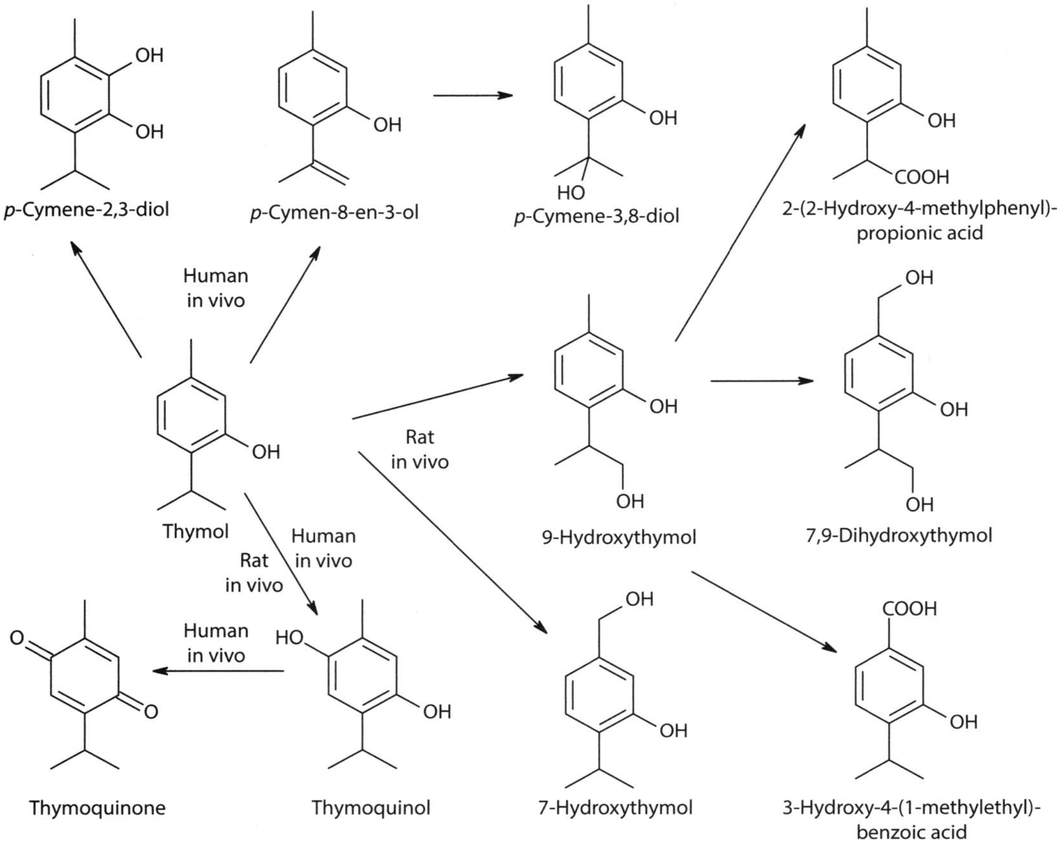

FIGURE 9.27 Proposed metabolism of thymol in humans and rats. (Adapted from Austgulen, L.T. et al., *Pharmacol. Toxicol.*, 61, 98, 1987; Thalhamer, B. et al., *J. Pharm. Biomed. Anal.*, 56, 64, 2011.)

9.3 METABOLISM OF SESQUITERPENES

9.3.1 CARYOPHYLLENE

(−)-β-Caryophyllene is a common sesquiterpene with a clove- or turpentine-like odor. It can be found in the essential oils of *Syzygium aromaticum*, *P. nigrum*, and *H. lupulus*. It is used as flavoring substance, for example, for chewing gums (Bornscheuer et al., 2014; O'Neil, 2006). The biotransformation of (−)-β-caryophyllene in rabbits yielded the main metabolite [10*S*]-(−)-14-hydroxycaryophyllene-5,6-oxide and the minor biotransformation product cayrophyllene-5,6-oxide-2,12-diol. The formation of the minor metabolites is easily explained via the regioselective hydroxylation of the diepoxide intermediate (Asakawa et al., 1981, 1986). Based on in vivo data from rabbits, extensive biotransformation in humans is highly suggested after oral administration (Figure 9.28).

9.3.2 FARNESOL

Farnesol is present in many essential oils such as *Cymbopogon* species and neroli. It is used in perfumery and soaps to emphasize the odors of sweet floral perfumes and due to its fixative and antibacterial properties (Bornscheuer et al., 2014; O'Neil, 2006). Interestingly, it is also produced in humans where it acts on numerous nuclear receptors (Joo and Jetten, 2010). In vitro studies using

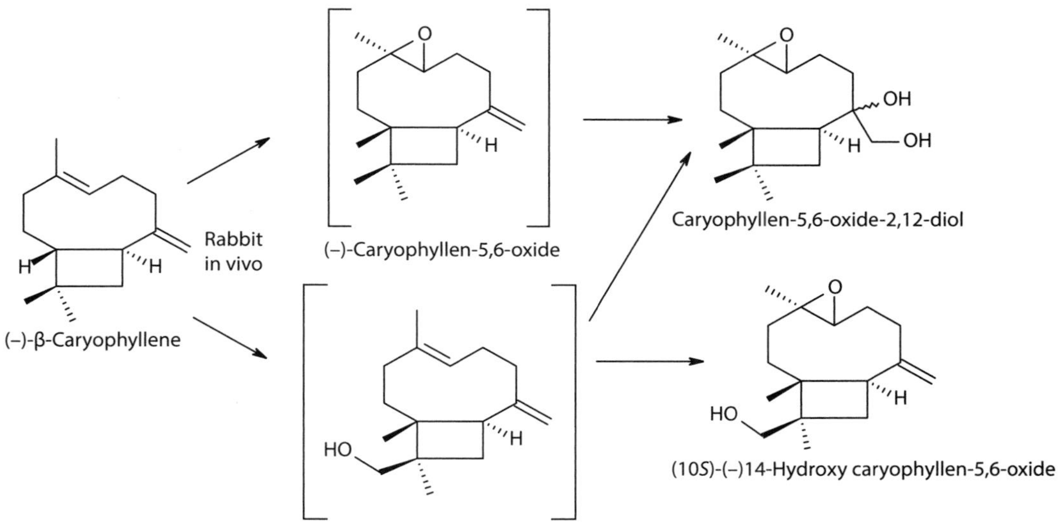

FIGURE 9.28 Proposed metabolism of β-caryophyllene in rabbits. (Adapted from Asakawa, Y. et al., *J. Pharm. Sci.*, 70, 710, 1981; Asakawa, Y. et al., *Xenobiotica*, 16, 753, 1986.)

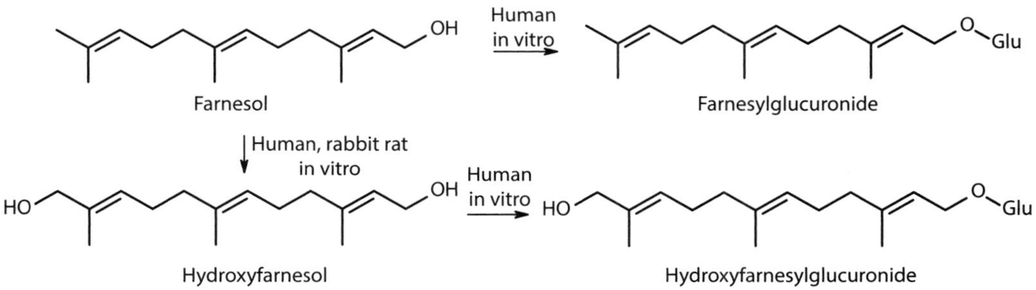

FIGURE 9.29 Proposed metabolism of farnesol in human liver microsomes. (Adapted from DeBarber, A.E. et al., *Biochim. Biophys. Acta*, 1682, 18, 2004; Staines, A.G. et al., *Biochem. J.*, 384, 637, 2004.)

recombinant drug metabolizing enzymes and human liver microsomes have shown that CYP isoenzymes participate in the metabolism of farnesol to 12-hydroxyfarnesol (Figure 9.29). Subsequently, farnesol and its metabolite are glucuronidated (DeBarber et al., 2004; Staines et al., 2004).

9.3.3 Longifolene

Longifolene is primarily found in Indian turpentine oil, which is commercially extracted from *Pinus roxburghii* (chir pine) (Bornscheuer et al., 2014). In rabbits, longifolene is metabolized as follows: attack on the *exo*-methylene group from the *endo*-face to form its epoxide followed by isomerization of the epoxide to a stable *endo*-aldehyde. Then rapid CYP-catalyzed hydroxylation of this *endo*-aldehyde occurs (Asakawa et al., 1986) (Figure 9.30).

9.3.4 Patchoulol

(−)-Patchoulol is the major active ingredient and the most odor-intensive component of patchouli oil, the volatile oil of *Pogostemon cablin*, one of the most important raw materials of perfumery,

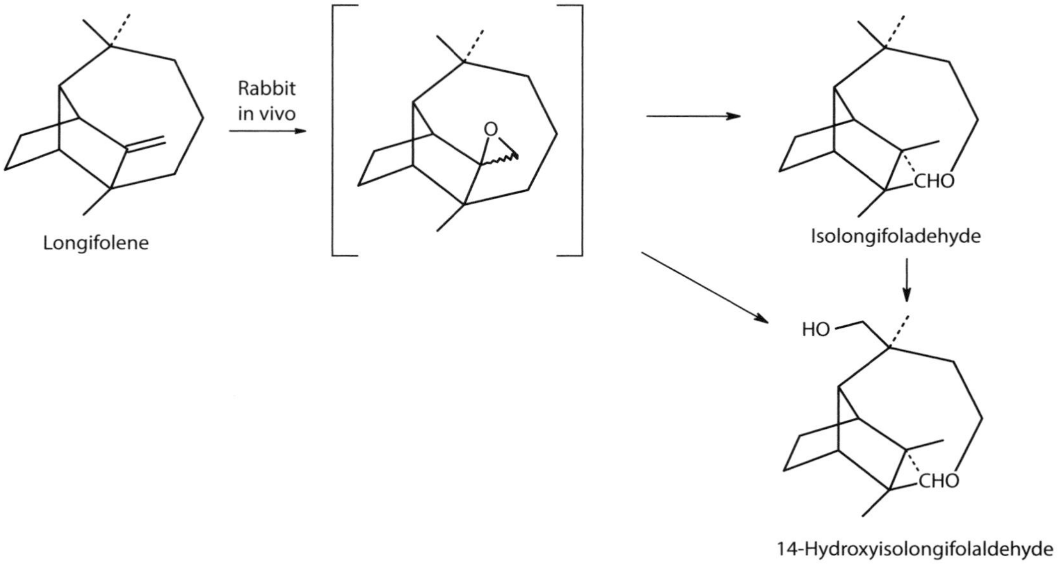

FIGURE 9.30 Proposed metabolism of (+)-longifolene in rabbits. (Adapted from Asakawa, Y. et al., *Xenobiotica*, 16, 753, 1986.)

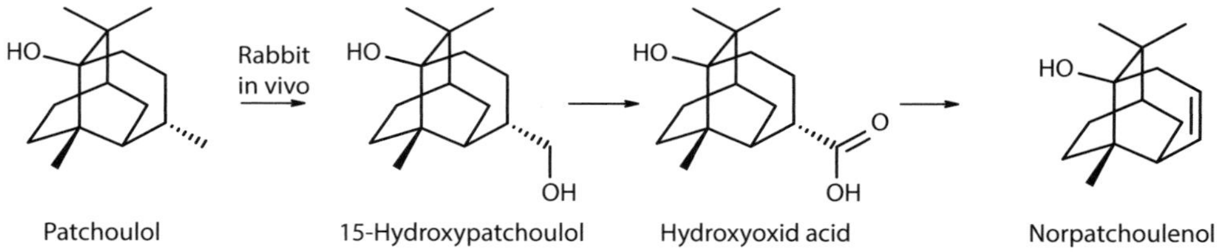

FIGURE 9.31 Proposed metabolism of patchoulol in rabbits and dogs. (Adapted from Bang, L. et al., *Tetrahedron Lett.*, 26, 2211, 1975; Ishida, T., *Chem. Biodivers.*, 2, 569, 2005.)

which is also used for its insect repellent activity. In rabbits and dogs, patchoulol is hydroxylated at the C-15 yielding a diol that is subsequently oxidized to a hydroxyl acid. After decarboxylation and oxidation, the 3,4-unsaturated norpatchoulen-1-ol is formed, which also has a characteristic odor (Figure 9.31). All these urinary metabolites are also found as glucuronides, explaining their excellent water solubility (Bang et al., 1975; Ishida, 2005).

REFERENCES

Abass, K., P. Reponen, S. Mattila, and O. Pelkonen, 2011. Metabolism of α-thujone in human hepatic preparations in vitro. *Xenobiotica*, 41: 101–111.

Asakawa, Y., T. Ishida, M. Toyota, and T. Takemoto, 1986. Terpenoid biotransformation in mammals. IV. Biotransformation of (+)-longifolene, (−)-caryophyllene, (−)-caryophyllene oxide, (−)-cyclocolorenone, (+)-nootkatone, (−)-elemol, (−)-abietic acid and (+)-dehydroabietic acid in rabbits. *Xenobiotica*, 16: 753–767.

Asakawa, Y., Z. Taira, T. Takemoto, T. Ishida, M. Kido, and Y. Ichikawa, 1981. X-ray crystal structure analysis of 14-hydroxycaryophyllene oxide, a new metabolite of (−)-caryophyllene, in rabbits. *J. Pharm. Sci.*, 70: 710–711.

Asakawa, Y., M. Toyota, and T. Ishida, 1988. Biotransformation of 1,4-cineole, a monoterpene ether. *Xenobiotica*, 18: 1129–1134.

Austgulen, L. T., E. Solheim, and R. R. Scheline, 1987. Metabolism in rats of p-cymene derivatives: Carvacrol and thymol. *Pharmacol. Toxicol.*, 61: 98–102.

Bang, L., G. Ourisson, and P. Teisseire, 1975. Hydroxylation of patchoulol by rabbits. Hemi-synthesis of nor-patchoulenol, the odour carrier of patchouli oil. *Tetrahedron Lett.*, 16: 2211–2214.

Bickers, D., P. Calow, H. Greim, J. M. Hanifin, A. E. Rogers, J. H. Saurat, I. G. Sipes, R. L. Smith, and H. Tagami, 2003. A toxicologic and dermatologic assessment of linalool and related esters when used as fragrance ingredients. *Food Chem. Toxicol. Int. J. Publ. Br. Ind. Biol. Res. Assoc.*, 41: 919–942.

Bornscheuer, U et al. (eds.), 2014. *Römpp Online 4.0*. Stuttgart, Germany: Georg Thieme Verlag KG. Retrieved on February 12, 2014 from https://roempp.thieme.de/.

Boyle, R., S. McLean, W. Foley, N. W. Davies, E. J. Peacock, and B. Moore, 2001. Metabolites of dietary 1,8-cineole in the male koala (*Phascolarctos cinereus*). *Comp. Biochem. Physiol. C*, 129: 385–395.

Chadha, A. and K. M. Madyastha, 1984. Metabolism of geraniol and linalool in the rat and effects on liver and lung microsomal enzymes. *Xenobiotica*, 14: 365–374.

Chen, L.-J., E. H. Lebetkin, and L. T. Burka, 2001. Metabolism of (R)-(+)-pulegone in F344 rats. *Drug Metab. Dispos.*, 29: 1567–1577.

Crowell, P. L., S. Lin, E. Vedejs, and M. N. Gould, 1992. Identification of metabolites of the antitumor agent d-limonene capable of inhibiting protein isoprenylation and cell growth. *Cancer Chemother. Pharmacol.*, 31: 205–212.

DeBarber, A. E., L. A. Bleyle, J.-B. O. Roullet, and D. R. Koop, 2004. Omega-hydroxylation of farnesol by mammalian cytochromes p450. *Biochim. Biophys. Acta*, 1682: 18–27.

Diliberto, J. J., P. Srinivas, D. Overstreet, G. Usha, L. T. Burka, and L. S. Birnbaum, 1990. Metabolism of citral, an alpha,beta-unsaturated aldehyde, in male F344 rats. *Drug Metab. Dispos. Biol. Fate Chem.*, 18: 866–875.

Dong, R.-H., Z.-Z. Fang, L.-L. Zhu, G.-B. Ge, Y.-F. Cao, X.-B. Li, C.-M. Hu, L. Yang, and Z.-Y. Liu, 2012a. Identification of CYP isoforms involved in the metabolism of thymol and carvacrol in human liver microsomes (HLMs). *Pharmazie*, 67: 1002–1006.

Dong, R.-H., Z.-Z. Fang, L.-L. Zhu, G.-B. Ge, L. Yang, and Z.-Y. Liu, 2012b. Identification of UDP-glucuronosyltransferase isoforms involved in hepatic and intestinal glucuronidation of phytochemical carvacrol. *Xenobiotica*, 42: 1009–1016.

Duisken, M., D. Benz, T. H. Peiffer, B. Blömeke, and J. Hollender, 2005. Metabolism of Delta(3)-carene by human cytochrome p450 enzymes: Identification and characterization of two new metabolites. *Curr. Drug Metab.*, 6: 593–601.

Engel, W., 2001. In vivo studies on the metabolism of the monoterpenes S-(+)- and R-(−)-carvone in humans using the metabolism of ingestion-correlated amounts (MICA) approach. *J. Agric. Food Chem.*, 49: 4069–4075.

Engel, W., 2003. In vivo studies on the metabolism of the monoterpene pulegone in humans using the metabolism of ingestion-correlated amounts (MICA) approach: Explanation for the toxicity differences between (S)-(−)- and (R)-(+)-pulegone. *J. Agric. Food Chem.*, 51: 6589–6597.

Eriksson, K. and J. O. Levin, 1996. Gas chromatographic-mass spectrometric identification of metabolites from alpha-pinene in human urine after occupational exposure to sawing fumes. *J. Chromatogr. B: Biomed. Appl.*, 677: 85–98.

Gelal, A., P. Jacob 3rd, L. Yu, and N. L. Benowitz, 1999. Disposition kinetics and effects of menthol. *Clin. Pharmacol. Ther.*, 66: 128–135.

Gyoubu, K. and M. Miyazawa, 2007. In vitro metabolism of (−)-camphor using human liver microsomes and CYP2A6. *Biol. Pharm. Bull.*, 30: 230–233.

Hagvall, L., J. M. Baron, A. Börje, L. Weidolf, H. Merk, and A.-T. Karlberg, 2008. Cytochrome P450-mediated activation of the fragrance compound geraniol forms potent contact allergens. *Toxicol. Appl. Pharmacol.*, 233: 308–313.

Haigou, R. and M. Miyazawa, 2012. Metabolism of (+)-terpinen-4-ol by cytochrome P450 enzymes in human liver microsomes. *J. Oleo Sci.*, 61: 35–43.

Höld, K. M., N. S. Sirisoma, and J. E. Casida, 2001. Detoxification of alpha- and beta-thujones (the active ingredients of absinthe): Site specificity and species differences in cytochrome P450 oxidation in vitro and in vivo. *Chem. Res. Toxicol.*, 14: 589–595.

Höld, K. M., N. S. Sirisoma, T. Ikeda, T. Narahashi, and J. E. Casida, 2000. Alpha-thujone (the active component of absinthe): Gamma-aminobutyric acid type A receptor modulation and metabolic detoxification. *Proc. Natl. Acad. Sci. USA*, 97: 3826–3831.

Horst, K. and M. Rychlik, 2010. Quantification of 1,8-cineole and of its metabolites in humans using stable isotope dilution assays. *Mol. Nutr. Food Res.*, *54*: 1515–1529.

Ishida, T., 2005. Biotransformation of terpenoids by mammals, microorganisms, and plant-cultured cells. *Chem. Biodivers.*, *2*: 569–590.

Ishida, T., Y. Asakawa, T. Takemoto, and T. Aratani, 1979. Terpenoid biotransformation in mammals. II: Biotransformation of dl-camphene in rabbits. *J. Pharm. Sci.*, *68*: 928–930.

Ishida, T., Y. Asakawa, T. Takemoto, and T. Aratani, 1981. Terpenoids biotransformation in mammals. III: Biotransformation of alpha-pinene, beta-pinene, pinane, 3-carene, carane, myrcene, and p-cymene in rabbits. *J. Pharm. Sci.*, *70*: 406–415.

Ishida, T., M. Toyota, and Y. Asakawa, 1989. Terpenoid biotransformation in mammals. V. Metabolism of (+)-citronellal, (+)-7-hydroxycitronellal, citral, (–)-perillaldehyde, (–)-myrtenal, cuminaldehyde, thujone, and (+)-carvone in rabbits. *Xenobiotica*, *19*: 843–855.

Jäger, W., M. Mayer, P. Platzer, G. Reznicek, H. Dietrich, and G. Buchbauer, 2000. Stereoselective metabolism of the monoterpene carvone by rat and human liver microsomes. *J. Pharm. Pharmacol.*, *52*: 191–197.

Jäger, W., M. Mayer, G. Reznicek, and G. Buchbauer, 2001. Percutaneous absorption of the montoterpene carvone: Implication of stereoselective metabolism on blood levels. *J. Pharm. Pharmacol.*, *53*: 637–642.

Jiang, Y., and P. R. Ortiz de Montellano, 2009. Cooperative effects on radical recombination in CYP3A4-catalyzed oxidation of the radical clock β-thujone. *ChemBioChem*, *10*: 650–653.

Joo, J. H. and A. M. Jetten, 2010. Molecular mechanisms involved in farnesol-induced apoptosis. *Cancer Lett.*, *287*: 123–135.

Khojasteh, S. C., S. Oishi, and S. D. Nelson, 2010. Metabolism and toxicity of menthofuran in rat liver slices and in rats. *Chem. Res. Toxicol.*, *23*: 1824–1832.

Kohlert, C., G. Schindler, R. W. März, G. Abel, B. Brinkhaus, H. Derendorf, E.-U. Gräfe, and M. Veit, 2002. Systemic availability and pharmacokinetics of thymol in humans. *J. Clin. Pharmacol.*, *42*: 731–737.

Leibman, K. C. and E. Ortiz, 1973. Mammalian metabolism of terpenoids. I. Reduction and hydroxylation of camphor and related compounds. *Drug Metab. Dispos.*, *1*: 543–551.

Madyastha, K. M. and V. Srivatsan, 1988a. Studies on the metabolism of l-menthol in rats. *Drug Metab. Dispos. Biol. Fate Chem.*, *16*: 765–772.

Madyastha, K. M. and V. Srivatsan, 1988b. Biotransformations of alpha-terpineol in the rat: Its effects on the liver microsomal cytochrome P-450 system. *Bull. Environ. Contam. Toxicol.*, *41*: 17–25.

Matsumoto, T., T. Ishida, T. Yoshida, H. Terao, Y. Takeda, and Y. Asakawa, 1992. The enantioselective metabolism of p-cymene in rabbits. *Chem. Pharm. Bull.* (Tokyo), *40*: 1721–1726.

Meesters, R. J. W., M. Duisken, and J. Hollender, 2007. Study on the cytochrome P450-mediated oxidative metabolism of the terpene alcohol linalool: Indication of biological epoxidation. *Xenobiotica*, *37*: 604–617.

Meesters, R. J. W., M. Duisken, and J. Hollender, 2009. Cytochrome P450-catalysed arene-epoxidation of the bioactive tea tree oil ingredient p-cymene: Indication for the formation of a reactive allergenic intermediate? *Xenobiotica*, *39*: 663–671.

Miyazawa, M. and K. Gyoubu, 2006. Metabolism of (+)-fenchone by CYP2A6 and CYP2B6 in human liver microsomes. *Biol. Pharm. Bull.*, *29*: 2354–2358.

Miyazawa, M. and K. Gyoubu, 2007. Metabolism of (–)-fenchone by CYP2A6 and CYP2B6 in human liver microsomes. *Xenobiotica*, *37*: 194–204.

Miyazawa M. and R. Haigou, 2011. Determination of cytochrome P450 enzymes involved in the metabolism of (–)-terpinen-4-ol by human liver microsomes. *Xenobiotica*, *41*: 1056–1062.

Miyazawa, M., S. Marumoto, T. Takahashi, H. Nakahashi, R. Haigou, and K. Nakanishi, 2011. Metabolism of (+)- and (–)-menthols by CYP2A6 in human liver microsomes. *J. Oleo Sci.*, *60*: 127–132.

Miyazawa, M. and K. Nakanishi, 2006. Biotransformation of (–)-menthone by human liver microsomes. *Biosci. Biotechnol. Biochem.*, *70*: 1259–1261.

Miyazawa, M., M. Shindo, and T. Shimada, 2001a. Roles of cytochrome P450 3A enzymes in the 2-hydroxylation of 1,4-cineole, a monoterpene cyclic ether, by rat and human liver microsomes. *Xenobiotica*, *31*: 713–723.

Miyazawa, M., M. Shindo, and T. Shimada, 2001b. Oxidation of 1,8-cineole, the monoterpene cyclic ether originated from *Eucalyptus polybractea*, by cytochrome P450 3A enzymes in rat and human liver microsomes. *Drug Metab. Dispos.*, *29*: 200–205.

Miyazawa, M., M. Shindo, and T. Shimada, 2002. Metabolism of (+)- and (–)-limonenes to respective carveols and perillyl alcohols by CYP2C9 and CYP2C19 in human liver microsomes. *Drug Metab. Dispos.*, *30*: 602–607.

Nakahashi, H., N. Yagi, and M. Miyazawa, 2013. Biotransformation of (+)-fenchone by *Salmonella typhimurium* OY1002/2A6 expressing human CYP2A6 and NADPH-P450 reductase. *J. Oleo Sci.*, 62: 293–296.

O'Neil, M. J. (ed.), 2006. *The Merck Index: An Encyclopedia of Chemicals, Drugs, and Biologicals.* Whitehouse Station, NJ: Wiley.

Pass, G. J., S. McLean, I. Stupans, and N. W. Davies, 2002. Microsomal metabolism and enzyme kinetics of the terpene p-cymene in the common brushtail possum (*Trichosurus vulpecula*), koala (*Phascolarctos cinereus*) and rat. *Chem. Biol. Interact.*, 32: 383–397.

Schmidt, L., V. N. Belov, and T. Göen, 2013. Sensitive monitoring of monoterpene metabolites in human urine using two-step derivatisation and positive chemical ionisation-tandem mass spectrometry. *Anal. Chim. Acta*, 793: 26–36.

Shimada, T., M. Shindo, and M. Miyazawa, 2002. Species differences in the metabolism of (+)- and (−)-limonenes and their metabolites, carveols and carvones, by cytochrome P450 enzymes in liver microsomes of mice, rats, guinea pigs, rabbits, dogs, monkeys, and humans. *Drug Metab. Pharmacokinet.*, 17: 507–515.

Shipley, L. A., E. M. Davis, L. A. Felicetti, S. McLean, and J. S. Forbey, 2012. Mechanisms for eliminating monoterpenes of sagebrush by specialist and generalist rabbits. *J. Chem. Ecol.*, 38: 1178–1189.

Spatzenegger, M. and W. Jäger, 1995. Clinical importance of hepatic cytochrome P450 in drug metabolism. *Drug Metab. Rev.*, 27: 397–417.

Spichiger, M. et al., 2004. Determination of menthol in plasma and urine of rats and humans by headspace solid phase microextraction and gas chromatography–mass spectrometry. *J. Chromatogr. B*, 799, 111–117.

Spichiger, M., R. Brenneisen, R. Felix, and R. C. Mühlbauer, 2006. The inhibition of bone resorption in rats treated with (−)-menthol is due to its metabolites. *Planta Med.*, 72: 1290–1295.

Staines, A. G., P. Sindelar, M. W. H. Coughtrie, and B. Burchell, 2004. Farnesol is glucuronidated in human liver, kidney and intestine in vitro, and is a novel substrate for UGT2B7 and UGT1A1. *Biochem. J.*, 384: 637–645.

Sun, X.-M., Q.-F. Liao, Y.-T. Zhou, X.-J. Deng, and Z.-Y. Xie, 2014. Simultaneous determination of borneol and its metabolite in rat plasma by GC-MS and its application to pharmacokinetics study. *J. Pharm. Anal.*, 4: 345–350.

Thalhamer, B., W. Buchberger, and M. Waser, 2011. Identification of thymol phase I metabolites in human urine by headspace sorptive extraction combined with thermal desorption and gas chromatography mass spectrometry. *J. Pharm. Biomed. Anal.*, 56: 64–69.

Zhang, R., C. Liu, T. Huang, N. Wang, and S. Mi, 2008. In vitro characterization of borneol metabolites by GC–MS upon incubation with rat liver microsomes. *J. Chromatogr. Sci.*, 46: 419–423.

Zhao, J., Y. Lu, S. Du, X. Song, J. Bai, and Y. Wang, 2012. Comparative pharmacokinetic studies of borneol in mouse plasma and brain by different administrations. *J. Zhejiang Univ. Sci. B*, 13: 990–996.

10 Biological Activities of Essential Oils
An Update

Gerhard Buchbauer and Ramona Bohusch

CONTENTS

10.1 Introduction 281
10.2 Antibacterial Activity 281
10.3 Antiviral Activity..285
10.4 Antinociceptive Activity...286
10.5 Anti-Inflammatory Activity..289
10.6 Vasodilatory Activity..296
10.7 Cytotoxic Activity...297
10.8 Penetration-Enhancing Activity 303
References...307

10.1 INTRODUCTION

This chapter deals with an overview of the research results of antibacterial, antiviral, antinociceptive, anti-inflammatory, vasodilatory, and penetration enhancing activity of essential oils (EO) and covers the literature from 2009 to March 2014. It is well known that EOs exert many effects in the human and animal body. Many authors before have described several biological activities. This treatise should afford an overview of the research results concerning the aforementioned topics from 2009 to 2014 following earlier chapters. The treatment of common disease and indispositions, rheumatoid arthritis, pain, high blood pressure, cancer, and many more, with alternative methods such as EOs, is still current. Because of a high rate of resistances against common treatment, science is in the quest of alternative or at least complementary treatment opportunities.

10.2 ANTIBACTERIAL ACTIVITY

In a study about antibiofilm activity of lemongrass and grapefruit EO against *Staphylococcus aureus*, Adukwu et al. (2012) studied the antibacterial activity of lemongrass (*Cymbopogon flexuosus* Stapf.) (Poaceae) and grapefruit (*Citrus paradisi MACF*) (Rutaceae). They used five different strains of *S. aureus* (three methicillin-sensitive *S. aureus* strains and two methicillin-resistant *S. aureus* strains) for testing the antibacterial activity, with the disc diffusion test and some biofilm methods. Tests showed that lemongrass and citral, the main component of lemongrass EO possessed the highest activity against all strains of *S. aureus*. Concentrations between 0.03% and 0.06% (v/v) were inhibiting growth of all five *S. aureus* and at 0.125% (v/v) lemongrass EO operated as bactericide. Grapefruit EO showed higher minimum inhibitory concentration (MIC) than lemongrass EO for all the strains of *S. aureus*. Between 0.5% and 2% (v/v) the grapefruit EO indicated bacterial growth inhibiting, bactericidal activity was observed between 2% and 4% (v/v). Valencia orange (*Citrus sinensis* L.) (Rutaceae) EO was studied for antibacterial activity against methicillin-susceptible *S. aureus* and vancomycin intermediate resistant strains. The authors used terpeneless cold-pressed Valencia orange oil (CPV). Keratinocytes infected with *S. aureus* as a model for soft part and skin infections were treated with CPV (0.1% or 0.2%). After 3 h of incubation with 0.2% CPV, *S. aureus* was undetectable (Muthaiyan et al., 2012). The EO of *Ruta* species were tested against their antimicrobial activity by the presence of inhibition zones, zone diameters and MIC values. Results exhibited low in vitro antimicrobial activity for *Enterococcus cloacae, Enterococcus faecalis, Acinetobacter baumannii, Bacillus cereus, Escherichia coli, Citrobacter freundii, Pseudomonas aeruginosa, Klebsiella pneumonia, Listeria monocytogenes, Proteus mirabilis, S. aureus, Salmonella typhi* (Haddouchi et al., 2013).

Artemisia spicigera C. Koch (Asteraceae) EO was found to be active against Gram-positive and Gram-negative bacteria. The aerial parts of *A. spicigera* were collected from different parts of East Azerbaijan. Eight different EOs were tested with the agar gel diffusion method against *E. coli, Enterobacter aerogenes, Serratia marcescens, B. cereus, Citrobacter amalonaticus, S. aureus, Staphylococcus saprophyticus,* and *Bacillus megaterium*. Antibacterial tests showed good activity against the studied bacterial strains. Quality and quantity of the EO depend on ecological conditions and the way the EO was extracted. Different populations showed different activities against tested bacterial strains. While the EO of population Nr. 1 (geographical character: N: 37°41′5″; E: 45°52′29″) showed higher inhibition zones against *E. coli, E. aerogenes, S. marcescens,* and *S. aureus,* the EO of population Nr. 5 (geographical characters: N: 33°09′24.9″; E: 46°39′0.2″) indicated higher inhibition zones against *St. saprophyticus, B. megaterium,* and *B. cereus*. Their effects are also more outstanding than studied standard antibiotics (Chehregani et al., 2013). The EO from *Launaea resedifolia* L. (Asteraceae) was tested for its antibacterial activity against *E. coli, S. aureus, St. intermedius, P. mirabilis, K. pneumoniae,* and *P. aeruginosa*. The authors used the disc diffusion test and MIC for testing the EO. The efficacy of the EO was shown to depend on the size of the inhibiting zone. It was found that *S. aureus* (inhibition zone of 37 mm) was the most sensitive bacterial strain for *L. resedifolia* EO. *St. intermedius, K. pneumoniae, St. pyogenes,* and *P. mirabilis* were found to be sensitive as well, as they showed inhibition zones of 29, 27, 23, and 20 mm. *E. coli* and *P. aeruginosa* also reacted moderately sensitive on the treatment with EO (inhibition zone of 15 and 12 mm). (Zellagui et al., 2012a) Tunisian *Conyza sumatrensis* (Retz.) E. Walker (Asteraceae) EO was found to have highest antimicrobial activity against *E. faecalis* and *S. aureus*. Generally, EO of the aerial parts, especially the leaf oil, showed higher inhibiting activity than root oil (Mabrouk et al., 2013). Two different EOs of *Chromolaena laevigata* (Lam.) R. M. King and H. Rob. (Asteraceae) (blooming and fruiting stage) were investigated for their antibacterial activity. Used bacterial strains were *S. aureus, P. aeruginosa,* and *E. coli*. Both EOs showed strong antibacterial activity against all three bacteria. The EO from plants in their fruiting stage showed higher antibacterial activity (Murakami et al., 2013).

Kavoosi et al. (2013) found that the EO of *Carum copticum* L. (Apiaceae) significantly reduced the growth of Gram-positive and Gram-negative bacteria. MIC of carum oil for *S. typhi, E. coli, S. aureus,* and *Bacillus subtilis* were 78 ± 8, 65 ± 7, 14 ± 3, and 5 ± 2 µL/mL, respectively. The EO of *Ferula assa-foetida* Linn. (Apiaceae) oleo-gum resin was examined for its antibacterial activity in dependence of collection time. Three different samples of EO, of plants, collected at different times, were produced and investigated. All EOs significantly decreased the growth of Gram-positive and Gram-negative bacteria. The various EOs showed different activity in relation to growth inhibition of bacterial strains (Kavoosi and Rowshan, 2013). *Daucus muricatus* L. (Apiaceae) EO was studied for the antibacterial activity against nine pathogenic microorganisms (*B. subtilis, L. monocytogenes, B. cereus, S. aureus, P. aeruginosa, E. faecalis, K. pneumoniae, E. coli*) using the paper disc agar-diffusion technique. The EO showed highest activity against *S. aureus* (zone of inhibition: 22 mm), while the lowest activity was exhibited against *E. coli*. *B. cereus* and *L. monocytogenes* were also sensitive against *D. muricatus* EO with zones of inhibition ranging from 12 to 16 mm. In general, root oil of *D. muricatus* showed higher activity against the tested microorganism than the EO of the aerial parts (Bendiabdellah et al., 2012). An antibacterial activity test was run on EOs of leaves of *Ferula vesceritensis* Coss et Dur. (Apiaceae) using the disc diffusion method. Investigated were four human pathogens (*E. coli*, *K. pneumonia*, *S. aureus*, *P. aeruginosa*). *F. vesceritensis* inhibited growth of all four microorganisms. Inhibiting Zone Diameters increased proportionally with the increase of concentrations (Zellagui et al., 2012b).

The antibacterial activities of the EOs of three ecotypes of *Zataria multiflora* Boiss. (Lamiaceae) were tested against several Gram-positive and Gram-negative bacterial strains. At a concentration of 0.12–4 µL/mL, the EOs operated as growth inhibiting on Gram-positive cocci. Besides, the EOs showed bactericidal activity at concentrations of 1–8 µL/mL for all tested Gram-positive cocci. In addition, all used strains of *E. coli* and about half of the isolated *P. aeruginosa* strains were also sensitive to the EOs of *Z. multiflora* (Zomorodian et al., 2011). The same author group studied the antimicrobial activity of *Nepeta cataria* L. (Lamiaceae) EO against *E. coli*, *S. aureus*, *Shigella flexneri*, *B. cereus*, and *P. aeruginosa* (Zomorodian et al., 2012). The EO inhibited growth of all Gram-positive bacteria, and it further exhibited bactericidal activity against Gram-positive bacteria at a concentration ranging from 0.5 to 8 µL/mL. The EO also showed bacteriostatic and bactericidal activity against *Shigella* and *Salmonella* species at concentrations of 0.25–2 µL/mL and 0.5–8 µL/mL. Ten different strains of Gram-negative (*E. coli*, *Salmonella* sp., *E. cloacae*, *K. pneumoniae*, *P. aeruginosa*) and Gram-positive bacteria (*B. subtilis*, *B. cereus*, *Micrococcus luteus*, *S. aureus*) were tested to be sensitive against *Thymus maroccanus* Ball and *Thymus broussonetii* Boiss (Lamiaceae) EO. The antibacterial activity of the mentioned EOs were settled using the disc diffusion agar method. Furthermore, MIC was determined using the macrodilution broth method. The tested EOs showed a wide antibacterial spectrum as they produced inhibition zone diameters ranging from 9 to 42 mm. The inhibition zone diameters were sometimes even higher than those achieved with general antibiotics. Gram-positive bacteria were found to be more sensitive against the aforementioned EOs than Gram-negative, while *P. aeruginosa* proved to be the most resistant. Synergistic interaction of antibiotics with two *Thymus* EOs was determined by the checkerboard test. Eighty combinations of the EOs and four antibiotics (ciprofloxacin, gentamicin, pristinamycin, and cefixime) were studied. The best antibacterial effect was achieved with the combination of *T. maroccanus* and ciprofloxacin. The total synergistic effect of this combination was observed for Gram-positive and Gram-negative bacteria (Fadli et al., 2012). De Azeredo et al. (2012) investigated the cytotoxic effect of EOs from *Origanum vulgare* L. and *Rosmarinus officinalis* L. (both Lamiaceae) on *Aeromonas hydrophila*. *Mentha suaveolens* ssp. *Timija* (Briq.) Harley (Lamiaceae) is known as "Mint Timija" in Moroccan folk medicine and has been used for the treatment of coughs, bronchitis, ulcerative colitis, and much more. Kasrati et al. (2013) cultivated and tested the species for its antibacterial activity. They used disc diffusion and MIC tests against several pathogenic bacteria. The EO of cultivated mint timija featured strong antibacterial activity (MIC 0.93 mg/mL). Gram-positive and Gram-negative bacteria reacted equally sensitive to the EO.

The EO of *Eucalyptus globulus* Labill. (Myrtaceae) was tested against two clinical isolated strains of *E. coli* and *S. aureus*. The agar disc diffusion method was used to determine the antimicrobial activities. The addition of *E. globulus* EO to the bacterial strains caused growth inhibition on both microorganisms. Furthermore, the inhibition was greater on *E. coli* than observed on *S. aureus* (Bachir Raho and Benali, 2012). *Hypericum rochelii* Griseb. and Schenk and *Hypericum umbellatum* A. Kern. (Hypericaceae) were studied for their antibacterial activity against five bacterial strains (*B. subtilis*, *S. aureus*, *E. coli*, *Salmonella abony*, *P. aeruginosa*) using broth microdilution assay. *H. rochelii* EO was the most active against *S. aureus*. *H. umbellatum* EO also indicated moderate antibacterial activity (Dordevic et al., 2013). The EOs used alone and in combination severely inhibited cell viability of *A. hydrophila*. *Melaleuca* species were found to have several antibiotic activities. The EOs exhibited inhibitory activity against *St. epidermidis,* which is a Gram-positive pathogen. Gram-negative bacteria, such as *E. coli*, were also affected by the oil of *Melaleuca armillaris* SM and *M. acuminate* F. Muell. (Myrtaceae), the latter was more active than *M. armillaris* oil (Amri et al., 2012). *E. coli*, two different strains of *S. aureus*, *P. aeruginosa*, *Neisseria gonorrhoeae*, *B. subtilis,* and *E. faecalis* were investigated for their sensitivity against *Syzygium cumini*

Linn. (Myrtaceae) leaves EO. The EO contained antibacterial activity against both Gram-positive and Gram-negative bacterial strains (Mohamed et al., 2013).

The antibacterial activity of the EO of *Pelargonium graveolens* L'Her (Geraniaceae) and *Vitex agnus-castus* L. (Verbenaceae) was investigated against six bacterial species (*L. monocytogenes*, *S. enteritidis*, *P. aeruginosa*, *E. coli*, *S. aureus* and *B. subtilis*). The disc diffusion method showed inhibition zones ranging from 42 ± 3.2 to 9.5 ± 0.7 mm for the EO of *P. graveolens* except for *S. enteritidis* and *L. monocytogenes*, which showed no sensitivity against Pelargonium EO. *V. agnus-castus* EO exhibited no activity against *E. coli* and *L. monocytogenes* but for the other four strains inhibition zones ranged from 50.0 ± 0.0 to 9.0 ± 0.0. *S. aureus* was exposed to be the most sensitive strain for *P. graveolens* and *V. agnus-castus* EO (Ghannadi et al., 2012). Cunha et al. (2013) studied the inhibitory effect of the EO of different parts of *Cassia bakeriana* Craib. (Fabaceae) on aerobic and anaerobic pathogens. The leaves and bark oils showed strong antibacterial activity against both aerobic and anaerobic bacteria (MIC ranging from 62.5 to 125 µg/mL). The EOs of *C. bakeriana* leaves and barks have indicated promising activity against oral pathogens, including *Streptococcus mutans*, the main etiological agent of dental caries. *B. subtilis*, *E. coli*, *Raoultella planticola*, *K. pneumoniae*, *M. luteus*, *Salmonella typhimurium*, two strains of *S. aureus*, *St. mutans*, and *P. aeruginosa* were studied for their sensitivity against fruit EOs of two Cameroonian *Zanthoxylum* species (Rutaceae). The in vitro screening was performed by disc diffusion and MIC was determined. *Zanthoxylum zanthoxyloides* Lam. inhibited growth of 6 of the 10 tested bacterial strains. Inhibition zone diameters ranged from 8 to 12 mm. The most sensitive bacterial strain was Gram-negative *K. pneumonia* (12 mm), whereas other species showed no antibacterial interaction. The EO of *Z. zanthoxyloides* also showed significant antimicrobial activity against Gram-negative *S. typhimurium* and Gram-positive *B. subtilis*. Overall, the EO of *Z. zanthoxyloides* was more active against Gram-positive than Gram-negative bacteria (Misra et al., 2013).

The EO of *Trachyspermum copticum* L., an annual plant of the Iranian flora, and *Satureja hortensis* L. were investigated for their antimicrobial activity against 15 microbial strains. MIC and minimal lethal concentration (MLC) were performed with microbroth dilution assay. The MIC values of the two oils ranged from 0.06 to 8 µL/mL, whereas *K. pneumonia* was the Gram-positive bacteria most sensitive to both oils. *Streptococcus sanguinis*, *Streptococcus salivarius*, and *S. aureus* were more susceptible than *S. flexneri* and *Sh. dysantri*. In general, the antimicrobial activity of *T. copticum* was higher than that of *S. hortensis* oil (Mahboubi and Kazempour, 2011). An experimental in vitro study was carried out to evaluate the antibacterial activity against of *R. officinalis* L., *Origanum syriacum* L., *Thymus syriacus* Boiss, *Salvia palaestina* Beth., *Mentha piperita* L. and *Lavandula stoechas* L. *Brucella melitensis*. The EOs of these plants have been used alone and in combination with some antibiotics. *O. syriacum* and *T. syriacus* demonstrated the highest activity against *B. melitensis*. The MIC of the EOs was 3.125 and 6.25 µL/mL, respectively. The combination of levofloxacin with *T. syriacus* increased the bactericidal activity of the EO (Al-Mariri and Safi, 2013). *O. syriacum*, *Majorana hortensis* Moench, *R. officinalis*, *Cymbopogon citratus* Stapf., *Thymus vulgaris*, and *Artemisia annua* were examined for their antibacterial activity against *Listeria innocua* (CECT 910). The agar disc diffusion method was used for investigation. Pure *C. citratus* EO showed the highest activity against *L. innocua* with diameters of the inhibition zone of 49.00 ± 5.66 mm. Except for *A. annua* all EOs showed antibacterial activity with diameters of inhibition zones ranging from 13.00 ± 0.00 to 40.00 ± 0.00 mm. Even concentrations of 6.25% EOs showed promising antibacterial activity except for *R. officinalis* and *A. annua* (Viuda-Martos et al., 2010). Eugenol is a component of many EOs especially of clove oil, nutmeg, cinnamon, and bay leaf. Devi et al. (2013) found that eugenol acts growth inhibiting and bactericidal on *P. mirabilis*, a nosocomial pathogen. The antibacterial activity of eugenol against *P. mirabilis* was examined by the disc diffusion method. Ciprofloxacin acted as positive control. Eugenol, at concentrations of 1% and 10% (v/v) with zones of inhibition of 8 and 11 mm, respectively, showed promising bactericidal activity. In comparison, ciprofloxacin (500 ng/mL) produced zones of inhibition of 13 mm. Sfeir et al. (2013) found that the EOs of *Cinnamomum verum* J.S. Presl (Lauraceae), *C. citratus* Stapf.

(Poaceae), *T. vulgaris* CT thymol, *Origanum compactum* and *Stagmomantis montana* have high activity against *St. pyogenes* with diameters of inhibition zones ranging from 48.0 to 35.0 mm. *St. pyogenes* was also sensitive against the EO of *Eugenia caryophyllus* (Spreng.) Bullock and S.G. Harrison (Myrtaceae) and *Cymbopogon martinii* (Roxb. Wats.) with zones of inhibition of 18.3 and 15.3 mm, respectively. Moderate activity was found for the EOs of *Cinnamomum camphora* CT linalool, *M. piperita*, *L. stoechas* L., *Melaleuca cajeput*, and *Melaleuca alternifolia* Cheel with diameters of inhibition zones ranging from 13.0 to 9.0 mm (Sfeir et al., 2013).

The antimicrobial activity of cinnamon bark, lavender, marjoram, tea tree, peppermint EOs combined with ampicillin, piperacillin, cefazolin, cefuroxime, carbenicillin, ceftazidime, and meropenem was investigated against β-lactamase-producing *E. coli*. Fractional inhibitory concentration (FIC) values were the results of the trial. It was found that peppermint EO improved the activity of piperacillin against multidrug-resistant *E. coli* strains (FIC: 0.31) compared with piperacillin alone. Also the used dose of meropenem can be reduced when combined with peppermint oil (FIC: 0.26) (Xi Yap et al., 2013).

10.3 ANTIVIRAL ACTIVITY

The inhibition of the replication of herpes simplex virus 1 (HSV-1) was tested with different concentrations of *M. suaveolens* Ehrh. (Lamiaceae) EO and its major constituent piperitenone oxide. African green monkey kidney (VERO) cells were incubated with HSV-1 and afterwards treated with *M. suaveolens* EO and piperitenone oxide. The inhibiting concentration 50% (IC_{50}) values were 5.1 ± 0.48 and 1.4 ± 0.07 µg/mL for the EO and piperitenone oxide, respectively. The most reduction of viral replication of HSV-1 was achieved at a concentration of 30 µg/mL of the EO. At concentrations of up to 10 µg/mL, the viral replication was nearly obliterated. The EO and piperitenone oxide also showed synergistic effects with acyclovir (Civitelli et al., 2014). The EO of *Lippia graveolens* Kunth. and its main compound carvacrol were also investigated against several human and animal viruses. It was shown that the EO had inhibitory effects against human herpes virus 1 (HHV-1), acyclovir-resistant HHV-1, Bovine herpes virus 2, HRSV (human respiratory syncytial virus), and bovine viral diarrhea virus. Carvacrol was also active against these human and animal viruses except for HHV-1. The effective concentration 50% (EC_{50}) was higher for carvacrol than for the EO. This may be due to the synergistic effect of many components in the EO (Pilau et al., 2011). Lai et al. (2012) found that thymol and carvacrol had significant antiviral activities against HSV-1 with an IC_{50} of 7 µM. Ninety percent of HSV-1 was inactivated after the treatment with thymol or carvacrol. Zhu et al. (2012) found that the EO of *Paederia scandens* Lour. (Rubiaceae) showed good inhibitory effects against hepatitis B surface antigen and hepatitis B e-Antigen. The EOs of tea tree, eucalyptus and thyme and their major monoterpenes were investigated for their antiviral effects on HSV-1 too. The EOs and the monoterpenes were able to reduce the viral infection by >96% and >80%, respectively (Astani et al., 2010).

The EO of *Cinnamomi ramulus* L. (Lauraceae) and its main compound cinnamaldehyde were investigated for their anti-influenza virus (A7PR/8/34) activity. Madin–Darby canine kidney cells were infected with influenza. Both the EO and cinnamaldehyde showed virucidal effects on the influenza virus. The IC_{50} values were $1.85 * 10^{-7}$ and $5.77 * 10^{-7}$ g/mL. The virucidal effects are related to the activation of toll-like receptor-7 signaling pathway and interleukin receptor associated kinase-4 (Liu et al., 2013). Also the EO of *Schizonepeta tenuifolia* Briq. (Lamiaceae) and the EO compounds menthone and pulegone were tested for their anti-influenza activity by Tang et al. (2012). The EO and pulegone showed significant anti-influenza effects in the chick embryo test, while all three agents showed antiviral effects on infected mice. Ichihashi and Jimbo (2012) found that patchouli alcohol has antiviral effects on influenza virus A too. Moreover, the EO of *Mosla dianthera* Thunb. (Lamiaceae) was evaluated for its antiviral, anti-influenza activity. Influenza A/PR/8/34 virus (H1N1) infected mice were used for the test. The infected mice were treated with the EO for 5 successive days at doses of 90–360 mg/kg. The EO-treated mice showed significant decrease in lung viral titer it inhibited pneumonia and reduced the level of interferon-γ (IFN-γ) and interleukin-4 (IL-4). Thus, the EO of *M. dianthera* exhibited potent anti-influenza effects (Wu et al., 2012b).

Eryngium alpinum L. and *Eryngium amethystinum* L. (Apiaceae) EO were evaluated for their antiviral activity. It was found that they were able to reduce the infection with mosaic virus, and an associated satellite (Dunkic et al., 2013). Liu et al. (2012b) found that the EO of *Elsholtzia densa* Benth. (Lamiaceae) had antiviral activities. Moreover, the EO of *Lippia citriodora* Paláu, *Lippia alba* (Miller) N.E. Brown (Verbenaceae) and the compounds citral and limonene exhibited antiviral activities against yellow fever virus with IC_{50} values ranging from 4.3 to 25 μg/mL (Gomez et al., 2013). *L. alba* and *L. citroideae* were also investigated for their virucidal activity against dengue virus (DENV). Both *Lippia* species showed significant virucidal effects on the tested virus with IC_{50} values of 1.4–33 μg/mL (*L. alba*) and 1.9–33.7 μg/mL (*L. citroideae*). Among the different dengue virus strains, *L. citroideae* showed stronger activity against DENV-1, 2 und 3. The EO exhibited lower activity against DENV-4 (Ocazionez et al., 2010). Furthermore, porcine epidemic diarrhea virus (PDEV) infected VERO cells were treated with the EO of *Artemisia dubia* Wall. (Asteraceae). It showed inhibiting effects of the replication of PDEV with an IC_{50} value of 43.7 μ^3/mL (Kim, 2012).

Eight species of *Eucalyptus* (Myrtaceae) EO were tested for their antiviral activity. To evaluate the antiviral activity, cells and virus were pretreated with the EO before the cells were infected. The EO of *Eucalyptus bicostata* and *Eucalyptus astringens* showed most significant antiviral effects with IC_{50} values of 0.7–4.8 and 8.4 mg/mL, respectively. *Eucalyptus cinerea* and *Eucalyptus maidennii* also showed moderate antiviral activity with IC_{50} values of 102.131 and 136.5–233.5 mg/mL (Elissi et al., 2012).

Orhan et al. (2012) evaluated the antiviral activity of different Umbelliferae and Lamiaceae plants (*Anethum graveolens* L., *Foeniculum vulgare* Mill., *M. piperita* L., *Mentha spicata* L., *Lavandula officinalis* Chaix ex Villars, *Ocimum basilicum* L., *Origanum onites* L., *O. vulgare* L., *Origanum munitiflorum* Haussкн., *Origanum majorana* L., *R. officinalis* L., *Salvia officinalis* L., and *Satureja cuneifolia* Ten.) All of the EOs showed significant inhibitory effects on HSV-1(representative for DNA virus) at concentrations ranging from 0.8 to 0.012 μg/mL. Against PI-3 (a representative for RNA virus) the EOs exhibited less inhibition activities. *A. graveolens*, *F. vulgare M. piperita*, *M. spicata*, *O. munitiflorum*, *O. vulgaris*, and *S. cuneifolia* were found to be active against this virus too.

The EO of *Thymus capitatus* (L.) Hoffmans and Link (Lamiaceae) was tested for its antiviral effects on HSV-1, ECV 11 (Echovirus 11), and ADV (Adenovirus). HSV-1 and ECV 11 were sensitive to the EO when the virus was pretreated, while ADV was affected after penetration (Salah-Fatnassi et al., 2010). Garcia et al. (2010) evaluated the virucidal activities of several aromatic plants of west Argentina. It was found that the EO of *Lantana grisebachii* (Seckt.) var. *grisebachii* showed good inhibitory effects on HSV-1 and DENV-2 with IC_{50} values of 21.1 and 26.1 ppm, respectively. The EO of *L. grisebachii* was also effective against HSV-2 and acyclovir-resistant herpes viruses.

10.4 ANTINOCICEPTIVE ACTIVITY

Formalin-induced nociception was significantly inhibited by *Piper aleyreanum* C.DC (Piperaceae) EO. At concentrations of 30–100 mg/kg the EO inhibited both phases of formalin licking in rats; however, the antinociceptive effects were more distinct in the second phase. Naloxone, a nonselective opioid antagonist, did not reverse the antinociceptive effect of *P. aleyreanum* EO (Lima et al., 2012b). Other authors found similar results for the EOs of Piperaceae family. Pretreatment with *Peperomia serpens* (Sw.) Loud (Piperaceae) EO (1 h before) caused a significant inhibition of acetic-acid-induced abdominal writhes in mice. The effective concentration 50% (ED_{50}) value was 188.8 mg/kg (Pinheiro et al., 2011). Jeena et al. (2014) found that the EO of *Piper nigrum* L. has antinociceptive effects too. The EO significantly reduced the number of abdominal writhes at concentration of 500 and 1000 mg/kg by 47.94% and 46%, respectively.

Lippia gracilis Schauer (Verbenaceae) EO significantly inhibited acetic-acid-induced writhes in Wistar rats. Concentrations ranging from 50 to 200 mg/kg peroral (PO), 1 h before treatment with acetic acid, reduced symptoms by 17.4%, 31.4%, and 42.8% dose dependently (Mendes et al. 2010).

The EO of *Heracleum persicum* Desf. ex Fisch.(Apiaceae) showed analgesic activity in acetic acid writhes tests and in formalin tests. The EO inhibited the number of acetic-acid-induced writhes at concentrations of 50 and 100 mg/kg by 52% and 65%. Formalin-induced licking was reduced at concentrations of 50 and 100 mg/kg by 12% and 24%, respectively (Hajhashemi et al., 2009). The same author group (2011) also found that *Bunium persicum* (Boiss) B. Fedtsh. (Apiaceae) fruit EO (100–400 μL/kg, PO) showed good antinociceptive activity. The EO significantly reduced acetic-acid-induced writhing as well as pain response of formalin tests in both phases. Two species endemic on Balkans were evaluated for their antinociceptive effects in a rat model: *Laserpitium zerny* L. and *Laserpitium achridanum* L. (Apiaceae). Both EOs showed significant antinociceptive effects (Popovic et al., 2014). *Ligusticum porteri* Coulter and Rose (Apiaceae) EO was also found to have significant antinociceptive activity. It showed potent antinociceptive effects in the writhing test. The main compounds, Z-ligustilide and Z-3-butylidenephthalide, exhibited antinociceptive effects in the hot plate test, and diligustilide showed a potent reduction in pain in both tests (Juarez-Reyes et al., 2014).

Zingiber zerumbet Smith (Zingiberaceae) EO was investigated for its antinociceptive properties using the formalin test. The results showed that *Z. zerumbet* EO had distinctive analgesic activity. Formalin-induced pain and discomfort were reduced by the EO in both phases. Concentrations of 100 mg/kg produced an antinociceptive activity similar to 100 mg/kg acetyl salicylic acid (Zakaria et al., 2011). *Z. zerumbet* EO was also evaluated for its antinociceptive effects using the acetic acid abdominal writhes test, capsaicin-induced nociception, glutamate-induced nociception, and phorbol-12-myristate-13-acetate-induced nociception. The EO showed significant reduction of abdominal writhes at doses of 50, 100, 200, and 300 mg/kg with a percentage of 23.02, 53.89, 83.63, and 98.57, respectively. In the capsaicin-induced nociception test, the EO also showed significant inhibition of pain in a dose-dependent manner (Khalid et al., 2011). Sulaiman et al. (2010) also found that the EO of *Z. zerumbet* has antinociceptive effects. It significantly increased the latency of paw removing in the hot plate test. It also produced significant inhibition of acetic-acid-induced writhing at concentrations of 30, 100, and 300 mg/kg. The effects are the same as those of acetyl salicylic acid (100 mg/kg). Formalin-induced paw licking was reduced in both phases too. Other species of the Zingiberaceae family showed similar effects. So Liju et al. (2011) found that *Curcuma longa* L. (Zingiberaceae) EO (turmeric oil) showed significant antinociceptive activities using the acetic-acid-induced writhing movement test in mice.

Carrageenan and tumor necrosis factor-α (TNF-α) induced mechanical hypernociception in mice. These animals were treated with α-terpineol. Results demonstrated antinociceptive properties. Pretreatment with α-terpineol at concentrations of 25, 50, and 100 mg/kg intraperitoneal (IP) inhibited mechanical hypernociception. Also prostaglandine E_2 (PGE_2) and dopamine-induced hypernociception were significantly reduced by α-terpineol at concentrations of 25, 50, and 100 mg/kg (de Oliveira et al., 2012). As well the antinociceptive effects of α-bisabolol, another monoterpene, were investigated using formalin test and acetic-acid-induced writhes. In the second phase of formalin-induced paw licking, α-bisabolol showed a significant reduction. It also exhibited less writhing induced by acetic acid by 54.68% and 64.29% at concentrations of 25 and 50 mg/kg PO, respectively. The positive control indomethacin (10 mg/kg) showed 48.52% of writhing reduction (Rocha et al., 2011). Furthermore, visceral pain was induced by mustard oil. Here too, α-bisabolol significantly reduced nociceptive behavior (Leite et al., 2011). Other monoterpenes showed similar effects. For example, the monoterpene α-phellandrene was evaluated for its antinociceptive effects. In the formalin test, α-phellandrene, at a concentration of 50 mg/kg, showed inhibition effects of both phases. It also reduced carrageenan-induced hyperalgesia at a dose of 50 mg/kg (Lima et al., 2012a). (R)-(+)-Pulegone was found to have antinociceptive effect too. It showed significant reduction of nociception in both phases of the formalin test and in the hot plate test (de Sousa et al., 2011).

(−)-Linalool, a monoterpene alcohol that is present in many EOs, was evaluated by Batista et al. (2010) for its antinociceptive effects using a neuropathic pain model. It significantly decreased the paw withdrawal response, and so it showed potent antinociceptive effects.

The analgesic effect of the EO of *C. citratus* Stapf. and *Eucalyptus citriodora* (Hook.) K.D. Hill and L.A.S. Johnson (Myrtaceae) was tested using tail immersion test. The tails of Wistar rats were kept in hot water (50°C). Animals treated with EOs were able to keep their tails longer in the hot water bath than untreated rats. The results showed that *E. citriodora* EO was more effective than the EO of *C. citratus* (Gbenou et al., 2013). The EO of *Pimenta pseudocaryophyllus* (Gomes) L.R. Landrum (Myrtaceae) at concentrations of 60, 200, and 600 mg/kg exhibited potent antinociceptive effects in the acetic acid writhing test (de Paula et al., 2012)

Carvacrol was investigated for its antinociceptive effects using carrageenan-induced mouse paw mechanical hypernociception. At doses of 50 and 100 mg/kg, applied 30 min before carrageenan injection, carvacrol significantly inhibited the hypernociceptive response. Carvacrol at these doses had similar effects as indomethacin (10 mg/kg) (Guimaraes et al., 2012b). Furthermore, male mice were pretreated with carvacrol (IP) at doses of 25, 50, and 100 mg/kg. IP treatment with carvacrol reduces pain behavior in both phases of the formalin test. It also had significant antinociceptive effects in the capsaicin and glutamate test at all concentrations (Guimaraes et al., 2012). Cavalcante et al. (2012) also found that carvacrol at doses of 50 and 100 mg/kg showed high reduction of nociception in the acetic acid writhing test and the hot plate test. The antinociceptive effects of carvacrol were not reversible by naloxone or L-arginine.

The EO of *Illicium lanceolatum* A.C. Smith (Illiciaceae) showed distinct antinociceptive properties in acetic acid writhing tests. The EO was able to reduce writhing responses at ratios of 26.6%, 31.73%, and 35.9%, respectively. The highest ratio was achieved at a concentration of 0.8 g/kg (Liang et al., 2012).

Many new research results concerning the antinociceptive activity were found within the Lamiaceae family. Tail-flicking tests showed significant acute pain reduction at the treatment of *Nepeta crispa* Willd. (Lamiaceae) EO (100 and 200 mg/kg). In the first phase of the formalin test, 30, 100, and 200 mg/kg EO exhibited strong reduction of licking duration as well as it did in the second phase except for 30 mg/kg, which did not show any effect in the second phase of formalin test. According to these results, authors suggest that *N. crispa* EO has central and peripheral antinociceptive effects (Ali et al., 2012). *Ocimum micranthum* Willd. EO, another species of the Lamiaceae family, was studied for its antinociceptive activity too. The EO (15–100 mg/kg PO) showed significant reduction in writhing response and the nociceptive response induced by acetic acid. Licking time induced by formalin was also significantly reduced (de Pinho et al., 2012). Moreover, *Ocimum gratissimum* L. (Lamiaceae) EO was tested for its antinociceptive effects in mice. In the hot plate test, the EO was able to increase the latency of paw withdrawal at doses of 40 mg/kg. Eugenol (5 and 100 mg/kg), the main compound of the EO, also showed significant reduction of nociception in the hot plate test. The EO at doses of 40 mg/kg also exhibited reducing of paw licking time in the first and second phase of nociception (Paula-Freire et al., 2013). Basil EO (*O. basilicum* L. Lamiaceae) was found to have significant antinociceptive effects. The EO was able to decrease acetic-acid-induced abdominal writhes by 48%–78%, dose dependently. In the hot plate test, the latency of paw removing was also increased by the EO. In the formalin-licking test, the EO showed significant reduction of licking in both phases (Venancio et al., 2011). *Teucrium stocksianum* Boiss. (Lamiaceae) EO also showed significant antinociceptive effects. The inhibition of writhes was 93% at a concentration of 80 mg/kg (Shah et al., 2012). Moreover, the antinociceptive activity of *Hyptis pectinata* (L.) Poit (Lamiaceae) was evaluated using acetic acid, formalin, and heat as pain stimuli. The EO at concentrations of 10, 30, and 100 mg/kg demonstrated reduced time, and the animals spend licking their formalin-injected paw (Raymundo et al., 2011). Acetic-acid-induced writhing tests showed that *S. hortensis* L. (Lamiaceae) EO significantly reduced abdominal writhes at concentrations of 100 and 200 μL/kg. The formalin test showed inhibition of licking behavior only in the chronic phase (Hajhashemi et al., 2010).

Biological Activities of Essential Oils

The bark EO of *Vanillosmopsis arborea* Baker (Asteraceae) exhibited potent antinociceptive activities. The topical pretreatment with the EO at concentrations of 25, 50, 100, and 200 mg/kg decreased the symptoms of both phases of the formalin test. It also decreased the number of eye wipes induced by 5 M NaCl solution, topically applied. The oral application of the EO also reduced the number of eye wipes (Inocencio et al., 2014). Maham et al. (2014) evaluated the antinociceptive effects of the EO of *Artemisia dracunculus* L. (Asteraceae). It showed significant reduction of pain on both phases of the formalin test by 91.4% and 83.3%, respectively. Also in the hot plate test, the EO showed good analgesic activity. The EO of *Centipeda minima* (L.) A. Braun and Asch. (Asteraceae) was also tested for its antinociceptive effects using the hot plate test. The EO showed significant reduction of pain induced by hot plate (Zhang et al., 2013).

Citronellal is a main compound of the EO of *Cymbopogon winterianus* J. (Poaceae). This agent showed significant antinociceptive activity in carrageenan, TNF-α, PGE_2 or dopamine-induced Swiss mice. Citronellal inhibited mechanical nociception at all doses. The effects of citronellal were even higher than those of L-NAME and glibenclamide (de Santana et al., 2013). Citronellal was also investigated by Quintans-Junior et al. (2011). It was found that citronellal showed significant reduction of face-rubbing behavior induced by capsaicin and glutamate in a dose-dependent manner. Brito et al. (2013) also found that citronellol showed significant antinociceptive effects. Swiss mice were pretreated with citronellol at concentrations of 25, 50, and 100 mg/kg IP The EO compound showed significant reduction in face-rubbing behavior induced by capsaicin at all doses. The pain was decreased in both phases of nociception.

The EO of *Croton adamantinus* Müll. Arg. (Euphorbiaceae) showed mild antinociceptive activity in the first phase of the formalin-licking test and even higher effects than morphine on the second phase of this test at concentrations of 50 and 100 mg/kg. Also, the number of abdominal contortions was decreased by the EO at both doses (Ximens et al., 2013).

The analgesic effect of *Croton tiglium* L. (Euphorbiaceae) was investigated by Liu et al. (2011). At doses of 25, 50, and 100 mg/kg, the EO showed significant reduction of acetic-acid-induced writhing. The results exhibited that the EO had low analgesic effects compared to aspirin. *Citrus lemon* (L.) Burm (Rutaceae) EO showed significant antinociceptive effects in acetic acid writhing tests and the formalin test. The acetic-acid-induced writhing was significantly reduced at EO doses of 50, 100, and 150 mg/kg. Licking response after treatment with formalin was also decreased by the EO in a dose-dependent manner (Campelo et al., 2011).

10.5 ANTI-INFLAMMATORY ACTIVITY

In a rat study, Mueller et al. (2013) found that turmeric oil, thyme oil, and rosemary oil acted as anti-inflammatory substances. In the trial, mild colitis was induced by 4% dextran sulfate sodium (DSS) in 6 weeks old rats. The animals were feed with 1494 mg/kg *C. longa* L. (Zingiberaceae) oil (turmeric oil), 618 mg/kg *T. vulgaris* L. (Lamiaceae) oil, or 680 mg/kg *R. officinalis* L. (Lamiaceae) oil. These EO food additives possess anti-inflammatory properties, because they reduce proinflammatory mediators, which are increased after the DSS treatment, like nuclear factor kappa B (NFκB), vascular cell adhesion molecule 1 (VCAM-1), monocyte chemotactic protein-1 (MCP-1), and cyclo-oxygenase-2 (COX-2). They also reduce proinflammatory adhesions molecules (like VCAM-1 and Clnd3) in non-DSS-treated animals. which lead to an improved gut barrier. Of all three EOs, rosemary oil showed the highest activity (Mueller et al., 2013). Turmeric Oil was also found to block SCW-induced (peptidoglycan–polysaccharide polymer isolated from *St. Pyogenes*) leukocytosis and the granulomatous inflammatory response that occurs at sites of hepatic SCW deposition. The anti-inflammatory effect of the EO was dose dependent and was only present at doses ≥28 mg/kg/day. While IP doses adjustable in therapeutic height were associated with extremely high mortality, oral application of doses at 560 mg/kg/day showed no increased mortality in the test animals. The conclusion was drawn that oral application of turmeric oil had promising anti-inflammatory effects with no toxic risks (Funk et al., 2010). Moreover, turmeric oil showed significant inhibitory effects in rat paw thickness induced by carrageenan or dextran too. Formalin acute inflammation was also decreased when treated with the EO (Liju et al., 2011).

Z. zerumbet Smith. (Zingiberaceae) at doses of 30, 100, and 300 mg/kg significantly reduced carrageenan-induced paw edema in rats dose dependently. The cotton pellet–induced granuloma test also showed a significant anti-inflammatory activity of *Z. zerumbet* EO in a dose-dependent manner (Zakaria et al., 2011). Furthermore, *Zingiber officinale* Roscoe (Zingiberaceae) was found to have high anti-inflammatory activity. Lipoxygenase (LOX) was inhibited by 50.9% at a concentration of 0.4 mg/mL. At a dose of 8 mg/mL LOX inhibition was complete (by 100%). *O. basilicum* L. (Lamiaceae), *Eucalyptus camaldulensis* Dehnhardt (Myrtaceae), *Lippia multiflora* L. (Verbenaceae), *Hyptis spicigera* Lamarck (Lamiaceae), *Ageratum conyzoides* L. (Asteraceae). *Ocimum americanum* L. (Lamiaceae) also showed inhibitory effects at a concentration of 8 mg/mL by 98.2%, 96.4%, 96.4%, 75.1%, 48.3%, and 31.6%, respectively. These EOs showed no significant inhibitory effects at a concentration of 4 mg/mL (Bayala et al., 2014).

Cupressus sempervirens L. var. *horizontalis*, *C. sempervirens* var. *pyramidalis*, *Juniperus communis* L., *Juniperus excelsa* M. B., *Juniperus foetidissima* Wild., *Juniperus oxycedrus* L., and *Juniperus phoenicea* L. (all Cupressaceae) were evaluated for their potential anti-inflammatory activity in rats. Increased capillary permeability was induced by acetic acid in mice. *J. oxycedrus* ssp.*oxycedrus* and *J. phoenicea* were found to be potential anti-inflammatory substances at a dose of 200 mg/kg with highest inhibitory values of 33.1% and 27.3%, respectively. At a concentration of 100 mg/kg, both species showed no significant anti-inflammatory activity (Tumen et al., 2012). The EO of *Juniperus virginiana* L. and *Juniperus occidentalis* Hook. exerted anti-inflammatory effects at a concentration of 200 mg/kg using the Whittle method (Tumen et al., 2013) as well. Moreover, *Juniperus macrocarpa* Sibth. et Sm. EO was found to inhibit the activity of cyclooxygenase-1 (COX-1) and LOX-12 as well as aspirin® and quercetin, well known as potent COX-1 and LOX-12 inhibitors (Lesjak et al., 2014).

S. cumini Skeels and *Psidium guajava* L. (Myrtaceae) were found to inhibit effectively the migration of eosinophils- and neutrophil-stimulated lipopolysaccharides (LPS). Therefore, these EOs were confirmed to be useful for the treatment of inflammatory diseases (Siani et al., 2013). *M. alternifolia* Cheel another species of the Myrtaceae family was found to have potent anti-inflammatory properties. The major compound of the EO is terpinen-4-ol, which exhibits anti-inflammatory activity too. It decreases the production of TNF-α, interleukin 1 (IL-1), IL-8, and prostaglandine E_2 (PGE2), which are potent proinflammatory agents (Pazyar and Yaghoobi, 2012). *Cleistocalyx operculatus* (Roxb.) Merr and Perry (Myrtaceae) EO was also investigated for its anti-inflammatory activity. LPS-induced production of TNF-α and IL-1β in RAW 264.7 cells (mouse leukemic monocyte macrophage cell line) was significantly inhibited by *C. operculatus* EO. Treatment with 10 and 20 µg/mL caused 33% and 45% inhibition of TNF-α production, respectively. These effects are attributable to the suppression of LPS-induced messenger RNA expression of TNF-α and IL-1β. These results showed that the reduction of proinflammatory agents occurs mainly at transcriptional level. EO treatment also decreased LPS-induced NFκB activity in RAW 264.7 cells. The in vivo tests exhibited potent anti-inflammatory effects as well. Therapy of inflamed mouse ears, pretreated with 12-*O*-tetradecanoylphorbol-13-acetate (TPA) with *C. operculatus* EO completely blocked skin inflammation. EO at 0.1% and 1% was more effective than 1% hydrocortisone (Dung et al., 2009). In vivo tests with carrageenan-induced paw edema in rats showed that *Myrciaria tenella* (DC.) Berg (Myrtaceae) EO (50 mg/kg) significantly reduced the symptoms like indomethacin (10 mg/kg). In vitro tests proved that *Calycorectes sellowianus* O. Berg and *M. tenella* EO (Myrtaceae) reduced leucocytes migration of the inflamed tissue (Apel et al., 2010). Furthermore, *Myrtus communis* L. (Myrtaceae) EO exhibited an effective decrease of croton oil-induced ear edema in rats as well as myeloperoxidase activity. Moreover, the oil blocked cotton pellet-induced granuloma and the level of TNF-α and IL-6 in serum (Maxia et al., 2011). El-Ahmady et al. (2013) also found that the leaf and fruit EO of *P. guajava* L. (Myrtaceae) inhibited the LOX-5 with IC_{50} values of 32.53 and 49.76 µg/mL, respectively.

Linalool, a major compound in several EOs of aromatic plants, was found to have anti-inflammatory activity. Tests showed inhibition of LPS induction of IL-6, TNF-α, and NFκB. Also, mitogen-activated protein (MAP) kinase pathway was blocked by linalool. The in vivo study also exhibited significant reduction of IL-6 and TNF-α as well as a significant decrease of inflammatory cells like neutrophils and macrophages. Histological changes in infected lung tissue could be alleviated with linalool treatment (Huo et al., 2013).

Moreover, many anti-inflammatory effects were found within the Lauraceae family. The EO of *Ocotea quixos* Lam. (Lauraceae) and its main constituent trans-cinnamaldehyde showed promising anti-inflammatory activity. They inhibited NO release from J774 macrophages in a dose-dependent manner. The EO also increased the level of cyclic adenosine monophosphate (cAMP) in Forskolin-stimulated SK-N-MC cells. Moreover, *Ocotea* EO (1–10 μg/mL) reduced LPS-induced COX-2 expression, dose dependently. In vivo tests showed significant reduction of carrageenan-induced swelling in rat paws. Maximum inhibition (about 20% compared to vehicle treated rats) was achieved with *Ocotea* EO at concentrations of 30 and 10 mg/kg of trans-cinnamaldehyde, respectively (Ballabeni et al., 2010). The anti-inflammatory potency of *Neolitsea sericea* (Blume) Koidz. (Lauraceae) leaf oil exhibited excellent dose-dependent inhibitory activities on nitric oxide (NO), PGE_2, TNF-α, IL-1β, and IL-6, which were LPS induced in RAW 264.7 macrophages. COX-2 and MAP kinase pathway as well as NFκB in association of IκB-α were suppressed by the EO (Yoon et al., 2010a). *Lindera umbellata* Thunberg (Lauraceae) EO and its main compound linalool were tested for their inhibitory activity on LPS-induced NO production. Cells treated with *L. umbellata* EO and linalool (25, 50 μg/mL) significantly reduced the NO production. The EO also showed decrease of IL-6 mRNA expression. Linalool (50 μg/mL) suppressed TNF-α mRNA expression. Furthermore, NO synthetase (iNOS) and COX-2 mRNA expression was diminished (Maeda et al., 2013). The leaf-EO of *Cinnamomum osmophloeum* Kaneh. (Lauraceae) and its main compounds were found to have excellent anti-inflammatory activities. The leaf oil strongly inhibited NO production, with inhibiting concentration 50% (IC_{50}) ranging from 9.7–15.5 μg/mL (Tung et al., 2010).

Anti-inflammatory activity of *Zizyphus jujuba* Meikl (Rhamnaceae) seeds EO was investigated. The authors used the TPA-induced skin inflammation model. Topical application of *Z. jujuba* oil (1% and 10%) definitely inhibited the increase of ear thickness, which was measured to be 0.30 ± 0.02 and 0.35 ± 0.01 mm. Therefore, EO treatment showed improvement of 51% and 53%, respectively. Hydrocortisone, the positive control, only caused 39% of decreased ear thickness (Al-Reza et al., 2010).

Artemisia fukudo Makino (Asteraceae) EO was evaluated for its anti-inflammatory activity. It exhibited dose-dependent inhibition of the production of NO and PGE2. The results indicate that the inhibitory effects are dependent on the suppression of iNOS and COX-2. The EO also suppressed the production of TNF-α, IL-1β, and IL-6. LPS-induced NFκB and IκB-α production was also reduced by *A. fukudo* EO (Yoon et al., 2010b)

Abies koreana Wilson (Pinaceae) EO was investigated for its anti-inflammatory activity using LPS-induced RAW 264.7 macrophages. The EO of *A. koreana* at doses of 12.5, 25, and 50 μg/mL showed dose-dependent inhibition of LPS-induced NO production. LPS-induced COX-2 activity was also suppressed by the EO, and consequently, the PGE_2 concentration was reduced. The results exhibited promising anti-inflammatory properties of the EO of *A. koreana* (Yoon et al., 2009).

Five selected herbs were investigated for their anti-inflammatory activity. *T. vulgaris* EO showed stronger inflammatory inhibition than the positive control (tea tree oil, α-bisabolol). In the lipoxygenase-5 (LOX-5) inhibition assay as well, *Lindernia anagallis* (Burm.f.) Pennell (Linderniaceae) and *Pelargonium fragrans* Willd. showed anti-inflammatory activity (Tsai et al., 2011). *T. vulgaris* EO is rich in thymol and carvacrol. The EO and both constituents were tested for their anti-inflammatory activity using a pleurisy model. *T. vulgaris* EO (250, 500, and 750 mg/kg) reduced inflammatory response. At a concentration of 750 mg/kg, the EO also reduced the number of cell migration. Thymol and carvacrol (400 mg/kg) reduced the volume of pleural exudates by 47.3% and 34.2% as well as cell migration. Carvacrol also suppressed COX-2. The efficacy of the topical application of thymol and carvacrol was evaluated using croton oil-induced ear edema. Croton oil induces increased weight of the ears. Carvacrol at doses of 10 mg/ear reduced ear edema similar to dexamethasone (Fachini-Queiroz et al., 2012). Carvacrol was found to decrease significantly paw edema induced by histamine or dextran. The anti-inflammatory activity only occurred at a concentration of 50 mg/kg (46% and 35%). Paw edema induced by substance P was reduced by 100 mg/kg carvacrol for 46%. TPA and arachidonic-acid-induced edema were also significantly reduced (Silva et al., 2012). α-Terpineol is a monoterpene and a compound of many EOs. Pretreatment with α-terpineol reduced the number of neutrophils in CG-induced mouse pleurisy. NO production in LPS-induced murine macrophages was also significantly reduced, when treated with α-terpineol (1, 10, and 100 μg/mL) (de Oliveira et al., 2012). Carvacrol is a phenolic monoterpene and a main compound of EOs of *Origanum* and *Thymus* species. Carrageenan-induced paw edema were significantly reduced by carvacrol at concentrations of 50 and 100 mg/kg, as did dexamethasone, the positive control. Carvacrol inhibited paw edema by 49.7%, 65.3%, and 76.7% for 25, 50, and 100 mg/kg, respectively. Leukocyte migration to mouse pleural cavity, dependent on intrapleural carrageenan injection, was also inhibited by carvacrol. Moreover, the production of TNF-α was significantly reduced by carvacrol at doses of 25, 50, and 100 mg/kg. LPS-induced NO production by murine macrophages was inhibited by carvacrol (1, 10, and 100 μg/mL) too (Guimaraes et al., 2012). Thyme oil (*T. vulgaris*) also showed strong inhibition of LPS-induced COX-2 promoter activity in a dose-dependent manner (Hotta et al., 2010). Thymol, a compound of the EO of *L. gracilis* Schauer (Verbenaceae) was evaluated for its anti-inflammatory effects. It was found that thymol reduced carrageenan-induced paw edema in rats at a concentration of 100 mg/kg. Furthermore, thymol reduced symptoms of carrageenan-induced peritonitis in rats at concentrations of 10, 30, and 100 mg/kg as well as dexamethasone. The thymol-treated rats showed an improvement in histological analysis of rat paws, for histological changes were limited to the superficial areas (Riella et al., 2012). Mendes et al. (2010) also investigated the anti-inflammatory effects of *L. gracilis* EO. They found that concentrations of 200 and 300 mg/kg of the EO were able to reduce rat paw edema induced by carrageenan. *L. gracilis* EO also inhibited leukocyte migration into peritoneal cavern after intraperitoneal injection of carrageenan. Croton oil–, arachidonic acid–, capsaicin-, phenol-, and histamine-induced edema were treated with the EO of *Lippia sidoides* Cham. The EO decreased the formation of the croton oil–induced ear edema at a concentration of 2 mg/ear twice daily for 4 days. Also edema induced by arachidonic acid was reduced by application of *L. sidoides* for 45% (Veras et al., 2013).

Candida-infected tongues were treated with terpinen-4-ol to evaluate the anti-inflammatory activity of this agent. Terpinen-4-ol is the main constituent of tea tree oil. Tongue treatment with terpinen-4-ol at a concentration of 10 mg/mL significantly reduced signs of inflammation. The addition of terpinen-4-ol to macrophage monolayer, which was cultivated with *Candida albicans*, significantly suppressed the production of TNF-α dose dependently. IC_{50} of terpinen-4-ol was 800 μg/mL (Ninomiya et al., 2013).

A collagen-induced arthritis (CIA) model showed anti-inflammatory effects of eugenol. Arthritic mice showed lower clinical scores of CIA when treated with eugenol than treated with the vehicle. Eugenol also reduced leukocyte recruitment to the knee joints of arthritic mice and reduced TNF-α, IFN-γ and transforming growth factor-β (TGF-β) in paw samples of arthritic mice (Grespan et al., 2012). *Syzygium aromaticum* EO (clove oil) (Myrtaceae) and its main component eugenol exhibited good anti-inflammatory properties. Clove oil (100 μg/well) inhibited production of IL-1β, IL-6, and IL-10. Moreover, it did not make any difference whether clove oil was administered either before or after LPS induction. Eugenol also inhibited IL-6 and IL-10 but only for IL-10 it did not matter whether eugenol was administered before or after LPS induction (Bachiega et al., 2012). Lemongrass oil was found to suppress LPS-induced COX-2 promoter activity at concentrations of 0.002% and 0.004%–20% and 58%, respectively. Moreover, citral, the main compound of lemongrass oil, showed LPS-induced COX-2 mRNA and COX-2 protein suppressing activity. However, the suppressing effect of LPS-induced COX-2 protein was more potent than the suppressing effect on LPS-induced COX-2 mRNA (Katsukawa et al., 2010). Anethole is the main constituent of star anise EO. It was found to have promising anti-inflammatory effects in LPS-induced acute lung injury in mice. At a concentration of 250 mg/kg anethole decreased the number of inflamed cells, including neutrophils and macrophages. The production of inflammatory mediators (TNF-α, matrix metallopeptidase 9 [MMP-9] and NO) was also reduced by anethole. They also suppressed the activation of NFκB by blocking the IκB-α-degradation (Kang et al., 2013). Regarding anethole, other authors found similar effects with other major compounds of the EO of *Illicium verum* Hook. f. (Illiciaceae). Ear edema induced by croton oil was used as inflammatory model. Anethole (250 and 400 mg/kg PO) significantly reduced the repercussions of topical application of croton oil. Carrageenan-induced pleurisy was also inhibited by anethole. Moreover, the EO compounds decreased the number of leukocytes at—concentrations of—250 and 500 mg/kg—by 48.78%. NO levels were also significantly reduced (Domiciano et al., 2013).

P. aleyreanum C.DC (Piperaceae) EO showed potent anti-inflammatory activity. The intrapleural injection of carrageenan induced acute inflammation, which was significantly decreased when the animals were treated with *P. aleyreanum* EO 1 h prior. The number of neutrophils and mononuclear cells was reduced at EO concentration of 53.6, 21.7, and 43.5 mg/kg of 54% ± 13%, 66% ± 10%, and 60% ± 8%, respectively (Lima et al., 2012b). *P. serpens* (Sw.) Loud (Piperaceae) showed significant inhibition of carrageenan-induced paw edema in rats when applied 1 h before injection of carrageenan as well as indomethacin. Also, dextran-induced edema in rat paws were inhibited with treatment of the EO 1 h before the dextran injection. Pretreatment with *P. serpens* EO (188.8 mg/kg PO) inhibited carrageenan-induced migration of leucocytes and neutrophils by 58.5% and 63.1%. (Pinheiro et al., 2011). The EO of *P. nigrum* L. (Piperaceae) was evaluated for its anti-inflammatory effects. Carrageenan-induced abdominal writhes were reduced significantly by 72% at a concentration of 500 mg/kg. Acute inflammation induced by dextran was also reduced by the EO at concentrations of 100, 500, and 1000 mg/kg by 33.3%, 53.3%, and 73.4%, respectively. Chronic inflammation, induced by formalin was even more reduced than acute inflammation (Jeena et al., 2014).

Pistacia lentiscus L. (Anacardiaceae) EO was evaluated for its anti-inflammatory effects on carrageenan-induced rat paw edema and cotton pellet–induced granuloma. The oil significantly reduced the production of TNF-α and IL-6 as well as the cotton pellet–induced granuloma (Maxia et al., 2010).

The anti-inflammatory effects of *Echinacea purpurea* L. (Asteraceae) were investigated by using xylene-induced mouse ear edema, egg-white-induced rat paw edema, and cotton-induced granuloma tissue proliferating in mice. Low, medium, and high dose of *E. purpurea* EO showed inhibition of 39.24%, 47.22%, and 44.79% in xylene-induced ear edema, respectively. The high dose of *E. purpurea* EO also showed significant inhibition of paw edema (48.51%) and cotton-induced granuloma (28.52%). Moreover the proinflammatory mediators such as IL-2, IL-6, and TNF-α were reduced in the treated group (Yu et al., 2013). *Artemisia capillaris* Thunb. (Asteraceae) EO significantly inhibited the secretion on NO in LPS-stimulated RAW 264.7 macrophages. It also suppressed the production of PGE2 and the expression of COX-2 in these cells. The anti-inflammatory effects may be correlated to the blockade of MAP kinase pathway and to the inhibition of NFκB (Cha et al., 2009a). The EO of *Anthemis wiedemanniana* Fisch. and C.A. Meyer (Asteraceae) showed inhibition effects of NO production in RAW 264.7 macrophages. These macrophages were pretreated with LPS, which induced NO production. At a concentration of 200 µg/mL, the EO exhibited an inhibitory effect of 93%. The IC_{50} value was more significant than the IC_{50} of the reference drug indomethacin (Conforti et al., 2012). The main constituents of the volatile oil of *C. minima* (L.) A. Braun and Asch. (Asteraceae) showed significant suppressing effects on rat's paw edema induced by fresh egg white and ear edema induced by xylene (Zhang et al., 2013).

The EO of *Chamaecyparis obtusa* (Siebold and Zucc.) Endl. (Cupressaceae) showed promising decrease of the production of LPS-induced PGE-2 and peripheral blood mononuclear cells. The expression of TNF-α and COX-2 was also inhibited. These results showed promising anti-inflammatory activity (An et al., 2013).

Leaf EO of *C. winterianus* J. (Poaceae) reduced carrageenan-induced neutrophil migration to the peritoneal cavity. The reduction of this inflammatory symptom was dose dependent with 35.5%, 42.8%, and 66.1% at concentrations of 50, 100, and 200 mg/kg, respectively (Leite et al., 2012). The anti-inflammatory activity of the EO of *C. citratus* Stapf., and *E. citriodora* (Hook.) K.D. Hill and L.A.S. Johnson (Myrtaceae) was evaluated. *C. citratus* EO was found to have a reducing effect on formol-induced edema dose dependently (Gbenou et al., 2013). Myrtol a constituent of eucalyptus EO was tested for its anti-inflammatory activity on alveolar macrophages from patients with chronic obstructive pulmonary disease. These cells were precultured with eucalyptus oil and myrtol for 1 h. Afterward, they were stimulated with LPS. Myrtol showed 37.3% reduction of TNF-α in a 0.0001% dilution. It also decreased the quantity of IL-8 and granulocyte macrophage colony-stimulating factor by 12.3% and 35.8%, respectively (Rantzsch et al., 2009).

Pterodon polygalaeflorus (Benth.) (Fabaceae) EO was investigated for its anti-inflammatory effects with the air pouch model. The EO reduced acute inflammatory symptoms and inhibited lymphocyte proliferation. It also diminished neutrophil migration to the inflamed tissue (Velozo et al., 2013).

The injection of 1 mL carrageenan into pouches of mice induced increased exudate volume. Pretreatment with *H. pectinata* (L.) Poit (Lamiaceae) EO significantly reduced these symptoms. Furthermore, *H. pectinata* EO reduced IL-6, TNF, PGE2, and NO concentration in the exudates (Raymundo et al., 2011). Supercritical fluid extraction of oregano (*O. vulgare* L. Lamiaceae) EO (S1 and S2) was also evaluated for its anti-inflammatory properties by using oxidated low-density lipoprotein-activated THP-1 macrophages. These cells showed increased release of cytokine, such as TNF-α, IL-1β, IL-6, and IL-10. Treatment with S1 and S2 showed decrease of all proinflammatory cytokines. This happened in a dose-dependent manner. Tests exhibited that the reduced TNF-α production is attributed to the gene expression (Ocana-Fuentes et al., 2010). The anti-inflammatory properties of *O. americanum* L. (Lamiaceae) EO were investigated using zymosan-induced arthritis. The EO suppressed leukocyte migration and reduced paw edema induced by zymosan. Also, IF-γ and TGF-β levels were decreased by the EO (Yamada et al., 2013). Paw edema were induced by formalin and treated with *N. crispa* Willd. (Lamiaceae) EO. The paw edema was significantly reduced by 69%, 74%, and 43% at concentration of 30, 100, and 200 mg/kg (Ali et al., 2012). Different concentrations of *R. officinalis* L. (Lamiaceae) EO were tested by in vitro chemotaxis assay and on in vivo leukocyte migration. In vitro and in vivo tests showed significant reduction of leukocyte migration (de Melo et al., 2011). The anti-inflammatory effect of the EO of *Nepeta glomerata* Montbret et Aucher ex Bentham (Lamiaceae) was investigated. LPS-induced RAW 264.7 macrophages were treated with different concentrations of the EO. LPS-induced NO production of the macrophages was decreased by the EO. The IC_{50} value was 78.1 μg/mL (Rigano et al., 2011). Neves et al. (2010) evaluated the anti-inflammatory effects of *M. x piperita* L. (Lamiaceae) *Origanum virens* L. (Lamiaceae), *Lavandula luisieri* L. (Lamiaceae) and *J. oxycedrus* L. ssp. oxycedrus (Cupressaceae). Four of these five EO showed reduction of IL-1 induced NO production. The greatest effect was obtained with *J. oxycedrus* EO with inhibition of 80% ± 8% at a concentration of 0.02% (V/V).

Distacco Selinum tenuifolium (Lag.) Garcia Martin and Silvestre (Apiaceae Umbelliferae) showed inhibition effects in LPS-induced NO production. At concentrations of 0.64 and 1.25 μL/kg the EO reduced NO production after LPS stimulation by 117.7% ± 3.7% and 125.6% ± 3.2%, respectively (Tavares et al., 2010). Oral administration of *H. persicum* Desf. ex Fisch. (Apiaceae Umbelliferae) EO at concentrations of 100 and 200 mg/kg reduced carrageenan-induced paw edema by 33% and 48%, respectively (Hajhashemi et al., 2009). *S. hortensis* EO was investigated for its anti-inflammatory activities by using carrageenan-induced paw edema. The EO (400 μL/kg) significantly inhibited the carrageenan-induced paw edema (Hajhashemi et al., 2010). *B. persicum* (Boiss) B. Fedtsh. (Apiaceae) fruit EO significantly reduced the inflammatory response of carrageenan-induced edema. Also, croton oil–induced ear edema was inhibited of *B. persicum* EO (Hajhashemi et al., 2011). Popovic et al. (2014) evaluated the anti-inflammatory and antiedematous activities of two Balkan endemic *Laserpitium* L. species (*L. zernyi* and *L. ochridanum*) (Apiaceae). Both EOs significantly reduced the paw edema at concentrations of 25.50 and 100 mg/kg. The anti-inflammatory activity increased with an increasing dose.

Acute dermatitis was induced by croton oil, arachidonic acid, phenol, and capsaicin. The dermatitis was treated with α-bisabolol. The EO compound showed good anti-inflammatory activity (Leite et al., 2011). (−)-α-Bisabolol is a sesquiterpene alcohol and the main compound of the EO of *Matricaria chamomilla* L. (Asteraceae). It was evaluated for its anti-inflammatory activity using carrageenan-induced mouse paw edema. α-Bisabolol was able to reduce the paw edema at concentrations of—100 or 200 mg/kg—at pre-treatment and so did indomethacin, the positive control. It also had significant reduction effect in edema induced by dextran. Rat treatment with sesquiterpene alcohol (100 and 200 mg/kg) showed a significant decrease of leukocyte migration on carrageenan-induced peritonitis. The level of TNF-α in the peritoneal fluid was definitely reduced in rats when treated with α-bisabolol (Rocha et al., 2011).

In vitro tests of the leaves of the EO of *Cedrelopsis grevei* Bill. and Courchet. (Rutaceae) showed good 5-LOX inhibitor activity. IC_{50} values are 21.33 ± 0.5 mg/L; therefore, authors concluded good 5-LOX inhibitor properties (Afoulous et al., 2013). *Citrus sunki* Hort. ex. Tan. and *Fortunella japonica* var. *margarita* Swingle EO (Rutaceae) were tested for their anti-inflammatory activity. The EOs reduced LPS-induced production of NO in RAW 264.7 cells. These findings indicate that the EOs have anti-inflammatory effects (Yang et al., 2010).

The topical application of the EO of *Myrica esculenta* Buch. Ham. Ex D. Don. (Myricaceae) exhibited strong suppression of mouse ear edema, induced by xylene. The anti-inflammatory activity of the EO was found to be 85.3% at topical application. Diclofenac exposed 87.5% of inhibition, which is somewhat more than for the EO. Nevertheless *M. esculenta* EO showed potent anti-inflammatory properties (Agnihotri et al., 2012).

I. lanceolatum A.C. Smith EO (Illiciaceae) was evaluated for its anti-inflammatory activity using xylene-induced ear edema test. In a dose-dependent manner, the EO of the roots of *I. lanceolatum* at doses of 0.4, 0.6, and 0.8 g/kg significantly reduced the effects of xylene in mouse ears (29.8%, 30.9%, and 35.3%). These results are slightly lower than those of dexamethasone (49.1%) (Liang et al., 2012). The EO of *Illicium anisatum* L. (Illiciaceae) was tested for its anti-inflammatory effect using Raw 264.7 cells. LPS-induced NO production in these cells was significantly decreased in the presence of *I. anisatum* EO. The EO also suppressed the production of PGE2 in LPS-induced cells in a dose-dependent manner. It was found that the EO interferes with the expression of iNOS and COX-2 mRNA (Kim et al., 2009).

Limonene was found to decrease monocyte chemoattractant protein-1 (MCP-1) production in df-HL-60 cells at concentrations greater than 14.68 mmol/L. MCP-1 is one of the C–C chemokines that are spontaneously produced by eosinophile. Diesel exhaust particles–induced MCP-1 production was also reduced by limonene at a concentration of 14.68 mmol/L. Furthermore, the activity of NFκB in relation to the treatment of limonene was significantly decreased. MAP kinase pathway was also suppressed by limonene. These tests showed promising anti-inflammatory activity of limonene (Hirota et al., 2010).

The EO of *Annona sylvatica* A. St.-Hil. (Annonaceae) was evaluated for its anti-inflammatory activity using carrageenan-induced paw edema and CFA (a mixture of oils and water with killed *Mycobacterium tuberculosis*), which induced intense inflammation. The results exhibited inhibition of carrageenan-induced paw edema at doses of 2, 20, and 200 mg/kg for 19% ± 3%, 27% ± 7%, and 35% ± 2%. In the CFA model, the EO also showed significant anti-inflammatory activity against persistent inflammation (Formagio et al., 2013). *Xylopia parviflora* (A. Rich.) Benth. (Annonaceae). EO showed anti-inflammatory effects. The NO production in LPS-induced RAW 264.7 macrophages was significantly decreased by the EO (Woguem et al., 2014).

Rose geranium (Geraniaceae) EO was evaluated for its anti-inflammatory activity using several in vivo tests. The EO showed significant reduction of carrageenan-induced paw edema in mice at concentrations of 100, 200, and 400 mg/kg for 30%, 38%, and 73%, respectively. The effects were not significantly different than those of diclofenac (50 mg/kg). The formation of croton oil–induced ear edema in mice was also enhanced using *R. geranium* EO. Histopathology analyses showed that edematous ear thickness, and substantial inflammatory cell infiltration in the dermis with associated connective tissue disturbance were significantly decreased after using the EO (Boukhatem et al., 2013).

The seed EO of *Ceiba pentandra* (L.) Gaertn. (Malvaceae) was found to have potent antiinflammatory activity by (Ravi and Raghava, 2014).

Curcumol is a main compound of the EO of *C. longa* L. (Zingiberaceae). The anti-inflammatory effect was investigated using LPS-induced RAW 264.7 macrophages. Curcumol inhibited NO production in LPS-induced RAW 264.7 cells in a dose-dependent manner. iNOS was not affected by The EO component (Chen et al., 2014).

10.6 VASODILATORY ACTIVITY

Pinto et al. (2009) found that *Alpinia zerumbet* (Pers.) Burtt. et Smith (Zingiberaceae) and its main constituent 1,8-cineole showed vasorelaxant effects. The effects are fully attributed to the integrity of a functional vascular endothelium. The methanolic fraction of the EO of *A. zerumbet* (Pers.) Burtt. et Smith (MFEOAz) was investigated in in vitro and in vivo experiments. MFEOAz at concentrations ranging from 0.1 to 3000 µL/mL reduced contraction in porphyrin-induced aortic rings, extracted from Wistar rats. MFEOAz also significantly reduced $CaCl_2$-induced contraction. To support the in vitro vasorelaxing properties, in vivo tests were conducted with hypertensive rats. These tests also showed good antihypertensive properties of MFEOAz (da Cunha et al., 2013). *A. Zerumbet* fruit oil was also investigated for its vasodilating activity. It was found that the EO had relaxing effects on endothelium with aortic rings precontracted with norepinephrine (1.0 µM) or potassium chloride (60 µM). The vasodilatory effect was stronger on norepinephrine precontracted endothelium with aortic rings than on KCl precontracted ones. The relaxing effect on endothelium intact aortic rings was found to be fully related to nitric oxide (NO)/guanylate cyclase system (Tao et al., 2013).

The EO of *Citrus bergamia* Risso (bergamot oil) (Rutaceae) was investigated for its vasodilatory activity. Bergamot oil induced relaxation of mouse aortic rings contracted with 3 µM prostaglandine Fα (PGFα). The effective concentration of 50% (EC_{50}) was determined to be 0.047% (v/v). Maximum relaxation was obtained at a concentration of 0.2% (v/v). Maybe the activation of K^+-channels and the inhibition of Ca^+-influx are involved in the vasorelaxing effect of bergamot oil (Kang et al., 2012). The EO of *Croton argyrophylloides* Muell. Arg. (Euphorbiaceae) relaxed isolated endothelium with aortic rings. These were precontracted with phenylephrine (Franca-Neto et al., 2012). Yvon et al. (2012) evaluated the vasorelaxant effects of six EOs using phenylephrine precontracted rat aorta. *Thymus capitatus* L. (Lamiaceae) and *Laurus nobilis* L. (Lauraceae) showed relaxing effects on rat aorta of 96.9% and 83%, respectively.

Pectus brevipedunculata Sch. Bip. (Asteraceae) EO and its main component citral were investigated for their vasodilating activity in aortic rings of Wistar Kyoto rats. Both induced relaxation of aortic rings, precontracted with phenylephrine. EC_{50} of *P. brevipedunculata* for phenylephrine-induced contraction of endothelium intact and denuded rings were 0.044% ± 0.006% and 0.093% ± 0.015%, respectively. The EC_{50} of citral were 1.42 ± 0.26 and 1.33 ± 0.18 mM, respectively. Also K^+-induced contraction of the aortic rings was inhibited by citral. The vasodilatory effect on KCl-induced contraction of citral was more effective than on phenylephrine-induced contraction. Moreover, citral alleviates the contractile response of Ca^{2+}. Maximal contraction induced by $CaCl_2$ was reduced to 53.38% ± 5.33% compared to control or completely removed at citral concentration of 0.6 and 6 mM (Lopes Pereira et al., 2013).

O. gratissimum L. (Lamiaceae) EO was determined for its vasorelaxant effects in aortas and mesenteric vascular beds of rats. The EO relaxed aortas precontracted by phenylephrine in dependence of concentration. The interaction is completely dependent on endothelial nitric oxide (NO) release in mesenteric vascular beds and partly related on NO release in the aortas (Pires et al., 2012).

The EO of *Aniba canelilla* (H.B.K.) Mez (Lauraceae) and its main compound 1-nitro-2-phenylethane (NP) were investigated for their vasodilating activity in an in vivo study with spontaneously hypertensive rats. The hypotensive and bradycardic effects were significant at concentration of 1 mg/kg. At 5 and 10 mg/kg, the effect response to NP, and at 10 and 20 mg/kg to the EO were biphasic. There occurred a first rapid effect 1–1.5 and 1.5–2.5 s after injection and a second hypertensive and bradycardic effect after 3.6 and 4–8 s, respectively. There were no significant differences between the EO and NP detected. The in vitro study also showed promising vasorelaxant effects of aorta preparations with intact endothelium. All these effects of the EO were attributed to the main compound NP (Interaminense et al., 2011). *A. canelilla* EO and NP were tested for their vasorelaxant activity on mesenteric arteries. The EO and NP showed relaxing activity in endothelium-containing mesenteric arteries precontracted with 75 mM KCl at concentration of 20 μg/mL. The EO and NP also showed significant relaxing activity in phenylephrine and CaCl precontracted arteries. All vasodilating effects of the EO an NP were reversible (Interaminense et al., 2013).

C. winterianus J. (Poaceae) exhibited hypotension and vasorelaxing effects, which are mainly caused by Ca^{2+}-channel blocking. In nonanaesthetized rats, an intravenous injection (1, 5, 10, and 20 mg/kg) induced dose-dependent hypotension associated with tachycardia. A dose of 20 mg/kg induced transitory bradycardia before tachycardia. Moreover, the EO of *C. winterianus* showed vasodilating effects in rat mesenteric artery, which appeared to be mainly related to Ca^{2+}-influx (De Menezes et al., 2010). Citronellol, a monoterpene alcohol and compound of the EOs of *C. citratus* Stapf, *C. winterianus*, and *L. alba* (Miller) N.E. Brown (Verbenaceae), was found to have hypotensive activity. Bolus injections of citronellol (1, 5, 10, and 20 mg/kg) induced hypotension linked with tachycardia in nonanaesthetized rats. These effects were not dependent on the dose. The in vitro tests also showed significant vasodilating activity. Citronellol induced relaxation in phenylephrine and KCl precontracted rings of mesenteric artery with and without functional endothelium. In endothelium-denuded rings, contracted with $CaCl_2$, citronellol strongly inhibited vasoconstriction in doses of $1.9 * 10^{-1}$, $6.4 * 10^{-1}$, and 1.9 M (Bastos et al., 2009).

Thymol and carvacrol, two monoterpenic phenol isomers, which are constituents of EOs, induced endothelium-independent relaxation of rat aorta. The effects seem to be related to Ca^{2+}-release from the sarcoplasmic reticulum and/or the Ca^{2+}-sensitivity of the smooth muscle cell. Furthermore, it is entirely possible that thymol and carvacrol, at low concentration inhibits the Ca^{2+}-influx through the membrane (Peixoto-Neves et al., 2010).

The vasorelaxant effect of the EO of *Mentha pulegium* L. (Lamiaceae) was tested in tracheal and bladder smooth muscles isolated from rats. The EO induced relaxation in precontracted smooth muscles. Relaxing effects are likely to be correlated with the inhibition of Ca^{2+}-entry. The effects are likely to be mediated with the main component pulegone (Soares et al., 2012).

Trans-caryophyllene, the major compound of many EOs showed significant relaxing effects on the basal tone and on precontracted isolated rat trachea. Basal tone was only affected when the endothelium was intact. Trans-caryophyllene demonstrated significant relaxing effects on tracheal muscle, dependent on an induced blocked of Ca^{2+}-influx through voltage-dependent Ca^{2+}-channels (Pinho-da-Silva et al., 2012).

10.7 CYTOTOXIC ACTIVITY

C. grevei Bill. and Courchet. (Rutaceae) EO exhibited cytotoxic effects against MCF-7 cell line (human breast cancer cells) with IC_{50} (inhibitory concentration 50%) values of 21.5 ± 2 mg/L. The cytotoxic activity of the EO may be attributed to synergistic effects of all terpenes (Afoulous et al., 2013). Aydin et al. (2013b) tested the cytotoxic effect of terpinolene, the main compound of many aromatic plants. It showed cytotoxic effects at concentrations of 100, 200, and 400 mg/L in primary rat neurons. In N2a (neuroblastoma) cells terpinolene showed cytotoxic effects already at a concentration of 50 mg/L. The antiproliferative effects of carvone were also tested. High doses of carvone (100, 200, and 400 mg/L) showed high cytotoxic effects on N2a cell lines (Aydin et al.,

2013a). The monoterpene myrtenal showed significant anticancer activity. Mice were treated with diethylnitrosamine–phenobarbital, and which furnished an increase of liver weight. Treatment with myrtenal decreased relative liver weight to near normal. α-Fetoprotein and carcinoembryonic antigen were also significant decreased when treated with myrtenal (Babu et al., 2012). Eugenol, orally applied 15 days prior, showed significant decrease in the numbers of skin cancer, induced by croton oil (topical application) in Swiss mice. The count of developed tumors in mice was reduced by 42%. Eugenol was also able to inhibit the proliferation of osteosarcoma cells and human leukemia cells (Jaganathan and Supriyanto, 2012). Furthermore, *M. alternifolia* Cheel (tea tree oil) EO and its main compound terpinen-4-ol was found to have anticancer activity on murine tumor cell lines. Inhibiting concentration 50% (IC_{50}) values were 0.03% for the EO and 0.02% for terpinen-4-ol. Both substances induced cell death by necrosis and low level apoptosis (Greay et al., 2010).

The EO of *Eugenia caryophyllata* Thunb. (Myrtaceae) and its main compound eugenol were tested against Hela (human cervix carcinoma) cell lines. Eugenol emphasized to be a potent cytotoxic agent against Hela cells (Chang et al., 2011). *E. caryophyllata* clove EO showed decrease of cell viability at concentrations ranging from 12.5 to 500 µg/mL. The most sensitive cell lines against the EO were MRC-5 (noncancer human fibroblasts). Among the cancer cell lines murine leukemia macrophages RAW 264.7 were the most vulnerable cells. The IC_{50} value was 18.8 ± 2.4 µg/mL (Kouidhi et al., 2010).

The EO of *A. sylvatica* A. St.-Hil. (Annonaceae) was found to have antiproliferative activity. The EO was evaluated for its antiproliferative activity against nine human tumor cell lines. It was found to have anticancer activity with GI_{50} values (concentration that elicit inhibition by 50% of the cell growth) of 36.04–45.37 µg/mL. At the highest concentration, cytotoxic effects were reached for all cell lines (Formagio et al., 2013). Furthermore, the leaf EO of *Porcelia macrocarpa* R.E. Fries (Annonaceae) was tested against six cancer cell lines. The EO reduced more than 50% of viability of the different cancer cells at a concentration of 100 µg/mL. The anticancer effects are possibly due to the compound germacrene D (Da Silva et al., 2013). The EO of *Xylopia frutescens* Aubl. (Annonaceae) was evaluated for its anticancer activity on OVCAR-8 (ovarian adenocarcinoma), NCI-H358M (bronchoalveolar lung carcinoma) and PC-3M (metastatic prostate carcinoma) human tumor cell lines. The EO exhibited IC_{50} values from 24.6 to 40.0 µg/mL for NCI-H358M and PC-3M cells. In vivo tests also showed significant antitumor activity of the leaf EO. On day 8 the average tumor weight was reduced by 63.2% in EO-treated mice compared to the control group (Ferraz et al., 2013b).

The EO of *Neolitsea variabillima* (Hayata) Kaneh. and Sasaki (Lauraceae) also showed cytotoxic effects against human oral, liver, lung, colon, melanoma, and leukemic cancer cells, as the results of Su et al. (2013) showed. They also evaluated the anticancer effects of heartwood EO *of Cunninghamia lanceolata* Lamb. var. *konishii* (Cupressaceae). The EO showed anticancer activity against human lung, liver, and oral cancer cell lines (Su et al., 2012). Kuramoji, the EO of *Lindera umbellate* Thunberg (Lauraceae) showed inhibitory effects in cell proliferation in in vitro tests. The growth of HL-60 cells, treated with 5 or 50 µg/mL EO for 24 h, was significantly reduced in a dose-dependent manner. It is to suggest that Kuromoji EO suppressed cell proliferation by induction of apoptosis (Hayato et al., 2012). Kim et al. (2009) found that the EO of *I. anisatum* L. (Illicaceae) showed low cytotoxic activity against fibroblasts and keratinocytes at a concentration of 100 µL/mL.

A. capillaris Thunb. (Asteraceae) EO was found to have cytotoxic effects on KB cells in a dose-dependent manner. Addition of 0.5 µg/mL EO reduced cell viability by 80%. The tests also showed that *A. capillaris* EO induced apoptosis at concentrations of 0.5 and 0.5 µg/mL by 24.3% ± 2.8% and 73.2% ± 2.9% (Cha et al., 2009a). Twenty EOs from herbal plants and citrus fruits were evaluated for their potential antitumor activity. Chamomile (*M. chamomilla*, Asteraceae) showed the strongest inhibitory effect against α and λ mammalian pol (mammalian DNA polymerase) (Mitoshi et al., 2012). The leaf and root oils of *Pulicaria jaubertii* E. Gamal-Eldin (Asteraceae) were tested for their antitumor activity against MCF-7 and HEPG-2 (human liver carcinoma cell line) carcinoma cell lines. Both EOs exhibited antitumor effects against both cell lines. The IC_{50} of the leaf oil for MCF-7 cells was 3.8 µg/mL, for HEPG-2 cells IC50 was 5.1 µg/mL. The root oil showed lower activity (Fawzy et al., 2013). An in vitro cytotoxicity study showed 70.23% cancer cell (B16F-10 mouse melanoma cells) death when treated with the EO of *Tridax procumbens* L. (Asteraceae). The percentage of tumor nodule formation was also decreased by 71.67% with the EO when compared to the control group (Manjamalai et al., 2012). Manjamalai and Grace (2013) also found that the EO of *Plectranthus amboinicus* (Lour) Spreng (Lamiaceae) had cytotoxic effects on B16F-10 melanoma cells. It showed potent chemotherapeutic and chemoprotective effects over lung metastasis. The EO of *Artemisia indica* Willd (Asteraceae) was evaluated for its anticancer activity against THP-1 (leukemia), A-549 (Lung), Hep-2, and Caco-2 (coloreactal adenocarcinoma) cancer cell lines. Concentrations from 10 to 100 µg/mL decreased the cell viability of all four cell lines. At a concentration of 100 µg/mL growth inhibition was found to be 95% (THP-1), 72% (Hep-2), 82% (Caco-2), and 78% (A-549) (Rashid et al., 2013).

The EO of the aerial parts of *Artemisia herba-alba* Asso. (Asteraceae) exhibited significant antiproliferative effects on acute lymphocytic leukemia cancer cells. The cytotoxic activity started at very low doses (less than 0.5 µg/mL). At a high concentration (50 µg/mL), the cell viability was decreased by 80%. The IC_{50} value was 3 µg/mL (Tilaoui et al., 2011).

β-Elemene is a natural sesquiterpene isolated from the EO of *Curcuma aromatica* Salisb. (Zingiberaceae). They showed antiproliferative effects on several cell lines. The results of in vitro tests exhibited growth inhibition of laryngeal cancer cells. Hep-2-cells were transplanted into nude mice for in vivo tests. β-Elemene was also able to inhibit the growth of those induced tumors. The inhibitory rate of β-elemene (40 µg/mL) was 73.7% ± 4.4%. The antiproliferative effect was found to be related to the cell cycle arrest, the induction of apoptosis, and the inhibition of metastasis (Dai et al., 2013).

Misharina et al. (2013) found that the EO *O. vulgare* L. (Lamiaceae) showed significant anticancer activity in in vivo experiments. Engraftment and development of the tumors in F1 DBA C57 black hybrid mice were significantly decreased. The EO of *Marrubium vulgare* L. (Lamiaceae) exhibited significant anticancer activity against HeLa cell lines. Different concentrations (3.91–3000 µg/mL) of *M. vulgare* EO were used. Concentrations up to 250 µg/mL EO showed deletion by 27%, at higher concentrations (up to 500 µg/mL) all HeLa cells were destroyed. The *M. vulgare* EO showed an IC_{50} value of 0.258 µg/mL (Zarai et al., 2011). *S. officinalis* L. (Lamiaceae) EO was tested for its anticancer activity against oral epithelial cell carcinoma cell line (UMSCC1). Cell viability was decreased at a concentration of 135 µg/mL to 3%. IC_{50} value was 135 µg/mL (Sertel et al., 2011a). *T. vulgaris* also showed cytotoxic effects against UMSCC1 cell line with an IC_{50} value of 369.55 µg/mL (Sertel et al., 2011c). The main compound of the EO of *Nepeta sibirica* L. (Lamiaceae) was found to be nepetalactone. These agents showed cytotoxic activities against HL60 (melanoma) and Kato III (stomach carcinoma) cell lines (Tsuruoka et al., 2012). Two different EO of *H. spicigera* Lam (Lamiaceae) showed cytotoxic effect against MCF-7 cell line. IC_{50} values were 170 and 84 µg/mL. The cytotoxic effect is based on the compounds of the EO (Bogninou-Agbidinoukoun et al., 2013).

Xylopia laevigata (Mart.) R.E. Fries (Annonaceae) EO was tested for its anticancer activity in in vitro and in vivo tests. The IC_{50} values were ranging from 14.4 to 31.6 µg/mL in SF-295 (glioblastoma) and OVCAR-8 cell lines. The EO was also cytotoxic to normal cells. In vivo tests were performed with mice, which were transplanted with sarcoma180 cells. These mice were treated with the EO in 5% DMSO intraperitoneally once a day for 7 days. On day 8, the tumor growth inhibition rates were 37.3%–42.5% (Quintans et al., 2013).

Pinus densiflora Sieb. et Zucc. (Pinaceae) EO inhibited proliferation and induced apoptosis in YD-8 cells (human oral squamous carcinoma). Cell proliferation was inhibited by 30% and 60% at concentrations of 40 and 60 µg/mL EO. The EO reduced cancer cells by 70% when treated with 60 µg/mL EO. Apoptosis was induced by activation of caspase-9 (Jo et al., 2012). Soeur et al. (2011) found that *Aniba rosaeodora* Ducke (Lauraceae) had cytotoxic effects against A431 (human epidermoid carcinoma cell line) and on HaCaT (human keratinocytes) cells. At a concentration up to 400 nL/mL the EO showed significant reduction of cell viability of these cell lines. The cytotoxic activity is due to induction of caspase-dependent apoptosis in the aforementioned cell lines. Nonsmall cell lung cancer cells were treated with the EO of *L. cubeba*. The vapor of the EO induced apoptosis in the cancer cells. Cell death was caused through induction of caspase-9 (Seal et al., 2012). *Boswellia sacra* (Burseraceae) EO was found to have antitumor activity against human breast cancer cells. Cell viability in all tested cancer cells was decreased, dependent on the hydrodistillation temperatures of the EOs. Immortalized normal breast epithelial cells (MCF-10-2A) were insensitive to the EOs. Therefore, the EOs induces tumor cell-specific death. *B. sacra* EO induced apoptosis in dependence on activation of caspase-8 and caspase-9 (Suhail et al., 2011). KB cells were treated with the EO of *Cryptomeria japonica* D. Don (Cupressaceae). Cell viability was significantly inhibited when treated with the EO. The treatment with the EO arrested cells in gap-0 (G0) phase, and proliferation was impossible. The EO also activates caspase-8 and induces apoptosis (Cha and Kim, 2012). De Martino et al. (2011) found that the EO of *Verbena officinalis* L. (Verbenaceae) was able to induce apoptosis in chronic lymphocytic leukemia cells, dependent on the activation of caspase-3. The EO of *Kadsura longipedunculata* Finet and Gagnep. (Schisandraceae) exhibited moderate cytotoxic activity against MIA PaCa-2, HepG2, and SW-480 cell lines with IC_{50} values of 133.53, 136.96, and 136.62 µg/mL, respectively. Cytotoxic activity may be due to the activation of caspase-3 and caspase-7 (Mulyaningsih et al., 2010)

L. gracilis Schauer (Verbenaceae) EO was tested for its in vitro cytotoxic effects against HepG2, K562 (human chronic myelocytic leukemia), and B16-F10 (mouse melanoma) cell lines. These cell lines were treated with different concentrations of the EO for 72 h. The major constituents (thymol, *p*-cymene, γ-terpinene, and myrcene) were tested against the cell lines. The EO exhibited IC_{50} values ranging from 4.93 to 22.92 µg/mL for HepG2 and K562 cells. The compound myrcene was the most cytotoxic agent among the compounds. IC_{50} values ranging from 9.23 to 12.2 µg/mL for HepG2 and B16-F10 cell lines. *p*-Cymene and γ-terpinene only showed cytotoxic effects on B16-F10. Due to these results it can be concluded that the EO of *L. gracilis* is a potent cytotoxic agent (Ferraz et al., 2013a). Other different EOs of Verbenaceae and Asteraceae family were tested for their cytotoxic activity. *Achillea ligustica* All. (Asteraceae) displayed cytotoxic activity against the B16-F1 cell line and *Ambrosia arborescens* Mill. against Hela cells. *L. alba* and *L. micromera* Schauer (Verbenaceae) also showed cytotoxic effects on Hela cells (Zapata et al., 2014). The EO of *C. lanceolata* var. konishii (Lamb.) Hook. (Cupressaceae) showed potent cytotoxic effects on human lung, liver, and oral cancer cells. The active agent seemed to be cedrol (Su et al., 2012). The EOs of *Erigeron acris* L. and *Erigeron annuus* (L.) Pers (Asteraceae) were investigated for their anticancer activity against cultured fibroblasts, MCF-7, MDA-MBA-231, endometrial adenocarcinoma (Ishikawa) and colon adenocarcinoma (DLD-1) cells. The EO of *E. acris* showed strong activities on MCF-7 cell line, with an IC_{50} value of 14.5 µg/mL. The EOs had no effects on noncancer cells (Nazaruk et al., 2010).

X. laevigata L. (Apiaceae) EO exhibited potent antiproliferative effects against U251 (glioma, CNS), UACC-62 (melanoma), MCF-7 (mamma), NCI-ADR/RES (ovarian-resistant), 786-0 (kidney), NCI-H460 (lung, no small cells), PC-3 (prostate), OVCAR-3 (ovarian), HT-29 (colon), and VERO (African green monkey kidney) cell lines in in vitro tests with total growth inhibition (TGI) of 4.03–36.71 µg/mL. UACC-62, NCI-ADR/RES and NCI-H460 were the most sensitive cells against the EO with TGI lower than 10 µg/mL (Costa et al., 2013a).

The EO mandarin peel (*Citrus reticulata* Blanco Rutaceae) and its main compound, limonene, were tested against A549 and HepG2 cell lines. Both agents showed dose-dependent growth inhibitory effects. The IC_{50} values of limonene were 0.150 µL/mL in HepG2 cells and 0.098 µL/mL in A549 cells. For the mandarin peel oil, the IC_{50} values were 0.063 and 0.036 µL/mL, respectively (Manassero et al., 2013).

Costa et al. (2013b) found that the EO of *Annona pickelii* (Diels) H. Rainer and *Annona salzmannii* A. DC (Annonaceae) exhibited potent antitumor effects on U251 (glioma, CNS), UACC-62 (melanoma), MCF-7, NCI-H460 (lung, nonsmall cells), OVCAR-03 (ovarian), PC-3 (prostate),

HT-29 (colon), 786-0 (kidney), K562 (leukemia) and VERO (noncancer cell line) human tumor cell lines. TGI was lower than 100 µg/mL.

The EO of *Solanum erianthum* D. Don and *Solanum macranthum* Dunal (Solanaceae) were tested for their in vitro cytotoxic effects. *S. erianthum* showed inhibitory effects against Hs 578T (human breast ductal carcinoma) cells and PC-3 (human prostate carcinoma) cells. Fruit oil of *S. macranthum* exhibited anticancer activity against Hs 578T cells by 78% (Essien et al., 2012).

The EOs of *Zanthoxylum zanthoxyloides* Lam. and *Zanthoxylum leprieurii* Guill. and Perr (Rutaceae) were evaluated for the growth inhibiting activity on four different human cancer cell lines. Both EOs showed potent against T98G (human glioblastoma), MDA-MB231 (breast adenocarcinoma), A375 (melanoma), and HCT116 (colon carcinoma) cell lines, but the EO of *Z. zanthoxyloides* had stronger growth inhibitory effects than *Z. leprieurii*. The highest activity was achieved on the MDA-MB231cells with IC_{50} values of 18.2 ± 1.5 and 76.0 ± 11.1 µg/mL for *Z. zanthoxyloides* and *Z. leprieurii*, respectively (Fogang et al., 2012).

Origanum marjoram L. and *O. vulgare* L. (Lamiaceae) EOs were evaluated for their cytotoxic activity on two human cancer cell lines (MCF-7 and LNCaP) and one fibroblast cell line (NIH-3T3). The EO decreased the cell viability ranging from 79% to 88% at a concentration of 0.5 µg/mL. The EO of *O. marjoram* showed notable stronger activity than the EO of *O. vulgare* (Hussain et al., 2011). *R. officinalis* L. EO was tested against MCF-7 and NIH-3T3 (mouse embryonic fibroblasts) cells. The EO inhibited the cell viability by 81%–89% at a concentration of 500 µg/mL. The IC_{50} values were 190.1 and 180.9 µg/mL for MCF-7 and NIH-3T3 cell lines, respectively (Hussain et al., 2010). *Thymus caespititius* Hoffmanns and Link (Lamiaceae) EOs was tested for its cytotoxic effects on the cell line ACC201 (adenocarcinoma gastric cells). The cell viability was decreased to 45%, 30 min after the treatment with the EO. At the end of the first hour, cell viability was recovered to 65% at a concentration of 0.08 mg/mL. Higher concentrations (0.5 and 1.0 mg/mL) were more harmful to the tumor cells. The cell viability decreased to $30\% \pm 0.05\%$ and $20\% \pm 0.05\%$, respectively (Dandlen et al., 2011).

Different pancreatic cancer cell lines (KLM1, KP4, Panc1, MIA Paca2) and pancreatic epithelial cells (ACBRI515) were treated with 0–250 µM α-bisabolol, a compound of many EOs. α-Bisabolol significantly suppressed the proliferation of KLM1 and KP4 cells at a concentration of 5 µM. To suppress the proliferation of Panc1 and MIA Paca2 cells a higher concentration (6.25 µM) was needed. It was found that α-bisabolol induced apoptosis in consequence of Akt (serine/threonine-specific protein kinase) activation (Seki et al., 2011).

Murraya koenigii (L.) Spreng (Rutaceae) was evaluated for its in vitro cytotoxic effects against MCF-7, P388 (murine leukemia) and Hela cells. The results showed that both, the EO and its main compound carbazole alkaloids have significant antiproliferative effects on all three tested cell lines in a dose- and time-dependent manner (Nagappan et al., 2011).

The cytotoxic effects of the EOs of *M. communis* and *Eugenia supraaxillaris* Spreng. (Myrtaceae) were evaluated. Both EO samples exhibited strong cytotoxic activity against cervices, colon, larynx, liver, and breast cancer cell lines. The activity may be attributed to the presence of limonene, cineole eugenol, and methyl eugenol as compounds of the EOs (Aboutabl et al., 2011). *Myrcia laruotteana* Camb. (Myrtaceae) EO was tested against U251, UACC-62 (melanoma), MCF-7, NCI-ADR/RES, 786-0 (kidney), NCI-H460, PC-3, OVCAR-3, HT-29, K562, and VERO cells. The EO showed antiproliferative effects on all tested cancer cell lines, accept for NCI-ADR/RES. TGI was ranging from 20.46 to 88.31 µg/mL. The most sensitive cell lines against *M. laruotteana* EO were U251, 786-0, UACC-62 and PC-3, with TGI less than 30 µg/mL. The EO also exhibited toxicity against VERO cells (Stefanello et al., 2011).

Ceratonia siliqua L. (Fabaceae) EO was tested for its cytotoxic effects on Hela, MCF-7 and HUVEC (human umbilical vein endothelial) cells using the MTT test. The EO exhibited, dose dependently, growth inhibition effects on all tested cell lines, but inhibition was significantly stronger in Hela cells than in MCF-7 cells. It also showed lower cytotoxic activity against normal human cells (HUVEC) (Hsouna et al., 2011).

Patharakorn et al. (2010) found that the leaf EO of *Morus rotundiloba* Koidz exhibited cytotoxic activity against Hep2 and SW620 cell lines. The EO showed no toxic effects on VERO cells at concentrations of 0.1–100 μg/mL.

The EO of four different plants growing in Benin (*Ozoroa insignis* Delile, Anacardiaceae; *Pentadesma butyracea* Sabine, Clusiaceae; *Siphonochilus aethiopicus* (Schweinf.) B.L. Burtt, Zingiberaceae; and *Xylopia aethiopica* (Dunal) A. Rich., Annonaceae) were evaluated for their cytotoxic effects. All four EOs inhibited the proliferation of breast cancer cells. Seventy percent of all cancer cells were inhibited at the highest test concentration of *P. butyracea* EO. The IC_{50} value was 133.5 ± 2.6 μg/mL. *S. aethiopicus* at a concentration of 178.0 μg/mL also displayed antiproliferative effects on MCF-7 cell lines. The IC_{50} values for *X. aethiopica* EO was 127.9 ± 28.8 μg/mL, for *O. insignis* EO the IC_{50} value was >327.6 μg/mL. Due to these results, all four EOs, except for *O. insignis*, showed potent cytotoxic activities (Noudogbessi et al., 2013). The EO of *Casearia lasiophylla* Eichler (Salicaceae) exhibited antiproliferative activity on U251 (glioma), UACC-62 (melanoma), MCF-7, NCI-ADR/RES (ovarian resistant), PC-3, OVCAR-3, HT-29, K562, and VERO cell lines. The most sensitive cell lines were UACC-62 (TGI: 7.30 μg/mL) and K562 (TGI: 7.56 μg/mL). The cytotoxic effect is possibly due to the attendance of germacrene D, (E)-caryophyllene, and α-cardinol in the EO (Salvador et al., 2011). Flower EO of *Convolvulus althaeoides* L. (Concolvulaceae) exhibited significant cytotoxic effects against human breast cancer cells. The cytotoxic effects of the EO may be due to the synergistic effects of the active compounds (α-humolene, β-caryophyllene and germacrene D) (Hassine et al., 2014).

Patchouli alcohol is a main compound of the EO of *Pogostemon cablin* (Blanco) Benth. (Lamiaceae). The treatment with 50, 75, and 100 μM patchouli alcohol reduced the proliferation of HCT116 cells by 22%, 35%, and 56% at 24 h. The cell proliferation of MCF-7 cells was reduced by 24% and 59% at concentrations of 75 and 100 μM patchouli alcohol at 24 h. The antiproliferative effects may be due to the induction of apoptosis (Jeong et al., 2013).

P. guajava L. (Myrtaceae) EO of fruits and leaves were tested against HepG2 and MCF-7 cell lines. IC_{50} values of leaf EO were ranging from 130.69 to 351 μg/mL, IC_{50} values of fruit EO were ranging from 196.45 to 544.38 μg/mL. The cytotoxic effect against MCF-7 cells was much lower than against HepG2 cells. The EO concentration used was 0.11% (El-Ahmady et al., 2013).

The cytotoxic effects of *Guatteria pogonopus* Maritus (Annonaceae) was evaluated by Fontes et al. (2013). The used cell lines were OVCAR-8, NCI-H358M, and PC-3M. The EO showed cytotoxic effects against all three cell lines. The most sensitive cells were OVCAR-8 cells; the lowest cytotoxic effects were reached on NCI-H358M. The in vivo antitumor activity was tested on mice, transplanted with sarcoma180 cells. The EO exhibited a growth inhibition of 25.3% and 42.6% at concentrations of 50 and 100 mg/kg/day.

Marigold (*Tagetes minuta* L.) and basil (*O. basilicum*) (Lamiaceae) EO showed in vitro cytotoxic effects at concentrations of 25–200 μg/mL. Both EOs reduced the cell viability of HL-60 and NB4 (human leukemia cell line) cells. 82.33% of cell death of HL-60 cells was reached at a concentration of 200 μg/mL basil oil. The highest rate of cell death on NB4 cells (81.87%) was reached with marigold oil at a concentration of 200 μg/mL. Both EOs showed no cytotoxic effects in the in vivo tests (Mahmoud, 2013).

Vepris macrophylla (Baker) I. Verd. (Rutaceae) EO was tested for its anticancer effects by Maggi et al. (2013). The EO showed inhibitory effects on tumor cell growth on T98G (human glioblastoma), MDA-MB-231, A375, and HCT116 cell lines. The highest activity of the EO was reached against MDA-MB-231 and HCT116 cell lines. The IC_{50} values for these two cell lines were 3.14 ± 0.21 and 3.21 ± 0.22 mg/mL. The IC50 value for T98G was 28.4 ± 2.2 mg/mL. These cells were the most resistant against the EO.

Abietane, a new type of natural diterpene, was found in the EO of *Tetradenia riparia* (Hochstetter) Codd. (Lamiaceae). The EO and the diterpene were tested against their cytotoxic activity. The EO and its compounds (abietane) showed prominent anticancer activity against SF-295 (tumoral) cell line by 78.06% and 94.80%, respectively. Inhibitory effects for HCT-8 cells were 85.00% and

86.54%. MDA-MB-435 cells were inhibited by 59.48% and 45.43%. These results indicate that the new found agent exhibits promising anticancer activities (Gazim et al., 2014).

Thymus serpyllum, *Thymus algeriensis* Boiss. And Reut and *T. vulgaris* EO were tested against MCF-7, NCI-H460, HCT-15, Hela and HepG2 cells. *T. serpyllum* showed the most potent cytotoxic activity against the tested cell lines. The GI50 values were ranging from 7.02 to 52.69 µg/mL. All three EOs exhibited no toxicity against noncancer cells. MCF-7 cells were the most sensitive cells human cancer cells (Nicolic et al., 2014).

The EO of *Monarda citriodora* Cerv. ex Lag. (Lamiaceae) showed chemotherapeutic potential in HL-60 cell lines. It induces apoptosis and disrupted the PI3K/AKT/mTOR signaling cascade. The EO and its main compound thymol also inhibited the cell proliferation of MCF-7, PC-3, A-549, and MDA-MB-231 cells (Pathania et al., 2013). *Thymus citriodorus* (Pers.) Schreb. ex Schweigg. and Körte (Lamiaceae) showed toxic effects on HepG2 cells. It induced apoptosis associated to the expression of NFκB65. The IC_{50} value was 0.34% (Wu et al., 2013).

Lindera strychnifolia (Sieb. and Zucc.) (Lauraceae) leaf EO was investigated for its cytotoxic effects on eight human cancer cell lines. The anticancer activity against Eca-109 (human esophageal carcinoma), HepG2, MDA-MB-231, PC-3, and SGC-7901 (human gastric carcinoma) of the EO was more significant than those of cisplatin. HepG2 was the most sensitive cell line of the leaf EO. It induced apoptosis in the tumor cells (Yan et al., 2013).

Bou et al. (2013) evaluated the cytotoxic activity of the EO of *Casearia sylvestris* L. (Salicaceae). It exhibited antitumor effects on all tested cancer cells. With IC50 values ranging from 12 to 42 µg/mL. The main compound α-Zingiberene showed cytotoxic effects against Hela, U-87 (glioblastome), Siha (uterus carcinoma) and HL60 cell lines.

Further research results concerning the cytotoxic effects of EOs are summarized in Table 10.1.

10.8 PENETRATION-ENHANCING ACTIVITY

The volatile oils of *Rhizoma Zingiberis* (Zingiberaceae), *Flos Caryophyllata* (Myrtaceae), and *Fructus Litseae* (Piperaceae) were tested for their penetration-enhancing effects. EOs at 3%, 5%, 7%, and 10% and mixtures in ratio of 1:1 were tested in an in vitro transdermal mouse skin patch model. The best penetration-enhancing effects were observed with 10% *R. zingiberis* EO, 5% *F. Litseae* EO and 5% *F. caryophyllata* EO. These oils were able to enhance the penetration volume from patch to mouse skin significantly (Li et al., 2013a). The volatile oils of *R. Zingiberis*, *Flos Magnoliae,* and *F. Litseae* were found to have good penetration-enhancing activities of rotundine permeation. 2.5% *F. litseae* EO plus 2.5% azone showed the pest effects on the penetration of rotundine using the Franz cell diffusion assay (Cui et al., 2011).

Yuan et al. (2014) found that the EO of *Asarum mialaicum* Hook.f. *and* Thomson *ex* Klotzsch (Aristolochiaceae) has penetration-enhancing activities. Patches impregnated with 15%–35% EO increased penetration by 10%–20%.

The EO of *Z. bungeanum* Maxim. (Rutaceae) was evaluated for its penetration enhancer effects in rat skin. The EO was able to increase effectively the percutaneous absorption of the applied drugs in a dose-dependent manner (Lan et al., 2014). The EO of *Cyperus rotundus* L. (Cyperaceae) exhibited potent penetration-enhancing activities for nitrazepam in a rat skin model. After treatment with more than 1% EO the percutaneous permeability of nitrazepam was significantly increased (Zhou et al., 2012). The EO of *Caulis sinomenii* showed significant increase of percutaneous penetration of sinomenine and triptolide (Deng et al., 2011). Eucalyptus EO (*E. globulus* Labill. (Myrtaceae) showed penetration enhancing activities on chlorhexidine (Karpanen et al., 2010).

The volatile oils of *Flos caryophylli*, *Semen myristicae,* and *Fructus anisi stellati* were found to exhibit penetration-enhancing activities on ligustrazine phosphate. EOs at 3%, 5%, and 7% were compared to 3% of azone, a prominent permeation enhancer. The EO of *F. caryophylli* showed the highest enhancing effect (Luo et al., 2013).

TABLE 10.1
Cytotoxic Effects of Essential Oils

Name of Plant and Family	Sensitive Cell Line	References
Eucalyptus benthamii Maiden et Cambage (Myrtaceae)	Jurkat (T leukemia cells) J774A (murine macrophage tumor) HeLa (cervical cancer)	Döll-Boscardin et al. (2012)
Myrrh (*Commiphora myrrha* Engl. Burseraceae) and frankincense (*Boswellia carterii* and *Commiphora pyracatoides* Engl. Burseraceae)	MCF-7 (human breast cancer), HS-1 (human skin)	Chen et al. (2013c)
Myristica fragrans Houtt (Myristicaceae)	Human colon adenocarcinoma	Piras et al. (2012)
R. officinalis L. (Lamiaceae)	Human ovarian cancer cell lines (SK-OV-3 and HO-8910) and human hepatocellular liver carcinoma cell line (Bel-7402)	Wang et al. (2012b)
O. basilicum L. (Lamiaceae)	HeLa (human cervical cancer cell line), Hep-2 (human laryngeal epithelial carcinoma cells) and NIH 3T3 (mouse embryonic fibroblasts)	Kathirvel and Ravi (2012)
M. armillaris (Sol Ex Gateau) Sm (Myrtaceae)	MCF7	Chabir et al. (2011)
Levisticum officinale L. (Apiaceae)	UMSCC1	Sertel et al. (2011b)
Genista tinctoria L. (Leguminosae), *Genista sessilifolia* DC	Malignant melanoma cells	Rigano et al. (2010)
Schinus molle L. and *Schinus terebinthifolius* Raddi (Anacardiaceae)	MCF-7	Bendaoud et al. (2010)
Litsea cubeba (lour.) pers.	Human lung, liver, and oral cancer cells	Ho et al. (2010)
Ducrosia anethifolia Boiss. and *Ducrosia flabellifolia* Boiss. (Apiaceae)	K562, LS180 (human colon adenocarcinoma) and MCF-7	Shahabipour et al. (2013)
Citrus limettioides Tan. (Rutaceae)	SW480 cells	Jayaprakasha et al. (2013)
Smyrnium olusatrum L. (Apiaceae)	T98G (human glioblastoma multiform), MDA-MB 231 (human breast adenocarcinoma) and HCT116 (human colon carcinoma)	Quassinti et al. (2013a)
The anticancer activity of *Angelica acutiloba* Sieb. and Zucc. (Apiaceae)	HeLa and MCF-7	Roh et al. (2012)
The EO of *Daucus carota* Linn. (Apiaceae)	HuH7 (hepatocellular carcinoma), NCI-H446 (human lung cancer)	Li et al. (2012b)
Pterodon emarginatus Vogel (Fabaceae)	C6 (rat glioma), MeWo (human melanoma), CT26-WT (mouse colon carcinoma), MDA-MB 231 (human breast cancer), A549 (human lung carcinoma), B16-F1 (mouse melanoma), CHO-K1 (hamster ovary cell), and BHK-21 (normal hamster kidney fibroblasts)	Dutra et al. (2012)
Acanthopanax leucorrhizus (Oliv.) Harm (Araliaceae)	Hela, A-549, Bel-7402 (Liver carcinoma), and Hep G2	Hu et al. (2012)
Myrothamnus moschatus (Baillon) Niedenzu (Myrothamnaceae)	Four human glioblastoma cell lines (T98G, U-87 MG, GL15, and U251) and one human breast cancer cell line (MDA-MB 231)	Nicoletti et al. (2012)
Citrus aurantium L. (Rutaceae) peel	Colorectal cell line (Lim1863)	Fadi et al. (2012)
C. citratus EO	Human colorectal (HCT-116), MCF-7	Piaru et al. (2012b)
M. fragrans and *Morinda citrifolia* L. (Rubiaceae)	HCT-116 and MCF-7	Piaru et al. (2012a)

(Continued)

TABLE 10.1 (*Continued*)

Name of Plant and Family	Sensitive Cell Line	References
Cleidion javanicum Bl. (Euphorbiaceae) EO	KB oral cavity cancer, MCF-7, and H187-small cell lung cancer.	Sanseera et al. (2012)
Satureja sahendica Bornm. (Lamiaceae)	MCF-7, Vero, SW480 (human colon adenocarcinoma) and JET 3 (choriocarcinoma cell)	Yousefzadi et al. (2012)
Fortunella margarita Swingle (Rutaceae)	LNCaP cell (prostate cancer cells)	Jayaprakasha et al. (2012)
Pinus wallichiana P. strobus (Pinaceae)	THP-1, A-549, HEP-2, PC-3 (prostate cancer cells), and IGR-OV-1 (ovarian carcinoma)	Dar et al. (2012)
Cinnamomum longepaniculatum N. Chao ex H.W. Li leave EO	BEL-7402 (hepatocellular carcinoma)	Ye et al. (2012)
Micromeria fruticosa L. Druce ssp. serpyllifolia	HCT (human colon tumor) and MCF-7 cells.	Shehab and Abu-Gharbieh (2012)
M. spicata EO	HeLa	Liu et al. (2012a)
Senecio leucanthemifolius Poiret (Asteraceae)	COR-L23 (large cell carcinoma) Caco-2, C32 (amelanotic melanoma), HepG-2 and MRC-5 (human fetal lung).	Ouchbani et al. (2011)
A. wiedemanniana EO	COR-L23 and C23 cell	Conforti et al. (2012)
Anoectochilus roxburghii (Wall.) Lindl. (Orchidaceae)	NCI-H446 (human lung cancer)	Chen et al. (2012)
Chromolaena odorata (L.) R.M. King and H. Rob (Asteraceae)	Hela, Hep-2, and NIH 3T3 (mouse embryonic fibroblasts)	Prabhu et al. (2011)
Euphorbia macrorrhiza C.A. Mey (Euphorbiaceae)	Caco-2 cell line	Lin et al. (2012)
Ligusticum chuanxiong Hort. (Umbelliferae)	MCF-7, Hela and SK-Hep-1 (liver adenocarcinoma)	Sim and Shin (2011)
Z. officinale Roscoe (Zingiberaceae)	HO-8910 and Bel-7402	Wang et al. (2012c)
Hypericum hircinum L. ssp. majus (Aiton) N. Robson	T98G (human glioblastoma), PC3 (human prostatic adenocarcinoma), A431 (human squamous carcinoma), and B16-F1 cell lines.	Quassinti et al. (2013b)
Pulicaria undulata Gamal Eddin (Asteraceae)	MCF-7	Awadh et al. (2012)
Solidago canadensis L. (Asteraceae)	Hela, PLC (human liver carcinoma) and SMMC-7721 (human hepatocellular carcinoma)	Li et al. (2012a)
Capparis spinosa L. (Capparaceae)	HT-29 cancer cells	Kulisic-Bilusic et al. (2012)
Helichrysum gymnocephalum (DC.) Humb. (Asteraceae)	MCF-7	Afoulous et al. (2011)
Cyperus kyllingia Endl. (Cyperaceae)	NCI-H187	Khamsan et al. (2011)
A. campestris	HT-29	Akrout et al. (2011)
Lycopus lucidus Turcz. var. hirtus Regel (Lamiaceae)	Bel-7402, Hep-G2, MDA-MB-435S, and ZR-75-30	Yu et al. (2011)
Mangifera indica var. coquinho	K562, NCI-ADR/RES, OVCAR-3, NCI-H460, 786-0 and UACC-62, HT-29, PC3 and MCF-7	Simionatto et al. (2010)
Peristrophe bicalyculata (Retz) Nees (Acanthaceae)	MCF-7 and MDA-MB-468 (human breast cancer) cells	Ogunwande et al. (2010)
Flower and Fruit EO of *Datura metel* L. (Solanaceae)	Hs 578T (human breast ductal carcinoma)	Essien et al. (2010)
Salvia pisidica Boiss. et Heldr. (Lamiaceae)	HEP-G2	Ozkan et al. (2010)

(*Continued*)

TABLE 10.1 (*Continued*)

Name of Plant and Family	Sensitive Cell Line	References
Cinnamomum zeylanicum Blume (Lauraceae)	F2408 (normal rat embryonic fibroblasts) and 5RP7 (c-H-ras transformed rat embryonic fibroblasts)	Unlu et al. (2010)
Annona senegalensis Pers. leaves (Annonaceae)	A549, HT29, MCF-7, RPMI, and U251	Ahmed et al. (2010)
Croton matourensis Aubl. and *Croton micans* Müll. Arg.	LoVo (colon carcinoma), Hela and X-17 (colon carcinoma)	Campagnone et al. (2010)
Heracleum transcaucasica (Manden.) (Umbelliferae)	Hela, LS180	Firuzi et al. (2010)
Glycyrrhiza glabra L. (Fabaceae) and *M. chamomilla*	MCF-7	Ali (2013)
Tarchonanthus comphoratus L. (Asteraceae)	HT29 tumor	Awadh et al. (2013)
Caesalpinia peltophoroides Benth. (Caesalpiniaceae)	U87, HCT, and A2058	de Carvalho et al. (2013b)
Pylorae herba	SW1353	Cai et al. (2013)
Oxandra sessiliflora R.E. Fries (Annonaceae)	HL-60 (human leukemia cells) B16F10-Nex-2, MCF-7 and Hela	de Carvalho et al. (2013a)
Senecio grandiflorus DC EO (Asteraceae)	A-549, THP-1, PC-3, and HCT-116	Lone et al. (2014)
Zanthoxylum bungeanum L. (Rutaceae)	Hela	Li et al. (2013b)
Artabotrys hexapetalus (L.) Bhandari (Annonaceae)	Liver carcinoma cell lines (BEL-7402)	Wang et al. (2013b)
Cercis chinensis Bunge	K-562 (human leukemia)	Wang et al. (2013a)
Wild pepper (*Piper carpense* L. Piperaceae)	MDA-MB 231, A375, and HCT116	Woguem et al. (2013)
Oxytropis falcata Bunge (Fabaceae)	SMMC-7721	Yang et al. (2013)
Stachys rupestris Montbret et Aucher ex Benth *Salvia heldreichiana* Boiss. ex Benth (Lamiaceae)	PC-3 and MCF-7	Erdogan et al. (2013)
Curcuma zedoaria Roscoe (Zingiberaceae)	H1299, A549, and H23	Chen et al. (2013a)
Atractylodes lancea Thunb.	MKN-45 cancer cells.	Chen et al. (2013b)
Artemisia anomala S. Moore	A549, BRO (human melanoma), MCF-7 and PC-3	Zhao et al. (2013)
Fissistigma cavaleriei (Lev.) Rehd (Annonaceae)	K562, S-180, and A549	Zhang et al. (2012)

The percutaneous penetration enhancement of the EO of *E. caryophyllata* Thunb (Myrtaceae) was investigated on ligustrazine phosphate patch. The cumulative permeation volumes (Q_{12}) were 3.0, 4.1, 1.8, and 1.2 mg/cm^2 for concentrations of 3%, 5%, 7%, and 10%, respectively. Due to these results the EO of *E. caryophyllata* showed potent permeation enhancer activities (Luo et al., 2012).

Wang et al. (2012a) found that the EO of *Angelica sinensis* (Oliv.) Diels (Apiaceae) is able to increase the percutaneous penetration of resveratrol. At concentrations of 1.0% and 2%, the cumulative permeation was 48.11 ± 15.93 and 16.74 ± 1.30 μg/m^2/h. The EO of *A. sinensis* was also found to raise the percutaneous permeation of nimodipine in transdermal preparations (Wang Q. et al., 2010).

The permeation contingents were evaluated by Han et al. (2011) using ligustrazine phosphate transdermal patches. The cumulative permeation significantly increased after treatment with the

EO. At concentrations of 3%, 5%, and 7% of the EO the cumulative penetration quantities were 720.11, 470.13, and 115.41 µg/cm^2/h, respectively. Thus, the concentration of 5% EO had the highest penetration-enhancing effects.

Black cumin seed oil (*Nigella sativa* L., Ranunculaceae) was tested for its penetration enhancer activity on carvedilol using excised albino Wistar rat abdominal skin. At a concentration of 5% the EO showed the most potent enhancing effects. The transdermal flux and the permeability coefficient were higher than in the control group (Saima et al., 2010).

Volatile oils of *Alpinia katsumadai* Hayata (Zingiberaceae), *Amomum tsao-ko*, and *Amomum kravanh* Pierre (both Cornaceae) were tested for their penetration-enhancing effects on strychnine through mouse skin. At a concentration of 7% the EO of *A. katsumadai* and 5% of *A. tsao-ko* EO were able to increase the percutaneous permeation of strychnine in the in vitro study (Shen et al., 2010).

REFERENCES

Aboutabl, E. A., Meselhy, K. M., Elkhreisy, E. M., Nassar, M. I., and Fawzi, R. (2011). Composition and bioactivity of essential oils from leaves and fruits of *Myrtus communis* and *Eugenia supraxillaris* (Myrtaceae) grown in Egypt. *J. Ess. Oil Baer. Plants* 14(*2*), 192–200.

Adukwu, E., Allen, S., and Phillips, C. (2012). The anti-biofilm activity of lemongrass (*Cymbopogon flexuosus*) and grapefruit (*Citrus paradisi*) essential oils against five strains of *Staphylococcus aureus*. *J. Appl. Microbiol.* 113, 1217–1227.

Afoulous, S., Ferhout, H., Raoelison, E. G., Valentin, A., Moukarzel, B., Coudere, F. et al. (2011). *Helichrysum gymnocephalum* essential oil: Chemical composition and cytotoxic, antimalaria and antioxidant activities, attribution of the activity origin by correlation. *Molecules* 16, 8273–8291.

Afoulous, S., Ferhout, H., Raoelison, E. G., Valentin, A., Moukarzel, B., Couderc, F. et al. (2013). Chemical composition and anticancer, antiinflammatory, antioxidant and antimalaria activities of leaves essential oil of *Cedrelopsis grevei*. *Food Chem. Toxicol.* 56, 352–362.

Agnihotri, S., Wakode, S., and Ali, M. (December 2012). Essential oil of *Myrica esculenta* Buch. Ham.: Composition, antimicrobial and topical anti-inflammatory activities. *Nat. Prod. Res.* 26(*23*), 2266–2269.

Ahmed, A. L., Bassem, S. E., Mohamed, Y. H., and Gamila, M. W. (2010). Cytotoxic essential oil from *Annona sengalensis* Pers. leaves. *Pharmacog. Res.* 2(*4*), 211–214.

Akrout, A., Gonzalez, L. A., Jani, H. E., and Madrid, P. C. (2011). Antioxidant and antitumor activities of artemisia campestris and *Thymelaea hirsuta* from southern Tunesia. *Food Chem. Toxicol.* 49, 342–347.

Ali, E. M. (2013). Phytochemical composition, antifungal, antiaflotoxigenic, antioxidant and anticancer activities of *Glycorrhiza glabra* L. and *Matricaria chamomilla* L. essential oils. *Acad. J.* 7(*29*), 2197–2207.

Ali, T., Javan, M., Sonboli, A., and Semnanian, S. (2012). Antinociceptive and anti-inflammatory activities of the essential oil of *Nepeta crispa* Willd. in experimental rat model. *Nat. Prod. Res.* 26(*16*), 1529–1534.

Al-Mariri, A. and Safi, M. (March 2013). The antimicrobial activity of selected Labiatae (Lamiaceae) essential oils against *Brucella melitensis*. *IJMS* 38(*1*), 44–50.

Al-Reza, S. M., Yoon, H. J., Kim, J.-S., and Kang, S. C. (2010). Anti-inflammatory activity of seed essential oil from *Zizyphus jujuba*. *Food Chem. Toxicol.* 48, 639–643.

Amri, I., Mancini, E., De Martino, L., Marandino, A., Lamia, H., Mohsen, H. et al. (2012). Chemical composition and biological activity of the essential oils from three Melaleuca species grown in Tunisia. *Int.J. Mol. Sci.* 13, 16580–16591.

An, B. S., Kang, J. H., Yang, H., Jung, E. M., Kang, H. S., Choi, I. G. et al. (2013). Anti-inflammatory effects of essential oils from *Chamaecyparis obtusa* via the cyclooxygenase-2 pathway in rats. *Mol. Med. Rep.* 8(*1*), 255–259.

Apel, M. A., Lima, M. E., Sorbal, M., Young, M. C., Cordeiro, I., Schapoval, E. E. et al. (2010). Anti-inflammatory activities of essential oils from leaves of *Myrciaria tenella* and *Calycorectes sellowianus*. *Pharm. Biol.* 48(*4*), 433–438.

Astani, A., Reichling, J., and Schnitzler, P. (2010). Comparative study on the antiviral activity of selected monoterpenes derived from essential oils. *Phytother. Res.* 24, 673–679.

Awadh, A. N., Al-Fatima, M. A., Crouch, R. A., Denkert, A., Setzer, W. N., and Wessjohann, L. (2013). Antimicrobial, antioxidant and cytotoxic activities of the essential oil of *Tarchonanthus camphoratus*. *Nat. Prod. Commun.* 8(*5*), 679–682.

Awadh, A. N., Sharopov, F. S., Alhaj, M., Hill, G. M., Porzel, A., Arnold, N. et al. (2012). Chemical composition and biological activity of essential oil from *Pulcaria undulata* from Yemen. *Nat. Prod. Med. 7*(2), 257–260.

Aydin, E., Türkez, H., and Keles, M. S. (2013a). Potential anticancer activity of carvone in N2a neuroblastoma cell lines. *Toxicol. Ind. Health* 1–9.

Aydin, E., Türkez, H., and Tasdemir, S. (2013b). Anticancer and antioxidant properties of terpinolene in rat brain cells. *Arh. Hig. Toksikol 64*, 415–424.

Babu, L. H., Perumal, S., and Balasubramanian, M. P. (2012). Myrtenal, a natural monoterpene, downregulates TNF-a expression and suppresses carcinogen-induced hepatocellular carcinoma in rats. *Mol. Cell Biochem. 339*, 183–193.

Bachiega, T. F., de Sousa, J. P., Bastos, J. K., and Sforcin, J. M. (2012). Clove and eugenol in noncytotoxic concentration exert immunomodulatory/anti-inflammatory action on cytokine by murine macrophages. *J. Pharm. Pharmacol. 64*(4), 610–616.

Bachir Raho, G. and Benali, M. (2012). Antibacterial activity of the essential oils from the leaves of *Eucalyptus globulus* against *Escherichia coli* and *Staphylococcus aureus*. *Asian Pac. J. Trop. Biomed. 2*, 739–742.

Ballabeni, V., Tognolini, M., Giorgio, C., Bertoni, S., Bruni, R., and Baricelli, E. (2010). *Ocotea quixos* Lam. essential oil: In vitro and in vivo investigation on its anti-inflammatory properties. *Fitoterapia 81*, 289–295.

Bastos, J. F., Moreira, I. J., Ribeiro, T. P., Medeiros, I. A., Antoniolli, A. R., De Sousa, D. P. et al. (2009). Hypertensive and vasorelaxant effects of citronellol, a monoterpene alcohol, in rats. *Basic Clin. Pharmacol. Toxicol. 106*, 331–337.

Batista, P. A., Werner, M. F., Oliveira, E. C., Burgos, L., Pereira, P., Brum, L. F. et al. (2010). The antinociceptive effect of (−)-linalool in models of chronic inflammatory and neuropathic hypersensitivity in mice. *J. Pain 11*(11), 1222–1229.

Bayala, B., Bassole, I. H., Gnoula, C., Nebie, R., Yonli, A., Morel, L. et al. (2014). Chemical composition, antioxidant, anti-inflammatory and anti-proliferative activities of essential oils of plants from Burkina Faso. *PLoS ONE 9*(3), E-ISSN:1932-6203.

Bendaoud, H., Romdhane, M., Souchard, J. P., Cazaux, S., and Bouajila, J. (2010). Chemical composition and anticancer and antioxidant activities of *Schinus molle* L. and *Schinus terebinthifolius* Raddi berries essential oils. *J. Food Chem. 75*(6), C466–C472.

Bendiabdellah, A., El Amine Dib, M., Djabou, N., Allali, H., Tabti, B., Muselli, A. et al. (2012). Biological activities and volatile constituents of *Daucus muricatus* L. from Algeria. *Chem. Cent. J. 6*, 48.

Bogninou-Agbidinoukoun, G. S., Yedomonhan, H., Avlessi, F., Sohounhloue, D., Chalard, P., Chalchat, J.-C. et al. (2013). Volatile oil composition and antiproliferative activity of *Hyptis spicigera* Lam against human breast adenocarcinoma cells MCF-7. *Res. J. Chem. Sci. 3*(1), 27–31.

Bou, D. D., Lago, J. H., Figueiredo, C. R., Matsuo, A. L., Guadagnin, R., Soares, M. G. et al. (2013). Chemical composition and cytotoxic evaluation of essential oil from leaves of *Casearia sylvestris*, its main compound α-zingiberene and derivatives. *Molecules 18*, 9477–9487.

Boukhatem, M. N., Kameli, A., Ferhat, M. A., Saidi, F., and Mekarnia, M. (2013). *Rose geranium* essential oil as a source of new and safe anti-inflammatory drugs. *Libyan J. Med. 8*, 22520.

Brito, R. G., Santos, P. L., Prado, D. S., Santana, M. T., Arauja, A. A., Bonjardim, L. R. et al. (2013). Citronellol reduces orofacial nociceptive behavior in mice—Evidence of involvement of retrosplenial cortex and periaqueductal grey areas. *Basic Clin. Pharmacol. Toxicol. 112*, 215–221.

Cai, L., Ye, H., Li, X., Lin, Y., Yu, F., Chen, J. et al. (2013). Chemical constituents of volatile oil from Pylorae herba and antiproliferative activity against SW1353 human chondrosarcoma cells. *Int. J. Oncol. 42*(4), 1452–1458.

Campagnone, R. S., Chavez, K., Mateu, E., Orsini, G., Arvelo, F., and Suarez, A. I. (2010). Composition and cytotoxic activity of essential oils from *Croton matourensis* and *Croton micans* from Venezuela. *Rec. Nat. Prod. 4*(2), 101–108.

Campelo, L. M., de Almeida, A. A., de Freitas, R. L., Cerqueira, G. S., de Sousa, G. F., Saldanha, G. B. et al. (2011). Antioxidant and antinociceptive effects of citrus limon essential oil in mice. *J. Biomed. Biotechnol. 2011*, 678673.

Cavalcante, M. F., Rios, E. R., Rocha, N. F., Cito, M. C., Fernandes, M. L., de Sousa, D. P. et al. (2012). Antinociceptive activity of carvacrol (5-isopropyl-2-methylphenol) in mice. *J. Pharm. Parmakol. 64*(12), 1772–1729.

Cha, J.-D. and Kim, J.-Y. (2012). Essential oil from *Cryptomeria japonica* induces apoptosis in human oral epidermoid carcinoma cells via mitochondrial stress and activation of caspases. *Molecules 17*, 3890–3901.

Cha, J.-D., Moon, S.-E., Kim, H.-Y., Cha, I.-H., and Lee, K.-Y. (2009a). Essential oil of *Artemisia capillaris* induces apoptosis in KB cells via mitochondrial stress and caspase activation mediated by MAPK-stimulated signaling pathway. *J. Food Sci. 75*(9), T75–T81.

Cha, J. D., Moon, S. E., Kim, H. Y., Lee, J. C., and Lee, K. Y. (2009b). The essential oil isolated from *Artemisia capillaris* prevents LPS-induced production of NO and PGE(2) by inhibiting MAPK-mediated pathways in RAW 264.7 macrophages. *Immunol. Invest. 38*(6), 483–497.

Chabir, N., Romdhane, M., Valentin, A., Moukarzel, B., Marzoug, H. N., Brahim, N. B. et al. (2011). Chemical study and antimalaria, antioxidant and anticancer activities of *Melaleuca armillaris* (sol Ex Gateau) Sm essential oil. *J. Med. Food 14*(11), 1383–1388.

Chang, W.-C., Hsiao, M.-W., Wu, H.-C., Chang, Y.-Y., Hung, Y.-C., and Ye, J.-C. (2011). The analysis of eugenol from the essential oil of *Eugenia caryophyllata* by HPLC and against the proliferation of cervical cancer cells. *J. Med. Plants Res. 5*(7), 1121–1127.

Chehregani, A., Atri, M., Yousefi, S., Albooyeh, Z., and Mohsenzadeh, F. (2013). Essential oil variations in the populations of *Artemisia spicigera* from northwest Iran: Chemical composition and antibacterial activity. *Pharm. Biol. 51*(2), 246–252.

Chen, C.-C., Chen, Y., Hsi, Y.-T., Chang, C.-S., Huang, L.-F., Ho, C.-T. et al. (2013a). Chemical constituents and anticancer activity of *Curcuma zedoaria* Roscoe essential oils against non-small cell lung cancer cells in vitro and in vivo. *J. Agr. Food Chem. 61*, 11418–11427.

Chen, H.-N., Shao, C., Zhao, J.-Z., Guan, X.-W., Ding, J., and Shao, S.-H. (2013b). Essential oil extracted from rhizoma of *Atractylode lancea* induces oncosis in human MKN-45 cancer cells. *Eur. J. Inflam. 11*(2), 397–403.

Chen, X., Zong, C., Gao, Y., Cai, R., Fang, L., Lu, J. et al. (2014). Curcumol exhibits anti-inflammatory properties by interfering with the JNK-mediated AP-1 pathway in lipopolysaccharide-activated RAW 264.7 cells. *Eur. J. Pharm. 723*, 339–345.

Chen, Y., Chen, X., Que, W., and Yang, J. (2012). Extraction of volatile oil from *Anoectochilus roxburghii* and the antitumour effect. *Zhongguo Yaoye 21*(6), 21–22.

Chen, Y., Zhou, C., Ge, Z., Liu, Y., Liu, Y., Feng, W. et al. (2013c). Composition and potential anticancer activities of essential oils obtained from myrrh and frankincense. *Oncol. Lett. 6*, 1140–1146.

Civitelli, L., Panella, S., Marcocci, M. E., De Petris, A., Garzoli, S., Pepi, F. et al. (2014). Invitro inhibition of herpes simplex virus type 1 replication by *Mentha suaveolens* essential oil and its main component piperitenone oxide. *Phytomedicine 21*, 857–865. doi:10.1016/j.phymed.2014.01.013.

Conforti, F., Menichini, F., Formisano, C., Rigano, D., Senatore, F., Bruno, M. et al. (2012). *Anthemis wiedemanniana* essential oil prevents LPS-induced production of NO an RAW 264.7 macrophages and exerts antiproliferative and antibacterial activities in vitro. *Nat. Prod. Med. 26*(17), 1594–1601.

Costa, E. V., da Silva, T. B., Menezes, L. R., Ribeiro, L. H., Gadelha, F. R., de Carvalho, J. E. et al. (2013a). Biological activities of the essential oil from the leaves of *Xylpia laevigata* (Annonaceae). *J. Ess. Oil Res. 25*(3), 179–185.

Costa, E. V., Dutra, L. M., Salvador, M. J., Ribeiro, L. H., Gadelha, F. R., and de Carvalho, J. E. (2013b). Chemical composition of the essential oils of *Annona pickelii* and *Annona salzmanii* (Annonaceae), and their antitumour and trypanocidal activities. *Nat. Prod. Res. 27*(11), 997–1001.

Cui, L., Ma, Y., and Han, H. (2011). Enhancing effect of volatile oil of *Rhizoma Zingiberis*, *Flos Magnoliae* and *Fructus Litseae* on permeation of rotundine in vitro. *Zhong Yao Cai. 34*(5), 753–757.

Cunha, L. C., de Morais, S. A., Martins, C. H., Martins, M. M., Chang, R., de Aquino, F. J. et al. (2013). Chemical composition, cytotoxic and antimicrobial activity of essential oils from *Cassia bakariana* Craib. against aerobic and anaerobic oral pathogens. *Molecules, 18*, 4588–4598.

da Cunha, G. H., de Moraes, M. O., Fechine, F. V., Frota Bezerra, F. A., Rocha Silveira, E., Marques Canuto, K. et al. (2013). Vasorelaxant and antihypertensive effects of methanolic fraction of essential oil of *Alpinia zerumbet*. *Vasc. Pharmakol. 58*, 337–345.

Da Silva, E. B., Matsuo, A. L., Figueiredo, C. R., Chaves, M. H., Sartorelli, P., and Lago, J. H. (2013). Chemical constituent and cytotoxic evaluation of essential oil from leaves of *Porcelia macrocarpa* (Annonaceae). *Nat. Prod. Commun. 8*(2), 277–279.

Dai, Z.-J., Tang, W., Lu, W.-F., Goa, J., Kang, H.-F., Ma, X.-B. et al. (2013). Antiproloferative and apoptotic effects of b-elemene on human hepatoma HepG2 cells. *Cancer Cell Int. 13*, 27.

Dandlen, S. A., Lima, S., Mendes, M. D., Miguel, M. G., Faleiro, M. L., Sousa, M. J. et al. (2011). Antimicrobial activity, cytotoxic and intracellular growth inhibition of *Portuguese Thymus* essential oils. *Braz. J. Pharm. 21*(6), 1012–1024.

Dar, M. Y., Shah, W. A., Mubashir, S., and Rather, M. A. (2012). Chromatographic analysis, anti-proliferative and radical scavenging activity of *Pinus wallichina* essential oil growing in high altitude areas of Kashmir, India. *Phytomedicine 19*, 1228–1233.

de Azeredo, G. A., Montenegro Stamford, T. L., de Figueiredo, R. C., and de Souza, E. (2012). The cytotoxic effect of essential Ois from *Origanum vulgare* L. and/or *Rosmarinus officinalis* L. an *Aeromonas hydrophila*. *Foodborne Path. Dis. 9(4)*, 298–304.

de Carvalho, A. A., da Silva, A., de Sousa, E. A., Matsuo, A. L., Lago, J. H., and Chaves, M. H. (2013a). Intraspecific variation and cytotoxic evaluation of the essential oils from *Oxandra sessilifolia* R. E. Fries. *J. Med. Plants Res. 7(9)*, 504–508.

de Carvalho, B. A., Domingos, O. S., Massoni, M., dos santos, M. H., Ionta, M., Lago, J. H. et al. (2013b). Essential oils from *Caesalpina petophoroides* flowers—Chemical composition and in vitro cytotoxic evaluation. *Nat. Prod. Commun. 8(5)*, 679–682.

De Martino, L., D'Arena, G., Minervini, M. M., Deaglio, S., Sinisi, N., Cascavilla, N. et al. (2011). Active caspase-3 detection to evaluate apoptosis induced by *Verbena officinalis* essential oil and citral in chronic lymphicytic leukaemia cells. *Braz. J. Pharm. 21(5)*, 869–873.

de Melo, G. A., Grespan, R., Fonseca, J. P., Farinha, T. O., Silva, E. L., Romero, A. L. et al. (2011). *Rosmarinus officinalis* L. essential oil inhibits in vivo and in vitro leukocyte migration. *J. Med. Food 14 (9)*, 944–949.

De Menezes, I. A., Moreira, I. J., De Paula, J. W., Blank, A. F., anatoniolli, A. R., Quintans-Junior, L. J. et al. (2010). Cardiovascular effects induced by *Cymbopogon winterianus* essential oil in rats: Involvement of calcium channels and vagal pathway. *J. Pharm. Phamakol. 62*, 215–221.

de Oliveira, M. G., Marques, R. B., de Santana, M. F., Santos, A. B., Brito, F. A., Barreto, E. O. et al. (2012). α-terpineol reduces mechanical hypernociception and inflammatory response. *Basic Clin. Pharmakol. Toxikol. 111*, 120–125.

de Paula, J. A., Silva, M. D., Costa, M. P., Diniz, D. G., Sa, F. A., Alves, S. F. et al. (2012). Phytochemical analysis and antimicrobial, antinociceptive, and anti-inflammatory activities of two chemotypes of *Pimenta pseudocaryophyllus* (Myrtaceae). *Evid. Based Complement. Alternat. Med. 2012*, 420715.

de Pinho, J. P., Silva, A. S., Pinheiro, B. G., Sombra, I., Bayma, J. d., Lahlou, S. et al. (2012). Antinociceptive and antispasmodic effects of the essential oil of *Ocimum micranthum*: Potential anti-inflammatory properties. *Planta Med. 78(7)*, 681–685.

de Santana, M. T., de Oliveira, M. G., Santana, M. F., De Sousa, D. P., Santana, D. G., Camargo, E. A. et al. (2013). Citronellal, a monoterpene present in java citronella oil, attenuates mechanical nociception response in mice. *Pharm. Biol. 51(9)*, 1144–1149.

de Sousa, D. P., Nobrega, F. F., de Lima, M. R., and de Almeida, R. N. (2011). Pharmacologocal activity of ®-(+)-pulegone, a chemical constituent of essential oils. *Z. Naturforsch. C 66(7–8)*, 353–359.

Deng, Y.-L., Zhou, L.-L., and Chen, D.-Y. (2011). In vivo and in vitro evaluation of essential oil from *Caulis sinomenii* as enhancers to accelerate transdermal drug delivery. *Lia. Zhong. Daxue Xue. 13(4)*, 23–25.

Devi, K. P., Sakthivel, R., Nisha, S. A., Suganthy, N., and Pandian, S. K. (2013). Eugenol alters the integrity of cell membrane and acts against the nosocomial pathogen *Proteus mirabilis*. *Arch. Pharm. Res. 36*, 282–292.

Doll-Boscardin, P. M., Sartoratto, A., Maia, B. H., de Paula, J. P., Nakashima, T., Farago, P. V. et al. (2012). In vitro cytotoxic potential of essential oils of *Eucalyptus benthamii* and its related terpenes on tumor cell lines. *Evid. Based Complement. Alternat. Med. 2012*, 342652.

Domiciano, T. P., Dalalio, M. M., Silva, E. L., Ritter, A. M., Estevao-Silva, C. F., Ramos, F. S. et al. (2013). Inhibitory effect of anethole in nonimmune acute inflammation. *Naunyn-Schmiedenberg's Arch Pharmakol. 386*, 331–338.

Dordevic, A., Lazarevic, J., Smelcerovic, A., and Stojanovic, G. (2013). The case of *Hypericum rochelii* Griseb. & Schenk and *Hyperikum umbellatum* A. Kern. essential oils: Chemical composition and antimicrobial activity. *J. Pharm. Biomed. Anal. 77*, 145–148.

Dung, N. T., Bajpai, V. K., Yoon, J. I., and Kang, S. C. (2009). Anti-inflammatory effects of essential oil isolated from the buds of *Cleistocalyx operculatus* (Roxb.) Merr and Perry. *Food Chem. Toxicol. 47*, 449–453.

Dunkic, V., Vuko, E., Bezic, N., Kremer, D., and Ruscic, M. (2013). Composition and antiviral activity of the essential oils of *Eryngium alpinum* and *E. amethystinum*. *Chem. Biodivers. 10*, 1894–1902.

Dutra, R. C., Pittella, F., Dittz, D., Marcon, R., Pimenta, D. S., Lopes, M. T. et al. (2012). Chemical composition and cytotoxic activity of the essential oil of *Pterodon emarginatus*. *Braz. J. Pharmakog. 22(5)*, 971–978.

El-Ahmady, S., Ashour, M. L., and Wink, M. (2013). Chemical composition and anti-inflammatory activity of the essential oils of *Psidium guayava* fruits and leaves. *J. Ess. Oil Res. 25(6)*, 475–481.

Elissi, A., Rouis, Z., Salem, N. A., Mabrouk, S., Salem, Y. B., Salah, K. B. et al. (2012). Chemical composition of 8 eucalyptus species essential oils and the evaluation of their antibacterial, antifungal and antiviral activities. *BMC Complement. Alternat. Med. 12*, 81.

Erdogan, E. A., Everest, A., Martino, L. D., Mancini, E., Festa, M., and Feo, V. D. (2013). Chemical composition and in vitro cytotoxic activity of the essential oil of stachys ruestris and *Salvia heldreichiana*, two endemic plants of turkey. *Nat. Prod. Commun. 8(11)*, 1637–1640.

Essien, E. E., Ogunwande, I. A., Setzer, W. N., and Ekundayo, O. (2012). Chemical composition, antibacterial, and cytotoxic studies on *S. erianthum* and *S. macranthum* essential oil. *Pharm. Biol. 50(4)*, 474–480.

Essien, E. E., Walker, T. M., Ogunwande, L. A., Bansal, A., Setzer, W. N., and Ekundayo, O. (2010). Essential oil composition, cytotoxic and antimicrobial activities of *Dutra metel* L. from Nigeria. *Int. J. Ess. Oil Ther. 4(1–2)*, 69–72.

Fachini-Queiroz, F. C., Kummer, R., Estevao-Silva, C. F., Carvalho, M. D., Cunha, J. M., Grespan, R. et al. (2012). Effects of thymol and carvacrol, constituents of *Thymus vulgaris* L. essential oil, on the inflammatory response. *Evid. Based Complement. Alternat. Med. 2012*, 657026.

Fadi, O., Abdulkader, R., Samir, A. A., and Ayad, C. M. (2012). The cytotoxic effect of essential oils *Citrus aurantium* peels on human colorectal carcinoma cell line (Lim1863). *J. Microbiol. Biotechnol. Food Sci. 1(6)*, 1476–1487.

Fadli, M., Saad, A., Sayadi, S., Chevalier, J., Mezarioui, N., Pages, J.-M. et al. (2012). Antibacterial activity of *Thymus maroccanus* and *Thymus broussonetii* essential oils against nosocomial infection-bacteria and their synergistic potential with antibiotics. *Phytomedicine 19*, 464–471.

Fawzy, G. A., Ati, H. Y., and Gamal, A. A. (2013). Chemical composition and biological evaluation of essential oils of *Pulicaria jaubertii*. *Pharmacogn. Mag. 9(33)*, 28–32.

Ferraz, R. P., Bomfim, D. S., Carvalho, N. C., Soares, M. B., da Silva, T. B., Machadeo, W. J. et al. (2013a). Cytotoxic effects of leaf essential oil of *Lippia gracilis* Schauer (Verbenaceae). *Phytomedicine 20*, 615–621.

Ferraz, R. P., Cardoso, G. M., da Silva, T. B., Fontes, J. E., Prata, A. P., Carvalho, A. A. et al. (2013b). Antitumor properties of the leaf essential oil of *Xylopia frutescens* Aubl. (Annonaceae). *Food Chem. 141*, 296–200.

Firuzi, O., Asadollahi, M., Gholami, M., and Javidnia, K. (2010). Composition and biological activities of essential oils from four *Heracleum* species. *Food Chem. 122*, 117–122.

Fogang, H. P., Tapondjou, L. A., Womeni, H. M., Quassinti, L., Bramucci, M., Vitali, L. A. et al. (2012). Characterization and biological activity of essential oils from fruits of *Zanthoxylum xanthoxyloides* Lam. and *Z. leurieurii* Guill & Perr., two culinary plants from Cameroon. *Flav. Frag. J. 27(2)*, 171–179.

Fontes, J. E., Ferraz, R. P., Britto, A. C., Carvalho, A. A., Moraes, M. O., Pessoa, C. et al. (2013). Antitumor effect of the essential oil from leaves of *Guatteria pogonopus* (Annonaceae). *Chem. Biodiv. 10*, 722–729.

Formagio, A. S., Vieira, M. C., dos Santos, L. A., Cardoso, C. A., Foglio, M. A., de Carvalho, J. E. et al. (2013). Composition and evaluation of the anti-inflammatory and anticancer activities of the essential oil from *Annona sylvatica* A. St.-Hil. *J. Med. Food 16(1)*, 20–25.

Franca-Neto, A., Cardoso-Teixeira, A. C., Medeiros, T. C., Quinto-Farias, M. S., Sampaio, C. M., Celjo-de-Souza, A. N. et al. (2012). Essential oil of *Croton argyrophylloides*: Toxicological aspects and vasorelaxant activity in rats. *Nat. Prod. Commun. 7(10)*, 1397–1400.

Funk, J. L., Frye, J. B., Oyarzo, J. N., Zhang, H., and Timmermann, B. N. (2010). Anti-arthritic and toxicity of the essential oils of tumeric (*Curcuma longa* L.). *J. Agric. Food Chem. 58*, 842–849.

Garcia, C. C., Acosta, E. G., Carro, A. C., Belmonte, M. C., Bomben, R., Duschatzky, C. B. et al. (2010). Virucidal activity and chemical composition of essential oils from aromatic plants of central west Argentine. *Nat. Prod. Commun. 5(8)*, 1307–1310.

Gazim, Z. C., Rodrigues, F., Amorin, A. C., de Rezende, C. M., Sokovic, M., Tesevic, V. et al. (2014). New natural diterpene-type abietane from *Tetradenia riparia* essential oil with cytotoxic and antioxidant activities. *Molecules 19*, 514–524.

Gbenou, J. D., Ahounou, J. F., Akakpo, H. B., Laleye, A., Yaxi, E., Gbaguidi, F. et al. (2013). Phytochemical composition of *Cymbopogon citratus* and *Eucalyptus citiodora* essential oils and their anti-inflammatory and analgesic properties on Wistar rats. *Mol. Biol. Rep. 40*, 1127–1134.

Ghannadi, A., Bagherinejad, M. R., Abedi, D., Jalali, M., Absalan, B., and Sadeghi, N. (2012). Antibacterial activity and composition of essential oils from *Pelargonum graveolens* L'Her and *Vitex agus-castus* L. *Iran. J. Microbiol. 4*, 171–176.

Gomez, L. A., Stashenko, E., and Ocazionez, R. E. (2013). Comparative study on in vitro activities of citral, limonene and essential oil from *Lippia citriodora* and *L. alba* on yellowq fever virus. *Nat. Prod. Commun. 8(2)*, 249–252.

Greay, S. J., Ireland, D. J., Kissick, H. T., Levy, A., Beilharz, M. W., Riley, T. V. et al. (2010). Induction of necrosis and cell cycle arrest in murine cancer cell lines by *Melaleuca alternifolia* (tea tree) oil and terpinen-4-ol. *Cancer Chemother. Pharmacol. 65*, 877–888.

Grespan, R., Paludo, M., Lemos, H. P., Barbosa, C. P., Bersani-Amado, C. A., Dalalio, M. M. et al. (2012). Anti-arthritic effect of eugenol an collagen-induced arthritis experimental model. *Biol. Pharm. Bull. 35(10)*, 1818–1820.

Guimaraes, A. G., Silva, F. V., Xavier, M., Santos, M. R., Oliveira, R. C., Oliveira, M. G. et al. (2012a). Orofacial analgesic-like activity of carvacrol in rodent. *Z. Naturforschung 67(9–10)*, 481–485.

Guimaraes, A. G., Xavier, M. A., de Santana, M. T., Camargo, E. A., Santos, C. A., Brito, F. A. et al. (2012b). Carvacrol attenuates mechanical hypernociception and inflammatory response. *Naunyn-Schmiedberger's Arch. Pharmacol. 385*, 253–263.

Haddouchi, F., Chaouche, T. M., Zaouali, Y., Ksouri, R., Attou, A., and Benmansour, A. (2013). Chemical composition and antimicrobial activity of the essential oils from four Ruta species growing in Algeria. *Food Chem. 141*, 253–258.

Hajhashemi, V., Sajjadi, S. E., and Heshmati, M. (2009). Anti-inflammatory and analgesic properties of *Heracleum persicum* essential oil and hydroalcoholic extract in animal model. *J. Ethnopharmacol. 124*, 475–480.

Hajhashemi, V., Sajjadi, S. E., and Zomorodkia, M. (2011). Antinociceptive and anti-inflammatory activities of *Bunium persicum* essential oil, hydroalcoholic and polyphenolic extracts in animal models. *Pharm. Biol. 498(2)*, 146–151.

Hajhashemi, V., Zolfaghari, B., and Yousefi, A. (2010). Antinociceptive and anti-inflammatory activiteis of *Satureja hortensis* seed essential oil, hydroalcoholic and polyphenolic extracts in animal models. *Med. Princ. Pract. 21*, 178–182.

Han, H., Ma, Y., Cui, L., Zhang, H., and Kang, S. (2011). Primarily exploration of preparation of ligustrazine phosphate transdermal patches and effect on permeation enhancing of volatile oil of *Flos magnoliae*. *Zhong. Yaoxue Zazhi 46(24)*, 1915–1918.

Hassine, M., Zardi-Berguaoui, A., Znati, M., Flamini, G., Jannet, H. B., and Hamza, M. A. (2014). Chemical composition, antibacterial and cytotoxic activities of the essential oil from the flowers of Tunisian *Convolvulus althaeoides* L. *Nat. Prod. Res. 28(11)*, 769–775.

Hayato, M., Yamazaki, M., and Katagata, Y. (2012). Kuramoji (*Lindera umbellata*) essential oil-induced apoptosis and differentiation in human leukemia HL-60 cells. *Exp. Ther. Med. 3*, 49–52.

Hirota, R., Roger, N. N., Nakamura, H., Song, H.-S., Suwamura, M., and Suganuma, N. (2010). Anti-inflammatory effects of limonenen from Yuzu (Citrus junos Tanaka) essential oil and eosinophils. *J. Food Sci. 75(3)*, H87–H92.

Ho, C. L., Jie-Pinge, O., Liu, C. P., Hung, C. P., Tsai, M. C., Liao, P. C. et al. (2010). Composition and in vitro anticancer activities of the leaf and fruit oils of *Litsea cubeba* from Taiwan. *Nat. Prod. Med. 5(4)*, 617–620.

Hotta, M., Nakata, R., Katsukawa, M., Hori, K., Takahshi, S., and Inoue, H. (2010). Carvacrol, a component of thyme oil, activates PPARa and y and suppresses COX-2 expression. *J. Lipid Res. 51*, 132–139.

Hsouna, A. B., Trigui, M., Mansour, R. B., Jarraya, R. M., Damak, M., and Jaoua, S. (2011). Chemical composition, cytotoxicity effect and antimicrobial activity of *Ceratonia siliqua* essential oil with preservative effects against Listeria inoculated in minced beef meat. *Int. J. Food Microbiol. 148*, 66–72.

Hu, H., Zheng, X., and Hu, H. (2012). Chemical composition, antimicrobial, antioxidant and cytotoxic activity of the essential oil from the leaves of *Acanthopanax leucorrhizus* (Oliv.) Harms. *Environ. Toxicol. Pharmacol. 34*, 618–623.

Huo, M. D., Cui, X., Xue, J., Chi, G., Gao, R., Deng, X. et al. (2013). Anti-inflammatory effects of linalool in Raw 264.7 macrophages and lipopolysaccharide-induced lung injury model. *J. Surg. Res. 180*, E47–E54.

Hussain, A. I., Anwar, F., Chatha, S. A., Jabbar, A., Mahboob, S., and Nigam, P. S. (2010). *Rosmarinus officinalis* essential oil: Antiproliferative, antioxidant and antibacterial activities. *Braz. J. Microbiol. 41*, 1070–1078.

Hussain, A. I., Anwar, F., Rasheed, S., Nigam, P. S., Janneh, O., and Sarker, S. D. (2011). Composition, antioxidant and chemotherapeutic properties of the essential oils from two Origanum species growing in Pakistan. *Braz. J. Pharm. 21(6)*, 943–952.

Ichihashi, K. and Jimbo, D. (2012). Latest trend in study of patchuli essential oil and its prospects. *Aromatopia 111*, 29–32.

Inocencio, L. L., Leite, G. O., Silva, C. T., de Sousa, S. D., Sampaio, R. S., da Costa, J. G. et al. (2014). Topical antinociceptive effect of *Vanillosmopsis arboea* Baker on acute corneal pain in mice. *Evid. Based Complement. Alternat. Med.* Article ID 708636, 6pp.

Interaminense, L. D., dos Ramos-Alves, F. E., de Siqueira, R. J., Xavier, F. E., Duarte, G. P., Magalhaes, P. J. et al. (2013). Vasorelaxant effects of 1-nitro-2-phenylethane, the main constituent of the essential oil of *Aniba canelilla*, in superior mesenteric arteries from spontaneously hypertensive rats. *Eur. J. Pharm. Sci. 48*, 709–716.

Interaminense, L. F., de Siqueira, R. J., Xavier, F. E., Duarte, G., Magalhaes, P. J., da siva, J. K. et al. (2011). Cardiovascular effects of 1-nitro-2phenylethane, the main constituent of the essential oil of *Aniba canelilla*, in spontaneously hypertensive rats. *Fund. Clin. Pharmacol. 25*, 661–669.

Jaganathan, S. K. and Supriyanto, E. (2012). Antiproliferative and molecular mechanism of eugenol-induced apoprosis in cancer cells. *Molecules 17*, 6290–6304.

Jayaprakasha, G. K., Murthy, C., Uckoo, R. M., and Patil, B. S. (2013). Chemical composition of volatile oil from *Citrus limettioides* and their inhibition of colon cancer cell proliferation. *Indust. Crops Prod. 45*, 200–207.

Jayaprakasha, G. K., Murthy, K. N., Demarais, R., and Patil, B. S. (2012). Inhibition of prostate cancer (LNCaP) cell proliferation by volatile components from Nagami kumquats. *Planta Med. 78*, 974–980.

Jeena, K., Vijayasteltar, B., Umadevi, N. P., and Kuttan, R. (2014). Antioxidant, anti-inflammatory and antinociceptive properties of black pepper essential oil (*Piper nigrum* Linn). *J. Ess. Oil Bear. Plants 17(1)*, 1–12.

Jeong, J. B., Choi, J., Lou, Z., Jiang, X., and Lee, S.-H. (2013). Patchuli alcohol, an essential oil of *Pogostemon cablin*, exhibits anti-tumorgenic activity in human colorectal cancer cells. *Int. Immunopharmacol. 16*, 184–190.

Jo, J.-R., Park, J. S., Park, Y.-K., Chae, Y. Z., Lee, G.-Y., Park, G.-Y. et al. (2012). *Pinus desiflora* leaf essential oil induces apoptosis via ROS generation and activation of caspases in YD-8 human oral cancer cells. *Int. J. Oncol. 40*, 1238–1245.

Juarez-Reyes, K., angeles-Lopez, G. E., Rivero-Cruz, I., Bye, R., and Mata, R. (2014). Antinociceptive activity of Ligusticum preparations and compounds. *Pharm. Biol. 52(1)*, 14–20.

Kang, P., Kim, K. Y., Lee, H. S., Min, S. S., and Seol, G. H. (2013). Anti-inflammatory effects of anethole in lipopolysaccharide-induced acute lung injury in mice. *Life Sci. 93(5)*, 955–961.

Kang, P., Suh, S. H., Min, S. S., and Seol, G. (2012). The essential oil of *Citrus bergamia* Risso induces vasorelaxation of the mouse aorta by activation K^+ channels and inhibiting Ca^+ influx. *J. Pharm. Pharmacol. 65(5)*, 745–749.

Karpanen, T. J., Conway, B. R., Worthington, T., Hilton, A. C., Elliott, T. S., and Lambert, P. A. (2010). Enhanced chlorhexidine skin penetration with eucalyptus oil. *BMC Inf. Dis. 10*, 278.

Kasrati, A., Alaoui Jamal, C., Bekkouche, K., Lahcen, H., Markouk, M., Wohlmuth, H. et al. (2013). Essential oil composition and antimicrobial activity of wild cultivated mint timija (*Mantha suaveolens* subsp. timija (Briq.) Harley), an endemic and threatened medicinal species in Morocco. *Nat. Prod. Res. 27(12)*, 1119–1122.

Kathirvel, P. and Ravi, S. (2012). Chemical composition of the essential oil from basil (*Ocimum basilicum* Linn.) and its in vitro cytotoxicity against HeLa and HEp-2 human cancer cell lines and NIH 3T3 mouse embryonic fibroblasts. *Nat. Prod. Res. 26(12)*, 1112–1118.

Katsukawa, M., Nakata, R., Takizawa, Y., Hori, K., Takahashi, S., and Inoue, H. (2010). Citral, a component of lemongrass oil, activates PPARa and y and suppresses COX-2 expression. *Biochim. et Pyioph. Acta 1801*, 1214–1220.

Kavoosi, G. and Rowshan, V. (2013). Chemical composition, antioxidant and antimicrobial activities of essential oil obtained from *Ferula assa-foetida* oleo-gum-rasin: Effect of collecting time. *Food Chem. 138*, 2180–2187.

Kavoosi, G., Tafsiry, A., Ebdam, A., and Rowshan, V. (2013). Evaluation of antioxidant and antimicrobial activities of essential oils from *Carum copticum* seed and *Ferula assafoetida* Latex. *J. Food Sci. 78(2)*, T356–T361.

Khalid, M. H., Akthar, M. N., Mohamad, A. S., Periml, E. K., Akira, A., Israf, D. A. et al. (2011). Antinociceptive effects of the essentiaoil of *Zingiber zerumbet* in mice: Possible mechanism. *J. Ethnopharmacol. 137*, 345–351.

Khamsan, S., Liawruangrath, B., Liawruangrath, S., Teerawutkulrag, A., Pyne, S. G., and Garson, M. J. (2011). Antimalaria, anticancer, antimicrobial activities and chemical constituents of essential oil from the aerial parts of *Cyperus kyllingia* Endl. *Rec. Nat. Prod. 5(4)*, 324–327.

Kim, J.-I. (2012). Anti-porcine epidemic diarrhea virus (PEDV) activity and antimicrobial activities of *Artemisia dubia* essential oil. *Han'guk Misa. Saeng. Hak. 40(4)*, 396–402.

Kim, J.-Y., Kim, S.-S., Oh, T.-H., Baik, J., Song, G., Lee, N. H. et al. (2009). Chemical composition, antioxidant, anti-elastace, and anti-inflammatory activities of *Illicium anisatum* essential oil. *Acta Pharm. 59*, 289–300.

Kouidhi, B., Zmantar, T., and Bakhrouf, A. (2010). Anticarcinogenic and cytotoxic activity of clove essential oil (*Eugenia caryophyllata*) against a large number of oral pathogens. *Ann. Microbiol. 60*, 599–604.

Kulisic-Bilusic, T., Schmöller, I., Schnäbele, K., Siracusa, L., and Ruberto, G. (2012). The anticarcinogenic potential of essential oil and aqueous infusion from caper (*Capparis spinosa* L.). *Food Chem. 132*, 261–267.

Lai, W.-L., Chuang, H.-S., Lee, M.-H., Wei, C.-L., Lin, C.-F., and Tsai, Y.-C. (2012). Inhibition of herpes simplex virus type 1 by thymol-related monoterpenoids. *Planta Med. 78*, 1636–1638.

Lan, Y., Wu, Q., Mao, Y.-Q., Wang, Q., An, J., Chen, Y.-Y. et al. (2014). Cytotoxic and enhancer activity of essential oil from *Zanthoxylum bungeanum* Mixim. as a natural transdermal penetration enhancer. *J. Zheij. Univ. Sci. B* 15(2), 153–164.

Leite, B. L., Bonfim, R. R., Antoniolli, A. R., Thomazzi, S. M., Arauja, A. A., Blank, A. F. et al. (2012). Assessment of antinociceptive, anti-inflammatory and antioxidant properties of *Cymbopogon winterianus* leaf essential oil. *Pharm. Biol.* 48(10), 1164–1169.

Leite, G. D., Leite, L. H., Sampaio, R. D., Araruna, M. K., de Menezes, I. R., da Costa, J. G. et al. (2011). (−)-α-bisabolol attenuates visceral nociception and inflammation in mice. *Fitoterapia 82*, 208–211.

Lesjak, M. M., Beara, I. N., Orcic, D. Z., Petar, K. N., Simin, N. D., Emilija, S. D. et al. (2014). Phytochemical composition and antioxidant anti-inflammatory and antimicrobial activities of *Juniperus macrocarpa* Sibeth. et Sm. *J. Funct. Food 7*, 257–268.

Li, D., Pan, S., Zhu, X., and Tan, L. C. (2012a). Anticancer activity and chemical composition of leaf essential oil from *Solidago canadensis* L. in China. *Adv. Mater. Res. 347–353*, 1584–1589.

Li, M., Shao, L., Xu, L., Cheng, W., Gao, H., and Ding, Q. (2012b). Constituents and biological activity of volatile oil from umbels of *Daucus carota* Linn. *Zhong. Lian. Xue.* 27(9), 112–115.

Li, Q., Ma, Y.-S., Yang, F., Zhang, H.-L., and Wang, B. (2013a). Percutaneous enhancement effect of volatile oil of *Rhizoma Zingiberis*, *Flos Caryophyllata* and *Fructus Litsea* on total alkaloids of semen strychni transdermal patch. *Schiz. Guoyi Guoy.* 24(10), 2321–2324.

Li, Z.-D., Han, S.-N., Jiang, J.-L., Zhang, X.-H., Li, Y., Chen, H. et al. (2013b). Antitumor compound identification from *Zanthoxylum bungeanum* essential oil based an composition-activity relationship. *Chem. Res. Chin. Univ.* 29(6), 1068–1071.

Liang, J., Huang, B., and Wang, G. (2012). Chemical composition, antinociceptive and anti-inflammatory properties of essential oil from the roots of *Illicium lanceolatum*. *Nat. Prod. Res.* 26(18), 1712–1714.

Liju, V. B., Jeena, K., and Kuttan, R. (2011). An evaluation of antioxidamt, anti-inflammatory and anti-nociceptive activities of essential oil from *Curcuma longa* L. *Ind. J. Pharm.* 43(5), 526–531.

Lima, D. F., Brandaos, M. S., Moura, J. B., Leitao, J. M., Carvalho, F. A., Miura, L. M. et al. (2012a). Antinociceptive activity of the monoterpene α-phellandrene in rodents: Possible mechanism of action. *J. Pharm. Pharmaocol.* 64(2), 283–292.

Lima, D. K., Ballico, L. J., Lapa, F. R., Goncalves, H. P., de Souza, L. M., Iacomini, M. et al. (2012b). Evaluation of the antinociceptive, anti-inflammatory and gastric antiulcer activities of the essential oil from *Pipier aleyreanum* C.DC in rodent. *J. Ethnopharmacol.* 142, 274–282.

Lin, J., Dou, J., Xu, J., and Aisa, H. A. (2012). Chemical composition, antimicrobial and antitumor activities of the essential oil and crude extracts of *Euphorbia macrorrhiza*. *Molecules* 17, 5030–5039.

Liu, K., Zuh, Q., Zhang, J., Xu, J., and Wang, X. (2012a). Chemical composition and biological activities of the essential oil of *Mentha spicata* Lamiaceae. *Adv. Mater. Res.* 524–527, 2269–2272.

Liu, R., He, T., Zeng, N., Chen, T., Gou, L., and Liu, J.-W. (2013). Mechanism of anti-influenza virus of volatile oil in *Cinnamomi ramulus* and cinnamaldehyde. *Chin. Trad. Herb. Drugs* 44(11), 1460–1464.

Liu, Y., Cao, L., Jia, X.-G., Li, X.-J., and Liu, J.-Y. (2012b). Chemical composition, antimicrobial and antiviral activities of the essential oil of *Elsholtzia densa* Benth. *Tian. Chan. Yan. Yu Kaifa* 24(8), 1070–1074.

Liu, Z., Gao, W., Zhang, J., and Hu, J. (2011). Antinociceptive and smooth muscle relaxant activity of *Croton tiglium* L. Seed: An in-vitro and in-vivo study. *Iran. J. Pharm. Res.* 11(2), 611–620.

Lone, S. H., Bhat, K. A., Bhat, H. M., Majeed, R., Anand, R., Hamid, A. et al. (2014). Essential oil composition of *Senecio graciliflorus* DC: Comperative analysis of different parts and evaluation of antioxidant and cytotoxic activities. *Phytomedicine* 21(6), 919–925.

Lopes Pereira, S., Mesquita, A., Takashi, R., Coelho, A. A., and Zapata-Sudo, G. (2013). Vasodilator activity of the essential oil from aerial parts of *Pectis brevipedunculata* and its mail constituent citral in rat aorta. *Molecules 18*, 3072–3085.

Luo, H.-M., Ma, Y.-S., Huang, J.-E., Zhang, G.-H., and Kang, S.-J. (2012). Percutanous enhancement function of volatile oil from *Eugenia caryophyllata* on ligustracine phoshate patch. *Zhon. Shi. Fang. Zazhi* 18(11), 40–43.

Luo, H.-M., Ma, Y.-S., Huang, J.-E., Zhang, G.-H., and Kang, S.-J. (2013). Enhancing effect of three kinds of volatile oil on permeation of ligustrazine phosphate patches trough hairless skin mice. *Zhong. Yiyuan yao. Zathi* 33(13), 1032–1035.

Mabrouk, S., Bel Hadj Salah, K., Elaissi, A., Jlaiel, L., Ben Jannet, H., Aouni, M. et al. (2013). Chemical composition and antimicrobial activity of Iunesian Conyza sumatrensis (RETZ.) e. Walker essential oils. *Chem. Biodiv.* 10, 209–223.

Maeda, H., Yamazaki, M., and Katagata, Y. (2013). Kurumoij (*Lindera umbellata*) essential oil inhibits LPS-induced inflammation in RAW 264.7 cells. *Biosci. Biotechnol. Biochem.* 77(3), 482–486.

Maggi, F., Randriana, R. F., Rasoanaivo, P., Nicoletti, M., Quassinti, L., Bramucci, M. et al. (2013). Chemical composition and in vitro biological activites of the essential oil of *Vepris macrophylla* (Baker) I. Verd. endemic to Madagascar. *Chem. Biodiv. 10*, 356–366.

Maham, M., Moslemzadeh, H., and Jalilzadeh-Amin, G. (2014). Antinociceptive effect of the essential oil of tarragon (*Artemisia dranunculus*). *Pharm. Biol. 52(2)*, 208–212.

Mahboubi, M. and Kazempour, N. (2011). Chemical composition and antimicrobial activity of *Satureja hortensis* and *Trachyspermum copticom* essential oil. *Iran. J. Microbiol. 3(4)*, 194–200.

Mahmoud, G. I. (2013). Biological effects, antioxidant and anticancer activites of marigold and Basil essential oil. *J. Med. Plant Res. 7(10)*, 561–572.

Manassero, C. A., Girotti, J. R., Mijailovsky, S., de Bravo, M. G., and Polo, M. (2013). In vitro comparative analysis of antiproliferative activity of essential oil from mandarin peel and its principal component limonene. *Nat. Prod. Res. 27(16)*, 1475–1478.

Manjamalai, A. and Grace, V. M. (2013). The chemotherapeutic effect of essential oil of *Plectranthus amboinicus* (Lour) on lung metastasis developed by B16F-10 cell line in C57Bl/6 mice. *Cancer Invest. 31(1)*, 74–82.

Manjamalai, A., Kumar, M. J., and Grace, V. M. (2012). Essential oil *Tridax procumbens* L induces apoptosis and suppresses angiogenesis and lung metastasis af the B16F-10 cell line in C57BL/6 mice. *Asian Pacif. J. Cancer Prev. 13*, 5887–5895.

Maxia, A., Frau, M. A., Falconieri, D., Karchili, M. S., and Kasture, S. (2011). Essential oil of *Myrthus communis* inhibits inflammation in rats by reducing serum IL-6 and TNF-alpha. *Nat. Prod. Commun. 6(10)*, 1545–1548.

Maxia, A., Sanna, C., Frau, M. A., Piras, A., Karchuli, M. S., and Kasture, V. (2010). Anti-inflammatory activity of *Pistacia lentiscus* essential oil: Involvement of IL-6 and TNF-alpha. *Nat. Prod. Commun. 6(10)*, 1543–1544.

Mendes, S. S., Bomfim, R. R., Jesus, H. C., Alves, P. B., Blank, A. F., Estevam, C. S. et al. (2010). Evaluation of the analgesic and anti-inflammatory effects of the essential oil of *Lippia gracilis* leaves. *J. Ethnopharmakol. 129*, 391–397.

Misharina, T. A., Burlakova, E. B., Fatkullina, L. D., Alinkina, E. S., Vorob'eva, A. K., Medvedeva, A. K. et al. (2013). Effects of oregano essential oils on the engraftment and development of Lewis carcinoma in F1 DBA C57 black hybrid mice. *Prikl. Biokhim. Mikrobiol. 49(4)*, 423–428.

Misra, L. N., Vyry Wouatsa, N. A., Kumar, S., Venkatesh Kumar, R., and Tchoumbougnang, F. (2013). Antibacterial, cytotoxic activities and chemical composition of fruits of two Cameroonian Zanthoxylum species. *J. Ethnopharmcol. 148*, 74–80.

Mitoshi, M., Kuriyama, I., Nakayama, H., Miyazato, H., Sugimoto, K., Kobayashi, Y. et al. (2012). Effects of essential oils from herbal plants and citrus fruits on DNA polymerase inhibitory, cancer cell growth inhibitory, antiallergic, and antioxidant activity. *J. Agric. Food Chem. 60*, 11343–11350.

Mohamed, A., Ali, S. I., and El-Baz, F. (2013). Antioxidant and antibacterial activities of crude extracts and essential oils of *Syzygium cumini* leaves. *PLoS ONE 8(4)*, e60269.

Mueller, K., Blum, N. M., and Mueller, A. S. (2013). Examination of the anti-inflammatory, antioxidant and xenobiotic-inducing potential of broccoli extract and various essential oils during mild DSS-induced colitis in rats. *ISRN Gastroenterol. 2013*, 710–856.

Mulyaningsih, S., Youns, M., El-Readi, M. Z., ashour, M. L., Nibret, E., Sporer, F. et al. (2010). Biological activity of the essential oil of *Kadsura longipendiculata* (Schisandraceae) and its major components. *J. Pharm. Pharmacol. 62(8)*, 1037–1044.

Murakami, C., Lago, J. H., Perazzo, F. F., Ferreira, K. S., Lima, M. E., Moreno, P. R. et al. (2013). Chemical composition and antimicrobial activity of essential oils from *Chromolaena laevigata* during flowering and fruiting stages. *Chem. Biodiv. 10*, 621–627.

Muthaiyan, A., Biswas, D., Crandall, P. G., Wilkinson, B. J., and Ricke, S. C. (2012). Application of orange essential oil as an antistaphylococcal agent in a dressing model. *BMC Complement. Alternat. Med. 12*, 125.

Nagappan, T., Ramasamy, P., Wahid, M. E., Segaran, T. C., and Vairappan, C. S. (2011). Biological activity of carbazole alkaloids and essential oil of *Murraya koenigii* against antibiotic resistant microbes and cancer cell lines. *Molecules 16*, 9651–9664.

Nazaruk, J., Karna, E., Wieczorek, P., Sacha, P., and Tryniszewska, E. (2010). In vitro antiproliferative and antifungal activity of essential oils from *Erigeron acris* L. and *Erigeron annuus* (L.) Pers. *J. Biosci. 65(11/12)*, 642–646.

Neves, A., Rosa, S., Goncalves, J., Rufino, A., Judas, F., Salgueiro, L. et al. (2010). Screening of five essential oils for identification of potential inhibitors of IL-1-induced NF-kB activation and NO production in human chondrocytes: Characterisation of the activity of α-pinene. *Planta Med. 76*, 303–308.

Nicoletti, M., Maggi, F., Papa, F., Vittori, S., Quassinti, L., Bramucci, M. et al. (2012). In vitro biological activities of the essential oil from the resurrection plant *Myrozhamnus moschatus* (Baillon) Niedenzu endemic in Madagaskar. *Nat. Prod. Res. 26(24)*, 2291–2300.

Nicolic, M., Glamoclija, J., Ferreira, I. C., Calhelha, R. C., Fernandes, A., Markovic, T. et al. (2014). Chemical composition, antimicrobial, antioxidant and antitumor activity of *Thymus serpyllum* L., *Thymus algeriensis* Boiss. and Reut and *Thymus vulgaris* L. essential oil. *Ind. Crops Prod. 52*, 183–190.

Ninomiya, K., Hayama, K., Ishijima, S. A., Maruyama, N., Irie, H., Kurihara, J. et al. (2013). Suppression of inflammatory reactions by terpinen-4.ol, a main constituent of tea tree oil, in a murine model of oral candidiasis and its suppressive activity to cytokine production of macrophages in vitro. *Biol. Pharm. Bull. 36(5)*, 838–844.

Noudogbessi, J. P., Delort, L. P., Billard, H., Figueredo, G., Ruiz, N., Chalchat, J. C. et al. (2013). Antiproliferative activity of four aromatic plants of Benin. *J. Nat. Prod. 6*, 123–131.

Ocana-Fuentes, A., Arranz-Gutierrez, E., Senorans, F. J., and Reglero, G. (2010). Supercritical fluid extraction of oregano (*Origanum vulgare*) essential oils: Anti-inflammatory properties based an cytokine response an THP-1 macrophages. *Food Chem. Toxicol. 48*, 1568–1575.

Ocazionez, R. E., Meneses, R., Torrese, F. A., and Stashenko, E. (2010). Virucidal activity of Colombian Lippia essential oil on dengue virus replication in vitro. *Mem. Inst. Oswald Cruz, Rio de Janeiro 105/3*, 304–309.

Ogunwande, I. A., Walker, T. M., Bansal, A., Setzer, W. N., and Essien, E. E. (2010). Essential oil constituent and biological activities of *Peristrophe bicalyculata* and *Borreria verticillata*. *Nat. Prod. Med. 5(11)*, 1815–1818.

Orhan, I. E., Özcelik, B., Kartal, M., and Kan, Y. (2012). Antimicrobial and antiviral effects of essential oils from selected Umbelliferae and Labiatea plants and individual essential oil components. *Turk. J. Biol. 36*, 239–246.

Ouchbani, T., Ouchbani, S., Bouhfid, R., Merghoub, N., Guessous, A. R., Mzibri, M. E. et al. (2011). Chemical composition and antiproliferative activity of *Senecio leucanthemifolius* poiret essential oil. *J. Ess. Oil Bear. Plants 14(6)*, 815–819.

Ozkan, A., Erdogan, A., Sokmen, M., Tugrulay, S., and Unal, O. (2010). Antitumoral and antioxidant effect of essential oil and in vitro antioxidant properties of essential oil and aqueous extracts from *Salvia pisidica*. *Biologia 65(6)*, 990–996.

Pathania, A. S., Guru, S. K., Verma, M. K., Sharma, C., Abdullah, S. T., Malik, F. et al. (2013). Disruption of the PI3K/AKT/mTOR signaling cascade and induction of apoptosis in HL-60 cells by essential oil from *Monarda citriodora*. *Food Chem. Toxicol. 62*, 246–254.

Patharakorn, T., Arpornsuwan, T., Wetprasit, N., Promboon, A., and Ratanapo, S. (2010). Antibacterial activity and cytotoxicity of the leaf essential oil of *Morus rotunbiloba* Koidz. *J. Med. Plants Res. 4(9)*, 837–843.

Paula-Freire, L. I., Andersen, M. L., Molska, G. R., Köhn, D. O., and Carlini, E. L. (2013). Evaluation of the antinociceptive activity of *Ocimum gratissimum* L. (Lamiaceae) essential oil and its isolated active principles in mice. *Phytoth. Res. 27*, 1220–1224.

Pazyar, N. and Yaghoobi, R. (2012). Tea tree oil as a novel antipsoriasis weapon. *Skin Pharmakol. Physiol. 25*, 162–163.

Peixoto-Neves, D., Silva-Alves, K. S., Gomes, M. D., Lima, F. C., Lahlou, S., Magalhaes, P. J. et al. (2010). Vasorelaxant effects of the monoterpenic phenol isomers, carvacrol and thymol on rat isolated aorta. *Fundam. Clinic. Pharmakol. 24*, 341–350.

Piaru, S. P., Mahmud, R., Majid, A. M., Ismail, S., and Man, C. N. (2012a). Chemical composition, antioxidant and cytotoxic activities of the essential oils of *Myritica fragrans* and *Mornda citrifolia*. *J. Sci. Food Agric. 92*, 593–597.

Piaru, S. P., Perumal, S., Cai, L. W., Mahmud, R., Majid, A. M., and Man, C. N. (2012b). Chemical composition, anti-angiogenic and cytotoxic actiities of the essential oils of *Cymbopogon citratus* (lemon grass) against colorectal and breast cancer cell lines. *J. Ess. Oil Res. 24(5)*, 453–459.

Pilau, M. R., Alves, S. H., Weilen, R., Arenhart, S., Cueto, A. P., and Lovato, L. T. (2011). Antiviral activity of the *Lippia graveolens* (Mexican oregano) essential oil and its main compound carvacrol against human and animal virus. *Braz. J. Microbiol. 423*, 1616–1624.

Pinheiro, B. G., Silva, A. S., Souza, G. E., Figueiredo, J. G., Cunha, F. Q., Lahlou, S. et al. (2011). Chemical composition, antinociceptive and antiinflammatory effects in rodents of the essential oil of *Peperomia serpens* (Sw.) Loud. *J. Ethnopharmacol. 138*, 479–486.

Pinho-da-Silva, L., Mendes-Maia, P. V., Teofilo, T. M., Barbosa, R., Caccatto, V. M., Coelho e Souza, A. N. et al. (2012). Trans-Caryophyllene, a natural sesquiterpene, causes tracheal smooth muscle relaxation through blockade of voltage-dependent Ca^{2+} channels. *Molecules 17*, 11965–11977.

Pinto, N. V., Assreuy, A. M., Coelho-de Souza, A. N., Ceccatto, M., Magalhaes, P. J., Lahlou, S. et al. (2009). Endothelium-dependent vasorelaxant effects of the essential oil from aerial parts of *Alpina zerumbet* and its main constituent 1,8-cineol in rats. *Phytomedicine 16*, 1151–1155.

Piras, A., Rosa, A., Marongiu, B., Atzeri, A., Dessi, M. A., Falconieri, D. et al. (2012). Extractn and seperation of volatile and fixed oils from seed of *Mysterica fragrans* by supercritical CO_2: Chemical composition and cytotoxic activity on Caco-2 cancer cells. *J. Food Sci. 77(4)*, C448–C453.

Pires, A. F., Frota Madeira, S. V., Soares, P. M., Montenegro, C. M., Souza, E. P., Resende, A. C. et al. (2012). The role of endothelium in the vasorelaxant effects of the essential oil of *Ocimum gratissimum* in aorta and mesenteric vascular bed of rats. *Can. J. Phys. Pharmacol. 90(10)*, 1380–1385.

Popovic, V., Petrivic, S., Tomic, M., Stepanovic-Petrovic, R., Micovic, A., Pavlovic-Drobac, M. et al. (2014). Antinociceptive and anti-edematous activities of the essential oils of two Balkan endemic Laserpitium species. *Nat. Prod. Commun. 9(1)*, 125–128.

Prabhu, V., Sujina, I., Hemal, H., and Ravi, S. (2011). Essential oil composition, antimicrobila, MRSA and in-vitro cytotoxic activity of fresh leaves of *Chromolaena odorata*. *J. Parm. Res. 4(12)*, 4609–4611.

Quassinti, L., Bramucci, M., Lupidi, G., Barboni, L., Ricciutelli, M., Sagratini, G. et al. (2013a). In vitro biological activity of essential oils and isolated furanosesquiterpenes from the neglected vegetable *Smyrnium olusatrum* L. (Apiaceae). *Food Chem. 138*, 808–813.

Quassinti, L., Lupidi, G., Maggi, F., Sagratini, G., Papa, F., Vittori, S. et al. (2013b). Antioxidant and antiproliferative activity of *Hypericum hircinum* L. subsp. majus (Aiton) N. Robson essential oil. *Nat. Prod. Res. 27(10)*, 862–868.

Quintans, J. D., Soares, B. M., Ferraz, R. P., Oliveira, A. C., da Silva, T. B., Menezes, L. R. et al. (2013). Chemical constituents and anticancer effects of the essential oil from leaves of *Xylopia laevigata*. *Planta Med. 79*, 123–130.

Quintans-Junior, L., da Rocha, R. F., Caregnato, F. F., Moreira, J. C., da Silva, F. A., de Souza Araujo, A. A. et al. (2011). Antinociceptive action and redox properties of citronellal an essential oil present in Lemongrass. *J. Med. Food 14(6)*, 630–639.

Rantzsch, U., Vacca, G., Dück, R., and Gillissen, A. (2009). Anti-inflammatory effects of myrtol standardized and other essential oils on alveolar macrophages from patients with chronic obstructive pulmonary disease. *Eur. J. Med. Res. 14(IV)*, 205–209.

Rashid, S., Rather, M. A., Shad, W. A., and Bhat, B. A. (2013). Chemical composition, antimicrobial, cytotoxic and antioxidant activities of the essential oil of *Artemisia indiaca* Wild. *Food Chem. 138*, 693–700.

Ravi, K. C. and Raghava, R. T. (2014). Lipid profiling by GC-MS and anti-inflammatory activities of *Ceiba pentandra* seed oil. *J. Biol. Active Prod. Nat. 4(1)*, 62–70. doi:10.1080/22311866.2014.890064.

Raymundo, L. J., Guilhon, C. C., Alviano, D. S., Matheus, M. E., Antoniolli, A. R., Cavalcanti, S. C. et al. (2011). Characterisation of anti-inflammatory and antinociceptiv activities of the *Hyptis pectinata* (L.) Poit essential oil. *J. Ethnopharmacol. 134*, 725–732.

Riella, K. R., Marinho, R. R., Santos, J. S., Peireira-Filho, R. N., Cardoso, J. C., Albuquerque-Junior, R. L. et al. (2012). Anti-inflammatory and cicatrizing activities of thymol, a monoterpene of essentiol oil from *Lippia gracilis*, in rodent. *J. Ethnopharmacol. 143*, 656–663.

Rigano, D., Arnold, N. A., Conforti, F., Menichini, F., Formisano, C., Piozzi, F. et al. (2011). Characterisation of the essential oil of *Nepeta glomerata* Montbret et Aucher ex Bentham from Lebanon and its biological activities. *Nat. Prod. Med. 25(6)*, 614–626.

Rigano, D., Russo, A., Formisano, C., Cardile, V., and Senatore, F. (2010). Antiproliferative and cytotoxic effects on malignant melanoma cells of essential oils from aereal parts of *Genista sesilifolia* and *G. tinctoria*. *Nat. Prod. Commun. 5(7)*, 1127–1132.

Rocha, N. F., Rios, E. R., Carvalho, A. M., Cerqueira, G. S., Lopes, A. D., Leal, L. K. et al. (2011). Antinociceptive and anti-inflammatory activites of (−)-α-bisabolol in rodent. *Naunyn-Schmiedberger's Arch Pharmacol. 384*, 525–533.

Roh, J., Lim, H., and Shin, S. (2012). Biological activities of the essential oil from *Angelica acutiloba*. *Nat. Prod. Sci. 18(4)*, 244–249.

Saima, A., Showkat, R., Kanchan, K., Babar, A., and Mohad, A. (2010). A study of the chemical composition of black cumin oil and its effect on penetratin enhancement from transdermal formulation. *Nat. Prod. Res. 24(12)*, 1151–1157.

Salah-Fatnassi, P. B., Slim-Bannour, A., Harzallah-Skhiri, F., Mahjoub, M. A., Mighri, Z., Chaumont, J.-P. et al. (2010). Activities antivirale es antioydante in vitro d'huiles essentielles de *Thymus capitatus* (L.) Hoffmans.& Link de Tunisie. *Acta Bot. Gallica 157(3)*, 433–444.

Salvador, M. J., de Carvalho, J. E., Wisniewski Jr., A., Kassuya, C. A., Santos, E. P., Riva, D. et al. (2011). Chemical composition and cytotoxic activity of the essential oil from the leaves of *Casearia lasiophylla*. *Braz. J. Pharm. 21(5)*, 864–868.

Sanseera, D., Niwatananun, W., Liawruangrath, B., Liawruangrath, S., Baramee, A., and Pyne, S. G. (2012). Chemical composition and biological activities of the essential oils from leaves of *Cleidion javancum* Bl. *J. Ess. Oil Ber. Plants 15(29*, 186–194.

Seal, S., Chatterjee, P., Bhattacharya, S., Pal, D., Dasgupta, S., Kundu, R. et al. (2012). Vapor of volatile oils from *Litsea cubeba* seed induces apoptosis and causes call cycle arrest in lung cancer cells. *PLoS ONE 7(10)*, e47014. doi:10.1371/journal.pone.0047014.

Seki, T., Kokuryo, T., Yokoyama, Y., Suzuki, H., Itatsu, K., Nakagawa, A. et al. (2011). Antitumor effects of α-bisabolol against pancreatic cancer. *Off. J. Jap. Cancer Assoc. 102(12)*, 2199–2205.

Sertel, S., Eichhorn, T., Plinkert, P. K., and Efferth, T. (2011a). Krebshemmende Wirkung des ätherischen salbei-Öls auf eine Plattenepithelzellkarzinom-Zelllinie der Mundhöhle (UMSCC1). *HNO 59*, 1203–1208.

Sertel, S., Eichhorn, T., Plinkert, P. K., and Efferth, T. (2011b). Chemical composition and antiproliferative activity of essential oil from the leaves of a medicinal herb, *Levisticum officinale*, against UMSCC1 head and neck squamous carcinoma cells. *Anticancer Res. 31*, 185–192.

Sertel, S., Eichhorn, T., Plinkert, P. K., and Efferth, T. (2011c). Cytotoxicity of *Thymus vulgaris* essential oil towards human oral vavity squamous cell carcinoma. *Anticancer Res. 31*, 81–88.

Sfeir, J., Lefrancois, C., Baudoux, D., Derbre, S., and Licznar, P. (2013). In vitro antibacterial activity of essential oils against *Streprococcus pyogenes*. *Evid. Based Complement. Alternat. Med. 2013*, 269161.

Shah, S. M., Ullah, F., Shah, S. M., Zahoor, M., and Sadiq, A. (2012). Analysis of chemical constituents and antinociceptive potential of essential oil of *Teucrium stocksianum* bioss collected from the North West of Pakistan. *BMC Complement. Alternat. Med. 12*, 244.

Shahabipour, S., Firuzi, O., Asadollahi, M., Faghihmirzaei, E., and Javidnia, K. (2013). Essential oil composition and cytotoxic activity of *Ducrosia anethifolia* and *Ducrosia flabellifolia* from Iran. *J. Ess. Oil Res. 25(2)*, 160–163.

Shehab, N. G. and Abu-Gharbieh, E. (2012). Constituents and biological activites of the essential oil and the aqueous extract of *Micromeria fruticosa* (L.) Druce subsp. serpyllifolia. *Park. J. Pharm. Sci. 25(3)*, 687–692.

Shen, L.-Y., Yang, Z.-Y., Zhang, Y., Jiang, Y.-F., and Ma, Y.-S. (2010). Enhancing effects of 3 kinds of volatile oils on percutaneous penetration of strychnine through mouse skin in vitro. *Huaxi Yao. Zazhi 25(1)*, 4–6.

Siani, A. C., Souza, M. C., Henriques, M. G., and Ramos, M. F. (2013). Anti-inflammatory activity of essential oils from *Syzygium cumini* and *Psidium guajava*. *Pharm. Biol. 51*, 881–887.

Silva, F. V., Guimaraes, A. G., Silva, E. R., Sousa-Neto, B. P., Machado, F. D., Quintans-Junior, L. J. et al. (2012). Anti-inflammatory and anti-ulcer activities of carvacrol, a monoterpene present in the essential oil of oregano. *J. Med. Food 15(11)*, 984–991.

Sim, Y. and Shin, S. (2011). Study on cytotoxic activities of the essential oil compounds from *Ligusticum chuanxiong* against some human cancer strains. *Yakhak Hoechi 55(5)*, 398–403.

Simionatto, E., Peres, M. T., Hess, S. C., da Silva, C. B., Chagas, M. O., Poppi, N. R. et al. (2010). Chemical composition and cytotoxic activity of leaves essential oil from *Mangifera indica* var. coquinho (Anacardiaceae). *J. Ess. Oil Res. 22*, 596–599.

Soares, P. M., de Freitas, A. P., de Souza, E. P., Assreuy, A. M., and Criddle, D. N. (2012). Relaxant effects of the essential oil of *Mentha pulegium* L. in rat isolated trachea and urinary bladder. *J. Pharm. Pharmacol. 64(12)*, 1777–1784.

Soeur, J., Marrot, L., Perez, P., Iraqui, I., Kienda, G., Dardalhon, M. et al. (2011). Selective cytotoxicity of *Aniba rosaeodora* essential oil toward epidermoid cancer cells through induction of apoptosis. *Mutat. Res. 718*, 24–32.

Stefanello, M. E., Riva, D., Simionatto, E. L., de Carvalho, J. E., Gois Ruiz, A. L., and Salvador, M. J. (2011). Chemical composition and cytotoxic activity of essential oil from *Myrica laruotteana* fruits. *J. Ess. Oil Res. 23(5)*, 7–10.

Su, Y. C., Hsu, K. P., Wang, E. I., and Ho, C. L. (2012). Composition, anticancer, and antimicrobial activities in vitro of the heartwood essential oil of *Cunninghamia lanceolata* var. konishii from Taiwn. *Nat. Prod. Med. 7(9)*, 1245–1247.

Su, Y. C., Hsu, K. P., Wang, E. I., and Ho, C. L. (2013). Composition and in vitro ancancer activities of the leaf essential oil of *Neolitsea variabillima* from Taiwan. *Nat. Prod. Commun. 8(4)*, 531–532.

Suhail, M. M., Wu, W., Cao, A., Mondalek, F. G., Fung, K.-M., Shih, P.-T. et al. (2011). *Boswellia sacra* essential oil induces tumor cell-specific apoptosis and suppresses tumor aggressiveness in cultured human breast cancer cells. *BMC Complement. Altern. Med. 11*, 129.

Sulaiman, M. R., Tengku, M. T., Shaik, M. W., Moin, S., Yusuf, M., Mokhtar, A. F. et al. (2010). Antinociceptive activity of the essential oil of *Zingiber zerumbet*. *Planta Med. 76(2)*, 107–112.

Tang, Q., Yang, F., Zeng, N., He, T., Yu, L., Gou, L. et al. (2012). Study on anti-influenza virus effects of *Schizonepeta tenuifolia* volatile oil, menthone and pulegone. *Zhong. Yaoli Yu Linch. 28(2)*, 28–31.

Tao, L., Hu, H. S., and Shen, X. C. (2013). Endothelium-dependant vasodilatation effects of the essential oil from Fructus *Alpinae Zerumbet* (EOFAZ) an rat thoracic aortic rings in vitro. *Phytomedicine 20*, 387–393.

Tavares, A. C., Goncalves, M. J., Cruz, M. T., Cavaleiro, C., Lopes, M. C., Canhoto, J. et al. (2010). Essential oils from *Distichoselinum tenuifolium*: Chemical composition, cytotoxic, antifungal and anti-inflammatory properties. *J. Ethnopharmacol. 130*, 593–598.

Tilaoui, M., Mouse, H. A., Jaafari, A., Aboufatima, R., Chait, A., and Zyad, A. (2011). Chemical composition and antiproloferatice activity of essential oil from aerial parts a medicinal herb *Artemisia herb-alba*. *Braz. J. Pharm. 21(4)*, 781–785.

Tsai, M.-L., Lin, C.-C., Lin, W.-C., and Yang, C.-H. (2011). Antimicrobial, antioxidant and antiinflammatory activities of essential oils from five selected herbs. *Biosci. Biotechnol. Biochem. 75*, 1977–1983.

Tsuruoka, T., Bekh-Ochir, D., Kato, F., Snduin, S., Shataryn, A., Ayurzana, A. et al. (2012). The essential oil of Mongolian *Nepeta sibirica*: A single component and its biological activity. *J. Ess. Oil Res. 24(6)*, 555–559.

Tumen, I., Süntar, I., Eller, F. J., Keles, H., and Akkol, E. K. (2013). Topical wound-healing and phytochemical composition of heartwood essential oil of *Juniperus virdiniana* L., *Juniperus occidentalis* Hook., and *Juniperus ashei* J. Buchholz. *J. Med. Food 16(1)*, 48–55.

Tumen, I., Süntar, I., Keles, H., and Akkol, E. K. (2012). A therapeutic approach for wound healing by using essential oils of Cypressus and Juniperus species growing in Turkey. *Evid. Based Complement. Alternat. Med. 2012*, 728281.

Tung, Y. T., Yen, P. L., Lin, C. Y., and Chang, S. T. (2010). Anti-inflammatory activities of the essential oils and their constituents from different provenances of indigenous cinnamon (*Cinnamomum osmophloeum*) leaves. *Pharm. Biol. 48(10)*, 1130–1136.

Unlu, M., Ergene, E., Unlu, G. V., Zeytinoglu, H. S., and Vural, N. (2010). Composition, antimicrobial activity and in vitro cytotoxic of essential oil from *Cinnamomum zeylanicum* Blume (Lauraceae). *Food Chem. Toxicol. 48*, 3274–3280.

Velozo, L. S., Martino, T., Vigliano, M. V., Pinto, F. A., Silva, G. P., Justo Mda, G. et al. (2013). *Pterodon polygalaeflorus* essential oil modulates acute inflammation and B and T lymphocyte activation. *Am. J. Clin. Med. 41(3)*, 545–563.

Venancio, A. M., Onofre, A. S., Lira, A. F., Alves, P. B., Blank, A. F., Antoniolli, A. R. et al. (2011). Chemical composition, acute toxicity and antinociceptive activity of the essential oil of a plant breeding cultivar of basil (*Ocimum basilicum* L.). *Planta Med. 77(8)*, 825–829.

Veras, H. N., Araruna, M. K., Costa, J. G., Continho, H. D., Kerntopf, M. R., Botelho, M. A. et al. (2013). Topical antiinflammatory activity of essential oil of *Lippia sidoides* Cham: Possible mechanism of action. *Phytother. Res. 27*, 179–185.

Viuda-Martos, M., El.Nasser, A., El Gendy, G. S., Sendra, E., Fernandez-Lopez, J., Abd El Razik, K. A. et al. (2010). Chemical composition and antioxidant and anti-listeria activities of essential oils obtained from some Egyptian plants. *J. Agric. Food Chem. 58*, 9063–9070.

Wang, G.-X., Zhang, H., Geng, Z.-L., and Wang, Q.-W. (2012a). Enhancement of resveratrol percutaneous penetration by essential oil of *Angelica sinensis* (Oliv.) Diels. *Zhon. Zhongy. Zazhi 27(1)*, 117–120.

Wang, Q., Zhang, J., Liu, X., Li, X., Xu, Y., and Qi, Z. (2010). Effects of essential oil from *Angelica sinensis* (Oliv.) Diels on percutaneous permeability of nimodipine. *Yi. Dao. 29(11)*, 1397–1400.

Wang, W., Li, N., Luo, M., Zu, Y., and Efferth, T. (2012b). Antibacterial activity and anticancer activity of *Rosmarinus officinalis* L. essential oil compared to that of its main components. *Molecules 17*, 2704–2713.

Wang, W., Zhang, L., and Zu, Y. (2012c). Chemical composition and in vitro antioxidant, cytotoxic activities of *Zingiber officinale* Roscoe essential oil. *Afr. J. Biochem. Res. 6(6)*, 75–80.

Wang, Y., Chen, G.-Y., Chen, W.-H., Zhang, D.-S., Ping, Y.-Y., and Hu, X.-Y. (2013a). Chemical composition and antitumor actvities of the essential oil from leaves of *Cecris chinesis* Bunee. *Shiz. Guo. Guoy. 24(8)*, 1830–1832.

Wang, Y., Chen, W.-H., Chen, G.-Y., Song, X.-P., Zhang, D.-S., and Ping, Y.-Y. (2013b). GC-MS analysis and bioactivity of essential oil from *Artabotrys hexapetalus*. *Zhong. Shi. Fang. Zaz. 19(17)*, 100–103.

Woguem, V., Fogang, H. P., Maggi, F., Tapondjou, L. A., Womeni, H. M., Quassinti, L. et al. (2014). Volatile oil from striped African pepper (*Xylopia parvifolium*, Annonaceae) possesses notable chemoprotective, anti-inflammatory and antimicrobial potential. *Food Chem. 149*, 183–189.

Woguem, V., Maggi, F., Fogang, H. P., Tapondjou, L. A., Womeni, H. M., Quassinti, L. et al. (2013). Antioxidant, antiproliferative and antimicrobial activities of the volatile oil from wild pepper piper capense used in Cameroon as a culinary spice. *Nat. Prod. Commun. 8(12)*, 1791–1796.

Wu, Q.-F., Wamg, W., Dai, X.-Y., Wang, Z.-Y., Shen, Z.-H., Ying, H.-Z. et al. (2012b). Chemical composition and anti-influenza activities of essential oils from *Mosla dianthera*. *J. Ethnopharmacol. 139*, 668–671.

Wu, S., Wei, F. X., Li, H. Z., Liu, X. G., Zhang, J. H., and Liu, J. X. (2013). Chemical composition of essential oil from *Thymus citriodorus* and its toxic effects on liver cancer cells. *J. Chin. Med. Mat. 36(5)*, 756–759.

Xi Yap, P. S., Erin Lim, S., Ping Hu, C., and Chin Yiap, B. (2013). Combination of essential oils and antibiotics reduce antibiotic resistance in plasmid-conferred multidrug resistant bacteria. *Phytomedicine 20*, 710–713.

Xia, H., Liang, W., Song, Q., Chen, X., Chen, X., and Hong, J. (2013). The in vitro study of apoptosis in NB4 cell induced by cital. *Cytotechnology 65*, 49–57.

Ximens, R. M., Nogueira, L. D., Cassunde, N. M., Jorge, R. J., dos Santos, S. M., Magalhaes, L. P. et al. (2013). Antinociceptive and wound healing activities od *Croton adamantinus* Müll. Arg. essential oil. *J. Nat. Med. 67*, 758–734.

Yamada, A. N., Grespan, R., Yamada, A. T., Silva, E. L., Silva-Filho, S. E., Damiano, M. J. et al. (2013). Anti-inflammatory activity of *Ocimum americanum* L. essential oil in experimental model of zymosan-induced arthritis. *Am. J. Chin. Med. 41(4)*, 913.

Yan, R., Yang, Y., and Zou, G. (2013). Cytotoxic and apoptotoc effects of *Lindera strychnifolia* leaf essential oil. *J. Ess. Oil Res. 26(4)*: 308–314.

Yang, E. J., Kim, S. S., Moon, J. Y., Oh, T. H., Baik, J. S., Lee, N. H. et al. (2010). Inhibitory effects of *Fortunella japonica* var. margarita and *Citrus sunki* essential oils on nitric oxide productions and skin pathogens. *Acta Microbial. Immunol. Hung. 57 (1)*, 15–27.

Yang, G.-M., Yan, R., Wang, Z.-X., Zhang, F.-F., Pan, Y., and Cai, B.-C. (2013). Antitumor effects of two extracts from *Oxytropis falcata* on hepatocellular carcinoma in vitro and in vivo. *Chin. J. Nat. Med. 11(5)*, 519–524.

Ye, K., Yin, Z., Wei, Q., Zhou, L., Jia, R., Xu, j. et al. (2012). Anticancer activity of essential oil from *Cinnamomum longepaniculatum* leaves and its major components against human BEL-7402. *Jiepou Xuebao 43(3)*, 381–386.

Yoon, W., Moon, J. Y., Kang, J. Y., Kim, G. O., Lee, N. H., and Hyun, C. G. (2010a). *Neolitsea sericea* essential oil attenuates LPS-induced inflammation in RAW264.7 macrophages by suppressing NF-kappaB and MAPK activation. *Nat. Prod. Commun. 5(8)*, 1311–1316.

Yoon, W.-J., Kim, S.-S., Oh, T.-H., Lee, N. H., and Hyun, C.-G. (2009). *Abies koreana* essential oil inhibits drug-resistant skin pathogen growth and LPS-induced inflammatory effects of murine macrophage. *Lipids 44*, 471–476.

Yoon, W. J., Moon, J. Y., Song, G., Lee, Y. K., Han, M. S., Lee, J. S. et al. (2010b). *Artemisia fukudo* essential oil attenuates LPS-induced inflammation by suppressing NF-kB and MAPK activation in Raw 264.7 macrophages. *Food Chem. Toxicol. 48*, 1222–1229.

Yousefzadi, M., Riahi-Madvar, A., Hadian, J., Rezaee, F., and Rafiee, R. (2012). In vitro cytotoxic and antimicrobial activity of the essential oil from *Satureja sahendica*. *Toxicol. Environ. Chem. 94(9)*, 1735–1745.

Yu, D., Yuan, Y., Jiang, L., Tai, Y., Yang, X., Hu, F. et al. (2013). Anti-inflammatory effects of essential oil in *Echinacea purpurea* L. *Pak. J. Pharm. Sci. 26(2)*, 403–408.

Yu, J.-Q., Lei, J.-C., Zhang, X.-Q., Yu, H.-D., Tian, D.-Z., Liao, Z.-X. et al. (2011). Anticancer, antioxidant and antimicrobial activites of the essential oil of *Lycopus lucidus* Turcz. var. hirtus Regel. *Food Chem. 126*, 1593–1598.

Yuan, D., Xiao, F. W., Wang, L., Wen, H., and Kong, D. (2014). Patch containing asarum himalaicum volatile oil, preparation method and medical uses. *Faming Zhuanli Shenqing*. CN 103690520 A 20140402.

Yvon, Y., Raoelison, E. G., Razafindrazaka, R., Randriantsoa, A., Romdhane, M., Chabir, N. et al. (2012). Relation between chemical composition or antioxidant activity and antihypertensive activity for six essential oils. *J. Food Sci. 77(8)*, H184–H191.

Zakaria, Z. A., Mohamad, A. S., Ahmad, M. S., Mokhtar, A. F., Israf, D. A., Lajis, N. H. et al. (2011). Preliminary analysis of the anti-inflammatory activity of the essential oils of *Zingiber zerumbet*. *Biol. Res. Nursing 13(4)*, 425–432.

Zapata, B., Betancur-Galvis, L., Duran, C., and Stashenko, E. (2014). Cytotoxic activity of Asteraceae and Verbenaceae family essential oils. *J. Ess. Oil Res. 26(1)*, 50–57.

Zarai, Z., Kadri, A., Chobba, I. B., Mansour, R. B., Bekir, A., Mejdoub, H. et al. (2011). The in-vitro evaluation of antibacterial, antifungal and cytotoxic properties of *Marrubium vulgare* L. essential oil grown in Tunesia. *Lipids Health Dis. 10*, 161.

Zellagui, A., Gherraf, N., Ladjel, S., and Hameurlaine, S. (2012a). Chemical composition and antibacterial activity of the essential oils from *Launaea resedifolia* L. *Org. Med. Chem. Lett. 2(1)*: 2–5.

Zellagui, A., Gherraf, N., and Rhouati, S. (2012b). Chemical composition and antibacterial activity of the essential oils of *Ferula vesceritensis* Coss et Dur. leaves, endemic in Algeria. *Org. Med. Chem. Lett. 2*, 31.

Zhang, G.-Y., Gong, L.-I., and Dang, R.-M. (2012). Analysis on main components and antitumor activities of essential oil from *Fissistigma cavaleriei* leaves. *Gui. Non. Kex. 40(9)*, 67–69.

Zhang, H., Lin, Y., Liu, X., Hou, M.-Y., and Wu, J.-J. (2013). Anti-inflammatory and anti-nociceptive active chemical compounds of the volatile oil of *Centipeda minima*. *Guangpu Shiyan. 30(4)*, 1913–1921.

Zhao, J., Zheng, X., Newman, R. A., Zhong, Y., Liu, Z., and Nam, P. (2013). Chemical composition and bioactivity of the essential oil of *Artemisia anomala* from Chinese. *J. Ess. Oil Res. 25(6)*, 520–525.

Zhou, X.-W., Jiang, R., Wang, Q.-W., Li, X., Xu, Y., and Li, X.-Y. (2012). Enhancer percutaneous permeability of nitrazepam through rat skin by essential oil from *Cyprus rotundus* L. *Yiyao Dao. 31(7)*, 867–870.

Zhu, N., Huang, D., Hou, G., and Ge, Q. (2012). Antiviral activity of *Paederia scandens* essential oil in HBV. *Shiz. Guoyi Guoy. 21(11)*, 2754–2756.

Zomorodian, K., Saharkhiz, M. J., Rahimi, M. J., Bandegi, A., Shekarkhar, G., Bandegani, A. et al. (January–March 2011). Chemical composition and antimicrobial activities of the essential oils from three ecotypes of *Zataria multiflora*. *Pharmak. Mag. 7(25)*, 53–59.

Zomorodian, K., Saharkhiz, M. J., Shariati, S., Pakshir, K., Rahimi, M. J., and Khashei, R. (2012). Chemical composition and antimicrobial activities of essential oil from *Nepeta cataria* L. against common causes of food-borne infections. *Int. Schol. Res. Net. 2012*: 591953.

11 Antioxidative Properties of Essential Oils and Single Fragrance Compounds

Gerhard Buchbauer and Marina Erkic

CONTENTS

11.1 Introduction 323
11.2 Free Radicals 324
    11.2.1 ROS 324
    11.2.2 Reactive Nitrogen Species 325
    11.2.3 Synthesis of ROS 325
    11.2.4 Positive and Side Effects of Free Radicals 325
11.3 Oxidative Stress 326
11.4 Antioxidants 326
11.5 Antioxidant Activity 326
11.6 New Test Methods 327
    11.6.1 FRAP 327
    11.6.2 FRAP Assay 327
    11.6.3 CUPRAC Assay 327
    11.6.4 DMPD Assay 328
11.7 Antioxidant Properties of Essential Oils 328
11.8 Essential Oil Effect on Different Enzyme Functions and Free Radical Formation 337
11.9 Conclusion 339
References 340

11.1 INTRODUCTION

Essential oils have found widespread use in various domains of everyday human life (Bassolé and Juliani, 2012; Mimica-Dukić et al., 2010). They are applied as a flavoring agent (Prakash et al., 2013) in medicinal and industrial purposes, and their usage is still increasing because of their safe status, great consumer compliance, and multifunctional administration. Most of them are famous for their antimicrobial, carminative, spasmolytic, antiviral, and many other properties. Essential oils show radical scavenging activity, which qualifies them as natural antioxidants. When compared with synthetic ones, essential oils show less toxic side effects, which make them an important substitution in food preservation (Aidi Wannes et al., 2010; Bassolé and Juliani, 2012; Miguel, 2010; Mimica-Dukić et al., 2010).

Essential oils are a heterogeneous group of organic substances produced by plants. They have low molecular weight, have an oily consistency, and can be liquid or solid at room temperature, and can be obtained by steam distillation or squeezing out the peels of citrus fruits. They are synthesized as secondary metabolites by all plant organs, flowers, stems, seeds, leaves, fruits, roots, etc., and can possess different colors like yellow, blue, brownish red, or green. Their chemical composition and quality vary from different environmental conditions (Aidi Wannes et al., 2010; Bassolé and Juliani, 2012; Karakaya et al., 2011; Miguel, 2010).

11.2 FREE RADICALS

Free radicals are a highly unstable species with a single electron in their atomic electron shell. Because of their attempt to establish the balance, they bind electrons from other atoms or molecules, where they trigger biochemical instability that mostly ends up with successive chain reaction (Ríos-Arrabal et al., 2013) and oxidation of lots of biomolecules such as proteins, lipids, and DNA. Under physiological condition, free radicals are formed during ATP production in mitochondria or as metabolic products in liver cells. They are also formed in muscle cells during exercise or in phagocytes during immune reaction. Physiologically, they show an essential function in protecting an organism from infection or in keeping the regular metabolism of cells; however, their occurrence in high concentrations represents a significant risk to human health (Juranek et al., 2013; Ríos-Arrabal et al., 2013).

11.2.1 ROS

A *superoxide* free radical (O_2^{-}) is produced from an enzyme called NADPH oxidase mostly in mitochondrion (Ríos-Arrabal et al., 2013). It is less reactive than a hydroxyl radical ($\cdot$OH), but much more selective. Its lifetime is not longer than few seconds in biological systems, and it reacts with another superoxide molecule (self-dismutation reaction) to form a hydrogen peroxide. Superoxide also reacts with a *nitric oxide* to form a *peroxynitrite*, a very potent oxidant that belongs to *reactive nitrogen species* (RNS) (Juranek et al., 2013; Kalyanaraman, 2013; Miguel, 2010; Sahin Basak and Candan, 2013).

A radical–radical reaction between two superoxide radicals leads to formation of oxygen and *hydrogen peroxide* by superoxide dismutase (SOD). Hydrogen peroxide is rapidly destroyed in the presence of UV radiation or ferric or cupric ions resulting in formation of *hydroxyl radical* ($\cdot$OH). On the other hand, antioxidant enzymes such as catalase (CAT) and glutathione peroxidase degrade hydrogen peroxide (Kalyanaraman, 2013; Miguel, 2010; Narotzki et al., 2013). Hydroxyl radical is highly reactive with a lifetime of about 10^{-9} s, and there is no selectivity for this radical that reacts with the nearest cell compounds (Kalyanaraman, 2013). Although not being a free radical, *singlet oxygen* belongs to reactive oxygen species (ROS), because it rapidly reacts with cell proteins, nucleic acids, and unsaturated lipids causing oxidative damage. One of the most frequent side effects is increased photosensitivity in the presence of light. Singlet oxygen has a half-life of only 10 μs, and it has no unpaired electron in comparison to oxygen that we breathe, leaving an empty orbital to be occupied by electrons of other molecules (Kalyanaraman, 2013). In addition, binding of a proton to an oxygen molecule forms a *hydroperoxide* radical (HO_2^{-}) (Ríos-Arrabal et al., 2013). Eventually, myeloperoxidase-/H_2O_2-dependent oxidation of chloride anion forms *hypochlorous acid* (HOCl, or bleach), which is also an ROS but less active than hydroxyl radicals (Kalyanaraman, 2013; Miguel, 2010) (Table 11.1).

TABLE 11.1
Reactive Oxygen Species

Superoxide anion	$O_2 \xrightarrow{\text{NADPH Oxidase}} O_2^{-\bullet}$
Hydrogen peroxide	$O_2^{-\bullet} \xrightarrow[\text{SOD}]{O_2^{\bullet}} H_2O_2$
Hydroxyl radical	$H_2O_2 \xrightarrow{\text{Fenton Reaction}} HO^{\bullet}$
Hydroperoxide radical	$O_2 \xrightarrow{H^+} HO_2^{-\bullet}$

Source: Ríos-Arrabal, S. et al., *SpringerPlus*, 2, 404, 2013.

TABLE 11.2
Reactive Nitrogen Species

Nitric oxide	L-Arginine $\xrightarrow{NOS}$ NO$^{\bullet}$
Peroxynitrite	NO$^{\bullet}$ $\xrightarrow{O_2^{\bullet-}}$ ONOO$^-$
Anhydride nitrous	NO$^{\bullet}$ → N$_2$O$_3$
Nitrogen dioxide	NO$_2^{\bullet}$

Source: Ríos-Arrabal, S. et al., *SpringerPlus*, 2, 404, 2013.

11.2.2 Reactive Nitrogen Species

Free radicals such as *nitric oxide* (NO$^{\bullet}$), *peroxynitrite* (ONOO$^-$), the radical *nitrogen dioxide* (NO$_2^{\bullet}$), and *nitrite* (NO$_2^-$) belong to the *RNS* group. Nitric oxide synthetase (NOS) enzyme produces nitric oxide from L-arginine. Peroxynitrite is formed by the reaction of nitric oxide with a molecule of superoxide. An intermediator in this reaction is nitrogen dioxide (NO$_2^{\bullet}$), which reacts with nitric oxide forming the *anhydride nitrous* acid (N$_2$O$_3$) (Miguel, 2010; Ríos-Arrabal et al., 2013).

Endothelial, neuronal, mitochondrial, and inducible (iNOS) forms belong to the NOS family, whose activities are based on NADPH and calmodulin (Ríos-Arrabal et al., 2013). eNOS modulates cancer-related processes, such as cell death, angiogenesis, and invasion, so it possesses significant importance in tumor development. Cyclic guanylate monophosphate and guanylyl cyclase mediate many biochemical reactions of nitric oxide (Miguel, 2010; Ríos-Arrabal et al., 2013) (Table 11.2).

11.2.3 Synthesis of ROS

Physiologically, free radicals are produced during a lot of metabolic reactions in human body, where 80% of oxygen that we breathe is consumed by mitochondria, while only 5% of oxygen is converted into hydroxyl radicals and superoxide. In contrast to this, lack of oxygen inhibits mitochondrial respiration and ATP production, resulting in increased free radical synthesis. ROS is also formed by the cytochrome P-450 system and oxidative enzymes such as NAD(P)H oxidases, endothelial xanthine oxidase, myeloperoxidases, and arachidonate oxygenases (Juranek et al., 2013; Serviddio et al., 2013). In hypoxia stages and cytosolic calcium overload, xanthine oxidase can be activated resulting in elevated ROS synthesis. The consequence of this is mostly an irreversible tissue damage. On the other hand, elevated ROS synthesis activates neutrophils that increase further ROS formation. Leucocytes and macrophages also produce ROS, protecting the body against virus and bacteria. Hydroxyl radicals are the most synthesized radicals during metabolism of prostaglandins and lipids. Metabolism of drugs and flavoring agents also increase ROS synthesis (Juranek et al., 2013).

11.2.4 Positive and Side Effects of Free Radicals

Under physiological conditions, many of the biochemical reactions are forced by free radicals. Endogenous substances such as cholesterol, steroidal hormones, and prostaglandins are mediated by free radicals. Hydroxyl radicals, for example, are involved in hydroxylation of amino acids needed for collagen biosynthesis. ROS regulate gene transcription, maintain cell homeostasis, activate apoptosis, and protect against cancer. They also increase skeletal muscle endurance after physical exercise. When speaking of immune system, one of the most important functions is protecting the body from diverse infections and tissue insults. Moderate ROS concentrations guarantee maintenance of human health (Juranek et al., 2013). However, sustained exposure to free radicals can trigger various tissue injuries, where the most frequent targets are proteins, DNA, and lipids (Juranek et al., 2013;

TABLE 11.3
Role of Free Radicals in Tumor Development

- DNA damage
- Genetic instability
- Cell cycle/repair/cell death
- Inflammation
- Cellular transformation
- Proliferation
- Apoptosis
- Tumor promoting
- Angiogenesis
- Immune response
- Metastasis

Source: Ríos-Arrabal, S. et al., *SpringerPlus*, 2, 404, 2013.

Kannan et al., 2013; Ríos-Arrabal et al., 2013). They also generate tumor development and contribute to cancer metastasis (Ríos-Arrabal et al., 2013) (Table 11.3).

11.3 OXIDATIVE STRESS

Conditions of increased free radical formation and their decreased degradation are defined as oxidative stress. These kind of cytosolic conditions contribute to great disorders and damages in humans resulting in degradation of basic biomolecules such as lipids, nucleic acids, and proteins, which can end up with genotoxic effects (Kalyanaraman, 2013; Mimica-Dukić et al., 2010; Narotzki et al., 2013; Ríos-Arrabal et al., 2013; Saleh et al., 2010).

The imbalance in free radical concentration mostly comes in stages of hypoxia events. On the other side, ROS formation elevates with age. And finally, defects in expressing antioxidant genes, reduced activity of antioxidant enzymes, or lack of dietary antioxidants increase free radical levels leading to oxidative stress (Juranek et al., 2013; Kalyanaraman, 2013; Narotzki et al., 2013). Prolonged oxidative stress may lead to the formation of many diseases such as diabetes, heart and liver diseases, arthritis, and Alzheimer's disease (Bhalla et al., 2013; Fu et al., 2010; Narotzki et al., 2013).

11.4 ANTIOXIDANTS

Antioxidants degrade free radicals and protect the body from oxidation. There are three different types of defense mechanism, to which antioxidative enzymes, low-molecular nonenzymatic antioxidants, and repair mechanisms belong. CAT, SOD, glutathione peroxidase, and peroxidase belong to antioxidative enzymes, while nonenzymatic antioxidants include ascorbic acid, tocopherols, glutathione, ubiquinone, and β-carotene. Repair enzymes degrade impaired biomolecules and produce new ones. To this group, enzymes such as proteases, ligases, polymerases, and phospholipases belong (Juranek et al., 2013; Kalyanaraman, 2013; Kannan et al., 2013; Ríos-Arrabal et al., 2013). Glutathione maintains redox balance in hypoxia conditions or in case of increased ROS and NO· formation during tumor invasion. The reduced state of glutathione (GSH) catches ROS, which is lately recycled by the glutathione-reductase enzyme (GRd) (Ríos-Arrabal et al., 2013). Lipid peroxyl radicals are formed in lipid peroxidation processes, and a lipophilic vitamin E easily scavenges these radicals terminating the chain reaction. On the other hand, vitamin C is water soluble and scavenges superoxide and hydroxyl radicals. This cytosolic antioxidant recycles membrane-bounded vitamin E, increasing its total antiradical activity (Kalyanaraman, 2013; Lebold and Traber, 2013).

11.5 ANTIOXIDANT ACTIVITY

Antioxidant activity is defined as a property of antioxidants to neutralize any free radicals. Generally speaking of antioxidants, their activity is required to act as a hydrogen donor. On the other hand, the aromatic ring plays a significant role and especially phenolic hydroxyl groups enhance the inhibition of oxidation (Gülcin, 2011).

Antioxidant activity is also a measure for substance ability to prevent free radical concentration increment, oxidative stress, and risk for development-related diseases. And for the purpose of measuring antioxidant activity, many assays are applied such as 2,2-diphenyl-1-picrylhydrazyl (DPPH) assay, 2,2′-azino-bis(3-ethylbenzthiazoline-6)-sulfonic acid (ABTS) assay, β-carotene bleaching test, ferric and cupric reducing power, and linoleic acid assay (Akrout et al., 2011; Chabir et al., 2011; Jia et al., 2010; Serrano et al., 2011).

Many studies assess the antioxidant activity of the essential oil to be used as a natural source of antioxidants. Generally, plant species are rich in phenolic compounds, flavonoids, tannins, lignans, etc., which are all able to protect human health. Single compounds of essential oils donate protons to highly reactive radicals, inactivate it, and prevent possible damage. Various studies propagate essential oils as a great source of antioxidants and an ideal substitution of synthetic ones, such as butylated hydroxytoluene and butylated hydroxyanisole (BHA), which are defined as highly potent antioxidants but also with high carcinogenicity and other toxic properties as side effects (Aidi Wannes et al., 2010; Bhalla et al., 2013; Fu et al., 2010; Iqbal et al., 2013; Messaoud and Boussaid, 2011; Mimica-Dukić et al., 2010).

11.6 NEW TEST METHODS

These new test methods are presented in addition to those already described by Buchbauer (2010, Chapter 9, p. 257), there are new methods that need to be mentioned: FRAP, FRAO assay, CUPRAC assay and DMPD assay.

11.6.1 FRAP

The ferric ion reducing antioxidant power (FRAP) assay represents reducing antioxidant activity of substances, which reduce colorless oxidized Fe^{3+} to a blue-colored Fe^{2+} ion. FRAP reagent is made from tripyridyltriazine (TPTZ) and ferric chloride and sodium acetate (pH 3.6) and is used as a solvent in this test. To assess the reducing power of essential oils, FRAP reagent is mixed with essential oil diluted in ethanol, and after the reaction at room temperature during 4 min, its absorbance is measured at 593 nm with UV–visible spectrophotometer. Fifty percent ethanol is used as a negative control. The reducing capacity of the essential oil is compared with a standard solution, which is prepared from ferrous sulfate and TPTZ. The ferric reducing power is presented as μmol Fe^{2+}/g of sample (Serrano et al., 2011; Teixeira et al., 2013).

11.6.2 FRAP Assay

This assay measures the electron transfer reaction where the ferric salt represents the oxidant, while the reducing power is investigated with a mixture of the diluted essential oil, potassium hexacyanoferrate (III) (2.5 mL, 10 mg/mL), and phosphate buffer (2.5 mL, 0.2 mol/L, pH 6.6). Then, trichloroacetic acid (2.5 mL, 0.1 g/L) should be added at room temperature after 30 min of incubation at 50°C. Of this solution, 2.5 mL needs to be added to a mixture of 0.5 mL of ferric chloride and 2.5 mL of water, finally to be measured at 700 nm in a spectrophotometer (Gülcin, 2011; Mossa and Nawwar, 2011; Teixeira et al., 2013). The positive control is ascorbic acid, and a negative is ethanol (Teixeira et al., 2013). Trolox can also be used as a positive control (Mossa and Nawwar, 2011). The ferric reducing power is expressed in μmol ascorbic acid/g of the sample (Teixeira et al., 2013). Elevated absorbance represents higher reducing power of antioxidants (Gülcin, 2011).

11.6.3 CUPRAC Assay

The reduction of Cu^{2+} to Cu^+ is used in this assay to investigate the reducing power of an essential oil. For this reason, the essential oil sample is mixed with a mixture made from 0.25 mL ethanolic neocuproine solution ($7.5 \cdot 10^{-3}$ M), 0.25 mL $CuCl_2$ solution (0.01 M), and 0.25 mL CH_3COONH_4

buffer solution (1 M). This mixture is filled up to 2 mL with distilled water and measured against a blank solution at 450 nm. Results were compared with those of positive controls, BHA and α-tocopherol. As in the FRAP assay, the increase of absorbance indicates a high reduction ability (Gülcin, 2011; Tel et al., 2010).

11.6.4 DMPD Assay

Also, the dimethylphenylendiamin (DMPD) assay is used for the measurement of the radical scavenging activity of essential oils, because DMPD radicals can be neutralized by taking protons from single compounds of the applied essential oil. DMPD reagent contains dissolved DMPD in water and acetate buffer (pH 5.3). After adding 0.05 M ferric chloride ($FeCl_3$) and essential oil sample to DMPD reagent, the absorbance was measured at 505 nm. Acetate buffer is to be used as a blank solution (Gülcin, 2011).

11.7 ANTIOXIDANT PROPERTIES OF ESSENTIAL OILS

Thyme species (Lamiaceae) are well known aromatic herbs used not only as spice but also as treatment of different diseases (Jamali et al., 2012). *Thymus vulgaris* L. has been applied for different indications such as dry cough, bronchitis, and digestive problems (Grosso et al., 2010; Tsai et al., 2011). Thyme essential oil contains thymol, carvacrol, *p*-cymene, γ-terpinene, and linalool, where the thymol and carvacrol are present in relatively high percentage, up to 41.6% and 7.9%, respectively (Aazza et al., 2011; Asbaghian et al., 2011; Grosso et al., 2010). Thymol and carvacrol are phenolic compounds with estimated antioxidant activity (Aazza et al., 2011). They are considered to be responsible not only for antioxidant features of thyme oil but also for antimicrobial and anesthetic (Jamali et al., 2012). Thymol showed an IC_{50} of 0.051 mg/mL using DPPH measurement and an IC_{50} of 0.172 mg/mL using the thiobarbituric acid reactive substances (TBARS) assay, whereas the corresponding IC_{50} values for carvacrol were 0.052 mg/mL and 0.276, respectively (Aazza et al., 2011). *T. vulgaris* essential oil exhibited the IC_{50} value for scavenging the DPPH radical of 0.1 µg/mL (Tsai et al., 2011), and the antioxidant activity for preventing lipid oxidation in TBARS assay was 0.116 mg/mL (Aazza et al., 2011). The β-carotene/linoleic acid assay showed an IC_{50} value of 0.14 µg/mL and revealed an even better activity than the synthetic antioxidant butylhydroxytoluene (BHT) (Aazza et al., 2011; Tsai et al., 2011; Viuda-Martos et al., 2010).

The essential oils of *Thymus serpyllum* L., *Thymus maroccanus* Ball., *Thymus pallidus* Batt., *Thymus broussonetii* Boiss., *Thymus ciliatus* Desf., *Thymus satureioides* Coss., and *Thymus leptobotrys* Murb. contain monoterpenes, and thymol and carvacrol are present with the highest percentage. These compounds are followed by *p*-cymene, γ-terpinene, linalool, and borneol. Samples were collected in southern and southwestern Morocco and were used for isolation of essential oil by hydrodistillation (HD). While the oil of *T. serpyllum* had no activity in the DPPH assay, *T. leptobotrys* and *T. maroccanus* oils showed the highest radical scavenging activity. In comparison to this, hydroxytoluene (BHT) and quercetin showed higher radical scavenging. Keeping step with this, *T. leptobotrys* oil obtained the highest ferric reducing capacity with a half maximal effective concentration (EC_{50}) of 19.24 mg/mL. The second potent oil was *T. maroccanus* with EC_{50} of 139.5 mg/mL. Taking the oil composition into account, a strong linear correlation between the antioxidant activity and the phenolic (r = 0.980, p = 0.001) or carvacrol content was observed (r = 0.911, p = 0.011). For *Thymus capitatus* L., its oil is reported to be rich with carvacrol (62%–83%), and it was estimated to have a ferric reducing capacity of EC_{50} 312.5 mg/mL (Jamali et al., 2012).

Thymol, *p*-cymene, and γ-terpinene were major components of *Thymus proximus* Serg. and *Thymus marschallianus* Will. essential oils, growing wild in Xinjiang. In a dose-dependent manner, both essential oils showed results between 12,000 and 20,000 µg/mL in reducing power assays, while the results of the positive controls BHT and thymol were better than from the oil. Results from the DPPH assay were also concentration dependent. *T. proximus* essential oil exhibited an

IC_{50} value of 3326 μg/mL, while the *T. marschallianus* essential oil had a slightly lower IC_{50}. Thymol showed the best DPPH radical scavenging activity better than BHT (Jia et al., 2010). Ichrak et al. (2011) investigated the antioxidant and antibacterial activities of *T. pallidus* Coss. ex Batt. and *T. satureioides* Coss. essential oils obtained by steam HD. *T. pallidus* essential oil revealed to have camphor (29.8%) as a major compound, followed by dihydrocarvone (17.6%), borneol (7.6%), and camphene (7.5%), respectively. *T. satureioides* contains borneol, carvacrol, and β-caryophyllene as most content (29.5%, 9.1%, and 8.2%, respectively). *T. pallidus* and *T. satureioides* essential oils exhibited strong antioxidant activity (Ichrak et al., 2011).

Artemisia arborescens L. (Asteraceae) is a Mediterranean plant used in a traditional medicine especially in skin diseases. Areal parts collected in February were used for HD to obtain an essential oil that was consisting mostly of chamazulene (52%). ABTS and DPPH assay were applied, where the ABTS assay showed significantly better results than the DPPH test. IC_{50} value for *A. arborescens* essential oil was 21.9 μg/mL, while for Trolox, it was 15.1 μg/mL. DPPH assay showed the IC_{50} for *A. arborescens* essential oil to be higher than 200.0 μg/mL, while for Trolox, it was only 4.36 μg/mL. FRAP assay was also applied and showed the essential oil to possess 147.7 mmol TE/g of essential oil (Ornano et al., 2013). *Artemisia campestris* L. possesses antiseptic and antioxidant activities. β-Pinene and limonene were the major constituents analyzed by gas chromatography/mass spectrometry (GC–MS). Three tests were applied for investigating antioxidant activity of essential oil obtained from areal parts of this plant: ABTS, DPPH, and β-carotene bleaching method. All tests showed moderate antioxidant activity (Akrout et al., 2011).

Melaleuca (Myrtaceae) is a native genus of Australia and has found widespread use in medicinal and cosmetic purposes. Essential oils extracted by steam distillation from *Melaleuca* species are mostly composed of 1,8-cineole, α-pinene, β-pinene, and terpinen-4-ol, and they possess antimicrobial and antioxidant properties. Melaleuca oil is a commercial name for oil extracted from leaves of *Melaleuca armillaris* Sm. with 1,8-cineole (85.8%) as main compound, followed by camphene and α-pinene as constituents also in major concentrations, but to a lower extent. This oil showed a better effect on radical scavenging in the ABTS than in the DPPH assay. Vitamin C was a reference with a higher antioxidant activity for ABTS and DPPH assays (Chabir et al., 2011).

Pino et al. (2010) investigated the chemical composition and antioxidant features of leaf and fruit essential oil of *Melaleuca lucandendra* L. GC–MS revealed 41 volatile compounds for the leaf oil, where the main compounds were 1,8-cineole (43.0%), α-pinene (5.3%), α-terpineol (7.0%), limonene (4.8%), and viridiflorol (24.2%), while for fruit oil, 64 compounds were identified with viridiflorol (47.6%) as the most present compound, followed by globulol (5.8%), guaiol (5.3%), and α-pinene (4.5%) also as major compounds.

The DPPH radical scavenging capacity of leaf essential oil solutions with concentrations of 0.3 mg/mL was between 9.9% and 76.6% with an EC_{50} value of 2.4 mg/mL, while 0.2 mg/mL fruit essential oil had a scavenging capacity between 6.1% and 78.8% and an EC_{50} value of 2.3 mg/mL. Ascorbic acid was used as a positive control. The second method was the TBARS assay, which showed the extent of lipid degradation inhibition in a dose-dependent manner. BHT was used as a positive control. Concentrations of the volatile leaf oils from 20 to 250 mg/mL lead to 7.54% to 60.66% inhibition of lipid peroxidation. Fruit volatile oil showed 9.46%–61.89% inhibition of lipid peroxidation for the same concentrations as for the leaf oil. The third test was the ABTS assay with Trolox as a reference. The results were 448 and 565 mM for the leaf and fruit volatile oils, respectively. In comparison to this, volatile oils of *Origanum vulgare* L. and *Ocimum basilicum* L. had lower antioxidant power with 1105 and 997 mM, respectively. In all three tests, both *Melaleuca leucadendra* essential oils (obtained from leaves and fruits) showed a significant antioxidant capacity, while the oil obtained from fruits had the higher one. Although major compounds such as 1,8-cineole, α-terpineol, α-pinene, limonene, globulol, and guaiol may contribute to the wanted activities, it is hard to tell whether only those among the whole mixture are responsible for the results. It is possible that the trace constituents also contribute to the antioxidant activity (Pino et al., 2010).

Eucalyptus camaldulensis Dehnh. belongs to a eucalyptus species mostly common in Sardinia, Italy. Its aerial parts are used for isolation of essential oil by steam distillation, with their chemical composition investigated using GC–ITMS (ion trap mass spectrometry) and GC–FID (flame ionization detection). Thirty-seven compounds were found to be present in the essential oil, where the major compounds were *p*-cymene, 1,8-cineole, spathulenol, β-phellandrene, and cryptone, with chemical presence that varied during the vegetative stage of single samples. DPPH radical scavenging assay was carried out, showing the values in between 0.5 and 5.8 mmol/L (Barra et al., 2010). Herzi et al. (2013) found in this oil 1,8-cineole (45.7%) and *p*-cymene (17.1%) as major components. He also proved by ABTS assay an antioxidant activity of 183 mg/L as IC_{50} value, whereas in the DPPH assay revealed an IC_{50} of 1146 mg/L. GC–MS and GC–FID analyses showed that the main compounds of *Eucalyptus cinerea* F. Muell. essential oil were 1,8-cineole (64.9%) and *p*-menth-1-en-8-ol (88.2%), respectively. DPPH and ABTS assays were also applied (Herzi et al., 2013). The antioxidant capacity and chemical content of the essential oils obtained from *E. cinerea* and *E. camaldulensis* were influenced by the extraction technique, whereas the supercritical carbon dioxide extract furnished better results than the essential oil obtained by HD. In the oil of *Eucalyptus gracilis* F. Muell., 26 compounds were found with the major components 1,8-cineole (77.9%) and then *cis*-sabinol (4.2%), *p*-cymene (4.5%), and α-pinene (2.3%), respectively. This essential oil showed a low DDPH activity and the ABTS assay furnished better results (Yvon et al., 2012).

DPPH and ABTS tests were carried out for the essential oil of *Eucalyptus oleosa* F. Muell. too. Using steam distillation, essential oils were obtained from different aerial parts, stems, leaves, flowers, and fruits, and the main compounds were 1,8-cineole, spathulenol, and γ-eudesmol. DPPH showed a moderate to low antioxidant capacity, where the fruit essential oil had the highest activity. The second active oil was the one obtained from flowers. The ABTS assay showed the best activity for the leaves essential oil, followed by the stem essential oil. Fruits and flowers also showed only moderate ABTS scavenging activity (Ben Marzoug et al., 2011). Ben Hassine et al. (2012) showed for the first time the antioxidant activity of *Eucalyptus gilli* Maid. essential oil. For this reason, the IC_{50} was demonstrated to be 163.5 and 94.7 mg/L in DPPH and ABTS measurements, respectively. This activity may be due to the presence of oxygenated monoterpene hydrocarbons and monoterpenes, which possess moderate activity compared to phenols and ascorbic acid. In comparison to *E. gilli*, which had good results, *Eucalyptus radiata* Benth. possesses only low antioxidative property in the ABTS assay (Ben Hassine et al., 2012).

Aazza et al. (2011) tested the antioxidative activity of some commercial essential oils. Various in vitro tests were applied for investigating essential oils of *T. vulgaris*, *Eucalyptus globules* Labill., *Foeniculum vulgare* Mill., *Cupressus sempervirens* L., and *Citrus aurantium* L. (Aazza et al., 2011). *T. vulgaris* showed the best lipid peroxidation inhibition among other oils having IC_{50} value of 0.116 mg/mL in TBARS method, whereas *Citrus limon* oil had the lowest one. The activity of the *T. vulgaris* oil is due to its phenolic compounds thymol and carvacrol, which showed the greatest antioxidant activity among all other essential oil compounds used in test as standards. In contrast to that, limonene possessed a low activity per se, IC_{50} 3.346 mg/mL, explaining the low results of *C. limon* essential oil, which is rich in limonene. *Eucalyptus globulus* essential oil showed that its activity increased with the increment concentration. It also showed not so higher IC_{50} values than other essential oils, with the best IC_{50} result of 1.109 mg/mL. *E. globulus* oil contained 1,8-cineole (38%) and limonene (55%), respectively, and both of these compounds showed lower activity in comparison to *E. globulus* oil. However, 1,8-cineole yielded an IC_{50} value of 9.360 mg/mL explaining that a synergistic effect is responsible for the lower IC_{50} value of this essential oil. *F. vulgare* essential oil was better than *E. globulus* essential oil. On the other hand, *C. aurantium* oil showed no difference in lipid peroxidation inhibition than *F. vulgare* essential oil. *Trans*-anethol, an essential oil single compound characterized by high antioxidant activity, was present up to 70% in *F. vulgare* oil, which explains its inhibitory activity, as compared with *C. aurantium* that mostly contains linalool. Keeping in step with thymol, carvacrol, and *trans*-anethol, a compound with also good activity is δ-3-carene.

δ-3-Carene is highly present in *C. sempervirens*, explaining its inhibitory activity of 0.766 mg/mL (Aazza et al., 2011). *T. vulgaris* also showed the best results in DPPH scavenging assay, whereas the essential oils of *E. globulus* and *F. vulgare* showed the weakest ones, not even able to reach the IC_{50} value (Aazza et al., 2011; Miguel et al., 2010; Senatore et al., 2013). The ferric reducing features of the oils of *T. vulgaris* were also the best among other essential oils, in contrast to *F. vulgare* with the weakest one. Following the *T. vulgaris* oil, *C. limon* oil also showed great reductive performance of Fe(III) ion, in comparison to other assays where it furnished mostly the weakest one (Aazza et al., 2011; Senatore et al., 2013). The distillation time influences the essential oil quality and the antioxidant capacity of *F. vulgare* (Zheljazkov et al., 2013).

Citrus genus is one of the largest subunit of Rutaceae family with widespread use, mostly applied for its antibacterial, antifungal, and antioxidant effects (Campelo et al., 2011). *Citrus maxima* Merr. and *Citrus sinensis* L. essential oils are used because of their antifungal and antiaflatoxigenic properties, having moderate antioxidant activity in DPPH assay (Singh et al., 2010a,b). *C. aurantium* L. (bitter orange) belongs to the *Citrus* species commonly used as an aromatic plant, and its essential oil is mostly popular because of its antimicrobial and antioxidant features. Fresh flowers of this species collected from northeast of Tunisia were analyzed for their chemical composition and antioxidant activity. Limonene was most present up to 27%, while (*E*)-nerolidol, α-terpineol, α-terpinyl acetate, and (*E*)-farnesol were also present, but in lower concentrations (Ammar et al., 2012; Ben Hsouna et al., 2013). Ability to scavenge DPPH radicals was highly present (IC_{50} of 1.8 µg/mL), whereas β-carotene bleaching inhibition showed an IC_{50} value of 15.3 µg/mL. These results approved its wide usage, proposing it to be used in food preservation and pharmaceutical purposes as well (Ben Hsouna et al., 2013). In ABTS assay, the IC_{50} value was determined to be 672 mg/L (Ammar et al., 2012). Sarrou et al. (2013) investigated the antioxidant activity of *C. aurantium* too, using peel, flower, and leaves of plants growing in Greece. Upon 31 different constituents, most common were limonene, linalool, and α-terpineol, whereas *trans*-β-ocimene and β-pinene were also present, but in lower concentrations. The essential oil of old leaves (94.36%) had the highest antiradical activity in DPPH assay. Flower oil had double lower value with 53.98%, while young leaves and peel oil had similar results, 22.79% and 19.29%, respectively (Sarrou et al., 2013). The antioxidant activity of *C. aurantium* leaves oil was also measured according to the season's variations of its content on essential oil single compounds. Fresh leaves were collected in Nabel, Tunisia, during January, April, July, September, and November between 8 and 10 a.m. Among the 46 compounds of the essential oil obtained by HD, mostly present were linalool, linalyl acetate, and α-terpineol. The tests showed that the best activities in DPPH and ABTS assays were registered for the essential oil obtained in July (Ellouze et al., 2012). *C. limon* L. leaves oil showed significant scavenging effects in TBARS assay, representing itself as a potent inhibitor of lipid oxidation against NO and hydroxyl radicals. Trolox was used as a reference. Although concentrations increment exhibited increase of *C. limon* essential oil activity, low concentrations inhibited NO scavenging effect. In comparison to NO radicals, the essential oil yields a strong antiradical activity of hydroxyl radicals, which is probably due to high concentrations of limonene (52.8%). Geranyl acetate and *trans*-limonene-oxide were also present in the oil and contributed to its activity (Campelo et al., 2011).

Cuminum cyminum L. (Apiaceae) is a commonly used spice with widespread distribution in Asia and Mediterranean countries (Hajlaoui et al., 2010). Tunisian *C. cyminum* had been hydrodistillated, and its essential oil was analyzed by means of GC and GC–MS. Four tests were applied for investigating the antiradical features of this essential oil. The first one was DPPH assay with an antioxidative value of IC_{50} 31 mg/mL. BHT was used as a reference with a lower IC_{50} of 11.5 mg/mL. The second applied system used to assess the superoxide scavenging activity was phenazine methosulfate (PMS)–NADH–nitroblue tetrazolium (NBT). The coupling reaction of PMS and nicotinamide adenine dinucleotide (NAD) produces a superoxide anion, which is able to reduce NBT. An absorbance decrease suggests consumption of superoxide anion. The third method was used for measuring the reducing power. The essential oil showed a better activity in comparison to the BHT. The fourth test applied was the β-carotene bleaching test, and here *C. cyminum* essential oil showed a better inhibition than the control (20 and 75 µg/mL, respectively). All four tests confirm the capacity of *C. cyminum* oil, rich with cuminaldehyde and γ-terpinene, to be able to neutralize the reactive species and to prevent lipid oxidation (Hajlaoui et al., 2010).

O. vulgare L. (oregano, Lamiaceae) is an aromatic herb traditionally used as spice in Mediterranean countries. It is also applied to treat gastrointestinal and respiratory disorders. Because of its main essential oil compounds such as thymol, carvacrol, γ-terpinene, and β-fenchyl, oregano possesses antioxidative and antibacterial properties. Its antioxidant capacity was measured with FRAP and DPPH assays. *O. vulgare* essential oil exhibited moderate ferric reducing power with 38.5 µmol/g in FRAP assay, while it revealed only a poor antioxidant activity in DPPH assay. Even ethanolic, hot water and cold water extracts, showed better results in comparison to the essential oil in the DPPH assay, where the hot water extract had the best one. In contrast to this, the essential oil furnished the best results in the FRAP test. Oregano essential oil showed 16.3 mg/g of total phenolic content (Teixeira et al., 2013). The DPPH test of *O. vulgare* L. spp. *vulgare* essential oil dissolved in extra virgin oil showed the antioxidative capacity of 65.93%, mostly due to its antioxidative essential oil constituents thymol and carvacrol (Asensio et al., 2011). *Origanum majorana* L. essential oil mostly composed of 4-terpineol, γ-terpinene, α-terpinene, and *trans*-sabinene showed radical scavenging activity in a concentration-dependent manner (10–160 mg/mL). The DPPH assay of it had the IC_{50} value of 58.67 mg/mL, in comparison to the reference Trolox with a double lower value of 23.95 mg/mL. The Fenton reaction was used for measuring the hydroxyl radical scavenging effect of this essential oil in comparison to Trolox as a reference, where the 160 mg/mL of the *O. majorana* essential oil showed an IC_{50} value of 67.11 mg/mL. The reference was still better in inhibiting activity. At the same concentration of 160 mg/mL, the essential oil was able to scavenge H_2O_2 radicals (IC_{50} value of 91.25 mg/mL), while Trolox had 51.30 mg/mL. The reducing power of Fe^{3+} was measured too and showed the concentration of 160 mg/mL from essential oil and reference values of 78.67 and 42.22 mg/mL, respectively. Lipid peroxidation was induced with H_2O_2, and various essential oil concentrations were also used in this test for measurement of malondialdehyde (MDA) inhibition, 10–160 mg/mL. Higher concentrations showed better activity, with IC_{50} of 68.75 and 52.72 mg/mL for the essential oil and Trolox, respectively, but in all tests, the essential oil possessed lower radical scavenging activity than its reference (Mossa and Nawwar, 2011). *O. vulgare* L. ssp. *hirtum* rich with thymol, carvacrol, and *p*-cymene possesses significant antioxidative properties. The total antioxidant activity in ABTS assay of the essential oil, obtained by conventional HD, was 97.45%, and it yielded higher values than Trolox. Its antioxidant capacity was also measured by the DPPH assay where it showed an 87.09% activity, lower than the oil obtained by solvent-free microwave extraction with a great activity of 91.97%. Keeping in step with this, linoleic peroxidation inhibition was also high with 90.43% (Karakaya et al., 2011). The essential oil of *O. vulgare* L. subsp. *glandulosum*, collected from three different localities of Tunisia (Krib, Bargou, and Nefza), consisted mostly of *p*-cymene, thymol, γ-terpinene, and carvacrol. The antioxidative property for oils from all three localities ranged from 59 to 80 mg/L in DPPH assay (Mechergui et al., 2010). The essential oil of *Origanum acutidens* L. consisted mostly of carvacrol, *p*-cymene, γ-terpinene, and β-caryophyllene. It showed strong antioxidant activity in the DPPH test and 65% of inhibition in the β-carotene linoleic acid assay. BHT was the positive reference with 125 mg/mL and 100%, respectively (Goze et al., 2010). Oregano leaves oil of *Origanum compactum* L. contained carvacrol, thymol, and *p*-cymene as major components among its 46 identified constituents, with 36.4%, 29.7%, and 24.3% respectively. In the ABTS test, oregano essential oils revealed an IC_{50} value of 2.0 mg/L, showing an antioxidant activity as strong as vitamin C. On the other hand, in DPPH assay, oregano oil revealed an IC_{50} value of 60.1 mg/L, while carvacrol and thymol yielded 0.4 and 0.7 mg/L, respectively (El Babili et al., 2010). *Origanum syriacum* L. tested within TBARS assay showed a significant percentage of inhibition with 85.79% (Viuda-Martos et al., 2010).

Fresh and dry rhizomes of *Curcuma longa* L. (turmeric, Zingiberaceae) possess potent antiarthritic effects (Funk et al., 2010) and were used for isolation of the essential oil by HD followed by testing its antioxidant activity. GS–MS revealed a slight difference in the composition between fresh and dry rhizome essential oil. Fresh rhizome essential oil was mostly composed of β-turmerone, aromatic-turmerone, and α-turmerone, while the dry rhizome oil contained aromatic curcumene, α-santalene, and aromatic turmerone (Singh et al., 2010a,b). DPPH, metal chelating, and lipid peroxidation assays were used to assess the antiradical capacity of turmeric oil. BHT and BHA were used as positive controls. Turmeric essential oil showed in all tests better results than references, and fresh rhizome oil showed higher antioxidant activity than dry rhizome oil. The presence of curcumin, ar-turmerone, and α-turmerone explains a good radical scavenging activity of these essential oils. α- and β-Turmerones are particularly present in fresh rhizome oil, explaining its higher antioxidative capacity. This explains that drying of the rhizomes leads to a loss of its active compounds. Therefore, it is advisable that fresh rhizomes should be applied in pharmaceuticals and other purposes (Singh et al., 2010a,b). Liju et al. (2011) showed mild DPPH scavenging activity of turmeric essential oil (IC_{50} 1000 mg/mL). On the other side, lipid peroxidation inhibition was better with an IC_{50} value of 400 mg/mL. Capacities to scavenge superoxide and hydroxyl radicals were 135 and 200 mg/mL, while the ferric reducing one was 5 mM for 50 mg of turmeric oil (Liju et al., 2011). Essential oils of *Curcuma wenyujin* Y.H. Chen and C. Ling and *C. longa* L. were tested for their activities too, and in DPPH assay, it was found that they possessed strong activities, showing *C. wenyujin* oil with a higher potential than *C. longa* oil (Zhao et al., 2010). *Curcuma aromatica* Salisb. essential oil was able to scavenge DPPH radicals with an IC_{50} value of 14.45 µg/mL (Al-Reza et al., 2010).

Satureja khuzestanica Jamzad (Lamiaceae) (SKE) essential oil possesses many beneficial properties for human health against diabetes or cardiovascular diseases. The essential oil of this endemic plant from Iran showed a strong antioxidant activity (Bagheri et al., 2013; Hashemi et al., 2012), with good results in the DPPH assay possessing an IC_{50} of 5.30 ng/mL. BHT as a positive control had an IC_{50} value of 3.89 ng/mL. The total antiradical capacity of SKE essential oil was 3.20 nmol of ascorbic acid equivalents/g SKE by means of the phosphomolybdenum method (Bagheri et al., 2013). LDL oxidation may lead to arterial cell damage, platelet aggregation, and even diabetes mellitus deterioration. SKE oil is supposed to decrease LDL oxidation, so it was tested trough continuous monitoring of formation of conjugated dienes (CDs) in LDL. $CuSO_4$ was used for induction of LDL oxidation and it was estimated that it induces formation of CDs for about sixfold in comparison to the control LDL. Vitamin E is used as a positive control for inhibition of the oxidation. For SKE, it is found to have a dose-dependent activity. A concentration of 5 µg/mL increased the lag time for about 33.33%, while for concentrations of 100 and 200 µg/mL, the lag time increment was 67% and 100%, respectively, which points out the strong peroxidation inhibition capacity of SKE. For the positive control, the lag time increase was found to be 111% for a concentration of 100 µmol/L (Bagheri et al., 2013). Lipid peroxidation inhibition was tested using TBARS assay. Secondary metabolite concentrations of the lipid oxidation were measured by MDA formation analysis. Tests showed that MDA production was significantly inhibited by SKE and vitamin E, where the SKE showed a dose-dependent activity in the dosage of 50, 100, and 200 µg/mL (Bagheri et al., 2013). *S. khuzestanica* essential oil inhibited oxidation of sunflower oil stored ad 60°C, mostly due to its main compound carvacrol (Hadian et al., 2011). All concentrations of SKE (0.02%, 0.04%, 0.06%, and 0.08%) showed antioxidant activity on peroxide value (PV), while the DPPH assay revealed an IC_{50} of 31.5 µg/mL (Hashemi et al., 2012). *Satureja thymbra* L. essential oil exhibited high antioxidative capacity, having the ability to scavenge DPPH radicals by IC_{50} 0.0967 mg/mL. The essential oil composition of this plant, growing wild in Libya, mostly consisted of γ-terpinene, *p*-cymene, carvacrol, and thymol. *S. thymbra* essential oil also showed better results than thymol and γ-terpinene. On the other hand, its results were more or less equal to BHA. *O. vulgare* essential oil, also containing much γ-terpinene and thymol, had shown better results than the positive control BHA (Giweli et al., 2012). Mediterranean *Satureja montana* L. is famous because of its fungicidal, antiviral, and antimicrobial properties. This annual plant yields an essential oil rich of carvacrol, thymol, and carvacrol methyl ether, all famous for their antioxidant activity. In comparison to ethanol and water extract, the essential oil of *S. montana* showed the lowest antioxidant activity in DPPH assay. In the FRAP assay, this essential oil showed better results (Serrano et al., 2011).

Kavoosi et al. (2013) tested two essential oils of different plants for their antioxidant activity, *Carum copticum* L. (Ajowain, Apiaceae) and *Ferula assafoetida* L. (Apiaceae). Both of them are used in traditional medicine for their sedative, analgesic, antitussive, antihypertensive, digestive, and other properties. Both are potent antioxidants in a concentration-dependent manner (Kavoosi et al., 2013). Scavenging activities for ROS and RNS are demonstrated by IC_{50} values of 40 and 45 µg/mL for *C. copticum* essential oil and IC_{50} 130 and 150 µg/mL for *F. assafoetida* essential oil, respectively. IC_{50} values for hydrogen peroxide radicals were 60 µg/mL for *C. copticum* essential oil (EO) and 160 µg/mL for *F. assafoetida* EO. TBARS assay was also applied with results of IC_{50} values of 50 and 155 µg/mL for *C. copticum* and *F. assafoetida* essential oils, respectively. Ascorbic acid was used as a positive control (Kavoosi et al., 2013). Carum oil contains thymol, *p*-cymene, and γ-terpinene as major components, whereas the ferula oil possessed β-pinene, 1,2-dithiolane, and α-pinene as major constituents. These differences may explain the different antioxidant behaviors of these two essential oils, particularly showing carum more effective than ferula essential oil (Kavoosi et al., 2013). Inhibitory concentrations (IC_{50}) for total radical scavenging capacity were between 40 and 60 µg/mL of carum oil and 130 and 160 µg/mL of ferula oil. Carum and ferula essential oils displayed potent and concentration-dependent ROS, RNS, H_2O_2, and TBARS scavenging activities (Kavoosi et al., 2013). Another paper reported on the DPPH assay of *Carum carvi* L. and *Coriandrum sativum* L. essential oils to have good antioxidant property (Samojlik et al., 2010).

Zataria multiflora Boiss. (Lamiaceae) has been used in traditional medicine for many years and is endemic in Iran (Sharififar et al., 2012). Five ecotypes of *Z. multiflora* were used for testing the antioxidant activity of their essential oils. According to the different phytogeographic occurrence in Iran, names given to these ecotypes were names of towns where they were collected, including Hajiabad, Farashband, Yazd, Najafabad, and Poldokhtar. Using GC-MS, it was found that these essential oils have small differences in the concentration of essential oil constituents. Among all, thymol was the most abundant compound in all essential oils ranging from 27.1% to 64.8%, followed by carvacrol, γ-terpinene, thymol methyl ether, and carvacrol methyl ether (Karimian et al., 2012; Saei-Dehkordi et al., 2010). The best activity in DPPH assay was assessed with the Najafabad sample (IC_{50} 19.7 µg/mL). The Yazd sample was the second one and then Farashband and Hajiabad samples followed, while the Poldokhtar sample showed the lowest activity due to the fact that thymol and carvacrol were less present. Ascorbic acid and BHT were used as references, showing that Najafabad sample possessed similar activity as BHT (Saei-Dehkordi et al., 2010).

Pakistan *Rosmarinus officinalis* L. essential oil (Lamiaceae) contains 1,8-cineole (38.5%) as a main compound. This essential oil, famous for its antioxidant and antibacterial (Beretta et al., 2011; Sirocchi et al., 2013) properties, was able to reduce DPPH radical and to inhibit lipid peroxidation in linoleic acid system. On the other hand, 1,8-cineole showed poor activity, suggesting that the effects of the essential oil were due to the synergistic effect of other compounds contained in the oil, such as camphor, limonene, camphene, and linalool (Hussain et al., 2010). Zaouali et al. (2010), on the other hand, tested the antioxidant activity of two *R. officinalis* varieties: *typicus* Batt. and *troglodytorum* Maire., endemic to Tunisia. The essential oils of both varieties yielded positive results in DPHH assay, ferric reducing (FRAP) assay, and β-carotene bleaching test. *R. officinalis* var. *troglodytorum* showed higher activity in DPPH assay compared with the var. *typicus*. The IC_{50} values in β-carotene bleaching test were 1.1 µL/mL for var. *troglodytorum* and 3.2 µL/mL for var. *typicus* essential oils, showing that these essential oils possess a significant lipid peroxidation inhibition capacity. Although the presence of α-thujene, camphor, and linalyl acetate as essential oil constituents may contribute to the obtained results, it is believed that the synergistic effect of all compounds is responsible for the antioxidative results (Zaouali et al., 2010).

D-Borneol is a major compound of *Cinnamomum burmannii* Nees leaves oil (Lauraceae), being present up to 78.6%. Other constituents are bornyl acetate, (−)-spathulenol, and eucalyptol. The oil was investigated for its antiradical properties in DPPH and ABTS assays and showed moderate scavenging activity, however weaker than the positive control BHT. Although concentration dependent, maximal scavenging rates were assessed 58.89% for ABTS radicals and 21.71% for DPPH radicals (Al-Dhubiab, 2012). *Cinnamomum glaucescens* Hand.-Mazz. is used in antifungal, insecticidal, and antiaflatoxin purposes. DPPH test was applied to assess the antioxidant activity of *C. glaucescens* essential oil and BHT was used as a positive control (Prakash et al., 2013).

Clove oil from *Eugenia caryophyllata* Thunb. or *Syzygium aromaticum* L. (Myrtaceae) consists mostly eugenol (Merchán Arenas et al., 2011), which is used for food flavoring and posseses antiseptic, antibacterial, sedative, and antifungal properties. The total antioxidant activity of 15 µg/mL eugenol was 96.7%, whereas among the positive controls, the highest activity was BHT with 99.7%, while the other controls, Trolox, α-tocopherol, BHA, were less effective. Eugenol showed high reducing capacity among the ferric and cupric ions. In both tests, eugenol furnished better results than controls; however, all results were dose dependent. Eugenol also showed a significant scavenging activity for DPPH stable radicals with an IC_{50} value of 16.06 µg/mL, representing itself as the best among positive controls compared with the IC_{50} values for BHA, BHT, α-tocopherol, and Trolox. Keeping in step with the DPPH assay, eugenol also showed the best results in comparison to controls among DMPD radicals (*N,N*-dimethyl-*p*-phenylenediamine) (Gülcin, 2011). GS–MS analysis showed for the *Syzygium cumini* L. (Myrtaceae) leaves essential oil to contain α-pinene, β-pinene, and *trans*-caryophyllene as major compounds. The total phenolic content of essential oil in the DPPH assay showed 55.87% scavenging activity at 50 µg/mL. On the other hand, ferric reducing power in FRAP method was 0.47 µg/100 mg of essential oil (Mohamed et al., 2013).

For investigation of the antioxidant activity also was applied *Leucas virgata* Balf.f. (Lamiaceae) essential oil, collected from Island Soqotra (Yemen). Using GC–MS, this oil was analyzed and 43 constituents, mostly oxygenated monoterpenes (camphor and fenchon, followed by borneol), were detected. Among oxygenated sesquiterpenes, β-eudesmol and caryophyllene oxide were found. *L. virgata* essential oil showed 31% of scavenging activity in DPPH assay for 1 mg/mL of essential oil (Mothana et al., 2013).

Laurel essential oil was distilled from *Laurus nobilis* L. (Lamiaceae) leaves, which is mostly used for flavoring and preservative purposes, and it was identified to have 29 different compounds, where the major compounds were 1,8-cineole, 1-(*S*)-α-pinene, and *R*-(+)-limonene (Ramos et al., 2012; Sahin Basak and Candan, 2013). Laurel essential oil has a low antioxidant activity (Ramos et al. 2012; Sahin Basak and Candan, 2013). Sahin Basak and Candan (2013) investigated the potential of laurel essential oil to inhibit hydrogen peroxide, superoxide, and hydroxyl radicals. The DPPH scavenging activity and lipid peroxidation inhibition were also measured. In vitro Fe^{3+}–ascorbate–EDTA–H_2O_2 system was applied for determination of hydroxyl radicals, which are one of the most presented reactive oxygen species (ROS). Tests showed for the oil to possess low IC_{50} values, representing it as a good scavenger for hydroxyl radicals. Curcumin and BHT were positive controls. 1,8-Cineole, *R*-(+)-limonene, and 1-(*S*)-α-pinene also proved to be hydroxyl radical scavengers, among which 1,8-cineole had the major activity. The superoxide radical inhibiting test presented laurel essential oil as the strongest superoxide scavenger among the controls curcumin, ascorbic acid, and BHT. Upon testing some single compounds of this essential oil, 1,8-cineole showed the highest scavenging activity. Laurel oil was also the most active antioxidant against hydrogen peroxide. Positive controls were curcumin, BHT, and ascorbic acid in the hydrogen peroxide scavenging test. In lipid peroxidation assay and in DPPH test, it was assessed for laurel oil to be better than the controls (Sahin Basak and Candan, 2013).

Corsican *Limbarda crithmoides* L. ssp. *longifolia* (Asteraceae) essential oil was investigated, and it was found to contain 65 constituents, among which the major compounds were *p*-cymene, α-pinene, thymol-methyl-ether, 2,5-dimethoxy-*p*-cymenene, 3-methoxy-*p*-cymenene, and α-phellandrene. In DPPH and ABTS assays, this essential oil showed positive results. A moderate reducing power of ferric ion was measured with FRAP assay and was attributed to the essential oil (Andreani et al., 2013).

Mentha longifolia L. (Lamiaceae), also called wild mint, has been used against different infections and gastrointestinal disorders for many years. Using GC–MS, 19 compounds were identified, with piperitenone oxide, borneol, germacrene D, and β-caryophyllene as major constituents. Different tests such as linoleic acid assay, β-carotene bleaching test, and DPPH assay were applied in order to investigate the antioxidant property of mint, growing wild in the dry region of Pakistan. In DPPH assay, *M. longifolia* essential oil showed a scavenging activity (IC_{50} 21.8 µg/mL), while the linoleic acid peroxidation inhibition test had revealed an antioxidant activity of 37.3%. On the other hand, *M. longifolia* essential oil showed poor activity than in the linoleic acid assay (Iqbal et al., 2013).

Essential oils obtained from needles of *Pinus densiflora* Siebold et Zucc. and *Pinus thunbergii* Parl. (Pinaceae) possess anti-inflammatory, anticarcinogenic, and antioxidant properties. *P. densiflora* essential oil is rich in camphene, α-limonene, and α-pinene. The *P. thunbergii* essential oil consists of δ-3-carene, α-terpinolene, and β-phellandrene. For investigation of the antioxidant activity, the DPPH and nitrite-scavenging activity assays were proposed. The DPPH assay of *P. densiflora* and *P. thunbergii* essential oils exerted antioxidant activity of 120 and 30 µg/mL (IC_{50}), respectively. On the other hand, *P. densiflora* and *P. thunbergii* EO revealed a nitrite-scavenging activity higher than 50% (Park and Lee, 2011).

Piper officinarum C. DC. (Piperaceae) essential oils were obtained from leaves and stems by steam distillation. Leaf oil mostly consisted of β-caryophyllene, α-pinene, limonene, β-selinene, and sabinene. The stem oil was mostly composed of β-caryophyllene, limonene, and α-pinene, and in comparison to the leaf oil, it possessed higher concentrations of α-phellandrene and linalool. In the DPPH assay, the essential oils showed low antioxidant activity with an IC_{50} value of 777.4 µg/mL. The β-carotene linoleic acid test yielded better results for the essential oil with peroxidation inhibition of 88.9%. In comparison to this essential oil, BHT revealed 95.5% peroxidation inhibition (Salleh et al., 2012). Another pepper species, *Piper capense* Piperaceae, a wild pepper, is applied as a spice in western Cameroon. Its essential oil is rich in monoterpene hydrocarbons (mostly β-pinene, sabinene, and α-pinene) and showed to be an effective scavenger in ABTS and DPPH assays (Woguem et al., 2013). Using the same assays, the essential oil of *Ampelopsis megalophylla* showed a moderate antioxidant activity. This essential oil comprises mostly β-elemene, α-pinene, and borneol (Xie et al, 2014). Also, the in vitro antioxidative properties of essential oils obtained from seeds of *Afrostyrax lepidophyllus* and tropical *Scorodophloeus zenkeri* were measured using the DPPH and ABTS assays. Compared with Trolox, it is estimated for these essential oils to possess a good antioxidant activity where the essential oil obtained from *A. lepidophyllus* showed the better one (Fogang et al., 2014).

PV, anisidine value, Totox value, and thiobarbituric acid (TBA) value measurements and DPPH assay were used for investigation of the oxidative stability of sunflower oil mixed with *C. copticum* essential oil. The stabilization factor (F) determination also proved the efficacy of this essential oil to be an antioxidant in sunflower oil. *C. copticum* essential oil, rich with thymol, γ-terpinene, and *p*-cymene, therefore can be used as a natural antioxidant in food lipids (Hashemi et al., 2014). *Trifolium pratense* essential oil of three growth stages (TP1, TP2, and TP3) was used for the measurement of antioxidant and antimicrobial activities. Essential oil was mostly composed of β-myrcene, *p*-cymene, limonene, and tetrahydroionone. The scavenging capacity for the free radicals $NO^•$, $DPPH^•$, and $O_2^{•-}$ was measured. Furthermore, the effects of lipid peroxidation were also investigated. Among all tests, *T. pratense* essential oils showed good results, where the TP1 showed the best one. IC_{50} values for TP1 were 27.61 µg/mL for $DPPH^•$ radicals, 16.03 µg/mL for $NO^•$, and 16.62 µg/mL for $O_2^{•-}$. In the Fe^{2+}/ascorbate induction system, TP1 had the IC_{50} value of 9.35 µg/mL (Vlaisavljevic et al., 2014). *Athanasia brownii* (Asteraceae) essential oil is rich with oxygenated sesquiterpenes, such as selin-11-en-4α-ol, caryophyllene oxide, humulene epoxide II, and (*E*)-nerolidol. DPPH, ABTS, and FRAP assays represented the essential oil to have a mild antioxidant ability (Rasoanaivo et al., 2013).

Dracocephalum multicaule essential oil is reported to have an antioxidative effect on K562 cells. In DDPH assay, it was estimated to possess an IC_{50} value of 438.2 μg/mL. H_2O_2-induced oxidative damage was decreased for 49.5% after cell pretreatment with this essential oil. It consists mostly of perilla aldehyde and limonene. It also increases the activities of antioxidant enzymes and glutathione content in these cells (Esmaeili et al., 2014). Hypericum species are traditionally used for treating wounds, burns, neuralgia, stomach diseases, depression, and hysteria. The essential oils obtained from dried flowering aerial parts of *Hypericum helianthemoides* (Spach) Boiss., *Hypericum perforatum* L., and *Hypericum scabrum* L. (Hypericaceae) are rich in monoterpene and sesquiterpene hydrocarbons, among them α-pinene, β-pinene, β-caryophyllene, (E)-β-ocimene, and germacrene-D as the major constituents. DPPH assay was applied for the measurement of antioxidant activity. All three essential oils possessed antioxidant activity, the essential oil of *H. scabrum* showed the highest activity (Pirbalouti et al., 2014). Another plant, *Schinus molle* L., has been applied in traditional medicine because of its antimicrobial, antioxidant, anticancer, anti-inflammatory, analgesic, and spasmolytic features. *S. molle* leaf and fruit essential oils with the major constituents α-pinene, α-phellandrene, β-phellandrene, β-myrcene, and limonene were used for determination of antioxidant activity by using DPPH and β-carotene/linoleic acid assays. Here, the results proved the expected antioxidant activity, where the essential oils showed a better activity in β-carotene/linoleic acid method than in DPPH assay (Mdo et al., 2014).

11.8 ESSENTIAL OIL EFFECT ON DIFFERENT ENZYME FUNCTIONS AND FREE RADICAL FORMATION

Many studies have shown that essential oils and their single compounds exert an effect on various enzymes, mostly linked to free radical production. Infections, different kind of inflammations, tumor cells, and macrophages are the greatest cause of reactive species production and oxidative stress. They all exhibit their effect through activation of different enzymes and cytokines such as cytochrome P-450 system and oxidative enzymes such as NAD(P)H oxidases, endothelial xanthine oxidase, myeloperoxidases, and arachidonate oxygenases (Juranek et al., 2013; Serviddio et al., 2013), or iNOS, COX-2, TNF-α, IL-6, etc. (Chou et al., 2013; Valente et al., 2013). On the other hand, antioxidative enzyme expression will be induced during oxidative stress too. To this group of enzymes belong CAT, SOD, glutathione peroxidase, and peroxidase (Juranek et al., 2013; Kalyanaraman, 2013; Kannan et al., 2013; Ríos-Arrabal et al., 2013). It is assessed that many essential oils can exhibit their efficacy not only by their antioxidative properties but also by their ability to inhibit oxidative enzymes and cytokines or to enhance antioxidative enzyme reactions. For example, the essential oil compound (−)-α-bisabolol, obtained by direct distillation of the essential oil from *Vanillosmopsis erythropappa* Sch. Bip. (Asteraceae), is able to increase the SOD activity, which is responsible for forming hydrogen peroxide (Rocha et al., 2011).

Achillea millefolium L. (Asteraceae) is a popular traditional plant in many countries. Also called yarrow, this plant is applied for ages in folk medicine to treat different kinds of inflammations, skin diseases, and gastrointestinal disorders. *A. millefolium* essential oil (AM-EO) is rich in artemisia ketone, linalyl acetate, camphor, and 1,8-cineole, all famous for their antioxidative, antimicrobial, and anticancer activities. Various infections, tissue damages, or other health disorders stimulate an inflammatory response, followed by free radical formation. Macrophages are mostly responsible for immune response and thought to be one of the main sources of free radical production and inflammation. For this reason, RAW 264.7 macrophages were applied for investigation of AM-EO activity and its capacity to inhibit production of free radicals and inflammation, when pretreated with lipopolysaccharide (LPS), a bacterial toxic molecule responsible for activation of macrophages and inflammation processes. Firstly, it was assessed that essential oils have no effect on RAW 264.7 macrophages viability. When pretreated with 80 μg/mL of AM-EP, NO production decreased to 35%. LPSs are significant agents of oxidative stress. They not only activate NO production but also activate formation of superoxide anion and MDA. They influence GSH concentration and induce DNA damage. AM-EO decreased superoxide anion formation at any concentration added. Generally, superoxide promotes oxidative stress and cell death of healthy cells during inflammation too. A concentration of 20 µg/mL is able to inhibit 58%, while 40 and 80 µg/mL concentrations were able to decrease superoxide formation in that manner, similar to untreated macrophages with LPS. AM-EO was able to inhibit lipid peroxidation too, being able to decrease formations of the MDA and GS. Also, the DNA damage was less present in AM-EO-treated RAW 264.7 macrophages. The effect of this essential oil on antioxidant enzymes was also measured, which are activated from cells in inflammation conditions. SOD, CAT, and GPx activities were reduced at all concentrations of AM-EO added, proving that AM-EO activity is obtained from its antioxidative property and not from antioxidative enzyme activation. INOS, COX-2, TNF-α, and IL-6 also influence an increase of free radicals and inflammation. iNOS activates NO synthesis and COX-2 converts arachidonic acid into prostaglandins, while proinflammatory cytokines IL-6 and TNF-α are expressed during inflammation. Expression of these inflammatory agents is highly decreased after pretreatment of macrophages with AM-EO, indicating that this essential oil has significant protective properties during oxidative stress. On the other hand, HO-1 expression levels, which suppress ROS formation and protect the body from inflammation by inhibiting NO function, are low after cell pretreatment with this essential oil, which is due to the fact that AM-EO suppresses oxidation via its own antioxidant property (Chou et al., 2013).

The aerial parts of *Oenanthe crocata* L. (Apiaceae) were used for HD to obtain an essential oil, which mostly contained sabinene, *trans*-β-ocimene, and *cis*-β-ocimene. Although the root of *O. crocata* is highly toxic, its leaves and flowers are used in traditional medicine for dermatological- and brain-related diseases. Nitric oxide scavenging activity was measured and the essential oil presented itself as a potent NO scavenger, and it also inhibited the expression of NOS. Already 0.04 µL/mL of essential oil is able to decrease nitrite oxidation by 30%. NO scavenging activity of sabinene was also measured, and it was less potent than that of the essential oil. High levels of NO during some inflammation are mostly due to iNOS. After pretreatment of macrophages with this essential oil, nitrite production was reduced from 211.1% to 103%, indicating the high potency of *O. crocata* essential oil to inhibit the iNOS (Valente et al., 2013).

Macrophages were used for generation of superoxide radicals by phorbol-12-myristate-13-acetate and for measurement of turmeric oil scavenging activity. It was found for turmeric oil to inhibit an increase of superoxide radical concentration. Turmeric oil effect also was tested on antioxidant enzymes in blood and serum of mice. After administration of turmeric oil, it was found for enzymes such as CAT, SOD, glutathione, and glutathione reductase to be significantly increased, particularly by a concentration of 500 mg/kg after treatment over 30 days. Mice liver tissues were also used for estimation of turmeric oil activity on antioxidant enzyme induction. Increased enzyme levels were found to be for glutathione peroxidase, SOD. Even glutathione and glutathione-*S*-transferase levels were elevated. CAT and glutathione reductase remained unaltered (Liju et al., 2011). *Z. multiflora* Boiss. (Lamiaceae) is a traditional plant used in the Middle East for carminative, analgesic, and antimicrobial purposes. Only growing in warm parts of Afghanistan, Pakistan, and Iran, *Z. multiflora* essential oil has a high presence of thymol and carvacrol, explaining its strong antioxidant property. It is also proven to be able to reduce hydrogen peroxide (H_2O_2) and nitric oxide (NO) production, and it is believed to react as radical scavenger and inhibitor of related oxidative enzymes. It is also believed that carvacrol and thymol alone inhibit NOS and NADH oxidase (NOX). H_2O_2 and NO are formed in human monocytes via glucose degradation, especially when high glucose concentrations are present. Major compounds of *Z. multiflora* essential oil determined by GS–MS were carvacrol, thymol, and *p*-cymene with percentage. Linalool and γ-terpinene were also present but to lower extent of 9.6% and 8%, respectively. The ABTS assay was used to identify reactive oxygen inhibitory capacity of this essential oil, showing an IC_{50} value of 5.7 µg/mL. Reactive nitrogen scavenging activity was also measured in the NO scavenging assay. Thymol and carvacrol showed a better activity, while linalool, γ-terpinene, and *p*-cymene were inactive. Human monocytes were used for testing NO and H_2O_2 production. When incubated with glucose, the significant increase was found not only for these reactive species but also for NOS and NOX. After incubation of cells with 20 mM of glucose, the NO production was 263 nM, and after pretreatment with 1, 10, and 100 ng/mL of *Z. multiflora* essential oil, the NO production significantly decreased to 100 ng/mL. Using the same concentrations by carvacrol and thymol, pretreatment of monocytes showed reduction of NO in similar manner. Thymol was slightly better in comparison to carvacrol with 152 nM. Linalool, γ-terpinene, and *p*-cymene revealed no difference to NO concentration after pretreatment of monocytes before incubation. The same affects were measured on NOS activity changes after pretreatment with *Z. multiflora* essential oil and its single constituents. Linalool, γ-terpinene, and *p*-cymene caused no change in NOS activity. *Z. multiflora* essential oil also showed a significant reduction of H_2O_2 production in glucose-stimulated monocytes. At a concentration of 100 ng/mL, thymol extended similar result with 63 nM, while carvacrol showed itself as the better one with 59 nM at the same concentration. Linalool, γ-terpinene, and *p*-cymene had no effect on H_2O_2 production. Effects on NOS production in monocytes were also present after pretreatment with *Z. multiflora* essential oil, thymol, and carvacrol. Like in all tests, *p*-cymene, linalool, and γ-terpinene did not alter NOS activity (Kavoosi and Teixeira da Silva, 2012). This author group investigated *Z. multiflora* essential oil activity on nitric oxide and hydrogen peroxide production in macrophages, stimulated with LPS. *Z. multiflora* essential oil reduces NO concentration in macrophages. After pretreatment, it showed a clear reduction of different concentrations of this essential oil. Pretreatment of macrophages with 1, 10, and 100 ng/mL essential oil leads also to a significant reduction of H_2O_2 radicals. Reduction of RNS and NOX was also present. Concentrations of 1, 10, and 100 ng/mL of this essential oil inactivated RNS (NOS) and furnished the reduction of NOX for the same concentrations as for other radicals. Thymol and carvacrol were positive controls and showed in all tests significant reduction too. *p*-Cymene on the other hand had no effects on radical and enzyme concentration change after its pretreatment of macrophages (Kavoosi et al., 2012).

Eugenol and methyl eugenol are the main compounds of the essential oil of *Ocimum sanctum* L., which are able to decrease serum lipids. *O. sanctum* leaves oil is used for the measurement of the antioxidative and antihyperlipidamic properties in rats, which were on cholesterol diet. Lipid peroxide concentration was assessed in TABRS assay, and it was found for this essential oil to decrease TBA levels in heart and liver. In heart tissue, it is also shown for the oil to exert a decreasing effect on SOD and glutathione peroxidase (GPx) without effecting CAT. On the other hand GPx, SOD, and CAT were not impacted in liver tissue (Suanarunsawat et al., 2010).

11.9 CONCLUSION

Speaking of an antioxidant generally, its activity is required to act as a hydrogen donor and neutralize free radicals. On the other hand, aromatic rings play a significant role and phenolic hydroxyl groups enhance the inhibition of oxidation (Gülcin, 2011).

Essential oils comprise many different single constituents in different concentrations. An antioxidant effect is mostly attributed to compounds with highest presence. However, investigations have shown that minor constituents also contribute to the essential oil activity, which concludes that single compounds are on one hand highly active substances and show already in small concentration great activity and on the other hand they contribute by their synergistic effect to the property of the essential oil (Victoria et al., 2012). For example, essential oils obtained from *M. leucadendra* leaves and fruits showed significant antioxidant capacity in DPPH, TBARS, and ABTS assays. However, it is hard to tell whether the antioxidant activity results from the presence of 1,8-cineole, α-terpineol, α-pinene, limonene, globulol, and guaiol as major constituents or from a synergistic effect. On the other hand, trace constituents could also contribute to the activity (Pino et al., 2010).

Different kinds of methods, such as DPPH assay, ABTS assay, or β-carotene bleaching test, already mentioned in previous text, are used to assess the antioxidative properties of essential oils and their single compounds. Being one of the main topics in many studies for many years, essential oils have shown themselves as very potent substances in decreasing high free radical concentrations and in decreasing their formations. Many scientific articles have shown how antioxidant activity is of special importance for protecting cells against DNA damage and mutagenesis, cancer invasion and inhibition of bacterial growth and infections (Gülcin, 2011). Following this, they are good in preventing different human diseases and food spoilage. Some of them have pleasant aroma and are used as a flavor or as fragrant agents. Especially because of their low or even no side-effects properties, their popularity is high, and they show themselves as a potent replacement for synthetic antioxidants having a bright application spectrum in pharmacy and food industry (Aidi Wannes et al., 2010; Bassolé and Juliani, 2012; Mimica-Dukić et al., 2010; Prakash et al., 2013).

REFERENCES

Aazza, S., B. Lyoussi, and M.G. Miguel, 2011. Antioxidant and antiacetylcholinesterase activities of some commercial essential oils and their major compounds. *Molecules*, 16(9):7672–7690.

Aidi Wannes, W., B. Mhamdi, J. Sriti et al., 2010. Antioxidant activities of the essential oils and methanol extracts from myrtle (*Myrtus communis* var. *italica* L.) leaf, stem and flower. *Food Chem. Toxicol.*, 48(5):1362–1370.

Akrout, A., L.A. Gonzalez, H. El Jani, and P.C. Madrid, 2011. Antioxidant and antitumor activities of *Artemisia campestris* and *Thymelaea hirsuta* from southern Tunisia. *Food Chem. Toxicol.*, 49(2):342–347.

Al-Dhubiab, B.E. 2012. Pharmaceutical applications and phytochemical profile of *Cinnamomum burmannii*. *Pharmacogn. Rev.*, 6(12):125–131.

Al-Reza, S.M., A. Rahman, M.A. Sattar et al., 2010. Essential oil composition and antioxidant activities of *Curcuma aromatica* Salisb. *Food Chem. Toxicol.*, 48(6):1757–1760.

Ammar, A.H., J. Bouajila, A. Lebrihi et al., 2012. Chemical composition and in vitro antimicrobial and antioxidant activities of *Citrus aurantium* L. flowers essential oil (Neroli oil). *Pak. J. Biol. Sci.*, 15(21):1034–1040.

Andreani, S., M.C. De Cian, J. Paolini et al., 2013. Chemical variability and antioxidant activity of *Limbarda crithmoides* L. essential oil from Corsica. *Chem. Biodivers.*, 10(11):2061–2077.

Asbaghian, S., A. Shafaghat, K. Zarea et al., 2011. Comparison of volatile constituents, and antioxidant and antibacterial activities of the essential oils of *Thymus caucasicus*, *T. kotschyanus* and *T. vulgaris*. *Nat. Prod. Commun.*, 6(1):137–140.

Asensio, C.M., V. Nepote, and N.R. Grosso, 2011. Chemical stability of extra-virgin olive oil added with oregano essential oil. *J. Food Sci.*, 76(7):S445–S450.

Bagheri, S., H. Ahmadvand, A. Khosrowbeygi et al., 2013. Antioxidant properties and inhibitory effects of *Satureja khozestanica* essential oil on LDL oxidation induced-$CuSO_4$ in vitro. *Asian Pac. J. Trop. Biomed.*, 3(1):22–27.

Barra, A., V. Coroneo, S. Dessi et al., 2010. Chemical variability, antifungal and antioxidant activity of *Eucalyptus camaldulensis* essential oil from Sardinia. *Nat. Prod. Commun.*, 5(2):329–335.

Bassolé, I.H. and H.R. Juliani, 2012. Essential oils in combination and their antimicrobial properties. *Molecules*, 17(4):3989–4006.

Ben Hassine, D., M. Abderrabba, Y. Yvon et al., 2012. Chemical composition and in vitro evaluation of the antioxidant and antimicrobial activities of *Eucalyptus gilli* essential oil and extracts. *Molecules*, 17(8):9540–9558.

Ben Hsouna, A., N. Hamdi, N. Ben Halima, and S. Abdelkafi, 2013. Characterization of essential oil from *Citrus aurantium* L. flowers: Antimicrobial and antioxidant activities. *J. Oleo Sci.*, 62(10):763–772.

Ben Marzoug H.N, M. Romdhane, A. Lebrihi et al., 2011. *Eucalyptus oleosa* essential oils: Chemical composition and antimicrobial and antioxidant activities of the oils from different plant parts (stems, leaves, flowers and fruits). *Molecules*, 16(2):1695–1709.

Beretta, G., R. Artali, R.M. Facino, and F. Gelmini, 2011. An analytical and theoretical approach for the profiling of the antioxidant activity of essential oil: The case of *Rosmarinus officinalis* L. *J. Pharm. Biomed. Anal.*, 55(5):1255–1264.

Bhalla, Y., V.K. Gupta, and V. Jaitak, 2013. Anticancer activity of essential oils: A review. *J. Sci. Food Agric.*, 93(15): 3643–3653

Buchbauer, G. (2010). Biological activities of essential oils. In: Baser, K.H.C. and Buchbauer, G., eds. *Handbook of Essential Oils—Science, Technology and Applications*. Boca Raton, FL: Taylor & Francis, pp. 235–280.

Campêlo, L.M., A.A. de Almeida, R.L. de Freitas et al., 2011. Antioxidant and antinociceptive effects of *Citrus limon* essential oil in mice. *J. Biomed. Biotechnol.*, 2011:678673.

Chabir, N., M. Romdhane, A. Valentin et al., 2011. Chemical study and antimalarial, antioxidant and anticancer activities of *Melaleuca armillaris* (Sol Ex Gateau) Sm essential oil. *J. Med. Food*, 14(11):1383–1388.

Chou, S.T., H.Y. Peng, J.C. Hsu et al., 2013. *Achillea millefolium* L. essential oil inhibits LPS-induced oxidative stress and nitric oxide production in RAW 264.7 macrophages. *Int. J. Mol. Sci.*, 14(7):12978–12993.

El Babili, F., J. Bouajila, J.P. Souchard et al., 2010. Oregano: Chemical analysis and evaluation of its antimalarial, antioxidant, and cytotoxic activities. *J. Food Sci.*, 76(3):C512–C518.

Ellouze, I., M. Abderrabba, N. Sabaou et al., 2012. Season's variation impact on *Citrus aurantium* leaves essential oil: Chemical composition and biological activities. *J. Food Sci.*, 77(9):T173–T180.

Esmaeili, M.A., A. Sonboli, and M.H. Mirjalili, 2014. Oxidative stress protective effect of *Dracocephalum multicaule* essential oil against human cancer cell line. *Nat. Prod. Res.*, 28(11): 848–852

Fogang, H.P., F. Maggi, L.A. Tapondjou et al., 2014. In vitro biological activities of seed essential oils from the Cameroonian spices *Afrostyrax lepidophyllus* Mildbr. and *Scorodophloeus zenkeri* Harms rich in sulfur-containing compounds. *Chem. Biodivers.*, 11(1):161–169.

Fu, L., B.T. Xu, X.R. Xu et al., 2010. Antioxidant capacities and total phenolic contents of 56 wild fruits from South China. *Molecules*, 15(12):8602–8617.

Funk, J.L., J.B. Frye, J.N. Oyarzo et al., 2010. Anti-arthritic effects and toxicity of the essential oil of turmeric (*Curcuma longa* L.). *J. Agric. Food Chem.*, 58(2):842–849.

Giweli, A., A.M. Džamić, M. Soković et al., 2012. Antimicrobial and antioxidant activities of essential oils of *Satureja thymbra* growing wild in Libya. *Molecules*, 17(5):4836–4850.

Goze, I., A. Alim, S.A. Cetinus et al., 2010. In vitro antimicrobial, antioxidant, and antispasmodic activities and the composition of the essential oil of *Origanum acutidens* (Hand.-Mazz.) Ietswaart. *J. Med. Food*, 13(3):705–709.

Grosso, C, A.C. Figueiredo, J. Burillo et al., 2010. Composition and antioxidant activity of *Thymus vulgaris* volatiles: Comparison between supercritical fluid extraction and hydrodistillation. *J. Sep. Sci.*, 33(14):2211–2218.

Gülcin, I., 2011. Antioxidant activity of eugenol: A structure–activity relationship study. *J. Med. Food*, 14(9):975–985.

Hadian, J., M. Hossein Mirjalili, M. Reza Kanani et al., 2011. Phytochemical and morphological characterization of *Satureja khuzestanica* Jamzad populations from *Iran*. *Chem. Biodivers.*, 8(5):902–915.

Hajlaoui, H., H. Mighri, E. Noumi et al., 2010. Chemical composition and biological activities of Tunisian *Cuminum cyminum* L. essential oil: A high effectiveness against *Vibrio* spp. strains. *Food Chem. Toxicol.*, 48(8–9):2186–2192.

Hashemi, M.B., M. Niakousari, M.J. Saharkhiz et al., 2014. Stabilization of sunflower oil with *Carum copticum* Benth & Hook essential oil. *J. Food Sci. Technol.*, 51(1):142–147.

Hashemi, M.B., M. Niakousari, M.J. Saharkhiz, and M.H. Eskandari, 2012. Effect of *Satureja khuzestanica* essential oil on oxidative stability of sunflower oil during accelerated storage. *Nat. Prod. Res.*, 26(15):1458–1463.

Herzi, N., J. Bouajila, S. Camy et al., 2013. Comparison between supercritical CO_2 extraction and hydrodistillation for two species of *Eucalyptus*: Yield, chemical composition, and antioxidant activity. *J. Food Sci.*, 78(5):C667–C672.

Hussain, A.I., F. Anwar, S.A. Chatha et al., 2010. *Rosmarinus officinalis* essential oil: Antiproliferative, antioxidant and antibacterial activities. *Braz. J. Microbiol.*, 41(4):1070–1078.

Ichrak, G., B. Rim, A.S. Loubna et al., 2011. Chemical composition, antibacterial and antioxidant activities of the essential oils from *Thymus satureioides* and *Thymus pallidus*. *Nat. Prod. Commun.*, 6(10):1507–1510.

Iqbal, T., A.I. Hussain, S.A. Chatha et al., 2013. Antioxidant activity and volatile and phenolic profiles of essential oil and different extracts of wild mint (*Mentha longifolia*) from the Pakistani Flora. *J. Anal. Methods Chem.*, 2013:536490.

Jamali, C.A., L. El Bouzidi, K. Bekkouche et al., 2012. Chemical composition and antioxidant and anticandidal activities of essential oils from different wild Moroccan *Thymus* species. *Chem. Biodivers.*, 9(6):1188–1197.

Jia, H.L., Q.L. Ji, S.L. Xing et al., 2010. Chemical composition and antioxidant, antimicrobial activities of the essential oils of *Thymus marschallianus* Will. and *Thymus proximus* Serg. *J. Food Sci.*, 75(1):E59–E65.

Juranek, I., D. Nikitovic, D. Kouretas et al., 2013. Biological importance of reactive oxygen species in relation to difficulties of treating pathologies involving oxidative stress by exogenous antioxidants. *Food Chem. Toxicol.*, 61:240–247.

Kalyanaraman, B., 2013. Teaching the basics of redox biology to medical and graduate students: Oxidants, antioxidants and disease mechanisms. *Redox. Biol.*, 1(1):244–257.

Kannan, P.R., S. Deepa, S.V. Kanth, and R. Rengasamy, 2013. Growth, osmolyte concentration and antioxidant enzymes in the leaves of *Sesuvium portulacastrum* L. under salinity stress. *Appl. Biochem. Biotechnol.*, 171(8):1925–1932.

Karakaya, S., S.N. El, N. Karagözlü, and S. Sahin, 2011. Antioxidant and antimicrobial activities of essential oils obtained from oregano (*Origanum vulgare* ssp. hirtum) by using different extraction methods. *J. Med. Food.*, 14(6):645–652.

Karimian, P., G. Kavoosi, and M.J. Saharkhiz, 2012. Antioxidant, nitric oxide scavenging and malondialdehyde scavenging activities of essential oils from different chemotypes of *Zataria multiflora*. *Nat. Prod. Res.*, 26(22):2144–2147.

Kavoosi, G., A. Tafsiry, A.A. Ebdam, and V. Rowshan, 2013. Evaluation of antioxidant and antimicrobial activities of essential oils from *Carum copticum* seed and *Ferula assafoetida* latex. *J. Food Sci.*, 78(2):T356–T361.

Kavoosi, G. and J.A. Teixeira da Silva, 2012. Inhibitory effects of *Zataria multiflora* essential oil and its main components on nitric oxide and hydrogen peroxide production in glucose-stimulated human monocytes. *Food Chem. Toxicol.*, 50(9):3079–3085.

Kavoosi, G., J.A. Teixeira da Silva, and M.J. Saharkhiz, 2012. Inhibitory effects of *Zataria multiflora* essential oil and its main components on nitric oxide and hydrogen peroxide production in lipopolysaccharide-stimulated macrophages. *J. Pharm. Pharmacol.*, 64(10):1491–1500.

Lebold, K.M. and M.G. Traber, 2013. Interactions between α-tocopherol, polyunsaturated fatty acids, and lipoxygenases during embryogenesis. *Free Radic. Biol. Med.*, 66:13–19.

Liju, V.B., K. Jeena, and R. Kuttan, 2011. An evaluation of antioxidant, anti-inflammatory, and antinociceptive activities of essential oil from *Curcuma longa* L. *Indian J. Pharmacol.*, 43(5):526–531.

Mdo, M.R., S. Arantes, F. Candeias et al., 2014. Antioxidant, antimicrobial and toxicological properties of *Schinus molle* L. essential oils. *Ethnopharmacology*, 151(1):485–492.

Mechergui, K., J.A. Coelho, M.C. Serra et al., 2010. Essential oils of *Origanum vulgare* L. subsp. *glandulosum* (Desf.) Ietswaart from Tunisia: Chemical composition and antioxidant activity. *J. Sci. Food Agric.*, 90(10):1745–1749.

Merchán Arenas, D.R., A.M. Acevedo, L.Y. Vargas Méndez, and V.V. Kouznetsov, 2011. Scavenger activity evaluation of the clove bud essential oils (*Eugenia caryophyllus*) and eugenol derivatives employing ABTS decolorization. *Sci. Pharm.*, 79(4):779–791.

Messaoud, C. and M. Boussaid, 2011. *Myrtus communis* Berry Color Morphs: A comparative analysis of essential oils, fatty acids, phenolic compounds, and antioxidant activities. *Chem. Biodivers.*, 8(2):300–310.

Miguel, M.G., 2010. Antioxidant and anti-inflammatory activities of essential oils: A short review. *Molecules*, 15(12):9252–9287.

Miguel, M.G., C. Cruz, L. Faleiro et al., 2010. *Foeniculum vulgare* essential oils: Chemical composition, antioxidant and antimicrobial activities. *Nat. Prod. Commun.*, 5(2):319–328.

Mimica-Dukić, N., D. Bugarin, S. Grbović et al. 2010. Essential oil of *Myrtus communis* L. as a potential antioxidant and antimutagenic agents. *Molecules*, 15(4):2759–2770.

Mohamed, A.A., S.I. Ali, and F.K. El-Baz, 2013. Antioxidant and antibacterial activities of cude extracts and essential oils of *Syzygium cumini* leaves. *PLoS One*, 8(4):e60269.

Mossa, A.T. and G.A. Nawwar, 2011. Free radical scavenging and antiacetylcholinesterase activities of *Origanum majorana* L. essential oil. *Hum. Exp. Toxicol.*, 30(10):1501–1513.

Mothana, R.A., M.S. Al-Said, M.A. Al-Yahya et al., 2013. GC and GC/MS analysis of essential oil composition of the endemic soqotraen leucas virgata Balf.f. and its antimicrobial and antioxidant activities. *Int. J. Mol. Sci.*, 14(11):23129–23139.

Narotzki, B., A.Z. Reznick, and D. Navot-Mintzer, 2013. Green tea and vitamin E enhance exercise-induced benefits in body composition, glucose homeostasis, and antioxidant status in elderly men and women. *J. Am. Coll. Nutr.*, 32(1):31–40.

Ornano, L., A. Vendittin, M. Ballero et al., 2013. Chemopreventive and antioxidant activity of the chamazulene-rich essential oils obtained from *Artemisia arborescens* L. growing on the Isle of La Maddalena, Sardinia, Italy. *Chem. Biodivers.*, 10(8):1464–1474.

Park, J.S. and G.H. Lee, 2011. Volatile compounds and antimicrobial and antioxidant activities of the essential oil of the needles of *Pinus densiflora* and *Pinus thunbergii*. *J. Sci. Food. Agric.*, 91(4):703–709.

Pino, J.A., E.L. Regalado, J.L. Rodríguez, and M.D. Fernández, 2010. Phytochemical analysis and in vitro free-radical-scavenging activities of the essential oils from leaf and fruit of *Melaleuca leucadendra* L. *Chem. Biodivers.*, 7(9):2281–2288.

Pirbalouti, G., M. Fatahi-Vanani, L. Craker, and H. Shirmardi, 2014. Chemical composition and bioactivity of essential oils of *Hypericum helianthemoides*, *Hypericum perforatum* and *Hypericum scabrum*. *Pharm Biol.*, 52(2):175–181.

Prakash, B., P. Singh, S. Yadav et al., 2013. Safety profile assessment and efficacy of chemically characterized *Cinnamomum glaucescens* essential oil against storage fungi, insect, aflatoxin secretion and as antioxidant. *Food Chem. Toxicol.*, 53:160–167.

Ramos, C., B. Teixeira, I. Batista et al., 2012. Antioxidant and antibacterial activity of essential oils and extracts of bay laurel *Laurus nobilis* Linnaeus (Lauraceae) from Portugal. *Nat. Prod. Res.*, 26(6):518–529.

Rasoanaivo P., R. Fortuné Randriana, F. Maggi et al., 2013. Chemical composition and biological activities of the essential oil of *Athanasia brownii* Hochr. (Asteraceae) endemic to Madagascar. *Chem. Biodivers*, 10(10):1876–1886.

Ríos-Arrabal, S., F. Artacho-Cordón, J. León et al., 2013. Involvement of free radicals in breast cancer. *Springerplus*, 2:404.

Rocha, N.F., G.V. Oliveira, F.Y. Araújo et al., 2011. (−)-α-Bisabolol-induced gastroprotection is associated with reduction in lipid peroxidation, superoxide dismutase activity and neutrophil migration. *Eur. J. Pharm. Sci.*, 44(4):455–461.

Saei-Dehkordi, S.S., H. Tajik, M. Moradi, and F. Khalighi-Sigaroodi, 2010. Chemical composition of essential oils in *Zataria multiflora* Boiss. from different parts of Iran and their radical scavenging and antimicrobial activity. *Food Chem. Toxicol.*, 48(6):1562–1567.

Sahin Basak, S. and F. Candan, 2013. Effect of *Laurus nobilis* L. essential oil and its main components on α-lucosidase and reactive oxygen species scavenging activity. *Iran. J. Pharm. Res.*, 12(2):367–379.

Saleh, M.A., S. Clark, B. Woodard, and S.A. Deolu-Sobogun, 2010. Antioxidant and free radical scavenging activities of essential oils. *Ethn. Dis.*, 20(1 Suppl. 1):S1-78–S1-82.

Salleh, W.M., F. Ahmad, K.H. Yen, and H.M. Sirat, 2012. Chemical compositions, antioxidant and antimicrobial activity of the essential oils of *Piper officinarum* (Piperaceae). *Nat. Prod. Commun.*, 7(12):1659–1662.

Samojlik, I., N. Lakić, N. Mimica-Dukić, K. Daković-Svajcer, and B. Bozin, 2010. Antioxidant and hepatoprotective potential of essential oils of coriander (*Coriandrum sativum* L.) and caraway (*Carum carvi* L.) (Apiaceae). *J. Agric. Food Chem.*, 58(15):8848–8853.

Sarrou, E., P. Chatzopoulou, K. Dimassi-Theriou, and I. Therios, 2013. Volatile constituents and antioxidant activity of peel, flowers and leaf oils of *Citrus aurantium* L. growing in Greece. *Molecules*, 18(9):10639–10647.

Senatore, F., F. Oliviero, E. Scandolera et al., 2013. Chemical composition, antimicrobial and antioxidant activities of anethole-rich oil from leaves of selected varieties of fennel [*Foeniculum vulgare* Mill. ssp. *vulgare* var. *azoricum* (Mill.) Thell]. *Fitoterapia*, 90:214–219.

Serrano, C., O. Matos, B. Teixeira et al., 2011. Antioxidant and antimicrobial activity of *Satureja montana* L. extracts. *J. Sci. Food Agric.*, 91(9):1554–1560.

Serviddio, G., F. Bellanti, and G. Vendemiale, 2013. Free radical biology for medicine: Learning from nonalcoholic fatty liver disease. *Free Radic. Biol. Med.*, 65C:952–968.

Sharififar, F., A. Derakhshanfar, G. Dehghan-Nudeh et al., 2012. In vivo antioxidant activity of *Zataria multiflora* Boiss essential oils. *Pak. J. Pharm. Sci.*, 24(2):221–225.

Singh, G., I.P. Kapoor, P. Singh et al., 2010a. Comparative study of chemical composition and antioxidant activity of fresh and dry rhizomes of turmeric (*Curcuma longa* Linn.). *Food Chem. Toxicol.*, 48(4):1026–1031.

Singh, P., R. Shukla, B. Prakash et al., 2010b. Chemical profile, antifungal, antiaflatoxigenic and antioxidant activity of *Citrus maxima* Burm. and *Citrus sinensis* (L.) Osbeck essential oils and their cyclic monoterpene, DL-limonene. *Food Chem. Toxicol.*, 48(6):1734–1740.

Sirocchi, V., G. Caprioli, C. Cecchini et al., 2013. Biogenic amines as freshness index of meat wrapped in a new active packaging system formulated with essential oils of *Rosmarinus officinalis*. *Int. J. Food Sci. Nutr.*, 64(8):921–928.

Suanarunsawat, T., W. Devakul Na Ayutthaya, T. Songsak et al., 2010. Antioxidative activity and lipid lowering effect of essential oils extracted from *Ocimum sanctum* L. leaves in rats fed with a high cholesterol diet. *J. Clin. Biochem. Nutr.*, 46(1):52–59.

Teixeira, B., A. Marques, C. Ramos et al., 2013. Chemical composition and bioactivity of different oregano (*Origanum vulgare*) extracts and essential oil. *J. Sci. Food Agric.* doi: 10.1002/jsfa.6089.

Tel, G., M. Oztürk, M.E. Duru et al., 2010. Chemical composition of the essential oil and hexane extract of *Salvia chionantha* and their antioxidant and anticholinesterase activities. *Food Chem. Toxicol.*, 48(11):3189–3193.

Tsai, M.L, C.C. Lin, W.C. Lin, and C.H. Yang, 2011. Antimicrobial, antioxidant, and anti-inflammatory activities of essential oils from five selected herbs. *Biosci. Biotechnol. Biochem.*, 75(10):1977–1983.

Valente, J., M. Zuzarte, M.J. Gonçalves et al., 2013. Antifungal, antioxidant and anti-inflammatory activities of *Oenanthe crocata* L. essential oil. *Food Chem. Toxicol.*, 62:349–354.

Victoria, F.N., E.J. Lenardão, L. Savegnago et al., 2012. Essential oil of the leaves of *Eugenia uniflora* L.: Antioxidant and antimicrobial properties. *Food Chem. Toxicol.*, 50(8):2668–2674.

Viuda-Martos, M., A.N.G.S. El Gendy, E. Sendra et al., 2010. Chemical composition and antioxidant and anti-Listeria activities of essential oils obtained from Egyptian plants. *J. Agric. Food Chem.*, 58(16):9063–9092.

Vlaisavljevic, S., B. Kaurinovic, M. Popovic et al., 2014. *Trifolium pratense* L. as a potential natural antioxidant. *Molecules*, 19(1):713–725.

Woguem, V., F. Maggi, H.P. Fogang et al., 2013. Antioxidant, antiproliferative and antimicrobial activities of the volatile oil from the wild pepper *Piper capense* used in Cameroon as a culinary spice. *Nat. Prod. Commun.*, 8(12):1791–1796.

Xie, X.F., J.W. Wang, H.P. Zhang et al., 2014. Chemical composition, antimicrobial and antioxidant activities of essential oil from *Ampelopsis megalophylla*. *Nat. Prod. Res.* doi: 10.1080/14786419.2014.886208

Yvon, Y., E.G. Raoelison, R. Razafindrazaka et al., 2012. Relation between chemical composition or antioxidant activity and antihypertensive activity for six essential oils. *J. Food Sci.*, 77(8):H184–H191.

Zaouali, Y., T. Bouzaine, and M. Boussaid, 2010. Essential oils composition in two *Rosmarinus officinalis* L. varieties and incidence for antimicrobial and antioxidant activities. *Food Chem. Toxicol.*, 48(11):3144–3152.

Zhao, J., J.S. Zhang, B. Yang et al., 2010. Free radical scavenging activity and characterization of sesquiterpenoids in four species of curcuma using a TLC bioautography assay and GC-MS analysis. *Molecules*, 15(11):7547–7557.

Zheljazkov, V.D., T. Horgan, T. Astatkie, and V. Schlegel, 2013. Distillation time modifies essential oils yield, composition, and antioxidant capacity of fennel (*Foeniculum vulgare* Mill). *J. Oleo Sci.*, 62(9):665–672.

12 Effects of Essential Oils in the Central Nervous System

CONTENTS

12.1 Central Nervous System Effects of Essential Oils in Humans...345
Eva Heuberger
    12.1.1 Introduction 345
    12.1.2 Activation and Arousal: Definition and Neuroanatomical Considerations 348
    12.1.3 Influence of Essential Oils and Fragrances on Brain Potentials Indicative of Arousal....349
        12.1.3.1 Spontaneous Activity in the Electroencephalogram (EEG)......................349
        12.1.3.2 Contingent Negative Variation..353
    12.1.4 Effects of Essential Oils and Fragrances on Selected Basic and Higher Cognitive Functions...........................354
        12.1.4.1 Alertness and Attention 355
        12.1.4.2 Learning and Memory 358
        12.1.4.3 Other Cognitive Tasks...360
    12.1.5 Conclusions..362
12.2 Psychopharmacology of Essential Oils 362
Domingos Sávio Nunes, Viviane de Moura Linck, Adriana Lourenço da Silva, Micheli Figueiró, and Elaine Elisabetsky
    12.2.1 Aromatic Plants Used in Traditional Medical Systems as Sedatives or Stimulants......363
    12.2.2 Effects of Essential Oils in Animal Models..365
        12.2.2.1 Effects of Individual Components 366
        12.2.2.2 Effects of Inhaled Essential Oils 367
    12.2.3 Mechanism of Action Underlying Psychopharmacological Effects of Essential Oils....368
    12.2.4 Chemical Structures of Mentioned CNS Active Compounds 369
References..372

12.1 CENTRAL NERVOUS SYSTEM EFFECTS OF ESSENTIAL OILS IN HUMANS

Eva Heuberger

12.1.1 INTRODUCTION

A number of attempts has been made to unravel the effects of natural essential oils (EOs) and fragrances on the human central nervous system (CNS). Among these attempts, two major lines of research have been followed to identify psychoactive, particularly stimulating and sedative, effects of fragrances. On the one hand, researchers have investigated the influence of EO and fragrances on brain potentials that are indicative of the arousal state of the humans by means of neurophysiological methods. On the other hand, behavioral studies have elucidated the effects of EOs and fragrances on basic and higher cognitive functions, such as alertness and attention, learning and memory, or problem solving. The scope of the following section, though not claiming to be complete, is to give an overview about the current knowledge in these fields. Much of the research reviewed has been carried out in healthy populations, and only recently investigators have started to focus on clinical aspects of the administration of fragrances and EOs. However, since this topic is covered in Chapter 13, it is omitted here due to space constraints.

Olfaction differs from other senses in several ways. First, in humans and many other mammals, the information received by peripheral olfactory receptor cells is mainly processed in brain areas located ipsilaterally to the stimulated side of the body, whereas in the other sensory systems, it is transferred to the contralateral hemisphere. Second, in contrast to the other sensory systems, olfactory information reaches a number of cortical areas without being relayed in the thalamus (Kandel et al. 1991; Wiesmann et al. 2001; Zilles and Rehkämpfer 1998) (Figure 12.1). Owing to this missing thalamic control as well as to the fact that the olfactory system presents anatomical connections and overlaps with brain areas involved in emotional processing, such as the amygdala, hippocampus, and prefrontal cortex of the limbic system (Bermpohl et al. 2006; Davidson and Irwin 1999; Phan et al. 2002; Reiman et al. 1997), the effects of odorants on the organisms are supposedly exerted not only via pharmacological but also via psychological mechanisms. In humans and probably also in other mammals, psychological factors may be based on certain stimulus features, such as odor valence (Baron and Thomley 1994); on semantic cues, for example, memories and experiences associated with a particular odor; or on placebo effects related to the expectation of certain effects

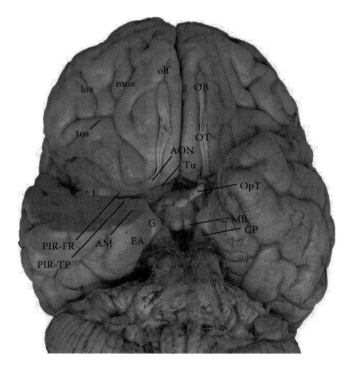

FIGURE 12.1 (See color insert.) Macroscopic view of the human ventral forebrain and medial temporal lobes, depicting the olfactory tract, its primary projections, and surrounding nonolfactory structures. The right medial temporal lobe has been resected horizontally through the midportion of the amygdala (AM) to expose the olfactory cortex. AON, anterior olfactory nucleus; CP, cerebral peduncle; EA, entorhinal area; G, gyrus ambiens; L, limen insula; los, lateral olfactory sulcus; MB, mammillary body; mos, medial olfactory sulcus; olf, olfactory sulcus; PIR-FR, frontal piriform cortex; OB, olfactory bulb; OpT, optic tract; OT, olfactory tract; tos, transverse olfactory sulcus; Tu, olfactory tubercle; PIR-TP, temporal piriform cortex. Figure prepared with the help of Dr. Eileen H. Bigio, Dept. of Pathology, Northwestern University Feinberg School of Medicine, Chicago, IL. (Taken with permission from Gottfried, J.A. and Zald, D.A., *Brain Res. Rev.*, 50, 287, 2005.)

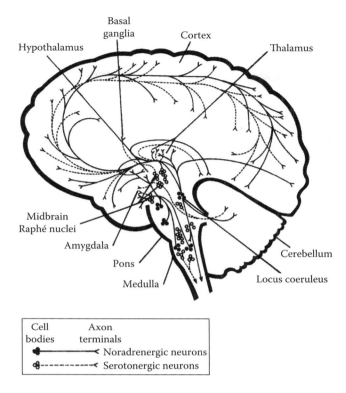

FIGURE 12.2 Schematic of the reticular activating system (RAS) with noradrenergic and serotonergic connections. (Taken with permission from Grilly, D.M., *Drugs & Human Behavior*, Allyn and Bacon, Boston, MA, 2002.)

(Jellinek 1997). None of the latter mechanisms is substance, that is, odorant, specific, but their effectiveness depends on cognitive mediation and control.

Many odorants stimulate not only the olfactory system via the first cranial nerve (*N. olfactorius*) but also the trigeminal system via the fifth cranial nerve (*N. trigeminus*) that enervates the nasal mucosa. The trigeminal system is part of the body's somatosensory system and mediates mechanical and temperature-related sensations, such as itching and burning or warmth and cooling sensations. Trigeminal information reaches the brain via the trigeminal ganglion and the ventral posterior nucleus of the thalamus. The primary cortical projection area of the somatosensory system is the contralateral postcentral gyrus of the parietal lobe (Zilles and Rehkämpfer 1998). The reticular formation in the brain stem that is part of the reticular activating system (RAS) (Figure 12.2) receives collaterals from the trigeminal system. Thus, trigeminal stimuli have direct effects on arousal. Utilizing this direct connection, highly potent trigeminal stimulants, such as ammonia and menthol, have been used in the past in smelling salts to awaken people who fainted.

It has been shown in experimental animals that, due to their lipophilic properties, fragrances do not only penetrate the skin (Hotchkiss 1998) but also the blood–brain barrier (Buchbauer et al. 1993a). Also, odorants have been found to bind to several types of brain receptors (Aoshima and Hamamoto 1999; Elisabetsky et al. 1999; Okugawa et al. 2000), and it has been suggested that these odorant–receptor interactions are responsible for psychoactive effects of fragrances in experimental animals. With regard to these findings, it is important to note that Heuberger and coworkers (2001) have observed differential effects of fragrances as a function of chirality. It seems likely that such differences in effectiveness are related to the enantiomeric selectivity of receptor proteins. However, the question remains to be answered whether effects of fragrances on human arousal and cognition rely on a similar psychopharmacological mechanism.

12.1.2 ACTIVATION AND AROUSAL: DEFINITION AND NEUROANATOMICAL CONSIDERATIONS

Activation, or arousal, refers to the ability of an organism to adapt to internal and external challenges (Schandry 1989). Activation is an elementary process that serves in the preparation for overt activity. Nevertheless, it does not necessarily result in overt behavior (Duffy 1972). Activation varies in degree and can be described along a continuum from deep sleep to overexcitement. Early theoretical accounts of activation have emphasized physiological responses as the sole measurable correlate of arousal. Current models, however, consider physiological, cognitive, and emotional activity as observable consequences of activation processes. It has been shown that arousal processes within each of these three systems, that is, physiological, cognitive, and emotional, can occur to varying degrees so that the response of one system need not be correlated linearly to that of the other systems (Baltissen and Heimann 1995).

It has long been established that the RAS that comprises the reticular formation with its sensory afferents and widespread hypothalamic, thalamic, and cortical projections plays a crucial role in the control of both phasic and tonic activation processes (Becker-Carus 1981; Schandry 1989). Pribram and McGuinness (1975) distinguish three separate but interacting neural networks in the control of activation (Figure 12.3). The arousal network involves amygdalar and related frontal cortical structures and regulates phasic physiological responses to novel incoming information. The activation network centers on the basal ganglia of the forebrain and controls the tonic physiological readiness to respond. Finally, the effort network, which comprises hippocampal circuits, coordinates the arousal and activation networks. Noradrenergic projections from the locus ceruleus, which is located within the dorsal wall of the rostral pons, are particularly important in the regulation of circadian alertness, the sleep–wake rhythm, and the sustainment of alertness (alerting) (Aston-Jones et al. 2001; Pedersen et al. 1998). On the other hand, tonic alertness seems to be dependent on cholinergic (Baxter and Chiba 1999; Gill et al. 2000) frontal and inferior parietal thalamic structures of the right hemisphere (Sturm et al. 1999). Other networks that are involved in the control of arousal and attentional functions are found in posterior parts of the brain, for example, the parietal cortex, superior colliculi, and posterolateral thalamus, as well as in anterior regions, for example, the cingular and prefrontal cortex (Paus 2001; Posner and Petersen 1990).

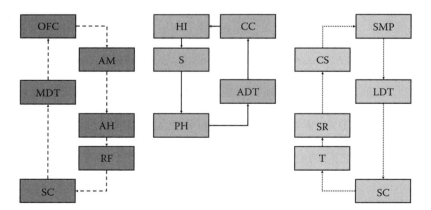

FIGURE 12.3 (See color insert.) Control of activation processes. OFC, orbitofrontal cortex; AM, amygdala; MDT, medial dorsal thalamus; AH, anterior hypothalamus; RF, reticular formation; SC, spinal cord; HI, hippocampus; CC, cingulate cortex; S, septum; ADT, anterior dorsal thalamus; PH, posterior hypothalamus; SMP, sensory-motor projections; CS, corpus striatum; LDT, lateral dorsal thalamus; SR, subthalamic regions; T, tectum. *Orange*, structures of the arousal network; *green*, structures of the effort network; *blue*, structures of the activation network. (Adapted from Pribram, K.H. and McGuinness, D., *Psychol. Rev.*, 82(2), 116, 1975.)

12.1.3 Influence of Essential Oils and Fragrances on Brain Potentials Indicative of Arousal

12.1.3.1 Spontaneous Activity in the Electroencephalogram (EEG)

Recordings of spontaneous EEG activity during the administration of EOs and fragrances have widely been used to assess stimulant and sedative effects of these substances. Particular attention has been paid to changes within the alpha and beta bands, sometimes also the theta band of the EEG in response to olfactory stimulation since these bands are thought to be most indicative of central arousal processes. Alpha waves are slow brain waves within a frequency range of 8–13 Hz and amplitudes between 5 and 100 μV that typically occur over posterior areas of the brain in an awake but relaxed state, especially with closed eyes. The alpha rhythm disappears immediately when subjects open their eyes and when cognitive activity is required, for example, when external stimuli are processed or tasks are solved. This phenomenon is often referred to as alpha block or desynchronization. Simultaneously with the alpha block, faster brain waves occur, such as beta waves with smaller amplitudes (2–20 μV) and frequencies between 14 and 30 Hz. The beta rhythm that is most evident frontally is characteristic of alertness, attention, and arousal. In contrast, theta waves are very slow brain waves occurring in frontotemporal areas with amplitudes between 5 and 100 μV in the frequency range between 4 and 7 Hz. Although the theta rhythm is most commonly associated with drowsiness and light sleep, some researchers found theta activity to correlate with memory processes (Grunwald et al. 1999; Hoedlmoser et al. 2007) and creativity (Razumnikova 2007). Other authors found correlations between theta activity and ratings of anxiety and tension (Lorig and Schwartz 1988). With regard to animal olfaction, it has been proposed that the theta rhythm generated by the hippocampus is concomitant to sniffing and allows for encoding and integration of olfactory information with other cognitive and motor processes (Kepecs et al. 2006).

A large number of measures can be derived from recordings of the spontaneous EEG. Time (index)- or voltage (power)-based rates of typical frequency bands, as well as ratios between certain frequency bands (e.g., between the alpha and the beta band or the theta and the beta band) within a selected time interval, are most commonly used to quantify EEG patterns. Period analysis quantifies the number of waves that occur in the various frequency bands within a distinct time interval of the EEG record and is supposed to be more sensitive to task-related changes than spectral analysis (Lorig 1989). Other parameters that describe the covariation of a given signal at different electrodes are coherence and neural synchrony. These measures inform on the functional link between brain areas (Oken et al. 2006).

Generally speaking, the pattern of the spontaneous EEG varies with the arousal level of the CNS. Thus, different states of consciousness, such as sleep, wakefulness, or meditation, can be distinguished by their characteristic EEG patterns. For instance, an increase of central activation is typically characterized by a decrease in alpha and an increase in beta activity (Schandry 1989). More precisely, a decline of alpha and beta power together with a decrease of alpha index and an increase of beta and theta activity has been observed under arousing conditions, that is, in a mental calculation and a psychosocial stress paradigm (Walschburger 1976). Also when subjects are maximally attentive, frequencies in the alpha band are attenuated and activity in the beta and even higher frequency bands can be observed. Fatigue and performance decrements in situations requiring high levels of attention are often associated with increases in theta and decreases in beta activity (Oken et al. 2006). On the other hand, drowsiness and the onset of sleep are characterized by an increase in slow and a decrease in fast EEG waves. However, high activity in the alpha band, particularly in the range between 7 and 10 Hz, is not indicative of low arousal states of the brain, such as relaxation, drowsiness, and the onset of sleep, but rather seems to be a component of selective neural inhibition processes that are necessary for a number of cognitive processes, such as perception, attention, and memory (Miller and O'Callaghan 2006; Palva and Palva 2007).

Because changes of spontaneous EEG activity accompany a wide range of cognitive as well as emotional brain processes and "EEG measurements […] do not tell investigators what the brain is doing" (Lorig 1989, p. 93), it is somewhat naïve to interpret changes that are induced by the application of an odorant as a result of a single and specific process, particularly when no other correlates of the process of interest are assessed. Nevertheless, this is exactly the approach that has been taken by many researchers to identify stimulant or relaxing effects of odors. The simplest, but probably most problematic, set up for such experiments in terms of interpretation of the results is the comparison of spontaneous EEG activity in response to odorants with a no-odor baseline. Using this design, Sugano (1992) observed increased EEG alpha activity after inhalation of α-pinene (**1**), 1,8-cineole (**2**), lavender, sandalwood, musk, and eucalyptus odors. Considering that traditional aromatherapy discriminates these fragrances by their psychoactive effects, for example, lavender is assigned relaxing properties while eucalyptus is supposedly stimulant (Valnet 1990), these findings are at least rather curious. Also, Ishikawa and coworkers recorded the spontaneous EEG in 13 Japanese subjects while drinking either lemon juice with a supplement of lemon odor or lemon juice without it (Ishikawa et al. 2002). It was shown that alpha power, indicative of increased relaxation, was enhanced by supplementation with lemon odor. Again, with regard to aromatherapeutical accounts of the lemon EO, this result is somewhat counterintuitive, even more so as in the same study the juice supplemented with lemon odor increased spontaneous locomotion in experimental animals. Haneyama and Kagatani tested a fragrant spray made from extracts of Chinese spikenard roots (*Nardostachys chinensis* Batalin (Valerianaceae)) in butylene glycol and found increased alpha activity in subjects under stress (Haneyama and Kagatani 2007). This finding was interpreted by the authors as demonstrating a sedative effect of the extract. However, it is unknown how stress was induced in the subjects and how it was measured. Similarly, Ishiyama (2000) concluded from measurements of the frequency fluctuation patterns of the alpha band that smelling a blend of terpene compounds typically found in forests induced feelings of refreshment and relaxation in human subjects without proper description of how these feelings were assessed.

Inconclusive findings as those described earlier are not unexpected with such simple experimental designs as there are several problems associated with this kind of experiments. First, a no-odor baseline is often inappropriate as it does not control for cognitive activity of the subject. For instance, subjects might be puzzled by the fact that they do not smell anything eventually focusing attention to the search for an odor. This may lead to quite high arousal levels rather than the intended resting brain state. This was the case with Lorig and Schwartz (1988) who tested changes of spontaneous EEG in response to spiced apple, eucalyptus, and lavender fragrances diluted in an odorless base; however, contrary to the authors' expectations it was found that alpha activity in the no-odor condition was less than during odor presentation.

Considering the various mechanisms outlined by Jellinek (1997) by which fragrances influence human arousal and behavior, another inherent problem of such simple designs is that little is known about how subjects process stimulus-related information, for example, the pleasantness or intensity of an odorant, and whether or not higher cognitive processes related to the odorant are initiated by the stimulation. For instance, subjects might be able to identify and label some odors but might fail to do so with others; similarly, some odors might trigger the recall of associated memories while others might not. In order to assess psychoactive effects of EOs and fragrances, it seems necessary to control for these factors, for instance, by assessing additional variables that inform on the subject's perception of primary and secondary stimulus features and that correlate to the subject's cognitive or emotional arousal state. In the previously mentioned study, Lorig and Schwartz (1988) collected ratings of intensity and pleasantness of the tested fragrances as well as subjective ratings of a number of affective states in addition to the EEG recordings. Analysis of the amount of EEG theta activity revealed that the spiced apple odor produced more relaxation than lavender and eucalyptus; the analysis of the secondary variables suggested that this relaxing effect was correlated with subjective estimates of anxiety and tension. As to the EEG patterns, similar results were observed when subjects imagined food odors and practiced relaxation techniques. Thus, the authors conclude that the relaxing effect of spiced apple was probably related to its association with food. These cognitive influences also seem to be a plausible explanation for the increase in alpha power by lemon odor in the study by Ishikawa et al. (Ishikawa et al. 2002). Other studies related to food odors were conducted by Kaneda and colleagues (2005, 2006, 2011). These authors investigated the influence of smelling beer flavors on the frequency fluctuation of alpha waves in frontal areas of the brain. The results showed relaxing effects of the aroma of hop extracts as well as of linalool (**3**), geraniol (**39**) ethyl acetate, and isoamyl acetate, but not of humulene and myrcene (**15**). In addition, in the right hemisphere these fluctuations were correlated with subjective estimates of arousal and with the intensity of the hop aroma. Lee et al. (1994) also found evidence for differential EEG patterns as a function of odor intensity for citrus, lavender, and a floral odor. A 10-min exposure to the weaker intensity of the citrus fragrance in comparison to lavender odor increased the rate of occipital alpha. Moreover, there was a general trend for citrus to be rated as more comfortable than the other fragrances. In contrast, the higher intensity of the floral fragrance increased the rate of occipital beta more than the lavender odor.

Several authors have shown influences of odor pleasantness and familiarity on changes of the spontaneous EEG. For instance, Kaetsu and colleagues (1994) reported that pleasant odors increased alpha activity, while unpleasant ones decreased it. In a study on the effects of lavender and jasmine odor on electrical brain activity (Yagyu 1994), it was shown that changes in the alpha, beta, and theta bands in response to these fragrances were similar when subjects rated them as pleasant, while lavender and jasmine odor led to distinct patterns when they were rated as unpleasant. Increases of alpha activity in response to pleasant odors might be explained by altered breathing patterns since it has been demonstrated that pleasant odors induce deeper inhalations and exhalations than unpleasant odors and that this form of breathing by itself increases activity in the alpha band (Lorig 2000). Masago and coworkers (2000) tested the effects of lavender, chamomile, sandalwood, and eugenol (**4**) fragrances on ongoing EEG activity and self-ratings of comfort and found a significant positive correlation between the degree of comfort and the odorants' potency to decrease alpha activity in parietal and posterior temporal regions. In relation to the previously described investigations, this finding is rather difficult to explain, although it differs from the other studies in that it differentiated between electrode sites rather than reporting merely global changes in electrical brain activity. Therefore, this result suggests that topographical differences in electrical brain activity induced by fragrances may be important and need further investigation. In fact, differences in hemispheric localization of spontaneous EEG activity in response to pleasant and unpleasant fragrances, respectively, seem to be quite consistent. While pleasant odors induced higher activation in left frontal brain regions, unpleasant ones led to bilateral and widespread activation (Kim and Watanuki 2003) or no differences were observed when an unpleasant odor (valerian) was compared to a no-odor control condition (Kline et al. 2000). Another interesting finding in the study of Kim and Watanuki (2003) was that EEG activity in response to the tested fragrances was observed when subjects were at rest but vanished after performing a mental task. The importance of distinguishing EEG activity arising from different areas of the brain is highlighted by an investigation by Van Toller and coworkers (1993). These authors recorded alpha wave activity at 28 sites of the scalp immediately after the exposure to a number of fragrances covering a range of different odor types and hedonic tones at isointense concentrations; the odorants had to be rated in terms of pleasantness, familiarity, and intensity. It was shown that in posterior regions of the brain, changes in alpha activity in response to these odors compared to an odorless blank were organized in distinct topographical maps. Moreover, alpha activity in a set of electrodes at frontal and temporal sites correlated with the psychometric ratings of the fragrances.

More recently, a research group from Thailand screened a variety of EOs used in aromatherapy for their effects on electrical brain activity and mood after inhalation. All experiments were conducted in healthy human subjects and employed the same A–B–C design, that is, a no-odor control condition (A) was followed by a condition (B) in which a carrier oil without EO was presented and finally by the experimental condition (C) in which the EO of interest diluted in the carrier was administered. The EOs under study were jasmine (Sayowan et al. 2013), rosemary (Sayorwan et al. 2013), citronella (Sayowan et al. 2012), and lavender (Sayorwan et al. 2012a,b). In light of the previously mentioned influence of hedonic odor evaluation on EEG activity, it is important to mention that in all of these investigations, subjects were only allowed to participate if they rated the tested odors as highly to moderately pleasant. The authors were able to confirm the effect described by traditional aromatherapy for every single EO. In regard to the inconsistent findings reported by others, the perfect conformity between observed effects and aromatherapeutical claims is quite remarkable. The selection of subjects based on their hedonic preferences does not seem to offer a fully satisfactory explanation since there should not have emerged differences between the EOs if the effects relied solely on mechanisms involving hedonic odor valence. Perhaps other factors, such as familiarity with the odors and, even more importantly, expectations of their effects (Jellinek 1997; Lorig and Roberts 1990), have influenced the physiological response of the study participants. Thus, it would have been worthwhile for the authors to assess these potentially confounding influences.

As to influences of the familiarity of or experience with fragrances, Kawano (2001) reported that the odors of lemon, lavender, patchouli, marjoram, rosemary, and sandalwood increased alpha activity over occipital electrode sites in subjects to whom these fragrances were well known. On the one hand, lag times between frontal and occipital alpha phases were shorter in subjects less experienced with the fragrances indicating that these subjects were concentrating more on smelling—and probably identifying—the odorants. These findings were confirmed in an investigation comparing professional perfume researchers, perfume salespersons, and general workers (Min et al. 2003). This study showed that measures of corticocortical connectivity, that is, the averaged cross mutual information content, in odor processing were more pronounced in frontal areas with perfume researchers, whereas with perfume salespersons and general workers, a larger network of posterior temporal, parietal, and frontal regions was activated. These results could result from a greater involvement of orbitofrontal cortex neurons in perfume researches who exhibit high sophistication in discriminating and identifying odors. Moreover, it was shown that the value of the averaged cross mutual information content was inversely related to preference in perfume researchers and perfume salespersons, but not in general workers.

As pointed out earlier, the administration of EO and fragrances to naïve subjects can lead to cognitive processes that are unknown to the investigator and sometimes even the subject. Nevertheless, these processes will affect spontaneous EEG activity. Some researchers have sought to solve this problem by engaging subjects in a secondary task while the influence of the odorant of interest is assessed. This procedure does not only draw the subject's attention away from the odor stimulus but also provides the desired information about his or her arousal state. Another benefit of such experimental designs is that the task may control for the subject's arousal state if a certain amount of attention is required to perform it. Measurement of changes of alpha and theta activity in the presence or absence of 1,8-cineole (**2**), methyl jasmonate (**5**), and *trans*-jasmin lactone (**6**) in subjects who performed a simple visual task showed that the increase in slow-wave activity was attenuated by 1,8-cineole (**2**) and methyl jasmonate (**5**), while augmented by *trans*-jasmin lactone (**6**) (Nakagawa et al. 1992). At least in the case of 1,8-cineole (**2**), these findings are supported by results from experimental animals and humans indicating activating effects of this odorant (Bensafi et al. 2002; Kovar et al. 1987; Nasel et al. 1994). An investigation of the effects of lemon odor EEG alpha, beta, and theta activity during the administration of lemon odor showed that the odor reduced power in the lower alpha range while it increased power in the higher alpha, lower beta, and lower theta bands (Krizhanovs'kii et al. 2004). These findings of increased arousal were in agreement with better performance in a cognitive task. In addition, it was shown that inhalation of the lemon fragrance was most effective during rest and in the first minutes of the cognitive task but wore off after less than 10 min. In several experiments, the group of Sugawara demonstrated complex interactions between electrical brain activity induced by the exposure to fragrances, sensory profiling, and various types of tasks (Satoh and Sugawara 2003; Sugawara et al. 2000). In the first study they showed that the odor of peppermint, in contrast to basil, was rated less favorable on a number of descriptors and reduced the magnitude of beta waves after as compared to before performance of a cognitive task. In a similar investigation, it was shown that the sensory evaluation as well as changes

//
Effects of Essential Oils in the Central Nervous System

in spontaneous EEG activity in response to the odors of the linalool enantiomers differed as a function of the molecular structure and the kind of task. For instance, R-(−)-linalool (**7**) was rated more favorable and led to larger decreases of beta activity after listening to natural sounds than before. In contrast, after as compared to before cognitive effort, R-(−)-linalool (**7**) was rated as less favorable and tended to increase beta power. A similar pattern was found for RS-(±)-linalool (**3**), whereas for S-(+)-linalool (**8**) the pattern was different, particularly with regard to EEG activity.

In the study of Yagyu (1994), the effects of lavender and jasmine fragrances on the performance in a critical flicker fusion and an auditory reaction time task were assessed in addition to changes of the ongoing EEG. It was demonstrated that in contrast to the EEG findings, lavender decreased performance in both tasks independent of its hedonic evaluation. Jasmine, however, had no effect on task performance. The EEG changes in response to these odorants might well explain their effects on performance: lavender induced decreases of activity in the beta band that is associated with states of low attention regardless of being rated as pleasant or unpleasant; jasmine, on the other hand, increased EEG beta activity when it was judged unpleasant but lowered beta activity when judged pleasant, so that overall its effect on performance leveled out. The effects of lavender and rosemary fragrances on electrical brain activity, mood states, and math computations were investigated by Diego, Field, and coworkers (Diego et al. 1998; Field et al. 2005). These investigations showed that the exposure to lavender increased beta power, elevated feelings of relaxation, reduced feelings of depression, and improved both speed and accuracy in the cognitive task. In contrast, rosemary odor decreased frontal alpha and beta power, decreased feelings of anxiety, increased feelings of relaxation and alertness, and increased speed in math computations. The EEG results were interpreted as indicating increased drowsiness in the lavender group and increased alertness in the rosemary group; however, the behavioral data showed performance improvement and similar mood ratings in both groups. These findings suggest that different electrophysiological arousal patterns may still be associated with similar behavioral arousal patterns emphasizing the importance of collecting additional endpoints to evaluate psychoactive effects of EO and fragrances.

12.1.3.2 Contingent Negative Variation

The contingent negative variation (CNV) is a slow, negative event-related brain potential that is generated when an imperative stimulus is preceded by a warning stimulus and reflects expectancy and preparation (Walter et al. 1964). The amplitude of the CNV is correlated to attention and arousal (Tecce 1972). Since changes of the magnitude and latency of CNV components have long been associated with the effects of psychoactive drugs (Ashton et al. 1977; Kopell et al. 1974), measurement of the CNV has also been used to evaluate psychostimulant and sedating effects of EOs and fragrances. In a pioneering investigation, Torii and colleagues (Torii et al. 1988) measured CNV magnitude changes evoked by a variety of EOs, such as jasmine, lavender, and rose oil, in male subjects. CNV was recorded at frontal, central, and parietal sites after the presentation of an odorous or blank stimulus in the context of a cued reaction time paradigm. In addition, physiological markers of arousal, that is, skin potential level and heart rate, were simultaneously measured. Results showed that at frontal sites the amplitude of the early negative shift of the CNV was significantly altered after the presentation of odor stimuli, and that these changes were mostly congruent with stimulating and sedative properties reported for the tested oils in the traditional aromatherapy literature. In contrast to other psychoactive substances, such as caffeine or benzodiazepines, presentation of the EOs neither affected physiological parameters nor reaction times. The authors concluded that the EOs tested influenced brain waves *almost exclusively* while having no effects on other indicators of arousal.

Subsequently, CNV recordings have been used by a number of researchers on a variety of EOs and fragrances to establish effects of odors on the human brain along the activation–relaxation continuum. For instance, Sugano (1992) in the aforementioned study demonstrated that α-pinene (**1**), sandalwood, and lavender odor increased the magnitude of the CNV in healthy young adults, whereas eucalyptus reduced it. It is interesting to note, however, that all of these odors—despite their differential influence on the CNV—increased spontaneous alpha activity in the same experiment. An increase of CNV magnitude was also observed with the EO from pine needles (Manley 1993) that was interpreted as a stimulating effect. Aoki (1996) investigated the influence of odors from several coniferous woods, that is, hinoki (*Chamaecyparis obtusa* (Siebold & Zucc.) Endl. (Cupressaceae)), sugi (*Cryptomeria japonica* D. Don (Cupressaceae)), akamatsu (*Pinus densiflora* Siebold & Zucc. (Pinaceae)), hiba (*Thujopsis dolabrata* var. *hondai* Siebold & Zucc. (Cupressaceae)), Alaska cedar (*Chamaecyparis nootkatensis* (D. Don) Spach (Cupressaceae)), Douglas fir (*Pseudotsuga menziesii* (Mirbel) Franco (Pinaceae)), and Western red cedar (*Thuja plicata* Donn (Cupressaceae)), on the CNV and found conflicting effects: The amplitude of the early CNV component at central sites was decreased by these wood odors, and the alpha/beta-wave ratio of the EEG increased. Moreover, the decrease of CNV magnitude was correlated with the amount of α-pinene (**1**) in the tested wood odors. Also, Sawada and coworkers (2000) measured changes of the early component of the CNV in response to stimulation with terpenes found in the EO of woods and leaves. These authors noticed a reduction of the CNV magnitude after the administration of α-pinene (**1**), Δ-3-carene (**9**), and bornyl acetate (**10**). However, a more recent investigation (Hiruma et al. 2002) showed that hiba (*T. dolabrata* Siebold & Zucc. (Cupressaceae)) odor increased the CNV magnitude at frontal and central sites and shortened reaction times to the imperative stimulus in female subjects. These authors thus concluded that the odor of hiba heightened the arousal level of the CNS.

Although the CNV is believed to be largely independent of individual differences, such as age, sex, or race (Manley 1997), there seem to be cognitive influences that must not be neglected when interpreting the effects of odor stimuli on the CNV. Lorig and Roberts (1990) repeated the study by Torii and colleagues (1988) and investigated cognitive factors by introducing a manipulation of their subjects as an additional variable into the paradigm. In this experiment subjects were exposed to the original two odors, that is, lavender and jasmine, referred to as odor A and odor B, respectively, as well as to a mixture of the two fragrances. However, in half of the trials in which the mixture was administered, subjects were led to believe that they received a low concentration of odor A, while in the other half of trials they thought that they would be exposed to a low concentration of odor B. In none of the four conditions, subjects were given the correct odor names. As in the study by Torii et al., lavender reduced the amplitude of the CNV, whereas jasmine increased it. When the mixture was administered, however, the CNV magnitude decreased when subjects believed to receive a low concentration of lavender but increased when they thought they were inhaling a low concentration of jasmine. This means that the alteration of the CNV amplitude was not solely related to the substance that had been administered but also to the expectation of the subjects. Another point made by Lorig and Roberts (1990) is that in their study self-report data indicated that lavender was actually rated as more arousing than jasmine. Since low CNV amplitudes are not only associated with low arousal but also with high arousal in the context of distraction (Travis and Tecce 1998), the lavender odor might in fact have led to higher arousal levels than jasmine even though the CNV magnitude was smaller with lavender. Other authors have noted that CNV changes might not only reflect effects of odor stimuli but also the anticipation, expectancy, and emotional state of the subjects who are exposed to these odorants (Hiruma et al. 2005). The involvement of these and other cognitive factors might well explain why the findings of CNV changes in response to odorants are rather inconsistent.

12.1.4 Effects of Essential Oils and Fragrances on Selected Basic and Higher Cognitive Functions

Psychoactive effects of odorants at the cognitive level have been explored in humans using a large number of methods. A variety of testing procedures ranging from simple alertness or mathematical tasks to tests that assess higher cognitive functions, such as memory or creativity, have been employed to study stimulant or relaxing/sedating effects of EOs and fragrances.

Nevertheless, the efficiency of odorants is commonly defined by changes in performance in such tasks as a function of the exposure to fragrances.

12.1.4.1 Alertness and Attention

A number of studies is available on the influence of fragrances on attentional functions. The integrity and the level of the processing efficiency of the attentional systems are fundamental prerequisites of all higher cognitive functions. Attentional functions can be divided into four categories, that is, alertness, selective and divided attention, and vigilance (Keller and Groemminger 1993; Posner and Rafal 1987; Sturm 1997). Alertness is the most basic form of attention and is intrinsically dependent of the general level of arousal. Selective attention describes the ability to focus on relevant stimulus information while nonrelevant features are neglected; divided attention describes the ability to concomitantly process several stimuli from different sensory modalities. Vigilance refers to the sustainment of attention over longer periods of time. Since the critical stimuli typically occur only rarely in time, vigilance can be seen as a counterforce against increasing fatigue in boring situations that is crucial in everyday life, in situations like long-distance driving (particularly at night), working in assembly lines, or monitoring a radar screen (e.g., in air traffic control).

In a pioneering study, Warm and colleagues (1991) investigated the influence of peppermint and muguet odors on human visual vigilance. Peppermint that was rated stimulant was expected to increase task performance, while muguet rated as relaxing was expected to impair it. After intermittent inhalation, none of these fragrances increased processing speed in the task, but subjects in both odorant conditions detected more targets than a control group receiving unscented air. On the other hand, neither fragrance influenced subjective mood or judgments of workload. Gould and Martin (2001) studied the effects of bergamot and peppermint EOs on human sustained attention, where again peppermint was expected to improve performance, while bergamot characterized as relaxing by an independent sample of subjects was expected to have a deteriorating effect on vigilance performance. However, only bergamot had a significant influence in the anticipated direction, that is, subjects in this condition detected fewer targets than subjects in the peppermint or a no-odor control condition, which was probably related to the subjects' expectation of a relaxing effect.

The influence of the inhalation of a number of EOs and fragrances on performance in basic attentional tasks was assessed by Heuberger and Ilmberger (2010) and Ilmberger and coworkers (2001). In the first experiment, several EOs with presumed activating effects were inhaled by human subjects during an alertness task. Contrary to the authors' hypotheses, the results suggested that these EOs when compared to an odorless control did not increase the speed of information processing. Even more unexpectedly, motor learning was impaired in the groups that received EOs; this effect is likely a consequence of distraction induced by the strong odor stimuli, an explanation that was supported by the reaction times that tended to be higher in the EO-treated groups than in the corresponding control groups. Alternatively, these authors argued that a ceiling effect might be responsible for the observed effects. Given that healthy subjects with intact attentional systems already perform at optimal levels of information processing in such basic tasks, it seems likely that activating EOs cannot enhance performance any further. Similarly, performance of healthy subjects may be too robust to be influenced by deactivating fragrances. In the second study, subjects inhaled several activating and sedating fragrances while engaging in a vigilance task. Again the expected effects were not observed, but for one compound, that is, linalyl acetate (**44**), even the opposite effect on vigilance performance was demonstrated. Further analyses showed that this effect was strongly correlated to subjective estimates of odor pleasantness.

The observation that the effectiveness of fragrances is dependent on task complexity has been supported by investigations on the EO of peppermint (Ho and Spence 2005), lavender, and rosemary (Moss et al. 2003). These fragrances rather affected performance in difficult tasks or in tasks testing higher cognitive functions than in simple ones testing basic functions. Another interesting finding of the study of Ilmberger et al. (2001) was that changes in performance were correlated with subjective ratings of characteristic odor properties, particularly with pleasantness and efficiency. Similar results were obtained in another study for the EO of peppermint (Sullivan et al. 1998) that showed that in a vigilance task subjects benefited most from the effects of this fragrance when they experienced the task as quite difficult and thought that the EO had a stimulant effect. In addition, the studies of Ilmberger and coworkers clearly demonstrated effects of expectation, that is, a placebo effect, as correlations between individual task performance and odor ratings were not only revealed in the EO groups but also in the no-odor control groups. Nevertheless, a recent study by Moss and Oliver (2012) strongly supports the existence of pharmacological factors involved in the effects of fragrances on human attentional performance after inhalation. These authors studied the influence of the EO of rosemary on several tasks with varying cognitive load and found significant correlations between a number of performance measures and plasma levels of 1,8-cineole (**2**).

The effects of a pleasant and an unpleasant blend of fragrances on selective attention were studied by Gilbert and colleagues (1997). No influence of either fragrance blend was found on attentional performance, but the authors observed a sex-specific effect of suggesting the presence of ambient odors. In the presence of a pleasant or no odor in the testing room, male subjects performed better when they were led to believe that no odor was present. Female subjects, however, performed better when they thought that they were exposed to an odorant under the same conditions. No such interaction was found in the unpleasant fragrance condition. These data again emphasize that, in addition to hedonic preferences, expectation of an effect may crucially influence the effects of odorants on human performance and that these factors may affect women and men differently.

Millot and coworkers (2002) evaluated the influence of pleasant (lavender oil) and unpleasant (pyridine, **11**) ambient odors on performance in a visual or auditory alertness task and in a divided attention task. The results showed that in the alertness task, irrespective of the tested modality, both fragrances independent of their hedonic valence improved performance by shortening reaction times compared to an unscented control condition. However, none of the odorants exerted any influence on performance in the selective attention task in which subjects had to attend to auditory stimuli while neglecting visual ones. The authors conclude that pleasant odors enhance task performance by decreasing subjective feelings of stress, that is, by reducing overarousal, while unpleasant fragrances increase activation from suboptimal to optimal levels, thus having the same beneficial effects on cognitive performance. With this explanation the authors, however, presume that subjects in their experimental groups started from dissimilar arousal levels that seems rather unlikely given that subjects were assigned to these groups at random. Moreover, cognitive performance should be affected by alterations of the arousal level more readily with increasing task difficulty. Thus, the interpretation given by the authors does not thoroughly explain why reaction times were influenced by the odorants in the simple alertness task but not in the more sophisticated selective attention task.

Degel and Köster (1999) exposed healthy subjects to either lavender, jasmine, or no fragrance with subjects being unaware of the odorants. Subjects had to perform a mathematical test, a letter counting (i.e., selective attention) task, and a creativity test. The authors expected a negative effect on performance of lavender and a positive effect of jasmine. The results, however, showed that lavender decreased the error rate in the selective attention task, whereas jasmine increased the number of errors in the mathematical test. Ratings of odor valence collected after testing demonstrated that lavender was judged more pleasant than jasmine, independent of which odor had been presented during the testing. Although subjects did not know that a fragrance had been administered, implicit evaluation of odor pleasantness has probably influenced their performance. This relation is supported by the fact that subjects who were not able to correctly identify the odors preferentially associated pictures of the room they had been tested in with the odor that had been present during testing. Improvement of performance as a result of the inhalation of lavender EO has also been reported in another investigation (Sakamoto et al. 2005). In this study, subjects were exposed to lavender, jasmine, or no aroma during phases of rest in between sessions in which they completed a visual vigilance task involving tracking of a moving target. In the penultimate of five sessions, when fatigue was highest and arousal lowest as estimated from the decrement in performance between sessions of the control group, tracking speed increased and tracking error decreased in the lavender group when compared to the no-aroma group. Jasmine had no effect on task performance. The authors argued that lavender aroma may have decreased arousal during the resting period and hence helped to achieve optimal levels for the following task period. Since no secondary variables indicative of arousal or of subjective evaluation of aroma quality were assessed in this investigation, no inferences can be made on the mechanisms underlying the observed effects. Diego and coworkers in the aforementioned investigation studied the influence of lavender and rosemary EOs after a 3 min inhalation period on a mathematical task (Diego et al. 1998). In contrast to the authors' expectations, both odorants positively affected performance by increasing calculation speed, although only lavender improved calculation accuracy. In addition, subjects in both fragrance groups reported to be more relaxed. Those in the lavender group had less depressed mood, while those in the rosemary group felt more alert and had lower state anxiety scores. These findings were interpreted as indicating overarousal caused by rosemary EO that led to an increase of calculation speed at the cost of accuracy. In contrast, lavender EO seemed to have reduced the subjects' arousal level and thus led to better performance than rosemary EO. However, since subjects in both fragrance groups felt more relaxed—but obviously only the lavender group benefited from this increase in relaxation—this is a somewhat unsatisfying explanation for the observed results.

Evidence for the influence of physicochemical odorant properties on visual information processing was supplied by Michael and colleagues (2005). These authors found that the exposure to both allyl isothiocyanate (AIC, **12**), a mixed olfactory/trigeminal stimulus, and 2-phenyl ethyl alcohol (2-PEA, **13**), a pure olfactory stimulant, impaired performance in a highly demanding visual attention task involving reaction to a target stimulus when a distractor appeared at different intervals after presentation of the target, although different mechanisms were responsible for these effects. In trials without a distractor, only 2-PEA (**13**) significantly increased the reaction times of healthy subjects; in trials with a distractor, subjects reacted more slowly in both odor conditions as compared to the no-odor control condition. However, AIC impaired performance independent of the interval between distractor and target, whereas 2-PEA (**13**) only had a negative effect when the interval between target and distractor was short. While 2-PEA (**13**) seemed to have led to performance decrements by decreasing subjects' arousal level, AIC as a strong trigeminal irritant seemed to have affected the shift of attention toward the distractor stimuli, so that they were considered more important than in the other conditions. A similar observation has also been made for the annoying odor of propionic acid (Hey et al. 2009). In this study, the error rate in a response-inhibition task increased as a function of odorant concentration suggesting a relationship between cognitive distraction and sensory annoyance.

Differences in effectiveness of fragrances as a function of the route of administration were explored by Heuberger and coworkers (2008). In several experiments, these authors investigated the influence of two monoterpenes, that is, 1,8-cineole (**2**) and ($\pm$)-linalool (**3**), on performance in a visual sustained attention task after 20 min inhalation and dermal application, respectively. 1,8-Cineole (**2**) was expected to induce activation and improve task performance, while ($\pm$)-linalool (**3**) was considered sedating/relaxing thus impairing performance. Since one of the aims of the study was to assess fragrance effects that were not mediated by stimulation of the olfactory system, inhalation of the odorants was prevented in the dermal application conditions. In each condition, subjects rated their mood and well being. In addition, ratings of odor pleasantness, intensity, and effectiveness were assessed in the inhalation conditions. In regard to performance on the vigilance task, the results showed no difference between the fragrance groups compared to a control group that had received odorless air. However, 1,8-cineole (**2**) increased feelings of relaxation and calmness, whereas ($\pm$)-linalool (**3**) led to increased vigor and mood. In addition, individual performance was correlated to the pleasantness of the odor and of expectations of its effect. In contrast, in the dermal application conditions, subjects having received 1,8-cineole (**2**) performed faster than those having received ($\pm$)-linalool (**3**). These findings were interpreted as indicating the involvement of different mechanisms after inhalation and nonolfactory administration of fragrances. It seems that psychological effects are predominant when fragrances are applied by means of inhalation, that is, when the sense of smell is stimulated. On the other hand, pharmacological effects of odorants that might be overridden when fragrances are inhaled are evident when processing of odor information is prevented.

Along the same lines, two studies investigated the effects of Spanish sage (*Salvia lavandulifolia* Vahl (Lamiaceae)) on the performance in a number of cognitive tasks after oral application or inhalation. In the first study by Kennedy and colleagues (2011), participants who received a single oral dose of 50 µL of the EO of *S. lavandulifolia* reacted faster in a simple reaction time task than subjects who received a placebo. By contrast, this facilitating effect on attention speed was not detected in the second study when subjects inhaled the EO of *S. lavandulifolia* (Moss et al. 2010).

12.1.4.2 Learning and Memory

Effects of EOs and fragrances on memory functions and learning have less frequently been explored than influences on more basic cognitive functions. While learning can briefly be defined as "a process through which experience produces a lasting change in behavior or mental processes" (Zimbardo et al. 2003, p. 206), memory is a cognitive system composed of three separate subsystems or stages that cooperate closely to encode, store, and retrieve information. Sensory memory constitutes the first of the three memory stages and is responsible for briefly retaining sensory information. The second stage, working memory, transitorily preserves recent events and experiences. Long-term memory, the third subsystem, has the highest capacity of all stages and stores information based on meaning associated with the information (Zimbardo et al. 2003). A basic form of learning that has been identified as a potent mediator of fragrance effects in humans (Jellinek 1997) is conditioning, that is, the (conscious or unconscious) association of a stimulus with a specific response or behavior. For instance, Epple and Herz (1999) demonstrated that children who were exposed to an odorant during the performance of an insolvable task performed worse on other solvable tasks when the same odorant was presented again. In contrast, no such impairment was observed when no odor or a different odor was presented. These results were interpreted as demonstrating negative olfactory conditioning. Along the same lines, Chu (2008) was able to show positive olfactory conditioning. In his study, children successfully performed a cognitive task that they believed was insolvable in the presence of an ambient odor. When they were reexposed to the same odorant, performance on other tasks improved significantly in comparison to another group of children who receive a different odor. Since the children in Chu's investigation were described in school reports as underachieving and lacking self-confidence, one might speculate that only children with these specific attributes benefit of the influence of fragrances. This seems to be confirmed by a study of Kerl (1997) who found that, in general, ambient odors of lavender and jasmine did not improve memory functions in school children. However, lavender tended to increase performance in the memory task in children with high anxiety levels, which may have been consequent to the stress-relieving properties of lavender. On the other hand, jasmine impaired performance in lethargic children and this impairment in performance was negatively correlated with the children's rating of the odor. This result might indicate that lethargic children were distracted by the presence of an odorant they liked.

Several studies examined the influence of EOs on a number of memory-related variables in adults (Moss et al. 2003, 2008, 2010; Tildesley et al. 2005). These authors reported that lavender reduced the quality of memory and rosemary increased it, while both EOs reduced the speed of memory when compared to a no-odor control condition. At the same time, rosemary increased alertness in comparison to both the control and the lavender group, but exposure to the odorants led to higher contentedness than no scent. Similarly, peppermint enhanced memory quality while ylang-ylang impaired it and reduced processing speed. Ratings of mood showed that peppermint increased alertness while ylang-ylang decreased it and increased subjective calmness. In a third experiment, oral administration of Spanish sage (*S. lavandulifolia* Vahl (Lamiaceae)) improved both quality and speed of memory and increased subjective ratings of alertness, calmness, and contentedness. In a fourth study, the effects of inhalation of two species of sage, that is,

S. officinalis L. (Lamiaceae) and *S. lavandulifolia* Vahl (Lamiaceae), were compared. This investigation demonstrated differences between the two species in regard to their impact on the quality of memory factor, that is, in comparison to a no-odor control, *S. officinalis* aroma improved quality of memory-related endpoints, whereas *S. lavandulifolia* aroma did not. In contrast, the impact of the two species on subjective measures of mood was identical: compared to the no-odor control both increased alertness but had no effect on calmness and contendedness. Also, there were no differences between the two species with respect to odor hedonics or intensity. More interestingly, the comparison between oral administration and inhalation of *S. lavandulifolia* EO also highlights differences in effectiveness in relation to the route of administration that are in accordance with the authors' own observations (Friedl et al. 2010; Heuberger et al. 2006, 2008; Hongratanaworakit et al. 2004).

The beneficial influence of an odorant being present in the learning phase on successive retrieval of information was shown by Morgan (1996). In this study, subjects were exposed or not exposed to a fragrance during the encoding of words unrelated to odor. Recall of the learned material was tested in three unannounced sessions 15 min apart, as well as 5 days after the learning phase. The results showed that performance in those groups that had not been exposed to an odorant in the learning phase declined continuously over time, whereas it remained stable in those groups that had learned with ambient odor present. In addition, subjects who had learned under odor exposure performed significantly better when the odor was present during recall than those who had not received an odorant during the learning phase. These findings show that odorants in the encoding phase may serve as cues for later recall of the stored information. Recently, similar results have been reported in 3-month-old infants (Suss et al. 2012). Schwabe and Wolf (2009) showed that the detrimental effects of stress experienced before memory retrieval can be attenuated by odor context cues presented during the encoding phase. Enhancement of declarative memory by olfactory context cues has even been found in adults who were presented with odorants while asleep (Rasch et al. 2007). Moreover, a recent study showed that memory consolidation during sleep is accelerated by odors (Diekelmann et al. 2012). On the other hand, fear memory can be extinguished selectively when odors are presented during sleep that have served as contextual cues during a learning phase in the awake state (Hauner et al. 2013).

According to a study by Walla and coworkers, it seems to be crucial whether an odorant in the encoding phase of a mnemonic task is consciously perceived and processed (Walla et al. 2002). These authors found differences in brain activation in a word recognition task as a function of conscious versus unconscious olfactory processing in the encoding phase. In other words, when odorants were presented during the learning phase and consciously perceived, word recognition was more likely negatively affected than when the odor was not consciously processed. In addition, the same group of researchers demonstrated that word recognition performance was significantly poorer when the odorants were presented simultaneously with the words as opposed to continuously during the encoding phase and when semantic (deep) as opposed to nonsemantic (shallow) encoding was required. These effects can be explained by a competition of processing resources in brain areas engaged in both language and odor processing (Walla et al. 2003a). Similar results were also observed in an experiment involving the encoding of faces with and without odorants present in the learning phase (Walla et al. 2003b). Again, recognition accuracy was impaired when an odor was simultaneously presented during encoding.

As discussed earlier, subjective experience of valence seems to modulate the influence of fragrances on cognition. For instance, Danuser and coworkers (2003) found no effects of pleasant olfactory stimuli on short-term memory, whereas unpleasant odorants reduced the performance of healthy subjects, probably by distracting them. Habel and coworkers (2007) studied the effect of neutral and unpleasant olfactory stimulation on the performance of a working memory task and found that malodors significantly deteriorated working memory in only about half of the subjects. It was also shown that subjects in the affected group differed significantly in brain activation patterns from those in the unaffected group, that is, the latter showing stronger activation in fronto-parieto-cerebellar networks associated with working memory. In contrast, subjects whose performance was impaired by the unpleasant odor showed greater activation in areas associated with emotional processing, such as the temporal and medial frontal cortex. The authors concluded that individual differences exist for the influence of fragrances on working memory and that unaffected subjects were better able to counteract the detrimental effect of unpleasant odor stimuli.

Leppanen and Hietanen (2003) tested the recognition speed of happy and disgusted facial expressions when pleasant or unpleasant odorants were presented during the recognition task. This study showed that pleasant olfactory stimuli had no particular influence on the speed of recognition of emotional facial expressions, that is, happy faces were recognized faster than disgusted faces. This result was also observed when no odorant was administered. In the unpleasant condition, however, the advantage for recognizing happy faces disappeared. These findings were interpreted as evidence for the modulation of emotion-related brain structures that form the perceptual representation of facial expressions by unpleasant odorants. Walla and colleagues supplied evidence that performance in a face recognition task was only affected when conscious odor processing took place (Walla et al. 2005). In this study, two olfactory stimuli, that is, 2-PEA (**13**) and dihydrogen sulfide (H_2S); a trigeminal stimulus, that is, carbon dioxide (CO_2); and no odor were presented briefly and simultaneously to the presentation of faces. The results showed that the pure odorants irrespective of their valence improved recognition performance, whereas CO_2 decreased it, and only CO_2 that is associated with painful sensations was processed consciously by the participants of this investigation.

The effect of expectancy on implicit learning was demonstrated in a recent study by Colagiuri et al. (2011). Expectancy was manipulated in this experiment by providing to the subjects positive, negative, or no information about a possible effect of an odor present during a visual search task. Unknown to the subjects, the search task comprised a contingency, that is, on half of the trials the location of the target was cued by a distinct spatial configuration of the distractors. While neither the mere presence of the odor nor the manipulation had any influence on the participants' awareness of the contingency, reaction time on cued trials was affected by the information subjects had received on the effect of the odor, that is, participants who had received the negative information reacted more slowly than those who had received no information, whereas those who had received the positive information reacted faster than the other groups. These findings strongly resemble the effects of expectation observed on EEG measures such as the CNV and basic cognitive tasks.

With regard to the content of memory, some researchers have claimed a special relationship between autobiographical, that is, personally meaningful episodic, memories and fragrances. As a result of this special link, it has been observed that memories evoked by olfactory cues are often older, more vivid, more detailed, and more affectively toned than those cued by other sensory stimuli (Chu and Downes 2002; Goddard et al. 2005; Willander and Larsson 2007). This phenomenon has been explained by the peculiar neuroanatomical connection of the memory systems with the emotional systems. Evidence for this hypothesis has, for instance, been supplied by the group of Herz (Herz 2004; Herz and Cupchik 1995; Herz et al. 2004) who demonstrated that presentation of odorants resulted in more emotional memories than presentation of the same cue in auditory or visual form. The authors also showed that if the odor cue was hedonically congruent with the item that had to be remembered, memory for associated emotional experience was improved. Moreover, personally salient fragrance cues were associated with higher functional activity in emotion-related brain regions, such as the amygdala and the hippocampus.

12.1.4.3 Other Cognitive Tasks

The study of Degel and Köster described earlier (Degel and Köster 1999) showed that under certain conditions odorants may influence attentional performance even without subjects being aware of their presence. According to an investigation by Holland and coworkers, the unnoticed presence of odorants may also affect everyday behavior and higher cognitive functions (Holland et al. 2005). The authors reported that subliminal concentrations of a citrus-scented cleaning product increased identification of cleaning-related words in a lexical decision task. Moreover, subjects exposed to the subliminal odor listed cleaning-related activities more frequently when asked to describe planned activities during the day and kept their environment tidier during an eating task.

The effects of pleasant suprathreshold fragrances on other higher cognitive functions were, for instance, investigated by Baron (1990). In this study, subjects were exposed to pleasant or neutral ambient odors while solving a clerical coding task and negotiating about monetary issues with a fellow participant. Before performing these tasks subjects indicated self-set goals and self-efficacy. Following the tasks subjects rated the experimental rooms in terms of pleasantness and comfort, as well as their mood. In addition, they were asked which conflict management strategies they would adopt in the future. Although neither fragrance had a direct effect on performance in the clerical task, subjects in the pleasant odor condition set higher goals and adopted a more efficient strategy in the task than those in the neutral odor condition. In addition, male subjects in the pleasant odor condition rated themselves as more efficient than those in the neutral odor condition. In the negotiation task, subjects in the pleasant odor condition set higher monetary goals and made more concessions than those in the neutral odor condition. Moreover, subjects in the pleasant odor condition were in better mood and reported planning to handle future conflicts less often through confrontation and avoidance. Thus, this study showed that pleasant ambient fragrances offer a potential to create a more comfortable work environment and diminish aggressive behavior in situations involving competition. Gilbert and coworkers (1997) examined the effect of a pleasant and an unpleasant blend of fragrances on the same clerical coding task but were not able to show any influence of either fragrance on task performance. However, subjects exposed to unpleasant odorants believed that these odorants had negative effects on their performance in simple and difficult mathematical and verbal tasks (Knasko 1993).

Also, Ludvigson and Rottman (1989) studying the influence of lavender and clove EOs in ambient air in comparison to a no-odor control condition found that lavender impaired performance in a mathematical reasoning task, while clove was devoid of effects. However, the lavender effect was only observed in the first of two sessions held 1 week apart. Also, subjects in the lavender condition rated the experimental conditions more favorably, while clove odor decreased subjects' willingness to return to the second session. Moreover, subjects who were exposed to an odor in one of the two sessions were generally less willing to return and had worse mood than subjects who never received an odor. To further complicate things, these odorant by session interactions were related to personality factors in a highly complex manner.

Another study that demonstrated rather complex effects of the administration order of olfactory stimuli was reported by Gaygen and Hedge (2009). These authors found detrimental effects of a commercial air freshener containing fragrances rated as pleasant on certain aspects of word recognition in a lexical decision task only when the odorants were applied in the second of two independent sessions that was preceded by a no-odor condition. In contrast, no effects were found when the odor was applied in the first session and the no-odor condition in the second. Probably, the observed effect is attributable to the influence of a novel cue (odor in session two) in an otherwise familiar context (experimental context without odor) that allocated attentional resources.

More recently, Finkelmeyer and coworkers (2010) investigated the effect of mood states induced by olfactory stimuli on inhibitory control performance. In their study, subjects performed the Stroop color–word interference task in the presence of either an unpleasant odor (H_2S) generating a negative affective state or an emotionally neutral odor (eugenol (**4**)). In this task, color names are written in colored ink and the task of the subject is to attend to the color of the ink and ignore the color name. Typically, naming of the ink color is more difficult and reaction times are higher when the color of the ink and the color name are incongruent. Contrary to the authors' expectations presentation of the negative odor facilitated cognitive processing and reduced Stroop interference while the neutral odor had no effects on performance. In the opinion of the authors this finding can be explained by increasing cognitive control as a result of mood-congruent processing.

12.1.5 CONCLUSIONS

The presented review of the literature on the effects of EOs and fragrances on human arousal and cognition demonstrates that coherent findings are quite scarce and that we are still far from painting a detailed picture of which effects can be achieved by administering a particular EO and how these effects are precisely exerted. One reason for the inconsistency in results may be that in most studies in humans, clear associations between constituents of EOs and observed effects are missing. While exact specifications about the origin, composition, and concentration of the tested oils would be necessary to establish clear pharmacological profiles and dose–response curves for specific oils, in many investigations no such details are given. This renders attempts to compare the results from different studies and to generalize findings rather difficult.

Another aspect that clearly contributes to inconclusive findings is the involvement of a variety of mechanisms of action in the effects of EOs on human arousal and cognition. In a very valuable review on the assessment of olfactory processes with electrophysiological techniques, Lorig (2000) points out that EEG changes induced by odorants have to be interpreted with great care, since other than direct odor effects may be responsible for changes, particularly when relaxing effects reflected by the induction of slow-wave activity are concerned. Physiological processes, such as altered breathing patterns in response to pleasant versus unpleasant odors; cognitive factors, for example, as a consequence of expectancy or the processing of secondary stimulus features; or an inappropriate baseline condition can lead to changes of the EEG pattern that are quite unrelated to any psychoactive effect of the tested odorant. Well-designed paradigms are thus necessary to control for cognitive influences that might mask substance-specific effects of fragrances. Also, while EEG and other electrophysiological techniques are highly efficient to elucidate fragrance effects on the CNS in the time domain, we know only little about spatial aspects of such effects. Brain imaging techniques, such as functional magnetic resonance imaging (fMRI), will prove valuable to address this issue. Using fMRI, preliminary results from the author's group (Friedl et al. 2010) have shown that, after prolonged exposure, fragrances alter neuronal activity in distinct regions of the brain in a time-dependent manner and that this influence is sex specific.

When evaluating psychoactive effects of EOs and fragrances on human cognitive functions, the results should be interpreted just as cautiously as those of electrophysiological studies as similar confounding factors, ranging from influences of stimulus-related features, for example, pleasantness, to expectation of fragrance effects, and even personality traits may be influencing the observed outcome. In regard to higher cognitive functioning, such as language or emotional processing, conscious as opposed to sub- or unconscious processing of odor information seems to differentially affect performance due to differences in the utilization of shared neuronal resources. Again, it seems worthwhile to measure additional parameters that are indicative of (subjective) stimulus information processing and emotional arousal if hypotheses are being built on direct (pharmacological) and cognitively mediated (psychological) odor effects on human behavior. Moreover, comparisons of different forms of application that involve or exclude stimulation of the olfactory system, such as inhalative versus noninhalative dermal administration, have proven useful in the distinction of pharmacological from psychological mechanisms and will serve to enlarge our understanding of psychoactive effects of EO and fragrances in humans.

12.2 PSYCHOPHARMACOLOGY OF ESSENTIAL OILS

Domingos Sávio Nunes, Viviane de Moura Linck, Adriana Lourenço da Silva, Micheli Figueiró, and Elaine Elisabetsky

The use of aromas goes back to ancient times, as implied by the nearly 200 references in the Bible relating the use of aromas for "mental, spiritual and physical healing" (Perry and Perry 2006). It is currently accepted that aromas and some of its individual components may in fact possess pharmacological and/or psychological properties, and in many instances the overall effect is likely to result from a combination of both. The following sections review the psychopharmacology of EOs and/or its individual components, as well as its mechanisms of action.

12.2.1 Aromatic Plants Used in Traditional Medical Systems as Sedatives or Stimulants

EOs are generally products of rather complex compositions used contemporaneously in aromatherapy and for centuries as aromatic medicinal plant species in traditional systems of medicine. Aromatic formulas are used for the treatment of a variety of illnesses, including those that affect the CNS (Almeida et al. 2004). Volatile compounds presenting sedative or stimulatory properties have been and continue to be identified in EOs from aromatic medicinal species spread into different families and genera. The majority of these substances have small structures with less than 12 carbons and present low-polarity chemical functions, being therefore quite volatile. Since most natural EOs are formed by complex mixtures, their bioactivity(ies) is obviously dependent on the contribution of their various components. Therefore, studies failing to characterize at least the main components of the EO studied are not discussed in this chapter.

Several *Citrus* EOs contain high proportions of limonene (**14**) as its major component. Orange peels are used as sedative in several countries, and EOs obtained from *Citrus aurantium* L. (Rutaceae) fruit peels can contain as much as 97.8% of limonene (**14**). The anxiolytic and sedative properties of *Citrus* EO suggested by traditional uses have been assessed in mice (Carvalho-Freitas and Costa 2002; Pultrini et al. 2006) and also shown in a clinical setting (Lehrner et al. 2000). The relaxant effects observed in female patients in a dental office were produced with a *C. sinensis* (L.) Osbeck (Rutaceae) EO composed of 88.1% limonene (**14**) and 3.77% myrcene (**15**).

The common and large variability in the composition of natural EOs poses difficulties for the evaluation and the safe and effective use of aromatic medicinal plants. Genetic variations lead to the occurrence of chemotypes, as in the case of *Lippia alba* (P. Mill.) N.E. Br. ex Britt. & Wilson (Verbenaceae). Analyses revealed three monoterpenic chemotypes characterized by the prevalence of myrcene (**15**) and citral (**16**) in the chemotype I, limonene (**14**) and citral (**16**) in the chemotype II, and limonene (**14**) and carvone (**17**) in the chemotype III (Matos 1996). This species is known in Brazil as *cidreira*; the aromatic tea from its leaves is traditionally used as a tranquilizer, being one of the most widely known homemade remedies. Pharmacological assays showed anxiolytic (Vale et al. 1999) and anticonvulsant (Viana et al. 2000) effects of EO samples from all three chemotypes. Anticonvulsive and sedative effects in mice were also demonstrated for the three isolated principal constituents of *L. alba* oils: limonene (**14**), myrcene (**15**), and citral (**16**) (Vale et al. 2002).

Various *Ocimum* species are used traditionally for sedative and anticonvulsive purposes, and its EOs seem to play an important role for these properties. However, besides the normal variability in the composition of the EOs, the occurrence of chemotypes seems to be generalized in this genus (Grayer et al. 1996; Vieira et al. 2001). The comparison of EO samples from different accessions of *O. basilicum* L. (Lamiaceae) pointed to the occurrence of linalool (**3**) and methylchavicol (**18**) types, and it has been suggested that such data should be taken into account for an intraspecific classification of the taxon (Grayer et al. 1996). An EO of *O. basilicum* L. of the linalool-type containing 44.18% linalool (**3**), 13.65% 1,8-cineole (**2**), and 8.59% eugenol (**4**) as main constituents presented anticonvulsant and hypnotic activities (Ismail 2006). There are two varieties of *O. gratissimum* L. (Lamiaceae) (*O. gratissimum* L. var. *gratissimum* and *O. gratissimum* var. *macrophyllum* Briq.) that form a polymorphic complex very difficult to differentiate by morphological traits (Vieira et al. 2001). It was clearly demonstrated that the genetic variations of *O. gratissimum* lead to three different chemotypes (Vieira et al. 2001): eugenol (**4**), thymol (**19**), and geraniol (**20**). EOs obtained from *O. gratissimum* of the eugenol-type collected during the 4 year seasons contained eugenol (**4**) (44.89%–56.10%) and 1,8-cineole (**2**) (16.83%–33.67%) as main components and presented sedative and anticonvulsant activities slightly altered by the composition of each sample (Freire et al. 2006). A dose-dependent sedative effect was observed in mice and rats (Orafidiya et al. 2004) treated with

O. gratissimum thymol-type EO containing 47.0% thymol (**19**), 16.2% *p*-cymene (**21**), and 6.2% α-terpinene (**22**) as major constituents.

Among Amazonian traditional communities, a widespread recipe that includes *Cissus sicyoides* L. (Vitaceae), *Aeolanthus suaveolens* Mart. ex Spreng. (Lamiaceae), *Ruta graveolens* L. (Rutaceae), and *Sesamum indicum* L. (Pedaliaceae) was identified as the most frequently indicated for the management of epilepsy-like symptoms. The results of pharmacological studies on the traditional preparations and the EO obtained from *A. suaveolens* and its principal components led to the suggestion that the volatile lactones could be interesting target compounds in the search for new anticonvulsant agents (de Souza et al. 1997). Linalool (**3**) was identified as one of the principal active components of the *A. suaveolens* EO (Elisabetsky et al. 1995a) and proved to play a key role in the central activities of the traditional preparation (Brum et al. 2001a,b; Elisabetsky et al. 1995b, 1999). (*R*)-(−)-Linalool (**7**) is recognized today as the sedative and calming component of numerous traditional and commercial preparations or their isolated natural EOs (Heuberger et al. 2004; Kuroda et al. 2005; Shaw et al. 2007; Sugawara et al. 1998). Epinepetalactone (**23**) is a volatile apolar compound and major component in the EO from *Nepeta sibthorpii* Benth. (Lamiaceae) responsible for the EO anticonvulsant activity (Galati et al. 2004).

Nature continues to reveal its inventiveness in combining different monoterpenes and arylpropenoids to achieve special nuances regarding central activities. An EO obtained from *Artemisia dracunculus* L. (Asteraceae) containing 21.1% *trans*-anethole (**24**), 20.6% α-*trans*-ocimene (**25**), 12.4% limonene (**14**), 5.1% α-pinene (**1**), 4.8% *allo*-ocimene (**26**), and 2.2% methyleugenol (**27**) shows anticonvulsant activity likely to be assigned to a combination of these various monoterpenes (Sayyah et al. 2004). β-Asarone (**28**) and its isomers are recognized as important sedative and anticonvulsive active components as in, for example, drugs based on *Acorus* species in which the total proportion of isomers can reach 90% (Koo et al. 2003; Mukherjee et al. 2007). In the EO from *Acorus tatarinowii* Schott (Acoraceae) rhizome, traditionally used for epilepsy, the combination of monoterpenes and arylpropenoids was found to be ~25% 1,8-cineole (**2**), ~12% linalool (**3**), and ~10% β-asarone (**28**) and isomers (Ye et al. 2006).

Traditional Chinese medicine (TCM) is known to use especially complex preparations, more often than not composed of several plant materials including therefore a number of active principles pertaining to different chemical classes. The TCM prescription SuHeXiang Wan combines different proportions of as much as 15 crude drugs and is used orally for the treatment of seizure, infantile convulsion, and other conditions affecting the CNS (Koo et al. 2004). The EO obtained by hexane extraction at room temperature from a SuHeXiang Wan composed of nine crude drugs proved to have a relatively simple composition: 21.4% borneol (**29**), 33.3% isoborneol (**30**), 5.9% eugenol (**4**), and other minor components (Koo et al. 2004). The inhalation of this volatile mixture delayed the appearance of pentylenetetrazole (PTZ)-induced convulsions suggesting GABAergic (gamma-amino-butyric acid) modulation (Koo et al. 2004).

Other monoterpenoid or arylpropenoid derivatives identified as the active components of traditional sedatives include methyleugenol (**27**) (Norte et al. 2005), isopulegol (**31**) (Silva et al. 2007), and α-terpineol (**32**) (de Sousa et al. 2007). It is worth mentioning the monoterpene thujone (**33**), the dangerous principle of the ancient Absinthii herba, *Artemisia absinthium* L. (Asteraceae), that induces marked central stimulatory effects, especially when used in the form of liqueur. Frequent and excessive use of this drug can cause intoxicated states accompanied by clonic convulsions among other serious consequences (Bielenberg 2007).

Only a few volatile sesquiterpenes presenting important central activities are currently known. β-Eudesmol (**34**) was found to be one of the volatile active principles of the Chinese medicinal herb *Atractylodes lancea* DC. (Asteraceae) with alleged antagonist properties useful in intoxication by anticholinesterase agents of the organophosphorus type (Chiou et al. 1997). Experimental data show that β-eudesmol (**34**) prevents convulsions and lethality induced by electroshock but not those induced by PTZ or picrotoxin (Chiou et al. 1995). With a very similar chemical structure, α-eudesmol (**35**) protects the development of postischemic brain injury in rats by blocking ϖ-Aga-IVA-sensitive Ca^{2+} channels (Asakura et al. 2000).

The sesquiterpenes caryophyllene oxide (**36**) and β-selinene (= β-eudesmene) (**37**) isolated from the hexane extract from leaves of *Psidium guajava* var. *minor* Mattos (Myrtaceae) potentiated pentobarbital sleep and increased the latency for PTZ-induced convulsions in mice; blockade of extracellular Ca^{2+} was observed in isolated guinea-pig ileum with the hexane extract and its fractions containing both sesquiterpenes (Meckes et al. 1997). The similarity between the chemical structures of the sesquiterpenes (**34**), (**35**), and (**37**) is noteworthy and could indicate relevant characteristic patterns required for central activity.

12.2.2 Effects of Essential Oils in Animal Models

Pure compounds isolated from aromas and complex EOs have been proved to induce a variety of effects on human and other mammalian species. Biological properties such as antispasmodic (Carvalho-Freitas et al. 2002) or other autonomic nervous system-related activities (Haze et al. 2002; Sadraei et al. 2003) are outside the scope of this chapter. Central effects have been extensively documented and fall more often than not into the sedative (Buchbauer et al. 1991, 1993; Elisabetsky et al. 1995a; Lehrner et al. 2000), anxiolytic (Carvalho-Freitas et al. 2002; Cooke and Ernst 2000; Diego et al. 1998; Lehrner et al. 2000), antidepressant (Komori et al. 1995a,b), and hypnotic (Diego et al. 1998) categories. As earlier stated, it is likely that the overall effects of EOs in humans result from a combination of physiological (in this case psychopharmacological) and psychological effects. Even when accepting the limitations of the validity of animal models in regard to assessing the effects of drugs on complex human emotional states, in the case of EOs the usefulness of experiments with rodent models is largely limited to clarify the physiological part of the potential effects in humans. Again, the reproducibility of the data compiled in this section would be highly dependent on the EO composition, unfortunately not always adequately reported.

Relevant to the data that follow, it has been shown that individual components of EOs administered orally, by means of intraperitoneal or subcutaneous injections, dermally, or by inhalation do reach and adequately cross the blood–brain barrier (Buchbauer et al. 1993; Fujiwara et al. 1998; Jirovetz et al. 1990; Kovar et al. 1987; Moreira et al. 2001; Perry et al. 2002). The question of whether psychopharmacological effects in animals are dependent of olfactory functions is surprisingly not yet entirely clarified. Cedrol (from pine EO) was shown to be sedative in normal rats and rats made anosmic with zinc sulfate (Kagawa et al. 2003). In contrast, a mix of chamomile and lavender oils reduced pentobarbital-induced sleeping time in normal but not anosmic rats and mice (Kagawa et al. 2003).

Rose and lavender oil, administered intraperitoneally (ip) to mice, showed anticonflict effects (Umezu 1999; Umezu et al. 2002), suggesting anxiolytic properties. Unequivocal anxiolytic properties were also demonstrated for lavender EO (ip, gerbils) with the elevated plus maze test (Bradley et al. 2007). Chen et al. (2004) reported that *Angelica sinensis* (Oliv.) Diels (Apiaceae) EO (orally administered, mice) also has an anxiolytic effect in the elevated plus maze, the light/dark model, and the stress-induced hyperthermia paradigms.

The dry rhizomes of *Acorus gramineus* Soland (Acoraceae) are officially listed in the Korean pharmacopeia for sedative, digestive, analgesic, diuretic, and antifungal effects (Koo et al. 2003). Various CNS effects have been characterized for *A. gramineus* EO, including antagonism of PTZ-induced convulsion, potentiation of pentobarbital-induced sleeping time, sedation, and decreased spontaneous locomotion with the water and methanol extracts (mice, ip) (Liao et al. 1998; Vohora et al. 1990). Unexpectedly, given that sedative drugs usually impair cognition in animals, the oral administration of *A. gramineus* rhizoma EO, with eugenol (**4**) as the principal component, improved cognitive function in aged rats and mice; based on brain amine analyses, the authors suggest that such effects may be related to increased norepinephrine (NE), dopamine (DA), and serotonin relative levels and to decreased activity of brain acetylcholinesterase (Zhang et al. 2007).

Cymbopogon winterianus Jowitt (Poaceae), popularly known as "citronella" and "java grass," is an important EO yielding aromatic grass, mostly cultivated in India and Brazil. *C. winterianus* EO is rich in citronellal (**38**), geraniol (**39**), and citronellol (**40**) (Cassel and Vargas 2006) and has demonstrated anticonvulsant effects (ip, mice) (Quintans-Júnior et al. 2007). Pharmacological studies with *Cymbopogon citratus* Stapf EO (presenting high percentage of citral (**16**) in its composition) revealed anxiolytic, hypnotic, and anticonvulsant properties when orally administered to mice (Blanco et al. 2007).

The EO of *Eugenia caryophyllata* Thunb (Myrtaceae), used in Iranian traditional medicine, exhibits anticonvulsant activity against tonic seizures induced by maximal electroshock (MES) (ip, mice) (Pourgholami et al. 1999a). The leaf EO of *Laurus nobilis* L. (Lauraceae), which has been used as an antiepileptic remedy in Iranian traditional medicine, demonstrated more anticonvulsant activity against experimental seizures induced by PTZ (ip, in mice) than MES-induced seizures. Components responsible for this effect may include methyleugenol, eugenol, and pinene present in the EO (Sayyah et al. 2002). In the same manner, EO obtained from fruits of *Pimpinella anisum* S.G.Gmel. (Umbelliferae) demonstrated anticonvulsant activity against seizures induced by PTZ or MES in mice (ip) (Pourgholami et al. 1999b). As mentioned earlier, since no chemical analysis is reported for the studied EOs, one can only speculate on the components that may be responsible for the activities observed with *E. caryophyllata*, *L. nobilis*, and *P. anisum*.

Various *Thymus* species of the Lamiaceae family (including *T. fallax* Fisch. & Mey., *T. kotschyanus* Boiss. & Hohen., *T. pubescens* Boiss. & Kotschy ex Celak, *T. vulgaris* M. Bieb.), which are widely distributed as aromatic and medicinal plants in many regions of Iran, have shown CNS effects (Duke et al. 2002). Pharmacological studies in mice (ip) with *T. fallax*, *T. kotschyanus*, and *T. pubescens* containing, respectively, 30.2% of carvacrol (**41**), 18.7% of pulegone (**42**), and 32.1% of carvacrol (**41**) demonstrated that the *T. fallax* EO has more antidepressant activity than *T. kotschyanus* and *T. pubescens* during the forced swimming test (Morteza-Semnani et al. 2007). These results illustrate that minor components of EOs can modify the activity of main components, reaffirming the importance of chemically characterizing EOs in order to understand its overall bioactivity.

An EO fraction obtained from powdered seeds of *Licaria puchury-major* (Mart.) Kosterm. (Lauraceae) containing 51.3% of safrol (**43**), 3.3% of eugenol (**4**), and 2.9% of methyleugenol (**27**) reduced locomotion and anesthetized mice, as well as affording protection against electroshock-induced convulsions (Carlini et al. 1983).

12.2.2.1 Effects of Individual Components

Section 12.2.2 listed the effects of several EO components found in traditionally used species, including α-pinene (**1**), eugenol (**4**), limonene (**14**), myrcene (**15**), citral (**16**), epinepetalactone (**23**), *trans*-anethole (**24**), α-*trans*-ocimene (**25**), *allo*-ocimene (**26**), methyleugenol (**27**), borneol (**29**), isoborneol (**30**), citronellal (**38**), geraniol (**39**), and citronellol (**40**); anticonvulsive, muscle relaxants, anxiolytic, and/or hypnotic properties were observed with these compounds when given ip to mice (Blanco et al. 2007; Cassel and Vargas 2006; Galati et al. 2004; Koo et al. 2004; Quintans-Júnior et al. 2007; Sayyah et al. 2004; Vale et al. 1999, 2002).

Linalool (**3**) is a monoterpene commonly found as a major volatile component of EOs in several aromatic plant species, such as *Lavandula angustifolia* Mill. (Lamiaceae), *Rosa damascene* Mill. (Rosaceae), *Citrus bergamia* Risso (Rutaceae), *Melissa officinalis* L. (Lamiaceae), *Rosmarinus officinalis* L. (Lamiaceae), *C. citratus* DC ex Nees (Poaceae), and *Mentha piperita* L. (Lamiaceae). Interestingly, many linalool-producing species are traditionally used as sedative, analgesic, hypnotic, or anxiolytic remedies in traditional medicine and some as well in aromatherapy (Elisabetsky et al. 1995a).

As mentioned earlier, *A. suaveolens* Mar. ex Spreng. (Lamiaceae) is used as an anticonvulsant through the Brazilian Amazon. The EO obtained from *A. suaveolens* and its main component linalool (**3**) proved to be anticonvulsant against several types of experimental convulsions, including those induced by PTZ and transcorneal electroshock (Elisabetsky et al. 1995a), intracerebrally injected quinolinic acid, and ip N-methyl-D-aspartate (NMDA) (Elisabetsky et al. 1999). Moreover, psychopharmacological evaluation of linalool (**3**) showed dose-dependent marked sedative effects, including hypnotic, hypothermic, increased sleeping time, and decreased spontaneous locomotion in mice (ip) (Elisabetsky et al. 1995a; Linck et al. 2009). Decreased motor activity was also reported in mice by Buchbauer's group (Buchbauer et al. 1991). Indicating anxiolytic properties, linalool (**3**) was reported to have anticonflict effects (mice, ip) in the Geller and Vogel tests, and similar findings were reported for lavender oil (Umezu et al. 2006). Analgesic properties were observed against chemical (rats, po; Barocelli et al. 2004) (mice, sc; Peana et al. 2004a,b) and thermal (sc mice; Peana et al. 2003) nociceptive stimuli. Since anti-inflammatory activity (rats, sc; Peana et al. 2002) has also been reported, it is not clear if linalool-induced analgesia is of central origin. Nevertheless, these experimental data are relevant to clinical studies indicating that aromatherapy with lavender can reduce the demand for opioids during the immediate postoperative period (Kim et al. 2007) and deserve further investigation. Linalool local anesthetic effects were observed in vivo by the conjunctival reflex test and in vitro by phrenic nerve–hemidiaphragm preparation (Ghelardini et al. 1999).

12.2.2.2 Effects of Inhaled Essential Oils

Despite the wide and growing use of aromatherapy in the treatment of a diversity of ailments, including those of central origin (Perry and Perry 2006), and the alleged effects of incenses and other means of ambient aromas, experimental data on psychopharmacological properties of inhaled EOs are surprisingly scarce. Moreover, few of the studies control for inhalation flow and it is difficult to estimate the actual concentration of whatever is being inhaled. However, evidences that EOs and its components are absorbed by inhalation are available. After mice were exposed to a cage loaded with 27 mg of linalool (**3**) for 30, 60, or 90 min, plasma linalool concentrations of, respectively, 1.0, 2.7, and 3.0 ng/mL were found (Buchbauer et al. 1991); these exposure schedules resulted in an exposure-dependent decrease in locomotion. Moreover, detectable plasma concentrations at the nanogram level were reported for as many as 40 different aromatic compounds after 60 min of inhalation (Buchbauer et al. 1993). Therefore, despite the lack of precise measures of inhaled and/or absorbed quantities, it is arguable that animal studies with inhaled aromas are nevertheless informative.

Inhalation of citrus-based aromas (*Citrus sinensis* (L.) Osbeck (Rutaceae)) or fragrances were found to restore stress-induced immunosuppression (Shibata et al. 1990) and antidepressant-like effects in rats (2 mL/min EO in the air flux) (Komori et al. 1995a). A clinical study with depressed patients revealed that inhaling a mixture of citrus oils was capable of reducing the needed antidepressant doses; moreover, inhalation of the oil by itself was antidepressive and normalized neuroendocrine hormone levels (cortisol and DA) in depressive patients (Komori et al. 1995b). Relevant to these findings, inhaled lemon oil (*Citrus limonum* Risso (Rutaceae)) has been shown to increase the turnover of DA and serotonin after inhalation in mice (Komiya et al. 2006).

Anxiolytic proprieties were observed in rats with the plus maze model after 7 min rose oil inhalation; this is the only report of inhaled effects in animals after a short period of inhalation, although the procedure of leaving four cotton balls embedded with 2 mL of EO lacks standardization (Almeida et al. 2004). As previously mentioned, inhaled cedrol was shown to be sedative in rats (Kagawa et al. 2003), and the inhalation of a volatile mixture from the TCM SuHeXiang Wan composed of 21.4% borneol (**29**), 33.3% isoborneol (**30**), 5.9% eugenol (**4**), and other minor components delayed the appearance of PTZ-induced convulsions suggesting GABAergic modulation (Koo et al. 2004).

Inhaled lavender oil (0–1–2 mL, 25% of linalool (**3**) and 46% of linalyl acetate (**44**)) demonstrated anxiolytic effects in rats, as shown by decreased peripheral movement and defecation in an open field, after at least 30 min of inhalation (Shaw et al. 2007). Similar effects were observed with inhaled lavender containing 38.47% of linalool (**3**) and 43.98% of linalyl acetate (**44**) in gerbils, showing increased exploratory behavior in the elevated plus maze test after 1 or 14 days inhalation (Bradley et al. 2007a). In agreement with studies with inhaled lavender EO, anxiolytic activity was observed in mice after inhaling (±)-linalool (**3**) at 1% or 3% inhaled for 60 min using the light/dark and immobilization-induced stress paradigms; moreover, after the same inhalation procedure, isolated mice exhibited decreased aggression toward an intruder, and increased social interaction was also observed (Linck et al. 2010).

Finally, antinociceptive (mice) and gastroprotective effects (rats) of orally given or inhaled (60 min in a camera saturated with 2.4 µL/L) *Lavandula hybrida* Rev. (Lamiaceae) EO and its principal constituents linalool (**3**) and linalyl acetate (**44**) have also been reported (Barocelli et al. 2004).

12.2.3 Mechanism of Action Underlying Psychopharmacological Effects of Essential Oils

The mechanisms of action underlying the effects of fragrances as complex mixtures as found *in natura*, or even for isolated components, are far from being clarified, but seem to differ among different fragrances (Komiya et al. 2006). If the overall pharmacological effects of an EO depend on the contribution of its various components, the mechanism of action of a complex mixture is far more complex than a simple sum of each component's physiological consequence. Interaction among the various substances present in EOs can modify each other's pharmacodynamic and pharmacokinetic properties; nevertheless, studying the pharmacodynamic basis of isolated components is helpful for a comprehensive understanding of the basis of EO physiological and psychopharmacological effects.

Rats treated (ip) with the isolated components of lemon EO, such as *R*-(+)-limonene (**14**), *S*-(−)-limonene (**45**), and citral, did not show the cold stress-induced elevation in NE and DA (Fukumoto et al. 2008); the authors suggested that these effects are related to changes in monoamine release in rat brain slices induced by these compounds (Fukumoto el al. 2003). Complementary to these findings, inhaled lemon oil (cage with a cotton ball with 1 mL, 90 min) has also been reported to increase the turnover of DA and serotonin in mice (Komiya et al. 2006).

Inhalation (2 g of fragrance/day, 2 × 3 h/day, for 7, 14, or 30 days at home cages) of *A. gramineus* Solander (Acoraceae) EO inhibited the activity of GABA transaminase (an enzyme critical for metabolizing GABA at synapses), thereby significantly increasing GABA levels; a decrease in glutamate levels was also reported in this study (Koo et al. 2003). Both of these alterations in the inhibitory and excitatory neurotransmitter systems are compatible with and relevant to the sedative and anticonvulsant effects earlier mentioned. Neuroprotective effects on cultured neurons were also reported for *A. gramineus* EO, apparently attained through the blockade of NMDA receptors given that this oil inhibits [^{3}H]-MK801 binding (Cho et al. 2001).

Other examples of EO actions on amino acid neurotransmitters are available. Morrone et al. (2007) using in vivo microdialysis showed that bergamot EO (ip) increased levels of aspartate, glycine, taurine, GABA, and glutamate in a Ca^{2+}-dependent manner in rat hippocampus; the authors suggested that these same effects may be relevant for other monoterpenes affecting the CNS. Inhibition of glutamate-mediated neurotransmission is in part responsible for the mechanisms of action underlying the previously mentioned anticonvulsant effects of linalool (**3**). Neurochemical assays reveal that (±)-linalool (**3**) acts as a competitive antagonist of L-[^{3}H]-glutamate binding (Elisabetsky et al. 1999) and also shows a dose-dependent noncompetitive inhibition of [^{3}H]-MK801 binding (IC_{50} = 2.97 mM) indicating antagonism of NMDA glutamate receptors (Brum et al. 2001a). (±)-Linalool (**3**) also decreases the potassium-stimulated glutamate release and uptake in mice cortical synaptosomes, without affecting basal glutamate release (Brum et al. 2001b). This neurochemical profile explains, for instance, the linalool-induced delay in NMDA-induced convulsions and blockade of quinolinic acid-induced convulsions (Elisabetsky et al. 1999). Eugenol (**4**) was also found to inhibit excitotoxic neuronal effects induced by NMDA, apparently involving modulation of NMDA glutamate receptor and inhibition of Ca^{2+} uptake (Wie et al. 1997; Won et al. 1998).

Because (±)-linalool (**3**) is also able to protect against PTZ- and picrotoxin-induced convulsions (Elisabetsky et al. 1999), a GABAergic modulation could also be in place. Nevertheless, (±)-linalool (**3**) did not alter [^{3}H]-muscimol binding (Brum 2001a) suggesting that, if existing at all, linalool modulation of the GABAergic system is not mediated by $GABA_A$ receptors.

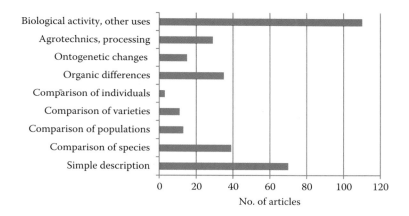

FIGURE 4.1 Distribution of topics of publications in *Journal of Essential Oil Research* between 2010 and 2014.

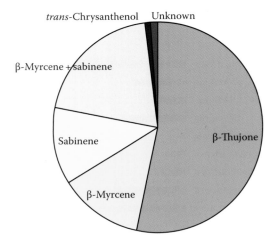

FIGURE 4.3 Distribution of individuals of different main compounds in their essential oil in a commercial wormwood (*Artemisia absinthium*) population. (From Zámboriné, N.É. et al., Individual variability of wormwood [*Artemisia absinthium* L.] essential oil composition, In: *Program and Book of Abstracts, 43rd ISEO*, Lisbon, Portugal, 2012, p. 105.)

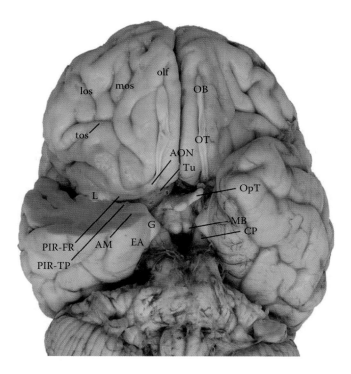

FIGURE 12.1 Macroscopic view of the human ventral forebrain and medial temporal lobes, depicting the olfactory tract, its primary projections, and surrounding nonolfactory structures. The right medial temporal lobe has been resected horizontally through the midportion of the amygdala (AM) to expose the olfactory cortex. AON, anterior olfactory nucleus; CP, cerebral peduncle; EA, entorhinal area; G, gyrus ambiens; L, limen insula; los, lateral olfactory sulcus; MB, mammillary body; mos, medial olfactory sulcus; olf, olfactory sulcus; PIR-FR, frontal piriform cortex; OB, olfactory bulb; OpT, optic tract; OT, olfactory tract; tos, transverse olfactory sulcus; Tu, olfactory tubercle; PIR-TP, temporal piriform cortex. Figure prepared with the help of Dr. Eileen H. Bigio, Dept. of Pathology, Northwestern University Feinberg School of Medicine, Chicago, IL. (Taken with permission from Gottfried, J.A. and Zald, D.A., *Brain Res. Rev.*, 50, 287, 2005.)

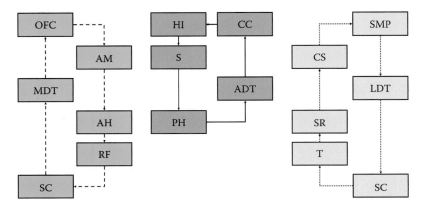

FIGURE 12.3 Control of activation processes. OFC, orbitofrontal cortex; AM, amygdala; MDT, medial dorsal thalamus; AH, anterior hypothalamus; RF, reticular formation; SC, spinal cord; HI, hippocampus; CC, cingulate cortex; S, septum; ADT, anterior dorsal thalamus; PH, posterior hypothalamus; SMP, sensory-motor projections; CS, corpus striatum; LDT, lateral dorsal thalamus; SR, subthalamic regions; T, tectum. *Orange*, structures of the arousal network; *green*, structures of the effort network; *blue*, structures of the activation network. (Adapted from Pribram, K.H. and McGuinness, D., *Psychol. Rev.*, 82(2), 116, 1975.)

Potentially relevant to the pharmacology profile of linalool (**3**), patch clamp techniques demonstrated that the monoterpene (no isomer specified) suppressed the Ca^{2+} current in rat sensory neurons and in rat cerebellar Purkinje cells (Narusuye et al. 2005). Based on their study with lavender EO, Aoshima and Hamamoto (1999) suggested that the GABAergic transmission may be of relevance for the mechanism of action of other monoterpenes, such as α-pinene (**1**), eugenol (**4**), citronellal (**38**), citronellol (**40**), and hinokitiol (**46**) (Aoshima and Hamamoto 1999). Additionally, the antinociceptive activity of *R*-(−)-linalool (**7**) seems to involve several receptors, including opioids, cholinergic M_2, DA D_2, and adenosine A_1 and A_{2A}, as well as changes in K^+ channels (Peana et al. 2003, 2004a, 2006).

Relevant to studies on cognition, *S. lavandulifolia* Vahl (Lamiaceae) EO and isolated monoterpene constituents were shown to inhibit brain acetylcholinesterase and to present antioxidant properties (Perry et al. 2001, 2002).

Overall, it seems reasonable to argue that the modulation of glutamate and GABA neurotransmitter systems is likely to be the critical mechanisms responsible for the sedative, anxiolytic, and anticonvulsant proprieties of linalool and EOs containing linalool (**3**) in significant proportions.

12.2.4 CHEMICAL STRUCTURES OF MENTIONED CNS ACTIVE COMPOUNDS

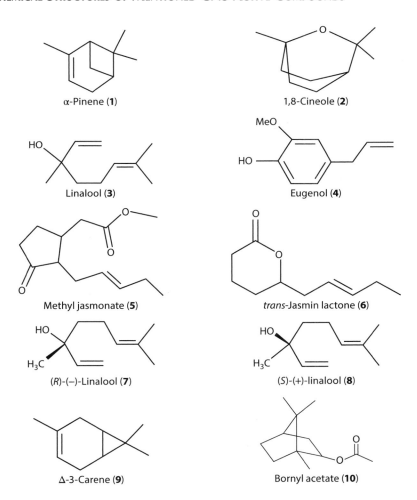

α-Pinene (**1**)

1,8-Cineole (**2**)

Linalool (**3**)

Eugenol (**4**)

Methyl jasmonate (**5**)

trans-Jasmin lactone (**6**)

(*R*)-(−)-Linalool (**7**)

(*S*)-(+)-linalool (**8**)

Δ-3-Carene (**9**)

Bornyl acetate (**10**)

Pyridine (**11**)

Allyl isothiocyanate (**12**)

2-Phenyl ethyl alcohol (**13**)

Limonene (**14**)

Myrcene (**15**)

Citral (**16**)

Carvone (**17**)

Methylchavicol (**18**)

Thymol (**19**)

Geraniol (**20**)

p-Cymene (**21**)

α-Terpinene (**22**)

Epinepetalactone (**23**)

trans-Anethole (**24**)

trans-Ocimene (**25**)

allo-Ocimene (**26**)

Methyl eugenol (**27**)

β-Asarone (**28**)

Effects of Essential Oils in the Central Nervous System

Borneol (**29**)

Isoborneol (**30**)

Isopulegol (**31**)

α-Terpineol (**32**)

Thujone (**33**)

β-Eudesmol (**34**)

α-Eudesmol (**35**)

Caryophyllene oxide (**36**)

β-Eudesmene (**37**)

Citronellal (**38**)

Geraniol (**39**)

Citronellol (**40**)

Carvacrol (**41**)

Pulegone (**42**)

Safrol (**43**)

Linalyl acetate (**44**)

S-(−)-Limonene (**45**)

Hinokitiol (**46**)

REFERENCES

Almeida, R. N., Motta, S. C., Faturi, C. B., Cattallini, B., and Leite, J. R. 2004. Anxiolytic-like effects of rose oil inhalation on the elevated plus-maze test in rats. *Pharmacology, Biochemistry and Behavior* 77:361–364.

Aoki, H. 1996. Effect of odors from coniferous woods on contingent negative variation (CNV). *Zairyo* 45(4):397–402.

Aoshima, H. and Hamamoto, K. 1999. Potentiation of GABAA receptors expressed in Xenopus oocytes by perfume and phytoncid. *Bioscience, Biotechnology, and Biochemistry* 63(4):743–748.

Asakura, K., Matsuo, Y., Oshima, T. et al. 2000. Ω-Agatoxin IVA-sensitive Ca^{2+} channel blocker, α-eudesmol, protects against brain injury after focal ischemia in rats. *European Journal of Pharmacology* 394:57–65.

Ashton, H., Millman, J. E., Telford, R., and Thompson, J. W. 1977. The use of event-related slow potentials of the human brain as an objective method to study the effects of centrally acting drugs [proceedings]. *Neuropharmacology* 16(7–8):531–532.

Aston-Jones, G., Chen, S., Zhu, Y., and Oshinsky, M. L. 2001. A neural circuit for circadian regulation of arousal. *Nature Neuroscience* 4(7):732–738.

Baltissen, R. and Heimann, H. 1995. Aktivierung, Orientierung und Habituation bei Gesunden und psychisch Kranken. In *Biopsychologie von Streß und emotionalen Reaktionen—Ansätze interdisziplinärer Forschung*, edited by F. Debus, G. Erdmann, and K. W. Dallus. Göttingen, Germany: Hogrefe-Verlag für Psychologie.

Barocelli, E., Calcina, F., Chiavarini, M. et al. 2004. Antinociceptive and gastroprotective effects of inhaled and orally administered *Lavandula hybrida* Reverchon "Grosso" essential oil. *Life Science* 76(2):213–223.

Baron, R. A. 1990. Environmentally induced positive affect: Its impact on self-efficacy, task performance, negotiation, and conflict. *Journal of Applied Social Psychology* 20(5, Pt 2):368–384.

Baron, R. A. and Thomley, J. 1994. A whiff of reality: Positive affect as a potential mediator of the effects of pleasant fragrances on task performance and helping. *Environment and Behavior* 26(6):766–784.

Baxter, M. G. and Chiba, A. A. 1999. Cognitive functions of the basal forebrain. *Current Opinion in Neurobiology* 9(2):178–183.

Becker-Carus, C. 1981. *Grundriß der Physiologischen Psychologie*. Heidelberg, Germany: Quelle & Meyer.

Bensafi, M., Rouby, C., Farget, V. et al. 2002. Autonomic nervous system responses to odours: The role of pleasantness and arousal. *Chemical Senses* 27(8):703–709.

Bermpohl, F., Pascual-Leone, A., Amedi, A. et al. 2006. Dissociable networks for the expectancy and perception of emotional stimuli in the human brain. *Neuroimage* 30(2):588–600.

Bielenberg, J. 2007. Zentralnervöse Effekte durch Thujon. *Medizinische Monatsschrift für Pharmazeuten* 30(9):322–326.

Blanco, M. M., Costa, C. A. R. A., Freire, A. O., Santos Jr., J. G., and Costa, M. 2007. Neurobehavioral effect of essential oil of *Cymbopogon citratus* in mice. *Phytomedicine* 16(2–3):265–270. doi:10.1016/j.phymed.2007.04.007.

Bradley, B. F., Starkey, N. J., Brown, S. L., and Lea, R. W. 2007. Anxiolytic effects of *Lavandula angustifolia* odour on the Mongolian gerbil elevated plus maze. *Journal of Ethnopharmacology* 111(3):517–525.

Brum, L. F. S., Elisabetsky, E., and Souza, D. 2001a. Effects of linalool on [^{3}H] MK801 and [^{3}H] muscimol binding in mouse cortical membranes. *Phytotherapy Research* 15(5):422–425.

Brum, L. F. S., Emanuelli, T., Souza, D., and Elisabetsky, E. 2001b. Effects of linalool on glutamate release and uptake in mouse cortical synaptosomes. *Neurochemical Research* 25(3):191–194.

Buchbauer, G., Jirovetz, L., Czejka, M., Nasel, C., and Dietrich, H. 1993a. New results in aromatherapy research. Paper read at *24th International Symposium on Essential Oils*, July 21–24, 1993, TU Berlin, Germany.

Buchbauer, G., Jirovetz, L., Jäger, W., Dietrich, H., and Plank, C. 1991. Aromatherapy: Evidence for sedative effects of the essential oil of lavender after inhalation. *Journal of Bioscience* 46:1067–1072.

Buchbauer, G., Jirovetz, L., Jäger, W., Plank, C., and Dietrich, H. 1993b. Fragrance compounds and essential oil with sedative effects upon inhalation. *Journal of Pharmacological Science* 82:660–664.

Carlini, E. A., de Oliveira, A. B., and de Oliveira, G. G. 1983. Psychopharmacological effects of the essential oil fraction and of the hydrolate obtained from the seeds of *Licaria puchury-major*. *Journal of Ethnopharmacology* 8(2):225–236.

Carvalho-Freitas, M. I. R. and Costa, M. 2002. Anxiolytic and sedative effects of extracts and essential oils from *Citrus aurantium* L. *Biological and Pharmaceutical Bulletin* 25(12):1629–1633.

Cassel, E. and Vargas, R. M. F., 2006. Experiments and modeling of the *Cymbopogon winterianus* essential oil extraction by steam distillation. *Journal of Mexican Chemical Society* 50(3):126–129.

Chen, S. W., Min, L., Li, W. J. et al. 2004. The effects of angelica essential oil in three murine tests of anxiety. *Pharmacology, Biochemistry, and Behavior* 79:377–382.

Chiou, L. C., Ling, J. Y., and Chang, C. C. 1995. β-Eudesmol as an antidote for intoxication from organophosphorus anticholinesterase agents. *European Journal of Pharmacology* 292:151–156.

Chiou, L. C., Ling, J. Y., and Chang, C. C. 1997. Chinese herb constituent β-eudesmol alleviated the electroshock seizures in mice and electrographic seizures in rat hippocampal slices. *Neuroscience Letters* 231:171–174.

Cho, J., Kong, J. Y., Jeong, D. Y. et al. 2001. NMDA receptor-mediated neuroprotection by essential oils from the rhizomes of *Acorus gramineus*. *Life Science* 68:1567–1573.

Chu, S. 2008. Olfactory conditioning of positive performance in humans. *Chemical Senses* 33(1):65–71.

Chu, S. and Downes, J. J. 2002. Proust nose best: Odors are better cues of autobiographical memory. *Memory and Cognition* 30(4):511–518.

Colagiuri, B., Livesey, E. J., and Harris, J. A. 2011. Can expectancies produce placebo effects for implicit learning? *Psychonomic Bulletin & Review* 18(2):399–405.

Cooke, B. and Ernst, E. 2000. Aromatherapy: A systematic review. *British Journal of General Practice* 50:493–496.

Danuser, B., Moser, D., Vitale-Sethre, T., Hirsig, R., and Krueger, H. 2003. Performance in a complex task and breathing under odor exposure. *Human Factors* 45(4):549–562.

Davidson, R. J. and Irwin, W. 1999. The functional neuroanatomy of emotion and affective style. *Trends in Cognitive Sciences* 3(1):11–21.

de Sousa, D. P., Quintans, L., and de Almeida, R. N. 2007. Evolution of the anticonvulsant activity of α-terpineol. *Pharmaceutical Biology* 45(1):69–70.

de Souza, G. P. C., Elisabetsky, E., Nunes, D. S., Rabelo, S. K. L., and da Silva, M. N. 1997. Anticonvulsant properties of γ-decanolactone in mice. *Journal of Ethnopharmacology* 58(3):175–181.

Degel, J. and Köster, E. P. 1999. Odors: Implicit memory and performance effects. *Chemical Senses* 24(3):317–325.

Diego, M. A., Jones, N. A., Field, T. et al. 1998. Aromatherapy positively affects mood, EEG patterns of alertness and math computations. *International Journal of Neuroscience* 96(3–4):217–224.

Diekelmann, S., Biggel, S., Rasch, B., and Born, J. 2012. Offline consolidation of memory varies with time in slow wave sleep and can be accelerated by cuing memory reactivations. *Neurobiology of Learning and Memory* 98(2):103–111.

Duffy, E. 1972. Activation. In *Handbook of Psychophysiology*, edited by N. S. Greenfield and R. A. Sternbach. New York: Holt, Rinehart and Winston, Inc.

Duke, J. A., Bogenschutz-Godwin, M. J., De Cellier, J., and Duke, P. A. K. 2002. *Handbook of Medicinal Herbs*, 2nd edn. Boca Raton, FL: CRC Press.

Elisabetsky, E., Brum, L. F., and Souza, D. O. 1999. Anticonvulsant properties of linalool in glutamate-related seizure models. *Phytomedicine* 6(2):107–113.

Elisabetsky, E., de Souza, G. P. C., Santos, M. A. C. et al. 1995a. Sedative properties of linalool. *Fitoterapia* 66:407–414.

Elisabetsky, E., Marschner, J., and Souza, D. O. 1995b. Effects of linalool on glutamatergic system in the rat cerebral-cortex. *Neurochemical Research* 20(4):461–465.

Epple, G. and Herz, R. S. 1999. Ambient odors associated to failure influence cognitive performance in children. *Developmental Psychobiology* 35(2):103–107.

Field, T., Diego, M., Hernandez-Reif, M. et al. 2005. Lavender fragrance cleansing gel effects on relaxation. *International Journal of Neuroscience* 115(2):207–222.

Finkelmeyer, A., Kellermann, T., Bude, D. et al. 2010. Effects of aversive odour presentation on inhibitory control in the Stroop colour-word interference task. *BMC Neuroscience* 11:131.

Freire, C. M., Marques, M. O. M., and Costa, M. 2006. Effects of seasonal variation on the central nervous system activity of *Ocimum gratissimum* L. essential oil. *Journal of Ethnopharmacology* 105:161–166.

Friedl, S. M., Oedendorfer, K., Kitzer, S. et al. 2010. Comparison of liquid-liquid partition, HS-SPME and static HS GC/MS analysis for the quantification of (−)-linalool in human whole blood samples. *Natural Product Communications* 5(9):1447–1452.

Fujiwara, R., Komori, T., Noda, Y. et al. 1998. Effects of a long-term inhalation of fragrances on the stress-induced immunosuppression in mice. *Neuroimmunomodulation* 5(6):318–322.

Fukumoto, S., Morishita, A., Furutachi, K. et al. 2008. Effect of flavors components in lemon essential oil on physical or psychological stress. *Stress and Health* 24:3–12.

Fukumoto, S., Sawasaki, E., Okuyama, S., Miyake, Y., and Yokogoshi, H. 2003. Flavor components of monoterpenes in citrus essential oils enhance the release of monoamines from rat brain slices. *Nutritional Neuroscience* 9(1–2):73–80.

Galati, E. M., Miceli, N., Galluzzo, M., Taviano, M. F., and Tzakou, O. 2004. Neuropharmacological effects of epinepetalactone from *Nepeta sibthorpii* behavioral and anticonvulsant activity. *Pharmaceutical Biology* 42(6):391–395.

Gaygen, D. E. and Hedge, A. 2009. Effect of acute exposure to a complex fragrance on lexical decision performance. *Chemical Senses* 34(1):85–91.

Ghelardini, C., Galeotti, N., Salvatore, G., and Mazzanti, G. 1999. Local anesthetic activity of the essential oil of *Lavandula angustifolia*. *Planta Medica* 65(8):700–703.

Gilbert, A. N., Knasko, S. C., and Sabini, J. 1997. Sex differences in task performance associated with attention to ambient odor. *Archives of Environmental Health* 52(3):195–199.

Gill, T. M., Sarter, M., and Givens, B. 2000. Sustained visual attention performance-associated prefrontal neuronal activity: Evidence for cholinergic modulation. *Journal of Neuroscience* 20(12):4745–4757.

Goddard, L., Pring, L., and Felmingham, N. 2005. The effects of cue modality on the quality of personal memories retrieved. *Memory* 13(1):79–86.

Gottfried, J. A. and Zald, D. H. 2005. On the scent of human olfactory orbitofrontal cortex: Meta-analysis and comparison to non-human primates. *Brain Research Reviews* 50(2):287–304.

Gould, A. and Martin, G. N. 2001. 'A good odour to breathe?' The effect of pleasant ambient odour on human visual vigilance. *Applied Cognitive Psychology* 15:225–232.

Grayer, R. J., Kite, G. C., Goldstone, F. J. et al. 1996. Intraspecific taxonomy and essential oil chemotypes in sweet basil, *Ocimum basilicum*. *Phytochemistry* 43(5):1033–1039.

Grilly, D. M. 2002. *Drugs & Human Behavior*. Boston, MA: Allyn and Bacon.

Grunwald, M., Weiss, T., Krause, W. et al. 1999. Power of theta waves in the EEG of human subjects increases during recall of haptic information. *Neuroscience Letters* 260(3):189–192.

Habel, U., Koch, K., Pauly, K. et al. 2007. The influence of olfactory-induced negative emotion on verbal working memory: Individual differences in neurobehavioral findings. *Brain Research* 1152:158–170.

Haneyama, H. and Kagatani, M. 2007. Sedatives containing Nardostachys chinensis extracts. *CAN* 147:197215.

Hauner, K. K., Howard, J. D., Zelano, C., and Gottfried, J. A. 2013. Stimulus-specific enhancement of fear extinction during slow-wave sleep. *Nature Neuroscience* 16(11):1553–1555.

Haze, S., Sakai, K., and Gozu, Y. 2002. Effects of fragrance inhalation on sympathetic activity in normal adults. *Japanese Journal of Pharmacology* 90:247–253.

Herz, R. S. 2004. A naturalistic analysis of autobiographical memories triggered by olfactory visual and auditory stimuli. *Chemical Senses* 29(3):217–224.

Herz, R. S. and Cupchik, G. C. 1995. The emotional distinctiveness of odor-evoked memories. *Chemical Senses* 20(5):517–528.

Herz, R. S., Eliassen, J., Beland, S., and Souza, T. 2004. Neuroimaging evidence for the emotional potency of odor-evoked memory. *Neuropsychologia* 42(3):371–378.

Heuberger, E., Hongratanaworakit, T., Bohm, C., Weber, R., and Buchbauer, G. 2001. Effects of chiral fragrances on human autonomic nervous system parameters and self-evaluation. *Chemical Senses* 26(3):281–292.

Heuberger, E., Hongratanaworakit, T., and Buchbauer, G. 2006. East Indian Sandalwood and α-Santalol odor increase physiological and self-rated arousal in humans. *Planta Medica* 72(9):792–800.

Heuberger, E. and Ilmberger, J. 2010. The influence of essential oils on human vigilance. *Natural Product Communications* 5(9):1441–1446.

Heuberger, E., Ilmberger, J., Hartter, E., and Buchbauer, G. 2008. Physiological and behavioral effects of 1,8-cineol and (±)-linalool: A comparison of inhalation and massage aromatherapy. *Natural Product Communications* 3(7):1103–1110.

Heuberger, E., Redhammer, S., and Buchbauer, G. 2004. Transdermal absorption of (-)-linalool induces autonomic deactivation but has no impact on ratings of well-being in humans. *Neuropsychopharmacology* 29(10):1925–1932.

Hey, K., Juran, S., Schaeper, M. et al. 2009. Neurobehavioral effects during exposures to propionic acid—An indicator of chemosensory distraction? *NeuroToxicology* 30(6):1223–1232.

Hiruma, T., Matuoka, T., Asai, R. et al. 2005. Psychophysiological basis of smells. *Seishin shinkeigaku zasshi* 107(8):790–801.

Hiruma, T., Yabe, H., Sato, Y., Sutoh, T., and Kaneko, S. 2002. Differential effects of the hiba odor on CNV and MMN. *Biological Psychology* 61(3):321–331.

Ho, C. and Spence, C. 2005. Olfactory facilitation of dual-task performance. *Neuroscience Letters* 389(1):35–40.

Hoedlmoser, K., Schabus, M., Stadler, W., et al. 2007. EEG Theta-Aktivität während deklarativem Lernen und anschließendem REM-Schlaf im Zusammenhang mit allgemeiner Gedächtnisleistung. *Klinische Neurophysiologie* 38(01):P304.

Holland, R. W., Hendriks, M., and Aarts, H. 2005. Smells like clean spirit. Nonconscious effects of scent on cognition and behavior. *Psychological Science* 16(9):689–693.

Hongratanaworakit, T., Heuberger, E., and Buchbauer, G. 2004. Evaluation of the effects of East Indian sandalwood oil and a-santalol on humans after transdermal absorption. *Planta Medica* 70(1):3–7.

Hotchkiss, S. A. M. 1998. Absorption of fragrance ingredients using in vitro models with human skin. In *Fragrances: Beneficial and Adverse Effects*, edited by P. J. Frosch, J. D. Johansen, and I. R. White. Berlin, Germany: Springer-Verlag.

Ilmberger, J., Heuberger, E., Mahrhofer, C. et al. 2001. The influence of essential oils on human attention. I: Alertness. *Chemical Senses* 26(3):239–245.

Ishikawa, S., Miyake, Y., and Yokogoshi, H. 2002. Effect of lemon odor on brain neurotransmitters in rat and electroencephalogram in human subject. *Aroma Research* 3(2):126–130.

Ishiyama, S. 2000. Aromachological effects of volatile compounds in forest. *Aroma Research* 1(4):15–21.

Ismail, M. 2006. Central properties and chemical composition of *Ocimum basilicum* essential oil. *Pharmaceutical Biology* 44(8):619–626.

Jellinek, J. S. 1997. Psychodynamic odor effects and their mechanisms. *Cosmetics & Toiletries* 112(9):61–71.

Jirovetz, L., Buchbauer, G., Jäger, W. et al. 1990. Determination of lavender oil fragrance compounds in blood samples. *Fresenius' Journal of Analytical Chemistry* 338:922–923.

Kaetsu, I., Tonoike, T., Uchida, K. et al. 1994. Effect of controlled release of odorants on electroencephalogram during mental activity. *Proceedings of the International Symposium on Controlled Release of Bioactive Materials*, Nice, France, Vol. 21ST, pp. 589–590.

Kagawa, D., Jokura, H., Ochiai, R., Tokimitsu, I., and Tsubone, H. 2003. The sedative effects and mechanism of action of cedrol inhalation with behavioral pharmacological evaluation. *Planta Medica* 69(7):637–641.

Kandel, E. R., Schwartz, J. H., and Jessel, T. M. 1991. *Principles of Neural Science*, 3rd edn. Englewood Cliffs, NJ: Prentice-Hall International, Inc.

Kaneda, H., Kojima, H., Takashio, M., and Yoshida, T. 2005. Relaxing effect of hop aromas on human. *Aroma Research* 6(2):164–170.

Kaneda, H., Kojima, H., and Watari, J. 2006. Effect of beer flavors on changes in human feelings. *Aroma Research* 7(4):342–347.

Kaneda, H., Kojima, H., and Watari, J. 2011. Novel psychological and neurophysiological significance of beer aromas. Part I: Measurement of changes in human emotions during the smelling of hop and ester aromas using a measurement system for brainwaves. *Journal of the American Society of Brewing Chemists* 69(2):67–74.

Kawano, K. 2001. Meditation-like effects of aroma observed in EEGs with consideration of each experience. *Aroma Research* 2(1):30–46.

Keller, I. and Groemminger, O. 1993. Aufmerksamkeit. In *Neuropsychologische Diagnostik*, edited by D. Y. von Cramon, N. Mai, and W. Ziegler. Weinheim, Germany: VCH.

Kennedy, D. O., Dodd, F. L., Robertson, B. C. et al. 2011. Monoterpenoid extract of sage (*Salvia lavandulaefolia*) with cholinesterase inhibiting properties improves cognitive performance and mood in healthy adults. *Journal of Psychopharmacology* 25(8):1088–1100.

Kepecs, A., Uchida, N., and Mainen, Z. F. 2006. The sniff as a unit of olfactory processing. *Chemical Senses* 31(2):167–179.

Kerl, S. 1997. Zur olfaktorischen Beeinflussbarkeit von Lernprozessen. *Dragoco Report* 44:45–59.

Kim, J. T., Ren, C. J., Fielding, G. A. et al. 2007. Treatment with lavender aromatherapy in post-anesthesia care unit reduces opioid requirements of morbidly obese patients undergoing laparoscopic adjustable gastric banding. *Obesity Surgery* 17(7):920–925.

Kim, Y.-K. and Watanuki, S. 2003. Characteristics of electroencephalographic responses induced by a pleasant and an unpleasant odor. *Journal of Physiological Anthropology and Applied Human Science* 22(6):285–291.

Kline, J. P., Blackhart, G. C., Woodward, K. M., Williams, S. R., and Schwartz, G. E. 2000. Anterior electroencephalographic asymmetry changes in elderly women in response to a pleasant and an unpleasant odor. *Biological Psychology* 52(3):241–250.

Knasko, S. C. 1993. Performance, mood, and health during exposure to intermittent odors. *Archives of Environmental Health* 48(5):305–308.

Komiya, M., Takeuchi, T., and Harada, E. 2006. Lemon oil vapor causes an anti-stress effect via modulating the 5-HT and DA activities in mice. *Behavioural Brain Research* 172:240–249.

Komori, T., Fujiwara, R., Tanida, M., and Nomura, J. 1995a. Potential antidepressant effects of lemon odor in rats. *European Neuropsychopharmacology* 5(4):477–480.

Komori, T., Fujiwara, R., Tanida, M., Nomura, J., and Yokoyama, M. M. 1995b. Effects of citrus fragrance on immune function and depressive states. *Neuroimmunomodulation* (2):174–180.

Koo, B.-S., Lee, S.-I., Ha, J.-H., and Lee, D.-U. 2004. Inhibitory effects of the essential oil from SuHeXiang Wa non the central nervous system after inhalation. *Biological and Pharmaceutical Bulletin* 27(4):515–519.

Koo, B.-S., Park, K.-S., Ha, J.-H. et al. 2003. Inhibitory effects of the fragrance inhalation of essential oil from *Acorus gramineus* on central nervous system. *Biological and Pharmaceutical Bulletin* 26(7):978–982.

Kopell, B. S., Wittner, W. K., Lunde, D. T., Wolcott, L. J., and Tinklenberg, J. R. 1974. The effects of methamphetamine and secobarbital on the contingent negative variation amplitude. *Psychopharmacology* 34(1):55–62.

Kovar, K. A., Gropper, B., Friess, D., and Ammon, H. P. 1987. Blood levels of 1,8-cineole and locomotor activity of mice after inhalation and oral administration of rosemary oil. *Planta Medica* 53(4):315–318.

Krizhanovs'kii, S. A., Zima, I. H., Makarchuk, M. Y., Piskors'ka, N. H., and Chernins'kii, A. O. 2004. Effect of citrus essential oil on the attention level and electrophysical parameters of human brain. *Physics of the Alive* 12(1):111–120.

Kuroda, K., Inoue, N., Ito, Y. et al. 2005. Sedative effects of the jasmine tea odor and (R)-(−)-linalool, one of its major odor components, on autonomic nerve activity and mood sates. *European Journal of Applied Physiology* 95:107–114.

Lee, C. F., Katsuura, T., Shibata, S. et al. 1994. Responses of electroencephalogram to different odors. *Annals of Physiological Anthropology* 13(5):281–291.

Lehrner, J., Eckersberger, C., Walla, P., Postsch, G., and Deecke, L. 2000. Ambient odor of orange in a dental office reduces anxiety and improves mood in female patients. *Physiology and Behaviour* 71:83–86.

Leppanen, J. M. and Hietanen, J. K. 2003. Affect and face perception: Odors modulate the recognition advantage of happy faces. *Emotion* 3(4):315–326.

Liao, J. F., Huang, S. Y., Jan, Y. M., Yu, L. L., and Chen, C. F. 1998. Central inhibitory effects of water extract of *Acori graminei* rhizoma in mice. *Journal of Ethnopharmacology* 61:185–193.

Linck V. M., da Silva A. L., Figueiró M., et al. 2009. Inhaled linalool-induced sedation in mice. *Phytomedicine* 16(4):303–307.

Linck V. M., da Silva A. L., Figueiró M., et al. 2010. Effects of inhaled Linalool in anxiety, social interaction and aggressive behavior in mice. *Phytomedicine* 17(8–9):679–683.

Lorig, T. S. 1989. Human EEG and odor response. *Progress in Neurobiology* 33(5–6):387–398.

Lorig, T. S. 2000. The application of electroencephalographic techniques to the study of human olfaction: A review and tutorial. *International Journal of Psychophysiology* 36(2):91–104.

Lorig, T. S. and Roberts, M. 1990. Odor and cognitive alteration of the contingent negative variation. *Chemical Senses* 15(5):537–545.

Lorig, T. S. and Schwartz, G. E. 1988. Brain and odor: I. Alteration of human EEG by odor administration. *Psychobiology* 16(3):281–284.

Ludvigson, H. W. and Rottman, T. R. 1989. Effects of ambient odors of lavender and cloves on cognition, memory, affect and mood. *Chemical Senses* 14(4):525–536.

Manley, C. H. 1993. Psychophysiological effect of odor. *Critical Reviews in Food Science and Nutrition* 33(1):57–62.

Manley, C. H. 1997. Psychophysiology of odor. *Rivista Italiana EPPOS* (Spec. Num., 15th Journees Internationales Huiles Essentielles, 1996):375–386.

Masago, R., Matsuda, T., Kikuchi, Y. et al. 2000. Effects of inhalation of essential oils on EEG activity and sensory evaluation. *Journal of Physiological Anthropology and Applied Human Science* 19(1):35–42.

Matos, F. J. A. 1996. As ervas cidreiras do Nordeste do Brasil: Estudo de três quimiotipos de *Lippia alba* (Mill.) N. E. Brown (Verbenaceae). Parte II—Farmacoquímica. *Revista Brasileira de Farmácia* 77:137–141.

Meckes, M., Calzada, F., Tortoriello, J., González, J. L., and Martinez, M. 1997. Terpenoids isolated from *Psidium guajava* hexane extract with depressant activity on central nervous system. *Phytotherapy Research* 10(7):600–603.

Michael, G. A., Jacquot, L., Millot, J. L., and Brand, G. 2005. Ambient odors influence the amplitude and time course of visual distraction. *Behavioral Neuroscience* 119(3):708–715.

Miller, D. B. and O'Callaghan, J. P. 2006. The pharmacology of wakefulness. *Metabolism* 55(Suppl. 2): S13–S19.

Millot, J. L., Brand, G., and Morand, N. 2002. Effects of ambient odors on reaction time in humans. *Neuroscience Letters* 322(2):79–82.

Min, B.-C., Jin, S.-H., Kang, I.-H. et al. 2003. Analysis of mutual information content for EEG responses to odor stimulation for subjects classified by occupation. *Chemical Senses* 28(9):741–749.

Moreira, M. R., Cruz, G. M., Lopes, M. S. et al. 2001. Effects of terpineol on the compound action potential of the rat sciatic nerve. *Brazilian Journal of Medical and Biological Research* 34:1337–1340.

Morgan, C. L. 1996. Odors as cues for the recall of words unrelated to odor. *Perceptual and Motor Skills* 83(3 Pt 2):1227–1234.

Morrone, L. A., Romboià, L., Pelle, C. et al. 2007. The essential oil of bergamot enhances the levels of amino acid neurotransmitters in the hippocampus of rat: Implication of monoterpene hydrocarbons. *Pharmacology Research* 55(4):255–262.

Morteza-Semnani, K., Mahmoudi, M., and Riahi, G. 2007. Effects of essential oil and extracts from certain *Thymus* species on swimming performance in mice. *Pharmaceutical Biology* 45(6):464–467.

Moss, L., Rouse, M., Wesnes, K. A., and Moss, M. 2010. Differential effects of the aromas of Salvia species on memory and mood. *Human Psychopharmacology* 25(5):388–396.

Moss, M., Cook, J., Wesnes, K., and Duckett, P. 2003. Aromas of rosemary and lavender essential oils differentially affect cognition and mood in healthy adults. *International Journal of Neuroscience* 113(1):15–38.

Moss, M., Hewitt, S., Moss, L., and Wesnes, K. 2008. Modulation of cognitive performance and mood by aromas of peppermint and ylang-ylang. *International Journal of Neuroscience* 118(1):59–77.

Moss, M. and Oliver, L. 2012. Plasma 1,8-cineole correlates with cognitive performance following exposure to rosemary essential oil aroma. *Therapeutic Advances in Psychopharmacology* 2(3):103–113.

Mukherjee, P. K., Kumar, V., Mal, M., and Houghton, P. J. 2007. *Acorus calamus*: Scientific validation of Ayurvedic tradition from natural resources. *Pharmaceutical Biology* 45(8):651–666.

Nakagawa, M., Nagai, H., and Inui, T. 1992. Evaluation of drowsiness by EEGs. Odors controlling drowsiness. *Fragrance Journal* 20(10):68–72.

Narusuye, K., Kawai, F., Mtsuzaki, K., and Miyachi, E. 2005. Linalool suppresses voltage-gated currents in sensory neurons and cerebellar Purkinje cells. *Journal of Neural Transmission* 112:193–203.

Nasel, C., Nasel, B., Samec, P., Schindler, E., and Buchbauer, G. 1994. Functional imaging of effects of fragrances on the human brain after prolonged inhalation. *Chemical Senses* 19(4):359–364.

Norte, M. C., Cosentino, R. M., and Lazarini, C. A. 2005. Effects of methyleugenol administration on behavior models related to depression and anxiety in rats. *Phytomedicine* 12(4):294–298.

Oken, B. S., Salinsky, M. C., and Elsas, S. M. 2006. Vigilance, alertness, or sustained attention: Physiological basis and measurement. *Clinical Neurophysiology* 117(9):1885–1901.

Okugawa, H., Ueda, R., Matsumoto, K., Kawanishi, K., and Kato, K. 2000. Effects of sesquiterpenoids from "Oriental incenses" on acetic acid-induced writhing and D_2 and $5-HT_{2A}$ receptors in rat brain. *Phytomedicine* 7(5):417–422.

Orafidiya, L. O., Agbani, E. O., Iwalewa, E. O., Adelusola, K. A., and Oyedapo, O. O. 2004. Studies on the acute and sub-chronic toxicity of the essential oil of *Ocimum gratissimum* L. leaf. *Phytomedicine* 11:71–76.

Palva, S. and Palva, J. M. 2007. New vistas for [alpha]-frequency band oscillations. *Trends in Neurosciences* 30(4):150–158.

Paus, T. 2001. Primate anterior cingulate cortex: Where motor control, drive and cognition interface. *Nature Reviews Neuroscience* 2(6):417–424.

Peana, A. T., D'Aquila, P. S., Chessa, M. L. et al. 2003. (−)-Linalool produces antinociception in two experimental models of pain. *European Journal of Pharmacology* 460:37–41.

Peana, A. T., D'Aquila, P. S., Panin, F. et al. 2002. Anti-inflammatory activity of linalool and linalyl acetate constituents of essential oils. *Phytomedicine* 9(8):721–726.

Peana, A. T., De Montis, M. G., Nieddu, E. et al. 2004a. Profile of spinal and supra-spinal antinociception of (−)-linalool. *European Journal of Pharmacology* 485:165–174.

Peana, A. T., De Montis, M. G., Sechi, S. et al. 2004b. Effects of (−)-linalool in the acute hyperalgesia induced by carrageenan, L-glutamate and prostaglandin E2. *European Journal of Pharmacology* 497(3):279–284.

Peana, A. T., Rubattu, P., Piga, G. G. et al. 2006. Involvement of adenosine A1 and A2A receptors in (−)-linalool-induced antinociception. *Life Science* 78:2471–2474.

Pedersen, C. A., Gaynes, B. N., Golden, R. N., Evans, D. L., and Haggerty Jr., J. J. 1998. Neurobiological aspects of behavior. In *Human Behavior: An Introduction for Medical Students*, edited by A. Stoudemire. Philadelphia, PA: Lippincott-Raven Publishers.

Perry, N. and Perry, E. 2006. Aromatherapy in the management of psychiatric disorders: Clinical and neuropharmacological perspectives. *CNS Drugs* 20(4):257–280.

Perry, N. S., Hougthon, P. J., Jenner, P., Keith, A., and Perry, E. K. 2002. *Salvia lavandulaefolia* essential oil inhibits cholinesterase in vivo. *Phytomedicine* 9(1):48–51.

Perry, N. S., Hougthon, P. J., Sampson, J. et al. 2001. In-vitro activity of *S. lavandulaefolia* (Spanish sage) relevant to treatment of Alzheimer's disease. *Journal of Pharmacy and Pharmacology* 53(10):1347–1356.

Phan, K. L., Wager, T., Taylor, S. F., and Liberzon, I. 2002. Functional neuroanatomy of emotion: A meta-analysis of emotion activation studies in PET and fMRI. *Neuroimage* 16(2):331–348.

Posner, M. I. and Petersen, S. E. 1990. The attention system of the human brain. *Annual Review of Neuroscience* 13:25–42.

Posner, M. I. and Rafal, R. D. 1987. Cognitive theories of attention and the rehabilitation of attentional deficits. In *Neuropsychological Rehabilitation*, edited by M. J. Meier, A. L. Benton, and L. Diller. London, U.K.: Churchill-Livingston.

Pourgholami, M. H., Kamalinejad, M., Javadi, M., Majzoob, S., and Sayyah, M. 1999a. Evaluation of the anticonvulsant activity of the essential oil of *Eugenia caryophyllata* in male mice. *Journal of Ethnopharmacology* 64:167–171.

Pourgholami, M. H., Majzoob, S., Javadi, M. et al. 1999b. The fruit essential oil of *Pimpinella anisum* exerts anticonvulsant effects in mice. *Journal of Ethnopharmacology*. 66(2):211–215.

Pribram, K. H. and McGuinness, D. 1975. Arousal, activation, and effort in the control of attention. *Psychological Review* 82(2):116–149.

Pultrini, A. M., Galindo, L. A., and Costa, M. 2006. Effects of essential oil from *Citrus aurantium* L. in experimental anxiety models in mice. *Life Sciences* 78:1720–1725.

Quintans-Júnior, L. J., Souza, T. T., Leite, B. S. et al. 2007. Phythochemical screening and anticonvulsant activity of *Cymbopogon winterianus* Jowitt (Poaceae) leaf essential oil in rodents. *Phytomedicine* 15(8):619–624.

Rasch, B., Buchel, C., Gais, S., and Born, J. 2007. Odor cues during slow-wave sleep prompt declarative memory consolidation. *Science* 315(5817):1426–1429.

Razumnikova, O. M. 2007. Creativity related cortex activity in the remote associates task. *Brain Research Bulletin* 73(1–3):96–102.

Reiman, E. M., Lane, R. D., Ahern, G. L. et al. 1997. Neuroanatomical correlates of externally and internally generated human emotion. *American Journal of Psychiatry* 154(7):918–925.

Sadraei, H., Ghannadi, A., and Malekshahi, K. 2003. Relaxant effect of essential oil of *Melissa officinalis* and citral on rat ileum contractions. *Fitoterapia* 74:445–452.

Sakamoto, R., Minoura, K., Usui, A., Ishizuka, Y., and Kanba, S. 2005. Effectiveness of aroma on work efficiency: Lavender aroma during recesses prevents deterioration of work performance. *Chemical Senses* 30(8):683–691.

Satoh, T. and Sugawara, Y. 2003. Effects on humans elicited by inhaling the fragrance of essential oils: Sensory test, multi-channel thermometric study and forehead surface potential wave measurement on basil and peppermint. *Analytical Sciences* 19(1):139–146.

Sawada, K., Komaki, R., Yamashita, Y., and Suzuki, Y. 2000. Odor in forest and its physiological effects. *Aroma Research* 1(3):67–71.

Sayorwan, W., Ruangrungsi, N., Piriyapunyporn, T. et al. 2013. Effects of inhaled rosemary oil on subjective feelings and activities of the nervous system. *Scientia Pharmaceutica* 81(2):531–542.

Sayowan, W., Siripornpanich, V., Hongratanaworakit, T., Kotchabhakdi, N., and Ruangrungsi, N. 2013. The effects of jasmine oil inhalationon brainwave activies and emotions. *Journal of Health Research* 27(2):73–77.

Sayorwan, W., Siripornpanich, V., Piriyapunyaporn, T. et al. 2012a. The effects of lavender oil inhalation on emotional states, autonomic nervous system, and brain electrical activity. *Journal of the Medical Association of Thailand* 95(4):598–606.

Sayowan, W., Siripornpanich, V., Piriyapunyaporn, T. et al. 2012b. The harmonizing effects of citronella oil on mood states and brain activities. *Journal of Health Research* 26(2):69–75.

Sayyah, M., Nadjafnia, L., and Kamalinejad, M. 2004. Anticonvulsant activity and chemical composition of *Artemisia dracunculus* L. essential oil. *Journal of Etnopharmacology* 94:283–287.

Sayyah, M., Valizadehl, J., and Kamalinejad, M. 2002. Anticonvulsant activity of the leaf essential oil of *Laurus nobilis* against pentylenetetrazole- and maximal electroshock-induced seizures. *Phytomedicine* 9:212–216.

Schandry, R. 1989. *Lehrbuch der Psychophysiologie*. 2. Auflage ed. Weinheim, Germany: Psychologie Verlags Union.

Schwabe, L. and Wolf, O. T. 2009. The context counts: Congruent learning and testing environments prevent memory retrieval impairment following stress. *Cognitive, Affective, & Behavioral Neuroscience* 9(3):229–236.

Shaw, D., Annett, J. M., Doherty, B., and Leslle, J. C. 2007. Anxiolytic effects of lavender oil inhalation on open-field behaviour in rats. *Phytomedicine* 14(9):613–620.

Shibata, H., Fujiwara, R., Iwamoto, M., Matsuoka, H., and Yokoyama, M. M. 1990. Recovery of PFC in mice exposed to high pressure stress by olfactory stimulation with fragrance. *International Journal of Neuroscience* 51:245–247.

Silva, M. I., Aquino Neto, M. R., Teixeira Neto, P. F. et al. 2007. Central nervous system activity of acute administration of isopulegol in mice. *Pharmacology Biochemistry and Behavior* 88:141–147.

Sturm, W. 1997. Aufmerksamkeitsstoerungen. In *Klinische Neuropsychologie*, edited by W. Hartje and K. Poeck. Stuttgart, Germany: Thieme.

Sturm, W., de Simone, A., Krause, B. J. et al. 1999. Functional anatomy of intrinsic alertness: Evidence for a fronto-parietal-thalamic-brainstem network in the right hemisphere. *Neuropsychologia* 37(7):797–805.

Sugano, H. 1992. Psychophysiological studies of fragrance. In *Fragrance: The Psychology and Biology of Perfume*, edited by S. Van Toller and G. H. Dodd. Barking, U.K.: Elsevier Science Publishers Ltd.

Sugawara, Y., Hara, C., Aoki, T., Sugimoto, N., and Masujima, T. 2000. Odor distinctiveness between enantiomers of linalool: Difference in perception and responses elicited by sensory test and forehead surface potential wave measurement. *Chemical Senses* 25(1):77–84.

Sugawara, Y., Hara, C., Tamura, K. et al. 1998. Sedative effect on humans of inhalation of linalool: Sensory evaluation and physiological measurements using optically active linalools. *Analitica Chimica Acta* 365:293–299.

Sullivan, T. E., Warm, J. S., Schefft, B. K. et al. 1998. Effects of olfactory stimulation on the vigilance performance of individuals with brain injury. *Journal of Clinical and Experimental Neuropsychology* 20(2):227–236.

Suss, C., Gaylord, S., and Fagen, J. 2012. Odor as a contextual cue in memory reactivation in young infants. *Infant Behavior and Development* 35(3):580–583.

Tecce, J. J. 1972. Contingent negative variation (CNV) and psychological processes in man. *Psychological Bulletin* 77(2):73–108.

Tildesley, N. T. J., Kennedy, D. O., Perry, E. K. et al. 2005. Positive modulation of mood and cognitive performance following administration of acute doses of *Salvia lavandulaefolia* essential oil to healthy young volunteers. *Physiology & Behavior* 83(5):699–709.

Torii, S., Fukada, H., Kanemoto, H. et al. 1988. Contingent negative variation (CNV) and the psychological effects of odour. In *Perfumery—The Psychology and Biology of Fragrance*, edited by S. Van Toller and G. H. Dodd. London, U.K.: Chapman and Hall.

Travis, F. and Tecce, J. J. 1998. Effects of distracting stimuli on CNV amplitude and reaction time. *International Journal of Psychophysiology* 31(1):45–50.

Umezu, T. 1999. Anticonflict effects of plant-derived essential oils. *Pharmacology, Biochemistry and Behavior* 64:35–40.

Umezu, T., Ito, H., Nagano, K. et al. 2002. Anticonflict effects of rose oil and identification of its active constituents. *Life Sciences* 72:91–102.

Umezu, T., Nagano, K., Ito, H. et al. 2006. Anticonflict effects of lavender oil and identification of its active constituents. *Pharmacology Biochemistry Behavior* 85(4):713–721.

Vale, T. G., Furtado, E. C., Santos Jr., J. G., and Viana, G. S. B. 2002. Central effects of citral, myrcene and limonene, constituents of essential oil chemotypes from *Lippia alba* (Mill.) N. E. Brown. *Phytomedicine* 9:709–714.

Vale, T. G., Matos, F. J. A., Lima, T. C., and Viana, G. S. B. 1999. Behavioral effects of essential oils from *Lippia alba* (Mill.) N.E.Brown chemotypes. *Journal of Ethnopharmacology* 67:127–133.

Valnet, J. 1990. *The Practice of Aromatherapy*. Rochester, NY: Inner Traditions.

Van Toller, S., Behan, J., Howells, P., Kendal-Reed, M., and Richardson, A. 1993. An analysis of spontaneous human cortical EEG activity to odors. *Chemical Senses* 18(1):1–16.

Veira, R. F., Grayer, R. J., Paton, A., and Simon, J. E. 2001. Genetic diversity of *Ocimum gratissimum* L. based on volatile oil constituents, flavonoids and RAPD markers. *Biochemical Systematics and Ecology* 29:287–304.

Viana, G. S. B., Vale, T. G., and Matos, F. J. A. 2000. Anticonvulsant activity of essential oils and active principles from chemotypes of *Lippia alba* (Mill.) N. E. Brown. *Biological and Pharmaceutical Bulletin* 23:1314–1317.

Vohora, S. B., Shah, S. A., and Dandiya, P. C. 1990. Central nervous system studies on an ethanol extract of *Acorus calamus* rhizomes. *Journal of Ethnopharmacology* 28:53–62.

Walla, P., Hufnagl, B., Lehrner, J. et al. 2002. Evidence of conscious and subconscious olfactory information processing during word encoding: A magnetoencephalographic (MEG) study. *Brain Research, Cognitive Brain Research* 14(3):309–316.

Walla, P., Hufnagl, B., Lehrner, J. et al. 2003a. Olfaction and depth of word processing: A magnetoencephalographic study. *Neuroimage* 18(1):104–116.

Walla, P., Hufnagl, B., Lehrner, J. et al. 2003b. Olfaction and face encoding in humans: A magnetoencephalographic study. *Brain Research, Cognitive Brain Research* 15(2):105–115.

Walla, P., Mayer, D., Deecke, L., and Lang, W. 2005. How chemical information processing interferes with face processing: A magnetoencephalographic study. *Neuroimage* 24(1):111–117.

Walschburger, P. 1976. Zur Beschreibung von Aktivierungsprozessen: Eine Methodenstude zur psychophysiologischen Diagnostik. Dissertation, Philosophische Fakultät, Albert-Ludwigs-Universität, Freiburg (Breisgau), Germany.

Walter, W. G., Cooper, R., Aldridge, V. J., McCallum, W. C., and Winter, A. L. 1964. Contingent negative variation: An electric sign of sensorimotor association and expectancy in the human brain. *Nature* 203:380–384.

Warm, J. S., Dember, W. N., and Parasuraman, R. 1991. Effects of olfactory stimulation on performance and stress in a visual sustained attention task. *Journal of the Society of Cosmetic Chemists* 42:199–210.

Wie, M. B., Won, M. H., Lee, K. H. et al. 1997. Eugenol protects neuronal cells form excitotoxic and oxidative injure in primary cortical cultures. *Neuroscience Letters* 225:93–96.

Wiesmann, M., Yousry, I., Heuberger, E. et al. 2001. Functional magnetic resonance imaging of human olfaction. *Neuroimaging Clinics of North America* 11(2):237–250.

Willander, J. and Larsson, M. 2007. Olfaction and emotion: The case of autobiographical memory. *Memory and Cognition* 35(7):1659–1663.

Won, M. H., Lee, K. H., Kim, Y. H. et al. 1998. Postischemic hypothermia induced by eugenol protects hippocampal neurons form global ischemia in gerbils. *Neuroscience Letters* 254:101–104.

Yagyu, T. 1994. Neurophysiological findings on the effects of fragrance: Lavender and jasmine. *Integrative Psychiatry* 10:62–67.

Ye, H., Ji, J., Deng, C. et al. 2006. Rapid analysis of the essential oil of *Acorus tatarinowii* Schott by microwave distillation, SPME, and GC-MS. *Chromatographia* 63:591–594.

Zhang, H., Han, T., Yu, C. H. et al. 2007. Ameliorating effects of essential oil from *Acori graminei* rhizoma on learning and memory in aged rats and mice. *Journal of Pharmacy and Pharmacology* 59(2):301–309.

Zilles, K. and Rehkämpfer, G. 1998. *Funktionelle Neuroanatomie*. 3. Aufl. Berlin, Germany: Springer-Verlag.

Zimbardo, P. G., Weber, A. L., and Hohnson, R. L. 2003. *Psychology—Core Concepts*, 4th edn. Boston, MA: Allyn and Bacon.

13 Phytotherapeutic Uses of Essential Oils

Robert Harris

CONTENTS

- 13.1 Introduction 382
- 13.2 Allergic Rhinopathy 383
- 13.3 Antihypotensive 383
- 13.4 Anticarcinogenic 384
- 13.5 Anti-Inflammatory 384
- 13.6 Antimicrobial 385
    - 13.6.1 Antibacterial 385
        - 13.6.1.1 Acne 385
        - 13.6.1.2 Bacterial Vaginosis 385
        - 13.6.1.3 MRSA 386
    - 13.6.2 Antifungal 386
    - 13.6.3 Antiviral 388
    - 13.6.4 Microbes of the Oral Cavity 389
        - 13.6.4.1 Activity of Listerine® against Plaque and/or Gingivitis 389
        - 13.6.4.2 Antiviral Listerine® 390
        - 13.6.4.3 Activity of Essential Oils 390
    - 13.6.5 Controlling Microflora in Atopic Dermatitis 392
    - 13.6.6 Odor Management for Fungating Wounds 392
- 13.7 Anxiety 393
- 13.8 Cognitive Performance 394
- 13.9 Dissolution of Hepatic and Renal Stones 394
    - 13.9.1 Gall and Biliary Tract Stones 394
    - 13.9.2 Renal Stones 395
- 13.10 Functional Dyspepsia 396
- 13.11 Gastroesophageal Reflux 397
- 13.12 Hyperlipoproteinemia 397
- 13.13 Insecticidal 398
    - 13.13.1 Acaricidal Activity 398
    - 13.13.2 Pediculicidal Activity 400
- 13.14 Irritable Bowel Syndrome 401
- 13.15 Medical Examinations 402
- 13.16 Mucositis 403
- 13.17 Nausea 403
- 13.18 Pain Relief 404
    - 13.18.1 Dysmenorrhea 404
    - 13.18.2 Headache 405
    - 13.18.3 Infantile Colic 405
    - 13.18.4 Joint Physiotherapy 405

        13.18.5 Muscle Strain..406
        13.18.6 Nipple Pain 406
        13.18.7 Osteoarthritis...406
        13.18.8 Postherpetic Neuralgia 406
13.19 Postoperative Pain..407
        13.19.1 Breast Biopsy...407
        13.19.2 Cesarean Section 407
        13.19.3 Episiotomy...407
        13.19.4 Gastric Banding...408
        13.19.5 Prostatitis...408
13.20 Pruritus 408
13.21 Recurrent Aphthous Stomatitis..408
13.22 Respiratory Tract 409
        13.22.1 Menthol..409
                13.22.1.1 Antitussive 409
                13.22.1.2 Nasal Decongestant 410
                13.22.1.3 Inhibition of Respiratory Drive and Respiratory Comfort..................410
                13.22.1.4 Bronchodilation and Airway Hyperresponsiveness...................411
                13.22.1.5 Summary 411
        13.22.2 1,8-Cineole...411
                13.22.2.1 Antimicrobial...412
                13.22.2.2 Antitussive 412
                13.22.2.3 Bronchodilation 413
                13.22.2.4 Mucolytic and Mucociliary Effects..413
                13.22.2.5 Anti-Inflammatory Activity 414
                13.22.2.6 Pulmonary Function...415
                13.22.2.7 Summary 417
        13.22.3 Treatment with Blends Containing Both Menthol and 1,8-Cineole......................417
13.23 Snoring...418
13.24 Swallowing Dysfunction..419
13.25 Conclusion 419
References..419

13.1 INTRODUCTION

For many, the term "aromatherapy" originally became associated with the concept of the holistic use of essential oils to promote health and well-being. As time has progressed and the psychophysiological effects of essential oils have been explored further, their uses to reduce anxiety and aid sedation have also become associated with the term. This is especially so since the therapy has moved into the field of nursing, where such activities are of obvious benefit to patients in a hospital environment. More importantly, the practice of aromatherapy (in English-speaking countries) is firmly linked to the inhalation of small doses of essential oils and their application to the skin in high dilution as part of an aromatherapy massage.

This chapter is concerned with the medical use of essential oils, given to the patient by all routes of administration to treat specific conditions and in comparably concentrated amounts. Studies that use essential oils in an aromatherapy-like manner, for example, to treat anxiety by essential oil massage, are therefore excluded here. It is true that such studies have value in a medical setting and the use of aromatherapy to treat the symptoms of dementia is a good example (11 clinical trials), but they are detailed in Chapter 15.

Of the literature published in peer-reviewed journals over the last 30 years, only a small percentage concerns the administration of essential oils or their components to humans in order to treat disease processes. These reports are listed in the succeeding text in alphabetical order of their activity. The exception is the section on the respiratory tract, where the many activities of the two principal components (menthol and 1,8-cineole) are discussed and related to respiratory pathologies.

All of the references cited are from peer-reviewed publications; a minority is open to debate regarding methodology and/or interpretation of results but this is not the purpose of this compilation. Reports of individual case studies have been omitted.

13.2 ALLERGIC RHINOPATHY

Allergic rhinitis is an inflammatory condition of the nasal airways caused by the inhalation of an allergen by someone with a sensitized immune system. When caused by grass pollens, it is known as hay fever. Immunoglobulin E production is triggered, and the antibody binds to mast cells and basophils, leading to the release of inflammatory mediators. Systemic, intranasal, and topical corticosteroids, along with antihistamines, are used in treatment regimes, but there are concerns about side effects.

In a proof of concept study, a nasal spray was made from the essential oil of *Artemisia abrotanum* L. (4 mg/mL) and flavonoid extracts (2.5 µg/mL) from the same plant. The essential oil consisted primarily of 1,8-cineole and davanone at approximately 40% and 50%, respectively. Apart from a spasmolytic activity (Perfumi et al., 1995), little is known about the biological activity of davanone. The flavonoids present were thought to inhibit histamine release and interfere with arachidonic acid metabolism. The nasal spray was self-administered by 12 patients with allergic rhinitis, allergic conjunctivitis, and/or bronchial obstructive disease. They were instructed to use 1–2 puffs in each nostril at the first sign of symptoms, to a maximum of six treatments per day. All patients experienced rapid and significant relief of nasal symptoms, and for those with allergic conjunctivitis, a significant relief of subjective eye symptoms was also experienced. Three of the six patients with bronchial obstructive disease experienced rapid and clinically significant bronchial relief (Remberg et al., 2004).

A novel nasal spray formulation was prepared using a lemon extract, *Aloe barbadensis* juice, propoli, and the essential oils of *Ravensara aromatica* and *Melaleuca quinquenervia*. One hundred patients suffering from vasomotor allergic rhinopathy with no ongoing treatment were recruited for a clinical trial. Basic nasal cytology was performed, and then the treatment group was administered with the nasal spray three times daily for 10 days. A small control group of 10 patients was given a saline spray. Cytological examination was performed again after 15 and 30 days. In all test subjects, a marked regression of symptoms was noted that reduced to zero symptoms at the end of treatment. Rhinocytograms demonstrated the absence of neutrophils and lymphocytes and a reduction of more than 50% of eosinophils and mast cells when compared to results before treatment. In the control group, the inflammatory cell count remained unchanged, as did the allergic symptoms. Although the trial was small in size, it was concluded that the spray was a good alternative to conventional therapies for treatment of perennial and seasonal allergic rhinopathy (Patel et al., 2007; Ferrara et al., 2012).

13.3 ANTIHYPOTENSIVE

Several papers have reported the hypotensive action of inhaled essential oils due to relief of stress and/or anxiety. There are many reports of essential oils and individual components producing a hypotensive effect through intravenous (IV) injection in animals, for example, *Mentha x villosa* (Lahlou et al., 2001), *Aniba canelilla* (Lahlou et al., 2005), *Croton zehntneri* (de Siqueira et al., 2006), and *Cymbopogon winterianus* (de Menezes et al., 2010). A review of hypotensive components administered by IV found that terpene alcohols were more effective than terpene hydrocarbons (Menezes et al., 2010). There is little evidence, other than anecdotal, that essential oils may possess an antihypotensive effect.

The essential oil of *Rosmarinus officinalis* has been listed in old herbal publications as indicated in states of "cardiovascular weakness." A Spanish prospective study used the 1,8-cineole chemotype of rosemary essential oil with 32 patients diagnosed with hypotension. Evaluation was performed by regular measurement of systolic and diastolic blood pressure levels and the completion of a Short Form (SF)-36 Health Survey to determine various physical and mental parameters. In the first 12 weeks, patients were given 1 mL of placebo (aromatic olive oil) every 8 h. During the treatment phase (44 weeks), the patients were given 1 mL of rosemary essential oil every 8 h. In a posttreatment phase, the placebo was again administered for 12 weeks. Both blood pressure variables demonstrated a significant clinical antihypotensive effect of the essential oil during the whole treatment period. After the treatment, pressure levels returned to those prior to commencing, with no rebound effect. There were significant differences in pre- and posttreatment values of both physical and mental summaries, which were directly indicative of improvements relating to the variation in blood pressure. No adverse effects were noted during the treatment (Fernandez et al., 2014).

13.4 ANTICARCINOGENIC

Despite the popularity of in vitro experimentation concerning the cellular mechanisms of carcinogenic prevention by essential oil components (mainly by inducing apoptosis), there is no evidence that the direct administration of essential oils can cure cancer. There is evidence to suggest that the mevalonate pathway of cancer cells is sensitive to the inhibitory actions of dietary plant isoprenoids (e.g., Elson and Yu, 1994; Duncan et al., 2005). Animal testing has shown that some components can cause a significant reduction in the incidence of chemically induced cancers when administered before and during induction (e.g., Reddy et al., 1997; Uedo et al., 1999).

Phase II clinical trials have all involved perillyl alcohol. Results demonstrated that despite preclinical evidence, there appeared to be no anticarcinogenic activity in cases of advanced ovarian cancer (Bailey et al., 2002), metastatic colorectal cancer (Meadows et al., 2002), and metastatic breast cancer (Bailey et al., 2008). Only one trial has demonstrated antitumor activity as evidenced by a reduction of tumor size in patients with recurrent malignant gliomas (Orlando da Fonseca et al., 2008).

13.5 ANTI-INFLAMMATORY

The topical use of essential oils to treat inflammatory conditions has been investigated for many years, but with few exceptions, always using in vitro or animal models. The activity of monoterpenes in particular has been summarized (de Cassia da Silveira et al., 2013).

The potential topical anti-inflammatory activity of *Coriandrum sativum* essential oil has been investigated in human subjects. Using a monocentric, double-blind trial, erythema was initiated on the backs of 40 volunteers using ultraviolet B (UVB) radiation. The test areas were subsequently treated under occlusion with a lipolotion containing 0.5% or 1% coriander essential oil for 47 h. Betamethasone valerate and hydrocortisone were used as controls, with the vehicle alone as a placebo. The erythema was measured photometrically after 48 h. The 0.5% coriander essential oil significantly reduced the UV-induced erythema in comparison to the placebo but was not as effective as hydrocortisone. A mild anti-inflammatory effect was thus demonstrated that could be used in the concomitant treatment of inflammatory skin diseases (Reuter et al., 2008).

A randomized, double-blind trial investigated the clinical effects of a spray containing the essential oils of *Eucalyptus citriodora*, *E. globulus*, *Mentha piperita*, *Origanum syriacum*, and *R. officinalis* (3% in a polysorbate solvent). Sixty adults with upper respiratory tract infection (URTI) symptoms and a clinical diagnosis of pharyngotonsillitis, viral laryngitis, or tracheitis were given either the test spray or a placebo of polysorbate and lemon additive. Four sprays to the throat were given every 5 min for 20 min, and then the major symptoms of sore throat, hoarseness, and cough were appraised. After this, the patients applied the spray at home five times a day for 3 days. The symptoms were again assessed. Twenty minutes after spray use, the test group reported a significant improvement in the severity of their URTI symptoms in comparison to the placebo group. Both groups reported significantly improved symptoms after 3 days, and this was attributed to the natural course of URTI symptoms over time (Ben-Arye et al., 2011). The anti-inflammatory and analgesic effects of the essential oils explained the rapid benefits, although a direct antiviral activity cannot be excluded.

13.6 ANTIMICROBIAL

Considering that the majority of essential oil research is directed toward antimicrobial activity, there is a surprising lack of corresponding in vivo human trials. This is disappointing since the topical and systemic application of essential oils to treat infection is a widespread practice among therapists with (apparently) good results.

13.6.1 ANTIBACTERIAL

13.6.1.1 Acne

Antibiotics that affect *Propionibacterium acnes* are a standard treatment for acne, but antibiotic resistance is becoming prevalent. A preliminary study of 126 patients showed that topical 2% essential oil of *Ocimum gratissimum* (thymol chemotype) in a hydrophilic cream base was more effective than 10% benzyl peroxide lotion at reducing the number of lesions when applied twice daily for 4 weeks (Orafidiya et al., 2002).

In a randomized, single-blind, parallel group-controlled trial, the same group examined the effects of the addition of aloe vera gel at varying concentrations to the *O. gratissimum* cream and compared its activity with 1% clindamycin phosphate. In the 84 patients with significant acne, it was found that increasing the aloe gel content improved efficacy; the essential oil preparations formulated with undiluted or 50% aloe gels were more effective at reducing lesions than the reference product. The aloe vera gels alone had minimal activity (Orafidiya et al., 2004).

A later report judged the efficacy of a 5% *Melaleuca alternifolia* gel in the amelioration of mild to moderate acne, since a previous study (Raman et al., 1995) had demonstrated the effectiveness of tea tree oil components against *P. acnes*. The randomized, double-blind, placebo-controlled trial used 60 patients who were given the tea tree oil gel or the gel alone twice daily for 45 days. The total acne lesion count was significantly reduced by 43.64%, and the acne severity index was significantly reduced by 40.49% after the tea tree oil treatment, as compared to the placebo scores of 12.03% and 7.04%, respectively (Enshaieh et al., 2007).

A gel containing 1% essential oil of *Copaifera langsdorffii* Desf. was used in a small, double-blind study using 10 patients. The copaiba gel or placebo gel was applied to different regions of the face twice daily for 21 days. These regions were photographed before and after treatment, the affected areas occupied by acne pustules being measured by computer software. Although the placebo itself produced a decreasing effect, the copaiba gel caused a highly significant decrease in the surface area affected with acne. It was postulated that copaiba essential oil might be used as a treatment for mild acne (da Silva et al., 2012).

13.6.1.2 Bacterial Vaginosis

An imbalance of the normal vaginal flora, with overgrowth of anaerobic bacteria and a reduction of lactobacilli, causes bacterial vaginosis. In women of child-bearing age, it is the most common vaginal infection. Metronidazole is effective in the management of bacterial vaginosis, but it does have side effects including discharge, vulvovaginal irritation, and candidiasis.

The first mention of the use of an essential oil in the treatment of bacterial vaginosis was made when *M. alternifolia* was suggested instead of nitroimidazoles during pregnancy (Blackwell, 1991). The antibacterial activity of tea tree oil was examined in vitro against the microorganisms associated with bacterial vaginosis (e.g., *Bacteroides*, *Prevotella*, *Fusobacteria*, *Gardnerella*, and anaerobic Gram-positive cocci). It was found that these bacteria were susceptible to the essential oil while the commensal lactobacilli were relatively resistant (Hammer et al., 1999). *Artemisia princeps* essential oil was found to ameliorate bacterial vaginosis in animal models (Trinh et al., 2011), but no human trials were conducted with either essential oils.

A commercial vaginal cream containing the essential oil of *Zataria multiflora* (Lucorex®) was compared to metronidazole gel in the treatment of 90 women diagnosed with bacterial vaginosis by Gram stain and the presence of 3 of Amsel's criteria (vaginal pH higher than 4.5, vaginal epithelial cells in vaginal fluid, amine odor after the addition of KOH to the fluid, and a homogenous discharge. Both groups were given the cream or gel to apply for 5 consecutive nights. Except for itching, all signs and symptoms were significantly decreased in the *Z. multiflora* group. All signs and symptoms were significantly decreased in the metronidazole group. There was a significant improvement in Amsel's criteria for both groups. In the essential oil group, 14% of users reported a vaginal burning sensation. Overall, the therapeutic effects of *Z. multiflora* essential oil were considered similar to metronidazole. It could be used as an alternative treatment, especially in those patients suffering the side effects of the antibiotic (Simbar et al., 2008).

13.6.1.3 MRSA

A number of papers have demonstrated the in vitro effects of various essential oils against methicillin-resistant *Staphylococcus aureus* (MRSA), for example, *Lippia origanoides* (Dos Santos et al., 2004), *Backhousia citriodora* (Hayes and Markovic, 2002), *M. piperita*, *M. arvensis* and *Mentha spicata* (Imai et al., 2001), and *M. alternifolia* (Carson et al., 1995). There have been no trials involving the use of essential oils to combat active MRSA infections, although there have been two studies involving the use of tea tree oil as a topical decolonization agent for MRSA carriers.

A pilot study compared the use of 2% mupirocin nasal ointment and triclosan body wash (routine care) with 4% *M. alternifolia* essential oil nasal ointment and 5% tea tree oil body wash in 30 MRSA patients. The interventions lasted a minimum of 3 days, and screening for MRSA was undertaken 48 and 96 h posttreatment from sites previously colonized by the bacteria. There was no correlation between length of treatment and outcome in either group. Of the tea tree oil group, 33% were initially cleared of MRSA carriage, while 20% remained chronically infected at the end of treatment; this was in comparison to routine care group of 13% and 53%, respectively. The trial was too small to provide significant results (Caelli et al., 2000).

A randomized, controlled trial compared the use of a standard regime for MRSA decolonization with *M. alternifolia* essential oil. The 5-day study involved 236 patients. The standard treatment group was given 2% mupirocin nasal ointment three times daily, 4% chlorhexidine gluconate soap as a body wash once daily, and 1% silver sulfadiazine cream for skin lesions, wounds, and leg ulcers once daily. The tea tree oil group received 10% essential oil cream three times daily to the nostrils and to specific skin sites and 5% essential oil body wash at least once daily. In the tea tree oil group, 41% were cleared of MRSA as compared to 49% using the standard regime; however, this was not a significant difference. Tea tree oil cream was significantly less effective at clearing nasal carriage than mupirocin (47% compared to 78%) but was more effective at clearing superficial sites than chlorhexidine or silver sulfadiazine (Dryden et al., 2004).

13.6.2 ANTIFUNGAL

The essential oil of *Citrus aurantium* var. *amara* was used to treat 60 patients with tinea corporis, cruris, or pedis. One group received a 25% bitter orange oil emulsion (BO) thrice daily, a second group was treated with 20% bitter orange oil in alcohol (BOa) thrice daily, and a third group used undiluted bitter orange oil once daily. The trial lasted 4 weeks, and clinical and mycological examinations were performed every week until cure, which was defined as an elimination of signs and symptoms. In the BO group, 80% of patients were cured in 1–2 weeks and the rest within

2–3 weeks. By using BOa, 50% of patients were cured in 1–2 weeks, 30% in 2–3 weeks, and 20% in 3–4 weeks. With the undiluted essential oil, 25% of patients did not continue treatment, 33.3% were cured in 1 week, 60% in weeks 1–2, and 6.7% in 2–3 weeks.

A double-blind, randomized, placebo-controlled trial investigated the efficacy of 2% butenafine hydrochloride cream with added 5% *M. alternifolia* essential oil in 60 patients with toenail onychomycosis. After 16 weeks, 80% of patients in the treatment group were cured, as opposed to none in the control group (Syed et al., 1999). However, butenafine hydrochloride is a potent antimycotic in itself, and the results were not compared with this product when used alone.

After an initial in vitro study, which showed that the essential oil of *Eucalyptus pauciflora* had a strong fungicidal activity against *Epidermophyton floccosum*, *Microsporum canis*, *M. nanum*, *M. gypseum*, *Trichophyton mentagrophytes*, *T. rubrum*, *T. tonsurans*, and *T. violaceum*, an in vivo trial was commenced. Fifty patients with confirmed dermatophytosis were treated with 1% v/v essential oil twice daily for 3 weeks. At the end of treatment, a cure was demonstrated in 60% of patients with the remaining 40% showing significant improvement (Shahi et al., 2000).

On the surmise that infection with *Pityrosporum ovale* is a major contributing factor to dandruff and that anti-*Pityrosporum* drugs such as nystatin were proven effective treatments, the use of 5% *M. alternifolia* essential oil was investigated. In this randomized, single-blind, parallel group study, tea tree oil shampoo or placebo shampoo was used daily for 4 weeks by 126 patients with mild to moderate dandruff. In the treatment group, the dandruff severity score showed an improvement of 41%, as compared to 11% in the placebo group. The area involvement and total severity scores also demonstrated a statistically significant improvement, as did itchiness and greasiness. Scaliness was not greatly affected. The condition resolved for one patient in each group and so ongoing application of tea tree oil shampoo was recommended for dandruff control (Satchell et al., 2002a).

Tinea versicolor (pityriasis versicolor) is a pigmentation disorder with macular lesions that occurs especially in individuals with decreased immunity. It can affect all regions of the body. The causative agents are *Malassezia* spp. (syn. *Pityrosporum*). A randomized trial was conducted using *Cymbopogon citratus* essential oil at 1.25 µL/mL concentration in a cream and shampoo. Patients were divided into two groups: 29 were treated with the essential oil, and the same number with similar formulations but using 2% ketoconazole as the active ingredient. The shampoo was applied three times a week and the cream twice a day, both for 40 days. The mycological cure rate was 60% for the essential oil group compared to over 80% for the ketoconazole group. Nevertheless, the results proved the potential of *C. citratus* essential oil in the treatment of such infections (Carmo et al., 2013).

For inclusion in a randomized, double-blind, controlled trial, 158 patients with the clinical features of intertriginous tinea pedis and confirmed dermatophyte infection were recruited. They were administered 25% or 50% *M. alternifolia* essential oil (in an ethanol and polyethylene glycol vehicle) or the vehicle alone, twice daily for 4 weeks. There was an improvement in the clinical severity score, falling by 68% and 66% in the 25% and 50% tea tree oil groups, in comparison to 41% for the placebo. There was an effective cure in the 25% and 50% tea tree oil and placebo groups of 48%, 50%, and 13%, respectively. The essential oil was less effective than standard topical treatments (Satchell et al., 2002b).

The anti-*Candida* properties of *Z. multiflora* essential oil and its active components (thymol, carvacrol, and eugenol) were demonstrated in vitro by Mahmoudabadi in 2006. A randomized, clinical trial was conducted using 86 patients with acute vaginal candidiasis. They were treated with a cream containing 0.1% *Z. multiflora* essential oil or 1% clotrimazole once daily for 7 days. Statistically significant decreases in vulvar pruritus (80.9%), vaginal pruritus (65.5%), vaginal burning (73.95), urinary burning (100%), and vaginal secretions (90%) were obtained by the essential oil treatment as compared to the clotrimazole treatment of 73.91%, 56.7%, 82.1%, 100%, and 70%, respectively. In addition, the *Z. multiflora* cream reduced erythema and satellite vulvar lesions in

100% of patients, vaginal edema in 100%, vaginal edema in 83.3%, and vulvovaginal excoriation and fissures in 92%. The corresponding results for clotrimazole were 100%, 100%, 76%, and 88%. In terms of overall efficacy, the rates of improvement were 90% and 74.8% for the *Z. multiflora* and clotrimazole groups, respectively. Use of the cream alone provided no significant changes (Khosravi et al., 2008).

13.6.3 Antiviral

The majority of in vitro studies that have been conducted indicate that many essential oils possess antiviral properties, but they affect only enveloped viruses and only when they are in the free state, that is, before the virus is attached to or has entered the host cell (e.g., Schnitzler et al., 2008; Astani et al., 2010). This is in contrast to the majority of synthetic antiviral agents, which either bar the complete penetration of viral particles into the host cell or interfere with viral replication once the virus is inside the cell.

However, recent investigations indicate the possibility that antiviral essential oil activity may involve other processes, such as the following:

1. Inhibition of viral protein synthesis at the posttranscription level (Wu et al., 2010)
2. Causing dysfunction of the cellular cytoskeleton thereby disrupting transport of viruses in caveolae and other pH-independent vesicles to their target organelles (Morise et al., 2010)
3. Prevention of the release of viruses from endosomes within the host cell by interference with the acidification process, thus halting infection (Garozzo et al., 2011)
4. Inhibition of the early (Liao et al., 2013) and late (Paulpandi et al., 2012) stages of viral ribonucleic acid (RNA) replication

A randomized, investigator-blinded, placebo-controlled trial used 6% *M. alternifolia* essential oil gel to treat recurrent herpes labialis. It was applied five times daily and continued until reepithelialization occurred, and the polymerase chain reaction (PCR) for *Herpes simplex* virus was negative for 2 consecutive days. The median time to reepithelialization after treatment with tea tree oil was 9 days as compared to 12.5 days with the placebo, which is similar to reductions caused by other topical therapies. The median duration of PCR positivity was the same for both groups (6 days) although the viral titers appeared slightly lower in the oil group on days 3 and 4. None of the differences reached statistical significance, probably due to the small group size (Carson et al., 2001).

Children under 5 were enrolled in a randomized trial to test a 10% v/v solution of the essential oil of *B. citriodora* against molluscum contagiosum (caused by *Molluscipoxvirus*). Of the 31 patients, 16 were assigned to the treatment group and the rest to the control of olive oil. The solutions were applied directly to the papules once daily at bedtime for 21 days or until the lesions had resolved. In the essential oil group, five children had a total resolution of lesions, and four had reductions of greater than 90% at the end of 21 days. In contrast, none of the control group had any resolution or reduction of lesions by the end of the study period (Burke et al., 2004).

A study was conducted on 60 patients who were chronic carriers of hepatitis B or C. The essential oils of *Cinnamomum camphora* ct 1,8-cineole, *Daucus carota*, *Ledum groelandicum*, *Laurus nobilis*, *Helichrysum italicum*, *Thymus vulgaris* ct thujanol, and *M. quinquenervia* were used orally in various combinations. They were used as a monotherapy or as a complement to allopathic treatment. The objectives of treatment were normalization of transaminase levels, reduction of viral load, and stabilization or regression of fibrosis. In patients with hepatitis C given bitherapy and essential oils, there was an improvement to treatment of 100%. With essential oil monotherapy, improvements were noted in 64% of patients with hepatitis C, and there were two cures of hepatitis B (Giraud-Robert, 2005).

13.6.4 MICROBES OF THE ORAL CAVITY

The activities of essential oils against disease-producing microbes in the oral cavity have been documented separately because there are numerous reports of relevance. The easy administration of essential oils in mouth rinses, gargles, and toothpastes, and the success of such commercial preparations, has no doubt led to the popularity of this research.

The in vitro activities of essential oils against the oral microflora are well documented, and these include effects on cariogenic and periodontopathic bacteria. One example is the in vitro activity of *Leptospermum scoparium*, *M. alternifolia*, *E. radiata*, *Lavandula officinalis*, and *R. officinalis* against *Porphyromonas gingivalis*, *Actinobacillus actinomycetemcomitans*, *Fusobacterium nucleatum*, and *Streptococcus mutans*. The essential oils inhibited all of the test bacteria, acting bactericidally except for *L. officinalis*. In addition, significant adhesion-inhibiting activity was shown against *S. mutans* by all essential oils and against *P. gingivalis* by tea tree and manuka (Takarada et al., 2004).

There have been at least six in vivo studies concerning the activity of individual essential oils against the microflora of the oral cavity. In addition, a review of the literature finds a surprising number of in vivo papers that detail the activities of "an essential oil mouth rinse." Closer examination reveals that the essential oil mouth rinse is the commercial product Listerine®. This product was originally formulated in 1879 and contained the essential oils of peppermint, thyme, eucalyptus, and wintergreen. The active ingredients were later changed to the individual components of 1,8-cineole (0.092%), menthol (0.042%), methyl salicylate (0.06%), and thymol (0.64%). Although Listerine contains 21% or 26% alcohol (depending on the exact product), a 6-month study has shown that it contributes nothing to the efficaciousness of the mouth rinse (Lamster et al., 1983). For this reason, a small random selection of such papers is included in the succeeding text. Sometime, just prior to or during 2012, menthol and methyl salicylate were replaced by synthetic derivatives and so Listerine can no longer be described as a phytomedicine (Vlachojannis et al., 2013). For this reason, more recent papers have been omitted from this chapter.

13.6.4.1 Activity of Listerine® against Plaque and/or Gingivitis

An observer-blind, 4-day plaque regrowth, crossover study compared the use of Listerine® with a triclosan mouth rinse and two placebo controls in 32 volunteers. All normal hygiene procedures were suspended except for the rinses. The triclosan product produced a 45% reduction in plaque area and a 12% reduction in plaque index against its placebo, in comparison to 52% and 17%, respectively, for the essential oil rinse. The latter was thus deemed more effective (Moran et al., 1997).

A similar protocol was used to compare the effects of Listerine against an amine fluoride/stannous fluoride-containing mouth rinse (Meridol®) and a 0.1% chlorhexidine mouth rinse (Chlorhexamed®) in inhibiting the development of supragingival plaque. On day 5 of each treatment, the results from 23 volunteers were evaluated. In comparison to their placebos, the median plaque reductions were 12.2%, 23%, and 38.2% for the fluoride, essential oil, and chlorhexidine rinses, respectively. The latter two results were statistically significant (Riep et al., 1999).

After assessment for the presence of gingivitis and target pathogens (*P. gingivalis*, *F. nucleatum*, *Veillonella* spp.) and total anaerobes, 37 patients undertook a twice daily mouth rinse with Listerine for 14 days. After a washout period, the study was conducted again using a flavored hydroalcoholic placebo. The results of this randomized, double-blind, crossover study showed that the essential oil rinse significantly lowered the number of all target pathogens by 66.3%–79.2%, as compared to the control (Fine et al., 2007).

The effect of adding Listerine mouth rinse to a standard oral hygiene regime in 50 orthodontic patients was examined. The control group brushed and flossed twice daily, while the test group also used the mouth rinse twice daily. Measurements of bleeding, gingival, and plaque indices were conducted at 3 and 6 months. All three indices were significantly lowered in the test group as compared to the control at both time intervals (Tufekci et al., 2008).

The same fixed combination of essential oils that is found in Listerine mouth rinse has been incorporated into a dentifrice.

Such a dentifrice was used in a 6-month double-blind study to determine its effect on the microbial composition of dental plaque as compared to an identical dentifrice without essential oils. Supragingival plaque and saliva samples were collected at baseline and their microbial content characterized, after which the study was conducted for 6 months. The essential oil dentifrice did not significantly alter the microbial flora, and opportunistic pathogens did not emerge, nor was there any sign of developing resistance to the essential oils in tested bacterial species (Charles et al., 2000).

The same dentifrice was examined for antiplaque and antigingivitis properties in a blinded, randomized, controlled trial. Before treatment, 200 patients were assessed using a plaque index, modified gingival index, and a bleeding index. The dentifrice was used for 6 months, after which another assessment was made. It was found that the essential oil dentifrice had a statistically significant lower whole-mouth and interproximal plaque index (18.3% and 18.1%), mean gingival index (16.2% and 15.5%), and mean bleeding index (40.5% and 46.9%), as compared to the control. It was therefore proven to be an effective antiplaque and antigingivitis agent (Coelho et al., 2000).

13.6.4.2 Antiviral Listerine®

A trial was conducted to examine whether a mouth rinse could decrease the risk of viral cross contamination from oral fluids during dental procedures. Forty patients with a perioral outbreak of recurrent herpes labialis were given a 30 s mouth rinse with either water or Listerine. Salivary samples were taken at baseline, immediately following the rinse and 30 min after the rinse and evaluated for the viral titer. Infectious virions were reduced immediately to zero postrinse, and there was a continued significant reduction 30 min postrinse. The reduction by the control was not significant (Meiller et al., 2005).

13.6.4.3 Activity of Essential Oils

The antibacterial activity of the essential oil of *Lippia multiflora* was first examined in vitro for antimicrobial activity against American type culture collection (ATCC) strains and clinical isolates of the buccal flora. A significant activity was found, with an minimum bactericidal concentration (MBC) of 1/1400 for streptococci and staphylococci, 1/800 for enterobacteria and neisseria, and 1/600 for candida. A mouthwash was prepared with the essential oil at a 1/500 dilution, and this was used in two clinical trials.

The buccodental conditions of 26 French children were documented by measuring the percentage of dental surface free of plaque, gum inflammation (GI), and the papillary bleeding index (PBI). After 7 days of rinsing with the mouthwash for 2 min, the test group was found to have a reduction of dental plaque in 69% of cases and a drop in PBI with a clear improvement of GI in all cases. The second trial was conducted in the Cote d'Ivoire with 60 adult patients with a variety of conditions. After using the mouthwash after every meal for 5 days, it was found that candidiasis had disappeared in most cases, gingivitis was resolved in all patients, and 77% of dental abscesses had resorbed (Pélissier et al., 1994).

Fluconazole-refractory oropharyngeal candidiasis is a common condition in human immunodeficiency virus (HIV)-positive patients. Twelve such patients were treated with 15 mL of a *M. alternifolia* oral solution (Breath-Away) four times daily for 2 weeks, in a single center, open-label clinical trial. The solution was swished in the mouth for 30–60 s and then expelled, with no rinsing for at least 30 min. Clinical assessment was carried out on days 7 and 14, and also on days 28 and 42 of the follow-up. Two patients were clinically cured and six were improved after the therapy; however, four remained unchanged and one deteriorated. The overall clinical response rate was thus 67% and was considered as a possible alternative antifungal treatment in such cases (Jandourek et al., 1998).

A clinical pilot study compared the effect of 0.34% *M. alternifolia* essential oil solution with 0.1% chlorhexidine on supragingival plaque formation and vitality. Eight subjects participated, with a 10-days washout period between each treatment regime of 1 week. The plaque area was calculated using a stain, and plaque vitality was estimated using a fluorescence technique. Neither of these parameters was reduced by the tea tree oil treatment (Arweiler et al., 2000).

A gel containing 2.5% *M. alternifolia* essential oil was used in a double-blind, longitudinal noncrossover trial and compared to a chlorhexidine gel positive control and a placebo gel in the treatment of plaque and chronic gingivitis. The gels were applied as a dentifrice twice daily by 49 subjects for 8 weeks, and treatment was assessed using a gingival index (GI), a PBI, and a plaque-staining score. The tea tree group showed a significant reduction in PBI and GI scores, although plaque scores were not reduced. It was apparent that the tea tree gel decreased the level of gingival inflammation more than the positive or negative controls (Soukoulis and Hirsch, 2004).

A mouth care solution consisting of an essential oil mixture of *M. alternifolia*, *M. piperita*, and *Citrus limon* in a 2:1:2 ratio diluted in water to a 0.125% solution was used to treat oral malodor in 32 intensive care unit patients, 13 of whom were ventilated. The solution was used to clean the teeth, tongue, and oral cavity twice during day 1. The level of malodor was assessed by a nurse using a visual analog scale (VAS), and volatile sulfur compounds (VSC) were measured via a probe in the mouth, before, 5, and 60 min after treatment. On the second day, the procedure was repeated using benzydamine hydrochloride (BH), which is normally used for oral hygiene, instead of essential oil solution. The perception of oral malodor was significantly lowered after the essential oil treatment but not after the BH treatment. There was a decrease in VSC levels at 60 min for both treatment groups, but not after 5 min for the oil mixture. The results suggested that just one session with the essential oil mixture could improve oral malodor and VSC in intensive care patients (Hur et al., 2007).

Halitosis is a common problem and originates from the oral cavity due to conditions such as poor oral hygiene, deep caries, endodontic lesions, and periodontitis. The malodor occurs as a result of bacterial metabolism of amino acids, producing compounds such as hydrogen sulfide and methyl mercaptan. Hinokitiol (β-thujaplicin), a component found in Cupressaceae essential oils, has demonstrated in vitro antibacterial activity against oral microorganisms (Saeki et al., 1989) and has demonstrated in vivo lipopolysaccharide-induced nitric oxide and tumor necrosis factor-α production (Shih et al., 2012). It has also been used to treat periodontal disease in students (Suzuki et al., 2008).

Eighteen subjects complaining of halitosis took part in a randomized, open-label pilot study. The test group cleaned their mouths three times daily for 4 weeks with an oral care gel containing 0.01%–0.2% hinokitiol. The control group used a gel containing cetylpyridinium chloride. Malodor was assessed by an organoleptic test and gas chromatography of breath. Major clinical outcomes evaluated were plaque control, periodontal health, and tongue coating. In the hinokitiol group, the organoleptic test, the level of VSC, frequency of bleeding, and the plaque index all significantly improved, whereas only the organoleptic test scores improved in the control group (Iha et al., 2013). The antimicrobial and anti-inflammatory activities of hinokitiol were credited with the improvement in periodontal parameters.

The essential oil of *Lippia sidoides* (rich in thymol and carvacrol) was used in a double-blind, randomized, parallel-armed study against gingival inflammation and bacterial plaque. Fifty-five patients used a 1% essential oil solution as a mouth rinse twice daily for 7 days, and the results were compared to a positive control, 0.12% chlorhexidine. Clinical assessment demonstrated decreased plaque index and gingival bleeding scores as compared to the baseline, with no significant difference between test and control. The essential oil of *L. sidoides* was considered a safe and effective treatment.

Dental caries can quickly occur in young children and lead to mass destruction of primary dentition. The condition is linked to early contamination with *Mutans streptococci* and treatment is focused on controlling the bacteria. The essential oil of *L. sidoides* was used in a randomized trial involving 48 children with dental caries. They applied different concentrations of a gel or mouth rinse containing 0.8%–1.4% or 0.6%–1.2% essential oil, respectively. Efficacy was compared to a mouth rinse and gel containing a thymol/carvacrol mixture isolated from the essential oil at a 7:3 ratio. The rinse and gel formulations of *L. sidoides* all produced a significant reduction in salivary *Mutans streptococci*. The phenolic mixture was also highly effective but was not superior to the whole essential oil. Rinses above 0.8% were associated with an intraoral burning sensation (Lobo et al., 2011).

13.6.5 Controlling Microflora in Atopic Dermatitis

Rarely found on healthy skin, *Staphylococcus aureus* is usually present in dry skin and is one of the factors that can worsen atopic dermatitis. Toxins and enzymes deriving from this bacteria cause skin damage and form a biofilm from fibrin and glycocalyx, which aids adhesion to the skin and resistance to antibiotics. An initial in vitro study found that a mixture of xylitol (a sugar alcohol) and farnesol was an effective agent against *S. aureus*; xylitol inhibited the formation of glycocalyx, while farnesol dissolved fibrin and suppressed *S. aureus* growth without affecting *S. epidermidis* (Masako et al., 2005a).

The same mixture of xylitol and farnesol was used in a double-blind, randomized, placebo-controlled study of 17 patients with mild to moderate atopic dermatitis to their arms. A skin care cream containing 0.02% farnesol and 5% xylitol or the cream alone was applied to either the left or right arms for 7 days. The ratio of *S. aureus* to other aerobic skin microflora was significantly decreased in the test group compared to placebo, from 74% to 41%, while the numbers of coagulase-negative staphylococci increased. In addition, skin conductance (indicating hydration of skin surface) significantly increased at the test cream sites when compared to before application and to the placebo (Masako et al., 2005b).

13.6.6 Odor Management for Fungating Wounds

Fungating wounds may be caused by primary skin carcinomas, underlying tumors, or via spread from other tissues. The malodor associated with such necrosis is caused by the presence of aerobic and anaerobic bacteria. The wounds rarely heal and require constant palliative treatment, leading to social isolation of the patients and poor quality of life.

Smell reduction with essential oils was first reported in 2004 by Warnke et al. in 25 malodorous patients with inoperable squamous cell carcinoma of the head and neck. A commercial product containing eucalyptus, grapefruit, and tea tree essential oils (Megabac®) was applied topically to the wounds twice daily. Normal medication apart from Betadine disinfection was continued. The smell disappeared completely within 2–3 days and signs of superinfection and pus secretion were reduced in the necrotic areas.

Megabac has also been used in a small pilot study (10 patients) to treat gangrenous areas, being applied via spray thrice daily until granulation tissue formed. The treatment was then continued onto newly formed split skin grafts. All wounds healed within 8 weeks and no concurrent antibiotics were used (Sherry et al., 2003).

Use of essential oils to reduce the smell of fungating wounds in 13 palliative care patients was detailed by another group the following year. *Lavandula angustifolia*, *M. alternifolia*, and *Pogostemon cablin* essential oils were used alone or in combinations at 2.5%–5% concentrations in a cream base. The treatments were effective (Mercier and Knevitt, 2005).

A further study was conducted with 30 patients suffering incurable head and neck cancers with malodorous necrotic ulcers. A custom made product (Klonemax®) containing eucalyptus, tea tree, lemongrass, lemon, clove, and thyme essential oils was applied topically (5 mL) twice daily. All patients had a complete resolution of the malodor; in addition to the antibacterial activity, an anti-inflammatory effect was also noted (Warnke et al., 2006).

The use of essential oils to treat malodorous wounds in cancer patients is becoming widespread in many palliative care units although no formal clinical trials have been conducted as yet.

13.7 ANXIETY

There have been a number of clinical trials involving the inhalation of lavender essential oil that indicate a reduction in anxiety and stress (e.g., Motomura et al., 2001; Toda and Morimoto, 2008; Kritsidima et al., 2010). One study employing oral lavender essential oil in 100 and 200 µL capsules found that heart rate variation significantly increased compared to placebo while watching an anxiety-provoking film. This suggested that lavender had anxiolytic effects under conditions of low anxiety (Bradley et al., 2009).

Silexan is a commercial preparation containing 80 mg of the essential oil of *L. angustifolia* in an immediate release soft gelatin capsule. It is authorized in Germany for the treatment of anxiety and accompanying restlessness states. A number of clinical trials have been performed.

The primary study investigated the anxiolytic effects of Silexan on 221 patients with an otherwise unspecified anxiety disorder. They were given either one capsule of Silexan daily or a placebo for 10 weeks. Outcome measures used the Hamilton Anxiety Rating Scale (HAMA), Clinical Global Impressions Scale (CGI), the SF-36 Health Survey Questionnaire (HSQ), the Pittsburgh Sleep Quality Index (PSQI), and the Zung Self-Rating Anxiety Scale (SAS). Silexan patients showed total score decreases of 59.3% and 44.7% for the HAMA and PSQI, respectively. Overall, the preparation improved general mental and physical health and provided a significant beneficial effect on the quality and duration of sleep. There were no sedative or other side effects (Kasper et al., 2010).

A general and persistent anxiety with nervousness and other symptoms is termed generalized anxiety disorder (GAD). It can lead to a chronic condition exacerbated by stressful events. Benzodiazepines are commonly used in treatment but they cause sedation, and long-term use may induce dependence or abuse. A second study (Woelk and Schläfke, 2010) compared the effects of Silexan with lorazepam in 78 patients with GAD. Over a 6-week period, they were administered daily either one Silexan capsule with a lorazepam placebo or a 0.5 mg lorazepam capsule and a Silexan placebo. Assessments were based on HAMA, HSQ, SAS, CGI, and the Penn State Worry Questionnaire. Results showed that the total HAMA score decreased by 45% and 46% in the Silexan and lorazepam groups, respectively. Other assessment scores improved to a similar extent for both groups. It was concluded that both Silexan and lorazepam had comparable positive effects in adults with GAD (Woelk and Schläfke, 2010).

A phase II trial was conducted to assess the use of Silexan in treating subthreshold anxiety in relation to somatization disorder, neurasthenia, and posttraumatic stress disorder. Fifty patients were given one Silexan capsule daily for 6 weeks. Outcomes were measured using the State-Trait Anxiety Inventory, von Zerssen's Depression Scale (D-S), the Maslach Burnout Inventory, and the HSQ. Pretreatment, 96%, 98%, 92%, and 72% of patients suffered from restlessness, depressed mood, sleep disturbances, or anxiety, respectively. During treatment, 62%, 57%, 51%, and 62% of patients, respectively, showed improvements. For all patients, the D-S score decreased by 32.7% and the HSQ increased by 48.2%. Sleep and mood improved, while morning tiredness and wake-up duration decreased. Patients demonstrated statistically significant and clinically meaningful improvements in symptoms affecting their mental health status and quality of life (Uehleke et al., 2012).

A case series was performed on eight patients suffering from major depressive disorder (MDD) and symptoms of psychomotor agitation, insomnia, and anxiety. The majority were prescribed one capsule daily of Lasea® (the active substance is Silexan) with concomitant antidepressant medication for 3 weeks. One patient received two capsules daily. The HAMA was used to measure effectiveness. The treatment reduced MDD in six cases, agitation in five, anxiety in four, and sleep onset and maintenance in three cases. It was concluded that Lasea reduces some of the anxiety-related symptoms and sleep disturbances in MDD patients (Fissler and Quante, 2014).

13.8 COGNITIVE PERFORMANCE

Acetylcholinesterase inhibitors are used for the symptomatic treatment of Alzheimer's disease. The primary symptom is a loss of memory, and the consistent neurological finding associated with this cognitive dysfunction is a cholinergic deficit. The leaves and extracts of *Salvia officinalis* and *S. lavandulifolia* have been used for millennia to improve cognitive function and combat cognitive decline (Kennedy and Scholey, 2006). In vitro research documenting the acetylcholinesterase inhibition by *S. lavandulifolia* essential oil monoterpenoids was begun by Perry et al. (2000, 2001, 2002). Since then, no clinical trials involving Alzheimer's patients have taken place, but the effect of Spanish sage essential oil on the cognitive function of healthy adults has been investigated.

The first trial involved 20 students who were given daily oral capsules containing either 50 μL each of essential oil and sunflower oil (1–3 capsules) or 100 μL of sunflower oil as placebo. Students therefore received 0, 50, 100, or 150 μL of essential oil. The Cognitive Drug Research (CDR)-computerized assessment battery was used predose and up to 6 h posttreatment. Performance was found to be enhanced at 1 and 2.5 h for both immediate and delayed memory recall. A similar second test using 25 and 50 μL of essential oil found that improved memory performance was maintained from 1 to 4 h posttreatment for the 50 μL dose only. Overall, the 25 and 150 μL doses had no significant effect (Tildesley et al., 2003).

Using daily 25 or 50 μL essential oil capsules and sunflower oil placebo, another trial of 24 volunteers over 4 days used the CDR test battery and global "quality of memory" measure to assess cognitive performance pretest and up to 6 h posttreatment. Both doses gave an improvement of the "speed memory" factor and the 25 μL dose only for the "secondary memory" factor. The higher dose produced significant positive effects in the mood factors of alertness, contentedness, and calmness, while the 25 μL dose was associated with improvements in the calmness factor only. The results suggested that the essential oil was capable of acute modulation of cognition and mood in healthy young adults (Tildesley et al., 2005).

A double-blind crossover trial administered a single 50 μL capsule of essential oil in olive oil to 36 individuals. One week later, an olive oil placebo capsule was given to the same subjects. Cognitive function was assessed by various computerized memory and attention tasks and the Cognitive Demand Battery pretest and 1 and 4 h posttest. There was improved performance of attention and secondary memory tasks that were more pronounced at 1 h postdose. Increased alertness and reduced mental fatigue were more noticeable at 4 h postdose. A separate in vitro test demonstrated the potent acetylcholinesterase inhibition of the *S. lavandulifolia* essential oil. As with the previous trials, a beneficial modulation of cholinergic function was suggested with an improved cognitive performance (Kennedy et al., 2011).

13.9 DISSOLUTION OF HEPATIC AND RENAL STONES

13.9.1 Gall and Biliary Tract Stones

Rowachol and Rowatinex are two commercial products that have been marketed for many years and are based on essential oil components. They are sometimes thought of as being the same product but in fact they are different. The compositions have changed slightly over the years and the most recently disclosed are shown in Table 13.1.

Rowachol has been in use for over 50 years for the dissolution of gallstones and biliary tract stones. There have been many published works as to its effects and at least one double-blind trial (Lamy, 1967). It has been stated that although the dissolution rate of Rowachol is not impressive, it is still much greater than could occur spontaneously (Doran and Bell, 1979). It has been employed alone as a useful therapy for common duct stones (Ellis and Bell, 1981), although improved results were demonstrated when Rowachol was used in conjunction with bile acid therapy (Ellis et al., 1981).

TABLE 13.1
Composition of Rowachol and Rowatinex Declared by the Manufacturers

Component	Rowachol	Rowatinex
α-Pinene	20.0	37.0
β-Pinene	5.0	9.0
Camphene	8.0	22.0
1,8-Cineole	3.0	4.0
Fenchone	—	6.0
Menthone	9.0	—
Borneol	8.0	15.0
Menthol	48.0	—
Anethole	—	6.0

Source: Sybilska, D. and Asztemborska, M., *J. Biochem. Biophys. Methods.*, 54, 187, 2002.

Rowachol has been shown to inhibit hepatic cholesterol synthesis mediated by a decreased hepatic S-3-hydroxy-3-methylgutaryl-CoA reductase activity (Middleton and Hui, 1982); the components mostly responsible for this activity were menthol and 1,8-cineole, with pinene and camphene having no significant effect (Clegg et al., 1980). A reduction in cholesterol crystal formation in the bile of gallstone patients has been demonstrated in a small trial using Rowachol (von Bergmann, 1987).

Two early uncontrolled trials reported that Rowachol significantly increased plasma high-density lipoprotein (HDL) cholesterol when administered to patients with low HDL cholesterol; a twofold increase was found in 10 subjects after 6 weeks of treatment (Hordinsky and Hordinsky, 1979), while a progressive increase in HDL of 14 subjects was noted, >100% after 6 months (Bell et al., 1980). This was interesting as low plasma concentrations of HDLs are associated with an elevated risk of coronary heart disease. However, a double-blind, placebo-controlled trial that administered six capsules of Rowachol daily for 24 weeks to 19 men found that there were no significant HDL-elevating effects of the treatment (Cooke et al., 1998). It is currently thought that monoterpenes have no HDL-elevating potential that is useful for disease prevention.

In vitro, a solution of 97% d-limonene was found to be 100-fold better at solubilizing cholesterol than sodium cholate. A small trial followed with 15 patients, whereby 20 mL of the d-limonene preparation, was introduced into the gallbladder via a catheter on alternate days for up to 48 days. Treatment was successful in 13 patients with gallstone dissolution sometimes occurring after three infusions. Side effects included vomiting and diarrhea (Igimi et al., 1976).

A further study was conducted in 1991 (Igimi et al.) using the same technique with 200 patients. Treatments lasted from 3 weeks to 4 months. Complete or partial dissolution of gallstones was achieved in 141, with complete disappearance of stones in 48% of cases. Epigastric pain was experienced by 168 patients and 121 suffered nausea and vomiting. Further trials have not been conducted.

13.9.2 RENAL STONES

While Rowachol is used as a measure against gallstones and biliary tract stones, Rowatinex is used in the treatment of renal stones. The mechanism of action is not known, but is thought to involve an improved renal blood supply thereby stimulating an increased urine secretion, and an antispasmodic effect to allow passage of the calculi.

The first double-blind, randomized trial was conducted in 1987 by Mukamel et al. on 40 patients with acute renal colic. In the Rowatinex group, there was a significantly higher expulsion rate of stones ≥ 3 mm in diameter in comparison to the placebo (61% and 28%, respectively). There was also a higher overall success rate in terms of spontaneous stone expulsion and/or disappearance of ureteral dilatation in the treatment group compared to placebo (78%–52%), but the difference was not statistically significant.

A second double-blind, randomized trial was conducted on 87 patients with ureterolithiasis. Four Rowatinex capsules were prescribed four times a day, the average treatment time being 2 weeks. The overall stone expulsion rate was significantly higher in the Rowatinex group as compared to placebo, 81% and 51%, respectively. Mild to moderate gastrointestinal disturbances were noted in 7 patients. It was concluded that the early treatment of ureteral stones with Rowatinex is preferable before more aggressive measures were considered (Engelstein et al., 1992).

Rowatinex has also been used with success in the removal of residual stone fragments after extracorporeal shock wave lithotripsy (ESWL), a situation that occurs in up to 72% of patients given this therapy. These fragments are important to remove since they may lead to calculus regrowth, infection, or obstruction. With 50 patients, it was found that Rowatinex decreased the number of calculi debris, reducing the number of late complications and further interventions. By day 28, 82% of patients were free of calculi, whereas this situation is normally reached after 3 months without Rowatinex treatment (Siller et al., 1998).

After undergoing ESWL, 100 patients were randomly assigned to receive either Rowatinex (100 mg 3× daily) or placebo. The results demonstrated a total calculus clearance rate of 24% for the Rowatinex group as compared to 18% for the control group, 1 month post-ESWL. There were no significant differences in other outcome measurements such as signs and symptoms (Djaladat et al., 2009). It was concluded that Rowatinex accelerated calculus passage but had no effect on overall outcome.

A minor study examined the use of Rowatinex in the management of childhood urolithiasis. Six children aged from 4 months to 5 years were administered varying doses of the preparation from 10 days to 12 weeks. All patients became stone-free with no side effects although a definite conclusion as to the efficacy of treatment could not be established due to the small patient number involved (Al-Mosawi, 2005).

A comparison of the effects of an α-blocker (tamsulosin) and Rowatinex for the spontaneous expulsion of ureter stones and pain control was undertaken using 192 patients. They were divided into three groups: analgesics only, Rowatinex with analgesics, and tamsulosin with analgesics. For ureter stones less than 4 mm in diameter, both Rowatinex and tamsulosin accelerated their excretion. The use of these two treatments also decreased the amount of analgesics required, and it was concluded that they should be considered as adjuvant regimes (Bak et al., 2007).

13.10 FUNCTIONAL DYSPEPSIA

Several essential oils have been used in the treatment of functional (nonulcer) dyspepsia. All of the published trials have concerned the commercial preparation known as Enteroplant®, an enteric-coated capsule containing 90 mg of *M. piperita* and 50 mg of *Carum carvi* essential oils.

The combination of peppermint and caraway essential oils has been shown to act locally in the gut as an antispasmodic (Micklefield et al., 2000, 2003) and to have a relaxing effect on the gallbladder (Goerg and Spilker, 2003). The antispasmodic effect of peppermint is well documented and that of caraway essential oil has also been demonstrated (Reiter and Brandt, 1985). The latter alone has also been shown to inhibit gallbladder contractions in healthy volunteers, increasing gallbladder volume by 90% (Goerg and Spilker, 1996).

One of the first studies involved 45 patients in a double-blind, placebo-controlled multicenter trial with the administration of Enteroplant thrice daily for 4 weeks. It was found to be superior to placebo with regard to pain frequency, severity, efficacy, and medical prognosis. Clinical Global Impressions were improved for 94.5% of patients using the essential oil combination (May et al., 1996).

The activity of Enteroplant (twice daily) was compared to that of cisapride (30 mg daily), a serotonin 5-HT$_4$ agonist that stimulates upper gastrointestinal tract motility, over a 4-week period. This double-blind, randomized trial found that both products had comparable efficacy in terms of pain severity and frequency, Dyspeptic Discomfort Score, and Clinical Global Impressions (Madisch et al., 1999).

Another double-blind, randomized trial administered either Enteroplant or placebo twice daily for 28 days. Pain intensity and pressure, heaviness, and fullness were reduced in the test group by 40% and 43% as compared to 22% for both in the placebo group, respectively. In addition, Clinical Global Impressions were improved by 67% for the peppermint/caraway combination, while the placebo scored 21% (May et al., 2000).

Holtmann et al. (2001) were the first to investigate the effect of Enteroplant (twice daily) on disease-specific quality of life as measured by the Nepean Dyspepsia Index. All scores were significantly improved compared to the placebo. In 2002, the same team also demonstrated that patients suffering with severe pain or severe discomfort both responded significantly better in comparison to the placebo.

Approximately 50% of patients suffering from functional dyspepsia are infected with *Helicobacter pylori* (Freidman, 1998). The *Helicobacter* status of 96 patients and the efficacy of Enteroplant were compared by May et al. in 2003. They found that patients with *H. pylori* infection demonstrated a substantially better treatment response than those who were not infected. However, a previous study found no efficacy differences between infected and noninfected functional dyspepsia patients (Madisch et al., 2000) and so the effect of the presence of the bacterium on Enteroplant treatment has yet to be elucidated.

A short review of the literature concluded that treatment with the fixed peppermint/caraway essential oil combination had demonstrated significant efficacy in placebo-controlled trials, had good tolerability and safety, and could thus be considered for the long-term management of functional dyspepsia patients (Holtmann et al., 2003).

13.11 GASTROESOPHAGEAL REFLUX

d-Limonene has been found to be effective in the treatment of gastroesophageal reflux disorder. Nineteen patients took one capsule of 1000 mg *d*-limonene every day and rated their symptoms using a severity/frequency index. After 2 days, 32% of patients had significant relief, and by day 14, 89% of patients had complete relief of symptoms (Wilkins, 2002).

A double-blind, placebo-controlled trial was conducted with 13 patients who were administered one 1000 mg capsule of D-limonene daily or on alternate days. By day 14, 86% of patients were asymptomatic compared to 29% in the placebo group (Wilkins, 2002).

The mechanism of action of D-limonene has not been fully elucidated in this regard, but it is thought that it may coat the mucosal lining and offer protection against gastric acid and/or promote healthy peristalsis.

13.12 HYPERLIPOPROTEINEMIA

Girosital is a Bulgarian encapsulated product consisting of rose essential oil (68 mg) and vitamin A in sunflower vegetable oil. Initial animal studies found that rose oil administered at 0.01 and 0.05 mL/kg had a hepatoprotective effect against ethanol. Dystrophy and lipid infiltration were lowered and glycogen tended to complete recovery, suggesting a beneficial effect of rose oil on lipid metabolism (Kirov et al., 1988a).

Girosital was administered to 33 men with long-standing alcohol abuse, twice daily for 3 months. It significantly reduced serum triglycerides and low-density lipoprotein and increased the level of HDL-cholesterin; it was particularly effective for the treatment of hyperlipoproteinemia types IIb and IV. Liver lesions relating to alcohol intoxication improved, and subjective complaints such as dyspeptic symptoms and pain were reduced (Konstantinova et al., 1988).

The hypolipidemic effect of Girosital was again studied by giving a capsule once daily for 20 days in 35 patients with hyperlipoproteinemia. In type IIa hyperlipoproteinemia cases, the total lipids were reduced by 23.91% and total cholesterol by 10.64%. For type IIb patients, the total lipid reduction was 15.93%, triglycerides fell by 25.45%, and cholesterol by 14.06%; in type IV cases, the reductions were 33.83%, 25.33%, and 36%, respectively. Girosital was more effective in comparison to treatment with bezalip and clofibrate (Stankusheva, 1988).

Thirty-two patients with hyperlipoproteinemia and arterial hypertension were administered one Girosital capsule twice daily for 110 days. A marked reduction in hyperlipoproteinemia was demonstrated in all patients. The hypocholesterolemic effect manifested first in type IIa patients after 20 days and later in type IIb cases. Reduction of serum triglycerides in type IIb began 50 days after commencement of treatment (Kirov et al., 1988b).

A further study (Mechkov et al., 1988) examined the effect of Girosital capsules twice daily for 110 days in 30 patients with cholelithiasis, liver steatosis, and hyperlipoproteinemia. Total cholesterol decreased after 20 days of treatment although it tended to rise slightly later in the test period. The triglycerides were most affected in hyperlipoproteinemia types IIb and IV. The beta-lipoprotein values were not altered by the treatment.

13.13 INSECTICIDAL

There are protozoal and viral diseases worldwide affecting humans that are transmitted by insect vectors. In many cases, the treatment of such diseases is difficult and sometimes impossible. The most efficacious approach to minimize incidence is to prevent transmission into new hosts by controlling the insect vectors. Their larvae and adults can be killed and the chances of being bitten lowered by the use of repellents. Synthetic insecticides may have toxic side effects and generally harm the environment. Moreover, insects are continually developing resistance to them. A viable alternative is the employment of botanical insecticides that are biodegradable and environmentally safe. For many years, essential oils have been investigated with success for their repellent, larvicidal, and adulticidal activities against various insect vectors. Examples of research are given in Table 13.2.

Essential oils have also been examined for their direct effect, both in vitro and in vivo, against *Herpetomonas*, *Leishmania*, *Plasmodium*, and *Trypanosoma* species. Recent examples are given in Table 13.3. Unfortunately, the promising results have not encouraged human clinical trials as yet.

13.13.1 Acaricidal Activity

A number of essential oils have been found to have effective acaricidal activity against infections in the animal world. Recent examples include *Origanum onites* against cattle ticks (Coskun et al., 2008) and *Cinnamomum zeylanicum* against rabbit mange mites (Fichi et al., 2007). In comparison to veterinary research, there have been few investigations into human acaricidal infections.

The scabies mite, *Sarcoptes scabiei* var. *hominis*, is becoming increasingly resistant to existing acaricidal compounds such as lindane, benzyl benzoate, permethrin, and oral ivermectin. The potential use of a 5% *M. alternifolia* essential oil solution to treat scabies infections was investigated in vitro. It was found to be highly effective at reducing mite survival times, and the main active component was terpinen-4-ol. However, the in vivo effectiveness was only tested on one individual, in combination with benzyl benzoate and ivermectin (Walton et al., 2004).

A successful in vitro study using *L. multiflora* Moldenke essential oil against scabies led to an in vivo trial using concentrations of essential oil diluted in liquid paraffin ranging from 10% to 50% (eight patients per concentration). A 20% concentration applied once daily to affected areas for 3 and 5 days produced a cure rate of 75% and 100%, respectively (Oladimeji et al., 2000). A second trial was conducted with larger patient numbers in 2005.

TABLE 13.2
Examples of Repellent, Larvicidal, and Adulticidal Activities of Essential Oils against Insect Vectors of Disease

Insect Vector	Essential Oil	Origin	Disease	References
Rhodnius neglectus *Triatoma infestans*	*Minthostachys andina* *Hedeoma mandonianum*	Bolivia	Chaga's disease	Fournet et al. (1996)
Aedes aegypti *Anopheles annularis* *A. culicifacies* *A. stephensi* *Culex quinquefasciatus*	*Mentha piperita*	India	Dengue fever Yellow fever Malaria	Ansari et al. (2000)
A. gambiae	*Conyza newii* *Plectranthus marruboides*	Kenya	Malaria	Omolo et al. (2005)
Phlebotomus papatasi	*Myrtus communis*	Iran	Cutaneous leishmaniasis	Yabhoobi-Ershadi et al. (2006)
C. quinquefasciatus	*Zingiber officinalis*	India	Filariasis	Pushpanathan et al. (2008)
C. pipiens	*Oenanthe pimpinelloides*	Greece	West Nile virus	Evergetis et al. (2009)
C. pipiens pallens	*Coriandrum sativum* *Elettaria cardamomum* *Rosmarinus officinalis* *Santalum album*	South Korea	Dengue fever Encephalitis Filariasis Malaria	Kang et al. (2009)
C. pipiens	*Chrysanthemum coronarium* *Hypericum scabrum* *Pistacia terebinthus* *Vitex agnus castus*	Turkey	West Nile virus	Cetin et al. (2011)
A. anthropophagus	*Blumea densiflora*	China	Malaria	Zhu and Tian (2011)
Aedes albopictus	*Citrus limon* *Citrus sinensis* *Citrus paradisi*	Greece	Dengue fever Chikungunya	Giatropoulos et al. (2012)
A. aegypti	*Croton spp.* *Lippia sidoides*	Brazil	Dengue fever	de Lima et al. (2013)
A. albopictus	*Coriandrum sativum*		Chikungunya Dengue fever Encephalitis Filariasis Malaria	Benelli et al. (2013)
C. quinquefasciatus	*Pimpinella anisum*		Filariasis Rift Valley fever virus St. Louis encephalitis West Nile virus	Pavela (2014)

TABLE 13.3
Examples of Recent Research into the Antiprotozoal Effects of Essential Oils

Essential Oil	Protozoal Species	References
Citrus sinensis	*Trypanosoma brucei*	Habila et al. (2010)
Cymbopogon citratus	*T. evansi*	
Eucalyptus camaldulensis		
E. citriodora		
Lippia alba	*Leishmania chagasi*	Escobar et al. (2010)
L. origanoides	*T. cruzi*	
Carum carvi	*Plasmodium berghei*	Fujisaki et al. (2012)
Monarda fistulosa	*P. falciparum*	
Myristica fragrans		
Santalum album		
Thujopsis dolabrata		
Annona crassiflora	*L. amazonensis*	Oliani et al. (2013)
	L. braziliensis	
	L. chagasi	
	L. major	
	T. cruzi	
Keetia leucantha	*T. brucei brucei*	Bero et al. (2013)
Cymbopogon citratus	*P. falciparum*	Kpoviessi et al. (2014)
C. giganteus	*T. brucei brucei*	
C. nardus		
C. schoenanthus		

A double-blind, randomized, parallel group study was used to compare the effects of 25% w/w benzyl benzoate emulsion with 20% w/w *L. multiflora* essential oil emulsion in the treatment of scabies infection in 105 patients. Applied daily, the cure rate for the oil emulsion was 50%, 80%, and 80% for 3, 5, and 7 days, respectively, as compared to 30%, 60%, and 70% for the benzyl benzoate emulsion. There were also less adverse reactions to the oil emulsion, leading it to be considered as an additional formulation for the treatment of scabies (Oladimeji et al., 2005).

Although not an infection, the lethal activity of essential oils toward the house dust mite (*Dermatophagoides farinae* and *D. pteronyssinus*) is important as these mites are a major cause of respiratory allergies and an etiologic agent in the sensitization and triggering of asthma in children. Numerous studies have been conducted, including the successful inclusion of *E. globulus* in blanket-washing solutions (Tovey and McDonald, 1997), the high acaricidal activity of clove, rosemary, eucalyptus, and caraway (El-Zemity et al., 2006), and of tea tree and lavender (Williamson et al., 2007).

13.13.2 Pediculicidal Activity

Over the years, a number of in vitro investigations have indicated the potential of individual essential oils as pediculicidal agents against *Pediculus humanus capitis* (head louse). These include *Myrtus communis* (Gauthier et al., 1989); *C. zeylanicum* fol., *M. alternifolia, Pimpinella anisum*, and *Thymus vulgaris* (Veal 1996); *Eugenia caryophyllata* (Yang et al., 2003); *Origanum majorana, R. officinalis*, and *Elettaria cardamomum* (Yang et al., 2004); *C. zeylanicum* cort. (Yang et al., 2005); *L. angustifolia* (Williamson et al., 2007); and *Eucalyptus sideroxylon, E. globulus* ssp.

globulus, and *E. globulus* ssp. *maidenii* (Toloza et al., 2010). The action of *L. multiflora* essential oil was investigated in vitro against both head and body lice (*P. humanus corporis*). The head lice were more susceptible than body lice (Oladimeji et al., 2000).

Three clinical trials have been documented. The first employed a spray formulation (Chick-Chack) of coconut oil, *Illicium verum*, and *Cananga odorata* essential oils, comparing the effect with a commercial pediculicide containing permethrin and malathion. The 119 children were randomly divided into two groups and treated with either product three times at 5-day intervals. Treatments were successful (i.e., absence of living lice or eggs) in 92.3% and 92.2% for the natural and commercial products, respectively (Mumcuoglu et al., 2002).

Again, another spray product (Paranix®) containing coconut oil, *I. verum,* and *C. odorata* essential oils was compared to the effects of malathion in 24 subjects. Four days after treatment, 11/12 subjects in both groups had no lice or eggs present (Scanni and Bonifazi, 2005). In a small study, the same authors applied the product once to five children. Viable lice were not found after 7 days (Scanni and Bonifazi, 2006).

A third trial used Paranix applied to 50 infected children twice over 9 days. The results were compared to a control group of equal number using a standard permethrin lotion. The spray was more successful than permethrin, with cure rates of 82% and 42% (Burgess et al., 2010).

13.14 IRRITABLE BOWEL SYNDROME

The essential oil of *M. piperita* has been used for many years as a natural carminative of the gastrointestinal tract. This effect is principally due to the antispasmodic activity of menthol, which acts as a calcium channel antagonist of the intestinal smooth muscle (Taylor et al., 1984, 1985). Secondary effects include a reduction of gastrointestinal foam by peppermint oil (Harries et al., 1978) and a choleretic activity that is attributed to menthol (Rangelov et al., 1988). The reduction of intestinal hydrogen production caused by bacterial overgrowth has also been demonstrated in patients by enteric-coated peppermint oil (Logan and Beaulne, 2002).

The first clinical trial of peppermint for the treatment of irritable bowel syndrome was conducted in 1979 by Rees et al. They prescribed 0.2 mL of peppermint oil in enteric-coated capsules (1–2 capsules depending on symptom severity) thrice daily. Patient assessment considered the oil to be superior to the placebo in relieving abdominal symptoms.

Since then, a further 15 double-blind and two open trials have been conducted; examples of these can be seen in Table 13.4. Eight studies used the commercial preparation known as Colpermin® and two used Mintoil®, the capsules of which contain 187 and 225 mg of peppermint oil, respectively. The other studies used enteric-coated capsules usually containing 0.2 mL of the essential oil.

The latest trial (Cappello et al., 2007) used a randomized, double-blind, placebo-controlled design to test the efficacy of two capsules of Mintoil twice daily for 4 weeks. The symptoms evaluated before treatment and at 4 and 8 weeks posttreatment were abdominal bloating, pain or discomfort, diarrhea, constipation, and incomplete or urgency of defecation and the passage of gas or mucus. The frequency and intensity of these symptoms was used to calculate the total irritable bowel syndrome symptoms score. At 4 weeks, 75% of patients in the peppermint oil group demonstrated a >50% reduction of the symptoms score as compared to 38% in the placebo group. At 4 and 8 weeks in the peppermint oil group compared to that before treatment, there was a statistically significant reduction of the total irritable bowel syndrome symptoms score, whereas there was no change with the placebo.

A critical review and meta-analysis of the use of peppermint oil for irritable bowel syndrome was published in 1998 (Pittler and Ernst). They examined five double-blind, placebo-controlled trials; there was a significant difference between peppermint oil and placebo in three cases and no significant difference in two cases. It was concluded that although a beneficial effect of peppermint oil was demonstrated, its role in treatment was not established.

TABLE 13.4
Examples of Clinical Trials of Peppermint Oil in the Treatment of Irritable Bowel Syndrome

Patients	Treatment	Outcome	References
18	0.2–0.4 mL thrice daily for 3 weeks	Superior to placebo in relieving abdominal symptoms.	Rees et al. (1979)
29	0.2–0.4 mL thrice daily for 2 weeks	Superior to placebo in relieving abdominal symptoms.	Dew et al. (1984)
25	0.2 mL thrice daily for 2 weeks	No significant change in symptoms as compared to placebo.	Lawson et al. (1988)
35	1 Colpermin thrice daily for 24 weeks	Effective in relieving symptoms.	Shaw et al. (1991)
110	1 Colpermin 3–4 times daily for 2 weeks	Significant improvement in symptoms as compared to placebo.	Lui et al. (1997)
42	1–2 Colpermin thrice daily for 2 weeks	75% of children had reduced pain severity.	Kline et al. (2001)
178	2 Mintoil thrice daily for 3 months	Significant improvement in gastroenteric symptoms as compared to placebo (97% vs. 33%, respectively).	Capanni et al. (2005)
57	2 Mintoil twice daily for 4 weeks	Significant reduction in overall symptom score.	Cappello et al. (2007)

A review of 16 trials was conducted by Grigoleit and Grigoleit in 2005. They concluded that there was reasonable evidence that the administration of enteric-coated peppermint oil (180–200 mg) thrice daily was an effective treatment for irritable bowel syndrome when compared to placebo or the antispasmodic drugs investigated (mebeverine, hyoscyamine, and alverine citrate).

A comparison between two commercial delayed release peppermint oil preparations found that there were differences in the pharmacokinetics in relation to bioavailability times and release site. A capsule that is more effective in delivering the peppermint oil to the distal small intestine and ascending colon would be more beneficial in the treatment of irritable bowel syndrome (White et al., 1987). It has also been suggested that the conflicting results in some trials may be due to the inclusion of patients suffering from lactose intolerance, syndrome of small intestinal bacterial overgrowth, and celiac disease, all of which have symptoms similar to irritable bowel disease (Cappello et al., 2007).

13.15 MEDICAL EXAMINATIONS

Although not employed in a treatment context, the antispasmodic activity of peppermint essential oil has been used to facilitate examinations of the upper and lower gastrointestinal tract. A few examples are highlighted in the succeeding text.

Peppermint oil has also been used during double-contrast barium enemas. The study comprised 383 patients in four groups, two being no-treatment and a Buscopan groups. The preparation, consisting of 8 mL of essential oil, 0.2 mL of Tween 80 in 100 mL water, was administered in 30 mL quantities via the enema tube or mixed in with the barium meal. Peppermint oil had the same spasmolytic effect as systemic Buscopan in the transverse and descending colon and a stronger effect in the cecum and ascending colon. Both methods of peppermint oil administration were equally effective (Asao et al., 2003).

Orally administered peppermint oil was used in a randomized trial in 430 patients undergoing a double-contrast barium meal examination, without other antispasmodics. A reduction in spasms of the esophagus, lower stomach, and duodenal bulb was found, along with an inhibition of barium flow to the distal duodenum and an improvement of diagnostic quality (Mizuno et al., 2006).

During endoscopic retrograde cholangiopancreatography, Buscopan or glucagon is used to inhibit duodenal motility but produce adverse effects. Various concentrations of peppermint oil were introduced into the upper gastrointestinal tract of 40 patients undergoing the procedure. Duodenal relaxation was obtained with 20 mL of 1.6% peppermint oil solution, and the procedure was performed successfully in 91.4% of patients. The inhibitory effect of peppermint oil appeared to be identical to that of glucagon but without side effects (Yamamoto et al., 2006).

13.16 MUCOSITIS

High-dose chemotherapy was used to treat 80 patients with breast cancer; all developed severe oral mucositis. The test group received a mouthwash consisting of one drop each of the essential oils of *M. alternifolia*, *Citrus bergamia*, and *Pelargonium graveolens* blended together in warm water. The control group used a proprietary mouthwash. Both mouthwashes were used five times daily during chemotherapy. The results showed no difference in outcomes between the two groups although the proprietary mouthwash was considerably more expensive than the essential oil treatment (Gravett, 2000).

Between 80% and 100% of patients suffer radiation-induced mucositis of the oropharyngeal area as a result of radiotherapy for head and neck cancers. Nineteen patients participated in a randomized trial. One group was given two drops of a 1:1 mixture of the essential oils of *L. scoparium* and *Kunzea ericoides* in water as a gargle; the placebo group gargled with water, and a third group received "usual care." During radiotherapy days, patients were gargling up to five times. Treatment was continued 1-week postradiotherapy, with 3–5 gargles per day. After gargling for 15 s and then spitting out the fluid, the essential oil group then swallowed the same quantity of mixture. This was to cover as much of the pharyngeal area as possible. Those patients in the essential oil group had a delayed onset of mucositis, reduced pain, and oral symptoms compared to the placebo (Maddocks-Jennings et al., 2009). The small sample size was a significant limitation to this study.

13.17 NAUSEA

A small study examined a variety of aromatherapy treatments to 25 patients suffering from nausea in a hospice and palliative care facility. Patients were offered the essential oils of *Foeniculum vulgare* var. *dulce*, *Chamaemelum nobile*, and *M. piperita*, either singly or in blends, depending on individual preferences. Delivery methods included abdominal compress or massage, personal air spritzer or scentball diffuser. Only 32% of patients reported no response to the treatments, and they had all just finished heavy courses of chemotherapy. Using a visual-numeric analog scale, the remainder of patients experienced an improvement in their nausea symptoms when using the aromatherapy interventions. All patients were also taking antiemetic drugs, and so the essential oils were regarded as successful complements to standard medications (Gilligan, 2005).

About 80% of all cancer patients that undertake chemotherapy experience nausea and vomiting, the former being ranked first on the list of distressing symptoms. A double-blind clinical trial examined the antiemetic effects of ingested essential oils of *M. spicata* or *M. piperita* during chemotherapy. Patients suffering from any form of cancer were divided into groups ($n = 50$). The control group received their normal antiemetic drugs. The test groups were given spearmint or peppermint capsules (two drops of essential oil in sugar) administered 30 min before commencement of chemotherapy and twice again every 4 h. They also received their normal antiemetic drugs. Both essential oils were found to produce a significant reduction of emetic events during 24 h of treatment, with 0.6 and 0.7 emetic events for *M. spicata* and *M. piperita*, respectively, as compared to 1.8 for the control. The intensity of nausea was also significantly reduced by both essential oils as compared to the control. It was concluded that spearmint and peppermint essential oils were a safe and effective option for the treatment of chemotherapy-induced nausea and emesis (Tayarani-Najaran et al., 2013).

A 6-month trial investigated the effect of inhaled 5% *Zingiber officinale* essential oil in the prevention of postoperative nausea and vomiting (PONV). All patients were at high risk for PONV and all used similar combinations of prophylactic intravenous antiemetics. The test group had the essential oil applied to the volar aspects of both wrists via a rollerball immediately prior to surgery. In the recovery room, patients were questioned as to their feelings of nausea. Any patient that felt that they required further medication was considered a "failure." Prevention of PONV by ginger essential oil was effective in 80% of cases, as measured by no complaint of nausea during the recovery period. In those patients that did not receive the essential oil, 50% experienced nausea (Geiger, 2005).

Another experiment used essential oils to prevent PONV, but they were applied after surgery if the patient complained of nausea. An undiluted mixture of *Z. officinale*, *E. cardamomum*, and *Artemisia dracunculus* essential oils in equal parts was applied with light friction to the sternocleidomastoid area and carotid–jugular axis of the neck. Of the 73 cases treated, 50 had a positive response, that is, a complete block of nausea and vomiting within 30 min. It was found that the best response (75%) was with patients that had received a single analgesic/anesthetic (de Pradier, 2006).

The use of essential oils to alleviate motion sickness has also been investigated. A blend of *Z. officinale*, *L. angustifolia*, *M. spicata*, and *M. piperita* essential oils in an inhalation dispenser (QueaseEase™) was given to 55 ocean boat passengers with a history of motion sickness. The oil blend was inhaled as needed during the trip, and queasiness was assessed using a linear analogue scale. The product was more effective than the placebo in lowering sensations of nausea when the seas were roughest but was not significant at other times (Post-White and Nichols, 2007).

13.18 PAIN RELIEF

There follows a number of differing conditions that have been treated with essential oils with varying biological activities, such as antispasmodic and anti-inflammatory. They all share a common effect, that of pain relief.

13.18.1 DYSMENORRHEA

The seeds of *F. vulgare* have been used in traditional remedies for the treatment of dysmenorrhea, an action attributed to the antispasmodic effect of the essential oil. An in vitro experiment demonstrated that fennel essential oil inhibited oxytocin- and prostaglandin E_2-induced contractions of isolated uterus; the former was considered to have a similar activity to diclofenac, a nonsteroidal anti-inflammatory drug. The overall mechanism of action is still unknown (Ostad et al., 2001).

A randomized, double-blind crossover study examined the effect of oral fennel essential oil at 1% or 2% concentration as compared to placebo for the treatment of 60 women with mild to moderate dysmenorrhea. Up to 1 mL of the solution was taken as required for the pain at intervals of not less than 4 h. In the treatment groups, the severity of the pain was significantly decreased; the efficacy of the 2% fennel oil was 67.4%, which was comparable to the efficacy of nonsteroidal anti-inflammatory drugs (Khorshidi et al., 2003).

Thirty patients with moderate to severe dysmenorrhea took part in a study to compare the activity of mefenamic acid with the essential oil of *F. vulgare* var. *dulce*. The evaluation was done during the first 5 days of three consecutive menstrual cycles. In the first cycle, no intervention was given (control); during the second cycle, 250 mg of mefenamic acid every 6 h was prescribed; and in the third, 25 drops of a 2% solution of fennel essential oil were given every 4 h. A self-scoring linear analog technique was used to determine effect and potency. Both interventions effectively relieved menstrual pain as compared to the control. Mefenamic acid was more potent on the second and third days, but the result was not statistically significant. It was concluded that fennel essential oil was a safe and effective remedy but was probably less effective than mefenamic acid at the dosage used (Jahromi et al., 2003).

A third study used aromatherapy massage for the relief of the symptoms of dysmenorrhea in 67 students. The essential oils of *L. officinalis*, *Salvia sclarea*, and *Rosa centifolia* (2:1:1 ratio) were diluted to 3% in 5 mL of almond oil and applied in a 15 min abdominal massage daily, 1 week before the start of menstruation and stopping on the first day of menstruation. The control group received no treatment, and the placebo group received massage with almond oil only. The results showed a significant improvement of dysmenorrhea as assessed by a verbal multidimensional scoring system for the essential oil group when compared to the other two groups (Han et al., 2006).

13.18.2   Headache

The effect of peppermint and eucalyptus essential oils on the neurophysiological, psychological, and experimental algesimetric parameters of headache mechanisms were investigated using a double-blind placebo-controlled trial with 32 healthy subjects. Measurements included sensitivity to mechanical, thermal, and ischemically induced pain. Four preparations consisting of varying amounts of peppermint and/or eucalyptus oils in ethanol were applied to the forehead and temples. Eucalyptus alone had no effect on the parameters studied. A combination of both oils (10% peppermint and 5% eucalyptus) increased cognitive performance and had a muscle-relaxing and mentally relaxing effect but did not influence pain sensitivity. Peppermint alone (10%) had a significant analgesic effect with reduction in sensitivity to headache. It was shown to exert significant effects on the pathophysiological mechanisms of clinical headache syndromes (Göbel et al., 1995a).

A second study used the same essential oils when investigating the skin perfusion of the head in healthy subjects and migraine patients. In the former, capillary flow was increased by 225% in comparison to baseline by peppermint oil, while eucalyptus decreased the flow by 16%. In migraine patients, neither essential oil had any effect. It was suggested that the absence of capillary vasodilation (normally caused by menthol) was due to impaired calcium channel function in migraine patients (Göbel et al., 1995b).

13.18.3   Infantile Colic

Since animal studies had demonstrated that the essential oil of *F. vulgare* reduced intestinal spasm and increased the motility of the small intestine, it was used in a double-blind, randomized, placebo-controlled trial in the treatment of infantile colic. The 125 infants were all 2–12 weeks of age and those in the treatment group received a water emulsion of 0.1% fennel essential oil and 0.4% polysorbate (5–20 mL) up to four times a day. The dose was estimated to provide about 12 mg/kg/day of fennel essential oil. The control group received the polysorbate only. The treatment provided a significant improvement of colic, eliminating symptoms in 65% of infants as compared to 23.7% for the control. No side effects were noted (Alexandrovich et al., 2003).

13.18.4   Joint Physiotherapy

Six sports physiotherapists treated 30 patients suffering from knee or ankle pathologies of traumatic or surgical origin. Two commercial products were used simultaneously, Dermasport® and Solution Cryo®. The former was a gel consisting of the essential oils of *Betula alba*, *Melaleuca leucadendron*, *C. camphora*, *Syzygium aromaticum*, *E. globulus*, and *Gaultheria procumbens*. Solution Cryo contained the same essential oils minus *G. procumbens* but with the addition of *C. nobile*, *C. limon*, and *Cupressus sempervirens*. Both products were at an overall concentration of 6%. Thirty minutes after application, a net reduction in movement pain and joint circumference were demonstrated, along with an increase in articular flexion and extension of both joints in all patients (Le Faou et al., 2005).

13.18.5 Muscle Strain

Various over-the-counter preparations containing methyl salicylate and menthol have been used to treat musculoskeletal pain for many years, but few have been the subjects of clinical trials. Menthol is classified as a counterirritant, producing a cooling sensation via activation of specific thermoreceptors (Patel et al., 2007), while methyl salicylate has analgesic, anti-inflammatory, and vasodilatory effects (Green and Flammer, 1989).

A topical patch containing 10% methyl salicylate and 3% *l*-menthol or placebo was applied to the site of muscle sprains on various parts of the bodies of 208 individuals. Pain intensity was assessed at rest and with movement, the primary efficacy endpoint being the summed pain intensity difference score through 8 h with movement. Compared to placebo, the test group experienced significantly greater pain relief (~40%) resulting from mild to moderate muscle strain (Higashi et al., 2010).

13.18.6 Nipple Pain

Nipple cracks and pain are a common cause of breastfeeding cessation. In a randomized trial, 196 primiparous women were studied during the first 2 weeks postpartum. The test group applied peppermint water (essential oil in water, concentration not given) to the nipple and areola after each breastfeed, while the control group applied expressed breast milk. The overall nipple crack rate at the end of the period in the peppermint group was 7% as compared to 23% for the control. Only 2% of peppermint group experienced severe nipple pain in contrast to 23% of the control, with 93% and 71% experiencing no pain, respectively (Melli et al., 2007).

13.18.7 Osteoarthritis

A blend of *Z. officinale* (1%) and *Citrus sinensis* (0.5%) essential oils was used in an experimental double-blind study using 59 patients with moderate to severe knee pain caused by osteoarthritis. The treatment group received six massage sessions over a 3-week period; the placebo received the same massage sessions but without the essential oils and the control had no intervention. Assessment of pain intensity, stiffness, and physical functioning was done at baseline and at post 1 and 4 weeks. There were improvements in pain and function for the intervention group in comparison to the placebo and control at week post 1, but this was not sustained to week 4. The treatment was suggested for the relief of short-term knee pain (Yip and Tam, 2008).

13.18.8 Postherpetic Neuralgia

A double-blind crossover study examined the effect of the essential oil of *Pelargonium* spp. on moderate to severe postherpetic pain in 30 subjects. They were assigned to groups receiving 100%, 50%, or 10% geranium essential oil (in mineral oil), mineral oil placebo, or capsaicin control. Pain relief was measured using a VAS from 0–60 min after treatment. Mean values for the time integral of spontaneous pain reduction was 21.3, 12.7, and 8.0 for the 100%, 50%, and 10% geranium oils, and evoked pain reduction values were 15.8, 7.7, and 5.9, respectively. Both evoked and spontaneous pains were thus significantly reduced in a dose-dependent manner (Greenway et al., 2003).

The result is interesting because topical capsaicin cream (one of the standard treatments for this condition) relieves pain gradually over 2 weeks, while the essential oil acted within minutes. Geranium essential oil applied cutaneously in animal studies has suppressed cellular inflammation and neutrophil accumulation in inflammatory sites (Maruyama et al., 2006), but postherpetic neuralgia normally occurs after the inflammation has subsided. One of the main components of the essential oil, geraniol, and the minor components of geranial, nerol, and neral has been shown to interact with the transient receptor potential channel, TRPV1, as does capsaicin (Stotz et al., 2008). This sensory inhibition may explain the efficacy of topical geranium oil.

13.19 POSTOPERATIVE PAIN

13.19.1 Breast Biopsy

A similar study in the previous year with 50 patients who had undergone breast biopsy surgery had found that lavender essential oil had no significant effect on postoperative pain or analgesic requirements. However, a significantly higher satisfaction with pain control was noted by patients in the lavender group (Kim et al., 2006).

13.19.2 Cesarean Section

Caesarian section is a common operation and, in some countries, accounts for nearly 50% of deliveries (Ganji et al., 2006), with more than half the patients reporting severe pain. This is managed by opiates and derivatives, anesthesia, and nonsteroidal anti-inflammatory drugs. Their use is limited due to efficacy, safety, unavailability, and potential side effects (Goodman and Limbird, 2001).

For a single-blind clinical trial, 200 women who underwent planned elective cesarean section were divided into two groups. At least 3 h after receiving intravenous analgesics, one group inhaled 2% lavender essential oil via a face mask for 3 min, while the control inhaled an aromatic placebo. The severity of pain was documented using a VAS both before and after inhalation (30 min later). The same process was repeated after 8 and 16 h. The mean VAS decreased significantly in both groups. The amelioration of pain was significantly more prominent in the lavender group when compared to the control at all three intervention stages. It was concluded that the use of lavender was a successful and safe therapy for reducing post cesarean pain (Hadi and Hanid, 2011).

13.19.3 Episiotomy

About 33% of women with vaginal deliveries undergo an episiotomy, and it is the most common perineal incision in obstetrics and midwifery. Wound care usually involves regular antiseptic sitz baths using aqueous solutions of povidone-iodine (Betadine). However, this substance has been demonstrated to impair wound healing and to suppress the function of lymphocytes and fibroblasts (Cooper et al., 1991).

The use of *L. angustifolia* essential oil for perineal wound care was first investigated in 1994. The 635 patients were divided into three groups; those receiving lavender essential oil, a synthetic lavender oil, or 2-methyl 3-isobutyl pyrazine, an inert but aromatic placebo. Six drops were added to a daily bath for 10 consecutive days. The results provided no statistically significant differences between the three groups, although the essential oil group reported lower mean discomfort scores (Dale and Cornwell, 1994).

A second clinical trial was conducted using 120 primiparous women who had undergone episiotomy. The test group used a sitz bath containing 5–7 drops of lavender essential oil in 4 L of water, twice daily for 10 days. The control group received routine wound care with povidone-iodine. Twenty-five patients in the lavender group reported no pain as compared to 17 of the control group; this was not considered significant. There was a significant difference between the two groups regarding redness of the wound, and lavender was considered to be suitable for episiotomy wound care (Vakilian et al., 2011).

Sixty primiparous women who had undergone episiotomy were divided into two groups of 30; the control group was given 30 min sitz baths (containing 10 mL Betadine in 4 L water) twice daily for 5 days, while the test group received the same sitz bath routine but with 0.25 mL *L. angustifolia* essential oil per 5 L of water. A significant difference was found between the groups in the prescriptions of analgesics, with a consumption mean of 2.07 and 1.93 at days two and five, respectively, for the control group and 0.8 and 0.4 for the lavender group. Moreover, 70% of the lavender group required no analgesics in contrast to 33% of the control group. The redness, edema, ecchymosis, discharge, and approximation scale was used to assess episiotomy pain and discomfort. This was significantly lower in the lavender group after 5 days. It was concluded that lavender essential oil can assist in episiotomy pain relief and healing (Sheikhan et al., 2012).

13.19.4 GASTRIC BANDING

A randomized, placebo-controlled clinical trial was conducted to determine whether the inhalation of lavender essential oil could reduce opioid requirements after laparoscopic adjustable gastric banding. In the postanesthesia care unit, 54 patients were given either lavender (2 drops of a 2% dilution) or nonscented oil in a face mask. It was found that patients in the lavender group required significantly less morphine postoperatively than the placebo group (2.38 and 4.26 mg, respectively). Moreover, significantly more patients in the placebo group required analgesics in comparison to the lavender group; 82% compared to 46% (Kim et al., 2007).

13.19.5 PROSTATITIS

One study has evaluated the use of Rowatinex for the treatment of chronic prostatitis/chronic pelvic pain syndrome, the rationale being based on the known anti-inflammatory properties of the product. A 6-week, randomized single-blind trial compared the use of Rowatinex 200 mg thrice daily with ibuprofen 600 mg thrice daily in 50 patients. Efficacy was measured by the National Institute of Health Chronic Prostatitis Symptom Index (NIH-CPSI) that was completed by the patients on four occasions. The decrease in the NIH-CPSI was significant in both groups at the end of treatment, and a 25% improvement in the total score was superior in the Rowatinex group (68%) compared to the ibuprofen group (40%). Although the symptomatic response was significant, no patients became asymptomatic (Lee et al., 2006).

13.20 PRURITUS

Pruritus is one of the most common complications of patients undergoing hemodialysis. Thirteen such patients were given an arm massage with lavender and tea tree essential oils (5% dilution in sweet almond and jojoba oil) three times a week for 4 weeks. A control group received no intervention. Pruritus score, pruritus-related biochemical markers, skin pH, and skin hydration were measured before and after the study. There was a significant decrease in the pruritus score and blood urea nitrogen level for the test group. The control group showed a decreased skin hydration between pre- and posttest, whereas for the essential oil group, it was significantly increased (Ro et al., 2002). The lack of a massage only group in the study meant that the effects could not be definitely associated with the essential oils.

13.21 RECURRENT APHTHOUS STOMATITIS

Recurrent aphthous stomatitis (RAS), also known as canker sores, is the most common oral mucosal lesions, and although the process is sometimes self-limiting, the ulcer activity is mostly continuous and some forms may last for 20 years. Predisposing agents include bacteria and fungi, stress, mouth trauma, certain medications, and food allergies. Two essential oils both endemic to Iran have been investigated for treatment of this condition; *Z. multiflora*, a thyme-like plant containing thymol, carvacrol, and linalool as major components and *Satureja khuzestanica* containing predominantly carvacrol.

In a double-blind, randomized study, 60 patients with RAS received either 30 mL of an oral mouthwash composed of 60 mg of *Z. multiflora* essential oil in an aqueous-alcoholic solution or placebo thrice daily for 4 weeks. In the treatment group, 83% of patients responded well, while 17% reported a deterioration of their condition. This was compared to 13% and 87% for the placebo group, respectively. A significant clinical improvement with regard to less pain and shorter duration of the condition was found in the essential oil group (Mansoori et al., 2002).

S. khuzestanica essential oil 0.2% v/v was prepared in a hydroalcoholic solution and used in double-blind, randomized trial with 60 RAS patients. Its activity was compared to a 25% hydroalcoholic extract of the same plant and a hydroalcoholic placebo. A cotton pad was impregnated with five drops of preparation and placed on the ulcers for 1 min (fasting for 30 min afterward) four times a day. The results of the extract and the essential oil groups were similar, with a significantly lower time for both pain elimination and complete healing of the ulcers in comparison to the placebo (Amanlou et al., 2007). The reported antibacterial, analgesic, antioxidant, and anti-inflammatory activities of this essential oil (Abdollahi et al., 2003; Amanlou et al., 2004, 2005) were thought responsible for the result.

13.22 RESPIRATORY TRACT

Given the volatile nature of essential oils, it should come as no surprise that their ability to directly reach the site of intended activity via inhalation therapy has led to their use in the treatment of a range of respiratory conditions. Moreover, a number of components are effective when taken internally, since they are bioactive at the level of bronchial secretions during their excretion. With the exception of one report, all of the research has used the individual components of either 1,8-cineole or menthol or has employed them in combination with several other isolated essential oil components within commercial preparations.

13.22.1 MENTHOL

Menthol-containing essential oils have been used in the therapy of respiratory conditions for many years and the individual component is present in a wide range of over-the-counter medications. Of the eight optical isomers of menthol, *l*-(-)-menthol is the most abundant in nature and imparts a cooling sensation to the skin and mucous membranes.

Menthol is known to react with a temperature-sensitive (8°C–28°C range) transient receptor potential channel, leading to an increase in intracellular calcium, depolarization, and initiation of an action potential (Jordt et al., 2003). This channel, known as TRPM8, is expressed in distinct populations of afferent neurons; primarily thinly myelinated Aδ cool fibers and to a lesser extent, unmyelinated C-fiber nociceptors (Thut et al., 2003). It is the interaction with the TRPM8 thermoreceptor that is responsible for the cooling effect of menthol when it is applied to the skin. This activity is not confined to the dermis, since the presence of TRPM8 has been demonstrated by animal experimentation in the squamous epithelium of the nasal vestibule (Clarke et al., 1992), the larynx (Sant'Ambrogio et al., 1991), and lung tissue (Wright et al., 1998). Thus, the activation of cold receptors via inhaled menthol leads to a number of beneficial effects.

13.22.1.1 Antitussive

Despite being used as a component in cough remedies since the introduction of a "VapoRub" in 1890, there are few human trials of menthol used alone as being effective. In a citric acid-induced cough model in healthy subjects, Packman and London (1980) found that menthol was effective, although 1,8-cineole was more efficacious. The use of an aromatic unction rather than direct inhalation may have affected the results, since the inhalation of menthol has been shown in animal models to be significantly more effective at cough frequency reduction (28% and 56% at 10 and 30 μg/L, respectively) when compared to 1,8-cineole (Laude et al., 1994).

A single-blind pseudo randomized crossover trial in 42 healthy children was used to compare the effect of an inhalation of either menthol or placebo on citric acid-induced cough. It was found that cough frequency was reduced in comparison to the baseline but not to that of the placebo (Kenia et al., 2008). However, the placebo chosen was eucalyptus oil, whose main component is 1,8-cineole and known to have similar antitussive properties to menthol.

Along with other ion channel modulators, menthol is recognized as a potential "novel therapy" for the treatment of chronic cough (Morice et al., 2004). It is not clear whether the antitussive activity of menthol is due solely to its stimulation of airway cold receptors; it may also involve pulmonary C-fibers (a percentage of which also express TRPM8) or there may be a specific interaction with the neuronal cough reflex.

13.22.1.2 Nasal Decongestant

Menthol is often thought of as a decongestant, but this effect is a sensory illusion. Burrow et al. (1983) and Eccles et al. (1988) showed that there was no change in nasal resistance to airflow during inhalation of menthol, although the sensation of nasal airflow was enhanced. In the former experiment, 1,8-cineole and camphor were also shown to enhance the sensation of airflow but to a lesser extent than menthol.

In a double-blind, randomized trial, subjects suffering from the common cold were given lozenges containing 11 mg of menthol. Posterior rhinomanometry could detect no change in nasal resistance to airflow after 10 min; however, there were significant changes in the nasal sensation of airflow (Eccles et al., 1990).

A single-blind pseudo randomized crossover trial compared the effect of an inhalation of either menthol or placebo. The main outcome measures were nasal expiratory and inspiratory flows and volumes, as measured by a spirometer and the perception of nasal patency, assessed with a VAS. It was found that there was no effect of menthol on any of the spirometric measurements, although there was a significant increase in the perception of nasal patency (Kenia et al., 2008).

Thus, it has been demonstrated that menthol is not a nasal decongestant. However, it is useful in therapy since stimulation of the cold receptors causes a subjective sensation of nasal decongestion and so relieves the feeling of a blocked nose. In commercial preparations that include menthol, a true decongestant such as oxymetazoline hydrochloride is often present.

13.22.1.3 Inhibition of Respiratory Drive and Respiratory Comfort

When cold air was circulated through the nose in human breath-hold experiments, subjects were able to hold their breath longer (McBride and Whitelaw, 1981), and inhaling cold air was shown to inhibit normal breathing patterns (Burgess and Whitelaw, 1988). This indicated that cold receptors could be one source of monitoring inspiratory flow rate and volume. Several animal experiments demonstrated that the inhalation of cold air, warm air plus menthol, or menthol alone (390 ng/mL) significantly enhanced ventilator inhibition (Orani et al., 1991; Sant'Ambrogio et al., 1992).

In 1992, Sloan et al. conducted breath-hold experiments with 20 healthy volunteers. The ingestion of a lozenge containing 11 mg of menthol significantly increased the hold time, indicating a depression of the ventilatory drive. It was later postulated by Eccles (2000) that in addition to chemoreceptors detecting oxygen and carbon dioxide in the blood, cold receptors in the respiratory tract may also modulate the drive to breathe.

Eleven healthy subjects breathed through a device that had either an elastic load or a flow-resistive load. Sensations of respiratory discomfort were compared using a VAS before, during, and after inhalation of menthol. It was found that the discomfort associated with loaded breathing was significantly reduced and was more effective during flow-resistive loading than elastic loading. Inhalation of another fragrance had no effect, and so the result was attributed to a direct stimulation of cold receptors by menthol, a reduction in respiratory drive being perhaps responsible (Nishino et al., 1997).

During an investigation of dyspnea, the effect of menthol inhalation on respiratory discomfort during loaded breathing was found to be inconsistent. Further tests found that the effect of menthol was most important during the first few minutes of inhalation and in the presence of high loads (Peiffer et al., 2001). The therapeutic application of menthol in the alleviation of dyspnea has yet to be described.

13.22.1.4 Bronchodilation and Airway Hyperresponsiveness

The spasmolytic activity of menthol on airway smooth muscle has been demonstrated in vitro (Taddei et al., 1988). To examine the bronchodilatory effects of menthol, a small trial was conducted on six patients with mild to moderate asthma. A poultice containing menthol was applied daily for 4 weeks, and it was found that bronchoconstriction was decreased and airway hyperresponsiveness improved (Chiyotani et al., 1994b).

A randomized, placebo-controlled trial examined the effects of menthol (10 mg nebulized twice daily for 4 weeks) on airway hyperresponsiveness in 23 patients with mild to moderate asthma. The diurnal variation in the peak expiratory flow rate (a value reflecting airway hyperexcitability) was decreased but the forced expiratory volume was not significantly altered. This indicated an improvement of airway hyperresponsiveness without affecting airflow limitation (Tamaoki et al., 1995). Later in vivo research examined the effect of menthol on airway resistance caused by capsaicin- and neurokinin-induced bronchoconstriction; there was a significant decrease in both cases by inhalation of menthol at 7.5 µg/L air concentration. The in vitro effect of menthol on bronchial rings was also studied. It was concluded that menthol attenuated bronchoconstriction by a direct action on bronchial smooth muscle (Wright et al., 1997).

In cases of asthma, the beneficial effects of menthol seem to be mainly due to its bronchodilatory activity on smooth muscle; interaction with cold receptors and the respiratory drive may also play an important role.

Recent in vitro studies have shown that a subpopulation of airway vagal afferent nerves express TRPM8 receptors and that activation of these receptors by cold and menthol excite these airway autonomic nerves. Thus, activation of TRPM8 receptors may provoke an autonomic nerve reflex to increase airway resistance. It was postulated that this autonomic response could provoke menthol- or cold-induced exacerbation of asthma and other pulmonary disorders (Xing et al., 2008). Direct cold stimulation or inhalation of menthol can cause immediate airway constriction and asthma in some people; perhaps, the TRPM8 receptor expression is up regulated in these subjects. The situation is far from clear.

13.22.1.5 Summary

The respiratory effects of menthol that have been demonstrated are the following:

Antitussive at low concentration
Increases the sensation of nasal airflow giving the impression of decongestion
No physical decongestant activity
Depresses ventilation and the respiratory drive at comparatively higher concentration
Reduces respiratory discomfort and sensations of dyspnea

A number of in vitro and animal experiments have demonstrated the bronchomucotropic activity of menthol (Boyd and Sheppard, 1969; Welsh et al., 1980; Chiyotani et al., 1994a), while there have been conflicting reports as to whether menthol is a mucociliary stimulant (Das et al., 1970) or is ciliotoxic (Su et al., 1993). Apart from the inclusion of relatively small quantities of menthol in commercial preparations that have known beneficial mucociliary effects, there are no documented human trials to support the presence of these activities.

13.22.2 1,8-Cineole

This oxide has a number of biological activities that make it particularly useful in the treatment of the respiratory tract. 1,8-Cineole has been registered as a licensed medication in Germany for over 20 years and is available as enteric-coated capsules (Soledum®). It is therefore not surprising that the majority of trials originate from this country and use oral dosing of 1,8-cineole instead of inhalation. Rather than discuss specific pathologies, the individual activities are examined and their relevance (alone or in combination) in treatment regimes should become apparent.

13.22.2.1 Antimicrobial

The antiinfectious properties of essential oils high in 1,8-cineole content may warrant their inclusion into a treatment regime, but other components are more effective in this regard. 1,8-Cineole is often considered to have marginal or no antibacterial activity (Kotan et al., 2007), although it is very effective at causing leakage of bacterial cell membranes (Carson et al., 2002). It may thus allow more active components to enter the bacteria by permeabilizing their membranes.

1,8-Cineole does possess noted antiviral properties when compared to the common essential oil components of borneol, citral, geraniol, limonene, linalool, menthol, and thymol; only that of eugenol was greater (Bourne et al., 1999). However, in comparison to the potent thujone, the antiviral potential of 1,8-cineole was considered relatively low (Sivropoulou et al., 1997).

A placebo-controlled, double-blind, randomized parallel group trial examined the long-term treatment of 246 chronic bronchitis during winter with myrtol-standardized Gelomyrtol® forte. This established German preparation consists mainly of 15% a-pinene, 35% limonene, and 47% 1,8-cineole and was administered 3× daily in 300 mg capsules. It was found to reduce the requirement for antibiotics during acute exacerbations; 51.6% compared to 61.2% under placebo. Of those patients needing antibiotics, 62.5% in the test group required them for ≤7 days, whereas 76.7% of patients in the placebo group needed antibiotics for >7 days. Moreover, 72% of patients remained without acute exacerbations in the test group as compared to 53% in the placebo group (Meister et al., 1999).

Although emphasis was given to antibiotic reduction, a significant antimicrobial effect by the preparation is unlikely to have paid an important contribution. Indeed, the paper refers to reduced health impairment due to sputum expectoration and cough and notes other beneficial properties of 1,8-cineole that is discussed in the succeeding text.

13.22.2.2 Antitussive

The antitussive effects of 1,8-cineole were first proven by Packman and London in 1980, who induced coughing in 32 healthy human subjects via the use of an aerosol spray containing citric acid. This single-blind crossover study examined the effect of a commercially available chest rub containing, among others, eucalyptus essential oil; the rub was applied to the chest in a 7.5 mg dose and massaged for 10–15 s, after which the frequency of the induced coughing was noted. It was found that the chest rub produced a significant decrease in the induced cough counts, and that eucalyptus oil was the most active component of the rub.

1,8-Cineole interacts with TRPM8, the cool-sensitive thermoreceptor that is primarily affected by menthol. In comparison to menthol, the effect of 1,8-cineole on TRPM8 (as measured by Ca^{2+} influx kinetics) is much slower and declines more rapidly (Behrendt et al., 2004). In a similar manner to menthol, the antitussive activity of 1,8-cineole may be due in part to its stimulation of airway cold receptors.

In a double-blind clinical trial, Gelomyrtol forte (300 mg 4× daily) or a placebo was administered to 413 patients with acute bronchitis for 2 weeks. Evaluation of efficacy was based primarily on the frequency of coughing fits, ability to expel mucus, and the bronchitis severity score, as noted by patient daily records and independent assessment. The results demonstrated significantly less daytime coughing fits, less difficulty in expectoration, less sleep disturbance due to coughing, and less symptomatic impairment. Gelomyrtol forte was thus significantly better than placebo in treating acute bronchitis (Gillissen et al., 2013). This trial used the product in a new capsule coating that reduced the incidence of gastrointestinal side effects, resulting in a relative risk reduction of 58%.

13.22.2.3 Bronchodilation

In vitro tests using guinea pig trachea determined that the essential oil of *Eucalyptus tereticornis* had a myorelaxant, dose-dependent effect (10–1000 µg/mL) on airway smooth muscle, reducing tracheal basal tone and K^+-induced contractions, and attenuating acetylcholine-induced contractions at higher concentrations (Coelho-de-Sousa et al., 2005). This activity was found to be mainly due to 1,8-cineole, although the overall effect was thought due to a synergistic relationship between the oxide and α- and β-pinene. Similar results were obtained using the essential oil of *Croton nepetaefolius*, whose major component was also 1,8-cineole (Magalhães et al., 2003).

A double-blind, randomized clinical trial over 7 days compared oral pure 1,8-cineole (3 × 200 mg/day) to Ambroxol (3 × 30 mg/day) in 29 patients with chronic obstructive pulmonary disease (COPD). Vital capacity, airway resistance, and specific airway conductance improved significantly for both drugs, while the intrathoracic gas volume was reduced by 1,8-cineole but not Ambroxol. All parameters of lung function, peak flow, and symptoms of dyspnea were improved by 1,8-cineole therapy but were not statistically significant in comparison to Ambroxol due to the small number of patients. In addition to other properties, it was noted that the oxide seemed to have bronchodilatory effects (Wittman et al., 1998).

13.22.2.4 Mucolytic and Mucociliary Effects

Mucolytics break down or dissolve mucus and thus facilitate the easier removal of these secretions from the respiratory tract by the ciliated epithelium, a process known as mucociliary clearance. Some mucolytics also have a direct action on the mucociliary apparatus itself.

Administered via steam inhalation to rabbits, 1,8-cineole in concentrations that produced a barely detectable scent (1–9 mg/kg) augmented the volume output of respiratory tract fluid from 9.5% to 45.3% (Boyd and Sheppard, 1971), an effect that they described as "mucotropic." Interestingly, in the same experiment, fenchone at 9 mg/kg increased the output by 186.2%, thus confirming the strong effects of some ketones in this regard. Also using rabbits, Zanker and Blumel (1983) found that oxygenated monoterpenoids reduced mucus deposition and partially recovered the activity of ciliated epithelium.

Since these early animal experiments, the beneficial effects of 1,8-cineole on mucociliary clearance have been clearly demonstrated in a number of human trials. Dorow et al. (1987) examined the effects of a 7-day course of either Gelomyrtol forte (4 × 300 mg/day) or Ambroxol (3 × 30 mg/day) in 20 patients with chronic obstructive bronchitis. Improved mucociliary clearance was observed in both groups, although improvement in lung function was not detected.

Twelve patients with chronic obstructive bronchitis were given a 4-day treatment with 1,8-cineole (4 × 200 mg/day). By measuring the reduction in percentage radioactivity of an applied radioaerosol, significant improvements in mucociliary clearance were demonstrated at the 60 and 120 min after each administration, (Dorow, 1989).

In a small double-blind study, the expectorant effect of Gelomyrtol forte (1 × 300 mg/day, 14 days) was examined in 20 patients with chronic obstructive bronchitis. The ability to expectorate, frequency of coughing attacks, and shortness of breath were all improved by the therapy, as was sputum volume and color. Both patients and physicians rated the effects of Gelomyrtol forte as better than the placebo, but due to the small group size, statistically significant differences could not be demonstrated (von Ulmer and Schött, 1991).

A randomized, double-blind, placebo-controlled trial was used to investigate the use of mucolytics to alleviate acute bronchitis (Mattys et al., 2000). They compared Gelomyrtol forte (4 × 300 mg, days 1–14), with Ambroxol (3 × 30 mg, days 1–3; 2 × 30 mg, days 4–14) and Cefuroxime (2 × 250 mg, days 1–6), in 676 patients. By monitoring cough frequency data, regression of the frequency of abnormal auscultation, hoarseness, headache, joint pain, and fatigue, it was shown that Gelomyrtol forte was very efficacious and comparable to the other active treatments. Overall, it scored slightly more than Ambroxol and Cefuroxime and was therefore considered to be a well-evidenced alternative to antibiotics for acute bronchitis.

Several studies have demonstrated a direct effect of 1,8-cineole on the ciliated epithelium itself. Kaspar et al. (1994) conducted a randomized, double-blind three-way crossover 4-day study of the effects of 1,8-cineole (3 × 200 mg/day) or Ambroxol (3 × 30 mg/day) on mucociliary clearance in 30 patients with COPD. Treatment with the oxide resulted in a statistically significant increase in the ciliary beat frequency of nasal cilia, a phenomenon that did not occur with the use of Ambroxol (an increase of 8.2% and 1.1%, respectively). A decrease of "saccharine time" was clinically relevant and significant after 1,8-cineole therapy (241 s) but not after Ambroxol (48 s). Lung function parameters were significantly improved equally by both drugs.

After the ingestion of Gelomyrtol forte (3 × 1 capsule/day for 4 days) by four healthy persons and one person after sinus surgery, there was a strong increase in mucociliary transport velocity, as detected by movement of a radiolabeled component (Behrbohm et al., 1995).

In sinusitis, the ciliated beat frequency is reduced and 30% of ciliated cells convert to mucus-secreting goblet cells. The impaired mucociliary transport, excessive secretion of mucus, and edema block drainage sites lead to congestion, pain, and pressure.

To demonstrate the importance of drainage and ventilation of sinuses as a therapeutic concept, Federspil et al. (1997) conducted a double-blind, randomized, placebo-controlled trial using 331 patients with acute sinusitis. The secretolytic effects of Gelomyrtol forte (300 mg) over a 6-day period proved to be significantly better than the placebo.

Kehrl et al. (2004) used the known stimulatory effects of 1,8-cineole on ciliated epithelium and its mucolytic effect as a rationale for treating 152 acute rhinosinusitis patients in a randomized, double-blind, placebo-controlled study. The treatment group received 3 × 200 mg 1,8-cineole daily for 7 days. There was a clinically relevant and significant improvement in frontal headache, headache on bending, pressure point sensitivity of the trigeminal nerve, nasal obstruction, and rhinological secretions in the test group, as compared to the control group. It was concluded that 1,8-cineole was a safe and effective treatment for acute nonpurulent rhinosinusitis before antibiotics are indicated.

13.22.2.5 Anti-Inflammatory Activity

The effects of 1,8-cineole on stimulated human monocyte mediator production were studied in vitro and compared to that of budesonide, a corticosteroid agent with anti-inflammatory and immunosuppressive effects (Juergens et al., 1998a). At therapeutic levels, both substances demonstrated a similar inhibition of the inflammatory mediators leukotriene B_4 (LTB_4), prostaglandin E_2 (PGE_2), and interleukin-1β (IL-1β). This was the first evidence of a steroid-like inhibition of arachidonic acid metabolism and IL-1β production by 1,8-cineole.

Later that year, the same team (Juergens et al., 1998b) reported a dose-dependent and highly significant inhibition of tumor necrosis factor-α (TNF-α), IL-1β, thromboxane B_2, and LTB_4 production by 1,8-cineole from stimulated human monocytes in vitro.

A third experiment combined ex vivo and in vivo testing; 10 patients with bronchial asthma were given 3 × 200 mg of 1,8-cineole daily for 3 days. Lung function was measured before the first dose, at the end of the third dose, and 4 days after discontinuation of the therapy. At the same time, blood samples were taken from which monocytes were collected and stimulated ex vivo for LTB_4 and PGE_2 production. Twelve healthy volunteers also underwent the treatment, and their blood was taken for testing. It was found that by the end of treatment and 4 days after, the production of LTB_4 and PGE_2 from the monocytes of both asthmatics and healthy individuals was significantly inhibited. Lung function parameters of asthmatic patients were significantly improved (Juergens et al., 1998c).

These three reports suggested a strong anti-inflammatory activity of 1,8-cineole via both the cyclooxygenase and 5-lipooxygenase pathways, and the possibility of a new, well-tolerated treatment of airway inflammation in obstructive airway disease.

In 2003, Juergens et al. conducted a double-blind, placebo-controlled clinical trial involving 32 patients with steroid-dependent severe bronchial asthma. The subjects were randomly assigned to receive either a placebo or 3 × 200 mg 1,8-cineole daily for 12 weeks. Oral glucocorticosteroids were reduced by 2.5 mg increments every 3 weeks with the aim of establishing the glucocorticosteroid-sparing capacity of 1,8-cineole. The majority of asthma patients receiving oral 1,8-cineole remained clinically stable despite a mean reduction of oral prednisolone dosage of 36%, equivalent to 3.8 mg/day. In the placebo group, where only four patients could tolerate a steroid decrease, the mean reduction was 7%, equivalent to 0.9 mg/day. Compared to the placebo group, 1,8-cineole recipients maintained their lung function four times longer despite receiving lower doses of prednisolone.

Increased mucus secretion often appears as an initial symptom in exacerbated COPD and asthma, where stimulated mediator cells migrate to the lungs to produce cytokines; of particular importance are TNF-α, IL-1β, IL-6, and IL-8, and those known to induce IgE antibody synthesis and maintain allergic eosinophilic inflammation (IL-4 and IL-5). Therefore, a study was conducted to investigate the role of 1,8-cineole in inhibiting cytokine production in stimulated human monocytes and lymphocytes in vitro (Juergens et al., 2004). It was shown that 1,8-cineole is a strong inhibitor of TNF-α and IL-1β in both cell types. At known therapeutic blood levels, it also had an inhibitory effect on the production of the chemotactic cytokines IL-8 and IL-5 and may possess additional antiallergic activity by blocking IL-4 production.

A clinically relevant anti-inflammatory activity of 1,8-cineole has thus been proven for therapeutic use in airway diseases.

13.22.2.6 Pulmonary Function

An inhaler was used to apply 1,8-cineole (Soledum Balm) to 24 patients with asthma or chronic bronchitis in a 8-day-controlled trial. In all but one patient, an objective rise in expiratory peak flow values was demonstrated. The subjective experience of their illness was significantly improved for all subjects (Grimm, 1987).

In an open trial of 100 chronic bronchitis using both inhaled (4 × 200 mg) and oral (3 × 200 mg) 1,8-cineole over 7 days, the clinical parameters of forced vital capacity, forced expiratory volume, peak expiratory flow, and residual volume were all significantly improved when compared to initial values before treatment (Mahlo, 1990).

In a randomized, double-blind, placebo-controlled study of 51 patients with COPD, 1,8-cineole (3 × 200 mg/day) was given for 8 weeks. For the objective lung functions of "airway resistance" and "specific airway resistance," there was a clinically significant reduction of 21% and 26%, respectively. The improvement was attributed to a positive influence on disturbed breathing patterns, mucociliary clearance, and anti-inflammatory effects (Habich and Repges, 1994).

As 1,8-cineole reduces the exacerbation rate in patients with COPD, it was thought that asthma patients would also benefit from the anti-inflammatory, bronchodilating, and mucolytic effects. A double-blind, placebo-controlled study administered 200 mg of 1,8-cineole or placebo three times a day for 6 months to 247 patients with confirmed asthma. This was done concomitantly with normal medication. The outcome measures of lung function, asthma symptoms, and quality of life were all significantly improved in the test group as compared to the placebo. It was concluded that concomitant therapy with 1,8-cineole in asthma patients could lead to substantial improvement in lung function and health condition and reduce dyspnea (Worth and Dethlefsen, 2012).

The majority of the in vivo trials involving 1,8-cineole report good, if not significant, changes in lung function parameters, whether the investigation concerns the common cold, asthma, or COPD. This is not a convenient, accidental side effect of treatment but is a direct result of one or more of the factors already discussed that have direct effects on the pathophysiology of the airways. The ability to breathe more effectively and easily is an important consequence of therapy that is sometimes minimized when dealing with the specific complexities of infection, inflammation, etc.

A compilation of human trials with 1,8-cineole is given in Table 13.5.

TABLE 13.5
Summary of Human Trials Demonstrating the Beneficial Effects of 1,8-cineole in Various Respiratory Conditions

Patients	Treatment	Outcome	References
Asthma			
11	Cineole inhalation, 8 days	Objective rise in expiratory peak flow found. Subjective experiences of illness significantly improved.	Grimm (1987)
10	Cineole 3 × 200 mg daily, 3 days	LTB_4 and PGE_2 production by monocytes was significantly inhibited. Lung functions were significantly improved.	Juergens et al. (1998)
32	Cineole 3 × 200 mg daily, 12 weeks	12 of 16 patients in cineole group remained stable despite a 36% reduction in oral steroid dosage.	Juergens et al. (2003)
247	Cineole 3 × 200 mg daily, 6 months	All test patients showed significant improvement in lung function, asthma symptoms, and quality of life.	Worth and Dethlefsen (2012)
Acute bronchitis			
60	VapoRub 3 min	Decreased breathing frequency, suggesting "easier breathing."	Berger et al. (1978)
676	Gelomyrtol 4 × 300 mg daily, 14 days	Coughing, sputum consistency, well-being, bronchial hyperreactivity, and associated symptoms improved similarly by Gelomyrtol, Ambroxol, and Cefuroxime.	Mathys et al. (2000)
413	Gelomyrtol 4 × 300 mg daily, 14 days	Significantly less daytime coughing fits, less difficulty in expectoration, less sleep disturbance due to coughing, and less symptomatic impairment.	Gillissen et al. (2013)
Chronic bronchitis			
9	Cineole inhalation, 8 days	Objective rise in expiratory peak flow found. Subjective experiences of illness significantly improved.	Grimm (1987)
100	Cineole 3 × 200 mg daily, 7 days	All lung function parameters significantly improved.	Mahlo (1990)
246	Gelomyrtol 3 × 300 mg daily, 6 months	Reduced acute exacerbations; reduced requirement for antibiotics; reduced treatment times when antibiotics taken. Well-being significantly improved.	Meister et al. (1999)
Chronic obstructive pulmonary disease			
20	Gelomyrtol 4 × 0.3 g daily, 7 days	Improved mucociliary clearance.	Dorow et al. (1987)
20	Gelomyrtol 1 × 0.3 g daily, 14 days	All parameters relating to coughing improved. Sputum volume increased.	von Ulmer and Schött (1991)
12	Cineole 4 × 200 mg, 4 days	Significant improvement of mucociliary clearance.	Dorow (1989)
51 inc. 16 asthmatics	Cineole 3 × 200 mg daily, 8 weeks	Significant improvement in airway resistance (21%); positive effects on sputum output and dyspnea.	Habich and Repges (1994)

(Continued)

TABLE 13.5 (*Continued*)
Summary of Human Trials Demonstrating the Beneficial Effects of 1,8-cineole in Various Respiratory Conditions

Patients	Treatment	Outcome	References
30	Cineole 1 × 200 mg daily, 4 days	Significant improvements in lung functions of FVC and FEV_1 (ambroxol and cineole equieffective); significant increase in ciliary beat frequency.	Kaspar et al. (1994)
29	Cineole 3 × 200 mg daily, 7 days	All lung function parameters, peak flow, and dyspnea improved from day 1 onward.	Wittmann et al. (1998)
Common cold			
24	Eucalyptus oil (9% of a mixture)	Reversed lung function abnormalities in small and large airways.	Cohen and Dressler (1982)
Sinusitis			
331	Gelomyrtol 300 mg, 6 days	Effective treatment instead of antibiotics.	Federspil et al. (1997)
152	Cineole 3 × 200 mg daily, 7 days	Effective reduction of symptoms without the need for antibiotics.	Kehrl et al. (2004)

13.22.2.7 Summary

Although discussed separately, the multifaceted activities of 1,8-cineole perform together in harmony to provide an effective intervention that can inherently adapt to the needs of the individual patient. As already described, 1,8-cineole is known to possess the following properties:

- Antimicrobial
- Antitussive
- Bronchodilatory
- Mucolytic
- Ciliary transport promotion
- Anti-inflammatory
- Lung function improvement

Therefore, it may be seen that a diverse range of respiratory conditions of varying complexities will benefit from the use of pure 1,8-cineole or from essential oils containing this oxide as a major component.

13.22.3 Treatment with Blends Containing Both Menthol and 1,8-Cineole

A study measured transthoracic impedance pneumographs of 60 young children (2–40 months) with acute bronchitis before and after a 3 min application of VapoRub® to the back and chest. The data showed an early increase in amplitude up to 33%, which slowly descended during the 70-min posttreatment period to slightly above the control. Breathing frequency progressively decreased during the same period by 19.4%. Clinical observations combined with these results suggested a condition of "easier breathing" (Berger et al., 1978a). Currently, the active ingredients of VapoRub are camphor 4.8%, 1,8-cineole 1.2%, and menthol 2.6, but these components and percentages may have changed over the years.

The same team employed a similar experiment but used the pneumographic data to examine the quiet periods, that is, parts of the pneumogram where changes in the baseline were at least half of the average amplitude in more than five consecutive breathing excursions. It was found that the application of VapoRub increased quiet periods by up to 213.8%, whereas the controls (petroleum jelly application or rubbing only) never exceeded 62.4%. Thus, the breathing restlessness of children with bronchitis was diminished, and this was confirmed by clinical observations (Berger et al., 1978b).

By the measurement of lung and forced expiratory volumes, nasal, lower and total airway resistances, closing volume data, the phase III slope of the alveolar plateau, and the maximum expiratory flow volume, peripheral airway dysfunction was confirmed in 24 adults with common colds. In a randomized, controlled trial, an aromatic mixture of menthol, eucalyptus oil, and camphor (56%, 9%, and 35% w/w, respectively) were vaporized in a room where the subjects were seated. Respiratory function measurements were made at baseline, 20 and 60 min after exposure. After the last measurement, phenylephrine was sprayed into the nostrils and the measurements taken again 5–10 min later to determine potential airway responsiveness. The control consisted of tap water. The results showed significant changes in forced vital capacity, forced expiratory volume, closing capacity, and the phase III slope after aromatic therapy as compared to the control. It was concluded that the aromatic inhalation favorably modified the peripheral airway dysfunction (Cohen and Dressler, 1982).

In a randomized, placebo-controlled trial of citric acid-induced cough in 20 healthy subjects, the inhalation of a combination of menthol and eucalyptus oil (75% and 25%, respectively) significantly decreased the cough frequency (Morice et al., 1994).

The effect of an aromatic inunction (VapoRub) was studied by the inhalation of a radioaerosol in a randomized, single-blinded, placebo-controlled crossover trial with 12 chronic bronchitis. It was found that after application of 7.5 g of the product to the chest, removal of the tracheobronchial deposit was significantly enhanced at 30 and 60 min postinhalation, although further effects could not be demonstrated during the following 5 h, despite further application of the rub. During the first hour, mucociliary clearance was correlated with the concentration level of the aromatics (Hasani et al., 2003).

Another commercial preparation Pinimenthol®, a mixture of eucalyptus and pine needle oils plus menthol, reduced bronchospasm and demonstrated significant secretolytic effects when insufflated through the respiratory tract and when applied to the epilated skin of animals (Schäfer and Schäfer, 1981). In addition to the known effects of menthol and 1,8-cineole, pine needle oil is considered to be weakly antiseptic and secretolytic (Commission E, 1998).

In a randomized, double-blind 14-day trial, 100 patients with chronic obstructive bronchitis received a combination of theophylline with β-adrenergica 2–3 times daily. The test group also received Pinimenthol. The parameters that were investigated were objective (measurement of lung function and sputum) and subjective (cough, respiratory insufficiency, and pulmonary murmur). All differences in the subjective evaluations were statistically significant and of clinical importance; secretolysis was clearly shown. The addition of Pinimenthol showed a clear superiority to the basic combination therapy alone (Linsenmann and Swoboda, 1986).

A postmarketing survey was conducted of 3060 patients prescribed Pinimenthol suffering from cold, acute or chronic bronchitis, bronchial catarrh, or hoarseness. The product was given by inunction (29.6%), inhalation (17.3%), or inunction and inhalation (53.1%). Only 22 patients reported adverse effects, and the efficacy of the product was judged as excellent or good by 88.3% of physicians and 88.1% of patients (Kamin and Kieser, 2007).

13.23 SNORING

A blend of 15 essential oils was developed into a commercial product called "Helps stop snoring," and 140 adult snorers were recruited into a randomized trial using the product as a spray or gargle. VAS was completed by the snorers' partners relating to sleep disturbance each night. The treatment lasted 14 days and results were compared to a pretrial period of the same length. The partners of

82% of the patients using the spray and 71% of patients using the gargle reported a reduction in snoring. This was compared to 44% of placebo users. The mode of action was postulated as being antispasmodic to the soft palate and pharynx (Pritchard, 2004).

13.24 SWALLOWING DYSFUNCTION

A delayed triggering of the swallowing reflex, mainly in elderly people, predisposes to aspiration pneumonia. To improve dysphagia, two different approaches using essential oils have been tried with success.

As black pepper is a strong appetite stimulant, it was postulated that nasal inhalation of the essential oil may stimulate cerebral blood flow in the insular cortex, the dysfunction of which has been reported to play a role in dysphagia. A randomized, controlled study of 105 elderly patients found that the inhalation of black pepper oil for 1 min significantly shortened the delayed swallowing time and increased the number of swallowing movements. Emission-computed tomography demonstrated activation of the anterior cingulate cortex by the treatment. The inhalation of lavender essential oil or water had no effects (Ebihara et al., 2006a).

A second study used the established stimulating effects of menthol on cold receptors, since cold stimulation was known to restore sensitivity to trigger the swallowing reflex in dysphagic patients. Menthol was introduced into the pharynx of patients with mild to moderate dysphagia via a nasal catheter. The latent time of swallowing reflex was reduced significantly by menthol in a concentration-dependent manner; 10^{-2} menthol reduced the time to 9.4 s as compared to 13.8 s for distilled water. The use of a menthol lozenge before meals was thought appropriate (Ebihara et al., 2006b).

13.25 CONCLUSION

It is apparent from the diverse range of conditions that have benefitted from the administration of essential oils that their therapeutic potential is vast and yet underdeveloped. Moreover, since they are not composed of a single "magic bullet" with one target, they often have multiple effects that have additive or synergistic properties within a treatment regime.

A great many research papers investigating the bioactivity of essential oils conclude that the results are very encouraging and that clinical trials are the next step. For the majority, this step is never taken. The expense is one limiting factor, and it is not surprising that clinical trials are mostly conducted once the essential oils have been formulated into a commercial product that has financial backing.

It is evident that many of the claims made for essential oils in therapeutic applications have not been substantiated and an evidence base is clearly lacking. However, there is similarly a lack of research to demonstrate that essential oils are not effective interventions.

With the continuing search for new medicaments from natural sources, especially in the realm of antimicrobial therapy, it is hoped that future research into the efficacy of essential oils will be both stimulated and funded.

REFERENCES

Abdollahi, M., Salehnia, A., Mortazavi, S. H. et al. 2003. Antioxidant, antidiabetic, antihyperlipidaemic, reproduction stimulatory properties and safety of essential oil of Satureja khuzestanica in rat in vivo: A toxicopharmacological study. *Med. Sci. Monit.* 9:BR331–BR335.

Alexandrovich, I., Rakovitskaya, O., Kolmo, E., Sidorova, T., Shushunov, S. 2003. The effect of fennel (*Foeniculum vulgare*) seed oil emulsion in infantile colic: A randomised, placebo-controlled study. *Altern. Ther. Health Med.* 9:58–61.

Al-Mosawi, A. J. 2005. A possible role of essential oil terpenes in the management of childhood urolithiasis. *Therapy* 2:243–247.

Amanlou, M., Babaee, N., Saheb-Jamee, M. et al. 2007. Efficacy of Satureja khuzistanica extract and essential oil preparations in the management of recurrent aphthous stomatitis. *Daru* 15:231–234.

Amanlou, M., Dadkhah, F., Salehnia, A., Farsam, H., Dehpour, A. R. 2005. An anti-inflammaory and anti-nociceptive effects of hydroalcoholic extract of Satureja khuzistanica Jamzad extract. *J. Pharm. Pharmacol. Sci.* 8:102–106.

Amanlou, M., Fazeli, M. R., Arvin, A., Amin, H. G., Farsam, H. 2004. Antimicrobial activity of crude methanolic extract of Satureja khuzistanica. *Fitoterapia* 75:768–770.

Ansari, M. A., Vasudevan, P., Tandon, M., Razdan, R. K. 2000. Larvicidal and mosquito repellent action of peppermint (*Mentha piperita*) oil. *Bioresour. Technol.* 71(3):267–271.

Arweiler, N. B., Donos, N., Netuschil, L., Reich, E., Sculean, A. 2000. Clinical and antibacterial effect of tea tree oil—A pilot study. *Clin. Oral Investig.* 4:70–73.

Asao, T., Kuwano, H., Ide, M. et al. 2003. Spasmolytic effect of peppermint oil in barium during double-contrast barium enema compared with buscopan. *Clin. Radiol.* 58:301–305.

Astani, A. M., Reichling, J., Schnitzler, P. 2010. Comparative study on the antiviral activity of selected monoterpenes derived from essential oils. *Phytother. Res.* 24(5):673–679.

Bailey, H. H., Attia, S., Love, R. R. et al. 2008. Phase II trial of daily oral perillyl alcohol (NSC 641066) in treatment-refractory metastatic breast cancer. *Cancer Chemother. Pharmacol.* 62:149–157.

Bailey, H. H., Levy, D., Harris, L. S. et al. 2002. A phase II trial of daily perillyl alcohol in patients with advanced ovarian cancer: Eastern cooperative oncology group study E2E96. *Gynecol. Oncol.* 85:464–468.

Bak, C. W., Yoon, S. J., Chung, H. 2007. Effects of an α-blocker and terpene mixture for pain control and spontaneous expulsion of ureter stone. *Korean J. Urol.* 48:517–521.

Behrbohm, H., Kaschke, O., Sydnow, K. 1995. Der Einfluß des pflanzlichen Sekretolytikums Gelomyrtol® forte auf die mukozililäre Clearance der Kieferhöhle. *Laryngo-Rhino-Otol.* 74:733–737.

Behrendt, H-J., Germann, T., Gillen, C., Hatt, H., Jostock, R. 2004. Characterisation of the mouse cold-menthol receptor TRPM8 and vanilloid receptor type-1 VR1 using a fluorometric imaging plate reader (FLIPR) assay. *Br. J. Pharmacol.* 141:737–745.

Bell, G. D., Bradshaw, J. P., Burgess, A. et al. 1980. Elevation of serum high density lipoprotein cholesterol by Rowachol, a proprietary mixture of six pure monoterpenes. *Atherosclerosis* 36:47–54.

Ben-Arye, E., Dudai, N., Eini, A., Torem, M., Schiff, E., Rakover, Y. 2011. Treatment of upper respiratory tract infections in primary care: A randomized study using aromatic herbs. *Evid. Based Complement. Alternat. Med.* 2011:690346, doi:10.1155/2011/690346.

Benelli, G., Flamini, G., Fiore, G., Cioni, P. L., Conti, B. 2013. Larvicidal and repellent activity of the essential oil of *Coriandrum sativum* L. (Apiaceae) fruits against the filariasis vector *Aedes albopictus* Skuse (Diptera: Culicidae). *Parasitol. Res.* 112(3):1155–1161.

Berger, H., Jarosch, E., Madreiter, H. 1978a. Effect of vaporub and petroleum on frequency and amplitude of breathing in children with acute bronchitis. *J. Int. Med. Res.* 6:483–486.

Berger, H., Jarosch, E., Madreiter, H. 1978b. Effect of vaporub on the restlessness of children with acute bronchitis. *J. Int. Med. Res.* 6(6):491–493.

Bero, J., Beaufay, C., Hannaert, V., Herent, M. F., Michels, P. A., Quetin-Leclercq, J. 2013. Antitrypanosomal compounds from the essential oil and extracts of Keetia leucantha leaves with inhibitor activity on *Trypanosoma brucei* glyceraldehyde-3-phosphate dehydrogenase. *Phytomedicine* 20(3–4):270–274.

Blackwell, A. L. 1991. Tea tree oil and anaerobic (bacterial) vaginosis. *Lancet* 337(8736):300.

Botelho, M. A., Bezerra Filho, J. G., Correa, L. L. et al. 2007. Effect of a novel essential oil mouthrinse without alcohol on gingivitis: A double-blinded randomised controlled trial. *J. Appl. Oral Sci.* 15:175–180.

Bourne, K. Z., Bourne, N., Reising, S. F., Stanberry, L. R. 1999. Plant products as topical microbicide candidates: Assessment of in vitro and in vivo activity against *Herpes simplex* virus type 2. *Antiviral Res.* 42:219–226.

Boyd, E. M., Sheppard, E. P. 1969. A bronchomucotropic action in rabbits from inhaled menthol and thymol. *Arch. Int. Pharmacodyn. Ther.* 182:206–213.

Boyd, E. M., Sheppard, E. P. 1971. An autumn-enhanced mucotropic action of inhaled terpenes and related volatile agents. *Pharmacology* 6:65–80.

Bradley, B. F., Brown, S. L., Chu, S., Lea, R. W. 2009. Effects of orally administered lavender essential oil on responses to anxiety-provoking film clips. *Hum. Psychopharmacol.* 24(4):319–330.

Burgess, I. F., Brunton, E. R., Burgess, N. A. 2010. Clinical trial showing superiority of a coconut and anise spray over permethrin 0.43% lotion for head louse infestation, ISRCTN96469780. *Eur. J. Pediatr.* 169(1):55–62.

Burgess, K. R., Whitelaw, W. A. 1988. Effects of nasal cold receptors on patterns of breathing. *J. Appl. Physiol.* 64:371–376.

Burke, B. E., Baillie, J-E., Olson, R. D. 2004. Essential oil of Australian lemon myrtle (*Backhousia citriodora*) in the treatment of molluscum contagiosum in children. *Biomed. Pharmacother.* 58:245–247.

Burrow, A., Eccles, R., Jones, A. S. 1983. The effects of camphor, eucalyptus and menthol vapours on nasal resistance to airflow and nasal sensation. *Acta Otolaryngol.* 96:157–161.

Caelli, M., Porteous, J., Carson, C. F., Heller, R., Riley, T. V. 2000. Tea tree oils an alternative topical decolonization agent for methicillin-resistant *Staphylococcus aureus*. *J. Hosp. Infect.* 46:236–237.

Capanni, M., Surrenti, E., Biagini, M. R., Milani, S., Surrenh, C., Galli, A. 2005. Efficacy of peppermint oil in the treatment of irritable bowel syndrome: A randomized, controlled trial. *Gazz. Med. Ital. Arch. Sci. Med.* 164:119–126.

Cappello, G., Spezzaferro, M., Grossi, L., Manzoli, L., Marzio, L. 2007. Peppermint oil (Mintol®) in the treatment of irritable bowel syndrome: A prospective double blind placebo-controlled randomized trial. *Digest. Liver Dis.* 39:530–536.

Carmo, E. S., Pereira Fde, O., Cavalcante, N. M., Gayoso, C. W., Lima Ede, O. 2013. Treatment of pityriasis versicolor with topical application of essential oil of Cymbopogon citratus (DC) Stapf - therapeutic pilot study. *An. Bras. Dermatol.* 88(3):381–385.

Carson, C. F., Ashton, L., Dry, L., Smith, D. W., Riley, T. V. 2001. Melaleuca alternifolia (tea tree) oil gel (6%) for the treatment of recurrent herpes labialis. *J. Antimicrob. Chemother* 48:445–446.

Carson, C.F., Cookson, B.D., Farrelly, H.D, Riley, T.V. 1995. Susceptibility of methicillin-resistant Staphylococcus aureus to the essential oil of *Melaleuca alternifolia*. *J. Antimicrob. Chemother.* 35:421–424.

Carson, C. F., Mee, B. J., Rilet, T. V. 2002. Mechanism of action of *Melaleuca alternifoila* (tea tree) oil on *Staphylococcus aureus* determined by time-kill, lysis, leakage and salt tolerance assays and electron microscopy. *Antimicrob Agents Chemother* 48:1914–1920.

Cetin, H., Yanikoglu, A., Cilek, J. E. 2011. Larvicidal activity of selected plant hydrodistillate extracts against the house mosquito, *Culex pipiens*, a West Nile virus vector. *Parasitol. Res.* 108(4):943–948.

Charles, C. H., Vincent, J. W., Borycheski, L. et al. 2000. Effect of an essential oil-containing dentifrice on dental plaque microbial composition. *Am. J. Dent.* 13:26C–30C.

Chiyotani, A., Tamaoki J., Sakai, N. 1994b. Effect of menthol on peak expiratory flow in patients with bronchial asthma. *Jpn. J. Chest Dis.* 53:949–953.

Chiyotani, A., Tamaoki, J., Takeuchi, S., Kondo, M., Isono, K., Konno, K. 1994a. Stimulation by menthol of Cl secretion via a Ca^{2+}-dependent mechanism in canine airway epithelium. *Br. J. Pharmacol.* 112:571–575.

Clarke, R. W., Jones, A. S., Charters, P., Sherman, I. 1992. The role of mucosal receptors in the nasal sensation of airflow. *Clin. Otolaryngol.* 17:383–387.

Clegg, R. J., Middleton, B., Bell, G. D., White, D. A. 1980. Inhibition of hepatic cholesterol synthesis and S-3-hydroxy-3-methylglutaryl-CoA reductase by mono and bicyclic monoterpenes administered in vivo. *Biochem. Pharmacol.* 29:2125–2127.

Coelho, J., Kohut, B. E., Mankodi, S., Parikh, R., Wu, M-M. 2000. Essential oils in an antiplaque and antigingivitis dentifrice: A 6-month study. *Am. J. Dent.* 13:5C–10C.

Coelho-de-Souza, L. N., Leal-Cardoso, J. H., de Abreu Matos, F. J., Lahlou, S., Magalhães, P. J. C. 2005. Relaxant effects of the essential oil of *Eucalyptus tereticornis* and its main constituent 1,8-cineole on guinea-pig tracheal smooth muscle. *Planta Med.* 71:1173–1175.

Cohen, B. M., Dressler, W. E. 1982. Acute aromatics inhalation modifies the airways. Effects of the common cold. *Respiration* 43:285–293.

Commission E Monographs. 1998. Pine needle oil. Blumenthal, E., Busse, W. R., Goldberg, A. et al. (eds.). Boston, MA: American Botanical Council. Chapter 2, p. 185.

Cooke, C. J., Nanjee, M. N., Dewey, P., Cooper, J. A., Miller, G. J., Miller, N. E. 1998. Plant monoterpenes do not raise plasma high-density-lipoprotein concentrations in humans. *Am. J. Clin. Nutr.* 68:1042–1045.

Cooper, M. L., Laxer, J. A., and Hansbrought, J. F. 1991. The cytotoxic effects of commonly used topical antimicrobial agents on human fibroblasts and keratinocytes. *J. Trauma.* 31:775–782.

Coskun, S., Girisgin, O., Kürkcüoglu, M. et al. 2008. Acaricidal efficacy of *Origanum onites* L. essential oil against *Rhipicephalus turanicus* (Ixodidae). *Parasitol. Res.* 103:259–261.

da Silva, A. G., Puziol, P. F., Leitão, R. N., Gomes, T. R., Scherer, R., Martins, M. L. L., Cavalcanti, A. S. S., and Cavalcanti, L. C. 2012. Application of the essential oil from copaiba (*Copaifera langsdorffii* Desf.) for acne vulgaris: A double-blind, placebo controlled clinical trial. *Altern. Med. Rev.* 17:69–75.

Dale, A., Cornwell, S. 1994. The role of lavender oil in relieving perineal discomfort following childbirth: A blind randomised clinical trial. *J. Adv. Nurs.* 19(1):89–96.

Das, P. K., Rathor, R. S., San Yal, A. K. 1970. Effect on ciliary movements of some agents which come into contact with the respiratory tract. *Ind. J. Physiol. Pharmacol.* 14:297–303.

de Cassia da Silveira, E. S. R., Andrade, L. N., de Sousa, D. P. 2013. A review on anti-inflammatory activity of monoterpenes. *Molecules* 18(1):1227–1254.

de Lima, G. P. C., de Souza, T. M., Feire, G. P., Farais, D. F., Cunha, A. P., Ricardo, N. M. P. S., de Morais, S. M., Carvalho, A. F. U. 2013. Further insecticidal activities of essential oils from *Lippia sidoides* and Croton species against *Aedes aegypti* L. *Parasitol. Res.* 112(5):1953–1958.

de Menezes, I. A. C., Moreira, I. J. A., de Paula, J. W. A., Blank, A. F., Antoniolli, A. R., Quintans-Júnior, L. J., Santos, M. R. V. 2010. Cardiovascular effects induced by Cymbopogon winterianus essential oil in rats: Involvement of calcium channels and vagal pathway. *J. Pharm. Pharmacol.* 62(2):215–221.

de Pradier, E. 2006. A trial of a mixture of three essential oils in the treatment of postoperative nausea and vomiting. *Int. J. Aromather.* 16:15–20.

de Siqueira, R., Magalhães, P., Leal-Cardoso, J., Duarte, G., Lahlou, S. 2006. Cardiovascular effects of the essential oil of *Croton zehntneri* leaves and its main constituents, anethole and estragole, in normotensive conscious rats. *Life Sci.* 78(20):2365–2372.

Dew, M. J., Evans, B. K., Rhodes, J. 1984. Peppermint oil for the irritable bowel syndrome: A multicentre trial. *Br. J. Clin. Pract.* 38:394, 398.

Djaladat, H., Mahouri, K., Shooshtary, F. K., Ahmadieh, A. 2009. Effect of Rowatinex on calculus clearance after extracorporeal shock wave lithotripsy. *Urol. Res.* 6(1):9–13.

Doran, J., Bell, G. D. 1979. Gallstone dissolution in man using an essential oil preparation. *Br. Med. J.* 1:24.

Dorow, P. 1989. Welchen Einfluß hat Cineol auf die mukoziliare Clearance? *Therapiewoche* 39:2652–2654.

Dorow, P., Weiss, Th., Felix, R., Schmutzler, H. 1987. Einfluß eines Sekretolytikums und einer Kombination von Pinen, Limonen und Cineol auf die mukoziliäre Clearance bei Patienten mit chronisch obstruktiver Atemwegserkrankung. *Arzneim.-Forsch.* 37:1378–1381.

DosSantos, F.J.B., Lopes, J.A.D., Cito, A.M.G.L., DeOliveira, E. H., DeLima, S. G., Reis, F.D.A.M. 2004. Composition and biological activity of essential oils from Lippia origanoides H.B.K. *J. Essent. Oil Res.* 16:504–506.

Dryden, M. S., Dailly, S., Crouch, M. 2004. A randomised, controlled trial of tea tree topical preparations versus a standard topical regimen for the clearance of MRSA colonization. *J. Hosp. Infect.* 56:283–286.

Duncan, R. E., El-Sohemy, A., Archer, M. C. et al. 2005. Dietary factors and the regulation of 3-hydroxy-3-methylglutaryl coenzyme A reductase: Implications for breast cancer development. *Mol. Nutr. Food Res.* 49:93–100.

Ebihara, T., Ebihara, S., Maruyama, M. et al. 2006a. A randomised trial of olfactory stimulation using black pepper oil in older people with swallowing dysfunction. *J. Am. Geriatr. Soc.* 54:1401–1406.

Ebihara, T., Ebihara, S., Watando, A. et al. 2006b. Effects of menthol on the triggering of the swallowing reflex in elderly patients with dysphagia. *Br. J. Clin. Pharmacol.* 62:369–371.

Eccles, R. 2000. Role of cold receptors and menthol in thirst, the drive to breathe and arousal. *Appetite.* 34:29–35.

Eccles, R., Griffiths, D. H., Newton, C. G., Tolley, N. S. 1988. The effects of menthol isomers on nasal sensation of airflow. *Clin. Otolaryngol.* 13:25–29.

Eccles, R., Jawad, M. S., Morris, S. 1990. The effects of oral administration of (-)-menthol on nasal resistance to airflow and nasal sensation of airflow in subjects suffering from nasal congestion associated with the common cold. *J. Pharm. Pharmacol.* 42:652–654.

El-Zemity, S., Rezk, H., Farok, S., Zaitoon, A. 2006. Acaricidal activities of some essential oils and their monoterpenoidal constituents against house dust mite, dermatophagoides pteronyssinus (Acari: Pyroglyphidae). *J. Zhejiang Uni. Sci. B.* 7:957–962.

Ellis, W. R., Bell, G. D. 1981. Treatment of biliary duct stones with a terpene preparation. *Br. Med. J.* 282:611.

Ellis, W. R., Bell, G. D., Middleton, B., White, D. A. 1981. An adjunct to bile acid therapy for gallstone dissolution—Combination of low dose chenodeoxycholic acid with a terpene preparation. *Br. Med. J.* 1:611–612.

Elson, C. E., Yu, S. G. 1994. The chemoprevention of cancer by mevalonate-derived constituents of fruits and vegetables. *J. Nutr.* 124:607–614.

Engelstein, D., Kahan, E., Servadio, C. 1992. Rowatinex for the treatment of ureterolithiasis. *J. Urol.* 98:98–100.

Enshaieh, S., Jooya, A., Siadat, A. H., Iraji, F. 2007. The efficacy of 5% topical tea tree oil gel in mild to moderate acne vulgaris: A randomised, double-blind-placebo-controlled study. *Indian J. Dermatol. Venereol. Leprol.* 73:22–25.

Escobar, P., Leal, S. M., Herrera, L. V., Martinez, J. R., Staskenko, E. 2010. Chemical composition and antiprotozoal activities of *Colombian Lippia* spp essential oils and their major components. *Mem. Instit. Oswaldo Cruz.* 105(2):184–190.

Evergetis, E., Michaelakis, A., Kioulos, E., Koliopoulos, G., Haroutounian, S. A. 2009. Chemical composition and larvicidal activity of essential oils from six Apiaceae family taxa against the West Nile virus vector *Culex pipiens. Parasitol. Res.* 105(1):117–124.

Federspil, P., Wulkow, R., Zimmermann, Th. 1997. Wirkung von Myrtol standardisiert bei der Therapie der akuten Sinusitis—Ergebnisse einer doppelblinden, randomisierten Multicenterstudie gegen Plazebo. *Laryngo-Rhino-Otol.* 76:23–27.

Fernandez, L. F., Palomino, O. M., Frutos, G. 2014. Effectiveness of *Rosmarinus officinalis* essential oil as antihypotensive agent in primary hypotensive patients and its influence on health-related quality of life. *J. Ethnopharmacol.* 151(1):509–516.

Ferrara, L., Naviglio, D., Caruso, A. A. 2012. Cytological aspects on the effects of a nasal spray consisting of standardized extract of citrus lemon and essential oils in allergic rhinopathy. *ISRN Pharm.* 2012: doi:10.5402/2012/404606.

Fichi, G., Flamini, G., Zaralli, L.J., Perrucci, S. 2007. Efficacy of an essentifal oil of *Cinnamomum zeylanicum* against *Psoroptes cuniculi*. *Phytomedicine* 14:227–231.

Fine, D. H., Markowitz, K., Furgang, D. et al. 2007. Effect of rinsing with an essential oil-containing mouthrinse on subgingival periodontopathogens. *J. Periodontol.* 78:1935–1942.

Fissler, M., Quante, A. 2014. A case series on the use of lavendula oil capsules in patients suffering from major depressive disorder and symptoms of psychomotor agitation, insomnia and anxiety. *Complement. Ther. Med.* 22(1):63–69.

Fournet, A., Rojas de Arias, A., Charles, B., Bruneton, J. 1996. Chemical constituents of essential oils of Muna, Bolivian plants traditionally used as pesticides, and their insecticidal properties against Chagas' disease vectors. *J. Ethnopharmacol.* 52(3):145–149.

Freidman, L. S. 1998. Helicobacter pylori and nonulcer dyspepsia. *N. Engl. J. Med.* 339:1928–1930.

Fujisaki, R., Kamei, K., Yamamura, M., Nishiya, H., Inouye, S., Takahashi, M., Abe, S. 2012. In vitro and in vivo anti-plasmodial activity of essential oils, including hinokitiol. *Southeast Asian J. Trop. Med. Public Health.* 43(2):270–279.

Ganji, F. H., Yusefi, H., Baradaran, A. 2006. Effect of a participatory intervention to reduce the number of unnecessary caesarean sections performed in Shahrekord of Iran. *J. Med Sci.* 6:690–692.

Garozzo, A., Timpanaro, R., Stivala, A., Bisignano, G., Castro, A. 2011. Activity of *Melaleuca alternifolia* (tea tree) oil on influenza virus A/PR/8: Study on the mechanism of action. *Antivir. Res.* 89(1):83–88.

Gauthier, R., Agoumi, A., Gourai, M. 1989. The activity of extracts of *Myrtus communis* against *Pediculus humanus capitis*. *Plant. Med. Phytother.* 23(2):95–108.

Geiger, J. L. 2005. The essential oil of ginger, *Zingiber officinale*, and anaesthesia. *Int. J. Aromather.* 15:7–14.

Giatropoulos, A., Papachristos, D. P., Kimbaris, A., Koliopoulos, G., Polissiou, M. G., Emmanouel, N., Michaelakis, A. 2012. Evaluation of bioefficacy of three *Citrus* essential oils against the dengue vector *Aedes albopictus* (Diptera: Culicidae) in correlation to their components enantiomeric distribution. *Parasitol. Res.* 111(6):2253–2263.

Gilligan, N. P. 2005. The palliation of nausea in hospice and palliative care patients with essential oils of *Pimpinella anisum* (aniseed), *Foeniculum vulgare* var. dulce (sweet fennel), *Anthemis nobilis* (Roman chamomile) and *Mentha x piperita* (peppermint). *Int. J. Aromather.* 15:163–167.

Gillissen, A., Wittig, T., Ehman, M., Krezdorn, H. G., de Mey, C. 2013. A multi-centre, randomised, double-blind, placebo-controlled clinical trial on the efficacy and tolerability of GeloMyrtol® forte in acute bronchitis. *Arzneimittel-Forschung.* 63(1):19–27.

Giraud-Robert, A. M. 2005. The role of aromatherapy in the treatment of viral hepatitis. *Int. J. Aromather.* 15:183–192.

Göbel, H., Dworschak, M., Ardabili, A., Stolze, H., Soyka, D. 1995b. Effect of volatile oils on the flow of skin-capillaries of the head in healthy people and migraine patients. *Cephalalgia* 15:93.

Göbel, H., Schmidt, G., Dworschak, M., Stolze, H., Heuss, D. 1995a. Essential plant oils and headache mechanisms. *Phytomedicine* 2(2):93–102.

Goerg, K. J., Spilker, T. 1996. Simultane sonographische Messung der Magen- und Gallenblasenentleerung mit gleichzeitiger Bestimmung der orozökalen Transitzeit mittels H_2-Atemtest. In: Low, D. and Rietbrock, N. (eds.) *Phytopharmaka II. Forschung und klinische Anwendung.* Damstadt, Germany: Steinkopff.

Goerg, K. J., Spilker, Th. 2003. Effect of peppermint oil and caraway oil on gastrointestinal motility in healthy volunteers: A pharmacodynamic study using simultaneous determination of gastric and gall-bladder emptying and orocaecal transit time. *Aliment. Pharmacol. Ther.* 17:445–451.

Goodman, L. S., Limbird, L. E. 2001. *Pharmacological Basis of Therapeutics,* 10th edn. McGraw Hill, New York.

Gravett, P., 2000. Aromatherapy treatment of severe oral mucositis. *Int. J. Aromather.* 10(1–2):52–53.

Green, B. G., Flammer, L. J. 1989. Methyl salicylate as a cutaneous stimulus: A psychophysical analysis. *Somatosens. Mot. Res.* 6(3):253–274.

Greenway, F. L., Frome, B. M., Engels, T. M., McLellan, A. 2003. Temporary relief of postherpetic neuralgia pain with topical geranium oil [letter]. *Am. J. Med.* 115:586–587.
Grigoleit, H-G., Grigoleit, P. 2005. Peppermint oil in irritable bowel syndrome. *Phytomedicine* 12:601–606.
Grimm, H. 1987. Antiobstruktive Wirksamkeit von Cineol bei Atemwegserkrankungen. *Therapiewoche* 37:4306–4311.
Habich, G., Repges, R. 1994. Chronisch obstruktive Atemwegserkrankungen. Cineol als Medikation sinnvoll und bewährt! *Therapiewoche* 44:356–365.
Habila, N., Agbaji, A. S., Ladan, Z., Bello, I. A., Haruna, E., Dakare, M. A., Atolagbe, T. O. 2010. Evaluation of in vitro activity of essential oils against *Trypanosoma brucei brucei* and *Trypanosoma evansi*. *J. Parasitol. Res.* 2010: doi: 10.1155/2010/534601.
Hadi, N., Hanid, A. A. 2011. Lavender essence for post-cesarean pain. *Pakistan J. Biol. Sci.* 14(11):664–667.
Hammer, K. A., Carson, C. F., Riley, T. V. 1999. In vitro susceptibilities of lactobacilli and organisms associated with bacterial vaginosis to *Melaleuca alternifolia* (tea tree) oil. *Antimicrob. Agents Chemother.* 43(1):196.
Han, S-H., Hur, M-H., Buckle, J., Choi, J., Lee, M. S. 2006. Effect of aromatherapy on symptoms of dysmenorrhea in college students: A randomised placebo-controlled clinical trial. *J. Altern. Complement. Med.* 12:535–541.
Harries, N., James, K. C., Pugh, W. K. 1978. Antifoaming and carminative actions of volatile oils. *J. Clin. Pharmacy.* 2:171–177.
Hasani, A., Demetri, P., Toms, N., Dilworth, P., Agnew, J. E. 2003. Effect of aromatics on lung mucociliary clearance in patients with chronic airways obstruction. *J. Altern. Comp. Med.* 9:243–249.
Hayes, A. J., Markovic, B. 2002. Toxicity of Australian essential oil *Backhousia citriodora* (Lemon myrtle). Part 1. Antimicrobial activity and in vitro cytotoxicity. *Food Chem. Toxicol.* 40:535–543.
Higashi, Y., Kiuchi, T., Furuta, K. 2010. Efficacy and safety profile of a topical methyl salicylate and menthol patch in adult patients with mild to moderate muscle strain: A randomized, double-blind, parallel-group, placebo-controlled, multicenter study. *Clin. Ther.* 32(1):34–43.
Holtmann, G., Gschossmann, J., Buenger, L., Wieland, V., Heydenreich, C.-J. 2001. Effects of a fixed peppermint oil/caraway oil combination (PCC) on symptoms and quality of life in functional dyspepsia. A multicentre, placebo-controlled, double-blind, randomised trial. *Gastroenterology* 120(Suppl 1): A-237.
Holtmann, G., Gschossmann, J., Buenger, L., Wieland, V., Heydenreich, C.-J. 2002. Effects of a fixed peppermint oil/caraway oil combination (FPCO) on symptoms of functional dyspepsia accentuated by pain or discomfort. *Gastroenterology* 122(Suppl 1):A-471.
Holtmann, G., Haag, S., Adam, B., Funk, P., Wieland, V., Heydenreich, C.-J. 2003. Effects of a fixed combination of peppermint oil and caraway oil on symptoms and quality of life in patients suffering from functional dyspepsia. *Phytomedicine* 10:56–57.
Hordinsky B. Z., Hordinsky, W. 1979. Rowachol—Its use in hyperlipidemia. *Arch. Arzneither.* 1:45–51.
Hur, M-H., Park, J., Maddock-Jennings, W., Kim, D. O., Lee, M. S. 2007. Reduction of mouth malodour and volatile sulphur compounds in intensive care patients using an essential oil mouthwash. *Phytother. Res.* 21:641–643.
Igimi, H., Histasugo,T., Nishimura, M. 1976. The use of d-limonene preparation as a dissolving agent of gallstones. *Am. J. Dig. Dis.* 21:926–939.
Igimi, H., Tamura, R., Toraishi, K. et al. 1991. Medical dissolution of gallstones. Clinical experience of d-limonene as a simple, safe and effective solvent. *Dig. Dis. Sci.* 36:329–332.
Iha, K., N. Suzuki, M. Yoneda, T. Takeshita and T. Hirofuji, 2013. Effect of mouth cleaning with hinokitiol-containing gel on oral malodor: A randomized, open-label pilot study. *Oral Surg Oral Med Oral Pathol Oral Radiol*, 116(4):433–439.
Imai, H., Osawa, K., Yasuda, H., Hamashima, H., Arai, T., Sasatsu, M. 2001. Inhibition by the essential oils of peppermint and spearmint of the growth of pathogenic bacteria. *Microbios.* 106:31–39.
Jahromi, B. N., Tartifizadeh, A., Khabnadideh, S. 2003. Comparison of fennel and mefenamic acid for the treatment of primary dysmenorrhea. *Int. J. Gynecol. Obstet.* 80:153–157.
Jandourek, A., Viashampayan, J. K., Vazquez, J. A. 1998. Efficacy of melaleuca oral solution for the treatment of fluconazole refractory oral candidiasis in AIDS patients. *Aids.* 12:1033–1037.
Jordt, S-E., McKemmy, D. D., Julius, D. 2003. Lessons from peppers and peppermint: The molecular logic of thermosensation. *Curr. Opin. Neurobiol.* 13:73–77.
Juergens, U. R., Dethlefsen, U., Steinkamp, G., Gillissen, A., Repges, R., Vetter, H. 2003. Anti-inflammatory activity of 1,8-cineol (eucalyptol) in bronchial asthma: A double-blind placebo-controlled trial. *Resp. Med.* 97:250–256.

Juergens, U. R., Engelen, T., Racké, K., Stöber, M., Gillissen, A. S., Vetter, H. 2004. Inhibitory activity of 1,8-cineole (eucalyptol) on cytokine production in cultured human lymphocytes and monocytes. *Pulmon. Pharmacol. Therapeut.* 17:281–287.

Juergens, U. R., Stöber, M., Schmidt-Schilling, L., Kleuver, T., Vetter, H. 1998c. Antiinflammatory effects of eucalyptol (1,8-cineole) in bronchial asthma: Inhibition of arachidonic acid metabolism in human blood monocytes ex vivo. *Eur. J. Med. Res.* 3:407–412.

Juergens, U. R., Stöber, M., Vetter, H. 1998a. Steroidartige Hemmung des monozytären Arachidonsäuremetabolismus und der IL-1β-Produktion durch 1,8-Cineol. *Atemwegs- Lungenkrank.* 24(1):3–11.

Juergens, U. R., Stöber, M., Vetter, H. 1998b. Inhibition of cytokine production and arachidonic acid metabolism by eucalyptol (1,8-cineole) in human blood monocytes in vitro. *Eur. J. Med. Res.* 3:508–510.

Kamin, W., Kieser, M. 2007. Pinimenthol® ointment in patients suffering from upper respiratory tract infections—A post-marketing observational study. *Phytomedicine.* 14:787–791.

Kang, S. H., Kim, M. K., Noh, D. J., Yoon, C., Kim, G. H. 2009. Spray adulticidal effects of plant oils against house mosquito, *Culex pipiens pallens* (Diptera: Culicidae). *J. Pest. Sci.* 34(2):100–106.

Kaspar, P., Repges, R., Dethlefsen, U., Petro, W. 1994. Sekretolytika im Vergleich. Änderung der Ziliarfrequenz und Lungenfunktion nach Therapie mit Cineol und Ambroxol. *Atemw.-Lungenkrank.* 20:605–614.

Kasper, S., Gastpar, M., Müller, W. E., Volz, H. P., Möller, H. J., Dienel, A., Schläfke, S. 2010. Silexan, an orally administered Lavandula oil preparation, is effective in the treatment of "subsyndromal" anxiety disorder: A randomized, double-blind, placebo controlled trial. *Int. Clin. Psychpharmacol.* 25(5):277–287.

Kehrl, W., Sonnemann, U., Dethlefson, U. 2004. Therapy for acute nonpurulent rhinosinusitis with cineole: Results of a double-blind, randomised, placebo-controlled trial. *Laryngoscope* 114:738–742.

Kenia, P., Houghton, T., Beardsmore, C. 2008. Does inhaling menthol affect nasal patency or cough? *Pediatr. Pulmonol.* 43:532–537.

Kennedy, D. L., Scholey, A. B. 2006. The psychopharmacology of European herbs with cognitive-enhancing properties. *Curr. Pharm. Des.* 12:4613–4623.

Kennedy, D. O., Dodd, F. L., Robertson, B. C., Okello, E. J., Reay, J. L., Scholey, A. B., Haskell, C. F. 2011. Monoterpenoid extract of sage (*Salvia lavandulaefolia*) with cholinesterase inhibiting properties improves cognitive performance and mood in healthy adults. *J. Psychopharmacol.* 25(8):1088–1100.

Khorshidi, N., Ostad, S. N., Mopsaddegh, M., Soodi, M. 2003. Clinical effects of fennel essential oil on primary dysmenorrhea. *Iranian J. Pharmaceut. Res.* 2:89–93.

Khosravi, A. R., Eslami, A. R., Shokri, H., Kashanian, M. 2008. *Zataria multiflora* cream for the treatment of acute vaginal candidiasis. *Int. J. Gynecol. Obstet.* 101:201–202.

Kim, H.-K., Yun, Y.-K., Ahn, Y.-J. 2008. Fumigant toxicity of cassia bark and cassia and cinnamon oil compounds to *Dermatophagoides farinae* and *Dermatophagoides pteronyssinus* (Acari: Pyroglyphidae). *Exp. Appl. Acarol.* 44:1–9.

Kim, J. T., Ren, C. J., Fielding, G. A. et al. 2007. Treatment with lavender aromatherapy in the post-anaesthesia care unit reduces opioid requirements of morbidly obese patients undergoing laparoscopic adjustable gastric banding. *Obes. Surg.* 17:920–925.

Kim, J. T., Wadja, M., Cuff, G. et al. 2006. Evaluation of aromatherapy in treating postoperative pain: Pilot study. *Pain Pract.* 6:273–277.

Kirov, M., Burkova, T., Kapurdov, V., Spasovski, M. 1988a. Rose oil. Lipotropic effect in modelled fatty dystrophy of the liver. *Med. Biol. Info.* 3:18–22.

Kirov, M., Koev, P., Popiliev, I., Apostolov, I., Marinova, V. 1988b. Girositol. Clinical trial in primary hyperlipoproteinemia. *Med. Biologic Info.* 3:30–34.

Kline, R. M., Kline, J. J., Di Palma, J., Barbaro, G J. 2001. Enteric-coated, pH-dependent peppermint oil capsules for the treatment of irritable bowel syndrome in children. *J. Pediatr.* 138:125–128.

Konstantinova, L., Kirov, M., Petrova, M., Ortova, M., Marinova, V. 1988. Girosital. Continuous administration in hyperlipoproteinemia and chronic alcoholic lesions. *Med. Biol. Info.* 3:35–38.

Kotan, R., Kordali, S., Cakir, A. 2007. Screening of antibacterial activities of twenty-one oxygenated monoterpenes. *Zeit. Naturforschung C.* 62:507–513.

Kpoviessi, S., Bero, J., Agbani, P., Gbaguidi, F., Kpadonou-Kpoviessi, B., Sinsin, B., Accrombessi, G., Frederich, M., Moudachirou, M., Quetin-Leclercq, J. 2014. Chemical composition, cytotoxicity and in vitro antitrypanosomal and antiplasmodial activity of the essential oils of four Cymbopogon species from Benin. *J. Ethnopharmacol.* 151(1):652–659.

Kritsidima, M., Newton, T., Asimakopoulou, K. 2010. The effects of lavender scent on dental patient anxiety levels: A cluster randomised-controlled trial. *Community Dent. Oral Epidemiol.* 38(1):83–87.

Lahlou, S., Carneiro-Leao, R. F. L., Leãl-Cardaso, J. H., Toscano, C. F. 2001. Cardiovascular effects of the essential oil of *Mentha x villosa* and its main constituent, piperitone oxide, in normotensive anaesthetised rats: Role of the autonomic nervous system. *Planta Med.* 67(7):638–643.

Lahlou, S., Magalhães, P. C. G., de Siqueira, R. J. B., Figueiredo, A. F., Interaminense, L. F. L., Maia, J. G. S., Sousa, P. J. C. 2005. Cardiovascular effects of the essential oil of Aniba canelilla bark in normotensive rats. *J. Cardiovasc. Surg.* 46(4):412–421.

Lamster, I. B., Alfano, M. C., Seiger, M. C., Gordon, J. M. 1983. The effect of Listerine antiseptic on reduction of existing plaque and gingivitis. *Clin. Prev. Dent.* 5:12–16.

Lamy, J. 1967. Essais cliniques de dérivés terpeniques en therapeutique hépato-biliare. *L'Information Therapeutique* 5:39–45.

Laude, E. A., Morice, A. H., Grattan, T J. 1994. The antitussive effects of menthol, camphor and cineole in conscious guinea pigs. *Pulmon. Pharmacol.* 7:179–184.

Lawson, M. J., Knight, R. E., Walker, T. G., Roberts-Thomson, I. C. 1988. Failure of enteric-coated peppermint oil in the irritable bowel syndrome: A randomised, double-blind crossover study. *J. Gastroenterol. Hepatol.* 3:235–238.

Le Faou, M., Beghe, T., Bourguignon, E. et al. 2005. The effects of the application of Dermasport® plus Solution Cryo® in physiotherapy. *Int. J. Aromather.* 15:123–128.

Lee, C. B., Ha, U-S., Lee, S. J., Kim, S. W., Cho, Y-H. 2006. Preliminary experience with a terpene mixture versus ibuprofen for treatment of category III chronic prostatitis/chronic pelvic pain syndrome. *World J.Urol.* 24:55–60.

Liao, Q., Qian, Z., Liu, R., An, L.,Chen, X. 2013. Germacrone inhibits early stages of influenza virus infection. *Antivir. Res.* 100(3):578–588.

Linsenmann, P., Swoboda, M. 1986. Therapeutic efficacy of volatile oils in chronic obstructive bronchitis. *Therapiewoche* 36:1162–1166.

Liu, J-H., Chen, G-H., Yeh, H-Z., Huang, C-K., Poon, S-K. 1997. Enteric-coated peppermint-oil capsules in the treatment of irritable bowel syndrome: A prospective, randomised trial. *J. Gastroenterol.* 32:765–768.

Lobo, P. L., Fonteles, C. S., de Carvalho, C. B., do Nascimento, D. F., da Cruz Fonseca, S. G., Jamacaru, F. V., de Moraes, M. E. 2011. Dose-response evaluation of a novel essential oil against *Mutans streptococci* in vivo. *Phytomedicine* 18(7):551–556.

Logan, A. C., Beaulne, T. M. 2002. The treatment of small intestinal bacterial overgrowth with enteric-coated peppermint oil: A case report. *Altern. Med. Rev.* 7:410–417.

Maddocks-Jennings, W., Wilkinson, J. M., Cavanagh, H. M., Shillington, D. 2009. Evaluating the effects of the essential oils *Leptospermum scoparium* (manuka) and *Kunzea ericoides* (kanuka) on radiotherapy induced mucositis: A randomized, placebo controlled feasibility study. *Eur. J. Oncol. Nurs.* 13(2):87–93.

Madisch, A., Heydenreich, C-J., Wieland, V., Hufnagel, R., Hotz, J. 1999. Treatment of functional dyspepsia with a fixed peppermint oil and caraway oil combination preparation as compared to cisapride. *Arzneim-Forsch.* 49:925–932.

Madisch, A., Heydenreich, C-J., Wieland, V., Hufnagel, R., Hotz, J. 2000. Equivalence of a fixed herbal combination preparation as compared with cisapride in functional dyspepsia—Influence of *H. pylori* status [Abstract]. *Gut* 47(Suppl 1):A111.

Magalhães, P. J., Lahlou, S., Vasconcelos dos Santos, M. A., Pradines, T. L., Leal-Cardoso, J. H. 2003. Myorelaxant effects of the essential oil of *Croton nepetaefolius* on the contractile activity of the guinea pig tracheal smooth muscle. *Planta Med.* 69:74–77.

Mahlo, D-H. 1990. Obstruktive Atemwegserkrankungen. Mit Cineol die Lungen-funktionsparameter verbessern. *Therapiewoche* 40:3157–3162.

Mahmoudabadi, A. Z., Dabbagh, M. A., Fouladi, Z. 2006. In vitro anti-candida activity of *Zataria multiflora* Boiss. *Evid. Based Complement. Alternat Med.* 4:351–353.

Mansoori, P., Hadjiakhondi, A., Ghavami, R., Shafiee, A. 2002. Clinical evaluation of *Zataria multiflora* essential oil mouthwash in the management of recurrent aphthous stomatitis. *Daru* 10:74–77.

Marchand, S., Arsenault, P. 2002. Odors modulate pain perception. A gender-specific effect. *Physiol. Behav.* 76:251–256.

Maruyama, N., Hiroko Ishibashi, H., Weimin Hu, W. et al. 2006. Suppression of carrageenan- and collagen II-induced inflammation in mice by geranium oil. *Mediat. Inflamm.* 2006(3). doi: 10.1155/MI/2006/62537.

Masako, K., Hideyki, I., Shigeyuki O., Zenro, I. 2005a. A novel method to control the balance of skin microflora. Part 1. Attack on biofilm of Staphylococcus aureus without antibiotics. *J. Dermatol. Sci.* 38:197–205.

Masako, K., Yusuke, K., Hideyuki, I. et al. 2005b. A novel method to control the balance of skin microflora. Part 2. A study to assess the effect of a cream containing farnesol and xylitol on atopic dry skin. *J. Dermatol. Sci.* 38:207–213.

Mattys, H., de Mey, C., Carls, C., Ryś, A., Geib, A., Wittig, T. 2000. Efficacy and tolerability of myrtol standardised in acute bronchitis. A multi-centre, randomised, double-blind, placebo-controlled parallel group clinical trial vs. cefuroxime and ambroxol. *Arzneim-Forsch.* 50(8):700–711.

May, B., Funk, P., Schneider, B. 2003. Peppermint and caraway oil in functional dyspepsia—Efficacy unaffected by *H. pylori* [letter]. *Aliment. Pharmacol. Ther.* 17:975–976.

May, B., Köhler, S., Schneider, B. 2000. Efficacy and tolerability of a fixed combination of peppermint oil and caraway oil in patients suffering from functional dyspepsia. *Aliment. Pharmacol. Ther.* 14:1671–1677.

May, B., Kuntz, H.-D., Kieser, M., Köhler, S. 1996. Efficacy of a fixed peppermint oil/caraway oil combination in non-ulcer dyspepsia. *Arzneim-Forsch.* 46:1149–1153.

McBride, B., Whitelaw, W. A. 1981. Physiological stimulus to upper airway receptors in humans. *J. Appl. Physiol.* 57:1189–1197.

Meadows, S. M., Mulkerin, D., Berlin et al. 2002. Phase II trial of perillyl alcohol in patients with metastatic colorectal cancer. *Int. J. Gastrointest. Cancer.* 32:125–128.

Mechkov, G., Kirov, M., Yankov, B., Georgiev, P., Marinova, V., Doncheva, T. 1988. Girosital. Hypolipidemic effect in cholelithiasis and liver steatosis. *Med. Biol. Info.* 3:26–29.

Meiller, T. F., Silva, A., Ferreira, S. M., Jabra-Rizk, M. A., Kelley, J. I., DePaola, L. G. 2005. Efficacy of Listerine® antiseptic in reducing viral contamination of saliva. *J. Clin. Periodontol.* 32:341–346.

Meister, R., Wittig, T., Beuscher, N., de Mey, C. 1999. Efficacy and tolerability of myrtol standardized in long-term treatment of chronic bronchitis. A double-blind, placebo-controlled study. *Arzneim.-Forsch.* 49:351–358.

Melli, M. S., Rashidi, M R., Delazar, A. et al. 2007. Effect of peppermint water on prevention of nipple cracks in lactating primiparous women: A randomised controlled trial. *Int. Breastfeed. J.* 2:7. http://internationalbreastfeedingjournal.com/content/2/1/7.

Menezes, I. G. A., Barreto, C. M. N., Antoniolli, A. R., Santos, M. R. V., de Sousa, D. P. 2010. Hypotensive activity of terpenes found in essential oils. *Z. Naturforsch.* C 65(9–10):562–566.

Mercier, D., Knevitt, A. 2005. Using topical aromatherapy for the management of fungating wounds in a palliative care unit. *J. Wound Care* 14:497–501.

Mickelfield, G., Greving I, May B. 2000. Effects of peppermint oil and caraway oil on gastroduodenal motility. *Phytother. Res.* 14:20–23.

Micklefield, G., Jung, O., Greving, I., May, B. 2003. Effects of intraduodenal application of peppermint oil (WS® 1340) and caraway oil (WS® 1520) on gastrointestinal motility in healthy volunteers. *Phytother. Res.* 17:135–140.

Middleton, B., Hui, K-P. 1982. Inhibition of hepatic S-3-hydrox-3-methylglutaryl-CoA redsuctase and in vivo rates of lipogenesis by a mixture of pure cyclic monoterpenes. *Biochem. Pharmacol.* 31:2897–2901.

Mizuno, S., Kimitoshi, K., Yoshiki, O. et al. 2006. Oral peppermint oil is a useful antispasmodic for double-contrast barium meal examination. *J. Gastroenterol. Hepatol.* 21:1297–1301.

Moran, J., Addy, M., Newcombe, R. 1997. A 4-day plaque regrowth study comparing an essential oil mouthrinse with a triclosan mouthrinse. *J. Clin. Periodontol.* 24:636–639.

Morice, A. H., Fontana, G. A., Sovijarvi, A. R. et al. 2004. The diagnosis and management of chronic cough. *Eur. Resp. J.* 24:481–492.

Morice, A. H., Marshall, A. E., Higgins, K. S., Grattan, T. J. 1994. Effect of menthol on citric acid induced cough in normal subjects. *Thorax* 49:1024–1026.

Morise, M., Ito, Y., Matsuno, T., Hibino, Y., Mizutani, T., Ito, S., Hashimoto, N., Kondo, M., Imaizumi, K. and Hasegawa, Y. 2010. Heterologous regulation of anion transporters by menthol in human airway epithelial cells. *Eur. J. Pharmacol.* 635:204–211.

Motomura, N., Sakurai, A., Yotsuya, Y. 2001. Reduction of mental stress with lavender odorant. *Percept. Motor Skill.* 93(3):713–718.

Mumcuoglu, K. Y., Miller, J. A., Zamir, C., Zentner, G., Helbin, V., Ingber, A. 2002. The in vivo pediculicidal efficacy of a natural remedy. *Isr. Med. Assoc. J.* 4(10):790–793.

Mukamel, E., Engelstein, D., Simon, D., Servadio, C. 1987. The value of Rowatinex in the treatment of ureterolithiasis. *J. Urol.* 93:31–33.

Nishino, T, Tagaito, Y., Sakurai, Y. 1997. Nasal inhalation of l-menthol reduces respiratory discomfort associated with loaded breathing. *Am. J. Respir. Crit. Med.* 155:309–313.

Oladimeji, F. A., Orafidiya, O. O., Ogunniyi, T. A. B., Adewunmi, T. A. 2000. Pediculocidal and scabicidal properties of *Lippia multiflora* essential oil. *J. Ethnopharmacol.* 72(1–2):305–311.

Oladimeji, F. A., Orafidiya, L. O., Ogunniyi, T. A. B. Adewunmi, T. A., Onayemi, O. 2005. A comparative study of the scabicidal activities of formulations of essential oil of *Lippia multiflora* Moldenke and benzyl benzoate emulsion BP. *Int. J. Aromather.* 15:87–93.

Olando da Fonseca, C., Schwartsmann, G., Fischer, J. et al. 2008. Preliminary results from a phase I/II study of perillyl alcohol intranasal administration in adults with recurrent malignant gliomas. *Surg. Neurol.* 70:259–267.

Oliani, J., Siqueira, C. A. T., Sartoratto, A., Queiroga, C. L., Moreno, P. H., Reimão, J. Q. R., Tempone, A. G., Diaz, I. E. C., Fischer, D. C. H. 2013. Chemical composition and in vitro antiprotozoal activity of the volatile oil from leaves of *Annona crassiflora* Mart. (Annonaceae). *Pharmacologyonline* 3:8–15.

Omolo, M. O., Okinyo, D., Ndiege, I. O., Lwande, W., Hassanali, A. 2005. Fumigant toxicity of the essential oils of some African plants against *Anopheles gambiae* sensu stricto. *Phytomedicine* 12(3):241–246.

Orafidiya, L. O., Agbani, E. O., Oyedele, A. O., Babalola, O. O., Onayemi, O. 2002. Preliminary clinical tests on topical preparations of *Ocimum gratissimum* Linn leaf essential oil for the treatment of acne vulgaris. *Clin. Drug Investig.* 22(5):313–319.

Orafidiya, L. O., Agbani, E. O., Oyedele, A. O., Babalola, O. O., Onayemi, O., Aiyedun, F. F. 2004. The effect of aloe vera gel on the anti-acne properties of the essential oil of *Ocimum gratissimum* Linn—A preliminary clinical investigation. *Int. J. Aromather.* 14:15–21.

Orani, G. P., Anderson, J. W., Sant'Ambrogio, G., Sant'Ambrogio, F. B. 1991. Upper airway cooling and l-menthol reduce ventilation in the guinea pig. *J. Appl. Physiol.* 70:2080–2086.

Ostad, S.N., Soodi, M., Sharifzadeh, M., Khorsidi, N. 2001. The effect of fennel essential oil on uterine contraction as a model for dysmenorrhoea, pharmacology and toxicology study. *J. Ethnopharmacol.* 76:299–304.

Packman, E. W., London, S. J. 1980. The utility of artificially induced cough as a clinical model for evaluating the antitussive effects of aromatics delivered by inunction. *Eur. J. Resp. Dis.* 61(Suppl 10):101–109.

Patel, T., Ishiuji, Y., Yosipovitch, G. 2007. Menthol: A refreshing look at this ancient compound. *J. Am. Acad. Dermatol.* 57(5):873–878.

Paulpandi, M., Kannan, S., Thangam, R., Kaveri, K., Gunasekaran, P., Rejeeth, C. 2012. In vitro anti-viral effect of β-santalol against influenza viral replication. *Phytomedicine* 19(3–4):231–235.

Pavela, R. 2014. Insecticidal properties of *Pimpinella anisum* essential oils against the *Culex quinquefasciatus* and the non-target organism *Daphnia magna*. *J. Asia Pac. Entomol.* 17(3):287–293.

Peiffer, C., Pline, J-B., Thivard, L., Aubier, M., Samson, Y. 2001. Neural substrates for the perception of acutely induced dyspnea. *Am. J. Respir. Crit. Care* 163:951–957.

Pélissier, Y., Marion, C., Casadebaig, J. et al. 1994. A chemical bacteriological, toxicological and clinical study of the essential oil of *Lippia multiflora* Mold. (Verbenaceae). *J. Essent. Oil Res.* 6:623–630.

Perfumi, M., Paparelli, F., Cingolani, M. L. 1995. Spasmolytic activity of essential oil of *Artemisia thuscula* Cav. from the Canary Islands. *J. Essent. Oil Res.* 7:387–392.

Perry, N. S. L., Houghton, P. J., Jenner, P., Keith, A., Perry, E. K. 2002. *Salvia lavandulaefolia* essential oil inhibits cholinesterase in vivo. *Phytomedicine* 9(1):48–51.

Perry, N. S. L., Houghton, P. J., Sampson, J., Theobald, A. E., Hart, S., Lis-Balchin, M., Hoult, J. R. S. et al. 2001. In vitro activity of *S. lavandulaefolia* (Spanish sage) relevant to treatment of Alzheimer's disease. *J. Pharm. Pharmacol.* 53(10):1347–1356.

Perry, N. S. L., Houghton, P. J., Theobald, A., Jenner, P. M., Perry, E. K. 2000. In vitro inhibition of human erythrocyte acetylcholinesterase by *Salvia lavandulaefolia* essential oil and constituent terpenes. *J. Pharm. Pharmacol.* 52(7):895–902.

Pittler, M. H., Ernst, E. 1998. Peppermint oil for irritable bowel syndrome: A critical review and meta-analysis. *Am. J. Gastroenterol.* 93:1131–1135.

Post-White, J., Nichols, W. 2007. Randomised rrial testing of QueaseEase essential oil for motion sickness. *Int. J. Essent. Oil Ther.* 1:143–152.

Pritchard, A. J. N. 2004. The use of essential oils to treat snoring. *Phytother. Res.* 18:696–699.

Pushpanathan, T., Jebanesan, A., Govindarajan, M. 2008. The essential oil of *Zingiber officinalis* Linn (Zingiberaceae) as a mosquito larvicidal and repellent agent against the filarial vector *Culex quinquefasciatus* Say (Diptera: Culicidae). *Parasitol. Res.* 102(6):1289–1291.

Ramadan, W., Mourad, B., Ibrahim, S., Sonbol, F. 1996. Oil of bitter orange: New topical antifungal agent. *Int. J. Dermatol.* 35:448–449.

Raman, A., Weir, U., Bloomfield, S.F. 1995. Antimicrobial effects of tea tree oil and its major components on *Staphylococcus aureus*, Staph. epidermidis and Propionibacterium acnes. *Lett. Appl. Microbiol.* 21:242–245.

Rangelov, A. M., Toreva D., Kosev, R. 1988. Experimental study of the cholagogic and choleretic action of some of the basic ingredients of essential oils on laboratory animals. *Folia Medica (Plovdiv).* 30:30–38.

Reddy, B. S., Wang, C-X., Samaha, H. et al. 1997. Chemoprevention of colon carcinogenesis by dietary perillyl alcohol. *Cancer Res.* 57:420–425.

Rees, W. D. W., Evans, B. K., Rhodes, J. 1979. Treating irritable bowel syndrome with peppermint oil. *Br. Med. J.* 2:835–836.

Reiter, M., Brandt, W. 1985. Relaxant effects on tracheal and ileal smooth muscles of the guinea pig. *Arzneim-Forsch.* 35:408–414.

Remberg, P., Björk, L., Hedner, T., Sterner, O. 2004. Characteristics, clinical effect profile and tolerability of a nasal spray preparation of *Artemisia abrotanum* L. for allergic rhinitis. *Phytomedicine* 11:36–42.

Reuter, J., Huyke, C., Casetti, F., Theek, C., Frank, U., Augustin, M., Schempp, C. 2008. Anti-inflammatory potential of a lipolotion containing coriander oil in the ultraviolet erythema test. *J. German Soc. Dermatol.* 6(10):847–851.

Riep, B. G., Bernimoulin, J-P., Barnett, M. L. 1999. Comparative antiplaque effectiveness of an essential oil and an amione fluoride/stannous fluoride mouthrinse. *J. Clin. Periodontol.* 26:164–168.

Ro, Y-J., Ha, H-C., Kim, C-G., Yeom, H-A. 2002. The effects of aromatherapy on pruritis in patients undergoing hemodialysis. *Dermatol. Nurs.* 14:231–238.

Saeki, Y., Ito, Y., Shibata, M., Sato, Y., Okuda, K., Takazoe, I. 1989. Antimicrobial action of natural substances on oral bacteria. *Bull. Tokyo Dent. College* 30(3):129–135.

Sant'Ambrogio, F. B., Anderson, J. W., Sant'Ambrogio, G. 1991. Effect of l-menthol on laryngeal receptors. *J. Appl. Physiol.* 70:788–793.

Sant'Ambrogio, F. B., Anderson, J. W., Sant'Ambrogio, G. 1992. Menthol in upper airway depresses ventilation in newborne dogs. *Resp. Physiol.* 89:299–307.

Satchell, A. C., Saurajen, A., Bell, C., Barnetson, R. StC. 2002a. Treatment of dandruff with 5% tea tree oil shampoo. *J. Am. Acad. Dermatol.* 47:852–855.

Satchell, A. C., Saurajen, A., Bell, C., Barnetson, R. StC. 2002b. Treatment of interdigital tinea pedis with 25% and 50% tea tree oil solution: A randomised, placebo-controlled, blinded study. *Austr. J. Dermatol.* 45:175–178.

Scanni, G., Bonifazi, E. 2005. Efficacy and safety of a new non-pesticide lice removal product. *Eur. J. Pediat. Dermatol.* 15:249–252.

Scanni, G., Bonifazi, E. 2006. Efficacy of a single application of a new natural lice removal product. Preliminary data. *Eur. J. Pediatr. Dermatol.* 16:231–234.

Schäfer, D., Schäfer, W. 1981. Pharmakologische Untersuchungen zur broncholytischen und sekretolytisch-expektorierenden Wirksamkeit einer Salbe auf der Basis von Menthol, Camphor und ätherischen Ölen. *Arzneim-Forsch.* 31(1):82–86.

Schnitzler, P., Schumacher, A., Astani, A., Reichling, J. 2008. Melissa officinalis oil affects infectivity of enveloped herpesviruses. *Phytomedicine* 15:734–740.

Shahi, S. K., Shukla, A. C., Bajaj, A. K. et al. 2000. Broad spectrum herbal therapy against superficial fungal infections. *Skin Pharmacol. Appl. Skin Physiol.* 13:60–64.

Shaw, G., Srivastava, E.D., Sadlier, M., Swann, P., James, J.Y., Rhodes, J. 1991. Stress management for irritable bowel syndrome: A controlled trial. *Digestion* 50:36–42.

Sheikhan, F., Jahdi, F., Khoei, E. M., Shamsalizadeh, N., Sheikhan, M., Haghani, H. 2012. Episiotomy pain relief: Use of Lavender oil essence in primiparous Iranian women. *Complement. Ther. Clin. Pract.* 18(1):66–70.

Sherry, E., Sivananthan, S., Warnke, P., Eslick, G. D. 2003. Topical phytochemicals used to salvage the gangrenous lower limbs of type I diabetic patients [letter]. *Diabetes Res. Clin. Pract.* 62:65–66.

Shih, M. F., Chen, L. Y., Tsai, P. J., Cherng, J. Y. 2012. In vitro and in vivo therapeutics of β-thujaplicin on LPS-induced inflammation in macrophages and septic shock in mice. *Int. J. Immunopathol. Pharmacol.* 25:39–48.

Siller, G., Kottasz, S., Palfi, Z. 1998. Effect of Rowatinex on expulsion of post ESWL residual calculi. *MagyarUrologia* 10:139–146.

Simbar, M., Azarbad, Z., Mojab, F., Alavi Majd, H. 2008. A comparative study of the therapeutic effects of the Zataria multiflora vaginal cream and metronidazole vaginal gel on bacterial vaginosis. *Phytomedicine* 15(12):1025–1031.

Sivropoulou, A., Nikolaou, C., Papanikolaou, E., Kokkini, S., Lanaras, T., Arsenakis, M. 1997. Antimicrobial, cytotoxic and antiviral activities of *Salvia fruticosa* essential oil. *J. Agric. Food Chem.* 45:3197–3201.

Sloane, A., De Cort, S. C., Eccles, R. 1993. Prolongation of breath-hold time following treatment with an l-menthol lozenge in healthy man. *J. Physiol.* 473:53.

Soukoulis, S., Hirsch, R. 2004. The effects of a tea tree oil-containing gel on plaque and chronic gingivitis. *Aust. Dent. J.* 49:78–83.

Stankusheva, T., Balabanski, L., Narmova, R. Girosital. 1988. Hypolipidemic effect. *Med. Biol. Inform.* 3:23–25.

Stotz, S. C., Vtiens, J., Martyn, D., Clardy, J., Clapham, D. E. 2008. Citral sensing by transient receptor potential channels in dorsal root ganglions. *PLoS ONE.* 3:e2082. doi:10.1371/journal.pone.0002082.

Su, XC. Y., Wan Po, A., Millership, J. S. 1993. Ciliotoxicity of intranasal formulations: Menthol enantiomers. *Chirality.* 5:58–60.

Suzuki, J., Tokiwa, T., Mochizuki, M. 2008. Effects of a newly designed toothbrush for the application of periodontal disease treatment medicine (Hinopron) on the plaque removal and the improvement of gingivitis. *J. Jpn. Soc. Periodontal.* 50:30–38.

Sybilska, D., Asztemborska, M. 2002. Chiral recognition of terpenoids in some pharmaceuticals derived from natural sources. *J. Biochem. Biophys. Methods.* 54:187–195.

Syed, T. A., Qureshi, Z. A., Ali, S. M., Ahmed, S., Ahmed, S. A. 1999. Treatment of toenail onychomycosis with 2% butenafine and 5% Melaleuca alternifolia (tea tree) oil in cream. *Trop. Med. Int. Health* 4:284–287.

Taddei, I., Giachetti, D., Taddei, E., Mantovani, P. 1988. Spasmolytic activity of peppermint, sage and rosemary essences and their major constituents. *Fitoterapia* 59:463–468.

Takarada, K., Kimizuka, R., Takahashi, N., Honma, K., Okuda, K., Kato, T. 2004 A comparison of the antibacterial efficacies of essential oils against oral pathogens. *Oral Microbiol. Immunol.* 19:61–64.

Tamaoki, J., Chiyotani, A., Sakai, A., Takemura, H., Konno, K. 1995. Effect of menthol vapour on airway hyperresponsiveness in patients with mild asthma. *Resp. Med.* 89:503–504.

Tayarani-Najaran, Z., Talasaz-Firoozi, E., Nasiri, R., Jalali, N., Hassanzadeh, M. 2013. Antiemetic activity of volatile oil from Mentha spicata and Mentha x piperita in chemotherapy-induced nausea and vomiting. *Ecancermedicalscience* 7: doi: 10.3332/ecancer.2013.3290.

Taylor, B. A., Duthie, H. L., Luscombe, D. K. 1985. Calcium antagonist activity of menthol on gastrointestinal smooth muscle. *Br. J. Clin. Pharm.* 20:293–294.

Taylor, B. A., Luscombe, D. K., Duthie, H. L. 1984. Inhibitory effects of peppermint oil and menthol on human isolated coli. *Gut.* 25:A1168–A1169.

Thut, P. D., Wrigley, D., Gold, M. S. 2003. Cold transduction in rat trigeminal ganglia neurons in vitro. *Neurosci.* 119:1071–1083.

Tildesley, N. T. J., Kennedy, D. O., Perry, E. K., Ballard, C. G., Savelev, S., Wesnes, K. A., Scholey, A. B. 2003. *Salvia lavandulaefolia* (Spanish sage) enhances memory in healthy young volunteers. *Pharmacol. Biochem. Behav.* 75(3):669–674.

Tildesley, N. T. J., Kennedy, D. O., Perry, E. K., Ballard, C. G., Wesnes, K. A., Scholey, A. B. 2005. Positive modulation of mood and cognitive performance following administration of acute doses of *Salvia lavandulaefolia* essential oil to healthy young volunteers. *Physiol. Behav.* 83(5):699–709.

Toda, M., Morimoto, K. 2008. Effect of lavender aroma on salivary endocrinological stress markers. *Arch. Oral Biol.* 53(10):964–968.

Toloza, A. C., Lucía, A., Zerba, E., Masuh, H., Picollo, M. I. 2010. Eucalyptus essential oil toxicity against permethrin-resistant *Pediculus humanus capitis* (Phthiraptera: Pediculidae). *Parasitol. Res.* 106(2):409–414.

Tovey, E. R., McDonald, L. G. 1997. A simple washing procedure with eucalyptus oil for controlling house dust mites and their allergens in clothing and bedding. *J.Allergy Clin. Immunol.* 100:464–466.

Trinh, H.-T., Lee, I.-A., Hyun, Y.-J., Kim, D.-H. 2011. Artemisia princeps Pamp. essential oil and its constituents eucalyptol and α-terpineol ameliorate bacterial vaginosis and vulvovaginal candidiasis in mice by inhibiting bacterial growth and NF-κB activation. *Planta Med.* 77(18):1996–2002.

Tufekci, E., Casagrande, Z. A., Lindauer, S. J., Fowler, C. E., Williams, K. T. 2008. Effectiveness of an essential oil mouthrinse in improving oral health in orthodontic patients. *Angle Orthod.* 78:294–298.

Uedo, N., Tatsuta, M., Iishi, H. et al. 1999. Inhibition by d-limonene of gastric carcinogenesis induced by N-methyl-N'-nitro-N-nitroguanidine in Wistar rats. *Cancer Lett.* 137:131–136.

Uehleke, B., Schaper, S., Dienel, A., Schlaefke, S., Stange, R. 2012. Phase II trial on the effects of Silexan in patients with neurasthenia, post-traumatic stress disorder or somatization disorder. *Phytomedicine* 19(8–9):665–671.

von Bergmann, K., Beck, A., Engel, C., Leiß, O. 1987. Administration of a terpene mixture inhibits cholesterol nucleation in bile from patients with cholesterol gallstones. *J. Mol. Med.* 65:1432–1440.

Vakilian, K., Atarha, M., Bekhradi, R., Chaman, R. 2011. Healing advantages of lavender essential oil during episiotomy recovery: A clinical trial. *Complement. Ther. Clin. Pract.* 17(1):50–53.

Veal, L., 1996. The potential effectiveness of essential oils as a treatment for head lice, Pediculus humanus capitis. *Complement. Ther. Nurs. Midwif.*, 2(4):97–101.

Vlachojannis, C., Winsauer, H., Chrubasik, S. 2013. Effectiveness and safety of a mouthwash containing essential oil ingredients. *Phytother. Res.* 27(5):685–691.

von Ulmer, W. T., Schött, D. 1991. Chronisch-obstruktive Bronchitis. Wirkung von Gelomyrtol forte in einer plazebokontrollierten Doppelblindstudie. *Fortschr. Med.* 109:547–550.

Walton, S. F., McKinnon, M., Pizzutto, S., Dougall, A., Williams, E., Currie, B. J. 2004. Acaricidal activity of Melaleuca alternifolia (tea tree) oil: In vitro sensitivity of Sarcoptes scabiei var hominis to terpinen-4-ol. *Arch. Dermatol.* 140:563–566.

Warnke, P., Sherry, E., Russo, P. A. J. et al. 2006. Antibacterial essential oils in malodorous cancer patients: Clinical observations in 30 patients. *Phytomedicine* 13:463–467.

Warnke, P., Terheyden, H., Açil, Y., Springer, I. N. 2004. Tumor smell reduction with antibacterial essential oils [letter]. *Cancer* 100:879–880.

Welsh, M. J., Widdiscombe, J. H., Nadel, J. A. 1980. Fluid transport across canine tracheal epithelium. *J. Appl. Physiol.* 49:905–909.

White, D. A., Thompson, S. P., Wilson, C. G., Bell, G. D. 1987. A pharmacokinetic comparison of two delayed-release peppermint oil preparations, Colpermin and Mintec, for treatment of the irritable bowel syndrome. *Int. J. Pharmaceut.* 40:151–155.

Wilkins, J. 2002. Method for treating gastrointestinal disorder. US Patent 642045.

Willimason, E. M., Priestley, C. M., Burgess, I. F. 2007. An investigation and comparison of the bioactivity of selected essential oil on human lice and house dust mites. *Fitoterapia* 78:521–525.

Wittmann, M., Petro, W., Kaspar, P., Repges, R., Dethlefsen, U. 1998. Zur therapie chronisch obstruktiver atemwegserkrankungen mit sekretolytika. *Atemw.-Lungenkrank.* 24:67–74.

Worth, H., Dethlefsen, U. 2012. Patients with asthma benefit from concomitant therapy with cineole: A placebo-controlled, double-blind trial. *J. Asthma* 49(8):849–853.

Woelk, H., Schläfke, S. 2010. A multi-center, double-blind, randomised study of the Lavender oil preparation Silexan in comparison to Lorazepam for generalized anxiety disorder. *Phytomedicine* 17(2):94–99.

Wright, C. E., Bowen, W. P., Grattan, T. J., Morice, A. H. 1998. Identification of the l-menthol binding site in guinea pig lung membranes. *Br. J. Pharmacol.* 123:481–486.

Wright, C. E., Laude, E. A., Grattan, T. J., Morice, A. H. 1997. Capsaicin and neurokinin A-induced bronchoconstriction in the anaesthetised guinea pig: Evidence for a direct action of menthol on isolated bronchial smooth muscle. *Br. J. Pharmacol.* 121:1645–1650.

Wu, S., Patel, K. B., Booth, L. J., Metcalf, J. P., Lin, H.-K., Wu, W. 2010. Protective essential oil attenuates influenza virus infection: An in vitro study in MDCK cells. *BMC Complement. Altern. Med.*, 10(69): doi:10.1186/1472-6882-1110-1169.

Xing, H., Ling, J. X., Chen, M. et al. 2008. TRPM8 mechanism of autonomic nerve response to cold in respiratory airway. *Mol. Pain.* 4:22.

Yabhoobi-Ershadi, M. R., Akhavan, A. A., Jahanifard, E., Vatandoost, H., Amin, G., Moosavi, L., Zahraei Ramazani, A. R., Abdoli, H., Arandian, M. H. 2006. Repellency effect of myrtle essential oil and DEET against Phlebotomus papatasi, under laboratory conditions. *Iran. J. Public Health.* 35(3):7–13.

Yamamoto, N., Nakai, Y., Sasahira, N. et al. 2006. Efficacy of peppermint oil as an antispasmodic during endoscopic retrograde cholangiopancreatography. *J. Gastroenterol. Hepatol.* 21:1394–1398.

Yang, Y. C., Lee, H. S., Clark, J. M., Ahn, Y. J. 2004. Insecticidal activity of plant essential oils against *Pediculus humanus capitis* (Anoplura: Pediculidae). *J. Med. Entomol.* 41:699–704.

Yang, Y.-C., Lee, H.-S., Clark, J. M., Ahn, Y.-J. 2005. Ovicidal and adulticidal activities of *Cinnamomum zeylanicum* bark essential oil compounds and related compounds against *Pediculus humanus capitis* (Anoplura: Pediculicidae). *Int. J. Parasitol.* 35(14):1595–1600.

Yang, Y. C., Lee, S. H., Lee, W. J., Choi, D. H., Ahn, Y. J. 2003. Ovicidal and adulticidal effects of *Eugenia caryophyllata* bud and leaf oil compounds on *Pediculus capitis*. *J. Agric. Food Chem.* 51(17):4884–4888.

Yip, Y. B., Tam, A. C. Y. 2008. An experimental study on the effectiveness of massage with aromatic ginger and orange essential oil for moderate-to-severe knee pain among the elderly in Hong Kong. *Compl. Ther. Med.* 16:131–138.

Zanker, K. S., Blumel, G. 1983. Terpene-induced lowering of surface tension in vitro: A rationale for surfactant substitution. *Res. Exp. Med.* 182(1):33–38.

Zhu, L., Tian, Y. 2011. Chemical composition and larvicidal activity of Blumea densiflora essential oils against Anopheles anthropophagus: A malarial vector mosquito. *Parasitol. Res.* 109(5):1417–1422.

14 In Vitro Antimicrobial Activities of Essential Oils Monographed in the *European Pharmacopoeia 8th Edition*

Alexander Pauli and Heinz Schilcher

CONTENTS

14.1 Introduction 433
    14.1.1 Agar Diffusion Test 434
    14.1.2 Dilution Test 434
    14.1.3 Vapor Phase Test 435
14.2 Results 435
14.3 Discussion 608
References 612

14.1 INTRODUCTION

One of the first systematic in vitro examinations of the antimicrobial activity of essential oils dates back to the late nineteenth century when Buchholtz studied the growth inhibitory properties of caraway oil, thyme oil, phenol, and thymol on bacteria having been cultivated in a tabac decoction. In this examination, thymol turned out to be 10-fold stronger than phenol (Buchholtz, 1875), which was in use as surgical antiseptic at that time (Ashhurst, 1927). The German pharmacopoeia "Deutsches Arzneibuch 6" (DAB 6) issued in 1926 and later supplements (1947, 1959) listed together 26 essential oils, and by this, it has become obvious that essential oils have a long history in pharmaceutical practice due to their pharmacological activities. The *European Pharmacopoeia 8th Edition* (Council of Europe, 2014) lists 32 essential oils. Among them are 20 oils already present in DAB 6 (anise, bitter fennel, caraway, cassia, cinnamon bark, citronella, clove, coriander, eucalyptus, juniper, lavender, lemon, matricaria, neroli, peppermint, pine needle, pumilio pine, rosemary, thyme, and turpentine), three oils have been previously listed in the British Pharmacopoeia in the year 1993 (dementholized mint, nutmeg, and sweet orange), one in the French Pharmacopoeia X (star anise), and five oils were added later (cinnamon leaf, clary sage, mandarin, star anise, and tea tree).

Since essential oils are subject for pharmacological studies, tests on their antimicrobial activities have been done frequently. In consequence, a comprehensive data material exists, which, however, has never been compiled together for important essential oils, like such listed actually in the *European Pharmacopoeia*. Therefore, an attempt was done to collect and examine such information from scientific literature to obtain an insight into the variability of test parameters, data variation, and the significance of such factors in the interpretation of results.

Among the testing methods used to characterize in vitro antimicrobial activity of essential oils, the three main methods turned out to be agar diffusion test, serial broth or agar dilution test, and vapor phase (VP) test. Further tests comprise various kill-time studies, for example, the activity of a compound relative to phenol after 15 min (phenol coefficient) (Rideal and Walker, 1903), killing time determination after contact to a test compounds using contaminated silk threads (Koch, 1881), recording of growth curves and determination of the amount of a compound being effective to inhibit growth of 50% of test organisms (Friedman et al., 2004), poisoned food techniques in which the delay of microbial growth is determined in presence of growth inhibitors (Reiss, 1982; Kurita and Koike, 1983), spore germination, and short contact time studies in fungi (Smyth and Smyth, 1932; Mikhlin et al., 1983). Other studies monitor presence or absence of growth by measuring metabolic CO_2 in yeast (Belletti et al., 2004) or visualize growth by indicators, such as sulfur salts from sulfur-supplemented cow milk as growth medium (Geinitz, 1912), 2,3,5-triphenyltetrazolium chloride (Canillac et al., 1996), or *p*-iodonitrophenyltetrazolium violet (Weseler et al., 2005; Al-Bayati, 2008). The bioautographic assay on thin-layer chromatography plates has been developed for identification of active compounds in plant extracts (Rahalison et al., 1994) but was later also taken for the examination of essential oils (Iscan et al., 2002).

14.1.1 Agar Diffusion Test

In the agar diffusion test, the essential oil to be tested is placed on top of an agar surface. Two techniques exist: In the first one, the essential oil is placed onto a paper disk; in the second, a hole is made into the agar surface and the essential oil is put into the hole. In the following, the essential oil diffuses from its reservoir through the agar medium, which is seeded with microorganism. Antimicrobially active oils cause an inhibition zone around the reservoir after incubation, respectively, and normally the size of inhibition zone is regarded as measure for the antimicrobial potency of an essential oil. However, lipophilic compounds such as farnesol cause only small inhibition zones against *Bacillus subtilis* in the agar diffusion test (Weis, 1986), although the compound resulted in a strong inhibition in the serial dilution test (DIL) (minimum inhibitory concentration [MIC] = 12.5 µg/mL) (Kubo et al., 1983). Thus, strong inhibitors having low water solubility gave a poor or even negative result in the agar diffusion test. It is therefore, wrong to conclude that an essential oil without resulting in an inhibition zone in the agar diffusion test is without any antimicrobially active constituents, or in other words, antimicrobially active compounds are easily overlooked by this method. Another problem is the interpretation of the size of inhibition zones, which depend on both, the diffusion coefficient plus antimicrobial activity of every compound present in an essential oil. Beside generally unknown diffusion coefficients of essential oil constituents, the size of inhibition zones is influenced by several other factors: volatilization of essential oil, disk or hole sizes, amount of compound given to disk or into hole, adsorption by the disk, agar type, agar–agar content, pH, volume of agar, microbial strains tested (Janssen et al., 1987). Taken together, this test method can be used as a pretest, but the results should not be overrated. In the following data schedule, the experimental conditions are briefly given (culture medium, incubation temperature, incubation time, disk or hole size, amount of essential oil on disk or hole). The amount of compound in microgram added to a reservoir (disk or hole) is recalculated from microliter with a density of 1 for all essential oils.

14.1.2 Dilution Test

In the DIL, the essential oil to be tested is incorporated in a semisolid agar medium or liquid broth in several defined amounts. Absence of growth in agar plates or test tubes is determined with the naked eye after incubation. The MIC is the concentration of essential oil present in the ungrown agar plate or test tube with the highest amount of test material. When essential oils are tested, the main difficulty is caused by their low water solubility. The addition of solvents (e.g., dimethylsulfoxide and ethanol) or detergents (e.g., Tween 20) to the growth medium are unavoidable, which however, influences the MIC (Remmal et al., 1993; Hili et al., 1997; Hammer et al., 1999c). Another problem is the volatilization of essential oils during incubation. Working in a chamber with saturated moistened atmosphere (Pauli, 2006) or high water activity levels (Guynot et al., 2005) improved the situation. The MIC is additionally influenced by the selection of growth media, for example, in RPMI 1640, the MIC toward yeast is about 15 times lower than in Sabouraud medium (McCarthy et al., 1992; Jirovetz et al., 2007). Further, MIC-influencing test parameters are size of inoculum, pH of growth medium, and incubation time. Nevertheless, the serial DIL in liquid broth was recommended for natural substances (Boesel, 1991; Hadacek and Greger, 2000; Pauli, 2007) and is standardized for the testing of antibacterial and antifungal drugs in liquid broth and agar plates (Clinical and Laboratory Standards Institute, 2015). Its use enables a link to data of pharmaceutical drugs and an easier interpretation of test results. In the data section, all concentrations are recalculated in µg/mL. To distinguish between the agar dilution test and the serial DIL, the growth medium is abbreviated as either agar (A) or broth (B). Test parameters—exceptionally citations—are systematically given and comprise growth medium, incubation time, incubation temperature, and MIC in µg/mL or ppm.

14.1.3 Vapor Phase Test

In the VP, a standardized method does not exist among tests to study antimicrobial activity of essential oils. In most of the examinations, a reservoir (paper disk, cup, and glass) contains the sample of essential oil, and a seeded agar plate was inverted and covered the reservoir. After inoculation, an inhibition zone is formed, which is the measure of activity. Most of the data listed in the following tables have been worked out by such methods. Because these methods allow only the creation of relative values, a few examiners defined the MIC in atmosphere (MIC_{air}) by using airtight boxes (Inouye et al., 2001; Nakahara et al., 2003), which contained a seeded agar plate and the essential oil on the glass or paper. Otherwise, the results were estimated with +++ = normal growth, ++ = reduced growth, + = visible growth, and NG = no growth. The test parameters given in the following tables are growth medium, incubation time, incubation temperature, and activity evaluation or MIC_{air} in µg/mL or ppm.

To give detailed information, the following abbreviations are used in Tables 14.1 through 14.80: (h), essential oil was given into a hole; BA, Bacto agar; BHA, brain–heart infusion agar; BlA, blood agar; CA, Czapek's agar; CAB, Campylobacter agar base; CDA, Czapek Dox agar; DMSO, dimethylsulfoxide; EtOH, ethanol; EYA, Emerson's Ybss agar; germ., germination; HIB, heart infusion broth; HS, horse serum; inh., inhibition; ISA, Iso-sensitest agar; KBA, King's medium B agar; LA, Laury agar; LSA, Listeria selective agar; MA, malt agar; MAA, medium A agar; MBA, mycobiotic agar; MCA, MacConkey agar; MEB, malt extract broth; MHA, Mueller–Hinton agar; MHA, Mueller–Hinton broth; MIA, minimum inhibitory amount in µg per disk; MIC in µg/mL or ppm; MIC_{air}, minimal inhibitory concentration in µg/mL or ppm in the VP; MYA, malt extract–yeast extract–peptone–glucose agar; MYB, malt extract–yeast extract–glucose–peptone broth; NA, nutrient agar; NB, nutrient broth; OA, oat agar; PB, Penassay broth; Ref., reference number; SA, Sabouraud agar; SB, Sabouraud broth; sd, saturated disk; SDA, Sabouraud dextrose agar; SGB, Sabouraud glucose broth; SMA, Sabouraud maltose agar; sol., solution; TYA, tryptone–glucose–yeast extract agar; THB, Todd–Hewitt broth; TSA, trypticase soy agar; TSB, trypticase soy broth; TYB, tryptone–glucose–yeast extract broth; VC, various conditions: not given in detail; WA, sucrose–peptone agar; WFA, wheat flour agar; and YPB, yeast extract–peptone–dextrose broth.

14.2 RESULTS

The results of antimicrobial testing of the 28 essential oils listed in the *European Pharmacopoeia 6th Edition* (Council of Europe, 2008) are shown in Tables 14.1 through 14.80. Although literature is not fully covered, the quantity of available information was in part unexpected.

Anise oil, *Anisi Aetheroleum*
Definition: Essential oil obtained by steam distillation of the dry ripe fruits of *Pimpinella anisum* L.
Content: trans-anethole 87%–94%.

Bitter-fennel fruit oil, *Foeniculi amari fructus aetheroleum*
Definition: Essential oil obtained by steam distillation of the ripe fruits of *Foeniculum vulgare* Miller, ssp. *vulgare* var. *vulgare*.
Content: fenchone 12%–25%, *trans*-anethole 55%–75%.

Bitter-fennel herb oil, *Foeniculi amari herbal aetheroleum*
Definition: Essential oil obtained by steam distillation of the aerial parts of *F. vulgare* Miller, ssp. *vulgare* var. *vulgare*.
Content: Spanish type: *trans*-anethole 15%–40%, limonene 8%–30%, α-phellandrene 1%–25%, fenchone 7%–16%; α-pinene 2%–8%.
Tasmanian type: *trans*-anethole 45%–78%, fenchone 10%–25%, α-pinene 2%–11%.

Caraway oil, *Carvi aetheroleum*
Definition: Essential oil obtained by steam distillation of the dry fruits of *Carum carvi* L.
Content: carvone 50%–65%, limonene 30%–45%.

Cassia oil, *Cinnamomi cassia aetheroleum*
Definition: Cassia oil is obtained by steam distillation of the leaves and young branches of *Cinnamomum cassia* Blume (*Cinnamomum aromaticum* Nees).
Content: *trans*-cinnamic aldehyde 70%–90%, *trans*-2-methoxycinnamaldehyde 3%–15%.

Ceylon cinnamon bark oil, *Cinnamon zeylanicii corticis aetheroleum*
Definition: Ceylon cinnamon bark oil is obtained by steam distillation of the bark of the shoots of *Cinnamomum zeylanicum* Nees (*Cinnamomum verum* J.S. Presl.).
Content: *trans*-cinnamic aldehyde 55%–75%, *trans*-2-methoxycinnamaldehyde 0.1%–1.0%.

Ceylon cinnamon leaf oil, *Cinnamomi zeylanici folii aetheroleum*
Definition: Essential oil obtained by steam distillation of the leaves of *Cinnamomum verum* J.S. Presl.
Content: eugenol 70%–85%.

Citronella oil, *Citronella aetheroleum*
Definition: Essential oil obtained by steam distillation of the fresh or partially dried aerial parts of *Cymbopogon winterianus* Jowitt.
Content: citronellal 30%–45%, geraniol 20%–25%, citronellol 9%–15%.

Clary sage oil, *Salvia sclarae aetheroleum*
Definition: Essential oil obtained by steam distillation of the fresh or dried flowering stems of *Salvia sclarea* L.
Content: linalyl acetate 56%–78%, linalool 6.5%–24%, sclareol 0.4%–2.6%.

Clove oil, *Caryophylli floris aetheroleum*
Definition: Clove oil is obtained by steam distillation of the dried flower buds of *Syzygium aromaticum* (L.) Merill et L. M. Perry (*Eugenia caryophyllus* C. Spreng. Bull. et Harr.).
Content: eugenol 75%–88%, acetyleugenol 4%–15%, β-caryophyllene 5%–14%.

Coriander oil, *Coriandri aetheroleum*
Definition: Essential oil obtained by steam distillation of the fruits of *Coriandrum sativum* L.
Content: linalool 65%–78%.

Dwarf-pine oil, *Pini pumilionis aetheroleum*
Definition: Essential oil obtained by steam distillation of the fresh needles, twigs, and branches of *Pinus mugo* ssp. *mugo* ZENARI and/or *Pinus mugo* ssp. *pumilio* (HAENKE) FRANCO.
Pinus mugo Turra [*Pinus mugo* var. *pumilio* (Haenke) Zenari], plant part: leaves and twigs.
Content: a-pinene 10%–30%, δ-3-carene 10%–20%, β-phellandrene 10%–19%.

Eucalyptus oil, *Eucalypti aetheroleum*
Definition: Eucalyptus oil is obtained by steam distillation and rectification of the fresh leaves or the fresh terminal branchlets of various species of *Eucalyptus* rich in 1,8-cineole. The species mainly used are *Eucalyptus globulus* Labill., *Eucalyptus polybractea* R.T. Baker, and *Eucalyptus smithii* R.T. Baker.
Content: 1,8-cineole min. 70%.

Juniper oil, *Juniperi aetheroleum*
Definition: Essential oil obtained by steam distillation of the ripe, nonfermented berry cones of *Juniperus communis* L.
Content: α-pinene 20%–50%, β-myrcene 1%–35%, sabinene less than 20%.

Lavender oil, *Lavandulae aetheroleum*
Definition: Essential oil obtained by steam distillation of the flowering tops of *Lavandula angustifolia* Miller (*Lavandula officinalis* Chaix).
Content: linalyl acetate 25%–46%, linalool 20%–45%.

Lemon oil, *Lemonis aetheroleum*
Definition: Essential oil obtained by suitable mechanical means, without the aid of heat, from the fresh peel of *Citrus limon* (L.) Burman fil.
Content: limonene 56%–78%, β-pinene 7%–17%, γ-terpinene 6%–12%.

Mandarin oil, *Citri reticulatae aetheroleum*
Definition: Essential oil obtained without heating, by suitable mechanical treatment of the fresh peel of the fruit of *Citrus reticulata* Blanco var. mandarin.
Content: limonene 65%–75%, γ-terpinene 16%–22%, methyl N-methylanthranilate 0.30%–0.60%.

Matricaria oil, *Matricariae aetheroleum*
Definition: Blue essential oil obtained by steam distillation of the fresh or dried flower heads or flowering tops of *Matricaria recutita* L. (*Chamomilla recutita* L. Rauschert). There are two types of matricaria oils, which are characterized as rich in bisabolol oxides or rich in (−)-α-bisabolol.
Content: 1. Matricaria oil rich in bisabolol oxides: bisabolol oxides 29%–81%; chamazulene ≥ 1%.
2. Matricaria oil rich in (−)- a-bisabolol: (−)-a-bisabolol 10%–65%, chamazulene ≥ 1%, total of bisabolol oxides and (−)-a-bisabolol ≥ 20%.

Mint oil, *Menthae arvensis aetheroleum partim mentholum depletum*
Definition: Essential oil obtained by steam distillation of the fresh, flowering aerial parts, recently gathered from *Mentha canadensis* L. [syn. *Mentha arvensis* L. var. *glabrata* (Benth) Fern., *Mentha arvensis* var. *piperascens* Malinv. ex Holmes], followed by partial separation of menthol by crystallization.
Content: menthol 30%–50%, menthone 17%–35%, isomenthone 5%–13%.

Neroli oil, *Neroli aetheroleum*
Definition: Neroli oil is obtained by steam distillation of the fresh flowers of *Citrus aurantium* L. subsp. *aurantium* L. (*C. aurantium* L. subsp. *amara* Engl.).
Content: linalool 28%–44%, limonene 9%–18%, β-pinene 7%–17%, linalyl acetate 2%–15%.

Niaouli oil, Cineole type, *Niaouli typo cineolo aetheroleum*
Definition: Essential oil obtained by steam distillation from young leafy branches of *Melaleuca quinquenervia* (Cav.) S.T. Blake.
Content: 1,8-cineole 45%–65%.

Nutmeg oil, *Myristicae fragrantis aetheroleum*
Definition: Nutmeg oil is obtained by steam distillation of the dried and crushed kernels of *Myristica fragrans* Houtt.
Content: sabinene 14%–29%, α-pinene 15%–28%, β-pinene 13%–18%.

Peppermint oil, *Menthae piperitae aetheroleum*
Definition: Essential oil obtained by steam distillation of the fresh aerial parts of the flowering plant of *Mentha x piperita* L.
Content: menthol 30%–55%, menthone 14%–32%, isomenthone 1.5%–10%.

Pinus sylvestris oil, *Pini sylvestris aetheroleum*
Definition: Essential oil obtained by steam distillation of the fresh needles, twigs, and branches of *Pinus sylvestris* L.
Content: α-pinene 32%–60%, β-pinene 5%–22%, δ-3-carene 6%–18%.

Turpentine oil, *Pinus pinaster* type, *Terebinthi aetheroleum ab pinum pinastrum*
Definition: Essential oil obtained by steam distillation, followed by rectification at a temperature below 180°C, from the oleoresin obtained by tapping *Pinus pinaster* Aiton.
Content: α-pinene 70%–85%, β-pinene 11%–20%.

Rosemary oil, *Rosmarini aetheroleum*
Definition: Essential oil obtained by steam distillation of the flowering aerial parts of *Rosmarinus officinalis* L.
Content: Spanish type: α-pinene 18%–26%, 1,8-cineole 16%–25%, camphor 13%–21%
Moroccan, Tunisian type: 1,8-cineole 38%–55%, camphor 5%–15%, α-pinene 9%–14%

Spanish sage oil, *Salvage lavandulifolia aetheroleum*
Definition: Essential oil obtained by steam distillation from the aerial parts of *Salvia lavandulifolia* Vahl. collected at the flowering stage.
Content: camphor 11%–36%, 1,8-cineole 10%–30.5%, α-pinene 4%–11%.

Spike lavender oil, *Spicae aetheroleum*
Definition: Essential oil obtained by steam distillation of the flowering tops of *Lavandula latifolia* Medik.
Content: linalool 34%–50%, 1,8-cineole 16%–39%, camphor 8%–16%.

Star anise oil, *Anisi stellati aetheroleum*
Definition: Essential oil obtained by steam distillation of the dry ripe fruits of *Illicium verum* Hook. fil.
Content: *trans*-anethole 86%–93%.

Sweet orange oil, *Aurantii dulcis aetheroleum*
Definition: Essential oil obtained without heating, by suitable mechanical treatment of the fresh peel of the fruit of *Citrus sinensis* (L.) Osbeck (*C. aurantium* L. var. *dulcis* L.).

Tea tree oil, *Melaleucae aetheroleum*
Definition: Essential oil obtained by steam distillation of the foliage and terminal branchlets of *Melaleuca alternifolia* (Maiden and Betche) Cheel, *Melaleuca linariifolia* Smith, *Melaleuca dissitiflora* F. Mueller, and/or other species of *Melaleuca*.
Content: terpinen-4-ol min. 30%, γ-terpinene 10%–28%.

Thyme oil thymol type, *Thymi typo thymolo aetheroleum*
Definition: Essential oil obtained by steam distillation of the fresh flowering aerial parts of *Thymus vulgaris* L., *T. zygis* Loefl. ex L. or a mixture of both species.
Content: thymol 37%–55%, p-cymene 14%–28%.

TABLE 14.1
Inhibitory Data of Anise Oil Obtained in the Agar Diffusion Test

Microorganism	MO Class	Test Parameters	Disk Size (mm), Quantity (µg)	Inhibition Zone (mm)	Reference
Acinetobacter calcoaceticus	Bac–	ISA, 48 h, 25°C	4 (h), 10,000	0	Deans and Ritchie (1987)
Aerobacter aerogenes	Bac–	NA, 24 h, 37°C	—, sd	0	Maruzzella and Lichtenstein (1956)
Aeromonas hydrophila	Bac–	ISA, 48 h, 25°C	4 (h), 10,000	6	Deans and Ritchie (1987)
Alcaligenes faecalis	Bac–	ISA, 48 h, 25°C	4 (h), 10,000	0	Deans and Ritchie (1987)
Beneckea natriegens	Bac–	ISA, 48 h, 25°C	4 (h), 10,000	0	Deans and Ritchie (1987)
Campylobacter jejuni	Bac–	TSA, 24 h, 42°C	4 (h), 25,000	4.5	Smith-Palmer et al. (1998)
Citrobacter freundii	Bac–	ISA, 48 h, 25°C	4 (h), 10,000	0	Deans and Ritchie (1987)
Enterobacter aerogenes	Bac–	ISA, 48 h, 25°C	4 (h), 10,000	0	Deans and Ritchie (1987)
Enterococcus faecalis	Bac–	NA, 24 h, 37°C	5,000 on agar	0	Di Pasqua et al. (2005)
Erwinia carotovora	Bac–	ISA, 48 h, 25°C	4 (h), 10,000	0	Deans and Ritchie (1987)
Escherichia coli	Bac–	NA, 24 h, 37°C	—, sd	0	Maruzzella and Lichtenstein (1956)
Escherichia coli	Bac–	NA, 24 h, 37°C	10 (h), 100,000	0	Narasimha Rao and Nigam (1970)
Escherichia coli	Bac–	ISA, 48 h, 25°C	4 (h), 10,000	0	Deans and Ritchie (1987)
Escherichia coli	Bac–	Cited	15, 2,500	1	Pizsolitto et al. (1975)
Escherichia coli	Bac–	Cited, 18 h, 37°C	6, 2,500	7	Janssen et al. (1986)
Escherichia coli	Bac–	TSA, 24 h, 35°C	4 (h), 25,000	4.5	Smith-Palmer et al. (1998)
Escherichia coli	Bac–	NA, 18 h, 37°C	6 (h), pure	11	Yousef and Tawil (1980)
Flavobacterium suaveolens	Bac–	ISA, 48 h, 25°C	4 (h), 10,000	6	Deans and Ritchie (1987)
Klebsiella aerogenes	Bac–	NA, 24 h, 37°C	10 (h), 100,000	0	Narasimha Rao and Nigam (1970)
Klebsiella pneumoniae	Bac–	ISA, 48 h, 25°C	4 (h), 10,000	0	Deans and Ritchie (1987)
Klebsiella sp.	Bac–	Cited	15, 2,500	0	Pizsolitto et al. (1975)
Moraxella sp.	Bac–	ISA, 48 h, 25°C	4 (h), 10,000	0	Deans and Ritchie (1987)
Neisseria perflava	Bac–	NA, 24 h, 37°C	—, sd	1	Maruzzella and Lichtenstein (1956)
Proteus sp.	Bac–	Cited	15, 2,500	1	Pizsolitto et al. (1975)
Proteus vulgaris	Bac–	ISA, 48 h, 25°C	4 (h), 10,000	0	Deans and Ritchie (1987)
Proteus vulgaris	Bac–	NA, 24 h, 37°C	—, sd	1.5	Maruzzella and Lichtenstein (1956)
Pseudomonas aeruginosa	Bac–	NA, 24 h, 37°C	—, sd	0	Maruzzella and Lichtenstein (1956)
Pseudomonas aeruginosa	Bac–	Cited	15, 2,500	0	Pizsolitto et al. (1975)
Pseudomonas aeruginosa	Bac–	ISA, 48 h, 25°C	4 (h), 10,000	0	Deans and Ritchie (1987)

(Continued)

TABLE 14.1 (Continued)
Inhibitory Data of Anise Oil Obtained in the Agar Diffusion Test

Microorganism	MO Class	Test Parameters	Disk Size (mm), Quantity (μg)	Inhibition Zone (mm)	Reference
Pseudomonas aeruginosa	Bac–	NA, 18 h, 37°C	6 (h), pure	0	Yousef and Tawil (1980)
Pseudomonas aeruginosa	Bac–	Cited, 18 h, 37°C	6, 2,500	6.7	Janssen et al. (1986)
Pseudomonas sp.	Bac–	NA, 24 h, 37°C	5,000 on agar	0	Di Pasqua et al. (2005)
Salmonella enteritidis	Bac–	TSA, 24 h, 35°C	4 (h), 25,000	4.5	Smith-Palmer et al. (1998)
Salmonella pullorum	Bac–	ISA, 48 h, 25°C	4 (h), 10,000	0	Deans and Ritchie (1987)
Salmonella sp.	Bac–	Cited	15, 2,500	0	Pizsolitto et al. (1975)
Salmonella typhi	Bac–	NA, 24 h, 37°C	10 (h), 100,000	16	Narasimha Rao and Nigam (1970)
Serratia marcescens	Bac–	ISA, 48 h, 25°C	4 (h), 10,000	0	Deans and Ritchie (1987)
Serratia marcescens	Bac–	NA, 24 h, 37°C	—, sd	1	Maruzzella and Lichtenstein (1956)
Serratia sp.	Bac–	Cited	15, 2,500	1	Pizsolitto et al. (1975)
Shigella sp.	Bac–	Cited	15, 2,500	1	Pizsolitto et al. (1975)
Vibrio cholerae	Bac–	NA, 24 h, 37°C	10 (h), 100,000	17	Narasimha Rao and Nigam (1970)
Yersinia enterocolitica	Bac–	ISA, 48 h, 25°C	4 (h), 10,000	0	Deans and Ritchie (1987)
Bacillus mesentericus	Bac+	NA, 24 h, 37°C	—, sd	3	Maruzzella and Lichtenstein (1956)
Bacillus sp.	Bac+	Cited	15, 2,500	1	Pizsolitto et al. (1975)
Bacillus subtilis	Bac+	ISA, 48 h, 25°C	4 (h), 10,000	0	Deans and Ritchie (1987)
Bacillus subtilis	Bac+	NA, 24 h, 37°C	—, sd	4	Maruzzella and Lichtenstein (1956)
Bacillus subtilis	Bac+	Cited, 18 h, 37°C	6, 2,500	8	Janssen et al. (1986)
Bacillus subtilis	Bac+	NA, 18 h, 37°C	6 (h), pure	12.5	Yousef and Tawil (1980)
Brevibacterium linens	Bac+	ISA, 48 h, 25°C	4 (h), 10,000	9	Deans and Ritchie (1987)
Brochothrix thermosphacta	Bac+	NA, 24 h, 37°C	5,000 on agar	0	Di Pasqua et al. (2005)
Brochothrix thermosphacta	Bac+	ISA, 48 h, 25°C	4 (h), 10,000	5.5	Deans and Ritchie (1987)
Clostridium sporogenes	Bac+	ISA, 48 h, 25°C	4 (h), 10,000	0	Deans and Ritchie (1987)
Corynebacterium diphtheriae	Bac+	NA, 24 h, 37°C	10 (h), 100,000	0	Narasimha Rao and Nigam (1970)
Lactobacillus delbrueckii	Bac+	NA, 24 h, 37°C	5,000 on agar	0	Di Pasqua et al. (2005)
Lactobacillus plantarum	Bac+	ISA, 48 h, 25°C	4 (h), 10,000	0	Deans and Ritchie (1987)
Lactobacillus plantarum	Bac+	NA, 24 h, 37°C	5,000 on agar	0	Di Pasqua et al. (2005)
Lactococcus garvieae	Bac+	NA, 24 h, 37°C	5,000 on agar	0	Di Pasqua et al. (2005)
Lactococcus lactis subsp. *lactis*	Bac+	NA, 24 h, 37°C	5,000 on agar	0	Di Pasqua et al. (2005)

(Continued)

TABLE 14.1 (Continued)
Inhibitory Data of Anise Oil Obtained in the Agar Diffusion Test

Microorganism	MO Class	Test Parameters	Disk Size (mm), Quantity (μg)	Inhibition Zone (mm)	Reference
Leuconostoc cremoris	Bac+	ISA, 48 h, 25°C	4 (h), 10,000	5	Deans and Ritchie (1987)
Listeria monocytogenes	Bac+	ISA, 48 h, 25°C	4 (h), 10,000	0	Lis-Balchin et al. (1998)
Listeria monocytogenes	Bac+	NA, 24 h, 37°C	5,000 on agar	0	Di Pasqua et al. (2005)
Listeria monocytogenes	Bac+	TSA, 24 h, 35°C	4 (h), 25,000	4	Smith-Palmer et al. (1998)
Micrococcus luteus	Bac+	ISA, 48 h, 25°C	4 (h), 10,000	0	Deans and Ritchie (1987)
Mycobacterium phlei	Bac+	NA, 18 h, 37°C	6 (h), pure	18	Yousef and Tawil (1980)
Sarcina lutea	Bac+	NA, 24 h, 37°C	—, sd	0	Maruzzella and Lichtenstein (1956)
Staphylococcus aureus	Bac+	NA, 24 h, 37°C	—, sd	0	Maruzzella and Lichtenstein (1956)
Staphylococcus aureus	Bac+	NA, 24 h, 37°C	10 (h), 100,000	0	Narasimha Rao and Nigam (1970)
Staphylococcus aureus	Bac+	ISA, 48 h, 25°C	4 (h), 10,000	0	Deans and Ritchie (1987)
Staphylococcus aureus	Bac+	NA, 24 h, 37°C	5,000 on agar	0	Di Pasqua et al. (2005)
Staphylococcus aureus	Bac+	Cited	15, 2,500	1	Pizsolitto et al. (1975)
Staphylococcus aureus	Bac+	TSA, 24 h, 35°C	4 (h), 25,000	4	Smith-Palmer et al. (1998)
Staphylococcus aureus	Bac+	Cited, 18 h, 37°C	6, 2,500	8.3	Janssen et al. (1986)
Staphylococcus aureus	Bac+	NA, 18 h, 37°C	6 (h), pure	14.5	Yousef and Tawil (1980)
Staphylococcus epidermidis	Bac+	Cited	15, 2,500	4	Pizsolitto et al. (1975)
Staphylococcus epidermidis	Bac+	NA, 24 h, 37°C	10 (h), 100,000	12	Narasimha Rao and Nigam (1970)
Streptococcus faecalis	Bac+	ISA, 48 h, 25°C	4 (h), 10,000	0	Deans and Ritchie (1987)
Streptococcus sp.	Bac+	NA, 24 h, 37°C	10 (h), 100,000	16	Narasimha Rao and Nigam (1970)
Streptococcus viridans	Bac+	Cited	15, 2,500	0	Pizsolitto et al. (1975)
Streptomyces venezuelae	Bac+	SMA, 2–7 days, 20°C	6.35, sd	7	Maruzzella and Liguori (1958)
Alternaria porri	Fungi	PDA, 72 h, 28°C	5, 5,000	35	Pawar and Thaker (2007)
Alternaria solani	Fungi	SMA, 2–7 days, 20°C	6.35, sd	9	Maruzzella and Liguori (1958)
Aspergillus flavus	Fungi	Cited	—, pure	28	Gangrade et al. (1991)
Aspergillus fumigatus	Fungi	SMA, 2–7 days, 20°C	6.35, sd	9	Maruzzella and Liguori (1958)
Aspergillus niger	Fungi	PDA, 48 h, 28°C	5, 5,000	0	Pawar and Thaker (2006)
Aspergillus niger	Fungi	SMA, 2–7 days, 20°C	6.35, sd	13	Maruzzella and Liguori (1958)
Aspergillus niger	Fungi	Cited	—, pure	27	Gangrade et al. (1991)

(Continued)

TABLE 14.1 (*Continued*)
Inhibitory Data of Anise Oil Obtained in the Agar Diffusion Test

Microorganism	MO Class	Test Parameters	Disk Size (mm), Quantity (μg)	Inhibition Zone (mm)	Reference
Aspergillus niger	Fungi	SDA, 8 days, 30°C	6 (h), pure	60	Yousef and Tawil (1980)
Fusarium oxysporum	Fungi	Cited	—, pure	32	Gangrade et al. (1991)
Fusarium oxysporum f.sp. *ciceri*	Fungi	PDA, 72 h, 28°C	5, 5,000	15	Pawar and Thaker (2007)
Geotrichum sp.	Fungi	SDA, 2–7 days, 20°C	5 (h), 60,000	17	Kosalec et al. (2005)
Helminthosporium sativum	Fungi	SMA, 2–7 days, 20°C	6.35, sd	10	Maruzzella and Liguori (1958)
Microsporum canis	Fungi	SDA, 2–7 days, 20°C	5 (h), 60,000	27	Kosalec et al. (2005)
Microsporum gypseum	Fungi	SDA, 2–7 days, 20°C	5 (h), 60,000	21	Kosalec et al. (2005)
Mucor mucedo	Fungi	SMA, 2–7 days, 20°C	6.35, sd	6	Maruzzella and Liguori (1958)
Mucor sp.	Fungi	SDA, 8 days, 30°C	6 (h), pure	12	Yousef and Tawil (1980)
Nigrospora panici	Fungi	SMA, 2–7 days, 20°C	6.35, sd	12	Maruzzella and Liguori (1958)
Penicillium chrysogenum	Fungi	SDA, 8 days, 30°C	6 (h), pure	60	Yousef and Tawil (1980)
Penicillium digitatum	Fungi	SMA, 2–7 days, 20°C	6.35, sd	15	Maruzzella and Liguori (1958)
Rhizopus nigricans	Fungi	SMA, 2–7 days, 20°C	6.35, sd	3	Maruzzella and Liguori (1958)
Rhizopus sp.	Fungi	SDA, 8 days, 30°C	6 (h), pure	13	Yousef and Tawil (1980)
Trichophyton mentagrophytes	Fungi	SDA, 2–7 days, 20°C	5 (h), 60,000	23	Kosalec et al. (2005)
Trichophyton rubrum	Fungi	SDA, 2–7 days, 20°C	5 (h), 60,000	25	Kosalec et al. (2005)
Candida albicans	Yeast	SMA, 2–7 days, 20°C	6.35, sd	2	Maruzzella and Liguori (1958)
Candida albicans	Yeast	Cited, 18 h, 37°C	6, 2,500	12	Janssen et al. (1986)
Candida albicans	Yeast	SDA, 18 h, 30°C	6 (h), pure	15	Yousef and Tawil (1980)
Candida albicans	Yeast	SDA, 2–7 days, 20°C	5 (h), 60,000	29	Kosalec et al. (2005)
Candida glabrata	Yeast	SDA, 2–7 days, 20°C	5 (h), 60,000	21	Kosalec et al. (2005)
Candida krusei	Yeast	SDA, 2–7 days, 20°C	5 (h), 60,000	[a]	Kosalec et al. (2005)
Candida krusei	Yeast	SMA, 2–7 days, 20°C	6.35, sd	0	Maruzzella and Liguori (1958)
Candida parapsilosis	Yeast	SDA, 2–7 days, 20°C	5 (h), 60,000	30	Kosalec et al. (2005)
Candida pseudotropicalis	Yeast	SDA, 2–7 days, 20°C	5 (h), 60,000	[a]	Kosalec et al. (2005)
Candida tropicalis	Yeast	SDA, 2–7 days, 20°C	5 (h), 60,000	[a]	Kosalec et al. (2005)
Candida tropicalis	Yeast	SMA, 2–7 days, 20°C	6.35, sd	2	Maruzzella and Liguori (1958)
Cryptococcus neoformans	Yeast	SMA, 2–7 days, 20°C	6.35, sd	9	Maruzzella and Liguori (1958)
Cryptococcus rhodobenhani	Yeast	SMA, 2–7 days, 20°C	6.35, sd	0	Maruzzella and Liguori (1958)
Saccharomyces cerevisiae	Yeast	SMA, 2–7 days, 20°C	6.35, sd	0	Maruzzella and Liguori (1958)

[a] Fungicidal.

TABLE 14.2
Inhibitory Data of Anise Oil Obtained in the Dilution Test

Microorganism	MO Class	Test Parameters	MIC (μg/mL)	Reference
Campylobacter jejuni	Bac–	TSB, 24 h, 42°C	>10,000	Smith-Palmer et al. (1998)
Escherichia coli	Bac–	TSB, 24 h, 35°C	>10,000	Smith-Palmer et al. (1998)
Escherichia coli	Bac–	TSB, 24 h, 37°C	>10,000	Di Pasqua et al. (2005)
Escherichia coli	Bac–	MHB, DMSO, 24 h, 37°C	>500	Al-Bayati (2008)
Escherichia coli	Bac–	NB, Tween 20, 18 h, 37°C	50,000	Yousef and Tawil (1980)
Helicobacter pylori	Bac–	Cited, 20 h, 37°C	294.7–589.4	Weseler et al. (2005)
Klebsiella pneumoniae	Bac–	MHB, DMSO, 24 h, 37°C	>500	Al-Bayati (2008)
Proteus mirabilis	Bac–	MHB, DMSO, 24 h, 37°C	125	Al-Bayati (2008)
Proteus vulgaris	Bac–	MHB, DMSO, 24 h, 37°C	62.5	Al-Bayati (2008)
Pseudomonas aeruginosa	Bac–	MHB, DMSO, 24 h, 37°C	>500	Al-Bayati (2008)
Pseudomonas aeruginosa	Bac–	NB, Tween 20, 18 h, 37°C	>50,000	Yousef and Tawil (1980)
Salmonella enteritidis	Bac–	TSB, 24 h, 35°C	>10,000	Smith-Palmer et al. (1998)
Salmonella typhi	Bac–	MHB, DMSO, 24 h, 37°C	500	Al-Bayati (2008)
Salmonella typhimurium	Bac–	TSB, 24 h, 37°C	>10,000	Di Pasqua et al. (2005)
Salmonella typhimurium	Bac–	MHB, DMSO, 24 h, 37°C	250	Al-Bayati (2008)
Bacillus cereus	Bac+	MHB, DMSO, 24 h, 37°C	62.5	Al-Bayati (2008)
Bacillus subtilis	Bac+	NB, Tween 20, 18 h, 37°C	1,600	Yousef and Tawil (1980)
Brochothrix thermosphacta	Bac+	M17, 24 h, 20°C	>10,000	Di Pasqua et al. (2005)
Listeria monocytogenes	Bac+	TSB, 24 h, 35°C	>10,000	Smith-Palmer et al. (1998)
Mycobacterium phlei	Bac+	NB, Tween 20, 18 h, 37°C	200	Yousef and Tawil (1980)
Staphylococcus aureus	Bac+	TSB, 24 h, 35°C	>10,000	Smith-Palmer et al. (1998)
Staphylococcus aureus	Bac+	MHB, DMSO, 24 h, 37°C	125	Al-Bayati (2008)
Staphylococcus aureus	Bac+	NB, Tween 20, 18 h, 37°C	25,000	Yousef and Tawil (1980)
Alternaria alternata	Fungi	Cited	100% inh. 600	Shukla and Tripathi (1987)
Alternaria tenuissima	Fungi	Cited	100% inh. 600	Shukla and Tripathi (1987)
Aspergillus awamori	Fungi	Cited	100% inh. 600	Shukla and Tripathi (1987)
Aspergillus flavus	Fungi	PDA, 5 days, 27°C	>1,000	Thompson and Cannon (1986)
Aspergillus flavus	Fungi	PDA, 7–14 days, 28°C	500	Soliman and Badeaa (2002)
Aspergillus flavus	Fungi	PDA, 6–8 h 20°C, spore germ. inh.	50–100	Thompson (1986)
Aspergillus fumigatus	Fungi	Cited	100% inh. 600	Shukla and Tripathi (1987)
Aspergillus niger	Fungi	Cited	100% inh. 600	Shukla and Tripathi (1987)
Aspergillus niger	Fungi	NB, Tween 20, 8 days, 30°C	1,600	Yousef and Tawil (1980)

(*Continued*)

TABLE 14.2 (Continued)
Inhibitory Data of Anise Oil Obtained in the Dilution Test

Microorganism	MO Class	Test Parameters	MIC (µg/mL)	Reference
Aspergillus niger	Fungi	Yeast extract (YES) broth, 10 days	83% inh. 10,000	Lis-Balchin et al. (1998)
Aspergillus ochraceus	Fungi	Cited	100% inh. 600	Shukla and Tripathi (1987)
Aspergillus ochraceus	Fungi	PDA, 7–14 days, 28°C	500	Soliman and Badeaa (2002)
Aspergillus ochraceus	Fungi	YES broth, 10 days	82% inh. 10,000	Lis-Balchin et al. (1998)
Aspergillus oryzae	Fungi	Cited	250	Okazaki and Oshima (1953)
Aspergillus parasiticus	Fungi	PDA, 5 days, 27°C	>1,000	Thompson and Cannon (1986)
Aspergillus parasiticus	Fungi	Cited	100% inh. 600	Shukla and Tripathi (1987)
Aspergillus parasiticus	Fungi	PDA, 7–14 days, 28°C	500	Soliman and Badeaa (2002)
Aspergillus parasiticus	Fungi	PDA, 6–8 h 20°C, spore germ. inh.	50–100	Thompson (1986)
Aspergillus sydowii	Fungi	Cited	100% inh. 600	Shukla and Tripathi (1987)
Aspergillus tamari	Fungi	Cited	100% inh. 600	Shukla and Tripathi (1987)
Aspergillus terreus	Fungi	Cited	100% inh. 600	Shukla and Tripathi (1987)
Botryodiplodia theobromae	Fungi	Cited	100% inh. 600	Shukla and Tripathi (1987)
Cephalosporium sacchari	Fungi	OA, EtOH, 3 days, 20°C	20,000	Narasimba Rao et al. (1971)
Ceratocystis paradoxa	Fungi	OA, EtOH, 3 days, 20°C	1,000	Narasimba Rao et al. (1971)
Cladosporium herbarum	Fungi	Cited	100% inh. 600	Shukla and Tripathi (1987)
Colletotrichum capsici	Fungi	Cited	100% inh. 600	Shukla and Tripathi (1987)
Curvularia lunata	Fungi	Cited	100% inh. 600	Shukla and Tripathi (1987)
Curvularia lunata	Fungi	OA, EtOH, 3 days, 20°C	4,000	Narasimba Rao et al. (1971)
Curvularia pallescens	Fungi	Cited	100% inh. 600	Shukla and Tripathi (1987)
Epicoccum nigrum	Fungi	Cited	100% inh. 600	Shukla and Tripathi (1987)
Epidermophyton floccosum	Fungi	SA, Tween 80, 21 days, 20°C	>300	Janssen et al. (1988)
Fusarium accuminatum	Fungi	Cited	100% inh. 600	Shukla and Tripathi (1987)
Fusarium culmorum	Fungi	YES broth, 10 days	69% inh. 10,000	Lis-Balchin et al. (1998)

(*Continued*)

TABLE 14.2 (Continued)
Inhibitory Data of Anise Oil Obtained in the Dilution Test

Microorganism	MO Class	Test Parameters	MIC (μg/mL)	Reference
Fusarium equiseti	Fungi	Cited	100% inh. 600	Shukla and Tripathi (1987)
Fusarium moniliforme	Fungi	Cited	100% inh. 600	Shukla and Tripathi (1987)
Fusarium moniliforme	Fungi	PDA, 7–14 days, 28°C	500	Soliman and Badeaa (2002)
Fusarium moniliforme	Fungi	PDA, 7 days, 23.5°C	52% inh. 10,000	Mueller-Ribeau et al. (1995)
Fusarium moniliforme var. subglutinans	Fungi	OA, EtOH, 3 days, 20°C	2,000	Narasimba Rao et al. (1971)
Fusarium oxysporum	Fungi	Cited	100% inh. 600	Shukla and Tripathi (1987)
Fusarium semitectum	Fungi	Cited	100% inh. 600	Shukla and Tripathi (1987)
Fusarium udum	Fungi	Cited	100% inh. 600	Shukla and Tripathi (1987)
Geotrichum sp.	Fungi	SGB, 3–7 days, 25°C	15,600	Kosalec et al. (2005)
Helminthosporium sacchari	Fungi	OA, EtOH, 3 days, 20°C	4,000	Narasimba Rao et al. (1971)
Macrophomina phaseoli	Fungi	Cited	100% inh. 600	Shukla and Tripathi (1987)
Microsporum canis	Fungi	SGB, 3–7 days, 25°C	1,000	Kosalec et al. (2005)
Microsporum gypseum	Fungi	SGB, 3–7 days, 25°C	2,000	Kosalec et al. (2005)
Mucor hiemalis	Fungi	PDA, 5 days, 27°C	>1,000	Thompson and Cannon (1986)
Mucor mucedo	Fungi	PDA, 5 days, 27°C	>1,000	Thompson and Cannon (1986)
Mucor racemosus	Fungi	Cited	250	Okazaki and Oshima (1953)
Mucor racemosus f. racemosus	Fungi	PDA, 5 days, 27°C	>1,000	Thompson and Cannon (1986)
Mucor sp.	Fungi	NB, Tween 20, 8 days, 30°C	400	Yousef and Tawil (1980)
Nigrospora oryzae	Fungi	Cited	100% inh. 600	Shukla and Tripathi (1987)
Penicillium chrysogenum	Fungi	Cited	100% inh. 600	Shukla and Tripathi (1987)
Penicillium chrysogenum	Fungi	NB, Tween 20, 8 days, 30°C	1,600	Yousef and Tawil (1980)
Penicillium chrysogenum	Fungi	Cited	250	Okazaki and Oshima (1953)
Penicillium citrinum	Fungi	Cited	100% inh. 600	Shukla and Tripathi (1987)
Physalospora tucumanensis	Fungi	OA, EtOH, 3 days, 20°C	2,000	Narasimba Rao et al. (1971)
Phytophthora capsici	Fungi	PDA, 7 days, 23.5°C	36% inh. 10,000	Mueller-Ribeau et al. (1995)

(Continued)

TABLE 14.2 (*Continued*)
Inhibitory Data of Anise Oil Obtained in the Dilution Test

Microorganism	MO Class	Test Parameters	MIC (μg/mL)	Reference
Rhizoctonia solani	Fungi	PDA, 7 days, 23.5°C	100% inh. 10,000	Mueller-Ribeau et al. (1995)
Rhizopus 66-81-2	Fungi	PDA, 5 days, 27°C	>1,000	Thompson and Cannon (1986)
Rhizopus arrhizus	Fungi	PDA, 5 days, 27°C	>1,000	Thompson and Cannon (1986)
Rhizopus chinensis	Fungi	PDA, 5 days, 27°C	>1,000	Thompson and Cannon (1986)
Rhizopus circinans	Fungi	PDA, 5 days, 27°C	>1,000	Thompson and Cannon (1986)
Rhizopus japonicus	Fungi	PDA, 5 days, 27°C	>1,000	Thompson and Cannon (1986)
Rhizopus kazanensis	Fungi	PDA, 5 days, 27°C	>1,000	Thompson and Cannon (1986)
Rhizopus nigricans	Fungi	Cited	100% inh. 600	Shukla and Tripathi (1987)
Rhizopus oryzae	Fungi	PDA, 5 days, 27°C	>1,000	Thompson and Cannon (1986)
Rhizopus pymacus	Fungi	PDA, 5 days, 27°C	>1,000	Thompson and Cannon (1986)
Rhizopus sp.	Fungi	NB, Tween 20, 8 days, 30°C	3200	Yousef and Tawil (1980)
Rhizopus stolonifer	Fungi	PDA, 5 days, 27°C	>1,000	Thompson and Cannon (1986)
Rhizopus tritici	Fungi	PDA, 5 days, 27°C	>1,000	Thompson and Cannon (1986)
Sclerotina sclerotiorum	Fungi	PDA, 7 days, 23.5°C	100% inh. 10,000	Mueller-Ribeau et al. (1995)
Sclerotium rolfsii	Fungi	OA, EtOH, 6 days, 20°C	2,000	Narasimba Rao et al. (1971)
Trichophyton mentagrophytes	Fungi	SA, Tween 80, 21 days, 20°C	300–625	Janssen et al. (1988)
Trichophyton mentagrophytes	Fungi	SGB, 3–7 days, 25°C	7,800	Kosalec et al. (2005)
Trichophyton rubrum	Fungi	SA, Tween 80, 21 days, 20°C	>300	Janssen et al. (1988)
Trichophyton rubrum	Fungi	SGB, 3–7 days, 25°C	2,000	Kosalec et al. (2005)
Candida albicans	Yeast	SGB, 3–7 days, 25°C	10,000	Kosalec et al. (2005)
Candida albicans	Yeast	NB, Tween 20, 18 h, 37°C	800	Yousef and Tawil (1980)
Candida glabrata	Yeast	SGB, 3–7 days, 25°C	1,000	Kosalec et al. (2005)
Candida krusei	Yeast	SGB, 3–7 days, 25°C	2,000	Kosalec et al. (2005)
Candida parapsilosis	Yeast	SGB, 3–7 days, 25°C	1,000	Kosalec et al. (2005)
Candida pseudotropicalis	Yeast	SGB, 3–7 days, 25°C	1,000	Kosalec et al. (2005)
Candida tropicalis	Yeast	SGB, 3–7 days, 25°C	1,000	Kosalec et al. (2005)

TABLE 14.3
Inhibitory Data of Anise Oil Obtained in the Vapor Phase Test

Microorganism	MO Class	Test Parameters	Quantity (µg)	Activity	Reference
Escherichia coli	Bac−	NA, 24 h, 37°C	~20,000	+++	Kellner and Kober (1954)
Neisseria sp.	Bac−	NA, 24 h, 37°C	~20,000	+	Kellner and Kober (1954)
Salmonella typhi	Bac−	NA, 24 h, 37°C	~20,000	++	Kellner and Kober (1954)
Salmonella typhi	Bac−	NA, 24 h, 37°C	6.35, sd	+++	Maruzzella and Sicurella (1960)
Bacillus megaterium	Bac+	NA, 24 h, 37°C	~20,000	++	Kellner and Kober (1954)
Bacillus subtilis var. aterrimus	Bac+	NA, 24 h, 37°C	6.35, sd	+++	Maruzzella and Sicurella (1960)
Corynebacterium diphtheriae	Bac+	NA, 24 h, 37°C	~20,000	+	Kellner and Kober (1954)
Mycobacterium avium	Bac+	NA, 24 h, 37°C	6.35, sd	+++	Maruzzella and Sicurella (1960)
Staphylococcus aureus	Bac+	NA, 24 h, 37°C	~20,000	+++	Kellner and Kober (1954)
Staphylococcus aureus	Bac+	NA, 24 h, 37°C	6.35, sd	+++	Maruzzella and Sicurella (1960)
Streptococcus faecalis	Bac+	NA, 24 h, 37°C	~20,000	++	Kellner and Kober (1954)
Streptococcus faecalis	Bac+	NA, 24 h, 37°C	6.35, sd	+++	Maruzzella and Sicurella (1960)
Streptococcus pyogenes	Bac+	NA, 24 h, 37°C	~20,000	+	Kellner and Kober (1954)
Aspergillus flavus	Fungi	WFA, 42 days, 25°C	Disk, 50,000	+++	Guynot et al. (2003)
Aspergillus niger	Fungi	WFA, 42 days, 25°C	Disk, 50,000	+++	Guynot et al. (2003)
Eurotium amstelodami	Fungi	WFA, 42 days, 25°C	Disk, 50,000	+++	Guynot et al. (2003)
Eurotium herbarum	Fungi	WFA, 42 days, 25°C	Disk, 50,000	+++	Guynot et al. (2003)
Eurotium repens	Fungi	WFA, 42 days, 25°C	Disk, 50,000	+++	Guynot et al. (2003)
Eurotium rubrum	Fungi	WFA, 42 days, 25°C	Disk, 50,000	+++	Guynot et al. (2003)
Penicillium corylophilum	Fungi	WFA, 42 days, 25°C	Disk, 50,000	+	Guynot et al. (2003)
Candida albicans	Yeast	NA, 24 h, 37°C	~20,000	+	Kellner and Kober (1954)

TABLE 14.4
Inhibitory Data of Bitter Fennel Obtained in the Agar Diffusion Test

Microorganism	MO Class	Test Parameters	Disk Size (mm), Quantity (µg)	Inhibition Zone (mm)	Reference
Acinetobacter calcoaceticus	Bac−	ISA, 48 h, 25°C	4 (h), 10,000	0	Deans and Ritchie (1987)
Aeromonas hydrophila	Bac−	ISA, 48 h, 25°C	4 (h), 10,000	9.5	Deans and Ritchie (1987)
Agrobacterium tumefaciens	Bac−	WA, 48 h, 25°C	6, 8,000	MIA 2880	Cantore et al. (2004)
Alcaligenes faecalis	Bac−	ISA, 48 h, 25°C	4 (h), 10,000	0	Deans and Ritchie (1987)
Beneckea natriegens	Bac−	ISA, 48 h, 25°C	4 (h), 10,000	0	Deans and Ritchie (1987)
Burkholderia gladioli pv. *agaricicola*	Bac−	WA, 48 h, 25°C	6, 8,000	MIA 7680	Cantore et al. (2004)
Citrobacter freundii	Bac−	ISA, 48 h, 25°C	4 (h), 10,000	0	Deans and Ritchie (1987)
Enterobacter aerogenes	Bac−	ISA, 48 h, 25°C	4 (h), 10,000	0	Deans and Ritchie (1987)
Enterobacter aerogenes	Bac−	MHA, 48 h, 27°C	6, 15,000	10	Ertürk et al. (2006)
Erwinia carotovora	Bac−	ISA, 48 h, 25°C	4 (h), 10,000	0	Deans and Ritchie (1987)
Erwinia carotovora subsp. *atroseptica*	Bac−	WA, 48 h, 25°C	6, 8,000	MIA 7680	Cantore et al. (2004)
Erwinia carotovora subsp. *carotovora*	Bac−	WA, 48 h, 25°C	6, 8,000	MIA 7680	Cantore et al. (2004)
Escherichia coli	Bac−	ISA, 48 h, 25°C	4 (h), 10,000	0	Deans and Ritchie (1987)
Escherichia coli	Bac−	NA, 18 h, 37°C	6 (h), pure	0	Yousef and Tawil (1980)
Escherichia coli	Bac−	NA, 18 h, 37°C	5 (h), −30,000	14	Schelz et al. (2006)
Escherichia coli	Bac−	MHA, 48 h, 27°C	6, 15,000	25	Ertürk et al. (2006)
Escherichia coli	Bac−	WA, 48 h, 25°C	6, 8,000	MIA >7680	Cantore et al. (2004)
Flavobacterium suaveolens	Bac−	ISA, 48 h, 25°C	4 (h), 10,000	6	Deans and Ritchie (1987)
Klebsiella pneumoniae	Bac−	ISA, 48 h, 25°C	4 (h), 10,000	0	Deans and Ritchie (1987)
Moraxella sp.	Bac−	ISA, 48 h, 25°C	4 (h), 10,000	6.5	Deans and Ritchie (1987)
Proteus vulgaris	Bac−	ISA, 48 h, 25°C	4 (h), 10,000	0	Deans and Ritchie (1987)
Pseudomonas aeruginosa	Bac−	ISA, 48 h, 25°C	4 (h), 10,000	0	Deans and Ritchie (1987)
Pseudomonas aeruginosa	Bac−	NA, 18 h, 37°C	6 (h), pure	0	Yousef and Tawil (1980)
Pseudomonas aeruginosa	Bac−	MHA, 48 h, 27°C	6, 15,000	18	Ertürk et al. (2006)

(*Continued*)

TABLE 14.4 (*Continued*)
Inhibitory Data of Bitter Fennel Obtained in the Agar Diffusion Test

Microorganism	MO Class	Test Parameters	Disk Size (mm), Quantity (µg)	Inhibition Zone (mm)	Reference
Pseudomonas agarici	Bac–	KBA, 48 h, 25°C	6, 8,000	MIA >7680	Cantore et al. (2004)
Pseudomonas chichori	Bac–	KBA, 48 h, 25°C	6, 8,000	MIA >7680	Cantore et al. (2004)
Pseudomonas corrugate	Bac–	KBA, 48 h, 25°C	6, 8,000	MIA >7680	Cantore et al. (2004)
Pseudomonas reactants	Bac–	KBA, 48 h, 25°C	6, 8,000	MIA >7680	Cantore et al. (2004)
Pseudomonas syringae pv. *aptata*	Bac–	KBA, 48 h, 25°C	6, 8,000	MIA >7680	Cantore et al. (2004)
Pseudomonas syringae pv. *apii*	Bac–	KBA, 48 h, 25°C	6, 8,000	MIA >7680	Cantore et al. (2004)
Pseudomonas syringae pv. *atrofaciens*	Bac–	KBA, 48 h, 25°C	6, 8,000	MIA 3840	Cantore et al. (2004)
Pseudomonas syringae pv. *Glycinea*	Bac–	KBA, 48 h, 25°C	6, 8,000	MIA 960	Cantore et al. (2004)
Pseudomonas syringae pv. *lachrymans*	Bac–	KBA, 48 h, 25°C	6, 8,000	MIA >7680	Cantore et al. (2004)
Pseudomonas syringae pv. *maculicola*	Bac–	KBA, 48 h, 25°C	6, 8,000	MIA >7680	Cantore et al. (2004)
Pseudomonas syringae pv. *phaseolicola*	Bac–	KBA, 48 h, 25°C	6, 8,000	MIA >7680	Cantore et al. (2004)
Pseudomonas syringae pv. *pisi*	Bac–	KBA, 48 h, 25°C	6, 8,000	MIA >7680	Cantore et al. (2004)
Pseudomonas syringae pv. *syringae*	Bac–	KBA, 48 h, 25°C	6, 8,000	MIA >7680	Cantore et al. (2004)
Pseudomonas syringae pv. *tomato*	Bac–	KBA, 48 h, 25°C	6, 8,000	MIA >7680	Cantore et al. (2004)
Pseudomonas tolaasii	Bac–	KBA, 48 h, 25°C	6, 8,000	MIA 7680	Cantore et al. (2004)
Pseudomonas viridiflava	Bac–	KBA, 48 h, 25°C	6, 8,000	MIA >7680	Cantore et al. (2004)
Salmonella pullorum	Bac–	ISA, 48 h, 25°C	4 (h), 10,000	0	Deans and Ritchie (1987)
Salmonella typhimurium	Bac–	MHA, 48 h, 27°C	6, 15,000	8	Ertürk et al. (2006)
Serratia marcescens	Bac–	ISA, 48 h, 25°C	4 (h), 10,000	0	Deans and Ritchie (1987)
Xanthomonas campestris pv. *campestris*	Bac–	WA, 48 h, 25°C	6, 8,000	MIA 5760	Cantore et al. (2004)
Xanthomonas campestris pv. *phaesoli* var. *fuscans*	Bac–	WA, 48 h, 25°C	6, 8,000	MIA 720	Cantore et al. (2004)
Xanthomonas campestris pv. *phaseoli* var. *phaseoli*	Bac–	WA, 48 h, 25°C	6, 8,000	MIA 480	Cantore et al. (2004)
Xanthomonas campestris pv. *Vesicatoria*	Bac–	WA, 48 h, 25°C	6, 8,000	MIA 1440	Cantore et al. (2004)

(*Continued*)

TABLE 14.4 (*Continued*)
Inhibitory Data of Bitter Fennel Obtained in the Agar Diffusion Test

Microorganism	MO Class	Test Parameters	Disk Size (mm), Quantity (µg)	Inhibition Zone (mm)	Reference
Yersinia enterocolitica	Bac−	ISA, 48 h, 25°C	4 (h), 10,000	0	Deans and Ritchie (1987)
Bacillus megaterium	Bac+	WA, 48 h, 25°C	6, 8,000	MIA >7680	Cantore et al. (2004)
Bacillus subtilis	Bac+	ISA, 48 h, 25°C	4 (h), 10,000	0	Deans and Ritchie (1987)
Bacillus subtilis	Bac+	NA, 18 h, 37°C	6 (h), pure	18	Yousef and Tawil (1980)
Brevibacterium linens	Bac+	ISA, 48 h, 25°C	4 (h), 10,000	7.5	Deans and Ritchie (1987)
Brochothrix thermosphacta	Bac+	ISA, 48 h, 25°C	4 (h), 10,000	0	Deans and Ritchie (1987)
Clavibacter michiganensis subsp. *michiganensis*	Bac+	WA, 48 h, 25°C	6, 8,000	MIA 7680	Cantore et al. (2004)
Clavibacter michiganensis subsp. *sepedonicus*	Bac+	WA, 48 h, 25°C	6, 8,000	MIA 960	Cantore et al. (2004)
Clostridium sporogenes	Bac+	ISA, 48 h, 25°C	4 (h), 10,000	0	Deans and Ritchie (1987)
Curtobacterium flaccumfaciens pv. *betae*	Bac+	WA, 48 h, 25°C	6, 8,000	MIA >7680	Cantore et al. (2004)
Curtobacterium flaccumfaciens pv. *flaccumfaciens*	Bac+	WA, 48 h, 25°C	6, 8,000	MIA >7680	Cantore et al. (2004)
Lactobacillus plantarum	Bac+	ISA, 48 h, 25°C	4 (h), 10,000	0	Deans and Ritchie (1987)
Leuconostoc cremoris	Bac+	ISA, 48 h, 25°C	4 (h), 10,000	9.5	Deans and Ritchie (1987)
Micrococcus luteus	Bac+	ISA, 48 h, 25°C	4 (h), 10,000	0	Deans and Ritchie (1987)
Mycobacterium phlei	Bac+	NA, 18 h, 37°C	6 (h), pure	20	Yousef and Tawil (1980)
Rhodococcus fascians	Bac+	WA, 48 h, 25°C	6, 8,000	MIA 1920	Cantore et al. (2004)
Staphylococcus aureus	Bac+	ISA, 48 h, 25°C	4 (h), 10,000	7.5	Deans and Ritchie (1987)
Staphylococcus aureus	Bac+	MHA, 48 h, 27°C	6, 15,000	16	Ertürk et al. (2006)
Staphylococcus aureus	Bac+	NA, 18 h, 37°C	6 (h), pure	17	Yousef and Tawil (1980)
Staphylococcus epidermidis	Bac+	MHA, 48 h, 27°C	6, 15,000	12	Ertürk et al. (2006)
Staphylococcus epidermidis	Bac+	NA, 18 h, 37°C	5 (h), −30,000	15	Schelz et al. (2006)
Streptococcus faecalis	Bac+	ISA, 48 h, 25°C	4 (h), 10,000	0	Deans and Ritchie (1987)
Streptomyces venezuelae	Bac+	SMA, 2–7 days, 20°C	6.35, sd	10	Maruzzella and Liguori (1958)
Absidia corymbifera	Fungi	EYA, 48 h, 45°C	5, sd	0	Nigam and Rao (1979)

(*Continued*)

TABLE 14.4 (*Continued*)
Inhibitory Data of Bitter Fennel Obtained in the Agar Diffusion Test

Microorganism	MO Class	Test Parameters	Disk Size (mm), Quantity (µg)	Inhibition Zone (mm)	Reference
Alternaria solani	Fungi	SMA, 2–7 days, 20°C	6.35, sd	6	Maruzzella and Liguori (1958)
Alternaria sp.	Fungi	PDA, 18 h, 37°C	6, sd	12	Sharma and Singh (1979)
Aspergillus candidus	Fungi	PDA, 18 h, 37°C	6, sd	0	Sharma and Singh (1979)
Aspergillus flavus	Fungi	PDA, 18 h, 37°C	6, sd	7	Sharma and Singh (1979)
Aspergillus fumigatus	Fungi	PDA, 18 h, 37°C	6, sd	0	Sharma and Singh (1979)
Aspergillus fumigatus	Fungi	SMA, 2–7 days, 20°C	6.35, sd	12	Maruzzella and Liguori (1958)
Aspergillus nidulans	Fungi	PDA, 18 h, 37°C	6, sd	9	Sharma and Singh (1979)
Aspergillus niger	Fungi	SMA, 2–7 days, 20°C	6.35, sd	0	Maruzzella and Liguori (1958)
Aspergillus niger	Fungi	PDA, 18 h, 37°C	6, sd	12	Sharma and Singh (1979)
Aspergillus niger	Fungi	MHA, 48 h, 27°C	6, 15,000	12	Ertürk et al. (2006)
Aspergillus niger	Fungi	SDA, 8 days, 30°C	6 (h), pure	31	Yousef and Tawil (1980)
Cladosporium herbarum	Fungi	PDA, 18 h, 37°C	6, sd	12.5	Sharma and Singh (1979)
Cunninghamella echinulata	Fungi	PDA, 18 h, 37°C	6, sd	21	Sharma and Singh (1979)
Fusarium oxysporum	Fungi	PDA, 18 h, 37°C	6, sd	0	Sharma and Singh (1979)
Helminthosporium sacchari	Fungi	PDA, 18 h, 37°C	6, sd	16	Sharma and Singh (1979)
Helminthosporium sativum	Fungi	SMA, 2–7 days, 20°C	6.35, sd	6	Maruzzella and Liguori (1958)
Humicola grisea var. *thermoidea*	Fungi	EYA, 48 h, 45°C	5, sd	0	Nigam and Rao (1979)
Microsporum gypseum	Fungi	PDA, 18 h, 37°C	6, sd	14.5	Sharma and Singh (1979)
Mucor mucedo	Fungi	SMA, 2–7 days, 20°C	6.35, sd	0	Maruzzella and Liguori (1958)
Mucor mucedo	Fungi	PDA, 18 h, 37°C	6, sd	12	Sharma and Singh (1979)
Mucor sp.	Fungi	SDA, 8 days, 30°C	6 (h), pure	20	Yousef and Tawil (1980)

(*Continued*)

TABLE 14.4 (Continued)
Inhibitory Data of Bitter Fennel Obtained in the Agar Diffusion Test

Microorganism	MO Class	Test Parameters	Disk Size (mm), Quantity (µg)	Inhibition Zone (mm)	Reference
Nigrospora panici	Fungi	SMA, 2–7 days, 20°C	6.35, sd	7	Maruzzella and Liguori (1958)
Penicillium aculeatum	Fungi	CA, 48 h, 27°C	5, sd	0	Nigam and Rao (1979)
Penicillium chrysogenum	Fungi	CA, 48 h, 27°C	5, sd	15	Nigam and Rao (1979)
Penicillium chrysogenum	Fungi	SDA, 8 days, 30°C	6 (h), pure	60	Yousef and Tawil (1980)
Penicillium digitatum	Fungi	PDA, 18 h, 37°C	6, sd	7	Sharma and Singh (1979)
Penicillium digitatum	Fungi	SMA, 2–7 days, 20°C	6.35, sd	7	Maruzzella and Liguori (1958)
Penicillium javanicum	Fungi	CA, 48 h, 27°C	5, sd	25	Nigam and Rao (1979)
Penicillium jensenii	Fungi	CA, 48 h, 27°C	5, sd	10	Nigam and Rao (1979)
Penicillium lividum	Fungi	CA, 48 h, 27°C	5, sd	0	Nigam and Rao (1979)
Penicillium notatum	Fungi	CA, 48 h, 27°C	5, sd	0	Nigam and Rao (1979)
Penicillium obscurum	Fungi	CA, 48 h, 27°C	5, sd	15	Nigam and Rao (1979)
Penicillium sp. I	Fungi	CA, 48 h, 27°C	5, sd	15	Nigam and Rao (1979)
Penicillium sp. II	Fungi	CA, 48 h, 27°C	5, sd	10	Nigam and Rao (1979)
Penicillium sp. III	Fungi	CA, 48 h, 27°C	5, sd	0	Nigam and Rao (1979)
Rhizopus nigricans	Fungi	SMA, 2–7 days, 20°C	6.35, sd	3	Maruzzella and Liguori (1958)
Rhizopus nigricans	Fungi	PDA, 18 h, 37°C	6, sd	7	Sharma and Singh (1979)
Rhizopus sp.	Fungi	SDA, 8 days, 30°C	6 (h), pure	20	Yousef and Tawil (1980)
Sporotrichum thermophile	Fungi	EYA, 48 h, 45°C	5, sd	25	Nigam and Rao (1979)
Thermoascus aurantiacus	Fungi	EYA, 48 h, 45°C	5, sd	10	Nigam and Rao (1979)
Thermomyces lanuginosus	Fungi	EYA, 48 h, 45°C	5, sd	15	Nigam and Rao (1979)
Thielava minor	Fungi	EYA, 48 h, 45°C	5, sd	0	Nigam and Rao (1979)
Trichophyton rubrum	Fungi	PDA, 18 h, 37°C	6, sd	8	Sharma and Singh (1979)
Trichothecium roseum	Fungi	PDA, 18 h, 37°C	6, sd	9	Sharma and Singh (1979)
Brettanomyces anomalus	Yeast	MYA, 4 days, 30°C	5, 10% sol. sd	0	Conner and Beuchat (1984)
Candida albicans	Yeast	SMA, 2–7 days, 20°C	6.35, sd	0	Maruzzella and Liguori (1958)
Candida albicans	Yeast	MHA, 48 h, 27°C	6, 15,000	12	Ertürk et al. (2006)

(*Continued*)

TABLE 14.4 (*Continued*)
Inhibitory Data of Bitter Fennel Obtained in the Agar Diffusion Test

Microorganism	MO Class	Test Parameters	Disk Size (mm), Quantity (µg)	Inhibition Zone (mm)	Reference
Candida albicans	Yeast	SDA, 18 h, 30°C	6 (h), pure	24	Yousef and Tawil (1980)
Candida krusei	Yeast	SMA, 2–7 days, 20°C	6.35, sd	0	Maruzzella and Liguori (1958)
Candida lipolytica	Yeast	MYA, 4 days, 30°C	5, 10% sol. sd	0	Conner and Beuchat (1984)
Candida tropicalis	Yeast	SMA, 2–7 days, 20°C	6.35, sd	0	Maruzzella and Liguori (1958)
Cryptococcus neoformans	Yeast	SMA, 2–7 days, 20°C	6.35, sd	15	Maruzzella and Liguori (1958)
Cryptococcus rhodobenhani	Yeast	SMA, 2–7 days, 20°C	6.35, sd	20	Maruzzella and Liguori (1958)
Debaryomyces hansenii	Yeast	MYA, 4 days, 30°C	5, 10% sol. sd	0	Conner and Beuchat (1984)
Geotrichum candidum	Yeast	MYA, 4 days, 30°C	5, 10% sol. sd	0	Conner and Beuchat (1984)
Hansenula anomala	Yeast	MYA, 4 days, 30°C	5, 10% sol. sd	0	Conner and Beuchat (1984)
Kloeckera apiculata	Yeast	MYA, 4 days, 30°C	5, 10% sol. sd	10	Conner and Beuchat (1984)
Kluyveromyces fragilis	Yeast	MYA, 4 days, 30°C	5, 10% sol. sd	0	Conner and Beuchat (1984)
Lodderomyces elongisporus	Yeast	MYA, 4 days, 30°C	5, 10% sol. sd	0	Conner and Beuchat (1984)
Metschnikowia pulcherrima	Yeast	MYA, 4 days, 30°C	5, 10% sol. sd	0	Conner and Beuchat (1984)
Pichia membranifaciens	Yeast	MYA, 4 days, 30°C	5, 10% sol. sd	0	Conner and Beuchat (1984)
Rhodotorula rubra	Yeast	MYA, 4 days, 30°C	5, 10% sol. sd	7	Conner and Beuchat (1984)
Saccharomyces cerevisiae	Yeast	MYA, 4 days, 30°C	5, 10% sol. sd	0	Conner and Beuchat (1984)
Saccharomyces cerevisiae	Yeast	SMA, 2–7 days, 20°C	6.35, sd	3	Maruzzella and Liguori (1958)
Saccharomyces cerevisiae	Yeast	NA, 24 h, 20°C	5 (h), –30,000	11	Schelz et al. (2006)
Torula glabrata	Yeast	MYA, 4 days, 30°C	5, 10% sol. sd	7	Conner and Beuchat (1984)
Torula thermophila	Yeast	EYA, 48 h, 45°C	5, sd	20	Nigam and Rao (1979)

TABLE 14.5
Inhibitory Data of Bitter Fennel Oil Obtained in the Dilution Test

Microorganism	MO Class	Test Parameters	MIC (µg/-mL)	Reference
Enterobacter aerogenes	Bac–	MHB, 24 h, 37°C	4,880	Ertürk et al. (2006)
Escherichia coli	Bac–	MHB, 24 h, 37°C	1,220	Ertürk et al. (2006)
Escherichia coli	Bac–	TYB, 18 h, 37°C	3,000	Schelz et al. (2006)
Escherichia coli	Bac–	NB, Tween 20, 18 h, 37°C	12,500	Yousef and Tawil (1980)
Pseudomonas aeruginosa	Bac–	MHB, 24 h, 37°C	9,760	Ertürk et al. (2006)
Pseudomonas aeruginosa	Bac–	NB, Tween 20, 18 h, 37°C	25,000	Yousef and Tawil (1980)
Salmonella typhimurium	Bac–	MHB, 24 h, 37°C	19,560	Ertürk et al. (2006)
Bacillus subtilis	Bac+	NB, Tween 20, 18 h, 37°C	800	Yousef and Tawil (1980)
Mycobacterium phlei	Bac+	NB, Tween 20, 18 h, 37°C	400	Yousef and Tawil (1980)
Staphylococcus aureus	Bac+	MHB, 24 h, 37°C	4,880	Ertürk et al. (2006)
Staphylococcus aureus	Bac+	NB, Tween 20, 18 h, 37°C	12,500	Yousef and Tawil (1980)
Staphylococcus epidermidis	Bac+	TYB, 18 h, 37°C	3,000	Schelz et al. (2006)
Staphylococcus epidermidis	Bac+	MHB, 24 h, 37°C	9,760	Ertürk et al. (2006)
Alternaria alternata	Fungi	MEB, 72 h, 28°C	1,500	Mimica-Dukic et al. (2003)
Aspergillus flavus	Fungi	MEB, 72 h, 28°C	1,800–2,700	Mimica-Dukic et al. (2003)
Aspergillus niger	Fungi	MEB, 72 h, 28°C	1,700–2,200	Mimica-Dukic et al. (2003)
Aspergillus niger	Fungi	NB, Tween 20, 8 days, 30°C	200	Yousef and Tawil (1980)
Aspergillus niger	Fungi	MHB, 24 h, 37°C	9,760	Ertürk et al. (2006)
Aspergillus ochraceus	Fungi	MEB, 72 h, 28°C	1,800–2,000	Mimica-Dukic et al. (2003)
Aspergillus oryzae	Fungi	Cited	250	Okazaki and Oshima (1953)
Aspergillus terreus	Fungi	MEB, 72 h, 28°C	1,800–2,200	Mimica-Dukic et al. (2003)
Aspergillus versicolor	Fungi	MEB, 72 h, 28°C	2,000–2,200	Mimica-Dukic et al. (2003)
Cladosporium cladosporioides	Fungi	MEB, 72 h, 28°C	1,200–1,500	Mimica-Dukic et al. (2003)
Epidermophyton floccosum	Fungi	MEB, 72 h, 28°C	1,200–1,300	Mimica-Dukic et al. (2003)
Fusarium tricinctum	Fungi	MEB, 72 h, 28°C	800–1,500	Mimica-Dukic et al. (2003)
Microsporum canis	Fungi	MEB, 72 h, 28°C	1,000–1,200	Mimica-Dukic et al. (2003)
Mucor racemosus	Fungi	Cited	500	Okazaki and Oshima (1953)
Mucor sp.	Fungi	NB, Tween 20, 8 days, 30°C	3,200	Yousef and Tawil (1980)
Penicillium chrysogenum	Fungi	NB, Tween 20, 8 days, 30°C	400	Yousef and Tawil (1980)
Penicillium chrysogenum	Fungi	Cited	500	Okazaki and Oshima (1953)
Penicillium funiculosum	Fungi	MEB, 72 h, 28°C	2,800–3,000	Mimica-Dukic et al. (2003)
Penicillium ochroyloron	Fungi	MEB, 72 h, 28°C	2,500–2,800	Mimica-Dukic et al. (2003)
Phomopsis helianthi	Fungi	MEB, 72 h, 28°C	800–1,300	Mimica-Dukic et al. (2003)
Rhizopus sp.	Fungi	NB, Tween 20, 8 days, 30°C	12,500	Yousef and Tawil (1980)
Trichoderma viride	Fungi	MEB, 72 h, 28°C	2,700–3,200	Mimica-Dukic et al. (2003)
Trichophyton mentagrophytes	Fungi	MEB, 72 h, 28°C	1,000–1,200	Mimica-Dukic et al. (2003)
Candida albicans	Yeast	NB, Tween 20, 18 h, 37°C	800	Yousef and Tawil (1980)
Candida albicans	Yeast	MHB, 24 h, 37°C	4,880	Ertürk et al. (2006)
Saccharomyces cerevisiae	Yeast	YPB, 24 h, 20°C	800	Schelz et al. (2006)

TABLE 14.6
Inhibitory Data of Bitter Fennel Oil Obtained in the Vapor Phase Test

Microorganism	MO Class	Test Parameters	Quantity (µg)	Activity	Reference
Escherichia coli	Bac−	NA, 24 h, 37°C	~20,000	++	Kellner and Kober (1954)
Neisseria sp.	Bac−	NA, 24 h, 37°C	~20,000	NG	Kellner and Kober (1954)
Salmonella typhi	Bac−	NA, 24 h, 37°C	~20,000	+	Kellner and Kober (1954)
Salmonella typhi	Bac−	NA, 24 h, 37°C	6.35, sd	+++	Maruzzella and Sicurella (1960)
Bacillus megaterium	Bac+	NA, 24 h, 37°C	~20,000	+	Kellner and Kober (1954)
Bacillus subtilis var. *aterrimus*	Bac+	NA, 24 h, 37°C	6.35, sd	+++	Maruzzella and Sicurella (1960)
Corynebacterium diphtheriae	Bac+	NA, 24 h, 37°C	~20,000	+	Kellner and Kober (1954)
Mycobacterium avium	Bac+	NA, 24 h, 37°C	6.35, sd	+++	Maruzzella and Sicurella (1960)
Staphylococcus aureus	Bac+	NA, 24 h, 37°C	~20,000	+	Kellner and Kober (1954)
Streptococcus faecalis	Bac+	NA, 24 h, 37°C	~20,000	+	Kellner and Kober (1954)
Streptococcus faecalis	Bac+	NA, 24 h, 37°C	6.35, sd	+++	Maruzzella and Sicurella (1960)
Streptococcus pyogenes	Bac+	NA, 24 h, 37°C	~20,000	NG	Kellner and Kober (1954)
Candida albicans	Yeast	NA, 24 h, 37°C	~20,000	NG	Kellner and Kober (1954)

TABLE 14.7
Inhibitory Data of Caraway Oil Obtained in the Agar Diffusion Test

Microorganism	MO Class	Test Parameters	Disk Size (mm), Quantity (μg)	Inhibition Zone (mm)	Reference
Acinetobacter calcoaceticus	Bac−	ISA, 48 h, 25°C	4 (h), 10,000	30	Deans and Ritchie (1987)
Aerobacter aerogenes	Bac−	NA, 24 h, 37°C	—, sd	0	Maruzzella and Lichtenstein (1956)
Aeromonas hydrophila	Bac−	ISA, 48 h, 25°C	4 (h), 10,000	8.5	Deans and Ritchie (1987)
Alcaligenes faecalis	Bac−	ISA, 48 h, 25°C	4 (h), 10,000	5	Deans and Ritchie (1987)
Beneckea natriegens	Bac−	ISA, 48 h, 25°C	4 (h), 10,000	7	Deans and Ritchie (1987)
Citrobacter freundii	Bac−	ISA, 48 h, 25°C	4 (h), 10,000	10.5	Deans and Ritchie (1987)
Enterobacter aerogenes	Bac−	ISA, 48 h, 25°C	4 (h), 10,000	0	Deans and Ritchie (1987)
Enterococcus faecalis	Bac−	NA, 24 h, 37°C	5,000 on agar	0	Di Pasqua et al. (2005)
Erwinia carotovora	Bac−	ISA, 48 h, 25°C	4 (h), 10,000	7	Deans and Ritchie (1987)
Escherichia coli	Bac−	NA, 24 h, 37°C	—, sd	0	Maruzzella and Lichtenstein (1956)
Escherichia coli	Bac−	TYA, 18–24 h, 37°C	9.5, 2,000	0	Morris et al. (1979)
Escherichia coli	Bac−	NA, 24 h, 37°C	4,—	4	El-Gengaihi and Zaki (1982)
Escherichia coli	Bac−	Cited, 18 h, 37°C	6, 2,500	10.3	Janssen et al. (1986)
Escherichia coli	Bac−	ISA, 48 h, 25°C	4 (h), 10,000	9.5	Deans and Ritchie (1987)
Escherichia coli	Bac−	NA, 18 h, 37°C	6 (h), pure	18	Yousef and Tawil (1980)
Flavobacterium suaveolens	Bac−	ISA, 48 h, 25°C	4 (h), 10,000	0	Deans and Ritchie (1987)
Klebsiella pneumoniae	Bac−	ISA, 48 h, 25°C	4 (h), 10,000	5.5	Deans and Ritchie (1987)
Moraxella sp.	Bac−	ISA, 48 h, 25°C	4 (h), 10,000	14.5	Deans and Ritchie (1987)
Neisseria perflava	Bac−	NA, 24 h, 37°C	—, sd	0	Maruzzella and Lichtenstein (1956)
Proteus vulgaris	Bac−	NA, 24 h, 37°C	—, sd	0	Maruzzella and Lichtenstein (1956)
Proteus vulgaris	Bac−	ISA, 48 h, 25°C	4 (h), 10,000	9.5	Deans and Ritchie (1987)
Pseudomonas aeruginosa	Bac−	NA, 18 h, 37°C	6 (h), pure	0	Yousef and Tawil (1980)
Pseudomonas aeruginosa	Bac−	ISA, 48 h, 25°C	4 (h), 10,000	0	Deans and Ritchie (1987)
Pseudomonas aeruginosa	Bac−	NA, 24 h, 37°C	—, sd	2	Maruzzella and Lichtenstein (1956)
Pseudomonas aeruginosa	Bac−	Cited, 18 h, 37°C	6, 2,500	7.3	Janssen et al. (1986)
Pseudomonas fluorescens	Bac−	NA, 24 h, 37°C	4,—	4	El-Gengaihi and Zaki (1982)
Salmonella pullorum	Bac−	ISA, 48 h, 25°C	4 (h), 10,000	6.5	Deans and Ritchie (1987)
Salmonella sp.	Bac−	NA, 24 h, 37°C	4,—	10	El-Gengaihi and Zaki (1982)
Serratia marcescens	Bac−	NA, 24 h, 37°C	—, sd	2	Maruzzella and Lichtenstein (1956)
Serratia marcescens	Bac−	ISA, 48 h, 25°C	4 (h), 10,000	13	Deans and Ritchie (1987)
Yersinia enterocolitica	Bac−	ISA, 48 h, 25°C	4 (h), 10,000	8.5	Deans and Ritchie (1987)
Actinomyces sp.	Bac+	NA, 24 h, 37°C	4,—	10	El-Gengaihi and Zaki (1982)
Bacillus mesentericus	Bac+	NA, 24 h, 37°C	—, sd	2	Maruzzella and Lichtenstein (1956)

(*Continued*)

TABLE 14.7 (*Continued*)
Inhibitory Data of Caraway Oil Obtained in the Agar Diffusion Test

Microorganism	MO Class	Test Parameters	Disk Size (mm), Quantity (μg)	Inhibition Zone (mm)	Reference
Bacillus subtilis	Bac+	NA, 24 h, 37°C	4,—	4	El-Gengaihi and Zaki (1982)
Bacillus subtilis	Bac+	ISA, 48 h, 25°C	4 (h), 10,000	6	Deans and Ritchie (1987)
Bacillus subtilis	Bac+	NA, 24 h, 37°C	—, sd	9	Maruzzella and Lichtenstein (1956)
Bacillus subtilis	Bac+	Cited, 18 h, 37°C	6, 2,500	8.3	Janssen et al. (1986)
Bacillus subtilis	Bac+	NA, 18 h, 37°C	6 (h), pure	20	Yousef and Tawil (1980)
Brevibacterium linens	Bac+	ISA, 48 h, 25°C	4 (h), 10,000	0	Deans and Ritchie (1987)
Brochothrix thermosphacta	Bac+	ISA, 48 h, 25°C	4 (h), 10,000	0	Deans and Ritchie (1987)
Clostridium butyricum	Bac+	NA, 24 h, 37°C	4,—	0	El-Gengaihi and Zaki (1982)
Clostridium sporogenes	Bac+	ISA, 48 h, 25°C	4 (h), 10,000	0	Deans and Ritchie (1987)
Corynebacterium sp.	Bac+	TYA, 18–24 h, 37°C	9.5, 2,000	0	Morris et al. (1979)
Lactobacillus delbrueckii	Bac+	NA, 24 h, 37°C	5,000 on agar	0	Di Pasqua et al. (2005)
Lactobacillus plantarum	Bac+	NA, 24 h, 37°C	5,000 on agar	0	Di Pasqua et al. (2005)
Lactobacillus plantarum	Bac+	ISA, 48 h, 25°C	4 (h), 10,000	5	Deans and Ritchie (1987)
Lactococcus garvieae	Bac+	NA, 24 h, 37°C	5,000 on agar	0	Di Pasqua et al. (2005)
Lactococcus lactis subsp. *lactis*	Bac+	NA, 24 h, 37°C	5,000 on agar	0	Di Pasqua et al. (2005)
Leuconostoc cremoris	Bac+	ISA, 48 h, 25°C	4 (h), 10,000	0	Deans and Ritchie (1987)
Listeria monocytogenes	Bac+	NA, 24 h, 37°C	5,000 on agar	0	Di Pasqua et al. (2005)
Micrococcus luteus	Bac+	ISA, 48 h, 25°C	4 (h), 10,000	8.5	Deans and Ritchie (1987)
Micrococcus sp.	Bac+	NA, 48 h, 37°C	4,—	4	El-Gengaihi and Zaki (1982)
Mycobacterium phlei	Bac+	NA, 24 h, 37°C	4,—	10	El-Gengaihi and Zaki (1982)
Mycobacterium phlei	Bac+	NA, 18 h, 37°C	6 (h), pure	28	Yousef and Tawil (1980)
Sarcina lutea	Bac+	NA, 24 h, 37°C	—, sd	0	Maruzzella and Lichtenstein (1956)
Sarcina lutea	Bac+	NA, 48 h, 37°C	4,—	4	El-Gengaihi and Zaki (1982)
Staphylococcus aureus	Bac+	NA, 24 h, 37°C	—, sd	0	Maruzzella and Lichtenstein (1956)
Staphylococcus aureus	Bac+	TYA, 18–24 h, 37°C	9.5, 2,000	0	Morris et al. (1979)
Staphylococcus aureus	Bac+	NA, 24 h, 37°C	4,—	4	El-Gengaihi and Zaki (1982)
Staphylococcus aureus	Bac+	ISA, 48 h, 25°C	4 (h), 10,000	7	Deans and Ritchie (1987)
Staphylococcus aureus	Bac+	Cited, 18 h, 37°C	6, 2,500	7.7	Janssen et al. (1986)
Staphylococcus aureus	Bac+	NA, 18 h, 37°C	6 (h), pure	17	Yousef and Tawil (1980)
Streptococcus faecalis	Bac+	ISA, 48 h, 25°C	4 (h), 10,000	0	Deans and Ritchie (1987)
Streptomyces venezuelae	Bac+	SMA, 2–7 days, 20°C	6.35, sd	7	Maruzzella and Liguori (1958)
Absidia corymbifera	Fungi	EYA, 48 h, 45°C	5, sd	0	Nigam and Rao (1979)
Alternaria porri	Fungi	PDA, 72 h, 28°C	5, 5,000	7.4	Pawar and Thaker (2007)
Alternaria solani	Fungi	SMA, 2–7 days, 20°C	6.35, sd	15	Maruzzella and Liguori (1958)
Alternaria sp.	Fungi	PDA, 18 h, 37°C	6, sd	0	Sharma and Singh (1979)
Aspergillus candidus	Fungi	PDA, 18 h, 37°C	6, sd	13.5	Sharma and Singh (1979)
Aspergillus flavus	Fungi	PDA, 18 h, 37°C	6, sd	0	Sharma and Singh (1979)

(*Continued*)

TABLE 14.7 (Continued)
Inhibitory Data of Caraway Oil Obtained in the Agar Diffusion Test

Microorganism	MO Class	Test Parameters	Disk Size (mm), Quantity (μg)	Inhibition Zone (mm)	Reference
Aspergillus fumigatus	Fungi	SMA, 2–7 days, 20°C	6.35, sd	10	Maruzzella and Liguori (1958)
Aspergillus fumigatus	Fungi	PDA, 18 h, 37°C	6, sd	12	Sharma and Singh (1979)
Aspergillus nidulans	Fungi	PDA, 18 h, 37°C	6, sd	8	Sharma and Singh (1979)
Aspergillus niger	Fungi	PDA, 18 h, 37°C	6, sd	0	Sharma and Singh (1979)
Aspergillus niger	Fungi	PDA, 48 h, 28°C	5, 5,000	7	Pawar and Thaker (2006)
Aspergillus niger	Fungi	SMA, 2–7 days, 20°C	6.35, sd	13	Maruzzella and Liguori (1958)
Aspergillus niger	Fungi	SDA, 8 days, 30°C	6 (h), pure	42	Yousef and Tawil (1980)
Cladosporium herbarum	Fungi	PDA, 18 h, 37°C	6, sd	0	Sharma and Singh (1979)
Cunninghamella echinulata	Fungi	PDA, 18 h, 37°C	6, sd	21	Sharma and Singh (1979)
Fusarium oxysporum	Fungi	PDA, 18 h, 37°C	6, sd	6	Sharma and Singh (1979)
Fusarium oxysporum f.sp. *ciceri*	Fungi	PDA, 72 h, 28°C	5, 5,000	11	Pawar and Thaker (2007)
Helminthosporium sacchari	Fungi	PDA, 18 h, 37°C	6, sd	6.5	Sharma and Singh (1979)
Helminthosporium sativum	Fungi	SMA, 2–7 days, 20°C	6.35, sd	12	Maruzzella and Liguori (1958)
Humicola grisea var. *thermoidea*	Fungi	EYA, 48 h, 45°C	5, sd	0	Nigam and Rao (1979)
Microsporum gypseum	Fungi	PDA, 18 h, 37°C	6, sd	11	Sharma and Singh (1979)
Mucor mucedo	Fungi	SMA, 2–7 days, 20°C	6.35, sd	9	Maruzzella and Liguori (1958)
Mucor mucedo	Fungi	PDA, 18 h, 37°C	6, sd	15	Sharma and Singh (1979)
Mucor sp.	Fungi	SDA, 8 days, 30°C	6 (h), pure	28	Yousef and Tawil (1980)
Nigrospora panici	Fungi	SMA, 2–7 days, 20°C	6.35, sd	15	Maruzzella and Liguori (1958)
Penicillium aculeatum	Fungi	CA, 48 h, 27°C	5, sd	0	Nigam and Rao (1979)
Penicillium chrysogenum	Fungi	CA, 48 h, 27°C	5, sd	0	Nigam and Rao (1979)
Penicillium chrysogenum	Fungi	SDA, 8 days, 30°C	6 (h), pure	60	Yousef and Tawil (1980)
Penicillium digitatum	Fungi	SMA, 2–7 days, 20°C	6.35, sd	7	Maruzzella and Liguori (1958)
Penicillium digitatum	Fungi	PDA, 18 h, 37°C	6, sd	78	Sharma and Singh (1979)
Penicillium javanicum	Fungi	CA, 48 h, 27°C	5, sd	10	Nigam and Rao (1979)
Penicillium jensenii	Fungi	CA, 48 h, 27°C	5, sd	15	Nigam and Rao (1979)
Penicillium lividum	Fungi	CA, 48 h, 27°C	5, sd	0	Nigam and Rao (1979)
Penicillium notatum	Fungi	CA, 48 h, 27°C	5, sd	15	Nigam and Rao (1979)
Penicillium obscurum	Fungi	CA, 48 h, 27°C	5, sd	20	Nigam and Rao (1979)
Penicillium sp. I	Fungi	CA, 48 h, 27°C	5, sd	10	Nigam and Rao (1979)
Penicillium sp. II	Fungi	CA, 48 h, 27°C	5, sd	15	Nigam and Rao (1979)
Penicillium sp. III	Fungi	CA, 48 h, 27°C	5, sd	0	Nigam and Rao (1979)
Rhizopus nigricans	Fungi	PDA, 18 h, 37°C	6, sd	0	Sharma and Singh (1979)
Rhizopus nigricans	Fungi	SMA, 2–7 days, 20°C	6.35, sd	6	Maruzzella and Liguori (1958)

(Continued)

TABLE 14.7 (*Continued*)
Inhibitory Data of Caraway Oil Obtained in the Agar Diffusion Test

Microorganism	MO Class	Test Parameters	Disk Size (mm), Quantity (µg)	Inhibition Zone (mm)	Reference
Rhizopus sp.	Fungi	SDA, 8 days, 30°C	6 (h), pure	0	Yousef and Tawil (1980)
Sporotrichum thermophile	Fungi	EYA, 48 h, 45°C	5, sd	10	Nigam and Rao (1979)
Thermoascus aurantiacus	Fungi	EYA, 48 h, 45°C	5, sd	10	Nigam and Rao (1979)
Thermomyces lanuginosus	Fungi	EYA, 48 h, 45°C	5, sd	10	Nigam and Rao (1979)
Thielava minor	Fungi	EYA, 48 h, 45°C	5, sd	0	Nigam and Rao (1979)
Trichophyton rubrum	Fungi	PDA, 18 h, 37°C	6, sd	7	Sharma and Singh (1979)
Trichothecium roseum	Fungi	PDA, 18 h, 37°C	6, sd	20	Sharma and Singh (1979)
Brettanomyces anomalus	Yeast	MYA, 4 days, 30°C	5, 10% sol. sd	0	Conner and Beuchat (1984)
Candida albicans	Yeast	SMA, 2–7 days, 20°C	6.35, sd	0	Maruzzella and Liguori (1958)
Candida albicans	Yeast	TYA, 18–24 h, 37°C	9.5, 2,000	0	Morris et al. (1979)
Candida albicans	Yeast	NA, 24 h, 37°C	4,—	0	El-Gengaihi and Zaki (1982)
Candida albicans	Yeast	Cited, 18 h, 37°C	6, 2,500	9.7	Janssen et al. (1986)
Candida albicans	Yeast	SDA, 18 h, 30°C	6 (h), pure	30	Yousef and Tawil (1980)
Candida krusei	Yeast	SMA, 2–7 days, 20°C	6.35, sd	0	Maruzzella and Liguori (1958)
Candida lipolytica	Yeast	MYA, 4 days, 30°C	5, 10% sol. sd	0	Conner and Beuchat (1984)
Candida tropicalis	Yeast	SMA, 2–7 days, 20°C	6.35, sd	0	Maruzzella and Liguori (1958)
Candida utilis	Yeast	NA, 24 h, 37°C	4,—	0	El-Gengaihi and Zaki (1982)
Cryptococcus neoformans	Yeast	SMA, 2–7 days, 20°C	6.35, sd	14	Maruzzella and Liguori (1958)
Cryptococcus rhodobenhani	Yeast	SMA, 2–7 days, 20°C	6.35, sd	0	Maruzzella and Liguori (1958)
Debaryomyces hansenii	Yeast	MYA, 4 days, 30°C	5, 10% sol. sd	0	Conner and Beuchat (1984)
Geotrichum candidum	Yeast	MYA, 4 days, 30°C	5, 10% sol. sd	0	Conner and Beuchat (1984)
Hansenula anomala	Yeast	MYA, 4 days, 30°C	5, 10% sol. sd	0	Conner and Beuchat (1984)
Kloeckera apiculata	Yeast	MYA, 4 days, 30°C	5, 10% sol. sd	0	Conner and Beuchat (1984)
Kluyveromyces fragilis	Yeast	MYA, 4 days, 30°C	5, 10% sol. sd	0	Conner and Beuchat (1984)
Lodderomyces elongisporus	Yeast	MYA, 4 days, 30°C	5, 10% sol. sd	0	Conner and Beuchat (1984)
Metschnikowia pulcherrima	Yeast	MYA, 4 days, 30°C	5, 10% sol. sd	0	Conner and Beuchat (1984)
Pichia membranifaciens	Yeast	MYA, 4 days, 30°C	5, 10% sol. sd	0	Conner and Beuchat (1984)
Rhodotorula rubra	Yeast	MYA, 4 days, 30°C	5, 10% sol. sd	0	Conner and Beuchat (1984)
Saccharomyces cerevisiae	Yeast	SMA, 2–7 days, 20°C	6.35, sd	0	Maruzzella and Liguori (1958)
Saccharomyces cerevisiae	Yeast	NA, 24 h, 37°C	4,—	0	El-Gengaihi and Zaki (1982)
Saccharomyces cerevisiae	Yeast	MYA, 4 days, 30°C	5, 10% sol. sd	0	Conner and Beuchat (1984)
Torula glabrata	Yeast	MYA, 4 days, 30°C	5, 10% sol. sd	0	Conner and Beuchat (1984)
Torula thermophila	Yeast	EYA, 48 h, 45°C	5, sd	10	Nigam and Rao (1979)

TABLE 14.8
Inhibitory Data of Caraway Oil Obtained in the Dilution Test

Microorganism	MO Class	Test Parameters	MIC (µg/mL)	Reference
Bacteria	Bac	5% Glucose, 9 days, 37°C	2,000	Buchholtz (1875)
Escherichia coli	Bac−	TYB, 18–24 h, 37°C	>1,000	Morris et al. (1979)
Escherichia coli	Bac−	TSB, 24 h, 37°C	>10,000	Di Pasqua et al. (2005)
Escherichia coli	Bac−	NA, 1–3 days, 30°C	2,000	Farag et al. (1989)
Escherichia coli	Bac−	NB, Tween 20, 18 h, 37°C	6,400	Yousef and Tawil (1980)
Helicobacter pylori	Bac−	Cited, 20 h, 37°C	273.1	Weseler et al. (2005)
Pseudomonas aeruginosa	Bac−	NB, Tween 20, 18 h, 37°C	>50,000	Yousef and Tawil (1980)
Pseudomonas fluorescens	Bac−	NA, 1–3 days, 30°C	2,000	Farag et al. (1989)
Pseudomonas sp.	Bac−	TSB, 24 h, 37°C	6,000	Di Pasqua et al. (2005)
Salmonella typhimurium	Bac−	TSB, 24 h, 37°C	>10,000	Di Pasqua et al. (2005)
Serratia marcescens	Bac−	NA, 1–3 days, 30°C	2,500	Farag et al. (1989)
Bacillus subtilis	Bac+	NA, 1–3 days, 30°C	1,000	Farag et al. (1989)
Bacillus subtilis	Bac+	NB, Tween 20, 18 h, 37°C	3,200	Yousef and Tawil (1980)
Brochothrix thermosphacta	Bac+	M17, 24 h, 20°C	10,000	Di Pasqua et al. (2005)
Corynebacterium sp.	Bac+	TYB, 18–24 h, 37°C	500	Morris et al. (1979)
Micrococcus sp.	Bac+	NA, 1–3 days, 30°C	1,000	Farag et al. (1989)
Mycobacterium phlei	Bac+	NA, 1–3 days, 30°C	750	Farag et al. (1989)
Mycobacterium phlei	Bac+	NB, Tween 20, 18 h, 37°C	200	Yousef and Tawil (1980)
Sarcina sp.	Bac+	NA, 1–3 days, 30°C	1,250	Farag et al. (1989)
Staphylococcus aureus	Bac+	TSB, 24 h, 37°C	10,000	Di Pasqua et al. (2005)
Staphylococcus aureus	Bac+	NA, 1–3 days, 30°C	1,250	Farag et al. (1989)
Staphylococcus aureus	Bac+	TYB, 18–24 h, 37°C	500	Morris et al. (1979)
Staphylococcus aureus	Bac+	NB, Tween 20, 18 h, 37°C	6,400	Yousef and Tawil (1980)
Alternaria citri	Fungi	PDA, 8 days, 22°C	>1,000	Arras and Usai (2001)
Aspergillus flavus	Fungi	PDA, 5 days, 27°C	>1,000	Thompson and Cannon (1986)
Aspergillus flavus	Fungi	PDA, 7–14 days, 28°C	2,000	Soliman and Badeaa (2002)
Aspergillus flavus	Fungi	PDA, 6–8 h 20°C, spore germ. inh.	50–100	Thompson (1986)
Aspergillus flavus	Fungi	CA, 7 days, 28°C	84%–96% inh. 500	Kumar et al. (2007)
Aspergillus niger	Fungi	NB, Tween 20, 8 days, 30°C	3,200	Yousef and Tawil (1980)
Aspergillus ochraceus	Fungi	PDA, 7–14 days, 28°C	3,000	Soliman and Badeaa (2002)
Aspergillus parasiticus	Fungi	PDA, 5 days, 27°C	>1,000	Thompson and Cannon (1986)
Aspergillus parasiticus	Fungi	PDA, 7–14 days, 28°C	2,000	Soliman and Badeaa (2002)
Aspergillus parasiticus	Fungi	PDA, 6–8 h 20°C, spore germ. inh.	50–100	Thompson (1986)
Botrytis cinerea	Fungi	PDA, 8 days, 22°C	>1,000	Arras and Usai (2001)
Epidermophyton floccosum	Fungi	SA, Tween 80, 21 days, 20°C	300–625	Janssen et al. (1988)
Fusarium moniliforme	Fungi	PDA, 7–14 days, 28°C	3,000	Soliman and Badeaa (2002)

(*Continued*)

TABLE 14.8 (*Continued*)
Inhibitory Data of Caraway Oil Obtained in the Dilution Test

Microorganism	MO Class	Test Parameters	MIC (µg/mL)	Reference
Mucor hiemalis	Fungi	PDA, 5 days, 27°C	>1,000	Thompson and Cannon (1986)
Mucor mucedo	Fungi	PDA, 5 days, 27°C	>1,000	Thompson and Cannon (1986)
Mucor racemosus f. *racemosus*	Fungi	PDA, 5 days, 27°C	>1,000	Thompson and Cannon (1986)
Mucor sp.	Fungi	NB, Tween 20, 8 days, 30°C	800	Yousef and Tawil (1980)
Penicillium chrysogenum	Fungi	NB, Tween 20, 8 days, 30°C	1,600	Yousef and Tawil (1980)
Penicillium digitatum	Fungi	PDA, 8 days, 22°C	>1,000	Arras and Usai (2001)
Penicillium italicum	Fungi	PDA, 8 days, 22°C	>1,000	Arras and Usai (2001)
Rhizopus 66-81-2	Fungi	PDA, 5 days, 27°C	>1,000	Thompson and Cannon (1986)
Rhizopus arrhizus	Fungi	PDA, 5 days, 27°C	>1,000	Thompson and Cannon (1986)
Rhizopus chinensis	Fungi	PDA, 5 days, 27°C	>1,000	Thompson and Cannon (1986)
Rhizopus circinans	Fungi	PDA, 5 days, 27°C	>1,000	Thompson and Cannon (1986)
Rhizopus japonicus	Fungi	PDA, 5 days, 27°C	>1,000	Thompson and Cannon (1986)
Rhizopus kazanensis	Fungi	PDA, 5 days, 27°C	>1,000	Thompson and Cannon (1986)
Rhizopus oryzae	Fungi	PDA, 5 days, 27°C	>1,000	Thompson and Cannon (1986)
Rhizopus pymacus	Fungi	PDA, 5 days, 27°C	>1,000	Thompson and Cannon (1986)
Rhizopus sp.	Fungi	NB, Tween 20, 8 days, 30°C	3,200	Yousef and Tawil (1980)
Rhizopus stolonifer	Fungi	PDA, 5 days, 27°C	>1,000	Thompson and Cannon (1986)
Rhizopus tritici	Fungi	PDA, 5 days, 27°C	>1,000	Thompson and Cannon (1986)
Trichophyton mentagrophytes	Fungi	SA, Tween 80, 21 days, 20°C	<300	Janssen et al. (1988)
Trichophyton rubrum	Fungi	SA, Tween 80, 21 days, 20°C	<300	Janssen et al. (1988)
Candida albicans	Yeast	NB, Tween 20, 18 h, 37°C	1,600	Yousef and Tawil (1980)
Candida albicans	Yeast	TYB, 18–24 h, 37°C	500	Morris et al. (1979)
Saccharomyces cerevisiae	Yeast	NA, 1–3 days, 30°C	1,000	Farag et al. (1989)

TABLE 14.9
Inhibitory Data of Caraway Oil Obtained in the Vapor Phase Test

Microorganism	MO Class	Test Parameters	Quantity (µg)	Activity	Reference
Escherichia coli	Bac−	NA, 24 h, 37°C	~20,000	NG	Kellner and Kober (1954)
Neisseria sp.	Bac−	NA, 24 h, 37°C	~20,000	NG	Kellner and Kober (1954)
Salmonella typhi	Bac−	NA, 24 h, 37°C	~20,000	NG	Kellner and Kober (1954)
Salmonella typhi	Bac−	NA, 24 h, 37°C	6.35, sd	+++	Maruzzella and Sicurella (1960)
Bacillus megaterium	Bac+	NA, 24 h, 37°C	~20,000	NG	Kellner and Kober (1954)
Bacillus subtilis var. *aterrimus*	Bac+	NA, 24 h, 37°C	6.35, sd	+++	Maruzzella and Sicurella (1960)
Corynebacterium diphtheriae	Bac+	NA, 24 h, 37°C	~20,000	NG	Kellner and Kober (1954)
Mycobacterium avium	Bac+	NA, 24 h, 37°C	6.35, sd	+++	Maruzzella and Sicurella (1960)
Staphylococcus aureus	Bac+	NA, 24 h, 37°C	~20,000	NG	Kellner and Kober (1954)
Streptococcus faecalis	Bac+	NA, 24 h, 37°C	~20,000	NG	Kellner and Kober (1954)
Streptococcus faecalis	Bac+	NA, 24 h, 37°C	6.35, sd	+++	Maruzzella and Sicurella (1960)
Streptococcus pyogenes	Bac+	NA, 24 h, 37°C	~20,000	NG	Kellner and Kober (1954)
Candida albicans	Yeast	NA, 24 h, 37°C	~20,000	NG	Kellner and Kober (1954)

TABLE 14.10
Inhibitory Data of Cassia Oil Obtained in the Agar Diffusion Test

Microorganism	MO Class	Test Parameters	Disk Size (mm), Quantity (µg)	Inhibition Zone (mm)	Reference
Enterobacter aerogenes	Bac−	MHA, 24 h, 30°C	6, 15,000	18	Rossi et al. (2007)
Escherichia coli	Bac−	BA, 24–48 h, 37°C	12.7, sd	14	Goutham and Purohit (1974)
Escherichia coli	Bac−	MHA, 24 h, 30°C	6, 15,000	21	Rossi et al. (2007)
Proteus vulgaris	Bac−	BA, 24–48 h, 37°C	12.7, sd	17.5	Goutham and Purohit (1974)
Pseudomonas aeruginosa	Bac−	BA, 24–48 h, 37°C	12.7, sd	18	Goutham and Purohit (1974)
Pseudomonas aeruginosa	Bac−	MHA, 24 h, 30°C	6, 15,000	19	Rossi et al. (2007)
Salmonella pullorum	Bac−	BA, 24–48 h, 37°C	12.7, sd	13.5	Goutham and Purohit (1974)
Bacillus subtilis	Bac+	BA, 24–48 h, 37°C	12.7, sd	13.5	Goutham and Purohit (1974)
Corynebacterium pyogenes	Bac+	BA, 24–48 h, 37°C	12.7, sd	13.5	Goutham and Purohit (1974)
Listeria monocytogenes	Bac+	ISA, 48 h, 25°C	4 (h), 10,000	20	Lis-Balchin et al. (1998)
Pasteurella multocida	Bac+	BlA, 24–48 h, 37°C	12.7, sd	19	Goutham and Purohit (1974)
Staphylococcus aureus	Bac+	BA, 24–48 h, 37°C	12.7, sd	0	Goutham and Purohit (1974)
Staphylococcus aureus	Bac+	MHA, 24 h, 37°C	6, 15,000	30	Rossi et al. (2007)
Alternaria porri	Fungi	PDA, 72 h, 28°C	5, 5,000	45	Pawar and Thaker (2007)
Aspergillus niger	Fungi	PDA, 48 h, 28°C	5, 5,000	40	Pawar and Thaker (2006)
Fusarium oxysporum f.sp. ciceri	Fungi	PDA, 72 h, 28°C	5, 5,000	37	Pawar and Thaker (2007)

TABLE 14.11
Inhibitory Data of Cassia Oil Obtained in the Dilution Test

Microorganism	MO Class	Test Parameters	MIC (µg/mL)	Reference
Escherichia coli	Bac–	NA, cited	75–600	Ooi et al. (2006)
Escherichia coli O157:H7	Bac–	BHI, 48 h, 35°C	500	Oussalah et al. (2006)
Pseudomonas aeruginosa	Bac–	NA, cited	75–600	Ooi et al. (2006)
Salmonella typhimurium	Bac–	NA, cited	75–600	Ooi et al. (2006)
Salmonella typhimurium	Bac–	BHI, 48 h, 35°C	250	Oussalah et al. (2006)
Vibrio cholerae	Bac–	NA, cited	75–600	Ooi et al. (2006)
Vibrio parahaemolyticus	Bac–	NA, cited	75–600	Ooi et al. (2006)
Yersinia enterocolitica	Bac–	MHA, Tween 20, 24 h, 37°C	300	Rossi et al. (2007)
Listeria monocytogenes	Bac+	BHI, 48 h, 35°C	300	Oussalah et al. (2006)
Staphylococcus aureus	Bac+	NA, cited	75–600	Ooi et al. (2006)
Staphylococcus aureus	Bac+	BHI, 48 h, 35°C	250	Oussalah et al. (2006)
Alternaria alternata	Fungi	PDA, 7 days, 28°C	100% inh. 300	Feng and Zheng (2007)
Aspergillus niger	Fungi	YES broth, 10 d	87% inh. 10,000	Lis-Balchin et al. (1998)
Aspergillus ochraceus	Fungi	YES broth, 10 d	89% inh. 10,000	Lis-Balchin et al. (1998)
Aspergillus oryzae	Fungi	Cited	500	Okazaki and Oshima (1953)
Aspergillus sp.	Fungi	NA, cited	75–150	Ooi et al. (2006)
Fusarium culmorum	Fungi	YES broth, 10 d	54% inh. 10,000	Lis-Balchin et al. (1998)
Fusarium sp.	Fungi	NA, cited	75–150	Ooi et al. (2006)
Microsporum gypseum.	Fungi	NA, cited	18.8–37.5	Ooi et al. (2006)
Mucor racemosus	Fungi	Cited	500	Okazaki and Oshima (1953)
Penicillium chrysogenum	Fungi	Cited	500	Okazaki and Oshima (1953)
Trichophyton mentagrophytes	Fungi	NA, cited	18.8–37.5	Ooi et al. (2006)
Trichophyton rubrum	Fungi	NA, cited	18.8–37.5	Ooi et al. (2006)
Candida albicans	Yeast	NA, cited	100–450	Ooi et al. (2006)
Candida glabrata	Yeast	NA, cited	100–450	Ooi et al. (2006)
Candida krusei	Yeast	NA, cited	100–450	Ooi et al. (2006)
Candida tropicalis	Yeast	NA, cited	100–450	Ooi et al. (2006)

TABLE 14.12
Inhibitory Data of Cassia Oil Obtained in the Vapor Phase Test

Microorganism	MO Class	Test Parameters	Quantity (µg)	Activity	Reference
Escherichia coli	Bac−	NA, 24 h, 37°C	~20,000	NG	Kellner and Kober (1954)
Neisseria sp.	Bac−	NA, 24 h, 37°C	~20,000	NG	Kellner and Kober (1954)
Salmonella typhi	Bac−	NA, 24 h, 37°C	~20,000	NG	Kellner and Kober (1954)
Salmonella typhi	Bac−	NA, 24 h, 37°C	6.35, sd	++	Maruzzella and Liguori (1958)
Salmonella typhi	Bac−	NA, 24 h, 37°C	6.35, sd	++	Maruzzella and Sicurella (1960)
Bacillus megaterium	Bac+	NA, 24 h, 37°C	~20,000	NG	Kellner and Kober (1954)
Bacillus subtilis var. aterrimus	Bac+	NA, 24 h, 37°C	6.35, sd	++	Maruzzella and Liguori (1958)
Bacillus subtilis var. aterrimus	Bac+	NA, 24 h, 37°C	6.35, sd	++	Maruzzella and Sicurella (1960)
Corynebacterium diphtheriae	Bac+	NA, 24 h, 37°C	~20,000	NG	Kellner and Kober (1954)
Mycobacterium avium	Bac+	NA, 24 h, 37°C	6.35, sd	NG	Maruzzella and Liguori (1958)
Mycobacterium avium	Bac+	NA, 24 h, 37°C	6.35, sd	+	Maruzzella and Sicurella (1960)
Staphylococcus aureus	Bac+	NA, 24 h, 37°C	~20,000	NG	Kellner and Kober (1954)
Staphylococcus aureus	Bac+	NA, 24 h, 37°C	6.35, sd	++	Maruzzella and Liguori (1958)
Staphylococcus aureus	Bac+	NA, 24 h, 37°C	6.35, sd	++	Maruzzella and Sicurella (1960)
Streptococcus faecalis	Bac+	NA, 24 h, 37°C	~20,000	NG	Kellner and Kober (1954)
Streptococcus faecalis	Bac+	NA, 24 h, 37°C	6.35, sd	+	Maruzzella and Liguori (1958)
Streptococcus faecalis	Bac+	NA, 24 h, 37°C	6.35, sd	+	Maruzzella and Sicurella (1960)
Streptococcus pyogenes	Bac+	NA, 24 h, 37°C	~20,000	NG	Kellner and Kober (1954)
Candida albicans	Yeast	NA, 24 h, 37°C	~20,000	NG	Kellner and Kober (1954)

TABLE 14.13
Inhibitory Data of Ceylon Cinnamon Bark Oil Obtained in the Agar Diffusion Test

Microorganism	MO Class	Test Parameters	Disk Size (mm), Quantity (µg)	Inhibition Zone (mm)	Reference
Acinetobacter calcoaceticus	Bac–	ISA, 48 h, 25°C	4 (h), 10,000	16.5	Deans and Ritchie (1987)
Aerobacter aerogenes	Bac–	NA, 24 h, 37°C	—, sd	0	Maruzzella and Lichtenstein (1956)
Aeromonas hydrophila	Bac–	ISA, 48 h, 25°C	4 (h), 10,000	15	Deans and Ritchie (1987)
Alcaligenes faecalis	Bac–	ISA, 48 h, 25°C	4 (h), 10,000	18	Deans and Ritchie (1987)
Beneckea natriegens	Bac–	ISA, 48 h, 25°C	4 (h), 10,000	27	Deans and Ritchie (1987)
Campylobacter jejuni	Bac–	TSA, 24 h, 42°C	4 (h), 25,000	8.9	Smith-Palmer et al. (1998)
Citrobacter freundii	Bac–	ISA, 48 h, 25°C	4 (h), 10,000	12	Deans and Ritchie (1987)
Enterobacter aerogenes	Bac–	ISA, 48 h, 25°C	4 (h), 10,000	13	Deans and Ritchie (1987)
Erwinia carotovora	Bac–	ISA, 48 h, 25°C	4 (h), 10,000	16	Deans and Ritchie (1987)
Escherichia coli	Bac–	NA, 24 h, 37°C	—, sd	5	Maruzzella and Lichtenstein (1956)
Escherichia coli	Bac–	TSA, 24 h, 35°C	4 (h), 25,000	10.1	Smith-Palmer et al. (1998)
Escherichia coli	Bac–	Cited	15, 2,500	13	Pizsolitto et al. (1975)
Escherichia coli	Bac–	ISA, 48 h, 25°C	4 (h), 10,000	13	Deans and Ritchie (1987)
Escherichia coli	Bac–	NA, 18 h, 37°C	6 (h), pure	15.5	Yousef and Tawil (1980)
Escherichia coli	Bac–	Cited, 18 h, 37°C	6, 2,500	14.7	Janssen et al. (1986)
Escherichia coli	Bac–	NA, 24 h, 37°C	10 (h), 2,000	35	Singh et al. (2007)
Escherichia coli	Bac–	NA, 24 h, 30°C	Drop, 5,000	45	Hili et al. (1997)
Flavobacterium suaveolens	Bac–	ISA, 48 h, 25°C	4 (h), 10,000	22	Deans and Ritchie (1987)
Klebsiella pneumoniae	Bac–	ISA, 48 h, 25°C	4 (h), 10,000	18	Deans and Ritchie (1987)
Klebsiella sp.	Bac–	Cited	15, 2,500	11	Pizsolitto et al. (1975)
Moraxella sp.	Bac–	ISA, 48 h, 25°C	4 (h), 10,000	0	Deans and Ritchie (1987)
Neisseria perflava	Bac–	NA, 24 h, 37°C	—, sd	0	Maruzzella and Lichtenstein (1956)
Proteus sp.	Bac–	Cited	15, 2,500	17	Pizsolitto et al. (1975)
Proteus vulgaris	Bac–	NA, 24 h, 37°C	—, sd	12	Maruzzella and Lichtenstein (1956)
Proteus vulgaris	Bac–	ISA, 48 h, 25°C	4 (h), 10,000	15.5	Deans and Ritchie (1987)
Pseudomonas aeruginosa	Bac–	Cited	15, 2,500	2	Pizsolitto et al. (1975)
Pseudomonas aeruginosa	Bac–	NA, 24 h, 37°C	—, sd	4	Maruzzella and Lichtenstein (1956)
Pseudomonas aeruginosa	Bac–	ISA, 48 h, 25°C	4 (h), 10,000	13	Deans and Ritchie (1987)
Pseudomonas aeruginosa	Bac–	Cited, 18 h, 37°C	6, 2,500	7.7	Janssen et al. (1986)
Pseudomonas aeruginosa	Bac–	NA, 18 h, 37°C	6 (h), pure	17.5	Yousef and Tawil (1980)
Pseudomonas aeruginosa	Bac–	NA, 24 h, 30°C	Drop, 5,000	25	Hili et al. (1997)
Pseudomonas aeruginosa	Bac–	NA, 24 h, 37°C	10 (h), 2,000	60	Singh et al. (2007)
Salmonella enteritidis	Bac–	TSA, 24 h, 35°C	4 (h), 25,000	10.9	Smith-Palmer et al. (1998)
Salmonella pullorum	Bac–	ISA, 48 h, 25°C	4 (h), 10,000	21.2	Deans and Ritchie (1987)
Salmonella sp.	Bac–	Cited	15, 2,500	12	Pizsolitto et al. (1975)
Salmonella typhi	Bac–	NA, 24 h, 37°C	10 (h), 2,000	41	Singh et al. (2007)
Serratia marcescens	Bac–	NA, 24 h, 37°C	—, sd	0	Maruzzella and Lichtenstein (1956)
Serratia marcescens	Bac–	ISA, 48 h, 25°C	4 (h), 10,000	12	Deans and Ritchie (1987)

(Continued)

TABLE 14.13 (*Continued*)
Inhibitory Data of Ceylon Cinnamon Bark Oil Obtained in the Agar Diffusion Test

Microorganism	MO Class	Test Parameters	Disk Size (mm), Quantity (µg)	Inhibition Zone (mm)	Reference
Serratia sp.	Bac–	Cited	15, 2,500	12	Pizsolitto et al. (1975)
Shigella sp.	Bac–	Cited	15, 2,500	13	Pizsolitto et al. (1975)
Yersinia enterocolitica	Bac–	ISA, 48 h, 25°C	4 (h), 10,000	20	Deans and Ritchie (1987)
Bacillus cereus	Bac+	NA, 24 h, 37°C	10 (h), 2,000	27	Singh et al. (2007)
Bacillus mesentericus	Bac+	NA, 24 h, 37°C	—, sd	16	Maruzzella and Lichtenstein (1956)
Bacillus sp.	Bac+	Cited	15, 2,500	25	Pizsolitto et al. (1975)
Bacillus subtilis	Bac+	Cited, 18 h, 37°C	6, 2,500	16	Janssen et al. (1986)
Bacillus subtilis	Bac+	NA, 24 h, 37°C	—, sd	21	Maruzzella and Lichtenstein (1956)
Bacillus subtilis	Bac+	NA, 18 h, 37°C	6 (h), pure	25	Yousef and Tawil (1980)
Bacillus subtilis	Bac+	ISA, 48 h, 25°C	4 (h), 10,000	23.5	Deans and Ritchie (1987)
Bacillus subtilis	Bac+	NA, 24 h, 37°C	10 (h), 2,000	56	Singh et al. (2007)
Brevibacterium linens	Bac+	ISA, 48 h, 25°C	4 (h), 10,000	35	Deans and Ritchie (1987)
Brochothrix thermosphacta	Bac+	ISA, 48 h, 25°C	4 (h), 10,000	5	Deans and Ritchie (1987)
Clostridium sporogenes	Bac+	ISA, 48 h, 25°C	4 (h), 10,000	0	Deans and Ritchie (1987)
Lactobacillus plantarum	Bac+	ISA, 48 h, 25°C	4 (h), 10,000	19	Deans and Ritchie (1987)
Leuconostoc cremoris	Bac+	ISA, 48 h, 25°C	4 (h), 10,000	8	Deans and Ritchie (1987)
Listeria monocytogenes	Bac+	TSA, 24 h, 35°C	4 (h), 25,000	6.8	Smith-Palmer et al. (1998)
Micrococcus luteus	Bac+	ISA, 48 h, 25°C	4 (h), 10,000	18	Deans and Ritchie (1987)
Mycobacterium phlei	Bac+	NA, 18 h, 37°C	6 (h), pure	40	Yousef and Tawil (1980)
Sarcina lutea	Bac+	NA, 24 h, 37°C	—, sd	0	Maruzzella and Lichtenstein (1956)
Staphylococcus aureus	Bac+	NA, 24 h, 37°C	—, sd	0	Maruzzella and Lichtenstein (1956)
Staphylococcus aureus	Bac+	Cited, 18 h, 37°C	6, 2,500	10	Janssen et al. (1986)
Staphylococcus aureus	Bac+	ISA, 48 h, 25°C	4 (h), 10,000	10	Deans and Ritchie (1987)
Staphylococcus aureus	Bac+	TSA, 24 h, 35°C	4 (h), 25,000	7.5	Smith-Palmer et al. (1998)
Staphylococcus aureus	Bac+	Cited	15, 2,500	25	Pizsolitto et al. (1975)
Staphylococcus aureus	Bac+	NA, 18 h, 37°C	6 (h), pure	27.3	Yousef and Tawil (1980)
Staphylococcus aureus	Bac+	NA, 24 h, 30°C	Drop, 5,000	45	Hili et al. (1997)
Staphylococcus aureus	Bac+	NA, 24 h, 37°C	10 (h), 2,000	57	Singh et al. (2007)
Staphylococcus epidermidis	Bac+	Cited	15, 2,500	20	Pizsolitto et al. (1975)
Streptococcus faecalis	Bac+	ISA, 48 h, 25°C	4 (h), 10,000	9	Deans and Ritchie (1987)
Streptococcus viridans	Bac+	Cited	15, 2,500	8	Pizsolitto et al. (1975)
Streptomyces venezuelae	Bac+	SMA, 2–7 days, 20°C	6.35, sd	15	Maruzzella and Liguori (1958)
Alternaria porri	Fungi	PDA, 72 h, 28°C	5, 5,000	50	Pawar and Thaker (2007)
Alternaria solani	Fungi	SMA, 2–7 days, 20°C	6.35, sd	12	Maruzzella and Liguori (1958)
Aspergillus fumigatus	Fungi	SDA, 3 days, 28°C	6, sd	18	Saksena and Saksena (1984)
Aspergillus fumigatus	Fungi	SMA, 2–7 days, 20°C	6.35, sd	19	Maruzzella and Liguori (1958)

(*Continued*)

TABLE 14.13 (*Continued*)
Inhibitory Data of Ceylon Cinnamon Bark Oil Obtained in the Agar Diffusion Test

Microorganism	MO Class	Test Parameters	Disk Size (mm), Quantity (µg)	Inhibition Zone (mm)	Reference
Aspergillus niger	Fungi	SMA, 2–7 days, 20°C	6.35, sd	16	Maruzzella and Liguori (1958)
Aspergillus niger	Fungi	PDA, 48 h, 28°C	5, 5,000	43	Pawar and Thaker (2006)
Aspergillus niger	Fungi	SDA, 8 days, 30°C	6 (h), pure	60	Yousef and Tawil (1980)
Fusarium oxysporum f.sp. *ciceri*	Fungi	PDA, 72 h, 28°C	5, 5,000	40	Pawar and Thaker (2007)
Helminthosporium sativum	Fungi	SMA, 2–7 days, 20°C	6.35, sd	15	Maruzzella and Liguori (1958)
Keratinomyces afelloi	Fungi	SDA, 3 days, 28°C	6, sd	18	Saksena and Saksena (1984)
Keratinophyton terreum	Fungi	SDA, 3 days, 28°C	6, sd	12	Saksena and Saksena (1984)
Microsporum gypseum	Fungi	SDA, 3 days, 28°C	6, sd	21	Saksena and Saksena (1984)
Mucor mucedo	Fungi	SMA, 2–7 days, 20°C	6.35, sd	12	Maruzzella and Liguori (1958)
Mucor sp.	Fungi	SDA, 8 days, 30°C	6 (h), pure	40	Yousef and Tawil (1980)
Nigrospora panici	Fungi	SMA, 2–7 days, 20°C	6.35, sd	15	Maruzzella and Liguori (1958)
Penicillium chrysogenum	Fungi	SDA, 8 days, 30°C	6 (h), pure	60	Yousef and Tawil (1980)
Penicillium digitatum	Fungi	SMA, 2–7 days, 20°C	6.35, sd	17	Maruzzella and Liguori (1958)
Rhizopus nigricans	Fungi	SMA, 2–7 days, 20°C	6.35, sd	7	Maruzzella and Liguori (1958)
Rhizopus sp.	Fungi	SDA, 8 days, 30°C	6 (h), pure	19	Yousef and Tawil (1980)
Trichophyton equinum	Fungi	SDA, 3 days, 28°C	6, sd	28	Saksena and Saksena (1984)
Trichophyton rubrum	Fungi	SDA, 3 days, 28°C	6, sd	28	Saksena and Saksena (1984)
Brettanomyces anomalus	Yeast	MYA, 4 days, 30°C	5, 10% sol. sd	18	Conner and Beuchat (1984)
Candida albicans	Yeast	SDA, 3 days, 28°C	6, sd	14	Saksena and Saksena (1984)
Candida albicans	Yeast	SMA, 2–7 days, 20°C	6.35, sd	15	Maruzzella and Liguori (1958)
Candida albicans	Yeast	Cited, 18 h, 37°C	6, 2,500	27	Janssen et al. (1986)
Candida albicans	Yeast	NA, 24 h, 30°C	Drop, 5,000	39	Hili et al. (1997)
Candida albicans	Yeast	SDA, 18 h, 30°C	6 (h), pure	48	Yousef and Tawil (1980)
Candida krusei	Yeast	SMA, 2–7 days, 20°C	6.35, sd	4	Maruzzella and Liguori (1958)
Candida lipolytica	Yeast	MYA, 4 days, 30°C	5, 10% sol. sd	21	Conner and Beuchat (1984)

(*Continued*)

TABLE 14.13 (*Continued*)
Inhibitory Data of Ceylon Cinnamon Bark Oil Obtained in the Agar Diffusion Test

Microorganism	MO Class	Test Parameters	Disk Size (mm), Quantity (µg)	Inhibition Zone (mm)	Reference
Candida tropicalis	Yeast	SMA, 2–7 days, 20°C	6.35, sd	8	Maruzzella and Liguori (1958)
Candida tropicalis	Yeast	SDA, 3 days, 28°C	6, sd	21	Saksena and Saksena (1984)
Cryptococcus neoformans	Yeast	SMA, 2–7 days, 20°C	6.35, sd	18	Maruzzella and Liguori (1958)
Cryptococcus rhodobenhani	Yeast	SMA, 2–7 days, 20°C	6.35, sd	13	Maruzzella and Liguori (1958)
Debaryomyces hansenii	Yeast	MYA, 4 days, 30°C	5, 10% sol. sd	29	Conner and Beuchat (1984)
Geotrichum candidum	Yeast	MYA, 4 days, 30°C	5, 10% sol. sd	18	Conner and Beuchat (1984)
Hansenula anomala	Yeast	MYA, 4 days, 30°C	5, 10% sol. sd	16	Conner and Beuchat (1984)
Kloeckera apiculata	Yeast	MYA, 4 days, 30°C	5, 10% sol. sd	15	Conner and Beuchat (1984)
Kluyveromyces fragilis	Yeast	MYA, 4 days, 30°C	5, 10% sol. sd	8	Conner and Beuchat (1984)
Lodderomyces elongisporus	Yeast	MYA, 4 days, 30°C	5, 10% sol. sd	17	Conner and Beuchat (1984)
Metschnikowia pulcherrima	Yeast	MYA, 4 days, 30°C	5, 10% sol. sd	28	Conner and Beuchat (1984)
Pichia membranifaciens	Yeast	MYA, 4 days, 30°C	5, 10% sol. sd	11	Conner and Beuchat (1984)
Rhodotorula rubra	Yeast	MYA, 4 days, 30°C	5, 10% sol. sd	17	Conner and Beuchat (1984)
Saccharomyces cerevisiae	Yeast	SMA, 2–7 days, 20°C	6.35, sd	6	Maruzzella and Liguori (1958)
Saccharomyces cerevisiae	Yeast	MYA, 4 days, 30°C	5, 10% sol. sd	22	Conner and Beuchat (1984)
Saccharomyces cerevisiae	Yeast	NA, 24 h, 30°C	Drop, 5,000	53	Hili et al. (1997)
Schizosaccharomyces pombe	Yeast	NA, 24 h, 30°C	Drop, 5,000	43	Hili et al. (1997)
Torula glabrata	Yeast	MYA, 4 days, 30°C	5, 10% sol. sd	20	Conner and Beuchat (1984)
Torula utilis	Yeast	NA, 24 h, 30°C	Drop, 5,000	42	Hili et al. (1997)

TABLE 14.14
Inhibitory Data of Ceylon Cinnamon Bark Oil Obtained in the Dilution Test

Microorganism	MO Class	Test Parameters	MIC (µg/mL)	Reference
Campylobacter jejuni	Bac–	TSB, 24 h, 42°C	500	Smith-Palmer et al. (1998)
Escherichia coli	Bac–	NB, 16 h, 37°C	200	Lens-Lisbonne et al. (1987)
Escherichia coli	Bac–	NB, Tween 20, 18 h, 37°C	400	Yousef and Tawil (1980)
Escherichia coli	Bac–	TSB, 24 h, 35°C	500	Smith-Palmer et al. (1998)
Escherichia coli	Bac–	MYB, DMSO, 40 h, 30°C	67% inh. 500	Hili et al. (1997)
Escherichia coli O157:H7	Bac–	BHI, 48 h, 35°C	250	Oussalah et al. (2006)
Pseudomonas aeruginosa	Bac–	NB, 16 h, 37°C	300	Lens-Lisbonne et al. (1987)
Pseudomonas aeruginosa	Bac–	NB, Tween 20, 18 h, 37°C	400	Yousef and Tawil (1980)
Pseudomonas aeruginosa	Bac–	MYB, DMSO, 40 h, 30°C	85% inh. 500	Hili et al. (1997)
Salmonella enteritidis	Bac–	TSB, 24 h, 35°C	500	Smith-Palmer et al. (1998)
Salmonella typhimurium	Bac–	BHI, 48 h, 35°C	500	Oussalah et al. (2006)
Bacillus subtilis	Bac+	NB, Tween 20, 18 h, 37°C	200	Yousef and Tawil (1980)
Listeria monocytogenes	Bac+	TSB, 24 h, 35°C	300	Smith-Palmer et al. (1998)
Listeria monocytogenes	Bac+	TSB, 10 days, 4°C	300	Smith-Palmer et al. (1998)
Listeria monocytogenes	Bac+	BHI, 48 h, 35°C	500	Oussalah et al. (2006)
Mycobacterium phlei	Bac+	NB, Tween 20, 18 h, 37°C	320	Yousef and Tawil (1980)
Staphylococcus aureus	Bac+	BHI, 48 h, 35°C	250	Oussalah et al. (2006)
Staphylococcus aureus	Bac+	NB, 16 h, 37°C	350	Lens-Lisbonne et al. (1987)
Staphylococcus aureus	Bac+	MHB, Tween 80, 24 h, 37°C	390	Bastide et al. (1987)
Staphylococcus aureus	Bac+	NB, Tween 20, 18 h, 37°C	400	Yousef and Tawil (1980)
Staphylococcus aureus	Bac+	TSB, 24 h, 35°C	400	Smith-Palmer et al. (1998)
Staphylococcus aureus	Bac+	MYB, DMSO, 40 h, 30°C	70% inh. 500	Hili et al. (1997)
Streptococcus faecalis	Bac+	NB, 16 h, 37°C	200	Lens-Lisbonne et al. (1987)
Aspergillus flavus	Fungi	PDA, 6–8 h 20°C, spore germ. inh.	50–100	Thompson (1986)
Aspergillus flavus	Fungi	PDA, 5 days, 27°C	1000	Thompson and Cannon (1986)
Aspergillus flavus	Fungi	PDA, 7–14 days, 28°C	1000	Soliman and Badeaa (2002)
Aspergillus niger	Fungi	NB, Tween 20, 8 days, 30°C	100	Yousef and Tawil (1980)
Aspergillus ochraceus	Fungi	PDA, 7–14 days, 28°C	1000	Soliman and Badeaa (2002)
Aspergillus parasiticus	Fungi	PDA, 6–8 h 20°C, spore germ. inh.	50–100	Thompson (1986)
Aspergillus parasiticus	Fungi	PDA, 5 days, 27°C	1000	Thompson and Cannon (1986)
Aspergillus parasiticus	Fungi	PDA, 7–14 days, 28°C	1000	Soliman and Badeaa (2002)
Botrytis cinerea	Fungi	PDA, Tween 20, 7 days, 24°C	25% inh. 1000	Bouchra et al. (2003)
Colletotrichum musae	Fungi	SMKY, EtOH, 7 days, 28°C	300	Ranasinghe et al. (2002)
Epidermophyton floccosum	Fungi	SA, Tween 80, 21 days, 20°C	300–625	Janssen et al. (1988)
Fusarium moniliforme	Fungi	PDA, 7–14 days, 28°C	1000	Soliman and Badeaa (2002)
Fusarium proliferatum	Fungi	SMKY, EtOH, 7 days, 28°C	500	Ranasinghe et al. (2002)
Geotrichum citri-aurantii	Fungi	PDA, Tween 20, 7 days, 24°C	30% inh. 1000	Bouchra et al. (2003)

(Continued)

TABLE 14.14 (*Continued*)
Inhibitory Data of Ceylon Cinnamon Bark Oil Obtained in the Dilution Test

Microorganism	MO Class	Test Parameters	MIC (µg/mL)	Reference
Lasiodiplodia theobromae	Fungi	SMKY, EtOH, 7 days, 28°C	350	Ranasinghe et al. (2002)
Mucor hiemalis	Fungi	PDA, 5 days, 27°C	500	Thompson and Cannon (1986)
Mucor mucedo	Fungi	PDA, 5 days, 27°C	1000	Thompson and Cannon (1986)
Mucor racemosus f. *racemosus*	Fungi	PDA, 5 days, 27°C	1000	Thompson and Cannon (1986)
Mucor sp.	Fungi	NB, Tween 20, 8 days, 30°C	100	Yousef and Tawil (1980)
Penicillium chrysogenum	Fungi	NB, Tween 20, 8 days, 30°C	100	Yousef and Tawil (1980)
Penicillium digitatum	Fungi	PDA, Tween 20, 7 days, 24°C	32% inh. 1000	Bouchra et al. (2003)
Phytophthora citrophthora	Fungi	PDA, Tween 20, 7 days, 24°C	38% inh. 1000	Bouchra et al. (2003)
Rhizopus 66-81-2	Fungi	PDA, 5 days, 27°C	1000	Thompson and Cannon (1986)
Rhizopus arrhizus	Fungi	PDA, 5 days, 27°C	500	Thompson and Cannon (1986)
Rhizopus chinensis	Fungi	PDA, 5 days, 27°C	500	Thompson and Cannon (1986)
Rhizopus circinans	Fungi	PDA, 5 days, 27°C	500	Thompson and Cannon (1986)
Rhizopus japonicus	Fungi	PDA, 5 days, 27°C	1000	Thompson and Cannon (1986)
Rhizopus kazanensis	Fungi	PDA, 5 days, 27°C	500	Thompson and Cannon (1986)
Rhizopus oryzae	Fungi	PDA, 5 days, 27°C	1000	Thompson and Cannon (1986)
Rhizopus pymacus	Fungi	PDA, 5 days, 27°C	500	Thompson and Cannon (1986)
Rhizopus sp.	Fungi	NB, Tween 20, 8 days, 30°C	400	Yousef and Tawil (1980)
Rhizopus stolonifer	Fungi	PDA, 5 days, 27°C	500	Thompson and Cannon (1986)
Rhizopus tritici	Fungi	PDA, 5 days, 27°C	1000	Thompson and Cannon (1986)
Trichophyton mentagrophytes	Fungi	SA, Tween 80, 21 days, 20°C	<300	Janssen et al. (1988)
Trichophyton rubrum	Fungi	SA, Tween 80, 21 days, 20°C	<300	Janssen et al. (1988)
Candida albicans	Yeast	NB, Tween 20, 18 h, 37°C	80	Yousef and Tawil (1980)
Candida albicans	Yeast	MYB, DMSO, 40 h, 30°C	72% inh. 500	Hili et al. (1997)
Saccharomyces cerevisiae	Yeast	MYB, DMSO, 40 h, 30°C	59% inh. 500	Hili et al. (1997)
Schizosaccharomyces pombe	Yeast	MYB, DMSO, 40 h, 30°C	96% inh. 500	Hili et al. (1997)
Torula utilis	Yeast	MYB, DMSO, 40 h, 30°C	100% inh. 500	Hili et al. (1997)

TABLE 14.15
Inhibitory Data of Ceylon Cinnamon Bark Oil Obtained in the Vapor Phase Test

Microorganism	MO Class	Test Parameters	Quantity (µg)	Activity	Reference
Escherichia coli	Bac–	NA, 24 h, 37°C	~20,000	NG	Kellner and Kober (1954)
Escherichia coli	Bac–	BIA, 18 h, 37°C	MIC_{air}	12.5	Inouye et al. (2001)
Haemophilus influenzae	Bac–	MHA, 18 h, 37°C	MIC_{air}	3.13	Inouye et al. (2001)
Neisseria sp.	Bac–	NA, 24 h, 37°C	~20,000	NG	Kellner and Kober (1954)
Salmonella typhi	Bac–	NA, 24 h, 37°C	~20,000	NG	Kellner and Kober (1954)
Salmonella typhi	Bac–	NA, 24 h, 37°C	6.35, sd	++	Maruzzella and Sicurella (1960)
Bacillus megaterium	Bac+	NA, 24 h, 37°C	~20,000	NG	Kellner and Kober (1954)
Bacillus subtilis var. *aterrimus*	Bac+	NA, 24 h, 37°C	6.35, sd	++	Maruzzella and Sicurella (1960)
Corynebacterium diphtheriae	Bac+	NA, 24 h, 37°C	~20,000	NG	Kellner and Kober (1954)
Mycobacterium avium	Bac+	NA, 24 h, 37°C	6.35, sd	+	Maruzzella and Sicurella (1960)
Staphylococcus aureus	Bac+	NA, 24 h, 37°C	~20,000	NG	Kellner and Kober (1954)
Staphylococcus aureus	Bac+	NA, 24 h, 37°C	6.35, sd	++	Maruzzella and Sicurella (1960)
Staphylococcus aureus	Bac+	MHA, 18 h, 37°C	MIC_{air}	6.25	Inouye et al. (2001)
Streptococcus faecalis	Bac+	NA, 24 h, 37°C	~20,000	NG	Kellner and Kober (1954)
Streptococcus faecalis	Bac+	NA, 24 h, 37°C	6.35, sd	+	Maruzzella and Sicurella (1960)
Streptococcus pneumoniae	Bac+	MHA, 18 h, 37°C	MIC_{air}	1.56–3.13	Inouye et al. (2001)
Streptococcus pyogenes	Bac+	NA, 24 h, 37°C	~20,000	NG	Kellner and Kober (1954)
Streptococcus pyogenes	Bac+	MHA, 18 h, 37°C	MIC_{air}	6.25	Inouye et al. (2001)
Aspergillus flavus	Fungi	CDA, 6 d	12, 6,000	+++	Singh et al. (2007)
Aspergillus niger	Fungi	CDA, 6 d	12, 6,000	+	Singh et al. (2007)
Aspergillus ochraceus	Fungi	CDA, 6 d	12, 6,000	++	Singh et al. (2007)
Aspergillus terreus	Fungi	CDA, 6 d	12, 6,000	NG	Singh et al. (2007)
Fusarium graminearum	Fungi	CDA, 6 d	12, 6,000	NG	Singh et al. (2007)
Fusarium moniliforme	Fungi	CDA, 6 d	12, 6,000	NG	Singh et al. (2007)
Penicillium citrinum	Fungi	CDA, 6 d	12, 6,000	NG	Singh et al. (2007)
Penicillium viridicatum	Fungi	CDA, 6 d	12, 6,000	NG	Singh et al. (2007)
Candida albicans	Yeast	NA, 24 h, 37°C	~20,000	NG	Kellner and Kober (1954)

TABLE 14.16
Inhibitory Data of Ceylon Cinnamon Leaf Oil Obtained in the Agar Diffusion Test

Microorganism	MO Class	Test Parameters	Disk Size (mm), Quantity (µg)	Inhibition Zone (mm)	Reference
Escherichia coli	Bac–	TYA, 18–24 h, 37°C	9.5, 2,000	17	Morris et al. (1979)
Escherichia coli	Bac–	NA, 24 h, 37°C	10 (h), 2,000	25	Singh et al. (2007)
Pseudomonas aeruginosa	Bac–	NA, 24 h, 37°C	10 (h), 2,000	90	Singh et al. (2007)
Salmonella typhi	Bac–	NA, 24 h, 37°C	10 (h), 2,000	17	Singh et al. (2007)
Bacillus cereus	Bac+	NA, 24 h, 37°C	10 (h), 2,000	32	Singh et al. (2007)
Bacillus subtilis	Bac+	NA, 24 h, 37°C	10 (h), 2,000	90	Singh et al. (2007)
Corynebacterium sp.	Bac+	TYA, 18–24 h, 37°C	9.5, 2,000	12	Morris et al. (1979)
Listeria monocytogenes	Bac+	ISA, 48 h, 25°C	4 (h), 10,000	20	Lis-Balchin et al. (1998)
Staphylococcus aureus	Bac+	TYA, 18–24 h, 37°C	9.5, 2,000	18	Morris et al. (1979)
Staphylococcus aureus	Bac+	NA, 24 h, 37°C	10 (h), 2,000	48	Singh et al. (2007)
Alternaria porri	Fungi	PDA, 72 h, 28°C	5, 5,000	50	Pawar and Thaker (2007)
Aspergillus niger	Fungi	PDA, 48 h, 28°C	5, 5,000	30	Pawar and Thaker (2006)
Fusarium oxysporum f.sp. *ciceri*	Fungi	PDA, 72 h, 28°C	5, 5,000	35	Pawar and Thaker (2007)
Candida albicans	Yeast	TYA, 18–24 h, 37°C	9.5, 2,000	14	Morris et al. (1979)

TABLE 14.17
Inhibitory Data of Ceylon Cinnamon Leaf Oil Obtained in the Dilution Test

Microorganism	MO Class	Test Parameters	MIC (µg/mL)	Reference
Escherichia coli	Bac−	TYB, 18–24 h, 37°C	1,000	Morris et al. (1979)
Escherichia coli O157:H7	Bac−	BHI, 48 h, 35°C	1,000	Oussalah et al. (2006)
Salmonella typhimurium	Bac−	BHI, 48 h, 35°C	1,000	Oussalah et al. (2006)
Corynebacterium sp.	Bac+	TYB, 18–24 h, 37°C	500	Morris et al. (1979)
Listeria monocytogenes	Bac+	BHI, 48 h, 35°C	2,000	Oussalah et al. (2006)
Staphylococcus aureus	Bac+	TYB, 18–24 h, 37°C	500	Morris et al. (1979)
Staphylococcus aureus	Bac+	BHI, 48 h, 35°C	500	Oussalah et al. (2006)
Aspergillus flavus	Fungi	PDA, 6–8 h 20°C, spore germ. inh.	50–100	Thompson (1986)
Aspergillus flavus	Fungi	PDA, 5 days, 27°C	500	Thompson and Cannon (1986)
Aspergillus niger	Fungi	YES broth, 10 d	95% inh. 10,000	Lis-Balchin et al. (1998)
Aspergillus ochraceus	Fungi	YES broth, 10 d	94% inh. 10,000	Lis-Balchin et al. (1998)
Aspergillus parasiticus	Fungi	PDA, 6–8 h 20°C, spore germ. inh.	50–100	Thompson (1986)
Aspergillus parasiticus	Fungi	PDA, 5 days, 27°C	500	Thompson and Cannon (1986)
Colletotrichum musae	Fungi	SMKY, EtOH, 7 days, 28°C	500	Ranasinghe et al. (2002)
Fusarium culmorum	Fungi	YES broth, 10 days	73% inh. 10,000	Lis-Balchin et al. (1998)
Fusarium proliferatum	Fungi	SMKY, EtOH, 7 days, 28°C	500	Ranasinghe et al. (2002)
Lasiodiplodia theobromae	Fungi	SMKY, EtOH, 7 days, 28°C	600	Ranasinghe et al. (2002)
Mucor hiemalis	Fungi	PDA, 5 days, 27°C	500	Thompson and Cannon (1986)
Mucor mucedo	Fungi	PDA, 5 days, 27°C	500	Thompson and Cannon (1986)
Mucor racemosus f. *racemosus*	Fungi	PDA, 5 days, 27°C	500	Thompson and Cannon (1986)
Rhizopus 66-81-2	Fungi	PDA, 5 days, 27°C	500	Thompson and Cannon (1986)
Rhizopus arrhizus	Fungi	PDA, 5 days, 27°C	500	Thompson and Cannon (1986)
Rhizopus chinensis	Fungi	PDA, 5 days, 27°C	100	Thompson and Cannon (1986)
Rhizopus circinans	Fungi	PDA, 5 days, 27°C	500	Thompson and Cannon (1986)
Rhizopus japonicus	Fungi	PDA, 5 days, 27°C	500	Thompson and Cannon (1986)
Rhizopus kazanensis	Fungi	PDA, 5 days, 27°C	500	Thompson and Cannon (1986)
Rhizopus oryzae	Fungi	PDA, 5 days, 27°C	500	Thompson and Cannon (1986)
Rhizopus pymacus	Fungi	PDA, 5 days, 27°C	500	Thompson and Cannon (1986)
Rhizopus stolonifer	Fungi	PDA, 5 days, 27°C	500	Thompson and Cannon (1986)
Rhizopus tritici	Fungi	PDA, 5 days, 27°C	500	Thompson and Cannon (1986)
Candida albicans	Yeast	TYB, 18–24 h, 37°C	500	Morris et al. (1979)

TABLE 14.18
Inhibitory Data of Ceylon Cinnamon Leaf Oil Obtained in the Vapor Phase Test

Microorganism	MO Class	Test Parameters	Quantity (µg)	Activity	Reference
Escherichia coli	Bac−	NA, 24 h, 37°C	~20,000	NG	Kellner and Kober (1954)
Neisseria sp.	Bac−	NA, 24 h, 37°C	~20,000	NG	Kellner and Kober (1954)
Salmonella typhi	Bac−	NA, 24 h, 37°C	~20,000	NG	Kellner and Kober (1954)
Bacillus megaterium	Bac+	NA, 24 h, 37°C	~20,000	NG	Kellner and Kober (1954)
Corynebacterium diphtheriae	Bac+	NA, 24 h, 37°C	~20,000	NG	Kellner and Kober (1954)
Staphylococcus aureus	Bac+	NA, 24 h, 37°C	~20,000	NG	Kellner and Kober (1954)
Streptococcus faecalis	Bac+	NA, 24 h, 37°C	~20,000	NG	Kellner and Kober (1954)
Streptococcus pyogenes	Bac+	NA, 24 h, 37°C	~20,000	NG	Kellner and Kober (1954)
Aspergillus flavus	Fungi	WFA, 42 days, 25°C	Disk, 50,000	NG	Guynot et al. (2003)
Aspergillus flavus	Fungi	Bread, 14 days, 25°C	30,000	+++	Suhr and Nielsen (2003)
Aspergillus flavus	Fungi	CDA, 6 days	12, 6,000	NG	Singh et al. (2007)
Aspergillus niger	Fungi	WFA, 42 days, 25°C	Disk, 50,000	NG	Guynot et al. (2003)
Aspergillus niger	Fungi	CDA, 6 days	12, 6,000	NG	Singh et al. (2007)
Aspergillus ochraceus	Fungi	CDA, 6 days	12, 6,000	+	Singh et al. (2007)
Aspergillus terreus	Fungi	CDA, 6 days	12, 6,000	+	Singh et al. (2007)
Endomyces fibuligera	Fungi	Bread, 14 days, 25°C	30,000	+++	Suhr and Nielsen (2003)
Eurotium amstelodami	Fungi	WFA, 42 days, 25°C	Disk, 50,000	NG	Guynot et al. (2003)
Eurotium herbarum	Fungi	WFA, 42 days, 25°C	Disk, 50,000	NG	Guynot et al. (2003)
Eurotium repens	Fungi	WFA, 42 days, 25°C	Disk, 50,000	NG	Guynot et al. (2003)
Eurotium repens	Fungi	Bread, 14 days, 25°C	30,000	+++	Suhr and Nielsen (2003)
Eurotium rubrum	Fungi	WFA, 42 days, 25°C	Disk, 50,000	NG	Guynot et al. (2003)
Fusarium graminearum	Fungi	CDA, 6 days	12, 6,000	NG	Singh et al. (2007)
Fusarium moniliforme	Fungi	CDA, 6 days	12, 6,000	NG	Singh et al. (2007)
Penicillium citrinum	Fungi	CDA, 6 days	12, 6,000	NG	Singh et al. (2007)
Penicillium corylophilum	Fungi	WFA, 42 days, 25°C	Disk, 50,000	NG	Guynot et al. (2003)
Penicillium corylophilum	Fungi	Bread, 14 days, 25°C	30,000	+++	Suhr and Nielsen (2003)
Penicillium roqueforti	Fungi	Bread, 14 days, 25°C	30,000	+++	Suhr and Nielsen (2003)
Penicillium viridicatum	Fungi	CDA, 6 days	12, 6,000	NG	Singh et al. (2007)
Candida albicans	Yeast	NA, 24 h, 37°C	~20,000	NG	Kellner and Kober (1954)

TABLE 14.19
Inhibitory Data of Citronella Oil Obtained in the Agar Diffusion Test

Microorganism	MO Class	Conditions		Inhibition Zone (mm)	Reference
Aerobacter aerogenes	Bac−	NA, 24 h, 37°C	—, sd	0[a]	Maruzzella and Lichtenstein (1956)
Aerobacter aerogenes	Bac−	NA, 24 h, 37°C	5 × 20, 1,000	0[a]	Möse and Lukas (1957)
Brucella abortus	Bac−	NA, 24 h, 37°C	5 × 20, 1,000	2–5[a]	Möse and Lukas (1957)
Campylobacter jejuni	Bac−	Cited	6, 15,000	40[a]	Wannissorn et al. (2005)
Escherichia coli	Bac−	NA, 24 h, 37°C	—, sd	0[a]	Maruzzella and Lichtenstein (1956)
Escherichia coli	Bac−	NA, 24 h, 37°C	5 × 20, 1,000	0–1[a]	Möse and Lukas (1957)
Escherichia coli	Bac−	Cited	6, 15,000	10.5[a]	Wannissorn et al. (2005)
Escherichia coli	Bac−	Cited	20,000	0	Lemos et al. (1992)
Escherichia coli	Bac−	NA, 18 h, 37°C	6 (h), pure	0	Yousef and Tawil (1980)
Escherichia coli	Bac−	TGA, 18–24 h, 37°C	9.5, 2,000	0	Morris et al. (1979)
Escherichia coli	Bac−	Agar, 24 h, 37°C	6, 6,000	7	Jirovetz et al. (2006)
Klebsiella pneumonia	Bac−	NA, 24 h, 37°C	5 × 20, 1,000	2–5[a]	Möse and Lukas (1957)
Klebsiella pneumonia subsp. ozaenae	Bac−	NA, 24 h, 37°C	5 × 20, 1,000	2–5[a]	Möse and Lukas (1957)
Klebsiella pneumoniae	Bac−	Agar, 24 h, 37°C	6, 6,000	0	Jirovetz et al. (2006)
Neisseria perflava	Bac−	NA, 24 h, 37°C	—, sd	0[a]	Maruzzella and Lichtenstein (1956)
Proteus OX19	Bac−	NA, 24 h, 37°C	5 × 20, 1,000	2–5[a]	Möse and Lukas (1957)
Proteus vulgaris	Bac−	NA, 24 h, 37°C	—, sd	0[a]	Maruzzella and Lichtenstein (1956)
Proteus vulgaris	Bac−	NA, 24 h, 37°C	5 × 20, 1,000	0[a]	Möse and Lukas (1957)
Proteus vulgaris	Bac−	Agar, 24 h, 37°C	6, 6,000	10	Jirovetz et al. (2006)
Pseudomonas aeruginosa	Bac−	NA, 24 h, 37°C	—, sd	0[a]	Maruzzella and Lichtenstein (1956)
Pseudomonas aeruginosa	Bac−	NA, 24 h, 37°C	5 × 20, 1,000	0[a]	Möse and Lukas (1957)
Pseudomonas aeruginosa	Bac−	NA, 18 h, 37°C	6 (h), pure	0	Yousef and Tawil (1980)
Pseudomonas aeruginosa	Bac−	Agar, 24 h, 37°C	6, 6,000	0	Jirovetz et al. (2006)
Pseudomonas fluorescens	Bac−	NA, 24 h, 37°C	5 × 20, 1,000	0–1[a]	Möse and Lukas (1957)
Pseudomonas mangiferae indica	Bac−	NA, 36–48 h, 37°C	6, sd	11	Garg and Garg (1980)
Salmonella enteritidis	Bac−	Cited	6, 15,000	12.8[a]	Wannissorn et al. (2005)
Salmonella enteritidis	Bac−	NA, 24 h, 37°C	5 × 20, 1,000	6–10[a]	Möse and Lukas (1957)
Salmonella paratyphi	Bac−	NA, 36–48 h, 37°C	6, sd	14	Garg and Garg (1980)
Salmonella paratyphi B	Bac−	NA, 24 h, 37°C	5 × 20, 1,000	1[a]	Möse and Lukas (1957)
Salmonella sp.	Bac−	Agar, 24 h, 37°C	6, 6,000	0	Jirovetz et al. (2006)
Salmonella typhi	Bac−	NA, 24 h, 37°C	5 × 20, 1,000	2–5[a]	Möse and Lukas (1957)

(Continued)

TABLE 14.19 (*Continued*)
Inhibitory Data of Citronella Oil Obtained in the Agar Diffusion Test

Microorganism	MO Class	Conditions		Inhibition Zone (mm)	Reference
Salmonella typhi	Bac−	NA, 36–48 h, 37°C	6, sd	18	Garg and Garg (1980)
Salmonella typhimurium	Bac−	Cited	6, 15,000	21[a]	Wannissorn et al. (2005)
Serratia marcescens	Bac−	NA, 24 h, 37°C	5 × 20, 1,000	0[a]	Möse and Lukas (1957)
Serratia marcescens	Bac−	NA, 24 h, 37°C	—, sd	0	Maruzzella and Lichtenstein (1956)
Vibrio albicans	Bac−	NA, 24 h, 37°C	5 × 20, 1,000	6–10[a]	Möse and Lukas (1957)
Vibrio cholera	Bac−	NA, 24 h, 37°C	5 × 20, 1,000	6–10[a]	Möse and Lukas (1957)
Vibrio cholera	Bac−	NA, 36–48 h, 37°C	6, sd	0	Garg and Garg (1980)
Bacillus anthracis	Bac+	NA, 24 h, 37°C	5 × 20, 1,000	1[a]	Möse and Lukas (1957)
Bacillus cereus	Bac+	Cited	20,000	20	Lemos et al. (1992)
Bacillus mesentericus	Bac+	NA, 24 h, 37°C	—, sd	3[a]	Maruzzella and Lichtenstein (1956)
Bacillus mycoides	Bac+	NA, 36–48 h, 37°C	6, sd	12	Garg and Garg (1980)
Bacillus pumilus	Bac+	NA, 36–48 h, 37°C	6, sd	12	Garg and Garg (1980)
Bacillus subtilis	Bac+	NA, 24 h, 37°C	5 × 20, 1,000	1[a]	Möse and Lukas (1957)
Bacillus subtilis	Bac+	NA, 24 h, 37°C	—, sd	8[a]	Maruzzella and Lichtenstein (1956)
Bacillus subtilis	Bac+	NA, 36–48 h, 37°C	6, sd	12	Garg and Garg (1980)
Bacillus subtilis	Bac+	NA, 18 h, 37°C	6 (h), pure	15.5	Yousef and Tawil (1980)
Clostridium perfringens	Bac+	Cited	6, 15,000	39.5[a]	Wannissorn et al. (2005)
Corynebacterium diphtheria	Bac+	NA, 24 h, 37°C	5 × 20, 1,000	6–10[a]	Möse and Lukas (1957)
Enterococcus faecalis	Bac+	Agar, 24 h, 37°C	6, 6,000	10	Jirovetz et al. (2006)
Listeria monocytogenes	Bac+	NA, 24 h, 37°C	5 × 20, 1,000	1[a]	Möse and Lukas (1957)
Mycobacterium phlei	Bac+	NA, 18 h, 37°C	6 (h), pure	16	Yousef and Tawil (1980)
Sarcina alba	Bac+	NA, 24 h, 37°C	5 × 20, 1,000	6–10[a]	Möse and Lukas (1957)
Sarcina beige	Bac+	NA, 24 h, 37°C	5 × 20, 1,000	16–23[a]	Möse and Lukas (1957)
Sarcina citrea	Bac+	NA, 24 h, 37°C	5 × 20, 1,000	0[a]	Möse and Lukas (1957)
Sarcina lutea	Bac+	NA, 24 h, 37°C	—, sd	0[a]	Maruzzella and Lichtenstein (1956)
Sarcina lutea	Bac+	NA, 36–48 h, 37°C	6, sd	0	Garg and Garg (1980)
Sarcina rosa	Bac+	NA, 24 h, 37°C	5 × 20, 1,000	11–15[a]	Möse and Lukas (1957)
Sporococcus sarc.	Bac+	NA, 24 h, 37°C	5 × 20, 1,000	26–33[a]	Möse and Lukas (1957)
Staphylococcus albus	Bac+	NA, 24 h, 37°C	5 × 20, 1,000	6–10[a]	Möse and Lukas (1957)
Staphylococcus aureus	Bac+	NA, 24 h, 37°C	—, sd	0[a]	Maruzzella and Lichtenstein (1956)
Staphylococcus aureus	Bac+	Cited	20,000	0[a]	Lemos et al. (1992)

(*Continued*)

TABLE 14.19 (*Continued*)
Inhibitory Data of Citronella Oil Obtained in the Agar Diffusion Test

Microorganism	MO Class	Conditions		Inhibition Zone (mm)	Reference
Staphylococcus aureus	Bac+	NA, 24 h, 37°C	5 × 20, 1,000	2–5[a]	Möse and Lukas (1957)
Staphylococcus aureus	Bac+	Agar, 24 h, 37°C	6, 6,000	10	Jirovetz et al. (2006)
Staphylococcus aureus	Bac+	TGA, 18–24 h, 37°C	9.5, 2,000	11	Morris et al. (1979)
Staphylococcus aureus	Bac+	NA, 36–48 h, 37°C	6, sd	12	Garg and Garg (1980)
Staphylococcus aureus	Bac+	NA, 18 h, 37°C	6 (h), pure	12.6	Yousef and Tawil (1980)
Staphylococcus epidermidis	Bac+	Cited	20,000	12[a]	Lemos et al. (1992)
Streptococcus haemolyticus	Bac+	NA, 24 h, 37°C	5 × 20, 1,000	6–10[a]	Möse and Lukas (1957)
Streptococcus viridians	Bac+	NA, 24 h, 37°C	5 × 20, 1,000	1[a]	Möse and Lukas (1957)
Streptomyces venezuelae	Bac+	SMA, 2–7 days, 20°C	sd	4	Maruzzella and Liguori (1958)
Alternaria solani	Fungi	SMA, 2–7 days, 20°C	sd	6	Maruzzella and Liguori (1958)
Aspergillus fumigatus	Fungi	SMA, 2–7 days, 20°C	sd	6	Maruzzella and Liguori (1958)
Aspergillus niger	Fungi	SDA, 8 days, 30°C	6 (h), pure	15	Yousef and Tawil (1980)
Aspergillus niger	Fungi	SMA, 2–7 days, 20°C	sd	7	Maruzzella and Liguori (1958)
Aspergillus niger	Fungi	SDA, 7–10 days, 28°C	5, 5,000	8	Saikia et al. (2001)
Helminthosporium sativum	Fungi	SMA, 2–7 days, 20°C	sd	2	Maruzzella and Liguori (1958)
Microsporum gypseum	Fungi	SDA, 7–10 days, 28°C	5, 5,000	34	Saikia et al. (2001)
Mucor mucedo	Fungi	SMA, 2–7 days, 20°C	sd	6	Maruzzella and Liguori (1958)
Mucor sp.	Fungi	SDA, 8 days, 30°C	6 (h), pure	18	Yousef and Tawil (1980)
Nigrospora panici	Fungi	SMA, 2–7 days, 20°C	sd	6	Maruzzella and Liguori (1958)
Penicillium chrysogenum	Fungi	SDA, 8 days, 30°C	6 (h), pure	40	Yousef and Tawil (1980)
Penicillium digitatum	Fungi	SMA, 2–7 days, 20°C	sd	5	Maruzzella and Liguori (1958)
Rhizopus nigricans	Fungi	SMA, 2–7 days, 20°C	sd	1	Maruzzella and Liguori (1958)
Rhizopus sp.	Fungi	SDA, 8 days, 30°C	6 (h), pure	12	Yousef and Tawil (1980)
Sporothrix schenckii	Fungi	SDA, 7–10 days, 28°C	5, 5,000	12	Saikia et al. (2001)
Candida albicans	Yeast	TGA, 18–24 h, 37°C	9.5, 2,000	1	Morris et al. (1979)

(*Continued*)

TABLE 14.19 (*Continued*)
Inhibitory Data of Citronella Oil Obtained in the Agar Diffusion Test

Microorganism	MO Class	Conditions		Inhibition Zone (mm)	Reference
Candida albicans	Yeast	Cited	20,000	17	Lemos et al. (1992)
Candida albicans	Yeast	SDA, 18 h, 30°C	6 (h), pure	17	Yousef and Tawil (1980)
Candida albicans	Yeast	Agar, 24 h, 37°C	6, 6,000	28	Jirovetz et al. (2006)
Candida albicans	Yeast	SMA, 2–7 days, 20°C	sd	6	Maruzzella and Liguori (1958)
Candida albicans	Yeast	SDA, 7–10 days, 28°C	5, 5,000	7	Saikia et al. (2001)
Candida krusei	Yeast	SMA, 2–7 days, 20°C	sd	0	Maruzzella and Liguori (1958)
Candida tropicalis	Yeast	Cited	20,000	0	Lemos et al. (1992)
Candida tropicalis	Yeast	SMA, 2–7 days, 20°C	sd	3	Maruzzella and Liguori (1958)
Cryptococcus neoformans	Yeast	SMA, 2–7 days, 20°C	sd	2	Maruzzella and Liguori (1958)
Cryptococcus rhodobenhani	Yeast	SMA, 2–7 days, 20°C	sd	7	Maruzzella and Liguori (1958)
Saccharomyces cerevisiae	Yeast	SMA, 2–7 days, 20°C	sd	2	Maruzzella and Liguori (1958)

[a] Citronella oil obtained from *Cymbopogon nardus* (Ceylon type).

TABLE 14.20
Inhibitory Data of Citronella Oil Obtained in the Dilution Test

Microorganism	MO Class	Conditions	MIC	Reference
Escherichia coli	Bac−	MHB, 24 h, 36°C	2,000–8,000	Duarte et al. (2006)
Escherichia coli	Bac−	NB, Tween 20, 18 h, 37°C	50,000	Yousef and Tawil (1980)
Escherichia coli O157:H7	Bac−	BHI, 48 h, 35°C	>8,000	Oussalah et al. (2006)
Pseudomonas aeruginosa	Bac−	NA, Tween 80, 24 h, 37°C	>500[a]	Koba et al. (2004)
Pseudomonas aeruginosa	Bac−	NB, Tween 20, 18 h, 37°C	50,000	Yousef and Tawil (1980)
Pseudomonas cepacia	Bac−	NA, Tween 80, 24 h, 37°C	>500[a]	Koba et al. (2004)
Salmonella typhimurium	Bac−	BHI, 48 h, 35°C	4,000	Oussalah et al. (2006)
Bacillus subtilis	Bac+	NB, Tween 20, 18 h, 37°C	400	Yousef and Tawil (1980)
Listeria monocytogenes	Bac+	BHI, 48 h, 35°C	4,000	Oussalah et al. (2006)
Mycobacterium phlei	Bac+	NB, Tween 20, 18 h, 37°C	1,250	Yousef and Tawil (1980)
Staphylococcus aureus	Bac+	BHI, 48 h, 35°C	500	Oussalah et al. (2006)
Staphylococcus aureus	Bac+	NB, Tween 20, 18 h, 37°C	6,400	Yousef and Tawil (1980)
Staphylococcus intermedius	Bac+	NA, Tween 80, 24 h, 37°C	>500[a]	Koba et al. (2004)
Aspergillus candidus	Fungi	Cited	>250[a]	Nakahara et al. (2003)
Aspergillus flavus	Fungi	Cited	>250[a]	Nakahara et al. (2003)
Aspergillus flavus	Fungi	PDA, 28 days, 21°C	4,000–10,000	Thanaboripat et al. (2004)
Aspergillus fumigatus	Fungi	NA, Tween 80, 24 h, 37°C	200[a]	Koba et al. (2004)
Aspergillus niger	Fungi	SDB, 7–10 days, 28°C	2,500	Saikia et al. (2001)
Aspergillus niger	Fungi	NB, Tween 20, 8 days, 30°C	6,400	Yousef and Tawil (1980)
Aspergillus oryzae	Fungi	Cited	250	Okazaki and Oshima (1953)
Aspergillus versicolor	Fungi	Cited	>250[a]	Nakahara et al. (2003)
Eurotium amstelodami	Fungi	Cited	>250[a]	Nakahara et al. (2003)
Eurotium chevalieri	Fungi	Cited	>250[a]	Nakahara et al. (2003)
Microsporum canis	Fungi	NA, Tween 80, 24 h, 37°C	200[a]	Koba et al. (2004)
Microsporum gypseum	Fungi	NA, Tween 80, 24 h, 37°C	500[a]	Koba et al. (2004)
Microsporum gypseum	Fungi	SDB, 7–10 days, 28°C	625	Saikia et al. (2001)
Mucor racemosus	Fungi	Cited	500	Okazaki and Oshima (1953)
Mucor sp.	Fungi	NB, Tween 20, 8 days, 30°C	6,400	Yousef and Tawil (1980)
Penicillium adametzii	Fungi	Cited	>250[a]	Nakahara et al. (2003)
Penicillium chrysogenum	Fungi	NB, Tween 20, 8 days, 30°C	12,500	Yousef and Tawil (1980)
Penicillium chrysogenum	Fungi	Cited	250	Okazaki and Oshima (1953)
Penicillium citrinum	Fungi	Cited	>250[a]	Nakahara et al. (2003)
Penicillium griseofulvum	Fungi	Cited	>250[a]	Nakahara et al. (2003)
Penicillium islansicum	Fungi	Cited	>250[a]	Nakahara et al. (2003)
Rhizopus sp.	Fungi	NB, Tween 20, 8 days, 30°C	12,500	Yousef and Tawil (1980)
Sporothrix schenckii	Fungi	SDB, 7–10 days, 28°C	1,250	Saikia et al. (2001)
Trichophyton mentagrophytes	Fungi	NA, Tween 80, 24 h, 37°C	150[a]	Koba et al. (2004)
Candida albicans	Yeast	NA, Tween 80, 24 h, 37°C	>500[a]	Koba et al. (2004)
Candida albicans	Yeast	SDB, 7–10 days, 28°C	1,250	Saikia et al. (2001)
Candida albicans	Yeast	NB, Tween 20, 18 h, 37°C	6,400	Yousef and Tawil (1980)
Cryptococcus neoformans	Yeast	NA, Tween 80, 24 h, 37°C	500[a]	Koba et al. (2004)
Malassezia pachydermatis	Yeast	NA, Tween 80, 24 h, 37°C	150[a]	Koba et al. (2004)

[a] Citronella oil obtained from *Cymbopogon nardus* (Ceylon type).

TABLE 14.21
Inhibitory Data of Citronella Oil Obtained in the Vapor Phase Test

Microorganism	MO Class	Conditions		Activity	Reference
Escherichia coli	Bac−	NA, 24 h, 37°C	~20,000	+	Kellner and Kober (1954)
Neisseria sp.	Bac−	NA, 24 h, 37°C	~20,000	NG	Kellner and Kober (1954)
Salmonella typhi	Bac−	NA, 24 h, 37°C	sd	+++[a]	Maruzzella and Sicurella (1960)
Salmonella typhi	Bac−	NA, 24 h, 37°C	~20,000	NG	Kellner and Kober (1954)
Bacillus megaterium	Bac+	NA, 24 h, 37°C	~20,000	+	Kellner and Kober (1954)
Bacillus subtilis var. *aterrimus*	Bac+	NA, 24 h, 37°C	sd	+[a]	Maruzzella and Sicurella (1960)
Corynebacterium diphtheriae	Bac+	NA, 24 h, 37°C	~20,000	NG	Kellner and Kober (1954)
Mycobacterium avium	Bac+	NA, 24 h, 37°C	sd	NG[a]	Maruzzella and Sicurella (1960)
Staphylococcus aureus	Bac+	NA, 24 h, 37°C	sd	+++[a]	Maruzzella and Sicurella (1960)
Staphylococcus aureus	Bac+	NA, 24 h, 37°C	~20,000	NG	Kellner and Kober (1954)
Streptococcus faecalis	Bac+	NA, 24 h, 37°C	sd	+++[a]	Maruzzella and Sicurella (1960)
Streptococcus faecalis	Bac+	NA, 24 h, 37°C	~20,000	NG	Kellner and Kober (1954)
Streptococcus pyogenes	Bac+	NA, 24 h, 37°C	~20,000	NG	Kellner and Kober (1954)
Aspergillus candidus	Fungi	PDA, 3–5 days, 27°C	250 mg/L air	NG[b]	Nakahara et al. (2003)
Aspergillus flavus	Fungi	PDA, 3–5 days, 27°C	250 mg/L air	NG[b]	Nakahara et al. (2003)
Aspergillus flavus	Fungi	PDA, 10 days, 25°C	50	++	Sarbhoy et al. (1978)
Aspergillus fumigatus	Fungi	PDA, 10 days, 25°C	50	NG	Sarbhoy et al. (1978)
Aspergillus sulphureus	Fungi	PDA, 10 days, 25°C	50	NG	Sarbhoy et al. (1978)
Aspergillus versicolor	Fungi	PDA, 3–5 days, 27°C	250 mg/L air	NG[b]	Nakahara et al. (2003)
Eurotium amstelodami	Fungi	PDA, 3–5 days, 27°C	250 mg/L air	NG[b]	Nakahara et al. (2003)
Eurotium chevalieri	Fungi	PDA, 3–5 days, 27°C	250 mg/L air	NG[b]	Nakahara et al. (2003)
Mucor fragilis	Fungi	PDA, 10 days, 25°C	50	NG	Sarbhoy et al. (1978)
Penicillium adametzii	Fungi	PDA, 3–5 days, 27°C	250 mg/L air	NG[b]	Nakahara et al. (2003)
Penicillium citrinum	Fungi	PDA, 3–5 days, 27°C	250 mg/L air	NG[b]	Nakahara et al. (2003)
Penicillium griseofulvum	Fungi	PDA, 3–5 days, 27°C	250 mg/L air	NG[b]	Nakahara et al. (2003)
Penicillium islansicum	Fungi	PDA, 3–5 days, 27°C	250 mg/L air	NG[b]	Nakahara et al. (2003)
Rhizopus stolonifer	Fungi	PDA, 10 days, 25°C	50	++	Sarbhoy et al. (1978)
Candida albicans	Yeast	NA, 24 h, 37°C	~20,000	NG	Kellner and Kober (1954)

[a] Formosan citronella oil.
[b] Citronella oil obtained from *Cymbopogon nardus* (Ceylon type).

TABLE 14.22
Inhibitory Data of Clary Sage Oil Obtained in the Agar Diffusion Test

Microorganism	MO Class	Conditions		Inhibition Zone (mm)	Reference
Escherichia coli	Bac−	MHA, 24 h, 37°C	6, 10,000	14	Yousefzadi et al. (2007)
Klebsiella pneumoniae	Bac−	MHA, 24 h, 37°C	6, 10,000	10	Yousefzadi et al. (2007)
Pseudomonas aeruginosa	Bac−	MHA, 24 h, 37°C	6, 10,000	0	Yousefzadi et al. (2007)
Bacillus pumilus	Bac+	MHA, 24 h, 37°C	6, 10,000	16	Yousefzadi et al. (2007)
Bacillus subtilis	Bac+	MHA, 24 h, 37°C	6, 10,000	17	Yousefzadi et al. (2007)
Enterococcus faecalis	Bac+	MHA, 24 h, 37°C	6, 10,000	9	Yousefzadi et al. (2007)
Listeria monocytogenes	Bac+	ISA, 48 h, 25°C	4 (h), 10,000	9–15	Lis-Balchin et al. (1998)
Staphylococcus aureus	Bac+	MHA, 24 h, 37°C	6, 10,000	15	Yousefzadi et al. (2007)
Staphylococcus epidermidis	Bac+	MHA, 24 h, 37°C	6, 10,000	15	Yousefzadi et al. (2007)
Alternaria porri	Fungi	PDA, 72 h, 28°C	5, 5,000	0	Pawar and Thaker (2007)
Aspergillus niger	Fungi	MHA, 48 h, 30°C	6, 10,000	0	Yousefzadi et al. (2007)
Aspergillus niger	Fungi	PDA, 48 h, 28°C	5, 5,000	0	Pawar and Thaker (2006)
Fusarium oxysporum f.sp. *ciceri*	Fungi	PDA, 72 h, 28°C	5, 5,000	9	Pawar and Thaker (2007)
Candida albicans	Yeast	MHA, 48 h, 30°C	6, 10,000	0	Yousefzadi et al. (2007)
Saccharomyces cerevisiae	Yeast	MHA, 48 h, 30°C	6, 10,000	0	Yousefzadi et al. (2007)

TABLE 14.23
Inhibitory Data of Clary Sage Oil Obtained in the Dilution Test

Microorganism	MO Class	Conditions	MIC	Reference
Acinetobacter baumannii	Bac−	MHA, Tween 20, 48 h, 35°C	>20,000	Hammer et al. (1999)
Aeromonas sobria	Bac−	MHA, Tween 20, 48 h, 35°C	>20,000	Hammer et al. (1999)
Erwinia carotovora	Bac−	NA, 48 h, 22°C	>2,000	Maruzzella (1963)
Escherichia coli	Bac−	NA, 48 h, 37°C	>2,000	Maruzzella (1963)
Escherichia coli	Bac−	MHA, Tween 20, 48 h, 35°C	>20,000	Hammer et al. (1999)
Escherichia coli	Bac−	MHB, Tween 80, 24 h, 37°C	15,000	Yousefzadi et al. (2007)
Escherichia coli	Bac−	Cited	1,500–2,000	Peana et al. (1999)
Klebsiella pneumoniae	Bac−	MHB, Tween 80, 24 h, 37°C	>15,000	Yousefzadi et al. (2007)
Klebsiella pneumoniae	Bac−	MHA, Tween 20, 48 h, 35°C	>20,000	Hammer et al. (1999)
Proteus vulgaris	Bac−	NA, 48 h, 37°C	>2,000	Maruzzella (1963)
Pseudomonas aeruginosa	Bac−	MHA, Tween 20, 48 h, 35°C	>20,000	Hammer et al. (1999)
Salmonella typhi	Bac−	NA, 48 h, 37°C	>2,000	Maruzzella (1963)
Salmonella typhimurium	Bac−	MHA, Tween 20, 48 h, 35°C	>20,000	Hammer et al. (1999)
Serratia marcescens	Bac−	MHA, Tween 20, 48 h, 35°C	>20,000	Hammer et al. (1999)
Bacillus cereus	Bac+	NA, 48 h, 37°C	250	Maruzzella (1963)
Bacillus circulans	Bac+	NA, 48 h, 37°C	100	Maruzzella (1963)
Bacillus megaterium	Bac+	NA, 48 h, 37°C	250	Maruzzella (1963)
Bacillus pumilus	Bac+	MHB, Tween 80, 24 h, 37°C	7,500	Yousefzadi et al. (2007)
Bacillus subtilis	Bac+	MHB, Tween 80, 24 h, 37°C	7,500	Yousefzadi et al. (2007)
Bacillus subtilis var. aterrimus	Bac+	NA, 48 h, 37°C	100	Maruzzella (1963)
Enterococcus faecalis	Bac+	MHB, Tween 80, 24 h, 37°C	>15,000	Yousefzadi et al. (2007)
Enterococcus faecalis	Bac+	MHA, Tween 20, 48 h, 35°C	>20,000	Hammer et al. (1999)
Staphylococcus aureus	Bac+	NA, 48 h, 37°C	>2,000	Maruzzella (1963)
Staphylococcus aureus	Bac+	MHA, Tween 20, 48 h, 35°C	>20,000	Hammer et al. (1999)
Staphylococcus aureus	Bac+	Cited	1,500–2,000	Peana et al. (1999)
Staphylococcus aureus	Bac+	MHB, Tween 80, 24 h, 37°C	7,500	Yousefzadi et al. (2007)
Staphylococcus epidermidis	Bac+	Cited	1,500–2,000	Peana et al. (1999)
Staphylococcus epidermidis	Bac+	MHB, Tween 80, 24 h, 37°C	7,500	Yousefzadi et al. (2007)
Alternaria alternata	Fungi	SDA, 6–8 h, 20°C	500, 62% inh.	Dikshit et al. (1986)
Aspergillus flavus	Fungi	SDA, 6–8 h, 20°C	500, 52% inh.	Dikshit et al. (1986)
Aspergillus niger	Fungi	YES broth, 10 d	−92% inh. 10,000	Lis-Balchin et al. (1998)
Aspergillus ochraceus	Fungi	YES broth, 10 d	−>10,000	Lis-Balchin et al. (1998)
Fusarium culmorum	Fungi	YES broth, 10 d	−69% inh. 10,000	Lis-Balchin et al. (1998)
Fusarium oxysporum f.sp. dianthi	Fungi	PDA, Tween 20, 4 days, 23°C	72% inh. 2,000	Pitarokili et al. (2002)
Geotrichum candidum	Fungi	NA, 5 days, 22°C	>2,000	Maruzzella (1963)
Gibberella fujikuroi	Fungi	NA, 5 days, 22°C	>2,000	Maruzzella (1963)
Helminthosporium truicicum	Fungi	NA, 5 days, 22°C	>2,000	Maruzzella (1963)
Microsporum gypseum	Fungi	SDA, 7 days, 30°C	400, 56% inh.	Dikshit and Husain (1984)
Microsporum gypseum	Fungi	SDA, 6–8 h, 20°C	500, 56% inh.	Dikshit et al. (1986)
Penicillium italicum	Fungi	SDA, 6–8 h, 20°C	500, 59% inh.	Dikshit et al. (1986)
Phoma betae	Fungi	NA, 5 days, 22°C	>2,000	Maruzzella (1963)

(Continued)

TABLE 14.23 (*Continued*)
Inhibitory Data of Clary Sage Oil Obtained in the Dilution Test

Microorganism	MO Class	Conditions	MIC	Reference
Pityrosporum ovale	Fungi	NA, 5 days, 22°C	>2,000	Maruzzella (1963)
Sclerotina cepivorum	Fungi	PDA, Tween 20, 4 days, 23°C	94% inh. 1,000	Pitarokili et al. (2002)
Sclerotina sclerotiorum	Fungi	PDA, Tween 20, 4 days, 23°C	1,000	Pitarokili et al. (2002)
Trichophyton equinum	Fungi	SDA, 7 days, 30°C	400, 30% inh.	Dikshit and Husain (1984)
Trichophyton mentagrophytes	Fungi	SDA, 6–8 h, 20°C	500, 40% inh.	Dikshit et al. (1986)
Trichophyton rubrum	Fungi	SDA, 7 days, 30°C	400, 43% inh.	Dikshit and Husain (1984)
Trichophyton rubrum	Fungi	SDA, 6–8 h, 20°C	500, 43% inh.	Dikshit et al. (1986)
Candida albicans	Yeast	NA, 48 h, 37°C	>2,000	Maruzzella (1963)
Candida albicans	Yeast	MHA, Tween 20, 48 h, 35°C	>20,000	Hammer et al. (1999)
Candida albicans	Yeast	Cited	1,500–2,000	Peana et al. (1999)

TABLE 14.24
Inhibitory Data of Clary Sage Oil Obtained in the Vapor Phase Test

Microorganism	MO Class	Conditions		Activity	Reference
Escherichia coli	Bac–	NA, 48 h, 37°C	500 µL in cover	+++	Maruzzella (1963)
Proteus vulgaris	Bac–	NA, 48 h, 37°C	500 µL in cover	NG	Maruzzella (1963)
Salmonella typhi	Bac–	NA, 24 h, 37°C	sd	+++	Maruzzella and Sicurella (1960)
Bacillus subtilis var. *aterrimus*	Bac+	NA, 48 h, 37°C	500 µL in cover	+	Maruzzella (1963)
Bacillus subtilis var. *aterrimus*	Bac+	NA, 24 h, 37°C	sd	+++	Maruzzella and Sicurella (1960)
Mycobacterium avium	Bac+	NA, 24 h, 37°C	sd	NG	Maruzzella and Sicurella (1960)
Staphylococcus aureus	Bac+	NA, 48 h, 37°C	500 µL in cover	NG	Maruzzella (1963)
Staphylococcus aureus	Bac+	NA, 24 h, 37°C	sd	+++	Maruzzella and Sicurella (1960)
Streptococcus faecalis	Bac+	NA, 24 h, 37°C	sd	+++	Maruzzella and Sicurella (1960)
Botrytis cinerea	Fungi	PDA, 3 days, 25°C	1,000	+++	Lee et al. (2007)
Colletotrichum gloeosporioides	Fungi	PDA, 3 days, 25°C	1,000	+++	Lee et al. (2007)
Fusarium oxysporum	Fungi	PDA, 3 days, 25°C	1,000	+++	Lee et al. (2007)
Geotrichum candidum	Fungi	NA, 5 days, 22°C	500 µL in cover	+++	Maruzzella (1963)
Gibberella fujikuroi	Fungi	NA, 5 days, 22°C	500 µL in cover	+++	Maruzzella (1963)
Helminthosporium truicicum	Fungi	NA, 5 days, 22°C	500 µL in cover	+++	Maruzzella (1963)
Phoma betae	Fungi	NA, 5 days, 22°C	500 µL in cover	+++	Maruzzella (1963)
Pityrosporum ovale	Fungi	NA, 5 days, 22°C	500 µL in cover	+++	Maruzzella (1963)
Pythium ultimum	Fungi	PDA, 3 days, 25°C	1,000	+++	Lee et al. (2007)
Rhizoctonia solani	Fungi	PDA, 3 days, 25°C	1,000	+++	Lee et al. (2007)
Candida albicans	Yeast	NA, 48 h, 37°C	500 µL in cover	+++	Maruzzella (1963)

Annotation: Clary sage absolute tested (Maruzzella, 1963).

TABLE 14.25
Inhibitory Data of Clove Oil Obtained in the Agar Diffusion Test

Microorganism	MO Class	Conditions		Inhibition Zone (mm)	Reference
Acinetobacter calcoaceticus	Bac–	ISA, 48 h, 25°C	4 (h), 10,000	16	Deans and Ritchie (1987)
Aerobacter aerogenes	Bac–	NA, 24 h, 37°C	—, sd	0	Maruzzella and Lichtenstein (1956)
Aeromonas hydrophila	Bac–	ISA, 48 h, 25°C	4 (h), 10,000	16.5	Deans and Ritchie (1987)
Alcaligenes faecalis	Bac–	ISA, 48 h, 25°C	4 (h), 10,000	1	Deans and Ritchie (1987)
Beneckea natriegens	Bac–	ISA, 48 h, 25°C	4 (h), 10,000	13	Deans and Ritchie (1987)
Campylobacter jejuni	Bac–	TSA, 24 h, 42°C	4 (h), 25,000	9	Smith-Palmer et al. (1998)
Campylobacter jejuni	Bac–	Cited	6, 15,000	22.5	Wannissorn et al. (2005)
Citrobacter freundii	Bac–	ISA, 48 h, 25°C	4 (h), 10,000	9	Deans and Ritchie (1987)
Citrobacter sp.	Bac–	NA, 24 h, 37°C	11, sd	0	Prasad et al. (1986)
Enterobacter aerogenes	Bac–	ISA, 48 h, 25°C	4 (h), 10,000	8.5	Deans and Ritchie (1987)
Enterobacter sp.	Bac–	NA, 24 h, 37°C	11, sd	16	Prasad et al. (1986)
Erwinia carotovora	Bac–	ISA, 48 h, 25°C	4 (h), 10,000	14.5	Deans and Ritchie (1987)
Escherichia coli	Bac–	NA, 24 h, 37°C	—, sd	0	Maruzzella and Lichtenstein (1956)
Escherichia coli	Bac–	NA, 24 h, 37°C	11, sd	0	Prasad et al. (1986)
Escherichia coli	Bac–	Cited	15, 2,500	6	Pizsolitto et al. (1975)
Escherichia coli	Bac–	ISA, 48 h, 25°C	4 (h), 10,000	8.5	Deans and Ritchie (1987)
Escherichia coli	Bac–	NA, 18 h, 37°C	6 (h), pure	16	Yousef and Tawil (1980)
Escherichia coli	Bac–	TSA, 24 h, 35°C	4 (h), 25,000	9.7	Smith-Palmer et al. (1998)
Escherichia coli	Bac–	TGA, 18–24 h, 37°C	9.5, 2,000	19	Morris et al. (1979)
Escherichia coli	Bac–	Cited, 18 h, 37°C	6, 2,500	13.7	Janssen et al. (1986)
Escherichia coli	Bac–	Cited	6, 15,000	16.5	Wannissorn et al. (2005)
Escherichia coli	Bac–	NA, 24 h, 30°C	Drop, 5,000	38	Hili et al. (1997)
Flavobacterium suaveolens	Bac–	ISA, 48 h, 25°C	4 (h), 10,000	15	Deans and Ritchie (1987)
Klebsiella pneumoniae	Bac–	ISA, 48 h, 25°C	4 (h), 10,000	7	Deans and Ritchie (1987)
Klebsiella sp.	Bac–	Cited	15, 2,500	3	Pizsolitto et al. (1975)
Moraxella sp.	Bac–	ISA, 48 h, 25°C	4 (h), 10,000	0	Deans and Ritchie (1987)
Neisseria perflava	Bac–	NA, 24 h, 37°C	—, sd	0	Maruzzella and Lichtenstein (1956)
Proteus sp.	Bac–	Cited	15, 2,500	4	Pizsolitto et al. (1975)
Proteus vulgaris	Bac–	NA, 24 h, 37°C	—, sd	2	Maruzzella and Lichtenstein (1956)
Proteus vulgaris	Bac–	ISA, 48 h, 25°C	4 (h), 10,000	9.5	Deans and Ritchie (1987)
Pseudomonas aeruginosa	Bac–	NA, 18 h, 37°C	6 (h), pure	0	Yousef and Tawil (1980)
Pseudomonas aeruginosa	Bac–	NA, 24 h, 37°C	—, sd	0	Maruzzella and Lichtenstein (1956)
Pseudomonas aeruginosa	Bac–	Cited	15, 2,500	1	Pizsolitto et al. (1975)
Pseudomonas aeruginosa	Bac–	Cited, 18 h, 37°C	6, 2,500	9	Janssen et al. (1986)
Pseudomonas aeruginosa	Bac–	ISA, 48 h, 25°C	4 (h), 10,000	11	Deans and Ritchie (1987)

(*Continued*)

TABLE 14.25 (*Continued*)
Inhibitory Data of Clove Oil Obtained in the Agar Diffusion Test

Microorganism	MO Class	Conditions		Inhibition Zone (mm)	Reference
Pseudomonas aeruginosa	Bac−	NA, 24 h, 30°C	Drop, 5,000	23	Hili et al. (1997)
Pseudomonas sp.	Bac−	NA, 24 h, 37°C	11, sd	0	Prasad et al. (1986)
Salmonella enteritidis	Bac−	TSA, 24 h, 35°C	4 (h), 25,000	11.1	Smith-Palmer et al. (1998)
Salmonella enteritidis	Bac−	Cited	6, 15,000	14.3	Wannissorn et al. (2005)
Salmonella pullorum	Bac−	ISA, 48 h, 25°C	4 (h), 10,000	16	Deans and Ritchie (1987)
Salmonella saintpaul	Bac−	NA, 24 h, 37°C	11, sd	22	Prasad et al. (1986)
Salmonella sp.	Bac−	Cited	15, 2,500	7	Pizsolitto et al. (1975)
Salmonella sp. B	Bac−	NA, 24 h, 37°C	11, sd	26	Prasad et al. (1986)
Salmonella typhimurium	Bac−	Cited	6, 15,000	19.5	Wannissorn et al. (2005)
Salmonella weltevreden	Bac−	NA, 24 h, 37°C	11, sd	22	Prasad et al. (1986)
Serratia marcescens	Bac−	NA, 24 h, 37°C	—, sd	0	Maruzzella and Lichtenstein (1956)
Serratia marcescens	Bac−	ISA, 48 h, 25°C	4 (h), 10,000	25	Deans and Ritchie (1987)
Serratia sp.	Bac−	Cited	15, 2,500	2	Pizsolitto et al. (1975)
Shigella sp.	Bac−	Cited	15, 2,500	6	Pizsolitto et al. (1975)
Yersinia enterocolitica	Bac−	ISA, 48 h, 25°C	4 (h), 10,000	7.5	Deans and Ritchie (1987)
Bacillus anthracis	Bac+	NA, 24 h, 37°C	11, sd	25	Prasad et al. (1986)
Bacillus mesentericus	Bac+	NA, 24 h, 37°C	—, sd	4	Maruzzella and Lichtenstein (1956)
Bacillus saccharolyticus	Bac+	NA, 24 h, 37°C	11, sd	20	Prasad et al. (1986)
Bacillus sp.	Bac+	Cited	15, 2,500	9	Pizsolitto et al. (1975)
Bacillus stearothermophilus	Bac+	NA, 24 h, 37°C	11, sd	20	Prasad et al. (1986)
Bacillus subtilis	Bac+	NA, 24 h, 37°C	—, sd	7	Maruzzella and Lichtenstein (1956)
Bacillus subtilis	Bac+	Cited, 18 h, 37°C	6, 2,500	16	Janssen et al. (1986)
Bacillus subtilis	Bac+	ISA, 48 h, 25°C	4 (h), 10,000	12.5	Deans and Ritchie (1987)
Bacillus subtilis	Bac+	NA, 24 h, 37°C	11, sd	22	Prasad et al. (1986)
Bacillus subtilis	Bac+	NA, 18 h, 37°C	6 (h), pure	22.5	Yousef and Tawil (1980)
Bacillus thuringiensis	Bac+	NA, 24 h, 37°C	11, sd	19	Prasad et al. (1986)
Brevibacterium linens	Bac+	ISA, 48 h, 25°C	4 (h), 10,000	14	Deans and Ritchie (1987)
Brochothrix thermosphacta	Bac+	ISA, 48 h, 25°C	4 (h), 10,000	5.5	Deans and Ritchie (1987)
Clostridium perfringens	Bac+	Cited	6, 15,000	20.5	Wannissorn et al. (2005)
Clostridium sporogenes	Bac+	ISA, 48 h, 25°C	4 (h), 10,000	0	Deans and Ritchie (1987)
Corynebacterium sp.	Bac+	TGA, 18–24 h, 37°C	9.5, 2,000	15	Morris et al. (1979)
Lactobacillus casei	Bac+	NA, 24 h, 37°C	11, sd	22	Prasad et al. (1986)
Lactobacillus plantarum	Bac+	ISA, 48 h, 25°C	4 (h), 10,000	11	Deans and Ritchie (1987)
Lactobacillus plantarum	Bac+	NA, 24 h, 37°C	11, sd	50	Prasad et al. (1986)
Leuconostoc cremoris	Bac+	ISA, 48 h, 25°C	4 (h), 10,000	13	Deans and Ritchie (1987)
Listeria monocytogenes	Bac+	TSA, 24 h, 35°C	4 (h), 25,000	8.4	Smith-Palmer et al. (1998)
Listeria monocytogenes	Bac+	ISA, 48 h, 25°C	4 (h), 10,000	20	Lis-Balchin et al. (1998)
Micrococcus glutamicus	Bac+	NA, 24 h, 37°C	11, sd	44	Prasad et al. (1986)
Micrococcus luteus	Bac+	ISA, 48 h, 25°C	4 (h), 10,000	9	Deans and Ritchie (1987)
Mycobacterium phlei	Bac+	NA, 18 h, 37°C	6 (h), pure	20	Yousef and Tawil (1980)

(*Continued*)

TABLE 14.25 (*Continued*)
Inhibitory Data of Clove Oil Obtained in the Agar Diffusion Test

Microorganism	MO Class	Conditions		Inhibition Zone (mm)	Reference
Sarcina lutea	Bac+	NA, 24 h, 37°C	—, sd	4	Maruzzella and Lichtenstein (1956)
Sarcina lutea	Bac+	NA, 24 h, 37°C	11, sd	32	Prasad et al. (1986)
Staphylococcus aureus	Bac+	NA, 24 h, 37°C	—, sd	0	Maruzzella and Lichtenstein (1956)
Staphylococcus aureus	Bac+	Cited	15, 2,500	5	Pizsolitto et al. (1975)
Staphylococcus aureus	Bac+	ISA, 48 h, 25°C	4 (h), 10,000	7	Deans and Ritchie (1987)
Staphylococcus aureus	Bac+	TSA, 24 h, 35°C	4 (h), 25,000	8	Smith-Palmer et al. (1998)
Staphylococcus aureus	Bac+	Cited, 18 h, 37°C	6, 2,500	11	Janssen et al. (1986)
Staphylococcus aureus	Bac+	TGA, 18–24 h, 37°C	9.5, 2,000	14	Morris et al. (1979)
Staphylococcus aureus	Bac+	NA, 18 h, 37°C	6 (h), pure	14.3	Yousef and Tawil (1980)
Staphylococcus aureus	Bac+	NA, 24 h, 30°C	Drop, 5,000	21	Hili et al. (1997)
Staphylococcus aureus	Bac+	NA, 24 h, 37°C	11, sd	30	Prasad et al. (1986)
Staphylococcus epidermidis	Bac+	Cited	15, 2,500	10	Pizsolitto et al. (1975)
Staphylococcus sp.	Bac+	NA, 24 h, 37°C	11, sd	26	Prasad et al. (1986)
Streptococcus faecalis	Bac+	ISA, 48 h, 25°C	4 (h), 10,000	8.5	Deans and Ritchie (1987)
Streptococcus viridans	Bac+	Cited	15, 2,500	1	Pizsolitto et al. (1975)
Streptomyces venezuelae	Bac+	SMA, 2–7 days, 20°C	sd	9	Maruzzella and Liguori (1958)
Alternaria porri	Fungi	PDA, 72 h, 28°C	5, 5,000	40.5	Pawar and Thaker (2007)
Alternaria solani	Fungi	SMA, 2–7 days, 20°C	sd	9	Maruzzella and Liguori (1958)
Aspergillus fumigatus	Fungi	SMA, 2–7 days, 20°C	sd	10	Maruzzella and Liguori (1958)
Aspergillus niger	Fungi	SMA, 2–7 days, 20°C	sd	10	Maruzzella and Liguori (1958)
Aspergillus niger	Fungi	PDA, 48 h, 28°C	5, 5,000	28	Pawar and Thaker (2006)
Aspergillus niger	Fungi	SDA, 8 days, 30°C	6 (h), pure	34	Yousef and Tawil (1980)
Fusarium oxysporum f.sp. *ciceri*	Fungi	PDA, 72 h, 28°C	5, 5,000	12	Pawar and Thaker (2007)
Helminthosporium sativum	Fungi	SMA, 2–7 days, 20°C	sd	7	Maruzzella and Liguori (1958)
Mucor mucedo	Fungi	SMA, 2–7 days, 20°C	sd	7	Maruzzella and Liguori (1958)
Mucor sp.	Fungi	SDA, 8 days, 30°C	6 (h), pure	20	Yousef and Tawil (1980)
Nigrospora panici	Fungi	SMA, 2–7 days, 20°C	sd	8	Maruzzella and Liguori (1958)
Penicillium chrysogenum	Fungi	SDA, 8 days, 30°C	6 (h), pure	47	Yousef and Tawil (1980)
Penicillium digitatum	Fungi	SMA, 2–7 days, 20°C	sd	9	Maruzzella and Liguori (1958)
Rhizopus nigricans	Fungi	SMA, 2–7 days, 20°C	sd	7	Maruzzella and Liguori (1958)
Rhizopus sp.	Fungi	SDA, 8 days, 30°C	6 (h), pure	20	Yousef and Tawil (1980)

(*Continued*)

TABLE 14.25 (Continued)
Inhibitory Data of Clove Oil Obtained in the Agar Diffusion Test

Microorganism	MO Class	Conditions		Inhibition Zone (mm)	Reference
Brettanomyces anomalus	Yeast	MPA, 4 days, 30°C	5, 10% sol. sd	18	Conner and Beuchat (1984)
Candida albicans	Yeast	SMA, 2–7 days, 20°C	sd	3	Maruzzella and Liguori (1958)
Candida albicans	Yeast	TGA, 18–24 h, 37°C	9.5, 2,000	19	Morris et al. (1979)
Candida albicans	Yeast	SDA, 18 h, 30°C	6 (h), pure	20	Yousef and Tawil (1980)
Candida albicans	Yeast	Cited, 18 h, 37°C	6, 2,500	28.3	Janssen et al. (1986)
Candida albicans	Yeast	NA, 24 h, 30°C	Drop, 5,000	40	Hili et al. (1997)
Candida krusei	Yeast	SMA, 2–7 days, 20°C	sd	5	Maruzzella and Liguori (1958)
Candida lipolytica	Yeast	MPA, 4 days, 30°C	5, 10% sol. sd	27	Conner and Beuchat (1984)
Candida tropicalis	Yeast	SMA, 2–7 days, 20°C	sd	3	Maruzzella and Liguori (1958)
Cryptococcus neoformans	Yeast	SMA, 2–7 days, 20°C	sd	8	Maruzzella and Liguori (1958)
Cryptococcus rhodobenhani	Yeast	SMA, 2–7 days, 20°C	sd	13	Maruzzella and Liguori (1958)
Debaryomyces hansenii	Yeast	MPA, 4 days, 30°C	5, 10% sol. sd	26	Conner and Beuchat (1984)
Geotrichum candidum	Yeast	MPA, 4 days, 30°C	5, 10% sol. sd	19	Conner and Beuchat (1984)
Hansenula anomala	Yeast	MPA, 4 days, 30°C	5, 10% sol. sd	19	Conner and Beuchat (1984)
Kloeckera apiculata	Yeast	MPA, 4 days, 30°C	5, 10% sol. sd	21	Conner and Beuchat (1984)
Kluyveromyces fragilis	Yeast	MPA, 4 days, 30°C	5, 10% sol. sd	14	Conner and Beuchat (1984)
Lodderomyces elongisporus	Yeast	MPA, 4 days, 30°C	5, 10% sol. sd	15	Conner and Beuchat (1984)
Metschnikowia pulcherrima	Yeast	MPA, 4 days, 30°C	5, 10% sol. sd	29	Conner and Beuchat (1984)
Pichia membranifaciens	Yeast	MPA, 4 days, 30°C	5, 10% sol. sd	13	Conner and Beuchat (1984)
Rhodotorula rubra	Yeast	MPA, 4 days, 30°C	5, 10% sol. sd	18	Conner and Beuchat (1984)
Saccharomyces cerevisiae	Yeast	SMA, 2–7 days, 20°C	sd	5	Maruzzella and Liguori (1958)
Saccharomyces cerevisiae	Yeast	MPA, 4 days, 30°C	5, 10% sol. sd	19	Conner and Beuchat (1984)
Saccharomyces cerevisiae	Yeast	NA, 24 h, 30°C	Drop, 5,000	50	Hili et al. (1997)
Schizosaccharomyces pombe	Yeast	NA, 24 h, 30°C	Drop, 5,000	34	Hili et al. (1997)
Torula glabrata	Yeast	MPA, 4 days, 30°C	5, 10% sol. sd	15	Conner and Beuchat (1984)
Torula utilis	Yeast	NA, 24 h, 30°C	Drop, 5,000	39	Hili et al. (1997)

TABLE 14.26
Inhibitory Data of Clove Oil Obtained in the Dilution Test

Microorganism	MO Class	Conditions	MIC	Reference
Acinetobacter baumannii	Bac−	MHA, Tween 20, 48 h, 35°C	2,500	Hammer et al. (1999)
Brucella abortus	Bac−	Cited	250	Okazaki and Oshima (1952a)
Brucella melitensis	Bac−	Cited	500	Okazaki and Oshima (1952a)
Brucella suis	Bac−	Cited	15	Okazaki and Oshima (1952a)
Campylobacter jejuni	Bac−	TSB, 24 h, 42°C	500	Smith-Palmer et al. (1998)
Escherichia coli	Bac−	LA, 18 h, 37°C	500–1,000	Remmal et al. (1993)
Escherichia coli	Bac−	TSB, 24 h, 35°C	400	Smith-Palmer et al. (1998)
Escherichia coli	Bac−	Cited	500	Okazaki and Oshima (1952)
Escherichia coli	Bac−	MPB, DMSO, 40 h, 30°C	74% inh. 500	Hili et al. (1997)
Escherichia coli	Bac−	TGB, 18–24 h, 37°C	1,000	Morris et al. (1979)
Escherichia coli	Bac−	NA, 1–3 days, 30°C	1,250	Farag et al. (1989)
Escherichia coli	Bac−	MHA, Tween 20, 48 h, 35°C	2,500	Hammer et al. (1999)
Escherichia coli	Bac−	NB, Tween 20, 18 h, 37°C	50,000	Yousef and Tawil (1980)
Escherichia coli O157:H7	Bac−	BHI, 48 h, 35°C	1,000	Oussalah et al. (2006)
Klebsiella pneumoniae	Bac−	Cited	>500	Okazaki and Oshima (1952a)
Klebsiella pneumoniae	Bac−	MHA, Tween 20, 48 h, 35°C	2,500	Hammer et al. (1999)
Proteus vulgaris	Bac−	Cited	250	Okazaki and Oshima (1952a)
Pseudomonas aeruginosa	Bac−	Cited	>500	Okazaki and Oshima (1952a)
Pseudomonas aeruginosa	Bac−	MPB, DMSO, 40 h, 30°C	75% inh. 500	Hili et al. (1997)
Pseudomonas aeruginosa	Bac−	MHA, Tween 20, 48 h, 35°C	>20,000	Hammer et al. (1999)
Pseudomonas aeruginosa	Bac−	NB, Tween 20, 18 h, 37°C	50,000	Yousef and Tawil (1980)
Pseudomonas fluorescens	Bac−	NA, 1–3 days, 30°C	1,500	Farag et al. (1989)
Salmonella enteritidis	Bac−	TSB, 24 h, 35°C	400	Smith-Palmer et al. (1998)
Salmonella hadar	Bac−	LA, 18 h, 37°C	500	Remmal et al. (1993)
Salmonella paratyphi A	Bac−	Cited	500	Okazaki and Oshima (1952a)
Salmonella paratyphi B	Bac−	Cited	>500	Okazaki and Oshima (1952a)
Salmonella typhimurium	Bac−	BHI, 48 h, 35°C	1,000	Oussalah et al. (2006)
Salmonella typhimurium	Bac−	MHA, Tween 20, 48 h, 35°C	>20,000	Hammer et al. (1999)
Serratia marcescens	Bac−	NA, 1–3 days, 30°C	1,500	Farag et al. (1989)
Serratia marcescens	Bac−	MHA, Tween 20, 48 h, 35°C	2,500	Hammer et al. (1999)
Shigella dysenteriae I	Bac−	Cited	500	Okazaki and Oshima (1952a)
Shigella dysenteriae II	Bac−	Cited	500	Okazaki and Oshima (1952a)
Bacillus megaterium	Bac+	LA, 18 h, 37°C	500–1,000	Remmal et al. (1993)
Bacillus subtilis	Bac+	NA, 1–3 days, 30°C	500	Farag et al. (1989)
Bacillus subtilis	Bac+	NB, Tween 20, 18 h, 37°C	50,000	Yousef and Tawil (1980)
Corynebacterium sp.	Bac+	TGB, 18–24 h, 37°C	500	Morris et al. (1979)
Enterococcus faecalis	Bac+	MHA, Tween 20, 48 h, 35°C	5,000	Hammer et al. (1999)
Listeria monocytogenes	Bac+	TSB, 10 days, 4°C	200	Smith-Palmer et al. (1998)
Listeria monocytogenes	Bac+	TSB, 24 h, 35°C	300	Smith-Palmer et al. (1998)
Listeria monocytogenes	Bac+	BHI, 48 h, 35°C	2,000	Oussalah et al. (2006)
Micrococcus sp.	Bac+	NA, 1–3 days, 30°C	250	Farag et al. (1989)
Mycobacterium phlei	Bac+	NA, 1–3 days, 30°C	500	Farag et al. (1989)
Mycobacterium phlei	Bac+	NB, Tween 20, 18 h, 37°C	1,600	Yousef and Tawil (1980)
Mycobacterium tuberculosis	Bac+	Cited	125	Okazaki and Oshima (1952a)
Sarcina sp.	Bac+	NA, 1–3 days, 30°C	500	Farag et al. (1989)

(*Continued*)

TABLE 14.26 (Continued)
Inhibitory Data of Clove Oil Obtained in the Dilution Test

Microorganism	MO Class	Conditions	MIC	Reference
Staphylococcus aureus	Bac+	TSB, 24 h, 35°C	400	Smith-Palmer et al. (1998)
Staphylococcus aureus	Bac+	Cited	500	Okazaki and Oshima (1952a)
Staphylococcus aureus	Bac+	TGB, 18–24 h, 37°C	500	Morris et al. (1979)
Staphylococcus aureus	Bac+	BHI, 48 h, 35°C	500	Oussalah et al. (2006)
Staphylococcus aureus	Bac+	MPB, DMSO, 40 h, 30°C	83% inh. 500	Hili et al. (1997)
Staphylococcus aureus	Bac+	NA, 1–3 days, 30°C	750	Farag et al. (1989)
Staphylococcus aureus	Bac+	LA, 18 h, 37°C	1,000	Remmal et al. (1993)
Staphylococcus aureus	Bac+	MHA, Tween 20, 48 h, 35°C	2,500	Hammer et al. (1999)
Staphylococcus aureus	Bac+	NB, Tween 20, 18 h, 37°C	6,400	Yousef and Tawil (1980)
Staphylococcus citreus	Bac+	Cited	500	Okazaki and Oshima (1952a)
Achorion gypseum	Fungi	Cited, 15 d	125	Okazaki and Oshima (1952b)
Alternaria alternata	Fungi	RPMI, 1.5% EtOH, 7 days, 30°C	156–312	Tullio et al. (2006)
Alternaria citri	Fungi	PDA, 8 days, 22°C	500	Arras and Usai (2001)
Aspergillus flavus	Fungi	PDA, 8 h, 20°C, spore germ. inh.	50–100	Thompson (1986)
Aspergillus flavus	Fungi	PDA, 5 days, 27°C	100	Thompson and Cannon (1986)
Aspergillus flavus	Fungi	RPMI, 1.5% EtOH, 7 days, 30°C	2,500	Tullio et al. (2006)
Aspergillus flavus var. *columnaris*	Fungi	RPMI, 1.5% EtOH, 7 days, 30°C	1,250	Tullio et al. (2006)
Aspergillus fumigatus	Fungi	RPMI, 1.5% EtOH, 7 days, 30°C	5,000	Tullio et al. (2006)
Aspergillus niger	Fungi	RPMI, 1.5% EtOH, 7 days, 30°C	2,500	Tullio et al. (2006)
Aspergillus niger	Fungi	NB, Tween 20, 8 days, 30°C	3,200	Yousef and Tawil (1980)
Aspergillus niger	Fungi	YES broth, 10 d	95% inh. 10,000	Lis-Balchin et al. (1998)
Aspergillus ochraceus	Fungi	YES broth, 10 d	94% inh. 10,000	Lis-Balchin et al. (1998)
Aspergillus oryzae	Fungi	Cited	250	Okazaki and Oshima (1953b)
Aspergillus parasiticus	Fungi	PDA, 8 h, 20°C, spore germ. inh.	50–100	Thompson (1986)
Aspergillus parasiticus	Fungi	PDA, 5 days, 27°C	100	Thompson and Cannon (1986)
Botrytis cinerea	Fungi	PDA, 8 days, 22°C	500	Arras and Usai (2001)
Cladosporium cladosporioides	Fungi	RPMI, 1.5% EtOH, 7 days, 30°C	78–156	Tullio et al. (2006)
Colletotrichum musae	Fungi	SMKY, EtOH, 7 days, 28°C	400	Ranasinghe et al. (2002)
Epidermophyton floccosum	Fungi	RPMI, 1.5% EtOH, 7 days, 30°C	125	Tullio et al. (2006)
Epidermophyton floccosum	Fungi	SA, Tween 80, 21 days, 20°C	<300	Janssen et al. (1988)
Epidermophyton inguinale	Fungi	Cited, 15 days	125	Okazaki and Oshima (1952b)

(Continued)

TABLE 14.26 (*Continued*)
Inhibitory Data of Clove Oil Obtained in the Dilution Test

Microorganism	MO Class	Conditions	MIC	Reference
Fusarium culmorum	Fungi	YES broth, 10 days	73% inh. 10,000	Lis-Balchin et al. (1998)
Fusarium oxysporum	Fungi	RPMI, 1.5% EtOH, 7 days, 30°C	625	Tullio et al. (2006)
Fusarium proliferatum	Fungi	SMKY, EtOH, 7 days, 28°C	500	Ranasinghe et al. (2002)
Lasiodiplodia theobromae	Fungi	SMKY, EtOH, 7 days, 28°C	450	Ranasinghe et al. (2002)
Microsporum canis	Fungi	RPMI, 1.5% EtOH, 7 days, 30°C	125–500	Tullio et al. (2006)
Microsporum gypseum	Fungi	RPMI, 1.5% EtOH, 7 days, 30°C	125–250	Tullio et al. (2006)
Mucor hiemalis	Fungi	PDA, 5 days, 27°C	100	Thompson and Cannon (1986)
Mucor mucedo	Fungi	PDA, 5 days, 27°C	100	Thompson and Cannon (1986)
Mucor racemosus	Fungi	Cited	2	Okazaki and Oshima (1953b)
Mucor racemosus f. *racemosus*	Fungi	PDA, 5 days, 27°C	100	Thompson and Cannon (1986)
Mucor sp.	Fungi	NB, Tween 20, 8 days, 30°C	1,600	Yousef and Tawil (1980)
Mucor sp.	Fungi	RPMI, 1.5% EtOH, 7 days, 30°C	>10,000	Tullio et al. (2006)
Penicillium chrysogenum	Fungi	Cited	250	Okazaki and Oshima (1953b)
Penicillium chrysogenum	Fungi	NB, Tween 20, 8 days, 30°C	1,600	Yousef and Tawil (1980)
Penicillium digitatum	Fungi	PDA, 8 days, 22°C	500	Arras and Usai (2001)
Penicillium frequentans	Fungi	RPMI, 1.5% EtOH, 7 days, 30°C	1,250	Tullio et al. (2006)
Penicillium italicum	Fungi	PDA, 8 days, 22°C	500	Arras and Usai (2001)
Penicillium lanosum	Fungi	RPMI, 1.5% EtOH, 7 days, 30°C	5,000	Tullio et al. (2006)
Rhizopus 66-81-2	Fungi	PDA, 5 days, 27°C	100	Thompson and Cannon (1986)
Rhizopus arrhizus	Fungi	PDA, 5 days, 27°C	100	Thompson and Cannon (1986)
Rhizopus chinensis	Fungi	PDA, 5 days, 27°C	100	Thompson and Cannon (1986)
Rhizopus circinans	Fungi	PDA, 5 days, 27°C	100	Thompson and Cannon (1986)
Rhizopus japonicus	Fungi	PDA, 5 days, 27°C	100	Thompson and Cannon (1986)
Rhizopus kazanensis	Fungi	PDA, 5 days, 27°C	100	Thompson and Cannon (1986)
Rhizopus oryzae	Fungi	PDA, 5 days, 27°C	100	Thompson and Cannon (1986)
Rhizopus pymacus	Fungi	PDA, 5 days, 27°C	100	Thompson and Cannon (1986)
Rhizopus sp.	Fungi	RPMI, 1.5% EtOH, 7 days, 30°C	>10,000	Tullio et al. (2006)

(*Continued*)

TABLE 14.26 (*Continued*)
Inhibitory Data of Clove Oil Obtained in the Dilution Test

Microorganism	MO Class	Conditions	MIC	Reference
Rhizopus sp.	Fungi	NB, Tween 20, 8 days, 30°C	50,000	Yousef and Tawil (1980)
Rhizopus stolonifer	Fungi	PDA, 5 days, 27°C	100	Thompson and Cannon (1986)
Rhizopus tritici	Fungi	PDA, 5 days, 27°C	100	Thompson and Cannon (1986)
Scopulariopsis brevicaulis	Fungi	RPMI, 1.5% EtOH, 7 days, 30°C	10,000	Tullio et al. (2006)
Trichophyton asteroides	Fungi	Cited, 15 days	125	Okazaki and Oshima (1952b)
Trichophyton interdigitale	Fungi	Cited, 15 days	125	Okazaki and Oshima (1952b)
Trichophyton mentagrophytes	Fungi	RPMI, 1.5% EtOH, 7 days, 30°C	250–500	Tullio et al. (2006)
Trichophyton mentagrophytes	Fungi	SA, Tween 80, 21 days, 20°C	<300	Janssen et al. (1988)
Trichophyton purpureum	Fungi	Cited, 15 days	125	Okazaki and Oshima (1952b)
Trichophyton rubrum	Fungi	SA, Tween 80, 21 days, 20°C	<300	Janssen et al. (1988)
Trichophyton schoenleinii	Fungi	Cited, 15 days	125	Okazaki and Oshima (1952b)
Candida albicans	Yeast	TGB, 18–24 h, 37°C	500	Morris et al. (1979)
Candida albicans	Yeast	MPB, DMSO, 40 h, 30°C	61% inh. 500	Hili et al. (1997)
Candida albicans	Yeast	MHA, Tween 20, 48 h, 35°C	1,200	Hammer et al. (1999)
Candida albicans	Yeast	MHB, Tween 80, 48 h, 35	1,200	Hammer et al. (1998)
Candida albicans	Yeast	NB, Tween 20, 18 h, 37°C	50,000	Yousef and Tawil (1980)
Saccharomyces cerevisiae	Yeast	MPB, DMSO, 40 h, 30°C	99% inh. 500	Hili et al. (1997)
Saccharomyces cerevisiae	Yeast	NA, 1–3 days, 30°C	750	Farag et al. (1989)
Schizosaccharomyces pombe	Yeast	MPB, DMSO, 40 h, 30°C	94% inh. 500	Hili et al. (1997)
Torula utilis	Yeast	MPB, DMSO, 40 h, 30°C	95% inh. 500	Hili et al. (1997)

TABLE 14.27
Inhibitory Data of Clove Oil Obtained in the Vapor Phase Test

Microorganism	MO Class	Conditions		Activity	Reference
Escherichia coli	Bac−	NA, 24 h, 37°C	~20,000	NG	Kellner and Kober (1954)
Neisseria sp.	Bac−	NA, 24 h, 37°C	~20,000	NG	Kellner and Kober (1954)
Salmonella typhi	Bac−	NA, 24 h, 37°C	~20,000	NG	Kellner and Kober (1954)
Salmonella typhi	Bac−	NA, 24 h, 37°C	sd	++	Maruzzella and Sicurella (1960)
Bacillus megaterium	Bac+	NA, 24 h, 37°C	~20,000	NG	Kellner and Kober (1954)
Bacillus subtilis var. aterrimus	Bac+	NA, 24 h, 37°C	sd	+++	Maruzzella and Sicurella (1960)
Corynebacterium diphtheriae	Bac+	NA, 24 h, 37°C	~20,000	NG	Kellner and Kober (1954)
Mycobacterium avium	Bac+	NA, 24 h, 37°C	sd	+	Maruzzella and Sicurella (1960)
Staphylococcus aureus	Bac+	NA, 24 h, 37°C	~20,000	NG	Kellner and Kober (1954)
Staphylococcus aureus	Bac+	NA, 24 h, 37°C	sd	+++	Maruzzella and Sicurella (1960)
Streptococcus faecalis	Bac+	NA, 24 h, 37°C	~20,000	NG	Kellner and Kober (1954)
Streptococcus faecalis	Bac+	NA, 24 h, 37°C	sd	+++	Maruzzella and Sicurella (1960)
Streptococcus pyogenes	Bac+	NA, 24 h, 37°C	~20,000	NG	Kellner and Kober (1954)
Alternaria alternata	Fungi	RPMI, 7 days, 30°C	MIC_{air}	312–625	Tullio et al. (2006)
Aspergillus flavus	Fungi	WFA, 42 days, 25°C	Disk, 50,000	NG	Guynot et al. (2003)
Aspergillus flavus	Fungi	RPMI, 7 days, 30°C	MIC_{air}	625–1250	Tullio et al. (2006)
Aspergillus fumigatus	Fungi	RPMI, 7 days, 30°C	MIC_{air}	625–1250	Tullio et al. (2006)
Aspergillus niger	Fungi	WFA, 42 days, 25°C	Disk, 50,000	NG	Guynot et al. (2003)
Aspergillus niger	Fungi	RPMI, 7 days, 30°C	MIC_{air}	625–1250	Tullio et al. (2006)
Cladosporium cladosporioides	Fungi	RPMI, 7 days, 30°C	MIC_{air}	78–156	Tullio et al. (2006)
Eurotium amstelodami	Fungi	WFA, 42 days, 25°C	Disk, 50,000	NG	Guynot et al. (2003)
Eurotium herbarum	Fungi	WFA, 42 days, 25°C	Disk, 50,000	NG	Guynot et al. (2003)
Eurotium repens	Fungi	WFA, 42 days, 25°C	Disk, 50,000	NG	Guynot et al. (2003)
Eurotium rubrum	Fungi	WFA, 42 days, 25°C	Disk, 50,000	NG	Guynot et al. (2003)
Fusarium oxysporum	Fungi	RPMI, 7 days, 30°C	MIC_{air}	312	Tullio et al. (2006)
Microsporum canis	Fungi	RPMI, 7 days, 30°C	MIC_{air}	312–1250	Tullio et al. (2006)
Microsporum gypseum	Fungi	RPMI, 7 days, 30°C	MIC_{air}	156–312	Tullio et al. (2006)
Mucor sp.	Fungi	RPMI, 7 days, 30°C	MIC_{air}	625	Tullio et al. (2006)
Penicillium corylophilum	Fungi	WFA, 42 days, 25°C	Disk, 50,000	NG	Guynot et al. (2003)
Penicillium frequentans	Fungi	RPMI, 7 days, 30°C	MIC_{air}	>10,000	Tullio et al. (2006)
Penicillium lanosum	Fungi	RPMI, 7 days, 30°C	MIC_{air}	>10,000	Tullio et al. (2006)
Rhizopus sp.	Fungi	RPMI, 7 days, 30°C	MIC_{air}	125	Tullio et al. (2006)
Scopulariopsis brevicaulis	Fungi	RPMI, 7 days, 30°C	MIC_{air}	312–1250	Tullio et al. (2006)
Trichophyton mentagrophytes	Fungi	RPMI, 7 days, 30°C	MIC_{air}	78–156	Tullio et al. (2006)
Candida albicans	Yeast	NA, 24 h, 37°C	~20,000	NG	Kellner and Kober (1954)

TABLE 14.28
Inhibitory Data of Coriander Oil Obtained in the Agar Diffusion Test

Microorganism	MO Class	Conditions		Inhibition Zone (mm)	Reference
Acinetobacter calcoaceticus	Bac–	ISA, 48 h, 25°C	4 (h), 10,000	0	Deans and Ritchie (1987)
Aerobacter aerogenes	Bac–	NA, 24 h, 37°C	—, sd	0	Maruzzella and Lichtenstein (1956)
Aeromonas hydrophila	Bac–	ISA, 48 h, 25°C	4 (h), 10,000	8.5	Deans and Ritchie (1987)
Agrobacterium tumefaciens	Bac–	WA, 48 h, 25°C	6, 8,000	MIA 435	Cantore et al. (2004)
Alcaligenes faecalis	Bac–	ISA, 48 h, 25°C	4 (h), 10,000	6	Deans and Ritchie (1987)
Beneckea natriegens	Bac–	ISA, 48 h, 25°C	4 (h), 10,000	0	Deans and Ritchie (1987)
Burkholderia gladioli pv. *agaricicola*	Bac–	WA, 48 h, 25°C	6, 8,000	MIA 3480	Cantore et al. (2004)
Citrobacter freundii	Bac–	ISA, 48 h, 25°C	4 (h), 10,000	8.5	Deans and Ritchie (1987)
Enterobacter aerogenes	Bac–	ISA, 48 h, 25°C	4 (h), 10,000	7.5	Deans and Ritchie (1987)
Enterobacter aerogenes	Bac–	MHA, 48 h, 27°C	6, 15,000	12	Ertürk et al. (2006)
Erwinia carotovora	Bac–	ISA, 48 h, 25°C	4 (h), 10,000	8	Deans and Ritchie (1987)
Erwinia carotovora subsp. *atroseptica*	Bac–	WA, 48 h, 25°C	6, 8,000	MIA 435	Cantore et al. (2004)
Erwinia carotovora subsp. *carotovora*	Bac–	WA, 48 h, 25°C	6, 8,000	MIA 435	Cantore et al. (2004)
Escherichia coli	Bac–	NA, 24 h, 37°C	—, sd	0	Maruzzella and Lichtenstein (1956)
Escherichia coli	Bac–	TGA, 18–24 h, 37°C	9.5, 2,000	0	Morris et al. (1979)
Escherichia coli	Bac–	Cited, 18 h, 37°C	6, 2,500	10	Janssen et al. (1986)
Escherichia coli	Bac–	ISA, 48 h, 25°C	4 (h), 10,000	7.6	Deans and Ritchie (1987)
Escherichia coli	Bac–	NA, 24 h, 30°C	Drop, 5,000	13	Hili et al. (1997)
Escherichia coli	Bac–	NA, 18 h, 37°C	6 (h), pure	10.5	Yousef and Tawil (1980)
Escherichia coli	Bac–	MHA, 48 h, 27°C	6, 15,000	20	Ertürk et al. (2006)
Escherichia coli	Bac–	WA, 48 h, 25°C	6, 8,000	MIA 870	Cantore et al. (2004)
Flavobacterium suaveolens	Bac–	ISA, 48 h, 25°C	4 (h), 10,000	13	Deans and Ritchie (1987)
Klebsiella pneumoniae	Bac–	ISA, 48 h, 25°C	4 (h), 10,000	8	Deans and Ritchie (1987)
Moraxella sp.	Bac–	ISA, 48 h, 25°C	4 (h), 10,000	0	Deans and Ritchie (1987)
Neisseria perflava	Bac–	NA, 24 h, 37°C	—, sd	0	Maruzzella and Lichtenstein (1956)
Proteus vulgaris	Bac–	NA, 24 h, 37°C	—, sd	0	Maruzzella and Lichtenstein (1956)
Proteus vulgaris	Bac–	ISA, 48 h, 25°C	4 (h), 10,000	6	Deans and Ritchie (1987)
Pseudomonas aeruginosa	Bac–	NA, 18 h, 37°C	6 (h), pure	0	Yousef and Tawil (1980)
Pseudomonas aeruginosa	Bac–	NA, 24 h, 37°C	—, sd	0	Maruzzella and Lichtenstein (1956)
Pseudomonas aeruginosa	Bac–	ISA, 48 h, 25°C	4 (h), 10,000	0	Deans and Ritchie (1987)
Pseudomonas aeruginosa	Bac–	NA, 24 h, 30°C	Drop, 5,000	6	Hili et al. (1997)

(*Continued*)

TABLE 14.28 (Continued)
Inhibitory Data of Coriander Oil Obtained in the Agar Diffusion Test

Microorganism	MO Class	Conditions		Inhibition Zone (mm)	Reference
Pseudomonas aeruginosa	Bac–	Cited, 18 h, 37°C	6, 2,500	9	Janssen et al. (1986)
Pseudomonas aeruginosa	Bac–	MHA, 48 h, 27°C	6, 15,000	12	Ertürk et al. (2006)
Pseudomonas agarici	Bac–	KBA, 48 h, 25°C	6, 8,000	MIA 3480	Cantore et al. (2004)
Pseudomonas cichorii	Bac–	KBA, 48 h, 25°C	6, 8,000	MIA 6960	Cantore et al. (2004)
Pseudomonas corrugate	Bac–	KBA, 48 h, 25°C	6, 8,000	MIA 3480	Cantore et al. (2004)
Pseudomonas reactants	Bac–	KBA, 48 h, 25°C	6, 8,000	MIA >6960	Cantore et al. (2004)
Pseudomonas syringae pv. aptata	Bac–	KBA, 48 h, 25°C	6, 8,000	MIA 3480	Cantore et al. (2004)
Pseudomonas syringae pv. atrofaciens	Bac–	KBA, 48 h, 25°C	6, 8,000	MIA 6960	Cantore et al. (2004)
Pseudomonas syringae pv. glycinea	Bac–	KBA, 48 h, 25°C	6, 8,000	MIA 870	Cantore et al. (2004)
Pseudomonas syringae pv. lachrymans	Bac–	KBA, 48 h, 25°C	6, 8,000	MIA >6960	Cantore et al. (2004)
Pseudomonas syringae pv. maculicola	Bac–	KBA, 48 h, 25°C	6, 8,000	MIA 870	Cantore et al. (2004)
Pseudomonas syringae pv. phaseolicola	Bac–	KBA, 48 h, 25°C	6, 8,000	MIA 2610	Cantore et al. (2004)
Pseudomonas syringae pv. pisi	Bac–	KBA, 48 h, 25°C	6, 8,000	MIA 2610	Cantore et al. (2004)
Pseudomonas syringae pv. syringae	Bac–	KBA, 48 h, 25°C	6, 8,000	MIA 3480	Cantore et al. (2004)
Pseudomonas syringae pv. tomato	Bac–	KBA, 48 h, 25°C	6, 8,000	MIA 3480	Cantore et al. (2004)
Pseudomonas tolaasii	Bac–	KBA, 48 h, 25°C	6, 8,000	MIA >6960	Cantore et al. (2004)
Pseudomonas viridiflava	Bac–	KBA, 48 h, 25°C	6, 8,000	MIA >6960	Cantore et al. (2004)
Salmonella pullorum	Bac–	ISA, 48 h, 25°C	4 (h), 10,000	11	Deans and Ritchie (1987)
Salmonella typhimurium	Bac–	MHA, 48 h, 27°C	6, 15,000	7	Ertürk et al. (2006)
Serratia marcescens	Bac–	NA, 24 h, 37°C	—, sd	3	Maruzzella and Lichtenstein (1956)
Serratia marcescens	Bac–	ISA, 48 h, 25°C	4 (h), 10,000	7.5	Deans and Ritchie (1987)
Xanthomonas campestris pv. campestris	Bac–	WA, 48 h, 25°C	6, 8,000	MIA 217	Cantore et al. (2004)
Xanthomonas campestris pv. phaseoli var. fuscans	Bac–	WA, 48 h, 25°C	6, 8,000	MIA 217	Cantore et al. (2004)
Xanthomonas campestris pv. phaseoli var. phaseoli	Bac–	WA, 48 h, 25°C	6, 8,000	MIA 217	Cantore et al. (2004)

(Continued)

TABLE 14.28 (*Continued*)
Inhibitory Data of Coriander Oil Obtained in the Agar Diffusion Test

Microorganism	MO Class	Conditions		Inhibition Zone (mm)	Reference
Xanthomonas campestris pv. *vesicatoria*	Bac–	WA, 48 h, 25°C	6, 8,000	MIA 217	Cantore et al. (2004)
Yersinia enterocolitica	Bac–	ISA, 48 h, 25°C	4 (h), 10,000	7	Deans and Ritchie (1987)
Bacillus megaterium	Bac+	WA, 48 h, 25°C	6, 8,000	MIA 435	Cantore et al. (2004)
Bacillus mesentericus	Bac+	NA, 24 h, 37°C	—, sd	4	Maruzzella and Lichtenstein (1956)
Bacillus subtilis	Bac+	Cited, 18 h, 37°C	6, 2,500	7.3	Janssen et al. (1986)
Bacillus subtilis	Bac+	NA, 24 h, 37°C	—, sd	11	Maruzzella and Lichtenstein (1956)
Bacillus subtilis	Bac+	ISA, 48 h, 25°C	4 (h), 10,000	7.5	Deans and Ritchie (1987)
Bacillus subtilis	Bac+	NA, 18 h, 37°C	6 (h), pure	27	Yousef and Tawil (1980)
Brevibacterium linens	Bac+	ISA, 48 h, 25°C	4 (h), 10,000	9	Deans and Ritchie (1987)
Brochothrix thermosphacta	Bac+	ISA, 48 h, 25°C	4 (h), 10,000	14	Deans and Ritchie (1987)
Clavibacter michiganensis subsp. *michiganensis*	Bac+	WA, 48 h, 25°C	6, 8,000	MIA 374	Cantore et al. (2004)
Clavibacter michiganensis subsp. *sepedonicus*	Bac+	WA, 48 h, 25°C	6, 8,000	MIA 435	Cantore et al. (2004)
Clostridium sporogenes	Bac+	ISA, 48 h, 25°C	4 (h), 10,000	0	Deans and Ritchie (1987)
Corynebacterium sp.	Bac+	TGA, 18–24 h, 37°C	9.5, 2,000	0	Morris et al. (1979)
Curtobacterium flaccumfaciens pv. *betae*	Bac+	WA, 48 h, 25°C	6, 8,000	MIA 632	Cantore et al. (2004)
Curtobacterium flaccumfaciens pv. *flaccumfaciens*	Bac+	WA, 48 h, 25°C	6, 8,000	MIA 435	Cantore et al. (2004)
Lactobacillus plantarum	Bac+	ISA, 48 h, 25°C	4 (h), 10,000	9	Deans and Ritchie (1987)
Leuconostoc cremoris	Bac+	ISA, 48 h, 25°C	4 (h), 10,000	7.5	Deans and Ritchie (1987)
Micrococcus luteus	Bac+	ISA, 48 h, 25°C	4 (h), 10,000	7	Deans and Ritchie (1987)
Mycobacterium phlei	Bac+	NA, 18 h, 37°C	6 (h), pure	19	Yousef and Tawil (1980)
Rhodococcus fascians	Bac+	WA, 48 h, 25°C	6, 8,000	MIA 435	Cantore et al. (2004)
Sarcina lutea	Bac+	NA, 24 h, 37°C	—, sd	0	Maruzzella and Lichtenstein (1956)
Staphylococcus aureus	Bac+	NA, 24 h, 37°C	—, sd	0	Maruzzella and Lichtenstein (1956)
Staphylococcus aureus	Bac+	TGA, 18–24 h, 37°C	9.5, 2,000	0	Morris et al. (1979)
Staphylococcus aureus	Bac+	ISA, 48 h, 25°C	4 (h), 10,000	5	Deans and Ritchie (1987)
Staphylococcus aureus	Bac+	NA, 24 h, 30°C	Drop, 5,000	12	Hili et al. (1997)
Staphylococcus aureus	Bac+	NA, 18 h, 37°C	6 (h), pure	14	Yousef and Tawil (1980)
Staphylococcus aureus	Bac+	Cited, 18 h, 37°C	6, 2,500	11.3	Janssen et al. (1986)
Staphylococcus aureus	Bac+	MHA, 48 h, 27°C	6, 15,000	18	Ertürk et al. (2006)

(*Continued*)

TABLE 14.28 (*Continued*)
Inhibitory Data of Coriander Oil Obtained in the Agar Diffusion Test

Microorganism	MO Class	Conditions		Inhibition Zone (mm)	Reference
Staphylococcus epidermidis	Bac+	MHA, 48 h, 27°C	6, 15,000	10	Ertürk et al. (2006)
Streptococcus faecalis	Bac+	ISA, 48 h, 25°C	4 (h), 10,000	0	Deans and Ritchie (1987)
Streptomyces venezuelae	Bac+	SMA, 2–7 days, 20°C	sd	11	Maruzzella and Liguori (1958)
Alternaria porri	Fungi	PDA, 72 h, 28°C	5, 5,000	12	Pawar and Thaker (2007)
Alternaria solani	Fungi	SMA, 2–7 days, 20°C	sd	18	Maruzzella and Liguori (1958)
Alternaria sp.	Fungi	PDA, 18 h, 37°C	6, sd	9.5	Sharma and Singh (1979)
Aspergillus candidus	Fungi	PDA, 18 h, 37°C	6, sd	0	Sharma and Singh (1979)
Aspergillus flavus	Fungi	PDA, 18 h, 37°C	6, sd	0	Sharma and Singh (1979)
Aspergillus fumigatus	Fungi	PDA, 18 h, 37°C	6, sd	12	Sharma and Singh (1979)
Aspergillus fumigatus	Fungi	SMA, 2–7 days, 20°C	sd	21	Maruzzella and Liguori (1958)
Aspergillus nidulans	Fungi	PDA, 18 h, 37°C	6, sd	0	Sharma and Singh (1979)
Aspergillus niger	Fungi	PDA, 48 h, 28°C	5, 5,000	0	Pawar and Thaker (2006)
Aspergillus niger	Fungi	MHA, 48 h, 27°C	6, 15,000	0	Ertürk et al. (2006)
Aspergillus niger	Fungi	PDA, 18 h, 37°C	6, sd	16	Sharma and Singh (1979)
Aspergillus niger	Fungi	SMA, 2–7 days, 20°C	sd	21	Maruzzella and Liguori (1958)
Aspergillus niger	Fungi	SDA, 8 days, 30°C	6 (h), pure	60	Yousef and Tawil (1980)
Cladosporium herbarum	Fungi	PDA, 18 h, 37°C	6, sd	21.5	Sharma and Singh (1979)
Cunninghamella echinulata	Fungi	PDA, 18 h, 37°C	6, sd	20	Sharma and Singh (1979)
Fusarium oxysporum	Fungi	PDA, 18 h, 37°C	6, sd	0	Sharma and Singh (1979)
Fusarium oxysporum f.sp. *ciceri*	Fungi	PDA, 72 h, 28°C	5, 5,000	11.5	Pawar and Thaker (2007)
Helminthosporium sacchari	Fungi	PDA, 18 h, 37°C	6, sd	13	Sharma and Singh (1979)
Helminthosporium sativum	Fungi	SMA, 2–7 days, 20°C	sd	11	Maruzzella and Liguori (1958)
Microsporum gypseum	Fungi	PDA, 18 h, 37°C	6, sd	8	Sharma and Singh (1979)
Mucor mucedo	Fungi	PDA, 18 h, 37°C	6, sd	12	Sharma and Singh (1979)

(*Continued*)

TABLE 14.28 (*Continued*)
Inhibitory Data of Coriander Oil Obtained in the Agar Diffusion Test

Microorganism	MO Class	Conditions		Inhibition Zone (mm)	Reference
Mucor mucedo	Fungi	SMA, 2–7 days, 20°C	sd	25	Maruzzella and Liguori (1958)
Mucor sp.	Fungi	SDA, 8 days, 30°C	6 (h), pure	21	Yousef and Tawil (1980)
Nigrospora panici	Fungi	SMA, 2–7 days, 20°C	sd	28	Maruzzella and Liguori (1958)
Penicillium chrysogenum	Fungi	SDA, 8 days, 30°°C	6 (h), pure	60	Yousef and Tawil (1980)
Penicillium digitatum	Fungi	PDA, 18 h, 37°C	6, sd	12	Sharma and Singh (1979)
Penicillium digitatum	Fungi	SMA, 2–7 days, 20°C	sd	14	Maruzzella and Liguori (1958)
Rhizopus nigricans	Fungi	PDA, 18 h, 37°C	6, sd	11	Sharma and Singh (1979)
Rhizopus nigricans	Fungi	SMA, 2–7 days, 20°C	sd	13	Maruzzella and Liguori (1958)
Rhizopus sp.	Fungi	SDA, 8 days, 30°C	6 (h), pure	19	Yousef and Tawil (1980)
Trichophyton rubrum	Fungi	PDA, 18 h, 37°C	6, sd	10	Sharma and Singh (1979)
Trichothecium roseum	Fungi	PDA, 18 h, 37°C	6, sd	8	Sharma and Singh (1979)
Brettanomyces anomalus	Yeast	MPA, 4 days, 30°C	5, 10% sol. sd	0	Conner and Beuchat (1984)
Candida albicans	Yeast	TGA, 18–24 h, 37°C	9.5, 2,000	0	Morris et al. (1979)
Candida albicans	Yeast	SMA, 2–7 days, 20°C	sd	5	Maruzzella and Liguori (1958)
Candida albicans	Yeast	MHA, 48 h, 27°C	6, 15,000	10	Ertürk et al. (2006)
Candida albicans	Yeast	Cited, 18 h, 37°C	6, 2,500	11	Janssen et al. (1986)
Candida albicans	Yeast	SDA, 18 h, 30°C	6 (h), pure	28	Yousef and Tawil (1980)
Candida albicans	Yeast	NA, 24 h, 30°C	Drop, 5,000	29	Hili et al. (1997)
Candida krusei	Yeast	SMA, 2–7 days, 20°C	sd	0	Maruzzella and Liguori (1958)
Candida lipolytica	Yeast	MPA, 4 days, 30°C	5, 10% sol. sd	0	Conner and Beuchat (1984)
Candida tropicalis	Yeast	SMA, 2–7 days, 20°C	sd	9	Maruzzella and Liguori (1958)
Cryptococcus neoformans	Yeast	SMA, 2–7 days, 20°C	sd	5	Maruzzella and Liguori (1958)
Cryptococcus rhodobenhani	Yeast	SMA, 2–7 days, 20°C	sd	20	Maruzzella and Liguori (1958)

(*Continued*)

TABLE 14.28 (*Continued*)
Inhibitory Data of Coriander Oil Obtained in the Agar Diffusion Test

Microorganism	MO Class	Conditions		Inhibition Zone (mm)	Reference
Debaryomyces hansenii	Yeast	MPA, 4 days, 30°C	5, 10% sol. sd	0	Conner and Beuchat (1984)
Geotrichum candidum	Yeast	MPA, 4 days, 30°C	5, 10% sol. sd	0	Conner and Beuchat (1984)
Hansenula anomala	Yeast	MPA, 4 days, 30°C	5, 10% sol. sd	12	Conner and Beuchat (1984)
Kloeckera apiculata	Yeast	MPA, 4 days, 30°C	5, 10% sol. sd	8	Conner and Beuchat (1984)
Kluyveromyces fragilis	Yeast	MPA, 4 days, 30°C	5, 10% sol. sd	0	Conner and Beuchat (1984)
Lodderomyces elongisporus	Yeast	MPA, 4 days, 30°C	5, 10% sol. sd	0	Conner and Beuchat (1984)
Metschnikowia pulcherrima	Yeast	MPA, 4 days, 30°C	5, 10% sol. sd	0	Conner and Beuchat (1984)
Pichia membranifaciens	Yeast	MPA, 4 days, 30°C	5, 10% sol. sd	0	Conner and Beuchat (1984)
Rhodotorula rubra	Yeast	MPA, 4 days, 30°C	5, 10% sol. sd	0	Conner and Beuchat (1984)
Saccharomyces cerevisiae	Yeast	MPA, 4 days, 30°C	5, 10% sol. sd	0	Conner and Beuchat (1984)
Saccharomyces cerevisiae	Yeast	SMA, 2–7 days, 20°C	sd	7	Maruzzella and Liguori (1958)
Saccharomyces cerevisiae	Yeast	NA, 24 h, 30°C	Drop, 5,000	32	Hili et al. (1997)
Schizosaccharomyces pombe	Yeast	NA, 24 h, 30°C	Drop, 5,000	33	Hili et al. (1997)
Torula glabrata	Yeast	MPA, 4 days, 30°C	5, 10% sol. sd	7	Conner and Beuchat (1984)
Torula utilis	Yeast	NA, 24 h, 30°C	Drop, 5,000	37	Hili et al. (1997)

TABLE 14.29
Inhibitory Data of Coriander Oil Obtained in the Dilution Test

Microorganism	MO Class	Conditions	MIC	Reference
Enterobacter aerogenes	Bac–	MHB, 24 h, 37°C	4,315	Ertürk et al. (2006)
Escherichia coli	Bac–	MPB, DMSO, 40 h, 30°C	>500	Hili et al. (1997)
Escherichia coli	Bac–	TGB, 18–24 h, 37°C	>1,000	Morris et al. (1979)
Escherichia coli	Bac–	MHB, 24 h, 37°C	2,150	Ertürk et al. (2006)
Escherichia coli	Bac–	NB, Tween 20, 18 h, 37°C	3,200	Yousef and Tawil (1980)
Escherichia coli O157:H7	Bac–	BHI, 48 h, 35°C	2,000	Oussalah et al. (2006)
Pseudomonas aeruginosa	Bac–	MPB, DMSO, 40 h, 30°C	74% inh. 500	Hili et al. (1997)
Pseudomonas aeruginosa	Bac–	MHB, 24 h, 37°C	4,350	Ertürk et al. (2006)
Pseudomonas aeruginosa	Bac–	NB, Tween 20, 18 h, 37°C	12,500	Yousef and Tawil (1980)
Salmonella typhimurium	Bac–	BHI, 48 h, 35°C	2,000	Oussalah et al. (2006)
Salmonella typhimurium	Bac–	MHB, 24 h, 37°C	17,260	Ertürk et al. (2006)
Bacillus subtilis	Bac+	NB, Tween 20, 18 h, 37°C	800	Yousef and Tawil (1980)
Corynebacterium sp.	Bac+	TGB, 18–24 h, 37°C	500	Morris et al. (1979)
Listeria monocytogenes	Bac+	BHI, 48 h, 35°C	>8,000	Oussalah et al. (2006)
Mycobacterium phlei	Bac+	NB, Tween 20, 18 h, 37°C	200	Yousef and Tawil (1980)
Staphylococcus aureus	Bac+	MPB, DMSO, 40 h, 30°C	44% inh. 500	Hili et al. (1997)
Staphylococcus aureus	Bac+	TGB, 18–24 h, 37°C	1,000	Morris et al. (1979)
Staphylococcus aureus	Bac+	MHB, 24 h, 37°C	1,070	Ertürk et al. (2006)
Staphylococcus aureus	Bac+	BHI, 48 h, 35°C	2,000	Oussalah et al. (2006)
Staphylococcus aureus	Bac+	MHB, Tween 80, 24 h, 37°C	3,120	Bastide et al. (1987)
Staphylococcus aureus	Bac+	NB, Tween 20, 18 h, 37°C	3,200	Yousef and Tawil (1980)
Staphylococcus epidermidis	Bac+	MHB, 24 h, 37°C	8,630	Ertürk et al. (2006)
Aspergillus flavus	Fungi	PDA, 8 h, 20°C, spore germ. inh.	50–100	Thompson (1986)
Aspergillus flavus	Fungi	PDA, 5 days, 27°C	>1,000	Thompson and Cannon (1986)
Aspergillus flavus	Fungi	CDA, cited	3,000	Dubey et al. (1990)
Aspergillus niger	Fungi	NB, Tween 20, 8 days, 30°C	3,200	Yousef and Tawil (1980)
Aspergillus niger	Fungi	MHB, 24 h, 37°C	>20,000	Ertürk et al. (2006)
Aspergillus parasiticus	Fungi	PDA, 8 h, 20°C, spore germ. inh.	50–100	Thompson (1986)
Aspergillus parasiticus	Fungi	PDA, 5 days, 27°C	>1,000	Thompson and Cannon (1986)
Epidermophyton floccosum	Fungi	SA, Tween 80, 21 days, 20°C	<300	Janssen et al. (1988)
Mucor hiemalis	Fungi	PDA, 5 days, 27°C	>1,000	Thompson and Cannon (1986)
Mucor mucedo	Fungi	PDA, 5 days, 27°C	>1,000	Thompson and Cannon (1986)
Mucor racemosus f. *racemosus*	Fungi	PDA, 5 days, 27°C	>1,000	Thompson and Cannon (1986)
Mucor sp.	Fungi	NB, Tween 20, 8 days, 30°C	800	Yousef and Tawil (1980)
Penicillium chrysogenum	Fungi	NB, Tween 20, 8 days, 30°C	800	Yousef and Tawil (1980)
Rhizopus 66-81-2	Fungi	PDA, 5 days, 27°C	>1,000	Thompson and Cannon (1986)
Rhizopus arrhizus	Fungi	PDA, 5 days, 27°C	>1,000	Thompson and Cannon (1986)

(Continued)

TABLE 14.29 (*Continued*)
Inhibitory Data of Coriander Oil Obtained in the Dilution Test

Microorganism	MO Class	Conditions	MIC	Reference
Rhizopus chinensis	Fungi	PDA, 5 days, 27°C	>1,000	Thompson and Cannon (1986)
Rhizopus circinans	Fungi	PDA, 5 days, 27°C	>1,000	Thompson and Cannon (1986)
Rhizopus japonicus	Fungi	PDA, 5 days, 27°C	>1,000	Thompson and Cannon (1986)
Rhizopus kazanensis	Fungi	PDA, 5 days, 27°C	>1,000	Thompson and Cannon (1986)
Rhizopus oryzae	Fungi	PDA, 5 days, 27°C	>1,000	Thompson and Cannon (1986)
Rhizopus pymacus	Fungi	PDA, 5 days, 27°C	>1,000	Thompson and Cannon (1986)
Rhizopus sp.	Fungi	NB, Tween 20, 8 days, 30°C	1,600	Yousef and Tawil (1980)
Rhizopus stolonifer	Fungi	PDA, 5 days, 27°C	>1,000	Thompson and Cannon (1986)
Rhizopus tritici	Fungi	PDA, 5 days, 27°C	>1,000	Thompson and Cannon (1986)
Trichophyton mentagrophytes	Fungi	SA, Tween 80, 21 days, 20°C	300,625	Janssen et al. (1988)
Trichophyton rubrum	Fungi	SA, Tween 80, 21 days, 20°C	<300	Janssen et al. (1988)
Candida albicans	Yeast	MPB, DMSO, 40 h, 30°C	75% inh. 500	Hili et al. (1997)
Candida albicans	Yeast	NB, Tween 20, 18 h, 37°C	800	Yousef and Tawil (1980)
Candida albicans	Yeast	TGB, 18–24 h, 37°C	>1,000	Morris et al. (1979)
Candida albicans	Yeast	MHB, Tween 80, 48 h, 35°C	2,500	Hammer et al. (1998)
Candida albicans	Yeast	MHB, 24 h, 37°C	4,310	Ertürk et al. (2006)
Saccharomyces cerevisiae	Yeast	MPB, DMSO, 40 h, 30°C	68% inh. 500	Hili et al. (1997)
Schizosaccharomyces pombe	Yeast	MPB, DMSO, 40 h, 30°C	18% inh. 500	Hili et al. (1997)
Torula utilis	Yeast	MPB, DMSO, 40 h, 30°C	87% inh. 500	Hili et al. (1997)

TABLE 14.30
Inhibitory Data of Coriander Oil Obtained in the Vapor Phase Test

Microorganism	MO Class	Conditions		Activity	Reference
Escherichia coli	Bac−	NA, 24 h, 37°C	~20,000	NG	Kellner and Kober (1954)
Escherichia coli	Bac−	BLA, 18 h, 37°C	MIC$_{air}$	250	Inouye et al. (2001)
Haemophilus influenzae	Bac−	MHA, 18 h, 37°C	MIC$_{air}$	12.5	Inouye et al. (2001)
Neisseria sp.	Bac−	NA, 24 h, 37°C	~20,000	NG	Kellner and Kober (1954)
Salmonella typhi	Bac−	NA, 24 h, 37°C	~20,000	NG	Kellner and Kober (1954)
Salmonella typhi	Bac−	NA, 24 h, 37°C	sd	++	Maruzzella and Sicurella (1960)
Bacillus megaterium	Bac+	NA, 24 h, 37°C	~20,000	NG	Kellner and Kober (1954)
Bacillus subtilis var. aterrimus	Bac+	NA, 24 h, 37°C	sd	+++	Maruzzella and Sicurella (1960)
Corynebacterium diphtheriae	Bac+	NA, 24 h, 37°C	~20,000	NG	Kellner and Kober (1954)
Mycobacterium avium	Bac+	NA, 24 h, 37°C	sd	NG	Maruzzella and Sicurella (1960)
Staphylococcus aureus	Bac+	NA, 24 h, 37°C	~20,000	NG	Kellner and Kober (1954)
Staphylococcus aureus	Bac+	NA, 24 h, 37°C	sd	+++	Maruzzella and Sicurella (1960)
Staphylococcus aureus	Bac+	MHA, 18 h, 37°C	MIC$_{air}$	50	Inouye et al. (2001)
Streptococcus faecalis	Bac+	NA, 24 h, 37°C	~20,000	NG	Kellner and Kober (1954)
Streptococcus faecalis	Bac+	NA, 24 h, 37°C	sd	+++	Maruzzella and Sicurella (1960)
Streptococcus pneumoniae	Bac+	MHA, 18 h, 37°C	MIC$_{air}$	25	Inouye et al. (2001)
Streptococcus pyogenes	Bac+	NA, 24 h, 37°C	~20,000	NG	Kellner and Kober (1954)
Streptococcus pyogenes	Bac+	MHA, 18 h, 37°C	MIC$_{air}$	25	Inouye et al. (2001)
Candida albicans	Yeast	NA, 24 h, 37°C	~20,000	NG	Kellner and Kober (1954)

TABLE 14.31
Inhibitory Data of Dwarf Pine Oil Obtained in the Agar Diffusion Test

Microorganism	MO Class	Conditions		Inhibition Zone (mm)	Reference
Aerobacter aerogenes	Bac–	NA, 24 h, 37°C	—, sd	2	Maruzzella and Lichtenstein (1956)
Escherichia coli	Bac–	NA, 24 h, 37°C	—, sd	0	Maruzzella and Lichtenstein (1956)
Escherichia coli	Bac–	Cited, 18 h, 37°C	6, 2,500	11	Janssen et al. (1986)
Neisseria perflava	Bac–	NA, 24 h, 37°C	—, sd	0	Maruzzella and Lichtenstein (1956)
Proteus vulgaris	Bac–	NA, 24 h, 37°C	—, sd	5	Maruzzella and Lichtenstein (1956)
Pseudomonas aeruginosa	Bac–	NA, 24 h, 37°C	—, sd	0	Maruzzella and Lichtenstein (1956)
Pseudomonas aeruginosa	Bac–	Cited, 18 h, 37°C	6, 2,500	8.3	Janssen et al. (1986)
Serratia marcescens	Bac–	NA, 24 h, 37°C	—, sd	0	Maruzzella and Lichtenstein (1956)
Bacillus mesentericus	Bac+	NA, 24 h, 37°C	—, sd	0	Maruzzella and Lichtenstein (1956)
Bacillus subtilis	Bac+	NA, 24 h, 37°C	—, sd	5	Maruzzella and Lichtenstein (1956)
Bacillus subtilis	Bac+	Cited, 18 h, 37°C	6, 2,500	16.7	Janssen et al. (1986)
Sarcina lutea	Bac+	NA, 24 h, 37°C	—, sd	0	Maruzzella and Lichtenstein (1956)
Staphylococcus aureus	Bac+	NA, 24 h, 37°C	—, sd	3	Maruzzella and Lichtenstein (1956)
Staphylococcus aureus	Bac+	Cited, 18 h, 37°C	6, 2,500	8.7	Janssen et al. (1986)
Streptomyces venezuelae	Bac+	SMA, 2–7 days, 20°C	sd	2	Maruzzella and Liguori (1958)
Alternaria solani	Fungi	SMA, 2–7 days, 20°C	sd	0	Maruzzella and Liguori (1958)
Aspergillus fumigatus	Fungi	SMA, 2–7 days, 20°C	sd	0	Maruzzella and Liguori (1958)
Aspergillus niger	Fungi	SMA, 2–7 days, 20°C	sd	0	Maruzzella and Liguori (1958)
Helminthosporium sativum	Fungi	SMA, 2–7 days, 20°C	sd	0	Maruzzella and Liguori (1958)
Mucor mucedo	Fungi	SMA, 2–7 days, 20°C	sd	0	Maruzzella and Liguori (1958)
Nigrospora panici	Fungi	SMA, 2–7 days, 20°C	sd	0	Maruzzella and Liguori (1958)
Penicillium digitatum	Fungi	SMA, 2–7 days, 20°C	sd	0	Maruzzella and Liguori (1958)
Rhizopus nigricans	Fungi	SMA, 2–7 days, 20°C	sd	0	Maruzzella and Liguori (1958)
Candida albicans	Yeast	SMA, 2–7 days, 20°C	sd	0	Maruzzella and Liguori (1958)
Candida albicans	Yeast	Cited, 18 h, 37°C	6, 2,500	32.3	Janssen et al. (1986)
Candida krusei	Yeast	SMA, 2–7 days, 20°C	sd	0	Maruzzella and Liguori (1958)
Candida tropicalis	Yeast	SMA, 2–7 days, 20°C	sd	0	Maruzzella and Liguori (1958)
Cryptococcus neoformans	Yeast	SMA, 2–7 days, 20°C	sd	0	Maruzzella and Liguori (1958)
Cryptococcus rhodobenhani	Yeast	SMA, 2–7 days, 20°C	sd	0	Maruzzella and Liguori (1958)
Saccharomyces cerevisiae	Yeast	SMA, 2–7 days, 20°C	sd	0	Maruzzella and Liguori (1958)

TABLE 14.32
Inhibitory Data of Dwarf Pine Oil Obtained in the Dilution Test

Microorganism	MO Class	Conditions	MIC	Reference
Epidermophyton floccosum	Fungi	SA, Tween 80, 21 days, 20°C	>1250	Janssen et al. (1988)
Trichophyton mentagrophytes	Fungi	SA, Tween 80, 21 days, 20°C	>1250	Janssen et al. (1988)
Trichophyton rubrum	Fungi	SA, Tween 80, 21 days, 20°C	>1250	Janssen et al. (1988)

TABLE 14.33
Inhibitory Data of Dwarf Pine Oil Obtained in the Vapor Phase Test

Microorganism	MO Class	Conditions		Activity	Reference
Escherichia coli	Bac−	NA, 24 h, 37°C	~20,000	+++	Kellner and Kober (1954)
Neisseria sp.	Bac−	NA, 24 h, 37°C	~20,000	NG	Kellner and Kober (1954)
Salmonella typhi	Bac−	NA, 24 h, 37°C	~20,000	+++	Kellner and Kober (1954)
Bacillus megaterium	Bac+	NA, 24 h, 37°C	~20,000	NG	Kellner and Kober (1954)
Corynebacterium diphtheriae	Bac+	NA, 24 h, 37°C	~20,000	NG	Kellner and Kober (1954)
Staphylococcus aureus	Bac+	NA, 24 h, 37°C	~20,000	+	Kellner and Kober (1954)
Streptococcus faecalis	Bac+	NA, 24 h, 37°C	~20,000	NG	Kellner and Kober (1954)
Streptococcus pyogenes	Bac+	NA, 24 h, 37°C	~20,000	NG	Kellner and Kober (1954)
Candida albicans	Yeast	NA, 24 h, 37°C	~20,000	++	Kellner and Kober (1954)

TABLE 14.34
Inhibitory Data of Eucalyptus Oil Obtained in the Agar Diffusion Test

Microorganism	MO Class	Conditions		Inhibition Zone (mm)	Reference
Acinetobacter calcoaceticus	Bac−	ISA, 48 h, 25°C	4 (h), 10,000	7	Deans and Ritchie (1987)
Aerobacter aerogenes	Bac−	NA, 24 h, 37°C	—, sd	0	Maruzzella and Lichtenstein (1956)
Aeromonas hydrophila	Bac−	ISA, 48 h, 25°C	4 (h), 10,000	7.5	Deans and Ritchie (1987)
Alcaligenes faecalis	Bac−	ISA, 48 h, 25°C	4 (h), 10,000	0	Deans and Ritchie (1987)
Beneckea natriegens	Bac−	ISA, 48 h, 25°C	4 (h), 10,000	0	Deans and Ritchie (1987)
Campylobacter jejuni	Bac−	TSA, 24 h, 42°C	4 (h), 25,000	8.3	Smith-Palmer et al. (1998)
Citrobacter freundii	Bac−	ISA, 48 h, 25°C	4 (h), 10,000	6	Deans and Ritchie (1987)
Enterobacter aerogenes	Bac−	ISA, 48 h, 25°C	4 (h), 10,000	0	Deans and Ritchie (1987)
Enterobacter aerogenes	Bac−	MHA, 24 h, 30°C	6, 15,000	7	Rossi et al. (2007)
Erwinia carotovora	Bac−	ISA, 48 h, 25°C	4 (h), 10,000	6	Deans and Ritchie (1987)
Escherichia coli	Bac−	TGA, 18–24 h, 37°C	9.5, 2,000	0	Morris et al. (1979)
Escherichia coli	Bac−	ISA, 48 h, 25°C	4 (h), 10,000	0	Deans and Ritchie (1987)
Escherichia coli	Bac−	NA, 24 h, 37°C	—, sd	3	Maruzzella and Lichtenstein (1956)
Escherichia coli	Bac−	MHA, 24 h, 30°C	6, 15,000	6	Rossi et al. (2007)
Escherichia coli	Bac−	Cited, 18 h, 37°C	6, 2,500	8	Janssen et al. (1986)
Escherichia coli	Bac−	TSA, 24 h, 35°C	4 (h), 25,000	10.3	Smith-Palmer et al. (1998)
Escherichia coli	Bac−	Cited	15, 2,500	16	Pizsolitto et al. (1975)
Escherichia coli	Bac−	NA, 18 h, 37°C	6 (h), pure	20	Yousef and Tawil (1980)
Escherichia coli	Bac−	NA, 18 h, 37°C	5 (h), −30,000	22	Schelz et al. (2006)
Flavobacterium suaveolens	Bac−	ISA, 48 h, 25°C	4 (h), 10,000	10	Deans and Ritchie (1987)
Klebsiella pneumoniae	Bac−	ISA, 48 h, 25°C	4 (h), 10,000	0	Deans and Ritchie (1987)
Klebsiella sp.	Bac−	Cited	15, 2,500	1	Pizsolitto et al. (1975)
Moraxella sp.	Bac−	ISA, 48 h, 25°C	4 (h), 10,000	0	Deans and Ritchie (1987)
Neisseria perflava	Bac−	NA, 24 h, 37°C	—, sd	4	Maruzzella and Lichtenstein (1956)
Proteus sp.	Bac−	Cited	15, 2,500	10	Pizsolitto et al. (1975)
Proteus vulgaris	Bac−	ISA, 48 h, 25°C	4 (h), 10,000	0	Deans and Ritchie (1987)
Proteus vulgaris	Bac−	NA, 24 h, 37°C	—, sd	17	Maruzzella and Lichtenstein (1956)
Pseudomonas aeruginosa	Bac−	Cited	15, 2,500	0	Pizsolitto et al. (1975)
Pseudomonas aeruginosa	Bac−	ISA, 48 h, 25°C	4 (h), 10,000	0	Deans and Ritchie (1987)
Pseudomonas aeruginosa	Bac−	NA, 24 h, 37°C	—, sd	4	Maruzzella and Lichtenstein (1956)
Pseudomonas aeruginosa	Bac−	MHA, 24 h, 30°C	6, 15,000	6	Rossi et al. (2007)
Pseudomonas aeruginosa	Bac−	NA, 18 h, 37°C	6 (h), pure	13	Yousef and Tawil (1980)
Pseudomonas aeruginosa	Bac−	Cited, 18 h, 37°C	6, 2,500	9.7	Janssen et al. (1986)

(Continued)

TABLE 14.34 (Continued)
Inhibitory Data of Eucalyptus Oil Obtained in the Agar Diffusion Test

Microorganism	MO Class	Conditions		Inhibition Zone (mm)	Reference
Salmonella enteritidis	Bac−	TSA, 24 h, 35°C	4 (h), 25,000	7.5	Smith-Palmer et al. (1998)
Salmonella pullorum	Bac−	ISA, 48 h, 25°C	4 (h), 10,000	6	Deans and Ritchie (1987)
Salmonella sp.	Bac−	Cited	15, 2,500	10	Pizsolitto et al. (1975)
Serratia marcescens	Bac−	NA, 24 h, 37°C	—, sd	7	Maruzzella and Lichtenstein (1956)
Serratia marcescens	Bac−	ISA, 48 h, 25°C	4 (h), 10,000	5.5	Deans and Ritchie (1987)
Serratia sp.	Bac−	Cited	15, 2,500	2	Pizsolitto et al. (1975)
Shigella sp.	Bac−	Cited	15, 2,500	13	Pizsolitto et al. (1975)
Yersinia enterocolitica	Bac−	ISA, 48 h, 25°C	4 (h), 10,000	6	Deans and Ritchie (1987)
Bacillus mesentericus	Bac+	NA, 24 h, 37°C	—, sd	10	Maruzzella and Lichtenstein (1956)
Bacillus sp.	Bac+	Cited	15, 2,500	4	Pizsolitto et al. (1975)
Bacillus subtilis	Bac+	ISA, 48 h, 25°C	4 (h), 10,000	7	Deans and Ritchie (1987)
Bacillus subtilis	Bac+	Cited, 18 h, 37°C	6, 2,500	8.3	Janssen et al. (1986)
Bacillus subtilis	Bac+	NA, 18 h, 37°C	6 (h), pure	23.5	Yousef and Tawil (1980)
Bacillus subtilis	Bac+	NA, 24 h, 37°C	—, sd	34	Maruzzella and Lichtenstein (1956)
Brevibacterium linens	Bac+	ISA, 48 h, 25°C	4 (h), 10,000	4.5	Deans and Ritchie (1987)
Brochothrix thermosphacta	Bac+	ISA, 48 h, 25°C	4 (h), 10,000	0	Deans and Ritchie (1987)
Clostridium sporogenes	Bac+	ISA, 48 h, 25°C	4 (h), 10,000	0	Deans and Ritchie (1987)
Corynebacterium sp.	Bac+	TGA, 18–24 h, 37°C	9.5, 2,000	0	Morris et al. (1979)
Lactobacillus plantarum	Bac+	ISA, 48 h, 25°C	4 (h), 10,000	0	Deans and Ritchie (1987)
Leuconostoc cremoris	Bac+	ISA, 48 h, 25°C	4 (h), 10,000	9.5	Deans and Ritchie (1987)
Listeria monocytogenes	Bac+	ISA, 48 h, 25°C	4 (h), 10,000	6–20	Lis-Balchin et al. (1998)
Listeria monocytogenes	Bac+	TSA, 24 h, 35°C	4 (h), 25,000	5.4	Smith-Palmer et al. (1998)
Micrococcus luteus	Bac+	ISA, 48 h, 25°C	4 (h), 10,000	0	Deans and Ritchie (1987)
Mycobacterium phlei	Bac+	NA, 18 h, 37°C	6 (h), pure	29	Yousef and Tawil (1980)
Sarcina lutea	Bac+	NA, 24 h, 37°C	—, sd	4	Maruzzella and Lichtenstein (1956)
Staphylococcus aureus	Bac+	NA, 24 h, 37°C	—, sd	0	Maruzzella and Lichtenstein (1956)
Staphylococcus aureus	Bac+	TGA, 18–24 h, 37°C	9.5, 2,000	0	Morris et al. (1979)
Staphylococcus aureus	Bac+	ISA, 48 h, 25°C	4 (h), 10,000	0	Deans and Ritchie (1987)
Staphylococcus aureus	Bac+	Cited	15, 2,500	5	Pizsolitto et al. (1975)
Staphylococcus aureus	Bac+	Cited, 18 h, 37°C	6, 2,500	8	Janssen et al. (1986)
Staphylococcus aureus	Bac+	TSA, 24 h, 35°C	4 (h), 25,000	8.5	Smith-Palmer et al. (1998)
Staphylococcus aureus	Bac+	MHA, 24 h, 37°C	6, 15,000	16	Rossi et al. (2007)
Staphylococcus aureus	Bac+	NA, 18 h, 37°C	6 (h), pure	30	Yousef and Tawil (1980)
Staphylococcus epidermidis	Bac+	Cited	15, 2,500	15	Pizsolitto et al. (1975)
Staphylococcus epidermidis	Bac+	NA, 18 h, 37°C	5 (h), −30,000	15	Schelz et al. (2006)

(Continued)

TABLE 14.34 (Continued)
Inhibitory Data of Eucalyptus Oil Obtained in the Agar Diffusion Test

Microorganism	MO Class	Conditions		Inhibition Zone (mm)	Reference
Streptococcus faecalis	Bac+	ISA, 48 h, 25°C	4 (h), 10,000	0	Deans and Ritchie (1987)
Streptococcus viridans	Bac+	Cited	15, 2,500	0	Pizsolitto et al. (1975)
Streptomyces venezuelae	Bac+	SMA, 2–7 days, 20°C	sd	13	Maruzzella and Liguori (1958)
Alternaria solani	Fungi	SMA, 2–7 days, 20°C	sd	10	Maruzzella and Liguori (1958)
Aspergillus fumigatus	Fungi	SMA, 2–7 days, 20°C	sd	13	Maruzzella and Liguori (1958)
Aspergillus niger	Fungi	SMA, 2–7 days, 20°C	sd	7	Maruzzella and Liguori (1958)
Aspergillus niger	Fungi	SDA, 8 days, 30°C	6 (h), pure	10	Yousef and Tawil (1980)
Helminthosporium sativum	Fungi	SMA, 2–7 days, 20°C	sd	9	Maruzzella and Liguori (1958)
Mucor mucedo	Fungi	SMA, 2–7 days, 20°C	sd	9	Maruzzella and Liguori (1958)
Mucor sp.	Fungi	SDA, 8 days, 30°C	6 (h), pure	10	Yousef and Tawil (1980)
Nigrospora panici	Fungi	SMA, 2–7 days, 20°C	sd	10	Maruzzella and Liguori (1958)
Penicillium chrysogenum	Fungi	SDA, 8 days, 30°C	6 (h), pure	13	Yousef and Tawil (1980)
Penicillium digitatum	Fungi	SMA, 2–7 days, 20°C	sd	4	Maruzzella and Liguori (1958)
Rhizopus nigricans	Fungi	SMA, 2–7 days, 20°C	sd	3	Maruzzella and Liguori (1958)
Rhizopus sp.	Fungi	SDA, 8 days, 30°C	6 (h), pure	0	Yousef and Tawil (1980)
Candida albicans	Yeast	TGA, 18–24 h, 37°C	9.5, 2,000	0	Morris et al. (1979)
Candida albicans	Yeast	SMA, 2–7 days, 20°C	sd	5	Maruzzella and Liguori (1958)
Candida albicans	Yeast	Cited, 18 h, 37°C	6, 2,500	11.7	Janssen et al. (1986)
Candida albicans	Yeast	SDA, 18 h, 30°C	6 (h), pure	31	Yousef and Tawil (1980)
Candida krusei	Yeast	SMA, 2–7 days, 20°C	sd	5	Maruzzella and Liguori (1958)
Candida tropicalis	Yeast	SMA, 2–7 days, 20°C	sd	3	Maruzzella and Liguori (1958)
Cryptococcus neoformans	Yeast	SMA, 2–7 days, 20°C	sd	8	Maruzzella and Liguori (1958)
Cryptococcus rhodobenhani	Yeast	SMA, 2–7 days, 20°C	sd	7	Maruzzella and Liguori (1958)
Saccharomyces cerevisiae	Yeast	NA, 24 h, 20°C	5 (h), –30,000	16–21	Schelz et al. (2006)
Saccharomyces cerevisiae	Yeast	SMA, 2–7 days, 20°C	sd	8	Maruzzella and Liguori (1958)

TABLE 14.35
Inhibitory Data of Eucalyptus Oil Obtained in the Dilution Test

Microorganism	MO Class	Conditions	MIC	Reference
Campylobacter jejuni	Bac–	TSB, 24 h, 42°C	>10,000	Smith-Palmer et al. (1998)
Citrobacter freundii	Bac–	ISB, Tween 80, 20–24 h, 37°C	10,000	Harkenthal et al. (1999)
Enterobacter aerogenes	Bac–	ISB, Tween 80, 20–24 h, 37°C	20,000	Harkenthal et al. (1999)
Escherichia coli	Bac–	TGB, 18–24 h, 37°C	>1,000	Morris et al. (1979)
Escherichia coli	Bac–	TSB, 24 h, 35°C	>10,000	Smith-Palmer et al. (1998)
Escherichia coli	Bac–	ISB, Tween 80, 20–24 h, 37°C	>40,000	Harkenthal et al. (1999)
Escherichia coli	Bac–	TGB, 18 h, 37°C	2,800	Schelz et al. (2006)
Escherichia coli	Bac–	NB, Tween 20, 18 h, 37°C	6,400	Yousef and Tawil (1980)
Klebsiella pneumoniae	Bac–	ISB, Tween 80, 20–24 h, 37°C	5,000	Harkenthal et al. (1999)
Proteus mirabilis	Bac–	ISB, Tween 80, 20–24 h, 37°C	20,000	Harkenthal et al. (1999)
Pseudomonas aeruginosa	Bac–	ISB, Tween 80, 20–24 h, 37°C	>40,000	Harkenthal et al. (1999)
Pseudomonas aeruginosa	Bac–	NB, Tween 20, 18 h, 37°C	12,500	Yousef and Tawil (1980)
Salmonella choleraesuis	Bac–	ISB, Tween 80, 20–24 h, 37°C	20,000	Harkenthal et al. (1999)
Salmonella enteritidis	Bac–	TSB, 24 h, 35°C	>10,000	Smith-Palmer et al. (1998)
Shigella flexneri	Bac–	ISB, Tween 80, 20–24 h, 37°C	2,500	Harkenthal et al. (1999)
Bacillus subtilis	Bac+	ISB, Tween 80, 20–24 h, 37°C	10,000	Harkenthal et al. (1999)
Bacillus subtilis	Bac+	NB, Tween 20, 18 h, 37°C	3,200	Yousef and Tawil (1980)
Corynebacterium pseudodiphtheriae	Bac+	ISB, Tween 80, 20–24 h, 37°C	20,000	Harkenthal et al. (1999)
Corynebacterium sp.	Bac+	TGB, 18–24 h, 37°C	>1,000	Morris et al. (1979)
Enterococcus durans	Bac+	ISB, Tween 80, 20–24 h, 37°C	20,000	Harkenthal et al. (1999)
Enterococcus faecalis	Bac+	ISB, Tween 80, 20–24 h, 37°C	20,000	Harkenthal et al. (1999)
Enterococcus faecium	Bac+	ISB, Tween 80, 20–24 h, 37°C	20,000	Harkenthal et al. (1999)
Listeria monocytogenes	Bac+	ISB, Tween 80, 20–24 h, 37°C	2,500	Harkenthal et al. (1999)
Listeria monocytogenes	Bac+	TSB, 24 h, 35°C	750	Smith-Palmer et al. (1998)
Mycobacterium phlei	Bac+	NB, Tween 20, 18 h, 37°C	3,200	Yousef and Tawil (1980)
Staphylococcus aureus	Bac+	TGB, 18–24 h, 37°C	>1,000	Morris et al. (1979)
Staphylococcus aureus	Bac+	TSB, 24 h, 35°C	1,000	Smith-Palmer et al. (1998)
Staphylococcus aureus	Bac+	ISB, Tween 80, 20–24 h, 37°C	20,000	Harkenthal et al. (1999)
Staphylococcus aureus	Bac+	NB, Tween 20, 18 h, 37°C	6,400	Yousef and Tawil (1980)
Staphylococcus epidermidis	Bac+	TGB, 18 h, 37°C	2,800	Schelz et al. (2006)
Staphylococcus epidermidis	Bac+	ISB, Tween 80, 20–24 h, 37°C	5,000	Harkenthal et al. (1999)

(Continued)

TABLE 14.35 (Continued)
Inhibitory Data of Eucalyptus Oil Obtained in the Dilution Test

Microorganism	MO Class	Conditions	MIC	Reference
Staphylococcus saprophyticus	Bac+	ISB, Tween 80, 20–24 h, 37°C	5,000	Harkenthal et al. (1999)
Staphylococcus xylosus	Bac+	ISB, Tween 80, 20–24 h, 37°C	5,000	Harkenthal et al. (1999)
Alternaria alternata	Fungi	PDA, 7 days, 28°C	0% inh. 500	Feng and Zheng (2007)
Aspergillus niger	Fungi	NB, Tween 20, 8 days, 30°C	50,000	Yousef and Tawil (1980)
Aspergillus niger	Fungi	YES broth, 10 d	−87% inh. 10,000	Lis-Balchin et al. (1998)
Aspergillus ochraceus	Fungi	YES broth, 10 d	−61% inh. 10,000	Lis-Balchin et al. (1998)
Aspergillus oryzae	Fungi	Cited	500	Okazaki and Oshima (1953)
Botrytis cinerea	Fungi	PDA, Tween 20, 7 days, 24°C	2% inh. 1,000	Bouchra et al. (2003)
Epidermophyton floccosum	Fungi	SA, Tween 80, 21 days, 20°C	625–1,250	Janssen et al. (1988)
Fusarium culmorum	Fungi	YES broth, 10 d	−78% inh. 10,000	Lis-Balchin et al. (1998)
Geotrichum citri-aurantii	Fungi	PDA, Tween 20, 7 days, 24°C	0% inh. 1,000	Bouchra et al. (2003)
Mucor racemosus	Fungi	Cited	>500	Okazaki and Oshima (1953)
Mucor sp.	Fungi	NB, Tween 20, 8 days, 30°C	12,500	Yousef and Tawil (1980)
Penicillium chrysogenum	Fungi	NB, Tween 20, 8 days, 30°C	25,000	Yousef and Tawil (1980)
Penicillium chrysogenum	Fungi	Cited	500	Okazaki and Oshima (1953)
Penicillium digitatum	Fungi	PDA, Tween 20, 7 days, 24°C	2% inh. 1,000	Bouchra et al. (2003)
Phytophthora citrophthora	Fungi	PDA, Tween 20, 7 days, 24°C	38% inh. 1,000	Bouchra et al. (2003)
Rhizopus sp.	Fungi	NB, Tween 20, 8 days, 30°C	6,400	Yousef and Tawil (1980)
Trichophyton mentagrophytes	Fungi	SA, Tween 80, 21 days, 20°C	>1,250	Janssen et al. (1988)
Trichophyton rubrum	Fungi	SA, Tween 80, 21 days, 20°C	625–1,250	Janssen et al. (1988)
Candida albicans	Yeast	TGB, 18–24 h, 37°C	1,000	Morris et al. (1979)
Candida albicans	Yeast	NB, Tween 20, 18 h, 37°C	1,600	Yousef and Tawil (1980)
Saccharomyces cerevisiae	Yeast	YPB, 24 h, 20°C	700	Schelz et al. (2006)

TABLE 14.36
Inhibitory Data of Eucalyptus Oil Obtained in the Vapor Phase Test

Microorganism	MO Class	Conditions		Activity	Reference
Escherichia coli	Bac−	NA, 24 h, 37°C	~20,000	NG	Kellner and Kober (1954)
Neisseria sp.	Bac−	NA, 24 h, 37°C	~20,000	NG	Kellner and Kober (1954)
Salmonella typhi	Bac−	NA, 24 h, 37°C	~20,000	NG	Kellner and Kober (1954)
Salmonella typhi	Bac−	NA, 24 h, 37°C	sd	+++	Maruzzella and Sicurella (1960)
Bacillus megaterium	Bac+	NA, 24 h, 37°C	~20,000	NG	Kellner and Kober (1954)
Bacillus subtilis var. aterrimus	Bac+	NA, 24 h, 37°C	sd	+++	Maruzzella and Sicurella (1960)
Corynebacterium diphtheriae	Bac+	NA, 24 h, 37°C	~20,000	NG	Kellner and Kober (1954)
Mycobacterium avium	Bac+	NA, 24 h, 37°C	sd	NG	Maruzzella and Sicurella (1960)
Staphylococcus aureus	Bac+	NA, 24 h, 37°C	~20,000	NG	Kellner and Kober (1954)
Staphylococcus aureus	Bac+	NA, 24 h, 37°C	sd	+++	Maruzzella and Sicurella (1960)
Streptococcus faecalis	Bac+	NA, 24 h, 37°C	~20,000	NG	Kellner and Kober (1954)
Streptococcus faecalis	Bac+	NA, 24 h, 37°C	sd	++	Maruzzella and Sicurella (1960)
Streptococcus pyogenes	Bac+	NA, 24 h, 37°C	~20,000	NG	Kellner and Kober (1954)
Botrytis cinerea	Fungi	PDA, 3 days, 25°C	1,000	+++	Lee et al. (2007)
Colletotrichum gloeosporioides	Fungi	PDA, 3 days, 25°C	1,000	+++	Lee et al. (2007)
Fusarium oxysporum	Fungi	PDA, 3 days, 25°C	1,000	+++	Lee et al. (2007)
Pythium ultimum	Fungi	PDA, 3 days, 25°C	1,000	+++	Lee et al. (2007)
Rhizoctonia solani	Fungi	PDA, 3 days, 25°C	1,000	+++	Lee et al. (2007)
Candida albicans	Yeast	NA, 24 h, 37°C	~20,000	NG	Kellner and Kober (1954)

TABLE 14.37
Inhibitory Data of Juniper Oil Obtained in the Agar Diffusion Test

Microorganism	MO Class	Conditions		Inhibition Zone (mm)	Reference
Aerobacter aerogenes	Bac−	NA, 24 h, 37°C	—, sd	0	Maruzzella and Lichtenstein (1956)
Enterobacter aerogenes	Bac−	MHA, 24 h, 30°C	6, 15,000	8	Rossi et al. (2007)
Escherichia coli	Bac−	Cited, 18 h, 37°C	6, 2,500	0	Janssen et al. (1986)
Escherichia coli	Bac−	NA, 18 h, 37°C	5 (h), −30,000	0	Schelz et al. (2006)
Escherichia coli	Bac−	NA, 24 h, 37°C	—, sd	1	Maruzzella and Lichtenstein (1956)
Escherichia coli	Bac−	MHA, 24 h, 30°C	6, 15,000	11	Rossi et al. (2007)
Neisseria perflava	Bac−	NA, 24 h, 37°C	—, sd	0	Maruzzella and Lichtenstein (1956)
Proteus vulgaris	Bac−	NA, 24 h, 37°C	—, sd	10	Maruzzella and Lichtenstein (1956)
Pseudomonas aeruginosa	Bac−	NA, 24 h, 37°C	—, sd	0	Maruzzella and Lichtenstein (1956)
Pseudomonas aeruginosa	Bac−	MHA, 24 h, 30°C	6, 15,000	6	Rossi et al. (2007)
Pseudomonas aeruginosa	Bac−	Cited, 18 h, 37°C	6, 2,500	8	Janssen et al. (1986)
Serratia marcescens	Bac−	NA, 24 h, 37°C	—, sd	4	Maruzzella and Lichtenstein (1956)
Bacillus mesentericus	Bac+	NA, 24 h, 37°C	—, sd	3	Maruzzella and Lichtenstein (1956)
Bacillus subtilis	Bac+	NA, 24 h, 37°C	—, sd	3	Maruzzella and Lichtenstein (1956)
Bacillus subtilis	Bac+	Cited, 18 h, 37°C	6, 2,500	7.7	Janssen et al. (1986)
Sarcina lutea	Bac+	NA, 24 h, 37°C	—, sd	5	Maruzzella and Lichtenstein (1956)
Staphylococcus aureus	Bac+	NA, 24 h, 37°C	—, sd	3	Maruzzella and Lichtenstein (1956)
Staphylococcus aureus	Bac+	Cited, 18 h, 37°C	6, 2,500	9	Janssen et al. (1986)
Staphylococcus aureus	Bac+	MHA, 24 h, 37°C	6, 15,000	17	Rossi et al. (2007)
Staphylococcus epidermidis	Bac+	NA, 18 h, 37°C	5 (h), −30,000	0	Schelz et al. (2006)
Streptomyces venezuelae	Bac+	SMA, 2–7 days, 20°C	sd	0	Maruzzella and Liguori (1958)
Alternaria porri	Fungi	PDA, 72 h, 28°C	5, 5,000	11	Pawar and Thaker (2007)
Alternaria solani	Fungi	SMA, 2–7 days, 20°C	sd	2	Maruzzella and Liguori (1958)
Aspergillus flavus	Fungi	SDA, 72 h, 26°C	8, 25,000	0	Shin (2003)
Aspergillus fumigatus	Fungi	SMA, 2–7 days, 20°C	sd	13	Maruzzella and Liguori (1958)
Aspergillus niger	Fungi	SDA, 72 h, 26°C	8, 25,000	0	Shin (2003)
Aspergillus niger	Fungi	PDA, 48 h, 28°C	5, 5,000	5	Pawar and Thaker (2006)
Aspergillus niger	Fungi	SMA, 2–7 days, 20°C	sd	6	Maruzzella and Liguori (1958)

(Continued)

TABLE 14.37 (*Continued*)
Inhibitory Data of Juniper Oil Obtained in the Agar Diffusion Test

Microorganism	MO Class	Conditions		Inhibition Zone (mm)	Reference
Botrytis cinerea	Fungi	PDA, few days, 24°C	12.7, sd	0	Angioni et al. (2003)
Cercospora beticola	Fungi	PDA, few days, 24°C	12.7, sd	0	Angioni et al. (2003)
Fusarium graminearum	Fungi	PDA, few days, 24°C	12.7, sd	0	Angioni et al. (2003)
Fusarium oxysporum f.sp. *ciceri*	Fungi	PDA, 72 h, 28°C	5, 5,000	7.5	Pawar and Thaker (2007)
Fusarium oxysporum lycopersici	Fungi	PDA, few days, 24°C	12.7, sd	0	Angioni et al. (2003)
Helminthosporium oryzae	Fungi	PDA, few days, 24°C	12.7, sd	0	Angioni et al. (2003)
Helminthosporium sativum	Fungi	SMA, 2–7 days, 20°C	sd	0	Maruzzella and Liguori (1958)
Mucor mucedo	Fungi	SMA, 2–7 days, 20°C	sd	0	Maruzzella and Liguori (1958)
Nigrospora panici	Fungi	SMA, 2–7 days, 20°C	sd	0	Maruzzella and Liguori (1958)
Penicillium digitatum	Fungi	SMA, 2–7 days, 20°C	sd	0	Maruzzella and Liguori (1958)
Phytophthora capsici	Fungi	PDA, few days, 24°C	12.7, sd	0	Angioni et al. (2003)
Pyricularia oryzae	Fungi	PDA, few days, 24°C	12.7, sd	0	Angioni et al. (2003)
Pythium ultimum	Fungi	PDA, few days, 24°C	12.7, sd	0	Angioni et al. (2003)
Rhizoctonia solani	Fungi	PDA, few days, 24°C	12.7, sd	0	Angioni et al. (2003)
Rhizopus nigricans	Fungi	SMA, 2–7 days, 20°C	sd	6	Maruzzella and Liguori (1958)
Sclerotium rolfsii	Fungi	PDA, few days, 24°C	12.7, sd	0	Angioni et al. (2003)
Septoria tritici	Fungi	PDA, few days, 24°C	12.7, sd	0	Angioni et al. (2003)
Candida albicans	Yeast	SMA, 2–7 days, 20°C	sd	0	Maruzzella and Liguori (1958)
Candida albicans	Yeast	Cited, 18 h, 37°C	6, 2,500	15	Janssen et al. (1986)
Candida krusei	Yeast	SMA, 2–7 days, 20°C	sd	0	Maruzzella and Liguori (1958)
Candida tropicalis	Yeast	SMA, 2–7 days, 20°C	sd	0	Maruzzella and Liguori (1958)
Cryptococcus neoformans	Yeast	SMA, 2–7 days, 20°C	sd	11	Maruzzella and Liguori (1958)
Cryptococcus rhodobenhani	Yeast	SMA, 2–7 days, 20°C	sd	6	Maruzzella and Liguori (1958)
Saccharomyces cerevisiae	Yeast	NA, 24 h, 20°C	5 (h), –30,000	15–17	Schelz et al. (2006)
Saccharomyces cerevisiae	Yeast	SMA, 2–7 days, 20°C	sd	0	Maruzzella and Liguori (1958)

TABLE 14.38
Inhibitory Data of Juniper Oil Obtained in the Dilution Test

Microorganism	MO Class	Conditions	MIC	Reference
Escherichia coli	Bac−	NB, DMSO, 24 h, 37°C	>900	Angioni et al. (2003)
Escherichia coli	Bac−	NB, 24 h, 37°C	>900	Cosentino et al. (2003)
Escherichia coli	Bac−	TGB, 18 h, 37°C	5,400	Schelz et al. (2006)
Escherichia coli O157:H7	Bac−	NB, 24 h, 37°C	>900	Cosentino et al. (2003)
Pseudomonas aeruginosa	Bac−	NB, DMSO, 24 h, 37°C	>900	Angioni et al. (2003)
Pseudomonas aeruginosa	Bac−	NB, 24 h, 37°C	>900	Cosentino et al. (2003)
Salmonella typhimurium	Bac−	NB, 24 h, 37°C	>900	Cosentino et al. (2003)
Yersinia enterocolitica	Bac−	NB, 24 h, 37°C	>900	Cosentino et al. (2003)
Yersinia enterocolitica	Bac−	MHA, Tween 20, 24 h, 37°C	2,500	Rossi et al. (2007)[a]
Bacillus cereus	Bac+	NB, 24 h, 37°C	>900	Cosentino et al. (2003)
Enterococcus faecalis	Bac+	NB, 24 h, 37°C	>900	Cosentino et al. (2003)
Enterococcus faecium VRE	Bac+	HIB, Tween 80, 18 h, 37°C	>20,000	Nelson (1997)
Listeria monocytogenes	Bac+	NB, 24 h, 37°C	>900	Cosentino et al. (2003)
Staphylococcus aureus	Bac+	NB, DMSO, 24 h, 37°C	>900	Angioni et al. (2003)
Staphylococcus aureus	Bac+	NB, 24 h, 37°C	>900	Cosentino et al. (2003)
Staphylococcus aureus MRSA	Bac+	HIB, Tween 80, 18 h, 37°C	>20,000	Nelson (1997)
Staphylococcus epidermidis	Bac+	TGB, 18 h, 37°C	>11,300	Schelz et al. (2006)
Alternaria alternate	Fungi	SDA, 6–8 h, 20°C	500, 13% inh.	Dikshit et al. (1986)
Aspergillus flavus	Fungi	MYB, 72 h, 26°C	>25,000	Shin (2003)
Aspergillus flavus	Fungi	RPMI, 24 h, 37°C	>900	Cosentino et al. (2003)
Aspergillus flavus	Fungi	RPMI, DMSO, 72 h, 37°C	20,000	Cavaleiro et al. (2006)[a]
Aspergillus flavus	Fungi	SDA, 6–8 h, 20°C	500, 11% inh.	Dikshit et al. (1986)
Aspergillus fumigatus	Fungi	RPMI, DMSO, 72 h, 37°C	10,000	Cavaleiro et al. (2006)[a]
Aspergillus niger	Fungi	MYB, 72 h, 26°C	>25,000	Shin (2003)
Aspergillus niger	Fungi	RPMI, DMSO, 72 h, 37°C	10,000–20,000	Cavaleiro et al. (2006)[a]
Epidermophyton floccosum	Fungi	SA, Tween 80, 21 days, 20°C	>1,250	Janssen et al. (1988)
Epidermophyton floccosum	Fungi	RPMI, DMSO, 72 h, 37°C	1,250	Cavaleiro et al. (2006)[a]
Microsporum canis	Fungi	RPMI, DMSO, 72 h, 37°C	1,250	Cavaleiro et al. (2006)[a]
Microsporum gypseum	Fungi	RPMI, DMSO, 72 h, 37°C	2,500	Cavaleiro et al. (2006)[a]
Microsporum gypseum	Fungi	SDA, 6–8 h, 20°C	500, 48% inh.	Dikshit et al. (1986)
Penicillium italicum	Fungi	SDA, 6–8 h, 20°C	500, 20% inh.	Dikshit et al. (1986)
Trichophyton mentagrophytes	Fungi	SA, Tween 80, 21 days, 20°C	>1,250	Janssen et al. (1988)
Trichophyton mentagrophytes	Fungi	RPMI, DMSO, 72 h, 37°C	1,250	Cavaleiro et al. (2006)[a]
Trichophyton mentagrophytes	Fungi	SDA, 6–8 h, 20°C	500, 48% inh.	Dikshit et al. (1986)
Trichophyton rubrum	Fungi	SA, Tween 80, 21 days, 20°C	>1,250	Janssen et al. (1988)
Trichophyton rubrum	Fungi	RPMI, DMSO, 72 h, 37°C	1,250	Cavaleiro et al. (2006)[a]
Trichophyton rubrum	Fungi	SDA, 6–8 h, 20°C	500, 53% inh.	Dikshit et al. (1986)
Candida albicans	Yeast	NB, DMSO, 24 h, 30°C	>900	Angioni et al. (2003)
Candida albicans	Yeast	RPMI, DMSO, 24 h, 37°C	1,250–10,000	Cavaleiro et al. (2006)[a]
Candida glabrata	Yeast	RPMI, DMSO, 24 h, 37°C	5,000	Cavaleiro et al. (2006)[a]

(*Continued*)

TABLE 14.38 (*Continued*)
Inhibitory Data of Juniper Oil Obtained in the Dilution Test

Microorganism	MO Class	Conditions	MIC	Reference
Candida krusei	Yeast	RPMI, DMSO, 24 h, 37°C	5,000	Cavaleiro et al. (2006)[a]
Candida parapsilosis	Yeast	RPMI, DMSO, 24 h, 37°C	5,000	Cavaleiro et al. (2006)[a]
Candida tropicalis	Yeast	RPMI, DMSO, 24 h, 37°C	20,000	Cavaleiro et al. (2006)[a]
Saccharomyces cerevisiae	Yeast	SB, 24 h, 37°C	>900	Cosentino et al. (2003)
Saccharomyces cerevisiae	Yeast	YPB, 24 h, 20°C	2,700	Schelz et al. (2006)

[a] *Juniper communis ssp. alpine.*

TABLE 14.39
Inhibitory Data of Juniper Oil Obtained in the Vapor Phase Test

Microorganism	MO Class	Conditions		Activity	Reference
Escherichia coli	Bac−	NA, 24 h, 37°C	~20,000	++	Kellner and Kober (1954)
Neisseria sp.	Bac−	NA, 24 h, 37°C	~20,000	NG	Kellner and Kober (1954)
Salmonella typhi	Bac−	NA, 24 h, 37°C	~20,000	+	Kellner and Kober (1954)
Bacillus megaterium	Bac+	NA, 24 h, 37°C	~20,000	NG	Kellner and Kober (1954)
Corynebacterium diphtheriae	Bac+	NA, 24 h, 37°C	~20,000	NG	Kellner and Kober (1954)
Staphylococcus aureus	Bac+	NA, 24 h, 37°C	~20,000	+	Kellner and Kober (1954)
Streptococcus faecalis	Bac+	NA, 24 h, 37°C	~20,000	+	Kellner and Kober (1954)
Streptococcus pyogenes	Bac+	NA, 24 h, 37°C	~20,000	NG	Kellner and Kober (1954)
Botrytis cinerea	Fungi	PDA, 3 days, 25°C	1,000	+++	Lee et al. (2007)
Colletotrichum gloeosporioides	Fungi	PDA, 3 days, 25°C	1,000	+++	Lee et al. (2007)
Fusarium oxysporum	Fungi	PDA, 3 days, 25°C	1,000	+++	Lee et al. (2007)
Pythium ultimum	Fungi	PDA, 3 days, 25°C	1,000	+++	Lee et al. (2007)
Rhizoctonia solani	Fungi	PDA, 3 days, 25°C	1,000	+++	Lee et al. (2007)
Candida albicans	Yeast	NA, 24 h, 37°C	~20,000	NG	Kellner and Kober (1954)

TABLE 14.40
Inhibitory Data of Lavender Oil Obtained in the Agar Diffusion Test

Microorganism	MO Class	Conditions		Inhibition Zone (mm)	Reference
Acinetobacter calcoaceticus	Bac–	ISA, 48 h, 25°C	4 (h), 10,000	10	Deans and Ritchie (1987)
Aerobacter aerogenes	Bac–	NA, 24 h, 37°C	—, sd	2	Maruzzella and Lichtenstein (1956)
Aeromonas hydrophila	Bac–	ISA, 48 h, 25°C	4 (h), 10,000	8	Deans and Ritchie (1987)
Alcaligenes faecalis	Bac–	ISA, 48 h, 25°C	4 (h), 10,000	0	Deans and Ritchie (1987)
Beneckea natriegens	Bac–	ISA, 48 h, 25°C	4 (h), 10,000	0	Deans and Ritchie (1987)
Citrobacter freundii	Bac–	ISA, 48 h, 25°C	4 (h), 10,000	8.5	Deans and Ritchie (1987)
Enterobacter aerogenes	Bac–	ISA, 48 h, 25°C	4 (h), 10,000	7.5	Deans and Ritchie (1987)
Enterococcus faecalis	Bac–	NA, 24 h, 37°C	5,000 on agar	0	Di Pasqua et al. (2005)
Erwinia carotovora	Bac–	ISA, 48 h, 25°C	4 (h), 10,000	6	Deans and Ritchie (1987)
Escherichia coli	Bac–	NA, 18 h, 37°C	6 (h), pure	0	Yousef and Tawil (1980)
Escherichia coli	Bac–	NA, 24 h, 37°C	—, sd	0	Maruzzella and Lichtenstein (1956)
Escherichia coli	Bac–	TGA, 18–24 h, 37°C	9.5, 2,000	0	Morris et al. (1979)
Escherichia coli	Bac–	ISA, 48 h, 25°C	4 (h), 10,000	7.5	Deans and Ritchie (1987)
Escherichia coli	Bac–	Cited, 24 h, 37°C	—	<12	Rota et al. (2004)
Escherichia coli	Bac–	Cited, 18 h, 37°C	6, 2,500	7.7	Janssen et al. (1986)
Flavobacterium suaveolens	Bac–	ISA, 48 h, 25°C	4 (h), 10,000	0	Deans and Ritchie (1987)
Klebsiella pneumoniae	Bac–	ISA, 48 h, 25°C	4 (h), 10,000	0	Deans and Ritchie (1987)
Moraxella sp.	Bac–	ISA, 48 h, 25°C	4 (h), 10,000	9	Deans and Ritchie (1987)
Neisseria perflava	Bac–	NA, 24 h, 37°C	—, sd	2	Maruzzella and Lichtenstein (1956)
Proteus vulgaris	Bac–	NA, 24 h, 37°C	—, sd	0	Maruzzella and Lichtenstein (1956)
Proteus vulgaris	Bac–	ISA, 48 h, 25°C	4 (h), 10,000	9	Deans and Ritchie (1987)
Pseudomonas aeruginosa	Bac–	NA, 18 h, 37°C	6 (h), pure	0	Yousef and Tawil (1980)
Pseudomonas aeruginosa	Bac–	NA, 24 h, 37°C	—, sd	0	Maruzzella and Lichtenstein (1956)
Pseudomonas aeruginosa	Bac–	Cited, 18 h, 37°C	6, 2,500	0	Janssen et al. (1986)
Pseudomonas aeruginosa	Bac–	ISA, 48 h, 25°C	4 (h), 10,000	0	Deans and Ritchie (1987)

(*Continued*)

TABLE 14.40 (Continued)
Inhibitory Data of Lavender Oil Obtained in the Agar Diffusion Test

Microorganism	MO Class	Conditions		Inhibition Zone (mm)	Reference
Pseudomonas sp.	Bac−	NA, 24 h, 37°C	5,000 on agar	0	Di Pasqua et al. (2005)
Salmonella enteritidis	Bac−	Cited, 24 h, 37°C	—	<12	Rota et al. (2004)
Salmonella pullorum	Bac−	ISA, 48 h, 25°C	4 (h), 10,000	7	Deans and Ritchie (1987)
Salmonella typhimurium	Bac−	Cited, 24 h, 37°C	—	<12	Rota et al. (2004)
Serratia marcescens	Bac−	NA, 24 h, 37°C	—, sd	2	Maruzzella and Lichtenstein (1956)
Serratia marcescens	Bac−	ISA, 48 h, 25°C	4 (h), 10,000	7.5	Deans and Ritchie (1987)
Shigella flexneri	Bac−	Cited, 24 h, 37°C	—	<12	Rota et al. (2004)
Yersinia enterocolitica	Bac−	ISA, 48 h, 25°C	4 (h), 10,000	9	Deans and Ritchie (1987)
Yersinia enterocolitica	Bac−	Cited, 24 h, 37°C	—	<12	Rota et al. (2004)
Bacillus mesentericus	Bac+	NA, 24 h, 37°C	—, sd	4	Maruzzella and Lichtenstein (1956)
Bacillus subtilis	Bac+	NA, 24 h, 37°C	—, sd	17	Maruzzella and Lichtenstein (1956)
Bacillus subtilis	Bac+	ISA, 48 h, 25°C	4 (h), 10,000	12.5	Deans and Ritchie (1987)
Bacillus subtilis	Bac+	Cited, 18 h, 37°C	6, 2,500	13.7	Janssen et al. (1986)
Bacillus subtilis	Bac+	NA, 18 h, 37°C	6 (h), pure	43	Yousef and Tawil (1980)
Brevibacterium linens	Bac+	ISA, 48 h, 25°C	4 (h), 10,000	8.5	Deans and Ritchie (1987)
Brochothrix thermosphacta	Bac+	ISA, 48 h, 25°C	4 (h), 10,000	5.5	Deans and Ritchie (1987)
Brochothrix thermosphacta	Bac+	NA, 24 h, 37°C	5,000 on agar	0	Di Pasqua et al. (2005)
Clostridium sporogenes	Bac+	ISA, 48 h, 25°C	4 (h), 10,000	0	Deans and Ritchie (1987)
Corynebacterium sp.	Bac+	TGA, 18–24 h, 37°C	9.5, 2,000	0	Morris et al. (1979)
Lactobacillus delbrueckii	Bac+	NA, 24 h, 37°C	5,000 on agar	0	Di Pasqua et al. (2005)
Lactobacillus plantarum	Bac+	NA, 24 h, 37°C	5,000 on agar	0	Di Pasqua et al. (2005)
Lactobacillus plantarum	Bac+	ISA, 48 h, 25°C	4 (h), 10,000	7.5	Deans and Ritchie (1987)
Lactobacillus sp.	Bac+	MRS, cited	9, 20,000	20 to >90	Pellecuer et al. (1980)
Lactococcus garvieae	Bac+	NA, 24 h, 37°C	5,000 on agar	0	Di Pasqua et al. (2005)
Lactococcus lactis subsp. *lactis*	Bac+	NA, 24 h, 37°C	5,000 on agar	0	Di Pasqua et al. (2005)
Leuconostoc cremoris	Bac+	ISA, 48 h, 25°C	4 (h), 10,000	0	Deans and Ritchie (1987)
Listeria monocytogenes	Bac+	ISA, 48 h, 25°C	4 (h), 10,000	0–18	Lis-Balchin et al. (1998)
Listeria monocytogenes	Bac+	NA, 24 h, 37°C	5,000 on agar	0	Di Pasqua et al. (2005)
Listeria monocytogenes	Bac+	Cited, 24 h, 37°C	—	<12	Rota et al. (2004)

(*Continued*)

TABLE 14.40 (*Continued*)
Inhibitory Data of Lavender Oil Obtained in the Agar Diffusion Test

Microorganism	MO Class	Conditions		Inhibition Zone (mm)	Reference
Micrococcus luteus	Bac+	MHA, cited	9, 20,000	0–22	Pellecuer et al. (1980)
Micrococcus luteus	Bac+	ISA, 48 h, 25°C	4 (h), 10,000	9.5	Deans and Ritchie (1987)
Micrococcus ureae	Bac+	MHA, cited	9, 20,000	22	Pellecuer et al. (1980)
Mycobacterium phlei	Bac+	NA, 18 h, 37°C	6 (h), pure	22	Yousef and Tawil (1980)
Sarcina lutea	Bac+	NA, 24 h, 37°C	—, sd	0	Maruzzella and Lichtenstein (1956)
Sarcina ureae	Bac+	MHA, cited	9, 20,000	>90	Pellecuer et al. (1980)
Staphylococcus aureus	Bac+	TGA, 18–24 h, 37°C	9.5, 2,000	0	Morris et al. (1979)
Staphylococcus aureus	Bac+	NA, 24 h, 37°C	5,000 on agar	0	Di Pasqua et al. (2005)
Staphylococcus aureus	Bac+	NA, 24 h, 37°C	—, sd	3	Maruzzella and Lichtenstein (1956)
Staphylococcus aureus	Bac+	ISA, 48 h, 25°C	4 (h), 10,000	8	Deans and Ritchie (1987)
Staphylococcus aureus	Bac+	Cited, 18 h, 37°C	6, 2,500	12.3	Janssen et al. (1986)
Staphylococcus aureus	Bac+	NA, 18 h, 37°C	6 (h), pure	14.3	Yousef and Tawil (1980)
Staphylococcus epidermidis	Bac+	MHA, cited	9, 20,000	14	Pellecuer et al. (1980)
Streptococcus D	Bac+	MHA, cited	9, 20,000	0–29	Pellecuer et al. (1980)
Streptococcus faecalis	Bac+	ISA, 48 h, 25°C	4 (h), 10,000	14	Deans and Ritchie (1987)
Streptococcus micros	Bac+	MHA, cited	9, 20,000	26	Pellecuer et al. (1980)
Streptomyces venezuelae	Bac+	SMA, 2–7 days, 20°C	sd	15	Maruzzella and Liguori (1958)
Alternaria porri	Fungi	PDA, 72 h, 28°C	5, 5,000	13.3	Pawar and Thaker (2007)
Alternaria solani	Fungi	SMA, 2–7 days, 20°C	sd	0	Maruzzella and Liguori (1958)
Aspergillus flavus	Fungi	SDA, 72 h, 26°C	8, 25,000	4	Shin (2003)
Aspergillus fumigatus	Fungi	SMA, 2–7 days, 20°C	sd	7	Maruzzella and Liguori (1958)
Aspergillus niger	Fungi	SDA, 72 h, 26°C	8, 25,000	2	Shin (2003)
Aspergillus niger	Fungi	SMA, 2–7 days, 20°C	sd	8	Maruzzella and Liguori (1958)
Aspergillus niger	Fungi	PDA, 48 h, 28°C	5, 5,000	10	Pawar and Thaker (2006)
Aspergillus niger	Fungi	SDA, 8 days, 30°C	6 (h), pure	32	Yousef and Tawil (1980)
Fusarium oxysporum f.sp. *ciceri*	Fungi	PDA, 72 h, 28°C	5, 5,000	9.8	Pawar and Thaker (2007)
Helminthosporium sativum	Fungi	SMA, 2–7 days, 20°C	sd	7	Maruzzella and Liguori (1958)

(*Continued*)

TABLE 14.40 (*Continued*)
Inhibitory Data of Lavender Oil Obtained in the Agar Diffusion Test

Microorganism	MO Class	Conditions		Inhibition Zone (mm)	Reference
Mucor mucedo	Fungi	SMA, 2–7 days, 20°C	sd	10	Maruzzella and Liguori (1958)
Mucor sp.	Fungi	SDA, 8 days, 30°C	6 (h), pure	21	Yousef and Tawil (1980)
Nigrospora panici	Fungi	SMA, 2–7 days, 20°C	sd	5	Maruzzella and Liguori (1958)
Penicillium chrysogenum	Fungi	SDA, 8 days, 30°C	6 (h), pure	16	Yousef and Tawil (1980)
Penicillium digitatum	Fungi	SMA, 2–7 days, 20°C	sd	3	Maruzzella and Liguori (1958)
Rhizopus nigricans	Fungi	SMA, 2–7 days, 20°C	sd	8	Maruzzella and Liguori (1958)
Rhizopus sp.	Fungi	SDA, 8 days, 30°C	6 (h), pure	14	Yousef and Tawil (1980)
Candida albicans	Yeast	TGA, 18–24 h, 37°C	9.5, 2,000	0	Morris et al. (1979)
Candida albicans	Yeast	SMA, 2–7 days, 20°C	sd	7	Maruzzella and Liguori (1958)
Candida albicans	Yeast	SDA, 18 h, 30°C	6 (h), pure	11.5	Yousef and Tawil (1980)
Candida albicans	Yeast	Cited, 18 h, 37°C	6, 2,500	14.3	Janssen et al. (1986)
Candida krusei	Yeast	SMA, 2–7 days, 20°C	sd	2	Maruzzella and Liguori (1958)
Candida tropicalis	Yeast	SMA, 2–7 days, 20°C	sd	6	Maruzzella and Liguori (1958)
Cryptococcus neoformans	Yeast	SMA, 2–7 days, 20°C	sd	0	Maruzzella and Liguori (1958)
Cryptococcus rhodobenhani	Yeast	SMA, 2–7 days, 20°C	sd	7	Maruzzella and Liguori (1958)
Saccharomyces cerevisiae	Yeast	SMA, 2–7 days, 20°C	sd	3	Maruzzella and Liguori (1958)

TABLE 14.41
Inhibitory Data of Lavender Oil Obtained in the Dilution Test

Microorganism	MO Class	Conditions	MIC	Reference
Bordetella bronchiseptica	Bac−	Cited	2,000	Pellecuer et al. (1976)
Escherichia coli	Bac−	Cited	500	Pellecuer et al. (1976)
Escherichia coli	Bac−	TGB, 18–24 h, 37°C	>1,000	Morris et al. (1979)
Escherichia coli	Bac−	TSB, 24 h, 37°C	>10,000	Di Pasqua et al. (2005)
Escherichia coli	Bac−	NB, Tween 20, 18 h, 37°C	50,000	Yousef and Tawil (1980)
Haemophilus influenza	Bac−	Cited	2,000	Pellecuer et al. (1976)
Klebsiella pneumoniae	Bac−	Cited	1,000	Pellecuer et al. (1976)
Moraxella glucidolytica	Bac−	Cited	2,000	Pellecuer et al. (1976)
Neisseria catarrhalis	Bac−	Cited	250	Pellecuer et al. (1976)
Neisseria flava	Bac−	Cited	500	Pellecuer et al. (1976)
Pseudomonas aeruginosa	Bac−	NB, Tween 20, 18 h, 37°C	>50,000	Yousef and Tawil (1980)
Salmonella typhimurium	Bac−	TSB, 24 h, 37°C	10,000	Di Pasqua et al. (2005)
Bacillus subtilis	Bac+	NB, Tween 20, 18 h, 37°C	400	Yousef and Tawil (1980)
Bacillus subtilis	Bac+	Cited	1,000	Pellecuer et al. (1976)
Corynebacterium pseudodiphtheriae	Bac+	Cited	1,000	Pellecuer et al. (1976)
Corynebacterium sp.	Bac+	TGB, 18–24 h, 37°C	1,000	Morris et al. (1979)
Enterococcus faecium VRE	Bac+	HIB, Tween 80, 18 h, 37°C	5,000–10,000	Nelson (1997)
Lactobacillus sp.	Bac+	MRS, cited	5	Pellecuer et al. (1980)
Micrococcus flavus	Bac+	Cited	1,000	Pellecuer et al. (1976)
Micrococcus luteus	Bac+	MHB, cited	1.25–2.5	Pellecuer et al. (1980)
Micrococcus ureae	Bac+	MHB, cited	2.5	Pellecuer et al. (1980)
Mycobacterium phlei	Bac+	NB, Tween 20, 18 h, 37°C	100	Yousef and Tawil (1980)
Sarcina lutea	Bac+	Cited	2,000	Pellecuer et al. (1976)
Sarcina ureae	Bac+	MHB, cited	0.31–0.62	Pellecuer et al. (1980)
Staphylococcus aureus	Bac+	TGB, 18–24 h, 37°C	>1,000	Morris et al. (1979)
Staphylococcus aureus	Bac+	Cited	2,000	Pellecuer et al. (1976)
Staphylococcus aureus	Bac+	TSB, 3% EtOH, 24 h, 37°C	10,000	Rota et al. (2004)
Staphylococcus aureus	Bac+	NB, Tween 20, 18 h, 37°C	12,500	Yousef and Tawil (1980)
Staphylococcus aureus	Bac+	MHB, Tween 80, 24 h, 37°C	12,500	Bastide et al. (1987)
Staphylococcus aureus MRSA	Bac+	HIB, Tween 80, 18 h, 37°C	5,000	Nelson (1997)
Staphylococcus epidermidis	Bac+	MHB, cited	5	Pellecuer et al. (1980)
Staphylococcus epidermidis	Bac+	Cited	2,000	Pellecuer et al. (1976)
Streptococcus D	Bac+	MHB, cited	0.62–5	Pellecuer et al. (1980)
Streptococcus micros	Bac+	MHB, cited	>5	Pellecuer et al. (1980)
Streptococcus pyogenes	Bac+	Cited	2,000	Pellecuer et al. (1976)
Alternaria alternata	Fungi	RPMI, 1.5% EtOH, 7 days, 30°C	5,000	Tullio et al. (2006)

(Continued)

TABLE 14.41 (Continued)
Inhibitory Data of Lavender Oil Obtained in the Dilution Test

Microorganism	MO Class	Conditions	MIC	Reference
Aspergillus flavus	Fungi	CA, 7 days, 28	5%–8% inh. 500	Kumar et al. (2007)
Aspergillus flavus	Fungi	MYB, 72 h, 26°C	3,120	Shin (2003)
Aspergillus flavus	Fungi	RPMI, 1.5% EtOH, 7 days, 30°C	10,000	Tullio et al. (2006)
Aspergillus flavus var. *columnaris*	Fungi	RPMI, 1.5% EtOH, 7 days, 30°C	>10,000	Tullio et al. (2006)
Aspergillus fumigatus	Fungi	RPMI, 1.5% EtOH, 7 days, 30°C	10,000	Tullio et al. (2006)
Aspergillus niger	Fungi	Cited	1,000	Pellecuer et al. (1976)
Aspergillus niger	Fungi	MYB, 72 h, 26°C	3,120	Shin (2003)
Aspergillus niger	Fungi	YES broth, 10 d	−93% inh. 10,000	Lis-Balchin et al. (1998)
Aspergillus niger	Fungi	RPMI, 1.5% EtOH, 7 days, 30°C	10,000	Tullio et al. (2006)
Aspergillus niger	Fungi	NB, Tween 20, 8 days, 30°C	25,000	Yousef and Tawil (1980)
Aspergillus ochraceus	Fungi	YES broth, 10 d	−90% inh. 10,000	Lis-Balchin et al. (1998)
Cladosporium cladosporioides	Fungi	RPMI, 1.5% EtOH, 7 days, 30°C	2,500–10,000	Tullio et al. (2006)
Epidermophyton floccosum	Fungi	SA, Tween 80, 21 days, 20°C	<300	Janssen et al. (1988)
Epidermophyton floccosum	Fungi	Cited	1,000	Pellecuer et al. (1976)
Epidermophyton floccosum	Fungi	RPMI, 1.5% EtOH, 7 days, 30°C	5,000	Tullio et al. (2006)
Fusarium culmorum	Fungi	YES broth, 10 d	−89% inh. 10,000	Lis-Balchin et al. (1998)
Fusarium oxysporum	Fungi	RPMI, 1.5% EtOH, 7 days, 30°C	156	Tullio et al. (2006)
Microsporum canis	Fungi	RPMI, 1.5% EtOH, 7 days, 30°C	2,500–5,000	Tullio et al. (2006)
Microsporum canis	Fungi	MBA, Tween 80, 10 days, 30°C	75 to >300	Perrucci et al. (1994)
Microsporum gypseum	Fungi	RPMI, 1.5% EtOH, 7 days, 30°C	5,000–10,000	Tullio et al. (2006)
Microsporum gypseum	Fungi	MBA, Tween 80, 10 days, 30°C	50 to >300	Perrucci et al. (1994)
Microsporum gypseum	Fungi	SDA, 7 days, 30°C	400, 24% inh.	Dikshit and Husain (1984)
Mucor sp.	Fungi	RPMI, 1.5% EtOH, 7 days, 30°C	>10,000	Tullio et al. (2006)
Mucor sp.	Fungi	NB, Tween 20, 8 days, 30°C	50,000	Yousef and Tawil (1980)
Penicillium chrysogenum	Fungi	NB, Tween 20, 8 days, 30°C	12,500	Yousef and Tawil (1980)
Penicillium frequentans	Fungi	RPMI, 1.5% EtOH, 7 days, 30°C	5,000	Tullio et al. (2006)
Penicillium lanosum	Fungi	RPMI, 1.5% EtOH, 7 days, 30°C	10,000	Tullio et al. (2006)
Rhizopus sp.	Fungi	RPMI, 1.5% EtOH, 7 days, 30°C	>10,000	Tullio et al. (2006)

(Continued)

TABLE 14.41 (*Continued*)
Inhibitory Data of Lavender Oil Obtained in the Dilution Test

Microorganism	MO Class	Conditions	MIC	Reference
Rhizopus sp.	Fungi	NB, Tween 20, 8 days, 30°C	12,500	Yousef and Tawil (1980)
Scopulariopsis brevicaulis	Fungi	RPMI, 1.5% EtOH, 7 days, 30°C	10,000	Tullio et al. (2006)
Trichophyton equinum	Fungi	SDA, 7 days, 30°C	400, 10% inh.	Dikshit and Husain (1984)
Trichophyton interdigitale	Fungi	Cited	1,000	Pellecuer et al. (1976)
Trichophyton mentagrophytes	Fungi	RPMI, 1.5% EtOH, 7 days, 30°C	5,000–10,000	Tullio et al. (2006)
Trichophyton mentagrophytes	Fungi	SA, Tween 80, 21 days, 20°C	300–625	Janssen et al. (1988)
Trichophyton rubrum	Fungi	SA, Tween 80, 21 days, 20°C	<300	Janssen et al. (1988)
Trichophyton rubrum	Fungi	SDA, 7 days, 30°C	400, 21% inh.	Dikshit and Husain (1984)
Candida albicans	Yeast	TGB, 18–24 h, 37°C	1,000	Morris et al. (1979)
Candida albicans	Yeast	Cited	2,000	Pellecuer et al. (1976)
Candida albicans	Yeast	MHB, Tween 80, 48 h, 35	5,000	Hammer et al. (1998)
Candida albicans	Yeast	NB, Tween 20, 18 h, 37°C	6,400	Yousef and Tawil (1980)
Candida mycoderma	Yeast	Cited	2,000	Pellecuer et al. (1976)
Candida parapsilosis	Yeast	Cited	4,000	Pellecuer et al. (1976)
Candida pelliculosa	Yeast	Cited	1,000	Pellecuer et al. (1976)
Candida tropicalis	Yeast	Cited	4,000	Pellecuer et al. (1976)
Geotrichum asteroides	Yeast	Cited	2,000	Pellecuer et al. (1976)
Hansenula sp.	Yeast	Cited	2,000	Pellecuer et al. (1976)
Saccharomyces carlsbergensis	Yeast	Cited	2,000	Pellecuer et al. (1976)

TABLE 14.42
Inhibitory Data of Lavender Oil Obtained in the Vapor Phase Test

Microorganism	MO Class	Conditions		Activity	Reference
Escherichia coli	Bac–	NA, 24 h, 37°C	~20,000	+	Kellner and Kober (1954)
Escherichia coli	Bac–	BLA, 18 h, 37°C	MIC_{air}	>1600	Inouye et al. (2001)
Haemophilus influenzae	Bac–	MHA, 18 h, 37°C	MIC_{air}	25	Inouye et al. (2001)
Neisseria sp.	Bac–	NA, 24 h, 37°C	~20,000	NG	Kellner and Kober (1954)
Salmonella typhi	Bac–	NA, 24 h, 37°C	~20,000	NG	Kellner and Kober (1954)
Salmonella typhi	Bac–	NA, 24 h, 37°C	sd	+++	Maruzzella and Sicurella (1960)
Bacillus megaterium	Bac+	NA, 24 h, 37°C	~20,000	+	Kellner and Kober (1954)
Bacillus subtilis var. *aterrimus*	Bac+	NA, 24 h, 37°C	sd	+++	Maruzzella and Sicurella (1960)
Corynebacterium diphtheriae	Bac+	NA, 24 h, 37°C	~20,000	NG	Kellner and Kober (1954)
Lactobacillus sp.	Bac+	MRS, cited	Disk, 20,000?	+++	Pellecuer et al. (1980)
Micrococcus luteus	Bac+	MHB, cited	Disk, 20,000?	+++	Pellecuer et al. (1980)
Micrococcus ureae	Bac+	MHB, cited	Disk, 20,000?	++	Pellecuer et al. (1980)
Mycobacterium avium	Bac+	NA, 24 h, 37°C	sd	NG	Maruzzella and Sicurella (1960)
Sarcina ureae	Bac+	MHB, cited	Disk, 20,000?	+	Pellecuer et al. (1980)
Staphylococcus aureus	Bac+	NA, 24 h, 37°C	~20,000	NG	Kellner and Kober (1954)
Staphylococcus aureus	Bac+	NA, 24 h, 37°C	sd	+++	Maruzzella and Sicurella (1960)
Staphylococcus aureus	Bac+	MHA, 18 h, 37°C	MIC_{air}	100	Inouye et al. (2001)
Staphylococcus epidermidis	Bac+	MHB, cited	Disk, 20,000?	++	Pellecuer et al. (1980)
Streptococcus D	Bac+	MHB, cited	Disk, 20,000?	+++	Pellecuer et al. (1980)
Streptococcus faecalis	Bac+	NA, 24 h, 37°C	~20,000	NG	Kellner and Kober (1954)
Streptococcus faecalis	Bac+	NA, 24 h, 37°C	sd	+++	Maruzzella and Sicurella (1960)
Streptococcus micros	Bac+	MHB, cited	Disk, 20,000?	+++	Pellecuer et al. (1980)
Streptococcus pneumoniae	Bac+	MHA, 18 h, 37°C	MIC_{air}	50	Inouye et al. (2001)
Streptococcus pyogenes	Bac+	NA, 24 h, 37°C	~20,000	NG	Kellner and Kober (1954)
Streptococcus pyogenes	Bac+	MHA, 18 h, 37°C	MIC_{air}	50	Inouye et al. (2001)
Alternaria alternata	Fungi	RPMI, 7 days, 30°C	MIC_{air}	625–2500	Tullio et al. (2006)
Aspergillus flavus	Fungi	RPMI, 7 days, 30°C	MIC_{air}	2500	Tullio et al. (2006)

(*Continued*)

TABLE 14.42 (Continued)
Inhibitory Data of Lavender Oil Obtained in the Vapor Phase Test

Microorganism	MO Class	Conditions		Activity	Reference
Aspergillus fumigatus	Fungi	RPMI, 7 days, 30°C	MIC_{air}	1250–2500	Tullio et al. (2006)
Aspergillus niger	Fungi	RPMI, 7 days, 30°C	MIC_{air}	1250	Tullio et al. (2006)
Cladosporium cladosporioides	Fungi	RPMI, 7 days, 30°C	MIC_{air}	156–312	Tullio et al. (2006)
Fusarium oxysporum	Fungi	RPMI, 7 days, 30°C	MIC_{air}	5000	Tullio et al. (2006)
Microsporum canis	Fungi	RPMI, 7 days, 30°C	MIC_{air}	312–1250	Tullio et al. (2006)
Microsporum gypseum	Fungi	RPMI, 7 days, 30°C	MIC_{air}	312	Tullio et al. (2006)
Mucor sp.	Fungi	RPMI, 7 days, 30°C	MIC_{air}	1250	Tullio et al. (2006)
Penicillium frequentans	Fungi	RPMI, 7 days, 30°C	MIC_{air}	625	Tullio et al. (2006)
Penicillium lanosum	Fungi	RPMI, 7 days, 30°C	MIC_{air}	625	Tullio et al. (2006)
Rhizopus sp.	Fungi	RPMI, 7 days, 30°C	MIC_{air}	2500	Tullio et al. (2006)
Scopulariopsis brevicaulis	Fungi	RPMI, 7 days, 30°C	MIC_{air}	125	Tullio et al. (2006)
Trichophyton mentagrophytes	Fungi	RPMI, 7 days, 30°C	MIC_{air}	312–625	Tullio et al. (2006)
Candida albicans	Yeast	NA, 24 h, 37°C	~20,000	NG	Kellner and Kober (1954)

TABLE 14.43
Inhibitory Data of Lemon Oil Obtained in the Agar Diffusion Test

Microorganism	MO Class	Conditions		Inhibition Zone (mm)	Reference
Acinetobacter calcoaceticus	Bac−	ISA, 48 h, 25°C	4 (h), 10,000	0	Deans and Ritchie (1987)
Aerobacter aerogenes	Bac−	NA, 24 h, 37°C	5 × 20, 1,000	1–10	Möse and Lukas (1957)
Aerobacter aerogenes	Bac−	NA, 24 h, 37°C	—, sd	0	Maruzzella and Lichtenstein (1956)
Aeromonas hydrophila	Bac−	ISA, 48 h, 25°C	4 (h), 10,000	6.5	Deans and Ritchie (1987)
Alcaligenes faecalis	Bac−	ISA, 48 h, 25°C	4 (h), 10,000	0	Deans and Ritchie (1987)
Beneckea natriegens	Bac−	ISA, 48 h, 25°C	4 (h), 10,000	0	Deans and Ritchie (1987)
Brucella abortus	Bac−	NA, 24 h, 37°C	5 × 20, 1,000	2–5	Möse and Lukas (1957)
Campylobacter jejuni	Bac−	TSA, 24 h, 42°C	4 (h), 25,000	4.6	Smith-Palmer et al. (1998)
Campylobacter jejuni	Bac−	CAB, 24 h, 42°C	Disk, 10,000	41	Fisher and Phillips (2006)
Citrobacter freundii	Bac−	ISA, 48 h, 25°C	4 (h), 10,000	0	Deans and Ritchie (1987)
Enterobacter aerogenes	Bac−	ISA, 48 h, 25°C	4 (h), 10,000	0	Deans and Ritchie (1987)
Erwinia carotovora	Bac−	ISA, 48 h, 25°C	4 (h), 10,000	6	Deans and Ritchie (1987)
Escherichia coli	Bac−	NA, 24 h, 37°C	5 × 20, 1,000	0–10	Möse and Lukas (1957)
Escherichia coli	Bac−	Cited	15, 2,500	0	Pizsolitto et al. (1975)
Escherichia coli	Bac−	NA, 24 h, 37°C	10 (h), 100,000	0	Narasimha Rao and Nigam (1970)
Escherichia coli	Bac−	NA, 24 h, 37°C	—, sd	0	Maruzzella and Lichtenstein (1956)
Escherichia coli	Bac−	TGA, 18–24 h, 37°C	9.5, 2,000	0	Morris et al. (1979)
Escherichia coli	Bac−	ISA, 48 h, 25°C	4 (h), 10,000	0	Deans and Ritchie (1987)
Escherichia coli	Bac−	TSA, 24 h, 35°C	4 (h), 25,000	4	Smith-Palmer et al. (1998)
Escherichia coli	Bac−	NA, 24 h, 37°C	Disk, 10,000	21	Fisher and Phillips (2006)
Escherichia coli	Bac−	NA, 18 h, 37°C	6 (h), pure	18.5	Yousef and Tawil (1980)
Flavobacterium suaveolens	Bac−	ISA, 48 h, 25°C	4 (h), 10,000	13.5	Deans and Ritchie (1987)
Klebsiella aerogenes	Bac−	NA, 24 h, 37°C	10 (h), 100,000	0	Narasimha Rao and Nigam (1970)
Klebsiella pneumonia	Bac−	NA, 24 h, 37°C	5 × 20, 1,000	6–10	Möse and Lukas (1957)
Klebsiella pneumonia subsp. *ozaenae*	Bac−	NA, 24 h, 37°C	5 × 20, 1,000	2–5	Möse and Lukas (1957)
Klebsiella pneumoniae	Bac−	ISA, 48 h, 25°C	4 (h), 10,000	0	Deans and Ritchie (1987)
Klebsiella sp.	Bac−	Cited	15, 2,500	0	Pizsolitto et al. (1975)
Moraxella sp.	Bac−	ISA, 48 h, 25°C	4 (h), 10,000	0	Deans and Ritchie (1987)
Neisseria perflava	Bac−	NA, 24 h, 37°C	—, sd	0	Maruzzella and Lichtenstein (1956)
Proteus OX19	Bac−	NA, 24 h, 37°C	5 × 20, 1,000	2–5	Möse and Lukas (1957)
Proteus sp.	Bac−	Cited	15, 2,500	0	Pizsolitto et al. (1975)
Proteus vulgaris	Bac−	NA, 24 h, 37°C	5 × 20, 1,000	2–5	Möse and Lukas (1957)

(Continued)

TABLE 14.43 (Continued)
Inhibitory Data of Lemon Oil Obtained in the Agar Diffusion Test

Microorganism	MO Class	Conditions		Inhibition Zone (mm)	Reference
Proteus vulgaris	Bac−	NA, 24 h, 37°C	—, sd	0	Maruzzella and Lichtenstein (1956)
Proteus vulgaris	Bac−	ISA, 48 h, 25°C	4 (h), 10,000	0	Deans and Ritchie (1987)
Pseudomonas aeruginosa	Bac−	Cited	15, 2,500	0	Pizsolitto et al. (1975)
Pseudomonas aeruginosa	Bac−	NA, 18 h, 37°C	6 (h), pure	0	Yousef and Tawil (1980)
Pseudomonas aeruginosa	Bac−	NA, 24 h, 37°C	—, sd	0	Maruzzella and Lichtenstein (1956)
Pseudomonas aeruginosa	Bac−	ISA, 48 h, 25°C	4 (h), 10,000	0	Deans and Ritchie (1987)
Pseudomonas aeruginosa	Bac−	NA, 24 h, 37°C	5 × 20, 1,000	1	Möse and Lukas (1957)
Pseudomonas fluorescens	Bac−	NA, 24 h, 37°C	5 × 20, 1,000	1	Möse and Lukas (1957)
Pseudomonas mangiferae indica	Bac−	NA, 36–48 h, 37°C	6, sd	14	Garg and Garg (1980a,b)
Salmonella enteritidis	Bac−	NA, 24 h, 37°C	5 × 20, 1,000	6–10	Möse and Lukas (1957)
Salmonella enteritidis	Bac−	TSA, 24 h, 35°C	4 (h), 25,000	4	Smith-Palmer et al. (1998)
Salmonella paratyphi	Bac−	NA, 36–48 h, 37°C	6, sd	16	Garg and Garg (1980a,b)
Salmonella paratyphi B	Bac−	NA, 24 h, 37°C	5 × 20, 1,000	6–10	Möse and Lukas (1957)
Salmonella pullorum	Bac−	ISA, 48 h, 25°C	4 (h), 10,000	7	Deans and Ritchie (1987)
Salmonella sp.	Bac−	Cited	15, 2,500	0	Pizsolitto et al. (1975)
Salmonella typhi	Bac−	NA, 24 h, 37°C	5 × 20, 1,000	2–5	Möse and Lukas (1957)
Salmonella typhi	Bac−	NA, 24 h, 37°C	10 (h), 100,000	0	Narasimha Rao and Nigam (1970)
Salmonella typhi	Bac−	NA, 36–48 h, 37°C	6, sd	28	Garg and Garg (1980)
Serratia marcescens	Bac−	NA, 24 h, 37°C	—, sd	0	Maruzzella and Lichtenstein (1956)
Serratia marcescens	Bac−	ISA, 48 h, 25°C	4 (h), 10,000	0	Deans and Ritchie (1987)
Serratia marcescens	Bac−	NA, 24 h, 37°C	5 × 20, 1,000	1	Möse and Lukas (1957)
Serratia sp.	Bac−	Cited	15, 2,500	0	Pizsolitto et al. (1975)
Shigella sp.	Bac−	Cited	15, 2,500	0	Pizsolitto et al. (1975)
Vibrio albicans	Bac−	NA, 24 h, 37°C	5 × 20, 1,000	11–13	Möse and Lukas (1957)
Vibrio cholera	Bac−	NA, 36–48 h, 37°C	6, sd	12	Garg and Garg (1980a,b)
Vibrio cholerae	Bac−	NA, 24 h, 37°C	5 × 20, 1,000	6–10	Möse and Lukas (1957)
Vibrio cholerae	Bac−	NA, 24 h, 37°C	10 (h), 100,000	22	Narasimha Rao and Nigam (1970)
Yersinia enterocolitica	Bac−	ISA, 48 h, 25°C	4 (h), 10,000	0	Deans and Ritchie (1987)
Bacillus anthracis	Bac+	NA, 24 h, 37°C	5 × 20, 1,000	6–10	Möse and Lukas (1957)
Bacillus cereus	Bac+	BHA, 24 h, 30°C	Disk, 10,000	19	Fisher and Phillips (2006)

(Continued)

TABLE 14.43 (Continued)
Inhibitory Data of Lemon Oil Obtained in the Agar Diffusion Test

Microorganism	MO Class	Conditions		Inhibition Zone (mm)	Reference
Bacillus mesentericus	Bac+	NA, 24 h, 37°C	—, sd	4	Maruzzella and Lichtenstein (1956)
Bacillus mycoides	Bac+	NA, 36–48 h, 37°C	6, sd	12	Garg and Garg (1980a,b)
Bacillus pumilus	Bac+	NA, 36–48 h, 37°C	6, sd	16	Garg and Garg (1980a,b)
Bacillus sp.	Bac+	Cited	15, 2,500	0	Pizsolitto et al. (1975)
Bacillus subtilis	Bac+	NA, 24 h, 37°C	5 × 20, 1,000	1	Möse and Lukas (1957)
Bacillus subtilis	Bac+	NA, 24 h, 37°C	—, sd	5	Maruzzella and Lichtenstein (1956)
Bacillus subtilis	Bac+	ISA, 48 h, 25°C	4 (h), 10,000	6	Deans and Ritchie (1987)
Bacillus subtilis	Bac+	NA, 36–48 h, 37°C	6, sd	16	Garg and Garg (1980a,b)
Bacillus subtilis	Bac+	NA, 18 h, 37°C	6 (h), pure	19.5	Yousef and Tawil (1980)
Brevibacterium linens	Bac+	ISA, 48 h, 25°C	4 (h), 10,000	6	Deans and Ritchie (1987)
Brochothrix thermosphacta	Bac+	ISA, 48 h, 25°C	4 (h), 10,000	8	Deans and Ritchie (1987)
Clostridium sporogenes	Bac+	ISA, 48 h, 25°C	4 (h), 10,000	0	Deans and Ritchie (1987)
Corynebacterium diphtheria	Bac+	NA, 24 h, 37°C	5 × 20, 1,000	1	Möse and Lukas (1957)
Corynebacterium diphtheria	Bac+	NA, 24 h, 37°C	10 (h), 100,000	0	Narasimha Rao and Nigam (1970)
Corynebacterium sp.	Bac+	TGA, 18–24 h, 37°C	9.5, 2,000	0	Morris et al. (1979)
Lactobacillus plantarum	Bac+	ISA, 48 h, 25°C	4 (h), 10,000	0	Deans and Ritchie (1987)
Leuconostoc cremoris	Bac+	ISA, 48 h, 25°C	4 (h), 10,000	7	Deans and Ritchie (1987)
Listeria monocytogenes	Bac+	NA, 24 h, 37°C	5 × 20, 1,000	2–5	Möse and Lukas (1957)
Listeria monocytogenes	Bac+	ISA, 48 h, 25°C	4 (h), 10,000	3	Lis-Balchin et al. (1998)
Listeria monocytogenes	Bac+	TSA, 24 h, 35°C	4 (h), 25,000	5.3	Smith-Palmer et al. (1998)
Listeria monocytogenes	Bac+	LSA, 24 h, 37°C	Disk, 10,000	41	Fisher and Phillips (2006)
Micrococcus luteus	Bac+	ISA, 48 h, 25°C	4 (h), 10,000	0	Deans and Ritchie (1987)
Mycobacterium phlei	Bac+	NA, 18 h, 37°C	6 (h), pure	24	Yousef and Tawil (1980)
Sarcina alba	Bac+	NA, 24 h, 37°C	5 × 20, 1,000	2–5	Möse and Lukas (1957)
Sarcina beige	Bac+	NA, 24 h, 37°C	5 × 20, 1,000	6–10	Möse and Lukas (1957)
Sarcina citrea	Bac+	NA, 24 h, 37°C	5 × 20, 1,000	2–5	Möse and Lukas (1957)
Sarcina lutea	Bac+	NA, 24 h, 37°C	—, sd	0	Maruzzella and Lichtenstein (1956)
Sarcina lutea	Bac+	NA, 36–48 h, 37°C	6, sd	17	Garg and Garg (1980a,b)
Sarcina rosa	Bac+	NA, 24 h, 37°C	5 × 20, 1,000	6–10	Möse and Lukas (1957)
Sporococcus sarc.	Bac+	NA, 24 h, 37°C	5 × 20, 1,000	6–10	Möse and Lukas (1957)
Staphylococcus albus	Bac+	NA, 24 h, 37°C	5 × 20, 1,000	6–10	Möse and Lukas (1957)
Staphylococcus aureus	Bac+	NA, 24 h, 37°C	5 × 20, 1,000	11–15	Möse and Lukas (1957)
Staphylococcus aureus	Bac+	Cited	15, 2,500	0	Pizsolitto et al. (1975)

(Continued)

TABLE 14.43 (Continued)
Inhibitory Data of Lemon Oil Obtained in the Agar Diffusion Test

Microorganism	MO Class	Conditions		Inhibition Zone (mm)	Reference
Staphylococcus aureus	Bac+	NA, 24 h, 37°C	10 (h), 100,000	0	Narasimha Rao and Nigam (1970)
Staphylococcus aureus	Bac+	NA, 24 h, 37°C	—, sd	0	Maruzzella and Lichtenstein (1956)
Staphylococcus aureus	Bac+	TGA, 18–24 h, 37°C	9.5, 2,000	0	Morris et al. (1979)
Staphylococcus aureus	Bac+	ISA, 48 h, 25°C	4 (h), 10,000	0	Deans and Ritchie (1987)
Staphylococcus aureus	Bac+	TSA, 24 h, 35°C	4 (h), 25,000	6	Smith-Palmer et al. (1998)
Staphylococcus aureus	Bac+	NA, 36–48 h, 37°C	6, sd	14	Garg and Garg (1980)
Staphylococcus aureus	Bac+	BHA, 24 h, 37°C	Disk, 10,000	14	Fisher and Phillips (2006)
Staphylococcus aureus	Bac+	NA, 18 h, 37°C	6 (h), pure	22	Yousef and Tawil (1980a,b)
Staphylococcus epidermidis	Bac+	Cited	15, 2,500	0	Pizsolitto et al. (1975)
Staphylococcus epidermidis	Bac+	NA, 24 h, 37°C	10 (h), 100,000	18	Narasimha Rao and Nigam (1970)
Streptococcus faecalis	Bac+	ISA, 48 h, 25°C	4 (h), 10,000	7	Deans and Ritchie (1987)
Streptococcus haemolyticus	Bac+	NA, 24 h, 37°C	5 × 20, 1,000	6–10	Möse and Lukas (1957)
Streptococcus sp.	Bac+	NA, 24 h, 37°C	10 (h), 100,000	0	Narasimha Rao and Nigam (1970)
Streptococcus viridans	Bac+	Cited	15, 2,500	0	Pizsolitto et al. (1975)
Streptococcus viridians	Bac+	NA, 24 h, 37°C	5 × 20, 1,000	1	Möse and Lukas (1957)
Streptomyces venezuelae	Bac+	SMA, 2–7 days, 20°C	sd	11	Maruzzella and Liguori (1958)
Alternaria porri	Fungi	PDA, 72 h, 28°C	5, 5,000	7.3	Pawar and Thaker (2007)
Alternaria solani	Fungi	SMA, 2–7 days, 20°C	sd	6	Maruzzella and Liguori (1958)
Aspergillus fumigatus	Fungi	SMA, 2–7 days, 20°C	sd	9	Maruzzella and Liguori (1958)
Aspergillus niger	Fungi	PDA, 48 h, 28°C	5, 5,000	7	Pawar and Thaker (2006)
Aspergillus niger	Fungi	SMA, 2–7 days, 20°C	sd	9	Maruzzella and Liguori (1958)
Aspergillus niger	Fungi	SDA, 8 days, 30°C	6 (h), pure	18	Yousef and Tawil (1980)
Fusarium oxysporum f.sp. ciceri	Fungi	PDA, 72 h, 28°C	5, 5,000	13.2	Pawar and Thaker (2007)
Helminthosporium sativum	Fungi	SMA, 2–7 days, 20°C	sd	16	Maruzzella and Liguori (1958)
Mucor mucedo	Fungi	SMA, 2–7 days, 20°C	sd	11	Maruzzella and Liguori (1958)
Mucor sp.	Fungi	SDA, 8 days, 30°C	6 (h), pure	14	Yousef and Tawil (1980)
Nigrospora panici	Fungi	SMA, 2–7 days, 20°C	sd	10	Maruzzella and Liguori (1958)

(Continued)

TABLE 14.43 (Continued)
Inhibitory Data of Lemon Oil Obtained in the Agar Diffusion Test

Microorganism	MO Class	Conditions		Inhibition Zone (mm)	Reference
Penicillium chrysogenum	Fungi	SDA, 8 days, 30°C	6 (h), pure	17	Yousef and Tawil (1980)
Penicillium digitatum	Fungi	SMA, 2–7 days, 20°C	sd	11	Maruzzella and Liguori (1958)
Rhizopus nigricans	Fungi	SMA, 2–7 days, 20°C	sd	0	Maruzzella and Liguori (1958)
Rhizopus sp.	Fungi	SDA, 8 days, 30°C	6 (h), pure	0	Yousef and Tawil (1980)
Brettanomyces anomalus	Yeast	MPA, 4 days, 30°C	5, 10% sol. sd	0	Conner and Beuchat (1984)
Candida albicans	Yeast	TGA, 18–24 h, 37°C	9.5, 2,000	0	Morris et al. (1979)
Candida albicans	Yeast	SMA, 2–7 days, 20°C	sd	2	Maruzzella and Liguori (1958)
Candida albicans	Yeast	SDA, 18 h, 30°C	6 (h), pure	33.5	Yousef and Tawil (1980)
Candida krusei	Yeast	SMA, 2–7 days, 20°C	sd	3	Maruzzella and Liguori (1958)
Candida lipolytica	Yeast	MPA, 4 days, 30°C	5, 10% sol. sd	0	Conner and Beuchat (1984)
Candida tropicalis	Yeast	SMA, 2–7 days, 20°C	sd	1	Maruzzella and Liguori (1958)
Cryptococcus neoformans	Yeast	SMA, 2–7 days, 20°C	sd	5	Maruzzella and Liguori (1958)
Cryptococcus rhodobenhani	Yeast	SMA, 2–7 days, 20°C	sd	4	Maruzzella and Liguori (1958)
Debaryomyces hansenii	Yeast	MPA, 4 days, 30°C	5, 10% sol. sd	0	Conner and Beuchat (1984)
Geotrichum candidum	Yeast	MPA, 4 days, 30°C	5, 10% sol. sd	0	Conner and Beuchat (1984)
Hansenula anomala	Yeast	MPA, 4 days, 30°C	5, 10% sol. sd	0	Conner and Beuchat (1984)
Kloeckera apiculata	Yeast	MPA, 4 days, 30°C	5, 10% sol. sd	0	Conner and Beuchat (1984)
Kluyveromyces fragilis	Yeast	MPA, 4 days, 30°C	5, 10% sol. sd	0	Conner and Beuchat (1984)
Lodderomyces elongisporus	Yeast	MPA, 4 days, 30°C	5, 10% sol. sd	0	Conner and Beuchat (1984)
Metschnikowia pulcherrima	Yeast	MPA, 4 days, 30°C	5, 10% sol. sd	0	Conner and Beuchat (1984)
Pichia membranifaciens	Yeast	MPA, 4 days, 30°C	5, 10% sol. sd	0	Conner and Beuchat (1984)
Rhodotorula rubra	Yeast	MPA, 4 days, 30°C	5, 10% sol. sd	0	Conner and Beuchat (1984)
Saccharomyces cerevisiae	Yeast	MPA, 4 days, 30°C	5, 10% sol. sd	0	Conner and Beuchat (1984)
Saccharomyces cerevisiae	Yeast	SMA, 2–7 days, 20°C	sd	7	Maruzzella and Liguori (1958)
Torula glabrata	Yeast	MPA, 4 days, 30°C	5, 10% sol. sd	0	Conner and Beuchat (1984)

TABLE 14.44
Inhibitory Data of Lemon Oil Obtained in the Dilution Test

Microorganism	MO Class	Conditions	MIC	Reference
Aerobacter aerogenes	Bac−	NA, pH 7	>2,000	Subba et al. (1967)
Campylobacter jejuni	Bac−	CAB, 24 h, 42°C	>40,000	Fisher and Phillips (2006)
Escherichia coli	Bac−	TGB, 18–24 h, 37°C	>1,000	Morris et al. (1979)
Escherichia coli	Bac−	NA, 24 h, 37°C	10,000	Fisher and Phillips (2006)
Escherichia coli	Bac−	NB, Tween 20, 18 h, 37°C	25,000	Yousef and Tawil (1980)
Escherichia coli	Bac−	NB, 24–72 h, 37°C	98% inh. 10,000	Dabbah et al. (1970)
Pseudomonas aeruginosa	Bac−	NB, Tween 20, 18 h, 37°C	50,000	Yousef and Tawil (1980)
Pseudomonas aeruginosa	Bac−	NB, 24–72 h, 37°C	90% inh. 10,000	Dabbah et al. (1970)
Salmonella schottmuelleri	Bac−	NA, pH 7	>2,000	Subba et al. (1967)
Salmonella senftenberg	Bac−	NB, 24–72 h, 37°C	98% inh. 10,000	Dabbah et al. (1970)
Serratia marcescens	Bac−	NA, pH 7	>2,000	Subba et al. (1967)
Bacillus cereus	Bac+	BHA, 24 h, 30°C	10,000	Fisher and Phillips (2006)
Bacillus subtilis	Bac+	NA, pH 7	2,000	Subba et al. (1967)
Bacillus subtilis	Bac+	NB, Tween 20, 18 h, 37°C	3,200	Yousef and Tawil (1980)
Corynebacterium sp.	Bac+	TGB, 18–24 h, 37°C	>1,000	Morris et al. (1979)
Lactobacillus plantarum	Bac+	NA, pH 7	1,000	Subba et al. (1967)
Listeria monocytogenes	Bac+	LSA, 24 h, 37°C	2,500	Fisher and Phillips (2006)
Micrococcus sp.	Bac+	NA, pH 7	2,000	Subba et al. (1967)
Mycobacterium phlei	Bac+	NB, Tween 20, 18 h, 37°C	800	Yousef and Tawil (1980)
Staphylococcus aureus	Bac+	BHA, 24 h, 37°C	>40,000	Fisher and Phillips (2006)
Staphylococcus aureus	Bac+	NB, 24–72 h, 37°C	10,000	Dabbah et al. (1970)
Staphylococcus aureus	Bac+	TGB, 18–24 h, 37°C	500	Morris et al. (1979)
Staphylococcus aureus	Bac+	NB, Tween 20, 18 h, 37°C	6,400	Yousef and Tawil (1980)
Streptococcus faecalis	Bac+	NA, pH 7	1,000	Subba et al. (1967)
Aspergillus awamori	Fungi	PDA, pH 4.5	>2,000	Subba et al. (1967)
Aspergillus flavus	Fungi	PDA, 5 days, 27°C	>1,000	Thompson and Cannon (1986)
Aspergillus flavus	Fungi	PDA, pH 4.5	>2,000	Subba et al. (1967)
Aspergillus flavus	Fungi	PDA, 8 h, 20°C, spore germ. inh.	50–100	Thompson (1986)
Aspergillus niger	Fungi	PDA, pH 4.5	>2,000	Subba et al. (1967)
Aspergillus niger	Fungi	YES broth, 10 d	4% inh. 10,000	Lis-Balchin et al. (1998)
Aspergillus niger	Fungi	NB, Tween 20, 8 days, 30°C	800	Yousef and Tawil (1980)
Aspergillus ochraceus	Fungi	YES broth, 10 d	22% inh. 10,000	Lis-Balchin et al. (1998)
Aspergillus oryzae	Fungi	Cited	>500	Okazaki and Oshima (1953)
Aspergillus parasiticus	Fungi	PDA, 5 days, 27°C	>1,000	Thompson and Cannon (1986)
Aspergillus parasiticus	Fungi	PDA, 8 h, 20°C, spore germ. inh.	50–100	Thompson (1986)
Botrytis cinerea	Fungi	PDA, Tween 20, 7 days, 24°C	4% inh. 1,000	Bouchra et al. (2003)
Fusarium culmorum	Fungi	YES broth, 10 d	0% inh. 10,000	Lis-Balchin et al. (1998)

(*Continued*)

TABLE 14.44 (Continued)
Inhibitory Data of Lemon Oil Obtained in the Dilution Test

Microorganism	MO Class	Conditions	MIC	Reference
Geotrichum citri-aurantii	Fungi	PDA, Tween 20, 7 days, 24°C	0% inh. 1,000	Bouchra et al. (2003)
Mucor hiemalis	Fungi	PDA, 5 days, 27°C	>1,000	Thompson and Cannon (1986)
Mucor mucedo	Fungi	PDA, 5 days, 27°C	>1,000	Thompson and Cannon (1986)
Mucor racemosus	Fungi	Cited	500	Okazaki and Oshima (1953)
Mucor racemosus f. *racemosus*	Fungi	PDA, 5 days, 27°C	>1,000	Thompson and Cannon (1986)
Mucor sp.	Fungi	NB, Tween 20, 8 days, 30°C	25,000	Yousef and Tawil (1980)
Penicillium chrysogenum	Fungi	NB, Tween 20, 8 days, 30°C	1,600	Yousef and Tawil (1980)
Penicillium chrysogenum	Fungi	Cited	500	Okazaki and Oshima (1953)
Penicillium digitatum	Fungi	PDA, Tween 20, 7 days, 24°C	0% inh. 1,000	Bouchra et al. (2003)
Penicillium digitatum	Fungi	SDB, 5 days, 20°C, MIC = ED50	500–1,000	Caccioni et al. (1998)
Penicillium italicum	Fungi	SDB, 5 days, 20°C, MIC = ED50	1,000–2,500	Caccioni et al. (1998)
Phytophthora citrophthora	Fungi	PDA, Tween 20, 7 days, 24°C	20% inh. 1,000	Bouchra et al. (2003)
Rhizopus 66-81-2	Fungi	PDA, 5 days, 27°C	>1,000	Thompson and Cannon (1986)
Rhizopus arrhizus	Fungi	PDA, 5 days, 27°C	>1,000	Thompson and Cannon (1986)
Rhizopus chinensis	Fungi	PDA, 5 days, 27°C	>1,000	Thompson and Cannon (1986)
Rhizopus circinans	Fungi	PDA, 5 days, 27°C	>1,000	Thompson and Cannon (1986)
Rhizopus japonicus	Fungi	PDA, 5 days, 27°C	>1,000	Thompson and Cannon (1986)
Rhizopus kazanensis	Fungi	PDA, 5 days, 27°C	>1,000	Thompson and Cannon (1986)
Rhizopus oryzae	Fungi	PDA, 5 days, 27°C	>1,000	Thompson and Cannon (1986)
Rhizopus pymacus	Fungi	PDA, 5 days, 27°C	>1,000	Thompson and Cannon (1986)
Rhizopus sp.	Fungi	NB, Tween 20, 8 days, 30°C	12,500	Yousef and Tawil (1980)
Rhizopus stolonifer	Fungi	PDA, 5 days, 27°C	>1,000	Thompson and Cannon (1986)
Rhizopus tritici	Fungi	PDA, 5 days, 27°C	>1,000	Thompson and Cannon (1986)
Candida albicans	Yeast	MHB, Tween 80, 48 h, 35°C	20,000	Hammer et al. (1998)
Candida albicans	Yeast	TGB, 18–24 h, 37°C	500	Morris et al. (1979)
Candida albicans	Yeast	NB, Tween 20, 18 h, 37°C	6,400	Yousef and Tawil (1980)
Saccharomyces cerevisiae	Yeast	PDA, pH 4.5	500	Subba et al. (1967)
Torula utilis	Yeast	PDA, pH 4.5	1,000	Subba et al. (1967)
Zygosaccharomyces mellis	Yeast	PDA, pH 4.5	>2,000	Subba et al. (1967)

TABLE 14.45
Inhibitory Data of Lemon Oil Obtained in the Vapor Phase Test

Microorganism	MO Class	Conditions		Activity	Reference
Campylobacter jejuni	Bac−	CAB, 24 h, 42°C	Disk, 10,000	+++	Fisher and Phillips (2006)
Escherichia coli	Bac−	NA, 24 h, 37°C	~20,000	++	Kellner and Kober (1954)
Escherichia coli	Bac−	NA, 24 h, 37°C	Disk, 10,000	+++	Fisher and Phillips (2006)
Escherichia coli	Bac−	BLA, 18 h, 37°C	MIC_{air}	>1600	Inouye et al. (2001)
Haemophilus influenzae	Bac−	MHA, 18 h, 37°C	MIC_{air}	200	Inouye et al. (2001)
Neisseria sp.	Bac−	NA, 24 h, 37°C	~20,000	NG	Kellner and Kober (1954)
Salmonella typhi	Bac−	NA, 24 h, 37°C	~20,000	NG	Kellner and Kober (1954)
Bacillus cereus	Bac+	BHA, 24 h, 30°C	Disk, 10,000	+++	Fisher and Phillips (2006)
Bacillus megaterium	Bac+	NA, 24 h, 37°C	~20,000	+	Kellner and Kober (1954)
Corynebacterium diphtheriae	Bac+	NA, 24 h, 37°C	~20,000	NG	Kellner and Kober (1954)
Listeria monocytogenes	Bac+	LSA, 24 h, 37°C	Disk, 10,000	+++	Fisher and Phillips (2006)
Staphylococcus aureus	Bac+	NA, 24 h, 37°C	~20,000	NG	Kellner and Kober (1954)
Staphylococcus aureus	Bac+	BHA, 24 h, 37°C	Disk, 10,000	+++	Fisher and Phillips (2006)
Staphylococcus aureus	Bac+	MHA, 18 h, 37°C	MIC_{air}	800	Inouye et al. (2001)
Streptococcus faecalis	Bac+	NA, 24 h, 37°C	~20,000	NG	Kellner and Kober (1954)
Streptococcus pneumoniae	Bac+	MHA, 18 h, 37°C	MIC_{air}	400	Inouye et al. (2001)
Streptococcus pyogenes	Bac+	NA, 24 h, 37°C	~20,000	NG	Kellner and Kober (1954)
Streptococcus pyogenes	Bac+	MHA, 18 h, 37°C	MIC_{air}	200	Inouye et al. (2001)
Aspergillus flavus	Fungi	WFA, 42 days, 25°C	Disk, 50,000	+++	Guynot et al. (2003)
Aspergillus niger	Fungi	WFA, 42 days, 25°C	Disk, 50,000	+++	Guynot et al. (2003)
Botrytis cinerea	Fungi	PDA, 3 days, 25°C	1,000	+++	Lee et al. (2007)
Colletotrichum gloeosporioides	Fungi	PDA, 3 days, 25°C	1,000	+++	Lee et al. (2007)
Eurotium amstelodami	Fungi	WFA, 42 days, 25°C	Disk, 50,000	+++	Guynot et al. (2003)
Eurotium herbarum	Fungi	WFA, 42 days, 25°C	Disk, 50,000	+++	Guynot et al. (2003)
Eurotium repens	Fungi	WFA, 42 days, 25°C	Disk, 50,000	+++	Guynot et al. (2003)
Eurotium rubrum	Fungi	WFA, 42 days, 25°C	Disk, 50,000	+++	Guynot et al. (2003)
Fusarium oxysporum	Fungi	PDA, 3 days, 25°C	1,000	+++	Lee et al. (2007)
Penicillium corylophilum	Fungi	WFA, 42 days, 25°C	Disk, 50,000	++	Guynot et al. (2003)
Pythium ultimum	Fungi	PDA, 3 days, 25°C	1,000	+++	Lee et al. (2007)
Rhizoctonia solani	Fungi	PDA, 3 days, 25°C	1,000	+++	Lee et al. (2007)
Candida albicans	Yeast	NA, 24 h, 37°C	~20,000	NG	Kellner and Kober (1954)

TABLE 14.46
Inhibitory Data of Mandarin Oil Obtained in the Agar Diffusion Test

Microorganism	MO Class	Conditions		Inhibition Zone (mm)	Reference
Acinetobacter calcoaceticus	Bac−	ISA, 48 h, 25°C	4 (h), 10,000	0	Deans and Ritchie (1987)
Aerobacter aerogenes	Bac−	NA, 24 h, 37°C	—, sd	1	Maruzzella and Lichtenstein (1956)
Aeromonas hydrophila	Bac−	ISA, 48 h, 25°C	4 (h), 10,000	6	Deans and Ritchie (1987)
Alcaligenes faecalis	Bac−	ISA, 48 h, 25°C	4 (h), 10,000	0	Deans and Ritchie (1987)
Beneckea natriegens	Bac−	ISA, 48 h, 25°C	4 (h), 10,000	0	Deans and Ritchie (1987)
Campylobacter jejuni	Bac−	TSA, 24 h, 42°C	4 (h), 25,000	4	Smith-Palmer et al. (1998)
Campylobacter jejuni	Bac−	Cited	6, 15,000	11	Wannissorn et al. (2005)
Citrobacter freundii	Bac−	ISA, 48 h, 25°C	4 (h), 10,000	0	Deans and Ritchie (1987)
Enterobacter aerogenes	Bac−	ISA, 48 h, 25°C	4 (h), 10,000	0	Deans and Ritchie (1987)
Enterobacter aerogenes	Bac−	MHA, 24 h, 30°C	6, 15,000	7	Rossi et al. (2007)
Erwinia carotovora	Bac−	ISA, 48 h, 25°C	4 (h), 10,000	0	Deans and Ritchie (1987)
Escherichia coli	Bac−	NA, 24 h, 37°C	—, sd	0	Maruzzella and Lichtenstein (1956)
Escherichia coli	Bac−	TGA, 18–24 h, 37°C	9.5, 2,000	0	Morris et al. (1979)
Escherichia coli	Bac−	ISA, 48 h, 25°C	4 (h), 10,000	0	Deans and Ritchie (1987)
Escherichia coli	Bac−	Cited	6, 15,000	0	Wannissorn et al. (2005)
Escherichia coli	Bac−	TSA, 24 h, 35°C	4 (h), 25,000	4	Smith-Palmer et al. (1998)
Escherichia coli	Bac−	MHA, 24 h, 30°C	6, 15,000	8	Rossi et al. (2007)
Flavobacterium suaveolens	Bac−	ISA, 48 h, 25°C	4 (h), 10,000	0	Deans and Ritchie (1987)
Klebsiella pneumoniae	Bac−	ISA, 48 h, 25°C	4 (h), 10,000	0	Deans and Ritchie (1987)
Moraxella sp.	Bac−	ISA, 48 h, 25°C	4 (h), 10,000	0	Deans and Ritchie (1987)
Neisseria perflava	Bac−	NA, 24 h, 37°C	—, sd	0	Maruzzella and Lichtenstein (1956)
Proteus vulgaris	Bac−	ISA, 48 h, 25°C	4 (h), 10,000	0	Deans and Ritchie (1987)
Proteus vulgaris	Bac−	NA, 24 h, 37°C	—, sd	9	Maruzzella and Lichtenstein (1956)
Pseudomonas aeruginosa	Bac−	NA, 24 h, 37°C	—, sd	0	Maruzzella and Lichtenstein (1956)
Pseudomonas aeruginosa	Bac−	ISA, 48 h, 25°C	4 (h), 10,000	0	Deans and Ritchie (1987)
Pseudomonas aeruginosa	Bac−	MHA, 24 h, 30°C	6, 15,000	6	Rossi et al. (2007)

(*Continued*)

TABLE 14.46 (Continued)
Inhibitory Data of Mandarin Oil Obtained in the Agar Diffusion Test

Microorganism	MO Class	Conditions		Inhibition Zone (mm)	Reference
Salmonella enteritidis	Bac–	Cited	6, 15,000	0	Wannissorn et al. (2005)
Salmonella enteritidis	Bac–	TSA, 24 h, 35°C	4 (h), 25,000	4	Smith-Palmer et al. (1998)
Salmonella pullorum	Bac–	ISA, 48 h, 25°C	4 (h), 10,000	0	Deans and Ritchie (1987)
Salmonella typhimurium	Bac–	Cited	6, 15,000	9	Wannissorn et al. (2005)
Serratia marcescens	Bac–	NA, 24 h, 37°C	—, sd	0	Maruzzella and Lichtenstein (1956)
Serratia marcescens	Bac–	ISA, 48 h, 25°C	4 (h), 10,000	0	Deans and Ritchie (1987)
Yersinia enterocolitica	Bac–	ISA, 48 h, 25°C	4 (h), 10,000	0	Deans and Ritchie (1987)
Bacillus mesentericus	Bac+	NA, 24 h, 37°C	—, sd	3	Maruzzella and Lichtenstein (1956)
Bacillus subtilis	Bac+	NA, 24 h, 37°C	—, sd	3	Maruzzella and Lichtenstein (1956)
Bacillus subtilis	Bac+	ISA, 48 h, 25°C	4 (h), 10,000	5	Deans and Ritchie (1987)
Brevibacterium linens	Bac+	ISA, 48 h, 25°C	4 (h), 10,000	0	Deans and Ritchie (1987)
Brochothrix thermosphacta	Bac+	ISA, 48 h, 25°C	4 (h), 10,000	7	Deans and Ritchie (1987)
Clostridium perfringens	Bac+	Cited	6, 15,000	35	Wannissorn et al. (2005)
Clostridium sporogenes	Bac+	ISA, 48 h, 25°C	4 (h), 10,000	0	Deans and Ritchie (1987)
Corynebacterium sp.	Bac+	TGA, 18–24 h, 37°C	9.5, 2,000	0	Morris et al. (1979)
Lactobacillus plantarum	Bac+	ISA, 48 h, 25°C	4 (h), 10,000	0	Deans and Ritchie (1987)
Leuconostoc cremoris	Bac+	ISA, 48 h, 25°C	4 (h), 10,000	7	Deans and Ritchie (1987)
Listeria monocytogenes	Bac+	TSA, 24 h, 35°C	4 (h), 25,000	4.2	Smith-Palmer et al. (1998)
Micrococcus luteus	Bac+	ISA, 48 h, 25°C	4 (h), 10,000	0	Deans and Ritchie (1987)
Sarcina lutea	Bac+	NA, 24 h, 37°C	—, sd	15	Maruzzella and Lichtenstein (1956)
Staphylococcus aureus	Bac+	NA, 24 h, 37°C	—, sd	0	Maruzzella and Lichtenstein (1956)
Staphylococcus aureus	Bac+	TGA, 18–24 h, 37°C	9.5, 2,000	0	Morris et al. (1979)
Staphylococcus aureus	Bac+	ISA, 48 h, 25°C	4 (h), 10,000	0	Deans and Ritchie (1987)
Staphylococcus aureus	Bac+	TSA, 24 h, 35°C	4 (h), 25,000	4.1	Smith-Palmer et al. (1998)
Staphylococcus aureus	Bac+	MHA, 24 h, 37°C	6, 15,000	17	Rossi et al. (2007)

(Continued)

TABLE 14.46 (Continued)
Inhibitory Data of Mandarin Oil Obtained in the Agar Diffusion Test

Microorganism	MO Class	Conditions		Inhibition Zone (mm)	Reference
Streptococcus faecalis	Bac+	ISA, 48 h, 25°C	4 (h), 10,000	0	Deans and Ritchie (1987)
Streptomyces venezuelae	Bac+	SMA, 2–7 days, 20°C	sd	7	Maruzzella and Liguori (1958)
Alternaria solani	Fungi	SMA, 2–7 days, 20°C	sd	7	Maruzzella and Liguori (1958)
Aspergillus fumigatus	Fungi	SMA, 2–7 days, 20°C	sd	11	Maruzzella and Liguori (1958)
Aspergillus niger	Fungi	SMA, 2–7 days, 20°C	sd	12	Maruzzella and Liguori (1958)
Helminthosporium sativum	Fungi	SMA, 2–7 days, 20°C	sd	22	Maruzzella and Liguori (1958)
Mucor mucedo	Fungi	SMA, 2–7 days, 20°C	sd	3	Maruzzella and Liguori (1958)
Nigrospora panici	Fungi	SMA, 2–7 days, 20°C	sd	6	Maruzzella and Liguori (1958)
Penicillium digitatum	Fungi	SMA, 2–7 days, 20°C	sd	7	Maruzzella and Liguori (1958)
Rhizopus nigricans	Fungi	SMA, 2–7 days, 20°C	sd	2	Maruzzella and Liguori (1958)
Candida albicans	Yeast	SMA, 2–7 days, 20°C	sd	0	Maruzzella and Liguori (1958)
Candida albicans	Yeast	TGA, 18–24 h, 37°C	9.5, 2,000	0	Morris et al. (1979)
Candida krusei	Yeast	SMA, 2–7 days, 20°C	sd	0	Maruzzella and Liguori (1958)
Candida tropicalis	Yeast	SMA, 2–7 days, 20°C	sd	2	Maruzzella and Liguori (1958)
Cryptococcus neoformans	Yeast	SMA, 2–7 days, 20°C	sd	10	Maruzzella and Liguori (1958)
Cryptococcus rhodobenhani	Yeast	SMA, 2–7 days, 20°C	sd	7	Maruzzella and Liguori (1958)
Saccharomyces cerevisiae	Yeast	SMA, 2–7 days, 20°C	sd	5	Maruzzella and Liguori (1958)

TABLE 14.47
Inhibitory Data of Mandarin Oil Obtained in the Dilution Test

Microorganism	MO Class	Conditions	MIC	Reference
Escherichia coli	Bac−	TGB, 18–24 h, 37°C	>1,000	Morris et al. (1979)
Escherichia coli	Bac−	NB, 24–72 h, 37°C	98% inh. 10,000	Dabbah et al. (1970)
Pseudomonas aeruginosa	Bac−	NB, 24–72 h, 37°C	87% inh. 10,000	Dabbah et al. (1970)
Salmonella senftenberg	Bac−	NB, 24–72 h, 37°C	>10,000	Dabbah et al. (1970)
Yersinia enterocolitica	Bac−	MHA, Tween 20, 24 h, 37°C	2,500	Rossi et al. (2007)
Corynebacterium sp.	Bac+	TGB, 18–24 h, 37°C	>1,000	Morris et al. (1979)
Staphylococcus aureus	Bac+	TGB, 18–24 h, 37°C	>1,000	Morris et al. (1979)
Staphylococcus aureus	Bac+	NB, 24–72 h, 37°C	10,000	Dabbah et al. (1970)
Candida albicans	Yeast	TGB, 18–24 h, 37°C	1,000	Morris et al. (1979)

Annotation: Terpeneless mandarin oil tested (Maruzzella and Liguori, 1958).

TABLE 14.48
Inhibitory Data of Mandarin Oil Obtained in the Vapor Phase Test

Microorganism	MO Class	Conditions		Activity	Reference
Aspergillus flavus	Fungi	WFA, 42 days, 25°C	Disk, 50,000	+++	Guynot et al. (2003)
Aspergillus niger	Fungi	WFA, 42 days, 25°C	Disk, 50,000	+++	Guynot et al. (2003)
Eurotium amstelodami	Fungi	WFA, 42 days, 25°C	Disk, 50,000	+++	Guynot et al. (2003)
Eurotium herbarum	Fungi	WFA, 42 days, 25°C	Disk, 50,000	+++	Guynot et al. (2003)
Eurotium repens	Fungi	WFA, 42 days, 25°C	Disk, 50,000	+++	Guynot et al. (2003)
Eurotium rubrum	Fungi	WFA, 42 days, 25°C	Disk, 50,000	+++	Guynot et al. (2003)
Penicillium corylophilum	Fungi	WFA, 42 days, 25°C	Disk, 50,000	+	Guynot et al. (2003)

TABLE 14.49
Inhibitory Data of Matricaria Oil Obtained in the Agar Diffusion Test

Microorganism	MO Class	Conditions		Inhibition Zone (mm)	Reference
Acinetobacter calcoaceticus	Bac−	ISA, 48 h, 25°C	4 (h), 10,000	7	Deans and Ritchie (1987)
Aerobacter aerogenes	Bac−	NA, 24 h, 37°C	5 × 20, 1,000	0	Möse and Lukas (1957)
Aeromonas hydrophila	Bac−	ISA, 48 h, 25°C	4 (h), 10,000	0	Deans and Ritchie (1987)
Alcaligenes faecalis	Bac−	ISA, 48 h, 25°C	4 (h), 10,000	0	Deans and Ritchie (1987)
Beneckea natriegens	Bac−	ISA, 48 h, 25°C	4 (h), 10,000	0	Deans and Ritchie (1987)
Brucella abortus	Bac−	NA, 24 h, 37°C	5 × 20, 1,000	0	Möse and Lukas (1957)
Citrobacter freundii	Bac−	ISA, 48 h, 25°C	4 (h), 10,000	0	Deans and Ritchie (1987)
Enterobacter aerogenes	Bac−	ISA, 48 h, 25°C	4 (h), 10,000	0	Deans and Ritchie (1987)
Erwinia carotovora	Bac−	ISA, 48 h, 25°C	4 (h), 10,000	0	Deans and Ritchie (1987)
Escherichia coli	Bac−	NA, 18 h, 37°C	6 (h), pure	0	Yousef and Tawil (1980)
Escherichia coli	Bac−	TGA, 18–24 h, 37°C	9.5, 2,000	0	Morris et al. (1979)
Escherichia coli	Bac−	NA, 24 h, 37°C	5 × 20, 1,000	0	Möse and Lukas (1957)
Escherichia coli	Bac−	ISA, 48 h, 25°C	4 (h), 10,000	0	Deans and Ritchie (1987)
Flavobacterium suaveolens	Bac−	ISA, 48 h, 25°C	4 (h), 10,000	0	Deans and Ritchie (1987)
Klebsiella pneumonia	Bac−	NA, 24 h, 37°C	5 × 20, 1,000	2–5	Möse and Lukas (1957)
Klebsiella pneumonia subsp. *ozaenae*	Bac−	NA, 24 h, 37°C	5 × 20, 1,000	2–5	Möse and Lukas (1957)
Klebsiella pneumoniae	Bac−	ISA, 48 h, 25°C	4 (h), 10,000	0	Deans and Ritchie (1987)
Moraxella sp.	Bac−	ISA, 48 h, 25°C	4 (h), 10,000	0	Deans and Ritchie (1987)
Proteus OX19	Bac−	NA, 24 h, 37°C	5 × 20, 1,000	0	Möse and Lukas (1957)
Proteus vulgaris	Bac−	NA, 24 h, 37°C	5 × 20, 1,000	0	Möse and Lukas (1957)
Proteus vulgaris	Bac−	ISA, 48 h, 25°C	4 (h), 10,000	0	Deans and Ritchie (1987)
Pseudomonas aeruginosa	Bac−	NA, 18 h, 37°C	6 (h), pure	0	Yousef and Tawil (1980)
Pseudomonas aeruginosa	Bac−	NA, 24 h, 37°C	5 × 20, 1,000	0	Möse and Lukas (1957)
Pseudomonas aeruginosa	Bac−	ISA, 48 h, 25°C	4 (h), 10,000	0	Deans and Ritchie (1987)
Pseudomonas fluorescens	Bac−	NA, 24 h, 37°C	5 × 20, 1,000	0	Möse and Lukas (1957)
Salmonella enteritidis	Bac−	NA, 24 h, 37°C	5 × 20, 1,000	2–5	Möse and Lukas (1957)
Salmonella paratyphi B	Bac−	NA, 24 h, 37°C	5 × 20, 1,000	0	Möse and Lukas (1957)
Salmonella pullorum	Bac−	ISA, 48 h, 25°C	4 (h), 10,000	0	Deans and Ritchie (1987)
Salmonella typhi	Bac−	NA, 24 h, 37°C	5 × 20, 1,000	2–5	Möse and Lukas (1957)
Serratia marcescens	Bac−	NA, 24 h, 37°C	5 × 20, 1,000	0	Möse and Lukas (1957)
Serratia marcescens	Bac−	ISA, 48 h, 25°C	4 (h), 10,000	0	Deans and Ritchie (1987)
Vibrio albicans	Bac−	NA, 24 h, 37°C	5 × 20, 1,000	1	Möse and Lukas (1957)
Vibrio cholerae	Bac−	NA, 24 h, 37°C	5 × 20, 1,000	2–5	Möse and Lukas (1957)
Yersinia enterocolitica	Bac−	ISA, 48 h, 25°C	4 (h), 10,000	0	Deans and Ritchie (1987)
Bacillus anthracis	Bac+	NA, 24 h, 37°C	5 × 20, 1,000	2–5	Möse and Lukas (1957)
Bacillus subtilis	Bac+	NA, 24 h, 37°C	5 × 20, 1,000	0	Möse and Lukas (1957)
Bacillus subtilis	Bac+	ISA, 48 h, 25°C	4 (h), 10,000	0	Deans and Ritchie (1987)
Bacillus subtilis	Bac+	NA, 18 h, 37°C	6 (h), pure	13.5	Yousef and Tawil (1980)
Brevibacterium linens	Bac+	ISA, 48 h, 25°C	4 (h), 10,000	0	Deans and Ritchie (1987)
Brochothrix thermosphacta	Bac+	ISA, 48 h, 25°C	4 (h), 10,000	0	Deans and Ritchie (1987)
Clostridium sporogenes	Bac+	ISA, 48 h, 25°C	4 (h), 10,000	0	Deans and Ritchie (1987)

(*Continued*)

TABLE 14.49 (Continued)
Inhibitory Data of Matricaria Oil Obtained in the Agar Diffusion Test

Microorganism	MO Class	Conditions		Inhibition Zone (mm)	Reference
Corynebacterium diphtheria	Bac+	NA, 24 h, 37°C	5 × 20, 1,000	1	Möse and Lukas (1957)
Corynebacterium sp.	Bac+	TGA, 18–24 h, 37°C	9.5, 2,000	0	Morris et al. (1979)
Lactobacillus plantarum	Bac+	ISA, 48 h, 25°C	4 (h), 10,000	0	Deans and Ritchie (1987)
Leuconostoc cremoris	Bac+	ISA, 48 h, 25°C	4 (h), 10,000	0	Deans and Ritchie (1987)
Listeria monocytogenes	Bac+	ISA, 48 h, 25°C	4 (h), 10,000	0–11	Lis-Balchin et al. (1998)
Listeria monocytogenes	Bac+	NA, 24 h, 37°C	5 × 20, 1,000	2–5	Möse and Lukas (1957)
Micrococcus luteus	Bac+	ISA, 48 h, 25°C	4 (h), 10,000	0	Deans and Ritchie (1987)
Mycobacterium phlei	Bac+	NA, 18 h, 37°C	6 (h), pure	15.5	Yousef and Tawil (1980)
Sarcina alba	Bac+	NA, 24 h, 37°C	5 × 20, 1,000	0	Möse and Lukas (1957)
Sarcina beige	Bac+	NA, 24 h, 37°C	5 × 20, 1,000	6–10	Möse and Lukas (1957)
Sarcina citrea	Bac+	NA, 24 h, 37°C	5 × 20, 1,000	0	Möse and Lukas (1957)
Sarcina rosa	Bac+	NA, 24 h, 37°C	5 × 20, 1,000	6–10	Möse and Lukas (1957)
Sporococcus sarc.	Bac+	NA, 24 h, 37°C	5 × 20, 1,000	6–10	Möse and Lukas (1957)
Staphylococcus albus	Bac+	NA, 24 h, 37°C	5 × 20, 1,000	2–5	Möse and Lukas (1957)
Staphylococcus aureus	Bac+	TGA, 18–24 h, 37°C	9.5, 2,000	0	Morris et al. (1979)
Staphylococcus aureus	Bac+	ISA, 48 h, 25°C	4 (h), 10,000	0	Deans and Ritchie (1987)
Staphylococcus aureus	Bac+	NA, 24 h, 37°C	5 × 20, 1,000	2–5	Möse and Lukas (1957)
Staphylococcus aureus	Bac+	NA, 18 h, 37°C	6 (h), pure	11.5	Yousef and Tawil (1980)
Streptococcus faecalis	Bac+	ISA, 48 h, 25°C	4 (h), 10,000	0	Deans and Ritchie (1987)
Streptococcus haemolyticus	Bac+	NA, 24 h, 37°C	5 × 20, 1,000	2–5	Möse and Lukas (1957)
Streptococcus viridians	Bac+	NA, 24 h, 37°C	5 × 20, 1,000	2–5	Möse and Lukas (1957)
Alternaria porri	Fungi	PDA, 72 h, 28°C	5, 5,000	9.7	Pawar and Thaker (2007)
Aspergillus niger	Fungi	SDA, 8 days, 30°C	6 (h), pure	10	Yousef and Tawil (1980)
Aspergillus niger	Fungi	PDA, 48 h, 28°C	5, 5,000	0	Pawar and Thaker (2006)
Fusarium oxysporum f.sp. ciceri	Fungi	PDA, 72 h, 28°C	5, 5,000	9	Pawar and Thaker (2007)
Mucor sp.	Fungi	SDA, 8 days, 30°C	6 (h), pure	10	Yousef and Tawil (1980)
Penicillium chrysogenum	Fungi	SDA, 8 days, 30°C	6 (h), pure	11	Yousef and Tawil (1980)
Rhizopus sp.	Fungi	SDA, 8 days, 30°C	6 (h), pure	0	Yousef and Tawil (1980)
Candida albicans	Yeast	TGA, 18–24 h, 37°C	9.5, 2,000	0	Morris et al. (1979)
Candida albicans	Yeast	SDA, 18 h, 30°C	6 (h), pure	9.5	Yousef and Tawil (1980)

TABLE 14.50
Inhibitory Data of Matricaria Oil Obtained in the Dilution Test

Microorganism	MO Class	Conditions	MIC	Reference
Escherichia coli	Bac−	TGB, 18–24 h, 37°C	>1,000	Morris et al. (1979)
Escherichia coli	Bac−	NB, Tween 80, 24 h, 37°C	>8,000	Aggag and Yousef (1972)
Escherichia coli	Bac−	Agar, cited	10,000	Kedzia (1991)
Escherichia coli	Bac−	NB, Tween 20, 18 h, 37°C	>50,000	Yousef and Tawil (1980)
Helicobacter pylori	Bac−	Cited, 20 h, 37°C	35.7–70.4	Weseler et al. (2005)
Klebsiella pneumoniae	Bac−	Agar, cited	10,000	Kedzia (1991)
Pseudomonas aeruginosa	Bac−	Agar, cited	7,500	Kedzia (1991)
Pseudomonas aeruginosa	Bac−	NB, Tween 20, 18 h, 37°C	>50,000	Yousef and Tawil (1980)
Bacillus subtilis	Bac+	NB, Tween 20, 18 h, 37°C	400	Yousef and Tawil (1980)
Bacillus subtilis	Bac+	NB, Tween 80, 24 h, 37°C	6,000	Aggag and Yousef (1972)
Corynebacterium sp.	Bac+	TGB, 18–24 h, 37°C	>1,000	Morris et al. (1979)
Mycobacterium phlei	Bac+	NB, Tween 20, 18 h, 37°C	1,600	Yousef and Tawil (1980)
Pseudomonas aeruginosa	Bac+	NB, Tween 80, 24 h, 37°C	>8,000	Aggag and Yousef (1972)
Staphylococcus aureus	Bac+	TGB, 18–24 h, 37°C	1,000	Morris et al. (1979)
Staphylococcus aureus	Bac+	Agar, cited	2,500	Kedzia (1991)
Staphylococcus aureus	Bac+	NB, Tween 80, 24 h, 37°C	7,000	Aggag and Yousef (1972)
Staphylococcus aureus	Bac+	NB, Tween 20, 18 h, 37°C	25,000	Yousef and Tawil (1980)
Streptococcus faecalis	Bac+	Agar, cited	2,500	Kedzia (1991)
Aspergillus flavus	Fungi	PDA, 7–14 days, 28°C	>3,000	Soliman and Badeaa (2002)
Aspergillus niger	Fungi	YES broth, 10 d	−63% inh. 10,000	Lis-Balchin et al. (1998)
Aspergillus niger	Fungi	NB, Tween 20, 8 days, 30°C	50,000	Yousef and Tawil (1980)
Aspergillus ochraceus	Fungi	PDA, 7–14 days, 28°C	>3,000	Soliman and Badeaa (2002)
Aspergillus ochraceus	Fungi	YES broth, 10 d	−56% inh. 10,000	Lis-Balchin et al. (1998)
Aspergillus parasiticus	Fungi	PDA, 7–14 days, 28°C	>3,000	Soliman and Badeaa (2002)
Botrytis cinerea	Fungi	PDA, Tween 20, 7 days, 24°C	0% inh. 1,000	Bouchra et al. (2003)
Fusarium culmorum	Fungi	YES broth, 10 d	−75% inh. 10,000	Lis-Balchin et al. (1998)
Fusarium moniliforme	Fungi	PDA, 7–14 days, 28°C	>3,000	Soliman and Badeaa (2002)
Microsporum gypseum	Fungi	Agar, cited	1,000	Kedzia (1991)
Mucor sp.	Fungi	NB, Tween 20, 8 days, 30°C	50,000	Yousef and Tawil (1980)
Penicillium chrysogenum	Fungi	NB, Tween 20, 8 days, 30°C	50,000	Yousef and Tawil (1980)
Phytophthora citrophthora	Fungi	PDA, Tween 20, 7 days, 24°C	2% inh. 1,000	Bouchra et al. (2003)
Rhizopus sp.	Fungi	NB, Tween 20, 8 days, 30°C	50,000	Yousef and Tawil (1980)
Trichophyton mentagrophytes	Fungi	SDA, 21 days, 20°C	1000	Szalontai et al. (1977)
Trichophyton mentagrophytes	Fungi	SDA, 21 days, 20°C	1,000	Szalontai et al. (1977)
Candida albicans	Yeast	TGB, 18–24 h, 37°C	500	Morris et al. (1979)
Candida albicans	Yeast	SDA, 7 days, 37°C	1,000	Szalontai et al. (1977)
Candida albicans	Yeast	Agar, cited	5,000	Kedzia (1991)
Candida albicans	Yeast	NB, Tween 80, 24 h, 37°C	7,000	Aggag and Yousef (1972)
Candida albicans	Yeast	NB, Tween 20, 18 h, 37°C	50,000	Yousef and Tawil (1980)

TABLE 14.51
Inhibitory Data of Matricaria Oil Obtained in the Vapor Phase Test

Microorganism	MO Class	Conditions		Activity	Reference
Escherichia coli	Bac−	NA, 24 h, 37°C	~20,000	+++	Kellner and Kober (1954)
Neisseria sp.	Bac−	NA, 24 h, 37°C	~20,000	+++	Kellner and Kober (1954)
Salmonella typhi	Bac−	NA, 24 h, 37°C	~20,000	+++	Kellner and Kober (1954)
Bacillus megaterium	Bac+	NA, 24 h, 37°C	~20,000	++	Kellner and Kober (1954)
Corynebacterium diphtheriae	Bac+	NA, 24 h, 37°C	~20,000	+	Kellner and Kober (1954)
Staphylococcus aureus	Bac+	NA, 24 h, 37°C	~20,000	+	Kellner and Kober (1954)
Streptococcus faecalis	Bac+	NA, 24 h, 37°C	~20,000	++	Kellner and Kober (1954)
Streptococcus pyogenes	Bac+	NA, 24 h, 37°C	~20,000	+	Kellner and Kober (1954)
Botrytis cinerea	Fungi	PDA, 3 days, 25°C	1,000	+++	Lee et al. (2007)
Colletotrichum gloeosporioides	Fungi	PDA, 3 days, 25°C	1,000	+++	Lee et al. (2007)
Fusarium oxysporum	Fungi	PDA, 3 days, 25°C	1000	+++	Lee et al. (2007)
Pythium ultimum	Fungi	PDA, 3 days, 25°C	1,000	+++	Lee et al. (2007)
Rhizoctonia solani	Fungi	PDA, 3 days, 25°C	1,000	+++	Lee et al. (2007)
Candida albicans	Yeast	NA, 24 h, 37°C	~20,000	+++	Kellner and Kober (1954)

TABLE 14.52
Inhibitory Data of Mint Oil Obtained in the Agar Diffusion Test

Microorganism	MO Class	Conditions		Inhibition Zone (mm)	Reference
Acinetobacter calcoaceticus	Bac–	ISA, 48 h, 25°C	4 (h), 10,000	9	Deans and Ritchie (1987)
Aeromonas hydrophila	Bac–	ISA, 48 h, 25°C	4 (h), 10,000	8.5	Deans and Ritchie (1987)
Alcaligenes faecalis	Bac–	ISA, 48 h, 25°C	4 (h), 10,000	5	Deans and Ritchie (1987)
Beneckea natriegens	Bac–	ISA, 48 h, 25°C	4 (h), 10,000	9.5	Deans and Ritchie (1987)
Campylobacter jejuni	Bac–	Cited	6, 15,000	90	Wannissorn et al. (2005)
Citrobacter freundii	Bac–	ISA, 48 h, 25°C	4 (h), 10,000	8.5	Deans and Ritchie (1987)
Enterobacter aerogenes	Bac–	ISA, 48 h, 25°C	4 (h), 10,000	6.5	Deans and Ritchie (1987)
Erwinia carotovora	Bac–	ISA, 48 h, 25°C	4 (h), 10,000	0	Deans and Ritchie (1987)
Escherichia coli	Bac–	ISA, 48 h, 25°C	4 (h), 10,000	0	Deans and Ritchie (1987)
Escherichia coli	Bac–	Cited	15, 2,500	1	Pizsolitto et al. (1975)
Escherichia coli	Bac–	Cited	6, 15,000	14.5	Wannissorn et al. (2005)
Escherichia coli O157	Bac–	Cited	6, 15,000	13.5	Wannissorn et al. (2005)
Flavobacterium suaveolens	Bac–	ISA, 48 h, 25°C	4 (h), 10,000	9	Deans and Ritchie (1987)
Klebsiella pneumoniae	Bac–	ISA, 48 h, 25°C	4 (h), 10,000	6	Deans and Ritchie (1987)
Klebsiella sp.	Bac–	Cited	15, 2,500	2	Pizsolitto et al. (1975)
Moraxella sp.	Bac–	ISA, 48 h, 25°C	4 (h), 10,000	8	Deans and Ritchie (1987)
Proteus sp.	Bac–	Cited	15, 2,500	2	Pizsolitto et al. (1975)
Proteus vulgaris	Bac–	ISA, 48 h, 25°C	4 (h), 10,000	7	Deans and Ritchie (1987)
Pseudomonas aeruginosa	Bac–	Cited	15, 2,500	0	Pizsolitto et al. (1975)
Pseudomonas aeruginosa	Bac–	ISA, 48 h, 25°C	4 (h), 10,000	0	Deans and Ritchie (1987)
Salmonella agona	Bac–	Cited	6, 15,000	12	Wannissorn et al. (2005)
Salmonella braenderup	Bac–	Cited	6, 15,000	13	Wannissorn et al. (2005)
Salmonella derby	Bac–	Cited	6, 15,000	10	Wannissorn et al. (2005)
Salmonella enteritidis	Bac–	Cited	6, 15,000	13.5	Wannissorn et al. (2005)
Salmonella gallinarum	Bac–	Cited	6, 15,000	52.5	Wannissorn et al. (2005)
Salmonella hadar	Bac–	Cited	6, 15,000	13.5	Wannissorn et al. (2005)
Salmonella mbandaka	Bac–	Cited	6, 15,000	11.5	Wannissorn et al. (2005)
Salmonella montevideo	Bac–	Cited	6, 15,000	12	Wannissorn et al. (2005)
Salmonella pullorum	Bac–	ISA, 48 h, 25°C	4 (h), 10,000	6	Deans and Ritchie (1987)
Salmonella saintpaul	Bac–	Cited	6, 15,000	12	Wannissorn et al. (2005)
Salmonella schwarzengrund	Bac–	Cited	6, 15,000	11.5	Wannissorn et al. (2005)
Salmonella senftenberg	Bac–	Cited	6, 15,000	12	Wannissorn et al. (2005)
Salmonella sp.	Bac–	Cited	15, 2,500	4	Pizsolitto et al. (1975)
Salmonella typhimurium	Bac–	Cited	6, 15,000	47.5	Wannissorn et al. (2005)
Serratia marcescens	Bac–	ISA, 48 h, 25°C	4 (h), 10,000	4.5	Deans and Ritchie (1987)
Serratia sp.	Bac–	Cited	15, 2,500	6	Pizsolitto et al. (1975)
Shigella sp.	Bac–	Cited	15, 2,500	2	Pizsolitto et al. (1975)
Yersinia enterocolitica	Bac–	ISA, 48 h, 25°C	4 (h), 10,000	6.5	Deans and Ritchie (1987)
Bacillus sp.	Bac+	Cited	15, 2,500	10	Pizsolitto et al. (1975)
Bacillus subtilis	Bac+	ISA, 48 h, 25°C	4 (h), 10,000	14.5	Deans and Ritchie (1987)
Brevibacterium linens	Bac+	ISA, 48 h, 25°C	4 (h), 10,000	0	Deans and Ritchie (1987)

(Continued)

TABLE 14.52 (*Continued*)
Inhibitory Data of Mint Oil Obtained in the Agar Diffusion Test

Microorganism	MO Class	Conditions		Inhibition Zone (mm)	Reference
Brochothrix thermosphacta	Bac+	ISA, 48 h, 25°C	4 (h), 10,000	6	Deans and Ritchie (1987)
Clostridium perfringens	Bac+	Cited	6, 15,000	90	Wannissorn et al. (2005)
Clostridium sporogenes	Bac+	ISA, 48 h, 25°C	4 (h), 10,000	0	Deans and Ritchie (1987)
Lactobacillus plantarum	Bac+	ISA, 48 h, 25°C	4 (h), 10,000	9	Deans and Ritchie (1987)
Leuconostoc cremoris	Bac+	ISA, 48 h, 25°C	4 (h), 10,000	0	Deans and Ritchie (1987)
Micrococcus luteus	Bac+	ISA, 48 h, 25°C	4 (h), 10,000	8.5	Deans and Ritchie (1987)
Staphylococcus aureus	Bac+	Cited	15, 2,500	2	Pizsolitto et al. (1975)
Staphylococcus aureus	Bac+	ISA, 48 h, 25°C	4 (h), 10,000	12	Deans and Ritchie (1987)
Staphylococcus epidermidis	Bac+	Cited	15, 2,500	2	Pizsolitto et al. (1975)
Streptococcus faecalis	Bac+	ISA, 48 h, 25°C	4 (h), 10,000	14	Deans and Ritchie (1987)
Streptococcus viridans	Bac+	Cited	15, 2,500	1	Pizsolitto et al. (1975)
Absidia corymbifera	Fungi	EYA, 48 h, 45°C	5, sd	10	Nigam and Rao (1979)
Aspergillus flavus	Fungi	PDA, 10 days, 25°C	50	NG	Sarbhoy et al. (1978)
Aspergillus fumigatus	Fungi	PDA, 10 days, 25°C	50	NG	Sarbhoy et al. (1978)
Aspergillus sulphureus	Fungi	PDA, 10 days, 25°C	50	NG	Sarbhoy et al. (1978)
Humicola grisea var. *thermoidea*	Fungi	EYA, 48 h, 45°C	5, sd	15	Nigam and Rao (1979)
Mucor fragilis	Fungi	PDA, 10 days, 25°C	50	NG	Sarbhoy et al. (1978)
Rhizopus stolonifer	Fungi	PDA, 10 days, 25°C	50	NG	Sarbhoy et al. (1978)
Sporotrichum thermophile	Fungi	EYA, 48 h, 45°C	5, sd	20	Nigam and Rao (1979)
Thermoascus aurantiacus	Fungi	EYA, 48 h, 45°C	5, sd	10	Nigam and Rao (1979)
Thermomyces lanuginosus	Fungi	EYA, 48 h, 45°C	5, sd	0	Nigam and Rao (1979)
Thielava minor	Fungi	EYA, 48 h, 45°C	5, sd	0	Nigam and Rao (1979)
Torula thermophila	Yeast	EYA, 48 h, 45°C	5, sd	10	Nigam and Rao (1979)

TABLE 14.53
Inhibitory Data of Mint Oil Obtained in the Dilution Test

Microorganism	MO Class	Conditions	MIC	Reference
Escherichia coli	Bac−	Cited, 24 h	400	Imai et al. (2001)
Helicobacter pylori	Bac−	Cited, 48 h	100	Imai et al. (2001)
Salmonella enteritidis	Bac−	Cited, 24 h	800	Imai et al. (2001)
Enterococcus faecium VRE	Bac+	HIB, Tween 80, 18 h, 37°C	5,000–10,000	Nelson (1997)
Staphylococcus aureus MRSA	Bac+	Cited, 24 h	400	Imai et al. (2001)
Staphylococcus aureus MRSA	Bac+	HIB, Tween 80, 18 h, 37°C	5,000	Nelson (1997)
Staphylococcus aureus MSSA	Bac+	Cited, 24 h	400	Imai et al. (2001)
Alternaria alternate	Fungi	SDA, 6–8 h, 20°C	63% inh. 500	Dikshit et al. (1986)
Aspergillus flavus	Fungi	CA, 7 days, 28°C	500	Kumar et al. (2007)
Aspergillus flavus	Fungi	Cited	1,000	Kumar et al. (2007)
Aspergillus flavus	Fungi	SDA, 6–8 h, 20°C	65% inh. 500	Dikshit et al. (1986)
Aspergillus niger	Fungi	CA, 7 days, 28°C	81% inh. 500	Kumar et al. (2007)
Botryodiplodia theobromae	Fungi	CA, 7 days, 28°C	87% inh. 500	Kumar et al. (2007)
Cladosporium cladosporioides	Fungi	CA, 7 days, 28°C	>500	Kumar et al. (2007)
Fusarium oxysporum	Fungi	CA, 7 days, 28°C	83% inh. 500	Kumar et al. (2007)
Helminthosporium oryzae	Fungi	CA, 7 days, 28°C	500	Kumar et al. (2007)
Helminthosporium oryzae	Fungi	MA, cited[a]	2,000	Dikshit et al. (1979)
Helminthosporium oryzae	Fungi	MA, pH 5.0, 6 days, 28°C[a,b]	2,000	Dikshit et al. (1982)
Helminthosporium oryzae	Fungi	MA, pH 4.0, 6 days, 28°C[a,b]	500	Dikshit et al. (1982)
Macrophomina phaseoli	Fungi	CA, 7 days, 28°C	94% inh. 500	Kumar et al. (2007)
Microsporum gypseum	Fungi	SDA, 6–8 h, 20°C	64% inh. 500	Dikshit et al. (1986)
Penicillium italicum	Fungi	SDA, 6–8 h, 20°C	65% inh. 500	Dikshit et al. (1986)
Sclerotium rolfsii	Fungi	CA, 7 days, 28°C	500	Kumar et al. (2007)
Trichophyton mentagrophytes	Fungi	SDA, 6–8 h, 20°C	69% inh. 500	Dikshit et al. (1986)
Trichophyton rubrum	Fungi	SDA, 6–8 h, 20°C	36% inh. 500	Dikshit et al. (1986)
Candida albicans	Yeast	Cited, 48 h, 36°C[a]	1,100	Duarte et al. (2005)

[a] *Mentha arvensis* var. *piperascens*.
[b] Dementholized oil.

TABLE 14.54
Inhibitory Data of Neroli Oil Obtained in the Agar Diffusion Test

Microorganism	MO Class	Conditions		Inhibition Zone (mm)	Reference
Aerobacter aerogenes	Bac−	NA, 24 h, 37°C	—, sd	0	Imai et al. (2001)
Escherichia coli	Bac−	Cited	15, 2,500	0	Imai et al. (2001)
Escherichia coli	Bac−	NA, 24 h, 37°C	—, sd	2	Imai et al. (2001)
Escherichia coli	Bac−	NA, 18 h, 37°C	6 (h), pure	14	Nelson (1997)
Escherichia coli	Bac−	NA, 24 h, 37°C	10 (h), 100,000	22	Imai et al. (2001)
Klebsiella aerogenes	Bac−	NA, 24 h, 37°C	10 (h), 100,000	0	Nelson (1997)
Klebsiella sp.	Bac−	Cited	15, 2,500	0	Imai et al. (2001)
Neisseria perflava	Bac−	NA, 24 h, 37°C	—, sd	2	Dikshit et al. (1986)
Proteus sp.	Bac−	Cited	15, 2,500	0	Kumar et al. (2007)
Proteus vulgaris	Bac−	NA, 24 h, 37°C	—, sd	2	Kumar et al. (2007)
Pseudomonas aeruginosa	Bac−	Cited	15, 2,500	0	Dikshit et al. (1986)
Pseudomonas aeruginosa	Bac−	NA, 24 h, 37°C	—, sd	0	Kumar et al. (2007)
Pseudomonas aeruginosa	Bac−	NA, 18 h, 37°C	6 (h), pure	12	Kumar et al. (2007)
Salmonella sp.	Bac−	Cited	15, 2,500	0	Kumar et al. (2007)
Salmonella typhi	Bac−	NA, 24 h, 37°C	10 (h), 100,000	19	Kumar et al. (2007)
Serratia marcescens	Bac−	NA, 24 h, 37°C	—, sd	0	Kumar et al. (2007)
Serratia sp.	Bac−	Cited	15, 2,500	0	Dikshit et al. (1979)
Shigella sp.	Bac−	Cited	15, 2,500	1	Dikshit et al. (1982)
Vibrio cholerae	Bac−	NA, 24 h, 37°C	10 (h), 100,000	0	Dikshit et al. (1982)
Bacillus mesentericus	Bac+	NA, 24 h, 37°C	—, sd	7	Kumar et al. (2007)
Bacillus sp.	Bac+	Cited	15, 2,500	1	Dikshit et al. (1986)
Bacillus subtilis	Bac+	NA, 24 h, 37°C	—, sd	19	Dikshit et al. (1986)
Bacillus subtilis	Bac+	NA, 18 h, 37°C	6 (h), pure	22	Kumar et al. (2007)
Corynebacterium diphtheriae	Bac+	NA, 24 h, 37°C	10 (h), 100,000	18	Dikshit et al. (1986)
Listeria monocytogenes	Bac+	ISA, 48 h, 25°C	4 (h), 10,000	11–19	Dikshit et al. (1986)
Mycobacterium phlei	Bac+	NA, 18 h, 37°C	6 (h), pure	20	Duarte et al. (2005)
Sarcina lutea	Bac+	NA, 24 h, 37°C	—, sd	0	Imai et al. (2001)
Staphylococcus aureus	Bac+	NA, 24 h, 37°C	10 (h), 100,000	0	Imai et al. (2001)
Staphylococcus aureus	Bac+	NA, 24 h, 37°C	—, sd	0	Imai et al. (2001)
Staphylococcus aureus	Bac+	Cited	15, 2,500	2	Nelson (1997)
Staphylococcus aureus	Bac+	NA, 18 h, 37°C	6 (h), pure	13.6	Imai et al. (2001)
Staphylococcus epidermidis	Bac+	Cited	15, 2,500	1	Nelson (1997)
Staphylococcus epidermidis	Bac+	NA, 24 h, 37°C	10 (h), 100,000	22	Imai et al. (2001)
Streptococcus sp.	Bac+	NA, 24 h, 37°C	10 (h), 100,000	0	Dikshit et al. (1986)
Streptococcus viridans	Bac+	Cited	15, 2,500	0	Kumar et al. (2007)
Streptomyces venezuelae	Bac+	SMA, 2–7 days, 20°C	sd	9	Kumar et al. (2007)
Alternaria solani	Fungi	SMA, 2–7 days, 20°C	sd	9	Dikshit et al. (1986)
Aspergillus fumigatus	Fungi	SMA, 2–7 days, 20°C	sd	5	Kumar et al. (2007)
Aspergillus niger	Fungi	SMA, 2–7 days, 20°C	sd	9	Kumar et al. (2007)

(Continued)

TABLE 14.54 (Continued)
Inhibitory Data of Neroli Oil Obtained in the Agar Diffusion Test

Microorganism	MO Class	Conditions		Inhibition Zone (mm)	Reference
Aspergillus niger	Fungi	SDA, 8 days, 30°C	6 (h), pure	26	Kumar et al. (2007)
Helminthosporium sativum	Fungi	SMA, 2–7 days, 20°C	sd	9	Kumar et al. (2007)
Mucor mucedo	Fungi	SMA, 2–7 days, 20°C	sd	9	Kumar et al. (2007)
Mucor sp.	Fungi	SDA, 8 days, 30°C	6 (h), pure	22	Dikshit et al. (1979)
Nigrospora panici	Fungi	SMA, 2–7 days, 20°C	sd	5	Dikshit et al. (1982)
Penicillium chrysogenum	Fungi	SDA, 8 days, 30°C	6 (h), pure	56	Dikshit et al. (1982)
Penicillium digitatum	Fungi	SMA, 2–7 days, 20°C	sd	9	Kumar et al. (2007)
Rhizopus nigricans	Fungi	SMA, 2–7 days, 20°C	sd	6	Dikshit et al. (1986)
Rhizopus sp.	Fungi	SDA, 8 days, 30°C	6 (h), pure	21	Dikshit et al. (1986)
Candida albicans	Yeast	SMA, 2–7 days, 20°C	sd	8	Kumar et al. (2007)
Candida albicans	Yeast	SDA, 18 h, 30°C	6 (h), pure	30	Dikshit et al. (1986)
Candida krusei	Yeast	SMA, 2–7 days, 20°C	sd	0	Dikshit et al. (1986)
Candida tropicalis	Yeast	SMA, 2–7 days, 20°C	sd	0	Duarte et al. (2005)
Cryptococcus neoformans	Yeast	SMA, 2–7 days, 20°C	sd	6	Imai et al. (2001)
Cryptococcus rhodobenhani	Yeast	SMA, 2–7 days, 20°C	sd	6	Imai et al. (2001)
Saccharomyces cerevisiae	Yeast	SMA, 2–7 days, 20°C	sd	4	Imai et al. (2001)

TABLE 14.55
Inhibitory Data of Neroli Oil Obtained in the Dilution Test

Microorganism	MO Class	Conditions	MIC	Reference
Escherichia coli	Bac−	NB, Tween 20, 18 h, 37°C	25,000	Yousef and Tawil (1980)
Pseudomonas aeruginosa	Bac−	NB, Tween 20, 18 h, 37°C	25,000	Yousef and Tawil (1980)
Bacillus subtilis	Bac+	NB, Tween 20, 18 h, 37°C	3,200	Yousef and Tawil (1980)
Mycobacterium phlei	Bac+	NB, Tween 20, 18 h, 37°C	50,000	Yousef and Tawil (1980)
Staphylococcus aureus	Bac+	NB, Tween 20, 18 h, 37°C	6400	Yousef and Tawil (1980)
Aspergillus niger	Fungi	NB, Tween 20, 8 days, 30°C	3,200	Yousef and Tawil (1980)
Aspergillus niger	Fungi	YES broth, 10 days	−86% inh. 10,000	Lis-Balchin et al. (1998)
Aspergillus ochraceus	Fungi	YES broth, 10 days	−90% inh. 10,000	Lis-Balchin et al. (1998)
Aspergillus oryzae	Fungi	Cited	250	Okazaki and Oshima (1953)
Cephalosporium sacchari	Fungi	OA, EtOH, 3 days, 20°C	20,000	Narasimba Rao et al. (1971)
Ceratocystis paradoxa	Fungi	OA, EtOH, 3 days, 20°C	20,000	Narasimba Rao et al. (1971)
Curvularia lunata	Fungi	OA, EtOH, 3 days, 20°C	4,000	Narasimba Rao et al. (1971)
Fusarium culmorum	Fungi	YES broth, 10 days	−71% inh. 10,000	Lis-Balchin et al. (1998)
Fusarium moniliforme var. subglutinans	Fungi	OA, EtOH, 3 days, 20°C	2,000	Narasimba Rao et al. (1971)
Helminthosporium sacchari	Fungi	OA, EtOH, 3 days, 20°C	2,000	Narasimba Rao et al. (1971)
Mucor racemosus	Fungi	Cited	500	Okazaki and Oshima (1953)
Mucor sp.	Fungi	NB, Tween 20, 8 days, 30°C	3,200	Yousef and Tawil (1980)
Penicillium chrysogenum	Fungi	Cited	250	Okazaki and Oshima (1953)
Penicillium chrysogenum	Fungi	NB, Tween 20, 8 days, 30°C	6,400	Yousef and Tawil (1980)
Physalospora tucumanensis	Fungi	OA, EtOH, 3 days, 20°C	20,000	Narasimba Rao et al. (1971)
Rhizopus sp.	Fungi	NB, Tween 20, 8 days, 30°C	1,600	Yousef and Tawil (1980)
Sclerotium rolfsii	Fungi	OA, EtOH, 6 days, 20°C	20,000	Narasimba Rao et al. (1971)
Candida albicans	Yeast	NB, Tween 20, 18 h, 37°C	3,200	Yousef and Tawil (1980)

TABLE 14.56
Inhibitory Data of Neroli Oil Obtained in the Vapor Phase Test

Microorganism	MO Class	Conditions		Activity	Reference
Escherichia coli	Bac−	NA, 24 h, 37°C	~20,000	+++	Kellner and Kober (1954)
Neisseria sp.	Bac−	NA, 24 h, 37°C	~20,000	NG	Kellner and Kober (1954)
Salmonella typhi	Bac−	NA, 24 h, 37°C	~20,000	+++	Kellner and Kober (1954)
Salmonella typhi	Bac−	NA, 24 h, 37°C	sd	+++	Maruzzella and Sicurella (1960)
Bacillus megaterium	Bac+	NA, 24 h, 37°C	~20,000	NG	Kellner and Kober (1954)
Bacillus subtilis var. *aterrimus*	Bac+	NA, 24 h, 37°C	sd	+++	Maruzzella and Sicurella (1960)
Corynebacterium diphtheriae	Bac+	NA, 24 h, 37°C	~20,000	NG	Kellner and Kober (1954)
Mycobacterium avium	Bac+	NA, 24 h, 37°C	sd	NG	Maruzzella and Sicurella (1960)
Staphylococcus aureus	Bac+	NA, 24 h, 37°C	~20,000	NG	Kellner and Kober (1954)
Staphylococcus aureus	Bac+	NA, 24 h, 37°C	sd	+++	Maruzzella and Sicurella (1960)
Streptococcus faecalis	Bac+	NA, 24 h, 37°C	~20,000	NG	Kellner and Kober (1954)
Streptococcus faecalis	Bac+	NA, 24 h, 37°C	sd	+++	Maruzzella and Sicurella (1960)
Streptococcus pyogenes	Bac+	NA, 24 h, 37°C	~20,000	NG	Kellner and Kober (1954)
Candida albicans	Yeast	NA, 24 h, 37°C	~20,000	NG	Kellner and Kober (1954)

TABLE 14.57
Inhibitory Data of Nutmeg Oil Obtained in the Agar Diffusion Test

Microorganism	MO Class	Conditions		Inhibition Zone (mm)	Reference
Acinetobacter calcoaceticus	Bac−	ISA, 48 h, 25°C	4 (h), 10,000	10.5	Deans and Ritchie (1987)
Aeromonas hydrophila	Bac−	ISA, 48 h, 25°C	4 (h), 10,000	11.5	Deans and Ritchie (1987)
Alcaligenes faecalis	Bac−	ISA, 48 h, 25°C	4 (h), 10,000	8	Deans and Ritchie (1987)
Beneckea natriegens	Bac−	ISA, 48 h, 25°C	4 (h), 10,000	7	Deans and Ritchie (1987)
Campylobacter jejuni	Bac−	TSA, 24 h, 42°C	4 (h), 25,000	5.5	Smith-Palmer et al. (1998)
Citrobacter freundii	Bac−	ISA, 48 h, 25°C	4 (h), 10,000	13.5	Deans and Ritchie (1987)
Enterobacter aerogenes	Bac−	ISA, 48 h, 25°C	4 (h), 10,000	10.5	Deans and Ritchie (1987)
Erwinia carotovora	Bac−	ISA, 48 h, 25°C	4 (h), 10,000	9	Deans and Ritchie (1987)
Escherichia coli	Bac−	TGA, 18–24 h, 37°C	9.5, 2,000	0	Morris et al. (1979)
Escherichia coli	Bac−	Cited	15, 2,500	3	Pizsolitto et al. (1975)
Escherichia coli	Bac−	Cited, 18 h, 37°C	6, 2,500	8	Janssen et al. (1986)
Escherichia coli	Bac−	ISA, 48 h, 25°C	4 (h), 10,000	10.5	Deans and Ritchie (1987)
Escherichia coli	Bac−	TSA, 24 h, 35°C	4 (h), 25,000	6.9	Smith-Palmer et al. (1998)
Escherichia coli	Bac−	NA, 18 h, 37°C	6 (h), pure	18	Yousef and Tawil (1980)
Flavobacterium suaveolens	Bac−	ISA, 48 h, 25°C	4 (h), 10,000	13	Deans and Ritchie (1987)
Klebsiella pneumoniae	Bac−	ISA, 48 h, 25°C	4 (h), 10,000	9	Deans and Ritchie (1987)
Klebsiella sp.	Bac−	Cited	15, 2,500	2	Pizsolitto et al. (1975)
Moraxella sp.	Bac−	ISA, 48 h, 25°C	4 (h), 10,000	0	Deans and Ritchie (1987)
Proteus sp.	Bac−	Cited	15, 2,500	5	Pizsolitto et al. (1975)
Proteus vulgaris	Bac−	ISA, 48 h, 25°C	4 (h), 10,000	6	Deans and Ritchie (1987)
Pseudomonas aeruginosa	Bac−	Cited	15, 2,500	0	Pizsolitto et al. (1975)
Pseudomonas aeruginosa	Bac−	NA, 18 h, 37°C	6 (h), pure	0	Yousef and Tawil (1980)
Pseudomonas aeruginosa	Bac−	Cited, 18 h, 37°C	6, 2,500	0	Janssen et al. (1986)
Pseudomonas aeruginosa	Bac−	ISA, 48 h, 25°C	4 (h), 10,000	0	Deans and Ritchie (1987)
Salmonella enteritidis	Bac−	TSA, 24 h, 35°C	4 (h), 25,000	7.6	Smith-Palmer et al. (1998)
Salmonella pullorum	Bac−	ISA, 48 h, 25°C	4 (h), 10,000	17	Deans and Ritchie (1987)
Salmonella sp.	Bac−	Cited	15, 2,500	2	Pizsolitto et al. (1975)
Serratia marcescens	Bac−	ISA, 48 h, 25°C	4 (h), 10,000	18	Deans and Ritchie (1987)
Serratia sp.	Bac−	Cited	15, 2,500	3	Pizsolitto et al. (1975)
Shigella sp.	Bac−	Cited	15, 2,500	2	Pizsolitto et al. (1975)
Yersinia enterocolitica	Bac−	ISA, 48 h, 25°C	4 (h), 10,000	10.5	Deans and Ritchie (1987)
Bacillus sp.	Bac+	Cited	15, 2,500	5	Pizsolitto et al. (1975)
Bacillus subtilis	Bac+	ISA, 48 h, 25°C	4 (h), 10,000	8	Deans and Ritchie (1987)
Bacillus subtilis	Bac+	Cited, 18 h, 37°C	6, 2,500	8.7	Janssen et al. (1986)
Bacillus subtilis	Bac+	NA, 18 h, 37°C	6 (h), pure	34	Yousef and Tawil (1980)

(*Continued*)

TABLE 14.57 (Continued)
Inhibitory Data of Nutmeg Oil Obtained in the Agar Diffusion Test

Microorganism	MO Class	Conditions		Inhibition Zone (mm)	Reference
Brevibacterium linens	Bac+	ISA, 48 h, 25°C	4 (h), 10,000	7.5	Deans and Ritchie (1987)
Brochothrix thermosphacta	Bac+	ISA, 48 h, 25°C	4 (h), 10,000	5.5	Deans and Ritchie (1987)
Clostridium sporogenes	Bac+	ISA, 48 h, 25°C	4 (h), 10,000	0	Deans and Ritchie (1987)
Corynebacterium sp.	Bac+	TGA, 18–24 h, 37°C	9.5, 2,000	0	Morris et al. (1979)
Lactobacillus plantarum	Bac+	ISA, 48 h, 25°C	4 (h), 10,000	7	Deans and Ritchie (1987)
Leuconostoc cremoris	Bac+	ISA, 48 h, 25°C	4 (h), 10,000	9	Deans and Ritchie (1987)
Listeria monocytogenes	Bac+	ISA, 48 h, 25°C	4 (h), 10,000	0–12	Lis-Balchin et al. (1998)
Listeria monocytogenes	Bac+	TSA, 24 h, 35°C	4 (h), 25,000	7.7	Smith-Palmer et al. (1998)
Micrococcus luteus	Bac+	ISA, 48 h, 25°C	4 (h), 10,000	0	Deans and Ritchie (1987)
Mycobacterium phlei	Bac+	NA, 18 h, 37°C	6 (h), pure	27	Yousef and Tawil (1980)
Staphylococcus aureus	Bac+	TGA, 18–24 h, 37°C	9.5, 2,000	0	Morris et al. (1979)
Staphylococcus aureus	Bac+	ISA, 48 h, 25°C	4 (h), 10,000	0	Deans and Ritchie (1987)
Staphylococcus aureus	Bac+	Cited	15, 2,500	3	Pizsolitto et al. (1975)
Staphylococcus aureus	Bac+	TSA, 24 h, 35°C	4 (h), 25,000	6.5	Smith-Palmer et al. (1998)
Staphylococcus aureus	Bac+	Cited, 18 h, 37°C	6, 2,500	8.7	Janssen et al. (1986)
Staphylococcus aureus	Bac+	NA, 18 h, 37°C	6 (h), pure	15.5	Yousef and Tawil (1980)
Staphylococcus epidermidis	Bac+	Cited	15, 2,500	1	Pizsolitto et al. (1975)
Streptococcus faecalis	Bac+	ISA, 48 h, 25°C	4 (h), 10,000	0	Deans and Ritchie (1987)
Streptococcus viridans	Bac+	Cited	15, 2,500	0	Pizsolitto et al. (1975)
Alternaria porri	Fungi	PDA, 72 h, 28°C	5, 5,000	8.3	Pawar and Thaker (2007)
Aspergillus niger	Fungi	PDA, 48 h, 28°C	5, 5,000	5	Pawar and Thaker (2006)
Aspergillus niger	Fungi	SDA, 8 days, 30°C	6 (h), pure	40	Yousef and Tawil (1980)
Fusarium oxysporum f.sp. *ciceri*	Fungi	PDA, 72 h, 28°C	5, 5,000	0	Pawar and Thaker (2007)
Mucor sp.	Fungi	SDA, 8 days, 30°C	6 (h), pure	28	Yousef and Tawil (1980)
Penicillium chrysogenum	Fungi	SDA, 8 days, 30°C	6 (h), pure	60	Yousef and Tawil (1980)
Rhizopus sp.	Fungi	SDA, 8 days, 30°C	6 (h), pure	19	Yousef and Tawil (1980)
Brettanomyces anomalus	Yeast	MPA, 4 days, 30°C	5, 10% sol. sd	0	Conner and Beuchat (1984)
Candida albicans	Yeast	TGA, 18–24 h, 37°C	9.5, 2,000	0	Morris et al. (1979)
Candida albicans	Yeast	Cited, 18 h, 37°C	6, 2,500	13	Janssen et al. (1986)
Candida albicans	Yeast	SDA, 18 h, 30°C	6 (h), pure	24	Yousef and Tawil (1980)
Candida lipolytica	Yeast	MPA, 4 days, 30°C	5, 10% sol. sd	0	Conner and Beuchat (1984)

(Continued)

TABLE 14.57 (Continued)
Inhibitory Data of Nutmeg Oil Obtained in the Agar Diffusion Test

Microorganism	MO Class	Conditions		Inhibition Zone (mm)	Reference
Debaryomyces hansenii	Yeast	MPA, 4 days, 30°C	5, 10% sol. sd	0	Conner and Beuchat (1984)
Geotrichum candidum	Yeast	MPA, 4 days, 30°C	5, 10% sol. sd	0	Conner and Beuchat (1984)
Hansenula anomala	Yeast	MPA, 4 days, 30°C	5, 10% sol. sd	0	Conner and Beuchat (1984)
Kloeckera apiculata	Yeast	MPA, 4 days, 30°C	5, 10% sol. sd	0	Conner and Beuchat (1984)
Kluyveromyces fragilis	Yeast	MPA, 4 days, 30°C	5, 10% sol. sd	0	Conner and Beuchat (1984)
Lodderomyces elongisporus	Yeast	MPA, 4 days, 30°C	5, 10% sol. sd	0	Conner and Beuchat (1984)
Metschnikowia pulcherrima	Yeast	MPA, 4 days, 30°C	5, 10% sol. sd	0	Conner and Beuchat (1984)
Pichia membranifaciens	Yeast	MPA, 4 days, 30°C	5, 10% sol. sd	0	Conner and Beuchat (1984)
Rhodotorula rubra	Yeast	MPA, 4 days, 30°C	5, 10% sol. sd	0	Conner and Beuchat (1984)
Saccharomyces cerevisiae	Yeast	MPA, 4 days, 30°C	5, 10% sol. sd	0	Conner and Beuchat (1984)
Torula glabrata	Yeast	MPA, 4 days, 30°C	5, 10% sol. sd	0	Conner and Beuchat (1984)

TABLE 14.58
Inhibitory Data of Nutmeg Oil Obtained in the Dilution Test

Microorganism	MO Class	Conditions	MIC	Reference
Campylobacter jejuni	Bac–	TSB, 24 h, 42°C	>10,000	Smith-Palmer et al. (1998)
Escherichia coli	Bac–	NB, Tween 20, 18 h, 37°C	800	Yousef and Tawil (1980)
Escherichia coli	Bac–	TGB, 18–24 h, 37°C	>1,000	Morris et al. (1979)
Escherichia coli	Bac–	TSB, 24 h, 35°C	>10,000	Smith-Palmer et al. (1998)
Pseudomonas aeruginosa	Bac–	NB, Tween 20, 18 h, 37°C	>50,000	Yousef and Tawil (1980)
Salmonella enteritidis	Bac–	TSB, 24 h, 35°C	>10,000	Smith-Palmer et al. (1998)
Bacillus subtilis	Bac+	NB, Tween 20, 18 h, 37°C	6,400	Yousef and Tawil (1980)
Corynebacterium sp.	Bac+	TGB, 18–24 h, 37°C	1,000	Morris et al. (1979)
Listeria monocytogenes	Bac+	TSB, 24 h, 35°C	<100	Smith-Palmer et al. (1998)
Listeria monocytogenes	Bac+	TSB, 10 days, 4°C	500	Smith-Palmer et al. (1998)
Mycobacterium phlei	Bac+	NB, Tween 20, 18 h, 37°C	3,200	Yousef and Tawil (1980)
Staphylococcus aureus	Bac+	TGB, 18–24 h, 37°C	500	Morris et al. (1979)
Staphylococcus aureus	Bac+	TSB, 24 h, 35°C	>10,000	Smith-Palmer et al. (1998)
Staphylococcus aureus	Bac+	NB, Tween 20, 18 h, 37°C	25,000	Yousef and Tawil (1980)
Alternaria alternata	Fungi	PDA, 7 days, 28°C	0% inh. 500	Feng and Zheng (2007)
Aspergillus flavus	Fungi	PDA, 8 h, 20°C, spore germ. inh.	50–100	Thompson (1986)
Aspergillus flavus	Fungi	PDA, 5 days, 27°C	>1,000	Thompson and Cannon (1986)
Aspergillus niger	Fungi	NB, Tween 20, 8 days, 30°C	1,600	Yousef and Tawil (1980)
Aspergillus niger	Fungi	YES broth, 10 days	−88% inh. 10,000	Lis-Balchin et al. (1998)
Aspergillus ochraceus	Fungi	YES broth, 10 days	−86% inh. 10,000	Lis-Balchin et al. (1998)
Aspergillus parasiticus	Fungi	PDA, 8 h, 20°C, spore germ. inh.	50–100	Thompson (1986)
Aspergillus parasiticus	Fungi	PDA, 5 days, 27°C	>1,000	Thompson and Cannon (1986)
Epidermophyton floccosum	Fungi	SA, Tween 80, 21 days, 20°C	300–625	Janssen et al. (1988)
Fusarium culmorum	Fungi	YES broth, 10 days	−>10,000	Lis-Balchin et al. (1998)
Mucor hiemalis	Fungi	PDA, 5 days, 27°C	>1,000	Thompson and Cannon (1986)
Mucor mucedo	Fungi	PDA, 5 days, 27°C	>1,000	Thompson and Cannon (1986)
Mucor racemosus f. *racemosus*	Fungi	PDA, 5 days, 27°C	>1,000	Thompson and Cannon (1986)
Mucor sp.	Fungi	NB, Tween 20, 8 days, 30°C	3,200	Yousef and Tawil (1980)
Penicillium chrysogenum	Fungi	NB, Tween 20, 8 days, 30°C	400	Yousef and Tawil (1980)
Rhizopus 66-81-2	Fungi	PDA, 5 days, 27°C	>1,000	Thompson and Cannon (1986)
Rhizopus arrhizus	Fungi	PDA, 5 days, 27°C	>1,000	Thompson and Cannon (1986)
Rhizopus chinensis	Fungi	PDA, 5 days, 27°C	>1,000	Thompson and Cannon (1986)
Rhizopus circinans	Fungi	PDA, 5 days, 27°C	>1,000	Thompson and Cannon (1986)

(*Continued*)

TABLE 14.58 (*Continued*)
Inhibitory Data of Nutmeg Oil Obtained in the Dilution Test

Microorganism	MO Class	Conditions	MIC	Reference
Rhizopus japonicus	Fungi	PDA, 5 days, 27°C	>1,000	Thompson and Cannon (1986)
Rhizopus kazanensis	Fungi	PDA, 5 days, 27°C	>1,000	Thompson and Cannon (1986)
Rhizopus oryzae	Fungi	PDA, 5 days, 27°C	>1,000	Thompson and Cannon (1986)
Rhizopus pymacus	Fungi	PDA, 5 days, 27°C	>1,000	Thompson and Cannon (1986)
Rhizopus sp.	Fungi	NB, Tween 20, 8 days, 30°C	50,000	Yousef and Tawil (1980)
Rhizopus stolonifer	Fungi	PDA, 5 days, 27°C	>1,000	Thompson and Cannon (1986)
Rhizopus tritici	Fungi	PDA, 5 days, 27°C	>1,000	Thompson and Cannon (1986)
Trichophyton mentagrophytes	Fungi	SA, Tween 80, 21 days, 20°C	625–1,250	Janssen et al. (1988)
Trichophyton rubrum	Fungi	SA, Tween 80, 21 days, 20°C	300–625	Janssen et al. (1988)
Candida albicans	Yeast	TGB, 18–24 h, 37°C	500	Morris et al. (1979)
Candida albicans	Yeast	NB, Tween 20, 18 h, 37°C	3,200	Yousef and Tawil (1980)

TABLE 14.59
Inhibitory Data of Nutmeg Oil Obtained in the Vapor Phase Test

Microorganism	MO Class	Conditions		Activity	Reference
Salmonella typhi	Bac–	NA, 24 h, 37°C	sd	+++	Maruzzella and Sicurella (1960)
Bacillus subtilis var. *aterrimus*	Bac+	NA, 24 h, 37°C	sd	+++	Maruzzella and Sicurella (1960)
Mycobacterium avium	Bac+	NA, 24 h, 37°C	sd	++	Maruzzella and Sicurella (1960)
Staphylococcus aureus	Bac+	NA, 24 h, 37°C	sd	+++	Maruzzella and Sicurella (1960)
Streptococcus faecalis	Bac+	NA, 24 h, 37°C	sd	+++	Maruzzella and Sicurella (1960)

TABLE 14.60
Inhibitory Data of Peppermint Oil Obtained in the Agar Diffusion Test

Microorganism	MO Class	Conditions		Inhibition Zone (mm)	Reference
Acinetobacter calcoaceticus	Bac−	ISA, 48 h, 25°C	4 (h), 10,000	14.5	Deans and Ritchie (1987)
Aerobacter aerogenes	Bac−	NA, 24 h, 37°C	—, sd	0	Maruzzella and Lichtenstein (1956)
Aeromonas hydrophila	Bac−	ISA, 48 h, 25°C	4 (h), 10,000	13	Deans and Ritchie (1987)
Alcaligenes faecalis	Bac−	ISA, 48 h, 25°C	4 (h), 10,000	8.5	Deans and Ritchie (1987)
Beneckea natriegens	Bac−	ISA, 48 h, 25°C	4 (h), 10,000	16.5	Deans and Ritchie (1987)
Campylobacter jejuni	Bac−	TSA, 24 h, 42°C	4 (h), 25,000	7.1	Smith-Palmer et al. (1998)
Citrobacter freundii	Bac−	ISA, 48 h, 25°C	4 (h), 10,000	8	Deans and Ritchie (1987)
Enterobacter aerogenes	Bac−	ISA, 48 h, 25°C	4 (h), 10,000	9	Deans and Ritchie (1987)
Erwinia carotovora	Bac−	ISA, 48 h, 25°C	4 (h), 10,000	11	Deans and Ritchie (1987)
Escherichia coli	Bac−	NA, 18 h, 37°C	6 (h), pure	0	Yousef and Tawil (1980)
Escherichia coli	Bac−	TGA, 18–24 h, 37°C	9.5, 2,000	0	Morris et al. (1979)
Escherichia coli	Bac−	NA, 24 h, 37°C	—, sd	3	Maruzzella and Lichtenstein (1956)
Escherichia coli	Bac−	Cited	15, 2,500	5	Pizsolitto et al. (1975)
Escherichia coli	Bac−	ISA, 48 h, 25°C	4 (h), 10,000	9	Deans and Ritchie (1987)
Escherichia coli	Bac−	NA, 18 h, 37°C	5 (h), −30,000	23	Schelz et al. (2006)
Escherichia coli	Bac−	Cited, 18 h, 37°C	6, 2,500	9.7	Janssen et al. (1986)
Escherichia coli	Bac−	TSA, 24 h, 35°C	4 (h), 25,000	6.8	Smith-Palmer et al. (1998)
Flavobacterium suaveolens	Bac−	ISA, 48 h, 25°C	4 (h), 10,000	12	Deans and Ritchie (1987)
Klebsiella pneumoniae	Bac−	ISA, 48 h, 25°C	4 (h), 10,000	11	Deans and Ritchie (1987)
Klebsiella sp.	Bac−	Cited	15, 2,500	0	Pizsolitto et al. (1975)
Moraxella sp.	Bac−	ISA, 48 h, 25°C	4 (h), 10,000	0	Deans and Ritchie (1987)
Neisseria perflava	Bac−	NA, 24 h, 37°C	—, sd	2	Maruzzella and Lichtenstein (1956)
Proteus sp.	Bac−	Cited	15, 2,500	10	Pizsolitto et al. (1975)
Proteus vulgaris	Bac−	NA, 24 h, 37°C	—, sd	0	Maruzzella and Lichtenstein (1956)
Proteus vulgaris	Bac−	ISA, 48 h, 25°C	4 (h), 10,000	8	Deans and Ritchie (1987)
Pseudomonas aeruginosa	Bac−	Cited	15, 2,500	0	Pizsolitto et al. (1975)
Pseudomonas aeruginosa	Bac−	NA, 18 h, 37°C	6 (h), pure	0	Yousef and Tawil (1980)
Pseudomonas aeruginosa	Bac−	NA, 24 h, 37°C	—, sd	0	Maruzzella and Lichtenstein (1956)
Pseudomonas aeruginosa	Bac−	Cited, 18 h, 37°C	6, 2,500	0	Janssen et al. (1986)
Pseudomonas aeruginosa	Bac−	ISA, 48 h, 25°C	4 (h), 10,000	9	Deans and Ritchie (1987)

(Continued)

TABLE 14.60 (Continued)
Inhibitory Data of Peppermint Oil Obtained in the Agar Diffusion Test

Microorganism	MO Class	Conditions		Inhibition Zone (mm)	Reference
Salmonella enteritidis	Bac−	TSA, 24 h, 35°C	4 (h), 25,000	6.3	Smith-Palmer et al. (1998)
Salmonella paratyphi	Bac−	NA, 24 h, 37°C	6, sd	18	Dube and Rao (1984)
Salmonella pullorum	Bac−	ISA, 48 h, 25°C	4 (h), 10,000	15	Deans and Ritchie (1987)
Salmonella sp.	Bac−	Cited	15, 2,500	1	Pizsolitto et al. (1975)
Serratia marcescens	Bac−	NA, 24 h, 37°C	—, sd	2	Maruzzella and Lichtenstein (1956)
Serratia marcescens	Bac−	ISA, 48 h, 25°C	4 (h), 10,000	20	Deans and Ritchie (1987)
Serratia sp.	Bac−	Cited	15, 2,500	3	Pizsolitto et al. (1975)
Shigella sp.	Bac−	Cited	15, 2,500	3	Pizsolitto et al. (1975)
Vibrio cholera	Bac−	NA, 24 h, 37°C	6, sd	16	Dube and Rao (1984)
Yersinia enterocolitica	Bac−	ISA, 48 h, 25°C	4 (h), 10,000	13	Deans and Ritchie (1987)
Bacillus anthracis	Bac+	NA, 24 h, 37°C	6, sd	17	Dube and Rao (1984)
Bacillus mesentericus	Bac+	NA, 24 h, 37°C	—, sd	3	Maruzzella and Lichtenstein (1956)
Bacillus pumilus	Bac+	NA, 24 h, 37°C	6, sd	13	Dube and Rao (1984)
Bacillus sp.	Bac+	Cited	15, 2,500	10	Pizsolitto et al. (1975)
Bacillus subtilis	Bac+	NA, 24 h, 37°C	—, sd	8	Maruzzella and Lichtenstein (1956)
Bacillus subtilis	Bac+	ISA, 48 h, 25°C	4 (h), 10,000	10	Deans and Ritchie (1987)
Bacillus subtilis	Bac+	NA, 18 h, 37°C	6 (h), pure	21.5	Yousef and Tawil (1980)
Bacillus subtilis	Bac+	Cited, 18 h, 37°C	6, 2,500	14.7	Janssen et al. (1986)
Brevibacterium linens	Bac+	ISA, 48 h, 25°C	4 (h), 10,000	12	Deans and Ritchie (1987)
Brochothrix thermosphacta	Bac+	ISA, 48 h, 25°C	4 (h), 10,000	7	Deans and Ritchie (1987)
Clostridium sporogenes	Bac+	ISA, 48 h, 25°C	4 (h), 10,000	0	Deans and Ritchie (1987)
Corynebacterium sp.	Bac+	TGA, 18–24 h, 37°C	9.5, 2,000	0	Morris et al. (1979)
Lactobacillus plantarum	Bac+	ISA, 48 h, 25°C	4 (h), 10,000	11	Deans and Ritchie (1987)
Leuconostoc cremoris	Bac+	ISA, 48 h, 25°C	4 (h), 10,000	10	Deans and Ritchie (1987)
Listeria monocytogenes	Bac+	ISA, 48 h, 25°C	4 (h), 10,000	13–20	Lis-Balchin et al. (1998)
Listeria monocytogenes	Bac+	TSA, 24 h, 35°C	4 (h), 25,000	5.3	Smith-Palmer et al. (1998)
Micrococcus luteus	Bac+	ISA, 48 h, 25°C	4 (h), 10,000	9	Deans and Ritchie (1987)
Mycobacterium phlei	Bac+	NA, 18 h, 37°C	6 (h), pure	16	Yousef and Tawil (1980)
Sarcina lutea	Bac+	NA, 24 h, 37°C	—, sd	5	Maruzzella and Lichtenstein (1956)
Staphylococcus aureus	Bac+	TGA, 18–24 h, 37°C	9.5, 2,000	0	Morris et al. (1979)
Staphylococcus aureus	Bac+	NA, 24 h, 37°C	—, sd	3	Maruzzella and Lichtenstein (1956)
Staphylococcus aureus	Bac+	Cited	15, 2,500	8	Pizsolitto et al. (1975)

(Continued)

TABLE 14.60 (*Continued*)
Inhibitory Data of Peppermint Oil Obtained in the Agar Diffusion Test

Microorganism	MO Class	Conditions		Inhibition Zone (mm)	Reference
Staphylococcus aureus	Bac+	ISA, 48 h, 25°C	4 (h), 10,000	8	Deans and Ritchie (1987)
Staphylococcus aureus	Bac+	NA, 18 h, 37°C	6 (h), pure	12	Yousef and Tawil (1980)
Staphylococcus aureus	Bac+	NA, 24 h, 37°C	6, sd	13	Dube and Rao (1984)
Staphylococcus aureus	Bac+	Cited, 18 h, 37°C	6, 2,500	19	Janssen et al. (1986)
Staphylococcus aureus	Bac+	TSA, 24 h, 35°C	4 (h), 25,000	6.4	Smith-Palmer et al. (1998)
Staphylococcus epidermidis	Bac+	Cited	15, 2,500	10	Pizsolitto et al. (1975)
Staphylococcus epidermidis	Bac+	NA, 18 h, 37°C	5 (h), −30,000	15	Schelz et al. (2006)
Streptococcus faecalis	Bac+	ISA, 48 h, 25°C	4 (h), 10,000	0	Deans and Ritchie (1987)
Streptococcus viridans	Bac+	Cited	15, 2,500	0	Pizsolitto et al. (1975)
Streptomyces venezuelae	Bac+	SMA, 2–7 days, 20°C	sd	3	Maruzzella and Liguori (1958)
Absidia corymbifera	Fungi	EYA, 48 h, 45°C	5, sd	0	Nigam and Rao (1979)
Alternaria porri	Fungi	PDA, 72 h, 28°C	5, 5,000	10.6	Pawar and Thaker (2007)
Alternaria solani	Fungi	SMA, 2–7 days, 20°C	sd	0	Maruzzella and Liguori (1958)
Aspergillus flavus	Fungi	PDA, 10 days, 25°C	10	—	Sarbhoy et al. (1978)
Aspergillus fumigatus	Fungi	PDA, 10 days, 25°C	10	—	Sarbhoy et al. (1978)
Aspergillus fumigatus	Fungi	SMA, 2–7 days, 20°C	sd	4	Maruzzella and Liguori (1958)
Aspergillus fumigatus	Fungi	SDA, 3 days, 28°C	6, sd	10	Saksena and Saksena (1984)
Aspergillus niger	Fungi	PDA, 48 h, 28°C	5, 5,000	0	Pawar and Thaker (2006)
Aspergillus niger	Fungi	SMA, 2–7 days, 20°C	sd	6	Maruzzella and Liguori (1958)
Aspergillus niger	Fungi	SDA, 8 days, 30°C	6 (h), pure	60	Yousef and Tawil (1980)
Aspergillus sulphureus	Fungi	PDA, 10 days, 25°C	10	—	Sarbhoy et al. (1978)
Fusarium oxysporum f.sp. *ciceri*	Fungi	PDA, 72 h, 28°C	5, 5,000	10	Pawar and Thaker (2007)
Helminthosporium sativum	Fungi	SMA, 2–7 days, 20°C	sd	7	Maruzzella and Liguori (1958)
Humicola grisea var. *thermoidea*	Fungi	EYA, 48 h, 45°C	5, sd	30	Nigam and Rao (1979)
Keratinomyces afelloi	Fungi	SDA, 3 days, 28°C	6, sd	18	Saksena and Saksena (1984)
Keratinophyton terreum	Fungi	SDA, 3 days, 28°C	6, sd	0	Saksena and Saksena (1984)

(*Continued*)

TABLE 14.60 (Continued)
Inhibitory Data of Peppermint Oil Obtained in the Agar Diffusion Test

Microorganism	MO Class	Conditions		Inhibition Zone (mm)	Reference
Microsporum gypseum	Fungi	SDA, 3 days, 28°C	6, sd	10	Saksena and Saksena (1984)
Mucor fragilis	Fungi	PDA, 10 days, 25°C	10	—	Sarbhoy et al. (1978)
Mucor mucedo	Fungi	SMA, 2–7 days, 20°C	sd	8	Maruzzella and Liguori (1958)
Mucor sp.	Fungi	SDA, 8 days, 30°C	6 (h), pure	23	Yousef and Tawil (1980)
Nigrospora panici	Fungi	SMA, 2–7 days, 20°C	sd	8	Maruzzella and Liguori (1958)
Penicillium chrysogenum	Fungi	SDA, 8 days, 30°C	6 (h), pure	60	Yousef and Tawil (1980)
Penicillium digitatum	Fungi	SMA, 2–7 days, 20°C	sd	8	Maruzzella and Liguori (1958)
Rhizopus nigricans	Fungi	SMA, 2–7 days, 20°C	sd	7	Maruzzella and Liguori (1958)
Rhizopus sp.	Fungi	SDA, 8 days, 30°C	6 (h), pure	22	Yousef and Tawil (1980)
Rhizopus stolonifer	Fungi	PDA, 10 days, 25°C	10	—	Sarbhoy et al. (1978)
Sporotrichum thermophile	Fungi	EYA, 48 h, 45°C	5, sd	0	Nigam and Rao (1979)
Thermoascus aurantiacus	Fungi	EYA, 48 h, 45°C	5, sd	0	Nigam and Rao (1979)
Thermomyces lanuginosus	Fungi	EYA, 48 h, 45°C	5, sd	0	Nigam and Rao (1979)
Thielava minor	Fungi	EYA, 48 h, 45°C	5, sd	15	Nigam and Rao (1979)
Trichophyton equinum	Fungi	SDA, 3 days, 28°C	6, sd	14	Saksena and Saksena (1984)
Trichophyton rubrum	Fungi	SDA, 3 days, 28°C	6, sd	16	Saksena and Saksena (1984)
Brettanomyces anomalus	Yeast	MPA, 4 days, 30°C	5, 10% sol. sd	0	Conner and Beuchat (1984)
Candida albicans	Yeast	TGA, 18–24 h, 37°C	9.5, 2,000	0	Morris et al. (1979)
Candida albicans	Yeast	SMA, 2–7 days, 20°C	sd	4	Maruzzella and Liguori (1958)
Candida albicans	Yeast	Cited, 18 h, 37°C	6, 2,500	10	Janssen et al. (1986)
Candida albicans	Yeast	SDA, 18 h, 30°C	6 (h), pure	16	Yousef and Tawil (1980)
Candida albicans	Yeast	SDA, 3 days, 28°C	6, sd	18	Saksena and Saksena (1984)
Candida krusei	Yeast	SMA, 2–7 days, 20°C	sd	1	Maruzzella and Liguori (1958)
Candida lipolytica	Yeast	MPA, 4 days, 30°C	5, 10% sol. sd	0	Conner and Beuchat (1984)
Candida tropicalis	Yeast	SMA, 2–7 days, 20°C	sd	3	Maruzzella and Liguori (1958)

(*Continued*)

TABLE 14.60 (*Continued*)
Inhibitory Data of Peppermint Oil Obtained in the Agar Diffusion Test

Microorganism	MO Class	Conditions		Inhibition Zone (mm)	Reference
Candida tropicalis	Yeast	SDA, 3 days, 28°C	6, sd	23	Saksena and Saksena (1984)
Cryptococcus neoformans	Yeast	SMA, 2–7 days, 20°C	sd	3	Maruzzella and Liguori (1958)
Cryptococcus rhodobenhani	Yeast	SMA, 2–7 days, 20°C	sd	2	Maruzzella and Liguori (1958)
Debaryomyces hansenii	Yeast	MPA, 4 days, 30°C	5, 10% sol. sd	0	Conner and Beuchat (1984)
Geotrichum candidum	Yeast	MPA, 4 days, 30°C	5, 10% sol. sd	7	Conner and Beuchat (1984)
Hansenula anomala	Yeast	MPA, 4 days, 30°C	5, 10% sol. sd	0	Conner and Beuchat (1984)
Kloeckera apiculata	Yeast	MPA, 4 days, 30°C	5, 10% sol. sd	0	Conner and Beuchat (1984)
Kluyveromyces fragilis	Yeast	MPA, 4 days, 30°C	5, 10% sol. sd	0	Conner and Beuchat (1984)
Lodderomyces elongisporus	Yeast	MPA, 4 days, 30°C	5, 10% sol. sd	0	Conner and Beuchat (1984)
Metschnikowia pulcherrima	Yeast	MPA, 4 days, 30°C	5, 10% sol. sd	9	Conner and Beuchat (1984)
Pichia membranifaciens	Yeast	MPA, 4 days, 30°C	5, 10% sol. sd	0	Conner and Beuchat (1984)
Rhodotorula rubra	Yeast	MPA, 4 days, 30°C	5, 10% sol. sd	7	Conner and Beuchat (1984)
Saccharomyces cerevisiae	Yeast	MPA, 4 days, 30°C	5, 10% sol. sd	0	Conner and Beuchat (1984)
Saccharomyces cerevisiae	Yeast	SMA, 2–7 days, 20°C	sd	2	Maruzzella and Liguori (1958)
Saccharomyces cerevisiae	Yeast	NA, 24 h, 20°C	5 (h), −30,000	12–15	Schelz et al. (2006)
Torula glabrata	Yeast	MPA, 4 days, 30°C	5, 10% sol. sd	8	Conner and Beuchat (1984)
Torula thermophila	Yeast	EYA, 48 h, 45°C	5, sd	0	Nigam and Rao (1979)

TABLE 14.61
Inhibitory Data of Peppermint Oil Obtained in the Dilution Test

Microorganism	MO Class	Conditions	MIC	Reference
Treponema denticola	Bac	HS, 72 h, 37°C	1,000	Shapiro et al. (1994)
Treponema vincentii	Bac	HS, 72 h, 37°C	2,000	Shapiro et al. (1994)
Actinobacillus actinomycetemcomitans	Bac–	HS, 72 h, 37°C	3,000	Shapiro et al. (1994)
Campylobacter jejuni	Bac–	TSB, 24 h, 42°C	1,000	Smith-Palmer et al. (1998)
Capnocytophaga sp.	Bac–	HS, 72 h, 37°C	3,000	Shapiro et al. (1994)
Eikenella corrodens	Bac–	HS, 72 h, 37°C	2,000	Shapiro et al. (1994)
Escherichia coli	Bac–	Cited, 24 h	800	Imai et al. (2001)
Escherichia coli	Bac–	TGB, 18–24 h, 37°C	>1,000	Morris et al. (1979)
Escherichia coli	Bac–	NB, Tween 20, 18 h, 37°C	1,600	Yousef and Tawil (1980)
Escherichia coli	Bac–	TGB, 18 h, 37°C	5,700	Schelz et al. (2006)
Escherichia coli	Bac–	MHB, 24 h, 36°C	>10,000	Duarte et al. (2006)
Escherichia coli	Bac–	TSB, 24 h, 35°C	>10,000	Smith-Palmer et al. (1998)
Fusobacterium nucleatum	Bac–	HS, 72 h, 37°C	2,000	Shapiro et al. (1994)
Helicobacter pylori	Bac–	Cited, 48 h	100	Imai et al. (2001)
Helicobacter pylori	Bac–	Cited, 20 h, 37°C	135.6	Weseler et al. (2005)
Porphyromonas gingivalis	Bac–	HS, 72 h, 37°C	2,000	Shapiro et al. (1994)
Prevotella buccae	Bac–	HS, 72 h, 37°C	2,000	Shapiro et al. (1994)
Prevotella intermedia	Bac–	HS, 72 h, 37°C	3000	Shapiro et al. (1994)
Prevotella nigrescens	Bac–	HS, 72 h, 37°C	2,000	Shapiro et al. (1994)
Pseudomonas aeruginosa	Bac–	NB, Tween 20, 18 h, 37°C	>50,000	Yousef and Tawil (1980)
Salmonella enteritidis	Bac–	Cited, 24 h	400	Imai et al. (2001)
Salmonella enteritidis	Bac–	TSB, 24 h, 35°C	>10,000	Smith-Palmer et al. (1998)
Selenomonas artemidis	Bac–	HS, 72 h, 37°C	1,000	Shapiro et al. (1994)
Actinomyces viscosus	Bac+	HS, 16–24, 37°C	5,000	Shapiro et al. (1994)
Bacillus subtilis	Bac+	NB, Tween 20, 18 h, 37°C	1,600	Yousef and Tawil (1980)
Corynebacterium sp.	Bac+	TGB, 18–24 h, 37°C	>1,000	Morris et al. (1979)
Listeria monocytogenes	Bac+	TSB, 24 h, 35°C	300	Smith-Palmer et al. (1998)
Mycobacterium phlei	Bac+	NB, Tween 20, 18 h, 37°C	400	Yousef and Tawil (1980)
Peptostreptococcus anaerobius	Bac+	HS, 72 h, 37°C	2,000	Shapiro et al. (1994)
Staphylococcus aureus	Bac+	TSB, 24 h, 35°C	400	Smith-Palmer et al. (1998)
Staphylococcus aureus	Bac+	TGB, 18–24 h, 37°C	1,000	Morris et al. (1979)
Staphylococcus aureus	Bac+	NB, Tween 20, 18 h, 37°C	6,400	Yousef and Tawil (1980)
Staphylococcus aureus MRSA	Bac+	Cited, 24 h	200	Imai et al. (2001)
Staphylococcus aureus MSSA	Bac+	Cited, 24 h	200	Imai et al. (2001)
Staphylococcus epidermidis	Bac+	TGB, 18 h, 37°C	5,700	Schelz et al. (2006)

(*Continued*)

TABLE 14.61 (Continued)
Inhibitory Data of Peppermint Oil Obtained in the Dilution Test

Microorganism	MO Class	Conditions	MIC	Reference
Streptococcus sanguinis	Bac+	HS, 16–24, 37°C	6,000	Shapiro et al. (1994)
Streptococcus sobrinus	Bac+	HS, 16–24, 37°C	3,000	Shapiro et al. (1994)
Aspergillus flavus	Fungi	PDA, 8 h, 20°C, spore germ. inh.	50–100	Thompson (1986)
Aspergillus flavus	Fungi	PDA, 5 days, 27°C	>1,000	Thompson and Cannon (1986)
Aspergillus niger	Fungi	NB, Tween 20, 8 days, 30°C	800	Yousef and Tawil (1980)
Aspergillus niger	Fungi	YES broth, 10 days	−98% inh. 10,000	Lis-Balchin et al. (1998)
Aspergillus ochraceus	Fungi	YES broth, 10 days	−93% inh. 10,000	Lis-Balchin et al. (1998)
Aspergillus oryzae	Fungi	Cited	500	Okazaki and Oshima (1953)
Aspergillus parasiticus	Fungi	PDA, 8 h, 20°C, spore germ. inh.	50–100	Thompson (1986)
Aspergillus parasiticus	Fungi	PDA, 5 days, 27°C	>1,000	Thompson and Cannon (1986)
Botrytis cinerea	Fungi	PDA, Tween 20, 7 days, 24°C	0% inh. 1,000	Bouchra et al. (2003)
Epidermophyton floccosum	Fungi	SA, Tween 80, 21 days, 20°C	300–625	Janssen et al. (1988)
Fusarium culmorum	Fungi	YES broth, 10 days	−>10,000	Lis-Balchin et al. (1998)
Geotrichum citri-aurantii	Fungi	PDA, Tween 20, 7 days, 24°C	0% inh. 1,000	Bouchra et al. (2003)
Microsporum gypseum	Fungi	SDA, 7 days, 30°C	400, 37% inh.	Dikshit and Husain (1984)
Mucor hiemalis	Fungi	PDA, 5 days, 27°C	>1,000	Thompson and Cannon (1986)
Mucor mucedo	Fungi	PDA, 5 days, 27°C	>1,000	Thompson and Cannon (1986)
Mucor racemosus	Fungi	Cited	500	Okazaki and Oshima (1953)
Mucor racemosus f. racemosus	Fungi	PDA, 5 days, 27°C	>1,000	Thompson and Cannon (1986)
Mucor sp.	Fungi	NB, Tween 20, 8 days, 30°C	1,600	Yousef and Tawil (1980)
Penicillium chrysogenum	Fungi	Cited	500	Okazaki and Oshima (1953)
Penicillium chrysogenum	Fungi	NB, Tween 20, 8 days, 30°C	3,200	Yousef and Tawil (1980)
Penicillium digitatum	Fungi	PDA, Tween 20, 7 days, 24°C	0% inh. 1,000	Bouchra et al. (2003)
Phytophthora citrophthora	Fungi	PDA, Tween 20, 7 days, 24°C	14% inh. 1000	Bouchra et al. (2003)
Rhizopus 66-81-2	Fungi	PDA, 5 days, 27°C	>1,000	Thompson and Cannon (1986)
Rhizopus arrhizus	Fungi	PDA, 5 days, 27°C	>1,000	Thompson and Cannon (1986)
Rhizopus chinensis	Fungi	PDA, 5 days, 27°C	>1,000	Thompson and Cannon (1986)
Rhizopus circinans	Fungi	PDA, 5 days, 27°C	>1,000	Thompson and Cannon (1986)
Rhizopus japonicus	Fungi	PDA, 5 days, 27°C	>1,000	Thompson and Cannon (1986)
Rhizopus kazanensis	Fungi	PDA, 5 days, 27°C	>1,000	Thompson and Cannon (1986)
Rhizopus oryzae	Fungi	PDA, 5 days, 27°C	>1,000	Thompson and Cannon (1986)

(Continued)

TABLE 14.61 (Continued)
Inhibitory Data of Peppermint Oil Obtained in the Dilution Test

Microorganism	MO Class	Conditions	MIC	Reference
Rhizopus pymacus	Fungi	PDA, 5 days, 27°C	>1,000	Thompson and Cannon (1986)
Rhizopus sp.	Fungi	NB, Tween 20, 8 days, 30°C	6,400	Yousef and Tawil (1980)
Rhizopus stolonifer	Fungi	PDA, 5 days, 27°C	>1,000	Thompson and Cannon (1986)
Rhizopus tritici	Fungi	PDA, 5 days, 27°C	>1,000	Thompson and Cannon (1986)
Trichophyton equinum	Fungi	SDA, 7 days, 30°C	400, 51% inh.	Dikshit and Husain (1984)
Trichophyton mentagrophytes	Fungi	SA, Tween 80, 21 days, 20°C	625–1,250	Janssen et al. (1988)
Trichophyton rubrum	Fungi	SDA, 7 days, 30°C	400, 61% inh.	Dikshit and Husain (1984)
Trichophyton rubrum	Fungi	SA, Tween 80, 21 days, 20°C	300–625	Janssen et al. (1988)
Candida albicans	Yeast	NB, Tween 20, 18 h, 37°C	800	Yousef and Tawil (1980)
Candida albicans	Yeast	TGB, 18–24 h, 37°C	>1,000	Morris et al. (1979)
Candida albicans	Yeast	MHB, Tween 80, 48 h, 35°C	5,000	Hammer et al. (1998)
Saccharomyces cerevisiae	Yeast	YPB, 24 h, 20°C	400	Schelz et al. (2006)

TABLE 14.62
Inhibitory Data of Peppermint Oil Obtained in the Vapor Phase Test

Microorganism	MO Class	Conditions		Activity	Reference
Escherichia coli	Bac–	NA, 24 h, 37°C	~20,000	++	Kellner and Kober (1954)
Escherichia coli	Bac–	BLA, 18 h, 37°C	MIC_{air}	>1,600	Inouye et al. (2001)
Haemophilus influenzae	Bac–	MHA, 18 h, 37°C	MIC_{air}	12.5	Inouye et al. (2001)
Neisseria sp.	Bac–	NA, 24 h, 37°C	~20,000	NG	Kellner and Kober (1954)
Salmonella typhi	Bac–	NA, 24 h, 37°C	~20,000	+	Kellner and Kober (1954)
Salmonella typhi	Bac–	NA, 24 h, 37°C	sd	+++	Maruzzella and Sicurella (1960)
Bacillus megaterium	Bac+	NA, 24 h, 37°C	~20,000	+	Kellner and Kober (1954)
Bacillus subtilis var. *aterrimus*	Bac+	NA, 24 h, 37°C	sd	++	Maruzzella and Sicurella (1960)
Corynebacterium diphtheriae	Bac+	NA, 24 h, 37°C	~20,000	+	Kellner and Kober (1954)
Mycobacterium avium	Bac+	NA, 24 h, 37°C	sd	NG	Maruzzella and Sicurella (1960)
Staphylococcus aureus	Bac+	NA, 24 h, 37°C	~20,000	+	Kellner and Kober (1954)
Staphylococcus aureus	Bac+	NA, 24 h, 37°C	sd	++	Maruzzella and Sicurella (1960)
Staphylococcus aureus	Bac+	MHA, 18 h, 37°C	MIC_{air}	25	Inouye et al. (2001)
Streptococcus faecalis	Bac+	NA, 24 h, 37°C	~20,000	+	Kellner and Kober (1954)
Streptococcus faecalis	Bac+	NA, 24 h, 37°C	sd	+++	Maruzzella and Sicurella (1960)
Streptococcus pneumoniae	Bac+	MHA, 18 h, 37°C	MIC_{air}	25	Inouye et al. (2001)
Streptococcus pyogenes	Bac+	NA, 24 h, 37°C	~20,000	NG	Kellner and Kober (1954)
Streptococcus pyogenes	Bac+	MHA, 18 h, 37°C	MIC_{air}	25	Inouye et al. (2001)
Aspergillus flavus	Fungi	WFA, 42 days, 25°C	Disk, 50,000	+++	Guynot et al. (2003)
Aspergillus niger	Fungi	WFA, 42 days, 25°C	Disk, 50,000	+++	Guynot et al. (2003)
Botrytis cinerea	Fungi	PDA, 3 days, 25°C	1,000	+++	Lee et al. (2007)
Colletotrichum gloeosporioides	Fungi	PDA, 3 days, 25°C	1,000	+++	Lee et al. (2007)
Eurotium amstelodami	Fungi	WFA, 42 days, 25°C	Disk, 50,000	++	Guynot et al. (2003)
Eurotium herbarum	Fungi	WFA, 42 days, 25°C	Disk, 50,000	+++	Guynot et al. (2003)
Eurotium repens	Fungi	WFA, 42 days, 25°C	Disk, 50,000	++	Guynot et al. (2003)
Eurotium rubrum	Fungi	WFA, 42 days, 25°C	Disk, 50,000	+++	Guynot et al. (2003)
Fusarium oxysporum	Fungi	PDA, 3 days, 25°C	1,000	+++	Lee et al. (2007)
Penicillium corylophilum	Fungi	WFA, 42 days, 25°C	Disk, 50,000	+	Guynot et al. (2003)
Pythium ultimum	Fungi	PDA, 3 days, 25°C	1,000	+++	Lee et al. (2007)
Rhizoctonia solani	Fungi	PDA, 3 days, 25°C	1,000	+++	Lee et al. (2007)
Candida albicans	Yeast	NA, 24 h, 37°C	~20,000	+	Kellner and Kober (1954)

TABLE 14.63
Inhibitory Data of *Pinus sylvestris* Oil Obtained in the Agar Diffusion Test

Microorganism	MO Class	Conditions		Inhibition Zone (mm)	Reference
Aerobacter aerogenes	Bac–	NA, 24 h, 37°C	—, sd	0	Maruzzella and Lichtenstein (1956)
Bordetella bronchiseptica	Bac–	MHA, 18 h, 37°C	6, 17,500	0	Schales et al. (1993)
Citrobacter freundii	Bac–	MHA, 18 h, 37°C	6, 17,500	0	Schales et al. (1993)
Escherichia coli	Bac–	NA, 24 h, 37°C	—, sd	1	Maruzzella and Lichtenstein (1956)
Escherichia coli	Bac–	Cited, 18 h, 37°C	6, 2,500	11.7	Janssen et al. (1986)
Escherichia coli 1	Bac–	MHA, 18 h, 37°C	6, 17,500	8	Schales et al. (1993)
Escherichia coli 2	Bac–	MHA, 18 h, 37°C	6, 17,500	0	Schales et al. (1993)
Klebsiella pneumoniae	Bac–	MHA, 18 h, 37°C	6, 17,500	0	Schales et al. (1993)
Neisseria perflava	Bac–	NA, 24 h, 37°C	—, sd	3	Maruzzella and Lichtenstein (1956)
Proteus mirabilis	Bac–	MHA, 18 h, 37°C	6, 17,500	0	Schales et al. (1993)
Proteus vulgaris	Bac–	NA, 24 h, 37°C	—, sd	4	Maruzzella and Lichtenstein (1956)
Pseudomonas aeruginosa	Bac–	NA, 24 h, 37°C	—, sd	0	Maruzzella and Lichtenstein (1956)
Pseudomonas aeruginosa	Bac–	MHA, 18 h, 37°C	6, 17,500	0	Schales et al. (1993)
Pseudomonas aeruginosa	Bac–	Cited, 18 h, 37°C	6, 2,500	0	Janssen et al. (1986)
Serratia marcescens	Bac–	NA, 24 h, 37°C	—, sd	0	Maruzzella and Lichtenstein (1956)
Bacillus mesentericus	Bac+	NA, 24 h, 37°C	—, sd	2	Maruzzella and Lichtenstein (1956)
Bacillus sp.	Bac+	MHA, 18 h, 37°C	6, 17,500	11	Schales et al. (1993)
Bacillus subtilis	Bac+	NA, 24 h, 37°C	—, sd	3	Maruzzella and Lichtenstein (1956)
Bacillus subtilis	Bac+	Cited, 18 h, 37°C	6, 2,500	8.7	Janssen et al. (1986)
Clostridium perfringens	Bac+	MHA, 18 h, 37°C	6, 17,500	23	Schales et al. (1993)
Enterococcus sp.	Bac+	MHA, 18 h, 37°C	6, 17,500	15	Schales et al. (1993)
Sarcina lutea	Bac+	NA, 24 h, 37°C	—, sd	0	Maruzzella and Lichtenstein (1956)
Staphylococcus aureus	Bac+	NA, 24 h, 37°C	—, sd	2	Maruzzella and Lichtenstein (1956)
Staphylococcus aureus	Bac+	Cited, 18 h, 37°C	6, 2,500	7.7	Janssen et al. (1986)
Staphylococcus sp.	Bac+	MHA, 18 h, 37°C	6, 17,500	6	Schales et al. (1993)
Streptomyces venezuelae	Bac+	SMA, 2–7 days, 20°C	sd	7	Maruzzella and Liguori (1958)
Alternaria solani	Fungi	SMA, 2–7 days, 20°C	sd	8	Maruzzella and Liguori (1958)
Aspergillus fumigatus	Fungi	SMA, 2–7 days, 20°C	sd	9	Maruzzella and Liguori (1958)
Aspergillus niger	Fungi	SMA, 2–7 days, 20°C	sd	12	Maruzzella and Liguori (1958)
Helminthosporium sativum	Fungi	SMA, 2–7 days, 20°C	sd	12	Maruzzella and Liguori (1958)

(Continued)

TABLE 14.63 (Continued)
Inhibitory Data of *Pinus sylvestris* Oil Obtained in the Agar Diffusion Test

Microorganism	MO Class	Conditions		Inhibition Zone (mm)	Reference
Mucor mucedo	Fungi	SMA, 2–7 days, 20°C	sd	15	Maruzzella and Liguori (1958)
Nigrospora panici	Fungi	SMA, 2–7 days, 20°C	sd	12	Maruzzella and Liguori (1958)
Penicillium digitatum	Fungi	SMA, 2–7 days, 20°C	sd	10	Maruzzella and Liguori (1958)
Rhizopus nigricans	Fungi	SMA, 2–7 days, 20°C	sd	5	Maruzzella and Liguori (1958)
Candida albicans	Yeast	SMA, 2–7 days, 20°C	sd	5	Maruzzella and Liguori (1958)
Candida albicans	Yeast	Cited, 18 h, 37°C	6, 2,500	16.3	Janssen et al. (1986)
Candida krusei	Yeast	SMA, 2–7 days, 20°C	sd	5	Maruzzella and Liguori (1958)
Candida tropicalis	Yeast	SMA, 2–7 days, 20°C	sd	10	Maruzzella and Liguori (1958)
Cryptococcus neoformans	Yeast	SMA, 2–7 days, 20°C	sd	6	Maruzzella and Liguori (1958)
Cryptococcus rhodobenhani	Yeast	SMA, 2–7 days, 20°C	sd	14	Maruzzella and Liguori (1958)
Saccharomyces cerevisiae	Yeast	SMA, 2–7 days, 20°C	sd	18	Maruzzella and Liguori (1958)

TABLE 14.64
Inhibitory Data of *Pinus sylvestris* Oil Obtained in the Dilution Test

Microorganism	MO Class	Conditions	MIC	Reference
Acinetobacter baumannii	Bac−	MHA, Tween 20, 48 h, 35°C	20,000	Hammer et al. (1999)
Aeromonas sobria	Bac−	MHA, Tween 20, 48 h, 35°C	20,000	Hammer et al. (1999)
Escherichia coli	Bac−	MHB, Tween 80, 24 h, 37°C	>29,000	Chalchat et al. (1989)
Escherichia coli	Bac−	MHA, Tween 20, 48 h, 35°C	20,000	Hammer et al. (1999)
Escherichia coli	Bac−	MHB, Tween 80, 24 h, 37°C	64,300	Bastide et al. (1987)
Klebsiella pneumonia	Bac−	MHB, Tween 80, 24 h, 37°C	3,500	Chalchat et al. (1989)
Klebsiella pneumoniae	Bac−	MHA, Tween 20, 48 h, 35°C	>20,000	Hammer et al. (1999)
Proteus mirabilis	Bac−	MHB, Tween 80, 24 h, 37°C	>29,000	Chalchat et al. (1989)
Pseudomonas aeruginosa	Bac−	MHA, Tween 20, 48 h, 35°C	>20,000	Hammer et al. (1999)
Pseudomonas aeruginosa	Bac−	MHB, Tween 80, 24 h, 37°C	>29,000	Chalchat et al. (1989)
Salmonella typhimurium	Bac−	MHA, Tween 20, 48 h, 35°C	>20,000	Hammer et al. (1999)
Serratia marcescens	Bac−	MHA, Tween 20, 48 h, 35°C	>20,000	Hammer et al. (1999)
Bacillus sp.	Bac+	CA, 7 days, 25°C	5,000	Motiejunaite and Peciulyte (2004)
Enterococcus faecalis	Bac+	MHA, Tween 20, 48 h, 35°C	>20,000	Hammer et al. (1999)
Rhodococcus sp.	Bac+	CA, 7 days, 25°C	5,000	Motiejunaite and Peciulyte (2004)
Staphylococcus aureus	Bac+	MHA, Tween 20, 48 h, 35°C	>20,000	Hammer et al. (1999)
Staphylococcus aureus	Bac+	MHB, Tween 80, 24 h, 37°C	3,500	Chalchat et al. (1989)
Staphylococcus aureus	Bac+	MHB, Tween 80, 24 h, 37°C	4000	Bastide et al. (1987)
Alternaria alternata	Fungi	RPMI, 1.5% EtOH, 7 days, 30°C	10,000	Tullio et al. (2006)
Aspergillus flavus	Fungi	RPMI, 1.5% EtOH, 7 days, 30°C	5,000	Tullio et al. (2006)
Aspergillus flavus var. *columnaris*	Fungi	RPMI, 1.5% EtOH, 7 days, 30°C	1,250	Tullio et al. (2006)
Aspergillus fumigatus	Fungi	RPMI, 1.5% EtOH, 7 days, 30°C	5,000–10,000	Tullio et al. (2006)
Aspergillus niger	Fungi	RPMI, 1.5% EtOH, 7 days, 30°C	5,000	Tullio et al. (2006)
Aspergillus niger	Fungi	CA, 7 days, 25°C	7,500–15,000	Motiejunaite and Peciulyte (2004)
Aspergillus versicolor	Fungi	CA, 7 days, 25°C	7,500–15,000	Motiejunaite and Peciulyte (2004)
Aureobasidium pullulans	Fungi	CA, 7 days, 25°C	5,000–7,500	Motiejunaite and Peciulyte (2004)
Chaetomium globosum	Fungi	CA, 7 days, 25°C	5,000	Motiejunaite and Peciulyte (2004)
Cladosporium cladosporioides	Fungi	RPMI, 1.5% EtOH, 7 days, 30°C	1,250–2,500	Tullio et al. (2006)
Cladosporium cladosporioides	Fungi	CA, 7 days, 25°C	5,000	Motiejunaite and Peciulyte (2004)
Epidermophyton floccosum	Fungi	RPMI, 1.5% EtOH, 7 days, 30°C	1,250	Tullio et al. (2006)
Fusarium oxysporum	Fungi	RPMI, 1.5% EtOH, 7 days, 30°C	312	Tullio et al. (2006)
Microsporum canis	Fungi	RPMI, 1.5% EtOH, 7 days, 30°C	1,250–5,000	Tullio et al. (2006)

(*Continued*)

TABLE 14.64 (*Continued*)
Inhibitory Data of *Pinus sylvestris* Oil Obtained in the Dilution Test

Microorganism	MO Class	Conditions	MIC	Reference
Microsporum gypseum	Fungi	RPMI, 1.5% EtOH, 7 days, 30°C	2,500–5,000	Tullio et al. (2006)
Mucor sp.	Fungi	RPMI, 1.5% EtOH, 7 days, 30°C		Tullio et al. (2006)
Paecilomyces variotii	Fungi	CA, 7 days, 25°C	>10,000	Thanaboripat et al. (2004), Motiejunaite and Peciulyte (2004)
Penicillium chrysogenum	Fungi	CA, 7 days, 25°C	10,000–25,000	Motiejunaite and Peciulyte (2004)
Penicillium frequentans	Fungi	RPMI, 1.5% EtOH, 7 days, 30°C	1,250	Tullio et al. (2006)
Penicillium lanosum	Fungi	RPMI, 1.5% EtOH, 7 days, 30°C	10,000	Tullio et al. (2006)
Phoma glomerata	Fungi	CA, 7 days, 25°C	10,000–25,000	Motiejunaite and Peciulyte (2004)
Phoma sp.	Fungi	CA, 7 days, 25°C	7,500	Motiejunaite and Peciulyte (2004)
Rhizopus sp.	Fungi	RPMI, 1.5% EtOH, 7 days, 30°C	>10,000	Tullio et al. (2006)
Rhizopus stolonifer	Fungi	CA, 7 days, 25°C	5,000–7,500	Motiejunaite and Peciulyte (2004)
Scopulariopsis brevicaulis	Fungi	RPMI, 1.5% EtOH, 7 days, 30°C	10,000	Tullio et al. (2006)
Stachybotrys chartarum	Fungi	CA, 7 days, 25°C	10,000–15,000	Motiejunaite and Peciulyte (2004)
Trichoderma viride	Fungi	CA, 7 days, 25°C	5,500	Motiejunaite and Peciulyte (2004)
Trichophyton mentagrophytes	Fungi	RPMI, 1.5% EtOH, 7 days, 30°C	2,500–5,000	Tullio et al. (2006)
Candida albicans	Yeast	MHB, Tween 80, 24 h, 37°C	14,000	Chalchat et al. (1989)
Candida albicans	Yeast	MHA, Tween 20, 48 h, 35°C	20,000	Hammer et al. (1999)
Candida lipolytica	Yeast	CA, 7 days, 25°C	5,000	Motiejunaite and Peciulyte (2004)
Geotrichum candida	Yeast	CA, 7 days, 25°C	3,500–5,000	Motiejunaite and Peciulyte (2004)

TABLE 14.65
Inhibitory Data of *Pinus sylvestris* Oil Obtained in the Vapor Phase Test

Microorganism	MO Class	Conditions		Activity	Reference
Botrytis cinerea	Fungi	PDA, 3 days, 25°C	1,000	+++	Lee et al. (2007)
Colletotrichum gloeosporioides	Fungi	PDA, 3 days, 25°C	1,000	+++	Lee et al. (2007)
Fusarium oxysporum	Fungi	PDA, 3 days, 25°C	1,000	+++	Lee et al. (2007)
Pythium ultimum	Fungi	PDA, 3 days, 25°C	1,000	+++	Lee et al. (2007)
Rhizoctonia solani	Fungi	PDA, 3 days, 25°C	1,000	+++	Lee et al. (2007)

TABLE 14.66
Inhibitory Data of Rosemary Oil Obtained in the Agar Diffusion Test

Microorganism	MO Class	Conditions		Inhibition Zone (mm)	Reference
Acinetobacter calcoaceticus	Bac−	ISA, 48 h, 25°C	4 (h), 10,000	15	Deans and Ritchie (1987)
Aerobacter aerogenes	Bac−	NA, 24 h, 37°C	—, sd	3	Maruzzella and Lichtenstein (1956)
Aerobacter tumefaciens	Bac−	Cited	(h) 20,000	—	Hethenyi et al. (1989)
Aeromonas hydrophila	Bac−	ISA, 48 h, 25°C	4 (h), 10,000	7	Deans and Ritchie (1987)
Alcaligenes faecalis	Bac−	ISA, 48 h, 25°C	4 (h), 10,000	0	Deans and Ritchie (1987)
Beneckea natriegens	Bac−	ISA, 48 h, 25°C	4 (h), 10,000	0	Deans and Ritchie (1987)
Campylobacter jejuni	Bac−	TSA, 24 h, 42°C	4 (h), 25,000	9.3	Smith-Palmer et al. (1998)
Citrobacter freundii	Bac−	ISA, 48 h, 25°C	4 (h), 10,000	8	Deans and Ritchie (1987)
Enterobacter aerogenes	Bac−	MHA, 24 h, 30°C	6, 15,000	7	Rossi et al. (2007)
Enterobacter aerogenes	Bac−	ISA, 48 h, 25°C	4 (h), 10,000	8.5	Deans and Ritchie (1987)
Erwinia carotovora	Bac−	ISA, 48 h, 25°C	4 (h), 10,000	7.5	Deans and Ritchie (1987)
Escherichia coli	Bac−	Cited	(h) 20,000	− to +++	Hethenyi et al. (1989)
Escherichia coli	Bac−	Cited	15, 2,500	1	Pizsolitto et al. (1975)
Escherichia coli	Bac−	NA, 24 h, 30°C	Drop, 5,000	1	Hili et al. (1997)
Escherichia coli	Bac−	NA, 24 h, 37°C	—, sd	3	Maruzzella and Lichtenstein (1956)
Escherichia coli	Bac−	ISA, 48 h, 25°C	4 (h), 10,000	12	Deans and Ritchie (1987)
Escherichia coli	Bac−	Cited, 24 h, 37°C	—	<12	Rota et al. (2004)
Escherichia coli	Bac−	MHA, 24 h, 30°C	6, 15,000	12	Rossi et al. (2007)
Escherichia coli	Bac−	Cited, 18 h, 37°C	6, 2,500	8.7	Janssen et al. (1986)
Escherichia coli	Bac−	TSA, 24 h, 35°C	4 (h), 25,000	8.7	Smith-Palmer et al. (1998)
Escherichia coli	Bac−	NA, 18 h, 37°C	5 (h), −30,000	15	Schelz et al. (2006)
Escherichia coli	Bac−	NA, 18 h, 37°C	6 (h), pure	13.5	Yousef and Tawil (1980)
Escherichia coli	Bac−	NA, 24 h, 37°C	10 (h), 100,000	18	Narasimha Rao and Nigam (1970)
Flavobacterium suaveolens	Bac−	ISA, 48 h, 25°C	4 (h), 10,000	8.5	Deans and Ritchie (1987)
Haemophilus influenza	Bac−	Cited	(h) 20,000	—	Hethenyi et al. (1989)
Klebsiella aerogenes	Bac−	NA, 24 h, 37°C	10 (h), 100,000	20	Narasimha Rao and Nigam (1970)

(Continued)

TABLE 14.66 (*Continued*)
Inhibitory Data of Rosemary Oil Obtained in the Agar Diffusion Test

Microorganism	MO Class	Conditions		Inhibition Zone (mm)	Reference
Klebsiella pneumoniae	Bac–	ISA, 48 h, 25°C	4 (h), 10,000	6	Deans and Ritchie (1987)
Klebsiella sp.	Bac–	Cited	15, 2,500	0	Pizsolitto et al. (1975)
Moraxella sp.	Bac–	ISA, 48 h, 25°C	4 (h), 10,000	9.5	Deans and Ritchie (1987)
Neisseria perflava	Bac–	NA, 24 h, 37°C	—, sd	3	Maruzzella and Lichtenstein (1956)
Pectobacterium carotovorum	Bac–	Cited	(h) 20,000	+++	Hethenyi et al. (1989)
Proteus sp.	Bac–	Cited	15, 2,500	0	Pizsolitto et al. (1975)
Proteus vulgaris	Bac–	Cited	w 20,000	+++	Hethenyi et al. (1989)
Proteus vulgaris	Bac–	NA, 24 h, 37°C	—, sd	0	Maruzzella and Lichtenstein (1956)
Proteus vulgaris	Bac–	ISA, 48 h, 25°C	4 (h), 10,000	8.5	Deans and Ritchie (1987)
Pseudomonas aeruginosa	Bac–	Cited	(h) 20,000	+++	Hethenyi et al. (1989)
Pseudomonas aeruginosa	Bac–	Cited	15, 2,500	0	Pizsolitto et al. (1975)
Pseudomonas aeruginosa	Bac–	ISA, 48 h, 25°C	4 (h), 10,000	0	Deans and Ritchie (1987)
Pseudomonas aeruginosa	Bac–	NA, 24 h, 30°C	Drop, 5,000	0	Hili et al. (1997)
Pseudomonas aeruginosa	Bac–	NA, 24 h, 37°C	—, sd	2	Maruzzella and Lichtenstein (1956)
Pseudomonas aeruginosa	Bac–	MHA, 24 h, 30°C	6, 15,000	6	Rossi et al. (2007)
Pseudomonas aeruginosa	Bac–	NA, 18 h, 37°C	6 (h), pure	11	Yousef and Tawil (1980)
Pseudomonas aeruginosa	Bac–	Cited, 18 h, 37°C	6, 2,500	11	Janssen et al. (1986)
Pseudomonas pisi	Bac–	Cited	(h) 20,000	++	Hethenyi et al. (1989)
Pseudomonas tabaci	Bac–	Cited	(h) 20,000	+++	Hethenyi et al. (1989)
Salmonella enteritidis	Bac–	TSA, 24 h, 35°C	4 (h), 25,000	9.3	Smith-Palmer et al. (1998)
Salmonella enteritidis	Bac–	Cited, 24 h, 37°C	—	<12	Rota et al. (2004)
Salmonella pullorum	Bac–	ISA, 48 h, 25°C	4 (h), 10,000	5	Deans and Ritchie (1987)
Salmonella sp.	Bac–	Cited	15, 2,500	0	Pizsolitto et al. (1975)
Salmonella typhi	Bac–	NA, 24 h, 37°C	10 (h), 100,000	18	Narasimha Rao and Nigam (1970)
Serratia marcescens	Bac–	NA, 24 h, 37°C	—, sd	4	Maruzzella and Lichtenstein (1956)
Serratia marcescens	Bac–	ISA, 48 h, 25°C	4 (h), 10,000	7.5	Deans and Ritchie (1987)
Serratia sp.	Bac–	Cited	15, 2,500	0	Pizsolitto et al. (1975)
Shigella sonnei	Bac–	Cited	(h) 20,000	++	Hethenyi et al. (1989)
Shigella sp.	Bac–	Cited	15, 2,500	3	Pizsolitto et al. (1975)

(*Continued*)

TABLE 14.66 (*Continued*)
Inhibitory Data of Rosemary Oil Obtained in the Agar Diffusion Test

Microorganism	MO Class	Conditions		Inhibition Zone (mm)	Reference
Vibrio cholerae	Bac–	NA, 24 h, 37°C	10 (h), 100,000	22	Narasimha Rao and Nigam (1970)
Xanthomonas versicolor	Bac–	Cited	(h) 20,000	+++	Hethenyi et al. (1989)
Yersinia enterocolitica	Bac–	ISA, 48 h, 25°C	4 (h), 10,000	0	Deans and Ritchie (1987)
Bacillus mesentericus	Bac+	NA, 24 h, 37°C	—, sd	3	Maruzzella and Lichtenstein (1956)
Bacillus sp.	Bac+	Cited	15, 2,500	0	Pizsolitto et al. (1975)
Bacillus subtilis	Bac+	Cited	(h) 20,000	—	Hethenyi et al. (1989)
Bacillus subtilis	Bac+	NA, 24 h, 37°C	—, sd	8	Maruzzella and Lichtenstein (1956)
Bacillus subtilis	Bac+	Cited, 18 h, 37°C	6, 2,500	8	Janssen et al. (1986)
Bacillus subtilis	Bac+	ISA, 48 h, 25°C	4 (h), 10,000	20	Deans and Ritchie (1987)
Bacillus subtilis	Bac+	NA, 18 h, 37°C	6 (h), pure	22.5	Yousef and Tawil (1980)
Brevibacterium linens	Bac+	ISA, 48 h, 25°C	4 (h), 10,000	5.5	Deans and Ritchie (1987)
Brochothrix thermosphacta	Bac+	ISA, 48 h, 25°C	4 (h), 10,000	6	Deans and Ritchie (1987)
Clostridium sporogenes	Bac+	ISA, 48 h, 25°C	4 (h), 10,000	13.5	Deans and Ritchie (1987)
Corynebacterium diphtheriae	Bac+	NA, 24 h, 37°C	10 (h), 100,000	30	Narasimha Rao and Nigam (1970)
Corynebacterium fascians	Bac+	Cited	(h) 20,000	++	Hethenyi et al. (1989)
Lactobacillus plantarum	Bac+	ISA, 48 h, 25°C	4 (h), 10,000	8	Deans and Ritchie (1987)
Lactobacillus sp.	Bac+	MRS, cited	9, 20,000	16.5 to >90	Pellecuer et al. (1980)
Leuconostoc cremoris	Bac+	ISA, 48 h, 25°C	4 (h), 10,000	8.5	Deans and Ritchie (1987)
Listeria monocytogenes	Bac+	TSA, 24 h, 35°C	4 (h), 25,000	7.1	Smith-Palmer et al. (1998)
Listeria monocytogenes	Bac+	ISA, 48 h, 25°C	4 (h), 10,000	16	Lis-Balchin et al. (1998)
Micrococcus luteus	Bac+	MHA, cited	9, 20,000	18–26	Pellecuer et al. (1980)
Micrococcus luteus	Bac+	ISA, 48 h, 25°C	4 (h), 10,000	6	Deans and Ritchie (1987)
Micrococcus ureae	Bac+	MHA, cited	9, 20,000	21	Pellecuer et al. (1980)
Mycobacterium phlei	Bac+	NA, 18 h, 37°C	6 (h), pure	23	Yousef and Tawil (1980)
Pneumococcus sp.	Bac+	Cited	(h) 20,000	++	Hethenyi et al. (1989)
Sarcina lutea	Bac+	NA, 24 h, 37°C	—, sd	0	Maruzzella and Lichtenstein (1956)
Sarcina ureae	Bac+	MHA, cited	9, 20,000	>90	Pellecuer et al. (1980)
Staphylococcus aureus	Bac+	Cited	(h) 20,000	—	Hethenyi et al. (1989)
Staphylococcus aureus	Bac+	NA, 24 h, 37°C	10 (h), 100,000	0	Narasimha Rao and Nigam (1970)

(*Continued*)

TABLE 14.66 (*Continued*)
Inhibitory Data of Rosemary Oil Obtained in the Agar Diffusion Test

Microorganism	MO Class	Conditions		Inhibition Zone (mm)	Reference
Staphylococcus aureus	Bac+	NA, 24 h, 30°C	Drop, 5,000	0	Hili et al. (1997)
Staphylococcus aureus	Bac+	NA, 24 h, 37°C	—, sd	2	Maruzzella and Lichtenstein (1956)
Staphylococcus aureus	Bac+	Cited	15, 2,500	10	Pizsolitto et al. (1975)
Staphylococcus aureus	Bac+	ISA, 48 h, 25°C	4 (h), 10,000	12	Deans and Ritchie (1987)
Staphylococcus aureus	Bac+	TSA, 24 h, 35°C	4 (h), 25,000	5.9	Smith-Palmer et al. (1998)
Staphylococcus aureus	Bac+	MHA, 24 h, 37°C	6, 15,000	16	Rossi et al. (2007)
Staphylococcus aureus	Bac+	Cited, 18 h, 37°C	6, 2,500	10.7	Janssen et al. (1986)
Staphylococcus aureus	Bac+	NA, 18 h, 37°C	6 (h), pure	17.9	Yousef and Tawil (1980)
Staphylococcus epidermidis	Bac+	Cited	15, 2,500	0	Pizsolitto et al. (1975)
Staphylococcus epidermidis	Bac+	NA, 18 h, 37°C	5 (h), −30,000	0	Schelz et al. (2006)
Staphylococcus epidermidis	Bac+	MHA, cited	9, 20,000	12	Pellecuer et al. (1980)
Staphylococcus epidermidis	Bac+	NA, 24 h, 37°C	10 (h), 100,000	22	Narasimha Rao and Nigam (1970)
Streptococcus equi	Bac+	Cited	(h) 20,000	+	Hethenyi et al. (1989)
Streptococcus D	Bac+	MHA, cited	9, 20,000	0–30	Pellecuer et al. (1980)
Streptococcus faecalis	Bac+	Cited	(h) 20,000	++	Hethenyi et al. (1989)
Streptococcus faecalis	Bac+	ISA, 48 h, 25°C	4 (h), 10,000	15	Deans and Ritchie (1987)
Streptococcus haemolyticus	Bac+	Cited	(h) 20,000	+	Hethenyi et al. (1989)
Streptococcus micros	Bac+	MHA, cited	9, 20,000	18	Pellecuer et al. (1980)
Streptococcus sp.	Bac+	Cited	(h) 20,000	+	Hethenyi et al. (1989)
Streptococcus sp.	Bac+	NA, 24 h, 37°C	10 (h), 100,000	30	Narasimha Rao and Nigam (1970)
Streptococcus viridans	Bac+	Cited	15, 2,500	0	Pizsolitto et al. (1975)
Streptomyces venezuelae	Bac+	SMA, 2–7 days, 20°C	sd	10	Maruzzella and Liguori (1958)
Alternaria porri	Fungi	PDA, 72 h, 28°C	5, 5,000	21	Pawar and Thaker (2007)
Alternaria solani	Fungi	SMA, 2–7 days, 20°C	sd	7	Maruzzella and Liguori (1958)
Aspergillus flavus	Fungi	SDA, 72 h, 26°C	8, 25,000	2	Shin (2003)
Aspergillus fumigatus	Fungi	SMA, 2–7 days, 20°C	sd	6	Maruzzella and Liguori (1958)
Aspergillus niger	Fungi	SDA, 72 h, 26°C	8, 25,000	0	Shin (2003)
Aspergillus niger	Fungi	SMA, 2–7 days, 20°C	sd	3	Maruzzella and Liguori (1958)

(*Continued*)

TABLE 14.66 (*Continued*)
Inhibitory Data of Rosemary Oil Obtained in the Agar Diffusion Test

Microorganism	MO Class	Conditions		Inhibition Zone (mm)	Reference
Aspergillus niger	Fungi	PDA, 48 h, 28°C	5, 5,000	7	Pawar and Thaker (2006)
Aspergillus niger	Fungi	SDA, 8 days, 30°C	6 (h), pure	10	Yousef and Tawil (1980)
Fusarium moniliforme	Fungi	Cited	(h) 20,000	—	Hethenyi et al. (1989)
Fusarium oxysporum f.sp. *ciceri*	Fungi	PDA, 72 h, 28°C	5, 5,000	11	Pawar and Thaker (2007)
Fusarium solani	Fungi	Cited	(h) 20,000	—	Hethenyi et al. (1989)
Helminthosporium sativum	Fungi	SMA, 2–7 days, 20°C	sd	9	Maruzzella and Liguori (1958)
Mucor mucedo	Fungi	SMA, 2–7 days, 20°C	sd	5	Maruzzella and Liguori (1958)
Mucor sp.	Fungi	SDA, 8 days, 30°C	6 (h), pure	18	Yousef and Tawil (1980)
Nigrospora panici	Fungi	SMA, 2–7 days, 20°C	sd	5	Maruzzella and Liguori (1958)
Ophiobolus graminis	Fungi	Cited	(h) 20,000	—	Hethenyi et al. (1989)
Penicillium chrysogenum	Fungi	SDA, 8 days, 30°C	6 (h), pure	20	Yousef and Tawil (1980)
Penicillium digitatum	Fungi	SMA, 2–7 days, 20°C	sd	6	Maruzzella and Liguori (1958)
Rhizopus nigricans	Fungi	SMA, 2–7 days, 20°C	sd	6	Maruzzella and Liguori (1958)
Rhizopus sp.	Fungi	SDA, 8 days, 30°C	6 (h), pure	0	Yousef and Tawil (1980)
Brettanomyces anomalus	Yeast	MPA, 4 days, 30°C	5, 10% sol. sd	0	Conner and Beuchat (1984)
Candida albicans	Yeast	Cited	(h) 20,000	—	Hethenyi et al. (1989)
Candida albicans	Yeast	SMA, 2–7 days, 20°C	sd	2	Maruzzella and Liguori (1958)
Candida albicans	Yeast	NA, 24 h, 30°C	Drop, 5,000	7	Hili et al. (1997)
Candida albicans	Yeast	Cited, 18 h, 37°C	6, 2,500	13	Janssen et al. (1986)
Candida albicans	Yeast	SDA, 18 h, 30°C	6 (h), pure	24	Yousef and Tawil (1980)
Candida krusei	Yeast	SMA, 2–7 days, 20°C	sd	1	Maruzzella and Liguori (1958)
Candida lipolytica	Yeast	MPA, 4 days, 30°C	5, 10% sol. sd	0	Conner and Beuchat (1984)
Candida tropicalis	Yeast	SMA, 2–7 days, 20°C	sd	2	Maruzzella and Liguori (1958)
Cryptococcus neoformans	Yeast	SMA, 2–7 days, 20°C	sd	6	Maruzzella and Liguori (1958)
Cryptococcus rhodobenhani	Yeast	SMA, 2–7 days, 20°C	sd	6	Maruzzella and Liguori (1958)

(*Continued*)

TABLE 14.66 (Continued)
Inhibitory Data of Rosemary Oil Obtained in the Agar Diffusion Test

Microorganism	MO Class	Conditions		Inhibition Zone (mm)	Reference
Debaryomyces hansenii	Yeast	MPA, 4 days, 30°C	5, 10% sol. sd	0	Conner and Beuchat (1984)
Geotrichum candidum	Yeast	MPA, 4 days, 30°C	5, 10% sol. sd	0	Conner and Beuchat (1984)
Hansenula anomala	Yeast	MPA, 4 days, 30°C	5, 10% sol. sd	0	Conner and Beuchat (1984)
Kloeckera apiculata	Yeast	MPA, 4 days, 30°C	5, 10% sol. sd	0	Conner and Beuchat (1984)
Kluyveromyces fragilis	Yeast	MPA, 4 days, 30°C	5, 10% sol. sd	0	Conner and Beuchat (1984)
Lodderomyces elongisporus	Yeast	MPA, 4 days, 30°C	5, 10% sol. sd	0	Conner and Beuchat (1984)
Metschnikowia pulcherrima	Yeast	MPA, 4 days, 30°C	5, 10% sol. sd	0	Conner and Beuchat (1984)
Pichia membranifaciens	Yeast	MPA, 4 days, 30°C	5, 10% sol. sd	0	Conner and Beuchat (1984)
Rhodotorula rubra	Yeast	MPA, 4 days, 30°C	5, 10% sol. sd	8	Conner and Beuchat (1984)
Saccharomyces cerevisiae	Yeast	Cited	(h) 20,000	—	Hethenyi et al. (1989)
Saccharomyces cerevisiae	Yeast	MPA, 4 days, 30°C	5, 10% sol. sd	0	Conner and Beuchat (1984)
Saccharomyces cerevisiae	Yeast	SMA, 2–7 days, 20°C	sd	7	Maruzzella and Liguori (1958)
Saccharomyces cerevisiae	Yeast	NA, 24 h, 20°C	5 (h), –30,000	10	Schelz et al. (2006)
Saccharomyces cerevisiae	Yeast	NA, 24 h, 30°C	Drop, 5,000	12	Hili et al. (1997)
Schizosaccharomyces pombe	Yeast	NA, 24 h, 30°C	Drop, 5,000	16	Hili et al. (1997)
Torula glabrata	Yeast	MPA, 4 days, 30°C	5, 10% sol. sd	0	Conner and Beuchat (1984)
Torula utilis	Yeast	NA, 24 h, 30°C	Drop, 5,000	10	Hili et al. (1997)

TABLE 14.67
Inhibitory Data of Rosemary Oil Obtained in the Dilution Test

Microorganism	MO Class	Conditions	MIC	Reference
Acinetobacter baumannii	Bac–	MHA, Tween 20, 48 h, 35°C	10,000	Hammer et al. (1999)
Aeromonas sobria	Bac–	MHA, Tween 20, 48 h, 35°C	5,000	Hammer et al. (1999)
Bordetella bronchiseptica	Bac–	Cited	2,500	Pellecuer et al. (1976)
Campylobacter jejuni	Bac–	TSB, 24 h, 42°C	500	Smith-Palmer et al. (1998)
Escherichia coli	Bac–	TSB, Tween 80, 48 h, 35°C	40	Panizzi et al. (1993)
Escherichia coli	Bac–	MPB, DMSO, 40 h, 30°C	25% inh. 500	Hili et al. (1997)
Escherichia coli	Bac–	NB, DMSO, 24 h, 37°C	>900	Angioni et al. (2004)
Escherichia coli	Bac–	Cited	2,500	Pellecuer et al. (1976)
Escherichia coli	Bac–	NA, 1–3 days, 30°C	3,500	Farag et al. (1989)
Escherichia coli	Bac–	TSB, 24 h, 35°C	>10,000	Smith-Palmer et al. (1998)
Escherichia coli	Bac–	MHA, Tween 20, 48 h, 35°C	10,000	Hammer et al. (1999)
Escherichia coli	Bac–	NB, Tween 20, 18 h, 37°C	25,000	Yousef and Tawil (1980)
Haemophilus influenza	Bac–	Cited	2,500	Pellecuer et al. (1976)
Helicobacter pylori	Bac–	Cited, 20 h, 37°C	137	Weseler et al. (2005)
Klebsiella pneumoniae	Bac–	Cited	2,500	Pellecuer et al. (1976)
Klebsiella pneumoniae	Bac–	MHA, Tween 20, 48 h, 35°C	20,000	Hammer et al. (1999)
Moraxella glucidolytica	Bac–	Cited	2,500	Pellecuer et al. (1976)
Neisseria catarrhalis	Bac–	Cited	1,250	Pellecuer et al. (1976)
Neisseria flava	Bac–	Cited	1,250	Pellecuer et al. (1976)
Pseudomonas aeruginosa	Bac–	TSB, Tween 80, 48 h, 35°C	>40	Panizzi et al. (1993)
Pseudomonas aeruginosa	Bac–	MPB, DMSO, 40 h, 30°C	54% inh. 500	Hili et al. (1997)
Pseudomonas aeruginosa	Bac–	NB, DMSO, 24 h, 37°C	>900	Angioni et al. (2004)
Pseudomonas aeruginosa	Bac–	MHA, Tween 20, 48 h, 35°C	>20,000	Hammer et al. (1999)
Pseudomonas aeruginosa	Bac–	NB, Tween 20, 18 h, 37°C	50,000	Yousef and Tawil (1980)
Pseudomonas fluorescens	Bac–	NA, 1–3 days, 30°C		Thanaboripat et al. (2004), Farag et al. (1989)
Salmonella enteritidis	Bac–	TSB, 24 h, 35°C	>10,000	Smith-Palmer et al. (1998)
Salmonella typhimurium	Bac–	TSB, 3% EtOH, 24 h, 37°C	10,000–13,000	Rota et al. (2004)
Salmonella typhimurium	Bac–	MHA, Tween 20, 48 h, 35°C	>20,000	Hammer et al. (1999)
Serratia marcescens	Bac–	NA, 1–3 days, 30°C	11,000	Farag et al. (1989)
Serratia marcescens	Bac–	MHA, Tween 20, 48 h, 35°C	>20,000	Hammer et al. (1999)
Shigella flexneri	Bac–	TSB, 3% EtOH, 24 h, 37°C	20,000	Rota et al. (2004)
Yersinia enterocolitica	Bac–	TSB, 3% EtOH, 24 h, 29°C	10,000–15,000	Rota et al. (2004)
Yersinia enterocolitica	Bac–	MHA, Tween 20, 24 h, 37°C	1250	Rossi et al. (2007)
Actinomyces viscosus	Bac+	HS, 16–24, 37°C	>6,000	Shapiro et al. (1994)
Bacillus subtilis	Bac+	TSB, Tween 80, 48 h, 35°C	10	Panizzi et al. (1993)
Bacillus subtilis	Bac+	NA, 1–3 days, 30°C	750	Farag et al. (1989)
Bacillus subtilis	Bac+	NB, Tween 20, 18 h, 37°C	800	Yousef and Tawil (1980)
Bacillus subtilis	Bac+	Cited	1,250	Pellecuer et al. (1976)
Corynebacterium pseudodiphtheriae	Bac+	Cited	1,250	Pellecuer et al. (1976)
Enterococcus faecalis	Bac+	MHA, Tween 20, 48 h, 35°C	>20,000	Hammer et al. (1999)
Lactobacillus sp.	Bac+	MRS, cited	5	Pellecuer et al. (1980)
Listeria monocytogenes	Bac+	TSB, 3% EtOH, 24 h, 37°C	7,000–10,000	Rota et al. (2004)
Listeria monocytogenes	Bac+	TSB, 24 h, 35°C	200	Smith-Palmer et al. (1998)
Micrococcus flavus	Bac+	Cited	1,250	Pellecuer et al. (1976)
Micrococcus luteus	Bac+	MHB, cited	2.5–5	Pellecuer et al. (1980)

(Continued)

TABLE 14.67 (Continued)
Inhibitory Data of Rosemary Oil Obtained in the Dilution Test

Microorganism	MO Class	Conditions	MIC	Reference
Micrococcus sp.	Bac+	NA, 1–3 days, 30°C	1,500	Farag et al. (1989)
Micrococcus ureae	Bac+	MHB, cited	5	Pellecuer et al. (1980)
Mycobacterium phlei	Bac+	NB, Tween 20, 18 h, 37°C	400	Yousef and Tawil (1980)
Mycobacterium phlei	Bac+	NA, 1–3 days, 30°C	1,250	Farag et al. (1989)
Sarcina lutea	Bac+	Cited	1,250	Pellecuer et al. (1976)
Sarcina sp.	Bac+	NA, 1–3 days, 30°C	2,000	Farag et al. (1989)
Sarcina ureae	Bac+	MHB, cited	2.5	Pellecuer et al. (1980)
Staphylococcus aureus	Bac+	TSB, 3% EtOH, 24 h, 37°C	30,000–50,000	Rota et al. (2004)
Staphylococcus aureus	Bac+	TSB, Tween 80, 48 h, 35°C	20	Panizzi et al. (1993)
Staphylococcus aureus	Bac+	TSB, 24 h, 35°C	400	Smith-Palmer et al. (1998)
Staphylococcus aureus	Bac+	MPB, DMSO, 40 h, 30°C	1% inh. 500	Hili et al. (1997)
Staphylococcus aureus	Bac+	NB, DMSO, 24 h, 37°C	>900	Angioni et al. (2004)
Staphylococcus aureus	Bac+	NA, 1–3 days, 30°C	1,000	Farag et al. (1989)
Staphylococcus aureus	Bac+	Cited	1,250	Pellecuer et al. (1976)
Staphylococcus aureus	Bac+	MHA, Tween 20, 48 h, 35°C	10,000	Hammer et al. (1999)
Staphylococcus aureus	Bac+	NB, Tween 20, 18 h, 37°C	12,500	Yousef and Tawil (1980)
Staphylococcus epidermidis	Bac+	MHB, cited	5	Pellecuer et al. (1980)
Staphylococcus epidermidis	Bac+	NB, DMSO, 24 h, 37°C	>900	Angioni et al. (2004)
Staphylococcus epidermidis	Bac+	Cited	1,250	Pellecuer et al. (1976)
Streptococcus D	Bac+	MHB, cited	2.5–5	Pellecuer et al. (1980)
Streptococcus micros	Bac+	MHB, cited	5	Pellecuer et al. (1980)
Streptococcus pyogenes	Bac+	Cited	625	Pellecuer et al. (1976)
Streptococcus sanguinis	Bac+	HS, 16–24, 37°C	>6,000	Shapiro et al. (1994)
Streptococcus sobrinus	Bac+	HS, 16–24, 37°C	>6,000	Shapiro et al. (1994)
Absidia glauca	Fungi	Cited	2,000	Pellecuer et al. (1976)
Aspergillus chevalieri	Fungi	Cited	2,000	Pellecuer et al. (1976)
Aspergillus clavatus	Fungi	Cited	2,000	Pellecuer et al. (1976)
Aspergillus flavus	Fungi	PDA, 8 h, 20°C, spore germ. inh.	50–100	Thompson (1986)
Aspergillus flavus	Fungi	PDA, 5 days, 27°C	>1,000	Thompson and Cannon (1986)
Aspergillus flavus	Fungi	Cited	2,000	Pellecuer et al. (1976)
Aspergillus flavus	Fungi	MYB, 72 h, 26°C	12,500	Shin (2003)
Aspergillus giganteus	Fungi	Cited	2,000	Pellecuer et al. (1976)
Aspergillus niger	Fungi	Cited	2,000	Pellecuer et al. (1976)
Aspergillus niger	Fungi	YES broth, 10 days	12% inh. 10,000	Lis-Balchin et al. (1998)
Aspergillus niger	Fungi	MYB, 72 h, 26°C	12,500	Shin (2003)
Aspergillus niger	Fungi	NB, Tween 20, 8 days, 30°C	50,000	Yousef and Tawil (1980)
Aspergillus ochraceus	Fungi	YES broth, 10 days	14% inh. 10,000	Lis-Balchin et al. (1998)
Aspergillus oryzae	Fungi	Cited	250	Okazaki and Oshima (1953)
Aspergillus oryzae	Fungi	Cited	2,000	Pellecuer et al. (1976)

(*Continued*)

TABLE 14.67 (*Continued*)
Inhibitory Data of Rosemary Oil Obtained in the Dilution Test

Microorganism	MO Class	Conditions	MIC	Reference
Aspergillus parasiticus	Fungi	PDA, 8 h, 20°C, spore germ. inh.	50–100	Thompson (1986)
Aspergillus parasiticus	Fungi	PDA, 5 days, 27°C	>1,000	Thompson and Cannon (1986)
Aspergillus repens	Fungi	Cited	2,000	Pellecuer et al. (1976)
Cephalosporium sacchari	Fungi	OA, EtOH, 3 days, 20°C	20,000	Narasimba Rao et al. (1971)
Ceratocystis paradoxa	Fungi	OA, EtOH, 3 days, 20°C	4,000	Narasimba Rao et al. (1971)
Cladosporium herbarum	Fungi	Cited	2,000	Pellecuer et al. (1976)
Curvularia lunata	Fungi	OA, EtOH, 3 days, 20°C	20,000	Narasimba Rao et al. (1971)
Epidermophyton floccosum	Fungi	SA, Tween 80, 21 days, 20°C	625–1,250	Janssen et al. (1988)
Epidermophyton floccosum	Fungi	Cited	>4,000	Pellecuer et al. (1976)
Fusarium culmorum	Fungi	YES broth, 10 days	0% inh. 10,000	Lis-Balchin et al. (1998)
Fusarium moniliforme var. *subglutinans*	Fungi	OA, EtOH, 3 days, 20°C	20,000	Narasimba Rao et al. (1971)
Helminthosporium sacchari	Fungi	OA, EtOH, 3 days, 20°C	20,000	Narasimba Rao et al. (1971)
Microsporum canis	Fungi	MBA, Tween 80, 10 days, 30°C	300 to >300	Perrucci et al. (1994)
Microsporum gypseum	Fungi	MBA, Tween 80, 10 days, 30°C	>300	Perrucci et al. (1994)
Mucor hiemalis	Fungi	PDA, 5 days, 27°C	>1,000	Thompson and Cannon (1986)
Mucor mucedo	Fungi	Cited	1,000	Pellecuer et al. (1976)
Mucor mucedo	Fungi	PDA, 5 days, 27°C	>1,000	Thompson and Cannon (1986)
Mucor racemosus	Fungi	Cited	500	Okazaki and Oshima (1953)
Mucor racemosus f. *racemosus*	Fungi	PDA, 5 days, 27°C	>1,000	Thompson and Cannon (1986)
Mucor sp.	Fungi	NB, Tween 20, 8 days, 30°C	6,400	Yousef and Tawil (1980)
Penicillium chrysogenum	Fungi	Cited	500	Okazaki and Oshima (1953)
Penicillium chrysogenum	Fungi	Cited	2,000	Pellecuer et al. (1976)
Penicillium chrysogenum	Fungi	NB, Tween 20, 8 days, 30°C	12,500	Yousef and Tawil (1980)
Penicillium lilacinum	Fungi	Cited	2,000	Pellecuer et al. (1976)
Penicillium rubrum	Fungi	Cited	2,000	Pellecuer et al. (1976)
Physalospora tucumanensis	Fungi	OA, EtOH, 3 days, 20°C	20,000	Narasimba Rao et al. (1971)
Rhizopus 66-81-2	Fungi	PDA, 5 days, 27°C	>1,000	Thompson and Cannon (1986)
Rhizopus arrhizus	Fungi	PDA, 5 days, 27°C	>1,000	Thompson and Cannon (1986)
Rhizopus chinensis	Fungi	PDA, 5 days, 27°C	>1,000	Thompson and Cannon (1986)
Rhizopus circinans	Fungi	PDA, 5 days, 27°C	>1,000	Thompson and Cannon (1986)

(*Continued*)

TABLE 14.67 (*Continued*)
Inhibitory Data of Rosemary Oil Obtained in the Dilution Test

Microorganism	MO Class	Conditions	MIC	Reference
Rhizopus japonicus	Fungi	PDA, 5 days, 27°C	>1,000	Thompson and Cannon (1986)
Rhizopus kazanensis	Fungi	PDA, 5 days, 27°C	>1,000	Thompson and Cannon (1986)
Rhizopus nigricans	Fungi	Cited	2,000	Pellecuer et al. (1976)
Rhizopus oryzae	Fungi	PDA, 5 days, 27°C	>1,000	Thompson and Cannon (1986)
Rhizopus pymacus	Fungi	PDA, 5 days, 27°C	>1,000	Thompson and Cannon (1986)
Rhizopus sp.	Fungi	NB, Tween 20, 8 days, 30°C	25,000	Yousef and Tawil (1980)
Rhizopus stolonifer	Fungi	PDA, 5 days, 27°C	>1,000	Thompson and Cannon (1986)
Rhizopus tritici	Fungi	PDA, 5 days, 27°C	>1,000	Thompson and Cannon (1986)
Sclerotium rolfsii	Fungi	OA, EtOH, 6 days, 20°C	4,000	Narasimba Rao et al. (1971)
Scopulariopsis brevicaulis	Fungi	Cited	2,000	Pellecuer et al. (1976)
Syncephalastrum racemosum	Fungi	Cited	2,000	Pellecuer et al. (1976)
Trichophyton interdigitale	Fungi	Cited	>4,000	Pellecuer et al. (1976)
Trichophyton mentagrophytes	Fungi	SA, Tween 80, 21 days, 20°C	300–625	Janssen et al. (1988)
Trichophyton rubrum	Fungi	SA, Tween 80, 21 days, 20°C	<300	Janssen et al. (1988)
Candida albicans	Yeast	SDB, Tween 80, 48 h, 35	10	Panizzi et al. (1993)
Candida albicans	Yeast	MPB, DMSO, 40 h, 30°C	8% inh. 500	Hili et al. (1997)
Candida albicans	Yeast	NB, DMSO, 24 h, 30°C	>900	Angioni et al. (2004)
Candida albicans	Yeast	Cited	1,000	Pellecuer et al. (1976)
Candida albicans	Yeast	NB, Tween 20, 18 h, 37°C	3,200	Yousef and Tawil (1980)
Candida albicans	Yeast	MHA, Tween 20, 48 h, 35°C	10,000	Hammer et al. (1999)
Candida mycoderma	Yeast	Cited	2,000	Pellecuer et al. (1976)
Candida parapsilosis	Yeast	Cited	1,000	Pellecuer et al. (1976)
Candida pelliculosa	Yeast	Cited	2,000	Pellecuer et al. (1976)
Candida tropicalis	Yeast	Cited	2,000	Pellecuer et al. (1976)
Geotrichum asteroides	Yeast	Cited	4,000	Pellecuer et al. (1976)
Geotrichum candidum	Yeast	Cited	2,000	Pellecuer et al. (1976)
Hansenula sp.	Yeast	Cited	2,000	Pellecuer et al. (1976)
Saccharomyces carlsbergensis	Yeast	Cited	4,000	Pellecuer et al. (1976)
Saccharomyces cerevisiae	Yeast	SDB, Tween 80, 48 h, 35°C	5	Panizzi et al. (1993)
Saccharomyces cerevisiae	Yeast	MPB, DMSO, 40 h, 30°C	88% inh. 500	Hili et al. (1997)
Saccharomyces cerevisiae	Yeast	NA, 1–3 days, 30°C	2,000	Farag et al. (1989)
Schizosaccharomyces pombe	Yeast	MPB, DMSO, 40 h, 30°C	86% inh. 500	Hili et al. (1997)
Torula utilis	Yeast	MPB, DMSO, 40 h, 30°C	2% inh. 500	Hili et al. (1997)
Escherichia coli	Bac−	TGB, 18 h, 37°C	11,300	Schelz et al. (2006)
Staphylococcus epidermidis	Bac+	TGB, 18 h, 37°C	11,300	Schelz et al. (2006)
Saccharomyces cerevisiae	Yeast	YPB, 24 h, 20°C	2,800	Schelz et al. (2006)

TABLE 14.68
Inhibitory Data of Rosemary Oil Obtained in the Vapor Phase Test

Microorganism	MO Class	Conditions		Activity	Reference
Aerobacter tumefaciens	Bac−	Cited	Disk, 20,000	NG	Hethenyi et al. (1989)
Escherichia coli	Bac−	NA, 24 h, 37°C	~20,000	++	Kellner and Kober (1954)
Escherichia coli	Bac−	Cited	Disk, 20,000	NG to +++	Hethenyi et al. (1989)
Haemophilus influenza	Bac−	Cited	Disk, 20,000	NG	Hethenyi et al. (1989)
Neisseria sp.	Bac−	NA, 24 h, 37°C	~20,000	NG	Kellner and Kober (1954)
Pectobacterium carotovorum	Bac−	Cited	Disk, 20,000	+++	Hethenyi et al. (1989)
Proteus vulgaris	Bac−	Cited	Disk, 20,000	+++	Hethenyi et al. (1989)
Pseudomonas aeruginosa	Bac−	Cited	Disk, 20,000	+++	Hethenyi et al. (1989)
Pseudomonas pisi	Bac−	Cited	Disk, 20,000	+++	Hethenyi et al. (1989)
Pseudomonas tabaci	Bac−	Cited	Disk, 20,000	+++	Hethenyi et al. (1989)
Salmonella typhi	Bac−	NA, 24 h, 37°C	~20,000	+	Kellner and Kober (1954)
Salmonella typhi	Bac−	NA, 24 h, 37°C	sd	+++	Maruzzella and Sicurella (1960)
Shigella sonnei	Bac−	Cited	Disk, 20,000	+++	Hethenyi et al. (1989)
Xanthomonas versicolor	Bac−	Cited	Disk, 20,000	+++	Hethenyi et al. (1989)
Bacillus megaterium	Bac+	NA, 24 h, 37°C	~20,000	+	Kellner and Kober (1954)
Bacillus subtilis	Bac+	Cited	Disk, 20,000	NG	Hethenyi et al. (1989)
Bacillus subtilis var. aterrimus	Bac+	NA, 24 h, 37°C	sd	+++	Maruzzella and Sicurella (1960)
Corynebacterium diphtheriae	Bac+	NA, 24 h, 37°C	~20,000	+	Kellner and Kober (1954)
Corynebacterium fascians	Bac+	Cited	Disk, 20,000	+++	Hethenyi et al. (1989)
Lactobacillus sp.	Bac+	MRS, cited	Disk, 20,000?	++	Pellecuer et al. (1980)
Micrococcus luteus	Bac+	MHB, cited	Disk, 20,000?	++	Pellecuer et al. (1980)
Micrococcus ureae	Bac+	MHB, cited	Disk, 20,000?	+++	Pellecuer et al. (1980)
Mycobacterium avium	Bac+	NA, 24 h, 37°C	sd	++	Maruzzella and Sicurella (1960)
Pneumococcus sp.	Bac+	Cited	Disk, 20,000	NG	Hethenyi et al. (1989)
Sarcina ureae	Bac+	MHB, cited	Disk, 20,000?	++	Pellecuer et al. (1980)
Staphylococcus aureus	Bac+	NA, 24 h, 37°C	~20,000	NG	Kellner and Kober (1954)
Staphylococcus aureus	Bac+	Cited	Disk, 20,000	NG	Hethenyi et al. (1989)
Staphylococcus epidermidis	Bac+	MHB, cited	Disk, 20,000?	+++	Pellecuer et al. (1980)
Streptococcus equi	Bac+	Cited	Disk, 20,000	NG	Hethenyi et al. (1989)
Streptococcus D	Bac+	MHB, cited	Disk, 20,000?	++	Pellecuer et al. (1980)
Streptococcus faecalis	Bac+	NA, 24 h, 37°C	~20,000	+	Kellner and Kober (1954)
Streptococcus faecalis	Bac+	Cited	Disk, 20,000	NG	Hethenyi et al. (1989)
Streptococcus faecalis	Bac+	NA, 24 h, 37°C	sd	+++	Maruzzella and Sicurella (1960)
Streptococcus haemolyticus	Bac+	Cited	Disk, 20,000	NG	Hethenyi et al. (1989)
Streptococcus pyogenes	Bac+	NA, 24 h, 37°C	~20,000	NG	Kellner and Kober (1954)
Streptococcus sp.	Bac+	Cited	Disk, 20,000	NG	Hethenyi et al. (1989)
Aspergillus flavus	Fungi	WFA, 42 days, 25°C	Disk, 50,000	+++	Guynot et al. (2003)

(Continued)

TABLE 14.68 (Continued)
Inhibitory Data of Rosemary Oil Obtained in the Vapor Phase Test

Microorganism	MO Class	Conditions		Activity	Reference
Aspergillus flavus	Fungi	Bread, 14 days, 25°C	30,000	+++	Suhr and Nielsen (2003)
Aspergillus niger	Fungi	WFA, 42 days, 25°C	Disk, 50,000	+++	Guynot et al. (2003)
Botrytis cinerea	Fungi	PDA, 3 days, 25°C	1,000	+++	Lee et al. (2007)
Colletotrichum gloeosporioides	Fungi	PDA, 3 days, 25°C	1,000	+++	Lee et al. (2007)
Endomyces fibuligera	Fungi	Bread, 14 days, 25°C	30,000	+++	Suhr and Nielsen (2003)
Eurotium amstelodami	Fungi	WFA, 42 days, 25°C	Disk, 50,000	+++	Guynot et al. (2003)
Eurotium herbarum	Fungi	WFA, 42 days, 25°C	Disk, 50,000	+++	Guynot et al. (2003)
Eurotium repens	Fungi	WFA, 42 days, 25°C	Disk, 50,000	+++	Guynot et al. (2003)
Eurotium repens	Fungi	Bread, 14 days, 25°C	30,000	+++	Suhr and Nielsen (2003)
Eurotium rubrum	Fungi	WFA, 42 days, 25°C	Disk, 50,000	+++	Guynot et al. (2003)
Fusarium moniliforme	Fungi	Cited	Disk, 20,000	NG	Hethenyi et al. (1989)
Fusarium oxysporum	Fungi	PDA, 3 days, 25°C	1,000	+++	Lee et al. (2007)
Fusarium solani	Fungi	Cited	Disk, 20,000	NG	Hethenyi et al. (1989)
Ophiobolus graminis	Fungi	Cited	Disk, 20,000	NG	Hethenyi et al. (1989)
Penicillium corylophilum	Fungi	WFA, 42 days, 25°C	Disk, 50,000	++	Guynot et al. (2003)
Penicillium corylophilum	Fungi	Bread, 14 days, 25°C	30,000	+++	Suhr and Nielsen (2003)
Penicillium roqueforti	Fungi	Bread, 14 days, 25°C	30,000	+++	Suhr and Nielsen (2003)
Pythium ultimum	Fungi	PDA, 3 days, 25°C	1,000	+++	Lee et al. (2007)
Rhizoctonia solani	Fungi	PDA, 3 days, 25°C	1,000	+++	Lee et al. (2007)
Candida albicans	Yeast	NA, 24 h, 37°C	~20,000	NG	Kellner and Kober (1954)
Candida albicans	Yeast	Cited	Disk, 20,000	NG	Hethenyi et al. (1989)
Saccharomyces cerevisiae	Yeast	Cited	Disk, 20,000	NG	Hethenyi et al. (1989)

TABLE 14.69
Inhibitory Data of Star Anise Oil Obtained in the Agar Diffusion Test

Microorganism	MO Class	Conditions		Inhibition Zone (mm)	Reference
Acinetobacter calcoaceticus	Bac−	ISA, 48 h, 25°C	4 (h), 10,000	0	Deans and Ritchie (1987)
Aeromonas hydrophila	Bac−	ISA, 48 h, 25°C	4 (h), 10,000	6	Deans and Ritchie (1987)
Alcaligenes faecalis	Bac−	ISA, 48 h, 25°C	4 (h), 10,000	5	Deans and Ritchie (1987)
Beneckea natriegens	Bac−	ISA, 48 h, 25°C	4 (h), 10,000	5	Deans and Ritchie (1987)
Citrobacter freundii	Bac−	ISA, 48 h, 25°C	4 (h), 10,000	0	Deans and Ritchie (1987)
Enterobacter aerogenes	Bac−	ISA, 48 h, 25°C	4 (h), 10,000	0	Deans and Ritchie (1987)
Erwinia carotovora	Bac−	ISA, 48 h, 25°C	4 (h), 10,000	0	Deans and Ritchie (1987)
Escherichia coli	Bac−	ISA, 48 h, 25°C	4 (h), 10,000	0	Deans and Ritchie (1987)
Escherichia coli	Bac−	Cited, 18 h, 37°C	6, 2,500	7	Janssen et al. (1986)
Escherichia coli	Bac−	NA, 24 h, 37°C	10 (h), 2,000	18.5	Singh et al. (2006)
Flavobacterium suaveolens	Bac−	ISA, 48 h, 25°C	4 (h), 10,000	9	Deans and Ritchie (1987)
Klebsiella pneumoniae	Bac−	ISA, 48 h, 25°C	4 (h), 10,000	0	Deans and Ritchie (1987)
Moraxella sp.	Bac−	ISA, 48 h, 25°C	4 (h), 10,000	0	Deans and Ritchie (1987)
Proteus vulgaris	Bac−	ISA, 48 h, 25°C	4 (h), 10,000	0	Deans and Ritchie (1987)
Pseudomonas aeruginosa	Bac−	ISA, 48 h, 25°C	4 (h), 10,000	0	Deans and Ritchie (1987)
Pseudomonas aeruginosa	Bac−	Cited, 18 h, 37°C	6, 2,500	6.7	Janssen et al. (1986)
Pseudomonas aeruginosa	Bac−	NA, 24 h, 37°C	10 (h), 2,000	20.3	Singh et al. (2006)
Salmonella pullorum	Bac−	ISA, 48 h, 25°C	4 (h), 10,000	6	Deans and Ritchie (1987)
Salmonella typhi	Bac−	NA, 24 h, 37°C	10 (h), 2,000	30.1	Singh et al. (2006)
Serratia marcescens	Bac−	ISA, 48 h, 25°C	4 (h), 10,000	5.5	Deans and Ritchie (1987)
Yersinia enterocolitica	Bac−	ISA, 48 h, 25°C	4 (h), 10,000	9	Deans and Ritchie (1987)
Bacillus cereus	Bac+	NA, 24 h, 37°C	10 (h), 2,000	0	Singh et al. (2006)
Bacillus subtilis	Bac+	ISA, 48 h, 25°C	4 (h), 10,000	0	Deans and Ritchie (1987)
Bacillus subtilis	Bac+	Cited, 18 h, 37°C	6, 2,500	6.7	Janssen et al. (1986)
Bacillus subtilis	Bac+	NA, 24 h, 37°C	10 (h), 2,000	26.2	Singh et al. (2006)
Brevibacterium linens	Bac+	ISA, 48 h, 25°C	4 (h), 10,000	0	Deans and Ritchie (1987)
Brochothrix thermosphacta	Bac+	ISA, 48 h, 25°C	4 (h), 10,000	0	Deans and Ritchie (1987)
Clostridium sporogenes	Bac+	ISA, 48 h, 25°C	4 (h), 10,000	0	Deans and Ritchie (1987)
Lactobacillus plantarum	Bac+	ISA, 48 h, 25°C	4 (h), 10,000	0	Deans and Ritchie (1987)
Leuconostoc cremoris	Bac+	ISA, 48 h, 25°C	4 (h), 10,000	0	Deans and Ritchie (1987)
Micrococcus luteus	Bac+	ISA, 48 h, 25°C	4 (h), 10,000	6	Deans and Ritchie (1987)

(Continued)

TABLE 14.69 (*Continued*)
Inhibitory Data of Star Anise Oil Obtained in the Agar Diffusion Test

Microorganism	MO Class	Conditions		Inhibition Zone (mm)	Reference
Staphylococcus aureus	Bac+	NA, 24 h, 37°C	10 (h), 2,000	0	Singh et al. (2006)
Staphylococcus aureus	Bac+	ISA, 48 h, 25°C	4 (h), 10,000	0	Deans and Ritchie (1987)
Staphylococcus aureus	Bac+	Cited, 18 h, 37°C	6, 2,500	9	Janssen et al. (1986)
Streptococcus faecalis	Bac+	ISA, 48 h, 25°C	4 (h), 10,000	0	Deans and Ritchie (1987)
Brettanomyces anomalus	Yeast	MPA, 4 days, 30°C	5, 10% sol. sd	0	Conner and Beuchat (1984)
Candida albicans	Yeast	Cited, 18 h, 37°C	6, 2,500	8.7	Janssen et al. (1986)
Candida lipolytica	Yeast	MPA, 4 days, 30°C	5, 10% sol. sd	7	Conner and Beuchat (1984)
Debaryomyces hansenii	Yeast	MPA, 4 days, 30°C	5, 10% sol. sd	0	Conner and Beuchat (1984)
Geotrichum candidum	Yeast	MPA, 4 days, 30°C	5, 10% sol. sd	0	Conner and Beuchat (1984)
Hansenula anomala	Yeast	MPA, 4 days, 30°C	5, 10% sol. sd	0	Conner and Beuchat (1984)
Kloeckera apiculata	Yeast	MPA, 4 days, 30°C	5, 10% sol. sd	8	Conner and Beuchat (1984)
Kluyveromyces fragilis	Yeast	MPA, 4 days, 30°C	5, 10% sol. sd	0	Conner and Beuchat (1984)
Lodderomyces elongisporus	Yeast	MPA, 4 days, 30°C	5, 10% sol. sd	0	Conner and Beuchat (1984)
Metschnikowia pulcherrima	Yeast	MPA, 4 days, 30°C	5, 10% sol. sd	0	Conner and Beuchat (1984)
Pichia membranifaciens	Yeast	MPA, 4 days, 30°C	5, 10% sol. sd	0	Conner and Beuchat (1984)
Rhodotorula rubra	Yeast	MPA, 4 days, 30°C	5, 10% sol. sd	0	Conner and Beuchat (1984)
Saccharomyces cerevisiae	Yeast	MPA, 4 days, 30°C	5, 10% sol. sd	0	Conner and Beuchat (1984)
Torula glabrata	Yeast	MPA, 4 days, 30°C	5, 10% sol. sd	0	Conner and Beuchat (1984)

TABLE 14.70
Inhibitory Data of Star Anise Oil Obtained in the Dilution Test

Microorganism	MO Class	Conditions	MIC	Reference
Epidermophyton floccosum	Fungi	SA, Tween 80, 21 days, 20°C	300–625	Janssen et al. (1988)
Trichophyton mentagrophytes	Fungi	SA, Tween 80, 21 days, 20°C	300–625	Janssen et al. (1988)
Trichophyton rubrum	Fungi	SA, Tween 80, 21 days, 20°C	300–625	Janssen et al. (1988)

TABLE 14.71
Inhibitory Data of Star Anise Oil Obtained in the Vapor Phase Test

Microorganism	MO Class	Conditions		Activity	Reference
Escherichia coli	Bac–	NA, 24 h, 37°C	~20,000	+	Kellner and Kober (1954)
Neisseria sp.	Bac–	NA, 24 h, 37°C	~20,000	NG	Kellner and Kober (1954)
Salmonella typhi	Bac–	NA, 24 h, 37°C	~20,000	NG	Kellner and Kober (1954)
Bacillus megaterium	Bac+	NA, 24 h, 37°C	~20,000	+	Kellner and Kober (1954)
Corynebacterium diphtheriae	Bac+	NA, 24 h, 37°C	~20,000	NG	Kellner and Kober (1954)
Staphylococcus aureus	Bac+	NA, 24 h, 37°C	~20,000	NG	Kellner and Kober (1954)
Streptococcus faecalis	Bac+	NA, 24 h, 37°C	~20,000	+	Kellner and Kober (1954)
Streptococcus pyogenes	Bac+	NA, 24 h, 37°C	~20,000	NG	Kellner and Kober (1954)
Aspergillus flavus	Fungi	CDA, 6 days	12, 6,000	+	Singh et al. (2006)
Aspergillus niger	Fungi	CDA, 6 days	12, 6,000	++	Singh et al. (2006)
Aspergillus ochraceus	Fungi	CDA, 6 days	12, 6,000	+++	Singh et al. (2006)
Aspergillus terreus	Fungi	CDA, 6 days	12, 6,000	+++	Singh et al. (2006)
Fusarium graminearum	Fungi	CDA, 6 days	12, 6,000	+++	Singh et al. (2006)
Fusarium moniliforme	Fungi	CDA, 6 days	12, 6,000	NG	Singh et al. (2006)
Penicillium citrinum	Fungi	CDA, 6 days	12, 6,000	+	Singh et al. (2006)
Penicillium viridicatum	Fungi	CDA, 6 days	12, 6,000	++	Singh et al. (2006)
Candida albicans	Yeast	NA, 24 h, 37°C	~20,000	NG	Kellner and Kober (1954)

TABLE 14.72
Inhibitory Data of Sweet Orange Oil Obtained in the Agar Diffusion Test

Microorganism	MO Class	Conditions		Inhibition Zone in mm	Reference
Acinetobacter calcoaceticus	Bac–	ISA, 48 h, 25°C	4 (h), 10,000	0	Deans and Ritchie (1987)
Aerobacter aerogenes	Bac–	NA, 24 h, 37°C	—, sd	0	Maruzzella and Lichtenstein (1956)
Aeromonas hydrophila	Bac–	ISA, 48 h, 25°C	4 (h), 10,000	0	Deans and Ritchie (1987)
Alcaligenes faecalis	Bac–	ISA, 48 h, 25°C	4 (h), 10,000	0	Deans and Ritchie (1987)
Beneckea natriegens	Bac–	ISA, 48 h, 25°C	4 (h), 10,000	0	Deans and Ritchie (1987)
Campylobacter jejuni	Bac–	CAB, 24 h, 42°C	Disk, 10,000	0	Fisher and Phillips (2006)
Campylobacter jejuni	Bac–	TSA, 24 h, 42°C	4 (h), 25,000	4	Smith-Palmer et al. (1998)
Citrobacter freundii	Bac–	ISA, 48 h, 25°C	4 (h), 10,000	0	Deans and Ritchie (1987)
Enterobacter aerogenes	Bac–	ISA, 48 h, 25°C	4 (h), 10,000	0	Deans and Ritchie (1987)
Enterobacter aerogenes	Bac–	MHA, 24 h, 30°C	6, 15,000	6	Rossi et al. (2007)
Erwinia carotovora	Bac–	ISA, 48 h, 25°C	4 (h), 10,000	0	Deans and Ritchie (1987)
Escherichia coli	Bac–	NA, 24 h, 37°C	—, sd	0	Maruzzella and Lichtenstein (1956)
Escherichia coli	Bac–	ISA, 48 h, 25°C	4 (h), 10,000	0	Deans and Ritchie (1987)
Escherichia coli	Bac–	NA, 18 h, 37°C	5 (h), –30,000	0	Schelz et al. (2006)
Escherichia coli	Bac–	TSA, 24 h, 35°C	4 (h), 25,000	4	Smith-Palmer et al. (1998)
Escherichia coli	Bac–	Cited, 18 h, 37°C	6, 2,500	7	Janssen et al. (1986)
Escherichia coli	Bac–	MHA, 24 h, 30°C	6, 15,000	8	Rossi et al. (2007)
Escherichia coli	Bac–	Cited	15, 2,500	16	Pizsolitto et al. (1975)
Escherichia coli	Bac–	NA, 24 h, 37°C	Disk, 10,000	18	Fisher and Phillips (2006)
Escherichia coli	Bac–	NA, 18 h, 37°C	6 (h), pure	19.5	Yousef and Tawil (1980)
Escherichia coli	Bac–	TGA, 18–24 h, 37°C	9.5, 2,000	20	Morris et al. (1979)
Flavobacterium suaveolens	Bac–	ISA, 48 h, 25°C	4 (h), 10,000	10	Deans and Ritchie (1987)
Klebsiella pneumoniae	Bac–	ISA, 48 h, 25°C	4 (h), 10,000	0	Deans and Ritchie (1987)
Klebsiella sp.	Bac–	Cited	15, 2,500	2	Pizsolitto et al. (1975)
Moraxella sp.	Bac–	ISA, 48 h, 25°C	4 (h), 10,000	0	Deans and Ritchie (1987)
Neisseria perflava	Bac–	NA, 24 h, 37°C	—, sd	0	Maruzzella and Lichtenstein (1956)
Proteus sp.	Bac–	Cited	15, 2,500	3	Pizsolitto et al. (1975)
Proteus vulgaris	Bac–	NA, 24 h, 37°C	—, sd	0	Maruzzella and Lichtenstein (1956)
Proteus vulgaris	Bac–	ISA, 48 h, 25°C	4 (h), 10,000	0	Deans and Ritchie (1987)
Pseudomonas aeruginosa	Bac–	NA, 18 h, 37°C	6 (h), pure	0	Yousef and Tawil (1980)

(*Continued*)

TABLE 14.72 (*Continued*)
Inhibitory Data of Sweet Orange Oil Obtained in the Agar Diffusion Test

Microorganism	MO Class	Conditions		Inhibition Zone in mm	Reference
Pseudomonas aeruginosa	Bac−	Cited, 18 h, 37°C	6, 2,500	0	Janssen et al. (1986)
Pseudomonas aeruginosa	Bac−	ISA, 48 h, 25°C	4 (h), 10,000	0	Deans and Ritchie (1987)
Pseudomonas aeruginosa	Bac−	Cited	15, 2,500	2	Pizsolitto et al. (1975)
Pseudomonas aeruginosa	Bac−	NA, 24 h, 37°C	—, sd	2	Maruzzella and Lichtenstein (1956)
Pseudomonas aeruginosa	Bac−	MHA, 24 h, 30°C	6, 15,000	6	Rossi et al. (2007)
Pseudomonas mangiferae indica	Bac−	NA, 36–48 h, 37°C	6, sd	0	Garg and Garg (1980a,b)
Pseudomonas mangiferae indica	Bac−	NA, 24 h, 28°C	6, sd	9	Kindra and Satyanarayana (1978)
Salmonella enteritidis	Bac−	TSA, 24 h, 35°C	4 (h), 25,000	4	Smith-Palmer et al. (1998)
Salmonella paratyphi	Bac−	NA, 36–48 h, 37°C	6, sd	14	Garg and Garg (1980a,b)
Salmonella paratyphi	Bac−	NA, 24 h, 28°C	6, sd	19	Kindra and Satyanarayana (1978)
Salmonella pullorum	Bac−	ISA, 48 h, 25°C	4 (h), 10,000	0	Deans and Ritchie (1987)
Salmonella sp.	Bac−	Cited	15, 2,500	15	Pizsolitto et al. (1975)
Salmonella typhi	Bac−	NA, 36–48 h, 37°C	6, sd,	24	Garg and Garg (1980a,b)
Serratia marcescens	Bac−	NA, 24 h, 37°C	—, sd	0	Maruzzella and Lichtenstein (1956)
Serratia marcescens	Bac−	ISA, 48 h, 25°C	4 (h), 10,000	0	Deans and Ritchie (1987)
Serratia sp.	Bac−	Cited	15, 2,500	2	Pizsolitto et al. (1975)
Shigella sp.	Bac−	Cited	15, 2,500	13	Pizsolitto et al. (1975)
Vibrio cholera	Bac−	NA, 36–48 h, 37°C	6, sd	13	Garg and Garg (1980a,b)
Vibrio cholera	Bac−	NA, 24 h, 28°C	6, sd	13	Kindra and Satyanarayana (1978)
Xanthomonas campestris	Bac−	NA, 24 h, 28°C	6, sd	20.5	Kindra and Satyanarayana (1978)
Yersinia enterocolitica	Bac−	ISA, 48 h, 25°C	4 (h), 10,000	0	Deans and Ritchie (1987)
Bacillus anthracis	Bac+	NA, 24 h, 28°C	6, sd	23	Kindra and Satyanarayana (1978)
Bacillus cereus	Bac+	BHA, 24 h, 30°C	Disk, 10,000	36	Fisher and Phillips (2006)
Bacillus mesentericus	Bac+	NA, 24 h, 37°C	—, sd	2	Maruzzella and Lichtenstein (1956)
Bacillus mycoides	Bac+	NA, 36–48 h, 37°C	6, sd	0	Garg and Garg (1980a,b)
Bacillus mycoides	Bac+	NA, 24 h, 28°C	6, sd	17	Kindra and Satyanarayana (1978)

(*Continued*)

TABLE 14.72 (*Continued*)
Inhibitory Data of Sweet Orange Oil Obtained in the Agar Diffusion Test

Microorganism	MO Class	Conditions		Inhibition Zone in mm	Reference
Bacillus pumilus	Bac+	NA, 36–48 h, 37°C	6, sd	0	Garg and Garg (1980a,b)
Bacillus pumilus	Bac+	NA, 24 h, 28°C	6, sd	16	Kindra and Satyanarayana (1978)
Bacillus sp.	Bac+	Cited	15, 2,500	7	Pizsolitto et al. (1975)
Bacillus subtilis	Bac+	NA, 24 h, 37°C	—, sd	4	Maruzzella and Lichtenstein (1956)
Bacillus subtilis	Bac+	ISA, 48 h, 25°C	4 (h), 10,000	10	Deans and Ritchie (1987)
Bacillus subtilis	Bac+	Cited, 18 h, 37°C	6, 2,500	10.3	Janssen et al. (1986)
Bacillus subtilis	Bac+	NA, 36–48 h, 37°C	6, sd	14	Garg and Garg (1980)
Bacillus subtilis	Bac+	NA, 18 h, 37°C	6 (h), pure	19	Yousef and Tawil (1980a,b)
Bacillus subtilis	Bac+	NA, 24 h, 28°C	6, sd	21.5	Kindra and Satyanarayana (1978)
Brevibacterium linens	Bac+	ISA, 48 h, 25°C	4 (h), 10,000	0	Deans and Ritchie (1987)
Brochothrix thermosphacta	Bac+	ISA, 48 h, 25°C	4 (h), 10,000	8	Deans and Ritchie (1987)
Clostridium sporogenes	Bac+	ISA, 48 h, 25°C	4 (h), 10,000	0	Deans and Ritchie (1987)
Corynebacterium sp.	Bac+	TGA, 18–24 h, 37°C	9.5, 2,000	22	Morris et al. (1979)
Lactobacillus plantarum	Bac+	ISA, 48 h, 25°C	4 (h), 10,000	0	Deans and Ritchie (1987)
Leuconostoc cremoris	Bac+	ISA, 48 h, 25°C	4 (h), 10,000	7.5	Deans and Ritchie (1987)
Listeria monocytogenes	Bac+	TSA, 24 h, 35°C	4 (h), 25,000	4	Smith-Palmer et al. (1998)
Listeria monocytogenes	Bac+	ISA, 48 h, 25°C	4 (h), 10,000	10	Lis-Balchin et al. (1998)
Listeria monocytogenes	Bac+	LSA, 24 h, 37°C	Disk, 10,000	>90	Fisher and Phillips (2006)
Micrococcus luteus	Bac+	ISA, 48 h, 25°C	4 (h), 10,000	0	Deans and Ritchie (1987)
Mycobacterium phlei	Bac+	NA, 18 h, 37°C	6 (h), pure	23.5	Yousef and Tawil (1980)
Sarcina lutea	Bac+	NA, 24 h, 37°C	—, sd	0	Maruzzella and Lichtenstein (1956)
Sarcina lutea	Bac+	NA, 36–48 h, 37°C	6, sd	0	Garg and Garg (1980a,b)
Staphylococcus aureus	Bac+	NA, 24 h, 37°C	—, sd	0	Maruzzella and Lichtenstein (1956)
Staphylococcus aureus	Bac+	TGA, 18–24 h, 37°C	9.5, 2,000	0	Morris et al. (1979)
Staphylococcus aureus	Bac+	NA, 36–48 h, 37°C	6, sd	0	Garg and Garg (1980a,b)
Staphylococcus aureus	Bac+	ISA, 48 h, 25°C	4 (h), 10,000	0	Deans and Ritchie (1987)
Staphylococcus aureus	Bac+	TSA, 24 h, 35°C	4 (h), 25,000	4	Smith-Palmer et al. (1998)
Staphylococcus aureus	Bac+	Cited	15, 2,500	6	Pizsolitto et al. (1975)

(*Continued*)

TABLE 14.72 (Continued)
Inhibitory Data of Sweet Orange Oil Obtained in the Agar Diffusion Test

Microorganism	MO Class	Conditions		Inhibition Zone in mm	Reference
Staphylococcus aureus	Bac+	Cited, 18 h, 37°C	6, 2,500	7	Janssen et al. (1986)
Staphylococcus aureus	Bac+	MHA, 24 h, 37°C	6, 15,000	17	Rossi et al. (2007)
Staphylococcus aureus	Bac+	NA, 18 h, 37°C	6 (h), pure	20.6	Yousef and Tawil (1980)
Staphylococcus aureus	Bac+	BHA, 24 h, 37°C	Disk, 10,000	46	Fisher and Phillips (2006)
Staphylococcus epidermidis	Bac+	NA, 18 h, 37°C	5 (h), –30,000	0	Schelz et al. (2006)
Staphylococcus epidermidis	Bac+	Cited	15, 2,500	12	Pizsolitto et al. (1975)
Streptococcus faecalis	Bac+	ISA, 48 h, 25°C	4 (h), 10,000	0	Deans and Ritchie (1987)
Streptococcus viridans	Bac+	Cited	15, 2,500	0	Pizsolitto et al. (1975)
Streptomyces venezuelae	Bac+	SMA, 2–7 days, 20°C	sd	5	Maruzzella and Liguori (1958)
Alternaria porri	Fungi	PDA, 72 h, 28°C	5, 5,000	10	Pawar and Thaker (2007)
Alternaria solani	Fungi	SMA, 2–7 days, 20°C	sd	13	Maruzzella and Liguori (1958)
Aspergillus fumigatus	Fungi	SMA, 2–7 days, 20°C	sd	8	Maruzzella and Liguori (1958)
Aspergillus niger	Fungi	SMA, 2–7 days, 20°C	sd	4	Maruzzella and Liguori (1958)
Aspergillus niger	Fungi	PDA, 48 h, 28°C	5, 5,000	6	Pawar and Thaker (2006)
Aspergillus niger	Fungi	SDA, 8 days, 30°C	6 (h), pure	28	Yousef and Tawil (1980)
Fusarium oxysporum f.sp. ciceri	Fungi	PDA, 72 h, 28°C	5, 5,000	10	Pawar and Thaker (2007)
Helminthosporium sativum	Fungi	SMA, 2–7 days, 20°C	sd	7	Maruzzella and Liguori (1958)
Mucor mucedo	Fungi	SMA, 2–7 days, 20°C	sd	4	Maruzzella and Liguori (1958)
Mucor sp.	Fungi	SDA, 8 days, 30°C	6 (h), pure	18	Yousef and Tawil (1980)
Nigrospora panici	Fungi	SMA, 2–7 days, 20°C	sd	6	Maruzzella and Liguori (1958)
Penicillium chrysogenum	Fungi	SDA, 8 days, 30°C	6 (h), pure	25	Yousef and Tawil (1980)
Penicillium digitatum	Fungi	SMA, 2–7 days, 20°C	sd	7	Maruzzella and Liguori (1958)
Rhizopus nigricans	Fungi	SMA, 2–7 days, 20°C	sd	0	Maruzzella and Liguori (1958)
Rhizopus sp.	Fungi	SDA, 8 days, 30°C	6 (h), pure	0	Yousef and Tawil (1980)
Brettanomyces anomalus	Yeast	MPA, 4 days, 30°C	5, 10% sol. sd	0	Conner and Beuchat (1984)

(Continued)

TABLE 14.72 (Continued)
Inhibitory Data of Sweet Orange Oil Obtained in the Agar Diffusion Test

Microorganism	MO Class	Conditions		Inhibition Zone in mm	Reference
Candida albicans	Yeast	TGA, 18–24 h, 37°C	9.5, 2,000	0	Morris et al. (1979)
Candida albicans	Yeast	SMA, 2–7 days, 20°C	sd	5	Maruzzella and Liguori (1958)
Candida albicans	Yeast	Cited, 18 h, 37°C	6, 2,500	13	Janssen et al. (1986)
Candida albicans	Yeast	SDA, 18 h, 30°C	6 (h), pure	22.5	Yousef and Tawil (1980)
Candida krusei	Yeast	SMA, 2–7 days, 20°C	sd	5	Maruzzella and Liguori (1958)
Candida lipolytica	Yeast	MPA, 4 days, 30°C	5, 10% sol. sd	7	Conner and Beuchat (1984)
Candida tropicalis	Yeast	SMA, 2–7 days, 20°C	sd	4	Maruzzella and Liguori (1958)
Cryptococcus neoformans	Yeast	SMA, 2–7 days, 20°C	sd	6	Maruzzella and Liguori (1958)
Cryptococcus rhodobenhani	Yeast	SMA, 2–7 days, 20°C	sd	4	Maruzzella and Liguori (1958)
Debaryomyces hansenii	Yeast	MPA, 4 days, 30°C	5, 10% sol. sd	0	Conner and Beuchat (1984)
Geotrichum candidum	Yeast	MPA, 4 days, 30°C	5, 10% sol. sd	0	Conner and Beuchat (1984)
Hansenula anomala	Yeast	MPA, 4 days, 30°C	5, 10% sol. sd	0	Conner and Beuchat (1984)
Kloeckera apiculata	Yeast	MPA, 4 days, 30°C	5, 10% sol. sd	0	Conner and Beuchat (1984)
Kluyveromyces fragilis	Yeast	MPA, 4 days, 30°C	5, 10% sol. sd	0	Conner and Beuchat (1984)
Lodderomyces elongisporus	Yeast	MPA, 4 days, 30°C	5, 10% sol. sd	0	Conner and Beuchat (1984)
Metschnikowia pulcherrima	Yeast	MPA, 4 days, 30°C	5, 10% sol. sd	0	Conner and Beuchat (1984)
Pichia membranifaciens	Yeast	MPA, 4 days, 30°C	5, 10% sol. sd	0	Conner and Beuchat (1984)
Rhodotorula rubra	Yeast	MPA, 4 days, 30°C	5, 10% sol. sd	0	Conner and Beuchat (1984)
Saccharomyces cerevisiae	Yeast	NA, 24 h, 20°C	5 (h), −30,000	7–8	Schelz et al. (2006)
Saccharomyces cerevisiae	Yeast	MPA, 4 days, 30°C	5, 10% sol. sd	0	Conner and Beuchat (1984)
Saccharomyces cerevisiae	Yeast	SMA, 2–7 days, 20°C	sd	18	Maruzzella and Liguori (1958)
Torula glabrata	Yeast	MPA, 4 days, 30°C	5, 10% sol. sd	10	Conner and Beuchat (1984)

TABLE 14.73
Inhibitory Data of Sweet Orange Oil Obtained in the Dilution Test

Microorganism	MO Class	Conditions	MIC	Reference
Aerobacter aerogenes	Bac–	NA, pH 7	2,000	Subba et al. (1967)
Escherichia coli	Bac–	NB, Tween 20, 18 h, 37°C	3,200	Yousef and Tawil (1980)
Escherichia coli	Bac–	TGB, 18–24 h, 37°C	>1,000	Morris et al. (1979)
Escherichia coli	Bac–	NA, 24 h, 37°C	10,000	Fisher and Phillips (2006)
Escherichia coli	Bac–	NB, 24–72 h, 37°C	93% inh. 10,000	Dabbah et al. (1970)
Escherichia coli	Bac–	TGB, 18 h, 37°C	>11,300	Schelz et al. (2006)
Helicobacter pylori	Bac–	Cited, 20 h, 37°C	65.1	Weseler et al. (2005)
Pseudomonas aeruginosa	Bac–	NB, Tween 20, 18 h, 37°C	>50,000	Yousef and Tawil (1980)
Pseudomonas aeruginosa	Bac–	NB, 24–72 h, 37°C	87% inh. 10,000	Dabbah et al. (1970)
Salmonella heidelberg	Bac–	NB, 1–2 days, 35–37°C	90% inh., 1,000	Dabbah et al. (1970)
Salmonella montevideo	Bac–	NB, 1–2 days, 35–37°C	90% inh., 1,000	Dabbah et al. (1970)
Salmonella oranienburg	Bac–	NB, 1–2 days, 35–37°C	90% inh., 1,000	Dabbah et al. (1970)
Salmonella schottmuelleri	Bac–	NA, pH 7	1,000	Subba et al. (1967)
Salmonella senftenberg	Bac–	NB, 24–72 h, 37°C	93% inh. 10,000	Dabbah et al. (1970)
Salmonella typhimurium	Bac–	NB, 1–2 days, 35–37°C	90% inh., 1,000	Dabbah et al. (1970)
Serratia marcescens	Bac–	NA, pH 7	>2,000	Subba et al. (1967)
Yersinia enterocolitica	Bac–	MHA, Tween 20, 24 h, 37°C	1,250	Rossi et al. (2007)
Bacillus cereus	Bac+	BHA, 24 h, 30°C	>40,000	Fisher and Phillips (2006)
Bacillus subtilis	Bac+	NA, pH 7	2,000	Subba et al. (1967)
Bacillus subtilis	Bac+	NB, Tween 20, 18 h, 37°C	3,200	Yousef and Tawil (1980)
Corynebacterium sp.	Bac+	TGB, 18–24 h, 37°C	500	Morris et al. (1979)
Lactobacillus plantarum	Bac+	NA, pH 7	1,000	Subba et al. (1967)
Listeria monocytogenes	Bac+	LSA, 24 h, 37°C	2,500	Fisher and Phillips (2006)
Micrococcus sp.	Bac+	NA, pH 7	1,000	Subba et al. (1967)
Mycobacterium phlei	Bac+	NB, Tween 20, 18 h, 37°C	800	Yousef and Tawil (1980)
Staphylococcus aureus	Bac+	NB, Tween 20, 18 h, 37°C	12,500	Yousef and Tawil (1980)
Staphylococcus aureus	Bac+	TGB, 18–24 h, 37°C	500	Morris et al. (1979)
Staphylococcus aureus	Bac+	NB, 24–72 h, 37°C	10,000	Dabbah et al. (1970)
Staphylococcus epidermidis	Bac+	TGB, 18 h, 37°C	>11,300	Schelz et al. (2006)
Streptococcus faecalis	Bac+	NA, pH 7	1,000	Subba et al. (1967)
Aspergillus awamori	Fungi	PDA, pH 4.5	2,000	Subba et al. (1967)
Aspergillus flavus	Fungi	PDA, pH 4.5	2,000	Subba et al. (1967)
Aspergillus flavus	Fungi	PDA, 8 h, 20°C, spore germ. inh.	50–100	Thompson (1986)
Aspergillus flavus	Fungi	PDA, 5 days, 27°C	>1,000	Thompson and Cannon (1986)
Aspergillus niger	Fungi	PDA, pH 4.5	2,000	Subba et al. (1967)
Aspergillus niger	Fungi	NB, Tween 20, 8 days, 30°C	50,000	Yousef and Tawil (1980)
Aspergillus niger	Fungi	YES broth, 10 days	0% inh. 10,000	Lis-Balchin et al. (1998)
Aspergillus ochraceus	Fungi	YES broth, 10 days	34% inh. 10,000	Lis-Balchin et al. (1998)
Aspergillus oryzae	Fungi	Cited	500	Okazaki and Oshima (1953)
Aspergillus parasiticus	Fungi	PDA, 8 h, 20°C, spore germ. inh.	50–100	Thompson (1986)

(Continued)

TABLE 14.73 (*Continued*)
Inhibitory Data of Sweet Orange Oil Obtained in the Dilution Test

Microorganism	MO Class	Conditions	MIC	Reference
Aspergillus parasiticus	Fungi	PDA, 5 days, 27°C	>1,000	Thompson and Cannon (1986)
Botrytis cinerea	Fungi	PDA, Tween 20, 7 days, 24°C	4% inh. 1,000	Bouchra et al. (2003)
Cephalosporium sacchari	Fungi	OA, EtOH, 3 days, 20°C	>20,000	Narasimba Rao et al. (1971)
Ceratocystis paradoxa	Fungi	OA, EtOH, 3 days, 20°C	20,000	Narasimba Rao et al. (1971)
Curvularia lunata	Fungi	OA, EtOH, 3 days, 20°C	20,000	Narasimba Rao et al. (1971)
Epidermophyton floccosum	Fungi	SA, Tween 80, 21 days, 20°C	>1,250	Janssen et al. (1988)
Fusarium culmorum	Fungi	YES broth, 10 days	84% inh. 10,000	Lis-Balchin et al. (1998)
Fusarium moniliforme var. *subglutinans*	Fungi	OA, EtOH, 3 days, 20°C	>20,000	Narasimba Rao et al. (1971)
Geotrichum citri-aurantii	Fungi	PDA, Tween 20, 7 days, 24°C	7% inh. 1,000	Bouchra et al. (2003)
Helminthosporium sacchari	Fungi	OA, EtOH, 3 days, 20°C	20,000	Narasimba Rao et al. (1971)
Mucor hiemalis	Fungi	PDA, 5 days, 27°C	>1,000	Thompson and Cannon (1986)
Mucor mucedo	Fungi	PDA, 5 days, 27°C	>1,000	Thompson and Cannon (1986)
Mucor racemosus	Fungi	Cited	500	Okazaki and Oshima (1953)
Mucor racemosus f. *racemosus*	Fungi	PDA, 5 days, 27°C	>1,000	Thompson and Cannon (1986)
Mucor sp.	Fungi	NB, Tween 20, 8 days, 30°C	12,500	Yousef and Tawil (1980)
Penicillium chrysogenum	Fungi	Cited	500	Okazaki and Oshima (1953)
Penicillium chrysogenum	Fungi	NB, Tween 20, 8 days, 30°C	25,000	Yousef and Tawil (1980)
Penicillium digitatum	Fungi	SDB, 5 days, 20°C, MIC = ED50	1,000–2,400	Caccioni et al. (1998)
Penicillium digitatum	Fungi	PDA, Tween 20, 7 days, 24°C	32% inh. 1,000	Bouchra et al. (2003)
Penicillium italicum	Fungi	SDB, 5 days, 20°C, MIC = ED50	3,000–5,500	Caccioni et al. (1998)
Physalospora tucumanensis	Fungi	OA, EtOH, 3 days, 20°C	4,000	Narasimba Rao et al. (1971)
Phytophthora citrophthora	Fungi	PDA, Tween 20, 7 days, 24°C	13% inh. 1,000	Bouchra et al. (2003)
Rhizopus 66-81-2	Fungi	PDA, 5 days, 27°C	>1,000	Thompson and Cannon (1986)
Rhizopus arrhizus	Fungi	PDA, 5 days, 27°C	>1,000	Thompson and Cannon (1986)
Rhizopus chinensis	Fungi	PDA, 5 days, 27°C	>1,000	Thompson and Cannon (1986)
Rhizopus circinans	Fungi	PDA, 5 days, 27°C	>1,000	Thompson and Cannon (1986)

(*Continued*)

TABLE 14.73 (*Continued*)
Inhibitory Data of Sweet Orange Oil Obtained in the Dilution Test

Microorganism	MO Class	Conditions	MIC	Reference
Rhizopus japonicus	Fungi	PDA, 5 days, 27°C	>1,000	Thompson and Cannon (1986)
Rhizopus kazanensis	Fungi	PDA, 5 days, 27°C	>1,000	Thompson and Cannon (1986)
Rhizopus oryzae	Fungi	PDA, 5 days, 27°C	>1,000	Thompson and Cannon (1986)
Rhizopus pymacus	Fungi	PDA, 5 days, 27°C	>1,000	Thompson and Cannon (1986)
Rhizopus sp.	Fungi	NB, Tween 20, 8 days, 30°C	3,200	Yousef and Tawil (1980)
Rhizopus stolonifer	Fungi	PDA, 5 days, 27°C	>1,000	Thompson and Cannon (1986)
Rhizopus tritici	Fungi	PDA, 5 days, 27°C	>1,000	Thompson and Cannon (1986)
Sclerotium rolfsii	Fungi	OA, EtOH, 6 days, 20°C	20,000	Narasimba Rao et al. (1971)
Trichophyton mentagrophytes	Fungi	SA, Tween 80, 21 days, 20°C	>1,250	Janssen et al. (1988)
Trichophyton rubrum	Fungi	SA, Tween 80, 21 days, 20°C	>1,250	Janssen et al. (1988)
Candida albicans	Yeast	NB, Tween 20, 18 h, 37°C	1,600	Yousef and Tawil (1980)
Candida albicans	Yeast	TGB, 18–24 h, 37°C	500	Morris et al. (1979)
Saccharomyces cerevisiae	Yeast	PDA, pH 4.5	1,000	Subba et al. (1967)
Saccharomyces cerevisiae	Yeast	YPB, 24 h, 20°C	2,800	Schelz et al. (2006)
Torula utilis	Yeast	PDA, pH 4.5	1,000	Subba et al. (1967)
Zygosaccharomyces mellis	Yeast	PDA, pH 4.5	1,000	Subba et al. (1967)

TABLE 14.74
Inhibitory Data of Sweet Orange Oil Obtained in the Vapor Phase Test

Microorganism	MO Class	Conditions		Activity	Reference
Campylobacter jejuni	Bac–	CAB, 24 h, 42°C	Disk, 10,000	+++	Fisher and Phillips (2006)
Escherichia coli	Bac–	NA, 24 h, 37°C	~20,000	+	Kellner and Kober (1954)
Escherichia coli	Bac–	NA, 24 h, 37°C	Disk, 10,000	+++	Fisher and Phillips (2006)
Neisseria sp.	Bac–	NA, 24 h, 37°C	~20,000	NG	Kellner and Kober (1954)
Salmonella typhi	Bac–	NA, 24 h, 37°C	~20,000	NG	Kellner and Kober (1954)
Bacillus cereus	Bac+	BHA, 24 h, 30°C	Disk, 10,000	+++	Fisher and Phillips (2006)
Bacillus megaterium	Bac+	NA, 24 h, 37°C	~20,000	NG	Kellner and Kober (1954)
Corynebacterium diphtheriae	Bac+	NA, 24 h, 37°C	~20,000	NG	Kellner and Kober (1954)
Listeria monocytogenes	Bac+	LSA, 24 h, 37°C	Disk, 10,000	+++	Fisher and Phillips (2006)
Staphylococcus aureus	Bac+	NA, 24 h, 37°C	~20,000	NG	Kellner and Kober (1954)
Staphylococcus aureus	Bac+	BHA, 24 h, 37°C	Disk, 10,000	+++	Fisher and Phillips (2006)
Streptococcus faecalis	Bac+	NA, 24 h, 37°C	~20,000	NG	Kellner and Kober (1954)
Streptococcus pyogenes	Bac+	NA, 24 h, 37°C	~20,000	NG	Kellner and Kober (1954)
Aspergillus flavus	Fungi	Bread, 14 days, 25°C	30,000	NG	Suhr and Nielsen (2003)
Aspergillus flavus	Fungi	WFA, 42 days, 25°C	Disk, 50,000	+++	Guynot et al. (2003)
Aspergillus niger	Fungi	WFA, 42 days, 25°C	Disk, 50,000	+++	Guynot et al. (2003)
Botrytis cinerea	Fungi	PDA, 3 days, 25°C	1,000	+++	Lee et al. (2007)
Colletotrichum gloeosporioides	Fungi	PDA, 3 days, 25°C	1,000	+++	Lee et al. (2007)
Endomyces fibuligera	Fungi	Bread, 14 days, 25°C	30,000	+++	Suhr and Nielsen (2003)
Eurotium amstelodami	Fungi	WFA, 42 days, 25°C	Disk, 50,000	+++	Guynot et al. (2003)
Eurotium herbarum	Fungi	WFA, 42 days, 25°C	Disk, 50,000	+++	Guynot et al. (2003)
Eurotium repens	Fungi	Bread, 14 days, 25°C	30,000	+	Suhr and Nielsen (2003)
Eurotium repens	Fungi	WFA, 42 days, 25°C	Disk, 50,000	+++	Guynot et al. (2003)
Eurotium rubrum	Fungi	WFA, 42 days, 25°C	Disk, 50,000	+++	Guynot et al. (2003)
Fusarium oxysporum	Fungi	PDA, 3 days, 25°C	1,000	+++	Lee et al. (2007)
Penicillium corylophilum	Fungi	WFA, 42 days, 25°C	Disk, 50,000	+	Guynot et al. (2003)
Penicillium corylophilum	Fungi	Bread, 14 days, 25°C	30,000	+++	Suhr and Nielsen (2003)
Penicillium roqueforti	Fungi	Bread, 14 days, 25°C	30,000	+++	Suhr and Nielsen (2003)
Pythium ultimum	Fungi	PDA, 3 days, 25°C	1,000	+++	Lee et al. (2007)
Rhizoctonia solani	Fungi	PDA, 3 days, 25°C	1,000	+++	Lee et al. (2007)
Candida albicans	Yeast	NA, 24 h, 37°C	~20,000	NG	Kellner and Kober (1954)

TABLE 14.75
Inhibitory Data of Tea Tree Oil Obtained in the Agar Diffusion Test

Microorganism	MO Class	Conditions		Inhibition Zone (mm)	Reference
Escherichia coli	Bac−	NA, 24 h, 30°C	Drop, 5,000	14	Hili et al. (1997)
Escherichia coli	Bac−	NA, 18 h, 37°C	5 (h), −30,000	17	Schelz et al. (2006)
Pseudomonas aeruginosa	Bac−	NA, 24 h, 30°C	Drop, 5,000	5	Hili et al. (1997)
Listeria monocytogenes	Bac+	ISA, 48 h, 25°C	4 (h), 10,000	20	Lis-Balchin et al. (1998)
Staphylococcus aureus	Bac+	NA, 24 h, 30°C	Drop, 5,000	7	Hili et al. (1997)
Staphylococcus aureus MRSA	Bac+	MHA, 24 h, 37°C	12.7, 30,000	21–33	Carson et al. (1995)
Staphylococcus epidermidis	Bac+	NA, 18 h, 37°C	5 (h), −30,000	14	Schelz et al. (2006)
Alternaria porri	Fungi	PDA, 72 h, 28°C	5, 5,000	15.3	Pawar and Thaker (2007)
Aspergillus flavus	Fungi	SDA, 72 h, 26°C	8, 25,000	9	Shin (2003)
Aspergillus niger	Fungi	PDA, 48 h, 28°C	5, 5,000	7	Pawar and Thaker (2006)
Aspergillus niger	Fungi	SDA, 72 h, 26°C	8, 25,000	8	Shin (2003)
Fusarium oxysporum f.sp. ciceri	Fungi	PDA, 72 h, 28°C	5, 5,000	10.5	Pawar and Thaker (2007)
Candida albicans	Yeast	NA, 24 h, 30°C	Drop, 5,000	11	Hili et al. (1997)
Saccharomyces cerevisiae	Yeast	NA, 24 h, 20°C	5 (h), −30,000	19–21	Schelz et al. (2006)
Saccharomyces cerevisiae	Yeast	NA, 24 h, 30°C	Drop, 5,000	13	Hili et al. (1997)
Schizosaccharomyces pombe	Yeast	NA, 24 h, 30°C	Drop, 5,000	20	Hili et al. (1997)
Torula utilis	Yeast	NA, 24 h, 30°C	Drop, 5,000	47	Hili et al. (1997)

TABLE 14.76
Inhibitory Data of Tea Tree Oil Obtained in the Dilution Test

Microorganism	MO Class	Conditions	MIC	Reference
Mycoplasma fermentans	Bac	Cited	100–600	Furneri et al. (2006)
Mycoplasma hominis	Bac	Cited	600–1,200	Furneri et al. (2006)
Mycoplasma pneumoniae	Bac	Cited	100	Furneri et al. (2006)
Acinetobacter baumannii	Bac–	HIB, Tween 80, 24 h, 35°C	600–10,000	Hammer et al. (1996)
Actinobacillus actinomycetemcomitans	Bac–	HS, 72 h, 37°C	1,100	Shapiro et al. (1994)
Aeromonas sobria	Bac–	MHA, Tween 20, 48 h, 35°C	2,500	Hammer et al. (1999)
Bacteroides sp.	Bac–	VC, Tween 80	300–5,000	Hammer et al. (1999)
Citrobacter freundii	Bac–	ISB, Tween 80, 20–24 h, 37°C	5,000	Harkenthal et al. (1999)
Coliform bacilli	Bac–	MHB, Tween 80, 24 h, 37°C	10,000–20,000	Banes-Marshall et al. (2001)
Coliform bacilli	Bac–	BA, 24 h, 37°C	5,000	Banes-Marshall et al. (2001)
Enterobacter aerogenes	Bac–	ISB, Tween 80, 20–24 h, 37°C	2,500	Harkenthal et al. (1999)
Enterococcus faecalis	Bac–	ISB, Tween 20, 24 h, 37°C	5,000–7,500	Griffin et al. (2000)
Escherichia coli	Bac–	ISB, Tween 20, 24 h, 37°C	2,000	Griffin et al. (2000)
Escherichia coli	Bac–	MHA, Tween 20, 48 h, 35°C	2,500	Hammer et al. (1999)
Escherichia coli	Bac–	ISB, Tween 80, 16–20 h, 37°C	2,500	Gustafson et al. (1998)
Escherichia coli	Bac–	ISB, Tween 80, 20–24 h, 37°C	2,500	Harkenthal et al. (1999)
Escherichia coli	Bac–	MPB, DMSO, 40 h, 30°C	28% inh. 500	Hili et al. (1997)
Escherichia coli	Bac–	TGB, 18 h, 37°C	5,600	Schelz et al. (2006)
Fusobacterium nucleatum	Bac–	HS, 72 h, 37°C	>6,000	Shapiro et al. (1994)
Fusobacterium sp.	Bac–	VC, Tween 80	600–2,500	Hammer et al. (1999)
Gardnerella vaginalis	Bac–	VC, Tween 80	600	Hammer et al. (1999)
Klebsiella pneumoniae	Bac–	HIB, Tween 80, 24 h, 35°C	1,200–50,000	Hammer et al. (1996)
Klebsiella pneumoniae	Bac–	ISB, Tween 80, 20–24 h, 37°C	2,500	Harkenthal et al. (1999)
Klebsiella pneumoniae	Bac–	ISB, Tween 20, 24 h, 37°C	3,000	Griffin et al. (2000)
Klebsiella pneumoniae	Bac–	MHA, Tween 20, 48 h, 35°C	5,000	Hammer et al. (1999)
Porphyromonas gingivalis	Bac–	HS, 72 h, 37°C	1,100	Shapiro et al. (1994)
Prevotella sp.	Bac–	VC, Tween 80	300–2,500	Hammer et al. (1999)
Proteus mirabilis	Bac–	ISB, Tween 80, 20–24 h, 37°C	2,500	Harkenthal et al. (1999)
Proteus vulgaris	Bac–	ISB, Tween 20, 24 h, 37°C	3,000	Griffin et al. (2000)
Pseudomonas aeruginosa	Bac–	MCA, 24 h, 37	>20,000	Banes-Marshall et al. (2001)
Pseudomonas aeruginosa	Bac–	ISB, Tween 80, 20–24 h, 37°C	>40,000	Harkenthal et al. (1999)
Pseudomonas aeruginosa	Bac–	ISB, Tween 20, 24 h, 37°C	10,000 to >20,000	Griffin et al. (2000)
Pseudomonas aeruginosa	Bac–	MHB, Tween 80, 24 h, 37°C	10,000–80,000	Banes-Marshall et al. (2001)
Pseudomonas aeruginosa	Bac–	HIB, Tween 80, 24 h, 35°C	20,000–50,000	Hammer et al. (1996)
Pseudomonas aeruginosa	Bac–	MHB, 18–24 h, 37°C	40,000	Papadopoulos et al. (2006)

(*Continued*)

TABLE 14.76 (Continued)
Inhibitory Data of Tea Tree Oil Obtained in the Dilution Test

Microorganism	MO Class	Conditions	MIC	Reference
Pseudomonas aeruginosa	Bac–	MHA, Tween 20, 48 h, 35°C	5,000	Hammer et al. (1999)
Pseudomonas aeruginosa	Bac–	MPB, DMSO, 40 h, 30°C	75% inh. 500	Hili et al. (1997)
Pseudomonas fluorescens	Bac–	MHB, 18–24 h, 37°C	40,000	Papadopoulos et al. (2006)
Pseudomonas putida	Bac–	ISB, Tween 20, 24 h, 37°C	>20,000	Griffin et al. (2000)
Pseudomonas putida	Bac–	MHB, 18–24 h, 37°C	10,000	Papadopoulos et al. (2006)
Salmonella choleraesuis	Bac–	ISB, Tween 80, 20–24 h, 37°C	2,500	Harkenthal et al. (1999)
Salmonella typhimurium	Bac–	MHA, Tween 20, 48 h, 35°C	5,000	Hammer et al. (1999)
Serratia marcescens	Bac–	ISB, Tween 20, 24 h, 37°C	1,000–3,000	Griffin et al. (2000)
Serratia marcescens	Bac–	HIB, Tween 80, 24 h, 35°C	2,500–50,000	Hammer et al. (1996)
Serratia marcescens	Bac–	MHA, Tween 20, 48 h, 35°C	5,000	Hammer et al. (1999)
Shigella flexneri	Bac–	ISB, Tween 80, 20–24 h, 37°C	2,500	Harkenthal et al. (1999)
Actinomyces viscosus	Bac+	HS, 16–24, 37°C	6,000	Shapiro et al. (1994)
Anaerobic cocci	Bac+	VC, Tween 80	600–2,500	Hammer et al. (1999)
Bacillus cereus	Bac+	ISB, Tween 20, 24 h, 37°C	3,000	Griffin et al. (2000)
Bacillus subtilis	Bac+	ISB, Tween 80, 20–24 h, 37°C	2,500	Harkenthal et al. (1999)
Bacillus subtilis	Bac+	ISB, Tween 20, 24 h, 37°C	3,000	Griffin et al. (2000)
Corynebacterium pseudodiphtheriae	Bac+	ISB, Tween 80, 20–24 h, 37°C	5,000	Harkenthal et al. (1999)
Corynebacterium sp.	Bac+	ISB, Tween 20, 24 h, 37°C	2,000–3,000	Griffin et al. (2000)
Corynebacterium sp.	Bac+	HIB, Tween 80, 24 h, 35°C	600–20,000	Hammer et al. (1996)
Enterococcus durans	Bac+	ISB, Tween 80, 20–24 h, 37°C	1,000	Harkenthal et al. (1999)
Enterococcus faecalis	Bac+	ISB, Tween 80, 20–24 h, 37°C	1,000	Harkenthal et al. (1999)
Enterococcus faecalis	Bac+	MHA, Tween 20, 48 h, 35°C	10,000	Hammer et al. (1999)
Enterococcus faecalis	Bac+	MHB, Tween 80, 24 h, 37°C	80,000	Banes-Marshall et al. (2001)
Enterococcus faecium	Bac+	ISB, Tween 80, 20–24 h, 37°C	1,000	Harkenthal et al. (1999)
Enterococcus faecium VRE	Bac+	HIB, Tween 80, 18 h, 37°C	5,000–10,000	Nelson (1997)
Listeria monocytogenes	Bac+	ISB, Tween 80, 20–24 h, 37°C	2,500	Harkenthal et al. (1999)
Micrococcus luteus	Bac+	ISB, Tween 20, 24 h, 37°C	2,000–3,000	Griffin et al. (2000)
Micrococcus luteus	Bac+	HIB, Tween 80, 24 h, 35°C	600–5,000	Hammer et al. (1996)
Micrococcus sp.	Bac+	HIB, Tween 80, 24 h, 35°C	600–5,000	Hammer et al. (1996)
Micrococcus varians	Bac+	HIB, Tween 80, 24 h, 35°C	5,000–10,000	Hammer et al. (1996)
Mobiluncus sp.	Bac+	VC, Tween 80	300–600	Hammer et al. (1999)
Peptostreptococcus anaerobius	Bac+	HS, 72 h, 37°C	2,000	Shapiro et al. (1994)
Peptostreptococcus anaerobius	Bac+	VC, Tween 80	600–2,500	Hammer et al. (1999)
Propionibacterium acnes	Bac+	Agar, 24 h, 37°C	3,100–6,300	Raman et al. (1995)
Propionibacterium acnes	Bac+	ISB, Tween 20, 24 h, 37°C	5,000	Griffin et al. (2000)
Staphylococcus aureus	Bac+	BA, 24 h, 37°C	10,000	Banes-Marshall et al. (2001)

(*Continued*)

TABLE 14.76 (Continued)
Inhibitory Data of Tea Tree Oil Obtained in the Dilution Test

Microorganism	MO Class	Conditions	MIC	Reference
Staphylococcus aureus	Bac+	HIB, Tween 80, 24 h, 35°C	1,200–5,000	Hammer et al. (1996)
Staphylococcus aureus	Bac+	ISB, Tween 20, 24 h, 37°C	2,000	Griffin et al. (2000)
Staphylococcus aureus	Bac+	MHB, Tween 80, 24 h, 37°C	20,000	Banes-Marshall et al. (2001)
Staphylococcus aureus	Bac+	ISB, Tween 80, 20–24 h, 37°C	2,500	Harkenthal et al. (1999)
Staphylococcus aureus	Bac+	MPB, DMSO, 40 h, 30°C	43% inh. 500	Hili et al. (1997)
Staphylococcus aureus	Bac+	MHA, Tween 20, 48 h, 35°C	5,000	Hammer et al. (1999)
Staphylococcus aureus	Bac+	MHB, Tween 80, 24 h, 37°C	5,000	Banes-Marshall et al. (2001)
Staphylococcus aureus	Bac+	Agar, 24 h, 37°C	6,300–12,500	Raman et al. (1995)
Staphylococcus aureus MRSA	Bac+	BA, 24 h, 37°C	10,000	Banes-Marshall et al. (2001)
Staphylococcus aureus MRSA	Bac+	MHB, Tween 80, 24 h, 37°C	20,000–40,000	Banes-Marshall et al. (2001)
Staphylococcus aureus MRSA	Bac+	ISB, Tween 20, 24 h, 37°C	2,000–3,000	Griffin et al. (2000)
Staphylococcus aureus MRSA	Bac+	HIB, Tween 80, 18 h, 37°C	2,500	Nelson (1997)
Staphylococcus aureus MRSA	Bac+	Cited	2,500	Carson and Messager (2005)
Staphylococcus aureus MRSA	Bac+	HIB, Tween 80, 24 h, 37°c	2,500–5,000	Carson et al. (1995)
Staphylococcus capitis	Bac+	HIB, Tween 80, 24 h, 35°C	10,000–100,000	Hammer et al. (1996)
Staphylococcus capitis	Bac+	ISB, Tween 80, 20–24 h, 37°C	1,200–2,500	Harkenthal et al. (1999)
Staphylococcus epidermidis	Bac+	HIB, Tween 80, 24 h, 35°C	1,200–40,000	Hammer et al. (1996)
Staphylococcus epidermidis	Bac+	ISB, Tween 80, 20–24 h, 37°C	2,500	Harkenthal et al. (1999)
Staphylococcus epidermidis	Bac+	ISB, Tween 20, 24 h, 37°C	5,000	Griffin et al. (2000)
Staphylococcus epidermidis	Bac+	TGB, 18 h, 37°C	5,600	Schelz et al. (2006)
Staphylococcus epidermidis	Bac+	Agar, 24 h, 37°C	6,300–12,500	Raman et al. (1995)
Staphylococcus haemolyticus	Bac+	HIB, Tween 80, 24 h, 35°C	10,000–40,000	Hammer et al. (1996)
Staphylococcus haemolyticus	Bac+	ISB, Tween 80, 20–24 h, 37°C	2,500–5,000	Harkenthal et al. (1999)
Staphylococcus hominis	Bac+	HIB, Tween 80, 24 h, 35°C	10,000–40,000	Hammer et al. (1996)
Staphylococcus hominis	Bac+	ISB, Tween 80, 20–24 h, 37°C	1,200	Harkenthal et al. (1999)
Staphylococcus saprophyticus	Bac+	HIB, Tween 80, 24 h, 35°C	20,000–30,000	Hammer et al. (1996)
Staphylococcus saprophyticus	Bac+	ISB, Tween 80, 20–24 h, 37°C	2,500–5,000	Harkenthal et al. (1999)

(Continued)

TABLE 14.76 (*Continued*)
Inhibitory Data of Tea Tree Oil Obtained in the Dilution Test

Microorganism	MO Class	Conditions	MIC	Reference
Staphylococcus warneri	Bac+	HIB, Tween 80, 24 h, 35°C	20,000–80,000	Hammer et al. (1996)
Staphylococcus xylosus	Bac+	HIB, Tween 80, 24 h, 35°C	10,000–30,000	Hammer et al. (1996)
Staphylococcus xylosus	Bac+	ISB, Tween 80, 20–24 h, 37°C	2,500	Harkenthal et al. (1999)
Streptococci beta-haemolytic	Bac+	BA, 24 h, 37°C	1,200–5,000	Banes-Marshall et al. (2001)
Streptococci beta-haemolytic	Bac+	MHB, Tween 80, 24 h, 37°C	80,000	Banes-Marshall et al. (2001)
Streptococci beta-haemolytic Gp.D	Bac+	MHB, Tween 80, 24 h, 37°C	5,000–20,000	Banes-Marshall et al. (2001)
Streptococci, faecal	Bac+	MHB, Tween 80, 24 h, 37°C	>80,000	Banes-Marshall et al. (2001)
Streptococci, faecal	Bac+	BA, 24 h, 37°C	10,000	Banes-Marshall et al. (2001)
Streptococcus equi	Bac+	THB, Tween 80, 24 h, 35°C	1,200	Carson et al. (1996)
Streptococcus equisimilis	Bac+	THB, Tween 80, 24 h, 35°C	1,200	Carson et al. (1996)
Streptococcus pyogenes	Bac+	THB, Tween 80, 24 h, 35°C	1,200	Carson et al. (1996)
Streptococcus pyogenes	Bac+	MHB, Tween 80, 24 h, 37°C	20,000	Banes-Marshall et al. (2001)
Streptococcus sobrinus	Bac+	HS, 16–24, 37°C	6,000	Shapiro et al. (1994)
Streptococcus sp. group G	Bac+	THB, Tween 80, 24 h, 35°C	1,200	Carson et al. (1996)
Streptococcus zooepidemicus	Bac+	THB, Tween 80, 24 h, 35°C	600	Carson et al. (1996)
Alternaria alternata	Fungi	PDA, 7 days, 28°C	62% inh. 500	Feng and Zheng (2007)
Alternaria sp.	Fungi	Cited data	160–1,200	Carson et al. (2006)
Alternaria sp.	Fungi	RPMI, Tween 80, 48 h, 30°C	160–1,200	Hammer et al. (2002)
Aspergillus flavus	Fungi	Cited data	3,100–7,000	Carson et al. (2006)
Aspergillus flavus	Fungi	MYB, 72 h, 26°C	3,120	Shin (2003)
Aspergillus flavus	Fungi	MA, Tween 20, 24 h, 30°C	4,000–5,000	Griffin et al. (2000)
Aspergillus flavus	Fungi	MA, Tween 20, 24 h, 30°C	5,000–7,000	Griffin et al. (2000)
Aspergillus flavus	Fungi	RPMI, Tween 80, 48 h, 35°C	600–1,200	Hammer et al. (2002)
Aspergillus fumigatus	Fungi	RPMI, Tween 80, 48 h, 35°C	600–1,200	Hammer et al. (2002)
Aspergillus niger	Fungi	Cited data	160–4,000	Carson et al. (2006)
Aspergillus niger	Fungi	MA, Tween 20, 24 h, 30°C	3,000–4,000	Griffin et al. (2000)
Aspergillus niger	Fungi	MYB, 72 h, 26°C	3,120	Shin (2003)
Aspergillus niger	Fungi	RPMI, Tween 80, 48 h, 35°C	600–1,200	Hammer et al. (2002)
Aspergillus niger	Fungi	YES broth, 10 days	>10,000	Lis-Balchin et al. (1998)
Aspergillus ochraceus	Fungi	YES broth, 10 days	91% inh. 10,000	Lis-Balchin et al. (1998)
Blastoschizomyces capitatus	Fungi	Cited data	2,500	Carson et al. (2006)
Cladosporium sp.	Fungi	RPMI, Tween 80, 72 h, 30°C	160–1,200	Hammer et al. (2002)
Cladosporium sp.	Fungi	Cited data	80–1,200	Carson et al. (2006)
Epidermophyton floccosum	Fungi	RPMI, Tween 80, 96 h, 30°C	80–300	Hammer et al. (2002)

(*Continued*)

TABLE 14.76 (*Continued*)
Inhibitory Data of Tea Tree Oil Obtained in the Dilution Test

Microorganism	MO Class	Conditions	MIC	Reference
Epidermophyton floccosum	Fungi	Cited data	80–7,000	Carson et al. (2006)
Fusarium culmorum	Fungi	YES broth, 10 days	76% inh. 10,000	Lis-Balchin et al. (1998)
Fusarium sp.	Fungi	Cited data	80–2,500	Carson et al. (2006)
Fusarium sp.	Fungi	RPMI, Tween 80, 48 h, 35°C	80–2,500	Hammer et al. (2002)
Malassezia sympodialis	Fungi	Cited data	160–1,200	Carson et al. (2006)
Microsporum canis	Fungi	Cited data	300–5,000	Carson et al. (2006)
Microsporum canis	Fungi	RPMI, Tween 80, 96 h, 30°C	40–300	Hammer et al. (2002)
Microsporum gypseum	Fungi	RPMI, Tween 80, 96 h, 30°C	160–300	Hammer et al. (2002)
Penicillium sp.	Fungi	Cited data	300–600	Carson et al. (2006)
Penicillium sp.	Fungi	RPMI, Tween 80, 48 h, 35°C	300–600	Hammer et al. (2002)
Pleurotus ferulae	Fungi	SDA, 7 days, 25°C	72%–82% inh. 1,000	Angelini et al. (2008)
Pleurotus nebrodensis	Fungi	SDA, 7 days, 25°C	64%–88% inh. 1,000	Angelini et al. (2008)
Pleurotus nebrodensis	Fungi	SDA, 7 days, 25°C	83%–88% inh. 1,000	Angelini et al. (2008)
Trichophyton interdigitale	Fungi	RPMI, Tween 80, 96 h, 30°C	80–300	Hammer et al. (2002)
Trichophyton mentagrophytes	Fungi	Cited data	1,100–4,400	Carson et al. (2006)
Trichophyton mentagrophytes	Fungi	MA, Tween 20, 24 h, 30°C	3,000–4,000	Griffin et al. (2000)
Trichophyton mentagrophytes	Fungi	RPMI, Tween 80, 96 h, 30°C	80–600	Hammer et al. (2002)
Trichophyton rubrum	Fungi	MA, Tween 20, 24 h, 30°C	10,000	Banes-Marshall et al. (2001)
Trichophyton rubrum	Fungi	Cited data	300–6,000	Carson et al. (2006)
Trichophyton rubrum	Fungi	RPMI, Tween 80, 96 h, 30°C	80–300	Hammer et al. (2002)
Trichophyton tonsurans	Fungi	Cited data	40–160	Carson et al. (2006)
Trichophyton tonsurans	Fungi	RPMI, Tween 80, 96 h, 30°C	40–160	Hammer et al. (2002)
Candida albicans	Yeast	RPMI, Tween 80, 48 h, 30°C	1,250	Oliva et al. (2003)
Candida albicans	Yeast	MEB, Tween 20, 24 h, 37	2,000	Griffin et al. (2000)
Candida albicans	Yeast	RPMI, Tween 80, 48 h, 35°C	2,500	Mondello et al. (2006)
Candida albicans	Yeast	MHB, Tween 80, 48 h, 35°C	2,500–5,000	Hammer et al. (1998)
Candida albicans	Yeast	MPB, DMSO, 40 h, 30°C	37% inh. 500	Hili et al. (1997)
Candida albicans	Yeast	MHA, Tween 20, 48 h, 35°C	5,000	Hammer et al. (1999)
Candida albicans	Yeast	SDA, 24 h, 37°C	5,000–10,000	Banes-Marshall et al. (2001)
Candida capitatus	Yeast	RPMI, Tween 80, 48 h, 30°C	1,250–2,500	Oliva et al. (2003)
Candida famata	Yeast	SDA, 24 h, 37°C	2,500	Banes-Marshall et al. (2001)

(*Continued*)

TABLE 14.76 (Continued)
Inhibitory Data of Tea Tree Oil Obtained in the Dilution Test

Microorganism	MO Class	Conditions	MIC	Reference
Candida glabrata	Yeast	MHB, Tween 80, 48 h, 35°C	1,200–5,000	Hammer et al. (1998)
Candida glabrata	Yeast	SDA, 24 h, 37°C	2,500–5,000	Banes-Marshall et al. (2001)
Candida glabrata	Yeast	RPMI, Tween 80, 48 h, 30°C	300–1,250	Oliva et al. (2003)
Candida glabrata	Yeast	Cited data	300–8,000	Carson et al. (2006)
Candida glabrata	Yeast	RPMI, Tween 80, 48 h, 35°C	600	Mondello et al. (2006)
Candida guilliermondii	Yeast	RPMI, Tween 80, 48 h, 30°C	1,250	Oliva et al. (2003)
Candida inconspicua	Yeast	RPMI, Tween 80, 48 h, 30°C	300	Oliva et al. (2003)
Candida krusei	Yeast	RPMI, Tween 80, 48 h, 30°C	1,250	Oliva et al. (2003)
Candida krusei	Yeast	RPMI, Tween 80, 48 h, 35°C	2,500	Mondello et al. (2006)
Candida krusei	Yeast	SDA, 24 h, 37°C	5,000	Banes-Marshall et al. (2001)
Candida lipolytica	Yeast	RPMI, Tween 80, 48 h, 30°C	600–1,250	Oliva et al. (2003)
Candida lusitaniae	Yeast	RPMI, Tween 80, 48 h, 30°C	1,250	Oliva et al. (2003)
Candida parapsilosis	Yeast	RPMI, Tween 80, 48 h, 35°C	1,250	Mondello et al. (2006)
Candida parapsilosis	Yeast	MHB, Tween 80, 48 h, 35°C	2,500–5,000	Hammer et al. (1998)
Candida parapsilosis	Yeast	Cited data	300–5,000	Carson et al. (2006)
Candida parapsilosis	Yeast	RPMI, Tween 80, 48 h, 30°C	600–1,250	Oliva et al. (2003)
Candida sp.	Yeast	MHB, Tween 80, 48 h, 35°C	1,200–5,000	Hammer et al. (1998)
Candida sp.	Yeast	MHB, Tween 80, 24 h, 37°C	5,000	Banes-Marshall et al. (2001)
Candida tropicalis	Yeast	Cited data	1,200–20,000	Carson et al. (2006)
Candida tropicalis	Yeast	RPMI, Tween 80, 48 h, 35°C	600	Mondello et al. (2006)
Cryptococcus neoformans	Yeast	Cited data	150–600	Carson et al. (2006)
Cryptococcus neoformans	Yeast	RPMI, Tween 80, 48 h, 35°C	300	Mondello et al. (2006)
Malassezia furfur	Yeast	MMA, Tween 20, 7 days	1,200–2,500	Hammer et al. (2000)
Malassezia furfur	Yeast	Cited data	300–1,200	Carson et al. (2006)
Malassezia globosa	Yeast	MMA, Tween 20, 7 days	300–1,200	Hammer et al. (2000)
Malassezia obtusa	Yeast	MMA, Tween 20, 7 days	1,200	Hammer et al. (2000)
Malassezia slooffiae	Yeast	MMA, Tween 20, 7 days	1,200–2,500	Hammer et al. (2000)
Malassezia sympodialis	Yeast	MMA, Tween 20, 7 days	160–2,500	Hammer et al. (2000)
Rhodotorula rubra	Yeast	Cited data	600	Carson et al. (2006)
Saccharomyces cerevisiae	Yeast	MEB, Tween 20, 24 h, 37°C	2,000	Griffin et al. (2000)
Saccharomyces cerevisiae	Yeast	Cited data	2,500	Carson et al. (2006)
Saccharomyces cerevisiae	Yeast	YPB, 24 h, 20°C	2,800	Schelz et al. (2006)
Saccharomyces cerevisiae	Yeast	MPB, DMSO, 40 h, 30°C	69% inh. 500	Hili et al. (1997)
Schizosaccharomyces pombe	Yeast	MPB, DMSO, 40 h, 30°C	74% inh. 500	Hili et al. (1997)
Torula utilis	Yeast	MPB, DMSO, 40 h, 30°C	33% inh. 500	Hili et al. (1997)
Trichophyton tonsurans	Yeast	Cited data	1,200–2,200	Carson et al. (2006)

TABLE 14.77
Inhibitory Data of Tea Tree Oil Obtained in the Vapor Phase Test

Microorganism	MO Class	Conditions		Activity	Reference
Escherichia coli	Bac−	BLA, 18 h, 37°C	MIC_{air}	50	Inouye et al. (2001)
Haemophilus influenzae	Bac−	MHA, 18 h, 37°C	MIC_{air}	25	Inouye et al. (2001)
Salmonella typhi	Bac−	NA, 24 h, 37°C	sd	+++	Maruzzella and Sicurella (1960)
Bacillus subtilis var. aterrimus	Bac+	NA, 24 h, 37°C	sd	+++	Maruzzella and Sicurella (1960)
Mycobacterium avium	Bac+	NA, 24 h, 37°C	sd	NG	Maruzzella and Sicurella (1960)
Staphylococcus aureus	Bac+	NA, 24 h, 37°C	sd	+++	Maruzzella and Sicurella (1960)
Staphylococcus aureus	Bac+	MHA, 18 h, 37°C	MIC_{air}	50	Inouye et al. (2001)
Streptococcus faecalis	Bac+	NA, 24 h, 37°C	sd	+++	Maruzzella and Sicurella (1960)
Streptococcus pneumoniae	Bac+	MHA, 18 h, 37°C	MIC_{air}	50	Inouye et al. (2001)
Streptococcus pyogenes	Bac+	MHA, 18 h, 37°C	MIC_{air}	50	Inouye et al. (2001)
Botrytis cinerea	Fungi	PDA, 3 days, 25°C	1,000	+++	Lee et al. (2007)
Colletotrichum gloeosporioides	Fungi	PDA, 3 days, 25°C	1,000	+++	Lee et al. (2007)
Fusarium oxysporum	Fungi	PDA, 3 days, 25°C	1,000	+++	Lee et al. (2007)
Pythium ultimum	Fungi	PDA, 3 days, 25°C	1,000	+++	Lee et al. (2007)
Rhizoctonia solani	Fungi	PDA, 3 days, 25°C	1,000	+++	Lee et al. (2007)

TABLE 14.78
Inhibitory Data of Thyme Oil Obtained in the Agar Diffusion Test

Microorganism	MO Class	Conditions		Inhibition Zone (mm)	Reference
Acinetobacter calcoaceticus	Bac–	ISA, 48 h, 25°C	4 (h), 10,000	19	Deans and Ritchie (1987)
Aerobacter aerogenes	Bac–	NA, 24 h, 37°C	—, sd	0	Maruzzella and Lichtenstein (1956)
Aeromonas hydrophila	Bac–	ISA, 48 h, 25°C	4 (h), 10,000	22.5	Deans and Ritchie (1987)
Alcaligenes faecalis	Bac–	ISA, 48 h, 25°C	4 (h), 10,000	20	Deans and Ritchie (1987)
Beneckea natriegens	Bac–	ISA, 48 h, 25°C	4 (h), 10,000	20	Deans and Ritchie (1987)
Campylobacter jejuni	Bac–	TSA, 24 h, 42°C	4 (h), 25,000	10.4	Smith-Palmer et al. (1998)
Citrobacter freundii	Bac–	ISA, 48 h, 25°C	4 (h), 10,000	21.5	Deans and Ritchie (1987)
Enterobacter aerogenes	Bac–	ISA, 48 h, 25°C	4 (h), 10,000	25.5	Deans and Ritchie (1987)
Erwinia carotovora	Bac–	ISA, 48 h, 25°C	4 (h), 10,000	21.5	Deans and Ritchie (1987)
Escherichia coli	Bac–	NA, 24 h, 37°C	—, sd	4	Maruzzella and Lichtenstein (1956)
Escherichia coli	Bac–	TSA, 24 h, 35°C	4 (h), 25,000	8.3	Smith-Palmer et al. (1998)
Escherichia coli	Bac–	NA, 24 h, 37°C	8 (h), pure	16	Fawzi (1991)
Escherichia coli	Bac–	Cited	15, 2,500	21	Pizsolitto et al. (1975)
Escherichia coli	Bac–	Cited, 18 h, 37°C	6, 2,500	19.3	Janssen et al. (1986)
Escherichia coli	Bac–	NA, 18 h, 37°C	5 (h), –30,000	26	Schelz et al. (2006)
Escherichia coli	Bac–	TGA, 18–24 h, 37°C	9.5, 2,000	27	Morris et al. (1979)
Escherichia coli	Bac–	ISA, 48 h, 25°C	4 (h), 10,000	22.5	Deans and Ritchie (1987)
Escherichia coli	Bac–	NA, 24 h, 30°C	Drop, 5,000	41	Hili et al. (1997)
Flavobacterium suaveolens	Bac–	ISA, 48 h, 25°C	4 (h), 10,000	38	Deans and Ritchie (1987)
Klebsiella pneumoniae	Bac–	ISA, 48 h, 25°C	4 (h), 10,000	19	Deans and Ritchie (1987)
Klebsiella sp.	Bac–	Cited	15, 2,500	12	Pizsolitto et al. (1975)
Moraxella sp.	Bac–	ISA, 48 h, 25°C	4 (h), 10,000	24	Deans and Ritchie (1987)
Neisseria perflava	Bac–	NA, 24 h, 37°C	—, sd	3	Maruzzella and Lichtenstein (1956)
Proteus sp.	Bac–	Cited	15, 2,500	16	Pizsolitto et al. (1975)
Proteus vulgaris	Bac–	NA, 24 h, 37°C	—, sd	0	Maruzzella and Lichtenstein (1956)
Proteus vulgaris	Bac–	ISA, 48 h, 25°C	4 (h), 10,000	20	Deans and Ritchie (1987)
Pseudomonas aeruginosa	Bac–	Cited	15, 2,500	0	Pizsolitto et al. (1975)
Pseudomonas aeruginosa	Bac–	NA, 24 h, 37°C	—, sd	0	Maruzzella and Lichtenstein (1956)
Pseudomonas aeruginosa	Bac–	NA, 24 h, 37°C	8 (h), pure	8	Fawzi (1991)
Pseudomonas aeruginosa	Bac–	NA, 24 h, 30°C	Drop, 5,000	14	Hili et al. (1997)
Pseudomonas aeruginosa	Bac–	Cited, 18 h, 37°C	6, 2,500	8.7	Janssen et al. (1986)

(*Continued*)

TABLE 14.78 (*Continued*)
Inhibitory Data of Thyme Oil Obtained in the Agar Diffusion Test

Microorganism	MO Class	Conditions		Inhibition Zone (mm)	Reference
Pseudomonas aeruginosa	Bac−	ISA, 48 h, 25°C	4 (h), 10,000	22.5	Deans and Ritchie (1987)
Salmonella enteritidis	Bac−	TSA, 24 h, 35°C	4 (h), 25,000	11.1	Smith-Palmer et al. (1998)
Salmonella pullorum	Bac−	ISA, 48 h, 25°C	4 (h), 10,000	26	Deans and Ritchie (1987)
Salmonella sp.	Bac−	Cited	15, 2,500	10	Pizsolitto et al. (1975)
Serratia marcescens	Bac−	NA, 24 h, 37°C	—, sd	4	Maruzzella and Lichtenstein (1956)
Serratia marcescens	Bac−	ISA, 48 h, 25°C	4 (h), 10,000	20.5	Deans and Ritchie (1987)
Serratia sp.	Bac−	Cited	15, 2,500	5	Pizsolitto et al. (1975)
Shigella sp.	Bac−	Cited	15, 2,500	8	Pizsolitto et al. (1975)
Yersinia enterocolitica	Bac−	ISA, 48 h, 25°C	4 (h), 10,000	23	Deans and Ritchie (1987)
Bacillus cereus	Bac+	NA, 24 h, 37°C	8 (h), pure	21	Fawzi (1991)
Bacillus mesentericus	Bac+	NA, 24 h, 37°C	—, sd	0	Maruzzella and Lichtenstein (1956)
Bacillus sp.	Bac+	Cited	15, 2,500	26	Pizsolitto et al. (1975)
Bacillus subtilis	Bac+	ISA, 48 h, 25°C	4 (h), 10,000	12.5	Deans and Ritchie (1987)
Bacillus subtilis	Bac+	NA, 24 h, 37°C	—, sd	25	Maruzzella and Lichtenstein (1956)
Bacillus subtilis	Bac+	Cited, 18 h, 37°C	6, 2,500	33.2	Janssen et al. (1986)
Brevibacterium linens	Bac+	ISA, 48 h, 25°C	4 (h), 10,000	24.5	Deans and Ritchie (1987)
Brochothrix thermosphacta	Bac+	ISA, 48 h, 25°C	4 (h), 10,000	16	Deans and Ritchie (1987)
Clostridium sporogenes	Bac+	ISA, 48 h, 25°C	4 (h), 10,000	0	Deans and Ritchie (1987)
Corynebacterium sp.	Bac+	TGA, 18–24 h, 37°C	9.5, 2,000	16	Morris et al. (1979)
Lactobacillus plantarum	Bac+	ISA, 48 h, 25°C	4 (h), 10,000	18	Deans and Ritchie (1987)
Lactobacillus sp.	Bac+	MRS, cited	9, 20,000	25 to >90	Pellecuer et al. (1980)
Leuconostoc cremoris	Bac+	ISA, 48 h, 25°C	4 (h), 10,000	0	Deans and Ritchie (1987)
Listeria monocytogenes	Bac+	ISA, 48 h, 25°C	4 (h), 10,000	6–20	Lis-Balchin et al. (1998)
Listeria monocytogenes	Bac+	TSA, 24 h, 35°C	4 (h), 25,000	10	Smith-Palmer et al. (1998)
Micrococcus luteus	Bac+	ISA, 48 h, 25°C	4 (h), 10,000	32	Deans and Ritchie (1987)
Micrococcus luteus	Bac+	MHA, cited	9, 20,000	40	Pellecuer et al. (1980)
Micrococcus ureae	Bac+	MHA, cited	9, 20,000	40	Pellecuer et al. (1980)
Sarcina lutea	Bac+	NA, 24 h, 37°C	—, sd	0	Maruzzella and Lichtenstein (1956)
Sarcina ureae	Bac+	MHA, cited	9, 20,000	>90	Pellecuer et al. (1980)
Staphylococcus aureus	Bac+	NA, 24 h, 37°C	—, sd	0	Maruzzella and Lichtenstein (1956)
Staphylococcus aureus	Bac+	TSA, 24 h, 35°C	4 (h), 25,000	8.5	Smith-Palmer et al. (1998)
Staphylococcus aureus	Bac+	Cited	15, 2,500	14	Pizsolitto et al. (1975)

(*Continued*)

TABLE 14.78 (*Continued*)
Inhibitory Data of Thyme Oil Obtained in the Agar Diffusion Test

Microorganism	MO Class	Conditions		Inhibition Zone (mm)	Reference
Staphylococcus aureus	Bac+	NA, 24 h, 37°C	8 (h), pure	19	Fawzi (1991)
Staphylococcus aureus	Bac+	NA, 18 h, 37°C	5 (h), –30,000	24	Schelz et al. (2006)
Staphylococcus aureus	Bac+	TGA, 18–24 h, 37°C	9.5, 2,000	25	Morris et al. (1979)
Staphylococcus aureus	Bac+	ISA, 48 h, 25°C	4 (h), 10,000	29	Deans and Ritchie (1987)
Staphylococcus aureus	Bac+	Cited, 18 h, 37°C	6, 2,500	33.7	Janssen et al. (1986)
Staphylococcus aureus	Bac+	NA, 24 h, 30°C	Drop, 5,000	51	Hili et al. (1997)
Staphylococcus epidermidis	Bac+	Cited	15, 2,500	16	Pizsolitto et al. (1975)
Staphylococcus epidermidis	Bac+	MHA, cited	9, 20,000	40	Pellecuer et al. (1980)
Streptococcus D	Bac+	MHA, cited	9, 20,000	17 to >90	Pellecuer et al. (1980)
Streptococcus faecalis	Bac+	ISA, 48 h, 25°C	4 (h), 10,000	19.5	Deans and Ritchie (1987)
Streptococcus micros	Bac+	MHA, cited	9, 20,000	>90	Pellecuer et al. (1980)
Streptococcus viridans	Bac+	Cited	15, 2,500	3	Pizsolitto et al. (1975)
Streptomyces venezuelae	Bac+	SMA, 2–7 days, 20°C	sd	21	Maruzzella and Liguori (1958)
Alternaria porri	Fungi	PDA, 72 h, 28°C	5, 5,000	26.6	Pawar and Thaker (2007)
Alternaria solani	Fungi	SMA, 2–7 days, 20°C	sd	22	Maruzzella and Liguori (1958)
Aspergillus fumigatus	Fungi	SMA, 2–7 days, 20°C	sd	21	Maruzzella and Liguori (1958)
Aspergillus niger	Fungi	PDA, 48 h, 28°C	5, 5,000	12	Pawar and Thaker (2006)
Aspergillus niger	Fungi	SMA, 2–7 days, 20°C	sd	22	Maruzzella and Liguori (1958)
Aspergillus niger	Fungi	SDA, 3 days, 30	8 (h), pure	35	Fawzi (1991)
Fusarium oxysporum f.sp. *ciceri*	Fungi	PDA, 72 h, 28°C	5, 5,000	10.5	Pawar and Thaker (2007)
Helminthosporium sativum	Fungi	SMA, 2–7 days, 20°C	sd	20	Maruzzella and Liguori (1958)
Mucor mucedo	Fungi	SMA, 2–7 days, 20°C	sd	22	Maruzzella and Liguori (1958)
Nigrospora panici	Fungi	SMA, 2–7 days, 20°C	sd	16	Maruzzella and Liguori (1958)
Penicillium digitatum	Fungi	SMA, 2–7 days, 20°C	sd	17	Maruzzella and Liguori (1958)
Rhizopus nigricans	Fungi	SMA, 2–7 days, 20°C	sd	14	Maruzzella and Liguori (1958)
Brettanomyces anomalus	Yeast	MPA, 4 days, 30°C	5, 10% sol. sd	31	Conner and Beuchat (1984)

(*Continued*)

TABLE 14.78 (*Continued*)
Inhibitory Data of Thyme Oil Obtained in the Agar Diffusion Test

Microorganism	MO Class	Conditions		Inhibition Zone (mm)	Reference
Candida albicans	Yeast	SMA, 2–7 days, 20°C	sd	14	Maruzzella and Liguori (1958)
Candida albicans	Yeast	TGA, 18–24 h, 37°C	9.5, 2,000	14	Morris et al. (1979)
Candida albicans	Yeast	SDA, 24 h, 37°C	8 (h), pure	37	Fawzi (1991)
Candida albicans	Yeast	Cited, 18 h, 37°C	6, 2,500	40.7	Janssen et al. (1986)
Candida albicans	Yeast	NA, 24 h, 30°C	Drop, 5,000	61	Hili et al. (1997)
Candida krusei	Yeast	SMA, 2–7 days, 20°C	sd	13	Maruzzella and Liguori (1958)
Candida lipolytica	Yeast	MPA, 4 days, 30°C	5, 10% sol. sd	18	Conner and Beuchat (1984)
Candida tropicalis	Yeast	SMA, 2–7 days, 20°C	sd	12	Maruzzella and Liguori (1958)
Cryptococcus neoformans	Yeast	SMA, 2–7 days, 20°C	sd	25	Maruzzella and Liguori (1958)
Cryptococcus rhodobenhani	Yeast	SMA, 2–7 days, 20°C	sd	14	Maruzzella and Liguori (1958)
Debaryomyces hansenii	Yeast	MPA, 4 days, 30°C	5, 10% sol. sd	15	Conner and Beuchat (1984)
Geotrichum candidum	Yeast	MPA, 4 days, 30°C	5, 10% sol. sd	34	Conner and Beuchat (1984)
Hansenula anomala	Yeast	MPA, 4 days, 30°C	5, 10% sol. sd	18	Conner and Beuchat (1984)
Kloeckera apiculata	Yeast	MPA, 4 days, 30°C	5, 10% sol. sd	19	Conner and Beuchat (1984)
Kluyveromyces fragilis	Yeast	MPA, 4 days, 30°C	5, 10% sol. sd	17	Conner and Beuchat (1984)
Lodderomyces elongisporus	Yeast	MPA, 4 days, 30°C	5, 10% sol. sd	16	Conner and Beuchat (1984)
Metschnikowia pulcherrima	Yeast	MPA, 4 days, 30°C	5, 10% sol. sd	38	Conner and Beuchat (1984)
Pichia membranifaciens	Yeast	MPA, 4 days, 30°C	5, 10% sol. sd	34	Conner and Beuchat (1984)
Rhodotorula rubra	Yeast	MPA, 4 days, 30°C	5, 10% sol. sd	21	Conner and Beuchat (1984)
Saccharomyces cerevisiae	Yeast	NA, 24 h, 20°C	5 (h), –30,000	23–25	Schelz et al. (2006)
Saccharomyces cerevisiae	Yeast	SMA, 2–7 days, 20°C	sd	16	Maruzzella and Liguori (1958)
Saccharomyces cerevisiae	Yeast	MPA, 4 days, 30°C	5, 10% sol. sd	25	Conner and Beuchat (1984)
Saccharomyces cerevisiae	Yeast	NA, 24 h, 30°C	Drop, 5,000	80	Hili et al. (1997)
Schizosaccharomyces pombe	Yeast	NA, 24 h, 30°C	Drop, 5,000	69	Hili et al. (1997)
Torula glabrata	Yeast	MPA, 4 days, 30°C	5, 10% sol. sd	15	Conner and Beuchat (1984)
Torula utilis	Yeast	NA, 24 h, 30°C	Drop, 5,000	67	Hili et al. (1997)

TABLE 14.79
Inhibitory Data of Thyme Oil Obtained in the Dilution Test

Microorganism	MO Class	Conditions	MIC	Reference
Bacteria	Bac	5% Glucose, 9 days, 37	500–1,000	Buchholtz (1875)
Acinetobacter baumannii	Bac–	MHA, Tween 20, 48 h, 35°C	1,200	Hammer et al. (1999)
Aeromonas sobria	Bac–	MHA, Tween 20, 48 h, 35°C	1,200	Hammer et al. (1999)
Bordetella bronchiseptica	Bac–	Cited	250	Pellecuer et al. (1976)
Campylobacter jejuni	Bac–	TSB, 24 h, 42°C	400	Smith-Palmer et al. (1998)
Enterococcus faecalis	Bac–	M17, 24 h, 20°C	>10,000	Di Pasqua et al. (2005)
Escherichia coli	Bac–	MHA, Tween 20, 48 h, 35°C	1,200	Hammer et al. (1999)
Escherichia coli	Bac–	TGB, 18 h, 37°C	1,500	Schelz et al. (2006)
Escherichia coli	Bac–	TSB, Tween 80, 48 h, 35°C	2	Panizzi et al. (1993)
Escherichia coli	Bac–	NB, 16 h, 37°C	200	Lens-Lisbonne et al. (1987)
Escherichia coli	Bac–	TSB, 3% EtOH, 24 h, 37°C	3,000–8,000	Rota et al. (2004)
Escherichia coli	Bac–	LA, 18 h, 37°C	375–500	Remmal et al. (1993)
Escherichia coli	Bac–	NB, 24 h, Tween 20, 37	400	Fawzi (1991)
Escherichia coli	Bac–	TGB, 18–24 h, 37°C	500	Morris et al. (1979)
Escherichia coli	Bac–	Cited	500	Pellecuer et al. (1976)
Escherichia coli	Bac–	TSB, 24 h, 35°C	500	Smith-Palmer et al. (1998)
Escherichia coli	Bac–	MHB, DMSO, 24 h, 37°C	62.5	Al-Bayati (2008)
Escherichia coli	Bac–	TSB, 24 h, 37°C	700	Di Pasqua et al. (2005)
Escherichia coli	Bac–	MPB, DMSO, 40 h, 30°C	75% inh. 500	Hili et al. (1997)
Escherichia coli	Bac–	NA, 1–3 days, 30°C	750	Farag et al. (1989)
Escherichia coli	Bac–	MHB, 24 h, 36°C	>10,000	Oussalah et al. (2006)
Escherichia coli O157:H7	Bac–	BHI, 48 h, 35°C	>8,000	Oussalah et al. (2006)
Escherichia coli O157:H7	Bac–	BHI, 48 h, 35°C	500	Oussalah et al. (2006)
Escherichia coli O157:H7	Bac–	BHI, 48 h, 35°C	500	Oussalah et al. (2006)
Escherichia coli O157:H7	Bac–	BHI, 48 h, 35°C	8,000	Oussalah et al. (2006)
Haemophilus influenza	Bac–	Cited	1,000	Pellecuer et al. (1976)
Helicobacter pylori	Bac–	Cited, 20 h, 37°C	275.2	Weseler et al. (2005)
Klebsiella pneumoniae	Bac–	MHA, Tween 20, 48 h, 35°C	2,500	Hammer et al. (1999)
Klebsiella pneumoniae	Bac–	MHB, DMSO, 24 h, 37°C	500	Al-Bayati (2008)
Klebsiella pneumoniae	Bac–	Cited	5,000	Pellecuer et al. (1976)
Moraxella glucidolytica	Bac–	Cited	1,000	Pellecuer et al. (1976)
Neisseria catarrhalis	Bac–	Cited	125	Pellecuer et al. (1976)
Neisseria flava	Bac–	Cited	500	Pellecuer et al. (1976)
Proteus mirabilis	Bac–	MHB, DMSO, 24 h, 37°C	62.5	Al-Bayati (2008)
Proteus vulgaris	Bac–	MHB, DMSO, 24 h, 37°C	31.2	Al-Bayati (2008)
Pseudomonas aeruginosa	Bac–	MHA, Tween 20, 48 h, 35°C	>20,000	Hammer et al. (1999)
Pseudomonas aeruginosa	Bac–	TSB, Tween 80, 48 h, 35°C	>40	Panizzi et al. (1993)
Pseudomonas aeruginosa	Bac–	MHB, DMSO, 24 h, 37°C	>500	Al-Bayati (2008)
Pseudomonas aeruginosa	Bac–	NB, 24 h, Tween 20, 37°C	>50,000	Fawzi (1991)
Pseudomonas aeruginosa	Bac–	MPB, DMSO, 40 h, 30°C	77% inh. 500	Hili et al. (1997)
Pseudomonas aeruginosa	Bac–	NB, 16 h, 37°C	800	Lens-Lisbonne et al. (1987)
Pseudomonas fluorescens	Bac–	NA, 1–3 days, 30°C	1,000	Farag et al. (1989)

(Continued)

TABLE 14.79 (*Continued*)
Inhibitory Data of Thyme Oil Obtained in the Dilution Test

Microorganism	MO Class	Conditions	MIC	Reference
Pseudomonas sp.	Bac−	TSB, 24 h, 37°C	1,500	Di Pasqua et al. (2005)
Salmonella enteritidis	Bac−	TSB, 24 h, 35°C	400	Smith-Palmer et al. (1998)
Salmonella enteritidis	Bac−	TSB, 3% EtOH, 24 h, 37°C	4,000–8,000	Rota et al. (2004)
Salmonella hadar	Bac−	LA, 18 h, 37°C	500	Remmal et al. (1993)
Salmonella typhi	Bac−	MHB, DMSO, 24 h, 37°C	250	Al-Bayati (2008)
Salmonella typhimurium	Bac−	MHA, Tween 20, 48 h, 35°C	>20,000	Hammer et al. (1999)
Salmonella typhimurium	Bac−	BHI, 48 h, 35°C	1,000	Oussalah et al. (2006)
Salmonella typhimurium	Bac−	MHB, DMSO, 24 h, 37°C	125	Al-Bayati (2008)
Salmonella typhimurium	Bac−	BHI, 48 h, 35°C	2,000	Oussalah et al. (2006)
Salmonella typhimurium	Bac−	TSB, 24 h, 37°C	300	Di Pasqua et al. (2005)
Salmonella typhimurium	Bac−	BHI, 48 h, 35°C	4,000	Oussalah et al. (2006)
Salmonella typhimurium	Bac−	BHI, 48 h, 35°C	500	Oussalah et al. (2006)
Salmonella typhimurium	Bac−	TSB, 3% EtOH, 24 h, 37°C	5,000	Rota et al. (2004)
Serratia marcescens	Bac−	NA, 1–3 days, 30°C	1,250	Farag et al. (1989)
Serratia marcescens	Bac−	MHA, Tween 20, 48 h, 35°C	2,500	Hammer et al. (1999)
Shigella flexneri	Bac−	TSB, 3% EtOH, 24 h, 37°C	3,000–8,000	Rota et al. (2004)
Yersinia enterocolitica	Bac−	TSB, 3% EtOH, 24 h, 29°C	3,000–5,000	Rota et al. (2004)
Bacillus cereus	Bac+	NB, 24 h, Tween 20, 37°C	>50,000	Fawzi (1991)
Bacillus cereus	Bac+	MHB, DMSO, 24 h, 37°C	15.6	Al-Bayati (2008)
Bacillus megaterium	Bac+	LA, 18 h, 37°C	375–500	Remmal et al. (1993)
Bacillus subtilis	Bac+	TSB, Tween 80, 48 h, 35°C	2	Panizzi et al. (1993)
Bacillus subtilis	Bac+	NA, 1–3 days, 30°C	250	Farag et al. (1989)
Bacillus subtilis	Bac+	Cited	500	Pellecuer et al. (1976)
Brochothrix thermosphacta	Bac+	M17, 24 h, 20°C	2,500	Di Pasqua et al. (2005)
Corynebacterium pseudodiphtheriae	Bac+	Cited	125	Pellecuer et al. (1976)
Corynebacterium sp.	Bac+	TGB, 18–24 h, 37°C	500	Morris et al. (1979)
Enterococcus faecalis	Bac+	MHA, Tween 20, 48 h, 35°C	5,000	Hammer et al. (1999)
Enterococcus faecium VRE	Bac+	HIB, Tween 80, 18 h, 37°C	500–10,000	Nelson (1997)
Lactobacillus delbrueckii	Bac+	MRS, 24 h, 37°C	>10,000	Di Pasqua et al. (2005)
Lactobacillus plantarum	Bac+	MRS, 24 h, 30°C	>10,000	Di Pasqua et al. (2005)
Lactobacillus sp.	Bac+	MRS, cited	310–620	Pellecuer et al. (1980)
Lactococcus garvieae	Bac+	M17, 24 h, 20°C	>10,000	Di Pasqua et al. (2005)
Lactococcus lactis subsp. *lactis*	Bac+	M17, 24 h, 20°C	>10,000	Di Pasqua et al. (2005)
Listeria monocytogenes	Bac+	TSB, 3% EtOH, 24 h, 37°C	<1,000–5,000	Rota et al. (2004)
Listeria monocytogenes	Bac+	BHI, 48 h, 35°C	>8,000	Oussalah et al. (2006)
Listeria monocytogenes	Bac+	BHI, 48 h, 35°C	>8,000	Oussalah et al. (2006)
Listeria monocytogenes	Bac+	TSB, 24 h, 37°C	1,000	Di Pasqua et al. (2005)
Listeria monocytogenes	Bac+	BHI, 48 h, 35°C	1,000	Oussalah et al. (2006)
Listeria monocytogenes	Bac+	TSB, 24 h, 35°C	200	Smith-Palmer et al. (1998)
Listeria monocytogenes	Bac+	TSB, 10 days, 4°C	200	Smith-Palmer et al. (1998)
Listeria monocytogenes	Bac+	BHI, 48 h, 35°C	2,000	Oussalah et al. (2006)

(*Continued*)

TABLE 14.79 (*Continued*)
Inhibitory Data of Thyme Oil Obtained in the Dilution Test

Microorganism	MO Class	Conditions	MIC	Reference
Micrococcus flavus	Bac+	Cited	500	Pellecuer et al. (1976)
Micrococcus luteus	Bac+	MHB, cited	150	Pellecuer et al. (1980)
Micrococcus sp.	Bac+	NA, 1–3 days, 30°C	250	Farag et al. (1989)
Micrococcus ureae	Bac+	MHB, cited	150	Pellecuer et al. (1980)
Mycobacterium phlei	Bac+	NA, 1–3 days, 30°C	125	Farag et al. (1989)
Sarcina lutea	Bac+	Cited	250	Pellecuer et al. (1976)
Sarcina sp.	Bac+	NA, 1–3 days, 30°C	125	Farag et al. (1989)
Sarcina ureae	Bac+	MHB, cited	78	Pellecuer et al. (1980)
Staphylococcus aureus	Bac+	TSB, 3% EtOH, 24 h, 37°C	<1,000–5,000	Rota et al. (2004)
Staphylococcus aureus	Bac+	Cited	1,000	Pellecuer et al. (1976)
Staphylococcus aureus	Bac+	BHI, 48 h, 35°C	1,000	Oussalah et al. (2006)
Staphylococcus aureus	Bac+	NB, 16 h, 37°C	150	Lens-Lisbonne et al. (1987)
Staphylococcus aureus	Bac+	TSB, 24 h, 37°C	1,700	Di Pasqua et al. (2005)
Staphylococcus aureus	Bac+	TSB, 24 h, 35°C	200	Smith-Palmer et al. (1998)
Staphylococcus aureus	Bac+	BHI, 48 h, 35°C	250	Oussalah et al. (2006)
Staphylococcus aureus	Bac+	BHI, 48 h, 35°C	250	Oussalah et al. (2006)
Staphylococcus aureus	Bac+	MHA, Tween 20, 48 h, 35°C	2,500	Hammer et al. (1999)
Staphylococcus aureus	Bac+	MHB, DMSO, 24 h, 37°C	31.2	Al-Bayati (2008)
Staphylococcus aureus	Bac+	NB, 24 h, Tween 20, 37°C	400	Fawzi (1991)
Staphylococcus aureus	Bac+	BHI, 48 h, 35°C	4,000	Oussalah et al. (2006)
Staphylococcus aureus	Bac+	TSB, Tween 80, 48 h, 35°C	5	Panizzi et al. (1993)
Staphylococcus aureus	Bac+	TGB, 18–24 h, 37°C	500	Morris et al. (1979)
Staphylococcus aureus	Bac+	NA, 1–3 days, 30°C	500	Farag et al. (1989)
Staphylococcus aureus	Bac+	LA, 18 h, 37°C	500	Remmal et al. (1993)
Staphylococcus aureus	Bac+	MPB, DMSO, 40 h, 30°C	95% inh. 500	Hili et al. (1997)
Staphylococcus aureus MRSA	Bac+	HIB, Tween 80, 18 h, 37°C	5,000	Nelson (1997)
Staphylococcus epidermidis	Bac+	TGB, 18 h, 37°C	1,500	Schelz et al. (2006)
Staphylococcus epidermidis	Bac+	Cited	500	Pellecuer et al. (1976)
Staphylococcus epidermidis	Bac+	MHB, cited	620	Pellecuer et al. (1980)
Streptococcus D	Bac+	MHB, cited	78–620	Pellecuer et al. (1980)
Streptococcus faecalis	Bac+	NB, 16 h, 37°C	150	Lens-Lisbonne et al. (1987)
Streptococcus pyogenes	Bac+	Cited	250	Pellecuer et al. (1976)
Absidia glauca	Fungi	Cited	500	Pellecuer et al. (1976)
Alternaria alternata	Fungi	RPMI, 1.5% EtOH, 7 days, 30°C	5,000	Tullio et al. (2006)
Alternaria alternata	Fungi	PDA, 7 days, 28°C	62% inh. 500	Feng and Zheng (2007)
Alternaria citri	Fungi	PDA, 8 days, 22°C	500	Arras and Usai (2001)
Aspergillus chevalieri	Fungi	Cited	250	Pellecuer et al. (1976)
Aspergillus clavatus	Fungi	Cited	1,000	Pellecuer et al. (1976)
Aspergillus flavus	Fungi	Cited	1,000	Pellecuer et al. (1976)
Aspergillus flavus	Fungi	PDA, 7–14 days, 28°C	250	Soliman and Badeaa (2002)
Aspergillus flavus	Fungi	RPMI, 1.5% EtOH, 7 days, 30°C	2,500	Tullio et al. (2006)

(*Continued*)

TABLE 14.79 (*Continued*)
Inhibitory Data of Thyme Oil Obtained in the Dilution Test

Microorganism	MO Class	Conditions	MIC	Reference
Aspergillus flavus	Fungi	PDA, 5 days, 27°C	500	Thompson and Cannon (1986)
Aspergillus flavus	Fungi	PDA, 8 h, 20°C, spore germ. inh.	50–100	Thompson (1986)
Aspergillus flavus	Fungi	CA, 7 days, 28°C	74%–76% inh. 500	Kumar et al. (2007)
Aspergillus flavus var. *columnaris*	Fungi	RPMI, 1.5% EtOH, 7 days, 30°C	2,500	Tullio et al. (2006)
Aspergillus fumigatus	Fungi	RPMI, 1.5% EtOH, 7 days, 30°C	2,500	Tullio et al. (2006)
Aspergillus giganteus	Fungi	Cited	1,000	Pellecuer et al. (1976)
Aspergillus niger	Fungi	Cited	1,000	Pellecuer et al. (1976)
Aspergillus niger	Fungi	SB, 72 h, Tween 20, 37°C	200	Fawzi (1991)
Aspergillus niger	Fungi	RPMI, 1.5% EtOH, 7 days, 30°C	2,500–5,000	Tullio et al. (2006)
Aspergillus niger	Fungi	YES broth, 10 days	−96% inh. 10,000	Lis-Balchin et al. (1998)
Aspergillus ochraceus	Fungi	PDA, 7–14 days, 28°C	500	Soliman and Badeaa (2002)
Aspergillus ochraceus	Fungi	YES broth, 10 days	−92% inh. 10,000	Lis-Balchin et al. (1998)
Aspergillus oryzae	Fungi	Cited	2,000	Pellecuer et al. (1976)
Aspergillus parasiticus	Fungi	PDA, 5 days, 27°C	500	Thompson and Cannon (1986)
Aspergillus parasiticus	Fungi	PDA, 7–14 days, 28°C	500	Soliman and Badeaa (2002)
Aspergillus parasiticus	Fungi	PDA, 8 h, 20°C, spore germ. inh.	50–100	Thompson (1986)
Aspergillus repens	Fungi	Cited	250	Pellecuer et al. (1976)
Botrytis cinerea	Fungi	PDA, 8 days, 22°C	500	Arras and Usai (2001)
Cladosporium cladosporioides	Fungi	RPMI, 1.5% EtOH, 7 days, 30°C	1,250–2,500	Tullio et al. (2006)
Cladosporium herbarum	Fungi	Cited	1,000	Pellecuer et al. (1976)
Epidermophyton floccosum	Fungi	SA, Tween 80, 21 days, 20°C	<300	Janssen et al. (1988)
Epidermophyton floccosum	Fungi	Cited	1,000	Pellecuer et al. (1976)
Epidermophyton floccosum	Fungi	RPMI, 1.5% EtOH, 7 days, 30°C	625	Tullio et al. (2006)
Fusarium culmorum	Fungi	YES broth, 10 days	−86% inh. 10,000	Lis-Balchin et al. (1998)
Fusarium moniliforme	Fungi	PDA, 7–14 days, 28°C	250	Soliman and Badeaa (2002)
Fusarium oxysporum	Fungi	RPMI, 1.5% EtOH, 7 days, 30°C	1,250	Tullio et al. (2006)
Microsporum canis	Fungi	MBA, Tween 80, 10 days, 30°C	12.5 to >300	Perrucci et al. (1994)
Microsporum canis	Fungi	RPMI, 1.5% EtOH, 7 days, 30°C	1,250–2,500	Tullio et al. (2006)
Microsporum gypseum	Fungi	RPMI, 1.5% EtOH, 7 days, 30°C	1,250–2,500	Tullio et al. (2006)

(*Continued*)

TABLE 14.79 (*Continued*)
Inhibitory Data of Thyme Oil Obtained in the Dilution Test

Microorganism	MO Class	Conditions	MIC	Reference
Microsporum gypseum	Fungi	MBA, Tween 80, 10 days, 30°C	25 to >300	Perrucci et al. (1994)
Mucor hiemalis	Fungi	PDA, 5 days, 27°C	500	Thompson and Cannon (1986)
Mucor mucedo	Fungi	PDA, 5 days, 27°C	100	Thompson and Cannon (1986)
Mucor mucedo	Fungi	Cited	1,000	Pellecuer et al. (1976)
Mucor racemosus f. *racemosus*	Fungi	PDA, 5 days, 27°C	500	Thompson and Cannon (1986)
Mucor sp.	Fungi	RPMI, 1.5% EtOH, 7 days, 30°C	>10,000	Tullio et al. (2006)
Penicillium chrysogenum	Fungi	Cited	1,000	Pellecuer et al. (1976)
Penicillium digitatum	Fungi	PDA, 8 days, 22°C	1,000	Arras and Usai (2001)
Penicillium frequentans	Fungi	RPMI, 1.5% EtOH, 7 days, 30°C	2,500	Tullio et al. (2006)
Penicillium italicum	Fungi	PDA, 8 days, 22°C	1,000	Arras and Usai (2001)
Penicillium lanosum	Fungi	RPMI, 1.5% EtOH, 7 days, 30°C	2,500	Tullio et al. (2006)
Penicillium lilacinum	Fungi	Cited	500	Pellecuer et al. (1976)
Penicillium rubrum	Fungi	Cited	1,000	Pellecuer et al. (1976)
Rhizopus 66-81-2	Fungi	PDA, 5 days, 27°C	500	Thompson and Cannon (1986)
Rhizopus arrhizus	Fungi	PDA, 5 days, 27°C	500	Thompson and Cannon (1986)
Rhizopus chinensis	Fungi	PDA, 5 days, 27°C	500	Thompson and Cannon (1986)
Rhizopus circinans	Fungi	PDA, 5 days, 27°C	500	Thompson and Cannon (1986)
Rhizopus japonicus	Fungi	PDA, 5 days, 27°C	500	Thompson and Cannon (1986)
Rhizopus kazanensis	Fungi	PDA, 5 days, 27°C	1,000	Thompson and Cannon (1986)
Rhizopus nigricans	Fungi	Cited	500	Pellecuer et al. (1976)
Rhizopus oryzae	Fungi	PDA, 5 days, 27°C	500	Thompson and Cannon (1986)
Rhizopus pymacus	Fungi	PDA, 5 days, 27°C	500	Thompson and Cannon (1986)
Rhizopus sp.	Fungi	RPMI, 1.5% EtOH, 7 days, 30°C	>10,000	Tullio et al. (2006)
Rhizopus stolonifer	Fungi	PDA, 5 days, 27°C	500	Thompson and Cannon (1986)
Rhizopus tritici	Fungi	PDA, 5 days, 27°C	500	Thompson and Cannon (1986)
Scopulariopsis brevicaulis	Fungi	Cited	1,000	Pellecuer et al. (1976)
Scopulariopsis brevicaulis	Fungi	RPMI, 1.5% EtOH, 7 days, 30°C	10,000	Tullio et al. (2006)

(*Continued*)

TABLE 14.79 (*Continued*)
Inhibitory Data of Thyme Oil Obtained in the Dilution Test

Microorganism	MO Class	Conditions	MIC	Reference
Syncephalastrum racemosum	Fungi	Cited	500	Pellecuer et al. (1976)
Trichophyton interdigitale	Fungi	Cited	1,000	Pellecuer et al. (1976)
Trichophyton mentagrophytes	Fungi	SA, Tween 80, 21 days, 20°C	<300	Janssen et al. (1988)
Trichophyton mentagrophytes	Fungi	RPMI, 1.5% EtOH, 7 days, 30°C	2,500	Tullio et al. (2006)
Trichophyton rubrum	Fungi	SA, Tween 80, 21 days, 20°C	<300	Janssen et al. (1988)
Candida albicans	Yeast	SDB, Tween 80, 48 h, 35°C	1	Panizzi et al. (1993)
Candida albicans	Yeast	MPB, DMSO, 40 h, 30°C	500	Hili et al. (1997)
Candida albicans	Yeast	MHA, Tween 20, 48 h, 35°C	1,200	Hammer et al. (1999)
Candida albicans	Yeast	NB, 24 h, Tween 20, 37°C	200	Fawzi (1991)
Candida albicans	Yeast	Cited	2,000	Pellecuer et al. (1976)
Candida albicans	Yeast	Cited, 48 h, 36°C	2,000	Duarte et al. (2005)
Candida albicans	Yeast	TGB, 18–24 h, 37°C	500	Morris et al. (1979)
Candida albicans	Yeast	RPMI, DMSO, 48 h, 35°C	7.5–10.5	12,166
Candida guilliermondii	Yeast	RPMI, DMSO, 48 h, 35°C	7.5	12,166
Candida mycoderma	Yeast	Cited	2,000	Pellecuer et al. (1976)
Candida parapsilosis	Yeast	Cited	4,000	Pellecuer et al. (1976)
Candida pelliculosa	Yeast	Cited	1,000	Pellecuer et al. (1976)
Candida tropicalis	Yeast	Cited	4,000	Pellecuer et al. (1976)
Candida tropicalis	Yeast	RPMI, DMSO, 48 h, 35°C	8.5	12,166
Candida utilis	Yeast	RPMI, DMSO, 48 h, 35°C	7.5	12,166
Geotrichum asteroides	Yeast	Cited	2,000	Pellecuer et al. (1976)
Geotrichum candidum	Yeast	Cited	2,000	Pellecuer et al. (1976)
Hansenula sp.	Yeast	Cited	2,000	Pellecuer et al. (1976)
Saccharomyces carlsbergensis	Yeast	Cited	2,000	Pellecuer et al. (1976)
Saccharomyces cerevisiae	Yeast	MPB, DMSO, 40 h, 30°C	500	Hili et al. (1997)
Saccharomyces cerevisiae	Yeast	SDB, Tween 80, 48 h, 35°C	2	Panizzi et al. (1993)
Saccharomyces cerevisiae	Yeast	YPB, 24 h, 20°C	400–700	Schelz et al. (2006)
Saccharomyces cerevisiae	Yeast	NA, 1–3 days, 30°C	500	Farag et al. (1989)
Schizosaccharomyces pombe	Yeast	MPB, DMSO, 40 h, 30°C	500	Hili et al. (1997)
Torula utilis	Yeast	MPB, DMSO, 40 h, 30°C	500	Hili et al. (1997)

Annotation: (1) *Thymus vulgaris carvacroliferum*, (2) *Thymus vulgaris linaloliferum*, (3) *Thymus vulgaris thuyanoliferum*, and (4) *Thymus vulgaris thymoliferum*.

TABLE 14.80
Inhibitory Data of Thyme Oil Obtained in the Vapor Phase Test

Microorganism	MO Class	Conditions		Activity	Reference
Escherichia coli	Bac–	NA, 24 h, 37°C	~20,000	NG	Kellner and Kober (1954)
Escherichia coli	Bac–	BLA, 18 h, 37°C	MIC_{air}	12.5	Inouye et al. (2001)
Haemophilus influenzae	Bac–	MHA, 18 h, 37°C	MIC_{air}	3.13	Inouye et al. (2001)
Neisseria sp.	Bac–	NA, 24 h, 37°C	~20,000	NG	Kellner and Kober (1954)
Salmonella typhi	Bac–	NA, 24 h, 37°C	~20,000	NG	Kellner and Kober (1954)
Salmonella typhi	Bac–	NA, 24 h, 37°C	sd	++	Maruzzella and Sicurella (1960)
Bacillus megaterium	Bac+	NA, 24 h, 37°C	~20,000	NG	Kellner and Kober (1954)
Bacillus subtilis var. *aterrimus*	Bac+	NA, 24 h, 37°C	sd	++	Maruzzella and Sicurella (1960)
Corynebacterium diphtheriae	Bac+	NA, 24 h, 37°C	~20,000	NG	Kellner and Kober (1954)
Lactobacillus sp.	Bac+	MRS, cited	Disk, 20,000?	++	Pellecuer et al. (1980)
Micrococcus luteus	Bac+	MHB, cited	Disk, 20,000?	++	Pellecuer et al. (1980)
Micrococcus ureae	Bac+	MHB, cited	Disk, 20,000?	++	Pellecuer et al. (1980)
Mycobacterium avium	Bac+	NA, 24 h, 37°C	sd	+	Maruzzella and Sicurella (1960)
Sarcina ureae	Bac+	MHB, cited	Disk, 20,000?	+	Pellecuer et al. (1980)
Staphylococcus aureus	Bac+	NA, 24 h, 37°C	~20,000	NG	Kellner and Kober (1954)
Staphylococcus aureus	Bac+	NA, 24 h, 37°C	sd	++	Maruzzella and Sicurella (1960)
Staphylococcus aureus	Bac+	MHA, 18 h, 37°C	MIC_{air}	6.25–12.5	Inouye et al. (2001)
Staphylococcus epidermidis	Bac+	MHB, cited	Disk, 20,000?	++	Pellecuer et al. (1980)
Streptococcus D	Bac+	MHB, cited	Disk, 20,000?	++	Pellecuer et al. (1980)
Streptococcus faecalis	Bac+	NA, 24 h, 37°C	~20,000	NG	Kellner and Kober (1954)
Streptococcus faecalis	Bac+	NA, 24 h, 37°C	sd	++	Maruzzella and Sicurella (1960)
Streptococcus pneumoniae	Bac+	MHA, 18 h, 37°C	MIC_{air}	3.13–6.25	Inouye et al. (2001)
Streptococcus pyogenes	Bac+	NA, 24 h, 37°C	~20,000	NG	Kellner and Kober (1954)
Streptococcus pyogenes	Bac+	MHA, 18 h, 37°C	MIC_{air}	3.13–6.25	Inouye et al. (2001)
Alternaria alternata	Fungi	RPMI, 7 days, 30°C	MIC_{air}	78	Tullio et al. (2006)
Aspergillus flavus	Fungi	WFA, 42 days, 25°C	Disk, 50,000	NG	Guynot et al. (2003)
Aspergillus flavus	Fungi	Bread, 14 days, 25°C	30,000	+++	Suhr and Nielsen (2003)
Aspergillus flavus	Fungi	RPMI, 7 days, 30°C	MIC_{air}	156–312	Tullio et al. (2006)
Aspergillus fumigatus	Fungi	RPMI, 7 days, 30°C	MIC_{air}	78–312	Tullio et al. (2006)
Aspergillus niger	Fungi	WFA, 42 days, 25°C	Disk, 50,000	NG	Guynot et al. (2003)
Aspergillus niger	Fungi	RPMI, 7 days, 30°C	MIC_{air}	78–156	Tullio et al. (2006)
Botrytis cinerea	Fungi	PDA, 3 days, 25°C	1,000	+++	Lee et al. (2007)

(Continued)

TABLE 14.80 (Continued)
Inhibitory Data of Thyme Oil Obtained in the Vapor Phase Test

Microorganism	MO Class	Conditions		Activity	Reference
Cladosporium cladosporioides	Fungi	RPMI, 7 days, 30°C	MIC_{air}	78	Tullio et al. (2006)
Colletotrichum gloeosporioides	Fungi	PDA, 3 days, 25°C	1,000	+	Lee et al. (2007)
Endomyces fibuligera	Fungi	Bread, 14 days, 25°C	30,000	+++	Suhr and Nielsen (2003)
Eurotium amstelodami	Fungi	WFA, 42 days, 25°C	Disk, 50,000	NG	Guynot et al. (2003)
Eurotium herbarum	Fungi	WFA, 42 days, 25°C	Disk, 50,000	NG	Guynot et al. (2003)
Eurotium repens	Fungi	WFA, 42 days, 25°C	Disk, 50,000	NG	Guynot et al. (2003)
Eurotium repens	Fungi	Bread, 14 days, 25°C	30,000	NG	Suhr and Nielsen (2003)
Eurotium rubrum	Fungi	WFA, 42 days, 25°C	Disk, 50,000	NG	Guynot et al. (2003)
Fusarium oxysporum	Fungi	PDA, 3 days, 25°C	1,000	+	Lee et al. (2007)
Fusarium oxysporum	Fungi	RPMI, 7 days, 30°C	MIC_{air}	156	Tullio et al. (2006)
Microsporum canis	Fungi	RPMI, 7 days, 30°C	MIC_{air}	78–156	Tullio et al. (2006)
Microsporum gypseum	Fungi	RPMI, 7 days, 30°C	MIC_{air}	78	Tullio et al. (2006)
Mucor sp.	Fungi	RPMI, 7 days, 30°C	MIC_{air}	156	Tullio et al. (2006)
Penicillium corylophilum	Fungi	WFA, 42 days, 25°C	Disk, 50,000	NG	Guynot et al. (2003)
Penicillium corylophilum	Fungi	Bread, 14 days, 25°C	30,000	+++	Suhr and Nielsen (2003)
Penicillium frequentans	Fungi	RPMI, 7 days, 30°C	MIC_{air}	156	Tullio et al. (2006)
Penicillium lanosum	Fungi	RPMI, 7 days, 30°C	MIC_{air}	312	Tullio et al. (2006)
Penicillium roqueforti	Fungi	Bread, 14 days, 25°C	30,000	+++	Suhr and Nielsen (2003)
Pythium ultimum	Fungi	PDA, 3 days, 25°C	1,000	+++	Lee et al. (2007)
Rhizoctonia solani	Fungi	PDA, 3 days, 25°C	1,000	+	Lee et al. (2007)
Rhizopus sp.	Fungi	RPMI, 7 days, 30°C	MIC_{air}	312	Tullio et al. (2006)
Scopulariopsis brevicaulis	Fungi	RPMI, 7 days, 30°C	MIC_{air}	78	Tullio et al. (2006)
Trichophyton mentagrophytes	Fungi	RPMI, 7 days, 30°C	MIC_{air}	39–78	Tullio et al. (2006)
Candida albicans	Yeast	NA, 24 h, 37°C	~20,000	NG	Kellner and Kober (1954)

14.3 DISCUSSION

The in vitro most antimicrobially active essential oils regularly (or normally) contain substances as main components, which are themselves known to exhibit pronounced antimicrobial properties. These are cinnamic aldehyde (cinnamon bark and cassia oil) and the phenolic compounds eugenol (clove and cinnamon leaf oil) and thymol (thyme oil) (Pauli, 2001). All these essential oils reveal a broadband spectrum of activity in various in vitro test systems (agar diffusion, dilution, and VP) due to their considerable water solubility and volatility. The evaluated antimicrobial inhibitory data of the essential oils obtained in agar dilution tests, serial DIL, and VP tests are summarized in Table 14.81.

A few essential oils exhibit limited activities and act against a class of microorganism, for example, anise and bitter fennel oil specifically inhibit the growth of filamentous fungi (Table 14.81), or act well against microbial species belonging to an identical genus, for example, caraway oil inhibits *Trichophyton* species. Some essential oils are active only in a specific test system against a defined group of microorganism, for example, dwarf-pine oil and juniper oil preferably inhibit gram-positive

TABLE 14.81
Summary of Results of Antimicrobial Activities of Essential Oils Listed in the *European Pharmacopoeia* 6th Edition

Essential Oil	Bac−			Bac+			Fungi			Yeast		
Test Method	ADT	DIL	VP	ADT	DIL	VP	ADT	DIL	VP	ADT	DIL	VP
Anise	0–1	0–3	0–2	0–1	0–3	0–2	2–3	0–3	0–2	0–1	1–2	0
Bitter fennel	0–2	1–2	0–3	0–2	1–3	0–2	0–2	2–3	ND	0–2	2	3
Caraway	0–1	0–3	0–3	0–1	1–3	0–3	1–3	0–3	ND	0–1	2	3
Cassia	2	2–3	1–3	2	3	1–3	3	1–3	ND	ND	3	3
Cinnamon bark	1–2	3	1–3	1–3	3	1–3	2–3	2–3	0–3	2–3	2–3	3
Cinnamon leaf	2–3	2	3	2–3	2	3	3	1–3	0–3	2	2	3
Citronella	0–1	1–2	0–3	0–1	2–3	0–3	1–3	1–3	3	1–2	1–3	3
Clary sage	0–1	0–1	0–3	1–2	0–3	0–3	0–1	0–2	0	0	0–2	0
Clove	0–2	0–3	1–3	1–3	0–3	0–3	1–3	0–3	3	1–3	0–2	3
Coriander	0–2	0–2	3	0–2	1–3	0–3	0–2	0–3	ND	0–2	2	3
Dwarf pine	0–1	ND	0–3	0–1	ND	2–3	0	ND	ND	0–1	0	1
Eucalyptus	0–1	0–1	0–3	0–1	1	0–3	0–1	0–2	0	1	2	3
Juniper	0–1	0–2	1–3	0–1	0	2–3	0–1	0–2	0	0–1	2	3
Lavender	0–1	0–3	0–3	0–3	1–3	0–3	0–2	0–3	3	0–1	2	3
Lemon	0–1	0–1	0–3	0–1	0–2	0–3	0–1	0–3	0–1	0–1	2	3
Mandarin	0–1	0–2	ND	0–1	0–1	ND	1	ND	0–2	0–1	2	ND
Matricaria	0–1	0–3	0	0–1	0–3	1–2	0–1	0–2	0	0–1	0–2	0
Mint	0–3	2–3	ND	0–1	2–3	ND	0–3	2	ND	1	2	ND
Neroli	0–1	0	0–3	0–1	1–2	0–3	1–2	1–3	ND	0–1	2	3
Nutmeg	0–1	0–2	0	0–1	0–3	0–1	0–3	0–3	ND	0–1	2	ND
Peppermint	0–1	0–3	0–3	0–1	1–3	2–3	0–3	0–3	1–3	0–1	2–3	2
Pinus sylvestris	0–1	0–1	ND	1	0–2	ND	1	0–2	0	1	1–2	ND
Pinus pinaster	0–1	0–2	1–3	0–1	2	1–3	0–1	2	ND	0–1	0–3	2
Rosemary	0–1	0–3	0–3	0–2	0–3	0–3	0–2	0–2	0–3	0–1	0–2	3
Star anise	0–1	ND	2–3	0–1	ND	2–3	ND	ND	0–3	0–1	2–3	3
Sweet orange	0–2	0–3	0–3	0–2	0–2	0–3	0–2	0–3	0–3	0–1	2	3
Tea tree	1	0–3	0–3	1–2	0–3	0–3	1	1–3	0	1–3	1–3	ND
Thyme	1–2	1–3	2–3	0–3	0–3	1–3	1–3	1–3	0–3	1–3	2–3	3

Evaluation: Agar diffusion test (ADT): 0 = inactive, 1 = weak (most of the inhibition zones up to ~15 mm), 2 = moderate (most of the inhibition zones between 16 and 30 mm), 3 = strong inhibitory (most of the inhibition zones >30 mm); dilution test (DIL): 0 = MIC > 20,000 µg/mL, 1 = >5,000–20,000 µg/mL, 2 = 500–5,000 µg/mL, 3 = <500 µg/mL; vapor phase test (VP): 0 = no, 1 = weak, 2 = moderate, 3 strong growth reduction.

bacteria in the VP, but both were of low activity in the agar diffusion or DILs. Similarly, citronella oil was not inhibitory toward nine different fungal species in the agar dilution test but inhibited all of them in the VP test (Nakahara et al., 2003). Yeasts turned out to be susceptible toward essential oil vapors, which is in agreement to observations made with volatile esters and monoterpenes. Possibly, in yeast, the biosynthesis of chitin is inhibited by volatile compounds emitted from plants (Pauli, 2006).

The inhibitory data itself differ considerably from each other when results of different examiners are compared as it can be seen, for example, by the activities of lemon oil toward *Escherichia coli* (five tests inactive and four tests active) in the agar diffusion test (Table 14.43) or by the MIC values of rosemary oil against *Staphylococcus aureus* (Table 14.67), which cover a range from 20 to 50,000 µg/mL in nine examinations. Even if the unit of the low value of 20 was confused—this might had happened in references Panizzi et al. (1993) and Pellecuer et al. (1980)—the activity range of rosemary oil is 400–50,000 μg/mL. Taken together, exceptionally the aforementioned relatively strong-acting essential oils, the results are not coordinated and cover frequently the complete activity spectrum from inactive to strong active (evaluation from 0 to 3 in Table 14.81). In part, the results are contradictory and no general rules can be raised from the data shown in Tables 14.1 through 14.80.

Several reasons can be made responsible for this undesirable but at least normal situation:

- Natural variability in the composition of essential oils
- Natural variability in the susceptibility of microorganism
- Different parameters in microbiological testing methods
- Unknown history of the tested essential oils: production, storage, and age
- Insufficient knowledge about exact phytochemical composition

The composition of essential oils depends on several factors: plant part used, place of growth, climate, natural variation (varieties, subspecies, and chemotypes), harvesting time, production, storage conditions, and analysis parameters in compound identification. Because some of these influencing factors differ from year to year, no constant composition of an essential can be expected, even when it is grown and produced at the same place. Chemotypes possess in part a completely different composition, for example, the MICs of four thyme oil chemotypes (linalool, thuyanol, carvacrol, and thymol type) toward *S. aureus* (Table 14.79) differ from 250 to 4000 μg/mL and low MIC values depended on the presence of thymol (Oussalah et al., 2006). In some literature works, the botanical description of investigated plant material is not defined exactly. The characterization of "fennel oil" is not sufficient to decide between sweet or bitter fennel. The same is true for orange oil, where sweet orange (*C. sinensis*) and bitter orange (*C. aurantium*) refer to two different botanical species. The characterization of essential oils with physical properties alone (e.g., density and solubility) is not sufficient for an unequivocal definition of its constituents. At best distillation, analysis, and pharmacological examination of an essential oil are done in a close-time relationship.

To find out the most appropriate method for antimicrobial testing of essential oils or components thereof, literature inhibitory data of eugenol against *E. coli* were compared with each other, and it was concluded that the variation of data as it is obtained in the serial DIL is tolerable (MIC range from 250 to 600 μg/mL) (Pauli and Kubeczka, 1996). The same is true for clove oil, which was found to be active in DILs against *E. coli* in the range from 400 to 1250 μg/mL in seven of nine examinations (Table 14.26).

Questions concerning disinfectant activity of essential oils, for example, the minimum time needed to kill a given microbial species or the determination of microbial survivors after short time contact, are not answered by agar diffusion or DILs. In older literature, the killing concentration relative to phenol was determined after 15 or 30 min exposure of the respective microbials species to the compound to be tested. The so-called carboxylic acid coefficient or phenol coefficient was introduced in 1903 (Rideal and Walker, 1903) and was also taken for the characterization of the killing activity of essential oils toward microorganism (Martindale, 1910).

Differences in the susceptibility exist between organisms of the identical species as it was shown in experiments with three *E. coli* strains. One of them was inhibited by eugenol methyl ether at a low concentration (MIC = 550 μg/mL), while two other strains tolerated still 8000 μg/mL without any visible growth reduction. Remarkably, the MIC of eugenol toward all three strains was almost equal (550–600 μg/mL) (Pauli, 1994). Because these *E. coli* strains never had had the opportunity to develop a specific resistance against eugenol methyl ether before, it is evident that a natural variation of susceptibility toward natural antimicrobials exists.

To improve the data situation of in vitro antimicrobial data of essential oils, all aforementioned biological and experimental parameters should be controlled as best as possible. An appropriate microbiological test system should be taken, which allows comparison of inhibitory data with drugs used in the therapy of human infectious diseases. Such worked out and standardized serial DILs (Clinical & Laboratory Standards Institute 2008) are already utilized in the examination of natural substances, for example, Hostettmann and Schaller (1999) and Jirovetz et al. (2007). To avoid complications by strains with unknown susceptibilities toward antimicrobials and to make the results from different laboratories comparable to each other, available standard strains from collections (e.g., American Type Culture Collection, Deutsche Sammlung für Mikroorganismen und Zellkulturen, and Institute for Fermentation Osaka) should be taken in the routine analysis of antimicrobial activities of natural compounds and essential oils. Antimicrobials tests should include a greater number of different organisms belonging to the groups: gram-positive, gram-negative bacteria, filamentous fungi, and yeasts.

Two principal reasons for performing the in vitro tests are as follows:

1. Identification of antimicrobially active compounds
2. Control of microbial susceptibilities toward approved antibiotics and antimycotics

The procedure from identification of antimicrobially active compounds to their use in humans to treat infectious diseases is a multistep pathway, which includes pharmacological (concentration of the active compound at the site of action, half-life time, serum levels, dose–response relationship, etc.) and toxicological (e.g., toxicity, allergic responses, and interactions) aspects.

Essential oils consist frequently of over 100 individual compounds, which themselves plus their metabolic transformation products cannot be followed up in the living body. This fact may explain why pharmacological studies with entire essential oils never have been in the focus of pharmaceutical research.

In animal experiments, it was demonstrated by Imura over 80 years ago that the in vivo protection against tuberculosis was not parallel with results in vitro (Table 14.82).

The superior protection of tuberculosis-infected guinea pigs by anethole and lemon oil is contradictory to their weak in vitro antitubercular activity. By means of these results obtained parallelly in vivo and in vitro, it is obvious that in vitro inhibitory data alone cannot be used as an information basis for the treatment of infectious diseases in humans (Table 14.81). Therefore, the recording of so-called aromatograms with patient isolates and a greater number of essential oils—as it is

TABLE 14.82
Comparison of In Vitro Growth Inhibition of Human Type of *Mycobacterium tuberculosis* with the In Vivo Protection of Tuberculosis-Infected Guinea Pigs by Essential Oils and Components Thereof

Test Materials	In Vivo[a]		In Vitro[b]
	Lymph Nodes	Viscera	
Anethole	33	50	2500
Lemon oil	49	55	1250
Terpineol	59	57	625
Nutmeg oil, expressed	61	73	312
Geraniol	96	71	625
Eugenol	89	84	1250

Source: Imura, K., *J. Shanghai Sci. Inst./Sec.*, 4, 1, 235, 1935.

[a] Grade of tuberculous change in percent was observed after 10 weeks in lymph nodes and viscera in groups of five infected guinea pigs fed with ~250 mg/kg (recalculated from experimental section) test material per day for 1 week, respectively.

[b] Growth reduction (dry weight tubercle bacilli 5 mg) by a given concentration in µg/mL determined in glycerin-bouillon after 3 weeks incubation at 37°C.

common in "aromatherapy"—is critical (DHZ-Spektrum, 2007). The selection of the most active essential oil by using in vitro testing methods for the therapy of infectious diseases may have success, but at least there is no rational relationship between in vitro testing and in vivo success (Table 14.81). Besides that, many factors influence the results obtained in the agar diffusion test and may lead to wrong interpretation of the results concerning the antimicrobial strength of an essential oil.

It seems to be suitable to discuss a special status of essential oils in concern of their pharmacological activities, since essential oils cannot be followed up by classical pharmacological methods due to their complex nature. Recently, successes of wound treatment with essential oils have been demonstrated under clinical conditions, which cannot be realized with pharmaceuticals (Warnke et al., 2005). In another case, a deep infection of a hip arthroplasty was successfully treated with a pure compound occurring in matricaria oil: (−)-⟨-bisabolol (Pauli et al., 2007, 2009). Due to its lipophilic nature (−)-⟨-bisabolol is thought to be taken up by the skin and it enters the blood circulatory system, which should be measureable with pharmacological methods. Interestingly, the toxicity of essential oils toward mammalians decreases significantly with increase of average lipophilicity of their components (Pauli, 2008), while simultaneously the toxicity toward bacteria and fungi increases significantly with increasing lipophilicity (Pauli, 2007a), which points to the extraordinary role of essential oils among natural compounds, especially of their highly lipophilic constituents.

REFERENCES

Aggag, M.E. and R.T. Yousef, 1972. Study of antimicrobial activity of chamomile oil. *Planta Med.*, 22: 140–144.

Al-Bayati, F., 2008. Synergistic antibacterial activity between *Thymus vulgaris* and *Pimpinella anisum* essential oils and methanol extracts. *J. Ethnopharmacol.*, 116: 403–406.

Angelini, P., R. Pagiotti, and B. Granetti, 2008. Effect of antimicrobial activity of *Melaleuca alternifolia* essential oil on antagonistic potential of *Pleurotus* species against *Trichoderma harzianum* in dual culture. *World J. Microbiol. Biotechnol.*, 24: 197–202.

Angioni, A., A. Barra, W. Cereti et al., 2004. Chemical composition, plant genetic differences, antimicrobial and antifungal activity investigation of the essential oil of *Rosmarinus officinalis* L. *J. Agric. Food Chem.*, 52: 3530–3535.

Angioni, A., A. Barra, M.T. Russo et al., 2003. Chemical composition of the essential oils of *Juniperus* from ripe and unripe berries and leaves and their antimicrobial activity. *J. Agric. Food Chem.*, 51: 3073–3078.

Arras, G. and M. Usai, 2001. Fungitoxic activity of 12 essential oils against four postharvest citrus pathogens: Chemical analysis of *Thymus capitatus* oil and its effect in subatmospheric pressure conditions, *J. Food Prot.*, 64: 1025–1029.

Ashhurst, A.P., 1927. The centenary of Lister (1827–1927). A tale of sepsis and anti-sepsis. *Ann. Med. History*, 9: 205–211.

Banes-Marshall, L., P. Cawley, and C.A. Phillips, 2001. In vitro activity of *Melaleuca alternifolia* (tea tree) oil against bacterial and *Candida* spp. isolates from clinical specimens. *Br. J. Biomed. Sci.*, 58: 139–145.

Bastide, P., R. Malhuret, J.C. Chalchat et al., 1987. Correlation of chemical composition and antimicrobial activity. II. Activity of three resinous essential oils on two bacterial strains. *Plantes Medicinales et Phytotherapie*, 21: 209–217.

Belletti, N., M. Ndagijimana, C. Sisto et al., 2004. Evaluation of the antimicrobial activity of citrus essences on *Saccharomyces cerevisiae*. *J. Agric. Food Chem.*, 52: 6932–6938.

Boesel, R. 1991. Pharmakognostische, phytochemische und mikrobiologische Untersuchung von Queckenwurzelstock Rhizoma graminis (Erg.-B. 6) (*Agropyron repens* (L.) PALISOT DE BEAUVOIS, Poaceae), Dissertation, FU Berlin, Germany.

Bouchra, C., A. Mohamed, I.H. Mina et al., 2003. Antifungal activity of essential oil from several medicinal plants against four postharvest citrus pathogens. *Phytopathol. Mediterranea*, 42: 251–256.

Buchholtz, L., 1875. Antiseptika und Bakterien. *Archiv experim. Pathologie Pharmakologie*, 4: 1–82.

Caccioni, D.R., M. Guizzardi, D.M. Biondi et al., 1998. Relationship between volatile components of citrus fruit essential oils and antimicrobial action on *Penicillium digitatum* and *Penicillium italicum*. *Int. J. Food Microbiol.*, 43: 73–79.

Canillac, N. and A. Mourey, 1996. Compartement de Listeria en presence d'huiles essentielles de sapin et de pin. *Sciences des Aliments*, 16: 403–411.

Cantore, P.L., N.S. Iacobellis, A. DeMarco et al., 2004. Antibacterial activity of *Coriandrum sativum* L. and *Foeniculum vulgare* Miller var. *vulgare* (Miller) essential oils. *J. Agric. Food Chem.*, 52: 7862–7866.

Carson, C.F., B.D. Cookson, H.D. Farrelly et al., 1995. Susceptibility of methicillin-resistant *Staphylococcus aureus* to the essential oil of *Melaleuca alternifolia*. *J. Antimicrob. Chemother.*, 35: 421–424.

Carson, C.F., K.A. Hammer, and T.V. Riley, 1996. In-vitro activity of the essential oil of *Melaleuca alternifolia* against *Streptococcus* spp. *J. Antimicrob. Chemother.*, 37: 1177–1178.

Carson, C.F., K.A. Hammer, and T.V. Riley, 2006. *Melaleuca alternifolia* (tea tree) oil: A review of antimicrobial and other medicinal properties. *Clin. Microbiol. Rev.*, 19: 50–62.

Carson C.F. and S. Messager, 2005. Tea tree oil: A potential alternative for the management of methicillin-resistant *Staphylococcus aureus* (MRSA). *Aust. Infect. Control.*, 10: 32–34.

Cavaleiro, C., E. Pinto, M.J. Goncalves et al., 2006. Antifungal activity of *Juniperus* essential oils against dermatophyte, *Aspergillus* and *Candida* strains. *J. Appl. Microbiol.*, 100: 1333–1338.

Chalchat, J.C., R.P. Garry, A. Michet et al., 1989. Chemical composition/antimicrobial activity correlation: IV. Comparison of the activity of natural and oxygenated essential oils against six strains. *Plantes Medicinales et Phytotherapie*, 23: 305–314.

Clinical and Laboratory Standards Institute, 2012. Antimicrobial Susceptibility Testing Standards. http://clsi.org/blog/2012/01/13/clsi-publishes-2012-antimicrobial-susceptibility-testing-standards/ (accessed June 24, 2015).

Conner, D.E. and L.R. Beuchat, 1984. Effect of essential oils from plants on growth of food spoilage yeasts. *J. Food Sci.*, 49: 429–434.

Cosentino, S., A. Barra, B. Pisano et al., 2003. Composition and antimicrobial properties of Sardinian *Juniperus* essential oils against foodborne pathogens and spoilage microorganisms. *J. Food Prot.*, 66: 1288–1291.

Council of Europe, 2008. *European Pharmacopoeia 6th Edition*. Council of Europe, European Directorate for the Quality of Medicines and Healthcare, Strasbourg, France.

Council of Europe, 2014. *European Pharmacopoeia 8th Edition*. Council of Europe, European Directorate for the Quality of Medicines and Healthcare, Strasbourg, France.

Dabbah, R., V.M. Edwards, and W.A. Moats, 1970. Antimicrobial action of some citrus fruit oils on selected food-borne bacteria. *Appl. Microbiol.*, 19: 27–31.

Deans, S.G. and G. Ritchie, 1987. Antibacterial properties of plant essential oils. *Int. J. Food Microbiol.*, 5: 165–180.

DHZ-Spektrum, 2007. Aromatogramm als Schlüssel zur "antibiotischen Aromatherapie". *Dtsch. Heilpraktiker-Zeitschr.* 2: 22–25.

Di Pasqua, R., V. De Feo, F. Villani et al., 2005. In vitro antimicrobial activity of essential oils from Mediterranean Apiaceae, Verbenaceae and Lamiaceae against foodborne pathogens and spoilage bacteria. *Ann. Microbiol.*, 55: 139–145.

Dikshit, A. and A. Husain, 1984. Antifungal action of some essential oils against animal pathogens. *Fitoterapia*, 55: 171–176.

Dikshit, A., A.A. Naqvi, and A. Husain, 1986. *Schinus molle*: A new source of natural fungitoxicant. *Appl. Environ. Microbiol.*, 51: 1085–1088.

Dikshit, A., A.K. Singh, and S.N. Dixit, 1982. Effect of pH on fungitoxic activity of some essential oils. *Bokin Bobai*, 10: 9–10.

Dikshit, A., A.K. Singh, R.D. Tripathi et al., 1979. Fungitoxic and phytotoxic studies of some essential oils. *Biolog. Bull. India*, 1: 45–51.

Duarte, M.C., G.M. Figueira, A. Sartoratto et al., 2005. Anti-*Candida* activity of Brazilian medicinal plants. *J. Ethnopharmacol.*, 97: 305–311.

Duarte, M.C.T., C. Leme-Delarmelina, G.M. Figueira et al., 2006. Effects of essential oils from medicinal plants used in Brazil against epec and etec *Escherichia coli*. *Revista Brasileira de Plantas Medicinais*, 8: 129–143.

Dube, K.G. and T.S.S. Rao, 1984. Antibacterial efficacy of some Indian essential oils. *Chemicals Petro-Chemicals J.*, 15: 13–14.

Dubey, P., S. Dube, and S.C. Tripathi, 1990. Fungitoxic properties of essential oil of *Anethum graveolens* L. *Proc. Indian Natl. Sci. Acad.— Part B: Biol. Sci.*, 60: 179–184.

El-Gengaihi, S. and D. Zaki, 1982. Biological investigation of some essential oils isolated from Egyptian plants. *Herba Hungarica*, 21: 107–111.

Ertürk, Ö., T.B. Özbucak, and A. Bayrak, 2006. Antimicrobial activities of some medicinal essential oils. *Herba Polonica*, 52: 58–66.

Farag, R.S., Z.Y. Daw, F.M. Hewedi et al., 1989. Antimicrobial activity of some Egyptian spice essential oils. *J. Food Prot.*, 52: 665–667.

Fawzi, M.A., 1991. Studies on the antimicrobial activity of the volatile oil of *Thymus vulgaris*. *Alexandria J. Pharm. Sci.*, 5: 113–115.

Feng, W. and X. Zheng, 2007. Essential oils to control *Alternaria alternata* in vitro and in vivo. *Food Control*, 18: 1126–1130.

Fisher, K. and C.A. Phillips, 2006. The effect of lemon, orange and bergamot essential oils and their components on the survival of *Campylobacter jejuni, Escherichia coli* O157, *Listeria monocytogenes, Bacillus cereus* and *Staphylococcus aureus* in vitro and in food systems. *J. Appl. Microbiol.*, 101: 1232–1240.

Friedman, M., P.R. Henika, C.E. Levin et al., 2004. Antibacterial activities of plant essential oils and their components against *Escherichia coli* O157:H7 and *Salmonella enterica* in apple juice. *J. Agric. Food Chem.*, 52: 6042–6048.

Furneri, P.M., D. Paolino, A. Saija et al., 2006. In vitro antimycoplasmal activity of *Melaleuca alternifolia* essential oil. *J. Antimicrob. Chemother.*, 58: 706–707.

Gangrade, S.K., S.K. Shrivastave, O.P. Sharma et al. 1991. In vitro antifungal effect of the essential oils. *Indian Perf.*, 35: 46–48.

Garg, S.C. and D.C. Garg, 1980a. Antibakterielle Wirksamkeit in vitro einiger etherischer Öle. Teil 1. *Parfuem. Kosmet.*, 61: 219–220.

Garg, S.C. and D.C. Garg, 1980b. Antibakterielle Wirksamkeit in vitro einiger etherischer Öle. Teil 2. *Parfuem. Kosmet.*, 61: 255–256.

Geinitz, R., 1912. Vergleichende Versuche über die narkotischen und desinfizierenden Wirkungen der gangbarsten ätherischen Öle und deren wirksamen Bestandteile. *Sitzungsberichte und Abhandlungen der naturforschenden Gesellschaft zu Rostock.* Neue Folge. IV: 33–98.

Goutham, M.P. and R.M. Purohit, 1974. Überprüfung einiger ätherischer Öle auf ihre antibakteriellen Eigenschaften. *Riechstoffe, Aromen, Körperpflegemittel*, 3: 70–71.

Griffin, S.G., J.L. Markham, and D.N. Leach, 2000. An agar dilution method for the determination of the minimum inhibitory concentration of essential oils. *J. Essent. Oil Res.*, 12: 249–255.

Gustafson, J.E., Y.C. Liew, S. Chew et al., 1998. Effects of tea tree oil on *Escherichia coli*. *Lett. Appl. Microbiol.*, 26: 194–198.

Guynot, M.E., S. Marin, L. Seto et al. 2005. Screening for antifungal activity of some essential oils against common spoilage fungi of bakery products. *Food Sci. Technol. Int.*, 11: 25–32.

Guynot, M.E., A.J. Ramos, L. Seto et al., 2003. Antifungal activity of volatile compounds generated by essential oils against fungi commonly causing deterioration of bakery products. *J. Appl. Microbiol.*, 94: 893–899.

Hadacek, F. and H. Greger. 2000. Testing of antifungal natural products: Methodologies, comparability of results and assay of choice. *Phytochem. Anal.*, 11: 137–147.

Hammer, K.A., C.F. Carson, and T.V. Riley, 1996. Susceptibility of transient and commensal skin flora to the essential oil of *Melaleuca alternifolia* (tea tree oil). *Am. J. Infect. Control*, 24: 186–189.

Hammer, K.A., C.F. Carson, and T.V. Riley, 1998. In-vitro activity of essential oils in particular *Melaleuca alternifolia* (tea tree) oil and tea tree oil products, against *Candida* spp. *J. Antimicrob. Chemother.*, 42: 591–595.

Hammer, K.A., C.F. Carson, and T.V. Riley, 1999a. Antimicrobial activity of essential oils and other plant extracts. *J. Appl. Microbiol.*, 86: 985–990.

Hammer, K.A., C.F. Carson, and T.V. Riley, 1999b. In vitro susceptibilities of lactobacilli and organisms associated with bacterial vaginosis to *Melaleuca alternifolia* (tea tree) oil. *Antimicrob. Agents Chemother.*, 43: 196.

Hammer, K.A., C.F. Carson, and T.V. Riley, 1999c. Influence of organic matter, cations and surfactants on the antimicrobial activity of *Melaleuca alternifolia* (tea tree) oil in vitro. *J. Appl. Microbiol.*, 86: 446–452.

Hammer, K.A., C.F. Carson, and T.V. Riley, 2000. In vitro activities of ketoconazole, econazole, miconazole, and *Melaleuca alternifolia* (tea tree) oil against *Malassezia* species. *Antimicrob. Agents Chemother.*, 44: 467–469.

Hammer, K.A., C.F. Carson, and T.V. Riley, 2002. In vitro activity of *Melaleuca alternifolia* (tea tree) oil against dermatophytes and other filamentous fungi. *J. Antimicrob. Chemother.*, 50: 195–199.

Harkenthal, M., J. Reichling, H.K. Geiss et al., 1999. Comparative study on the in vitro antibacterial activity of Australian tea tree oil, cajuput oil, niaouli oil, manuka oil, kanuka oil, and eucalyptus oil. *Pharmazie*, 54: 460–463.

Hethenyi, E., I. Koczka, and P. Tetenyi, 1989. Phytochemical and antimicrobial analysis of essential oils. *Herba Hungarica*, 28: 99–115.

Hili, P., C.S. Evans, and R.G. Veness, 1997. Antimicrobial action of essential oils: The effect of dimethylsulphoxide on the activity of cinnamon oil. *Lett. Appl. Microbiol.*, 24: 269–275.

Hostettmann, K. and F. Schaller, 1999. Antimicrobial diterpenes. U.S. Patent 5,929,124.

Imai, H., K. Osawa, H. Yasuda et al., 2001. Inhibition by the essential oils of peppermint and spearmint of the growth of pathogenic bacteria. *Microbios*, 106(Suppl. 1): 31–39.

Imura, K., 1935. On the influence of the ethereal oils upon the culture of tubercle bacilli and upon development of experimental tuberculosis in animals. *J. Shanghai Sci. Instit./Sect.*, 4, 1: 235–270.

Inouye, S., T. Takizawa, and H. Yamaguchi, 2001. Antibacterial activity of essential oils and their major constituents against respiratory tract pathogens by gaseous contact. *J. Antimicrob. Chemother.*, 47: 565–573.

Iscan, G., N. Kirimer, M. Kurkcuoglu et al., 2002. Antimicrobial screening of *Mentha piperita* essential oils. *J. Agric. Food Chem.*, 50: 3943–3946.

Janssen, A.M., N.L.J. Chin, J.J.C. Scheffer et al., 1986. Screening for antimicrobial activity of some essential oils by the agar overlay technique. *Pharmacy World Sci.*, 8: 289–292.

Janssen, A.M., J.J.C. Scheffer, and A. Baerheim-Svendsen, 1987. Antimicrobial activities of essential oils. A 1976–1986 literature review on possible applications. *Pharm. Weekblad Sci. Ed.*, 9: 193–197.

Janssen, A.M., J.J.C. Scheffer, A.W. Parhan-van Atten et al., 1988. Screening of some essential oils on their activities on dermatophytes. *Pharm. Weekblad Sci. Ed.*, 10: 277–280.

Jirovetz, L., G. Eller, G. Buchbauer et al., 2006. Chemical composition and antimicrobial activities and odor description of some essential oils with characteristic flory-rosy scent and of their principal aroma compounds. *Recent Res. Dev. Agron. Hort.*, 2: 1–12.

Jirovetz, L., K. Wlcek, G. Buchbauer et al., 2007. Antifungal activity of various Lamiaceae essential oils rich in thymol and carvacrol against clinical isolates of pathogenic fungi. *Int. J. Essent. Oil Therap.*, 1: 153–157.

Kedzia, B., 1991. Antimicrobial activity of chamomile oil and its components. *Herba Polonica*, 37: 29–38.

Kellner, W. and W. Kober, 1954. Möglichkeiten der Verwendung ätherischer Öle zur Raumdesinfektion. 1. Mitteilung. Die Wirkung gebräuchlicher ätherischer Öle auf Testkeime. *Arzneim.-Forsch.*, 4: 319–325.

Kindra, K.J. and T. Satyanarayana, 1978. Inhibitory activity of essential oils of some plants against pathogenic bacteria. *Indian Drugs*, 16: 15–17.

Koba, K., K. Sanda, C. Raynaud et al., 2004. Activites antimicrobiennes d'huiles essentielles de trois *CMYBopogon* sp. africains vis-avis de germes pathogenes d'animaux de compagnie. *Annales de medecine veterinaire*, 148: 202–206.

Koch, R., 1881. Über Desinfektion. *Mittheilungen aus dem Kaiserlichen Gesundheitsamte*, 1: 234–282.

Kosalec, I., S. Pepeljnjak, and D. Kustrak, 2005. Antifungal activity of fluid extract and essential oil from anise fruits (*Pimpinella anisum* L., Apiaceae). *Acta Pharm.* (Zagreb)., 55: 377–385.

Kubo, I., H. Muroi, and A. Kubo, 1993. Antibacterial activity of long chain alcohols against *Streptococcus mutans*. *J. Agric. Food Chem.*, 41: 2447–2450.

Kumar, R., N.K. Dubey, O.P. Tiwari et al., 2007. Evaluation of some essential oils as botanical fungitoxicants for the protection of stored food commodities from fungal infestation. *J. Sci. Food Agric.*, 87: 1737–1742.

Kurita, N. and S. Koike, 1983. Synergistic antimicrobial effect of ethanol, sodium chloride, acetic acid and essential oil components. *J. Agric. Biol. Chem.*, 47: 67–75.

Lee, S.O., G.J. Choi, K.S. Jang et al., 2007. Antifungal activity of five plant essential oils as fumigant against postharvest and soilborne plant pathogenic fungi. *Plant Pathol. J.*, 23: 97–102.

Lemos, T.L.G., F.J.Q. Monte, F.J.A. Matos et al., 1992. Chemical composition and antimicrobial activity of essential oils from Brazilian plants. *Fitoterapia*, 63: 266–268.

Lens-Lisbonne, C., A. Cremieux, C. Maillard et al., 1987. Methods d'evaluation de l'activite antibacterienne des huiles essentieles. Application aux essences de thym et de canelle. *J. Pharmacie Belg.*, 42: 297–302.

Lis-Balchin, M., S.G. Deans, and E. Eaglesham, 1998. Relationship between bioactivity and chemical composition of commercial essential oils. *Flav. Fragr. J.*, 13: 98–104.

Martindale, W.H., 1910. Essential oils in relation to their antiseptic powers as determined by their carbolic acid coefficients. *Perf. Essent. Oil Rec.*, 1: 266.

Maruzzella, J.C., 1963. An investigation of the antimicrobial properties of absolutes. *Am. Perfum. Cosmet.*, 78: 19–20.

Maruzzella, J.C. and M.B. Lichtenstein, 1956. The in vitro antibacterial activity of oils. *J. Am. Pharm. Assoc. Sci. Ed.*, 45: 378–381.

Maruzzella, J.C. and L. Liguori, 1958. The in vitro antifungal activity of essential oils. *J. Am. Pharm. Assoc. Sci. Ed.*, 47: 250–255.

Maruzzella, J.C. and N.A. Sicurella. 1960. Antibacterial activity of essential oil vapores. *J. Am. Pharm. Assoc., Sci. Ed.*, 49: 692–694.

McCarthy, P.J., T.P. Pitts, G.P. Gunawardana et al., 1992. Antifungal activity of meridine, a natural product from the marine sponge *Corticium* sp. *J. Nat. Prod.*, 55: 1664–1668.

Mikhlin, E.D., V.P. Radina, A.A. Dmitrovskii et al., 1983. Antifungal and antimicrobial activity of beta-ionone and vitamin A derivatives. *Prikladnaja biochimija i mikrobiologija*, 19: 795–803.

Mimica-Dukic, N., S. Kujundzic, M. Sokovic et al., 2003. Essential oils composition and antifungal activity of *Foeniculum vulgare* Mill. obtained by different distillation conditions. *Phytother. Res.*, 17: 368–371.

Mondello, F., F. De Bernardis, A. Girolamo et al., 2006. In vivo activity of terpinen-4-ol, the main bioactive component of *Melaleuca alternifolia* Cheel (tea tree) oil against azole-susceptible and -resistant human pathogenic *Candida* species. *BMC Infect. Dis.*, 6: 91.

Morris, J.A., A. Khettry, and E.W. Seitz, 1979. Antimicrobial activity of aroma chemicals and essential oils. *J. Am. Oil Chem. Soc.*, 56: 595–603.

Möse, J.R. and G. Lukas, 1957. Zur Wirksamkeit einiger ätherischer Öle und deren Inhaltsstoffe auf Bakterien. *Arzneim.-Forsch.*, 7: 687–692.

Motiejunaite, O. and D. Peciulyte, 2004. Fungicidal properties of *Pinus sylvestris* L. for improvement of air quality. *Medicina* (Kaunas), 40: 787–794.

Mueller-Ribeau, F., B. Berger, and O. Yegen, 1995. Chemical composition and fungitoxic properties to phytopathogenic fungi of essential oils of selected aromatic plants growing wild in Turkey. *J. Agric. Food Chem.*, 43: 2262–2266.

Nakahara, K., N.S. Alzoreky, T. Yoshihashi et al., 2003. Chemical composition and antifungal activity of essential oils from *CMYBopogon nardus* (citronella grass). *Jpn. Agric. Res. Quart.*, 37: 249–252.

Narasimba Rao, B.G.V. and P.L. Joseph, 1971. Die Wirksamkeit einiger ätherischer Öle gegenüber phytopathogenen Fungi. *Riechstoffe, Aromen, Körperpflegemittel*, 21: 405–410.

Narasimha Rao, B.G.V. and S.S. Nigam, 1970. The in vitro antimicrobial efficiency of some essential oils. *Flav. Ind.*, 1: 725–729.

Nelson, R.R., 1997. In-vitro activities of five plant essential oils against methicillin-resistant *Staphylococcus aureus* and vancomycin-resistant *Enterococcus faecium*. *J. Antimicrob. Chemother.*, 40: 305–306.

Nigam, S.S. and J.T. Rao, 1979. Efficacy of some Indian essential oils against thermophilic fungi and *Penicillium* species. *7th Int. Congr. Essent. Oils*, 1977 (pub. 1979). 7: 485–487.

Okazaki, K. and S. Oshima, 1952a. Antibacterial activity of higher plants. XX. Antimicrobial effect of essential oils. 1). Clove oil and eugenol. *Yakugaku Zasshi*, 72: 558–560.

Okazaki, K. and S. Oshima, 1952b. Antibacterial activity of higher plants. XXII. Antimicrobial effect of essential oils. 1. Fungistatic effect of clove oil and eugenol. *Yakugaku Zasshi*, 72: 664–667.

Okazaki, K. and S. Oshima, 1953. Studies on the fungicides for drug preparation. *Yakugaku Zasshi*, 73: 692–696.

Oliva, B., E. Piccirilli, T. Ceddia et al., 2003. Antimycotic activity of *Melaleuca alternifolia* essential oil and its major components. *Lett. Appl. Microbiol.*, 37: 185–187.

Ooi, L.S., Y. Li, S.L. Kam et al., 2006. Antimicrobial activities of cinnamon oil and cinnamaldehyde from the Chinese medicinal herb *Cinnamon cassia* Blume. *Am. J. Chin. Med.*, 34: 511–522.

Oussalah, M., S. Caillet, L. Saucier et al., 2006. Inhibitory effects of selected plant essential oils on the growth of four pathogenic bacteria: *Escherichia coli* O157:H7, *Salmonella typhimurium*, *Staphylococcus aureus* and *Listeria monocytogenes*. *Meat Science*, 73: 236–244.

Panizzi, L., G. Flamini, P.L. Cioni et al., 1993. Composition and antimicrobial properties of essential oils of four mediterannean Lamiaceae. *J. Ethnopharmacol.*, 39: 167–170.

Papadopoulos, C.J., C.F. Carson, K.A. Hammer et al., 2006. Susceptibility of pseudomonads to *Melaleuca alternifolia* (tea tree) oil and components. *J. Antimicrob. Chemother.*, 58: 449–451.

Pauli, A., 1994. Chemische, physikalische und antimikrobielle Eigenschaften von in ätherischen Ölen vorkommenden Phenylpropanen. *Dissertation*, University Würzburg, Germany.

Pauli, A., 2001. Antimicrobial properties of essential oil constituents. *Int. J. Aromather.*, 11: 126–133.

Pauli, A., 2006. Anticandidal low molecular compounds from higher plants with special reference to compounds from essential oils. *Med. Res. Rev.*, 26: 223–268.

Pauli, A., 2007a. Identification strategy of mechanism-based lipophilic antimicrobials. In *New Biocides Development: The Combined Approach of Chemistry and Microbiology* (ACS Symposium Series), P. Zhu, ed. pp. 213–268. Corby, U.K.: Oxford University Press.

Pauli, A., 2007b. Kritische Anmerkungen zur keimhemmenden Wirkung ätherischer Öle und deren Bestandteile. *Kompl. Integr. Med.*, 48: 20–23.

Pauli, A., 2008. Relationship between lipophilicity and toxicity of essential oils. *Int. J. Essent. Oil Ther.*, 2: 60–68.

Pauli, A. and K.-H. Kubeczka, 1996. Evaluation of inhibitory data of essential oil constituents obtained with different microbiological testing methods. In *Essential Oils: Basic and Applied Research. Proceedings of 27th International Symposium on Essential Oils*, C. Franz, A. Mathe, and G. Buchbauer, eds, pp. 33–36. Carol Stream, IL: Allured Publishing Corporation.

Pauli, A. and H. Schilcher, 2009. Anwendungsbeobachtungen mit dem Kamilleninhaltsstoff (−)-α-Bisabolol. Infektionen, Entzündungen, Wunden: 18 Fallbeispiele zu (−)-alpha-Bisabolol. *Kompl. Integr. Med.*, 50: 27–31.

Pauli, A., R. Wölfel, and H. Schilcher, 2007. Fallbeispiel mit (−)-alpha-Bisabolol zur Behandlung einer infizierten Hüft-Endoprothese. *Kompl. Integr. Med.*, 48: 35–38.

Pawar, V.C. and V.S. Thaker, 2006. In vitro efficacy of 75 essential oils against *Aspergillus niger*. *Mycoses*, 49: 316–323.

Pawar, V.C. and V.S. Thaker, 2007. Evaluation of the anti-*Fusarium oxysporum* f. sp. *cicer* and anti-*Alternaria porri* effects of some essential oils. *World J. Microbiol. Biotechnol.*, 23: 1099–1106.

Peana A., Moretti M., and C. Julidano, 1999. Chemical composition and antimicrobial action of the essential oils of *Salvia desoleana* and *S. sclarea*. *Planta Med.*, 65: 751–754.

Pellecuer, J., J. Allegrini, and M.S. DeBuochberg, 1976. Bactericidal and fungicidal essential oils. *Revue de l'Institut Pasteur de Lyon*, 9: 135–159.

Pellecuer, J., M. Jacob, S.M. DeBuochberg et al., 1980. Tests on the use of the essential oils of Mediterranean aromatic plants in conservative odontology. *Plantes medicinales et phytotherapie*, 14: 83–98.

Perrucci, S., F. Mancianti, P.-L. Cioni et al., 1994. In vitro antifungal activity of essential oils against some isolates of *Mycrosporum canis* and *Microsporum gypseum*. *Planta Med.*, 60: 184–186.

Pitarokili, D. and M. Couladis, N. Petsikos-Panayotarou et al., 2002. Composition and antifungal activity on soil-borne pathogens of the essential oil of Salvia sclarea from Greece. *J. Agric. Food Chem.* 50: 6688–6691.

Pizsolitto, A.C., B. Mancini, S.E. Longo Fracalanzza et al., 1975. Determination of antibacterial activity of essential oils officialized by the Brazilian pharmacopeia, 2nd Edition. *Revista da Faculdade de Farmacia e Odontologia de Araraquara.*, 9: 55–61.

Prasad, G., A. Kumar, A.K. Singh et al., 1986. Antimicrobial activity of essential oils of some *Ocimum* species and clove oil. *Fitoterapia*, 57: 429–432.

Rahalison, L., M. Hamburger, M. Monod et al., 1994. Antifungal tests in phytochemical investigations: Comparison of bioautographic methods using phytopathogenic and human pathogenic fungi. *Planta Med.*, 60: 41–44.

Raman, A., U. Weir, and S.F. Bloomfield, 1995. Antimicrobial effects of tea-tree oil and its major components on *Staphylococcus aureus, Staph. epidermidis* and *Propionibacterium acnes*. *Lett. Appl. Microbiol.*, 21: 242–245.

Ranasinghe, L., B. Jayawardena, and K. Abeywickrama, 2002. Fungicidal activity of essential oils of *Cinnamomum zeylanicum* (L.) and *Syzygium aromaticum* (L.) Merr et L.M. Perry against crown rot and anthracnose pathogens isolated from banana. *Lett. Appl. Microbiol.*, 35: 208–211.

Reiss, J., 1982. Einfluss von Zimtrinde auf Wachstum und Toxinbildung von Schimmelpilzen auf Brot und auf die Bildung von Mykotoxinen. *Getreide, Mehl, Brot*, 36: 50–53.

Remmal, A., T. Bouchikhi, K. Rhayour et al., 1993. Improved method for the determination of antimicrobial activity of essential oils in agar medium. *J. Essent. Oil Res.*, 5: 179–184.

Rideal, S. and J.T.A. Walker, 1903. Standardisation of disinfectants. *J. Roy. Sanitary Soc.*, 24: 424–441.

Rossi, P.-G., L. Berti, J. Panighi et al., 2007. Antibacterial action of essential oils from Corsica. *J. Essent. Oil Res.*, 19: 176–182.

Rota, C., J.J. Carraminana, J. Burillo et al., 2004. In vitro antimicrobial activity of essential oils from aromatic plants against selected foodborne pathogens. *J. Food Prot.*, 67: 1252–1256.

Saikia, D., S.P.S. Khanuja, A.P. Kahol et al. 2001. Comparative antifungal activity of essential oils and constituents from three distinct genotypes of *CMYBopogon* sp. *Curr. Sci.*, 80: 1264–1266.

Saksena, N.K. and S. Saksena, 1984. Enhancement in the antifungal activity of some essential oils in combination against some dermatophytes. *Indian Perf.*, 28: 42–45.

Sarbhoy, A.K., J.L. Varshney, M.L. Maheshwari et al., 1978. Effect of some essential oils and their constituents on few ubiquitous moulds. *Zentralbl. Bakteriologie, Parasitenkunde Infektionskrankheiten*, 133: 732–734.

Schales, C., H. Gerlach, and Kosters, J., 1993. Investigations on the antibacterial effect of conifer needle oils on bacteria isolated from the feces of captive Capercaillies (*Tetrao urogallus* L., 1758). *J. Veter. Med. Ser. B*, 40: 381–390.

Schelz, Z., J. Molnar, and J. Hohmann. 2006. Antimicrobial and antiplasmid activities of essential oils. *Fitoterapia*, 77: 279–285.

Shapiro, S., A. Meier, and B. Guggenheim, 1994. The antimicrobial activity of essential oils and essential oil components towards oral bacteria. *Oral Microbiol. Immunol.*, 9: 202–208.

Sharma, S.K. and V.P. Singh, 1979. The antifungal activity of some essential oils. *Indian Drugs Pharm. Ind.*, 14: 3–6.

Shin, S., 2003. Anti-*Aspergillus* activities of plant essential oils and their combination effects with ketoconazole or amphotericin B. *Arch. Pharmacol. Res.*, 26: 389–393.

Shukla, H.S. and S.C. Tripathi, 1987. Antifungal substance in the essential oil of anise (*Pimpinella anisum* L.). *Agric. Biol. Chem.*, 51: 1991–1993.

Singh, G., S. Maurya, M.P. deLampasona et al., 2006. Chemical constituents, antimicrobial investigations and antioxidative potential of volatile oil and acetone extract of star anise fruits. *J. Sci. Food Agric.*, 86: 111–121.

Singh, G., S. Maurya, M.P. DeLampasona et al., 2007. A comparison of chemical, antioxidant and antimicrobial studies of cinnamon leaf and bark volatile oils, oleoresins and their constituents. *Food Chem. Toxicol.*, 45: 1650–1661.

Smith-Palmer, A., J. Stewart, and L. Fyfe, 1998. Antimicrobial properties of plant essential oils and essences against five important food-borne pathogens. *Lett. Appl. Microbiol.*, 26: 118–122.

Smyth, H.F. and H.F. Jr. Smyth, 1932. Action of pine oil on some fungi of the skin. *Arch. Dermat. Syphiol.*, 26: 1079–1085.

Soliman, K.M. and R.I. Badeaa, 2002. Effect of oil extracted from some medicinal plants on different mycotoxigenic fungi. *Food Chem. Toxicol.*, 40: 1669–1675.

Subba, M.S., T.C. Soumithri, and R. Suryanarayana Rao, 1967. Antimicrobial action of citrus oils. *J. Food Sci.*, 32: 225–227.

Suhr, K.I. and P.V. Nielsen, 2003. Antifungal activity of essential oils evaluated by two different application techniques against rye bread spoilage fungi. *J. Appl. Microbiol.*, 94: 665–674.

Szalontai, M., G. Verzar-Petri, and E. Florian, 1977. Beitrag zur Untersuchung der antimykotischen Wirkung biologisch aktiver Komponenten der *Matricaria chamomilla*. *Parfum. Kosmet.*, 58: 121–127.

Thanaboripat, D., N. Mongkontanawut, Y. Suvathi et al., 2004. Inhibition of aflatoxin production and growth of *Aspergillus flavus* by citronella oil. *KMITL Sci. Technol.*, 4: 1–8.

Thompson, D.P., 1986. Effect of some essential oils on spore germination of *Rhizopus*, *Mucor* and *Aspergillus* species. *Mycologia*, 78: 482.

Thompson, D.P. and C. Cannon, 1986. Toxicity of essential oils on toxigenic and nontoxigenic fungi. *Bull. Environ. Contam. Toxicol.*, 36: 527–532.

Tullio, V., A. Nostro, N. Mandras et al., 2006. Antifungal activity of essential oils against filamentous fungi determined by broth microdilution and vapour contact methods. *J. Appl. Microbiol.*, 102: 1544–1550.

Wannissorn, B., S. Jarikasem, T. Siriwangchai et al., 2005. Antibacterial properties of essential oils from Thai medicinal plants. *Fitoterapia*, 76: 233–236.

Warnke, P.H., E. Sherry, P.A.J. Russo et al., 2005. Antibacterial essential oils reduce tumor smell and inflammation in cancer patients. *J. Clinic. Oncol.*, 23: 1588–1589.

Weis, N., 1986. Zur Wirkweise von Terpenoiden auf den Energiestoffwechsel von Bakterien: Wirkung auf respiratorischen Elektronentransport und oxidative Phoshorylierung. *Dissertation*. Erlangen, Germany: Friedrich-Alexander Universität.

Weseler, A., H.K. Geiss, R. Saller et al., 2005. A novel colorimetric broth microdilution method to determine the minimum inhibitory concentration (MIC) of antibiotics and essential oils against *Helicobacter pylori*. *Pharmazie*, 60: 498–502.

Yousef, R.T. and G.G. Tawil, 1980. Antimicrobial activity of volatile oils. *Pharmazie*, 35: 698–701.

Yousefzadi, M., A. Sonboli, F. Karimic et al., 2007. Antimicrobial activity of some *Salvia* species essential oils from Iran. *Zeitschr. Naturforsch. C.*, 62: 514–518.

15 Aromatherapy with Essential Oils

Maria Lis-Balchin

CONTENTS

15.1 Introduction 620
    15.1.1 Aromatherapy Practice in the United Kingdom and the United States 620
15.2 Definitions of Aromatherapy 620
15.3 Introduction to Aromatherapy Concepts 621
    15.3.1 Aromatherapy, Aromatology, and Aromachology 621
    15.3.2 Scientifically Accepted Benefits of Essential Oils versus the Lack
            of Evidence for Aromatherapy 621
15.4 Historical Background to Aromatherapy 623
    15.4.1 Scented Plants Used as Incense in Ancient Egypt 624
15.5 Perfume and Cosmetics: Precursors of Cosmetological Aromatherapy 624
    15.5.1 Three Methods of Producing Perfumed Oils by the Egyptians 625
15.6 Medicinal Uses: Precursors of Aromatology or "Clinical" Aromatherapy 625
    15.6.1 Middle Ages: Use of Aromatics and Quacks 626
15.7 Modern Perfumery 627
15.8 Aromatherapy Practice 627
    15.8.1 Methods of Application of Aromatherapy Treatment 628
15.9 Massage Using Essential Oils 629
    15.9.1 Massage Techniques 629
15.10 Aromatherapy: Blending of Essential Oils 631
    15.10.1 Fixed Oils 631
15.11 Internal Usage of Essential Oils by Aromatherapists 631
15.12 Use of Pure or Synthetic Components 632
15.13 Therapeutic Claims for the Application of Essential Oils 632
    15.13.1 False Claims Challenged in Court 633
15.14 Physiological and Psychological Responses to Essential Oils and Psychophysiology 633
15.15 Placebo Effect of Aromatherapy 634
15.16 Safety Issue in Aromatherapy 635
15.17 Toxicity in Humans 636
    15.17.1 Increase in Allergic Contact Dermatitis in Recent Years 636
    15.17.2 Photosensitizers 637
    15.17.3 Commonest Allergenic Essential Oils and Components 637
    15.17.4 Toxicity in Young Children: A Special Case 638
    15.17.5 Selected Toxicities of Common Essential Oils and Their Components 639
15.18 Clinical Studies of Aromatherapy 640

15.19 Recent Clinical Studies...640
    15.19.1 Aromatherapy in Dementia..640
15.20 Past Clinical Studies 641
    15.20.1 Critique of Selected Clinical Trials..642
15.21 Use of Essential Oils Mainly as Chemical Agents and Not for Their Odor 645
    15.21.1 Single-Case Studies 646
15.22 Conclusion 647
References..647

15.1 INTRODUCTION

15.1.1 Aromatherapy Practice in the United Kingdom and the United States

Aromatherapy has become more of an art than a science. This is mostly due to the health and beauty industries, which have taken over the original concept as a money-spinner in the United Kingdom, United States, and almost all other parts of the world. There are virtually thousands of "aromatherapy" products in pharmacies, high street shops, supermarkets, hair salons, and beauty salons. The products are supposedly made with "essential oils" (which are usually perfumes) and include skin creams, hair shampoos, shower gels, moisturizers, bath salts, lotions, candles, and essential oils themselves.

Many aromatherapy products, such as perfumes, are also linked with sexual attractiveness. There are numerous "health and beauty" salons or clinics that offer aromatherapy as part of their *treatments* together with waxing, electrolysis, massage (of various types, including "no-hands massage"), facial treatments including botox, manicures and pedicures, eyes and eyebrow shaping, ear piercing, tanning, and makeup application. Often hundreds of these *therapies* are offered in one small shop, with aromatherapy thrown in. Most people, especially men, consider aromatherapy to be a sensual massage with some perfumes given all over the body by a young lady. This is often the case, although aromatherapy massage is often provided just on the back or even just on the face and hands for busy people. The use of pure essential oils both in such beauty massage and all the aromatherapy products on sale everywhere is very doubtful (because of the cost), but the purchaser believes the advertisements assuring pure oil usage. Beauty consultants/therapists use massage skills and a nice odor simply for relaxation; they sometimes include beautifying treatments using specific essential oils as initiated by Marguerite Maury (1989). Aromatherapy has thus become an art.

However, aromatherapists (who have studied the "science" for 3 h, a week, a year, or even did a 3-year degree) are keen to bring science into this alternative treatment. The multitude of books written on the subject, aromatherapy journals, and the websites consider that there has been enough proof of the scientific merit of aromatherapy. They quote studies that have shown *no* positive or statistically significant effects as proof that aromatherapy works. The actual validity of these claims will be discussed later and several publications criticized this on scientific grounds. Aromatherapy is often combined with *counseling* by a *qualified* therapist, with no counseling qualifications. Massaging is carried out using very diluted plant essential oils (2–5 drops per 10 mL of carrier oil, such as almond oil) on the skin—that is, in almost homeopathic dilutions. But they believe that the essential oils are absorbed and go straight to the target organ where they exert the healing effect. Many aromatherapists combine their practice with cosmology, crystals, colors, music, and so on. These may also be associated with a commercial sideline in selling "own trademark" essential oils and associated items, including diffusers, scented candles, and scented jewelry.

15.2 DEFINITIONS OF AROMATHERAPY

Aromatherapy is defined as "the use of aromatic plant extracts and essential oils in massage and other treatment" (Concise Oxford Dictionary, 1995). However, there is no mention of massage or the absorption of essential oils through the skin and their effect on the target organ (which is the mainframe of

Aromatherapy with Essential Oils

aromatherapy in the United Kingdom and the United States) in *Aromatherapie* (Gattefossé, 1937/1993). This was where the term "aromatherapy" was coined after all, by the "father of aromatherapy"—but was actually based on the odor of essential oils and perfumes and their antimicrobial, physiological, and cosmetological properties (Gattefossé, 1928, 1952, 1937/1993). *Pure* essential oils were of no concern to Gattefossé. Recently, definitions have begun to encompass the effects of aromatherapy on the mind as well as on the body (Lawless, 1994; Worwood, 1996, 1998; Hirsch, 1998).

15.3 INTRODUCTION TO AROMATHERAPY CONCEPTS

The original concept of modern aromatherapy was based on the assumption that the volatile, fat-soluble essential oil was equivalent in bioactivity to that of the whole plant when inhaled or massaged onto the skin. Information about the medicinal and other properties of the plants was taken from old English herbals (e.g., Culpeper, 1653/1995), combined with some more esoteric nuances involving the planets and astrology (Tisserand, 1977).

This notion is clearly flawed. As an example, a whole orange differs from just the essential oil (extracted from the rind alone) as the water-soluble vitamins (thiamine, riboflavin, nicotinic acid, and vitamins C and A) are excluded, as are calcium, iron, proteins, carbohydrates, and water. Substantial differences in bioactivity are found in different fractions of plants, for example, the essential oils of Pelargonium species produced a consistent relaxation of the smooth muscle of the guinea pig in vitro, whereas the water-soluble extracts did not (Lis-Balchin, 2002b). Botanical misinterpretations are also common in many aromatherapy books, for example, "geranium oil" bioactivity is based on the herb Robert, a hardy Geranium species found widely in European hedgerows, whereas geranium oil is distilled from species of the South African genus Pelargonium (Lis-Balchin, 2002a).

15.3.1 AROMATHERAPY, AROMATOLOGY, AND AROMACHOLOGY

Aromatherapy can now be divided into three "sciences": aromatherapy, aromatology, and aromachology.

Aromachology (coined by the Sense of Smell Institute [SSI], USA, 1982) is based on the interrelationship of psychology and odor, that is, its effect on specific feelings (e.g., relaxation, exhilaration, sensuality, happiness, and achievement) by its direct effect on the brain.

Aromatherapy is defined by the SSI as "the therapeutic effects of aromas on physical conditions (such as menstrual disorders, digestive problems, etc.) as well as psychological conditions (such as chronic depression)." The odor being composed of a mixture of fat-soluble chemicals may thus, have an effect on the brain via inhalation, skin absorption, or even directly via the nose.

Aromatology is concerned with the internal use of oils (SSI). This is similar to the use of aromatherapy in most of Europe, excluding the United Kingdom; it includes the effect of the chemicals in the essential oils via oral intake or via the anus, vagina, or any other possible opening by medically qualified doctors or at least herbalists, using essential oils as internal medicines.

There is a vast difference between aromatherapy in the United Kingdom and that in continental Europe (aromatology): the former is *alternative*, while the latter is *conventional*. The alternative aromatherapy is largely based on *healing*, which is largely based on belief (Millenson, 1995; Benson and Stark, 1996; Lis-Balchin, 1997). This is credited with a substantial placebo influence. However, the placebo effect can be responsible for results in both procedures.

15.3.2 SCIENTIFICALLY ACCEPTED BENEFITS OF ESSENTIAL OILS VERSUS THE LACK OF EVIDENCE FOR AROMATHERAPY

There is virtually no scientific evidence, as yet, regarding the direct action of essential oils, applied through massage on the skin, on specific internal organs—rather than through the odor pathway leading into the midbrain's "limbic system" and then through the normal sympathetic and parasympathetic pathways. This is despite some evidence that certain components of essential oils can be absorbed either through the skin or lungs (Buchbauer et al., 1992; Jager et al., 1992; Fuchs et al., 1997).

Many fragrances have been shown to have an effect on mood and, in general, pleasant odors generate happy memories, more positive feelings, and a general sense of well-being (Knasko et al., 1990; Knasko, 1992; Warren and Warrenburg, 1993) just like perfumes. Many essential oil vapors have been shown to depress contingent negative variation (CNV) brain waves in human volunteers, and these are considered to be sedative (Torii et al., 1988). Others increase CNV and are considered stimulants (Kubota et al., 1992). An individual with anosmia showed changes in cerebral blood flow on inhaling certain essential oils, just as in people able to smell (Buchbauer et al., 1993c), showing that the oil had a positive brain effect despite the patient's inability to smell it. There is some evidence that certain essential oils (e.g., nutmeg) can lower high blood pressure (Warren and Warrenburg, 1993). Externally applied essential oils (e.g., tea tree) can reduce/eliminate acne (Bassett et al., 1990) and athlete's foot (Tong et al., 1992). This happens, however, using conventional chemical effects of essential oils rather than aromatherapy.

Most clients seeking out aromatherapy are suffering from some stress-related conditions, and improvement is largely achieved through relaxation. An alleviation of suffering and possibly pain, due to gentle massage and the presence of someone who cares and listens to the patient, could be beneficial in such cases as in cases of terminal cancer; the longer the time spent by the therapist with the patient, the stronger the belief imparted by the therapist, and the greater the willingness of the patient to believe in the therapy, the greater the effect achieved (Benson and Stark, 1996). There is a need for this kind of healing contact, and aromatherapy with its added power of odor fits this niche, as the main action of essential oils is probably on the primitive, unconscious, limbic system of the brain (Lis-Balchin, 1997), which is not under the control of the cerebrum or higher centers and has a considerable subconscious effect on the person. However, as mood and behavior can be influenced by odors, and memories of past odor associations could also be dominant, aromatherapy should not be used by aromatherapists, unqualified in psychology, and so on in the treatment of Alzheimer's or other diseases of aging (Lis-Balchin, 2006).

Proven uses of essential oils and their components are found in industry, for example, foods, cosmetic products, and household products. They impart the required odor or flavor to food, cosmetics and perfumery, tobacco, and textiles. Essential oils are also used in the paint industry, which capitalizes on the exceptional *cleaning* properties of certain oils. This, together with their embalming properties, suggests that essential oils are very potent and dangerous chemicals—not the sort of natural products to massage into the skin!

Why, therefore, should essential oils be of great medicinal value? They are, after all, just chemicals. However, essential oils have many functions in everyday life ranging from their use in dentistry (e.g., cinnamon and clove oils), as decongestants (e.g., *Eucalyptus globulus*, camphor, peppermint, and cajuput) to their use as mouthwashes (e.g., thyme), also external usage as hyperemics (e.g., rosemary, turpentine, and camphor) and anti-inflammatories (e.g., German chamomile and yarrow). Some essential oils are used internally as stimulants of digestion (e.g., anise, peppermint, and cinnamon) and as diuretics (e.g., buchu and juniper oils) (Lis-Balchin, 2006).

Many plant essential oils are extremely potent antimicrobials in vitro (Deans and Ritchie, 1987; Bassett et al., 1990; Lis-Balchin, 1995; Lis-Balchin et al., 1996a; Deans, 2002). Many are also strong antioxidant agents and have recently been shown to stop some of the symptoms of aging in animals (Dorman et al., 1995a,b). The use of camphor, turpentine oils, and their components as rubefacients, causing increased blood flow to a site of pain or swelling when applied to the skin, is well known and is the basis of many well-known medicaments such as Vicks VapoRub and Tiger Balm. Some essential oils are already used as orthodox medicines: peppermint oil is used for treating irritable bowel syndrome and some components of essential oils, such as pinene, limonene, camphene, and borneol, given orally, have been found to be effective against certain internal ailments, such as gallstones (Somerville et al., 1985) and ureteric stones (Engelstein et al., 1992). Many essential oils have been shown to be active on many different animal tissues, for example, skeletal and/or smooth muscles, in vitro (Lis-Balchin et al., 1997b). There are many examples of the benefits of using essential oils by topical application for acne, alopecia areata, and athlete's foot (discussed in Section 15.21), but this is a treatment using chemicals rather than aromatherapy treatment.

Future scientific studies, such as those on Alzheimer's syndrome (Perry et al., 1998, 1999), may reveal the individual benefits of different essential oils for different ailments, but in practice, this may not be of utmost importance as aromatherapy massage for relief from stress. Aromatherapy has had very little scientific evaluation to date. As with so many alternative therapies, the placebo effect may provide the largest percentage benefit to the patient (Benson and Stark, 1996). Many aromatherapists have not been greatly interested in scientific research, and some have even been antagonistic to any such research (Vickers, 1996; Lis-Balchin, 1997). Animal experiments, whether maze studies using mice or pharmacology using isolated tissues, are considered unacceptable and only essential oils that are "untested on animals" are acceptable, despite all essential oils having been already tested on animals (denied by assurances of essential oil suppliers) because this is required by law before they can be used in foods.

The actual mode of action of essential oils in vivo is still far from clear, and clinical studies to date have been scarce and mostly rather negative (Stevenson, 1994; Dunn et al., 1995; Brooker et al., 1997; Anderson et al., 2000). The advent of scientific input into the clinical studies, rather than aromatherapist-led studies, has recently yielded some more positive and scientifically acceptable data (Burns et al., 2000; Smallwood et al., 2001; Ballard et al., 2002; Holmes et al., 2002; Kennedy et al., 2002). The main difficulty in clinical studies is that it is virtually impossible to do randomized double-blind studies involving different odors as it is almost impossible to provide an adequate control as this would have to be either odorless or else of a different odor, neither of which is satisfactory. In aromatherapy, as practiced, there is a variation in the treatment for each client, based on *holistic* principles, and each person can be treated by an aromatherapist with one to five or more different essential oil mixtures on subsequent visits, involving one to four or more different essential oils in each mixture. This makes scientific evaluation almost useless, as seen by studies during childbirth (Burns and Blaney, 1994; see also Section 15.19). There is also the belief among alternative medicine practitioners that if the procedure *works* in one patient, there is no need to study it using scientific double-blind procedures. There is therefore, a great bias when clinical studies in aromatherapy are conducted largely by aromatherapists.

Recent European regulations (the seventh Amendment to the European Cosmetic Directive 76/768/EEC, 2002; see Appendices 27 and 28) have listed 26 sensitizers found in most of the common essential oils used: this could be a problem for aromatherapists as well as clients, both in possibly causing sensitization and also resulting in legal action regarding such an eventuality in the case of the client. Care must be taken regarding the sensitization potential of the essential oils, especially when massaging patients with cancer or otherwise sensitive skin. It should also be borne in mind when considering the use of essential oils during childbirth and in other clinical studies (Burns and Blaney, 1994; Burns et al., 2000) that studies in animals have indicated that some oils cause a decrease in uterine contractions (Lis-Balchin and Hart, 1997a).

15.4 HISTORICAL BACKGROUND TO AROMATHERAPY

The advent of "aromatherapy" has been attributed to both the Ancient Egyptians and Chinese over 4500 years ago, as scented plants and their products were used in religious practices, as medicines, perfumes, and embalming agents (Manniche, 1989, 1999), and to bring out greater sexuality (Schumann Antelme and Rossini, 2001). But essential oils as such were unlikely to have been used. In Ancient Egypt, crude plant extracts of frankincense, myrrh, or galbanum, and so on were used in an oily vegetable or animal fat that was massaged onto the bodies of workers building the pyramids or the rich proletariat after their baths (Manniche, 1999). These contained essential oils, water-soluble extractives, and pigments. Incense smoke from resinous plant material provided a more sacrosanct atmosphere for making sacrifices, both animal and human, to the gods. The incense was often mixed with narcotics like cannabis to anesthetize the sacrificial animals, especially with humans (Devereux, 1997). The frankincense extract in oils (citrusy odor) was entirely different to that burnt (church-like) in chemical composition (Arctander, 1960), and therefore, would have entirely different functions.

15.4.1 SCENTED PLANTS USED AS INCENSE IN ANCIENT EGYPT

Frankincense (*Boswellia carterii, Boswellia thurifera*) (Burseraceae), myrrh (*Commiphora myrrha, Balsamodendron myrrha, Balsamodendron opobalsamum*) (Burseraceae), labdanum (*Cistus ladanifer*), galbanum (*Ferula galbaniflua*), styrax (*Styrax officinalis*), or *Liquidambar orientalis*, Balm of Gilead (*Commiphora opobalsamum*), sandalwood (*Santalum album*), and opopanax (*Opopanax chironium*).

Uses included various concoctions of kyphi, burnt three times a day to the sun god Ra, morning, noon, and sunset, in order for him to come back. The ingredients included raisins, juniper, cinnamon, honey, wine, frankincense, myrrh, burnt resins, cyperus, sweet rust, sweet flag, and aspalathus in a certain secret proportion (Forbes, 1955; Loret, 1887; Manniche, 1989), as shown on the walls of the laboratory in the temples of Horus at Edfu and Philae. Embalming involved odorous plants such as juniper, cassia, cinnamon, cedar wood, and myrrh, together with natron to preserve the body and ensure safe passage to the afterlife. The bandages in which the mummy was wrapped were drenched in stacte (oil of myrrh) and sprinkled with other spices (for further descriptions and uses, see Lis-Balchin, 2006).

The Chinese also used an incense, *hsiang*, meaning "aromatic," made from a variety of plants, with sandalwood being particularly favored by Buddhists. In India, fragrant flowers including jasmine and the root of spikenard giving a sweet scent were used. The Hindus obtained cassia from China and were the first to organize trading routes to Arabia where frankincense was exclusively found. The Hebrews traditionally used incense for purification ceremonies. The use of incense probably spread to Greece from Egypt around the eighth century B.C. The Indians of Mesoamerica used copal, a hard, lustrous resin, obtained from pine trees and various other tropical trees by slicing the bark (*Olibanum americanum*). Copal pellets bound to corn-husk tubes would be burnt in hollows on the summits of holy hills and mountains, and these places, blackened by centuries of such usage, are still resorted to by today's Maya in Guatemala (Janson, 1997) and used medicinally to treat diseases of the respiratory system and the skin.

Anointing also involves incense (Unterman, 1991). Queen Elizabeth II underwent the ritual in 1953 at her coronation, with a composition of oils originated by Charles I: essential oils of roses, orange blossom, jasmine petals, sesame seeds, and cinnamon combined with gum benzoin, musk, civet, and ambergris were used (Ellis, 1960). Similarly, musk, sandalwood, and other fragrances were used by the Hindus to wash the effigies of their gods, and this custom was continued by the early Christians. This probably accounts for the divine odor frequently reported when the tombs of early Christians were opened (Atchley and Cuthbert, 1909). The Christian Church was slow to adopt the use of incense until medieval times, when it was used for funerals (Genders, 1972). The reformation reversed the process as it was considered to be of pagan origin, but it still survives in the Roman Catholic Church. Aromatic substances were also widely used in magic (Pinch, 1994).

15.5 PERFUME AND COSMETICS: PRECURSORS OF COSMETOLOGICAL AROMATHERAPY

The word "perfume" is derived from the Latin term *per fumare*: "by smoke." The preparation of perfumes in Ancient Egypt was done by the priests, who passed on their knowledge to new priests (Manniche, 1989, 1999). Famous people like Nefertiti and Cleopatra and the workers building the great pyramids, who even went on strike when they were denied their allocation of "aromatherapy massage oil," used huge amounts of fragranced materials as unguents, powders, and perfumes (Manniche, 1999).

15.5.1 Three Methods of Producing Perfumed Oils by the Egyptians

Enfleurage involved steeping the flowers or aromatics in oils or animal fats (usually goat) until the scent from the materials was imparted to the fat. The impregnated fat was often molded into cosmetic cones and used for perfuming hair wigs, worn on festive occasions, which could last for 3 days; the fat would soften and start melting, spreading the scented grease not only over the wig, but also over the clothes and body—more pleasing than the stench of stale wine, food, and excrement (Manniche, 1999).

Maceration was used principally for skin creams and perfumes: flowers, herbs, spices, or resins were chopped up and immersed in hot oils. The oil was strained and poured into alabaster (calcite) containers and sealed with wax. These scented fatty extracts were also massaged onto the skin (Manniche, 1999).

Expression involved putting flowers or herbs into bags or presses, which extracted the aromatic oils. Expression is now only used for citrus fruit oils (Lis-Balchin, 1995). Wine was often included in the process, and the resulting potent liquid was stored in jars. These methods are still used today.

Megaleion, an Ancient Greek perfume described by Theophrastus who believed it to be good for wounds, was made of burnt resins and balanos oil and boiled for 10 days before adding cassia, cinnamon, and myrrh (Groom, 1992). Rose, marjoram, sage, lotus flower, and galbanum perfumes were also made. Apart from these, aromatic oils from basil, celery, chamomile, cumin, dill, fenugreek, fir, henna, iris, juniper, lily, lotus, mandrake, marjoram, myrtle, pine, rose, rue, and sage were sometimes used in perfumes or as medicines taken internally and externally.

Dioscorides, in his *De Materia Medica*, discussed the components of perfumes and their medicinal properties, providing detailed perfume formulae. Alexandrian chemists were divided into three schools, one of which was the school of Maria the Jewess, which produced pieces of apparatus for distillation and sublimation, such as the *bain-marie*, useful for extracting the aromatic oils from plant material. Perfumes became more commonly known in medieval Europe as knights returning from the Crusades brought back musk, floral waters, and a variety of spices.

15.6 MEDICINAL USES: PRECURSORS OF AROMATOLOGY OR "CLINICAL" AROMATHERAPY

The ancient use of plants, not essential oils, can be found in fragments of Egyptian herbals. The names of various plants, their habitats, characteristics, and the purposes for which they were used are included in the following: *Veterinary papyrus* (ca. 2000 B.C.), *Gynaecological papyrus* (ca. 2000 B.C.), *Edwin Smith Papyrus* (an army surgeon's manual, ca. 1600 B.C.), *Papyrus Ebers* (includes remedies for health, beauty, and the home, ca. 1600 B.C.), *Hearst Papyrus* (with prescriptions and spells, ca. 1400 B.C.), and *Demotic medical papyri* (second century B.C. to first century A.D.).

Magic was often used as part of the treatment and gave the patient the expectation of a cure and thus, provided a placebo effect (Pinch, 1994). The term "placing the hand" appears frequently in a large number of medical papyri; this probably alludes to the manual examination in order to reach a diagnosis but could also imply cure by the "laying on of hands," or even both (Nunn, 1997). This could be the basis of modern massage (with or without aromatherapy). It is certainly the basis of many alternative medicine practices at present (Lis-Balchin, 1997).

Plants were used in numerous ways. Onions were made into a paste with wine and inserted into the vagina to stop a woman menstruating. Garlic ointment was used to keep away serpents and snakes, heal dog bites, and bruises; raw garlic was given to asthmatics; fresh garlic and coriander in wine was a purgative and an aphrodisiac! Juniper mixed with honey and beer was used orally to encourage defecation; and origanum was boiled with hyssop for a sick ear (Manniche, 1989).

Egyptians also practiced inhalation by using a double-pot arrangement whereby a heated stone was placed in one of the pots and a liquid herbal remedy poured over it. The second pot, with a hole in the bottom through which a straw was inserted, was placed on top of the first pot, allowing the patient to breathe in the steaming remedy (Manniche, 1989), that is, aromatherapy by inhalation.

15.6.1 MIDDLE AGES: USE OF AROMATICS AND QUACKS

In the twelfth century, the Benedictine Abbess Hildegard of Bingen (1098–1179) was authorized by the Church to publish her visions on medicine (*Causae et Curae*), dealing with the causes and remedies for illness (Brunn and Epiney-Burgard, 1989). The foul smell of refuse in European towns in the seventeenth century was thought to be the major cause of disease, including the plague (Classen et al., 1994), and aromatics were used for both preventing and in some cases curing diseases; herbs such as rosemary were in great demand and sold for exorbitant prices as a prophylactic against the plague (Wilson, 1925). People forced to live near victims of the plague would carry a pomander, which contained a mixture of aromatic plant extracts. Medical practitioners carried a small cassolette or "perfume box" on the top of their walking sticks, when visiting contagious patients, which was filled with aromatics (Rimmel, 1865). Some physicians wore a device filled with herbs and spices over their nose when they examined plague patients (Wilson, 1925). These became known as "beaks," and it is from this word that the term "quack" developed.

Apothecaries were originally wholesale merchants and spice importers, and in 1617, the Worshipful Society of Apothecaries was formed, under the control of the London Royal College of Physicians, which produced an *official* pharmacopoeia specifying the drugs the apothecaries were allowed to dispense. The term "perfumer" occurs in some places instead of "apothecary" (Rimmel, 1865).

John Gerard (1545–1612) and Nicholas Culpeper (1616–1654) were two of the better-known apothecaries of their time. Nicholas Culpeper combined healing herbs with astrology as he believed that each plant, like each part of the body, and each disease, was governed or under the influence of one of the planets: rosemary was believed to be ruled by the Sun, lavender by Mercury, and spearmint by Venus. Culpeper also adhered to the doctrine of signatures, introduced by Paracelsus in the sixteenth century, and mythology played a role in many of the descriptive virtues in Culpeper's herbal. This astrological tradition is carried through by many aromatherapists today, together with other innovations such as yin and yang, crystals, and colors.

Culpeper's simple or distilled waters and oils (equivalent to the present hydrosols) were prepared by the distillation of herbs in water in a pewter still, and then fractionating them to separate out the essential or "chymical" oil from the scented plants. The plant waters were the weakest of the herbal preparations and were not regarded as being beneficial. Individual plants such as rose or elderflower were used to make the corresponding waters, or else mixtures of herbs were used to make compound waters (Culpeper, 1826/1981; Tobyn, 1997). Essential oils of single herbs were regarded by Culpeper as too strong to be taken alone, due to their vehement heat and burning, but had to be mixed with other medicinal preparations. Two or three drops were used in this way at a time. Culpeper mentioned the oils of wormwood, hyssop, marjoram, mints, oregano, pennyroyal, rosemary, rue, sage, thyme, chamomile, lavender, orange, and lemon. Spike lavender, not *Lavandula angustifolia*, is used in aromatherapy nowadays. Herbs such as dried wormwood and rosemary were also steeped in wine and set in the sun for 30–40 days to make a "physical wine." The "herbal extracts" mentioned in the herbals were mostly water soluble and at best, alcoholic extracts, none of which are equivalent to essential oils, which contain many potent chemical components are not found in essential oils.

15.7 MODERN PERFUMERY

In the fourteenth century, alcohol was used for the extraction and preservation of plants, and oleum mirable, an alcoholic extract of rosemary and resins, was later popularized as "Hungary water," without the resins (Müller et al., 1984).

In the sixteenth century, perfumes were made using animal extracts, which were the base notes or fixatives, and made the scent last longer (Piesse, 1855). Among these ingredients were ambergris, musk, and civet.

Perfumes came into general use in England during the reign of Queen Elizabeth (1558–1603). Many perfumes, such as rose water, benzoin, and storax, were used for sweetening the heavy ornate robes of the time, which were impossible to wash. Urinals were treated with orris powder, damask rose powder, and rose water. Bags of herbs, musk, and civet were used to perfume bath water.

Elizabeth I carried a pomander filled with ambergris, benzoin, civet, damask rose, and other perfumes (Rimmel, 1865) and used a multitude of perfumed products in later life. Pomanders, from the French *pomme d'amber* ("ball of ambergris"), were originally hung in silver perforated balls from the ceiling to perfume the room. The ingredients such as benzoin, amber, labdanum, storax, musk, civet, and rose buds could be boiled with gum tragacanth and kneaded into balls; the small ones were made into necklaces.

Various recipes were used for preparing aromatic waters, oils, and perfumes. Some of these were for perfumes and some undoubtedly for alcoholic beverages, as one of the major ingredients for many concoctions was a bottle or two of wine, which when distilled produced a very alcoholic brew.

Ambergris, musk, and civet went out of fashion, as the excremental odors could not be reconciled with modesty (Corbin, 1986). The delicate floral perfumes became part of the ritual of bodily hygiene, gave greater variety, and allowed Louis XV a different perfume every day. Today, the sentiment "odours are carried in bottles, for fear of annoying those who do not like them" (Dejeans, 1764) is reemerging, as more and more people are becoming sensitive to odors, giving them headaches, asthma, and migraines.

The Victorians liked simple perfumes made of individual plant extracts. Particular favorites were rose, lavender, and violet. These would be steam distilled or extracted with solvents. The simple essential oils produced would often be blended together to produce perfumes like eau de cologne (1834).

The first commercial scent production was produced in the United Kingdom, in Mitcham, Surrey, in the seventeenth century, using lavender (Festing, 1989). In 1865, cinnamaldehyde, the first synthetic, was made. Adulteration and substitution by the essential oil or component of another plant species became rampant. Aroma chemicals synthesized from coal, petroleum by-products, and terpenes are much cheaper than the equivalent plant products, so perfumes became cheap.

The way was now open for the use of scent in the modern era. It seems therefore, a retrograde step to use pure essential oils in "aromatherapy," especially as the "father of aromatherapy," René-Maurice Gattefossé, used scents or deterpenated essential oils.

15.8 AROMATHERAPY PRACTICE

Aromatherapists usually treat their clients (patients) after an initial full consultation, which usually involves taking down a full medical case history. The aromatherapist then decides what treatment to give, which usually involves massage with three essential oils, often one each chosen from those with top, middle, and base perfumery notes, which balances the mixture. Sometimes, only *specific* essential oils for the *disease* are used. Most aromatherapists arrange to see the client three to five times and the mixture will often be changed on the next visit, if not on each visit, in order to treat all the possible symptoms presented by the client (holistically), or simply as a substitute when no improvement was initially obtained. Treatment may involve other alternative medicine procedures, including chakras.

Many aromatherapists offer to treat any illness, as they are convinced that essential oils have great powers. They embark on the treatment of endometriosis, infertility, asthma, diabetes, and arthritis, even cancer, as they are convinced of the therapeutic nature of essential oils, but are often without the necessary scientific and medical knowledge. "Psychoneuroimmunology" treatment is the current buzzword.

Although aromatherapists consider themselves professionals, there is no Hippocratic oath involved. The aromatherapist, being nonmedically qualified, may not even be acquainted with most of the illnesses or symptoms, so there could be a very serious mistake made as potentially serious illnesses could be adversely affected by being *treated* by a layperson. Some, but not all, aromatherapists ask the patients to tell their doctor of the aromatherapy treatment. Counseling is greatly recommended by aromatherapy schools. Aromatherapists are not necessarily, however, trained in counseling and, with few exceptions, could do more damage than good, especially when dealing with psychiatric illness, cases of physical or drug abuse, people with learning difficulties, and so on, where their *treatment* should only be complementary and under a doctor's control (Lis-Balchin, 2006).

15.8.1 Methods of Application of Aromatherapy Treatment

Various methods are used to apply the treatment in aromatherapy. The most usual methods are the following:

- A diffuser, usually powered by electricity, giving out a fine mist of the essential oil.
- A burner, with water added to the fragrance to prevent burning of the essential oil. About 1–4 drops of essential oil are added to about 10 mL water. The burner can be warmed by candles or electricity. The latter would be safer in a hospital or a children's room or even a bedroom.
- Ceramic or metal rings, placed on an electric light bulb with a drop or two of essential oil. This results in a rapid burnout of the oil and lasts for a very short time due to the rapid volatilization of the essential oil in the heat.
- A warm bath with drops of essential oil added. This results in the slow volatilization of the essential oil, and the odor is inhaled via the mouth and nose. Any effect is not likely to be through the absorption of the essential oil through the skin as stated in aromatherapy books, as the essential oil does not mix with water. Droplets either form on the surface of the water, often coalescing, or else the essential oil sticks to the side of the bath. Pouring in an essential oil mixed with milk serves no useful purpose as the essential oil still does not mix with water, and the premixing of the essential oil in a carrier oil, as for massage, just results in a nasty oily scum around the bath.
- A bowl of hot water with drops of essential oil, often used for soaking feet or used as a bidet. Again, the essential oil does not mix with the water. This is, however, a useful method for inhaling essential oils in respiratory conditions and colds; the essential oil can be breathed in when the head is placed over the container and a towel placed over the head and container. This is an established method of treatment and has been used successfully with Vicks VapoRub, olbas oil, and *Eucalyptus* oils for many years, so it is not surprising that it works with aromatherapy essential oils!
- Compresses using essential oil drops on a wet cloth, either hot or cold, to relieve inflammation, treat wounds, and so on. Again, the essential oil is not able to mix with water and can be concentrated in one or two areas, making it a possible health hazard.
- Massage of hands, feet, back, or all over the body using 2–4 drops of essential oil (single essential oil or mixture) diluted in 10 mL carrier oil (fixed, oily), for example, almond oil or jojoba oil, grape-seed, wheat-germ oils, and so on. The massage applied is usually by gentle effleurage with some petrissage (kneading), with and without some

shiatsu, lymph drainage in some cases, and is more or less vigorous, according to the aromatherapist's skills and beliefs.
- Oral intake is more like a conventional than an alternative usage of essential oils. Although it is practiced by a number of aromatherapists, this is not to be condoned unless the aromatherapist is medically qualified. Essential oil drops are *mixed* in a tumbler of hot water or presented on a sugar cube or *mixed* with a teaspoonful of honey and taken internally. The inability of the essential oil to mix with aqueous solutions presents a health hazard, as do the other methods, as such strong concentrations of essential oils are involved.

15.9 MASSAGE USING ESSENTIAL OILS

The most popular method of using aromatherapy is through massage. The first written records referring to massage date back to its practice in China more than 4000 years and in Egypt. Hippocrates, the father of modern medicine, wrote "the physician must be experienced in many things, but most assuredly in rubbing."

Massage has been used for centuries in Ayurvedic medicine in India as well as in China and shiatsu, acupressure, reflexology, and many other contemporary techniques have their roots in these sources. Massage was used for conventional therapeutic purposes in hospitals before World War II and is still used by physiotherapists for various conditions including sports injuries.

René-Maurice Gattefossé, credited as being the founding father of modern aromatherapy, never made a connection between essential oils and massage. It was Marguerite Maury who advocated the external use of essential oils combined with carrier oils (Maury, 1989). She used carefully selected essential oils for cleansing the skin, including that in acne, using a unique blend of oils for each client created specifically for the person's temperament and health situation. Maury's main focus was on rejuvenation; she was convinced that aromas could be used to slow down the aging process if the correct oils were chosen. In recent experiments on animals, it has been shown that the oral intake of some antioxidant essential oils can appear to defer aging, as indicated by the composition of membranes in various tissues (Youdim and Deans, 2000).

Massage *per se* can be a relaxing experience and can help to alleviate the stresses and strains of daily life. In a review of the literature on massage, Vickers (1996) found that in most studies massage had no psychological effect, in a few studies there was arousal, and in an even smaller number of studies there was sedation; some massage has both local and systemic effects on blood flow and possibly on lymph flow and reduction of muscle tension.

It may be that these variable responses are directly related to the variability of massage techniques, of which there are over 200. Massage can be given over the whole body or limited to the face, neck, or just hands, feet, legs—depending on the patient and his or her condition or illness, for example, patients with learning disabilities, and many psychiatric patients are often only able to have limited body contact for a short time.

15.9.1 MASSAGE TECHNIQUES

Massage is customarily defined as the manual manipulation of the soft tissues of the body for therapeutic purposes, using strokes that include gliding, kneading, pressing, tapping, and/or vibrating (Tisserand, 1977; Price and Price, 1999). Massage therapists may also cause movement within the joints, apply heat or cold, use holding techniques, and/or advise clients on exercises to improve muscle tone and range of motion. Some common massage techniques include Swedish massage, acupressure, craniosacral therapy, deep tissue massage, infant massage, lymph system massage, polarity therapy, reflexology, reiki, rolfing, shiatsu, and therapeutic touch.

Massage usually involves the use of a lubricating oil to help the practitioner's hands glide more evenly over the body. The addition of perfumed essential oils further adds to its potential to relax. In most English-speaking countries, massage is nowadays seen as an alternative or complementary treatment. However, before World War II, it was regarded as a conventional treatment (Goldstone, 1999, 2000), as it is now in continental Europe. In Austria, for example, most patients with back pain receive (and are usually reimbursed for) massage treatment (Ernst, 2003a).

Not all massage treatments are free of risk. Too much force can cause fractures of osteoporotic bones, and even rupture of the liver and damage to nerves have been associated with massage (Ernst, 2003b). These events are rarities, however, and massage is relatively safe, provided that well-trained therapists observe the contraindications: phlebitis, deep vein thrombosis, burns, skin infections, eczema, open wounds, bone fractures, and advanced osteoporosis (Ernst et al., 2001).

It is not known exactly how massage works, although many theories abound (Vickers, 1996; Ernst et al., 2001). The mechanical action of the hands on cutaneous and subcutaneous structures enhances circulation of blood and lymph, resulting in increased supply of oxygen and removal of waste products or mediators of pain (Goats, 1994). Certain massage techniques have been shown to increase the threshold for pain (Dhondt et al., 1999). Also, most importantly from the standpoint of aromatherapy, a massage can relax the mind and reduce anxiety, which could positively affect the perception of pain (Vickers, 1996; Ernst, 2003a). Many studies have been carried out, most of which are unsatisfactory. It appears that placebo-controlled, double-blind trials may not be possible, yet few randomized clinical trials have been forthcoming.

Different client groups require proper recognition before aromatherapy trials are started or aromatherapy massage is given. For example, for cancer patients, guidelines must be observed (Wilkinson et al., 1999): special care must be taken for certain conditions such as autoimmune disease (where there are tiny bruises present); low blood cell count, which makes the patient lethargic and needing nothing more than very gentle treatment; and lymphedema, which should not be treated unless the therapist has special knowledge and where enfleurage toward the lymph nodes should not be used.

Recent individual studies to investigate the benefit of massage for certain complaints have given variable results. Many are positive, although the standard of the studies has, in general, been poor (Vickers, 1996). The most successful applications of massage or aromatherapy massage have been in cancer care, and about a third of patients with cancer use complementary/alternative medicine during their illness (Ernst and Cassileth, 1998). Massage is commonly provided within UK cancer services (Kohn, 1999), and although only anecdotal and qualitative evidence is available, it is considered by patients to be beneficial. Only a few small-scale studies among patients with cancer have identified short-term benefits from a course of massage, mainly in terms of reduced anxiety (Corner et al., 1995; Kite et al., 1998; Wilkinson et al., 1999). These studies have been criticized by scientists, however, as they were either nonrandomized, had inadequate control groups, or were observational in design (Cooke and Ernst, 2000). Complementary therapy practitioners have criticized medical research for not being sufficiently holistic in approach, focusing on efficacy of treatments in terms of tumor response and survival, rather than quality of life (Wilkinson, 2003).

A general study of the clinical effectiveness of massage by Ernst (1994) used numerous trials, with and without control groups. A variety of control interventions were used in the controlled studies including placebo, analgesics, transcutaneous electrical nerve stimulation, and so on. There were some positive effects of vibrational or manual massage, assessed as improvements in mobility, Doppler flow, expiratory volume, and reduced lymphedema in controlled studies. Improvements in musculoskeletal and phantom limb pain, but not cancer pain, were recorded in controlled studies. Uncontrolled studies were invariably positive. Adverse effects included thrombophlebitis and local inflammation or ulceration of the skin.

Different megastudies included massage for delayed-onset muscle soreness—seven trials were included with 132 patients in total (Ernst, 1998); effleurage backrub for relaxation—nine trials were included with a total of 250 patients (Labyak and Metzger, 1997), and massage for low-back pain (Ernst, 1999a,b). All gave positive and negative outcomes.

15.10 AROMATHERAPY: BLENDING OF ESSENTIAL OILS

There are numerous suggestions for the use of particular essential oils for treating specific illnesses in books on aromatherapy. However, when collated, each essential oil can treat each illness (Vickers, 1996; compare also individual essential oil monographs in Lis-Balchin, 2006).

A few drops of the essential oil or oils chosen are always mixed with a carrier oil before being applied to the skin for an aromatherapy massage. The exact dilution of the essential oils in the carrier oil is often controversial and can be anything from 0.5% to 20% and more. Either 5, 10, or 20 mL of carrier oil is first poured into a (usually brown) bottle with a stoppered dropper. The essential oil is then added dropwise into the carrier oil, either as a single essential oil or as a mixture of 2–3 different essential oils, and then stoppered.

Volumes of essential oils used for dilutions vary widely in different aromatherapies and the fact that even the size of a "dropper" varies raised the question of possible safety problems (Lis-Balchin, 2006), and a recent article in a nursing journal makes a request for standardization of the measurement of the dropper size (Ollevant et al., 1999).

15.10.1 FIXED OILS

Many fixed oils are used for dilution and all provide a lubricant; many have a high vitamin E and A content. By moistening the skin, they can assist in a variety of mild skin conditions especially where the skin is rough, cracked, or dry (Healey and Aslam, 1996).

Almond (*Prunus amygdalus* var. *dulcis*) is sweet, cheapest, and most commonly used. Others include apricot kernel (*Prunus armeniaca*), borage seed (*Borago officinalis*), calendula (*Calendula officinalis*), coconut oil (*Cocos nucifera*), evening primrose (*Oenothera biennis*), grape seed (*Vitis vinifera*), macadamia nut (*Macadamia integrifolia*), olive (*Olea europaea*), rose hip seed (*Rosa mosqueta*), soya bean (*Glycine soja*), sunflower (*Helianthus annuus*), wheat germ (*Triticum vulgare*), and jojoba (*Simmondsia californica*). The latest oil in vogue is emu oil (*Dromaius novaehollandiae*), which comes from a thick pad of fat on the bird's back. For centuries, the aborigines of Australia have been applying emu oil to their wounds with excellent results. It is now found in muscle pain relievers, skin care products, and natural soaps.

The exact method of mixing is controversial, but most aromatherapists are taught not to shake the bottle containing the essential oil(s) and the diluent fixed oils, but to gently mix the contents by turning the bottle in the hand. Differences in the actual odor and thereby presumable benefits of the diluted oils made by different aromatherapists can just be due to the different droppers (Lis-Balchin, 2006).

15.11 INTERNAL USAGE OF ESSENTIAL OILS BY AROMATHERAPISTS

Oral intake of essential oils is not true "aromatherapy" as the odor has virtually no effect past the mouth, and the effect of the chemical components takes over as odors cannot influence the internal organs (Lis-Balchin, 1998a). Most aromatherapists consider that essential oils should only be prescribed by primary care practitioners such as medical doctors or medical herbalists who have intimate knowledge of essential oil toxicology (Tisserand and Balacs, 1995). In the United Kingdom, such "clinical aromatherapy" is rare, unlike on the continent. Maladies treated include arthritis, bronchitis, rheumatism, chilblains, eczema, high blood pressure, and venereal diseases. In clinical aromatherapy, there is a real risk of overdosage due to variable droppers on bottles, which can differ by as much as 200% (Lis-Balchin, 2006); this may be the cause of asphyxiation of a baby, as already shown by peppermint oil (Bunyan, 1998). It is possible that aromatherapists would not be covered by their insurance if there were adverse effects. However, most of us ingest small amounts of essential oils and their components daily almost all processed foods and drinks, but it does not make us all healthy.

Conventional drugs involving essential oils and their components have been used internally for a long time, for example, decongestants containing menthol, camphor, and pine, and various throat drops containing components from essential oils such as lemon, thyme, peppermint, sage, and hyssop.

Essential oils in processed foods are used in very minute amounts of 10 ppm but can be 1000 ppm in mint confectionery or chewing gum (Fenaroli, 1997). This contrasts greatly with the use of drops of undiluted essential oils on sugar lumps for oral application or on suppositories in anal or vaginal application. Damage to mucous membranes could result due to the high concentration of the essential oils in certain areas of the applicator.

Essential oils and their components are incorporated into enterically coated capsules to prevent damage and used for treating irritable bowel syndrome (peppermint in Colpermin), a mixture of monoterpenes for treating gallstones (Rowachol) and ureteric stones (Rowatinex); these are under product licenses as medicines (Somerville et al., 1984, 1985; Engelstein et al., 1992).

Some aromatherapists support the use of essential oils in various venereal conditions. However, aromatherapists are qualified either to treat venereal disease conditions or to make an accurate diagnosis in the first place, unless they are also medically qualified. Tea tree oil (2–3 drops undiluted) was used on a tampon for candidiasis with apparently very encouraging results (Zarno, 1994). *Candida* treatments also include chamomile, lavender, bergamot, and thyme (Schnaubelt, 1999). Essential oils used in this way, sometimes for months, often produced extremely painful reactions and putrid discharges due to damage to delicate mucosal membranes.

15.12 USE OF PURE OR SYNTHETIC COMPONENTS

Does it really matter whether the essential oil is pure or a synthetic mixture as long as the odor is the same? The perfumers certainly do not see any difference and even prefer the synthetics as they remain constant. Many of the so-called pure essential oils used today are, however, adulterated (Lis-Balchin et al., 1996, 1998; Which Report, 2001). There is often a difference in the proportion of different enantiomers of individual components that often have different odors and different biological properties (Lis-Balchin, 2002a,b). This was not, however, appreciated by Gattefosse (1937/1993), who worked with perfumes and not with the "pure plant essential oils" (*Formulaires de Parfumerie Gattefossé*, 1906). He studied the antimicrobial and wound-healing properties of essential oils on soldiers during World War I (Arnould-Taylor, 1981). He later worked in hospitals on the use of perfumes and essential oils as antiseptics and other (unstated) applications, and also in dermatology, which led to advances in the development of beauty products and treatments and the publication of *Physiological Aesthetics and Beauty Products* in 1936 (Gattefosse, 1992).

Gattefossé promoted the deterpenization of essential oils because, being a perfumer, he was aware that his products must be stable, have a long shelf life, and will not go cloudy when diluted in alcohol. Terpenes also oxidize rapidly, often giving rise to toxic oxidation products (e.g., limonene of citrus essential oils). But this goes against the use of pure essential oils, as their wholeness or natural synergy is apparently destroyed (Price, 1993). Bergamot and other citrus essential oils obtained by expression are therefore recommended, despite their phototoxicity (Price and Price, 1999). There is no reason why a toxic essential oil should be preferentially used if the nontoxic and furanocoumarin-free (FCF) alternative is available. If adverse effects resulted, it is possible that there could be legal implications for the therapist.

15.13 THERAPEUTIC CLAIMS FOR THE APPLICATION OF ESSENTIAL OILS

There are a wide range of properties ascribed to each essential oil in aromatherapy books, without any scientific proof of effectiveness (Vickers, 1996; Lis-Balchin, 2006). The following are a few examples.

Aromatherapy with Essential Oils

Diabetes can be treated by eucalyptus, geranium, and juniper (Tisserand, 1977); clary sage, eucalyptus, geranium, juniper, lemon, pine, red thyme, sweet thyme, vetiver, and ylang ylang (Price, 1993); eucalyptus, geranium, juniper, and onion (Valnet, 1982); and eucalyptus, geranium, cypress, lavender, hyssop, and ginger (Worwood, 1991).

Allergies can be treated by immortelle, chamomile, balm, and rose (Fischer-Rizzi, 1990); lemon balm, chamomile (German and Roman), helichrysum, true lavender, and spikenard (Lawless, 1992); and chamomile, jasmine, neroli, and rose (Price, 1983).

No botanical names are, however, given in the lists, even when there are several possible species. No indication is provided as to why these particular essential oils are used and how they are supposed to affect the condition. Taking the case of diabetes, where there is a lack of the hormone insulin, it is impossible to say how massage with any given essential oil could cure the condition, without giving the hormone itself in juvenile-type diabetes or some blood glucose–decreasing drugs in late-onset diabetes. Unfortunately, constant repetition of a given statement often lends it credence—at least to the layperson, who does not require scientific evidence of its validity.

15.13.1 False Claims Challenged in Court

The false promotion of products for treating not only medical conditions but also well-being generally is now being challenged in the law courts. For example, in 1997, Los Angeles Attorney Morsé Mehrban charged that Lafabre and Aroma Vera had violated the California Business and Professions Code by advertising that their products could promote health and well-being, relax the body, relax the mind, enhance mood, purify the air, are antidotes to air pollution, relieve fatigue, tone the body, nourish the skin, promote circulation, alleviate feminine cramps, and do about 50 other things (Barrett, 2000). In September 2000, the case was settled out of court with a $5700 payment to Mehrban and a court-approved stipulation prohibiting the defendants from making 57 of the disputed claims in advertising within California (Horowitz, 2000).

15.14 PHYSIOLOGICAL AND PSYCHOLOGICAL RESPONSES TO ESSENTIAL OILS AND PSYCHOPHYSIOLOGY

Many examples of essential oil effects abound in animal studies, for example, the sedative action of lavender on the overall activity of mice decreased when exposed to lavender vapor (*L. angustifolia* P. Miller); its components linalool and linalyl acetate showed a similar effect (Buchbauer et al., 1992). A possible explanation for the observed sedative effects was shown by linalool, which produced a dose-dependent inhibition of the binding of glutamate (an excitatory neurotransmitter in the brain) to its receptors on membranes of the rat cerebral cortex (Elisabetsky et al., 1995). More recently, this action was related to an anticonvulsant activity of linalool in rats (Elisabetsky et al., 1999). Other oils with sedative activity were found to be neroli and sandalwood; active components included citronellal, phenylethyl acetate, linalool, linalyl acetate, benzaldehyde, terpineol, and isoeugenol (in order of decreasing activity).

Stimulant oils included jasmine, patchouli, ylang ylang, basil, and rosemary; active components included fenchone, 1,8-cineole, isoborneol, and orange terpenes (Lis-Balchin, 2006). There was considerable similarity in the sedative and stimulant effects of some essential oils studied physiologically (e.g., their effect on smooth muscle of the guinea pig in vitro) and in various psychological assessments, mostly on humans (Lis-Balchin, 2006).

1,8-Cineole when inhaled showed a decreased blood flow through the brain (measured using computerized tomography), although no changes were found with lavender oil or linalyl acetate (Buchbauer et al., 1993c). Changing electrical activity, picked up by scalp electrodes, in response to lavender odors was considered a measure of brain activity by electro encephalogram (EEG) (Van Toller et al., 1993). The most consistent responses to odors were in the theta band (Klemm et al., 1992). Many essential oil vapors have been shown to depress CNV brain waves (an upward shift in EEG waves that occurs when people are expecting something to happen) in human volunteers, and these are considered to be sedatives; others increase CNV and are considered stimulants: lavender was found to have a sedative effect on humans (Torii et al., 1988; Kubota et al., 1992; Manley, 1993) and had a *positive* effect on mood, EEG patterns, and math computations (Diego et al., 1998). It also caused reduced motility in mice (Kovar et al., 1987; Ammon, 1989; Buchbauer et al., 1992, 1993a,b,c; Jager et al., 1992). However, Karamat et al. (1992) found that lavender had a stimulant effect on decision times in human experiments.

A large workplace in Japan with odorized air via the whole building showed that citrus smells refreshed the workers first thing in the morning and after the lunch break, and floral smells improved their concentration in between. In the lunch break and during late afternoon, woodland scents were circulated to relax the workers, and this increased productivity (Van Toller and Dodd, 1991). It is also possible that the use of a general regime of odorants could have very negative effects on some members of the workforce or on patients in hospital wards, where the use of pleasant odors could mask the usual unpleasant odors providing the smell of fear. Ambient odors have an effect on creativity, mood, and perceived health (Knasko, 1992, 1993) and so does feigned odor (Knasko et al., 1990).

It is very difficult to make simple generalizations concerning the effects of any fragrance on psychological responses, which are based on the immediate perceptual effects, rather than the longer-term pharmacological effects because the pharmacological effect is likely to affect people similarly, but the additional psychological mechanisms will create complex effects at the individual level. Odors are perceptible even during sleep, as shown in another experiment; college students were tested with fragrances during the night and the day (Badia, 1991).

Various nonscientific studies have been published in perfumery journals on the treatment of psychiatric patients by psychoaromatherapy in the 1920s (Gatti and Cajola, 1923a,b, 1929; Tisserand, 1997), but there was virtually no information on their exact illnesses. Sedative essential oils or essences were identified as chamomile, melissa, neroli, petitgrain, opopanax, asafoetida, and valerian. Stimulants were angelica, cardamom, lemon, fennel, cinnamon, clove, and ylang ylang. Many aromatherapists have also written books on the effect of essential oils on the mind, giving directives for the use of specific plant oils for treating various conditions, without any scientific proof (Lawless, 1994; Worwood, 1996, 1998; Hirsch, 1998).

15.15 PLACEBO EFFECT OF AROMATHERAPY

The placebo effect is an example of a real manifestation of mind over matter. It does not confine itself to alternative therapies, but there is a greater likelihood of the placebo effect accounting for over 90% of the effect in the latter (Millenson, 1995). Reasons for the potency of the placebo effect are either the patient's belief in the method, the practitioner's belief in the method, or the patient and practitioner's belief in each other, that is, the strength of their relationship (Weil, 1983).

Placebo effects have been shown to relieve postoperative pain, induce sleep, or mental awareness, bring about drastic remission in both symptoms and objective signs of chronic diseases, initiate the rejection of warts and other abnormal growths, and so on (Weil, 1983). Placebo affects headaches, seasickness, and coughs, as well as have beneficial effects on pathological conditions such as rheumatoid and degenerative arthritis, blood cell count, respiratory rates, vasomotor function, peptic ulcers, hay fever, and hypertension (Cousins, 1979). There can also be undesirable side effects, such as nausea, headaches, skin rashes, allergic reactions, and even addiction, that is, a nocebo effect. This is almost akin to voodoo death threats or when patients are mistakenly told that their illness is hopeless—both are said to cause death soon after.

Rats were found to have increased levels of opioids in their brains after inhaling certain essential oils. Opioids are a factor in pain relief (Lis-Balchin, 1998b) and can be increased in the body by autosuggestion, relaxation, belief, and so on.

Aromatherapy with Essential Oils

The use of aromatherapy for pain relief is best achieved through massage, personal concern and touch of the patient, and also listening to their problems. The extra benefit of real *healers* found among aromatherapists is an added advantage.

15.16 SAFETY ISSUE IN AROMATHERAPY

Many aromatherapists and laymen consider natural essential oils to be completely safe. This is based on the misconception that all herbs are safe—because they are *natural*, which is a fallacy. The toxicity of essential oils can also be entirely different to that of the herb, not only because of their high concentration, but also because of their ability to pass across membranes very efficiently due to their lipophilicity.

Some aromatherapists erroneously believe that aromatherapy is self-correcting, unlike conventional therapy with medicines, and if errors are made in aromatherapy, they may be resolved through discontinuation of the wrongful application of the oil (e.g., Schnaubelt, 1999).

Many essential oils are inherently toxic at very low concentrations due to very toxic components; these are not normally used in aromatherapy. Many essential oils that are considered to be nontoxic can have a toxic effect on some people; this can be influenced by previous sensitization to a given essential oil, a group of essential oils containing similar components, or some adulterant in the essential oil. It can also be influenced by the age of the person; babies and young children are especially vulnerable and so are very old people. The influence of other medicaments, both conventional and herbal, is still in the preliminary stages of being studied. It is possible that these medicaments, and also probably household products, including perfumes and cosmetics, can influence the adverse reactions to essential oils.

Aromatherapists themselves have also been affected by sensitization (Crawford et al., 2004); in a 12-month period under study, prevalence of hand dermatitis in a sample of massage therapists was 15% by self-reported criteria and 23% by a symptom-based method and included the use of aromatherapy products in massage oils, lotions, or creams. In contrast, the suggestion that aromatherapists have any adverse effects to long-term usage of essential oils was apparently disproved by a nonscientific survey (Price and Price, 1999).

As most essential oils were tested over 30 years ago, the toxicity data may now be meaningless, as different essential oils are now used, some of which contain different quantities of many different synthetic components (Lis-Balchin, 2006).

The major drawbacks of trying to extrapolate toxicity studies in animals to humans concern feelings—from headaches to splitting migraines; feeling sick, vertigo, profound nausea; tinnitus; sadness, melancholia, suicidal thoughts; feelings of hate—which are clearly impossible to measure in animals (Lis-Balchin, 2006). The toxicity of an individual essential oil/component is also tested in isolation in animals and disregards the possibility of modification by other substances, including food components and food additive chemicals, the surrounding atmosphere with gaseous and other components, fragrances used in perfumes, domestic products, in the car, in public transport (including the people), workplace, and so on. These could cause modification of the essential oil/component, its bioavailability, and possibly the enhancement or loss of its function. The detoxification processes in the body are all directed to the production of a more polar product(s), which can be excreted mainly by the kidneys regardless of whether this/these are more toxic or less toxic than the initial substance and differ in different animals.

Most essential oils have generally recognized as safe (GRAS) status granted by the Flavor and Extract Manufacturers Association (FEMA) and approved by the U.S. Food and Drug Administration for food use, and many appear in the food chemical codex. This was reviewed in 1996 after evaluation by the expert panel of the FEMA. The assessment was based on data of exposure, and as most flavor ingredients are used at less than 100 ppm, predictions regarding their safety can be assessed from the data on their structurally related group(s) (Munro et al., 1996). The no-observed-adverse-effect levels are more than 100,000 times their exposure levels from use as flavor ingredients (Adams et al., 1996). Critical to GRAS assessment are data of metabolic fate and chronic studies rather than acute toxicity. Most essential oils and components have an LD50 of 1–20 g/kg body weight or roughly 1–20 mL/kg, with a few exceptions as follows: boldo leaf oil 0.1/0.9 (oral/dermal), calamus 0.8–9/5, *Chenopodium* 0.2/0.4, pennyroyal 0.4/4, and *Thuja* 0.8/4.

Research Institute for Fragrance Materials (RIFM) testing is generally limited to acute oral and dermal toxicity, irritation and dermal sensitization, and phototoxicity of individual materials, and there is little effort to address synergistic and modifying effects of materials in combination (Johansen et al., 1998).

Many materials that were widely used for decades in the past had severe neurotoxic properties and accumulated in body tissues (Spencer et al., 1979; Furuhashi et al., 1994), but most fragrance materials have never been tested for neurological effects, despite the fact that olfactory pathways provide a direct route to the brain (Hastings et al., 1991).

15.17 TOXICITY IN HUMANS

The most recent clinical review of the adverse reactions to fragrances (de Groot and Frosch, 1997) showed many examples of cutaneous reactions to essential oils reported elsewhere (Guin, 1982, 1995). In the United States, about 6 million people have a skin allergy to fragrance, and this has a major impact on their quality of life. Symptoms include headaches, dizziness, nausea, fatigue, shortness of breath, and difficulty in concentrating. Fragrance materials are readily absorbed into the body via the respiratory system and once absorbed they cause systemic effects.

Migraine headaches are frequently triggered by fragrances that can act on the same receptors in the brain as alcohol and tobacco, altering mood and function [Institute of Medicine (United States), sponsored by the Environmental Protection Agency]. Perfumes and fragrances are recognized as triggers for asthma by the American Lung Association. The vast majority of materials used in fragrances are respiratory irritants, and there are a few that are known to be respiratory sensitizers. Most have *not* been evaluated for their effects on the lungs and the respiratory system.

Respiratory irritants are known to make the airways more susceptible to injury and allergens, as well as to trigger and exacerbate conditions such as asthma, allergies, sinus problems, and other respiratory disorders. In addition, there is a subset of asthmatics that is specifically triggered by fragrances (Shim and Williams, 1986; Bell et al., 1993; Baldwin et al., 1999), which suggests that fragrances not only trigger asthma, but they may also cause it in some cases (Millqvist and Lowhagen, 1996). Placebo-controlled studies using perfumes to challenge people with asthma-like symptoms showed that asthma could be elicited with perfumes without the presence of bronchial obstruction, and these were not transmitted by the olfactory nerve as the patients were unaware of the smell (Millqvist and Lowhagen, 1996).

Adverse reactions to fragrances are difficult or even impossible to link to a particular chemical—often due to secrecy rules of the cosmetic/perfumery companies and the enormous range of synthetic components, constituting about 90% of flavor and fragrance ingredients (Larsen, 1998). The same chemicals are used in foods and cosmetics—there is, therefore, a greater impact due to the three different modes of entry: oral, inhalation, and skin.

15.17.1 Increase in Allergic Contact Dermatitis in Recent Years

A study of 1600 adults in 1987 showed that 12% reacted adversely to cosmetics and toiletries, 4.3% of which were used for their odor (i.e., they contained high levels of fragrances). Respiratory problems worsened with prolonged fragrance exposure (e.g., at cosmetic/perfumery counters) and even in churches. In another study, 32% of the women tested had adverse reactions, and 80% of these had positive skin tests for fragrances (deGroot and Frosch, 1987). Problems with essential oils have also been increasing. For example, contact dermatitis and allergic contact dermatitis (ACD) caused by tea tree oil has been reported, which was previously considered to be safe (Carson and Riley, 1995).

Aromatherapy with Essential Oils

It is unclear whether eucalyptol was responsible for the allergenic response (Southwell, 1997); out of seven patients sensitized to tea tree oil, six reacted to limonene, five to α-terpinene and aromadendrene, two to terpinen-4-ol, and one to *p*-cymene and α-phellandrene (Knight and Hausen, 1994).

Many studies on ACD have been done in different parts of the world (deGroot and Frosch, 1997), and the more recent studies are listed here:

- *Japan* (Sugiura et al., 2000): The patch test with lavender oil was found to be positive in increased numbers and above that of other essential oils in 10 years.
- *Denmark* (Johansen et al., 2000): There was an 11% increase in the patch test in the last year, and of 1537 patients, 29% were allergic to scents.
- *Hungary* (Katona and Egyud, 2001): Increased sensitivity to balsams and fragrances was noted.
- *Switzerland* (Kohl et al., 2002): ACD incidence has increased over the years, and recently, 36% of 819 patch tests were positive to cosmetics.
- *Belgium* (Kohl et al., 2002): Increased incidence of ACD has been noted.

Occupational increases have also been observed. Two aromatherapists developed ACD: one to citrus, neroli, lavender, frankincense, and rosewood and the other to geraniol, ylang ylang, and angelica (Keane et al., 2000). Allergic airborne contact dermatitis from the essential oils used in aromatherapy was also reported (Schaller and Korting, 1995). ACD occurred in an aromatherapist due to French marigold essential oil, *Tagetes* (Bilsland and Strong, 1990). A physiotherapist developed ACD to eugenol, cloves, and cinnamon (Sanchez-Perez and Garcia Diez, 1999).

There is also the growing problem that patients with eczema are frequently treated by aromatherapists using massage with essential oils. A possible allergic response to a variety of essential oils was found in children with atopic eczema, who were massaged with or without the oils. At first, both massages proved beneficial, though not significantly different; but on reapplying the essential oil massage after a month's break, there was a notable adverse effect on the eczema, which could suggest sensitization (Anderson et al., 2000).

15.17.2 Photosensitizers

Berloque dermatitis is frequently caused by bergamot or other citrus oil applications on the skin (often due to their inclusion in eau de cologne) followed by exposure to UV light. This effect is caused by psoralens or furanocoumarins (Klarmann, 1958). Citrus essential oils labeled FCF have no phototoxic effect but are suspected carcinogens (Young et al., 1990). Other phototoxic essential oils include yarrow and angelica, neroli, petitgrain, cedar wood, rosemary, cassia, calamus, cade, eucalyptus (species not stated), orange, anise, bay, bitter almond, ylang ylang, carrot seed, and linaloe (the latter probably due to linalool, which, like citronellol, has a sensitizing methylene group exposed) (Guin, 1995). Photosensitizer oils include cumin, rue, dill, sandalwood, lemon (oil and expressed), lime (oil and expressed), opopanax, and verbena (the latter being frequently adulterated) (Klarmann, 1958). Even celery soup eaten before UV irradiation has been known to cause severe sunburn (Boffa et al., 1996).

Many of these photosensitizers are now banned or restricted. New International Fragrance Research Association (IFRA) proposals for some phototoxic essential oils include rue oil to be 0.15% maximum in consumer products, marigold oil and absolute to be 0.01%, and petitgrain mandarin oil to be 0.165%.

15.17.3 Commonest Allergenic Essential Oils and Components

The most common fragrance components causing allergy are cinnamic alcohol, hydroxycitronellal, musk ambrette, isoeugenol, and geraniol (Scheinman, 1996). These are included in the eight commonest markers used to check for ACD, usually as a 2% mix. Other components considered allergenic are benzyl salicylate, sandalwood oil, anisyl alcohol, benzyl alcohol, and coumarin.

IFRA and RIFM have forbidden the use of several essential oils and components, including costus root oil, dihydrocoumarin, musk ambrette, and balsam of Peru (Ford, 1991); a concentration limit is imposed on the use of isoeugenol, cold-pressed lemon oil, bergamot oil, angelica root oil, cassia oil, cinnamic alcohol, hydroxycitronellal, and oakmoss absolute. Cinnamic aldehyde, citral, and carvone oxide can only be used with a quenching agent.

Photosensitivity and phototoxicity occur with some allergens such as musk ambrette and 6-methyl coumarin that are now removed from skin care products. Children were often found to be sensitive to Peru balsam, probably due to the use of baby-care products containing this (e.g., talcum powder used on nappy rash).

Fragrance materials have been found to interact with food flavorings, for example, a "balsam of Peru-free diet" has been devised in cases where cross reactions are known to occur (Veien et al., 1985). *Newer* sensitizers include ylang ylang (Romaguera and Vilplana, 2000), sandalwood oil (Sharma et al., 1994) but much of this essential oil is adulterated or completely synthetic, lyral (Frosch et al., 1999; Hendriks et al., 1999), and eucalyptol (Vilaplana and Romaguera, 2002).

Some sensitizers have been shown to interact with other molecules. For example, cinnamaldehyde interacts with proteins (Weibel et al., 1989), indicating how the immunogenicity occurs.

There have been very few published reports on neurotoxic aroma chemicals such as musk ambrette (Spencer et al., 1984), although many synthetic musks took over as perfume ingredients when public opinion turned against the exploitation of animal products. Musk ambrette was found to have neurotoxic properties in orally fed mice in 1967 and was readily absorbed through the skin. A similar story occurred with acetyl-ethyl-tetramethyl-tetralin, another synthetic musk, also known as versalide, patented in the early 1950s. During routine tests for irritancy in 1975, it was noted that with repeated applications, the skin of the mice turned bluish, and they exhibited signs of neurotoxicity. The myelin sheath was damaged irreversibly in a manner similar to that which occurs with multiple sclerosis. Musk xylene, one of the commonest fragrance materials, is found in blood samples from the general population (Kafferlein et al., 1998) and bound to human hemoglobin (Riedel et al., 1999). These musk products have been found to have an effect on the life stages of experimental animals such as the frog, *Xenopus laevis*, the zebra fish, *Danio rerio* (Chou and Dietrich, 1999), and the rat (Christian et al., 1999). The hepatotoxic effect of musks is under constant study (Steinberg et al., 1999).

15.17.4 Toxicity in Young Children: A Special Case

Many aromatherapy books give dangerous advice on the treatment of babies and children, for example, 5–10 drops of "chamomile oil" three times a day in a little warmed milk given to their babies to treat colic with no indication as to which of the three commercially available chamomile oils is to be used and because, depending on the dropper size, the dose could easily approach the oral LD50 for the English and German chamomile oils, this could result in a fatality. Peppermint, often mentioned, could possibly be given by mothers in the form of oil and has been known to kill a 1-week-old baby (*Evening Standard*, 1998). Dosages given in terms of drops can vary widely according to the size of the dropper in an essential oil.

Many "cosmetics" designed for use by children contain fragrance allergens (Rastogi et al., 1999). In Denmark, samples of children's cosmetics were found to contain geraniol, hydroxycitronellol, isoeugenol, and cinnamic alcohol (Rastogi et al., 1999). Children are more susceptible than adults to any chemical, so the increase in childhood asthma reported in recent years could be caused by fragrance components also found in fast foods. Aromatherapy therefore, could be dangerous.

15.17.5 Selected Toxicities of Common Essential Oils and Their Components

Limonene and linalool are found in a multitude of the commonest aromatherapy oils.

Limonene is a common industrial cleaner and is also the main citrus oil component, which causes ACD, particularly when aged (Chang et al., 1997; Karlberg and Dooms-Goossens, 1997). The major volatile component of lactating mothers' milk in the United States was found to contain *d*-limonene, and the component is used as a potential skin penetration promoter for drugs such as indomethacin, especially when mixed with ethanol (Falk-Filipsson et al., 1993). Lastly, cats and dogs are very susceptible to insecticides and baths containing *d*-limonene, giving rise to neurological symptoms including ataxia, stiffness, apparent severe CNS depression, tremors, and coma (von Burg, 1995; see also Beasley, 1999).

Linalool, when oxidized for just 10 weeks, fell to 80%, and the remaining 20% consisted of a range of breakdown chemicals including linalool hydroperoxide, which was confirmed as a sensitizing agent. The fresh linalool was not a sensitizer; therefore, the European Commission (EC) regulations that are warnings about sensitization potential are looking for potential harm even on storage (Skoeld et al., 2002a,b).

Most cosmetics and perfumes are tested on human "guinea pigs" using similar tests to those described for animals. These are demanded by the RIFM as a final test before marketing a product. Further data are accumulated from notifications from disgruntled consumers who report dermatitis, itching, or skin discoloration after use. These notifications can result in legal claims, although most cases are probably settled out of court and not reported to the general public.

The Internet has made it possible for a trusting, although often ill informed, public to purchase a wide range of dubious plant extracts and essential oils. Even illegal essential oils can now be obtained. Furthermore, unqualified people can offer potentially dangerous advice, such as internal usage or the use of undiluted essential oils on the skin for "mummification" or in order to rid the body of toxic waste. The latter can result in excruciating pain from the burns produced and the subsequent loss of layers of skin.

There is a recipe for suntan oil, including bergamot, carrot seed, and lemon essential oils (all phototoxic) in an aromatherapy book (Fischer-Rizzi, 1990). The author then advises that bergamot oil is added to suntan lotion to get the bonus of the substance called "furocoumarin," which lessens the skin's sensitivity to the sun while it helps one to tan quickly. This could cause severe burns. Elsewhere, sassafras (*Ocotea pretiosa*) was said to be only toxic for rats, due to its metabolism and not dangerous to humans (Pénoel, 1991a,b), and a 10% solution in oil was suggested for treating muscular and joint pain and sports injuries. Safrole (and sassafras oil) is, however, controlled under the Controlled Drugs Regulations (1993) and listed as a Category 1 substance, as it is a precursor to the illicit manufacture of hallucinogenic, narcotic, and psychotropic drugs like ecstasy.

French practitioners and other therapists have apparently become *familiar* with untested oils (Guba, 2000). The use of toxicologically untested Nepalese essential oils and the like includes lichen resinoids, sugandha kokila oil, jatamansi oil, and Nepalese lemongrass (*Cymbopogon flexuosus*), also *Tagetes* oil (Basnyet, 1999). *Melaleuca rosalina* (*Melaleuca ericifolia*), 1,8-cineole 18%–26%, is apparently especially useful for the respiratory system (Pénoel, 1998), but it is untested and could be a sensitizer.

The Medicines and Healthcare Products Regulatory Agency in the United Kingdom may bring about changes in aromatherapy practice similar to their threat on herbal remedies. Aromatherapists are now using some potentially harmful products in their therapy. This immediately places them at serious risk if there is any untoward reaction to their specific treatment. It virtually means that bottles and containers of essential oils now rank with domestic bleach for labeling purposes and companies are now obliged to self-classify their essential oils on their labels and place them in suitable containers; this applies both to large distributing companies as well as individual aromatherapists reselling essential oils under their own name. Finally, new legislation has gone to the Council of Ministers and may imply that only qualified people will be able to use essential oils, and retail outlets for oils will be pharmacies. Their definition of "qualified" is limited to academic qualifications—doctors or pharmacists.

15.18 CLINICAL STUDIES OF AROMATHERAPY

Very few scientific clinical studies on the effectiveness of aromatherapy have been published to date. Perhaps, the main reason is that until recently, scientists were not involved, and people engaging in aromatherapy clinical studies had accepted the aromatherapy doctrine in its entirety, precluding any possibility of a nonbiased study. This has been evident in the design and execution of the studies; the main criterion has usually been the use of massage with essential oils and not the effect of the odorant itself. The latter is considered by most aromatherapists as irrelevant to clinical aromatherapy, which implies that it is simply the systemic action of essential oils absorbed through the skin that exerts an effect on specific organs or tissues. Odorant action is considered to be just "aromachology," despite its enormous psychological and physiological impact (Lis-Balchin, 2006). In some studies, attempts are even made to bypass the odorant effect entirely by making the subjects wear oxygen masks throughout (Dunn et al., 1995).

The use of particular essential oils for certain medical conditions is also adhered to, despite the wide assortment of supposed functions for each essential oil claimed by different aromatherapy source materials. In many studies, it is even unclear exactly which essential oil was used; as often the correct nomenclature, chemical composition, and exact purity are not given.

Many aromatherapists feel that they know that aromatherapy works as they have enormous numbers of case studies to prove it. But the production of lists of *positive* results on diverse clients, with diverse ailments, using diverse essential oils in the treatments, and diverse methods of application (which also frequently change from visit to visit for the same client) does not satisfy scientific criteria.

Negative results must surely be among the positive ones, due to the change in essential oils during the course of the treatment, which suggests that they did not work, but these are never stated. There are also no controls in case studies and no attempt to control the bias of the individual aromatherapist and clients.

Double-blind studies are *not* possible in *individual case* studies. Physiological or psychological changes due to the treatment are not properly defined, and loose phrases such as "the client felt better" or "happier" are inappropriate for a scientific study.

These faults in the design and interpretation of results of aromatherapy research have been pointed out many times, for example, in Vickers (1996) Kirk-Smith (1996), Nelson (1997), and Lis-Balchin (2002b). However, the lack of statistically significant results does not prevent many aromatherapists from accepting vaguely positive clinical research results, and numerous poor-grade clinical studies are now quoted as factual confirmations that aromatherapy works.

It is almost impossible to do a double-blind study using odorants, as the patient and treatment provider would experience the odor differences and would inevitably react knowingly or unknowingly to that factor alone. The psychological effect(s) could be very diverse, as recall of odors can bring about very acute reactions in different people, depending on the individual's past experiences and on the like (Lis-Balchin, 2006). Lastly, there is potential bias as patients receiving aromatherapy treatment could be grateful for the attention given to them and, not wanting to upset the givers of such attention, would state that they were better and happier than before.

15.19 RECENT CLINICAL STUDIES

15.19.1 Aromatherapy in Dementia

A meticulously conducted double-blind study involved 72 dementia patients with clinically significant agitation treated with melissa oil (Ballard et al., 2002). Agitation included anxiety and irritability, motor restlessness, and abnormal vocalization—symptoms that often lead to disturbed behaviors such as pacing, wandering, aggression, shouting, and nighttime disturbance, all characterized by appropriate inventories.

Ten percent (by weight) melissa oil (active) or sunflower oil (placebo), combined with a base lotion (*Prunus dulcis* oil, glycerine, stearic acid, cetearyl alcohol, and tocopheryl acetate), was dispensed in metered doses and applied to the face and both arms twice daily for 4 weeks by a care assistant, the process taking 1–2 min. The patients also received neuroleptic treatment and other conventional treatments when necessary; this was therefore, a study of complementary aromatherapy treatment—not an alternative treatment.

The "melissa group" showed a higher significant improvement in reducing aggression than the control group by week 4; the total Cohen–Mansfield Agitation Inventory scores had decreased significantly in both groups, from a mean of 68 to 45 (35%; $P < .0001$) in the treatment group and from 61 to 53 (11%; $P < .005$) in the placebo group. Clinically, significant reduction in agitation occurred in 60% of the melissa group compared with 14% of placebo responders ($P < .0001$). Neuropsychiatric Inventory scores also declined with melissa treatment, and quality of life was improved, with less social isolation and more involvement in activities. The latter was in contrast to the usual neuroleptic treatment effects.

The authors concluded that the melissa treatment was successful but pointed out that there was also a significant, but lower, improvement in the control group and suggested that a stronger odor should have been used.

The effect of the melissa oil was probably on cholinergic receptors as shown by previous in vitro studies (Perry et al., 1999; Wake et al., 2000). The authors also concluded that as most people with severe dementia have lost any meaningful sense of smell, a direct placebo effect due to a pleasant-smelling fragrance, although possible, is an unlikely explanation for the positive effects of melissa in this study, but others may disagree with this conclusion as it has been shown that subliminal odors can have an effect. The fragrance may have had some impact upon the care staff and influenced ratings to some degree on the informant schedules.

A further recent study found no support for the use of a purely olfactory form of aromatherapy to decrease agitation in severely demented patients using lavender and thyme oil (Snow et al., 2004).

Other research (Burns et al., 2002) suggested that aromatherapy and light therapy were more effective and gentler alternatives to the use of neuroleptics in patients with dementia. Three studies were analyzed in each category; in the aromatherapy section, it included the aforementioned study, plus the use of 2% lavender oil via inhalation in a double-blind study for 10 days (Holmes et al., 2002) and a 2-week single-blind study using either aromatherapy plus massage, aromatherapy plus conversation, or massage alone (Smallwood et al., 2001). All of the interventions in the aromatherapy groups proved significantly beneficial. However, so did the light treatment, where patients sat in front of a light box that beamed out 10,000 lux of artificial light, which adjusts the body's melatonin levels, affects the body clock, and is used in the treatment of seasonal affective disorder.

15.20 PAST CLINICAL STUDIES

In contrast to more recent studies, past clinical trials were often very defective in design and also outcomes. In a recent review, Cooke and Ernst (2000) included only those aromatherapy trials that were randomized and included human patients; they excluded those with no control group or if only local effects (e.g., antiseptic effects of tea tree oil) or preclinical studies on healthy volunteers occurred. The six trials included massage with or without aromatherapy (Buckle, 1993; Stevenson, 1994; Corner et al., 1995; Dunn et al., 1995; Wilkinson, 1995; Wilkinson et al., 1999) and were based on their relaxation outcomes. The authors concluded that the effects of aromatherapy were probably not strong enough for it to be considered for the treatment of anxiety or for any other indication.

A further study included trials with no replicates and contained six studies. It showed that in five out of six cases, the main outcomes were positive; however, these were limited to very specific criteria, such as small airways resistance for common colds (Cohen and Dressler, 1982), prophylaxis of bronchi for bronchitis (Ferley et al., 1989), lessening smoking withdrawal symptoms

(Rose and Behm, 1994), relief of anxiety (Morris et al., 1995), and treatment of alopecia areata (Hay et al., 1998). The alleviation of perineal discomfort (Dale and Cornwell, 1994) was not significant.

Psychological effects, which include inhalation of essential oils and behavioral changes, were already discussed.

15.20.1 Critique of Selected Clinical Trials

The following clinical studies attempted to show that aromatherapy was more efficient than massage alone, but they showed mainly negative results; however, in some cases, the authors clearly emphasized some very small positive results, and this was then accepted and the report was welcomed in aromatherapy journals as a positive trial that supported aromatherapy.

Massage, aromatherapy massage, or a period of rest in 122 patients in an intensive care unit (ICU) (Dunn et al., 1995) showed no difference between massage with or without lavender oil and no treatment in the physiological parameters and all psychological parameters showed no effects throughout, bar a significantly greater improvement in mood and in anxiety levels between the rest group and essential oil massage group after the first session. The trial had a large number of changeable parameters: it involved patients in the ICU for about 5 days (age range 2–92 years), who received 1–3 therapy sessions in 24 h given by six different nurses. Massage was performed on the back or outside of limbs or scalp for 15–30 min with lavender (*Lavandula vera* at 1% in grape-seed oil, which was the only constant parameter). The patients wore oxygen masks, for some of the time. It seems unlikely that confused patients in ICU could remember the massage or its effects, and a child of 2 years could not be expected to answer any pertinent questions.

Massage with and without Roman chamomile in 51 palliative care patients (Wilkinson, 1995) showed that both groups experienced the same decrease in symptoms and severity after three full body massages in 3 weeks. There was, however, a statistically significant difference between the two groups after the first aromatherapy massage and also an improvement in the "quality of life" from pre- to postmassage. German chamomile was likely to have been used, not Roman chamomile as stated, according to the chemical composition and potential bioactivity given.

Aromatherapy with and without massage, and massage alone on disturbed behavior in four patients with severe dementia (Brooker et al., 1997), was an unusual single-case study evaluating the use of *true* aromatherapy (using inhaled lavender oil) for 10 treatments of each, randomly given to each patient over a 3-month period and assessed against 10 no-treatment periods. Two patients became more agitated following their treatment sessions, and only one patient seemed to have benefited. According to the staff providing the treatment, however, the use of all the treatments seemed to have been beneficial to the patients, suggesting pronounced bias.

An investigation of the psychophysiological effects of aromatherapy massage following cardiac surgery (Stevenson, 1994) showed experimenter bias due to the statement that "neroli is also especially valuable in the relief of anxiety, it calms palpitations, has an antispasmodic effect and an anti-inflammatory effect ... it is useful in the treatment of hysteria, as an antidepressant and a gentle sedative." None of this has been scientifically proven, but as this was not a double-blind study and presumably the author did the massaging, communicating, and collating information alone, bias is probable. Statistical significances were not shown, nor the age ranges of the 100 patients, and no differences between the aromatherapy-only and massage-only groups were shown, except for an immediate increase in respiratory rate when the two control groups (20 min chat or rest) were compared with the aromatherapy massage and massage-only groups.

Atopic eczema in 32 children treated by massage with and without essential oils (Anderson et al., 2000) in a single-case experimental design across subjects showed that this complementary therapy provided no statistically significant differences between the two groups after 8 weeks of treatment. This indicated that massage and thereby regular parental contact and attention showed positive results, which was expected in these children. However, a continuation of the study, following a 3-month period of rest, using only the essential oil massage group showed a possible sensitization effect, as the symptoms worsened.

Massage using two different types of lavender oil on postcardiotomy patients (Buckle, 1993) was proclaimed to be a *double-blind* study but had no controls, and the results by the author did not appear to be assessed correctly (Vickers, 1996). The author attempted to show that the *real* lavender showed significant benefits in the state of the patients compared with the other oil. However, outcome measures were not described, and the chemical composition and botanical names of the *real* and *not real* lavender remains a mystery, as three lavenders were stated in the text. Although the results were insignificant, this paper is quoted widely as proof that only *real* essential oils work through aromatherapy massage.

Aromatherapy trails in childbirth have been of dubious design and low scientific merit and, not surprisingly, have yielded confusing results (Burns and Blaney, 1994), mainly due to the numerous parameters incorporated. In the study by Burns and Blaney (1994), many different essential oils were used in various uncontrolled ways during childbirth and assessed using possibly biased criteria as to their possible benefits to the mother and midwife. The first pilot study used 585 women in a delivery suite over a 6-month period using lavender, clary sage, peppermint, eucalyptus, chamomile, frankincense, jasmine, lemon, and mandarin. These oils were either used singly or as part of a mixture where they could be used as the first, second, third, or fourth essential oil. The essential oils were applied in many different ways and at different times during parturition, for example, sprayed in a *solution* in water onto a face flannel, pillow, or bean bag, in a bath, foot bath, an absorbent card for inhalation, or in almond oil for massage. Peppermint oil was applied as an undiluted drop on the forehead and frankincense onto the palm.

Midwives and mothers filled in a form as to the effects of the essential oils including their relaxant value, effect on nausea and vomiting, analgesic action, mood enhancer action, accelerator, or not of labor. The results were inconclusive, and there was a bias toward the use of a few oils, for example, lavender was stated to be "estrogenic and used to calm down uterine tightenings if a woman was exhausted and needed sleep," and clary sage was given to "encourage the establishment of labor." This shows complete bias and a belief in unproven clinical attributes by the authors and presumably those carrying out the study. Which of the lavender, peppermint, eucalyptus, chamomile, or frankincense species were used remains a mystery.

The continuation of this study (Burns et al., 2000) on 8058 mothers during childbirth was intended to show that aromatherapy would "relieve anxiety, pain, nausea and/or vomiting, or strengthen contractions." Data from the unit audit were compared with those of 15,799 mothers not given aromatherapy treatment. The results showed that 50% of the aromatherapy group mothers found the intervention "helpful" and only 14% "unhelpful." The use of pethidine over the year declined from 6% to 0.2% by women in the aromatherapy group. The study also (apparently) showed that aromatherapy may have the potential to augment labor contractions for women in dysfunctional labor, in contrast to scientific data showing that the uterine contractions decrease due to administration of any common essential oils (Lis-Balchin and Hart, 1997a).

It is doubtful whether a woman would in her first labor, or in subsequent ones, be able to judge whether the contractions were strengthened or the labor shortened due to aromatherapy. It seems likely that there was some placebo effect (itself a very powerful effector) due to the bias of the experimenters and the *suggestions* made to the aromatherapy group regarding efficacy of essential oils, which were obviously absent in the case of the control group.

Lavender oil (volatilized from a burner during the night in their hospital room) has been successful in replacing medication to induce sleep in three out of four geriatrics (Hardy et al., 1995). There was a general deterioration in the sleep patterns when the medication was withdrawn, but lavender oil seemed to be as good as the original medication. However, the deterioration in the sleep patterns (due to "rebound insomnia"?) may simply have been due to recovery of normal sleep patterns when lavender was given (Vickers, 1996).

The efficacy of peppermint oil was studied on postoperative nausea in 18 women after gynecological operations (Tate, 1997) using peppermint oil or a control, peppermint essence (obviously of similar odor). A statistically significant difference was found between the controls and the test group. The test group required less antiemetics and received less opioid analgesia. However, the use of a peppermint essence as a control seems rather like having two test groups as inhalation was used.

A group of 313 patients undergoing radiotherapy were randomly assigned to receive either carrier oil with fractionated oils, carrier oil only, or pure essential oils of lavender, bergamot, and cedar wood administered by inhalation concurrently with radiation treatment. There were no significant differences in Hospital Anxiety and Depression Scale (HADS) scores and other scores between the randomly assigned groups. Aromatherapy, as administered in this study, was not found to be beneficial (Graham et al., 2003).

Heliotropin, a sweet, vanilla-like scent, reduced anxiety during magnetic resonance imaging (Redd and Manne, 1991), which causes distress to many patients as they are enclosed in a "coffin"-like apparatus. Patients experienced approximately 63% less overall anxiety than a control group of patients.

A double-blind randomized trial was conducted on 66 women undergoing abortions (Wiebe, 2000). Ten minutes were spent sniffing a numbered container with either a mixture of the essential oils (vetivert, bergamot, and geranium) or a hair conditioner (placebo). Aromatherapy involving essential oils was no more effective than having patients sniff other pleasant odors in reducing preoperative anxiety.

An audit into the effects of aromatherapy in palliative care (Evans, 1995) showed that the most frequently used oils were lavender, marjoram, and chamomile. These were applied over a period of 6 months by a therapist available for 4 h on a weekly basis in the ward. Relaxing music was played throughout each session to allay fears of the hands-on massage. The results revealed that 81% of the patients stated that they either felt *better* or *very relaxed* after the treatment; most appreciated the music greatly. The researchers themselves confessed that it is uncertain whether the benefits were the result of the patient being given individual attention, talking with the therapist, the effects of touch and massage, the effects of the aromatherapy essential oils, or the effects of the relaxation music.

Aromatherapy massage studied in eight cancer patients did not show any psychological benefit. However, there was a statistically significant reduction in all of the four physical parameters, which suggests that aromatherapy massage affects the autonomic nervous system, inducing relaxation. This finding was supported by the patients themselves, all of whom stated during interview that they felt *relaxed* after aromatherapy massage (Hadfield, 2001).

Forty-two cancer patients were randomly allocated to receive weekly massages with lavender essential oil in carrier oil (aromatherapy group), carrier oil only (massage group), or no intervention for 4 weeks (Soden et al., 2004). Outcome measures included a visual analogue scale of pain intensity, the Verran and Snyder–Halpern Sleep Scale, the HADS, and the Rotterdam Symptom Checklist. No significant long-term benefits of aromatherapy or massage in terms of improving pain control, anxiety, or quality of life were shown. However, sleep scores improved significantly in both the massage and the combined massage (aromatherapy and massage) groups. There were also statistically significant reductions in depression scores in the massage group. In this study of patients with advanced cancer, the addition of lavender essential oil did not appear to increase the beneficial effects of massage.

A randomized controlled pilot study was carried out to examine the effects of adjunctive aromatherapy massage on mood, quality of life, and physical symptoms in patients with cancer attending a specialist unit (Wilcock et al., 2004). Patients were randomized to conventional day care alone or day care plus weekly aromatherapy massage using a standardized blend of oils for 4 weeks. At baseline and at weekly intervals, patients rated their mood, quality of life, and the intensity and bother of two symptoms most important to them. However, although 46 patients were recruited to the study, only 11 of 23 (48%) patients in the aromatherapy group and 18 of 23 (78%) in the control group completed all 4 weeks. Mood, physical symptoms, and quality of life improved in both groups but there was *no* statistically significant difference between groups, but all patients were satisfied with the aromatherapy and wished to continue it.

Aromatherapy sessions in deaf and deaf–blind people became an accepted, enjoyable, and therapeutic part of the residents' lifestyle in an uncontrolled series of case studies. It appeared that this gentle, noninvasive therapy could benefit deaf and deaf–blind people, especially as their intact senses can be heightened (Armstrong and Heidingsfeld, 2000).

A scientifically unacceptable study of the effect of aromatherapy on endometriosis, reported only at an aromatherapy conference (Worwood, 1996), involved 22 aromatherapists who treated a total of 17 women in two groups over 24 weeks. One group was initially given massage with essential oils and then not *touched* for the second period, while the second group had the two treatments reversed. Among the many parameters measured were constipation, vaginal discharge, fluid retention, abdominal and pelvic pain, degree of feeling well, renewed vigor, depression, and tiredness. The data were presented as means (or averages, possibly, as this was not stated) but without standard errors of mean and lacked any statistical analyses. Unfortunately, the study has been accepted by many aromatherapists as being a conclusive proof of the value in treating endometriosis using aromatherapy.

In all the aforementioned trials, there was a more positive outcome for aromatherapy if there were no stringent scientific double-blind and randomized control measures, suggesting that in the latter case, bias is removed.

15.21 USE OF ESSENTIAL OILS MAINLY AS CHEMICAL AGENTS AND NOT FOR THEIR ODOR

The efficacy and safety of capsules containing peppermint oil (90 mg) and caraway oil (50 mg), when studied in a double-blind, placebo-controlled, multicenter trial in patients with nonulcer dyspepsia, was shown by May et al. (1996). Intensity of pain was significantly improved for the experimental group compared with the placebo group after 4 weeks.

Six drops of pure lavender oil included in the bath water for 10 days following childbirth was assessed against *synthetic* lavender oil and a placebo (distilled water containing an unknown GRAS additive) for perineal discomfort (Cornwell and Dale, 1995). No significant differences between groups were found for discomfort, but lower scores in discomfort means for days 3 and 5 for the lavender group were seen. This was very unsatisfactory as a scientific study, mainly because essential oils do not mix with water and there was no proof whether the lavender oil itself was pure.

Alopecia areata was treated in a randomized trial using "aromatherapy" carried out over 7 months. The test group massaged a mixture of 2 drops of *Thymus vulgaris*, 3 drops of *L. angustifolia*, 3 drops of *Rosmarinus officinalis*, and 2 drops of *Cedrus atlantica* in 3 mL of jojoba and 20 mL grape-seed oil into the scalp for 2 min minimum every night. The control group massaged the carrier oils alone (Hay et al., 1998). There was a significant improvement in the test group (44%) compared with the control group (15%). The smell of the essential oils (psychological/physiological) and/or their chemical nature on the scalp may have achieved these long-term results. On the other hand, the scalp may have healed naturally anyway after 7 months.

Ureterolithiasis was treated with Rowatinex, a mixture of terpenes smelling like Vicks VapoRub in 43 patients against a control group treated with a placebo. The overall expulsion rate of the ureteric stones was greater in the Rowatinex group (Engelstein et al., 1992). Similar mixes have shown both positive and negative results on gallstones over the years.

In a double-blind, placebo-controlled, randomized crossover study involving 332 healthy subjects, four different preparations were used to treat headaches (Gobel et al., 1994). Peppermint oil, eucalyptus oil (species not stated), and ethanol were applied to large areas of the forehead and temples. A combination of the three increased cognitive performance, muscle relaxation, and mental relaxation but had no influence on pain. Peppermint oil and ethanol decreased the headache. The reason for the success could have been the intense coldness caused by the application of the latter mixture, which was followed by a warming up as the peppermint oil caused counterirritation on the skin; the essential oils were also inhaled.

A clinical trial on 124 patients with acne, randomly distributed to a group treated with 5% tea tree oil gel or a 5% benzoyl peroxide lotion group (Bassett et al., 1990), showed improvement in both groups and fewer side effects in the tea tree oil group. The use of tea tree oil has, however, had detrimental effects in some people (Lis-Balchin, 2006).

A 10% tea tree oil was used on 104 patients with athlete's foot (*Tinea pedis*) in a randomized double-blind study against 1% tolnaftate and placebo creams. The tolnaftate group showed a better effect; tea tree oil was as effective in improving the condition but was no better than the placebo at curing it (Tong et al., 1992). Surprisingly, tea tree oil is sold as a *cure* for athlete's foot.

15.21.1 SINGLE-CASE STUDIES

In the past few years, the theme of the case studies (reported mainly in aromatherapy journals) has started to change, and most of the aromatherapists are no longer announcing that they are *curing* cancer and other serious diseases. Emphasis has swung toward real complementary treatment, often in the area of palliative care. However, the so-called clinical aromatherapists persist in attempting to cure various medical conditions using high doses of oils mainly by mouth, vagina, anus, or on the skin. Many believe that healing wounds using essential oils is also classed as aromatherapy (Guba, 2000) despite the evidence that odor does not kill germs and any effect is due to the chemical activity alone.

Because of the lack of scientific evidence in many studies, we could assume that aromatherapy is mainly based on faith; it works because the aromatherapist believes in the treatment and because the patient believes in the supposed action of essential oils, that is, the placebo effect.

Decreased smoking withdrawal symptoms in 48 cigarette smokers were achieved by black pepper oil puffed out of a special instrument for 3 h after an overnight cigarette deprivation against mint/menthol or nothing (Rose and Behm, 1994).

Chronic respiratory infection was successfully treated when the patient was massaged with tea tree, rosemary, and bergamot oils while on her second course of antibiotics and taking a proprietary cough medicine. She also used lavender and rosemary oils in her bath, a drop of eucalyptus oil and lavender oil on her tissue near the pillow at night, three drops of eucalyptus and ginger for inhalations daily, and reduced her dairy products and starches. In a week, her cough was better, and by 3 weeks, it was gone (Laffan, 1992). It is unclear which treatment actually helped the patient, and as it took a long time, the infection may well have gone away by then, or sooner, without any medicinal aid.

After just one treatment of aromatherapy massage using rose oil, bergamot, and lavender at 2.5% in almond oil, a 36-year-old woman managed to get pregnant after being told she was possibly infertile following the removal of her right fallopian tube (Rippon, 1993)!

Aromatherapy can apparently help patients with multiple sclerosis, especially for relaxation, in association with many other changes in the diet and also use of conventional medicines (Barker, 1994). French basil, black pepper, and true lavender in evening primrose oil with borage oil was used to counteract stiffness and also to stimulate; this mixture was later changed to include relaxing and sedative oils such as Roman chamomile, ylang ylang, and melissa.

Specific improvements in clients given aromatherapy treatment in dementia include increased alertness, self-hygiene, contentment, initiation of toileting, sleeping at night, and reduced levels of agitation, withdrawal, and wandering. Family carers reported less distress, improved sleeping patterns, and calmness (Kilstoff and Chenoweth, 1998). Other patients with dementia were monitored over a period of 2 months and then for a further 2 months during which they received aromatherapy treatments in a clinical trial; they showed a significant improvement in motivational behavior during the period of aromatherapy treatment (MacMahon and Kermode, 1998).

15.22 CONCLUSION

Aromatherapy, using essential oils as an odorant by inhalation or massage onto the skin, has not been shown to work better than massage alone or a control. No failures have, however, been reported, although treatment is invariably changed on each visit. Many patients feel better, even if their disease is getting worse, due to their belief in an alternative therapist, and this is a good example of "mind over matter," that is, the placebo effect. This effect has been recommended by some members of the House of Lords Select Committee on Science and Technology, Sixth Report (2000), as a good basis for retaining complementary and alternative medicine, but other members argued that scientific proof of effects is necessary.

It is hoped that aromatherapists do not try to convince their patients of a cure, especially in the case of serious ailments such as cancer, which often recede naturally for a time on their own. Conventional treatment should always be advised in the first instance and retained during aromatherapy treatment with the consent of the patient's primary health-care physician or consultant. Aromatherapy can provide a useful complementary medical service both in healthcare settings and in private practice and should not be allowed to become listed as a bogus cure in alternative medicine.

REFERENCES

Adams, T.B. et al., 1996. The FEMA GRAS assessment of alicyclic substances used as flavour ingredients. *Food Chem. Toxicol.*, 34: 763–828.
Ammon, H.P.T., 1989. Phytotherapeutika in der Kneipp-therapie. *Therapiewoche*, 39: 117–127.
Anderson, C. et al., 2000. Evaluation of massage with essential oils on childhood atopic eczema. *Phytother. Res.*, 14: 452–456.
Arctander, S., 1960. *Perfume and Flavor Materials of Natural Origin*. Elizabeth, NJ: S Arctander.
Armstrong, F. and V. Heidingsfeld, 2000. Aromatherapy for deaf and deafblind people living in residential accommodation. *Complement Ther. Nurs. Midwifery*, 6: 180–188.
Arnould-Taylor, W.E., 1981. *A Textbook of Holistic Aromatherapy*. London, U.K.: Stanley Thornes.
Atchley, E.G. and F. Cuthbert, 1909. *A History of the Use of Incense in Divine Worship*. London, U.K.: Longmans, Green and Co.
Badia, P., 1991. Olfactory sensitivity in sleep: The effects of fragrances on the quality of sleep: A summary of research conducted for the fragrance research fund. *Perf. Flav.*, 16: 33–34.
Baldwin, C.M. et al., 1999. Odor sensitivity and respiratory complaint profiles in a community-based sample with asthma, hay fever, and chemical intolerance. *Toxicol. Ind. Health*, 15: 403–409.
Ballard, C. et al., 2002. Aromatherapy as a safe and effective treatment for the management of agitation in severe dementia: A double-blind, placebo-controlled trial with Melissa. *J. Clin. Psychiatry*, 63: 553–558.
Barker, A., 1994. Aromatherapy and multiple sclerosis. *Aromather. Q*, 4–6.
Barrett, S., 2000. Aromatherapy company sued for false advertising. *Quackwatch*, September 25, http://www.quackwatch.com/01QuackeryRelatedTopics/aroma.html, accessed August 8, 2009.
Basnyet, J., 1999. Tibetan essential oils. *IFA J.*, 43: 12–13.
Bassett, I.B. et al., 1990. A comparative study of tea tree oil versus benzoylperoxide in the treatment of acne. *Med. J. Aust.*, 153: 455–458.
Beasley, V., ed., 1999. Toxicants that cause central nervous system depression. In: *Veterinary Toxicology*. Ithaca, NY: International Veterinary Information Service, A2608.0899, http://www.ivis.org/ (accessed August 13, 2015).
Bell, I.R. et al., 1993. Self-reported illness from chemical odors in young adults without clinical syndromes or occupational exposures. *Arch. Environ. Health*, 48: 6–13.
Benson, H. and M. Stark, 1996. *Timeless Healing. The Power and Biology of Belief*. London, U.K.: Simon and Schuster.
Bilsland, D. and A. Strong, 1990. Allergic contact dermatitis from the essential oil of French Marigold (*Tagetes patula*) in an aromatherapist. *Contact Dermatitis*, 23: 55–56.
Boffa, M.J. et al., 1996. Celery soup causes severe phototoxicity during PUVA therapy. *Br. J. Dermatol.*, 135: 334.
Brooker, D.J.R. et al., 1997. Single case evaluation of the effects of aromatherapy and massage on disturbed behaviour in severe dementia. *Br. J. Clin. Psychol.*, 36: 287–296.
Brunn, E.Z. and G. Epiney-Burgard, 1989. *Women Mystics in Medieval Europe*. New York: Paragon House.

Buchbauer, G. et al., 1992. Passiflora and limeblossom: Motility effects after inhalation of the essential oils and of some of the main constituents in animal experiments. *Arch. Pharm. (Weinheim)*, 325: 247–248.

Buchbauer, G. et al., 1993a. Fragrance compounds and essential oils with sedative effects upon inhalation. *J. Pharm. Sci.*, 82: 660–664.

Buchbauer, G. et al., 1993b. Therapeutic properties of essential oils and fragrances. In: *Bioactive Volatile Compounds from Plants*, R. Teranishi et al., eds., pp. 159–165. Washington, DC: American Chemical Society.

Buchbauer, G. et al., 1993c. New results in aromatherapy research. Paper presented at the *24th International Symposium on Essential Oils*, Berlin, Germany.

Buckle, J., 1993. Does it matter which lavender oil is used? *Nurs. Times*, 89: 32–35.

Bunyan, N., 1998. Baby died after taking mint water for wind. *Daily Telegraph*, May 21.

Burns, E. and C. Blaney, 1994. Using aromatherapy in childbirth. *Nurs. Times*, 90: 54–58.

Burns, E. et al., 2000. An investigation into the use of aromatherapy in intrapartum midwifery practice. *J. Altern. Complement Med.*, 6: 141–147.

Carson, C.F. and T.V. Riley, 1995. Toxicity of the essential oil of *Melaleuca alternifolia* or tea tree oil. *J. Toxicol. Clin. Toxicol.*, 33: 193–194.

Chang, Y.-C. et al., 1997. Allergic contact dermatitis from oxidised d-limonene. *Contact Dermatitis*, 37: 308–309.

Chou, Y.J. and D.R. Dietrich, 1999. Toxicity of nitromusks in early life-stages of South African clawed frog, *Xenopus laevis* and zebra-fish, *Danio rerio. Toxicol. Lett.*, 111: 17–25.

Christian, M.S. et al., 1999. Developmental toxicity studies of four fragrances in rats. *Toxicol. Lett.*, 111: 169–174.

Classen, C. et al., 1994. *Aroma*. London, U.K.: Routledge.

Cohen, B.M. and W.E. Dressler, 1982. Acute aromatics inhalation modifies the airways. Effects of the common cold. *Respiration*, 43: 285–293.

Concise Oxford Dictionary, 9th edn., 1995. Thompson, D., ed. Oxford, U.K.: Oxford University Press.

Cooke, B. and E. Ernst, 2000. Aromatherapy: A systematic review. *Br. J. Gen. Pract.*, 50: 493–496.

Corbin, A., 1986. *The Foul and the Fragrant*. Cambridge, MA: Harvard University Press.

Corner, J. et al., 1995. An evaluation of the use of massage and essential oils on the well-being of cancer patients. *Int. J. Palliat. Nurs.*, 1: 67–73.

Cornwell, S. and A. Dale, 1995. Lavender oil and perineal repair. *Mod. Midwifery*, 5: 31–33.

Cousins, N., 1979. *Anatomy of an Illness*. New York: WW Norton.

Crawford, G.H. et al., 2004. Use of aromatherapy products and increased risk of hand dermatitis in massage therapists. *Arch. Dermatol.*, 140: 991–996.

Culpeper, N., 1653/1995. *Culpeper's Complete Herbal and The English Physician and Family Dispensatory*. Facsimile. Ware, U.K.: Wordsworth editions, 1995.

Culpeper, N., 1826/1981. *Culpeper's Complete Herbal and English Physician*. Facsimile. Bath, U.K.: Pitman Press Ltd., 1981.

Dale, A. and S. Cornwell, 1994. The role of lavender oil in relieving perineal discomfort following childbirth: A blind randomized clinical trial. *J. Adv. Nurs.*, 19: 89–96.

De Groot, A.C. and P.J. Frosch, 1997. Adverse reactions to fragrances. A clinical review. *Contact Dermatitis*, 36: 57–86.

Deans, S.G., 2002. Antimicrobial properties of lavender volatile oil. In: *Genus Lavandula, Aromatic and Medicinal Plants—Industrial Profiles*, M. Lis-Balchin, ed. London, U.K.: Taylor & Francis.

Deans, S.G. and G. Ritchie, 1987. Antibacterial properties of plant essential oils. *Int. J. Food Microbiol.*, 5: 165–180.

Dejeans, M. (Antoine de Hornot), 1764. *TraitŽ des Odeurs*. Paris, France: Nyon/Guillyn/Saugrain.

Devereux, P., 1997. *The Long Trip*. New York: Penguin.

Dhondt, W. et al., 1999. Pain threshold in patients with rheumatoid arthritis and effect of manual oscillations. *Scand. J. Rheumatol.*, 28: 88–93.

Diego, M.A. et al., 1998. Aromatherapy positively affects mood, EEG patterns of alertness and math computations. *Int. J. Neurosci.*, 96: 217–224.

Dorman, H.J.D. et al., 1995a. Antioxidant-rich plant volatile oils: In vitro assessment of activity. Paper presented at the *26th International Symposium on Essential Oils*, Hamburg, Germany, September 10–13, 1995.

Dorman, H.J.D. et al., 1995b. Evaluation in vitro of plant essential oils as natural antioxidants. *J. Essent. Oil Res.*, 7: 645–651.

Dunn, C. et al., 1995. Sensing an improvement: An experimental study to evaluate the use of aromatherapy, massage and periods of rest in an intensive care unit. *J. Adv. Nurs.*, 21: 34–40.

Elisabetsky, E. et al., 1995. Effects of linalool on glutamatergic system in the rat cerebral cortex. *Neurochem. Res.*, 20: 461–465.

Elisabetsky, E. et al., 1999. Anticonvulsant properties of linalool in glutamate-related seizure models. *Phytomedicine*, 6: 107–113.

Ellis, A., 1960. *The Essence of Beauty*. London, U.K.: Secker and Warburg.

Engelstein, E. et al., 1992. Rowatinex for the treatment of ureterolithiasis. *J. Urol.*, 98: 98–100.

Ernst, E., 1994. Clinical effectiveness of massage—A critical review. *Forsch Komplementärmed*, 1: 226–232.

Ernst, E., 1998. Does post-exercise massage treatment reduce delayed onset muscle soreness? *Br. J. Sports Med.*, 32: 212–214.

Ernst, E., 1999a. Abdominal massage for chronic constipation: A systematic review of controlled clinical trials. *Forsch Komplementärmed*, 6: 149–151.

Ernst, E., 1999b. Massage therapy for low back pain: A systematic review. *J. Pain Symptom. Manage.*, 17: 56–69.

Ernst, E., 2003a. Massage treatment for back pain. *BMJ*, 326: 562–563.

Ernst, E., 2003b. The safety of massage therapy. *Rheumatology*, 42: 1101–1106.

Ernst, E. and B.R. Cassileth, 1998. The prevalence of complementary/alternative medicine in cancer: A systematic review. *Cancer*, 83: 777–782.

Ernst, E. et al., 2001. *The Desktop Guide to Complementary and Alternative Medicine*. Edinburgh, U.K.: Mosby.

Evans, B., 1995. An audit into the effects of aromatherapy massage and the cancer patient in palliative and terminal care. *Complement Ther. Med.*, 3: 239–241.

Falk-Filipsson, A. et al., 1993. d-Limonene exposure to humans by inhalation: Uptake, distribution, elimination, and effects on the pulmonary function. *J. Toxicol. Environ. Health*, 38: 77–88.

Fenaroli, G., 1997. *Handbook of Flavour Ingredients*, 3rd edn., Vol. 1. London, U.K.: CRC Press.

Ferley, J.P. et al., 1989. Prophylactic aromatherapy for supervening infections in patients with chronic bronchitis. Statistical evaluation conducted in clinics against a placebo. *Phytother. Res.*, 3: 97–100.

Festing, S., 1989. *The Story of Lavender*. Guildford, U.K.: Heritage in Sutton Leisure.

Fischer-Rizzi, S., 1990. *Complete Aromatherapy Handbook*. New York: Sterling Publishing.

Forbes, R.J., ed., 1955. Cosmetics and perfumes in antiquity. In: *Studies in Ancient Technology*, Vol. III. Leiden, the Netherlands: EJ Brill.

Ford, R.A., 1991. The toxicology and safety of fragrances. In: *Perfumes, Art, Science and Technology*, P.M. Muller and D. Lamparsky, eds., pp. 442–463. New York: Elsevier.

Frosch, P.J. et al., 1999. Lyral is an important sensitizer in patients sensitive to fragrances. *Br. J. Dermatol.*, 141: 1076–1083.

Fuchs, N. et al., 1997. Systemic absorption of topically applied carvone: Influence of massage technique. *J. Soc. Cosmet. Chem.*, 48: 277–282.

Furuhashi, A. et al., 1994. Effects of AETT-induced neuronal ceroid lipofuscinosis on learning ability in rats. *Jpn. J. Psychiatry Neurol.*, 48: 645–653.

Gattefossé, M., 1992. Rene-Maurice Gattefossé, The father of modern aromatherapy. *Int. J. Aromather.*, 4: 18–20.

Gattefossé, R.M., 1928. *Formulaire du chimiste-Parfumeur et du Savonnier* [Formulary of cosmetics]. Paris, France: Librairie des Sciences.

Gattefossé, R.M., 1937/1993. In: *Gattefossé's Aromatherapy*, R. Tisserand, ed. Saffron Walden, U.K.: CW Daniel Co.

Gattefossé, R.M., 1952. *Formulary of Perfumery and of Cosmetology*. London, U.K.: Leonard Hill.

Gatti, G. and R. Cajola, 1923a. L'Azione delle essenze sul systema nervoso. *Riv Ital Della Essenze e Profumi*, 5: 133–135.

Gatti, G. and R. Cajola, 1923b. L'Azione terapeutica degli olii essenziali. *Riv Ital Della Essenze e Profumi*, 5: 30–33.

Gatti, G. and R. Cajola, 1929. L'essenza di valeriana nella cura delle malattie nervose. *Riv Ital Della Essenze e Profumi*, 2: 260–262.

Genders, R., 1972. *A History of Scent*. London, U.K.: Hamish Hamilton.

Goats, G.C., 1994. Massage—The scientific basis of an ancient art: Parts 1 and 2. *Br. J. Sports Med. (UK)*, 28: 149–152, 153–156.

Gobel, H. et al., 1994. Effect of peppermint and eucalyptus oil preparations on neurophysiological and experimental algesimetric headache parameters. *Cephalagia*, 14: 228–234.

Goldstone, L., 1999. From orthodox to complementary: The fall and rise of massage, with specific reference to orthopaedic and rheumatology nursing. *J. Orthop. Nurs.*, 3: 152–159.

Goldstone, L., 2000. Massage as an orthodox medical treatment, past and future. *Complement Ther. Nurs. Midwifery*, 6: 169–175.

Graham, P.H. et al., 2003. Inhalation aromatherapy during radiotherapy: Results of a placebo-controlled double-blind randomized trial. *J. Clin. Oncol.*, 15: 2372–2376.

Groom, N., 1992. *The Perfume Handbook*, 2nd edn. London, U.K.: Chapman & Hall.

Guba, R., 2000. Toxicity myths: The actual risks of essential oil use. *Perf. Flav.*, 25: 10–28.

Guin, J.D., 1982. History, manufacture and cutaneous reaction to perfumes. In: *Principles of Cosmetics for the Dermatologist*, P. Frost and S.W. Horwitz, eds. St. Louis, MO: The CV Mosby Co.

Guin, J.D., 1995. *Practical Contact Dermatitis*. New York: McGraw-Hill.

Hadfield, N., 2001. The role of aromatherapy massage in reducing anxiety in patients with malignant brain tumours. *Int. J. Palliat. Nurs.*, 7: 279–285.

Hardy, M. et al., 1995. Replacement of drug treatment for insomnia by ambient odour. *Lancet*, 346: 701.

Hastings, L. et al., 1991. Olfactory primary neurons as a route of entry for toxic agents into the CNS. *Neurotoxicology*, 12: 707–714.

Hay, I.C. et al., 1998. Randomized trial of aromatherapy. Successful treatment for alopecia areata. *Arch. Dermatol.*, 134: 1349–1352.

Healey, M.A. and M. Aslam, 1996. Aromatherapy. In: *Trease & Evans' Pharmacognosy*, 14th edn., W.C. Evans, ed. London, U.K.: WB Saunders.

Hendriks, S.A. et al., 1999. Allergic contact dermatitis from the fragrance ingredient Lyral in underarm deodorant. *Contact Dermatitis*, 41: 119.

Hirsch, A., 1998. *Sensational Sex: The Secret to Using Aroma for Arousal*. Boston, MA: Element.

Holmes, C. et al., 2002. Lavender oil as a treatment for agitated behaviour in severe dementia: A placebo controlled study. *Int. J. Geriatr. Psychiatry*, 17: 305–308.

Horowitz, D.A., 2000. Judgement (pursuant to stipulation). National Council against Health Fraud, Inc., v. Aroma Vera, Inc. et al., Superior Court No. BC183903, October 11.

Jager, W. et al., 1992. Percutaneous absorption of lavender oil from a massage oil. *J. Soc. Cosmet. Chem.*, 43: 49–54.

Janson, T., 1997. *Mundo Maya* (Editorial). Guatemala City, Guatemala: Artemis Edinter.

Johansen, J.D. et al., 1998. Allergens in combination have a synergistic effect on the elicitation response: A study of fragrance-sensitized individuals. *Br. J. Dermatol.*, 139: 264–270.

Johansen, J.D. et al., 2000. Rash related to use of scented products. A questionnaire study in the Danish population. Is the problem increasing? *Cont. Dermat.*, 42: 222–226.

Kafferlein, H.U. et al., 1998. Musk xylene: Analysis, occurrence, kinetics and toxicology. *Crit. Rev. Toxicol.*, 28: 431–476.

Karamat, R. et al., 1992. Excitatory and sedative effects of essential oils on human reaction time performance. *Chem. Senses*, 17: 847.

Karlberg, A.-T. and A. Dooms-Goossens, 1997. Contact allergy to oxidised d-limonene among dermatitis patients. *Contact Dermatitis*, 36: 201–206.

Katona, M. and K. Egyud, 2001. Increased sensitivity to balsams and fragrances among our patients. *Orv. Hetil.*, 142: 465–466.

Keane, F.M. et al., 2000. Occupational allergic contact dermatitis in two aromatherapists. *Contact Dermatitis*, 43: 49–51.

Kennedy, D.O. et al., 2002. Modulation of mood and cognitive performance following acute administration of *Melissa officinalis* (lemon balm). *Pharmacol. Biochem. Behav.*, 72: 953–964.

Kilstoff, K. and L. Chenoweth, 1998. New approaches to health and well-being for dementia day-care clients, family carers and day-care staff. *Int. J. Nurs. Pract.*, 4: 70–83.

Kirk-Smith, M., 1996. Clinical evaluation: Deciding what questions to ask. *Nurs. Times*, 92: 34–35.

Kite, S.M. et al., 1998. Development of an aromatherapy service at a Cancer Centre. *Palliat. Med.*, 12: 171–180.

Klarmann, E.G., 1958. Perfume dermatitis. *Ann. Allergy*, 16: 425–434.

Klemm, W.R. et al., 1992. Topographical EEG maps of human responses to odors. *Chem. Senses*, 17: 347–361.

Knasko, S.C., 1992. Ambient odours effect on creativity, mood and perceived health. *Chem. Senses*, 17: 27–35.

Knasko, S.C., 1993. Performance, mood and health during exposure to intermittent odours. *Arch. Environ. Health*, 48: 3058.

Knasko, S.C. et al., 1990. Emotional state, physical well-being and performance in the presence of feigned ambient odour. *J. Appl. Soc. Psychol.*, 20: 1345–1347.

Knight, T.E. and B.M. Hausen 1994. Melaleuca oil (tea tree oil) dermatitis. *J. Am. Acad. Dermatol.*, 30: 423–427.
Kohl, L. et al., 2002. Allergic contact dermatitis from cosmetics: Retrospective analysis of 819 patch-tested patients. *Dermatology*, 204: 334–337.
Kohn, M., 1999. Complementary therapies in cancer care. Macmillan Cancer Relief Report. New York: Macmillan.
Kovar, K.A. et al., 1987. Blood levels of 1,8-cineole and locomotor activity of mice after inhalation and oral administration of rosemary oil. *Planta Med.*, 3: 315–318.
Kubota, M. et al., 1992. Odor and emotion-effects of essential oils on contingent negative variation. In: *Proceedings of the 12th International Congress on Flavours, Fragrances and Essential Oils*, Vienna, Austria, October 4–8, 1992, pp. 456–461.
Labyak, S.E. and B.L. Metzger, 1997. The effects of effleurage backrub on the physiological components of relaxation: A meta-analysis. *Nurs. Res.*, 46: 59–62.
Laffan, G., 1992. Chronic respiratory infection. *Int. J. Aromather.*, 4: 17.
Larsen, W., 1998. A study of new fragrance mixtures. *Am. J. Contact Dermat.*, 9: 202–206.
Lawless, J., 1992. *The Encyclopedia of Essential Oils*. Shaftesbury, U.K.: Element Books.
Lawless, J., 1994. *Aromatherapy and the Mind*. London, U.K.: Thorsons.
Lis-Balchin, M., 1995. *Aroma Science: The Chemistry and Bioactivity of Essential Oils*. Guildford, U.K.: Amberwood Publishing Ltd.
Lis-Balchin, M., 1997. Essential oils and 'aromatherapy': Their modern role in healing. *J. R. Soc. Health*, 117: 324–329.
Lis-Balchin, M., 1998a. Aromatherapy versus the internal intake of odours. *NORA Newsletter*, 3: 17.
Lis-Balchin, M., 1998b. Aromatherapy for pain relief. *Pain Concern Summer*, pp. 12–15. (Now available as a pamphlet from the organisation.)
Lis-Balchin, M., ed., 2002a. *Geranium and Pelargonium Genera Geranium and Pelargonium: Medicinal and Aromatic Plants—Industrial Profiles*. London, U.K.: Taylor & Francis.
Lis-Balchin, M., ed., 2002b. *Genus Lavandula: Medicinal and Aromatic Plants—Industrial Profiles*. London, U.K.: Taylor & Francis.
Lis-Balchin, M., 2006. *Aromatherapy Science*. London, U.K.: Pharmaceutical Press.
Lis-Balchin, M. and S. Hart, 1997a. The effect of essential oils on the uterus compared to that on other muscles. In: *Proceedings of the 27th International Symposium on Essential Oils*, Ch. Franz, A. Mathé, and G. Buchbauer, eds. Vienna, Austria, September 8–11, 1996. Carol Stream, IL: Allured Publishing, pp. 29–32.
Lis-Balchin, M. and S. Hart, 1997b. A preliminary study of the effect of essential oils on skeletal and smooth muscles in vitro. *J. Ethnopharmacol.*, 58: 183–187.
Lis-Balchin, M., S. Hart, S.G. Deans, and E. Eaglesham, 1996a. Comparison of the pharmacological and antimicrobial action of commercial plant essential oils. *J. Herbs Spices Med. Plants*, 4: 69–86.
Lis-Balchin, M. et al., 1996b. Bioactivity of commercial geranium oil from different sources. *J. Essent. Oil Res.*, 8: 281–290.
Lis-Balchin, M. et al., 1998. Relationship between the bioactivity and chemical composition of commercial plant essential oils. *Flav. Fragr. J.*, 13: 98–104.
Loret, V., 1887. Le kyphi, parfum sacre des anciens egyptiens. *J. Asiatique*, 10: 76–132.
MacMahon, S. and S. Kermode 1998. A clinical trial of the effect of aromatherapy on motivational behaviour in a dementia care setting using a single subject design. *Aust. J. Holist. Nurs.*, 5: 47–49.
Mandy, R., 1993. Infertility and stress. *Int. J. Aromather.*, 5: 33.
Manley, C.H., 1993. Psychophysiological effect of odor. *Crit. Rev. Food Sci. Nutr.*, 33: 57–62.
Manniche, L., 1989. *An Ancient Egyptian Herbal*. London, U.K.: British Museum Publications.
Manniche, L., 1999. *Sacred Luxuries. Fragrance, Aromatherapy and Cosmetics in Ancient Egypt*. London, U.K.: Opus Publishing.
Maury, M., 1989. *Marguerite Maury's Guide to Aromatherapy*. Saffron Walden, U.K.: CW Daniel Co.
May, B. et al., 1996. Efficacy of a fixed peppermint oil/caraway combination in non-ulcer dyspepsia. *Arzneimettelforsch*, 146: 1149–1153.
Millenson, J.R., 1995. *Mind Matters: Psychological Medicine in Holistic Practice*. Seattle, WA: Eastland Press.
Millqvist, E. and O. Lowhagen 1996. Placebo-controlled challenges with perfume in patients with asthma-like symptoms. *Allergy*, 51: 434–439.
Morris, N. et al., 1995. Anxiety reduction by aromatherapy: Anxiolytic effects of inhalation of geranium and rosemary. *Int. J. Aromather.*, 7: 33–39.

Munro, I.C. et al., 1996. Correlation of structural class with no-observed-effect levels: A proposal for establishing a threshold of concern. *Food Cosmet. Toxicol.*, 34: 829–867.
Müller, J. et al., 1984. *The H&R Book of Perfume*. London, U.K.: Johnson Publications.
Nelson, N.J., 1997. Scents or nonsense: Aromatherapy's benefits still subject to debate. *J. Natl. Cancer Inst.*, 89: 1334–1336.
Nunn, J.F., 1997. *Ancient Egyptian Medicine*. London, U.K.: British Museum Press.
Ollevant, N.A. et al., 1999. How big is a drop? A volumetric assay of essential oils. *J. Clin. Nurs.*, 8: 299–304.
Pénoel, D., 1991a. *Médecine Aromatique, Médecine Planétaire*. Limoges, France: Roger Jollois.
Pénoel, D., 1991b. *Les travaux de J.-A. Giralt-Gonzalez*. Limoges, France: Roger Jollois.
Pénoel, D., 1998. *A Natural Home Health Care Using Essential Oils*. Hurricane, UT: Essential Science Publishing.
Perry, E.K. et al., 1998. Medicinal plants and Alzheimer's disease. *J. Altern. Complement. Med.*, 4: 419–428.
Perry, E.K. et al., 1999. Medicinal plants and Alzheimer's disease: From ethnobotany to phytotherapy. *J. Pharm. Pharmacol.*, 51: 527–534.
Piesse, S.G.W., 1855. *The Art of Perfumery*. London, U.K.: Longman, Brown, Green.
Pinch, G., 1994. *Magic in Ancient Egypt*. London, U.K.: British Museum Press.
Price, S., 1983. *Practical Aromatherapy*. Wellingborough, U.K.: Thorsons.
Price, S., 1993. *Aromatherapy Workbook*. London, U.K.: Thorsons.
Price, S. and L. Price, 1999. *Aromatherapy for Health Professionals*, 2nd edn. London, U.K.: Churchill Livingstone.
Rastogi, S.C. et al., 1999. Contents of fragrance allergens in children's cosmetics and cosmetic-toys. *Contact Dermatitis*, 41: 84–88.
Redd, W. and S. Manne, 1991. Fragrance reduces patient anxiety during stressful medical procedures. *Focus Fragr. Summ.*, 1: 1.
Riedel, J. et al., 1999. Haemoglobin binding of a musk xylene metabolite in man. *Xenobiotica*, 29: 573–582.
Rimmel, E., 1865. *The Book of Perfumes*. London, U.K.: Chapman & Hall.
Romaguera, C. and J. Vilplana, 2000. Occupational contact dermatitis from ylang-ylang oil. *Contact Dermatitis*, 43: 251.
Rose, J.E. and F.M. Behm, 1994. Inhalation of vapour from black pepper extract reduces smoking withdrawal symptoms. *Drug Alcohol Depend.*, 34: 225–229.
Sanchez-Perez, J. and A. Garcia-Diez, 1999. Occupational allergic contact dermatitis from eugenol, oil of cinnamon and oil of cloves in a physiotherapist. *Contact Dermatitis*, 41: 346–347.
Schaller, M. and H.C. Korting, 1995. Allergic airborne contact dermatitis from essential oils used in aromatherapy. *Clin. Exp. Dermatol.*, 20: 143–145.
Scheinman, P.L., 1996. Allergic contact dermatitis to fragrance. *Am. J. Contact Dermat.*, 7: 65–76.
Schnaubelt, K., 1999. *Medical Aromatherapy—Healing with Essential Oils*. Berkeley, CA: Frog Ltd.
Schumann Antelme, R. and S. Rossini, 2001. *Sacred Sexuality in Ancient Egypt*. Rochester, VT: Inner Traditions International.
Sharma, J.N. et al., 1994. Suppressive effects of eugenol and ginger oil on arthritic rats. *Pharmacology*, 49: 314–318.
Shim, C. and M.H. Williams Jr., 1986. Effect of odors in asthma. *Am. J. Med.*, 80: 18–22.
Skoeld, M. et al., 2002a. Sensitization studies on the fragrance chemical linalool, with respect to autooxidation. Abstract. *Contact Dermatitis*, 46(Suppl. 4): 20.
Skoeld, M. et al., 2002b. Studies on the auto-oxidation and sensitizing capacity of the fragrance chemical linalool, identifying a linalool hydroperoxide. *Contact Dermatitis*, 46: 267–272.
Smallwood, J. et al., 2001. Aromatherapy and behaviour disturbances in dementia: A randomized controlled trial. *Int. J. Geriatr. Psychiatry*, 16: 1010–1013.
Snow, A.L. et al., 2004. A controlled trial of aromatherapy for agitation in nursing home patients with dementia. *J. Altern. Complement. Med.*, 10: 431–437.
Soden, K. et al., 2004. A randomized controlled trial of aromatherapy massage in a hospice setting. *Palliat. Med.*, 18: 87–92.
Somerville, K.W. et al., 1984. Delayed release peppermint oil capsules (Colpermin) for the spastic colon syndrome—A pharmaco-kinetic study. *Br. J. Clin. Pharm.*, 18: 638–640.
Somerville, K.W. et al., 1985. Stones in the common bile duct: Experience with medical dissolution therapy. *Postgrad. Med. J.*, 61: 313–316.
Southwell, I.A., 1997. Skin irritancy of tea tree oil. *J. Essent. Oil Res.*, 9: 47–52.
Spencer, P.S. et al., 1979. Neurotoxic fragrance produces ceroid and myelin disease. *Science*, 204: 633–635.
Spencer, P.S. et al., 1984. Neurotoxic properties of musk ambrette. *Toxicol. Appl. Pharmacol.*, 75: 571–575.

Steinberg, P. et al., 1999. Acute hepatotoxicity of the polycyclic musk 7-acetyl-1,1,3,4,4,6-hexamethyl1,2,3,4-tetrahydronaphthaline (AHTN). *Toxicol. Lett.*, 111: 151–160.
Stevenson, C., 1994. The psychophysiological effects of aromatherapy massage following cardiac surgery. *Compl. Ther. Med.*, 2: 27–35.
Sugiura, M. et al., 2000. Results of patch testing with lavender oil in Japan. *Contact Dermatitis*, 43: 157–160.
Tate, S., 1997. Peppermint oil: A treatment for postoperative nausea. *J. Adv. Nurs.*, 26: 543–549.
Tisserand, R., 1977. *The Art of Aromatherapy*. Saffron Walden, U.K.: CW Daniel Co.
Tisserand, R. and T. Balacs, 1995. *Essential Oil Safety—A Guide for Health Care Professionals*. Edinburgh, U.K.: Churchill Livingstone.
Tobyn, G., 1997. *Culpeper's Medicine: A Practice of Holistic Medicine*. Shaftesbury, U.K.: Element Books.
Tong, M.M. et al., 1992. Tea tree oil in the treatment of tinea pedis. *Aust. J. Dermatol.*, 33: 145–149.
Torii, S. et al., 1988. Contingent negative variation and the psychological effects of odor. In: *Perfumery: The Psychology and Biology of Fragrance*, S. Toller and G.H. Dodd, eds. New York: Chapman & Hall.
Unterman, A., 1991. *Dictionary of Jewish Lore & Legend*. London, U.K.: Thames and Hudson.
Valnet, J., 1982. *The Practice of Aromatherapy*. Saffron Walden, U.K.: CW Daniel Co.
Van Toller, S. and G.H. Dodd, 1991. *Perfumery: The Psychology and Biology of Fragrance*. New York: Chapman & Hall.
Van Toller, S. et al., 1993. An analysis of spontaneous human cortical EEG activity to odours. *Chem. Senses*, 18: 1–16.
Veien, N.K. et al., 1985. Reduction of intake of balsams in patients sensitive to balsam of Peru. *Contact Dermatitis*, 12: 270–273.
Vickers, A., 1996. *Massage and Aromatherapy*. London, U.K.: Chapman & Hall.
Vilaplana, J. and C. Romaguera, 2002. Contact dermatitis from the essential oil of tangerine in fragrance. *Contact Dermatitis*, 46: 108.
von Burg, R., 1995. Toxicology update. *J. Appl. Toxicol.*, 15: 495–499.
Wake, G. et al., 2000. CNS acetylcholine receptor activity in European medicinal plants traditionally used to improve failing memory. *J. Ethnopharmacology*, 69: 105–114.
Warren, C. and S. Warrenburg, 1993. Mood benefits of fragrance. *Perf. Flav.*, 18: 9–16.
Weibel, H. et al., 1989. Cross-sensitization patterns in guinea pigs between cinnamaldehyde, cinnamyl alcohol and cinnamic acid. *Acta Dermatol. Venereol.*, 69: 302–307.
Weil, A., 1983. *Health and Healing*. Boston, MA: Houghton Mifflin.
Which?, 2001. Essential oils. Health Which? *February*, 17–19.
Wiebe, E., 2000. A randomized trial of aromatherapy to reduce anxiety before abortion. *Eff. Clin. Pract.*, 3: 166–169.
Wilcock, A. et al., 2004. Does aromatherapy massage benefit patients with cancer attending a specialist palliative care day centre? *Palliat. Med.*, 18: 287–290.
Wilkinson, S., 1995. Aromatherapy and massage in palliative care. *Int. J. Palliat. Care*, 1: 21–30.
Wilkinson, S. et al., 1999. An evaluation of aromatherapy massage in palliative care. *Palliat. Med.*, 13: 409–417.
Wilkinson, S.M., 2003. Evaluating the efficacy of massage in cancer care. *BMJ*, 326: 562–563.
Wilson, F.P., 1925. *The Plague Pamphlets of Thomas Dekker*. Oxford, U.K.: Clarendon Press.
Worwood, V., 1991. *The Fragrant Pharmacy*. London, U.K.: Bantam Books.
Worwood, V., 1996. *The Fragrant Mind*. London, U.K.: Doubleday.
Worwood, V., 1998. *The Fragrant Heavens. The Spiritual Dimension of Fragrance and Aromatherapy*. Novato, CA: New World Library.
Youdim, K.A. and S.G. Deans, 2000. Effect of thyme oil and thymol dietary supplementation on the antioxidant status and fatty acid composition of the ageing rat brain. *Br. J. Nutr.*, 83: 87–93.
Young, A.R. et al., 1990. Phototumorigenesis studies of 5-methoxypsoralen in bergamot oil: Evaluation and modification of risk of human use in an albino mouse skin model. *J. Phytochem. Photobiol.*, 7: 231–250.
Zarno, V., 1994. Candidiasis: A holistic view. *Int. J. Aromather.*, 6(2): 20–23.

16 Essential Oils Used in Veterinary Medicine

K. Hüsnü Can Başer and Chlodwig Franz

CONTENTS

16.1 Introduction 655
16.2 Oils Attracting Animals 657
16.3 Oils Repelling Animals 657
16.4 Oils against Pests 658
    16.4.1 Insecticidal, Pest Repellent, and Antiparasitic Oils 658
    16.4.2 Fleas and Ticks 658
    16.4.3 Mosquitoes 658
    16.4.4 Moths 659
    16.4.5 Aphids, Caterpillars, and Whiteflies 659
        16.4.5.1 Garlic Oil 659
    16.4.6 Ear Mites 659
    16.4.7 Antiparasitic 659
16.5 Essential Oils Used in Animal Feed 660
    16.5.1 Ruminants 660
    16.5.2 Poultry 661
        16.5.2.1 Studies with CRINA® Poultry 661
        16.5.2.2 Studies with Herbromix® 662
    16.5.3 Pigs 663
16.6 Essential Oils Used in Treating Diseases in Animals 664
References 665

16.1 INTRODUCTION

Essential oils are volatile constituents of aromatic plants. These liquid oils are generally complex mixtures of terpenoid and/or nonterpenoid compounds. Mono-, sesqui-, and sometimes diterpenoids, phenylpropanoids, fatty acids and their fragments, benzenoids, and so on may occur in various essential oils (Baser and Demirci, 2007).

Except for citrus oils obtained by cold pressing, all other essential oils are obtained by distillation. Products obtained by solvent extraction or supercritical fluid extraction are not technically considered as essential oils (Baser, 1995).

Essential oils are used in perfumery, food flavoring, pharmaceuticals, and sources of aromachemicals.

Essential oils exhibit a wide range of biological activities and 31 essential oils have monographs in the latest edition of the *European Pharmacopoeia* (Table 16.1).

Antimicrobial activities of many essential oils are well documented (Bakkali et al., 2008). Such oils may be used singly or in combination with one or more oils. For the sake of synergism this may be necessary.

TABLE 16.1
Essential Oil Monographs in the *European Pharmacopoeia* (6.5 Edition, 2009)

English Name	Latin Name	Plant Name
Anise oil	*Anisi aetheroleum*	*Pimpinella anisum* L. fruits
Bitter-fennel fruit oil	*Foeniculi amari fructus aetheroleum*	*Foeniculum vulgare* Miller subsp. *vulgare* var. *vulgare*
Bitter-fennel herb oil	*Foeniculi amari herba aetheroleum*	*Foeniculum vulgare* Miller subsp. *vulgare* var. *vulgare*
Caraway oil	*Carvi aetheroleum*	*Carum carvi* L.
Cassia oil	*Cinnamomi cassiae aetheroleum*	*Cinnamomum cassia* Blume (*Cinnamomum aromaticum* Nees)
Cinnamon bark oil, Ceylon	*Cinnamomi zeylanici corticis aetheroleum*	*Cinnamomum zeylanicum* Nees
Cinnamon leaf oil, Ceylon	*Cinnamomi zeylanici folium aetheroleum*	*Cinnamomum verum* J.S. Presl.
Citronella oil	*Citronellae aetheroleum*	*Cymbopogon winterianus* Jowitt
Clary sage oil	*Salviae sclareae aetheroleum*	*Salvia sclarea* L.
Clove oil	*Caryophylli aetheroleum*	*Syzygium* aromaticum (L.) Merill et L.M. Perry (*Eugenia caryophyllus* C.S. Spreng. Bull. et Harr.)
Coriander oil	*Coriandri aetheroleum*	*Coriandrum sativum* L.
Dwarf pine oil	*Pini pumilionis aetheroleum*	*Pinus mugo* Turra.
Eucalyptus oil	*Eucalypti aetheroleum*	*Eucalyptus globulus* Labill.
Juniper oil	*Juniperi aetheroleum*	*Juniperus communis* L. meyvesi
Lavender oil	*Lavandulae aetheroleum*	*Lavandula angustifolia* P. Mill. (*Lavandula officinalis* Chaix.)
Lemon oil	*Lemonis aetheroleum*	*Citrus limon* (L.) Burman fil.
Mandarin oil	*Citri reticulatae aetheroleum*	*Citrus reticulata* Blanco
Matricaria oil	*Matricariae aetheroleum*	*Matricaria recutita* L. (*Chamomilla recutita* (L.) Rauschert)
Mint oil, partly dementholized	*Menthae arvensis aetheroleum partim mentholi privum*	*Mentha canadensis* L. (*Mentha arvensis* L. var. *glabrata* (Benth.) Fern, *Mentha arvensis* L. var. *piperascens* Malinv. ex Holmes) *Japanese mint*
Neroli oil (formerly bitter-orange flower oil)	*Neroli aetheroleum* (formerly *Aurantii amari floris aetheroleum*)	*Citrus aurantium* L. subsp. *aurantium* (*Citrus aurantium* L. subsp. *amara* Engl.)
Nutmeg oil	*Myristicae fragrantis aetheroleum*	*Myristica fragrans* Houtt.
Peppermint oil	*Menthae piperitae aetheroleum*	*Mentha* × *piperita* L.
Pine silvestris oil	*Pini silvestris aetheroleum*	*Pinus silvestris* L.
Rosemary oil	*Rosmarini aetheroleum*	*Rosmarinus officinalis* L.
Spanish sage oil	*Salviae lavandulifoliae aetheroleum*	*Salvia lavandulifolia* Vahl.
Spike lavender oil	*Spicae aetheroleum*	*Lavandula latifolia* Medik.
Star anise oil	*Anisi stellati aetheroleum*	*Illicium verum* Hooker fil.
Sweet orange oil	*Aurantii dulcis aetheroleum*	*Citrus sinensis* (L.) Osbeck (*Citrus aurantium* L. var. *dulcis* L.)
Tea tree oil	*Melaleucae aetheroleum*	*Melaleuca alternifolia* (Maiden et Betch) Cheel, *Melaleuca linariifolia* Smith, *Melaleuca dissitiflora* F. Mueller, and other species
Thyme oil	*Thymi aetheroleum*	*Thymus vulgaris* L., *T. zygis* L.
Turpentine oil, *Pinus pinaster* type	*Terebinthini aetheroleum ab pinum pinastrum*	*Pinus pinaster* Aiton. (*Maritime pine*)

Essential Oils Used in Veterinary Medicine

Although many are generally regarded as safe, essential oils are generally not recommended for internal use. However, their much diluted forms (e.g., hydrosols) obtained during oil distillation as a by-product may be taken orally.

Topical applications of some essential oils (e.g., oregano and lavender) in wounds and burns bring about fast recovery without leaving any sign of cicatrix. By inhalation, several essential oils act as a mood changer and have effect especially on respiratory conditions.

Several essential oils (e.g., citronella oil) have been used as pest repellents or as insecticides and such uses are frequently encountered in veterinary applications.

In recent years, especially after the ban on the use of antibiotics in animal feed in the European Union since January 2006, essential oils have emerged as a potential alternative to antibiotics in animal feed.

Essential oils used in veterinary medicine may be classified as follows:

1. Oils attracting animals
2. Oils repelling animals
3. Insecticidal, pest repellent, and antiparasitic oils
4. Oils used in animal feed
5. Oils used in treating diseases in animals.

16.2 OILS ATTRACTING ANIMALS

Valeriana oils (and valerianic and isovalerianic acids) and nepeta oils (and nepetalactones) are well-known feline-attractant oils. Their odor attracts male cats.

Douglas fir oil and its monoterpenes have been claimed to attract deer and wild boar (Buchbauer et al., 1994).

Dogs are normally drawn to floral oils and usually choose to take these by inhalation only. Monoterpene-rich oils are usually too strong for dogs, with the exception of bergamot, *Citrus bergamia*.

Cats also usually select only floral oils for inhalation. Cats do not have metabolic mechanism to break down essential oils due to the lack of the enzyme glucuronidase. Therefore, they should not be taken by mouth and should not be generally applied topically (Ingraham, 2008).

16.3 OILS REPELLING ANIMALS

Peppermint oil (*Mentha piperita*) repels mice. It can be applied under the sink in the kitchen or applied in staples to prevent mice annoying horses and livestock. A few drops of peppermint oil in a bucket of water used to scrub out a stall and sprinkling a few drops around the perimeter and directly on straw or bedding are said to eliminate or severely curtail the habitation of mice (Scents & Sensibility, 2001).

A patent (U.S. Patent 4,961,929) claims that a mixture of methyl salicylate, birch oil, wintergreen oil, eucalyptus oil, pine oil, and pine-needle oil repels dogs.

Another patent (U.S. Patent 4,735,803) claims the same using lemon oil and α-terpinyl methyl ether.

Another similar formulation (U.S. Patent 4,847,292) claims that a mixture of citronellyl nitrile, citronellol, α-terpinyl methyl ether, and lemon oil repels dogs.

A mixture of black pepper and capsicum oils and the oleoresin of rosemary is claimed to repel animals (U.S. Patent 6,159,474).

Citronella oil repels cats and dogs (Moschetti, 2003).

Repellents alleged to repel cats include allyl isothiocyanate (oil of mustard), amyl acetate, anethole, capsaicin, cinnamaldehyde, citral, citronella, citrus oil, eucalyptus oil, geranium oil, lavender oil, lemongrass oil, menthol, methyl nonyl ketone, methyl salicylate, naphthalene, nicotine, paradichlorobenzene, and thymol. Oil of mustard, cinnamaldehyde, and methyl nonyl ketone are said to be the most potent.

Essential oils comprised of 10 g/L solutions of cedarwood, cinnamon, sage, juniper berry, lavender, and rosemary; all of these were potent snake irritants. Brown tree snakes exposed to a 2 s burst of aerosol of these oils exhibited prolonged, violent undirected locomotory behavior. In contrast, exposure to a 10 g/L concentration of ginger oil aerosol caused snakes to locomote, but in a deliberate, directed manner. The 10 g/L solutions delivered as aerosols of *m*-anisaldehyde, *trans*-anethole, 1,8-cineole, cinnamaldehyde, citral, ethyl phenylacetate, eugenol, geranyl acetate, or methyl salicylate acted as potent irritants for brown tree snakes (*Boiga irregularis*) (Clark and Shivik, 2002).

16.4 OILS AGAINST PESTS

16.4.1 INSECTICIDAL, PEST REPELLENT, AND ANTIPARASITIC OILS

The essential oil of bergamot (*C. bergamia*), anise (*Pimpinella anisum*), sage (*Salvia officinalis*), tea tree (*Melaleuca alternifolia*), geranium (*Pelargonium* sp.), peppermint (*M. piperita*), thyme (*Thymus vulgaris*), hyssop (*Hyssopus officinalis*), rosemary (*Rosmarinus officinalis*), and white clover (*Trifolium repens*) can be used to control certain pests on plants. They have been shown to reduce the number of eggs laid and the amount of feeding damage by certain insects, particularly lepidopteran caterpillars. Sprays made from tansy (*Tanacetum vulgare*) have demonstrated a repellent effect on imported cabbageworm on cabbage, reducing the number of eggs laid on the plants. Teas made from wormwood (*Artemisia absinthium*) or nasturtiums (*Nasturtium* spp.) are reputed to repel aphids from fruit trees, and sprays made from ground or blended catnip (*Nepeta cataria*), chives (*Allium schoenoprasum*), feverfew (*Tanacetum parthenium*), marigolds (*Calendula, Tagetes*, and *Chrysanthemum* spp.), or rue (*Ruta graveolens*) have also been used by gardeners against pests that feed on leaves (Moschetti, 2003).

16.4.2 FLEAS AND TICKS

Dogs, cats, and horses are plagued by fleas and ticks. One to two drops of citronella or lemongrass oils added to the shampoo will repel these pests. Alternatively, 4–5 drops of cedarwood oil and pine oil is added to a bowl of warm water, and a bristle hairbrush is soaked with this solution to brush the pet down with it. Eggs and parasites gathered in the brush are rinsed out. This is repeated several times. This solution can be used similarly for livestock after adding citronella and lemongrass oils to this mixture.

Flea collar can be prepared by a mixture of cedarwood (*Juniperus virginiana*), lavender (*Lavandula angustifolia*), citronella (*Cymbopogon winterianus* [Java]), thyme oils, and 4–5 garlic (*Allium sativum*) capsules. This mixture is thinned with a teaspoonful of ethanol and soaked with a collar or a cotton scarf. This is good for 30 days (Scents & Sensibility, 2001).

Ticks can be removed by applying 1 drop of cinnamon or peppermint oil on Q-tip by swabbing on it.

Carvacrol-rich oil (64%) of *Origanum onites* and carvacrol was found to be effective against the tick *Rhipicephalus turanicus*. Pure carvacrol killed all the ticks following 6 h of exposure, while 25% and higher concentrations of the oil were effective in killing the ticks by the 24 h posttreatment (Coskun et al., 2008).

16.4.3 MOSQUITOES

Catnip oil (*N. cataria*) containing nepetalactones can be used effectively as a mosquito repellent. It is said to be 10 times more effective than DEET (*N,N*-diethyl-meta-toluamide) (Moschetti, 2003).

Juniperus communis berry oil is a very good mosquito repellent. Ocimum volatile oils including camphor, 1,8-cineole, methyl eugenol, limonene, myrcene, and thymol strongly repelled mosquitoes (Regnault-Roger, 1997).

Citronella oil repels mosquitoes, biting insects, and fleas.

Essential oils of *Zingiber officinale* and *R. officinalis* were found to be ovicidal and repellent, respectively, toward three mosquito species (Prajapati et al., 2005). Root oil of *Angelica sinensis* and ligustilide was found to be mosquito repellent (Wedge et al., 2009).

16.4.4 Moths

Cedarwood oil is used in mothproofing. A large number of patents have been assigned to the preservation of cloths from moths and beetles: application of a solution containing clove (*Syzygium aromaticum*) essential oil on woolen cloth, filter paper containing *Juniperus rigida* oil, and tablets of *p*-dichlorobenzene mixed with essential oils to be placed in wardrobe.

16.4.5 Aphids, Caterpillars, and Whiteflies

16.4.5.1 Garlic Oil

Essential oils are effective in insect pest control (Regnault-Roger, 1997).

16.4.6 Ear Mites

Peppermint oil is applied to a Q-tip and swabbed inside of the ear.

16.4.7 Antiparasitic

There is a patent (U.S. Patent 6,800,294) on an antiparasitic formulation comprising eucalyptus oil (*Eucalyptus globulus*), cajeput oil (*Melaleuca cajuputi*), lemongrass oil, clove bud oil (*S. aromaticum*), peppermint oil (*M. piperita*), piperonyl, and piperonyl butoxide. The formulation can be used for treating an animal body, in the manufacture of a medicament for treating ectoparasitic infestation of an animal, or for repelling parasites.

Two essential oils derived from *L. angustifolia* and *Lavandula × intermedia* were investigated for any antiparasitic activity against the human protozoal pathogens *Giardia duodenalis* and *Trichomonas vaginalis* and the fish pathogen *Hexamita inflata*, all of which have significant infection and economic impacts. The study has demonstrated that low ($\leq$1%) concentrations of *L. angustifolia* and *Lavandula × intermedia* oil can completely eliminate *T. vaginalis*, *G. duodenalis*, and *H. inflata* in vitro. At 0.1% concentration, *L. angustifolia* oil was found to be slightly more effective than *Lavandula × intermedia* oil against *G. duodenalis* and *H. inflata* (Moon et al., 2006).

The antiparasitic properties of essential oils from *A. absinthium*, *Artemisia annua*, and *Artemisia scoparia* were tested on intestinal parasites *Hymenolepis nana*, *Lamblia intestinalis*, *Syphacia obvelata*, and *Trichocephalus muris* (*Trichuris muris*). Infested white mice were injected with 0.01 mL/g of the essential oils (6%) twice a day for 3 days. The effectiveness of the essential oils was observed in 70%–90% of the tested animals (Chobanov et al., 2004).

Parasites, such as head lice and scabies, as well as internal parasites, are repelled by oregano oil (86% carvacrol). The oil can be added to soaps, shampoos, and diluted in olive oil for topical applications. By taking a few drops daily under the tongue, one can gain protection from waterborne parasites, such as *Cryptosporidium* and *Giardia*. Internal dosages also are effective in killing parasites in the body (http://curingherbs.com/wild_oregano_oil.htm) (Foster, 2002).

Essential oils from *Pinus halepensis*, *Pinus brutia*, *Pinus pinaster*, *Pinus pinea*, and *Cedrus atlantica* were tested for molluscicidal activity against *Bulinus truncatus*. The oil from *C. atlantica* was found the most active (LC 50 = 0.47 ppm). Among their main constituents, α-pinene, β-pinene, and myrcene exhibited potent molluscicidal activity (LC 50 = 0.49, 0.54, and 0.56 ppm, respectively). These findings have important application of natural products in combating schistosomiasis (Lahlou, 2003).

Origanum essential oils have exhibited differential degrees of protection against myxosporean infections in gilthead and sharpsnout sea bream tested in land-based experimental facilities (Athanassopoulou et al., 2004a,b).

16.5 ESSENTIAL OILS USED IN ANIMAL FEED

Essential oils can be used in feed as appetite stimulant, stimulant of saliva production, gastric and pancreatic juice production enhancer, and antimicrobial and antioxidant to improve broiler performance. Antimicrobial effects of essential oils are well documented. Essential oils due to their potent nature should be used as low as possible levels in animal nutrition. Otherwise, they can lead to feed intake reduction, gastrointestinal tract (GIT) microflora disturbance, or accumulation in animal tissues and products. Odor and taste of essential oils may contribute to feed refusal; however, encapsulation of essential oils could solve this problem (Gauthier, 2005).

Generally, Gram-positive bacteria are considered more sensitive to essential oils than Gram-negative bacteria because of their less complex membrane structure (Lis-Balchin, 2003).

Carvacrol, the main constituent of oregano oils, is a powerful antimicrobial agent (Baser, 2008). It asserts its effect through the biological membranes of bacteria. It acts through inducing a sharp reduction of the intercellular adenosine triphosphate (ATP) pool through the reduction of ATP synthesis and increased hydrolysis. Reduction of the membrane potential (transmembrane electrical potential), which is the driving force of ATP synthesis, makes the membrane more permeable to protons. A high level of carvacrol (1 mM) decreases the internal pH of bacteria from 7.1 to 5.8 related to ion gradients across the cell membrane. One millimolar of carvacrol reduces the internal potassium (K) level of bacteria from 12 mmol/mg of cell protein to 0.99 mmol/mg in 5 min. K plays a role in the activation of cytoplasmic enzymes and in maintaining osmotic pressure and in the regulation of cytoplasmic pH. K efflux is a solid indication of membrane damage (Ultee et al., 1999).

It has been shown that the mode of action of oregano oils is related to an impairment of a variety of enzyme systems, mainly involved in the production of energy and the synthesis of structural components. Leakage of ions, ATP, and amino acids also explains the mode of action. Potassium and phosphate ion concentrations are affected at levels below the minimum inhibitory concentration (MIC) concentration (Lambert et al., 2001).

16.5.1 RUMINANTS

A recent review compiled information on botanicals including essential oils used in ruminant health and productivity (Rochfort et al., 2008). Unfortunately, there are few reports on the effects of essential oils and natural aromachemicals on ruminants. It was demonstrated that the consumption of terpene volatiles such as camphor and α-pinene in "tarbush" (*Flourensia cernua*) effected feed intake in sheep (Estell et al., 1998). In vitro and in vivo antimicrobial activities of essential oils have been demonstrated in ruminants (Cardozo et al., 2005; Elgayyar et al., 2001; Moreira et al., 2005; Wallace et al., 2002). Synergistic antinematodal effects of essential oils and lipids were demonstrated (Ghisalberti, 2002). Other nematocidal volatiles reported are as follows: benzyl isothiocyanate (goat), ascaridole (goat and sheep) (Ghisalberti, 2002; Githiori et al., 2006), geraniol, eugenol (Chitwood, 2002; Githiori et al., 2006), menthol, and 1,8-cineole (Chitwood, 2002).

Methylsalicylate, the main component of the essential oil of *Gaultheria procumbens* (wintergreen), is topically used as emulsion in cattle, horses, sheep, goats, and poultry in the treatment of muscular and articular pain. The recommended dose is 600 mg/kg bw twice a day. The duration of treatment is usually less than 1 week (EMA, 1999). It is included in Annex II of Council Regulation (EEC) No. 2377/90 as a substance that does not need an maximum residue limit (MRL) level. *G. procumbens* should not be used as flavoring in pet food since salicylates are toxic to dogs and cats. As cats metabolize salicylates much more slowly than other species, they are more likely to be overdosed. Use of methylsalicylate in combination with anticoagulants such as warfarin can result in adverse interactions and bleedings (Chow et al., 1989; Ramanathan, 1995; Tam et al., 1995; Yip et al., 1990).

The essential oil of *L. angustifolia* (*Lavandulae aetheroleum*) is used in veterinary medicinal products for topical use together with other plant extracts or essential oils for antiseptic and healing purposes. The product is used in horses, cattle, sheep, goats, rabbits, and poultry. It is included in Annex II of Council Regulation (EEC) No. 2377/90 as a substance that does not need an MRL level (EMEA, 1999; Franz et al., 2005).

The outcomes of in vitro studies investigating the potential of *P. anisum* essential oil as a feed additive to improve nutrient use in ruminants are inconclusive, and more and larger preferably in vivo studies are necessary for evaluation of efficacy (Franz et al., 2005).

Carvacrol, carvone, cinnamaldehyde, cinnamon oil, clove bud oil, eugenol, and oregano oil have resulted in a 30%–50% reduction in ammonia N concentration in diluted ruminal fluid with a 50:50 forage concentrate diet during the 24 h incubation (Busquet et al., 2006).

Carvacrol has been suggested as a potential modulator of ruminal fermentation (Garcia et al., 2007).

16.5.2 Poultry

16.5.2.1 Studies with CRINA® Poultry

Dietary addition of essential oils in a commercial blend (CRINA Poultry) showed a decreased *Escherichia coli* population in ileocecal digesta of broiler chickens. Furthermore, in high doses, a significant increase in certain digestive enzyme activities of the pancreas and intestine was observed in broiler chickens (Jang et al., 2007).

In another study, CRINA Poultry was shown to control the colonization of the intestine of broilers with *Clostridium perfringens*, and the stimulation of animal growth was put down to this development (Losa, 2001).

Commercial essential oil blends CRINA Poultry and CRINA Alternate were tested in broilers infected with viable oocysts of mixed *Eimeria* spp. It was concluded that these essential oil blends may serve as an alternative to antibiotics and/or ionophores in mixed *Eimeria* infections in non-cocci-vaccinated broilers, but no benefit of essential oil supplementation was observed for vaccinated broilers against coccidia (Oviedo-Rondon et al., 2006).

16.5.2.1.1 Other Studies

Supplementation of 200 ppm essential oil mixture (EOM) that included oregano, clove, and anise oils (no species name or composition given!) in broiler diets was said to significantly improve the daily live weight gain and feed conversion ratio (FCR) during a growing period of 5 weeks (Ertas et al., 2006). Similar results were obtained with 400 mg/kg anise oil (composition not known!) (Ciftci et al., 2005).

A total of 50 and 100 mg/kg of feed of oregano oil* were tested on broilers. No growth-promoting effect was observed. At 100 mg/kg of feed, antioxidant effect was detected on chicken tissues (Botsoglou et al., 2002a).

Positive results were also reported for oregano oil added in poultry feed (Bassett, 2000).

Antioxidant activities of rosemary and sage oils on lipid oxidation of broiler meat have been shown. Following dietary administration of rosemary and sage oils to the live birds, a significant

---

* Oregano essential oil was in the form of a powder called Orego-Stim. This product contains 5% oregano essential oil (Ecopharm Hellas S.A., Kilkis, Greece) and 95% natural feed grade inert carrier. The oil of *Origanum vulgare* subsp. *hirtum* used in this product contains 85% carvacrol and thymol.

inhibition of lipid peroxidation was reported in chicken meat stored for 9 days (Lopez-Bote et al., 1998). A dietary supplementation of oregano essential oil (300 mg/kg) showed a positive effect on the performance of broiler chickens experimentally infected with *Eimeria tenella*. Throughout the experimental period of 42 days, oregano essential oil exerted an anticoccidial effect against *E. tenella*, which was, however, lower than that exhibited by lasalocid. Supplementation with dietary oregano oil to *E. tenella*–infected chickens resulted in body weight gains and FCRs not differing from the noninfected group, but higher than those of the infected control group and lower than those of chickens treated with the anticoccidial lasalocid (Giannenas et al., 2003).

Inclusion of oregano oil at 0.005% and 0.01% in chicken diets for 38 days resulted in a significant antioxidant effect in raw and cooked breast and thigh muscle stored up to 9 days in refrigerator (Botsoglou et al., 2002b).

Oregano oil (55% carvacrol) exhibited a strong bactericidal effect against lactobacilli and following the oral administration of the oil MIC values of amikacin, apramycin, and streptomycin and neomycin against *E. coli* strains increased (Horosova et al., 2006).

An in vitro assay measuring the antimicrobial activity of essential oils of *Coridothymus capitatus*, *Satureja montana*, *Thymus mastichina*, *Thymus zygis*, and *Origanum vulgare* was carried out against poultry origin strains of *E. coli*, *Salmonella enteritidis*, and *Salmonella essen* and pig origin strains of enterotoxigenic *E. coli*, *Salmonella choleraesuis*, and *Salmonella typhimurium*. *O. vulgare* (MIC ≤ 1% v/v) oil showed the highest antimicrobial activity against the four strains of *Salmonella*. It was followed by *T. zygis* oil (MIC ≤ 2% v/v). *T. mastichina* oil inhibited all the microorganisms at the highest concentration, 4% (v/v). Monoterpenic phenols carvacrol and thymol showed higher inhibitory capacity than the monoterpenic alcohol linalool. The results confirmed potential application of such oils in the treatment and prevention of poultry and pig diseases caused by *Salmonella* (Penalver et al., 2005).

In another study, groups of male, 1-day-old Lohmann broilers were given maize–soya bean meal diets, with oils extracted from thyme, mace, and caraway or coriander, garlic, and onion (0, 20, 40, and 80 mg/kg) for 6 weeks. The average daily gain and FCR were not different between the broilers fed with the different oils; meat was not tainted with flavor or smell of the oils (Vogt and Rauch, 1991).

16.5.2.2 Studies with Herbromix®

Essential oils from oregano herb (*O. onites*), laurel leaf (*Laurus nobilis*), sage leaf (*Salvia fruticosa*), fennel fruit (*Foeniculum vulgare*), myrtle leaf (*Myrtus communis*), and citrus peel (rich in limonene) were mixed and formulated as feed additive after encapsulation. It is marketed in Turkey as poultry feed under the name Herbromix.

The following three in vivo experiments with this product were recently accomplished.

16.5.2.2.1 In Vivo Experiment 1

In this study, 1250 sexed 1-day-old broiler chicks obtained from a commercial hatchery were randomly divided into five treatment groups of 250 birds each (negative control, antibiotic, and essential oil combination [EOC] at 3 levels). Each treatment group was further subdivided into five replicates of 50 birds (25 males and 25 females) per replicate. Commercial EOC at three different levels (24, 48, and 72 mg) and antibiotic (10 mg avilamycin) per kg were added to the basal diet. There were significant effects of dietary treatments on body weight, feed intake (except at day 42), FCR, and carcass yield at 21 and 42 days. Body weights were significantly different between the treatments. Birds fed on diet containing 48 mg essential oil/kg being the highest, and this treatment was followed by chicks fed on the diet containing 72 mg essential oil/kg, antibiotic, negative control, and 24 mg essential oil/kg at day 42.

Supplementation with 48 mg EOC/kg to the broiler diet significantly improved the body weight gain, FCR, and carcass yield compared to other dietary treatments on 42 days of age. EOC may be considered as a potential growth promoter in the future of the new era, which agrees with producer needs for increased performance and today's consumer demands for environment-friendly broiler production. The EOC can be used cost-effectively when its cost is compared with antibiotics and other commercially available products in the market.

16.5.2.2.2 In Vivo Experiment 2

In this study, 1250 sexed 1-day-old broiler chicks were randomly divided into five treatment groups of 250 birds each (negative control, organic acid, probiotic, and EOC at two levels). Each treatment group was further subdivided into five replicates of 50 birds (25 males and 25 females) per replicate. The oils in the EOC were extracted from different herbs growing in Turkey. The organic acid at 2.5 g/kg diet, the probiotic at 1 g/kg diet, and the EOC at 36 and 48 mg/kg diet were added to the basal diet.

The results obtained from this study indicated that the inclusion of 48 mg EOC/kg broiler diet significantly improved the body weight gain, FCR, and carcass yield of broilers compared to organic acid and probiotic treatments after a growing period of 42 days. The EOC may be considered as a potential growth promoter like organic acids and probiotics for environment-friendly broiler production.

16.5.2.2.3 In Vivo Experiment 3

The aim of this study was to examine the effect of essential oils and breeder age on growth performance and some internal organs' weight of broilers. A total of 1008 unsexed 1-day-old broiler chicks (Ross-308) originating from young (30 weeks) and older (80 weeks) breeder flocks were randomly divided into three treatment groups of 336 birds each, consisting of control and two EOMs at a level of 24 and 48 mg/kg diet. There were no significant effects of dietary treatments on body weight gain of broilers at days 21 and 42.

On the other hand, there were significant differences on the feed intake at days 21 and 42. The addition of 24 or 48 mg/kg EOM to the diet reduced significantly the feed intake compared to the control. The groups fed with the added EOM had significantly better FCR than the control at days 21 and 42. Although there was no significant effect of broiler breeder age on body weight gain at day 21, significant differences were observed on body weight gain at 42 days of age. Broilers originating from young breeder flock had significantly higher body weight gain than those originating from old breeder flock at 42 days of age. No difference was noticed for carcass yield, liver, pancreas, proventriculus, gizzard, and small intestine weight. Supplementation with EOM to the diet in both levels significantly decreased mortality at days 21 and 42.

The results indicated that the Herbromix may be considered as a potential growth promoter. However, more trials are needed to determine the effect of essential oil supplementation to diet on the performance of broilers with regard to variable management conditions including different stress factors, essential oils and their optimal dietary inclusion levels, active substances of oils, dietary ingredients, and nutrient density (Alcicek et al., 2003, 2004; Bozkurt and Baser, 2002a,b; Cabuk et al., 2006a,b).

16.5.3 Pigs

CRINA® Pigs was tested on pigs. The results for the first 21-day period showed that males grew faster, ate less, and exhibited superior FCR compared to females. Although female carcass weight was higher, males had a significantly lower carcass fat than females (Losa, 2001).

The addition of fennel (*F. vulgare*) and caraway (*Carum carvi*) oils was not found beneficial for weaned piglets. In feed choice conditions, fennel oil caused feed aversion (Schoene et al., 2006).

Oregano oil was found to be beneficial for piglets (Molnar and Bilkei, 2005).

In a preliminary investigation, the effects of low-level dietary inclusion of rosemary, garlic, and oregano oils on pig performance and pork quality were carried out. Unfortunately, no information on the species from which the oils were obtained and their composition existed in the paper. The pigs appeared to prefer the garlic-treated diet, and the feed intake and the average daily gain were significantly increased although no difference in the feed efficiency was observed. Carcass and meat quality attributes were unchanged, although a slight reduction of lipid oxidation was noted in oregano-fed pork. Since the composition of the oils is not clear, it is not possible to evaluate the results (Janz et al., 2007).

A study revealed that the inclusion of essential oil of oregano in pigs' diet significantly improved the average daily weight gain and FCR of the pigs. Pigs fed with the essential oils had higher carcass weight, dressing percentage, and carcass length than those fed with the basal and antibiotic-supplemented diet. The pigs that received the essential oil supplementation had a significantly lower fat thickness. Also lean meat and ham portions from these pigs were significantly higher. Therefore, the use of *Origanum* essential oil as feed additive improves the growth of pigs and has greater positive effects on carcass composition than antibiotics (Onibala et al., 2001).

Ropadiar®, an essential oil of the oregano plant, was supplemented in the diet of weaning pigs as alternative for antimicrobial growth promoters (AMGPs), observing its efficacy on the performance of the piglets. Ropadiar liquid contains 10% oregano oil and has been designed to be added to water. Compared to the negative control (without AMGP), Ropadiar improved performance only during the first 14 days after weaning. Based on the results of this trial, it cannot be argued about the usefulness of Ropadiar as an alternative for AMGP in diets of weanling pigs. However, its addition in prestarter diets could improve performance of these animals (Krimpen and Binnendijk, 2001).

The objective of another trial was to ascertain the effect on nutrient digestibilities and N-balance, as well as on parameters of microbial activity in the GIT of weaned pigs after adding oregano oil to the feed. The apparent digestibility of crude nutrients (except fiber) and the N-balance of the weaned piglets in this study were not influenced by feeding piglets restrictively with this feed additive. By direct microbiological methods, no influence of the additive on the gut flora could be found (Moller, 2001).

The inclusion of essential oil of spices in the pigs' diet significantly improved the average daily weight gain and FCR of the pigs in Groups 3, 4, and 5, as compared to Groups 1 and 2 ($P < 0.01$). Furthermore, pigs fed with the essential oils had higher carcass weight ($P < 0.01$), dressing percentage ($P < 0.01$), and carcass length ($P < 0.01$) than those fed with the basal and antibiotic-supplemented diet. In Groups 3, 4, and 5, backfat thickness was significantly lower than those in Groups 1 and 2. Moreover, lean meat and ham portions from pigs in Groups 3, 4, and 5 were significantly higher than those from pigs in Groups 1 and 2. In conclusion, the use of essential oils as feed additives improves the growth of pigs and has greater positive effects on carcass composition than antibiotics (Onibala et al., 2001).

16.6 ESSENTIAL OILS USED IN TREATING DISEASES IN ANIMALS

There is scarce scientific information on the use of essential oils in treating diseases in animals. Generally, the oils used in treating diseases in humans are also recommended for animals.

Internet literature is abound with valid and/or suspicious information in this issue. We have tried to compile relevant information using the reachable resources. The information may not be concise or comprehensive but should be seen as an effort to combine the available information in a short period of time.

The oil of *Ocimum basilicum* has been reported as an expectorant in animals. The combined oils of *Ocimum micranthum* and *Chenopodium ambrosioides* are claimed to treat stomachache and colic in animals (Cornell University Animal Science, 2009).

Bad breath as a result of gum disease and bacterial buildup on the teeth of pets can be treated by brushing their teeth with a mixture of a couple of tablespoons of baking soda, 1 drop of clove oil, and 1 drop of aniseed oil. Lavender, myrrh, and clove oils can also be directly applied to their gums.

For wounds, abscesses, and burns, lavender and tea tree oils are used by topical application. Skin rashes can be treated with tea tree, lavender, and chamomile oils.

Earache of pets can be healed by dripping a mixture of lavender, chamomile, and tea tree oils (1 drop each) dissolved in a teaspoonful of grape-seed or olive oil in the infected ears.

Hoof rot in livestock can be treated with a hot compress made up of 10 drops of chamomile, 15 drops of thyme, and 5 drops of melissa oils diluted in about 100 mL of vegetable oil (e.g., grape-seed oil).

Intestinal worms of horses can be expelled by applying 3–4 drops of thyme oil and tansy leaves to each feed. Melissa oil can be added to feed to increase milk production of both cows and goats (http://scentsnsensibility.com/newsletter/Apr0601.htm).

Aromatic plants such as *Pimpinella isaurica*, *Pimpinella aurea*, and *Pimpinella corymbosa* are used as animal feed to increase milk secretion in Turkey (Tabanca et al., 2003).

To calm horses, chamomile oil is added to their feed. Pneumonia in young elephants caused by *Klebsiella* is claimed to be healed by *Lippia javanica* oil. Rose and yarrow oils bring about emotional release in donkeys by licking them. Wounds in horses are treated with *Achillea millefolium* oil; sweet itch is treated with peppermint oil. *Matricaria recutita* and *A. millefolium* oils are used to heal the skin with inflammatory conditions (Ingraham, 2008).

A study evaluated the effect of dietary oregano etheric oils as nonspecific immunostimulating agents in growth-retarded, low-weight growing–finishing pigs. A group of pigs were fed with commercial fattening diet supplemented with 3000 ppm oregano additive (Oregpig®, Pecs, Hungary), composed of dried leaf and flower of *O. vulgare*, enriched with 500 g/kg cold-pressed essential oils of the leaf and flower of *O. vulgare*, and composed of 60 g carvacrol and 55 g thymol/kg. Dietary oregano improved growth in growth-retarded growing–finishing pigs and had nonspecific immunostimulatory effects on porcine immune cells (Walter and Bilkei, 2004).

Menthol is often used as a repellent against insects and in lotions to cool legs (especially for horses) (Franz et al., 2005).

Milk cows become restless and aggressive each time a group of cows are separated and regrouped. This can last a few days putting cows in more stress resulting in a drop in milk production. Two Auburn University scientists could solve this problem by spraying anise oil (*P. anisum*) on the cows. Treated animals could not distinguish any differences among the cows in new or old groupings. They were mellower and kept their milk production up. Among many other oils tested but only anise seemed to work (HerbalGram, 1990).

Essential oils have been found effective in honeybee diseases (Ozkirim, 2006; Ozkirim et al., 2007).

In this review, we tried to give you an insight into the use of essential oils in animal health and nutrition. Due to the paucity of research in this important area, there is not much to report. Most information on usage exists in the form of not-so-well-qualified reports. We hope that this rather preliminary report can be of use as a starting point for more comprehensive reports.

REFERENCES

Alcicek, A., M. Bozkurt, and M. Cabuk, 2003. The effect of an essential oil combination derived from selected herbs growing wild in Turkey on broiler performance. *S. Afr. J. Anim. Sci.*, 33(2): 89–94.

Alcicek, A., M. Bozkurt, and M. Cabuk, 2004. The effect of a mixture of herbal essential oils, an organic acid or a probiotic on broiler performance. *S. Afr. J. Anim. Sci.*, 34(4): 217–222.

Athanassopoulou, F., E. Karagouni, E. Dotsika, V. Ragias, J. Tavla, and P. Christofilloyani, 2004a. Efficacy and toxicity of orally administered anticoccidial drugs for innovative treatments of *Polysporoplasma sparis* infection in *Sparus aurata* L. *J. Appl. Ichthyol.*, 20: 345–354.

Athanassopoulou, F., E. Karagouni, E. Dotsika, V. Ragias, J. Tavla, P. Christofilloyanis, and I. Vatsos, 2004b. Efficacy and toxicity of orally administered anticoccidial drugs for innovative treatments of *Myxobolus* sp. infection in *Puntazzo puntazzo*. *Dis. Aquat. Org.*, 62: 217–226.

Bakkali, F., S. Averbeck, D. Averbeck, and M. Idaomar, 2008. Biological effects of essential oils—A review. *Food Chem. Technol.*, 46: 446–475.

Baser, K.H.C., 1995. Analysis and quality assessment of essential oils. In: *A Manual on the Essential Oil Industry*, K.T. De Silva (ed.), pp. 155–177. Vienna, Austria: UNIDO.

Baser, K.H.C., 2008. Chemistry and biological activities of carvacrol and carvacrol-bearing essential oils. *Curr. Pharm. Des.*, 14: 3106–3120.

Baser, K.H.C. and F. Demirci, 2007. Chemistry of essential oils. In: *Flavours and Fragrances. Chemistry, Bioprocessing and Sustainability*, R.G. Berger (ed.), pp. 43–86. Berlin, Germany: Springer.

Bassett, R., 2000. Oreganos positive impact on poultry production. *World Poultry—Elsevier*, 16: 31–34.

Botsoglou, N.A., E. Christaki, D.J. Fletouris, P. Florou-Paneri, and A.B. Spais, 2002a. The effect of dietary oregano essential oil on lipid oxidation in raw and cooked chicken during refrigerated storage. *Meat Sci.*, 62: 259–265.

Botsoglou, N.A., P. Floron-Paneri, E. Christaki, D.J. Fletouris, and A.B. Spais, 2002b. Effect of dietary oregano essential oil on performance of chickens and on iron-induced lipid peroxidation of breast, thigh and abdominal fat tissues. *Br. Poult. Sci.*, 43(2): 223–230.

Bozkurt, M. and K.H.C. Baser, 2002a. The effect of antibiotic, Mannan oligosaccharide and essential oil mixture on the laying egg performance. In: *First European Symposium on Bioactive Secondary Plant Products in Veterinary Medicine*, Vienna, Austria, October 4–5, 2002.

Bozkurt, M. and K.H.C. Baser, 2002b. The effect of commercial organic acid, probiotic and essential oil mixture at two levels on the performance of broilers. In: *First European Symposium on Bioactive Secondary Plant Products in Veterinary Medicine*, Vienna, Austria, October 4–5, 2002.

Buchbauer, G., L. Jirovetz, M. Wasicky, and A. Nikiforov, 1994. Comparative investigation of Douglas fir headspace samples, essential oils, and extracts (needles and twigs) using GC-FID and GC-FTIR-MS. *J. Agric. Food Chem.*, 42: 2852–2854.

Busquet, M., S. Calsamiglia, A. Ferret, and C. Kamel, 2006. Plant extracts affect in vitro rumen microbial fermentation. *J. Dairy Sci.*, 89: 761–771.

Cabuk, M., M. Bozkurt, A. Alcicek, Y. Akbas, and K. Kucukyilmaz, 2006a. Effect of a herbal essential oil mixture on growth and internal organ weight of broilers from young and old breeder flocks. *S. Afr. J. Anim. Sci.*, 36(2): 135–141.

Cabuk, M., M. Bozkurt, A. Alcicek, A.U. Catli, and K.H.C. Baser, 2006b. Effect of dietary essential oil mixture on performance of laying hens in summer season. *S. Afr. J. Anim. Sci.*, 36(4): 215–221.

Cardozo, P.W., S. Calsamiglia, A. Ferret, and C. Kamel, 2005. Screening for the effects of natural plant extracts at different pH on in vitro Rumen microbial fermentation of a high-concentrate diet for beef cattle. *J. Anim. Sci.*, 83: 2572–2579.

Chitwood, D.J., 2002. Phytochemical based strategies for nematode control. *Annu. Rev. Phytopathol.*, 40: 221–249.

Chobanov, R.E., A.N. Aleskerova, S.N. Dzhanahmedova, and L.A. Safieva, 2004. Experimental estimation of antiparasitic properties of essential oils of some *Artemisia* (Asteraceae) species of Azerbaijan flora. *Rastitel'nye Resursy*, 40(4): 94–98.

Chow, W.H., K.L. Cheung, H.M. Ling, and T. See, 1989. Potentiation of warfarin anticoagulation by topical methylsalicylate ointment. *J. R. Soc. Med.*, 82(8): 501–502.

Ciftci, M., T. Guler, B. Dalkilic, and O.K. Ertas, 2005. The effect of anise oil (*Pimpinella anisum* L.) on broiler performance. *Int. J. Poultry Sci.*, 4(11): 851–855.

Clark, L. and J. Shivik, 2002. Aerosolized essential oils and individual natural product compounds as brown tree snake repellents. *Pest Manage. Sci.*, 58(8): 775–783.

Cornell University Animal Science, 2009. Treating livestock with medicinal plants: Beneficial or toxic?—*Ocimum basilicum, O. americanum* and *O. micranthum*. http://www.ansci.cornell.edu/plants/medicinal/basil.html. Accessed September 10, 2009.

Coskun, S., O. Grekin, M. Kurkcuoglu, H. Malyer, A.O. Grekin, N. Kirimer, and K.H.C. Baser, 2008. Acaricidal efficacy of *Origanum onites* L. essential oil against *Rhipicephalus turanicus* (Ixodidae). *Parasitol. Res.*, 103: 259–261.

Council Regulation (EEC) No. 2377/90, 1990. Laying down a community procedure for the establishment of maximum residue limits of veterinary medicinal products in foodstuffs of animal origin. http://ec.europa.eu/health/files/eudralex/vol-5/reg_1990_2377/reg_1990_2377_en.pdf. Accessed September 10, 2008.

Elgayyar, M., F.A. Draughon, D.A. Golden, and J.A. Mount, 2001. Antimicrobial activity of essential oils from plants against selected pathogenic and saprophytic microorganisms. *J. Food Prot.*, 64(7): 1019–1024.

EMA (European Medicines Agency), 1999. Salicylic acid, sodium salicylate, aluminium salicylate and methyl salicylate. Committee for Veterinary Medicinal Products (EMEA/MRL/696/99, 1999). www.ema.europa.eu/ema/pages/includes/document/open_document.jsp?webContentId=WC500015823. Accessed May 17, 2008.

Ertas, O.K., T. Guler, M. Ciftci, B. Dalkilic, and U.G. Simsek, 2006. The effect of an essential oil mixture from oregano, clove and anis on broiler performance. *Int. J. Poult. Sci.*, 4(11): 879–884.

Estell, R.E., E.L. Fredrickson, M.R. Tellez, K.M. Havstad, W.L. Shupe, D.M. Anderson, and M.D. Remmenga, 1998. Effects of volatile compounds on consumption of alfalfa pellets by sheep. *J. Anim. Sci.*, 76: 228–233.

Foster, S., 2002. The fighting power of Oregano: This versatile herb packs a powerful punch—Earth medicine. *Better Nutr.*, 1. http://www.oreganocures.com/articles74.html. Accessed May 17, 2015.

Franz, Ch., R. Bauer, R. Carle, D. Tedesco, A. Tubaro, and K. Zitterl-Eglseer, 2005. Study on the assessment of plants/herbs, plant/herb extracts and their naturally or synthetically produced components as "additives" for use in animal production. CFT/EFSA/FEEDAP/2005/01. http://www.agronavigator.cz/UserFiles/File/Agronavigator/Kvasnickova_2/EFSA_feedap_report_plantsherbs.pdf. Accessed September 15, 2008.

Garcia, V., P. Catala-Gregori, J. Madrid, F. Hernandez, M.D. Megias, and H.M. Andrade-Montemayor, 2007. Potential of carvacrol to modify in vitro rumen fermentation as compared with monensin. *Animal*, 1: 675–680.

Gauthier, R., 2005. Organic acids and essential oils, a realistic alternative to antibiotic growth promoters in poultry. *I Forum Internacional de Avicultura*. Foz do Iguaçu, PR, Brazil, August 17–19, 2005.

Ghisalberti, E.L., 2002. Secondary metabolites with antinematodal activity. In: *Studies in Natural Products Chemistry*, Atta-ur-Rahman (ed.), Vol. 26, pp. 425–506. Amsterdam, the Netherlands: Elsevier Science BV.

Giannenas, I., P.P. Florou, M. Papazahariadou, E. Christaki, N.A. Botsoglou, and A.B. Spais, 2003. Effect of dietary supplementation with oregano essential oil on performance of broilers after experimental infection with *Eimeria tenella*. *Arch. Anim. Nutr.*, 57(2): 99–106.

Githiori, J.B., S. Athanasiadou, and S.M. Thamsborg, 2006. Use of plants in novel approaches for control of gastrointestinal helminthes in livestock with emphasis on small ruminants. *Vet. Parasitol.*, 139: 308–320.

HerbalGram, 1990. Bovine aromatherapy: Common herb quells cowcophony. *HerbalGram*, 22: 8.

Horosova, K., D. Bujnakova, and V. Kmet, 2006. Effect of oregano essential oil on chicken lactobacilli and *E. coli*. *Folia Microbiol.*, 51(4): 278–280.

Ingraham, C, 2008. Zoopharmacognosy—Working with aromatic medicine. http://www.ingraham.co.uk/. Accessed May 17, 2008.

Jang, I.S., Y.H. Ko, S.Y. Kang, and C.Y. Lee, 2007. Effect of commercial essential oil on growth performance digestive enzyme activity and intestinal microflora population in broiler chickens. *Anim. Feed Sci. Technol.*, 134: 304–315.

Janz, J.A.M., P.C.H. Morel, B.H.P. Wilkinson, and R.W. Purchas, 2007. Preliminary investigation of the effects of low-level dietary inclusion of fragrant essential oils and oleoresins on pig performance and pork quality. *Meat Sci.*, 75: 350–355.

Krimpen, M.V. and G.P. Binnendijk, 2001. Ropadiar® as alternative for anti microbial growth promoter in diets of weanling pigs. Rapport Praktijkonderzoek Veehouderij, May 15, 2001. ISSN: 0169-3689.

Lahlou, M., 2003. Composition and molluscicidal properties of essential oils of five Moroccan Pinaceae. *Pharm. Biol.*, 41(3): 207–210.

Lambert, R.J.W., P.N. Skandamis, P.J. Coote, and G.-J.E. Nychas, 2001. A study of the minimum inhibitory concentration and mode of action of oregano essential oil, thymol and carvacrol. *J. Appl. Microbiol.*, 91: 453–462.

Lis-Balchin, M., 2003. Feed additives as alternatives to antibiotic growth promoters: Botanicals. In: *Proceedings of the Ninth International Symposium on Digestive Physiology in Pigs*, Vol. 1, pp. 333–352. Banff, Alberta, Canada: University of Alberta.

Lopez-Bote, L.J., J.I. Gray, E.A. Gomaa, and C.I. Flegal, 1998. Effect of dietary administration of oil extracts from rosemary and sage on lipid oxidation in broiler meat. *Br. Poult. Sci.*, 39: 235–240.

Losa, R., 2001. The use of essential oils in animal nutrition. In: *Feed Manufacturing in the Mediterranean Region. Improving Safety: From Feed to Food*, J. Brufau (ed.), pp. 39–44. Zaragoza, Spain: CIHEAM-IAMZ (Cahiers Options Méditerranéennes; v. 54), *Third Conference of Feed Manufacturers of the Mediterranean*, Reus, Spain, March 22–24, 2000.

Moller, T., 2001. Studies on the effect of an oregano-oil-addition to feed towards nutrient digestibilities, N-balance as well as towards the parameters of microbial activity in the alimentary tract of weanedpiglets. Thesis. http://www.agronavigator.cz/UserFiles/File/Agronavigator/Kvasnickova_2/EFSA_feedap_report_plantsherbs.pdf. Accessed July 15, 2009.

Molnar, C. and G. Bilkei, 2005. The influence of an oregano feed additive on production parameters and mortality of weaned piglets, Tieraerzliche Praxis, Ausgabe Grosstiere. *Nutztiere*, 33: 42–47.

Moon, T., J. Wilkinson, and H. Cavanagh, 2006. Antiparasitic activity of two Lavandula essential oils against *Giardia duodenalis*, *Trichomonas vaginalis* and *Hexamita inflata*. *Parasitol. Res.*, 99(6): 722–728.

Moreira, M.R., A.G. Ponze, C.E. del Valle, and S.I. Roura, 2005. Inhibitory parameters of essential oils to reduce a foodborne pathogen. *LWT*, 38: 565–570.

Moschetti, R., 2003. Pesticides made with botanical oils and extracts. http://www.plantoils.in/uses/other/other.html. Accessed September 2009.

Onibala, J.S.I.T., K.D. Gunther, and Ut. Meulen, 2001. Effects of essential oil of spices as feed additives on the growth and carcass characteristics of growing-finishing pigs. In: *Sustainable Development in the Context of Globalization and Locality: Challenges and Options for Networking in Southeast Asia*. EFSA. http://www.efsa.europa.eu/en/scdocs/doc/070828.pdf. Accessed September 10, 2008.

Oviedo-Rondon, E.O., S. Clemente-Hernandez, F. Salvador, R. Williams, and R. Losa, 2006. Essential oils on mixed coccidia vaccination and infection in broilers. *Int. J. Poult. Sci.*, 5(8): 723–730.

Ozkirim, A., 2006. The detection of antibiotic resistance in the American and European Foulbrood diseases of honey bees (*Apis mellifera* L.). PhD thesis, Hacettepe University, Ankara, Turkey.

Ozkirim, A., N. Keskin, M. Kurkcuoglu, and K.H.C. Baser, 2007. Screening alternative antibiotics-essential oils from *Seseli* spp. against *Paenibacillus larvae* subsp. *larvae* strains isolated from different regions of Turkey. In: *40th Apimondia International Apicultural Congress*, Melbourne, Victoria, Australia, September 9–14, 2007.

Penalver, P., B. Huerta, C. Borge, R. Astorga, R. Romero, and A. Perea, 2005. Antimicrobial activity of five essential oils against origin strains of the Enterobacteriaceae family. *APMIS*, 113: 1–6.

Prajapati, V., A.K. Tripathi, K.K. Aggarwal, and S.P.S. Khanuja, 2005. Insecticidal, repellent and oviposition-deterrent activity of selected essential oils against *Anopheles stephensi*, *Aedes aegypti* and *Culex quinquefasciatus*. *Bioresour. Technol.*, 96(16): 1749–1757.

Ramanathan, M., 1995. Warfarin–topical salicylate interactions: Case reports. *Med. J. Malaysia*, 50(3): 278–279.

Regnault-Roger, C., 1997. The potential of botanical essential oils for insect pest control. *Int. Pest Manage. Rev.*, 2: 25–34.

Rochfort, S., A.J. Parker, and F.R. Dunshea, 2008. Plant bioactives for ruminant health and productivity. *Phytochemistry*, 69: 299–322.

Scents & Sensibility, 2001. Pet care & pest control—Using essential oils. *Scents Sensibility Newsl.*, 2(12): 1. http://scentsnsensibility.com/newsletter/Apr0601.htm. Accessed May 17, 2008.

Schoene, F., A. Vetter, H. Hartung, H. Bergmann, A. Biertuempfel, G. Richter, S. Mueller, and G. Breitschuh, 2006. Effects of essential oils from fennel (*Foeniculi aetheroleum*) and caraway (*Carvi aetheroleum*) in pigs. *J. Anim. Physiol. Anim. Nutr.*, 90: 500–510.

Tabanca, N., E. Bedir, N. Kirimer, K.H.C. Baser, S.I. Khan, M.R. Jacob, and I.A. Khan, 2003. Antimicrobial compounds from *Pimpinella* species growing in Turkey. *Planta Med.*, 69: 933.

Tam, L.S., T.Y. Chan, W.K. Leung, and J.A. Critchley, 1995. Warfarin interactions with Chinese traditional medicines: Danshen and methyl salicylate medicated oil. *Aust. N. Z. J. Med.*, 25(3): 258.

Ultee, A., E.P.W. Kets, and E.J. Smid, 1999. Mechanisms of action of carvacrol in the food-borne pathogen *Bacillus cereus*. *Appl. Environ. Microbiol.*, 65: 4606–4610.

U.S. Patent 4,735,803. Naturally-odoriferous animal repellent. http://digitalcommons.unl.edu/cgi/viewcontent.cgi?article=1151&context=icwdm_usdanwrc. Accessed September 10, 2008.

U.S. Patent 4,847,292. Repelling animals with compositions comprising citronellyl nitrile, citronellol, alpha-terpinyl methyl ether and lemon oil. http://www.freepatentsonline.com/4847292.html. Accessed September 10, 2008.

U.S. Patent 4,961,929. Process of repelling dogs and dog repellent material. http://www.freepatentsonline.com/4961929.html. Accessed September 10, 2008.

U.S. Patent 6,159,474. Animal repellant containing oils of black pepper and/or capsicum. http://www.freepatentsonline.com/6159474.html. Accessed September 10, 2008.

Vogt, H. and H.W. Rauch, 1991. The use of several essential oils in broiler diets. *Landbauforschung Volkenrode*, 41: 94–97.

Wallace, R.J., N.R. McEwan, F.M. McIntosh, B. Teferedegne, and C.J. Newbold, 2002. Natural products as manipulators of rumen fermentation. *Asian-Aust. J. Anim. Sci.*, 15: 1458–1468.

Walter, B.M. and G. Bilkei, 2004. Immunostimulatory effect of dietary oregano etheric oils on lymphocytes from growth-retarded, low-weight growing-finishing pigs and productivity. *Tijdschrift voor Diergeneeskunde*, 129(6): 178–181.

Wedge, D.E., J.A. Klun, N. Tabanca, B. Demirci, T. Ozek, K.H.C. Baser, Z. Liu, S. Zhang, C.L. Cantrell, and J. Zhang, 2009. Bioactivity-guided fractionation and GC-MS fingerprinting of *Angelica sinensis* and *A. archangelica* root components for antifungal and mosquito deterrent activity. *J. Agric. Food Chem.*, 57: 464–470.

Yip, A.S., W.H. Chow, Y.T. Tai, and K.L. Cheung, 1990. Adverse effect of topical methylsalicylate ointment on warfarin anticoagulation: An unrecognized potential hazard. *Postgrad. Med. J.*, 66(775): 367–369.

17 Use of Essential Oils in Agriculture

Gerhard Buchbauer and Susanne Hemetsberger

CONTENTS

17.1 Introduction 670
17.2 Essential Oils as Antipests 671
    17.2.1 Health and Environmental Impact of Botanical Antipests 671
    17.2.2 Pesticidal and Repellent Action of Essential Oils 671
    17.2.3 Development and Commercialization of Botanicals 672
    17.2.4 Examples of Essential Oils Used as Antipests 676
        17.2.4.1 *Rosmarinus officinalis* 676
        17.2.4.2 *Thymus* sp. 676
        17.2.4.3 *Syzygium aromaticum* 679
        17.2.4.4 Muña 679
        17.2.4.5 *Eucalyptus* sp. 680
        17.2.4.6 *Satureja* sp. 681
        17.2.4.7 *Ocimum* sp. 681
        17.2.4.8 *Origanum* sp. 681
        17.2.4.9 *Artemisia* sp. 682
        17.2.4.10 *Mentha* sp. 682
        17.2.4.11 *Cinnamomum* sp. 682
        17.2.4.12 *Acorus* sp. 682
        17.2.4.13 *Foeniculum vulgare* 682
        17.2.4.14 *Lavandula* sp. 683
        17.2.4.15 *Carum* sp. 683
        17.2.4.16 *Chenopodium ambrosoides* 683
17.3 Essential Oils as Herbicides 683
    17.3.1 Phytotoxicity 683
    17.3.2 Prospects of Organic Weed Control 684
    17.3.3 Examples of Essential Oils in Weed Control 684
        17.3.3.1 *Thymus vulgaris* 684
        17.3.3.2 *Mentha* sp. 684
        17.3.3.3 *Cymbopogon* sp. 689
        17.3.3.4 *Eucalyptus* sp. 689
        17.3.3.5 *Lavandula* sp. 689
        17.3.3.6 *Origanum* sp. 689
        17.3.3.7 *Artemisia scoparia* 690
        17.3.3.8 *Zataria multiflora* 690
        17.3.3.9 *Tanacetum* sp. 690
17.4 Essential Oils as Inhibitors of Various Pests 690
    17.4.1 Effect on Bacteria 690
    17.4.2 Effect on Fungi 693

    17.4.3 Effect on Viruses 693
    17.4.4 Effect on Nematodes.. 693
17.5 Effect of Essential Oils on the Condition of the Soil 694
    17.5.1 Effects of Essential Oils on Microorganisms and Soil.. 694
    17.5.2 Examples of Essential Oils with an Effect on Soil Condition 696
        17.5.2.1 *Mentha spicata*... 697
        17.5.2.2 *Lavandula* sp. 697
        17.5.2.3 *Salvia* sp. 697
        17.5.2.4 *Myrtus communis*... 697
        17.5.2.5 *Laurus nobilis*.. 697
        17.5.2.6 *Cymbopogon* sp. 697
17.6 Essential Oils Used in Postharvest Disease Control 698
    17.6.1 Effects of Essential Oils on Stored-Product Pests 698
    17.6.2 Examples of Essential Oils Used on Stored Products 698
        17.6.2.1 *Thymus zygis*.. 698
        17.6.2.2 *Cinnamomum* sp. 700
        17.6.2.3 *Cymbopogon citratus*... 700
        17.6.2.4 *Laurus nobilis* 700
17.7 Conclusion 700
References... 701

17.1 INTRODUCTION

Essential oils play a major role in nature for the plants that produce them in order to protect the plants against bacteria, fungi, and viruses as well as against herbivores and pests (Bakkali et al., 2008). Since pollution of our environment and intoxication of mammalians with pesticides and herbicides is a severe problem all over the world, it is necessary to look for healthier alternatives. Especially natural products such as essential oils offer an enormous potential for agricultural usage since most of them are nontoxic to vertebrates and do not harm our ambience. Even though there are so many advantages of essential oils, some disadvantages must also be mentioned, like the slow action and short duration of effectiveness as well as the high quantities needed. The object of this compilation is to review the chances and problems of the use of essential oils in agriculture.

    Crop protection has always been an important topic in agriculture for thousands of years. Many Greek and Roman authors, for example, Theophrastus (371–287 BC), Cato the Censor (234–149 BC), Varro (116–27 BC), Vergil (70–19 BC), Columella (AD 4–70), and Pliny the Elder (AD 23–79), wrote about different substances, such as essential oils, to protect plants against pests. Also Chinese literature, such as a survey of the Shengnong Ben Tsao Jing era (AD 25–220), tells about pesticidal activity of plants. Famous Linnaeus (AD 1752) did research about natural pesticides that could be used on caterpillars. During the European agricultural revolution in the nineteenth century, pyrethrum- and rotenone-containing powders, which were made of Chrysanthemum flower heads and Derris root, were discovered. But due to highly synthetic insecticide usage, many pests nowadays are suspected to have a higher resistance against those products (Dayan et al., 2009).

    Also essential oils have a long history. The first written report on extraction of essential oils was written by Ibn al-Baitar in Andalusia in Spain in the thirteenth century, but they have been used since much longer (Regnault-Roger et al., 2012). They have been used since ancient times as cosmetics and pharmaceuticals and until today they still play a major role in our life (Ammon et al., 2010). They are also known for their ability to kill bacteria, viruses, and fungi, which allow their use as antiseptics (Bakkali et al., 2008). And due to their herbicidal, pesticidal, and antimicrobial activity, they also can be used in agriculture.

17.2 ESSENTIAL OILS AS ANTIPESTS

Insects form the largest population in the animal kingdom and many of them are harmful toward human beings, since they act as pathogenic vectors and devastate crops. That is why pesticides bear a huge market potential, which can be seen in the regular growth of 7%–10% per year. But due to the massive use of oil-based synthetic molecules as pesticides since the middle of the twentieth century, there have also been negative side effects reported, such as environmental pollution, toxicity toward mammalians, and increased insect resistance (Regnault-Roger, 1997). Every year million tons of pesticides are used and harm our environment, because of that it is important to find alternatives that do not damage our environment (Koul et al., 2008).

17.2.1 HEALTH AND ENVIRONMENTAL IMPACT OF BOTANICAL ANTIPESTS

Essential oils are natural products, which are excellent alternatives to synthetic products because they reduce negative impacts on human health and the environment (Koul et al., 2008). Even though there are a lot of botanical insecticides available, from 1980 to 2000, only one single product was registered in the United States and Europe, which is called neem. It is obtained by the seeds of the Indian tree *Azadirachta indica*, Meliaceae. In the last few years, essential oils also obtained influence as botanical insecticides in the United States. But due to the "relatively slow action, variable efficacy, lack of persistence and inconsistent availability" of natural products, they still cannot compete against synthetic pesticides. On the other hand, it is not necessary to kill harmful insects since botanical insecticides, like essential oils, can also be used as repellents against animals. Moreover, those natural pesticides can also be mixed with synthetic products that could lessen the needed quantities of pesticides and improve the environmental problems that are caused by excessive use of synthetic pesticides. Especially developing countries could benefit from the increased use of botanical pesticides, such as essential oils, since human poisoning from pesticides, which can even lead to death, is a severe problem there. The farmers, who often use pesticides, do not know about the dangers because many of them do not receive any information and they are often unable to read the official languages in which the instructions are written (Isman, 2008). In static water eugenol appears to be 1,500 times less toxic than pyrethrum, which is also a botanical insecticide, and not more than 15,000 times less toxic than the organophosphate insecticide azinphosmethyl. Moreover, eugenol is volatile, which means that its half time is extremely short. After about 2 days there will be no eugenol left, which also avoids rare side effects of essential oils (Isman, 2000). Also the tobacco cutworm, which is a severe problem to vegetable and tobacco crops in Asia, can be killed by compounds of essential oils (thymol, carvacrol, pulegone, eugenol, and trans-anethole) even though it is quite resistant (Hummelbrunner and Isman, 2001). Many monoterpenes like (+)-limonene, pinene, and Δ3-carene show acaricidal activity. Carvomenthenol and terpinen-4-ol proved to cause the highest toxicity against mites (Ibrahim et al., 2008). Eucalyptus essential oil was tested on several parasites, such as *Varroa destructor* (Varroa mite), *Tetranychus urticae*, *Phytoseiulus persimilis*, and *Boophilus microplus*. The studies concluded that several *Eucalyptus* sp. (Myrtaceae) essential oils can be used as acaricides (Batish et al., 2008). *T. urticae* can be killed by essential oils obtained from *Satureja hortensis* L. (Lamiaceae), *Ocimum basilicum* L. (Lamiaceae), and *Thymus vulgaris* L. (Lamiaceae) (Aslan et al., 2004). Essential oils do not only act as deterrents but also as attractants toward insects. Ethanolic extracts of *Rosmarinus officinalis* L. (Lamiaceae) essential oil attracted the moth *Lobesia botrana*, a pest of grape berries (Katerinopoulos et al., 2005).

17.2.2 PESTICIDAL AND REPELLENT ACTION OF ESSENTIAL OILS

Essential oils are neurotoxic to some pests as some of them interfere with neuromodulator octopamine and others with γ-aminobutyric acid (GABA)-gated chloride channel (Isman, 2006). Octopamine "controls and modulates neuronal development, circadian rhythm, locomotion 'fight or flight' responses, as well as learning and memory." It activates guanosine triphosphate (GTP)-binding protein G–coupled receptors that lead to cyclic adenosine monophosphate (cAMP) production or calcium release (Balfanz et al., 2005). Interrupting its function *results in total breakdown of the nervous system in insects* (Tripathi et al., 2009). GABA A receptors of insects showed similarities as well as differences to vertebrate ones. Activation of insect GABA channels can increase the chloride ion conductance across cell membranes. Insect GABA receptors are also sensitive to benzodiazepines and barbiturates but they are insensitive to antagonists like bicuculline and pitrazepin. The differences of insect and vertebrate GABA channels can be used as targets for pesticides (Anthony et al., 1993). At high concentrations of essential oils, they can also inhibit the acetylcholinesterase activity, but it cannot explain the low-dose pesticidal effect (Kostyukovsky et al., 2002). Moreover, the acetylcholinesterase inhibiting effect could not be seen in vivo for all monoterpenes (Rajendran and Sriranjini, 2008). It seems to be a very good idea to use nontoxic substances that only act as deterrents or antifeedants toward insects. This concept obtained influence with the discovery of the antifeedant azadirachtin in the 1970s and 1980s. But even though deterrents and antifeedants were hyped a lot during that time, a few problems remain, such as the variability of response—even closely related insects vary in their behavior toward these substances, and some may even get attracted by substances that other insects may find detesting. Furthermore, the deterrent action may change during a period of time and insects can get used to it (Isman, 2006). The exact mode of action of repellents still could not be explained completely. It is also unclear if there are common mechanisms in different arthropods. Just as insects do on the antennae, "tick detects repellents on the tarsi of the first pair of legs." It is known that hairs on the antennae of mosquitoes are used to detect temperature and moisture. Repellents target female mosquito olfactory receptors and block them. As for cockroaches' method of defense, it is just known that oleic and linoleic acid cause death recognition and aversion (Tripathi et al., 2009).

Even though single compounds of the essential oil can be isolated and tested for their repellent activity, it has been reported that the synergistic effect of the whole essential oil is more effective (Nerio et al., 2010) (Table 17.1).

17.2.3 Development and Commercialization of Botanicals

The utilization of insecticides, both botanicals and synthetics, are handled differently in various countries (Isman, 2006); in the United States many essential oils are registered for agricultural use. Canada has a slightly different law as there are not as many essential oils allowed. Mexico has a similar utilization of essential oils as the United States, but there is no exception of essential oils. The European Union allows the utilization of components of essential oils, even though the individual law differs for countries in the European Union. In India more biological pesticides are used than in other countries. In South America pesticide use varies in every region. In Africa there are no real restrictions by law for its local use of various pesticides. In general, countries tend to make law for use of pesticides stricter than they were, because the majority of the population and media are aware of the environmental and health impact the pesticides cause. This is especially the case in wealthier countries, but since developing countries are dependent on selling their products to so-called first-world countries, they also have to meet up with stricter criteria, even though it might not make sense for them (Isman, 2006). According to Isman three aspects are important for successful commercialization of pesticides: sustainability of the botanical resource, standardization of the chemically complex extracts, and regulatory approval. Sustainability means that the base of the product must be available every time and not only seasonal. Standardization means that there must be a method to investigate analytically the product so that the product obtained from plants will not vary. It will be a problem if the pesticidal action of a substance differs even though it has been cultivated under the same circumstances. Regulatory control is—among those—the most difficult aspect. The profit made by selling botanicals will not be very high in industrialized countries. That is why "green pesticides" might not even reach the market. In industrialized countries it is hard to

TABLE 17.1
Examples of Harmful Pests

Name and Family	Notes	Sources
Acanthoscelides obtectus Bean weevil Coleoptera: Bruchidae	Native to Africa, Europe, America; eats fruits of crops	Capinera (2008)
Acrolepiopsis assectella Leek moth Lepidoptera: Yponomeutidae	Native to Asia, Europe; eats foliage of crops	Capinera (2008)
Acyrthosiphon pisum Pea aphid Hemiptera: Aphididae	Pest of legumes, e.g., alfalfa, clover, and peas	Capinera (2008)
Aedes aegypti Yellow fever mosquito Diptera: Culicidae	Lives in tropics and subtropics; pathogenous	Capinera (2008)
Anopheles stephensi Diptera: Anophelinae	Vector of malaria disease	Prajapazi et al. (2005)
Bemisia tabaci Sweetpotato whitefly Homoptera: Aleyrodidae	Polyphagous, pathogenous vector (i.e., cassava mosaic disease); lives in America, Africa, Asia	Capinera (2008)
Boophilus microplus Ixodida, Ixodidae		Capinera (2008)
Callosobruchus chinensis Adzuki bean weevil Coleoptera: Bruchidae	Storage pest, i.e., in Syria	Capinera (2008)
Callosobruchus maculatus Cowpea weevil Coleoptera: Bruchidae	Damages cowpeas	Capinera (2008)
Cryptolestes pusillus Flat grain beetle Coleoptera: Laemophloeidae	Stored-grain pest; in temperate climate	Capinera (2008)
Culex pipiens House mosquito Diptera: Culicidae	Pathogenous vector of diseases	Capinera (2008)
Culex quinquefasciatus Diptera: Culicidae	Tropical regions; pathogenous vector of diseases	Capinera (2008)
Delia radicum Cabbage fly Diptera: Anthomyiidae	Native to Asia, Europe, and North America; damages crucifers	Capinera (2008)
Dendroctonus rufipennis Spruce beetle Coleoptera: Scolytidae	Damages spruce trees	Ibrahim et al. (2001)
Dendroctonus simplex Eastern larch beetle Coleoptera: Scolytidae	Damages spruce trees	Ibrahim et al. (2001)
Dermanyssus gallinae Roost mite Gamasida, Dermanyssoidea, Dermanyssidae	Worldwide; poultry pest; can reduce egg production	Capinera (2008)

(*Continued*)

TABLE 17.1 (Continued)
Examples of Harmful Pests

Name and Family	Notes	Sources
Drosophila auraria	Subgroup of *D. melanogaster*	Konstantopoulou et al. (1992)
Ephestia kuehniella Mediterranean flour moth Lepidoptera: Pyralidae	Pest of industrial flour mills; lives in temperate climate	Capinera (2008) and Ayvaz et al. (2010)
Helicoverpa armigera Cotton bollworm Lepidoptera: Noctuidae	Pest of maize and other crops; native to Asia and Africa	Capinera (2008)
Hyalomma marginatum Acari: Ixodidae	In Europe and Asia	Capinera (2008)
Hylobius abietis Large pine weevil Coleoptera: Curculionidae	Pest of conifers; in Europe and Asia	Capinera (2008)
Lasioderma serricorne Cigarette beetle Coleoptera: Anobiidae	Pest of stored food, household pest; native to America, Asia, and Africa	Capinera (2008)
Leptinotarsa decemlineata Colorado potato beetle Coleoptera: Chrysomelidae	Destroys solanaceous crops; in Europe, North America	Capinera (2008)
Lymantria dispar Gypsy moth Lepidoptera: Lymantriidae	Pest of cork oak forests	Moretti et al. (2002)
Listronotus oregonensis Carrot weevil Coleoptera: Curculionidae	Damages mainly carrots	Capinera (2008)
Lobesia botrana Grape berry moth Lepidoptera: Tortricidae	Pest of grapes; e.g., in Jordan and Iran	Capinera (2008)
Megastigmus pinus Hymenoptera: Torymidae	Destroys white fir trees	Ibrahim et al. (2001)
Megastigmus rafini Hymenoptera: Torymidae	Destroys white fir trees	Ibrahim et al. (2001)
Musca domestica Housefly Diptera: Muscidae	Feeds on garbage and feces; worldwide	Capinera (2008)
Oryzaephilus surinamensis Sawtoothed grain beetle Coleoptera: Silvanidae	Stored-product pest	Capinera (2008)
Papilio demoleus Common lime butterfly Lepidoptera: Papilionidae	Southeast Asia; pest of citrus	Capinera (2008)
Periplaneta americana American cockroach Blattaria: Blattidae	Household pest; omnivore	Capinera (2008)
Phytoseiulus persimilis Acarina: Phytoseiidae	Is used to control mite pests in greenhouses; predatory mite	Capinera (2008)

(Continued)

TABLE 17.1 (*Continued*)
Examples of Harmful Pests

Name and Family	Notes	Sources
Pissodes strobi White pine weevil Coleoptera: Curculionidae	Destroys pines, e.g., Eastern White pine	Capinera (2008) and Ibrahim et al. (2001)
Plodia interpunctella Indian meal moth Lepidoptera: Pyralidae	Stored-grain pest; worldwide	Capinera (2008)
Prays citri (Citrus blossom moth) Lepidoptera: Hyponomentidae	e.g., in Egypt	Capinera (2008)
Prostephanus truncatus Larger grain borer Coleoptera: Bostrichidae	Pest of maize; especially in tropical regions	Capinera (2008)
Rhyzopertha dominica Lesser grain borer Coleoptera: Bostrichidae	Native to America, Africa, Asia; pest of grain	Capinera (2008)
Sitophilus granarius Granary weevil Coleoptera: Curculionidae	Pest of stored grain	Capinera (2008)
Sitophilus oryzae Rice weevil Coleoptera: Curculionidae	Pest of stored grain	Capinera (2008)
Sitophilus zeamais Maize weevil Coleoptera: Curculionidae	Pest of stored grain	Capinera (2008)
Spodoptera frugiperda Fall armyworm Lepidoptera: Noctuidae	Pest of sugarcane fields and other crops; North and South America	Capinera (2008)
Spodoptera litura Rice cutworm Lepidoptera: Noctuidae	Native to Australia and Asia; damages legumes and solanaceous crops	Capinera (2008)
Stegobium paniceum Drugstore beetle Coleoptera: Anobiidae	Temperate and subtropical areas; eats pharmaceutical drugs	Capinera (2008)
Stephanitis pyri Azalea lace bug Heteroptera: Tingidae	Native to Europe; damages trees	Capinera (2008)
Tenebrio molitor Yellow mealworm Coleoptera: Tenebrionidae	Stored-product pest	Capinera (2008)
Tetranychus cinnabarinus Carmine mite Acari: Tetranychidae	Damages soybeans, cotton, plums, citrus, vegetables; in Syria, Saudi Arabia, Lebanon, Jordan	Capinera (2008)
Tetranychus urticae Twospotted spider mite Acari: Tetranychidae	Worldwide in temperate and subtropical areas; greenhouse pest	Capinera (2008)

(*Continued*)

TABLE 17.1 (*Continued*)
Examples of Harmful Pests

Name and Family	Notes	Sources
Thaumetopoea pityocampa Pine processionary Lepidoptera: Thaumetopoeidae	Pest of Pinaceae	Capinera (2008)
Triatoma infestans Kissing bug Hemiptera: Reduviidae: Triatominae	Pathogenous vectors of Chagas' disease in South America	Capinera (2008)
Tribolium castaneum Red flour beetle Coleoptera: Tenebrionidae	Native to America, Africa, and Asia; pest of stored food	Capinera (2008)
Tribolium confusum Confused flour beetle Coleoptera: Tenebrionidae	Stored food pest, especially flour; difficult to control	Capinera (2008)
Tyrophagus putrescentiae Mold mite Acaridida: Acaroidea: Acaridae	Eats processed cereals	Capinera (2008)
Varroa destructor (*Varroa jacobsoni*) Acari: Varroidae	Pest of honeybees	Capinera (2008)

imagine that botanicals such as essential oils will play a bigger role in the future since synthetic pesticides are affordable and very potent against various insects (Isman, 2006).

17.2.4 Examples of Essential Oils Used as Antipests

See Tables 17.2 through 17.4.

17.2.4.1 *Rosmarinus officinalis*

R. officinalis L. (Lamiaceae) can be found in the Mediterranean area. The used parts are the fresh flowering tops. The essential oil can be used as antipest due to its effect on beetles, caterpillar larvae, and many other insects (Dayan et al., 2009) like *Drosophila auraria* (Konstantopoulou et al., 1992), *Sitophilus oryzae* (Lee et al., 2004), or *Rhyzopertha dominica* (Shaaya et al., 1991). The microencapsulated essential oil also showed larvicidal effects on *Lymantria dispar* (Moretti et al., 2002). In *Acanthoscelides obtectus* males and females, *R. officinalis* volatile oil also caused high mortality rates (Papachristos and Stamopoulos, 2002). Repellent activity was reported against *Listronotus oregonensis* (Niepel, 2000).

17.2.4.2 *Thymus* sp.

T. vulgaris L. (Lamiaceae) also grows in the Mediterranean area. Parts used are partial dried or fresh aerial parts. The main components are "available for broad-spectrum insect control in organic farming" (Dayan et al., 2009). The essential oil was tested against *Bemisia tabaci* and caused its death (Aslan et al., 2004). *T. vulgaris* essential oil also showed toxicity against *Dermanyssus gallinae*, a pathogenic mite on hens (George et al., 2009a, 2010; Ghrabi-Gammar et al., 2009). It also caused high mortality in *Tenebrio molitor* (George et al., 2009b) and *Tyrophagus putrescentiae* (Kim et al., 2003a). The essential oil of *Thymus herba-barona* Loisel was microencapsulated and tested on *L. dispar*, on which it caused mortality (Moretti et al., 2002).

TABLE 17.2
Essential Oils That Can Be Used as Insecticide and Acaricide

Name and Family	Constituents	Notes	Sources
Acorus sp. Acoraceae	β-Asarone, acorenone, acoragermacrone		Teuscher et al. (2004)
Artemisia sp. Asteraceae	Limonene, myrcene, α-thujone, β-thujone, caryophyllene, sabinyl acetate		Chiasson et al. (2001)
Baccharis salicifolia (Ruiz & Parvon) Pers. Asteraceae	β-Pinene, pulegone, camphene, limonene, α-pinene, pulegone, pulegol, germacrol, germacrone	Insecticidal effect on *T. castaneum*	Garcia et al. (2005)
Carum sp. Apiaceae	D-Carvone, limonene		Teuscher et al. (2004)
Chenopodium ambrosoides L. Amaranthaceae	α-Terpinene, cymol, *cis*-β-farnesene, ascaridole, carvacrol		Tapondjou et al. (2002)
Cinnamomum sp. Lauraceae	Cinnamaldehyde, linalool, β-caryophyllene, eugenol		Teuscher et al. (2004)
Citrus sinensis (L.) Osbeck Rutaceae	Limonene, α-pinene, sabinene, α-terpinene	Repellent against *A. pisum*	Loebe (2001), Njoroge et al. (2005), and *Plant Encyclopedia* (2012)
Coriandrum sativum L. Apiaceae	Linalool, α-terpinyl acetate, 1,8-cineole, linalyl acetate, geranyl acetate, camphor	Active against *S. oryzae*, *R. dominica*, and *C. pusillus*	Teuscher et al. (2004) and Lopez et al. (2008)
Cuminum cyminum L. Umbelliferae	Cuminal, cuminic alcohol, γ-terpinene, safranal, *p*-cymene, β-pinene	Ovicidal effects on eggs of *E. kuehniella*	Li and Jiang (2004) and Tunc et al. (2000)
Cymbopogon sp. Poaceae	Citronellal, geraniol, citronellol, citral, limonene	Weed control, antipest; exogenous limonene against insect pests; acaricidal effect against *T. putrescentiae*	Teuscher et al. (2004), Dayan et al. (2009), Ibrahim et al. (2001), and Kim et al. (2003a)
Elettaria cardamomum (L.) Maton Zingiberaceae	α-Terpinyl acetate, 1,8-cineole, linalyl acetate	Active against *S. zeamais* and *T. castaneum*, acaricidal	Teuscher et al. (2004), Huang et al. (2000), and Kim et al. (2003a)
Eucalyptus sp. Myrtaceae	1,8-Cineole, limonene, α-pinene		Teuscher et al. (2004)
Evodia rutaecarpa Hook f. et Thomas Rutaceae	Limonene, β-elemene, linalool, myrcene, valencene, linalyl acetate, β-caryophyllene	Fumigant against *T. castaneum* and topical insecticide against *S. zeamais*	Pellati et al. (2005) and Liu and Ho (1999)
Foeniculum vulgare L. Apiaceae	Anethol, fenchone, estragole		Teuscher et al. (2004)
Hedeoma mandonianum Wedd. Lamiaceae	Eucalyptol, pulegone, menthone		Fournet et al. (1996)

(*Continued*)

TABLE 17.2 (*Continued*)
Essential Oils That Can Be Used as Insecticide and Acaricide

Name and Family	Constituents	Notes	Sources
Juniperus sp. Cupressaceae	α-Pinene, myrcene, β-pinene, terpinene-4-ol, germacrene D	Toxicity against *D. gallinae*; acaricide against *T. putrescentiae*; controls *A. stephensi*, *A. aegypti*, and *C. quinquefasciatus*	Teuscher et al. (2004), George et al. (2010), Kim et al. (2003a), and Prajapati et al. (2005)
Lavandula sp. Lamiaceae	Linalool, linalyl acetate, fenchone, camphor		Konstantopoulou et al. (1992)
Leptospermum scoparium J.R. et G. Forst. Myrtaceae	Trans-calamenene, δ-cadinene, β-caryophyllene, leptospermone	Highly effective against *D. gallinae*	Douglas et al. (2004) and George et al. (2009a,b)
Lippia sp. Verbenaceae	Piperitenone oxide, limonene, thymol, carvacrol, *p*-cymene, β-caryophyllene, and γ-terpinene	Acaricide toward *T. urticae*, repellent activity on *S. zeamais*	Duschatzky et al. (2004), Calvacanti et al. (2010), and Nerio et al. (2009)
Melaleuca sp. Myrtaceae	Terpinene-4-ol, γ-terpinene, α-terpinene	Fumigant action against *S. oryzae*, *R. dominica*, and *T. castaneum*	Teuscher et al. (2004) and Lee et al. (2004)
Mentha sp. Lamiaceae	Carvone, pulegone, menthol, menthone, methylacetate		Konstantopoulou et al. (1992)
Micromeria fruticosa L. Lamiaceae	Pulegone, β-caryophyllene, isomenthol, limonene	Effective against *B. tabaci* and *T. urticae*	Dudai et al. (1999) and Çalmaşur et al. (2006)
Minthostachys andina (Britton) Epling Lamiaceae	Menthone, pulegone, iso-menthone		Fournet et al. (1996)
Myristica fragrans Houtt. Myristicaceae	Sabinene, α-pinene, β-pinene, myristicin	Very high mortality rates on *T. putrescentiae*	Teuscher et al. (2004) and Kim et al. (2003a)
Myrtus communis Myrtaceae	1,8-Cineole, α-pinene, limonene, linalool, myrtenol	Highly effective against *E. kuehniella*, *P. interpunctella*, *A. obtectus*, *S. oryzae*, and *R. dominica*	Teuscher et al. (2004) and Ayvaz et al. (2010)
Nepeta racemosa L. Lamiaceae	Nepetalactone, myrtenol, terpinen-4-ol	Effective against *B. tabaci* and *T. urticae*	Mutlu et al. (2010) and Çalmaşur et al. (2006)
Ocimum sp. Lamiaceae	γ-Terpinene, α-terpinene, thymol, carvacrol, linalool, eugenol		Aslan et al. (2004)
Origanum sp. Lamiaceae	Carvacrol, thymol, γ-terpinene, *p*-cymene, linalool		Konstantopoulou et al. (1992)
Pelargonium graveolens L'Hér Geraniaceae	Geraniol, citronellol	Acaricide against *D. gallinae*	Ghrabi-Gammar et al. (2009)
Pimpinella anisum L. Apiaceae	*cis*-Anethol	Ovicidal effects against eggs of *T. confusum* and *E. kuehniella*	De Almeida et al. (2010) and Tunc et al. (2000)

(*Continued*)

TABLE 17.2 (Continued)
Essential Oils That Can Be Used as Insecticide and Acaricide

Name and Family	Constituents	Notes	Sources
Piper nigrum L. Piperaceae	β-Caryophyllene, α-pinene, sabinene, limonene	Effective against *T. molitor*	Teuscher et al. (2004) and George et al. (2009b)
Rosmarinus officinalis L. Lamiaceae	Limonene, cineol, borneol, terpineol		Konstantopoulou et al. (1992)
Salvia sp. Lamiaceae	Cineol, thujone, camphor, α-pinene, β-pinene	Insecticide against *D. auraria* and *R. dominica*	Konstantopoulou et al. (1992) and Shaaya et al. (1991)
Satureja sp. Lamiaceae	Carvacrol, thymol, γ-terpinene, *p*-cymene		Konstantopoulou et al. (1992)
Syzygium aromaticum (L.) Merr. et L.M. Perry Myrtaceae	Eugenol, aceteugenol, β-caryophyllene		Teuscher et al. (2004)
Tagetes sp. Asteraceae	Limonene, (Z)-β-ocimene, (Z)- and (E)-ocimenone	Toxic to *V. destructor*, no toxicity toward honeybees; causes repellence to *S. zeamais*	Eguaras et al. (2005) and Nerio et al. (2009)
Tanacetum vulgare L. Asteraceae	α-Pinene, α-terpinene, γ-terpinene, carvone, borneol, β-thujone, camphor	Antifeedant against *L. decemlineata* and acaricide toward *T. urticae*	Chiasson et al. (2001)
Thymbra sp. Lamiaceae	Carvacrol, thymol, γ-terpinene, *p*-cymene	High mortality rates in *D. gallinae* and *T. cinnabarinus*; insecticidal activity on *D. auraria*	Ghrabi-Gammar et al. (2009), Sertkaya et al. (2010), and Konstantopoulou et al. (1992)
Thymus sp. Lamiaceae	Thymol, carvacrol, *p*-cymene		Teuscher et al. (2004)

17.2.4.3 Syzygium aromaticum

Syzygium aromaticum (L.) Merr et L. M. Perry (Myrtaceae) (Teuscher et al., 2004) grows in tropical areas. Parts used are flower buds, which have been dried. Its field of application is insect pest management as eugenol is toxic to many common pests (Dayan et al., 2009). In another study conducted by Huang et al. (2002), toxicity of eugenol to *Sitophilus zeamais* and *Tribolium castaneum* could be shown. Acaricidal activity against *T. putrescentiae* (Kim et al., 2003a) and *D. gallinae* (Kim et al., 2007) could also be confirmed for clove essential oil.

17.2.4.4 Muña

Muña are Bolivian medical plants and include species of *Satureia*, *Minthostachys*, *Mentha*, and *Hedomea*. They derive their name muña by the Kechuas Indians, who live in the Andean mountains where these plants grow on an altitude between 2500 and 5000 m. Indians use these plants among others because of their insecticide and repellent activity in order to protect the crops of potatoes against insects and also to prevent Chagas' disease of which an insect acts as pathogenous vector. Plants used in the study against *Triatoma infestans* that was conducted by Fournet et al. (1996) were *Minthostachys andina* (Britton) Epling (Lamiaceae) and *Hedeoma mandonianum* Wedd. (Lamiaceae). Essential oil

TABLE 17.3
Effect of Limonene Application on Various Pests

Insect Species	Host Plant	Limonene Application	Response
		Reduced activity of pest insect	
Acrolepiopsis assectella (leek moth)	Leek and onion, all *Allium* crops	Olfactometer with two parallel air currents containing a Y-shaped nylon fiber	Limonene shows repellent activity.
Delia radicum (cabbage fly)	Brassicaceae	Olfactory stimuli for orientation behavior	Limonene from the surface part of plant host acts as a repellent.
Dendroctonus rufipennis, *D. simplex* (spruce beetle, eastern larch beetle)	*Spruce* spp.	Bioassayed for their toxicity	(+)-Limonene at 60 ppm kills 100% of the pests after 24 h.
Hylobius abietis (large pine weevil)	Pine, spruce	Exposure to limonene vapors	High limonene concentrations show signs of poisoning within a few hours.
Megastigmus pinus, *M. rafini* (silver fir seed wasp)	White fir	Olfactory responses to pure α-pinene and limonene	Limonene significantly acts as repellent.
Thaumetopoea pityocampa (pine processionary)	Pine	Emulsified with water and sprayed on the foliage of pine seedlings	(+)-Limonene reduces the number of egg cluster on plants sprayed with it.
		Increased activity of pest insect	
Helicoverpa armigera (cotton bollworm)	Polyphagous moth (e.g., cotton, tomatoes, conifers)	Electroantennography used to investigate electrophysiological responses	Attractive to 1–2-day-old moths.
Papilio demoleus (common lime butterfly)	Fabaceae	Orientation responses to different odors in olfactory	(−)-Limonene shows maximum attraction to the larvae.
Prays citri (citrus blossom moth)	*Citrus limonum, C. decuminata, C. aurantium*	Electroantennogram response to pure limonene	Limonene activates oviposition.
		No effect	
Hylobius abietis (large pine weevil)	Pine	Exposure to limonene vapors	Low limonene levels do not affect feeding activity.

Source: Ibrahim, M.A. et al., *Agric. Food Sci.*, 10, 243, 2008.

of *M. andina* could significantly decrease the number of insects: after 28 days, 10 out of 20 insects were found dead. *H. mandonianum* had no such effect (Fournet et al., 1996).

17.2.4.5 *Eucalyptus* sp.

Eucalyptus' essential oil is obtained from the Australian tree *Eucalyptus* sp. that belongs to the Myrtaceae family. It has been known for its antibacterial, antifungal, and antiseptic action for hundreds of years (Batish et al., 2008). Toxicity of eucalyptus essential oil toward *S. oryzae* was reported by Lee et al. (2001), and it proves to be a promising fumigant to control that pest. Lee et al. (2004) reported in another study that several Eucalyptus species were toxic to *S. oryzae*: *E. nicholii*, *E. codonocarpa*, and *E. blakelyi*. The same species cause mortality in *T. castaneum* and *R. dominica*.

TABLE 17.4
Effect of Eucalyptus Essential Oil on Various Pests

Eucalyptus sp.	Tested Organism
E. camaldulensis (river red gum E.)	Repels adult females of *Culex pipiens* (common house mosquito); mortality of eggs in *Tribolium confusum* and *Ephestia kuehniella*
E. citriodora (lemon scented E.)	Toxicity against *Sitophilus zeamais* (greater rice weevil)
E. globulus (Tasmanian blue gum E.)	Toxic to pupae of *Musca domestica* (housefly) as well as to female *Pediculus humanus* (louse)
E. intertexta	Causes death of adult *Callosobruchus maculatus*, *Sitophilus oryzae*, and *Tribolium castaneum*
E. sargentii	
E. camaldulensis	
E. saligna	Repelled *Sitophilus zeamais* and *Tribolium confusum* (confused flour beetle)
E. tereticornis	Pesticidal activity against *Anopheles stephensi*
Eucalyptus sp.	Toxic to *Thaumetopoea pityocampa* (pine processionary)

Source: Batish, D.R. et al., Forest Ecol. Manage., 256, 2166, 2008.

17.2.4.6 Satureja sp.

Aslan et al. tested the essential oil of *S. hortensis* L. (Lamiaceae) against *B. tabaci*. Of three oils (*T. vulgaris* and *O. basilicum*) tested, *S. hortensis* essential oil was most toxic (Aslan et al., 2004). *S. thymbra* L. (Lamiaceae) was tested against *Hyalomma marginatum* and was able to kill almost all of the ticks after 24 h (Cetin et al., 2010). It also showed toxicity to *D. auraria* (Konstantopoulou et al., 1992). According to Ayvaz et al., savory essential oil caused mortality against the pests *Ephestia kuehniella* and *Plodia interpunctella* (Ayvaz et al., 2010).

17.2.4.7 Ocimum sp.

Basil (*O. basilicum* L., Lamiaceae) essential oil was tested against *B. tabaci* and proved to be effective against it (Aslan et al., 2004). Keita et al. (2001) tested the fumigant and contact toxicity of *O. basilicum* and *O. gratissimum* L. on *Callosobruchus maculatus* beetles. Both essential oils showed a high toxicity toward the beetles compared to the control. Also the egg hatch rage was decreased. Another study reports methyleugenol, methylcinnamate, linalool, and estragole as main compounds. Estragole was identified as one of the effective compounds against *Cryptolestes pusillus* and *R. dominica*, while it was variable against *S. oryzae* (Lopez et al., 2008). Basil essential oil showed also very promising results in *Oryzaephilus surinamensis* (Shaaya et al., 1991). *O. kilimandscharicum* Wild. essential oil, with camphor as major compound, was proved to be toxic to four pests (*Sitophilus granarius*, *S. zeamais*, *T. castaneum*, and *Prostephanus truncatus*) while it also caused repellency in them. Compared with the other three, *T. castaneum* is the strongest among them in tolerating fumigation, as it has the lowest mortality rate (Obeng-Ofori et al., 1998).

17.2.4.8 Origanum sp.

Çalmaşur et al. tested the pesticidal effect of *Origanum vulgare* L. (Lamiaceae) essential oil on two pests: *T. urticae* and *B. tabaci*. They could show that after 120 h and at a concentration of 2 μL/L air, the mortality rate on *B. tabaci* is 100% and on *T. urticae* 95% (Çalmaşur et al., 2006). Oregano essential oil can also be used as acaricide against *T. putrescentiae* (Kim et al., 2003a) and as insecticide against *O. surinamensis* (Shaaya et al., 1991) and *D. auraria*. *O. dictamnus* L. and *O. majorana* L. were also effective against *D. auraria* (Konstantopoulou et al., 1992). *O. acutidens* essential oil can be used against *S. granarius* and *Tribolium confusum* (Kordali et al., 2008), and the oil of *Origanum onites* L. (with carvacrol, linalool, and thymol as main constituents) is successful in controlling *Tetranychus cinnabarinus* as reported by Sertkaya et al. (2010). Oregano essential oil showed high activity against *P. interpunctella* and *E. kuehniella* (Ayvaz et al., 2010).

17.2.4.9 Artemisia sp.

The main constituents of *Artemisia absinthium* L. (Asteraceae) essential oil are said to be effective against several flea, flies, and mosquitoes. In a study conducted by Chiasson et al. (2001), *A. absinthium* essential oil showed toxicity to *T. urticae* and *T. putrescentiae* (Kim et al., 2003a). When used as fumigant, *A. sieberi* Besser essential oil caused mortality in *C. maculatus, S. oryzae*, and *T. castaneum*. Of all three insects, it was most toxic to *C. maculatus*, as it caused 100% mortality at a concentration of 37 µL/L after 12 h (Negahban et al., 2007). In another study this author group tested the effect of *A. scoparia* Waldst et Kit essential oil on the same three stored-product pests as mentioned earlier. At a concentration of 37 µL/L, 100% mortality of the pests can be observed after 1 day. It also causes repellent effects on all three insects (Negahban et al., 2006).

17.2.4.10 Mentha sp.

Mentha pulegium L. (Lamiaceae) essential oil was tested on *D. gallinae*; it showed high toxicity as fumigant against the mite (George et al., 2009a). The same essential oil proved to be active as acaricide against *T. putrescentiae*. The same effect can be observed with *M. spicata* L. (Kim et al., 2003a). *M. pulegium* L. and *M. spicata* were very effective against *D. auraria* (Konstantopoulou et al., 1992). *Mentha arvensis* var. *piperascens* also showed high mortality rates on *D. gallinae* (Kim et al., 2007). *Mentha microphylla* and *M. viridis* proved to be toxic to *A. obtectus* and they also decreased the number of eggs (Papachristos and Stamopoulos, 2002).

17.2.4.11 Cinnamomum sp.

Cinnamomum aromaticum Nees (Lauraceae) proved to be a contact insecticide to *T. castaneum* and *S. zeamais* (Huang and Ho, 1998). *C. cassia* and *C. sieboldii* proved to have pesticidal action against *S. oryzae* and *Callosobruchus chinensis* (Kim et al., 2003b). Cinnamaldehyde causes acaricidal effects on *T. putrescentiae*, especially in closed containers (Kim et al., 2004). *C. camphora* (L.) J. Presl causes mortality to *D. gallinae* and can be used as acaricide against them (Kim et al., 2007).

17.2.4.12 Acorus sp.

Acorus calamus var. *angustatus* L. and *Acorus gramineus* Sol. showed 100% mortality at concentrations of 3.5 mg/cm^3 after 3 days treatment on *S. oryzae*. Essential oil obtained from *A. calamus* var. *angustatus* was also very potent against *C. chinensis*. A mortality rate of 100% after 1 day could be realized at a concentration of 3.5 mg/cm^3 (Kim et al., 2003b). Another pest that can be controlled by *A. calamus* var. *angustatus* is *D. gallinae* (Kim et al., 2007). *A. gramineus* essential oil shows insecticidal action against *S. oryzae* and *C. chinensis* when directly applied. It can also be used against *Lasioderma serricorne*, but it works more potent on the first two insects. As fumigant it worked best in closed containers (Park et al., 2003).

17.2.4.13 Foeniculum vulgare

Fennel essential oil caused 100% toxicity to *S. oryzae* after exposition for 3 days at concentrations of 3.5 mg/cm^3. The essential oil was even more potent against *C. chinensis* because 100% mortality can be realized at the same concentration after 1 day (Kim et al., 2003b) Compounds of fennel essential oil were also toxic, as proven in the previous study, to *S. oryzae* and *C. chinensis* and also to *L. serricorne*. At concentrations of 0.168 mg/cm^2, estragole was most toxic to *S. oryzae* after 1 day, followed by (+)-fenchone and (*E*)-anethol. Against *C. chinensis* (*E*)-anethole was most effective, followed by estragole and (+)-fenchone. After 1 day (*E*)-anethole was lethal to *L. serricorne*, whereas estragole and (+)-fenchone showed lower toxicity (Kim and Ahn, 2001). *Foeniculum vulgare* essential oil is also potent in controlling *D. gallinae* (Kim et al., 2007). Fennel essential oil also showed toxic effects to *T. putrescentiae* (Lee et al., 2006), which can possibly be attributed to fenchone, as the isolated compound also showed high mortality rates (Sánchez-Ramos and Castañera, 2000).

17.2.4.14 *Lavandula* sp.

Lavandula stoechas L. (Lamiaceae) essential oil is effective in killing *D. auraria* in a study conducted by Konstantopoulou et al. In the same study the insecticidal effect of *Lavandula angustifolia* Miller (Lamiaceae), with linalool and linalyl acetate as main compounds, could be confirmed as well (Konstantopoulou et al., 1992). *Lavandula hybrida* volatile oil caused fumigant toxicity to *A. obtectus* males and females and lower number of eggs (Papachristos and Stamopoulos, 2002). Lavender essential oil also showed high activity against *R. dominica* (Shaaya et al., 1991).

17.2.4.15 *Carum* sp.

Carum carvi L. (Apiaceae) essential oil is able to kill *S. oryzae*, and carvone seems to be the active compound against it. (*E*)-anethole is proven to be effective against *R. dominica*. Limonene and fenchone, for example, were active against *C. pusillus* (Lopez et al., 2008). *C. copticum* C.B. Clarke volatile oil constituents are thymol, α-terpineol, and *p*-cymene. Especially *S. oryzae* was weak against the fumigant action of the essential oil, but also mortality on *T. castaneum* can be observed (Sahaf et al., 2007).

17.2.4.16 *Chenopodium ambrosoides*

At a concentration of 0.4%, the essential oil of *Chenopodium ambrosoides* was able to kill *C. chinensis*, *C. maculatus*, and *A. obtectus* at a high percentage. *S. granarius* and *S. zeamais* were totally killed at 6.4%, and *P. truncatus* was least sensitive, as 56% survived (Tapondjou et al., 2002).

The use of essential oils as antipest is presented in this chapter. It could be shown that several essential oils, like *R. officinalis*, *T. vulgaris*, *Eucalyptus* species, and *Origanum* species, show great potential in controlling several pests, for example, *Sitophilus* sp. or *Tetranychus* sp., among many others, both on stored products and in protection of livestock, bees, and crops. Since synthetic pesticides bear a lot of risk toward environment and mammalian health, there is a need for natural and healthier alternatives, just like essential oils. Of course, the high volatility is a limiting factor in the use of volatile oils on the field, but in glasshouses or on stored food, they bear a great chance to change the current situation.

17.3 ESSENTIAL OILS AS HERBICIDES

Weed control in agriculture is very important as a high percentage of weed causes a high crop reduction, even more than pests. While in ancient times people were dependent on removing weed by hand, today's weed control is managed by synthetic and natural substances. Since synthetic substances for weed control cause the same environmental and health problems as synthetic substances for pest control, natural substances such as essential oils seem to be highly preferable (Dayan et al., 2009).

17.3.1 PHYTOTOXICITY

Phytotoxicity means the harmful impact of chemical substances on plants that can be seen in many different ways: Their parts can appear to be burned; they can suffer from chlorosis, which colors the leaves yellow; or they can also result in a lack of growth or even an excessive growth. When phytotoxicity of various components of essential oils was tested, pulegone was least and D-carvone most phytotoxic to maize plants. (+)-Limonene appeared most toxic to sugar beet seedlings. Limonene showed toxicity toward strawberry seedlings in preliminary screenings as well as toward cabbage and carrot seedlings (Ibrahim et al., 2008).

De Almeida et al. found out that essential oils of balm, caraway, hyssop, thyme, and vervain showed a 100% inhibition of germination of *Lepidium sativum* (Brassicaceae). *Raphanus sativus*'

(Brassicaceae) germination was inhibited by 100% by vervain oil at all concentrations. At low concentrations anise and basil essential oils promoted the germination and radicle growth, while caraway, sage, vervain, and marjoram essential oils inhibited the growth of radish. *Lactuca sativa*'s (Asteraceae) growth was inhibited by thyme essential oil. Vervain, balm, and caraway also had the ability to suppress the growth of lettuce. Generally said, a high level of oxygenated monoterpenes is most harmful toward weed. When it comes to inhibit the seed germination and subsequent growth, ketones and alcohol showed the highest activity followed by aldehydes and phenols (De Almeida et al., 2010).

Plants especially from the Lamiaceae family can inhibit the growth of several weeds by releasing phytotoxic monoterpenes (α-pinene, β-pinene, camphene, limonene, α-phellandrene, *p*-cymene, 1,8-cineole, borneol, pulegone, and camphor) (Angelini et al., 2003). The herbicide effect of 1,4-cineole and 1,8-cineole is also described by Dayan et al. (2012). Plants that are exposed to essential oils often metabolize them, and when citral was added geraniol, nerol and their acids appeared. When citronellal metabolization was tested, citronellol and citronellic acid were formed, and with pulegone (iso)-menthone, isopulegol and menthofuran were found (Dudai et al., 2000).

It was also reported that essential oils lead to accumulation of H_2O_2 in other plants. That way they increase oxidative stress, which leads to "disruption of metabolic activities in the cell." Another cause for phytotoxicity is that essential oil destroys the cell membrane of plants and inhibits their enzymes (Mutlu et al., 2010) (Table 17.5).

17.3.2 Prospects of Organic Weed Control

Organic weed control's problem is the huge quantities that must be applied to harm the weed. This may also cause a large negative impact on the environment and the microbes in the soil. Also the volatility is a problem of essential oils, on account of which they also have to be applied very often, which could be avoided by using microencapsulation of essential oils, because it will decrease their volatility. The third problem is that essential oils can also harm the desired crop and lead to a partly destruction of the culture. Altogether, organic weed control using essential oils does not seem too promising as a future replacement of synthetic products (Dayan et al., 2009).

17.3.3 Examples of Essential Oils in Weed Control

See Table 17.6.

17.3.3.1 *Thymus vulgaris*

T. vulgaris L. belongs to the *Lamiaceae* family as well (De Almeida et al., 2010; Angelini et al., 2003). The latter author group observed that thyme essential oil can inhibit the growth of the weeds *Portulaca oleracea*, *Echinochloa crus-galli*, and *Chenopodium album*, but also of lettuce and pepper crops. *T. vulgaris* essential oil also causes damage to dandelion (*Taraxacum* sp., Asteraceae) leaves, common lambsquarters (*C. album* L.), common ragweed (*Ambrosia artemisiifolia*), and Johnsongrass (*Sorghum halepense* L.) (Tworkowski, 2002).

17.3.3.2 *Mentha* sp.

2-Phenylethyl propionate can be obtained from peppermint oil (*Mentha x piperita* L., Lamiaceae); it also contains menthol and menthone (Dayan et al., 2009) *M. spicata* L. essential oil contains (−)-carvone and limonene. *M. spicata* shows good effects against several weeds. It is able to inhibit germination (under 15%) of *Amaranthus retroflexus*, *Centaurea solstitialis*, *Sinapis arvensis*, *Sonchus oleraceus*, *Raphanus raphanistrum*, and *Rumex nepalensis* Spreng (Azirak and Karaman, 2008). *Mentha longifolia* (L.) Huds and *Mentha officinalis* can be used as herbicide as they are able

TABLE 17.5
Examples of Tested Plants

Name	Common Name	Family	Sources
Achyranthes aspera	Prickly chaff flower	Amaranthaceae	*Plant Encyclopedia* (2012)
Ageratum conyzoides	Billygoat weed	Asteraceae	*Plant Encyclopedia* (2012)
Agrostis stolonifera	Creeping bentgrass	Poaceae	*Plant Encyclopedia* (2012)
Alcea pallida	Hollyhock	Malvaceae	*Plant Encyclopedia* (2012)
Amaranthus hybridus	Smooth amaranth	Amaranthaceae	*Plant Encyclopedia* (2012)
Amaranthus retroflexus	Red-root amaranth	Amaranthaceae	*Plant Encyclopedia* (2012)
Amaranthus viridis	Slender amaranth	Amaranthaceae	*Plant Encyclopedia* (2012)
Ambrosia artemisiifolia	Common ragweed	Asteraceae	*Plant Encyclopedia* (2012) and Tworkowski (2002)
Arabidopsis thaliana	Thale cress	Brassicaceae	*Plant Encyclopedia* (2012)
Bromus danthonia		Poaceae	*Plant Encyclopedia* (2012)
Bromus intermedius		Poaceae	*Plant Encyclopedia* (2012)
Cassia occidentalis			*Plant Encyclopedia* (2012)
Centaurea solstitialis			*Plant Encyclopedia* (2012)
Chenopodium album	Common lambsquarters	Amaranthaceae	*Plant Encyclopedia* (2012) and Tworkowski (2002)
Cirsium arvense	Creeping thistle	Asteraceae	*Plant Encyclopedia* (2012)
Cynodon dactylon	Bermuda grass	Poaceae	*Plant Encyclopedia* (2012)
Cyperus rotundus	Purple nut sedge	Cyperaceae	*Plant Encyclopedia* (2012)
Echinochloa crus-galli	Common barnyard grass	Poaceae	*Plant Encyclopedia* (2012)
Hordeum spontaneum	Wild barley	Poaceae	*Plant Encyclopedia* (2012)
Lactuca sativa	Lettuce	Asteraceae	*Plant Encyclopedia* (2012)
Lactuca serriola	Prickly lettuce	Asteraceae	*Plant Encyclopedia* (2012)
Lepidium sativum	Garden cress	Brassicaceae	*Plant Encyclopedia* (2012)
Lolium multiflorum	Annual ryegrass	Poaceae	*Plant Encyclopedia* (2012)
Lolium rigidum	Ryegrass	Poaceae	*Plant Encyclopedia* (2012)
Parthenium hysterophorus	Whitetop weed	Asteraceae	*Plant Encyclopedia* (2012)
Phalaris canariensis	Canary grass	Poaceae	*Plant Encyclopedia* (2012)
Phalaris minor	Littleseed canary grass	Poaceae	*Plant Encyclopedia* (2012)
Portulaca oleracea	Common purslane	Portulacaceae	*Plant Encyclopedia* (2012)
Ranunculus repens	Creeping buttercup	Ranunculaceae	*Plant Encyclopedia* (2012)
Raphanus raphanistrum	Wild radish	Brassicaceae	*Plant Encyclopedia* (2012)
Raphanus sativus	Radish	Brassicaceae	*Plant Encyclopedia* (2012)
Rumex crispus	Curled dock	Polygonaceae	*Plant Encyclopedia* (2012)
Rumex nepalensis		Polygonaceae	*Plant Encyclopedia* (2012)
Rumex obtusifolius	Broad-leaved dock	Polygonaceae	*Plant Encyclopedia* (2012)
Secale cereale	Rye	Poaceae	*Plant Encyclopedia* (2012)
Senecio jacobaea	Common ragwort	Asteraceae	*Plant Encyclopedia* (2012) and Clay et al. (2005)
Sinapis arvensis	Wild mustard	Brassicaceae	*Plant Encyclopedia* (2012)
Sonchus oleraceus	Common sowthistle	Asteraceae	*Plant Encyclopedia* (2012)
Sorghum halepense	Johnsongrass	Poaceae	*Plant Encyclopedia* (2012) and Tworkowski (2002)
Taraxacum sp.	Dandelion	Asteraceae	*Plant Encyclopedia* (2012)
Trifolium campestre	Hop trefoil	Fabaceae	*Plant Encyclopedia* (2012) and Teuscher et al. (2004)
Urtica dioica	Common nettle	Urticaceae	Teuscher et al. (2004)

TABLE 17.6
Essential Oils That Can Be Used in Weed Control

Name Family	Constituents	Notes	Sources
Achillea sp. Asteraceae	Camphor, 1,8-cineole, piperitone, borneol, α-terpineol	Inhibitory effect on germination and seedling growth of *A. retroflexus*, *C. arvense*, and *L. serriola*	Kordali et al. (2009)
Ageratum conyzoides L. Asteraceae	Precocene I and II, β-caryophyllene, γ-bisabolene, fenchyl acetate	Causes phytotoxic effects on radish, mungbean, and tomatoes	*Plant Encyclopedia* (2012) and Kong et al. (1999)
Anisomeles indica L. Lamiaceae	Isobornyl acetate, isothujone, nerolidol, camphene, eugenol	Herbicide against *P. minor*, positive effects on growth of wheat	Batish et al. (2007b) and Ushir et al. (2010)
Artemisia scoparia Waldst et Kit. Asteraceae	*p*-Cymene, β-myrcene, (+)-limonene		Kaur et al. (2010)
Callicarpa japonica Thunb. Verbenaceae	Spathulenol, germacrene B, viridiflorol, globulol	Toxic to *A. stolonifera*, but had no such effect on lettuce	Kobaisy et al. (2002)
Carum carvi L. Apiaceae	D-Carvone, limonene	Inhibits germination of *A. retroflexus*, *C. solstitialis*, *S. arvensis*, *S. oleraceus*, *R. raphanistrum* and *R. nepalensis*, and *A. pallida*	Teuscher et al. (2004), De Almeida et al. (2010), and Azirak et al. (2008)
Coriandrum sativum L. Apiaceae	Linalool, α-terpinyl acetate, 1,8-cineole, linalyl acetate	Effective against *C. solstitialis*, *S. arvensis*, *S. oleraceus*, *R. raphanistrum*, and *R. nepalensis*	Teuscher et al. (2004) and Azirak and Karaman (2008)
Cymbopogon sp. Poaceae	Citronellal, geraniol, citronellol, citral, limonene		Teuscher et al. (2004) and Dayan et al. (2009)
Eucalyptus sp. Myrtaceae	1,8-Cineole, limonene, α-pinene citronellal, citronellol, linalool, α-terpinene		Teuscher et al. (2004) and Batish et al. (2006)
Foeniculum vulgare L. Apiaceae	Anethol, fenchone, estragole	Reduces germination rate (under 25%) of *C. solstitialis*, *S. arvensis*, and *R. raphanistrum*	Teuscher et al. (2004) and Azirak et al. (2008)
Hibiscus cannabinus L. Malvaceae	α-Terpineol, myrtenol, limonene, trans-carveol, and γ-eudesmol	Controls various weeds, e.g., *A. retroflexus* and *L. multiflorum*, at higher concentration; effective against lettuce, bentgrass, and against one cyanobacterium	Kobaisy et al. (2001)
Hyssopus officinalis L. Lamiaceae	β-Pinene, iso-pinocamphone, trans-pinocamphone	Inhibitory effect on germination of wheat seeds	De Almeida et al. (2010) and Dudai et al. (1999)

(Continued)

TABLE 17.6 (Continued)
Essential Oils That Can Be Used in Weed Control

Name Family	Constituents	Notes	Sources
Juniperus sp. Cupressaceae	α-Pinene, myrcene, β-pinene, terpinene-4-ol, germacrene D	Effective against *S. arvensis*, *T. campestre*, *L. rigidum*, and *P. canariensis* causing electrolyte leakage in them	Teuscher et al. (2004) and Ismail et al. (2012)
Lavandula sp. Lamiaceae	Linalool, linalyl acetate, fenchone, camphor		Konstantopoulou et al. (1992) and De Almeida et al. (2010)
Majorana hortensis L. Lamiaceae	1,8-Cineole, α-pinene, limonene		De Almeida et al. (2010)
Melissa officinalis L. Lamiaceae	Citral, citronellal, carvacrol, iso-menthone	Inhibitory effect on germination of wheat seeds	Teuscher et al. (2004), De Almeida et al. (2010), and Dudai et al. (1999)
Mentha sp. Lamiaceae	Carvone, pulegone, menthol, menthone, menthylacetate		Teuscher et al. (2004) and Konstantopoulou et al. (1992)
Micromeria fruticosa Lamiaceae	Pulegone, β-caryophyllene, isomenthol, limonene	Inhibits germination of wheat seedlings	Dudai et al. (1999)
Nepeta meyeri Benth. Lamiaceae	Nepetalactone, myrtenol, terpinen-4-ol	Inhibits germination of seeds of *B. danthoniae*, *B. intermedius*, *A. retroflexus*, *C. dactylon*, *C. album*, and *L. serriola*	Mutlu et al. (2010)
Ocimum sp. Lamiaceae	γ-Terpinene, α-terpinene, thymol, carvacrol, linalool, eugenol, carvone	Inhibits germination of wheat seeds	Aslan et al. (2004), De Almeida et al. (2010), and Dudai et al. (1999)
Origanum sp. Lamiaceae	Carvacrol, thymol, γ-terpinene, *p*-cymene, linalool		Konstantopoulou et al. (1992) and De Almeida et al. (2010)
Peumus boldus Mol. Monimiaceae	Limonene, 1,8-cineole, *p*-cymene, ascaridole	Herbicidal activity against *A. hybridus* and *P. oleracea*	Teuscher et al. (2004), Ibrahim et al. (2008), and Verdeguer et al. (2011)
Pimpinella anisum L. Apiaceae	*cis*-Anethol	Herbicidal effect against *R. raphanistrum*	De Almeida et al. (2010) and Azirak and Karaman (2008)
Pinus sp. Pinaceae	α-Pinene, β-pinene	Herbicide against various weeds	Teuscher et al. (2004) and Dayan et al. (2009)
Pistacia sp. Anacardiaceae	α-Terpinene, limonene	Completely inhibits germination of dicotyles *S. arvensis* and *T. campestre*, partially germination of monocotyles *L. rigidum* and *P. canariensis*	Ismail et al. (2012)
Rosa damascena Mill. Rosaceae	Citronellol, geraniol, nerol, linalool		Teuscher et al. (2004) and Aridogan et al. (2002)

(*Continued*)

TABLE 17.6 (*Continued*)
Essential Oils That Can Be Used in Weed Control

Name Family	Constituents	Notes	Sources
Rosmarinus officinalis L. Lamiaceae	Limonene, 1,8-cineole, borneol, terpineol	Selective toxicity against *C. album, P. oleracea, E. crus-galli, C. annuum*; good inhibition rates against *C. solstitialis, S. arvensis*, and *R. raphanistrum* and wheat seeds	Konstantopoulou et al. (1992), Dudai et al. (1999), Azirak and Karaman (2008), and Angelini et al. (2003)
Ruta graveolens L. Rutaceae	α-Pinene, limonene, 1,8-cineole, nonan-2-one, undecan-2-one	Inhibits germination and radicle elongation of radish	De Feo et al. (2002)
Salvia officinalis L. Lamiaceae	1,8-Cineole, thujone, camphor, borneol	Lowers germination rate of *S. arvensis* and *R. raphanistrum*; inhibits germination of wheat seeds	Teuscher et al. (2004), De Almeida et al. (2010), Azirak and Karaman (2008), and Dudai et al. (1999)
Satureja sp. Lamiaceae	Carvacrol, thymol, γ-terpinene, *p*-cymene	Damages several weeds, e.g., dandelion leaves, *C. album, A. artemisiifolia, S. halepense*	Konstantopoulou et al. (1992), Angelini et al. (2003), and Tworkowski (2002)
Syzygium aromaticum (L.) Merr et L.M. Perry Myrtaceae	Eugenol, aceteugenol, β-caryophyllene	Damages *Taraxacum* sp., *S. halepense, C. album*, and *A. artemisiifolia*	Teuscher et al. (2004), Dayan et al. (2009), and Tworkowski (2002)
Tagetes minuta L. Asteraceae	Limonene, (Z)-β-ocimene, (Z)- and (E)-ocimenone	Inhibits germination of *E. crus-galli* and *C. rotundus*; inhibits root growth of *Zea mays*	Eguaras et al. (2005), Batish et al. (2007a), and Scrivanti et al. (2003)
Tanacetum sp. Asteraceae	α-Pinene, α-terpinene, γ-terpinene, carvone, borneol, β-thujone, camphor		Chiasson et al. (2001)
Thymbra sp. Lamiaceae	Thymol, carvacrol	Lowers germination rate to under 5% of *A. retroflexus, S. arvensis, S. oleraceus*, and *R. raphanistrum*	Ghrabi-Gammar et al. (2009) and Azirak and Karaman (2008)
Thymus vulgaris L. Lamiaceae	Thymol, *p*-cymene, carvacrol	Inhibits the growth of *P. oleracea, E. crus-galli*, and *C. album*, lettuce and pepper crops; damages *Taraxacum* sp., *C. album, A. artemisiifolia*, and *S. halepense*	Teuscher et al. (2004), De Almeida et al. (2010), Kim and Ahn (2001), and Tworkowski (2002)
Verbena officinalis L. Verbenaceae	Citral, isobornyl formate, linalool, geranial		De Almeida et al. (2010)
Zataria multiflora Boiss. Lamiaceae	Carvacrol, linalool, α-pinene, *p*-cymene		Saharkhiz et al. (2010)

to inhibit germination of wheat seeds. This must be considered if the crop happens to be wheat (Dudai et al., 1999).

17.3.3.3   *Cymbopogon* sp.

Besides its mosquito-repellent actions, lemongrass oil can also be used as herbicide and *Cymbopogon citratus* (DC) Stapf (Poaceae) as weed control and antipest. It is used as contact herbicide, since limonene is able to "remove the waxy cuticular layer from leaves" that leads to death by dehydration. Limonene can be used as pesticide as well as herbicide (Dayan et al., 2009). *Cymbopogon winterianus* essential oil shows toxicity toward several weeds: *Senecio jacobaea* L., *Ranunculus repens* L., *Rumex obtusifolius* L., and *Urtica dioica* L. When sprayed on trees it causes damage to the foliage but it does not inhibit their growth (Clay et al., 2005). The mode of action of citral is probably that it causes disruption of microtubule polymerization in weeds, as could be proved in *Arabidopsis thaliana* seedlings (Dayan et al., 2012). The effect of *Cy. citratus* essential oil was also proved by Dudai et al. The essential oil proves to be one of the most effective ones, as it is able to inhibit germination sooner (0% germination at 80 nL/mL) than most other essential oils tested (Dudai et al., 1999).

17.3.3.4   *Eucalyptus* sp.

Species of *Eucalyptus* (Myrtaceae), such as *Eucalyptus citriodora* and *Eucalyptus tereticornis*, can be used for weed control, because they could reduce growth and chlorophyll and water content when used as fumigant. It also reduced the cellular respiration of *Parthenium hysterophorus* (Asteraceae). After about 2 weeks injuries like necrosis or wilting could be observed (Batish et al., 2008; Dayan et al., 2009). Moreover, the essential oil of *E. citriodora* can also inhibit its seed germination. The suspected mode of action is the inhibition of mitosis (Sing et al., 2005). The essential oil obtained from *E. tereticornis* showed a higher toxicity than that of *E. citriodora*, because of a slightly different composition of the essential oil. In another study *E. citriodora* essential oil was tested on crops (*Triticum aestivum*, *Zea mays*, and *R. sativus*) and weeds (*Cassia occidentalis*, *Amaranthus viridis*, and *E. crus-galli*). It could be shown that the essential oil was more toxic to small-seeded crops such as *A. viridis*. When *Eucalyptus* essential oil was tested on *P. hysterophorus* and *Phalaris minor*, the herbicidal effect could also be confirmed (Batish et al., 2008). The herbicidal effect of *E. citriodora* on *P. minor* was examined in another study by Batish et al. They found out that the concentration of lemon-scented eucalyptus oil is important for inhibiting the growth. The inhibition is higher at higher concentrations. The mode of action is more on the inhibition of growth of seedlings than inhibition of germination; *E. citriodora* also lessens the chlorophyll content and the respiratory activity of the weed and causes an ion leakage in membranes (Batish et al., 2007c). The herbicidal effect of *E. citriodora* was also mentioned by Dudai et al. (1999) when they examined the inhibition of germination of wheat seeds. *E. camaldulensis* also showed promising effects in controlling *Amaranthus hybridus* and *P. oleracea*, because it inhibits seedling growth as well as germination in both of them. Its main compound is identified as spathulenol (Verdeguer et al., 2009).

17.3.3.5   *Lavandula* sp.

When tested on various weeds by Azirak and Karaman, *L. stoechas* was not able to significantly inhibit the germination (under 25%) of most of the weed plants. It showed an effect on a few of them: *C. solstitialis*, *S. arvensis*, and *R. raphanistrum* (Azirak and Karaman, 2008). *Lavandula x intermedia* cv. Grosso was tested on *Lolium rigidum* by Haig et al. It showed a high phytotoxicity when it comes to inhibiting root growth (Haig et al., 2009).

17.3.3.6   *Origanum* sp.

An author group found out that the composition of oregano was different compared to literature because other authors named *p*-cymene as main compound (De Almeida et al., 2010). *O. onites* L.

essential oil proved to be effective against several weeds (*A. retroflexus*, *C. solstitialis*, *S. arvensis*, *S. oleraceus*, and *R. raphanistrum*) as they lower germination rate to less than 20% (Azirak and Karaman, 2008). *O. majorana* L. is also able to significantly inhibit germination of wheat seeds (Dudai et al., 1999). *O. acutidens* is able to inhibit germination and seedling growth of *A. retroflexus*, *C. album*, and *Rumex crispus* (Kordali et al., 2008).

17.3.3.7 Artemisia scoparia

A. scoparia Waldst. & Kit. (Asteraceae) was tested on *Achyranthes aspera* L., *C. occidentalis* L., *E. crus-galli* (L.) P. Beauv., *P. hysterophorus* L., and *Ageratum conyzoides* L. The inhibitory effect on seedling growth was greatest in the latter two weeds. When the essential oil was used on the weeds, it caused wilting and necrosis of sprayed parts. Also a decreasing chlorophyll amount and ion leakage can be observed. This effect was greatest in *E. crus-galli* and *P. hysterophorus*. *A. scoparia* essential oil's main constituents are *p*-cymene, β-myrcene, and (+)-limonene (Kaur et al., 2010).

17.3.3.8 Zataria multiflora

Two different ecotypes of *Zataria multiflora* Boiss, Lamiaceae, were tested in the study conducted by Saharkhiz et al. (2010). Ecotype B had similar constituents but in lower concentrations; this also explains why ecotype A was more successful in inhibiting the germination and growth of the tested weeds, which were *Hordeum spontaneum* Koch, *Secale cereale*, *A. retroflexus*, and *Cynodon dactylon*.

17.3.3.9 Tanacetum sp.

Tanacetum aucheranum and *Tanacetum chiliophyllum* var. *chiliophyllum*, Asteraceae, are common in Europe and Western Asia. The essential oils were tested about their potential to inhibit germination and seedling growth of *A. retroflexus*, *C. album*, and *R. crispus*, in which they were very successful. Besides their herbicidal effect, they also have an antibacterial and antifungal activity (Salamci et al., 2007).

Weeds are a severe problem for agriculture today as they lead to decreased yields in crops. Essential oils, like those obtained from *Cymbopogon* sp., *Eucalyptus* sp., *Origanum* sp., and *T. vulgaris*, have been proven to be very potent inhibitors of both germination and seedling growth of various weeds, although two limiting factors must be mentioned: first the high volatility of essential oils and second the fact that some essential oils do not only inhibit the growth of weeds but also that of crops. More research needs to be done on which weeds and crops are influenced by different volatile oils.

17.4 ESSENTIAL OILS AS INHIBITORS OF VARIOUS PESTS

Volatile oils can also be used in controlling various other pests like bacteria, viruses, fungi, and nematodes. Their application and potential will be discussed in the following section (Table 17.7).

17.4.1 Effect on Bacteria

Essential oils also have the potential to kill bacteria. Since they are lipophilic themselves, they can pass through the membrane of cells and cause cytotoxicity (Bakkali et al., 2008). On the one hand they *inhibit the respiration and increase the permeability of bacterial cytoplasmic membranes*, and on the other hand they cause potassium leakage (Ibrahim et al., 2008). According to Burt (2004), other mechanisms that cause the bactericidal effects are depletion of proton motive force, damage to membrane proteins as well as to the cytoplasmic membrane, and degradation of the cell wall, for example, the essential oil of *Cymbopogon densiflorus* (Poaceae), whose main compounds are limonene, cymenene, *p*-cymene, carveol, and carvone, were active against Gram-negative as well as Gram-positive bacteria (Ibrahim et al., 2008). Another essential oil with effect against bacteria

Use of Essential Oils in Agriculture

TABLE 17.7
Essential Oils with Effect on Bacteria, Viruses, Fungi, and Nematodes

Name and Family	Constituents	Notes	Sources
Aloysia triphylla Verbenaceae	α-Thujone, *cis*-carveol, carvone, limonene	Nematicide	Duschatzky et al. (2004)
Artemisia sp. Asteraceae	Limonene, myrcene, α-thujone, β-thujone, caryophyllene, sabinyl acetate	Fungicide Nematicide	Chiasson et al. (2001)
Calamintha nepeta Lamiaceae	Limonene, menthone, pulegone, menthol	Bactericide	Ibrahim et al. (2008)
Carum carvi L. Apiaceae	D-Carvone, limonene	Nematicide	Teuscher et al. (2004)
Cinnamomum zeylanicum Lauraceae	Cinnamaldehyde, linalool, β-caryophyllene, eugenol	Fungicide	Teuscher et al. (2004)
Coridothymus sp. Lamiaceae	Carvacrol, thymol, γ-terpinene, *p*-cymene	Nematicide	Konstantopoulou et al. (1992)
Cotinus coggygria Scop. Anacardiaceae	Limonene, (Z)-β-ocimene, (E)-β-ocimene, β-caryophyllene	Bactericide	Demirci et al. (2003)
Cymbopogon sp. Poaceae	Citronellal, geraniol, citronellol	Virucide Bactericide Fungicide	Teuscher et al. (2004)
Ducrosia anethifolia (DC.) Boiss Apiaceae	α-Pinene, myrcene, limonene, terpinolene, (E)-β-ocimene	Bactericide	Ibrahim et al. (2008)
Echinophora tenuifolia Apiaceae	α-Phellandrene, eugenol, *p*-cymene	Bactericide	Aridogan et al. (2002)
Elettaria cardamomum (L.) Maton Zingiberaceae	α-Terpinyl acetate, 1,8-cineole, linalyl acetate	Fungicide	Teuscher et al. (2004)
Eucalyptus sp. Myrtaceae	1,8-Cineole, limonene, α-pinene	Bactericide Fungicide Nematicide	Teuscher et al. (2004)
Foeniculum vulgare L. Apiaceae	Anethol, fenchone, estragole	Bactericide Fungicide Nematicide	Teuscher et al. (2004)
Hibiscus cannabinus L. Malvaceae	α-Terpineol, myrtenol, limonene, *trans*-carveol, and γ-eudesmol	Fungicide	Kobaisy et al. (2001)
Juniperus sp. Cupressaceae	α-Pinene, myrcene, β-pinene, terpinene-4-ol, germacrene D	Bactericide	Teuscher et al. (2004)
Laurus nobilis L. Lauraceae	1,8-Cineole, linalool, terpineol acetate, methyleugenol, linalyl acetate, eugenol	Fungicide	De Corato et al. (2010)
Lavandula sp. Lamiaceae	Linalool, linalyl acetate, fenchone, camphor	Bactericide	Konstantopoulou et al. (1992)
Lippia sp. Verbenaceae	Piperitenone oxide, limonene	Nematicide	Duschatzky et al. (2004)
Melaleuca alternifolia (Maiden et Betche) Cheel Myrtaceae	Terpinene-4-ol, γ-terpinene, α-terpinene	Virucide	Teuscher et al. (2004)

(*Continued*)

TABLE 17.7 (*Continued*)
Essential Oils with Effect on Bacteria, Viruses, Fungi, and Nematodes

Name and Family	Constituents	Notes	Sources
Mentha sp. Lamiaceae	Carvone, pulegone, menthol, menthone, menthylacetate	Nematicide Bactericide	Teuscher et al. (2004) and Konstantopoulou et al. (1992)
Micromeria fruticosa Lamiaceae	Pulegone, β-caryophyllene, isomenthol, limonene	Nematicide	Dudai et al. (1999)
Ocimum sp. Lamiaceae	γ-Terpinene, α-terpinene, thymol, carvacrol, linalool, and eugenol	Nematicide	Aslan et al. (2004)
Origanum sp. Lamiaceae	Carvacrol, thymol, γ-terpinene, *p*-cymene, linalool	Fungicide Bactericide Nematicide	Konstantopoulou et al. (1992)
Pelargonium graveolens L'Hér. Geraniaceae	Geraniol, citronellol	Nematicide	Ghrabi-Gammar et al. (2009)
Peumus boldus Mol. Monimiaceae	Limonene, 1,8-cineole, *p*-cymene	Bactericide	Teuscher et al. (2004) and Ibrahim et al. (2008)
Pinus sp. Pinaceae	α-Pinene, β-pinene	Fungicide	Teuscher et al. (2004)
Rosa damascena Mill. Rosaceae	Citronellol, geraniol, nerol, linalool	Bactericide	Teuscher et al. (2004) and Aridogan et al. (2002)
Rosmarinus officinalis L. Lamiaceae	Limonene, cineol, borneol, terpineol	Bactericide	Konstantopoulou et al. (1992)
Salvia fruticosa Miller Lamiaceae	Cineol, thujone, camphor, α-pinene, β-pinene	Fungicide	Konstantopoulou et al. (1992)
Satureja thymbra L. Lamiaceae	Carvacrol, thymol, γ-terpinene, *p*-cymene	Fungicide	Konstantopoulou et al. (1992)
Syzygium aromaticum (L.) Merr. et L.M. Perry Myrtaceae	Eugenol, aceteugenol, β-caryophyllene	Fungicide	Teuscher et al. (2004)
Tagetes minuta L. Asteraceae	Limonene, (Z)-β-ocimene, (Z)- and (E)-ocimenone	Bactericide	Eguaras et al. (2005)
Thymbra sp. Lamiaceae	Thymol	Fungicide	Ghrabi-Gammar et al. (2009)
Thymus sp. Lamiaceae	Thymol, carvacrol, *p*-cymene	Bactericide Fungicide	Teuscher et al. (2004)
Xylopia longifolia A. DC. Annonaceae	α-Phellandrene, limonene, *p*-cymene, spathulenol	Bactericide	Fournier et al. (1993)

is that of *Eucalyptus* sp. (Myrtaceae), which mostly contains 1,8-cineole, linalool, citronellal, and limonene (Batish et al., 2008). The essential oil of *Calamintha nepeta* (Lamiaceae), which contains limonene, menthone, and pulegone, showed high activity against *Salmonella*. Of all the constituents, pulegone was the most effective. The essential oil from *Peumus boldus* (Monimiaceae) contains mainly monoterpenes, especially limonene, and was also highly efficient against Gram-negative and Gram-positive bacteria. Also the essential oils of *F. vulgare* (Apiaceae)—mostly sweet fennel—*R. officinalis* (Lamiaceae), and *T. vulgaris* (Lamiaceae) showed high activity against bacteria. Other essential oils with activity against bacteria are those obtained from *Tagetes minuta* (Asteraceae), *Xylopia longifolia* (Annonaceae), *Cotinus coggygria* Scop. (Anacardiaceae), and *Ducrosia anethifolia* (Ibrahim et al., 2008). Essential oils obtained from *O. onites* (Lamiaceae), *Rosa damascena* (Rosaceae), *M. piperita* (Lamiaceae), *Echinophora tenuifolia* (Apiaceae), *L. hybrida* (Lamiaceae), and *Juniperus exalsa* (Cupressaceae) were tested on various bacteria. *O. onites*' essential oil was reported to elicit the biggest effect (Aridogan et al., 2002).

17.4.2 Effect on Fungi

Volatile oils are also capable of killing fungi that harm crops, because they cause increased plasma membrane permeability. Several monoterpenes like isopulegol, (*R*)-carvone, and isolimonene seemed to be very efficient against *Candida albicans*. The essential oils obtained from *F. vulgare* (Apiaceae) proved to be very active against *Aspergillus niger*. Also essential oils of citrus fruits (other than limonene) were tested against several *Penicillium* species and they were highly efficient. Unlike many other essential oils, volatile compounds that can be found in tomato leaves were able to inhibit the growth of hyphae of *Alternaria alternata*. Cardamom oil, linalool, limonene, and cineole showed the highest toxicity against fungi (Ibrahim et al., 2008). Another essential oil that is able to kill harmful fungi like *Aspergillus* sp., *Penicillium* sp., *Fusarium* sp., and *Mucor* sp. is that of *Eucalyptus* sp., which mostly contains 1,8-cineole, citronellal, limonene, and linalool. It is able to inhibit the growth of mycel as well as spore production and germination (Batish et al., 2008). *Pinus roxburghii* (Pinaceae) essential oil was able to kill *Aspergillus* sp. and *Cymbopogon pendulus* (Poaceae) and can inhibit the mycelium growth of *Microsporum gypseum* and *Trichophyton mentagrophytes*. Moreover, essential oils of *Eucalyptus amygdalina* (Myrtaceae) and *Eucalyptus pauciflora* showed activity against *Erysiphe cichoracearum*. Moreover, essential oils obtained from *Thymbra spicata*, *S. thymbra*, *Salvia fruticosa*, *Eucalyptus* sp., and *Origanum minutiflorum* (Lamiaceae) were tested on several fungi. Most toxic to those fungi were the essential oils of *T. spicata*, *S. thymbra*, and *O. minutiflorum* (Blaeser et al., 2002). Origanum, cassia, and red thyme essential oils are able to inhibit fungi that destroy wood (Chao et al., 2000). *O. acutidens* (Lamiaceae) essential oil shows antifungal activity as well, for example, against several fungi of *Fusarium* and *Botrytis* family (Kordali et al., 2008). Against *Botrytis cinerea* essential oil from *Cymbopogon martini* (Poaceae), *Thymus zygis* (Lamiaceae), *Cinnamomum zeylanicum* (Lauraceae), and *S. aromaticum* (Myrtaceae) showed high activity as well. Especially *C. martini* and *T. zygis* were very effective in inhibiting spore germination, followed by *S. aromaticum* essential oil (Wilson et al., 1997). Laurel oil (*Laurus nobilis* L., Lauraceae) showed antifungal activity against *Rhizoctonia solani* and *Sclerotinia sclerotiorum* (De Corato et al., 2010). Essential oil of *Hibiscus cannabinus* L. (Malvaceae) was effective against fungi of *Colletotrichum* species, even though the antifungal effect was weak (Kobaisy et al., 2001). Essential oils from *Artemisia dracunculus*, *A. absinthium*, *Artemisia santonicum*, and *Artemisia spicigera*, all of them Asteraceae, were tested against various fungi and showed to be effective against them (Kordali et al., 2005).

17.4.3 Effect on Viruses

Essential oils were also found to be active against viruses. The tobacco mosaic virus, which is an important pest in agriculture, is weakly resistant to lemongrass essential oil (Chao et al., 2000). But also essential oil of *Melaleuca alternifolia* (Myrtaceae) resulted in less lesions caused by tobacco mosaic virus for at least 10 days (Bishop, 1995).

17.4.4 Effect on Nematodes

The population of nematodes (*Heterodera schachtii*) could be decreased down to less than 3% of the control within 3 months using (+)-limonene. Also menthol proved to be very efficient against nematodes (Ibrahim et al., 2008). Essential oil of *Ocimum sanctum* L. (Lamiaceae) with eugenol as main compound was tested on *Caenorhabditis elegans* and showed an anthelmintic effect (Asha et al., 2001). Another essential oil that is very promising in nematode control is *Eucalyptus* sp.

essential oil. It proved to be toxic to *Meloidogyne incognita* and *Meloidogyne exigua* (Batish et al., 2008). *O. basilicum* L. (Lamiaceae) and *O. sanctum* L. (Lamiaceae), as well as their main compounds linalool and eugenol, are effective in killing larvae of *M. incognita* (Chatterjee et al., 1982). Pandey et al. (2000) reported that several *Eucalyptus* and *Mentha* species as well as *O. basilicum*, as proven in the previous study, and *Pelargonium graveolens* L'Hér (Geraniaceae) volatile oils act nematicidal on *M. incognita*, too. Another study conducted on *Meloidogyne* sp. shows that *Aloysia triphylla* (Verbenaceae), *Lippia juneliana* (Verbenaceae), and *Lippia turbinata* (Verbenaceae) have a nematicidal effect. The main components are α-thujone, *cis*-carveol, carvone, and limonene in *A. triphylla*; piperitenone oxide, limonene, and camphor in *L. juneliana*; and limonene and piperitenone oxide in *L. turbinata* (Duschatzky et al., 2004). Oka et al. (2000) examined the effect of essential oil on *Meloidogyne javanica*. Essential oils that significantly inhibited its mobility (more than 80%) were *Artemisia judaica*, *C. carvi*, *Coridothymus capitatus*, *Coridothymus citratus*, *F. vulgare*, *Mentha rotundifolia*, *M. spicata*, *Micromeria fruticosa*, *Origanum syriacum*, and *O. vulgare*. Essential oils that lowered the egg hatching rate to less than 2% were *A. judaica*, *C. carvi*, *F. vulgare*, and *M. rotundifolia*. *O. gratissimum*, with eugenol and 1,8-cineole as main compounds, was reported to have anthelmintic activity on *Haemonchus contortus* as it decreases the egg hatch rate (Pessoa et al., 2002) (Table 17.8).

This section gave a short overview about various pests that play a role in agriculture. The bactericidal and fungicidal effects of various essential oils, like *Cymbopogon* sp., *Thymus* sp., *Eucalyptus* sp., and *F. vulgare* among many others, have been known for quite a long time. These effects have been studied in many different publications and many facts about the mode of action and the different effects are known already. But essential oils already show virucidal and nematicidal effects as well. An agrochemical relevant virus is the tobacco mosaic virus, for example, that causes big losses every year. An essential oil that could be of use in killing the virus is *Cymbopogon* sp. Also nematodes, like the root-knot nematode, can be controlled by essential oils, for example, by several *Mentha* species.

17.5 EFFECT OF ESSENTIAL OILS ON THE CONDITION OF THE SOIL

17.5.1 Effects of Essential Oils on Microorganisms and Soil

Successful biodegradation is influenced by several factors: soil water, oxygen, redox potential, pH, nutrient status, and temperature (Holden and Firestone, 1997). Essential oils are not only capable of destroying microorganisms but they can also be used as carbon and energy source if they have been exposed to it recently. That way some microorganisms can actually be activated by essential oils. This applies mostly to bacteria that have lived in Mediterranean areas where also many plants that contain essential oils are. That way they were able to make fast use of the offered essential oil that was added (Vokou and Liotiri, 1999). Especially bacteria like *Arthrobacter* sp. and *Nocardia* sp. have been reported to use essential oils as carbon source. Also several *Pseudomonas* species have been detected to use α-pinene as energy source (Hassiotis, 2010; Hassiotis and Lazari, 2010). Essential oils tested by Voukou et al. were *O. vulgare* subsp. *hirtum* (Lamiaceae), *R. officinalis* (Lamiaceae), *M. spicata* (Lamiaceae), and *C. capitatus* (Lamiaceae) as well as *L. angustifolia* (Lamiaceae). In another study Voukou et al. observed the effect of essential oils obtained from *L. stoechas* (Lamiaceae), *S. thymbra* (Lamiaceae), or fenchone (Vokou et al., 2002). They were all able to increase the CO_2 emission of the soil and the microbial population, even though some substances had not been in previous contact with the soil sample before (Vokou and Liotiri, 1999). According to Hassiotis and Dina, the deciding factor if essential oils act as inhibitors or not is the concentration of the essential oil applied. At lower concentrations they can induce bacterial growth; at higher concentrations most likely they inhibit bacterial action (Hassiotis and Dina, 2010). The only difference between the essential oils was the time after they showed an effect. While some were immediately

TABLE 17.8
List of Various Pests in Agriculture, Including Postharvest Food Pathogens and Soil Microbes

Name	Notes	Group	Sources
Aeromonas hydrophila	Spoils meat	Bacterium	Burt (2004)
Alternaria alternata		Fungus	Burt (2004)
Arthrobacter sp.	Can use essential oil as carbon source	Bacterium	Hassiotis (2010)
Aspergillus flavus	Storage fungus, can produce toxic aflatoxins	Fungus	Blaeser et al. (2002) and Razzaghi-Abyaneh et al. (2008)
Aspergillus niger	Postharvest pathogen	Fungus	Tzortzakis and Economakis (2007)
Aspergillus parasiticus	Can produce toxic aflatoxins	Fungus	Razzaghi-Abyaneh et al. (2008)
Bacillus cereus	Food pathogen	Bacterium	Burt (2004)
Botrytis cinerea	Postharvest pathogen	Fungus	Tzortzakis and Economakis (2007)
Caenorhabditis elegans		Nematode	Tzortzakis and Economakis (2007)
Candida albicans		Fungus	Tzortzakis and Economakis (2007)
Cladosporium herbarum	Postharvest pathogen	Fungus	Razzaghi-Abyaneh et al. (2008)
Clostridium botulinum	Spoils food, produces neurotoxins	Bacterium	Jobling (2000)
Colletotrichum coccodes	Postharvest pathogen	Fungus	Tzortzakis and Economakis (2007)
Erysiphe cichoracearum	Phytopathogenous fungus	Fungus	Blaeser et al. (2002)
Eurotium sp.	Spoils food, i.e., bakery products	Fungus	Guynot et al. (2005)
Fusarium oxysporum f sp. *dianthi*	Phytopathogenous fungus	Fungus	Pitarokili et al. (2002)
Haemonchus contortus	Gastrointestinal parasite of ruminants	Nematode	Pessoa et al. (2002)
Heterodera schachtii	Infects various plants, i.e., sugarbeet	Nematode	Ibrahim et al. (2008)
Listeria monocytogenes	Spoils food	Bacterium	Jobling (2000)
Meloidogyne exigua Root-knot nematode	Damages plant roots	Nematode	Batish et al. (2008)
Meloidogyne incognita Root-knot nematode	Damages plant roots	Nematode	Batish et al. (2008)
Meloidogyne javanica Root-knot nematode	Damages plant roots	Nematode	Oka et al. (2000)
Methylosinus trichosporium	Consumes methane in soils	Bacterium	Amaral and Knowles (1998)
Microsporum gypseum		Fungus	Amaral and Knowles (1998)
Monilinia laxa	Postharvest pathogen	Fungus	De Corato et al. (2010)
Mucor sp.		Fungus	De Corato et al. (2010)
Nitrosomonas sp.	Oxidize ammonia	Bacterium	Chalkos et al. (2010)
Nitrosospira sp.	Oxidize ammonia	Bacterium	Chalkos et al. (2010)
Nocardia sp.	Can use essential oils as carbon source	Bacterium	Hassiotis (2010)
Penicillium digitatum	Spoils food, i.e., bakery products	Fungus	Guynot et al. (2005)
Polysphondylium pallidum	Cellular slime mold, feeds on bacteria in forest soils	Fungus	Hwang and Kim (2004)
Pseudomonas sp.		Bacterium	Hwang and Kim (2004)
Rhizoctonia solani	Postharvest fungus	Fungus	De Corato et al. (2010)
Rhizopus stolonifer	Spoils food	Fungus	Tzortzakis and Economakis (2007)
Salmonella enteritidis	Spoils food, especially raw food	Bacteria	Walderhaug (2012)
Sclerotinia sclerotiorum	Phytopathogenous fungus	Fungus	Pitarokili et al. (2002)
Sclerotium cepivorum	Phytopathogenous fungus	Fungus	Pitarokili et al. (2002)
Trichophyton mentagrophytes		Fungus	Pitarokili et al. (2002)

effective, others needed more time to work. Gram-positive as well as Gram-negative bacteria were able to use monoterpenes or oxygenated products (Vokou et al., 2002). This effect is also enhanced by the ability of essential oils to kill harmful fungi and their spores. As one can see, plants that contain essential oils can influence the condition of the soil by activating the soil respiration through increasing the microbial growth. The plants must be native to the soil's place to achieve that effect. If essential oils are tested on soil where they do not belong to, they appear to be less or noneffective (Vokou and Liotiri, 1999). Amaral and Knowles (1998) found out that several monoterpenes, especially (−)-α-pinene, were able to inhibit the atmospheric methane consumption by forest soils. Essential oils can kill *Methylosinus trichosporium*, a methanotroph bacterium, and they are also said to be able to inhibit the nitrification of forest soils. In the same way as the nitrogen cycle, the carbon cycle is also affected by essential oils (Amaral and Knowles, 1998). Generally said, terpenes, alcohols, and esters can be degraded by bacteria in a very fast way, while it takes a lot longer for ketones (Hassiotis, 2010). Another interesting question is how essential oils affect *Polysphondylium*, a cellular slime mold that feeds on bacteria in forest soils. (*R*)-(−)-limonene, (−)-camphene, and (*S*)-(+)-carvone as well as (−)-menthone and (+)-β-pinene inhibited the growth of *Polysphondylium pallidum* (Hwang and Kim, 2004).

17.5.2 Examples of Essential Oils with an Effect on Soil Condition

See Table 17.9.

TABLE 17.9
Essential Oils That Can Influence Condition of the Soil

Name and Family	Constituents	Notes	Sources
Cymbopogon sp. Poaceae	Citronellal, geraniol, citronellol		Teuscher et al. (2004)
Laurus nobilis L. Lauraceae	1,8-Cineole, linalool, terpineol acetate, methyleugenol, linalyl acetate, eugenol		De Corato et al. (2010)
Lavandula sp. Lamiaceae	Linalool, linalyl acetate, fenchone, camphor		Konstantopoulou et al. (1992)
Mentha spicata L. Lamiaceae	Carvone, pulegone, menthol, menthone, menthylacetate		Konstantopoulou et al. (1992) and Teuscher et al. (2004)
Myrtus communis L. Myrtaceae	1,8-Cineole, α-pinene, limonene, linalool, myrtenol		Teuscher et al. (2004)
Origanum sp. Lamiaceae	Carvacrol, thymol, γ-terpinene, *p*-cymene, linalool	Activates soil respiration	Konstantopoulou et al. (1992), Vokou and Liotiri (1999), and Vokou et al. (2002)
Rosmarinus officinalis L. Lamiaceae	Limonene, 1,8-cineol, borneol, terpineol, camphor	Activates soil respiration	Konstantopoulou et al. (1992), Vokou and Liotiri (1999), and Vokou et al. (2002)
Salvia sp. Lamiaceae	Cineol, thujone, camphor, α-pinene, β-pinene		Konstantopoulou et al. (1992)
Satureja thymbra L. Lamiaceae	Carvacrol, thymol, γ-terpinene, *p*-cymene	Activates soil respiration	Konstantopoulou et al. (1992) and Vokou et al. (2002)
Thymbra capitata (L.) Cav. Lamiaceae	Carvacrol, thymol, γ-terpinene, *p*-cymene	Activates soil respiration	Ghrabi-Gammar et al. (2009) and Vokou et al. (2002)

17.5.2.1 *Mentha spicata*

M. spicata's (Lamiaceae) main compounds are (*R*)-(−)-carvone, limonene, and 1,8-cineole (Vokou et al., 2002). Chalkos et al. evaluated the ability of *M. spicata* composted plants to increase growth of tomato crops and avoid weed growth. It has also a positive effect on the bacterial and fungal population in the soil. The bacteria *Nitrosomonas* and *Nitrosospira* are still oxidizing ammonia in the soil despite adding the compost. Among all these effects, the compost of *M. spicata* was able to increase the pH. Upon adding *M. spicata* to the soil, compounds of its essential oil can be found even after a long time, although in a different composition than before. Carvone, for example, makes up 50% of the essential oil in the beginning, but after some time in the soil, the concentration lowers to 1%, while other monoterpenes are gone completely. Sesquiterpenes, which act as allelochemicals, on the other hand, can be found at the same or even higher concentrations than in the original oil (Chalkos et al., 2010).

17.5.2.2 *Lavandula* sp.

Lavender oil (*L. angustifolia* Mill., Lamiaceae) mainly contains linalool and linalyl acetate. Its natural habitat is also the Mediterranean area (Vokou et al., 2002). Spanish or French lavender oil (*L. stoechas* L., Lamiaceae), as it is also called, is obtained from *L. stoechas* and consists of fenchone (Vokou et al., 2002). Other compounds are camphor, *p*-cymene, and 1,8-cineole. In a study by Hassiotis and Dina, the bacterial growth stimulating effect could be shown (Hassiotis, 2010; Hassiotis and Dina, 2010). In another study *L. stoechas*' essential oil degradation was examined. The highest essential oil degradation was in October, November, and December. During the hot summer months, the essential oil decrease was not very high, which indicates that bacteria are the reason for the loss and not evaporation. After 17 months only camphene, 1,8-cineole, and camphor remained (Hassiotis, 2010).

17.5.2.3 *Salvia* sp.

Adding *S. fruticosa* Mill. (Lamiaceae) compost to soil decreases the high C/N rates and raises the pH of the soil. Moreover, it increases the number of microbes in the soil, such as bacteria and fungi (Chalkos et al., 2010). *S. sclarea* (Lamiaceae) volatile oil's compounds are linalyl acetate, linalool, geranyl acetate, and α-terpineol. The essential oil is able to cause fungicidal activity on *S. sclerotiorum*, a soilborne pathogen. It also inhibits growth of *Sclerotium cepivorum* and *Fusarium oxysporum* f.sp. *Dianthi* (Pitarokili et al., 2002).

17.5.2.4 *Myrtus communis*

Hassiotis and Lazari examined the bacterial population growth during degradation of *Myrtus communis* L. (Myrtaceae) essential oil. It could be shown that the bacterial population was able to grow and use myrtle essential oil, whose compounds were decreasing over the time, as energy source. At the end of the study, only low percentages of 1,8-cineole and camphene could be detected (Hassiotis and Lazari, 2010).

17.5.2.5 *Laurus nobilis*

L. nobilis showed inhibition of bacterial activity and lowered the number of colonies as well as soil respiration rate. This is because 1,8-cineole and eugenol are known for their bactericidal effects (Hassiotis and Dina, 2010).

17.5.2.6 *Cymbopogon* sp.

Malkomes tested the dehydrogenase activity in soil with and without dung and the addition of citronella (*Cymbopogon* sp., Poaceae) oil. A low concentration of citronella oil caused a decreasing stimulation of the activity of dehydrogenases. In the soil, in which no dung was added, citronella oil also stimulated the CO_2 production from the very beginning, and in the soil, in which dung was added, the CO_2 production decreased at first, but after 3 weeks it started to increase (Malkomes, 2006).

The bactericidal action of essential oils has been known for quite a long time, but they can also be used as energy source for bacteria and activate them. That way they can increase the quality of the soil because CO_2 production is stimulated and pathogenous microorganisms are killed. Several essential oils (*M. spicata, Lavandula* sp., and *Cymbopogon* sp.) bear great potential to be used as dung and control of soil microbes in the future.

17.6 ESSENTIAL OILS USED IN POSTHARVEST DISEASE CONTROL

Food safety is a very important topic. Even though there is an increased hygiene during food production, still 30% of people suffer from food poisoning every year and as a result two million people worldwide still die. A lot of people nowadays also want their food to have less synthetic food additives in it; therefore, there is a need to find natural substances that can replace them (Burt, 2004).

Insects and mites that damage food and act as pathogens are also a major problem in food storage, especially in developing countries. Essential oils that can be used as acaricides and insecticides are explained in Section 17.2.

17.6.1 Effects of Essential Oils on Stored-Product Pests

In order to keep the concentration of health-damaging effects low, it is important to find out on the benefits of essential oils on stored products. Essential oils that have proven themselves to be useful against postharvest pathogens (the fungi *B. cinerea*, are red thyme oil, clove oil, and cinnamon oil). Essential oils of *Monarda citriodora* (Lamiaceae) and *M. alternifolia* (Myrtaceae) showed an antifungal effect against many different fungi. Tea tree oil also works as inhibitor on bacteria (Jobling, 2000). The effect of anethole, carvacrol, cinnamaldehyde, eugenol, and safrole were tested against fungi of *Mucor, Aspergillus,* and *Rhizopus* species. Against most *Rhizopus* and *Mucor* species, carvacrol proved itself to be most effective. Against *Aspergillus* species, both carvacrol and eugenol showed the best results (Thompson, 1989). Carvacrol is also able to inhibit toxins that are produced by *Bacillus cereus* in various dishes (Burt, 2004). Against bacteria that damage stored products, cedar, eucalyptus, thyme, and chamomile oils proved themselves useful, especially when used against pathogenous bacteria *B. cereus, Clostridium botulinum,* and *Listeria monocytogenes* (Jobling, 2000). Generally said, essential oils with a high percentage of phenolic compounds seem to be more effective against pathogens that damage food. Moreover, essential oils have a bigger effect on Gram-positive bacteria than on Gram-negative (Burt, 2004).

But it must be mentioned that the effect of essential oils was tested in laboratories and not in the commercial field and that many essential oils only show a fungistatic effect, which means that the fungi will start growing again after the essential oil vapor is gone (Jobling, 2000). Moreover, a higher concentration of essential oils—up to 100 times as much in soft cheese—is needed to inhibit bacteria in food than in vitro. The chemical environment of the food is also important for the effect, for example, the bactericide effect increases when the pH decreases, and the concentration of the fat in the food also plays a major role because essential oils lose their ability to harm bacteria with increasing fat concentration (Burt, 2004).

17.6.2 Examples of Essential Oils Used on Stored Products

See Table 17.10.

17.6.2.1 *Thymus zygis*

The main components of thyme (*T. zygis* L., Lamiaceae) oil are thymol and *p*-cymene. Its home is the Mediterranean area (Jobling, 2000). Thyme oil can be used to protect meat products from two bacteria: *L. monocytogenes* and *Aeromonas hydrophila*. Thymol also showed activity against several *Salmonella* species (Burt, 2004).

TABLE 17.10
Essential Oils That Can Be Used to Preserve Stored Products

Name and Family	Constituents	Notes	Source
Artemisia sp. Asteraceae	Limonene, myrcene, α-thujone, β-thujone, caryophyllene, sabinyl acetate, camphor, 1,8-cineole, borneol, terpinen-4-ol, bornyl acetate	Very strong antifungal activity.	Chiasson et al. (2001) and Kordali et al. (2005)
Cinnamomum sp. Lauraceae	Cinnamaldehyde, linalool, β-caryophyllene, eugenol		Teuscher et al. (2004)
Coriandrum sp. Apiaceae	Linalool, α-terpinyl acetate, 1,8-cineole, linalyl acetate	Can be used to preserve meat products from bacteria.	Teuscher et al. (2004) and Burt (2004)
Cymbopogon sp. Poaceae	Citronellal, geraniol, citronellol		Teuscher et al. (2004)
Eucalyptus sp. Myrtaceae	1,8-Cineole, limonene, α-pinene	Against browning of mushrooms in plastic bags, antibacterial.	Teuscher et al. (2004) and Jobling (2000)
Laurus nobilis L. Lauraceae	1,8-Cineole, linalool, terpineol acetate, methyleugenol, linalyl acetate, eugenol		De Corato et al. (2010)
Melaleuca alternifolia (Maiden et Betche) Cheel Myrtaceae	Terpinene-4-ol, γ-terpinene, α-terpinene	At 500 ppm it is able to kill 100% of the fungi *Botrytis* sp. on grapes.	Teuscher et al. (2004) and Jobling (2000)
Mentha piperita L. Lamiaceae	Carvone, pulegone, menthol, menthone, menthylacetate	Is able to kill *Salmonella enteritidis* in several foods.	Konstantoploulou et al. (1992), Teuscher et al. (2004), and Tassou et al. (2000)
Monarda citriodora Lamiaceae	Thymol, thymol methylester, α-terpinene, *p*-cymene	Antifungal activity on most postharvest fungi, e.g., *Alternaria*, *Fusarium*, and *Rhizopus* species.	Rozzi et al. (2002), Jobling (2000), and Bishop and Thornton (1997)
Origanum sp. Lamiaceae	Carvacrol, thymol, γ-terpinene, *p*-cymene, linalool	Inhibits bacterial growth on meat, fish, and vegetable products.	Konstantopoulou et al. (1992) and Burt (2004)
Satureja hortensis Lamiaceae	Carvacrol, thymol	Inhibitor of *Aspergillus parasiticus* (can produce aflatoxins).	Razzaghi-Abyaneh et al. (2008)
Syzygium aromaticum (L.) Merr. et L.M. Perry Myrtaceae	Eugenol, aceteugenol, β-caryophyllene	Eugenol shows high activity against *A. flavus*; is antibacterial; inhibits *L. monocytogenes* and *A. hydrophila*.	Teuscher et al. (2004), Jobling (2000), Blaeser et al. (2002), and Burt (2004)
Thymus zygis L. Lamiaceae	Thymol, carvacrol, *p*-cymene	Protects meat products against *L. monocytogenes* and *A. hydrophila*; also against *Salmonella* species.	Teuscher et al. (2004), Jobling (2000), and Burt (2004)
Zizyphus jujuba Rhamnaceae	Eugenol, isoeugenol, caryophyllene, eucalyptol, caryophyllene oxide	Active against *Listeria monocytogenes*.	Al-Reza et al. (2009)

17.6.2.2 *Cinnamomum* sp.

Cinnamon oil (*C. zeylanicum* J. Presl, Lauraceae) consists mainly of cinnamaldehyde (Jobling, 2000). It can be used to preserve dairy products, but it also proved to be successful in defeating several *Salmonella* bacteria on vegetables (Burt, 2004). Guynot et al. (2005) tested cinnamon essential oil on *Eurotium* sp., *Aspergillus* sp., and *Penicillium* sp. to preserve bakery products.

17.6.2.3 *Cymbopogon citratus*

Cy. citratus (Poaceae) was able to kill *Aspergillus flavus* to 100%, a fungus that harms stored products. It has to be applied at about 1000 ppm to inhibit *A. flavus*. Besides *A. flavus*, it has the ability to kill many other fungi as well for about 210 days in storage. The essential oil of *Cy. citratus*, especially if it was produced between May and November, also appeared to be more successful in fighting fungi than several synthetic products (Mishra and Dubey, 1994). Lemongrass essential oil also showed activity against colony development of *Cladosporium herbarum* and *Rhizopus stolonifer*. A concentration of 500 ppm was lethal for *Colletotrichum coccodes*, *C. herbarum*, *R. stolonifer*, and *A. niger*. The fungal colony growth of *B. cinerea* was only inhibited up to 60%. The fungal spore production could be inhibited of all these five fungi (Tzortzakis and Economakis, 2007). Lemongrass essential oil can also be used against *A. flavus*, a fungus that can produce aflatoxins in parboiled rice. It causes fungistatic and fungicidal effects (Paraganama et al., 2003).

17.6.2.4 *Laurus nobilis*

Essential oil of *L. nobilis* L., Lauraceae, showed good effects in the protection of the harvested fruits of kiwi and peach against the postharvest fungi *Monilinia laxa*, *B. cinerea*, and *Penicillium digitatum*. The fungistatic effect is very high in *M. laxa* and *B. cinerea*, but *P. digitatum* cannot be completely killed (De Corato et al., 2010).

The protection of stored food against pathogens like bacteria, fungi, insects, and mites is a big challenge especially in developing countries. Especially volatile oils obtained from *Cinnamomum* sp., *Cymbopogon* sp., and *S. aromaticum* (earlier known as *Eugenia caryophyllata*) bear great potential. One thing that should be considered is that they have a strong taste themselves; therefore, not every essential oil can be used in every dish as the flavors might not match very well. But overall adding essential oils to food helps preserve it and seems to be a good idea since many essential oil plants are already used as spices throughout the world and most of them do not have any known toxic effects. But of course, studies must be conducted to make sure adding volatile oils do not lead to yet unknown side effects.

17.7 CONCLUSION

Essential oils can be used in various fields of application and one of them is in agriculture. Until now agriculture is dominated by synthetic products in order to control various pests, weed, and soil condition, but essential oils bear great potential to replace or complement those substances. In insect pest control, essential oils proved to be very promising as they have great potential for killing various harmful insects. Especially in developing countries they might play a bigger role in the future as they are very cheap and can be combined with synthetic products to minimize their toxicity. When it comes to repellent activity, essential oils might not be that useful as only some insects will be deterred while others can be attracted by them. When it comes to weed control, essential oils do not seem as promising because of the high quantities needed and the frequent application, which might also lead to negative environmental effects. Moreover, the possibility that the applied essential oils might damage the wanted crops cannot be excluded. Volatile oils are also able to kill various pests (bacteria, fungi, and viruses) as well as mites and nematodes. Especially in the control of tobacco mosaic virus, essential oils might become very important. More research needs to be done in this field of application. Furthermore, essential oils are also able to improve the condition of the soil because if exposed to the oils earlier, bacteria can use them as carbon source.

That way the respiration of the soil can be activated. Finally, a very interesting aspect of essential oils is to use them in postharvest pest management and against fungi and bacteria on stored products. Especially in developing countries, the insect and other pest management on stored products might become very useful. But also in industrialized countries there is some use for them, for example, in supermarkets on fruits and vegetables to make them less perishable.

REFERENCES

Al-Reza, S. M., V. K. Bajpai, and S. C. Kang, 2009. Antioxidant and antilisterial effect of seed essential oil and organic extracts from *Zizyphus jujuba*. *Food Chem. Toxicol.*, 47: 2374–2380.

Amaral, J. A. and R. Knowles, 1998. Inhibition of methane consumption in forest soils by monoterpenes. *J. Chem. Ecol.*, 24: 723–734.

Ammon, H. P. T., C. Hunnius, and A. Bihlmayer, 2010. *Hunnius pharmazeutisches Wörterbuch*. De Gruyter.

Angelini, L. G., G. Carpanese, P. L. Cioni, I. Morelli, M. Macchia, and G. Flamini, 2003. Essential oils from Mediterranean lamiaceae as weed germination inhibitors. *J. Agric. Food Chem.*, 51: 6158–6164.

Anthony, N. M., J. B. Harrison, and D. B. Sattelle, 1993. GABA receptor molecules of insects. *EXS*, 63: 172–209.

Aridoğan, B. C., H. Baydar, S. Kaya, M. Demirci, D. Ozbaşar, and E. Mumcu, 2002. Antimicrobial activity and chemical composition of some essential oils. *Arch. Pharm. Res.*, 25: 860–864.

Asha, M. K., D. Prashanth, B. Murali, R. Padmaja, and A. Amit, 2001. Anthelmintic activity of essential oil of *Ocimum sanctum* and eugenol. *Fitoterapia*, 72: 669–670.

Aslan, İ., H. Özbek, Ö. Çalmaşur, and F. Şahin, 2004. Toxicity of essential oil vapours to two greenhouse pests, *Tetranychus urticae* Koch and *Bemisia tabaci* Genn. *Indust. Crop Prod.*, 19: 167–173.

Ayvaz, A., O. Sagdic, S. Karaborklu, and I. Ozturk, 2010. Insecticidal activity of the essential oils from different plants against three stored-product insects. *J. Insect Sci.*, 10: 21.

Azirak, S. and S. Karaman, 2008. Allelopathic effect of some essential oils and components on germination of weed species. *Acta Agric. Scand. Sect. B—Soil Plant Sci.*, 58: 88–92.

Bakkali, F., S. Averbeck, D. Averbeck, and M. Idaomar, 2008. Biological effects of essential oils—A review. *Food Chem. Toxicol.*, 46: 446–475.

Balfanz, S., T. Strünker, S. Frings, and A. Baumann, 2005. A family of octopamine [corrected] receptors that specifically induce cyclic AMP production or Ca^{2+} release in *Drosophila melanogaster*. *J. Neurochem.*, 93: 440–451.

Batish, D. R., K. Arora, H. P. Singh, and R. K. Kohli, 2007a. Potential utilization of dried powder of *Tagetes minuta* as a natural herbicide for managing rice weeds. *Crop Prot.*, 26: 566–571.

Batish, D. R., M. Kaur, H. P. Singh, and R. K. Kohli, 2007b. Phytotoxicity of a medicinal plant, *Anisomeles indica*, against *Phalaris minor* and its potential use as natural herbicide in wheat fields. *Crop Prot.*, 26: 948–952.

Batish, D. R., H. P. Singh, R. K. Kohli, and S. Kaur, 2008. Eucalyptus essential oil as a natural pesticide. *Forest Ecol. Manage.*, 256: 2166–2174.

Batish, D. R., H. P. Singh, N. Setia, S. Kaur, and R. K. Kohli, 2006. Chemical composition and phytotoxicity of volatile essential oil from intact and fallen leaves of *Eucalyptus citriodora*. *Z. Naturforsch. C, J. Biosci.*, 61: 465–471.

Batish, D. R., H. P. Singh, N. Setia, R. K. Kohli, S. Kaur, and S. S. Yadav, 2007c. Alternative control of littleseed canary grass using eucalypt oil. *Agron. Sustain. Dev.*, 27: 171–177.

Bishop, C. D., 1995. Antiviral activity of the essential oil of *Melaleuca alternifolia* (Maiden amp; Betche) Cheel (Tea Tree) against tobacco mosaic virus. *J. Essent. Oil Res.*, 7: 641–644.

Bishop, C. D. and I. B. Thornton, 1997. Evaluation of the antifungal activity of the essential oils of *Monarda citriodora* var. *citriodora* and *Melaleuca alternifolia* on post-harvest pathogens. *J. Essent. Oil Res.*, 9: 77–82.

Blaeser, P., U. Steiner, and H. W. Dehne. 2002. Pflanzeninhaltsstoffe mit fungizider Wirkung. *Schriftenreihe des Lehr- und Forschungsschwerpunktes USL*, Vol. 97. Landwirtschaftliche Fakultät der Universität Bonn, Bonn, Germany, pp. 1–143.

Burt, S., 2004. Essential oils: Their antibacterial properties and potential applications in foods—A review. *Int. J. Food Microbiol.*, 94: 223–253.

Çalmaşur, Ö., İ. Aslan, and F. Şahin, 2006. Insecticidal and acaricidal effect of three Lamiaceae plant essential oils against *Tetranychus urticae* Koch and *Bemisia tabaci* Genn. *Indust. Crop Prod.*, 23: 140–146.

Capinera, J. L., 2008. *Encyclopedia of Entomology*. Springer Science & Business Media.

Cavalcanti, S. C. H., S. Niculau Edos, A. F. Blank, C. A. G. Câmara, I. N. Araújo, and P. B. Alves, 2010. Composition and acaricidal activity of *Lippia sidoides* essential oil against two-spotted spider mite (*Tetranychus urticae* Koch). *Bioresour. Technol.*, *101*: 829–832.

Cetin, H., J. E. Cilek, E. Oz, L. Aydin, O. Deveci, and A. Yanikoglu, 2010. Acaricidal activity of *Satureja thymbra* L. essential oil and its major components, carvacrol and γ-terpinene against adult *Hyalomma marginatum* (Acari: Ixodidae). *Vet. Parasitol.*, *170*: 287–290.

Chalkos, D., K. Kadoglidou, K. Karamanoli, C. Fotiou, A. S. Pavlatou-Ve, I. G. Eleftherohorinos, H.-I. A. Constantinidou, and D. Vokou, 2010. *Mentha spicata* and *Salvia fruticosa* composts as soil amendments in tomato cultivation. *Plant Soil*, *332*: 495–509.

Chao, S. C., D. G. Young, and C. J. Oberg, 2000. Screening for inhibitory activity of essential oils on selected bacteria, fungi and viruses. *J. Essent. Oil Res.*, *12*: 639–649.

Chatterjee, A., N. C. Sukul, S. Laskar, and S. Ghoshmajumdar, 1982. Nematicidal principles from two species of Lamiaceae. *J. Nematol.*, *14*: 118–120.

Chiasson, H., A. Bélanger, N. Bostanian, C. Vincent, and A. Poliquin, 2001. Acaricidal properties of *Artemisia absinthium* and *Tanacetum vulgare* (Asteraceae) essential oils obtained by three methods of extraction. *J. Econ. Entomol.*, *94*: 167–171.

Clay, D. V., F. L. Dixon, and I. Willoughby, 2005. Natural products as herbicides for tree establishment. *Forestry*, *78*: 1–9.

Dayan, F. E., C. L. Cantrell, and S. O. Duke, 2009. Natural products in crop protection. *Bioorg. Med. Chem.*, *17*: 4022–4034.

Dayan, F. E., D. K. Owens, and S. O. Duke, 2012. Rationale for a natural products approach to herbicide discovery. *Pest Manage. Sci.*, *68*: 519–528.

De Almeida, L. F. R., F. Frei, E. Mancini, L. De Martino, and V. De Feo, 2010. Phytotoxic activities of mediterranean essential oils. *Molecules*, *15*: 4309–4323.

De Corato, U., O. Maccioni, M. Trupo, and G. Di Sanzo, 2010. Use of essential oil of *Laurus nobilis* obtained by means of a supercritical carbon dioxide technique against post harvest spoilage fungi. *Crop Prot.*, *29*: 142–147.

De Feo, V., F. De Simone, and F. Senatore, 2002. Potential allelochemicals from the essential oil of *Ruta graveolens*. *Phytochemistry*, *61*: 573–578.

Demirci, B., F. Demirci, and K. H. C. Başer, 2003. Composition of the essential oil of *Cotinus coggygria* Scop. from Turkey. *Flavour Fragr. J.*, *18*: 43–44.

Douglas, M. H., J. W. van Klink, B. M. Smallfield, N. B. Perry, R. E. Anderson, P. Johnstone, and R. T. Weavers, 2004. Essential oils from New Zealand manuka: Triketone and other chemotypes of *Leptospermum scoparium*. *Phytochemistry*, *65*: 1255–1264.

Dudai, N., O. Larkov, E. Putievsky, H. R. Lerner, U. Ravid, E. Lewinsohn, and A. M. Mayer, 2000. Biotransformation of constituents of essential oils by germinating wheat seed. *Phytochemistry*, *55*: 375–382.

Dudai, N., A. Poljakoff-Mayber, A. M. Mayer, E. Putievsky, and H. R. Lerner, 1999. Essential oils as allelochemicals and their potential use as bioherbicides. *J.Chem. Ecol.*, *25*: 1079–1089.

Duschatzky, C. B., A. N. Martinez, N. V. Almeida, and S. L. Bonivardo, 2004. Nematicidal activity of the essential oils of several argentina plants against the root-knot nematode. *J. Essent. Oil Res.*, *16*: 626–628.

Eguaras, M. J., S. Fuselli, L. Gende, R. Fritz, S. R. Ruffinengo, G. Clemente, A. Gonzalez, P. N. Bailac, and M. I. Ponzi, 2005. An in vitro evaluation of *Tagetes minuta* essential oil for the control of the honeybee pathogens *Paenibacillus larvae* and *Ascosphaera apis*, and the parasitic mite *Varroa destructor*. *J. Essent. Oil Res.*, *17*: 336–340.

Fournet, A., A. Rojas de Arias, B. Charles, and J. Bruneton, 1996. Chemical constituents of essential oils of muña, Bolivian plants traditionally used as pesticides, and their insecticidal properties against Chagas' disease vectors. *J. Ethnopharmacol.*, *52*: 145–149.

Fournier, G., A. Hadjiakhoondi, M. Leboeuf, A. Cavé, J. Fourniat, and B. Charles, 1993. Chemical and biological studies of *Xylopia longifolia* A. DC. essential oils. *J. Essent. Oil Res.*, *5*: 403–410.

García, M., O. J. Donadel, C. E. Ardanaz, C. E. Tonn, and M. E. Sosa, 2005. Toxic and repellent effects of *Baccharis salicifolia* essential oil on *Tribolium castaneum*. *Pest Manage. Sci.*, *61*: 612–618.

George, D. R., G. Olatunji, J. H. Guy, and O. A. E. Sparagano, 2010. Effect of plant essential oils as acaricides against the poultry red mite, *Dermanyssus gallinae*, with special focus on exposure time. *Vet. Parasitol.*, *169*: 222–225.

George, D. R., T. J. Smith, R. S. Shiel, O. A. E. Sparagano, and J. H. Guy, 2009a. Mode of action and variability in efficacy of plant essential oils showing toxicity against the poultry red mite, *Dermanyssus gallinae*. *Vet. Parasitol.*, *161*: 276–282.

George, D. R., O. A. E. Sparagano, G. Port, E. Okello, R. S. Shiel, and J. H. Guy, 2009b. Repellence of plant essential oils to *Dermanyssus gallinae* and toxicity to the non-target invertebrate *Tenebrio molitor*. *Vet. Parasitol.*, 162: 129–134.

Ghrabi-Gammar, Z., D. R. George, A. Daoud-Bouattour, I. Ben Haj Jilani, S. B. Saad-Limam, and O. A. E. Sparagano, 2009. Screening of essential oils from wild-growing plants in Tunisia for their yield and toxicity to the poultry red mite, *Dermanyssus gallinae*. *Indust. Crop Prod.*, 30: 441–443.

Guynot, M. E., S. MarÍn, L. SetÚ, V. Sanchis, and A. J. Ramos, 2005. Screening for antifungal activity of some essential oils against common spoilage fungi of bakery products. *Food Sci. Technol. Int.*, 11: 25–32.

Haig, T. J., T. J. Haig, A. N. Seal, J. E. Pratley, M. An, and H. Wu, 2009. Lavender as a source of novel plant compounds for the development of a natural herbicide. *J. Chem. Ecol.*, 35: 1129–1136.

Hassiotis, C. N., 2010. Chemical compounds and essential oil release through decomposition process from *Lavandula stoechas* in Mediterranean region. *Biochem. System. Ecol.*, 38: 493–501.

Hassiotis, C. N. and E. I. Dina, 2010. The influence of aromatic plants on microbial biomass and respiration in a natural ecosystem. *Israel J. Ecol. Evol.*, 56: 181–196.

Hassiotis, C. N. and D. M. Lazari, 2010. Decomposition process in the Mediterranean region. Chemical compounds and essential oil degradation from *Myrtus communis*. *Int. Biodet. Biodegrad.*, 64: 356–362.

Holden, P. A. and M. K. Firestone, 1997. Soil microorganisms in soil cleanup: How can we improve our understanding? *J. Environ. Qual.*, 26: 32.

Huang, Y. and S. H. Ho, 1998. Toxicity and antifeedant activities of cinnamaldehyde against the grain storage insects, *Tribolium castaneum* (Herbst) and *Sitophilus zeamais* Motsch. *J. Stored Prod. Res.*, 34: 11–17.

Huang, Y., S.-H. Ho, H.-C. Lee, and Y.-L. Yap, 2002. Insecticidal properties of eugenol, isoeugenol and methyleugenol and their effects on nutrition of *Sitophilus zeamais* Motsch. (Coleoptera: Curculionidae) and *Tribolium castaneum* (Herbst) (Coleoptera: Tenebrionidae). *J. Stored Prod. Res.*, 38: 403–412.

Huang, Y., S. L. Lam, and S. H. Ho, 2000. Bioactivities of essential oil from *Elletaria cardamomum* (L.) Maton. to *Sitophilus zeamais* Motschulsky and: *Tribolium castaneum* (Herbst). *J. Stored Prod. Res.*, 36: 107–117.

Hummelbrunner, L. A. and M. B. Isman, 2001. Acute, sublethal, antifeedant, and synergistic effects of monoterpenoid essential oil compounds on the tobacco cutworm, *Spodoptera litura* (Lep., Noctuidae). *J. Agric. Food Chem.*, 49: 715–720.

Hwang, J.-Y. and J.-H. Kim, 2004. The effect of monoterpenoids on growth of a cellular slime mold, *Polysphondylium pallidum*. *J. Plant Biol.*, 47: 8–14.

Ibrahim, M. A., P. Kainulainen, and A. Aflatuni, 2008. Insecticidal, repellent, antimicrobial activity and phytotoxicity of essential oils: With special reference to limonene and its suitability for control of insect pests. *Agric. Food Sci.*, 10: 243–259.

Ismail, A., H. Lamia, H. Mohsen, and J. Bassem, 2012. Herbicidal potential of essential oils from three mediterranean trees on different weeds. *Curr. Bioactive Compd.*, 8: 3–12.

Isman, M. B., 2000. Plant essential oils for pest and disease management. *Crop Prot.*, 19: 603–608.

Isman, M. B., 2006. Botanical insecticides, deterrents, and repellents in modern agriculture and an increasingly regulated world. *Annu. Rev. Entomol.*, 51: 45–66.

Isman, M. B., 2008. Botanical insecticides: For richer, for poorer. *Pest Manage. Sci.*, 64: 8–11.

Jobling, J., 2000. Essential oils: A new idea for postharvest disease control. *Good Fruit Veget. Magaz.*, 11: 50.

Katerinopoulos, H. E., G. Pagona, A. Afratis, N. Stratigakis, and N. Roditakis, 2005. Composition and insect attracting activity of the essential oil of *Rosmarinus officinalis*. *J. Chem. Ecol.*, 31: 111–122.

Kaur, S., H. P. Singh, S. Mittal, D. R. Batish, and R. K. Kohli, 2010. Phytotoxic effects of volatile oil from *Artemisia scoparia* against weeds and its possible use as a bioherbicide. *Indust. Crop Prod.*, 32: 54–61.

Kéita, S. M., C. Vincent, J.-P. Schmit, J. T. Arnason, and A. Bélanger, 2001. Efficacy of essential oil of *Ocimum basilicum* L. and *O. gratissimum* L. applied as an insecticidal fumigant and powder to control *Callosobruchus maculatus* (Fab.) [Coleoptera: Bruchidae]. *J. Stored Prod. Res.*, 37: 339–349.

Kim, D.-H. and Y.-J. Ahn, 2001. Contact and fumigant activities of constituents of *Foeniculum vulgare* fruit against three coleopteran stored-product insects. *Pest. Manage. Sci.*, 57: 301–306.

Kim, E.-H., H.-K. Kim, and Y.-J. Ahn, 2003a. Acaricidal activity of plant essential oils against *Tyrophagus putrescentiae* (Acari: Acaridae). *J. Asia-Pacific Entomol.*, 6: 77–82.

Kim, H.-K., J.-R. Kim, and Y.-J. Ahn, 2004. Acaricidal activity of cinnamaldehyde and its congeners against *Tyrophagus putrescentiae* (Acari: Acaridae). *J. Stored Prod. Res.*, 40: 55–63.

Kim, S.-I., Y.-E. Na, J.-H. Yi, B.-S. Kim, and Y.-J. Ahn, 2007. Contact and fumigant toxicity of oriental medicinal plant extracts against *Dermanyssus gallinae* (Acari: Dermanyssidae). *Vet. Parasitol.*, 145: 377–382.

Kim, S.-I., J.-Y. Roh, D.-H. Kim, H.-S. Lee, and Y.-J. Ahn, 2003b. Insecticidal activities of aromatic plant extracts and essential oils against *Sitophilus oryzae* and *Callosobruchus chinensis*. *J. Stored Prod. Res.*, 39: 293–303.

Kobaisy, M., M. R. Tellez, F. E. Dayan, and S. O. Duke, 2002. Phytotoxicity and volatile constituents from leaves of *Callicarpa japonica* Thunb. *Phytochemistry*, 61: 37–40.

Kobaisy, M., M. R. Tellez, C. L. Webber, F. E. Dayan, K. K. Schrader, and D. E. Wedge, 2001. Phytotoxic and fungitoxic activities of the essential oil of kenaf (*Hibiscus cannabinus* L.) leaves and its composition. *J. Agric. Food Chem.*, 49: 3768–3771.

Kong, C., F. Hu, T. Xu, and Y. Lu, 1999. Allelopathic potential and chemical constituents of volatile oil from *Ageratum conyzoides*. *J. Chem. Ecol.*, 25: 2347–2356.

Konstantopoulou, I., L. Vassilopoulou, P. Mavragani-Tsipidou, and Z. G. Scouras, 1992. Insecticidal effects of essential oils. A study of the effects of essential oils extracted from eleven Greek aromatic plants on *Drosophila auraria*. *Experientia*, 48: 616–619.

Kordali, S., A. Cakir, T. A. Akcin, E. Mete, A. Akcin, T. Aydin, and H. Kilic, 2009. Antifungal and herbicidal properties of essential oils and n-hexane extracts of *Achillea gypsicola* Hub-Mor. and *Achillea biebersteinii* Afan. (Asteraceae). *Indust. Crop Prod.*, 29: 562–570.

Kordali, S., A. Cakir, H. Ozer, R. Cakmakci, M. Kesdek, and E. Mete, 2008. Antifungal, phytotoxic and insecticidal properties of essential oil isolated from Turkish *Origanum acutidens* and its three components, carvacrol, thymol and *p*-cymene. *Bioresour. Technol.*, 99: 8788–8795.

Kordali, S., R. Kotan, A. Mavi, A. Cakir, A. Ala, and A. Yildirim, 2005. Determination of the chemical composition and antioxidant activity of the essential oil of *Artemisia dracunculus* and of the antifungal and antibacterial activities of Turkish *Artemisia absinthium*, *A. dracunculus*, *Artemisia santonicum*, and *Artemisia spicigera* essential oils. *J. Agric. Food Chem.*, 53: 9452–9458.

Kostyukovsky, M., A. Rafaeli, C. Gileadi, N. Demchenko, and E. Shaaya, 2002. Activation of octopaminergic receptors by essential oil constituents isolated from aromatic plants: Possible mode of action against insect pests. *Pest Manage. Sci.*, 58: 1101–1106.

Koul, O., S. Walia, and G. S. Dhaliwal, 2008. Essential oils as green pesticides: Potential and constraints. *Biopest. Int.*, 4: 63–84.

Lee, B.-H., P. C. Annis, F. Tumaalii, and W.-S. Choi, 2004. Fumigant toxicity of essential oils from the Myrtaceae family and 1,8-cineole against 3 major stored-grain insects. *J. Stored Prod. Res.*, 40: 553–564.

Lee, B.-H., W.-S. Choi, S.-E. Lee, and B.-S. Park, 2001. Fumigant toxicity of essential oils and their constituent compounds towards the rice weevil, *Sitophilus oryzae* (L.). *Crop Prot.*, 20: 317–320.

Lee, C.-H., B.-K. Sung, and H.-S. Lee, 2006. Acaricidal activity of fennel seed oils and their main components against *Tyrophagus putrescentiae*, a stored-food mite. *J. Stored Prod. Res.*, 42: 8–14.

Li, R. and Z.-T. Jiang, 2004. Chemical composition of the essential oil of *Cuminum cyminum* L. from China. *Flavour Fragr. J.*, 19: 311–313.

Liu, Z. L. and S. H. Ho, 1999. Bioactivity of the essential oil extracted from *Evodia rutaecarpa* Hook f. et Thomas against the grain storage insects, *Sitophilus zeamais* Motsch. and *Tribolium castaneum* (Herbst). *J. Stored Prod. Res.*, 35: 317–328.

Loebe, L., 2001. Olfactory remedies for the evaluation of repellent and attractive properties of essential oils against the pea aphid *Acyrthosiphon pisum*, Master thesis. University of Vienna, Vienna, Austria.

López, M. D., M. J. Jordán, and M. J. Pascual-Villalobos, 2008. Toxic compounds in essential oils of coriander, caraway and basil active against stored rice pests. *J. Stored Prod. Res.*, 44: 273–278.

Malkomes, H.-P., 2006. Einfluss von herbizidem Citronella-Öl und Neem (Azadirachtin) auf mikrobielle Aktivitäten im Boden. *Gesunde Pflanzen*, 58: 205–212.

Mishra, A. K. and N. K. Dubey, 1994. Evaluation of some essential oils for their toxicity against fungi causing deterioration of stored food commodities. *Appl. Environ. Microbiol.*, 60: 1101–1105.

Moretti, M. D. L., G. Sanna-Passino, S. Demontis, and E. Bazzoni, 2002. Essential oil formulations useful as a new tool for insect pest control. *AAPS PharmSciTech*, 3: E13.

Mutlu, S., Ö. Atici, N. Esim, and E. Mete, 2010. Essential oils of catmint (*Nepeta meyeri* Benth.) induce oxidative stress in early seedlings of various weed species. *Acta Physiol. Plant*, 33: 943–951.

Negahban, M., S. Moharramipour, and F. Sefidkon, 2006. Chemical composition and insecticidal activity of *Artemisia scoparia* essential oil against three coleopteran stored-product insects. *J. Asia-Pacific Entomol.*, 9: 381–388.

Negahban, M., S. Moharramipour, and F. Sefidkon, 2007. Fumigant toxicity of essential oil from *Artemisia sieberi* Besser against three stored-product insects. *J. Stored Prod. Res.*, 43: 123–128.

Nerio, L. S., J. Olivero-Verbel, and E. Stashenko, 2010. Repellent activity of essential oils: A review. *Bioresour. Technol.*, 101: 372–378.

Nerio, L. S., J. Olivero-Verbel, and E. E. Stashenko, 2009. Repellent activity of essential oils from seven aromatic plants grown in Colombia against *Sitophilus zeamais* Motschulsky (Coleoptera). *J. Stored Prod. Res.*, 45: 212–214.

Niepel, D., 2000. Repellent properties of essential oils against the carrot weevil, Master thesis. University of Vienna, Vienna, Austria.

Njoroge, S. M., H. Koaze, P. N. Karanja, and M. Sawamura, 2005. Essential oil constituents of three varieties of Kenyan sweet oranges (*Citrus sinensis*). *Flavour Fragr. J.*, 20: 80–85.

Obeng-Ofori, D., C. H. Reichmuth, A. J. Bekele, and A. Hassanali, 1998. Toxicity and protectant potential of camphor, a major component of essential oil of *Ocimum kilimandscharicum*, against four stored product beetles. *Int. J. Pest Manage.*, 44: 203–209.

Oka, Y., S. Nacar, E. Putievsky, U. Ravid, Z. Yaniv, and Y. Spiegel, 2000. Nematicidal activity of essential oils and their components against the root-knot nematode. *Phytopathology*, 90: 710–715.

Pandey, R., A. Kalra, S. Tandon, N. Mehrotra, H. N. Singh, and S. Kumar, 2000. Essential oils as potent source of nematicidal compounds. *J. Phytopathol.*, 148: 501–502.

Papachristos, D. P. and D. C. Stamopoulos, 2002. Repellent, toxic and reproduction inhibitory effects of essential oil vapours on *Acanthoscelides obtectus* (Say) (Coleoptera: Bruchidae). *J. Stored Prod. Res.*, 38: 117–128.

Paranagama, P. A., K. H. T. Abeysekera, K. Abeywickrama, and L. Nugaliyadde, 2003. Fungicidal and anti-aflatoxigenic effects of the essential oil of *Cymbopogon citratus* (DC.) Stapf. (lemongrass) against *Aspergillus flavus* Link. isolated from stored rice. *Lett. Appl. Microbiol.*, 37: 86–90.

Park, C., S.-I. Kim, and Y.-J. Ahn, 2003. Insecticidal activity of asarones identified in *Acorus gramineus* rhizome against three coleopteran stored-product insects. *J. Stored Prod. Res.*, 39: 333–342.

Pellati, F., S. Benvenuti, F. Yoshizaki, D. Bertelli, and M. C. Rossi, 2005. Headspace solid-phase microextraction–gas chromatography–mass spectrometry analysis of the volatile compounds of *Evodia* species fruits. *J. Chromatogr. A*, 1087: 265–273.

Pessoa, L. M., S. M. Morais, C. M. L. Bevilaqua, and J. H. S. Luciano, 2002. Anthelmintic activity of essential oil of *Ocimum gratissimum* Linn. and eugenol against *Haemonchus contortus*. *Vet. Parasitol.*, 109: 59–63.

Pitarokili, D., M. Couladis, N. Petsikos-Panayotarou, and O. Tzakou, 2002. Composition and antifungal activity on soil-borne pathogens of the essential oil of *Salvia sclarea* from Greece. *J. Agric. Food Chem.*, 50: 6688–6691.

Plant Encyclopedia, 2012. Aden Earth Zone. http://www.theplantencyclopedia.org. Accessed August 16, 2014.

Prajapati, V., A. K. Tripathi, K. K. Aggarwal, and S. P. S. Khanuja, 2005. Insecticidal, repellent and oviposition-deterrent activity of selected essential oils against *Anopheles stephensi*, *Aedes aegypti* and *Culex quinquefasciatus*. *Bioresour. Technol.*, 96: 1749–1757.

Rajendran, S. and V. Sriranjini, 2008. Plant products as fumigants for stored-product insect control. *J. Stored Prod. Res.*, 44: 126–135.

Razzaghi-Abyaneh, M., M. Shams-Ghahfarokhi, T. Yoshinari, M.-B. Rezaee, K. Jaimand, H. Nagasawa, and S. Sakuda, 2008. Inhibitory effects of *Satureja hortensis* L. essential oil on growth and aflatoxin production by *Aspergillus parasiticus*. *Int. J. Food Microbiol.*, 123: 228–233.

Regnault-Roger, C., 1997. The potential of botanical essential oils for insect pest control. *Integr. Pest Manage. Rev.*, 2: 25–34.

Regnault-Roger, C., C. Vincent, and J. T. Arnason, 2012. Essential oils in insect control: Low-risk products in a high-stakes world. *Annu. Rev. Entomol.*, 57: 405–424.

Rozzi, N. L., W. Phippen, J. E. Simon, and R. K. Singh, 2002. Supercritical fluid extraction of essential oil components from lemon-scented botanicals. *LWT—Food Sci. Technol.*, 35: 319–324.

Sahaf, B. Z., S. Moharramipour, and M. H. Meshkatalsadat, 2007. Chemical constituents and fumigant toxicity of essential oil from *Carum copticum* against two stored product beetles. *Insect Sci.*, 14: 213–218.

Saharkhiz, M. J., S. Smaeili, and M. Merikhi, 2010. Essential oil analysis and phytotoxic activity of two ecotypes of *Zataria multiflora* Boiss. growing in Iran. *Nat. Prod. Res.*, 24: 1598–1609.

Salamci, E., S. Kordali, R. Kotan, A. Cakir, and Y. Kaya, 2007. Chemical compositions, antimicrobial and herbicidal effects of essential oils isolated from Turkish *Tanacetum aucheranum* and *Tanacetum chiliophyllum* var. *chiliophyllum*. *Biochem. System. Ecol.*, 35: 569–581.

Sánchez-Ramos, I. and P. Castañera, 2000. Acaricidal activity of natural monoterpenes on *Tyrophagus putrescentiae* (Schrank), a mite of stored food. *J. Stored Prod. Res.*, 37: 93–101.

Scrivanti, L. R., M. P. Zunino, and J. A. Zygadlo, 2003. *Tagetes minuta* and *Schinus areira* essential oils as allelopathic agents. *Biochem. System. Ecol.*, 31: 563–572.

Sertkaya, E., K. Kaya, and S. Soylu, 2010. Acaricidal activities of the essential oils from several medicinal plants against the carmine spider mite (*Tetranychus cinnabarinus* Boisd.) (Acarina: Tetranychidae). *Indust. Crop Prod.*, 31: 107–112.

Shaaya, E., U. Ravid, N. Paster, B. Juven, U. Zisman, and V. Pissarev, 1991. Fumigant toxicity of essential oils against four major stored-product insects. *J. Chem. Ecol.*, 17: 499–504.

Singh, H. P., D. R. Batish, N. Setia, and R. K. Kohli, 2005. Herbicidal activity of volatile oils from *Eucalyptus citriodora* against *Parthenium hysterophorus*. *Ann. Appl. Biol.*, *146*: 89–94.

Tapondjou, L. A., C. Adler, H. Bouda, and D. A. Fontem, 2002. Efficacy of powder and essential oil from *Chenopodium ambrosioides* leaves as post-harvest grain protectants against six-stored product beetles. *J. Stored Prod. Res.*, *38*: 395–402.

Tassou, C., K. Koutsoumanis, and G.-J. E. Nychas, 2000. Inhibition of *Salmonella enteritidis* and *Staphylococcus aureus* in nutrient broth by mint essential oil. *Food Res. Int.*, *33*: 273–280.

Teuscher, E., M. F. Melzig, and U. Lindequist, 2004. *Biogene Arzneimittel: Ein Lehrbuch der pharmazeutischen Biologie; 14 Tabellen*. Wiss. Verlag-Ges.

Thompson, D. P., 1989. Fungitoxic activity of essential oil components on food storage fungi. *Mycologia*, *81*: 151–153.

Tripathi, A. K., S. Upadhyay, M. Bhuiyan, and P. Bhattacharya, 2009. A review on prospects of essential oils as biopesticide in insect-pest management. *J. Pharmacog. Phytother.*, *1*: 52–63.

Tunç, İ., B. M. Berger, F. Erler, and F. Dağlı, 2000. Ovicidal activity of essential oils from five plants against two stored-product insects. *J. Stored Prod. Res.*, *36*: 161–168.

Tworkoski, T., 2002. Herbicide effects of essential oils. *Weed Sci.*, *50*: 425–431.

Tzortzakis, N. G. and C. D. Economakis, 2007. Antifungal activity of lemongrass (*Cympopogon citratus* L.) essential oil against key postharvest pathogens. *Innov. Food Sci. Emerg. Technol.*, *8*: 253–258.

Ushir, Y., A. Tatiya, S. Surana, and U. Patil, 2010. Gas chromatography-mass spectrometry analysis and antibacterial activity of essential oil from aerial parts and roots of *Anisomeles indica* Linn. *Int. J. Green Pharmacy*, *4*: 98.

Verdeguer, M., M. A. Blázquez, and H. Boira, 2009. Phytotoxic effects of *Lantana camara*, *Eucalyptus camaldulensis* and *Eriocephalus africanus* essential oils in weeds of Mediterranean summer crops. *Biochem. System. Ecol.*, *37*: 362–369.

Verdeguer, M., D. García-Rellán, H. Boira, E. Pérez, S. Gandolfo, and M. A. Blázquez, 2011. Herbicidal activity of *Peumus boldus* and *Drimys winterii* essential oils from Chile. *Molecules*, *16*: 403–411.

Vokou, D., D. Chalkos, G. Karamanlidou, and M. Yiangou, 2002. Activation of soil respiration and shift of the microbial population balance in soil as a response to *Lavandula stoechas* essential oil. *J. Chem. Ecol.*, *28*: 755–768.

Vokou, D. and S. Liotiri, 1999. Stimulation of soil microbial activity by essential oils. *Chemoecology*, *9*: 41–45.

Walderhaug, M., 2014. *Bad Bug Book: Foodborne Pathogenic Microorganisms and Natural Toxins Handbook*. Createspace Independent Pub.

Wilson, C. L., J. M. Solar, A. El Ghaouth, and M. E. Wisniewski, 1997. Rapid evaluation of plant extracts and essential oils for antifungal activity against *Botrytis cinerea*. *Plant Dis.*, *81*: 204–210.

18 Adulteration of Essential Oils

Erich Schmidt and Jürgen Wanner

CONTENTS

18.1 Introduction 709
    18.1.1 General Remarks 709
18.2 Definition and History 710
18.3 Adulteration 713
    18.3.1 Unintended Adulteration 713
    18.3.2 Intentional Adulteration 715
    18.3.3 Prices 715
    18.3.4 Availability 715
    18.3.5 Demand of Clients 716
    18.3.6 Regulations 716
    18.3.7 Aging 716
    18.3.8 Cupidity 717
    18.3.9 Simple Sports? 717
18.4 Possible Adulterations for Essential Oils 717
    18.4.1 Water 717
    18.4.2 Ethanol 717
    18.4.3 Fatty Oils or Mineral Oils 717
    18.4.4 High Boiling Glycols 718
    18.4.5 Oils from Other Parts of the Same Species or Other Species with Similar Essential Oil Composition 718
    18.4.6 Related Botanical Species 718
    18.4.7 Fractions of Essential Oils 719
    18.4.8 Natural Isolates 719
    18.4.9 Chemically Derived Synthetic Compounds, Which Are Proved to Appear in Nature 720
    18.4.10 Steam Distilled Residues from Expression 720
    18.4.11 Enzymatically Produced Chemicals (Natural by Law) 720
18.5 Methods to Detect Adulterations 721
    18.5.1 Organoleptic Methods 721
        18.5.1.1 Appearance and Color 721
        18.5.1.2 Odor 721
        18.5.1.3 Physical–Chemical Methods 721
        18.5.1.4 Calculation of Relationship Coefficient 722
    18.5.2 Analytical Methods 723
        18.5.2.1 General Tests 723
        18.5.2.2 Thin-Layer Chromatography 723
        18.5.2.3 Gas Chromatography (GC, GLC, HRGC, GC-FID, GC-MS) 724
        18.5.2.4 Chiral Analysis (Busch and Busch, 2006) 725
        18.5.2.5 GC-GC and GC×GC (Two-Dimensional Gas Chromatography, ^{2}D GC) 726
        18.5.2.6 ^{13}C NMR (Nuclear Magnetic Resonance) 726

18.6 Important Essential Oils and Their Possible Adulteration 727
    18.6.1 Ambrette Seed Oil 727
    18.6.2 Amyris Oil 727
    18.6.3 Angelica Oils 727
    18.6.4 Anise Fruit Oil 727
    18.6.5 Armoise Oil 727
    18.6.6 Basil Oils 727
    18.6.7 Bergamot Oil 727
    18.6.8 Bitter Orange Oil 728
    18.6.9 Bitter Orange Petitgrain Oil 728
    18.6.10 Cajeput Oil 729
    18.6.11 Camphor Oil 729
    18.6.12 Cananga Oil 729
    18.6.13 Caraway Oil 729
    18.6.14 Cardamom Oil 729
    18.6.15 Cassia Oil 730
    18.6.16 Cedar Leaf Oil 730
    18.6.17 Cedarwood Oils 730
    18.6.18 Celery Seed Oil 730
    18.6.19 Chamomile Oil Blue 730
    18.6.20 Chamomile Oil Roman 730
    18.6.21 Cinnamon Bark Oil 730
    18.6.22 Cinnamon Leaf Oil 731
    18.6.23 Citronella Oil 731
    18.6.24 Clary Sage Oil 731
    18.6.25 Clove Oils 731
    18.6.26 Coriander Fruit Oil 731
    18.6.27 Corymbia Citriodora Oil 731
    18.6.28 Corn Mint Oil 732
    18.6.29 Cumin Fruit Oil 732
    18.6.30 Cypress Oil 732
    18.6.31 Dill Oils 732
    18.6.32 Dwarf Pine Oil 732
    18.6.33 Elemi Oil 732
    18.6.34 Eucalyptus Oil 733
    18.6.35 Fennel Oil Sweet 733
    18.6.36 Fennel Oil Bitter 733
    18.6.37 Geranium Oils 733
    18.6.38 Grapefruit Oil 733
    18.6.39 Juniper Berry Oil 734
    18.6.40 Lavandin Oils 734
    18.6.41 Lavender Oil 734
    18.6.42 Lemon Oil 734
    18.6.43 Lemongrass Oil 735
    18.6.44 Lime Oil Distilled 735
    18.6.45 Lime Oil Expressed 735
    18.6.46 *Litsea cubeba* Oil 736
    18.6.47 Mandarin Oil 736
    18.6.48 Melissa Oil (Lemon Balm) 736

18.6.49 *Mentha citrata* Oil 736
18.6.50 Mountain Pine Oil 736
18.6.51 Neroli Oil 736
18.6.52 Nutmeg Oil 737
18.6.53 Orange Oil Sweet 737
18.6.54 Origanum Oil 737
18.6.55 Palmarosa Oil 737
18.6.56 Parsley Oil 737
18.6.57 Pine Oil Siberian 738
18.6.58 Patchouli Oil 738
18.6.59 Pepper Oil 738
18.6.60 Peppermint Oil 738
18.6.61 Petitgrain Oil Paraguay Type 738
18.6.62 Pimento Oils 738
18.6.63 Rose Oil 739
18.6.64 Rosemary Oil 739
18.6.65 Rosewood Oil 739
18.6.66 Sage Oil (*Salva officinalis*) 739
18.6.67 Sage Oil Spanish Type 739
18.6.68 Sandalwood Oil 740
18.6.69 Spearmint Oils 740
18.6.70 Spike Lavender Oil 740
18.6.71 Star Anise Oil 740
18.6.72 Tarragon Oil 740
18.6.73 Tea Tree Oil 741
18.6.74 Thyme Oil 741
18.6.75 Turpentine Oil 741
18.6.76 Vetiver Oil 741
18.6.77 Ylang-Ylang Oils 741
References 742

18.1 INTRODUCTION

18.1.1 GENERAL REMARKS

Requirements of governmental bodies in the use of natural products increased tremendously in the last years. Not only the directions by the flavor regulation but also the demand of the consumer organizations presses the industry and producers to supply solely natural products, often desired in "bio" quality. The green responsibility for natural products brought synthetic aroma chemicals in a poor light. Regarding the mass market for cosmetic products, genuineness in fragrances is very limited because of the sources of raw material. On the other side, prices for raw materials increase since years for natural flavor and fragrance ingredients. The acceptance of fragrance compounds containing essential oils and natural aroma chemicals is a way to fulfill the consumer demand for safe, effective, and too affordable finished products. But even today with that claims and in spite of fabulous analytical equipment and highly sophisticated methods, many adulterated essential oils are found on the market. The question arises, what are the motivations for such a serious condition? This and above all, the manner of adulteration and the matching analysis methods shall be treated within this chapter.

Essential oils are constituents of around 30,000 species of plants around the world. Only a few of them are used in today's flavor, cosmetic, animal feeding, and pharmaceutical industry as well as in aromatherapy. Observing the product range of producers and dealers, about 250–300 essential oils are offered in varying quantities. Within those oils, 150 can be characterized as important oils in quantity and price.

Consumption of these essential oils worldwide is tremendous. The data published by Lawrence (2009) show a total quantity of more than 120,000 metric tons. A monetary value can barely be calculated, as prices rise and fall over in a year; exchange rates in foreign currency and disposals within the dealer's community falsify the result. Gambling with natural raw materials and the finished essential oil takes place with all the important ones. Comparison of production figures and those of export and import statistic will vary sometimes obviously; discrepancy appears in higher amounts. Reunion Island's National Institute of Statistics and Economic Studies (INSEE, 2008–2009) reports for the year 2002 the low quantity of 0,4 to kg of "vetiver oil Bourbon" exported; the sold quantity in the market was more than ten times higher. From 2003 until today, no vetiver oil is produced on Reunion Island. The essential oils traded were proven pure and natural but were coming from other production areas.* However, "Bourbon" quality generated a much higher price. It must be assumed that the value of essential oils worldwide will be far above $10 billion and this dimension invites to make some extra money out of it.

Screening essential oil qualities from the market with high-end analytical equipment leads to detection of many adulterated essential oils. Alarming is the fact that these adulterations appear not only within the consumer market but also in industry and trade. The questions arise again, what is the explanation for such a behavior and what are the reasons? Adulterations are subject of many publications with high impact; most of them came out in the beginning of the nineties, when new analytical methods were developed and applied. However, adulteration of essential oils began a long time ago. This chapter deals with adulteration starting in history up to the present. High criminal energy must be responsible for such sacrileges. Financial advantages, market shares, monopole status, and sometimes simply a sportive action are only some reasons.

18.2 DEFINITION AND HISTORY

Observing the significant number of publications about essential oils using the Internet, the results show lack of knowledge, finding "extracted" essential oils, distilled ones by "supercritical fluids," distilled Jasmine oil, therapeutic essential oil, and so on. It is remarkable that in scientific publications of serious publishers, there is the same confusion and ignorance about the term "essential oil." Even in regulatory and institutional papers, this term is used falsely and unprofessionally. Essential oils are clearly defined by the International Standard Organization (ISO) in Draft International Standard (DIS) ISO 9235, 2013 (former 9235.2)—Aromatic natural raw materials—Vocabulary. This DIS war, revised 2013, tightened the strict definition for an essential oil:

> Essential oil: Product obtained from a natural vegetable raw material of plant origin, by steam distillation, by mechanical process from the epicarp of *Citrus* fruits, or by dry distillation, after separation of the aqueous phase—if any—by physical process.

Further it is mentioned that steam distillation can be carried out with or without added water in a still. No change has been made with the note that "the essential oil can undergo physical treatments (e.g., filtrations, decantation, centrifugation) which do not result in any significant change in its composition." This is a tightening of the old definition. Natural raw material also was updated in the definition: "Natural raw material of vegetal, animal or microbiologic origin, as such, obtained by physical, enzymatic or microbiological processes, or obtained by traditional preparation processes (e.g., extraction, distillation, heating, torrefaction, fermentation)." These definitions show clearly the importance of standards to avoid any adulteration during all stages of collection,

---

* Based on the authors' private information and experiences.

production, and trade. Terms like "pure or natural essential oils" now are misleading as an essential oil has to fulfill this demand. Other regulatory elements are used for standardization like the pharmacopoeias around the world. Furthermore medicinal authorities and animal feeding product industry settled own regulatory standards. In the field of aromatherapy, no standards are settled although just in that field, the existence of ISO standards or pharmacopoeias as base should be implemented. The use of natural cosmetics is reported to result in a higher demand for oils but the quantity available on the market was and is insufficient to satisfy this developing market. Today many aromatic chemicals from "natural" source, produced by enzymatic reaction from sometimes not natural starters by microorganisms, help the producers to fill up that shortage. But are these chemicals really "natural"?

Adulteration is defined as "making impure or inferior by adding foreign substances to something" or "is a legal term meaning that a product fails to meet international, federal or state standards" (The Free Dictionary n.d. [a]). In this context all kinds of adulterations are enclosed like poisonous, economical, and unethical ones. In the United States, FDA regulations for food and cosmetics are the basis of law; in Europe consumer protection and food regulations as well as the cosmetic regulation have to be applied. Adulterations must not always be deliberated with criminal or unethical background. Environmental pollution like herbicides or pesticides cannot be prevented; these are unavoidable "adulterations." Negligently adulteration, basing on the lack of knowledge, still happens today. By selection of the botanical raw material, very often mistakes are done. Another instance is the missing of suitable production facilities and again, insufficient professionalism of the staff working there. The vocabulary for adulteration of essential oils is fanciful: extending, stretching, cutting, bouquetting, and rounding up to even "sophistication."

The history of adulteration reaches back to the turn of nineteenth to twentieth century. It is ongoing hand in hand with the establishment of chemistry and the development of synthetic aromatic chemicals. Already in 1834 it was possible to isolate cinnamaldehyde from cinnamon bark oil and as a result in 1856, it was synthesized. It is worth to know that Otto Wallach, a basic scientist for terpene chemistry at the University of Göttingen, already started in 1884 with the clarification of terpene structures. In 1885 he defined the contribution for the structure of pinenes, limonene, 1,8-cineole, dipentene, terpineol, pulegol, caryophyllene, and cadinene. The development of synthetic aroma chemicals found in essential oils and extracts started with the synthesis of vanillin in 1874 by W. Haarmann and F. Tiemann. Literature gives only a few information about adulteration and the authors' research is going back to publications from 1919. The company Schimmel & Co. in Leipzig, Germany, was founded in 1829 under the name Spahn & Büttner. In 1879 Schimmel was the first company to establish an industrial laboratory for the production of essential oils. Famous well-known scientists like Wallach (Isolation of Muscon) and Eduard Gildemeister (the famous books of Gildemeister and Hofmann) were there at work. In 1909 Schimmel started publishing the famous "Schimmel-Berichte" (Schimmel reports). These reports about progress in research on essential oils, containing composition of essential oils, isolation of chemicals, cultivation of fragrance plants, and adulteration of essential oils, were discussed. In the report from Schimmel-Berichte (April/October 1919), the following adulterations were reported: "Bergamot oil with phthalates and terpenes, bitter almond oil with raw benzaldehyde, cassia oil with phthalates, cinnamon oil Ceylon with camphor oil, lemon oil with water, lavender oil with phthalates and terpineol, peppermint oil with glycerol esters, menthol, phthalates and spirit, sandalwood oil with benzyl alcohol, star anise oil with fatty oils and rose oil with palmarosa oil and spermaceti."

That points out clearly that already at that time adulteration was omnipresent. Schimmel was the first company to create a synthetic neroli oil on the market. The cause of that was the establishment of a syndicate comprising of the South French growers and, what is more, frost and storm damages during orange flower blooming. In 1920, again in one of the "Schimmel-chronicles" (Schimmel-Bericht April/Mai 1920), the adulteration of lemon oil is reported: "Skilled mingled, with terpenes, sesquiterpenes and citral, the determination with specific gravity and optical rotation are not sufficient to detect adulteration in lemon oil." As a solution for that problem, preparation of as much as possible fractions, tested for the solution in ethanol of 89% Vol., was proposed. Rochussen (1934) reported that several oils were adulterated, as other than the permitted parts of the plant were added. The utilization of turpentine oil was the most used component for adulteration of those essential oils, bearing large amounts of terpenes. Mineral oils (paraffin, petrolatum, benzene, and Vaseline [petrolatum]) were reported to adjust the change of physical data forced by other adulteration chemicals. For the same purpose cedarwood, gurjun balsam, and copaiba balsam oils were applied because of their weak odor.

At that time, physical constants like optical rotation, density, refraction, boiling point, freezing point, residue on evaporation, ester value, acid value, carbonyl value, and water content could give satisfying data about the examined oils. In addition the so-called wet chemistry could give some more answers. Detection reaction, titration, photometry, and gravimetric analysis were used to achieve a more precise result about the genuineness of an essential oil (detailed information will follow). Lawrence (2000) published in Perfumer & Flavorist an original text of E.W. Bovill, RC Treatt, London, about the essential oil market in 1934. One key sentence was as follows: "Essential oils, largely because they are liquids, are easy to adulterate in ways which are difficult to detect and – it must be admitted they very frequently adulterated." At that time it was obvious that adulteration was a fight likely as between a thief and a policeman. Previously, adulteration was done by producers and brokers and customers had to believe the words about purity and nativeness. The only key of trustfulness could be the relationship between producers/traders and clients by permanent control in the harvest and production time locally. Ohloff (1990) confirmed that through hydrodistillation in industrial scale, progress was made in synthesizing compounds with effective methods. Camphor was one of the important chemical components (not only for the flavor and fragrance use but as starting substance for important "plastic" products), together with borneol. Starting with the synthesis of important aromatic chemicals at accessible prices like linalool, geraniol, phenylethanol, and the esters, the adulteration of essential oils started progressively. In a flash, essential oil components were available in sufficient quantities and at prices much cheaper than the essential oil itself. Another event was the problem of the two world wars. Especially in World War II, the acquisition of raw materials of natural or synthetic kind was no more possible at all. The military machine used nearly all raw material resources, chemists were forced to establish substitutes for essential oils, and that resulted in several new synthetic aroma chemicals, which now could be used to stretch or to substitute an essential oil. Thus, the ingenuity of corruption by essential oil producers increased rapidly, and even with general application of chromatographic analysis in industrial laboratories beginning in the 1980s, falsifications could hardly be detected. The authors noticed in the course of a visit to Provence area and producers of lavender and lavandin oil that starting with the analysis by gas chromatography using packed columns in 1979 led to the result that compositions of samples of essential oils were not adequate to literature of that time. By informing the producers about the establishment of chromatographic equipment, the quality was immediately rising. But not all was satisfying. So a "touristic" tour to the facilities of the producers was decided during harvest and production time, end of August 1981. Starting in Puimoisson and with the help of a little blue empty perfume bottle ("flacon montre," used at that time for lavender perfumes), the staff was asked for a little bit of lavender oil. By giving them some Francs, he had the opportunity to watch around the distillation facility. Together with the oil sample, information about special observations was noticed. The same happened in facilities of Riez, Montguers, Simiane, Richerenches, Rosans, Remuzat, Banon, and Apt, and from all production units oil samples could be collected and these were analyzed in the laboratory by gas chromatography. As a result, only one from 10 samples was in accordance with the requirements. The others were adulterated mainly by synthetic linalool and linalyl acetate (recognized by dihydro- and dehydrolinalool as well as dihydro- and dehydrolinalyl acetate) but also with too high contents of limonene, borneol, and camphor (by adding lavandin oil), far too little amounts of lavandulol and lavandulyl acetate. Adulteration by adding those chemicals after distillation could not have happened, as the samples he received were filled directly from the Florentine flask. Finally he was in luck to observe that before closing the cover of the plant, linalyl acetate was spread by means of a portable pump with tank directly on the lavender. The cover was closed and the essential oil now running from the Florentine flask could be seriously called as natural. Interesting was the observation of the distillation facilities. In four cases drums with linalool and linalyl acetate from BASF (Badische Anilin- & Soda-Fabrik) could be detected. Concerning linalool and linalyl acetate, it is worth to mention that 1 year later other drums were occurring from a producer in Bulgaria produced by another synthesis pathway without de- and dihydro components.* At that time it was a customary practice, as reported by Touche et al. (1981), to use relationships of *cis*-β-ocimene to *trans*-β-ocimene (R1) and *trans*-β-ocimene to octanone-3 (R2) as well as linalool + linalyl acetate to lavandulol + lavandulyl acetate (R3). Statistically collected values over many years guaranteed the correct relations.

Another observation was the use of lavender bushels together with some lavandin bushels. This could be transparently seen; they were larger in size and the leaves were characteristic of Lavandin Grosso. At least, there were other chemicals in a rack, such as camphor and geraniol, borneol, terpinen-4-ol, and β-caryophyllene. What should these chemicals be for? He was told that these are for special clients that are so-called comunelles, meaning a mixture of several charges from different farmers but from the same plant was established. The last production unit had constrained not only a laboratory but also a production unit with many chemicals. From that time on the author did no more trust any producer or trade companies.*

Today such a way of working is no more possible. Ordering an essential oil means, with all possible parameters, the properties are scanned and tested and standards must be fulfilled. In spite of all loyalty of a producer/customer relationship, only analytical, physical, and sensory examination is inevitable.

18.3 ADULTERATION

18.3.1 UNINTENDED ADULTERATION

The term "adulteration" must be understood as a negative concept. It is in nearly all cases a criminal and unethical behavior. But there will be the possibility that it is an act of unwanted adulteration without any intent and has nothing to do with dishonesty. Reasons can be the lack of knowledge, bad equipment, wrong treatment of the plant material, not allowed methods for distillation and no good manufacturing practice. Of course, all these things should be avoided, but they happen.

Selection of correct plant material is the basic for the production of essential oils. It is clearly confirmed in every ISO standard, from which botanical source the oil has to be produced. When a cultivation of the biomass is possible, it must be guaranteed that the correct species is used. By using wild collected plants, the possible risk of collecting similar species or chemotypes cannot be excluded. As an example, essential oil of *Thymus serpyllum* L. from Turkey origin was sent for analysis. As a result, the oil was absolutely different from experience and literature. Thymol and carvacrol were too low, and linalool was higher as well as geraniol and geranyl acetate: What could have happened? People collecting the wild thyme took everything that look like a thyme plant without looking for the chemotypes (which is hardly possible by visual selection). Higher quantities of linalool and geraniol chemotypes were present. The oil was natural at all, but it is not to be used in flavor application or aromatherapy. A further example is the petitgrain oil: The petitgrain from Paraguay is not a bitter orange plant as such. *Citrus aurantium* L. is the true bitter orange plant used for production of the essential oil from peel, but also from leaves to obtain petitgrain oil "bigarade." The oil from Paraguayan plants is derived from *Citrus sinensis* L. Pers. × *Citrus aurantium* L. ssp. *amara* var. *pumila*. This is a hybrid from a sweet and bitter orange. The composition of that oil is different from the first because of its higher content of limonene, *trans*-β-ocimene, and mainly linalyl acetate (ISO/DIS 8901 and ISO/DIS 3064 2015). The Paraguayan

---

* Based on the authors' private information and experiences.

oil was sold as bitter orange oil; however, curiously, the oil of petitgrain from Paraguay was much cheaper. A further mistake can be the inappropriate treatment of the biomass. Fresh distilled plant material results in a different final product than the oil from dried or fermented material. The question arises, is it a desired procedure or a mistake or an omission? Good manufacturing practice (GMP) is a standard application procedure in technologically developed countries, but what about emerging nations or those without any technical progress? Can such essential oils be traded or used?

Hydrodistillation is nowadays a highly developed, automated and computerized method to produce essential oils. Applying computer programs for detection- and production parameters observing all conditions. Starting with the plant material (moisture content, degree of maturity, oil content) followed by temperature, steam quantity, steam pressure, cooling temperature, and subsequent treatment, all is handled electronically. In spite of such a technology, mistakes can happen as in the case of distilling *Melissa officinalis* L. from a biological culture. When analyzing that oil by gas chromatography–mass spectrometry (GC-MS), a content of more than 5% of thymol was detected and confirmed three times. After discussing the problem with the producer, he guaranteed that only balm, fresh cut, was used. Contamination in the analyzed sample could be excluded as the second sample showed identical results. After checking all distillation equipment, the reason was found. Before producing Melissa oil, thyme was distilled and he discovered a small chamber within his own constructed cooling unit. There remained about 50–60 mL of oil and it was mixed up with balm oil in the course of hydrodistillation. This is a great financial loss for the producer taking account of the price.* Adulteration can also be done by, for example, adding basic solution to the vessel while distilling. Sometimes it is used to avoid artifacts formation from the plant acids. By raw materials with high-grade acids, for example, barks like massoia, this is sometimes applied. The neutralization of the pH value inhibits the ester formation or degradation reactions in the course of the processing. The use of cohobation is another fact for discussion. This method uses the recovery of volatiles from distillation of water, remaining solved and without separation by Florentine flask. As long as this water is returned to the distillation vessel, there is no problem at all. For some oils like thyme oil, it is of importance to use cohobation, because of the high content of phenols, which dissolve not insignificantly in distilled water. Venskutonis (2002) reports that "cohobation minimizes the loss of oxygenated compounds but increases the risk of hydrolysis and degradation." Similar is the situation with rose oil. Here the cohobation is an important step to receive a genuine reflection of rose odor. Treating the distillation water with salt (NaCl) or pH adjusting to remove plant acids seems to be discussed, as it will still remain an essential oil. By extracting the distillation water with any solvents, which then will be removed by distillation in vacuum, will lead to an extract. This extract, added to the distilled essential oil will result in the loss of the term "essential oil" by definition.! Codistillation is reported for several "essential oils." Mainly sandalwood oils are mentioned to be extracted and redistilled. Valder (2003) reported that the production method of sandalwood oil is sometimes the hexane extraction followed by codistillation with propylene glycol and finally a rectification is done. Baldovini (2010) stated that this method is not used anymore as it does not longer comply with ISO rules. Rajeswara (2007) tested the effect of codistillation using crop weed with several plant materials like citronella, lemongrass, palmarosa, *Corymbia citriodora*, and basil. He observed the increase of yields and also changes in composition. The author was able to observe a similar effect, analyzing bio lavender oil. The oil differed in composition from standards with higher ester content, higher oxidation compounds, and less 3-octanone. The cultivation was bio, without removing any weed. The percentage of weed was esteemed to be more than 30%. The oil did not match ISO standard.* Contaminated containers for production and transport are another possibility of adulteration. Reused drums, not cleaned, are often the cause for worst damage of the essential oil.

---

* Based on the authors' private information and experiences.

Plastic containers contain phthalates as plasticizers and the hydrocarbons solve these. Phthalates contaminate even in smallest traces the oil for the use in natural finished products.

18.3.2 Intentional Adulteration

If anybody adulterates willfully and knowingly, it is a criminal act and never a wangle or minor offense. The target to cheat a customer by supplying an essential oil being not conform to any standard is a felony. There is no excuse for such an unethical behavior. Again, until now, adulteration can be found on essential oil market. Terms like cutting, stretching, blending, bouquetting, or sophistication try to moderate this act. The causes for adulteration are manifold and often related with economical and environmental reasons.

18.3.3 Prices

The cutting of essential oil happens in most cases because of the price. They are subject to falling and increasing prices, depending on crops and the demand but also speculation. The last is often the reason why essential oils come to market with adulteration in every dimension. All economical ambition is the increase of profit, the more the better to fulfill shareholders demand. Essential oil market makes no distinction as it is the same as the law of supply and demand. Speculation was always a reason to ameliorate the process. By hoarding raw materials, the profit can go up in high levels. Lavender oil is an example for that. Most of the farmers in France keep parts of the production in the cellars to wait as far as the new crop starts, until the price seems to be the best. Another example from the authors' experience was the competitive battle with cornmint oil in the 1980s. China and Brazil were the two countries with the highest production at that time. Both increased the cultivation area with the result that in the next season prices fall down tremendously because of oversupply. So Brazil decided to decrease the cultivations and in the next year, prices were going up like a rocket, as China too had decreased production, and demand of the market could not be fulfilled. At that time, most menthol was produced from cornmint and consumption was rising. Stabilization was achieved by supplement of sufficient synthetic menthol and India coming to the market as third production continent. Nowadays speculation has another dimension. Likely nearly all foodstuffs and plant crops are merchandized at the stocks and of course this has influence on the oil market. Increased prices in essential oil market cannot be transferred simply onto the industry of finished products. Mass market and global players will not accept any price advance, so the only solution is cutting or stretching. This situation leads to a competitive market, where adulterated essential oils with certificate of naturalness are sold.

18.3.4 Availability

The market changes from monopole situation for suppliers to buyers market within a few months. Shortage of biomass leads to a rapid increase of prices. The causes will be dryness in the production area, flood during harvest time, and ice and frost like in the case of citrus fruits in California or Florida. Tempests are crucial for destruction of stretches of land as well as deletion caused by insects or microbes. Annual rainfalls in tropic regions and lack of sunshine in Mediterranean region are reasons for partial failure. Unavailability or decreasing availability is a high risk for global players in the fine perfumery industry. The costs of launching a perfume worldwide exceed the amount of $100 million easily and one can imagine that the lack of an important natural raw material and therefore, unavailability of that perfume can lead to ruinous losses. In consequence of that, natural essential oils and also extracts are no more or only in small quantities added to the perfume compounds. It is easy to replace a natural product by a compound. Lemon oil, bergamot oil, rose oil, lavender oil, and many others can be substituted by basic synthetic compositions. Only some special oils like vetiver, sandalwood, patchouli, and orris are never perfect to compensate.

18.3.5 Demand of Clients

The various fields of applications sometimes cause the client industries to require special essential oils. One of the reasons is of course the price. The end-consumer industry settles the prices for the market. Competitive prices have to be adjusted and the problem for the use of essential oils is that prices can change from crop to crop or even from week to week.

18.3.6 Regulations

A huge number of regulations have been established for the protection of consumers worldwide. Cosmetic regulation, flavor regulation, and REACH (*R*egistration, *E*valuation, *A*uthorisation and Restriction of *C*hemicals; the European chemical law) extend to the Globally Harmonized System of Classification and Labeling of Chemicals (GHS). All these provisions cover essential oils, but not per se. They are dealt with their contents. Most oils contain one or more chemicals, classified as toxic, allergic, sensitizing, or carcinogenic. Consequently these oils are no more or only in very limited quantities applicable. As an example, the methyl eugenol must be mentioned. Methyl eugenol is a native ingredient of distilled rose oil. Between 1.5% and 3.5% is found in this essential oil. Considering the European Cosmetic Regulation, the maximum dose in finished leave-on products might not exceed 0.001%, which results in the fact that rose oil is no more applicable in cosmetic products. The same pertains to pimento oil from leaves and berries, oil of myrtle, magnolia flower oil, laurel leaf oil, and bay oil. The demand of the industry is rose oil with nearly no or only traces of methyl eugenol. Applying various methods like fractioning and remixing and chemical reactions it was tempted to fulfill the regulation. The result was not as desired, as the odor was not a rose anymore and the finished product could not be named as "essential oil" or "natural."

To avoid an increase of cases of allergic contact dermatitis in the public, many essential oils nowadays have to labeled as "sensitizing—can cause allergic reaction"! Cosmetic producers as well as discounters with private labels avoid naming and labeling and thus, force the replacement of natural products. Essential oil producers offer natural oils with reduced values of sensitizers and this is not possible. Maybe the essential oils are treated in some way, but then are no more "natural," according to ISO standard.

18.3.7 Aging

It is impossible to characterize aging as adulteration without intent. Changes over storing, either correct or false, always result in quantitative modification of the composition. The formation of oxidative compounds like epoxides, peroxides, and hyperoxides is a dangerous effect. Terpenoids show the properties to be volatile but also to be thermolabile. The mentioned oxidative substances show the negative reaction activities coming in contact with the skin. Sensitizing and allergic reactions are consecutive. Countless components of essential oils are subject to oxidation reaction with atmospheric oxygen. Citrus oils implicate such chemicals. Dugo and Mondello (2011) confirm that the reaction of degradation of sabinene, limonene, γ-terpinene, neral, and geranial in lemon oil results in the enlargement of p-cymene, *cis*-8,9-limonene oxide, (*E*)-dihydro carvone, (*Z*)-carveol, 2,3-epoxy geraniol, and carvone. Particularly the increase of p-cymene in citrus essential oils is a distinct detection for heterodyning and aging. Sawamura et al. (2004) demonstrated in a poster compositional changes in commercial mandarin essential oil and detected in a 12-month test the decrease of limonene (60%) and myrcene (nearly 100%) and the increase of (*Z*)-carveol (from 0.0% to 2.8%) and carvones (from 0.2% to 3.0%). Brophy et al. (1989) could observe degradation in terpinen-4-ol, γ-terpinene, and β-pinene but rise of p-cymene and α-terpineol. Examination of peroxide value seems not to be the suitable tool to recognize aging of essential oils, as some of the reactions are reversible. Eucalyptus oil, rosemary oil, and of course citrus oils have demonstrated that warming of these oils in a closed system at 40°C reduces obviously the peroxide value.

Adulteration of Essential Oils

The most suitable method of recognizing aging is analysis by GC-MS. It has to be the responsibility of producers and traders to take care for the safe quality of their essential oils.

18.3.8 Cupidity

This of course is the main reason for adulteration. Cupidity is defined as "strong desire, especially for possessions or money; greed" (The Free Dictionary n.d. [b]). Although this is the main reason, here is not the place to discuss anything about cupidity as it is human fault.

18.3.9 Simple Sports?

The knowledge about absence of qualified analytical equipment on the side of the client seduces producers as well as the trade to cut essential oils. With the development of new chromatographic methods and an increase in sensitivity as well as selectivity, the inventiveness of the producers shows high wealth of ideas. Again, the race between a thief and a police could have been stopped by the costs of investments in analytical systems, but there seems no end by observing bad qualities on the market.

18.4 POSSIBLE ADULTERATIONS FOR ESSENTIAL OILS

18.4.1 Water

Adding of water is very simple and cheap. This is not possible for all essential oils but is likely for those possessing compounds with high affinity of binding to water. Conifer varieties and citrus oils are examples. Siberian pine (*Abies sibirica* Ledeb.) was supplied in August at 25°C–30°C. Visually no water could be detected. In January the oil came from stock (−5°C) and the quantity of water could be esteemed to 8%. Is that adulteration or natural behavior? Rajeswara Rao et al. (2002) mentioned water contents up to 20%, but that level seems too high by following GMP. Responsible for this effect are monoterpenes, and like conifer oils, citrus oils contain higher quantities of these compounds. Citrus expression techniques use a lot of water to spray away the oil/wax emulsion. Centrifugation will separate water and waxes from essential oil. Unfortunately also aldehydes and alcohols contained in the water/wax phase are removed. Cotroneo et al. (1987) observed this effect by comparing oils from manual sponge method without water and the industrial technical methods. An easy method to detect higher value of water in citrus oil is the visible cloudiness and the deposit of waxes when cooled down to 10°C. A validated method to detect the water content in essential oils is the Karl Fischer titration (ISO 11021, 1999).

18.4.2 Ethanol

Mostafa (1990) reports that ethanol is the main alcohol used in moderate quantities to dilute essential oils. Ethanol is a component of rose oil. The process of water/steam distillation forms alcohol in certain quantities. In the chromatographic profile in ISO 9842, 2003, essential oil of rose, the value of ethanol is specified from 0% to 7%, depending on the origin (maxima: Bulgaria 2%, Turkey 7%, and Morocco 3%). The question arises whether the alcohol is coming directly from the process or if the distilled water is washed and extracted by alcohol and this extract is added to the essential oil. If it is the case it is no more an essential oil. Rose oil is a very expensive product and 1% or 2% of added ethanol increases the profit.

18.4.3 Fatty Oils or Mineral Oils

As mentioned in the introduction, adulteration with fatty oils in history is well known. Rochussen (1920) confirms that and mentions paraffin, petrolatum, and castor oil. The easiest way to detect was the blotting paper test. Essential oils containing fatty oils leave behind a lasting grease spot. By testing the solubility in ethanol of 90%, a blur occurs, showing the presence of fatty oils. Only castor oil is soluble in ethanol but can be detected by evaluation of nonvolatile residue content method. Since more than three decades, such an adulteration became no more apparent.

18.4.4 High Boiling Glycols

For a long time such adulterations remained undiscovered. High boiling materials like polyethylene glycols could not be detected with normal GC-MS method, even at 280°C column temperature. It takes many hours before these chemicals leave the column and give a long, small hill in chromatogram. Not only expensive essential oils like sandalwood, vetiver, and orris were cut with those chemicals but also extracts like rose absolute. By happenstance the author found the stretching of this rose absolute. Typically this product is of deep red color and its use in white cosmetic cream was impossible. A high vacuum distillation was performed to remove the color, which really seemed to be not a problem. The end temperature was 200°C in high vacuum but still about 20% remained in the retort. Never a component of rose absolute with that boiling point was detected or mentioned in the literature before.* At least in 1974, Peyron et al. reported about the use of hexylene glycol up to 40% in vetiver and cananga oils. By treatment with water, separation by rectification, and liquid gas chromatography, it could be detected. In 1991, John et al. reported the thin-layer chromatography method for detection. This method is simple and effective and can be used for routine analysis.

18.4.5 Oils from Other Parts of the Same Species or Other Species with Similar Essential Oil Composition

This is a very simple method to stretch. Clove leaves are easier to harvest and bring similar oil as the stem or bud oil. Clove bud oil is nearly three times higher in price than clove leaf oil. Therewith the concocting of that oil makes sense. Pimento berry oil and leaf oil are closely related in composition. The price of the berry oil is double the leaf oil; consequently, adulteration is interesting. By GC-MS method and regarding minor components, this immixture will be detected. Eucalyptus oil from China is derived from *Cinnamomum longepaniculatum* (Gamble) 1,8-cineole type. Either the whole oil sold as eucalyptus oil or *Eucalyptus* sp. added is clearly not allowed and must be seen as adulteration.

18.4.6 Related Botanical Species

Something of the kind can only happen in the origin, where biomass is growing and distillation is proceeded. Some plants have relatives, which then might have comparable oil compositions. Ylang oil is produced from the flowers of *Cananga odorata* (Lam.) Hook. F. Thomson forma *genuina*. The other form of *Cananga odorata* is cananga oil with *Cananga odorata* (Lam.) Hook. F. Thomson forma *macrophylla* as plant source. On the market, cananga oil is cheaper than ylang oil. Instead of mixing the distilled oils of both, the flowers are distilled together. Furthermore flowers of climbing ylang-ylang *Artabotrys uncinatus* (Lam.) Merill are sometimes added before distillation. When producing patchouli oil, the vessel is filled with leaves, and to avoid that they stick together during processing, branches of the gurjun tree (*Dipterocarpus alatus* Roxb. Ex G. Don and *Dipterocarpus turbinatus* C.F. Gaertn.) are added to avoid that. As a result, the oil of patchouli is contaminated with α-Gurjunene (up to 3%),which is not part of a pure patchouli oil (private information to the author).

---

* Based on the authors' private information.

18.4.7 Fractions of Essential Oils

This is one of the most applied adulteration methods. Fractions arise in all cases, when essential oils or extracts are concentrated, washed out, and rectified or are residues of removal from centrifugation and distillation as well as from recovery of water streams of expressed citrus peels. Heads and tails from rectification are added to similar essential oils, like peppermint terpenes to mint oil. Essential oils that are high in terpene content, cheap, and available on the market are citrus, eucalyptus, *Litsea cubeba*, cornmint and mint, petitgrain, spearmint, vetiver, lavender and lavandin, cedarwood, citronella, clove, and *Corymbia citriodora* terpenes. Especially citrus terpenes from various species are appropriate for stretching. Dugo (1993) recognized the contamination of bitter orange oil by the use of sweet orange and lemon oils and its terpenes. By applying liquid chromatography (LC), high-resolution gas chromatography (HRGC), and GC-MS methods, they confirmed the addition of less than 3% of these components. Limonene is the main component of many citrus oils. On the other side, limonene is a component of nearly all essential oils in variable quantities. Limonene is a chiral terpene; chirality describes a basic property of nature. Some molecules show the property to have three-dimensional structures that possess a chiral center. These molecules show a basic property of nature, to look like image and mirror image, but both cannot be aligned. It is a matter of common knowledge that chemical compounds in essential oils tend to be chiral and the preference of one form can be observed. In contrary synthetic molecules are racemic, meaning both forms of the molecule are balanced. This is a real fingerprint for detection of adulterations with terpenes owing different chiral properties. On normal GC columns a separation of chiral molecules is impossible. With the use of columns coated with modified cyclodextrins, those mirror-imaged molecules will be separated and appear as two different signals. In 1995 bergamot essential oils, sold in 10 mL bottles to end users, were subject for investigation of the German journal ÖKO Test. This magazine, dealing with naturality and safety of consumer products, charged the University of Frankfurt, Prof. Mosandl, to check the authenticity of the oils by chiral separation of linalool and linalyl acetate. Thirty-seven of fifty-three samples could be confirmed not being natural and "pure." In a private letter of Prof. König (University of Hamburg) to the author, he confirmed: "Linalool and linalyl acetate are present in bergamot oil only in pure R-enantiomers. That for bergamot oil—unnatural—(*S*)-linalyl acetate cannot be found even in traces." The invention of this technique by König, Mosandl, Casabianca, and Dugo, improved by Mondello, Bicchi, and Rubiolo, was a huge step to diminish the cases of adulterations in essential oils. Limonene occurs in D-, L-, and DL-form in nature. Limonene in sweet orange oil is always in D-form; most of the pine oils show pure L-forms. For the detection of stretching, other terpenes and terpene alcohols are used like α- and β-thujene, α- and β-pinene, α- and β-camphene, α- and β-sabinene, δ3-carene, linalool, β-borneol, myrtenol, linalyl acetate, menthol, *cis*- and *trans*-terpineol, 1-phenylethanol, nerolidol, terpinen-4-ol, and many more. Whenever the chiral distribution differs from normal proportion, a manual intervention was made. However, the last sentence has to be relativized carefully. Chiral composition of camphor and borneol in natural essential oil of rosemary is very variable, depending on the origin! Another method of stretching is the use of distilled residues of expressed citrus peels. After citrus production process, some quantities of essential oil components still can be found in the "waste products." In addition, Reeve et al. (2002) report of the use of distilled mandarin oil, produced from recovered water streams from spiking and pressing. Because of the contact with acidic juice, the composition of such an oil shows significant difference in composition from peel oil. This oil can be sold as "distilled" mandarin oil, but never be used to be blended with pressed oils. Such a handling must be seen as adulteration and is not covered by ISO definitions.

18.4.8 Natural Isolates

Single chemicals can be derived from essential oils by methods of fractioning and rectification. As this procedure is from a natural source, of course, this chemical must be characterized as natural.

Numerous components are offered on the market like citral from *Litsea cubeba* oil, geraniol from palmarosa oil, linalool from ho oil, coriander oil or lavandin, pinenes from different *Pinus* species, citronellal from *Corymbia citriodora*, cedrol from cedarwood, or even the santalols from different sandalwood species. All these are added to "finish" essential oils. As already mentioned before, synthetic chemicals can no more be applied as enantiomeric separation is a state of the art today and will convict the matter of fact of adulteration. By using synthesized, correct chiral compounds, the detection is hardly to be recognized but with NMR method, but this is an expensive analysis.

18.4.9 Chemically Derived Synthetic Compounds, Which Are Proved to Appear in Nature

Compounds formerly named "nature identical" are to be defined by law since some years as synthetic products. These molecules have been found in nature by analytical proof and are published in an authorized scientific journal. The term "nature identical" is no more valid and allowed in Europe in relation to flavor and fragrance substances. Such molecules are identical with those appearing in nature but are produced by a synthetic process. These processes contain undesired by-products. The use of such synthetic compounds is easy to detect, as by-products from manufacturing can easily be detected by GC-MS systems. On the other hand, chiral separation will help to confirm adulteration.

18.4.10 Steam Distilled Residues from Expression

After an expression process of citrus oils either in exhausted peels or in centrifugation residues, carryovers of the volatiles will remain. These can be removed by distillation with high-strung steam, and oils acquired are colorless and still have the smell of the starting material. Components are similar those of the expressed oils but contain for reason of the production method some oxidized chemicals. Nevertheless these oils are used to adulterate the expressed oils, and with that process the naturality per definition by ISO standard is lost. The main ingredient is limonene, followed by myrcene and γ-terpinene. Traces of aldehydes still can be found. Such adding can be detected by observing higher values of oxidation compounds.

18.4.11 Enzymatically Produced Chemicals (Natural by Law)

Enzyme is defined as "Any of numerous proteins or conjugated proteins produced by living organisms and functioning as biochemical catalysts" (The Free Dictionary n.d. [c]). Enzymes of microorganisms in a medium with added nutrients produced such molecules. Isolation has to be done by any physical process to isolate the desired molecule designed by the biosystem. Although this manufacturing process is authorized within the applicable legal requirements for the use of such enzymes in the EU, it is somewhat uncertain. These processes will as a rule not happen in nature. Within a plant, production of hydrocarbons takes place without microorganisms by conversion of a starting material and by enzymatic reaction within a cell system. The question arises if that molecule from microorganisms can really be named as natural, particularly if it is generated from "any" starters. In the authors' mind, adding such a compound must also be named as adulteration within the definition of essential oil, which does not allow such a process and any additives.

At least it must be mentioned that the mixture of essential oils from the same species but from different geographical sources cannot be called an adulteration. They fulfill the requirements of standards as long as the specific provenance is not laid down in the specification. Lavender oil from France can be mixed up with Chinese, Bulgarian, and Russian origin and is still lavender oil (*Lavandula angustifolia* Mill.), but may not be sold as "French lavender." Furthermore if an essential oil from the same species but from another geographical area is recognizably different in composition, it must be mentioned in standards and certificates. One example is geranium oil from North Africa compared to other origins. 10-epi-γ-Eudesmol is only present in North African oil. The cause for this is adaptation to different conditions in a longer period.

18.5 METHODS TO DETECT ADULTERATIONS

18.5.1 Organoleptic Methods

18.5.1.1 Appearance and Color

Appearance is the visual aspect of a thing or person. In this case it is the appearance of the "essential oil," starting with color, going on with mobility, and finally with the odor itself. The color of essential oils is dependent on the starting material. Citrus oils are colored weak yellowish with lemon oil, light green to darker green with bergamot oil, and orange to brownish red with orange and mandarin oils. The color is dependent on the degree of ripeness of the fruits but they alter with storage and influence of light and warmth. Hereby the age of such an essential oil can be detected. Colors too will appear in hydrodistilled essential oils. The normal case for these oils is colorlessness accompanied by mobile fluidness. Weak yellow color can appear in oils like cardamom, rosewood, tarragon, or turpentine. Oils of lavender and lavandin are pale yellow and mobile; sandalwood oil is almost colorless to golden yellow but viscous. Lemongrass and citronella Java-type oils are pale yellow to yellowish brown; the latter is slightly viscous. Geranium oil has various shades, starting from amber yellow to greenish yellow. Yellow to light brown is the rectified oil of clove leaf, but the crude oil is black. Vetiver and patchouli oils are yellow to reddish brown; both are viscous or even highly viscous. The oils of thyme (*Thymus vulgaris* L. and *Thymus zygis* (Loefl.) L.) are red. The color results from a reaction between thymol and iron of the still. Divergent colors have to be observed critical as these can be a result of aging. Citrus oils can be clear at ambient temperature but becomes cloudy at lower temperatures. Sometimes waxes can undergo precipitation by storage under 4°C.

18.5.1.2 Odor

The odor is the most important factor for the application of essential oils. The highest quantity of usage is in the flavor and fragrance industry. Hatt et al. (2011) confirmed that the human sense of scent is the most important, even if the visual is dominating today. Smelling influences the human being by affecting the limbic system and is responsible for feelings and instincts, to trigger hormonal effects and activate pheromone production. The fact, that the human species has 350 gene receptors for smelling but only 4 for seeing will demonstrate the vital necessity of this sense.

In times of high sophisticated analytical technique with highest resolutions, the human sense of smell is not replaceable. Especially small traces of aromatic chemicals with very low odor threshold are quickly identifiable. Examples are vanillin from vanilla beans and maltol, found in malt and caramel. For perfumer beginners, a single chemical is easy to recognize again. Essential oils are multicomponent mixtures, varying sometimes in wider limits. In spite of electronic noses a well-trained perfumer can recognize more nuances. Common human nose might be duped by adulterations, but not those of professional perfumers. At least the sense of taste is used to recognize adulteration. A good example is rose oil. Part of a drop applied to the tongue and mixture with synthetic geraniol or citronellol reflects in soapy and bitter taste.*

18.5.1.3 Physical–Chemical Methods

These are inevitable in spite of many analytical certificates according to regulations. In history these values have been the only reference points to confirm naturalness. Although the bandwidth is

---

* Based on the authors' private experience.

often too wide, the values can be used as simple and easy to establish proof-samples. Gross mistakes and adulteration can be detected as well as aging of the oil. A series of ISO standards have been established especially for the quality control of essential oils:

Relative density (ISO 279): This is defined as the relation between the mass of a defined volume of substance at 20°C and the mass of a comparable volume of water at the same temperature.

Refractive index (ISO 280): This is the ratio of the sine of the angle of refraction, when a ray of light is passing from one medium into another. Three decimals are obligatory for the result.

Optical rotation (ISO 592): This is the property of defined substances, to turn the plane of polarized light.

Residue on evaporation (ISO 4715), 1978: This is the residue of the essential oil after vaporization on the water quench, using defined conditions and is expressed in percent (*m/m*).

Determination of acid value (1242): This value shows the number of milligrams of potassium hydroxide required to neutralize free acid in one gram of essential oil.

Miscibility with ethanol (ISO 875), 1999: For that key figure, a defined quantity of an essential oil is added to an also defined mixture of distilled water and ethanol at 20°C. As result a blur occures visual. The value before obscuration is the value for a clear solution.

Determination of ethanol content (ISO 17494): This is the method of detecting the quantity of ethanol contained in an essential oil by GC analysis on a suitable column.

Determination of the carbonyl value—free hydroxylamine method (ISO 1271): The determination is done by converting the carbonyl compounds to oximes by reaction with free hydroxylamine liberated in a mixture of hydroxylammonium chloride and potassium hydroxide.

Determination of water content—Karl Fischer method (ISO 11021): A reagent (Karl Fischer reagent without pyridine) is used for the reaction of the absorbed water from an essential oil. The reagent is produced by titration, using a Karl Fischer apparatus. The final result of the reaction is obtained by an electronic method.

Determination of phenol content (ISO 1272), 2000: The transformation of phenolic compounds into their alkaline phenol esters, then soluble in aqueous phase, in a defined volume of essential oil is proceeded. The volume of unabsorbed portion is measured.

Determination of peroxide value (ISO 3960): The peroxide value is the quantity, expressed as oxygen, to oxidize potassium iodide under specific conditions, divided through the mass of testing substance.

Determination of freezing point (ISO 1041): This is the highest temperature during freezing of an undercooled liquid.

Iodine number: Kumar and Madaan (1979) report in their chapter the use of iodine number. This method is a measure of the unsaturation of a substance (as an oil or fat) expressed as the number of grams of iodine or equivalent halogen absorbed by 100 g of the substance (Merriam-Webster, n.d.). By using the iodine monobromide-mercuric acetate reagent for iodination, a much better result could be achieved, as reported by Kumar. With this method adulterations on essential oils can be successfully detected.

Thin-layer chromatography: This method belongs to wet chemistry (see Section 18.5.2).

All earlier-mentioned methods are still involved in the concerning standards and are useful tools to confirm naturalness and purity of essential oils.

18.5.1.4 Calculation of Relationship Coefficient

This method might be derided by some people, but once a relationship is recognized, it will help to detect mixtures with natural fractions or isolates. Touche et al. (1981) presented reference factors for identification of adulteration in French lavender. The ratio between (Z)-β-ocimene and (E)-β-ocimene (R_1), (E)-β-ocimene and 3-Octanone (R_2), and linalool + linalyl acetate to lavandulol + lavandulyl acetate had been determined using GC data of production values over more than 5 years. Schmidt (2003) showed in accordance to that the differentiation of the minor components of lavender oils as the possibility to detect even the provenance of the oil. Of course such a calculation of relationship can be done for many other essential oils. As long as typical minor substances are present, it is a valuable tool for genuineness control.

18.5.2 Analytical Methods

18.5.2.1 General Tests

Before the availability of sophisticated analytical instrumentation, only rather simple tests could be carried out to detect adulterations in essential oils (EOs), and it can be assumed that at that time many falsifications went undetected and adulterations were widespread. To determine the identity and purity of EOs, several elementary methods are described in the European Pharmacopoeia (Ph. Eur.) and ISO Standards (ISO/TC 54).

Physicochemical properties like relative density (Ph.Eur. 7.0/2.02.05.00, ISO 279), refractive index (Ph.Eur. 7.0/2.02.06.00, ISO 280), optical rotation (Ph.Eur. 7.0/2.02.07.00, ISO 592), and flash point (ISO/DIS 3679:2013-08) should be within certain limits and numerical values can be found in the literature, on the Internet, and in ISO standards. Solubility tests can indicate the presence of contaminants (solubility in ethanol ISO 875), for instance, a turbidity of a solution of an EO in carbon disulfide is an indication of water (Ph.Eur. 7.0/02.08.05, ISO 11021). To detect fatty oil adulterations, a drop of the EO is placed on a filter paper and after 24 h no visible grease spot must be left behind (Ph.Eur. 7.0/2.08.07.00). Determination of the evaporation residue serves the same purpose and detects nonvolatile or low volatile compounds (Ph.Eur. 7.0/2.08.09.00, ISO 4715). An instrumentally more demanding method would be thermogravimetry

Basic chemical tests can also provide useful information on the properties of EOs. Ester and saponification value (Ph.Eur. 7.0/2.05.02.00, ISO 7660, 1983), content of free and total alcohol (ISO 1241), carbonyl value (ISO 1279, 1996), and phenol content (ISO 1272, 2000) are helpful indicators of identity and purity of EOs.

Last but not least sensory evaluation is important especially in perfume houses and should be carried out by experts or a panel of fragrance professionals most effectively done with the help of reference samples of known status.

But some parameters alone determined by the earlier-mentioned methods tell us nothing about the composition of the tested EO. Therefore, separation of the oil into their individual components is necessary.

18.5.2.2 Thin-Layer Chromatography

One of the oldest and easiest separation methods is thin-layer chromatography (TLC), which nevertheless has the disadvantage of having a quite low separation efficiency. A drop of the EO is placed at one end of a plate (glass, aluminum foil) coated with a thin layer of an adsorbent (silica gel, diatomaceous earth, aluminum oxide, cellulose) as the stationary phase and placed in a glass chamber filled with a small amount of an appropriate solvent as a mobile phase. The solvent is drawn up the plate through capillary forces and separates the oil into several fractions. These fractions or spots can be made visible by derivatization with a chromophoric reagent or by inspection under UV light. A reference oil of known purity and composition must be analyzed simultaneously or under the same conditions for comparison (Ph. Eur. 7.0/2.02.26.00); (Jaspersen-Schib and Flueck, 1962; Phokas, 1965; Atal and Shah, 1966; Sen et al., 1974; Kubeczka and Bohn, 1985; Nova et al., 1986; John et al., 1991).

18.5.2.3 Gas Chromatography (GC, GLC, HRGC, GC-FID, GC-MS)

The analysis of natural products can be demanding especially in the field of EO research since an EO can contain 300 or more components at a concentration ranging from more than 90% to a few ppm or less of the total oil amount. A separation technique of high efficiency is needed and gas liquid chromatography (GLC) or simply GC is one of the most importantly used analytical methods for volatile compounds and is especially suited in EO and fragrance analysis. With the introduction of capillary columns, the separation performance increased dramatically and theoretical plate numbers of more than 1×10^5 are commonplace and provide gas chromatograms of high resolution (HRGC).

A small amount of the EO (nanograms to micrograms) is injected with a microliter syringe into a closed evaporation chamber (injector) held at elevated temperature (>200°C) where the sample evaporates and the vapor is transferred by an inert carrier gas (N_2, H_2, He, Ar) as the mobile phase to a quartz capillary column of an internal diameter ranging from 0.05 to 0.53 mm and a length between a few meters up to a hundred meters or more. The inner wall of the capillary is coated with a thin film of a liquid phase (polysiloxane, polyglycol) in which separation takes place through permanent absorption and desorption of the sample compounds between liquid and mobile phase. The column is mounted in an oven either held at constant temperature or more often programmed at a constant rising temperature rate and permanently flushed with the carrier gas. The substances pass through the column and leave at a reproducible time (retention time) and then need to be registered by an appropriate detector. The most common and versatile detector for organic compounds is the flame ionization detector (FID) in which the eluted substances are burnt in a tiny hot hydrogen/air flame producing charged carbon ions that are collected on an anode. The resulting current is registered after amplification by an electrometer. Amplitude and signal (peak) area are proportional to the amount of the corresponding compound. Retention times depend on many device parameters, and therefore, instrument-independent retention indices calculated according to Kovats (1958) or Van Den Dool and Kratz (1963) are used instead and can be compared with literature data for substance identification. To verify identification, a second analysis on a column of different polarity must be performed, and retention times or indices on both columns must match the data from the assumed compound or of those from a reference substance. This two-column approach is also useful to detect peak overlapping, which occurs quite often on single-column GC analysis of natural complex substances. Substance amounts are calculated from the individual peak areas as a percentage of the total peak areas of all substances multiplied by a substance-dependent correction factor, which accounts for the different sensitivities of each compound to FID detection.

These correction factors should be determined at least for the major compound found in an EO relative to a standard substance (preferably a *n*-paraffin) whose factor is typically set to 1. Unfortunately many EO compounds are not available commercially in pure form or easily synthesized, so for convenience all correction factors are set to 1 to a first approximation.

Chromatographic profiles and percentage composition of all important commercial essential oils and many oil-bearing plants can be found in the ISO standards and the literature of essential oil research (Formácek, 2002; Kubezka and Formácek, 2002; Adams, 2007), and genuine EOs should fit into this image within certain limits. Chemical profiling in this manner can detect adulterations if additional or missing peaks are found or percentage composition deviates substantially. Addition of synthetic compounds can in some cases be disclosed by the presence of by-products or impurities left behind from the synthesis pathway. Dihydro linalool, for example, is an intermediate in the industrial production of linalool and is always present in small amounts ($\ll 1\%$), and if a trace of this substance is found in an EO, it is a clear indicator of an adulteration with synthetic linalool.

However, fingerprinting of EOs by GC with FID detection alone cannot reveal the chemical identity of detected peaks if any deviations from the expected profiles occur. In that case further information is needed besides the chromatographic data, and mass spectrometry provides detailed information on the structure of the separated compounds. High-resolution gas chromatography coupled with a mass spectrometric detector (HRGC-MS or simply GC-MS), most commonly quadrupole ion trap detectors in EO analysis, together with sophisticated chromatographic software and special mass spectral libraries of essential oil components (Adams, 2007; König, Joulain, Hochmuth 2004, Mondello, 2011) separates and identifies most components of an EO. Since some classes of compounds show very similar mass spectra, like some groups of mono- and sesquiterpenes, retention indices must be taken into account as a second criterion for an unequivocal identification. Usually a GC-MS run is performed for identifying the EO components and a second GC-FID run for peak area, respectively, for percentage composition determination. Normally this is done on two different instruments. Identical capillary columns must be used in both GCs, and device parameters must be adjusted properly to obtain closely similar chromatographic profiles for both detectors to facilitate peak allocation between the two chromatograms. Peaks identified in the mass spectrometry (MS) chromatogram must be correctly assigned in the FID chromatogram for peak integration. Sometimes this is proving difficult since separation on two instruments is never exactly equal especially if one column ends up in a high vacuum (MS) and the other at atmospheric pressure (FID). Therefore, a series of closely eluting peaks may not be resolved in the same manner on both columns. To overcome this problem, an FID-MS splitter can be used since here the separation takes place on one GC column and the effluent is split to both detectors, MS and FID. To detect peak overlapping, the same procedure should be undertaken on another capillary column of different polarity. In EO analysis GC separations are preferably performed on 95% dimethylpolysiloxane/5% diphenylpolysiloxane and on Carbowax 20M columns. In the end you come up with two FID and two MS chromatograms, each of the two carried on different capillary columns. These results in two analyses that should for the most part coincide and accept the overlapping peaks on the other GC column. EO analysis performed in this way confirms the chemical composition of an EO and detects adulterations with exogenic substances if they are amenable by GC, that is, volatile. Adulterations with gas chromatographic undetectable substances like very high boiling vegetable or mineral oils can be disclosed by a change in the percentage composition or a too low total peak area. Specific marker compounds or a diastereomeric isomer distribution can also be used for authentication of EO (Teisseire, 1987).

In some unusual EOs not all detected peaks can be identified by MS library search because of missing entries but, with the necessary experience, it can often be estimated in which group of natural substances these peaks belong by looking at the mass spectrometric pattern. All in all an extensive experience in EO analysis and a thorough knowledge of the scientific literature is needed to evaluate the authenticity of an EO and detect adulterations. If this is the case the analyst should be in a position to disclose falsifications with cheaper or inappropriate EOs, synthetic natural substances, solvents, or other unnatural chemicals.

18.5.2.4 Chiral Analysis (Busch and Busch, 2006)

Many EOs contain substances with asymmetric carbon atoms, that is, there are two molecular forms of these molecules that behave like image und mirror image and cannot be brought into alignment. This is the reason why EOs rotate the oscillating plane of light and the resulting overall optical rotation is measured by a polarimeter. However, the enantiomeric distribution of single substances cannot be determined in this way because separation is needed in the first place. This will be enabled by GC, and additional separation of optically active compounds into their enantiomers can be achieved if a chiral selector is added to the stationary phase of the GC column. Cyclodextrin derivatives have proven to be very useful and chiral capillary columns with different optically active cyclodextrin selectors are commercially available. The enantiomeric separation should be done at rather low temperature since the interaction between the chiral molecules and the chiral selector is rather weak; therefore, the temperature gradient of the GC program should be low and the carrier gas flow is higher than the optimal flow. If the appropriate chiral GC column is used, the enantiomeric distribution of certain EO components can be determined, and since nature often prefers one enantiomer over the other, the determination of the enantiomeric excess is a valuable tool to detect adulterations by synthetic components that are racemic and change the enantiomeric ratio. The optical purity of authentic EOs can be found in the literature (Mosandl, 1998; Busch and Busch, 2006; Dugo and Mondello, 2011).

18.5.2.5 GC-GC and GC×GC (Two-Dimensional Gas Chromatography, ^{2}D GC)

Gas chromatography is one of the most widely deployed methods in analytical chemistry to investigate organic sample material due to its simple ease of use, the ready availability of sophisticated inexpensive instrumentation, and the large amount of qualitative and quantitative information that can be retrieved if the appropriate configuration is employed. Especially the high separation efficiency for volatiles makes GC very suitable to investigate complex mixtures and sample matrices. But for some applications, the separation performance is not sufficient when it comes to very complex mixtures like odors, flavors, crude oil products, and foodstuff. Co-elution with other analytes or sample matrix elements causes problems in detection and quantitation especially when the analytes differ greatly in their concentration. This problem can be solved by cutting out the co-eluting part of the chromatogram and a subsequent second chromatographic separation of the excised effluent preferably on a stationary phase of different polarity. This technique, called heart-cutting or two-dimensional GC (^{2}D GC or GC-GC), is done with the help of a diverting valve or a Deans switch. The sought-after substances are then resolved in the second GC column.

An even more elaborate technique recently developed not only cuts out one or several parts of the first GC column but virtually cuts the whole 1D chromatogram into small equal pieces (each several seconds long), refocuses each effluent, and separates it on a short second GC column within a very short time (several seconds). Refocusing, which here means stopping the effluent of the first column for several seconds onto a very small area and then releasing the focused substances into the second column, is done with a cryogenic modulator, and this procedure is repeated until all substances are eluted from the first GC column. The overall separation efficiency is calculated by multiplying the theoretical plate numbers of the two columns, which results in rather large figures and a separation efficiency, which cannot be achieved by simple GC alone. Instruments and the appropriate data acquisition software for this comprehensive ^{2}D GC (or GC×GC) are now commercially available. Instead of a 1D chromatogram, a 2D contour plot is generated, and since the "half-widths" of substances are very small, detection must proceed very fast, so in case a mass spectrometer is used as a detector, a fast scanning quadrupole or a time-of-flight MS (TOF-MS) must be chosen. Such an instrument arrangement like GC×GC-TOF-MS leaves almost nothing to be desired for EO analysis (Marsili, 2010).

18.5.2.6 ^{13}C NMR (Nuclear Magnetic Resonance)

Another useful method to validate the identity of an EO offers the spectroscopy of the magnetic properties of ^{13}C nuclei. ^{13}C NMR spectroscopy is a very useful technique regularly used to elucidate the structure of individual substances and has been applied by Kubeczka and Formácek (2002) to a large number of essential oils and individual reference compounds. Even though it is not very common to apply this technique to substance mixtures, as it is with an EO, the spectrum of a genuine oil is very distinct and characterized by the chemical shifts, signal multiplicities, and intensities of all components. Functional groups can be identified by chemical shifts and provide information on chemical structures of individual substances within the oil. No separation of the oil into its components is necessary, and therefore, the approach is simply measuring a solution of the oil in an NMR tube and comparing the spectrum with literature data or reference oils and compounds. Unfortunately NMR instruments are rarely found in traditional analytical laboratories, and thus, this method has not gained the significance in the field of EO analysis it deserves.

18.6 IMPORTANT ESSENTIAL OILS AND THEIR POSSIBLE ADULTERATION

18.6.1 Ambrette Seed Oil

This is a very expensive oil. The main ingredients are (*E*,*E*)-farnesyl acetate (≈60%) and ambrettolide (≈8.5%). Synthetic ambrettolide is used as a fixative in perfumery, is nearly odorless but has exalting properties. Compared to natural ambrette seed oil, the price of ambrettolide is only 10%. Detection of naturality can be done by isotope ratio mass spectrometry (IRMS).

18.6.2 Amyris Oil

ISO standard 3525 shows character and data for this oil. Blending (professional term within dealers and perfumers for adulterating) is done by Virginia cedarwood oil, α-terpineol, and copaiba balm. Also, elemol distilled from elemi resin is used. Detection is done by GC-MS.

18.6.3 Angelica Oils

The fruit oil contains up to 76% of β-phellandrene and α-pinene (13%) as main ingredients. As no chiral values for the β-phellandrene are described in literature, it must be assumed that adulteration is done by this compound. β-Phellandrene is naturally available by geranyl diphosphate cycling (BRENDA, BRaunschweig ENzyme DAtabase) 2007. For adulteration copaiva balm, gurjun balsam, lovage root oil, and amyris oil were used in the past. Cheap α-pinene can also be used to "improve" the composition.

18.6.4 Anise Fruit Oil

ISO standard 3475 shows character and data for that oil. This oil often is produced not only from the fruits but also from the whole aerial part. Values of *cis*- and *trans*-anethole are then reduced in smaller amount. Adulteration with star anise oil can be easily detected. Anise fruit oil does not contain any foeniculin, but star anise oil contains up to 3% (ISO 11016). On the other hand pseudo-isoeugenol-2-methylbutyrate is a component of anise fruit oil (0.3%–2%, ISO 3475) and does not appear in star anise oil. The midratio of *cis*-anethole to *trans*-anethole in anise seed oil in relation to star anise oil is 0.3%:0.6%. Synthetic anethole, fennel terpene limonene (80°), and terpineol were used for blending.

18.6.5 Armoise Oil

This oil is often blended with α- and β-thujones from cheaper sources like *Thuja orientalis*. Furthermore camphor or white camphor oil and camphene are used. GC-MS method is used to detect these adulterations.

18.6.6 Basil Oils

ISO standard 11043 shows character and data for the methyl chavicol–type oil. Synthetic methyl chavicol is used to adulterate that oil and can be detected by NMR. Basil oil from linalool type will be adulterated by synthetic linalool but can easily be detected by chiral separation. Casabianca (1996) found the minimum value *R*-(−)-linalool with 99.8% and is congruent with the authors' results.

18.6.7 Bergamot Oil

ISO standard 3520 shows character and data for that oil. As mentioned in the text, this oil was adulterated with synthetic linalool and linalyl acetate but also with bergamot terpenes and distilled oil from bergamot peel residues. Blending was done with limonene (80°), nerol, geranyl acetate, petitgrain oil Paraguay, rosewood oil, and citral from *Litsea cubeba* oil. This oil was subject to many studies with chiral separation. König, 1992, Mosandl, 1991, Dugo, 1992, and Casabianca, 1996a, reported similar values for linalool and linalyl acetate. (*R*)-(−)-linalyl acetate is present always in purity higher than 99.0% and (*R*)-(−)-linalool with a minimum of 99.0% too. (*S*-pinene value varies between α)-(−)- 68.0% and 71.1%, (*S*-pinene between 91.1% and 92.6%β)-(−)-, and (*R*)-(+)-limonene between 97.4% and 98.0% (Juchelka and Mosandl, 1996a). Blending is done by terpinyl acetate, citral synthetic, *n*-decanal, *n*-nonanal, nerol, limonene (80°), and sweet orange terpenes. Dugo and Mondello (2011) published the following data for chiral ratios: (*S*)-thujene (0.5%–1.3%):(α)-(+)-(*R*)-thujene (98.7%–99.5%); (α)-(−)-(*R*)-(+)-α-pinene (31.0%–36.1%):(*S*)-(−)-α-pinene (63.9–69.0); (1*S*,4*R*)-(−)-camphene (85.7%–92.7%):(1*R*,4*S*)-(+)-camphene (7.3%–14.3%); (*R*)-(+)-β-pinene (7.6%–10.3%):(*S*)-(−)-β-pinene (89.7%–92.4%); (*R*)-(+)-sabinene (13.7%–19.8%):(*S*)-(−) (80.2%–86.3%); (*R*)-(−)-α-phellandrene (43.1%–54.7%):(*S*)-(+)-α-phellandrene (45.3%–56.9%); (*R*)-(−)-β-phellandrene (24.1%–36.9%):(*S*)-(+)-β-phellandrene (63.1%–75.9%); (*S*)-(−)-limonene (1.2%–2.1%):(*R*)-(+)-limonene (97.9%–98.8%); (*R*)-(−)-linalool (97.8%–99.5%):(*S*)-(+)-linalool (0.5%–2.2%); (*S*)-(−)-citronellal (>98%):(*R*)-(+)-citronellal (<2%); (*R*)-(−)-linalyl acetate (99.1%–99.9%):(*S*)-(+)-linalyl acetate (0.1%–1.0%); (*S*)-(+)-terpinen-4-ol (22.4%–44.7%):(*R*)-(−)-terpinen-4-ol (55.3%–77.6%); (*S*)-(−)-α-terpineol (14.0%–68.5%):(*R*)-(+)-α-terpineol (31.5%–86.0%); (*S*)-(−)-citronellol (12.0%–20.0%):(*R*)-(+)-citronellol (80.0%–88.0%); (*S*)-(−)-terpinyl acetate (36.0%–44.0%):(*R*)-(+)-terpinyl acetate (56.0%–64.0%).

18.6.8 BITTER ORANGE OIL

ISO standard 3517 shows character and data for this oil. In the past, sweet orange oil was used as well as orange terpenes and distilled bitter orange residues from production for adulteration. Limonene of high purity from other citrus fruits was also applied. Dugo (2011) shows the similarity of the components between bitter and sweet orange oil. McHale et al. (1983) report the use of grapefruit oil, as it contains higher concentrations of coumarins and psoralens. Today the adding of purified components from other citrus sources is used. Mingling up with sweet orange oil can be detected by measuring the δ-3-carene and camphene content. As bitter orange oil contains only traces of δ-3-carene, the sweet oil goes up to 0.1%. Also, the ratio δ-3-carene/camphene can be used for detection (Dugo et al., 1992). Chiral values for components are reported by Dugo and Mondello (2011): (*R*)-(+)-α-pinene (89.7%–97.4%):(*S*)-(−)-α-pinene (2.6–10.3); (1*S*,4*R*)-(−)-camphene (35.8%–47.6%):(1*R*,4*S*)-(+)-camphene (52.4%–64.2%); (*R*)-(+)-β-pinene (6.1%–7.9%):(*S*)-(−)-β-pinene (92.1%–93.9%); (*R*)-(+)-sabinene (49.4%–80.6%):(*S*)-(−)-sabinene (19.4%–50.6%); (*R*)-(−)-α-phellandrene (60.1%–74.9%):(*S*)-(+)-α-phellandrene (25.1%–39.9%); (*R*)-(−)-β-phellandrene (0.6%–5.7%):(*S*)-(+)-β-phellandrene (94.3%–99.4%); (*S*)-(−)-limonene (0.5%–0.8%):(*R*)-(+)-limonene (99.2%–99.5%); (*R*)-(−)-linalool (61.2%–89.8%):(*S*)-(+)-linalool (10.2%–38.8%); (*S*)-(−)-citronellal 42.5%:(*R*)-(+)-citronellal (57.5%); (*R*)-(−)-linalyl acetate (99.2%–99.4%):(*S*)-(+)-linalyl acetate (0.6%–0.8%); (*S*)-(+)-terpinen-4-ol (67.5%–71.5%):(*R*)-(−)-terpinen-4-ol (28.5%–32.5%); (*S*)-(−)-α-terpineol (6.6%–29.8%):(*R*)-(+)-α-terpineol (93.4%–70.2%).

18.6.9 BITTER ORANGE PETITGRAIN OIL

ISO standard 8901, 2010 shows character and data for this oil. This is also known as "petitgrain oil bigarade." Adulteration is done by using petitgrain oil from Paraguay, heads and tails from fractioning of that oil (to receive the petitgrain oil "terpene free"), and rectified orange terpenes and limonene. Blending is done with synthetic linalool, linalyl acetate, geraniol, geranyl acetate, and α-terpineol. Detection is done by GC-MS as well as by multidimensional enantiomeric separation systems. Juchelka (1996) published the following chiral ratio values: (*R*)-(−)-linalyl acetate (98.0%–99.1%):(*S*)-(+)-linalyl acetate (0.2%–2.0%); (*R*)-(−)-linalool (66.4%–90.2%):(*S*)-(+)-linalool

(9.8%–33.6%); (S)-(−)-α-terpineol (26.4%–28.4%):(R)-(+)-α-terpineol (71.6%–73.6%). Mondello and Dugo (2011) reported following chiral ratios: (R)-(+)-α-pinene (6.7%–12.0%): (S)-(−)-α-pinene (93.3%–88.0%); (R)-(+)-β-pinene (0.1%–1.1%): (S)-(−)-β-pinene (98.8%–99.9%); (S)-(−)-limonene (29.2%–39.2%): (R)-(+)-limonene (60.8%–70.8%); (R)-(−)-linalool (66.4%–90.2%):(S)-(+)-linalool (9.8%–33.6%); (R)-(−)-linalyl acetate (93.4%–99.1%):(S)-(+)-linalyl acetate (0.9%–6.6%); (S)-(+)-terpinen-4-ol (47.9%–67.4%):(R)-(−)-terpinen-4-ol (32.6%–52.1%); (S)-(−)-α-terpineol (27.5%–28.4%):(R)-(+)-α-terpineol (71.6%–72.5%).

18.6.10 Cajeput Oil

The oil obtained from *Melaleuca cajuputi* Powell is the natural variety. To adulterate this oil, other species like *Eucalyptus* ssp. or *Cinnamomum camphora* (1,8-cineole type) are used. Eucalyptus terpenes, α-phellandrene, and α-terpineol were used for blending. Detection is done by GC/MS with the smaller components like α-selinene, α-humulene, and α-terpineol.

18.6.11 Camphor Oil

This oil is produced from *C. camphora* Sieb. wood in China and Taiwan. Many chemotypes and also varieties from *Cinnamomum* species exist, but the true camphor oil is meant. This oil is rarely adulterated because of the cheap price but, fractioned and enriched in camphor up to 90%, is used to mix up camphor containing oils like rosemary, Spanish sage, and spike lavender. Smallest traces of safrole (a carcinogenic compound) shows the use of *C. camphora* higher boiling fractions.

18.6.12 Cananga Oil

ISO standard 3523 shows character and data for the oil, produced from *C. odorata* (Lam.) Hook. F. et Thomson forma *macrophylla*. It can be adulterated by linalool from lavandin oil showing nearly equal chiral values and β-caryophyllene from clove oil fractions. Blending is done with benzyl acetate, linalool, α-terpineol, geraniol, α-terpinyl formate, methyl benzoate, and allo-ocimene, all synthetically produced further from Virginia cedarwood oil, copaiva balm oil, clove leaf oil, and gurjun balsam oil. Cananga oil is used to adulterate ylang-ylang oil, as yield is higher and the flowers have a higher distribution in nature. Detection is done by GC/MS and by multidimensional chiral separation. Bernreuther et al. (1991) mentions the ratio for linalool with (S)-(+) 2% and (R)-(−) 98%.

18.6.13 Caraway Oil

ISO standard 8896 shows technical data for that oil. Synthetic carvone is used for adulteration. European pharmacopoeia requires contents of (+)-limonene 35.5%–45.0%, (+)-carvone 50.0%–65.0%, and (−)-carvone with a maximum of 0.7% for a pure quality. ISO 8896 standard evaluated a content of *cis*-dihydrocarvone from 0.3% to 1.2%.

18.6.14 Cardamom Oil

ISO standard 4733 shows character and data for that oil. This oil is produced mainly in India, Sri Lanka, and Guatemala. Adulteration is done by 1,8-cineole from eucalyptus or camphor oil, α-terpinyl acetate, and linalool. The adding of 1,8-cineole is hard to be discovered; linalool ratio is reported by Dugo (1996) and Casabianca (2011) to be between 7% of (R)-(−)-linalool to 93% of (S)-(+)-linalool and 100.0% (R)-(−)-linalyl acetate to 0.0% (S)-(+)-linalyl acetate. The presence of δ-terpinyl acetate seems to show adulteration as to nowadays it is only detected in clary sage oil. δ-Terpinyl acetate is a side product in synthetic terpinyl acetate.

18.6.15 Cassia Oil

ISO standard 3216 shows character and data for the Chinese type oil. Main component is *trans*-cinnamaldehyde. Synthetic cinnamaldehyde as well as coumarin is used for adulteration. This oil is often used for the adulteration of cinnamon bark oil. That is easy to detect, as coumarin is not a component of that oil. If *o*-methoxy cinnamaldehyde is found in cinnamon bark oil, it is a sign for adulteration with cassia oil. The naturality of the cinnamaldehyde can be detected by the combination of GC–combustion– IRMS (GC-C-IRMS) and GC-P-IRMS.

18.6.16 Cedar Leaf Oil

Adulteration is done by adding thujones from other species. According to Ravid (1992), the chiral value of (−)-fenchone is 100%. The thujones were found to be (−)-α-thujone 90% and (+)-α-thujone 10% (Gnitka 2010).

18.6.17 Cedarwood Oils

ISO standard 9843 shows character and data for Virginia cedarwood oil. This is obtained from *Juniperus virginiana* L., whereas the Texas oil is distilled from *Juniperus ashei* J. Buchholz. If these two oils are mixed together, the status of essential oil per definition is lost. Another adulteration is adding cedrol of cheaper cedarwood from Chinese oil (*Chamaecyparis funebris* (Endl.) Franco). Production is done by cooling down the oil and separating the crystals by filtration. As by-products, the terpenes will also serve as adulterations for other cedarwood oils. This is only valid for cedarwood oils coming from the species *Juniperus*. Recognition of adulterations is done by GC-MS.

18.6.18 Celery Seed Oil

ISO standard 3760 shows character and data for this oil. Blending is done with α-terpineol, limonene from orange oil, rectified copaiba oil, lovage root oil, and amyris oil. Detection is best done by GC-MS.

18.6.19 Chamomile Oil Blue

ISO standard 19332 shows character and data for that oil. Adulteration is done mainly by synthetic α-bisabolol, chamazulene, and (*E*)-βfarnesene. The method developed by Carle et al. (1990) and Carle (1996) using $\delta^{13}C$ and δD isotopes with IRMS is a real tool for naturality assay.

18.6.20 Chamomile Oil Roman

This essential oil, distilled from *Chamaemelum nobile* (L.) All., possesses a series of angelates in various concentrations. Bail (2009) published the following values for angelates and specific esters: isobutyl angelate 32.1%, 2-methylbutyl angelate 16.2%, isobutyl isobutyrate 5.3%, methyl 2-methylbutyrate 1.9%, prenyl acetate 1.4%, 2-methylbutyl 2-methylbutyrate 1.2%, and 2-methylbutyl acetate 1.2%. As most of these compounds are available as synthetic chemicals, adulteration can be done easily. Detection is done by the combination of GC-C-IRMS.

18.6.21 Cinnamon Bark Oil

Only in some pharmacopoeia monographs this oil is dealt with a chromatic profile. Adulteration is done once by adding cassia oil. This can easily be detected by any quantity of coumarin, never to be found in pure cinnamon bark oil. On the other hand, adulteration by synthetic cinnamaldehyde will be detected by the combination of GC-C-IRMS and GC-P-IRMS as the assessment of synthetic and natural cinnamaldehyde is possible. Sewenig et al. (2003) reported that the chiral ratio of linalool is (R)-(−) 95% and (S)-(+) 5%.

18.6.22 Cinnamon Leaf Oil

ISO standard 3524 shows character and data for this oil. As eugenol is the main component, rectified clove leaf oil or isolated eugenol from this oil is used. Also β-caryophyllene, synthetic benzyl benzoate, and *trans*-cinnamaldehyde are added. Blending is done by α-terpineol. Casabianca (1998) reported the ratio of (R)-(−)-linalool and (S)-(+)-linalool to be 64.0% and 36.0%, respectively.

18.6.23 Citronella Oil

ISO standard 3849 shows character and data for the Sri Lanka–type oil and ISO standard 3848 for the Java-type oil. Synthetic citronellal, citronellol, and geraniol were used. In addition citronella oil terpenes were used to cover such adulterations. Lawrence (1996a) mentions the ratio of citronellal in Java type with R-(+) enantiomer is 90%. Casabianca (1996) found chiral ratio of (R)-(+)-citronellol and (S)-(−)-citronellol to be 75.0%–79.0% and 21.0%–25.0%, respectively.

18.6.24 Clary Sage Oil

The chiral ratio of (R)-(−)-linalool and (S)-(+)-linalool is 80.6%–94.0% and 6.0%–19.4%, respectively, and of (R)-(−)-linalyl acetate and (S)-(+)-linalyl acetate is 93.0%–98.1% and 1.9%–7.0%, respectively, dependent on fresh or ensilaged biomass, found by Casabianca (1996).

18.6.25 Clove Oils

ISO standards 3141, 3142, and 3143 dealing with the oil from buds, leaves, and stems show character and data for these oils. The best and most expensive oil is from buds. Adulterations with oils from leaves and stem are used for adulteration. For those other oils, residues from isolation of natural eugenol are applied. Sometimes β-caryophyllene from other sources was used too. Detection can be made by GC-MS analysis.

18.6.26 Coriander Fruit Oil

ISO standard 3516 shows character and data for this oil. It was subject to adulteration with linalool, as it is the main component. By using chiral separation, Braun and Franz (2001) found the chiral ratio for linalool as follows: (S)-(+) from 64.8% to 87.3% and (R)-(−) from 12.7% to 35.2%, while Casabianca (1996) found (S)-(+) 87% to (R)-(−) 13%. Chiral ratio in limonene according to Casabianca (1996) is (R)-(+) 62.0%–93.0% and (S)-(−) 7.0%–38.0%.

18.6.27 Corymbia Citriodora Oil

ISO standard 3044 shows character and data for this oil. Adulteration is done by synthetic citronellal as well as by citronellol. Detection is done by GC-MS but better by multidimensional chiral separation. The chiral ratio of linalool is (R)-(−) 100.0% and (S)-(+) 0.0%, citronellal (R)-(+) 55.6%–57.2% and (S)-(−) 42.8%–44.4%, and citronellol (R)-(+) 49.0%–54.0% and (S)-(−) 46.0%–51.0%, found by Casabianca (1996).

18.6.28 Corn Mint Oil

ISO standard 9776 shows character and data for this oil. Adulterations were carried out using synthetic menthol, menthyl acetate and menthone, terpenes from fractioning of corn mint oil, and limonene. Detection is done by enantiomeric separation by GC-MS. Mosandl (2000) reported the following chiral ratios: (1S)-(α)-pinene (56.5%–73.5%):(1R)-(α)-pinene (26.5%–43.5%); (+)-(1S)-(−)-β-pinene (49.1%–55.6%):(1R)-(+)-β-pinene (44.6%–50.9%); (1S)-(−)-limonene (98.1%–99.9%):(1R)-(+)-Limonene (0.1%–1.9%); (1R,3R,4S)-(−)-menthol (min 99.9%):(1S,3S,4R)-(+)-menthol (max 0.1%); and (4R)-(−)-piperitone 21.0%:(4S)-(+)-piperitone <0.1%.

18.6.29 Cumin Fruit Oil

ISO standard 9776 shows character and data for this oil. The main components are cumin aldehyde, p-mentha-1.3-dien-7al, p-mentha-1,4-dien-7al, γ-terpinene, and β-pinene. Blending is done by orange terpenes, p-cymene, and piperitone. Detection is done by GC-MS and by the combination of GC-C-IRMS and GC-P-IRMS as assessment of synthetic and natural cuminaldehyde is possible.

18.6.30 Cypress Oil

Adulteration is done by either turpentine oil or α- and β-pinene. Further, blending is done using δ-3-carene and cedrol from cedarwood Chinese type. AFNOR (1992) presents data in the standard NF T 75-254. Using chiral GC as his method of analysis, Casabianca (1996) determined that the enantiomeric ratio of α-thujene in cypress oil was as follows: (1R)-(+)-α-thujene (45%):(1S)-(−)-α-thujene (55%).

18.6.31 Dill Oils

Dill weed oil is dominated by α-phellandrene, limonene, and carvone. Dill ether and the absence of dill apiol are further criteria for that oil. Dill seed oil contains mainly carvone and dihydrocarvone. Adulteration is done using phellandrenes, distilled limonene coming from orange terpenes, synthetic carvone, and dihydrocarvone. Detection is done by 2D enantiomeric separation. Lawrence (1996) reports the following ratios for dill seed oil: (+)-limonene 98.4%:(−)-limonene 1.6%; (+)-carvone 98.7%:(−)-carvone 1.3%; (+)-trans-carveol 33.3%:(−)-trans-carveol 66.7%; and (+)-cis-carveol 100%:(−)-cis-carveol 0%. The authors own findings from biocultivated oil was (+)-carvone 98.4%:(−)-carvone 1.6%; (S)-(−)-α-pinene 4.0%:(R)-(+)-α-pinene 96.0%; (+)-limonene 95.4%:(−)-limonene 4.6%; (S)-(−)-β-phellandrene 0%:(S)-(+)-β-phellandrene 100%; and (R)-(−)-α-phellandrene 100%:(R)-(+)-α-phellandrene 0%.

18.6.32 Dwarf Pine Oil

ISO standard 21093 shows character and data for this oil. In the past, addition of turpentine oil was used for adulteration; later the essential oil of the needles of *Pinus maritima* was used, as the composition showed very close data of compounds. Chiral analysis gives helpful results. Kreis et al. (1991) reported chiral ratio in monoterpenes as follows: (1R,5R)-(+)-α-pinene (44%):(1S,5S)-(−)-α-pinene (56%); (1R,5R)-(+)-β-pinene (2%):(1S,5S)-(−)-β-pinene (98%); and (4R)-(+)-limonene (40%):(4S)-(−)-limonene (60%).

18.6.33 Elemi Oil

ISO standard 10624 shows character and data for this oil. Elemol and elemicin are the lead compounds. As minor component 10-epi-γ-eudesmol must be detected between 0.2% and 0.3% to ensure naturness. Adulteration is done by limonene, α-phellandrene, and sabinene. Detection is done by GC-MS.

18.6.34 Eucalyptus Oil

ISO standard 770 shows character and data for this oil. Schmidt (2010) reported the Chinese eucalyptus oil coming from Sichuan province and is derived from *C. longepaniculatum* (Gamble). This must be labeled and it is not correct to mix this up with true eucalyptus varieties. Adulteration is done by 1,8-cineole from various *Cinnamomum* varieties. The detection is not easy as all these adulterations are natural.

18.6.35 Fennel Oil Sweet

Although this oil should be produced solely from fruits, more often the whole aerial parts are used. Adulteration is done by synthetic anethole or from other sources like star anise oil. Blending is done too with star anise oil and limonene + 60°. Analysis is done by GC-MS or by multidimensional enantiomeric separation. According to Ravid (1992), the chiral ratio of (+)-fenchone is 100%–0%.

18.6.36 Fennel Oil Bitter

ISO standard 17412 shows character and data for this oil. It is often mixed up with the herb oil derived from bitter fennel and sweet fennel produced from the whole aerial parts of the plant. Blending is done by sweet fennel oil, star anise oil, and fennel terpenes. Analysis is done by GC-MS or by multidimensional enantiomeric separation. Ravid reported that the chiral ratio of (+)-fenchone is 100%:0%. Casabianca (1996b) confirms a chiral ratio of (4R)-(+)-α-phellandrene 100%:(4S)-(–)-α-phellandrene 0%.

18.6.37 Geranium Oils

ISO standard 4731 shows character and data for that oil. This important and high-price product is and was often the target for adulteration. Blending was done with synthetic geraniol, citronellol, limonene—60°, terpinyl formate, synthetic rhodinol, α-terpineol cristallin, and distilled bergamot oil. Lawrence (1996) showed results of chiral separation of citronellol, *cis-* and *trans-*rose oxide, menthone, isomenthone, and linalool for geranium oils from different origins. The chiral ratio of (R)-(–) linalool and (S)-(+)-linalool is 42.0%–55.0% and 45.0%–58.0% found by Casabianca (1996b). In 1996 he published the following chiral data: (from various origins) (R)-(–)-citronellol (18.0%–43.0%) and (S)-(+)-citronellol (50.0%–82.0%); (2R,4S)-(+)-*cis*-rose oxide (24.0%–38.0%):(2S,4R)-(–)-*cis*-rose oxide (62.0%–76.0%); (2R,4R)-(–) *trans*-rose oxide (70.0%–76.0%):(2S,4S)-(+)-*trans*-rose oxide (72.0%–76.0%); (1S,4R)-(+)-menthone (>99.0%):(1R,4S)-(–)-menthone (<1%); and (1S,4S)-(–)-isomenthone (>99.0%):(1R,4R)-(+)-isomenthone (<1.0%); Lawrence (1999) confirmed further the values: (2R,4S)-(+)-*cis*-rose oxide (35.5%–49.3%):(2S,4R)-(–)-*cis*-rose oxide (50.7%–64.5%); (2R,4R)-(–) *trans*-rose oxide (49.2%–62.4%):(2S,4S)-(+)-*trans*-rose oxide (37.6%–50.8%); and (2S,4R)-(–)-*cis*-rose oxide ketone (50.5%–62.6%):(2R,4S)-(+)-*cis*-rose oxide ketone (37.4%–49.5%).

18.6.38 Grapefruit Oil

ISO standard 3053 shows character and data for this oil. Pure oils possess as marker the compound nootkatone from traces up to 0.8%, depending on the fruit status. This compound is used for blending, together with *n*-octanal, *n*-nonanal, *n*-decanal, and synthetic citral. Adulteration is performed by orange terpenes and distilled grapefruit residues from expression and limonene—80°. Detection must be done exclusively by multidimensional enantiomeric separation. Dugo and Mondello (2011) published the following chiral data: (R)-(–)-α-pinene (0.3%–0.8%):(S)-(+)-α-pinene (99.2%–99.7%); (R)-(+)-β-pinene (62.0%–76.8%):(S)-(–)-β-pinene (23.2%–38.0%); (R)-(+)-sabinene (98.4%–98.5%):(S)-(–)-sabinene (1.5%–1.6%); (S)-(–)-limonene (0.5%–0.6%):(R)-(+)-limonene

(98.4%–98.5%); (R)-(–)-linalool (32.0%–43.0%):(S)-(+)-linalool (57.0%–68.0%); (S)-(–)-citronellal (16.6%–21.4%):(R)-(+)-citronellal (57.0%–68.0%); (S)-(–)-α-terpineol (1.2%–3.3%):(R)-(+)-α-terpineol (96.7%–99.8%); and (S)-(+)-carvone:(R)-(–)-carvone 34.8%.

18.6.39 Juniper Berry Oil

ISO standard 8897 shows character and data for this oil. *J. communis* oil is often mixed up with *J. oxycedrus*. As marker for that the myrcene content is rising up. Real markers are germacrene D- and δ-cadinene. The sesquiterpene fraction gives more information. Further on, addition of fractions of juniper berry oil from rectification as well as adding juniper branches oil is made. Kartnig et al. (1999) published some chiral data comparing self-distilled and commercial qualities of juniper berry α- and β-pinene, limonene, and terpinen-4-ol oils. Chirality was recognized as useful components for quality control of that oil. Mosandl et al. (1991) report a ratio for (S)-(–)-α-pinene 77%:(R-)-(+)-α-pinene 23%.

18.6.40 Lavandin Oils

ISO standard 3054 and 8902 shows character and data for Abrial and Grosso lavandin oil. This oil is adulterated by adding acetylated lavandin, lavandin distilled heads and tails, camphor oil white, Spanish sage oil, and spike lavender oil. Blending is done by terpinyl acetate, turpentine oil, methyl α-terpineol and ethyl amyl ketone, hexyl ketone, and geranyl acetate, all from synthetic source. Chiral ratio of linalool is from (R)-(–) 64.8% to 87.3% and (S)-(+) from 12.7% to 35.2%. Chiral ratio of linalyl acetate according to the findings of Casabianca (1996a) is (R)-(–) from 98.1% to 100.0% and (S)-(+) from 0.0% to 1.0%). Renaud (2001) reported chiral ratios for linalool and linalyl acetate as follows: (3R)-(–)-linalool (94.5%–98.2%):(3S)-(+)-linalool (1.8%–5.5%) and (3R)-(–)-linalyl acetate (99%–100%):(3S)-(+)-linalyl acetate (0%–1%). Lawrence (1996a) published the following chiral ratios: (3R)-(–)-linalyl acetate (98.3%–100%):(3S)-(+)-linalyl acetate (0.0%–1.7%); (3R)-(–)-linalool (95.0%–96.6%):(3S)-(+)-linalool (3.4%–5.0%). (2S,5S)-*trans*-linalool oxide (4.2%–23.3%):(2R,5R)-*trans*-linalool oxide (76.7%–95.8%); (2R,5S)-*cis*-linalool oxide (82.9%–95.8%):(2S,5R)-*trans*-linalool oxide (4.2%–17.1%); (R)-(–)-lavandulol (96.2%–99.0%):(S)-(+)-lavandulol (1.0%–3.8%); and (R)-(–)-terpinen-4-ol (1.6%–10.9%):(S)-(+)-terpinen-4-ol (89.1%–98.4%).

18.6.41 Lavender Oil

ISO standard 3515 shows character and data for oils from various origins. Most applied adulterations were already discussed in the text. Chiral separations showed the most effective results in analysis. The ratio of linalool is (R)-(–) 98.0%–100.0% and (S)-(+) 0.0%–2.0% found by Casabianca (1996a). He too found that the chiral ratio of linalyl acetate is (R)-(–) from 98.0% to 100.0% and (S)-(+) from 0.0% to 2.0%. Stoyanova reports the following data for Bulgarian lavender oil: (3R)-(–)-linalyl acetate (100%):(3S)-(+)-linalyl acetate (0%); (3R)-(–)-linalool (95.0%–96.6%):(3S)-(+)-linalool (3.4%–5.0%); and (3R)-(+)-camphor (27.4%–52.2%):(3S)-(–)-camphor (47.8%–78.6%). Kreis and Mosandl (1992) published the following data: (2S,5S)-*trans*-linalool oxide (3.9%–23.3%):(2R,5R)-*trans*-linalool oxide (76.7%–96.1%); (2R,5S)-*cis*-linalool oxide (82.9%–95.8%):(2S,5R)-*trans*-linalool oxide (4.2%–17.1%); (R)-(–)-lavandulol (89.8%–100%):(S)-(+)-lavandulol (0%–10.2%); (R)-(–)-terpinen-4-ol (0%–10.9%):(S)-(+)-terpinen-4-ol (89.1%–100%); and (3R)-(–)-linalool (95.1%–98.2%):(3S)-(+)-linalool (18%–4.9%).

18.6.42 Lemon Oil

ISO standard 855 shows character and data for that oil. Adulteration is done by distilled lemon oil from residues, orange terpenes or limonene from orange terpenes, and synthetic citral from

Litsea cubeba oil. Lemon oil washed as residues from production of terpene-free oil is preferably used, as these contain still all components of the pure lemon oil. Also lemon terpenes and heads of distilled grapefruit oils could be found. Blending is done by using synthetic decanal, nonanal, octanal, and citronellal from *Corymbia citriodora* oil. Detection is made by GC-MS and mainly by multidimensional enantiomeric separation with various methods (see part of methods). Mondello (1998) reports some constituents with chiral ratios as follows: (R)-(+)-β-pinene 6.3%:(S)-(−)-β-pinene 93.7%; (R)-(+)-sabinene 14.9%:(S)-(−)-sabinene 85.1%; (S)-(−)-limonene 1.6%:(R)-(+)-limonene 98.4%; (S)-(+)-terpinen-4-ol 24.7%:(R)-(−)-terpinen-4-ol 75.3%; and (S)-(−)-α-terpineol 75.2%:(R)-(+)-α-terpineol 75.2%. Further on, Dugo and Mondello (2011) gave the following data: (R)-(+)-α-pinene (25.5%–31.5%):(S)-(−)-α-pinene (68.5%–74.5%); (1S,4R)-(−)-camphene (86.2%–92.4%):(1R,4S)-(+)-camphene (7.6%–13.8%); (S)-(−)-β-pinene (93.2%–95.7%):(R)-(+)-β-pinene (4.3%–6.8%); (R)-(+)-sabinene (12.4%–15.0%):(S)-(−)-sabinene (85.0%–87.6%); (R)-(−)-α-phellandrene (46.9%–52.6%):(S)-(+)-α-phellandrene (47.4%–53.1%); (R)-(−)-β-phellandrene (31.1%–53.9%):(S)-(+)-β-phellandrene (46.1%–68.9%); (S)-(−)-limonene (1.4%–1.6%):(R)-(+)-limonene 98.4%–98.6%); (R)-(−)-linalool (52.0%–74.5%):(S)-(+)-linalool (25.5%–48.0%); (S)-(−)-citronellal (89.5%–94.8%):(R)-(+)-citronellal (5.2%–10.5%); (S)-(+)-terpinen-4-ol (12.0%–26.2%):(R)-(−)-terpinen-4-ol (73.8%–88.0%); and (S)-(−)-α-terpineol (66.4%–82.0%):(R)-(+)-α-terpineol (18.0%–33.6%).

18.6.43 Lemongrass Oil

ISO standard 4718 shows character and data for this oil. Adulteration is done by adding synthetic citral or citral from *Litsea cubeba* oil. Blending is done with addition of geranyl acetate and 6-methyl-5-heptene-2-one. Detecting is done by GC-MS and multidimensional chiral separation. Wang et al. (1995) reported chiral ratio for linalool to be (3S)-(+)-linalool 30.9%:(3R)-(−)-linalool 69.1% and (R)-(−)-linalool 58.0%:(S)-(+)-linalool 42.0%.

18.6.44 Lime Oil Distilled

ISO standard 3809 shows character and data for that oil. Adulteration is done by adding limonene from different sources, synthetic terpineol and γ-terpinene from lime terpenes as well as from heads of the production of terpene-free lemon oil. Detection must be done by multidimensional chiral separation. Dugo and Mondello (2011) report the following data for chiral ratios: (S)-(−)-β-pinene (96.0%–96.8%) (R)-(+)-β-pinene (3.2%–4.0%); (S)-(−)-limonene (5.5%–8.7%):(R)-(+)-limonene (91.3%–94.5%); (R)-(−)-linalool (49.8%–80.0%):(S)-(+)-linalool (50.0%–50.2%); (S)-(+)-terpinen-4-ol (42.3%–45.0%):(R)-(−)-terpinen-4-ol (55.0%–57.7%); and (S)-(−)-α-terpineol (53.3%–56.8%):(R)-(+)-α-terpineol (46.7%–43.2%).

18.6.45 Lime Oil Expressed

ISO standard 23954 shows character and data for this oil. Blending is done by adding limonene from different sources, citral from *Litsea cubeba* oil and γ-terpinene from lime terpenes as well as from heads of the production of terpene-free lemon oil. Detection is done by GC-MS, looking for absence of δ-3-carene (not even in traces) and multidimensional enantiomeric separation with various methods (see part of methods). Dugo and Mondello (2011) reported the following chiral ratios: (S-)-(−)-(β)-pinene (98.7%–90.9%):(R)-(+)-β-pinene (9.1%–10.3%); (R)-(+)-sabinene (18.2%–23.4%):(S)-(−)-sabinene (76.6%–81.8%); (S)-(−)-limonene (0.4%–2.7%):(R)-(+)-limonene (97.3%–99.6%); (R)-(−)-linalool (54.4%–69.3%):(S)-(+)-linalool (30.7%–45.6%); (S)-(+)-terpinen-4-ol (18.6%–24.9%):(R)-(−)-terpinen-4-ol (75.1%–81.4%); and (S)-(−)-α-terpineol (74.5%–80.8%):(R)-(+)-α-terpineol (19.2%–25.5%).

18.6.46 LITSEA CUBEBA OIL

ISO standard 3214 shows character and data for this oil. Adulteration is done by adding synthetic citral and can be detected by GC-MS by checking the values for geranial and neral but also for isogeranial and isoneral.

18.6.47 MANDARIN OIL

ISO standard 3528 shows character and data for this oil. α-Sinensal is a marker for mandarin essential oil. Adulteration is made by synthetic methyl-*n*-methyl anthranilate, as well as methyl anthranilate. Orange terpenes and limonene (80°), dipentene, citronellal, and citral are used for blending. Detection is done by GC-MS but improved results show multidimensional chiral separation. Dugo and Mondello (2011) reported the following chiral data: (*S*)-(+)-α-thujene (0.3%–1.9%):(*R*)-(−)-α-thujene (98.1%–99.7%); (*R*)-(+)-α-pinene (41.7%–54.5%):(*S*)-(−)-α-pinene (45.5–58.3); (1*S*,4*R*)-(−)-camphene (31.8%–72.6%):(1*R*,4*S*)-(+)-camphene (27.4%–68.2%); (*R*)-(+)-β-pinene (87.8%–99.1%):(*S*)-(−)-β-pinene(0.9%–12.2%);(*R*)-(+)-sabinene(71.3%–83.5%):(*S*)-(−)(16.5%–28.7%); (*R*)-(−)-α-phellandrene (44.3%–55.0%):(*S*)-(+)-α-phellandrene (45.0%–55.7%); (*R*)-(−)-β-phellandrene (0.4%–3.0%):(*S*)-(+)-β-phellandrene (97.0%–99.6%); (*S*)-(−)-limonene (1.5%–2.91%):(*R*)-(+)-limonene (97.1%–98.5%); (1*R*,4*R*)-(+)-camphor (17.0%–36.5%):(1*S*4*R*)-(−)-camphor (63.5%–83.0%); (*R*)-(−)-linalool (13.0%–22.7%):(*S*)-(+)-linalool (77.3%–87.0%); (*S*)-(−)-citronellal 3.9%–9.2%):(*R*)-(+)-citronellal 90.8%–96.1%); (*S*)-(+)-terpinen-4-ol (9.5%–23.8%):(*R*)-(−)-terpinen-4-ol (76.2%–90.5%); and (*S*)-(−)-α-terpineol (66.1%–75.9%):(*R*)-(+)-α-terpineol (24.1%–33.9%).

18.6.48 MELISSA OIL (LEMON BALM)

This expensive essential oil was and is often the target of adulteration: citronella oil (*Cymbopogon winterianus* or *Cymbopogon nardus*), lemongrass oil, lemon oil citral, and geraniol rose oxides, natural and synthetic. Lawrence (1996) reports the chiral values of (*R*)-(+)-citronellal in that oil being 97.2%–98.2%. Also the value for (*R*)-(+)-methyl citronellate is at a minimum of 99.0% By using GC analysis with different chiral GLC phases, Schultze (1993) confirmed citronellal from other sources as follows: lemongrass oil (*S*)-(−) 30%–55%, (*R*)-(+) 70%–45%; citronella oil (*S*)-(−) 10%–15%, (*R*)-(+) 90%–85%; and catnip oil (*S*)-(−) 98%–99.9%, (*R*)-(+) 2%–0.1%. Schultze also published δ-values to confirm naturality.

18.6.49 MENTHA CITRATA OIL

The chiral ratio of linalool is (*R*)-(−) 42.0%–55.0% and (*S*)-(+) 45.0%–58.0% found by Casabianca (1996a). Chiral ratio of linalyl acetate according to the findings of Casabianca (1996) is (*R*)-(−) from 98.6% to 99.0% and (*S*)-(+) from 1.0% to 1.4%.

18.6.50 MOUNTAIN PINE OIL

ISO standard 21093 shows character and data for that oil. *Pinus maritima* shows nearly identical values and adulteration is hardly to recognize. As long as the price of the oil of *P. maritima* was one-fifth of the *Pinus mugo*, it was sold in large quantities as mountain pine oil. Now prices are nearly identical and mixing up makes no sense anymore. Adulteration is done by α-pinene, β-pinene, δ-3-carene, (−)-limonene, myrcene, β-phellandrene, and *l*-bornyl acetate from various sources.

18.6.51 NEROLI OIL

ISO standard 3517 shows character and data for that oil. Adulteration is made by geraniol from palmarosa oil, linalool from rose wood oil, orange oil sweet, citral from *Litsea cubeba* oil, and petitgrain oil Paraguay. Blending is done by methyl-*n*-methyl anthranilate, methyl anthranilate, synthetic phenyl ethyl alcohol, and synthetic indol. Detection is done by GC-MS and multidimensional chiral separation. Casabianca (1996a) found the chiral ratio for linalool with (*S*)-(+) 12.0–13.8 and (*R*)-(−) 86.2%–88.0% and for linalyl acetate with (*S*)-(+) 1.8–5.0 and (*R*)-(−) 95.0%–98.2%. Dugo and Mondello (2011) reported the following chiral ratios: (*R*)-(+)-α-pinene (2.2%–13.6%):(*S*)-(−)-α-pinene (86.4%–97.8%); (*R*)-(+)-β-pinene (0.1%–0.8%):(*S*)-(−)-β-pinene (99.2%–99.9%); (*S*)-(−)-limonene (1.9%–6.9%):(*R*)-(+)-limonene (93.1%–98.1%); (*R*)-(−)-linalool (70.8%–81.5%):(*S*)-(+)-linalool (18.4%–29.2%); (*R*)-(−)-linalyl acetate (95.4%–98.2%):(*S*)-(+)-linalyl acetate (1.8%–4.6%); (*S*)-(+)-terpinen-4-ol (36.0%–37.6%):(*R*)-(−)-terpinen-4-ol (62.4%–64.0%); (*S*)-(−)-α-terpineol (28.1%–39.8%):(*R*)-(+)-α-terpineol (60.2%–71.9%); and (*R*)-(−)-(*E*)-nerolidol (0.4%–1.8%):(*S*)-(+)-(*E*)-nerolidol (98.2–99.6).

18.6.52 Nutmeg Oil

ISO standard 3215 shows character and data for this oil. Adulteration is done by monoterpenes α- and β-pinene, sabinene from different sources, and α- and β-phellandrene as well as synthetic linalool, terpinen-4-ol and α-terpineol. In the past, safrole from sassafras oil was used for blending. Detection is done by GC-MS, looking for the quantities of safrole and myristicin; in addition 2D chiral separation is recommended. König et al. (1992) reported the following chiral ratios: (+)-α-thujene 10.3%:(−)-α-thujene 89.7%; (−)-α-pinene 79.3:(+)-α-pinene 20.7%; (−)-camphene 100%:(+)-camphene 0%; (+)-β-pinene 41.9%:(−)-β-pinene 58.1%; (−)-α-phellandrene 0%:(+)-α-phellandrene 100%; (+)-sabinene 89.5%:(−)-sabinene 10.5%; (+)-δ-3-carene 0%:(−)-δ(−)-3-carene 100%; (−)-β-phellandrene 7.7%:(+)-β-phellandrene 92.3%; and (−)-limonene 60.9%:(+)-limonene 39.1%.

18.6.53 Orange Oil Sweet

ISO 3140 shows character and data for this oil. Adulteration is done by adding orange terpenes or purified limonene. Casabianca (1996) found chiral ratio for linalool with (*S*)-(+) 86 to (*R*)-(−) 14. Hara et al. (1999) reported chiral ratios of several hydrocarbon components as follows: (1*R*,5*R*)-(+)-α-pinene (99.6%–99.7%):(1*S*,5*S*)-(−)-α-pinene (0.3%–0.4%); (1*R*,5*R*)-(+)-β-pinene (66.1%–77.8%):(1*S*,5*S*)-(−)-β-pinene (22.2%–33.9%); (4*R*)-(+)-limonene (99.1%–99.4%):(4*S*)-(−)-limonene (0.6%–0.9%); (3*S*)-(+)-linalool (82.5%–96.3%):(3*R*)-(−)-linalool (3.7%–17.5%); (4*R*)-(+)-α-terpineol (97.0%–98.4%):(4*S*)-(−)-α-terpineol (1.9%–3.0%); and (3*R*)-(+)-citronellal (31.3%–87.8%):(3*S*)-(−)-citronellal (12.2%–68.7%).

18.6.54 Origanum Oil

This oil is from *Origanum vulgare* L. ssp. *hirtum*. Adulteration is done with synthetic thymol and carvacrol or with limonene from different sources. The chiral ratio of linalool is (*R*)-(−) 82.0% and (*S*)-(+) 18.0% found by Casabianca (1996).

18.6.55 Palmarosa Oil

ISO 4727 shows character and data for this oil. As the main component is geraniol, adulteration is done by synthetic geraniol. Detection must be done by the combination of GC-C-IRMS and GC-P-IRMS as the assessment of synthetic and natural geraniol is possible.

18.6.56 Parsley Oil

ISO standard 3527 shows character and data for this oil. Adulteration is done by turpentine oil or pure α-pinene, β-pinene, and elemicin from elemi resinoid. 1,2,3,4-Tetramethoxy-5-allyl benzene is a key compound up to 12% as well as apiol. Blending is done by celery grain oil, nutmeg oil, and carrot seed oil. Detection is done by GC-MS system.

18.6.57 Pine Oil Siberian

ISO 10869 shows character and data for this oil. Adulteration is done by turpentine oil, *l*-bornyl acetate synthetic, terpinyl acetate, and Virginia cedarwood oil. Detection is done either by GC-MS or chiral separation. Ochocka (2002) determined the enantiomeric ratios of four monoterpene hydrocarbons in *A. sibirica* oils (Korean source): (1R,5R)-(+)-α-pinene 25.4%:(1S,5S)-(−)-α-pinene 74.6%; (1R,5R)-(+)-β-pinene 13.2%:(1S,5S)-(−)-β-pinene 86.8%; (4R)-(+)-limonene 6.8%:(4S)-(−)-limonene 93.2%; and (3R)-(+)-camphene 4.8%:(3S)-(−)-camphene 95.2%.

18.6.58 Patchouli Oil

ISO standard 3757 shows character and data for this oil. This very complex oil is adulterated by gurjun balm oil (see text). Blending is done by patchouli terpenes, cedarwood oil, pepper oil, white camphor oil, and guaiac wood oil. Detection is done by GC-MS.

18.6.59 Pepper Oil

ISO standard 3061 shows character and data for this oil. Blending is done with turpentine oil, α-phellandrene from other sources, limonene from orange terpenes, and clove leaf oil terpenes. The chiral ratio of linalool is (R)-(−) 81.0%–89.0% and (S)-(+) 11.0%–19.0% found by Casabianca (1996a). König et al. (1992) reported the following chiral ratios: (+)-α-thujene 100%:(−)-α-thujene 0%; (−)-α-pinene 74.6%:(+)-α-pinene 25.4%; (−)-camphene 66.6%:(+)-camphene 33.4%; (+)-β-pinene 2.6%:(−)-β-pinene 97.4%; (+)-δ3-carene 2.6%:(−)-δ3-carene 97.4%; (−)-α-phellandrene 0%:(+)-α-phellandrene 100%; (−)-β-phellandrene 100%; and (−)-limonene 61.7%:(+)-limonene 38.3%.

18.6.60 Peppermint Oil

ISO standard 856 shows character and data for that oil. Adulteration is done by synthetic menthol, menthol from cornmint oil, and fractions of peppermint terpenes. Detection is done by chiral separation using 2D enantiomeric columns on GC-MS system. The chiral ratio of 3-octanol is (R)-(−) 94.1%–100.0% and (S)-(+) 0.0%–5.9% according to values found by Casabianca (1996b). Mosandl (2000) reports the following ratios: (1S)-(−)-α-pinene (45.1%–68.1%):(1R)-(+)-α-pinene (31.9%–54.9%); (1S)-(−)-β-pinene (41.7%–53.6%):(1R)-(+)-β-pinene (46.4%–58.3%); (4S)-(−)-limonene (74.4%–98.3%):(4R)-(+)-limonene (1.7%–25.6%); (1R,3R,4S)-(−)-menthol (min 99.9%):(1S,3S,4R)-(+)-menthol (max 0.1%); and (4R)-(−)-piperitone 2.0%–13.0%:(4S)-(+)-piperitone 87.0%–98.0%.

18.6.61 Petitgrain Oil Paraguay Type

ISO standard 25157 shows character and data for this oil. A marker for natural quality is (Z,Z)-farnesol with 2.0%–3.5%. Adulteration is done by limonene (80°) and rectified orange terpenes. The chiral ratio of linalool is (R)-(−) 71.4%–73.3% and (S)-(+) 26.7%–28.6% found by Casabianca (1996a).

18.6.62 Pimento Oils

ISO standard 4729 shows character and data for the oil from *Pimento dioica* (L.) Merr. (Pimento oil Jamaica type). ISO standard 3045 shows character and data for the oil of *Pimenta racemosa*

(Mill.) J.W. Moore (Bay oil). The difference between the oils is the content of eugenol and myrcenol. Pimento berry oil shows a content of more than 80% of eugenol and traces of myrcene, while bay oil contains maximum 56% of eugenol but up to 30% of myrcene. Eugenol from cinnamon leaf oil is added as adulteration for both oils and myrcene from other sources. Detection is done by GC-MS.

18.6.63 Rose Oil

ISO standard 9842 shows character and data for oils from various sources. Blending is done with synthetic phenyl ethyl alcohol, synthetic rhodinol, and geraniol from palmarosa oil. Geranium oil, ylang oil, rose absolute, and palmarosa oil are used for "finishing." Detection has to be done by GC-MS system and by chiral separation with multidimensional GC-MS. Kreis and Mosandl (1992) report the enantiomeric ratios for (S)-(−)-citronellol with >99%; (2S,4R)-cis-rose oxide as well as (2R,4R)-cis-rose oxide show a purity higher than 99.5%.

18.6.64 Rosemary Oil

ISO standard 1342 shows character and data for oils from various sources. Turpentine oil, synthetic camphor, and limonene from orange terpenes are used to blend this oil. Adulteration is done by 1,8-cineole from eucalyptus or white camphor oil. Detection is usually done by GC-MS but also by multidimensional chiral separation. Enantiomeric ratio of linalool is reported by Casabianca (1996a) to be (R)-(−) 23 to (S)-(+) 77. Kreis (1991a) published the following values of chiral separation: (−)-borneol (84.6%–97.8%):(+)-borneol (2.2%–15.4%) and (−)-isoborneol (29.1%–53.8%):(+)-isoborneol (46.2%–70.9%). König et al. (1992) reported the following chiral ratios: (+)-α-thujene 33.7%:(−)-α-thujene 66.3%; (−)-α-pinene 41.7%:(+)-α-pinene 58.3%; (−)-β-pinene 84.4%:(+)-β-pinene 15.6%; (−)-camphene 83.3%:(+)-camphene 16.7%; (−)-α-phellandrene 0%:(+)-α-phellandrene 100%; (−)-β-phellandrene 2.5%:(+)-β-phellandrene 97.5%; and (−)-limonene 64.8%:(+)-limonene 35.2%. The authors own results analyzing pure oils showed the following ratios: (+)-β-pinene 26.0%:(−)-β-pinene 74.0%; (−)-α-pinene 42.0%:(+)-α-pinene 58.0%; (S)-(−)-sabinene 66.5%:(R)-(+)-sabinene 33.5%; and (S)-(−)-camphene 31.4%:(R)-(+)-camphene 68.6%.

18.6.65 Rosewood Oil

ISO standard 3761 shows character and data for this oil. Blending is done by synthetic linalool, α-terpineol, geraniol, and heads of rosewood oil. Eremophilane is a marker according to Lawrence (1999) with values of 0.3%–0.9%. Detection is done by GC-MS on a chiral column. Casabianca (1996a) found chiral ratio for linalool with (S)-(+) 50.0%–51% and (R)-(−) 49.0%–50%. Lawrence (1999) reported the following chiral ratios: (3R)-(−)-linalool 10.0%:(3S)-(+)-linalool 90.0%. (For information, rosewood leaf oil ratio is (3R)-(−)-linalool 22.2%:(3S)-(+)-linalool 77.8%).

18.6.66 Sage Oil (*Salva officinalis*)

ISO standard 9909 shows character and data for this oil. As the main components α-thujone β-thujone are known, adulteration is done with thuja oil or cedar leaf oil. β-Caryophyllene, 1,8-cineole, and borneol from other sources are used. α-Humulene is a marker with up to 12% total content. Detection is done by GC-MS analysis.

18.6.67 Sage Oil Spanish Type

ISO standard 3526 shows character and data for this oil derived from *Salvia lavandulifolia* Vahl. Adulteration is done by eucalyptus oil, camphor oil chemotypes like 1,8-cineole and camphor, α-pinene, and limonene from different sources. Blending is done with synthetic linalyl acetate und terpinyl acetate. Detection can be achieved by using GC and GC-MS analytical equipment.

18.6.68 Sandalwood Oil

ISO standard 3518 shows character and data for the oil derived from *Santalum album* L. This very expensive oil with source from India is no longer sold on the market these days. Today other varieties from various sources like New Caledonia (*Santalum austrocaledonicum* Vieill. and *Santalum spicatum* (R.Br.) A. DC) from Australia are used in flavor and fragrance industry. α- and β-Santalols are the main components and responsible for the fragrance. Adulteration will take place with cheaper varieties, coming from Australia. This is very simple to detect as *Santalum spicatum* contains *cis*-nuciferol in high amounts. The same can be seen with *S. austrocaledonicum*, where the *cis*-lanceol can be found in high amounts compared to the other oils. Braun et al. (2005) found in analyzing *S. album* oil the following chiral substances and values: (1R,4R,5S)-α-acorenol (0.22%), (1R,4R,5R)-β-acorenol (0.11%), (1R,4S,5S)-epi-α-acorenol (0.13%), and (1R,4S,5R)-epi-β-acorenol (<0.01%).

18.6.69 Spearmint Oils

ISO 3033-1-3033.4 shows character and data for these oils from various varieties and hybrids. Adulteration is done by synthetic levo-carvone. Detection is done by enantiomeric separation by GC-MS. Coleman (2002), Lawrence and Cole (2011) report the chiral ratio of monoterpenes in native spearmint to be (1R,5R)-(+)-α-pinene-(40.3%):(1S,5S)-(−)-α-pinene (59.7%); (1R)-(+)-camphene (99.9%):(1S)-(−)-camphene (0.1%); (1R,5R)-(+)-β-pinene (48.7%):(1S,5S)-(−)-β-pinene (51.3%); and (4R)-(+)-limonene (1.9%):(4S)-(−)-limonene (98.1%). Further on, Nakamoto (1996) gave results about *cis*- and *trans*-carveol as follows: (2R,4S)-(+)-*trans*-carveol (15%):(2S,4R)-(−)-trans-carveol (85%) and (2S,4S)-(+)-cis-carveol (4%):(2R,4R)-(−)-*cis*-carveol (96%). Mosandl (2000) reports the following chiral ratios: (1S-)-(−)-α-pinene (59.7%–62.4%):(1R)-(+)-α-pinene (37.6%–40.3%); (1S)-(−)-β-pinene (51.3%–52.1%):(1R)-(+)-α-pinene (98.1%–98.8%); (−)-camphene (<0.1%):(+)-camphene (>99.9%), and (4S)-(+)-carvone (99.1%–99.9%):(4R)-(−)-carvone (0.1%–0.9%).

18.6.70 Spike Lavender Oil

ISO standard 4719 shows character and data for that oil. Adulteration is done by white camphor oil, 1,8-cineole distilled from eucalyptus oil, synthetic camphor, and linalool. Blending is done with terpenes from eucalyptus oil, turpentine oil, *n*-bornyl acetate, lavandin, rosemary oil, HO leaf oil, and α-terpineol. Detection can be made by GC-MS and by multidimensional chiral separation. Ravid (1992) mentions the chiral ratio of terpinen-4-ol as (4S)-(+)-terpinen-4-ol 93%:(4R)-(−)-terpinen-4-ol 7%.

18.6.71 Star Anise Oil

ISO standard 11016 shows character and data for the oil. This oil is not the target for adulteration but is itself an oil for the adulteration of oils containing *trans*-anethole. For the detection of possible adulteration by limonene and α-pinene trace components, *cis*- and *trans*-bergamotene and foeniculin are markers with quantities according to the standard.

18.6.72 Tarragon Oil

ISO standard 10115 shows character and data for this oil. Adulteration and blending are done by synthetic anethole, eugenol from cinnamon leaf oil, synthetic estragole, and *cis*- and *trans* ocimene from other sources. Detection is done either by GC-MS or by 2D chiral separation. The chiral ratio of linalool is (R)-(–) 80.0%–90.0% and (S)-(+) 10.0%–20.0% found by Casabianca (1996a).

18.6.73 TEA TREE OIL

ISO standard 4730 shows character and data for this oil. Adulteration is done by adding α-pinene and β-pinene from different sources and γ-terpinene from terpenes resulting from citrus concentration production. Synthetic terpinen-4-ol is also used. Detection is done by GC-MS and by using multidimensional chiral analytical equipment. Cornwell et al. (1995) reported the following chiral ratios for steam distillation of whole branches: (+)-sabinene 58%:(–)-sabinene 42%; (+)-α-pinene 91%:(–)-α-pinene 9%; (+)-α-phellandrene 49%:(–)-α-phellandrene 51%; (+)-β-phellandrene 71%:(–)-β-phellandrene 29%; (+)-limonene 62%:(–)-limonene 38%; (+)-terpinen-4-ol 65%:terpinen-4-ol 35%; and (+)-α-terpineol 80%:(–)-α (–)-terpineol 20%.

18.6.74 THYME OIL

ISO standard 14715 shows character and data for the oil of *T. zygis* (Loefl.) L. For real thyme oil (*T. vulgaris* L.), no ISO standard is available. Origanum oil and marjoram oil from Spain were used for adulteration in earlier times. Also synthetic thymol and carvacrol were applied components. Pure essential oil of thyme is red, as most of the steel vessels contain iron and this reacts with the thymol. White thyme oil, sold in high quantities, is mostly blended and synthetic. By using multidimensional chiral analytical equipment starting in 1990, Herner et al. 1990 reported the following enantiomeric ratios: (1S)-(–)-α-pinene 89%:(1R)-(+)-α-pinene 11%; (1S)-(–)-β-pinene 96%:(1R)-(+)-β-pinene 4%; and (4S)-(–)-limonene 70%:(4R)-(+)-limonene 30%. Kreis et al. (1991) reported the chiral ratio of borneol as follows: (–)-borneol (98.1%–99.6%):(+)-borneol (0.4%–1.9%). The chiral ratio of linalool is (R)-(–) 94.5%–99.0% and (S)-(+) 1.0%–5.5%. Casabianca (1996a) showed the following value for chiral ratios of linalyl acetate, although this is a minor component: (3R)-(–)-linalyl acetate (93.8%–99.2%):(3S)-(–)-linalyl acetate (0.8%–6.2%).

18.6.75 TURPENTINE OIL

ISO standard 21389 shows character and data for this oil. It is distilled from *Pinus massoniana* Lamb. and produced in China. Adulteration is rarely observed, but this oil is a source for the production of α-pinene and β-pinene. Both compounds are used for the adulteration of various essential oils containing these monoterpenes in higher value. Another type of turpentine oil is distilled from *Pinus pinaster* Aiton but is less of interest for adulterations as the quantities produced are minor.

18.6.76 VETIVER OIL

ISO standard 4716 shows character and data for this oil. Limonene chiral ratio is (R)-(–) with 100%. The identical value was reported by Möllenbeck (1997) for α-terpineol. He too found a ratio of linalool to be (R)-(–) 80% and (R)-(+) with 20% as well as terpinen-4-ol to be (R)-(–) 66% and (S)-(+) with 34%. Because of use GC×GC-TOF for detection it will be impossible to adulterate this essential oil.

18.6.77 YLANG-YLANG OILS

ISO standard 3063 shows character and data for this oil, distilled from the flowers of *C. odorata* (Lam.) Hook. F. Thomson forma genuina. Adulteration is done either by one of the fractions of cananga oil (extra, first, second, or third) or with various fractions like heads and tails. The mingling up of various fractions like extra, I, II, or III will not touch the purity or naturality of that oils as long as all are from the same botanical source. *C. odorata* (Lam.) Hook. F. Thomson forma *macrophylla* is used for adulteration. Blending is done by linalyl acetate, benzyl acetate, synthetic geraniol from various sources, methyl benzoate, benzyl alcohol, methyl salicylate, all synthetic and bay leaf oil, cedarwood terpenes, lavandin residues, and traces of ethyl vanillin (synthetic). Detection is done by GC/MS and by multidimensional chiral separation. Casabianca (1996a) reported the chiral ratio for linalool to be (S)-(+) 1.6–3.0 and (R)-(–) 97.0%–98.4%.

REFERENCES

Adams R.P., 2007, *Identification of Essential Oil Components by Gas Chromatography/Mass Spectrometry*, 4th edn., Allured Publishing Corp., Carol Stream, IL.

AFNOR (Association française de normalisation), 1992, NF T 75-254, Huile essentielle de cyprès, Tour Europe 92049 Paris La Défense Cedex.

Atal C.K., Shah K.C., 1966, TLC patterns of some volatile oils and crude drugs and their adulterants, *Indian J. Pharm.*, 28(6), 162–163.

Bail S., May/June 2009, Antimicrobial activities of Roman chamomile oil from France and its main compounds, *J. Essent. Oil Res.*, 21, 283–285.

Baldovini N., 2010, Phytochemistry of the heartwood from fragrant *Santalum* species: A review, *Flavour Fragr. J.*, 26, 7–26.

Bernreuther A. et al., 1991, Multidimensional gas chromatography/mass spectrometry: A powerful tool for the direct chiral evaluation of aroma compounds in plant tissue II. Linalool in essential oils and fruits, *Phytochem. Anal.*, 2, 167–170.

Braun M., Franz C., 2001, Chirale säulen decken Verfälschungen auf, *Pharm. Ztg.*, 146, 2493–2499.

Braun N. et al., 2005, *Santalum spicatum* (R.Br) DC. (*Santalaceae*)—Nor-helifolenal and acorenol isomers: Isolation and biogenetic considerations. *J. Essent. Oil Res.*, 15, 381–386.

BRENDA, The comprehensive enzyme information system, 2007, Technische Universität Braunschweig, Braunschweig, Germany. http://www.brenda-enzymes.org/php/ (accessed February 2014).

Brophy J.J. et al., 1989, Gas chromatographic quality control for oil of *Melaleuca* terpinen-4-ol type, *J. Agric. Food Chem.*, 37, 1330–1335.

Busch K.W., Busch M.A., 2006, *Chiral Analysis*, Elsevier B.V., Amsterdam, The Netherlands.

Carle R., 1996, Kamillenöl—Gewinnung und Qualitätsbeurteilung, Deutsche Apotherker Zeitung, 136. Jahrgang, Nr. 26, pp. 17–28.

Carle R. et al., 1990, Studies on the origin of (–)-α-bisabolol and chamazulene in chamomile preparations. Part 1. Investigations by isotope ratio mass spectrometry (IRMS), *Plant Med.*, 56, 456–460.

Casabianca H., 1996a, Le Point Sur L'Analyse Chirale Du Linalool Et De L'Acetate De Linalyle Dans Diverses Plantes, Rivista Italiana Eppos; 7, spi, pp. 227–243.

Casabianca H., 1996b, Analyses Chirales Et Isotopiques Des Principaux Constituants De Roses Et De Geraniums, Rivista Italiana Eppos; 7, spi, pp. 244–261.

Coleman B.M., Lawrence, Cole S.K., 2002, Semiquantitative determination of off-notes in mint oils by solid-phase microextraction. *J. Chromatogr. Sci.*, 40, 133–139.

Cornwell C.P. et al., November/December 1995, Incorporation of oxygen-18 into Terpinen-4-ol from the $H_2^{18}O$ steam distillates of *Melaleuca alternifolia* (Tea Tree), *J. Essent. Oil Res.*, 7, 613–620.

Cotroneo, A. et al., 1987. Sulla genuinità delle essenze agrumarie. Nota XVI. Differenze quantitative nella composizione di essenze di limone estratte a macchina e di essenze estratte manualmente senza uso di acqua, *Essenz. Deriv. Agrum.*, 57, 220–235.

Dugo G. et al., September/October 1992, High resolution gas chromatography for detection of adulterations of citrus cold-pressed essential oils, *Perf. Flavor.*, 17, 57–74.

Dugo G., 1993, On the genuineness of citrus essential oils, *Flavour Fragr. J.*, 8, 25–33.

Dugo D.L., 2011, *Citrus Oils, Advanced Analytical Techniques, Contaminant, and Biological Activity*, CRC Press, Taylor & Francis Group, Boca Raton, FL, p. 245.

Dugo G, Mondello L., 2011, *Composition of the Volatile Fraction of Citrus Peel Oils in Citrus Oils— Composition, Advanced Analytical Techniques, Contaminants, and Biological Activity*, Taylor & Francis Group, Boca Raton, FL, p. 147.

Formácek V., 2002, *Essential Oil Analysis by Capillary Gas Chromatography and Carbon-13 NMR Spectroscopy*, John Wiley & Sons, Ltd., Chichester, U.K.

Gnitka R., 2010, Efficient method of isolation of pure (−)-α- and (+)-β-thujone from *Thuja occidentalis* essential oil, Poster presented at the *Symposium on Essential Oils*, Wroclaw, Poland.

Hara F. et al., 1999, The analysis of some chiral components in citrus volatile compounds, *Proceedings 43rd TEAC Meeting*, Oita, Japan, pp. 360–362.

Hatt H. et al., 2011, Wo Düfte ihren Anfang nehmen, Spektrum der Wissenschaft, *Gehirn und Geist* (3), 38–41.

Herner U. et al., 1990, Enantiomeric distribution of α-pinene and β-pinene and limonene in essential oils and extracts. Part 2. Oils perfumes and cosmetics. *Flavour. Fragr. J.* 5, 201–205.

ISO, International Organization for Standardization TC 54, Standard catalogue, ISO/TC 54, 2015, http://www.iso.org/iso/home/store/catalogue_tc/catalogue_tc_browse.htm?commid=48956 (accessed April 2014).

ISO/DIS 875©, Essential oils—Evaluation of miscibility in ethanol, ISO, 1999, Geneva, Switzerland, http://www.iso.org (accessed January 2014).

ISO/DIS 1241©, Essential oils—Determination of ester values, before and after acetylation, and evaluation of the contents of free and total alcohols, ISO, 1996, Geneva, Switzerland, http://www.iso.org (accessed January 2014).

ISO/DIS 1272©, Essential oils—Determination of content of phenols, ISO, 2000, Geneva, Switzerland, http://www.iso.org (accessed January 2014).

ISO/DIS 1279©, Essential oils—Determination of carbonyl value—Potentiometric methods using hydroxylammonium chloride, ISO, 1996, Geneva, Switzerland, http://www.iso.org (accessed January 2014).

ISO/DIS 3064©, Essential oil of petitgrain, Paraguayan type (*Citrus aurantium* L. var. Paraguay (syn. *Citrus aurantium* var. bigaradia Hook f.)), ISO, 2010, Geneva, Switzerland, http://www.iso.org (accessed January 2014).

ISO/DIS 4715©, Essential oils—Quantitative evaluation of residue on evaporation, ISO, 1978, Geneva, Switzerland, http://www.iso.org (accessed January 2014).

ISO/DIS 7660©, Essential oils—Determination of ester value of oils containing difficult-to-saponify esters, ISO, 1983, Geneva, Switzerland, http://www.iso.org (accessed January 2014).

ISO/DIS 8901©, Oil of bitter orange petitgrain, cultivated (*Citrus aurantium* L.), ISO, 2010, Geneva, Switzerland, http://www.iso.org (accessed January 2014).

ISO/DIS 9235, Aromatic natural raw materials—Vocabulary, ISO, 2013, Geneva, Switzerland, www.iso.org.

ISO/DIS 9842, Oil of rose (Rosa × damascena Miller), 2003, ISO International.

ISO/DIS 11021©, Essential oils—Determination of water content—Karl Fischer method, ISO, 1999, Geneva, Switzerland, www.iso.org (accessed January 2014).

Jaspersen-Schib R., Flueck H., 1962, Identification and purity determination of essential oils by thin layer chromatography, *Congr. Sci. Farm. Conf. Commun.*, 21, Pisa, Italy, 1961, 608–614.

John, M.D. et al., 1991, Detection of adulteration of polyethylene glycols in oil of sandalwood, *Indian Perf.*, 35(4), 186–187.

Juchelka D., 1996a, Authenticity profiles of bergamot oil, *Pharmazie*, 51(6), 418.

Juchelka D., 1996b, Chiral compounds of essential oils. XX. Chirality evaluation and authenticity profiles of neroli and petitgrain oils. *J. Essent. Oil Res.*, 8, 487–497.

Kartnig T. et al., 1999, Gaschromatographische Untersuchungen an Ätheroleum Juniperi unter besonderer Berücksichtigung der Trennung enantiomerer Komponenten, *Sci. Pharm.*, 67, 77–82.

König W.A. et al., 1992, Enantiomeric composition of the chiral constituents in essential oils. Part 1: Monoterpene hydrocarbons, *J. High Res. Chromatogr.*

Konig W.A., Joulain D., Hochmuth D.H., Robertet S.A., Hochmuth G., 2004, Terpenoids and related constituents of essential oils. MassFinder 3: Convenient and Rapid Analysis of GCMS, Hamburg, Germany.

Kovats E., 1958, *Helv. Chim. Acta*, 41, 1915.

Kreis P. et al., 1990, Inhaltstoffe ätherischer Öle. III Stereodifferenzierung von α-Pinen, β-Pinen und Limonen in ätherischen Ölen, Drogen und Fertigarzneimitteln, *Dtsch. Apoth. Ztg.*, 130, 985–988.

Kreis P. et al., 1991, Chirale Inhaltsstoffe ätherischer Öle, Deutsche Apotheker Zeitung, Nr. 39 pp. 1984–1987.

Kreis P., Juchelka D., Motz C., Mosandl A., 1991a, Chirale Inhaltsstoffe aÅNtherischer OÅNle. IX: Stereodifferenzierung von Borneol, Isoborneol und Bornylacetat. *Dtsch. Apoth. Ztg.*, 131, 1984–1987.

Kreis P., Mosandl A., 1992, Chiral Compounds of Essential Oils. Part XI. Simultaneous Stereoanalysis of *Lavandula* Oil Constituents, *Flavour Fragr. J.*, 7, 187–193.

Kreis P., Mosandl A., 1992, Chiral Compounds of Essential oils. Part XII. Authenticity control of rose oils, using enantioselective multidimensional gas chromatography, *Flavour Fragr. J.*, 7, 199–201.

Kubeczka K.H., Bohn I., 1985, Pimpinella root and its adulteration. Detection of adulteration by thin-layer and gas chromatography. Structure revision of the principal components of the essential oils, *Deutsche Apotheker Zeitung*, 125(8), 399–402.

Kubezka K.H., Formácek V., 2002, *Essential Oil Analysis by Capillary Gas Chromatography and Carbon-13 NMR Spectroscopy*, John Wiley & Sons, Ltd., Chichester, U.K.

Sajilata M.G. and Singhal R.S., 2012, Quality indices for spice essential oils, in: *Handbook of Herbs and Sspices*, Second Edition, Woodhead Publishing Limited, Volume 1, Chapter 3, Section 3.3.5, p. 48.

Lawrence B., May/June 1996a, Progress in essential oils, Vol. 21, November/December 1996 *Perfum. Flavor.*, 21, 64.

Lawrence B., November/December 1996b, Progress in essential oils, Vol. 21, May/June *Perfum. Flavor.*, 21, 59.

Lawrence B., May/June 1999a, Progress in essential oils, *Perfum. Flavor.*, 24, 2–7.

Lawrence, B., January/February 1999b, Progress in essential oils, *Perfum. Flavor.*, 11.

Lawrenec B., Bovill E.W., 2000, The essential oil market in 1934, *Perfum. Flavor.*, 25, 22–32.

Lawrence B., 2009. A preliminary report on the world production of some selected essential oils and countries. *Perfum. Flavor.*, 34, 38–44.

Lawrence B.M., 2011, *Essential Oils 2008–2011*, Allured Business Media, Carol Stream, IL, p. 15.

Marsili R., 2010, *Flavor, Fragrance and Odor Analysis*, 2nd edn., Ray Marsili, CRC Press, Taylor & Francis Group, Boca Raton, FL.

McHale D. et al., February/March 1983 Detection of adulteration of cold-pressed bitter orange oil, *Perfum. Flavor.*, 8, 40–41.

Merriam-Webster, n.d., http://www.merriam-webster.com/dictionary/iodine%20number.

Möllenbeck S., König T, Schreier P., Schwab W., Rajaonarivony J., Ranarivelo, 1997, Chemical composition and analyses of enantiomeres of essential oils from Madagascar, *Flavour Frag. J.*, 12, 63–69.

Mondello L., April 1998, Multidimensional tandem capillary gas chromatography system for the analysis of real complex samples. Part I: Development of a fully automated tandem gas chromatography system, *J. Chromatogr. Sci.*, 36, 206.

Mondello L., 2011, *Flavour and Fragrance Natural and Synthetic Compounds 2*, 2nd edn. John Wiley & Sons, Ltd., Chichester, U.K.

Mosandl A., 1998, Enantioselective analysis, in: Ziegler E., Ziegler H. (eds.), *Flavourings*, Wiley-VCH Verlag GmbH & Co. KGaA, Weinheim, Germany.

Mosandl A., 2000, Authenticity assessment of essential oils—The current state and the future, Lecture hold on the *31st International Symposium on Essential Oils*, Hamburg, Germany.

Mosandl A. et al., 1991, Stereoisomeric flavor compounds 48. Chirospecific analysis of natural flavors and essential oils using multidimensional gas chromatography. *J. Agric. Food Chem.*, 39, 1131–1134.

Mostafa M.M., 1990, *Egypt. J. Food Sci.*, 16(1/2): 63–67.

Nakamoto et al., 1996, Enantiomeric distributions of carveols in grapefruit, orange and spearmint oils, in: *IFEAT Proceedings International Conference of Aromas and Essential Oils*, IFEAT Secretariat, Tel Aviv, Israel, pp. 36–50.

Nova D., Karmazin M., Buben I., 1986, Anatomical and chemical discrimination between the roots of various varieties of parsley (*Petroselinum crispum* Mill./A. W. Hill.) and parsnip (*Pastinaca sativa* L. ssp. sativa), *Cesko-Slovenska Farmacie*, 35(8), 363–366.

Ochocka J.R., 2002, Determination of enantiomers of terpene hydrocarbons in essential oils obtained from species of Pinus and Abies, *Pharm. Biol.*, 40, 395–399.

Ohloff G., 1990, *Riechstoffe und Geruchssinn: Die molekulare Welt der Düfte*, Springer-Verlag, Berlin, Germany.

ÖKO Test, 1995, Gepanschte Seelen, Oktober/95, ÖKO-TEST Verlag GmbH, Frankfurt, Germany, http://www.germany.ru/wwwthreads/files/1646–4014361-__196_t___246_le___214_kotest.pdf.

Peyron L. et al., 1974, Abnormal presence of hexylene glycol in commercial essential oils of vetiver and cananga, *Plantes Medicinales et Phytotherapie* (1975), 9(3), 192–203.

Ph.Eur, European Pharmacopoeia 7.4, 2012, European Directorate for the Quality of Medicines & HealthCare http://www.edqm.eu/en/edqm-homepage-628.html.

Phokas G., 1965, Thin-layer chromatography of the essential oil of *Melissa officinalis* and some of its adulterants, *Pharm. Deltion Epistemonike Ekdosis*, 5(1–2), 9–16.

Rajeswara Rao B.R. et al., 2002. Water soluble fractions of rose-scented geranium (*Pelargonium* species) essential oil, *Bioresource Technol.*, 84, 243–246.

Rajeswara Rao B.R., 2007, Effect of crop-weed mixed distillation on essential oil yield and composition of five aromatic crops, *JEOBP*, 10(2), 127–132.

Ravid U., 1992, Chiral GC analysis of enantiomerically pure Fenchone in essential oils, *Flav. Fragr. J.*, 7, 169–172.

Reeve D. et al., July/August 2002, Riding the citrus trail: When is a mandarin a tangerine, *Perf. Flavor.*, 27, 20–22.

Renaud E.N.C., 2001, Essential oil quantity and composition from 10 cultivars of organically grown lavender and lavandin, *J. Essent. Oil Res.*, 13, 269–273.

Reunion Islands National Institute of Statistics (INSEE), 2008–2009, 10.1: Revenus et production agricoles, 10.1.3—Production végétale, INSEE-RÉUNION—TER 2008–2009, p. 185, http://www.insee.fr/fr/insee_regions/reunion/themes/dossiers/ter/ter2008_production_vegetale.pdf.

Rochussen F., 1920, *Ätherische Öle und Riechstoffe*, Walter de Gruyter & Co., Berlin, Germany, pp. 19, 23, 41, 69–72.

Sawamura M. et al., 2004, Poster presented at the *35th International Symposium on Essential Oils*, Giardini Naxos, Italy.

Schimmel-Bericht & Co. Edition, April/October 1920, Schimmel & Co., Leipzig, Germany, pp. 30–31.

Schimmel-Bericht & Co. Edition, April/October 1919, Schimmel & Co., Leipzig, Germany, pp. 4, 13, 15, 22, 23, 36, 41–43, 48, 51, 52, 66.

Schmidt E., July/August 2003, The characteristics of lavender oils from Eastern Europe, *JEOR*, 28, 48–60.

Schmidt E., 2010, Production of essential oils, in: *Handbook of Essential Oils*, Edited by K.H.C. Baser and G. Buchbauer, CRC Press, Taylor & Francis Group, Boca Raton, FL.

Schultze W., 1993, Differentiation of original lemon balm oil (*Melissa officinalis*) from several lemonlike smelling oils by chirospecific GC analysis of citronellal and isotope ratio mass spectrometry, *Planta Med.*, 59(Suppl.), A635.

Sen A.R., Sen Gupta P., Ghose Dastidar N., 1974, Detection of *Curcuma zedoaria* and *C. aromatica* in *C. longa* (turmeric) by thin-layer chromatography, *Analyst* (Cambridge, United Kingdom), 99(1176), 153–155.

Sewenig S. et al., 2003, Online determination of $^2H/^1H$ and $^{13}C/^{12}C$ isotope ratios of cinnamaldehyde from different sources using gas chromatography isotope mass spectrometry, *Eur. Food Res. Technol.*, 217(5), 444–448.

Stoyanova A. et al., 2008, Traditional Bulgarian essential oil-bearing raw materials 2. Lavender (*Lavandula angustifolia* Mill.), *Indian Perfum.*, 52, 50–55.

Teisseire P., 1987, Industrial quality control of essential oils by capillary GC, in: Sandra P., Bicchi C. (eds.), *Capillary Gas Chromatography in Essential Oil Analysis*, Huethig, Basel, NY.

The Free Dictionary, n.d. (a), http://www.thefree dictionary.com/adulteration (accessed February 2014).

The Free Dictionary, n.d. (b), http://www.thefree dictionary.com/cupidity (accessed February 2014).

The Free Dictionary, n.d. (c),, http://www.thefree dictionary.com/enzymatic (accessed February 2014).

Touche J. et al., 1981, Maillettes et lavandes fines francaises, Rivista Italiana E.P.P.O.S., LXIII - n. 6 – settembre-ottobre, pp. 320–323.

Valder C., 2003, Western Australian sandalwood oil—New constituents of *Santalum spicatum* (R.Br) A DC. (Santalaceae), *J. Essent. Oil Res.*, 15, 178–186.

Van Den Dool H., Kratz P.D., *J. Chromatogr.*, 1963, 11, 463–471.

Venskutonis P.R., 2002, *Thyme—The Genus Thyme*, Taylor & Francis, New York, p. 226.

Wang X.-H. et al., 1995, The direct chiral separation of some optically active compounds in essential oils by multidimensional gas chromatography, *J. Chromatogr. Sci.*, 33, 22–25.

19 Biotransformation of Monoterpenoids by Microorganisms, Insects, and Mammals

Yoshiaki Noma and Yoshinori Asakawa

CONTENTS

19.1 Introduction 748
19.2 Metabolic Pathways of Acyclic Monoterpenoids 749
    19.2.1 Acyclic Monoterpene Hydrocarbons 749
        19.2.1.1 Myrcene 749
        19.2.1.2 Citronellene 750
    19.2.2 Acyclic Monoterpene Alcohols and Aldehydes 751
        19.2.2.1 Geraniol, Nerol, (+)- and (−)-Citronellol, Citral, and (+)- and (−)-Citronellal 751
        19.2.2.2 Linalool and Linalyl Acetate 759
        19.2.2.3 Dihydromyrcenol 765
19.3 Metabolic Pathways of Cyclic Monoterpenoids 766
    19.3.1 Monocyclic Monoterpene Hydrocarbon 766
        19.3.1.1 Limonene 766
        19.3.1.2 Isolimonene 776
        19.3.1.3 *p*-Menthane 778
        19.3.1.4 1-*p*-Menthene 778
        19.3.1.5 3-*p*-Menthene 780
        19.3.1.6 α-Terpinene 780
        19.3.1.7 γ-Terpinene 780
        19.3.1.8 Terpinolene 781
        19.3.1.9 α-Phellandrene 782
        19.3.1.10 *p*-Cymene 782
    19.3.2 Monocyclic Monoterpene Aldehyde 783
        19.3.2.1 Perillaldehyde 783
        19.3.2.2 Phellandral and 1,2-Dihydrophellandral 785
        19.3.2.3 Cuminaldehyde 786
    19.3.3 Monocyclic Monoterpene Alcohol 786
        19.3.3.1 Menthol 786
        19.3.3.2 Neomenthol 789
        19.3.3.3 (+)-Isomenthol 792
        19.3.3.4 Isopulegol 792
        19.3.3.5 α-Terpineol 792
        19.3.3.6 (−)-Terpinen-4-ol 796

   19.3.3.7 Thymol and Thymol Methyl Ether 796
   19.3.3.8 Carvacrol and Carvacrol Methyl Ether.. 798
   19.3.3.9 Carveol.. 798
   19.3.3.10 Dihydrocarveol 805
   19.3.3.11 Piperitenol 809
   19.3.3.12 Isopiperitenol 809
   19.3.3.13 Perillyl Alcohol... 811
   19.3.3.14 Carvomenthol.. 812
  19.3.4 Monocyclic Monoterpene Ketone... 814
   19.3.4.1 α, β-Unsaturated Ketone 814
   19.3.4.2 Saturated Ketone... 834
   19.3.4.3 Cyclic Monoterpene Epoxide... 837
19.4 Metabolic Pathways of Bicyclic Monoterpenoids 845
  19.4.1 Bicyclic Monoterpene 845
   19.4.1.1 α-Pinene.. 845
   19.4.1.2 β-Pinene 849
   19.4.1.3 (±)-Camphene... 856
   19.4.1.4 3-Carene and Carane 857
  19.4.2 Bicyclic Monoterpene Aldehyde 858
   19.4.2.1 Myrtenal and Myrtanal... 858
  19.4.3 Bicyclic Monoterpene Alcohol 858
   19.4.3.1 Myrtenol.. 858
   19.4.3.2 Myrtanol 859
   19.4.3.3 Pinocarveol 859
   19.4.3.4 Pinane-2,3-Diol... 861
   19.4.3.5 Isopinocampheol (3-Pinanol).. 862
   19.4.3.6 Borneol and Isoborneol... 864
   19.4.3.7 Fenchol and Fenchyl Acetate 866
   19.4.3.8 Verbenol 868
   19.4.3.9 Nopol and Nopol Benzyl Ether 868
  19.4.4 Bicyclic Monoterpene Ketones 870
   19.4.4.1 α-, β-Unsaturated Ketone 870
   19.4.4.2 Saturated Ketone... 870
19.5 Summary.. 878
  19.5.1 Metabolic Pathways of Monoterpenoids by Microorganisms....................................... 878
  19.5.2 Microbial Transformation of Terpenoids as Unit Reaction 890
References... 895

19.1 INTRODUCTION

A large number of monoterpenoids have been detected in or isolated from essential oils and solvent extracts of fungi, algae, liverworts, and higher plants, but the presence of monoterpenoids in fern is negligible. Vegetables, fruits, and spices contain monoterpenoids; however, their fate in human and other animal bodies has not yet been fully investigated systematically. The recent development of analytical instruments makes it easy to analyze the chemical structures of very minor components, and the essential oil chemistry field has dramatically developed.

  Since monoterpenoids, in general, show characteristic odor and taste, they have been used as cosmetic materials and food additives and often for insecticides, insect repellents, and attractant drugs. In order to obtain much more functionalized substances from monoterpenoids, various chemical reactions and microbial transformations of commercially available and cheap synthetic monoterpenoids have been carried out. On the other hand, insect larva and mammals have been used for direct biotransformations of monoterpenoids to study their fate and safety or toxicity in their bodies.

The biotransformation of α-pinene (**4**) by using the black fungus *Aspergillus niger* was reported by Bhattacharyya et al. (1960) half a century ago. During that period, many scientists studied the biotransformation of a number of monoterpenoids by using various kinds of bacteria, fungi, insects, mammals, and cultured cells of higher plants. In this chapter, the microbial transformation of monoterpenoids using bacteria and fungi is discussed. Furthermore, the biotransformation by using insect larva, mammals, microalgae, as well as suspended culture cells of higher plants is also summarized. In addition, several biological activities of biotransformed products are also represented. At the end of this chapter, the metabolite pathways of representative monoterpenoids for further development on biological transformation of monoterpenoids are demonstrated.

19.2 METABOLIC PATHWAYS OF ACYCLIC MONOTERPENOIDS

19.2.1 Acyclic Monoterpene Hydrocarbons

19.2.1.1 Myrcene

The microbial biotransformation of myrcene (**302**) was described with *Diplodia gossypina* ATCC 10936 (Abraham et al., 1985). The main reactions were hydroxylation, as shown in Figure 19.1. On oxidation, myrcene (**302**) gave the diol (**303**) (yield up to 60%) and also a side product (**304**) that possesses one carbon atom less than the parent compound, in yields of 1%–2%.

One of the publications dealing with the bioconversion of myrcene (Busmann and Berger, 1994) described its transformation to a variety of oxygenated metabolites, with *Ganoderma*

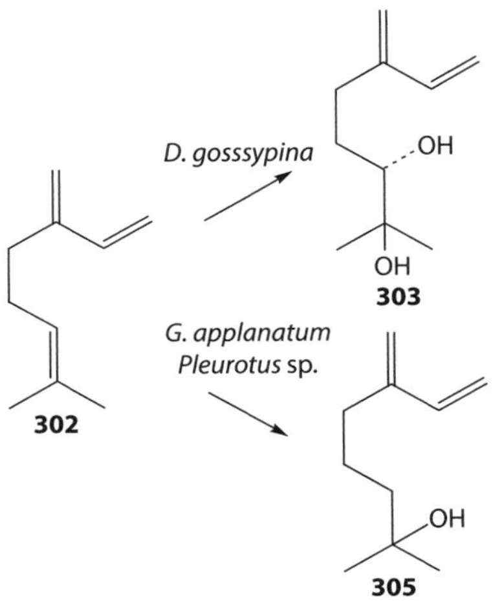

FIGURE 19.1 Biotransformation of myrcene (**302**) by *Diplodia gossypina* (Abraham et al., 1985), *Ganoderma applanatum*, and *Pleurotus* sp. (Modified from Busmann, D. and Berger, R.G., *J. Biotechol.*, 37, 39, 1994.)

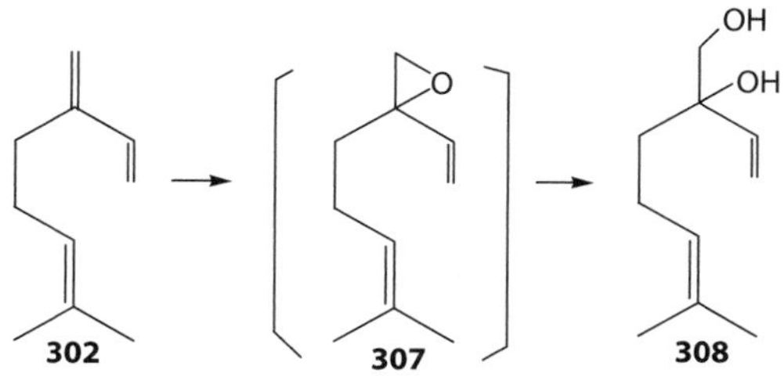

FIGURE 19.2 Biotransformation of myrcene (**302**) by *Spodoptera litura*. (Modified from Miyazawa, M. et al., Biotransformation of β-myrcene by common cutworm larvae, *Spodoptera litura* as a biocatalyst, *Proceedings of 42nd TEAC*, 1998, pp. 123–125.)

applanatum, *Pleurotus flabellatus*, and *Pleurotus sajor-caju* possessing the highest transformation activities. One of the main metabolites was myrcenol (**305**) (2-methyl-6-methylene-7-octen-2-ol), which gives a fresh, flowery impression and dominates the sensory impact of the mixture (see Figure 19.1).

β-Myrcene (**302**) was converted by common cutworm larvae, *Spodoptera litura*, to give myrcene-3,(10)-glycol (**308**) via myrcene-3,(10)-epoxide (**307**) (Figure 19.2) (Miyazawa et al., 1998).

19.2.1.2 Citronellene

(−)-Citronellene (**309**) and (+)-citronellene (**309′**) were biotransformed by the cutworm *S. litura* to give (3*R*)-3,7-dimethyl-6-octene-1,2-diol (**310**) and (3*S*)-3,7-dimethyl-6-octene-1,2-diol (**310′**), respectively (Takeuchi and Miyazawa, 2005) (Figure 19.3).

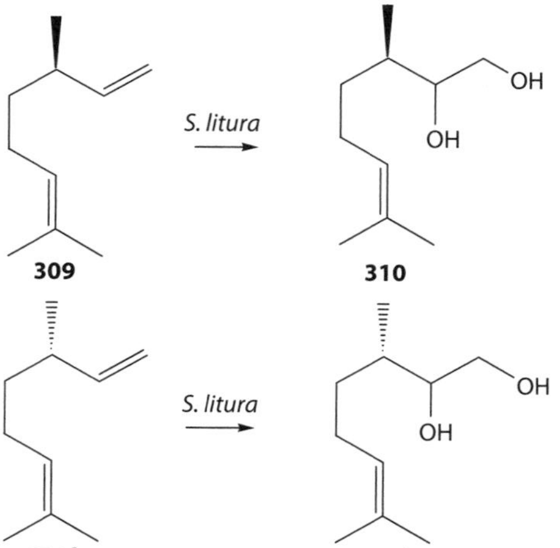

FIGURE 19.3 Biotransformation of (−)-citronellene (**309**) and (+)-citronellene (**309′**) by *Spodoptera litura*. (Modified from Takeuchi, H. and Miyazawa, M., Biotransformation of (−)- and (+)-citronellene by the larvae of common cutworm (*Spodoptera litura*) as biocatalyst, *Proceedings of 49th TEAC*, 2005, pp. 426–427.)

19.2.2 Acyclic Monoterpene Alcohols and Aldehydes

19.2.2.1 Geraniol, Nerol, (+)- and (−)-Citronellol, Citral, and (+)- and (−)-Citronellal

258 (R)-(+) **258′** (S)-(−)
Citronellol

261 (R)-(+) **261′** (S)-(−)
Citronellal

262 (R)-(+) **262′** (S)-(−)
Citronellic acid

271 Geraniol

272 Nerol

276 Geranial

275 Neral

278 Geranic acid

277 Neric acid

275 and 276 Citral

The microbial degradation of the acyclic monoterpene alcohols citronellol (**258**), nerol (**272**), geraniol (**271**), citronellal (**261**), and citral (equal mixture of **275** and **276**) was reported in the early part of 1960 (Seubert and Remberger, 1963; Seubert et al., 1963; Seubert and Fass, 1964a,b). *Pseudomonas citronellolis* metabolized citronellol (**258**), citronellal (**261**), geraniol (**271**), and geranic acid (**278**). The metabolism of these acyclic monoterpenes is initiated by the oxidation of the primary alcohol group to the carboxyl group, followed by the carboxylation of the C-10 methyl group (β-methyl) by a biotin-dependent carboxylase (Seubert and Remberger, 1963). The carboxymethyl group is eliminated at a later stage as acetic acid. Further degradation follows the β-oxidation pattern. The details of the pathway are shown in Figure 19.4 (Seubert and Fass, 1964a).

The microbial transformation of citronellal (**261**) and citral (**275 and 276**) was reported by way of *Pseudomonas aeruginosa* (Joglekar and Dhavlikar, 1969). This bacterium, capable of utilizing citronellal (**261**) or citral (**275 and 276**) as the sole carbon and energy source, has been isolated from soil by the enrichment culture technique. It metabolized citronellal (**261**) to citronellic acid (**262**) (65%), citronellol (**258**) (0.6%), dihydrocitronellol (**259**) (0.6%), 3,7-dimethyl-1,7-octanediol (**260**) (1.7%), and menthol (**137**) (0.75%) (Figure 19.5). The metabolites of citral (**275 and 276**) were geranic acid (**278**) (62%), 1-hydroxy-3,7-dimethyl-6-octen-2-one (**279**) (0.75%), 6-methyl-5- heptenoic acid (**280**) (0.5%), and 3-methyl-2-butenoic acid (**286**) (1%) (Figure 19.5). In a similar way, *Pseudomonas convexa* converted citral (**275 and 276**) to geranic acid (**278**) (Hayashi et al., 1967). The biotransformation of citronellol (**258**) and geraniol (**271**) by *P. aeruginosa*, *P. citronellolis*, and *Pseudomonas mendocina* was also reported by another group (Cantwell et al., 1978).

A research group in Czechoslovakia patented the cyclization of citronellal (**261**) with subsequent hydrogenation to menthol by *Penicillium digitatum* in 1952. Unfortunately the optical purities of the intermediates pulegol and isopulegol were not determined, and presumably the resulting menthol was a mixture of enantiomers. Therefore, it cannot be excluded that this extremely interesting cyclization is the result of a reaction primarily catalyzed by the acidic fermentation conditions and only partially dependent on enzymatic reactions (Babcka et al., 1956) (Figure 19.6).

Based on previous data (Madyastha et al., 1977; Rama Devi and Bhattacharyya, 1977a), two pathways for the degradation of geraniol (**271**) were proposed by Madyastha (1984) (Figure 19.7). Pathway

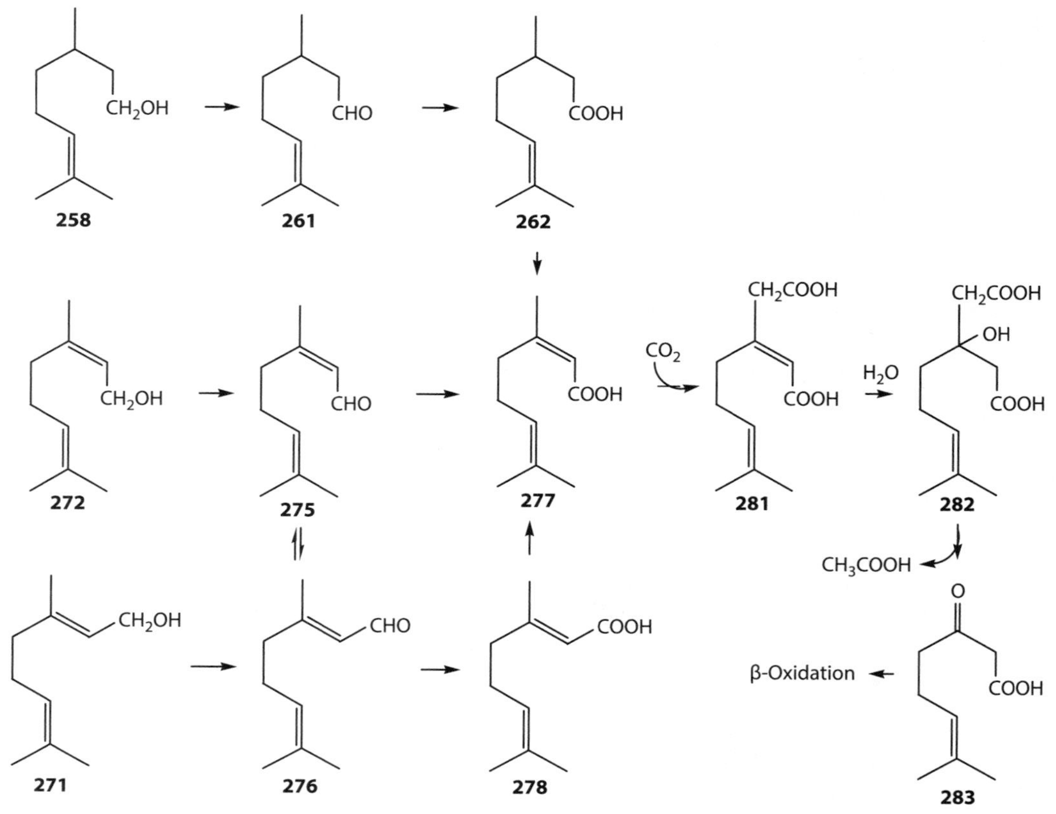

FIGURE 19.4 Biotransformation of citronellol (**258**), nerol (**272**), and geraniol (**271**) by *Pseudomonas citronellolis*. (Modified from Madyastha, K.M., *Proc. Indian Acad. Sci. (Chem. Sci.)*, 93, 677, 1984.)

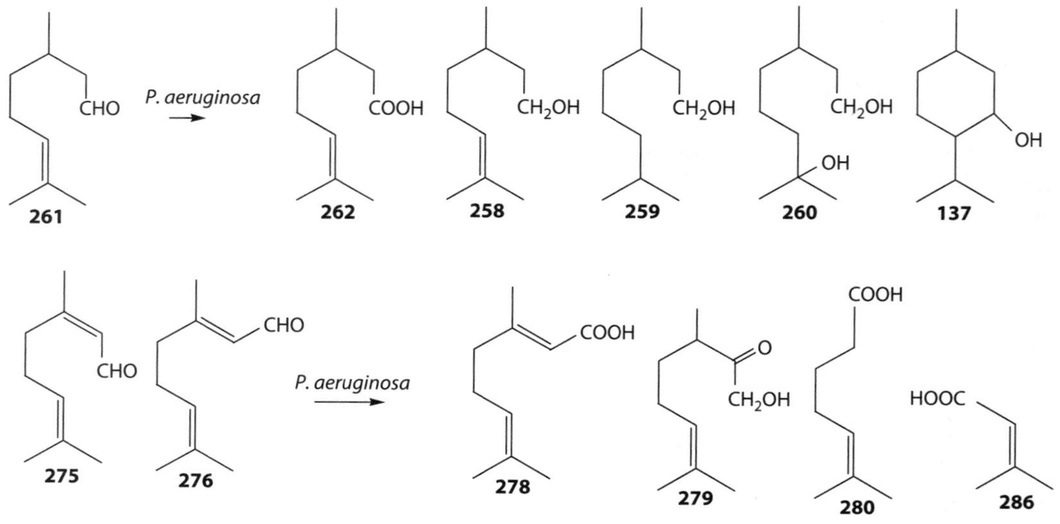

FIGURE 19.5 Biotransformation of citronellal (**261**) and citral (**275** and **276**) by *Pseudomonas aeruginosa*. (Modified from Joglekar, S.S. and Dhavlikar, R.S., *Appl. Microbiol.*, 18, 1084, 1969.)

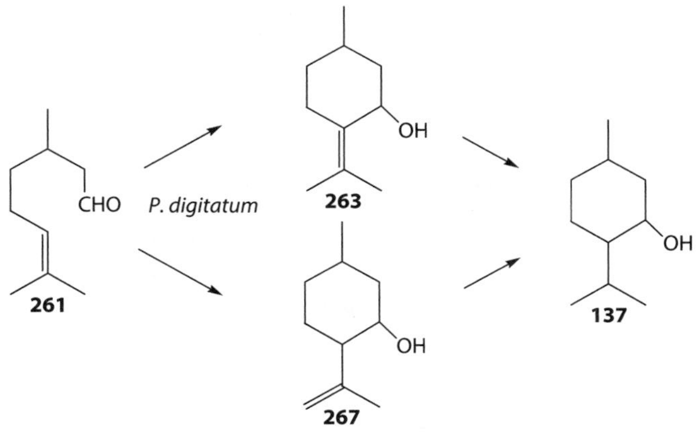

FIGURE 19.6 Biotransformation of citronellal to menthol by *Penicillium digitatum*. (Modified from Babcka, J. et al., Patent 56-9686b.)

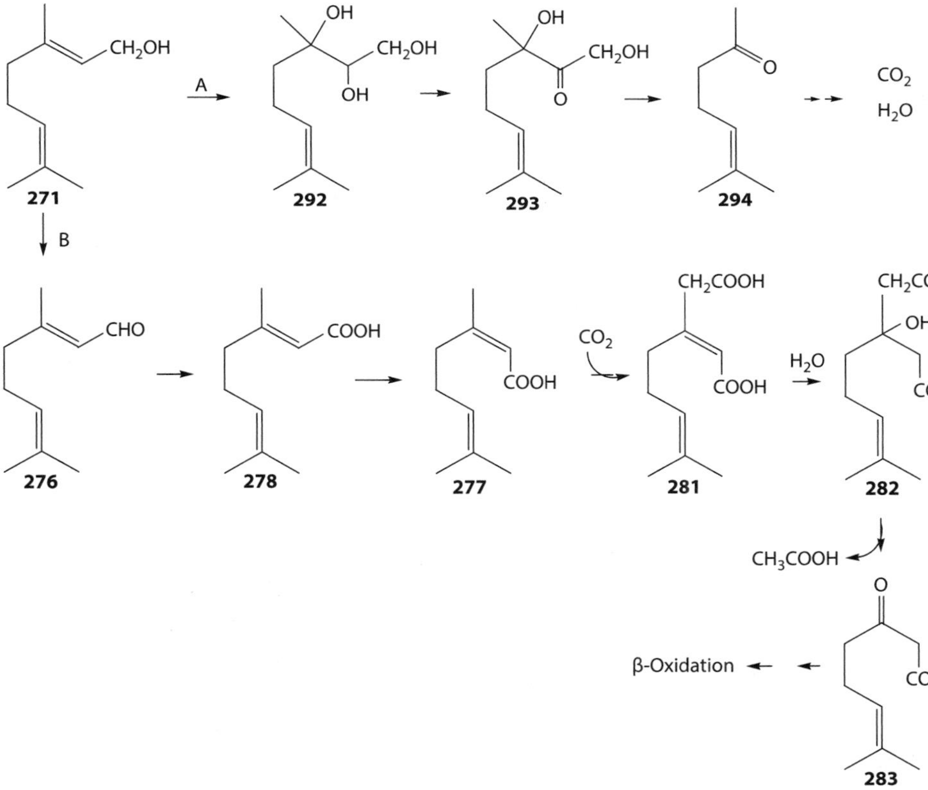

FIGURE 19.7 Metabolism of geraniol (**271**) by *Pseudomonas incognita*. (Modified from Madyastha, K.M., *Proc. Indian Acad. Sci. (Chem. Sci.)*, 93, 677, 1984.)

A involves an oxidative attack on the 2,3-double bond, resulting in the formation of an epoxide. Opening of the epoxide yields the 2,3-dihydroxygeraniol (**292**), which upon oxidation forms 2-oxo, 3-hydroxygeraniol (**293**). The ketodiol (**293**) is then decomposed to 6-methyl-5-hepten-2-one (**294**) by an oxidative process. Pathway B is initiated by the oxidation of the primary alcoholic group to geranic acid (**278**), and further metabolism follows the mechanism as proposed earlier for *P. citronellolis*

(Seubert and Remberger, 1963; Seubert et al., 1963). In the case of nerol (**272**), the Z-isomer of geraniol (**271**), degradative pathways analogous to pathways A and B as in geraniol (**271**) are observed. It was also noticed that *Pseudomonas incognita* metabolizes acetates of geraniol (**271**), nerol (**272**), and citronellol (**258**) much faster than their respective alcohols (Madyastha and Renganathan, 1983).

Euglena gracilis Z. converted citral (**275** and **276**, 56:44, peak area in gas chromatograph [GC]) to geraniol (**271**) and nerol (**272**), respectively, of which geraniol (**271**) was further transformed to (+)- and (−)-citronellol (**258** and **258′**). On the other hand, when either geraniol (**271**) or nerol (**272**) was added, both compounds were isomerized to each other, and then, geraniol (**271**) was transformed to citronellol. These results showed that *Euglena* could distinguish between the stereoisomers geraniol (**271**) and nerol (**272**) and hydrogenated geraniol (**271**) selectively. (+)-, (−)-, and (±)-Citronellal (**261**, **261′**, and **261** and **261′**) were also transformed to the corresponding (+)-, (−)-, and (±)-citronellol (**258**, **258′**, and **258** and **258′**) as the major products and (+)-, (−)-, and (±)-citronellic acids (**262**, **262′**, and **262** and **262′**) as the minor products, respectively (Noma et al., 1991a) (Figure 19.8).

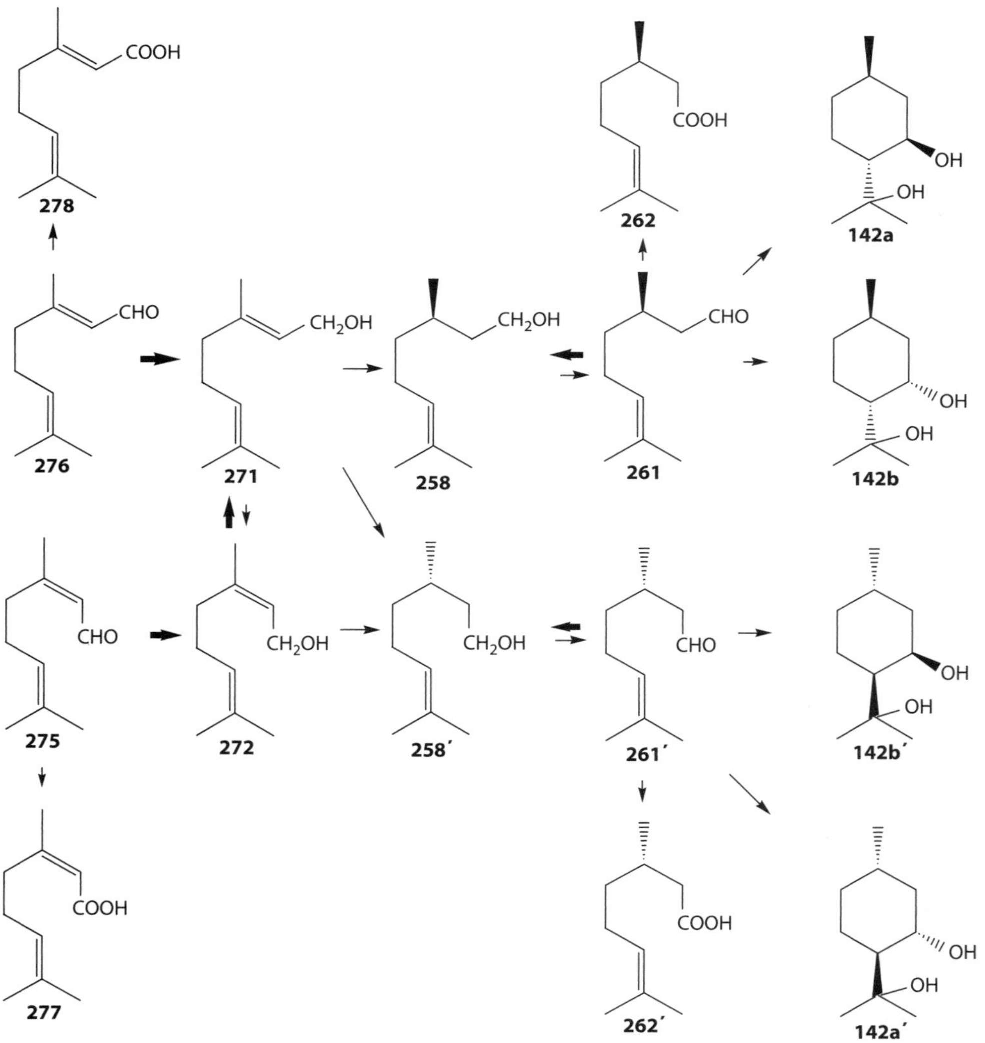

FIGURE 19.8 Metabolic pathways of citral (**275** and **276**) and its metabolites by *Euglena gracilis* Z. (Modified from Noma, Y. et al., *Phytochemistry*, 30, 1147, 1991a.)

Dunaliella tertiolecta also reduced citral (geranial (**276**) and neral (**275**) = 56:44) and (+)-, (−)-, and (±)-citronellal (**261**, **261′**, and **261** and **261′**) to the corresponding alcohols: geraniol (**271**), nerol (**272**), and (+)-, (−)-, and (±)-citronellol (**258**, **258′**, and **258** and **258′**) (Noma et al., 1991b, 1992a).

Citral (a mixture of geranial (**276**) and neral (**275**), 56:44 peak area in GC) is easily transformed to geraniol (**271**) and nerol (**272**), respectively, of which geraniol (**32**) is further hydrogenated to (+)-citronellol (**258**) and (−)-citronellol (**258′**). Geranic acid (**278**) and neric acid (**277**) as the minor products are also formed from **276** and **275**, respectively. On the other hand, when either **271** or **272** is used as a substrate, both compounds are isomerized to each other, and then **271** is transformed to citronellol (**258** or **258′**). These results showed the *Euglena* could distinguish between the stereoisomers **271** and **272** and hydrogenated selectively **271** to citronellol (**258** or **258′**). (+)-, (−)-, and (±)-Citronellal (**261**, **261′**, and equal mixture of **261** and **261′**) are also transformed to the corresponding citronellol and *p*-menthane-*trans*- and *cis*-3,8-diols (**142a, b, a′** and **b′**) as the major products, which are well known as mosquito repellents and plant growth regulators (Nishimura et al., 1982; Nishimura and Noma, 1996), and (+)-, (−)-, and (±)-citronellic acids (**262**, **262′**, and equal mixture of **262** and **262′**) as the minor products, respectively.

Streptomyces ikutamanensis, Ya-2-1, also reduced citral (geranial (**276**) and neral (**275**) = 56:44)) and (+)-, (−)-, and (±)-citronellal (**261**, **261′**, and **261** and **261′**) to the corresponding alcohols: geraniol (**271**), nerol (**272**), and (+)-, (−)-, and (±)-citronellol (**258**, **258′**, **258** and **258′**). Compounds **271** and **272** were isomerized to each other. Furthermore, terpene alcohols (**258′**, **272**, and **271**) were epoxidized to give 6,7-epoxygeraniol (**274**), 6,7-epoxynerol (**273**), and 2,3-epoxycitronellol (**268**). On the other hand, (+)- and (±)-citronellol (**258** and **258** and **258′**) were not converted at all (Noma et al., 1986) (Figure 19.9).

A strain of *A. niger*, isolated from garden soil, was able to transform geraniol (**271**), citronellol (**258** and **258′**), and linalool (**206**) to their respective 8-hydroxy derivatives. This reaction was called "ω-hydroxylation" (Madyastha and Krishna Murthy, 1988a,b).

Fermentation of citronellyl acetate with *A. niger* resulted in the formation of a major metabolite, 8-hydroxycitronellol, accounting for approximately 60% of the total transformation products, accompanied by 38% citronellol. Fermentation of geranyl acetate with *A. niger* gave geraniol and 8-hydroxygeraniol (50% and 40%, respectively, of the total transformation products).

One of the most important examples of fungal bioconversion of monoterpene alcohols is the biotransformation of citral by *Botrytis cinerea*. *B. cinerea* is a fungus of high interest in winemaking (Rapp and Mandery, 1988). In an unripe state of maturation, the infection of grapes by *B. cinerea* is very much feared, as the grapes become moldy (*gray rot*). With fully ripe grapes, however, the growth of *B. cinerea* is desirable; the fungus is then called "noble rot" and the infected grapes deliver famous sweet wines, such as Sauternes of France or Tokaji Aszu of Hungary (Brunerie et al., 1988).

One of the first reports in this area dealt with the biotransformation of citronellol (**258**) by *B. cinerea* (Brunerie et al., 1987a, 1988). The substrate was mainly metabolized by ω-hydroxylation. The same group also investigated the bioconversion of citral (**275** and **276**) (Brunerie et al., 1987b). A comparison was made between grape must and a synthetic medium. When using grape must, no volatile bioconversion products were found. With a synthetic medium, biotransformation of citral (**275** and **276**) was observed yielding predominantly nerol (**272**) and geraniol (**271**) as reduction products and some ω-hydroxylation products as minor compounds. Finally, the bioconversion of geraniol (**271**) and nerol (**272**) was described by the same group (Bock et al., 1988). When using grape must, a complete bioconversion of geraniol (**271**) was observed mainly yielding ω-hydroxylation products.

The most important metabolites from geraniol (**271**), nerol (**272**), and citronellol (**258**) are summarized in Figure 19.9. In the same year, the biotransformation of these monoterpenes by *B. cinerea* in model solutions was described by another group (Rapp and Mandery, 1988). Although the major metabolites found were ω-hydroxylation compounds, it is important to note that some new compounds that were not described by the previous group were detected (Figure 19.9). Geraniol (**271**) was mainly transformed to (2*E*,5*E*)-3,7-dimethyl-2,5-octadiene-1,7-diol (**318**), (*E*)-3,7-dimethyl-2,7-octadiene-1,6-diol (**319**), and (2*E*,6*E*)-2,6-dimethyl-2,6-octadiene-1,8-diol (**300**) and nerol (**272**) to (2*Z*,5*E*)-3,7-dimethyl-2,5-octadiene-1,7-diol (**314**), (*Z*)-3,7-dimethyl-2,6-octadiene-1,6-diol (**315**),

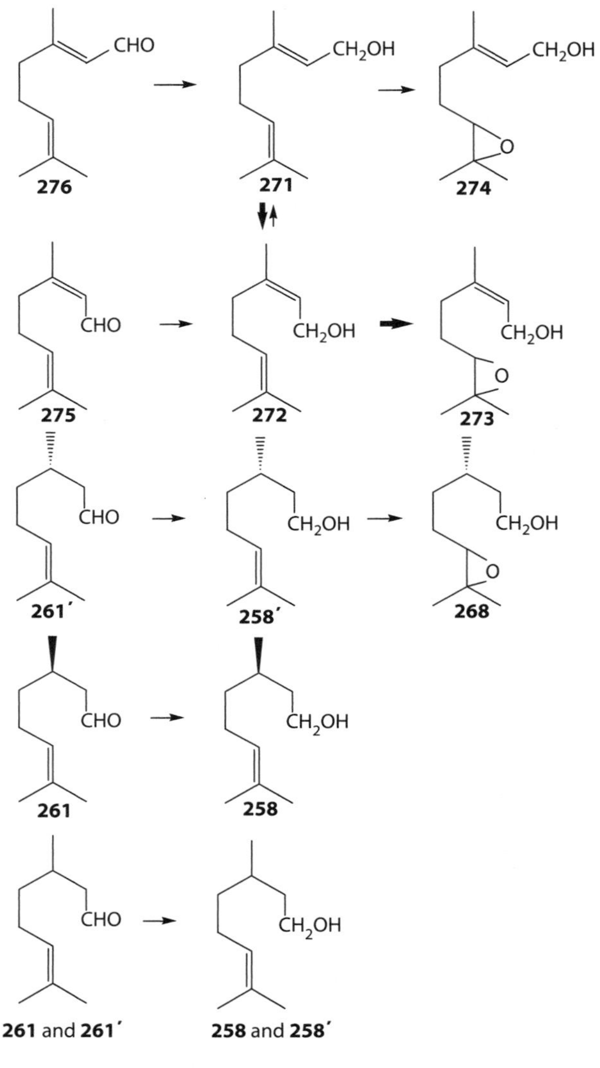

FIGURE 19.9 Reduction of terpene aldehydes and epoxidation of terpene alcohols by *Streptomyces ikutamanensis* Ya-2-1. (Modified from Noma, Y. et al., Reduction of terpene aldehydes and epoxidation of terpene alcohols by *S. ikutamanensis*, Ya-2-1, *Proceedings of 30th TEAC*, 1986, pp. 204–206.)

and (2E,6Z)-2,6-dimethyl-2,6-octadiene-1,8-diol (**316**). Furthermore, a cyclization product (**318**) that was not previously described was formed. Finally, citronellol (**258**) was converted to *trans*- (**312**) and *cis*-rose oxide (**313**) (a cyclization product not identified by the other group), (*E*)-3,7-dimethyl-5-octene-1,7-diol (**311**), 3,7-dimethyl-7-octene-1,6-diol (**260**), and (*E*)-2,6-dimethyl-2-octene-1,8-diol (**265**) (Miyazawa et al., 1996a) (Figure 19.10).

One of the latest reports in this area described the biotransformation of citronellol by the plant pathogenic fungus *Glomerella cingulata* to 3,7-dimethyl-1,6,7-octanetriol (Miyazawa et al., 1996a).

The ability of fungal spores of *P. digitatum* to biotransform monoterpene alcohols, such as geraniol (**271**) and nerol (**272**) and a mixture of the aldehydes, that is, citral (**276 and 275**), has only been discovered very recently by Demyttenaere and coworkers (Demyttenaere et al., 1996, 2000; Demyttenaere and De Pooter, 1996, 1998). Spores of *P. digitatum* were inoculated on solid media. After a short incubation period, the spores germinated and a mycelial mat was formed.

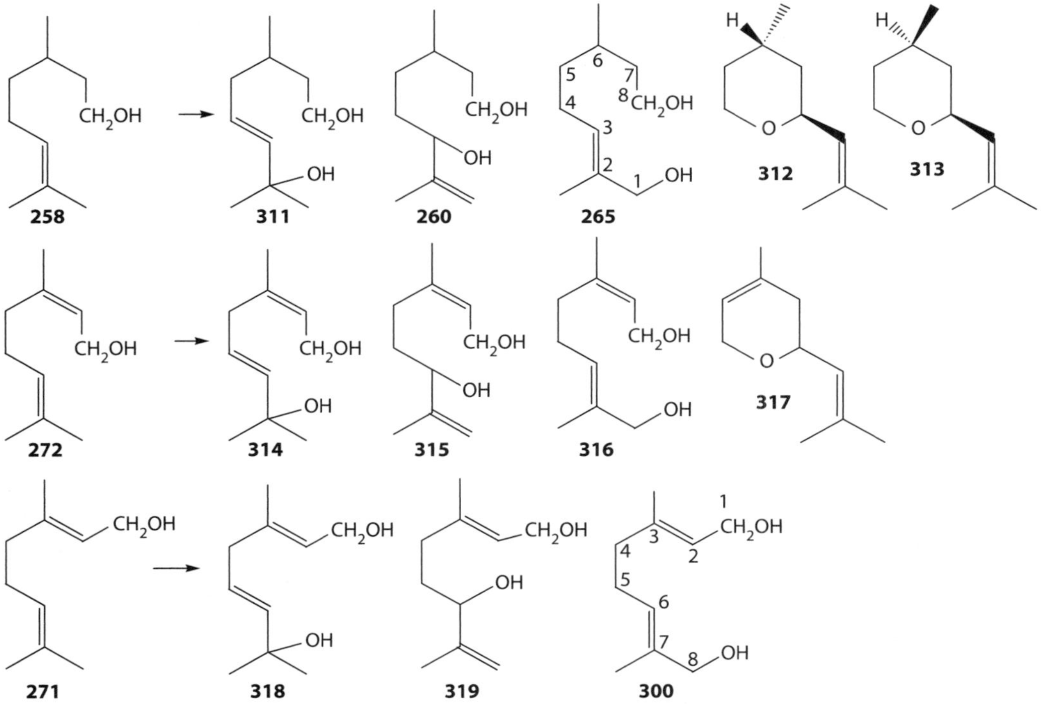

FIGURE 19.10 Biotransformation of geraniol (**271**), nerol (**272**), and citronellol (**258**) by *Botrytis cinerea*. (Modified from Miyazawa, M. et al., *Nat. Prod. Lett.*, 8, 303, 1996a.)

After 2 weeks, the culture had completely sporulated and bioconversion reactions were started. Geraniol (**271**), nerol (**272**), or citral (**276** and **275**) was sprayed onto the sporulated surface culture. After 1 or 2 days, the period during which transformation took place, the cultures were extracted. Geraniol and nerol were transformed into 6-methyl-5-hepten-2-one by sporulated surface cultures of *P. digitatum*. The spores retained their activity for at least 2 months. An overall yield of up to 99% could be achieved.

The bioconversion of geraniol (**271**) and nerol (**272**) was also performed with sporulated surface cultures of *A. niger*. Geraniol (**271**) was converted to linalool (**206**), α-terpineol (**34**), and limonene (**68**), and nerol (**272**) was converted mainly to linalool (**206**) and α-terpineol (**34**) (Demyttenaere et al., 2000).

The biotransformation of geraniol (**271**) and nerol (**272**) by *Catharanthus roseus* suspension cells was carried out. It was found that the allylic positions of geraniol (**271**) and nerol (**272**) were hydroxylated and reduced to double bond and ketones (Figure 19.11). Geraniol (**271**) and nerol (**272**) were isomerized to each other. Geraniol (**271**) and nerol (**272**) were hydroxylated at C10 to 8-hydroxygeraniol (**300**) and 8-hydroxynerol (**320**), respectively. 8-Hydroxygeraniol (**300**) was hydrogenated to 10-hydroxycitronellol (**265**). Geraniol (**271**) was hydrogenated to citronellol (**258**) (Hamada and Yasumune, 1995).

Cyanobacterium converted geraniol (**271**) to geranic acid (**278**) via geranial (**276**), followed by hydrogenation to give citronellic acid (**262**) via citronellal (**261**). Furthermore, the substrate **271** was isomerized to nerol (**272**), followed by oxidation, reduction, and further oxidation to afford neral (**275**), citronellal (**261**), citronellic acid (**262**), and nerolic acid (**277**) (Kaji et al., 2002; Hamada et al., 2004) (Figure 19.12).

Plant suspension cells of *C. roseus* converted geraniol (**271**) to 8-hydroxygeraniol (**300**). The same cells converted citronellol (**258**) to 8- (**265**) and 10-hydroxycitronellol (**264**) (Hamada et al., 2004) (Figure 19.13).

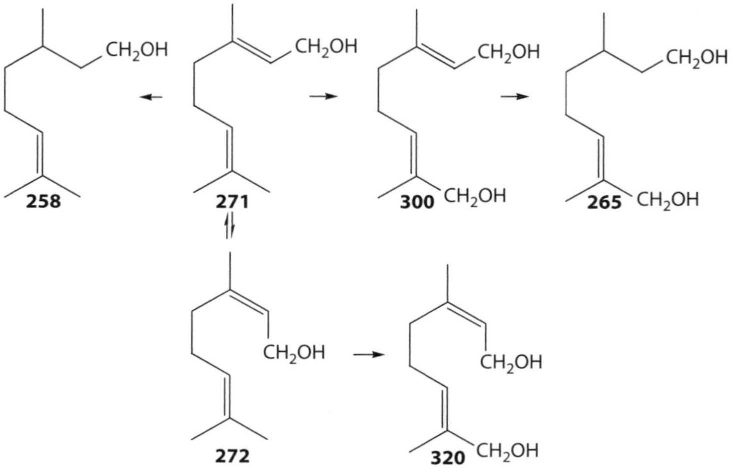

FIGURE 19.11 The biotransformation of geraniol (**271**) and nerol (**272**) by *Catharanthus roseus*. (Modified from Hamada, H. and Yasumune, H., The hydroxylation of monoterpenoids by plant cell biotransformation, *Proceedings of 39th TEAC*, 1995, pp. 375–377.)

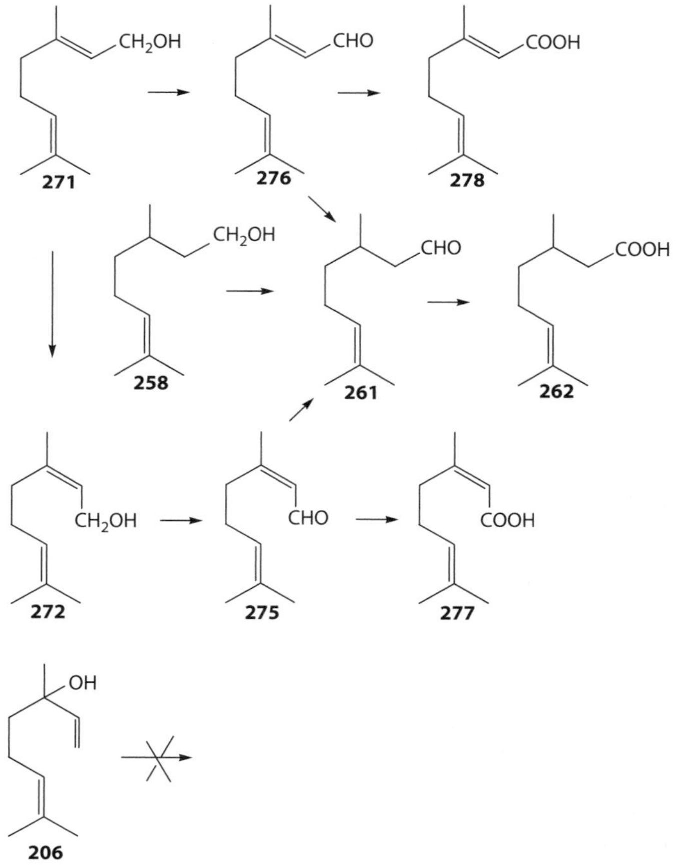

FIGURE 19.12 Biotransformation of geraniol (**271**) and citronellol (**258**) by cyanobacterium.

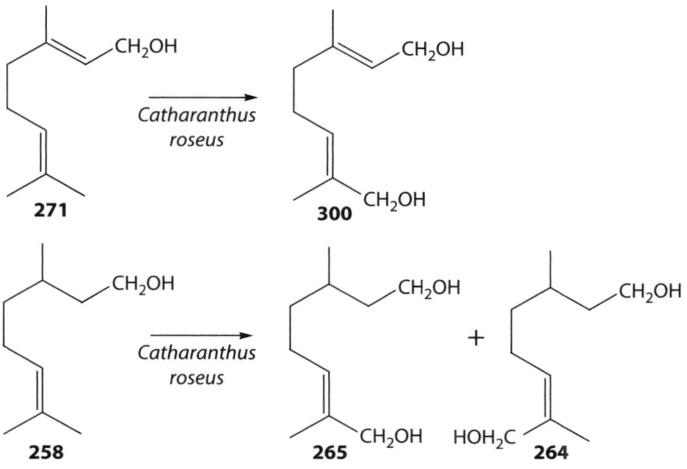

FIGURE 19.13 Biotransformation of geraniol (**271**), citronellol (**258**), and linalool (**206**) by plant suspension cells of *Catharanthus roseus*. (Modified from Hamada, H. et al., Biotransformation of acyclic monoterpenes by biocatalysts of plant cultured cells and *Cyanobacterium*, *Proceedings of 48th TEAC*, 2004, pp. 393–395.)

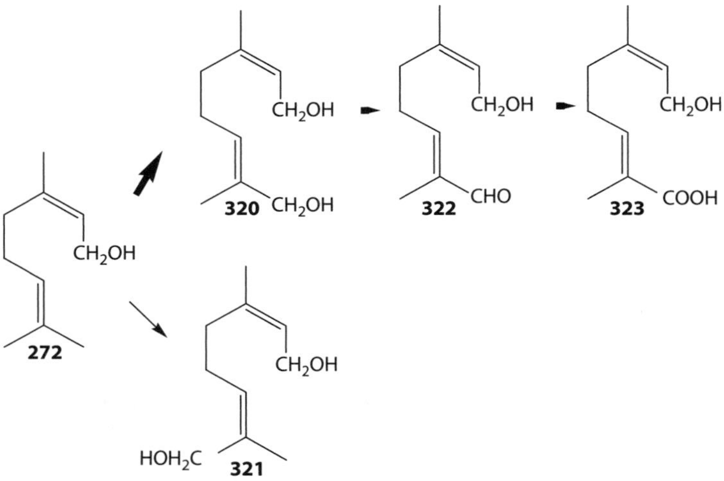

FIGURE 19.14 Biotransformation of nerol (**272**) by *Spodoptera litura*. (Modified from Takeuchi, H. and Miyazawa, M., Biotransformation of nerol by the larvae of common cutworm (*Spodoptera litura*) as a biocatalyst, *Proceedings of 48th TEAC*, 2004, pp. 399–400.)

Nerol (**272**) was converted by the insect larvae *S. litura* to give 8-hydroxynerol (**320**), 10-hydroxynerol (**321**), 1-hydroxy-3,7-dimethyl-(2*E*,6*E*)-octadienal (**322**), and 1-hydroxy-3,7-dimethyl-(2*E*,6*E*)-octadienoic acid (**323**) (Takeuchi and Miyazawa, 2004) (Figure 19.14).

19.2.2.2 Linalool and Linalyl Acetate

(+)-Linalool (**206**) [(*S*)-3,7-dimethyl-1,6-octadiene-3-ol] and its enantiomer (**206′**) [(*R*)-3,7-dimethyl-1,6-octadiene-3-ol] occur in many essential oils, where they are often the main component. (*S*)-(+)-Linalool (**206**) makes up 60%–70% of coriander oil. (*R*)-(−)-linalool (**206′**), for example, occurs at a concentration of 80%–85% in Ho oils from *Cinnamomum camphora*; rosewood oil contains ca. 80% (Bauer et al., 1990).

C. roseus converted (+)-linalool (**206**) to 8-hydroxylinalool (**219**) (Hamada et al., 2004) (Figure 19.15).

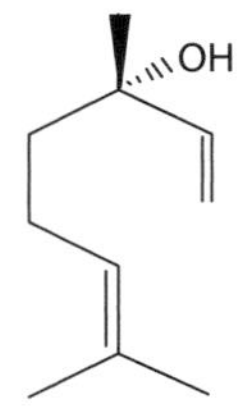

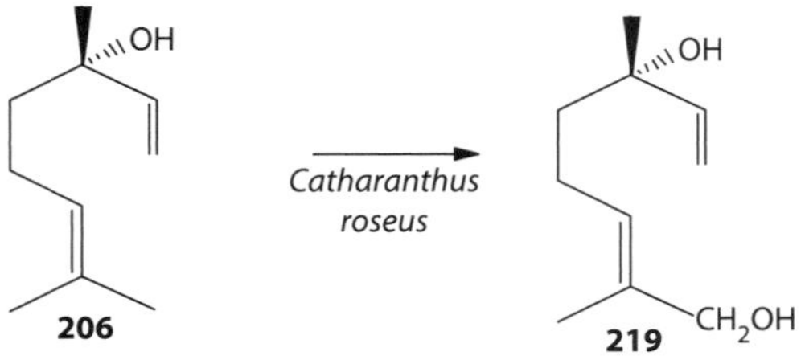

FIGURE 19.15 Biotransformation of linalool (**206**) by plant suspension cells of *Catharanthus roseus*. (Modified from Hamada, H. et al., Biotransformation of acyclic monoterpenes by biocatalysts of plant cultured cells and *Cyanobacterium*, Proceedings of 48th TEAC, pp. 393–395, 2004.)

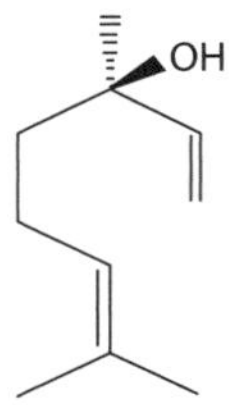

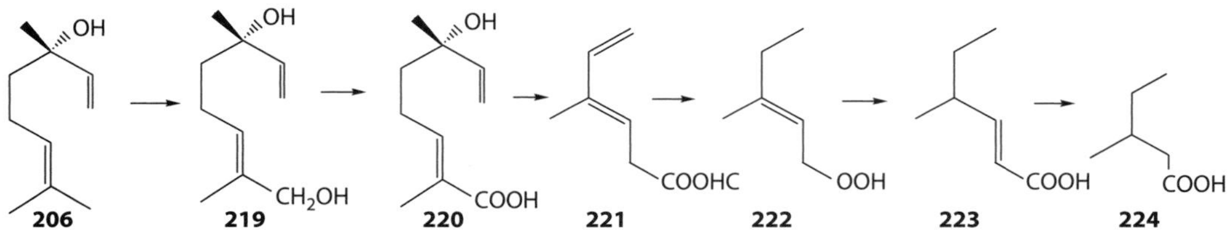

FIGURE 19.16 Degradative metabolic pathway of (+)-linalool (**206**) by *Pseudomonas pseudomallei*. (Modified from Murakami, T. et al., *Nippon Nogei Kagaku Kaishi*, 47, 699, 1973.)

The biodegradation of (+)-linalool (**206**) by *Pseudomonas pseudomallei* (strain A), which grows on linalool as the sole carbon source, was described in 1973 (Murakami et al., 1973) (Figure 19.16).

Madyastha et al. (1977) isolated a soil Pseudomonad, *P. incognita*, by the enrichment culture technique with linalool as the sole carbon source. This microorganism, the "linalool strain" as it was called, was also capable of utilizing limonene (**68**), citronellol (**258**), and geraniol (**271**) but failed to grow on citral (**275** and **276**), citronellal (**261**), and 1,8-cineole (**122**). Fermentation was carried out with shake cultures containing 1% linalool (**206**) as the sole carbon source. It was suggested by the authors that linalool (**206**) was metabolized by at least three different pathways of biodegradation (Figure 19.17). One of the pathways appeared to be initiated by the specific oxygenation of C-8 methyl group of linalool (**206**), leading to 8-hydroxylinalool (**219**), which was further oxidized to linalool-8-carboxylic acid (**220**). The presence of furanoid linalool oxide (**215**) and 2-methyl-2-vinyltetrahydrofuran-5-one (**216**) as the unsaturated lactone in the fermentation medium suggested another mode of utilization of linalool (**206**). The formation of these compounds was believed to proceed through the epoxidation of the 6,7-double bond giving rise to 6,7-epoxylinalool (**214**), which upon further oxidation yielded furanoid linalool oxide (**215**) and 2-methyl-2-vinyltetrahydrofuran-5-one (**216**) (Figure 19.17).

The presence of oleuropeic acid (**204**) in the fermentation broth suggested a third pathway. Two possibilities were proposed: (3a) water elimination giving rise to a monocyclic cation (**33**), yielding α-terpineol (**34**), which upon oxidation gave oleuropeic acid (**204**), and (3b) oxidation of the C-10

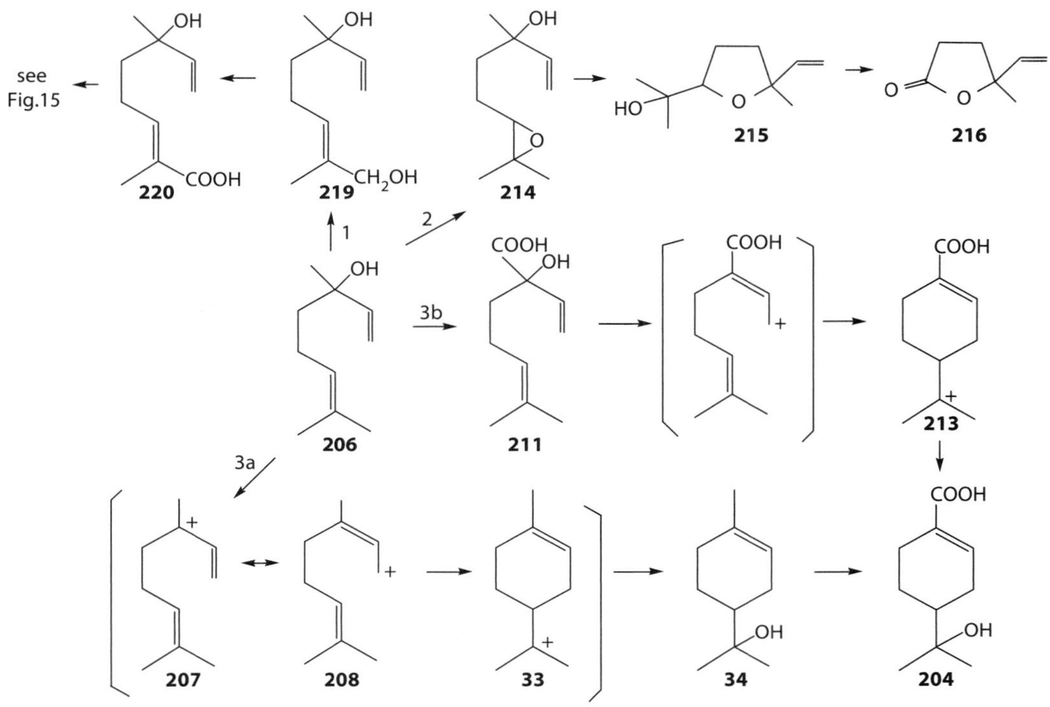

FIGURE 19.17 Biotransformation of linalool (**206**) by *Pseudomonas incognita* (Madyastha et al., 1977) and *Streptomyces albus* NRRL B1865. (Modified from David, L. and Veschambre, H., *Tetrahadron Lett.*, 25, 543, 1984.)

methyl group of linalool (**206**) before cyclization, giving rise to oleuropeic acid (**204**). This last pathway was also called the "prototropic cyclization" (Madyastha, 1984).

Racemic linalool (**206** and **206′**) is cyclized into *cis-* and *trans-*linalool oxides by various microorganisms such as *Streptomyces albus* NRRL B1865, *Streptomyces hygroscopicus* NRRL B3444, *Streptomyces cinnamonensis* ATCC 15413, *Streptomyces griseus* ATCC 10137, and *Beauveria sulfurescens* ATACC 7159 (David and Veschambre, 1984) (Figure 19.17).

A. niger isolated from garden soil biotransformed linalool and its acetates to give linalool (**206**), 2,6-dimethyl-2,7-octadiene-1,6-diol (8-hydroxylinalool **219a**), α-terpineol (**34**), geraniol (**271**), and some unidentified products in trace amounts (Madyastha and Krishna Murthy, 1988a,b).

The biotransformation of linalool (**206**) by *B. cinerea* was carried out and identified transformation products such as (*E*)-(**219a**) and (*Z*)-2,6-dimethyl-2,7-octadiene-1,6-diol (**219b**), *trans-* (**215a**) and *cis-*furanoid linalool oxide (**215b**), *trans-* (**217a**) and *cis-*pyranoid linalool oxide (**217b**) (Figure 19.18) and their acetates (**217a-Ac, 217b-Ac**), 3,9-epoxy-*p*-menth-1-ene (**324**) and 2-methyl-2-vinyltetrahydrofuran-5-one (**216**) (unsaturated lactone) (Bock et al., 1986) (Figure 19.19). Quantitative analysis, however, showed that linalool (**206**) was predominantly (90%) metabolized to (*E*)-2,6-dimethyl-2,7-octadiene-1,6-diol (**219a**) by *B. cinerea*. The other compounds were only found as by-products in minor concentrations.

The bioconversion of (*S*)-(+)-linalool (**206**) and (*R*)-(−)-linalool (**206′**) was investigated with *D. gossypina* ATCC 10936 (Abraham et al., 1990). The biotransformation of (±)-linalool (**206** and **206′**) by *A. niger* ATCC 9142 with submerged shaking culture yielded a mixture of *cis-* (**215b**) and *trans-*furanoid linalool oxide (**215a**) (yield 15%–24%) and *cis-* (**217b**) and *trans-*pyranoid linalool oxide (**217a**) (yield 5%–9%) (Demyttenaere and Willemen, 1998). The biotransformation of (*R*)-(−)-linalool (**206a**) with *A. niger* ATCC 9142 yielded almost pure *trans-*furanoid linalool oxide

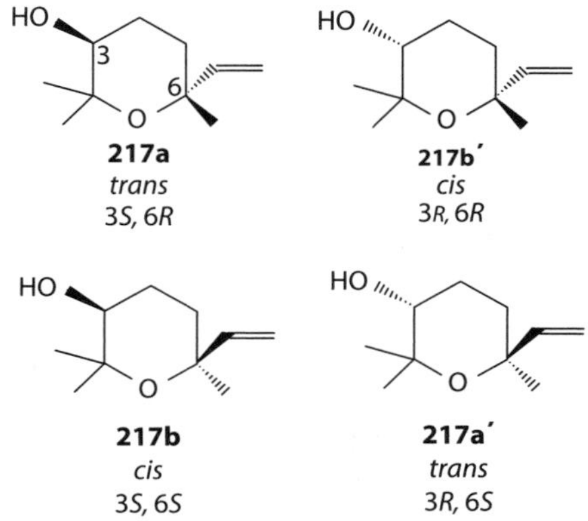

FIGURE 19.18 Four stereoisomers of pyranoid linalool oxides.

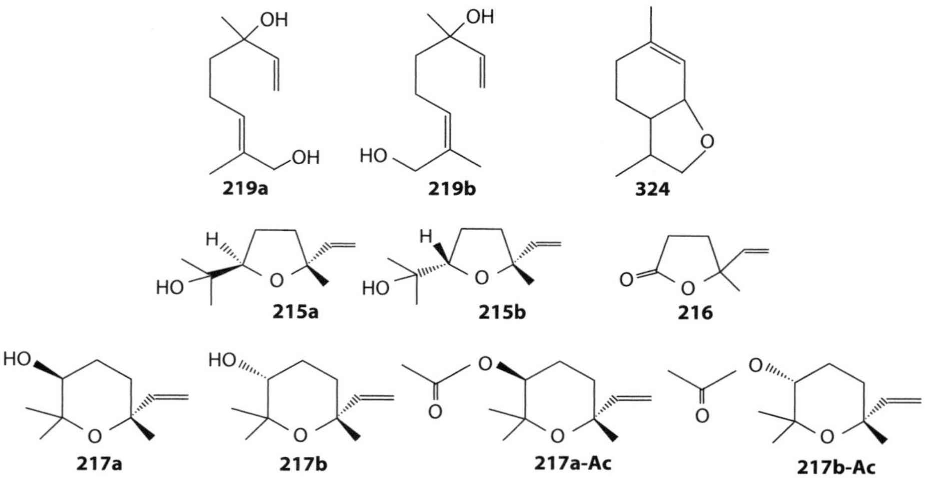

FIGURE 19.19 Biotransformation products of linalool (**206**) by *Botrytis cinerea*. (Modified from Bock, G. et al., *J. Food Sci.*, 51, 659, 1986.)

(**215a**) and *trans*-pyranoid linalool oxide (**217a**) (ee > 95) (Figure 19.20). These conversions were purely biocatalytic, since in acidified water (pH < 3.5) almost 50% linalool (**206**) was recovered unchanged, and the rest was evaporated. The biotransformation was also carried out with growing surface cultures.

S. ikutamanensis Ya-2-1 also converted (+)- (**206**), (−)- (**206′**), and racemic linalool (**206** and **206′**) via corresponding 2,3-epoxides (**214** and **214′**) to *trans*- and *cis*-furanoid linalool oxides (**215a, b, a′** and **b′**) (Noma et al., 1986) (Figure 19.21). The absolute configuration at C-3 and C-6 of *trans*- and *cis*-linalool oxides is shown in Figure 19.22.

Biotransformation of racemic *trans*-pyranoid linalool oxide (**217a** and **a′**) and racemic *cis*-linalool-pyranoid (**217b** and **b′**) has been carried out using fungus *G. cingulata* (Miyazawa et al., 1994b). *trans*- and *cis*-Pyranoid linalool oxide (**217a** and **217b**) were transformed to *trans*- (**217a′-1**) and *cis*-linalool oxide-3-malonate (**217b′-1**), respectively. In the biotransformation of racemic *cis*-linalool oxide-pyranoid, (+)-(3R,6R)-*cis*-pyranoid linalool oxide (**217a** and **a′**) was converted to (3R,6R)-pyranoid-*cis*-linalool oxide-3-malonate (**217a′-1**). (−)-(3S, 6S)-*cis*-Pyranoid linalool oxide-pyranoid

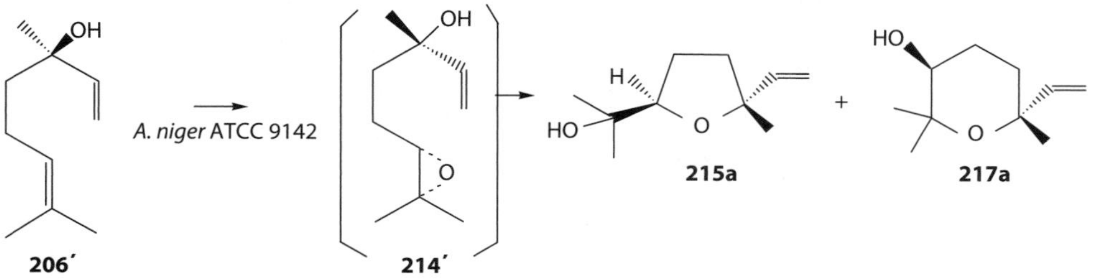

FIGURE 19.20 Biotransformation of (R)-(−)-linalool (**206′**) by *Aspergillus niger* ATCC 9142. (Modified from Demyttenaere, J.C.R. and Willemen, H.M., *Phytochemistry*, 47, 1029, 1998.)

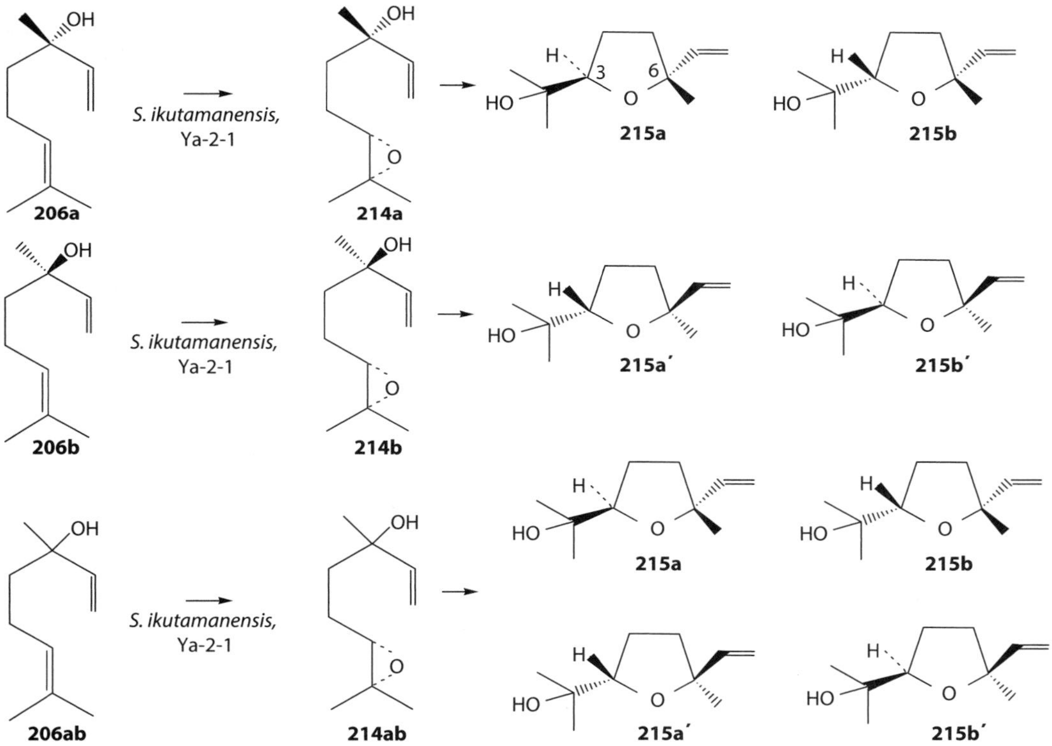

FIGURE 19.21 Metabolic pathway of (+)-(**206**), (−)-(**206′**), and racemic linalool (**206** and **206′**) by *Streptomyces ikutamanensis* Ya-2-1. (Modified from Noma, Y. et al., Reduction of terpene aldehydes and epoxidation of terpene alcohols by *S. ikutamanensis*, Ya-2-1, *Proceedings of 30th TEAC*, 1986, pp. 204–206.)

(**217a′**) was not metabolized. On the other hand, in the biotransformation of racemic *trans*-pyranoid linalool oxide (**217b** and **b′**), (−)-(3R,6S)-*trans*-linalool oxide (**217b′**) was transformed to (3R,6S)-*trans*-linalool oxide-3-malonate (**217b′-1**) (Figure 19.23). (+)-(3S,6S)-*trans*-Pyranoid-linalool oxide (**217b**) was not metabolized. These facts showed that *G. cingulata* recognized absolute configuration of the secondary hydroxyl group at C-3. On the basis of this result, it has become apparent that the optical resolution of racemic pyranoid linalool oxide proceeded in the biotransformation with *G. cingulata* (Miyazawa et al., 1994b).

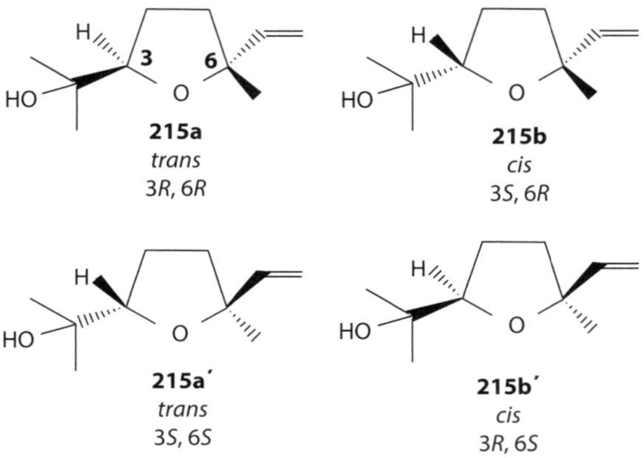

FIGURE 19.22 Four stereoisomers of furanoid linalool oxides. (Modified from Noma, Y. et al., Reduction of terpene aldehydes and epoxidation of terpene alcohols by *S. ikutamanensis*, Ya-2-1, *Proceedings of 30th TEAC*, 1986, pp. 204–206.)

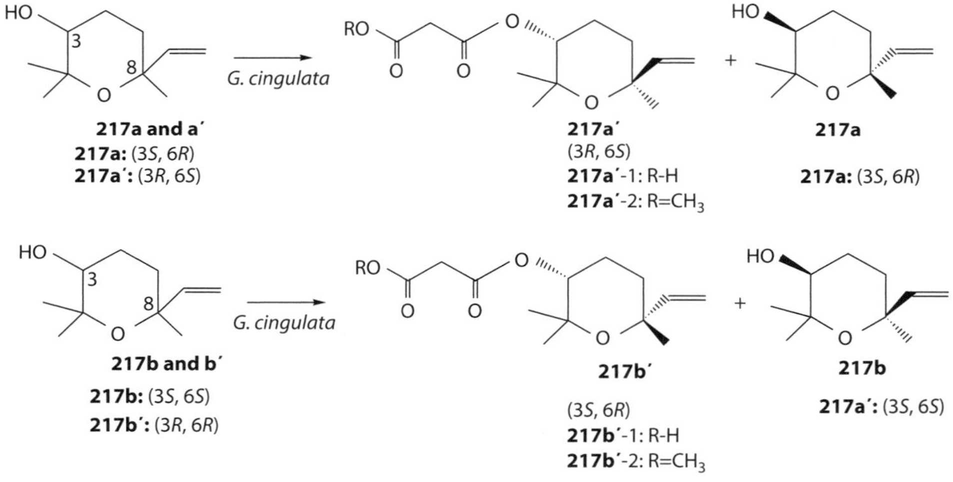

FIGURE 19.23 Biotransformation of racemic *trans*-linalool oxide-pyranoid (**217a** and **a′**) and racemic *cis*-linalool-pyranoid (**217b** and **b′**) by *Glomerella cingulata*. (Modified from Miyazawa, M. et al., Biotransformation of linalool oxide by plant pathogenic microorganisms, *Glomerella cingulata*, *Proceedings of 38th TEAC*, 1994a, pp. 101–102.)

Linalool (**206**) and tetrahydrolinalool (**325**) were converted by suspension cells of *C. roseus* to give 1-hydroxylinalool (**219**) from linalool (**206**) and 3,7-dimethyloctane-3,5-diol (**326**), 3,7-dimethyloctane-3,7-diol (**327**), and 3,7-dimethyloctane-3,8-diol (**328**) from tetrahydrolinalool (**325**) (Hamada and Furuya, 2000; Hamada et al., 2004) (Figure 19.24).

(±)-Linalyl acetate (**206-Ac**) was hydrolyzed to (+)-(*S*)-linalool (**206**) and (±)-linalyl acetate (**206-Ac**) by *Bacillus subtilis*, *Trichoderma S*, *Absidia glauca*, and *Gibberella fujikuroi* as shown

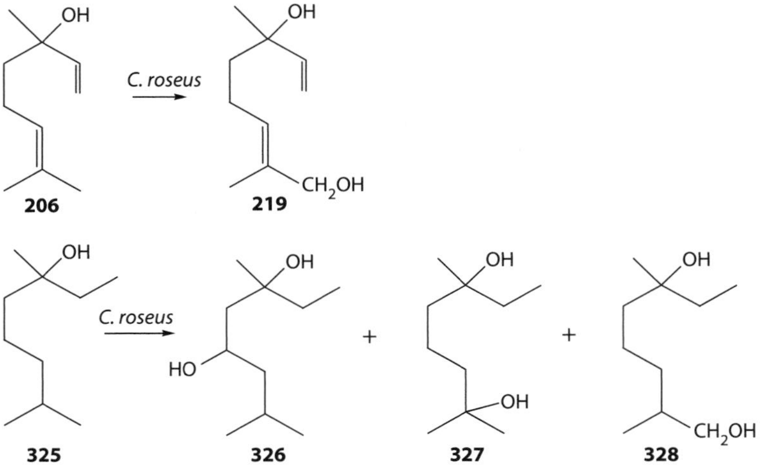

FIGURE 19.24 Biotransformation of linalool (**206**) and tetrahydrolinalool (**325**) by *Catharanthus roseus*. (Modified from Hamada, H. and Furuya, T., Hydroxylation of monoterpenes by plant suspension cells, *Proceedings of. 44th TEAC*, pp. 167–168, 2000; Hamada, H. et al.,. Biotransformation of acyclic monoterpenes by biocatalysts of plant cultured cells and *Cyanobacterium*, *Proceedings of 48th TEAC*, 2004, pp. 393–395.)

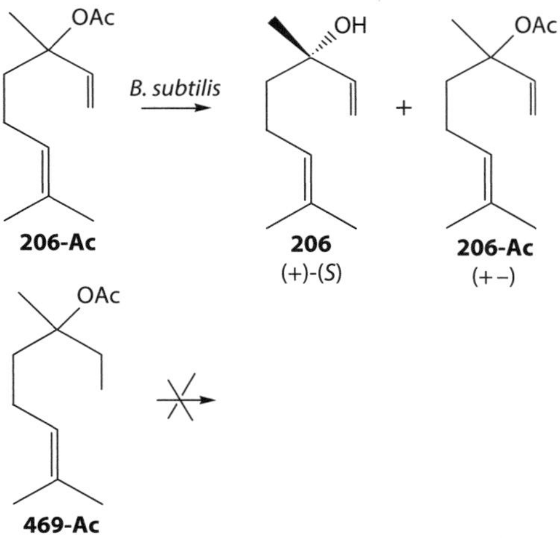

FIGURE 19.25 Hydrolysis of (±)-linalyl acetate (**206-Ac**) by microorganisms. (Modified from Oritani, T. and Yamashita, K., *Agric. Biol. Chem.*, 37, 1923, 1973a.)

in Figure 19.25. But (±)-dihydrolinalyl acetate (**469-Ac**) was not hydrolyzed by the aforementioned microorganisms (Oritani and Yamashita, 1973a).

19.2.2.3 Dihydromyrcenol

Dihydromyrcenol (**329**) was fed by *S. litura* to give 1,2-epoxydihydromyrcenol (**330**) as a main product and 3β-hydroxydihydromyrcenol (**331**) as a minor product. Dihydromyrcenyl acetate (**332**) was converted to 1,2-dihydroxydihydromyrcenol acetate (**333**) (Murata and Miyazawa, 1999) (Figure 19.26).

FIGURE 19.26 Biotransformation of dihydromyrcenol (**329**) and dihydromyrcenyl acetate (**332**) by *Spodoptera litura*. (Modified from Murata, T. and Miyazawa, M., Biotransformation of dihydromyrcenol by common cutworm larvae, *Spodoptera litura* as a biocatalyst, *Proceedings of 43rd TEAC*, 1999, pp. 393–394.)

19.3 METABOLIC PATHWAYS OF CYCLIC MONOTERPENOIDS

19.3.1 MONOCYCLIC MONOTERPENE HYDROCARBON

19.3.1.1 Limonene

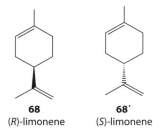

Limonene is the most widely distributed terpene in nature after α-pinene (**4**) (Krasnobajew, 1984). (4*R*)-(+)-Limonene (**68**) is present in citrus peel oils at a concentration of over 90%; a low concentration of the (4*S*)-(−)-limonene (**68′**) is found in oils from the *Mentha* species and conifers (Bauer et al., 1990). The first microbial biotransformation on limonene was carried out by using a soil Pseudomonad. The microorganism was isolated by the enrichment culture technique on limonene as the sole source of carbon (Dhavalikar and Bhattacharyya, 1966). The microorganism was also capable of growing on α-pinene (**4**), β-pinene (**1**), 1-*p*-menthene (**62**), and *p*-cymene (**178**). The optimal level of limonene for growth was 0.3%–0.6% (v/v) although no toxicity was observed at 2% levels. Fermentation of limonene (**68**) by this bacterium in a mineral salt medium resulted in the formation of a large number of neutral and acidic products such as dihydrocarvone (**64**), carvone (**61**), carveol (**60**), 8-*p*-menthene-1,2-*cis*-diol (**65b**), 8-*p*-menthen-1-ol-2-one (**66**), 8-*p*-menthene-1,2-*trans*-diol (**65a**), and 1-*p*-menthene-6,9-diol (**62**). Perillic acid (**69**), β-isopropenyl pimaric acid (**72**), 2-hydroxy-8-*p*-menthen-7-oic acid (**70**), and 4,9-dihydroxy-1-*p*-menthen-7-oic acid (**73**) were isolated and identified as acidic compounds. Based on these data, three distinct pathways for the catabolism of limonene (**68**) by the soil Pseudomonad were proposed by Dhavalikar et al. (1966), involving allylic oxygenation (pathway 1), oxygenation of the 1,2-double bond (pathway 2), and progressive oxidation

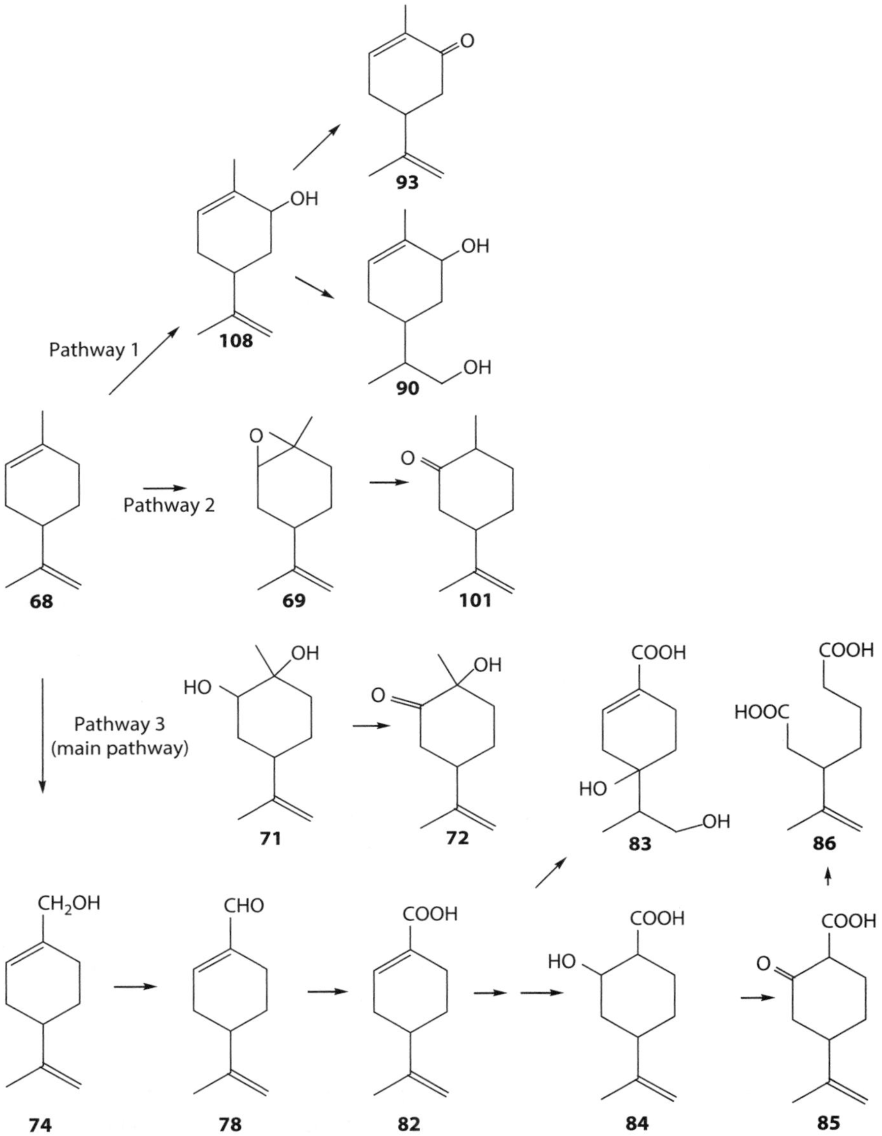

FIGURE 19.27 Pathways for the degradation of limonene (**68**) by a soil Pseudomonad sp. strain (L). (Modified from Krasnobajew, V., in: *Biotechnology*, Kieslich, K., ed., vol. 6a, Verlag Chemie, Weinheim, Germany, 1984, pp. 97–125.)

of the 7-methyl group to perillic acid (**82**) (pathway 3) (Figure 19.27) (Krasnobajew, 1984). Pathway 2 yields (+)-dihydrocarvone (**101**) via intermediate limonene epoxide (**69**) and 8-*p*-menthen-1-ol-2-one (**72**) as oxidation product of limonene-1,2-diol (**71**). The third and main pathway leads to perillyl alcohol (**74**), perillaldehyde (**78**), perillic acid (**82**), constituents of various essential oils and used in the flavor and fragrance industry (Fenaroli, 1975), 2-oxo-8-*p*-menthen-7-oic acid (**85**), β-isopropenyl pimaric acid (**86**), and 4,9-dihydroxy-1-*p*-menthene-7-oic acid (**83**).

(+)-Limonene (**68**) was biotransformed via limonene-1,2-epoxide (**69**) to 8-*p*-menthene 1,2-*trans*-diol (**71b**). On the other hand, (+)-carvone (**93**) was biotransformed via (−)-isodihydrocarvone (**101b**) and 1α-hydroxydihydrocarvone (**72**) to (+)-8-*p*-menthene-1,2-*trans*-diol (**71a**) (Noma et al., 1985a,b)

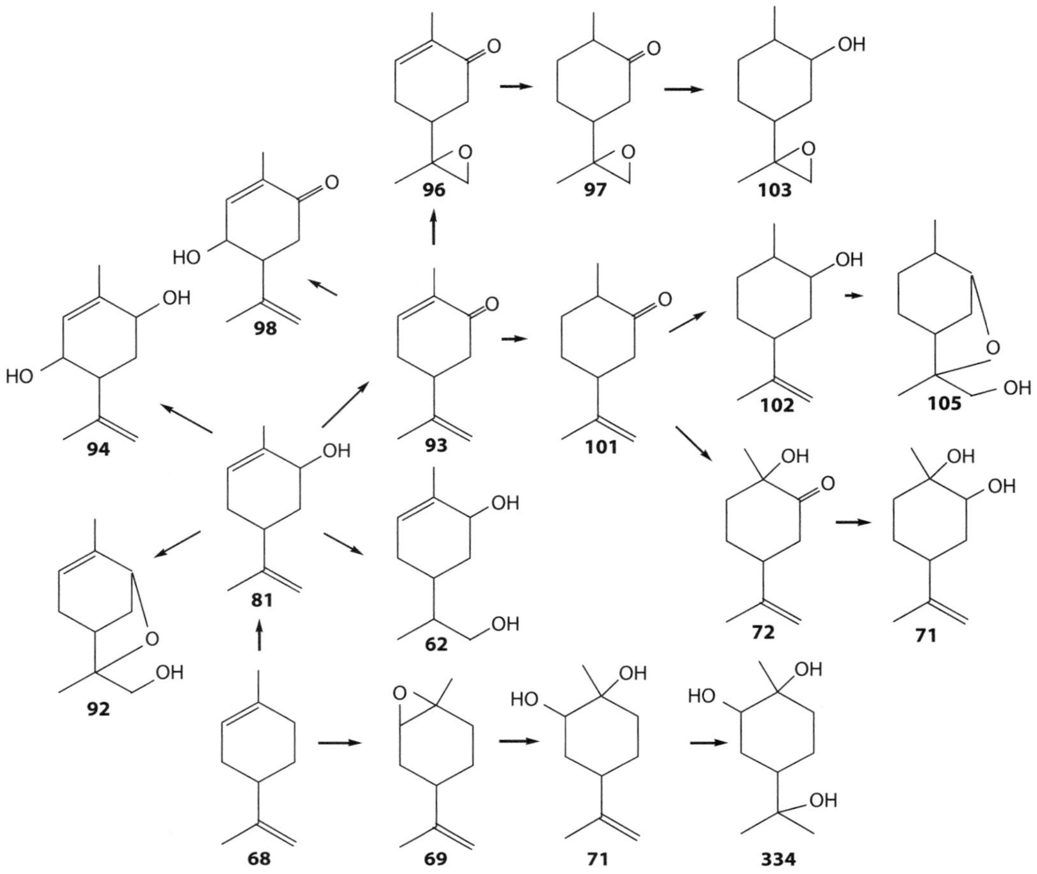

FIGURE 19.28 Formation of (+)-8-*p*-menthene-1,2-*trans*-diol (**71b**) in the biotransformation of (+)-limonene (**68**) and (+)-carvone (**93**) by *Aspergillus niger* TBUYN-2. (Modified from Noma, Y. et al., *Annual Meeting of Agricultural and Biological Chemistry*, Sapporo, p. 68; Noma, Y. et al., Biotransformation of carvone. 6. Biotransformation of (−)-carvone and (+)-carvone by a strain of *Aspergillus niger*, *Proceedings of 29th TEAC*, 1985a, pp. 235–237.)

(Figure 19.28). A soil Pseudomonad formed 1-hydroxydihydrocarvone (**72**) and 8-*p*-menthene-1,2-*trans*-diol (**71b**) from (+)-limonene (**68**). Dhavalikar and Bhattacharyya (1966) considered that the formation of 1-hydroxydihyFdrocarvone (**66**) is from dihydrocarvone (**64**).

Pseudomonas gladioli was isolated by an enrichment culture technique from pine bark and sap using a mineral salt broth with limonene as the sole carbon source (Cadwallander et al., 1989; Cadwallander and Braddock, 1992). Fermentation was performed during 4–10 days in shake flasks at 25°C using a pH 6.5 mineral salt medium and 1.0% (+)-limonene (**68**). Major products were identified as (+)-α-terpineol (**34**) and (+)-perillic acid (**82**). This was the first report of the microbial conversion of limonene to (+)-α-terpineol (**34**).

The first data on fungal bioconversion of limonene (**68**) date back to the late 1960s (Kraidman et al., 1969; Noma, 2007). Three soil microorganisms were isolated on and grew rapidly in mineral salt media containing appropriate terpene substrates as sole carbon sources. The microorganisms belonged to the class Fungi Imperfecti, and they had been tentatively identified as *Cladosporium* species. One of these strains, designated as *Cladosporium* sp. T$_7$, was isolated on (+)-limonene (**68a**). The growth medium of this strain contained 1.5 g/L of *trans*-limonene-1,2-diol (**71a**). Minor quantities of the corresponding *cis*-1,2-diol (**71b**) were also isolated. The same group isolated a

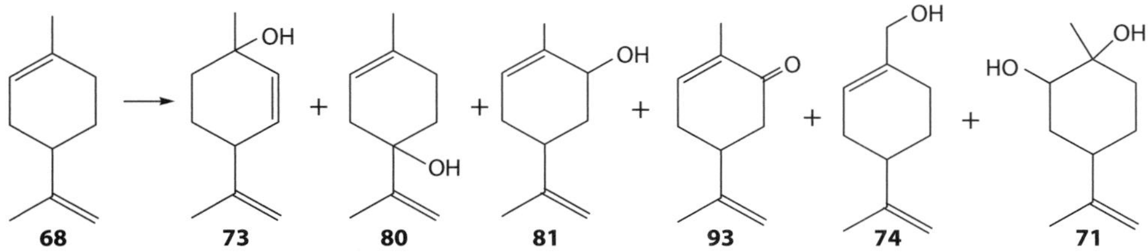

FIGURE 19.29 Biotransformation products of limonene (**68**) by *Penicillium digitatum* and *Penicillium italicum*. (Modified from Bowen, E.R., *Proc. Fla. State Hortic. Soc.*, 88, 304, 1975.)

fourth microorganism from a terpene-soaked soil on mineral salt media containing (+)-limonene as the sole carbon source (Kraidman et al., 1969). The strain, *Cladosporium*, designated T_{12}, was capable of converting (+)-limonene (**68a**) into an optically active isomer of α-terpineol (**34**) in yields of approximately 1.0 g/L.

α-Terpineol (**34**) was obtained from (+)-limonene (**68**) by fungi such as *P. digitatum*, *Penicillium italicum*, and *Cladosporium* and several bacteria (Figure 19.29). (+)-*cis*-Carveol (**81b**), (+)-carvone (**93**) (an important constituent of caraway seed and dill-seed oils) (Fenaroli, 1975; Bouwmester et al., 1995), and 1-*p*-menthene-6,9-diol (**90**) were also obtained by *P. digitatum* and *P. italicum*. (+)-(*S*)-Carvone (**93**) is a natural potato sprout–inhibiting, fungistatic, and bacteriostatic compound (Oosterhaven et al., 1995a,b). It is important to note that (−)-carvone (**93′**, the *spearmint flavor*) was not yet described in microbial transformation (Krasnobajew, 1984). However, the biotransformation of limonene to (−)-carvone (**93′**) was patented by a Japanese group (Takagi et al., 1972). *Corynebacterium* species grown on limonene was able to produce about 10 mg/L of 99% pure (−)-carvone (**93′**) in 24–48 h.

Mattison et al. (1971) isolated *Penicillium* sp. cultures from rotting orange rind that utilized limonene (**68**) and converted it rapidly to α-terpineol (**34**). Bowen (1975) isolated two common citrus molds, *P. italicum* and *P. digitatum*, responsible for the postharvest diseases of citrus fruits. Fermentation of *P. italicum* on limonene (**68**) yielded *cis*- (**81b**) and *trans*-carveol (**81a**) (26%) as the main products, together with *cis*- and *trans*-*p*-mentha-2,8-dien-1-ol (**73**) (18%), (+)-carvone (**93′**) (6%), *p*-mentha-1,8-dien-4-ol (**80**) (4%), perillyl alcohol (**74**) (3%), and 8-*p*-menthene-1,2-diol (**71**) (3%). Conversion of **68** by *P. digitatum* yielded the same products in lower yields (Figure 19.29).

The biotransformation of limonene (**68**) by *A. niger* is a very important example of fungal bioconversion. Screening for fungi capable of metabolizing the bicyclic hydrocarbon terpene α-pinene (**4**) yielded a strain of *A. niger* NCIM 612 that was also able to transform limonene (**68**) (Rama Devi and Bhattacharyya, 1978). This fungus was able to carry out three types of oxygenative rearrangements α-terpineol (**34**), carveol (**81**), and *p*-mentha-2,8-dien-1-ol (**73**) (Rama Devi and Bhattacharyya, 1978) (Figure 19.30). In 1985, Abraham et al. (1985) investigated the biotransformation of (*R*)-(+)-limonene (**68a**) by the fungus *P. digitatum*. A complete transformation for the substrate to α-terpineol (**34**) by *P. digitatum* DSM 62840 was obtained with 46% yield of pure product.

The production of glycols from limonene (**68**) and other terpenes with a 1-menthene skeleton was reported by *Corynespora cassiicola* DSM 62475 and *D. gossypina* ATCC 10936 (Abraham et al., 1984). Accumulation of glycols during fermentation was observed. An extensive overview on the microbial transformations of terpenoids with a 1-*p*-menthene skeleton was published by Abraham et al. (1986).

The biotransformation of (+)-limonene (**68**) was carried out by using *Aspergillus cellulosae* M-77 (Noma et al., 1992b) (Figures 19.31 and 19.32). It is important to note that (+)-limonene (**68a**) was mainly converted to (+)-isopiperitenone (**111**) (19%) as new metabolite, (1*S*,2*S*,4*R*)-(+)-limonene-1,2-*trans*-diol (**71a**) (21%), (+)-*cis*-carveol (**81b**) (5%), and (+)-perillyl alcohol (**74**) (12%) (Figure 19.32).

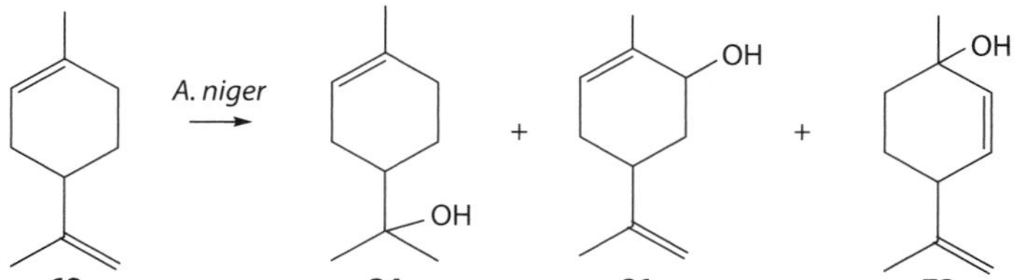

FIGURE 19.30 Biotransformation of limonene (**68**) by *Aspergillus niger* NCIM 612. (Modified from Rama Devi, J. and Bhattacharyya, P.K., *J. Indian Chem. Soc.*, 55, 1131, 1978.)

(+)-Limonene (**68**) was biotransformed by a kind of citrus pathogenic fungi, *P. digitatum* (Pers.; Fr.) Sacc. KCPYN., to isopiperitenone (**111**, 7% GC ratio), 2α-hydroxy-1,8-cineole (**125b**, 7%), (+)-limonene-1,2-*trans*-diol (**71a**, 6%), and (+)-*p*-menthane-1β,2α,8-triol (**334**, 45%) as main products and (+)-*trans*-sobrerol (**95a**, 2%), (+)-*trans*-carveol (**81a**), (+)-carvone (**93**), (−)-isodihydrocarvone (**101b**), and (+)-*trans*-isopiperitenol (**110a**) as minor products (Noma and Asakawa, 2006a, 2007a) (Figure 19.33). The metabolic pathways of (+)-limonene by *P. digitatum* are shown in Figure 19.34.

On the other hand, (−)-limonene (**68'**) was also biotransformed by a kind of citrus pathogenic fungi, *P. digitatum* (Pers.; Fr.) Sacc. KCPYN., to give isopiperitenone (**111'**), 2α-hydroxy-1,8-cineole (**125b'**), (−)-limonene-1,2-*trans*-diol (**71'**), and *p*-menthane-1,2,8-triol (**334'**) as main products together with (+)-*trans*-sobrerol (**80'**), (+)-*trans*-carveol (**81a'**), (−)-carvone (**93'**), (−)-dihydrocarvone (**101a'**), and (+)-isopiperitenol (**110a'**) as minor products (Noma and Asakawa, 2007b) (Figure 19.35.)

Newly isolated unidentified red yeast, *Rhodotorula* sp., converted (+)-limonene (**68**) mainly to (+)-limonene-1,2-*trans*-diol (**71a**), (+)-*trans*-carveol (**81a**), (+)-*cis*-carveol (**81b**), and (+)-carvone (**93'**) together with (+)-limonene-1,2-*cis*-diol (**71b**) as minor product (Noma and Asakawa, 2007b) (Figure 19.36).

Cladosporium sp. T₇ was cultivated with (+)-limonene (**68**) as the sole carbon source; it converted **68** to *trans-p*-menthane-1,2-diol (**71a**) (Figure 19.36) (Mukherjee et al., 1973).

On the other hand, the same red yeast converted (−)-limonene (**68'**) mainly to (−)-limonene-1,2-*trans*-diol (**71a'**), (−)-*trans*-carveol (**81a'**), (−)-*cis*-carveol (**81b'**), and (−)-carvone (**93'**) together with (−)-limonene-1,2-*cis*-diol (**71b'**) as minor product (Noma and Asakawa, 2007b) (Figure 19.37).

The biotransformation of (+)- and (−)-limonene (**68** and **68'**), (+)- and (−)-α-terpineol (**34** and **34'**), (+)- and (−)-limonene-1,2-epoxide (**69** and **69'**), and caraway oil was carried out by citrus pathogenic fungi *Penicillium* (Pers.; Fr.) Sacc. KCPYN and newly isolated red yeast, a kind of *Rhodotorula* sp. *P. digitatum* KCPYN converted limonenes (**68** and **68'**) to the corresponding isopiperitone (**111** and **111'**), 1α-hydroxy-1,8-cineole (**125b** and **125b'**), limonene-1,2-*trans*-diol (**71a** and **71a'**), *p*-menthane-1,2,8-triol (**334** and **334'**), and *trans*-sobrerol as main products. (+)- and (−)-α-Terpineol (**34** and **34'**) were the precursors of 2α-hydroxy-1,8-cineole (**125b** and **b'**) and *p*-menthane-1,2,8-triol (**334**). (+)- and (−)-Limonene-1,2-epoxide (**69** and **69'**) were also the precursor of limonene-1,2-*trans*-diol (**71a**). *Rhodotorula* sp. also biotransformed (+)- and (−)-limonene (**68** and **68'**) to the corresponding *trans*- and *cis*-carveols (**81a** and **b**) as main products. This microbe also converted caraway oil, equal mixture of (+)-limonene (**68**) and (+)-carvone (**93**). (+)-Limonene (**68**) disappeared and (+)-carvone (**93**) was produced and accumulated in the cultured broth (Noma and Asakawa, 2007b).

(4*S*)-(−)- (**68'**) and (4*R*)-(+)-Limonene (**68**) and their epoxides (**69** and **69'**) were incubated by cyanobacterium. It was found that the transformation was enantio- and regioselective. Cyanobacterium biotransformed only (4*S*)-limonene (**68'**) to (−)-*cis*- (**81b'**, 11.1%) and (−)-*trans*-carveol (**81a'**, 5%)

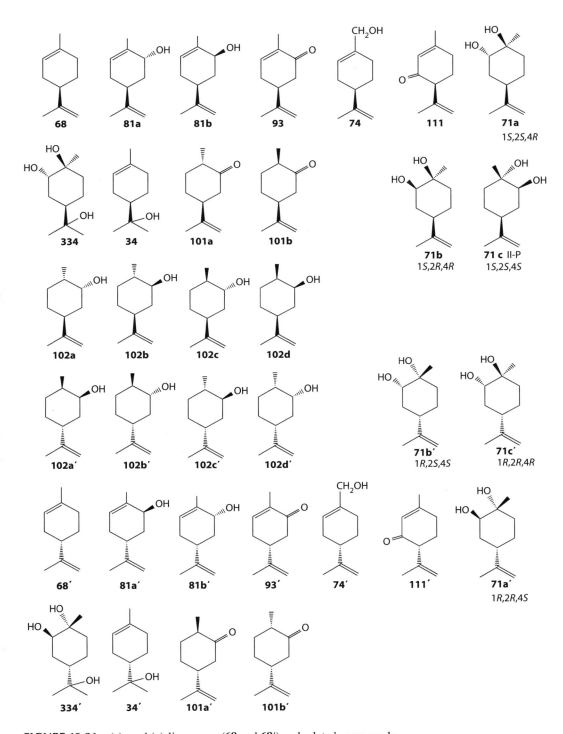

FIGURE 19.31 (+)- and (−)-limonenes (**68** and **68′**) and related compounds.

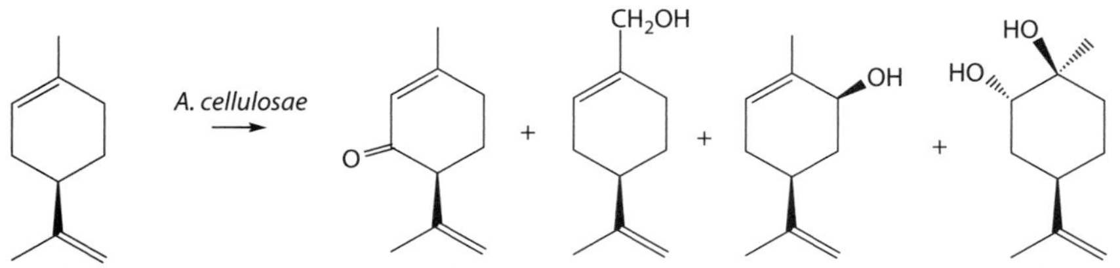

FIGURE 19.32 Biotransformation of (+)-limonene (**68**) by *Aspergillus cellulosae* IFO 4040. (Modified from Noma, Y. et al., *Phytochemistry*, 31, 2725, 1992b.)

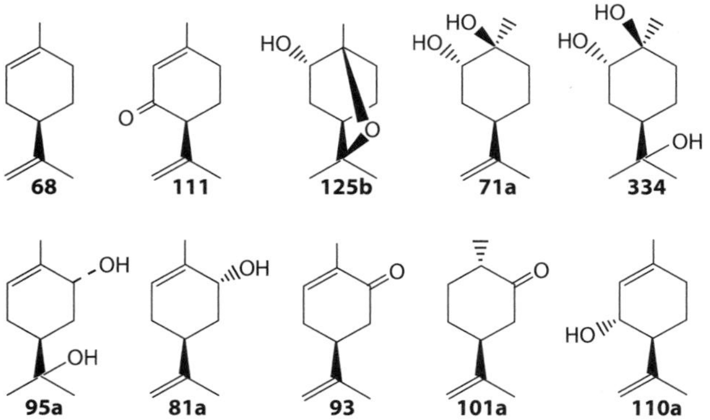

FIGURE 19.33 Metabolites of (+)-limonene (**68**) by a kind of citrus pathogenic fungi, *Penicillium digitatum* (Pers.; Fr.) Sacc. KCPYN. (Modified from Noma, Y. and Asakawa, Y., Biotransformation of (+)-limonene and related compounds by *Citrus* pathogenic fungi, *Proceedings of 50th TEAC*, pp. 431–433, 2006a; Noma, Y. and Asakawa, Y., Biotransformation of limonene and related compounds by newly isolated low temperature grown *citrus* pathogenic fungi and red yeast, *Book of Abstracts of the 38th ISEO*, 2007a, p. 7.)

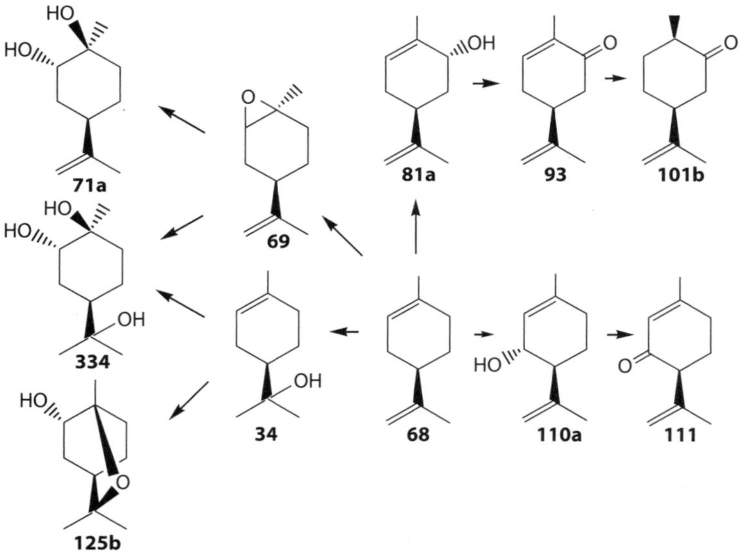

FIGURE 19.34 Biotransformation of (+)-limonene (**68**) by citrus pathogenic fungi, *Penicillium digitatum* (Pers.; Fr.) Sacc. KCPYN. (Modified from Noma, Y. and Asakawa, Y., Biotransformation of (+)-limonene and related compounds by *Citrus* pathogenic fungi, *Proceedings of 50th TEAC*, pp. 431–433, 2006a; Noma, Y. and Asakawa, Y., Biotransformation of limonene and related compounds by newly isolated low temperature grown *citrus* pathogenic fungi and red yeastm, *Book of Abstracts of the 38th ISEO*, 2007a, p. 7.)

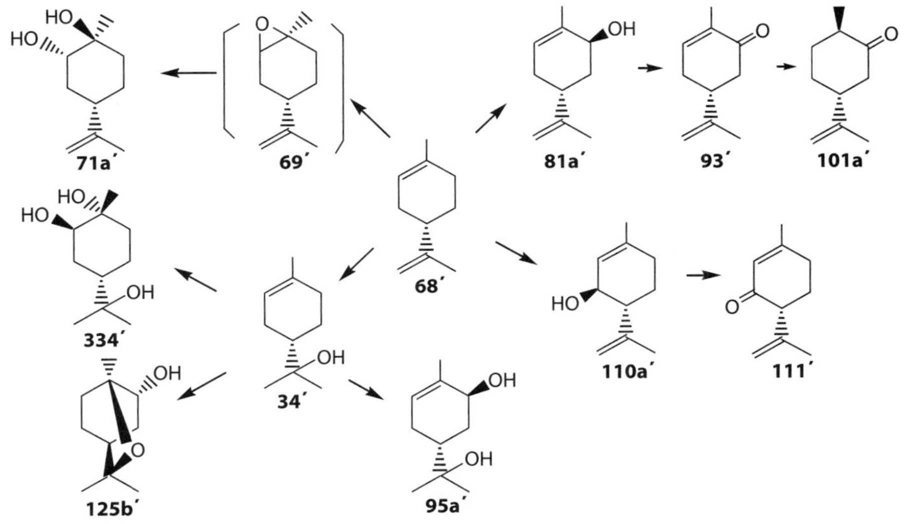

FIGURE 19.35 Biotransformation of (−)-limonene (**68′**) by citrus pathogenic fungi, *Penicillium digitatum* (Pers.; Fr.) Sacc. KCPYN. (Modified from Noma, Y. and Asakawa, Y., Microbial transformation of limonene and related compounds, *Proceedings of 51st TEAC*, 2007b, pp. 299–301.)

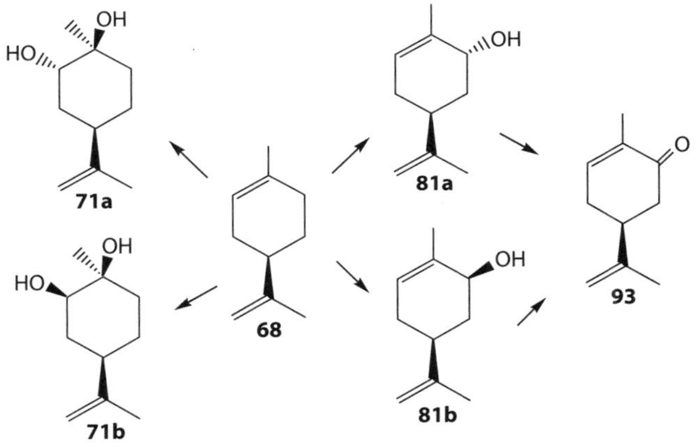

FIGURE 19.36 Biotransformation of (+)-limonene (**68**) by red yeast, *Rhodotorula* sp. and *Cladosporium* sp. T$_7$. (Modified from Mukherjee, B.B. et al., *Appl. Microbiol.*, 25, 447, 1973; Noma, Y. and Asakawa, Y., Microbial transformation of limonene and related compounds, *Proceedings of 51st TEAC*, 2007b, pp. 299–301.)

in low yield. On the other hand, (4*R*)-limonene oxide (**69**) was converted to limonene-1,2-*trans*-diol (**71a′**) and 1-hydroxy-(+)-dihydrocarvone (**72a′**). However, (4*R*)-(+)-limonene (**68**) and (4*S*)-limonene oxide (**69′**) were not converted at all (Figure 19.38) (Hamada et al., 2003).

(+)-Limonene (**68**) was fed by *S. litura* to give (+)-limonene-7-oic acid (**82**), (+)-limonene-9-oic acid (**70**), and (+)-limonene-8,9-diol (**79**); (−)-limonene (**68′**) was converted to (−)-limonene-7-oic acid (**82′**), (−)-limonene-9-oic acid (**70′**), and (−)-limonene-8,9-diol (**79′**) (Figure 19.39) (Miyazawa et al., 1995a).

Kieslich et al. (1985) found a nearly complete microbial resolution of a racemate in the biotransformation of (±)-limonene by *P. digitatum* (DSM 62840). The (*R*)-(+)-limonene (**68**) is converted to the optically active (+)-α-terpineol, [α]$_D$ = +99°, while the (*S*)-(−)-limonene (**68′**) is presumably adsorbed onto the mycelium or degraded via unknown pathways (Kieslich et al., 1985) (Figure 19.40).

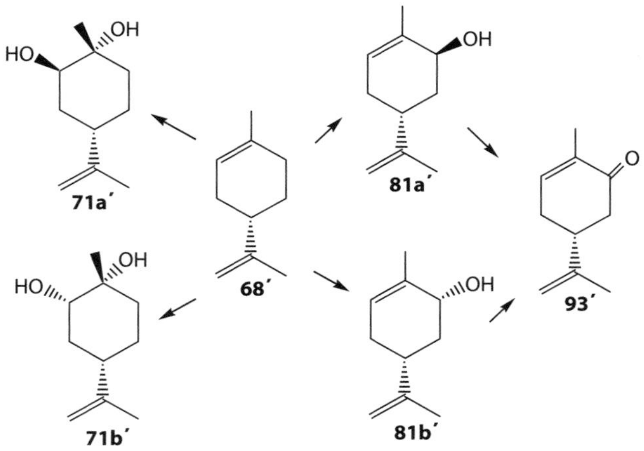

FIGURE 19.37 Biotransformation of (−)-limonene (**68′**) by a kind of *Rhodotorula* sp. (Modified from Noma, Y. and Asakawa, Y., Microbial transformation of limonene and related compounds, *Proceedings of 51st TEAC*, 2007b, pp. 299–301.)

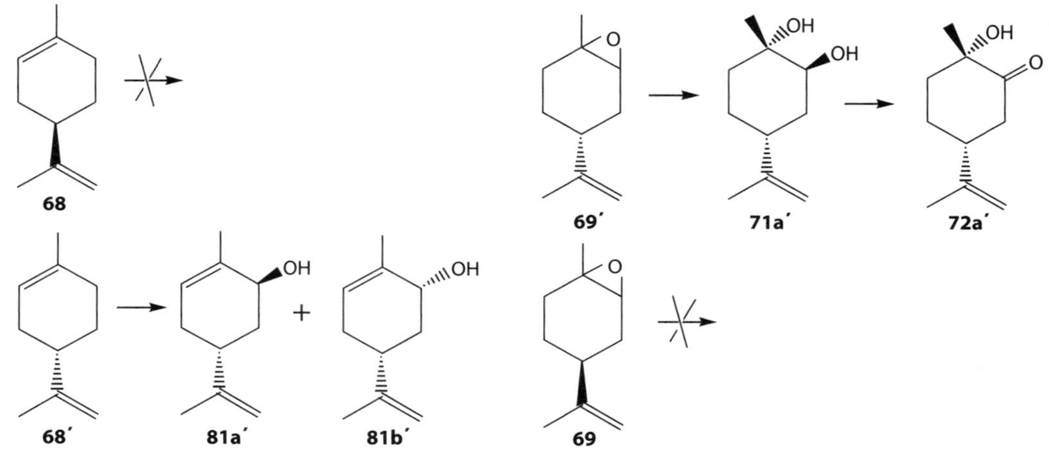

FIGURE 19.38 Biotransformation of (+)- and (−)-limonene (**68** and **68′**) and limonene epoxide (**69** and **69′**) by cyanobacterium. (Modified from Hamada, H. et al., Enantioselective biotransformation of monoterpenes by *Cyanobacterium*, *Proceedings of 47th TEAC*, 2003, pp. 162–163.)

(4*S*)- and (4*R*)-Limonene epoxides (**69a′** and **a**) were biotransformed by cyanobacterium to give 8-*p*-menthene-1α,2β-ol (**71a**, 68.4%) and 1α-hydroxy-8-*p*-menthen-2-one (**72**, 31.6%) (Hamada et al., 2003) (Figure 19.41).

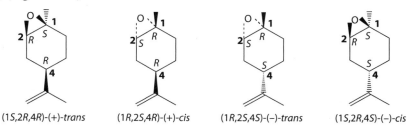

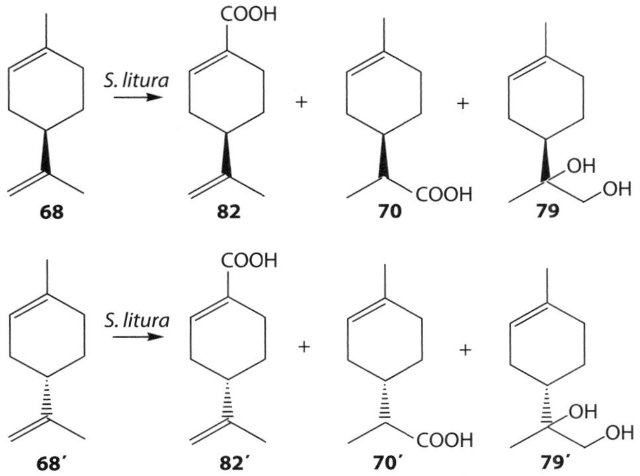

FIGURE 19.39 Biotransformation of (+)-limonene (**68**) and (−)-limonene (**68′**) by *Spodoptera litura*. (Modified from Miyazawa, M. et al., Biotransformation of terpinene, limonene and α-phellandrene in common cutworm larvae, *Spodoptera litura* Fabricius, *Proceedings of 39th TEAC*, 1995a, pp. 362–363.)

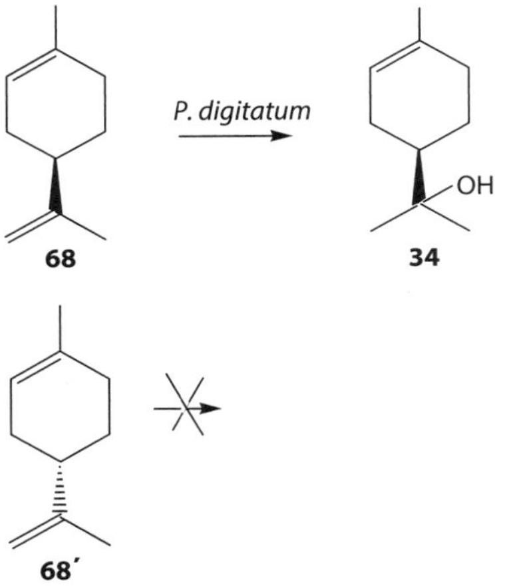

FIGURE 19.40 Microbial resolution of racemic limonene (**68** and **68′**) and the formation of optically active α-terpineol by *Penicillium digitatum*. (Modified from Kieslich, K. et al., In: *Topics in flavor research*, R.G. Berger, S. Nitz, and P. Schreier, eds., Marzling Hangenham, Eichborn, 1985, pp. 405–427.)

The mixture of (+)-*trans*- (**69a**) and *cis*- (**69b**) and the mixture of (−)-*trans*- (**69a′**) and *cis*-limonene-1,2-epoxide (**69b′**) were biotransformed by citrus pathogenic fungi *P. digitatum* (Pers.; Fr.) Sacc. KCPYN to give (1*R*,2*R*,4*R*)-(−)-*trans*-(**71a**) and (1*S*,2*S*,4*S*)-(+)-8-*p*-menthene-1,2-*trans*-diol (**71a′**) and (−)-*p*-menthane-1,2,8-triols (**334a** and **334a′**) (Noma and Asakawa, 2007b) (Figure 19.42).

Biotransformation of 1,8-cineole (**122**) by *A. niger* gave racemic 2α-hydroxy-1,8-cineole (**125b** and **b′**) (Nishimura et al., 1982). When racemic 2α-hydroxy-1,8-cineole (**125b** and **b′**) was biotransformed by *G. cingulata*, only (−)-2α-hydroxy-1,8-cineole (**125b′**) was selectively esterified with malonic acid to give its malonate (**125b′**-Mal). The malonate was hydrolyzed to give optical pure **125b′** (Miyazawa et al., 1995b). On the other hand, citrus pathogenic fungi *P. digitatum* biotransformed limonene (**68**) to give optical pure **125b** (Noma and Asakawa, 2007b) (Figure 19.43).

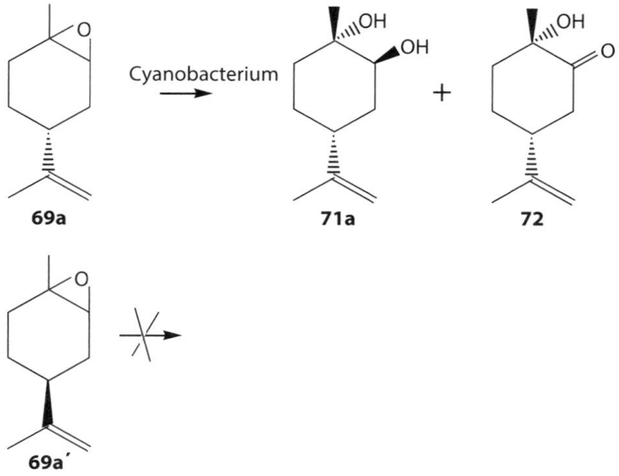

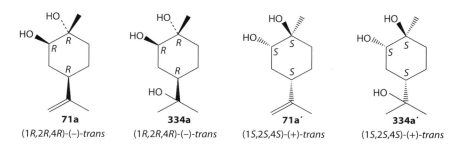

FIGURE 19.41 Enantioselective biotransformation of (4S)-(**69a′**) and (4R)-limonene epoxides (**69a**) by cyanobacterium. (Modified from Hamada, H. et al., Enantioselective biotransformation of monoterpenes by *Cyanobacterium*, *Proceedings of 47th TEAC*, 2003, pp. 162–163.)

FIGURE 19.42 Biotransformation of (+)-*trans*-(**69a**) and *cis*-(**69b**) and (−)-*trans*-(**69a′**) and *cis*-limonene-1,2-epoxide (**69b′**) by citrus pathogenic fungi, *Penicillium digitatum* (Pers.; Fr.) Sacc. KCPYN, and their metabolites. (Modified from Noma, Y. and Asakawa, Y., Microbial transformation of limonene and related compounds, *Proceedings of 51st TEAC*, 2007b, pp. 299–301.)

When monoterpenes, such as limonene (**68**), α-pinene (**4**), and 3-carene (**336**), were administered to the cultured cells of *Nicotiana tabacum*, they were converted to the corresponding epoxides enantio- and stereoselectively. The enzyme (p38) concerning with the epoxidation reaction was purified from the cultured cells by cation-exchange chromatography. The enzyme had not only epoxidation activity but also peroxidase activity. Amino acid sequence of p38 showed 89% homology in their 9 amino acid overlap with horseradish peroxidase (Yawata et al., 1998) (Figure 19.44). It was found that limonene and carene were converted to the corresponding epoxides in the presence of hydrogen peroxide and *p*-cresol by a radical mechanism with the peroxidase. (R)-limonene (**68**), (S)-limonene (**68′**), (1S,5R)-α-pinene (**4**), (1R,5R)-α-pinene (**4**), and (1R,6R)-3-carene (**336**) were oxidized by cultured cells of *N. tabacum* to give corresponding epoxides enantio- and stereoselectively (Yawata et al., 1998) (Figure 19.45).

19.3.1.2 Isolimonene

S. litura converted (1R)-*trans*-isolimonene (**338**) to (1R,4R)-*p*-menth-2-ene-8,9-diol (**339**) (Miyazawa et al., 1996b) (Figure 19.46).

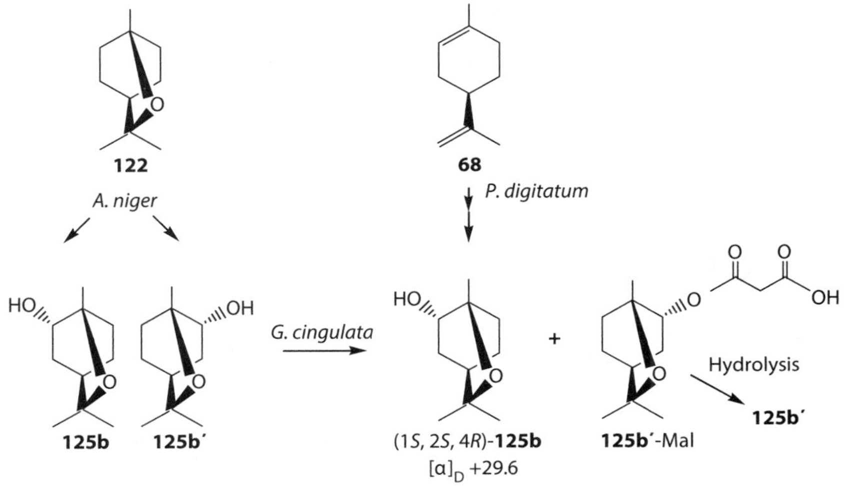

FIGURE 19.43 Formation of optical pure (+)- and (−)-2α-hydroxy-1,8-cineole (**125b** and **b′**) from the biotransformation of 1,8-cineole (**122**) and (+)-limonene (**68**) by citrus pathogenic fungi, *Penicillium digitatum* (Pers.; Fr.) Sacc. KCPYN and *Aspergillus niger* TBUYN-2. (Modified from Nishimura, H. et al., *Agric. Biol. Chem.*, 46, 2601, 1982; Miyazawa, M. et al., Biotransformation of 2-*endo*-hydroxy-1,4-cineole by plant pathogenic microorganism, *Glomerella cingulata*, *Proceedings of 39th TEAC*, pp. 352–353, 1995b; Noma, Y. and Asakawa, Y., Microbial transformation of limonene and related compounds, *Proceedings of 51st TEAC*, 2007b, pp. 299–301.)

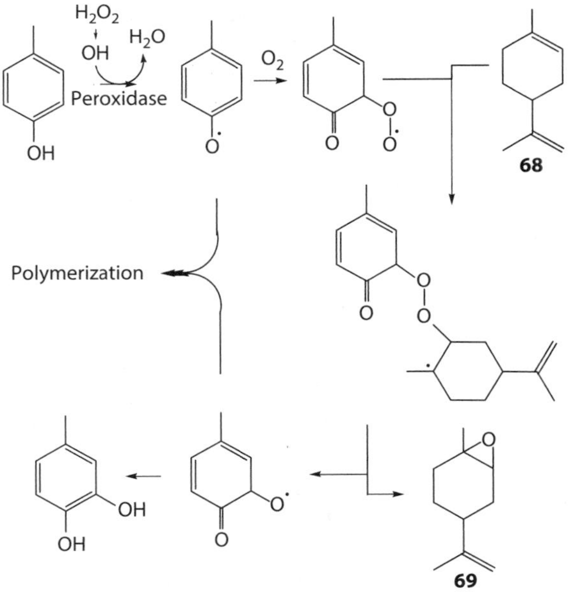

FIGURE 19.44 Proposed mechanism for the epoxidation of (+)-limonene (**68**) with p38 from the cultured cells of *Nicotiana tabacum*. (Modified from Yawata, T. et al., Epoxidation of monoterpenes by the peroxidase from the cultured cells of *Nicotiana tabacum*, *Proceedings of 42nd TEAC*, 1998, pp. 142–144.)

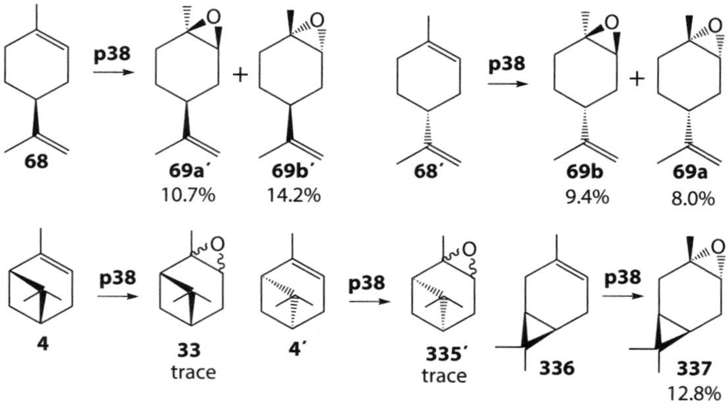

FIGURE 19.45 Epoxidation of limonene (**68**), α-pinene (**4**), and 3-carene (**336**) with p38 from the cultured cells of *Nicotiana tabacum*. (Modified from Yawata, T. et al., Epoxidation of monoterpenes by the peroxidase from the cultured cells of *Nicotiana tabacum*, *Proceedings of 42nd TEAC*, 1998, pp. 142–144.)

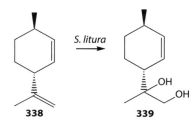

FIGURE 19.46 Biotransformation of (1*R*)-*trans*-isolimonene (**338**) by *Spodoptera litura*. (Modified from Miyazawa, M. et al., Biotransformation of *p*-menthanes using common cutworm larvae, *Spodoptera litura* as a biocatalyst, *Proceedings of 40th TEAC*, 1996b, pp. 80–81.)

19.3.1.3 *p*-Menthane

Hydroxylation of *trans*- and *cis*-*p*-menthane (**252a** and **b**) by microorganisms is also very interesting from the viewpoint of the formation of the important perfumes such as (−)-menthol (**137b′**) and (−)-carvomenthol (**49b′**), plant growth regulators, and mosquito repellents such as *p*-menthane-*trans*-3,8-diol (**142a**), *p*-menthane-*cis*-3,8-diol (**142b**) (Nishimura and Noma, 1996), and *p*-menthane-2,8-diol (**93**) (Noma, 2007). *P. mendocina* strain SF biotransformed **252b** stereoselectively to *p*-*cis*-menthan-1-ol (**253**) (Tsukamoto et al., 1975) (Figure 19.47).

On the other hand, the biotransformation of the mixture of *p*-*trans*- (**252a**) and *cis*-menthane (**252b**) (45:55, peak area in GC) by *A. niger* gave *p*-*cis*-menthane-1,9-diol (**254**) via *p*-*cis*-menthan-1-ol (**253**). No metabolite was obtained from **252a** at all (Noma et al., 1990) (Figure 19.47).

19.3.1.4 1-*p*-Menthene

Concentrated cell suspension of *Pseudomonas* sp. strain (PL) was inoculated to the medium containing 1-*p*-menthene (**62**) as the sole carbon source. It was degraded to give β-isopropyl pimelic acid (**248**) and methylisopropyl ketone (**251**) (Hungund et al., 1970) (Figure 19.48).

As shown in Figure 19.49, *S. litura* converted (4*R*)-*p*-menth-1-ene (**62**) at C-7 position to (4*R*)-phellandric acid (**65**) (Miyazawa et al., 1996b). On the other hand, when *Cladosporium* sp. T₁ was cultivated with (+)-limonene (**68**) as the sole carbon source, it converted **62′** to *trans*-*p*-menthane-1,2-diol (**54**) (Mukherjee et al., 1973).

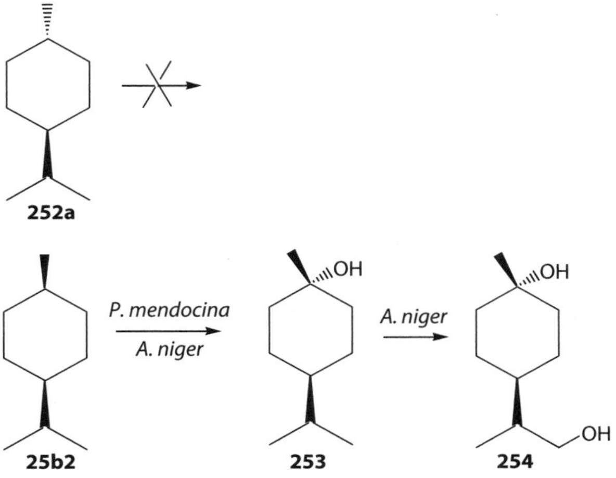

FIGURE 19.47 Biotransformation of the mixture of *trans*-(**252a**) and *cis*-*p*-menthane (**252b**) by *Pseudomonas mendocina* SF and *Aspergillus niger* TBUYN-2. (Modified from Tsukamoto, Y. et al., Microbiological oxidation of *p*-menthane 1. Formation of formation of *p*-*cis*-menthan-1-ol, *Proceedings of 18th TEAC*, pp. 24–26, 1974; Tsukamoto, Y. et al., *Agric. Biol. Chem.*, 39, 617, 1975; Noma, Y., *Aromatic Plants from Asia their Chemistry and Application in Food and Therapy*, L. Jiarovetz, N.X. Dung, and V.K. Varshney, Har Krishan Bhalla & Sons, Dehradun, India, pp. 169–186, 2007.)

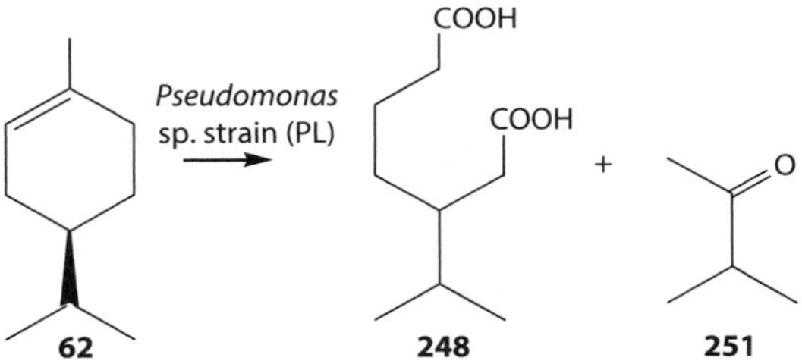

FIGURE 19.48 Biodegradation of (4R)-1-*p*-menthene (**62**) by *Pseudomonas* sp. *strain* (PL). (Modified from Hungund, B.L. et al., *Indian J. Biochem.*, 7, 80, 1970.)

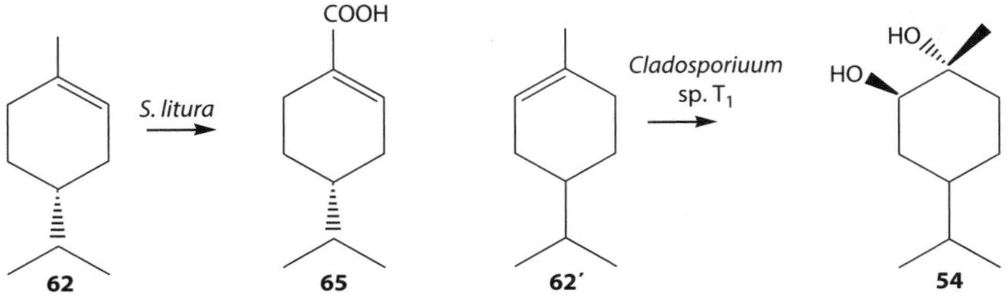

FIGURE 19.49 Biotransformation of (4R)-*p*-menth-1-ene (**62**) by *Spodoptera litura* and *Cladosporium* sp. T$_1$. (Modified from Miyazawa, M. et al., Biotransformation of *p*-menthanes using common cutworm larvae, *Spodoptera litura* as a biocatalyst, *Proceedings of 40th TEAC*, pp. 80–81, 1996b; Mukherjee, B.B. et al., *Appl. Microbiol.*, 25, 447, 1973.)

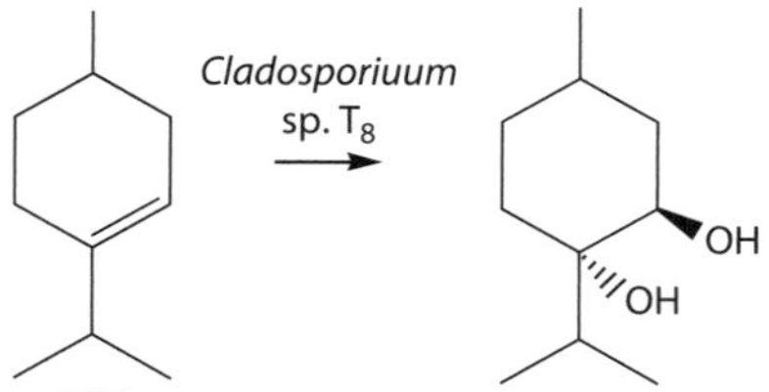

FIGURE 19.50 Biotransformation of *p*-menth-3-ene (**147**) by *Cladosporium* sp. T$_8$. (Modified from Mukherjee, B.B. et al., *Appl. Microbiol.*, 25, 447, 1973.)

19.3.1.5 3-*p*-Menthene

When *Cladosporium* sp. T$_8$ was cultivated with 3-*p*-menthene (**147**) as the sole carbon source, it was converted to *trans-p*-menthane-3,4-diol (**141**) as shown in Figure 19.50 (Mukherjee et al., 1973).

19.3.1.6 α-Terpinene

α-Terpinene (**340**) was converted by *S. litura* to give α-terpinene-7-oic acid (**341**) and *p*-cymene-7-oic acid (**194**, cuminic acid) (Miyazawa et al., 1995a) (Figure 19.51).

A soil Pseudomonad has been found to grow with *p*-mentha-1,3-dien-7-al (**463**) as the sole carbon source and to produce α-terpinene-7-oic acid (**341**) in a mineral salt medium (Kayahara et al., 1973) (Figure 19.51).

19.3.1.7 γ-Terpinene

γ-Terpinene (**344**) was converted by *S. litura* to give γ-terpinene-7-oic acid (**345**) and *p*-cymene-7-oic acid (**194**, cuminic acid) (Miyazawa et al., 1995a) (Figure 19.52).

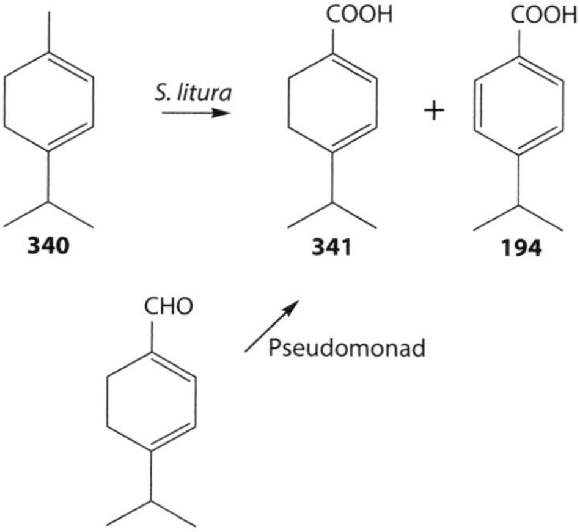

FIGURE 19.51 Biotransformation of α-terpinene (**340**) by *Spodoptera litura* and *p*-mentha-1,3-dien-7-al (**463**) by a soil Pseudomonad. (Modified from Kayahara, H. et al., *J. Ferment. Technol.*, 51, 254, 1973; Miyazawa, M. et al., Biotransformation of terpinene, limonene and α-phellandrene in common cutworm larvae, *Spodoptera litura* Fabricius, *Proceedings of 39th TEAC*, 1995a, pp. 362–363.)

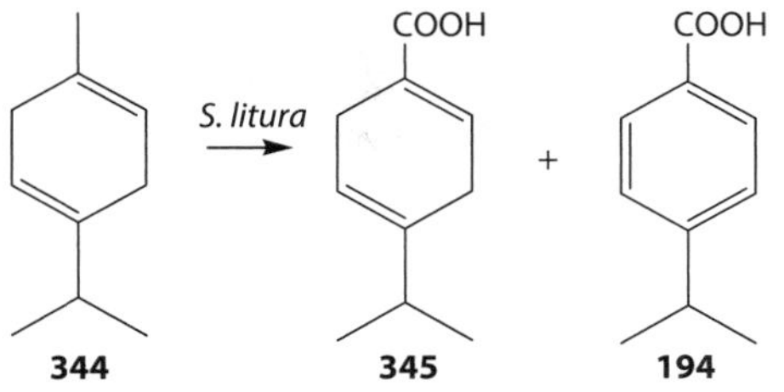

FIGURE 19.52 Biotransformation of γ-terpinene (**344**) by *Spodoptera litura*. (Modified from Miyazawa, M. et al., Biotransformation of terpinene, limonene and α-phellandrene in common cutworm larvae, *Spodoptera litura* Fabricius, *Proceedings of 39th TEAC*, 1995a, pp. 362–363.)

19.3.1.8 Terpinolene

Terpinolene (**346**) was converted by *A. niger* to give (1*R*)-8-hydroxy-3-*p*-menthen-2-one (**347**), (1*R*)-1,8-dihydroxy-3-*p*-menthen-2-one (**348**), and 5β-hydroxyfenchol (**350b′**). In the case of *C. cassiicola*, it was converted to terpinolene-1,2-*trans*-diol (**351**) and terpinolene-4,8-diol (**352**). Furthermore, in the case of rabbit, terpinolene-9-ol (**353**) and terpinolene-10-ol (**354**) were formed from **346** (Asakawa et al., 1983). *S. litura* also converted **346** to give 1-*p*-menthene-4,8-diol (**352**), cuminic acid (**194**, 29% main product), and terpinolene-7-oic acid (**357**) (Figure 19.53).

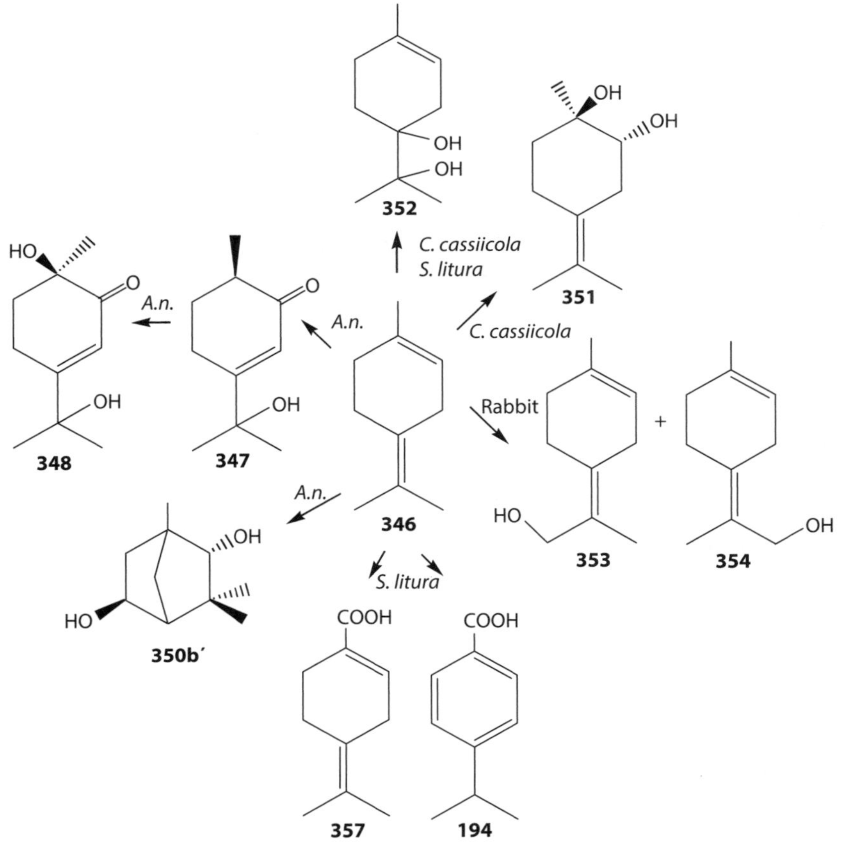

FIGURE 19.53 Biotransformation of terpinolene (**346**) by *Aspergillus niger* (Asakawa et al., 1991), *Corynespora cassiicola* (Abraham et al., 1985), rabbit (Asakawa et al., 1983), and *Spodoptera litura*. (Modified from Miyazawa, M. et al., Biotransformation of terpinene, limonene and α-phellandrene in common cutworm larvae, *Spodoptera litura* Fabricius, *Proceedings of 39th TEAC*, 1995a, pp. 362–363.)

19.3.1.9 α-Phellandrene

α-Phellandrene (**355**) was converted by *S. litura* to give α-phellandrene-7-oic acid (**356**) and *p*-cymene-7-oic acid (**194**, cuminic acid) (Miyazawa et al., 1995a) (Figure 19.54).

19.3.1.10 *p*-Cymene

Pseudomonas sp. strain (PL) was cultivated with *p*-cymene (**178**) as the sole carbon source to give cumyl alcohol (**192**), cumic acid (**194**), 3-hydroxycumic acid (**196**), 2,3-dihydroxycumic acid (**197**), 2-oxo-4-methylpentanoic acid (**201**), 9-hydroxy-*p*-cymene (**189**), and *p*-cymene-9-oic acid (**190**) as shown in Figure 19.55 (Madyastha and Bhattacharyya, 1968). On the other hand, *p*-cymene (**178**) was converted regiospecifically to cumic acid (**194**) by *Pseudomonas* sp., *Pseudomonas desmolytica*, and *Nocardia salmonicolor* (Madyastha and Bhattacharyya, 1968) (Figure 19.56).

p-Cymene (**178**) is converted to thymoquinone (**358**) and analogues, **179** and **180**, by various kinds of microorganisms (Demirci et al., 2007) (Figure 19.57).

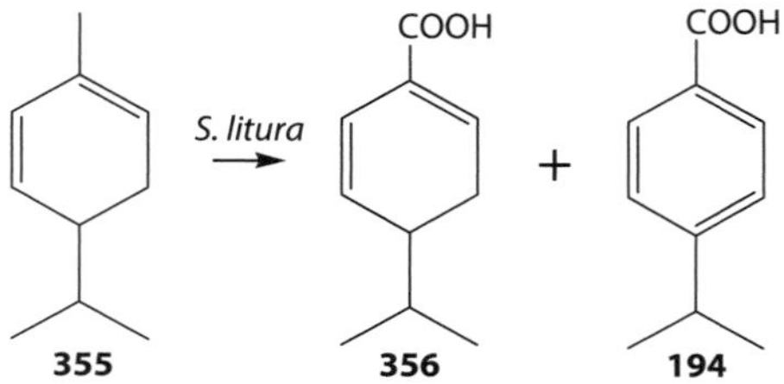

FIGURE 19.54 Biotransformation of α-phellandrene (**355**) by *Spodoptera litura*. (Modified from Miyazawa, M. et al., Biotransformation of terpinene, limonene and α-phellandrene in common cutworm larvae, *Spodoptera litura* Fabricius, *Proceedings of 39th TEAC*, 1995a, pp. 362–363.)

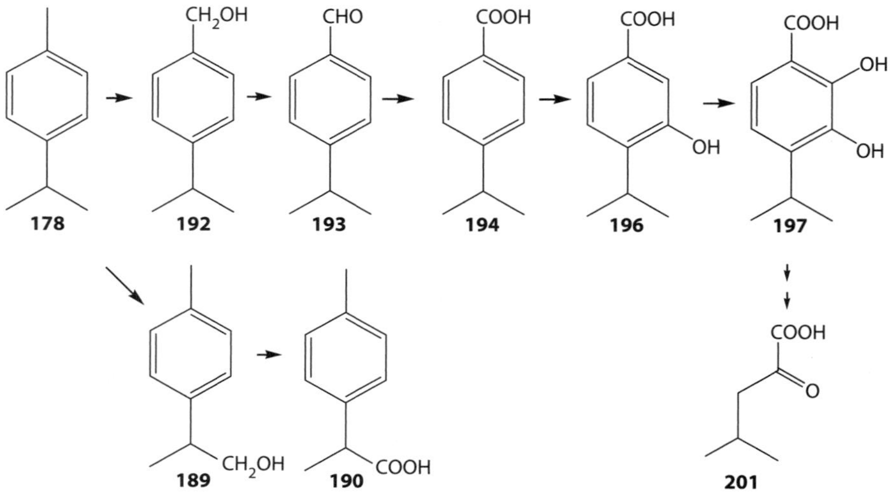

FIGURE 19.55 Biotransformation of *p*-cymene (**178**) by *Pseudomonas* sp. strain (PL). (Modified from Madyastha, K.M. and Bhattacharyya, P.K., *Indian J. Biochem.*, 5, 161, 1968.)

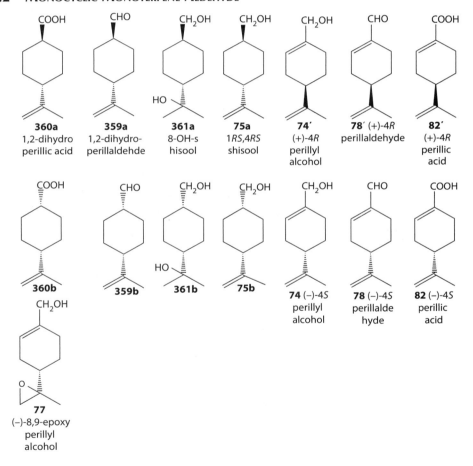

FIGURE 19.56 Biotransformation of *p*-cymene (**178**) to cumic acid (**194**) by *Pseudomonas* sp., *Pseudomonas desmolytica*, and *Nocardia salmonicolor*. (Modified from Yamada, K. et al., *Agric. Biol. Chem.*, 29, 943, 1965; Madyastha, K.M. and Bhattacharyya, P.K., *Indian J. Biochem.*, 5, 161, 1968; Noma, Y., 2000. unpublished data.)

19.3.2 Monocyclic Monoterpene Aldehyde

19.3.2.1 Perillaldehyde

Biotransformation of (−)-perillaldehyde (**78**), (+)-perillaldehyde (**78′**), (−)-perillyl alcohol (**74**), *trans*-1,2-dihydroperillaldehyde (**359a**) and *cis*-1,2-dihydroperillaldehyde (**359b**), and *trans*-shisoic acid (**360a**) and *cis*-shisoic acid (**360b**) was carried out by *E. gracilis* Z. (Noma et al., 1991a), *D. tertiolecta* (Noma et al., 1991b, 1992a), *Chlorella ellipsoidea* IAMC-27 (Noma et al., 1997), *S. ikutamanensis* Ya-2-1 (Noma et al., 1984, 1986), and other microorganisms (Kayahara et al., 1973) (Figure 19.58).

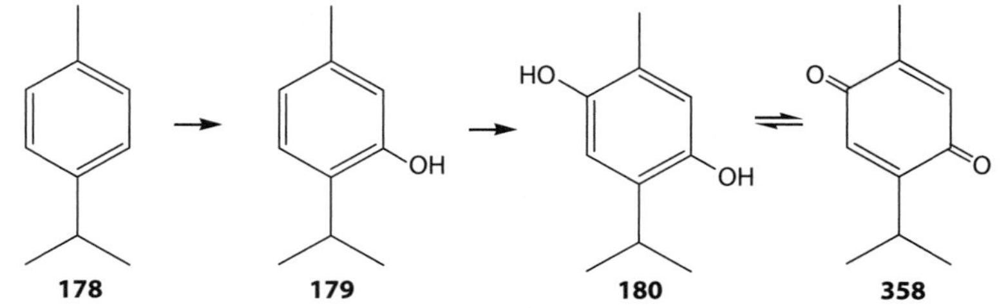

FIGURE 19.57 Biotransformation of *p*-cymene (**178**) to thymoquinone (**358**) and analogues by microorganisms. (Modified from Demirci, F. et al., Biotransformation of *p*-cymene to thymoquinone, *Book of Abstracts of the 38th ISEO*, SL-1, 2007, p. 6.)

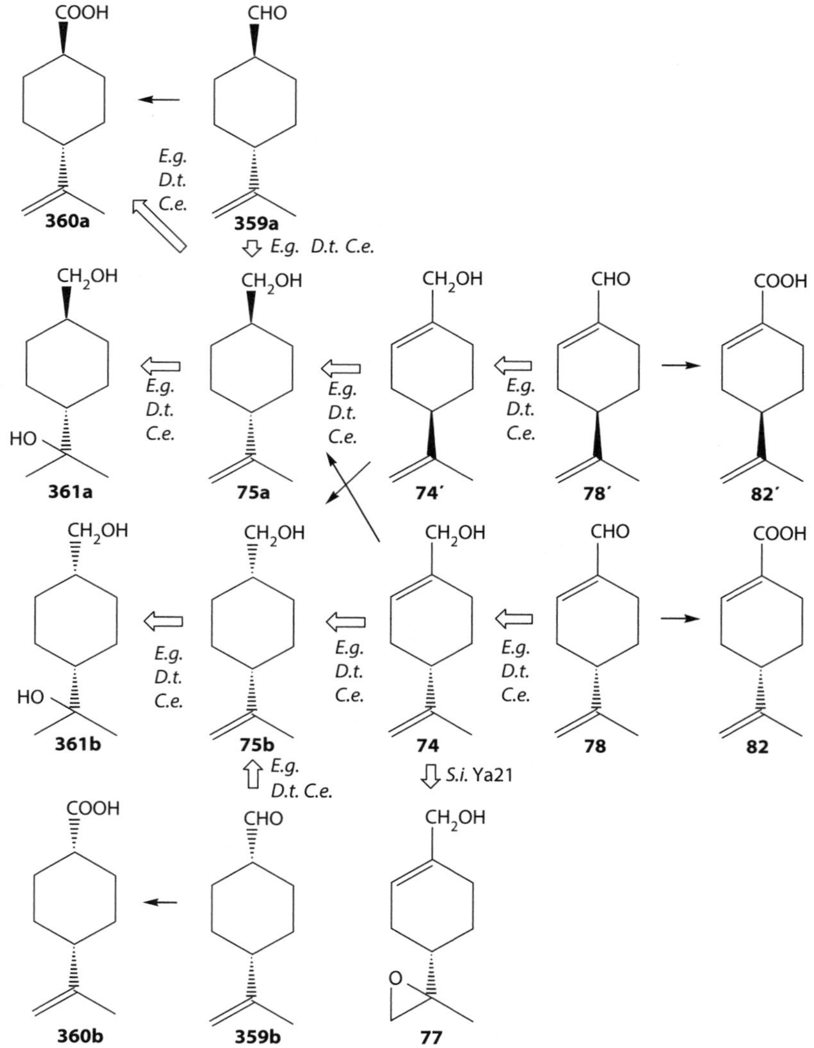

FIGURE 19.58 Metabolic pathways of perillaldehyde (**78** and **78′**) by *Euglena gracilis* Z (Noma et al., 1991a), *Dunaliella tertiolecta* (Noma et al., 1991b; 1992a), *Chlorella ellipsoidea* IAMC-27 (Noma et al., 1997), *Streptomyces ikutamanensis* Ya-2-1 (Noma et al., 1984, 1986), a soil Pseudomonad (Kayahara et al., 1973), and rabbit (Ishida et al., 1981a).

(−)-Perillaldehyde (**78**) is easily transformed to give (−)-perillyl alcohol (**74**) and *trans*-shisool (**75a**), which is well known as a fragrance, as the major product and (−)-perillic acid (**82**) as the minor product. (−)-Perillyl alcohol (**74**) is also transformed to *trans*-shisool (**75a**) as the major product with *cis*-shisool (**75b**) and 8-hydroxy-*cis*-shisool (**361b**). Furthermore, *trans*-shisool (**75a**) and *cis*-shisool (**75b**) are hydroxylated to 8-hydroxy-*trans*-shisool (**361a**) and 8-hydroxy-*cis*-shisool (**361b**), respectively. *trans*-1,2-Dihydroperillaldehyde (**359a**) and *cis*-1,2-dihydroperillaldehyde (**359b**) are also transformed to **75a** and **75b** as the major products and *trans*-shisoic acid (**360a**) and *cis*-shisoic acid (**360b**) as the minor products, respectively. Compound **360a** was also formed from **75a**. In the biotransformation of (±)-perillaldehyde (**74** and **74′**), the same results were obtained as described in the case of **74**. In the case of *S. ikutamanensis* Ya-2-1, (−)-perillaldehyde (**78**) was converted to (−)-perillic acid (**82**), (−)-perillyl alcohol (**74**), and (−)-perillyl alcohol-8,9-epoxide (**77**), which was the major product.

A soil Pseudomonad has been found to grow with (−)-perillaldehyde (**78**) as the sole carbon source and to produce (−)-perillic acid (**82**) in a mineral salt medium (Kayahara et al., 1973).

On the other hand, rabbit metabolized (−)-perillaldehyde (**78**) to (−)-perillic acid (**82**) along with minor shisool (**75a**) (Ishida et al., 1981a).

19.3.2.2 Phellandral and 1,2-Dihydrophellandral

Biotransformation of (−)-phellandral (**64**), *trans*-tetrahydroperillaldehyde (**362a**), and *cis*-tetrahydroperillaldehyde (**362b**) was carried out by microorganisms (Noma et al., 1986, 1991a,b, 1997). (−)-Phellandral (**64**) was metabolized mainly via (−)-phellandrol (**63**) to *trans*-tetrahydroperillyl alcohol (**66a**). *trans*-Tetrahydroperillaldehyde (**362a**) and *cis*-tetrahydroperillaldehyde (**362b**) were also transformed to *trans*-tetrahydroperillyl alcohol (**66a**) and *cis*-tetrahydroperillyl alcohol (**66b**) as the major products and *trans*-tetrahydroperillic acid (**363a**) and *cis*-tetrahydroperillic acid (**363b**) as the minor products, respectively (Figure 19.59).

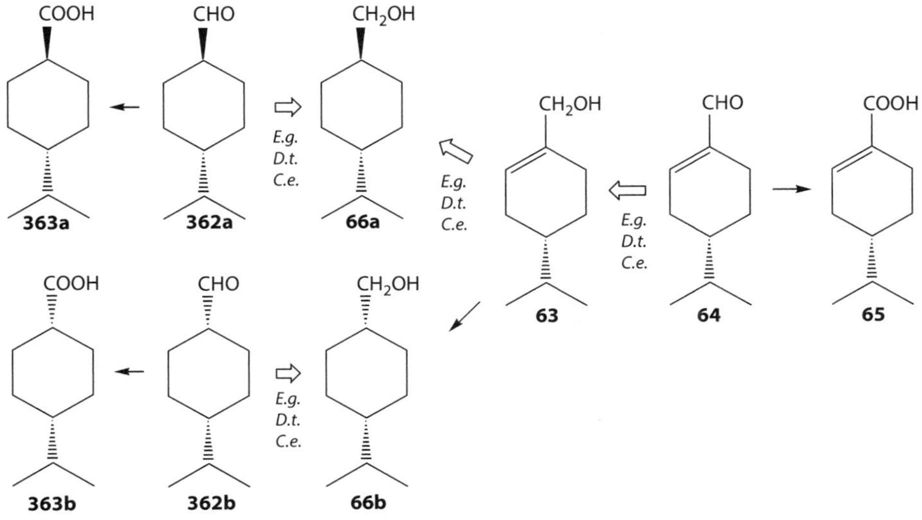

FIGURE 19.59 Metabolic pathways of (−)-phellandral (**64**) by microorganisms. (Modified from Noma, Y. et al., Reduction of terpene aldehydes and epoxidation of terpene alcohols by *S. ikutamanensis*, Ya-2-1, *Proceedings of 30th TEAC*, 1986, pp. 204–206; Noma, Y. et al., *Phytochemistry*, 30, 1147, 1991a; Noma, Y. et al., Biotransformation of monoterpenes by photosynthetic marine algae, *Dunaliella tertiolecta*, *Proceedings of 35th TEAC*, 1991b, pp. 112–114; Noma, Y. et al., Biotransformation of terpenoids and related compounds by *Chlorella* species, *Proceedings of 41st TEAC*, 1997, pp. 227–229.)

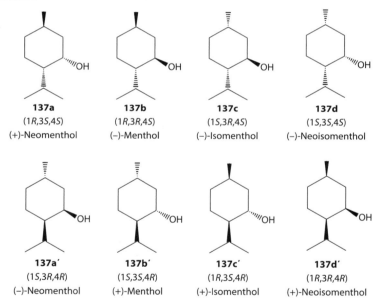

FIGURE 19.60 Metabolic pathway of cumin aldehyde (**193**) by microorganism. (Modified from Noma, Y. et al., Reduction of terpene aldehydes and epoxidation of terpene alcohols by *S. ikutamanensis*, Ya-2-1, *Proceedings of 30th TEAC*, 1986, pp. 204–206; Noma, Y. et al., *Phytochemistry*, 30, 1147, 1991a; Noma, Y. et al., Biotransformation of monoterpenes by photosynthetic marine algae, *Dunaliella tertiolecta*, *Proceedings of 35th TEAC*, 1991b, pp. 112–114.)

19.3.2.3 Cuminaldehyde

Cumin aldehyde (**193**) is transformed by *Euglena* (Noma et al., 1991a), *Dunaliella* (Noma et al., 1991b), and *S. ikutamanensis* (Noma et al., 1986) to give cumin alcohol (**192**) as the major product and cuminic acid (**194**) as the minor product (Figure 19.60).

19.3.3 Monocyclic Monoterpene Alcohol

19.3.3.1 Menthol

Menthol (**137**) is one of the rare naturally occurring monocyclic monoterpene alcohols that have not only various physiological properties, such as sedative, anesthetic, antiseptic, gastric, and antipruritic, but also characteristic fragrance (Bauer et al., 1990). There are in fact eight isomers with a menthol (*p*-menthan-3-ol) skeleton; (−)-menthol (**137b**) is the most important one, because of its cooling and refreshing effect. It is the main component of peppermint and cornmint oils obtained from the *Mentha piperita* and *Mentha arvensis* species. Many attempts have been made to produce (−)-menthol (**137b**) from inexpensive terpenoid sources, but these sources also unavoidably yielded the (±)-isomers (**137b** and **137b′**): isomenthol (**137c**), neomenthol (**137a**), and neoisomenthol (**137d**) (Krasnobajew, 1984). Japanese researchers have been active in this field, maybe because of the large demand for (−)-menthol (**137b**) in Japan itself, that is, 500 t/year (Janssens et al., 1992). Indeed, most

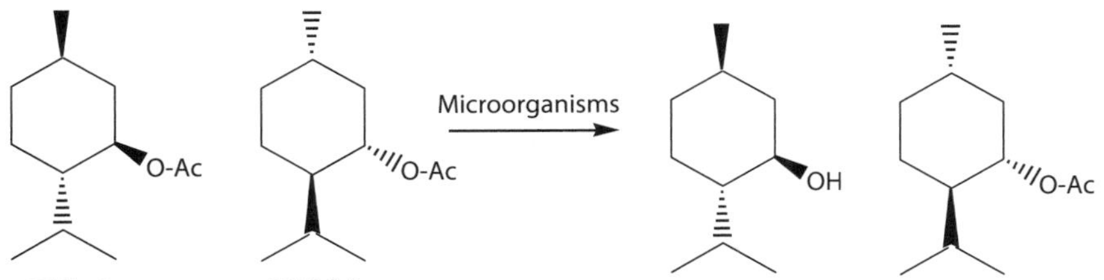

FIGURE 19.61 Asymmetric hydrolysis of racemic menthyl acetate (**137b-Ac** and **137b′-Ac**) to obtain pure (−)-menthol (**137b**). (Modified from Watanabe, Y. and Inagaki, T., Japanese Patent 77.12.989. No. 187696x, 1977a; Watanabe, Y. and Inagaki, T., Japanese Patent 77.122.690. No. 87656g, 1977b; Moroe, T. et al., Japanese Patent, 2.036. 875. no. 98195t, 1971; Oritani, T. and Yamashita, K., *Agric. Biol. Chem.*, 37, 1695, 1973b.)

literature deals with the enantiomeric hydrolysis of (±)-menthol (**137b** and **137b′**) esters to optically pure *l*-menthol (**137b**). The asymmetric hydrolysis of (±)-menthyl chloroacetate by an esterase of *Arginomonas non-fermentans* FERM-P-1924 has been patented by the Japanese Nippon Terpene Chemical Co. (Watanabe and Inagaki, 1977a,b). Investigators from the Takasago Perfumery Co. Ltd. claim that certain selected species of *Absidia, Penicillium, Rhizopus, Trichoderma, Bacillus, Pseudomonas*, and others asymmetrically hydrolyze esters of (±)-menthol isomers such as formates, acetates, propanoates, caproates, and esters of higher fatty acids (Moroe et al., 1971; Yamaguchi et al., 1977) (Figure 19.61).

Numerous investigations into the resolution of the enantiomers by selective hydrolysis with microorganisms or enzymes were carried out. Good results were described by Yamaguchi et al. (1977) with the asymmetric hydrolysis of (±)-methyl acetate by a mutant of *Rhodotorula mucilaginosa*, yielding 44 g of (−)-menthol (**137b**) from a 30% (±)-menthyl acetate mixture per liter of cultured medium for 24 h. The latest development is the use of immobilized cells of *Rhodotorula minuta* in aqueous saturated organic solvents (Omata et al., 1981) (Figure 19.62).

Besides the hydrolysis of menthyl esters, the biotransformation of menthol and its enantiomers has also been published (Shukla et al., 1987; Asakawa et al., 1991). The fungal biotransformation of (−)- (**137b**) and (+)-menthols (**137b′**) by *A. niger* and *A. cellulosae* was described (Asakawa et al., 1991). *A. niger* converted (−)-menthol (**137b**) to 1- (**138b**), 2- (**140b**), 6- (**139b**), 7- (**143b**), 9-hydroxymenthols (**144b**), and the mosquito repellent-active 8-hydroxymenthol (**142b**), whereas (+)-menthol (**137b′**) was smoothly biotransformed by the same microorganism to 7-hydroxymenthol

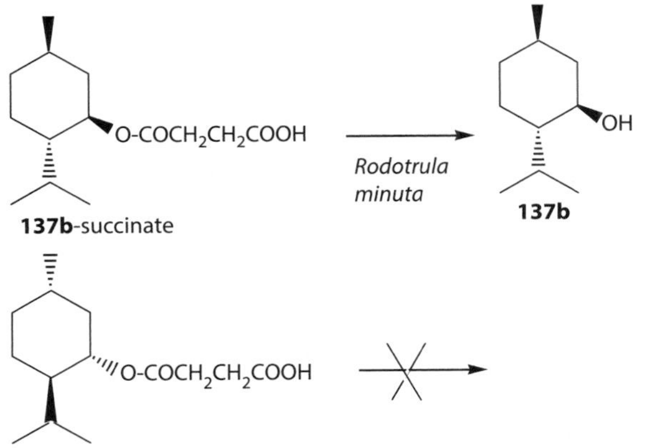

FIGURE 19.62 Asymmetric hydrolysis of racemic menthyl succinate (**137b-** and **137b′-succinates**) to obtain pure (−)-menthol (**137b**). (Modified from Yamaguchi, Y. et al., *J. Agric. Chem. Soc. Jpn.*, 51, 411, 1977.)

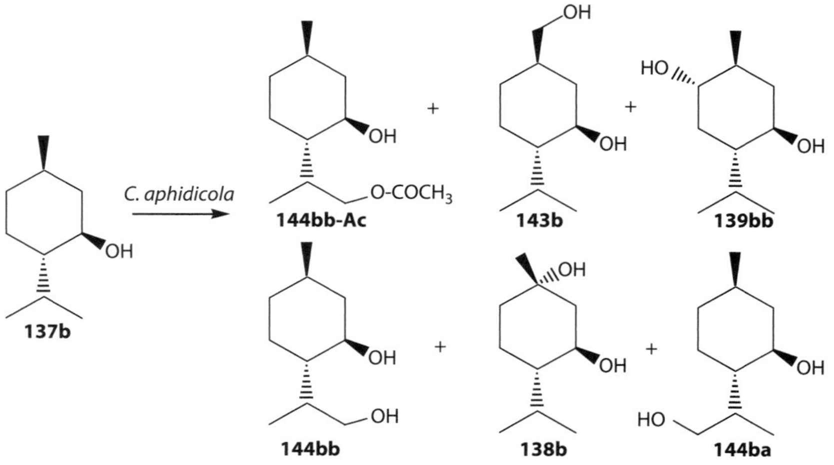

FIGURE 19.63 Biotransformation of (−)-menthol (**137b**) by *Cephalosporium aphidicola*. (Modified from Atta-ur-Rahman, M. et al., *J. Nat. Prod.*, 61, 1340, 1998.)

(**143b**). The bioconversion of (+)- (**137a′**) and (−)-neomenthol (**137a**) and (+)-isomenthol (**137c′**) by *A. niger* was studied later by Takahashi et al. (1994), mainly giving hydroxylated products. Noma and Asakawa (1995) reviewed the schematic menthol hydroxylation in detail.

Incubation of (−)-menthol (**137b**) with *Cephalosporium aphidicola* for 12 days yielded 10-acetoxymenthol (**144bb-Ac**), 1α-hydroxymenthol (**138b**), 6α-hydroxy-menthol (**139bb**), 7-hydroxymenthol (**143b**), 9-hydroxymenthol (**144ba**), and 10-hydroxymenthol (**144bb**) (Atta-ur-Rahman et al., 1998) (Figure 19.63).

A. niger TBUYN-2 converted (−)-menthol (**137b**) to 1α- (**138b**), 2α- (**140b**), 4β- (**141b**), 6α- (**139bb**), 7- (**143b**), 9-hydroxymenthols (**144ba**), and the mosquito repellent-active 8-hydroxymenthol (**142b**) (Figure 19.64). *A. cellulosae* M-77 biotransformed (−)-menthol (**137b**) to 4β-hydroxymenthol (**141b**) predominantly. The formation of **141b** is also observed in *A. cellulosae* IFO 4040 and *Aspergillus terreus* IFO 6123, but its yield is much less than that obtained from **137b** by *A. cellulosae* M-77 (Asakawa et al., 1991) (Table 19.1).

On the other hand, (+)-menthol (**137b′**) was smoothly biotransformed by *A. niger* to give 1β-hydroxymenthol (**138b′**), 6β-hydroxymenthol (**139ba′**), 2β-hydroxymenthol (**140ba′**), 4α-hydroxymenthol (**141b′**), 7-hydroxymenthol (**143b′**), 8-hydroxymenthol (**142b′**), and 9-hydroxymenthol (**144ba′**) (Figure 19.65) (Table 19.2).

S. litura converted (−)- and (+)-menthols (**137b** and **137b′**) that gave the corresponding 10-hydroxy products (**143b** and **143b′**) (Miyazawa et al., 1997a) (Figure 19.66).

(−)-Menthol (**137b**) was glycosylated by *Eucalyptus perriniana* suspension cells to (−)-menthol diglucoside (**364**, 26.6%) and another menthol glycoside. On the other hand, (+)-menthol (**137b′**) was glycosylated by *E. perriniana* suspension cells to (+)-menthol di- (**364′**, 44.0%) and triglucosides (**365**, 6.8%) (Hamada et al., 2002) (Figure 19.67).

(−)-Menthol (**137b**) and its enantiomer (**137b′**) were converted to their corresponding 8-hydroxy derivatives (**142b** and **142b′**) by human CYP2A6 (Nakanishi and Miyazawa, 2005) (Figure 19.68). By various assays, cytochrome P450 molecular species responsible for the metabolism of (−)- (**137b**) and (+)-menthol (**137b′**) was determined to be CYP2A6 and CYP2B1 in human and rat, respectively. Also, kinetic analysis showed that K and V_{max} values for the oxidation of (−)- (**137b**) and (+)-menthol (**137b′**) recombinant CYP2A6 and CYP2B1 were determined to be 28 μM and 10.33 nmol/min/nmol P450 and 27 μM and 5.29 nmol/min/nmol P450, 28 μM and 3.58 nmol/min/nmol P450, and 33 μM and 5.3 nmol/min/nmol P450, respectively (Nakanishi and Miyazawa, 2005) (Figure 19.68).

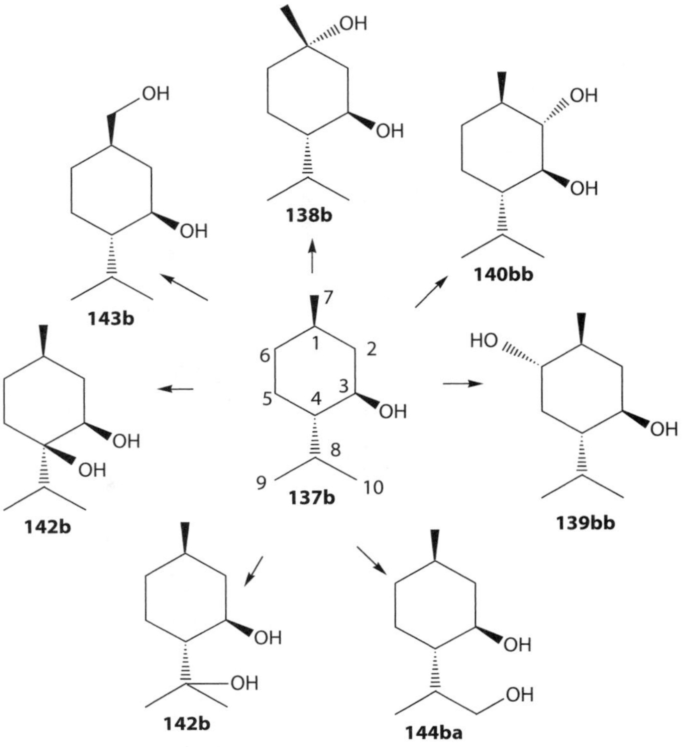

FIGURE 19.64 Metabolic pathways of (–)-menthol (**137b**) by *Aspergillus niger*. (Modified from Asakawa, Y. et al., *Phytochemistry*, 30, 3981, 1991.)

TABLE 19.1
Metabolites of (–)-Menthol (137b) by Various *Aspergillus* spp. (Static Culture)

Microorganisms	138b	142b	139bb	143b	139bb	144ba	141b
A. awamori IFO 4033	+[a]	++	−	+	++	+++	−
A. fumigatus IFO 4400	−	+	−	+	+	+	−
A. sojae IFO 4389	++	+	+	−	−	++++	−
A. usami IFO 4338	−	−	−	+	−	+++	−
A. cellulosae M-77	+	−	−	+	−	++	++++
A. cellulosae IFO 4040	−	+	−	−	−	++	++
A. terreus IFO 6123	+	+	+	−	+	+	−
A. niger IFO 4049	−	+	−	+	−	+++	−
A. niger IFO 4040	−	+	−	+++	−	+++	−
A. niger TBUYN-2	+	++	+	+	++	++	−

[a] Symbols +, ++, +++, etc., are relative concentrations estimated by GC-MS.

19.3.3.2 Neomenthol

(+)-Neomenthol (**137a**) is biotransformed by *A. niger* TBUYN-2 to give five kinds of diols (**138a**, **143a**, **144aa**, **144ab**, and **142a**) and two kinds of triols (**145a** and **146a**) as shown in Figure 19.69 (Takahashi et al., 1994).

(–)-Neomenthol (**137a′**) is biotransformed by *A. niger* to give six kinds of diols (**140a′**, **139a′**, **143a′**, **144aa′**, **144ab′**, and **142a′**) and a triol (**146a′**) as shown in Figure 19.70 (Takahashi et al., 1994).

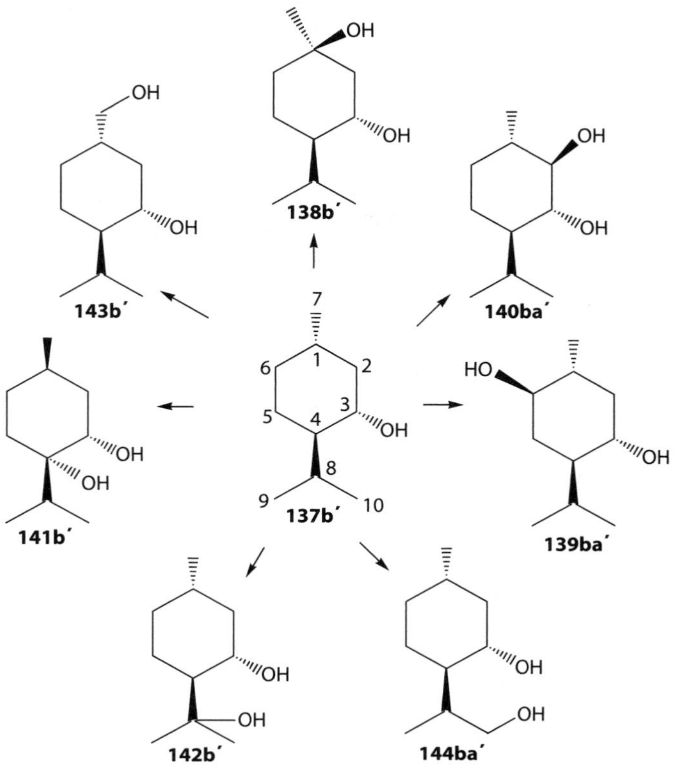

FIGURE 19.65 Metabolic pathways of (+)-menthol (**137b′**) by *Aspergillus niger*. (Modified from Noma, Y. et al., Microbiological conversion of menthol. Biotransformation of (+)-menthol by a strain of *Aspergillus niger*, *Proceedings of 33rd TEAC*, 1989, pp. 124–126; Asakawa, Y. et al., *Phytochemistry*, 30, 3981, 1991.)

TABLE 19.2
Metabolites of (+)-Menthol (137b′) by Various *Aspergillus* spp. (Static Culture)

Microorganisms	138b′	142b′	140ba′	143b′	139ba′	144ba′	141b′
A. awamori IFO 4033	+[a]	++	−	+++	−	+++	−
A. fumigates IFO 4400	+	++	−	+	−	++	−
A. sojae IFO 4389	+	++	−	−	−	+++	−
A. usami IFO 4338	+	−	−	+	−	+++	−
A. cellulosae M-77	−	+	−	−	−	++	++++
A. cellulosae IFO 4040	+	+	−	−	++	+	+
A. terreus IFO 6123	+	+++	+	+	+	++	−
A. niger IFO 4049	+	−	−	−	+	+++	−
A. niger IFO 4040	+	++	−	+	−	++	−
A. niger TBUYN-2	++	+	−	+++++	+	+	−

[a] Symbols +, ++, +++, etc., are relative concentrations estimated by GC-MS.

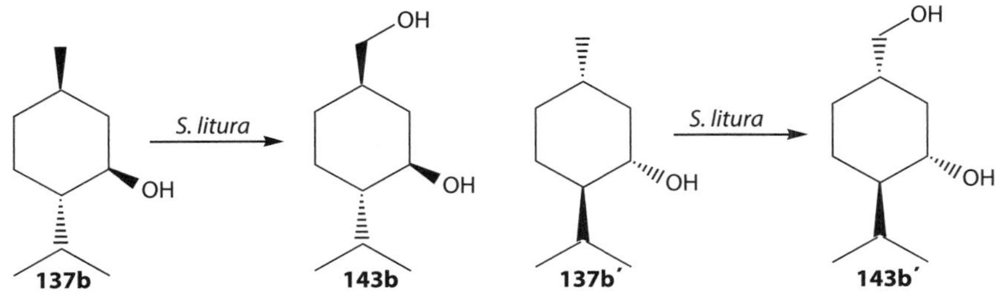

FIGURE 19.66 Biotransformation of (−)- (**137b**) and (+)-menthol (**137b′**) by *Spodoptera litura*. (Modified from Miyazawa, M. et al., Biotransformation of (−)-menthol and (+)-menthol by common cutworm Larvae, *Spodoptera litura* as a biocatalyst, *Proceedings of 41st TEAC*, 1997a, pp. 391–392.)

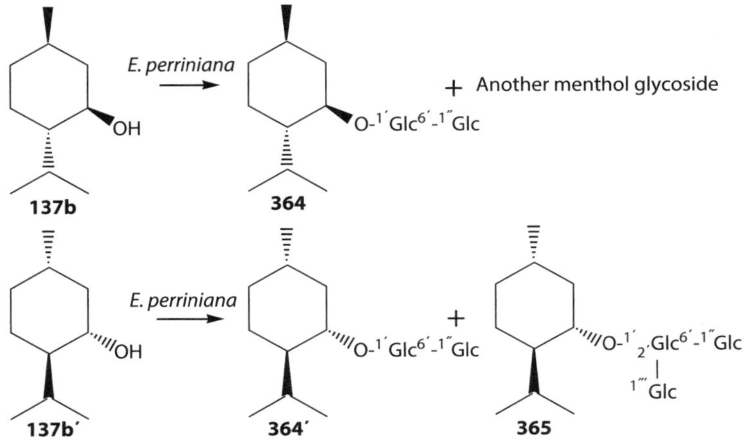

FIGURE 19.67 Biotransformation of (−)-(**137b**) and (+)-menthol (**137b′**) by *Eucalyptus perriniana* suspension cells. (Modified from Hamada, H. et al., Glycosylation of monoterpenes by plant suspension cells, *Proceedings of 46th TEAC*, 2002, pp. 321–322.)

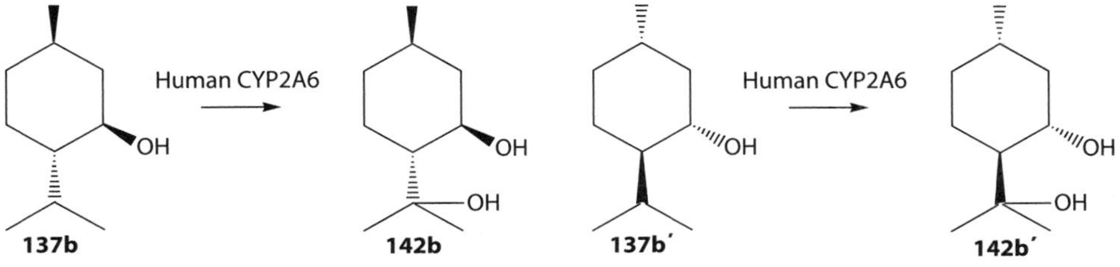

FIGURE 19.68 Biotransformation of (−)-menthol (**137b**) and its enantiomer (**137b′**) by human CYP2A6. (Modified from Nakanishi, K. and Miyazawa, M., Biotransformation of (+)- and (−)- menthol by liver microsomal humans and rats, *Proceedings of 49th TEAC*, 2005, pp. 423–425.)

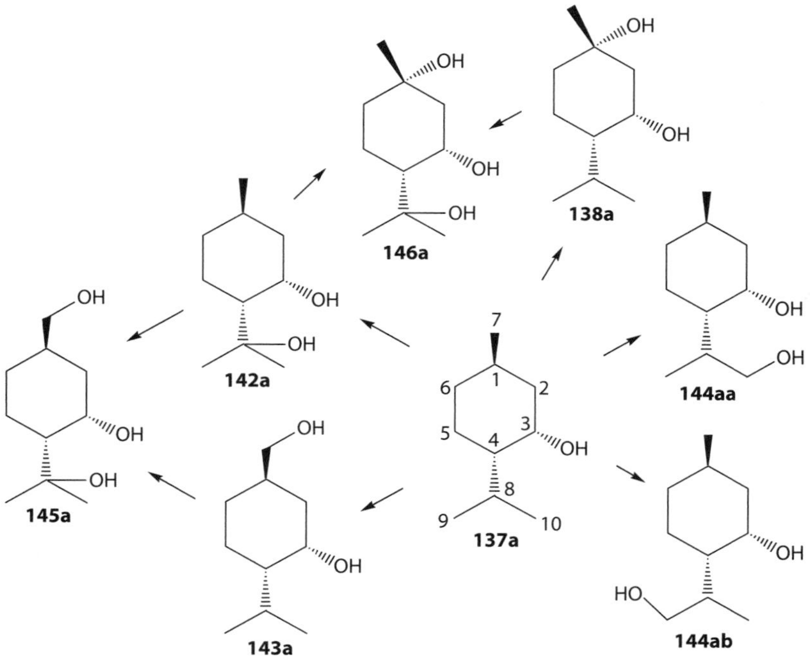

FIGURE 19.69 Metabolic pathways of (+)-neomenthol (**137a**) by *Aspergillus niger*. (Modified from Takahashi, H. et al., *Phytochemistry*, 35, 1465, 1994.)

19.3.3.3 (+)-Isomenthol

(+)-Isomenthol (**137c**) is biotransformed to give two kinds of diols such as 1β-hydroxy- (**138c**) and 6β-hydroxyisomenthol (**139c**) by *A. niger* (Takahashi et al., 1994) (Figure 19.71).

(±)-Isomenthyl acetate (**137c-Ac** and **137c′-Ac**) was asymmetrically hydrolyzed to (−)-isomenthol (**137c**) with (+)-isomenthol acetate (**137c′-Ac**) by many microorganisms and esterases (Oritani and Yamashita, 1973b) (Figure 19.72).

19.3.3.4 Isopulegol

(−)-Isopulegol (**366**) was biotransformed by *S. litura* larvae to give 7-hydroxy-(−)-isopulegol (**367**), 9-hydroxy-(−)-menthol (**144ba**), and 10-hydroxy-(−)-isopulegol (**368**). On the other hand, (+)-isopulegol (**366′**) was biotransformed by the same larvae in the same manner to give 7-hydroxy-(+)-isopulegol (**367′**), 9-hydroxy-(+)-menthol (**144ba′**), and 10-hydroxy-(+)-isopulegol (**368′**) (Ohsawa and Miyazawa, 2001) (Figure 19.73).

Microbial resolution of (±)-isopulegyl acetate (**366-Ac** and **366′-Ac**) was studied by microorganisms. (±)-Isopulegyl acetate (**366-Ac** and **366′-Ac**) was hydrolyzed asymmetrically to give a mixture of (−)-isopulegol (**366**) and (+)-isopulegyl acetate (**366′-Ac**) (Oritani and Yamashita, 1973c) (Figure 19.74).

19.3.3.5 α-Terpineol

P. pseudomallei strain T was cultivated with α-terpineol (**34**) as the sole carbon source to give 8,9-epoxy-*p*-menthan-1-ol (**58**) via epoxide (**369**) and diepoxide (**57**) as intermediates (Hayashi et al., 1972) (Figure 19.75).

(+)-α-Terpineol (**34**) was formed from (+)-limonene (**34**) by citrus pathogenic *P. digitatum* (Pers.; Fr.) Sacc. KCPYN, which was further biotransformed to *p*-menthane-1β,2α,8-triol (**334**), 2α-hydroxy-1,8-cineole (**125b**), and (+)-*trans*-sobrerol (**95a**) (Noma and Asakawa, 2006a, 2007a)

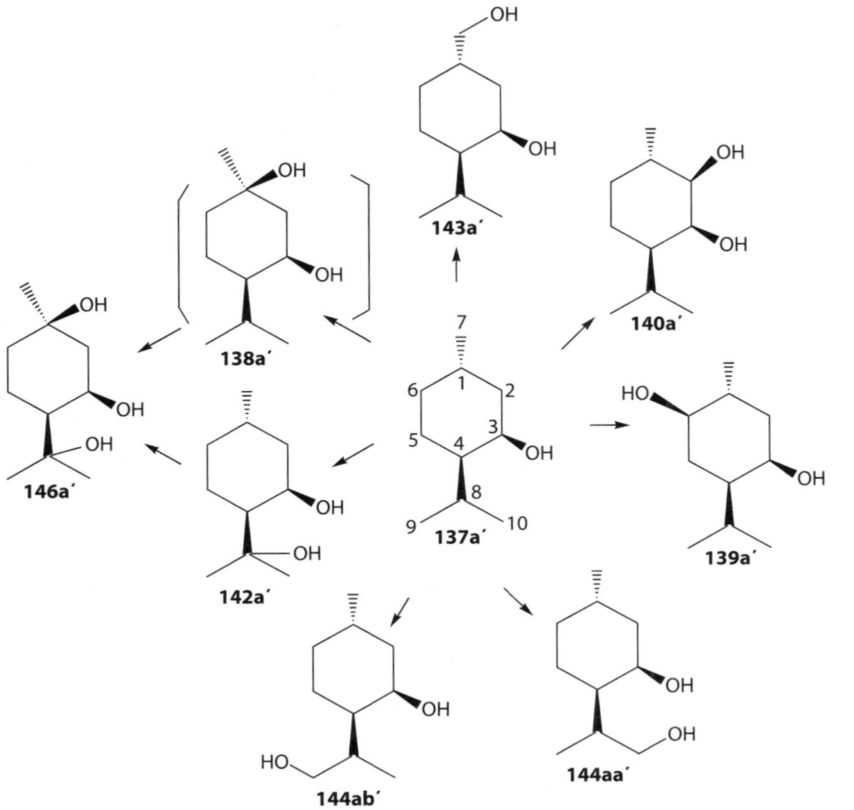

FIGURE 19.70 Metabolic pathways of (−)-neomenthol (**137a′**) by *Aspergillus niger*. (Modified from Takahashi, H. et al., *Phytochemistry*, 35, 1465, 1994.)

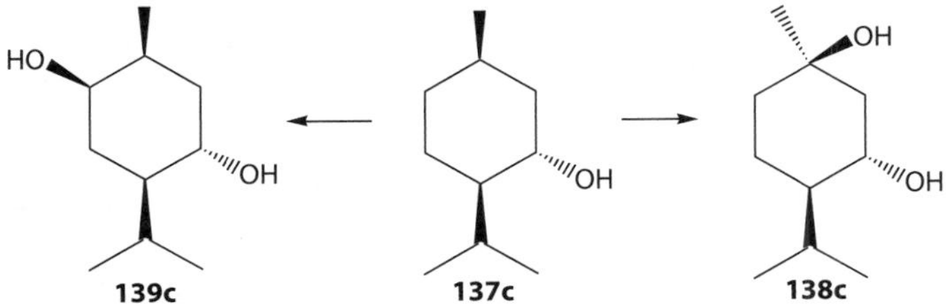

FIGURE 19.71 Metabolic pathways of (+)-isomenthol (**137c**) by *Aspergillus niger*. (Modified from Takahashi, H. et al., *Phytochemistry*, 35, 1465, 1994.)

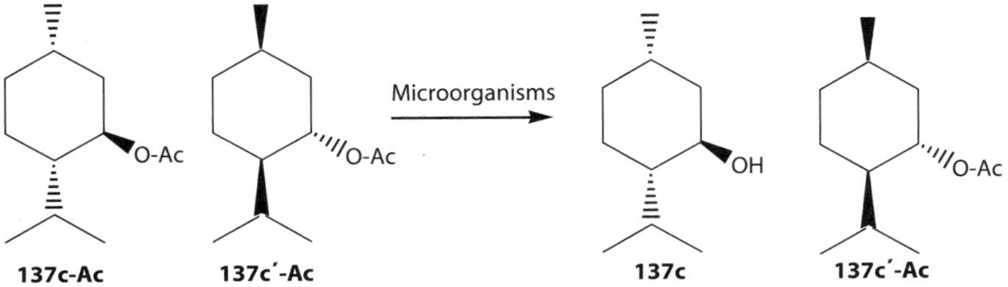

FIGURE 19.72 Microbial resolution of (±)-isomenthyl acetate (**137c-Ac** and **137c′-Ac**) by microbial esterase. (Modified from Oritani, T. and Yamashita, K., *Agric. Biol. Chem.*, 37, 1695, 1973b.)

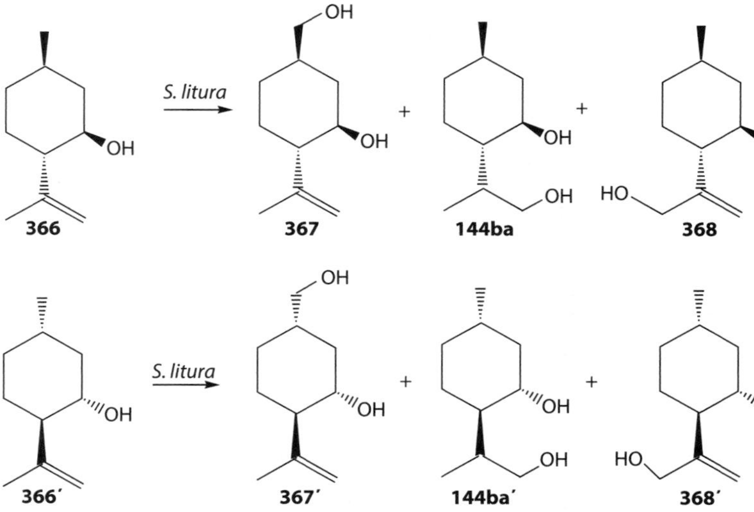

FIGURE 19.73 Biotransformation of (−)-(**366**) and (+)-isopulegol (**366′**) by *Spodoptera litura*. (Modified from Ohsawa, M. and Miyazawa, M., Biotransformation of (+)- and (−)-isopulegol by the larvae of common cutworm (*Spodoptera litura*) as a biocatalyst, *Proceedings of 45th TEAC*, 2001, pp. 375–376.)

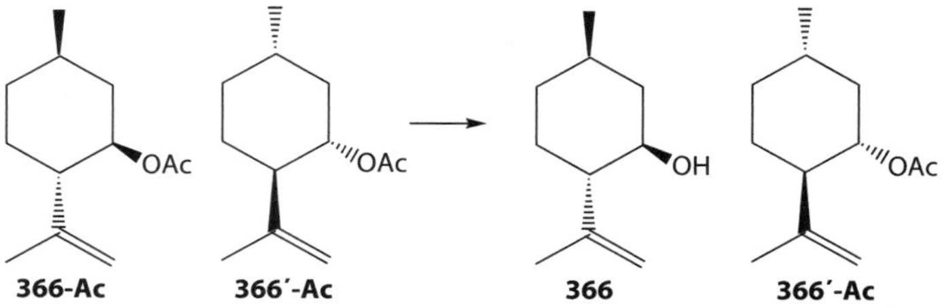

FIGURE 19.74 Microbial resolution of (±)-isopulegyl acetate (**366-Ac** and **366′-Ac**) by microorganisms. (Modified from Oritani, T. and Yamashita, K., *Agric. Biol. Chem.*, 37, 1687, 1973c.)

(Figure 19.76). *Penicillium* sp. YuzuYN also biotransformed **34** to **334**. Furthermore, *A. niger* Tiegh CBAYN and *C. roseus* biotransformed **34** to give **95a** and (+)-oleuropeyl alcohol (**204**), respectively (Hamada et al., 2001; Noma and Asakawa, 2006a, 2007a) (Figure 19.76).

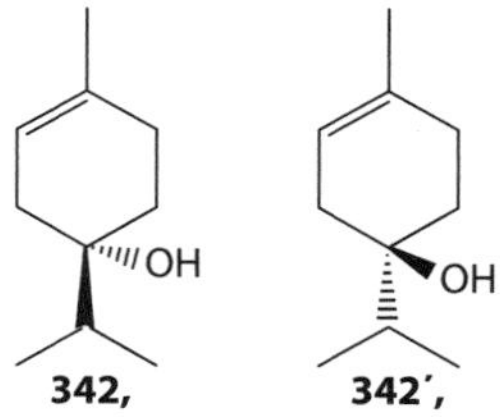

Gibberella cyanea DSM 62719 biotransformed (−)-α-terpineol (**34′**) to give *p*-menthane-1-β,2α,8-triol (**334′**), 2α-hydroxy-1,8-cineole (**125b′**), 1,2-epoxy-α-terpineol (**369′**), (−)-oleuropeyl alcohol (**204′**), (−)-*trans*-sobrerol (**95a′**), and *cis*-sobrerol (**95b′**) (Abraham et al., 1986) (Figure 19.76). In cases of *P. digitatum* (Pers. Fr.) Sacc. KCPYN, *Penicillium* sp. YuzuYN, and *A. niger* Tiegh CBAYN, **34′** was biotransformed to give **369′**, **95a′**, and **334′**, respectively (Noma and Asakawa, 2006a, 2007a) (Figure 19.77). *C. roseus* biotransformed **34′** to give **95a′** and **204′** (Hamada et al., 2001) (Figure 19.77).

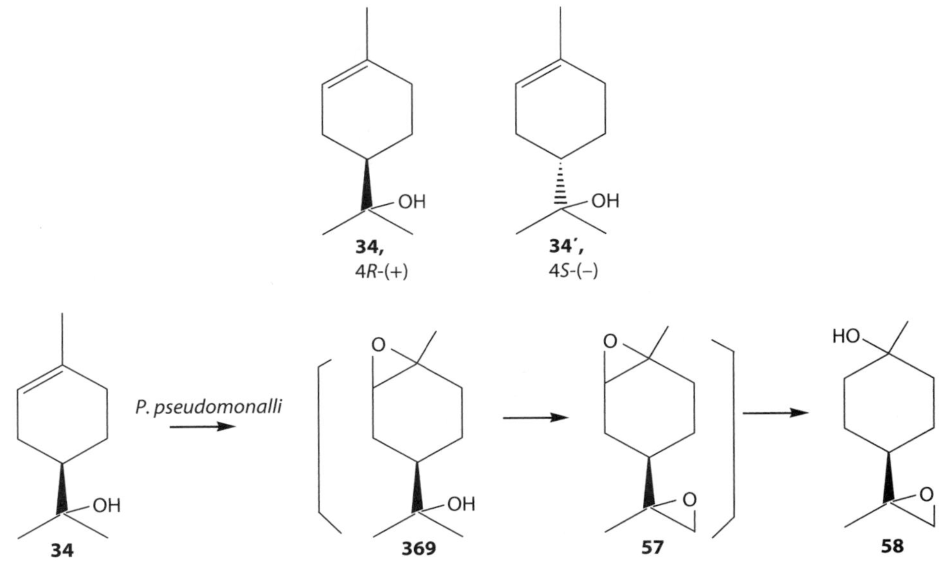

FIGURE 19.75 Biotransformation of (+)-α-terpineol (**34**) to 8,9-epoxy-*p*-menthan-1-ol (**58**) by *Pseudomonas pseudomallei* strain T. (Modified from Hayashi, T. et al., *Biol. Chem.*, 36, 690, 1972.)

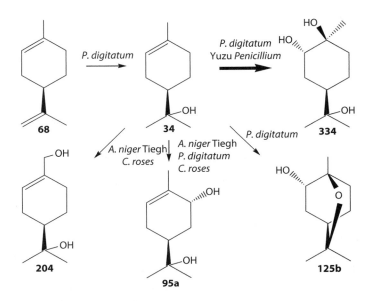

FIGURE 19.76 Biotransformation of (+)-α-terpineol (**34**) by citrus pathogenic fungi, *Penicillium digitatum* (Pers.; Fr.) Sacc. KCPYN, *Penicillium* sp. YuzuYN, and *Aspergillus niger* Tiegh CBAYN. (Modified from Noma, Y. and Asakawa, Y., Biotransformation of (+)-limonene and related compounds by *Citrus* pathogenic fungi, *Proceedings of 50th TEAC*, pp. 431–433, 2006a; Noma, Y. and Y. Asakawa, Biotransformation of limonene and related compounds by newly isolated low temperature grown *citrus* pathogenic fungi and red yeast, *Book of Abstracts of the 38th ISEO*, 2007a, p. 7.)

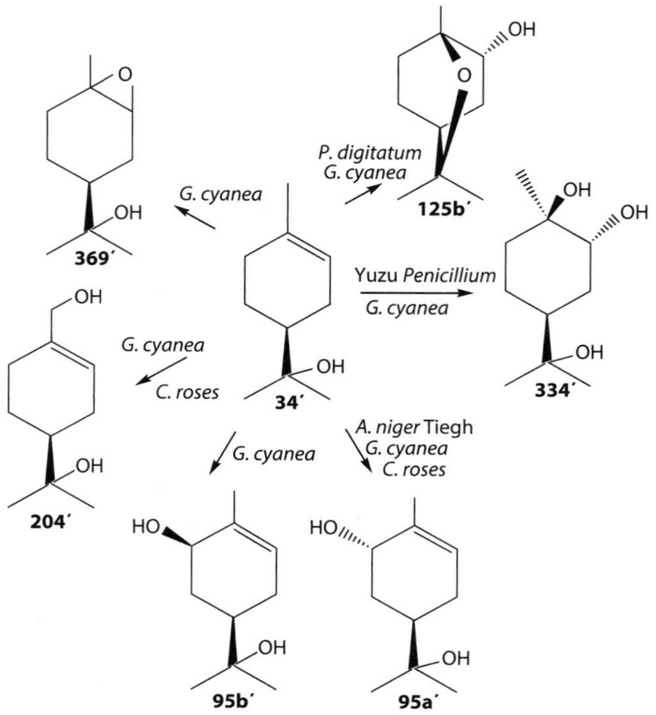

FIGURE 19.77 Biotransformation of (−)-α-terpineol (**34′**) by *Gibberella cyanea* DSM 62719, *Penicillium digitatum* (Pers. Fr.) Sacc. KCPYN, *Penicillium* sp. YuzuYN, and *Aspergillus niger* Tiegh CBAYN. (Modified from Abraham, W.-R. et al., *Appl. Microbiol. Biotechnol.*, 24, 24, 1986; Noma, Y. and Asakawa, Y., Biotransformation of (+)-limonene and related compounds by *Citrus* pathogenic fungi, *Proceedings of 50th TEAC*, 2006a, pp. 431–433; Noma, Y. and Asakawa, Y., Biotransformation of limonene and related compounds by newly isolated low temperature grown *citrus* pathogenic fungi and red yeast, *Book of Abstracts of the 38th ISEO*, 2007a, p. 7.)

19.3.3.6 (−)-Terpinen-4-ol

G. cyanea DSM 62719 biotransformed (*S*)-(−)-terpinen-4-ol (**342**) (1-*p*-menthen-4-ol) to give 2α-hydroxy-1,4-cineole (**132b**), 1-*p*-menthene-4α,6-diol (**372**), and *p*-menthane-1β,2α,4α-triol (**371**) (Abraham et al., 1986). On the other hand, *A. niger* TBUYN-2 also biotransformed (−)-terpinen-4-ol (**342**) to give 2α-hydroxy-1,4-cineole(**132b**) and (+)-*p*-menthane-1β,2α,4α-triol (**371**) (Noma and Asakawa, 2007b) (Figure 19.78). On the other hand, *S. litura* biotransformed (*R*)-terpinen-4-ol (**342′**) to (4*R*)-*p*-menth-1-en-4,7-diol (**373′**) (Kumagae and Miyazawa, 1999) (Figure 19.78).

19.3.3.7 Thymol and Thymol Methyl Ether

Thymol (**179**) was converted at the concentration of 14% by *Streptomyces humidus*, Tu-1 to give (1*R*,2*S*)- (**181a**) and (1*R*,2*R*)-2-hydroxy-3-*p*-menthen-5-one (**181b**) as the major products (Noma et al., 1988a) (Figure 19.79). On the other hand, in a *Pseudomonas*, thymol (**179**) was biotransformed to 6-hydroxy- (**180**), 7-hydroxy- (**479**), 9-hydroxy- (**480**), 7,9-dihydroxythymol (**482**), thymol-7-oic acid (**481**), and thymol-9-oic acid (**483**) (Chamberlain and Dagley, 1968) (Figure 19.79).

Thymol methyl ether (**459**) was converted by fungi *A. niger*, *Mucor ramannianus*, *Rhizopus arrhizus*, and *Trichothecium roseum* to give 7-hydroxy- (**460**) and 9-hydroxythymol methyl ether (**461**) (Demirci et al., 2001) (Figure 19.79).

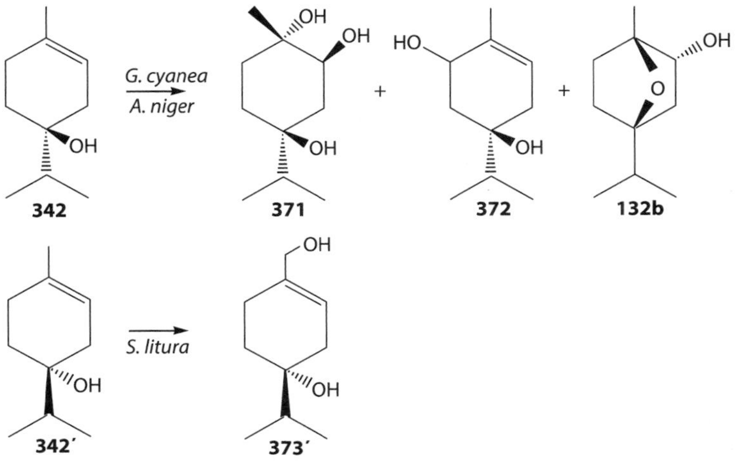

FIGURE 19.78 Biotransformation of (−)-terpinen-4-ol (**342**) by *Gibberella cyanea* DSM 62719, *Aspergillus niger* TBUYN-2, and *Spodoptera litura*. (Modified from Abraham, W.-R. et al., *Appl. Microbiol. Biotechnol.*, 24, 24, 1986; Kumagae, S. and Miyazawa, M., *Proceedings of 43rd TEAC*, 1999, pp. 389–390; Noma, Y. and Asakawa, Y., Microbial transformation of limonene and related compounds, *Proceedings of 51st TEAC*, 2007b, pp. 299–301.)

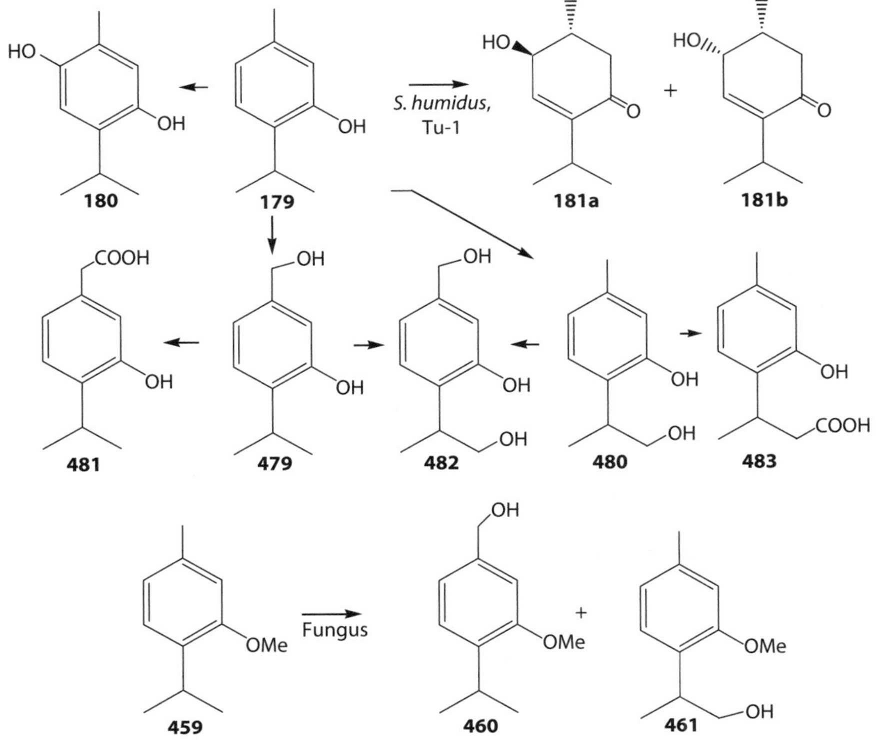

FIGURE 19.79 Biotransformation of thymol (**179**) and thymol methyl ether (**459**) by actinomycetes *Streptomyces humidus*, Tu-1, and fungi *Aspergillus niger*, *Mucor ramannianus*, *Rhizopus arrhizus*, and *Trichothecium roseum*. (Modified from Chamberlain, E.M. and Dagley, S., *Biochem. J.*, 110, 755, 1968; Noma, Y. et al., Microbial transformation of thymol formation of 2-hydroxy-3-*p*-menthen-5-one by *Streptomyces humidus*, Tu-1, *Proceedings of 28th TEAC*, 1988a, pp. 177–179; Demirci, F. et al., The biotransformation of thymol methyl ether by different fungi. *Book of Abstracts of the XII Biotechnology Congr.*, 2001, p. 47.)

19.3.3.8 Carvacrol and Carvacrol Methyl Ether

When cultivated in a liquid medium with carvacrol (**191**), as a sole carbon source, the bacterial isolated from savory and pine consumed the carvacrol in the range of 19%–22% within 5 days of cultivation. The fungal isolates grew much slower and after 13 days of cultivation consumed 7.1%–11.4% carvacrol (**191**). Pure strains belonging to the bacterial genera of bacterium, *Bacillus* and *Pseudomonas*, as well as fungal strain from *Aspergillus*, *Botrytis*, and *Geotrichum* genera, were also tested for their ability to grow in medium containing carvacrol (**191**). Among them, only in *Bacterium* sp. and *Pseudomonas* sp. carvacrol (**191**) uptake was monitored. Both *Pseudomonas* sp. 104 and 107 consumed the substrate in the amount of 19%. These two strains also exhibited the highest cell mass yield and the highest productivity (1.1 and 1.2 g/L/day) (Schwammle et al., 2001).

Carvacrol (**191**) was biotransformed to 3-hydroxy- (**470**), 9-hydroxy (**471**), 7-hydroxy- (**475**), and 8-hydroxycarvacrol (**474**), 8,9-dehydrocarvacrol (**473**), carvacrol-9-oic acid (**472**), carvacrol-7-oic acid (**476**), and 8,9-dihydroxycarvacrol (**477**) by rats (Ausgulen et al., 1987) and microorganisms (Demirci, 2000) including *T. roseum* and *Cladosporium* sp. (Figure 19.80). Furthermore, carvacrol methyl ether (**191-Me**) was converted by the same fungi to give 7-hydroxy- (**475-Ac**) and 9-hydroxycarvacrol methyl ether (**471-Me**) and 7,9-dihydroxycarvacrol methyl ether (**478**) (Demirci, 2000) (Figure 19.80).

19.3.3.9 Carveol

At first, soil Pseudomonad biotransformed (+)-limonene (**68**) to (+)-carvone (**93**) and (+)-1-*p*-menthene-6,9-diol (**90**) via (+)-*cis*-carveol (**81b**) as shown in Figure 19.81 (Dhavalikar and Bhattacharyya, 1966; Dhavalikar et al., 1966).

Second, *Pseudomonas ovalis* strain 6-1 (Noma, 1977) biotransformed the mixture of (−)-*cis*-carveol (**81b′**) and (−)-*trans*-carveol (**81a′**) (94:6, GC ratio) to (−)-carvone (**93′**) (Noma, 1977), which was further metabolized reductively to give (+)-dihydrocarvone (**101a′**), (+)-isodihydrocarvone (**101b′**), (+)-neodihydrocarveol (**102a**), and (−)-dihydrocarveol (**102b**) (Noma et al., 1984). Hydrogenation at C1, 2-position did not occur, but the dehydrogenation at C6-position occurred to give (−)-carvone (**93**) (Figure 19.82).

On the other hand, in *Streptomyces* A-5-1 and *Nocardia* 1-3-11, which were isolated from soil, (−)-carvone (**93′**) was reduced to give mainly (−)-*trans*-carveol (**81a′**) and (−)-*cis*-carveol (**81b′**), respectively. On the other hand, (−)-*trans*-carveol (**81a′**) and (−)-*cis*-carveol (**81b′**) were dehydrogenated to give **93′** by strain 1-3-11 and other microorganisms (Noma et al., 1986). The reaction between *trans*- and *cis*-carveols (**81a′** and **81b′**) and (−)-carvone (**93′**) is reversible (Noma, 1980) (Figure 19.82).

Third, the investigation for the biotransformation of the mixture of (−)-*trans*- (**81a′**) and (−)-*cis*-carveol (**81b′**) (60:40 in GC ratio) was carried out by using 81 strains of soil actinomycetes. All actinomycetes produced (−)-carvone (**93′**) from the mixture of (−)-*trans*- (**81a′**) and (−)-*cis*-carveol (**81b′**) (60:40 in GC ratio). However, 41 strains of actinomycetes converted (−)-*cis*-carveol

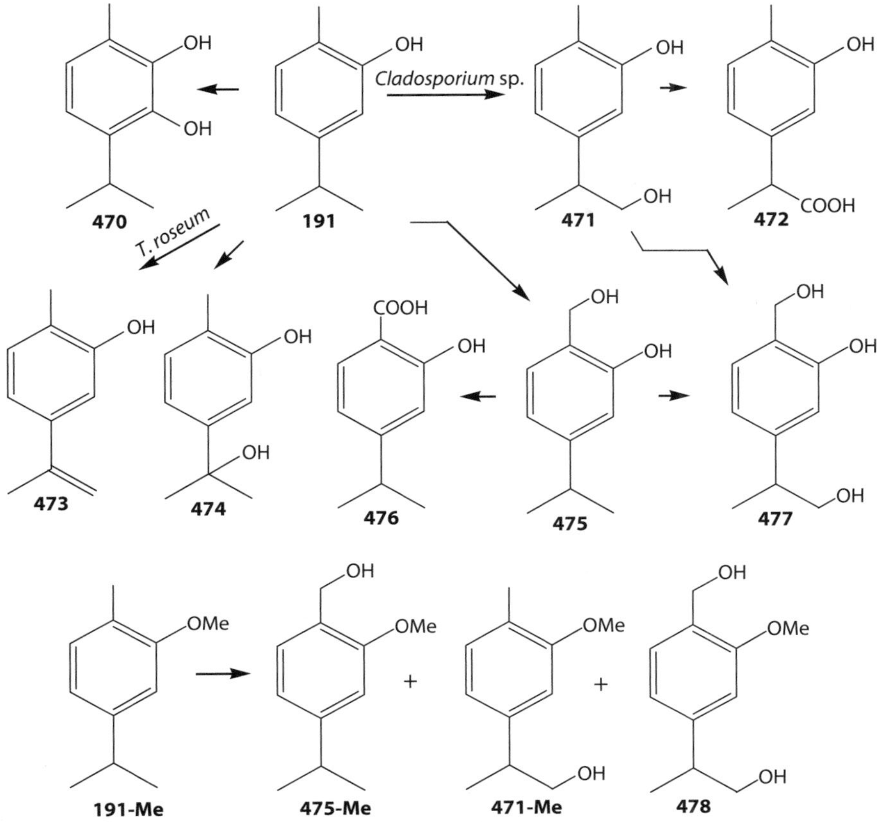

FIGURE 19.80 Biotransformation of carvacrol (**191**) and carvacrol methyl ether (**191-Me**) by rats. (Modified from Ausgulen, L.T. et al., *Pharmacol. Toxicol.*, 61, 98, 1987) and microorganisms (Modified from Demirci, F., *Microbial transformation of bioactive monoterpenes*. Ph.D. thesis, Anadolu University, Eskisehir, Turkey, 2000, pp. 1–137.)

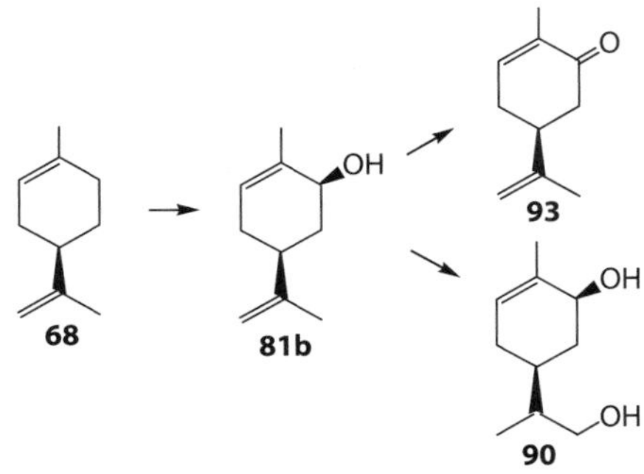

FIGURE 19.81 Proposed metabolic pathway of (+)-limonene (**68**) and (+)-*cis*-carveol (**81b**) by soil Pseudomonad. (Modified from Dhavalikar, R.S. and Bhattacharyya, P.K., *Indian J. Biochem.*, 3, 144, 1966; Dhavalikar, R.S. et al., *Indian J. Biochem.*, 3, 158, 1966.)

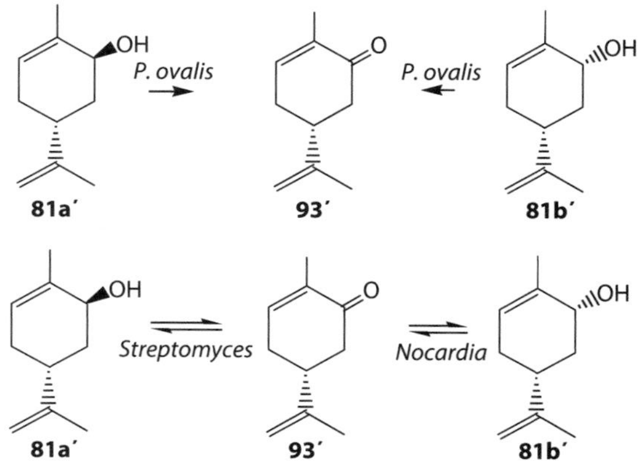

FIGURE 19.82 Biotransformation of (−)-*trans*-(**81a′**) and (−)-*cis*-carveol (**81b′**) (6:94, GC ratio) by *Pseudomonas ovalis* strain 6-1, *Streptomyces* A-5-1, and *Nocardia* 1-3-11. (Modified from Noma, Y., *Nippon Nogeikagaku Kaishi*, 51, 463, 1977; Noma, Y., *Agric. Biol. Chem.*, 44, 807, 1980.)

(**81b′**) to give (4*R*,6*R*)-(+)-6,8-oxidomenth-1-en-9-ol (**92a′**), which is named as bottrospicatol after the name of the microorganism *Streptomyces bottropensis* [Bottro], and (−)-*cis*-carveol (**81b′**) containing *Mentha spicata* [spicat] and alcohol [ol] (Nishimura et al., 1983a) (Figure 19.83).

(+)-Bottrospicatol (**92a′**) was prepared by epoxidation of (−)-carvone (**93′**) with *m*CPBA to (−)-carvone-8,9-epoxide (**96′**), followed by stereoselective reduction with NaBH₄ to alcohol, which was immediately cyclized with 0.1 N H₂SO₄ to give diastereomixture of bottrospicatol (**92a′** and **b′**) (Nishimura et al., 1983a) (Figure 19.84).

Further investigation showed *S. bottropensis* SY-2-1 (Noma and Iwami, 1994) has different metabolic pathways for (−)-*trans*-carveol (**81a′**) and (−)-*cis*-carveol (**81b′**). That is, *S. bottropensis* SY-2-1 converted (−)-*trans*-carveol (**81a′**) to (−)-carvone (**93′**), (−)-carvone-8,9-epoxide (**96′**), (−)-5β-hydroxycarvone (**98a′**), and (+)-5β-hydroxyneodihydrocarveol (**100aa′**) (Figure 19.85). On

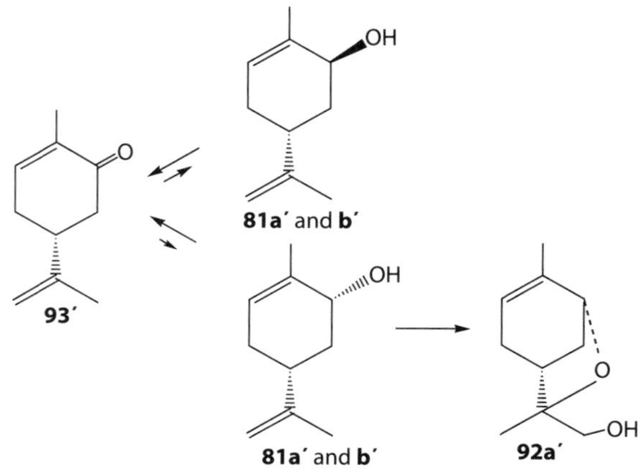

FIGURE 19.83 The metabolic pathways of *cis*-carveol (**81b′**) by *Pseudomonas ovalis* strain 6-1. (Modified from Noma, Y., *Nippon Nogeikagaku Kaishi*, 51, 463, 1977) and *Streptomyces bottropensis* SY-2-1 and other microorganisms (Modified from Noma, Y. et al., *Agric. Biol. Chem.*, 46, 2871, 1982; Nishimura, H. et al., Biological activity of bottrospicatol and related compounds produced by microbial transformation of (−)-*cis*-carveol towards plants, *Proceedings of 27th TEAC*, 1983a, pp. 107–109.)

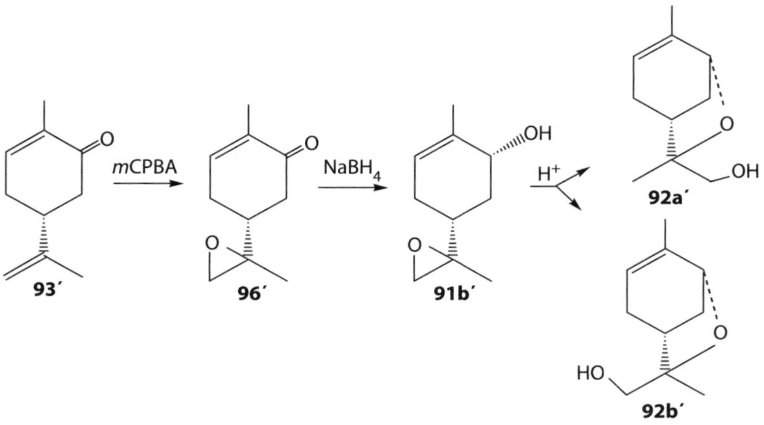

FIGURE 19.84 Preparation of (+)-bottrospicatol (**92a′**) and (+)-isobottrospicatol (**92b′**) from (−)-carvone (**93′**) with *m*CPBA. (Modified from Nishimura, H. and Noma, Y., *Biotechnology for Improved Foods and Flavors*, G.R. Takeoka, et al., ACS Symp. Ser. 637, pp. 173–187, 1996. American Chemical Society, Washington, DC.)

the other hand, *S. bottropensis* SY-2-1 converted (−)-*cis*-carveol (**81b′**) to give (+)-bottrospicatol (**92a′**) and (−)-5β-hydroxy-*cis*-carveol (**94ba′**) as main products together with (+)-isobottrospicatol (**92b′**) as the minor product as shown in Figure 19.85.

In the metabolism of *cis*-carveol by microorganisms, there are four pathways (pathways 1–4) as shown in Figure 19.86. At first, *cis*-carveol (**81**) is metabolized to carvone (**93**) by C2 dehydrogenation (Noma, 1977, 1980) (pathway 1). Second, *cis*-carveol (**81b**) is metabolized via epoxide as intermediate to bottrospicatol (**92**) by rearrangement at C2 and C8 (Noma et al., 1982; Nishimura et al., 1983a,b; Noma and Nishimura, 1987) (pathway 2). Third, *cis*-carveol (**81b**) is hydroxylated at C5 position to give 5-hydroxy-*cis*-carveol (**94**) (Noma and Nishimura, 1984) (pathway 3). Finally, *cis*-carveol

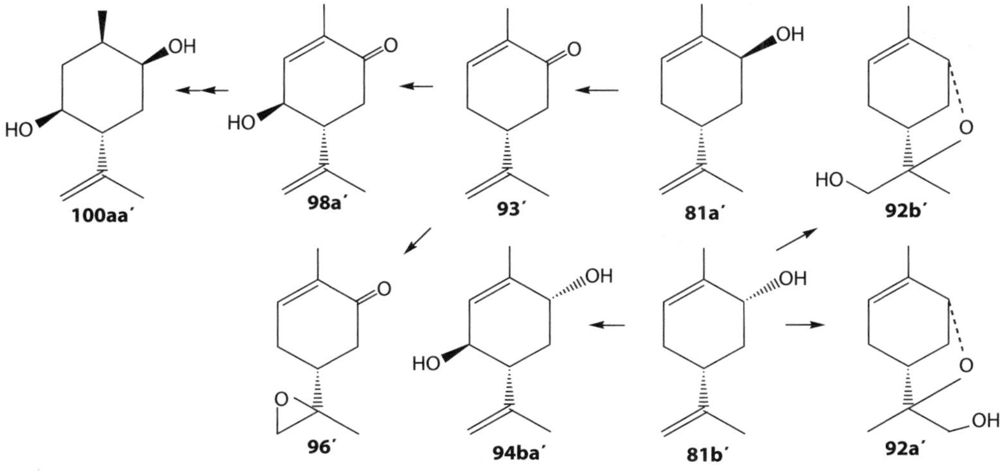

FIGURE 19.85 Biotransformation of (−)-*trans*- (**81a′**) and (−)-*cis*-carveol (**81b′**) by *Streptomyces bottropensis* SY-2-1 and *Streptomyces ikutamanensis* Ya-2-1. (Modified from Noma, Y. et al., *Agric. Biol. Chem.*, 46, 2871, 1982; Noma, Y. and Nishimura, H., Microbiological conversion of carveol. Biotransformation of (−)-*cis*-carveol and (+)-*cis*-carveol by *S. bottropensis*, Sy-2-1, *Proceedings of 28th TEAC*, 1984, pp. 171–173; Noma, Y. and H. Nishimura, *Agric. Biol. Chem.*, 51, 1845, 1987.)

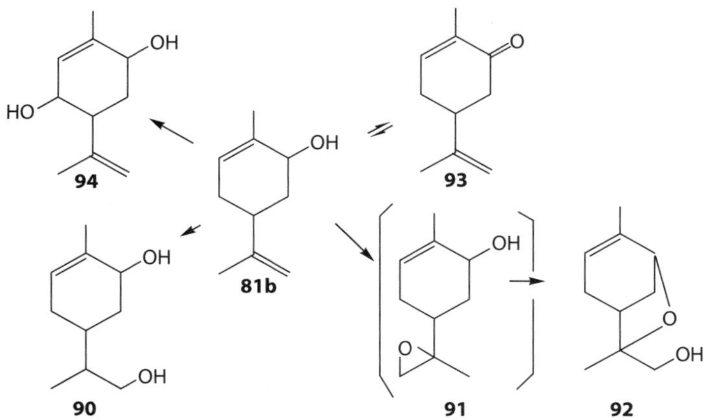

FIGURE 19.86 General metabolic pathways of carveol (**81**) by microorganisms. (Modified from Noma, Y. et al., *Agric. Biol. Chem.*, 46, 2871, 1982; Noma, Y. and Nishimura, H., Microbiological conversion of carveol. Biotransformation of (−)-*cis*-carveol and (+)-*cis*-carveol by *S. bottropensis*, Sy-2-1, *Proceedings of 28th TEAC*, pp. 171–173, 1984; Noma, Y. and Nishimura, H., *Agric. Biol. Chem.*, 51, 1845, 1987; Nishimura, H. and Noma, Y., *Biotechnology for Improved Foods and Flavors*, G.R. Takeoka, et al., ACS Symp. Ser. 637, pp.173–187. American Chemical Society, Washington, DC, 1996.)

(**81b**) is metabolized to 1-*p*-menthene-2,9-diol (**90**) by hydroxylation at C9 position (Dhavalikar and Bhattacharyya, 1966; Dhavalikar et al., 1966) (pathway 4).

Effects of (−)-*cis*- (**81b′**) and (−)-*trans*-carveol (**81a′**) conversion products by *S. bottropensis* SY-2-1 on the germination of lettuce seeds were examined, and the result is shown in Table 19.3. (+)-Bottrospicatol (**92′**) and (−)-carvone-8,9-epoxide (**96′**) showed strong inhibitory activity for the germination of lettuce seeds.

S. bottropensis SY-2-1 has also different metabolic pathways for (+)-*trans*-carveol (**81a**) and (+)-*cis*-carveol (**81b**) (Noma and Iwami, 1994). That is, *S. bottropensis* SY-2-1 converted (+)-*trans*-carveol (**81a**) to (+)-carvone (**93**), (+)-carvone-8,9-epoxide (**96**), and (+)-5α-hydroxycarvone (**98a**) (Noma and Nishimura, 1982, 1984) (Figure 19.87). On the other hand, *S. bottropensis* SY-2-1 converted (+)-*cis*-carveol (**81b**) to give (−)-isobottrospicatol (**92b**) and (+)-5-hydroxy-*cis*-carveol (**94b**)

TABLE 19.3
Effects of (−)-*cis*- (81b′) and (−)-*trans*-Carveol (81a′) Conversion Products by *Streptomyces bottropensis* SY-2-1 on the Germination of Lettuce Seeds

Compounds	Germination Rate (%)	
	24 h	48 h
(−)-Carvone (**93′**)	47	89
(+)-Bottrospicatol (**92′**)	3	48
(−)-Carvone-8,9-epoxide (**96′**)	2	77
5β-Hydroxyneodihydrocarveol (**102aa′**)	86	96
5β-Hydroxycarvone (**98a′**)	91	96
Control	95	96

Note: Concentration of each compound was adjusted at 200 ppm.

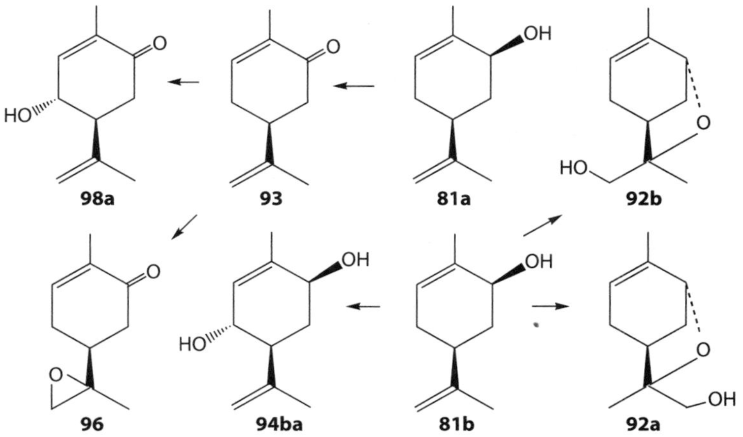

FIGURE 19.87 Metabolic pathways of (+)-*trans*- (**81a**) and (+)-*cis*-carveol (**81b**) by *Streptomyces bottropensis* SY-2-1. (Modified from Noma, Y. and Nishimura, H., *Agric. Biol. Chem.*, 51, 1845, 1987; Nishimura, H. and Noma, Y., *Biotechnology for Improved Foods and Flavors*, G.R. Takeoka, et al., ACS Symp. Ser. 637, pp.173–187. American Chemical Society, Washington, DC, 1996.)

as the main products and (−)-bottrospicatol (**92a**) as the minor product as shown in Figure 19.88 (Noma et al., 1980; Noma and Nishimura, 1987; Nishimura and Noma, 1996).

Biological activities of (+)-bottrospicatol (**92a′**) and related compounds for plant's seed germination and root elongation were examined toward barnyard grass, wheat, garden cress, radish, green foxtail, and lettuce (Nishimura and Noma, 1996).

Isomers and derivatives of bottrospicatol were prepared by the procedure shown in Figure 19.89. The chemical structure of each compound was confirmed by the interpretation of spectral data. The effects of all isomers and derivatives on the germination of lettuce seeds were compared. The germination inhibitory activity of (+)-bottrospicatol (**92a′**) was the highest of isomers. Interestingly, (−)-isobottrospicatol (**92b**) was not effective even in a concentration of 500 ppm. (+)-Bottrospicatol methyl ether (**92a′**-methyl ether) and esters [**92a′**-methyl (ethyl and *n*-propyl)

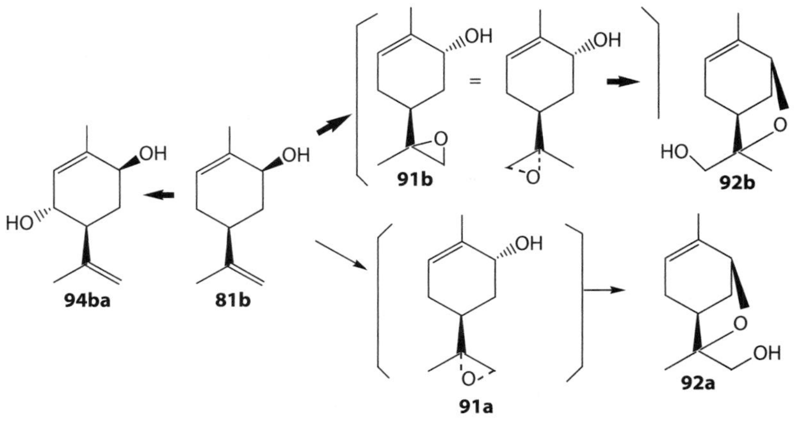

FIGURE 19.88 Metabolic pathways of (+)-*cis*-carveol (**81b**) by *Streptomyces bottropensis* SY-2-1 and *Streptomyces ikutamanensis* Ya-2-1. (Modified from Noma, Y. and Nishimura, H., *Agric. Biol. Chem.*, 51, 1845, 1987; Nishimura, H. and Noma, Y., *Biotechnology for Improved Foods and Flavors*, G.R. Takeoka et al., ACS Symp. Ser. 637, pp. 173–187. American Chemical Society, Washington, DC, 1996.)

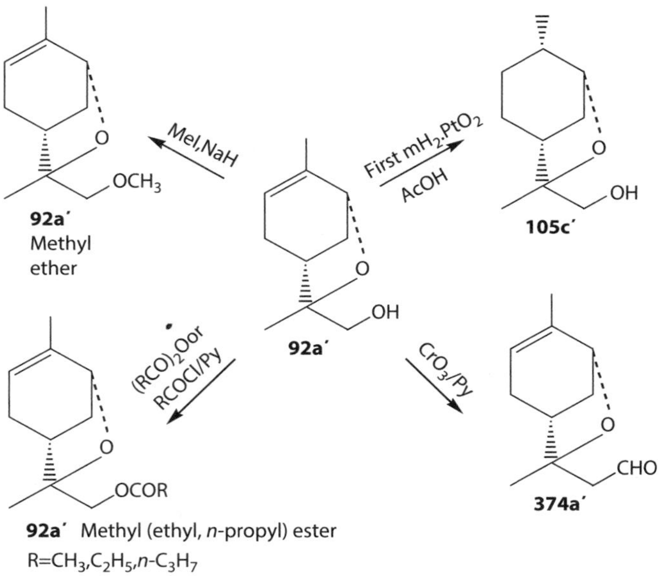

FIGURE 19.89 Preparation of (+)-bottrospicatol (**92a′**) derivatives. (Modified from Nishimura, H. and Noma, Y., *Biotechnology for Improved Foods and Flavors*, G.R. Takeoka et al., ACS Symp. Ser. 637, pp. 173–187. American Chemical Society, Washington, DC, 1996.)

ester] exhibited weak inhibitory activities. The inhibitory activity of (−)-isodihydrobottrospicatol (**105c′**) was as high as that of (+)-bottrospicatol (**92a′**). Furthermore, an oxidized compound, (+)-bottrospicatal (**374a′**), exhibited higher activity than (+)-bottrospicatol (**92a′**). So, the germination inhibitory activity of (+)-bottrospicatal (**374a′**) against several plant seeds, lettuce, green foxtail, radish, garden cress, wheat, and barnyard grass was examined. The result indicates that (+)-bottrospicatal (**374a′**) is a selective germination inhibitor as follows: lettuce > green foxtail > radish > garden cress > wheat > barnyard grass.

Enantio- and diastereoselective biotransformation of *trans*- (**81a** and **81a′**) and *cis*-carveols by *E. gracilis* Z. (Noma and Asakawa, 1992) and *Chlorella pyrenoidosa* IAM C-28 was studied (Noma et al., 1997).

In the biotransformation of racemic *trans*-carveol (**81a** and **81a′**), *C. pyrenoidosa* IAM C-28 showed high enantioselectivity for (−)-*trans*-carveol (**81a′**) to give (−)-carvone (**93′**), while (+)-*trans*-carveol (**81a**) was not converted at all. The same *C. pyrenoidosa* IAM C-28 showed high enantioselectivity for (+)-*cis*-carveol (**81b**) to give (+)-carvone (**93**) in the biotransformation of racemic *cis*-carveol (**81b** and **81b′**). (−)-*cis*-Carveol (**81b′**) was not converted at all. The same phenomenon was observed in the biotransformation of mixture of (−)-*trans*- and (−)-*cis*-carveol (**81a′** and **81b′**) and the mixture of (+)-*trans*- and (+)-*cis*-carveol (**81a** and **81b**) as shown in Figure 19.90. The high enantioselectivity and the high diastereoselectivity for the dehydrogenation of (−)-*trans*- and (+)-*cis*-carveols (**81a** and **81b′**) were shown in *E. gracilis* Z. (Noma and Asakawa, 1992), *C. pyrenoidosa* IAM C-28 (Noma et al., 1997), *N. tabacum*, and other *Chlorella* spp.

On the other hand, the high enantioselectivity for **81a′** was observed in the biotransformation of racemic (+)-*trans*-carveol (**81a**) and (−)-*trans*-carveol (**81a′**) by *Chlorella sorokiniana* SAG to give (−)-carvone (**93′**).

It was considered that the formation of (−)-carvone (**93′**) from (−)-*trans*-carveol (**81a′**) by diastereo- and enantioselective dehydrogenation is a very interesting phenomenon in order to produce mosquito repellent (+)-*p*-menthane-2,8-diol (**50a′**) (Noma, 2007).

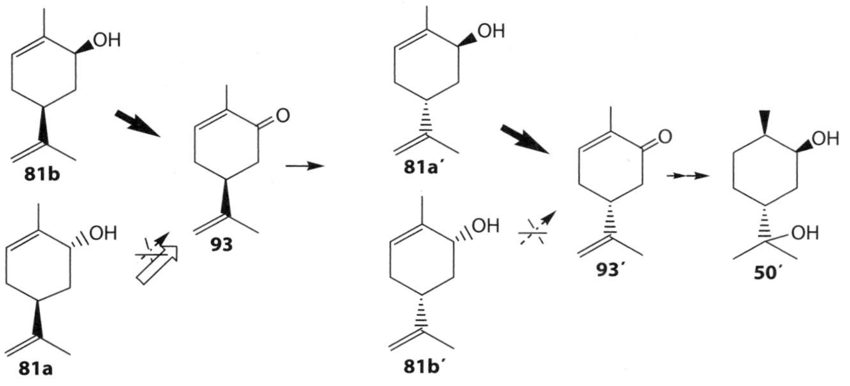

FIGURE 19.90 Enantio- and diastereoselective biotransformation of *trans*- (**81a** and **a′**) and *cis*-carveols (**81b** and **b′**) by *Euglena gracilis* Z and *Chlorella pyrenoidosa* IAM C-28. (Modified from Noma, Y. and Asakawa, Y., *Phytochemistry*, 31, 2009, 1992; Noma, Y. et al., Biotransformation of terpenoids and related compounds by *Chlorella* species, *Proceedings of 41st TEAC*, 1997, pp. 227–229.)

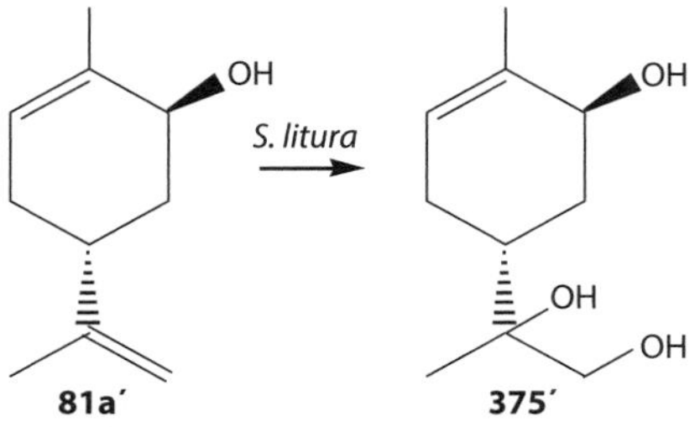

FIGURE 19.91 Biotransformation of (4R)-*trans*-carveol (**81a′**) by *Spodoptera litura*. (Modified from Miyazawa, M. et al., Biotransformation of *p*-menthanes using common cutworm larvae, *Spodoptera litura* as a biocatalyst, *Proceedings of 40th TEAC*, 1996b, pp. 80–81.)

(4R)-*trans*-Carveol (**81a′**) was converted by *S. litura* to give 1-*p*-menthene-6,8,9-triol (**375**) (Miyazawa et al., 1996b) (Figure 19.91).

19.3.3.10 Dihydrocarveol

(+)-Neodihydrocarveol (**102a′**) was converted to *p*-menthane-2,8-diol (**50a′**), 8-*p*-menthene-2,8-diol (**107a′**), and *p*-menthane-2,8,9-triols (**104a′** and **b′**) by *A. niger* TBUYN-2 (Noma et al., 1985a,b; Noma and Asakawa, 1995) (Figures 19.92 and 19.93). In the case of *E. gracilis* Z., mosquito repellent (+)-*p*-menthane-2,8-diol (**50a′**) was formed stereospecifically from (−)-carvone (**93′**) via (+)-dihydrocarvone (**101a′**) and (+)-neodihydrocarveol (**102a′**) (Noma et al., 1993; Noma, 2007). (−)-Neodihydrocarveol (**102a**) was also easily and stereospecifically converted by *E. gracilis* Z. to give (−)-*p*-menthane-2,8-diol (**50a**) (Noma et al., 1993).

On the other hand, *A. glauca* converted (−)-carvone (**93′**) stereospecifically to give (+)-8-*p*-menthene-2,8-diol (**107a′**) via (+)-dihydrocarvone (**101a′**) and (+)-neodihydrocarveol (**102a′**) (Demirci et al., 2004) (Figure 19.93).

(+)- (**102b**) and (−)-Dihydrocarveol (**102b′**) were converted by 10 kinds of *Aspergillus* spp. to give mainly (+)- (**107b′**) and (−)-10-hydroxydihydrocarveol (**107b**, 8-*p*-menthene-2,10-diol) and

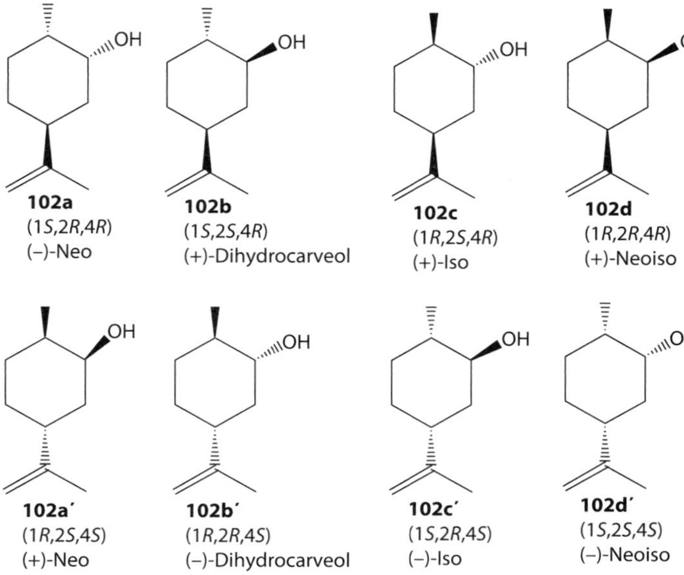

FIGURE 19.92 Chemical structure of eight kinds of dihydrocarveols.

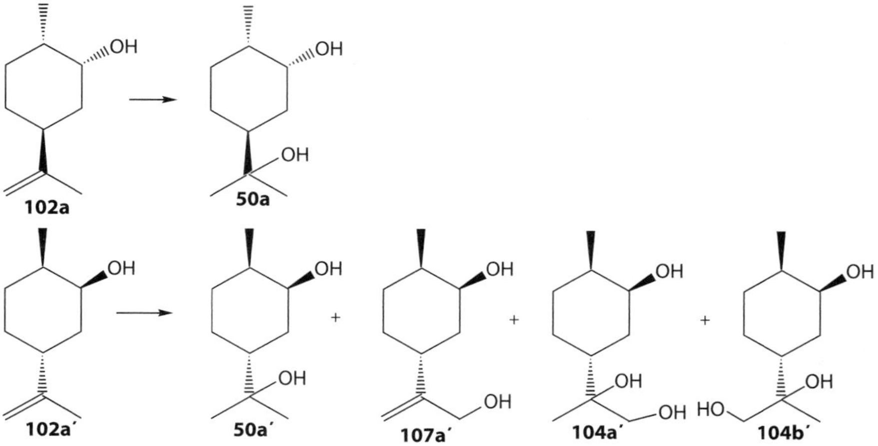

FIGURE 19.93 Biotransformation of (−)- and (+)-neodihydrocarveol (**102a** and **a′**) by *Euglena gracilis* Z, *Aspergillus niger* TBUYN-2, and *Absidia glauca*. (Modified from Noma, Y. et al., *Annual Meeting of Agricultural and Biological Chemistry*, Sapporo, 1985a, p. 68; Noma, Y. et al.,. Biotransformation of carvone. 6. Biotransformation of (−)-carvone and (+)-carvone by a strain of *Aspergillus niger*, *Proceedings of 29th TEAC*, 1985b, pp. 235–237; Noma, Y. et al., Formation of 8 kinds of *p*-menthane-2,8-diols from carvone and related compounds by *Euglena gracilis* Z. Biotransformation of monoterpenes by photosynthetic microorganisms. Part VIII, *Proceedings of 37th TEAC*, 1993, pp. 23–25; Noma, Y., *Aromatic Plants from Asia their Chemistry and Application in Food and Therapy*, L. Jiarovetz, N.X. Dung, and V.K. Varshney, pp. 169–186, Har Krishan Bhalla & Sons, India, 2007; Noma, Y. and Asakawa, Y., *Biotechnology in Agriculture and Forestry*, Vol. 33. Medicinal and Aromatic Plants VIII, Y.P.S. Bajaj, ed., pp. 62–96, Springer, Berlin, Germany, 1995; Demirci, F. et al., *Naturforsch.*, 59c, 389, 2004.)

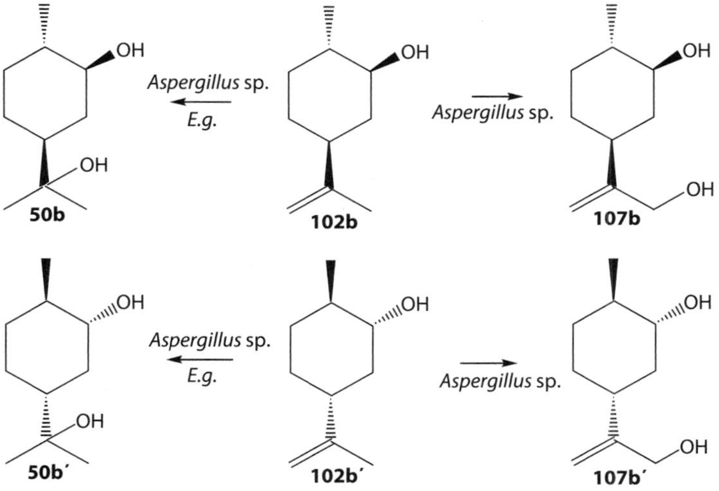

FIGURE 19.94 Biotransformation of (+)- (**102b**) and (−)-dihydrocarveol (**102b′**) by 10 kinds of *Aspergillus* spp. (Modified from Noma, Y., Formation of *p*-menthane-2,8-diols from (−)-dihydrocarveol and (+)-dihydrocarveol by *Aspergillus* spp., *The Meeting of Kansai Division of The Agricultural and Chemical Society of Japan*, Kagawa, 1988, p. 28) and *Euglena gracilis* Z (Modified from Noma, Y. et al., Formation of 8 kinds of *p*-menthane-2,8-diols from carvone and related compounds by *Euglena gracilis* Z. Biotransformation of monoterpenes by photosynthetic microorganisms. Part VIII, *Proceedings of 37th TEAC*, 1993, pp. 23–25.)

(+)- (**50b′**) and (−)-8-hydroxydihydrocarveol (**50b**, *p*-menthane-2,8-diol), respectively (Figure 19.94). The metabolic pattern of dihydrocarveols is shown in Table 19.4.

In the case of the biotransformation of *S. bottropensis* SY-2-1, (+)-dihydrocarveol (**102b**) was converted to (+)-dihydrobottrospicatol (**105aa**) and (+)-dihydroisobottrospicatol (**105ab**), whereas (−)-dihydrocarveol (**102b′**) was metabolized to (−)-dihydrobottrospicatol (**105aa′**) and

TABLE 19.4
Metabolic Pattern of Dihydrocarveols (102b and 102b′) by 10 Kinds of *Aspergillus* spp.

	Compounds					
Microorganisms	107b′	50b′	C.r. (%)	107b	50b	C.r. (%)
A. awamori, IFO 4033	0	98	99	3	81	94
A. fumigatus, IFO 4400	0	14	34	+	6	14
A. sojae, IFO 4389	0	47	59	1	50	85
A. usami, IFO 4338	0	32	52	+	5	7
A. cellulosae, M-77	0	27	52	+	7	14
A. cellulosae, IFO 4040	0	30	55	1	5	8
A. terreus, IFO 6123	0	79	92	+	18	46
A. niger, IFO 4034	0	29	49	+	8	12
A. niger, IFO 4049	4	50	67	9	34	59
A. niger, TBUYN-2	29	68	100	30	53	100

C.r.—conversion ratio.

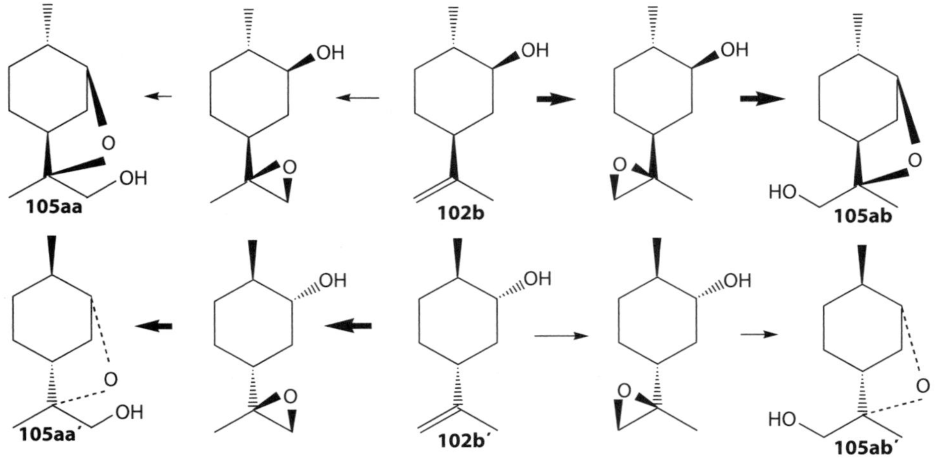

FIGURE 19.95 Biotransformation of (+)- (**102b**) and (−)-dihydrocarveol (**102b′**) by *Streptomyces bottropensis* SY-2-1. (Modified from Noma, Y., *Kagaku to Seibutsu*, 22, 742, 1984.)

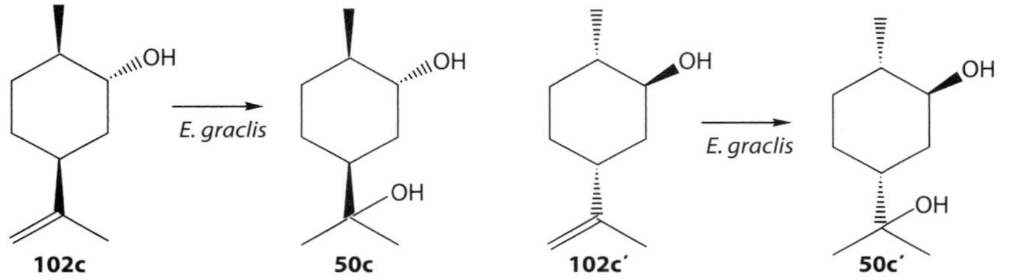

FIGURE 19.96 Biotransformation of (+)-iso- (**102c**) and (−)-dihydrocarveol (**102c′**) by *Euglena gracilis* Z. (Modified from Noma, Y. et al., Formation of 8 kinds of *p*-menthane-2,8-diols from carvone and related compounds by *Euglena gracilis* Z. Biotransformation of monoterpenes by photosynthetic microorganisms. Part VIII, *Proceedings of 37th TEAC*, 1993, pp. 23–25.)

(−)-dihydroisobottrospicatol (**105ab′**). (+)-Dihydroisobottrospicatol (**105ab**) and (−)-dihydrobottrospicatol (**105aa′**) are the major products (Noma, 1984) (Figure 19.95).

E. gracilis Z. converted (−)-iso- (**102c**) and (+)-isodihydrocarveol (**102c′**) to give the corresponding 8-hydroxyisodihydrocarveols (**50c** and **50c′**), respectively (Noma et al., 1993) (Figure 19.96).

In the case of the biotransformation of *S. bottropensis* SY-2-1, (−)-neoisodihydrocarveol (**102d**) was converted to (+)-isodihydrobottrospicatol (**105ba**) and (+)-isodihydroisobottrospicatol (**105bb**), whereas (+)-neoisodihydrocarveol (**102d′**) was metabolized to (−)-isodihydrobottrospicatol (**105ba′**) and (−)-isodihydroisobottrospicatol (**105bb′**). (+)-Isodihydroisobottrospicatol (**105bb**) and (−)-isodihydrobottrospicatol (**105ba′**) are the major products (Noma, 1984) (Figure 19.97).

E. gracilis Z. converted (−)- (**102d**) and (+)-neoisodihydrocarveol (**102d′**) to give the corresponding 8-hydroxyneoisodihydrocarveols (**50d** and **50d′**), respectively (Noma et al., 1993) (Figure 19.98).

Eight kinds of 8-hydroxydihydrocarveols (**50a–d** and **50a′–d′**; 8-*p*-menthane-2,8-diols) were obtained from carvone (**93** and **93′**), dihydrocarvones (**101a–b** and **101a′–b′**), and dihydrocarveols (**102a–d**, **102a′–d′**) by *E. gracilis* Z. as shown in Figure 19.99 (Noma et al., 1993).

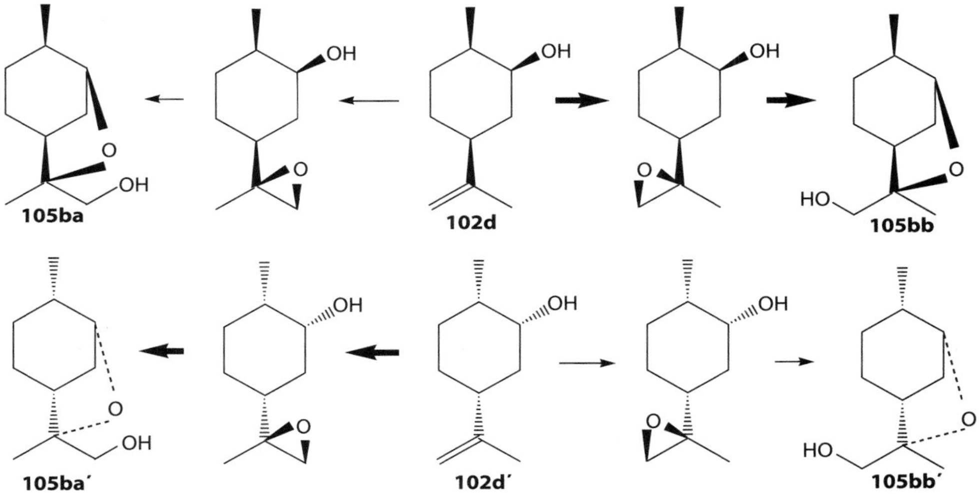

FIGURE 19.97 Formation of dihydroisobottrospicatols (**105**) from neoisodihydrocarveol (**102d** and **d′**) by *Streptomyces bottropensis* SY-2-1. (Modified from Noma, Y., *Kagaku to Seibutsu*, 22, 742, 1984.)

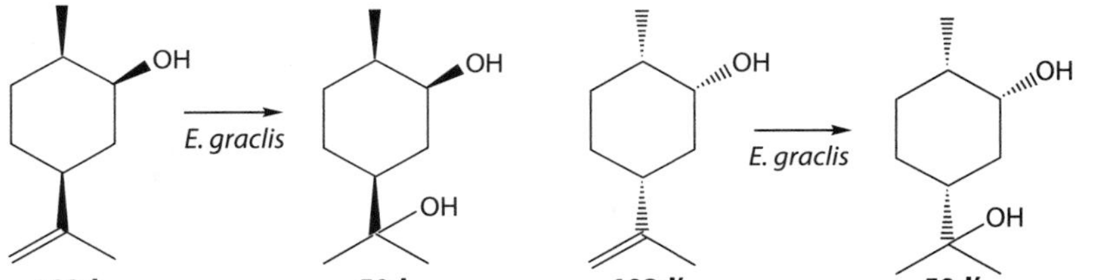

FIGURE 19.98 Biotransformation of (+)- (**102c**) and (−)-neoisodihydrocarveol (**102c′**) by *Euglena gracilis* Z. (Modified from Noma, Y. et al., Formation of 8 kinds of *p*-menthane-2,8-diols from carvone and related compounds by *Euglena gracilis* Z. Biotransformation of monoterpenes by photosynthetic microorganisms. Part VIII, *Proceedings of 37th TEAC*, 1993, pp. 23–25.)

19.3.3.11 Piperitenol

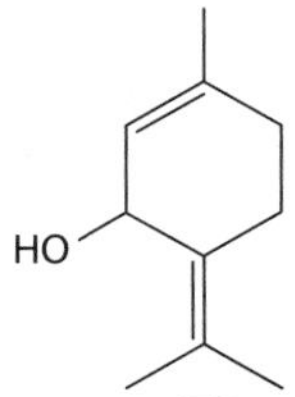

458

Incubation of piperitenol (**458**) with *A. niger* gave a complex metabolites whose structures have not yet been determined (Noma, 2000).

19.3.3.12 Isopiperitenol

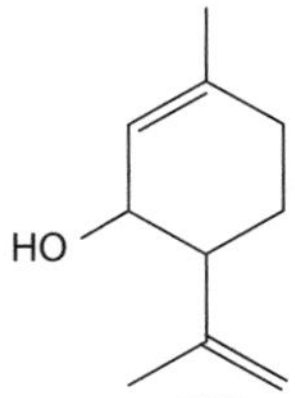

110

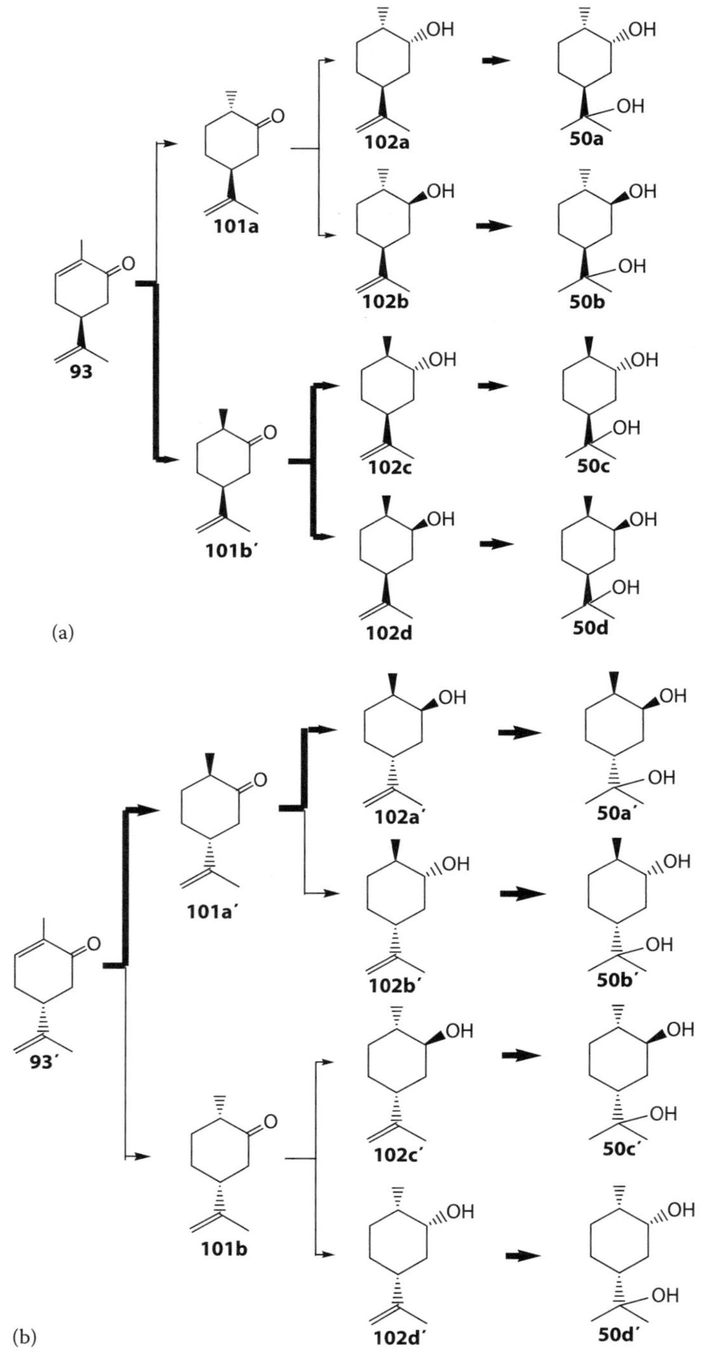

FIGURE 19.99 Formation of eight kinds of 8-hydroxydihydrocarveols (**50a–50d, 50a′–50d′**), dihydrocarvones (**101a–101b** and **101a′–101b′**), and dihydrocarveols (**102a–102d** and **102a′–102d′**) from (+)- (**93**) and (−)-carvone (**93′**) by *Euglena gracilis* Z.: (a) denotes 8-hydroxydihydrocarveols and (b) denotes dihydrocarveols. (Modified from Noma, Y. et al., Formation of 8 kinds of *p*-menthane-2,8-diols from carvone and related compounds by *Euglena gracilis* Z. Biotransformation of monoterpenes by photosynthetic microorganisms. Part VIII, *Proceedings of 37th TEAC*, 1993, pp. 23–25.)

Piperitenol (**458**) was metabolized by *A. niger* to give a complex alcohol mixtures whose structures have not yet been determined (Noma, 2000).

19.3.3.13 Perillyl Alcohol

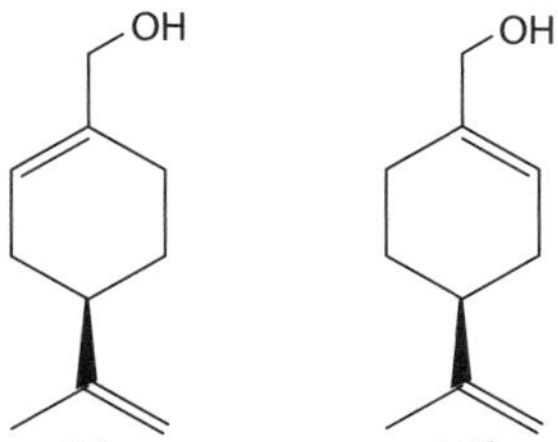

74
R-(+)-

74′
S-(−)-

(−)-Perillyl alcohol (**74′**) was epoxidized by *S. ikutamanensis* Ya-2-1 to give 8,9-epoxy-(−)-perillyl alcohol (**77′**) (Noma et al., 1986) (Figure 19.100).

(−)-Perillyl alcohol (**74′**) was glycosylated by *E. perriniana* suspension cells to (−)-perillyl alcohol monoglucoside (**376′**) and diglucoside (**377′**) (Hamada et al., 2002; Yonemoto et al., 2005) (Figure 19.101).

Furthermore, 1-perillyl-β-glucopyranoside (**376**) was converted into the corresponding oligosaccharides (**377–381**) using a cyclodextrin glucanotransferase (Yonemoto et al., 2005) (Figure 19.102).

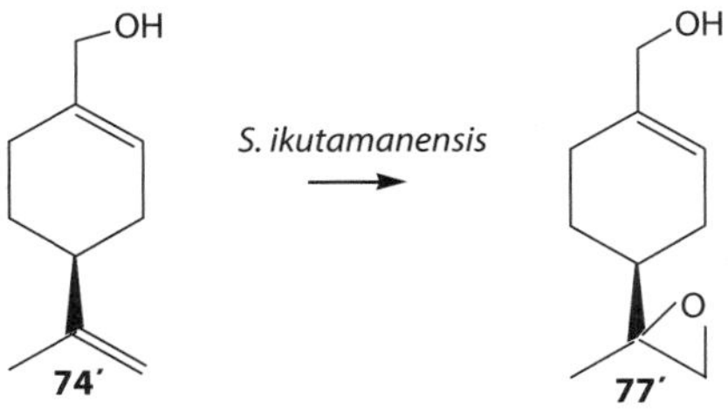

FIGURE 19.100 Biotransformation of (−)-perillyl alcohol (**74′**) by *Streptomyces ikutamanensis* Ya-2-1. (Modified from Noma, Y. et al., Reduction of terpene aldehydes and epoxidation of terpene alcohols by *S. ikutamanensis*, Ya-2-1, *Proceedings of 30th TEAC*, 1986, pp. 204–206.)

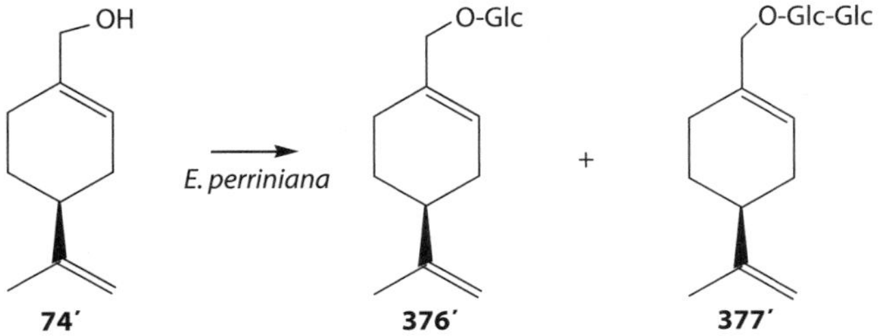

FIGURE 19.101 Biotransformation of (−)-perillyl alcohol (**74′**) by *Eucalyptus perriniana* suspension cell. (Modified from Hamada, H. et al., Glycosylation of monoterpenes by plant suspension cells, *Proceedings of 46th TEAC*, 2002, pp. 321–322; Yonemoto, N. et al., Preparation of (−)-perillyl alcohol oligosaccharides, *Proceedings of 49th TEAC*, 2005, pp. 108–110.)

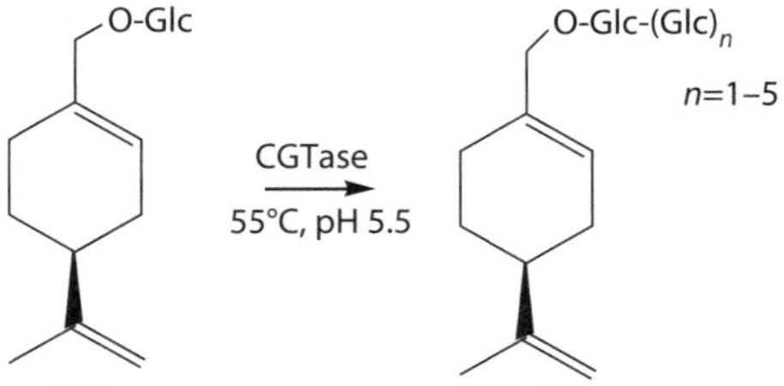

FIGURE 19.102 Biotransformation of (−)-perillyl alcohol monoglucoside (**376**) by CGTase. (Modified from Yonemoto, N. et al., Preparation of (−)-perillyl alcohol oligosaccharides, *Proceedings of 49th TEAC*, pp. 108–110, 2005.)

19.3.3.14 Carvomenthol

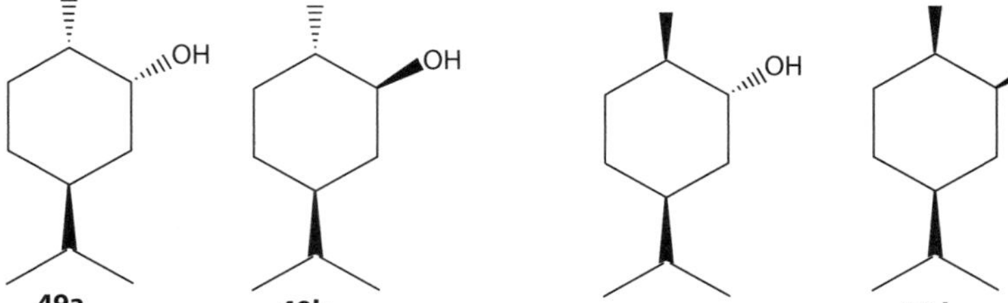

49a (1S,2R,4R) (−)-Neo

49b (1S,2S,4R) (+)-Carvomenthol

49c (1R,2S,4R) (−)-Iso

49d (1R,2R,4R) (−)-Neoiso

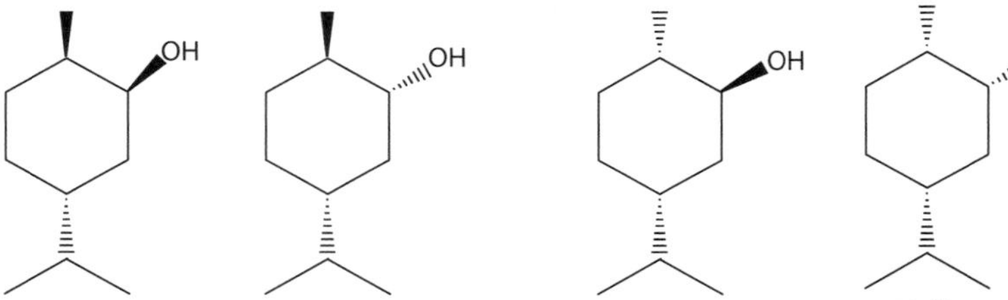

49a′ (1R,2S,4S) (+)-Neo

49b′ (1R,2R,4S) (−)-Carvomenthol

49c′ (1S,2R,4S) (+)-Iso

49d′ (1S,2S,4S) (+)-Neoiso

(+)-Iso- (**49c**) and (+)-neoisocarvomenthol (**49d**) were formed from (+)-carvotanacetone (**47**) via (−)-isocarvomenthone (**48b**) by *P. ovalis* strain 6-1, whereas (+)-neocarvomenthol (**49a′**) and (−)-carvomenthol (**49b′**) were formed from (−)-carvotanacetone (**47′**) via (+)-carvomenthone (**48a′**) by the same bacteria, of which **48b**, **48a′**, and **49d** were the major products (Noma et al., 1974a) (Figure 19.103).

Microbial resolution of carvomenthols was carried out by selected microorganisms such as *Trichoderma S* and *B. subtilis* var. *niger* (Oritani and Yamashita, 1973d). Racemic carvomenthyl acetate, racemic isocarvomenthyl acetate, and racemic neocarvomenthyl acetate were asymmetrically hydrolyzed to (−)-carvomenthol (**49b′**) with (+)-carvomenthyl acetate, (−)-isocarvomenthol (**49c**) with (+)-isocarvomenthyl acetate, and (+)-neoisocarvomenthol (**49d′**) with (−)-neoisocarvomenthyl acetate, respectively; racemic neocarvomenthyl acetate was not hydrolyzed (Oritani and Yamashita, 1973d) (Figure 19.104).

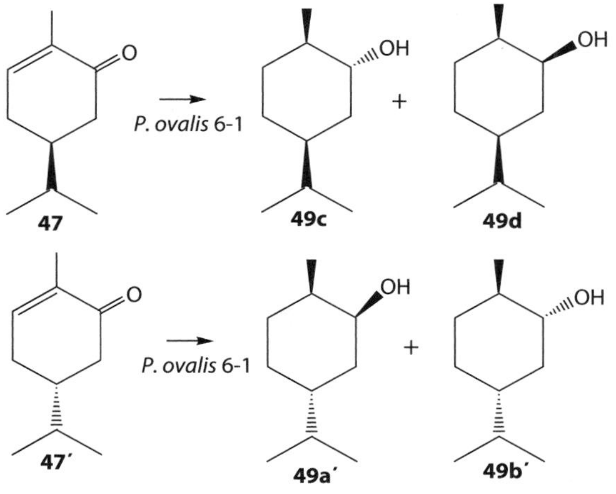

FIGURE 19.103 Formation of (−)-iso- (**49c**), (−)-neoiso- (**49d**), (+)-neo- (**49a′**), and (−)-carvomenthol (**49b′**) from (+)- (**47**) and (−)-carvotanacetone (**47′**) by *Pseudomonas ovalis* strain 6-1. (Modified from Noma, Y. et al., *Agric. Biol. Chem.*, 38, 1637, 1974a.)

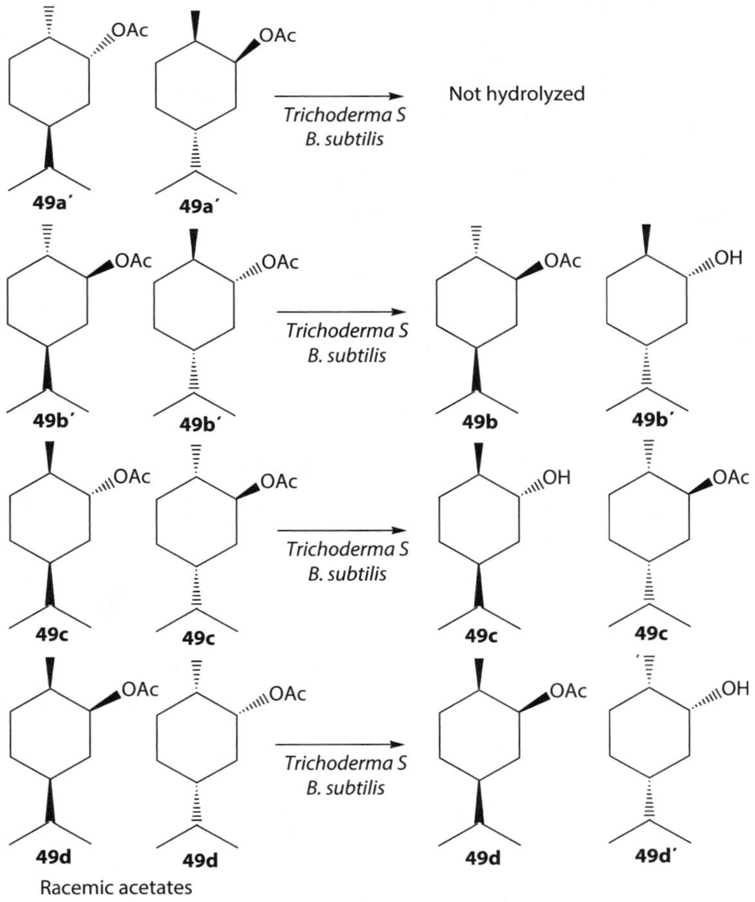

FIGURE 19.104 Microbial resolution of carvomenthols by *Trichoderma S* and *Bacillus subtilis* var. *niger*. (Modified from Oritani, T. and Yamashita, K., *Agric. Biol. Chem.*, 37, 1691, 1973d.)

19.3.4 Monocyclic Monoterpene Ketone

19.3.4.1 α, β-Unsaturated Ketone

19.3.4.1.1 Carvone

93 (4S)-(+)-form

93' (4R)-(−)-form

Carvone occurs as (+)-carvone (**93**), (−)-carvone (**93'**), or racemic carvone. (S)-(+)-Carvone (**93**) is the main component of caraway oil (*ca.* 60%) and dill oil and has a herbaceous odor reminiscent of caraway and dill seeds. (R)-(−)-Carvone (**93'**) occurs in spearmint oil at a concentration of 70%–80% and has a herbaceous odor similar to spearmint (Bauer et al., 1990).

The distribution of carvone convertible microorganisms is summarized in Table 19.5. When ethanol was used as a carbon source, 40% of bacteria converted (+)- (**93**) and (−)-carvone (**93'**). On the other hand, when glucose was used, 65% of bacteria converted carvone. In the case of yeasts, 75% converted (+)- (**93**) and (−)-carvone (**93'**). Of fungi, 90% and 85% of fungi converted **93** and **93'**, respectively. In actinomycetes, 56% and 90% converted **93** and **93'**, respectively.

Many microorganisms except for some strains of actinomycetes were capable of hydrogenating the C = C double bond at C-1, 2 position of (+)- (**93**) and (−)-carvone (**93'**) to give mainly (−)-isodihydrocarvone (**101b**) and (+)-dihydrocarvone (**101a'**), respectively (Noma and Tatsumi, 1973; Noma et al., 1974b; Noma and Nonomura, 1974; Noma, 1976, 1977) (Figure 19.105) (Tables 19.6 and 19.7).

Furthermore, it was found that (−)-carvone (**93'**) was converted via (+)-isodihydrocarvone (**101b'**) to (+)-isodihydrocarveol (**102c'**) and (+)-neoisodihydrocarveol (**102d'**) by some strains of actinomycetes (Noma, 1979a,b). (−)-Isodihydrocarvone (**101b**) was epimerized to

TABLE 19.5
Distribution of (+)- (93) and (−)-Carvone (93') Convertible Microorganisms

Microorganisms	Number of Microorganisms Used	Numbers of Carvone Convertible Microorganisms	Ratio (%)
Bacteria	40	16 (ethanol, **93**)	40
		16 (ethanol, **93'**)	40
		26 (glucose, **93**)	65
		26 (glucose, **93'**)	65
Yeasts	68	51 (**93**)	75
		51 (**93'**)	75
Fungi	40	34 (**93**)	85
		36 (**93'**)	90
Actinomycetes	48	27 (**93**)	56
		43 (**93'**)	90

Source: Noma, Y. et al., Formation of 8 kinds of *p*-menthane-2,8-diols from carvone and related compounds by *Euglena gracilis* Z. Biotransformation of monoterpenes by photosynthetic microorganisms. Part VIII, *Proceedings of 37th TEAC*, 1993, pp. 23–25.

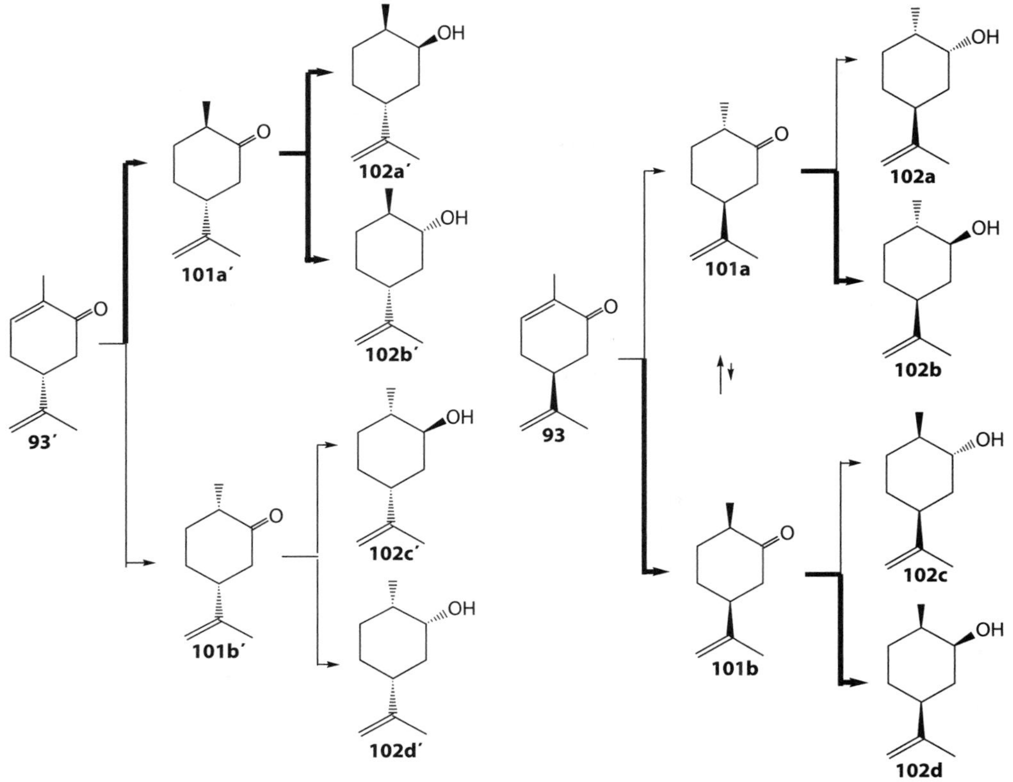

FIGURE 19.105 Biotransformation of (+)- (**93**) and (−)-carvone (**93′**) by various kinds of microorganisms. (Modified from Noma, Y. and Tatsumi, C., *Nippon Nogeikagaku Kaishi*, 47, 705, 1973; Noma, Y. et al., *Agric. Biol. Chem.*, 38, 735, 1974b; Noma, Y. et al., Microbial transformation of carvone, *Proceedings of 18th TEAC*, pp. 20–23, 1974c; Noma, Y. and Nonomura, S., *Agric. Biol. Chem.*, 38, 741, 1974; Noma, Y., *Ann. Res. Stud. Osaka Joshigakuen Junior College*, 20, 33, 1976; Noma, Y., *Nippon Nogeikagaku Kaishi*, 51, 463, 1977.)

TABLE 19.6
Ratio of Microorganisms That Carried Out the Hydrogenation of C = C Double Bond of Carvone by *Si* Plane Attack toward Microorganisms That Converted Carvone

Microorganisms	Ratio (%)
Bacteria	100[a]
	96[b]
Yeasts	74
Fungi	80
Actinomycetes	39

[a] When ethanol was used.
[b] When glucose was used.

TABLE 19.7
Summary of Microbial and Chemical Hydrogenation of
(−)-Carvone (93′) for the Formation of (+)-Dihydrocarvone
(101a′) and (+)-Isodihydrocarvone (101b′)

Microorganisms	Compounds	
	101a′	101b′
Amorphosporangium auranticolor	100	0
Microbispora rosea IFO 3559	86	0
Bacillus subtilis var. *niger*	85	13
Bacillus subtilis IFO 3007	67	11
Pseudomonas polycolor IFO 3918	75	15
Pseudomonas graveolens IFO 3460	74	17
Arthrobacter pascens IFO 121139	73	12
Pichia membranifaciens IFO 0128	70	16
Saccharomyces ludwigii IFO 1043	69	18
Alcaligenes faecalis IAM B-141-1	70	13
Zn-25% KOH–EtOH	73	27
Raney-10% NaOH	71	19

Source: Noma, Y., *Ann. Res. Stud. Osaka Joshigakuen Junior College*, 20, 33, 1976.

(−)-dihydrocarvone (**101a**) after the formation of (−)-isodihydrocarvone (**101b**) from (+)-carvone (**93**) by the growing cells, the resting cells, and the cell-free extracts of *Pseudomonas fragi* IFO 3458 (Noma et al., 1975).

Consequently, the metabolic pathways of carvone by microorganisms were summarized as the following eight groups (Figure 19.105):

Group 1: (−)-Carvone (**93′**)- (+)-dihydrocarvone (**101a′**)-(+)-neodihydrocarveol (**102a′**)
Group 2: **93′–101a′**-(−)-Dihydrocarveol (**102b′**)
Group 3: **93′–101a′-102a′** and **102b′**
Group 4: **93′**-(+)-Isodihydrocarvone (**101b′**)–**102c′** and **102d′**
Group 5: (+)-Carvone (**93**)-(−)-isodihydrocarvone (**101b**)-(−)-neoisodihydrocarveol (**102d**)
Group 6: **93–101b–102c**
Group 7: **93–101b–102c** and **102d**
Group 8: **93–101b–101a**

The result of the mode action of both the hydrogenation of carvone and the reduction for dihydrocarvone by microorganism is as follows. In bacteria, only two strains were able to convert (−)-carvone (**93′**) via (+)-dihydrocarvone (**101a′**) to (−)-dihydrocarveol (**102b′**) as the major product (group 3, when ethanol was used as a carbon source, 12.5% of (−)-carvone (**93′**) convertible microorganisms belonged to this group, and when glucose was used, 8% belonged to this group) (Noma and Tatsumi, 1973; Noma et al., 1975), whereas when (+)-carvone (**93**) was converted, one strain converted it to a mixture of (−)-isodihydrocarveol (**102c**) and (−)-neoisodihydrocarveol (**102d**) (group 7, 6% and 4% of **93** convertible bacteria belonged to this group, when ethanol and glucose were used, respectively), and four strains converted it via (−)-isodihydrocarvone (**101b**) to (−)-dihydrocarvone (**101a**) (group 8, 6% and 15% of (+)-carvone (**93′**) convertible bacteria belonged to this group, when ethanol and glucose were used, respectively.) (Noma et al., 1975). In yeasts, 43% of carvone convertible yeasts belong to group 1, 14% to group 2, and 33% to group 3 (of this group, three strains are close to group 1) and 12% to group 5, 4% to group 6, and 27% to group 7 (of this group, three strains are close to group 5 and one strain is close to group 6). In fungi, 51% of fungi metabolized (−)-carvone (**93′**) by way of group 1 and 3% via group 3, but there was no strain capable of metabolizing (−)-carvone (**93′**) via group 2, whereas 20% of fungi metabolized (+)-carvone (**93**) via group 5 and 29% via group 7, but there was no strain capable of metabolizing (+)-carvone (**93**) via group 6. In actinomycetes, (−)-carvone (**93′**) was converted to dihydrocarveols via group 1 (49%), group 2 (0%), group 3 (9%), and group 4 (28%), whereas (+)-carvone (**93**) was converted to dihydrocarveols via group 5 (7%), group 6 (0%), group 7 (19%), and group 8 (0%).

Furthermore, (+)-neodihydrocarveol (**102a′**) stereospecifically formed from (−)-carvone (**93′**) by *A. niger* TBUYN-2 was further biotransformed to mosquito repellent (1*R*,2*S*,4*R*)-(+)-*p*-menthane-2,8-diol (**50a′**), (1*R*,2*S*,4*R*)-(+)-8-*p*-menthene-2,10-diol (**107a′**), and the mixture of (1*R*,2*S*,4*R*,8*S/R*)-(+)-*p*-menthane-2,8,9-triols (**104aa′** and **104ab′**), while *A. glauca* ATCC 22752 gave **107a′** stereoselectively from **102a′** (Demirci et al., 2001) (Figure 19.106).

On the other hand, (−)-carvone (**93′**) was biotransformed stereoselectively to (+)-neodihydrocarveol (**102a′**) via (+)-dihydrocarvone (**101a′**) by a strain of *A. niger* (Noma and Nonomura, 1974), *E. gracilis* Z. (Noma et al., 1993), and *Chlorella miniata* (Gondai et al., 1999). Furthermore, in *E. gracilis* Z., mosquito repellent (1*R*,2*S*,4*R*)-(+)-*p*-menthane-2,8-diol (**50a′**) was obtained stereospecifically from (−)-carvone (**93′**) via **101a′** and **102a′** (Figure 19.107).

As the microbial method for the formation of mosquito repellent **50a′** was established, the production of (+)-dihydrocarvone (**101a′**) and (+)-neodihydrocarveol (**102a′**) as the precursor of mosquito repellent **50a′** was investigated by using 40 strains of bacteria belonging to *Escherichia*, *Aerobacter*, *Serratia*, *Proteus*, *Alcaligenes*, *Bacillus*, *Agrobacterium*, *Micrococcus*, *Staphylococcus*, *Corynebacterium*, *Sarcina*, *Arthrobacter*, *Brevibacterium*, *Pseudomonas*, and *Xanthomonas* spp.; 68 strains of yeasts belonging to *Schizosaccharomyces*, *Endomycopsis*, *Saccharomyces*, *Schwanniomyces*, *Debaryomyces*, *Pichia*, *Hansenula*, *Lipomyces*, *Torulopsis*, *Saccharomycodes*, *Cryptococcus*, *Kloeckera*, *Trigonopsis*, *Rhodotorula*, *Candida*, and *Trichosporon* spp.; 40 strains of fungi belonging to *Mucor*, *Absidia*, *Penicillium*, *Rhizopus*, *Aspergillus*, *Monascus*, *Fusarium*, *Pullularia*, *Keratinomyces*, *Oospora*, *Neurospora*, *Ustilago*, *Sporotrichum*, *Trichoderma*, *Gliocladium*, and *Phytophthora* spp.; and 48 strains of actinomycetes belonging to *Streptomyces*,

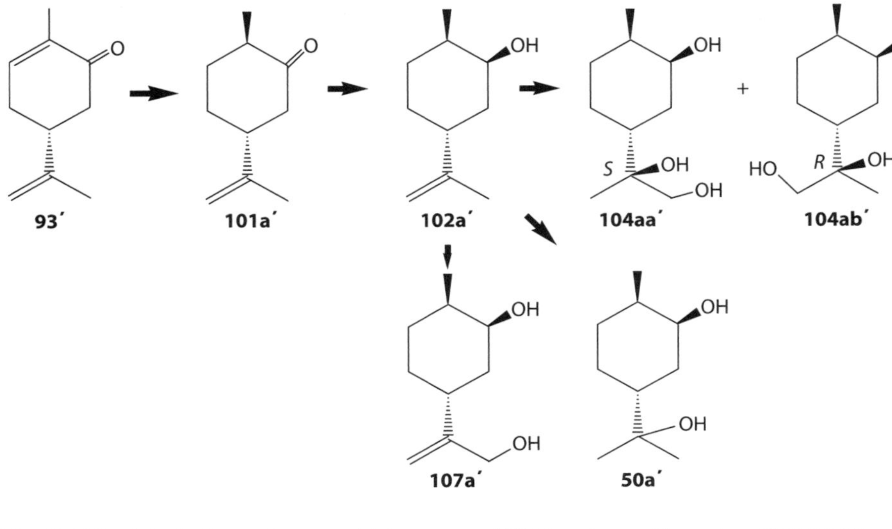

FIGURE 19.106 Metabolic pathways of (−)-carvone (**93′**) by *Aspergillus niger* TBUYN-2 and *Absidia glauca* ATCC 22752. (Modified from Demirci, F. et al., The biotransformation of thymol methyl ether by different fungi, *Book of Abstracts of the XII Biotechnology Congr.*, p. 47, 2001.)

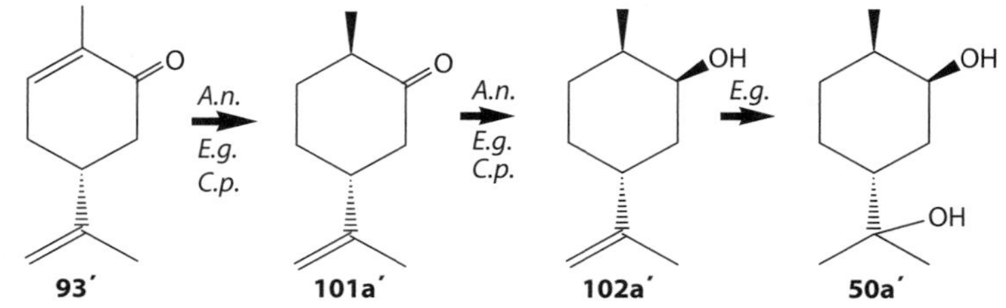

FIGURE 19.107 Metabolic pathway of (−)-carvone (**93′**) by *Aspergillus niger*, *Euglena gracilis* Z, and *Chlorella miniata*. (Modified from Noma, Y. and Nonomura, S., *Agric. Biol. Chem.*, 38, 741, 1974; Noma, Y. et al., Formation of 8 kinds of *p*-menthane-2,8-diols from carvone and related compounds by *Euglena gracilis* Z. Biotransformation of monoterpenes by photosynthetic microorganisms. Part VIII, *Proceedings of 37th TEAC*, 1993, pp. 23–25; Gondai, T. et al., Asymmetric reduction of enone compounds by *Chlorella miniata*, *Proceedings of 43rd TEAC*, 1999, pp. 217–219.)

Actinoplanes, Nocardia, Micromonospora, Microbispora, Micropolyspora, Amorphosporangium, Thermopolyspora, Planomonospora, and *Streptosporangium* spp.

As a result, 65% of bacteria, 75% of yeasts, 90% of fungi, and 90% of actinomycetes converted (−)-carvone (**93′**) to (+)-dihydrocarvone (**101a′**) or (+)-neodihydrocarveol (**102a′**) (Figures 19.105 and 19.108). Many microorganisms are capable of converting (−)-carvone (**93′**) to (+)-neodihydrocarveol (**102a′**) stereospecifically. Some of the useful microorganisms are listed in Tables 19.7 and 19.8. There is no good chemical method to obtain (+)-neodihydrocarveol (**102a′**) in large quantity. It was considered that the method utilizing microorganisms is a very useful means and better than the chemical synthesis for the production of mosquito repellent precursor (+)-neodihydrocarveol (**102a′**).

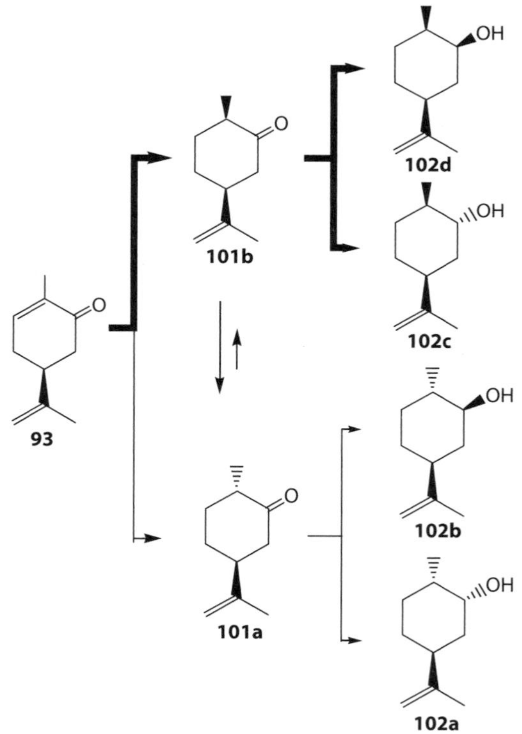

FIGURE 19.108 Metabolic pathways of (+)-carvone (**93**) by *Pseudomonas ovalis* strain 6-1 and other many microorganisms. (Modified from Noma, Y. et al., *Agric. Biol. Chem.*, 38, 735, 1974b.)

TABLE 19.8
Summary of Microbial and Chemical Reduction of (−)-Carvone (93′) for the Formation of (+)-Neodihydrocarveol (102a′)

Microorganisms	Compounds					
	101a′	101b′	102a′	102b′	102c′	102d′
Torulopsis xylinus IFO 454	0	0	100	0	0	0
Monascus anka var. *rubellus* IFO 5965	0	0	100	0	0	0
Fusarium anguioides Sherbakoff IFO 4467	0	0	100	0	0	0
Phytophthora infestans IFO 4872	0	0	100	0	0	0
Kloeckera magna IFO 0868	0	0	98	2	0	0
Kloeckera antillarum IFO 0669	19	4	72	0	0	0
Streptomyces rimosus	+	0	98	0	0	0
Penicillium notatum Westling IFO 464	6	2	92	0	0	0
Candida pseudotropicalis IFO 0882	17	4	79	0	0	0
Candida parapsilosis IFO 0585	16	4	80	0	0	0
$LiAlH_4$	0	0	17	67	2	13
Meerwein–Ponndorf–Verley reduction	0	0	29	55	9	5

Source: Noma, Y., *Ann. Res. Stud. Osaka Joshigakuen Junior College*, 20, 33, 1976.

(−)-Carvone (**93**) was biotransformed by *A. niger* TBUYN-2 to give mainly (+)-8-hydroxyneodihydrocarveol (**50a′**), (+)-8,9-epoxyneodihydrocarveol (**103a′**), and (+)-10-hydroxyneodihydrocarveol (**107a′**) via (+)-dihydrocarvone (**101a′**) and (+)-neodihydrocarveol (**102a′**). *A. niger* TBUYN-2 dehydrogenated (+)-*cis*-carveol (**81b**) to give (+)-carvone (**93**), which was further converted to (−)-isodihydrocarvone (**101b**). Compound **101b** was further metabolized by four pathways to give 10-hydroxy-(−)-isodihydrocarvone (**106b**), (1*S*,2*S*,4*S*)-*p*-menthane-1,2-diol (**71d**) via 1α-hydroxy-(−)-isodihydrocarvone (**72b**) as intermediate, (−)-isodihydrocarveol (**102c**), and (−)-neoisodihydrocarveol (**102d**). Compound **102d** was further converted to isodihydroisobottrospicatol (**105bb**) via 8,9-epoxy-(−)-neoisodihydrocarveol (**103d**); compound **105′** was a major product (Noma et al., 1985a) (Figure 19.109).

In the case of the plant pathogenic fungus, *A. glauca* (−)-carvone (**93′**) was metabolized to give the diol 10-hydroxy-(+)-neodihydrocarveol (**107a′**) (Nishimura et al., 1983b).

(+)-Carvone (**93**) was converted by five bacteria and one fungus (Verstegen-Haaksma et al., 1995) to give (−)-dihydrocarvone (**101a**), (−)-isodihydrocarvone (**101b**), and (−)-neoisodihydrocarveol (**102d**). Sensitivity of the microorganism to (+)-carvone (**93**) and some of the products prevented yields exceeding 0.35 g/L in batch cultures. The fungus *Trichoderma pseudokoningii* gave the highest yield of (−)-neoisodihydrocarveol (**102d**) (Figure 19.110). (+)-Carvone (**93**) is known to inhibit fungal growth of *Fusarium sulphureum* when it was administered via the gas phase (Oosterhaven et al., 1995a,b). Under the same conditions, the related fungus *Fusarium solani* var. *coeruleum* was not inhibited. In liquid medium, both fungi were found to convert (+)-carvone (**93**), with the same rate, mainly to (−)-isodihydrocarvone (**101b**), (−)-isodihydrocarveol (**102c**), and (−)-neoisodihydrocarveol (**102d**).

19.3.4.1.1.1 Biotransformation of Carvone to Carveols by Actinomycetes   The distribution of actinomycetes capable of reducing carbonyl group of carvone containing α, β-unsaturated ketone to (−)-*trans*- (**81a′**) and (−)-*cis*-carveol (**81b′**) was investigated. Of 93 strains of actinomycetes, 63 strains were capable of converting (−)-carvone (**93′**) to carveols. The percentage of microorganisms that produced carveols from (−)-carvone (**93′**) to total microorganisms was about 71%.

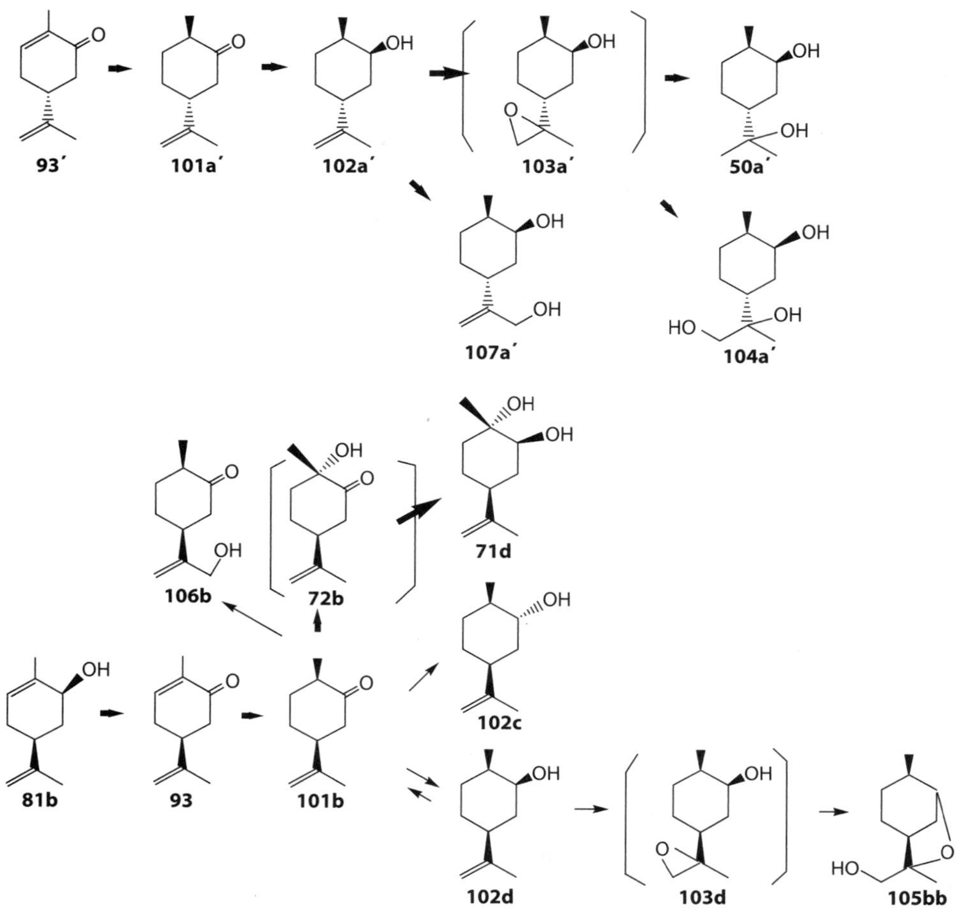

FIGURE 19.109 Possible main metabolic pathways of (−)-carvone (**93′**) and (+)-carvone (**93**) by *Aspergillus niger* TBUYN-2. (Modified from Noma, Y. et al., Biotransformation of (−)-carvone and (+)-carvone by *Aspergillus* spp., *Annual Meeting of Agricultural and Biological Chemistry*, Sapporo, Japan, 1985a, p. 68.)

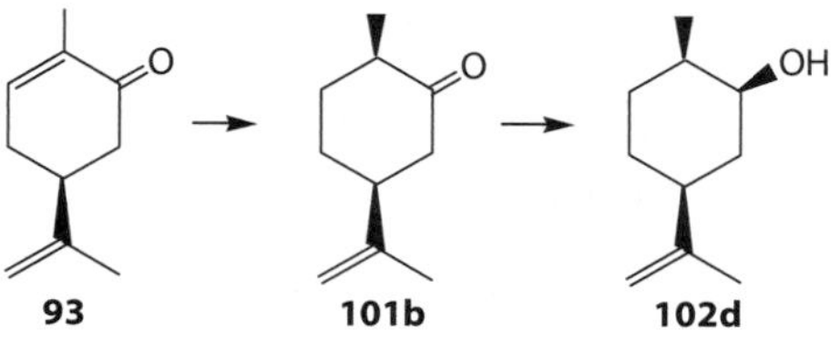

FIGURE 19.110 Biotransformation of (+)-carvone (**93**) by *Trichoderma pseudokoningii*. (Modified from Verstegen-Haaksma, A.A. et al., *Ind. Crops Prod.*, 4, 15, 1995.)

Microorganisms that produced carveols were classified into three groups according to the formation of (−)-*trans*-carveol (**81a′**) and (−)-*cis*-carveol (**81b′**): group 1, (−)-carvone-**81b′** only; group 2, (−)-carvone-**81a′** only; and group 3, (−)-carvone mixture of **81a′** and **81b′**. Three strains belonged to group 1 (4.5%), 34 strains belonged to group 2 (51.1%), and 29 strains belonged to group 3 (44%; of this group two strains were close to group 1 and 14 strains were close to group 2).

Streptomyces A-5-1 isolated from soil converted (−)-carvone (**93′**) to **101a′–102d′** and (−)-*trans*-carveol (**81a′**), whereas *Nocardia*, 1-3-11 converted (−)-carvone (**93′**) to (−)-*cis*-carveol (**81b′**)

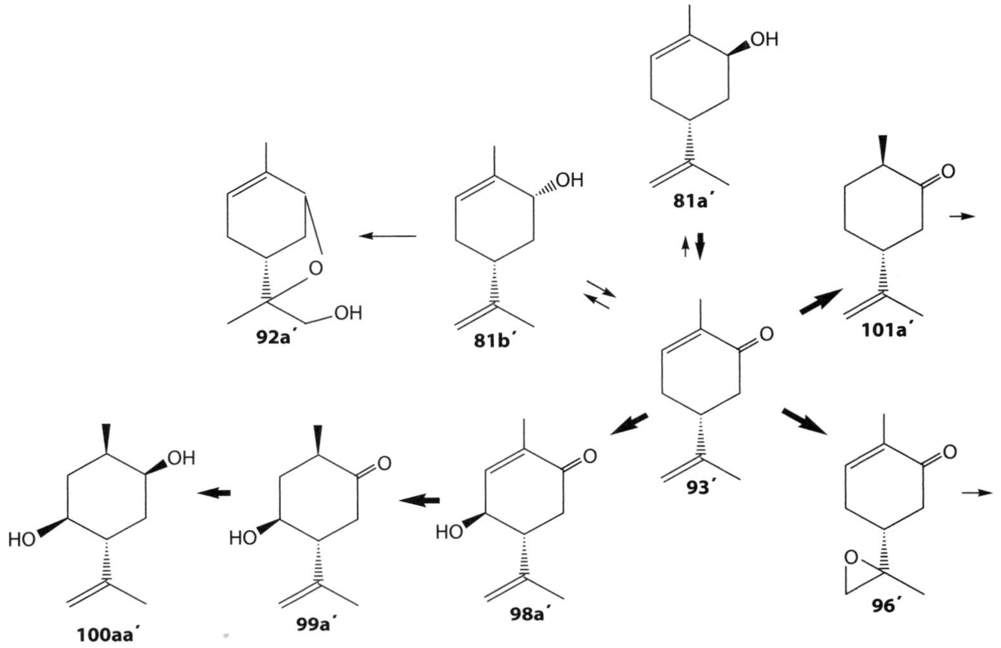

FIGURE 19.111 Metabolic pathways of (−)-carvone (**93′**) by *Streptomyces bottropensis* SY-2-, *Streptomyces ikutamanensis* Ya-2-1, *Streptomyces* A-5-1, and *Nocardia* 1-3-11. (Modified from Noma, Y., *Nippon Nogeikagaku Kaishi*, 53, 35, 1979a; Noma, Y., *Ann. Res. Stud. Osaka Joshigakuen Junior College*, 23, 27, 1979b; Noma, Y., *Agric. Biol. Chem.*, 44, 807, 1980; Noma, Y. and Nishimura, H., Biotransformation of (−)-carvone and (+)-carvone by *S. ikutamanensis* Ya-2-1, *Book of Abstracts of the Annual Meeting of Agricultural and Biological Chemical Society*, p. 390, 1983a; Noma, Y. and Nishimura, H., Biotransformation of carvone. 5. Microbiological transformation of dihydrocarvones and dihydrocarveols, *Proceedings of 27th TEAC*, 1983b, pp. 302–305.)

together with **101a′**–**81a′** (Noma, 1980). In the case of *Nocardia*, the reaction between **93′** and **81a′** was reversible and the direction from **81a′** to **93′** is predominantly (Noma, 1979a,b, 1980) (Figure 19.111).

(−)-Carvone (**93′**) was metabolized by actinomycetes to give (−)-*trans*- (**81a′**) and (−)-*cis*-carveol (**81b′**) and (+)-dihydrocarvone (**101a′**) as reduced metabolites. Compound **81b′** was further metabolized to (+)-bottrospicatol (**92a′**). Furthermore, **93′** was hydroxylated at C-5 position and C-8, 9 position to give 5β-hydroxy-(−)-carvone (**98a′**) and (−)-carvone-8,9-epoxide (**96′**), respectively. Compound **98a′** was further metabolized to 5β-hydroxyneodihydrocarveol (**100aa′**) via 5β-hydroxydihydrocarvone (**99a′**) (Noma, 1979a,b, 1980) (Figure 19.111).

Metabolic pattern of (+)-carvone (**93**) is similar to that of (−)-carvone (**93′**) in *S. bottropensis*. (+)-Carvone (**93**) was converted by *S. bottropensis* to give (+)-carvone-8,9-epoxide (**96**) and (+)-5α-hydroxycarvone (**98a**) (Figure 19.112). (+)-Carvone-8,9-epoxide (**96**) has light sweet aroma and has strong inhibitory activity for the germination of lettuce seeds (Noma and Nishimura, 1982).

The investigation of (−)-carvone (**93′**) and (+)-carvone (**93**) conversion pattern was carried out by using rare actinomycetes. The conversion pattern was classified as follows (Figure 19.113):

Group 1: Carvone (**93**), dihydrocarvones (**101**), dihydrocarveol (**102**), dihydrocarveol-8,9-epoxide (**103**), dihydrobottrospicatols (**105**), 5-hydroxydihydrocarveols (**100**)
Group 2: Carvone (**93**), carveols (**89**), bottrospicatols (**92**), 5-hydroxy-*cis*-carveols (**12**)
Group 3: Carvone (**93**), 5-hydroxycarvone (**98**), 5-hydroxyneodihydrocarveols (**15**)
Group 4: Carvone (**93**), carvone-8,9-epoxides (**96**)

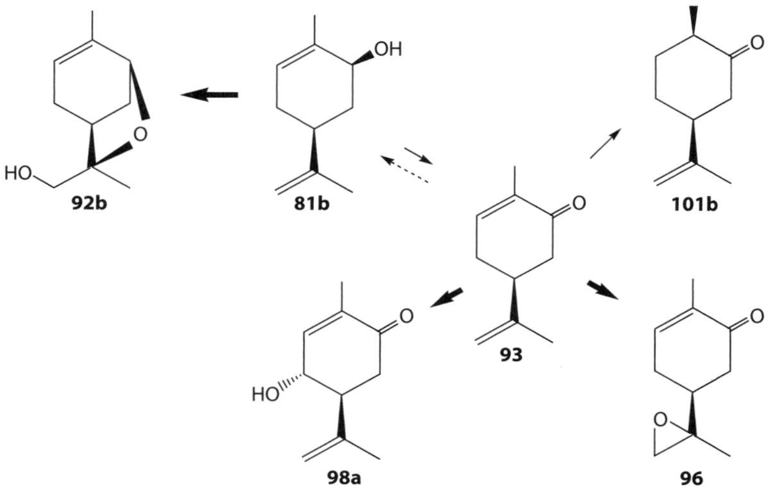

FIGURE 19.112 Metabolic pathways of (+)-carvone (**93′**) by *Streptomyces bottropensis* SY-2-1 and *Streptomyces ikutamanensis* Ya-2-1. (Modified from Noma, Y. and Nishimura, H., Biotransformation of carvone. 4. Biotransformation of (+)-carvone by *Streptomyces bottropensis*, SY-2-1, *Proceedings of 26th TEAC*, 1982, pp. 156–159l; Noma, Y. and Nishimura, H., Biotransformation of (−)-carvone and (+)-carvone by *S. ikutamanensis* Ya-2-1, *Book of Abstracts of the Annual Meeting of Agricultural and Biological Chemical Society*, 1983a, p. 390; Noma, Y. and Nishimura, H., Biotransformation of carvone. 5. Microbiological transformation of dihydrocarvones and dihydrocarveols, *Proceedings of 27th TEAC*, 1983b, pp. 302–305; Noma, Y., *Kagaku to Seibutsu*, 22, 742, 1984.)

Of 50 rare actinomycetes, 22 strains (44%) were capable of converting (−)-carvone (**93′**) to give (−)-carvone-8,9-epoxide (**96′**) via pathway 4 and (+)-5β-hydroxycarvone (**98a′**), (+)-5α-hydroxycarvone (**98b′**), and (+)-5β-hydroxyneodihydrocarveol (**100aa′**) via pathway 3 (Noma and Sakai, 1984).

On the other hand, in the case of (+)-carvone (**93**) conversion, 44% of rare actinomycetes were capable of converting (+)-carvone (**93**) to give (+)-carvone-8,9-epoxide (**96**) via pathway 4 and (−)-5α-hydroxycarvone (**98a**), (−)-5β-hydroxycarvone (**98b**), and (−)-5α-hydroxyneodihydrocarveol (**100aa**) via pathway 3 (Noma and Sakai, 1984).

19.3.4.1.1.2 Biotransformation of Carvone by Citrus Pathogenic Fungi, A. niger Tiegh TBUYN Citrus pathogenic *A. niger* Tiegh (CBAYN) and *A. niger* TBUYN-2 hydrogenated C = C double bond at C-1, 2 position of (+)-carvone (**93**) to give (−)-isodihydrocarvone (**101b**) as the major product together with a small amount of (−)-dihydrocarvone (**101a**), of which **101b** was further metabolized through two kinds of pathways as follows: One is the pathway to give (+)-1α-hydroxyneoisodihydrocarveol (**71**) via (+)-1α-hydroxyisodihydrocarvone (**72**) and the other one is the pathway to give (+)-4α-hydroxyisodihydrocarvone (**378**) (Noma and Asakawa, 2008) (Figure 19.114).

The biotransformation of enones such as (−)-carvone (**93′**) by the cultured cells of *C. miniata* was examined. It was found that the cells reduced stereoselectively the enones from *si* face at α-position of the carbonyl group and then the carbonyl group from *re* face (Figure 19.115).

Stereospecific hydrogenation occurs independent of the configuration and the kinds of the substituent at C-4 position, so that the methyl group at C-1 position is fixed mainly at *R*-configuration. [2-^{2}H]-(−)-Carvone ([2-^{2}H]-**93′**) was synthesized in order to clear up the hydrogenation mechanism at C-2 by microorganisms. Compound [2-^{2}H]-**93** was also easily biotransformed to [2-^{2}H]-8-hydroxy-(+)-neodihydrocarveol (**50a′**) via [2-^{2}H]-(+)-neodihydrocarveol (**102a′**). On the basis of ^{1}H-NMR spectral data of compounds **102a′** and **50a′**, the hydrogen addition of the carbon–carbon double bond at the C_1 and C_2 position by *A. niger* TBUYN-2, *E. gracilis* Z., and *D. tertiolecta* occurs from the *si* face and *re* face, respectively, that is, *anti* addition (Noma et al., 1995) (Figure 19.115) (Table 19.9).

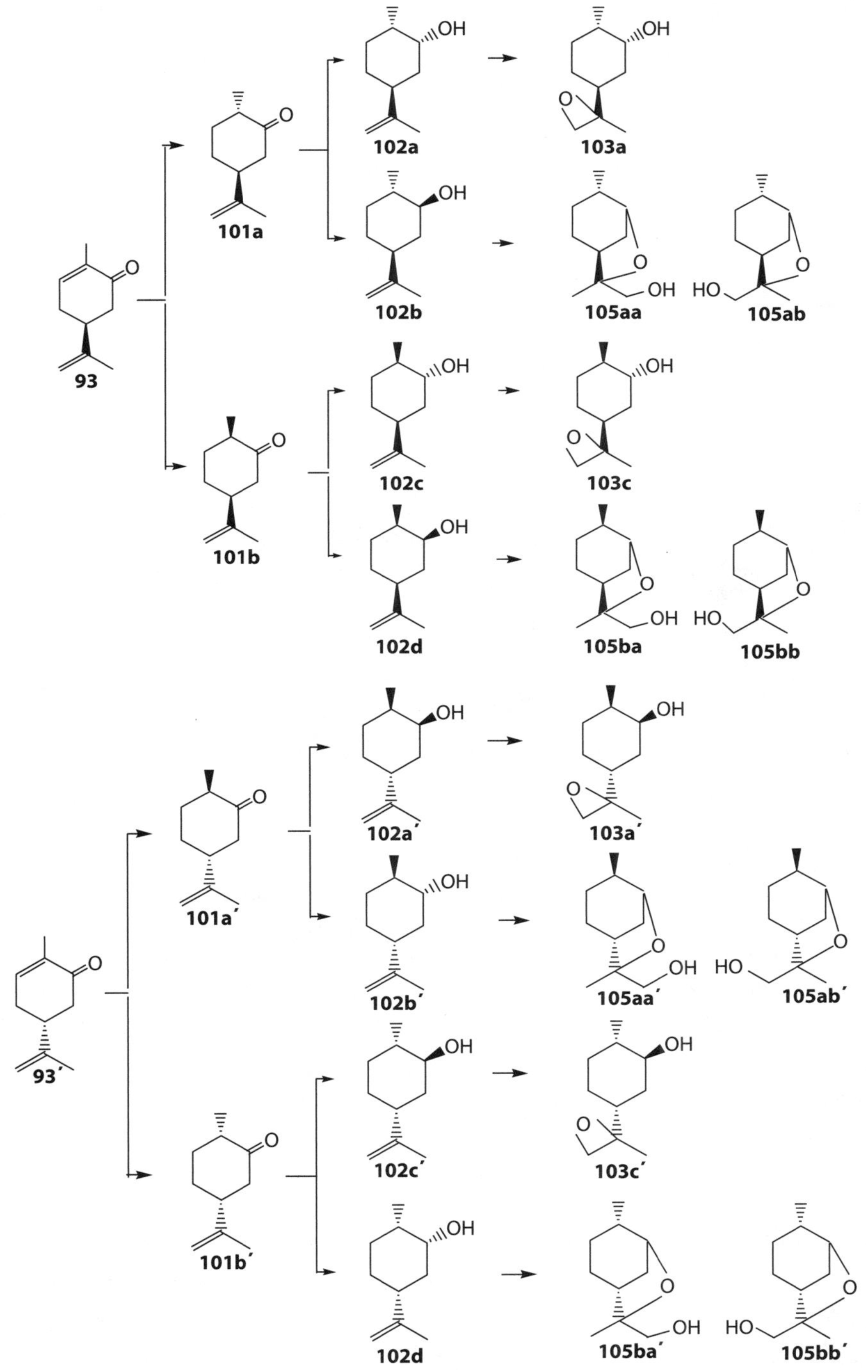

FIGURE 19.113 Metabolic pathways of (+)- (**93**) and (−)-carvone (**93′**) and dihydrocarveols (**102a-d** and **102a′-d′**) by *Streptomyces bottropensis* SY-2-1 and *Streptomyces ikutamanensis* Ya-2-1. (Modified from Noma, Y., *Kagaku to Seibutsu*, 22, 742, 1984.)

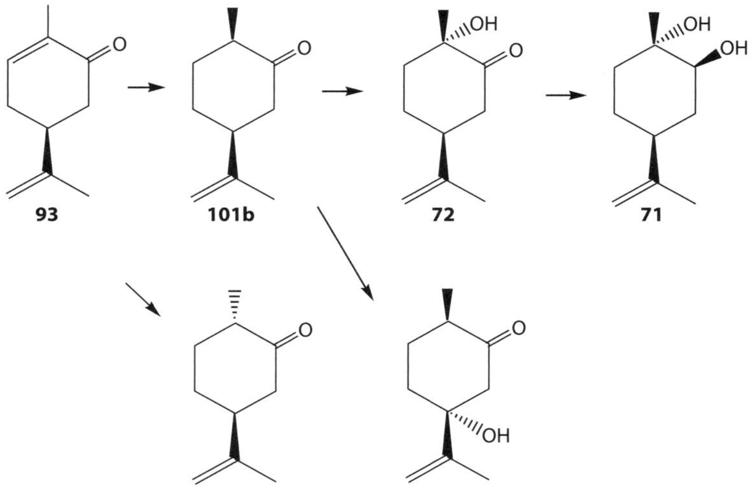

FIGURE 19.114 Metabolic pathways of (+)-carvone (**93**) by citrus pathogenic fungi, *Aspergillus niger* Tiegh CBAYN and *A. niger* TBUYN-2. (Modified from Noma, Y. and Asakawa, Y., New metabolic pathways of (+)-carvone by Citrus pathogenic *Aspergillus niger* Tiegh CBAYN and *A. niger* TBUYN-2, *Proceedings of 52nd TEAC*, 2008, pp. 206–208.)

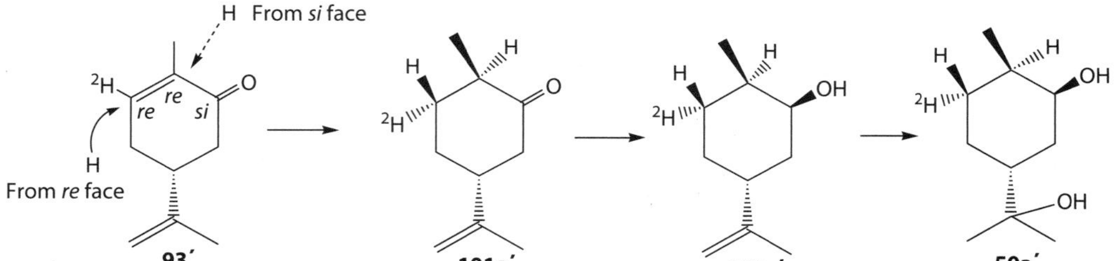

FIGURE 19.115 The stereospecific hydrogenation of the C = C double bond of α,β-unsaturated ketones, the reduction of saturated ketone, and the hydroxylation by *Euglena gracilis* Z. (Modified from Noma, Y. et al., Biotransformation of [6-^{2}H]-(−)-carvone by *Aspergillus niger*, *Euglena gracilis* Z and *Dunaliella tertiolecta*, *Proceedings of 39th TEAC*, 1995, pp. 367–368; Noma, Y. and Asakawa, Y., *Euglena gracilis* Z: Biotransformation of terpenoids and related compounds, in Bajaj, Y.P.S., (ed.), *Biotechnology in Agriculture and Forestry*, vol. 41, Medicinal and Aromatic Plants X, Springer, Berlin, Germany, 1998, pp. 194–237.)

TABLE 19.9
Summary for the Stereospecificity of the Reduction of the C = C Double Bond of [2-2H]-(−)-Carvone ([2-2H]-93) by Various Kinds of Microorganisms

	Stereochemistry at C-2H of Compounds	
Microorganisms	102a	50a
Aspergillus niger TBUYN-2	β	
Euglena gracilis Z	β	β
Dunaliella tertiolecta	β	
The cultured cells of *Nicotiana tabacum* (Suga et al., 1986)	β	

19.3.4.1.1.3 Hydrogenation Mechanisms of C = C Double Bond and Carbonyl Group In order to understand the mechanism of the hydrogenation of α-, β unsaturated ketone of (−)-carvone (**93′**) and the reduction of carbonyl group of dihydrocarvone (**101a′**) (−)-carvone (**93′**), (+)-dihydrocarvone (**101a′**) and the analogues of (−)-carvone (**93′**) were chosen and the conversion of the analogues was carried out by using *P. ovalis* strain 6-1. As the analogues of carvone (**93** and **93′**), (−)- (**47′**) and (+)-carvotanacetone (**47**), 2-methyl-2-cyclohexenone (**379**), the mixture of (−)-*cis*- (**81b′**) and (−)-*trans*-carveol (**81a′**), 2-cyclohexenone, racemic menthenone (**148**), (−)-piperitone (**156**), (+)-pulegone (**119**), and 3-methyl-2-cyclohexenone (**381**) were chosen. Of these analogues, (−)- (**47′**) and (+)-carvotanacetone (**47**) were reduced to give (+)-carvomenthone (**48a′**) and (−)-isocarvomenthone (**48b′**), respectively. 2-Methyl-2-cyclohexenone (**379**) was mainly reduced to (−)-2-methylcyclohexanone. But other compounds were not reduced.

The efficient formation of (+)-dihydrocarvone (**101a**), (−)-isodihydrocarvone (**101b′**), (+)-carvomenthone (**48a**), (−)-isocarvomenthone (**48b′**), and (−)-2-methylcyclohexanone from (−)-carvone (**93**), (+)-carvone (**93′**), (−)-carvotanacetone (**47**), (+)-carvotanacetone (**47′**), and 2-methyl-2-cyclohexenone (**379**) suggested at least that C = C double bond conjugated with carbonyl group may be hydrogenated from behind (*si* plane) (Noma et al., 1974b; Noma, 1977) (Figure 19.116).

19.3.4.1.1.4 What Is Hydrogen Donor in the Hydrogenation of Carvone to Dihydrocarvone? What Is Hydrogen Donor in Carvone Reductase? Carvone reductase prepared from *E. gracilis* Z., which catalyzes the NADH-dependent reduction of the C = C bond adjacent to the carbonyl group, was characterized with regard to the stereochemistry of the hydrogen transfer into the substrate. The reductase was isolated from *E. gracilis* Z. and was found to reduce stereospecifically the C=C double bond of carvone by *anti* addition of hydrogen from the *si* face at α-position to the carbonyl group and the *re* face at β-position (Tables 19.9 and 19.10). The hydrogen atoms participating

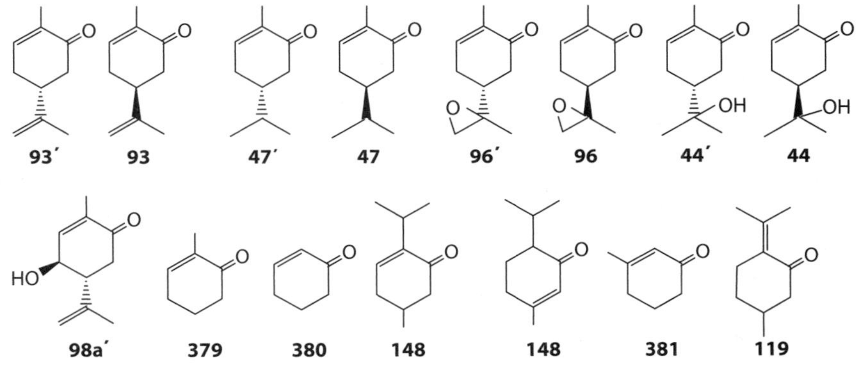

FIGURE 19.116 Substrates used for the hydrogenation of C = C double bond with *Pseudomonas ovalis* strain 6-1, *Streptomyces bottropensis* SY-2-1, *Streptomyces ikutamanensis* Ya-2-1, and *Euglena gracilis* Z.

TABLE 19.10
Purification of the Reductase from *Euglena gracilis* Z.

	Total Protein (mg)	Total Activity Unit × 104	Sp. Act. Units per Gram Protein	Fold
Crude extract	125	2.2	1.7	1
DEAE Toyopearl	7	1.5	21	12
AF-Blue Toyopearl	0.1	0.03	30	18

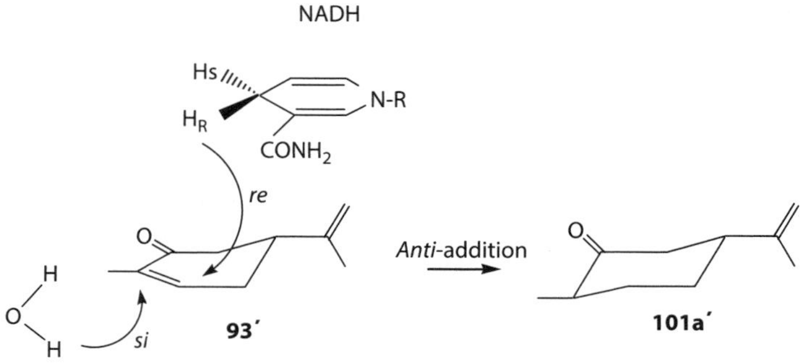

FIGURE 19.117 Stereochemistry in the reduction of (−)-carvone (**93′**) by the reductase from *Euglena gracilis* Z. (Modified from Shimoda, K. et al., *Phytochemistry*, 49, 49, 1998.)

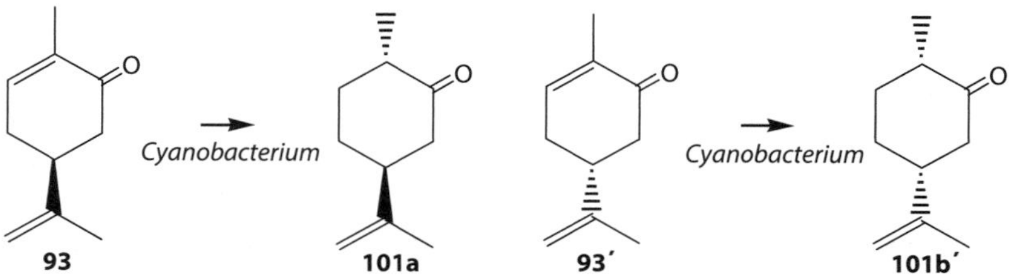

FIGURE 19.118 Biotransformation of (−)- and (+)-carvone (**93** and **93′**) by cyanobacterium. (Modified from Kaji, M. et al., Glycosylation of monoterpenes by plant suspension cells, *Proceedings of 46th TEAC*, 2002, pp. 323–325.)

in the enzymatic reduction at α- and β-position to the carbonyl group originate from the medium and the *pro*-4R hydrogen of NADH, respectively (Shimoda et al., 1998) (Figure 19.117).

In the case of biotransformation by using cyanobacterium, (+)- (**93**) and (−)-carvone (**93′**) were converted with a different type of pattern to give (+)-isodihydrocarvone (**101b′**, 76.6%) and (−)-dihydrocarvone (**101a**, 62.2%), respectively (Kaji et al., 2002) (Figure 19.118). On the other hand, *Catharanthus rosea*–cultured cell biotransformed (−)-carvone (**93′**) to give 5β-hydroxy-(+)-neodihydrocarveol (**100aa′**, 57.5%), 5α-hydroxy-(+)-neodihydrocarveol (**100ab′**, 18.4%), 5α-hydroxy-(−)-carvone (**98b′**), 4β-hydroxy-(−)-carvone (**384′**, 6.3%), 10-hydroxycarvone (**390′**), 5β-hydroxycarvone (**98′**), 5β-hydroxyneodihydrocarveol (**100ab′**), 5β-hydroxyneodihydrocarveol (**100aa′**), and 5α-hydroxydihydrocarvone (**99b′**) as the metabolites as shown in Figure 19.119, whereas (+)-carvone (**93**) gave 5α-hydroxy-(+)-carvone (**98a**, 65.4%) and 4α-hydroxy-(+)-carvone (**384**, 34.6%) (Hamada and Yasumune, 1995; Hamada et al., 1996; Kaji et al., 2002) (Figure 19.119) (Table 19.11).

(−)-Carvone (**93′**) was incubated with cyanobacterium, enone reductase (43 kDa) isolated from the bacterium, and microsomal enzyme to afford (+)-isodihydrocarvone (**101b′**) and (+)-dihydrocarvone (**101a′**). Cyclohexenone derivatives (**379** are **385**) were treated in the same enone reductase with microsomal enzyme to give the dihydro derivative (**382a, 386a**) with *R*-configuration in excellent *ee* (over 99%) and the metabolites (**382b, 386b**) with *S*-configuration in relatively high *ee* (85% and 80%) (Shimoda et al., 2003) (Figure 19.120).

In contrast, almost all the yeasts tested showed reduction of carvone, although the enzyme activity varied. The reduction of (−)-carvone (**93′**) was often much faster than the reduction of (+)-carvone (**93**). Some yeasts only reduced the carbon–carbon double bond to yield the dihydrocarvone isomers (**101a′** and **b′** and **101a** and **b**) with the stereochemistry at C-1 with *R*-configuration, while others also reduced the ketone to give the dihydrocarveols with the stereochemistry at C-2 always

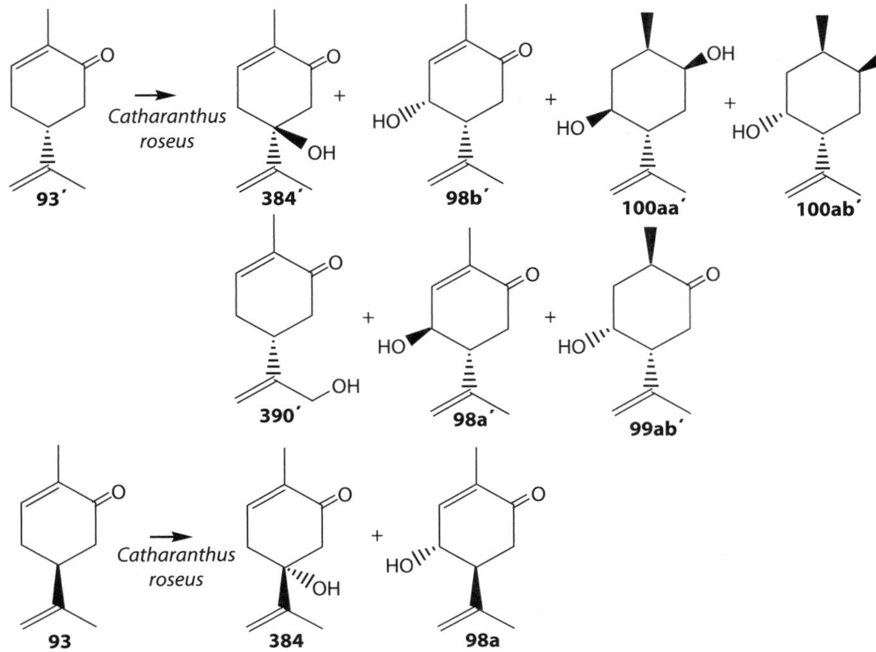

FIGURE 19.119 Biotransformation of (+)- and (−)-carvone (**93** and **93′**) by *Catharanthus roseus*. (Modified from Hamada, H. and Yasumune, H., The hydroxylation of monoterpenoids by plant cell biotransformation, *Proceedings of 39th TEAC*, pp. 375–377, 1995; Hamada, H. et al., The hydroxylation and glycosylation by plant catalysts, *Proceedings of 40th TEAC*, 1996, pp. 111–112; Kaji, M. et al., Glycosylation of monoterpenes by plant suspension cells, *Proceedings of 46th TEAC*, 2002, pp. 323–325.)

TABLE 19.11
Enantioselectivity in the Reduction of Enones (379 and 385) by Enone Reductase

Microsomal Enzyme	Substrate	Product	ee	Configuration[a]
−	379	382a	>99	R
−	385	386a	>99	R
+	379	382b	85	S
+	385	386b	80	S

[a] Preferred configuration at α-position to the carbonyl group of the products.

with S for (−)-carvone (**93′**), but sometimes S and sometimes R for (+)-carvone (**93**). In the case of (−)-carvone (**93′**) yields increased up to 90% within 2 h (van Dyk et al., 1998).

19.3.4.1.2 Carvotanacetone

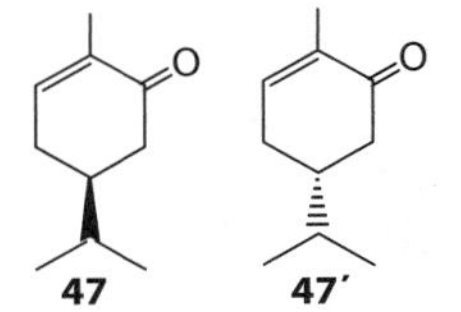

47
S-(+)-

47′
R-(−)-

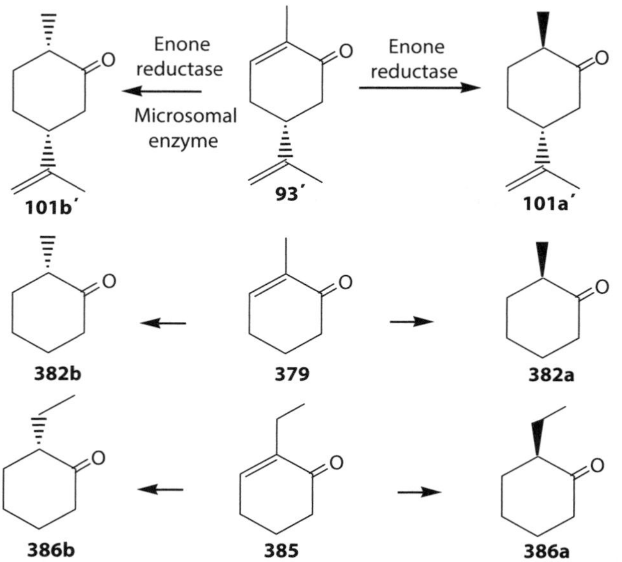

FIGURE 19.120 Biotransformation of 2-methyl-2-cyclohexenone (**379**) and 2-ethyl-2-cyclohexenone (**385**) by enone reductase.

In the conversion of (+)- (**47**) and (−)-carvotanacetone (**47′**) by *P. ovalis* strain 6-1, (−)-carvotanacetone (**47′**) is converted stereospecifically to (+)-carvomenthone (**48a′**) an the latter compound is further converted to (+)-neocarvomenthol (**49a′**) and (−)-carvomenthol (**49b′**) in small amounts, whereas (+)-carvotanacetone (**47**) is converted mainly to (−)-isocarvomenthone (**48b**) and (−)-neoisocarvomenthol (**49d**), forming (−)-carvomenthone (**48a**) and (−)-isocarvomenthol (**49c**) in small amounts as shown in Figure 19.121 (Noma et al., 1974a).

Biotransformation of (−)-carvotanacetone (**47**) and (+)-carvotanacetone (**47′**) by *S. bottropensis* SY-2-1 was carried out (Noma et al., 1985c).

As shown in Figure 19.122, (+)-carvotanacetone (**47**) was converted by *S. bottropensis* SY-2-1 to give 5β-hydroxy-(+)-neoisocarvomenthol (**139db**), 5α-hydroxy-(+)-carvotanacetone (**51a**), 5β-hydroxy-(−)-carvomenthone (**52ab**), 8-hydroxy-(+)-carvotanacetone (**44**), and 8-hydroxy-(−)-carvomenthone (**45a**), whereas (−)-carvotanacetone (**47′**) was converted to give 5β-hydroxy(−)-carvotanacetone (**51a′**) and 8-hydroxy-(−)-carvotanacetone (**44′**).

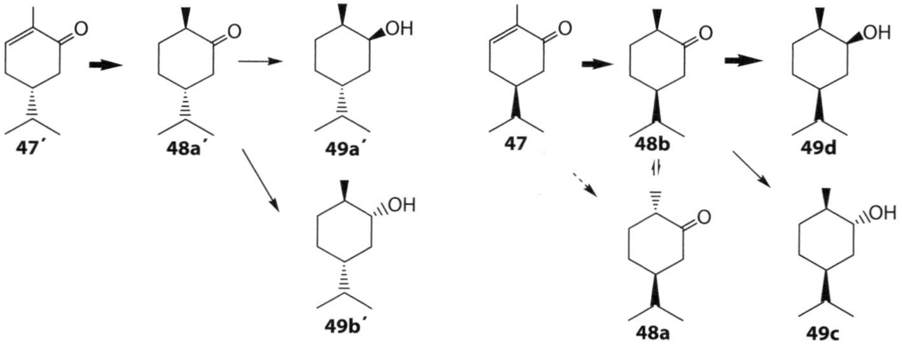

FIGURE 19.121 Metabolic pathways of (−)-carvotanacetone (**47′**) and (+)-carvotanacetone (**47**) by *Pseudomonas ovalis* strain 6-1. (Modified from Noma, Y. et al., *Agric. Biol. Chem.*, 38, 1637, 1974a.)

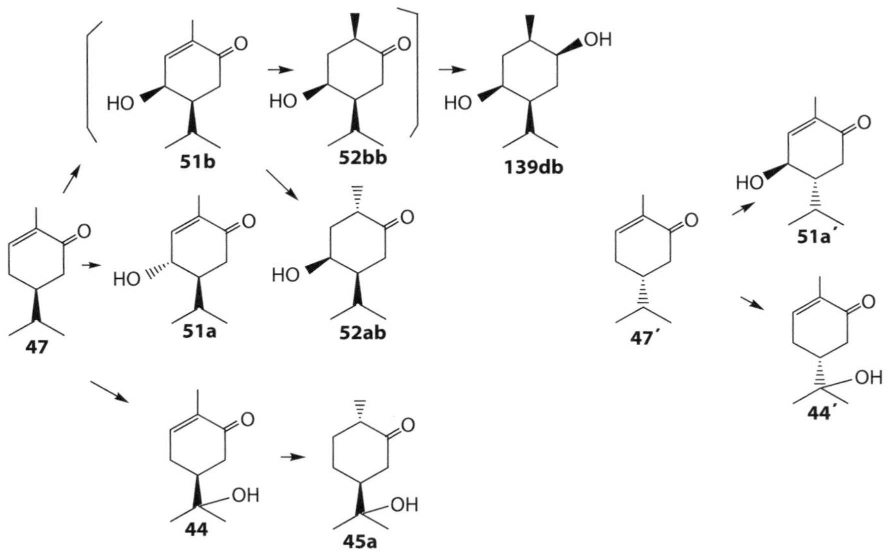

FIGURE 19.122 Proposed the metabolic pathways of (+)-carvotanacetone (**47**) and (−)-carvotanacetone (**47′**) by *Streptomyces bottropensis* SY-2-1. (Modified from Noma, Y. et al., Microbiological conversion of (−)-carvotanacetone and (+)-carvotanacetone by *S. bottropensis* SY-2-1, *Proceedings of 29th TEAC*, 1985c, pp. 238–240.)

A. niger TBUYN-2 converted (−)-carvotanacetone (**47′**) to (+)-carvomenthone (**48a′**), (+)-carvomenthone (**49a′**), diastereoisomeric *p*-menthane-2,9-diols [**55aa′** (8*R*) and **55ab′** (8*S*) in the ratio of 3:1], and 8-hydroxy-(+)-neocarvomenthol (**102a′**). On the other hand, the same fungus converted (+)-carvotanacetone (**47**) to (−)-isocarvomenthone (**48b**), 1α-hydroxy-(+)-neoisocarvomenthol (**54**) via 1α-hydroxy-(+)-isocarvomenthone (**53**), and 8-hydroxy-(−)-isocarvomenthone (**45b**) as shown in Figure 19.123 (Noma et al., 1988b).

19.3.4.1.3 Piperitone

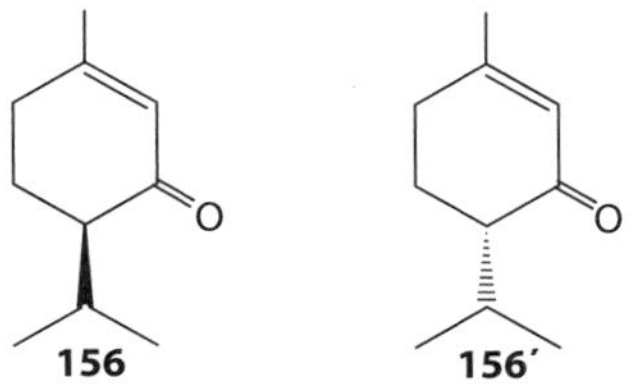

A large number of yeasts were screened for the biotransformation of (−)-piperitone (**156**). A relatively small number of yeasts gave hydroxylation products of (−)-piperitone (**156**). Products obtained from (−)-piperitone (**156**) were 7-hydroxypiperitone (**161**), *cis*-6-hydroxypiperitone (**158b**), *trans*-6-hydroxypiperitone (**158a**), and 2-isopropyl-5-methylhydroquinone (**180**). Yields for the hydroxylation reactions varied between 8% and 60%, corresponding to the product concentrations of 0.04–0.3 g/L. Not one of the yeasts tested reduced (−)-piperitone (**156**) (van Dyk et al., 1998). During the initial screen with (−)-piperitone (**156**), only hydroxylation products were obtained. The hydroxylation products (**161**, **158a**, and **158b**) obtained with nonconventional yeasts from the genera *Arxula*, *Candida*, *Yarrowia*, and *Trichosporon* have recently been described (van Dyk et al., 1998) (Figure 19.124).

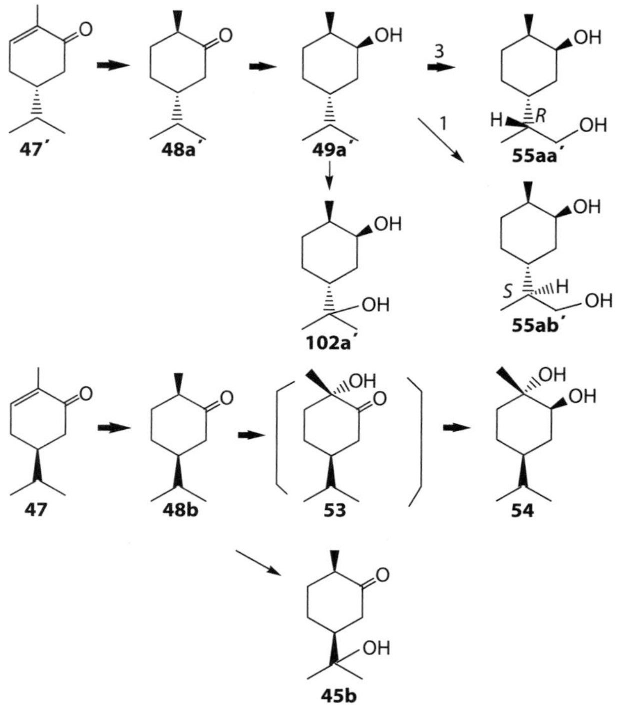

FIGURE 19.123 Proposed metabolic pathways of (−)-carvotanacetone (**47**) and (+)-carvotanacetone (**47′**) by *Aspergillus niger* TBUYN-2. (Modified from Noma, Y. et al., Microbiological conversion of (−)-carvotanacetone and (+)-carvotanacetone by a strain of *Aspergillus niger*, *Proceedings of 32nd TEAC*, 1988b, pp. 146–148.)

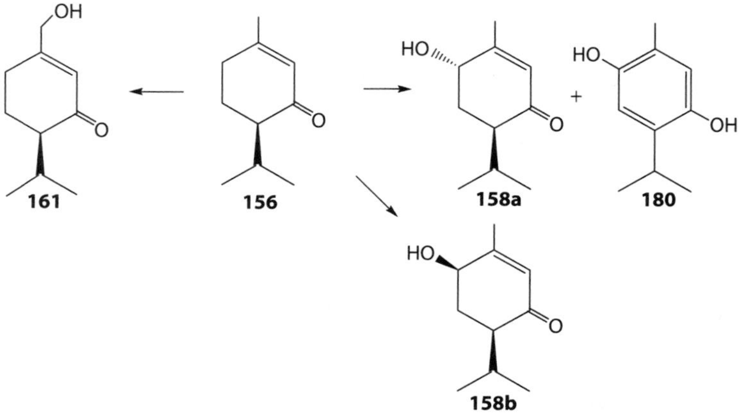

FIGURE 19.124 Hydroxylation products of (*R*)-(−)-piperitone (**156**) by yeast. (Modified from van Dyk, M.S. et al., *J. Mol. Catal. B: Enzym.*, 5, 149, 1998.)

19.3.4.1.4 Pulegone

(*R*)-(+)-Pulegone (**119**), with a mint-like odor monoterpene ketone, is the main component (up to 80%–90%) of *Mentha pulegium* essential oil (pennyroyal oil), which is sometimes used in beverages and food additive for human consumption and occasionally in herbal medicine as an abortifacient drug. The biotransformation of (+)-pulegone (**119**) by fungi was investigated (Ismaili-Alaoui et al., 1992). Most fungal strains grown in a usual liquid culture medium were able to metabolize (+)-pulegone (**119**) to some extent in a concentration range of 0.1–0.5 g/L; higher concentrations were generally toxic, except for a strain of *Aspergillus* sp. isolated from mint leaves infusion, which was able to survive to concentrations of up to 1.5 g/L. The predominant product was generally l-hydroxy-(+)-pulegone (**384**) (20%–30% yield). Other metabolites were present in lower amounts (5% or less) (see Figure 19.125). The formation of 1-hydroxy-(+)-pulegone (**387**) was explained by hydroxylation at a tertiary position. Its dehydration to piperitenone (**112**), even under the incubation conditions, during isolation or derivative reactions precluded any tentative determination of its optical purity and absolute configuration.

Botrytis allii converted (+)-pulegone (**119**) to (−)-(1*R*)-8-hydroxy-4-*p*-menthen-3-one (**121**) and piperitenone (**112**) (Miyazawa et al., 1991b,c). *Hormonema isolate* (UOFS Y-0067) quantitatively reduced (+)-pulegone (**119**) and (−)-menthone (**149a**) to (+)-neomenthol (**137a**) (van Dyk et al., 1998) (Figure 19.125).

Biotransformation by the recombinant reductase and the transformed *Escherichia coli* cells were examined with pulegone, carvone, and verbenone as substrates (Figure 19.126). The recombinant reductase catalyzed the hydrogenation of the exocyclic C=C double bond of pulegone (**119**) to give menthone derivatives (Watanabe et al., 2007) (Tables 19.12 and 19.13).

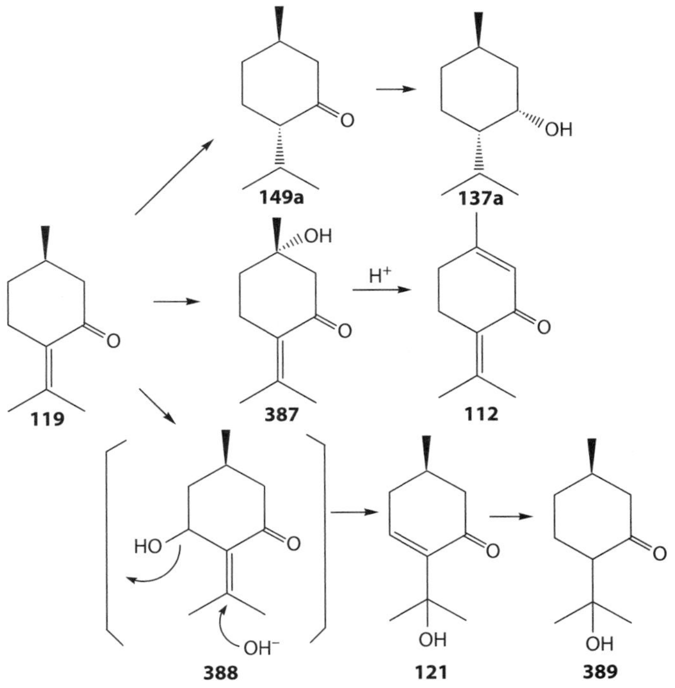

FIGURE 19.125 Biotransformation of (+)-pulegone (**119**) by *Aspergillus* sp., *Botrytis allii*, and *H. isolate* (UOFS Y-0067). (Modified from Miyazawa, M. et al., *Chem. Express*, 6, 479, 1991a; Miyazawa, M. et al., *Chem. Express*, 6, 873, 1991b; Ismaili-Alaoui, M. et al., *Tetrahedron Lett.*, 33, 2349, 1992; van Dyk, M.S. et al., *J. Mol. Catal. B: Enzym.*, 5, 149, 1998.)

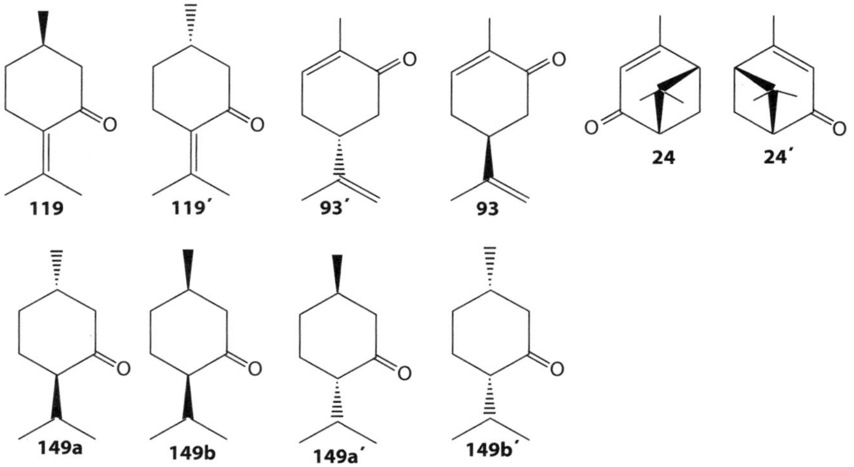

FIGURE 19.126 Chemical structures of substrate reduced by the recombinant pulegone reductase and the transformed *Escherichia coli* cells.

TABLE 19.12
Substrate Specificity in the Reduction of Enones with the Recombinant Pulegone Reductase

Entry No. (Reaction Time)	Substrates	Products	Conversions (%)
1 (3 h)	(R)-(+)-Pulegone (**119**)	(1R, 4R)-Isomenthone (**149b**)	4.4
2 (12 h)	(R)-Pulegone (**119**)	(1S, 4R)-Menthone (**149a**)	6.8
3 (3 h)	(S)-(−)-Pulegone (**119′**)	(1R, 4R)-Isomenthone (**149b**)	14.3
4 (12 h)	(S)-Pulegone (**119′**)	(1S, 4R)-Menthone (**149a**)	15.7
5 (12 h)	(R)-(−)-Carvone (**93′**)	(1S, 4S)-Isomenthone (**149b′**)	0.3
6 (12 h)	(S)-(+)-Carvone (**93**)	(1R, 4S)-Menthone (**149a′**)	0.5
7 (12 h)	(1S, 5S)-Verbenone (**24**)	(1S, 4S)-Isomenthone (**149b′**)	1.6
8 (12 h)	(1R, 5R)-Verbenone (**24′**)	(1R, 4S)-Menthone (**149a′**)	2.1
		—	N.d.
		—	N.d.
		—	N.d.
		—	N.d.

Note: N.d., denotes not detected.

19.3.4.1.5 Piperitenone and Isopiperitenone

Piperitenone (**112**) is metabolized to 5-hydroxypiperitenone (**117**), 7-hydroxypiperitenone (**118**), and 7,8-dihydroxypiperitenone (**157**). Isopiperitenol (**110**) is reduced to give isopiperitenone (**111**), which is further metabolized to piperitenone (**112**), 7-hydroxy- (**113**), 10-hydroxy- (**115**), 4-hydroxy- (**114**), and 5-hydroxyisopiperitenone (**116**). Compounds **111** and **112** are isomerized to each other. Pulegone (**119**) was metabolized to **112**, 8,9-dehydromenthone (**120**) and 8-hydroxymenthone (**121**) as shown in the biotransformation of the same substrate using *B. allii* (Miyazawa et al., 1991c) (Figure 19.127).

H. isolate (UOFS Y-0067) reduced (4S)-isopiperitenone (**111**) to (3R,4S)-isopiperitenol (**110**), a precursor of (−)-menthol (**137b**) (van Dyk et al., 1998) (Figure 19.128).

TABLE 19.13
Biotransformation of Pulegone (119 and 119′) with the Transformed *Escherichia coli* Cells[a]

Substrates	Products	Conversion (%)
(*R*)-(+)-Pulegone (**119**)	(1*R*, 4*R*)-Isomenthone (**149b**)	26.8
(*S*)-(−)-Pulegone (**119′**)	(1*S*, 4*R*)-Menthone (**149a**)	30.0
	(1*S*, 4*S*)-Isomenthone (**149b′**)	32.3
	(1*R*, 4*S*)-Menthone (**149a′**)	7.1

[a] Reaction times of the transformation reaction are 12 h.

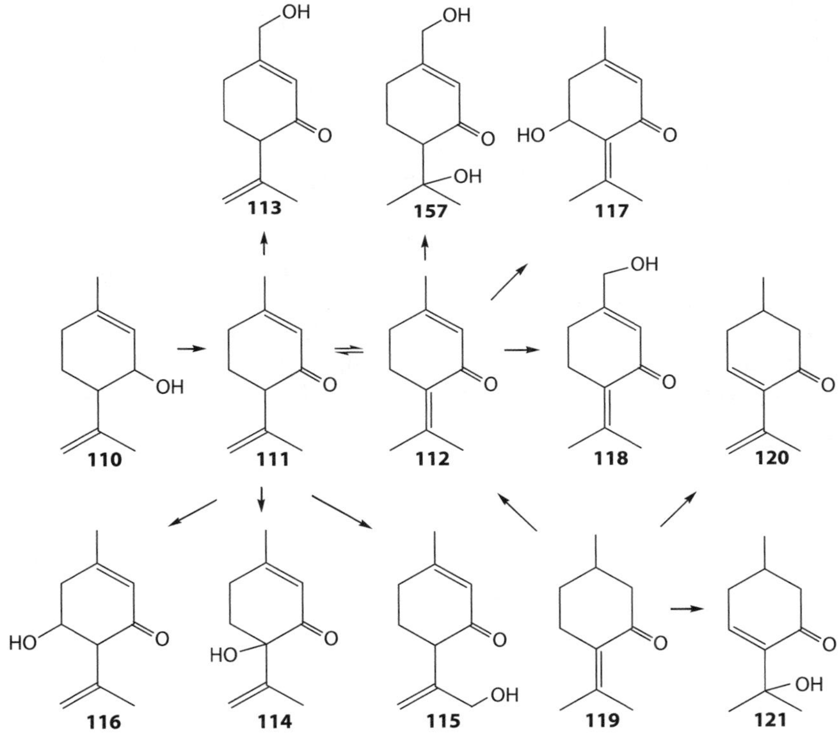

FIGURE 19.127 Biotransformation of isopiperitenone (**111**) and piperitenone (**112**) by *Aspergillus niger* TBUYN-2. (Modified from Noma, Y. et al., Biotransformation of isopiperitenone, 6-gingerol, 6-shogaol and neomenthol by a strain of *Aspergillus niger*, *Proceedings of 37th TEAC*, 1992c, pp. 26–28.)

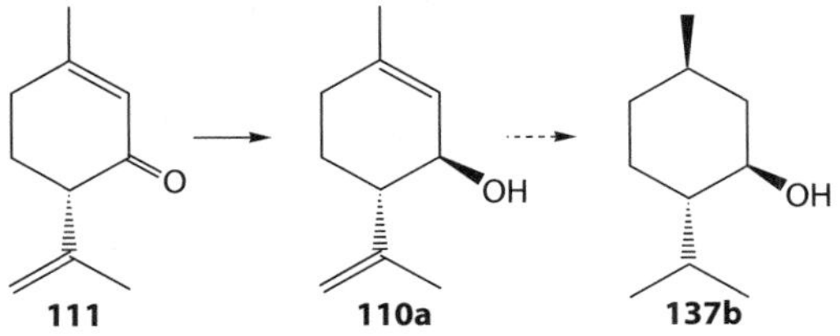

FIGURE 19.128 Biotransformation of isopiperitenone (**111**) by *H. isolate* (UOFS Y-0067). (Modified from van Dyk, M.S. et al., *J. Mol. Catal. B: Enzym.*, 5, 149, 1998.)

19.3.4.2 Saturated Ketone
19.3.4.2.1 Dihydrocarvone

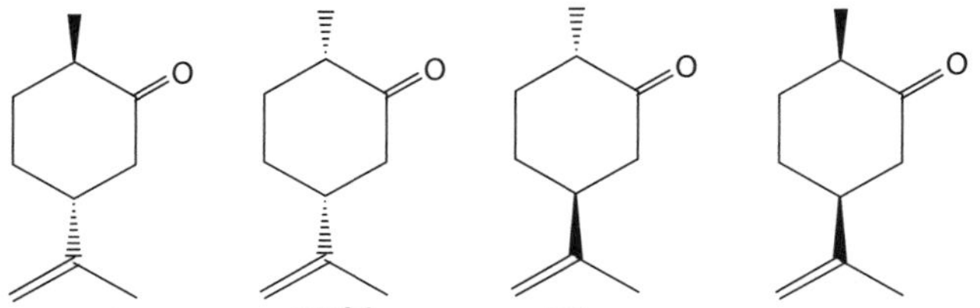

101a′	101b′	101a	101b
(1R,4S)	(1S,4S)	(1S,4R)	(1R,4R)
(+)	(+)-Iso	(−)	(−)-Iso

In the reduction of saturated carbonyl group of dihydrocarvone by microorganism, (+)-dihydrocarvone (**101a′**) is converted stereospecifically to either (+)-neodihydrocarveol (**102a′**) or (−)-dihydrocarveol (**102b′**) or nonstereospecifically to the mixture of **102a′** and **102b′**, whereas (−)-isodihydrocarvone (**101b**) is converted stereospecifically to either (−)-neoisodihydrocarveol (**102d**) or (−)-isodihydrocarveol (**102c**) or nonstereospecifically to the mixture of **102c** and **102d** by various microorganisms (Noma and Tatsumi, 1973; Noma et al., 1974c; Noma and Nonomura, 1974; Noma, 1976, 1977).

(+)-Dihydrocarvone (**101a′**) and (+)-isodihydrocarvone (**101b′**) are easily isomerized chemically to each other. In the microbial transformation of (−)-carvone (**93′**), the formation of (+)-dihydrocarvone (**101a′**) is predominant. (+)-Dihydrocarvone (**101a′**) was reduced to both/either (+)-neodihydrocarveol (**102a′**) and/or (−)-dihydrocarveol (**102b**), whereas in the biotransformation of (+)-carvone (**93**), (+)-isodihydrocarvone (**101b**) was formed predominantly. (+)-Isodihydrocarvone (**101b**) was reduced to both (+)-isodihydrocarveol (**102c**) and (+)-neoisodihydrocarveol (**102d**) (Figure 19.129).

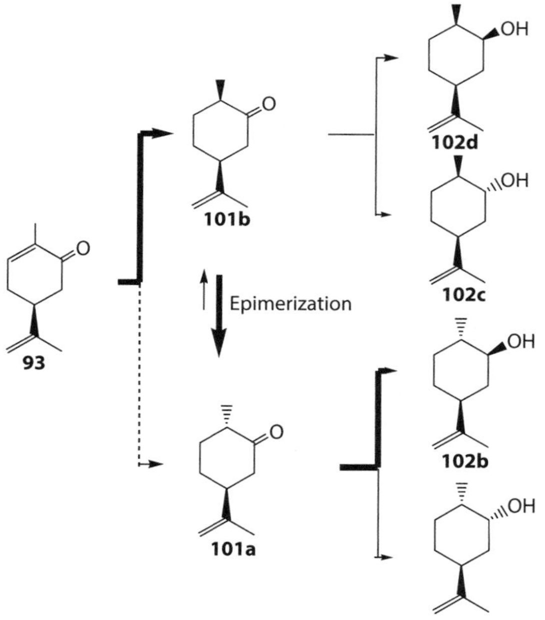

FIGURE 19.129 Proposed metabolic pathways of (+)-carvone (**93**) and (−)-isodihydrocarvone (**101b**) by *Pseudomonas fragi* IFO 3458. (Modified from Noma, Y. et al., *Agric. Biol. Chem.*, 39, 437, 1975.)

However, *P. fragi* IFO 3458, *Pseudomonas fluorescens* IFO 3081, and *Aerobacter aerogenes* IFO 3319 and IFO 12059 formed (−)-dihydrocarvone (**101a**) predominantly from (+)-carvone (**93**). In the time course study of the biotransformation of (+)-carvone (**93**), it appeared that predominant formation of (−)-dihydrocarvone is due to the epimerization of (−)-isodihydrocarvone (**101b′**) by epimerase of *P. fragi* IFO 3458 (Noma et al., 1975).

19.3.4.2.2 Isodihydrocarvone Epimerase

19.3.4.2.2.1 Preparation of Isodihydrocarvone Epimerase The cells of *P. fragi* IFO 3458 were harvested by centrifugation and washed five times with 1/100 M KH_2PO_4–Na_2HPO_4 buffer (pH 7.2). Bacterial extracts were prepared from the washed cells (20 g from 3 L medium) by sonic lysis (Kaijo Denki Co., Ltd., 20Kc., 15 min, at 5°C–7°C) in 100 mL of the same buffer. Sonic extracts were centrifuged at 25, 500 *g* for 30 min at −2°C. The opalescent yellow supernatant fluid had the ability to convert (−)-isodihydrocarvone (**101b**) to (−)-dihydrocarvone (**101a**). On the other hand, the broken cell preparation was incapable of converting (−)-isodihydrocarvone (**101b**) to (−)-dihydrocarvone (**101a**). The enzyme was partially purified from this supernatant fluid about 56-fold with heat treatment (95°C–97°C for 10 min), ammonium sulfate precipitation (0.4–0.7 saturation), and DEAE-Sephadex A-50 column chromatography.

The reaction mixture consisted of a mixture of (−)-isodihydrocarvone (**101b**) and (−)-dihydrocarvone (**101a**) (60:40 or 90:10), 1/30 M KH_2PO_4–Na_2HPO_4 buffer (pH 7.2), and the crude or partially purified enzyme solution. The reaction was started by the addition of the enzyme solution and stopped by the addition of ether. The ether extract was applied to analytical gas–liquid chromatography (GLC) (Shimadzu GC-4A 10% PEG-20M, 3 m × 3 mm, temperature 140°C–170°C at the rate of 1°C/min, N_2 35 mL/min), and epimerization was assayed by measuring the peak areas of (−)-isodihydrocarvone (**101b**) and (−)-dihydrocarvone (**101a**) in GLC before and after the reaction.

The crude extract and the partially purified preparation were found to be very stable to heat treatment; 66% and 36% of the epimerase activity remained after treatment at 97°C for 60 and 120 min, respectively (Noma et al., 1975).

A strain of *A. niger* TBUYN-2 hydroxylated at C-1 position of (−)-isodihydrocarvone (**101b**) to give 1α-hydroxyisodihydrocarvone (**72b**), which was easily and smoothly reduced to (1*S*, 2*S*, 4*S*)-(−)-8-*p*-menthene-1,2-*trans*-diol (**71d**), which was also obtained from the biotransformation of (−)-*cis*-limonene-1,2-epoxide (**69**) by microorganisms and decomposition by 20% HCl (Figure 19.127) (Noma et al., 1985a,b). Furthermore, *A. niger* TBUYN-2 and *A. niger* Tiegh (CBAYN) biotransformed (−)-isodihydrocarvone (**101b**) to give (−)-4α-hydroxyisodihydrocarvone (**378b**) and (−)-8-*p*-menthene-1,2-*trans*-diol (**71d**) as the major products together with a small amount of 1α-hydroxyisodihydrocarvone (**72b**) (Noma and Asakawa, 2008) (Figure 19.130).

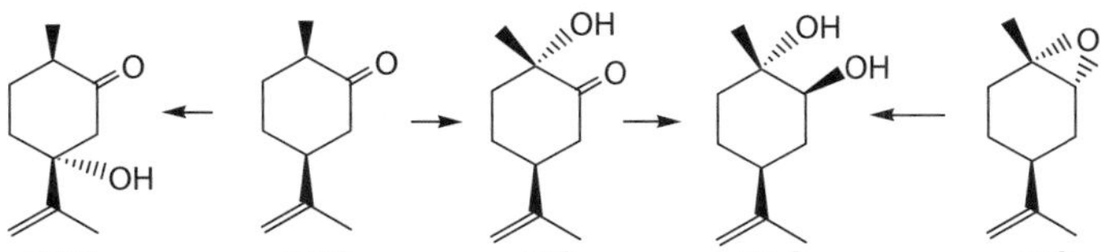

FIGURE 19.130 Biotransformation of (+)-carvone (**93**), (−)-isodihydrocarvone (**101b**), and (−)-*cis*-limonene-1,2-epoxide (**69b**) by *Aspergillus niger* TBUYN-2 and *A. niger* Tiegh (CBAYN). (Modified from Noma, Y. et al., Biotransformation of (−)-carvone and (+)-carvone by *Aspergillus* spp., *Annual Meeting of Agricultural and Biological Chemistry*, Sapporo, Japan, 1985a, p. 68; Noma, Y. and Asakawa, Y., New metabolic pathways of (+)-carvone by Citrus pathogenic *Aspergillus niger* Tiegh CBAYN and *A. niger* TBUYN-2, *Proceedings of 52nd TEAC*, 2008, pp. 206–208.)

19.3.4.2.3 Menthone and Isomenthone

149a
(1R,4S)
(−)-Menthone

149b
(1S,4S)
(−)-Isomenthone

149a′
(1S,4R)
(+)-Menthone

149b′
(1R,4R)
(+)-Isomenthone

The growing cells of *P. fragi* IFO 3458 epimerized 17% of racemic isomenthone (**149b** and **b′**) to menthone (**149a** and **a′**) (Noma et al., 1975). (−)-Menthone (**149a**) was converted by *P. fluorescens* M-2 to (−)-3-oxo-4-isopropyl-1-cyclohexanecarboxylic acid (**164a**), (+)-3-oxo-4-isopropyl-1-cyclohexanecarboxylic acid (**164b**), and (+)-3-hydroxy-4-isopropyl-1-cyclohexanecarboxylic acid (**165ab**). On the other hand, (+)-menthone (**149a′**) was converted to give (+)-3-oxo-4-isopropyl-1-cyclohexane carboxylic acid (**164a′**) and (−)-3-oxo-4-isopropyl-1-cyclohexane carboxylic acid (**164b′**). Racemic isomenthone (**149b** and **b′**) was converted to give racemic 1-hydroxy-1-methyl-4-isopropylcyclohexane-3-one (**150**), racemic piperitone (**156**), racemic 3-oxo-4-isopropyl-1-cyclohexene-1-carboxylic acid (**162**), 3-oxo-4-isopropyl-1-cyclohexane carboxylic acid (**164b**), 3-oxo-4-isopropyl-1-cyclohexane carboxylic acid (**164a**), and (+)-3-hydroxy-4-isopropyl-1-cyclohexane carboxylic acid (**165ab**) (Figure 19.131).

Soil plant pathogenic fungi *Rhizoctonia solani* 189 converted (−)-menthone (**149a**) to 4β-hydroxy(−)-menthone (**392**, 29%) and 1 α, 4 β-dihydroxy-(−)-menthone (**393**, 71%) (Nonoyama et al., 1999) (Figure 19.131). (−)-Menthone (**149a**) was transformed by *S. litura* to give 7-hydroxymenthone (**151a**), 7-hydroxyneomenthol (**165c**), and 7-hydroxy-9-carboxymenthone (**394a**) (Hagiwara et al., 2006) (Figure 19.132). (−)-Menthone (**149a**) gave 7-hydroxymenthone (**151a**) and (+)-neomenthol (**137c**) by human liver microsome (CYP2B6). Of 11 recombinant human P450 enzymes (expressed in *Trichoplusia ni* cells) tested, CYP2B6 catalyzed oxidation of (−)-menthone (**149a**) to 7-hydroxymenthone (**151a**) (Nakanishi and Miyazawa, 2004) (Figure 19.132).

19.3.4.2.4 Thujone

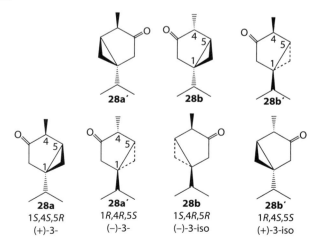

28a'

28b

28b'

28a
1S,4S,5R
(+)-3-

28a'
1R,4R,5S
(−)-3-

28b
1S,4R,5R
(−)-3-iso

28b'
1R,4S,5S
(+)-3-iso

β-Pinene (**1**) is metabolized to 3-thujone (**28**) via α-pinene (**4**) (Gibbon and Pirt, 1971). α-Pinene (**4**) is metabolized to give thujone (**28**). Thujone (**28**) was biotransformed to thujoyl alcohol (**29**) by

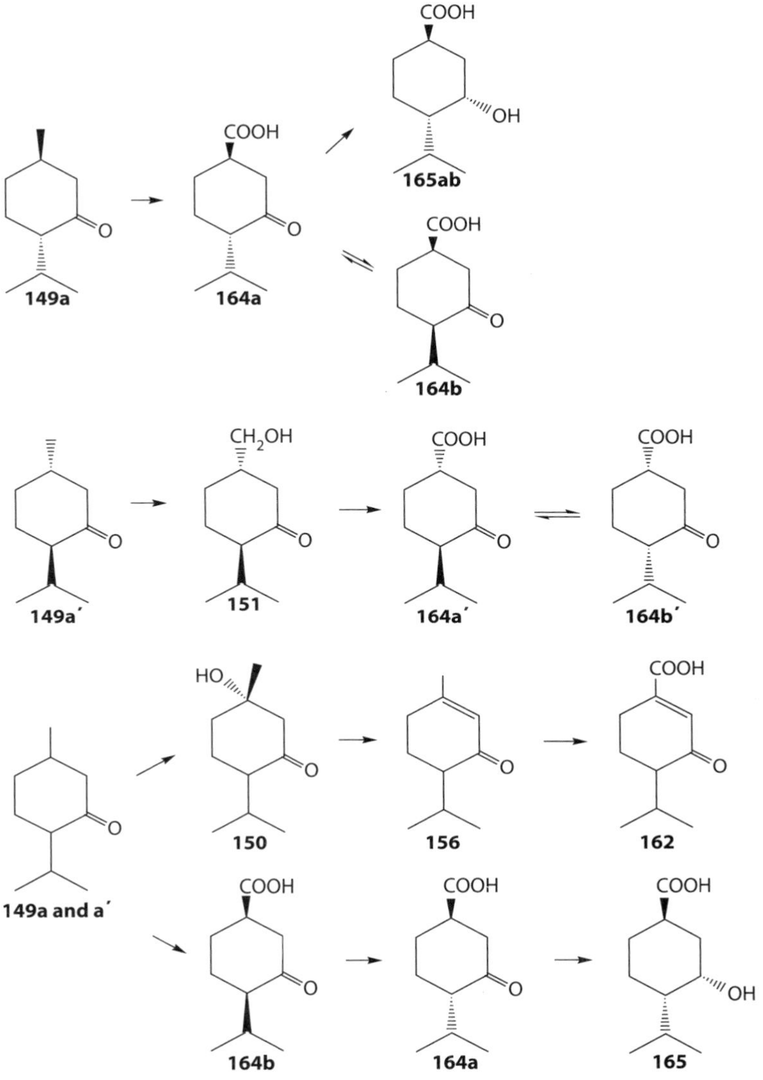

FIGURE 19.131 Biotransformation of (−)- (**149a**) and (+)-menthone (**149a′**) and racemic isomenthone (**149b** and **149b′**) by *Pseudomonas fluorescens* M-2. (Modified from Sawamura, Y. et al., Microbiological oxidation of *p*-menthane 1. Formation of formation of *p*-cis-menthan-1-ol, *Proceedings of 18th TEAC*, 1974, pp. 27–29.)

A. niger TBUYN-2 (Noma, 2000). Furthermore, (−)-3-isothujone (**28b**) prepared from *Armoise* oil was biotransformed by plant pathogenic fungus *B. allii* IFO 9430 to give 4-hydroxythujone (**30**) and 4,6-dihydroxythujone (**31**) (Miyazawa et al., 1992a) (Figure 19.133).

19.3.4.3 Cyclic Monoterpene Epoxide
19.3.4.3.1 1,8-Cineole

1,8-Cineole (**122**) is a main component of the essential oil of *Eucalyptus radiata* var. *australiana* leaves, comprising *ca*. 75% in the oil, which corresponds to 31 mg/g fr.wt. leaves (Nishimura et al., 1980).

The most effective utilization of **122** is very important in terms of renewable biomass production. It would be of interest, for example, to produce more valuable substances, such as plant growth regulators, by the microbial transformation of **122**. The first reported utilization of **122** was presented

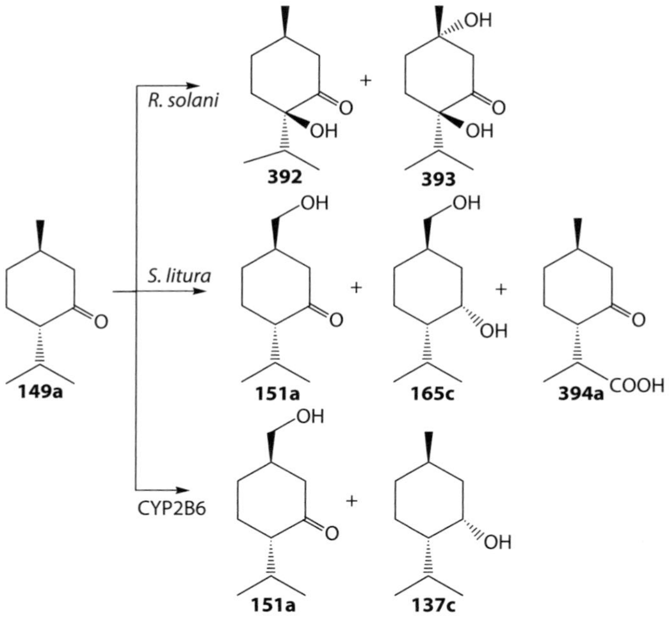

FIGURE 19.132 Metabolic pathway of (−)-menthone (**149a**) by *Rhizoctonia solani* 189, *Spodoptera litura*, and human liver microsome (CYP2B6). (Modified from Nonoyama, H. et al., Biotransformation of (−)-menthone using plant parasitic fungi, *Rhizoctonia solani* as a biocatalyst, *Proceedings of 43rd TEAC*, 1999, pp. 387–388; Nakanishi, K. and Miyazawa, M., Biotransformation of (−)-menthone by human liver microsomes, *Proceedings of 48th TEAC*, 2004, pp. 401–402; Hagiwara, Y. et al., Biotransformation of (+)-and (−)-menthone by the larvae of common cutworm (*Spodoptera litura*) as a biocatalyst, *Proceedings of 50th TEAC*, 2006, pp. 279–280.)

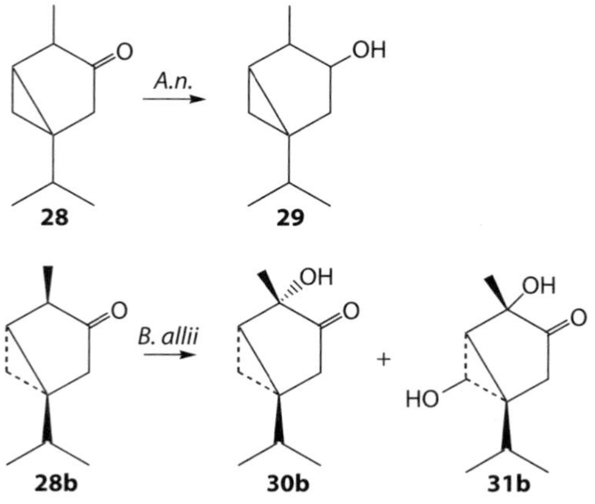

FIGURE 19.133 Biotransformation of (−)-3-isothujone (**28b**) by *Aspergillus niger* TBUYN-2 and plant pathogenic fungus *Botrytis allii* IFO 9430. (Modified from Gibbon, G.H. and Pirt, S.J., *FEBS Lett.*, 18, 103, 1971; Miyazawa, M. et al., Biotransformation of thujone by plant pathogenic microorganism, *Botrytis allii* IFO 9430, *Proceedings of 36th TEAC*, 1992a, pp. 197–198.)

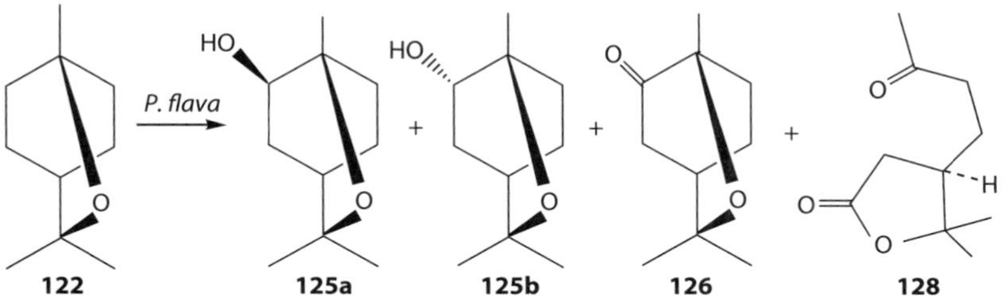

FIGURE 19.134 Biotransformation of 1,8-cineole (**84**) by *Pseudomonas flava*. (Modified from MacRae, I.C. et al., *Aust. J. Chem.*, 32, 917, 1979.)

by MacRae et al. (1979), who showed that it was a carbon source for *Pseudomonas flava* growing on *Eucalyptus* leaves. Growth of the bacterium in a mineral salt medium containing **122** resulted in the oxidation at the C-2 position of **122** to give the metabolites (1*S*,4*R*,6*S*)-(+)-2α-hydroxy-1,8-cineole (**225a**), (1*S*,4*R*,6*R*)-(−)-2β-hydroxy-1,8-cineole (**125a**), (1*S*,4*R*)-(+)-2-oxo-1,8-cineole (**126**), and (−)-(*R*)-5,5-dimethyl-4-(3′-oxobutyl)-4,5-dihydrofuran-2(3*H*)-one (**128**) (Figure 19.134).

S. *bottropensis* SY-2-1 biotransformed 1,8-cineole (**122**) stereochemically to (+)-2α-hydroxy-1,8-cineole (**125b**) as the major product and (+)-3α-hydroxy-1,8-cineole (**123b**) as the minor product. Recovery ratio of 1,8-cineole metabolites as ether extract was *ca.* 30% in *S. bottropensis* SY-2-1 (Noma and Nishimura, 1980, 1981) (Figure 19.135).

In the case of *S. ikutamanensis* Ya-2-1, 1,8-cineole (**122**) was biotransformed regioselectively to give (+)-3α-hydroxy-1,8-cineole (**123b**, 46%) and (+)-3β-hydroxy-1,8-cineole (**123b**, 29%) as the major product. Recovery ratio as ether extract was *ca.* 8.5% in *S. ikutamanensis*,Ya-2-1 (Noma and Nishimura, 1980, 1981) (Figure 19.135).

When (+)-3α-hydroxy-1,8-cineole (**123b**) was used as substrate in the cultured medium of *S. ikutamanensis* Ya-2-1, (+)-3β-hydroxy-1,8-cineole (**123a**, 32%) was formed as the major product together with a small amount of (+)-3-oxo-1,8-cineole (**126a**, 1.6%). When (+)-3β-hydroxy-1,8

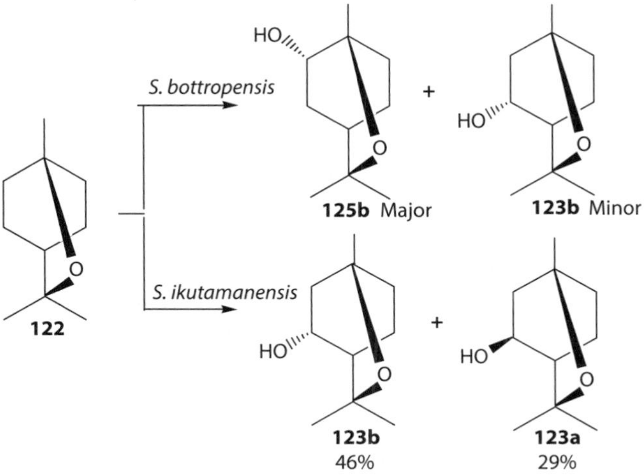

FIGURE 19.135 Biotransformation of 1,8-cineole (**122**) by *Streptomyces bottropensis* SY-2-1 and *Streptomyces ikutamanensis* Ya-2-1. (Modified from Noma, Y. and Nishimura, H., Microbiological transformation of 1,8-cineole. Oxidative products from 1,8-cineole by *S. bottropensis*, SY-2-1, *Book of abstracts of the Annual Meeting of Agricultural and Biological Chemical Society*, 1980, p. 28; Noma, Y. and Nishimura, H., Microbiological transformation of 1,8-cineole. Production of 3β-hydroxy-1,8-cineole from 1,8-cineole by *S. ikutamanensis*, Ya-2-1, *Book of Abstracts of the Annual Meeting of Agricultural and Biological Chemical Society*, 1981, p. 196.)

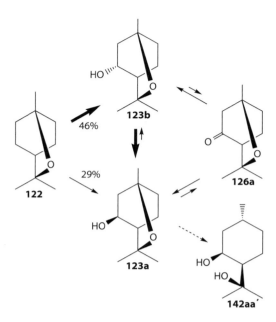

FIGURE 19.136 Biotransformation of 1,8-cineole (**122**), (+)-3α-hydroxy-1,8-cineole (**123b**), (+)-3β-hydroxy-1,8-cineole(**123a**), and (+)-3-oxo-1,8-cineole (**126a**) by *Streptomyces ikutamanensis* Ya-2-1. (Modified from Noma, Y. and Nishimura, H., Microbiological transformation of 1,8-cineole. Production of 3β-hydroxy-1,8-cineole from 1,8-cineole by *S. ikutamanensis*, Ya-2-1, *Book of Abstracts of the Annual Meeting of Agricultural and Biological Chemical Society*, 1981, p. 196.)

cineole (**123a**) was used, (+)-3-oxo-1,8-cineole (**126a**, 9.6%) and (+)-3α-hydroxy-1,8-cineole (**123b**, 2%) were formed. When (+)-3-oxo-1,8-cineole (**126a**) was used, (+)-3α-hydroxy- (**123b**, 19%) and (+)-3β-hydroxy-1,8-cineole (**123a**, 16%) were formed.

Based on the aforementioned results, it is obvious that (+)-3β-hydroxy-1,8-cineole (**123b**) is formed mainly in the biotransformation of 1,8-cineole (**122**), (+)-3α-hydroxy-1,8-cineole (**123b**), and (+)-3-oxo-1,8-cineole (**126a**) by *S. ikutamanensis* Ya-2-1. The production of (+)-3β-hydroxy-1,8-cineole (**123b**) is interesting, because it is a precursor of mosquito repellent, *p*-menthane-3,8-diol (**142aa′**) (Noma and Nishimura, 1981) (Figure 19.136).

When *A. niger* TBUYN-2 was cultured in the presence of 1,8-cineole (**122**) for 7 days, it was transformed to three alcohols [racemic 2α-hydroxy-1,8-cineoles (**125b** and **b′**), racemic 3α-hydroxy- (**123b** and **b′**), and racemic 3β-hydroxy-1,8-cineoles (**123a** and **123a′**)] and two ketones [racemic 2-oxo- (**126** and **126′**) and racemic 3-oxo-1,8-cineoles (**124** and **124′**)] (Figure 19.135). The formation of 3α-hydroxy- (**123b** and **b′**) and 3β-hydroxy-1,8-cineoles (**123a** and **123a′**) is of great interest not only due to the possibility of the formation of *p*-menthane-3,8-diol (**142** and **142′**), the mosquito repellents, and plant growth regulators that are synthesized chemically from 3α-hydroxy- (**123b** and **b′**) and 3β-hydroxy-1,8-cineoles (**123a** and **123a′**), respectively, but also from the viewpoint of the utilization of *E. radiata* var. *australiana* leaves oil as biomass. An Et$_2$O extract of the culture broth (products and **122** as substrate) was recovered in 57% of substrate (w/w) (Nishimura et al., 1982; Noma et al., 1996) (Figure 19.137).

Plant pathogenic fungus *Botryosphaeria dothidea* converted 1,8-cineole (**122**) to optical pure (+)-2α-hydroxy-1,8-cineole (**125b**) and racemic 3α-hydroxy-1,8-cineole (**123b** and **b′**), which were oxidized to optically active 2-oxo- (**126**) (100% ee) and racemic 3-oxo-1,8-cineole (**124** and **124′**), respectively (Table 19.14). The cytochrome P450 inhibitor 1-aminobenzotriazole inhibited the hydroxylation of the substrate (Noma et al., 1996) (Figure 19.138). *S. litura* also converted

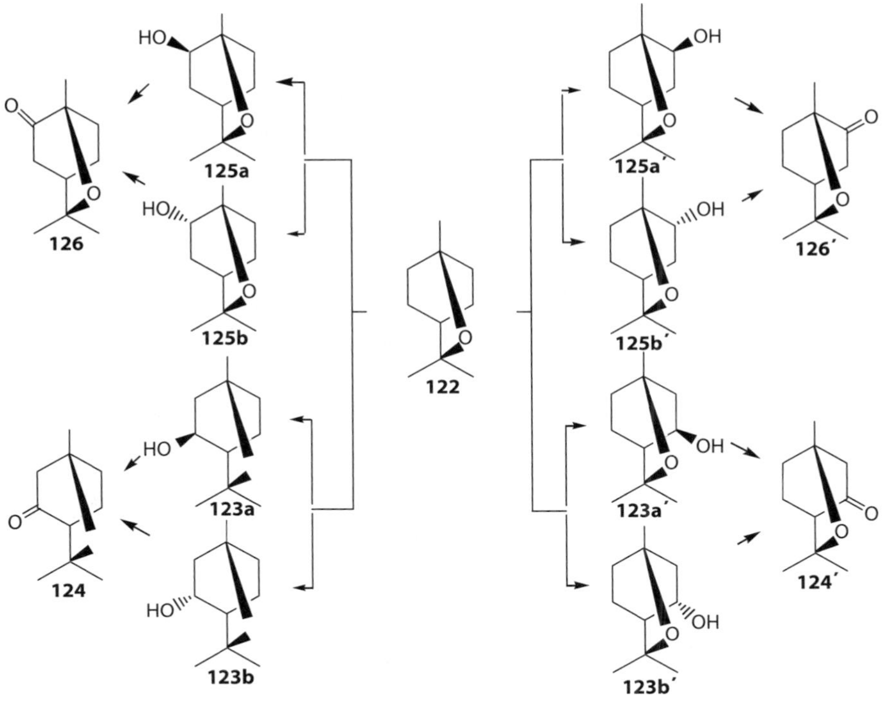

FIGURE 19.137 Biotransformation of 1,8-cineole (**122**) by *Aspergillus niger* TBUYN-2. (Modified from Nishimura, H. et al., *Agric. Biol. Chem.*, 46, 2601, 1982; Noma, Y. et al., Biotransformation of 1,8-cineole. Why do the biotransformed 2α- and 3α-hydroxy-1,8-cineole by *Aspergillus niger* have no optical activity? *Proceedings of 40th TEAC*, 1996, pp. 89–91.)

TABLE 19.14
Stereoselectivity in the Biotransformation of 1,8-Cineole (122) by *Aspergillus niger*, *Botryosphaeria dothidea*, and *Pseudomonas flava*

Microorganisms	Products 125a and a', 125b and b', 123b and b', 123a and a'
Aspergillus niger TBUYN-2	2:43:49:6
	50:50:41:59
Botryosphaeria dothidea	4:59:34:3
	100:0:53:47
Pseudomonas flava	
	29:71:0:0
	100:0

Source: Noma, Y. et al., Biotransformation of 1,8-cineole. Why do the biotransformed 2α- and 3α-hydroxy-1,8-cineole by *Aspergillus niger* have no optical activity? *Proceedings of 40th TEAC*, 1996, pp. 89–91.

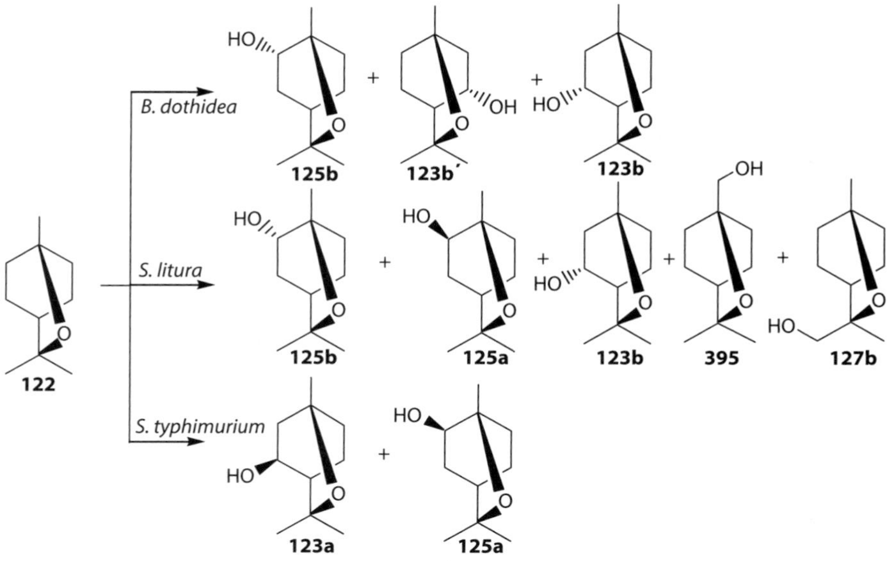

FIGURE 19.138 Biotransformation of 1,8-cineole (**122**) by *Botryosphaeria dothidea*, *Spodoptera litura*, and *Salmonella typhimurium*. (Modified from Noma, Y. et al., Biotransformation of 1,8-cineole. Why do the biotransformed 2α- and 3α-hydroxy-1,8-cineole by *Aspergillus niger* have no optical activity? *Proceedings of 40th TEAC*, 1996, pp. 89–91; Saito, H. and Miyazawa, M., Biotransformation of 1,8-cineole by *Salmonella typhimurium* OY1001/3A4, *Proceedings of 50th TEAC*, pp. 275–276, 2006; Hagiwara, Y. and Miyazawa, M., Biotransformation of cineole by the larvae of common cutworm (*Spodoptera litura*) as a biocatalyst, *Proceedings of 51st TEAC*, 2007, pp. 304–305.)

1,8-cineole (**122**) to give three secondary alcohols (**123b, 125a,** and **b**) and two primary alcohols (**395** and **127**) (Hagihara and Miyazawa, 2007). *Salmonella typhimurium* OY1001/3A4 and NADPH-P450 reductase hydroxylated 1,8-cineole (**122**) to 2β-hydroxy-1,8-cineole (**125a**, $[\alpha]_D$ + 9.3, 65.3% ee) and 3β-hydroxy-1 to 8-cineole (**123a**, $[\alpha]_D$ −27.8, 24.7% ee) (Saito and Miyazawa, 2006).

Extraction of the urinary metabolites from brushtail possums (*Trichosurus vulpecula*) maintained on a diet of fruit impregnated with 1,8-cineole (**122**) yielded *p*-cresol (**129**) and the novel C-9 oxidated products 9-hydroxy-1,8-cineole (**127a**) and 1,8-cineole-9-oic acid (**462a**) (Flynn and Southwell, 1979; Southwell and Flynn, 1980) (Figure 19.139).

1,8-Cineole (**122**) gave 2β-hydroxy-1,8-cineole (**125a**) by CYP450 human and rat liver microsome. Cytochrome P450 molecular species responsible for metabolism of 1,8-cineole (**122**) was determined to be CYP3A4 and CYP3A1/2 in human and rat, respectively. Kinetic analysis showed that K_m and V_{max} values for the oxidation of 1,8-cineole (**122**) by human and rat treated with pregnenolone-16α-carbonitrile recombinant CYP3A4 were determined to be 50 μM and 90.9 nmol/min/nmol P450, 20 μM and 11.5 nmol/min/nmol P450, and 90 μM and 47.6 nmol/min/nmol P450, respectively (Shindo et al., 2000).

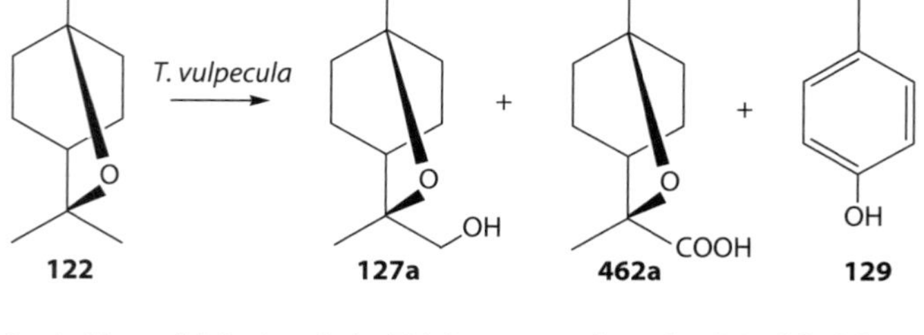

FIGURE 19.139 Metabolism of 1,8-cineole in *Trichosurus vulpecula*. (Modified from Southwell, I.A. and Flynn, T.M., *Xenobiotica*, 10, 17, 1980.)

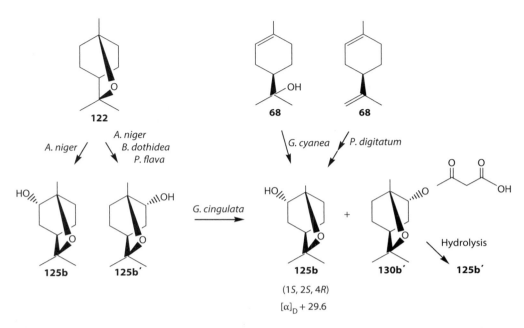

FIGURE 19.140 Formation of 2α-hydroxy-1,8-cineoles (**125b** and **b'**) from 1,8-cineole (**122**) and optical resolution by *Glomerella cingulata* and *Aspergillus niger* TBUYN-2 and **125b'** from (+)-limonene (**68**) by *Penicillium digitatum*. (Modified from Nishimura, H. et al., *Agric. Biol. Chem.*, 46, 2601, 1982; Abraham, W.-R. et al., *Appl. Microbiol. Biotechnol.*, 24, 24, 1986; Miyazawa, M. et al., Biotransformation of 2-endo-hydroxy-1,4-cineole by plant pathogenic microorganism, *Glomerella cingulata*, Proceedings of 39th TEAC, 1995b, pp. 352–353; Noma, Y. et al., Reduction of terpene aldehydes and epoxidation of terpene alcohols by *S. ikutamanensis*, Ya-2-1, Proceedings of 30th TEAC, 1986, pp. 204–206; Noma, Y. and Asakawa, Y., Biotransformation of limonene and related compounds by newly isolated low temperature grown *citrus* pathogenic fungi and red yeast, Book of Abstracts of the 38th ISEO, 2007a, p. 7.)

Microbial resolution of racemic 2α-hydroxy-1,8-cineoles (**125b** and **b'**) was carried out by using *G. cingulata*. The mixture of **125b** and **b'** was added to a culture of *G. cingulata* and esterified to give after 24 h (1*R*,2*R*,4*S*)-2α-hydroxy-1,8-cineole-2-yl-malonate (**130b'**) in 45% yield (ee 100%). The recovered alcohol showed 100% ee of the (1*S*,2*S*,4*R*)-enantiomer (**125b**) (Miyazawa et al., 1995c). On the other hand, optically active (+)-2α-hydroxy-1,8-cineole (**125b**) was also formed from (+)-limonene (**68**) by a strain of citrus pathogenic fungus *P. digitatum* (Saito and Miyazawa, 2006, Noma and Asakawa, 2007a) (Figure 19.140).

Esters of racemic 2α-hydroxy-1,8-cineole (**125b** and **b'**) were prepared by a convenient method (Figure 19.141). Their odors were characteristic. Then products were tested against antimicrobial activity and their microbial resolution was studied (Hashimoto and Miyazawa, 2001) (Table 19.15).

1,8-Cineole (**122**) was glucosylated by *E. perriniana* suspension cells to 2α-hydroxy-1,8 cineole monoglucoside (**404**, 16.0% and **404'**, 16.0%) and diglucosides (**405**, 1.4%) (Hamada et al., 2002) (Figure 19.142).

19.3.4.3.2 1,4-Cineole

Regarding the biotransformation of 1,4-cineole (**131**), *S. griseus* transformed it to 8-hydroxy-1,4-cineole (**134**), whereas *Bacillus cereus* transformed 1,4-cineole (**131**) to 2α-hydroxy-1,4 cineole (**132b**, 3.8%) and 2β-hydroxy-1,4-cineoles (**132a**, 21.3%) (Liu et al., 1988) (Figures 19.143 and 19.144). On the other hand, a strain of *A. niger* biotransformed 1,4-cineole (**131**) regiospecifically to 2α-hydroxy-1,4-cineole (**132b**) (Miyazawa et al., 1991e) and (+)-3α-hydroxy-1,4-cineole (**133b**) (Miyazawa et al., 1992d) along with the formation of 8-hydroxy-1,4-cineole (**134**) and 9-hydroxy-1,4 cineole (**135**) (Miyazawa et al., 1992e) (Figure 19.144).

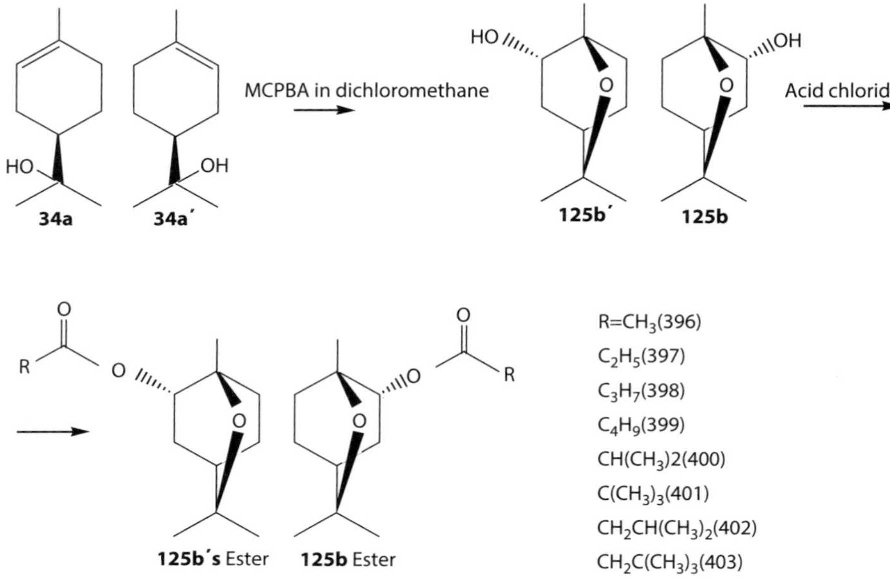

FIGURE 19.141 Chemical synthesis of esters of racemic 2α-hydroxy-1,8-cineole (**125b** and **b'**). (Modified from Hashimoto Y. and Miyazawa, M., Microbial resolution of esters of racemic 2-*endo*-hydroxy-1,8-cineole by *Glomerella cingulata*, Proceedings of 45th TEAC, 2001, pp. 363–365.)

TABLE 19.15
Yield and Enantiomer Excess of Esters of Racemic 2α-Hydroxy-1,8-Cineole (125b and b') on the Microbial Resolution by *Glomerella cingulata*

	0 h	24 h		48 h	
Compounds	% ee	% ee	Yield (%)	% ee	Yield (%)
396	(−)36.3	(+)85.0	24.0	(+)100	14.1
397	(−)36.9	(+)73.8	18.6	(+)100	8.6
398	(−)35.6	(+)33.2	13.7	(+)75.4	3.5
399	(−)36.8	(+)45.4	14.4	(+)100	2.3
400	(−)35.4	(−)21.4	25.2	(+)20.6	8.0
401	(−)36.7	(−)37.8	31.5	(−)40.6	15.2
402	(−)36.1	(−)29.8	46.8	(−)15.0	24.0
403	(−)36.3	(−)37.6	72.2	(−)39.0	36.9

Source: Hashimoto, Y. and Miyazawa, M., Microbial resolution of esters of racemic 2-*endo*-hydroxy-1,8-cineole by *Glomerella cingulata*, Proceedings of 45th TEAC, 2001, pp. 363–365.

Microbial optical resolution of racemic 2α-hydroxy-1,4-cineoles (**132b** and **b'**) was carried out by using *G. cingulata* (Liu et al., 1988). The mixture of 2α-hydroxy-1,4-cineoles (**132b** and **b'**) was added to a culture of *G. cingulata* and esterified to give after 24 h (1*R*,2*R*,4*S*)-2α-hydroxy-1,4-cineole-2-yl-malonate (**136'**) in 45% yield (ee 100%). The recovered alcohol showed an ee of 100% of the (1*S*,2*S*,4*R*)-enantiomer (**132b**). On the other hand, optically active (+)-2α-hydroxy-1,4-cineole (**132b**) was also formed from (−)-terpinen-4-ol (**342**) by *G. cyanea* DSM (Abraham et al., 1986) and *A. niger* TBUYN-2 (Noma and Asakawa, 2007b) (Figure 19.145).

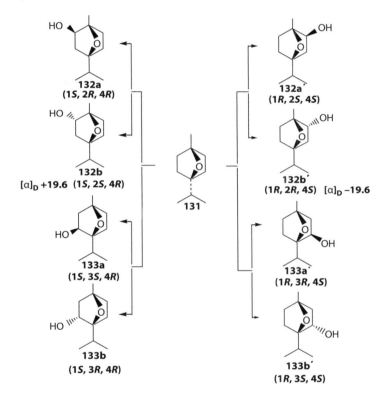

FIGURE 19.142 Biotransformation of 1,8-cineole (**122**) by *Eucalyptus perriniana* suspension cell. (Modified from Hamada, H. et al., Glycosylation of monoterpenes by plant suspension cells, *Proceedings of 46th TEAC*, 2002, pp. 321–322.)

FIGURE 19.143 Metabolic pathways of 1,4-cineole (**131**) by microorganisms.

19.4 METABOLIC PATHWAYS OF BICYCLIC MONOTERPENOIDS

19.4.1 Bicyclic Monoterpene

19.4.1.1 α-Pinene

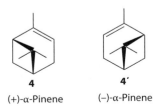

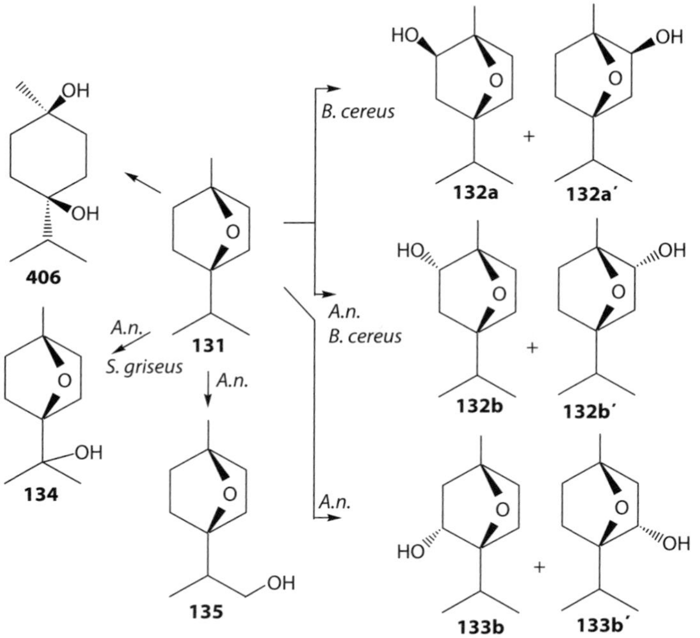

FIGURE 19.144 Metabolic pathways of 1,4-cineole (**131**) by *Aspergillus niger* TBUYN-2, *Bacillus cereus*, and *Streptomyces griseus*. (Modified from Liu, W. et al., *J. Org. Chem.*, 53, 5700, 1988; Miyazawa, M. et al., *Chem. Express*, 6, 771, 1991c; Miyazawa, M. et al., *Chem. Express*, 7, 305, 1992b; Miyazawa, M. et al., *Chem. Express*, 7, 125, 1992c; Miyazawa, M. et al., Biotransformation of 2-*endo*-hydroxy-1,4-cineole by plant pathogenic microorganism, *Glomerella cingulata*, Proceedings of 39th TEAC, 1995b, pp. 352–353.)

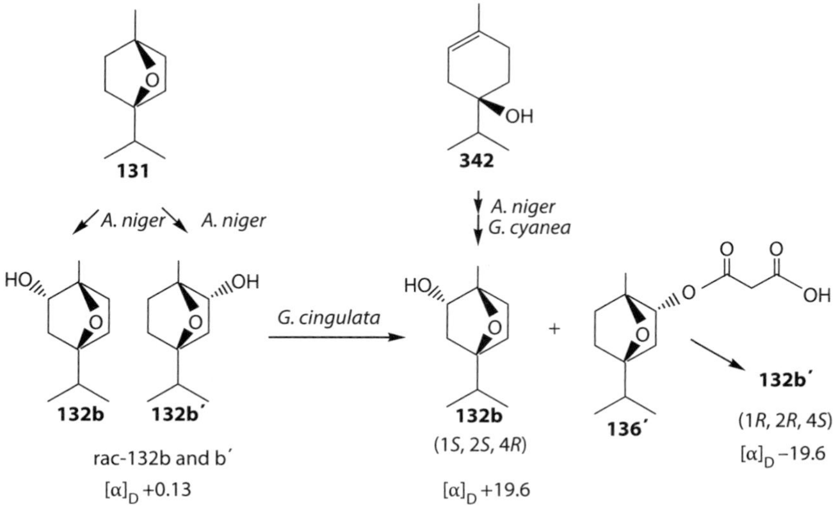

FIGURE 19.145 Formation of optically active 2α-hydroxycineole from 1,4-cineole (**131**) and terpinen-4-ol (**342**) by *Aspergillus niger* TBUYN-2, *Gibberella cyanea*, and *Glomerella cingulata*. (Modified from Abraham, W.-R. et al., *Appl. Microbiol. Biotechnol.*, 24, 24, 1986; Miyazawa, M. et al., *Chem. Express*, 6, 771, 1991c; Miyazawa, M. et al., Biotransformation of 2-*endo*-hydroxy-1,4-cineole by plant pathogenic microorganism, *Glomerella cingulata*, Proceedings of 39th TEAC, 1995b, pp. 352–353; Noma, Y. and Asakawa, Y., Microbial transformation of limonene and related compounds, Proceedings of 51st TEAC, 2007b, pp. 299–301.)

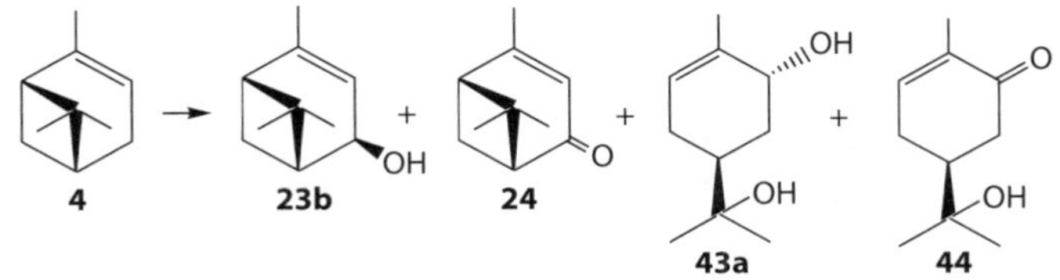

FIGURE 19.146 Biotransformation of (+)-α-pinene (**4**) by *Aspergillus niger* NCIM 612. (Modified from Bhattacharyya, P.K. et al., *Nature*, 187, 689, 1960; Prema, B.R. and Bhattachayya, P.K., *Appl. Microbiol.*, 10, 524, 1962.)

α-Pinene (**4** and **4′**) is the most abundant terpene in nature and obtained industrially by fractional distillation of turpentine (Krasnobajew, 1984). (+)-α-Pinene (**4**) occurs in oil of *Pinus palustris* Mill. at concentrations of up to 65% and in oil of *Pinus caribaea* at concentrations of 70% (Bauer et al., 1990). On the other hand, *P. caribaea* contains (−)-α-pinene (**4′**) at the concentration of 70%–80% (Bauer et al., 1990).

The biotransformation of (+)-α-pinene (**4**) was investigated by *A. niger* NCIM 612 (Bhattacharyya et al., 1960; Prema and Bhattacharyya, 1962). A 24 h shake culture of this strain metabolized 0.5% (+)-α-pinene (**4**) in 4–8 h. After the fermentation of the culture broth contained (+)-verbenone (**24**) (2%–3%), (+)-*cis*-verbenol (**23b**) (20%–25%), (+)-*trans*-sobrerol (**43a**) (2%–3%), and (+)-8-hydroxycarvotanacetone (**44**) (Bhattacharyya et al., 1960; Prema and Bhattacharyya, 1962) (Figure 19.146).

The degradation of (+)-α-pinene (**4**) by a soil *Pseudomonas* sp. (PL strain) was investigated by Hungund et al. (1970). A terminal oxidation pattern was proposed, leading to the formation of organic acids through ring cleavage. (+)-α-Pinene (**4**) was fermented in shake cultures by a soil *Pseudomonas* sp. (PL strain) that is able to grow on (+)-α-pinene (**4**) as the sole carbon source, and borneol (**36**), myrtenol (**5**), myrtenic acid (**84**), and α-phellandric acid (**65**) (Shukla and Bhattacharyya, 1968) (Figure 19.147) were obtained.

The degradation of (+)-α-pinene (**4**) by *P. fluorescens* NCIMB11671 was studied, and a pathway for the microbial breakdown of (+)-α-pinene (**4**) was proposed as shown in Figure 19.148 (Best et al., 1987; Best and Davis, 1988). The attack of oxygen is initiated by enzymatic oxygenation of the 1,2-double bond to form α-pinene epoxide (**38**), which then undergoes rapid rearrangement to produce a unsaturated aldehyde, occurring as two isomeric forms. The primary product of the reaction (Z)-2-methyl-5-isopropylhexa-2,5-dien-1-al (**39**, isonovalal) can undergo chemical isomerization to the *E*-form (novalal, **40**). Isonovalal (**39**), the native form of the aldehyde, possesses citrus, woody, and spicy notes, whereas novalal (**40**) has woody, aldehydic, and cyclone notes. The same biotransformation was also carried out by *Nocardia* sp. strain P18.3 (Griffiths et al., 1987a,b).

Pseudomonas PL strain and PIN 18 degraded α-pinene (**4**) by the pathway proposed in Figure 19.149 to give two hydrocarbon, limonene (**68**) and terpinolene (**346**), and neutral metabolite, borneol (**36**).

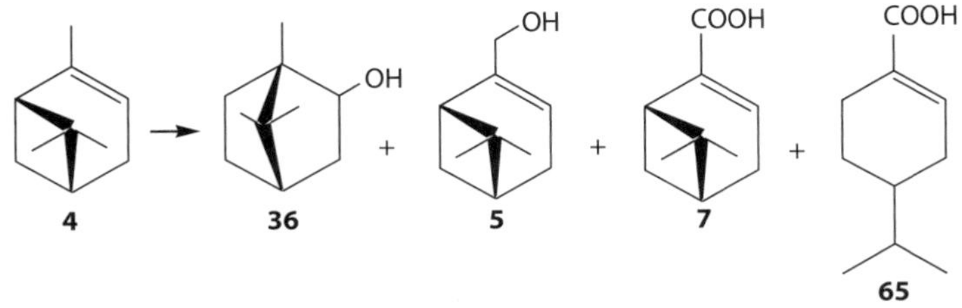

FIGURE 19.147 Biotransformation of (+)-α-pinene (**4**) by *Pseudomonas* sp. (PL strain). (Modified from Shukla, O.P., and Bhattacharyya, P.K., *Indian J. Biochem.*, 5, 92, 1968.)

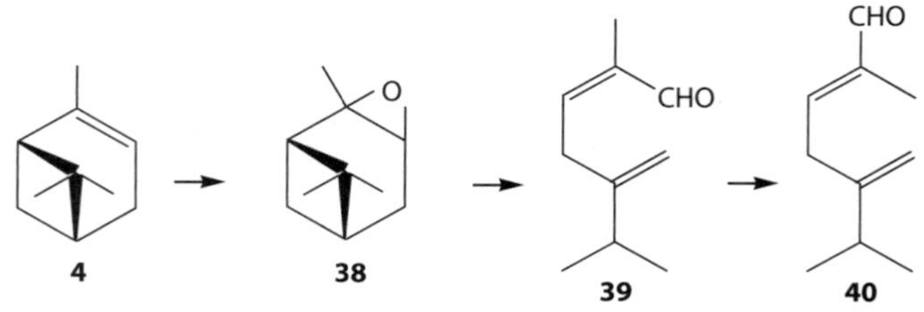

FIGURE 19.148 Biotransformation of (+)-α-pinene (**4**) by *Pseudomonas fluorescens* NCIMB 11671. (Modified from Best, D.J. et al., *Biocatal. Biotransform.*, 1, 147, 1987.)

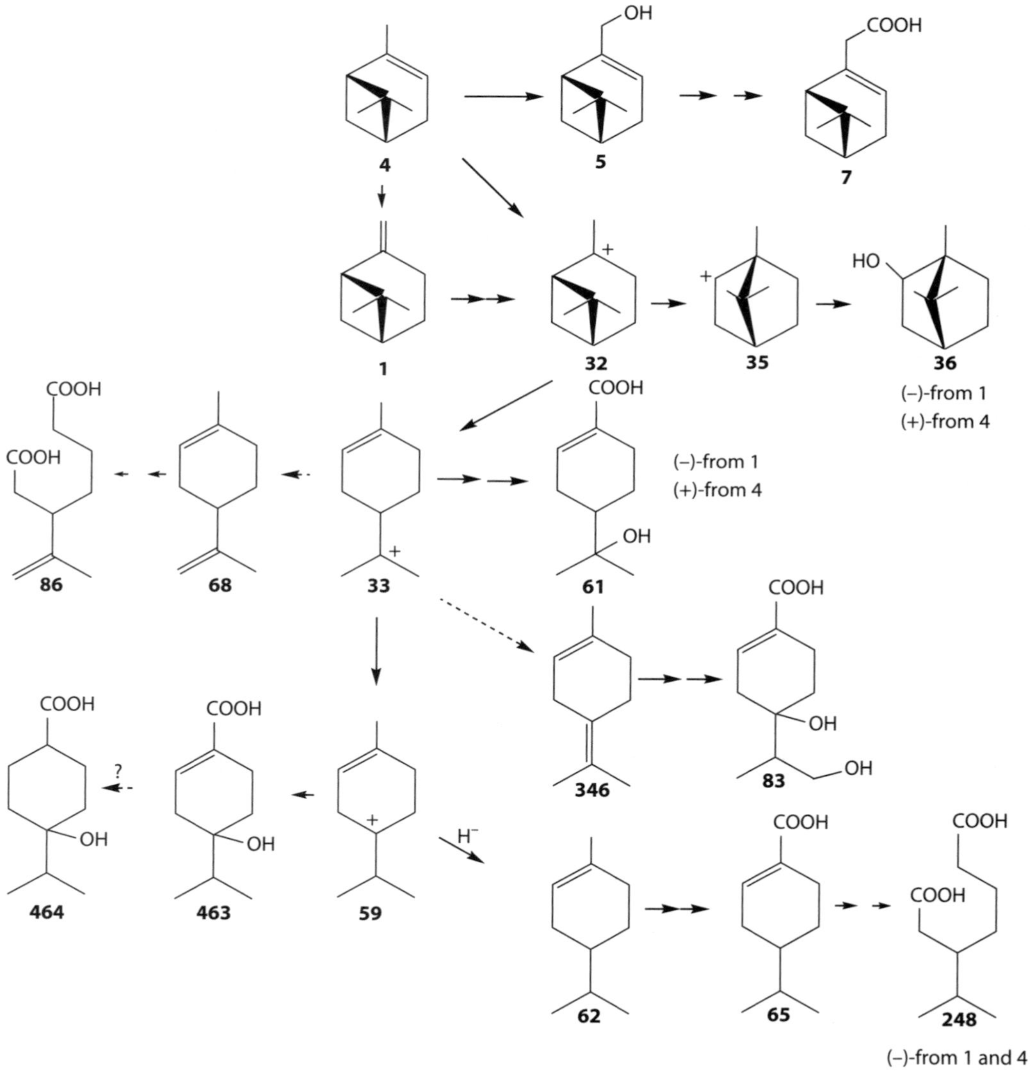

FIGURE 19.149 Metabolic pathways of degradation of α- and β-pinene by a soil Pseudomonad (PL strain) and *Pseudomonas* PIN 18. (Modified from Shukla, O.P. and Bhattacharyya, P.K., *Indian J. Biochem.*, 5, 92, 1968.)

A probable pathway has been proposed for the terminal oxidation of β-isopropylpimelic acid (**248**) in the PL strain and PIN 18 (Shukala and Bhattacharyya, 1968).

Pseudomonas PX 1 biotransformed (+)-α-pinene (**4**) to give (+)-*cis*-thujone (**29**) and (+)-*trans*-carveol (**81a**) as major compounds. Compounds **81a, 171, 173**, and **178** have been identified as fermentation products (Gibbon and Pirt, 1971; Gibbon et al., 1972) (Figure 19.150).

A. niger TBUYN-2 biotransformed (−)-α-pinene (**4′**) to give (−)-α-terpineol (**34′**) and (−)-*trans*-sobrerol (**43a′**) (Noma et al., 2001). The mosquitocidal (+)-(1*R*,2*S*,4*R*)-1-*p*-menthane-2,8-diol (**50a′**) was also obtained as a crystal in the biotransformation of (−)-α-pinene (**4′**) by *A. niger* TBUYN-2 (Noma et al., 2001; Noma, 2007) (Figure 19.151).

(1*R*)-(+)-α-Pinene (**4**) and its enantiomer (**4′**) were fed to *S. litura* to give the corresponding (+)- and (−)-verbenones (**24** and **24′**) and (+)- and (−)-myrtenols (**5** and **5′**) (Miyazawa et al., 1996c) (Figure 19.152).

(−)-α-Pinene (**4′**) was treated in human liver microsomes CYP2B6 to afford (−)-*trans*-verbenol (**23′**) and (−)-myrtenol (**5′**) (Sugie and Miyazawa, 2003) (Figure 19.153).

In rabbit, (+)-α-pinene (**4**) was metabolized to (−)-*trans*-verbenols (**23**) as the main metabolites together with myrtenol (**5**) and myrtenic acid (**7**). The purities of (−)-verbenol (**23**) from (−)- (**4′**), (+)- (**4**), and (+/−)-α-pinene (**4** and **4′**) were 99%, 67%, and 68%, respectively. This means that the biotransformation of (−)-**4′** in rabbit is remarkably efficient in the preparation of (−)-*trans*-verbenol (**23a**) (Ishida et al., 1981b) (Figure 19.154).

(−)-α-Pinene (**4′**) was biotransformed by the plant pathogenic fungus *B. cinerea* to afford 3α-hydroxy-(−)-β-pinene (**2a′**, 10%), 8-hydroxy-(−)-α-pinene (**434′**, 12%), 4β-hydroxy-(−)-pinene-6-one (**468′**, 16%), and (−)-verbenone (**24′**) (Farooq et al., 2002) (Figure 19.155).

19.4.1.2 β-Pinene

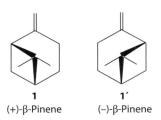

1
(+)-β-Pinene

1′
(−)-β-Pinene

(+)-β-Pinene (**1**) is found in many essential oils. Optically active and racemic β-pinenes are present in turpentine oils, although in smaller quantities than (+)-α-pinene (**4**) (Bauer et al., 1990).

Shukla et al. (1968) obtained a similarly complex mixture of transformation products from (−)-β-pinene (**1′**) through degradation by a *Pseudomonas* sp/(PL strain). On the other hand, Bhattacharyya and Ganapathy (1965) indicated that *A. niger* NCIM 612 acts differently and more specifically on the pinenes by preferably oxidizing (−)-β-pinene (**1′**) in the allylic position to form the interesting products pinocarveol (**2′**) and pinocarvone (**3′**), besides myrtenol (**5′**) (see Figure 19.156). Furthermore, the conversion of (−)-β-pinene (**1′**) by *Pseudomonas putida arvilla* (PL strain) gave borneol (**36′**) (Rama Devi and Bhattacharyya, 1978) (Figure 19.156).

P. pseudomallei isolated from local sewage sludge by the enrichment culture technique utilized caryophyllene as the sole carbon source (Dhavalikar et al., 1974). Fermentation of (−)-β-pinene (**1′**) by *P. pseudomallei* in a mineral salt medium (Seubert's medium) at 30°C with agitation and aeration for 4 days yielded camphor (**37′**), borneol (**36a′**), isoborneol (**36b′**), α-terpineol (**34′**), and β-isopropyl pimelic acid (**248′**) (see Figure 19.154). Using modified Czapek Dox medium and keeping the other conditions the same, the pattern of the metabolic products was dramatically changed. The metabolites were *trans*-pinocarveol (**2′**), myrtenol (**5′**), α-fenchol (**11′**), á-terpineol (**34′**), myrtenic acid (**7′**), and two unidentified products (see Figure 19.157).

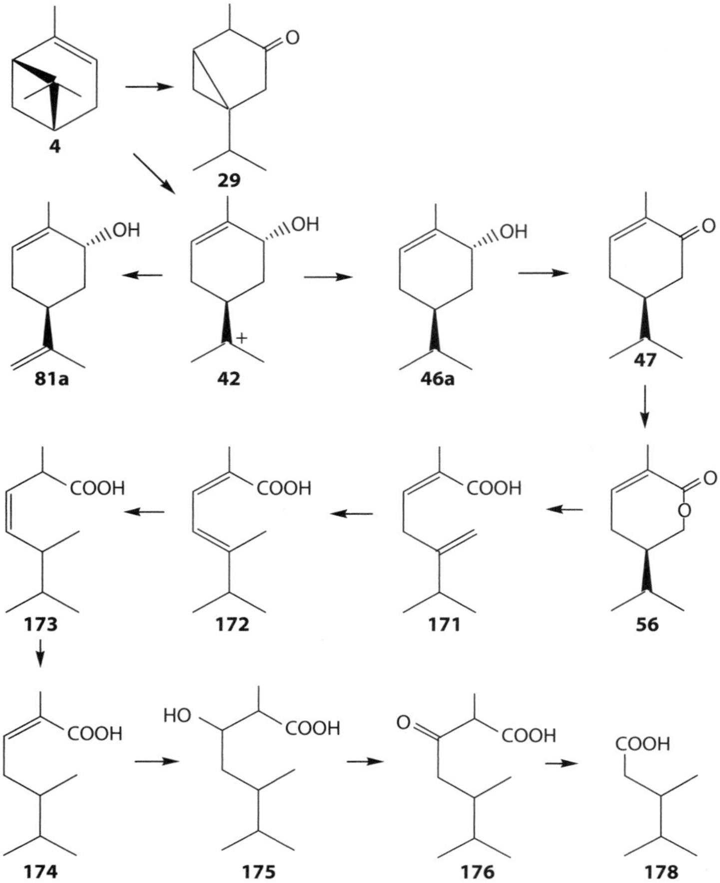

FIGURE 19.150 Proposed metabolic pathways for (+)-α-pinene (**4**) degradation by *Pseudomonas* PX 1. (Modified from Gibbon, G.H. and Pirt, S.J., *FEBS Lett.*, 18, 103, 1971; Gibbon, G.H. et al., Degradation of α-pinene by bacteria, *Proceedings of IV IFS: Fermentation Technology Today*, 1972, pp. 609–612.)

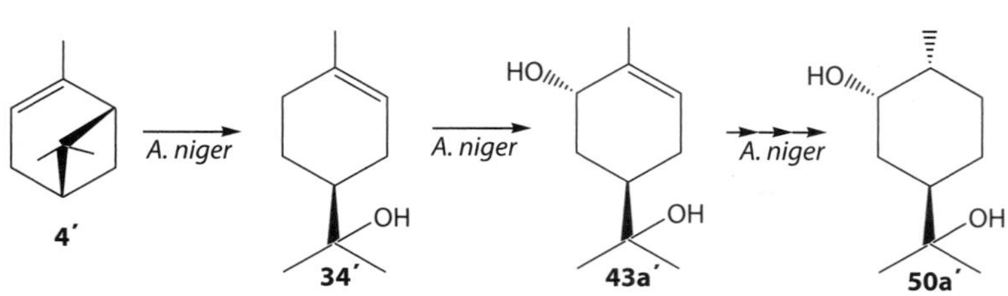

FIGURE 19.151 Biotransformation of (−)-α-pinene (**4**) by *Aspergillus niger* TBUYN-2. (Modified from Noma, Y. et al., Microbiological transformation of β-pinene, *Proceedings of 45th TEAC*, 2001, pp. 88–90.)

(−)-β-Pinene (**1′**) was converted by plant pathogenic fungi, *B. cinerea*, to give four new compounds such as (−)-pinane-2α,3α-diol (**408′**), (−)-6β-hydroxypinene (**409′**), (−)-4α,5-dihydroxypinene (**410′**), and (−)-4α-hydroxypinene-6-one (**411′**) (Figure 19.158).

This study progressed further biotransformation of (−)-pinane-2α,3α-diol (**408′**) and related compounds by microorganisms as shown in Figure 19.158.

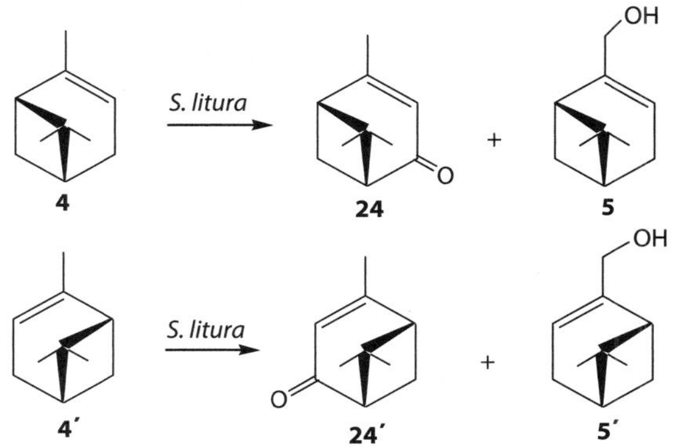

FIGURE 19.152 Biotransformation of (+)- (**4**) and (−)-α-pinene (**4′**) by *Spodoptera litura*. (Modified from Miyazawa, M. et al., Biotransformation of pinanes by common cutworm larvae, *Spodoptera litura* as a biocatalyst, *Proceedings of 40th TEAC*, 1996c, pp. 84–85.)

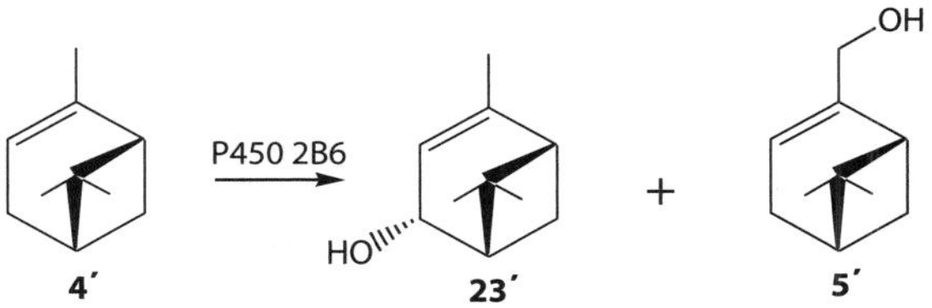

FIGURE 19.153 Biotransformation of (−)-α-pinene (**4′**) by human liver microsomes CYP2B6. (Modified from Sugie, A. and Miyazawa, M., Biotransformation of (−)-α-pinene by human liver microsomes, *Proceedings of 47th TEAC*, 2003, pp. 159–161.)

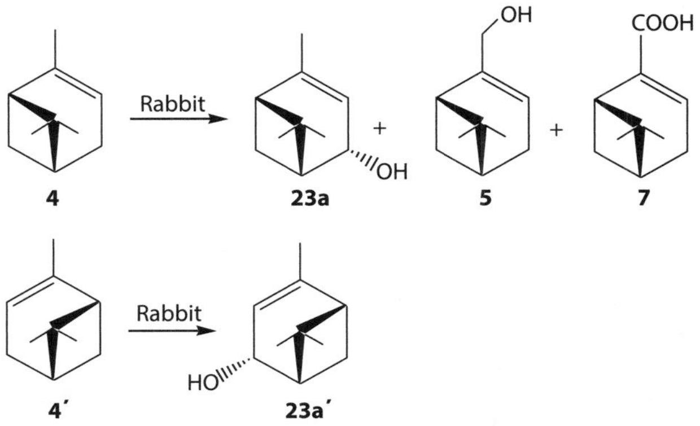

FIGURE 19.154 Biotransformation of α-pinene by rabbit. (Modified from Ishida, T. et al., *J. Pharm. Sci.*, 70, 406, 1981b.)

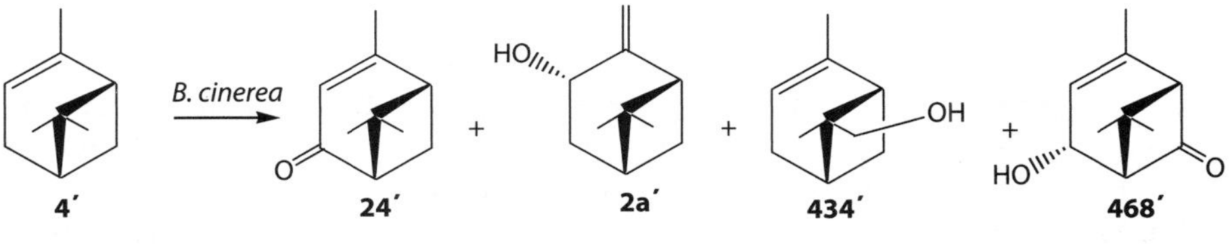

FIGURE 19.155 Microbial transformation of (−)-α-pinene (**4′**) by *Botrytis cinerea*. (Modified from Farooq, A. et al., *Z. Naturforsch.*, 57c, 686, 2002.)

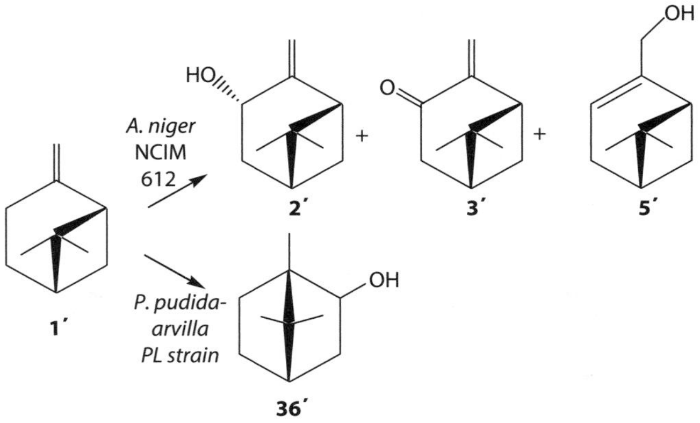

FIGURE 19.156 Biotransformation of (−)-β-pinene (**1′**) by *Aspergillus niger* NCIM 612 and *Pseudomonas putida arvilla* (PL strain). (Modified from Bhattacharyya, P.K. and Ganapathy, K., *Indian J. Biochem.*, 2, 137, 1965; Rama Devi, J. and Bhattacharyya, P.K., *J. Indian Chem. Soc.*, 55, 1131, 1978.)

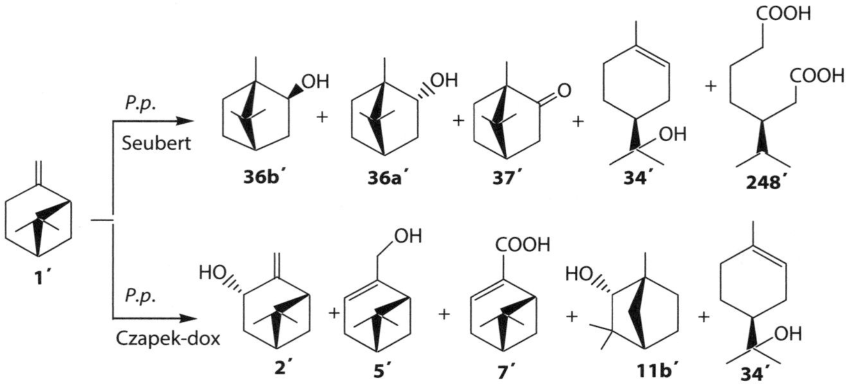

FIGURE 19.157 Biotransformation of (−)-β-pinene (**1′**) by *Pseudomonas pseudomallei*. (Modified from Dhavalikar, R.S. et al., *Dragoco Rep.*, 3, 47, 1974.)

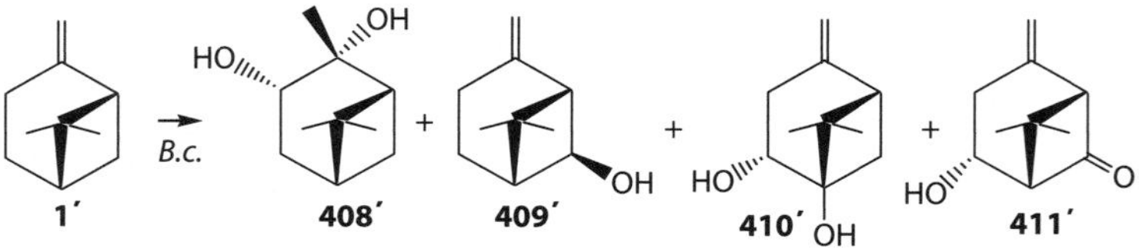

FIGURE 19.158 Biotransformation of (−)-β-pinene (**1′**) by *Botrytis cinerea*. (Modified from Farooq, A. et al., *Z. Naturforsch.*, 57c, 686, 2002.)

As shown in Figure 19.159, (+)- (**1**) and (−)-β-pinenes (**1′**) were biotransformed by *A. niger* TBUYN-2 to give (+)-α-terpineol (**34**) and (+)-oleuropeyl alcohol (**204**) and their antipodes (**34′** and **204′**), respectively. The hydroxylation process of α-terpineol (**34**) to oleuropeyl alcohol (**204**) was completely inhibited by 1-aminotriazole as a cytochrome P450 inhibitor.

(−)-β-Pinene (**1′**) was at first biotransformed by *A. niger* TBUYN-2 to give (+)-*trans*-pinocarveol (**2a′**) (**274**). (+)-*trans*-Pinocarveol (**2a′**) was further transformed by three pathways: First, (+)-*trans*-pinocarveol (**2a′**) was metabolized to (+)-pinocarvone (**3′**), (−)-3-isopinanone (**413′**),

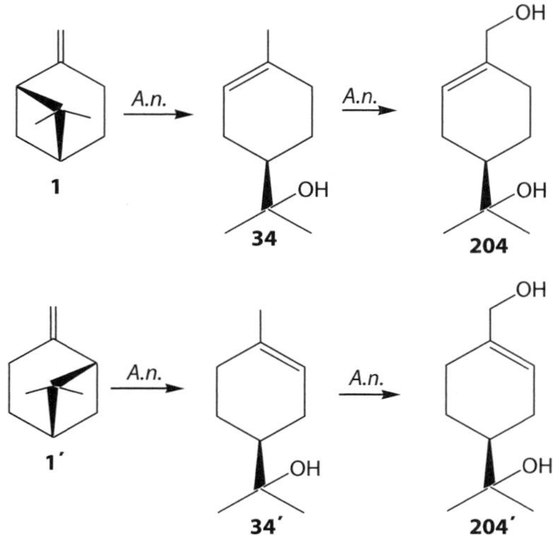

FIGURE 19.159 Biotransformation of (+)- (**1**) and (−)-β-pinene (**1′**) by *Aspergillus niger* TBUYN-2. (Modified from Noma, Y. et al., Microbiological transformation of β-pinene, *Proceedings of 45th TEAC*, 2001, pp. 88–90.)

(+)-2α-hydroxy-3-pinanone (**414′**), and (+)-2α,5-dihydroxy-3-pinanone (**415′**). Second, (+)-*trans*-pinocarveol (**2a′**) was metabolized to (+)-6β-hydroxyfenchol (**349ba′**), and third, (+)-*trans*-pinocarveol (**2a′**) was metabolized to (−)-6β,7-dihydroxyfenchol (**412ba′**) via epoxide and diol as intermediates (Noma and Asakawa, 2005a) (Figure 19.160).

(−)-β-Pinene (**1′**) was metabolized by *A. niger* TBUYN-2 with three pathways as shown in Figure 19.154 to give (−)-α-pinene (**4′**), (−)-α-terpineol (**34′**), and (+)-*trans*-pinocarveol (**2a′**). (−)-α-Pinene (**4′**) is further metabolized by three pathways. At first, (−)-α-pinene (**4′**) was metabolized via (−)-α-pinene epoxide (**38′**), *trans*-sobrerol (**43a′**), (−)-8-hydroxycarvotanacetone (**44′**), and (+)-8-hydroxycarvomenthone (**45a**) to (+)-*p*-menthane-2,8-diol (**50a′**), which was also metabolized in (−)-carvone (**93′**) metabolism. Second, (−)-α-pinene (**4′**) is metabolized to myrtenol (**83′**), which is metabolized by rearrangement reaction to give (−)-oleuropeyl alcohol (**204′**). (−)-α-Terpineol (**34′**), which is formed from β-pinene (**1′**), was also metabolized to (−)-oleuropeyl alcohol (**204′**), and (+)-*trans*-pinocarveol (**2a′**), formed

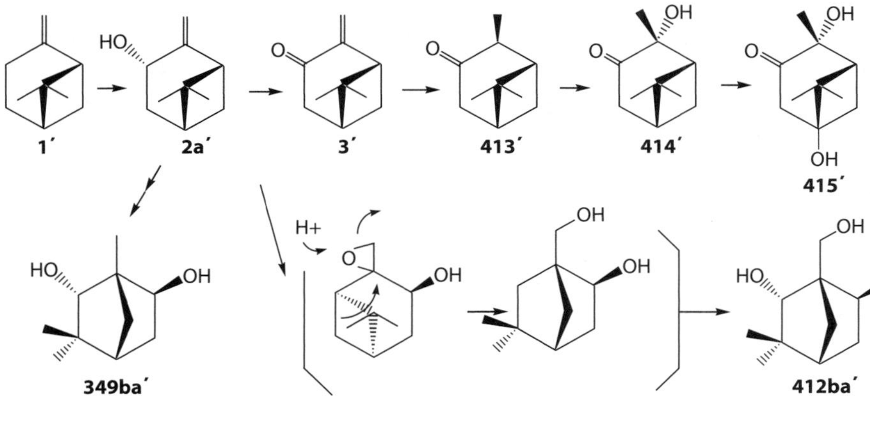

FIGURE 19.160 The metabolism of (−)-β-pinene (**1′**) and (+)-*trans*-pinocarveol (**2a′**) by *Aspergillus niger* TBUYN-2. (Modified from Noma, Y. and Asakawa, Y., New metabolic pathways of β-pinene and related compounds by *Aspergillus niger*, *Book of Abstracts of the 36th ISEO*, 2005a, p. 32.)

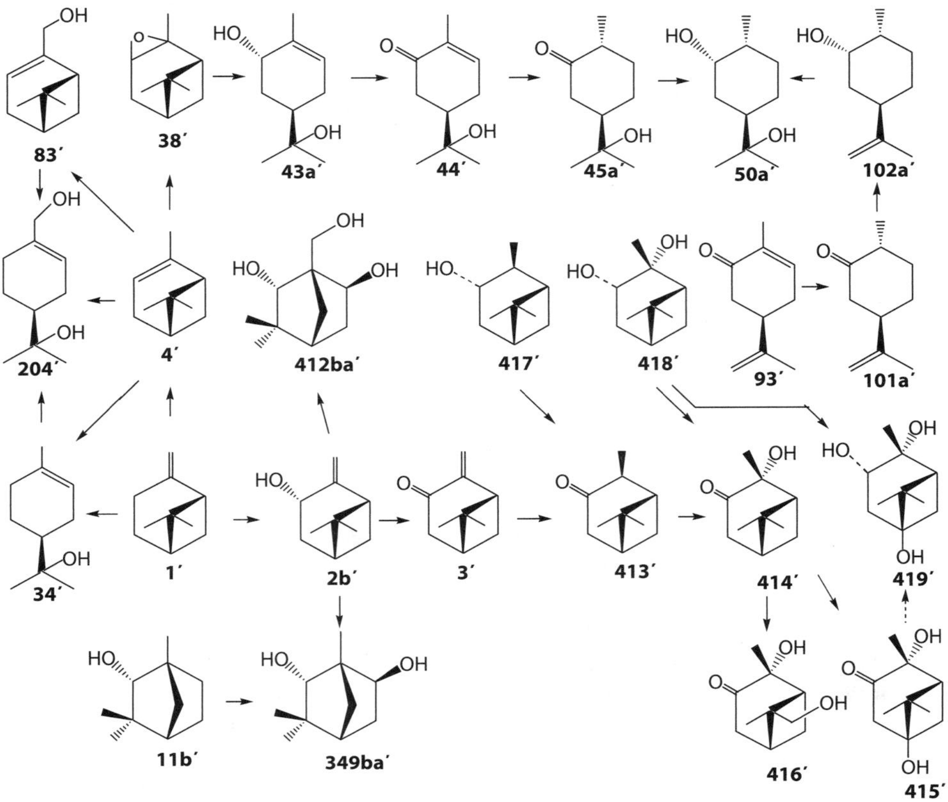

FIGURE 19.161 Biotransformation of (−)-β-pinene (**1′**), (−)-α-pinene (**4′**), and related compounds by *Aspergillus niger* TBUYN-2. (Modified from Noma, Y. and Asakawa, Y., New metabolic pathways of β-pinene and related compounds by *Aspergillus niger*, *Book of Abstracts of the 36th ISEO*, 2005a, p. 32.)

from (−)-β-pinene (**1′**), was metabolized to pinocarvone (**3′**), 3-pinanone (**413′**), 2α-hydroxy-3-pinanone (**414′**), 2α,5-dihydroxy-3-pinanone (**415′**), and 2α,9-dihydroxy-3-pinanone (**416′**). Furthermore, (+)-*trans*-pinocarveol (**2a′**) was metabolized by rearrangement reaction to give 6β-hydroxyfenchol (**349ba′**) and 6β,7-dihydroxyfenchol (**412ba′**) (Noma and Asakawa, 2005a) (Figure 19.161).

(−)-β-Pinene (**1′**) was metabolized by *A. niger* TBUYN-2 to give (+)-*trans*-pinocarveol (**2a′**), which was further metabolized to 6β-hydroxyfenchol (**349ba′**) and 6β, 7-dihydroxyfenchol (**412ba′**) by rearrangement reaction (Noma and Asakawa, 2005a) (Figure 19.162). 6β-Hydroxyfenchol (**349ba′**) was also obtained from (−)-fenchol (**11b′**). (−)-Fenchone was hydroxylated by the same fungus to give 6β- (**13a′**) and 6α-hydroxy-(−)-fenchone (**13b′**). There is a close relationship between the metabolism of (−)-β-pinene (**1′**) and those of (−)-fenchol (**11′**) and (−)-fenchone (**12′**).

(−)-β-Pinene (**1′**) and (−)-α-pinene (**4′**) were isomerized to each other. Both are metabolized via (−)-α-terpineol (**34'**) to (−)-oleuropeyl alcohol (**204′**) and (−)-oleuropeic acid (**61′**). (−)-Myrtenol (**5′**) formed from (−)-α-pinene (**1′**) was further metabolized via cation to (−)-oleuropeyl alcohol (**204′**) and (−)-oleuropeic acid (**61′**). (−)-α-Pinene (**4′**) is further metabolized by *A. niger* TBUYN-2 via (−)-α-pinene epoxide (**38′**) to *trans*-sobrerol (**43a′**), (−)-8-hydroxycarvotanacetone (**44′**), (+)-8-hydroxycarvomenthone (**45a**), and mosquitocidal (+)-*p*-menthane-2,8-diol (**50a′**) (Battacharyya et al., 1960; Noma et al., 2001, 2002, 2003) (Figure 19.163).

The major metabolites of (−)-β-pinene (**1′**) were *trans*-10-pinanol (myrtanol) (**8ba′**) (39%) and (−)-1-*p*-menthene-7,8-diol (oleuropeyl alcohol) (**204′**) (30%). In addition, (+)-*trans*-pinocarveol

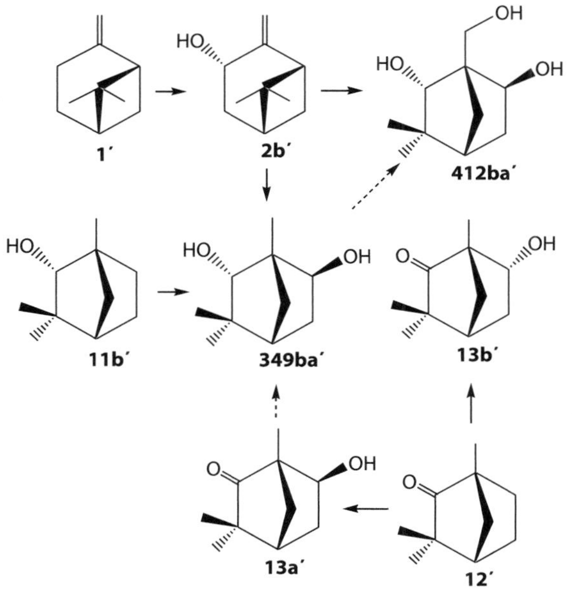

FIGURE 19.162 Relationship of the metabolism of (−)-β-pinene (**1′**), (+)-fenchol (**11′**), and (−)-fenchone (**12′**) by *Aspergillus niger* TBUYN-2. (Modified from Noma, Y. and Asakawa, Y., New metabolic pathways of β-pinene and related compounds by *Aspergillus niger*, *Book of Abstracts of the 36th ISEO*, 2005a, p. 32.)

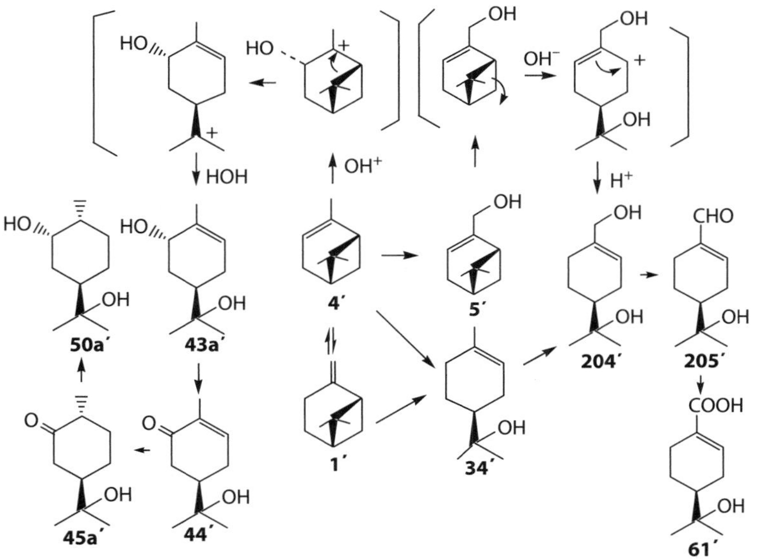

FIGURE 19.163 Metabolic pathways of (−)-β-pinene (**1′**) and related compounds by *Aspergillus niger* TBUYN-2. (Modified from Bhattacharyya, P.K. et al., *Nature*, 187, 689, 1960; Noma, Y. et al., Microbiological transformation of β-pinene, *Proceedings of 45th TEAC*, 2001, pp. 88–90; Noma, Y. et al., Stereoselective formation of (1R, 2S, 4R)-(+)-p-menthane-2,8-diol from α-pinene, *Book of Abstracts of the 33rd ISEO*, 2002, p. 142; Noma, Y. et al., Biotransformation of (+)- and (−)-pinane-2,3-diol and related compounds by *Aspergillus niger*, *Proceedings of 47th TEAC*, 2003, pp. 91–93.)

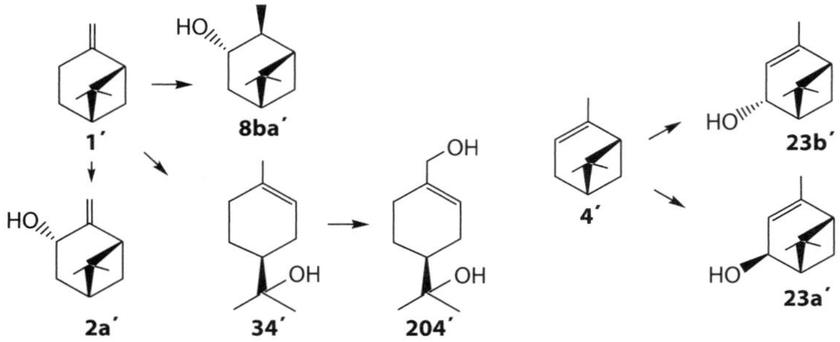

FIGURE 19.164 Metabolism of β-pinene (1) by bark beetle, *Dendroctonus frontalis*. (Modified from Ishida, T. et al., *J. Pharm. Sci.*, 70, 406, 1981b.)

(2a′) (11%) and (−)-α-terpineol (34′) (5%) and verbenol (23a and 23b) and pinocarveol (2a′) were oxidation products of α-(4) and β-pinene (1), respectively, in bark beetle, *Dendroctonus frontalis*. (−)-*cis*- (23b′) and (+)-*trans*-verbenols (23a′) have pheromonal activity in *Ips paraconfusus* and *Dendroctonus brevicomis*, respectively (Ishida et al., 1981b) (Figure 19.164).

19.4.1.3 (±)-Camphene

Racemate camphene (437 and 437′) is a bicyclic monoterpene hydrocarbon found in *Liquidambar* species, *Chrysanthemum*, *Zingiber officinale*, *Rosmarinus officinalis*, and among other plants. It was administered into rabbits. Six metabolites, camphene-2,10-glycols (438a, 438b), which were the major metabolites, together with 10-hydroxytricyclene (438c), 7-hydroxycamphene (438d), 6-exo-hydroxycamphene (438e), and 3-hydroxytricyclene (438f), were obtained (Ishida et al., 1979). On the basis of the production of the glycols (438a and 438b) in good yield, these alcohols might be formed through their epoxides as shown in Figure 19.165. The homoallyl camphene oxidation products (438c–f) apparently were formed through the nonclassical cation as the intermediate.

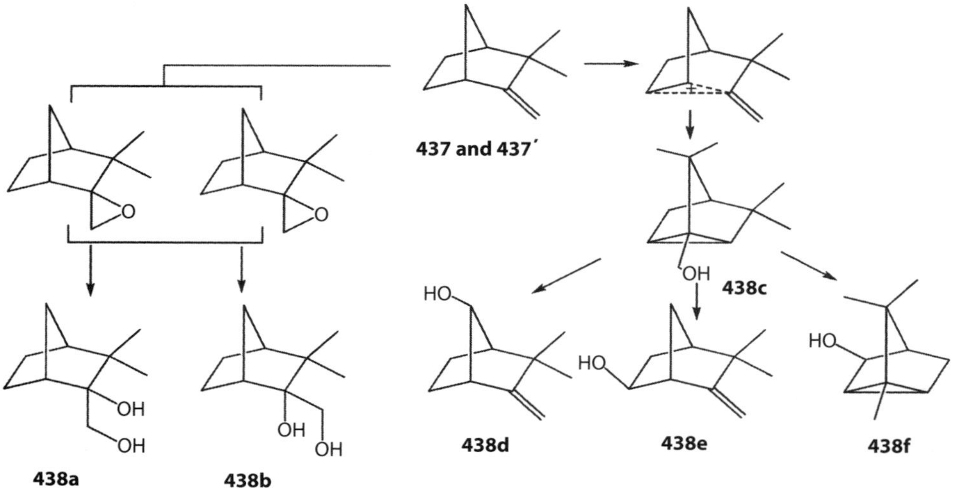

FIGURE 19.165 Biotransformation of (±)-camphene (437 and 437′) by rabbits. (Modified from Ishida, T. et al., *J. Pharm. Sci.*, 68, 928, 1979.)

19.4.1.4 3-Carene and Carane

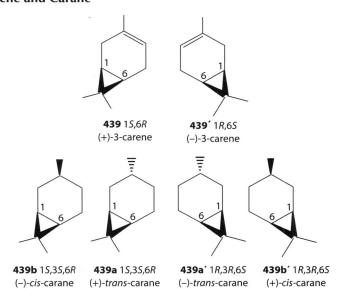

439 1S,6R
(+)-3-carene

439' 1R,6S
(−)-3-carene

439b 1S,3S,6R
(−)-*cis*-carane

439a 1S,3S,6R
(+)-*trans*-carane

439a' 1R,3R,6S
(−)-*trans*-carane

439b' 1R,3R,6S
(+)-*cis*-carane

(+)-3-Carene (**439**) was biotransformed by rabbits to give *m*-mentha-4,6-dien-8-ol (**440**) (71.6%) as the main metabolite together with its aromatized *m*-cymen-8-ol (**441**). The position of C-5 in the substrate is thought to be more easily hydroxylated than C-2 by enzymatic systems in the rabbit liver. In addition to the ring opening compound, 3-carene-9-ol (**442**), 3-carene-9-carboxylic acid (**443**), 3-carene-9,10-dicarboxylic acid (**445**), chamic acid, and 3-caren-10-ol-9-carboxylic acid (**444**) were formed. The formation of such compounds is explained by stereoselective hydroxylation and carboxylation of *gem*-dimethyl group (Ishida et al., 1981b) (Figure 19.166). In the case of (−)-*cis*-carane (**446**), two C-9 and C-10 methyl groups were oxidized to give dicarboxylic acid (**447**) (Ishida et al., 1981b) (Figure 19.166).

3-(+)-Carene (**439**) was converted by *A. niger* NC 1M612 to give either hydroxylated compounds of 3-carene-2-one or 3-carene-5-one, which was not fully identified (Noma et al., 2002) (Figure 19.167).

FIGURE 19.166 Metabolic pathways of (+)-3-carene (**439**) by rabbit (Modified from Ishida, T. et al., *J. Pharm. Sci.*, 70, 406, 1981b). 3-(+)-Carene (**439**) was converted by *Aspergillus niger* NC 1M612 to give either hydroxylated compounds of 3-carene-2-one or 3-carene-5-one, which was not fully identified (Figure 19.167). (Modified from Noma, Y. et al., Stereoselective formation of (1R, 2S, 4R)-(+)-*p*-menthane-2,8-diol from α-pinene, *Book of Abstracts of the 33rd ISEO*, 2002, p. 142.)

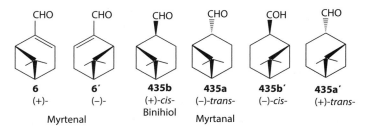

FIGURE 19.167 Metabolic pathways of (+)-3-Carene (**439**) by *Aspergillus niger* NC 1M612. (Modified from Noma, Y. et al., Stereoselective formation of (1*R*, 2*S*, 4*R*)-(+)-*p*-menthane-2,8-diol from α-pinene, *Book of Abstracts of the 33rd ISEO*, 2002, p. 142.)

19.4.2 Bicyclic Monoterpene Aldehyde

19.4.2.1 Myrtenal and Myrtanal

6	**6′**	**435b**	**435a**	**435b′**	**435a′**
(+)-	(−)-	(+)-*cis*-Binihiol	(−)-*trans*- Myrtanal	(−)-*cis*-	(+)-*trans*-

Myrtenal

E. gracilis Z. biotransformed (−)-myrtenal (**6′**) to give (−)-myrtenol (**5′**) as the major product and (−)-myrtenoic acid (**7′**) as the minor product. However, further hydrogenation of (−)-myrtenol (**5′**) to *trans*- and *cis*-myrtanol (**8a** and **8b**) did not occur even at a concentration less than ca. 50 mg/L. (*S*)-*trans*- and (*R*)-*cis*-myrtanal (**435a′** and **435b′**) were also transformed to *trans*- and *cis*-myrtanol (**8a′** and **8b′**) as the major products and (*S*)-*trans*- and (*R*)-*cis*-myrtanic acid (**436a′** and **436b′**) as the minor products, respectively (Noma et al., 1991a) (Figure 19.168).

In the case of *A. niger* TBUYN-2, *Aspergillus sojae*, and *Aspergillus usami*, (−)-myrtenol (**5′**) was further metabolized to 7-hydroxyverbenone (**25′**) as a minor product together with (−)-oleuropeyl alcohol (**204′**) as a major product (**279, 280**). (−)-Oleuropeyl alcohol (**204′**) is also formed from (−)-α-terpineol (**34**) by *A. niger* TBUYN-2 (Noma et al., 2001) (Figure 19.168).

Rabbits metabolized myrtenal (**6′**) to myrtenic acid (**7′**) as the major metabolite and myrtanol (**8a′** or **8b′**) as the minor metabolite (Ishida et al., 1981b) (Figure 19.168).

19.4.3 Bicyclic Monoterpene Alcohol

19.4.3.1 Myrtenol

5   **5′**

(−)-Myrtenol (**5′**) was biotransformed mainly to (−)-oleuropeyl alcohol (**204′**), which was formed from (−)-α-terpineol (**34′**) as a major product by *A. niger* TBUYN-2. In the case of *A. sojae* IFO 4389 and *A. usami* IFO 4338, (−)-myrtenol (**5′**) was metabolized to 7-hydroxyverbenone (**25′**) as a minor product together with (−)-oleuropeyl alcohol (**204′**) as a major product (Noma and Asakawa, 2005b) (Figure 19.169).

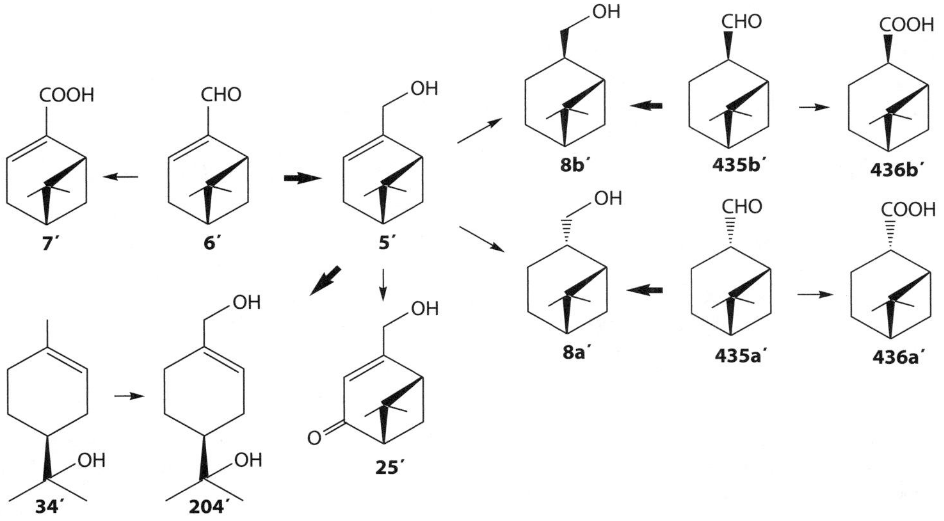

FIGURE 19.168 Biotransformation of (−)-myrtenal (**6′**) and (+)-*trans*- (**435a′**) and (−)-*cis*-myrtanal (**435b′**) by microorganisms. (Modified from Noma, Y. et al., *Phytochemistry*, 30, 1147, 1991a; Noma, Y. and Asakawa, Y., Microbial transformation of (−)-myrtenol and (−)-nopol, *Proceedings of 49th TEAC*, 2005b, pp. 78–80; Noma, Y. and Asakawa, Y., Biotransformation of β-pinene, myrtenol, nopol and nopol benzyl ether by *Aspergillus niger* TBUYN-2, *Book of Abstracts of the 37th ISEO*, 2006b, p. 144.)

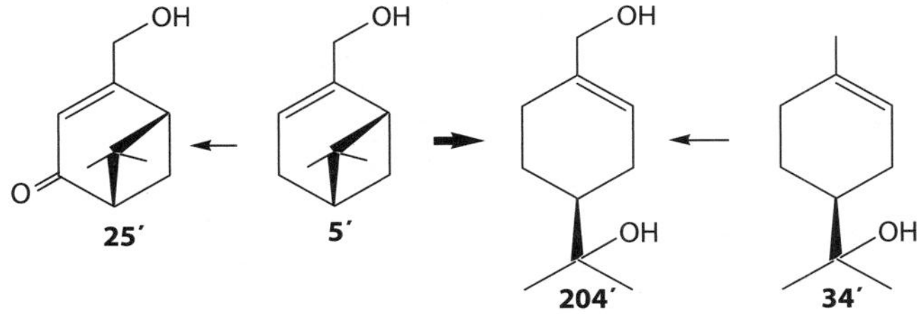

FIGURE 19.169 Biotransformation of (−)-myrtenol (**5′**) and (−)-α-terpineol (**34′**) by *Aspergillus niger* TBUYN-2. (Modified from Noma, Y. and Asakawa, Y., Microbial transformation of (−)-myrtenol and (−)-nopol, *Proceedings of 49th TEAC*, 2005b, pp. 78–80.)

19.4.3.2 Myrtanol

S. litura converted (−)-*trans*-myrtanol (**8a**) and its enantiomer (**8a′**) to give the corresponding myrtanic acid (**436** and **436′**) (Miyazawa et al., 1997b) (Figure 19.170).

19.4.3.3 Pinocarveol

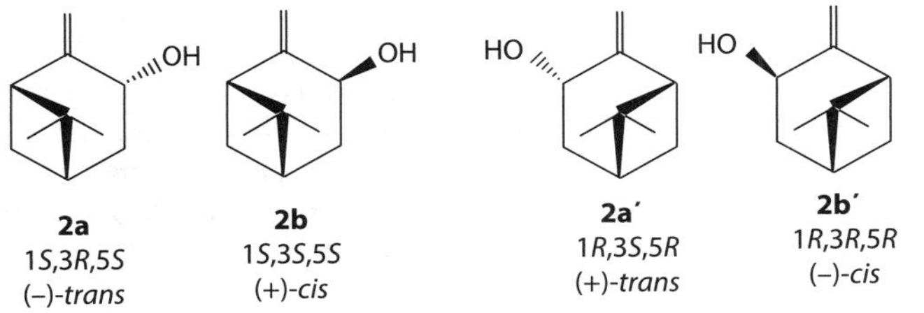

2a
1S,3R,5S
(−)-*trans*

2b
1S,3S,5S
(+)-*cis*

2a′
1R,3S,5R
(+)-*trans*

2b′
1R,3R,5R
(−)-*cis*

Pinocarveol

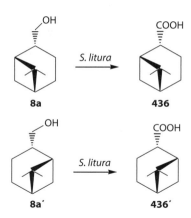

FIGURE 19.170 Biotransformation of (−)-*trans*-myrtanol (**8a**) and its enantiomer (**8a′**) by *Spodoptera litura*. (Modified from Miyazawa, M. et al., Biotransformation of (+)-*trans* myrtanol and (−)-*trans*-myrtanol by common cutworm Larvae, *Spodoptera litura* as a biocatalyst, *Proceedings of 41st TEAC*, 1997b, pp. 389–390.)

(+)-*trans*-Pinocarveol (**2a′**) was biotransformed by *A. niger* TBUYN-2 to the following two pathways. That is, (+)-*trans*-pinocarveol (**2a′**) was metabolized via (+)-pinocarvone (**3′**), (−)-3-isopinanone (**413′**), and (+)-2α-hydroxy-3-pinanone (**414′**) to (+)-2α,5-dihydroxy-3-pinanone (**415′**) (pathway 1). Furthermore, (+)-*trans*-pinocarveol (**2a′**) was metabolized to epoxide followed by rearrangement reaction to give 6β-hydroxyfenchol (**349ba′**) and 6β,7-dihydroxyfenchol (**412ba′**) (Noma and Asakawa, 2005a) (Figure 19.171). *S. litura* converted (+)-*trans*-pinocarveol (**2a′**) to (+)-pinocarvone (**3′**) as a major product (Miyazawa et al., 1995b) (Figure 19.171).

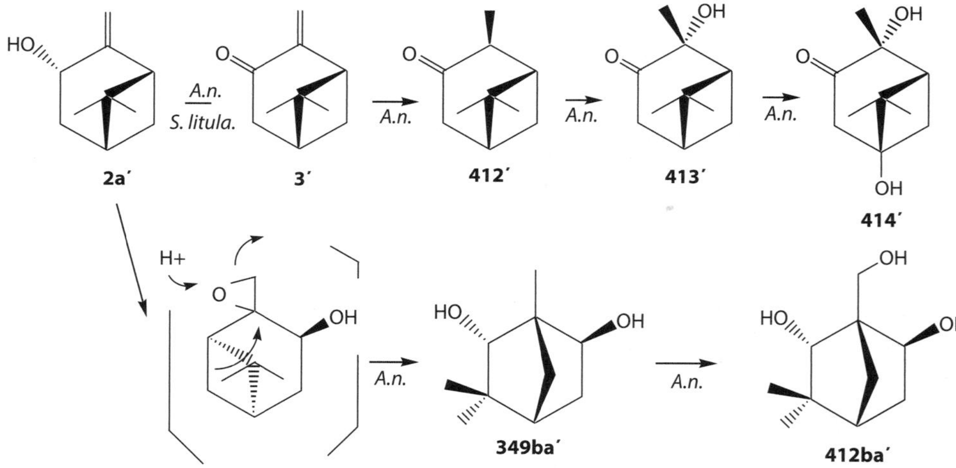

FIGURE 19.171 Biotransformation of (+)-*trans*-pinocarveol (**2a′**) by *Aspergillus niger* TBUYN-2 and *Spodoptera litura*. (Modified from Miyazawa, M. et al., Biotransformation of *trans*-pinocarveol by plant pathogenic microorganism, *Glomerella cingulata*, and by the larvae of common cutworm, *Spodoptera litura* Fabricius, *Proceedings of 39th TEAC*, 1995c, pp. 360–361; Noma, Y. and Asakawa, Y., New metabolic pathways of β-pinene and related compounds by *Aspergillus niger*, *Book of Abstracts of the 36th ISEO*, 2005a, p. 32.)

19.4.3.4 Pinane-2,3-Diol

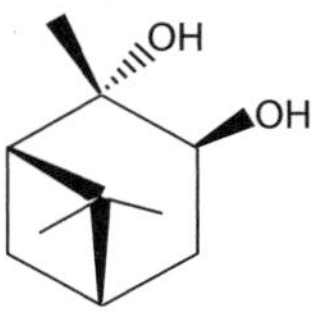

418aa
()-Pinane-2,3-diol
1S,2S,3S,5S

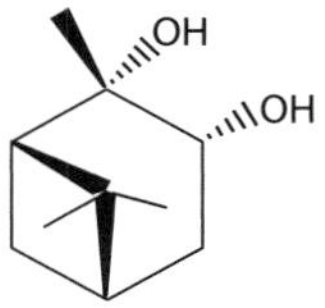

418ab
(+)-Pinane-2,3-diol
1S,2S,3R,5S

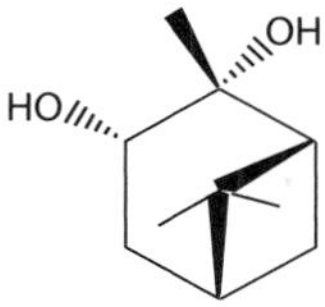

418ab′
(−)-Pinane-2,3-diol
1R,2R,3S,5R

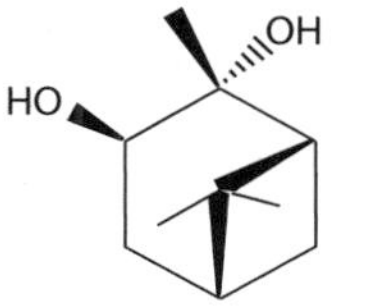

418aa′
()-Pinane-2,3-diol
1R,2R,3R,5R

This results led us to study the biotransformation of (−)-pinane-2,3-diol (**418ab′**) and (+)-pinane-2,3-diol (**418ab**) by *A. niger* TBUYN-2. (−)-Pinane-2,3-diol (**418ab′**) was easily biotransformed to give (−)-pinane-2,3,5-triol (**419ab′**) and (+)-2,5-dihydroxy-3-pinanone (**415a′**) as the major products and (+)-2-hydroxy-3-pinanone (**414a′**) as the minor product.

On the other hand, (+)-pinane-2,3-diol (**418ab**) was also biotransformed easily to give (+)-pinane-2,3,5-triol (**419ab**) and (−)-2,5-dihydroxy-3-pinanone (**415a**) as the major products and (−)-2-hydroxy-3-pinanone (**414a**) as the minor product (Noma et al., 2003) (Figure 19.172). *G. cingulata* transformed

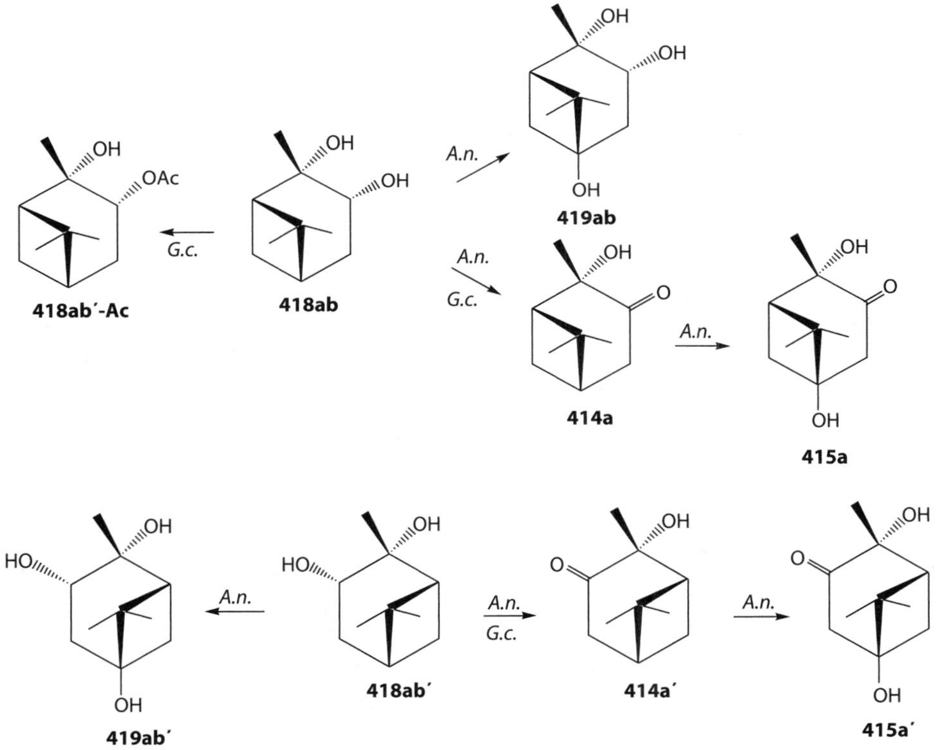

FIGURE 19.172 Biotransformation of (+)-pinane-2,3-diol (**418ab′**) and (−)-pinane-2,3-diol (**418ab′**) by *Aspergillus niger* TBUYN-2(**276**)] and *Glomerella cingulata*. (Modified from Noma, Y. et al., Biotransformation of (+)- and (−)-pinane-2,3-diol and related compounds by *Aspergillus niger*, *Proceedings of 47th TEAC*, 2003, pp. 91–93; Kamino, F. et al., Biotransformation of (1S,2S,3R,5S)-(+)-pinane-2,3-diol using plant pathogenic fungus, *Glomerella cingulata* as a biocatalyst, *Proceedings of 48th TEAC*, 2004, pp. 383–384; Kamino, F. and Miyazawa, M., Biotransformation of (+)-and (−)-pinane-2,3-diol using plant pathogenic fungus, *Glomerella cingulata* as a biocatalyst, *Proceedings of 49th TEAC*, 2005, pp. 395–396.)

(−)-pinane-2,3-diol (**418ab′**) to a small amount of (+)-2α-hydroxy-3-pinanone (**414ab′**, 5%) (Kamino and Miyazawa, 2005), whereas (+)-pinane-2,3-diol (**418ab**) was transformed to a small amount of (−)-2α-hydroxy-3-pinanone (**414ab**, 10%) and (−)-3-acetoxy-2α-pinanol (**433ab-Ac**, 30%) (Kamino et al., 2004) (Figure 19.172).

19.4.3.5 Isopinocampheol (3-Pinanol)

420 ba	**420bb**	**420aa**	**420ab**
(−)-isopino	(+)-neoiso	(−)-neo	(+)-pinocampherol
1R,2R,3R,5S	1R,2R,3S,5S	1R,2S,3R,5S	1R,2S,3S,5S

| **420ba′** (+)- | **420bb′** (−)- | **420aa′** (+)-neo | **420ab′** |
| 1S,2S,3S,5R | 1S,2S,3R,5R | 1S,2R,3S,5R | 1S,2R,3R,5R |

19.4.3.5.1 Chemical Structure of (−)-Isopinocampheol (420ba) and (+)-Isopinocampheol (420ba′)

Biotransformation of isopinocampheol (3-pinanol) with 100 bacterial and fungal strains yielded 1-, 2-, 4-, 5-, 7-, 8-, and 9-hydroxyisopinocampheol besides three rearranged monoterpenes, one of them bearing the novel isocarene skeleton. A pronounced enantioselectivity between (−)- (**420ba**) and (+)-isopinocampheol (**420ba′**) was observed. The phylogenetic position of the individual strains could be seen in their ability to form the products from (+)-isopinocampheol (**420ba′**). The formation of 1,3-dihydroxypinane (**421ba′**) is a domain of bacteria, while 3,5- (**415ba′**) or 3,6-dihydroxypinane (**428baa′**) was mainly formed by fungi, especially those of the phylum *Zygomycotina*. The activity of *Basidiomycotina* toward oxidation of isopinocampheol was rather low. Such informations can be used in a more effective selection of strains for screening (Wolf-Rainer, 1994) (Figure 19.173).

(+)-Isopinocampheol (**420ba′**) was metabolized to 4β-hydroxy-(+)-isopinocampheol (**424′**), 2β-hydroxy-(+)-isopinocampheol acetate (**425ba′-Ac**), and 2α-methyl,3-(2-methyl-2-hydroxypropyl)-cyclopenta-1β-ol (**432′**) (Wolf-Rainer, 1994) (Figure 19.174).

(−)-Isopinocampheol (**420ba**) was converted by *S. litura* to give (1R,2S,3R,5S)-pinane-2,3-diol (**418ba**) and (−)-pinane-3,9-diol (**423ba**), whereas (+)-isopinocampheol (**420ba′**) was converted to (+)-pinane-3,9-diol (**423ba′**) (Miyazawa et al., 1997c) (Figure 19.175).

(−)-Isopinocampheol (**420ba**) was biotransformed by *A. niger* TBUYN-2 to give (+)-(1S,2S,3S,5R)-pinane-3,5-diol (**422ba**, 6.6%), (−)-(1R,2R,3R,5S)-pinane-1,3-diol (**421ba**, 11.8%), and pinane-2,3-diol (**418ba**, 6.6%), whereas (+)-isopinocampheol (**420ba′**) was biotransformed by *A. niger* TBUYN-2 to give (+)-(1S,2S,3S,5R)-pinane-3,5-diol (**422ba′**, 6.3%) and

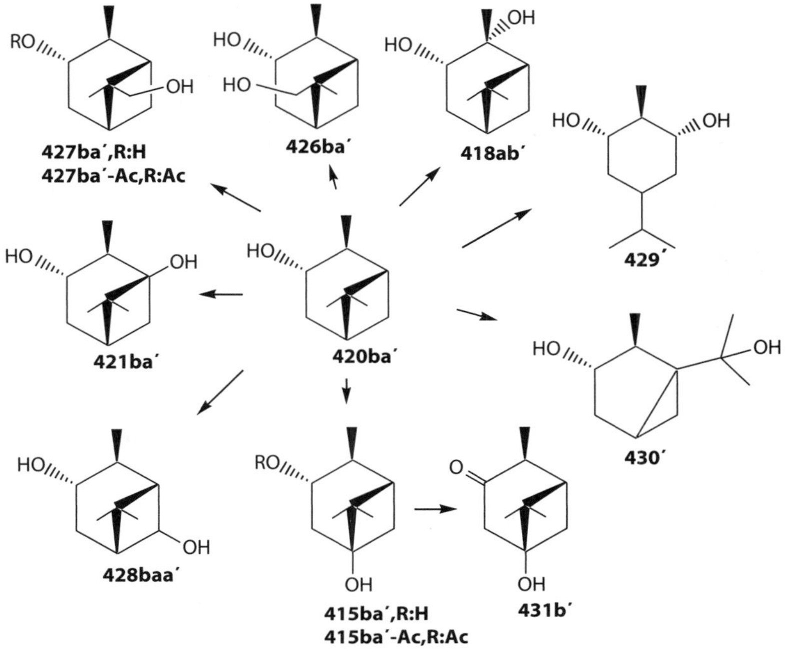

FIGURE 19.173 Metabolic pathways of (+)-isopinocampheol (**420ba′**) by microorganisms. (Modified from Wolf-Rainer, A., *Z. Naturforsch.*, 49c, 553, 1994.)

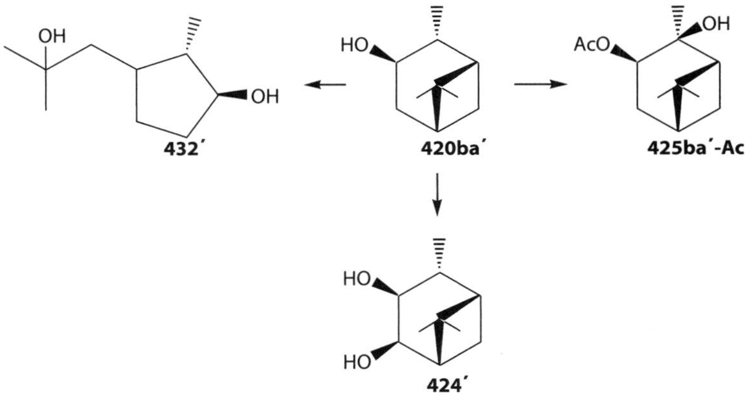

FIGURE 19.174 Metabolic pathways of (+)-isopinocampheol (**420ba′**) by microorganisms. (Modified from Wolf-Rainer, A., *Z. Naturforsch.*, 49c, 553, 1994.)

(−)-(1R,2R,3R,5S)-pinane-1,3,-diol (**421ba′**, 8.6%) (Noma et al., 2009) (Figure 19.176). On the other hand, *G. cingulata* converted (−)- (**420ba**) and (+)-isopinocampheol (**420ba′**) mainly to (1R,2R,3S,4S,5R)-3,4-pinanediol (**484ba**) and (1S,2S,3S,5R,6R)-3,6-pinanediol (**485ba′**), respectively, together with (**418ba**), (**422ba**), (**422ba′**), and (**486ba′**) as minor products (Miyazawa et al., 1997c) (Figure 19.176). Some similarities exist between the main metabolites with *G. cingulata* and *R. solani* (Miyazawa et al., 1997c) (Figure 19.176).

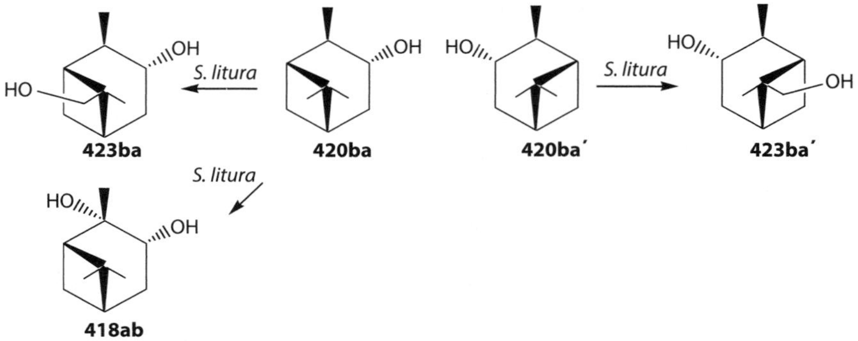

FIGURE 19.175 Biotransformation of (−)- (**420ba**) and (+)-isopinocampheol (**420ba′**) by *Spodoptera litura*. (Modified from Miyazawa, M. et al., *Phytochemistry*, 45, 945, 1997c.)

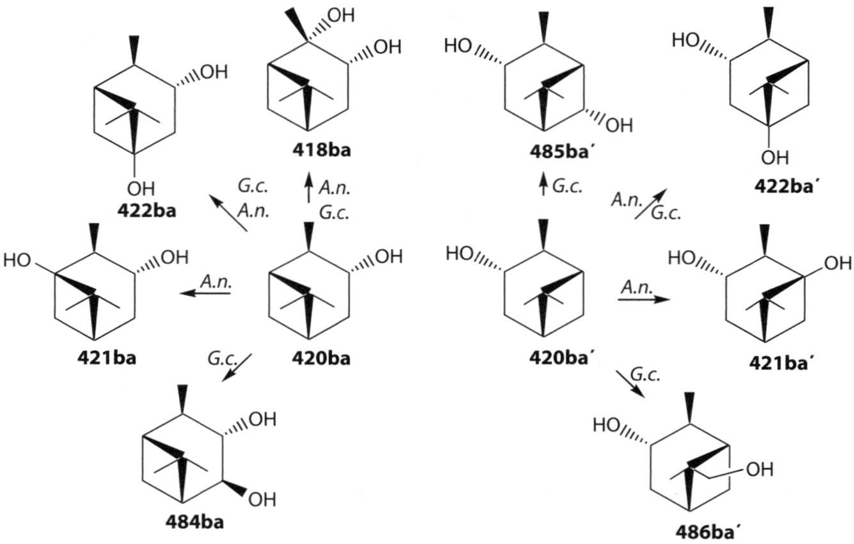

FIGURE 19.176 Biotransformation of (−)- (**420ba**) and (+)-isopinocampheol (**420ba′**) by *Aspergillus niger* TBUYN-2 and *Glomerella cingulata*. (Modified from Miyazawa, M. et al., *Phytochemistry*, 45, 945, 1997c; Noma, Y. et al., Unpublished data, 2009.)

19.4.3.6 Borneol and Isoborneol

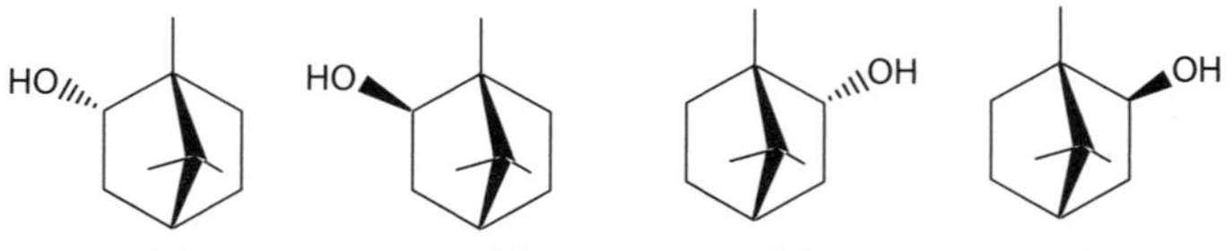

36a
(1R,2S)-(+)-
borneol
(**36a**)

36b′
(1R,2R)-(−)-
isoborneol
(**36b′**)

36b′
(1S,2R)-(−)-
borneol
(**36a′**)

36b
(1S,2S)-(+)-
isoborneol
(**36b**)

(−)-Borneol (**36a′**) was biotransformed by *P. pseudomallei* strain H to give (−)-camphor (**37′**), 6-hydroxycamphor (**228′**), and 2,6-diketocamphor (**229′**) (Hayashi et al., 1969) (Figure 19.177).

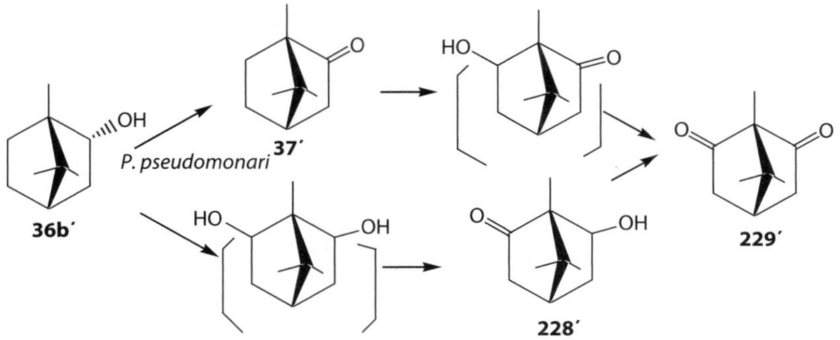

FIGURE 19.177 Biotransformation of (−)-borneol (**36a′**) by *Pseudomonas pseudomallei* strain. (Modified from Hayashi, T. et al., *J. Agric. Chem. Soc. Jpn.*, 43, 583, 1969.)

E. gracilis Z. showed enantio- and diastereoselectivity in the biotransformation of (+)- (**36a**), (−)- (**36a′**), and (±)-racemic borneols (equal mixture of **36a** and **36a′**) and (+)- (**36b**), (−)- (**36b′**), and (±)-isoborneols (equal mixture of **36b** and **36b′**). The enantio- and diastereoselective dehydrogenation for (−)-borneol (**36a′**) was carried out to give (−)-camphor (**37′**) at *ca.* 50% yield (Noma et al., 1992b; Noma and Asakawa, 1998). The conversion ratio was always *ca.* 50% even at different kinds of concentration of (−)-borneol (**36a′**). When (−)-camphor (**37′**) was used as a substrate, it was also converted to (−)-borneol (**36a′**) in 22% yield for 14 days. Furthermore, (+)-camphor (**37**) was also reduced to (+)-borneol (**36a**) in 4% and 18% yield for 7 and 14 days, respectively (Noma et al., 1992b; Noma and Asakawa, 1998) (Figure 19.178).

(+)- (**36a**) and (−)-Borneols (**36a′**) were biotransformed by *S. litura* to (+)- (**370a**) and (−)-bornane-2,8-diols (**370a′**), respectively (Miyamoto and Miyazawa, 2001) (Figure 19.179).

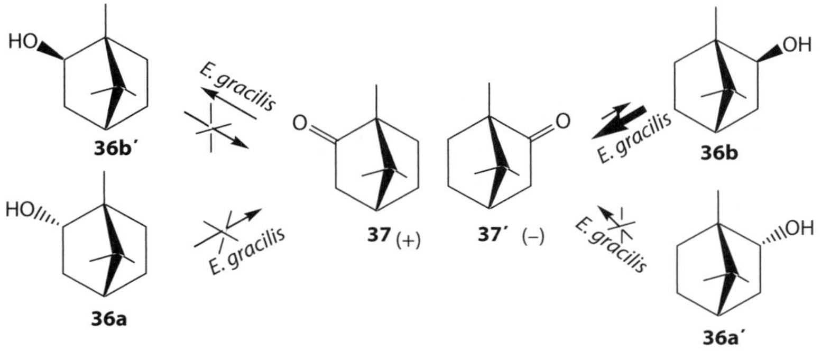

FIGURE 19.178 Enantio- and diastereoselectivity in the biotransformation of (+)- (**36a**) and (−)-borneols (**36a′**) by *Euglena gracilis* Z. (Modified from Noma, Y. et al., Biotransformation of terpenoids and related compounds, *Proceedings of 36th TEAC*, 1992d, pp. 199–201; Noma, Y. and Asakawa, Y., *Euglena gracilis* Z: Biotransformation of terpenoids and related compounds, in Bajaj, Y.P.S. (ed.), *Biotechnology in Agriculture and Forestry*, Vol. 41, Medicinal and Aromatic Plants X, Springer, Berlin, Germany, 1998, pp. 194–237.)

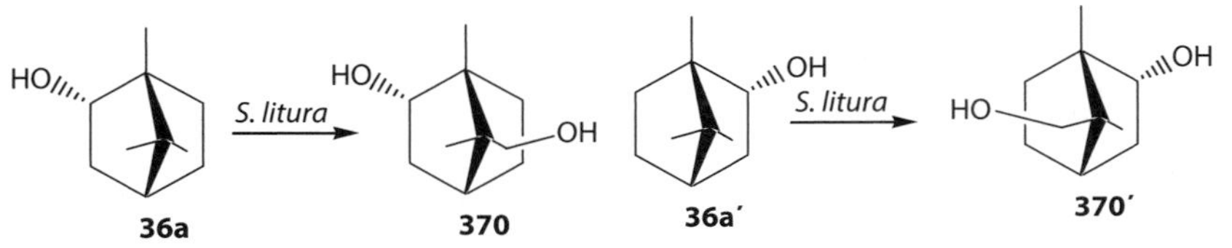

FIGURE 19.179 Biotransformation of (+)- (**36a**) and (−)-borneols (**36a′**) by *Spodoptera litura*. (Modified from Miyamoto, Y. and Miyazawa, M., Biotransformation of (+)- and (−)-borneol by the larvae of common cutworm (*Spodoptera litura*) as a biocatalyst, *Proceedings of 45th TEAC*, 2001, pp. 377–378.)

19.4.3.7 Fenchol and Fenchyl Acetate

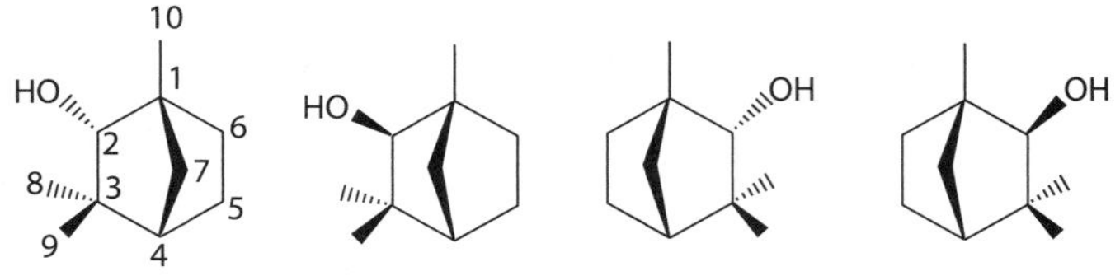

(1R,2R,4S)
(+)-endo
(+)-α-fenchol
11a

(1R,2S,4S)
(+)-exo
(+)-β-fenchol
11b

(1S,2S,4R)
(−)-endo
(−)-α-fenchol
11a′

(1S,2R,4R)
(−)-exo
(−)-β-fenchol
11b′

(1R,2R,4S)-(+)-Fenchol (**11a**) was converted by *A. niger* TBUYN-2 and *A. cellulosae* IFO 4040 to give (−)-fenchone (**12**), (+)-6β-hydroxyfenchol (**349ab**), (+)-5β-hydroxyfenchol (**350ab**), and 5α-hydroxyfenchol (**350aa**) (Noma and Asakawa, 2005a) (Figure 19.180). The larvae of common cutworm, *S. litura*, converted (+)-fenchol (**11a**) to (+)-10-hydroxyfenchol (**467a**), (+)-8-hydroxyfenchol (**465a**), (+)-6β-hydroxyfenchol (**349ab**), and (−)-9-hydroxyfenchol (**466a**) (Miyazawa and Miyamoto, 2004) (Figure 19.180).

(+)-*trans*-Pinocarveol (**2**), which was formed from (−)-β-pinene (**1**), was metabolized by *A. niger* TBUYN-2 to 6β-hydroxy-(+)-fenchol (**349ab**) and 6β,7-dihydroxy-(+)-fenchol (**412ba′**). (−)-Fenchone (**12**) was also metabolized to 6α-hydroxy- (**13b**) and 6β-hydroxy-(−)-fenchone (**13a**). (+)-Fenchol (**11**) was metabolized to 6β-hydroxy-(+)-fenchol (**349ab**) by *A. niger* TBUYN-2. The

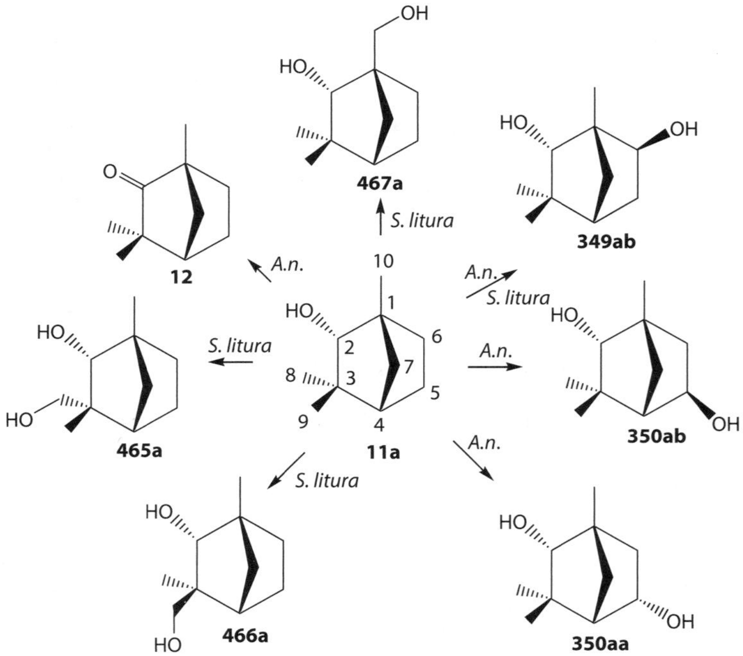

FIGURE 19.180 Biotransformation of (+)-fenchol (**11a**) by *Aspergillus niger* TBUYN-2, *Aspergillus cellulosae* IFO 4040, and the larvae of common cutworm, *Spodoptera litura*. (Modified from Miyazawa, M. and Miyamoto, Y., *Tetrahadron*, 60, 3091, 2004; Noma, Y. and Asakawa, Y., New metabolic pathways of β-pinene and related compounds by *Aspergillus niger*, *Book of Abstracts of the 36th ISEO*, 2005a, p. 32.)

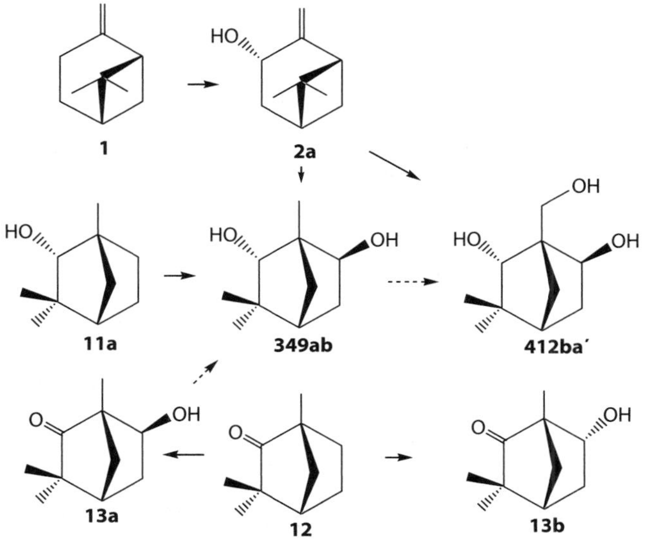

FIGURE 19.181 Metabolism of (+)-*trans*-pinocarveol (**2**), (−)-fenchone (**12**), and (+)-fenchol (**11**) by *Aspergillus niger* TBUYN-2. (Modified from Noma, Y. and Asakawa, Y., New metabolic pathways of β-pinene and related compounds by *Aspergillus niger*, *Book of Abstracts of the 36th ISEO*, 2005a, p. 32.)

relationship of the metabolisms of (+)-*trans*-pinocarveol (**2**), (−)-fenchone (**12**), and (+)-fenchol (**11**) by *A. niger* TBUYN-2 is shown in Figure 19.181 (Noma and Asakawa, 2005a).

(+)-α-Fencyl acetate (**11a-Ac**) was metabolized by *G. cingulata* to give (+)-5-β-hydroxy-α-fencyl acetate (**350a-Ac**, 50%) as the major metabolite and (+)-fenchol (**11a**, 20%) as the minor metabolite (Miyazato and Miyazawa, 1999). On the other hand, (−)-α-fencyl acetate (**11a′-Ac**) was metabolized to (−)-5-β-hydroxy-α-fencyl acetate (**350a′-Ac**, 70%) and (−)-fenchol (**11a′**, 10%) as the minor metabolite by *G. cingulata* (Miyazato and Miyazawa, 1999) (Figure 19.182).

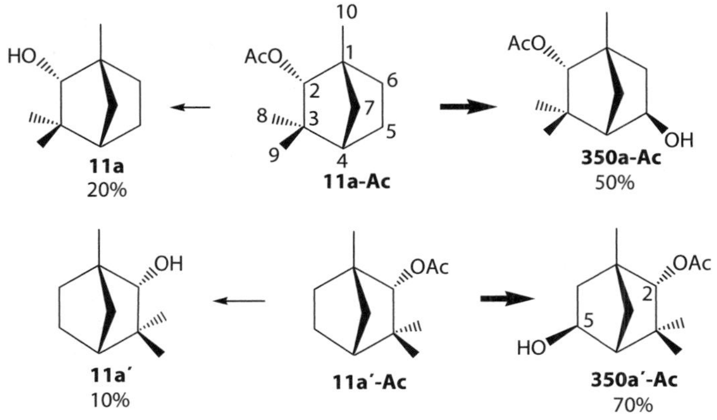

FIGURE 19.182 Biotransformation of (+)- (**11a-Ac**) and (−)-α-fencyl acetate (**11a′-Ac**) by *Glomerella cingulata*. (Modified from Miyazato, Y. and Miyazawa, M., Biotransformation of (+)- and (−)-α-fenchyl acetated using plant parasitic fungus, *Glomerella cingulata* as a biocatalyst, *Proceedings of 43rd TEAC*, 1999, pp. 213–214.)

19.4.3.8 Verbenol

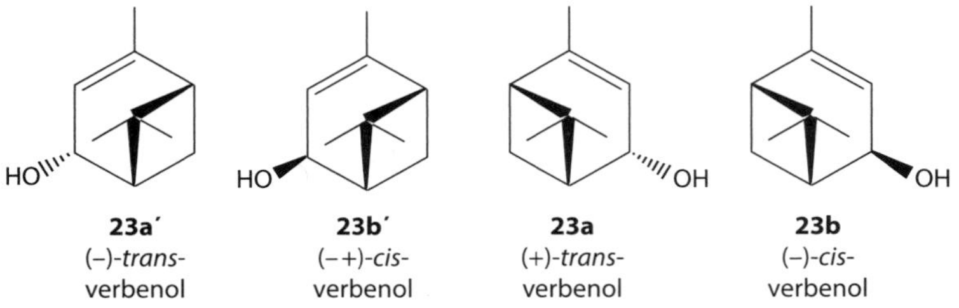

23a′
(−)-*trans*-
verbenol

23b′
(−+)-*cis*-
verbenol

23a
(+)-*trans*-
verbenol

23b
(−)-*cis*-
verbenol

(−)-*trans*-Verbenol (**23a′**) was biotransformed by *S. litura* to give 10-hydroxyverbenol (**451a′**). Furthermore, (−)-verbenone (**24′**) was also biotransformed in the same manner to give 10-hydroxyverbenone (**25′**) (Yamanaka and Miyazawa, 1999) (Figure 19.183).

19.4.3.9 Nopol and Nopol Benzyl Ether

Biotransformation of (−)-nopol (**452′**) was carried out at 30°C for 7 days at the concentration of 100 mg/200 mL medium by *A. niger* TBUYN-2, *A. sojae* IFO 4389, and *A. usami* IFO 4338. (−)-Nopol (**452′**) was incubated with *A. niger* TBUYN-2 to give 7-hydroxymethyl-1-*p*-menthen-8-ol (**453′**). In cases of *A. sojae* IFO 4389 and *A. usami* IFO 4338, (−)-nopol (**452′**) was metabolized to 3-oxonopol (**454′**) as a minor product together with 7-hydroxymethyl-1-*p*-menthen-8-ol (**453′**) as a major product (Noma and Asakawa, 2005b, 2006c) (Figure 19.184).

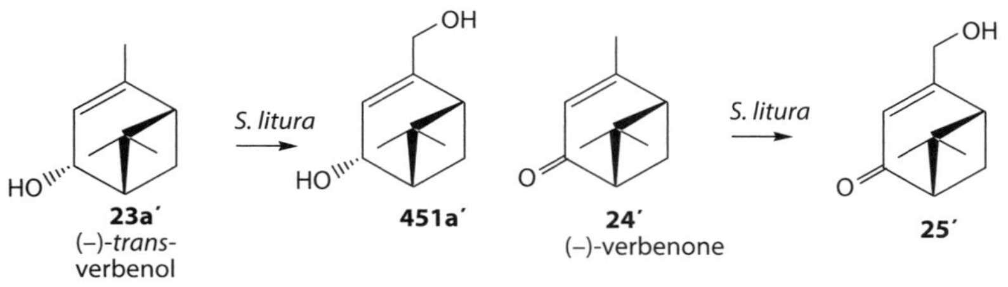

23a′
(−)-*trans*-
verbenol

451a′

24′
(−)-verbenone

25′

FIGURE 19.183 Metabolism of (−)-*trans*-verbenol (**23a′**) and (−)-verbenone (**24′**) by *Spodoptera litura*. (Modified from Yamanaka, T. and Miyazawa, M., Biotransformation of (−)-*trans*-verbenol by common cutworm larvae, *Spodoptera litura* as a biocatalyst, *Proceedings of 43rd TEAC*, 1999, pp. 391–392.)

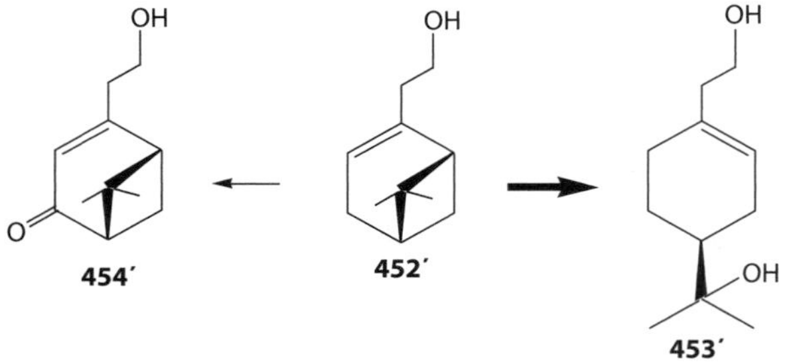

454′

452′

453′

FIGURE 19.184 Biotransformation of (−)-nopol (**452′**) by *Aspergillus niger* TBUYN-2, *Aspergillus sojae* IFO 4389, and *Aspergillus usami* IFO 4338. (Modified from Noma, Y. and Asakawa, Y., Microbial transformation of (−)-myrtenol and (−)-nopol, *Proceedings of 49th TEAC*, 2005b, pp. 78–80; Noma, Y. and Asakawa, Y., Microbial transformation of (−)-nopol benzyl ether, *Proceedings of 50th TEAC*, 2006c, pp. 434–436.)

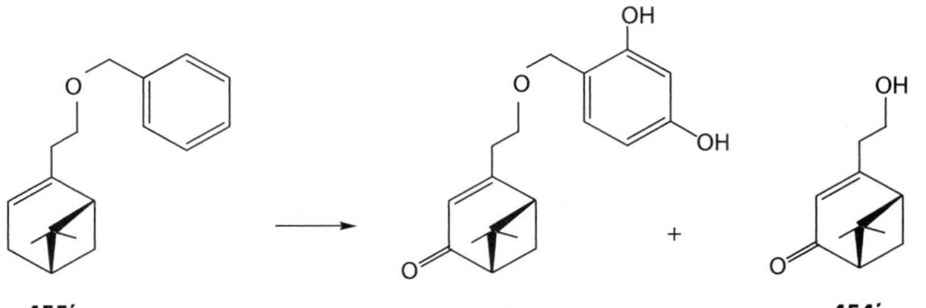

FIGURE 19.185 Biotransformation of (−)-nopol benzyl ether (**455′**) by *Aspergillus niger* TBUYN-2. (Modified from Noma, Y. and Asakawa, Y., Biotransformation of β-pinene, myrtenol, nopol and nopol benzyl ether by *Aspergillus niger* TBUYN-2, *Book of Abstracts of the 37th ISEO*, 2006b, p. 144; Noma, Y. and Asakawa, Y., Microbial transformation of (−)-nopol benzyl ether, *Proceedings of 50th TEAC*, 2006c, pp. 434–436.)

Biotransformation of (−)-nopol benzyl ether (**455′**) was carried out at 30°C for 8–13 days at the concentration of 277 mg/200 mL medium by *A. niger* TBUYN-2, *A. sojae* IFO 4389, and *A. usami* IFO 4338. (−)-Nopol benzyl ether (**455′**) was biotransformed by *A. niger* TBUYN-2 to give 4-oxonopol-2′, 4′-dihydroxy benzyl ether (**456′**), and (−)-oxonopol (**454′**). 7-Hydroxymethyl-1-*p*-menthen-8-ol benzyl ether (**457′**) was not formed at all (Figure 19.185). 4-Oxonopol-2′,4′-dihydroxybenzyl ether (**456′**) shows strong antioxidative activity (IC$_{50}$ 30.23 μM). The antioxidative activity of 4-oxonopol-2′,4′-dihydroxybenzyl ether (**456′**) is the same as that of butyl hydroxyl anisole (BHA) (Noma and Asakawa, 2006b,c).

Citrus pathogenic fungi *A. niger* Tiegh (CBAYN) also transformed (−)-nopol (**452′**) to (−)-oxonopol (**454′**) and 4-oxonopol-2′,4′-dihydroxybenzyl ether (**456′**) (Noma and Asakawa, 2006b,c) (Figure 19.186).

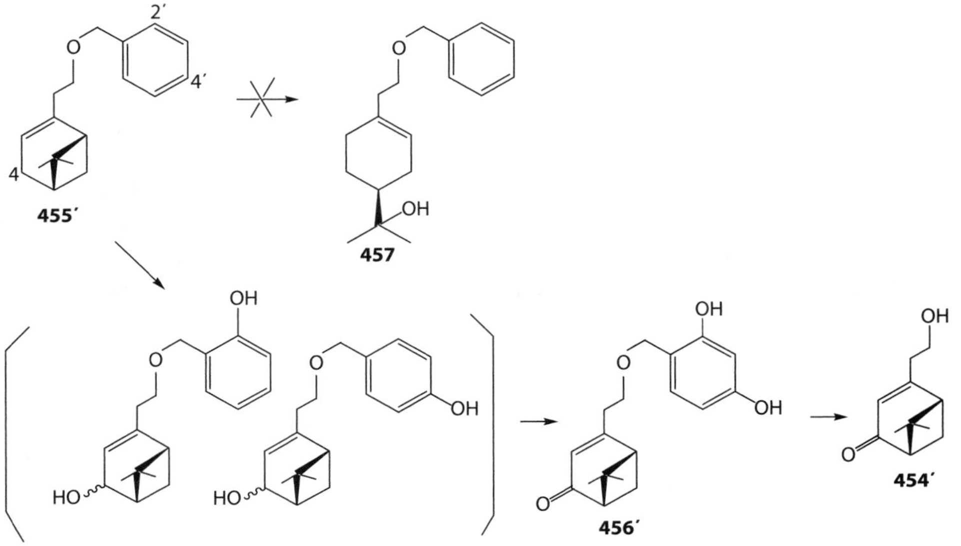

FIGURE 19.186 Proposed metabolic pathways of (−)-nopol benzyl ether (**455′**) by microorganisms. (Modified from Noma, Y. and Asakawa, Y., Biotransformation of β-pinene, myrtenol, nopol and nopol benzyl ether by *Aspergillus niger* TBUYN-2, *Book of Abstracts of the 37th ISEO*, 2006b, p. 144; Noma, Y. and Asakawa, Y., Microbial transformation of (−)-nopol benzyl ether, *Proceedings of 50th TEAC*, 2006c, pp. 434–436.)

19.4.4 Bicyclic Monoterpene Ketones

19.4.4.1 α-, β-Unsaturated Ketone

19.4.4.1.1 Verbenone

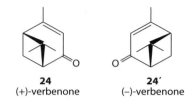

24
(+)-verbenone

24'
(−)-verbenone

(−)-Verbenone (**24'**) was hydrogenated by reductase of *N. tabacum* to give (−)-isoverbanone (**458b'**) (Suga and Hirata, 1990; Shimoda et al., 1996, 1998, 2002; Hirata et al., 2000) (Figure 19.187).

19.4.4.1.2 Pinocarvone

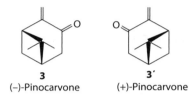

3
(−)-Pinocarvone

3'
(+)-Pinocarvone

A. niger TBUYN-2 transformed (+)-pinocarvone (**3'**) to give (−)-isopinocamphone (**413b'**), 2α-hydroxy-3-pinanone (**414b'**), and 2α, 5-dihydroxy-3-pinanone (**415b'**) together with small amounts of 2α, 10-dihydroxy-3-pinanone (**416b'**) (Noma and Asakawa, 2005a) (Figure 19.188).

19.4.4.2 Saturated Ketone

19.4.4.2.1 Camphor

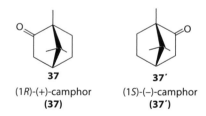

37
(1*R*)-(+)-camphor
(**37**)

37'
(1*S*)-(−)-camphor
(**37'**)

(+)- (**37**) and (−)-Camphor (**37'**) are found widely in nature, of which (+)-camphor (**37**) is more abundant. It is the main component of oils obtained from the camphor tree *C. camphora* (Bauer et al., 1990). The hydroxylation of (+)-camphor (**37**) by *P. putida* C_1 was described (Abraham et al., 1988). The substrate was hydroxylated exclusively in its 5-exo- (**235b**) and 6-exo- (**228b**) positions.

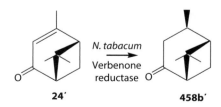

24' → **458b'**
N. tabacum Verbenone reductase

FIGURE 19.187 Hydrogenation of (−)-verbenone (**24'**) to (−)-isoverbanone (**458b'**) by verbenone reductase of *Nicotiana tabacum*. (Modified from Suga, T. and Hirata, T., *Phytochemistry*, 29, 2393, 1990; Shimoda, K. et al., *J. Chem. Soc., Perkin Trans.*, 1, 355, 1996; Shimoda, K. et al., *Phytochemistry*, 49, 49, 1998; Shimoda, K. et al., *Bull. Chem. Soc. Jpn.*, 75, 813, 2002; Hirata, T. et al., *Chem. Lett.*, 29, 850, 2000.)

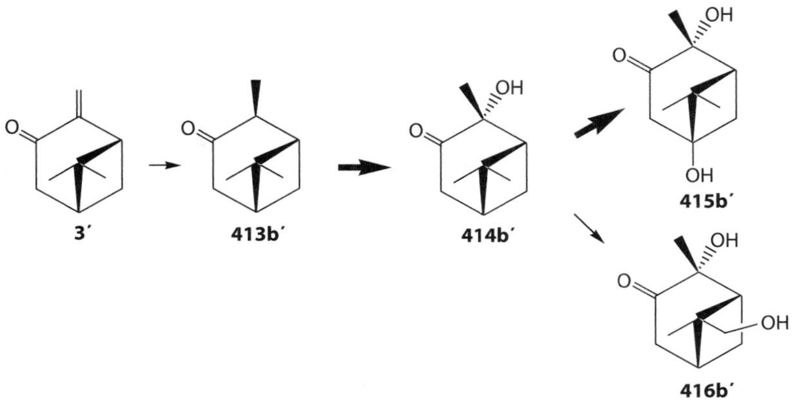

FIGURE 19.188 Biotransformation of (+)-pinocarvone (**3′**) by *Aspergillus niger* TBUYN-2. (Modified from Noma, Y. and Asakawa, Y., New metabolic pathways of β-pinene and related compounds by *Aspergillus niger*, *Book of Abstracts of the 36th ISEO*, 2005a, p. 32.)

Although only limited success was achieved in understanding the catabolic pathways of (+)-camphor (**37**), key roles for methylene group hydroxylation and biological Baeyer–Villiger monooxygenases in ring cleavage strategies were established (Trudgill, 1990). A degradation pathway of (+)-camphor (**37**) by *P. putida* ATCC 17453 and *Mycobacterium rhodochrous* T_1 was proposed (Trudgill, 1990).

The metabolic pathway of (+)-camphor (**37**) by microorganisms is shown in Figure 19.189. (+)-Camphor (**37**) is metabolized to 3-hydroxy- (**243**), 5-hydroxy- (**235**), 6-hydroxy- (**228**), and 9-hydroxycamphor (**225**) and 1,2-campholide (**237**). 6-Hydroxycamphor (**228**) is degradatively metabolized to 6-oxocamphor (**229**) and 4-carboxymethyl-2,3,3-trimethylcyclopentanone (**230**), 4-carboxymethyl-3,5,5-trimethyltetrahydro-2-pyrone (**231**), isohydroxycamphoric acid (**232**), isoketocamphoric acid (**233**), and 3,4,4-trimethyl-5-oxo-*trans*-2-hexenoic acid (**234**), whereas 1,2-campholide (**237**) is also degradatively metabolized to 6-hydroxy-1,2-campholide (**238**), 6-oxo-1,2-campholide (**239**), and 5-carboxymethyl-3,4,4-trimethyl-2-cyclopentenone (**240**), 6-carboxymethyl-4,5,5-trimethyl-5,6-dihydro-2-pyrone (**241**), and 5-carboxymethyl-3,4,4-trimethyl-2-heptene-1,7-dioic acid (**242**). 5-Hydroxycamphor (**235**) is metabolized to 6-hydroxy-1,2-campholide (**238**), 5-oxocamphor (**236**), and 6-oxo-1,2-campholide (**239**). 3-Hydroxycamphor (**243**) is also metabolized to camphorquinone (**244**) and 2-hydroxyepicamphor (**245**) (Bradshaw et al., 1959; Conrad et al., 1961, 1965a,b; Gunsalus et al., 1965; Chapman et al., 1966; Hartline and Gunsalus, 1971; Oritani and Yamashita, 1974) (Figure 19.189).

Human CYP2A6 converted (+)-camphor (**37**) and (−)-camphor (**37′**) to 6-*endo*-hydroxycamphor (**228a**) and 5-*exo*-hydroxycamphor (**235b**), while rat CYP2B1 did 5-*endo*- (**235a**), 5-*exo*- (**235b**), and 6-*endo*-hydroxycamphor (**228a**) and 8-hydroxycamphor (**225**) (Gyoubu and Miyazawa, 2006) (Figure 19.190).

(+)-Camphor (**37**) was glycosylated by *E. perriniana* suspension cells to (+)-camphor monoglycoside (3 new, 11.7%) (Hamada et al., 2002) (Figure 19.191).

19.4.4.2.2 Fenchone

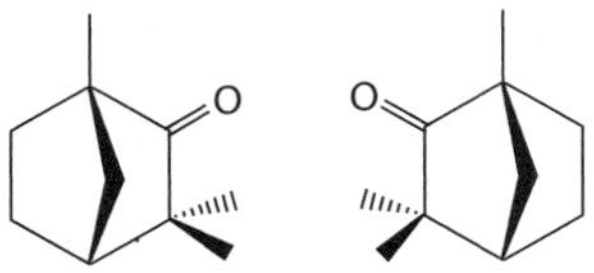

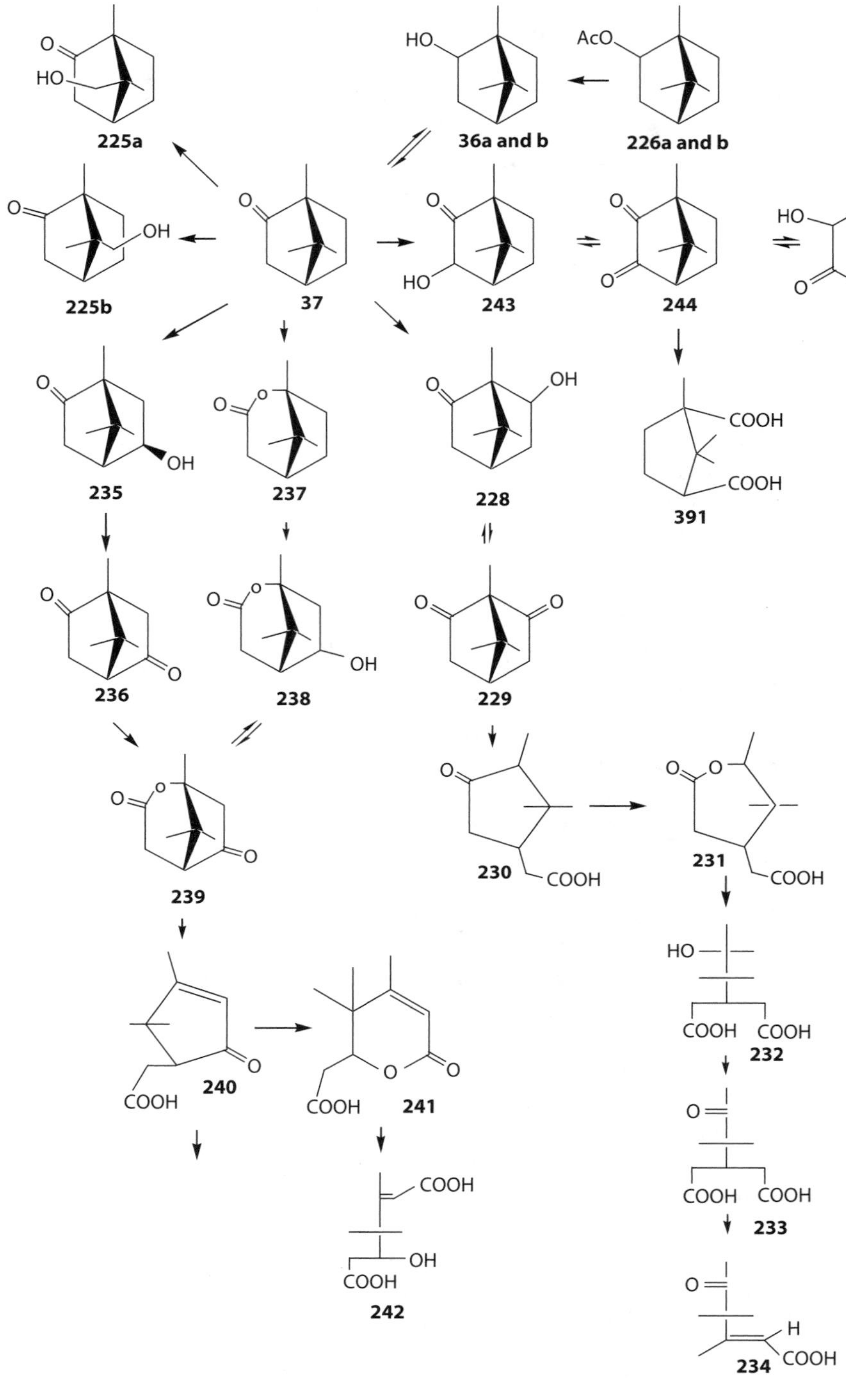

FIGURE 19.189 Metabolic pathways of (+)-camphor (**37**) by *Pseudomonas putida* and *Corynebacterium diphtheroides*. (Modified from Bradshaw, W.H. et al., *J. Am. Chem. Soc.*, 81, 5507, 1959; Conrad, H.E. et al., *Biochem. Biophys. Res. Commun.*, 6, 293, 1961; Conrad, H.E. et al., *J. Biol. Chem.*, 240, 495, 1965a; Conrad, H.E. et al., *J. Biol. Chem.*, 240, 4029, 1965b; Gunsalus, I.C. et al., *Biochem. Biophys. Res. Commun.*, 18, 924, 1965; Chapman, P.J. et al., *J. Am. Chem. Soc.*, 88, 618, 1966; Hartline, R.A. and Gunsalus, I.C., *J. Bacteriol.*, 106, 468, 1971; Oritani, T. and Yamashita, K., *Agric. Biol. Chem.*, 38, 1961, 1974.)

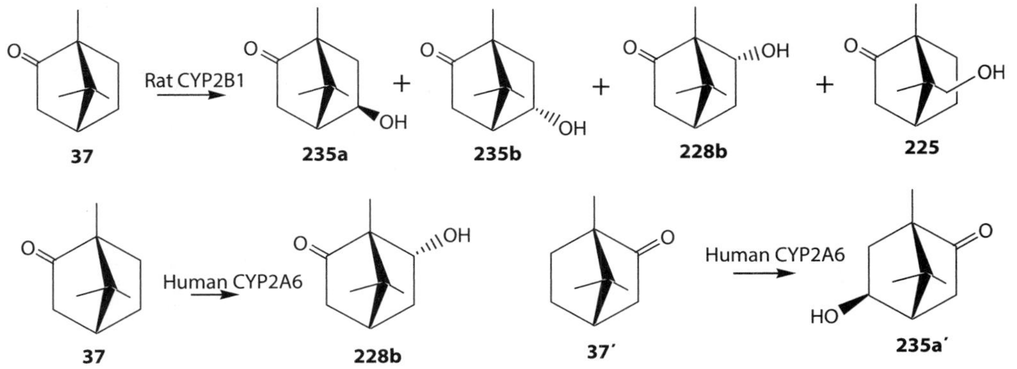

FIGURE 19.190 Biotransformation of (+)-camphor (**37**) by rat P450 enzyme (above) and (+)- (**37**) and (−)-camphor (**37′**) by human P450 enzymes.

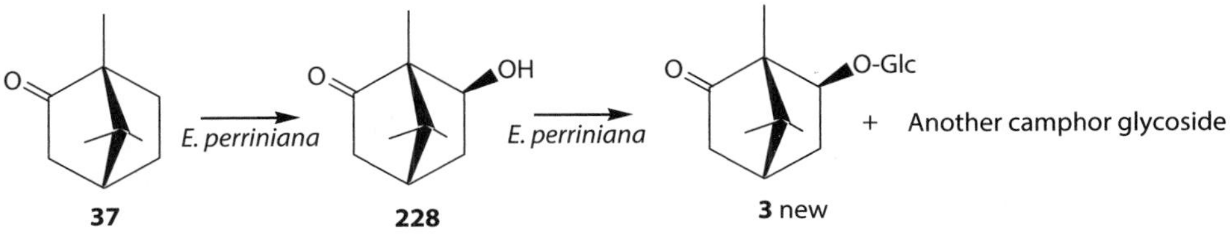

FIGURE 19.191 Biotransformation of (+)-camphor (**37**) by *Eucalyptus perriniana* suspension cell.

(+)-Fenchone (**12**) was incubated with *Corynebacterium* sp. (Chapman et al., 1965) and *Absidia orchidis* (Pfrunder and Tamm, 1969a) give 6β-hydroxy- (**13a**) and 5β-hydroxyfenchones (**14a**) (Figure 19.191). On the other hand, *A. niger* biotransformed (+)-fenchone (**12**) to (+)-6α- (**13b**) and (+)-5α-hydroxyfenchones (**14b**) (Miyazawa et al., 1990a,b) and 5-oxofenchone (**15**), 9-formylfenchone (**17b**), and 9-carboxyfenchone (**18b**) (Miyazawa et al., 1990a, 1990b) (Figure 19.192).

Furthermore, *A. niger* biotransformed (−)-fenchone (**12′**) to 5α-hydroxy- (**14b′**) and 6α-hydroxyfenchones (**13b′**) (Yamamoto et al., 1984) (Figure 19.193).

(+)- and (−)-Fenchone (**12** and **12′**) were converted to 6β-hydroxy- (**13a, 13a′**), 6α-hydroxyfenchone (**13b, 13b′**), and 10 hydroxyfenchone (**4, 4′**) by P450. Of the 11 recombinant human P450 enzymes tested, CYP2A6 and CYP2B6 catalyzed oxidation of (+)- (**12**) and (−)-fenchone (**12′**) (Gyoubu and Miyazawa, 2005) (Figure 19.194).

19.4.4.2.3 3-Pinanone (Pinocamphone and Isopinocamphone)

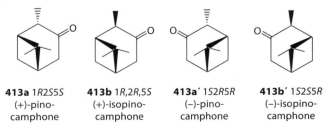

413a 1R2S5S
(+)-pino-
camphone

413b 1R,2R,5S
(+)-isopino-
camphone

413a′ 1S2R5R
(−)-pino-
camphone

413b′ 1S2S5R
(−)-isopino-
camphone

(+)- (**413**) and (−)-Isopinocamphone (**413′**) were biotransformed by *A. niger* to give (−)- (**414**) and (+)-2-hydroxy-3-pinanone (**414′**) as the main products, respectively, which inhibit strongly germination of lettuce seeds, and (−)- (**415**) and (+)-2,5-dihydroxy-3-pinanone (**415′**) as the minor components, respectively (Noma et al., 2003, 2004) (Figure 19.195).

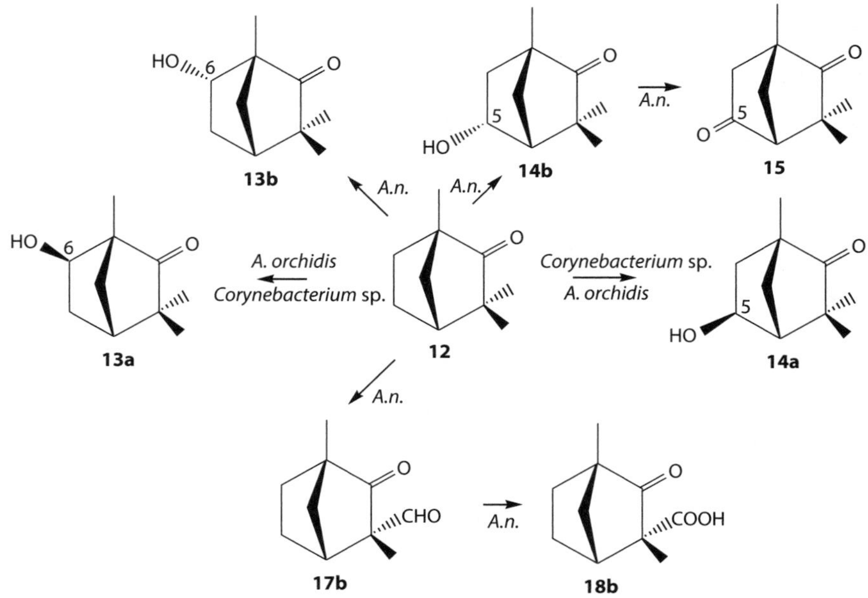

FIGURE 19.192 Metabolic pathways of (+)-fenchone (**12**) by *Corynebacterium* sp., *A. orchidis*, and *Aspergillus niger* TBUYN-2. (Modified from Chapman, P.J. et al., *Biochem. Biophys. Res. Commun.*, 20, 104, 1965; Pfrunder, B. and Tamm, Ch., *Helv. Chim. Acta.*, 52, 1643, 1969a; Miyazawa, M. et al., *Chem. Express*, 5, 237, 1990a; Miyazawa, M. et al., *Chem. Express*, 5, 407, 1990b.)

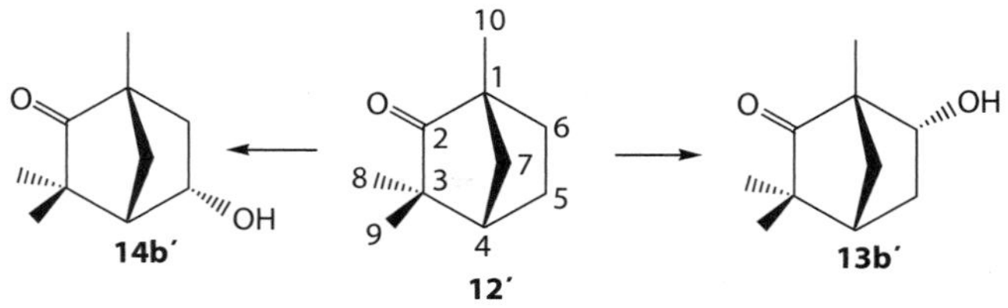

FIGURE 19.193 Metabolic pathways of (−)-fenchone (**12′**) by *Aspergillus niger* TBUYN-2. (Modified from Yamamoto, K. et al., Biotransformation of *d*- and *l*-fenchone by a strain of *Aspergillus niger*, *Proceedings of 28th TEAC*, 1984, pp. 168–170.)

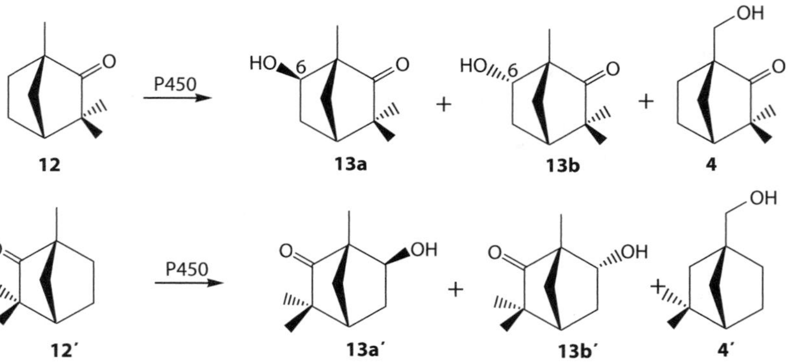

FIGURE 19.194 Biotransformation of (+)-fenchone (**12**) and (−)-fenchone (**12′**) by P450 enzymes. (Modified from Gyoubu, K. and Miyazawa, M., Biotransformation of (+)- and (−)-fenchone by liver microsomes, *Proceedings of 49th TEAC*, 2005, pp. 420–422.)

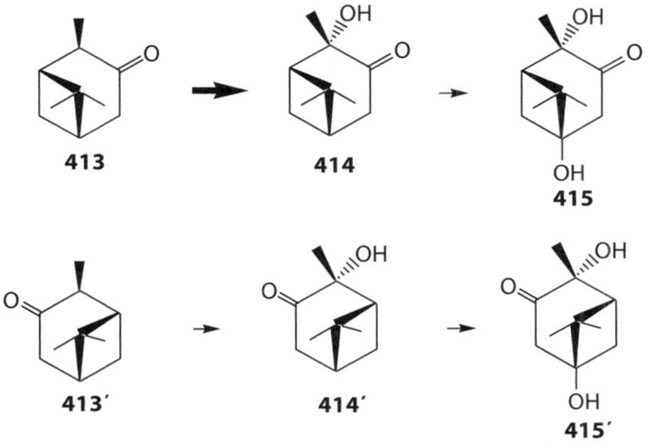

FIGURE 19.195 Biotransformation of (+)-isopinocamphone (**413b**) and (−)-isopinocamphone (**413b′**) by *Aspergillus niger* TBUYN-2. (Modified from Noma, Y. et al., Biotransformation of (+)- and (−)-pinane-2,3-diol and related compounds by *Aspergillus niger*, *Proceedings of 47th TEAC*, pp. 91–93, 2003; Noma, Y. et al., Biotransformation of (+)- and (−)-3-pinanone by *Aspergillus niger*, *Proceedings of 48th TEAC*, 2004, pp. 390–392.)

19.4.4.2.4 2-Hydroxy-3-Pinanone

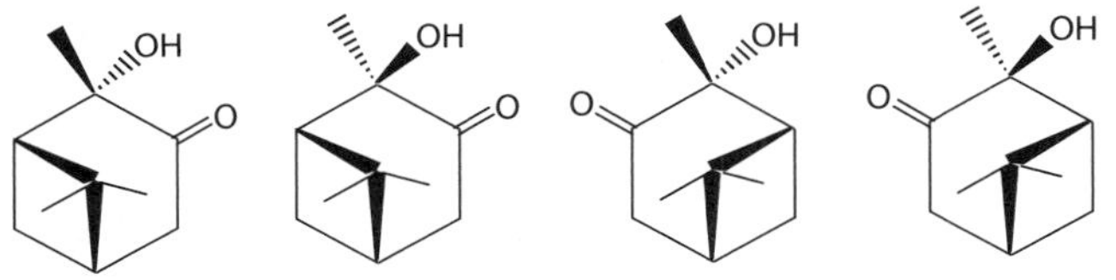

(1S,2R,5S)
(−)-2-OH-3-pinanone
414a

(1S,2R,5S)
(+)-2-OH-3-pinanone
414b

(1R,2R,5R)
(+)-2-OH-3-pinanone
414a′

(1R,2S,5R)
(+)-2-OH-3-pinanone
414b′

(−)-2α-Hydroxy-3-pinanone (**414**) was incubated with *A. niger* TBUYN-2 to give (−)-2α, 5-dihydroxy-3-pinanone (**415**) predominantly, whereas the fungus converted (+)-2α-hydroxy-3-pinanone (**414′**) mainly to 2α, 5-dihydroxy-3-pinanone (**415′**), 2α,9-dihydroxy-3-pinanone (**416′**), and (−)-pinane-2α,3α,5-triol (**419ba′**) (Noma et al., 2003, 2004) (Figure 19.196).

A. niger TBUYN-2 metabolized β-pinene (**1**), isopinocamphone (**414b**), 2α-hydroxy-3-pinanone (**414a**), and pinane-2,3-diol (**419ab**) as shown in Figure 19.197. On the other hand, *A. niger* TBUYN-2 and *B. cinerea* metabolized β-pinene (**1′**), isopinocamphone (**414b′**), 2α-hydroxy-3-pinanone (**414a′**), and pinane-2,3-diol (**419ab′**) as shown in Figure 19.198. The relationship of the metabolism of β-pinene (**1, 1′**), isopinocamphone (**414b, 414b′**), 2α-hydroxy-3-pinanone (**414a, 414a′**), and pinane-2,3-diol (**419ab, 419ab′**) in *A. niger* TBUYN-2 and *B. cinerea* is shown in Figures 19.197 and 19.198.

19.4.4.2.4.1 Mosquitocidal and Knockdown Activity Knockdown and mortality activity toward mosquito, *Culex quinquefasciatus*, was carried out for the metabolites of (+)- (**418ab**) and (−)-pinane-2,3-diols (**418ab′**) and (+)- and (−)-2-hydroxy-3-pinanones (**414** and **414′**) by Dr. Radhika Samarasekera, Industrial Technology Institute, Sri Lanka. (−)-2-Hydroxy-3-pinanone (**414′**) showed the mosquito knockdown activity and the mosquitocidal activity at the concentration of 1% and 2% (Table 19.16).

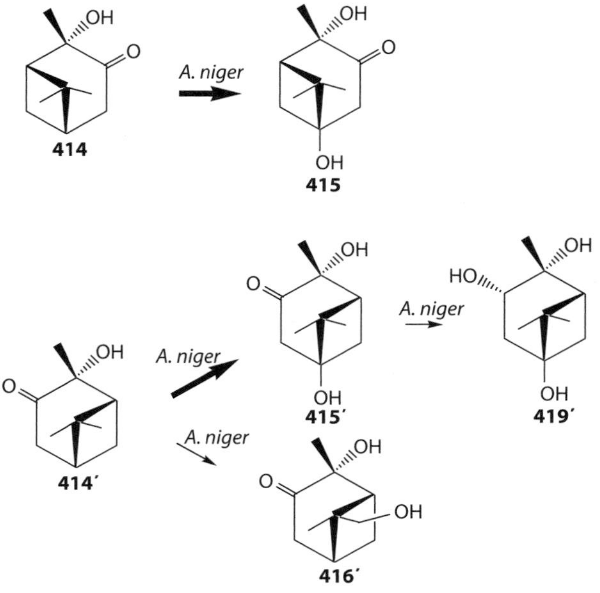

FIGURE 19.196 Biotransformation of (−)- (**414**) and (+)-2-hydroxy-3-pinanone (**414′**) by *Aspergillus niger* TBUYN-2. (Modified from Noma, Y. et al., Biotransformation of (+)- and (−)-pinane-2,3-diol and related compounds by *Aspergillus niger*, *Proceedings of 47th TEAC*, pp. 91–93, 2003; Noma, Y. et al., Biotransformation of (+)- and (−)-3-pinanone by *Aspergillus niger*, *Proceedings of 48th TEAC*, 2004, pp. 390–392.)

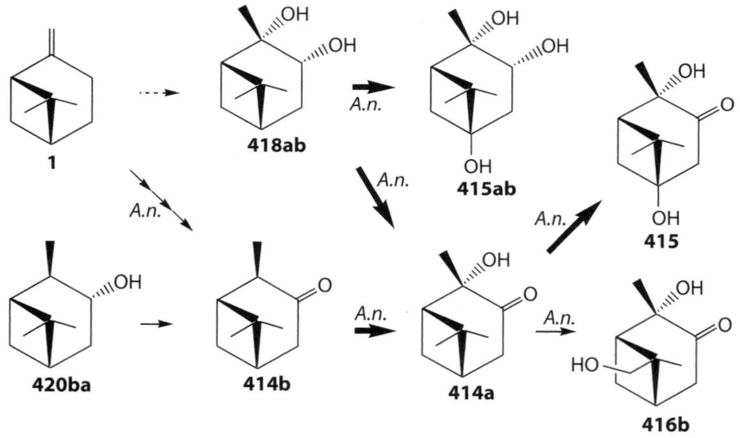

FIGURE 19.197 Relationship of the metabolism of β-pinene (**1**), isopinocamphone (**414b**), 2α-hydroxy-3-pinanone (**414a**), and pinane-2,3-diol (**419ab**) in *Aspergillus niger* TBUYN-2. (Modified from Noma, Y. et al., Biotransformation of (+)- and (−)-pinane-2,3-diol and related compounds by *Aspergillus niger*, *Proceedings of 47th TEAC*, 2003, pp. 91–93; Noma, Y. et al., Biotransformation of (+)- and (−)-3-pinanone by *Aspergillus niger*, *Proceedings of 48th TEAC*, 2004, pp. 390–392.)

19.4.4.2.4.2 Antimicrobial Activity   The microorganisms were refreshed in Mueller–Hinton broth (Merck) at 35°C–37°C and inoculated on Mueller–Hinton agar (Mast Diagnostics, Merseyside, United Kingdom) media for preparation of inoculum. *E. coli* (NRRL B-3008), *P. aeruginosa* (ATCC 27853), *Enterobacter aerogenes* (NRRL 3567), *S. typhimurium* (NRRL B-4420), *Staphylococcus epidermidis* (ATCC 12228), methicillin-resistant *Staphylococcus aureus* (MRSA, clinical isolate, Osmangazi University, Faculty of Medicine, Eskisehir, Turkey), and *Candida albicans* (clinical

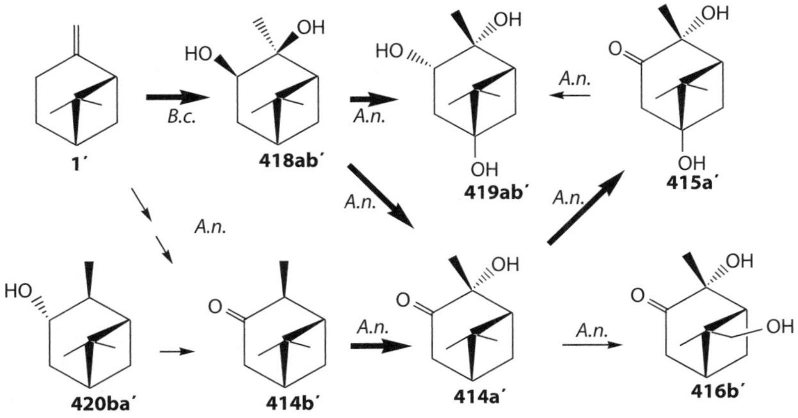

FIGURE 19.198 Relationship of the metabolism of β-pinene (**1′**), isopinocamphone (**414b′**), 2α-hydroxy-3-pinanone (**414a′**), and pinane-2,3-diol (**419ab′**) in *Aspergillus niger* TBUYN-2 and *Botrytis cinerea*. (Modified from Noma, Y. et al., Biotransformation of (+)- and (−)-pinane-2,3-diol and related compounds by *Aspergillus niger*, Proceedings of 47th TEAC, 2003, pp. 91–93; Noma, Y. et al., Biotransformation of (+)- and (−)-3-pinanone by *Aspergillus niger*, Proceedings of 48th TEAC, 2004, pp. 390–392.)

TABLE 19.16
Knockdown and Mortality Activity toward Mosquito[a]

Compounds	Knockdown (%)	Mortality (%)
(+)-2,5-Dihydroxy-3-pinanone (**415**, 2%)	27	20
(−)-2,5-Dihydroxy-3-pinanone (**415′**, 2%)	NT	7
(+)-2-Hydroxy-3-pinanone (**414**, 2%)	40	33
(−)-2-Hydroxy-3-pinanone (**414′**, 2%)	100	40
(−)-2-Hydroxy-3-pinanone (**414′**, 1%)	53	7
(+)-Pinane-2,3,5-triol (**419**, 2%)	NT	NT
(−)-Pinane-2,3,5-triol (**419**, 2%)	13	NT
(+)-Pinane-2,3-diol (**418**, 2%)	NT	NT
(−)-Pinane-2,3-diol (**418′**, 2%)	NT	NT

[a] The results are against *Culex quinquefasciatus*.

isolate, Osmangazi University, Faculty of Medicine, Eskisehir, Turkey) were used as pathogen test microorganisms. Microdilution broth susceptibility assay (*R1*, *R2*) was used for the antimicrobial evaluation of the samples. Stock solutions were prepared in DMSO (Carlo-Erba). Dilution series were prepared from 2 mg/mL in sterile distilled water in microtest tubes from where they were transferred to 96-well microtiter plates. Overnight grown bacterial and candial suspensions in double strength Mueller–Hilton broth (Merck) was standardized to approximately 10^8 CFU/mL using McFarland No. 0.5 (10^6 CFU/mL for *C. albicans*). A volume of 100 μL of each bacterial suspension was then added to each well. The last row containing only the serial dilutions of samples without microorganism was used as negative control. Sterile distilled water, medium, and microorganisms served as a positive growth control. After incubation at 37°C for 24 h, the first well without turbidity was determined as the minimal inhibition concentration (MIC), and chloramphenicol (sigma), ampicillin (sigma), and ketoconazole (sigma) were used as standard antimicrobial agents (Amsterdam, 1997; Koneman et al., 1997) (Table 19.17).

TABLE 19.17
Biological Activity of Pinane-2,3,5-Triol (419 and 419′), 2,5-Dihydroxy-3-Pinanone (415 and 415′), and 7-Hydroxymethyl-1-*p*-Menthene-8-ol (453′) toward MRSA

	MIC (mg/mL)							
	Compounds				Control			
Microorganisms	419	415′	415	419′	453′	ST1	ST2	ST3
Escherichia coli	0.5	0.5	0.25	0.5	0.25	0.007	0.0039	Nt
Pseudomonas aeruginosa	0.5	0.125	0.125	0.25	0.25	0.002	0.0078	Nt
Enterobacter aerogenes	0.5	0.5	0.25	0.5	1.00	0.007	0.0019	Nt
Salmonella typhimurium	0.25	0.125	0.125	0.25	0.25	0.01	0.0019	Nt
Candida albicans	0.5	0.125	0.125	0.25	1.00	NT	NT	0.0625
Staphylococcus epidermidis	0.5	0.5	0.25	0.5	1.00	0.002	0.0009	NT
MRSA	0.25	0.125	0.125	0.25	0.125	0.5	0.031	NT

Source: Iscan (2005, unpublished data).

Note: MRSA, methicillin-resistant Staphylococcus aureus; NT, not tested; ST1, ampicillin Na (sigma); ST2, chloramphenicol (sigma); ST3, ketoconazole (sigma).

19.5 SUMMARY

19.5.1 METABOLIC PATHWAYS OF MONOTERPENOIDS BY MICROORGANISMS

About 50 years is over since the hydroxylation of α-pinene (**4**) was reported by *A. niger* in 1960 (Bhattacharyya et al., 1960). During these years, many investigators have studied the biotransformation of a number of monoterpenoids by using various kinds of microorganisms. Now we summarize the microbiological transformation of monoterpenoids according to the literatures listed in the references including the metabolic pathways (Figures 19.199 through 19.206) for the further development of the investigation on microbiological transformation of terpenoids.

Metabolic pathways of β-pinene (**1**), α-pinene (**4**), fenchol (**11**), fenchone (**12**), thujone (**28**), carvotanacetone (**47**), and sobrerol (**43**) are summarized in Figure 19.199. In general, β-pinene (**1**) is metabolized by six pathways. At first, β-pinene (**1**) is metabolized via α-pinene (**4**) to many metabolites such as myrtenol (**5**) (Shukla et al., 1968; Shukla and Bhattacharyya, 1968), verbenol (**23**) (Bhattacharyya et al., 1960; Prema and Bhattacharyya, 1962), and thujone (**28**) (Gibbon and Pirt, 1971). Myrtenol (**5**) is further metabolized to myrtenal (**6**) and myrtenic acid (**7**). Verbenol (**23**) is further metabolized to verbenone (**24**), 7-hydroxyverbenone (**25**), 7-hydroxyverbenone (**26**), and 7-formyl verbanone (**27**). Thujone (**28**) is further metabolized to thujoyl alcohol (**29**), 1-hydroxythujone (**30**), and 1,3-dihydroxythujone (**31**). Second, β-pinene (**1**) is metabolized to pinocarveol (**2**) and pinocarvone (**3**) (Ganapathy and Bhattacharyya, unpublished data). Pinocarvone (**3**) is further metabolized to isopinocamphone (**413**), which is further hydroxylated to give 2-hydroxy-3-pinanone (**414**). Compound **414** is further metabolized to give pinane-2,3-diol (**419**), 2,5-dihydroxy- (**415**), and 2,9-dihydroxy-3-pinanone (**416**). Third, β-pinene (**1**) is metabolized to α-fenchol (**11**) and fenchone (**12**) (Dhavlikar et al., 1974), which are further metabolized to 6-hydroxy- (**13**) and 5-hydroxyfenchone (**14**), 5-oxofenchone (**15**), fenchone-9-al (**17**), fenchone-9-oic acid (**18**) via 9-hydroxyfenchone (**16**), 2,3-fencholide (**21**), and 1,2-fencholide (**22**) (Pfrunder and Tamm, 1969a,b; Yamamoto et al., 1984; Christensen and Tuthill, 1985; Miyazawa et al., 1990a, 1990b). Fenchol (**12**) is also metabolized to 9-hydroxyfenchol (**466**) and 7-hydroxyfenchol (**467**), 6-hydroxyfenchol (**349**), and 6,7-dihydroxyfenchol (**412**). Fourth, β-pinene (**1**) is metabolized via fenchoquinone (**19**) to 2-hydroxyfenchone (**20**) (Pfrunder and Tamm, 1969b; Gibbon et al., 1972). Fifth, β-pinene

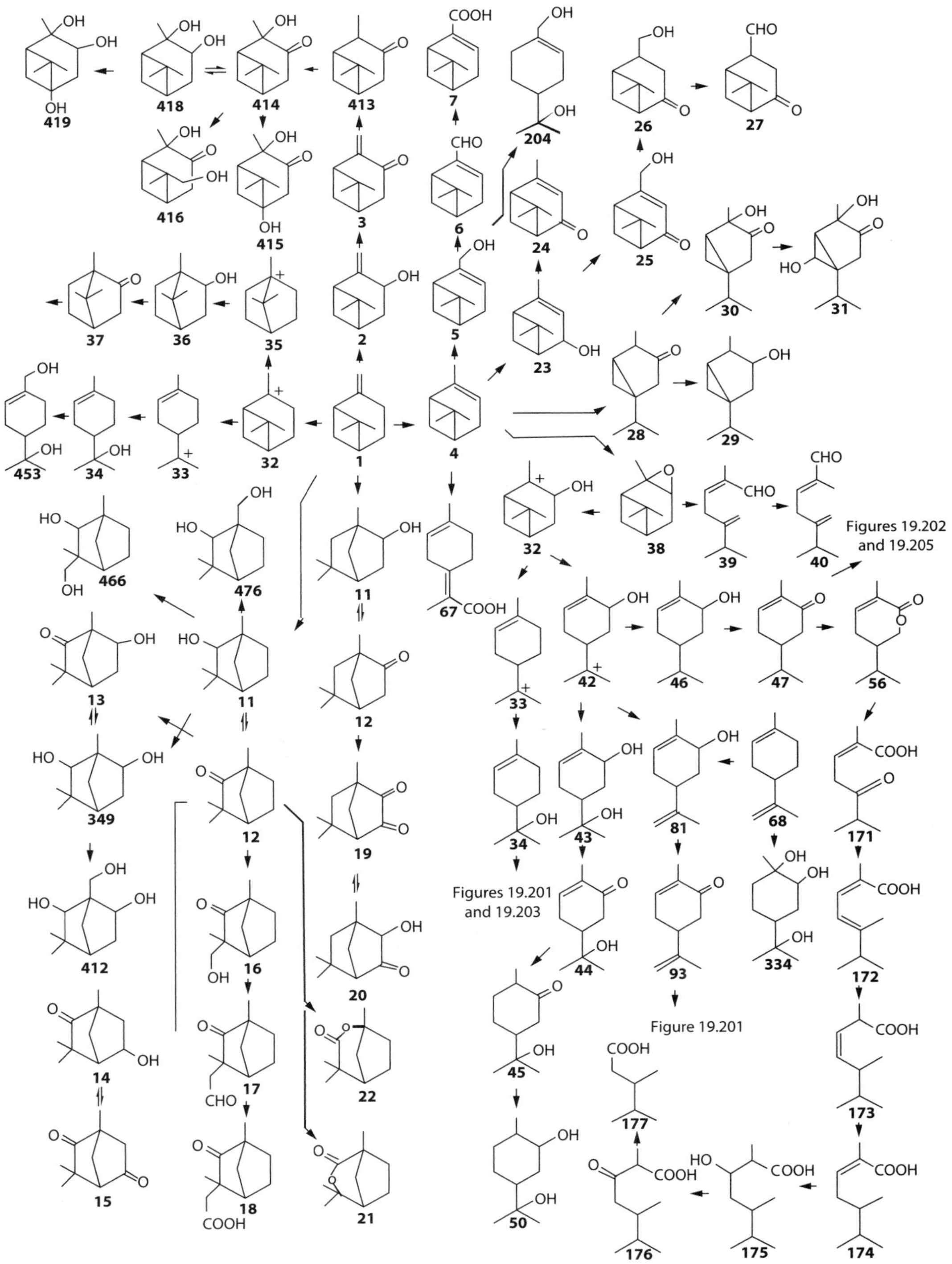

FIGURE 19.199 Metabolic pathways of β-pinene (**1**), α-pinene (**4**), fenchone (**9**), thujone (**28**), and carvotan-acetone (**44**) by microorganisms.

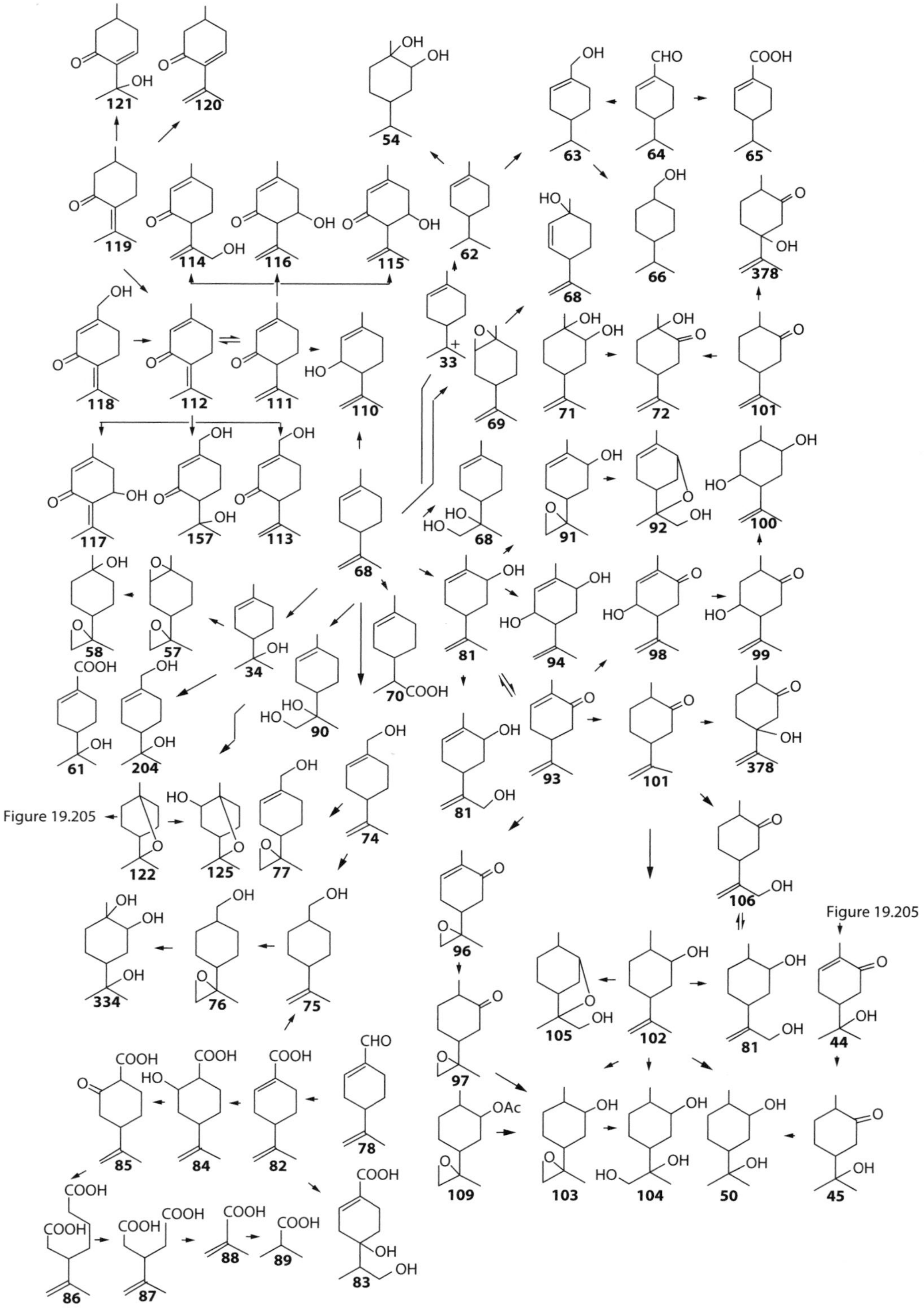

FIGURE 19.200 Metabolic pathways of limonene (**68**), perillyl alcohol (**74**), carvone (**93**), isopiperitenone (**111**), and piperitenone (**112**) by microorganisms.

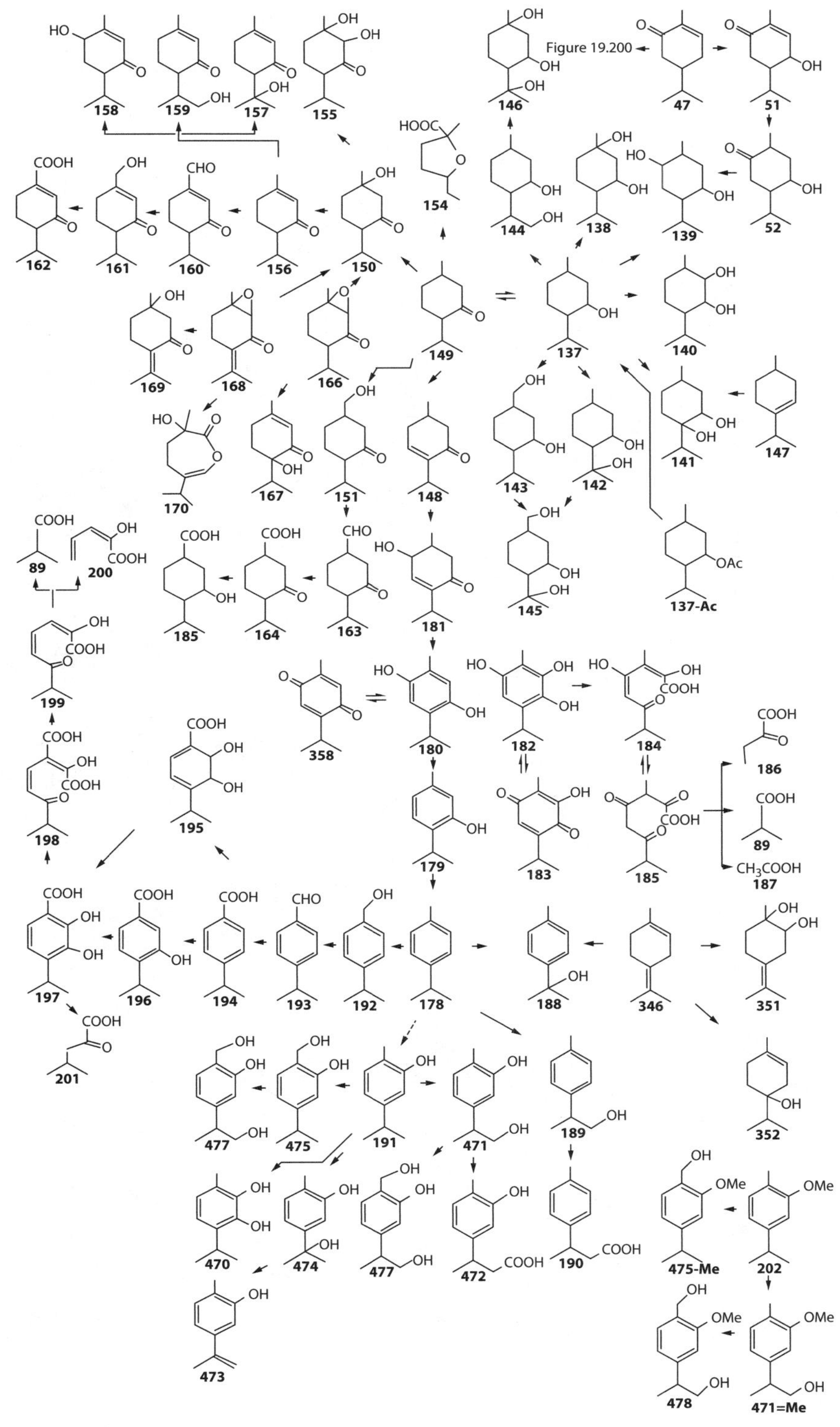

FIGURE 19.201 Metabolic pathways of menthol (**137**), menthone (**149**), *p*-cymene (**178**), thymol (**179**), carvacrol methyl ether (**201**), and carvotanacetone (**47**) by microorganisms and rabbit.

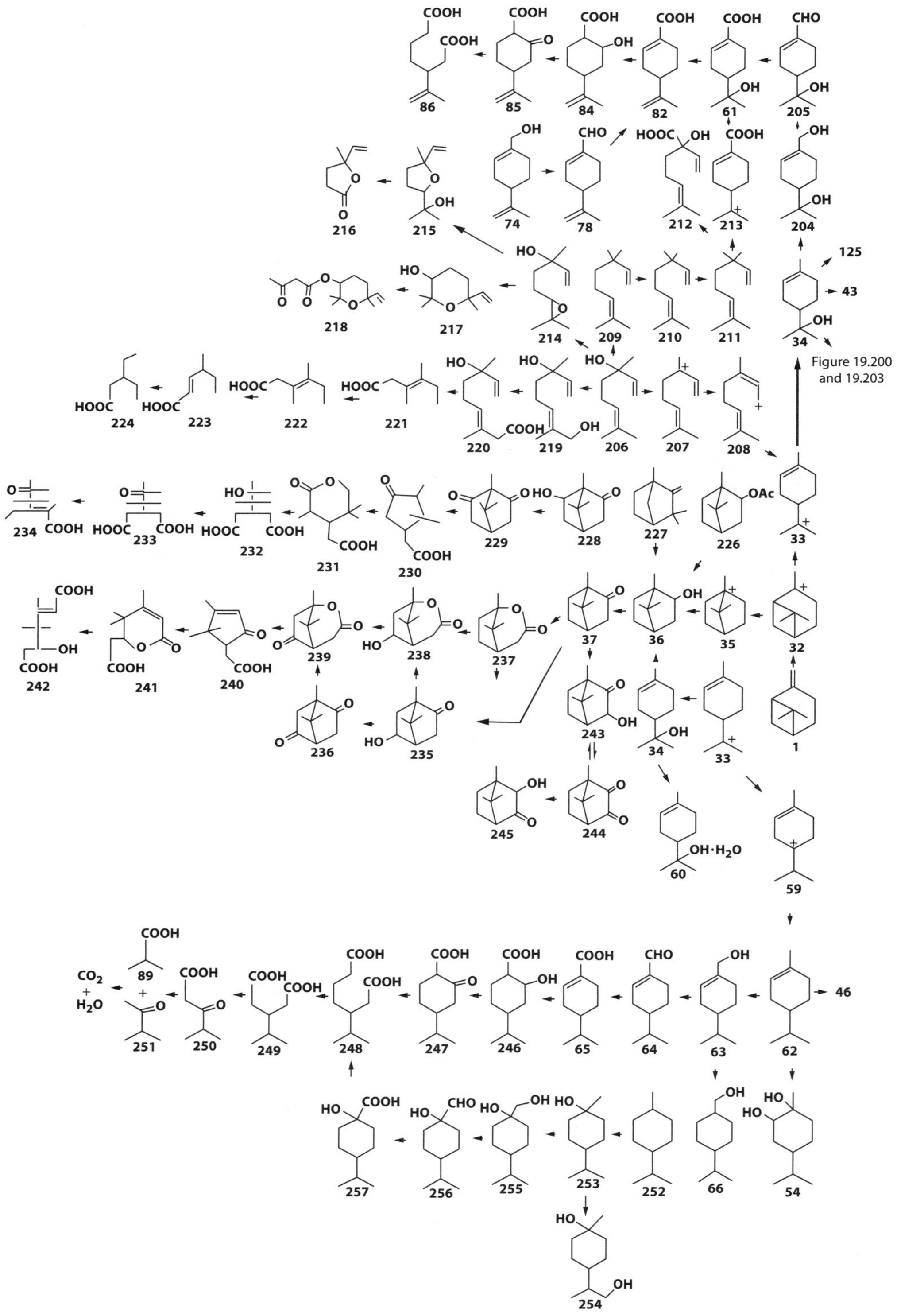

FIGURE 19.202 Metabolic pathway of borneol (**36**), camphor (**37**), phellandral (**64**), linalool (**206**), and *p*-menthane (**252**) by microorganisms.

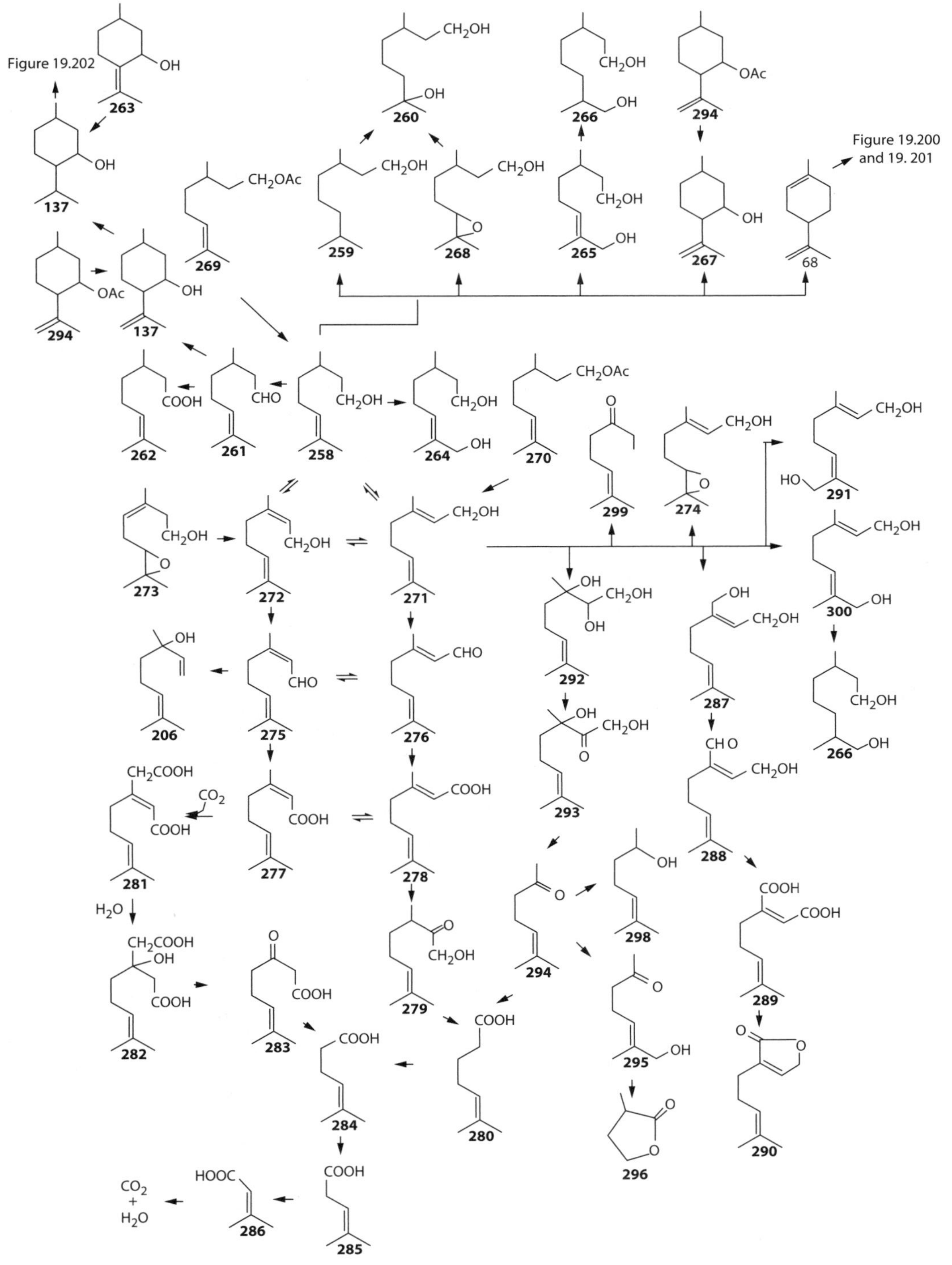

FIGURE 19.203 Metabolic pathways of citronellal (**258**), geraniol (**271**), nerol (**272**), and citral (**275** and **276**) by microorganisms.

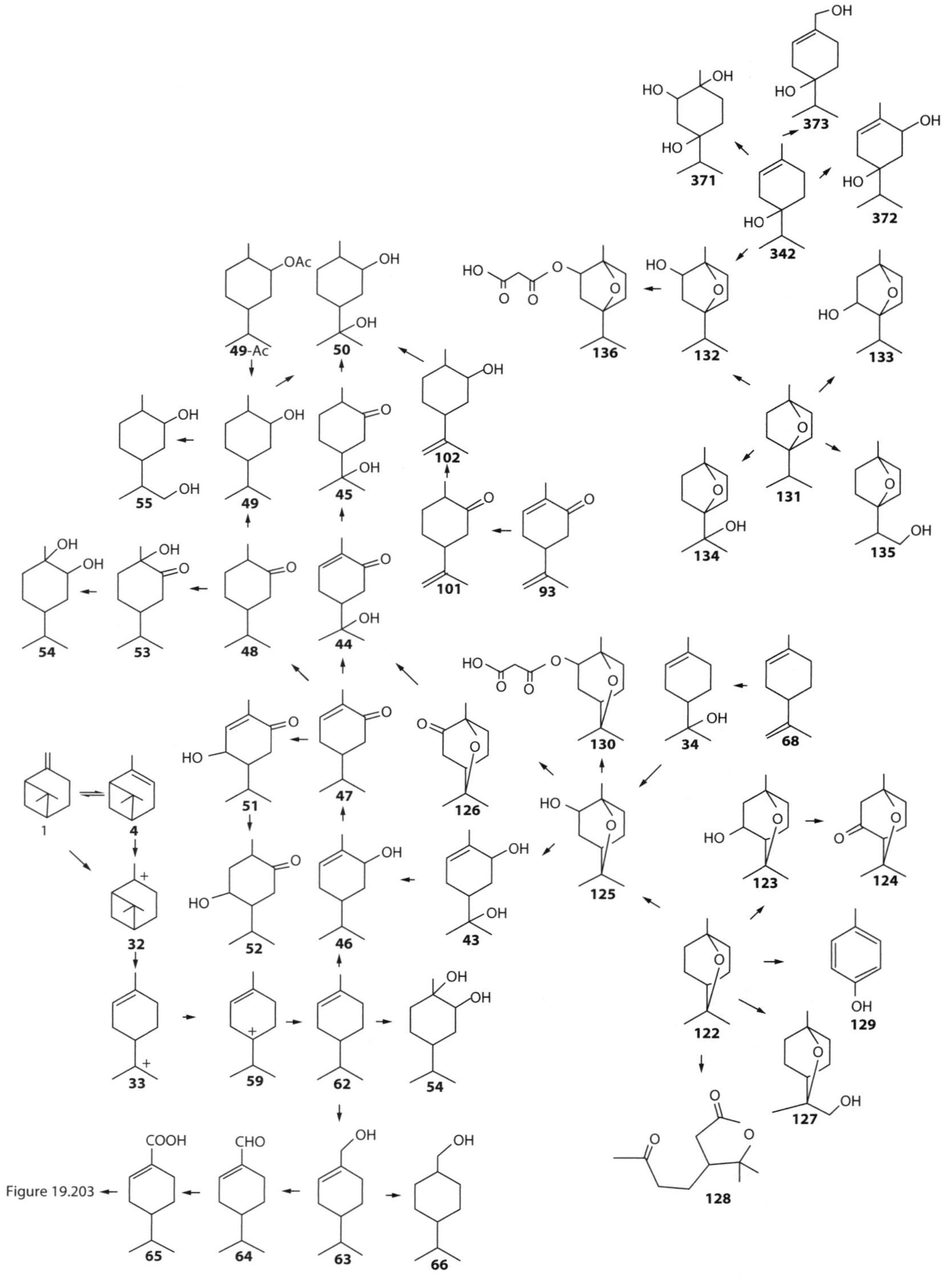

FIGURE 19.204 Metabolic pathways of 1,8-cineole (**122**), 1,4-cineole (**131**), phellandrene (**62**), and carvotanacetone (**47**) by microorganisms.

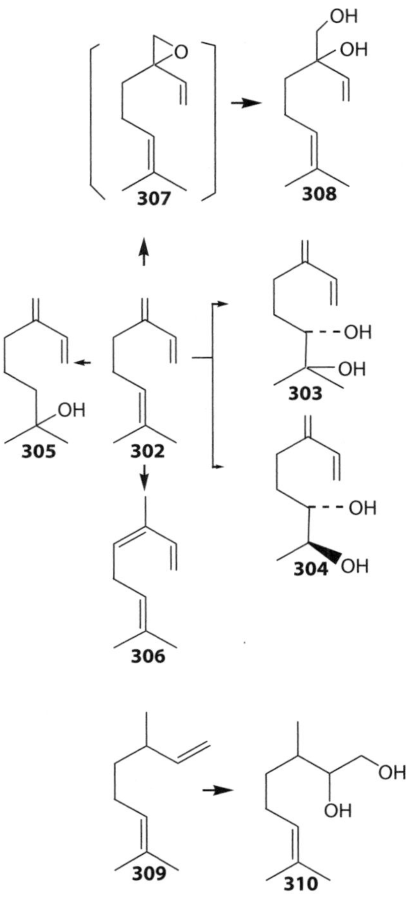

FIGURE 19.205 Metabolic pathways of myrcene (**302**) and citronellene (**309**) by rat and microorganisms.

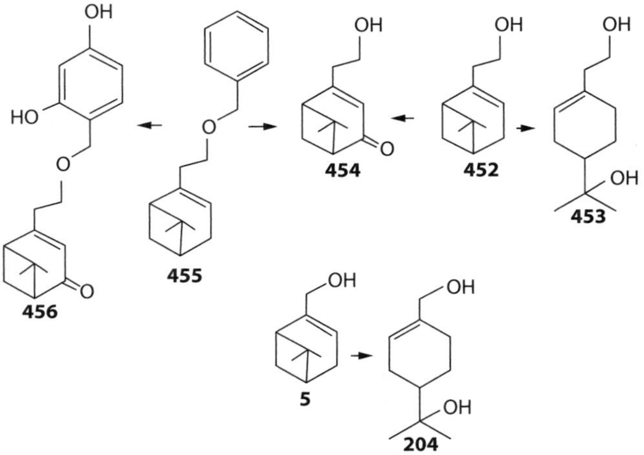

FIGURE 19.206 Metabolic pathways of nopol (**452**) and nopol benzyl ether (**455**) by microorganisms.

(**1**) is metabolized to α-terpineol (**34**) via pinyl cation (**32**) and 1-*p*-menthene-8-cation (**33**) (Saeki and Hashimoto, 1968, 1971; Hosler, 1969; Hayashi et al., 1972). α-Terpineol (**34**) is metabolized to 8,9-epoxy-1-*p*-methanol (**58**) via diepoxide (**57**), terpin hydrate (**60**), and oleuropeic acid (**204**) (Saeki and Hashimoto, 1968, 1971; Shukla et al., 1968; Shukla and Bhattacharyya, 1968; Hosler, 1969; Hungund et al., 1970; Hayashi et al., 1972). As shown in Figure 19.202, oleuropeic acid (**204**) is formed from linalool (**206**) and α-terpineol (**34**) via **204**, **205**, and **213** as intermediates (Shukla et al., 1968; Shukla and Bhattacharyya, 1968; Hungund et al., 1970) and degradatively metabolized to perillic acid (**82**), 2-hydroxy-8-*p*-menthen-7-oic acid (**84**), 2-oxo-8-*p*-menthen-7-oic acid (**84**), 2-oxo-8-*p*-menthen-1-oic acid (**85**), and β-isopropyl pimelic acid (**86**) (Shukla et al., 1968; Shukla and Bhattacharyya, 1968; Hungund et al., 1970). Oleuropeic acid (**204**) is also formed from β-pinene (**1**) via α-terpineol (**34**) as the intermediate (Noma et al., 2001). Oleuropeic acid (**204**) is also formed from myrtenol (**5**) by rearrangement reaction by *A. niger* TBUYN-2 (Noma and Asakawa, 2005b). Finally, β-pinene (**1**) is metabolized to borneol (**36**) and camphor (**37**) via two cations (**32** and **35**) and to 1-*p*-menthene (**62**) via two cations (**33** and **59**) (Shukla and Bhattacharyya, 1968). 1-*p*-Menthene (**62**) is metabolized to phellandric acid (**65**) via phellandrol (**63**) and phellandral (**64**), which is further degradatively metabolized through **246–251** and **89** to water and carbon dioxide as shown in Figure 19.204 (Shukla et al., 1968). Phellandral (**64**) is easily reduced to give phellandrol (**63**) by *Euglena* sp. and *Dunaliella* sp. (Noma et al., 1984, 1986, 1991a,b, 1992d). Furthermore, 1-*p*-menthene (**62**) is metabolized to 1-*p*-menthen-2-ol (**46**) and *p*-menthane-1,2-diol (**54**) as shown in Figure 19.204. Perillic acid (**82**) is easily formed from perillaldehyde (**78**) and perillyl alcohol (**74**) (Figure 19.19) (Swamy et al., 1965; Dhavalikar and Bhattacharyya, 1966; Dhavalikar et al., 1966; Ballal et al., 1967; Shima et al., 1972; Kayahara et al., 1973). α-Terpineol (**34**) is also formed from linalool (**206**). α-Pinene (**4**) is metabolized by five pathways as follows: First, α-pinene (**4**) is metabolized to myrtenol (**5**), myrtenal (**6**), and myrtenoic acid (**7**) (Shukla et al., 1968; Shukla and Bhattacharyya, 1968; Hungund et al., 1970; Ganapathy and Bhattacharyya, unpublished results). Myrtenal (**6**) is easily reduced to myrtenol (**5**) by *Euglena* and *Dunaliella* spp., (Noma et al., 1991a,b, 1992d). Myrtanol (**8**) is metabolized to 3-hydroxy- (**9**) and 4-hydroxymyrtanol (**10**) (Miyazawa et al., 1994a). Second, α-pinene (**4**) is metabolized to verbenol (**23**), verbenone (**24**), 7-hydroxyverbenone (**25**), and verbanone-7-al (**27**) (Bhattacharyya et al., 1960; Prema and Bhattacharyya, 1962; Miyazawa et al., 1991a). Third, α-pinene (**4**) is metabolized to thujone (**28**), thujoyl alcohol (**29**), 1-hydroxy- (**30**), and 1,3-dihydroxythujone (**31**) (Gibbon and Pirt, 1971; Miyazawa et al., 1992a; Noma, 2000). Fourth, α-pinene (**4**) is metabolized to sobrerol (**43**) and carvotanacetol (**46**, 1-*p*-menthen-2-ol) via α-pinene epoxide (**38**) and two cations (**41** and **42**). Sobrerol (**43**) is further metabolized to 8-hydroxycarvotanacetone (**44**, carvonhydrate), 8-hydrocarvomenthone (**45**), and *p*-menthane-2,8-diol (**50**) (Prema and Bhattacharyya, 1962; Noma, 2007). In the metabolism of sobrerol (**43**), 8-hydroxycarvotanacetone (**44**), and 8-hydroxycarvomenthone (**45**) by *A. niger* TBUYN-2, the formation of *p*-menthane-2,8-diol (**50**) is very highly enantio- and diastereoselective in the reduction of 8-hydroxycarvomenthone (Noma, 2007). 8-Hydroxycarvotanacetone (**44**) is a common metabolite from sobrerol (**43**) and carvotanacetone (**47**). That is, carvotanacetone (**47**) is metabolized to carvomenthone (**48**), carvomenthol (**49**), 8-hydroxycarvomenthol (**50**), 5-hydroxycarvotanacetone (**51**), 8-hydroxycarvotanacetone (**44**), 5-hydroxycarvomenthone (**52**), and 2,3-lactone (**56**) (Gibbon and Pirt, 1971; Gibbon et al., 1972; Noma et al., 1974a, 1985c, 1988b). Carvomenthone (**48**) is metabolized to **45**, 8-hydroxycarvomenthol (**50**), 1-hydroxycarvomenthone (**53**), and *p*-menthane-1,2-diol (**54**) (Noma et al., 1985b, 1988b). Compound **52** is metabolized to 6-hydroxymenthol (**139**), which is the common metabolite of menthol (**137**) (see Figure 19.201). Carvomenthol (**49**) is metabolized to 8-hydroxycarvomenthol (**50**) and *p*-menthane-2, 9-diol (**55**). Finally, α-pinene (**4**) is metabolized to borneol (**36**) and camphor (**37**) via **32** and **35** and to phellandrene (**62**) via **32** and two cations (**33** and **59**) as mentioned in the metabolism of β-pinene (**1**). Carvotanacetone (**47**) is also metabolized degradatively to 3,4-dimethylvaleric acid (**177**) via **56** and **158–163** as shown in Figure 19.201 (Gibbon and Pirt, 1971; Gibbon et al., 1972). α-Pinene (**4**) is also metabolized to 2-(4-methyl-3-cyclohexenylidene)-propionic acid (**67**) (Figure 19.199).

Metabolic pathways of limonene (**68**), perillyl alcohol (**74**), carvone (**93**), isopiperitenone (**111**), and piperitenone (**112**) are summarized in Figure 19.199. Limonene (**68**) is metabolized by eight pathways. That is, limonene (**68**) is converted into α-terpineol (**34**) (Savithiry et al., 1997), limonene-1,2-epoxide (**69**), 1-*p*-menthene-9-oic acid (**70**), perillyl alcohol (**74**), 1-*p*-menthene-8,9-diol (**79**), isopiperitenol (**110**), *p*-mentha-1,8-diene-4-ol (**80**, 4-terpineol), and carveol (**81**) (Dhavalikar and Bhattacharyya, 1966; Dhavalikar et al., 1966; Bowen, 1975; Noma et al., 1982, 1992d; Miyazawa et al., 1983; Savithrry et al., 1997; Van der Werf et al., 1997, 1998b; Van der Werf and de Bont, 1998a). Dihydrocarvone (**101**), limonene-1,2-diol (**71**), 1-hydroxy-8-*p*-menthene-2-one (**72**), and *p*-mentha-2,8-diene-1-ol (**73**) are formed from limonene (**68**) via limonene epoxide (**69**) as intermediate. Limonene (**68**) is also metabolized via carveol (**78**), limonene-1,2-diol (**71**), carvone (**93**), 1-*p*-menthene-6,9-diol (**95**), 8,9-dihydroxy-1-*p*-menthene (**90**), α-terpineol (**34**), 2α-hydroxy-1,8-cineole (**125**), and *p*-menthane-1,2,8-triol (**334**). Bottrospicatol (**92**) and 5-hydroxycarveol (**94**) are formed from *cis*-carveol by *S. bottropensis* SY-2-1 (Noma et al., 1982; Nishimura et al., 1983a; Noma and Asakawa, 1992; Noma and Nishimura, 1992). Carvyl acetate and carvyl propionate (both are shown as **106**) are hydrolyzed enantio- and diastereoselectively to carveol (**78**) (Oritani and Yamashita, 1980; Noma, 2000). Carvone (**93**) is metabolized through four pathways as follows: First, carvone (**93**) is reduced to carveol (**81**) (Noma, 1980). Second, it is epoxidized to carvone-8,9-epoxide (**96**), which is further metabolized to dihydrocarvone-8,9-epoxide (97), dihydrocarveol-8,9-epoxide (**103**), and menthane-2,8,9-triol (**104**) (Noma et al., 1980; Noma and Nishimura, 1982;Noma, 2000). Third, **93** is hydroxylated to 5-hydroxycarvone (**98**), 5-hydroxydihydrocarvone (**99**), and 5-hydroxydihydrocarveol (**100**) (Noma and Nishimura, 1982). Dihydrocarvone (**101**) is metabolized to 8-*p*-menthene-1,2-diol (**71**) via l-hydroxydihydrocarvone (**72**), 10-hydroxydihydrocarvone (**106**), and dihydrocarveol (**102**), which is metabolized to 10-hydroxydihydrocarveol (**107**), *p*-menthane-2,8-diol (**50**), dihydrocarveol-8,9-epoxide (**100**), *p*-menthane-2,8,9-triol (**104**), and dihydrobottrospicatol (**105**) (Noma et al., 1985a,b). In the biotransformation of (+)-carvone by plant pathogenic fungi, *A. niger* Tiegh, isodihydrocarvone (**101**) was metabolized to 4-hydroxyisodihydrocarvone (**378**) and 1-hydroxyisodihydrocarvone (**72**) (Noma and Asakawa, 2008). 8,9-Epoxydihydrocarvyl acetate (**109**) is hydrolyzed to 8,9-epoxydihydrocarveol (**103**). Perillyl alcohol (**74**) is metabolized through three pathways to shisool (**75**), shisool-8,9-epoxide (**76**), perillyl alcohol-8,9-epoxide (**77**), perillaldehyde (**78**), perillic acid (**82**), and 4,9-dihydroxy-1-*p*-menthen-7-oic acid (**83**). Perillic acid (**82**) is metabolized degradatively to **84–89** as shown in Figure 19.200 (Swamy et al., 1965; Dhavalikar and Bhattacharyya, 1966; Dhavalikar et al., 1966; Ballal et al., 1967; Shukla et al., 1968; Shukala and Bhattacharyya, 1968; Shima et al., 1972; Kayahara et al., 1973; Hungund et al., 1977). Isopiperitenol (**110**) is reduced to isopiperitenone (**111**), which is metabolized to 3-hydroxy- (**115**), 4-hydroxy- (**116**), 7-hydroxy- (**113**), and 10-hydroxyisopiperitenone (**114**) and piperitenone (**112**). Compounds isopiperitenone (**111**) and piperitenone (**112**) are isomerized to each other (Noma et al., 1992c). Furthermore, piperitenone (**112**) is metabolized to 8-hydroxypiperitone (**157**) and 5-hydroxy- (**117**) and 7-hydroxypiperitenone (**118**). Pulegone (**119**) is metabolized to **112**, 8-hydroxymenthenone (**121**), and 8,9-dehydromenthenone (**120**).

Metabolic pathways of menthol (**137**), menthone (**149**), thymol (**179**), and carvacrol methyl ether (**202**) are summarized in Figure 19.201. Menthol (**137**) is generally hydroxylated to give 1-hydroxy- (**138**), 2-hydroxy- (**140**), 4-hydroxy- (**141**), 6-hydroxy- (**139**), 7-hydroxy- (**143**), 8-hydroxy- (**142**), and 9-hydroxymenthol (**144**) and 1,8-dihydroxy- (**146**) and 7,8-dihydroxymenthol (**148**) (Asakawa et al., 1991; Takahashi et al., 1994; Van der Werf et al., 1997). Racemic menthyl acetate and menthyl chloroacetate are hydrolyzed asymmetrically by an esterase of microorganisms (Brit Patent, 1970; Moroe et al., 1971; Watanabe and Inagaki, 1977a,b). Menthone (**149**) is reductively metabolized to **137** and oxidatively metabolized to 3,7-dimethyl-6-hydroxyoctanoic acid (**152**), 3,7-dimethyl-6-oxooctanoic acid (**153**), 2-methyl-2,5-oxidoheptenoic acid (**154**), 1-hydroxymenthone (**150**), piperitone (**156**), 7-hydroxymenthone (**151**), menthone-7-al (**163**), menthone-7-oic acid (**164**), and 7-carboxylmenthol (**165**) (Sawamura et al., 1974). Compound **156** is metabolized to menthone-1,2-diol (**155**) (Miyazawa et al., 1991e, 1992d,e). Compound **148** is metabolized to 6-hydroxy- (**158**),

8-hydroxy- (**157**), and 9-hydroxypiperitone (**159**), piperitone-7-al (**160**), 7-hydroxypiperitone (**161**), and piperitone-7-oic acid (**162**) (Lassak et al., 1973). Compound **149** is also formed from menthenone (**148**) by hydrogenation (Mukherjee et al., 1973), which is metabolized to 6-hydroxymenthenone (**181**, 6-hydroxy-4-*p*-menthen-3-one). 6-Hydroxymenthenone (**181**) is also formed from thymol (**179**) via 6-hydroxythymol (**180**). 6-Hydroxythymol (**180**) is degradatively metabolized through **182–185** to **186**, **187**, and **89** (Mukherjee et al., 1974). Piperitone oxide (**166**) is metabolized to 1-hydroxymenthone (**150**) and 4-hydroxypiperitone (**167**) (Lassak et al., 1973; Miyazawa et al., 1991e). Piperitenone oxide (**168**) is metabolized to 1-hydroxymenthone (**150**), 1-hydroxypulegone (**169**), and 2,3-seco-*p*-menthalacetone-3-en-1-ol (**170**) (Lassak et al., 1973; Miyazawa et al., 1991e). *p*-Cymene (**178**) is metabolized to 8-hydroxy- (**188**) and 9-hydroxy-p-cymene (**189**), 2-(4-methylphenyl)-propanoic acid (**190**), thymol (**179**), and cumin alcohol (**192**), which is further converted degradatively to *p*-cumin aldehyde (**193**), cumic acid (**194**), *cis*-2,3-dihydroxy-2,3-dihydro-*p*-cumic acid (**195**), 2,3-dihydroxy-*p*-cumic acid (**197**), **198–200**, and **89** as shown in Figure 19.3 (Chamberlain and Dagley, 1968; DeFrank and Ribbons, 1977a,b; Hudlicky et al., 1999; Noma, 2000). Compound **197** is also metabolized to 4-methyl-2-oxopentanoic acid (**201**) (DeFrank and Ribbons, 1977a). Compound **193** is easily metabolized to 192 and **194** (Noma et al., 1991a, 1992). Carvacrol methyl ether (**202**) is easily metabolized to 7-hydroxycarvacrol methyl ether (**203**) (Noma, 2000).

Metabolic pathways of borneol (**36**), camphor (**37**), phellandral (**64**), linalool (**206**), and *p*-menthane (**252**) are summarized in Figure 19.202. Borneol (**36**) is formed from β-pinene (**1**), α-pinene (**4**), **34**, bornyl acetate (**226**), and camphene (**229**), and it is metabolized to **36**; 3-hydroxy- (**243**), 5-hydroxy- (**235**), 6-hydroxy- (**228**), and 9-hydroxycamphor (**225**); and 1,2-campholide (**23**). Compound **228** is degradatively metabolized to 6-oxocamphor (**229**) and **230–234**, whereas **237** is also degradatively metabolized to 6-hydroxy-1,2-campholide (**238**), 6-oxo-1,2-campholide (**239**), and **240–242**. 5-Hydroxycamphor (**235**) is metabolized to **238**, 5-oxocamphor (**236**), and 6-oxo-1,2-campholide (**239**). Compound **243** is also metabolized to camphorquinone (**244**) and 2-hydroxyepicamphor (**245**) (Bradshaw et al., 1959; Conrad et al., 1961, 1965a,b; Gunsalus et al., 1965; Chapman et al., 1966; Hartline and Gunsalus, 1971; Oritani and Yamashita, 1974). 1-*p*-Menthene (**62**) is formed **1** and **4** via three cations (**32**, **33**, and **59**) and metabolized to phellandrol (**63**) (Noma et al., 1991a) and *p*-menthane-1,2-diol (**54**). Compound **63** is metabolized to phellandral (**64**) and 7-hydroxy-*p*-menthane (**66**). Compound **64** is furthermore metabolized degradatively to CO_2 and water via phellandric acid (**65**), **246–251**, and **89** (Dhavalikar and Bhattacharyya, 1966; Dhavalikar et al., 1966; Bahhal et al., 1967; Shukla et al., 1968; Shukla and Bhattacharyya, 1968; Hungund et al., 1970). Compound **64** is also easily reduced to phellandrol (**63**) (Noma et al., 1991a, 1992a). *p*-Menthane (**252**) is metabolized via 1-hydroxy-*p*-menthane (**253**) to *p*-menthane-1,9-diol (**254**) and *p*-menthane-1,7-diol (**255**) (Tukamoto et al., 1974, 1975; Noma et al., 1990). Compound **255** is degradatively metabolized via **256–248** to CO_2 and water through the degradation pathway of phellandric acid (**65**, **246–251**, and **89**) as aforementioned. Linalool (**206**) is metabolized to α-terpineol (**34**), camphor (**37**), oleuropeic acid (**61**), 2-methyl-6-hydroxy-6-carboxy-2,7-octadiene (**211**), 2-methyl-6-hydroxy-6-carboxy-7-octene (**199**), 5-methyl-5-vinyltetrahydro-2-furanol (**215**), 5-methyl-5-vinyltetrahydro-2-furanone (**216**), and malonyl ester (**218**). 1-Hydroxylinalool (**219**) is metabolized degradatively to 2,6-dimethyl-6-hydroxy-*trans*-2,7-octadienoic acid (**220**), 4-methyl-*trans*-3, 5-hexadienoic acid (**221**), 4-methyl-*trans*-3,5-hexadienoic acid (**222**), 4-methyl-*trans*-2-hexenoic acid (**223**), and isobutyric acid (**224**). Compound **206** is furthermore metabolized via **213** to **61**, **82**, and **84–86** as shown in Figure 19.2 (Mizutani et al., 1971; Murakami et al., 1973; Madyastha et al., 1977; Rama Devi and Bhattacharyya, 1977a,b; Rama Devi et al., 1977; David and Veschambre, 1985; Miyazawa et al., 1994a,b).

Metabolic pathways of citronellol (**258**), citronellal (**261**), geraniol (**271**), nerol (**272**), citral [neral (**275**) and geranial (**276**)], and myrcene (**302**) are summarized in Figure 19.203 (Seubert and Fass, 1964; Hayashi et al., 1968; Rama Devi and Bhattacharyya, 1977a,b). Geraniol (**271**) is formed from citronellol (**258**), nerol (**272**), linalool (**206**), and geranyl acetate (**270**) and metabolized through

10 pathways. That is, compound **271** is hydrogenated to give citronellol (**258**), which is metabolized to 2,8-dihydroxy-2,6-dimethyl octane (**260**) via 6,7-epoxycitronellol (**268**), isopulegol (**267**), limonene (**68**), 3,7-dimethyloctane-1,8-diol (**266**) via 3,7-dimethyl-6-octene-1,8-diol (**265**), **267**, citronellal (**261**), dihydrocitronellol (**259**), and nerol (**272**). Citronellyl acetate (**269**) and isopulegyl acetate (**301**) are hydrolyzed to citronellol (**258**) and isopulegol (**267**), respectively. Compound **261** is metabolized via pulegol (**263**) and isopulegol (**267**) to menthol (**137**). Compound **271** and **272** are isomerized to each other. Compound **272** is metabolized to **271**, **258**, citronellic acid (**262**), nerol-6-,7-epoxide (**273**), and neral (**275**). Compound **272** is metabolized to neric acid (**277**). Compounds **275** and **276** are isomerized to each other. Compound **276** is completely decomposed to CO_2 and water via geranic acid (**278**), 2,6-dimethyl-8-hydroxy-7-oxo-2-octene (**279**), 6-methyl-5-heptenoic acid (**280**), 7-methyl-3-oxo-6-octenoic acid (**283**), 6-methyl-5-heptenoic acid (**284**), 4-methyl-3-heptenoic acid (**284**), 4-methyl-3-pentenoic acid (**285**), and 3-methyl-2-butenoic acid (**286**). Furthermore, compound **271** is metabolized via 3-hydroxymethyl-2,6-octadien-1-ol (**287**), 3-formyl-2,6-octadiene-1-ol (**288**), and 3-carboxy-2,6-octadiene-1-ol (**289**) to 3-(4-methyl-3-pentenyl)-3-butenolide (**290**). Geraniol (**271**) is also metabolized to 3,7-dimethyl-2,3-dihydroxy-6-octen-1-ol (**292**), 3,7-dimethyl-2-oxo-3-hydroxy-6-octen-1-ol (**293**), 2-methyl-6-oxo-2-heptene (**294**), 6-methyl-5-hepten-2-ol (**298**), 2-methyl-2-heptene-6-one-1-ol (**295**), and 2-methyl-γ-butyrolactone (**296**). Furthermore, **271** is metabolized to 7-methyl-3-oxo-6-octanoic acid (**299**), 7-hydroxymethyl-3-methyl-2,6-octadien-1-ol (**291**), 6,7-epoxygeraniol (**274**), 3,7-dimethyl-2,6-octadiene-1,8-diol (**300**), and 3,7-dimethyloctane-1,8-diol (**266**).

Metabolic pathways of 1,8-cineole (**122**), 1,4-cineole (**131**), phellandrene (**62**), carvotanacetone (**47**), and carvone (**93**) by microorganisms are summarized in Figure 19.204.

1,8-Cineole (**112**) is biotransformed to 2-hydroxy- (**125**), 3-hydroxy- (**123**), and 9-hydroxy-1,8-cineole (**127**), 2-oxo- (**126**) and 3-oxo-1,8-cineole (**124**), lactone [**128**, (*R*)-5,5-dimethyl-4-(3′-oxobutyl)-4,5-dihydrofuran-2-(3H)-one], and *p*-hydroxytoluene (**129**) (MacRae et al., 1979; Nishimura et al., 1982; Noma and Sakai, 1984). 2-Hydroxy-1,8-cineole (**125**) is further converted into 2-oxo-1,8-cineole (**126**), 1,8-cineole-2-malonyl ester (**130**), sobrerol (**43**), and 8-hydroxycarvotanacetone (**44**) (Miyazawa et al., 1995b). 2-Hydroxy-1,8-cineole (**125**) and 2-oxo-1,8-cineole (**126**) are also biodegraded to sobrerol (**43**) and 8-hydroxycarvotanacetone (**44**), respectively. 2-Hydroxy-1,8-cineole (**125**) was esterified to give malonyl ester (**130**). 2-Hydroxy-1,8-cineole (**125**) was formed from limonene (**68**) by citrus pathogenic fungi *P. digitatum* (Noma and Asakawa, 2007b). 1,4-Cineole (**131**) is metabolized to 2-hydroxy- (**132**), 3-hydroxy- (**133**), 8-hydroxy- (**134**), and 9-hydroxy-1,4-cineole (**135**). Compound **132** is also esterified to malonyl ester (**136**) as well as **125** (Miyazawa et al., 1995b). Terpinen-4-ol (342) is metabolized to 2-hydroxy-1,4-cineole (**132**), 2-hydroxy- (**372**) and 7-hydroxyterpinene-4-ol (**342**), and *p*-menthane-1,2,4-triol (**371**) (Abraham et al., 1986; Kumagae and Miyazawa, 1999; Noma and Asakawa, 2007a). Phellandrene (**62**) is metabolized to carvotanacetol (**46**) and phellandrol (**63**). Carvotanacetol (**46**) is further metabolized through the metabolism of carvotanacetone (**47**). Phelandrol (**63**) is also metabolized to give phellandral (**64**), phellandric acid (**65**), and 7-hydroxy-*p*-menthane (**66**). Phellandric acid (**65**) is completely degraded to carbon dioxide and water as shown in Figure 19.202.

Metabolic pathways of myrcene (**302**) and citronellene (**309**) by microorganisms and insects are summarized in Figure 19.205. β-Myrcene (**302**) was metabolized with *D. gossypina* ATCC 10936 (Abragam et al., 1985) to the diol (**303**) and a side product (**304**). β-Myrcene (**302**) was metabolized with *G. applanatum*, *P. flabellatus*, and *P. sajor-caju* to myrcenol (**305**) (2-methyl-6-methylene-7-octen-2-ol) and **306** (Busmann and Berger, 1994).

β-Myrcene (**302**) was converted by common cutworm larvae, *S. litura*, to give myrcene-3,(10)-glycol (**308**) via myrcene-3,(10)-epoxide (**307**) (Miyazawa et al., 1998). Citronellene (**309**) was metabolized by cutworm *S. litura* to give 3,7-dimethyl-6-octene-1,2-diol (**310**) (Takechi and Miyazawa, 2005). Myrcene (**302**) is metabolized to two kinds of diols (**303** and **304**), myrcenol (**305**), and ocimene (**306**) (Seubert and Fass, 1964; Abraham et al., 1985). Citronellene (**309**) was metabolized to (**310**) by *S. litura* (Takeuchi and Miyazawa, 2005).

Metabolic pathways of nopol (**452**) and nopol benzyl ether (**455**) by microorganisms are summarized in Figure 19.206. Nopol (**452**) is metabolized mainly to 7-hydroxyethyl-α-terpineol (453) by rearrangement reaction and 3-oxoverbenone (454) as minor metabolite by *Aspergillus* spp. including *A. niger* TBUYN-2 (Noma and Asakawa, 2006b,c). Myrtenol (**5**) is also metabolized to oleuropeic alcohol (**204**) by rearrangement reaction. However, nopol benzyl ether (**455**) was easily metabolized to 3-oxoverbenone (**454**) and 3-oxonopol-2′,4′-dihydroxybenzylether (**456**) as main metabolites without rearrangement reaction (Noma and Asakawa, 2006c).

19.5.2 Microbial Transformation of Terpenoids as Unit Reaction

Microbiological oxidation and reduction patterns of terpenoids and related compounds by fungi belonging to *Aspergillus* spp. containing *A. niger* TBUYN-2 and *E. gracilis* Z. are summarized in Tables 19.18 and 19.19, respectively. Dehydrogenation of secondary alcohols to ketones, hydroxylation of both nonallylic and allylic carbons, oxidation of olefins to form diols and triols via epoxides, reduction of both saturated and α,β-unsaturated ketones, and hydrogenation of olefin conjugated with the carbonyl group were the characteristic features in the biotransformation of terpenoids and related compounds by *Aspergillus* spp.

Compound names: **1**, β-pinene; **2**, pinocarveol; **3**, pinocarvone; **4**, α-pinene; **5**, myrtenol; **6**, myrtenal; **7**, myrtenoic acid; **8**, myrtanol; **9**, 3-hydroxymyrtenol; **10**, 4-hydroxymyrtanol; **11**, α-fenchol; **12**, fenchone; **13**, 6-hydroxyfenchone; **14**, 5-hydroxyfenchone; **15**, 5-oxofenchone; **16**, 9-hydroxyfenchone; **17**, fenchone-9-al; **18**, fenchone-9-oic acid; **19**, fenchoquinone; **20**, 2-hydroxyfenchone; **21**, 2,3-fencholide; **22**, 1,2-fencholide; **23**, verbenol; **24**, verbenone; **25**, 7-hydroxyverbenone; **26**, 7-hydroxyverbenone; **27**, verbanone-4-al; **28**, thujone; **29**, thujoyl alcohol; **30**, 1-hydroxythujone; **31**, 1,3-dihydroxythujone; **32**, pinyl cation; **33**, 1-*p*-menthene-8-cation; **34**, α-terpineol; **35**, bornyl cation; **36**, borneol; **37**, camphor; **38**, α-pinene epoxide; **39**, isonovalal; **40**, novalal; **41**, 2-hydroxypinyl cation; **42**, 6-hydroxy-1-*p*-menthene-8-cation; **43**, *trans*-sobrerol; **44**, 8-hydroxycarvotanacetone; (carvonehydrate); **45**, 8-hydrocarvomenthone; **46**, 1-*p*-menthen-2-ol; **47**, carvotanacetone; **48**, carvomenthone; **49**, carvomenthol; **50**, 8-hydroxycarvomenthol; **51**, 5-hydroxycarvotanacetone; **52**, 5-hydroxycarvomenthone; **53**, 1-hydroxycarvomenthone; **54**, *p*-menthane-1,2-diol; **55**, *p*-menthane-2,9-diol; **56**, 2,3-lactone; **57**, diepoxide; **58**, 8,9-epoxy-1-*p*-methanol; **59**, 1-*p*-menthene-4-cation; **60**, terpin hydrate; **61**, oleuropeic acid (8-hydroxyperillic acid); **62**, 1-*p*-menthene; **63**, phellandrol; **64**, phellandral; **65**, phellandric acid; **66**, 7-hydroxy-*p*-menthane; **67**, 2-(4-methyl-3-cyclohexenylidene)-propionic acid; **68**, limonene; **69**, limonene-1,2-epoxide; **70**, 1-*p*-menthene-9-oic acid; **71**, limonene-1,2-diol; **72**, 1-hydroxy-8-*p*-menthene-2-one; **73**, 1-hydroxy-*p*-menth-2,8-diene; **74**, perillyl alcohol; **75**, shisool; **76**, shisool-8,9-epoxide; **77**, perillyl alcohol-8,9-epoxide; **78**, perillaldehyde; **79**, 1-*p*-menthene-8,9-diol; **80**, 4-hydroxy-*p*-menth-1,8-diene (4-terpineol); **81**, carveol; **82**, perillic acid; **83**, 4,9-dihydroxy-1-*p*-menthene-7-oic acid; **84**, 2-hydroxy-8-*p*-menthen-7-oic acid; **85**, 2-oxo-8-*p*-menthen-7-oic acid; **86**, β-isopropyl pimelic acid; **87**, isopropenylglutaric acid; **88**, isobutenoic acid; **89**, isobutyric acid; **90**, 1-*p*-menthene-8,9-diol; **91**, carveol-8,9-epoxide; **92**, bottropsicatol; **93**, carvone; **94**, 5-hydroxycarveol; **95**, 1-*p*-menthene-6,9-diol; **96**, carvone-8,9-epoxide; **97**, dihydrocarvone-8,9-epoxide; **98**, 5-hydroxycarvone; **99**, 5-hydroxydihydrocarvone; **100**, 5-hydroxydihydrocarveol; **101**, dihydrocarvone; **102**, dihydrocarveol; **103**, dihydrocarveol-8,9-epoxide; **104**, *p*-menthane-2,8,9-triol; **105**, dihydrobottropsicatol; **106**, 10-hydroxydihydrocarvone; **107**, 10-hydroxydihydrocarveol; **108**, carvyl acetate and carvyl propionate; **109**, 8,9-epoxydihydrocarvyl acetate; **110**, isopiperitenol; **111**, isopiperitenone; **112**, piperitenone; **113**, 7-hydroxyisopiperitenone; **114**, 10-hydroxyisopiperitenone; **115**, 4-hydroxyisopiperitenone; **116**, 5-hydroxyisopiperitenone; **117**, 5-hydroxypiperitenone; **118**, 7-hydroxypiperitenone; **119**, pulegone; **120**, 8,9-dehydromenthenone; **121**, 8-hydroxymenthenone; **122**, 1,8-cineole; **123**, 3-hydroxy-1,8-cineole; **124**, 3-oxo-1,8-cineole; **125**, 2-hydroxy-1,8-cineole; **126**, 2-oxo-1,8-cineole; **127**,

TABLE 19.18
Microbiological Oxidation and Reduction Patterns of Monoterpenoids by *Aspergillus niger* TBUYN-2

Microbiological Oxidation

Oxidation of Alcohols	Oxidation of Primary Alcohols to Aldehydes and Acids		
	Oxidation of secondary alcohols to ketones	(−)-*trans*-Carveol (**81a′**), (+)-*trans*-carveol (**81a**),	
		(−)-*cis*-carveol (**81b′**),	
		(+)-*cis*-carveol (**81b**), 2α-hydroxy-1,8 cineole (**125b**), 3α-hydroxy-1,8-cineole (**123b**), 3β-hydroxy-1,8-cineole (**123a**)	
Oxidation of aldehydes to acids	Hydroxylation of nonallylic carbon	(−)-Isodihydrocarvone (**101c′**), (−)-carvotanacetone (**47′**), (+)-carvotanacetone (**47**), *cis*-*p*-menthane (**252**), 1α-hydroxy-*p*-menthane (**253**), 1,8-cineole (**122**), 1,4-cineole (**131**), (+)-fenchone (**12**), (−)-fenchone (**12′**), (−)-menthol (**137b′**), (+)-menthol (**137b**), (−)-neomenthol (**137a**), (+)-neomenthol (**137a**), (+)-isomenthol (**137c**)	
	Hydroxylation of allylic carbon	(−)-Isodihydrocarvone (**101b**), (+)-neodihydrocarveol (**102a′**), (−)-dihydrocarveol (**102b′**), (+)-dihydrocarveol (**102b**), (+)-limonene (**68**), (−)-limonene (**68′**)	
Oxidation of olefins	Formation of epoxides and oxides		
	Formation of diols	(+)-Neodihydrocarveol (**102a′**), (+)-dihydrocarveol (**102b**), (−)-dihydrocarveol (**102b′**), (+)-limonene (**68**), (−)-limonene (**68′**)	
	Formation of triols	(+)-Neodihydrocarveol (**102a′**)	
Lactonization			

Microbiological Reduction

Reduction of aldehydes to alcohols			
Reduction of ketones to alcohols	Reduction of saturated ketones	(+)-Dihydrocarvone (**101a′**), (−)-isodihydrocarvone (**101b**), (+)-carvomenthone (**48a′**), (−)-isocarvomenthone (**48b**)	
	Reduction of α,β-unsaturated ketones		
Hydrogenation of olefins	Hydrogenation of olefin conjugated with carbonyl group	(−)-Carvone (**93′**), (+)-carvone (**93**), (−)-carvotanacetone (**47′**), (+)-carvotanacetone (**47**)	
	Hydrogenation of olefin not conjugated with a carbonyl group		

9-hydroxy-1,8-cineole; **128**, lactone (*R*)-5,5-dimethyl-4-(3′-oxobutyl)-4,5-dihydrofuran-2-(3H)-one; **129**, *p*-hydroxytoluene; **130**, 1,8-cineole-2-malonyl ester; **131**, 1,4-cineole; **132**, 2-hydroxy-1,4-cineole; **133**, 3-hydroxy-1,4-cineole; **134**, 8-hydroxy-1,4-cineole; **135**, 9-hydroxy-1, 4-cineole; **136**, 1,4-cineole-2-malonyl ester; **137**, menthol; **138**, 1-hydroxymenthol; **139**, 6-hydroxymenthol; **140**, 2-hydroxymenthol; **141**, 4-hydroxymenthol; **142**, 8-hydroxymenthol; **143**, 7-hydroxymenthol; **144**, 9-hydroxymenthol; **145**, 7,8-dihydroxymenthol; **146**, 1,8-dihydroxymenthol; **147**, 3-*p*-menthene; **148**, menthenone; **149**, menthone; **150**, 1-hydroxymenthone; **151**, 7-hydroxymenthone; **152**, 3,7-dimethyl-6-hydroxyoctanoic acid; **153**, 3,7-dimethyl-6-oxooctanoic acid; **154**,

TABLE 19.19
Microbiological Oxidation, Reduction, and Other Reaction Patterns of Monoterpenoids by *Euglena gracilis Z*

Microbiological Oxidation

Oxidation of Alcohols	**Oxidation of Primary Alcohols to Aldehydes and Acids**	
	Oxidation of secondary alcohols to ketones	(−)-*trans*-Carveol (**81a′**), (+)-*cis*-carveol (**81b**), (+)-isoborneol (**36b**).[a]
Oxidation of aldehydes to acids		Myrtenal (**6**), myrtanal, (−)-perillaldehyde (**78**), *trans*- and *cis*-1,2-dihydroperillaldehydes (**261a** and **261b**), (−)-phellandral (**64**), *trans*- and *cis*-tetrahydroperillaldehydes, cuminaldehyde (**193**), (+)- and (−)-citronellal (**261** and **261′**).[b]
Hydroxylation	Hydroxylation of nonallylic carbon	(+)-Limonene (**68**), (−)-limonene (**68′**).
	Hydroxylation of allylic carbon	
Oxidation of olefins	Formation of epoxides and oxides	
	Formation of diols	
	Formation of triols	(+)- and (−)-Neodihydrocarveol (**102a′** and **a**), (−)- and (+)-dihydrocarveol (**102b′** and **b**), (+)- and (−)-isodihydrocarveol (**102c′** and **c**), (+)- and (−)-neoisodihydrocarveol (**102d′** and **d**).
Lactonization		

Microbiological Reduction

Reduction of aldehydes to alcohols	Reduction of terpene aldehydes to terpene alcohols	Myrtenal (**6**), myrtanal, (−)-perillaldehyde (**78**), *trans*- and *cis*-1,2-dihydroperillaldehydes (**261a** and **261b**), phellandral (**64**), *trans*- and *cis*-1,2-dihydroperillaldehydes (**261a** and **261b**), *trans*- and *cis*-tetrahydroperillaldehydes, cuminaldehyde (**193**), citral (**275** and **276**), (+)- (**261**) and (−)-citronellal (**261′**).
	Reduction of aromatic and related aldehydes to alcohols	
	Reduction of aliphatic aldehydes to alcohols	
Reduction of ketones to alcohols	Reduction of saturated ketones	(+)-Dihydrocarvone (**101a′**), (−)-isodihydrocarvone (**101b**), (+)-carvomenthone (**48a′**), (−)-isocarvomenthone (**48b**), (+)-dihydrocarvone-8,9-epoxides (**97a′**), (+)-isodihydrocarvone-8,9-epoxides (**97b′**), (−)-dihydrocarvone-8,9-epoxides (**97a**).
	Reduction of α,β-unsaturated ketones	
	Hydrogenation of olefins	(−)-Carvone (**93′**), (+)-carvone (**93**), (−)-carvotanacetone (**47′**), (+)-carvotanacetone (**47**), (−)-carvone-8,9-epoxides (**96′**), (+)-carvone-8,9-epoxides (**96**).
	Hydrogenation of olefin conjugated with carbonyl group	
	Hydrogenation of olefin not conjugated with a carbonyl group	

Hydrolysis

Hydrolysis	Hydrolysis of ester	(+)-*trans*- and *cis*-Carvyl acetates (**108a** and **b**), (−)-*cis*-carvyl acetate (**108b′**), (−)-*cis*-carvyl propionate, geranyl acetate (**270**).

(Continued)

TABLE 19.19 (Continued)
Microbiological Oxidation, Reduction, and Other Reaction Patterns of Monoterpenoids by *Euglena gracilis Z*

		Hydration
Hydration	Hydration of C = C bond in isopropenyl group to tertiary alcohol	(+)-Neodihydrocarveol (**102a′**), (−)-dihydrocarveol (**102b′**), (+)-isodihydrocarveol (**102c′**), (+)-neoisodihydrocarveol (**102d′**), (−)-neodihydrocarveol (**102a**), (+)-dihydrocarveol (**102b**), (−)-isodihydrocarveol (**102c**), (−)-neoisodihydrocarveol (**102d**), *trans*- and *cis*-shisools (**75a** and **75b**).
		Isomerization
Isomerization		Geraniol (**271**), nerol (**272**).

[a] Diastereo- and enantioselective dehydrogenation is observed in carveol, borneol, and isoborneol.
[b] Acids were obtained as minor products.

2-methyl-2,5-oxidoheptenoic acid; **155**, menthone-1,2-diol; **156**, piperitone; **157**, 8-hydroxypiperitone; **158**, 6-hydroxypiperitone; **159**, 9-hydroxypiperitone; **160**, piperitone-7-al; **161**, 7-hydroxypiperitone; **162**, piperitone-7-oic acid; **163**, menthone-7-al; **164**, menthone-7-oic acid; **165**, 7-carboxylmenthol; **166**, piperitone oxide; **167**, 4-hydroxypiperitone; **168**, piperitenone oxide; **169**, 1-hydroxypulegone; **170**, 2,3-seco-*p*-methylacetone-3-en-1-ol; **171**, 2-methyl-5-isopropyl-2,5-hexadienoic acid; **172**, 2,5,6-trimethyl-2,4-heptadienoic acid; **173**, 2,5,6-trimethyl-3-heptenoic acid; **174**, 2,5,6-trimethyl-2-heptenoic acid; **175**, 3-hydroxy-2,5,6-trimethyl-3-heptanoic acid; **176**, 3-oxo-2,5,6-trimethyl-3-heptanoic acid; **177**, 3,4-dimethylvaleric acid; **178**, *p*-cymene; **179**, thymol; **180**, 6-hydroxythymol **181**, 6-hydroxymenthenone, 6-hydroxy-4-*p*-menthen-3-one; **182**, 3-hydroxythymo-1,4-quinol; **183**, 2-hydroxythymoquinone; **184**, 2,4-dimethyl-6-oxo-3,7-dimethyl-2,4-octadienoic acid; **185**, 2,4,6-trioxo-3,7-dimethyl octanoic acid; **186**, 2-oxobutanoic acid; **187**, acetic acid; **188**, 8-hydroxy-*p*-cymene; **189**, 9-hydroxy-*p*-cymene; **190**, 2-(4-methylphenyl)-propanoic acid; **191**, carvacrol; **192**, cumin alcohol; **193**, *p*-cumin aldehyde; **194**, cumic acid; **195**, *cis*-2,3-dihydroxy-2, 3-dihydro-*p*-cumic acid; **196**, 3-hydroxycumic acid; **197**, 2,3-dihydroxy-*p*-cumic acid; **198**, 2-hydroxy-6-oxo-7-methyl-2,4-octadien-1,3-dioic acid; **199**, 2-methyl-6-hydroxy-6-carboxy-7-octene; **201**, 4-methyl-2-oxopentanoic acid; **202**, carvacrol methyl ether; **203**, 7-hydroxycarvacrol methyl ether; **204**, 8-hydroxyperillyl alcohol; **205**, 8-hydroxyperillaldehyde; **206**, linalool; **207**, linalyl-6-cation; **208**, linalyl-8-cation; **209**, 6-hydroxymethyl linalool; **210**, linalool-6-al; **211**, 2-methyl-6-hydroxy-6-carboxy-2,7-octadiene; **212**, 2-methyl-6-hydroxy-7-octen-6-oic acid; **213**, phellandric acid-8-cation; **214**, 2,3-epoxylinalool; **215**, 5-methyl-5-vinyltetrahydro-2-furanol; **216**, 5-methyl-5-vinyltetrahydro-2-furanone; **217**, 2,2,6-trimethyl-3-hydroxy-6-vinyltetrahydropyrane; **218**, malonyl ester; **219**, 1-hydroxylinalool (3,7-dimethyl-1,6-octadiene-8-ol); **220**, 2,6-dimethyl-6-hydroxy-*trans*-2,7-octadienoic acid; **221**, 4-methyl-*trans*-3,5-hexadienoic acid; **222**, 4-methyl-trans-3,5-hexadienoic acid; **223**, 4-methyl-*trans*-2-hexenoic acid; **224**, isobutyric acid; **225**, 9-hydroxycamphor; **226**, bornyl acetate; **228**, 6-hydroxycamphor; **229**, 6-oxocamphor; **230**, 4-carboxymethyl-2,3,3-trimethylcyclopentanone; **231**, 4-carboxymethyl-3,5,5-trimethyltetrahydro-2-pyrone; **232**, isohydroxycamphoric acid; **233**, isoketocamphoric acid; **234**, 3,4,4-trimethyl-5-oxo-trans-2-hexenoic acid; **235**, 5-hydroxycamphor; **236**, 5-oxocamphor; **237**, **238**, 6-hydroxy-1, 2-campholide; **239**, 6-oxo-1,2-campholide; **240**, 5-carboxymethyl-3,4,4-trimethyl-2-cyclopentenone; **241**, 6-carboxymethyl-4,5,5-trimethyl-5,6-dihydro-2-pyrone; **242**, 5-hydroxy-3,4,4-trimethyl-2-heptene-1,7-dioic acid; **243**, 3-hydroxycamphor; **244**, camphorquinone; **245**, 2-hydroxyepicamphor; **246**, 2-hydroxy-*p*-menthan-7-oic acid;

247, 2-oxo-*p*-menthan-7-oic acid; **248**, 3-isopropylheptane-1,7-dioic acid; **249**, 3-isopropylpentane-1,5-dioic acid; **250**, 4-methyl-3-oxopentanoic acid; **251**, methylisopropyl ketone; **252**, *p*-menthane; **253**, 1-hydroxy-*p*-menthane; **254**, *p*-menthane-1,9-diol; **255**, *p*-menthane-1,7-diol; **256**, 1-hydroxy-*p*-menthene-7-al; **257**, 1-hydroxy-*p*-menthene-7-oic acid; **258**, citronellol; **259**, dihydrocitronellol; **260**, 2,8-dihydroxy-2,6-dimethyl octane; **261**, citronellal; **262**, citronellic acid; **263**, pulegol; **264**, 7-hydroxymethyl-6-octene-3-ol; **265**, 3,7-dimethyl-6-octane-1,8-diol; **266**, 3,7-dimethyloctane-1,8-diol; **267**, isopulegol; **268**, 6,7-epoxycitronellol; **269**, citronellyl acetate; **270**, geranyl acetate; **271**, geraniol; **272**, nerol **273**, nerol-6,7-epoxide; **274**, 6,7-epoxygeraniol; **275**, neral; **276**, geranial; **277**, neric acid; **278**, geranic acid; **279**, 2,6-dimethyl-8-hydroxy-7-oxo-2-octene; **280**, 6-methyl-5-heptenoic acid; **281**, 7-methyl-3-carboxymethyl-2,6-octadiene-1-oic acid; **282**, 7-methyl-3-hydroxy-3-carboxymethyl-6-octen-1-oic acid; **283**, 7-methyl-3-oxo-6-octenoic acid; **284**, 6-methyl-5-heptenoic acid; **284**, 4-methyl-3-heptenoic acid; **285**, 4-methyl-3-pentenoic acid; **286**, 3-methyl-2-butenoic acid; **287**, 3-hydroxymethyl-2,6-octadiene-1-ol; **288**, 3-formyl-2,6-octadiene-1-ol; **289**, 3-carboxy-2,6-octadiene-1-ol; **290**, 3-(4-methyl-3-pentenyl)-3-butenolide; **291**, 7-hydroxymethyl-3-methyl-2,6-octadien-1-ol; **292**, 3,7-dimethyl-2,3-dihydroxy-6-octen-1-ol; **293**, 3,7-dimethyl-2-oxo-6-octene-1,3-diol; **294**, 6-methyl-5-hepten-2-one; **295**, 6-methyl-7-hydroxy-5-heptene-2-one; **296**, 2-methyl-γ-butyrolactone; **297**, 6-methyl-5-heptenoic acid; **298**, 6-methyl-5-hepten-2-ol; **299**, 7-methyl-3-oxo-6-octanoic acid; **300**, 3,7-dimethyl-2,6-octadiene-1,8-diol; **301**, isopulegyl acetate; **302**, myrcene; **303**, 2-methyl-6-methylene-7-octene-2,3-diol; **304**, 6-methylene-7-octene-2,3-diol; **305**, myrcenol; **306**, ocimene; **307**, myrcene-3,(10)-epoxide; **308**, myrcene-3,(10)-glycol; **309**, (−)-citronellene; **309′**, (+)-citronellene; **310**, (3*R*)-3,7-dimethyl-6-octene-1,2-diol; **310′**, (3*S*)-3,7-dimethyl-6-octene-1,2-diol; **311**, (*E*)-3,7-dimethyl-5-octene-1,7-diol; **312**, *trans*-rose oxide; **313**, *cis*-rose oxide; **314**, (2*Z*,5*E*)-3,7-dimethyl-2,5-octadiene-1,7-diol; **315**, (*Z*)-3,7-dimethyl-2,7-octadiene-1,6-diol; **316**, (2*E*,6*Z*)-2,6-dimethyl-2,6-octadiene-1,8-diol; **317**, a cyclization product; **318**, (2*E*,5*E*)-3,7-dimethyl-2,5-octadiene-1,7-diol; **319**, (*E*)-3,7-dimethyl-2,7-octadiene-1,6-diol; **320**, 8-hydroxynerol; **321**, 10-hydroxynerol; **322**, 1-hydroxy-3,7-dimethyl-2*E*,6*E*-octadienal; **323**, 1-hydroxy-3-,7-dimethyl-2*E*,6*E*-octadienoic acid; **324**, 3,9-epoxy-*p*-menth-1-ene; **325**, tetrahydrolinalool; **326**, 3,7-dimethyloctane-3,5-diol; **327**, 3,7-dimethyloctane-3,7-diol; **328**, 3,7-dimethyloctane-3,8-diol; **329**, dihydromyrcenol; **330**, 1,2-epoxydihydromyrcenol; **331**, 3β-hydroxydihydromyrcenol; **332**, dihydromyrcenyl acetate; **333**, 1,2-dihydroxydihydromyrcenyl acetate; **334**, (+)-*p*-menthane-1-β,2α,8-triol; **335**, α-pinene-1,2-epoxide; **336**, 3-carene; **337**, 3-carene-1,2-epoxide; **338**, (1*R*)-*trans*-isolimonene; **338**, (1*R*,4*R*)-*p*-menth-2-ene-8,9-diol; **339**, (1*R*,4*R*)-*p*-menth-2-ene-8,9-diol; **340**, α-terpinene; **341**, α-terpinene-7-oic acid; **342**, (−)-terpinen-4-ol; **343**, *p*-menthane-1,2,4-triol; **344**, γ-terpinene; **345**, γ-terpinene-7-oic acid; **346**, terpinolene; **347**, (1*R*)-8-hydroxy-3-*p*-menthen-2-one; **348**, (1*R*)-1,8-dihydroxy-3-*p*-menthen-2-one; **349**, 6β-hydroxyfenchol; **350**, 5β-hydroxyfenchol (a,5β,b,5α); **351**, terpinolene-1,2-*trans*-diol; **352**, terpinolene-4,8-diol; **353**, terpinolene-9-ol; **354**, terpinolene-10-ol; **355**, α-phellandrene; **356**, α-phellandrene-7-oic acid; **357**, terpinolene-7-oic acid; **358**, thymoquinone; **359**, 1,2-dihydroperillaldehyde; **360**, 1,2-dihydroperillic acid; **361**, 8-hydroxy-1,2-dihydroperillyl alcohol; **362**, tetrahydroperillaldehyde (a *trans*, b *cis*); **363**, tetrahydroperillic acid (a *trans*, b *cis*); **364**, (−)-menthol monoglucoside; **365**, (+)-menthol diglucoside; **366**, (+)-isopulegol; **367**, 7-hydroxy-(+)-isopulegol; **368**, 10-hydroxy-(+)-isopulegol; **369**, 1,2-epoxy-α-terpineol; **370**, bornane-2,8-diol; **371**, *p*-menthane-1α,2β,4β-triol; **372**, 1-*p*-menthene-4β,6-diol; **373**, 1-*p*-menthene-4a,7-diol; **374**, (+)-bottrospicatal; **375**, 1-*p*-menthene-2β,8,9-triol; **376**, (−)-perillyl alcohol monoglucoside; **377**, (−)-perillyl alcohol diglucoside; **378**, 4α-hydroxy-(−)-isodihydrocarvone; **379**, 2-methyl-2-cyclohexenone; **380**, 2-cyclohexenone; **381**, 3-methyl-2-cyclohexenone; **382**, 2-methylcyclohexanone; **383**, 2-methylcyclohexanol (a, *trans*; b, *cis*); **384**, 4-hydroxycarvone; **385**, 2-ethyl-2-cyclohexenone; **386**, 2-ethylcyclohexenone (a1*R*) (b1*S*); **387**, 1-hydroxypulegone; **388**, 5-hydroxypulegone; **389**, 8-hydroxymenthone; **390**, 10-hydroxy-(−)-carvone; **391**, 1,5,5-trimethylcyclopentane-1,4-dicarboxylic acid; **392**, 4β-hydroxy-(−)-menthone;

393, 1α,4β-dihydroxy-(−)-menthone; **394,** 7-hydroxy-9-carboxymenthone; **395,** 7-hydroxy-1,8-cineole; **396,** methyl ester of 2α-hydroxy-1,8-cineole; **397,** ethyl ester of 2α-hydroxy-1,8-cineole; **398,** *n*-propyl ester of 2α-hydroxy-1,8-cineole; **399,** *n*-butyl ester of 2α-hydroxy-1,8-cineole; **400,** isopropyl ester of 2α-hydroxy-1,8-cineole; **401,** tertiary butyl ester of 2α-hydroxy-1,8-cineole; **402,** methylisopropyl ester of 2α-hydroxy-1,8-cineole; **403,** methyl tertiary butyl ester of 2α-hydroxy-1,8-cineole; **404,** 2α-hydroxy-1,8-cineole monoglucoside (404 and 404'); **405,** 2α-hydroxy-1,8-cineole diglucoside; **406,** *p*-menthane-1,4-diol; **407,** 1-*p*-menthene-4β,6-diol; **408,** (−)-pinane-2α,3α-diol; **409,** (−)-6β-hydroxypinene; **410,** (−)-4α,5-dihydroxypinene; **411,** (−)-4α-hydroxypinene-6-one; **412,** 6β,7-dihydroxyfenchol; **413,** 3-oxo-pinane; **414,** 2α-hydroxy-3-pinanone; **415,** 2α, 5-dihydroxy-3-pinanone; **416,** 2α,10-dihydroxy-3-pinanone; **417,** *trans*-3-pinanol; **418,** pinane-2α,3α-diol; **419,** pinane-2α, 3α, 5-triol; **420,** isopinocampheol (3-pinanol); **421,** pinane-1,3α-diol; **422,** pinane-3α,5-diol; **423,** pinane-3β,9-diol; **424,** pinane-3β,4β,-diol; **425, 426,** pinane-3α,4β-diol; **427,** pinane-3α,9-diol; **428,** pinane-3α,6-diol; **429,** *p*-menthane-2α,9-diol; **430,** 2-methyl-3α-hydroxy-1-hydroxyisopropyl cyclohexane propane; **431,** 5-hydroxy-3-pinanone; **432,** 2α-methyl,3-(2-methyl-2-hydroxypropyl)-cyclopenta-1β-ol; **433,** 3-acetoxy-2α-pinanol; **434,** 8-hydroxy-α-pinene; **435, 436,** myrtanic acid; **437,** camphene; **438,** camphene glycol; **439,** (+)-3-carene; **440,** *m*-mentha-4,6-dien-8-ol; **441,** *m*-cymen-8-ol; **442,** 3-carene-9-ol; **443,** 3-carene-9-carboxylic acid; **444,** 3-caren-10-ol-9-carboxylic acid; **445,** 3-carene-9,10-dicarboxylic acid; **446,** (−)-*cis*-carane; **447,** dicarboxylic acid of (−)-*cis*-carane; **448,** (−)-6β-hydroxypinene; **449,** (−)-4α,5-dihydroxypinene; **450,** (−)-4α-hydroxypinene-6-one; **451,** 10-hydroxyverbenol; **452,** (−)-nopol; **453,** 7-hydroxymethyl-1-*p*-menthen-8-ol; **454,** 3-oxonopol; **455,** nopol benzyl ether; **456,** 4-oxonopol-2',4'-dihydroxybenzyl ether; **457,** 7-hydroxymethyl-1-*p*-menthen-8-ol benzyl ether; **458,** piperitenol; **459,** thymol methyl ether; **460,** 7-hydroxythymol methyl ether; **461,** 9-hydroxythymol methyl ether; **462,** 1,8-cineol-9-oic acid; **463,** 4-hydroxyphellandric acid; **464,** 4-hydroxydihydrophellandric acid; **465,** (+)-8-hydroxyfenchol; **466,** (−)-9-hydroxyfenchol; **467,** (+)-10-hydroxyfenchol; **468,** 4α-hydroxy-6-oxo-α-pinene; **469,** dihydrolinalyl acetate; **470,** 3-hydroxycarvacrol; **471,** 9-hydroxycarvacrol; **472,** carvacrol-9-oic acid; **473,** 8,9 dehydrocarvacrol; **474,** 8-hydroxycarvacrol; **475,** 7-hydroxycarvacrol; **476,** carvacrol-7-oic acid; **477,** 8,9-dihydroxycarvacrol; **478,** 7,9-dihydroxycarvacrol methyl ether; **479,** 7-hydroxythymol; **480,** 9-hydroxythymol; **481,** thymol-7-oic acid; **482,** 7,9-dihydroxythymol; **483,** thymol-9-oic acid; **484,** (1*R*,2*R*,3*S*,4*S*,5*R*)-3,4-pinanediol.

REFERENCES

Abraham, W.-R., H.-A. Arfmann, B. Stumpf, P. Washausen, and K. Kieslich, 1988. Microbial transformations of some terpenoids and natural compounds. In: *Bioflavour'87. Analysis—Biochemistry—Biotechnology*, P. Schreier, ed., pp. 399–414. Berlin, Germany: Walter de Gruyter and Co.

Abraham, W.-R., H.M.R. Hoffmann, K. Kieslich, G. Reng, and B. Stumpf, 1985. Microbial transformation of some monoterpenes and sesquiterpenoids. In: *Enzymes in Organic Synthesis*, R. Porter and S. Clark, ed., Ciba Foundation Symposium 111, pp. 146–160. London, U.K.: Pitman Press.

Abraham, W.-R., K. Kieslich, H. Reng, and B. Stumpf, 1984. Formation and production of 1,2-trans-glycols from various monoterpenes with 1-menthene skeleton by microbial transformations with *Diplodia gossypina*. In: *Third European Congress on Biotechnology*, Vol. 1, pp. 245–248. Verlag Chemie, Weinheim, Germany.

Abraham, W.-R., B. Stumpf, and H.-A. Arfmann, 1990. Chiral intermediates by microbial epoxidations. *J. Essent. Oil Res.*, 2: 251–257.

Abraham, W.-R., B. Stumpf, and K. Kieslich, 1986. Microbial transformation of terpenoids with 1-*p*-menthene skeleton. *Appl. Microbiol. Biotechnol.*, 24: 24–30.

Amsterdam, D. 1997. Susceptibility testing of antimicrobials in liquid media. In: *Antibiotics in Laboratory Medicine*, V. Lorian, ed., 4th ed. Baltimore, MD: Williams & Wilkins, Maple Press.

Asakawa, Y., H. Takahashi, M. Toyota, Y. and Noma, 1991. Biotransformation of monoterpenoids, (−)- and (+)-menthols, terpinolene and carvotanacetone by *Aspergillus* species. *Phytochemistry*, 30: 3981–3987.

Asakawa, Y., M. Toyota, T. Ishida, T. Takemoto, 1983. Metabolites in rabbit urine after terpenoid administration. *Proceedings of the 27th TEAC*, pp. 254–256.

Atta-ur-Rahman, M. Yaqoob, A. Farooq, S. Anjum, F. Asif, and M.I. Choudhary, 1998. Fungal transformation of (1R,2S,5R)-(−)-menthol by *Cephalosporium aphidicola*. *J. Nat. Prod.*, 61: 1340–1342.

Ausgulen, L.T., E. Solheim, and R.R. Scheline, 1987. Metabolism in rats of *p*-cymene derivatives: Carvacrol and thymol. *Pharmacol. Toxicol.*, 61: 98–102.

Babcka, J., J. Volf, J. Czchec, and P. Lebeda, 1956. Patent 56-9686b.

Ballal, N.R., P.K. Bhattacharyya, and P.N. Rangachari, 1967. Microbiological transformation of terpenes. Part XIV. Purification and properties of perillyl alcohol dehydrogenase. *Indian J. Biochem.*, 5: 1–6.

Bauer, K., D. Garbe, and H. Surburg (eds.), 1990. *Common Fragrance and Flavor Materials: Preparation, Properties and Uses*. 2nd revised ed., 218pp. New York: VCH Publishers.

Best, D.J. and K.J. Davis, 1988. *Soap, Perfumery Cosmetics* 4: 47.

Best, D.J., N.C. Floyd, A. Magalhaes, A. Burfield, and P.M. Rhodes, 1987. Initial enzymatic steps in the degradation of alpha-pinene by *Pseudomonas fluorescens* Ncimb 11671. *Biocatal. Biotransform.*, 1: 147–159.

Bhattacharyya, P.K. and K. Ganapathy, 1965. Microbial transformation of terpenes. VI. Studies on the mechanism of some fungal hydroxylation reactions with the aid of model systems. *Indian J. Biochem.*, 2: 137–145.

Bhattacharyya, P.K., B.R. Prema, B.D. Kulkarni, and S.K. Pradhan, 1960. Microbiological transformation of terpenes: Hydroxylation of α-pinene. *Nature*, 187: 689–690.

Bock, G., I. Benda, and P. Schreier, 1986. Biotransformation of linalool by *Botrytis cinerea*. *J. Food Sci.*, 51: 659–662.

Bock, G., I. Benda, and P. Schreier, 1988. Microbial transformation of geraniol and nerol by *Botrytis cinerea*. *Appl. Microbiol. Biotechnol.*, 27: 351–357.

Bouwmester, H.J., J.A.R. Davies, and H. Toxopeus, 1995. Enantiomeric composition of carvone, limonene, and carveols in seeds of dill and annual and biennial caraway varieties. *J. Agric. Food Chem.*, 43: 3057–3064.

Bowen, E.R., 1975. Potential by-products from microbial transformation of *d*-limonene, *Proc. Fla. State Hort. Soc.*, 88: 304–308.

Bradshaw, W.H., H.E. Conrad, E.J. Corey, I.C. Gunsalus, and D. Lednicer, 1959. Microbiological degradation of (+)-camphor. *J. Am. Chem. Soc.*, 81: 5507.

Brit Patent, 1970, No. 1,187,320.

Brunerie, P., I. Benda, G. Bock, and P. Schreier, 1987a. Bioconversion of citronellol by *Botrytis cinerea*. *Appl. Microbiol. Biotechnol.*, 27: 6–10.

Brunerie, P., I. Benda, G. Bock, and P. Schreier, 1987b. Biotransformation of citral by *Botrytis cinerea*. *Z. Naturforsch.*, 42C: 1097–1100.

Brunerie, P., I. Benda, G. Bock, and P. Schreier, 1988. Bioconversion of monoterpene alcohols and citral by *Botrytis cinerea*. In: *Bioflavour'87. Analysis—Biochemistry—Biotechnology*, P. Schreier, ed., pp. 435–444. Berlin, Germany: Walter de Gruyter and Co.

Busmann, D. and R.G. Berger, 1994. Conversion of myrcene by submerged cultured basidiomycetes. *J. Biotechnol.*, 37: 39–43.

Cadwallander, K.R. and R.J. Braddock, 1992. Enzymatic hydration of (4R)-(+)-limonene to (4R)-(+)-alpha-terpineol. *Dev. Food Sci.*, 29: 571–584.

Cadwallander, K.R., R.J. Braddock, M.E. Parish, and D.P. Higgins, 1989. Bioconversion of (+)-limonene by *Pseudomonas gladioli*. *J. Food Sci.*, 54: 1241–1245.

Cantwell, S.G., E.P. Lau, D.S. Watt, and R.R. Fall, 1978. Biodegradation of acyclic isoprenoids by *Pseudomonas* species. *J. Bacteriol.*, 135: 324–333.

Chamberlain, E.M. and S. Dagley, 1968. The metabolism of thymol by a *Pseudomonas*. *Biochem. J.*, 110: 755–763.

Chapman, P.J., G. Meerman, and I.C. Gunsalus, 1965. The microbiological transformation of fenchone. *Biochem. Biophys. Res. Commun.*, 20: 104–108.

Chapman, P.J., G. Meerman, I.C. Gunsalus, R. Srinivasan, and K.L. Rinehart Jr., 1966. A new acyclic acid metabolite in camphor oxidation. *J. Am. Chem. Soc.*, 88: 618–619.

Christensen, M. and D.E. Tuthill, 1985. *Aspergillus*: An overview. In: *Advances in Penicillium and Aspergillus Systematics*, R.A. Samson and J.I. Pitt, eds., pp. 195–209. New York: Plenum Press.

Conrad, H.E., R. DuBus, and I.C. Gunsalus, 1961. An enzyme system for cyclic ketone lactonization, *Biochem. Biophys. Res. Commun.*, 6: 293–297.

Conrad, H.E., R. DuBus, M.J. Mamtredt, and I.C. Gunsalus, 1965a. Mixed function oxidation II. Separation and properties of the enzymes catalyzing camphor ketolactonization. *J. Biol. Chem.*, 240: 495–503.

Conrad, H.E., K. Lieb, and I.C. Gunsalus, 1965b. Mixed function oxidation III. An electron transport complex in camphor ketolactonization. *J. Biol. Chem.*, 240: 4029–4037.
David, L. and H. Veschambre, 1984. Preparation d'oxydes de linalol par bioconversion. *Tetrahadron Lett.*, 25: 543–546.
David, L. and H. Veschambre, 1985. Oxidative cyclization of linalol by various microorganisms. *Agric. Biol. Chem.*, 49: 1487–1489.
DeFrank, J.J. and D.W. Ribbons, 1977a. p-Cymene pathway in *Pseudomonas putida*: Initial reactions. *J. Bacteriol.*, 129: 1356–1364.
DeFrank, J.J. and D.W. Ribbons, 1977b. p-Cymene pathway in *Pseudomonas putida*: Ring cleavage of 2,3-dihydroxy-p-cumate and subsequent reactions. *J. Bacteriol.*, 129: 1365–1374.
Demirci, F., 2000. Microbial transformation of bioactive monoterpenes. PhD thesis, pp. 1–137. Anadolu University, Eskisehir, Turkey.
Demirci, F., H. Berber, and K.H.C. Baser, 2007. Biotransformation of p-cymene to thymoquinone. *Book of Abstracts of the 38th ISEO*, SL-1, p. 6.
Demirci, F., N. Kirimer, B. Demirci, Y. Noma, and K.H.C. Baser, 2001. The biotransformation of thymol methyl ether by different fungi. *Book of Abstracts of the XII Biotechnology Congress*, p. 47.
Demirci, F., Y. Noma, N. Kirimer, and K.H.C. Baser, 2004. Microbial transformation of (−)-carvone. *Z. Naturforsch.*, 59c: 389–392.
Demyttenaere, J.C.R. and H.L. De Pooter, 1996. Biotransformation of geraniol and nerol by spores of *Penicillium italicum*. *Phytochemistry*, 41: 1079–1082.
Demyttenaere, J.C.R. and H.L. De Pooter, 1998. Biotransformation of citral and nerol by spores of *Penicillium digitatum*. *Flav. Fragr. J.*, 13: 173–176.
Demyttenaere, J.C.R., M. del Carmen Herrera, and N. De Kimpe, 2000. Biotransformation of geraniol, nerol and citral by sporulated surface cultures of *Aspergillus niger* and *Penicillium* sp. *Phytochemistry*, 55: 363–373.
Demyttenaere, J.C.R., I.E.I. Koninckx, and A. Meersman, 1996. Microbial production of bioflavours by fungal spores. In: *Flavour Science. Recent Developments*, A.J. Taylor and D.S. Mottram, eds., pp. 105–110. Cambridge, U.K.: The Royal Society of Chemistry.
Demyttenaere, J.C.R. and H.M. Willemen, 1998. Biotransformation of linalool to furanoid and pyranoid linalool oxides by *Aspergillus niger*. *Phytochemistry*, 47: 1029–1036.
Dhavalikar, R.S. and P.K. Bhattacharyya, 1966. Microbial transformation of terpenes. Part VIII. Fermentation of limonene by a soil Pseudomonad. *Indian J. Biochem.*, 3: 144–157.
Dhavalikar, R.S., A. Ehbrecht, and G. Albroscheit, 1974. Microbial transformations of terpenoids: β-pinene. *Dragoco Rep.*, 3: 47–49.
Dhavalikar, R.S., P.N. Rangachari, and P.K. Bhattacharyya, 1966. Microbial transformation of terpenes. Part IX. Pathways of degradation of limonene in a soil Pseudomonad. *Indian J. Biochem.*, 3: 158–164.
Farooq, A., M.I. Choudhary, S. Tahara, T.-U. Rahman, K.H.C. Baser, and F. Demirci, 2002. The microbial oxidation of (−)-p-pinene by *Botrytis cinerea*. *Z. Naturforsch.*, 57c: 686–690.
Fenaroli, G., 1975. Synthetic flavors. In: *Fenaroli's Handbook of Flavor Ingredients*, eds. T.E. Furia and N. Bellanca eds., Vol. 2, pp. 6–563. Cleveland, OH: CRC Press.
Flynn, T.M. and I.A. Southwell, 1979. 1,3-Dimethyl-2-oxabicyclo [2,2,2]-octane-3-methanol and 1,3-dimethyl-2-oxabicyclo[2,2,2]-octane-3-carboxylic acid, urinary metabolites of 1,8-cineole. *Aust. J. Chem.*, 32: 2093–2095.
Ganapathy, K. and P.K. Bhattacharyya, unpublished data.
Gibbon, G.H., N.F. Millis, and S.J. Pirt, 1972. Degradation of α-pinene by bacteria. *Proc. IV IFS, Ferment. Technol. Today*, pp. 609–612.
Gibbon, G.H. and S.J. Pirt, 1971. The degradation of a-pinene by *Pseudomonas* PX 1. *FEBS Lett.*, 18: 103–105.
Gondai, T., M. Shimoda, and T. Hirata, 1999. Asymmetric reduction of enone compounds by *Chlorella miniata*. *Proceedings of the 43rd TEAC*, pp. 217–219.
Griffiths, E.T., S.M. Bociek, P.C. Harries, R. Jeffcoat, D.J. Sissons, and P.W. Trudgill, 1987a. Bacterial metabolism of alpha-pinene: Pathway from alpha-pinene oxide to acyclic metabolites in *Nocardia* sp. strain P18.3. *J. Bacteriol.*, 169: 4972–4979.
Griffiths, E.T., P.C. Harries, R. Jeffcoat, and P.W. Trudgill, 1987b. Purification and properties of alpha-pinene oxide lyase from *Nocardia* sp. strain P18.3. *J. Bacteriol.*, 169: 4980–4983.
Gunsalus, I.C., P.J. Chapman, and J.-F. Kuo, 1965. Control of catabolic specificity and metabolism. *Biochem. Biophys. Res. Commun.*, 18: 924–931.
Gyoubu, K. and M. Miyazawa, 2005. Biotransformation of (+)- and (−)-fenchone by liver microsomes. *Proceedings of the 49th TEAC*, pp. 420–422.

Gyoubu, K. and M. Miyazawa, 2006. Biotransformation of (+)- and (−)-camphor by liver microsome. *Proceedings of the 50th TEAC*, pp. 253–255.
Hagiwara, Y. and M. Miyazawa, 2007. Biotransformation of cineole by the larvae of common cutworm (*Spodoptera litura*) as a biocatalyst. *Proceedings of the 51st TEAC*, pp. 304–305.
Hagiwara, Y., H. Takeuchi, and M. Miyazawa, 2006. Biotransformation of (+)- and (−)-menthone by the larvae of common cutworm (*Spodoptera litura*) as a biocatalyst. *Proceedings of the 50th TEAC*, pp. 279–280.
Hamada, H. and T. Furuya, 2000. Hydroxylation of monoterpenes by plant suspension cells. *Proceedings of the 44th TEAC*, pp. 167–168.
Hamada, H., T. Furuya, and N. Nakajima, 1996. The hydroxylation and glycosylation by plant catalysts. *Proceedings of the 40th TEAC*, pp. 111–112.
Hamada, H., T. Harada, and T. Furuya, 2001. Hydroxylation of monoterpenes by algae and plant suspension cells. *Proceedings of the 45th TEAC*, pp. 366–368.
Hamada, H., M. Kaji, T. Hirata, T. Furuya, 2003. Enantioselective biotransformation of monoterpenes by *Cyanobacterium*. *Proceedings of the 47th TEAC*, pp. 162–163.
Hamada, H., Y. Kondo, M. Kaji, and T. Furuta, 2002. Glycosylation of monoterpenes by plant suspension cells. *Proceedings of the 46th TEAC*, pp. 321–322.
Hamada, H., A. Matsumoto, and J. Takimura, 2004. Biotransformation of acyclic monoterpenes by biocatalysts of plant cultured cells and *Cyanobacterium*. *Proceedings of the 48th TEAC*, pp. 393–395.
Hamada, H. and H. Yasumune, 1995. The hydroxylation of monoterpenoids by plant cell biotransformation. *Proceedings of the 39th TEAC*, pp. 375–377.
Hartline, R.A. and I.C. Gunsalus, 1971. Induction specificity and catabolite repression of the early enzymes in camphor degradation by *Pseudomonas putida*. *J. Bacteriol.*, 106: 468–478.
Hashimoto Y. and M. Miyazawa, 2001. Microbial resolution of esters of racemic 2-*endo*-hydroxy-1,8-cineole by *Glomerella cingulata*. *Proceedings of the 45th TEAC*, pp. 363–365.
Hayashi, T., T. Kakimoto, H. Ueda, and C. Tatsumi, 1969. Microbiological conversion of terpenes. Part VI. Conversion of borneol. *J. Agric. Chem. Soc. Jpn.*, 43: 583–587.
Hayashi, T., H. Takashiba, S. Ogura, H. Ueda, and C. Tsutsumi, 1968. *Nippon Nogei-Kagaku Kaishi*, 42: 190–196.
Hayashi, T., H. Takashiba, H. Ueda, and C. Tsutsumi, 1967. *Nippon Nogei Kagaku Kaishi*, 41(254): 79878g.
Hayashi, T., S. Uedono, and C. Tatsumi, 1972. Conversion of a-terpineol to 8,9-epoxy-*p*-menthan-1-ol. *Agric. Biol. Chem.*, 36: 690–691.
Hirata, T., K. Shimoda, and T. Gondai, 2000. Asymmetric hydrogenation of the C–C double bond of enones with the reductases from *Nicotiana tabacum*. *Chem. Lett.*, 29: 850–851.
Hosler, P., 1969. U.S. Patent 3,458,399.
Hudlicky, T., D. Gonzales, and D.T. Gibson, 1999. Enzymatic dihydroxylation of aromatics in enantioselective synthesis: Expanding asymmetric methodology. *Aldrichim. Acta*, 32: 35–61.
Hungund, B.L., P.K. Bhattachayya, and P.N. Rangachari, 1970. Methylisopropyl ketone from a terpene fermentation by the soil Pseudomonad, PL-strain. *Indian J. Biochem.*, 7: 80–81.
Ishida, T., Y. Asakawa, and T. Takemoto, 1981a. Metabolism of myrtenal, perillaldehyde and dehydroabietic acid in rabbits. *Res. Bull. Hiroshima Inst. Technol.*, 15: 79–91.
Ishida, T., Y. Asakawa, T. Takemoto, and T. Aratani, 1979. Terpenoid biotransformation in mammals. II. Biotransformation of *dl*-camphene. *J. Pharm. Sci.*, 68: 928–930.
Ishida, T., Y. Asakawa, T. Takemoto, and T. Aratani, 1981b. Terpenoids biotransformation in mammals. III. Biotransformation of α-pinene, (3-pinene, pinane, 3-carene, carane, myrcene, and *p*-cymene in rabbits. *J. Pharm. Sci.*, 70: 406–415.
Iscan, G., 2005. Unpublished data.
Ismaili-Alaoui, M., B. Benjulali, D. Buisson, and R. Azerad, 1992. Biotransformation of terpenic compounds by fungi I. Metabolism of *R*-(+)-pulegone. *Tetrahedron Lett.*, 33: 2349–2352.
Janssens, L., H.L. De Pooter, N.M. Schamp, and E.J. Vandamme, 1992. Production of flavours by microorganisms. *Process Biochem.*, 27: 195–215.
Joglekar, S.S. and R.S. Dhavlikar, 1969. Microbial transformation of terpenoids. I. Identification of metabolites produced by a Pseudomonad from citronellal and citral. *Appl. Microbiol.*, 18: 1084–1087.
Kaji, M., H. Hamada, and T. Furuya, 2002. Biotransformation of monoterpenes by *Cyanobacterium* and plant suspension cells. *Proceedings of the 46th TEAC*, pp. 323–325.
Kamino, F. and M. Miyazawa, 2005. Biotransformation of (+)-and (−)-pinane-2,3-diol using plant pathogenic fungus, *Glomerella cingulata* as a biocatalyst. *Proceedings of the 49th TEAC*, pp. 395–396.
Kamino, F., Y. Noma, Y. Asakawa, and M. Miyazawa, 2004. Biotransformation of (1S,2S,3R,5S)-(+)-pinane-2,3-diol using plant pathogenic fungus, *Glomerella cingulata* as a biocatalyst. *Proceedings of the 48th TEAC*, pp. 383–384.

Kayahara, H., T. Hayashi, C. and Tatsumi, 1973. Microbiological conversion of (−)-perillaldehyde and *p*-mentha-1,3-dien-7-al. *J. Ferment. Technol.*, 51: 254–259.

Kieslich, K., W.-R. Abraham, and P. Washausen, 1985. Microbial transformations of terpenoids. In: *Topics in Flavor Research*, R.G. Berger, S. Nitz, and P. Schreier, eds., pp. 405–427. Marzling Hangenham, Germany: Eichborn.

Koneman, E.W., S.D. Allen, W.M. Janda, P.C. Schreckenberger, and W.C. Winn, 1997. *Color Atlas and Textbook of Diagnostic Microbiology*, Philadelphia, PA: Lippincott-Raven Publishers.

Kraidman, G., B.B. Mukherjee, and I.D. Hill, 1969. Conversion of limonene into an optically active isomer of α-terpineol by a *Cladosporium species*. *Bacteriol. Proc.*, 69: 63.

Krasnobajew, V., 1984. Terpenoids. In: *Biotechnology*, K. Kieslich, ed., Vol. 6a, pp. 97–125. Weinheim, Germany: Verlag Chemie.

Kumagae, S. and M. Miyazawa, 1999. Biotransformation of *p*-menthanes using common cutworm larvae, *Spodoptera litura* as a biocatalyst. *Proceedings of the 43rd TEAC*, pp. 389–390.

Lassak, E.V., J.T. Pinkey, B.J. Ralph, T. Sheldon, and J.J.H. Simes, 1973. Extractives of fungi. V. Microbial transformation products of piperitone. *Aust. J. Chem.*, 26: 845–854.

Liu, W., A. Goswami, R.P. Steffek, R.L. Chemman, F.S. Sariaslani, J.J. Steffens, and J.P.N. Rosazza, 1988. Stereochemistry of microbiological hydroxylations of 1,4-cineole. *J. Org. Chem.*, 53: 5700–5704.

MacRae, I.C., V. Alberts, R.M. Carman, and I.M. Shaw, 1979. Products of 1,8-cineole oxidation by a Pseudomonad. *Aust. J. Chem.*, 32: 917–922.

Madyastha, K.M. 1984. Microbial transformations of acyclic monoterpenes. *Proc. Indian Acad. Sci. (Chem. Sci.)*, 93: 677–686.

Madyastha, K.M. and P.K. Bhattacharyya, 1968. Microbiological transformation of terpenes. Part XIII. Pathways for degradation of *p*-cymene in a soil pseudomonad (PL-strain). *Indian J. Biochem.*, 5: 161–167.

Madyastha, K.M., P.K. Bhattacharyya, and C.S. Vaidyanathan, 1977. Metabolism of a monoterpene alcohol, linalool, by a soil pseudomonad. *Can. J. Microbiol.*, 23: 230–239.

Madyastha, K.M. and N.S.R. Krishna Murthy, 1988a. Regiospecific hydroxylation of acyclic monoterpene alcohols by *Aspergillus niger*. *Tetrahedron Lett.*, 29: 579–580.

Madyastha, K.M. and N.S.R. Krishna Murthy, 1988b. Transformations of acetates of citronellol, geraniol, and linalool by *Aspergillus niger*: Regiospecific hydroxylation of citronellol by a cell-free system. *Appl. Microbiol. Biotechnol.*, 28: 324–329.

Madyastha, K.M. and V. Renganathan, 1983. Bio-degradation of acetates of geraniol, nerol and citronellol by *P. incognita*: Isolation and identification of metabolites. *Indian J. Biochem. Biophys.*, 20: 136–140.

Mattison, J.E., L.L. McDowell, and R.H. Baum, 1971. Cometabolism of selected monoterpenoids by fungi associated with monoterpenoid-containing plants. *Bacteriol. Proc.*, 1971: 141.

Miyamoto, Y. and M. Miyazawa, 2001. Biotransformation of (+)- and (−)-borneol by the larvae of common cutworm (*Spodoptera litura*) as a biocatalyst. *Proceedings of the 45th TEAC*, pp. 377–378.

Miyazato, Y. and M. Miyazawa, 1999. Biotransformation of (+)- and (−)-a-fenchyl acetated using plant parasitic fungus, *Glomerella cingulata* as a biocatalyst. *Proceedings of the 43rd TEAC*, pp. 213–214.

Miyazawa, M., H. Furuno, and H. Kameoka, 1992a. Biotransformation of thujone by plant pathogenic microorganism, *Botrytis allii* IFO 9430. *Proceedings of the 36th TEAC*, pp. 197–198.

Miyazawa, M., H. Furuno, K. Nankai, and H. Kameoka, 1991a. Biotransformation of verbenone by plant pathogenic microorganism, *Rhizoctonia solani*. *Proceedings of the 35th TEAC*, pp. 274–275.

Miyazawa, M., H. Huruno, and H. Kameoka, 1991b. Biotransformation of (+)-pulegone to (−)-1*R*-8-hydroxy-4-*p*-menthen-3-one by *Botrytis allii*. *Chem. Express*, 6: 479–482.

Miyazawa, M., H. Huruno, and H. Kameoka, 1991c. *Chem. Express*, 6: 873.

Miyazawa, M., H. Kakita, M. Hyakumachi, and H. Kameoka, 1992b. Biotransformation of monoterpenoids having *p*-menthan-3-one skeleton by *Rhizoctonia solani*. *Proceedings of the 36th TEAC*, pp. 191–192.

Miyazawa, M., H. Kakita, M. Hyakumachi, K. Umemoto, and H. Kameoka, 1991d. Microbiological conversion of piperitone oxide by plant pathogenic fungi *Rhizoctonia solani*. *Proceedings of the 35th TEAC*, pp. 276–277.

Miyazawa, M., H. Kakita, M. Hyakumachi, K. Umemoto, and H. Kameoka, 1992c. Microbiological conversion of monoterpenoids containing *p*-menthan-3-one skeleton by plant pathogenic fungi *Rhizoctonia solani*. *Proceedings of the 36th TEAC*, pp. 193–194.

Miyazawa, M., S. Kumagae, H. Kameoka, 1997a. Biotransformation of (−)-menthol and (+)-menthol by common cutworm Larvae, *Spodoptera litura* as a biocatalyst. *Proceedings of the 41st TEAC*, pp. 391–392.

Miyazawa, M., S. Kumagae, H. Kameoka, 1997b. Biotransformation of (+)-*trans* myrtanol and (−)-*trans*-myrtanol by common cutworm Larvae, *Spodoptera litura* as a biocatalyst. *Proceedings of the 41st TEAC*, pp. 389–390.

Miyazawa, M. and Y. Miyamoto, 2004. Biotransformation of (+)-(1R, 2S)-fenchol by the larvae of common cutworm (*Spodoptera litura*). *Tetrahadron*, 60: 3091–3096.

Miyazawa, M., T. Murata, and H. Kameoka, 1998. Biotransformation of β-myrcene by common cutworm larvae, *Spodoptera litura* as a biocatalyst. *Proceedings of the 42nd TEAC*, pp. 123–125.

Miyazawa, M., H. Nankai, and H. Kameoka, 1996a. Microbial oxidation of citronellol by *Glomerella cingulata*. *Nat. Prod. Lett.*, 8: 303–305.

Miyazawa, M., Y. Noma, K. Yamamoto, and H. Kameoka, 1983. Microbiological conversion of d- and l-limonene, *Proceedings of the 27th TEAC*, pp. 147–149.

Miyazawa, M., Y. Noma, K. Yamamoto, and H. Kameoka, 1991e. Biotransformation of 1,4-cineole to 2-*endo*-hydroxy-1,4-cineole by *Aspergillus niger*. *Chem. Express*, 6: 771–774.

Miyazawa, M., Y. Noma, K. Yamamoto, and H. Kameoka, 1992d. Biohydroxylation of 1,4-cineole to 9-hydroxy-1,4-cineole by *Aspergillus niger*. *Chem. Express*, 7: 305–308.

Miyazawa, M., Y. Noma, K. Yamamoto, and H. Kameoka, 1992e. Biotransformation of 1,4-cineole to 3-*endo*-hydroxy-1,4-cineole by *Aspergillus niger*. *Chem. Express*, 7: 125–128.

Miyazawa, M., Y. Suzuki, and H. Kameoka, 1994a. Biotransformation of myrtanol by plant pathogenic microorganism, *Glomerella cingulata*, *Proceedings of the 38th TEAC*, pp. 96–97.

Miyazawa, M., Y. Suzuki, and H. Kameoka, 1997c. Biotransformation of (–)- and (+)-isopinocampheol by three fungi. *Phytochemistry*, 45: 945–950.

Miyazawa, M., T. Wada, and H. Kameoka, 1995a. Biotransformation of terpinene, limonene and α-phellandrene in common cutworm larvae, *Spodoptera litura* Fabricius, *Proceedings of the 39th TEAC*, pp. 362–363.

Miyazawa, M., T. Wada, and H. Kameoka, 1996b. Biotransformation of *p*-menthanes using common cutworm larvae, *Spodoptera litura* as a biocatalyst. *Proceedings of the 40th TEAC*, pp. 80–81.

Miyazawa, M., K. Yamamoto, Y. Noma, and H. Kameoka, 1990a. Bioconversion of (+)-fenchone to (+)-6-*endo*-hydroxyfenchone by *Aspergillus niger*. *Chem. Express*, 5: 237–240.

Miyazawa, M., K. Yamamoto, Y. Noma, and H. Kameoka, 1990b. Bioconversion of (+)-fenchone to 5-*endo*-hydroxyfenchone by *Aspergillus niger*. *Chem. Express*, 5: 407–410.

Miyazawa, M., H. Yanagihara, and H. Kameoka, 1996c. Biotransformation of pinanes by common cutworm larvae, *Spodoptera litura* as a biocatalyst. *Proceedings of the 40th TEAC*, pp. 84–85.

Miyazawa, M., H. Yanahara, and H. Kameoka, 1995b. Biotransformation of *trans*-pinocarveol by plant pathogenic microorganism, *Glomerella cingulata*, and by the larvae of common cutworm, *Spodoptera litura* Fabricius. *Proceedings of the 39th TEAC*, pp. 360–361.

Miyazawa, M., K. Yokote, and H. Kameoka, 1994b. Biotransformation of linalool oxide by plant pathogenic microorganisms, *Glomerella cingulata*. *Proceedings of the 38th TEAC*, pp. 101–102.

Miyazawa, M., K. Yokote, and H. Kameoka, 1995c. Biotransformation of 2-*endo*-hydroxy-1,4-cineole by plant pathogenic microorganism, *Glomerella cingulata*. *Proceedings of the 39th TEAC*, pp. 352–353.

Mizutani, S., T. Hayashi, H. Ueda, and C. Tstsumom, 1971. Microbiological conversion of terpenes. Part IX. Conversion of linalool. *Nippon Nogei Kagaku Kaishi*, 45: 368–373.

Moroe, T., S. Hattori, A. Komatsu, and Y. Yamaguchi, 1971. Japanese Patent, 2,036,875. No. 98195t.

Mukherjee, B.B., G. Kraidman, and I.D. Hill, 1973. Synthesis of glycols by microbial transformation of some monocyclic terpenes. *Appl. Microbiol.*, 25: 447–453.

Mukherjee, B.B., G. Kraidman, I.D. Hill, 1974. Transformation of 1-menthene by a *Cladosporium*: Accumulation of β-isopropyl glutaric acid in the growth medium. *Appl. Microbiol.*, 27: 1070–1074.

Murakami, T., I. Ichimoto, and C. Tstsumom, 1973. Microbiological conversion of linalool. *Nippon Nogei Kagaku Kaishi*, 47: 699–703.

Murata, T. and M. Miyazawa, 1999. Biotransformation of dihydromyrcenol by common cutworm larvae, *Spodoptera litura* as a biocatalyst. *Proceedings of the 43rd TEAC*, pp. 393–394.

Nakanishi, K. and M. Miyazawa, 2004. Biotransformation of (–)-menthone by human liver microsomes. *Proceedings of the 48th TEAC*, pp. 401–402.

Nakanishi, K. and M. Miyazawa, 2005. Biotransformation of (+)- and (–)- menthol by liver microsomal humans and rats. *Proceedings of the 49th TEAC*, pp. 423–425.

Nishimura, H., S. Hiramoto, and J. Mizutani, 1983a. Biological activity of bottrospicatol and related compounds produced by microbial transformation of (–)-*cis*-carveol towards plants. *Proceedings of the 27th TEAC*, pp. 107–109.

Nishimura, H., S. Hiramoto, J. Mizutani, Y. Noma, A. Furusaki, and T. Matsumoto, 1983b. Structure and biological activity of bottrospicatol, a novel monoterpene produced by microbial transformation of (–)-*cis*-carveol. *Agric. Biol. Chem.*, 47: 2697–2699.

Nishimura, H. and Y. Noma, 1996. Microbial transformation of monoterpenes: Flavor and biological activity. In: *Biotechnology for Improved Foods and Flavors*, G.R. Takeoka, R. Teranishi, P.J. Williams, and A. Kobayashi, A., eds., ACS Symposium Series 637, pp. 173–187. Washington, DC: American Chemical Society.

Nishimura, H., Y. Noma, and J. Mizutani, 1982. *Eucalyptus* as biomass. Novel compounds from microbial conversion of 1,8-cineole. *Agric. Biol. Chem.*, 46: 2601–2604.

Nishimura, H., D.M. Paton, and M. Calvin, 1980. *Eucalyptus radiata* oil as a renewable biomass. *Agric. Biol. Chem.*, 44: 2495–2496.

Noma, Y., 1976. Microbiological conversion of carvone. Biochemical reduction of terpenes, part VI. *Ann. Res. Stud. Osaka Joshigakuen Junior College*, 20: 33–47.

Noma, Y., 1977. Conversion of the analogues of carvone and dihydrocarvone by *Pseudomonas ovalis*, strain 6-1, Biochemical reduction of terpenes, part VII. *Nippon Nogeikagaku Kaishi*, 51: 463–470.

Noma, Y., 1979a. Conversion of (−)-carvone by *Nocardia lurida* A-0141 and *Streptosporangium roseum* IFO3776. Biochemical reduction of terpenes, part VIII. *Nippon Nogeikagaku Kaishi*, 53: 35–39.

Noma, Y., 1979b. On the pattern of reaction mechanism of (+)-carvone conversion by actinomycetes. Biochemical reduction of terpenes, part X, *Ann. Res. Stud. Osaka Joshigakuen Junior College*, 23: 27–31.

Noma, Y., 1980. Conversion of (−)-carvone by strains of *Streptomyces*, A-5-1 and *Nocardia*, 1-3-11. *Agric. Biol. Chem.*, 44: 807–812.

Noma, Y., 1984. Microbiological conversion of carvone, *Kagaku to Seibutsu*, 22: 742–746.

Noma, Y., 1988. Formation of *p*-menthane-2,8-diols from (−)-dihydrocarveol and (+)-dihydrocarveol by *Aspergillus* spp., *The Meeting of Kansai Division of The Agricultural and Chemical Society of Japan*, Kagawa, Japan, p. 28.

Noma, Y., 2000. unpublished data.

Noma, Y., 2007. Microbial production of mosquitocidal (1R,2S,4R)-(+)-menthane- 2,8-diol. In: *Aromatic Plants from Asia their Chemistry and Application in Food and Therapy*, L. Jiarovetz, N.X. Dung, and V.K. Varshney, eds., pp. 169–186. Dehradun, Uttarakhand: Har Krishan Bhalla & Sons.

Noma, Y., E. Akehi, N. Miki, and Y. Asakawa, 1992a. Biotransformation of terpene aldehyde, aromatic aldehydes and related compounds by *Dunaliella tertiolecta*. *Phytochemistry*, 31: 515–517.

Noma, Y. and Y. Asakawa, 1992. Enantio- and diastereoselectivity in the biotransformation of carveols by *Euglena gracilis* Z. *Phytochem.*, 31: 2009–2011.

Noma, Y. and Y. Asakawa, 1995. *Aspergillus* spp.: Biotransformation of terpenoids and related compounds. In: *Biotechnology in Agriculture and Forestry, Vol. 33. Medicinal and Aromatic Plants VIII*, Y.P.S. Bajaj, ed., pp. 62–96. Berlin, Germany: Springer.

Noma, Y. and Y. Asakawa, 1998. *Euglena gracilis* Z: Biotransformation of terpenoids and related compounds. In: *Biotechnology in Agriculture and Forestry, Vol. 41. Medicinal and Aromatic Plants X*, Y.P.S. Bajaj, ed., pp. 194–237. Berlin Heidelberg, Germany: Springer.

Noma, Y. and Y. Asakawa, 2005a. New metabolic pathways of β-pinene and related compounds by *Aspergillus niger*. *Book of Abstracts of the 36th ISEO*, p. 32.

Noma, Y. and Y. Asakawa, 2005b. Microbial transformation of (−)-myrtenol and (−)-nopol. *Proceedings of the 49th TEAC*, pp. 78–80.

Noma, Y. and Y. Asakawa, 2006a. Biotransformation of (+)-limonene and related compounds by *Citrus* pathogenic fungi. *Proceedings of the 50th TEAC*, pp. 431–433.

Noma, Y. and Y. Asakawa, 2006b. Biotransformation of β-pinene, myrtenol, nopol and nopol benzyl ether by *Aspergillus niger* TBUYN-2. *Book of Abstracts of the 37th ISEO*, p. 144.

Noma, Y. and Y. Asakawa, 2006c. Microbial transformation of (−)-nopol benzyl ether. *Proceedings of the 50th TEAC*, pp. 434–436.

Noma, Y. and Y. Asakawa, 2007a. Biotransformation of limonene and related compounds by newly isolated low temperature grown *citrus* pathogenic fungi and red yeast. *Book of Abstracts of the 38th ISEO*, p. 7.

Noma, Y. and Y. Asakawa, 2007b. Microbial transformation of limonene and related compounds. *Proceedings of the 51st TEAC*, pp. 299–301.

Noma, Y. and Y. Asakawa, 2008. New metabolic pathways of (+)-carvone by Citrus pathogenic *Aspergillus niger* Tiegh CBAYN and *A. niger* TBUYN-2, *Proceedings of the 52nd TEAC*, pp. 206–208.

Noma, Y., M. Furusawa, T. Hashimoto, and Y. Asakawa, 2002. Stereoselective formation of (1R, 2S, 4R)-(+)-*p*-menthane-2,8-diol from α-pinene. *Book of Abstracts of the 33rd ISEO*, p. 142.

Noma, Y., M. Furusawa, T. Hashimoto, and Y. Asakawa, 2004. Biotransformation of (+)- and (−)-3-pinanone by *Aspergillus niger*. *Proceedings of the 48th TEAC*, pp. 390–392.

Noma, Y., T. Hashimoto, S. Uehara, and Y. Asakawa, 2009. unpublished data.
Noma, Y., T. Higata, T. Hirata, Y. Tanaka, T. Hashimoto, and Y. Asakawa, 1995. Biotransformation of [6-^{2}H]-(−)-carvone by *Aspergillus niger*, *Euglena gracilis* Z and *Dunaliella tertiolecta*, *Proceedings of the 39th TEAC*, pp. 367–368.
Noma, Y., K. Hirata, and Y. Asakawa, 1996. Biotransformation of 1,8-cineole. Why do the biotransformed 2α- and 3α-hydroxy-1,8-cineole by *Aspergillus niger* have no optical activity? *Proceedings of the 40th TEAC*, pp. 89–91.
Noma, Y. and M. Iwami, 1994. Separation and identification of terpene convertible actinomycetes: *S. bottropensis* SY-2-1, *S. ikutamanensis* Ya-2-1 and *S. humidus* Tu-1. *Bull. Tokushima Bunri Univ.*, 47: 99–110.
Noma, Y., F. Kamino, T. Hashimoto, and Y. Asakawa, 2003. Biotransformation of (+)- and (−)-pinane-2,3-diol and related compounds by *Aspergillus niger*. *Proceedings of the 47th TEAC*, pp. 91–93.
Noma, Y., K. Matsueda, I. Maruyama, and Y. Asakawa, 1997. Biotransformation of terpenoids and related compounds by *Chlorella* species. *Proceedings of the 41st TEAC*, pp. 227–229.
Noma, Y., N. Miki, E. Akehi, E. Manabe, and Y. Asakawa, 1991b. Biotransformation of monoterpenes by photosynthetic marine algae, *Dunaliella tertiolecta*, *Proceedings of the 35th TEAC*, pp. 112–114.
Noma, Y., M. Miyazawa, K. Yamamoto, H. Kameoka, T. Inagaki, and H. Sakai, 1984. Microbiological conversion of perillaldehyde. Biotransformation of *l*- and *dl*-perillaldehyde by *Streptomyces ikutamanensis*, Ya-2-1, *Proceedings of the 28th TEAC*, pp. 174–176.
Noma, Y. and H. Nishimura, 1980. Microbiological transformation of 1,8-cineole. Oxidative products from 1,8-cineole by *S. bottropensis*, SY-2-1. *Book of Abstracts of the Annual Meeting of Agricultural and Biological Chemical Society*, p. 28.
Noma, Y. and H. Nishimura, 1981. Microbiological transformation of 1,8-cineole. Production of 3β-hydroxy-1,8-cineole from 1,8-cineole by *S. ikutamanensis*, Ya-2-1. *Book of Abstracts of the Annual Meeting of Agricultural and Biological Chemical Society*, p. 196.
Noma, Y. and H. Nishimura, 1982. Biotransformation of carvone. 4. Biotransformation of (+)-carvone by *Streptomyces bottropensis*, SY-2-1. *Proceedings of the 26th TEAC*, pp. 156–159.
Noma, Y. and H. Nishimura, 1983a. Biotransformation of (−)-carvone and (+)-carvone by *S. ikutamanensis* Ya-2-1. *Book of Abstracts of the Annual Meeting of Agricultural and Biological Chemical Society*, p. 390.
Noma, Y. and H. Nishimura, 1983b. Biotransformation of carvone. 5. Microbiological transformation of dihydrocarvones and dihydrocarveols, *Proceedings of the 27th TEAC*, pp. 302–305.
Noma, Y. and H. Nishimura, 1984. Microbiological conversion of carveol. Biotransformation of (−)-*cis*-carveol and (+)-*cis*-carveol by *S. bottropensis*, Sy-2-1. *Proceedings of the 28th TEAC*, pp. 171–173.
Noma, Y. and H. Nishimura, 1987. Bottrospicatols, novel monoterpenes produced on conversion of (−)- and (+)-*cis*-carveol by *Streptomyces*. *Agric. Biol. Chem.*, 51: 1845–1849.
Noma, Y., H. Nishimura, S. Hiramoto, M. Iwami, and C. Tstsumi, 1982. A new compound, (4R, 6R)-(+)-6,8-oxidomenth-1-en-9-ol produced by microbial conversion of (−)-*cis*-carveol. *Agric. Biol. Chem.*, 46: 2871–2872.
Noma, Y., H. Nishimura, and C. Tatsumi, 1980. Biotransformation of carveol by Actinomycetes. 1. Biotransformation of (−)-*cis*-carveol and (−)-*trans*-carveol by *Streptomyces bottropensis*, SY-2-1, *Proceedings of the 24th TEAC*, pp. 67–70.
Noma, Y. and S. Nonomura, 1974. Conversion of (−)-carvone and (+)-carvone by a strain of *Aspergillus niger*. *Agric. Biol. Chem.*, 38: 741–744.
Noma, Y., S. Nonomura, and H. Sakai, 1974a. Conversion of (−)-carvotanacetone and (+)-carvotanacetone by *Pseudomonas ovalis*, strain 6-1, *Agric. Biol. Chem.*, 38: 1637–1642.
Noma, Y., S. Nonomura, and H. Sakai, 1975. Epimerization of (−)-isodihydrocarvone to (−)-dihydrocarvone by *Pseudomonas fragi* IFO 3458. *Agric. Biol. Chem.*, 39: 437–441.
Noma, Y., S. Nonomura, H. Ueda, H. Sakai, and C. Tstusmi, 1974b. Microbial transformation of carvone. *Proceedings of the 18th TEAC*, pp. 20–23.
Noma, Y., S. Nonomura, H. Ueda, and C. Tatsumi, 1974c. Conversion of (+)-carvone by *Pseudomonas ovalis*, strain 6-1(1). *Agric. Biol. Chem.*, 38: 735–740.
Noma, Y. and H. Sakai, 1984. Investigation of the conversion of (−)-perillyl alcohol, 1,8-cineole, (+)-carvone and (−)-carvone by rare actinomycetes. *Ann. Res. Stud. Osaka Joshigakuen Junior College*, 28: 7–18.
Noma, Y., A. Sogo, S. Miki, N. Fujii, T. Hashimoto, and Y. Asakwawa, 1992d. Biotransformation of terpenoids and related compounds. *Proceedings of the 36th TEAC*, pp. 199–201.
Noma, Y., H. Takahashi, and Y. Asakawa, 1989. Microbiological conversion of menthol. Biotransformation of (+)-menthol by a strain of *Aspergillus niger*. *Proceedings of the 33rd TEAC*, pp. 124–126.

Noma, Y., H. Takahashi, and Y. Asakawa, 1990. Microbiological conversion of *p*-menthane 1. Formation of *p*-menthane-1,9-diol from *p*-menthane by a strain of *Aspergillus niger*. *Proceedings of the 34th TEAC*, pp. 253–255.

Noma, Y., H. Takahashi, and Y. Asakawa, 1991a. Biotransformation of terpene aldehyde by *Euglena gracilis* Z. *Phytochem.*, 30: 1147–1151.

Noma, Y., H. Takahashi, and Y. Asakawa, 1993. Formation of 8 kinds of *p*-menthane-2,8-diols from carvone and related compounds by *Euglena gracilis* Z. Biotransformation of monoterpenes by photosynthetic microorganisms. Part VIII. *Proceedings of the 37th TEAC*, pp. 23–25.

Noma, Y., H. Takahashi, M. Toyota, and Y. Asakawa, 1988b. Microbiological conversion of (−)-carvotanacetone and (+)-carvotanacetone by a strain of *Aspergillus niger*. *Proceedings of the 32nd TEAC*, pp. 146–148.

Noma, Y., H. Takahashi, T. Hashimoto, and Y. Asakawa, 1992c. Biotransformation of isopiperitenone, 6-gingerol, 6-shogaol and neomenthol by a strain of *Aspergillus niger*. *Proceedings of the 37th TEAC*, pp. 26–28.

Noma, Y. and C. Tatsumi, 1973. Conversion of (−)-carvone by *Pseudomonas ovalis*, strain 6-1(1), Microbial conversion of terpenes part XIII. *Nippon Nogeikagaku Kaishi*, 47: 705–711.

Noma, Y., M. Toyota, and Y. Asakawa, 1985a. Biotransformation of (−)-carvone and (+)-carvone by *Aspergillus* spp. *Annual Meeting of Agricultural and Biological Chemistry*, Sapporo, Japan, p. 68.

Noma, Y., M. Toyota, and Y. Asakawa, 1985b. Biotransformation of carvone. 6. Biotransformation of (−)-carvone and (+)-carvone by a strain of *Aspergillus niger*. *Proceedings of the 29th TEAC*, pp. 235–237.

Noma, Y., M. Toyota, Y. and Asakawa, 1985c. Microbiological conversion of (−)-carvotanacetone and (+)-carvotanacetone by *S. bottropensis* SY-2-1. *Proceedings of the 29th TEAC*, pp. 238–240.

Noma, Y., M. Toyota, and Y. Asakawa, 1986. Reduction of terpene aldehydes and epoxidation of terpene alcohols by *S. ikutamanensis*, Ya-2-1. *Proceedings of the 30th TEAC*, pp. 204–206.

Noma, Y., M. Toyota, and Y. Asakawa, 1988a. Microbial transformation of thymol formation of 2-hydroxy-3-*p*-menthen-5-one by *Streptomyces humidus*, Tu-1. *Proceedings of the 28th TEAC*, pp. 177–179.

Noma, Y., J. Watanabe, T. Hashimoto, and Y. Asakawa, 2001. Microbiological transformation of β-pinene. *Proceedings of the 45th TEAC*, pp. 88–90.

Noma, Y., S. Yamasaki, and Asakawa Y. 1992b. Biotransformation of limonene and related compounds by *Aspergillus cellulosae*. *Phytochemistry*, 31: 2725–2727.

Nonoyama, H., H. Matsui, M. Hyakumachi, and M. Miyazawa, 1999. Biotransformation of (−)-menthone using plant parasitic fungi, *Rhizoctonia solani* as a biocatalyst. *Proceedings of the 43rd TEAC*, pp. 387–388.

Ohsawa, M. and M. Miyazawa, 2001. Biotransformation of (+)- and (−)-isopulegol by the larvae of common cutworm (*Spodoptera litura*) as a biocatalyst. *Proceedings of the 45th TEAC*, pp. 375–376.

Omata, T., N. Iwamoto, T. Kimura, A. Tanaka, S. Fukui, 1981. Stereoselective hydrolysis of *dl*-menthyl succinate by gel-entrapped *Rhodotorula minuta* var. *texensis* cells in organic solvent. *Appl. Microbiol. Biotechnol.*, 11: 119–204.

Oosterhaven, K., K.J. Hartmans, and J.J.C. Scheffer, 1995a. Inhibition of potato sprouts growth by carvone enantiomers and their bioconversion in sprouts. *Potato Res.*, 38: 219–230.

Oosterhaven, K., B. Poolman, and E.J. Smid, 1995b. S-Carvone as a natural potato sprouts inhibiting, fungistatic and bacteriostatic compound. *Indian Crops Prod.*, 4: 23–31.

Oritani, T. and K. Yamashita, 1973a. Microbial DL-acyclic alcohols, *Agric. Biol. Chem.*, 37: 1923–1928.

Oritani, T. and K. Yamashita, 1973b. Microbial resolution of racemic 2- and 3-alkylcyclohexanols. *Agric. Biol. Chem.*, 37: 1695–1700.

Oritani, T. and K. Yamashita, 1973c. Microbial resolution of *dl*-isopulegol. *Agric. Biol. Chem.*, 37: 1687–1689.

Oritani, T. and K. Yamashita, 1973d. Microbial resolution of racemic carvomenthols. *Agric. Biol. Chem.*, 37: 1691–1694.

Oritani, T. and K. Yamashita, 1974. Microbial resolution of (±)-borneols. *Agric. Biol. Chem.*, 38: 1961–1964.

Oritani, T. and K. Yamashita, 1980. Optical resolution of *dl*-(β, γ-unsaturated terpene alcohols by biocatalyst of microorganism. *Proceedings of the 24th TEAC*, pp. 166–169.

Pfrunder, B. and Ch. Tamm, 1969a. Mikrobiologische Umwandlung von bicyclischen monoterpenen durch *Absidia orchidis* (Vuill.) Hagem. 2. Teil: Hydroxylierung von Fenchon und Isofenchon. *Helv. Chim. Acta.*, 52: 1643–1654.

Pfrunder, B. and Ch. Tamm, 1969b. Mikrobiologische Umwandlung von bicyclischen monoterpenen durch *Absidia orchidis* (Vuill.) Hagem. 1. Teil: Reduktion von Campherchinon und Isofenchonchinon. *Helv. Chim. Acta.*, 52: 1630–1642.

Prema, B.R. and P.K. Bhattachayya, 1962. Microbiological transformation of terpenes. II. Transformation of a-pinene. *Appl. Microbiol.*, 10: 524–528.

Rama Devi, J., S.G. Bhat, and P.K. Bhattacharyya, 1977. Microbiological transformations of terpenes. Part XXV. Enzymes involved in the degradation of linalool in the *Pseudomonas incognita*, linalool strain. *Indian J. Biochem. Biophys.*, 15: 323–327.

Rama Devi, J. and P.K. Bhattacharyya, 1977a. Microbiological transformations of terpenes. Part XXIV. Pathways of degradation of linalool, geraniol, nerol and limonene by *Pseudomonas incognita*, linalool strain. *Indian J. Biochem. Biophys.*, 14: 359–363.

Rama Devi, J. and P.K. Bhattacharyya, 1977b. Microbiological transformation of terpenes. Part XXIII. Fermentation of geraniol, nerol and limonene by soil Pseudomonad, *Pseudomonas incognita* (linalool strain). *Indian J. Biochem. Biophys.*, 14: 288–291.

Rama Devi, J. and P.K. Bhattacharyya, 1978. Molecular rearrangements in the microbiological transformations of terpenes and the chemical logic of microbial processes. *J. Indian Chem. Soc.*, 55: 1131–1137.

Rapp, A. and H. Mandery, 1988. Influence of *Botrytis cinerea* on the monoterpene fraction wine aroma. In: *Bioflavour'87. Analysis—Biochemistry—Biotechnology*, P. Schreier, ed., pp. 445–452. Berlin, Germany: Walter de Gruyter and Co.

Saeki, M. and N. Hashimoto, 1968. Microbial transformation of terpene hydrocarbons. Part I. Oxidation products of *d*-limonene and *d*-pentene. *Proceedings of the 12th TEAC*, pp. 102–104.

Saeki, M. and N. Hashimoto, 1971. Microorganism biotransformation of terpenoids. Part II. Production of *cis*-terpin hydrate and terpineol from *d*-limonene. *Proceedings of the 15th TEAC*, pp. 54–56.

Saito, H. and M. Miyazawa, 2006. Biotransformation of 1,8-cineole by *Salmonella typhimurium* OY1001/3A4. *Proceedings of the 50th TEAC*, pp. 275–276.

Savithiry, N., T.K. Cheong, and P. Oriel, 1997. Production of alpha-terpineol from *Escherichia coli* cells expressing thermostable limonene hydratase. *Appl. Biochem. Biotechnol.*, 63–65: 213–220.

Sawamura, Y., S. Shima, H. Sakai, and C. Tatsumi, 1974. Microbiological conversion of menthone. *Proceedings of the 18th TEAC*, pp. 27–29.

Schwammle, B., E. Winkelhausen, S. Kuzmanova, and W. Steiner, 2001. Isolation of carvacrol assimilating microorganisms. *Food Technol. Biotechnol.*, 39: 341–345.

Seubert, W. and E. Fass, 1964a. Studies on the bacterial degradation of isoprenoids. V. The mechanism of isoprenoid degradation. *Biochem. Z.*, 341: 35–44.

Seubert, W. and E. Fass, 1964b. Studies on the bacterial degradation of isoprenoids. IV. The purification and properties of beta-isohexenylglutaconyl-COA-hydratase and beta-hydroxy-beta-isohexenylglutaryl-COA-lyase. *Biochem. Z.*, 341: 23–34.

Seubert, W., E. Fass, and U. Remberger, 1963. Studies on the bacterial degradation of isoprenoid compounds. III. Purification and properties of geranyl carboxylase. *Biochem. Z.*, 338: 265–275.

Seubert, W. and U. Remberger, 1963. Studies on the bacterial degradation of isoprenoid compounds. II. The role of carbon dioxide. *Biochem. Z.*, 338: 245–246.

Shima, S., Y. Yoshida, Y. Sawamura, and C. Tstsumi, 1972. Microbiological conversion of perillyl alcohol. *Proceedings of the 16th TEAC*, pp. 82–84.

Shimoda, K., T. Hirata, and Y. Noma, 1998. Stereochemistry in the reduction of enones by the reductase from *Euglena gracilis*. *Z. Phytochem.*, 49: 49–53.

Shimoda, K., D.I. Ito, S. Izumi, and T. Hirata, 1996. Novel reductase participation in the syn-addition of hydrogen to the C=C bond of enones in the cultured cells of *Nicotiana tabacum*. *J. Chem. Soc. Perkin Trans.*, 1, 355–358.

Shimoda, K., S. Izumi, and T. Hirata, 2002. A novel reductase participating in the hydrogenation of an exocyclic C–C double bond of enones from *Nicotiana tabacum*. *Bull. Chem. Soc. Jpn.*, 75: 813–816.

Shimoda, K., N. Kubota, H. Hamada, and M. Kaji, 2003. *Cyanobacterium* catalyzed asymmetric reduction of enones. *Proceedings of the 47th TEAC*, pp. 164–166.

Shindo, M., T. Shimada, and M. Miyazawa, 2000. Metabolism of 1,8-cineole by cytochrome P450 enzymes in human and rat liver microsomes. *Proceedings of the 44th TEAC*, pp. 141–143.

Shukla, O.P., R.C. Bartholomeus, and I.C. Gunsalus, 1987. Microbial transformation of menthol and menthane-3,4-diol. *Can. J. Microbiol.*, 33: 489–497.

Shukla, O.P., and P.K. Bhattacharyya, 1968. Microbiological transformations of terpenes: Part XI—Pathways of degradation of α- & β-pinenes in a soil Pseudomonad (PL-strain). *Indian J. Biochem.*, 5: 92–101.

Shukla, O.P., M.N. Moholay, and P.K. Bhattacharyya, 1968. Microbiological transformation of terpenes: Part X—Fermentation of α- & β-pinenes by a soil Pseudomonad (PL-strain). *Indian J. Biochem.*, 5: 79–91.

Southwell, I.A. and T.M. Flynn, 1980. Metabolism of α- and β-pinene, *p*-cymene and 1,8-cineole in the brush tail possum. *Xenobiotica*, 10: 17–23.

Suga, T. and T. Hirata, 1990. Biotransformation of exogenous substrates by plant cell cultures. *Phytochemistry*, 29: 2393–2406.

Suga, T., T. Hirata, and H. Hamada, 1986. The stereochemistry of the reduction of carbon–carbon double bond with the cultured cells of *Nicotiana tabacum*. *Bull. Chem. Soc. Jpn.*, 59: 2865–2867.

Sugie, A. and M. Miyazawa, 2003. Biotransformation of (−)-a-pinene by human liver microsomes. *Proceedings of the 47th TEAC*, pp. 159–161.

Swamy, G.K., K.L. Khanchandani, and P.K. Bhattacharyya, 1965. *Symposium on Recent Advances in the Chemistry of Terpenoids*, Natural Institute of Sciences of India, New Delhi, India, p. 10.

Takagi, K., Y. Mikami, Y. Minato, I. Yajima, and K. Hayashi, 1972. Manufacturing method of carvone by microorganisms, *Japanese Patent* 7,238,998.

Takahashi, H., Y. Noma, M. Toyota, and Y. Asakawa, 1994. The biotransformation of (−)- and (+)-neomenthols and isomenthols by *Aspergillus niger*. *Phytochemistry*, 35: 1465–1467.

Takeuchi, H. and M. Miyazawa, 2004. Biotransformation of nerol by the larvae of common cutworm (*Spodoptera litura*) as a biocatalyst. *Proceedings of the 48th TEAC*, pp. 399–400.

Takeuchi, H. and M. Miyazawa, 2005. Biotransformation of (−)- and (+)-citronellene by the larvae of common cutworm (*Spodoptera litura*) as biocatalyst. *Proceedings of the 49th TEAC*, pp. 426–427.

Trudgill, P.W., 1990. Microbial metabolism of terpenes—Recent developments. *Biodegradation* 1: 93–105.

Tsukamoto, Y., S. Nonomura, and H. Sakai, 1975. Formation of *p-cis*-menthan-1-ol from *p*-menthane by *Pseudomonas mendocina* SF. *Agric. Biol. Chem.*, 39: 617–620.

Tsukamoto, Y., S. Nonomura, H. Sakai, and C. Tatsumi, 1974. Microbiological oxidation of *p*-menthane 1. Formation of formation of *p-cis*-menthan-1-ol. *Proceedings of the 18th TEAC*, pp. 24–26.

Van der Werf, M.J. and J.A.M. de Bont, 1998a. Screening for microorganisms converting limonene into carvone. In: *New Frontiers in Screening for Microbial Biocatalysts, Proceedings of the International Symposium*, Ede, the Netherlands, K. Kieslich, C.P. Beek, J.A.M. van der Bont, and W.J.J. van den Tweel, eds., Vol. 53, pp. 231–234. Amsterdam, the Netherlands: Studies in Organic Chemistry.

Van der Werf, M.J., J.A.M. de Bont, and D.J. Leak, 1997. Opportunities in microbial biotransformation of monoterpenes. *Adv. Biochem. Eng. Biotechnol.*, 55: 147–177.

Van der Werf, M.J., K.M. Overkamp, and J.A.M. de Bont, 1998b. Limonene-1,2-epoxide hydrolase from *Rhodococcus erythropolis* DCL14 belongs to a novel class of epoxide hydrolases. *J. Bacteriol.*, 180: 5052–5057.

van Dyk, M.S., E. van Rensburg, I.P.B. Rensburg, and N. Moleleki, 1998. Biotransformation of monoterpenoid ketones by yeasts and yeast-like fungi, *J. Mol. Catal. B: Enzym.*, 5: 149–154.

Verstegen-Haaksma, A.A., H.J. Swarts, B.J.M. Jansen, A. de Groot, N. Bottema-MacGillavry, and B. Witholt, 1995. Application of S-(+)-carvone in the synthesis of biologically active natural products using chemical transformations and bioconversions. *Indian Crops Prod.*, 4: 15–21.

Watanabe, Y. and T. Inagaki, 1977a. Japanese Patent 77,12,989. No. 187696x.

Watanabe, Y. and T. Inagaki, 1977b. Japanese Patent 77, 122,690. No. 87656g.

Watanabe, T., H. Nomura, T. Iwasaki, A. Matsushima, and T. Hirata, 2007. Cloning of pulegone reductase and reduction of enones with the recombinant reductase. *Proceedings of the 51st TEAC*, pp. 323–325.

Wolf-Rainer, A., 1994. Phylogeny and biotransformation. Part 5. Biotransformation of isopinocampheol. *Z. Naturforsch.*, 49c: 553–560.

Yamada, K., S. Horiguchi, and J. Tatahashi, 1965. Studies on the utilization of hydrocarbons by microorganisms. Part VI. Screening of aromatic hydrocarbon-assimilating microorganisms and cumic acid formation from *p*-cymene. *Agric. Biol. Chem.*, 29: 943–948.

Yamaguchi, Y., A. Komatsu, and T. Moroe, 1977. Asymmetric hydrolysis of *dl*-menthyl acetate by *Rhodotorula mucilaginosa*. *J. Agric. Chem. Soc. Jpn.*, 51: 411–416.

Yamamoto, K., M. Miyazawa, H. Kameoka, and Y. Noma, 1984. Biotransformation of *d*- and *l*-fenchone by a strain of *Aspergillus niger*. *Proceedings of the 28th TEAC*, pp. 168–170.

Yamanaka, T. and M. Miyazawa, 1999. Biotransformation of (−)-*trans*-verbenol by common cutworm larvae, *Spodoptera litura* as a biocatalyst. *Proceedings of the 43rd TEAC*, pp. 391–392.

Yawata, T., M, Ogura, K. Shimoda, S. Izumi, and T. Hirata, 1998. Epoxidation of monoterpenes by the peroxidase from the cultured cells of *Nicotiana tabacum*, *Proceedings of the 42nd TEAC*, pp. 142–144.

Yonemoto, N., S. Sakamoto, T. Furuya, and H. Hamada, 2005. Preparation of (−)-perillyl alcohol oligosaccharides. *Proceedings of the 49th TEAC*, pp. 108–110.

20 Biotransformation of Sesquiterpenoids, Ionones, Damascones, Adamantanes, and Aromatic Compounds by Green Algae, Fungi, and Mammals

Yoshinori Asakawa and Yoshiaki Noma

CONTENTS

20.1 Introduction 907
20.2 Biotransformation of Sesquiterpenoids by Microorganisms 908
    20.2.1 Highly Efficient Production of Nootkatone (2) from Valencene (1) 908
    20.2.2 Biotransformation of Valencene (1) by *Aspergillus niger* and *Aspergillus wentii*........910
    20.2.3 Biotransformation of Nootkatone (2) by *Aspergillus niger* 911
    20.2.4 Biotransformation of Nootkatone (2) by *Fusarium culmorum*
        and *Botryosphaeria dothidea* 913
    20.2.5 Biotransformation of (+)-1(10)-Aristolene (36) from the Crude Drug *Nardostachys
        chinensis* by *Chlorella fusca*, *Mucor* Species, and *Aspergillus niger* 916
    20.2.6 Biotransformation of Various Sesquiterpenoids by Microorganisms........................ 921
20.3 Biotransformation of Sesquiterpenoids by Mammals, Insects, and Cytochrome P-450...... 988
    20.3.1 Animals (Rabbits) and Dosing 988
    20.3.2 Sesquiterpenoids 989
20.4 Biotransformation of Ionones, Damascones, and Adamantanes 992
20.5 Biotransformation of Aromatic Compounds 996
References.. 1004

20.1 INTRODUCTION

Recently, environment-friendly green or clean chemistry is emphasized in the field of organic and natural product chemistry. Noyori's highly efficient production of (−)-menthol using (*S*)-BINAP-Rh catalyst is one of the most important green chemistries (Otsuka and Tani, 1991; Tani et al., 1982), and 1000 ton of (−)-menthol has been produced by this method in 1 year. On the other hand, enzymes of microorganisms and mammals are able to transform a huge variety of organic compounds, such as mono-, sesqui-, and diterpenoids, alkaloids, steroids, antibiotics, and amino acids from crude drugs and spore-forming green plants to produce pharmacologically and medicinally valuable substances.

Since Meyer and Neuberg (1915) studied the microbial transformation of citronellal, there are a great number of reports concerning biotransformation of essential oils, terpenoids, steroids, alkaloids, and acetogenins using bacteria, fungi, and mammals. In 1988, Mikami (1988) reported the review article of biotransformation of terpenoids entitled *Microbial Conversion of Terpenoids*. Lamare and Furstoss (1990) reviewed biotransformation of more than 25 sesquiterpenoids by microorganisms. In this chapter, the recent advances in the biotransformation of natural and synthetic compounds; sesquiterpenoids, ionones, α-damascone, and adamantanes; and aromatic compounds, using microorganisms including algae and mammals, are described.

20.2 BIOTRANSFORMATION OF SESQUITERPENOIDS BY MICROORGANISMS

20.2.1 HIGHLY EFFICIENT PRODUCTION OF NOOTKATONE (2) FROM VALENCENE (1)

The most important and expensive grapefruit aroma, nootkatone (**2**), decreases the somatic fat ratio (Haze et al., 2002), and therefore, its highly efficient production has been requested by the cosmetic and fiber industrial sectors. Previously, valencene (**1**) from the essential oil of Valencia orange was converted into nootkatone (**2**) by biotransformation using *Enterobacter* species only in 12% yield (Dhavlikar and Albroscheit, 1973), *Rhodococcus* KSM-5706 in 0.5% yield with a complex mixture (Okuda et al., 1994), and cytochrome P450 (CYP450) in 20% yield with other complex products (Sowden et al., 2005). Nootkatone (**2**) was chemically synthesized from valencene (**1**) with AcOOCMe$_3$ in three steps and chromic acid in low yield (Wilson and Saw, 1978) and using surface-functionalized silica supported by metal catalysts such as Co^{2+} and Mn^{2+} with *tert*-butyl hydroperoxide in 75% yield (Salvador and Clark, 2002). However, these synthetic methods are not safe because they involve toxic heavy metals. An environment-friendly method for the synthesis of nootkatone that does not use any heavy metals such as chromium and manganese must be designed. The commercially available and cheap sesquiterpene hydrocarbon (+)-valencene (**1**) ([α]$_D$ + 84.6°, c = 1.0) obtained from Valencia orange oil was very efficiently converted into nootkatone (**2**) by biotransformations using *Chlorella* (Hashimoto et al., 2003b), *Mucor* species (Hashimoto et al., 2003a), *Botryosphaeria dothidea*, and *Botryodiplodia theobromae* (Furusawa et al., 2005a,b; Noma et al., 2001b).

Chlorella fusca var. *vacuolata* IAMC-28 (Figure 20.1) was inoculated and cultivated while stationary under illumination in Noro medium MgCl$_2$·6H$_2$O (1.5 g), MgSO$_4$·7H$_2$O (0.5 g), KCl (0.2 g), CaCl$_2$·2H$_2$O (0.2 g), KNO$_3$ (1.0 g), NaHCO$_3$ (0.43 g), TRIS (2.45 g), K$_2$HPO$_4$ (0.045 g), Fe-EDTA

FIGURE 20.1 *Chlorella fusca* var. *vacuolata*.

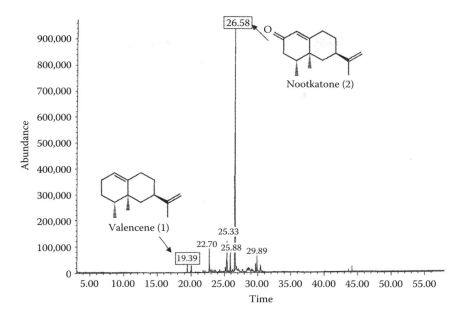

FIGURE 20.2 Total ion chromatogram of metabolites of valencene (**1**) by *Chlorella fusca* var. *vacuolata*.

(3.64 mg), EDTA-2Na (1.89 mg), $ZnSO_4 \cdot 7H_2O$ (1.5 g), H_3BO_2 (0.61 mg), $CoCl_2 \cdot 6H_2O$ (0.015 mg), $CuSO_4 \cdot 5H_2O$ (0.06 mg), $MnCl_2 \cdot 4H_2O$ (0.23 mg), and $(NH_4)_6Mo_7O_{24} \cdot 4H_2O$ (0.38 mg), in distilled H_2O 1 L (pH 8.0). Czapek-peptone medium (1.5% sucrose, 1.5% glucose, 0.5% polypeptone, 0.1% K_2HPO_4, 0.05% $MgSO_4 \cdot 7H_2O$, 0.05% KCl, and 0.001% $FeSO_4 \cdot 7H_2O$, in distilled water [pH 7.0]) was used for the biotransformation of substrate by microorganism. *Aspergillus niger* was isolated in our laboratories from soil in Osaka prefecture and was identified according to its physiological and morphological characters.

(+)-Valencene (**1**) (20 mg/50 mL) isolated from the essential oil of Valencia orange was added to the medium and biotransformed by *Chlorella fusca* for a further 18 days to afford nootkatone (**2**) (gas chromatography–mass spectrometry [GC-MS] peak area, 89%; isolated yield, 63%) (Figure 20.2) (Furusawa et al., 2005a,b; Noma et al., 2001b). The reduction of **2** with $NaBH_4$ and $CeCl_3$ gave 2α-hydroxyvalencene (**3**) in 87% yield, followed by Mitsunobu reaction with *p*-nitrobenzoic acid, triphenylphosphine, and diethyl azodicarboxylate to give nootkatol (2β-hydroxyvalencene) (**4**)—possessing calcium antagonistic activity—isolated from *Alpinia oxyphylla* (Shoji et al., 1984) in 42% yield. Compounds **3** and **4** thus obtained were easily biotransformed by *C. fusca* and *Chlorella pyrenoidosa* for only 1 day to give nootkatone (**2**) in good yield (80%–90%), respectively. The biotransformation of compound **1** was further carried out by *C. pyrenoidosa* and *Chlorella vulgaris* (Furusawa et al., 2005a,b) and soil bacteria (Noma et al., 2001a) to give nootkatone in good yield (Table 20.1).

TABLE 20.1
Conversion of Valencene (1) to Nootkatone (2) by *Chlorella* sp. for 14 Days

	Metabolites (% of the Total in GC-MS)				
Chlorella sp.	Valencene (1)	2α-Nootkatol (3)	2β-Nootkatol (4)	Nootkatone (2)	Conversion Ratio (%)
C. fusca	11	0	0	89	89
C. pyrenoidosa	7	0	0	93	93
C. vulgaris	0	0	0	100	100

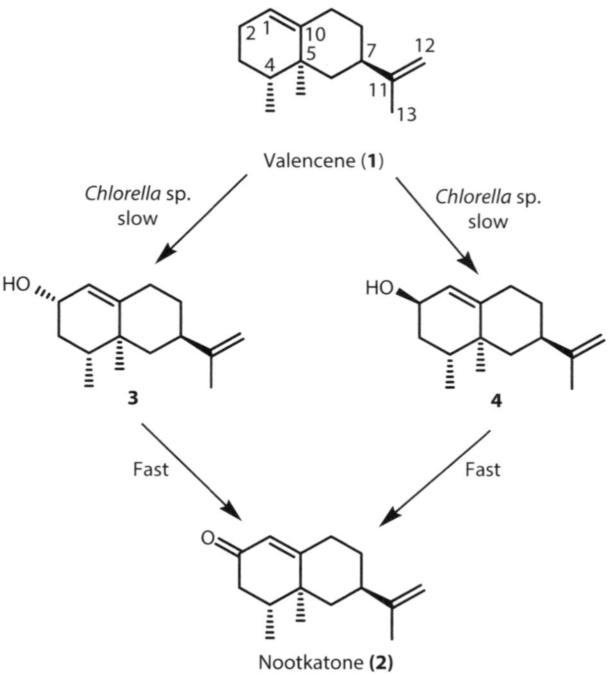

FIGURE 20.3 Biotransformation of valencene (**1**) by *Chlorella* species.

In the time course of the biotransformation of **1** by *C. pyrenoidosa*, the yield of nootkatone (**2**) and nootkatol (**4**) without 2α-hydroxyvalencene (**3**) increased with the decrease in that of **1**, and subsequently, the yield of **2** increased with decrease in that of **3**. In the metabolic pathway of valencene (**1**), **1** was slowly converted into nootkatol (**4**), and subsequently, **4** was rapidly converted into **2**, as shown in Figure 20.3.

A fungus strain from the soil adhering to the thalloid liverwort *Pallavicinia subciliata*, *Mucor* species, was inoculated and cultivated statically in Czapek-peptone medium (pH 7.0) at 30°C for 7 days. Compound **1** (20 mg/50 mL) was added to the medium and incubated for a further 7 days. Nootkatone (**2**) was then obtained in very high yield (82%) (Furusawa et al., 2005a; Noma et al., 2001b).

The biotransformation from **1** to **2** was also examined using the plant pathogenic fungi *B. dothidea* and *B. theobromae* (a total of 31 strains) separated from fungi infecting various types of fruit, and so on. *B. dothidea* and *B. theobromae* were both inoculated and cultivated while stationary in Czapek-peptone medium (pH 7.0) at 30°C for 7 days. The same size of the substrate **1** was added to each medium and incubated for a further 7 days to obtain nootkatone (42%–84%) (Furusawa et al., 2005a).

The expensive grapefruit aromatic nootkatone (**2**) used by cosmetic and fiber manufacturers was obtained in high yield by biotransformation of (+)-valencene (**1**), which can be cheaply obtained from Valencia orange, by *Chlorella* species, fungi such as *Mucor* species, *B. dothidea*, and *B. theobromae*. This is a very inexpensive and clean oxidation reaction, which does not use any heavy metals, and thus, this method is expected to find applications in the industrial production of nootkatone.

20.2.2 Biotransformation of Valencene (1) by *Aspergillus niger* and *Aspergillus wentii*

Valencene (**1**) from Valencia orange oil was cultivated by *A. niger* in Czapek-peptone medium, for 5 days to afford six metabolites: **5** (1.0%), **6** and **7** (13.5%), **8** (1.1%), **9** (1.5%), **10** (2.0%), and **11** (0.7%), respectively. Ratio of compounds **6** (11*S*) and **7** (11*R*) was determined as 1:3 by high-performance liquid chromatography (HPLC) analysis of their thiocarbonates (**12** and **13**) (Noma et al., 2001b) (Figure 20.4).

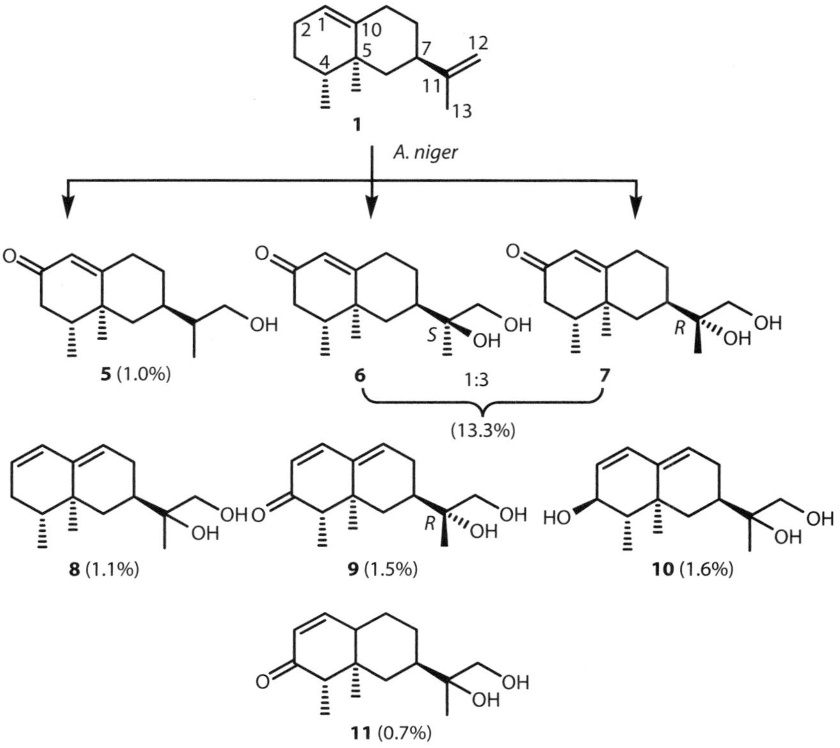

FIGURE 20.4 Biotransformation of valencene (**1**) by *Aspergillus niger*.

Compounds **8–11** could be biosynthesized by elimination of a hydroxy group of 2-hydroxyvalencenes (**3**, **4**). Compound **3** was biotransformed for 5 days by *A. niger* to give three metabolites: **6** and **7** (6.4%), **8** (34.6%), and **9** (5.5%), respectively. Compound **4** was biotransformed for 5 days by *A. niger* to give three metabolites: **6** and **7** (21.8%), **9** (5.5%), and **10** (10.4%), respectively (Figure 20.5).

Both ratios of **6** (11*S*) and **7** (11*R*) obtained from **3** and **4** were 1:3, respectively. From the aforementioned results, plausible metabolic pathways of valencene (**1**) and 2-hydroxyvalencene (**3**, **4**) by *A. niger* are shown in Figure 20.6 (Noma et al., 2001b).

Aspergillus wentii and *Epicoccum purpurascens* converted valencene (**1**) to 11,12-epoxide (**14a**) and the same diol (**6**, **7**) (Takahashi and Miyazawa, 2005) as well as nootkatone (**2**) and 2α-hydroxyvalencene (**3**) (Takahashi and Miyazawa, 2006).

Kaspera et al. (2005) reported that valencene (**1**) was incubated in submerged cultures of the ascomycete *Chaetomium globosum*, to give nootkatone (**2**), 2α-hydroxyvalencene (**3**), and valencene 11,12-epoxide (**14a**), together with a valencene ketodiol, valencene diols, a valencene triol, or valencene epoxydiol that were detected by liquid chromatography–MS (LC-MS) spectra and GC-MS of trimethylsilyl derivatives. These metabolites are accumulated preferably inside the fungal cells (Figure 20.7).

The metabolites of valencene, nootkatone (**2**), (**3**), and (**14a**), indicated grapefruit with sour and citrus with bitter odor, respectively. Nootkatone 11,12-epoxide (**14**) showed no volatile fragrant properties.

20.2.3 Biotransformation of Nootkatone (2) by *Aspergillus niger*

A. niger was inoculated and cultivated rotatory (100 rpm) in Czapek-peptone medium at 30°C for 7 days. (+)-Nootkatone (**2**) ($[\alpha]_D$ + 193.5°, c = 1.0) (80 mg/200 mL), which was isolated from the

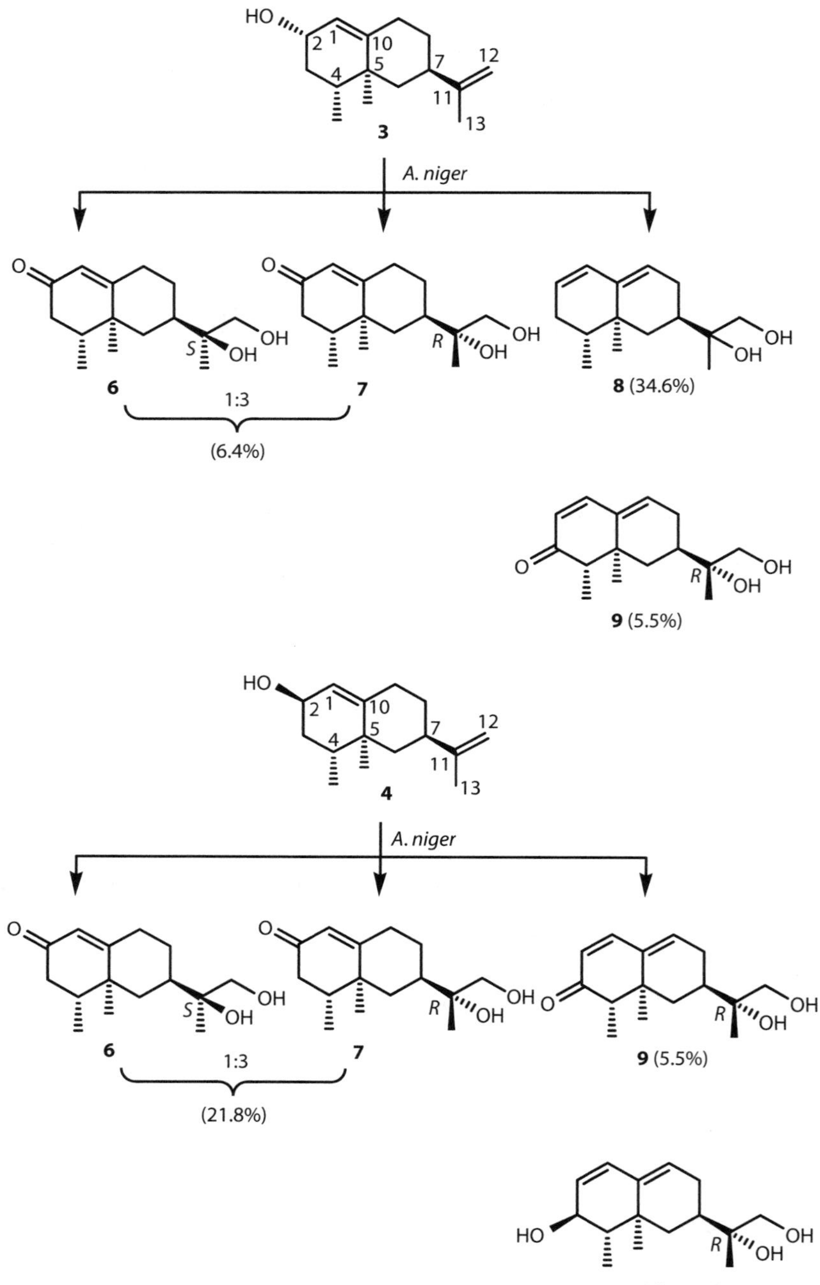

FIGURE 20.5 Biotransformation of 2α-hydroxyvalencene (**3**) and 2β-hydroxyvalencene (**4**) by *Aspergillus niger*.

essential oil of grapefruit, was added to the medium and further cultivated for 7 days to obtain two metabolites, 12-hydroxy-11,12-dihydronootkatone (**5**) (10.6%) and C11 stereo mixtures (51.5%), of nootkatone-11S,12-diol (**6**) and its 11R isomer (**7**) (11R:11S = 1:1) (Furusawa et al., 2003; Hashimoto et al., 2000b; Noma et al., 2001b) (Figure 20.8).

11,12-Epoxide (**14**) obtained by epoxidation of nootkatone (**2**) with meta-chloroperbenzoic acid (mCPBA) was biotransformed by *A. niger* for 1 day to afford **6** and **7** (11R:11S = 1:1) in good

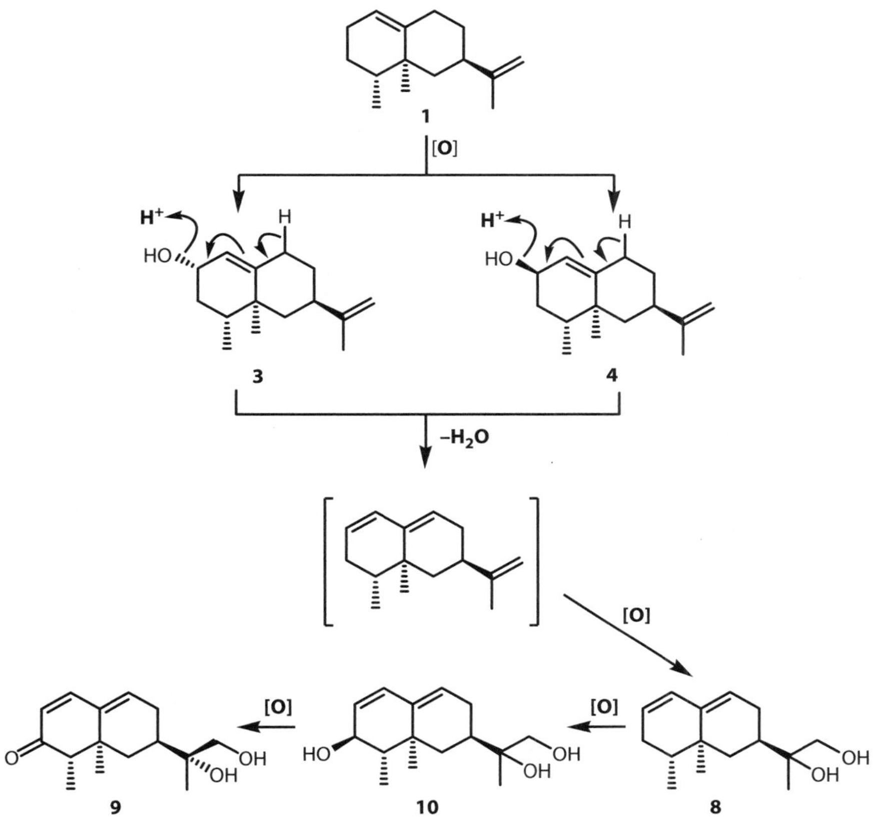

FIGURE 20.6 Possible pathway of biotransformation of valencene (**1**) by *Aspergillus niger*.

yield (81.4%). 1-Aminobenzotriazole, an inhibitor of CYP450, inhibited the oxidation process of **1** into compounds **5–7** (Noma et al., 2001b). From the aforementioned results, possible metabolic pathways of nootkatone (**2**) by *A. niger* might be considered as shown in Figure 20.9.

The same substrate was incubated with *A. wentii* to produce diol (**6, 7**) and 11,12-epoxide (**14**) (Takahashi and Miyazawa, 2005).

20.2.4 Biotransformation of Nootkatone (2) by *Fusarium culmorum* and *Botryosphaeria dothidea*

(+)-Nootkatone (**2**) was added to the same medium as mentioned earlier including *Fusarium culmorum* to afford nootkatone-11R,12-diol (**7**) (47.2%) and 9β-hydroxynootkatone (**15**) (14.9%) (Noma et al., 2001b).

Compound **7** was stereospecifically obtained at C11 by biotransformation of **1**. Purity of compound **7** was determined as ca. 95% by HPLC analysis of the thiocarbonate (**13**).

The biotransformation of nootkatone (**2**) was examined by the plant pathogenic fungus *B. dothidea* separated from the fungus that infected the peach. (+)-Nootkatone (**2**) was cultivated with *B. dothidea* (Peach PP8402) for 14 days to afford nootkatone diols (**6** and **7**) (54.2%) and 7α-hydroxynootkatone (**16**) (20.9%). Ratio of compounds **6** and **7** was determined as 3:2 by HPLC analysis of the thiocarbonates (**12, 13**) (Noma et al., 2001b). Nootkatone (**2**) was administered into rabbits to give the same diols (**6, 7**) (Asakawa et al., 1986; Ishida, 2005).

E. purpurascens also biotransformed nootkatone (**2**) to **5–7**, **14**, and **15a** (Takahashi and Miyazawa, 2006).

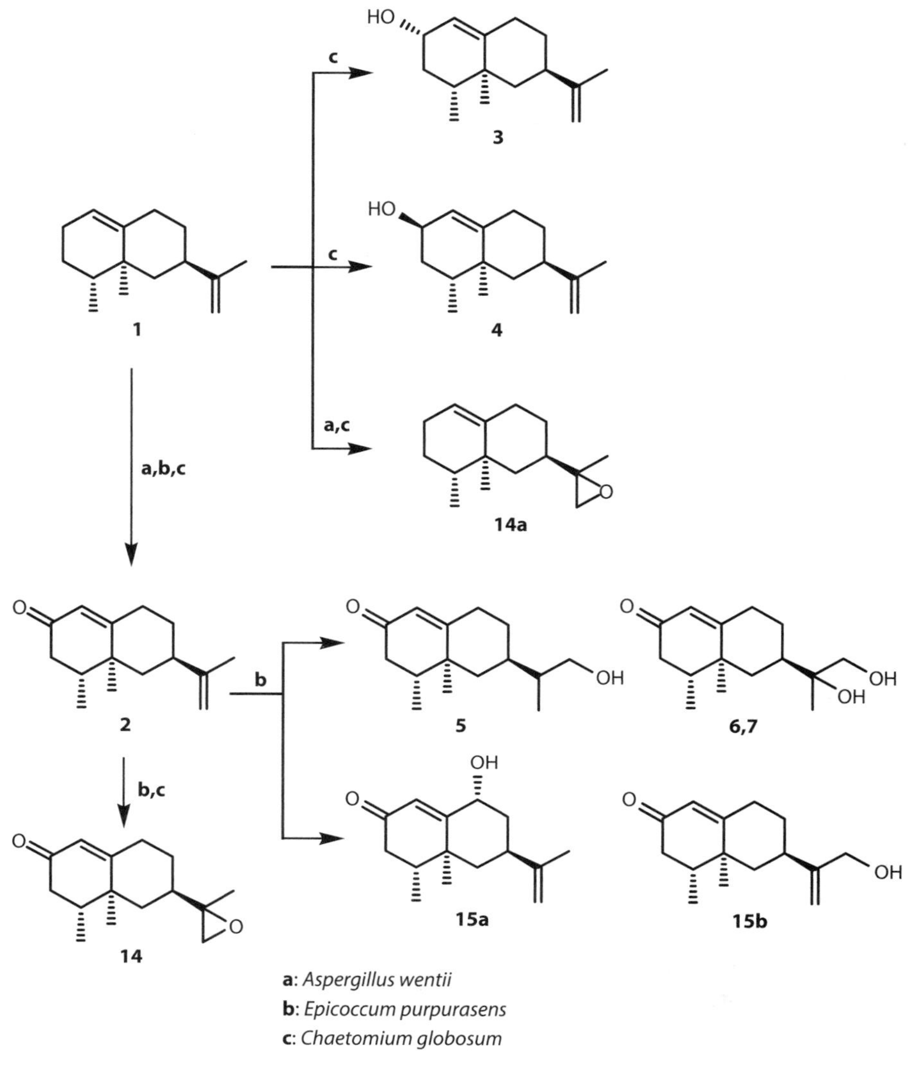

FIGURE 20.7 Biotransformation of valencene (**1**) and nootkatone (**2**) by *Aspergillus wentii*, *Epicoccum purpurascens*, and *Chaetomium globosum*.

The biotransformation of **2** by *A. niger* and *B. dothidea* resembled to that of the oral administration to rabbit since the ratio of the major metabolites 11*S*- (**6**) and 11*R*-nootkatone-11,12-diol (**7**) was similar. It is noteworthy that the biotransformation of **2** by *F. culmorum* affords stereospecifically nootkatone-11*R*,12-diol (**7**) (Noma et al., 2001b) (Figure 20.10).

Metabolites **3–5**, **12**, and **13** from (+)-nootkatone (**2**) and **14–17** from (+)-valencene (**1**) did not show an effective odor.

Dihydronootkatone (**17**), which shows that citrus odor possesses antitermite activity, was also treated in *A. niger* to obtain 11*S*-mono- (**18**) and 11*R*-dihydroxylated products (**19**) (the ratio 11*S* and 11*R* = 3:2). On the other hand, *Aspergillus cellulosae* reduced ketone group at C2 of **17** to give 2α- (**20**) (75.7%) and 2β-hydroxynootkatone (**21**) (0.7%) (Furusawa et al., 2003) (Figure 20.11).

Tetrahydronootkatone (**22**) also shows antitermite and mosquito-repellant activity. It was incubated with *A. niger* to give two similar hydroxylated compounds (**23**, 13.6% and **24**, 9.9%) to those obtained from **17** (Furusawa, 2006) (Figure 20.12).

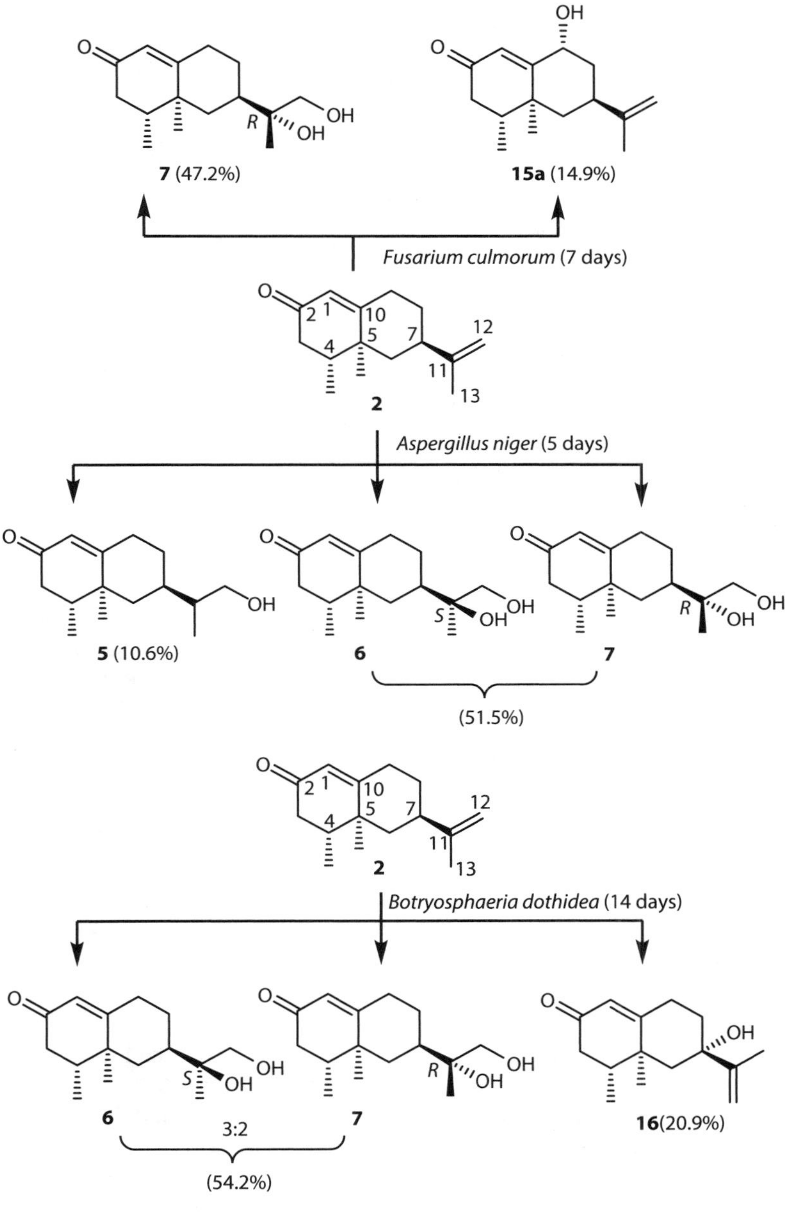

FIGURE 20.8 Biotransformation of nootkatone (**2**) by *Fusarium culmorum*, *Aspergillus niger*, and *Botryosphaeria dothidea*.

8,9-Dehydronootkatone (**25**) was incubated with *A. niger* to give four metabolites, a unique acetonide (**26**, 15.6%), monohydroxylated (**27**, 0.2%), dihydroxylated (**28**, 69%), and a carboxyl derivative (**29**, 0.8%) (Figure 20.13).

When the same substrate was treated in *Aspergillus sojae* IFO 4389, compound **25** was converted to the different monohydroxylated products (**30**, 15.8%) from that mentioned earlier. *A. cellulosae* is an interesting fungus since it did not give any same products as mentioned earlier; in place, it produced trinorsesquitepene ketone (**31**, 6%) and nitrogen-containing aromatic compound (**32**) (Furusawa et al., 2003) (Figure 20.14).

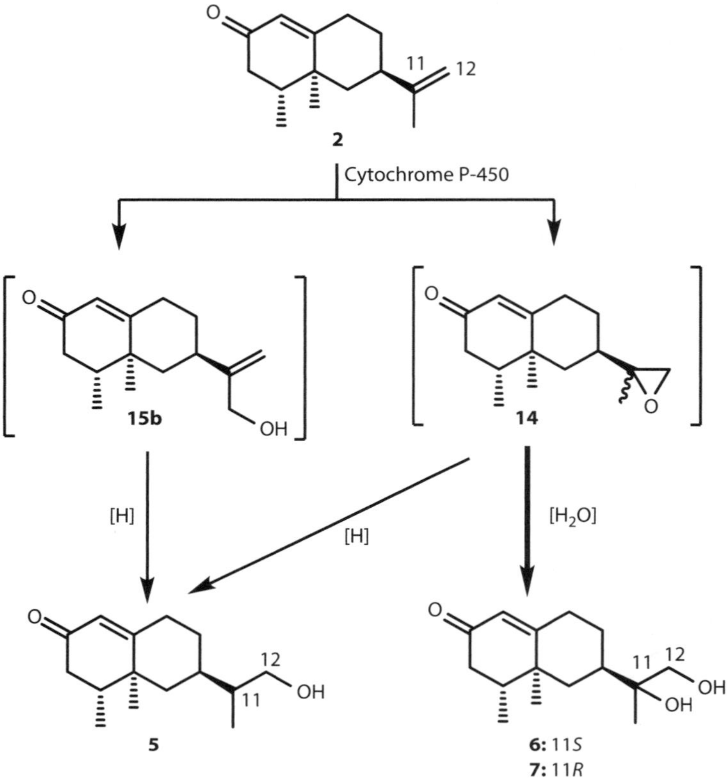

FIGURE 20.9 Possible pathway of biotransformation of valencene (**1**) by cytochrome P-450.

Mucor species also oxidized compound **25** to give three metabolites, 13-hydroxy-8,9-dehydronootkatone (**33**, 13.2%), an epoxide (**34**, 5.1%), and a diol (**35**, 19.9%) (Furusawa et al., 2003). The same substrate was investigated with cultured suspension cells of the liverwort *Marchantia polymorpha* to afford **33** (Hegazy et al., 2005) (Figure 20.15).

Although *Mucor* species could give nootkatone (**21**) from valencene (**1**), this fungus biotransformed the same substrate (**25**) to the same alcohol (**30**, 13.2%) obtained from the same starting compound (**25**) in *A. sojae*, a new epoxide (**34**, 5.1%), and a diol (**35**, 9.9%).

The metabolites (**3, 4, 20, 21**) inhibited the growth of lettuce stem, and **3** and **4** inhibited germination of the same plant (Hashimoto and Asakawa, 2007).

Valerianol (**35a**), from *Valeriana officinalis* whose dried rhizome is traditionally used for its carminative and sedative properties, was biotransformed by *Mucor plumbeus*, to produce three metabolites, a bridged ether (**35b**), and a triol (**35c**), which might be formed via C1–C10 epoxide, and **35d** arises from double dehydration (Arantes et al., 1999). In this case, allylic oxidative compounds have not been found (Figure 20.16).

20.2.5 Biotransformation of (+)-1(10)-Aristolene (36) from the Crude Drug *Nardostachys chinensis* by *Chlorella fusca*, *Mucor* Species, and *Aspergillus niger*

The structure of sesquiterpenoid (+)-1(10)-aristolene (=calarene) (**36**) from the crude drug *Nardostachys chinensis* was similar to that of nootkatone. 2-Oxo-1(10)-aristolene (**38**) shows antimelanin-inducing activity and excellent citrus fragrance. On the other hand, the enantiomer (**37**)

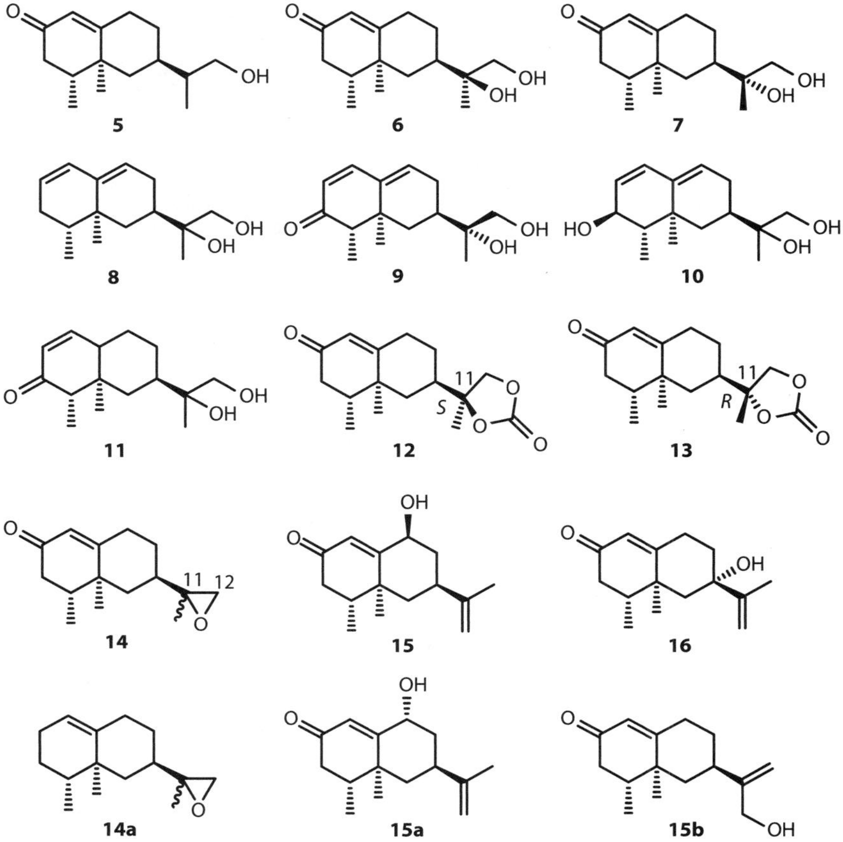

FIGURE 20.10 Metabolites (5–11, 14–15b) from valencene (1) and nootkatone (2) by various microorganisms.

of **36** and (+)-aristolone (**41**) were also found in the liverworts as the natural products. In order to obtain compound **38** and its analogues, compound **36** was incubated with *Chlorella fusca* var. *vacuolata* IAMC-28, *Mucor* species, and *A. niger* (Furusawa et al., 2006a) (Figure 20.17).

C. fusca was inoculated and cultivated stationary in Noro medium (pH 8.0) at 25°C for 7 days, and (+)-1(10)-aristolene (**36**) (20 mg/50 mL) was added to the medium and further incubated for 10–14 days and cultivated stationary under illumination (pH 8.0) at 25°C for 7 days to afford 1(10)-aristolen-2-one (**38**, 18.7%), (−)-aristolone (**39**, 7.1%), and 9-hydroxy-1(10)-aristolen-2-one (**40**). Compounds **38** and **40** were found in *Aristolochia* species (Figure 20.18).

Mucor species was inoculated and cultivated rotatory (100 rpm) in Czapek-peptone medium (pH 7.0) at 30°C for 7 days. (+)-1(10)-Aristolene (**36**) (100 mg/200 mL) was added to the medium and further for 7 days. The crude metabolites contained **38** (0.9%) and **39** (0.7%) as very minor products (Figure 20.19).

Although *Mucor* species produced a large amount of nootkatone (**2**) from valencene (**1**), however, only poor yield of similar products as those from valencene (**1**) was seen in the biotransformation of tricyclic substrate (**36**). Possible biogenetic pathway of (+)-1(10)-aristolene (**36**) is shown in Figure 20.20.

A. niger was inoculated and cultivated rotatory (100 rpm) in Czapek-peptone medium (pH 7.0) at 30°C for 3 days. (+)-1(10)-Aristolene (**36**) (100 mg/200 mL) was added to the medium and further for 7 days. From the crude metabolites, four new metabolic products **42**, 1.3%; **43**, 3.2%; **44**, 0.98%; and **45**, 2.8% were obtained in very poor yields (Figure 20.21). Possible metabolic pathways of **36** by *A. niger* are shown in Figure 20.22.

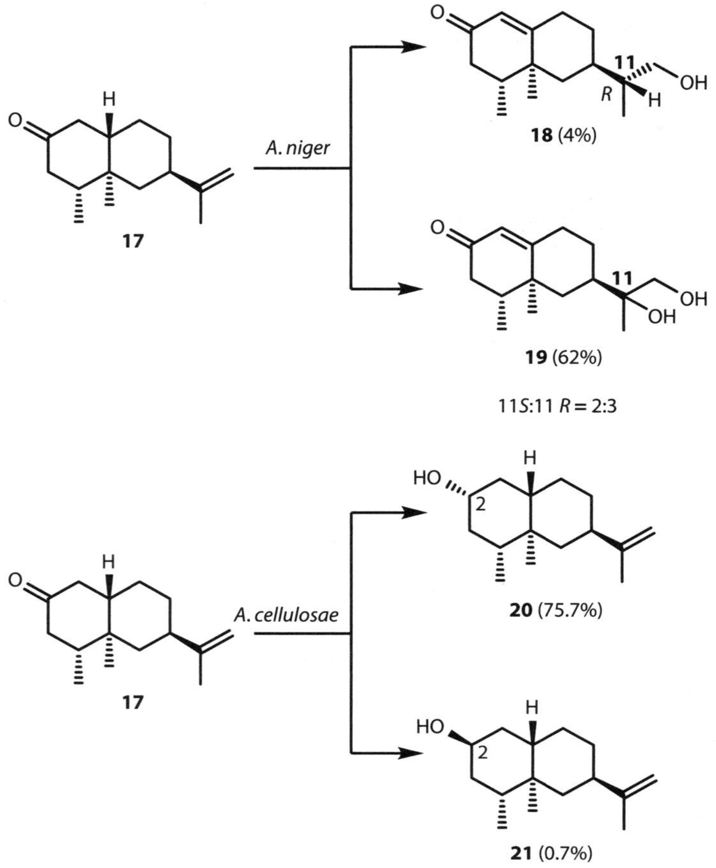

FIGURE 20.11 Biotransformation of dihydronootkatone (**17**) by *Aspergillus niger* and *Aspergillus cellulosae*.

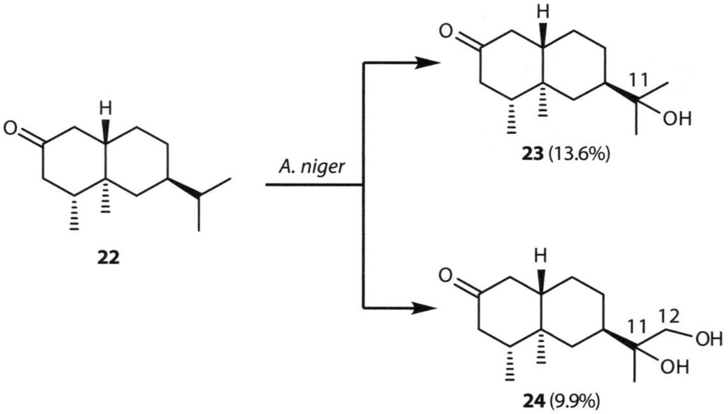

FIGURE 20.12 Biotransformation of tetrahydronootkatone (**22**) by *Aspergillus niger*.

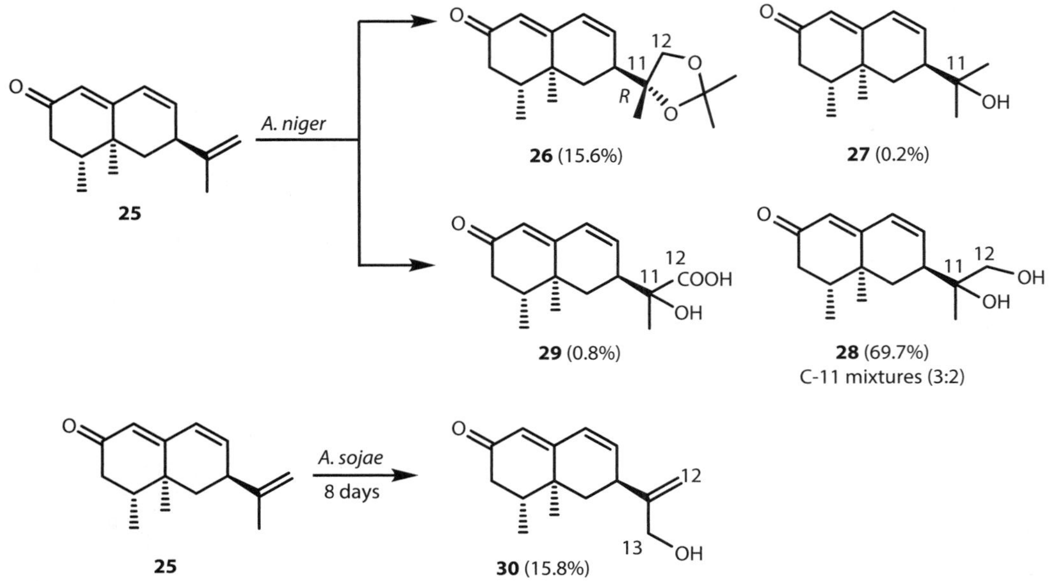

FIGURE 20.13 Biotransformation of 8,9-dehydronootkatone (**25**) by *Aspergillus sojae*.

Commercially available (+)-1(10)-aristolene (**36**) was treated with *Diplodia gossypina* and *Bacillus megaterium*. Both microorganisms converted **36-4** (**46–49**; 0.8%, 1.1%, 0.16%, 0.38%) and six metabolites (**40, 50–55**; 0.75%, 1.0%, 1.0%, 2.0%, 1.1%, 0.5%, 0.87%), together with **40** (0.75%), respectively (Abraham et al., 1992) (Figure 20.23).

It is noteworthy that *Chlorella* and *Mucor* species introduce hydroxyl group at C2 of the substrate (**36**) as seen in the biotransformation of valencene (**1**), while *D. gossypina* and *B. megaterium* oxidize C2, C8, C9, and/or 1,1-dimethyl group on a cyclopropane ring. *A. niger* oxidizes not only C2 but also stereoselectively oxidized one of the gem-dimethyl groups on cyclopropane ring. Stereoselective oxidation of one of gem-dimethyl of cyclopropane and cyclobutane derivatives is observed in biotransformation using mammals (see later).

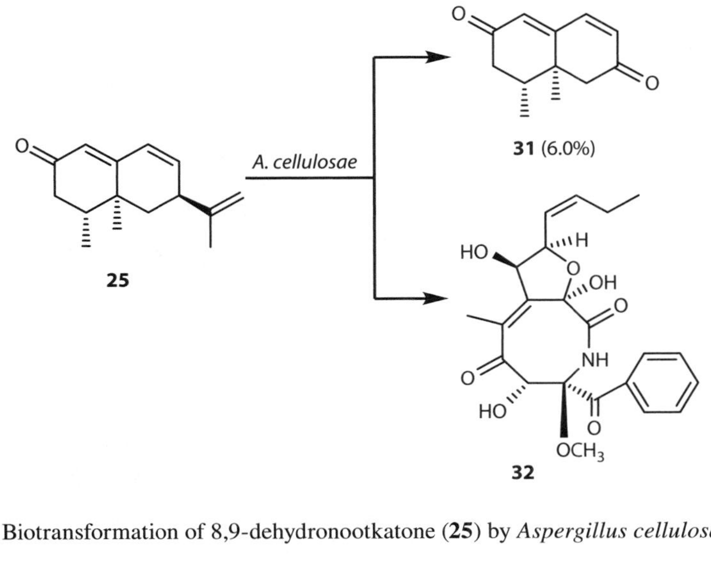

FIGURE 20.14 Biotransformation of 8,9-dehydronootkatone (**25**) by *Aspergillus cellulosae*.

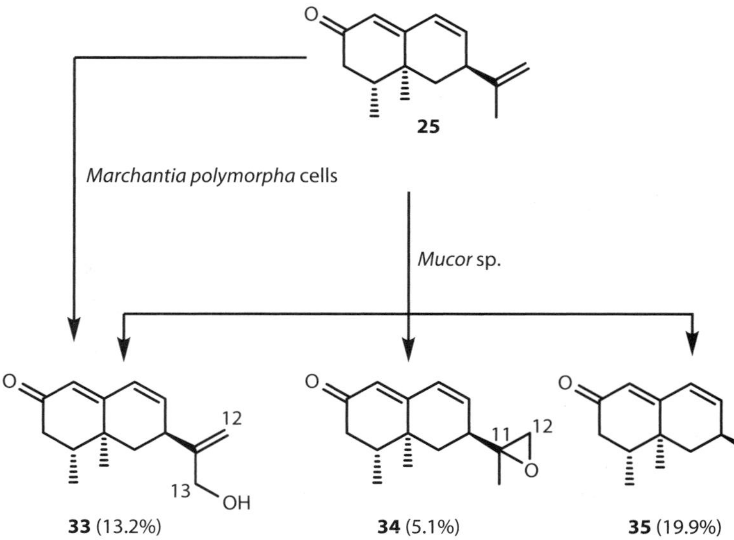

FIGURE 20.15 Biotransformation of 8,9-dehydronootkatone (**25**) by *Marchantia polymorpha* and *Mucor* species.

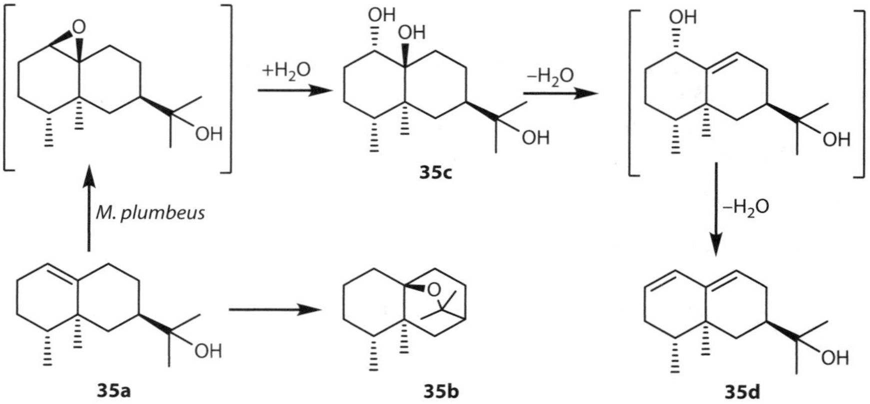

FIGURE 20.16 Biotransformation of valerianol (**35a**) by *Mucor plumbeus*.

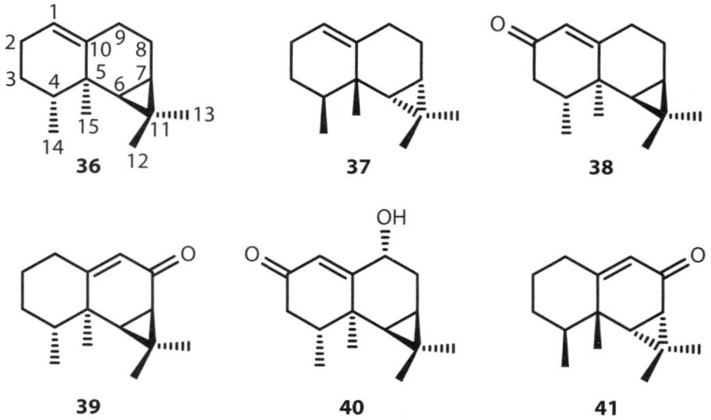

FIGURE 20.17 Naturally occurring aristolane sesquiterpenoids.

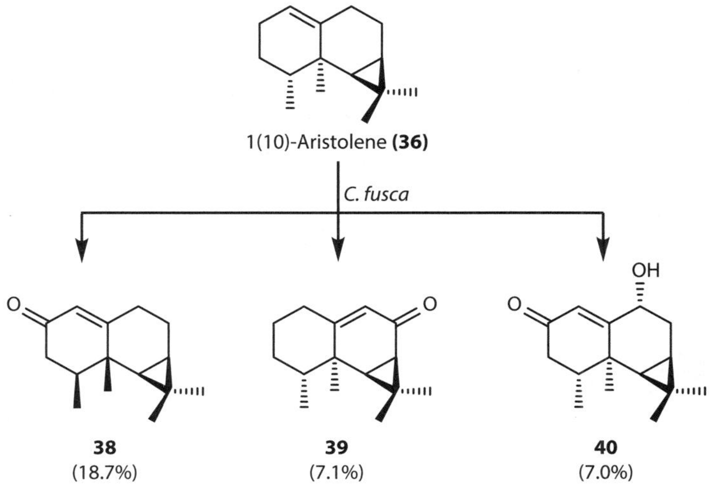

FIGURE 20.18 Biotransformation of 1(10)-aristolene (**36**) by *Chlorella fusca*.

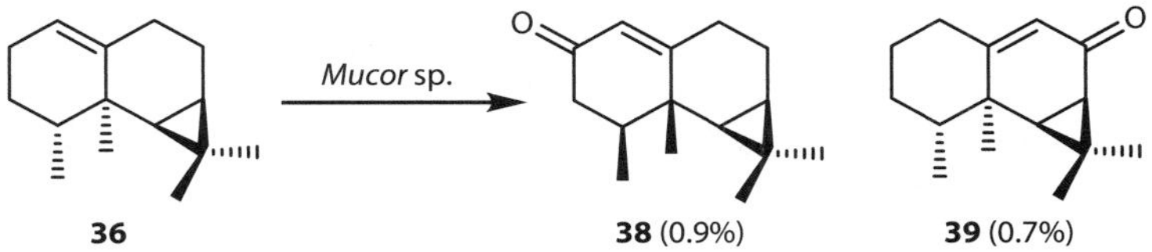

FIGURE 20.19 Biotransformation of 1(10)-aristolene (**36**) by *Mucor* species.

20.2.6 Biotransformation of Various Sesquiterpenoids by Microorganisms

Aromadendrene-type sesquiterpenoids have been found not only in higher plants but also in liverworts and marine sources. Three aromadendrenes (**56, 57, 58**) were biotransformed by *D. gossypina*, *B. megaterium*, and *Mycobacterium smegmatis* (Abraham et al., 1992). Aromadendrene (**56**) (800 mg) was converted by *B. megaterium* to afford a diol (**59**) and a triol (**60**) of which **59** (7 mg) was the major product. The triol (**60**) was also obtained from the metabolite of (+)-(1R)-aromadendrene (**56**) by the plant pathogen *Glomerella cingulata* (Miyazawa et al., 1995d). *allo*-Aromadendrene (**57**) (1.2 g) was also treated in *M. smegmatis* to afford **61** (10 mg) (Abraham et al., 1992) (Figure 20.24).

The same substrate was also incubated with *G. cingulata* to afford C10 epimeric triol (**62**) (Miyazawa et al., 1995d). Globulol (**58**) (400 mg) was treated in *M. smegmatis* to give only a carboxylic acid (**63**) (210 mg). The same substrate (**58**) (1 g) was treated in *D. gossypina* and *B. megaterium* to give two diols, **64** (182 mg) and **65**, and a triol (**66**) from the former and **67–69** from the latter organism among which **64** (60 mg) was predominant (Abraham et al., 1992). *G. cingulata* and *Botrytis cinerea* also bioconverted globulol (**58**) to diol (**64**) regio- and stereoselectively (Miyazawa et al., 1994) (Figures 20.25 and 20.26).

Globulol (**58**) (1.5 g) and 10-epiglobulol (**70**) (1.2 mL) were separately incubated with *Cephalosporium aphidicola* in shake culture for 6 days to give the same diol **64** (780 mg) as obtained from the same substrate by *B. megaterium* mentioned earlier and **71** (720 mg) (Hanson et al., 1994). *A. niger* also converted globulol (**58**) and epiglobulol (**70**) to a diol (**64**) and three 13-hydroxylated globulol (**71, 72, 74**) and 4α-hydroxylated product (**73**). The epimerization at C4 is very rare example (Hayashi et al., 1998).

Ledol (**75**), an epimer at C1 of globulol, was incubated with *G. cingulata* to afford C13 carboxylic acid (**76**) (Miyazawa et al., 1994) (Figure 20.27).

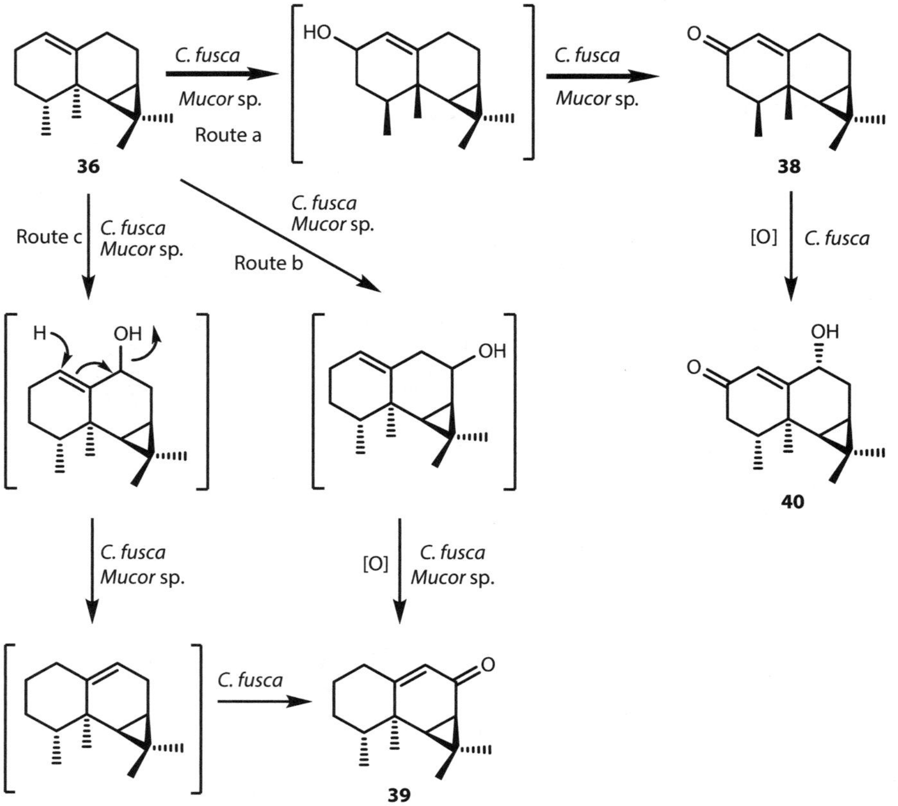

FIGURE 20.20 Possible pathway of biotransformation of 1(10)-aristolene (**36**) by *Chlorella fusca* and *Mucor* species.

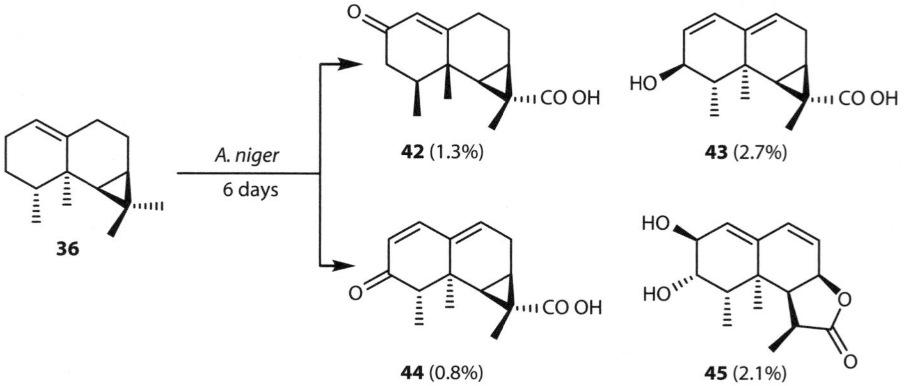

FIGURE 20.21 Biotransformation of 1(10)-aristolene (**36**) by *Aspergillus niger*.

Squamulosone (**77**), aromadendr-1(10)-en-9-one isolated from *Hyptis verticillata* (Labiatae), was reduced chemically to give **78–82**, which were incubated with the fungus *Curvularia lunata* in two different growth media (Figure 20.28).

From **78**, two metabolites **80** and **83** were obtained. Compound **79** and **80** were metabolized to give ketone **81** as the sole product and **78** and **83**, respectively. From compound **81**, two metabolites,

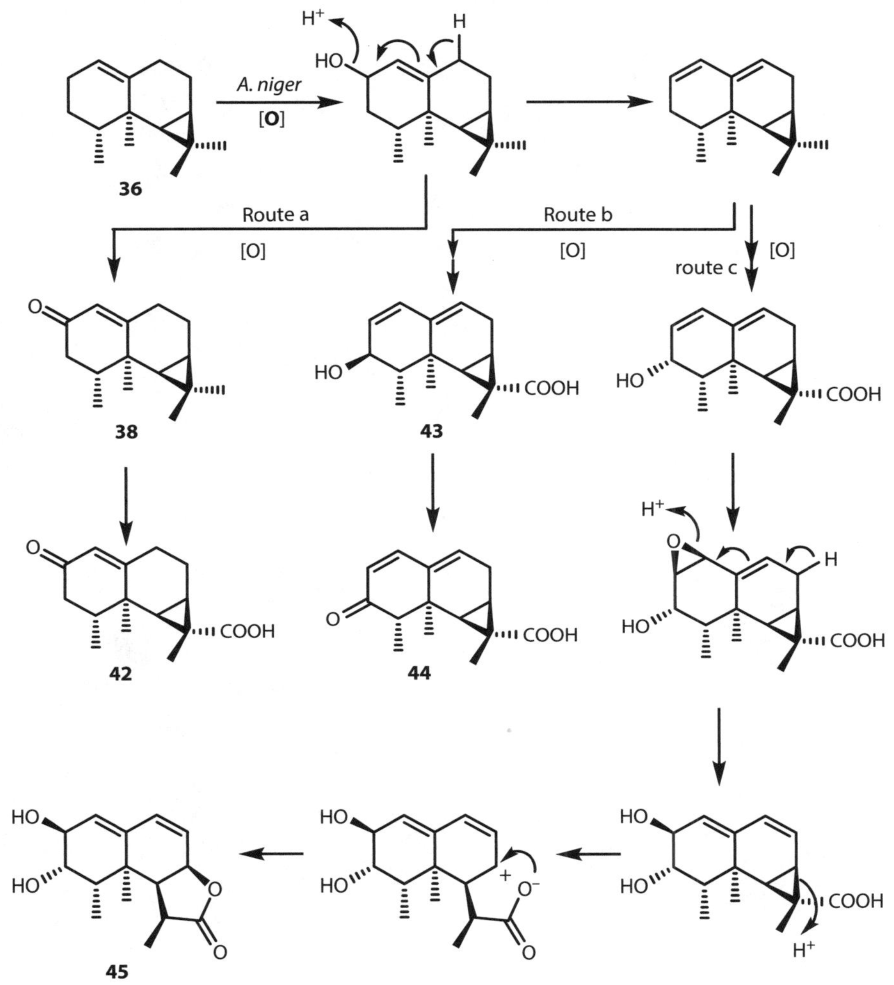

FIGURE 20.22 Possible pathway of biotransformation of 1(10)-aristolene (**36**) by *Aspergillus niger*.

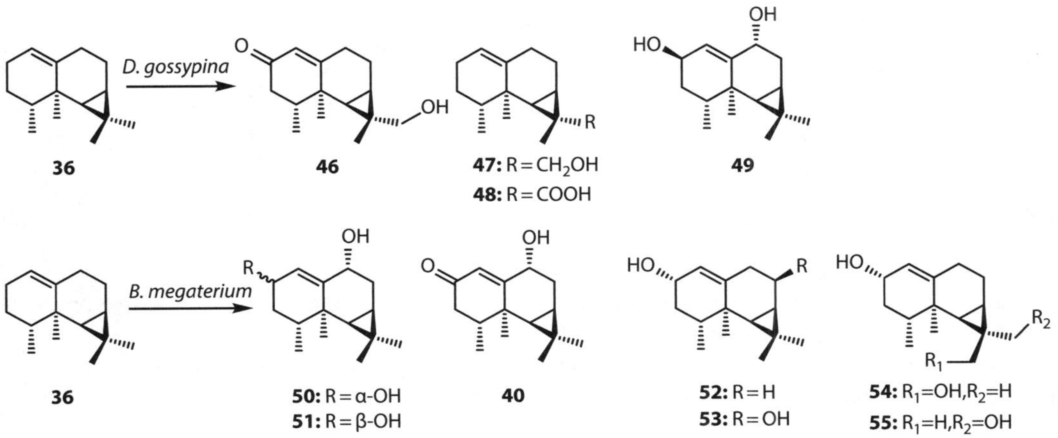

FIGURE 20.23 Biotransformation of 1(10)-aristolene (**36**) by *Diplodia gossypina* and *Bacillus megaterium*.

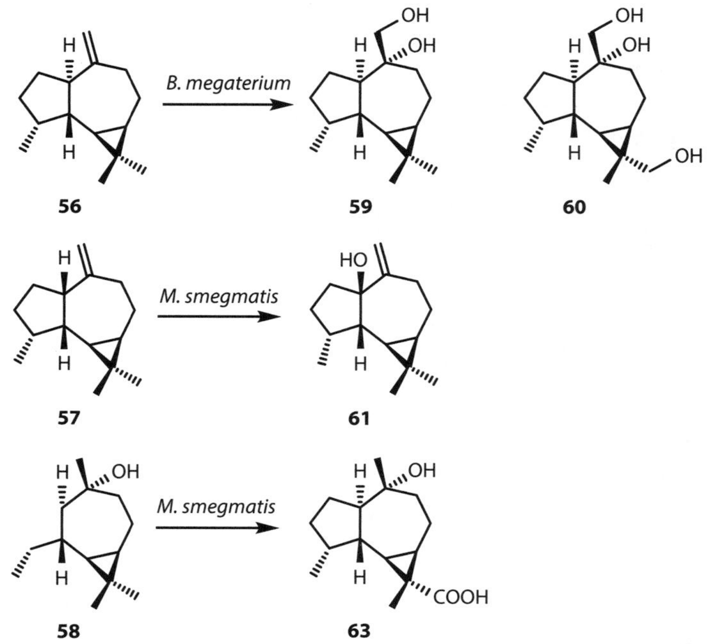

FIGURE 20.24 Biotransformation of aromadendrene (**56**), allo-aromadendrene (**57**), and globulol (**58**) by *Bacillus megaterium* and *Mycobacterium smegmatis*.

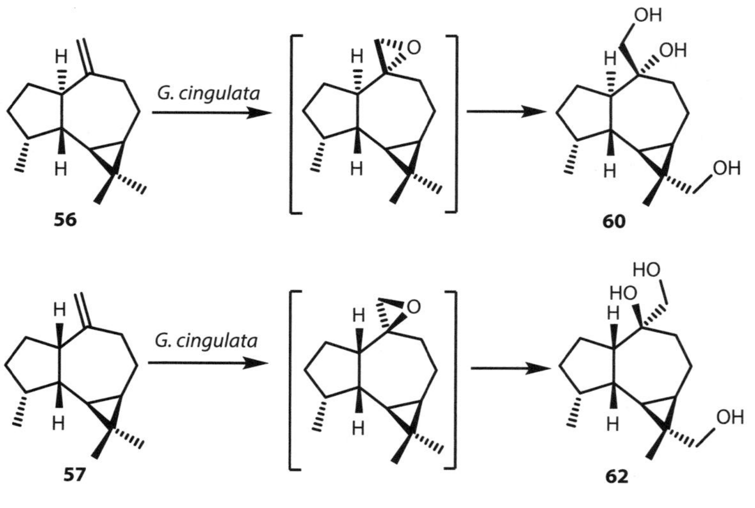

FIGURE 20.25 Biotransformation of aromadendrene (**56**) and allo-aromadendrene (**57**) by *Glomerella cingulata*.

79 and **84**, were obtained (Figure 20.29). From the metabolite of the substrate (**82**), five products (**84–88**) were isolated (Collins et al., 2002a) (Figure 20.30).

Squamulosone (**77**) was treated in the fungus *M. plumbeus* ATCC 4740 to give not only cyclopentanol derivatives (**89, 90**) but also C12-hydroxylated products (**91–93**) (Collins et al., 2002b) (Figure 20.31).

Spathulenol (**94**), which is found in many essential oils, was fed by *A. niger* to give a diol (**95**) (Higuchi et al., 2001). *ent*-10β-Hydroxycyclocolorenone (**96**) and myli-4(15)-en-9-one (**96a**) isolated

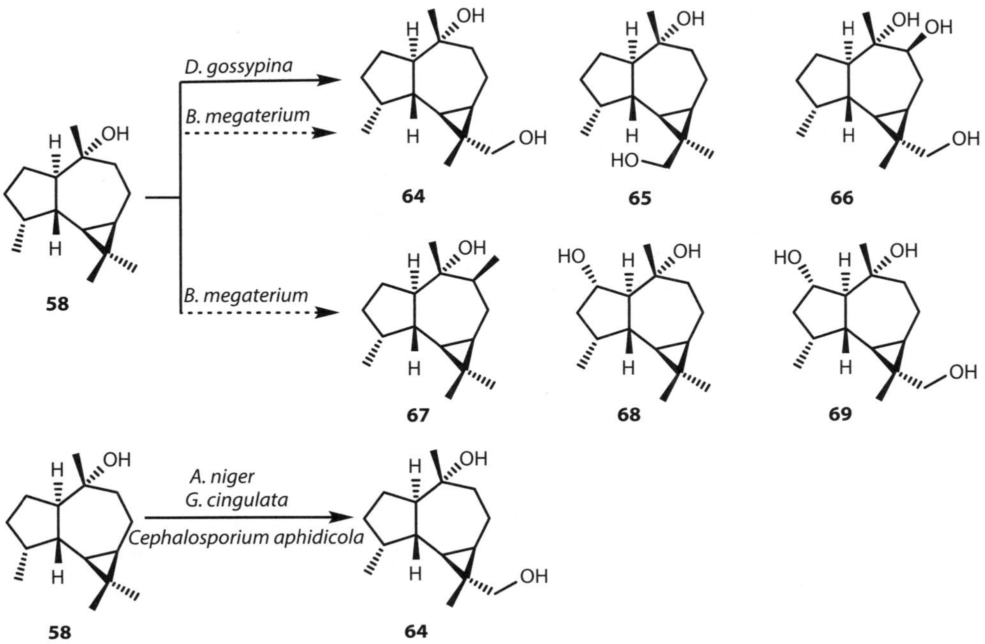

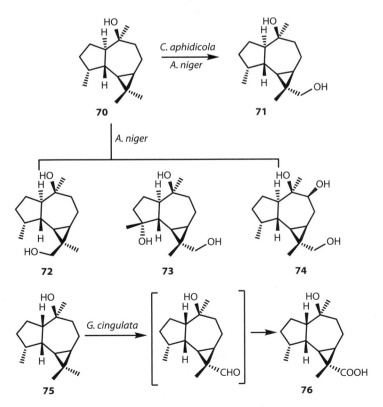

FIGURE 20.26 Biotransformation of globulol (**58**) by various microorganisms.

FIGURE 20.27 Biotransformation of 10-epi-glubulol (**70**) and ledol (**75**) by *Cephalosporium aphidicola*, *Aspergillus niger*, and *Glomerella cingulata*.

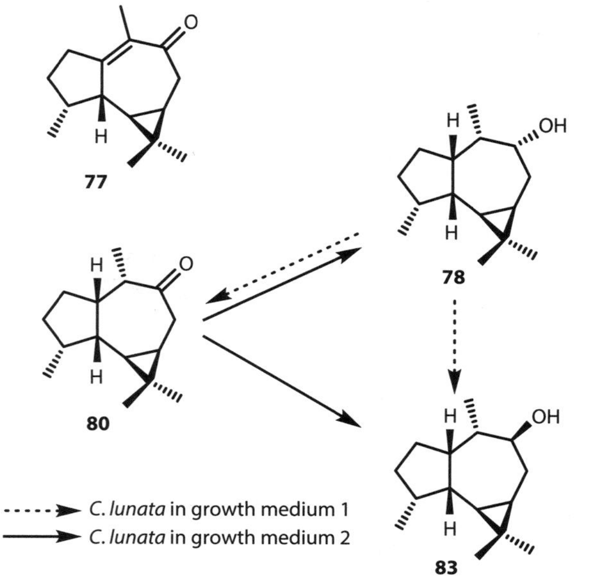

FIGURE 20.28 Biotransformation of aromadendra-9-one (**80**) by *Curvularia lunata*.

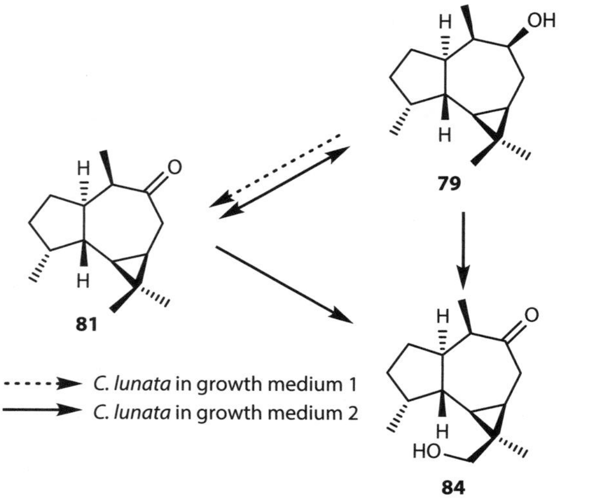

FIGURE 20.29 Biotransformation of 10-epi-aromadendra-9-one (**81**) by *Curvularia lunata*.

from the liverwort *Mylia taylorii* were incubated with *A. niger* IFO 4407 to give C10 epimeric product (**97**) (Hayashi et al., 1999) and 12-hydroxylated product (**96b**), respectively (Nozaki et al., 1996) (Figures 20.32 and 20.33).

(+)-*ent*-Cyclocolorenone (**98**) [α]$_D$ − 405° (c = 8.8, EtOH), one of the major compounds isolated from the liverwort *Plagiochila sciophila* (Asakawa, 1982, 1995), was treated by *A. niger* to afford three metabolites, 9-hydroxycyclocolorenone (**99**, 15.9%) 12-hydroxy-(+)-cyclocolorenone (**100**, 8.9%) and a unique cyclopropane-cleaved metabolite, 6β-hydroxy-4,11-guaiadien-3-one (**101**, 35.9%), and 6β,7β-dihydroxy-4,11-guaiadien-3-one (**102**, trace), of which **101** was the major component. The enantiomer (**103**) [α]$_D$ + 402° (c = 8.8, EtOH) of **98** isolated from *Solidago altissima* was biotransformed by the same organism to give 13-hydroxycyolorenone (**103a**, 65.5%), the enantiomer of **100**, 1β,13-dihydroxycyclocolorenone (**103b**, 5.0%), and its C11 epimer (**103c**)

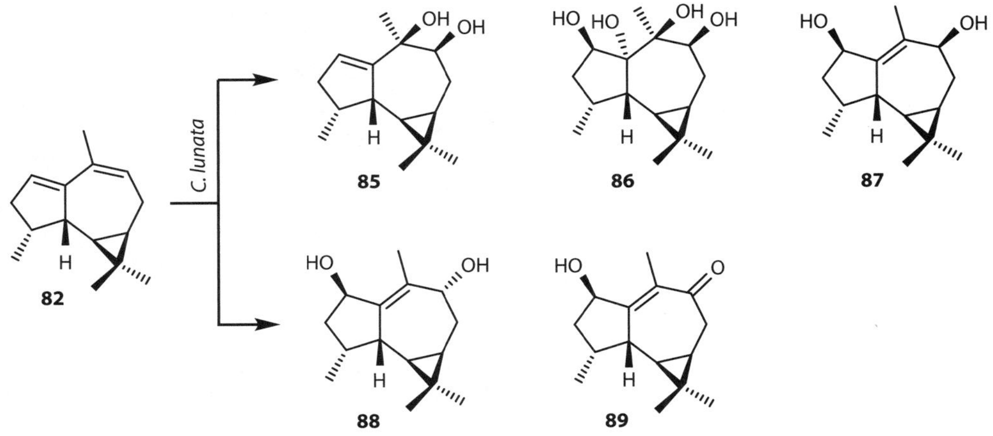

FIGURE 20.30 Biotransformation of aromadendr-1(10),9-diene (**82**) by *Curvularia lunata*.

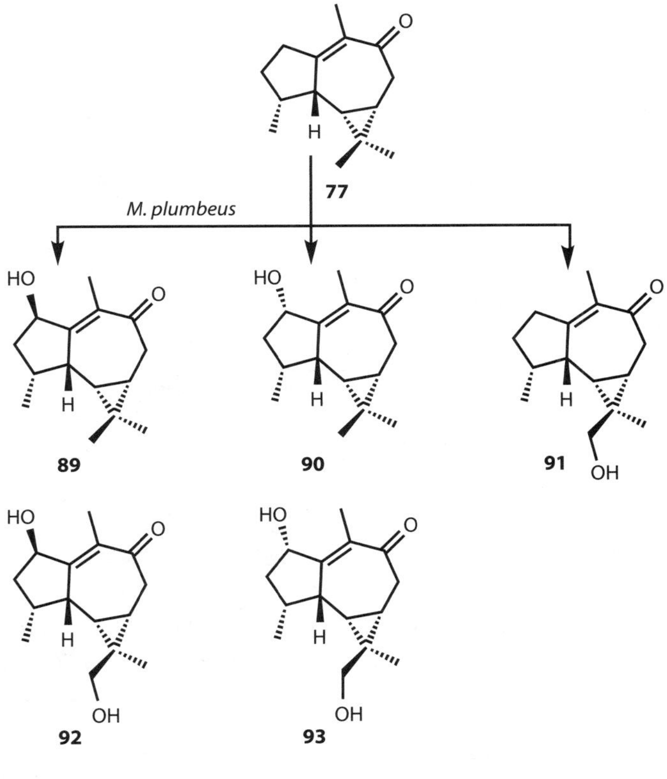

FIGURE 20.31 Biotransformation of squamulosone (**77**) by *Mucor plumbeus*.

(Furusawa et al., 2005c, 2006a). It is noteworthy that no cyclopropane-cleaved compounds from **103** have been detected in the crude metabolites even in GC-MS analysis (Figure 20.34).

Plagiochiline A (**104**) that shows potent insect antifeedant, cytotoxicity, and piscicidal activity is very pungent, 2,3-secoaromadendrane sesquiterpenoids having 1,1-dimethyl cyclopropane ring, isolated from the liverwort *Plagiochila fruticosa*. Plagiochilide (**105**) is the major component of this liverwort. In order to get more pungent component, the lactone (**105**, 101 mg) was incubated with *A. niger* to give two metabolites: **106** (32.5%) and **107** (9.7%). Compound **105** was incubated

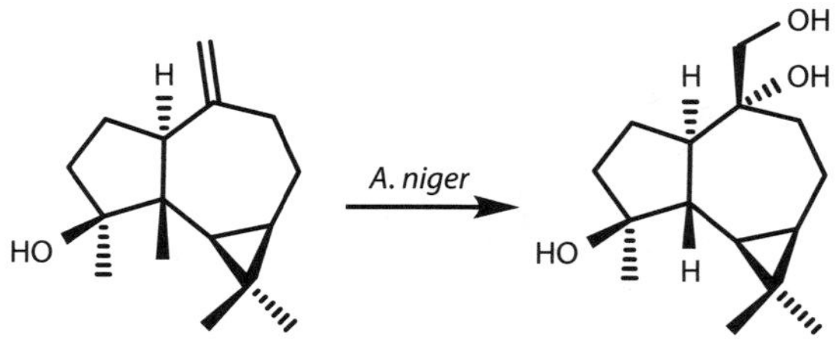

FIGURE 20.32 Biotransformation of spathulenol (**94**) by *Aspergillus niger*.

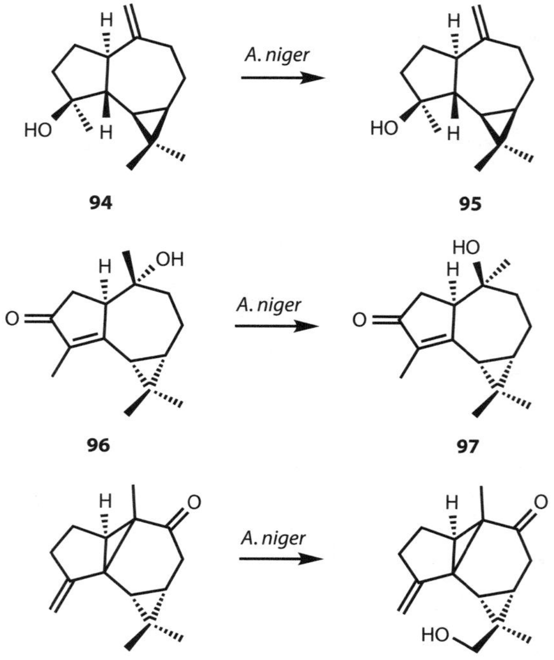

FIGURE 20.33 Biotransformation of spathulenol (**94**), *ent*-10β-hydroxycyclocolorenone (**96**) and myli-4(15)-en-9-one (**96a**) by *Aspergillus niger*.

in *A. niger* including 1-aminobenzotriazole, the inhibitor of CYP450, to produce only **106**, since this enzyme plays an important role in the formation of carboxylic acid (**107**) from primary alcohol (**106**). Unfortunately, two metabolites show no hot taste (Furusawa et al., 2006; Hashimoto et al., 2003c) (Figure 20.35).

Partheniol, 8α-hydroxybicyclogermacrene (**108**) isolated from *Parthenium argentatum* × *Parthenium tomentosum*, was cultured in the media of *Mucor circinelloides* ATCC 15242 to afford six metabolites, a humulane (**109**), three maaliane (**110, 112, 113**), an aromadendrane (**111**), and a tricylohumulane triol (**114**), the isomer of compound **111**. Compounds **110, 111**, and **114** were isolated as their acetates (Figure 20.36).

Compounds **110** might originate from the substrate by acidic transannular cyclization since the broth was pH 6.4 just before extraction (Maatooq, 2002a).

The same substrate (**108**) was incubated with the fungus *Calonectria decora* to afford six new metabolites (**108a–108f**). In these reactions, hydroxylation, epoxidation, and transannular cyclization were evidenced (Maatooq, 2002c) (Figure 20.37).

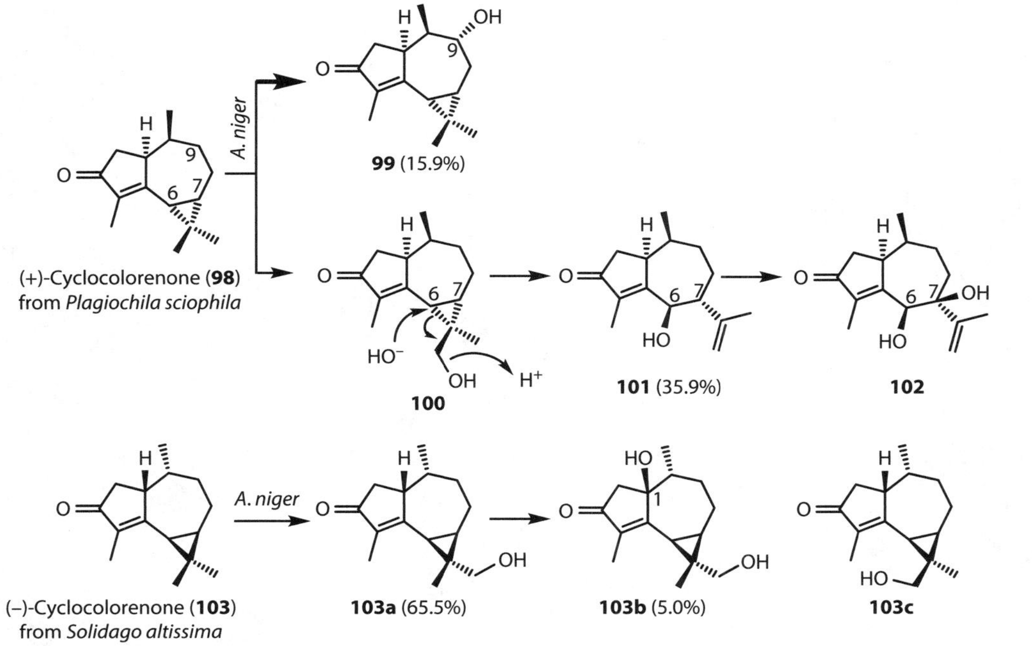

FIGURE 20.34 Biotransformation of (+)-cyclocolorenone (**98**) and (−)-cyclocolorenone (**103**) by *Aspergillus niger*.

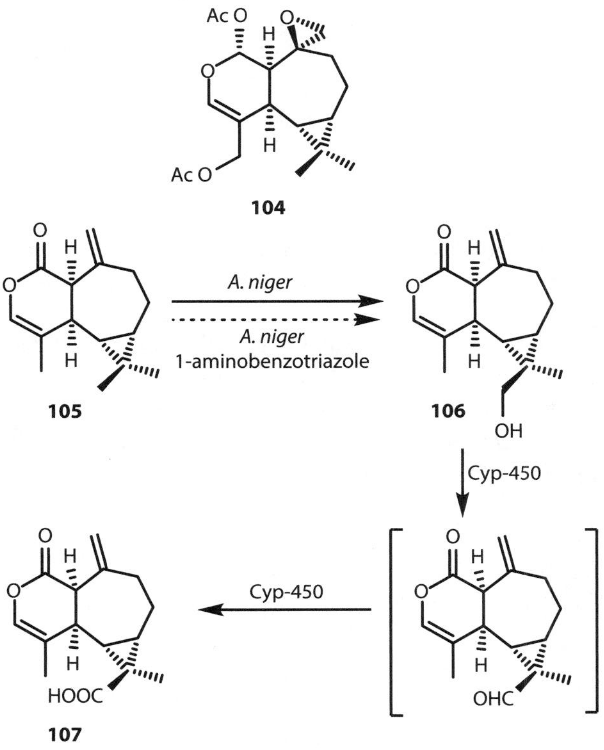

FIGURE 20.35 Biotransformation of plagiochiline C (**104**) by *Aspergillus niger*.

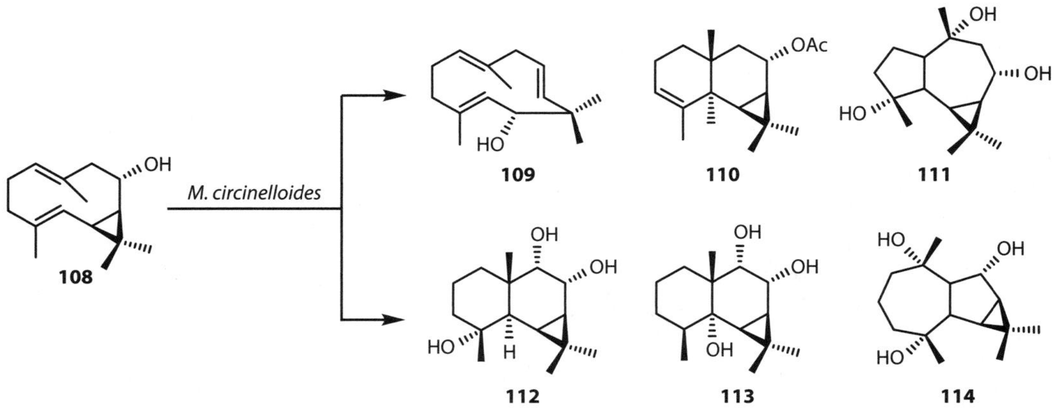

FIGURE 20.36 Biotransformation of 8α-hydroxybicyclogermacrene (**108**) by *Mucor circinelloides*.

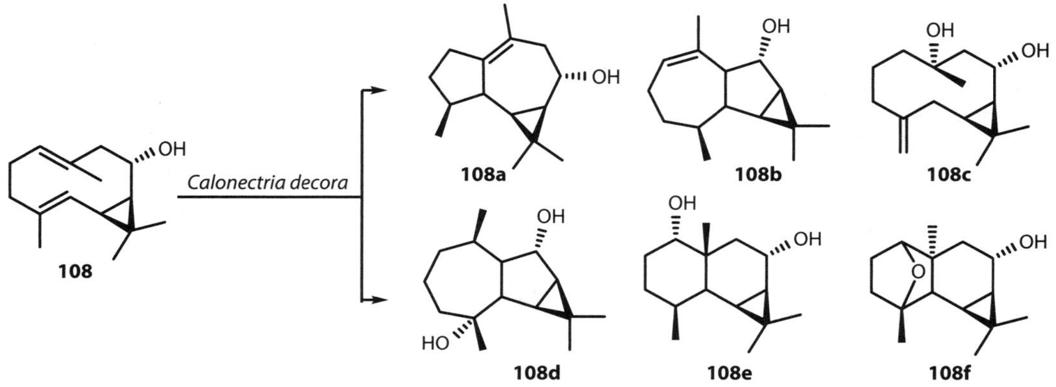

FIGURE 20.37 Biotransformation of 8α-hydroxybicyclogermacrene (**108**) by *Calonectria decora*.

ent-Maaliane-type sesquiterpene alcohol, 1α-hydroxymaaliene (**115**), isolated from the liverwort *M. taylorii*, was treated in *A. niger* to afford two primary alcohols (**116, 117**) (Morikawa et al., 2000). Such an oxidation pattern of 1,1-dimethyl group on the cyclopropane ring has been found in aromadendrane series as described earlier and mammalian biotransformation of a monoterpene hydrocarbon, D^3-carene (Ishida et al., 1981) (Figure 20.38).

9(15)-Africanene (**117a**), a tricyclic sesquiterpene hydrocarbon isolated from marine soft corals of *Simularia* species, was biotransformed by *A. niger* and *Rhizopus oryzae* for 8 days to give 10α-hydroxy (**117b**) and 9α,15-epoxy derivative (**117c**) (Venkateswarlu et al., 1999) (Figure 20.39).

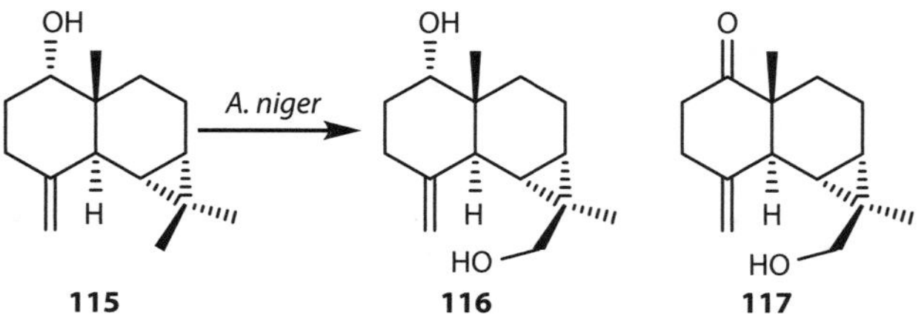

FIGURE 20.38 Biotransformation of 1α-hydroxymaaliene (**115**) *Aspergillus niger*.

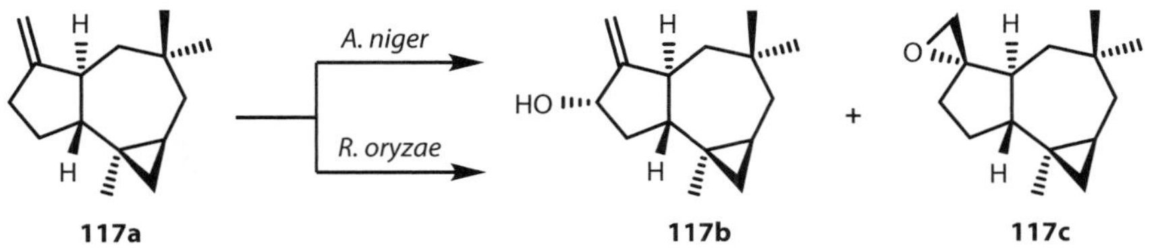

FIGURE 20.39 Biotransformation of 9(15)-africanene (**117a**) by *Aspergillus niger* and *Rhizopus oryzae*.

Germacrone (**118**), (+)-germacrone-4,5-epoxide (**119**), and curdione (**120**) isolated from *Curcuma aromatica*, which has been used as crude drug, were incubated with *A. niger*. From compound **119** (700 mg), two naturally occurring metabolites, zedoarondiol (**121**) and isozedoarondiol (**122**), were obtained (Takahashi, 1994). Compound **119** was cultured in callus of *Curcuma zedoaria* and *C. aromatica* to give the same secondary metabolites **121**, **122**, and **124** (Sakui et al., 1988) (Figures 20.40 and 20.41).

A. niger biotransformed germacrone (**118**, 3 g) to very unstable 3β-hydroxygermacrone (**123**) and 4,5-epoxygermacrone (**119**), which was further converted to two guaiane sesquiterpenoids (**121**) and (**122**) through transannular-type reaction (Takahashi, 1994). The same substrate was incubated in the microorganism *Cunninghamella blakesleeana* to afford germacrone-4,5-epoxide (**119**)

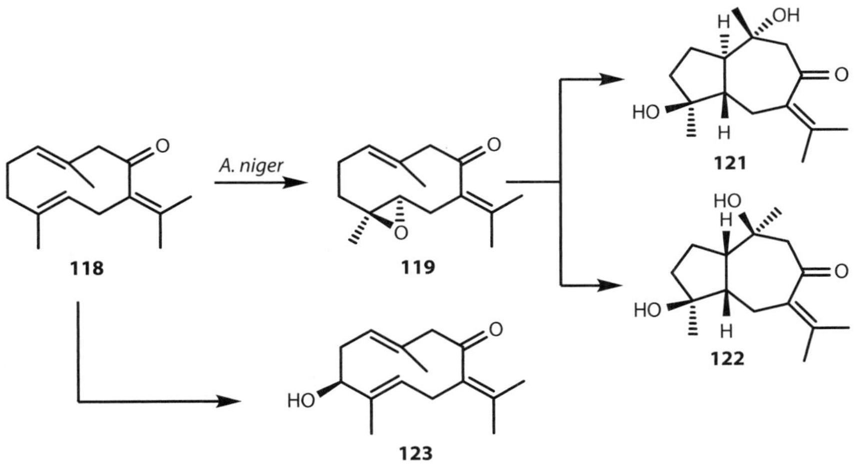

FIGURE 20.40 Biotransformation of germacrone (**118**) by *Aspergillus niger*.

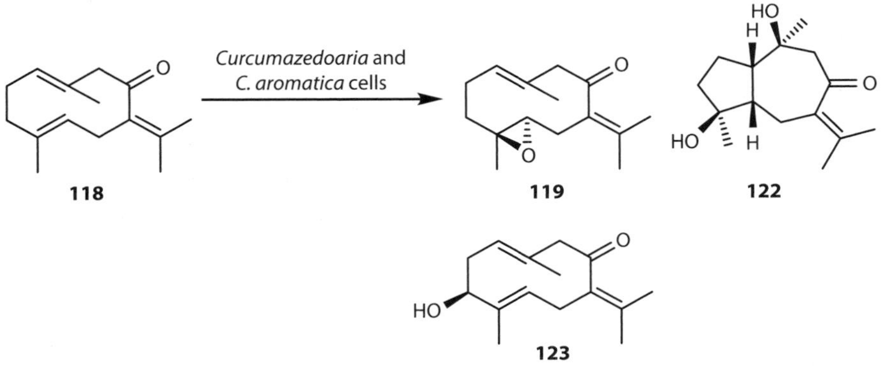

FIGURE 20.41 Biotransformation of germacrone (**118**) by *Curcuma zedoaria* and *Curcuma aromatica* cells.

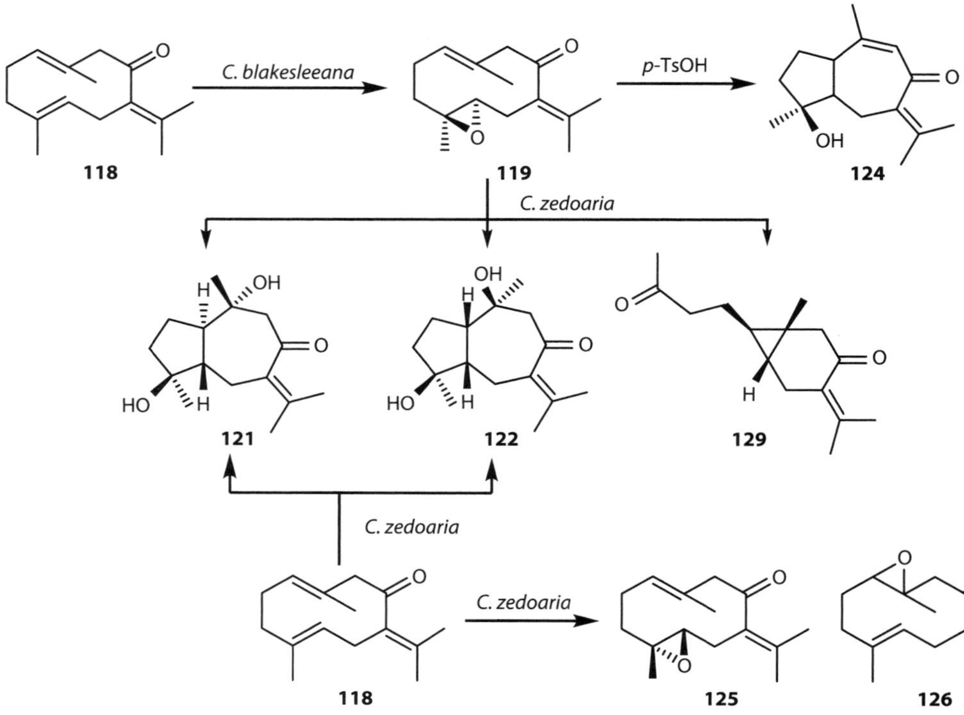

FIGURE 20.42 Biotransformation of germacrone (**118**) by *Cunninghamella blakesleeana* and *Curcuma zedoaria* cells.

(Hikino et al., 1971), while the treatment of **118** in the callus of *C. zedoaria* gave four metabolites **121**, **122**, **125**, and **126** (Sakamoto et al., 1994) (Figure 20.42).

The same substrate (**118**) was treated in plant cell cultures of *S. altissima* (Asteraceae) for 10 days to give various hydroxylated products (**121**, **127**, **125**, **128–132**) (Sakamoto et al., 1994). Guaiane (**121**) underwent further rearrangement C4–C5, cleavage, and C5–C10 transannular cyclization to the bicyclic hydroxyketone (**128**) and diketone (**129**) (Sakamoto et al., 1994) (Figure 20.43).

Curdione (**120**) was also treated in *A. niger* to afford two allylic alcohols (**133**, **134**) and a spirolactone (**135**). *C. aromatica* and *Curcuma wenyujin* produced spirolactone (**135**), which might be formed from curdione via transannular reaction in vivo and was biotransformed to spirolactone diol (**135**) (Asakawa et al., 1991; Sakui et al., 1992) (Figure 20.44).

A. niger also converted shiromodiol diacetate (**136**) isolated from *Neolitsea sericea* to 2β-hydroxy derivative (**137**) (Nozaki et al., 1996) (Figure 20.45).

Twenty strains of filamentous fungi and four species of bacteria were screened initially by thin-layer chromatography for their biotransformation capacity of curdione (**120**). *Mucor spinosus*, *Mucor polymorphosporus*, *Cunninghamella elegans*, and *Penicillium janthinellum* were found to be able to biotransform curdione (**120**) to more polar metabolites. Incubation of curdione with *M. spinosus*, which was most potent strain to produce metabolites, for 4 days using potato medium gave five metabolites (**134, 134a–134d**) among which compounds **134c** and **134d** are new products (Ma et al., 2006) (Figure 20.46).

Many eudesmane-type sesquiterpenoids have been biotransformed by several fungi and various oxygenated metabolites obtained.

β-Selinene (**138**) is ubiquitous sesquiterpene hydrocarbon of seed oil from many species of Apiaceae family, for example, *Cryptotaenia canadensis* var. *japonica*, which is widely used as vegetable for Japanese soup. β-Selinene was biotransformed by plant pathogenic fungus *G. cingulata*

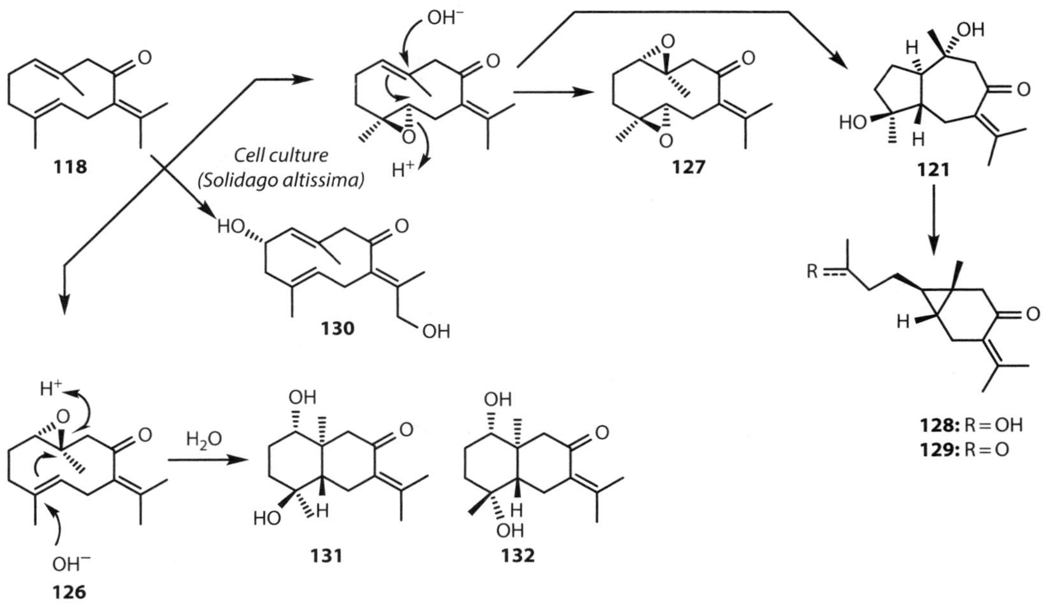

FIGURE 20.43 Biotransformation of germacrone (**118**) by *Solidago altissima* cells.

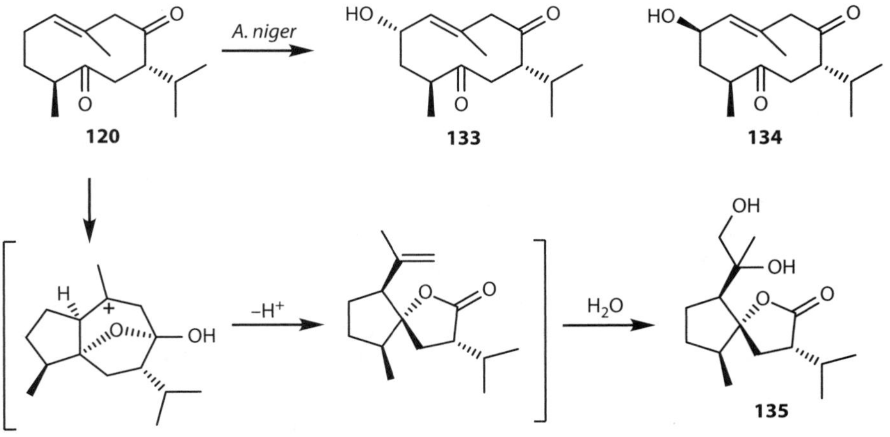

FIGURE 20.44 Biotransformation of curdione (**120**) by *Aspergillus niger*.

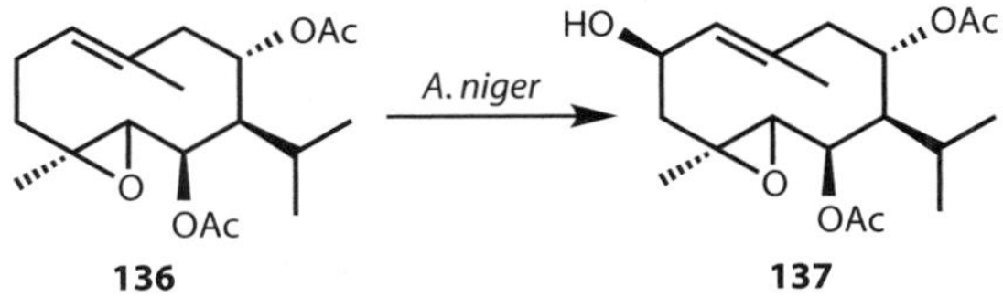

FIGURE 20.45 Biotransformation of shiromodiol diacetate (**136**) by *Aspergillus niger*.

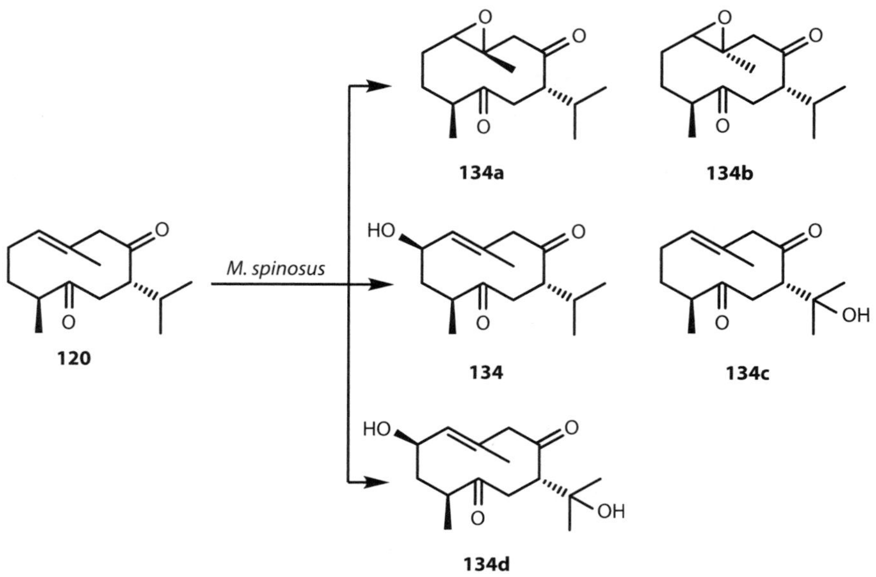

FIGURE 20.46 Biotransformation of curdione (**120**) by *Mucor spinosus*.

to give an epimeric mixtures (1:1) of 1β,11,12-trihydroxy product (**139**) (Miyazawa et al., 1997b). The same substrate was treated in *A. wentii* to give 2α,11,12-trihydroxy derivative (**140**) (Takahashi et al., 2007).

Eudesm-11(13)-en-4,12-diol (**141**) was biotransformed by *A. niger* to give 3β-hydroxy derivative (**142**) (Hayashi et al., 1999).

α-Cyperone (**143**) was fed by *Collectotrichum phomoides* (Lamare and Furstoss, 1990) to afford 11,12-diol (**144**) and 12-manool (**145**) (Higuchi et al., 2001) (Figure 20.47).

The filamentous fungi *Gliocladium roseum* and *Exserohilum halodes* were used as the bioreactors for 4β-hydroxyeudesmane-1,6-dione (**146**) isolated from *Sideritis varoi* subsp. *cuatrecasasii*. The former fungus transformed **146** to 7α-hydroxyl (**147**), 11-hydroxy (**148**), 7α,11-dihydroxy (**149**), 1α,11-dihydroxy (**150**), and 1α,8α-dihydroxy derivatives (**151**), while *E. halodes* gave only 1α-hydoxy product (**152**) (Garcia-Granados et al., 2001) (Figure 20.48).

Orabi (2000) reported that *Beauveria bassiana* is the most efficient microorganism to metabolize plectanthone (**152a**) among 20 microorganisms, such as *Absidia glauca*, *Aspergillus flavipes*, *B. bassiana*, *Cladosporium resinae*, and *Penicillium frequentans*. The substrate (**152a**) was incubated with *B. bassiana* to give metabolites **152b** (2.1%), **152c** (21.2%), **152d** (2.5%), **152e** (no data), and **152f** (1%) (Figure 20.49).

(−)-α-Eudesmol (**153**) isolated from the liverwort *Porella stephaniana* was treated by *A. cellulosae* and *A. niger* to give 2-hydroxy (**154**) and 2-oxo derivatives (**155**), among which the latter product was predominantly obtained. This bioconversion was completely blocked by 1-aminobenzotriazole, CYP450 inhibitor. Compound **155** has been known as natural product, isolated from *Pterocarpus santalinus* (Noma et al., 1996). Biotransformation of α-eudesmol (**153**) isolated from the dried *Atractylodes lancea* was reinvestigated by *A. niger* to give 2-oxo-11,12-dihydro-α-eudesmol (**156**) together with 2-hydroxy- (**154**) and 2-oxo-α-eudesmol (**155**). β-Eudesmol (**157**) was treated in *A. niger*, with the same culture medium, to afford 2α- (**158**) and 2β-hydroxy-α-eudesmol (**159**) and 2α,11,12-trihydroxy-β-eudesmol (**160**) and 2-oxo derivative (**161**), which was further isomerized to compound **162** (Noma et al., 1996, 1997a) (Figure 20.50).

Three new hydroxylated metabolites (**157b**–**157d**) along with a known **158** and **157e**–**157g** were isolated from the biotransformation reaction of a mixture of β- (**157**) and γ-eudesmols (**157a**) by *Gibberella suabinetii*. The metabolites proved a super activity of the hydroxylase, dehydrogenase,

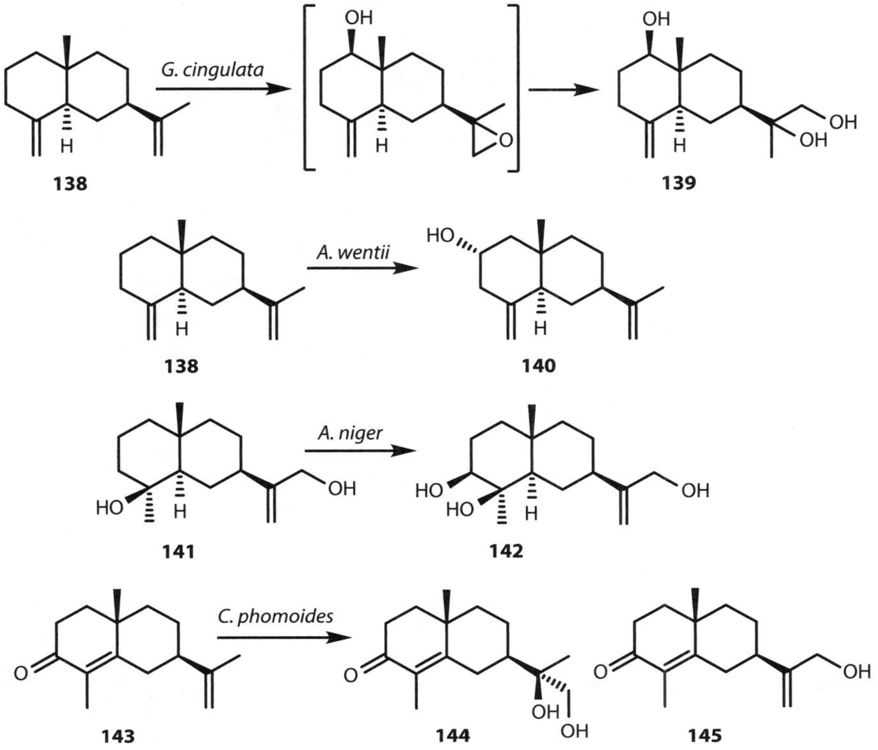

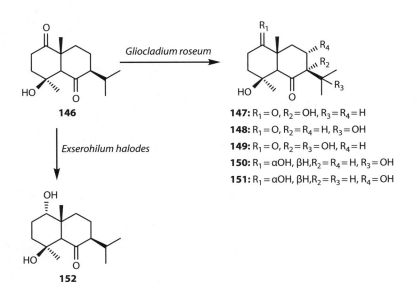

FIGURE 20.47 Biotransformation of eudesmenes (**138**, **141**, **143**) by *Aspergillus wentii*, *Glomerella cingulata*, and *Collectotrium phomoides*.

147: $R_1=O$, $R_2=OH$, $R_3=R_4=H$

148: $R_1=O$, $R_2=R_4=H$, $R_3=OH$

149: $R_1=O$, $R_2=R_3=OH$, $R_4=H$

150: $R_1=\alpha OH$, βH, $R_2=R_4=H$, $R_3=OH$

151: $R_1=\alpha OH$, βH, $R_2=R_3=H$, $R_4=OH$

FIGURE 20.48 Biotransformation of 4β-hydroxy-eudesmane-1,6-dione (**146**) by *Gliocladium roseum* and *Exserohilum halodes*.

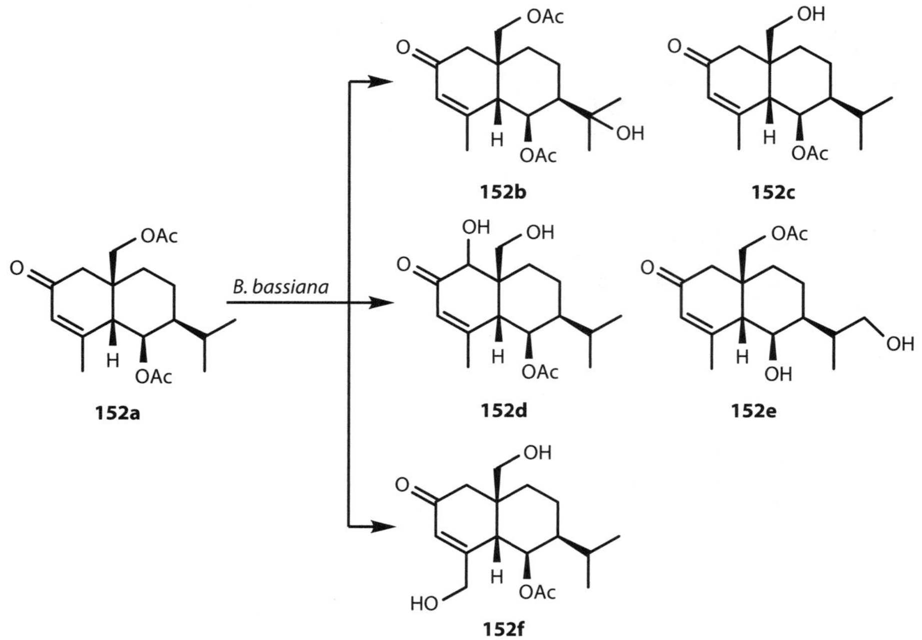

FIGURE 20.49 Biotransformation of eudesmenone (**152a**) by *Beauveria bassiana*.

and isomerase enzymes. The hydroxylation is a common feature; on the contrary, cyclopropyl ring formation like compound (**158d**) is very rare (Maatooq, 2002b) (Figure 20.51).

A furanosesquiterpene, atractylon (**163**) obtained from *Atractylodes* rhizoma, was treated with the same fungus to yield atractylenolide III (**164**) possessing inhibition of increased vascular permeability in mice induced by acetic acid (Hashimoto et al., 2001).

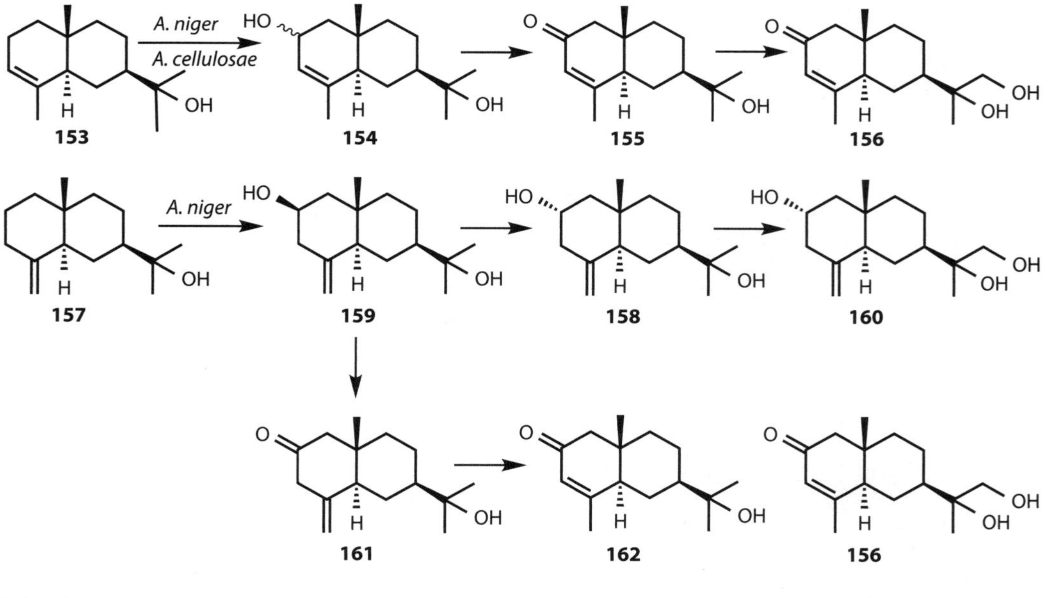

FIGURE 20.50 Biotransformation of α-eudesmol (**153**) and β-eudesmol (**157**) by *Aspergillus niger* and *Aspergillus cellulosae*.

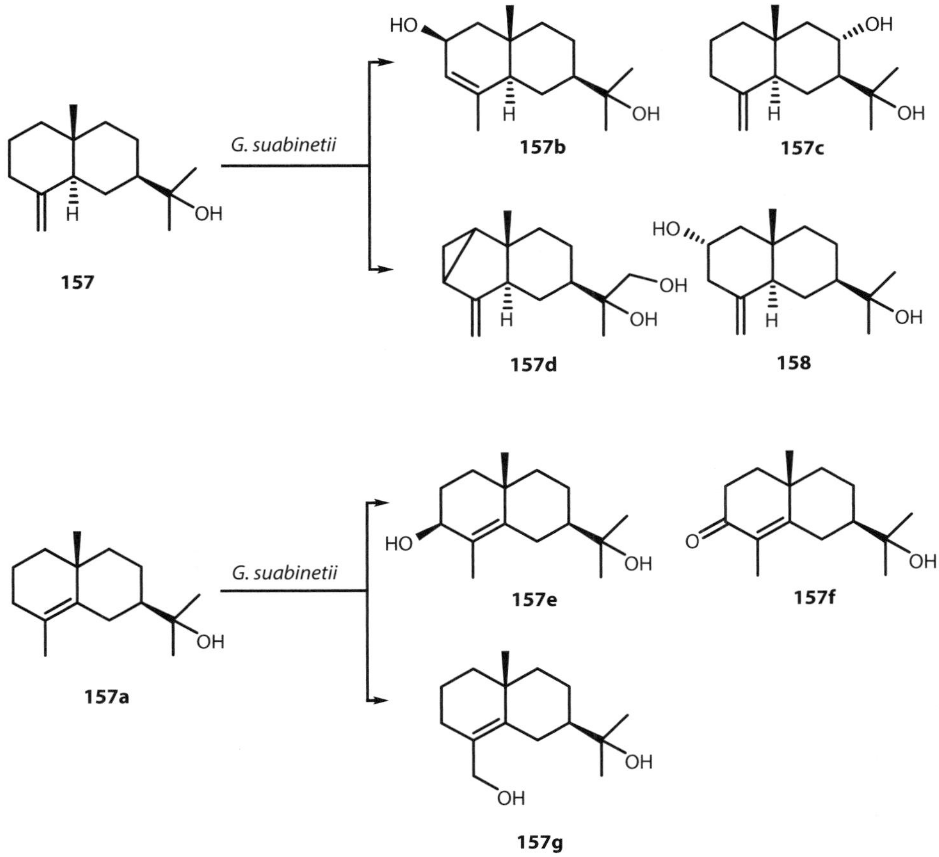

FIGURE 20.51 Biotransformation of β-eudesmol (**157**) and γ-eudesmol (**157a**) by *Gibberella suabinetii*.

The biotransformation of sesquiterpene lactones have been carried out by using different microorganisms.

Costunolide (**165**), a very unstable sesquiterpene γ-lactone, from *Saussurea radix*, was treated in *A. niger* to produce three dihydrocostunolides (**166–168**) (Clark and Hufford, 1979). Costunolide is easily converted into eudesmanolides (**169–172**) in diluted acid, thus, **166–168** might be biotransformed after being cyclized in the medium including the microorganisms. If the crude drug including costunolide (**165**) is orally administered, **165** will be easily converted into **169–172** by stomach juice (Figure 20.52).

(+)-Costunolide (**165**), (+)-cnicin (**172a**), and (+)-salonitgenolide (**172b**) were incubated with *Cunninghamella echinulata* and *R. oryzae*.

The former fungus converted compound **165**, to four metabolites, (+)-11β,13-dihydrocostunolide (**165a**), 1β-hydroxyeudesmanolide, (+)-santamarine (**166a**), (+)-reynosin (**166b**), and (+)-1β-hydroxyarbusculin A (**168a**), which might be formed from 1β,10α-epoxide (**166c**). Treatment of **172a** with *C. echinulata* gave (+)-salonitenolide (**172b**) (Barrero et al., 1999) (Figure 20.53).

α-Cyclocostunolide (**169**), β-cyclocostunolide (**170**), and γ-cyclocostunolide (**171**) prepared from costunolide were cultivated in *A. niger*. From the metabolite of **169**, four dihydro lactones (**173–176**) were obtained, among which sulfur-containing compound (**176**) was predominant (Figure 20.54).

The same substrate (**169**) was cultivated for 3 days by *A. cellulosae* to afford a sole metabolite, 11β,13-dihydro-α-cyclocostunolide (**177**). Possible metabolic pathways of **169** by both microorganisms were shown in Figure 20.55.

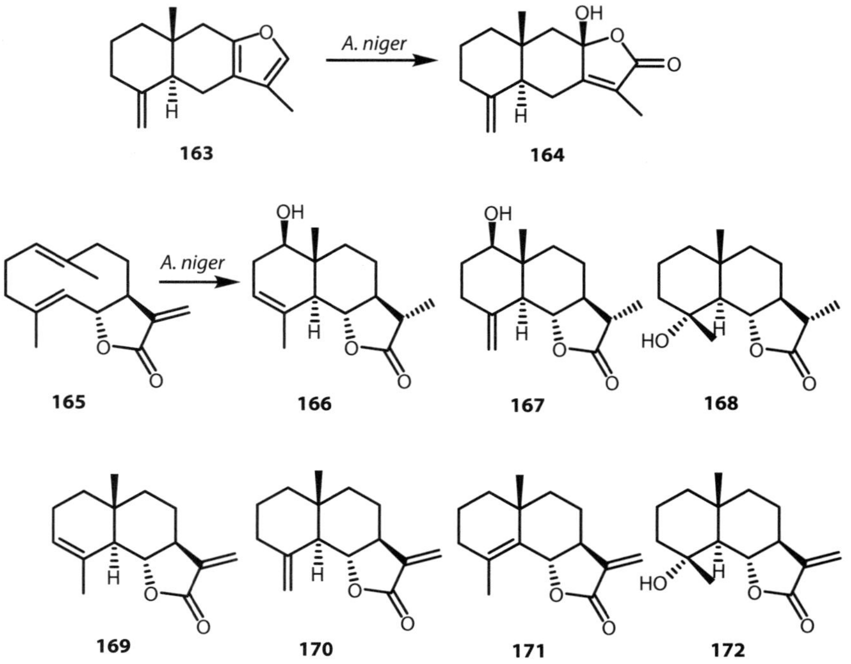

FIGURE 20.52 Biotransformation of atractylon (**163**) and costunolide (**165**) by *Aspergillus niger*.

A double bond at C11–C13 of **169** was firstly reduced stereoselectively to afford **177**, followed by oxidation at C2 to give **173**, and then further oxidation occurred to furnish two hydroxyl derivatives (**174, 175**) in *A. niger*. The sulfide compound (**176**) might be formed from **175** or by Michael condensation of ethyl 2-hydroxy-3-mercaptopropanate, which might originate from Czapek-peptone medium into exomethylene group of α-cyclocostunolide (Hashimoto et al., 1999c, 2001).

A. niger converted β-cyclocostunolide (**170**) to two oxygenated metabolites (**173, 174, 178–181**) of which **173** was predominant. It is suggested that compound **173** and **174** might be formed during biotransformation period since metabolite media after 7 days was acidic (pH 2.7). Surprisingly, *A. cellulosae* gave a sole 11β,13-dihydro-β-cyclocostunolide (**182**), which was abnormally folded in the mycelium of *A. cellulosae* as a crystal form after biotransformation of **170**. On the other hand, the metabolites were normally liberated in medium outside of the mycelium of *A. niger* and *B. dothidea* (Hashimoto et al., 1999c, 2001) (Figure 20.56).

B. dothidea has no stereoselectivity to reduce the C11–C13 double bond of β-cyclocostunolide (**170**) since this organism gave two dihydro derivatives, **182** (16.7%) and **183** (37.8%), respectively, as shown in Figure 20.57.

It is noteworthy that both α- and β-cyclocostunolides were biotransformed by *A. niger* to give the sulfur-containing metabolites (**176, 181**). Possible biogenetic pathway of **170** is shown in Figure 20.58.

When γ-cyclocostunolide (**171**) was cultivated in *A. niger* to give dihydro-α-santonin (**187**, 25%) and its related C11–C13, dihydro derivatives (**184–186, 188, 189**) were obtained as a small amount. Compound **186** was recultivated for 2 days by the same organism as mentioned earlier to afford **187** (25%) and 5β-hydroxy-α-cyclocostunolide (**189**, 54%). Recultivation of **185** for 2 days by *A. niger* afforded compound **187** as a sole metabolite. During the biotransformation of **171**, no sulfur-containing product was obtained. Both *A. cellulosae* and *B. dothidea* produced only dihydro-γ-cyclocostunolide (**184**) from the substrate (**171**) (Hashimoto et al., 1999c, 2001) (Figure 20.59).

Santonin (**190**) has been used as vermicide against roundworm. *C. blakesleeana* and *A. niger* converted **190**–**187** (Atta-ur-Rahman et al., 1998). When **187** was fed by *A. niger* for 1 week to

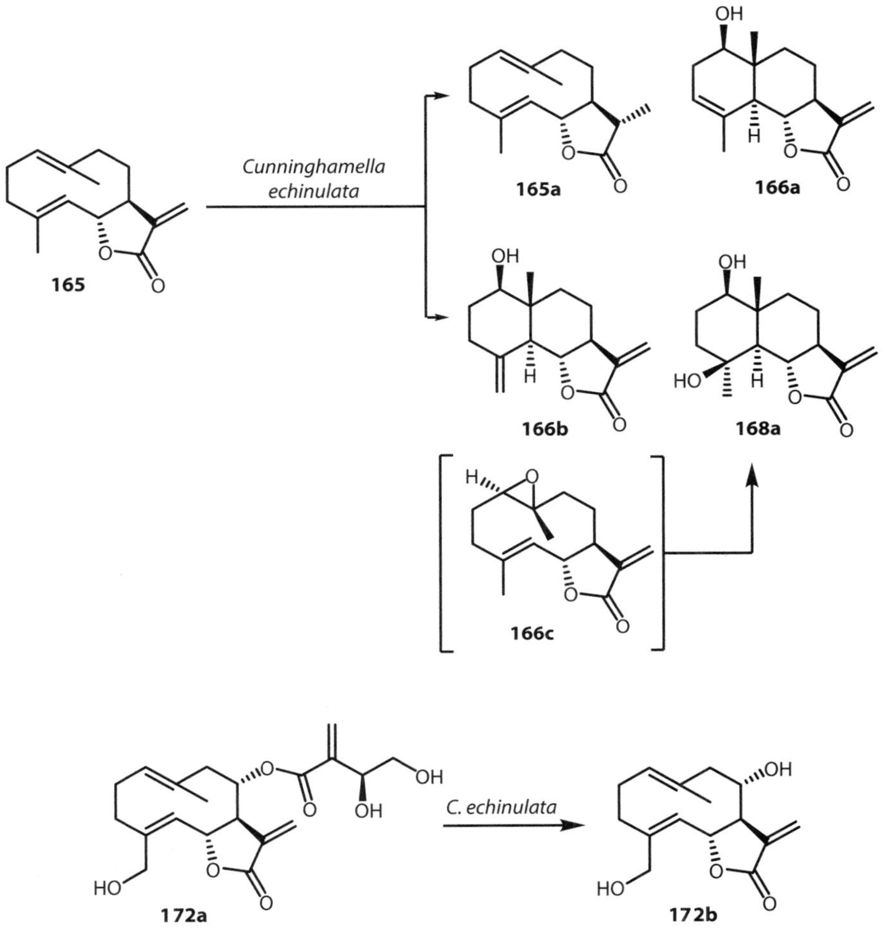

FIGURE 20.53 Biotransformation of costunolide (**165**) and its derivative (**172a**) by *Cunninghamella echinulata* and *Rhizopus oryzae*.

give 2β-hydroxy-1,2-dihydro-α-santonin (**188**, 39%) as well as 1β-hydroxy-1,2-dihydro-α-santonin (**195**, 6.5%), 9β-hydroxy-1,2-dihydro-α-santonin (**196**, 6.9%), and α-santonin (**190**, 5.4%), which might be obtained from dehydroxylation of **188**, as a minor component (Hashimoto et al., 2001), compound **188** was isolated from the crude metabolite of γ-cyclocostunolide (**171**) by *A. niger* as mentioned earlier (Figure 20.60).

It was treated with *A. niger* for 7 days to give **191** (18.3%), **192** (2.3%), **193** (19.3%), and **194** (3.5%) of which **193** was the major metabolite. Compound **191** was isolated from dog's urine after the oral administration of **190**. The structure of compound **194** was established as lumisantonin obtained by the photoreaction of **190**. α-Santonin **190** was not converted into 1,2-dihydro derivative by *A. niger*, whereas the other strain of *A. niger* gave a single product, 1,2-dihydro-α-santonin (**187**) (Hashimoto et al., 2001) (Figure 20.61).

Ata and Nachtigall (2004) reported that α-santonin (**190**) was incubated with *Rhizopus stolonifer* to give (**187a**) and (**183b**), while with *Cunninghamella bainieri*, *C. echinulata*, and *M. plumbeus* to afford the known 1,2-dihydro-α-santonin (**187**) (Figure 20.62).

α-Santonin (**190**) and 6-*epi*-α-santonin (**198**) were cultivated in *Absidia coerulea* for 2 days to give 11β-hydroxy- (**191**, 71.4%) and 8α-hydroxysantonin (**197**, 2.0%), while 6-*epi*-santonin (**198**) afforded four major products (**199–201, 206**) and four minor analogues (**202, 203–205**).

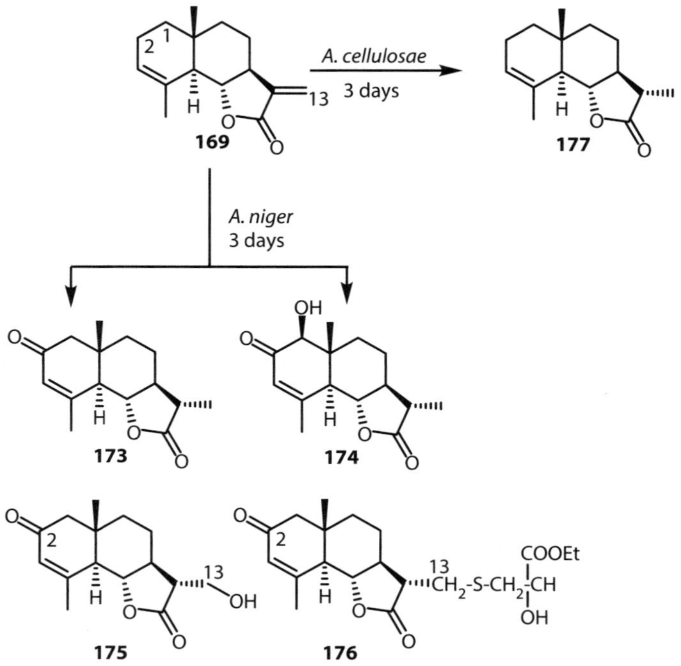

FIGURE 20.54 Biotransformation of α-cyclocostunolide (**169**) by *Aspergillus niger* and *Aspergillus cellulosae*.

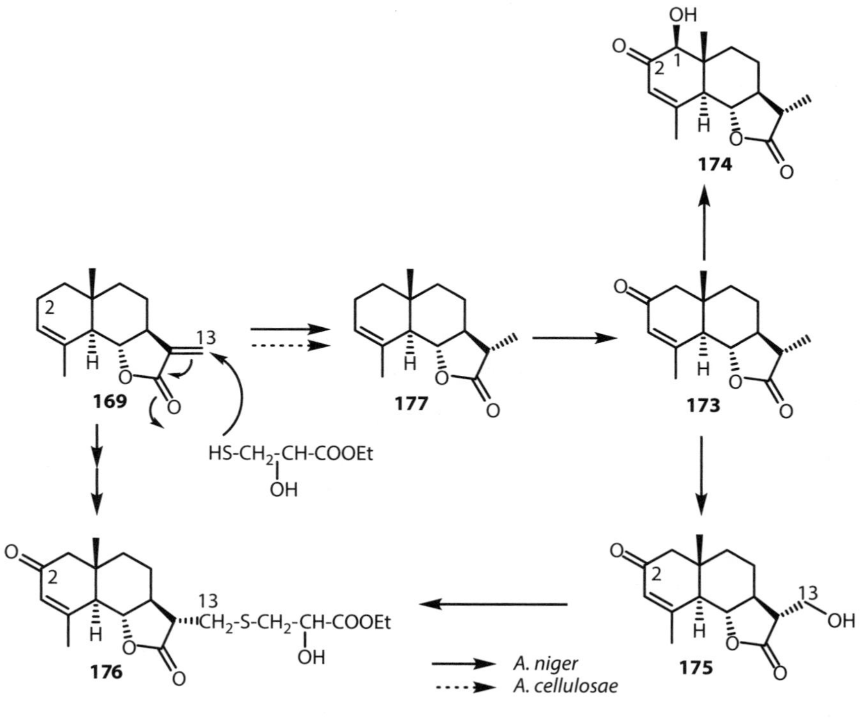

FIGURE 20.55 Possible pathway of biotransformation of α-cyclocostunolide (**169**) by *Aspergillus niger* and *Aspergillus cellulosae*.

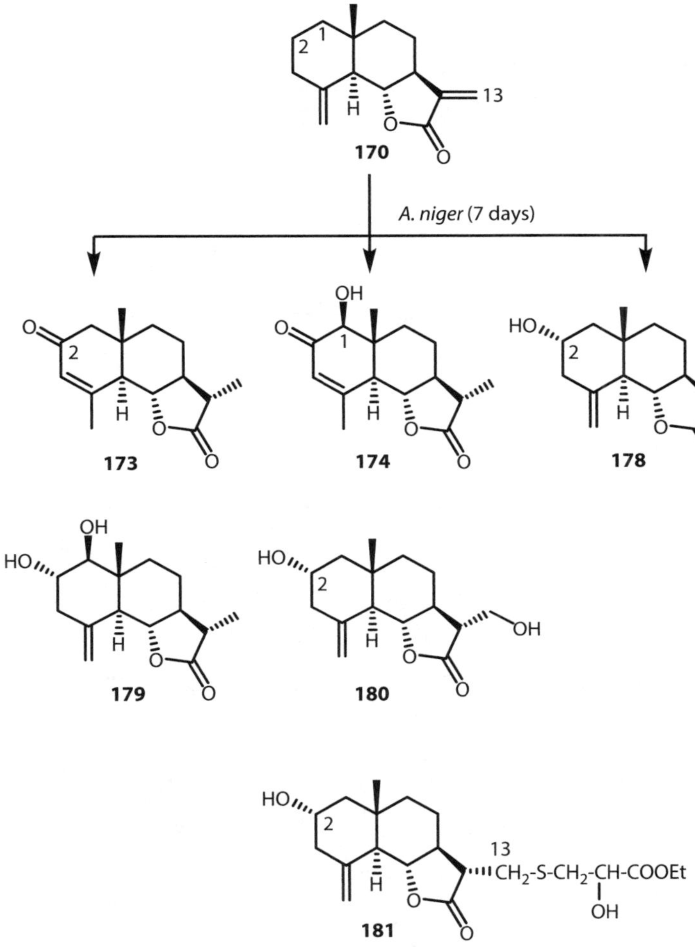

FIGURE 20.56 Biotransformation of β-cyclocostunolide (**170**) by *Aspergillus niger*.

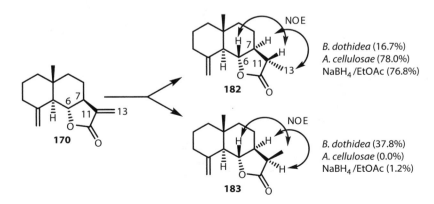

FIGURE 20.57 Biotransformation of β-cyclocostunolide (**170**) by *Aspergillus cellulosae* and *Botryosphaeria dothidea*.

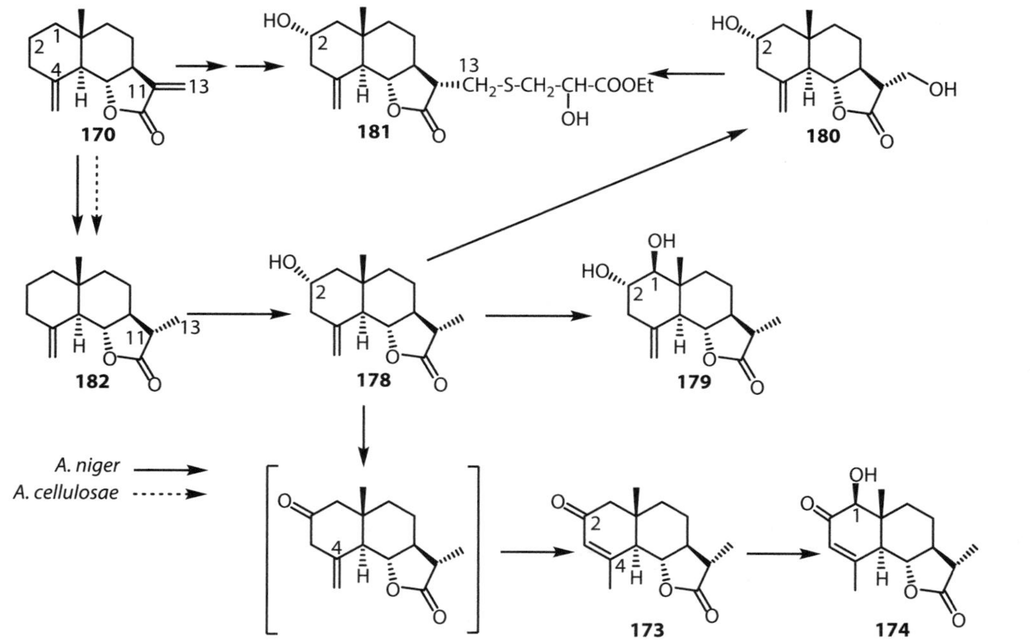

FIGURE 20.58 Possible pathway of biotransformation of β-cyclocostunolide (**170**) by *Aspergillus niger* and *Aspergillus cellulosae*.

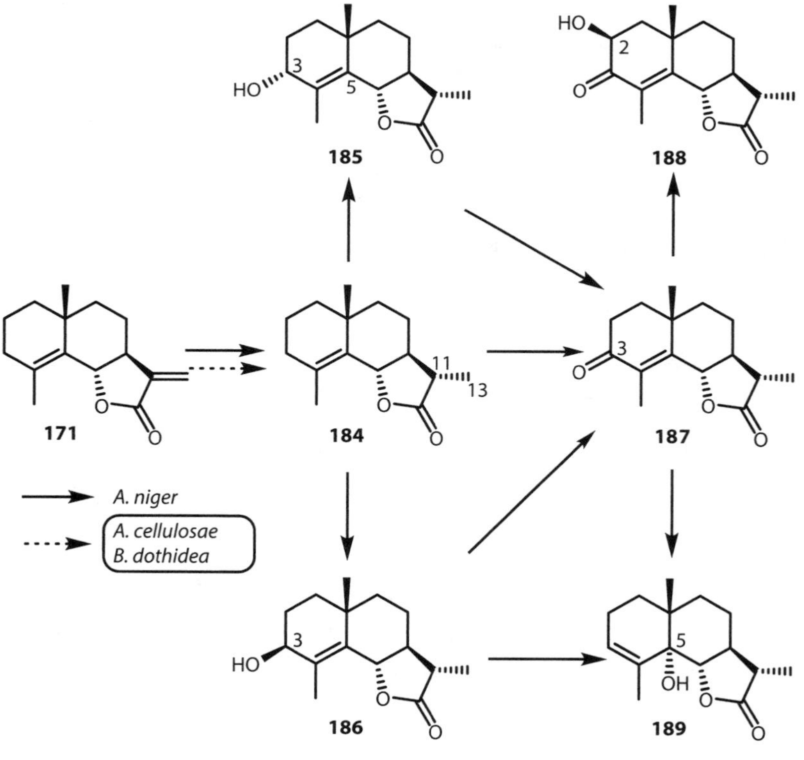

FIGURE 20.59 Biotransformation of γ-cyclocostunolide (**171**) by *Aspergillus niger*, *Aspergillus cellulosae*, and *Botryosphaeria dothidea*.

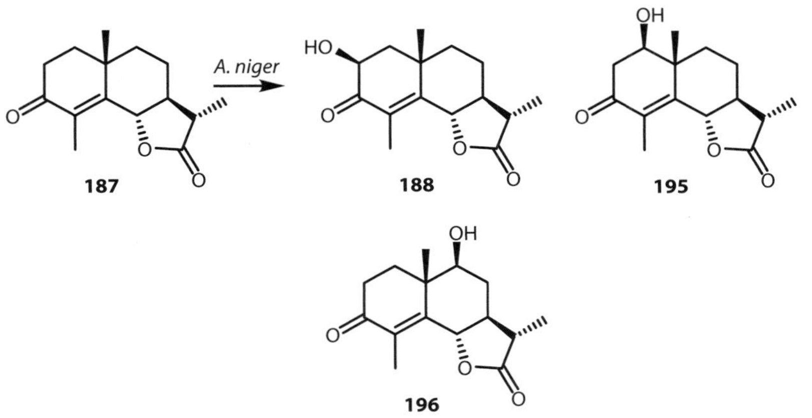

FIGURE 20.60 Biotransformation of dihydro-α-santonin (**187**) by *Aspergillus niger*.

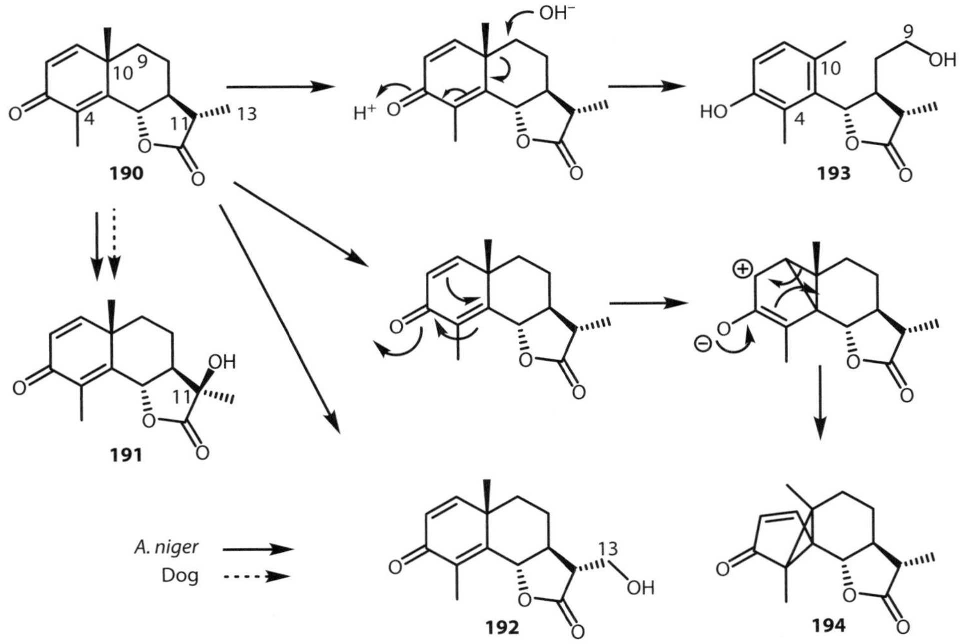

FIGURE 20.61 Biotransformation of α-santonin (**190**) by *Aspergillus niger* and dogs.

Asparagus officinalis also biotransformed α-santonin (**190**) into three eudesmanolides (**187, 207, 208**) and a guaianolide (**209**) in a small amount. 6-*epi*-Santonin (**198**) was also treated in the same bioreactor as mentioned earlier to give **199** and **206**, the latter of which was obtained as a major metabolite (44.7%) (Yang et al., 2003) (Figure 20.63).

α-Santonin (**190**) was incubated in the cultured cells of *Nicotiana tabacum* and the liverwort *M. polymorpha*. *N. tabacum* cells gave 1,2-dihydro-α-santonin (**187**) (50%) for 6 days. The latter cells also converted α-santonin to 1,2-dihydro-α-santonin, but conversion ratio was only 28% (Matsushima et al., 2004) (Figure 20.64).

6-*epi*-α-Santonin (**198**) and its tetrahydro analogue (**210**) were also incubated with fungus *Rhizopus nigricans* to give 2α-hydroxydihydro-α-santonin (**211**) (Amate et al., 1991), the epimer of **188** obtained from the biotransformation of dihydro-α-santonin (**187**) by *A. niger* (Hashimoto et al., 2001).

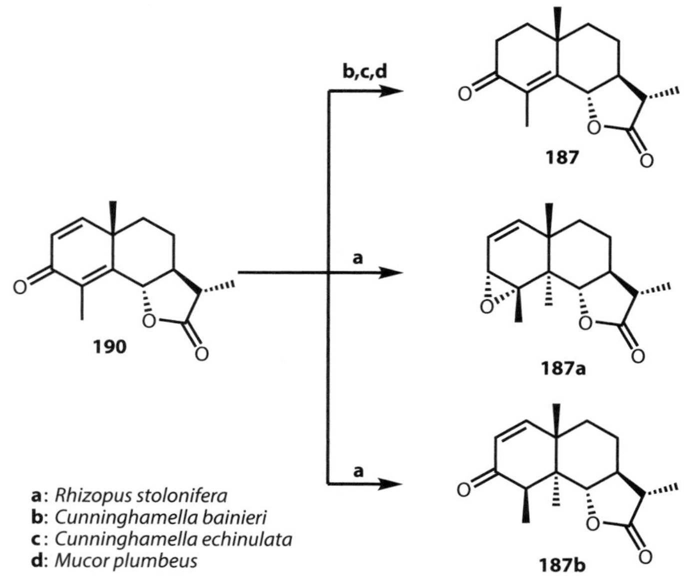

FIGURE 20.62 Biotransformation of α-santonin (**190**) by *Rhizopus stolonifer*, *Cunninghamella bainieri*, *Cunninghamella echinulata*, and *Mucor plumbeus*.

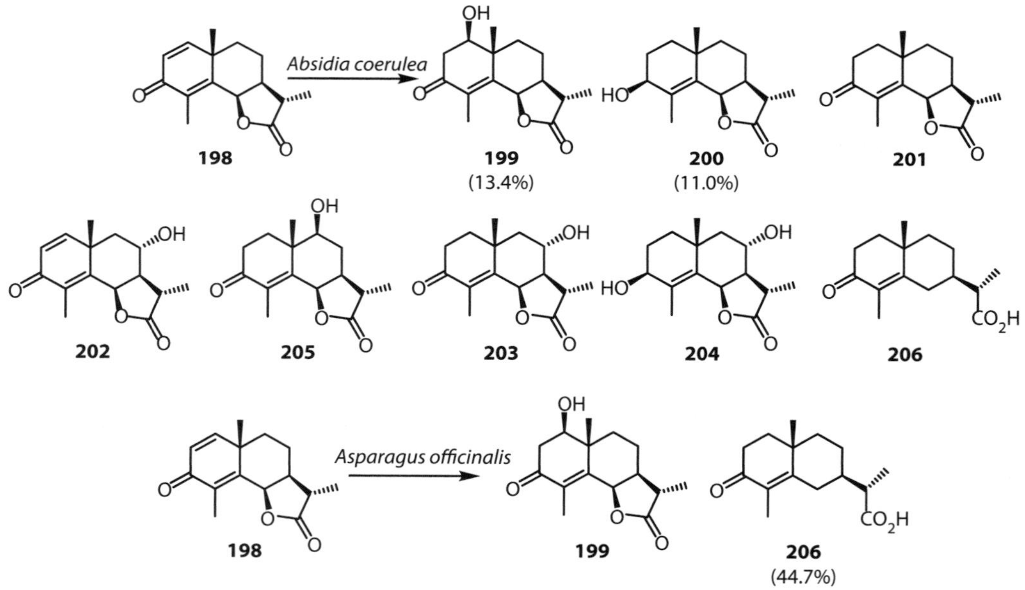

a: *Rhizopus stolonifera*
b: *Cunninghamella bainieri*
c: *Cunninghamella echinulata*
d: *Mucor plumbeus*

FIGURE 20.63 Biotransformation of α-epi-santonin (**198**) by *Absidia coerulea* and *Asparagus officinalis*.

The product **211** might be formed via 1,2-epoxide of **198**. Compound **210** was converted through carbonyl reduction to furnish **212** and **213** under epimerization at C4 (Amate et al., 1991) (Figure 20.65).

1,2,4β,5α-Tetrahydro-α-santonin (**214**) prepared from α-santonin (**190**) was treated with *A. niger* to afford six metabolites (**215–220**) of which **219** was the major product (21%). When the substrate (**214**) was treated with CYP450 inhibitor, 1-aminobenzotriazole, only **215** was obtained without its homologues, **216–220**, while the C4 epimer (**221**) of **214** was converted by the same microorganism to afford

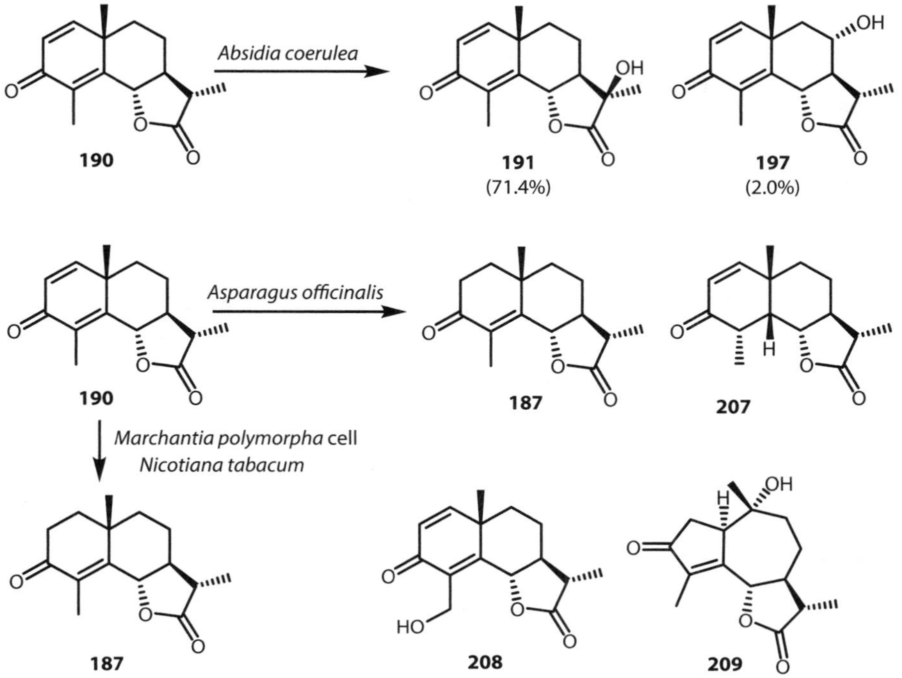

FIGURE 20.64 Biotransformation of 6-epi-α-santonin (**190**) by *Absidia coerulea*, *Asparagus officinalis*, *Marchantia polymorpha*, and *Nicotiana tabacum*.

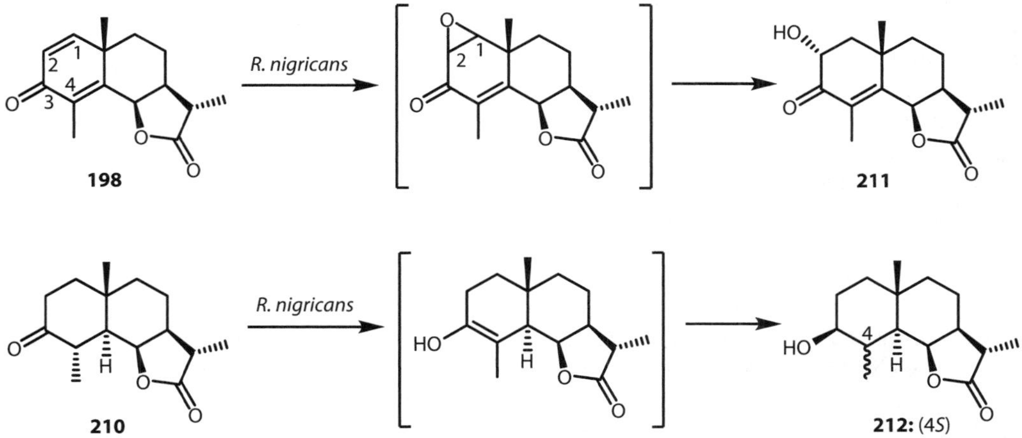

FIGURE 20.65 Biotransformation of α-epi-santonin (**198**) and tetrahydrosantonin (**210**) by *Rhizopus nigricans*.

a single metabolite (**222**) (73%). Further oxidation of **222** did not occur. This reason might be considered by the steric hindrance of β-(axial) methyl group at C4 (Hashimoto et al., 2001) (Figure 20.66).

7α-Hydroxyfrullanolide (**223**) possessing cytotoxicity and antitumor activity, isolated from *Sphaeranthus indicus* (Asteraceae), was bioconverted by *A. niger* to afford 13R-dihydro derivative (**224**). The same substrate was also treated in *Aspergillus quardilatus* (wild type) to give 13-acetyl product (**225**) (Atta-ur-Rahman et al., 1994) (Figure 20.67).

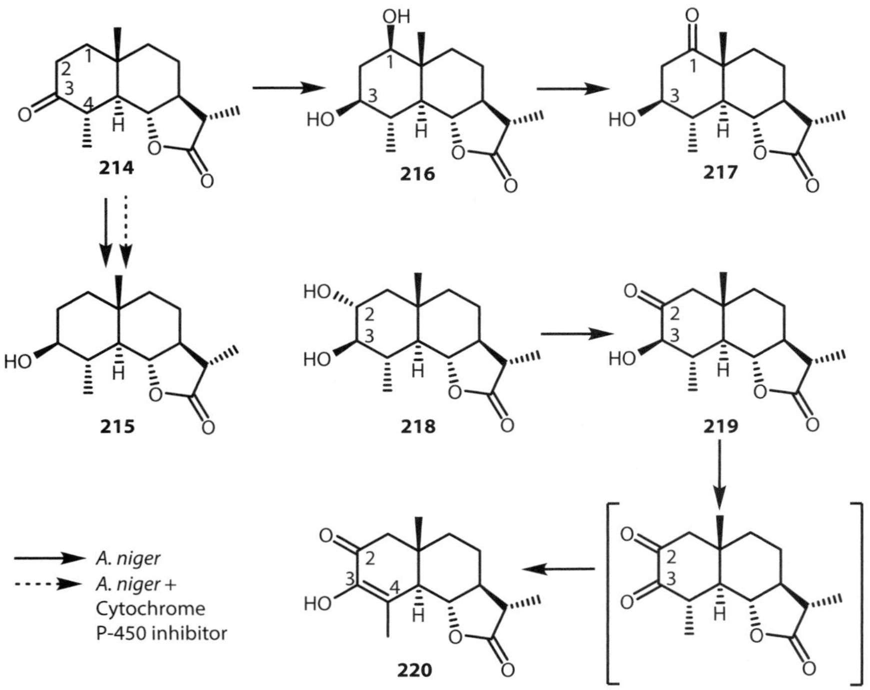

FIGURE 20.66 Biotransformation of 1,2,4β,5α-tetrahydro-α-santonin (**214**) by *Aspergillus niger*.

Incubation of (−)-frullanolide (**226**), obtained from the European liverwort *Frullania tamarisci* subsp. *tamarisci* that causes a potent allergenic contact dermatitis, was incubated by *A. niger* to give dihydrofrullanolide (**227**), nonallergenic compound in 31.8% yield. In this case, C11–C13 dihydro derivative was not obtained (Hashimoto et al., 2005).

Guaiane-type sesquiterpene hydrocarbon, (+)-γ-gurjunene (**228**), was treated in plant pathogenic fungus *G. cingulata* to give two diols, (1*S*,4*S*,7*R*,10*R*)-5-guaien-11,13-diol (**229**) and (1*S*,4*S*,7*R*,10*S*)-5-guaien-10,11,13-triol (**230**) (Miyazawa et al., 1997a, 1998a) (Figure 20.68).

G. cingulata converted guaiol (**231**) and bulnesol (**232**) to 5,10-dihydroxy (**233**) and 15-hydroxy derivative (**234**), respectively (Miyazawa et al., 1996a) (Figure 20.69).

When *Eurotium rubrum* was used as the bioreactor of guaiene (**235**), rotundone (**236**) was obtained (Sugawara and Miyazawa, 2004). Guaiol (**231**) was also transformed by *A. niger* to give a cyclopentane derivative, pancherione (**237**), and two dihydroxy guaiols (**238**, **239**) (Morikawa et al., 2000), of which **237** was obtained from the same substrate using *E. rubrum* for 10 days (Miyazawa and Sugawara, 2006; Sugawara and Miyazawa, 2004) (Figure 20.70).

Parthenolide (**240**), a germacrane-type lactone, isolated from the European feverfew (*Tanacetum parthenium*) as a major constituent, shows cytotoxic, antimicrobial, antifungal, anti-inflammatory, antirheumatic activity, apoptosis inducing, and NF-κB and DNA binding inhibitory activity. This substrate was incubated with *A. niger* in Czapek-peptone medium for 2 days to give six metabolites (**241**, 12.3%; **242**, 11.3%; **243**, 13.7%; **244**, 5.0%; **245**, 9.6%; **246**, 5.1%) (Hashimoto et al., 2005) (Figure 20.71). Compound **244** was a naturally occurring lactone from *Michelia champaca* (Jacobsson et al., 1995). The stereostructure of compound **243** was established by x-ray crystallographic analysis.

When parthenolide (**240**) was treated in *A. cellulosae* for 5 days, two new metabolites, 11β,13-dihydro-(**247**, 43.5%) and 11α,13-dihydroparthenolides (**248**, 1.6%), were obtained together with the same metabolites (**241**, 5.3%; **243**, 11.2%; **245**, 10.4%) as described earlier (Figure 20.72). Possible metabolic root of **240** has been shown in Figure 20.73 (Hashimoto et al., 2005).

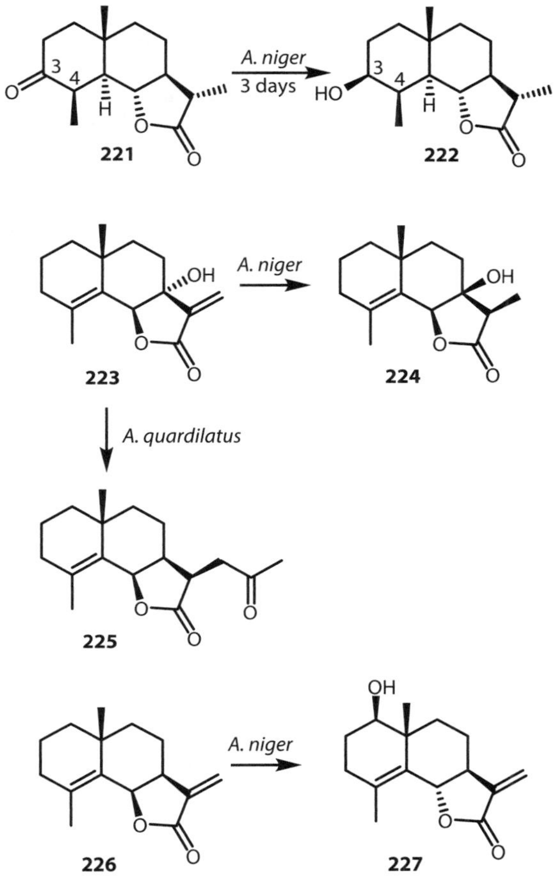

FIGURE 20.67 Biotransformation of C4-epimer (**221**) of **214**, 7α-hydroxyfrullanolide (**223**), and frullanolide (**226**) by *Aspergillus niger* and *Aspergillus quardilatus*.

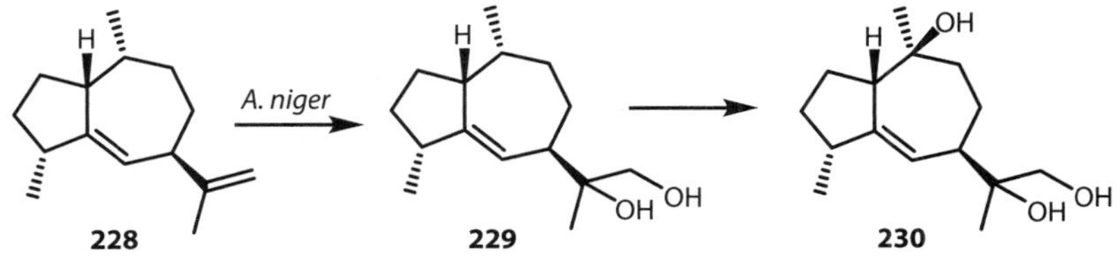

FIGURE 20.68 Biotransformation of (+)-γ-gurjunene (**228**) by *Glomerella cingulata*.

Galal et al. (1999) reported that *Streptomyces fulvissimus* or *R. nigricans* converted parthenolide (**240**) into 11α-methylparthenolide (**247**) in 20%–30% yield, while metabolite 11β-hydroxyparthenolide (**248**) was obtained by incubation of **240** with *R. nigricans* and *Rhodotorula rubra*. In addition to the metabolite **247**, *S. fulvissimus* gave a minor polar metabolite, 9β-hydroxy derivative (**248a**) in low yield (3%). The same metabolite (**248a**) was obtained from **247** by fermentation of *S. fulvissimus* as a minor constituent. Furthermore, 14-hydroxyparthenolide (**248b**) was obtained from **240** and **247** as a minor component (4%) by *R. nigricans* (Figure 20.74).

Pyrethrosin (**248c**), a germacranolide, was treated in the fungus *R. nigricans* to afford five metabolites (**248d**–**248h**). Pyrethrosin itself and metabolite **248e** displayed cytotoxic activity against

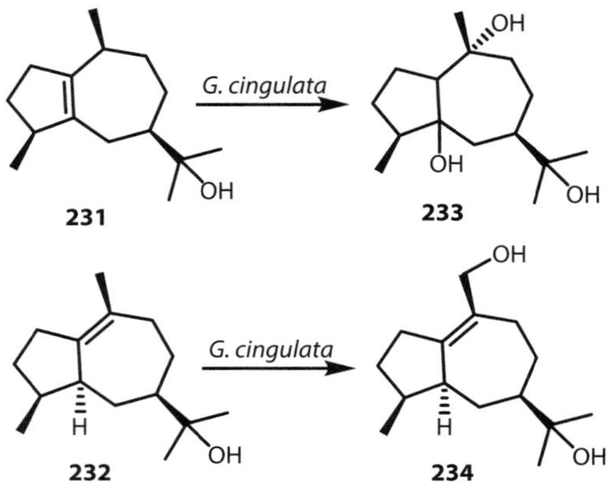

FIGURE 20.69 Biotransformation of guaiol (**221**) and bulnesol (**232**) by *Glomerella cingulata*.

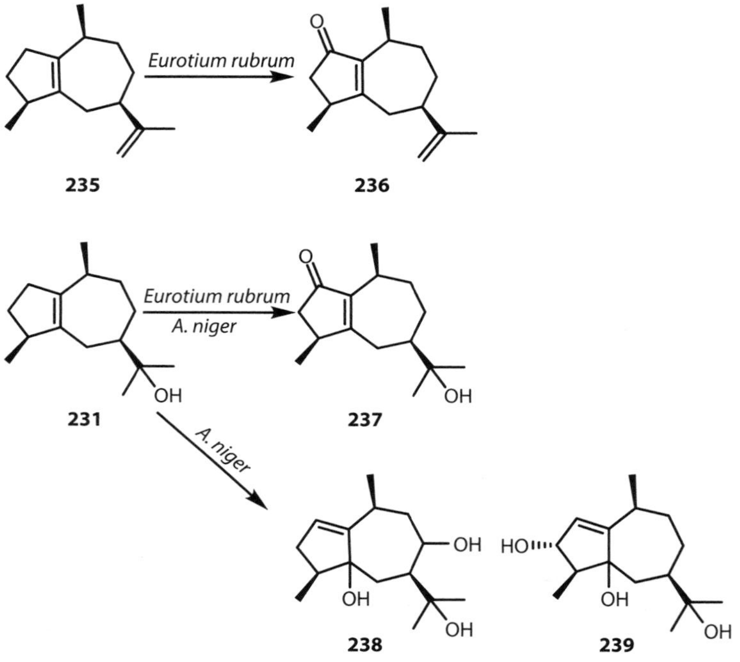

FIGURE 20.70 Biotransformation of guaiene (**235**) by *Eurotium rubrum* and guaiol (**231**) by *Aspergillus niger* and *Eurotium rubrum*.

human malignant melanoma with IC_{50} 4.20 and 7.5 mg/mL, respectively. Metabolite **248h** showed significant in vitro cytotoxic activity against human epidermoid carcinoma (KB cells) and against human ovary carcinoma with IC_{50} < 1.1 and 8.0 mg/mL, respectively. Compounds **248f** and **248i** were active against *Cryptococcus neoformans* with IC_{50} 35.0 and 25 mg/mL, respectively, while **248a** and **248g** showed antifungal activity against *Candida albicans* with IC_{50} 30 and 10 mg/mL. Metabolites **248g** and its acetate (**248i**), derived from **248g**, showed antiprotozoal activity against *Plasmodium falciparum* with IC_{50} 0.88 and 0.32 mg/mL, respectively, without significant toxicity. Compound **248i** also exhibited pronounced activity against the chloroquine-resistant strain of *P. falciparum* with IC_{50} 0.38 mg/mL (Galal, 2001) (Figure 20.75).

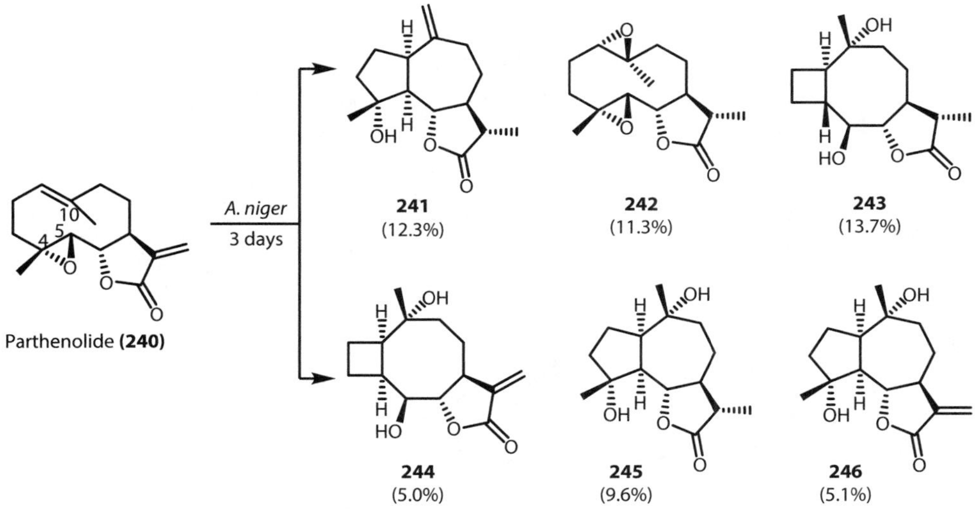

Biotransformation of Sesquiterpenoids, Ionones, Damascones

FIGURE 20.71 Biotransformation of parthenolide (**240**) by *Aspergillus niger*.

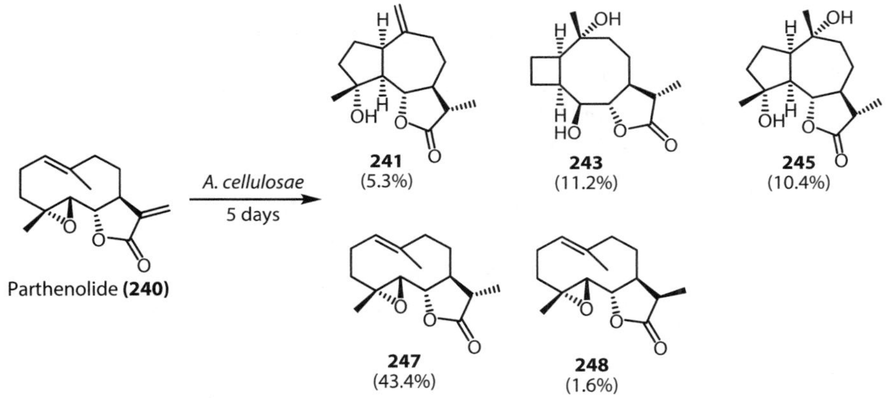

FIGURE 20.72 Biotransformation of parthenolide (**240**) by *Aspergillus cellulosae*.

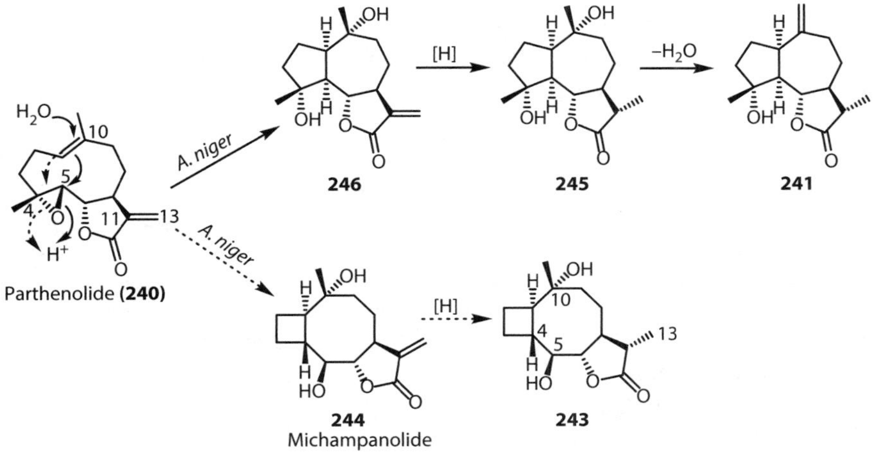

FIGURE 20.73 Possible pathway of biotransformation of parthenolide (**240**).

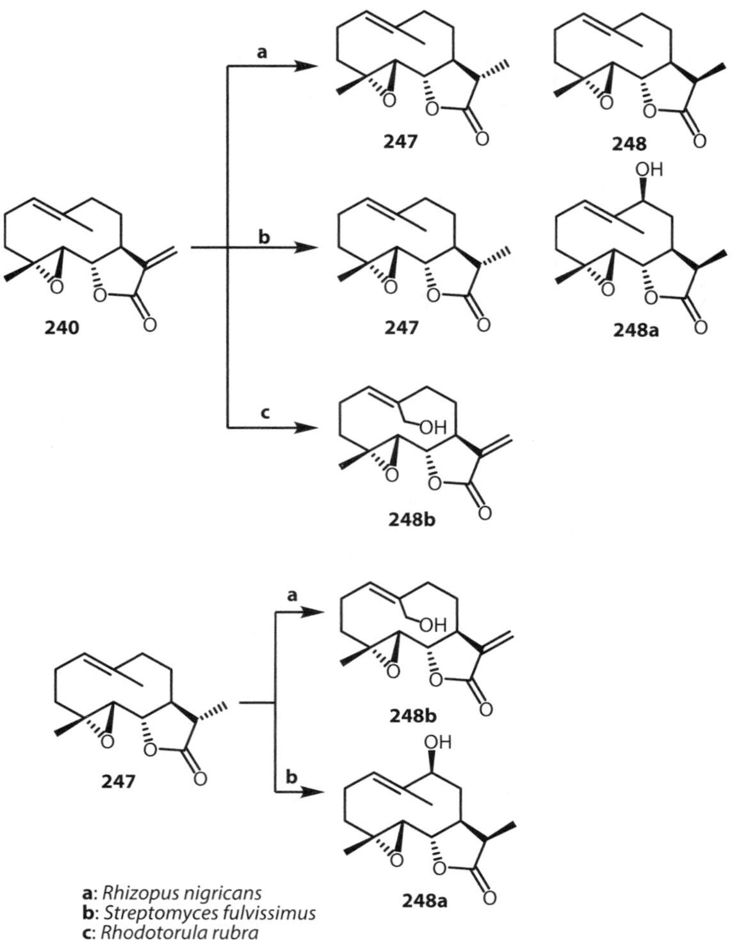

FIGURE 20.74 Biotransformation of parthenolide (**240**) and its dihydro derivative (**247**) by *Rhizopus nigricans*, *Streptomyces fulvissimus*, and *Rhodotorula rubra*.

(−)-Dehydrocostus lactone (**249**), inhibitors of nitric oxide synthases and TNF-α, isolated from *Saussurea* radix, was incubated with *C. echinulata* to afford (+)-11α,13-dihydrodehydrocostuslactone (**250a**). The epoxide (**251**) and a C11 reduced compound (**250**) were obtained by the aforementioned microorganisms (Galal, 2001).

C. echinulata and *R. oryzae* bioconverted **249** into C11/C13 dihydrogenated (**250**) and C10/C14 epoxidated product (**251**). Treatment of **252a** in *C. echinulata* and *R. oryzae* gave (−)-16-(1-methyl-1-propenyl)eremantholanolide (**252b**) (Galal, 2001) (Figure 20.76).

The same substrate (**249**) was fed by *A. niger* for 7 days to afford four metabolites, costus lactone (**250**), and their derivatives (**251–253**), of which **251** was the major product (28%); while the same substrate was cultivated with *A. niger* for 10 days, two minor metabolites (**254, 255**) were newly obtained in addition to **252** and **253** of which the latter lactone was predominant (20.7%) (Hashimoto et al., 2001) (Figure 20.77).

When compound **249** was treated with *A. niger* in the presence of 1-aminobenzotriazole, **249** was completely converted into 11β,13-dihydro derivative (**250**) for 3 days; however, further biodegradation did not occur for 10 days (Hashimoto et al., 1999a, 2001). The same substrate (**249**) was cultivated with *A. cellulosae* IFO to furnish 11,13-dihydro- (**250**) (82%) for only 1 day and then the

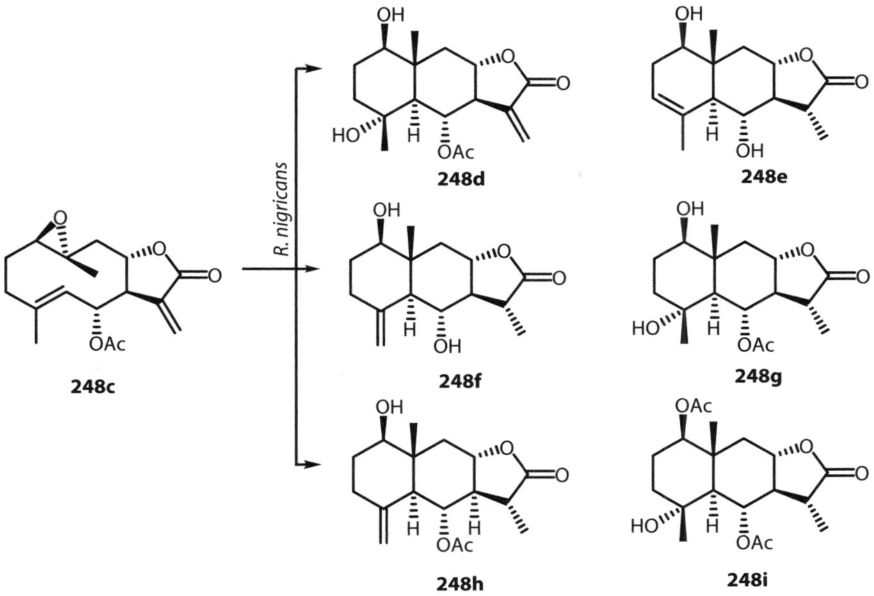

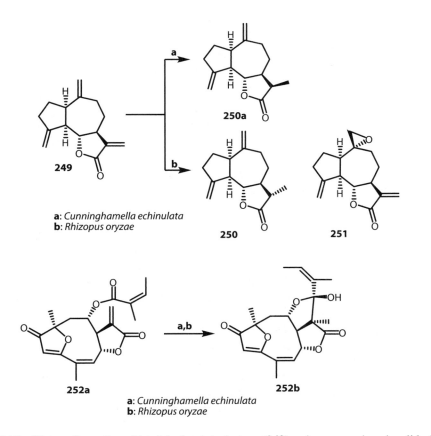

FIGURE 20.75 Biotransformation of pyrethrosin (**248c**) by *Rhizopus nigricans*.

FIGURE 20.76 Biotransformation of (−)-dehydrocostuslactone (**249**) and rearranged guaianolide (**252a**) by *Cunninghamella echinulata* and *Rhizopus oryzae*.

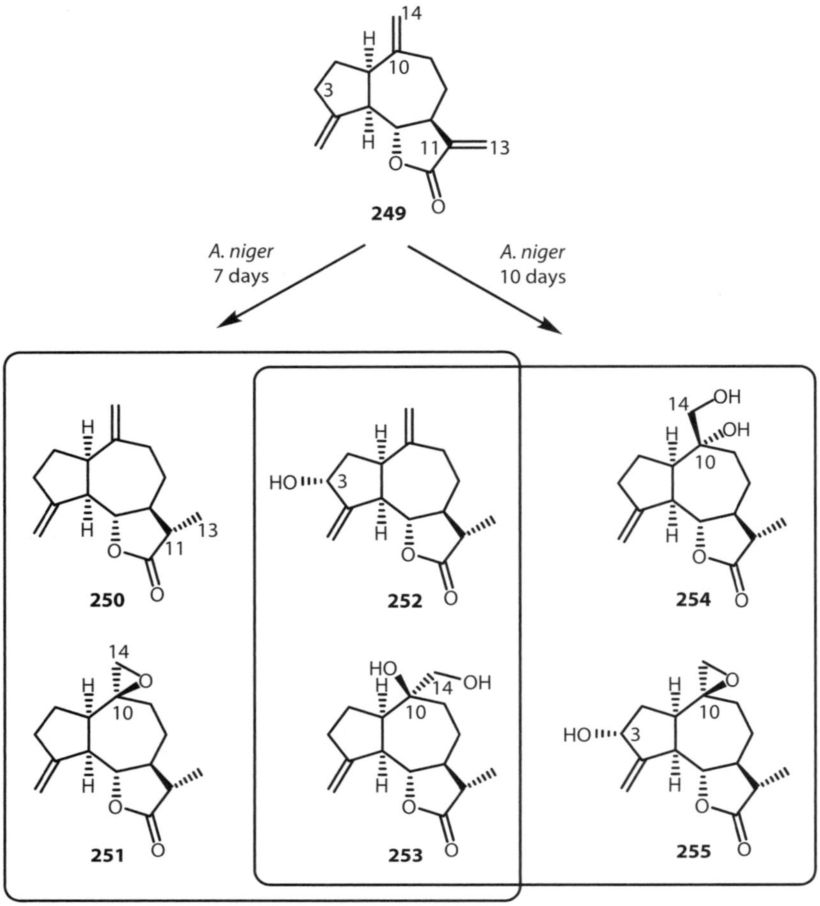

FIGURE 20.77 Biotransformation of (−)-dehydrocostuslactone (**249**) by *Aspergillus niger*.

product (**250**) slowly oxidized into 11,13-dihydro-8β-hydroxycostuslactone (**256**) (1.6%) for 8 days (Hashimoto et al., 1999a, 2001) (Figure 20.78).

The lactone (**249**) was biodegraded by the plant pathogen *B. dothidea* for 4 days to give the metabolites **250** (37.8%) and **257** (8.6%), while *A. niger* IFO-04049 (4 days) and *A. cellulosae* for 1 day gave only **250**. Thus, *B. dothidea* demonstrated low stereoselectivity to reduce the C11–C13 double bond (Hashimoto et al., 2001). Furthermore, three *Aspergillus* species, *A. niger* IFO 4034, *Aspergillus awamori* IFO 4033, and *Aspergillus terreus* IFO6123, were used as bioreactors for compounds **249**. *A. niger* IFO 4034 gave three products (**250–252**), of which **252** was predominant (56% in GC-MS). *A. awamori* IFO 4033 and *A. terreus* IFO 6123 converted **249** to give **250** (56% from *A. awamori*, 43% from *A. terreus*) and **252** (43% from *A. awamori*, 57% from *A. terreus*), respectively (Hashimoto et al., 2001) (Figure 20.79).

Vernonia arborea (Asteraceae) contains zaluzanin D (**258**) in high content. Ten microorganisms were used for the biotransformation of compound **258**. *B. cinerea* converted **258** into **259** and **260** (85%:15%) and *Fusarium equiseti* gave **259** and **260** (33%:66%). *C. lunata*, *Colletotrichum lindemuthianum*, *Alternaria alternata*, and *Phyllosticta capsici* produced **259** as the sole metabolite in good yield, while *Sclerotinia sclerotiorum* and *Rhizoctonia solani* gave deactyl product (**261**) as a sole product and **260** and **262–264**, among which **263** and **264** are the major products, respectively. Reduction of C11–C13 exocylic double bond is the common transformation of α-methylene-γ-butyrolactone (Kumari et al., 2003) (Figure 20.80).

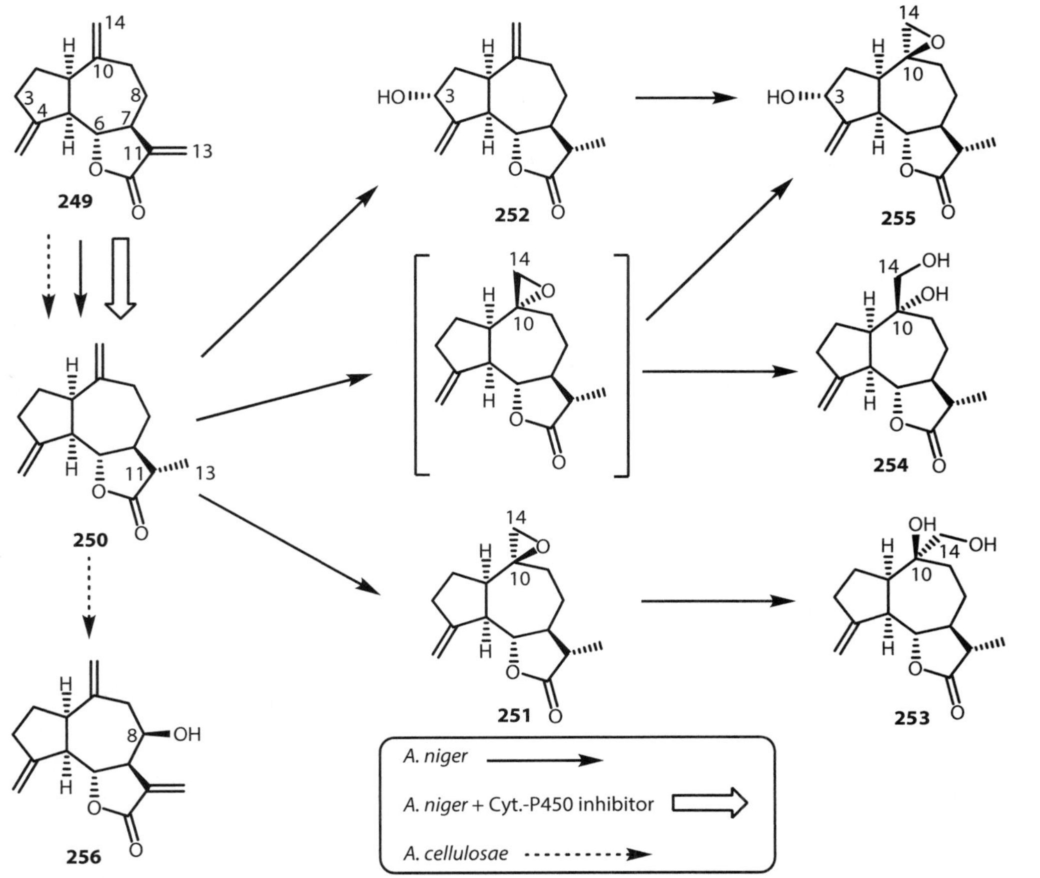

FIGURE 20.78 Possible pathway of biotransformation of (−)-dehydrocostuslactone (**249**) by *Aspergillus niger* and *Aspergillus cellulosae*.

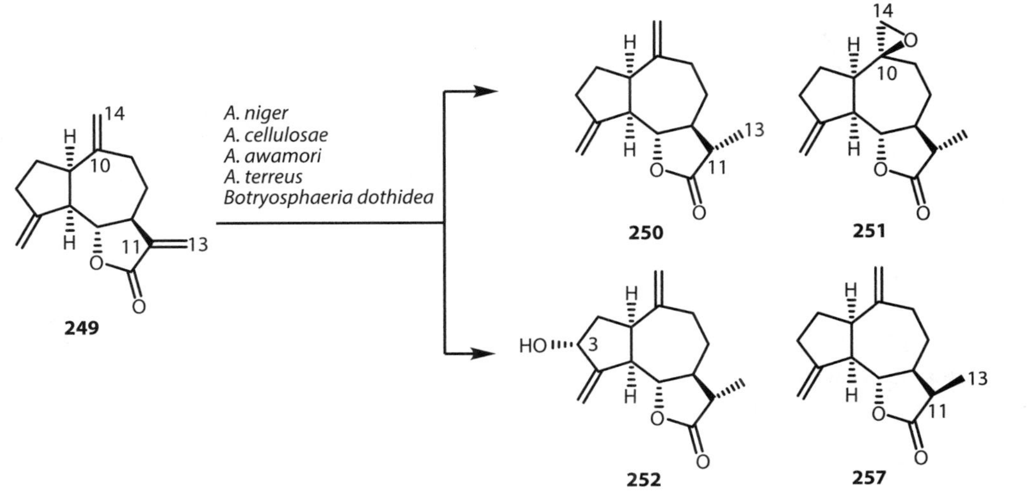

FIGURE 20.79 Biotransformation of (−)-dehydrocostuslactone (**249**) by *Aspergillus* species and *Botryosphaeria dothidea*.

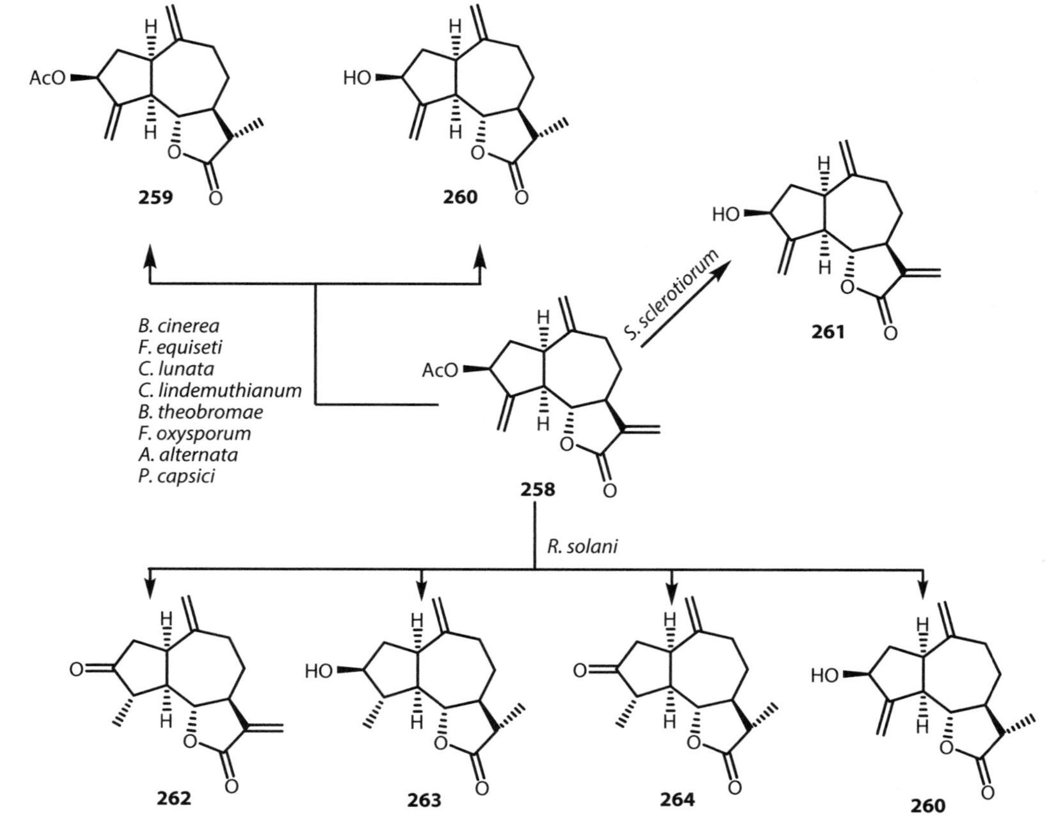

FIGURE 20.80 Biotransformation of zaluzanin D (**258**) by various fungi.

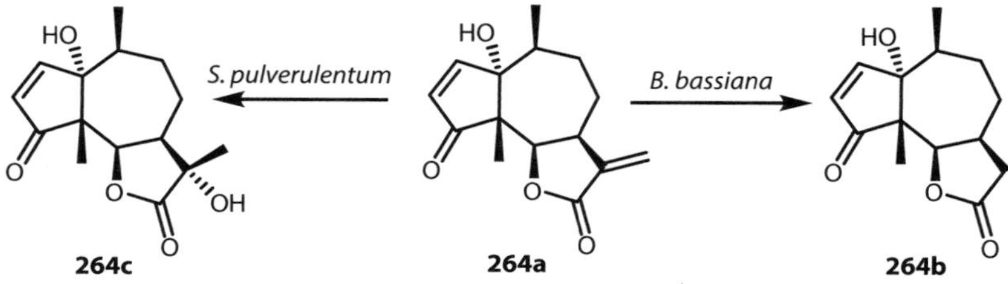

FIGURE 20.81 Biotransformation of parthenin (**264a**) by *Sporotrichum pulverulentum* and *Beauveria bassiana*.

Incubation of parthenin (**264a**) with the fungus *B. bassiana* in modified Richard's medium gave C11–C13 reduced product (**264b**) in 37% yield, while C11 α-hydroxylated product (**264c**) was obtained in 32% yield from the broth of the fungus *Sporotrichum pulverulentum* using the same medium (Bhutani and Thakur, 1991) (Figure 20.81).

Cadina-4,10(15)-dien-3-one (**265**) possessing insecticidal and ascaricidal activity, from the Jamaican medicinal plant *H. verticillata*, was metabolized by *C. lunata* ATCC 12017 in potato dextrose to give its 12-hydoxy- (**266**), 3α-hydroxycadina-4,10(15)-dien (**267**), and 3α-hydroxy-4,5-dihydrocadinenes (**268**), while **265** was incubated by the same fungus in peptone, yeast, and beef extracts and glucose medium, only **267** and **268** were obtained. Compound **267** derived synthetically was treated in the same fungus *C. lunata* to afford three metabolites (**269**–**271**) (Collins and Reese, 2002) (Figure 20.82).

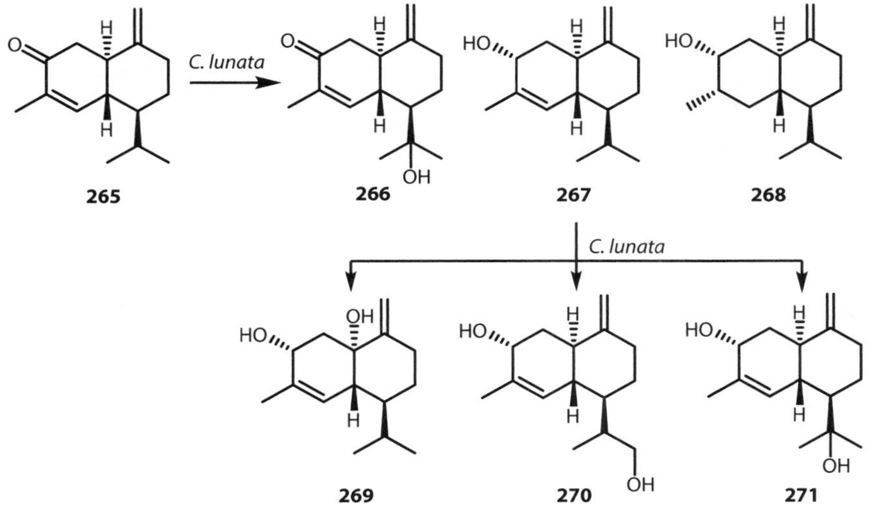

FIGURE 20.82 Biotransformation of cadina-4,10(15)-dien-3-one (**265**) by *Curvularia lunata*.

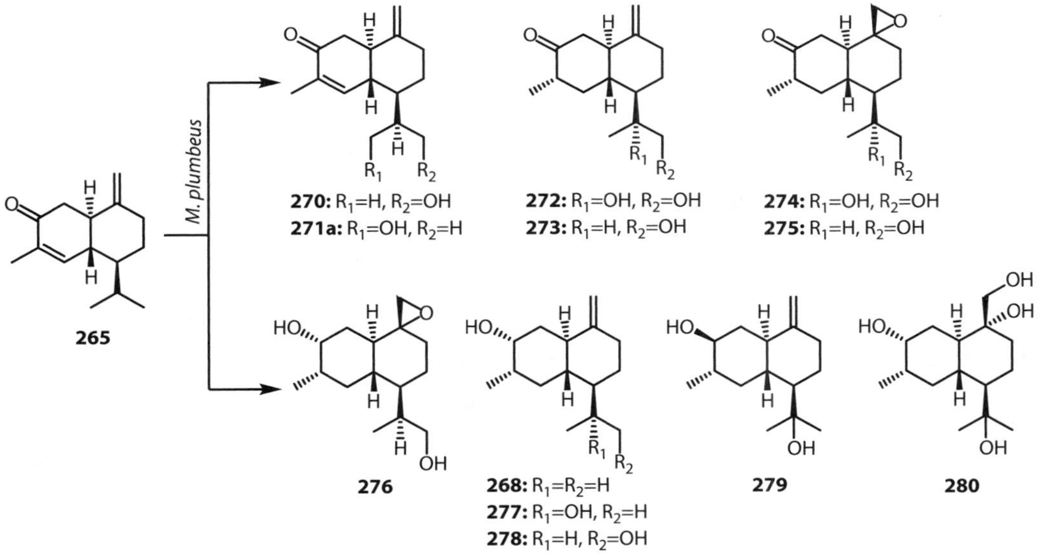

FIGURE 20.83 Biotransformation of cadina-4,10(15)-dien-3-one (**265**) by *Mucor plumbeus*.

The incubation of the same substrate (**265**) in *M. plumbeus* ATCC 4740 in high-iron-rich medium gave **270**, which was obtained from *C. lunata* mentioned earlier, **268, 272, 273, 277, 278**, and **279**. In low-iron medium, this fungus converted the same substrate **265** into three epoxides (**274–276**), a tetraol (**280**) with common metabolites (**268, 273, 277, 278**), and **271**, which was the same metabolite used by *C. lunata* (Collins et al., 2002a). It is interesting to note that only epoxides were obtained from the substrate (**265**) by *Mucor* fungus in low-iron medium (Figure 20.83).

The same substrate (**265**) was incubated with the deuteromycete fungus *B. bassiana*, which is responsible for the muscardine disease in insects, in order to obtain new functionalized analogues with improved biological activity. From compound **265**, nine metabolites were obtained. The insecticidal potential of the metabolites (**267, 268, 268a–268f**) were evaluated against *Cylas formicarius*. The metabolites (**273, 268, 268d**) showed enhanced activity compared with the substrate (**265**).

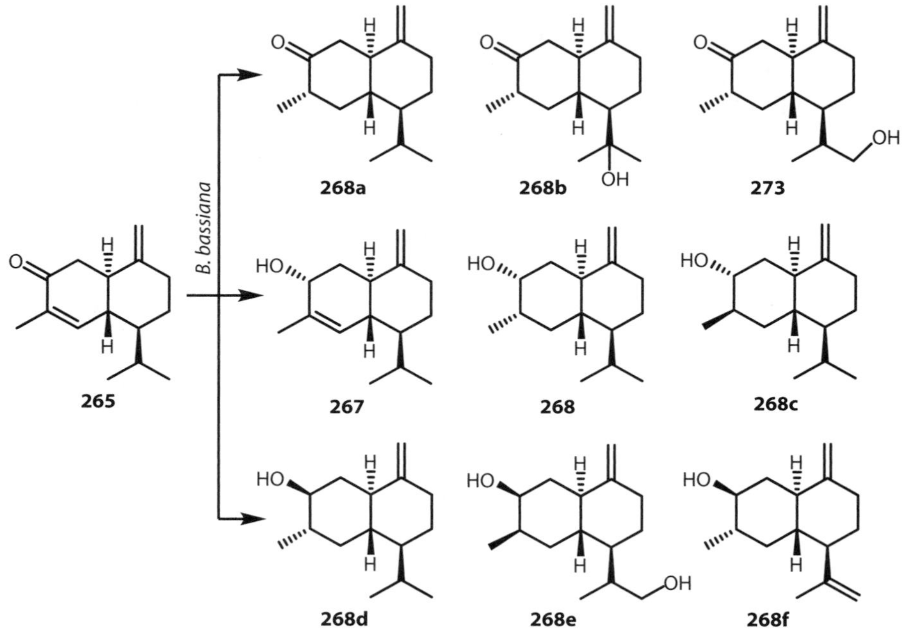

FIGURE 20.84 Biotransformation of cadina-4,10(15)-dien-3-one (**265**) by *Beauveria bassiana*.

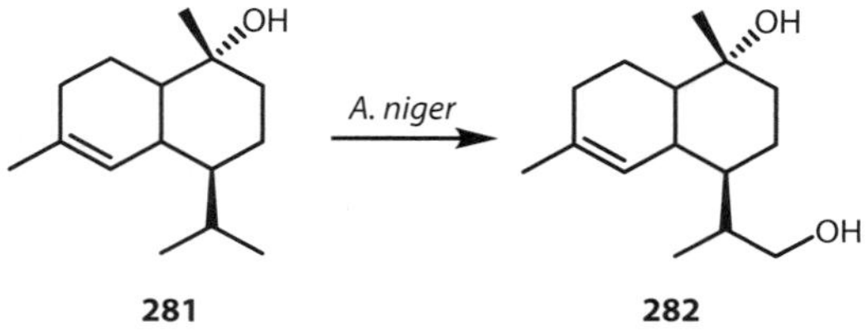

FIGURE 20.85 Biotransformation of cadinol (**281**) by *Aspergillus niger*.

The plant growth regulatory activity of the metabolites against radish seeds was tested. All the compounds showed inhibitory activity; however, their activity was less than colchicine (Buchanan et al., 2000) (Figure 20.84).

Cadinane-type sesquiterpene alcohol (**281**) isolated from the liverwort *M. taylorii* gave a primary alcohol (**282**) by *A. niger* treatment (Morikawa et al., 2000) (Figure 20.85).

Fermentation of (−)-α-bisabolol (**282a**) possessing an anti-inflammatory activity with the plant pathogenic fungus *G. cingulata* for 7 days yielded oxygenated products (**282b–282e**) of which compound **282e** was predominant. 3,4-Dihydroxy products (**282b**, **282d**, **282e**) could be formed by hydrolysis of the 3,4-epoxide from **282a** and **282c** (Miyazawa et al., 1995c) (Figure 20.86).

El Sayed et al. (2002) reported microbial and chemical transformation of (*S*)-(+)-curcuphenol (**282g**) and curcudiol (**282n**), isolated from the marine sponges *Didiscus axiata*. Incubation of compound **282g** with *Kluyveromyces marxianus* var. *lactis* resulted in the isolation of six metabolites (3–8, **282h–282j**). The same substrate was incubated with *Aspergillus alliaceus* to give the metabolites (**282p**, **282q**, **282s**) (Figure 20.87).

Compounds **282g** and **282n** were treated in *Rhizopus arrhizus* and *Rhodotorula glutinus* for 6 and 8 days to afford glucosylated metabolites, 1α-D-glucosides (**282o**) and **282r**, respectively. The substrate itself showed antimicrobial activity against *C. albicans*, *C. neoformans*, and MRSA-resistant *Staphylococcus aureus* and *S. aureus* with MIC and MFC/MBC ranges of 7.5–25 and 12.5–50 mg/mL, respectively. Compounds **282g** and **282h** also exhibited in vitro antimalarial

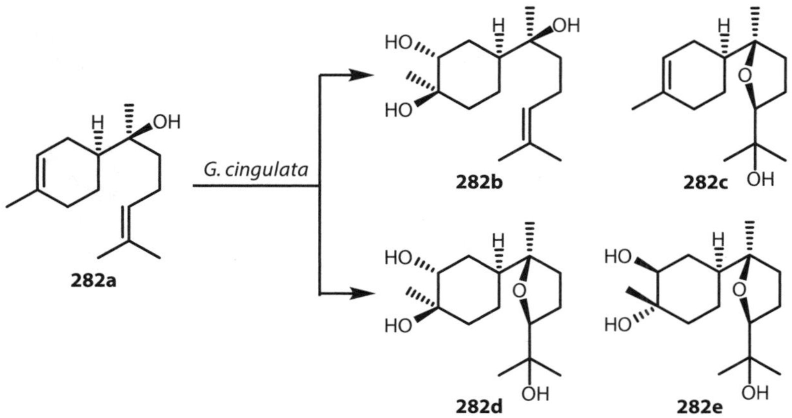

FIGURE 20.86 Biotransformation of β-bisabolol (**282a**) by *Glomerella cingulata*.

activity against *P. falciparum* (D6 clone) and *P. falciparum* (W2 clone) of 3600 and 3800 ng/mL (selective index [S.I.] > 1.3), 1800 (S.I. > 2.6), and 2900 (S.I. > 1.6), respectively (El Sayed et al., 2002) (Figure 20.87).

Artemisia annua is one of the most important Asteraceae species as antimalarial plant. There are many reports of microbial biotransformation of artemisinin (**283**), which is active antimalarial rearranged cadinane sesquiterpene endoperoxide, and its derivatives to give novel antimalarials with increased activities or differing pharmacological characteristics.

Lee et al. (1989) reported that deoxoartemisinin (**284**) and its 3α-hydroxy derivative (**285**) were obtained from the metabolites of artemisinin (**283**) incubated with *Nocardia corallina* and *Penicillium chrysogenum* (Figure 20.88).

Zhan et al. (2002) reported that incubation of artemisinin (**283**) with *C. echinulata* and *A. niger* for 4 days at 28°C resulted in the isolation of two metabolites, 10β-hydroxyartemisinin (**287a**) and 3α-hydroxydeoxoartemisinin (**285**), respectively.

Compound **283** was also biotransformed by *A. niger* to give four metabolites, deoxoartemisinin (**284**, 38%), 3α-hydroxydeoxoartemisinin (**285**, 15%), and two minor products (**286**, 8% and **287**, 5%) (Hashimoto et al., 2003d).

Artemisinin (**283**) was also bioconverted by *C. elegans*. During this process, 9β-hydroxyartemisinin (**287b**, 78.6%), 9β-hydroxy-8α-artemisinin (**287c**, 6.0%), 3α-hydroxydeoxoartemisinin (**285**, 5.4%), and 10β-hydroxyartemisinin (**287d**, 6.5%) have been formed. On the basis of quantitative structure–activity relationship and molecular modeling investigations, 9β-hydoxy derivatization of artemisinin skeleton may yield improvement in antimalarial activity and may potentially serve as an efficient means of increasing water solubility (Parshikov et al., 2004) (Figure 20.89).

Albicanal (**288**) and (−)-drimenol (**289**) are simple drimane sesquiterpenoids isolated from the liverwort, *Diplophyllum serrulatum*, and many other liverworts and higher plants. The latter compound was incubated with *M. plumbeus* and *R. arrhizus*. The former microorganism converted **289** to 6,7α-epoxy- (**290**), 3β-hydroxy- (**291**), and 6α-drimenol (**292**) in the yields of 2%, 7%, and 50%, respectively. On the other hand, the latter species produced only 3β-hydroxy derivative (**291**) in 60% yield (Aranda et al., 1992) (Figure 20.90).

(−)-Polygodial (**293**) possessing piscicidal, antimicrobial, and mosquito-repellant activity is the major pungent sesquiterpene dial isolated from *Polygonum hydropiper* and the liverwort *Porella vernicosa* complex. Polygodial was incubated with *A. niger*; however, because of its antimicrobial activity, no metabolite was obtained (Sekita et al., 2005). Polygodiol (**295**) prepared from polygodial (**293**) was also treated in the same manner as described earlier to afford 3β-hyrdoxy (**297**), which was isolated from *Marasmius oreades* as antimicrobial activity (Ayer and Craw, 1989), and 6α-hydroxypolygodiol (**298**) in 66%–70% and 5%–10% yields, respectively (Aranda et al., 1992).

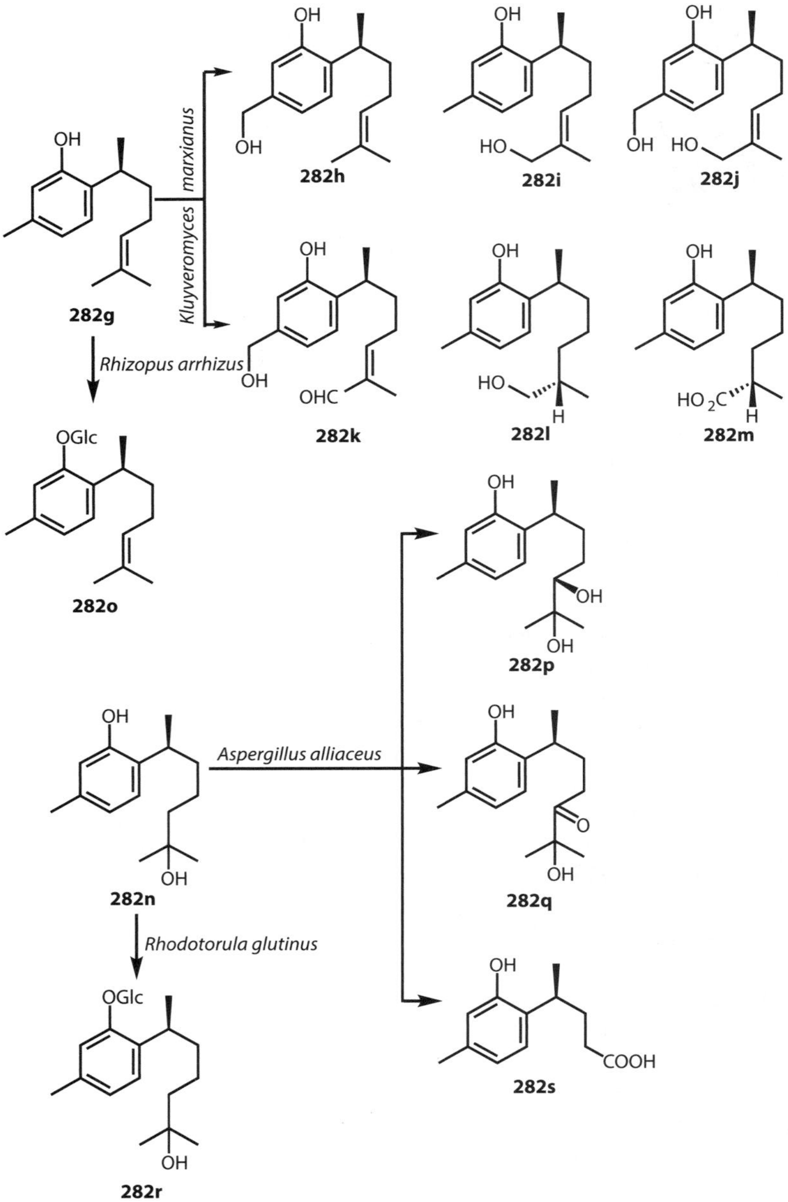

FIGURE 20.87 Biotransformation of (S)-(+)-curcuphenol (**282g**) by *Kluyveromyces marxianus* and *Rhizopus arrhizus* and curcudiol (282n) by *Aspergillus alliaceus* and *Rhodotorula glutinus*.

The same metabolite (**297**) was also obtained from polygodiol (**295**) as a sole metabolite from the culture broth of *A. niger* in Czapek-peptone medium for 3 days in 70.5% yield (Sekita et al., 2005), while the C9 epimeric product (**296**) from isopolygodial (**294**) was incubated with *M. plumbeus* to afford 3β-hydroxy- (**299**) and 6α-hydroxy derivative (**300**) in low yields, 7% and 13% (Aranda et al., 1992). Drim-9α-hydroxy-11β,12-diacetoxy-7-ene (**301**) derived from polygodiol (**295**) was treated in the same manner as described earlier to yield its 3β-hydroxy derivative (**302**, 42%) (Sekita et al., 2005) (Figures 20.91 and 20.92).

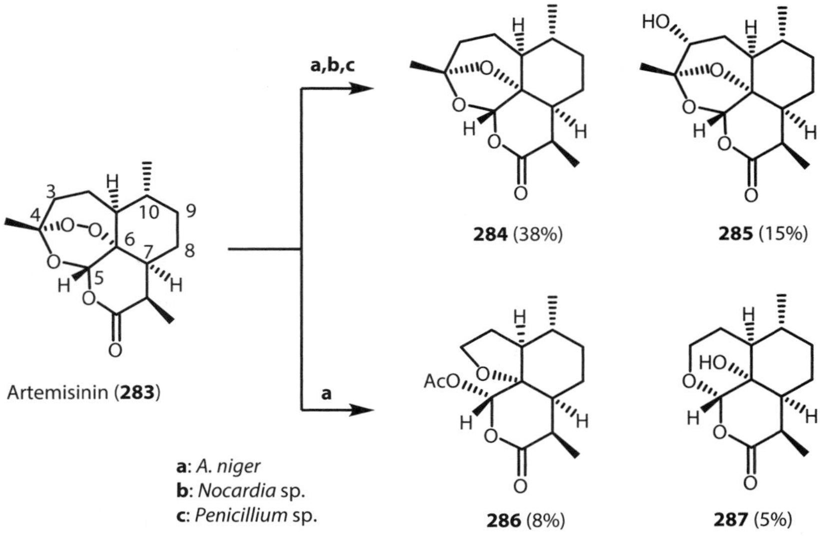

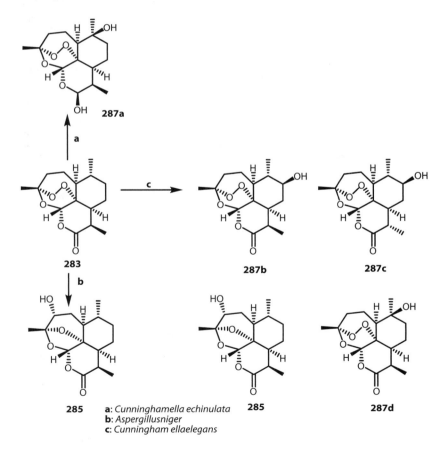

FIGURE 20.88 Biotransformation of artemisinin (**283**) by *Aspergillus niger*, *Nocardia corallina*, and *Penicillium chrysogenum*.

FIGURE 20.89 Biotransformation of artemisinin (**283**) by *Cunninghamella echinulata*, *Cunninghamella elegans*, and *Aspergillus niger*.

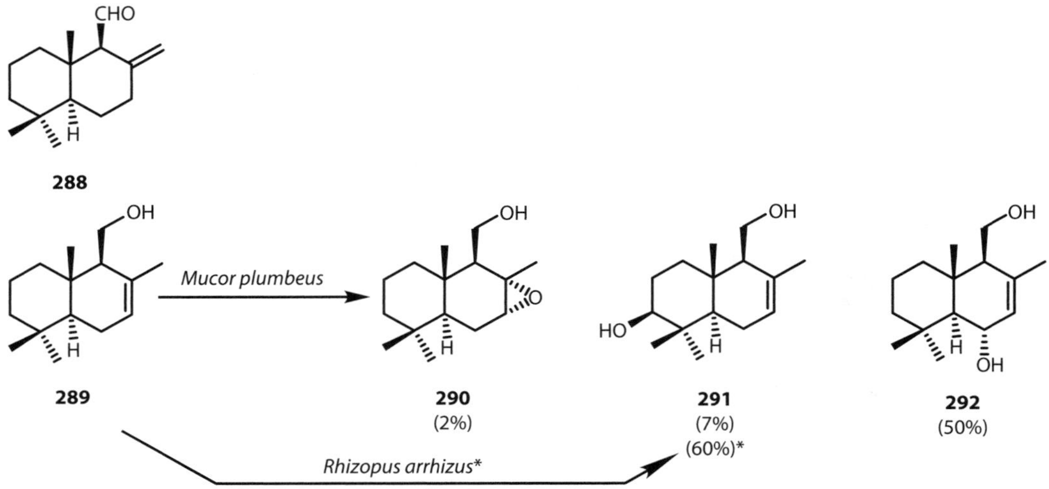

FIGURE 20.90 Biotransformation of drimenol (**289**) by *Mucor plumbeus* and *Rhizopus arrhizus*.

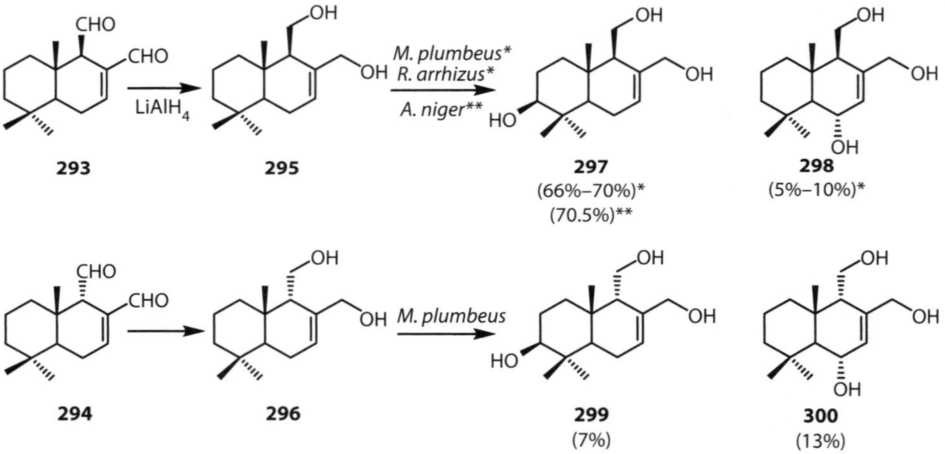

FIGURE 20.91 Biotransformation of polygodiol (**295**) by *Mucor plumbeus*, *Rhizopus arrhizus*, and *Aspergillus niger*.

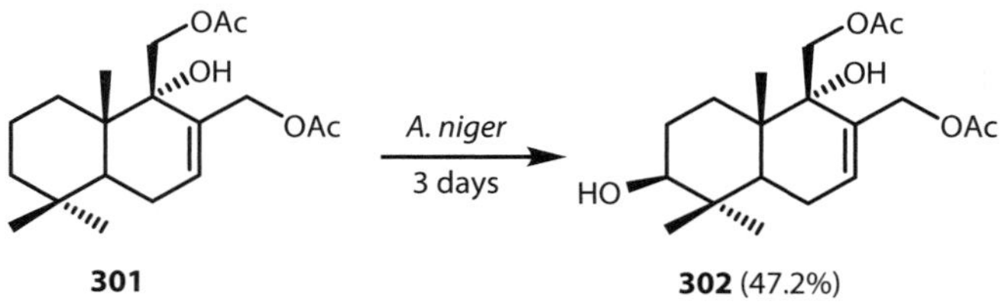

FIGURE 20.92 Biotransformation of drim-9α-hydroxy-11,12-diacetoxy-7-ene (**301**) by *Aspergillus niger*.

Cinnamodial (**303**) from the Malagasy medicinal plant, *Cinnamosma fragrans*, was also treated in the same medium including *A. niger* to furnish three metabolites, respectively, in very law yields (**304**, 2.2%; **305**, 0.05%; and **306**, 0.62%). Compound **305** and **306** are naturally occurring cinnamosmolide, possessing cytotoxicity and antimicrobial activity, and fragrolide. In this case, the introduction of 3β-hydroxy group was not observed (Sekita et al., 2006) (Figure 20.93).

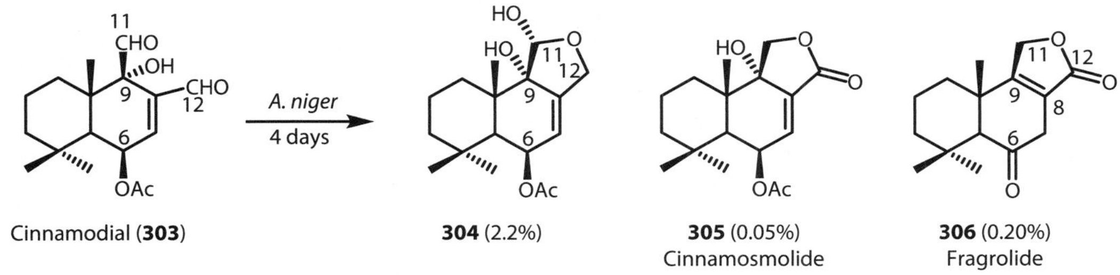

FIGURE 20.93 Biotransformation of cinnamodial (**303**) by *Aspergillus niger*.

Naturally occurring rare drimane sesquiterpenoids (**307–314**) were biosynthesized by the fungus *Cryptoporus volvatus* with isocitric acids. Among these compounds, in particular, cryptoporic acid E (**312**) possesses antitumor promoter, anticolon cancer, and very strong superoxide anion radical scavenging activities (Asakawa et al., 1992). When the fresh fungus allowed standing in moisture condition, olive fungus *Paecilomyces variotii* grows on the surface of the fruit body of this fungus. *C. volvatus* infected 2 kg of the fresh fungus for 1 month, followed by the extraction of methanol to give the crude extract and then purification using silica gel and Sephadex LH-20 to give five metabolites (**316, 318–321**), which were not found in the fresh fungus (Takahashi et al., 1993b). Compound **318** was also isolated from the liverworts, *Bazzania* and *Diplophyllum* species (Asakawa, 1982, 1995) (Figure 20.94).

Liverworts produce a large number of enantiomeric mono-, sesqui-, and diterpenoids to those found in higher plants and lipophilic aromatic compounds. It is also noteworthy that some liverworts produce both normal and its enantiomers. The more interesting phenomenon in the chemistry of liverworts is that the different species in the same genus, for example, *Frullania tamarisci* subsp. *tamarisci* and *Frullania dilatata*, produce totally enantiomeric terpenoids. Various sesqui- and diterpenoids, bibenzyls, and bisbibenzyls isolated from several liverworts show a

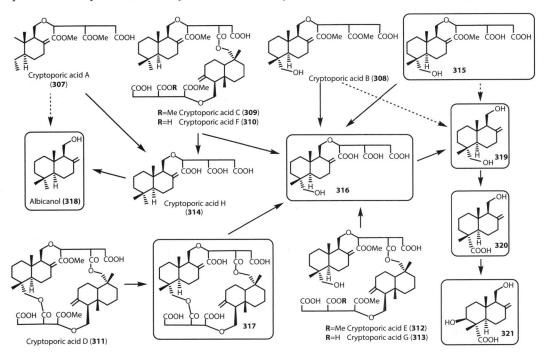

FIGURE 20.94 Biotransformation of cryptoporic acids (**307–317, 316**) by *Paecilomyces variotii*.

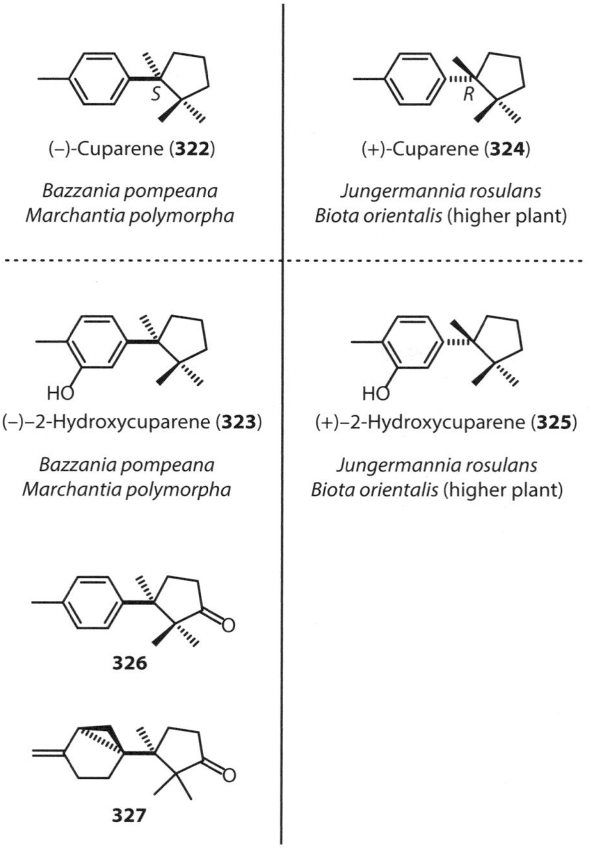

FIGURE 20.95 Naturally occurring cuparene sesquiterpenoids (**322–327**).

characteristic fragrant odor, intensely hot and bitter taste, muscle relaxation, allergenic contact dermatitis, and antimicrobial, antifungal, antitumor, insect antifeedant, superoxide anion release inhibitory, piscicidal, and neurotrophic activity (Asakawa, 1982, 1990, 1995, 1999, 2007, 2008; Asakawa and Ludwiczuk, 2008). In order to obtain the different kinds of biologically active products and to compare the metabolites of both normal and enantiomers of terpenoids, several secondary metabolites of specific liverworts were biotransformed by *Penicillium sclerotiorum*, *A. niger*, and *A. cellulosae*.

(−)-Cuparene (**322**) and (−)-2-hydroxycuparene (**323**) have been isolated from the liverworts *Bazzania pompeana* and *M. polymorpha*, while its enantiomer (+)-cuparene (**324**) and (+)-2-hydroxycuparene (**325**) from the higher plant *Biota orientalis* and the liverwort *Jungermannia rosulans*. (*R*)-(−)-α-Cuparenone (**326**) and grimaldone (**327**) demonstrate intense flagrance. In order to obtain such compounds from both cuparene and its hydroxy compounds, both enantiomers mentioned earlier were cultivated with *A. niger* (Hashimoto et al., 2001a) (Figure 20.95).

From (−)-cuparene (**322**), five metabolites (**328–332**) all of which contained cyclopentanediols or hydroxycyclopentanones were obtained. An aryl methyl group was also oxidized to give primary alcohol, which was further oxidized to afford carboxylic acids (**329–331**) (Hashimoto et al., 2001a) (Figure 20.96).

From (+)-cuparene, six metabolites (**333–338**) were obtained. These are structurally very similar to those found in the metabolites of (−)-cuparene, except for the presence of an acetonide (**336**), but they are not identical. All metabolites possess benzoic acid moiety.

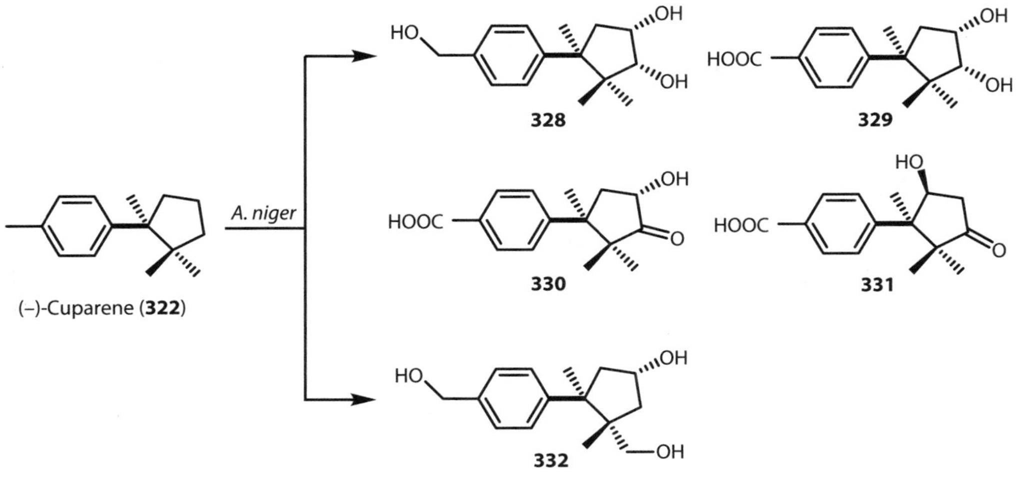

FIGURE 20.96 Biotransformation of (−)-cuparene (**322**) by *Aspergillus niger*.

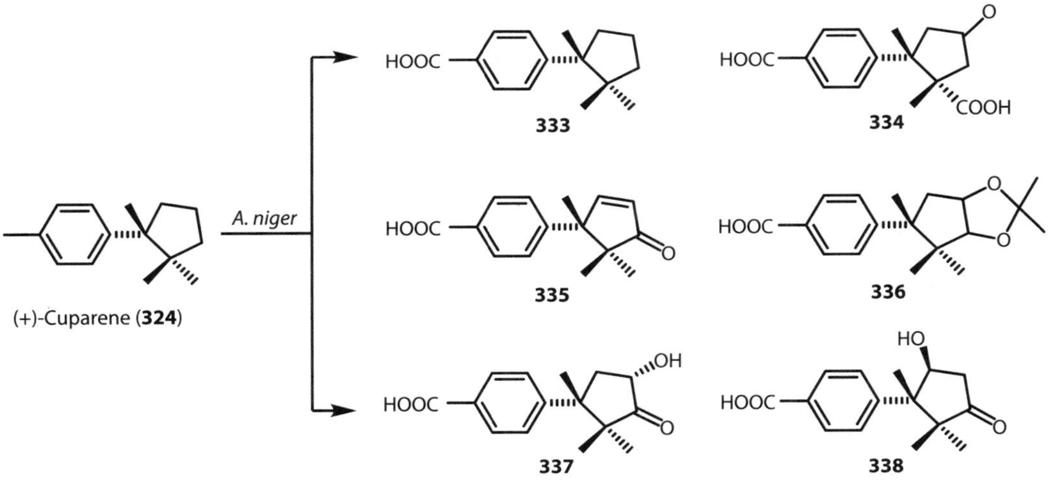

FIGURE 20.97 Biotransformation of (+)-cuparene (**324**) by *Aspergillus niger*.

The possible biogenetic pathways of (+)-cuparene (**324**) have been proposed in Figure 20.97. Unfortunately, none of the metabolites show strong mossy odor (Hashimoto et al., 2006). The presence of an acetonide in the metabolites has also been seen in those of dehydronootkatone (**25**) (Furusawa et al., 2003) (Figure 20.98).

The liverworts *Herbertus aduncus*, *Herbertus sakurai*, and *Mastigophora diclados* produce (−)-herbertene, the C3 methyl isomer of cuparene, with its hydroxy derivatives, for example, herbertanediol (**339**), which shows no production inhibitory activity (Harinantenaina et al., 2007), and herbertenol (**342**). Treatment of compound (**339**) in *P. sclerotiorum* in Czapek-polypeptone medium gave two dimeric products, mastigophorene A (**340**) and mastigophorene B (**341**), which showed neurotrophic activity (Harinantenaina et al., 2005).

When (−)-herbertenol (**342**) was biotransformed for 1 week by the same fungus, no metabolic product was obtained; however, five oxygenated metabolites (**344**–**348**) were obtained from its methyl ether (**343**). The possible metabolic pathway is shown in Figure 20.99. Except for the

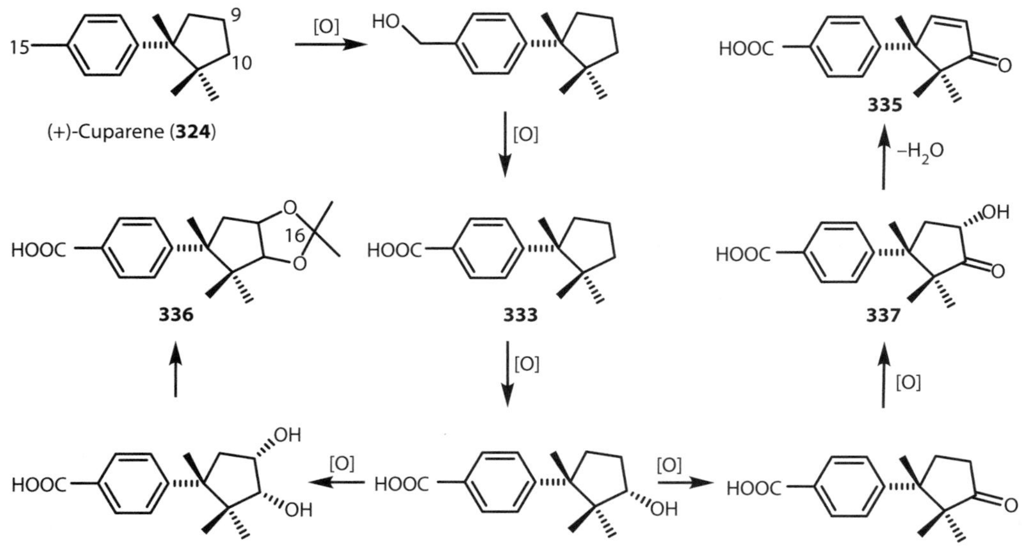

FIGURE 20.98 Possible pathway of biotransformation of (+)-cuparene (**324**) by *Aspergillus niger*.

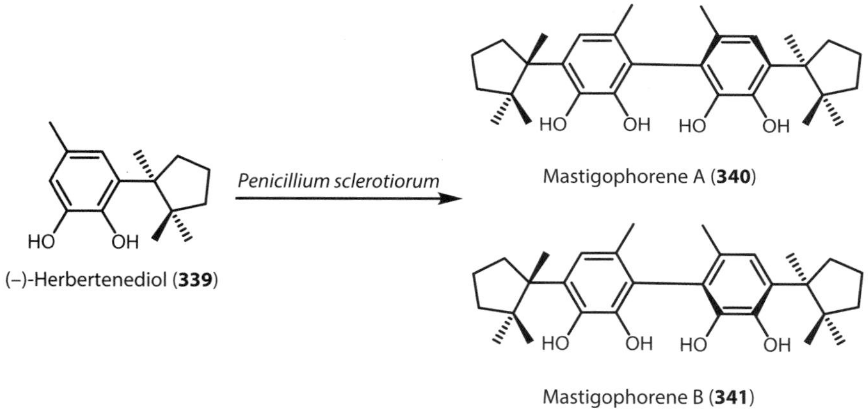

FIGURE 20.99 Biotransformation of (−)-herbertenediol (**339**) by *Penicillium sclerotiorum*.

presence of the ether (**348**), the metabolites from **342** resemble those found in (−)- and (+)-cuparene (Hashimoto et al., 2006) (Figures 20.100 and 20.101).

Maalioxide (**349**), mp 65°–66°, $[\alpha]_D^{21}$ − 34.4°, obtained from the liverwort *P. sciophila* was inoculated and cultivated rotatory (100 rpm) in Czapek-peptone medium (pH 7.0) at 30°C for 2 days. (−)-Maalioxide (**349**) (100 mg/200 mL) was added to the medium and further cultivated for 2 days to afford three metabolites, 1β-hydroxy- (**350**), 1β,9β-dihydroxy- (**351**), and 1β,12-dihydroxymaalioxides (**352**), of which **351** was predominant (53.6%). When the same substrate was cultured with *A. cellulosae* in the same medium for 9 days, 7β-hydroxymaalioxide (**353**) was obtained as a sole product in 30% yield. The same substrate (**349**) was also incubated with the fungus *M. plumbeus* to obtain a new metabolite, 9β-hydroxymaalioxide (**354**), together with two known hydroxylated products (**350**, **353**) (Wang et al., 2006).

Maalioxide (**349**) was oxidized by *m*-chloroperoxybenzoic acid to give a very small amount of **353** (1.2%), together with 2α-hydroxy- (**355**, 2%) and 8α-hydroxymaalioxide (**356**, 1.5%), which have not been obtained in the metabolite of **349** in *A. niger* and *A. cellulosae* (Tori et al., 1990) (Figure 20.102).

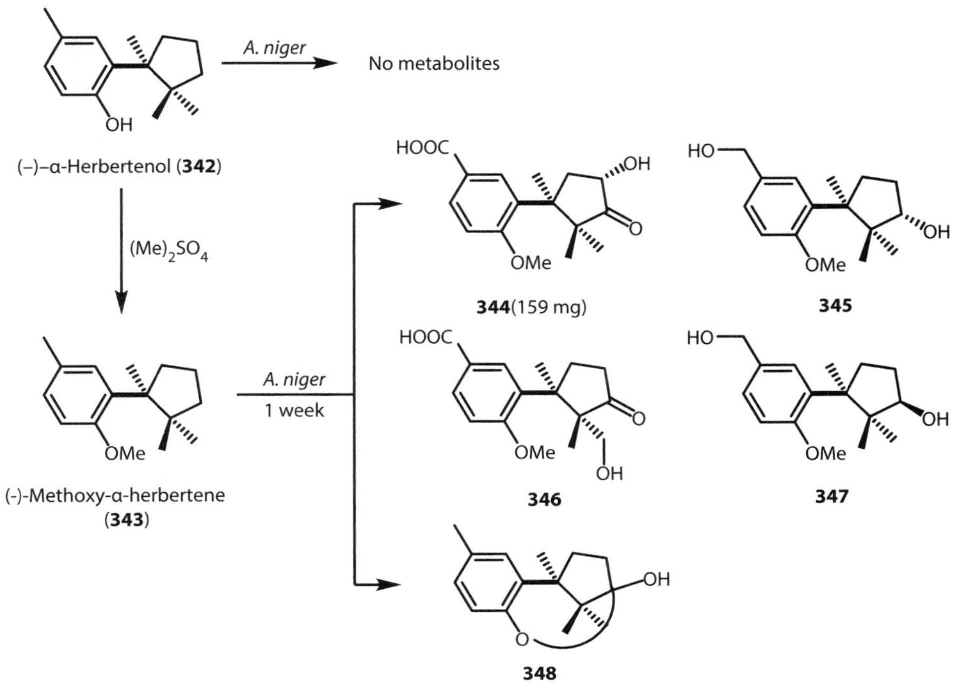

FIGURE 20.100 Biotransformation of (−)-methoxy-α-herbertene (**343**) by *Aspergillus niger*.

P. sciophila is one of the most important liverworts, since it produces bicyclohumulenone (**357**), which possesses strong mossy note and is expected to manufacture compounding perfume. In order to obtain much more strong scent, **357** was treated in *A. niger* for 4 days to give 4α,10β-dihydroxybicyclohumulenone (**358**, 27.4%) and bicyclohumurenone-12-oic acid (**359**). An epoxide (**360**) prepared by *m*-chloroperoxybenzoic acid was further treated in the same fungus as described earlier to give 10β-hydoxy derivative (**361**, 23.4%). Unfortunately, these metabolites possess only faint mossy odor (Hashimoto et al., 2003c) (Figure 20.103).

The liverwort *Reboulia hemisphaerica* biosynthesizes cyclomyltaylanoids like **362** and also *ent*-1α-hydroxy-β-chamigrene (**367**). Biotransformation of cyclomyltaylan-5-ol (**362**) in the same medium including *A. niger* gave four metabolites, 9β-hydroxy (**363**, 27%), 9β,15-dihydroxy (**364**, 1.7%), 10β-hydroxy (**365**, 10.3%), and 9β,15-dihydroxy derivative (**366**, 12.6%). In this case, the stereospecificity of alcohol was observed, but the regiospecificity of alcohol moiety was not seen in this substrate (Furusawa et al., 2005c, 2006b) (Figure 20.104).

The biotransformation of spirostructural terpenoids was not carried out. *ent*-1α-Hydroxy-β-chamigrene (**367**) was inoculated in the same manner as described earlier to give three new metabolites (**368**–**370**), of which **370** was the major product (46.2% in isolated yield). The hydroxylation of vinyl methyl group has been known to be very common in the case of microbial and mammalian biotransformation (Furusawa et al., 2005a, 2006) (Figure 20.105).

β-Barbatene (=gymnomitrene) (**4**), a ubiquitous sesquiterpene hydrocarbon, is from liverwort like *P. sciophila* and many others. Jungermanniales liverworts treated in the same manner using *A. niger* for 1 day gave a triol, 4β,9β,10β-trihydoxy-β-barbatene (**27**, 8%) (Hashimoto et al., 2003c).

Pinguisane sesquiterpenoids have been isolated from the Jungermanniales, Metzgeriales, and Marchantiales. In particular, the Lejeuneaceae and Porellaceae are rich sources of this unique type of sesquiterpenoids. One of the major furanosesquitepene (**373**) was biodegradated by *A. niger* to afford primary alcohol (**375**), which might be formed from **374** as shown in Figure 20.106 (Lahlou et al., 2000) (Figure 20.107).

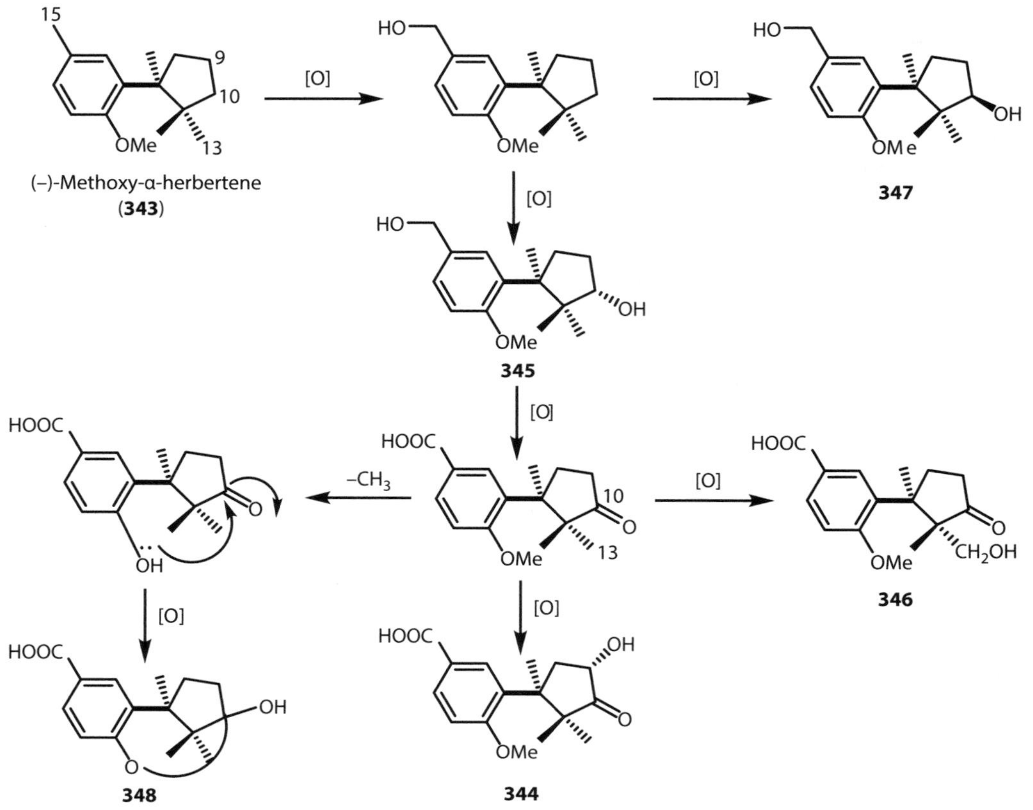

FIGURE 20.101 Possible pathway of biotransformation of (−)-methoxy-α-herbertene (**343**) by *Aspergillus niger*.

In order to obtain more pharmacologically active compounds, the secondary metabolites from crude drugs and animals, for example, nardosinone (**376**) isolated from the crude drug *N. chinensis*, which has been used for headache, stomachache, and diuresis, possesses antimalarial activity. Hinesol (**384**), possessing spasmolytic activity, obtained from *A. lancea* rhizoma, and animal perfume (−)-ambrox (**391**) from ambergris were biotransformed by *A. niger*, *A. cellulosae*, *B. dothidea*, and so on.

Nardosinone (**376**) was incubated in the same medium including *A. niger* as described earlier for 1 day to give six metabolites (**377**, 45%; **378**, 3%; **379**, 2%; **380**, 5%; **381**, 6%; and **382**, 3%). Compounds **380–382** are unique trinorsesquiterpenoids although their yields are very poor. Compound **380** might be formed by the similar manner to that of phenol from cumene (**383**) (Figures 20.108 and 20.109) (Hashimoto et al., 2003d).

From hinesol (**384**), two allylic alcohols (**386**, **387**) and their oxygenated derivative (**385**) and three unique metabolites (**388–390**) having oxirane ring were obtained. The metabolic pathway is very similar to that of oral administration of hinesol since the same metabolites (**395–387**) were obtained from the urine of rabbits (Hashimoto et al., 1998a, 1999b, 2001) (Figure 20.110).

To obtain a large amount of ambrox (**391**), a deterrence, labda-12,14-dien-7α,8-diol obtained from the liverwort *Porella pettottetiana* as a major component, was chemically converted into (−)-ambrox via six steps in relatively high yield (Hashimoto et al., 1998b). Ambrox was added to Czapek-peptone medium including *A. niger*, for 4 days, followed by chromatography of the crude extract to afford four oxygenated products (**392–395**), among which the carboxylic acid (**393**, 52.4%) is the major product (Hashimoto et al., 2001) (Figure 20.111).

When ambrox (**391**) was biotransformed by *A. niger* for 9 days in the presence of 1-aminobenzotriazole, an inhibitor of CYP450, compounds **396** and **397** were obtained instead of the

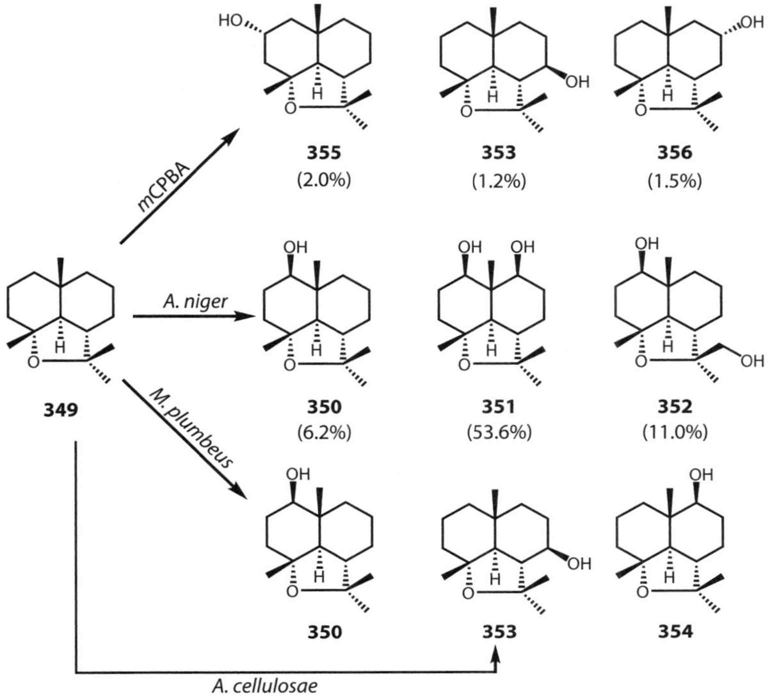

FIGURE 20.102 Biotransformation of maalioxide (**349**) by *Aspergillus niger*, *Aspergillus cellulosae*, and *Mucor plumbeus*.

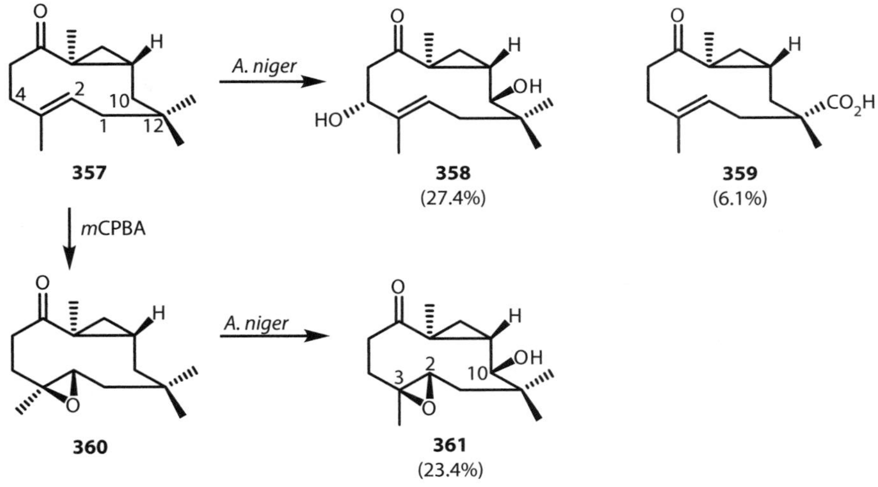

FIGURE 20.103 Biotransformation of bicyclohumulenone (**357**) by *Aspergillus niger*.

metabolites (**392–395**), which were obtained by incubation of ambrox in the absence of the inhibitor. Ambrox was cultivated by *A. cellulosae* for 4 days in the same medium to afford C1 oxygenated products (**398** and **399**), the former of which was the major product (41.3%) (Hashimoto et al., 2001) (Figure 20.112).

The metabolite pathways of ambrox are quite different between *A. niger* and *A. cellulosae*. Oxidation at C1 occurred in *A. cellulosae* to afford **398** and **399**, which was also afforded by John's

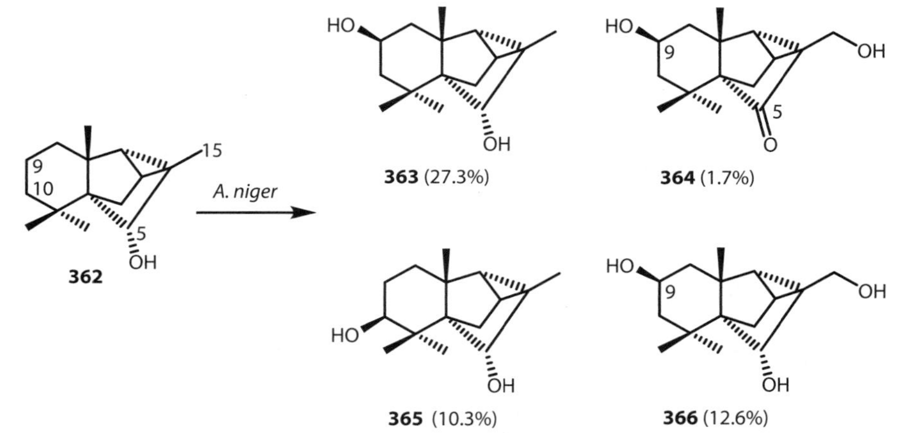

FIGURE 20.104 Biotransformation of cyclomyltaylan-5-ol (**362**) by *Aspergillus niger*.

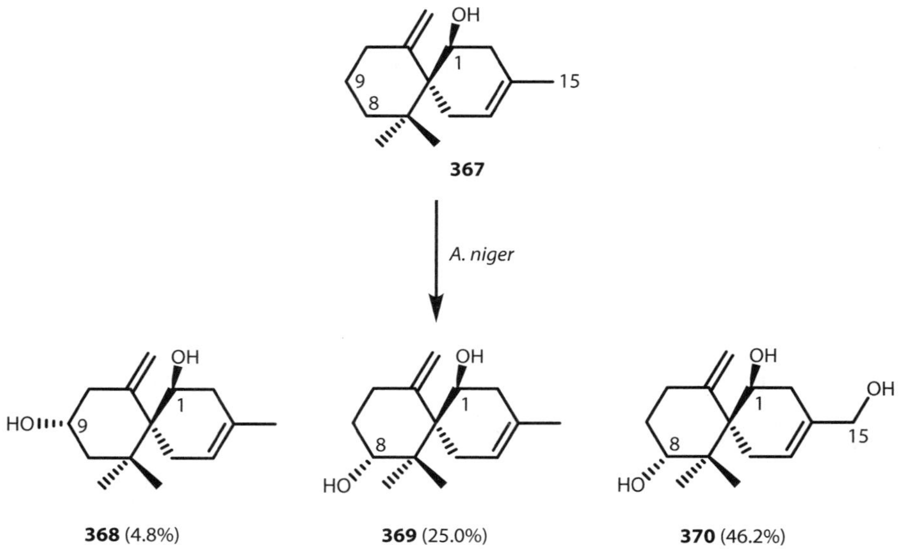

FIGURE 20.105 Biotransformation of *ent*-1α-hydroxy-β-chamigrene (**367**) by *Aspergillus niger*.

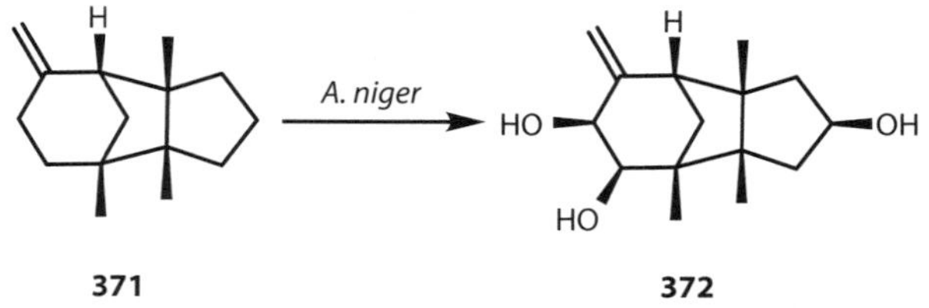

FIGURE 20.106 Biotransformation of β-barbatene (**371**) by *Aspergillus niger*.

oxidation of **398**, while oxidation at C3 and C18 and ether cleavage between C8 and C12 occurred in *A. niger* to give **392–395**. Ether cleavage seen in *A. niger* is very rare.

Fragrances of the metabolites (**392–395**) and 7α-hydroxy-(−)-ambrox (**400**) and 7-oxo-(−)-ambrox (**401**) obtained from labdane diterpene diol were estimated. Only **399** demonstrated a similar odor to ambrox (**391**) (Hashimoto et al., 2001) (Figure 20.113).

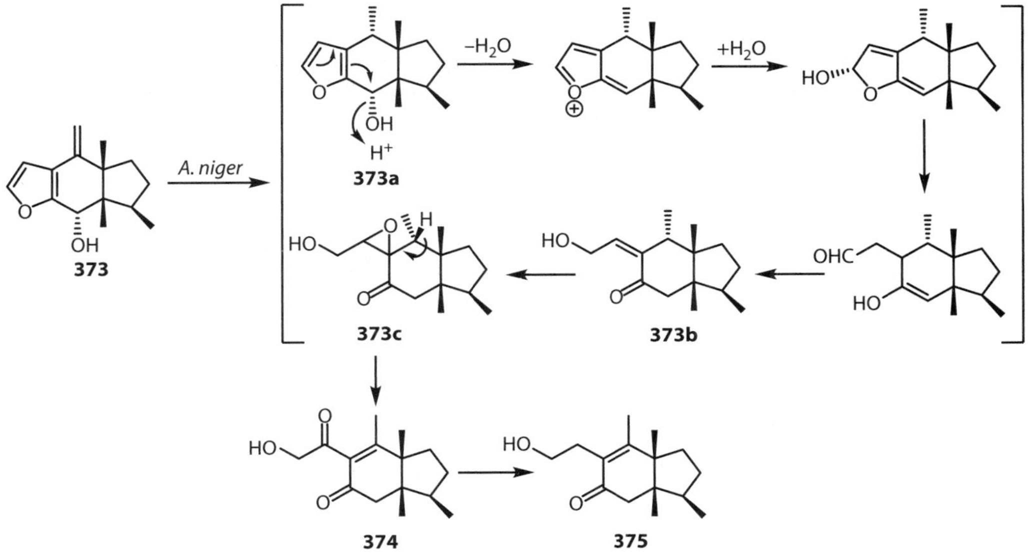

FIGURE 20.107 Biotransformation of pinguisanol (**373**) by *Aspergillus niger*.

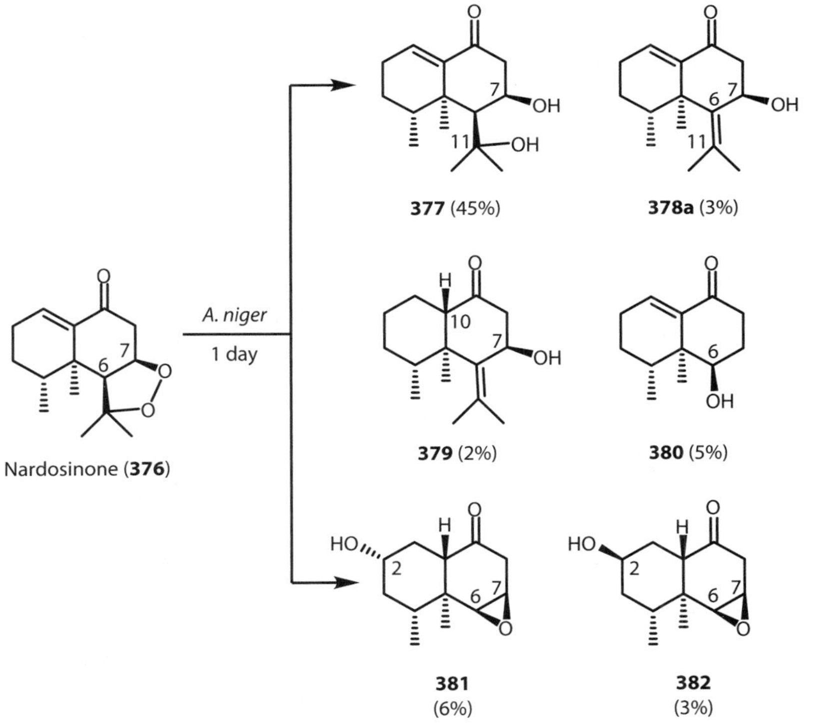

FIGURE 20.108 Biotransformation of nardosinone (**376**) by *Aspergillus niger*.

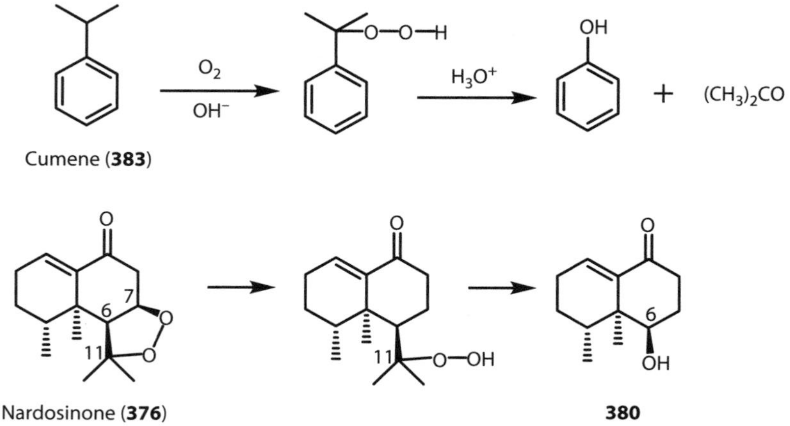

FIGURE 20.109 Possible pathway of biotransformation of nardosinone (**376**) to trinornardosinone (**380**) by *Aspergillus niger*.

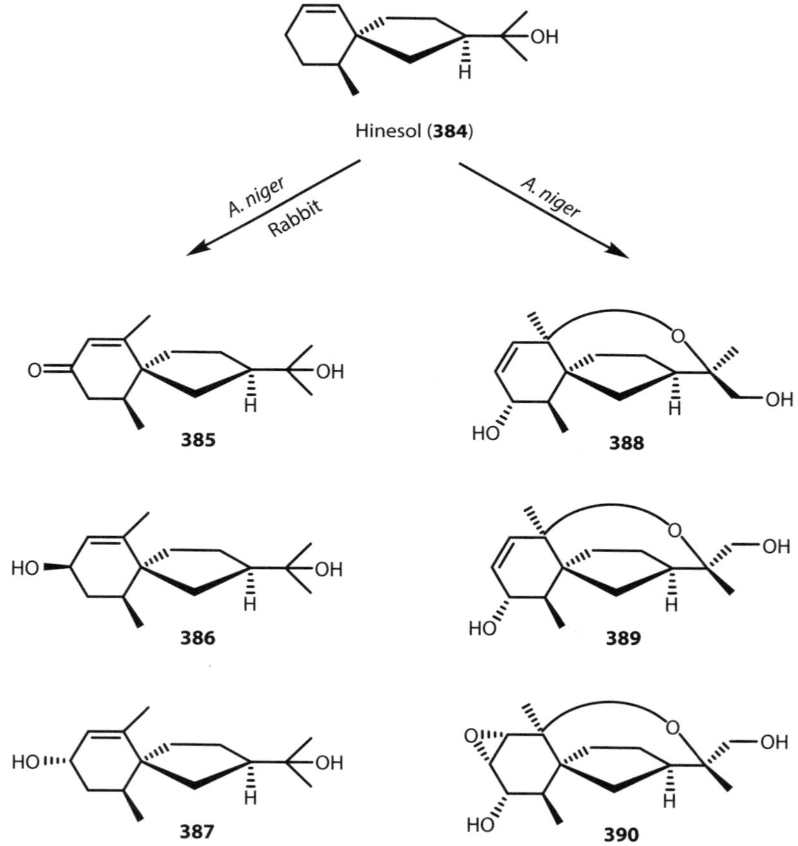

FIGURE 20.110 Biotransformation of hinesol (**384**) by *Aspergillus niger*.

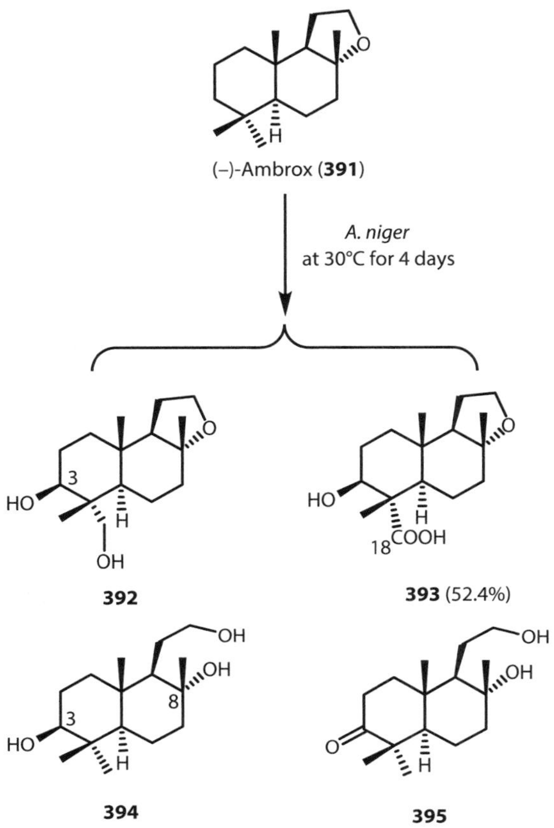

FIGURE 20.111 Biotransformation of (−)-ambrox (**391**) by *Aspergillus niger*.

(−)-Ambrox (**391**) was also microbiotransformed with *Fusarium lini* to give mono-, di-, and trihydroxylated metabolites (**401a–401d**), while incubation of the same substrate with *R. stolonifer* afforded two metabolites (**394, 396**), which were obtained from **391** by *A. niger* as mentioned earlier, together with **397** and **401e** (Choudhary et al., 2004) (Figure 20.114).

The sclareolide (**402**), which is C12 oxo derivative of ambrox, was incubated with *M. plumbeus* to afford three metabolites, 3β-hydroxy- (**403**, 7.9%), 1β-hydroxy- (**404**, 2.5%), and 3-ketosclareolide (**405**, 7.9%) (Aranda et al., 1991) (Figure 20.115).

A. niger in the same medium as mentioned earlier converted sclareolide (**402**) into two new metabolites (**406, 407**), together with known compounds (**403, 405**), of which 3β-hydroxy-sclareolide (**403**) is preferentially obtained (Hashimoto et al., 2007) (Figure 20.116).

From the metabolites of sclareolide (**402**) incubated with *C. lunata* and *A. niger*, five oxidized compounds (**403, 404, 405, 405a, 405b**) were obtained. Fermentation of **402** with *Gibberella fujikuroi* afforded **403, 404, 405**, and **405a**. Metabolites **403** and **405a** were formed from the same substrate by the incubation of *F. lini*. No microbial transformation of **402** was observed with *Pleurotus ostreatus* (Atta-ur-Rahman et al., 1997) (Figure 20.117).

Compound **391** treated in *C. lunata* gave metabolites **401e** and **396**, while *C. elegans* yielded compounds **401e** and **396** and (+)-sclareolide (**402**) (Figure 20.113). The metabolites (**401a–401e, 396**) from **391** do not release any effective aroma when compared to **391**. Compound **394** showed a strong sweet odor quite different from the amber-like odor (Choudhary et al., 2004).

Sclareolide (**402**) exhibited phytotoxic and cytotoxic activity against several human cancer cell lines. *C. elegans* gave new oxidized metabolites (**403, 404, 405a, 405c, 405d, 405e**), resulting from

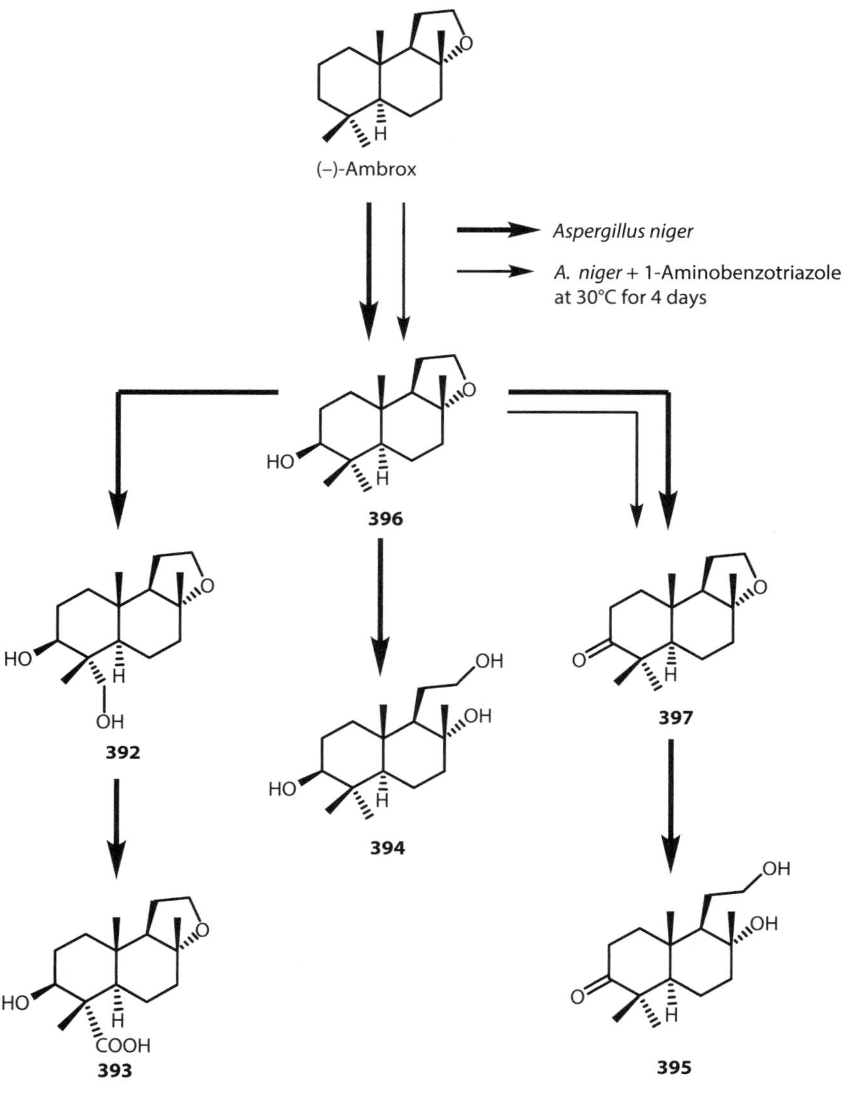

FIGURE 20.112 Possible pathway of biotransformation of (−)-ambrox (**391**) by *Aspergillus niger*.

the enantioselective hydroxylation. Metabolites **403**, **404**, and **405a** have been known as earlier as biotransformed products of **402** by many different fungi and have shown cytotoxicity against various human cancer cell lines. The metabolites (**403**, **404**, and **405a**) indicated significant phytotoxicity at higher dose against *Lemna minor* L. (Choudhary et al., 2004) (Figure 20.117).

Ambrox (**391**) and sclareolide (**402**) were incubated with the fungus *C. aphidicola* for 10 days in shake culture to give 3β-hydroxy- (**396**), 3β,6β-dihydroxy- (**401g**), 3β,12-dihydroxy- (**401h**), and sclareolide 3β,6β-diol (**401f**), and 3β-hydroxy- (**403**), 3-keto- (**405**), and sclareolide 3β,6β-diol (**401f**), respectively (Hanson and Truneh, 1996) (Figure 20.118).

Zerumbone (**408**), which is easily isolated from the wild ginger *Zingiber zerumbet*, and its epoxide (**409**) were incubated with *F. culmorum* and *A. niger* in Czapek-peptone medium, respectively. The former fungus gave (1R,2R)-(+)-2,3-dihydrozerumbol (**410**) stereospecifically via either 2,3-dihydrozerumbone (**408a**) or zerumbol (**408b**) or both and accumulated **410** in the mycelium. The facile production of optically active **410** will lead a useful material of woody fragrance,

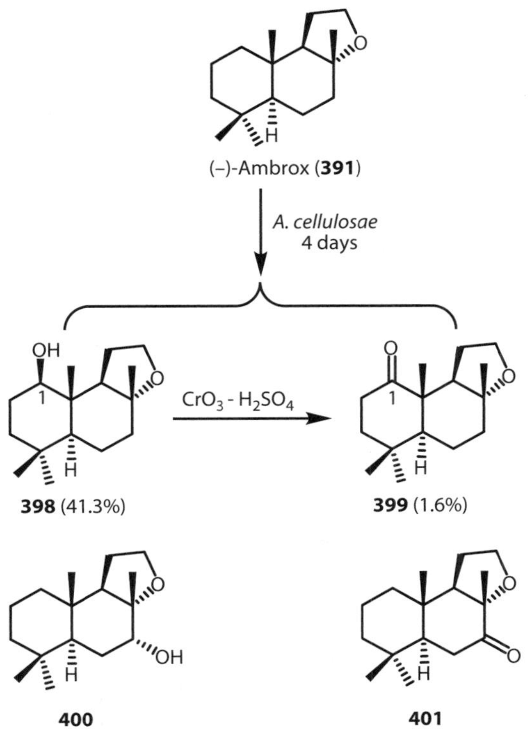

FIGURE 20.113 Biotransformation of (−)-ambrox (**391**) by *Aspergillus cellulosae*.

2,3-dihydrozerumbone. *A. niger* biotransformed **408** via epoxide (**409**) to several metabolites containing zerumbone-6,7-diol as a main product. The same fungus converted the epoxide (**409**) into three major metabolites containing (2*R*,6*S*,7*S*,10*R*,11*S*)-1-oxo7,9-dihydroxyisodaucane (**413**) via dihydro derivatives (**411**, **412**). However, *A. niger* biotransformed **409** only into **412** in the presence of the CYP450 inhibitor 1-aminobenzotriazole (Noma et al., 2002).

The same substrate was incubated in *A. niger*, *Aspergillus oryzae*, *Candida rugosa*, *Candida tropicalis*, *Mucor mucedo*, *Bacillus subtilis*, and *Schizosaccharomyces pombe*; however, any metabolites have been obtained. All microbes except for the last organism, zerumbone epoxide (**409**), prepared by *m*CPBA, bioconverted into two diastereoisomers, 2*R*,6*S*,7*S*-dihydro- (**411**) and 2*R*,6*R*,7*R*-derivative (**412**), whose ratio was determined by GC, and their enantio-excess was over 99% (Nishida and Kawai, 2007) (Figure 20.119).

Several microorganisms and a few mammals (see later) for the biotransformation of (+)-cedrol (**414**), which is widely distributed in the cedar essential oils, were used. Plant pathogenic fungus *G. cingulata* converted cedrol (**414**) into three diols (**415–417**) and 2α-hydroxycedrene (**418**) (Miyazawa et al., 1995a). The same substrate (**414**) was incubated with *A. niger* to give **416** and **417** together with a cyclopentanone derivative (**419**) (Higuchi et al., 2001). Human skin microbial flora, *Staphylococcus epidermidis* also converted (+)-cedrol into 2α-hydroxycedrol (**415**) (Itsuzaki et al., 2002) (Figure 20.120).

C. aphidicola bioconverted cedrol (**414**) into **417** (Hanson and Nasir, 1993). On the other hand, *Corynespora cassiicola* produced **419** in addition to **417** (Abraham et al., 1987). It is noteworthy that *B. cinerea* that damages many flowers, fruits, and vegetables biotransformed cedrol into different metabolites (**420–422**) from those mentioned earlier (Aleu et al., 1999a).

4α-Hydroxycedrol (**424**) was obtained from the metabolite of cedrol acetate (**423**) by using *G. cingulata* (Matsui et al., 1999) (Figure 20.121).

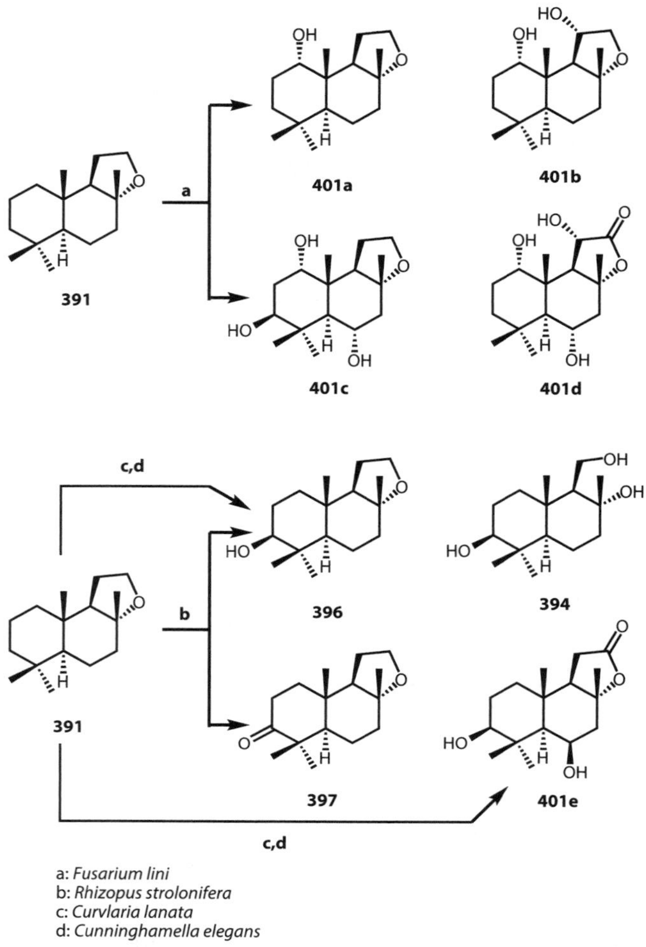

FIGURE 20.114 Biotransformation of (−)-ambrox (**391**) by *Fusarium lini* and *Rhizopus stolonifer*.

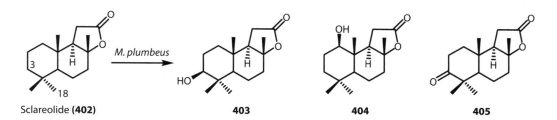

FIGURE 20.115 Biotransformation of (+)-sclareolide (**402**) by *Mucor plumbeus*.

Patchouli alcohol (**425**) was treated in *B. cinerea* to give three metabolites two tertiary alcohols (**426, 427**), four secondary alcohols (**428, 430, 430a**), and two primary alcohols (**430b, 430c**) of which compounds **425, 427**, and **428** are the major metabolites (Aleu et al., 1999a), while plant pathogenic fungus *G. cingulata* converted the same substrate to 5-hydroxy- (**426**) and 5,8-dihydroxy derivative (**429**) (Figure 20.122).

In order to confirm the formation of **429** from **426**, the latter product was reincubated in the same medium including *G. cingulata* to afford **429** (Miyazawa et al., 1997c) (Figure 20.123).

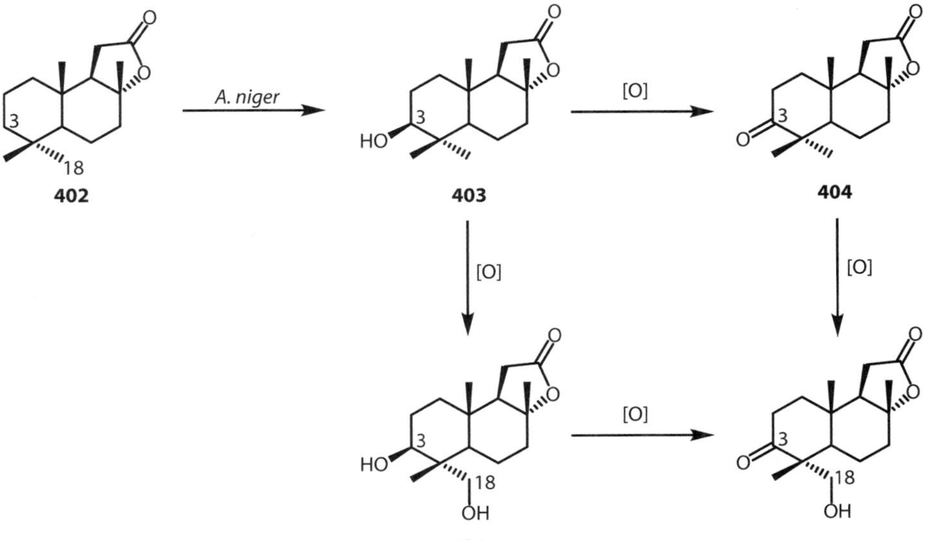

FIGURE 20.116 Biotransformation of (+)-sclareolide (**402**) by *Aspergillus niger*.

Patchouli acetate (**431**) was also treated in the same medium to give **426** and **429** (Matsui and Miyazawa, 2000). 5-Hydroxy-α-patchoulene (**432**) was incubated with *G. cingulata* to afford 1α-hydroxy derivative (**426**) (Miyazawa et al., 1998b).

(−)-α-Longipinene (**433**) was treated with *A. niger* to afford 12-hydroxylated product (**434**) (Sakata et al., 2007).

Ginsenol (**435**), which was obtained from the essential oil of *Panax ginseng*, was incubated with *B. cinerea* to afford four secondary alcohols (**436–439**) and two cyclohexanone derivatives (**440**) from **437** and **441** from **438** or **439**. Some of the oxygenated products were considered as potential antifungal agents to control *B. cinerea* (Aleu et al., 1999b) (Figures 20.124 and 20.125).

(+)-Isolongifolene-9-one (**442**), which was isolated from some cedar trees, was treated in *G. cingulata* for 15 days to afford two primary alcohols (**443**, **444**) and a secondary alcohol (**445**) (Sakata and Miyazawa, 2006) (Figure 20.126).

Choudhary et al. (2005) reported that fermentation of (−)-isolongifolol (**445a**) with *F. lini* resulted in the isolation of three metabolites, 10-oxo- (**445b**), 10α-hydroxy- (**445c**), and 9α-hydroxyisolongifolol (**445d**). Then the same substrate was incubated with *A. niger* to yield the products **445c** and **445d**. Both **445c** and **445d** showed inhibitory activity against butylcholinesterase enzyme in a concentration-dependent manner with IC_{50} 13.6 and 299.5 mM, respectively (Figure 20.127).

(+)-Cycloisolongifol-5β-ol (**445e**) was fermented with *C. elegans* to afford three oxygenated metabolites: 11-oxo (**445f**), 3β-hydroxy (**445g**), and 3β,11α-dihydroxy derivative (**445h**) (Choudhary et al., 2006a) (Figure 20.128).

A daucane-type sesquiterpene derivative, lancerroldiol *p*-hydroxybenzoate (**446**), was hydrogenated with cultured suspension cells of the liverwort *M. polymorpha* to give 3,4-dihydrolancerodiol (**447**) (Hegazy et al., 2005) (Figure 20.129).

Widdrane sesquiterpene alcohol (**448**) was incubated with *A. niger* to give an oxo and an oxy derivative (**449**, **450**) (Hayashi et al., 1999) (Figure 20.130).

(−)-β-Caryophyllene (**451**), one of the ubiquitous sesquiterpene hydrocarbons found not only in higher plants but also in liverworts, was biotransformed by *Pseudomonas cruciviae*, *D. gossypina*, and *Chaetomium cochliodes* (Lamare and Furstoss, 1990). *P. cruciviae* gave a ketoalcohol (**452**) (Devi, 1979), while the latter two species produced the 14-hydroxy-5,6-epoxide (**454**), its carboxylic (**455**), and 3α-hydroxy- (**456**) and norcaryophyllene alcohol (**457**), all of which might be formed

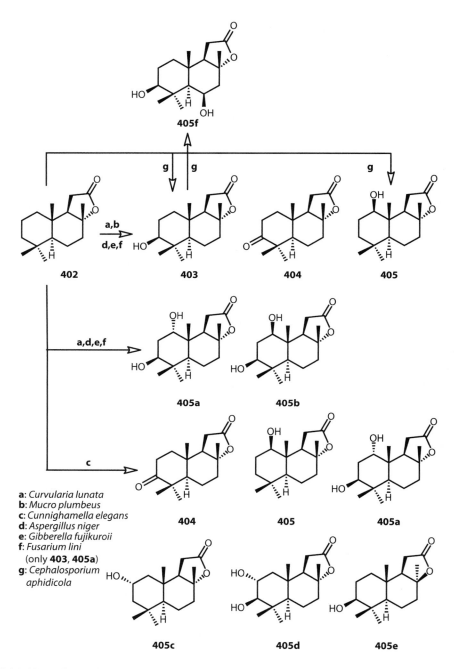

FIGURE 20.117 Biotransformation of (+)-sclareolide (**402**) by various fungi.

from caryophyllene C5,C6-epoxide (**453**). The oxidation pattern of (−)-β-caryophyllene by the fungi is very similar to that by mammals (see later) (Figure 20.131).

Fermentation of (−)-β-caryophyllene (**451**) with *D. gossypina* afforded **14** different metabolites (**453–457j**), among which 14-hydroxy-5,6-epoxide (**454**) and the corresponding acid (**455**) were the major metabolites. Compound **457j** is structurally very rare and found in *Poronia punctata*. The main reaction path is epoxidation at C5, C6 as mentioned earlier and selective hydroxylation at C4 (Abraham et al., 1990) (Figure 20.132).

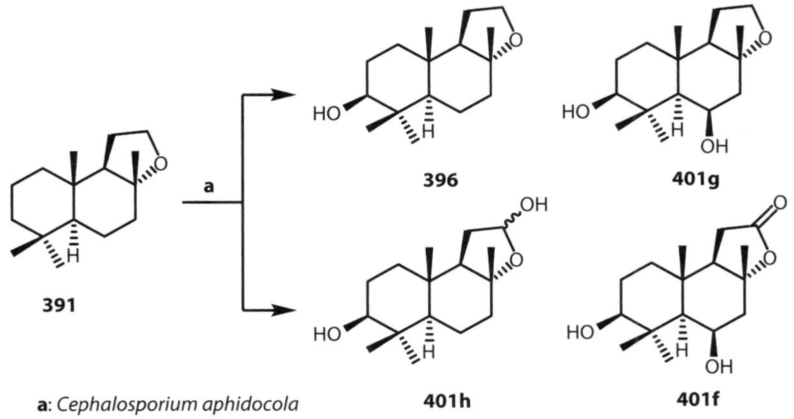

FIGURE 20.118 Biotransformation of (−)-ambrox (**391**) by *Cephalosporium aphidicola*.

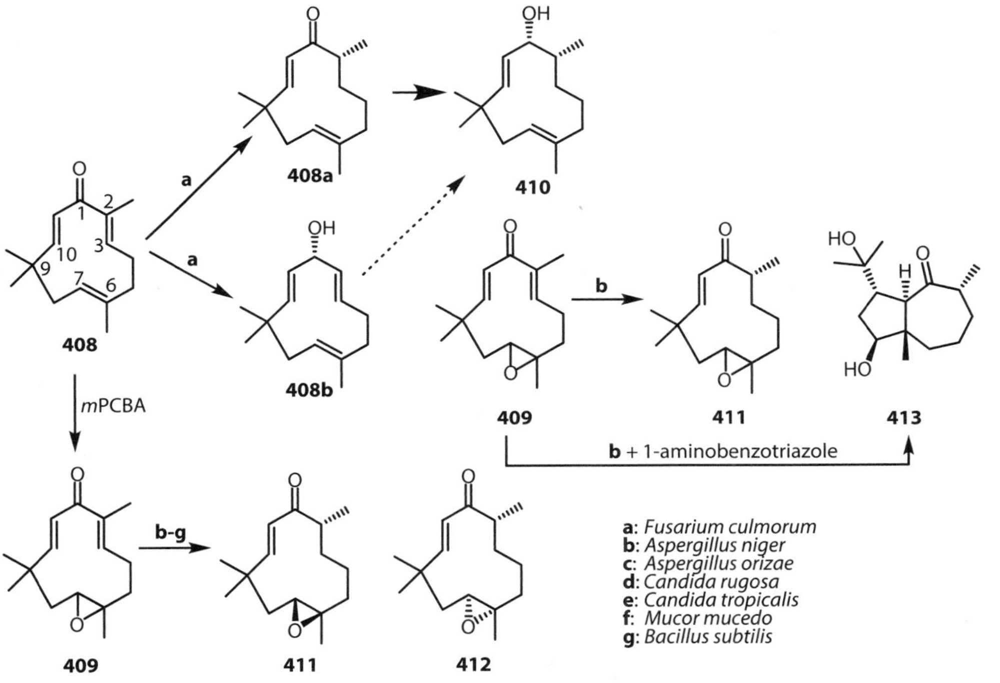

FIGURE 20.119 Biotransformation of zerumbone (**408**) by various fungi.

(−)-β-Caryophyllene epoxide (**453**) was incubated with *C. aphidicola* for 6 days to afford two metabolites (**457l, 457m**), while *Macrophomina phaseolina* biotransformed the same substrate to 14- (**454**) and 15-hydroxy derivatives (**457k**). The same substrate was treated in *A. niger*, *G. fujikuroi*, and *R. stolonifer* for 8 days and *F. lini* for 10 days to afford the metabolites **457n**, **457o**, **457p** and **457q**, and **457r**, respectively. All metabolites were estimated for butyrylcholine esterase inhibitory activity, and compound **457k** was found to show potency similar activity to galantamine HBr (IC_{50} 10.9 vs. 8.5 mM) (Choudhary et al., 2006b) (Figure 20.133).

The fermentation of (−)-β-caryophyllene oxide (**453**) using *B. cinerea* and the isolation of the metabolites were carried out by Duran et al. (1999). Kobuson (**457w**) was obtained with 14 products (**457s–457u, 457x**). Diepoxides **457t** and **457u** could be the precursors of epimeric alcohols **457q**

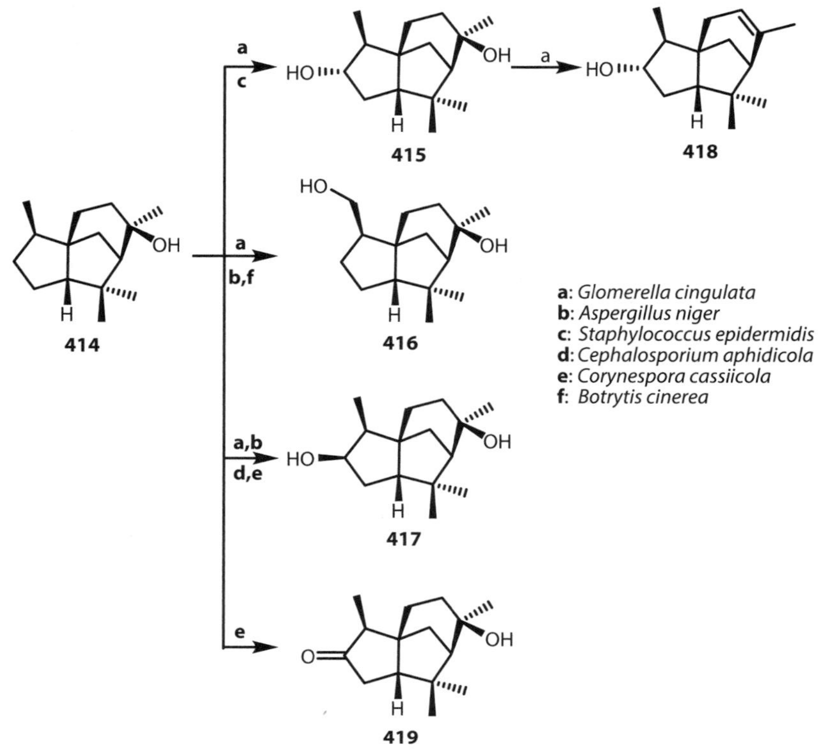

FIGURE 20.120 Biotransformation of cedrol (**414**) by various fungi.

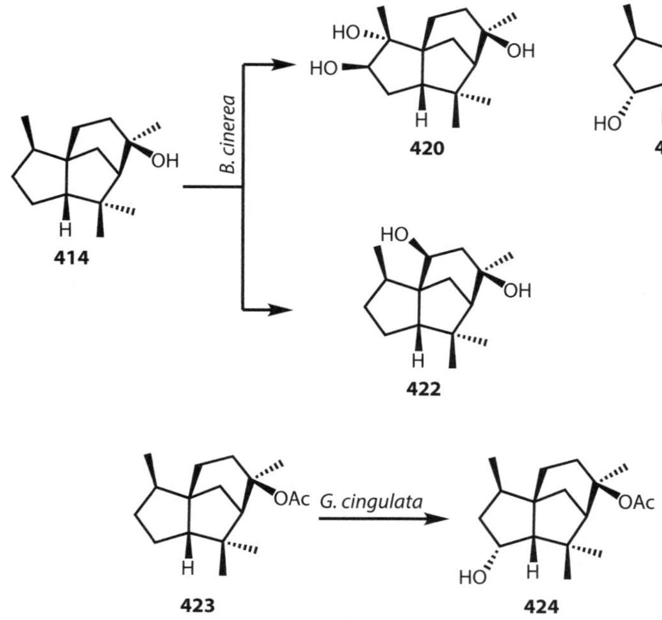

a: *Glomerella cingulata*
b: *Aspergillus niger*
c: *Staphylococcus epidermidis*
d: *Cephalosporium aphidicola*
e: *Corynespora cassiicola*
f: *Botrytis cinerea*

FIGURE 20.121 Biotransformation of cedrol (**414**) by *Botrytis cinerea* and *Glomerella cingulata*.

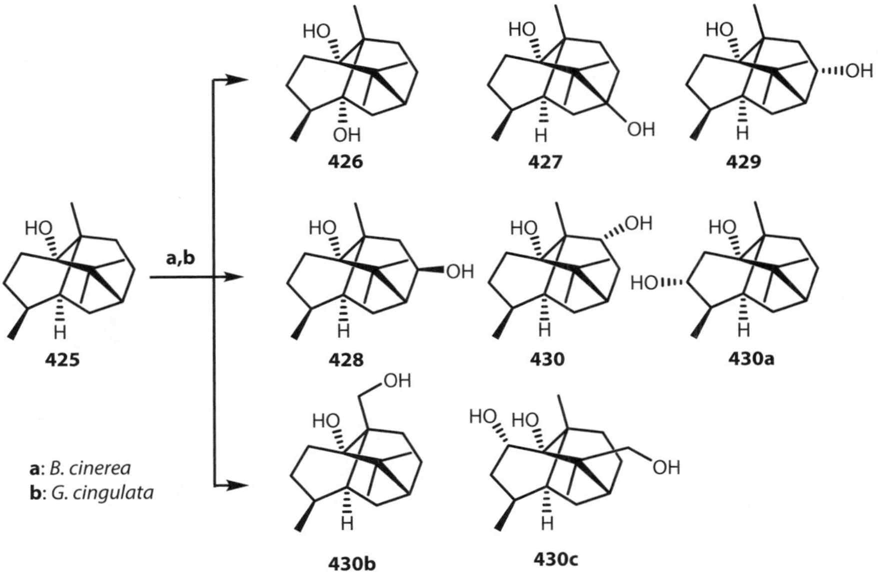

FIGURE 20.122 Biotransformation of patchoulol (**425**) by *Botrytis cinerea*.

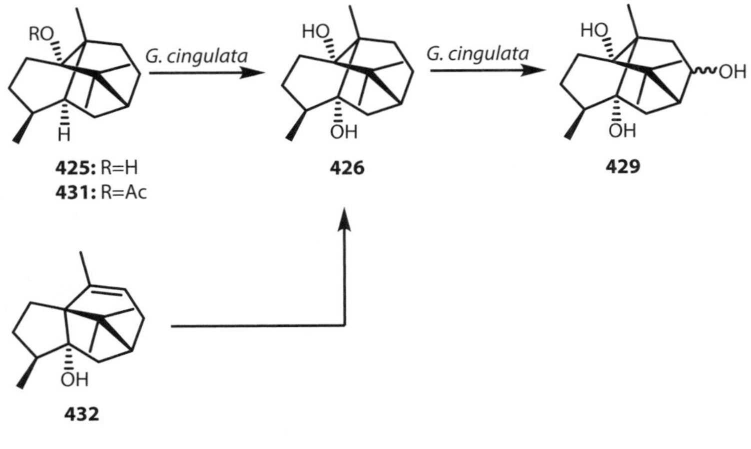

a: *B. cinerea*
b: *G. cingulata*

FIGURE 20.123 Biotransformation of patchoulol (**425**) by *Glomerella cingulata*.

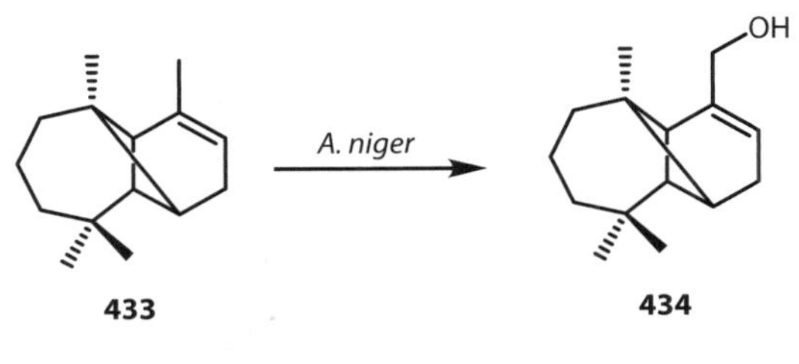

425: R=H
431: R=Ac

FIGURE 20.124 Biotransformation of α-longipinene (**433**) by *Aspergillus niger*.

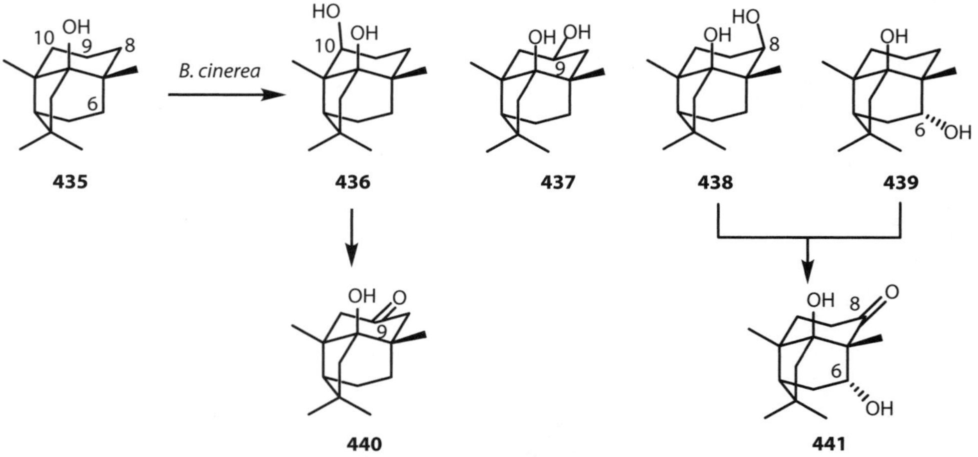

FIGURE 20.125 Biotransformation of ginsenol (**435**) by *Botrytis cinerea*.

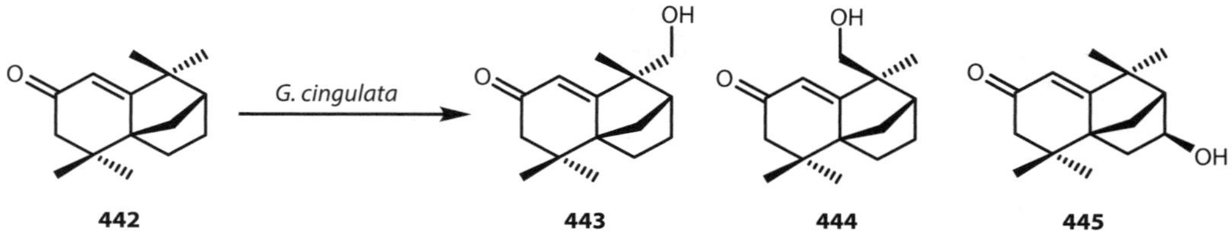

FIGURE 20.126 Biotransformation of (+)-isolongifolene-9-one (**442**) by *Glomerella cingulata*.

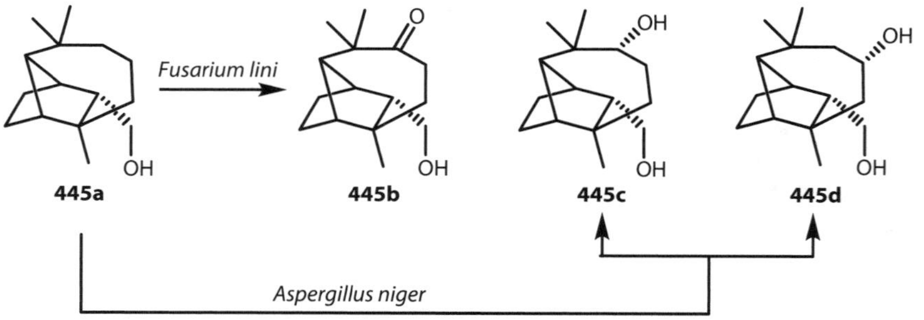

FIGURE 20.127 Biotransformation of (−)-isolongifolol (**445a**) by *Aspergillus niger* and *Fusarium lini*.

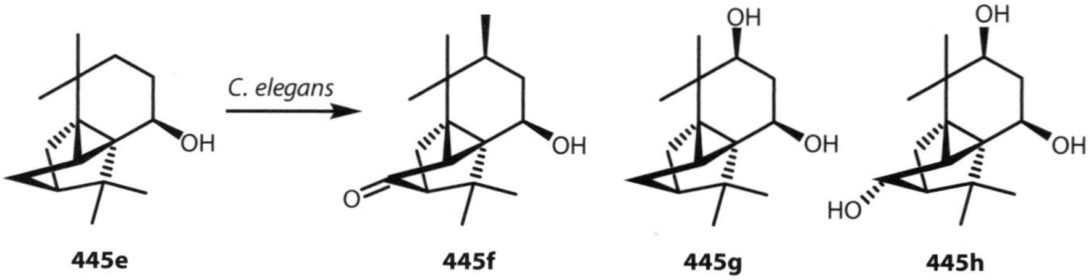

FIGURE 20.128 Biotransformation of (+)-cycloisolongifol-5β-ol (**445e**) by *Cunninghamella elegans*.

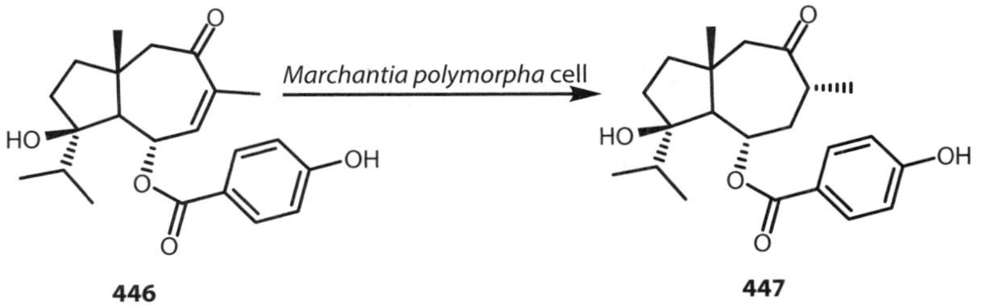

FIGURE 20.129 Biotransformation of lancerodiol *p*-hydroxybenzoate (**446**) by *Marchantia polymorpha* cells.

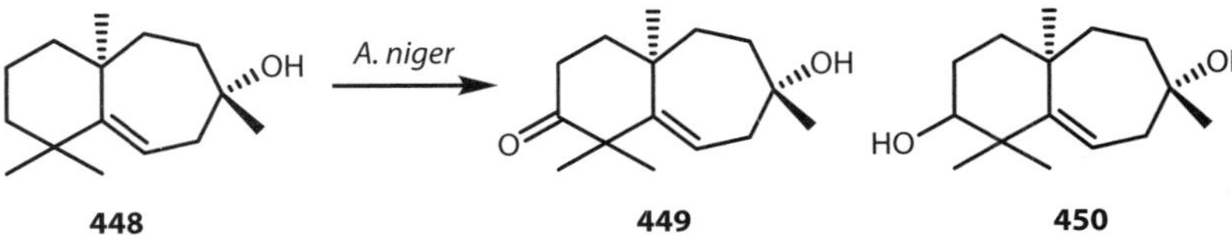

FIGURE 20.130 Biotransformation of widdrol (**448**) by *Aspergillus niger*.

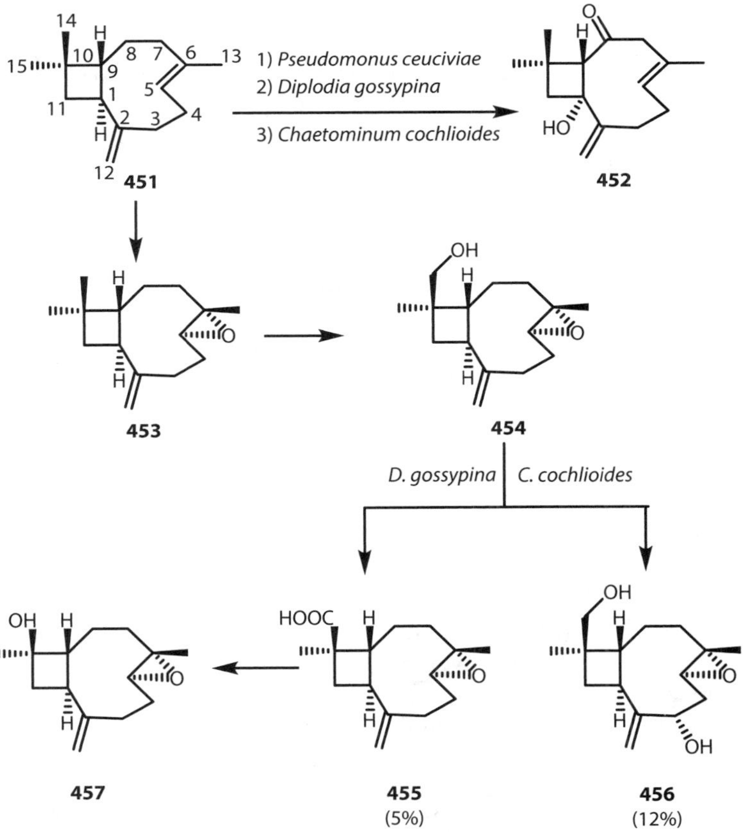

FIGURE 20.131 Biotransformation of (−)-β-caryophyllene (**451**) by *Pseudomonas cruciviae*, *Diplodia gossypina*, and *Chaetomium cochliodes*.

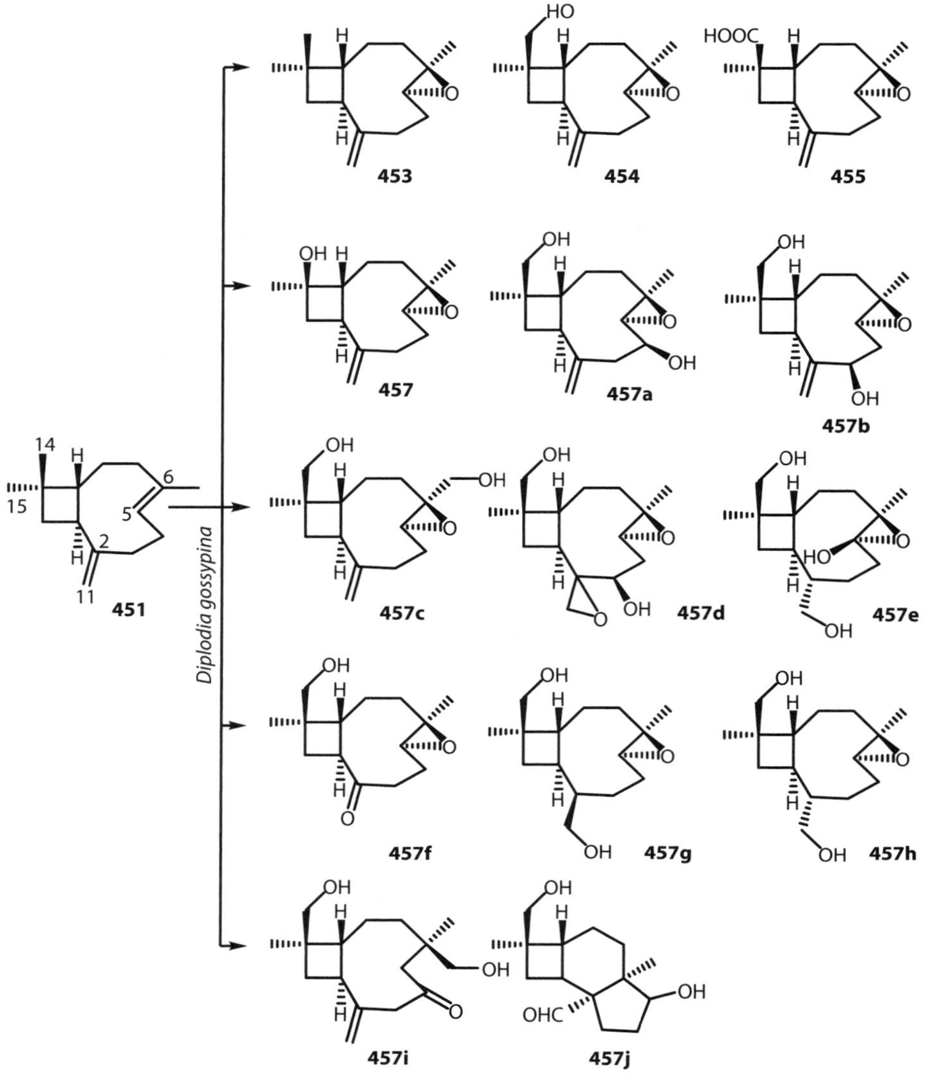

FIGURE 20.132 Biotransformation of (−)-β-caryophyllene (**451**) by *Diplodia gossypina*.

and **457y** obtained through reductive opening of the C2,C11-epoxide. The major reaction paths are stereoselective epoxidation and introduction of hydroxyl group at C3. Compound **457ae** has a caryolane skeleton (Figure 20.134).

When isoprobotryan-9α-ol (**458**) produced from isocaryophyllene was incubated with *B. cinerea*, it was hydroxylated at tertiary methyl groups to give three primary alcohols (**459–461**) (Aleu et al., 2002) (Figure 20.135).

Acyclic sesquiterpenoids, racemic *cis*-nerolidol (**462**), and nerylacetone (**463**) were treated by the plant pathogenic fungus *G. cingulata* (Miyazawa et al., 1995d). From the former substrate, a triol (**464**) was obtained as the major product. The latter was bioconverted to give the two methyl ketones (**465, 467**) and a triol (**468**), among which **465** was the predominant. The C10,C11 diols (**464, 465**) might be formed from both epoxides of the substrates, followed by the hydration although no C10,C11-epoxides were detected (Figure 20.136).

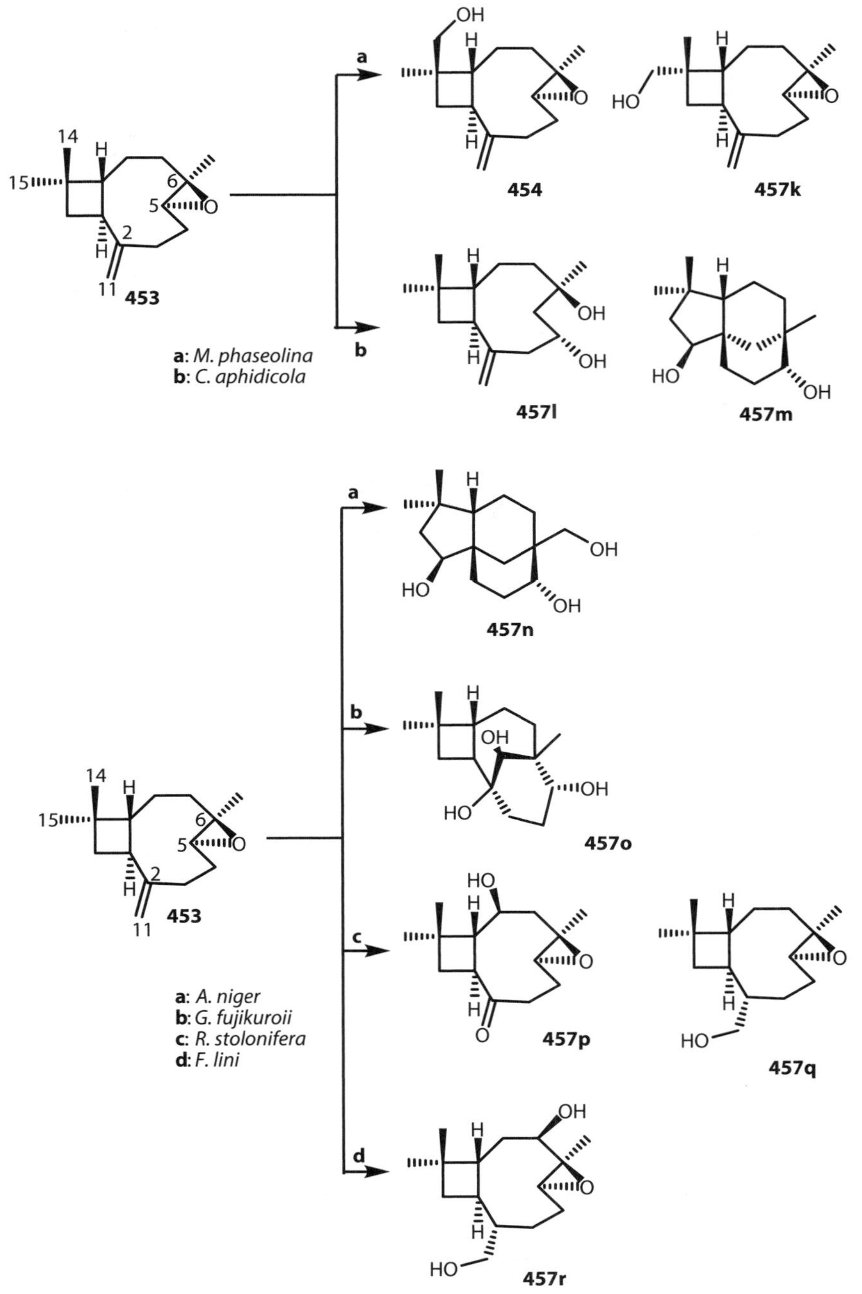

FIGURE 20.133 Biotransformation of (−)-β-caryophyllene epoxide (**453**) by various fungi.

Racemic *trans*-nerolidol (**469**) was also treated in the same fungus to afford w2-hydroxylated product (**471**) and C10,C11-hydroxylated compounds (**472**) as seen in racemic *cis*-nerolidol (**462**) (Miyazawa et al., 1996b) (Figure 20.137).

12-Hydroxy-*trans*-nerolidol (**472a**) is an important precursor in the synthesis of interesting flavor of α-sinensal. Hrdlicka et al. (2004) reported the biotransformation of *trans*- (**469**) and *cis*-nerolidol (**462**) and *cis*-/*trans*-mixture of nerolidol using repeated batch culture of *A. niger* grown in

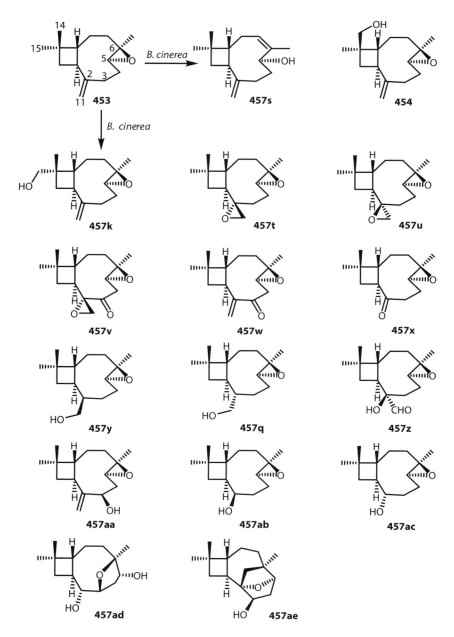

FIGURE 20.134 Biotransformation of (−)-β-caryophyllene epoxide (**453**) by *Botrytis cinerea*.

computer-controlled bioreactors. *Trans*-nerolidol (**469**) gave **472a** and **472** and *cis*-isomer (**462**) afforded **464a** and **464**. From a mixture of *cis*- and *trans*-nerolidol, 12-hydroxy-*trans*-neroridol **472a** (8%) was obtained in postexponential phase at high dissolved oxygen. At low dissolved oxygen condition, the mixture gave **472a** (7%) and **464a** (6%) (Figures 20.138 and 20.139).

From geranyl acetone (**470**) incubated with *G. cingulata*, four products (**473–477**) were formed. It is noteworthy that the major compounds from both substrates (**469, 470**) were w2-hydroxylated products, but not C10,C11-dihydroxylated products as seen in *cis*-nerolidol (**462**) and nerylacetone (**463**) (Miyazawa et al., 1995b) (Figure 20.136).

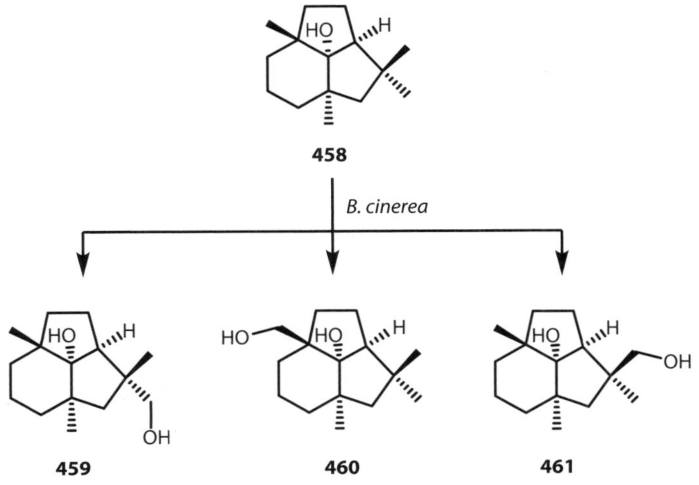

FIGURE 20.135 Biotransformation of isoprobotryan-9α-ol (**458**) by *Botrytis cinerea*.

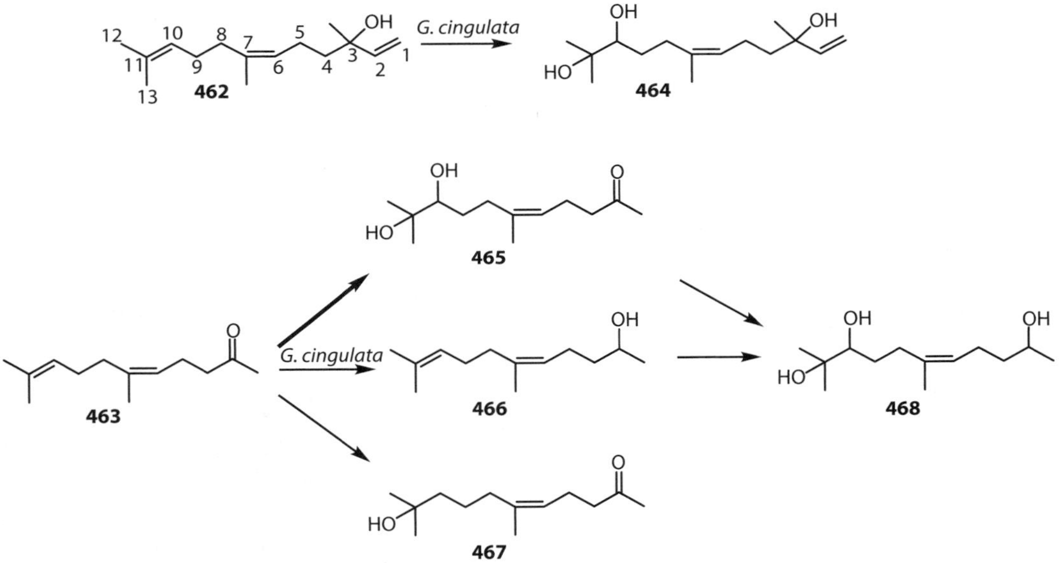

FIGURE 20.136 Biotransformation of *cis*-nerolidol (**462**) and *cis*-geranyl acetone (**463**) by *Glomerella cingulata*.

The same fungus bioconverted (2*E*,6*E*)-farnesol (**478**) to four products, w2-hydroxylated product (**479**), which was further oxidized to give C10,C11 dihydroxylated compound (**480**) and 5-hydroxy derivative (**481**), followed by isomerization at C2,C3 double bond to afford a triol (**482**) (Miyazawa et al., 1996c) (Figure 20.140).

The same substrate was bioconverted by *A. niger* to afford two metabolites, 10,11-dihydroxy (**480**) and 5,13-hydroxy derivative (**480a**) (Madyastha and Gururaja, 1993).

The same fungus also converted (2*Z*,6*Z*)-farnesol (**483**) to three hydroxylated products: 10,11-dihydroxy-(2*Z*,6*Z*)- (**484**), 10,11-dihydroxy-(2*E*,6*Z*)-farnesol (**485**), and (5*Z*)-9,10-dihydroxy-6,10-dimethyl-5-undecen-2-one (**486**) (Nankai et al., 1996) (Figures 20.140 and 20.141).

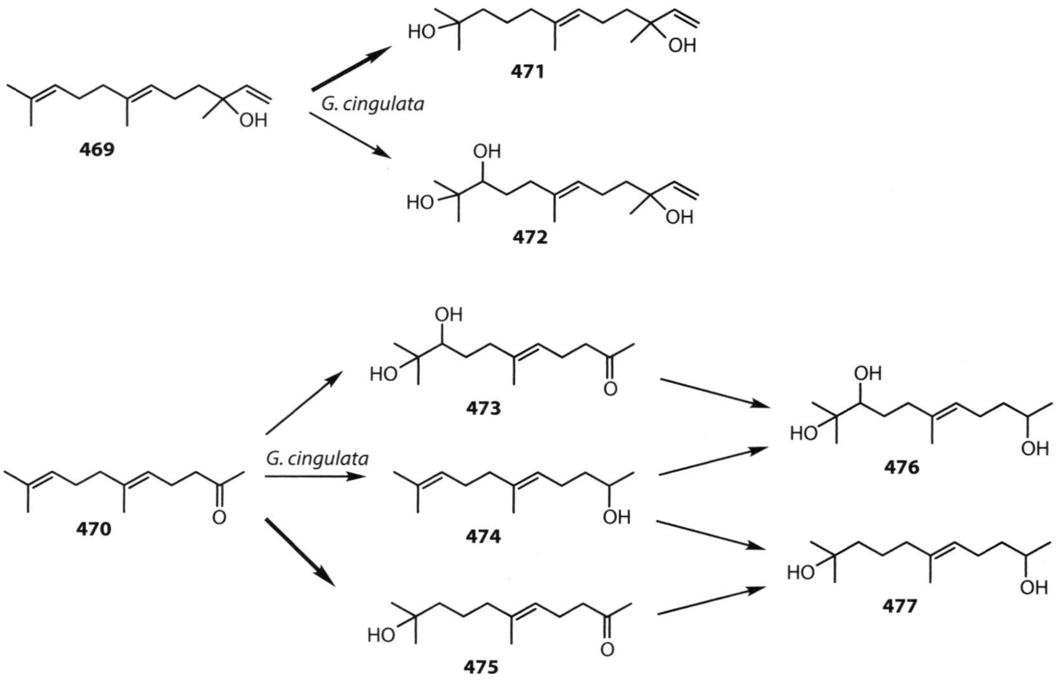

FIGURE 20.137 Biotransformation of *trans*-nerolidol (**469**) and *trans*-geranyl acetone (**470**) by *Glomerella cingulata*.

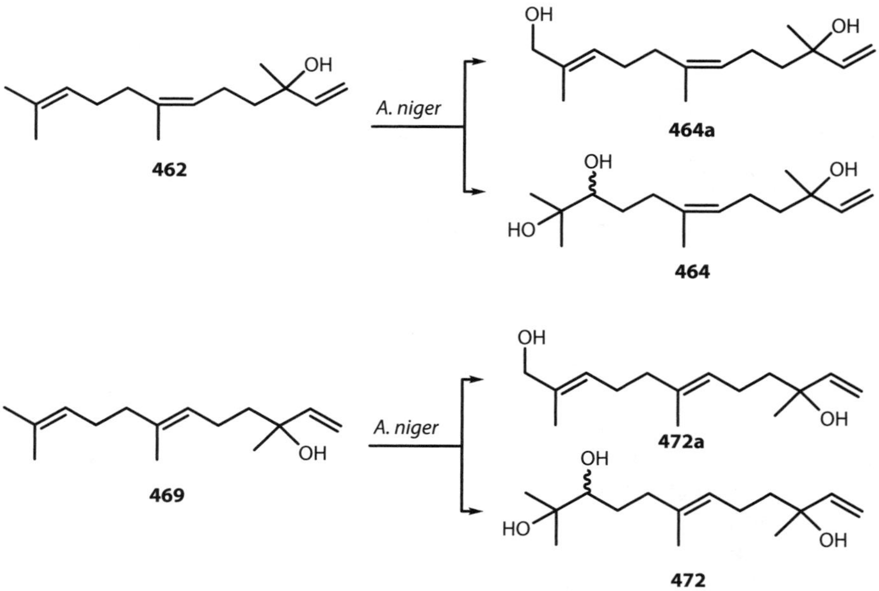

FIGURE 20.138 Biotransformation of *cis*- (**462**) and *trans*-nerolidol (**469**) by *Aspergillus niger*.

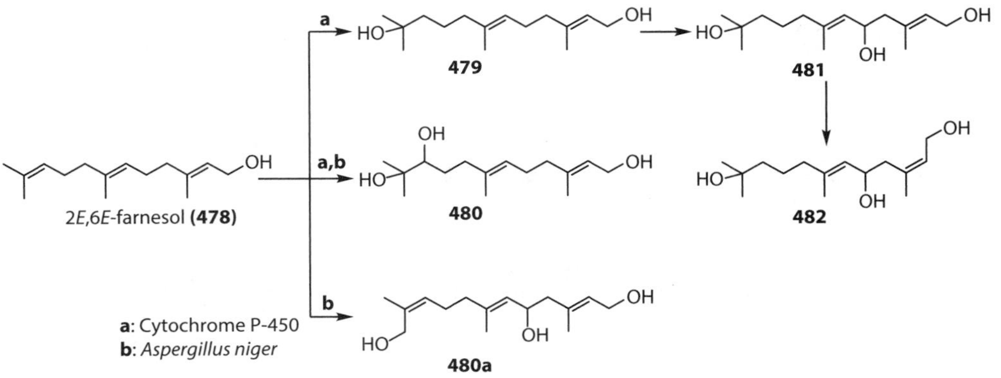

FIGURE 20.139 Biotransformation of 2E,6E-farnesol (**478**) by cytochrome P-450 and *Aspergillus niger*.

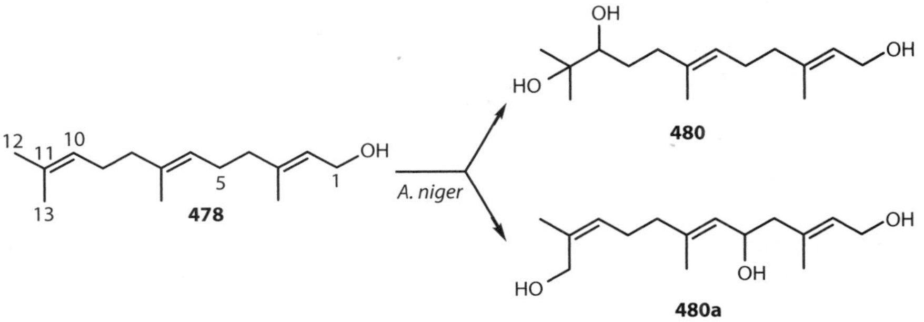

FIGURE 20.140 Biotransformation of 2E,6E-farnesol (**478**) by *Aspergillus niger*.

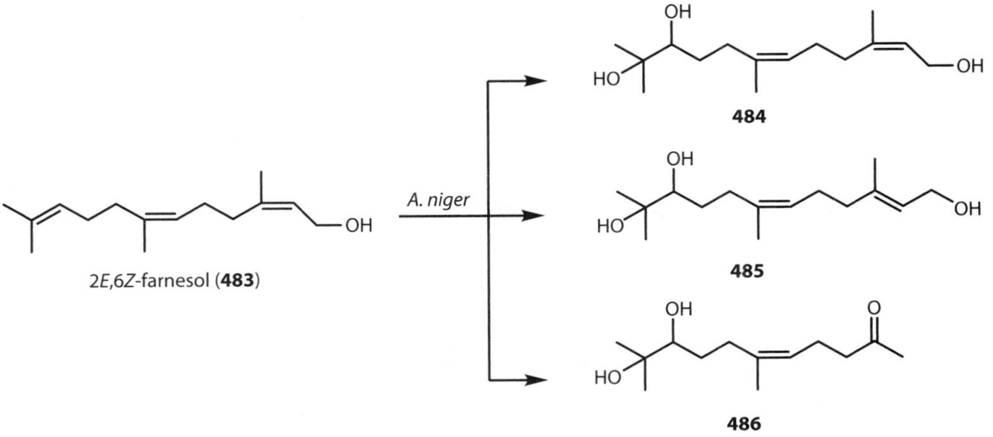

FIGURE 20.141 Biotransformation of 2Z,6Z-farnesol (**478**) by *Aspergillus niger*.

A linear sesquiterpene 9-oxonerolidol (**487**) was treated in *A. niger* to give w1-hydroxylated product (**488**) (Higuchi et al., 2001) (Figure 20.142).

Racemic diisophorone (**488a**) dissolved in ethanol was incubated with the Czapek–Dox medium of *A. niger* to afford 8α- (**488b**), 10β- (**488c**), and 17-hydroxydiisophorone (**488d**) (Kiran et al., 2004).

On the other hand, the same substrate was fed with *Neurospora crassa* and *C. aphidicola* to afford only 8β-hydroxydiisophorone (**488e**) in 20% and 10% yield, respectively (Kiran et al., 2005) (Figure 20.143).

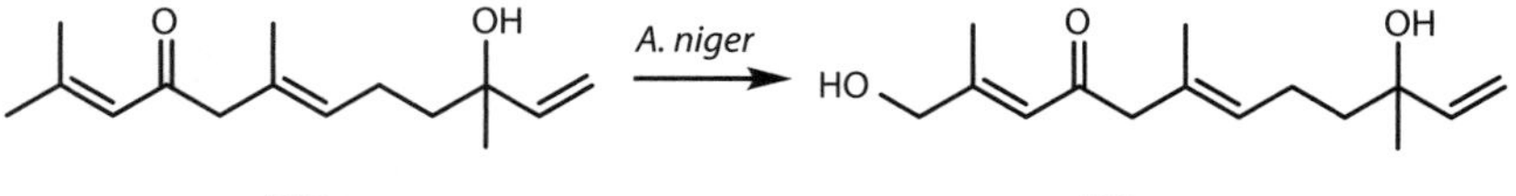

FIGURE 20.142 Biotransformation of 9-oxo-*trans*-nerolidol (**487**) by *Aspergillus niger*.

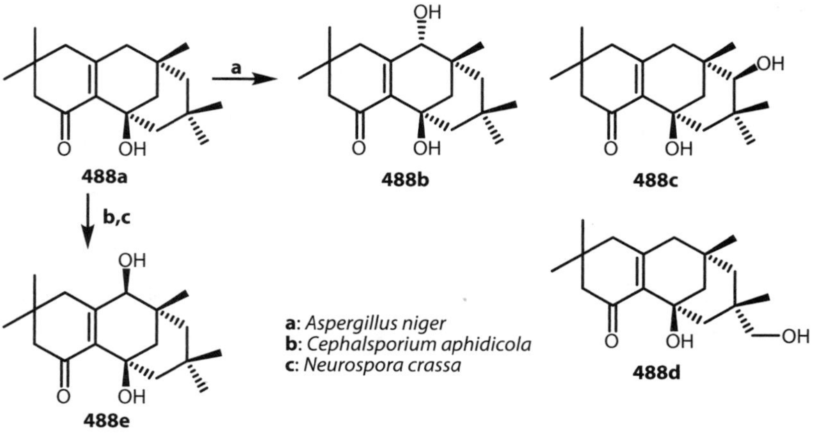

FIGURE 20.143 Biotransformation of diisophorone (**488a**) by *Aspergillus niger*, *Cephalosporium aphidicola*, and *Neurospora crassa*.

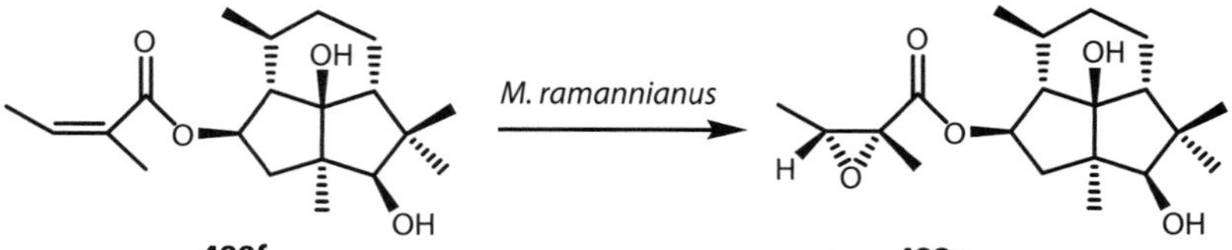

a: *Aspergillus niger*
b: *Cephalsporium aphidicola*
c: *Neurospora crassa*

FIGURE 20.144 Biotransformation of 5β,6β-dihydroxypresilphiperfolane 2β-angelate (**488f**) by *Mucor ramannianus*.

From the metabolites of 5β,6β-dihydroxypresilphiperfolane 2β-angelate (**488f**) using the fungus *Mucor ramannianus*, 2,3-epoxyangeloyloxy derivative (**488g**) was obtained (Orabi, 2001) (Figure 20.144).

20.3 BIOTRANSFORMATION OF SESQUITERPENOIDS BY MAMMALS, INSECTS, AND CYTOCHROME P-450

20.3.1 ANIMALS (RABBITS) AND DOSING

Six male albino rabbits (2–3 kg) were starved for 2 days before experiment. Monoterpenes were suspended in water (100 mL) containing polysorbate 80 (0.1 g) and were homogenized well. This solution (20 mL) was administered to each rabbit through a stomach tube followed by water (20 mL). This dose of sesquiterpenoids corresponds to 400–700 mg/kg. Rabbits were housed in stainless steel metabolism cages and were allowed rabbit food and water ad libitum. The urine was collected daily for 3 days after drug administration and stored at 0°C–5°C until the time of analysis. The urine was centrifuged to remove feces and hairs at 0°C and the supernatant was used for the experiments. The urine was adjusted to pH 4.6 with acetate buffer and incubated with β-glucuronidase–arylsulfatase

(3/100 mL of fresh urine) at 37°C for 48 h, followed by continuous ether extraction for 48 h. The ether extracts were washed with 5% NaHCO$_3$ and 5% NaOH to remove the acidic and phenolic components, respectively. The ether extract was dried over MgSO$_4$, followed by evaporation of the solvent to give the neutral crude metabolites (Ishida et al., 1981).

20.3.2 Sesquiterpenoids

Wild rabbits (hair) and deer damage the young leaves of *Chamaecyparis obtusa*, one of the most important furniture and house-constructing tree in Japan. The essential oil of the leaves contains a large amount of (−)-longifolene (**489**). Longifolene (36 g) was administered to 18 of rabbits to obtain the metabolites (3.7 g) from which an aldehyde (**490**) (35.5%) was isolated as pure state. In the metabolism of terpenoids having an exomethylene group, glycol formation was often found, but in the case of longifolene such as a diol was not formed. Introduction of an aldehydes group in biotransformation is very remarkable. Stereoselective hydroxylation of the gem-dimethyl group on a seven-membered ring is first time (Ishida et al., 1982) (Figure 20.145).

(−)-β-Caryophyllene (**451**) is one of the ubiquitous sesquiterpene hydrocarbons in the plant kingdom and the main component of beer hops and clove oil and is being used as a culinary ingredient and as a cosmetic in soaps and fragrances. (−)-β-Caryophyllene also has cytotoxic against breast carcinoma cells and its epoxide is toxic to *Planaria* worms. It contains unique 1,1-dimethylcyclobutane skeleton. (−)-β-Caryophyllene (3 g) was treated in the same manner as described earlier to afford the crude metabolite (2.27 g) from which (10*S*)-14-hydroxycaryophyllene-5,6-oxide (**491**) (80%) and a diol (**492**) were obtained (Asakawa et al., 1981). Later, compound **491** was isolated from the Polish mushroom *Lactarius camphoratus* (Basidiomycetes) as a natural product (Daniewski et al., 1981). 14-Hyroxy-β-callyophyllene and 1-hydroxy-8-keto-β-caryophyllene have been found in Asteraceae and *Pseudomonas* species, respectively. In order to confirm that caryophyllene epoxide (**453**) is the intermediate of both metabolites, it was treated in the same manner as described earlier to give the same metabolites, **491** and **492**, of which **491** was predominant (Asakawa et al., 1981, 1986) (Figure 20.146).

The grapefruit aroma, (+)-nootkatone (**2**), was administered into rabbits to give 11,12-diol (**6, 7**). The same metabolism has been found in that of biotransformation of nootkatone by microorganisms as mentioned in the previous paragraph. Compounds (**6, 7**) were isolated from the urine of hypertensive subjects and named urodiolenone. The endogenous production of **6** and **7** seem to occur interdentally from the administrative manner of nootkatone or grapefruit. Synthetic racemic nootkatone epoxide (**14**) was incubated with rabbit liver microsomes to give 11,12-diol (**6, 7**) (Ishida, 2005). Thus, the role of the epoxide was clearly confirmed as an intermediate of nootkatone (**2**).

(+)-*ent*-Cyclocolorenone (**98**) and its enantiomer (**103**) were biotransformed by *Aspergillus* species to give cyclopropane-cleaved metabolites as described in the previous paragraph.

In order to compare the metabolites between mammals and microorganisms, the essential oil (2 g/rabbit) containing (−)-cyclocolorenone (**103**) obtained from *S. altissima* was administered in rabbits to obtain two metabolites: 9β-hydroxycyclocolorenone (**493**) and 10-hydroxycyclocolorenone (**494**) (Asakawa et al., 1986). 10-Hydroxyaromadendrane-type compounds are well known as

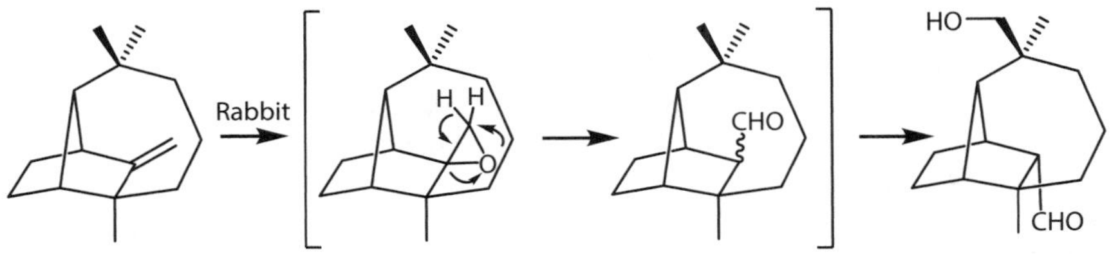

FIGURE 20.145 Biotransformation of longifolene (**489**) by rabbit.

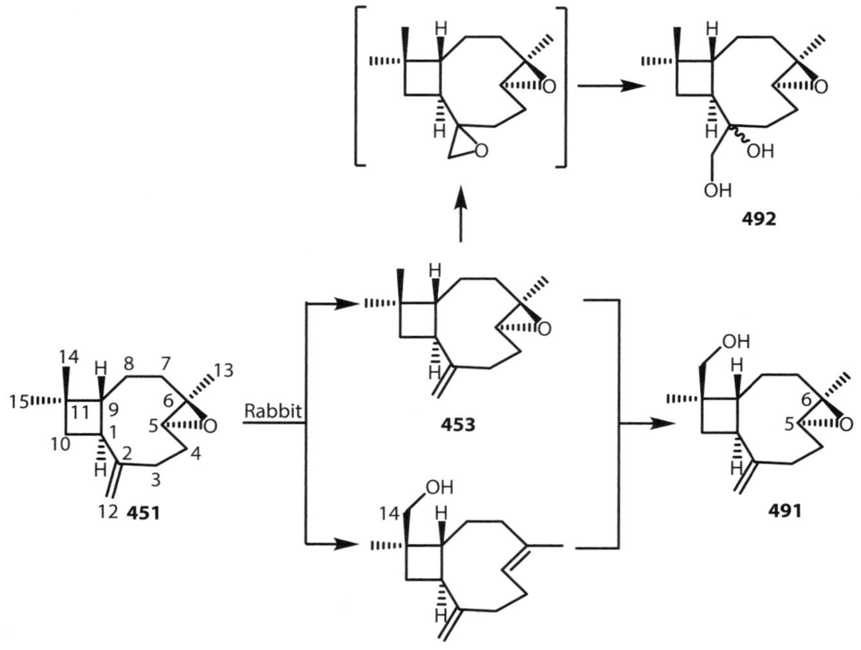

FIGURE 20.146 Biotransformation of (−)-β-caryophyllene (**451**) by rabbit.

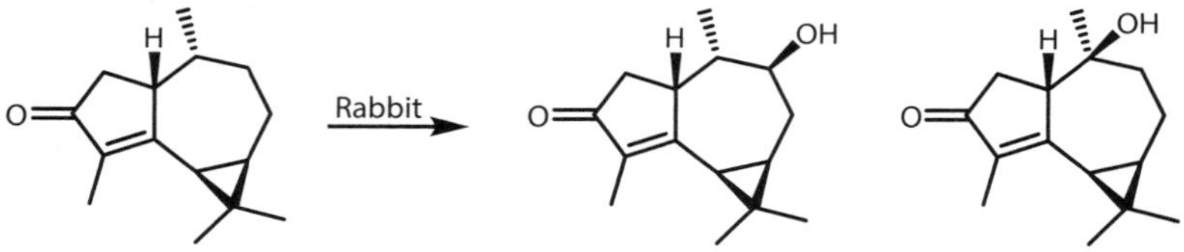

FIGURE 20.147 Biotransformation of (+)-*ent*-cyclocolorenone (**101**) by rabbit.

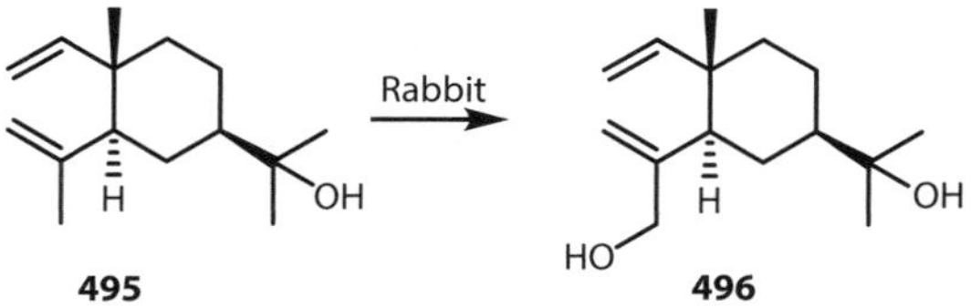

FIGURE 20.148 Biotransformation of elemol (**495**) by rabbit.

the natural products. No oxygenated compound of cyclopropane ring was found in the metabolites of cyclocolorenone in rabbit (Figure 20.147).

From the metabolites of elemol (**495**) possessing the same partial structures of monoterpene hydrocarbon, myrcene, and nootkatone, one primary alcohol (**496**) was obtained from rabbit urine after the administration of **495** (Asakawa et al., 1986) (Figure 20.148).

Components of cedar wood such as cedrol (**414**) and cedrene shorten the sleeping time of mice. In order to search for a relationship between scent, olfaction, and detoxifying enzyme induction, (+)-cedrol (**414**) was administered to rabbits and dogs. From the metabolites from rabbits, two

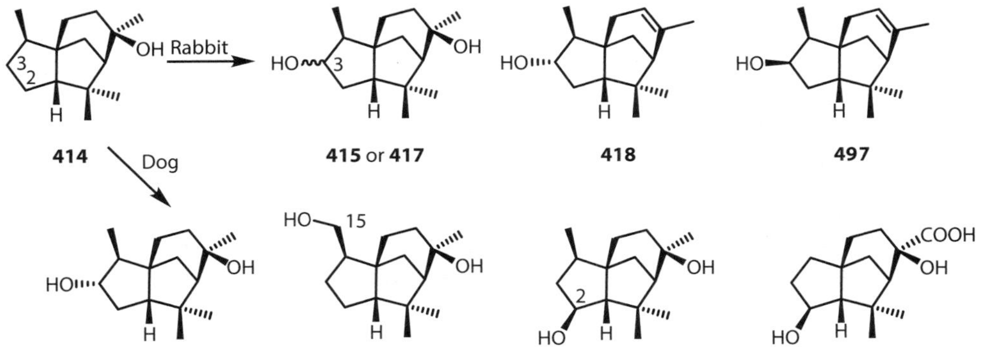

FIGURE 20.149 Biotransformation of cedrol (**414**) by rabbits or dogs.

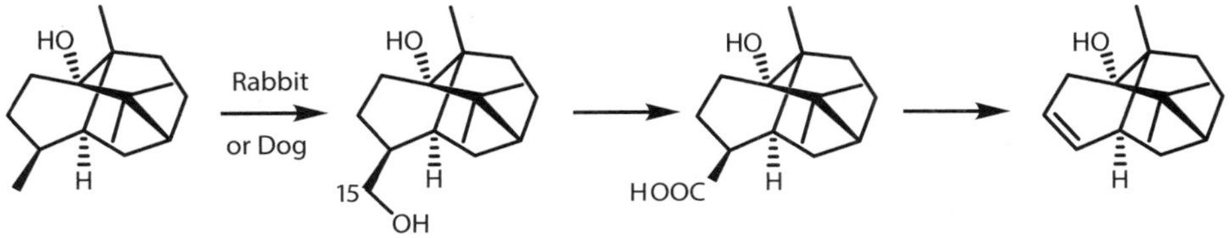

FIGURE 20.150 Biotransformation of patchouli alcohol (**425**) by rabbits or dogs.

C3-hydroxylated products (**418** and **497**) and a diol (**415** or **416**), which might be formed after the hydrogenation of double bond. Dogs converted cedrol (**414**) into different metabolite products, C2- (**498**) and C2/C14-hydroxylated products (**499**), together with the same C3- (**415**) and C15-hydroxylated products (**416**) as those found in the metabolites of microorganisms and rabbits. The aforementioned species-specific metabolism is very remarkable (Bang and Ourisson, 1975).

The microorganisms *C. aphidicola*, *C. cassiicola*, *B. cinerea*, and *G. cingulata* also biotransformed cedrol to various C2-, C3-, C4-, C6-, and C15-hydroxylated products as shown in the previous paragraph. The microbial metabolism of cedrol resembles that of mammals (Figure 20.149).

Patchouli alcohol (**425**) with fungi static properties is one of the important essential oils in perfumery industry. Rabbits and dogs gave two oxidative products (**500, 501**) and one norpatchoulen-1-one (**502**) possessing a characteristic odor. Plant pathogen *B. cinerea* causes many diseases for vegetables and flowers. This pathogen gave totally different five metabolites (**426–430**) from those found in the urine metabolites of mammals as described earlier (Bang et al., 1975) (Figure 20.150).

Sandalwood oil contains mainly α-santalol (**503**) and β-santalol. Rabbits converted α-santalol to three diastereomeric primary alcohols (**504–506**) and dogs did carboxylic acid (**507**) (Zundel, 1976) (Figure 20.151).

(2E,6E)-Farnesol (**478**) was treated in cockroach cytochrome P-450 (CYP4C7) to form regio- and diastereospecifically w-hydroxylated at the C12 methyl group to the corresponding diol (**508**) with 10E-configuration (Sutherland et al., 1998) (Figure 20.152).

Juvenile hormone III (**509**) was also treated in cockroach CYP4C7 to the corresponding 12-hydroxylated product (**510**) (Sutherland et al., 1998).

The African locust converted the same substrate (**509**) into a 7-hydroxy product (**511**) and a 13-hydroxylated product (**512**). It is noteworthy that the African locust and cockroach showed clear species specificity for introduction of oxygen function (Darrouzet et al., 1997).

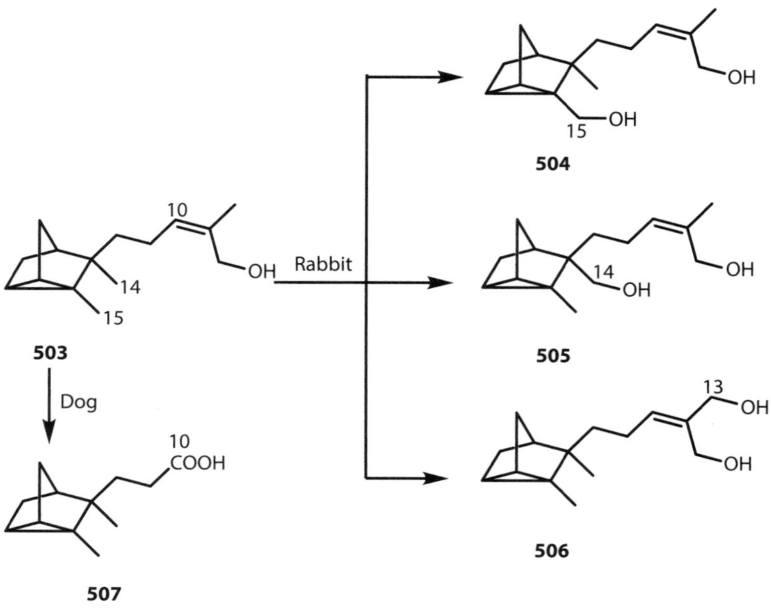

FIGURE 20.151 Biotransformation of santalol (**503**) by rabbits or dogs.

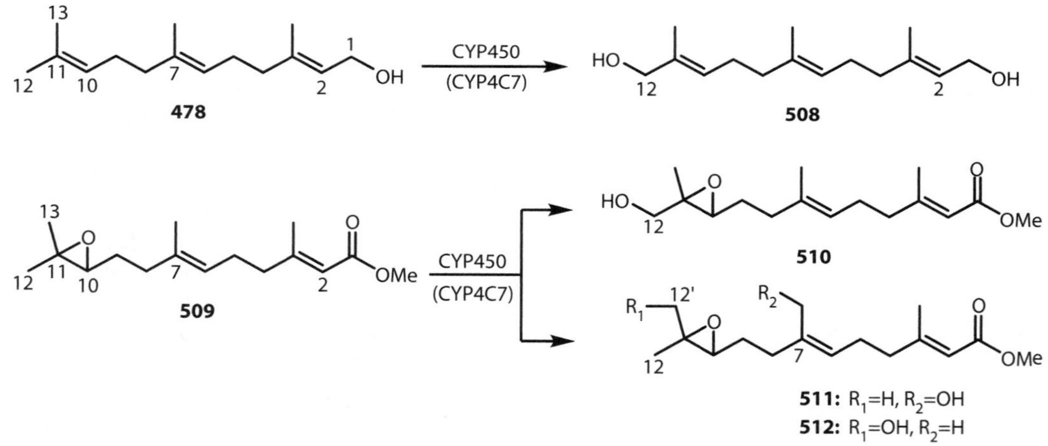

FIGURE 20.152 Biotransformation of 2E,6E-farnesol (**478**) by cockroach cytochrome P-450 and 10,11-epoxyfarnesic acid methyl ester (**509**) by African locust cytochrome P-450.

20.4 BIOTRANSFORMATION OF IONONES, DAMASCONES, AND ADAMANTANES

Racemic α-ionone (**513**) was converted to 4-hydroxy-α-ionone (**514**), which was further dehydrogenated to 4-oxo-α-ionone (**515**) by *Chlorella ellipsoidea* IAMC-27 and *C. vulgaris* IAMC-209. α-Ionone (**513**) was reduced preferentially to α-ionol (**516**) by *Chlorella sorokiniana* and *Chlorella salina* (Noma et al. 1997b).

α-Ionol (**516**) was oxidized by *C. pyrenoidosa* to afford 4-hydroxy-α-ionol (**524**). The same substrate was fed by the same microorganism and *A. niger* to furnish α-ionone (**513**) (Noma and Asakawa, 1998) (Figure 20.153).

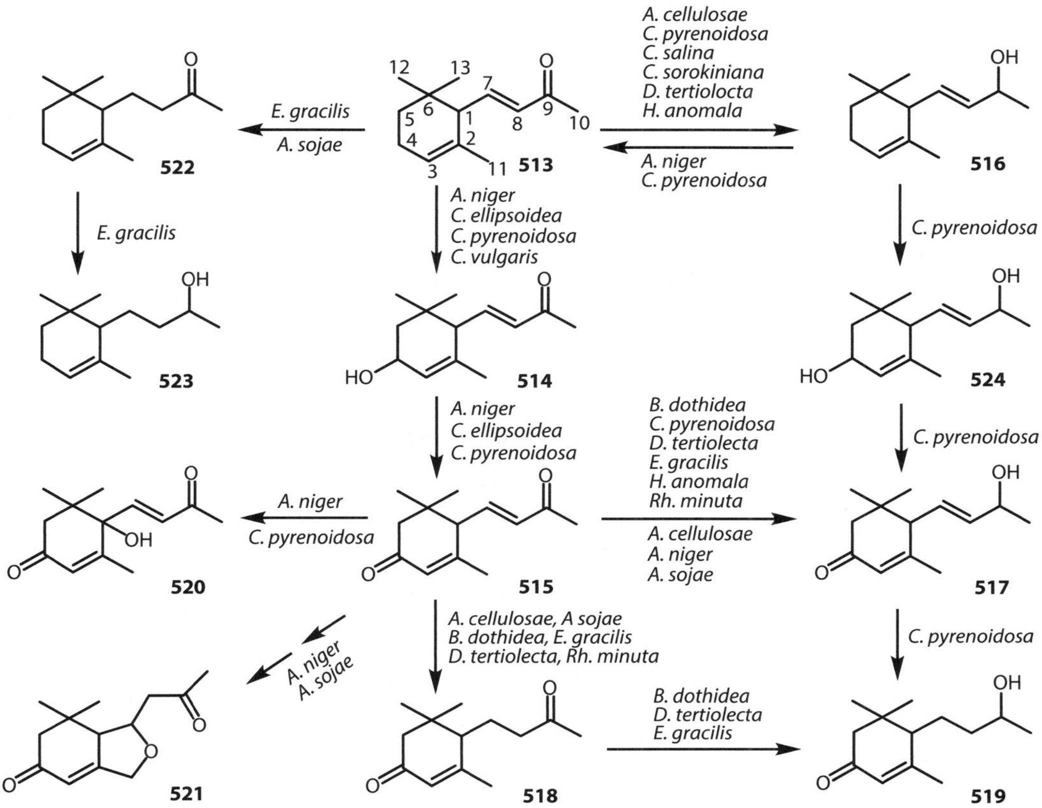

FIGURE 20.153 Biotransformation of α-ionone (**513**) by various microorganisms.

4-Oxo-α-ionone (**515**), which is one of the major product of α-ionone (**513**) by *A. niger*, was transformed reductively by *Hansenula anomala*, *Rhodotorula minuta*, *Dunaliella tertiolecta*, *Euglena gracilis*, *C. pyrenoidosa* C28 and other eight kinds of *Chlorella* species, *B. dothidea*, *A. cellulosae* IFO 4040, and *A. sojae* IFO 4389 to give 4-oxo-α-ionol (**517**), 4-oxo-7,8-dihydro-α-ionone (**518**), and 4-oxo-7,8-dihydro-α-ionol (**519**). Compound **515** was also oxidized by *A. niger* and *A. sojae* to give 1-hydroxy-4-oxo-α-ionone (**520**) and 7,11-oxido-4-oxo-7,8-dihydro-α-ionone (**521**). C7–C8 double bond of α-ionone (**513**), 4-oxo-α-ionone (**515**), and 4-oxo-α-ionol (**517**) were easily reduced to their corresponding dihydro products (**522**, **518**, **519**), respectively, by *Euglena*, *Aspergillus*, *Botryosphaeria*, and *Chlorella* species. The metabolite (**522**) was further reduced to **523** by *E. gracilis* (Noma and Asakawa, 1998).

Biotransformation of (+)-1R-α-ionone (**513a**), [α]$_D$ + 386.5°, 99% ee, and (−)-1S-α-ionone (**513a′**), [α]$_D$ 361.6°, 98% ee, which were obtained by optical resolution of racemic α-ionone (**513**), was fed by *A. niger* for 4 days in Czapek-peptone medium. From **513a**, 4α-hydroxy-α-ionone (**514a**), 4β-hydroxy-α-ionone (**514b**), and 4-oxo-α-ionone (**515a**) were obtained, while from compound **513a′**, the enantiomers (**514a′**, **514b′**, **515a′**) of the metabolites from **513a** were obtained; however, the difference of their yields was observed. In case of **513a**, 4α-hydroxy-α-ionone (**514a**) was obtained as the major product, while **515a′** was predominantly obtained from **513a′**. This oxidation was inhibited by 1-aminobenzotriazole, thus, CYP-450 is contributed to this oxidation process (Hashimoto et al., 2000a) (Figure 20.154).

α-Damascone (**525**) was incubated with *A. niger* and *A. terreus*, in Czapek-peptone medium to give *cis*- (**525**) and *trans*-3-hydroxy-α-damascones (**527**) and 3-oxo-α-damscone (**528**), while the latter *Aspergillus* species afforded 3-oxo-8,9-dihydro-α-damascone (**529**). The hydroxylation

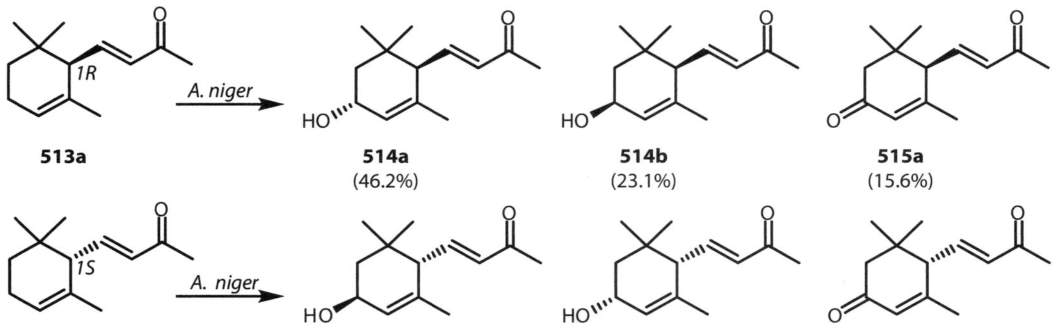

FIGURE 20.154 Biotransformation of (1R)-α-ionone (**513a**) and (1S)-α-ionone (**513a′**) by *Aspergillus niger*.

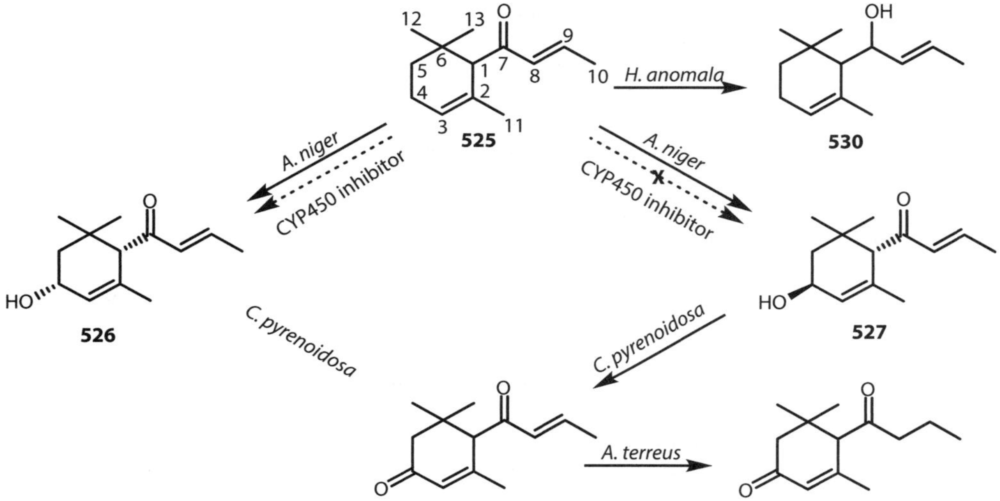

FIGURE 20.155 Biotransformation of α-damascone (**525**) by various microorganisms.

process of **525–527** was inhibited by CYP-450 inhibitor. *H. anomala* reduced α-damascone (**525**) to α-damascol (**530**). *Cis*- (**526**) and *trans*-4-hydroxy-α-damascone (**527**) were fed by *C. pyrenoidosa* in Noro medium to give 4-oxodamascone (**528**) (Noma et al., 2001b) (Figure 20.155).

β-Damascone (**531**) was also treated in *A. niger* to afford 5-hydroxy-β-damascone (**532**), 3-hydroxy-β-damascone (**533**), 5-oxo- (**534**), 3-oxo-β-damscone (**535**), and 3-oxo-1,9-dihydroxy-1,2-dihydro-β-damascone (**536**) as the minor components. In case of *A. terreus*, 3-hydroxy-8,9-dihydro-β-damascone (**537**) was also obtained (Figure 20.156).

Adamantane derivatives have been used as many medicinal drugs. In order to obtain the drugs, adamantanes were incubated by many microorganisms, such as *A. niger*, *A. awamori*, *A. cellulosae*, *Aspergillus fumigatus*, *A. sojae*, *A. terreus*, *B. dothidea*, *C. pyrenoidosa* IMCC-28, *C. sorokiniana*, *F. culmorum*, *E. gracilis*, and *H. anomala* (Figure 20.157).

Adamantane (**538**) was incubated with *A. niger*, *A. cellulosae*, and *B. dothidea* in Czapek-peptone medium. The same substrate was also treated in *C. pyrenoidosa* in Noro medium. Compound **538** was converted into both 1-hydroxy- (**539**) and 9α-hydroxyadamantane (**540**) by all four microorganisms, followed by oxidation oxidized to give 1,9α-dihydroxyadamantanol (**541**) by *A. niger*, which was further oxidized to 1-hydroxyadamantane-9-one (**542**), which was reduced to afford

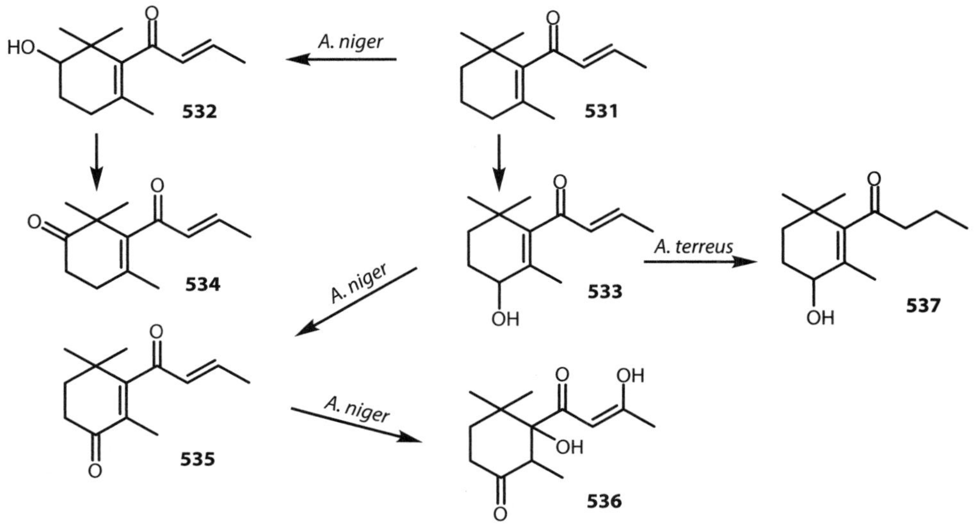

FIGURE 20.156 Biotransformation of β-damascone (**531**) by *Aspergillus niger* and *Aspergillus terreus*.

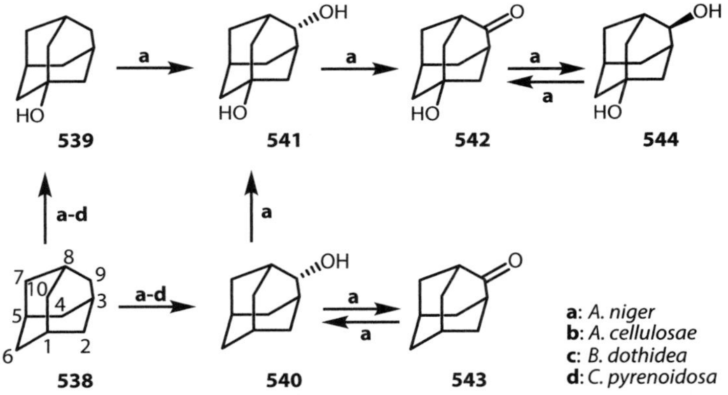

FIGURE 20.157 Biotransformation of adamantane (**538**) by various microorganisms.

a: *A. niger*
b: *A. cellulosae*
c: *B. dothidea*
d: *C. pyrenoidosa*

1,9β-hydroxyadamantane (**544**). *A. niger* gave the metabolite (**541**) as the major product in 80% yield. *A. cellulosae* converted **538**–**539** and **540** in the ratio of 81:19. *C. pyrenoidosa* gave **539**, **540**, and adamantane-9-one (**543**) in the ratio 74:16:10. 4α-Adamantanol (**540**) was directly converted by *C. pyrenoidosa*, *A. niger*, and *A. cellulosae* to afford **543**, which was also reduced to 9α-adamantanol (**540**) by *A. niger*. The biotransformation of adamantane, however, did not occur by the microorganisms *H. anomala*, *C. sorokiniana*, *D. tertiolecta*, and *E. gracilis* (Noma et al., 1999).

Adamantanes (**538**–**543**, **542**) were also incubated with various fungi including with *F. culmorum*. 1-Hydroxyadamantane-9-one (**542**) was reduced stereoselectively to **541** by *A. niger*, *A. cellulosae*, *B. dothidea*, and *F. culmorum*. On the contrary, *F. culmorum* reduced **541**–**542**. *A. cellulosae* and *B. dothidea* bioconverted **542** to 1,9β-hydroxyadamantane (**544**) stereoselectively. Adamantane-9-one (**543**) was treated by *A. niger* to give nonstereoselectively **545**–**547** that were further converted into diketone (**548**, **549**) and a diol (**550**). It is noteworthy that oxidation and reduction reactions were observed between ketoalcohol (**547**) and diols (**551**, **552**). The same phenomenon was also seen between **546** and **553**. The latter diol was also oxidized by *A. niger* to furnish diketone (**549**) (Noma et al., 2001c, 2003). Direct hydroxylation at C3 of 1-hydroxyadamantane-9-one (**542**) was seen in the incubation of **539** with *A. niger*.

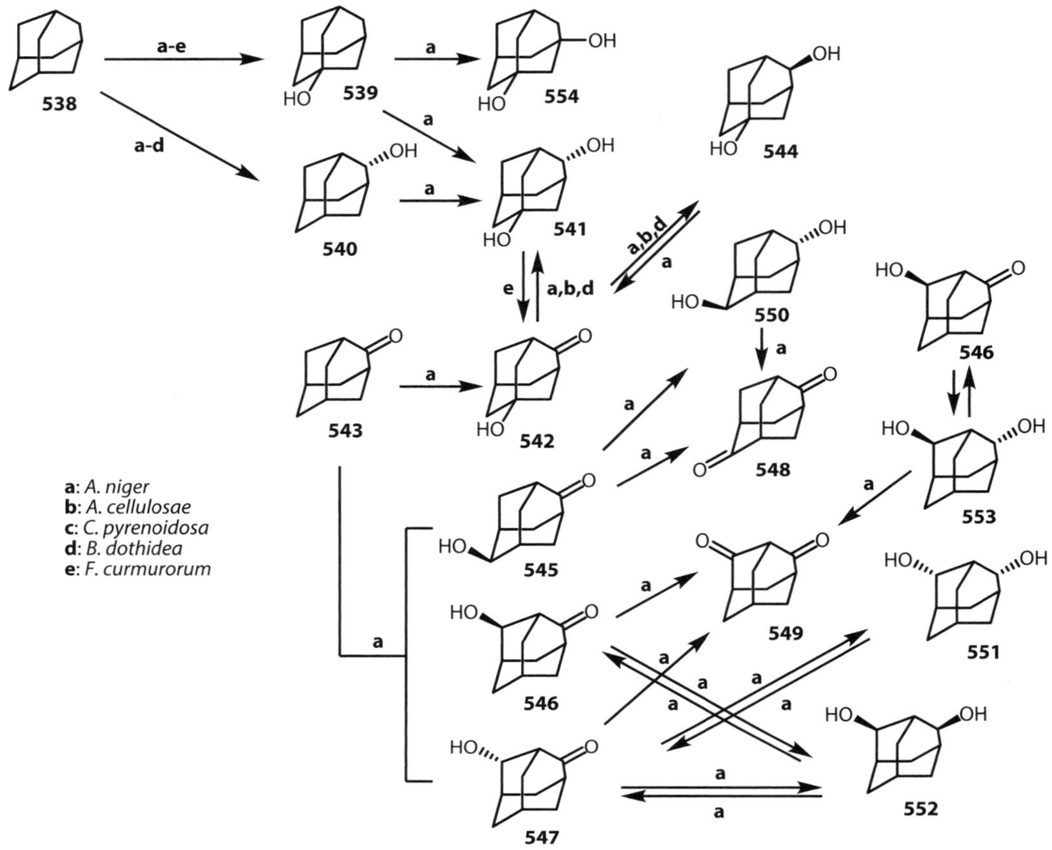

FIGURE 20.158 Biotransformation of adamantane (**538**) and adamantane-9-one by various microorganisms.

4-Adamantanone (**543**) showed promotion effect of cell division of the fungus, while 1-adamantanol (**539**) and adamantane-9-one (**543**) inhibited germination of lettuce seed. 1-Hydroxyadamantane-9-one (**542**) inhibited the elongation of root of lettuce, while adamantane-1,4-diol (**544**) and adamantane itself (**538**) promoted root elongation (Noma et al., 1999, 2001c) (Figure 20.158).

Stereoselective reduction of racemic bicyclo[3.3.1]nonane-2,6-dione (**555a**, **555a′**) was carried out by *A. awamori*, *A. fumigatus*, *A. cellulosae*, *A. sojae*, *A. terreus*, *A. niger*, *B. dothidea*, and *F. culmorum* in Czapek-peptone, *H. anomala* in yeast, *E. gracilis* in Hunter, and *D. tertiolecta* in Noro medium. All microorganisms reduced **555** and **555a′** to give corresponding monoalcohol (**556**, **556a**) and optical-active (–)-diol (**557a**) ($[\alpha]_D$ – 71.8° in the case of *A. terreus*), which was formed by enantioselective reduction of racemic monool, **556** and **556a** (Noma et al., 2003) (Figure 20.159).

20.5 BIOTRANSFORMATION OF AROMATIC COMPOUNDS

Essential oils contain aromatic compounds, such as *p*-cymene, carvacrol, thymol, vanillin, cinnamaldehyde, eugenol, chavicol, safrole, and asarone (**558**), among others.

Takahashi (1994) reported that simple aromatic compounds, propylbenzene, hexylbenzene, decylbenzene, *o*- and *p*-hydroxypropiophenones, *p*-methoxypropiophenone, 4-hexylresorcinol, and methyl 4-hexylresorcionol, were incubated with *A. niger*. From hexyl- and decylbenzenes, w1-hydroxylated products were obtained, whereas from propylbenzene, w2-hydroxylated metabolites were obtained (Takahashi, 1994).

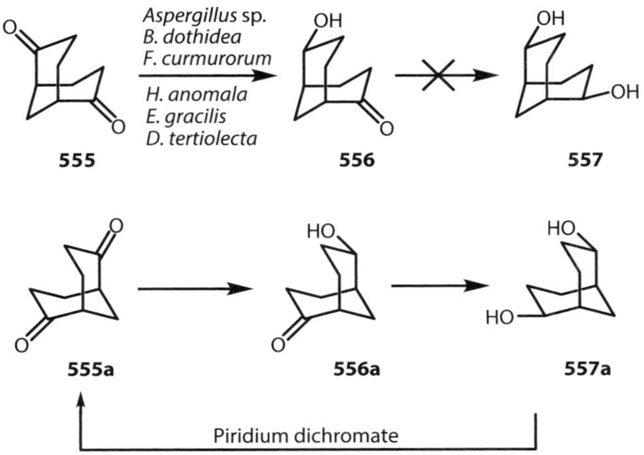

FIGURE 20.159 Biotransformation of bicyclo[3.3.1]nonane-2,6-dione (**555a**, **555a'**) by various microorganisms.

Asarone (**558**) and dihydroeugenol (**562**) were not biotransformed by *A. niger*. However, dihydroasarone (**559**) and methyl dihydroeugenol (**563**) were biotransformed by the same fungus to produce a small amount of 2-hydroxy (**560**, **561**) and 2-oxo derivatives (**564**, **565**), respectively. The chirality at C2 was determined to be *R* and *S* mixtures (1:2) by the modified Mosher method (Takahashi, 1994) (Figure 20.160).

Chlorella species are excellent microalgae as oxidation bioreactors as mentioned earlier. Treatment of monoterpene aldehydes and related aldehydes were reduced to the corresponding primary alcohols, indicating that these green algae possess reductase.

A microalgae *E. gracilis* Z. also contains reductase. The following aromatic aldehydes were treated in this organism: benzaldehyde; 2-cyanobenzaldehyde; *o*-, *m*-, and *p*-anisaldehyde; *o*-, *m*-, and *p*-salicylaldehyde; *o*-, *m*-, and *p*-tolualdehyde; *o*-chlorobenzaldehyde; *p*-hydroxybenzaldehyde; *o*-, *m*-, and *p*-nitrobenzaldehyde; 3-cyanobenzaldehyde; vanillin; isovanillin; *o*-vanillin; nicotine aldehyde; 3-phenylpropionaldehyde; and ethyl vanillin. Veratraldehyde, 3-nitrosalicylaldehde, phenylacetaldehyde, and 2-phenylproanaldehyde gave their corresponding primary alcohols. 2-Cyanobenzaldehyde gave its corresponding alcohol with phthalate. *m*- and *p*-Chlorobenzaldehyde gave its corresponding alcohols and *m*- and *p*-chlorobenzoic acids. *o*-Phthalaldehyde and *p*-phthalate and iso- and terephthalaldehydes gave their corresponding monoalcohols and dialcohols. When cinnamaldehyde and α-methyl cinnamaldehyde were incubated in *E. gracilis*, cinnamyl alcohol and 3-phenylpropanol, and 2-methylcinnamyl alcohol, and 2-methyo-3-phenylpropanol were obtained in good yield. *E. gracilis* could convert acetophenone to 2-phenylethanol; however, its enantio-excess is very poor (10%) (Takahashi, 1994).

Raspberry ketone (**566**) and zingerone (**574**) are the major components of raspberry (*Rubus idaeus*) and ginger (*Zingiber officinale*), and these are used as food additive and spice. Two substrates were incubated with the *Phytolacca americana* cultured cells for 3 days to produce two secondary alcohols (**567**, **568**) as well as five glucosides (**569–572**) from **566** and a secondary alcohol (**576**) and four glycoside products (**575**, **577–579**) from **574**. In the case of raspberry ketone, phenolic hydroxyl group was preferably glycosylated after the reduction of carbonyl group of the substrate occurred. It is interesting to note that one more hydroxyl group was introduced into the benzene ring to give **568**, which were further glycosylated by one of the phenolic hydroxyl groups, and no glycoside of the secondary alcohol at C2 were obtained (Figure 20.161).

On the other hand, zingerone (**574**) was converted into **576**, followed by glycosylation to give both glucosides (**577**, **578**) of phenolic and secondary hydroxyl groups and a diglucoside (**579**) of both phenolic and secondary hydroxyl group in the molecule. It is the first report on the introduction

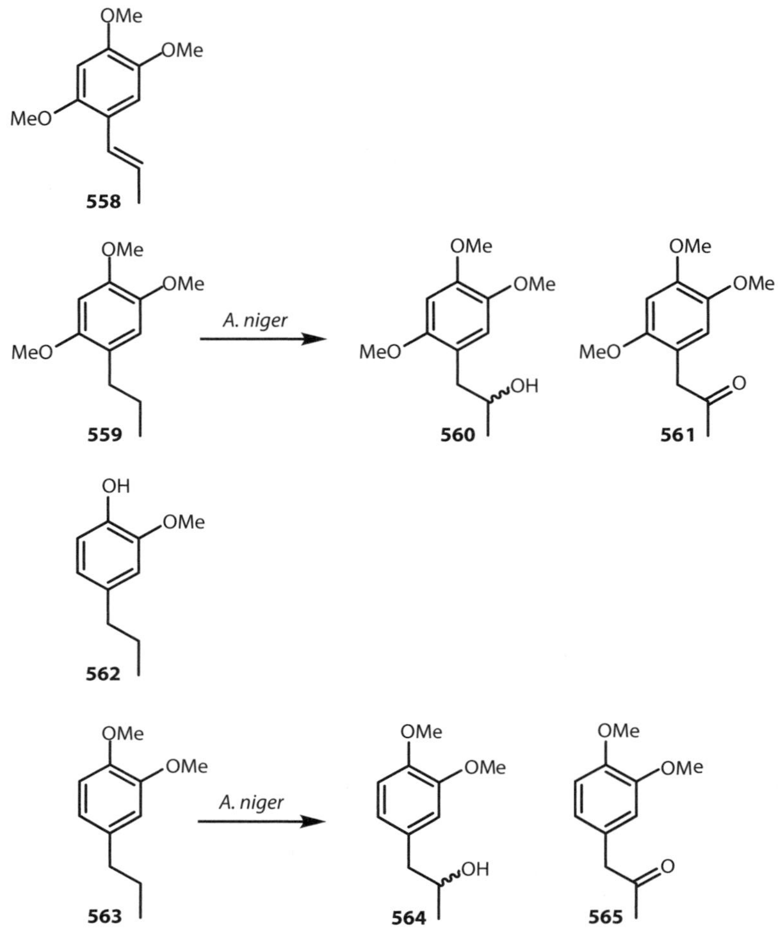

FIGURE 20.160 Biotransformation of dihydroasarone (**559**) and methyl dihydroeugenol (**563**) by *Aspergillus niger*.

of individual glucose residues onto both phenolic and secondary hydroxyl groups by cultured plant cells (Shimoda et al., 2007a) (Figure 20.162).

Thymol (**580**), carvacrol (**583**), and eugenol (**586**) were glucosylated by glycosyltransferase of cell-cultured *Eucalyptus perriniana* to each glucoside (**581**, 3%; **584**, 5%; **587**, 7%) and gentiobioside (**582**, 87%; **585**, 56%; **588**, 58%). The yield of thymol glycosides was 1.5 times higher than that of carvacrol and 4 times higher than that of eugenol. Such glycosylation is useful to obtain higher water-soluble products from natural and commercially available secondary metabolites for food additives and cosmetic fields (Shimoda et al., 2006) (Figure 20.163).

Hinokitiol (**589**), which is easily obtained from cell suspension cultures of *Thujopsis dolabrata* and possesses potent antimicrobial activity, was incubated with cultured cells of *E. perriniana* for 7 days to give its monoglucosides (**590**, **591**, 32%) and gentiobiosides (**592**, **593**) (Furuya et al., 1997; Hamada et al., 1998) (Figure 20.164).

(−)-Nopol benzyl ether (**594**) was smoothly biotransformed by *A. niger*, *A. cellulosae*, *A. sojae*, *Aspergillus usami*, and *Penicillium* species in Czapek-peptone medium to give (−)-4-oxonopol-2′,4′-dihydroxybenzyl ether (**595**, 23% in the case of *A. niger*), which demonstrated antioxidant activity (ID_{50} 30.23 m/M), together with a small amount of nopol (6.3% in *A. niger*). This is very rare direct introduction of oxygen function on the phenyl ring (Noma and Asakawa, 2006) (Figure 20.165).

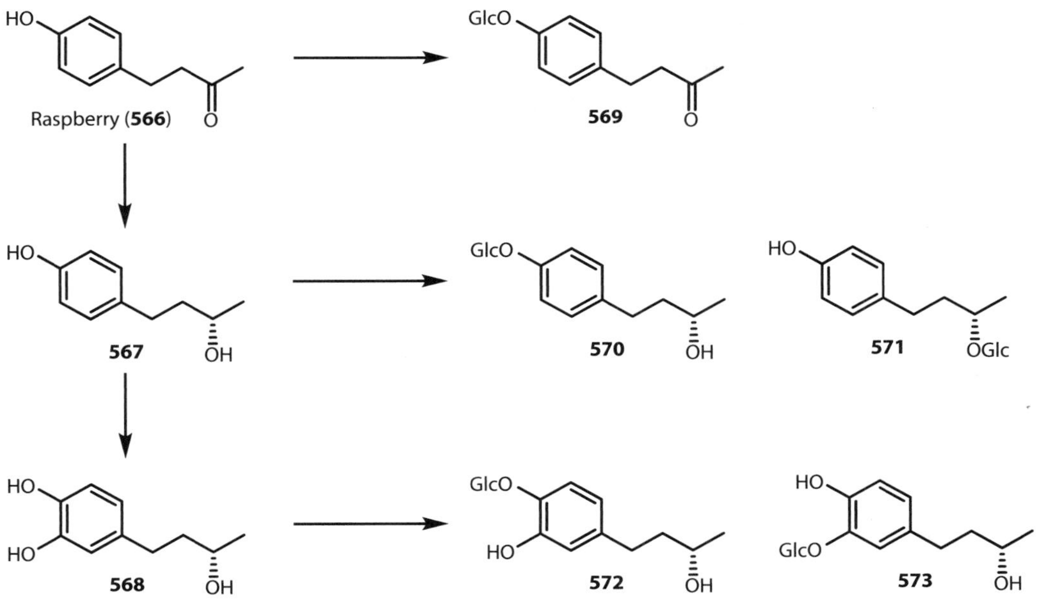

FIGURE 20.161 Biotransformation of raspberry ketone (**566**) by *Phytolacca americana* cells.

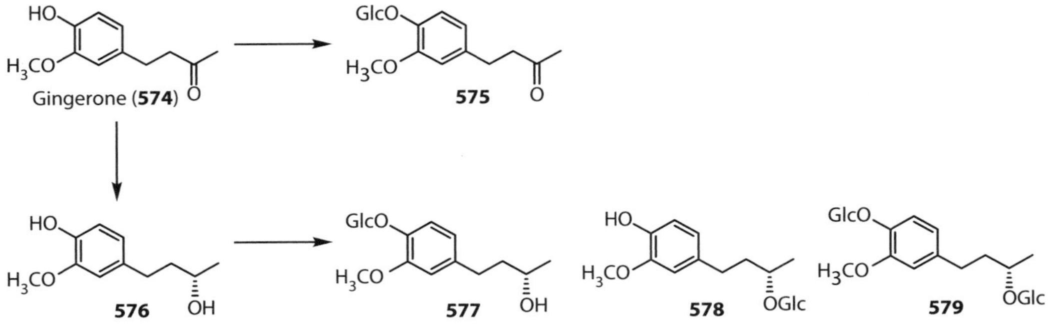

FIGURE 20.162 Biotransformation of zingerone (**574**) by *Phytolacca americana* cells.

Capsicum annuum contains capsaicin (**596**), and its homologues having an alkylvanillylamides possess various interesting biological activities such as anti-inflammatory, antioxidant, saliva- and stomach juice–inducing activity, analgesic, antigenotoxic, antimutagenic, anticarcinogenic, antirheumatoid arthritis, and diabetic neuropathy and are used as food additives. On the other hand, because of potent pungency and irritation on skin and mucous membrane, it has not yet been permitted as medicinal drug. In order to reduce this typical pungency and application of nonpungent capsaicin metabolites to the crude drug, capsaicin (**596**) (600 mg) including 30% of dihydrocapsaicin (**600**) was incubated in Czapek-peptone medium including *A. niger* for 7 days to give three metabolites, w1-hydroxylated capsaicin (**597**, 60.9%), 8,9-dihydro-w1-hydroxycapsaicin (**598**, 16%), and a carboxylic acid (**599**, 13.6%). All of the metabolites do not show pungency (Figure 20.166).

Dihydrocapsaicin (**600**) was also treated in the same manner as described earlier to afford w1-hydroxydihyrocpsaicin (**598**, 80.9%) in high yield and the carboxylic acid (**599**, 5.0%). Capsaicin itself showed carbachol-induced contraction of 60% in the bronchus at a concentration of 1 mmol/L. 11-Hydroxycapsicin (**85**) retained this activity of 60% at a concentration of 30 mmol/L. Dihydrocapsaicin (**600**) showed the same activity of contraction in the bronchus, at the same concentration as that used in capsaicin. However, the activity of contraction in the bronchus of 11-hydroxy

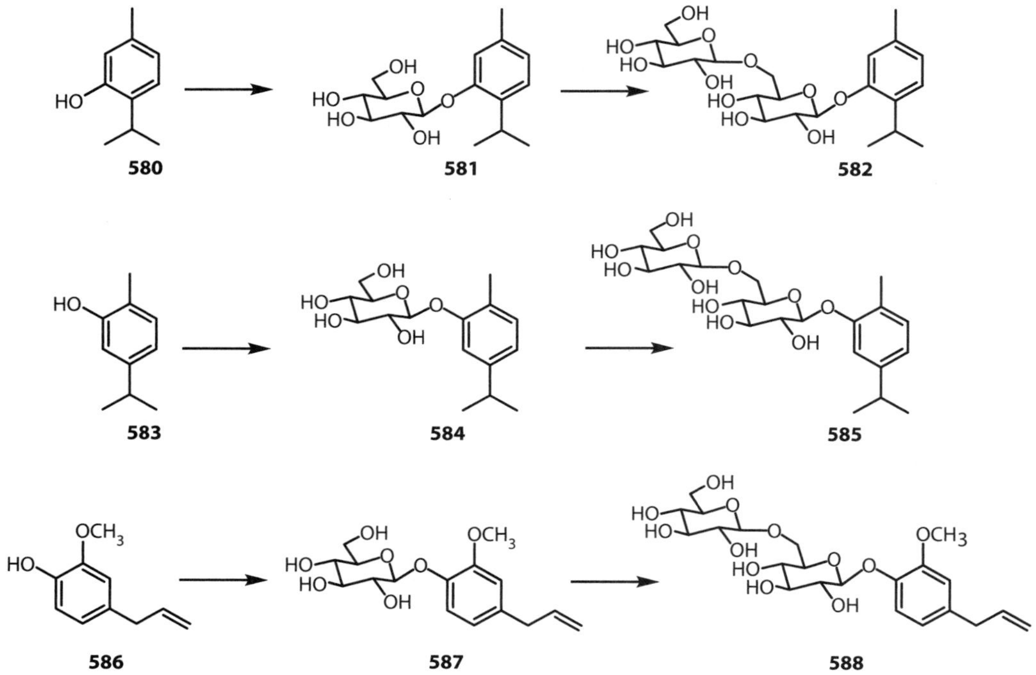

FIGURE 20.163 Biotransformation of thymol (**580**), carvacrol (**583**), and eugenol (**586**) by *Eucalyptus perriniana* cells.

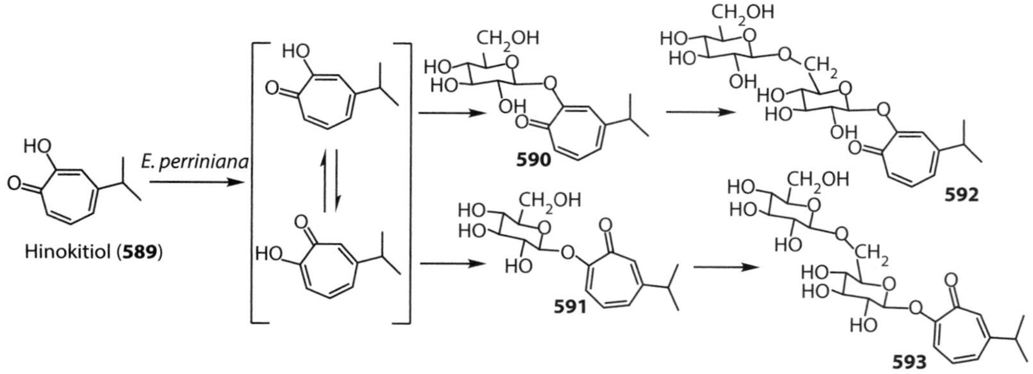

FIGURE 20.164 Biotransformation of hinokitiol (**589**) by *Eucalyptus perriniana* cells.

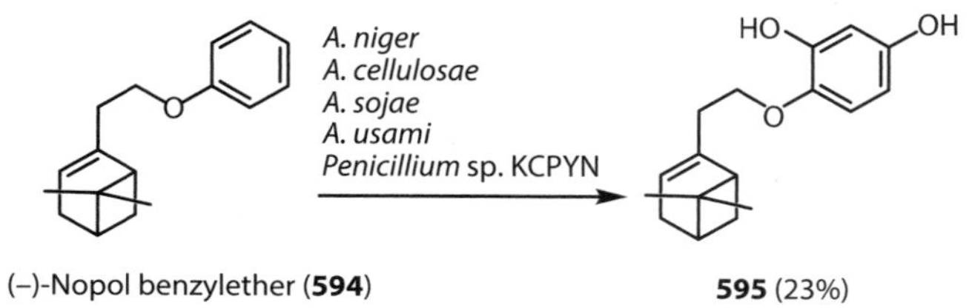

FIGURE 20.165 Biotransformation of nopol benzyl ether (**531**) by *Aspergillus* and *Penicillium* species.

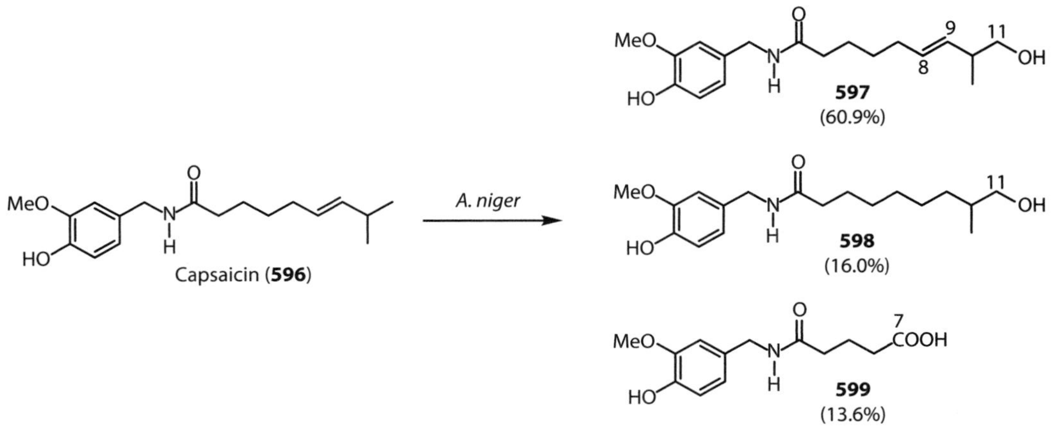

FIGURE 20.166 Biotransformation of capsaicin (**596**) by *Aspergillus niger*.

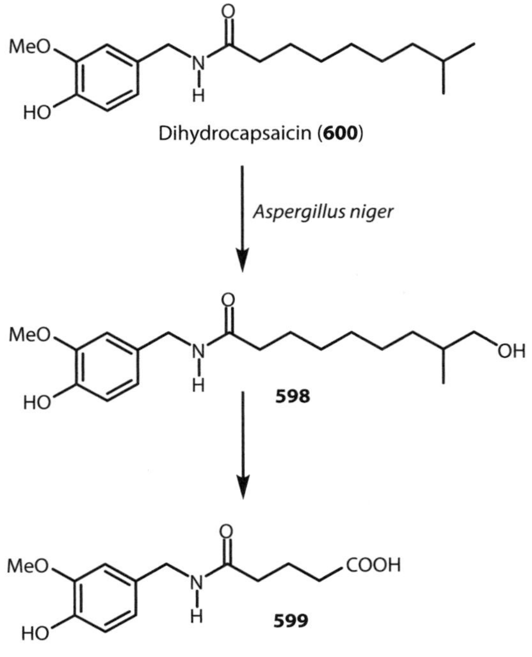

FIGURE 20.167 Biotransformation of dihydrocapsaicin (**600**) by *Aspergillus niger*.

derivative (**598**) showed weaker (50% at 30 mmol/L) than that of the substrate. Since both metabolites (**597** and **598**) are tasteless, these products might be valuable for the crude drug although the contraction in the bronchus is weak. 2,2-Diphenyl-1-picrylhydrazyl radical scavenging activity test of capsaicin and dihydrocapsaicin derivatives was carried out. 11-Hydroxycapsaicin (**597**), 11-dihydrocapsaicin (**598**), and capsaicin (**596**) showed higher activity than (±)-α-tocopherol, and 11-dihydroxycapsaicin (**598**) displayed a strong scavenging activity (IC_{50} 50 mmol/L) (Hashimoto and Asakawa, unpublished results) (Figure 20.167).

Shimoda et al. (2007b) reported the bioconversion of capsaicin (**596**) and 8-nor-dihydrocapsaicin (**601**) by the cultured cell of *Catharanthus roseus* to give more water-soluble capsaicin derivatives. From capsaicin, three glycosides, capsaicin 4-*O*-β-D-glucopyranoside (**602**), which was one of the

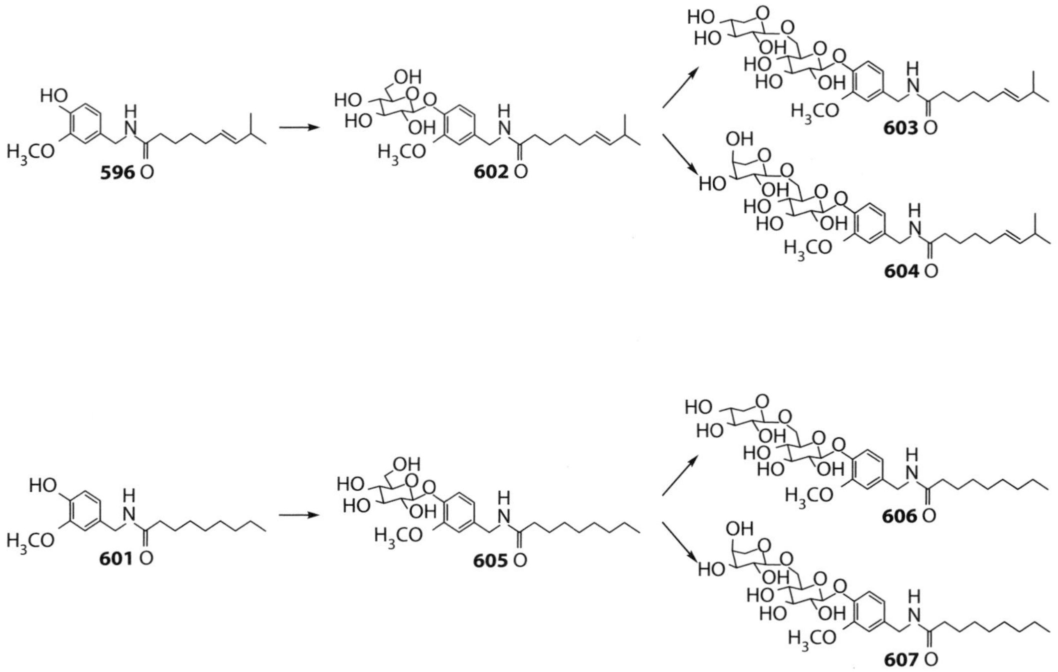

FIGURE 20.168 Biotransformation of capsaicin (**596**) and 8-nor-dihydrocapsaicin (**601**) by *Catharanthus roseus* cells.

capsaicinoids in the fruit of *Capsicum* and showed 1/100 weaker pungency than capsaicin, 4-*O*-(6-*O*-β-D-xylopyranosyl)-β-D-glucoside (**603**), and 4-*O*-(6-*O*-α-L-arbinosyl)-β-D-glucopyranoside (**604**), were obtained. 8-Nor-dihydrocapsaicin (**601**) was also incubated with the same cultured cell to afford the similar products (**605**–**607**) all of which reduced their pungency and enhanced water solubility. Since many synthetic capsaicin glycosides possess remarkable pharmacological activity, such as decrease of liver and serum lipids, the present products will be used for valuable prodrugs (Figure 20.168).

Z. officinale contains various sesquiterpenoids and pungent aromatic compounds such as 6-shogaol (**608**) and 6-gingerol (**613**), and their pungent compounds that possess cardio tonic and sedative activity. 6-Shogaol (**608**) was incubated with *A. niger* in Czapek-peptone medium for 2 days to afford w1-hydroxy-6-shagaol (**609**, 9.9%), which was further converted to 8-hydroxy derivative (**610**, 16.1%), a γ-lactone (**611**, 22.4%), and 3-methoxy-4-hydroxyphenylacetic acid (**612**, 48.5%) (Figure 20.169).

6-Gingerol (**613**) (1 g) was treated in the same condition as mentioned earlier to yield six metabolites, w1-hydroxy-6-gingerol (**614**, 39.8%), its carboxylic derivative (**616**, 14.5%), a γ-lactone (**618**) (16.9%) that might be formed from (**616**), its 8-hydoxy-γ-lactone (**619**, 12.1%), w2-hydroxy-6-gingerol (**615**, 19.9%), and 6-deoxy-gingerol (**617**, 14.5%) (Takahashi et al., 1993a).

The metabolic pathway of 6-gingerol (**613**) resembles that of 6-shagaol (**608**). That of 6-shogaol and dihydrocapsaicin (**600**) is also similar since both substrates gave carboxylic acids as the final metabolites (Takahashi et al., 1993a) (Figure 20.170).

In conclusion, a number of sesquiterpenoids were biotransformed by various fungi and mammals to afford many metabolites, several of which showed antimicrobial and antifungal, antiobesity, cytotoxic, neurotrophic, and enzyme inhibitory activity. Microorganisms introduce oxygen atom at allylic position to give secondary hydroxyl and keto groups. Double bond is also oxidized

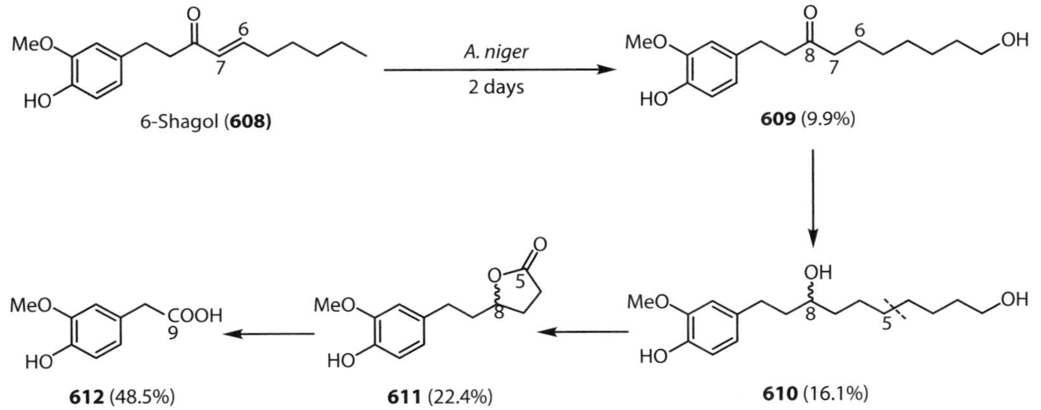

FIGURE 20.169 Biotransformation of 6-shogaol (**608**) by *Aspergillus niger*.

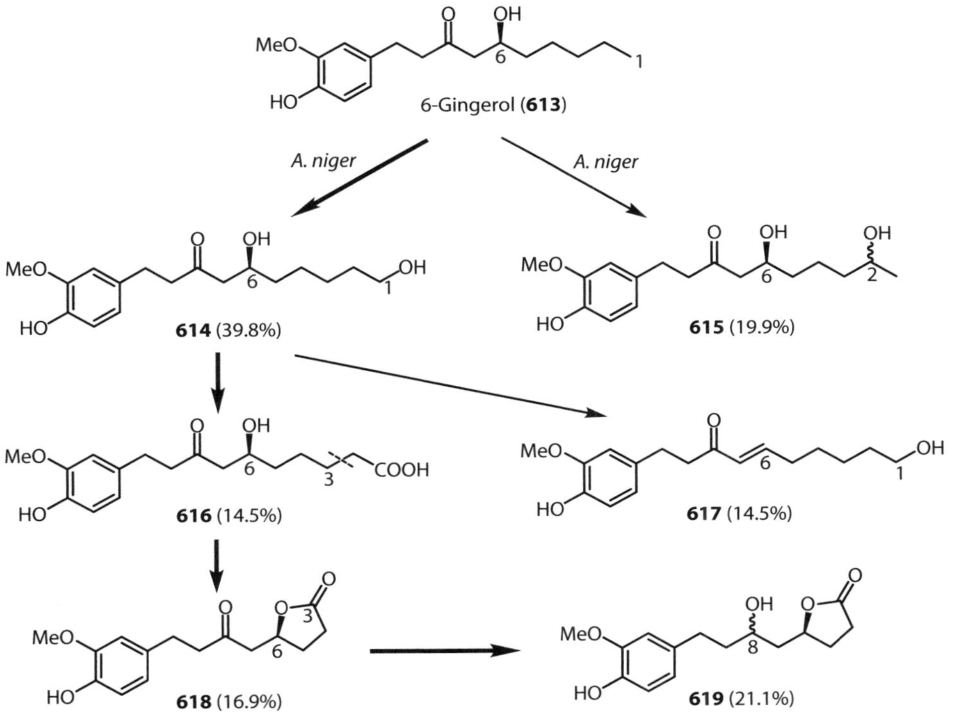

FIGURE 20.170 Biotransformation of 6-gingerol (**613**) by *Aspergillus niger*.

to give epoxide, followed by hydrolysis to afford a diol. These reactions precede stereo- and regio-specifically. Even at nonactivated carbon atom, oxidation reaction occurs to give primary alcohol. Some fungi like *A. niger* cleave the cyclopropane ring with a 1,1-dimethyl group. It is noteworthy that *A. niger* and *A. cellulosae* produce totally different metabolites from the same substrates. Some fungi occurs reduction of carbonyl group, oxidation of aryl methyl group, phenyl coupling, and cyclization of a 10-membered ring sesquiterpenoids to give C6/C6- and C5/C7-cyclic or spiro compounds. Cytochrome P-450 is responsible for the introduction of oxygen function into the substrates.

The present methods are very useful for the production of medicinal and agricultural drugs as well as fragrant components from commercially available cheap, natural, and unnatural terpenoids or a large amount of terpenoids from higher medicinal plants and spore-forming plants like liverworts and fungi.

The methodology discussed in this chapter is a very simple one-step reaction in water, nonhazard, and very cheap, and it gives many valuable metabolites possessing different properties from those of the substrates.

REFERENCES

Abraham, W.R., L. Ernst, and B. Stumpf, 1990. Biotransformation of caryophyllene by *Diplodia gossypina*. *Phytochemisrty*, 29: 115–120.

Abraham, W.-R., K. Kieslich, B. Stumpf, and L. Ernst, 1992. Microbial oxidation of tricyclic sesquiterpenoids containing a dimethylcyclopropane ring. *Phytochemistry*, 31: 3749–3755.

Abraham, W.-R., P. Washausen, and K. Kieslich, 1987. Microbial hydroxylation of cedrol and cedrene. *Z. Naturforsch.*, 42C: 414–419.

Aleu, J., J.R. Hanson, R. Hernandez-Galan, and I.G. Collado, 1999a. Biotransformation of the fungistatic sesquiterpenoid patchoulol by *Botrytis cinerea*. *J. Nat. Prod.*, 62: 437–440.

Aleu, J., R. Hernandez-Galan, and I.G. Collad, 2002. Biotransformation of the fungistatic sesquiterpenoid isobotryan-9α-ol by *Botrytis cinerea*. *J. Mol. Catal. B*, 16: 249–253.

Aleu, J., R. Hernandez-Galan, J.R. Hanson, P.B. Hitchcock, and I.G. Collado, 1999b. Biotransformation of the fungistatic sesquiterpenoid ginsenol by *Botrytis cinerea*. *J. Chem. Soc. Perkin Trans.*, 1: 727–730.

Amate, A., A. Garcia-Granados, A. Martinez et al., 1991. Biotransformation of 6α-eudesmanolides functionalized at C-3 with *Curvularia lunata* and *Rhizopus nigricans* cultures. *Tetrahedron*, 47: 5811–5818.

Aranda G., I. Facon, J.-Y. Lallemand, and M. Leclaire, 1992. Microbiological hydroxylation in the drimane series. *Tetrahedron Lett.*, 33, 7845–7848.

Aranda, G., M.S. Kortbi, J.-Y. Lallemand et al., 1991. Microbial transformation of diterpenes: Hydroxylation of sclareol, manool and derivatives by *Mucor plumbeus*. *Tetrahedron*, 47: 8339–8350.

Arantes, S.F., J.R. Hanson, and P.B. Hitchcok, 1999. The hydroxylation of the sesquiterpenoid valerianol by *Mucor plumbeus*. *Phytochemistry*, 52: 1063–1067.

Asakawa, Y., 1982. Chemical constituents of the Hepaticae. In: *Progress in the Chemistry of Organic Natural Products*, W. Herz, H. Grisebach, and G.W. Kirby (eds.), Vol. 42, pp. 1–285. Vienna, Austria: Springer.

Asakawa, Y., 1990. Terpenoids and aromatic compounds with pharmaceutical activity from bryophytes. In: *Bryophytes: Their Chemistry and Chemical Taxonomy*, D.H. Zinsmeister and R. Mues (eds.), pp. 369–410. Oxford, U.K.: Clarendon Press.

Asakawa, Y., 1995. Chemical constituents of the bryophytes. In: *Progress in the Chemistry of Organic Natural Products*, W. Herz, G.W. Kirby, R.E. Moore, W. Steglich, and Ch. Tamm (eds.), Vol. 65, pp. 1–618. Vienna, Austria: Springer.

Asakawa, Y., 1999. Phytochemistry of bryophytes. In: *Phytochemicals in Human Health Protection, Nutrition, and Plant Defense*, J. Romeo (ed.), Vol. 33, pp. 319–342. New York: Kluwer Academic, Plenum Publishers.

Asakawa, Y., 2007. Biologically active compounds from bryophytes. *Pure Appl. Chem.*, 75: 557–580.

Asakawa, Y., 2008. Recent advances of biologically active substances from the Marchantiophyta. *Nat. Prod. Commun.*, 3: 77–92.

Asakawa, Y., T. Hashimoto, Y. Mizuno, M. Tori, and Y. Fukuzawa, 1992. Cryptoporic acids A-G, drimane-type sesquiterpenoid ethers of isocitric acid from the fungus *Cryptoporus volvatus*. *Phytochemistry*, 31: 579–592.

Asakawa, Y., T. Ishida, M. Toyota, and T. Takemoto, 1986. Terpenoid biotransformation in mammals IV. Biotransformation of (+)-longifolene, (−)-caryophyllene, (−)-caryophyllene oxide, (−)-cyclocolorenone, (+)-nootkatone, (−)-elemol, (−)-abietic acid and (+)-dehydroabietic acid in rabbits. *Xenobiotica*, 6: 753–767.

Asakawa, Y. and A. Ludwiczuk, 2008. Bryophytes-Chemical diversity, bioactivity and chemosystematics. Part 1. Chemical diversity and bioactivity. *Med. Plants Poland World*, 14: 33–53.

Asakawa, Y., Z. Taira, T. Takemoto, T. Ishida, M. Kido, and Y. Ichikawa, 1981. X-ray crystal structure analysis of 14-hydroxycaryophyllene oxide, a new metabolite of (−)-caryophyllene in rabbits. *J. Pharm. Sci.*, 70: 710–711.

Asakawa, Y., H. Takahashi, and M. Toyota, 1991. Biotransformation of germacrane-type sesquiterpenoids by *Aspergillus niger*. *Phytochemistry*, 30: 3993–3997. Ata, A. and J.A. Nachtigall, 2004. Microbial transformation of α-santonin. *Z. Naturforsch.*, 59C: 209–214.

Atta-ur-Rahman, M.I. Choudhary, A. Ata et al., 1994. Microbial transformation of 7α-hydroxyfrullanolide. *J. Nat. Prod.*, 57: 1251–1255. Atta-ur-Rahman, M.I. Choudhary, F. Shaheen, A. Rauf, and A. Farooq, 1998. Microbial transformation of some bioactive natural products. *Nat. Prod. Lett.*, 12: 215–222.

Atta-ur-Rahman, A. Farooq, and M.I. Choudhary, 1997. Microbial transformation of sclareolide. *J. Nat. Prod.*, 60: 1038–1040.

Ayer, W.A. and P.A. Craw, 1989. Metabolites of fairy ring fungus, *Marasmius oreades*. Part 2. Norsesquiterpenes, further sesquiterpenes, and argocybin. *Can. J. Chem.*, 67: 1371–1380.

Bang, L. and G. Ourisson, 1975. Hydroxylation of cedrol by rabbits. *Tetrahedron Lett.*, 16: 1881–1884.

Bang, L., G. Ourisson, and P. Teisseire, 1975. Hydroxylation of patchoulol by rabbits. Hemisynthesis of norpatchoulenol, the odour carrier of patchouli oil. *Tetrahedron Lett.*, 16: 2211–2214.

Barrero, A.F., J.E. Oltra, D.S. Raslan, and D.A. Sade, 1999. Microbial transformation of sesquiterpene lactones by the fungi *Cunninghamella echinulata* and *Rhizopus oryzae*. *J. Nat. Prod.*, 62: 726–729.

Bhutani, K.K. and R.N. Thakur, 1991. The microbiological transformation of parthenin by *Beauveria bassiana* and *Sporotrichum pulverulentum*. *Phytochemistry*, 30: 3599–3600.

Buchanan, G.O., L.A.D. Williams, and P.B. Reese, 2000. Biotransformation of cadinane sesquiterpenes by *Beauveria bassiana* ATCC 7159. *Phytochemistry*, 54: 39–45.

Choudhary, M.I., W. Kausar, Z.A. Siddiqui, and Atta-ur-Rahman, 2006a. Microbial metabolism of (+)-cycloisolongifol-5β-ol. *Z. Naturforsch.*, 61B: 1035–1038.

Choudhary, M.I., S.G. Musharraf, S.A. Nawaz et al., 2005. Microbial transformation of (−)-isolongifolol and butyrylcholinesterase inhibitory activity of transformed products. *Bioorg. Med. Chem.*, 13: 1939–1944.

Choudhary, M.I., S.G. Musharraf, A. Sami, and Atta-ur-Rahman, 2004. Microbial transformation of sesquiterpenes, (−)-ambrox® and (+)-sclareolide. *Helv. Chim. Acta*, 87: 2685–2694.

Choudhary, M.I., Z.A. Siddiqui, S.A. Nawaz, and Atta-ur-Rahman, 2006b. Microbial transformation and butyrylcholinesterase inhibitory activity of (−)-caryophyllene oxide and its derivatives. *J. Nat. Prod.*, 69: 1429–1434.

Clark, A.M. and C.D. Hufford, 1979. Microbial transformation of the sesquiterpene lactone costunolide. *J. Chem. Soc. Perkin Trans.*, 1: 3022–3028.

Collins, D.O. and P.B. Reese, 2002. Biotransformation of cadina-4,10(15)-dien-3-one and 3α-hydroxycadina-4,10(15)-diene by *Curvularia lunata* ATCC 12017. *Phytochemistry*, 59: 489–492.

Collins, D.O., W.F. Reynold, and P.B. Reese, 2002a. Aromadendrane transformations by *Curvularia lunata* ATCC 12017. *Phytochemistry*, 60: 475–481.

Collins, D.O., P.L.D. Ruddock, J. Chiverton, C. de Grasse, W.F. Reynolds, and P.B. Reese, 2002b. Microbial transformation of cadina-4,10(15)-dien-3-one, aromadendr-1(10)-en-9-one and methyl ursolate by *Mucor plumbeus* ATCC 4740. *Phytochemistry*, 59: 479–488.

Daniewski, W.M., P.A. Grieco, J. Huffman, A. Rymkiewicz, and A. Wawrzun, 1981. Isolation of 12-hydroxycaryophyllene-4,5-oxide, a sesquiterpene from *Lactarius camphoratus*. *Phytochemistry*, 20: 2733–2734.

Darrouzet, E., B. Mauchamp, G.D. Prestwich, L. Kerhoas, I. Ujvary, and F. Couillaud 1997. Hydroxy juvenile hormones: New putative juvenile hormones biosynthesized by locust corpora allata *in vitro*. *Biochem. Biophys. Res. Commun.*, 240: 752–758.

Devi, J.R., 1979. Microbiological transformation of terpenes: Part XXVI. Microbial transformation of caryophyllene. *Ind. J. Biochem. Biophys.*, 16: 76–79.

Dhavlikar, R.S. and G. Albroscheit, 1973. Microbiologische Umsetzung von Terpenen: Valencen. *Dragoco Rep.*, 12: 250–258.

Duran, R., E. Corrales, R. Hernandez-Galan, and G. Collado, 1999. Biotransformation of caryophyllene oxide by *Botrytis cinerea*. *J. Nat. Prod.*, 62: 41–44.

El Sayed, K.A., M. Yousaf, M.T. Hamann, M.A. Avery, M. Kelly, and P. Wipf, 2002. Microbial and chemical transformation studies of the bioactive marine sesquiterpenes (S)-(+)-curcuphenol and -curcudiol isolated from a deep reef collection of the Jamaican sponge *Didiscus oxeata*. *J. Nat. Prod.*, 65: 1547–1553.

Furusawa, M., 2006. Microbial biotransformation of sesquiterpenoids from crude drugs and liverworts: Production of functional substances. PhD thesis. Tokushima Bunri University, Tokushima, Japan, pp. 1–156.

Furusawa, M., T. Hashimoto, Y. Noma, and Y. Asakawa, 2005a. Biotransformation of Citrus aromatics nootkatone and valencene by microorganisms. *Chem. Pharm. Bull.*, 53: 1423–1429.

Furusawa, M., T. Hashimoto, Y. Noma, and Y. Asakawa 2005b. Highly efficient production of nootkatone, the grapefruit aroma from valencene, by biotransformation. *Chem. Pharm. Bull.*, 53: 1513–1514.

Furusawa, M., T. Hashimoto, Y. Noma, and Y. Asakawa, 2005c. The structure of new sesquiterpenoids from the liverwort *Reboulia hemisphaerica* and their biotransformation. *Proceeding of 49th TEAC*, pp. 235–237.

Furusawa, M., T. Hashimoto, Y. Noma, and Y. Asakawa, 2006a. Biotransformation of aristolene- and 2,3-secoaromadendrane-type sesquiterpenoids having a 1,1-dimethylcyclopropane ring by *Chlorella fusca* var. *vacuolata, Mucor* species, and *Aspergillus niger*. *Chem. Pharm. Bull.*, 54: 861–868.

Furusawa, M., T. Hashimoto, Y. Noma, and Y. Asakawa, 2006b. Isolation and structures of new cyclomyltaylane and ent-chamigrane-type sesquiterpenoids from the liverwort *Reboulia hemisphaerica*. *Chem. Pharm. Bull.*, 54: 996–1003.

Furusawa, M., Y. Noma, T. Hashimoto, and Y. Asakawa, 2003. Biotransformation of Citrus oil nootkatone, dihydronootkatone and dehydronootkatone. *Proceedings of 47th TEAC*, pp. 142–144.

Furuya, T., Y. Asada, Y. Matsuura, S. Mizobata, and H. Hamada, 1997. Biotransformation of β-thujaplicin by cultured cells of *Eucalyptus perriniana*. *Phytochemistry*, 46: 1355–1358.

Galal, A.M., 2001. Microbial transformation of pyrethrosin. *J. Nat. Prod.*, 64: 1098–1099.

Galal, A.M., A.S. Ibrahim, J.S. Mossa, and F.S. El-Feraly, 1999. Microbial transformation of parthenolide. *Phytochemistry*, 51: 761–765.

Garcia-Granados, A., M.C. Gutierrez, F. Rivas, and J.M. Arias, 2001. Biotransformation of 4β-hydroxyeudesmane-1,6-dione by *Gliocladium roseum* and *Exserohilum halodes*. *Phytochemistry*, 58: 891–895.

Hamada, H., F. Murakami, and T. Furuya, 1998. The production of hinokitiol glycoside. *Proceedings of 42nd TEAC*, pp. 145–147.

Hanson, J.R. and H. Nasir, 1993. Biotransformation of the sesquiterpenoid, cedrol, by *Cephalosporium aphidicola*. *Phytochemistry*, 33: 835–837.

Hanson, J.R. and A. Truneh, 1996. The biotransformation of ambrox and sclareolide by *Cephalosporium aphidicola*. *Phytochemistry*, 42: 1021–1023.

Hanson, R.L., J.M. Wasylyk, V.B. Nanduri, D.L. Cazzulino, R.N. Patel, and L.J. Szarka, 1994. Site-specific enzymatic hydrolysis of tisanes at C-10 and C-13. *J. Biol. Chem.*, 269: 22145–22149.

Harinantenaina, L., Y. Noma, and Y. Asakawa, 2005. *Penicillium sclerotiorum* catalyzes the conversion of herbertenediol into its dimers: Mastigophorenes A and B. *Chem. Pharm. Bull.*, 53: 256–257.

Harinantenaina, L., D.N. Quang, T. Nishizawa et al., 2007. Bioactive compounds from liverworts: Inhibition of lipopolysaccharide-induced inducible NOS mRNA in RAW 264.7 cells by herbertenoids and cuparenoids. *Phytomedicine*, 14: 486–491.

Hashimoto, T. and Y. Asakawa, 2007. Biological activity of fragrant substances from Citrus and herbs, and production of functional substances using microbial biotransformation. In: *Development of Medicinal Foods*, M. Yoshikawa (ed.), pp. 168–184. Tokyo, Japan: CMC Publisher.

Hashimoto, T., Y. Asakawa, Y. Noma et al., 2003a. Production method of nootkatone. *Jpn. Kokai Tokkyo Koho*, 250591A.

Hashimoto, T., Y. Asakawa, Y. Noma et al., 2003b. Production method of nootkatone. *Jpn. Kokai Tokkyo Koho*, 70492A.

Hashimoto, T., M. Fujiwara, K. Yoshikawa, A. Umeyama, M. Tanaka, and Y. Noma, 2007. Biotransformation of sclareolide and sclareol by microorganisms. *Proceedings of 51st TEAC*, pp. 316–318.

Hashimoto, T., S. Kato, M. Tanaka, S. Takaoka, and Y. Asakawa, 1998a. Biotransformation of sesquiterpenoids by microorganisms (4): Biotransformation of hinesol by *Aspergillus niger*. *Proceedings of 42nd TEAC*, pp. 127–129.

Hashimoto, T., Y. Noma, Y. Akamatsu, M. Tanaka, and Y. Asakawa, 1999a. Biotransformation of sesquiterpenoids by microorganisms. (5): Biotransformation of dehydrocostuslactone. *Proceedings of 43rd TEAC*, pp. 202–204.

Hashimoto, T., Y. Noma, and Y. Asakawa, 2001a. Biotransformation of terpenoids from the crude drugs and animal origin by microorganisms. *Heterocycles*, 54: 529–559.

Hashimoto, T., Y. Noma, and Y. Asakawa, 2006. Biotransformation of cuparane- and herbertane-type sesquiterpenoids. *Proceedings of 50th TEAC*, pp. 263–265.

Hashimoto, T., Y. Noma, Y. Goto, S. Takaoka, M. Tanaka, and Y. Asakawa, 2003c. Biotransformation of sesquiterpenoids from the liverwort *Plagiochila* species. *Proceedings of 47th TEAC*, pp. 139–141.

Hashimoto, T., Y. Noma, Y. Goto, M. Tanaka, S. Takaoka, and Y. Asakawa, 2004. Biotransformation of (−)-maalioxide by *Aspergillus niger* and *Aspergillus cellulosae*. *Heterocycles*, 62: 655–666.

Hashimoto, T., Y. Noma, S. Kato, M. Tanaka, S. Takaoka, and Y. Asakawa, 1999b. Biotransformation of hinesol isolated from the crude drug *Atractylodes lancea* by *Aspergillus niger* and *Aspergillus cellulosae*. *Chem. Pharm. Bull.*, 47: 716–717. Hashimoto, T., Y. Noma, H. Matsumoto, Y. Tomita, M. Tanaka, and Y. Asakawa, 2000a. Microbial biotransformation of optically active (+)-α-ionone and (−)-α-ionone. *Proceedings of 44th TEAC*, pp. 154–156.

Hashimoto, T., Y. Noma, Y. Matsumoto, Y. Akamatsu, M. Tanaka, and Y. Asakawa, 1999c. Biotransformation of sesquiterpenoids by microorganisms. (6): Biotransformation of α-, β- and γ-cyclocostunolides. *Proceedings of 43rd TEAC*, pp. 205–207.

Hashimoto, T., Y. Noma, C. Murakami, N. Nishimatsu, M. Tanaka, and Y. Asakawa, 2001b. Biotransformation of valencene and aristolene. *Proceedings of 45th TEAC*, pp. 345–347.

Hashimoto, T., Y. Noma, C. Murakami, M. Tanaka, and Y. Asakawa, 2000b. Microbial transformation of α-santonin derivatives and nootkatone. *Proceedings of 44th TEAC*, pp. 157–159.

Hashimoto, T., Y. Noma, N. Nishimatsu, M. Sekita, M. Tanaka, and Y. Asakawa, 2003d. Biotransformation of antimalarial sesquiterpenoids by microorganisms. *Proceedings of 47th TEAC*, pp. 136–138.

Hashimoto, T., M. Sekita, M. Furusawa, Y. Noma, and Y. Asakawa, 2005. Biotransformation of sesquiterpene lactones, (−)-parthenolide and (−)-frullanolide by microorganisms. *Proceedings of 49th TEAC*, pp. 387–389.

Hashimoto, T., K. Shiki, M. Tanaka, S. Takaoka, and Y. Asakawa, 1998b. Chemical conversion of labdane-type diterpenoid isolated from the liverwort *Porella perrottetiana* into (−)-ambrox. *Heterocycles*, 49: 315–325.

Hayashi, K., H. Morikawa, A. Matsuo, D. Takaoka, and H. Nozaki, 1999. Biotransformation of sesquiterpenoids by *Aspergillus niger* IFO 4407. *Proceedings of 43rd TEAC*, pp. 208–210.

Hayashi, K., H. Morikawa, H. Nozaki, and D. Takaoka, 1998. Biotransformation of globulol and epiglubulol by *Aspergillus niger* IFO 4407. *Proceedings of 42nd TEAC*, pp. 136–138.

Haze, S., K. Sakai, and Y. Gozu, 2002. Effects of fragrance inhalation on sympathetic activity in normal adults. *Jpn. J. Pharmacol.*, 90: 247–253.

Hegazy, M.-E.F., C. Kuwata, Y. Sato et al., 2005. Research and development of asymmetric reaction using biocatalysts-biotransformation of enones by cultured cells of *Marchantia polymorpha*. *Proceedings of 49th TEAC*, pp. 402–404.

Higuchi, H., R. Tsuji, K. Hayashi, D. Takaoka, A. Matsuo, and H. Nozaki, 2001. Biotransformation of sesquiterpenoids by *Aspergillus niger*. *Proceedings of 45th TEAC*, pp. 354–355.

Hikino, H., T. Konno, T. Nagashima, T. Kohama, and T. Takemoto, 1971. Stereoselective epoxidation of germacrone by *Cunninghamella blakesleeana*. *Tetrahedron Lett.*, 12: 337–340.

Hrdlicka, P.J., A.B. Sorensen, B.R. Poulsen, G.J.G. Ruijter, J. Visser, and J.J.L. Iversen, 2004. Characterization of nerolidol biotransformation based on indirect on-line estimation of biomass concentration and physiological state in bath cultures of *Aspergillus niger*. *Biotechnol. Prog.*, 20: 368–376.

Ishida, T., 2005. Biotransformation of terpenoids by mammals, microorganisms, and plant-cultured cells. *Chem. Biodivers.*, 2: 569–590.

Ishida, T., Y. Asakawa, and T. Takemoto, 1982. Hydroxyisolongifolaldehyde: A new metabolite of (+)-longifolene in rabbits. *J. Pharm. Sci.*, 71: 965–966.

Ishida, T., Y. Asakawa, T. Takemoto, and T. Aratani, 1981. Terpenoid biotransformation in mammals. III. Biotransformation of α-pinene, β-pinene, pinane, 3-carene, carane, myrcene and p-cymene in rabbits. *J. Pharm. Sci.*, 70: 406–415.

Itsuzaki, Y., K. Ishisaka, and M. Miyazawa, 2002. Biotransformation of (+)-cedrol by using human skin microbial flora *Staphylococcus epidermidis*. *Proceedings of 46th TEAC*, pp. 101–102.

Jacobsson, U., V. Kumar, and S. Saminathan, 1995. Sesquiterpene lactones from *Michelia champaca*. *Phytochemistry*, 39: 839–843.

Kaspera, R., U. Krings, T. Nanzad, and R.G. Berger, 2005. Bioconversion of (+)-valencene in submerged cultures of the ascomycete *Chaetomium globosum*. *Appl. Microbiol. Biotechnol.*, 67: 477–583.Kiran, I., T. Akar, A. Gorgulu, and C. Kazaz, 2005. Biotransformation of racemic diisophorone by *Cephalosporium aphidicola* and *Neurospora crassa*. *Biotechnol. Lett.*, 27: 1007–1010.

Kiran, I., H.N. Yildirim, J.R. Hanson, and P.B. Hitchcock, 2004. The antifungal activity and biotransformation of diisophorone by the fungus *Aspergillus niger*. *J. Chem. Technol. Biotechnol.*, 79: 1366–1370.

Kumari, G.N.K., S. Masilamani, R. Ganesh, and S. Aravind, 2003. Microbial biotransformation of zaluzanin D. *Phytochemistry*, 62: 1101–1104.

Lahlou, E.L., Y. Noma, T. Hashimoto, and Y. Asakawa, 2000. Microbiotransformation of dehydropinguisenol by *Aspergillus* sp. *Phytochemistry*, 54: 455–460.

Lamare, V. and R. Furstoss, 1990. Bioconversion of sesquiterpenes. *Tetrahedron*, 46: 4109–4132.

Lee, I.-S., H.N. ElSohly, E.M. Coroom, and C.D. Hufford, 1989. Microbial metabolism studies of the antimalarial sesquiterpene artemisinin. *J. Nat. Prod.*, 52: 337–341.

Ma, X.C., M. Ye, L.J. Wu, and D.A. Guo, 2006. Microbial transformation of curdione by *Mucor spinosus*. *Enzyme Microb. Technol.*, 38: 367–371.

Maatooq, G.A., 2002b. Microbial transformation of a β- and γ-eudesmols mixture. *Z. Naturforsch.*, 57C: 654–659.
Maatooq, G.A., 2002c. Microbial conversion of partheniol by *Calonectria decora*. *Z. Naturforsch.*, 57C: 680–685.
Maatooq, G.T., 2002a. Microbial metabolism of partheniol by *Mucor circinelloides*. *Phytochemistry*, 59: 39–44.
Madyastha, K.M. and T.L. Gururaja, 1993. Utility of microbes in organic synthesis: Selective transformations of acyclic isoprenoids by *Aspergillus niger*. *Ind. J. Chem.*, 32B: 609–614.
Matsui, H., Y. Minamino, and M. Miyazawa, 1999. Biotransformation of (+)-cedryl acetate by *Glomerella cingulata*, parasitic fungus. *Proceedings of 43rd TEAC*, 215–216.
Matsui, H. and M. Miyazawa, 2000. Biotransformation of pathouli acetate using parasitic fungus *Glomerella cingulata* as a biocatalyst. *Proceedings of 44th TEAC*, pp.149–150.
Matsushima, A., M.-E.F. Hegazy, C. Kuwata, Y. Sato, M. Otsuka, and T. Hirata, 2004. Biotransformation of enones using plant cultured cells—The reduction of α-santonin. *Proceedings of 48th TEAC*, pp. 396–398.
Meyer, P. and C. Neuberg 1915. Phytochemische reduktionen. XII. Die umwandlung von citronellal in citronelol. *Biochem. Z*, 71: 174–179.
Mikami, Y., 1988. *Microbial Conversion of Terpenoids*. Biotechnology and Genetic Engineering Reviews, Vol. 6, pp. 271–320. Wimborne, UK: Intercept Ltd.
Miyazawa, M., S. Akazawa, H. Sakai, and H. Kameoka, 1997a. Biotransformation of (+)-γ-gurjunene using plant pathogenic fungus, *Glomerella cingulata* as a biocatalyst. *Proceedings of 41st TEAC*, pp. 218–219.
Miyazawa, M., Y. Honjo, and H. Kameoka, 1996a. Biotransformation of guaiol and bulnesol using plant pathogenic fungus *Glomerella cingulata* as a biocatalyst. *Proceedings of 40th TEAC*, pp. 82–83.
Miyazawa, M., Y. Honjo, and H. Kameoka, 1997b. Biotransformation of the sesquiterpenoid β-selinene using the plant pathogenic fungus *Glomerella cingulata*. *Phytochemistry*, 44: 433–436.
Miyazawa, M., Y. Honjo, and H. Kameoka, 1998a. Biotransformation of the sesquiterpenoid (+)-γ-gurjunene using a plant pathogenic fungus *Glomerella cingulata* as a biocatalyst. *Phytochemistry*, 49: 1283–1285.
Miyazawa, M., H. Matsui, and H. Kameoka, 1997c. Biotransformation of patchouli alcohol using plant parasitic fungus *Glomerella cingulata* as a biocatalyst. *Proceedings of 41st TEAC*, pp. 220–221.
Miyazawa, M., H. Matsui, and H. Kameoka, 1998b. Biotransformation of unsaturated sesquiterpene alcohol using plant parasitic fungus, *Glomerella cingulata* as a biocatalyst. *Proceedings of 42nd TEAC*, pp. 121–122.
Miyazawa, M., H. Nakai, and H. Kameoka, 1995a. Biotransformation of (+)-cedrol by plant pathogenic fungus, *Glomerella cingulata*. *Phytochemistry*, 40: 69–72.
Miyazawa, M., H. Nakai, and H. Kameoka, 1995b. Biotransformation of cyclic terpenoids, (±)-cis-nerolidol and nerylacetone, by plant pathogenic fungus, *Glomerella cingulata*. *Phytochemistry*, 40: 1133–1137.
Miyazawa, M., H. Nakai, and H. Kameoka, 1996b. Biotransformation of acyclic terpenoid (±)-*trans*-nerolidol and geranylacetone by *Glomerella cingulata*. *J. Agric. Food Chem.*, 44: 1543–1547.
Miyazawa, M., H. Nakai, and H. Kameoka, 1996c. Biotransformation of acyclic terpenoid (2E,6E)-farnesol by plant pathogenic fungus *Glomerella cingulata*. *Phytochemistry*, 43: 105–109.
Miyazawa, M., H. Nankai, and H. Kameoka, 1995c. Biotransformation of (−)-α-bisabolol by plant pathogenic fungus, *Glomerella cingulata*. *Phytochemistry*, 39: 1077–1080.
Miyazawa, M. and A. Sugawara, 2006. Biotransformation of (2)-guaiol by *Euritium rubrum*. *Nat. Prod. Res.*, 20: 731–734.
Miyazawa, M., T. Uemura, and H. Kameoka, 1994. Biotransformation of sesquiterpenoids, (−)-globulol and (+)-ledol by *Glomerella cingulata*. *Phytochemistry*, 37: 1027–1030.
Miyazawa, M., T. Uemura, and H. Kameoka, 1995d. Biotransformation of sesquiterpenoids, (+)-aromadendrene and (−)-alloaromadendrene by *Glomerella cingulata*. *Phytochemistry*, 40: 793–796.
Morikawa, H., K. Hayashi, K. Wakamatsu et al., 2000. Biotransformation of sesquiterpenoids by *Aspergillus niger* IFO 4407. *Proceedings of 44th TEAC*, pp. 151–153.
Nankai, H., M. Miyazawa, and H. Kameoka, 1996. Biotransformation of (Z,Z)-farnesol using plant pathogenic fungus, *Glomerella cingulata* as a biocatalyst. *Proceedings of 40th TEAC*, pp. 78–79.
Nishida, E. and Y. Kawai, 2007. Bioconversion of zerumbone and its derivatives. *Proceedings of 51st TEAC*, pp. 387–389.
Noma, Y. and Y. Asakawa, 1998. Microbiological transformation of 3-oxo-α-ionone. *Proceedings of 44th TEAC*, pp. 133–135.
Noma, Y. and Y. Asakawa, 2006. Biotransformation of (−)-nopol benzyl ether. *Proceedings of 50th TEAC*, pp. 434–436.
Noma, Y., M. Furusawa, C. Murakami, T. Hashimoto, and Y. Asakawa, 2001a. Formation of nootkatol and nootkatone from valencene by soil microorganisms. *Proceedings of 45th TEAC*, pp. 91–92.

Noma, Y., T. Hashimoto, and Y. Asakawa, 2001b. Microbiological transformation of damascone. *Proceedings of 45th TEAC*, pp. 93–95.
Noma, Y., T. Hashimoto, and Y. Asakawa, 2001c. Microbial transformation of adamantane. *Proceedings of 45th TEAC*, pp. 96–98.
Noma, Y., T. Hashimoto, Y. Akamatsu, S. Takaoka, and Y. Asakawa, 1999. Microbial transformation of adamantane (Part 1). *Proceedings of 43rd TEAC*, pp. 199–201.
Noma, Y., T. Hashimoto, S. Kato, and Y. Asakawa, 1997a. Biotransformation of (+)-β-eudesmol by *Aspergillus niger*. *Proceedings of 41st TEAC*, pp. 224–226.
Noma, Y., T. Hashimoto, A. Kikkawa, and Y. Asakawa, 1996. Biotransformation of (−)-α-eudesmol by *Asp. niger* and *Asp. cellulosae* M-77. *Proceedings of 40th TEAC*, pp. 95–97.
Noma, Y., T. Hashimoto, S. Sawada, T. Kitayama, and Y. Asakawa, 2002. Microbial transformation of zerumbone. *Proceedings of 46th TEAC*, pp. 313–315.
Noma, Y., K. Matsueda, I. Maruyama, and Y. Asakawa, 1997b. Biotransformation of terpenoids and related compounds by *Chlorella* species. *Proceedings of 41st TEAC*, pp. 227–229.
Noma, Y., Y. Takahashi, and Y. Asakawa, 2003. Stereoselective reduction of racemic bicycle[33.1]nonane-2,6-dione and 5-hydroxy-2-adamantanone by microorganisms. *Proceedings of 46th TEAC*, pp. 118–120.
Nozaki, H., K. Asano, K. Hayashi, M. Tanaka, A. Masuo, and D. Takaoka, 1996. Biotransformation of shiromodiol diacetate and myli-4(15)-en-9-one by *Aspergillus niger* IFO 4407. *Proceedings of 40th TEAC*, pp. 108–110.
Okuda, M., K. Sonohara, and H. Takikawa, 1994. Production of natural flavors by laccase catalysis. *Jpn. Kokai Tokkyo Koho*, 303967.
Orabi, K.Y., 2000. Microbial transformation of the eudesmane sesquiterpene plectranthone. *J. Nat. Prod.*, 63: 1709–1711.
Orabi, K.Y., 2001. Microbial epoxidation of the tricyclic sesquiterpene presilphiperfolane angelate ester. *Z. Naturforsch.*, 56C: 223–227.
Otsuka, S. and K. Tani, 1991. Catalytic asymmetric hydrogen migration of ally amines. *Synthesis*, 665–680.
Parshikov, I.A., K.M. Muraleedharan, and M.A. Avery, 2004. Transformation of artemisinin by Cunninghamella elegans. *Appl. Microbiol. Biotechnol.*, 64: 782–786.
Sakamoto, S., N. Tsuchiya, M. Kuroyanagi, and A. Ueno, 1994. Biotransformation of germacrone by suspension cultured cells. *Phytochemistry*, 35: 1215–1219.
Sakata, K., I. Horibe, and M. Miyazawa, 2007. Biotransformation of (+)-α-longipinene by microorganisms as a biocatalyst. *Proceedings of 51st TEAC*, 321–322.
Sakata, K. and M. Miyazawa, 2006. Biotransformation of (+)-isolongifolen-9-one by *Glomerella cingulata* as a biocatalyst. *Proceedings of 50th TEAC*, pp. 258–260.
Sakui, N., M. Kuroyamagi, Y. Ishitobi, M. Sato, and A. Ueno, 1992. Biotransformation of sesquiterpenes by cultured cells of *Curcuma zedoaria*. *Phytochemistry*, 31: 143–147.
Sakui, N., M. Kuroyanagi, M. Sato, and A. Ueno, 1988. Transformation of ten-membered sesquiterpenes by callus of *Curcuma*. *Proceedings of 32nd TEAC*, pp. 322–324.
Salvador, J.A.R. and J.H. Clark, 2002. The allylic oxidation of unsaturated steroids by *tert*-butyl hydroperoxide using surface functionalized silica supported metal catalysts. *Green Chem.*, 4: 352–356.
Sekita, M., M. Furusawa, T. Hashimoto, Y. Noma, and Y. Asakawa, 2005. Biotransformation of pungent tasting polygodial from *Polygonum hydropiper* and related compounds by microorganisms. *Proceedings of 49th TEAC*, pp. 380–381.
Sekita, M., T. Hashimoto, Y. Noma, and Y. Asakawa, 2006. Biotransformation of biologically active terpenoids, sacculatal and cinnamodial by microorganisms. *Proceedings of 50th TEAC*, pp. 406–408.
Shimoda, K., T. Harada, H. Hamada, N. Nakajima, and H. Hamda, 2007a. Biotransformation of raspberry ketone and zingerone by cultured cells of *Phytolacca americana*. *Phytochemistry*, 68: 487–492.
Shimoda, K., Y. Kondo, T. Nishida, H. Hamada, N. Nakajima, and H. Hamada, 2006. Biotransformation of thymol, carvacrol, and eugenol by cultured cells of *Eucalyptus perriniana*. *Phytochemistry*, 67: 2256–2261.
Shimoda, K., S. Kwon, A. Utsuki et al., 2007b. Glycosylation of capsaicin and 8-norhydrocapsaicin by cultured cells of *Catharanthus roseus*. *Phytochemistry*, 68: 1391–1396.
Shoji, N., A. Umeyama, Y. Asakawa, T. Takeout, K. Nocoton, and Y. Ohizumi, 1984. Structure determination of nootkatol, a new sesquiterpene isolated form *Alpinia oxyphylla* Miquel possessing calcium antagonist activity. *J. Pharm. Sci.*, 73: 843–844.
Sowden, R.J., S. Yasmin, N.H. Rees, S.G. Bell, and L.-L. Wong, 2005. Biotransformation of the sesquiterpene (+)-valencene by cytochrome $P450_{cam}$ and $P450_{BM-3}$. *Org. Biomol. Chem.*, 3: 57–64.
Sugawara, A. and M. Miyazawa, 2004. Biotransformation of guaiene using plant pathogenic fungus, *Eurotium rubrum* as a biocatalyst. *Proceedings of 48th TEAC*, pp. 385–386.

Sutherland, T.D., G.C. Unnithan, J.F. Andersen et al., 1998. A cytochrome P450 terpenoid hydroxylase linked to the suppression of insect juvenile hormone synthesis. *Proc. Natl. Acad. Sci. USA*, 95: 12884–12889.

Takahashi, H., 1994. Biotransformation of terpenoids and aromatic compounds by some microorganisms. PhD thesis. Tokushima Bunri University, Tokushima, Japan, pp. 1–115.

Takahashi, H., T. Hashimoto, Y. Noma, and Y. Asakawa, 1993a. Biotransformation of 6-gingerol, and 6-shogaol by *Aspergillus niger*. *Phytochemistry*, 34: 1497–1500.

Takahashi, T., I. Horibe, and M. Miyazawa, 2007. Biotransformation of β-selinene by *Aspergillus wentii*. *Proceedings of 51st TEAC*, pp. 319–320.

Takahashi, T. and M. Miyazawa, 2005. Biotransformation of (+)-nootkatone by *Aspergillus wentii*, as biocatalyst. *Proceedings of 49th TEAC*, pp. 393–394.

Takahashi, T. and M. Miyazawa, 2006. Biotransformation of sesquiterpenes which possess an eudesmane skeleton by microorganisms. *Proceedings of 50th TEAC*, pp. 256–257.

Takahashi, H., M. Toyota, and Y. Asakawa, 1993b. Drimane-type sesquiterpenoids from *Cryptoporus volvatus* infected by *Paecilomyces variotii*. *Phytochemistry*, 33: 1055–1059.

Tani, K., T. Yamagata, S. Otsuka et al., 1982. Cationic rhodium (I) complex-catalyzed asymmetric isomerization of allylamines to optically active enamines. *J. Chem. Soc. Chem. Commun.*, 600–601.

Tori, M., M. Sono, and Y. Asakawa, 1990. The reaction of three sesquiterpene ethers with *m*-chloroperbenzoic acid. *Bull. Chem. Soc. Jpn.*, 63: 1770–1776.

Venkateswarlu, Y., P. Ramesh, P.S. Reddy, and K. Jamil, 1999. Microbial transformation of $D^{9(15)}$-africane. *Phytochemistry*, 52: 1275–1277.

Wang, Y., T.-K. Tan, G.K. Tan, J.D. Connolly, and L.J. Harrison, 2006. Microbial transformation of the sesquiterpenoid (−)-maalioxide by *Mucor plumbeus*. *Phytochemistry*, 67: 58–61.

Wilson, C.W. III and P.E. Saw, 1978. Quantitative composition of cold-pressed grapefruit oil. *J. Agric. Food Chem.*, 26: 1430–1432.

Yang, L., K. Fujii, J. Dai, J. Sakai, and M. Ando, 2003. Biotransformation of α-santonin and its C-6 epimer by fungus and plant cell cultures. *Proceedings of 47th TEAC*, pp. 148–150.

Zhan, J., H. Guo, J. Dai, Y. Zhang, and D. Guo, 2002. Microbial transformation of artemisinin by *Cunninghamella echinulata* and *Aspergillus niger*. *Tetrahedron Lett.*, 43: 4519–4521.

Zundel, J.-L., 1976. PhD thesis. Universite Louis Pasteur. Strasbourg, France.

21 Industrial Uses of Essential Oils

W. S. Brud

CONTENTS

21.1 Introduction 1011
21.2 History 1011
21.3 Fragrances 1013
21.4 Flavors 1013
21.5 Production and Consumption 1014
21.6 Changing Trends 1017
21.7 Conclusions 1021
Acknowledgments 1021
References 1021
Further Readings 1021
Websites 1021

21.1 INTRODUCTION

The period when essential oils were used first on an industrial scale is difficult to identify. The nineteenth century is generally regarded as the commencement of the modern phase of industrial application of essential oils. However, the large-scale usage of essential oils dates back to ancient Egypt. In 1480 BC, Queen Hatshepsut of Egypt sent an expedition to the country of Punt (now Somalia) to collect fragrant plants, oils, and resins as ingredients for perfumes, medicaments, and flavors and for the mummification of bodies. Precious fragrances have been found in many Egyptian archeological excavations, as a symbol of wealth and social position.

If significant international trade of essential oil-based products is the criterion for industrial use, the *Queen of Hungary's Water* was the first alcoholic perfume in history. This fragrance, based on rosemary essential oil distillate, was created in the mid-fourteenth century for the Polish-born Queen Elisabeth of Hungary. Following a special presentation to King Charles V, The Wise of France in 1350, it became popular in all medieval European courts. The beginning of the eighteenth century saw the introduction of *eau de cologne*, based on bergamot and other citrus oils, which remains widely used to this day. This fresh citrus fragrance was the creation of Jean Maria Farina, a descendant of Italian perfumers who came to France with Catherine de Medici and settled in Grasse in the sixteenth century. According to the city of Cologne archives, Jean Maria Farina and Karl Hieronymus Farina, in 1749, established factory (Fabriek) of this water, which sounds very *industrial*. The *Kölnisch Wasser* became the first unisex fragrance rather than one simply for men, known and used all over Europe, and it has been repeated subsequently in innumerable countertypes as a fragrance for men.

21.2 HISTORY

The history of production of essential oils dates back to ca. 3500 BC when the oldest known water distillation equipment for essential oils was employed and may be seen today in the Taxila Museum in Pakistan. Ancient India, China, and Egypt were the locations where essential oils were produced and widely used as medicaments, flavors, and fragrances. Perfumes came to Europe most probably from the East at the time of the crusades, and perfumery was accorded a professional status by the approval of a

French guild of perfumers in Grasse by King Philippe Auguste in 1190. For centuries, Grasse remained the center of world perfumery and was also the home of the first ever officially registered essential oil–producing company—Antoine Chiris—in 1768. (It is worth noting that not much later, in 1798, the first American essential oil company—Dodge and Olcott Inc.—was established in New York.)

About 150 years earlier, in 1620, an Englishman, named Yardley, obtained a concession from King Charles I to manufacture soap for the London area. Details of this event are sparse, other than the high fee paid by Yardley for this privilege. Importantly, however, Yardley's soap was perfumed with English lavender, which remains the Yardley trademark today, and it was probably the first case of use of an essential oil as a fragrance in large-scale soap production.

The use of essential oils as food ingredients has a history dating back to ancient times. There are many examples of the use of citrus and other squeezed (manually or mechanically expressed) oils for sweets and desserts in ancient Egypt, Greece, and the Roman Empire. Numerous references exist to flavored ice creams in the courts of the Roman Emperor Nero and of China. The reintroduction of recipes in Europe is attributed to Marco Polo on his return from traveling to China. In other stories, Catherine de Medici introduced ice creams in France, whereas Charles I of England served the first dessert in the form of frozen cream. Ice was used for freezing drinks and food in many civilizations, and the Eastern practice of using spices and spice essential oils both as flavoring ingredients and as food conservation agents was adopted centuries ago in Europe.

Whatever may be regarded as the date of their industrial production, essential oils, together with a range of related products—pomades, tinctures, resins, absolutes, extracts, distillates, concretes, and so on—were the only ingredients of flavor and fragrance products until the late nineteenth century. At this stage, the growth in consumption of essential oils as odoriferous and flavoring materials stimulated the emergence of a great number of manufacturers in France, the United Kingdom, Germany, Switzerland, and the United States (Table 21.1).

TABLE 21.1
First Industrial Manufacturers of Essential Oils, Flavors, and Fragrances

Company Name	Country	Established
Antoine Chiris	France (Grasse)	1768
Cavallier Freres	France (Grasse)	1784
Dodge & Olcott Inc.	United States (New York)	1798
Roure Bertrand Fils and Justin Dupont	France (Grasse)	1820
Schimmel & Co.	Germany (Leipzig)	1829
J. Mero-Boyveau	France (Grasse)	1832
Stafford Allen and Sons	United Kingdom (London)	1833
Robertet et Cie	**France (Grasse)**	**1850**
W.J. Bush	United Kingdom (London)	1851
Payan–Bertrand et Cie	**France (Grasse)**	**1854**
A. Boake Roberts	United Kingdom (London)	1865
Fritzsche–Schimmel Co.	United States (New York)	1871
V. Mane et Fils	**France (Grasse)**	**1871**
Haarmann & Reimer	Germany (Holzminden)	1874
R.C. Treatt Co.	**United Kingdom (Bury)**	**1886**
N.V. Polak und Schwartz	Holland (Zaandam)	1889
Ogawa and Co.	**Japan (Osaka)**	**1893**
Firmenich and Cie	**Switzerland (Geneva)**	**1895**
Givaudan S.A.	**Switzerland (Geneva)**	**1895**
Maschmeijer Aromatics	Holland (Amsterdam)	1900

Note: Companies continuing to operate under their original name are printed in bold.

The rapid development of the fragrance and flavor industry in the nineteenth century was generally based on essential oils and related natural products. In 1876, however, Haarmann and Reimer started the first production of synthetic aroma chemicals—vanillin and then coumarin, anisaldehyde, heliotropin, and terpineol. Although aroma chemicals made a revolution in fragrances with top discoveries in the twentieth century, for many decades both flavors and fragrances were manufactured with constituents of natural origin, the majority of which were essential oils.

21.3 FRAGRANCES

The main reason for the expansion of the essential oil industry and the growing demand for products was the development of the food, soap, and cosmetics industries. Today's multinational companies, the main users of fragrances and flavors, have evolved directly from the developments during the mid-nineteenth century.

In 1806, William Colgate opened his first store for soaps, candles, and laundry starch on Dutch Street in New York. In 1864, B. J. Johnson in Milwaukee started the production of soap, which came to be known as Palmolive from 1898. In 1866, Colgate launched its first perfumed soaps and perfumes. In 1873, Colgate launched toothpaste in a glass jug on the market and in the tube first in 1896. In 1926, two soap manufacturers—Palmolive and Peet—merged to create Palmolive-Peet, which 2 years later merged with Colgate to establish the Colgate-Palmolive-Peet Company (renamed as the Colgate-Palmolive Company in 1953).

In October 1837, William Procter and James Gamble signed a formal partnership agreement to develop their production and marketing of soaps (Gamble) and candles (Procter). *Palm oil*, *rosin*, *toilet*, and *shaving* soaps were listed in their advertisements. An *oleine* soap was described as having a violet odor. Only 22 years later, Procter & Gamble (P&G) sales reached 1 million dollars. In 1879, a fine but inexpensive *ivory* white toilet soap was offered to the market with all-purpose applications as a toilet and laundry product. In 1890, P&G was selling more than 30 different soaps.

The story of a third player started in 1890 when William Hesketh Lever created his concept of the Sunlight Soap, which revolutionized the idea of cleanliness and hygiene in Victorian Britain.

The very beginning of twentieth century marked the next big event when the young French chemist Eugene Schueller prepared his first hair color in 1907 and established what is now L'Oreal. These were the flagships in hundreds of emerging (and disappearing by fusions, takeovers, or bankruptcy) manufacturers of perfumes, cosmetics, toiletries, detergents, household chemicals, and related products, the majority of which were and are perfumed with essential oils.

21.4 FLAVORS

Over the same time period, another group of users of essential oils entered the markets. In 1790, the term *soda water* for carbon dioxide saturated water as a new drink appeared for the first time in the United States, and in 1810, the first U.S. patent was issued for the manufacture of imitations of natural gaseous mineral waters. Only 9 years later, the soda fountain was patented by Samuel Fahnestock. In 1833, carbonated lemonade flavored with lemon juice and citric acid was on sale in England. In 1835, the first bottled soda water appeared in the United States. It is, however, interesting that the first flavored sparkling drink—ginger ale—was created in Ireland in 1851. The milestones in flavored soft drinks appeared 30 years later: in 1881, the first cola-flavored drink was introduced in the United States; in 1885, Dr. Pepper was invented by Charles Alderton in Waco, Texas; in 1886, Coca-Cola was developed by Dr. John S. Pemberton in Atlanta, Georgia; and in 1898, Pepsi-Cola was created by Caleb Bradham, which is known from 1893 as *Brad's Drink*.

Dr. Pepper was advertised as the king of beverages, free from caffeine (which was added to it later on), was flavored with black cherry artificial flavor, and was first sold in the Old Corner Drug Store owned by Wade Morrison. Its market success and position as one of the most popular U.S.

soft drinks started by a presentation during the St. Louis World's Fair, where some other important flavor-consuming products—ice cream cones, hot dog rolls, and hamburger buns—were also shown. All of them remain major users of natural flavors based on essential oils. Hundred years later after the merger with another famous lemon–lime drink 7UP in 1986, it finally became a part of Cadbury.

Dr. John Pemberton was a pharmacist and he mixed up a combination of lime, cinnamon, coca leaves, and cola to make the flavor for his famous drink, first used as a remedy against headache (Pemberton's French Wine Coca) and then reformulated according to the prohibition law and used it to add taste to soda water from his *soda fountain*. The unique name and logo were created by his bookkeeper Frank Robinson, and Coca-Cola was advertised as a delicious, exhilarating, refreshing, and invigorating temperance drink. Interestingly, the first year of sales resulted in $20 loss, as the cost of the flavor syrup used for the drink was higher than the total sales of $50. In 1887, another pharmacist, Asa Candler, bought the idea and with aggressive marketing in 10 years introduced his drink all over the United States and Canada by selling syrup to other companies licensed to manufacture and retail the drink. Until 1905, Coca-Cola was known as a tonic drink and contained the extract of cocaine and cola nuts and the flavoring of lime and sugar.

Like Pemberton, Caleb Bradham was a pharmacist and in his drugstore, he offered soda water from a soda fountain. To promote sales, he flavored the soda with sugar, vanilla, pepsin, cola, and *rare oils*—obviously the essential oils of lemon and lime—and started selling it as a cure for dyspepsia, under the name Brad's Drink rather than Pepsi-Cola.

The development of the soft drinks industry is of great importance because it is a major consumer of essential oils, especially those of citrus origin. It is enough to say that nowadays, according to their web pages, only Coca-Cola-produced beverages are consumed worldwide in a quantity exceeding 1 billion drinks per day. If we consider that the average content of the appropriate essential oil in the final drink is about 0.001%–0.002% and the standard drink is ca. 0.3 l (300 g), we approach a daily consumption of essential oils by this company alone at the level of 3–6 tons per day, which gives an annual usage well over 2000 tons. Although all other brands of the food industry use substantial quantities of essential oils in ice creams, confectionary, bakery, and a variety of fast foods (where spice oils are used), these together use less oils than the beverage manufacturers.

There is one special range of products that can be situated between the food and cosmetics–toiletries industry sectors, and it is a big consumer of essential oils, especially of all kinds of mint, eucalyptus, and some other herbal and fruity oils. These are oral care products, chewing gums, and all kinds of mouth refreshing confectioneries. As aforementioned, toothpastes appeared on the market in the late nineteenth century in the United States. Chewing gums or the custom of chewing certain plant secretions were known to the ancient Greeks (e.g., mastic tree resin) and to ancient Mayans (e.g., sapodilla tree gum). Chewing gum, as we know it now, started in America around 1850 when John B. Curtis introduced flavored chewing gum, which was first patented in 1859 by William Semple. In 1892, William Wrigley used chewing gum as a free gift with sales of baking powder in his business in Chicago, and very soon he realized that chewing gum has real potential. In 1893, Juicy Fruit gum came into market and was followed in the same year by Wrigley's Spearmint; today, both products are known and consumed worldwide and their names are global trademarks.

21.5 PRODUCTION AND CONSUMPTION

This brief and certainly incomplete look into the history of industrial usage of essential oils as flavor and fragrance ingredients shows that the real industrial scale of flavor and fragrance industry developed in the second half of the nineteenth century together with transformation of *manufacture* into *industry*.

There are no reliable data on the scale of consumption of essential oils in specific products. On the basis of different sources, it can be estimated that the world market for the flavors and fragrances has a value of 10–12 billion euros, being equally shared by each group of products. It is very difficult to estimate the usage of essential oils in each of the groups. More oils are used in flavors than in fragrances, which today are mainly based on aroma chemicals, especially in large-volume compounds used in detergents and household products. Table 21.2 presents estimated data on world consumption of major essential oils (each used over 500 tons per annum).

The following oils are used in quantities between 100 and 500 tons per annum: bergamot, cassia, cinnamon leaf, clary sage, dill, geranium, lemon petitgrain, lemongrass, petitgrain, pine, rosemary, tea tree, and vetivert. It must be emphasized that most of the figures given earlier on

TABLE 21.2
Estimated World Consumption of the Major Essential Oils

Oil Name	Consumption (Tons)	Approximate Value (€ Million)[a]	Major Applications[b]
Orange	50,000	275	Soft drinks, sweets, fragrances
Cornmint (*Mentha arvensis*)[c]	25,000	265	Oral care, chewing gum, confectionery, fragrances, menthol crystals
Peppermint	4,500	120	Oral care, chewing gum, confectionery, liquors, tobacco, fragrances
Eucalyptus (*Eucalyptus globulus*)	4,000	22	Oral care, chewing gum, confectionery, pharmaceuticals, fragrances
Lemon	3,500	21	Soft drinks, sweets, diary, fragrances, household chemicals
Citronella	3,000	33	Perfumery, toiletries, household chemicals
Eucalyptus (*Eucalyptus citriodora*)	2,100	10	Confectionery, oral care, chewing gum, pharmaceuticals, fragrances
Clove leaf	2,000	24	Condiments, sweets, pharmaceuticals, tobacco, toiletries, household chemicals
Spearmint (*Mentha spicata*)	2,000	46	Oral care, chewing gum, confectionery
Cedarwood (*Virginia*)	1,500	22	Perfumery, toiletries, household chemicals
Lime	1,500	66	Soft drinks, sweets, diary, fragrances
Lavandin	1,000	15	Perfumery, cosmetics, toiletries
Litsea cubeba	1,000	20	Citral for soft drinks, fragrances
Cedarwood (China)	800	11	Perfumery, toiletries, household chemicals
Camphor	700	3	Pharmaceuticals
Coriander	700	40	Condiments, pickles, processed food, fragrances
Grapefruit	700	9	Soft drinks, fragrances
Star anise	700	7	Liquors, sweets, bakery, household chemicals
Patchouli	600	69	Perfumery, cosmetics, toiletries
Basil	500	12	Condiments, processed food, perfumery, toiletries
Mandarin	500	30	Soft drinks, sweets, liquors, perfumery, toiletries

[a] Based on average prices offered in 2007.
[b] Almost all of the major oils are used in alternative medicine.
[c] Main source of natural menthol.

the production volume are probably underestimates because no reliable data are available on the domestic consumption of essential oils in major producing countries, such as China, India, and Indonesia. Therefore, quantities presented in various sources are sometimes very different. For example, consumption of *Mentha arvensis* is given as 5,000 and 25,000 tons per annum. The lower one probably relates to the direct usage of the oil, while the higher includes the oil used for the production of menthol crystals. In Table 21.2, the highest available figures are presented. Considering the aforementioned and general figures for flavors and fragrances, it can be estimated that the total value of essential oils used worldwide is somewhere between 2 and 3 billion euro. Price fluctuations (e.g., the patchouli oil price jump in mid-2007) and many other unpredictable changes cause any estimation of essential oil consumption value to be very risky and disputable. The figures given in Table 21.2 are based on average trade offers. Table 21.2 does not include turpentine, which is sometimes added into essential oil data. Being used mainly as a chemical solvent or a raw material in the aroma chemical industry, it has no practical application as an essential oil, except in some household chemicals.

As noted earlier, the largest world consumer of essential oils is the flavor industry, especially for soft drinks. However, this is limited to a few essential oils, mainly citrus (orange, lemon, grapefruit, mandarin, lime), ginger, cinnamon, clove, and peppermint. Similar oils are used in confectionery, bakery, desserts, and dairy products, although the range of oils may be wider and include some fruity products and spices. The spicy oils are widely used in numerous salted chips, which are commonly consumed along with beverages and long drinks. Also, the alcoholic beverage industry is a substantial user of essential oils, for example, anis in numerous specialties of the Mediterranean region, herbal oils in liqueurs, ginger in ginger beer, and peppermint in mint liquor and in many other flavored alcohols.

Next in importance to beverages in the food sector is the sweet, dairy, confectionery, dessert (fresh and powdered), sweet bakery, and cream manufacturing sector, for which the main oils used are citrus, cinnamon, clove, ginger, and anis. Many other oils are used in an enormous range of very different products in this category.

The fast-food and processed food industries are also substantial users of essential oils, although the main demand is for spicy and herbal flavors. Important oils here are coriander (especially popular in the United States), pepper, pimento, laurel, cardamom, ginger, basil, oregano, dill, and fennel, which are added to the spices with the aim of strengthening and standardizing the flavor.

The major users of essential oils are the big compounders—companies that emerged from the historical manufacturers of essential oils and fragrances and flavors and new ones established by various deals between old players in the market or, like International Flavors & Fragrances, were created by talented managers who left their parent companies and started on their own. Today's big 10 flavors and fragrances are listed in Table 21.3.

Out of the 20 companies listed in Table 21.1, seven were located in France. By 2007, of the 10 largest, only two were from France. Also, only four of today's big 10 are moe than a century old with two leaders: Givaudan and Firmenich from Switzerland and Mane and Robertet from France.

The flavor and fragrance industry is the one where the majority of oils are introduced into appropriate flavor and fragrance compositions. Created by flavorists and perfumers, an elite of professionals in the industry, the compositions, complicated mixtures of natural and nature-identical ingredients for flavoring, and natural and synthetic components for fragrances are offered to end users. The latter are the manufacturers of millions of very different products from luxurious *haute couture* perfumes and top-class-flavored liquors and chocolate pralines through cosmetics, household chemicals, sauces, condiments, cleaning products, air fresheners, and aroma marketing.

It is important to emphasize that a very wide range of essential oils are used in alternative or *natural* medicine with aromatherapy—treatment of many ailments with the use of essential oils as bioactive ingredients—being the leading outlet for the oils and products in which they are applied

TABLE 21.3
Leading Producers of Flavors and Fragrances

Position	Company Name (Headquarters)	Sales in Million (€)[a]
1	Givaudan S.A. (Vernier, Switzerland)	2550
2	Firmenich S.A. (Geneva, Switzerland)	1620
3	International Flavors & Fragrances (New York, United States)	1500
4	Symrise AG (Holzminden, Germany)	1160
5	Takasago International Corporation (Tokyo, Japan)	680
6	Sensient Technologies Flavors & Fragrances (Milwaukee, United States)	400
7	T. Hasegawa Co. Ltd. (Tokyo, Japan)	280
8	Mane S.A. (Le Bar-sur-Loup, France)	260
9	Frutarom Industries Ltd. (Haifa, Israel)	220
10	Robertet S.A. (Grasse, France)	210

[a] Estimated data based on web pages of the companies, various reports, and journals.

as major active components. The ideas of aromatherapy from a niche area dominated by lovers of nature and some kind of magic, although based on very old and clinically proved experience, came into mass production appearing as an advertising *hit* in many products including global ranges. Examples include Colgate-Palmolive liquid soaps; a variety of shampoos, body lotions, creams, and so on by many other producers; and fabric softeners emphasizing the benefits to users' mood and condition from the odors of essential oils (and other fragrant ingredients) remaining on fabrics. Aromatherapy and *natural* products, where essential oils are emphasized as *the natural* ingredients, are a very-fast-developing segment of the industry, and this is a return to what was a common practice in ancient and medieval times.

21.6 CHANGING TRENDS

Until the second half of the nineteenth century, formulas of perfumes and flavors (although much less data are available on flavoring products in history) were based on essential oils and some other naturals (musk, civet, amber, resins, pomades, tinctures, extracts, etc.). Now, some 150 years later, old formulations are being taken out of historical books and are advertised as the *back to nature* trend. Perfumery handbooks published until the early twentieth century listed essential oils and none or only one or two aroma chemicals (or isolates from essential oils). A very good illustration of the changes that affected the formulation of perfumes in the twentieth century is a comparison of rose fragrance as recorded in perfumery handbooks. Dr. Heinrich Hirzel in his *Die Toiletten-Chemie* (1892, p. 384) gave the following formula for high-quality white rose perfume:

400 g of rose extract
200 g of violet extract
150 g of acacia extract
100 g of jasmine extract
120 g of iris infusion
25 g of musk tincture
5 g of rose oil
10 drops of patchouli oil

Felix Cola's milestone work *Le Livre de Parfumeur* (1931, p. 192) recorded a white rose formula containing only 1% of rose oil, 2% of rose absolute, 7.5% of other oils, and aroma chemicals.

Rose Blanche	
Rose oil	10 g
Rose absolute	20 g
Patchouli oil	25 g
Bergamot oil	50 g
Linalool	60 g
Benzyl acetate	7 g
Phenylethyl acetate	75 g
Citronellol	185 g
Geraniol	200 g
Phenylethyl alcohol	300 g

In the mid-twentieth century, perfumers were educated to consider chemicals as the most convenient, stable, and useful ingredients for fragrance compositions. Several rose fragrance formulas with less than 2% rose oil or absolute can be found in F. V. Wells and M. Billot's *Perfumery Technology*, (1975), and rose fragrance without any natural rose product is nothing curious in a contemporary perfumers' notebook. However, looking through descriptions of new fragrances launched in the last few years, one can observe a very strong tendency to emphasize the presence of natural ingredients—oils, resinoids, and absolutes—in the fragrant mixture. The back to nature trend creates another area for essential oil usage in many products.

A very-fast-growing group of cosmetics and related products today are the so-called organic products. These are based on plant ingredients obtained from wild harvesting or from *organic cultivation* and are free of pesticides, herbicides, synthetic fertilizers, and other chemicals widely used in agriculture. According to different sources, sales of *organic* products in 2007 will reach 4–5 billion U.S. dollars. The same *organic raw materials* are becoming more and more popular in the food industry, which in consequence will increase the consumption of *organic flavors* based on *organic essential oils*. *Organic* certificates, available in many countries (in principle for agricultural products, although they are institutions that also certify cosmetics and related products), are product passports to a higher price level and selective shops or departments in supermarkets. The importance of that segment of essential oil consumption can be illustrated by comparison of the average prices for standard essential oils as listed in Table 21.4 and the same oils claimed as *organic*.

The consumption of essential oils in perfumed products varies according to the product (Table 21.5): from a very high level in perfumes (due to the high concentration of fragrance compounds in perfumes and the high content of natural ingredients in perfume fragrances) and in a wide range of *natural* cosmetics and toiletries to relatively low levels in detergents and household chemicals, in which fragrances are based on readily available low-priced aroma chemicals. However, it must be emphasized that although the concentration of essential oils in detergents and related products is low, the large-volume sales of these consumer products result in substantial consumption of the oils.

Average values given for fragrance dosage in products and for the content of oils in fragrances are based on literature data and private communications from the manufacturers. It should be noted that in many cases, the actual figures for individual products can be significantly different. *Eau Sauvage* from Dior is a very good example: analytical data indicate a content of essential oils (mainly bergamot) of over 70%. Toothpastes are exceptional in that the content of essential oils in the flavor is in some cases nearly 100% (mainly peppermint, spearmint cooled with natural menthol).

TABLE 21.4
Prices of Selected Standard and *Organic* Essential Oils

Oil Name	Standard Quality (€/kg)[a]	Organic Quality (€/kg)[a]
Orange	5.50	35
Cornmint (*M. arvensis*)	10.50	50
Peppermint	27.00	100
Eucalyptus (*E. globulus*)	5.50	26
Lemon	6.00	30
Citronella	11.00	23
Eucalyptus (*E. citriodora*)	5.00	34
Clove leaf	12.00	60
Spearmint (*M. spicata*)	23.00	40
Cedarwood (*Virginia*)	15.00	58
Lime	44.00	92
Lavandin	15.00	36
L. cubeba	20.00	44
Cedarwood (China)	14.00	53
Camphor	4.50	24
Coriander	57.00	143
Grapefruit	13.00	170
Patchouli	115.00	250

[a] Average prices based on commercial offers in 2007.

TABLE 21.5
Average Dosage of Fragrances in Consumer Products and Content of Essential Oils in Fragrance Compounds

Position	Product	Average Dosage of Fragrance Compound in Product (%)	Average Content of Essential Oils in Fragrance (%)
1	Perfumes	10.0–25.0	5–30[a]
2	Toilet waters	3.0–8.0	5–50[a]
3	Skin care cosmetics	0.1–0.6	0–10
4	Deodorants (inclusive deoparfum)	0.5–5.0	0–10
5	Shampoos	0.3–2.0	0–5
6	Body cleansing products (liquid soaps)	0.5–3.0	0–5
7	Bath preparations	0.5–6.0	0–10
8	Soaps	0.5–3.0	0–5
9	Toothpastes	0.5–2.5	10–50[b]
10	Air fresheners	0.5–30.0	0–20
11	Washing powders and liquids	0.1–0.5	0–5
12	Fabric softeners	0.1–0.5	0–10
13	Home care chemicals	0.5–5.0	0–5
14	Technical products	0.1–0.5	0–5
15	Aromatherapy and organic products	0.1–0.5	100

[a] Traditional perfumery products contained more natural oils than modern ones.
[b] Mainly mint oils.

TABLE 21.6
Average Content of Flavors in Food Products and of Essential Oils in Flavors

Position	Food Products	Flavor Dosage in Food Product (%)	Essential Oil Content in Flavor (%)
1	Alcoholic beverages	0.05–0.15	3–100
2	Soft drinks	0.10–0.15	2–5
3	Sweets (confectionery, chocolate, etc.)	0.15–0.25	1–100
4	Bakery (cakes, biscuits, etc.)	0.10–0.25	1–50
5	Ice creams	0.10–0.30	2–100
6	Diary products, desserts	0.05–0.25	1–50
7	Meat and fish products (also canned)	0.10–0.25	10–20
8	Sauces, ketchup, condiments	0.10–0.50	2–10
9	Food concentrates	0.10–0.50	1–25
10	Snacks	0.10–0.15	2–20

While the average dosage of fragrances in the final product can be very high, flavors in food products are used in very low dosages, well below 1%. The high consumption of essential oils by this sector results from the large volume of sales of flavored foods. Average dosages of flavors and the content of essential oils in the flavors are given in Table 21.6.

As in the case of fragrances, the average figures given in Table 21.6 vary in practice in individual cases, both in the flavor content in the product and much more in the essential oil percentage in the flavor. Again, *natural* or *organic* products contain only essential oils, since it is unacceptable to include any synthetic aroma chemicals or so-called nature-identical food flavors. It should be noted that a substantial number of flavorings are oleoresins: products that are a combination of essential oils and other plant-derived ingredients, which are especially common in hot spices (pepper, chili, pimento, etc.) containing organoleptically important pungent components that do not distill in steam. This group of oleoresin products must be included in the total consumption of essential oils.

For many years after World War II, aroma chemicals were considered the future for fragrance chemistry, and there was strong, if unsuccessful, pressure by the manufacturers to get approval for the wide introduction of synthetics (especially those regarded as *nature identical*) in food flavors. The very fast development of production and usage of aroma chemicals caused increasing concern over safety issues for the human health and for the environment. One by one certain products were found harmful either for human health (e.g., nitro musks) or for nature. This resulted in wide research on the safety of the chemicals and the development of new safe synthetics. Concurrently, the attention of perfumers and producers turned in the direction of essential oils, which as derived from natural sources and known and used for centuries were generally considered safe. According to recent research, however, this belief is not entirely true and some, fortunately very few, oils and other fragrance products obtained from plants have been found dangerous, and their use has been banned or restricted. However, these are exceptional cases and the majority of essential oils are found safe both for use on the human body as cosmetics and related products as well as for consumption as food ingredients.

It is important to appreciate that the market for *natural*, *organic*, and *ecological* products both in body care and food industries has changed from a niche area to a boom in recent years with the growth exceeding 30% per annum. The estimated value of sales for *organic* cosmetics and toiletries is 600–800 million euros in Europe, the United States, and Japan and will grow steadily together with organic foods. This creates a very sound future for the essential oil industry, which as such or as isolates derived from the oils will be widely used for fragrance compounds in cosmetics and related products as well as for flavors.

Furthermore, the modernization of agricultural techniques and the growth of plantation areas result in better economical factors for the production of essential oil–bearing plants, creating workplaces in developing countries of Southeast Asia, Africa, and South America as well as further development of modern farms in the United States and Europe (Mediterranean area, Balkans). Despite some regulatory restrictions (EU, REACH, FDA, etc.), essential oils are and will have an important and growing share in the fragrance and flavor industry. The same will be true for the usage of essential oils and other products of medicinal plants in pharmaceutical products. It is well known that the big pharmaceutical companies invest substantial resources in studies of folk and traditional medicine as well as in research on biologically active constituents of plant origin. Both of these areas cover applications of essential oils. The same is observed in cosmetics and toiletries using essential oils as active healing ingredients.

21.7 CONCLUSIONS

It can be concluded that the industrial use of essential oils is a very promising area and that regular growth shall be observed in future. Much research work will be undertaken both on the safety of existing products and on development of new oil-bearing plants that are used locally in different regions of the world both as healing agents and as food flavorings. Both directions are equally important. Global exchange of tastes and customs shall not lead to unification by Coca-Cola or McDonalds. With all the positive aspects of these products, there are many local specialties that can become world property, like basil–oregano-flavored pizza, curry dishes, spicy kebab, or the universal and always fashionable eau de cologne. With the growth of the usage of the commonly known essential oils, new ones coming from exotic flowers of the Amazon jungle or from Indian Ayurveda books can add new benefits to the flavor and fragrance industry.

ACKNOWLEDGMENTS

The author is most grateful to K. D. Protzen of Paul Kaders GmbH and Dr. C. Green for their help and assistance in preparation of this chapter.

REFERENCES

Cola, F., 1931. *Le Livre du Parfumeur.* Paris, France: Casterman.
Hirzel, H., 1892. *Die Toiletten Chemie.* Leipzig, Germany: J.J. Weber Verlag.

FURTHER READINGS

Dorland, W.F. and J.A. Rogers Jr., 1977. *The Fragrance and Flavor Industry.* Mendham, NJ: V.E. Dorland.
Lawrence, B.M., 2000. *Essential Oils 1995–2000.* Wheaton, IL: Allured Publishing.
Lawrence, B.M., 2004. *Essential Oils 2001–2004.* Wheaton, IL: Allured Publishing.
Lawrence, B.M., 2007. *Essential Oils 2005–2007.* Wheaton, IL: Allured Publishing.
Wells, F.V. and M. Billot, 1981. *Perfumery Technology.* London, U.K.: E. Horwood Ltd.

WEBSITES

American Beverage Association: http://www.ameribev.org.
The Coca-Cola Company: http://www.thecoca-colacompany.com/heritage/ourheritage.html.
Colgate-Palmolive: http://www.colgate.com/app/Colgate/US/Corp/History/1806.cvsp.
Pepsi Cola History: http://www.solarnavigator.net/sponsorship/pepsi_cola.htm.
Procter & Gamble: http://www.pg.com/company/who_we_are/ourhistory.shtml.
Unilever: http://www.unilever.com/aboutus.

22 Encapsulation and Other Programmed Release Techniques for Essential Oils and Volatile Terpenes

Jan Karlsen

CONTENTS

22.1 Introduction 1023
22.2 Controlled Release of Volatiles 1024
22.3 Use of Hydrophilic Polymers 1027
22.4 Alginate 1027
22.5 Stabilization of Essential Oil Constituents 1027
22.6 Controlled Release of Volatiles from Nonvolatile Precursors 1028
22.7 Cyclodextrin Complexation of Volatiles 1028
22.8 Enhanced Biological Effect by Prolonged Delivery of Volatiles and Essential Oils 1029
22.9 Methods for Producing Prolonged Delivery Units of Volatiles 1029
22.10 Concluding Remarks 1029
References 1030

22.1 INTRODUCTION

In order to widen the medical and industrial applications of volatiles (essential oils), it is necessary to lower the volatility of the compounds to obtain a longer shelf life of products containing volatiles. However, due to the many contradictory reports on the activity/nonactivity of volatile compounds toward a biological system, a prolonged-release formulation of volatiles would also be of great interest to basic research. The short contact time of the volatiles with a biological system may render it difficult to verify a biological effect, and this could be changed by, for instance, encapsulation of the volatile compounds before biological testing.

Volatile compounds like essential oils and terpenes have mainly been used in perfumery and fragrance industry for the impact on our sense of smelling. Formulations that can lower the volatility or prolong the release of volatiles obviously would be of interest not only to industry but also to scientists interested in the biological effects of volatile compounds. Encapsulation of special groups of volatile compounds in a mixture may change the smell of the original mixture of compounds and change the smell of a perfume even though the composition of the essential oils and volatiles are the same.

When the first edition of this *Handbook of Essential Oils* was published in 2010, there were several patents describing the encapsulation of essential oils, but the studies into the impact upon the biological effects of encapsulated (or "low volatile") volatiles were scarce. However, during recent years, a lot of publications describing the biological effect of encapsulated volatiles have been published. In the opinion of the author, we can expect many more studies to be carried out where the longer release of volatile compounds on the contact with a biological system may give rise to data on biological effects that until now have been bypassed.

There are many practical aspects of this use of volatile compounds where the limited contact time of volatiles can be significantly changed and therefore give rise to a range of new and better applications and industrial products. Encapsulation of essential oils can give a *dry* free-flowing powder easily incorporated into consumer products. Lowering the volatility can also allow incorporating these compounds into textiles; surface films; aerosols for spraying.

To lower the volatility, one needs to encapsulate the volatile into a polymer matrix, utilize a complex formation, and use a covalent bonding to a matrix—to mention a few techniques. We therefore need to formulate the volatiles and take many of the techniques from areas where controlled-release formulations have been in use for many years. Especially the area of controlled drug delivery has a large number of such formulations. Today, there exist a large number of sustained drug delivery formulations in both journal publications and patents (Deasy, 1984).

So far in the area of volatile terpenes/essential oils, we have seen a large number of investigations that focused on plant selection, volatile isolation techniques, separation of volatiles isolated, identification of isolated compounds, and the biochemical formation of terpenes. The formulation of volatile into products has been seen as an area of industrial research. This has naturally led to a large number of patents but very few scientific publications on the formulation of essential oils and lower terpenes.

The idea of this chapter has been to give an introduction to the area of making a controlled-release product of volatiles and, in particular, of essential oils and their constituents. It will be impossible to give a total survey of this area in a chapter, but hopefully some ideas are given that inspire the reader to use the Internet search in databases for more complete coverage. Several keywords for computer searching are given.

22.2 CONTROLLED RELEASE OF VOLATILES

The main interest of volatile encapsulation is the possibility to extend the biological effect of the compounds. For essential oils, we want to prolong the activity by lowering the evaporation of the volatile compounds (Baranaskiene et al., 2007). During the last 15 years, there are not many publications on this topic in the scientific literature compared to the number of patents that describe various ways of prolonging the effect of volatiles (Sair, 1980, Sair and Sair, 1980, Fulger and Popplewell, 1997, 1998, Zasypkin and Porzio, 2004, McIver, 2005, Porzio, 2008). One reason for this fact is that the prolonging of the evaporation of volatiles is regarded as so close to practical applications that further development of this area will immediately be blocked by a number of patents. However by 2013, there are signs that this attitude is changing. The change of the impact of a perfume by encapsulation of the ingredients is of importance. The possibility to change the top note, middle note, and end note in a perfume without changing the composition should be a tempting solution to new perfumes. This area should interest the perfume industry, but the cost of making well-controlled nanoparticles with the size less than 100 nm is for the moment too high to be applied to perfume production. In order to look into the lowering of the volatility of essential oils, we have to approach another active area of scientific research—controlled delivery of pharmacologically active ingredients.

The reasons for controlled release of volatiles may be the following:

- Changing the biological impact of volatiles
- Adding fragrance to textiles
- Stabilizing specific compounds
- Tailoring a fragrance to a specific application
- Lowering the volatility to increase the shelf life of a formulated product
- To improve the study of the biological effect of volatiles

Encapsulation and Other Programmed Release Techniques

The slow or controlled release of volatiles is achieved by

- Encapsulation
- Solution or dispersion into a polymer matrix
- Complex formation
- Covalent bonding to another molecule or matrix

For essential oils/terpenes and natural volatiles, the following techniques can be utilized depending upon the volatiles and the intended use of the final product:

- Microcapsule production
- Microparticle formation
- Melt extrusion
- Melt injection
- Complex formation
- Liposomes
- Micelles
- Covalent bonding to a matrix
- Combination of nanocapsules into larger microcapsules

The making of one of this type of products and techniques utilized will influence the activity toward the human biological membranes in one way or another. Therefore, the relevant sizes of biological units are listed in Table 22.1, and the average sizes of units produced in consumer products where volatiles are involved are listed in Table 22.2.

The introduction of encapsulation of volatiles, which 15 years ago saw very few publications, has now, by 2013, turned into a very active field of applied research. Encapsulation or prolonged delivery of volatiles gives us more predictable and long-lasting effect of the compounds. The areas of application are very varied, and the industry using essential oils and terpenes foresees many prospects for new microencapsulated products.

Application markets for encapsulated essential oils and volatiles are

- Applicated to the surface of various paper products
- Medicine
- Food, household items, and personal care
- Biotechnology
- Pharmaceuticals
- Electronics

TABLE 22.1
Size Diameters of Biological Entities

Human blood cells	7000–8000 nm
Bacteria	800–2000 nm
Human cell nucleus	1000 nm
Nanoparticles that can cross biomembranes	60 nm
Virus	17–300 nm
Hemoglobin molecules	3.3 nm
Nanoparticles that can cross blood–brain barrier	4 nm
DNA helix	3 nm
Water molecule	0.3 nm

TABLE 22.2
Average Size of Formulation Units in nm (Sizes below 150 nm May Be Invisible to the Naked Eye)

Solutions	0.1 nm
Micellar solutions	0.5 nm
Macromolecular solutions	0.5 nm
Microemulsions	5–20 nm
Liposomes SUV	20–150 nm
Nanospheres	100–500 nm
Nanocapsules	100–500 nm
Liposomes LUV	200–500 nm
Liposomes MLV	200–1,000 nm
Microcapsules	500–30,000 nm
Simple emulsions	500–5,000 nm
Multiple emulsions	10,000–100,000 nm

Abbreviations: SUV, Unilamellar vesicles; LUV, Large unilamellar vesicles; MLV, Multilamellar vesicles.

- Photography
- Chemical industry
- Textile industry
- Cosmetics

It is therefore easy to understand that the encapsulation procedures will open up a much larger and different market for essential oil/terpene products. Experience from all the areas mentioned previously can be applied to the study of volatile compounds in products.

In the area of essential oils and lower terpenes, simple encapsulation procedures from the area of drug delivery are applied. The essential oils or single active constituents are mixed with a hydrophilic polymer and spray dried using a commercial spray drier. Depending on whether we have an emulsion or a solution of the volatile fraction in the polymer, we obtain monolithic particles or normal microcapsules.

The most usual polymers used for encapsulation are

- Oligosaccharides from α-amylase
- Acacia gum
- Gum arabic
- Alginate
- Chitosan

Many different emulsifiers are used to dissolve the essential oils totally or partly| prior to the encapsulation process. This can result in a monolithic particle or a normal capsule, where the essential oil is surrounded by a hydrophilic coating. When the mixture of an essential oil and a hydrophilic polymer is achieved, the application of spray-drying procedure of the resulting mixture will result in the formation of microcapsules. The technique for achieving an encapsulated product in high efficiency will depend upon many technical parameters, and description of the procedures can be found in the patent literature. Often a mixture of essential oil/polymer (4:1) can be used, but a successful result will also depend on the type of apparatus being used for the production. The reader is advised to refer to the parameters given for the polymer used in the experiment. To achieve the encapsulated product, a mixture of low pressure and a slightly elevated temperature is used in the spray-drying equipment, and a certain loss of volatiles is inevitable. However, a recovery of more than 70% can be achieved by carefully monitoring the production parameters.

22.3 USE OF HYDROPHILIC POLYMERS

In product development, one tends to use cheap derivatives of starches or other low-grade quality polymers. Early studies with protein-based polymers such as gelatine, gelatine derivatives, soy proteins, and milk-derived proteins gave reasonable technical quality of the products. However, even if these materials show stable emulsification properties with essential oils, they have some unwanted side effects in products. We have seen that a more careful control of the polymer used can result in real high-tech products, where the predictability of the release of the volatiles can be assured like a programmed release of drug molecules in drug delivery devices. The polymer quality to be used will, of course, depend on the intention of the final product. In cosmetic industry, where one is looking for an essential oil product, free flowing and dry, to mix with a semisolid or solid matrix, the use of simple starch derivatives will be very good. For other applications, where the release of the volatile needs to be closely controlled or predicted more accurately, it is recommended that a more thorough selection of a well-characterized polymer is done. One example is the use of chitosan in prolonging the volatile profile of saffron by encapsulation (Chranioti et al., 2013). Another example of a very good and controllable hydrophilic polymer is alginate. This polymer is available in many qualities and can be tailored to any controlled-release product. The chemistry of alginate is briefly discussed later and will allow the reader to decide whether to opt for an alginate of technical quality, or if a high-tech product is the aim, to choose a better characterized hydrocolloid.

22.4 ALGINATE

Alginates are naturally occurring polycarbohydrates consisting of monomers of α-L-glucuronic acid (G) and β-D-mannuronic acid (M). The relative amounts of these two building blocks will influence the total chemistry of the polymer. The linear polymer is water soluble due to its polarity. Today alginate can be produced by bacteria that allow us to control the composition of the monomer (G/M) ratio. The chemical composition of the naturally occurring polymer is dependent upon the origin of the raw material. The marine species display seasonal differences in the composition and different parts of the plant produce different alginates and this may make standardization of the isolated polymer difficult. Alginates may undergo epimerization to obtain the preferred chemical composition. This composition (G/M ratio) will determine the diffusion rate through the swollen alginate gel, which surrounds the encapsulated essential oil (Ogston et al., 1973; Elias, 1997; Amsden, 1998a,b). An important structure parameter is also the distribution of the carboxyl groups along the polymer chain. The molecular weight of the polymer is equally important, and molecular weights between 12,000 and 250,000 are readily available in the market. The alginate polymer can form a swollen gel by hydrophobic interaction or by cross-linking with divalent ions like calcium. The G/M ratio determines the swelling rate and therefore also the release of encapsulated compounds. The diffusion of different substances has been studied and reference can be made to essential oil encapsulation. The size of alginate capsules can vary from 100 μm or more down to the nanometer range depending on the production procedure chosen (Draget et al., 1994, 1997; Tønnesen and Karlsen, 2002; Shilpa et al., 2003; Donati, et al., 2005).

22.5 STABILIZATION OF ESSENTIAL OIL CONSTITUENTS

The encapsulation of essential oils in a hydrophilic polymer may stabilize the constituents of the oil, but a better technique for this purpose will often be to use cyclodextrins in the encapsulation procedure. The use of cyclodextrins will lead to a complexation of the single compounds, which again will stabilize the complexed molecule. Complex formation with cyclodextrins is often used in drug delivery to promote solubility of lipophilic compounds; however, in the case of volatiles containing compounds that may oxidize, the complex formation will definitely prolong the shelf life of the finished product. A good review of the flavor encapsulation advantages is given by Risch and Reineccius (1988). The most important aspect of essential oil encapsulation in a hydrophilic polymer is that the volatility is lowered. Lowering the volatility will result in longer shelf life of products and a better stability of the finished product in this respect. A very interesting product is the development of microcapsules containing essential oils having adherence to keratinous surfaces. This should find applications in cosmetics and the personal care area (Alden-Danforth et al., 2013). By modifying the capsule surface, we can change the properties of the essential oil impact—in addition, lowering the volatility also changes the adherence of the capsules.

22.6 CONTROLLED RELEASE OF VOLATILES FROM NONVOLATILE PRECURSORS

The time-limited effect of volatiles for olfactive perception has led to the development of encapsulated volatiles and also to the development of covalent-bonded fragrance molecules to matrices. In this way, molecules release their fragrance components by the cleavage of the covalent bond. Mild reaction conditions met in practical life initiated by light, pH, hydrolysis, temperature, oxygen, and enzymes may release the flavors. The production of "profragrances" is a very active field for the industry and has led to numerous patents. The botanical plants producing essential oils have developed means by which the volatiles are produced, stored, and released into the atmosphere related to environmental factors. The making of a "profragrance" involves mimicking these natural procedures into flavor products. However, we are simplifying the process by using only one parameter in this release process, that is, the splitting of a covalent bond. In doing this, we mimic the formation of terpene glucosides in the plant that is then split and giving off the terpene. In theory, the making of a long-lasting biological impact and the breakdown of a constituent are contradictory reactions. However, for practical purposes, the use of a covalent bonding and thereafter the controlled splitting of this bond by the parameters mentioned earlier (temperature, humidity, etc) can be built into suitable flavor and fragrance products. This technique is only applicable to single essential oil constituents but constitutes a natural follow-up of essential encapsulation (Herrmann, 2004, 2007; Powell et al., 2006).

22.7 CYCLODEXTRIN COMPLEXATION OF VOLATILES

Cyclodextrin molecules are modified carbohydrates that have been used for many years to modify the solubility properties of drug molecules by complexation. The cyclodextrins can also be applied to volatiles to protect them against environmental hazards and thus prolong the shelf life of these compounds. Cyclodextrin complexation will also modify the volatility of essential oils and prolong the bioactivity (Han and Zhang, 2011; Na et al., 2012; Zhu and Di-jia, 2012). Recently, the activity of monoterpenes (linalool, S-carvone, camphor, geraniol, γ-terpinene, and fenchone) as insecticides has been shown (Don-Pedro, 1966). However, the efficient application of these compounds is difficult due to low stability and high evaporation properties. Product made by encapsulation into β-cyclodextrins show a better control of the release rate of the volatiles and therefore facilitate their use in products (Lopez and Pascual-Villalobos, 2010). The cyclodextrins will give a molecular encapsulation by the complexation reaction. There are many cyclodextrins on the market but β-cyclodextrin is the most popular in use for small molecules. Complexation of volatiles with cyclodextrins may improve the heat stability, improve the stability toward oxygen, and improve the stability against light-induced reactions (Szente and Szejtli, 1988). Cyclodextrin complexation will also protect enantiomers against racemization. A significant lowering of the volatility has been observed for the complexation with essential oils (Risch and Reineccius, 1988). The complexation with cyclodextrins will result in increased heat stability. This is in contrast with the stability of volatiles that has been adsorbed on a polymer matrix. Cyclodextrin complexation can protect against

- Loss of volatiles upon storage of finished product
- Light-induced instability
- Heat decomposition
- Oxidation
- Racemization of enantiomers

22.8 ENHANCED BIOLOGICAL EFFECT BY PROLONGED DELIVERY OF VOLATILES AND ESSENTIAL OILS

In cases where a biological effect has been indicated by introductory experiments, a prolonged contact time may give real evidence of the biological effect. This can then be applied to product development. The prolonged delivery has been introduced into a variety of commercial products like fruit juices (Fujimoto and Suehiro, 1992; Donsi et al., 2011) and used in the general improvement of microbial effect (Paluch et al., 2011) to mention a few applications. Several new patents and publications describe the encapsulation or prolonged delivery of essential oils (Trinh et al., 1994; Principato, 2007; Gaonkar et al., 2010; Behle et al., 2011; Bhala et al., 2012; Ortan et al., 2013).

22.9 METHODS FOR PRODUCING PROLONGED DELIVERY UNITS OF VOLATILES

The methods for prolonging the volatility of essential oils and other volatiles are varied. Some indications of useful excipient are given in this chapter. However, the efficiency of the encapsulation procedures shows great variation and a thorough investigation into suitable methods is needed. This is the reason for the many patents being published in this area. The surrounding matrix of the volatiles in a product will greatly influence the chosen method and polymer—whether the volatile will be incorporated into cosmetics, food, and pharmaceuticals or onto surfaces like paper, wood, lacquer, and textile fibers (Koike and Imai, 1992; Habar et al., 1996; Boh and Knez, 2006; Madene et al., 2006; Shah et al., 2012; Boardman and Stuart Lee, 2013; Vella and Marks, 2013; Zhang and Given, 2013; Jimenez et al., 2014).

22.10 CONCLUDING REMARKS

The encapsulation/complexation/covalent bonding of essential oils and single volatile compounds will result in a significant lowering of the volatility, stabilizing the compounds, improve the shelf life of finished products, and prolong the biological activity. It may also allow a better testing of the biological activity of volatiles and thereby improve basic research on the biological effect of volatile compounds. Until recently, most of the literature on prolonged flavor products was to be found in patent literature. Today, a series of papers describe the use of prolonged-effect formulations for volatile constituents, and the research activity in this field of volatile application is very high. The effect of controlled delivery of volatiles opens up different areas of applications that previously were limited due to the volatility of the essential oils and their constituents. The encapsulation or the lowering of the volatility of compounds like the essential oils will allow for more relevant studies into the biological effects of volatile compounds. The readers of this chapter should also be aware that the literature of essential oil encapsulation and other means of lowering the volatility of compounds cannot always be found in the traditional essential oil research journals. It may also be advisable to look into Chinese and Japanese journals and patents as scientists of these countries have a long tradition using hydrocolloids for industrial production among which many utilize the encapsulated volatiles.

REFERENCES

Alden-Danforth, E., W. Feuer, M. White, and N. Williams, 2013. Aqueous-based personal care product formula that combines friction controlled fragrance technology with a film-forming compound to improve adherence of capsules on keratinous surfaces. Patent WO 2013087549 A1 20130620.

Amsden, B., 1998a. Solute diffusion within hydrogels. *Macromolecules*, 31:8382–8395.

Amsden, B., 1998b. Solute diffusion in hydrogels. *Polym. Gels Netw.*, 6:13–43.

Baranauskiene, R., E. Bylaite, J. Zakauskaite, and R. Venskutonis, 2007. Flavour retention of peppermint (*Mentha piperita* L.) essential oil spray-dried in modified starches during encapsulation and storage. *J. Agric. Food Chem.*, 55:3027–3036.

Behle, R. W., L. B. Flor-Weiler, A. Bharadwaj, and C. S. Kirby, 2011. A formulation to encapsulate nootkatone for tick control. *J. Med. Entomol.*, 48:1120–1127.

Bhala, R., V. Dhandania, and A. P. Periyasamy, 2012. Bio-finishing of fabrics. *Asian Dyer*, 9:45–49.

Boardman, C. and K. Stuart Lee, 2013. Fabric treatment using encapsulated phase-change active materials. Patent WO 2013087550 A1 20130620.

Boh, B. and E. Knez, 2006. Microencapsulation of essential oils and phase change materials for application in textile products. *Ind. J. Fibre Text. Res.*, 31:72–82.

Chranioti, C., S. Popoutsakis, A. Stephanos, and C. Tzia, 2013. Special publication-Royal Society of Chemistry, section Nutrition. *Funct. Sens. Propert. Foods*, 344:111–116.

Deasy, P. B., 1984. *Microencapsulation and Related Drug Processes*. New York: Marcel Dekker.

Donati, I., S. Holtan, Y. A. Morch, M. Borgogna, M. Dentini, and G. Skjåk-Bræk, 2005. New hypothesis on the role of alternating sequences in calcium-alginate gels. *Biomacromolecules*, 6:1031–1040.

Don-Pedro, K. N., 1966. Fumigant toxicity is the major route of insecticidal activity of citrus peel essential oils. *Pest. Sci.*, 46:71–78.

Donsi, F., M. Annunziata, M. Sessa, and G. Ferrari, 2011. Nanoencapsulation of essential oils to enhance their microbial activity in foods. *Food Sci. Technol.*, 44:1908–1914.

Draget, K. I., G. Skjåk-Bræk, B. E. Christiansen, O. Gåserød, and O. Smidsrød, 1997. Swelling and partial solubilization of alginic acid beads in acids. *Carbohydr. Polym.*, 29:209–215.

Draget, K. I., G. Skjåk-Bræk, and O. Smidsrød, 1994. Alginic acid gels: The effect of alginate chemical composition and molecular weight. *Carbohydr. Polym.*, 25:31–38.

Elias, H.-G., 1997. *An Introduction to Polymer Science*. Weinheim, Germany: VCH.

Fujimoto, T. and K. Suehiro, 1992. Antimicrobial wax compositions containing essential oils from wood. JP Patent 04328182 A 19921117.

Fulger, C. and M. Popplewell, 1997. Flavour encapsulation. US Patent 5601845.

Fulger, C. and M. Popplewell, 1998. Flavour encapsulation. US Patent 5792505.

Gaonkar, A. G., A. Akashe, L. Lawrence, A. R. Lopez, R. L. Meibach, D. Sebesta, J. D. White, Y. Wang, and L. G. West, 2010. Delivery of essential oil esters or other functional compounds in an enteric matrix. US Patent 20100310666 A1 20101209.

Habar, G. L., A. L. Pape, and C. Descusse, 1996. Microcapsules containing terpene or abietic acid derivatives as biodegradable solvents for use on chemical copying paper and pressure sensitive papers. Patent EP 714786 A1 19960605.

Han, L. and Y. Zhang, 2011. Study on the preparation of the complex of β-cyclodextrin-clove oil inclusion. *Xibei Yaoxue Zazhi*, 26:447–449.

Herrmann, A., 2004. Photochemical fragrance delivery systems based on the Norrish type II reaction—A review. *Spectrum*, 17:10–13 and 19.

Herrmann, A., 2007. Controlled release of volatiles under mild reaction conditions. From nature to everyday products. *Angew. Chem. Int. Ed.*, 46:5836–5863.

Jimenez, A., L. Sanchez-Gonzales, S. Desobry, A. Chiralt, and E. A. Tehrany, 2014. Influence of nanoliposomes incorporation on properties of film forming dispersions and films based on corn starch and sodium caseinate. *Food Colloids*, 35:159–169.

Koike, S. and A. Imai, 1992. Microcapsules containing fragrant coatings. Patent JP 04351678 A 19921207.

Lopez, M. D. and M. J. Pascual-Villalobos, 2010. Analysis of monoterpenoids in inclusion complexes with β-cyclodextrin and study on ratio effect in these microcapsules. *10th International Working Conference on Stored Product Protection. Julius-Kahn-Archiv*, 425:705–709.

Madene, A., M. Jacquot, J. Scher, and S. Desobry, 2006. Flavour encapsulation and controlled release. *Int. J. Food Sci. Technol.*, 41:1–21.

McIver, B., 2005. Encapsulation of flavour and/or fragrance composition. US Patent 6932982.

Na, S., W. Rina, B. Ren, M. Ke, and G. Hexi, 2012. Mongolian medicinal compound preparation for treating bronchitis and its preparation method. Patent CN 102793840 A 20121128.

Ogston, A. G., B. N. Preston, and J. D. Wells, 1973. On the transport of compact particles through solutions of chain polymers. *Proc. R. Soc.(Lond.) A*, 333:297–309.

Ortan, A., M. Ferdes, S. Rodino, C. D. Pirvu, and D. Draganescu, 2013. Topical delivery system of liposomally encapsulated volatile oil of *Anethum graveolens*. *Farmacia (Bucharest, Roumania)*, 61:361–370.

Paluch, G., R. Bradbury, and S. Bessette, 2011. Development of botanical pesticides for public health. *J. ASTM Int.*, 8:JAI103468/1–JAI103468/7.

Porzio, M., 2008. Melt extrusion and melt injection. *Perfumer Flavorist*, 33:48–53.

Powell, K., J. Benkhoff, W. Fischer, and K. Fritsche, 2006. Secret sensations: Novel functionalities triggered by light—Part II: Photolatent fragrances. *Eur. Coat. J.*, 9:40–49.

Principato, M. A., 2007. Insecticidal/acaricidal and insectifungal/acarifungal formulation. Ital. Appl. Pat IT 2006BO0699 A1 20070110.

Risch, S. J. and G. A. Reineccius, 1988. Flavor encapsulation. *ACS Symposium*, Series 370. Washington, DC: American Chemical Society.

Sair, L., 1980b. Food supplement concentrate in a dense glass house extrudate. US Patent 4232047.

Sair, L. and R. Sair, 1980a. Encapsulation of active agents as microdispersions in homogenous natural polymers. US Patent 4230687.

Shah, B., M. P. Davidson, and Q. Zhong, 2012. Encapsulation of eugenol using Maillard-type conjugates to form transparent and heat-stable nanoscale dispersions. *Food Sci. Technol.*, 49:139–148.

Shilpa, A., S. S. Agarwal, and A. R. Ray, 2003. Controlled delivery of drugs from alginate matrix. *Macromol. Sci. Polym. Rev. C*, 43:187–221.

Szenta, L. and J. Szejtli, 1988. Stabilization of flavors by cyclodextrins. In: *Flavor Encapsulation*, S. J. Risch (Ed.), pp. 148–157. ACS Symposium 370. Washington, DC: American Chemical Society.

Trinh, T., G. F. Brunner, and T. A. Inglin, 1994. Adsorbent articles for odor control with positive scent signal. WO 9422500 A1 19941013.

Tønnesen, H. H. and J. Karlsen, 2002. Alginate in drug delivery systems. *Drug Dev. Ind. Pharm.*, 28:621–630.

Vella, J. and T. I. Marks, 2013. Micro-encapsulated chemical re-application method for laundering of fabrics. Patent US 20130239429 A1 20130919.

Zasypkin, D. and M. Porzio, 2004. Glass encapsulation of flavours with chemically modified starch blends. *J. Microencapsulation*, 21:385–397.

Zhang, N. and P. S. Given, 2013. Releasably encapsulated aroma. Patent WO 2013032631 A1 20130307.

Zhu, Y. and Y. Di-jia, 2012. Technology optimization of inclusion compound for volatile oil from *Cnidium monnieri* with hydroxypropyl-β-cyclodextrin. *Zhungguo Shiyan Fangjixue Zashi*, 18:28–31.

23 Trade of Essential Oils

Hugo Bovill

CONTENT

Reference 1039

The essential oil industry is highly complex and fragmented. There are at least 100 different producing countries, as can be seen from the map "Essential Oils of the World" (Figure 23.1). Many of these producing countries have been active in these materials for many decades. They are often involved in essential oils due to historical colonization; for example, clove oil from Madagascar has traditionally been purchased via France, nutmeg from Indonesia through Holland, and West Indian and Chinese products through Hong Kong and the United Kingdom. The main markets for essential oils are the United States (New Jersey), Germany, the United Kingdom, Japan, and France (Paris and Grasse). Within each producing country, there is often a long supply chain starting with the small peasant artisanal producer, producing just a few kilos, who then sells it to a collector who visits different producers and purchases the different lots that are then bulked together to form an export lot, which is then often exported by a firm based in the main capital or main seaport of that country. This exporter is equipped with the knowledge of international shipping regulations, in particular for hazardous goods, which applies to many essential oils. They also are able to quote in U.S.\$ or euros, which is often not possible for small local producers (Figure 23.2).

Producers of essential oils can vary from the very large, such as an orange juice factory where orange oil is a by-product, down to a small geranium distiller (Figures 23.3 and 23.4).

The business is commenced by sending type samples that are examples of the production from the supplier and should be typical of the production that can be made going forward. Lot samples are normally provided to the purchaser in the foreign country to enable them to chemically analyze the quality organoleptically both on odor and flavor. It is essential that the qualities remain constant as differing qualities are not acceptable and there is normally no such thing as a *better* quality; it is either the same or it is not good. This is the key to building close relationships between suppliers in the country of origin and the purchaser.

Many suppliers try to improve their processes by adapting their equipment and modernizing. In Paraguay, petitgrain distillers replaced wooden stills with stainless steel stills on the advice of overseas aid noncommercial organizations. This led to a change in quality and the declining usage of petitgrain oil. The quality issues made customers unhappy, and in fact the Paraguayan distillers reverted back to their traditional wooden stills (Figure 23.5).

Market information, as provided by the processor, is essential to developing long-term relationships. To enable the producer to understand market pricing, he should appreciate that when receiving more enquiries for an oil, it is likely that the price is moving upward and it is by these signs of demand that he can establish that there are potential shortages in the market (Figure 23.6).

Producers and dealers exporting oil should be prepared to commit to carry inventory to ensure carryover and adequate delivery reliability. It is important to note that with climate change, weather and market conditions are becoming increasingly important, and prior to planting, advice should be sought from the buyer as to their intentions, for short, medium, and long terms. Long- and medium-term contracts are unusual and it is becoming increasingly common for flavor and fragrance companies not to commit over 1 year but to buy hand to mouth and purely give estimated

FIGURE 23.1 (See color fold-out insert at the back of the book.) World map showing production centers of essential oils. (Courtesy of Treatt PLC, London, UK.)

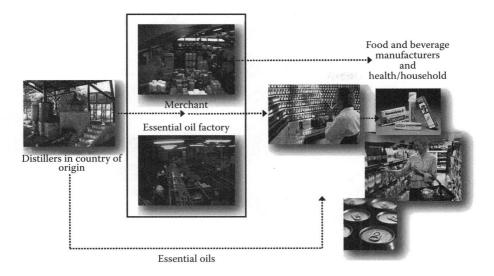

FIGURE 23.2 Flowchart showing the supply chain from distiller to finished product.

FIGURE 23.3 South American orange juice factory. (Photograph by kind permission of Sucocitrico Cutrale Ltd., Araraquara, Brazil.)

volume needs going forward. This strengthens the role of the essential oil dealers, of whom there are very few remaining in the main trading centers of the world, such as the United States, France, the United Kingdom, Germany, and Japan.

To quote from Ennever (1960),

> The dealer serves as a buffer between these two interests (producer and essential oil merchant house) by purchasing and carrying stocks of oils for his own account and risk when the producer and/or merchant house is unable to wait for the user's demand and hold stocks until the latter is ready to purchase. The risk of market fluctuations to the essential oil dealer or merchant in this practice, is quite considerable, but naturally, is reduced by his knowledge and experience of the trade. He is equipped to handle large or small quantities and a range of qualities, as a buyer or seller. Thus through the dealer's participation, the producer has a larger number of outlets for his production and the user can be reasonably certain of finding supplies of the oils required when he considers it necessary to purchase.

FIGURE 23.4 Copper still in East Africa.

FIGURE 23.5 Petitgrain still.

The dealer is aware of world markets and potential shortages that other producers may not be aware of, as these are happening in different continents. They can also have the knowledge of increasing demand and movements in consumer tastes.

Some essential oils are produced for their chemical constituents, whereas most are produced for their aromatic parts, and it is important that suppliers understand what is expected of them by their customer, whether it is chemical constituents naturally occurring or it is the aroma and flavor. Examples of this are turpentine oil, *Litsea cubeba* oil, sassafras oil, clove leaf oil, and coriander oil.

There is greater demand for ethical supplies, but it should be borne in mind that these surprisingly often do not receive a premium, and when entering the essential oil industry, it is important to note that it is not always the highest priced oils that give the best return as these are often those that are the most popular for new entrants to produce. Before entering into production of an essential oil, it is important to fully verify the market. It may be that there is good supply locally of the herb, for example, but maybe this is for a traditional purpose such as local medicinal use and production of local foodstuffs or liqueurs.

Origins are constantly changing and moving, as can be seen from the following: peppermint oil Mitcham production went from England to the United States; mint came from China and then went to Brazil and Paraguay, back to China, and now to India.

TREATT
PLC

Treatt Market Report
July 2007

Orange Oil Sweet

Strong demand currently for orange oil of all origins as we reach a period of the year where Florida plants are off season and Brazil is just beginning processing but oil of acceptable aldehyde is not yet available in volume for shipment from Brazil. The Brazilian crop this year is expected to be a very similar size to last season which is the first time the bi-annual cycle has been broken for 8 years. Better crop management including increased irrigation of groves and favourable weather conditions are two reasons cited for the better than expected crop in Brazil. Prices are moderately firm due to strong demand but this may subside as volume begins to come through in Brazil.

The 2006/07 crop in Florida was very low indeed at just 129 million boxes which contrasts markedly with the record crop of 1997/98 at 244 million boxes for example. As regularly reported in this column a significant hurricane event in Florida could result in a very firm market.

Lemon Oil

As volume availability improves, thanks to South American new crop the market price is showing signs of stabilising.

High quality oil continues to be in strong demand and discerning buyers are advised to carefully monitor the quality of their oils.

Lime Oil Distilled

Better fruit availability at the peak of the crop in Mexico has moved prices to lower levels as the market comes off the top of the cycle. However, strong fresh fruit demand is expected to keep the market firm compared with what we have seen in the last decade.

FIGURE 23.6 Market information.

Within the essential oil market, there are generally four different types of buyers: aromatherapy, the flavor and fragrance industries, and dealers. Many of these can be contacted through agents who would not pay for the goods themselves but would take a nominal commission of, say, 5%. The end users range from aromatherapists selling very small volumes of high- and fine-quality, natural essential oils to flavor and fragrance companies and, in a few cases, consumer product companies. The main markets are the essential oil dealers, of which there are probably 10 or 20 major companies remaining in the world, some of which are also involved in the manufacture of flavors or fragrances. To avoid conflicts of interest, it is perhaps better to work with those who concentrate solely on raw materials. Several of these companies have been established for many years and have a good trading history. Some information about them can be gained from their websites, but without meeting them in person, it is not easy to establish their credentials.

Conditions of trade are normally done on a FOB or a CIF basis, and the price should be given before samples are sent. With each sample, a material safety data sheet, a child labor certificate, and a certificate of analysis should be sent. It should be noted that the drums should be sealed and that the sample should be fully topped with nitrogen or be full to ensure that there is no oxygen present, in order to make sure that oxidation is avoided. The sample bottles should be made from glass and not from plastic to avoid contamination by phthalates. The lots should be bulked before sampling and a flashpoint test should be obtained to guarantee that it is within the law to send the sample by mail or by airfreight with the correct labeling.

Many customers are able to give advice on production, but dealers in particular are best placed to advise. To enable contact with such dealers, it is worthwhile attending international meetings such as the International Federation of Essential Oils and Aroma Trades annual conference or reading the *Perfumer & Flavorist* magazine, which gives full details of brokers, dealers, and essential oil suppliers. There is no reference site that is 100% reliable in pricing for essential oils; this information should be gained by working with a variety of buyers, and from this, a knowledge of the market can be acquired.

The essential oil industry is very traditional and even though there have been changes in analytical methods and demands, the knowledge required in 1950 by buyers such as W. A. Ennever of Treatt (as can be seen from his quotation from Ennever 1960, reprinted earlier in this chapter) is not too different from today. There is greater demand for organically certified, Kosher, Halal, and other standards. The market can change far quicker now than in the past, thanks to the World Wide Web. Producers are often their own worst enemies and can destroy their own successful markets by communicating with their neighboring farmers, thereby encouraging them to enter the market. This can depress prices as a result of increased supply, but on the other hand, it can sometimes be in the interest of a sole producer to have other producers participating in the supply, to ensure guarantees of supply, and to lower the costs of production, which in turn encourages buyers to use the oil. Oils such as patchouli and grapefruit have had significant changes in price, as can be seen in the price graphs in Figures 23.7 through 23.9.

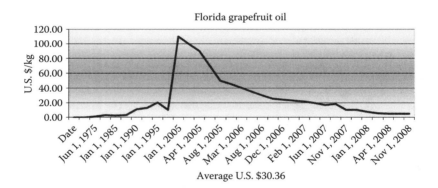

FIGURE 23.7 Price graph of grapefruit oil.

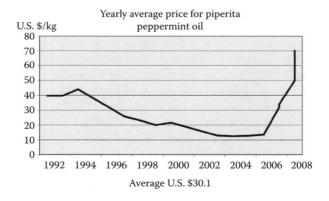

FIGURE 23.8 Price graph of peppermint oil (*Eucalyptus piperita*).

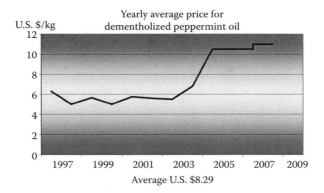

FIGURE 23.9 Price graph of peppermint oil (*Mentha arvensis*).

These price movements have reduced demand as major buyers of these products have had to look for alternatives to replace them as they are unable to cope with the massively increased prices from U.S.$10 to U.S.$100 for grapefruit and from U.S.$12.5 to U.S.$70 per kilo for peppermint oil. It can be seen, therefore, that stable pricing can lead to increased demand. Unstable pricing can lead to the death of essential oils. This is an important reason for holding inventory so that producers can enter into long-term associations with essential oil buyers to ensure good relationships.

In the 1970s, there was considerable fraud of millions of dollars, caused by the shipment of essential oils from Indonesia to the major buyers. The oils were in fact water, despite analysis certificates from Indonesian government laboratories showing them to be the named essential oil. Payment had been made by letters of credit and this fraudulent practice has discouraged buyers from opening letters of credit to suppliers today. Terms of trade should normally be cash against shipping documents or payment after receipt and quality control of goods.

The United States produces import statistics for essential oils, and these can often be useful sources of information, and the European Union (EU) also has such statistics. The EU statistics cover a wide range of essential oils in each tariff; therefore, the information is very vague and should not be used to make decisions. These statistics give no clues as to the quality of the product and it is that which can determine the price. The production of essential oils, as can be seen in the quotation by V. A. Beckley OBE, MC, Senior Agricultural Chemist, Kenya, during a meeting in 1931 in Nairobi, is perhaps more chancy than most farming propositions; it most certainly requires more attention and supervision than most, and, with certain rare exceptions, does not pay much more highly is still valid to this day, despite this being said in 1935.

The essential oil industry is a very small, tightly knit circle of traders, dealers, producers, and consumers, and apart from some notable exceptions, there is a very strong trade ethos. As it is a relatively small industry in terms of global commodities, statistics are not produced and it is by relationships with customers that information becomes available. Much that is on the Internet is misleading as it is for small quantities or is often written by consultants, and this information can be rapidly out of date as prices can move extremely quickly in either direction.

REFERENCE

Ennever, W.A. 1960. *Marketing Essential Oils*. London, U.K.: R.C. Treatt & Co. Ltd.

24 Storage, Labeling, and Transport of Essential Oils

Jens Jankowski, Jens-Achim Protzen, and Klaus-Dieter Protzen

CONTENTS

24.1 Marketing of Essential Oils: The Fragrant Gold of Nature Postulates Passion, Experience, and Knowledge 1041
24.2 Impact and Consequences on the Classification of Essential Oils as Natural but Chemical Substances in REACH 1043
24.3 Dangerous Substances and Dangerous Goods 1045
    24.3.1 Material Safety Data Sheet 1046
24.4 Packaging of Dangerous Goods 1047
24.5 Labeling 1048
24.6 List of Regulations for the Consideration of Doing Business in the EU 1050
Acronyms 1051
References 1052

24.1 MARKETING OF ESSENTIAL OILS: THE FRAGRANT GOLD OF NATURE POSTULATES PASSION, EXPERIENCE, AND KNOWLEDGE

Since the publication of the first edition of this book, quite a few changes have taken place regarding the regulations of handling and labeling of essential oils. These are considered and classified by regulatory authorities in most parts of the world not only as *natural* but also as *chemical substances*, abbreviated as NCS (the so-called natural complex substances).

The trade of essential oils is affected more and more by legal regulations related to safety aspects. The knowledge and the compliance with these superseding regulations that affect usual commercial aspects today have become a *conditio sine qua non* (precondition) to ensure trouble-free global business relations when placing essential oils on the market in the European Union (EU) (and other parts of the world) for use as natural flavors and fragrances in food, animal feed, cosmetics, pharmaceuticals, and aromatherapy.

Among others, the following regulations have to be observed (Dueshop, 2015):

- Regulation (EC) No. 1272/2008 on *c*lassification, *l*abeling, and *p*ackaging of substances and mixtures
- Regulation (EC) No. 1907/2006 on *REACH*
- Flavouring Regulation (EC) No. 1334/2008
- Regulation (EU) No. 1169/2011 on the provision of food information to the consumer
- Cosmetic Regulation (EC) No. 1223/2009
- Regulation (EU) No. 528/2012 on biocide products
- Regulation (EC) No. 1831/2003 on additives for use in animal nutrition
- EU Pharmaceutical Legislation—GMP and GDP aspects
- Regulation (EC) No. 648/2004 on detergents

- Regulation (EC) No. 178/2002 on food law
- Novel Food Regulation (EC) No. 258/97
- Regulation (EC) No. 396/2005 on maximum residue level of pesticides

Essential oils are natural substances mainly obtained from vegetable raw materials either by distillation with water or steam or by a mechanical process (expression) from the epicarp of citrus fruits. They are concentrated fragrance and flavor materials of complex composition, in general: volatile alcohols, aldehydes, ethers, esters, hydrocarbons, ketones, and phenols of the group of mono- and sesquiterpenes or phenylpropanes as well as nonvolatile lactones and waxes.

A definition of the term essential oils and related fragrance/aromatic substances is given in the ISO-Norm 9235 Aromatic Natural Raw Materials (International Standard Organization [ISO], Geneva, 1997).

In former times, essential oils were obtained from collected wild-growing plants—in these days, many of them are produced, however, from small-scale plantations and/or manufactured by small individual producers. The reason for this lies probably in the fact that a larger-scale production would require capital investment, which is rarely attracted as investors evidently realize that—if at all—no quick return of money is ensured. The negative factors influencing the market as follows:

- The dependency on weather as climatic conditions may affect the size of a crop over the whole vegetation period.
- Competing crops challenging the acreage.
- A keen global competition striving for market shares.

The general narrow margins do not compensate the involved risks—and last but not least, the adherence to comply with ever-changing administrative regulations is not necessarily the first target of investors of money.

All these aggravating factors also have an impact on the trade of essential oils. This is the reason that particularly the trade of the essential oils is dominated by small-scale and medium-sized family enterprises. Only entrepreneurs with passion, a personal engagement, and a persistent dedication, as well as a long-standing experience, nerve themselves to stay successfully in this business of the liquid gold of nature. A long-term philosophy, a lot of enthusiasm, and hard work together with a broad knowledge of the market situation and also the willingness of spending a lot of time and cost to investigate new ideas of state-of-the-art conditions of processing raw materials affecting the yield and quality and the return of investment.

In the EU, the classification for a chemical substance was laid down in the Council Directive 67/548 and subsequent amendments—Council Directive 79/831/EEC of 18-09-1979 (the famous sixth amendment) is the basis of all existing regulations for dangerous/hazardous chemicals and it earmarked the beginning of a new era that finally resulted in the EU in REACH—a regulation that is superseding and harmonizing the first attempts to regulate the safe handling of dangerous/hazardous materials.

REACH (*Registration, Evaluation, Authorization of Chemicals*) is the consistent continuation of rules to satisfy the EU administration with a perfect system to safeguard absolute security to protect humans and the environment regarding the use of chemicals. The topic of REACH will not be discussed in detail in this chapter because of its complexity and it touches only to a small extent the title of this contribution.

For the trade, that is, the industry and importers and dealers of essential oils, REACH is a heavy and costly burden demanding an unbelievable amount of time to furnish the required product information for an appropriate registration of essential oils as UVCB/NCS substances. In the following, however, a brief introduction to the historic development of the existing regulatory framework is given in order to help to understand the safety aspects, which are the background of the actual regulations and the impediments in connection with REACH.

24.2 IMPACT AND CONSEQUENCES ON THE CLASSIFICATION OF ESSENTIAL OILS AS NATURAL BUT CHEMICAL SUBSTANCES IN REACH

In the EU, the bell for the new era sounded when chemical substances in use within the EC during a reference period of 10 years had to be notified for the European Inventory of Existing Commercial Chemical Substances (EINECS).

At that time, EINECS enabled the EC administration not only to dispose of, for the first time, a survey of all chemical substances that had been in use in the EC between January 1971 and September 1981 but also to distinguish between "known substances" and "new substances."

"Known" substances are all chemicals notified for EINECS, whereas all chemical substances that were not notified (and subsequently had been registered as "known substances" in EINECS) were considered by the EU administration as "new chemicals."

"New" chemical substances could only be marketed in the EU after clearance according to uniform EC standards by competent authorities. Thus, from the beginning, all potential risks of a (new) chemical substance are ascertained for a proper labeling for handling and risks for humans and the environment could be minimized.

In a transitional phase, "known" chemical substances (notified for EINECS) enjoyed a temporary exemption from the obligation to furnish the same safety data required for new chemical substances. The assumption was that, based on the experience gathered during their use for decades and sometimes centuries, the temporary continuation of their use could be tolerated according to the hitherto used older standards of safety (Dueshop, 2007) as a short-term clearance of approximately 100,000 chemical substances registered in EINECS could not be effected overnight.

To make sure that the *known* substances which had been notified for EINECS—and thus became known to the regulative agencies in the EC—also did comply with the new safety standards, they were screened step by step according to the following volume bands either depending on their potential risk or according to the volumes produced or imported:

- >1000 tons
- 100–1000 tons
- 10–100 tons
- 1–10 tons
- 100–1000 kg
- <100 kg

To perform this task, the EU administration—as the United States already did many years ago—made use of the principles of the CAS system and arranged for the majority of essential oils and other "UVCB" (chemical substances of unknown or variable composition, complex reaction products and biological materials) and allocated (new) more precise CAS numbers, which were eventually also published in EINECS.

One should bear in mind, however, that, in principle, the CAS number is an identification number for a defined chemical substance that is allotted by a private enterprise in the United States must not be confused with the EINECS registration number, which is a registration number allocated by the EU administration, that is, ECB/JRC at ISPRA—CAS numbers are assigned by the (private) CAS organization in the United States with the purpose of identification of (defined) chemical substances.

In principle, a CAS number is allocated by the CAS organization to a new (defined) chemical substance only after thorough examination of the product as per the IUPAC rules to make sure that, irrespective of different chemical descriptions and/or coined names that have been given to a product, a substance can be clearly related by the allocated CAS number according to the (CAS) principle "one substance—one number."

Using the CAS number system to also register UVCB substances in EINECS, that is, products that are not defined chemicals, it made it necessary to extend the CAS system also to the so-called UVCBs. Essential oils as NCS eventually are registered by their botanical origin. As, for example,

- *Lavender oil*: lavender—*Lavandula angustifolia*
  - EINECS registration no. 289-995-2—CAS no. (Einecs) 90063-37-9: extractives and their physically modified derivatives such as tinctures, concretes, absolutes, essential oils, terpenes, terpene-free fractions, distillates, and residues from *Lavandula angustifolia*—Labiatae (Lamiaceae)
- *Lavender oil*: lavender—*Lavandula angustifolia*
  - EINECS registration no. 283-994-0—CAS no. (Einecs) 84776-65-8: extractives and their from *Lavandula angustifolia angustifolia*—Labiatae (Lamiaceae)
- *Lavender concrete/absolute*: lavender—*Lavandula angustifolia*
  - EINECS registration no. 289-995-2—CAS no. (Einecs) 90063-37-9 extractives and their physically modified derivatives such as tinctures, *concretes*, *absolutes*, essential oils, terpenes, terpene-free fractions, distillates, and residues from *Lavandula angustifolia*—Labiatae (Lamiaceae)
- *Lavandin oil*: *Lavandula hybrida*.
  - EINECS registration no. 294-470-6—CAS no. (Einecs) 91722-69-9 extractives and their physically modified derivatives such as tinctures, concretes, absolutes, essential oils, terpenes, terpene-free fractions, distillates, and residues from *Lavandula hybrida*—Labiatae (Lamiaceae)
- *Lavandin oil abrialis*: *Lavandula hybrida abrial*.
  - EINECS registration no. 297-384-7—CAS no. (Einecs) 93455-96-0 extractives and their from *Lavandula hybrida abrial*—Labiatae (Lamiaceae)
- *Lavandin oil grosso*: *Lavandula hybrida grosso* ext.
  - EINECS registration no. 297-385-2—CAS no. (Einecs) 93455-97-1 extractives and their from *Lavandula hybrida grosso*—Labiatae (Lamiaceae)

TABLE 24.1
Examples of Different CAS-Numbers Used in USA and EINECS in EU

Article	CAS No. USA	CAS No. EINECS	EC Registration No.
Eucalyptus oil	8000-48-4	84625-32-1	283-406-2
Eucalyptus globulus Lab.—Myrtaceae			
Lavender oil	8000-28-0	90063-37-9	289-995-2
Lavandula angustifolia—Labiatae			
Lavandula angustifolia angustifolia—Labiatae		84776-65-8	283-994-0
Lemon oil	8008-56-8	8028-48-6	284-515-8
Citrus limon L.—Rutaceae		84929-31-7	284-515-8
Orange oil	8008-52-9	8028-48-6	232-433-8
Citrus sinensis—Rutaceae			
Peppermint oil	8006-90-4	98306-02-6	308-770-2
Mentha piperita L.—Lamiaceae			

Storage, Labeling, and Transport of Essential Oils

The registration of essential oils under their botanical origin implicated that concretes/absolutes and other natural extractives of the same botanical origin also have the same EINECS and CAS numbers as the essential oil.

In this connection, it should be mentioned that when checking an EINECS number, it is important to investigate the correct number in the official original documentation, as in the secondary literature, there exist many inaccuracies.

Because of the lack of rules for a uniform classification of essential oils as UVCB, it happened that against the principles of the CAS organization in some cases, several CAS numbers had been allocated to essential oils of the same denomination and, in addition,

- An earlier (older) CAS number allocated for the product in the United States
- A new (and more precise) CAS number allocated for registration in the EC for EINECS, respectively (Table 24.1)

24.3 DANGEROUS SUBSTANCES AND DANGEROUS GOODS

There is a significant difference between the similar sounding words and regulations regarding "Dangerous Substances" and "Dangerous (Hazardous) Goods."

Both regulations are targeted to protect humans and the environment, but the term "Dangerous Substance" refers to the risks connected with the properties of the substance, that is, the potential risk of a direct contact with the product during production, packaging, and use.

"Dangerous Goods" refers to dangerous substances that are properly packed and labeled for storage and transport by road, rail, sea, or air (Figure 24.1).

In 2003, a working group of the United Nations (UN) that was trying to harmonize the hitherto often different national and international existing regulations on the classification and labeling of dangerous/hazardous products published the so-called *purple book*. It contained the results of their first attempts of a global uniform regulation. This publication is updated every 2 years.

In the EU *Regulation (EC) No. 1272 on Classification, Labelling and Packaging of substances and mixtures,* in general, often called as CLP-Regulation, was published in 2008. It is based on the Globally Harmonised System (GHS) and became binding in the EU on 2009.01.20 replacing the Regulations (EC) No. 67/548 as well as (EC) No. 1944/45.

The GHS distinguishes between 16 physical dangers, 10 health risks, and, in addition, the class "aquatic environment."

1. Physical hazards:
    a. Explosives/mixtures and products with explosive properties
    b. Flammable gases
    c. Flammable aerosols
    d. Oxidizing gases

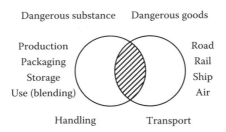

FIGURE 24.1 Interrelationship between dangerous substances and dangerous goods. (Courtesy of Paul Kaders, Hamburg/Germany).

e. Gases under pressure
  f. Flammable liquids
  g. Flammable solids
  h. Self-reactive substances and mixtures
  i. Pyrophoric liquids
  j. Pyrophoric solids
  k. Self-heating substances and mixtures
  l. Substances and mixtures which, in contact with water, emit flammable gases
  m. Oxidizing liquids
  n. Oxidizing solids
  o. Organic peroxides
  p. Corrosive to metals
2. Health hazards:
  a. Acute toxicity
  b. Skin corrosion/irritation
  c. Serious eye damage/eye irritation
  d. Respiratory or skin sensitization
  e. Germ cell mutagenicity
  f. Carcinogenicity
  g. Reproductive toxicity
  h. Specific target organ toxicity—single exposure
  i. Specific target organ toxicity—repeated exposure
  j. Aspiration hazard
3. Environmental hazards:
  a. Hazardous to the aquatic environment

In contrast to the CLP regulations—probably for historical reasons—the transport of dangerous goods is divided in nine classes only.

1. Class 1—Explosives
2. Class 2—Gases
3. Class 3—Flammable liquids
4. Class 4—Flammable solids
5. Class 5—Oxidizing substances and organic peroxides
6. Class 6—Toxic and infectious substances
7. Class 7—Radioactive material
8. Class 8—Corrosives
9. Class 9—Miscellaneous dangerous goods

Nowadays, for all ways of transport (road, rail, air, and sea), this uniform classification as per GHS is applicable with the requirement of a correspondent labeling.

24.3.1 Material Safety Data Sheet

As per GHS requirements, for each hazardous substance, a "Material Safety Data Sheet," as per ISO standard (Geneva) and/or REACH regulation, must be furnished.

In Europe, the REACH standard form is binding consisting of the following topics:

1. Identification of the substance/mixture and of the company
2. Hazard identification
3. Composition/information on ingredients

4. First aid measures
5. Fire-fighting measures
6. Accidental release measures
7. Handling and storage
8. Exposure controls/personal protection
9. Physical and chemical properties
10. Stability and reactivity
11. Toxicological information
12. Ecological information
13. Disposal considerations
14. Transport information
15. Regulatory information
16. Other information

The instructions that the supplier has to give in topic 14 of the MSDS permit a correct labeling for the transport of dangerous goods—in case of a doubt, however, the correct classification can also be verified by checking the remarks entered under section 2: "Hazard Identification" of the same MSDS.

In case that no MSDS is available, there exists the possibility to search the required information in the databank of the European Chemicals Agency (ECHA). In the C&L-Inventory, the CAS and/or EINECS numbers, the corresponding classification, required labels and the corresponding "H"-(hazard) and the P (precautionary) statements, and the GHS pictograms of all chemicals registered in the EU can be found. However, this ECHA databank is of help only for orientation, as, for example, there exist for Orange Oil sweet CAS No. 8028-48-6 a total of 56 entries.

Another source for information is the IFRA/IOFI Labelling Manual of the International Fragrance Association. The readings in this Manual are quite simple to understand and the desired information regarding the classification and thus the desired information is easier for use than in the C&L inventory of ECHA. In addition—besides the GHS classification—this manual also still contains a reference to the older Regulation (EC) No. 67/548, which sometimes is a valuable hint.

24.4 PACKAGING OF DANGEROUS GOODS

Those dangerous substances for which the international regulations for transport of dangerous goods apply have to be transported only in UN-approved container. All UN-approved packaging are marked with an immutable code number as, for example (Table 24.2).

According to the risk emanating from a substance, there exist three different packing groups (Table 24.3).

As illustrated, the packing codes "X", "Y," and "Z" refer to the risks emanating from a substance: PC "X" is required for substances with a high risk, and "Z" with a low risk (Table 24.4).

TABLE 24.2
Meaning of the Code Number on Container

UN1A1/Y/1,4/150/(06)/NL/VL824

1A1	Steel drum—nonremovable head
Y	Allowed for substances with packing group (PG) II and PG III
1.4	Maximum relative density at which the packing has been tested
150	Test pressure
(06)	Year of manufacture
(NL)	State
(VL123)	Code number of manufacturer

TABLE 24.3
Packing Group and Packing Code

Packing Group		Packing Code
PG III	Low risk	Z
PG II	Medium risk	Y
PG I	High risk	X

TABLE 24.4
Correlation between the Packing Code and Packing Group

Packing Code	Packing Group
X	PG I, PG II, PG III
Y	PG II and PG III
Z	PG III

24.5 LABELING

In addition to the correct selection of the packing, the CLP regulations and the guidelines for the transport of dangerous goods require labeling of each piece of packing piece together with "H" (hazard) and "P" (precautionary) statements. These statements must be given in English and national language(s) together with labels of pictograms according to the classification of the potential hazard emanating from the substance. In addition, if applicable, it should also contain the label/pictogram "environmentally hazardous" (Figure 24.2).

The introduction of the label/pictogram "environmentally hazardous" also became part of the present amendments of ADR/RID code/regulations for road and rail traffic in Europe and for the IMDG code for sea transport if the hazardous substance falls into the classification:

- aquatic acute 1
- aquatic chronic or
- aquatic chronic 2

FIGURE 24.2 "Environmentally Hazardous" label.

Storage, Labeling, and Transport of Essential Oils

This label is replacing the former label "Marine Pollutant" for dangerous goods shipped by sea.

For transport by air, as per IATA regulations, this new label is also compulsorily required for goods with the UN numbers UN 3077 and UN 3082—it may, however, also be fixed in those cases where it is required for the transport by other carriers.

In this connection, a few words are due on the so-called UN numbers for dangerous goods. These UN numbers are assigned to dangerous goods according to their hazard classification and composition. These UN numbers should not be confused with the (UN) number for packing. Labels with the UN numbers must be fixed distinctly and visibly on each packing in addition to the already mentioned other labels as per the CLP-Regulation.

UN numbers are listed in all codes for transport of dangerous goods and are identical for all types of transport. Most of the essential oils fall under the numbers:

- UN no. 1169 extracts, aromatic, and liquid
- UN no. 2319 terpene hydrocarbons
- UN no. 3082 environmentally hazardous substances, liquid. n.o.s.
- UN no. 1992 flammable liquid, toxic, n.o.s.
- UN no. 1272 pine oil(s)

N.O.S. (not otherwise specified) requires that the name of the hazardous substance has to be added in braces to the Proper Shipping Name (Figure 24.3).

The aim of dangerous goods regulations is not only to protect persons occupied with the conveyance of dangerous/hazardous substances but also to serve, for example, fire brigades, who in case of an accident or fire are called and have to be aware of special risks.

It should be noted that, at least in Europe, the transport police controls more and more transports of dangerous goods, the accompanying documentation even to the extent of the markings and correct labels.

For the transport of small quantities of hazardous substances, exceptions/exemptions exist that allow simplified procedures in two levels: the "excepted quantity" and the "limited quantity."

 (flame symbol, 3)	Label 3 Flammable liquids Symbol flame: Black or White Background: Red
 (flame symbol, 4)	Label 4.1 Flammable solid Symbol flame: Black Background: White with seven vertical red stripes
 (skull and crossbones, 6)	Label 6.1 Toxic Symbol skull and crossbones: Black Background: White
 (corrosive symbol, 8)	Label 8 Corrosive Symbol liquids spilling from two glass vessels and attacking a hand and a metal: black Background: upper half White, lower half Black with White border
 (stripes, 9)	Label 9 Miscellaneous Symbol seven vertical stripes in upper half: Black Background: white

FIGURE 24.3 Important hazard symbols for essential oils.

FIGURE 24.4 "Excepted Quantity" label with explanation of the printed*.

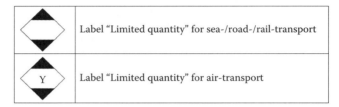

FIGURE 24.5 "Limited Quantity" label.

For the dispatch of samples of those essential oils classified as dangerous substances, the regulations for documentation are canceled and the requirements for packing have been reduced to a minimum and thus the "excepted quantity" is of invaluable help. However, the dispatch is only allowed in a combined packing and the packaging must be marked with the label "excepted quantity" (Figure 24.4).

In case the quantities to be sent exceed the exemptions for an "excepted quantity," there exists the possibility of a labeling as "limited quantity." As per (different) regulations of the carriers, these packages have to be labeled as shown in Figure 24.5.

For the transport of dangerous goods as "limited quantity," a list is required specifying the net weights of each class of hazardous substances planned for the shipment. This list should be furnished before hand to the forwarder and also has to be attached to the transport documentation. Furthermore, it should be noted that goods have to be packed in a combined packing.

Despite the adherence of aforementioned regulations, it can happen that a carrier might reject the transport of (certain) dangerous substance(s) under the provisions of a "limited quantity." In this case, there only remains the alternative to have the material transported as per the standard regulations of the respective carrier.

24.6 LIST OF REGULATIONS FOR THE CONSIDERATION OF DOING BUSINESS IN THE EU

- Regulation (EC) No 1272/2008 of the European Parliament and of the Council of December 16, 2008 on classification, labelling and packaging of substances and mixtures (Text with EEA relevance), OJ. L 353 published December 31, 2008, pp. 1–1355.
- Regulation (EC) No 1907/2006 of the European Parliament and of the Council of December 18, 2006 concerning the Registration, Evaluation, Authorisation and Restriction of Chemicals (REACH), establishing a European Chemicals Agency, OJ. L 396 published December 31, 2006, pp. 1–851.

- Regulation (EC) No. 1334/2008 of the European Parliament and of the Council of December 16, 2008 on flavorings and certain food ingredients with flavouring properties for use in and on foods, OJ L 354 published December 31, 2008, pp. 34–50.
- Regulation (EU) No. 1169/2011 of the European Parliament and of the Council of October 25, 2011 on the provision of food information to consumers, OJ L 304 published November 22, 2011, pp. 18–63.
- Regulation (EC) No. 1223/2009 of the European Parliament and of the Council of November 30, 2009 on cosmetic products, OJ L 342 published December 22, 2009, pp. 59–209.
- Regulation (EU) 528/2012 of the European Parliament and of the Council of May 22, 2012 concerning the making available on the market and use of biocidal products, OJ L 167 published June 27, 2012, pp. 1–123.
- Regulation (EC) No. 1831/2003 of the European Parliament and of the Council on additives for use in animal nutrition, OJ L 268 published October 18, 2003, pp. 29–43.
- Regulation (EC) No. 648/2004 of the European Parliament and of the Council of March 31, 2004 on detergents, OJ L 104 published April 8, 2004, pp. 1–35.
- Regulation (EC) No. 178/2002 of the European Parliament and of the Council of January 28, 2002 laying down the general principles and requirements of food law, establishing the European Food Safety Authority and laying down procedures in matters of food safety, OJ L 31 published February 1, 2002, pp. 1–24.
- Regulation (EC) No. 258/97 of the European Parliament and of the Council of January 27, 1997 concerning novel foods and novel food ingredients, OJ L 43 published February 14, 1997, pp. 1–6.
- Regulation (EC) No. 396/2005 of the European Parliament and of the Council of February 23, 2005 on maximum residue levels of pesticides in or on food and feed of plant and animal origin, OJ L 70 published March 16, 2005, pp. 1–16.
- Directive 2000/13/EC of the European Parliament and of the Council of March 20, 2000 on the approximation of the laws of the Member States relating to the labelling, presentation and advertising of foodstuffs, OJ L 106 published May 6, 2000, pp. 29–42.

ACRONYMS

ADR	Accord europeen relatif au transport international des marchandises dangereuses par route (European agreement for the transport of dangerous goods by road)
CAS	Chemical Abstracts Service
CLP	Classification, Labeling and Packaging of substances and mixtures
ECHA	European Chemicals Agency
EINECS	European Inventory of Existing Commercial Chemical Substances
GDP	Good Distribution Practice
GHS	Globally harmonized system
GMP	Good Manufacturing Practice
IFRA	International Fragrance Association
IMDG-Code	International Maritime Dangerous Goods Code
IOFI	International Organization of the Flavor Industry
ISO	International Standard Organization
MSDS	Material Safety Data Sheet
NOS	Not Otherwise Specified
RID	Règlement concernant le transport international ferroviaire des marchandises dangereuses (Order for the international transport of dangerous goods by train)
UVCB	Chemical Substances of Unknown or Variable Composition, Complex Reaction Products and Biological Materials

REFERENCES

C&L Inventory, European chemicals agency, Accessed October 20, 2013 from http://echa.europa.eu/de/regulations/clp/cl-inventory.

Council Directive 67/548/EEC of 27 June 1967 on the approximation of laws, regulations and administrative provisions relating to the classification, packaging and labeling of dangerous substances, OJ. 196 published August 16, 1967, pp. 1–98.

Council Directive 79/831/EEC of 19 September 1979 amending for the sixth time Directive 67/548/EEC on the approximation of the laws, regulations and administrative provisions relating on the classification, packaging and labelling of dangerous substances, OJ L 259, published October 15, 1979, pp. 10–28.

Dangerous Goods Regulations, 56th Edition, International Air Transport Association, Montreal, Quebec, Canada, September 2014.

Directive 1999/45/EC of the European Parliament and of the Council of 31 May 1999 concerning the approximation of the laws, regulations and administrative provisions of the Member States relating to the classification, packaging and labelling of dangerous preparations, OJ. L 200 published July 30, 1999, pp. 1–68.

Dueshop, L., 2007, Personal communications.

Dueshop, L., 2015, Personal communications.

Global harmonisiertes System zur Einstufung und Kennzeichnung, Bundesinstitut für Risikobewertung, Accessed June 17, 2013 from http://www.bfr.bund.de/de/global_harmonisiertes_system_zur_einstufung_und_kennzeichnung-61584.html.

IMDG-Code 2015, *Amdt.*, 37th edn., Storck Verlag, Hamburg, 2014, pp. 37–14.

Protzen, K.-D., October 1989, *Guideline for Classification and Labelling of Essential Oils for Transport and Handling*, distributed during *IFEAT Conference*, London, U.K.

Protzen, K.-D., November 8–12, 1998, Transportation/Safety Regulations Update, *International Conference on Essential Oils and Aromas*, London, U.K., IFRA/IOFI Labeling Manual 2014.

Regulation (EC) No. 178/2002 of the European Parliament and of the Council of 28 January 2002 laying down the general principles and requirements of food law, establishing the European Food Safety Authority and laying down procedures in matters of food safety, OJ L 31 published February 1, 2002, pp. 1–24.

Regulation (EC) No. 258/97 of the European Parliament and of the Council of 27 January 1997 concerning novel foods and novel food ingredients, OJ L 43 published February 14, 1997, pp. 1–6.

Regulation (EC) No. 396/2005 of the European Parliament and of the Council of 23 February 2005 on maximum residue levels of pesticides in or on food and feed of plant and animal origin, OJ L 70 published March 16, 2005, pp. 1–16.

Regulation (EU) 528/2012 of the European Parliament and of the Council of 22 May 2012 concerning the making available on the market and use of biocidal products, OJ L 167 published June 27, 2012, pp. 1–123.

Regulation (EC) No. 648/2004 of the European Parliament and of the Council of 31 March 2004 on detergents, OJ L 104 published April 8, 2004, pp. 1–35.

Regulation (EU) No. 1169/2011 of the European Parliament and of the Council of 25 October 2011 on the provision of food information to consumers, OJ L 304 published November 22, 2011, pp. 18–63.

Regulation (EC) No. 1223/2009 of the European Parliament and of the Council of 30 November 2009 on cosmetic products, OJ L 342 published November 22, 2009, pp. 59–209.

Regulation (EC) No. 1272/2008 of the European Parliament and of the council of 16 December 2008 on classification, labelling and packaging of substances and mixtures (Text with EEA relevance), OJ. L 353 published December 31, 2008, pp. 1–1355.

Regulation (EC) No. 1334/2008 of the European Parliament and of the Council of 16 December 2008 on flavourings and certain food ingredients with flavouring properties for use in and on foods, OJ L 354 published December 31, 2008, pp. 34–50.

Regulation (EC) No. 1831/2003 of the European Parliament and of the Council on additives for use in animal nutrition, OJ L 268 published October 18, 2003, pp. 29–43.

Regulation (EC) No. 1907/2006 of the European Parliament and of the council of 18 December 2006 concerning the Registration, Evaluation, Authorisation and Restriction of Chemicals (REACH), establishing a European Chemicals Agency, OJ. L 396 published December 31, 2006, pp. 1–851.

Verordnung zur Änderung der Anlagen A und B zum ADR-Übereinkommen (24. ADR-Änderungsverordnung—24. ADRÄndV) vom 06. Oktober 2014 (BGBl. II S. 722).

25 Aroma-Vital Cuisine
Healthy and Delightful Consumption by the Use of Essential Oils

Maria M. Kettenring and Lara-M. Vucemilovic-Geeganage

CONTENTS

25.1 Basic Principles of the Aroma-Vital Cuisine 1054
    25.1.1 The Heart of Culinary Arts Is Based on Exquisite Ingredients
            and an Accomplished Rounding 1054
    25.1.2 Quality Criteria and Specifics That Have to Be Adhered to While Handling
            Essential Oils for Food Preparation 1054
    25.1.3 Storage 1055
    25.1.4 Quantity 1055
    25.1.5 Emulsifiers and Forms of Administering 1055
    25.1.6 To Add Spice with Natural Aromas in a Balanced Way 1055
    25.1.7 Essential Oils Are Able to Lift Our Spirits As Well 1056
25.2 Small Culinary Trip: Aroma-Vital Cuisine Recipes and Introduction 1056
    25.2.1 Menu 1056
    25.2.2 Aroma-Vital Cuisine Recipes 1058
            25.2.2.1 Basics 1058
            25.2.2.2 Beverages 1059
            25.2.2.3 Entrees 1061
            25.2.2.4 Appetizer and Finger Food 1063
            25.2.2.5 Main Course 1065
            25.2.2.6 Dessert, Cakes, and Baked Goods 1068
            25.2.2.7 Résumé 1070

Your nourishment ought to be your remedies and your medicaments shall be your food.

—Hippocrates

Certainly, the value of our nutrition, in terms of nutritional physiology, is not only conditioned by its nutrient and calorie contents. Moreover, also health-conscious and constitutional eating habits require an adequate preparation of meals as well as an appropriate form of presentation. Early sophisticated civilizations and their health doctrines, like that of the traditional Chinese medicine, Ayurveda in Southeast Asia, and, for instance, the medical schools during the ancient Greek period, examined individuals and their reaction on life circumstances, habits, nutrition, and substances, to contribute to a long-lasting health. To support a person's balance, the aim was to develop a conscious way of using the senses and a balanced sensory perception.

Thus, fragrances are a kind of soul food, as the information of scents can be perceived in every section of our self, physical, energetic, and intellectual, from a holistic point of view. Adding spice with essential oils according to the aroma-vital cuisine combines sensuality with sanative potential.

People across continents and cultures have experimented with the healing virtues of "nature's bouquet" or just simply tried to enhance the flavor and vitality of their meals. The ancient Egyptian civilization reverted to an elaborated dinner ceremony by using the efficacy of essential oils to get the participants in the mood for the meal. Before the food was served, heated chalices with scented fats, enriched with a variety of herbs and spices were provided, not only to spread pleasant smells, rather as a kind of odorous aperitif to activate ones saliva to prepare for digestion. Meals that have been enriched with essential oils or expressed oils, rebound to a conscious awareness of consuming food, are well-nigh comparable, like going on a culinary expedition. This fare is perceived as a composition of tastes, which is not only tastefully ingenious, but also might be able to raise the food's virtue.

In this regard, the entropy rather than the potency of the condiment is significant. The abundance of nuances, the art of adding flavor on the cusp of being noticeable, becomes more important than giving aroma officiously. The scents hovering above the meals, almost like a slight breeze, compound the food's own natural flavor in a subtle manner. "Less is more" is the economic approach that in this context is indicative.

The sensation of satiety is taking place early on. Due to this desire to savor to the fullest, the taste is excited and leads to longer chewing. This in turn activates α-amylase (amylolytic enzyme, already working in the oral cavity). Conditionally on the high bioavailability, especially of the monoterpenes, which are significant and available in the paring of citrus fruits and some kind of herbs, in a sense, the aroma-vital cuisine shows aspects of the salutary genesis (Salutogenese). The savoriness of the food, pleasant smell, and appetizing appearance play a prominent role here; at last, the appetite regulates between physiological needs and pleasure, and thus variety- and vitally enhanced meals are in demand.

25.1 BASIC PRINCIPLES OF THE AROMA-VITAL CUISINE

25.1.1 THE HEART OF CULINARY ARTS IS BASED ON EXQUISITE INGREDIENTS AND AN ACCOMPLISHED ROUNDING

Natural aromas, from blossoms, herbs, seeds, and spices, extracted in artificial pure essential oils, delicately accompany the elaborate cuisine. They are not supposed to supersede fresh herbs, rather complementing them. If, however, herbs are not available, natural essences are delightfully suited to add nuances. They are giving impetus to and are flexible assistants for preparing last-minute menus. One should use this rich source to compile a first-aid assortment of condiments or even a mobile spice rack.

25.1.2 QUALITY CRITERIA AND SPECIFICS THAT HAVE TO BE ADHERED TO WHILE HANDLING ESSENTIAL OILS FOR FOOD PREPARATION

The regional legal regulations of the food chemical codex or the local food legislation might differ, and if one is going to use essential oils professionally, one has to be firm with them, but still there are certain basics that deserve attention and lead to a safe and healthy way of practicing this subtle culinary art.

For cooking, solely 100% pure essential oils from controlled organic cultivation should be used. With oils that are not available of controlled organic origin, particularly those that are cold-pressed, a residue check should be guaranteed by the manufacturer to ensure that the product does not contain harmful amounts of pesticides. The label should not only contain name, contents, and quantity but also should contain

- Latin definition
- Country of origin
- Description of used plant parts

Aroma-Vital Cuisine

- Used method of extraction
- Date of expiry
- If the oil has been thinned, the exact ratio of mixture
- Mentioning if solvents have been used

For the aroma-vital cuisine, the only acceptable solvent would be alcohol. As the oil is used in very small and thinned concentrations, it would not be harmful to children. Less qualitative oils from industrial origin sometimes might even contain other substances. It should be indicated that natural flavorings used in food production should be pure and free of animal by-products such as gelatin or glycerin, which has been obtained by saponification of animal fat.

25.1.3 Storage

Essential oils are very sensible to the disposure of light, air, and temperature; therefore, they should be stored adequately. In this way, long-lasting essential oils keep their aroma as well as their ingredients and might even develop their bouquet. Foods or processed foods with essential oils may not be stored in tin boxes. It is very important that essential oils should be kept away from children.

25.1.4 Quantity

The internal use of essential oils has to be practiced carefully. This subtle art is an amazing tool, but swallowed in too huge amounts, they are bad for one's health. One should never add the pure concentrate of essential oils to foods; it should not be forgotten that 1 drop is often comparable to a huge amount of plant material. Therefore, they ought to be always thinned, and the dilution should be used teaspoon by teaspoon.

25.1.5 Emulsifiers and Forms of Administering

Essential oils are not water soluble; therefore, emulsifiers are necessary to spread their aroma, they are, for example,

1. Basic oils, special oils, or macerated oils
2. Butter, milk, curd, egg yolk, and mayonnaise
3. Alcohol and vinegar
4. Syrups, molasses, honeys, treacles, and sugars
5. Salt
6. Tofu, soy sauce, and tamarind sauce
7. Avocado, lemon juice, and coconut
8. Sesame seeds, sunflower seeds, almonds, and walnuts

On the basis of these emulsifiers and a mixture of essential oils, a variety of "culinary assistants" can be conjured up: spiced oils, spiced butter or mayonnaises, spiced alcohols, spiced syrups, spiced sauces, or even spiced salts. These blends can be prepared in advance and stored to use them for everyday meals. Another nice variation is the use of hydrolates (a partial extract of plant material extracted by distillation) such as rose water, for food preparation.

25.1.6 To Add Spice with Natural Aromas in a Balanced Way

To know how food and essential oils interact is a great help to create a harmonic assembly of foods, which is nourishing us from a holistic point of view. In this manner, the sun-pervaded seed oils of anise, bay, dill, fennel, or caraway might be able to aerate the earthy corm and root vegetable. Salads can be enhanced and prepared to be more digestive by adding pure natural essential oils such as thyme, rosemary, and clementine to the marinade, or another rather Asian variation would be to add ginger, pepper, and lemon grass.

25.1.7 Essential Oils Are Able to Lift Our Spirits As Well

A condiment ensemble of orange, vanilla extract, cacao extract, and rose, for example, is able to support soul foods such as milk rice, milk shakes, and desserts in their attitude to supply security and confidence.

25.2 SMALL CULINARY TRIP: AROMA-VITAL CUISINE RECIPES AND INTRODUCTION

See Table 25.1.

25.2.1 Menu

Basics

    Crispy coconut flakes (flexible Asian spice variation)
    Gomasio (sesame sea-salt spice)
    Honey Provencal

Beverages

    Aroma shake with herbs
    Earl Grey at its best
    Lara's jamu
    Rose cider
    Syrup mint–orange

Entrees

1. Soups:
   a. Peppermint heaven
   b. Perky pumpkin soup
2. Salads:
   a. Melon–plum purple radish salad
   b. Salad with goat cheese and ricotta

Appetizer and Finger Food

    Crudities—flavored crispy raw vegetables
    Maria's dip
    Tapenade
    Tofu aromanaise
    Vegetable skewer

Main Course

    Celery—lemon grass patties
    Chèvre chaud (goat) cheese "provence" with pineapple
    Crispy wild rice (chapatis)
    Mango–dates–orange chutney
    Prawns bergamot

TABLE 25.1
Basic Spice Rack of Essential Oils: How to Prepare Essential Oil Mixtures and Essential Oil Seasonings

Basic Essential Oils	Mixtures	Emulsifier Seasonings		Recipes—Example
		Euro Asia		
Lime (*Citrus aurantifolia*)	5 drops	1. Oil	50 mL sesame oil	Asian style
		2. Dairy prod	50 mL mayonnaise	Eggs
Coriander seed (*Coriandrum sativum*)	1 drop	3. Vinegar	50 mL rice vinegar	Sushi
		4. Sweetener	50 mL agave syrup	Chutney
Ginger (*Zingiber officinalis*)	2 drops	5. Salt	50 mg sea-salt	Spice
		6. Tofu and co	50 mL soy sauce	Marinated fried tofu
Lemongrass (*Cymbopogon citratus*)	1 drop	7. Vegetables and fruits	50 mL coconut milk	Rice and curry
		8. Nuts and seeds	50 mg sesame seeds	Spice
Green pepper (*Piper nigrum*)	1 drop			
		O Sole Mio		
		1.	50 mL olive oil	Pasta
Thyme linalool (*Thymus vulgaris*)	1 drop	2.	50 mL egg yolk	Omelette
		3.	50 mL balmy vinegar	Salad
Rosemary cineole (*Rosmarinus officinalis*)	1/2 drop	4.	50 mL honey	Cuisine Provencal
		5.	50 mg sea-salt	Spice
Clementine (*Citrus deliciosa*)	5 drops	6.	50 mg tofu	Grilled tofu
		7.	50 mg avocado	Guacamole
		8.		Pesto
		Capri		
Orange (*Citrus sinensis*)	5 drops	1.	50 mL hazelnut oil	Desserts
		2.	50 mL buttermilk	Drink
Lemon (*Citrus limon*)	3 drops	3.	50 mL cider vinegar	Salad
		4.	100 mL maple syrup	Desserts
		5.	50 mg sea-salt	Spice
		6.	50 mL apple vinegar	Fruit salad
		7.	50 mg avocado	Sauce
		8.	50 mg walnuts	Cakes
		Bergamot-Grand Marnier		
Grapefruit (*Citrus paradisi*)	5 drops	1.	50 mL walnut oil	Salad
		2.	50 mg butter	Cake
Orange (*Citrus sinensis*)	5 drops	3.	1 L white wine	Beverage
		4.	50 mg raw sugar	Sweets
Limon (*Citrus limon*)	2 drops	5.	50 mg sea-salt	Spice
		6.	50 mL tamarind sauce	Thai cuisine
Bergamot (*Citrus bergamia*)	2 drops	7.	50 mL lemon juice	Drink
		8.	50 mg pumpkin seeds	Soup
		Magic Orange		
Orange (*Citrus sinensis*)	5 drops	1.	50 mL almond oil	Sweets
		2.	50 mg goat cheese	Oriental

(Continued)

TABLE 25.1 (*Continued*)
Basic Spice Rack of Essential Oils: How to Prepare Essential Oil Mixtures and Essential Oil Seasonings

Basic Essential Oils	Mixtures	Emulsifier Seasonings		Recipes—Example
Vanilla extract (*Vanilla planifolia*)	3 drops	3.	50 mL raspberry vinegar or balsamic vinegar	Fruit salad
		4.	100 mL honey/treacle	Sweets
Kakao extract (*Theobroma cacao*)	3 drops	5.	—	
		6.	50 mL seitan tofu	Oriental
Rose (*Rosa damascena*)	1/2 drop	7.	50 mg bananas	Desserts
		8.	50 mg almonds	Spice
Clary Sage and Bergamot				
Clary sage (*Salvia sclarea*)	2 drops			Spice
Bergamot (*Citrus bergamia*)	5 drops	5.	50 g sea-salt	
Peppermint				
Peppermint (*Mentha piperita*)	Rather less—2 drops per 100 mL/mg	4.	100 mL maple syrup	Drink
Lavender				
Lavender (*Lavandula officinalis*)	Rather less—2 drops per 100 mL/mg	4.	100 mL honey	Cuisine provencal

Dessert, Cakes, and Baked Goods

  Apple cake rose
  Chocolate fruits and leaves
  Homemade fresh berry jelly
  Rose semifreddo
  Sweet Florentines

(Chocolate should not be heated up more than 40°C. Essential oils are best at 40°C as well.)

25.2.2 AROMA-VITAL CUISINE RECIPES

25.2.2.1 Basics

25.2.2.1.1 Crispy Coconut Flakes (Flexible Asian Spice Variation)
Nice with Asian flavored dishes or sweet baked goods.
Ingredients:

- 50 g dried coconut flakes
- 10 drops Euro Asia intermixture (spicy variation) or 10 drops magic orange intermixture (sweet variation)
- 1 preserving jar

Preparation: Roast the coconut flakes in a frying pan. Lightly scatter the chosen essential oils into the empty jar. Spread the oil well, then fill in the roasted coconut rasps, and shake it well.

25.2.2.1.2 Gomasio (A Sesame Sea-Salt Spice)

Gomasio is a secret of the Middle Eastern cuisine, which completes your spice rack and gives a subtle salty flavor to the dish. It is nice to combine with soy sauce, fresh thyme leaves, or cumin.
Ingredients:

- 50 g sesame seeds
- 1 teaspoon Euro Asia seasoning salt no. 5
- 1 preserving jar

Preparation: Roast the sesame seeds in a frying pan, and then mix the seeds with the salt in a mortar. Crush them lightly with a pestle to release the flavor. Fill into a preserving jar and shake it well. If necessary, add a few more drops of Euro Asia intermixture.

25.2.2.1.3 Honey Provencal

A great basic for the cuisine Provencal.
Ingredients:

- 100 mL acacia honey
- 5 drops O Sole Mio intermixture
- 2 drops lavender pure essential oil
- 1 drop Euro Asia intermixture
- 1 drop clary sage and bergamot intermixture

Preparation: Emulsify the ingredients well. Use the honey to brush grilled vegetables, tofu, and goat and sheep cheese or to season gratins, to add a fabulous distinctly French flavor to a simple dish.

25.2.2.2 Beverages

25.2.2.2.1 Aroma Shake with Herbs

This green fruity flavored cleansing juice certainly is a great rejuvenator.
Ingredients:

- 500 mL organic buttermilk
- 100 mL organic soy milk
- 5 tablespoons sprouts (alfalfa, adzuki bean sprouts, and cress)
- 100 mL carrot juice
- 3 drops Capri intermixture
- 2 drops Euro Asia intermixture
- 1 tablespoon maple syrup
- 1 tablespoon parsley finely chopped

Preparation: Pour the buttermilk and soy milk in a blender and process for a few minutes until combined. Add the carrot juice, then emulsify the essential oils with the maple syrup, and stir it into the mixture. Fill into iced tall glasses and serve chilled. A decorative idea is to dive the top of the glasses into lemon juice and then into the finely chopped parsley, before filling in the shake.

25.2.2.2.2 Earl Grey at His Best

Ingredients:

- 1 preserving jar (100 g capacity)
- 100 g Darjeeling tea "first flush"
- 10 drops bergamot basic essential oil

Preparation: Lightly scatter the "bergamot" basic essential oil into the empty jar. Add the tea, close the jar, and shake it well. Repeat the procedure to shake the jug for the next 5–10 days; then this incredible sort of flavored tea will be ready to serve.

25.2.2.2.3 Lara's Jamu

Jamu is a kind of herbal tonic from Southeast Asia. Every country and family has their own recipes. This one is a tasty booster for the immune system. (Picture courtesy of Subash J. Geeganage.)
Ingredients:

- The rind of two limes in thin shreds
- Juice of two limes
- 2 tablespoons freshly grated ginger
- 1 handful fresh or dried nettle
- 50 mL maple treacle
- 2 teaspoons curcuma powder
- 500 mL water
- 750 mL of sparkling water (optional)
- 5 drops Euro Asia intermixture
- 2 drops peppermint basic essential oil
- 3 drops Capri intermixture

Preparation: Boil the mixture of lime, ginger, and nettle with 500 mL water for 10 min; then let it cool down a bit to be able to sieve it later into decorative chalices. Mix the curcuma powder with fresh lime juice and the Euro Asia basic essential oil, and stir it into the herbal mixture. Now, the maple treacle mixed with peppermint basic essential oil will be stirred in as a sweetener. Serve hot or chilled with sparkling water, fresh mint sprigs, and sliced lime.

25.2.2.2.4 Rose Cider

Refreshing and inspiring.
Ingredients:

- 1 L cider
- 1/2 drop rose basic essential oil or 1 tablespoon organic rose water

Preparation: Stir in the rose oil or rose water. Serve cold.

25.2.2.2.5 Syrup Mint–Orange
A refreshing hot summer drink.
Ingredients:

- 50 mL peppermint seasoning syrup no. 4
- 5 drops Capri intermixture

Preparation: Simply mix the ingredients and you have a refreshing basic syrup, which can be used for drinks and baked goods and to pour it into soda water, tea juices, or even ice cubes. To serve, garnish the drinks with some fresh peppermint leaves.

25.2.2.3 Entrees
25.2.2.3.1 Soups
25.2.2.3.1.1 Peppermint Heaven
Ingredients:

- 500 mL vegetable stock
- Fresh peppermint leaves for decoration
- 2–3 drops peppermint basic essential oil
- 1 drop bergamot basic essential oil
- 150 mL cream
- O sole mio salt no. 5 or regular salt to season to taste

Preparation: Whip the cream; then add the basic essential oils to it. Meanwhile, boil the vegetable stock; then stir in the cream. Ladle into soup bowls to serve and garnish each with a little bit whipped cream and fresh mint leaves.

25.2.2.3.1.2 Perky Pumpkin Soup   Warm and spicy—the perfect autumn dinner.
Ingredients:

- 2 drops Capri intermixture
- 1 large onion, finely chopped
- 2 carrots, sliced finely
- 1 tablespoon pumpkin seed oil or butter
- 500 g peeled pumpkin, finely chopped into cubes
- 200 mL vegetable stock
- 50 mL cream
- 1 teaspoon curry powder
- 1 tablespoon Euro Asia seasoning oil no. 1
- Fresh coriander to garnish
- 1 tablespoon Capri seasoning salt no. 5
- A little bit sherry

Preparation: Heat the pumpkin seed oil in a saucepan. Add the onion and carrots and cook over moderate heat until it softens. Stir in the pumpkin pieces and cook until the pumpkin is soft. Process the mixture in a blender and pour it to the pan. Stir in the vegetable stock and cream and season with the essential oils, salt, and sherry. Ladle into warm soup bowls and garnish each with some fresh coriander leaves.

25.2.2.3.2 Salads

25.2.2.3.2.1 Melon–Plum Purple Radish Salad   A refreshing hot summer party dish. (Picture courtesy of Ulla Mayer-Raichle.)
Ingredients:

- 1 midsize watermelon or 2 Galia melons
- 1 handful radishes rinsed and chopped
- 1 bell pepper rinsed and sliced
- 3 pears rinsed and chopped
- Juice of 1 lemon
- 1 tablespoon Capri or o sole mio seasoning oil no. 1
- 250 g sour cream
- 150 g curd
- Salt
- Freshly ground black pepper
- Some fresh summer herbs like thyme, cress, or lemon balm

Preparation: Half the melon in a zigzag manner, separate the halves, remove the seeds from the melon halves, and use a melon baller to scoop out even-sized balls. Place the half of the melon balls, radishes, bell pepper, and pears in a large salad bowl, and marinade the salad with lemon juice. Then store the melon halves and the salad in the fridge for at least half an hour. Meanwhile, mix the seasoning oil of your choice with sour cream and curd, and season with salt and pepper. Stir the mixture into the salad carefully and fill the salad into the melon halves. Garnish them with herbs and some of the extra melon balls.

25.2.2.3.2.2 Salad with Goat Cheese and Ricotta   A refreshing companion for spicy foods.
Ingredients:

- 1 red bell pepper rinsed, sliced
- 1 green bell pepper rinsed, sliced
- 1 scallion, chopped
- 1 head salad greens (arugula, sorrel, dandelion, etc.), rinsed, dried, and chopped

For the salad dressing:

- 1 drop o sole mio intermixture
- 3 drops Capri intermixture
- 4 tablespoons dark olive oil
- Juice of 1 lemon
- Sea salt
- 100 g goat cheese or ricotta, chopped
- 1/2 handful fresh eatable spring blossoms (daisies, primroses, etc.), rinsed
- 2 handfuls fresh herbs of your choice (coriander, parsley, basil, etc.), rinsed
- Roasted sesame

Preparation: Emulsify the essential oil intermixtures with the olive oil; add the lemon juice and season with salt. Place the dressing in a large bowl, marinade the cheese, and add the salad leaves, bell peppers, and scallion. Mix well and garnish with the herbs and blossoms and the roasted sesame.

25.2.2.4 Appetizer and Finger Food

25.2.2.4.1 Crudities—Flavored Crispy Raw Vegetables
Simple and delicious.
Ingredients:

- 750 g vegetables, well rinsed and cut into crudities (radishes, scallions, chicory, carrots, etc.)
- Juice of 1 lemon
- 5 drops Capri intermixture

Preparation: Emulsify the Capri intermixture into the lemon juice, fill it into a spray flacon, and spread it on top of the sliced vegetables. Serve with dip and breadsticks or baguette.

25.2.2.4.2 Maria's Dip
Ingredients:

- 3 drops Capri intermixture
- 1 tablespoon creme fraiche
- 1/2 teaspoon salt
- 250 g sour cream

Preparation: Emulsify the Capri essential oil intermixture into the creme fraiche. Stir in the salt and sour cream until combined. Ready to serve with bread, toast, and, for example, the flavored crudities.

25.2.2.4.3 Tapenade
An Italian secret simple to make and perfect for dipping or seasoning.
Ingredients:
For the olives:

- 200 g pitted green or black olives, rinsed and halved
- 100 mL dark olive oil
- 1 handful fresh rosemary
- 10 drops o sole mio intermixture

For the tapenade:

- 60 g capers
- 1 crushed garlic clove
- Freshly ground black pepper

Preparation: Marinate the olives in a mixture of olive oil, rosemary, and o sole mio intermixture for at least 1 h. Place the olives, capers, and garlic in a food processor or blender and process until combined. Gradually, add the flavored marinade and blend to a coarse paste; season with pepper. Keep stored in the fridge for up to 1 week.

25.2.2.4.4 Tofu Aromanaise

Served with the veggie skewers—a truly impressive dinner party dish. (Picture courtesy of Ulla Mayer-Raichle.)

Ingredients:

- 200 g organic pure tofu or smoked tofu
- 3 tablespoons sunflower oil
- 2 tablespoons Euro Asia seasoning oil no. 1
- Euro Asia seasoning salt no. 5
- A few chives

Preparation: Put the tofu in a blender and process it until the tofu is smooth. Transfer the creamy tofu to a bowl and stir in the sunflower oil very slowly, then add the Euro Asia seasoning oil, and season with Euro Asia salt. Garnish the top with chopped chives. Serve cold.

25.2.2.4.5 Veggie Skewers

A tasty idea for your next barbecue.
Ingredients:

- 20 skewers
- 1000 g fresh young vegetables

(Tomatoes, fennel, eggplants, carrots, bell peppers, scallions, etc.)
For the marinade:

- 5 tablespoons dark olive oil
- 3 tablespoons either o sole mio or Euro Asia seasoning oil no. 1
- Freshly ground pepper
- 1 handful fresh chopped herbs (basil, thyme, parsley, etc.) or dried herbs.

Preparation: Prepare the vegetables and cut them into cubes. Mix all the marinade ingredients in a shallow dish and add the vegetable cubes. Spoon the marinade over the vegetables and leave to marinate in the fridge for at least 1 h. Then thread the cubes onto skewers. Brush with the marinade and broil or grill until golden, turning occasionally. Serve with baguette, tofu aromannaise tapenade, or any other dip.

25.2.2.5 Main Course

25.2.2.5.1 Celery Lemon Grass Patties

Delicious, little, and flexible to combine.
Ingredients:

- 1–2 large celery
- 250 mL liquid vegetable stock
- 1 organic free range egg
- 4 lemon slices
- 1 pinch of bergamot and clary sage no. 5.

Asian variation:

- 3 tablespoons coconut flakes
- 2 tablespoons Euro Asia seasoning no. 1
- Coconut oil or roasted sesame oil to fry

Mediterranean variation:

- 2 tablespoons o sole mio seasoning no. 1
- 3 tablespoons sesame seeds
- Soy oil to fry

Preparation: Blanche the washed and sliced celery roots in the vegetable stock. Choose your favorite cookie cutter, like heart or star, and cut them out of the blanched celery. Whisk the egg and stir in the essential oil variation of your choice. Marinate the celery stars and hearts, then coat them with coconut flakes or sesame seeds and fry them until they have a delicious golden brown color. To serve, top them with a small amount of the essential oil seasoning. They are great to accompany salads, baked potatoes with sour cream, and other vegetarian dishes or, if you prefer, beef creations.

25.2.2.5.2 Chévre Chaud (Goat) Cheese "Provence" with Pineapple

Ingredients:

- 4 slices of fresh pineapple
- 1 tablespoon sunflower oil or butter or ghee
- 1 teaspoon "o sole mio honey" no. 4
- 1 tablespoon Capri honey no. 4
- 2 tablespoons honey Provencal (basics)
- 2–3 small goat or sheep cheese
- A little bit fresh or dried thyme to garnish
- Sour cream
- Salad or Parma ham (optional)

(Picture courtesy of Ulla Mayer-Raichle.)

Preparation: Halve the pineapple slices and fry them on both sides. Lower the heat and top them with Capri honey. Preheat the oven to 180°C. Halve the cheese and place them on top of each of the two pineapple slices. Drop a little bit of honey Provencal on each portion, and bake it shortly until the cheese starts to caramelize. Serve immediately with the rest of the aromatized honeys dispersed on the surface, fresh herbs above, the sour cream on top, and with Parma ham or fresh salad aside.

25.2.2.5.3 Crispy Wild Rice (Chapatis)
Ingredients:

- 200 g wild rice
- 400–500 mL warm water
- 1 laurel leaf
- 1 small onion or 3 scallions, finely chopped
- 1 teaspoon Euro Asia seasoning oil no. 1
- 1 tablespoon Euro Asia seasoning soy sauce no. 6
- 2 organic or free range eggs
- Curry powder
- Lemon juice as you like
- Around 2 tablespoons oil or ghee to fry

Preparation: Steam the wild rice briefly, then fill it up with the rest of the warm water, and add the laurel leave. Cook it for another 15–20 min, then turn the heat down and stir in the Euro Asia seasoning oil no. 1. Cover it, leave it, and let it chill until firm. Then stir all ingredients into the wild rice. Divide the mixture into walnut-sized balls; then flatten them slightly. Heat the oil or ghee in a pan and fry the chapatis until golden brown on each side. Drain on paper towels and serve at once. These crispy wild rice (chapatis) taste delicious with steamed vegetables and dips or even salads. They are ideal as a snack or a nice idea for the next picnic.

25.2.2.5.4 Mango–Dates–Orange Chutney
A spice dip trip to Asia.

Ingredients:
For 1000 g you need

- 250 g organic well-scrubbed oranges (e.g., sweet and juicy sorts like Valencia)
- 250 g onions
- 250 g sliced mangoes
- 350 mL acacia honey

(If this is not available, choose any other treacle or honeys that are neutral in taste and of organic origin.)

- 50 mL maple syrup
- 2 teaspoons Capri essential oil seasoning salt no. 5
- A little bit of chili powder or 1 fresh chili pepper
- 350 mL cider vinegar
- 250 mg chopped dates
- 50 mL of either Euro Asia or magic orange essential oil seasoning vinegar no. 3
- 2 tablespoons Capri essential oil seasoning syrup no. 4
- 5 drops pure Euro Asia condiment intermixture

Preparation: Remove long, thin shreds of orange rind, using a grater (zester). Scrape it firmly along the surface of the fruit. Remove the white layer of the oranges; then slice the oranges and remove the pits. Finely chop the onions. Peel the mangoes and cut them into small chunks. Mix honey, syrup, chili powder, and vinegar with 1 teaspoon of the Capri salt no. 5, and boil it in a huge saucepan until the honey melts, and stir it well. Add mangoes, onions, dates, oranges, and the half of the shredded orange rind. Then lower the heat, and let it simmer for 1 h, until the mixture has formed a thick mass. Stir in the rest of the shredded orange rind and the chosen essential oil vinegar no. 3. Then emulsify the pure Euro Asia condiment intermixture into the Capri syrup no. 4, and stir it in the chutney. Use the rest of the Capri salt no. 5 to add spice. Fill the mixture into sterilized warm preserving jars, and store them in cold and dark areas. It is nice to serve with the Chévre chaud or the crispy wild rice (chapatis) and veggie skewers.

25.2.2.5.5 Prawns Bergamot

Ingredients:

- 500 g large prawns

Marinade:

- 5 drops pure Capri essential oil intermixture
- 1 small onion
- 1/2 crushed garlic clove
- 1 handful flat leaf parsley
- 3 scallions
- Juice of a lemon
- 2 drops pure bergamot essential oil
- 1/2 teaspoon fennel seed
- 6 tablespoons olive oil
- Salt and fresh pepper
- 3 tablespoons bergamot–Grand Marnier vine no. 3

Preparation: Prepare and wash the prawns as usual. Slice the onions and garlic, chop the parsley finely, and cut the scallions into quarters. Take a teaspoon of lemon juice and emulsify the essential oils in it, and mix in the rest of the ingredients. Let the prawns soak in the marinade and keep it in the fridge for 1 h. Then separate the prawns from the marinade; filter the marinade and keep the parts separately. Fry the prawns inside of the liquid parts of the marinade, and then add the rest. Stir it well for another minute; season with salt, pepper, and bergamot vine; and let it simmer slowly. This is nice to serve with baguette or the crispy wild rice (chapatis) and vegetables like green asparagus tips.

25.2.2.6 Dessert, Cakes, and Baked Goods

25.2.2.6.1 Apple Cake Rose

This classic combination is an apple's favorite destiny and is suited even for diabetics.
Ingredients:

- 250 g spelt flour
- 120 g finely sliced cold butter
- 1 organic or free range egg
- 1 tablespoon Capri essential oil seasoning no. 1
- Salt
- 50–100 mL warm water
- 1000 g sweet ripe apples
- Juice of a half lemon
- 1 tablespoon organic rose water

For the topping:

- 250 mL cream
- 1 egg yolk of an organic or free range egg
- 5–7 drops magic orange pure seasoning intermixture
- 1 tablespoon organic rose water
- 50 g sliced almonds to garnish the top of the cake

Preparation: Sift the flour, butter, egg, warm water, and the Capri seasoning no. 1 into a large mixing bowl. Mix everything together until combined; then store the cake mixture in the fridge for a half hour. In the meanwhile, peel and core the apples, slice them into wedges, and slice the wedges thinly. Combine lemon juice with rose water and splash it over the apples. For the topping, beat the egg yolk with the cream and the pure essential oil intermixture magic orange. Then pour the cake mixture into the prepared pan, smooth the surface, and then make a shallow hollow in a ring around the edge of the mixture. Arrange the apple slices on top of the cake mixture. Pour the topping carefully above the apple slices, and garnish the sliced almonds above. Cover the cake with aluminum foil. Bake for 30–40 min, until firm, and the mixture comes away from the side of the pan. Lower the heat, remove the foil, and bake it for another 5 min. Serve warm.

25.2.2.6.2 Chocolate Fruits and Leaves

A delicious way to consume your favorite fruits, dried fruits, nuts, or even leaves like rose leaves. (Picture courtesy of Ulla Mayer-Raichle.)
Ingredients:

- 250 g organic chocolate couverture (bitter chocolate)
- 5 drops magic orange or bergamot–Grand Marnier or Capri intermixture or 2–3 drops peppermint, lavender, or ginger pure basic essential oil, depending on your taste—spicy, minty, or fruity

Aroma-Vital Cuisine

Preparation: Warm up the chocolate couverture until you have a creamy consistency. Stir in your choice of basic essential oils or intermixture. Dive in the fruits, and let them dry. Serve chilled.

25.2.2.6.3 Homemade Fresh Berry Jelly
Ingredients:

- 500 g mixed berries (blueberries; raspberries; red, white, and black currant; blackberries; strawberries; cranberries; cherries)
- 100 mL water
- 1 tablespoon agar or 2 tablespoons kuzu or sago (binding agent)
- 1–2 tablespoons cold water
- 12 drops magic orange intermixture
- 3 tablespoons maple syrup

Preparation: Take the clean fruits and boil them in the water. Stir the binding agent into the small amount of cold water, then add it to the warm fruits and let them boil for another 3–5 min before you lower the heat, and then leave the mixture to cool. Emulsify the essential oils intermixture with the maple honey; then stir it into the jelly. Serve cool with fresh berries or a spoonful of whipped cream with mint leaves.

25.2.2.6.4 Rose Semifreddo
Romantic and delicate aromatic dessert.
Ingredients:

- 150 g creme fraiche
- 75 g low fat quark
- 100 mL acacia honey
- 1 tablespoon rose water
- Rose leaves from 2 roses (organic farming)
- 2 tablespoons cognac
- Nonalcoholic alternative—1 drop pure magic orange intermixture in 2 tablespoons maple syrup
- 150 mL whipped cream
- 1 drop of pure magic orange intermixture

Preparation: Place the creme fraiche and the quark in a bowl and cream together. Keep some rose leaves for decoration aside, process the rest of the leaves in a food processor until smooth, and then transfer them into the bowl; add the acacia honey and stir to mix. Whisk in the rose water and either the cognac or the magic orange maple syrup. Fold in the whipped cream and the pure magic orange intermixture gently, being careful not to over mix. Pour the mixture into some small plastic containers, cover and freeze until the ice is firm. Transfer the ice to the refrigerator about 20 min before serving to allow it to soften a little. Serve in scoops decorated with rose leaves and berries.

25.2.2.6.5 Sweet Florentine

Sweet almond munchies.
Ingredients:

- 500 g butter
- 200 g sugar
- 2 packages organic bourbon vanilla sugar
- 250 mL cream
- 300 g sliced almonds
- 30 g spelt or wheat grain
- 15–20 drops magic orange or Capri intermixture emulsified in 1 tablespoon maple treacle
- 100 g chocolate couverture with 5–8 drops Capri or magic orange intermixture

Preparation: Caramelize the sugar, then stir into the bourbon, vanilla and butter until the sugar has been melted, and then stir in almonds and flour. Preheat the oven to 180°C, spoon the mixture on a baking tray, and bake for 10 min. Do not worry; it is in their nature to melt. To serve, just cut them into diamonds after cooling down and remove them from the pan. Dive them halfway into the chocolate couverture only (the lower smooth side) then let them dry. Serve chilled or iced.

25.2.2.7 Résumé

Aroma-vital cuisine is an aspect of aroma culture and therefore an art and cultivation of using the senses especially taste and smell.

26 Recent EU Legislation on Flavors and Fragrances and Its Impact on Essential Oils

Jan C.R. Demyttenaere

CONTENTS

26.1 Introduction 1071
26.2 Former Flavoring Directive and Current Flavoring Regulation: Impact on Essential Oils 1072
    26.2.1 Maximum Levels of "Restricted Substances" 1072
        26.2.1.1 (Restricted Substances under Former) Flavoring Directive 88/388/EC 1073
        26.2.1.2 (Restricted Substances under) New Flavoring Regulation 1334/2008/EC 1073
    26.2.2 Definition of "Natural" 1078
        26.2.2.1 (Definition of "Natural" under) Former Flavoring Directive 88/388/EC 1078
        26.2.2.2 (Definition of "Natural" under) New Flavoring Regulation 1334/2008/EC 1079
26.3 Hazard Classification and Labeling of Flavors and Fragrances 1081
    26.3.1 Orange Oil 1083
    26.3.2 Nutmeg Oil 1084
26.4 Conclusion 1086
26.A Appendix 1087
References 1087

26.1 INTRODUCTION

In the last two decades, several new European regulations and directives have been adopted and published in relation to flavors and fragrances. As essential oils and extracts are very important ingredients for flavoring and fragrance applications, these new regulations will have a major impact on their trade and use in commerce.

This chapter focuses on a major piece of legislation that is of particular importance for the flavor industry, namely, the new flavoring regulation [1] (part of the so-called Food Improvement Agents Package) comprising the flavoring, additives, and enzymes regulation and the so-called Common Authorisation Procedure (CAP). This regulation replaces the former flavoring directive 88/388/EEC [2].

The full title of this regulation is *Regulation (EC) No 1334/2008 of the European Parliament and of the Council on flavorings and certain food ingredients with flavoring properties for use in and on foods and amending Council Regulation (EEC) No 1601/91, Regulations (EC) No 2232/96 and (EC) No 110/2008 and Directive 2000/13/EC.*

In the wake of this new regulation, the former Council Directive 88/388/EEC of June 22, 1988, as well as its amendment directive 91/71/EEC [3] and the commission decision 88/389/EEC [4] have been repealed.

Especially, the impact on labeling, resulting from the difference between the new flavoring regulation and the former "flavor directive" 88/388/EC, is discussed from a B2B (business to business) perspective. Special focus is given to the labeling of "natural flavourings" (including essential oils) and changes in definitions (e.g., what is "natural?") in the new flavoring regulation.

In addition, the issue of hazard classification and labeling of dangerous substances and preparations, and essential oils containing hazardous components are discussed with examples. This relates, in particular, to the implementation of the GHS (Globally Harmonized System) in the EU through the publication of the CLP Regulation: classification, labeling, and packaging of substances and mixtures (Regulation (EC) No 1272/2008) [5]. This CLP regulation was entered into force on January 20, 2009, and the directives 67/548/EEC [6] (DSD: Dangerous Substances Directive) and Directive 1999/45/EC [7] (DPD: Dangerous Preparations Directive) were repealed on June 1, 2015.

26.2 FORMER FLAVORING DIRECTIVE AND CURRENT FLAVORING REGULATION: IMPACT ON ESSENTIAL OILS

In the EU, until 2008, for flavorings, the former flavoring directive 88/388/EC [2] was applied. This was the Council Directive of June 22, 1988 on the approximation of the laws of the Member States (MS) relating to flavorings for use in foodstuffs and to source materials for their production, as published in the Official Journal on July 15, 1988 (OJ L 184, p. 61). It has been amended once by the Commission Directive 91/71/EEC of 16/01/91 (OJ L 42, p. 25, 15/02/91) [3]. As this was a directive, it was up to the EU MS to take the necessary measures to ensure that flavorings may not be marketed or used if they do not comply with the rules stated in Art. 3 of this directive.

However, on December 31, 2008, the new flavoring regulation was published [1], which entered into force on January 20, 2009. This new regulation officially came into use on January 20, 2011, and the former flavoring directive was repealed. As many essential oils and extracts either contain flavoring substances or are regarded as "flavoring preparations" or "food ingredients with flavoring properties," this new flavoring regulation has an impact on essential oils and their use as flavoring ingredients for food products. Extracts and essential oils contain certain constituents (substances) that according to this regulation "should not be added as such to food" or should be added at a particular level. Especially, the application of maximum levels of these substances will have an impact on how and when extracts, essential oils, and herbs and spices may or can be applied to food. In addition, the definitions for "natural" have drastically changed. The difference between the former directive 88/388/EC and the new flavor regulation in this respect is outlined in the follwing.

26.2.1 MAXIMUM LEVELS OF "RESTRICTED SUBSTANCES"

Since 1999, the Scientific Committee on Food (SCF) and the European Food Safety Authority (EFSA)* have expressed their opinions on a number of substances found naturally in source materials for flavorings and food ingredients with flavoring properties, for example, some herbs and spices. According to the Committee of Experts on Flavoring Substances (CEFS) of the Council of Europe (CoE),[†,‡] these substances may raise toxicological concern. Although these substances

---

* Panel on Food Additives, Flavourings, Processing aids and Flavourings, and Materials in Contact with Food (AFC) and since July 2008 the EFSA Panel on Food Contact Materials, Enzymes, Flavourings and Processing Aids (CEF).
† Within CoE a Committee of Experts on Flavouring Substances has evaluated the safety-in-use of natural flavouring source materials since 1970. The results are published in the *"Blue Book"—Flavouring Substances and Natural Sources of Flavourings* [8].
‡ According to CoE "active principles" are chemically defined substances, which occur in certain natural flavoring source materials and preparations and which, on the basis of existing toxicological data, should not be used as flavoring substances in their own right [9].

have, in the past, commonly been referred to, within the food industry, as "biologically active principles" (BAPs), they are referred to as "restricted substances" (RS) in this chapter. Both the former flavoring directive and the new flavoring regulation use the undefined term "certain substances."

26.2.1.1 (Restricted Substances under Former) Flavoring Directive 88/388/EC

In the former flavoring directive 88/388/EC, Annex II set maximum levels (limits) for certain substances obtained from flavorings and other food ingredients with flavoring properties in foodstuffs as consumed in which flavorings have been used. Art. 4 (c) stipulated that

> (c) the use of flavourings and of other food ingredients with flavouring properties does not result in the presence of substances listed in Annex II in quantities greater than those specified therein.

The limits apply to foodstuffs and beverages (mg/kg), with some exceptions, for example, alcoholic beverages and confectionary. In Table 26.1, the maximum levels for these substances for foodstuffs and beverages are given (including the exceptions and/or special restrictions), as per Annex II to the former flavoring directive.

This means that for essential oils, extracts, complex mixtures containing these "restricted substances" (e.g., nutmeg, cinnamon, peppermint, sage oils, ...) and when added to food and flavorings, maximum levels applied. The same applies to herbs and spices containing these "restricted substances" as herbs and spices are also "food ingredients with flavoring properties."

26.2.1.2 (Restricted Substances under) New Flavoring Regulation 1334/2008/EC

Apart from the fact that the former directive 88/388/EC has now turned into a regulation, there are many changes that will have an impact on how essential oils and extracts will be used as source of flavors.

The most important issue is how the restricted substances are addressed.

According to the Regulation, substances for which toxicological concern was confirmed by SCF/EFSA should be regarded as naturally occurring undesirable substances, which should not be added as such to food. Due to their natural occurrence, these substances might be present in flavorings and certain food ingredients with flavoring properties.

This is addressed in Art. 6 of the Regulation: *Presence of certain substances,* which refers to Annex III with the same title. This article clearly states in the first paragraph that "Substances listed in Part A of Annex III shall not be added as such to food."

However, when it regards the levels of these substances coming from the use of flavorings and food ingredients with flavoring properties (such as extracts, essential oils, herbs, and spices) the Regulation further specifies (Art. 6.2) as follows:

> 2. Without prejudice to Regulation No 110/2008 maximum levels of certain substances, naturally present in flavourings and/or food ingredients with flavouring properties, in the compound foods listed in Part B of Annex III shall not be exceeded as a result of the use of flavourings and/or food ingredients with flavouring properties in and on those foods.

> The maximum levels of the substances set out in Annex III apply to foods as marketed, unless otherwise stated. By way of derogation from this principle, for dried and/or concentrated foods which need to be reconstituted the maximum levels apply to the food as reconstituted according to the instructions on the label, taking into account the minimum dilution factor.

This means that maximum levels of these substances also apply when the substances come from any type of food ingredients with flavoring properties; the only exception is dried and/or concentrated foods, which can have higher levels before they are diluted/reconstituted. Upon dilution/reconstitution, normal maximum levels apply.

The main difference between the former flavoring directive 88/388 and the new flavoring regulation is that in the directive 88/388, there is only one list (Annex II) of substances to which the

TABLE 26.1
Maximum Limits for Certain Substances Obtained from Flavorings and Other Food Ingredients with Flavoring Properties Present in Foodstuffs as Consumed in Which Flavorings Have Been Used (Annex II of 88/388/EC)

Substances	Foodstuffs (mg/kg)	Beverages (mg/kg)	Exceptions and/or Special Restrictions
Agaric acid[a]	20	20	100 mg/kg in alcoholic beverages and foodstuffs containing mushrooms
Aloin[a]	0.1	0.1	50 mg/kg in alcoholic beverages
Beta-Asarone[a]	0.1	0.1	1 mg/kg in alcoholic beverages and seasonings used in snack foods
Berberine[a]	0.1	0.1	10 mg/kg in alcoholic beverages
Coumarin[a]	2	2	10 mg/kg in certain types of caramel confectionery
			50 mg/kg in chewing gum
			10 mg/kg in alcoholic beverages
Hydrocyanic acid[a]	1	1	50 mg/kg in nougat, marzipan, or its substitutes or similar products
			1 mg/% volume of alcohol in alcoholic beverages
			5 mg/kg in canned stone fruit
Hypericine[a]	0.1	0.1	10 mg/kg in alcoholic beverages
			1 mg/kg in confectionery
Pulegone[a]	25	100	250 mg/kg in mint or peppermint-flavored beverages
			350 mg/kg in mint confectionery
Quassine[a]	5	5	10 mg/kg in confectionery in pastille form
			50 mg/kg in alcoholic beverages
Safrole and isosafrole[a]	1	1	2 mg/kg in alcoholic beverages with not more than 25% volume of alcohol
			5 mg/kg in alcoholic beverages with more than 25% volume of alcohol
			15 mg/kg in foodstuffs containing mace and nutmeg
Santonin[a]	0.1	0.1	1 mg/kg in alcoholic beverages with more than 25% volume of alcohol
Thujone (alpha and beta)[a]	0.5	0.5	5 mg/kg in alcoholic beverages with not more than 25% volume of alcohol
			10 mg/kg in alcoholic beverages with more than 25% volume of alcohol
			25 mg/kg in foodstuffs containing preparations based on sage
			35 mg/kg in bitters

[a] May not be added as such to foodstuffs or to flavorings. May be present in a foodstuff either naturally or following the addition of flavorings prepared from natural raw materials.

maximum levels apply—all those substances may not be added *as such* to food. In contrast, in the new flavoring regulation, the Annex III is split in two parts: Part A with "Substances which shall *not* be added *as such* to food" and Part B establishing "Maximum levels of certain substances, naturally present in flavourings and food ingredients with flavouring properties, in certain compound food as consumed to which flavourings and/or food ingredients with flavouring properties have been added."

Part A contains 15 substances, whereas Part B contains 11 substances.

TABLE 26.2
Annex III, Part A: Substances That May *Not* Be Added *As Such* to Food

Agaric Acid	Aloin	Capsaicin
1,2-Benzopyrone, coumarin	Hypericine	**Beta-Asarone**
1-Allyl-4-methoxybenzene, estragole	**Hydrocyanic acid**	**Menthofuran**
4-Allyl-1,2-dimethoxybenzene, methyleugenol	**Pulegone**	**Quassin**
1-Allyl-3,4-methylene dioxy benzene, safrole	**Teucrin A**	**Thujone (alpha and beta)**

Note: Substances in bold are those that are in both Part A and Part B of Annex III.

Table 26.2 lists the 15 substances of Part A of Annex III "which shall not be added as such to food," and Table 26.3 lists the 11 substances of Part B with their respective maximum levels in the various compound foods according to the new flavoring regulation.

An important statement in the new flavoring regulation was added only at a final stage of the political discussions of the draft text, as a footnote (asterisk) to the table, applying to three of the substances that are marked with an asterisk: estragol, safrol, and methyleugenol.

This means that the maximum levels do not apply to estragol, safrol, and methyleugenol when only fresh, dried, or frozen herbs and spices are added! However, when "food ingredients with flavoring properties" such as essential oils are added, or when essential oils and/or other flavorings are added in combination with herbs and spices, the levels *do* apply.

Moreover, as stipulated in the footnote under Annex III (applying to the three substances with an asterisk), amendments to the current derogations for herbs and spices can be expected (see Table 26.3).

Further compared to the former flavoring directive 88/388, some "restricted substances" are new, for example, methyleugenol, estragol, and menthofuran. This is because since the publication (and amendment) of the former directive, some new scientific evidence has become available that suggests that there would be some toxicological concern for these substances. As explained in Recital (8): "Since 1999, the Scientific Committee on Food and subsequently the European Food Safety Authority [EFSA]…have expressed opinions on a number of substances occurring naturally in source materials for flavourings and food ingredients with flavouring properties which, according to the Committee of Experts on Flavouring Substances of the Council of Europe, raise toxicological concern. Substances for which the toxicological concern was confirmed by the Scientific Committee on Food should be regarded as undesirable substances which should not be added as such to food."

It should be noted that for the same reason, as stipulated in Art. 22 of the Flavouring Regulation, the Annex can be amended "to reflect scientific and technical progress…following the opinion of the Authority" (EFSA).

An important example is methyleugenol (4-Allyl-1,2-dimethoxybenzene). In 1999, methyleugenol was evaluated by the CEFS of the council of Europe. The conclusions of this Committee were as follows:

> Available data show that methyleugenol is a naturally-occurring genotoxic carcinogen compound with a DNA-binding potency similar to that of safrole. Human exposure to methyleugenol may occur through the consumption of foodstuffs flavoured with aromatic plants and/or their essential oil fractions which contain methyleugenol. In view of the carcinogenic potential of methyleugenol, it is recommended that absence of methyleugenol in food products be ensured and checked with the most effective available analytical method [10].

TABLE 26.3
Maximum Levels of Certain Substances, Naturally Present in Flavorings and Food Ingredients with Flavoring Properties, in Certain Compound Foods Consumed to Which Flavorings and/or Food Ingredients with Flavoring Properties Have Been Added

Name of the Substance	Compound Food in which the Presence of the Substance is Restricted	Maximum Level (mg/kg)
Beta-Asarone	Alcoholic beverages	1.0
1-Allyl-4-methoxybenzene, estragol*	Dairy products	50
	Processed fruits, vegetables (including mushrooms, fungi, roots, tubers, pulses, and legumes), nuts, and seeds	50
	Fish products	50
	Nonalcoholic beverages	10
Hydrocyanic acid	Nougat, marzipan, or its substitutes or similar products	50
	Canned stone fruits	5
	Alcoholic beverages	35
Menthofuran	Mint-/peppermint-containing confectionery, except micro-breath freshening confectionery	500
	Micro-breath freshening confectionery	3000
	Chewing gum	1000
	Mint-/peppermint-containing alcoholic beverages	200
4-Allyl-1,2-dimethoxy-benzene, methyleugenol*	Dairy products	20
	Meat preparations and meat products, including poultry and game	15
	Fish preparations and fish products	10
	Soups and sauces	60
	Ready-to-eat savouries	20
	Nonalcoholic beverages	1
Pulegone	Mint-/peppermint-containing confectionery, except micro-breath freshening confectionery	250
	Micro-breath freshening confectionery	2000
	Chewing gum	350
	Mint-/peppermint-containing nonalcoholic beverages	20
	Mint-/peppermint-containing alcoholic beverages	100
Quassin	Nonalcoholic beverages	0.5
	Bakery wares	1
	Alcoholic beverages	1.5
1-Allyl-3,4-methylene dioxy benzene, safrole*	Meat preparations and meat products, including poultry and game	15
	Fish preparations and fish products	15
	Soups and sauces	25
	Nonalcoholic beverages	1
Teucrin A	Bitter-tasting spirit drinks or bitter[a]	5
	Liqueurs[b] with a bitter taste	5
	Other alcoholic beverages	2
Thujone (alpha and beta)	Alcoholic beverages, except those produced from *Artemisia* species	10
	Alcoholic beverages produced from *Artemisia* species	35
	Nonalcoholic beverages produced from *Artemisia* species	0.5

(*Continued*)

TABLE 26.3 (*Continued*)
Maximum Levels of Certain Substances, Naturally Present in Flavorings and Food Ingredients with Flavoring Properties, in Certain Compound Foods Consumed to Which Flavorings and/or Food Ingredients with Flavoring Properties Have Been Added

Name of the Substance	Compound Food in which the Presence of the Substance is Restricted	Maximum Level (mg/kg)
Coumarin	Traditional and/or seasonal bakery ware containing cinnamon in the labeling	50
	Breakfast cereals, including muesli	20
	Fine bakery ware with the exception of traditional and/or seasonal bakery ware containing cinnamon in the labeling	15
	Desserts	5

* The maximum levels shall not apply where a compound food contains no added flavorings and the only food ingredients with flavoring properties which have been added are fresh, dried or frozen herbs and spices. After consultation with the Member States and the Authority, based on data made available by the Member States and on the newest scientific information, and taking into account the use of herbs and spices and natural flavoring preparations, the Commission, if appropriate, proposes amendments to this derogation.
a As defined in Annex II, paragraph 30 of Regulation (EC) No 110/2008.
b As defined in Annex II, paragraph 32 of Regulation (EC) No 110/2008.

Methyleugenol was subsequently evaluated by the SCF, and an opinion on its safety was published in 2001 [11]. The conclusion of the SCF was as follows:

> Methyleugenol has been demonstrated to be genotoxic and carcinogenic. Therefore the existence of a threshold cannot be assumed and the Committee could not establish a safe exposure limit. Consequently, reductions in exposure and restrictions in use levels are indicated.

An equally important but similar example is estragol (1-Allyl-4-methoxybenzene) also known as methylchavicol. In 2000, the CEFS of the council of Europe evaluated estragol, and based on their findings (it was found to be a naturally occurring genotoxic carcinogen in experimental animals), a limit of 0.05 mg/kg (detection limit) was recommended [12].

Estragol was subsequently evaluated by the SCF, and an opinion on its safety was published in 2001 [13]. The conclusion of the SCF was

> Estragole has been demonstrated to be genotoxic and carcinogenic. Therefore the existence of a threshold cannot be assumed and the Committee could not establish a safe exposure limit. Consequently, reductions in exposure and restrictions in use levels are indicated.

As a consequence, both methyleugenol and estragol have been added to Annex III of the new flavoring regulation as "restricted substances."

A particular case is menthofuran, which had already been reviewed first by the SCF together with (*R*)-(+)-pulegone in 2002 [14] and which has more recently been evaluated/considered by EFSA and for which the EFSA CEF Panel raised some concerns [15]. Menthofuran was listed in the register of chemically defined flavoring substances laid down in Commission Decision 1999/217/EC [16], as amended [FL-no: 13.035], but due to these safety concerns, the substance has been introduced in Annex III of the new flavoring regulation and is no longer used/added as such in or on foods as a flavoring substance. It is also not part of the current EU "Union List of Flavourings and Source

Materials" Part A: "List of Flavouring Substances," which has been adopted and published recently as part of the *"Implementing Regulation"* [17].

Due to safety/toxicology concerns, some substances should be restricted not only when "added as such" (for which reason they appear in Annex III Part A) but also when they are naturally present in flavorings and food ingredients with flavoring properties (for which reason they appear in Annex III Part B). Others (e.g., capsaicin) are only restricted when added as such (as "chemically defined substance") and appear in Annex III Part A but not when naturally present (hence they do not appear in Annex III Part B). One of the reasons could be that the use of natural sources in which capsaicin is present (e.g., chilli peppers—*Capsicum*) is self-limiting (at least for consumers in the EU), and setting limits for the use of peppers would be extremely difficult from an implementation point of view and for control authorities (to check the maximum levels). An opinion on the safety of capsaicin was published by the SCF in 2002 [18].

It is also important to note that according to Art. 30 of the flavoring regulation (*entry into force*), which was applied 24 months after its entry into force (i.e., 20/01/2011), Art. 22 was applied from the date of its entry into force (i.e., 20/01/2009). Art. 22 concerns the Amendments to Annexes II–V. This means that even before the flavoring regulation was applied, the Annexes could be amended immediately, if necessary. However, between the entry into force and the application date of the regulation, the Annexes are amended, and it is not foreseen that they will be amended in the very near future.

Whereas Art. 6 relates to "certain substances," Art. 7 relates to the "Use of certain source materials," which is even more important in relation to herbs, spices, extracts, and essential oils. This article refers to Annex IV of the regulation, which is a new Annex that was not in the former flavoring directive 88/388/EC entitled: "List of source materials to which restrictions apply for their use in the production of flavourings and food ingredients with flavouring properties." Annex IV consists of two parts:

1. Part A: Source materials that shall not be used for the production of flavorings and food ingredients with flavoring properties.
2. Part B: Conditions of use for flavorings and food ingredients with flavoring properties produced from certain source materials.

The complete Annex IV to the flavoring regulation is given in Appendix 26.A to this chapter. Art. 7 stipulates the following:

1. Source materials listed in Part A of Annex IV shall not be used for the production of flavorings and/or food ingredients with flavoring properties.
2. Flavorings and/or food ingredients with flavoring properties produced from source materials listed in Part B of Annex IV may be used only under the conditions indicated in that Annex.

26.2.2 Definition of "Natural"

26.2.2.1 (Definition of "Natural" under) Former Flavoring Directive 88/388/EC

Also important is how the former flavoring directive addressed "naturalness" of flavors and how "natural" was defined for the purpose of labeling. This was stipulated by Art. 9a.2 (amending the original Art. 9.2 of 88/388/EC by 91/71/EEC) [3]:

> 2. the word 'natural', or any other word having substantially the same meaning, may be used only for flavourings in which the flavouring component contains exclusively flavouring substances as defined in Article 1 (2) (b) (i) and/or flavouring preparations as defined in Article 1 (2) (c). If the sales description of the flavourings contains a reference to a foodstuff or a flavouring source, the word 'natural' or any other word having substantially the same meaning, may not be used unless the flavouring component

has been isolated by appropriate physical processes, enzymatic or microbiological processes or traditional food-preparation processes solely or almost solely from the foodstuff or the flavouring source concerned.

How a "natural flavouring substance" could be obtained was thus defined in Art. 1.2 (b) (i):

(b) 'flavouring substance' means a defined chemical substance with flavouring properties that is obtained:

(i) by appropriate physical processes (including distillation and solvent extraction) or enzymatic or microbiological processes from material of vegetable or animal origin either in the raw state or after processing for human consumption by traditional food-preparation processes (including drying, torrefaction and fermentation),

How a 'flavouring preparation' could be obtained was defined in Art. 1.2 (c):

(c) 'flavouring preparation' means a product, other than the substances defined in (b) (i), whether concentrated or not, with flavouring properties, which is obtained by appropriate physical processes (including distillation and solvent extraction) or by enzymatic or microbiological processes from material of vegetable or animal origin, either in the raw state or after processing for human consumption by traditional food-preparation processes (including drying, torrefaction and fermentation);

This means that a "flavoring preparation" was by default always "natural" and that extracts and essential oils (obtained by appropriate physical processes such as distillation and solvent extraction) from the material of vegetable origin (e.g., plant material) could be considered as "flavoring preparation" and thus "natural."

26.2.2.2 (Definition of "Natural" under) New Flavoring Regulation 1334/2008/EC

Regarding "naturalness" of flavors and how "natural" is defined for the purpose of labeling the situation has drastically changed since the former flavoring directive 88/388/EC was repealed.

For example, in the past according to 88/388/EC, there were three categories of flavoring substances: natural, nature-identical (NI), and artificial. However, in the new flavoring regulation, there are only two categories: natural and not natural—meaning that the difference between NI and artificial have disappeared, and these two have merged into one category of "non-natural flavouring substances."

In the new Flavoring Regulation, "natural flavouring substance" is defined by Art. 3.2 (c):

(c) 'natural flavouring substance' shall mean a flavouring substance obtained by appropriate physical, enzymatic or microbiological processes from material of vegetable, animal or microbiological origin either in the raw state or after processing for human consumption by one or more of the traditional food preparation processes listed in Annex II. Natural flavouring substances correspond to substances that are naturally present and have been identified in nature;

Important is the last line stating (additional requirement) that a substance has to be identified in nature before it can be regarded as "natural," so it is not only sufficient to produce it "in a natural way" but it should also be identical to something that is present in nature. Of course, "present in nature" includes any natural substance present in any processed foods: for example, cured vanilla beans, roasted coffee, cooked meat, fermented cheese, wine, and beer. This is to avoid that when an enzymatic or microbial process would be developed by which a flavoring substance can be produced "by enzymatic or microbial processes from material of vegetable origin" (i.e., natural source materials) that up to then has never been identified in nature (and is not naturally occurring), such as ethylvanillin, such material would be labeled as a "natural flavouring substance."

Within the global flavor industry, there is a general principle and agreement that in order to determine that a flavoring substance has been "identified in nature," any identification needs to meet the

TABLE 26.4
List of Traditional Food Preparation Processes (Annex II of Flavouring Regulation (EC) No 1334/2008)

Chopping	Coating
Heating, cooking, baking, frying (up to 240°C at atmospheric pressure) and pressure cooking (up to 120°C)	Cooling
Cutting	Distillation/rectification
Drying	Emulsification
Evaporation	Extraction, including solvent extraction in accordance with Directive 88/344/EEC
Fermentation	Filtration
Grinding	
Infusion	Maceration
Microbiological processes	Mixing
Peeling	Percolation
Pressing	Refrigeration/freezing
Roasting/grilling	Squeezing
Steeping	

criteria for the validity of identifications in nature as described by IOFI (International Organization of the Flavor Industry) [19].

More important than the definition on "natural" as such (Art. 3.2 (c)) however is how "appropriate physical process" is defined.

Annex II to the flavoring regulation gives a list of "traditional food preparation processes" by which natural flavoring substances and (*natural*) flavoring preparations are obtained. The full list of *traditional food preparation processes* is given in Table 26.4.

The definition of "appropriate physical process" according to the flavor regulation is described in Art. 3.2 (k):

> (k) 'appropriate physical process' shall mean a physical process which does not intentionally modify the chemical nature of the components of the flavouring, without prejudice to the listing of traditional food preparation processes in Annex II, and does not involve, inter alinea, the use of singlet oxygen, ozone, inorganic catalysts, metal catalysts, organometallic reagents and/or UV radiation.

This definition refers to Annex II, which means that all processes listed in Annex II also fall under the definition of "appropriate physical processes."

'Flavoring preparations' (such as essential oils and extracts) are defined by Art. 3.2(d).

With respect to "natural labeling," according to Art. 16.2, the term "natural" may only be used for the description of a flavoring if the flavoring component comprises only flavoring preparations and/or natural flavoring substances, which means that a "flavoring preparation" is regarded as "natural" by definition. In other words, there is not such a thing as a "synthetic flavouring preparation." The definition for "flavouring preparation" reads as follows (Art. 3.2 (d)):

> (d) 'flavouring preparation' shall mean a product, other than a flavouring substance, obtained from:
>> i) food by appropriate physical, enzymatic or microbiological processes either in the raw state of the material or after processing for human consumption by one or more of the traditional food preparation processes listed in Annex II;

and/or

ii) material of vegetable, animal or microbiological origin, other than food, by appropriate physical, enzymatic or microbiological processes, the material being taken as such or prepared by one or more of the traditional food preparation processes listed in Annex II;

It can be concluded that according to this definition, essential oils and extracts obtained from plant material (*material of vegetable origin*) prepared by distillation (which is a *traditional food preparation process* listed in Annex II), followed by an *appropriate physical process,* can be considered as a "flavoring preparation" and thus natural, as long as the chemical nature of the components is not intentionally modified during the physical process.

However, it can also be argued that if the nature of the components is intentionally modified in the physical process (e.g., extraction, drying, evaporation, and condensation, dilution), which is often the case, then the end product can no longer be regarded as *flavouring preparation* and thus natural, according to the new Flavouring Regulation.

EFFA (the European Flavour Association) has developed a guidance document for the European Flavour Industry on the permissible processes to obtain natural flavoring ingredients (i.e., natural flavoring substances and (natural) flavoring preparations) [20], which is published as part of the General EFFA Guidelines on the flavoring regulation in the EFFA website [21].

26.3 HAZARD CLASSIFICATION AND LABELING OF FLAVORS AND FRAGRANCES

This section describes very briefly the general rules for hazard classification and labeling of F&F substances and preparations, including natural raw materials, such as extracts and essential oils, containing hazardous constituents, according to the EU regulations.

For the trade of F&F (including pure substances and mixtures or preparations thereof and natural raw materials) within the EU and also globally, certain rules apply within the flavur and fragrance industry, which are established by the IFRA/IOFI-Labeling Manual (IFRA/IOFI-LM), which is updated every year and shared with the IFRA and IOFI Membership (incl. EFFA-membership). The following general considerations are taken from the introductory note to the EFFA-Labeling Manual (available for EFFA members), which is based on the IFRA/IOFI-LM. It should be noted that the LM also takes into account the Globally Harmonized System of Classification and Labeling of Chemicals (UN-GHS). This GHS has, meanwhile, been implemented in the EU with the publication of the EU-GHS Regulation (CLP Regulation: Classification, Labeling and Packaging of substances and mixtures [Regulation (EC) No 1272/2008, OJ L 353/1, 31.12.2008]) [5]. It was entered into force on January 20, 2009, and the current Directive 67/548/EEC [6] (DSD—Dangerous Substance Directive) and Directive 1999/45/EC [7] (DPD—Dangerous Preparations Directive) was repealed June 1, 2015. However, Annex I of the DSD had already been repealed and transferred into Annex VI of the EU-CLP regulation.

The classification and labeling of substances is prescribed in Annex VI to the CLP Regulation (EC) No 1272/2008 (as it was formerly done by the supplier using the criteria of Annex VI of the Dangerous Substances Directive 67/548/EEC [EU DSD]). For preparations, like flavor and fragrance compounds, the rules of the Dangerous Preparations Directive 1999/45/EC (EU DPD) have to be followed.

Although the CLP regulation entered into force on January 20, 2009—which introduced the GHS into the EU—a staggered approach for that implementation was foreseen:

- *Substances*: since December 1, 2010: EU GHS (CLP) for labeling & packaging; DSD *and* EU GHS (CLP) for classification
- *Mixtures*: DPD for classification & labeling & packaging after June 1, 2015: EU GHS (CLP) for classification & labeling & packaging

(Also, a transition period of 24 months was provided for relabeling and repackaging, if mixtures were on the market before June 1, 2015; the transition period for substances placed on the market before December 1, 2010 expired on December 1, 2012.)

During the transition periods, ending on June 1, 2015, the indication of the classification following the DSD and DPD directives is still mandatory in the respective Safety Data Sheets (SDS).

Special emphasis is put on the classification of aspiration hazard (Xn; R65) of both substances that can easily reach the lungs upon ingestion and cause lung damage (substances with low viscosity and low surface tension) and mixtures/preparations with a high hydrocarbon (HC) content and low kinematic viscosity that will pose the same hazard.

Based on measurement results for a number of natural raw materials (e.g., extracts and essential oils) with HC contents between 10% and 90+% and on similar measurements of some F&F compounds, a dedicated working group of the F&F Industry has come to the conclusion that in practice, substances and preparations containing more than 10% of HC(s) fall within the criteria for viscosity and surface tension.

Therefore, the European F&F Industry through its EFFA LM recommends the following:

- To determine the HC content of substances (supplier information, analysis) and preparations (including extracts and essential oils) (calculation) and to classify as AH Cat 1 and Xn, R65, respectively if more than 10% HC is present.
- That nonclassification should only be possible if viscosity and/or surface tension measurement results are available for a specific substance or preparation (including extracts and essential oils).

In addition, classification and labeling of skin sensitizers are also addressed in the LM. Following the EFFA LM, skin sensitizers are labeled Xi, R43 (under former DSD rules), SS (skin sensitization), or "Danger" (GHS Signal Word) (under current CLP Regulation).

In the EFFA LM, special attention is paid to the hazard classification and labeling of natural raw materials, referred to as "natural complex substances" (NCSs) in the LM. The terminology "Natural Complex Substance" is used because in some cases the natural raw material (the complex) is regarded as a single substance, rather than a complex mixture.

NCSs (e.g., essential oils and extracts from botanical and animal sources) require special procedures due to the fact that they might have quite different chemical compositions (and therefore hazard classifications) under the same designation. This may occur even when this differentiates between species, cultivars, and chemotypes and different production procedures (e.g., absolutes, resinoids, and distilled oils).

There are two ways of classifying and labeling NCSs such as extracts and essential oils: either based on the data known and available on the natural raw material as such (NCS is regarded as a single "substance") or based on the hazardous constituents they are composed of (NCS is regarded as a complex mixture).

In the first case, an NCS may be classified on the basis of the data obtained by testing the NCS. The test results of an NCS, even if containing classified constituents, are evaluated in accordance with the DSD. The health and environmental hazard classifications derived following this approach are quality dependent, which is also indicated in the EFFA LM.

It must be noted that according to current practice of the F&F Industry, where hard test data are available, the F&F Industry takes them into account when modifying the "calculated" classification.

In the second case, for the grades of NCSs and for endpoints for which reliable test data are lacking, the EU's Labelling Guide (Annex VI to the DSD) incorporates a requirement introduced by Commission Directive 93/21/EEC, whereby the hazard classification of complex substances shall be evaluated on the basis of the levels of their known chemical constituents. Where knowledge about constituents exists, for example, on substances limited as per Annex III of Regulation 1334/2008/EC (the so-called Restricted Substances —detailed earlier) or on substances with sensitizing, toxic, harmful, corrosive, and environmentally hazardous properties, the classification and labeling of these NCSs according to the requirements of the EU should follow the rules for preparations (= mixtures) as prescribed by the DPD.

Recent EU Legislation on Flavors and Fragrances and Its Impact on Essential Oils

It should be underlined that the current practice of the F&F Industry is to evaluate NCSs based on their compositions (i.e., second case), *unless* very robust test data on the NCS as such are available.

One dedicated section of the EFFA LM also provides a list with the composition of the NCSs (extracts, essential oils, concretes, absolutes, etc.) in terms of the presence (content in %) of hazardous constituents and HCs in the NCSs that have to be taken into account for the classification and labeling of the NCSs or a preparation containing these NCSs, based on the DPD.

F&F compounds that are preparations (i.e., compounded mixtures, formulations, or compositions) should be classified and labeled according to the EU's DPD 1999/45/EC and its articles 6 and 7.

In practice, test data on the flavor or fragrance compounds (preparations) are not available or collected. Therefore, the classification of these preparations should be based on the chemical composition and should include the contributions of hazardous substances present as constituents in the NCSs present in the formulation. This is another reason why the composition of the NCSs in terms of presence of the hazardous constituents is also part of the EFFA LM.

Examples of important constituents to take into account for classification:

- Sensitizers (R43) → NCSs (essential oils and extracts) to be classified as R43 if the content (%) of the sensitizer (if one) or the content of their sum (if more than one) is greater than or equal to 1%.
- CMRs (carcinogenic, mutagenic, and reprotoxic materials: R45; R46; R68) → NCSs to be classified as CMR if the content of the CMR substance(s) is greater than or equal to 0.1%.

The final classification and labeling of an essential oil can be totally different depending on the approach used for the classification, either based on the data on the essential oil as such (the first case described earlier) or based on the hazardous constituents in the essential oil (the second case described earlier). This is illustrated with two examples: orange oil, containing mainly limonene [which is classified as both a sensitizer (R43) and very toxic for the environment (R50/53)], and nutmeg oil, containing safrole, which is a CMR (T; R45-22-68).

26.3.1 Orange Oil

Table 26.5 depicts the composition of orange oil with the labeling and classification of the main constituents.

The resulting classification and labeling of orange oil based on its constituents (according to the DPD) is as follows:

R10	Flammable
R38	Irritating to skin
R43	May cause sensitization by skin contact
R50/53	Very toxic to aquatic organisms (environment)
R65	Harmful: may cause lung damage if swallowed

TABLE 26.5
Composition of Orange Oil with Main Hazardous Constituents and Their Classification (according to DSD and GHS)

Constituent	Concentration (%)	Labeling and Classification (DSD)	GHS Hazard Categories
D-Limonene	96.2	Xi; R10; R38-43, N; R50/53	AH 1, EH A1, EH C1, FL 3, SCI 2, SS 1B
Linalool	0.5	Xi; R38	ATO 5(2790), EDI 2A, EH A3, FL 4(77), SCI 2
Citral	0.2	Xi; R38-43	ATD 5(2250), EDI 2A, EH A2, FL 4, SCI 2, SS 1B

From June 1, 2015, the GHS rules apply for classification of mixtures; consequently, the classification (based on GHS) of orange oil (as a mixture, based on its constituents) will then become as follows:

AH 1	(Aspiration hazard Cat. 1)
EH A1	(Acute aquatic toxicity Cat. 1)
EH C1	(Chronic aquatic toxicity Cat. 1)
FL 3	(Flammable liquid Cat. 3)
SCI 2	(Skin corrosion/irritation Cat. 2)
SS 1	(Skin sensitization Cat. 1).

The "global" classification (UN-GHS) and the "European classification" (EU-CLP) are equal in this example.

Labeling (pictograms) of orange oil based on its constituents (according to the DSD/DPD):

Xn:	Harmful	
N:	Dangerous for the environment	

The classification and labeling of orange oil-based on data on the essential oil as such may be different depending on the quality of the essential oil tested, but data are not provided here due to lack of sufficiently robust data. Therefore the F&F Industry perform C&L based on its constituents.

According to UN-GHS the following "Danger and Warning Pictograms" apply to limonene, the main constituent of orange oil (see Table 26.6).

From June 1, 2015, the pictograms for orange oil (based on its constituents) are as shown in Table 26.7.

26.3.2 NUTMEG OIL

Table 26.8 depicts the composition of nutmeg oil with the labeling and classification of the main constituents.

The resulting classification and labeling of nutmeg oil based on its constituents (according to the DPD) is as follows:

R10	Flammable
R43	May cause sensitization by skin contact
R45	May cause cancer
R51/53	Toxic to aquatic organisms (environment)
R65	Harmful: may cause lung damage if swallowed
R68	Possible risk of irreversible effects

TABLE 26.6
Labeling (Pictograms) of Limonene (according to CLP-Regulation)

GHS pictogram codes	GHS02	GHS07	GHS08	GHS09
GHS pictogram				

TABLE 26.7
Labeling (Pictograms) of Orange Oils (Based on Its Constituents, according to CLP Regulation)

GHS pictogram codes	GHS02	GHS08	GHS09
GHS pictogram			

TABLE 26.8
Composition of Nutmeg Oil with Main Hazardous Constituents and Their Classification (according to DSD and GHS)

Constituent	Concentration (%)	Labeling and Classification	GHS Hazard Categories
Pinenes (e.g., α-pinene)	40	Xi; R(10); R38-43-65, N; R50/53	AH 1,ATO 5(3500),FL 3,SCI 2,SS 1B
Limonene	7	Xi; R10; R38-43, N; R50/53	AH 1, EH A1, EH C1, FL 3, SCI 2, SS 1B
Safrole	2	T; R45, Xn; R22-68	ATO 4(1950),CAR 1B,MUT 2
Isoeugenol	1	Xn; R21/22, Xi; R36/38-43	ATD 4(1900),ATO 4(1500),EDI 2A,EH A2,SCI 2,SS 1A

From June 1, 2015, the GHS rules apply for classification of mixtures; consequently the "global" classification (based on GHS) of nutmeg oil (as a mixture, based on its constituents) will then become as follows:

AH 1	(Aspiration Hazard, Cat. 1)
ATO 5	(Acute toxicity, oral, Cat. 5* → "Not classified" according to EU-CLP)
CAR 1B	(Carcinogenicity, Cat. 1B)
EDI 2A	(Serious eye damage/eye irritation, Cat. 2A* → Cat. 2 according to EU-CLP)
EH A2	(Acute aquatic toxicity, Cat. A2* → "Not classified" according to EU-CLP)
EH C2	(Chronic aquatic toxicity, Cat. C2)
FL 3	(Flammable liquid, Cat. 3)
MUT 2	(Mutagenicity, Cat. 2)
SCI 2	(Skin corrosion/irritation, Cat. 2)
SS 1	(Skin sensitization, Cat. 1).

* The "global" classification (UN-GHS) and the "European classification" (EU-CLP) are not equal in this example. The difference in classification between UN-GHS and EU-CLP is caused by the difference of implementation of the building blocks. To derive from UN-GHS to EU-CLP is the subtraction of the EU-non applicable building blocks from UN-GHS classification. In that example "ATO5" and "EH A2" (GHS Hazard Categories) are not foreseen in the EU-CLP and since these are the lowest categories for the two resp. Hazard Classes (Acute toxicity, oral and Acute aquatic toxicity, resp.) according to UN-GHS for which no corresponding category is foreseen in EU-CLP the classification for these Hazard Classes becomes "Not classified."

TABLE 26.9
Labeling (Pictograms) of Nutmeg Oil (Based on Its Constituents, according to CLP-Regulation)

GHS pictogram codes	GHS02	GHS08	GHS09
GHS pictogram			

Labeling (pictograms) based on its constituents (according to the DSD/DPD):

T:	Toxic (CMR)	
N:	Dangerous for the environment	

As in the case above, the classification and labeling of nutmeg oil-based on data on the essential oil as such may be different depending on the quality of the essential oil tested, but data are not provided here due to lack of sufficiently robust data. Therefore the F&F Industry perform C&L based on its constituents.

According to UN-GHS the same "Danger and Warning Pictograms" apply to pinenes, the main constituent of Nutmeg oil, as to limonene (cfr example above—see Table 26.6).

Effective June 1, 2015, the pictograms for nutmeg oil (based on its constituents) are as shown in Table 26.9.

So with these examples, it can be demonstrated that the final Classification and Labeling (C&L) of essential oils can change significantly depending on the approach used based on existing data (for the various endpoints) on the essential oil as such or based on the hazardous constituents. It should however be underlined that according to the rules of the IFRA/IOFI and EFFA LM, C&L of natural raw materials or NCSs can *only* be done for the endpoints for which data (on the NCS as such) are available (e.g., skin irritation, sensitization, environmental toxicity, etc.)—if no data are available, then the constituents must be taken into account for the classification for these endpoints.

Thus, under the pressure of legislative bodies the F&F Industry has decided to evaluate NCSs based on their compositions. Only where very robust test data are available the F&F Industry takes them into account to modify the calculated classification. But the current practice of the F&F Industry is to evaluate and classify NCSs based on their constituents.

26.4 CONCLUSION

As described and outlined earlier, several new European regulations and directives have been adopted and published during the last two decades in relation to flavors and fragrances. NCSs or raw materials (such as essential oils and extracts) are very important ingredients for flavoring and fragrance applications. As a result, these new regulations have a major impact on the trade and use in commerce of these essential oils and extracts, in particular on labeling issues, as has been demonstrated with the new flavoring regulation (labeling of "natural"). The labeling issue is especially important because of its impact on consumer behavior: with respect to food, consumers prefer natural flavorings to synthetic ones. Good and pragmatic definitions in the flavoring regulation that has recently replaced the former flavoring directive are essential to ensure that all natural raw materials such as essential oils and extracts can continue to be labeled as natural.

26.A APPENDIX

List of source materials to which restrictions apply for their use in the production of flavorings and food ingredients with flavoring properties (Annex IV to Flavouring Regulation (EC) No 1334/2008):

Part A: Source materials which shall not be used for the production of flavorings and food ingredients with flavoring properties

Source Material	
Latin Name	**Common Name**
Tetraploid form of *Acorus calamus*	Tetraploid form of Calamus

Part B: Conditions of use for flavorings and food ingredients with flavoring properties produced from certain source materials

Source Material		Conditions of Use
Latin Name	**Common Name**	
Quassia amara L. and *Picrasma excelsa* (Sw)	Quassia	Flavorings and food ingredients with flavoring properties produced from the source material may only be used for the production of beverages and bakery wares
Laricifomes officinales (Vill.: Fr) Kotl. et Pouz or *Fomes officinalis*	White agaric mushroom	Flavorings and food ingredients with flavoring properties produced from the source material may only be used for the production of alcoholic beverages
Hypericum perforatum L.	St John's wort	
Teucrium chamaedrys L.	Wall germander	

REFERENCES

1. Regulation (EC) No 1334/2008 of the European Parliament and of the council on flavourings and certain food ingredients with flavouring properties for use in and on foods and amending Council Regulation (EEC) No 1601/91, Regulations (EC) No 2232/96 and (EC) No 110/2008 and Directive 2000/13/EC (OJ L 354/34, 31.12.2008).
2. Council Directive 88/388/EC of 22 June 1988 on the approximation of the laws of the Member States relating to flavourings for use in foodstuffs and to source materials for their production (OJ L 184, 15.7.1988, p. 61).
3. Commission Directive 91/71/EEC of 16 January, 1991 completing Council Directive 88/388/EEC on the approximation of the laws of the Member States relating to flavourings for use in foodstuffs and to source materials for their production (OJ L42, 15.2.1991, p. 25).
4. 88/389/EEC: Council Decision of 22 June, 1988 on the establishment, by the Commission, of an inventory of the source materials and substances used in the preparation of flavourings (OJ L184, 15.7.1988, p. 67).
5. Regulation (EC) No 1272/2008 of the European Parliament and of the Council on classification, labeling and packaging of substances and mixtures, amending and repealing Directives 67/548/EEC and 1999/45/EC, and amending Regulation (EC) No 1907/2006. OJ L 353, 31.12.2008, p. 1.
6. Dangerous Substances Directive: Council Directive 67/548/EEC of June 27, 1967 on the approximation of laws, regulations and administrative provisions relating to the classification, packaging and labelling of dangerous substances (OJ P 196, 16.8.1967, p. 1).
7. Dangerous Preparations Directive: Directive 1999/45/EC of the European Parliament and of the Council of May 31, 1999 concerning the approximation of the laws, regulations and administrative provisions of the Member States relating to the classification, packaging and labeling of dangerous preparations (OJ L 200, 30.7.1999, p. 1).
8. *Flavouring Substances and Natural Sources of Flavourings* ("*Blue Book*") Council of Europe (CoE), 1992.
9. Active principles (constituents of toxicological concern) contained in natural sources of flavorings. Approved by the Committee of Experts on Flavoring Substances, October 2005.

10. Council of Europe—Committee of Experts on Flavouring Substances, 1999. Publication datasheet on Methyleugenol. Document RD 4.14/2-45 submitted by the delegation of Italy for the 45th meeting in Zurich, October 1999.
11. Opinion of the Scientific Committee on Food on Methyleugenol (4-Allyl-1,2-dimethoxybenzene). *SCF/CS/FLAV/FLAVOUR/4 ADD1 FINAL*, 2001. European Commission, Health & Consumer Protection Directorate-General. http://ec.europa.eu/food/fs/sc/scf/out102_en.pdf (accessed July 14, 2015).
12. Council of Europe, Committee of Experts on Flavouring Substances, 2000. Final version of the publication datasheet on estragole. Document RD 4.5/1-47 submitted by Italy for the 47th meeting in Strasbourg, October 16–20, 2000.
13. Opinion of the Scientific Committee on Food on Estragole (1-Allyl-4-methoxybenzene). *SCF/CS/FLAV/FLAVOUR/6 ADD2 FINAL*, 2001. European Commission, Health & Consumer Protection Directorate-General. http://ec.europa.eu/food/fs/sc/scf/out104_en.pdf (accessed July 14, 2015).
14. Opinion of the Scientific Committee on Food on pulegone and menthofuran. *SCF/CS/FLAV/FLAVOUR/3 ADD2 Final*, 2002. European Commission, Health & Consumer Protection Directorate-General. http://ec.europa.eu/food/fs/sc/scf/out133_en.pdf (accessed July 14, 2015).
15. EFSA Scientific Opinion (FGE.57). Consideration of two structurally related pulegone metabolites and one ester thereof evaluated by JECFA (55th meeting). *The EFSA Journal*, 2009 ON-1079, 1–17.
16. EU Register: EC (European Commission), 1999. Commission Decision 1999/217/EC adopting a register of flavouring substances used in or on foodstuffs drawn up in application of Regulation (EC) No 2232/96 of the European Parliament and of the Council of 28 October 1996 (OJ L84/1-137, 27.3.1999).
17. Commission Implementing Regulation (EU) No 872/2012 of 1 October 2012 adopting the list of flavouring substances provided for by Regulation (EC) No 2232/96 of the European Parliament and of the Council, introducing it in Annex I to Regulation (EC) No 1334/2008 of the European Parliament and of the Council and repealing Commission Regulation (EC) No 1565/2000 and Commission Decision 1999/217/EC Text with EEA relevance (OJ L 267, 2.10.2012, pp. 1–161). http://eur-lex.europa.eu/legal-content/EN/ALL/?uri=CELEX:32012R0872 (accessed July 14, 2015).
18. Opinion of the Scientific Committee on Food on Capsaicin. *SCF/CS/FLAV/FLAVOUR/8 ADD1 Final*, 2002. European Commission, Health & Consumer Protection Directorate-General. http://ec.europa.eu/food/fs/sc/scf/out120_en.pdf (accessed July 14, 2015).
19. IOFI, Statement on the identification in nature of flavouring substances, made by the Working Group on Methods of Analysis of the International Organization of the Flavour Industry (IOFI). *Flavour and Fragrance Journal*, 2006, 21, 185.
20. EFFA Guidance Document for the Production of Natural Flavouring Substances and (Natural) Flavouring Preparations in the EU—1st Revision (V2.0). 2013. http://www.effa.eu/media/news/46/664/EFFAGuidanceDocumentproductionofnaturalFlavouringIngredientsRevvs20corr130513.pdf (accessed July 14, 2015).
21. EFFA Guidance Document on the EC Regulation on Flavourings—3rd Revision (V4.0). 2015. http://www.effa.eu/media/docs/guidance-document/EFFAGuidanceDocumentonthenewEC FlavouringRegulationV40revisedversion3rdrev300615.pdf (accessed July 14, 2015).

Index

A

Abbé-type refractometer, 197
Absidia
    A. glauca, 573, 603, 806, 817, 934
    A. orchidis, 873–874
Achillea millefolium, 45, 53, 59, 71, 92, 101, 103, 337
Acorus gramineus, 365, 602
Acyclic monoterpenoids
    alcohols and aldehydes
        citral, 751, 754–755
        citronellal, 751–753
        dihydromyrcenol, 765–766
        geraniol, 753, 757–759
        linalool (*see* Linalool)
        linalyl acetate, 764–765
        nerol, 757–759
    hydrocarbons
        citronellene, 750
        myrcene, 749–750
Adamantane, 994–996
4-Adamantanone, 996
Adenosine triphosphate (ATP), 166, 324–325, 660
Adulteration process
    adulterants
        enzymes, 720–721
        ethanol, 717
        fatty oils, 717–718
        fractions, 719
        glycols, 718
        natural isolates, 719–720
        residues, 720
        same species, 718
        synthetic compounds, 720
        water, 717
    analytical methods
        chiral analysis, 725–726
        GC, 724–726
        general test analysis, 723
        NMR spectroscopy, 726
        TLC, 723
    constraint analysis
        aging, 716–717
        client demands, 716
        cupidity, 717
        price, 715
        regulations, 716
        unavailability, 715
    definition and history, 711–713
    intentional, 715
    organoleptic method
        appearance, 721
        color, 721
        odor, 721
        quality control, 722
        relationship coefficient, 722–723
    unintentional, 713–715
AEDA. *see* Aroma extraction dilution analysis (AEDA)
9(15)-Africanene
    Aspergillus niger, 930–931
    Rhizopus oryzae, 930–931
Agar diffusion test
    anise oil, 439–442
    bitter-fennel oil, 448–453
    caraway oil, 456–459
    cassia oil, 463
    Ceylon cinnamon bark oil, 466–469
    Ceylon cinnamon leaf oil, 473
    citronella oil, 476–479
    clary sage oil, 482
    clove oil, 485–488
    coriander oil, 494–499
    dwarf-pine oil, 503
    eucalyptus oil, 505–507
    juniper oil, 511–512
    lavender oil, 515–518
    lemon oil, 524–528
    lipophilic compounds, 434
    Mandarin oil, 532–534
    matricaria oil, 537
    mint oil, 540–541
    neroli oil, 543–544
    peppermint oil, 552–556
    Pinus sylvestris oil, 561–562
    rosemary oil, 565–570
    star anise oil, 577–578
    sweet orange oil, 580–584
    tea tree oil, 589
    thyme oil, 597–600
Alginates, 1026–1027
Allergic rhinopathy, 383
Ambrox, 188, 966–968
    Aspergillus
        A. cellulosae, 973
        A. niger, 971–972
    Cephalosporium aphidicola, 977
    Fusarium lini, 974
    Rhizopus stolonifer, 974
Animals (rabbits) and dosing, 988–989
Anise oil
    content, 435
    definition, 435
    inhibitory data
        agar diffusion test, 439–442
        dilution test, 444–446
        vapor phase test, 447
Anisi
    A. Aetheroleum (*see* Anise oil)
    A. stellati aetheroleum (*see* Star anise oil)
Anointing, 624
Antibacterial activity
    acne, 385
    Artemisia spicigera C. Koch (Asteraceae), 282

bacterial strains, 282
bacterial vaginosis, 385–386
Carum copticum L. (Apiaceae), 282
ciprofloxacin, 284
Daucus muricatus L. (Apiaceae), 282
disc diffusion agar method, 283
Eucalyptus globulus Labill. (Myrtaceae), 283
eugenol, 284
E. Walker (Asteraceae) EO, 282
fractional inhibitory concentration values, 285
Gram-positive and Gram-negative bacterial strains., 283
grapefruit EO, 281
keratinocytes, 282
leaves and bark oils, 284
minimal lethal concentration (MLC), 284
"Mint Timija," 283
MRSA, 386
Pelargonium graveolens L'Her (Geraniaceae), 284
pure *C. citratus*, 284
S. aureus, 281
synergistic effects, 283
thymus EOs, 283
Trachyspermum copticum L., 284
Vitex agnus-castus L. (Verbenaceae), 284
Zataria multiflora Boiss. (Lamiaceae), 283
Anticarcinogenic acitivity, 384
Antifungal medications, 386–388
Antihypotensive effects, 383–384
Antioxidants
antioxidant activity, 326–327
Artemisia arborescens, 329
ascorbic acid, 329
D-Borneol, 334
Carum copticum, 334
citrus, 331
clove oil, 335
Cuminum cyminum, 331
DPPH and ABTS assays, 330
Dracocephalum multicaule, 337
Eucalyptus camaldulensis, 330
Fenton reaction, 332
fresh and dry rhizomes, 332
glutathione, 326
laurel essential oil, 335
Leucas virgata, 335
limonene, 331
lipid peroxidation inhibition, 333
lipid peroxyl radicals, 326
Melaleuca, 329
Mentha longifolia, 336
O. vulgare, 332
Pakistan *Rosmarinus officinalis*, 334
Pinus densiflora, 336
Piper officinarum, 336
Satureja khuzestanica, 333
S. khuzestanica, 333
thiobarbituric acid reactive substances (TBARS) assay, 328
Thyme species, 328
Trifolium pratense, 336
turmeric essential oil, 333
Zataria multiflora, 334

Antipests
acaricide, 677–679
Acorus sp., 682
Artemisia sp., 682
bacterial control, 690–693
Carum sp., 683
Chenopodium ambrosoides, 683
Cinnamomum sp., 682, 700
commercialization, 672
Cymbopogon citratus, 700
Eucalyptus sp., 680–681
examples, 673–676, 699
Foeniculum vulgare, 682–683
fungal control, 691–693
impacts of, 671
insecticide, 677–679
Laurus nobilis, 700
Lavandula sp., 683
Mentha sp., 682
Muña, 679–680
nematode control, 691–694
Ocimum sp., 681
Origanum sp., 681–682
repellent action, 671–672
Rosmarinus officinalis, 676
Satureja sp., 681
Syzygium aromaticum, 679
Thymus sp., 676, 698
viral control, 691–693
Antiviral Listeriner®, 390
Antiviral properties, 388, 412
Anxiety, 349–350, 353, 357, 393, 640–641, 644
(+)-1(10)-Aristolene, 916–919
Aspergillus
A. cellulosae, 917–918
A. niger, 917–918, 922
A. sojae, 919
Chlorella fusca, 917, 921–922
Diplodia gossypina, 919
Mucor species, 917, 922
Aroma extraction dilution analysis (AEDA), 206–207
Aromatherapy, 612
Alzheimer's syndrome, 623
antioxidant agents, 622
aromachology, 621
aromatherapy, 621
aromatology, 621
chronic respiratory infection, 646
"clinical" aromatherapy, 625–626
clinical studies, 640
autonomic nervous system, 644
deaf and deaf-blind people, 645
depression scores, 644
German chamomile, 641
heliotropin, 644
lavender oil, 643
midwives and mothers, 643
in palliative care, 644
peppermint oil, 644
randomized controlled pilot study, 644
sensitization effect, 642
contingent negative variation (CNV), 622
definition, 620–621
in dementia, 640–641, 646

Index

embalming properties, 622
ethanol, 646
fixed oils, 631
historical background, 623–624
internal usage, 631–632
limbic system, 622
massage, 630–631
modern perfumery, 627
mood and behavior, 622
with multiple sclerosis, 646
paint industry, 622
peppermint oil, 646
perfume and cosmetics
    enfleurage, 625
    expression, 625
    maceration, 625
physiological and psychological responses, 633–634
placebo effect of, 634–635
pure/synthetic components, 632
relaxation, 622
Rowatinex, 645
safety issue in, 635–636
sexual attractiveness, 620
stress-related conditions, 622
therapeutic claims, 632–633
Tiger Balm, 622
toxicity, in humans
    allergenic response, 636–637
    limonene, 639
    linalool, 639
    migraine headaches, 636
    photosensitizers, 637
    respiratory irritants, 636
    in young children, 638
treatment, 628–629
ureterolithiasis, 645
Vicks VapoRub, 622
Aroma-vital cuisine
    appetizer and finger food
        crudities, 1063
        Maria's dip, 1063
        tapenade, 1063–1064
        tofu aromanaise, 1064
        veggie skewers, 1064–1065
    beverages
        aroma shake, herbs, 1059
        earl grey, 1059–1060
        Lara's jamu, 1060
        rose cider, 1060–1061
        syrup mint–orange, 1061
    celery lemon grass patties, 1065
    chévre chaud cheese "provence," pineapple, 1065–1066
    crispy coconut flakes, 1058
    crispy wild rice, 1066
    dessert, cakes, and baked goods
        apple cake rose, 1068
        chocolate fruits and leaves, 1068–1069
        homemade fresh berry jelly, 1069
        rose semifreddo, 1069–1070
        sweet florentine, 1070
    emulsifiers, 1055
    entrees
        goat cheese and ricotta, 1062–1063
        melon–plum purple radish salad, 1062
        peppermint heaven, 1061
        perky pumpkin soup, 1061
    gomasio, 1059
    honey Provencal, 1059
    mango–dates–orange chutney, 1066–1067
    oil mixtures and seasonings, 1056–1058
    prawns bergamot, 1067–1068
    quality, 1054–1055
    quantity, 1055
    soul foods, 1056
    spice, 1055–1056
    storage, 1055
Artemisinin
    Artemisia annua, 957
    Aspergillus niger, 957, 959
    Cunninghamella
        C. echinulata, 957, 959
        C. elegans, 959
    Nocardia corallina, 957, 959
    Penicillium chrysogenum, 957, 959
Asarone, 997
Aspergillus
    A. cellulosae, 769–772
        ambrox, 973
        maalioxide, 967
        parthenolide, 949
    A. niger, 910–913, 1001
        (−)-ambrox, 971–972
        artemisinin, 959
        $1,2,4\beta,5\alpha$-tetrahydro-α-santonin, 946
        bicyclohumulenone, 967
        cadinol, 956
        capsaicin, 1001
        carvone, 817–818, 820
        carvotanacetone, 829–830
        1,8-cineole, 840–841
        cinnamodial, 961
        (+)-cuparene, 962–964
        (−)-cuparene, 962–963
        dihydroasarone, 998
        dihydrocapsaicin, 1001
        diisophorone, 988
        $2E,6E$-farnesol, 987
        fenchol, 866–867
        fenchone, 873–874
        6-Gingerol, 1003
        (−)-herbertenol, 963, 965
        hinesol, 970
        2-hydroxy-3-pinanone, 875–877
        (−)-isolongifolol, 980
        isomenthol, 792–793
        isopinocampheol, 863–864
        isopiperitenone, 832–833
        limonene, 769–770
        linalool, 761–763
        maalioxide, 967
        menthol, 788–790
        (−)-methoxy-α-herbertene, 963, 966
        methyl dihydroeugenol, 998
        microbiological oxidation and reduction, 890–891
        nardosinone, 969–970
        neomenthol, 789, 792–793
        parthenolide, 949
        3-pinanone, 873, 875

α-pinene, 847, 849–850
β-pinene, 849, 852–855
pinguisanol, 969
pinocarvone, 870–871
piperitenone, 832–833
polygodiol, 960
(+)-sclareolide, 975
6-shogaol, 1003
trans-nerolidol, 987
trinornardosinone, 970
widdrol, 981
A. sojae, 915
A. wentii, 910–911
Atopic dermatitis, 392
ATP. *see* Adenosine triphosphate (ATP)
Atractylon, 936, 938
Attenuated reflection (ATR) IR spectroscopy, 31
Aurantii dulcis aetheroleum. see Sweet orange oil

B

Bicyclic monoterpenoids
    alcohol
        borneol, 864–865
        fenchol and fenchyl acetate, 866–867
        isoborneol, 864–865
        isopinocampheol (3-pinanol), 862–864
        myrtanol, 859–860
        myrtenol, 858–859
        nopol, 868–869
        nopol benzyl ether, 868–869
        pinane-2,3-diol, 861–862
        pinocarveol, 859–860
        verbenol, 868
    β-pinene
        Aspergillus niger, 849, 852–855
        Botrytis cinerea, 850, 852
        Dendroctonus frontalis, 856
        Pseudomonas pseudomallei, 849, 852
        Pseudomonas putida arvilla, 849, 852
    camphene, 856
    carane, 857–858
    3-carene, 857–858
    myrtanal, 858–859
    myrtenal, 858–859
    α-pinene
        Aspergillus niger, 847
        Aspergillus niger TBUYN-2, 849–850
        Botrytis cinerea, 849, 851
        human liver microsomes CYP2B6, 849, 851
        Pseudomonas, 847
        Pseudomonas fluorescens, 847–848
        Pseudomonas PX 1, 849–850
        rabbit, 849, 851
        soil Pseudomonad, 848–849
        Spodoptera litura, 849, 851
    saturated ketone
        camphor, 870–873
        fenchone, 871, 873–874
        2-hydroxy-3-pinanone (*see* 2-hydroxy-3-pinanone)
        3-pinanone, 873, 875
    α-, β-unsaturated ketone
        pinocarvone, 870–871
        verbenone, 870

Bicyclohumulenone, 965, 967
BIOLANDES, 146, 155–156
β-Bisabolol, 957
Bitter-fennel oil
    content, 436
    definition, 436
    inhibitory data
        agar diffusion test, 448–453
        dilution test, 454
        vapor phase test, 455
Borneol, 254
    and isoborneol, 864–865
    metabolic pathways, 882, 886
    multiallelic genetic determination, 100
Botryosphaeria dothidea, 840–842, 913–916,
    941–942, 953
Botrytis allii, 831, 838
Botrytis cinerea, 755, 757, 761–762
    (−)-β-caryophyllene epoxide, 984
    citronellal, 756–757
    geraniol, 756–757
    ginsenol, 980
    isoprobotryan-9α-ol, 985
    patchoulol, 979
    α-Pinene, 849, 851
    β-pinene, 850, 852
Bronchodilation
    and airway hyperresponsiveness, 411
    1,8-cineole, 413
Brown oil extractors (BOEs), 141–142
Buscopan/glucagon, 402–403

C

Cadina-4,10(15)-dien-3-one, 954
    Curvularia lunata, 955
    Mucor plumbeus, 955
Cadinol, 956
Camphene, 7, 105, 214, 255, 336, 395, 697, 728, 856
Camphor, 255–256
    Corynebacterium diphtheroides, 871–872
    Eucalyptus perriniana, 873
    metabolic pathways, 882, 886
    P450 enzymes, 871, 873
    Pseudomonas putida, 871–872
Capsaicin, 289, 406, 999, 1078
    Aspergillus niger, 1001
    Catharanthus roseus, 1002
Carane, 94, 857–858
Caraway oil
    content, 436
    definition, 436
    inhibitory data
        agar diffusion test, 456–459
        dilution test, 460–461
        vapor phase test, 462
3-Carene, 256, 857–858
Carvacrol, 49–51, 72–73, 106–108, 257, 285, 288,
    291–292, 328, 332–334, 339, 660, 798–799,
    998, 1000
Carvacrol methyl ether, 333–334, 798–799, 881, 886–888
Carveol
    bottrospicatol, 800–801, 803–804
    Chlorella pyrenoidosa, 804–805

Index

E. gracilis Z., 804–805
enantio-and diastereoselective biotransformation, 804–805
isobottrospicatol, 800–801
metabolic pathways, 801–802
Pseudomonas ovalis, 798, 800
S. bottropensis, 800–803
soil Pseudomonad, 798–799
Spodoptera litura, 805
Carvi aetheroleum. see Caraway oil
Carvomenthols, 812–813, 886, 890
Carvone
    actinomycetes, 819
    Absidia glauca, 817
    Aspergillus
        A. niger, 817–818
        A.niger TBUYN-2., 818–820
    carvone reductase, 825–827
    Catharanthus roseus, 826–827
    Chlorella miniata, 817–818
    citrus pathogenic fungi, 822, 824
    convertible microorganisms, 814–815
    enone reductase, 826–827
    Euglena gracilis Z, 817–818, 825–826
    groups, 821–822
    hydrogenation mechanisms, 825
    metabolic pathways, 257–258, 880, 887
    microbial and chemical hydrogenation, 814, 816, 818–819
    microorganism groups, 816–817
    Pseudomonas ovalis, 818
    Streptomyces, 820–823
    Trichoderma pseudokoningii, 819–820
Carvotanacetone
    A. niger TBUYN-2, 829–830
    metabolic pathways, 878–879, 881, 884, 886–887, 889
    Pseudomonas ovalis, 828
    Streptomyces bottropensis, 828–829
Caryophyllene, 53, 72, 89, 182, 274–275, 336
(−)-β–caryophyllene
    Chaetomium cochliodes, 981
    Diplodia gossypina, 981–982
    Pseudomonas cruciviae, 981
(−)-β-caryophyllene epoxide, 977
    Botrytis cinerea, 984
    fungi, 983
Caryophylli floris aetheroleum. see Clove oil
Cassia oil
    content, 436
    definition, 436
    inhibitory data
        agar diffusion test, 463
        dilution test, 464
        vapor phase test, 465
Catharanthus roseus
    carvone, 826–827
    citronellal, 757, 759
    geraniol, 757–759
    linalool, 757, 759
    nerol, 757–758
CCC. *see* Countercurrent chromatography (CCC)
Cedrol, 973, 990–991
    Botrytis cinerea, 978
    fungi, 978
    Glomerella cingulata, 978

Central nervous system
    activation and arousal, 348
    CNS active compounds and chemical structures, 369–372
    contingent negative variation, 353–354
    fragrances (*see* Fragrances)
    lipophilic properties, 347
    olfaction differs, 346
    reticular activating system, 347
    spontaneous EEG activity
        alpha rhythm, 349
        aromatherapy, 350
        beta and theta activity, 349
        cognitive influence, 350
        cognitive processes, 352
        corticocortical connectivity, 352
        fatigue and performance decrements, 349
        frontal and occipital alpha phases, 352
        inconclusive finding, 350
        lavender and jasmine fragrance effects, 353
        mental task, 351
        odor pleasantness and familiarity, 351
        orbitofrontal cortex neurons, 352
        period analysis, 349
        topographical differences, 351
    trigeminal system, 347
Cephalosporium aphidicola
    ambrox, 977
    diisophorone, 988
    menthol, 787
Ceylon cinnamon bark oil
    content, 436
    definition, 436
    inhibitory data
        agar diffusion test, 466–469
        dilution test, 470–471
        vapor phase test, 472
Ceylon cinnamon leaf oil
    content, 436
    definition, 436
    inhibitory data
        agar diffusion test, 473
        dilution test, 474
        vapor phase test, 475
Chaetomium cochliodes, 975, 981
Chamomile
    Chamomilla recutita, 101
    ecological experiments, 73
    essential oil content and composition, 101
    frequency distribution, 54
    Matricaria recutita, 53
    morphological traits, 102
    synthetic cultivars, 67
CHARM. *see* Combined hedonic aroma response method (CHARM)
Chemical Abstracts Service (CAS) number, 1043–1044
Chemical substances of unknown/variable composition (UVCB), 1044–1045
Chemotaxonomy, 45–46
    essential oil-bearing plants, 45–46
    natural variability
        A. crithmifolia, 113
        Balkan *Achillea* species, 115

biological variability, 113
chemical compound, 111
chemotypes, 111
Journal of Essential Oil Research, 111
marjoram *(Majorana hortensis),* 112
M. citrata, 114
tansy *(T. vulgare),* 112
Chenopodium ambrosoides, 683
Chlorella
  C. fusca var. *vacuolata,* 908–909
  C. miniata, 817–818
  C.pyrenoidosa
    carveol, 804–805
1,4-Cineole, 843–846
  metabolic pathways, 884, 886, 889
  monoterpene, 259
1,8-Cineole
  2α-hydroxy-1,8-cineole, 843–844
  A. niger TBUYN-2, 840–841
  Botryosphaeria dothidea, 840, 842
  definition, 837
  Eucalyptus perriniana, 843, 845
  Glomerella cingulata, 842
  metabolic pathways, 884, 886, 889
  monoterpene, 259–260
  Pseudomonas flava, 839
  stereoselectivity, 840–841
  Streptomyces
    S. bottropensis, 839
    S. ikutamanensis, 839–840
  Trichosurus vulpecula, 842
Cinnamodial
  Aspergillus niger, 961
  Cinnamosma fragrans, 960
Cinnamomi
  C.cassia aetheroleum (see Cassia oil)
  C. ramulus, 285
  C. zeylanici folii aetheroleum (see Ceylon cinnamon leaf oil)
Cinnamon oil, 229–230
Cinnamon zeylanicii corticis aetheroleum.
  see Ceylon cinnamon bark oil
Cinnamosma fragrans, 960
cis-geranyl acetone, 984–985
cis-nerolidol, 984
  Aspergillus niger, 986
  Glomerella cingulata, 985
Citral
  Euglena gracilis Z., 754
  metabolic pathways, 883, 888
  monoterpene, 260–261
  Pseudomonas aeruginosa, 751–752
Citri reticulatae aetheroleum. see Mandarin oil
Citronella aetheroleum. see Citronella oil
Citronellal
  Botrytis cinerea, 756–757
  Catharanthus roseus, 757, 759
  cyanobacterium, 757–758
  metabolic pathways, 883, 888
  monoterpene, 260–261
  Penicillium digitatum, 751, 753
  Pseudomonas
    P. aeruginosa, 751–752
    P. citronellolis, 751–752

Citronella oil
  content, 436
  definition, 436
  inhibitory data
    agar diffusion test, 476–479
    dilution test, 480
    vapor phase test, 481
Citronellene
  metabolic pathways, 885, 889
  Spodoptera litura, 750
Cladosporium sp., 768–770, 773, 778–780, 798
Clary sage oil
  content, 436
  definition, 436
  inhibitory data
    agar diffusion test, 482
    dilution test, 483–484
    vapor phase test, 484
Clove oil
  content, 436
  definition, 436
  inhibitory data
    agar diffusion test, 485–488
    dilution test, 489–492
    vapor phase test, 493
^{13}C-NMR spectroscopy, 32–33, 726
Cognitive performance, 356, 394, 405, 645
Combined hedonic aroma response method (CHARM), 206–207
Committee of Experts on Flavoring Substances (CEFS), 1072, 1075, 1077
Complex reaction products and biological materials (UVCB), 1044–1045
Compositae (Asteraceae), 45, 71
Congeneric groups
  chemical composition and, 232–234
  chemical constituent, 240
  evaluation procedure, 240
  intake level, 239
  prioritization, 240–241
  toxic and carcinogenic potentials, 239
  toxicity
    corn mint oil, 244–247
    structural classes, 241–244
Contingent negative variation (CNV), 353–354, 622, 634
Controlled release mechanism, 1024–1025, 1028
Copal pellets, 624
Coriander, 105, 151, 173, 177, 236, 239
Coriander oil
  content, 436
  definition, 436
  inhibitory data
    agar diffusion test, 494–499
    dilution test, 500–501
    vapor phase test, 502
Coriandri aetheroleum. see Coriander oil
Coriandrum sativum. see Coriander
Corn mint oil, 241, 244–247
Corynebacterium diphtheroides, 871–872
Costunolide, 937–938
Council of Europe (CoE), 161, 1072, 1075
Countercurrent chromatography (CCC), 14, 23–24
Cryptoporic acids, 961
Cumin aldehyde, 786

Index

Cunninghamella
  C. echinulata
    artemisinin, 959
    (–)-dehydrocostuslactone, 951
  C. elegans, 693, 957, 959, 971, 975, 980
(+)-Cuparene, 962–964
(–)-Cuparene, 962–963
CUPRAC assay, 327–328
(S)-(+)-Curcuphenol, 958
Curdione, 931–933
Curvularia lunata, 954–955
Cyclic monoterpenoids
  cyclic monoterpene epoxide
    1,4-cineole, 843–846
    1,8-Cineole (*see* 1,8-Cineole)
  monocyclic monoterpene alcohol
    carvacrol, 798–799
    carvacrol methyl ether, 798–799
    carveol (*see* Carveol)
    carvomenthols, 812–813
    dihydrocarveol (*see* Dihydrocarveol)
    isomenthol, 792–793
    isopiperitenol, 809, 811
    isopulegol, 792, 794
    menthol (*see* Menthol)
    neomenthol, 789, 792–793
    perillyl alcohol, 811–812
    piperitenol, 809
    (–)-terpinen-4-ol, 796–797
    α-terpineol, 792, 794–796
    thymol, 796–797
    thymol methyl ether, 796–797
  monocyclic monoterpene aldehyde
    cumin aldehyde, 786
    1,2-Dihydrophellandral, 785
    perillaldehyde, 783–785
    phellandral, 785
  monocyclic monoterpene hydrocarbon
    isolimonene, 776, 778
    limonene (*see* Limonene)
    p-cymene, 782–784
    α-phellandrene, 782
    p-menthane, 778–779
    1-*p*-menthene, 778–779
    3-*p*-menthene, 780
    α-terpinene, 780
    γ-terpinene, 780–781
    terpinolene, 781
  saturated ketone
    dihydrocarvone, 834–835
    isodihydrocarvone epimerase, 835
    menthone and isomenthone, 836–838
    thujone, 836–838
  α, β-unsaturated ketone
    carvone (*see* Carvone)
    carvotanacetone
      (*see* Carvotanacetone)
    isopiperitenone, 832–833
    piperitenone, 832–833
    piperitone, 829–830
    pulegone, 831–833
α-Cyclocostunolide, 937, 940
  Aspergillus cellulosae, 940
  Aspergillus niger, 940
β-Cyclocostunolide, 937
γ-cyclocostunolide, 937–938
Cyclodextrin, 18, 719, 725, 1028–1029
(+)-cycloisolongifol-5α-ol, 980
Cymbopogon citratus, 284, 366, 387, 700
Cymbopogon winterianus, 289, 366, 658, 689
α-Cyperone, 934–935

D

α-Damascone, 993–994
β-Damascone, 994–995
DCCC. *see* Droplet countercurrent
      chromatography (DCCC)
(–)-Dehydrocostuslactone, 951
  Aspergillus
    A. cellulosae, 953
    A. niger, 952–953
  Botryosphaeria dothidea, 953
  Cunninghamella echinulata, 951
8,9-Dehydronootkatone
  Aspergillus
    A. cellulosae, 915, 919
    A. sojae, 915, 919
  Marchantia polymorpha, 916, 920
  Mucor species, 920
Dendroctonus frontalis, 856
Deoxoartemisinin, 957
Detection frequency method, 207–208
Deutsches Arzneibuch 6 (DAB 6), 433
Dihydroasarone, 997–998
Dihydrocapsaicin, 999, 1001
Dihydrocarveol
  chemical structure, 805–806
  E. gracilis Z., 808–809
  8-hydroxydihydrocarveols, 808, 810
  metabolic pathways, 807
  (–)- and (+)-neodihydrocarveol, 805–806
  S. bottropensis, 807–809
Dihydrocarvone, 732, 834–835
Dihydromyrcenol, 765–766
Dihydronootkatone
  antitermite activity, 914
  Aspergillus
    A. cellulosae, 918
    A. niger, 918
1,2-Dihydrophellandral, 785
Diisophorone, 987
  Aspergillus niger, 988
  Cephalosporium aphidicola, 988
  Neurospora crassa, 988
Dilution test (DIL)
  anise oil, 444–446
  bitter-fennel oil, 454
  caraway oil, 460–461
  cassia oil, 464
  Ceylon cinnamon bark oil, 470–471
  Ceylon cinnamon leaf oil, 474
  citronella oil, 480
  clary sage oil, 483–484
  clove oil, 489–492
  coriander oil, 500–501
  dwarf-pine oil, 504
  eucalyptus oil, 508–509

juniper oil, 513–514
lavender oil, 519–521
lemon oil, 529–530
Mandarin oil, 535
matricaria oil, 537–538
mint oil, 542
neroli oil, 545
peppermint oil, 557–559
Pinus sylvestris oil, 563–564
rosemary oil, 571–574
star anise oil, 579
sweet orange oil, 585–587
tea tree oil, 590–595
thyme oil, 601–606
Dimethylphenylendiamin (DMPD) assay, 328
2,3-di-*O*-ethyl-6-*O*-tert-butyldimethylsilyl-β-CD on polymethylphenylsiloxane (PS086) phase, 208
Diplodia gossypina
(–)-β–caryophyllene, 981–982
myrcene, 749
Distinctness, uniformity, and stability (DUS), 68–69
d-Limonene, 189, 395, 397, 639, 1083
Drimenol
Mucor plumbeus, 960
Rhizopus arrhizus, 960
Droplet countercurrent chromatography (DCCC), 23–24
Dunaliella tertiolecta, 755, 784, 786, 993
DUS. *see* Distinctness, uniformity, and stability (DUS)
Dwarf-pine oil
content, 436
definition, 436
inhibitory data
agar diffusion test, 503
dilution test, 504
vapor phase test, 504
Dysmenorrhea, 404–405

E

2*E*,6*E*-farnesol, 992
Aspergillus niger, 987
Elemol, 70, 92, 727, 732, 990
Enantioselective-GC (Es-GC), 18–20, 208–210, 212
(+)-*ent*-Cyclocolorenone
Aspergillus niger, 926
Solidago altissima, 926
Eryngium alpinum L., 286
Eryngium amethystinum L. (Apiaceae), 286
Es-GC. *see* Enantioselective-GC (Es-GC)
Es-MDGC analysis, 209, 212, 214–215
Essential oil (EO)
Abbé-type refractometer, 197
adulteration, 2–3
ambrette seed oil, 727
amyris oil, 727
angelica oil, 727
anise fruit oil, 727
armoise oil, 727
basil oil, 727
bergamot oil, 727–728
bitter orange oil, 728–729
cajeput oil, 729
camphor oil, 729
cananga oil, 729
caraway oil, 729
cardamom oil, 729
cassia oil, 730
cedar oil, 730
celery seed oil, 730
chamomile oil, 730
cinnamon oil, 730–731
citronella oil, 731
clary sage oil, 731
clove oil, 731
coriander oil, 731
corn mint oil, 732
corymbia citriodora oil, 731
cumin fruit oil, 732
cypress oil, 732
dill oil, 732
elemi oil, 732
eucalyptus oil, 733
fennel oil, 733
geranium oil, 733
grapefruit oil, 733–734
juniper berry oil, 734
lavandin oil, 734
lavender oil, 734
lemongrass oil, 735
lemon oil, 734–735
lime oil, 735
litsea cubeba oil, 736
mandarin oil, 736
melissa oil, 736
mentha citrata oil, 736
neroli oil, 736–737
nutmeg oil, 737
orange oil, 737
origanum oil, 737
palmarosa oil, 737
parsley oil, 737–738
patchouli oil, 738
pepper/peppermint oil, 738
petitgrain oil, 728, 738
pimento oil, 738–739
pine oil, 732, 736, 738
rosemary oil, 739
rose oil, 739
rosewood oil, 739
sage oil, 739–740
sandalwood oil, 740
spearmint oil, 740
spike lavender oil, 740
star anise oil, 740
tarragon oil, 740–741
tea tree oil, 741
thyme oil, 741
turpentine oil, 741
vetiver oil, 741
ylang-ylang oil, 741
adulterations, 196
agricultural crop establishment, 136–138
in agriculture, 3
antibacterial activity
Artemisia spicigera C. Koch (Asteraceae), 282
bacterial strains, 282
Carum copticum L. (Apiaceae), 282
ciprofloxacin, 284

Index

Daucus muricatus L. (Apiaceae), 282
disc diffusion agar method, 283
Eucalyptus globulus Labill. (Myrtaceae), 283
eugenol, 284
E. Walker (Asteraceae) EO, 282
fractional inhibitory concentration values, 285
Gram-positive and Gram-negative bacterial strains., 283
grapefruit EO, 281
keratinocytes, 282
leaves and bark oils, 284
minimal lethal concentration (MLC), 284
"Mint Timija," 283
Pelargonium graveolens L'Her (Geraniaceae), 284
pure *C. citratus*, 284
S.aureus, 281
synergistic effects, 283
thymus EOs, 283
Trachyspermum copticum L., 284
Vitex agnus-castus L. (Verbenaceae), 284
Zataria multiflora Boiss. (Lamiaceae), 283
anti-inflammatory activity
  Abies koreana, 291
  acute dermatitis, 295
  Annona sylvatica, 295
  Artemisia fukudo, 291
  Calycorectes sellowianus, 290
  candida-infected tongues, 292
  carvacrol, 292
  Ceiba pentandra, 296
  CFA model, 295
  Chamaecyparis obtusa, 293
  clove oil, 292
  collagen-induced arthritis model, 292
  croton oil, 292
  curcumol, 296
  C. winterianus, 294
  Distacco Selinum tenuifolium, 294
  ear edema, 293
  Echinacea purpurea, 293
  eugenol, 292
  IF-g and TGF-β levels, 294
  I. lanceolatum, 295
  limonene, 295
  linalool, 291
  Lindera umbellata, 291
  LPS-induced COX-2 protein, 292
  MAP kinase pathway, 293
  O. americanum, 294
  12-*O*-tetradecanoylphorbol-13-acetate, 290
  P. aleyreanum, 293
  paw edema, 294
  Pistacia lentiscus, 293
  Pterodon polygalaeflorus, 294
  Rose geranium, 295
  S. cumini Skeels and *Psidium guajava*, 290
  terpinen-4-ol, 290
  thymol, 292
  thymol and carvacrol, 291
  turmeric oil, 289
  T. vulgaris, 291
  Whittle method, 290
  Zingiber zerumbet, 290
  Zizyphus jujuba, 291

antimicrobial activities, 610
  in European Pharmacopoeia 6th Edition, 609
antinociceptive activity
  α-bisabolol, 287
  carrageenan, 287
  carvacrol, 288
  citronellal, 289
  Croton tiglium, 289
  Cymbopogon winterianus, 289
  Heracleum persicum, 287
  Hyptis pectinata, 288
  Illicium lanceolatum, 288
  Lippia gracilis Schauer (Verbenaceae), 287
  monoterpenes, 287
  Ocimum micranthum, 288
  Piper aleyreanum, 286
  prostaglandine E_2, 287
  tail-flicking tests, 288
  Teucrium stocksianum, 288
  tumor necrosis factor-α, 287
  Vanillosmopsis arborea, 289
  Zingiber zerumbet, 287
antiviral activity, 285–286
a-pinene, 189–190
benzene, 186–187
biomass, 135
biosynthetic pathways, 165–166
bisulfite method, 199
b-pinene, 191
chemical properties, 198
chromatographic procedures, 200–201
chromatographic separation techniques, 14–15
  CCC, 23–24
  GC (*see* Gas chromatography (GC))
  liquid column chromatography, 22–23
  supercritical fluid chromatography, 23
  TLC, 15
citral, 191–192
climate, 133–134
^{13}C-NMR spectroscopy, 32–33
Cupressaceae essential oils, 391
cytotoxic activity
  basil, 302
  Boswellia sacra, 300
  carvone, 297
  cell viability, 301
  Ceratonia siliqua, 301
  diterpene, 302
  β-elemene, 299
  Eugenia caryophyllata, 298
  Eugenia supraaxillaris, 301
  Guatteria pogonopus, 302
  in vitro cytotoxicity study, 299
  Kadsura longipedunculata, 300
  Lindera strychnifolia, 303
  Lippia gracilis, 300
  marigold, 302
  M. communis, 301
  monoterpene myrtenal, 298
  Morus rotundiloba, 302
  Neolitsea variabillima, 298
  nonsmall cell lung cancer cells, 300
  Origanum marjoram, 301
  pancreatic epithelial cells, 301

patchouli alcohol, 302
Pinus densiflora, 299
Solanum erianthum, 301
terpinolene, 297
Thymus caespititius, 301
T. vulgaris, 299
Vepris macrophylla, 302
X. laevigata, 300
Xylopia laevigata, 299
Zanthoxylum leprieurii, 301
Zanthoxylum zanthoxyloides, 301
definition, 129–130, 710
distillation and evaporation, 165
elements, 165
encapsulation
    (*see* Volatile encapsulation)
esters, 198
expression method
  BOE, 142
  Citrus fruit, 139–140
  FMC, 142–143
  peel oils classification, 141
  "pellatrici" method, 141–142
  "sfumatrici" method, 141
and fragrances (*see* Fragrances)
gas chromatographic analysis
  Es-GC, 208–210
  fast GC, 203–206
  GC-MS, 202–204
  GC-O, 206–208
  MDGC (*see* Multidimensional gas
    chromatographic (MDGC) techniques)
  retention index systems, 201–202
GC-IRMS, 221
halitosis, 391
history, 5–7, 130–132
HPLC, 210–211
HPLC-GC, 27–29
HPLC-MS, 29
HPLC-NMR spectroscopy, 29
industrial application
  consumption, 1014–1016, 1018–1019
  flavors, 1013–1014, 1020
  fragrances, 1013, 1018–1020
  history, 1011–1012
  manufacturers, 1012
  producers, 1016–1017
industrial processes, 8–9
insect stress, 134
IR spectroscopy, 31
laboratory-scale techniques, 9
Lippia
  L. multiflora, 390
  L. sidoides, 391
M. alternifolia, 391
MDLC
  DAD system, 220
  HPLC systems, 219
  LC x LC, 218
  NPLC x RPLC, 219
melting and congealing points, 198
L-menthol, 192–193
methyl anthranilate, 187–189
microorganisms, 134

microsampling techniques
  direct sampling, 10–11
  HS analysis, 11–13
  HSSE, 14
  microdistillation, 10
  SBSE, 14
  SPME, 13
MS, 31–32
natural extracts, 188, 190
NP-HPLC analysis, 210–211
oil cell location, 134–135
online coupled LC-GC, 221
optical rotation, 196–197
PC, 199–200
penetration-enhancing activity, 303–307
pesticidal action (*see* Antipests)
plant-derived feedstocks, 185–186
polyketides and lipids
  aldehydes, 168
  allylic oxidation, 168
  lipid-derived components, 168–169
  oakmoss components, 167–168
  prostaglandins and jasmines, 169–170
  structure, 167
production, 132–133
  countries, 129
  figures, 127–128
psychopharmacology
  in animal models, 365–366
  aromatic plants, 363–365
  individual components, 366–367
  inhaled essential oils, 367–368
  mechanisms, 368–369
refractive index, 197
RP-HPLC analysis, 210–211
seeds and clones, 138–139
sesquiterpenoids
  alcohols, 182
  carotenoid degradation products, 183–184
  components, 182–183
  farnesol, 179–181
  hydrocarbons, 182–183
  iripallidal and irones, 184–185
SFC-FTIR, 30
SFC-GC, 29
SFC-MS, 30
SFE-GC, 29
SG, 196
shikimates, 186–189
shikimic acid
  biosynthesis, 170
  biosynthetic intermediates, 170–171
  cinnamic acid, 172
  definition, 170
  essential oil components, 172
  ferulic acid derivatives, 173
  hydroxy-and aminobenzoic acid
    derivatives, 171
soil condition
  Cymbopogon sp., 697–698
  Laurus nobilis, 697
  Lavandula stoechas, 697
  Mentha spicata, 697
  microorganisms, 694–696

Index

Myrtus communis, 697
Salvia sp., 697
soil quality and preparation, 134
steam/water distillation
    (*see* Steam/water distillation)
Stillman-Reed method, 199
systematic investigations, 7–8
terpenes, 195
terpenoids
    C5 units coupling, 174
    definition, 173
    head-to-tail coupling, 174
    hemiterpenoids, 175
    isoprene units, 173–174
    monoterpenoids
        (*see* Monoterpenoids)
time of harvesting, 135–136
TLC, 199–200
trading
    buyers and dealers, 1037
    equipment, 1033, 1035–1036
    FOB and CIF, 1037
    international meetings, 1038
    market information, 1033, 1037
    price graph, 1038–1039
    producers and dealers, 1033, 1035–1036
    production centers, 1033–1034
    supply chain, 1033, 1035
turpentine, 189
UV spectroscopy, 31
vasodilatory activity, 296–297
water solubility, 197
water stress and drought, 134
yeasts, 609
Essential oil-bearing plants
    Asteraceae (Compositae)
        Achillea millefolium, 53
        chamomile, 53, 55–56
        chemotypes, 54–55
        phenylpropenes, 52
        proazulene frequency distribution, 53–54
        Tagetes lucida, 52
        Thymus vulgaris, 55, 57
    capitate trichomes, 44
    chemotaxonomy, 45–46
    contaminations, 75
    environmental influences, 73–74
    factors, 65
    GAP, 75–76
    harvesting, 75
    intellectual property rights, 68
    Lamiaceae (Labiatae) and Verbenaceae
        L. graveolens and *L. alba,* 50–51
        oregano, 47–50
        Perilla frutescens, 50
        Salvia officinalis, 46–47
        Salvia. fruticosa, 46–47
    monoterpene biosynthesis, 44
    peltate glands, 44
    phenotypic variation, 65
    phytochemical polymorphism, 44
    plant breeding
        patent protection, 68–69
        PVP, 68–69
    plant nutrition and fertilizing, 74
    plant parts *vs.* developmental stage
        Cinnamomum zeylanicum, 69
        Compositae (Asteraceae), 71
        Eucalyptus camaldulensis, 72
        F. vulgare Mill, 71
        Linalool, 72
        parsley seed oils, 70–71
        p-cymene, 72–73
        Pimpinella cumbrae, 70
        Pimpinella nigra, 70
        Rutaceae family, 70
    principal circumstances, 44
    protein engineering, 58–59
    quality and safety, 75
    salinity and salt stress, 75
    schizogenic oil ducts, 44
    secretory idioblasts, 44
    sustainability, 59
    targeted breeding
        artificially generated variability, 67–68
        hybrid breeding, 67
        natural variability, 65–67
        registered cultivars, 65–66
        synthetic varieties, 67
    wild collection
        common and botanical names, 59–63
        cultivation method, 64–65
        definition, 43
        DNA, 56–57
        domestication, 64
        FairWild standard, 77
        GACP, 76–77
        genetic engineering, 57–58
        GMOs, 57
        ISSC-MAP, 77
Eucalypti aetheroleum. See Eucalyptus oil
Eucalyptus
    E. camaldulensis, 72, 330, 381, 689
    E. citriodora, 178, 288, 294, 689
    E. globulus, 161, 283, 330–331, 400–401
    E. perriniana
        camphor, 871, 873
        1,8-Cineole, 843, 845
        menthol, 788, 791
        perillyl alcohol, 811
Eucalyptus oil
    content, 437
    definition, 437
    inhibitory data
        agar diffusion test, 505–507
        dilution test, 508–509
        vapor phase test, 510
α-Eudesmol, 365, 371, 934, 936
Eugenol, 50, 69, 173, 186, 284, 292, 298, 335, 671, 739, 998, 1000
Euglena gracilis Z, 754
    borneol, 865
    carveol, 804–805
    carvone, 817–818, 825–826
    citral, 754
    dihydrocarveol, 808–809
    microbiological oxidation and reduction, 890, 892–893

European Inventory of Existing Commercial Chemical Substances (EINECS)
  botanical origin, 1044–1045
  CAS number, 1043–1044
  known substances, 1043
  UVCB, 1044–1045
European Pharmacopoeia, 1–2, 9, 31, 75, 433, 609, 656, 723, 729

F

FairWild standard, 77
Farnesol, 179–182, 274–275, 392, 985, 987, 991–992
Fast GC analysis, 16–17, 203–206
FEMA. *see* Flavor and Extract Manufacturers Association (FEMA)
Fenchone, 262, 264
  Absidia orchidis, 873–874
  Aspergillus niger, 873–874
  Corynebacterium sp., 873–874
  metabolic pathways, 878–879
Ferric ion reducing antioxidant power (FRAP) assay, 327–329, 332, 334–336
FID. *see* Flame ionization detector (FID)
Flame ionization detector (FID), 15–16, 21, 23, 202, 206, 724–725
Flavor and Extract Manufacturers Association (FEMA), 234–235, 238, 242, 635
Flavors and fragrances (F&F)
  hazard classification and labeling
    constituents, 1083
    implementation, 1081–1082
    measurement results, 1082
    NCSs, 1082
    nutmeg oil, 1084–1086
    orange oil, 1083–1085
    transition periods, 1082
  natural flavouring substance
    flavoring regulation 1334/2008/EC, 1079–1081
    former flavoring directive 88/388/EC, 1078–1079
  properties, 1087
  restricted substances (*see* Restricted substances (RS))
Florentine flask, 6, 8, 145, 147, 149–152, 712–714
Fluconazole-refractory oropharyngeal candidiasis, 390
FMC. *see* Food machinery corporation (FMC)
Foeniculi amari fructus aetheroleum, 436, 656
Foeniculi amari herbal aetheroleum, 436
Foeniculum vulgare, 98, 262, 662, 682–683
Food machinery corporation (FMC), 8, 141–143
Fragrances
  alertness and attention
    cognitive performance, 356
    inhalation and nonolfactory administration, 357
    motor learning, 355
    peppermint and muguet odors, 355
    pharmacological factor, 356
    physicochemical odorant properties, 357
    Spanish sage, 358
    task complexity, 355
    vigilance, 355
  cognitive tasks, 360–361
  learning and memory, 358–360
  stress-induced immunosuppression, 367

Free radicals
  essential oil effects, 337–339
  reactive nitrogen species, 325
  ROS, 324
French Standards Association, 196
Full stochastic model (FSM), 235
Functional dyspepsia, 396–397
Fungating wounds, 392
Fusarium
  F. culmorum, 913–916, 995–996
  F. lini
    ambrox, 974
    (–)-isolongifolol, 980

G

GACP. *see* Guidelines on good agricultural and collection practices (GACP)
Gall and biliary tract stones, 394–395
GAP. *see* Good agricultural practices (GAP)
Gas chromatographic-isotope ratio mass spectrometry (GC-IRMS), 27, 221
Gas chromatographic-mass spectrometry (GC-MS), 24–25, 202–204, 714, 724–725, 731–732, 735–739, 909, 927
Gas chromatographic-olfactometry (GC-O), 206–208
Gas chromatography (GC), 724–726
  chiral GC
    cyclodextrin derivatives, 18–19
    cyclodextrins, 18
    enantioselective GC, 17–18
  comprehensive multidimensional GC, 21
  two2D GC, 19–21
  Es-GC, 208–210
  fast GC, 16–17, 203–206
  FID, 15–16
  GC-AES, 27
  GC-FTIR spectroscopy, 25–26
  GC-IRMS, 27
  GC-MS, 24–25, 202–204
  GC-O, 206–208
  GC-UV spectroscopy, 27
  MDGC (*see* Multidimensional gas chromatographic (MDGC) techniques)
  retention index systems, 201–202
  TCD, 15–16
  ultrafast GC, 16–17
Gas chromatography-atomic emission (GC-AES) spectroscopy, 15, 27
Gas chromatography-isotope ratio mass spectrometry (GC-IRMS), 15, 27–28
Gastroesophageal reflux, 397
GC-AES spectroscopy. *see* Gas chromatography-atomic emission (GC-AES) spectroscopy
GC-chemical ionization-MS (GC-CI-MS), 25
GC-FTIR spectroscopy, 15–16, 25–27
GC-IRMS. *see* Gas chromatographic-isotope ratio mass spectrometry (GC-IRMS)
GC-MS. *see* Gas chromatographic-mass spectrometry (GC-MS)
GC-O. *see* Gas chromatographic-olfactometry (GC-O)
GC-tandem MS (GC-MS-MS), 25–26

Index

GC-UV spectroscopy, 15, 27
Generally recognized as safe (GRAS), 230, 234, 247, 635–636, 645
Genetically modified organisms (GMOs), 57, 67
Genetic engineering, 57–58, 67
Geraniol, 262, 264
   Botrytis cinerea, 756–757
   Catharanthus roseus, 757–759
   cyanobacterium, 757–758
   metabolic pathways, 883, 888
   Pseudomonas incognita, 751, 753
Germacrone
   Aspergillus niger, 931
   Curcuma
      C. aromatica, 931
      C. zedoaria, 931
   Solidago altissima, 933
Gibberella cyanea, 794
6-Gingerol, 833, 1002–1003
Ginsenol, 975, 980
Globulol
   Bacillus megaterium, 921
   Cephalosporium aphidicola, 921
Glomerella cingulata
   1,8-Cineole, 842
   cis-geranyl acetate, 985
   cis-nerolidol, 985
   fencyl acetate, 867
   (+)-isolongifolene-9-one, 980
   patchoulol, 979
   trans-geranyl acetate, 986
   trans-nerolidol, 986
GMOs. *see* Genetically modified organisms (GMOs)
Good agricultural practices (GAP), 75–76
GRAS. *see* Generally recognized as safe (GRAS)
Guidelines on good agricultural and collection practices (GACP), 76–77

H

Headache, 405, 413–414, 646, 966, 1014
Headspace (HS) analysis
   dynamic methods, 11–12
   static methods, 11
Hemiterpenoids, 174–175
(–)-Herbertenediol
   Penicillium sclerotiorum, 963–964
(–)-Herbertenol
   Aspergillus niger, 963, 965
Herbicides
   Artemisia scoparia, 690
   Cymbopogon sp., 689
   essential oils, 686–688
   Eucalyptus sp., 689
   Lavandula sp., 689
   Mentha sp., 684, 689
   organic weed control, 684
   Origanum sp., 689–690
   phytotoxicity, 683–684
   Tanacetum sp., 690
   test specimen, 685
   Thymus vulgaris, 684
   Zataria multiflora, 690
High-performance LC (HPLC), 14–15, 22–23, 28–29, 209–211, 218–220, 913
High-performance liquid chromatography-gas chromatography (HPLC-GC), 15, 27–29
Hinesol, 966, 970
Hinokitiol, 369, 372, 391, 998, 1000
Hormonema. isolate, 831–833
Hospital Anxiety and Depression Scale (HADS) scores, 644
HPLC. *see* High-performance LC (HPLC)
HPLC-GC. *see* High-performance liquid chromatography-gas chromatography (HPLC-GC)
HPLC-MS spectroscopy, 15, 29
HPLC-NMR spectroscopy, 15, 29
HSSE. *see* HS sorptive extraction (HSSE)
HS sorptive extraction (HSSE), 14
Human herpes virus 1 (HHV-1), 285
8α-Hydroxybicyclogermacrene
   Calonectria decora, 928, 930
   Mucor circinelloides, 928
2α-Hydroxydihydro-α-santonin, 943
4β-Hydroxy-eudesmane-1,6-dione
   Exserohilum halodes, 934–935
   Gliocladium roseum, 934–935
1α-Hydroxymaaliene, 930
2-Hydroxy-3-pinanone
   antimicrobial activity, 876–878
   Aspergillus niger, 875–877
   mosquitocidal and knockdown activity, 875, 877
Hyperlipoproteinemia, 397–398

I

Infantile colic, 405
Insecticides
   acaricidal activity, 398, 401
   antiprotozoal effects, 400
   pediculicidal activity, 400–401
α-Ionol, 992–993
α-Ionone, 190, 992–994
Irritable bowel syndrome, 401–402, 622, 632
IR spectroscopy, 14, 25, 31
Isoborneol, 178, 214–215, 255, 366, 864–865
Isodihydrocarvone epimerase, 835
(+)-Isolongifolene-9-one, 975, 980
(–)-Isolongifolol
   Aspergillus niger, 980
   Fusarium lini, 980
Isomenthol, 208, 786, 788, 792–793
Isopinocampheol (3-pinanol), 88, 862–865
Isopiperitenol, 58, 108, 114, 770, 809, 811, 832, 887
Isopiperitenone, 115, 769–770, 832–833, 880, 887, 890
Isoprobotryan-9α-ol, 982, 985
Isopulegol, 178, 792, 794, 889, 894
ISSC-MAP, 59, 77

J

JECFA. *see* Joint Expert Committee on Food Additives (JECFA)
Joint Expert Committee on Food Additives (JECFA), 235–236, 242

Joint physiotherapy, 405
Juniperi aetheroleum. see Juniper oil
Juniper oil
   content, 437
   definition, 437
   inhibitory data
      agar diffusion test, 511–512
      dilution test, 513–514
      vapor phase test, 514

L

Laboratory-scale techniques, 9
Lactone, 106, 352, 760–761, 889, 946, 952
Lamiaceae (Labiatae)
   L. graveolens and *L. alba*, 50–51
   oregano, 47–50
   Perilla frutescens, 50
   Salvia
      S. fruticosa, 46–47
      S. officinalis, 46–47
Lancerodiol *p*-hydroxybenzoate
   Marchantia polymorpha, 981
Launaea resedifolia L. (Asteraceae), 282
Laurus nobilis, 700
Lavandula angustifolia, 138, 216
Lavandulae aetheroleum, 437, 656, 661
Lavender oil
   content, 437
   definition, 437
   inhibitory data
      agar diffusion test, 515–518
      dilution test, 519–521
      vapor phase test, 522–523
Ledol
   Aspergillus niger, 925
   Cephalosporium aphidicola, 925
   Glomerella cingulata, 921, 925
Lemonis aetheroleum. see Lemon oil
Lemon oil
   content, 437
   definition, 437
   inhibitory data
      agar diffusion test, 524–528
      dilution test, 529–530
      vapor phase test, 531
Limonene, 262, 265
   Aspergillus
      A. niger, 769–770
      A. cellulosae, 769–772
   Cladosporium sp., 770, 773
   cyanobacterium, 774, 776
   definition, 766
   Glomerella cingulata, 775, 777
   limonene epoxide, 773–774
   metabolic pathways, 767, 880, 887
   Nicotiana tabacum, 776–778
   Penicillium
      P. digitatum, 769–770, 772–773, 775–776
      P. italicum, 769
   8-*p*-menthene-1,2-*trans*-diol, 767–768
   Pseudomonas gladioli, 768
   Rhodotorula sp., 770, 773–774
   S. litura, 773, 775

Linalool, 262, 264, 266, 366
   A. niger, 761–763
   Botrytis cinerea, 761–762
   Catharanthus roseus, 757, 759, 764–765
   furanoid linalool oxides, 762, 764
   metabolic pathways, 882, 886
   Pseudomonas
      P. incognita, 760–761
      P. pseudomallei, 760
   pyranoid linalool oxides, 761–762
   S. ikutamanensis, 762–763
   tetrahydrolinalool, 764–765
   trans-pyranoid linalool oxide, 763–764
Linalyl acetate, 21, 266, 367–368, 372, 633, 697, 712–713, 719, 727–729, 764–765
Lippia. graveolens, 50–51
Listeriner®, 389–390
Longifolene, 180, 182, 275–276, 989
α-Longipinene, 975, 979
Lovage *(L. officinale)*, 91, 103

M

Maalioxide, 964
   Aspergillus
      A. cellulosae, 967
      A. niger, 967
   Mucor plumbeus, 967
Mandarin oil
   content, 437
   definition, 437
   inhibitory data
      agar diffusion test, 532–534
      dilution test, 535
      vapor phase test, 535
Marchantia polymorpha
   8,9-dehydronootkatone, 920
   6-epi-α-santonin, 945
   lancerodiol *p*-hydroxybenzoate, 981
   p-hydroxybenzoate, 981
Marjoram *(Majorana hortensis)*, 49–50, 67, 73, 112
Mass spectrometry (MS), 31–32
   gas chromatography, 24–25
   GC-IRMS, 27
Matricaria oil
   content, 437
   definition, 437
   inhibitory data
      agar diffusion test, 536–537
      dilution test, 538
      vapor phase test, 539
Maximized survey derived intake (MSDI), 235–236
MDGC. *see* Multidimensional gas chromatographic (MDGC) techniques
Melaleuca alternifolia
   MVA pathway, 108
   α-terpineol, 108
   tobacco mosaic virus, 693
Melaleucae aetheroleum. See Tea tree oil
Mentha arvensis, 45, 95, 178, 240, 245–246, 267, 1016, 1039
Menthae
   M. arvensis aetheroleum partim mentholum depletum
      (*See* Mint oil)
   M. piperitae aetheroleum (*see* Peppermint oil)

Index

Mentha piperita
  monoterpene biosynthesis, 44
  peppermint, 178, 214
  secretory structures, 10
Menthofuran, 58, 73, 107, 178, 245, 266–267, 1077
Menthol, 267–268
  A. niger, 788–790
  Cephalosporium aphidicola, 787
  definition, 786
  enantiomer, 788, 791
  E. perriniana, 788, 791
  isomers, 786
  menthyl acetate, 787
  menthyl succinate, 787
  metabolic pathways, 881, 886–887
  metabolites, 788–790
  S. litura, 788, 791
Menthone, 267–268, 881, 886–887
Menu-census approach, 235
(–)-Methoxy-α-herbertene, 963, 966
Methyl dihydroeugenol, 997–998
Methylerythritol phosphate (MEP) pathway, 58, 108
Mevalonic acid (MVA) pathway, 108
Microdistillation, 9–10, 136
Mint oil
  content, 437
  definition, 437
  inhibitory data
    agar diffusion test, 540–541
    dilution test, 542
Monoterpene biosynthesis, 44, 58, 108
Monoterpenoids
  acyclic (*see* Acyclic monoterpenoids)
  acyclic monoterpenoid alcohols, 177
  aldehydes, 178
  bicyclic (*see* Bicyclic monoterpenoids)
  cyclic (*see* Cyclic monoterpenoids)
  ethers, 178
  geraniol, 177
  geranyl carbocation, 175
  geranyl pyrophosphate, 175
  ketones, 179
  microbial transformation
    A. niger TBUYN-2, 890–891
    compound names, 890, 893–895
    Euglena gracilis Z, 890, 892–893
  monoterpenoid hydrocarbons, 175–176
  myrcene, 175
  structure formation, 175–176
MSDI. *see* Maximized surveyderived intake (MSDI)
Mucor
  M. plumbeus
    drimenol, 960
    maalioxide, 967
    polygodial, 960
    sclareolide, 974
  M. ramannianus, 796–797, 988
Mucositis, 403
Multidimensional gas chromatographic (MDGC) techniques, 212
  cold-pressed lemon oil, 212–213
  cryogenic system, 215–216
  Es-MDGC analysis, 212, 214
  fast MDGC method, 214–215
  GC x GC, 214–215, 217
  heart-cutting methods, 214
  Lavandula angustifolia, 216
  lavender essential oil, 217–218
  Melaleuca alternifolia, 216
  thermal modulator, 214, 216
Multidimensional liquid column chromatography (MDLC)
  DAD system, 220
  HPLC systems, 219
  LC x LC, 218
  NPLC x RPLC, 219
Muña, 679–680
Muscle strain, 406
Mycobacterium tuberculosis, 295, 611
Myrcene, 267, 269
  Diplodia gossypina, 749
  metabolic pathways, 885, 889
  Spodoptera litura, 750
Myrtanol, 859–860, 886
Myrtenal, 858–859
Myrtenol, 858–859

N

Nardosinone, 966, 969–970
Nasal decongestant, 410
Natural complex substances (NCS)
  classification, 1042
  dangerous goods
    definition, 1045
    "limited quantity" label, 1050
    packaging, 1047–1048
    purple book, 1045
    transport police control, 1049
    United Nations, 1049
  dangerous substances
    definition, 1045
    environmentally hazardous label, 1048–1049
    "excepted quantity" label, 1050
    GHS, 1045–1046
    material safety data sheet, 1046–1047
    purple book, 1045
  EINECS
    botanical origin, 1044–1045
    CAS number, 1043–1044
    known substances, 1043
    UVCB, 1044–1045
  fragrance/aromatic substances, 1042
  negative factors, 1042
  regulations, 1041–1042, 1050–1051
Natural complex substances (NCSs), 724, 1041, 1082
Natural flavourings
  flavoring regulation 1334/2008/EC, 1079–1081
  former flavoring directive 88/388/EC, 1078–1079
Natural variability
  chemotaxonomy
    A. crithmifolia, 113
    Balkan *Achillea* species, 115
    biological variability, 113
    chemical compound, 111
    chemotypes, 111
    Journal of Essential Oil Research, 111
    marjoram (*Majorana hortensis*), 112

M. citrata, 114
tansy *(T. vulgare),* 112
identification
   A. crithmifolia, 116
   agrotechnical methods, 117
   caraway, 117
   commercial samples, 116
   commercial seed, 116
   Mikania glauca, 116
   pool of samples, 115
   propagation method, 117–118
   sample misunderstanding, 116
   vitro methods, 118
Journal of Essential Oil Research, 87–88
lavender-odor type, 87
morphogenetic and ontogenetic manifestation
   A. absinthium, 104
   Achillea, 103
   biological variability, 108
   complex compartmentation system, 108
   coriander *(Coriandrum sativum),* 105
   fennel variety, 103
   Lovage *(L. officinale),* 103
   Mentha citrata, 106
   MEP pathway, 108
   MVA pathway, 108
   Origanum onites, 108
   peppermint oil, 104–107
   sage *(Salvia officinalis),* 105
   S. officinalis, 108
   tansy *(T. vulgare),* 106–107
   T. marschallianus, 107–108
   T. vulgaris, 107–108
origin of, 109–110
plant characteristics
   morphological characteristics, 101–102
   propagation and genetics, 99–100
populations
   Achillea collina populations, 96–97
   Achillea crithmifolia, 95–96
   intrapopulation variability, 97
   T. longicaulis ssp, 99
   wormwood *(A. absinthium),* 96, 98
rose-odor type, 87
species
   Achillea, 92
   Artemisia absinthium, 93
   caraway *(Carum carvi),* 89
   Carum carvi var. *annuum,* 89
   Carum carvi var. *biennis,* 89
   carvone accumulation, 89–90
   Dioryctria zimmermani, 88
   Fusarium, 88
   hyssop *(Hyssopus officinalis),* 89–90
   Mentha aquatica, 95
   Mentha arvensis, 95
   Mentha dumetorum, 95
   Mentha longifolia, 95
   Mentha piperita, 95
   Mentha pulegium, 95
   Mentha spicata, 95
   monoterpenes, 90
   phthalide components, 91
   Pinus sylvestris, 88
   root oil, 91
   Tagetes minuta, 88
   tansy *(Tanacetum vulgare),* 93–94
   Thymus, 94
   Thymus glabrescens Willd, 94–95
   wormwood *(Artemisia absinthium),* 92–93
thyme-odor type, 87
Nausea, 403–404
Neomenthol, 789, 792–793, 836
Nerol
   Botrytis cinerea, 756–757
   Catharanthus roseus, 757–758
   metabolic pathways, 883, 888
   Spodoptera litura, 759
Neroli oil *(Neroli aetheroleum)*
   content, 437
   definition, 437
   inhibitory data
      agar diffusion test, 543–544
      dilution test, 545
      vapor phase test, 546
Neurospora crassa, 987–988
Niaouli oil, 437
Niaouli typo cineolo aetheroleum, 437
Nicotiana tabacum, 776–778, 824, 870, 943
Nipple cracks, 406
Nocardia corallina, 957, 959
Nootkatone
   Aspergillus niger, 911–913
   Botryosphaeria dothidea, 913–916
   Fusarium culmorum, 913–916
   Valencene **(1),** 908–910
Nopol benzyl ether, 869, 885, 890, 998, 1000
8-nor-dihydrocapsaicin, 1001–1002
Normal-phase high-performance LC (NP-HPLC), 210–211

O

Ocimum species, 363
Online coupled LC-GC, 221
Oral cavity
   cariogenic and periodontopathic bacteria, 389
   Listeriner, 389–390
   microflora, 389
Oregano. majorana, 50
Organic certification, 1018
Organic weed control, 684
Osteoarthritis, 406
Oxidative stress, 326–327, 336–338, 684
4-Oxo-α-ionone, 993

P

Padova system, 154
Paecilomyces variotii, 564, 961
Paederia scandens Lour. (Rubiaceae), 285
Pain relief
   dysmenorrhea, 404–405
   headache, 405
   infantile colic, 405
   joint physiotherapy, 405
   muscle strain, 406
   nipple cracks and pain, 406

osteoarthritis, 406
  postherpetic neuralgia, 406–407
Paper chromatography (PC), 199–200
Papyrus Ebers, 6, 625
Parthenin, 954
Parthenolide, 946–947, 949
Patchouli acetate, 975
Patchouli alcohol, 59, 181, 285, 302, 974, 991
  rabbits/dogs, 991
Patchoulol, 275–276
  Botrytis cinerea, 979
  Glomerella cingulata, 979
PC. *see* Paper chromatography (PC)
p-cymene, 261–263, 782–783, 881, 886–887
PDMS. *see* Polydimethylsiloxane (PDMS)
"pellatrici" method, 141–142
Penicillium
  P. chrysogenum
    agar diffusion test, 442, 445, 452, 458, 468, 498, 555
    artemisinin, 959
    dilution test, 461, 471, 478, 480, 491, 544–545, 558, 564, 573, 586, 605
  P. digitatum
    citronellal, 751, 753
    limonene, 770, 772–773, 775
  P. sclerotiorum, 963–964
Peppermint oil, 402, 657
  content, 438
  definition, 438
  inhibitory data
    agar diffusion test, 552–556
    dilution test, 557–559
    vapor phase test, 560
Perilla frutescens, 50
Perillaldehyde, 140, 783–785, 886–887
Perillyl alcohol
  CGTase, 811–812
  Eucalyptus perriniana, 811
  metabolic pathways, 880, 887
  Streptomyces ikutamanensis, 811
Petitgrain oil, 139, 713, 728–729, 738, 1033, 1036
Phellandral, 785, 882
Phellandrene, 162, 732, 884, 886, 889
α-Phellandrene
  bitter orange oil, 728
  cajeput oil, 729
  dill oil, 732
  elemi oil, 732
  formalin test, 287
  mandarin oil, 736
  MDGC, 212
  mountain pine oil, 737
  pepper oil, 738
  rosemary oil, 739
  S. molle leaf, 337
  tea tree oil, 741
Phytochemical polymorphism, 44
Phytolacca americana
  raspberry ketone, 999
  zingerone, 999
Phytotoxicity, 683–684, 689, 972
Pinane-2,3-diol, 861–862
3-Pinanone, 873, 875

α-Pinene, 268–270
  A. niger, 847
  Aspergillus niger, 849–850
  Botrytis cinerea, 849, 851
  human liver microsomes CYP2B6, 849, 851
  metabolic pathways, 878–879
  Pseudomonas, 847
    P. fluorescens, 847–848
    P. PX 1, 849–850
  rabbit, 849, 851
  soil Pseudomonad, 848–849
  Spodoptera litura, 849, 851
β-Pinene, 268–270
  Aspergillus niger, 849, 852–855
  Botrytis cinerea, 850, 852
  Dendroctonus frontalis, 856
  metabolic pathways, 878–879
  Pseudomonas
    P. pseudomallei, 849, 852
    P. putida arvilla, 849, 852
Pinguisanol, 965, 969
Pini pumilionis aetheroleum. see Dwarf-pine oil
Pini sylvestris aetheroleum. see Pinus sylvestris oil
Pinocarveol, 859–860
Pinocarvone, 70, 90, 101, 849, 854, 860, 870–871
Pinus sylvestris oil
  content, 438
  definition, 438
  inhibitory data
    agar diffusion test, 561–562
    dilution test, 563–564
    vapor phase test, 564
Piperitenol, 809
Piperitenone
  Aspergillus niger, 832–833
  metabolic pathways, 880, 887
Piperitone, 115, 179, 829–830
Plagiochiline A
  Aspergillus niger, 928
  Plagiochila fruticosa, 927
Plant variety protection (PVP), 68–69
p-menthane
  metabolic pathways, 882, 886
  Pseudomonas mendocina, 778–779
1-*p*-menthene, 778–779
3-*p*-menthene, 780
Polydimethylsiloxane (PDMS), 13–14, 16, 21
Polygodiol
  Aspergillus niger, 960
  Mucor plumbeus, 960
  Rhizopus arrhizus, 960
Polyketides and lipids
  aldehydes, 168
  allylic oxidation, 168
  lipid-derived components, 168–169
  oakmoss components, 167–168
  prostaglandins and jasmines, 169–170
  structure, 167
Posterior intensity method, 207–208
Postherpetic neuralgia, 406–407
Postoperative pain
  breast biopsy, 407
  cesarean section, 407
  episiotomy, 407–408

gastric banding, 408
prostatitis, 408
Protein engineering, 58–59
Pruritus, 408
Pseudomonas
   P. aeruginosa, 751–752
   P. citronellolis, 751–752
   P. cruciviae, 981
   P. flava, 839
   P. fluorescens, 847–848
   P. fragi, 834–835
   P. gladioli, 768
   P. incognita
      geraniol, 751, 753–754
      linalool, 760–761
   P. ovalis
      carveol, 798, 800
      carvone, 818
      carvotanacetone, 828
   P. pseudomallei, 760
      borneol, 864–865
      linalool, 760
      β-pinene, 849, 852
   P. putida, 871–872
   P. putida arvilla, 849, 852
Psidium guajava, 365
Pulegone, 269–271, 831–833
Pyrethrosin, 947–948

R

Random amplification of polymorphic DNA (RAPD), 47, 57
Raspberry ketone, 997, 999
Reactive oxygen species (ROS), 325
   hydrogen peroxide, 324
   superoxide dismutase, 324
Recurrent aphthous stomatitis (RAS), 408–409
Renal stones, 395–396
Respiratory tract
   airway hyperresponsiveness, 411
   anti-inflammatory activity, 414–415
   antimicrobial effects, 412
   antitussive effects, 412
   antitussive properties, 409–410
   bronchodilation, 411, 413
   1,8-cineole, 411–412, 417–418
   menthol, 409, 417–418
   mucolytic and mucociliary effects, 413–414
   nasal decongestant, 410
   pulmonary function, 415–417
   respiratory drive and respiratory comfort, 410–411
Restricted substances (RS)
   CEFS, 1072
   CoE, 1072
   flavoring regulation 1334/2008/EC
      estragol, 1077
      maximum levels, 1073–1074, 1076–1077
      menthofuran, 1077–1078
      methyleugenol (4-Allyl-1,2-dimethoxybenzene), 1075, 1077
      Part A of Annex III, 1074–1075
      Presence of certain substances, 1073
      properties, 1078
      safety/toxicology, 1078

      former flavoring directive 88/388/EC, 1073–1074
   SCF, 1072–1073
Reversed-phase high-performance LC (NP-HPLC), 210–211
Rhizoctonia solani, 836, 838
Rhizopus
   R. arrhizus
      drimenol, 960
      polygodiol, 960
   R. nigricans, 950–951
   R. stolonifer, 974
Rhodotorula rubra, 950
Rhodotorula sp., 770, 773
Rosemary oil
   content, 438
   definition, 438
   inhibitory data
      agar diffusion test, 565–570
      dilution test, 571–574
      vapor phase test, 575–576
Rosmarini aetheroleum. See Rosemary oil
Rosmarinus officinalis, 45, 254, 334, 384, 676
Rotation locular countercurrent chromatography (RLCC), 24

S

Safety evaluation, essential oils
   chemical composition, 232–234
   chemical requirements
      analytical requirements, 236
      congeneric group, 236–237
      JECFA, 235–236
      menu-census approach, 235
      MSDI method, 235
      SPET, 235–236
      "threshold of regulation," 235
      volume-based approach, 235
   congeneric groups (*see* Congeneric groups)
   constituent-based evaluation, 231, 238–239
   corn mint oil, 241
   crude oil, 232
   factors, 231–232
   food relationship, 237–238
   GRAS, 230
   guide, 240
   multifaceted decision tree approach, 230
   plant sources, 231
   prioritization, 240–241
Sage (*Salvia officinalis*), 105
Salvage lavandulifolia aetheroleum, 438
Salvia
   S. fruticosa, 46–47
   S. lavandulifolia, 369
   S. officinalis, 46–47
   S. sclarae aetheroleum (*See* Clary sage oil)
Santalol, 182, 991–992
Santonin, 938–939
   Aspergillus niger, 943
   Cunninghamella
      C. bainieri, 944
      C. echinulata, 944
   Mucor plumbeus, 944
   Rhizopus stolonifer, 944

Index

6-*epi*-α-Santonin, 943
Schizonepeta tenuifolia Briq. (Lamiaceae), 285
Scientific Committee on Food (SCF), 1072–1073, 1075
Sclareolide, 972
    Aspergillus niger, 975
    Mucor plumbeus, 974
β-Selinene, 932, 934
Sesquiterpenoids
    alcohols, 182
    (–)-β-Caryophyllene, 989, 991
    carotenoid degradation products, 183–184
    cedrol, 990–991
    components, 182–183
    elemol, 990
    (+)-*ent*-Cyclocolorenone, 989–990
    farnesol, 179–181
    (2*E*,6*E*)-Farnesol, 991–992
    hydrocarbons, 182–183
    iripallidal and irones, 184–185
    (–)-longifolene, 989–990
    patchouli alcohol, 991
    santalol, 991–992
SFC. *see* Supercritical fluid chromatography (SFC)
SFC-FTIR, 30
SFC-GC. *see* Supercritical fluid chromatography-gas chromatography (SFC-GC)
SFC-MS, 30
SFE. *see* Supercritical fluid extraction (SFE)
"sfumatrici" method, 141
Shikimic acid
    biosynthesis, 170
    biosynthetic intermediates, 170–171
    cinnamic acid, 172
    definition, 170
    essential oil components, 172
    ferulic acid derivatives, 173
    hydroxy-and aminobenzoic acid derivatives, 171
6-Shogaol, 1002–1003
Single portion exposure technique (SPET), 235–236
Snoring, 418–419
Solid-phase microextraction (SPME), 11, 13–14, 20, 105, 196
Spanish sage oil, 438, 734
Spathulenol
    Aspergillus niger, 924
    Mylia taylorii, 926
Specific gravity (SG), 196, 234, 711
Spicae aetheroleum, 438, 656
Spike lavender oil, 438, 656, 734
Spodoptera litura
    borneol, 865
    carveol, 805
    citronellene, 750
    isolimonene, 776, 778
    isopinocampheol, 862, 864
    limonene, 773, 775
    menthol, 788, 791
    myrcene, 750
    nerol, 759
    α-Pinene, 849, 851
    1-*p*-menthene, 778–779
Squamulosone
    Curvularia lunata, 923, 926
    Hyptis verticillata, 923

Star anise oil
    content, 438
    definition, 438
    inhibitory data
        agar diffusion test, 577–578
        dilution test, 579
        vapor phase test, 579
Steam/water distillation
    battery, 146–147
    BIOLANDES, 155–156
    cooling unit, 147, 149
    cross section, 146–147
    definition, 143–144
    E. globulus, 161
    ester hydrolysis, 145
    Florentine flask, 147, 149–151
    folded/concentrated oils, 158
    fractionation, 157
    global trends, 159
    harvesting blue mallee, 152–154
    lavender oil, 159, 161
    loading kettle, 147, 149
    microwave-assisted hydrodistillation methods, 156
    old distillation apparatus, 145
    open kettle, 146, 148
    orris distillation, 151–152
    Padova system, 154
    rectification, 157–158
    rose oil, 146
    terpenic and sesquiterpenic hydrocarbons, 157
    texarome, 155
    turbo distillation, 151
    unloading kettle, 146–148
    vapor pressure, 144
    wild-growing plants, 160
Stillman-Reed method, 199
Stir bar sorptive extraction (SBSE), 14
Streptomyces
    S. bottropensis, 839
        carveol, 800–803
        carvotanacetone, 828–829
        dihydrocarveol, 807–809
    S. fulvissimus, 950
    S. humidus
        thymol, 796–797
        thymol methyl ether, 796–797
    S. ikutamanensis
        1,8-cineole, 839–840
        citronellol, 755–756
        geraniol, 755–756
        linalool, 762–763
        nerol, 755–756
        perillyl alcohol, 811
Supercritical fluid chromatography (SFC), 23
Supercritical fluid chromatography-gas chromatography (SFC-GC), 30
Supercritical fluid extraction (SFE), 9
Supercritical fluid extraction-gas chromatography (SFE-GC), 29
Swallowing dysfunction, 419
Sweet orange oil
    content, 438
    definition, 438
    inhibitory data

agar diffusion test, 580–584
   dilution test, 585–587
   vapor phase test, 588
Synthetic insecticides, 398, 670
Syzygium aromaticum, 44, 274, 292, 335, 405, 679

T

Tagetes lucida, 45, 52
Tansy *(T. vulgare),* 112
TCD. *see* Thermal conductivity detector (TCD)
Tea tree oil
   content, 438
   definition, 438
   inhibitory data
      agar diffusion test, 589
      dilution test, 590–595
      vapor phase test, 596
Terebinthi aetheroleum ab pinum pinastrum, 438
Terpenoids
   C5 units coupling, 174
   definition, 173
   head-to-tail coupling, 174
   hemiterpenoids, 175
   isoprene units, 173–174
   monoterpenoids (*see* Monoterpenoids)
Terpenoids metabolism
   monoterpenes metabolism
      α-and β-pinene, 268–270
      α-and β-thujone, 272–273
      borneol, 254
      camphene, 255
      camphor, 255–256
      3-carene, 256
      carvacrol, 257
      carvone, 257–258
      1,4-cineole, 259
      1,8-cineole, 259–260
      citral, 260–261
      citronellal, 260–261
      fenchone, 262, 264
      geraniol, 262, 264
      limonene, 262, 265
      linalool, 262, 264, 266
      linalyl acetate, 266
      menthofuran, 266–267
      menthol, 267–268
      menthone, 267–268
      myrcene, 267, 269
      p-cymene, 261–263
      pulegone, 269–271
      terpinen-4-ol, 272
      α-terpineol, 270, 272
      thymol, 273–274
   Phase I and Phase II reactions, 254
   sesquiterpenes metabolism
      caryophyllene, 274–275
      farnesol, 274–275
      longifolene, 275–276
      patchoulol, 275–276
α-Terpinene, 780
γ-Terpinene, 100, 300, 328, 332–334, 338–339, 735, 780–781
Terpinen-4-ol, 272

α-Terpineol, 270, 272
Terpinolene, 781
Tetrahydro analogue, 943
1,2,4β,5α-Tetrahydro-α-santonin, 943–946
Tetrahydronootkatone, 914, 918
Thermal conductivity detector (TCD), 15–16, 202
Thin-layer chromatography (TLC), 15, 199–200
Thujone
   metabolic pathways, 878–879
α-Thujone, 272–273
β-Thujone, 272–273
Thyme oil
   content, 438
   definition, 438
   inhibitory data
      agar diffusion test, 597–600
      dilution test, 601–606
      vapor phase test, 608
Thymol, 273–274, 998
   Eucalyptus perriniana, 1000
   metabolic pathways, 881, 886–887
Thymus
   T. vulgaris, 55, 57, 684
   T. zygis, 698
Time-intensity methods, 207
Tinea versicolor, 387
TLC. *see* Thin-layer chromatography (TLC)
Traditional Chinese medicine (TCM), 254, 364, 1053
Trans-geranyl acetone, 984, 986
Trans-nerolidol, 984
   Aspergillus niger, 987
   Glomerella cingulata, 986
Trichoderma pseudokoningii, 819–820
Trichosurus vulpecula, 842
Trinornardosinone, 970
Turpentine oil, 438
Two-dimensional GC, 19–21

U

Ultrafast module-GC (UFM-GC), 205
Ultraviolet (UV) spectroscopy, 7–8
Union for the Protection of New Varieties of Plants (UPOV), 68–69
UPOV. *see* Union for the Protection of New Varieties of Plants (UPOV)
UV spectroscopy, 31

V

Valencene
   Aspergillus
      A. niger, 910–911
      A. wentii, 910–911
   biotransformation, 909
Valerianol, 916, 920
Vapor phase test
   anise oil, 447
   bitter-fennel oil, 455
   caraway oil, 462
   cassia oil, 465
   Ceylon cinnamon bark oil, 472
   Ceylon cinnamon leaf oil, 475
   citronella oil, 481

Index

clary sage oil, 484
clove oil, 493
coriander oil, 502
dwarf-pine oil, 504
eucalyptus oil, 510
juniper oil, 514
lavender oil, 522–523
lemon oil, 531
mandarin oil, 535
matricaria oil, 539
neroli oil, 546
peppermint oil, 560
Pinus sylvestris oil, 564
rosemary oil, 575–576
star anise oil, 579
sweet orange oil, 588
tea tree oil, 596
thyme oil, 608
Verbenaceae
  Lippia
    L. alba, 50–51
    L. graveolens, 50–51
  oregano, 47–50
  Perilla frutescens, 50
  Salvia
    S. fruticosa, 46–47
    S. officinalis, 46–47
Verbenol, 868
Verbenone, 870
veterinary medicine
  antiparasitic formulation, 659–660
  antip arasitic oils, 658
  aromatic plants, 665
  black pepper and capsicum oils, 657
  citronella oil, 657
  dietary oregano etheric oils, 665
  douglas fir oil, 657
  ear mites, 659
  fleas and ticks, 658
  floral oils, 657
  garlic oil, 659
  hoof rot, 665
  horse intestinal worms, 665
  insecticidal and pest repellent, 658
  lemon oil, 657
  menthol, 665
  milk cows, 665
  mosquitoes, 658–659
  moths, 659
  Ocimum basilicum, 664
  peppermint oil, 657
  pneumonia, 665
  poultry
    CRINAr poultry, 661–662
    Herbromixr poultry, 662–663
    pigs, 663–624
  ruminants, 660–661
  α-terpinyl methyl, 657
  Valeriana oils, 657
Volatile encapsulation
  alginates, 1027
  application area, 1025–1026
  average size, 1025–1026
  biological entities, 1025
  controlled release, 1024–1025, 1028
  cyclodextrin complexation, 1028–1029
  emulsifiers, 1026
  hydrophilic polymers, 1026–1027
  prolonged delivery, 1029
  stabilization, 1027–1028
Volume-based approach, 235–236

W

Weed control. *see* Herbicides
Widdrol, 981

Z

Zaluzanin D, 952, 954
Zerumbone, 972, 977
Zingerone, 231, 997, 999